ALGEBRA

Arithmetic Operations

$$a(b + c) = ab + ac \qquad \frac{a}{b} + \frac{c}{d} = \frac{ad + bc}{bd}$$

$$\frac{a + c}{b} = \frac{a}{b} + \frac{c}{b} \qquad \frac{\frac{a}{b}}{\frac{c}{d}} = \frac{a}{b} \times \frac{d}{c} = \frac{ad}{bc}$$

Exponents and Radicals

$$x^m x^n = x^{m+n} \qquad \frac{x^m}{x^n} = x^{m-n}$$

$$(x^m)^n = x^{mn} \qquad x^{-n} = \frac{1}{x^n}$$

$$(xy)^n = x^n y^n \qquad \left(\frac{x}{y}\right)^n = \frac{x^n}{y^n}$$

$$x^{1/n} = \sqrt[n]{x} \qquad x^{m/n} = \sqrt[n]{x^m} = \left(\sqrt[n]{x}\right)^m$$

$$\sqrt[n]{xy} = \sqrt[n]{x}\sqrt[n]{y} \qquad \sqrt[n]{\frac{x}{y}} = \frac{\sqrt[n]{x}}{\sqrt[n]{y}}$$

Factoring Special Polynomials

$$x^2 - y^2 = (x + y)(x - y)$$
$$x^3 + y^3 = (x + y)(x^2 - xy + y^2)$$
$$x^3 - y^3 = (x - y)(x^2 + xy + y^2)$$

Binomial Theorem

$$(x + y)^2 = x^2 + 2xy + y^2 \qquad (x - y)^2 = x^2 - 2xy + y^2$$
$$(x + y)^3 = x^3 + 3x^2y + 3xy^2 + y^3$$
$$(x - y)^3 = x^3 - 3x^2y + 3xy^2 - y^3$$
$$(x + y)^n = x^n + nx^{n-1}y + \frac{n(n - 1)}{2}x^{n-2}y^2 + \cdots + \binom{n}{k}x^{n-k}y^k + \cdots + nxy^{n-1} + y^n$$

where $\binom{n}{k} = \dfrac{n(n - 1)\cdots(n - k + 1)}{1 \cdot 2 \cdot 3 \cdot \cdots \cdot k}$

Quadratic Formula

If $ax^2 + bx + c = 0$, then $x = \dfrac{-b \pm \sqrt{b^2 - 4ac}}{2a}$.

Inequalities and Absolute Value

If $a < b$ and $b < c$, then $a < c$.

If $a < b$, then $a + c < b + c$.

If $a < b$ and $c > 0$, then $ca < cb$.

If $a < b$ and $c < 0$, then $ca > cb$.

If $a > 0$, then

$|x| = a$ means $x = a$ or $x = -a$

$|x| < a$ means $-a < x < a$

$|x| > a$ means $x > a$ or $x < -a$

GEOMETRY

Geometric Formulas

Formulas for area A, circumfere

Triangle

$A = \frac{1}{2}bh$

$= \frac{1}{2}ab \sin\theta$

Circle

$A = \pi r^2$

$C = 2\pi r$

Sector of Circle

$A = \frac{1}{2}r^2\theta$

$s = r\theta$ (θ in radians)

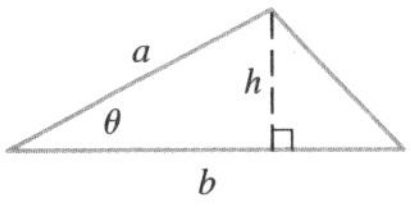

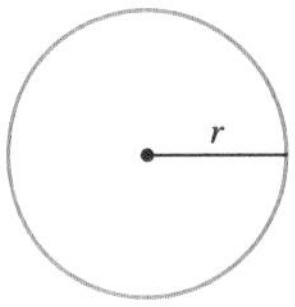

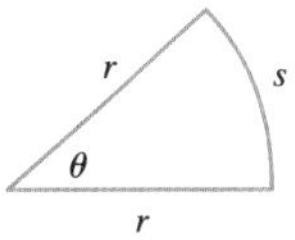

Sphere

$V = \frac{4}{3}\pi r^3$

$A = 4\pi r^2$

Cylinder

$V = \pi r^2 h$

Cone

$V = \frac{1}{3}\pi r^2 h$

$A = \pi r\sqrt{r^2 + h^2}$

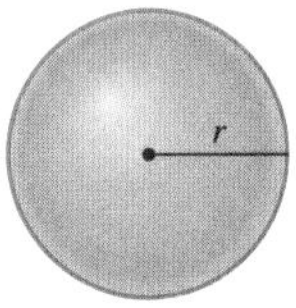

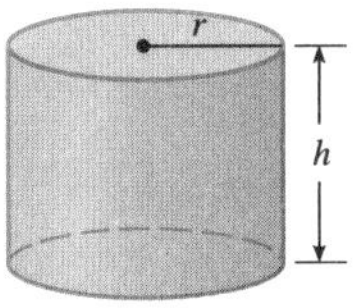

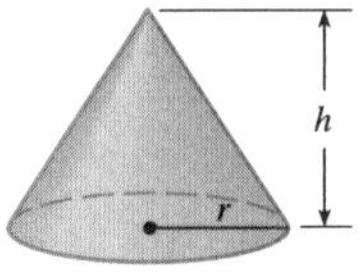

Distance and Midpoint Formulas

Distance between $P_1(x_1, y_1)$ and $P_2(x_2, y_2)$:

$$d = \sqrt{(x_2 - x_1)^2 + (y_2 - y_1)^2}$$

Midpoint of $\overline{P_1P_2}$: $\left(\dfrac{x_1 + x_2}{2}, \dfrac{y_1 + y_2}{2}\right)$

Lines

Slope of line through $P_1(x_1, y_1)$ and $P_2(x_2, y_2)$:

$$m = \frac{y_2 - y_1}{x_2 - x_1}$$

Point-slope equation of line through $P_1(x_1, y_1)$ with slope m:

$$y - y_1 = m(x - x_1)$$

Slope-intercept equation of line with slope m and y-intercept b:

$$y = mx + b$$

Circles

Equation of the circle with center (h, k) and radius r:

$$(x - h)^2 + (y - k)^2 = r^2$$

TRIGONOMETRY

Angle Measurement

π radians $= 180°$

$1° = \dfrac{\pi}{180}$ rad $\qquad$ 1 rad $= \dfrac{180°}{\pi}$

$s = r\theta$

(θ in radians)

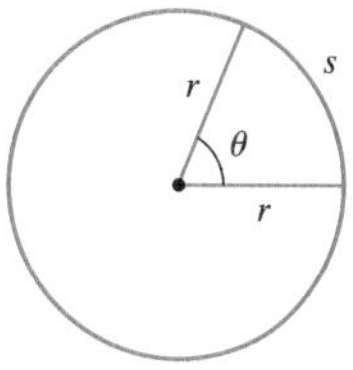

Right Angle Trigonometry

$\sin\theta = \dfrac{\text{opp}}{\text{hyp}}$ $\qquad$ $\csc\theta = \dfrac{\text{hyp}}{\text{opp}}$

$\cos\theta = \dfrac{\text{adj}}{\text{hyp}}$ $\qquad$ $\sec\theta = \dfrac{\text{hyp}}{\text{adj}}$

$\tan\theta = \dfrac{\text{opp}}{\text{adj}}$ $\qquad$ $\cot\theta = \dfrac{\text{adj}}{\text{opp}}$

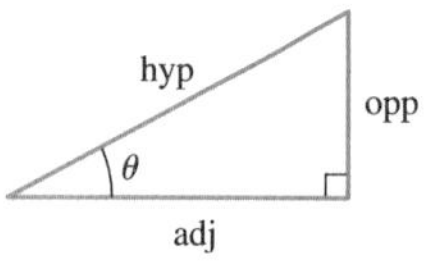

Trigonometric Functions

$\sin\theta = \dfrac{y}{r}$ $\qquad$ $\csc\theta = \dfrac{r}{y}$

$\cos\theta = \dfrac{x}{r}$ $\qquad$ $\sec\theta = \dfrac{r}{x}$

$\tan\theta = \dfrac{y}{x}$ $\qquad$ $\cot\theta = \dfrac{x}{y}$

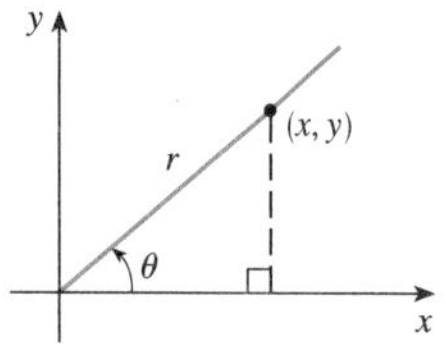

Graphs of Trigonometric Functions

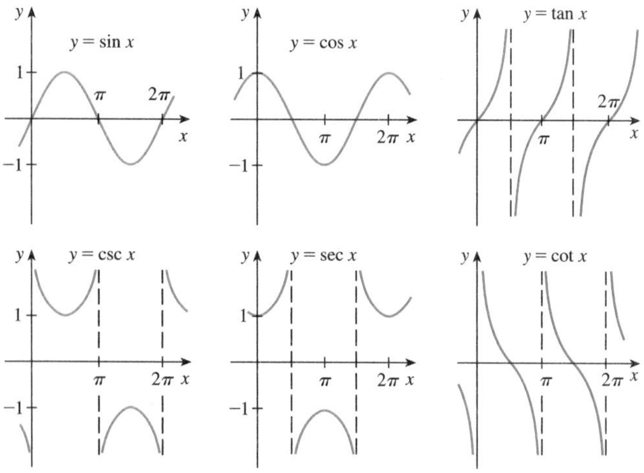

Trigonometric Functions of Important Angles

θ	radians	$\sin\theta$	$\cos\theta$	$\tan\theta$
$0°$	0	0	1	0
$30°$	$\pi/6$	$1/2$	$\sqrt{3}/2$	$\sqrt{3}/3$
$45°$	$\pi/4$	$\sqrt{2}/2$	$\sqrt{2}/2$	1
$60°$	$\pi/3$	$\sqrt{3}/2$	$1/2$	$\sqrt{3}$
$90°$	$\pi/2$	1	0	—

Fundamental Identities

$\csc\theta = \dfrac{1}{\sin\theta}$ $\qquad$ $\sec\theta = \dfrac{1}{\cos\theta}$

$\tan\theta = \dfrac{\sin\theta}{\cos\theta}$ $\qquad$ $\cot\theta = \dfrac{\cos\theta}{\sin\theta}$

$\cot\theta = \dfrac{1}{\tan\theta}$ $\qquad$ $\sin^2\theta + \cos^2\theta = 1$

$1 + \tan^2\theta = \sec^2\theta$ $\qquad$ $1 + \cot^2\theta = \csc^2\theta$

$\sin(-\theta) = -\sin\theta$ $\qquad$ $\cos(-\theta) = \cos\theta$

$\tan(-\theta) = -\tan\theta$ $\qquad$ $\sin\left(\dfrac{\pi}{2} - \theta\right) = \cos\theta$

$\cos\left(\dfrac{\pi}{2} - \theta\right) = \sin\theta$ $\qquad$ $\tan\left(\dfrac{\pi}{2} - \theta\right) = \cot\theta$

The Law of Sines

$$\frac{\sin A}{a} = \frac{\sin B}{b} = \frac{\sin C}{c}$$

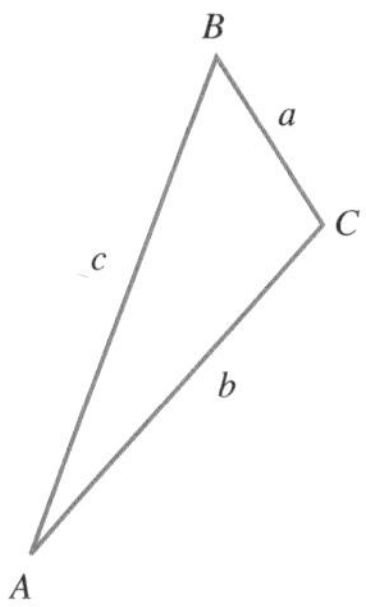

The Law of Cosines

$a^2 = b^2 + c^2 - 2bc\cos A$

$b^2 = a^2 + c^2 - 2ac\cos B$

$c^2 = a^2 + b^2 - 2ab\cos C$

Addition and Subtraction Formulas

$\sin(x + y) = \sin x\cos y + \cos x\sin y$

$\sin(x - y) = \sin x\cos y - \cos x\sin y$

$\cos(x + y) = \cos x\cos y - \sin x\sin y$

$\cos(x - y) = \cos x\cos y + \sin x\sin y$

$\tan(x + y) = \dfrac{\tan x + \tan y}{1 - \tan x\tan y}$

$\tan(x - y) = \dfrac{\tan x - \tan y}{1 + \tan x\tan y}$

Double-Angle Formulas

$\sin 2x = 2\sin x\cos x$

$\cos 2x = \cos^2 x - \sin^2 x = 2\cos^2 x - 1 = 1 - 2\sin^2 x$

$\tan 2x = \dfrac{2\tan x}{1 - \tan^2 x}$

Half-Angle Formulas

$\sin^2 x = \dfrac{1 - \cos 2x}{2}$ $\qquad$ $\cos^2 x = \dfrac{1 + \cos 2x}{2}$

Cut here and keep for reference

SPECIAL FUNCTIONS

Power Functions $f(x) = x^p$

(i) $f(x) = x^n$, n a positive integer

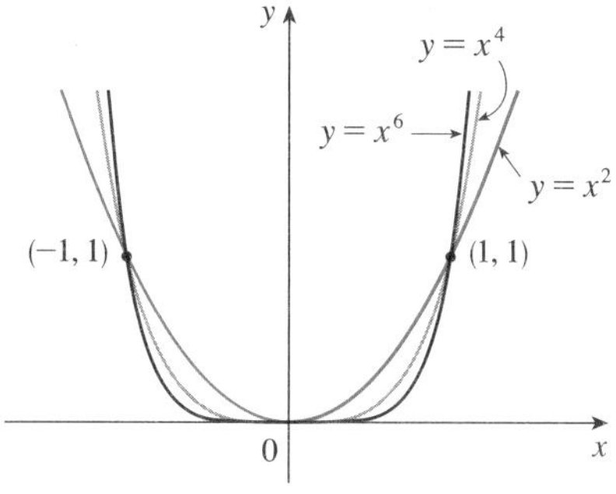

n even

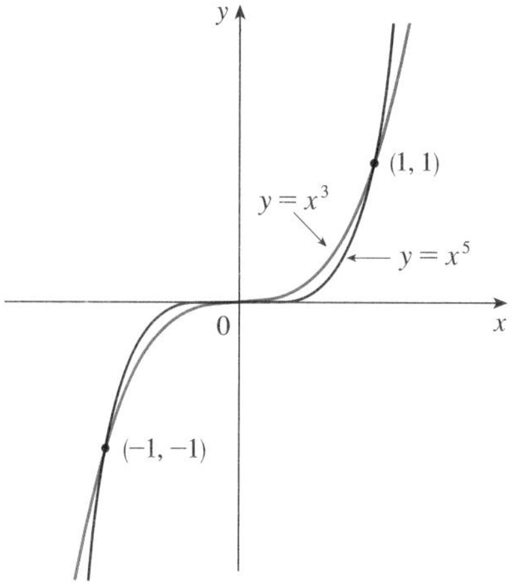

n odd

(ii) $f(x) = x^{1/n} = \sqrt[n]{x}$, n a positive integer

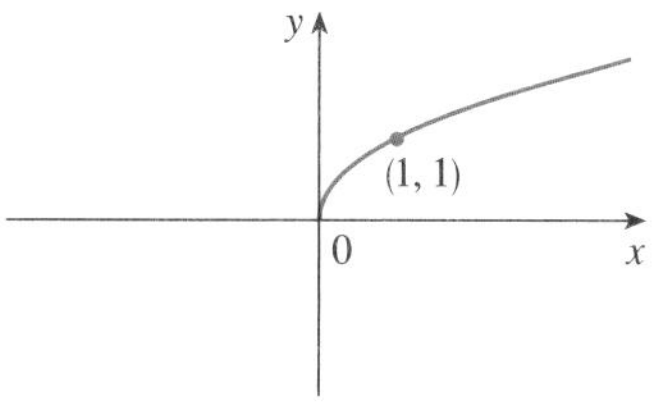

$f(x) = \sqrt{x}$

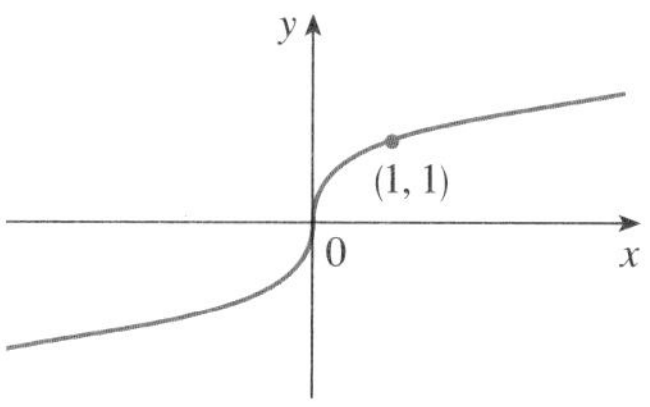

$f(x) = \sqrt[3]{x}$

(iii) $f(x) = x^{-1} = \dfrac{1}{x}$

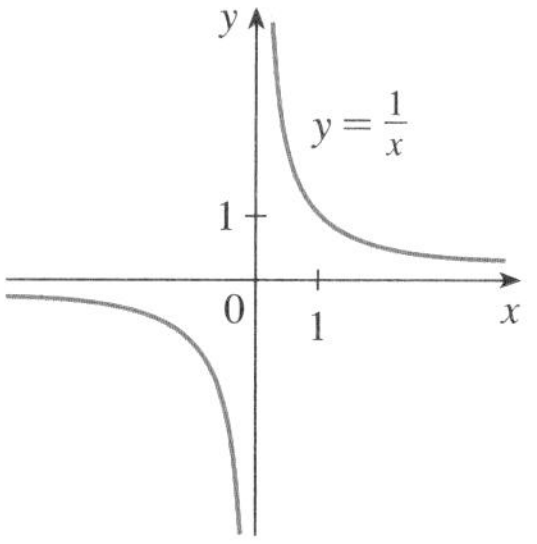

Inverse Trigonometric Functions

$$\arcsin x = \sin^{-1}x = y \iff \sin y = x \quad\text{and}\quad -\frac{\pi}{2} \leqslant y \leqslant \frac{\pi}{2}$$

$$\arccos x = \cos^{-1}x = y \iff \cos y = x \quad\text{and}\quad 0 \leqslant y \leqslant \pi$$

$$\arctan x = \tan^{-1}x = y \iff \tan y = x \quad\text{and}\quad -\frac{\pi}{2} < y < \frac{\pi}{2}$$

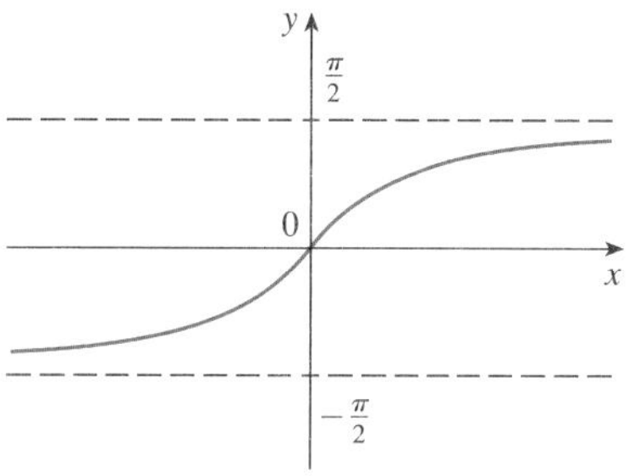

$y = \tan^{-1}x = \arctan x$

$$\lim_{x\to-\infty} \tan^{-1}x = -\frac{\pi}{2}$$

$$\lim_{x\to\infty} \tan^{-1}x = \frac{\pi}{2}$$

SPECIAL FUNCTIONS

Exponential and Logarithmic Functions

$\log_b x = y \iff b^y = x$

$\ln x = \log_e x$, where $\ln e = 1$

$\ln x = y \iff e^y = x$

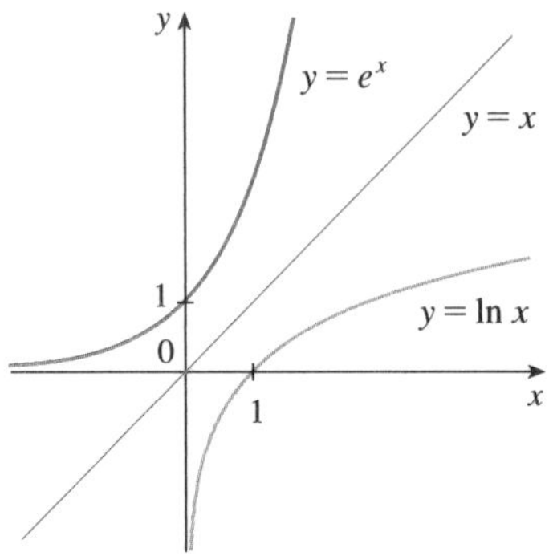

$\lim_{x \to -\infty} e^x = 0 \qquad \lim_{x \to \infty} e^x = \infty$

$\lim_{x \to 0^+} \ln x = -\infty \qquad \lim_{x \to \infty} \ln x = \infty$

Cancellation Equations

$\log_b(b^x) = x \qquad b^{\log_b x} = x$

$\ln(e^x) = x \qquad e^{\ln x} = x$

Laws of Logarithms

1. $\log_b(xy) = \log_b x + \log_b y$

2. $\log_b\left(\dfrac{x}{y}\right) = \log_b x - \log_b y$

3. $\log_b(x^r) = r \log_b x$

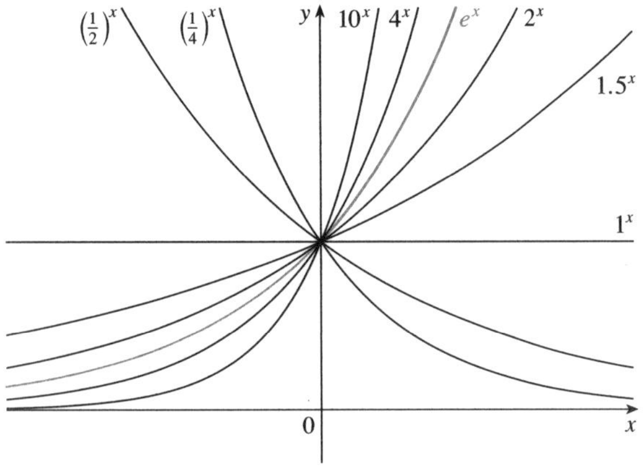

Exponential functions

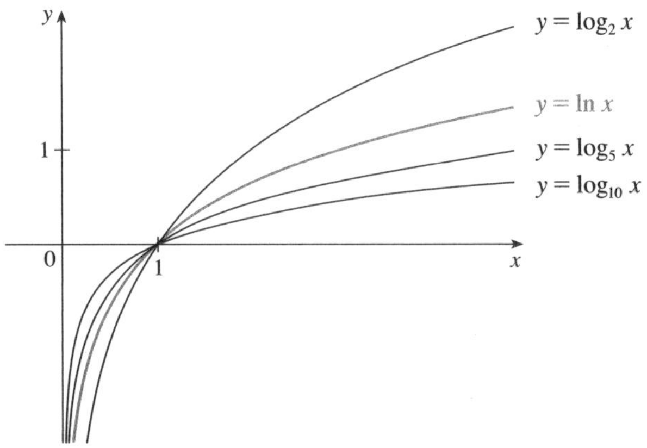

Logarithmic functions

Hyperbolic Functions

$\sinh x = \dfrac{e^x - e^{-x}}{2} \qquad \operatorname{csch} x = \dfrac{1}{\sinh x}$

$\cosh x = \dfrac{e^x + e^{-x}}{2} \qquad \operatorname{sech} x = \dfrac{1}{\cosh x}$

$\tanh x = \dfrac{\sinh x}{\cosh x} \qquad \coth x = \dfrac{\cosh x}{\sinh x}$

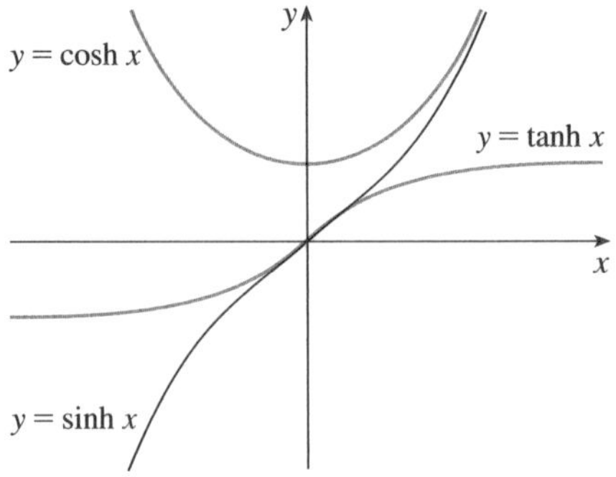

Inverse Hyperbolic Functions

$y = \sinh^{-1}x \iff \sinh y = x$

$y = \cosh^{-1}x \iff \cosh y = x$ and $y \geqslant 0$

$y = \tanh^{-1}x \iff \tanh y = x$

$\sinh^{-1}x = \ln\left(x + \sqrt{x^2 + 1}\right)$

$\cosh^{-1}x = \ln\left(x + \sqrt{x^2 - 1}\right)$

$\tanh^{-1}x = \frac{1}{2}\ln\left(\dfrac{1 + x}{1 - x}\right)$

Cut here and keep for reference

DIFFERENTIATION RULES

General Formulas

1. $\dfrac{d}{dx}(c) = 0$

2. $\dfrac{d}{dx}[cf(x)] = cf'(x)$

3. $\dfrac{d}{dx}[f(x) + g(x)] = f'(x) + g'(x)$

4. $\dfrac{d}{dx}[f(x) - g(x)] = f'(x) - g'(x)$

5. $\dfrac{d}{dx}[f(x)g(x)] = f(x)g'(x) + g(x)f'(x)$ (Product Rule)

6. $\dfrac{d}{dx}\left[\dfrac{f(x)}{g(x)}\right] = \dfrac{g(x)f'(x) - f(x)g'(x)}{[g(x)]^2}$ (Quotient Rule)

7. $\dfrac{d}{dx}f(g(x)) = f'(g(x))g'(x)$ (Chain Rule)

8. $\dfrac{d}{dx}(x^n) = nx^{n-1}$ (Power Rule)

Exponential and Logarithmic Functions

9. $\dfrac{d}{dx}(e^x) = e^x$

10. $\dfrac{d}{dx}(b^x) = b^x \ln b$

11. $\dfrac{d}{dx}\ln|x| = \dfrac{1}{x}$

12. $\dfrac{d}{dx}(\log_b x) = \dfrac{1}{x \ln b}$

Trigonometric Functions

13. $\dfrac{d}{dx}(\sin x) = \cos x$

14. $\dfrac{d}{dx}(\cos x) = -\sin x$

15. $\dfrac{d}{dx}(\tan x) = \sec^2 x$

16. $\dfrac{d}{dx}(\csc x) = -\csc x \cot x$

17. $\dfrac{d}{dx}(\sec x) = \sec x \tan x$

18. $\dfrac{d}{dx}(\cot x) = -\csc^2 x$

Inverse Trigonometric Functions

19. $\dfrac{d}{dx}(\sin^{-1}x) = \dfrac{1}{\sqrt{1 - x^2}}$

20. $\dfrac{d}{dx}(\cos^{-1}x) = -\dfrac{1}{\sqrt{1 - x^2}}$

21. $\dfrac{d}{dx}(\tan^{-1}x) = \dfrac{1}{1 + x^2}$

22. $\dfrac{d}{dx}(\csc^{-1}x) = -\dfrac{1}{x\sqrt{x^2 - 1}}$

23. $\dfrac{d}{dx}(\sec^{-1}x) = \dfrac{1}{x\sqrt{x^2 - 1}}$

24. $\dfrac{d}{dx}(\cot^{-1}x) = -\dfrac{1}{1 + x^2}$

Hyperbolic Functions

25. $\dfrac{d}{dx}(\sinh x) = \cosh x$

26. $\dfrac{d}{dx}(\cosh x) = \sinh x$

27. $\dfrac{d}{dx}(\tanh x) = \operatorname{sech}^2 x$

28. $\dfrac{d}{dx}(\operatorname{csch} x) = -\operatorname{csch} x \coth x$

29. $\dfrac{d}{dx}(\operatorname{sech} x) = -\operatorname{sech} x \tanh x$

30. $\dfrac{d}{dx}(\coth x) = -\operatorname{csch}^2 x$

Inverse Hyperbolic Functions

31. $\dfrac{d}{dx}(\sinh^{-1}x) = \dfrac{1}{\sqrt{1 + x^2}}$

32. $\dfrac{d}{dx}(\cosh^{-1}x) = \dfrac{1}{\sqrt{x^2 - 1}}$

33. $\dfrac{d}{dx}(\tanh^{-1}x) = \dfrac{1}{1 - x^2}$

34. $\dfrac{d}{dx}(\operatorname{csch}^{-1}x) = -\dfrac{1}{|x|\sqrt{x^2 + 1}}$

35. $\dfrac{d}{dx}(\operatorname{sech}^{-1}x) = -\dfrac{1}{x\sqrt{1 - x^2}}$

36. $\dfrac{d}{dx}(\coth^{-1}x) = \dfrac{1}{1 - x^2}$

TABLE OF INTEGRALS

Basic Form

1. $\int u\,dv = uv - \int v\,du$

2. $\int u^n\,du = \dfrac{u^{n+1}}{n+1} + C, \quad n \neq -1$

3. $\int \dfrac{du}{u} = \ln|u| + C$

4. $\int e^u\,du = e^u + C$

5. $\int b^u\,du = \dfrac{b^u}{\ln b} + C$

6. $\int \sin u\,du = -\cos u + C$

7. $\int \cos u\,du = \sin u + C$

8. $\int \sec^2 u\,du = \tan u + C$

9. $\int \csc^2 u\,du = -\cot u + C$

10. $\int \sec u \tan u\,du = \sec u + C$

11. $\int \csc u \cot u\,du = -\csc u + C$

12. $\int \tan u\,du = \ln|\sec u| + C$

13. $\int \cot u\,du = \ln|\sin u| + C$

14. $\int \sec u\,du = \ln|\sec u + \tan u| + C$

15. $\int \csc u\,du = \ln|\csc u - \cot u| + C$

16. $\int \dfrac{du}{\sqrt{a^2 - u^2}} = \sin^{-1}\dfrac{u}{a} + C, \quad a > 0$

17. $\int \dfrac{du}{a^2 + u^2} = \dfrac{1}{a}\tan^{-1}\dfrac{u}{a} + C$

18. $\int \dfrac{du}{u\sqrt{u^2 - a^2}} = \dfrac{1}{a}\sec^{-1}\dfrac{u}{a} + C$

19. $\int \dfrac{du}{a^2 - u^2} = \dfrac{1}{2a}\ln\left|\dfrac{u + a}{u - a}\right| + C$

20. $\int \dfrac{du}{u^2 - a^2} = \dfrac{1}{2a}\ln\left|\dfrac{u - a}{u + a}\right| + C$

Forms Involving $\sqrt{a^2 + u^2}$, $a > 0$

21. $\int \sqrt{a^2 + u^2}\,du = \dfrac{u}{2}\sqrt{a^2 + u^2} + \dfrac{a^2}{2}\ln\left(u + \sqrt{a^2 + u^2}\right) + C$

22. $\int u^2\sqrt{a^2 + u^2}\,du = \dfrac{u}{8}(a^2 + 2u^2)\sqrt{a^2 + u^2} - \dfrac{a^4}{8}\ln\left(u + \sqrt{a^2 + u^2}\right) + C$

23. $\int \dfrac{\sqrt{a^2 + u^2}}{u}\,du = \sqrt{a^2 + u^2} - a\ln\left|\dfrac{a + \sqrt{a^2 + u^2}}{u}\right| + C$

24. $\int \dfrac{\sqrt{a^2 + u^2}}{u^2}\,du = -\dfrac{\sqrt{a^2 + u^2}}{u} + \ln\left(u + \sqrt{a^2 + u^2}\right) + C$

25. $\int \dfrac{du}{\sqrt{a^2 + u^2}} = \ln\left(u + \sqrt{a^2 + u^2}\right) + C$

26. $\int \dfrac{u^2\,du}{\sqrt{a^2 + u^2}} = \dfrac{u}{2}\sqrt{a^2 + u^2} - \dfrac{a^2}{2}\ln\left(u + \sqrt{a^2 + u^2}\right) + C$

27. $\int \dfrac{du}{u\sqrt{a^2 + u^2}} = -\dfrac{1}{a}\ln\left|\dfrac{\sqrt{a^2 + u^2} + a}{u}\right| + C$

28. $\int \dfrac{du}{u^2\sqrt{a^2 + u^2}} = -\dfrac{\sqrt{a^2 + u^2}}{a^2 u} + C$

29. $\int \dfrac{du}{(a^2 + u^2)^{3/2}} = \dfrac{u}{a^2\sqrt{a^2 + u^2}} + C$

Cut here and keep for reference

TABLE OF INTEGRALS

Forms Involving $\sqrt{a^2 - u^2}$, $a > 0$

30. $\displaystyle\int \sqrt{a^2 - u^2}\,du = \frac{u}{2}\sqrt{a^2 - u^2} + \frac{a^2}{2}\sin^{-1}\frac{u}{a} + C$

31. $\displaystyle\int u^2\sqrt{a^2 - u^2}\,du = \frac{u}{8}(2u^2 - a^2)\sqrt{a^2 - u^2} + \frac{a^4}{8}\sin^{-1}\frac{u}{a} + C$

32. $\displaystyle\int \frac{\sqrt{a^2 - u^2}}{u}\,du = \sqrt{a^2 - u^2} - a\ln\left|\frac{a + \sqrt{a^2 - u^2}}{u}\right| + C$

33. $\displaystyle\int \frac{\sqrt{a^2 - u^2}}{u^2}\,du = -\frac{1}{u}\sqrt{a^2 - u^2} - \sin^{-1}\frac{u}{a} + C$

34. $\displaystyle\int \frac{u^2\,du}{\sqrt{a^2 - u^2}} = -\frac{u}{2}\sqrt{a^2 - u^2} + \frac{a^2}{2}\sin^{-1}\frac{u}{a} + C$

35. $\displaystyle\int \frac{du}{u\sqrt{a^2 - u^2}} = -\frac{1}{a}\ln\left|\frac{a + \sqrt{a^2 - u^2}}{u}\right| + C$

36. $\displaystyle\int \frac{du}{u^2\sqrt{a^2 - u^2}} = -\frac{1}{a^2u}\sqrt{a^2 - u^2} + C$

37. $\displaystyle\int (a^2 - u^2)^{3/2}\,du = -\frac{u}{8}(2u^2 - 5a^2)\sqrt{a^2 - u^2} + \frac{3a^4}{8}\sin^{-1}\frac{u}{a} + C$

38. $\displaystyle\int \frac{du}{(a^2 - u^2)^{3/2}} = \frac{u}{a^2\sqrt{a^2 - u^2}} + C$

Forms Involving $\sqrt{u^2 - a^2}$, $a > 0$

39. $\displaystyle\int \sqrt{u^2 - a^2}\,du = \frac{u}{2}\sqrt{u^2 - a^2} - \frac{a^2}{2}\ln\left|u + \sqrt{u^2 - a^2}\right| + C$

40. $\displaystyle\int u^2\sqrt{u^2 - a^2}\,du = \frac{u}{8}(2u^2 - a^2)\sqrt{u^2 - a^2} - \frac{a^4}{8}\ln\left|u + \sqrt{u^2 - a^2}\right| + C$

41. $\displaystyle\int \frac{\sqrt{u^2 - a^2}}{u}\,du = \sqrt{u^2 - a^2} - a\cos^{-1}\frac{a}{|u|} + C$

42. $\displaystyle\int \frac{\sqrt{u^2 - a^2}}{u^2}\,du = -\frac{\sqrt{u^2 - a^2}}{u} + \ln\left|u + \sqrt{u^2 - a^2}\right| + C$

43. $\displaystyle\int \frac{du}{\sqrt{u^2 - a^2}} = \ln\left|u + \sqrt{u^2 - a^2}\right| + C$

44. $\displaystyle\int \frac{u^2\,du}{\sqrt{u^2 - a^2}} = \frac{u}{2}\sqrt{u^2 - a^2} + \frac{a^2}{2}\ln\left|u + \sqrt{u^2 - a^2}\right| + C$

45. $\displaystyle\int \frac{du}{u^2\sqrt{u^2 - a^2}} = \frac{\sqrt{u^2 - a^2}}{a^2u} + C$

46. $\displaystyle\int \frac{du}{(u^2 - a^2)^{3/2}} = -\frac{u}{a^2\sqrt{u^2 - a^2}} + C$

TABLE OF INTEGRALS

Forms Involving $a + bu$

47. $\displaystyle\int \frac{u\,du}{a + bu} = \frac{1}{b^2}\big(a + bu - a \ln |a + bu|\big) + C$

48. $\displaystyle\int \frac{u^2\,du}{a + bu} = \frac{1}{2b^3}\Big[(a + bu)^2 - 4a(a + bu) + 2a^2 \ln |a + bu|\Big] + C$

49. $\displaystyle\int \frac{du}{u(a + bu)} = \frac{1}{a} \ln \left| \frac{u}{a + bu} \right| + C$

50. $\displaystyle\int \frac{du}{u^2(a + bu)} = -\frac{1}{au} + \frac{b}{a^2} \ln \left| \frac{a + bu}{u} \right| + C$

51. $\displaystyle\int \frac{u\,du}{(a + bu)^2} = \frac{a}{b^2(a + bu)} + \frac{1}{b^2} \ln |a + bu| + C$

52. $\displaystyle\int \frac{du}{u(a + bu)^2} = \frac{1}{a(a + bu)} - \frac{1}{a^2} \ln \left| \frac{a + bu}{u} \right| + C$

53. $\displaystyle\int \frac{u^2\,du}{(a + bu)^2} = \frac{1}{b^3}\left(a + bu - \frac{a^2}{a + bu} - 2a \ln |a + bu|\right) + C$

54. $\displaystyle\int u\sqrt{a + bu}\,du = \frac{2}{15b^2}(3bu - 2a)(a + bu)^{3/2} + C$

55. $\displaystyle\int \frac{u\,du}{\sqrt{a + bu}} = \frac{2}{3b^2}(bu - 2a)\sqrt{a + bu} + C$

56. $\displaystyle\int \frac{u^2\,du}{\sqrt{a + bu}} = \frac{2}{15b^3}(8a^2 + 3b^2u^2 - 4abu)\sqrt{a + bu} + C$

57. $\displaystyle\int \frac{du}{u\sqrt{a + bu}} = \frac{1}{\sqrt{a}} \ln \left| \frac{\sqrt{a + bu} - \sqrt{a}}{\sqrt{a + bu} + \sqrt{a}} \right| + C, \quad \text{if } a > 0$

$\displaystyle\qquad = \frac{2}{\sqrt{-a}} \tan^{-1}\sqrt{\frac{a + bu}{-a}} + C, \quad \text{if } a < 0$

58. $\displaystyle\int \frac{\sqrt{a + bu}}{u}\,du = 2\sqrt{a + bu} + a\int \frac{du}{u\sqrt{a + bu}}$

59. $\displaystyle\int \frac{\sqrt{a + bu}}{u^2}\,du = -\frac{\sqrt{a + bu}}{u} + \frac{b}{2}\int \frac{du}{u\sqrt{a + bu}}$

60. $\displaystyle\int u^n\sqrt{a + bu}\,du = \frac{2}{b(2n + 3)}\left[u^n(a + bu)^{3/2} - na\int u^{n-1}\sqrt{a + bu}\,du\right]$

61. $\displaystyle\int \frac{u^n\,du}{\sqrt{a + bu}} = \frac{2u^n\sqrt{a + bu}}{b(2n + 1)} - \frac{2na}{b(2n + 1)}\int \frac{u^{n-1}\,du}{\sqrt{a + bu}}$

62. $\displaystyle\int \frac{du}{u^n\sqrt{a + bu}} = -\frac{\sqrt{a + bu}}{a(n - 1)u^{n-1}} - \frac{b(2n - 3)}{2a(n - 1)}\int \frac{du}{u^{n-1}\sqrt{a + bu}}$

Cut here and keep for reference

TABLE OF INTEGRALS

Trigonometric Forms

63. $\int \sin^2 u\, du = \frac{1}{2}u - \frac{1}{4}\sin 2u + C$

64. $\int \cos^2 u\, du = \frac{1}{2}u + \frac{1}{4}\sin 2u + C$

65. $\int \tan^2 u\, du = \tan u - u + C$

66. $\int \cot^2 u\, du = -\cot u - u + C$

67. $\int \sin^3 u\, du = -\frac{1}{3}(2 + \sin^2 u)\cos u + C$

68. $\int \cos^3 u\, du = \frac{1}{3}(2 + \cos^2 u)\sin u + C$

69. $\int \tan^3 u\, du = \frac{1}{2}\tan^2 u + \ln|\cos u| + C$

70. $\int \cot^3 u\, du = -\frac{1}{2}\cot^2 u - \ln|\sin u| + C$

71. $\int \sec^3 u\, du = \frac{1}{2}\sec u \tan u + \frac{1}{2}\ln|\sec u + \tan u| + C$

72. $\int \csc^3 u\, du = -\frac{1}{2}\csc u \cot u + \frac{1}{2}\ln|\csc u - \cot u| + C$

73. $\int \sin^n u\, du = -\frac{1}{n}\sin^{n-1}u \cos u + \frac{n-1}{n}\int \sin^{n-2}u\, du$

74. $\int \cos^n u\, du = \frac{1}{n}\cos^{n-1}u \sin u + \frac{n-1}{n}\int \cos^{n-2}u\, du$

75. $\int \tan^n u\, du = \frac{1}{n-1}\tan^{n-1}u - \int \tan^{n-2}u\, du$

76. $\int \cot^n u\, du = \frac{-1}{n-1}\cot^{n-1}u - \int \cot^{n-2}u\, du$

77. $\int \sec^n u\, du = \frac{1}{n-1}\tan u \sec^{n-2}u + \frac{n-2}{n-1}\int \sec^{n-2}u\, du$

78. $\int \csc^n u\, du = \frac{-1}{n-1}\cot u \csc^{n-2}u + \frac{n-2}{n-1}\int \csc^{n-2}u\, du$

79. $\int \sin au \sin bu\, du = \frac{\sin(a-b)u}{2(a-b)} - \frac{\sin(a+b)u}{2(a+b)} + C$

80. $\int \cos au \cos bu\, du = \frac{\sin(a-b)u}{2(a-b)} + \frac{\sin(a+b)u}{2(a+b)} + C$

81. $\int \sin au \cos bu\, du = -\frac{\cos(a-b)u}{2(a-b)} - \frac{\cos(a+b)u}{2(a+b)} + C$

82. $\int u \sin u\, du = \sin u - u\cos u + C$

83. $\int u \cos u\, du = \cos u + u \sin u + C$

84. $\int u^n \sin u\, du = -u^n \cos u + n\int u^{n-1}\cos u\, du$

85. $\int u^n \cos u\, du = u^n \sin u - n\int u^{n-1}\sin u\, du$

86. $\int \sin^n u \cos^m u\, du = -\frac{\sin^{n-1}u \cos^{m+1}u}{n+m} + \frac{n-1}{n+m}\int \sin^{n-2}u \cos^m u\, du$

$= \frac{\sin^{n+1}u \cos^{m-1}u}{n+m} + \frac{m-1}{n+m}\int \sin^n u \cos^{m-2}u\, du$

Inverse Trigonometric Forms

87. $\int \sin^{-1}u\, du = u\sin^{-1}u + \sqrt{1-u^2} + C$

88. $\int \cos^{-1}u\, du = u\cos^{-1}u - \sqrt{1-u^2} + C$

89. $\int \tan^{-1}u\, du = u\tan^{-1}u - \frac{1}{2}\ln(1+u^2) + C$

90. $\int u\sin^{-1}u\, du = \frac{2u^2-1}{4}\sin^{-1}u + \frac{u\sqrt{1-u^2}}{4} + C$

91. $\int u\cos^{-1}u\, du = \frac{2u^2-1}{4}\cos^{-1}u - \frac{u\sqrt{1-u^2}}{4} + C$

92. $\int u\tan^{-1}u\, du = \frac{u^2+1}{2}\tan^{-1}u - \frac{u}{2} + C$

93. $\int u^n \sin^{-1}u\, du = \frac{1}{n+1}\left[u^{n+1}\sin^{-1}u - \int \frac{u^{n+1}\, du}{\sqrt{1-u^2}}\right], \quad n \neq -1$

94. $\int u^n \cos^{-1}u\, du = \frac{1}{n+1}\left[u^{n+1}\cos^{-1}u + \int \frac{u^{n+1}\, du}{\sqrt{1-u^2}}\right], \quad n \neq -1$

95. $\int u^n \tan^{-1}u\, du = \frac{1}{n+1}\left[u^{n+1}\tan^{-1}u - \int \frac{u^{n+1}\, du}{1+u^2}\right], \quad n \neq -1$

TABLE OF INTEGRALS

Exponential and Logarithmic Forms

96. $\displaystyle\int ue^{au}\,du = \frac{1}{a^2}(au - 1)e^{au} + C$

97. $\displaystyle\int u^n e^{au}\,du = \frac{1}{a}u^n e^{au} - \frac{n}{a}\int u^{n-1}e^{au}\,du$

98. $\displaystyle\int e^{au}\sin bu\,du = \frac{e^{au}}{a^2 + b^2}(a\sin bu - b\cos bu) + C$

99. $\displaystyle\int e^{au}\cos bu\,du = \frac{e^{au}}{a^2 + b^2}(a\cos bu + b\sin bu) + C$

100. $\displaystyle\int \ln u\,du = u\ln u - u + C$

101. $\displaystyle\int u^n \ln u\,du = \frac{u^{n+1}}{(n+1)^2}[(n+1)\ln u - 1] + C$

102. $\displaystyle\int \frac{1}{u\ln u}\,du = \ln|\ln u| + C$

Hyperbolic Forms

103. $\displaystyle\int \sinh u\,du = \cosh u + C$

104. $\displaystyle\int \cosh u\,du = \sinh u + C$

105. $\displaystyle\int \tanh u\,du = \ln\cosh u + C$

106. $\displaystyle\int \coth u\,du = \ln|\sinh u| + C$

107. $\displaystyle\int \operatorname{sech} u\,du = \tan^{-1}|\sinh u| + C$

108. $\displaystyle\int \operatorname{csch} u\,du = \ln\left|\tanh \tfrac{1}{2}u\right| + C$

109. $\displaystyle\int \operatorname{sech}^2 u\,du = \tanh u + C$

110. $\displaystyle\int \operatorname{csch}^2 u\,du = -\coth u + C$

111. $\displaystyle\int \operatorname{sech} u\tanh u\,du = -\operatorname{sech} u + C$

112. $\displaystyle\int \operatorname{csch} u\coth u\,du = -\operatorname{csch} u + C$

Forms Involving $\sqrt{2au - u^2}$, $a > 0$

113. $\displaystyle\int \sqrt{2au - u^2}\,du = \frac{u - a}{2}\sqrt{2au - u^2} + \frac{a^2}{2}\cos^{-1}\left(\frac{a - u}{a}\right) + C$

114. $\displaystyle\int u\sqrt{2au - u^2}\,du = \frac{2u^2 - au - 3a^2}{6}\sqrt{2au - u^2} + \frac{a^3}{2}\cos^{-1}\left(\frac{a - u}{a}\right) + C$

115. $\displaystyle\int \frac{\sqrt{2au - u^2}}{u}\,du = \sqrt{2au - u^2} + a\cos^{-1}\left(\frac{a - u}{a}\right) + C$

116. $\displaystyle\int \frac{\sqrt{2au - u^2}}{u^2}\,du = -\frac{2\sqrt{2au - u^2}}{u} - \cos^{-1}\left(\frac{a - u}{a}\right) + C$

117. $\displaystyle\int \frac{du}{\sqrt{2au - u^2}} = \cos^{-1}\left(\frac{a - u}{a}\right) + C$

118. $\displaystyle\int \frac{u\,du}{\sqrt{2au - u^2}} = -\sqrt{2au - u^2} + a\cos^{-1}\left(\frac{a - u}{a}\right) + C$

119. $\displaystyle\int \frac{u^2\,du}{\sqrt{2au - u^2}} = -\frac{(u + 3a)}{2}\sqrt{2au - u^2} + \frac{3a^2}{2}\cos^{-1}\left(\frac{a - u}{a}\right) + C$

120. $\displaystyle\int \frac{du}{u\sqrt{2au - u^2}} = -\frac{\sqrt{2au - u^2}}{au} + C$

DESCRIPTIVE STATISTICS

Let $x_1, x_2, \ldots, x_n$ be n data points.

The **average**, or **mean of the data**, denoted by $\bar{x}$, is

$$\bar{x} = \frac{x_1 + x_2 + \cdots + x_n}{n}$$

The **standard deviation of the data**, denoted by s.d., is

$$\text{s.d.} = \sqrt{\frac{1}{n}\sum_{k=1}^{n}(x_k - \bar{x})^2}$$

The **mode** of a data set is the number that appears most often.

The **median** of the data is the middle number when the data points are written in increasing order (if the number of data points is even, then the median is the average of the middle two numbers).

The **first quartile** of the data, denoted by Q_1, is the median of the lower half of the data. The **third quartile**, denoted by Q_3, is the median of the upper half of the data.

The **five-number summary** of a data set:

Minimum $\quad Q_1 \quad$ Median $\quad Q_3 \quad$ Maximum

Interquartile range, denoted by IQR:

$$\text{IQR} = Q_3 - Q_1$$

If $(x_1, y_1), (x_2, y_2), \ldots, (x_n, y_n)$ are the values of two numerical variables measured on n individuals, then the **least-squares line** $y = ax + b$ for the data has a and b given by

$$a = \frac{\overline{xy} - \bar{x}\,\bar{y}}{\overline{x^2} - \bar{x}^2} \qquad b = \bar{y} - a\bar{x}$$

PROBABILITY AND COUNTING

Counting

Number of permutations of n objects: $P_n = n!$

Number of permutations of n objects taken k at a time:

$$P_{n,k} = \frac{n!}{(n-k)!}$$

Number of combinations of n objects taken k at a time:

$$C_{n,k} = \frac{n!}{k!(n-k)!}$$

Probability

Terminology

An **experiment** is any procedure (such as rolling a die or flipping a coin) that can be repeated an indefinite number of times.

Each repeated instance of an experiment is called a **trial**.

An experiment has a set of distinct possible **outcomes**.

The set of all possible outcomes of an experiment is called the **sample space** (denoted by Ω).

An **event** is a subset of the sample space. A **simple event** consists of a single outcome. A **compound event** consists of more than one outcome.

The **complement** of event E, denoted by E^c, is the event consisting of all outcomes not in E.

The **union** of events E and F, denoted by $E \cup F$, is the event consisting of all outcomes that are in either E *or* F (or in both).

The **intersection** of events E and F, denoted by $E \cap F$, is the event consisting of all outcomes that are in both E *and* F.

The **probability** of an event E, denoted by $P(E)$, is equal to the proportion of times that event E occurs in n independent trials of an experiment as $n \to \infty$.

Formulas

If an experiment has a set of $n(\Omega)$ **equally likely outcomes**, and if E is an event consisting of $n(E)$ outcomes, then the **probability** of event E is

$$P(E) = \frac{n(E)}{n(\Omega)}$$

Complement Rule: For any event E and its complement E^c,

$$P(E^c) = 1 - P(E)$$

Union Rule: For any events E and F,

$$P(E \cup F) = P(E) + P(F) - P(E \cap F)$$

Conditional Probability: For any events E and F with $P(F) > 0$, the conditional probability of E given F, denoted by $P(E \mid F)$, is

$$P(E \mid F) = \frac{P(E \cap F)}{P(F)}$$

Bayes' Rule: For any two events E and F with $P(E) > 0$,

$$P(F \mid E) = \frac{P(E \mid F)\,P(F)}{P(E)}$$

PROBABILITY AND COUNTING

Random Variables

Definitions

	Discrete	*Continuous*
Probability density function	p_i such that $p_i = P(X = i)$	$f(x)$ such that $P(a \leqslant X \leqslant b) = \int_a^b f(x)\,dx$
Cumulative distribution function	F_i such that $F_i = P(X \leqslant i)$	$F(x)$ such that $F(x) = P(X \leqslant x)$
Mean or expected value (denoted by $E[X]$)	$E[X] = \sum_i i p_i$	$E[X] = \int_{-\infty}^{\infty} x f(x)\,dx$
Variance	$\mathrm{Var}[X] = \sum_i (i - E[X])^2 p_i$ $= \sum_i i^2 p_i - E[X]^2$	$\mathrm{Var}[X] = \int_{-\infty}^{\infty} (x - E[X])^2 f(x)\,dx$ $= \int_{-\infty}^{\infty} x^2 f(x)\,dx - E[X]^2$

Probability Distributions

	Parameters	*Possible values*	*PDF*	*Mean*	*Variance*
Discrete Random Variables					
Bernoulli	p	$X \in \{0, 1\}$	$P(X = 0) = 1 - p$ $P(X = 1) = p$	p	$p(1 - p)$
Binomial	p, n	$X \in \{0, 1, \dots, n\}$	$P(X = i) = \binom{n}{i} p^i (1 - p)^{n-i}$	np	$np(1 - p)$
Geometric	p	$X \in \{1, 2, 3, \dots\}$	$P(X = i) = p(1 - p)^{i-1}$	$\frac{1}{p}$	$\frac{1 - p}{p^2}$
Continuous Random Variables					
Exponential	c	$X \in [0, \infty)$	$f(t) = ce^{-ct}$	$\frac{1}{c}$	$\frac{1}{c^2}$
Normal	μ, σ	$X \in (-\infty, \infty)$	$f(x) = \frac{1}{\sigma\sqrt{2\pi}} e^{-(x-\mu)^2/(2\sigma^2)}$	μ	σ^2
Standard normal	none	$X \in (-\infty, \infty)$	$f(x) = \frac{1}{\sqrt{2\pi}} e^{-x^2/2}$	0	1

Cut here and keep for reference

INFERENTIAL STATISTICS

Let $x_1, x_2, \ldots, x_n$ be n data points.
Sample standard deviation (denoted by s):

$$s = \sqrt{\frac{1}{n-1} \sum_{k=1}^{n} (x_k - \bar{x})^2}$$

Confidence Intervals

Areas under a Standard Normal Curve

(a) Approximately 90% of the area under the curve lies between -1.64 and 1.64.

(b) Approximately 95% of the area under the curve lies between -1.96 and 1.96.

(c) Approximately 99% of the area under the curve lies between -2.58 and 2.58.

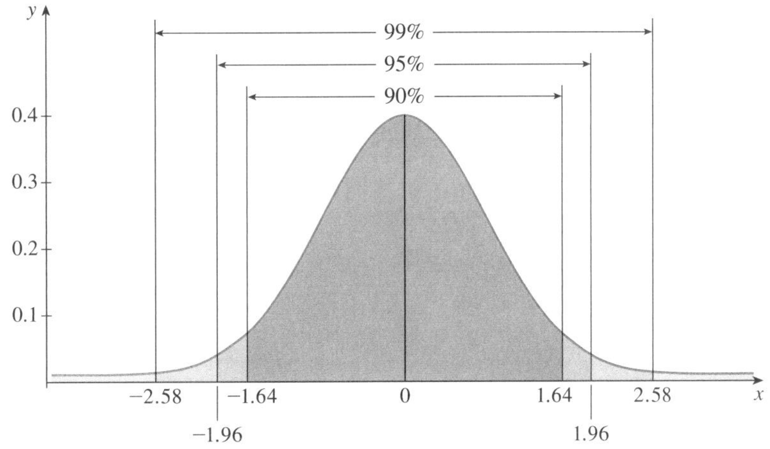

Confidence Intervals When the Population Standard Deviation σ Is Known

A $C\%$ confidence interval for the mean is constructed as follows.

(a) Calculate the sample mean Y_n, where n is the sample size.

(b) Determine the value z_C such that $P(-z_C < Z < z_C) = \dfrac{C}{100}$ from the standard normal distribution.

(c) Construct the interval whose endpoints are $Y_n \pm z_C \dfrac{\sigma}{\sqrt{n}}$.

These steps assume that the variable of interest is normally distributed.

Confidence Intervals When the Population Standard Deviation Is Unknown

A $C\%$ confidence interval for the mean is constructed as follows.

(a) Calculate the sample mean Y_n and the sample standard deviation s, where n is the sample size.

(b) Determine the value $t_{C,n-1}$ such that $P(-t_{C,n-1} < T_{n-1} < t_{C,n-1}) = \dfrac{C}{100}$ from Student's t-distribution, with $n - 1$ degrees of freedom.

(c) Construct the interval whose endpoints are $Y_n \pm t_{C,n-1} \dfrac{s}{\sqrt{n}}$.

These steps assume that the variable of interest is normally distributed.

INFERENTIAL STATISTICS

Hypothesis Testing

Definition

The two-tailed ***P*-value** associated with a specific value of the t-statistic, for example $T_{n-1} = t^*$, is the area in the tails of the Student's t-distribution with $n - 1$ degrees of freedom, beyond the values $-t^*$ and t^*.

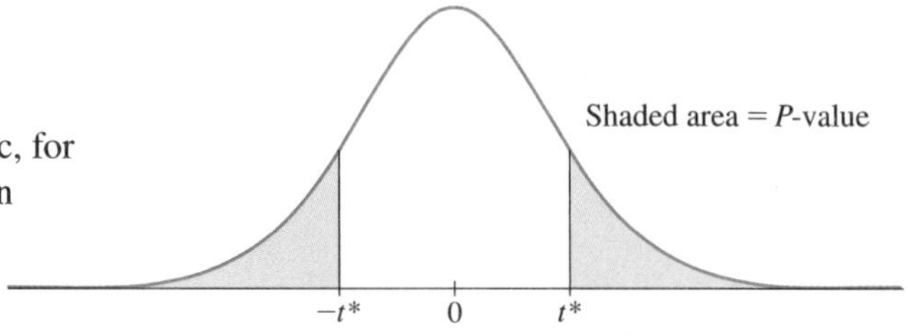

Hypothesis Testing for the Mean of a Normally Distributed Variable

(a) Specify the null and alternative hypotheses: H_0: $\mu = \mu_0$ H_A: $\mu \neq \mu_0$

(b) Choose a significance level α.

(c) The t-statistic has the form $T_{n-1} = \dfrac{Y_n - \mu_0}{s/\sqrt{n}}$.

Calculate the value of this statistic from the data, where n is the sample size.

(d) Determine the P-value associated with the t-statistic calculated in part (c).

(i) If $P \leq \alpha$, then reject H_0 (data provide evidence that $\mu \neq \mu_0$).

(ii) If $P > \alpha$, then do not reject H_0 (data are compatible with the assumption that $\mu = \mu_0$).

Contingency Table Analysis

Definition

The ***P*-value** associated with a specific value of the chi-squared statistic $X = x^*$ is the area in the tail of the probability density function (4) beyond this value of X.

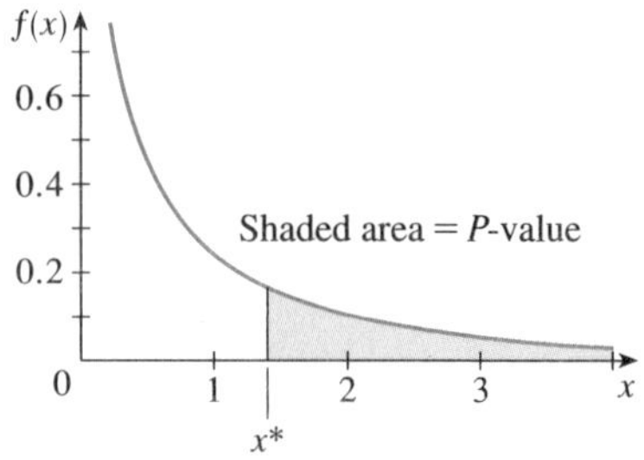

Testing For a Relationship Between Two Categorical Variables, Each with Two Possible Values

(a) Specify the null and alternative hypotheses, H_0 and H_A.

(b) Choose a significance level α.

(c) Fill in each cell of the contingency table with the observed and expected numbers, O_i and E_i, respectively. Check that the following conditions hold:

(i) None of the cells has an expected value less than 1.

(ii) No more than one cell has an expected value less than 5.

(d) The test statistic has the form

$$X = \sum_{i=1}^{4} \frac{(O_i - E_i)^2}{E_i}$$

Calculate the value of this statistic from the entries in the table.

(e) Determine the P-value associated with the test statistic calculated in part (d) by evaluating the integral

$$P = \int_X^{\infty} \frac{1}{\sqrt{2\pi}} x^{-1/2} e^{-x/2} \, dx$$

(i) If $P \leq \alpha$, then reject H_0 (the data provide evidence that the variables are associated).

(ii) If $P > \alpha$, then do not reject H_0 (the data are compatible with the assumption that the variables are independent).

Biocalculus

Calculus, Probability, and Statistics
for the Life Sciences

About the Cover Images

The sex ratio of barred owl offspring is studied using probability theory in Exercise 12.3.21.

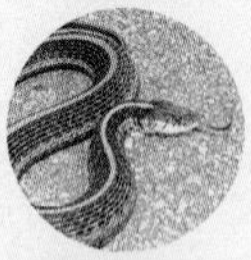
The fitness of a garter snake is a function of the degree of stripedness and the number of reversals of direction while fleeing a predator (Exercise 9.1.7).

The project on page 297 asks how birds can minimize power and energy by flapping their wings versus gliding.

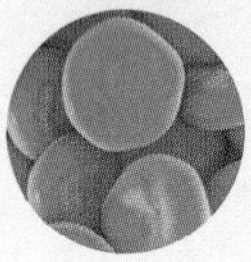
Example 13.3.7 uses hypothesis testing to determine if infection by malaria causes mice to become anemic.

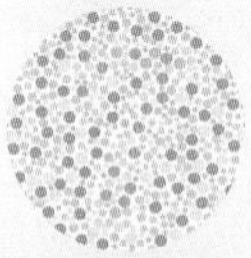
Color blindness is a genetically determined condition. Its inheritance in families is studied using conditional probability in Example 12.3.10.

Data from Gregor Mendel's famous genetic experiments with pea plants are used to introduce the techniques of descriptive statistics in Example 11.1.1.

The energy needed by an iguana to run is a function of two variables, weight and speed (Exercise 9.2.47).

Our study of probability theory in Chapter 12 forms the basis for predicting the inheritance of genetic diseases such as Huntington's disease.

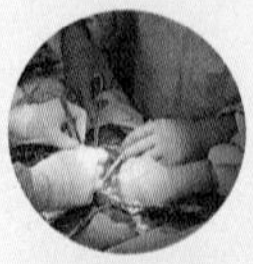
The project on page 222 illustrates how mathematics can be used to minimize red blood cell loss during surgery.

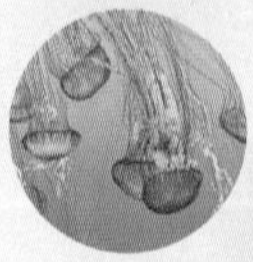
Jellyfish locomotion is modeled by a differential equation in Exercise 10.1.34.

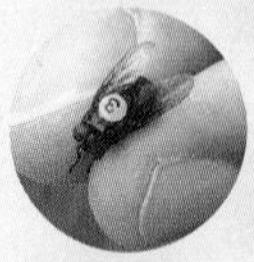
The screw-worm fly was effectively eliminated using the sterile insect technique (Exercise 5.6.24).

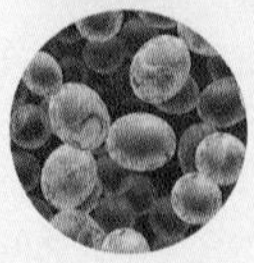
The growth of a yeast population leads naturally to the study of differential equations (Section 7.1).

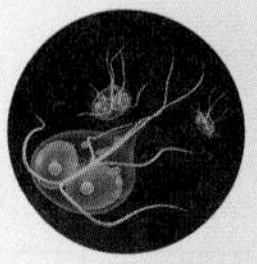
The doubling time of a population of the bacterium *G. lamblia* is determined in Exercise 1.4.29.

Experimental data on EPO injection by athletes for performance enhancement are used in Chapter 13 to illustrate techniques of inferential statistics.

Data on the wingspan of Monarch butterflies are used in Example 13.1.6 to illustrate the importance of sampling distributions in inferential statistics.

The optimal foraging time for bumblebees is determined in Example 4.4.2.

The vertical trajectory of zebra finches is modeled by a quadratic function (Figure 1.2.8).

The size of the gray-wolf population depends on the size of the food supply and the number of competitors (Exercise 9.4.21).

Courtship displays by male ruby-throated hummingbirds provide an interesting example of a geometric random variable in Exercise 12.4.72.

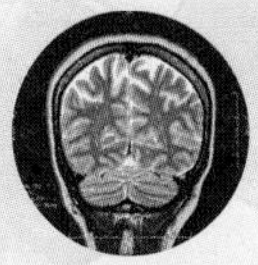
The area of a cross-section of a human brain is estimated in Exercise 6.Review.5.

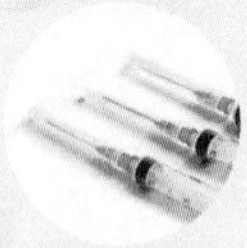
The project on page 479 determines the critical vaccination coverage required to eradicate a disease.

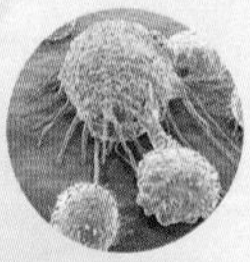
Natural killer cells attack pathogens and are found in two states described by a pair of differential equations developed in Section 10.3.

In Example 4.2.6 a junco has a choice of habitats with different seed densities and we determine the choice with the greatest energy reward.

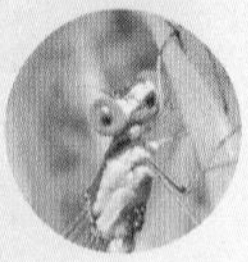
Data on the number of ectoparasites of damselflies are studied in Exercise 11.1.9.

Biocalculus

Calculus, Probability, and Statistics for the Life Sciences

James Stewart
McMaster University and University of Toronto

Troy Day
Queen's University

CENGAGE Learning®

Australia • Brazil • Mexico • Singapore • United Kingdom • United States

CENGAGE Learning

Biocalculus: Calculus, Probability, and Statistics for the Life Sciences

James Stewart, Troy Day

Product Manager: Neha Taleja

Senior Content Developer: Stacy Green

Associate Content Developer: Samantha Lugtu

Product Assistant: Stephanie Kreuz

Media Developer: Lynh Pham

Marketing Manager: Ryan Ahern

Content Project Manager: Cheryll Linthicum

Art Director: Vernon Boes

Manufacturing Planner: Becky Cross

Production Service and Composition: TECHarts

Text and Photo Researcher: Lumina Datamatics

Art and Copy Editor: Kathi Townes, TECHarts

Illustrator: TECHarts

Text and Cover Designer: Lisa Henry

Compositor: Stephanie Kuhns, TECHarts

Cover Images:
DNA strand © iStockphoto.com/Frank Ramspott; Junco © Steffen Foerster/Shutterstock.com; Owl © C. M. Corcoran/ Shutterstock.com; Snake © Matt Jeppson/Shutterstock.com; Bird in flight © Targn Pleiades/Shutterstock.com; Red blood cells © DTKUTOO/Shutterstock.com; Ishihara test for color blindness © Eveleen/Shutterstock.com; Pea plant in bloom © yuris/Shutterstock.com; Iguana © Ryan Jackson; DNA polymerase I © Leonid Andronov/Shutterstock.com; Surgery © Condor 36/Shutterstock.com; Jellyfish © Dreamframer/Shutterstock.com; Srew-worm fly Courtesy of U.S. Department of Agriculture; Yeast cells © Knorre/Shutterstock.com; *G. lamblia* © Sebastian Kaulitzki/Shutterstock.com; Cyclist © enciktat/ Shutterstock.com; Monarch butterfly © Lightspring/Shutterstock.com; Bumblebee foraging © Miroslav Halama/Shutterstock.com; Bacteria rods © Fedorov Oleksiy/Shutterstock.com; Zebra finches © Wang LiQuiang/Shutterstock.com; Wolf © Vladimir Gramagin/Shutterstock.com; Hummingbird © Steve Byland/Shutterstock.com; Brain MRI © Allison Herreid/ Shutterstock.com; Syringes © Tatik22/Shutterstock.com; NK cells © Juan Gaertner/Shutterstock.com; Junco © Steffen Foerster/Shutterstock.com; Damselfly © Laura Nagel; Wolves © Vladimir Gramagin/Shutterstock.com; Snake © Matt Jeppson/Shutterstock.com; Background gradient © ririro/ Shutterstock.com

Interior design images:
Pills © silver-john/Shutterstock.com;
Daphnia pulex © Lebendkulturen.de/Shutterstock.com

Library of Congress Control Number: 2015942360

ISBN-13: 978-1-305-11403-6

Cengage Learning
20 Channel Center Street
Boston, MA 02210
USA

Cengage Learning is a leading provider of customized learning solutions with employees residing in nearly 40 different countries and sales in more than 125 countries around the world. Find your local representative at **www.cengage.com**.

Cengage Learning products are represented in Canada by Nelson Education, Ltd.

To learn more about Cengage Learning Solutions, visit **www.cengage.com**. Purchase any of our products at your local college store or at our preferred online store **www.cengagebrain.com**.

K06T16

Printed in the United States of America
Print Number: 03 Print Year: 2016

To Dolph Schluter and Don Ludwig, for early inspiration

About the Authors

JAMES STEWART received the M.S. degree from Stanford University and the Ph.D. from the University of Toronto. After two years as a postdoctoral fellow at the University of London, he became Professor of Mathematics at McMaster University. His research has been in harmonic analysis and functional analysis. Stewart's books include a series of high-school textbooks as well as a best-selling series of calculus textbooks published by Cengage Learning. He is also coauthor, with Lothar Redlin and Saleem Watson, of a series of college algebra and precalculus textbooks. Translations of his books include those into Spanish, Portuguese, French, Italian, Korean, Chinese, Greek, Indonesian, and Japanese.

A talented violinist, Stewart was concertmaster of the McMaster Symphony Orchestra for many years and played professionally in the Hamilton Philharmonic Orchestra. He has given more than 20 talks worldwide on Mathematics and Music.

Stewart was named a Fellow of the Fields Institute in 2002 and was awarded an honorary D.Sc. in 2003 by McMaster University. The library of the Fields Institute is named after him. The James Stewart Mathematics Centre was opened in October, 2003, at McMaster University.

TROY DAY received the M.S. degree in biology from the University of British Columbia and the Ph.D. in mathematics from Queen's University. His first academic position was at the University of Toronto, before being recruited back to Queen's University as a Canada Research Chair in Mathematical Biology. He is currently Professor of Mathematics and Statistics and Professor of Biology. His research group works in areas ranging from applied mathematics to experimental biology. Day is also coauthor of the widely used book *A Biologist's Guide to Mathematical Modeling,* published by Princeton University Press in 2007.

Contents

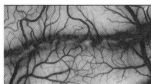

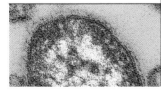

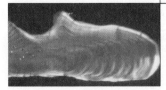

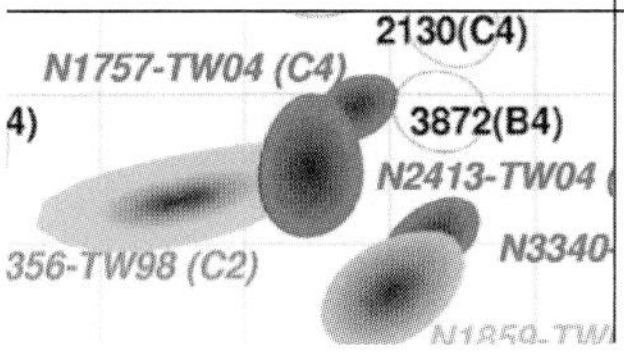
2130(C4)
N1757-TW04 (C4)
4)
3872(B4)
N2413-TW04
356-TW98 (C2)
N3340

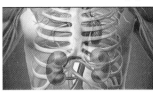

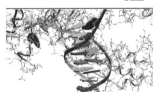

Preface

In recent years more and more colleges and universities have been introducing calculus courses specifically for students in the life sciences. This reflects a growing recognition that mathematics has become an indispensable part of any comprehensive training in the biological sciences.

Our chief goal in writing this textbook is to show students how calculus relates to biology. We motivate and illustrate the topics of calculus with examples drawn from many areas of biology, including genetics, biomechanics, medicine, pharmacology, physiology, ecology, epidemiology, and evolution, to name a few. We have paid particular attention to ensuring that all applications of the mathematics are genuine, and we provide references to the primary biological literature for many of these so that students and instructors can explore the applications in greater depth.

We strive for a style that maintains rigor without being overly formal. Although our focus is on the interface between mathematics and the life sciences, the logical structure of the book is motivated by the mathematical material. Students will come away from a course based on this book with a sound knowledge of mathematics and an understanding of the importance of mathematical arguments. Equally important, they will also come away with a clear understanding of how these mathematical concepts and techniques are central in the life sciences, just as they are in physics, chemistry, and engineering.

The book begins with a prologue entitled *Mathematics and Biology* detailing how the applications of mathematics to biology have proliferated over the past several decades and giving a preview of some of the ways in which calculus provides insight into biological phenomena.

Alternate Versions

There are two versions of this textbook. The first is entitled *Biocalculus: Calculus for the Life Sciences*; it focuses on calculus and some elements of linear algebra that are important in the life sciences. This is the second version, entitled *Biocalculus: Calculus, Probability, and Statistics for the Life Sciences*; it contains all of the content of the first version as well as three additional chapters titled *Descriptive Statistics*, *Probability*, and *Inferential Statistics* (see Content on page xviii).

Features

■ Real-World Data

We think it's important for students to see and work with real-world data in both numerical and graphical form. Accordingly, we have used data concerning biological phenomena to introduce, motivate, and illustrate the concepts of calculus. Many of the examples and exercises deal with functions defined by such numerical data or graphs. See, for example, Figure 1.1.1 (electrocardiogram), Figure 1.1.23 (malarial fever), Exercise 1.1.26 (blood alcohol concentration), Table 2 in Section 1.4 (HIV density), Table 3 in Section 1.5 (species richness in bat caves), Example 3.1.7 (growth of malarial parasites),

Exercise 3.1.42 (salmon swimming speed), Exercises 4.1.7–8 (influenza pandemic), Exercise 4.2.10 (HIV prevalence), Figure 5.1.17 (measles pathogenesis), Exercise 5.1.11 (SARS incidence), Figure 6.1.8 and Example 6.1.4 (cerebral blood flow), Table 1 and Figure 1 in Section 7.1 (yeast population), Figure 8.1.14 (antigenic cartography), Exercises 9.1.7, 9.2.48, and Examples 9.5.5 and 9.6.6 (snake reversals and stripes). And, of course, Chapters 11 and 13 are focused entirely on the analysis of biological data.

■ Graded Exercise Sets

Each exercise set is carefully graded, progressing from basic conceptual exercises and skill-development problems to more challenging problems involving applications and proofs.

■ Conceptual Exercises

One of the goals of calculus instruction is conceptual understanding, and the most important way to foster conceptual understanding is through the problems that we assign. To that end we have devised various types of problems. Some exercise sets begin with requests to explain the meanings of the basic concepts of the section. (See, for instance, the first few exercises in Sections 2.3, 2.5, 3.3, 4.1, and 8.2.) Similarly, all the review sections begin with a Concept Check and a True-False Quiz. Other exercises test conceptual understanding through graphs or tables (see Exercises 3.1.11, 5.2.41–43, 7.1.9–11, 9.1.1–2, and 9.1.26–32).

Another type of exercise uses verbal description to test conceptual understanding (see Exercises 2.5.12, 3.2.50, 4.3.47, and 5.8.29).

■ Projects

One way of involving students and making them active learners is to have them work (perhaps in groups) on extended projects that give a feeling of substantial accomplishment when completed. We have provided 24 projects in *Biocalculus: Calculus for the Life Sciences* and an additional four in *Biocalculus: Calculus, Probability, and Statistics for the Life Sciences*. *Drug Resistance in Malaria* (page 78), for example, asks students to construct a recursion for the frequency of the gene that causes resistance to an antimalarial drug. The project *Flapping and Gliding* (page 297) asks how birds can minimize power and energy by flapping their wings versus gliding. In *The Tragedy of the Commons: An Introduction to Game Theory* (page 298), two companies are exploiting the same fish population and students determine optimal fishing efforts. The project *Disease Progression and Immunity* (page 394) is a nice application of areas between curves. *DNA Supercoiling* (page 783) uses ideas from probability theory to predict how DNA is coiled and compacted into cells. We think that, even when projects are not assigned, students might well be intrigued by them when they come across them between sections.

■ Case Studies

We also provide two case studies: (1) *Kill Curves and Antibiotic Effectiveness* and (2) *Hosts, Parasites, and Time-Travel.* These are extended real-world applications from the primary literature that are more involved than the projects and that tie together multiple mathematical ideas throughout the book. An introduction to each case study is provided at the beginning of the book (page xli), and then each case study recurs in various chapters as the student learns additional mathematical techniques. The case studies can be used at the beginning of a course as motivation for learning the mathematics, and they can then be returned to throughout the course as they recur in the textbook. Alternatively, a case study may be assigned at the end of a course so students can work through all components of the case study in its entirety once all of the mathematical ideas are in place. Case studies might also be assigned to students as term projects. Additional case studies will be posted on the website www.stewartcalculus.com as they become available.

■ Biology Background

Although we give the biological background for each of the applications throughout the textbook, it is sometimes useful to have additional information about how the biological phenomenon was translated into the language of mathematics. In order to maintain a clear and logical flow of the mathematical ideas in the text, we have therefore included such information, along with animations, further references, and downloadable data on the website www.stewartcalculus.com. Applications for which such additional information is available are marked with the icon BB in the text.

■ Technology

The availability of technology makes it more important to clearly understand the concepts that underlie the images on the screen. But, when properly used, graphing calculators and computers are powerful tools for discovering and understanding those concepts. (See the section *Calculators, Computers, and Other Graphing Devices* on page xxvi for a discussion of these and other computing devices.) These textbooks can be used either with or without technology and we use two special symbols to indicate clearly when a particular type of machine is required. The icon indicates an exercise that definitely requires the use of such technology, but that is not to say that it can't be used on the other exercises as well. The symbol CAS is reserved for problems in which the full resources of a computer algebra system (like Maple, Mathematica, or the TI-89/92) are required. But technology doesn't make pencil and paper obsolete. Hand calculation and sketches are often preferable to technology for illustrating and reinforcing some concepts. Both instructors and students need to develop the ability to decide where the hand or the machine is appropriate.

■ Tools for Enriching Calculus (TEC)

TEC is a companion to the text and is intended to enrich and complement its contents. (It is now accessible in Enhanced WebAssign and CengageBrain.com. Selected Visuals and Modules are available at www.stewartcalculus.com.) Developed in collaboration with Harvey Keynes, Dan Clegg, and Hubert Hohn, TEC uses a discovery and exploratory approach. In sections of the book where technology is particularly appropriate, marginal icons TEC direct students to TEC Visuals and Modules that provide a laboratory environment in which they can explore the topic in different ways and at different levels. **Visuals are animations of figures in text; Modules are more elaborate activities and include exercises**. Instructors can choose to become involved at several different levels, ranging from simply encouraging students to use the Visuals and Modules for independent exploration, to assigning specific exercises from those included with each Module, to creating additional exercises, labs, and projects that make use of the Visuals and Modules.

■ Enhanced WebAssign

Technology is having an impact on the way homework is assigned to students, particularly in large classes. The use of online homework is growing and its appeal depends on ease of use, grading precision, and reliability. We have been working with the calculus community and WebAssign to develop a robust online homework system. Up to 50% of the exercises in each section are assignable as online homework, including free response, multiple choice, and multi-part formats.

The system also includes *Active Examples,* in which students are guided in step-by-step tutorials through text examples, with links to the textbook and to video solutions. The system features a customizable *YouBook*, a *Show My Work* feature, *Just in Time* review of precalculus prerequisites, an *Assignment Editor,* and an *Answer Evaluator* that accepts mathematically equivalent answers and allows for homework grading in much the same way that an instructor grades.

■ Website

The site www.stewartcalculus.com includes the following.

- Algebra Review
- Lies My Calculator and Computer Told Me
- History of Mathematics, with links to the better historical websites
- Additional Topics (complete with exercise sets): Approximate Integration: The Trapezoidal Rule and Simpson's Rule, First-Order Linear Differential Equations, Second-Order Linear Differential Equations, Double Integrals, Infinite Series, and Fourier Series
- Archived Problems (drill exercises and their solutions)
- Challenge Problems
- Links, for particular topics, to outside Web resources
- Selected Tools for Enriching Calculus (TEC) Modules and Visuals
- Case Studies
- Biology Background material, denoted by the icon BB in the text
- Data sets

Content

Diagnostic Tests The books begin with four diagnostic tests, in Basic Algebra, Analytic Geometry, Functions, and Trigonometry.

Prologue This is an essay entitled *Mathematics and Biology*. It details how the applications of mathematics to biology have proliferated over the past several decades and highlights some of the applications that will appear throughout the book.

Case Studies The case studies are introduced here so that they can be used as motivation for learning the mathematics. Each case study then recurs at the ends of various chapters throughout the book.

1 Functions and Sequences The first three sections are a review of functions from precalculus, but in the context of biological applications. Sections 1.4 and 1.5 review exponential and logarithmic functions; the latter section includes semilog and log-log plots because of their importance in the life sciences. The final section introduces sequences at a much earlier stage than in most calculus books. Emphasis is placed on recursive sequences, that is, difference equations, allowing us to discuss discrete-time models in the biological sciences.

2 Limits We begin with limits of sequences as a follow-up to their introduction in Section 1.6. We feel that the basic idea of a limit is best understood in the context of sequences. Then it makes sense to follow with the limit of a function at infinity, which we present in the setting of the Monod growth function. Then we consider limits of functions at finite numbers, first geometrically and numerically, then algebraically. (The precise definition is given in Appendix D.) Continuity is illustrated by population harvesting and collapse.

3 Derivatives Derivatives are introduced in the context of rate of change of blood alcohol concentration and tangent lines. All the basic functions, including the exponential and logarithmic functions, are differentiated here. When derivatives are computed in applied settings, students are asked to explain their meanings.

4 Applications of Derivatives The basic facts concerning extreme values and shapes of curves are deduced using the Mean Value Theorem as the starting point. In the section on l'Hospital's Rule we use it to compare rates of growth of functions. Among the applications of optimization, we investigate foraging by bumblebees and aquatic birds. The Stability Criterion for Recursive Sequences is justified intuitively and a proof based on the Mean Value Theorem is given in Appendix E.

5 Integrals The definite integral is motivated by the area problem, the distance problem, and the measles pathogenesis problem. (The area under the pathogenesis curve up to the time symptoms occur is equal to the total amount of infection needed to develop symptoms.) Emphasis is placed on explaining the meanings of integrals in various contexts and on estimating their values from graphs and tables. There is no separate chapter on techniques of integration, but substitution and parts are covered here, as well as the simplest cases of partial fractions.

6 Applications of Integrals The Kety-Schmidt method for measuring cerebral blood flow is presented as an application of areas between curves. Other applications include the average value of a fish population, blood flow in arteries, the cardiac output of the heart, and the volume of a liver.

7 Differential Equations Modeling is the theme that unifies this introductory treatment of differential equations. The chapter begins by constructing a model for yeast population size as a way to motivate the formulation of differential equations. We then show how phase plots allow us to gain considerable qualitative information about the behavior of differential equations; phase plots also provide a simple introduction to bifurcation theory. Examples range from cancer progression to individual growth, to ecology, to anesthesiology. Direction fields and Euler's method are then studied before separable equations are solved explicitly, so that qualitative, numerical, and analytical approaches are given equal consideration. The final two sections of this chapter explore systems of two differential equations. This brief introduction is given here because it allows students to see some applications of systems of differential equations without requiring any additional mathematical preparation. A more complete treatment is then given in Chapter 10.

8 Vectors and Matrix Models We start by introducing higher-dimensional coordinate systems and their applications in the life sciences including antigenic cartography and genome expression profiles. Vectors are then introduced, along with the dot product, and these are shown to provide insight ranging from influenza epidemiology, to cardiology, to vaccine escape, to the discovery of new biological compounds. They also provide some of the tools necessary for the treatment of multivariable calculus in Chapter 9. The remainder of this chapter is then devoted to the application of further ideas from linear algebra to biology. A brief introduction to matrix algebra is followed by a section where these ideas are used to model many different biological phenomena with the aid of matrix diagrams. The final three sections are devoted to the mathematical analysis of such models. This includes a treatment of eigenvalues and eigenvectors, which will also be needed as preparation for Chapter 10, and a treatment of the long-term behavior of matrix models using Perron-Frobenius Theory.

9 Multivariable Calculus Partial derivatives are introduced by looking at a specific column in a table of values of the heat index (perceived air temperature) as a function of the actual temperature and the relative humidity. Applications include body mass index, infectious disease control, lizard energy expenditure, and removal of urea from the blood in dialysis. If there isn't time to cover the entire chapter, then it would make sense to cover just sections 9.1 and 9.2 (preceded by 8.1) and perhaps 9.6. But if Section 9.5 is covered, then Sections 8.2 and 8.3 are prerequisites.

10 Systems of Linear Differential Equations Again modeling is the theme that unifies this chapter. Systems of linear differential equations enjoy very wide application in the life sciences and they also form the basis for the study of systems of nonlinear differential equations. To aid in visualization we focus on two-dimensional systems, and we begin with a qualitative exploration of the different sorts of behaviors that are possible in the context of population dynamics and radioimmunotherapy. The general solution to two-dimensional systems is then derived with the use of eigenvalues and eigenvectors. The third section then illustrates these results with four extended applications involving metapopulations, the immune system, gene regulation, and the transport of environmental pollutants. The chapter ends with a section that shows how the ideas from systems of linear differential equations can be used to understand local stability properties of equilibria in systems of nonlinear differential equations. To cover this chapter students will first need sections 8.1–8.4 and 8.6–8.7.

11 Descriptive Statistics Statistical analyses are central in most areas of biology. The basic ideas of descriptive statistics are presented here, including types of variables, measures of central tendency and spread, and graphical descriptions of data. Single variables are treated first, followed by an examination of the descriptive statistics for relationships between variables, including the calculus behind the least-square fit for scatter plots. A brief introduction to inferential statistics and its relationship to descriptive statistics is also given, including a discussion of causation in statistical analyses.

12 Probability Probability theory represents an important area of mathematics in the life sciences and it also forms the foundation for the study of inferential statistics. Basic principles of counting and their application are introduced first, and these are then used to motivate an intuitive definition of probability. This definition is then generalized to the axiomatic definition of probability in an accessible way that highlights the meanings of the axioms in a biological context. Conditional probability is then introduced with important applications to disease testing, handedness, color blindness, genetic disorders, and gender. The final two sections introduce discrete and continuous random variables and illustrate how these arise naturally in many biological contexts, from disease outbreaks to DNA supercoiling. They also demonstrate how the concepts of differentiation and integration are central components of probability theory.

13 Inferential Statistics The final chapter addresses the important issue of how one takes information from a data set and uses it to make inferences about the population from which it was collected. We do not provide an exhaustive treatment of inferential statistics, but instead present some of its core ideas and how they relate to calculus. Sampling distributions are explained, along with confidence intervals and the logic behind hypothesis testing. The chapter concludes with a simplified treatment of the central ideas behind contingency table analysis.

Student Resources

Enhanced WebAssign®

Printed Access Code ISBN: 978-1-285-85826-5
Instant Access Code ISBN: 978-1-285-85825-8

Enhanced WebAssign is designed to allow you to do your homework online. This proven and reliable system uses content found in this text, then enhances it to help you learn calculus more effectively. Automatically graded homework allows you to focus on your learning and get interactive study assistance outside of class. Enhanced WebAssign for *Biocalculus: Calculus, Probability, and Statistics for the Life Sciences* contains the Cengage YouBook, an interactive ebook that contains animated figures, video clips, highlighting and note-taking features, and more!

CengageBrain.com

To access additional course materials, please visit www.cengagebrain.com. At the CengageBrain.com home page, search for the ISBN of your title (from the back cover of your book) using the search box at the top of the page. This will take you to the product page where these resources can be found.

Stewart Website

www.stewartcalculus.com

This site includes additional biological background for selected examples, exercises, and projects, including animations, further references, and downloadable data files. In addition, the site includes the following:

- Algebra Review
- Additional Topics
- Drill exercises
- Challenge Problems
- Web Links
- History of Mathematics
- Tools for Enriching Calculus (TEC)

Student Solutions Manual

ISBN: 978-1-305-11406-7

Provides completely worked-out solutions to all odd-numbered exercises in the text, giving you a chance to check your answers and ensure you took the correct steps to arrive at an answer.

A Companion to Calculus

By Dennis Ebersole, Doris Schattschneider, Alicia Sevilla, and Kay Somers

ISBN 978-0-495-01124-8

Written to improve algebra and problem-solving skills of students taking a calculus course, every chapter in this companion is keyed to a calculus topic, providing conceptual background and specific algebra techniques needed to understand and solve calculus problems related to that topic. It is designed for calculus courses that integrate the review of precalculus concepts or for individual use. Order a copy of the text or access the eBook online at www.cengagebrain.com by searching the ISBN.

Linear Algebra for Calculus

by Konrad J. Heuvers, William P. Francis, John H. Kuisti, Deborah F. Lockhart, Daniel S. Moak, and Gene M. Ortner

ISBN 978-0-534-25248-9

This comprehensive book, designed to supplement a calculus course, provides an introduction to and review of the basic ideas of linear algebra. Order a copy of the text or access the eBook online at www.cengagebrain.com by searching the ISBN.

Instructor Resources

Enhanced WebAssign®

Printed Access Code ISBN: 978-1-285-85826-5
Instant Access Code ISBN: 978-1-285-85825-8

Exclusively from Cengage Learning, Enhanced WebAssign offers an extensive online program for *Biocalculus: Calculus, Probability, and Statistics for the Life Sciences* to encourage the practice that is so critical for concept mastery. The meticulously crafted pedagogy and exercises in our proven texts become even more effective in Enhanced WebAssign, supplemented by multimedia tutorial support and immediate feedback as students complete their assignments. Key features include:

- Thousands of homework problems that match your textbook's end-of-section exercises
- Opportunities for students to review prerequisite skills and content both at the start of the course and at the beginning of each section
- *Read It* eBook pages, *Watch It* videos, *Master It* tutorials, and *Chat About It* links
- A customizable Cengage Learning *YouBook* with highlighting, note-taking, and search features, as well as links to multimedia resources
- *Personal Study Plans* (based on diagnostic quizzing) that identify chapter topics that students will need to master
- A WebAssign *Answer Evaluator* that recognizes and accepts equivalent mathematical responses in the same way an instructor grades
- A *Show My Work* feature that gives instructors the option of seeing students' detailed solutions
- Lecture videos and more!

Cengage Customizable YouBook

YouBook is an eBook that is both interactive and customizable! Containing all the content from *Biocalculus: Calculus, Probability, and Statistics for the Life Sciences, YouBook* features a text edit tool that allows instructors to modify the textbook narrative as needed. With *YouBook,* instructors can quickly reorder entire sections and chapters or hide any content they don't teach to create an eBook that perfectly matches their syllabus. Instructors can further customize the text by adding instructor-created or YouTube video links. Additional media assets include animated figures, video clips, highlighting and note-taking features, and more! *YouBook* is available within Enhanced WebAssign.

Complete Solutions Manual

ISBN: 978-1-305-11407-4

Includes worked-out solutions to all exercises and projects in the text.

Instructor Companion Website (login.cengage.com)

This comprehensive instructor website contains all art from the text in both jpeg and PowerPoint formats.

Stewart Website

www.stewartcalculus.com

This comprehensive instructor website contains additional material to complement the text, marked by the logo BB. This material includes additional Biological Background for selected examples, exercises, and projects, including animations, further references, and downloadable data files. In addition, this site includes the following:

- Algebra Review
- Additional Topics
- Drill exercises
- Challenge Problems
- Web Links
- History of Mathematics
- Tools for Enriching Calculus (TEC)

Acknowledgments

We are grateful to the following reviewers and class testers for sharing their knowledge and judgment with us. We have learned something from each of them.

Reviewers

Anthony Barcellos, *American River College*

Frank Bauerle, *University of California–Santa Cruz*

Barbara Bendl, *University of the Sciences in Philadelphia*

Adam Bowers, *University of California–San Diego*

Richard Brown, *Johns Hopkins University*

Hannah Callender, *University of Portland*

Youn-Sha Chan, *University of Houston–Downtown*

Alberto Corso, *University of Kentucky*

Robert Crawford, *Sacramento City College*

Dwight Duffus, *Emory University*

Paula Federico, *Capital University*

Guillermo Goldsztein, *Georgia Institute of Technology*

Eli Goldwyn, *University of California–Davis*

Richard Gomulkiewicz, *Washington State University*

Genady Grabarnik, *St. John's University*

Mark Harbison, *Sacramento City College*

Jane Heffernan, *York University*

Sophia Jang, *Texas Tech University*

Yang Kuang, *Arizona State University*

Emile LeBlanc, *University of Toronto*

Glenn Ledder, *University of Nebraska–Lincoln*

Alun Lloyd, *North Carolina State University*

Melissa Macasieb, *University of Maryland, College Park*

Edward Migliore, *University of California–Santa Cruz*

Laura Miller, *University of North Carolina at Chapel Hill*

Val Mohanakumar, *Hillsborough Community College–Dale Mabry*

Douglas Norton, *Villanova University*

Michael Price, *University of Oregon*

Suzanne Robertson, *Virginia Commonwealth University*

Ayse Sahin, *DePaul University*

Asok Sen, *Indiana University–Purdue University*

Ellis Shamash, *Loyola Marymount University*

Patrick Shipman, *Colorado State University*

Jeff Suzuki, *Brooklyn College*

Nicoleta Tarfulea, *Purdue University Calumet*

Juan Tolosa, *Richard Stockton College of New Jersey*

Gerda de Vries, *University of Alberta*

Lindi Wahl, *Western University*

Saleem Watson, *California State University–Long Beach*

George Yates, *Youngstown State University*

Class Testers

Ayse Sahin and her students, *DePaul University,* 2012–2013 academic year

Ilie Ugarcovici and his students, *DePaul University,* 2013–2014 academic year

Michael Price and his students, *University of Oregon,* 2013–2014 academic year

In addition we would like to thank Alan Ableson, Daniel Ashlock, Kathy Davis, David Earn, Brian Gilbert, Lisa Hicks, Bob Montgomerie, Bill Nelson, Sarah Otto, Mary Pugh, Peter Taylor, Ron Wald, and Gail Wolkowicz for their advice and suggestions. We are particularly indebted to Jane Heffernan, for the contributions she made in Chapters 5 and 6 on measles pathogenesis, and to Lindi Wahl, for her careful editing of Chapters 11–13. Special thanks go to Alan Hastings and Saleem Watson for the extensive advice they gave us at the beginning of this writing project.

We also thank Kathi Townes, Stephanie Kuhns, Kira Abdallah, and Kristina Elliott at TECHarts for their production services; Josh Babbin for exercise solutions, Andrew Bulman-Fleming for solutions art, Lauri Semarne for solutions accuracy check; Lisa Henry for the cover image and text and cover design; and the following Cengage Learning staff: product manager Neha Taleja, senior content developer Stacy Green, associate content developer Samantha Lugtu, media developer Lynh Pham, product assistant Stephanie Kreuz, marketing manager Ryan Ahern, senior product development specialist Katherine Greig, content project manager Cheryll Linthicum, art director Vernon Boes, and manufacturing planner Rebecca Cross. They have all done an outstanding job.

JAMES STEWART

TROY DAY

To the Student

Reading a calculus textbook is different from reading a newspaper or a novel, or even a physics book. Don't be discouraged if you have to read a passage more than once in order to understand it. You should have pencil and paper and calculator at hand to sketch a diagram or make a calculation.

Some students start by trying their homework problems and read the text only if they get stuck on an exercise. We suggest that a far better plan is to read and understand a section of the text before attempting the exercises. In particular, you should look at the definitions to see the exact meanings of the terms. And before you read each example, we suggest that you cover up the solution and try solving the problem yourself. You'll get a lot more from looking at the solution if you do so.

Part of the aim of this course is to train you to think logically. Learn to write the solutions of the exercises in a connected, step-by-step fashion with explanatory sentences—not just a string of disconnected equations or formulas.

The answers to the odd-numbered exercises appear at the back of the book. Some exercises ask for a verbal explanation or interpretation or description. In such cases there is no single correct way of expressing the answer, so don't worry that you haven't found the definitive answer. In addition, there are often several different forms in which to express a numerical or algebraic answer, so if your answer differs from ours, don't immediately assume you're wrong. For example, if the answer given in the back of the book is $\sqrt{2} - 1$ and you obtain $1/(1 + \sqrt{2})$, then you're right and rationalizing the denominator will show that the answers are equivalent.

The icon indicates an exercise that definitely requires the use of either a graphing calculator or a computer with graphing software. (*Calculators, Computers, and Other Graphing Devices* discusses the use of these graphing devices and some of the pitfalls that you may encounter.) But that doesn't mean that graphing devices can't be used to check your work on the other exercises as well. The symbol CAS is reserved for problems in which the full resources of a computer algebra system (like Derive, Maple, Mathematica, or the TI-89/92) are required.

You will also encounter the symbol ⊘, which warns you against committing an error. We have placed this symbol in the margin in situations where we have observed that a large proportion of students tend to make the same mistake.

Applications with additional Biology Background available on www.stewartcalculus.com are marked with the icon BB in the text.

Tools for Enriching Calculus, which is a companion to this text, is referred to by means of the symbol TEC and can be accessed in Enhanced WebAssign (selected Visuals and Modules are available at www.stewartcalculus.com). It directs you to modules in which you can explore aspects of calculus for which the computer is particularly useful.

We recommend that you keep this book for reference purposes after you finish the course. Because you will likely forget some of the specific details of calculus, the book will serve as a useful reminder when you need to use calculus in subsequent courses. And, because this book contains more material than can be covered in any one course, it can also serve as a valuable resource for a working biologist.

Calculus is an exciting subject, justly considered to be one of the greatest achievements of the human intellect. We hope you will discover that it is not only useful but also intrinsically beautiful.

JAMES STEWART

TROY DAY

Calculators, Computers, and Other Graphing Devices

© Dan Clegg

Advances in technology continue to bring a wider variety of tools for doing mathematics. Handheld calculators are becoming more powerful, as are software programs and Internet resources. In addition, many mathematical applications have been released for smartphones and tablets such as the iPad.

Some exercises in this text are marked with a graphing icon, which indicates that the use of some technology is required. Often this means that we intend for a graphing device to be used in drawing the graph of a function or equation. You might also need technology to find the zeros of a graph or the points of intersection of two graphs. In some cases we will use a calculating device to solve an equation or evaluate a definite integral numerically. Many scientific and graphing calculators have these features built in, such as the Texas Instruments TI-84 or TI-Nspire CX. Similar calculators are made by Hewlett Packard, Casio, and Sharp.

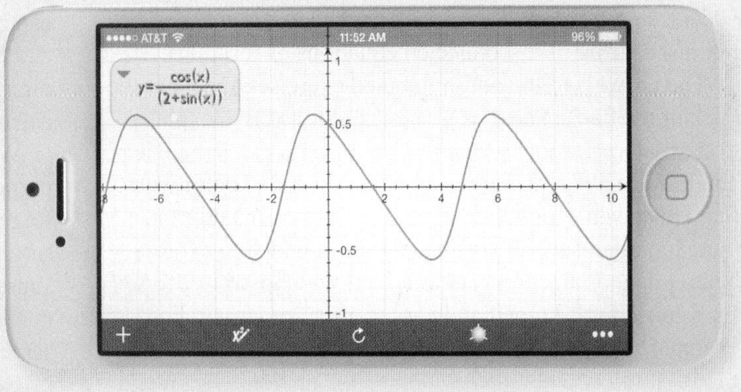

© Dan Clegg

You can also use computer software such as *Graphing Calculator* by Pacific Tech (www.pacifict.com) to perform many of these functions, as well as apps for phones and tablets, like Quick Graph (Colombiamug) or MathStudio (Pomegranate Apps). Similar functionality is available using a web interface at WolframAlpha.com.

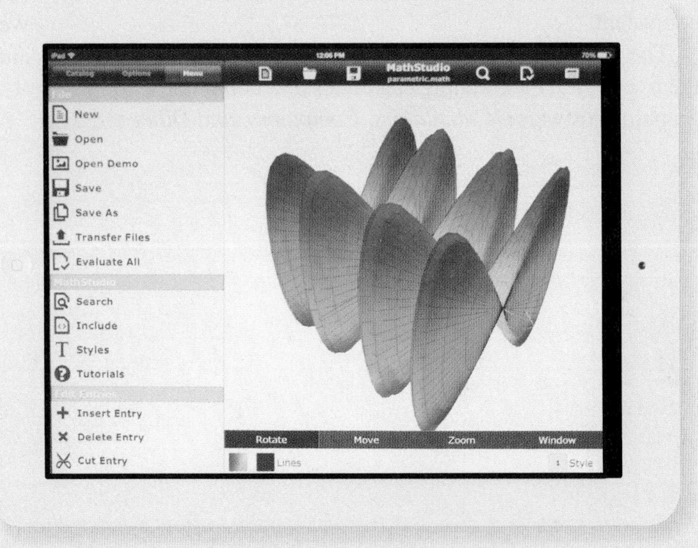

© Dan Clegg

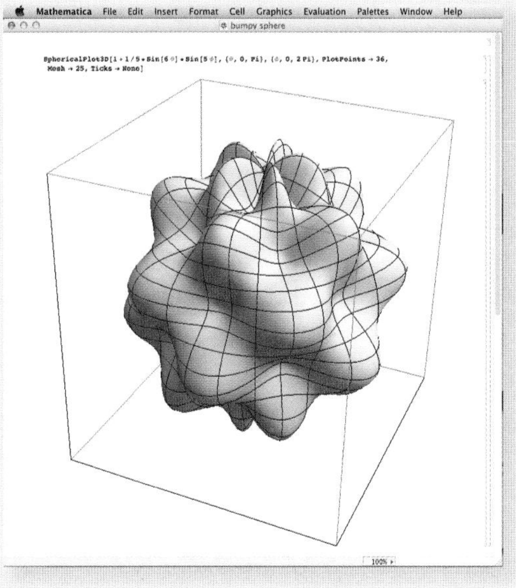

> In general, when we use the term "calculator" in this book, we mean the use of any of the resources we have mentioned.

The CAS icon is reserved for problems in which the full resources of a *computer algebra system* (CAS) are required. A CAS is capable of doing mathematics (like solving equations, computing derivatives or integrals) *symbolically* rather than just numerically.

Examples of well-established computer algebra systems are the computer software packages Maple and Mathematica. The WolframAlpha website uses the Mathematica engine to provide CAS functionality via the Web.

Many handheld graphing calculators have CAS capabilities, such as the TI-89 and TI-Nspire CX CAS from Texas Instruments. Some tablet and smartphone apps also provide these capabilities, such as the previously mentioned MathStudio.

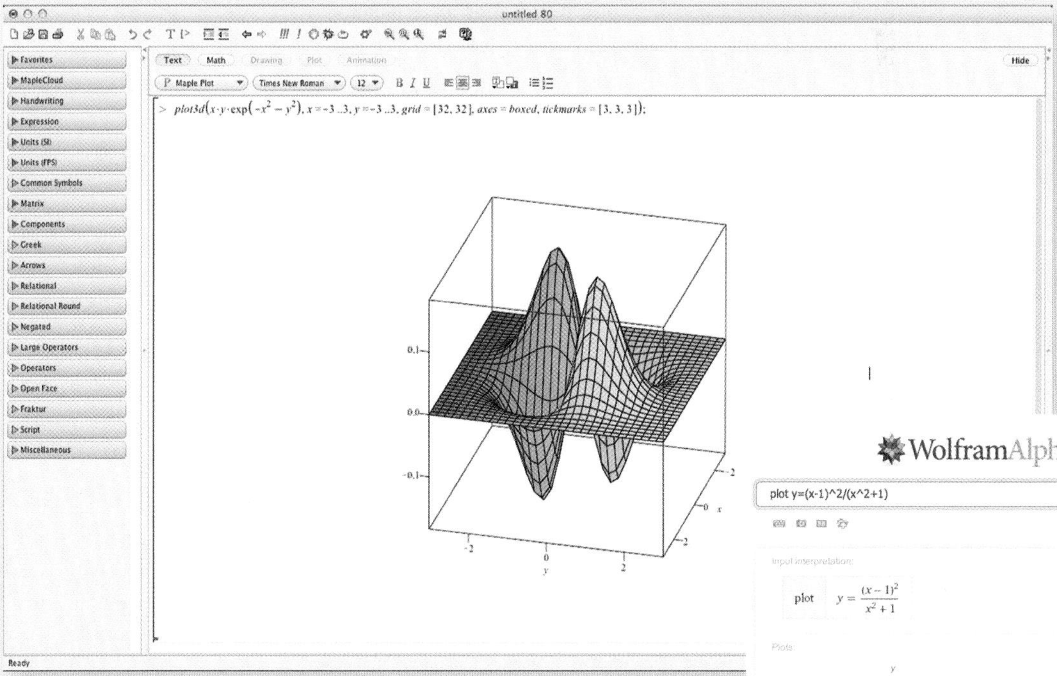

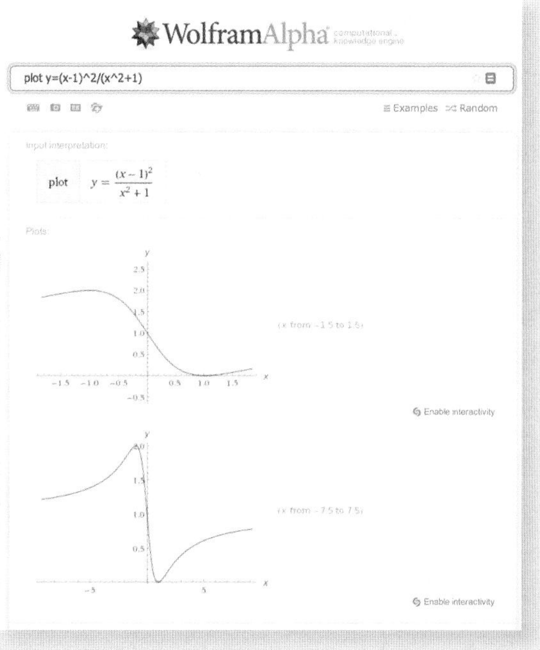

Diagnostic Tests

Success in calculus depends to a large extent on knowledge of the mathematics that precedes calculus. The following tests are intended to diagnose weaknesses that you might have. After taking each test you can check your answers against the given answers and, if necessary, refresh your skills by referring to the review materials that are provided.

A Diagnostic Test: Algebra

1. Evaluate each expression without using a calculator.

(a) $(-3)^4$ (b) -3^4 (c) 3^{-4}

(d) $\dfrac{5^{23}}{5^{21}}$ (e) $\left(\dfrac{2}{3}\right)^{-2}$ (f) $16^{-3/4}$

2. Simplify each expression. Write your answer without negative exponents.

(a) $\sqrt{200} - \sqrt{32}$

(b) $(3a^3b^3)(4ab^2)^2$

(c) $\left(\dfrac{3x^{3/2}y^3}{x^2y^{-1/2}}\right)^{-2}$

3. Expand and simplify.

(a) $3(x + 6) + 4(2x - 5)$ (b) $(x + 3)(4x - 5)$

(c) $\left(\sqrt{a} + \sqrt{b}\right)\left(\sqrt{a} - \sqrt{b}\right)$ (d) $(2x + 3)^2$

(e) $(x + 2)^3$

4. Factor each expression.

(a) $4x^2 - 25$ (b) $2x^2 + 5x - 12$

(c) $x^3 - 3x^2 - 4x + 12$ (d) $x^4 + 27x$

(e) $3x^{3/2} - 9x^{1/2} + 6x^{-1/2}$ (f) $x^3y - 4xy$

5. Simplify the rational expression.

(a) $\dfrac{x^2 + 3x + 2}{x^2 - x - 2}$ (b) $\dfrac{2x^2 - x - 1}{x^2 - 9} \cdot \dfrac{x + 3}{2x + 1}$

(c) $\dfrac{x^2}{x^2 - 4} - \dfrac{x + 1}{x + 2}$ (d) $\dfrac{\dfrac{y}{x} - \dfrac{x}{y}}{\dfrac{1}{y} - \dfrac{1}{x}}$

6. Rationalize the expression and simplify.

(a) $\dfrac{\sqrt{10}}{\sqrt{5}-2}$ (b) $\dfrac{\sqrt{4+h}-2}{h}$

7. Rewrite by completing the square.

(a) $x^2 + x + 1$ (b) $2x^2 - 12x + 11$

8. Solve the equation. (Find only the real solutions.)

(a) $x + 5 = 14 - \frac{1}{2}x$ (b) $\dfrac{2x}{x+1} = \dfrac{2x-1}{x}$

(c) $x^2 - x - 12 = 0$ (d) $2x^2 + 4x + 1 = 0$

(e) $x^4 - 3x^2 + 2 = 0$ (f) $3|x - 4| = 10$

(g) $2x(4-x)^{-1/2} - 3\sqrt{4-x} = 0$

9. Solve each inequality. Write your answer using interval notation.

(a) $-4 < 5 - 3x \leqslant 17$ (b) $x^2 < 2x + 8$

(c) $x(x-1)(x+2) > 0$ (d) $|x - 4| < 3$

(e) $\dfrac{2x-3}{x+1} \leqslant 1$

10. State whether each equation is true or false.

(a) $(p+q)^2 = p^2 + q^2$ (b) $\sqrt{ab} = \sqrt{a}\,\sqrt{b}$

(c) $\sqrt{a^2 + b^2} = a + b$ (d) $\dfrac{1 + TC}{C} = 1 + T$

(e) $\dfrac{1}{x-y} = \dfrac{1}{x} - \dfrac{1}{y}$ (f) $\dfrac{1/x}{a/x - b/x} = \dfrac{1}{a-b}$

ANSWERS TO DIAGNOSTIC TEST A: ALGEBRA

1. (a) 81 (b) -81 (c) $\frac{1}{81}$
(d) 25 (e) $\frac{9}{4}$ (f) $\frac{1}{8}$

2. (a) $6\sqrt{2}$ (b) $48a^5b^7$ (c) $\dfrac{x}{9y^7}$

3. (a) $11x - 2$ (b) $4x^2 + 7x - 15$
(c) $a - b$ (d) $4x^2 + 12x + 9$
(e) $x^3 + 6x^2 + 12x + 8$

4. (a) $(2x-5)(2x+5)$ (b) $(2x-3)(x+4)$
(c) $(x-3)(x-2)(x+2)$ (d) $x(x+3)(x^2-3x+9)$
(e) $3x^{-1/2}(x-1)(x-2)$ (f) $xy(x-2)(x+2)$

5. (a) $\dfrac{x+2}{x-2}$ (b) $\dfrac{x-1}{x-3}$
(c) $\dfrac{1}{x-2}$ (d) $-(x+y)$

6. (a) $5\sqrt{2} + 2\sqrt{10}$ (b) $\dfrac{1}{\sqrt{4+h}+2}$

7. (a) $\left(x + \frac{1}{2}\right)^2 + \frac{3}{4}$ (b) $2(x-3)^2 - 7$

8. (a) 6 (b) 1 (c) $-3, 4$
(d) $-1 \pm \frac{1}{2}\sqrt{2}$ (e) $\pm 1, \pm\sqrt{2}$ (f) $\frac{2}{3}, \frac{22}{3}$
(g) $\frac{12}{5}$

9. (a) $[-4, 3)$ (b) $(-2, 4)$
(c) $(-2, 0) \cup (1, \infty)$ (d) $(1, 7)$
(e) $(-1, 4]$

10. (a) False (b) True (c) False
(d) False (e) False (f) True

> If you had difficulty with these problems, you may wish to consult the Review of Algebra on the website **www.stewartcalculus.com**.

B | Diagnostic Test: Analytic Geometry

1. Find an equation for the line that passes through the point $(2, -5)$ and
(a) has slope -3
(b) is parallel to the x-axis
(c) is parallel to the y-axis
(d) is parallel to the line $2x - 4y = 3$

2. Find an equation for the circle that has center $(-1, 4)$ and passes through the point $(3, -2)$.

3. Find the center and radius of the circle with equation $x^2 + y^2 - 6x + 10y + 9 = 0$.

4. Let $A(-7, 4)$ and $B(5, -12)$ be points in the plane.
(a) Find the slope of the line that contains A and B.
(b) Find an equation of the line that passes through A and B. What are the intercepts?
(c) Find the midpoint of the segment AB.
(d) Find the length of the segment AB.
(e) Find an equation of the perpendicular bisector of AB.
(f) Find an equation of the circle for which AB is a diameter.

5. Sketch the region in the xy-plane defined by the equation or inequalities.
(a) $-1 \leqslant y \leqslant 3$
(b) $|x| < 4$ and $|y| < 2$
(c) $y < 1 - \frac{1}{2}x$
(d) $y \geqslant x^2 - 1$
(e) $x^2 + y^2 < 4$
(f) $9x^2 + 16y^2 = 144$

■ ANSWERS TO DIAGNOSTIC TEST B: ANALYTIC GEOMETRY

1. (a) $y = -3x + 1$ (b) $y = -5$
(c) $x = 2$ (d) $y = \frac{1}{2}x - 6$

2. $(x + 1)^2 + (y - 4)^2 = 52$

3. Center $(3, -5)$, radius 5

4. (a) $-\frac{4}{3}$
(b) $4x + 3y + 16 = 0$; x-intercept -4, y-intercept $-\frac{16}{3}$
(c) $(-1, -4)$
(d) 20
(e) $3x - 4y = 13$
(f) $(x + 1)^2 + (y + 4)^2 = 100$

5.

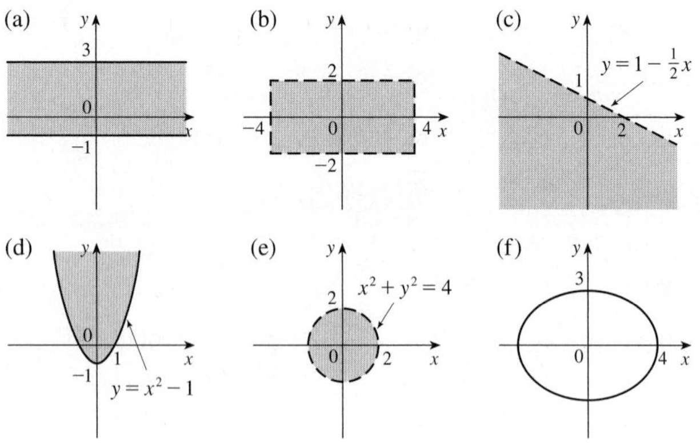

If you had difficulty with these problems, you may wish to consult the review of analytic geometry in Appendix B.

C | Diagnostic Test: Functions

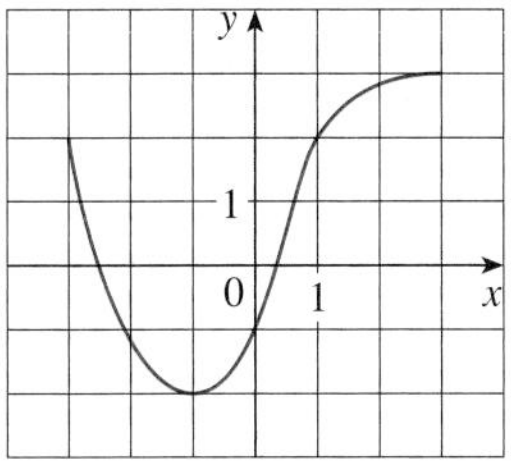

FIGURE FOR PROBLEM 1

1. The graph of a function f is given at the left.
  (a) State the value of $f(-1)$.
  (b) Estimate the value of $f(2)$.
  (c) For what values of x is $f(x) = 2$?
  (d) Estimate the values of x such that $f(x) = 0$.
  (e) State the domain and range of f.

2. If $f(x) = x^3$, evaluate the difference quotient $\dfrac{f(2 + h) - f(2)}{h}$ and simplify your answer.

3. Find the domain of the function.
  (a) $f(x) = \dfrac{2x + 1}{x^2 + x - 2}$ (b) $g(x) = \dfrac{\sqrt[3]{x}}{x^2 + 1}$ (c) $h(x) = \sqrt{4 - x} + \sqrt{x^2 - 1}$

4. How are graphs of the functions obtained from the graph of f?
  (a) $y = -f(x)$ (b) $y = 2f(x) - 1$ (c) $y = f(x - 3) + 2$

5. Without using a calculator, make a rough sketch of the graph.
  (a) $y = x^3$ (b) $y = (x + 1)^3$ (c) $y = (x - 2)^3 + 3$
  (d) $y = 4 - x^2$ (e) $y = \sqrt{x}$ (f) $y = 2\sqrt{x}$
  (g) $y = -2^x$ (h) $y = 1 + x^{-1}$

6. Let $f(x) = \begin{cases} 1 - x^2 & \text{if } x \leq 0 \\ 2x + 1 & \text{if } x > 0 \end{cases}$
  (a) Evaluate $f(-2)$ and $f(1)$. (b) Sketch the graph of f.

7. If $f(x) = x^2 + 2x - 1$ and $g(x) = 2x - 3$, find each of the following functions.
  (a) $f \circ g$ (b) $g \circ f$ (c) $g \circ g \circ g$

ANSWERS TO DIAGNOSTIC TEST C: FUNCTIONS

1. (a) -2 (b) 2.8
  (c) $-3, 1$ (d) $-2.5, 0.3$
  (e) $[-3, 3], [-2, 3]$

2. $12 + 6h + h^2$

3. (a) $(-\infty, -2) \cup (-2, 1) \cup (1, \infty)$
  (b) $(-\infty, \infty)$
  (c) $(-\infty, -1] \cup [1, 4]$

4. (a) Reflect about the x-axis
  (b) Stretch vertically by a factor of 2, then shift 1 unit downward
  (c) Shift 3 units to the right and 2 units upward

5.

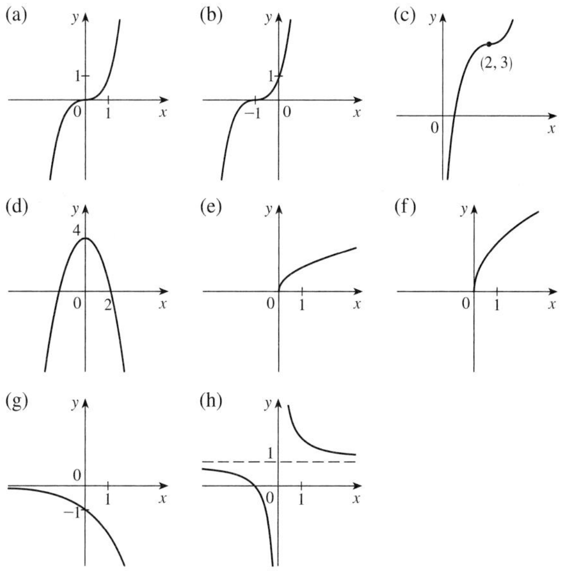

6. (a) $-3, 3$

(b)

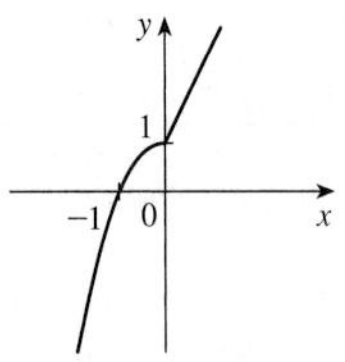

7. (a) $(f \circ g)(x) = 4x^2 - 8x + 2$

(b) $(g \circ f)(x) = 2x^2 + 4x - 5$

(c) $(g \circ g \circ g)(x) = 8x - 21$

> If you had difficulty with these problems, you should look at sections 1.1–1.3 of this book.

D | Diagnostic Test: Trigonometry

1. Convert from degrees to radians.

(a) $300°$ (b) $-18°$

2. Convert from radians to degrees.

(a) $5\pi/6$ (b) 2

3. Find the length of an arc of a circle with radius 12 cm if the arc subtends a central angle of $30°$.

4. Find the exact values.

(a) $\tan(\pi/3)$ (b) $\sin(7\pi/6)$ (c) $\sec(5\pi/3)$

5. Express the lengths a and b in the figure in terms of θ.

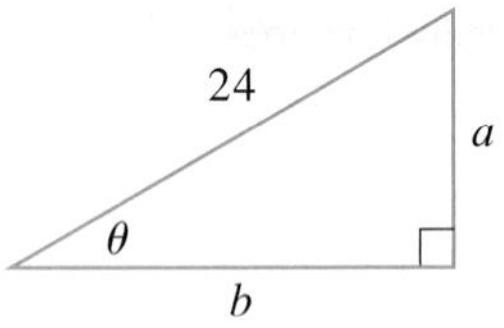

FIGURE FOR PROBLEM 5

6. If $\sin x = \frac{1}{3}$ and $\sec y = \frac{5}{4}$, where x and y lie between 0 and $\pi/2$, evaluate $\sin(x + y)$.

7. Prove the identities.

(a) $\tan\theta \sin\theta + \cos\theta = \sec\theta$ (b) $\dfrac{2\tan x}{1 + \tan^2 x} = \sin 2x$

8. Find all values of x such that $\sin 2x = \sin x$ and $0 \leq x \leq 2\pi$.

9. Sketch the graph of the function $y = 1 + \sin 2x$ without using a calculator.

ANSWERS TO DIAGNOSTIC TEST D: TRIGONOMETRY

1. (a) $5\pi/3$ (b) $-\pi/10$

2. (a) $150°$ (b) $360°/\pi \approx 114.6°$

3. 2π cm

4. (a) $\sqrt{3}$ (b) $-\frac{1}{2}$ (c) 2

5. (a) $24\sin\theta$ (b) $24\cos\theta$

6. $\frac{1}{15}\left(4 + 6\sqrt{2}\right)$

8. $0, \pi/3, \pi, 5\pi/3, 2\pi$

9.

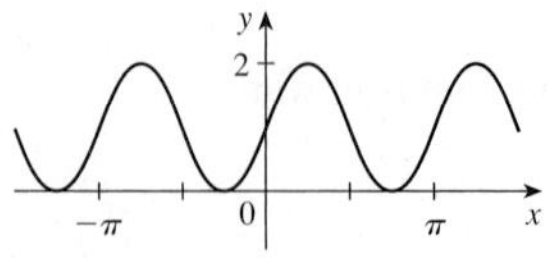

> If you had difficulty with these problems, you should look at Appendix C of this book.

Prologue: Mathematics and Biology

Galileo was keenly aware of the role of mathematics in the study of nature. In 1610 he famously wrote:

> Philosophy [Nature] is written in that great book which ever lies before our eye—I mean the universe—but we cannot understand it if we do not first learn the language and grasp the symbols in which it is written. The book is written in the language of mathematics and the symbols are triangles, circles, and other geometrical figures, without whose help it is impossible to comprehend a single word of it; without which one wanders in vain through a dark labyrinth.[1]

Indeed, in the seventeenth and later centuries Newton and other scientists employed mathematics in trying to explain physical phenomena. First physics and astronomy, and later chemistry, were investigated with the methods of mathematics. Most of the applications of mathematics to biology, however, occurred much later.

A connection between mathematics and biology that was noticed at an early stage was phyllotaxy, which literally means leaf arrangement. For some trees, such as the elm, the leaves occur alternately, on opposite sides of a branch, and we refer to $\frac{1}{2}$ phyllotaxis because the next leaf is half of a complete turn (rotation) beyond the first one. For beech trees each leaf is a third of a turn beyond the preceding one and we have $\frac{1}{3}$ phyllotaxis. Oak trees exhibit $\frac{2}{5}$ phyllotaxis, poplar trees $\frac{3}{8}$ phyllotaxis, and willow trees $\frac{5}{13}$ phyllotaxis. These fractions

$$\frac{1}{2} \quad \frac{1}{3} \quad \frac{2}{5} \quad \frac{3}{8} \quad \frac{5}{13} \quad \ldots$$

are related to the Fibonacci numbers

$$1 \quad 1 \quad 2 \quad 3 \quad 5 \quad 8 \quad 13 \quad 21 \quad 34 \quad \ldots$$

which we will study in Section 1.6. Each of the Fibonacci numbers is the sum of the two preceding numbers. Notice that each of the phyllotaxis fractions is a ratio of Fibonacci numbers spaced two apart. It has been suggested that the adaptive advantage of this arrangement of leaves comes from maximizing exposure to sunlight and rainfall.

The Fibonacci numbers also arise in other botanical examples of phyllotaxis: the spiral patterns of the florets of a sunflower, the scales of a fir cone, and the hexagonal cells of a pineapple. Shown are three types of spirals on a pineapple: 5 spirals sloping up gradually to the right, 8 spirals sloping up to the left, and 13 sloping up steeply.

5 parallel spirals 8 parallel spirals 13 parallel spirals

1. Galileo Galilei, *Le Opere di Galileo Galilei*, Edizione Nationale, 20 vols., ed. Antonio Favaro (Florence: G. Barbera, 1890–1909; reprinted 1929–39, 1964–66), vol. 4, p. 171.

Another early application of mathematics to biology was the study of the spread of smallpox by the Swiss mathematician Daniel Bernoulli in the 1760s. Bernoulli formulated a mathematical model of an epidemic of an infectious disease in the form of a differential equation. (Such equations will be studied in Chapter 7.) In particular, Bernoulli showed that, under the assumptions of his model, life expectancy would increase by more than three years if the entire population were inoculated at birth for smallpox. His work was the start of the field of mathematical epidemiology, which we will explore extensively in this book.

Aside from a few such instances, however, mathematical biology was slow to develop, probably because of the complexity of biological structures and processes. In the last few decades, however, the field has burgeoned. In fact, Ian Stewart has predicted that "Biology will be the great mathematical frontier of the twenty-first century."[2]

Already the scope of mathematical applications to biology is enormous, having led to important insights that have revolutionized our understanding of biological processes and spawned new fields of study. These successes have reached the highest levels of scientific recognition, resulting in Nobel Prizes to Ronald Ross in 1902 for his work on malaria transmission dynamics, to Alan Lloyd Hodgkin and Andrew Fielding Huxley in 1963 for their work on the transmission of nerve impulses, and to Alan Cormack and Godfrey Hounsfield in 1979 for the development of the methodology behind the now-common medical procedure of CAT scans. You will learn some of the mathematics behind each of these fundamental discoveries throughout this book.

Perhaps even more telling of the importance of mathematics to modern biology is the breadth of biological areas to which mathematics contributes. For example, mathematical analyses are central to our understanding of disease, from the function of immune molecules like natural killer cells and the occurrence of autoimmune diseases like lupus, to the spread of drug resistance. Likewise, modern medical treatments and techniques, from drug pharmacokinetics and dialysis, to the lung preoxygenation and hemodilution techniques used for surgery, have all been developed through the use of mathematical models.

The reach of mathematics in modern biology extends far beyond medicine, however, and is fundamental to virtually all areas of biology. Mathematical models and analyses are now routinely used in the study of physiology, from the growth and morphological structure of organisms, to photosynthesis, to the emergence of ordered patterns during cell division, to the dynamics of cell cycles and genome expression. Mathematics is used to understand organism movement, from humans to jellyfish, and to understand population and ecological processes, as well as the roles of habitat destruction and harvesting in the conservation of endangered species.

All of these applications are just a few of those explored in this book (a complete list can be found at the back of the book). But this book is just the beginning of the story. Modern biology and mathematics are now connected by a two-way street, with biological phenomena providing the impetus for advanced mathematical and computational analyses that go well beyond introductory calculus, probability, and statistics. High-tech research companies like Microsoft now have computational biology departments that examine the parallels between biological systems and computation. And these, in turn, are providing critical insight into a broad array of questions. From the dramatic failure and subsequent discontinuation of the breast cancer drug bevacizumab (Avastin) in 2011,[3] to the very nature of life itself, mathematics and biology are now moving for-

---

2. I. Stewart, *The Mathematics of Life* (New York: Basic Books, 2011).
3. N. Savage, "Computing Cancer," *Nature* (2012) 491: S62.

ward hand in hand. Techniques in advanced geometry are being developed to quantify similarities between different biological patterns, from electrical impulses in the neural cortex, to peptide sequences and patterns of protein folding. And these analyses have very close mathematical connections to other kinds of pattern matching as well, including those used by Web search engines like Google. Likewise, seemingly abstract topics from advanced algebra are being used in the statistical analysis of the reams of DNA sequence data that are now available and such biological questions are, in turn, reinvigorating these abstract areas of mathematics.[4]

This textbook provides the first steps into this exciting and fast-moving area that combines mathematics with biology. As motivation for our studies, we conclude this prologue with a brief description of some of the areas of application that will be covered.

Calculus and Biology

Living organisms change: they move, they grow, they reproduce. Calculus can be regarded as the mathematics of change. So it is natural that calculus plays a major role in mathematical biology. The following highlighted examples of applications are some of the recurring themes throughout the book. As we learn more calculus, we repeatedly return to these topics with increasing depth.

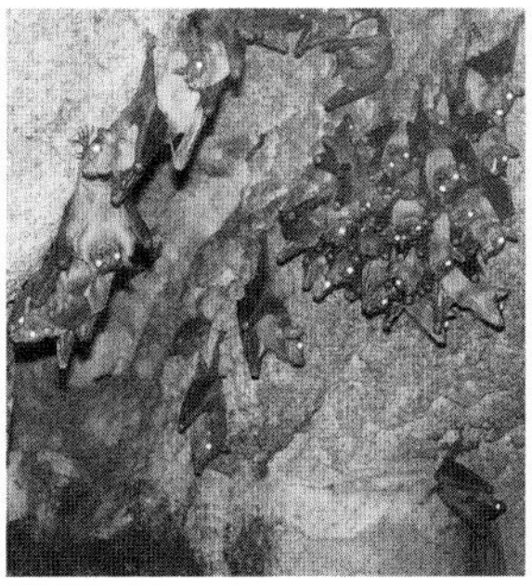

■ Species Richness

It seems reasonable that the larger the area of a region, the larger will be the number of species that inhabit that region. To make scientific progress, however, we need to describe this relationship more precisely. Can we describe such species–area relationships mathematically, and can we use mathematics to better understand the processes that give rise to these patterns?

In Examples 1.2.6 and 1.5.14 we show that the species–area relation for bats in Mexican caves is well modeled using functions called power functions. Later, in Exercise 3.3.48, we show the same is true for tree species in Malaysian forests and then use the model to determine the rate at which the number of species grows as the area increases. When we study differential equations in Chapter 7, we show how assumptions about rates of increase of species lead naturally to such power-function models. In Example 4.2.5 we also see, however, that for very large areas the power-function model is no longer appropriate.

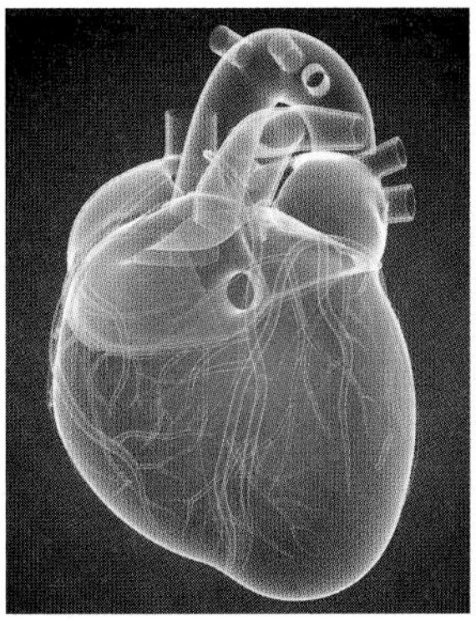

■ Vectorcardiography

Heartbeat patterns can be used to diagnose a variety of different medical conditions. These patterns are usually recorded by measuring the electrical potential on the surface of the body using several (often 12) wires, or "leads." How can we use the measurements from these leads to diagnose heart problems?

In Section 1.1 and Example 4.1.4 we introduce the idea of using functions to describe heartbeats. We then consider, in Exercises 4.1.5–6, how the shapes of their graphs are diagnostic of different heart conditions. In Chapter 8 we introduce vectors and show how the direction of the voltage vector created by a heartbeat can be measured with ECG leads using the dot product (Example 8.3.7) and how this can be used to diagnose spe-

4. L. Pachter and B. Sturmfels, *Algebraic Statistics for Computational Biology* (Cambridge: Cambridge University Press, 2005).

cific heart conditions (Exercises 8.2.39, 8.3.40, and 8.7.7). We also show how the techniques of matrix algebra can be used to model the change in the heartbeat voltage vector (Exercises 8.5.16, 8.6.30, and 8.6.35).

■ Drug and Alcohol Metabolism

Biomedical scientists study the chemical and physiological changes that result from the metabolism of drugs and alcohol after consumption. How does the level of alcohol in the blood vary over time after the consumption of a drink, and can we use mathematics to better understand the processes that give rise to these patterns?

In Exercise 1.1.26 we present some data that we use to sketch the graph of the blood alcohol concentration (BAC) function, illustrating the two stages of the reaction in the human body: absorption and metabolism. In Exercises 1.4.34 and 1.5.69 we model the second stage with a decaying exponential function to determine when the BAC will be less than the legal limit. In Chapter 3 we model the entire two-stage process with a surge function and use it to estimate the rate of increase of the BAC in the first stage and the rate of decrease in the second stage (Exercise 3.5.59). Later we find the maximum value of the BAC (Example 4.1.7), the limiting value (Example 4.3.9), and the average value (Exercise 6.2.16).

■ Population Dynamics

One of the central goals of population biology and ecology is to describe the abundance and distribution of organisms and species over time and space. Can we use mathematical models to describe the processes that alter these abundances, and can these models then be used to predict population sizes?

In Section 1.1 we begin by using different representations of functions to describe the human population. Section 1.4 then illustrates how exponential functions can be used to model population change, from humans to malaria. Section 1.6 introduces recursion equations, which are fundamental tools used to study population dynamics. Several examples and exercises in Chapters 3 and 4 use calculus to show how derivatives of functions can tell us important information about the rate of growth of populations, while Chapters 5 and 6 illustrate how integration can be used to quantify the size of populations. Chapters 7, 8 and 10 then use differential equations and techniques from matrix algebra to model populations and show that populations can even exhibit chaotic behavior (see the project on page 430).

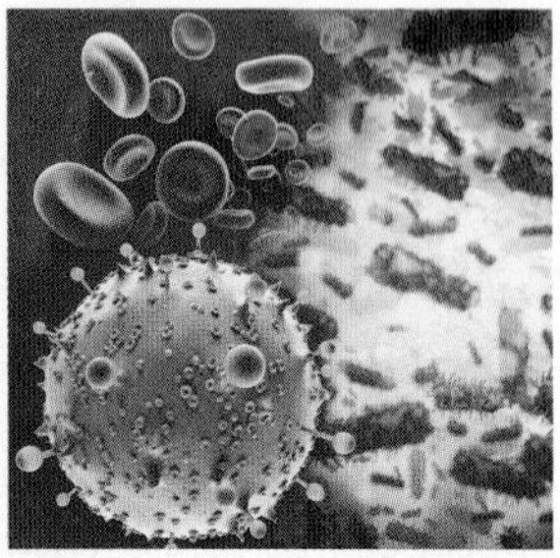

■ Antigenic Cartography and Vaccine Design

Cartography is the study of mapmaking. "Antigenic cartography" involves making maps of the antigenic properties of viruses. This allows us to better understand the changes that occur from year to year in viruses such as influenza. How can we describe these changes? Why is it that flu vaccines need to be updated periodically because of vaccine escape, and can we use mathematics to understand this process and to design new vaccines?

In Exercises 4.1.7 and 4.1.8 we use calculus to explore the epidemiological consequences of the antigenic change that occurs during an influenza pandemic. In the project on page 479 we model these processes using differential equations and determine the vaccine coverage needed to prevent an outbreak. Chapter 8 introduces the ideas of vectors and the geometry of higher-dimensional space and uses them in antigenic cartography (Examples 8.1.3, 8.1.6, and 8.1.8 and Exercise 8.1.39) and in vaccine design (Exercise 8.1.38). Vectors are then used to quantify antigenic evolution in Example 8.2.1 and Exercises 8.2.46, 8.3.37, 8.5.17, 8.6.31, and differential equations are used in the project on page 514 to understand vaccine escape.

© Maridav / Shutterstock.com

Biomechanics of Human Movement

When you walk, the horizontal force that the ground exerts on you is a function of time. Understanding human movement, and the energetic differences between walking, running, and other animal gaits, like galloping, requires an understanding of these forces. Can we quantify these processes using mathematical models?

The description of these forces when you are walking is investigated in Exercises 1.1.16 and 3.2.14. If you now start walking faster and faster and then begin to run, your gait changes. The metabolic power that you consume is a function of your speed and this is explored in Examples 1.1.10 and 3.2.7. In the project on page 40 we use trigonometric functions of time to model the vertical force that you exert on the ground with different gaits. In Chapter 8 we then introduce a three-dimensional coordinate system, enabling us to analyze the trajectory of the center of a human walking on a treadmill. Vectors are introduced in Section 8.2 and so we can then talk about the force vectors, such as those that sprinters exert on starting blocks (Example 8.2.6 and Exercise 8.2.38).

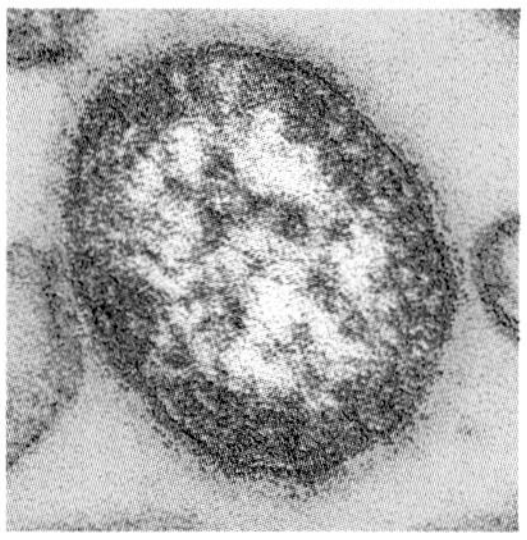
Scott Camazine / Alamy

Measles Pathogenesis

Infection with the measles virus results in symptoms and viral transmission in some patients and not in others. What causes these different outcomes, and can we predict when each is expected to occur?

The level of the measles virus in the bloodstream of a patient with no immunity peaks after about two weeks and can be modeled using a third-degree polynomial (Exercise 4.4.8). The area under this curve for the first 12 days turns out to be the total amount of infection needed for symptoms to develop (see the heading *Pathogenesis* on page 325 and Exercises 5.1.9 and 5.3.45). In the project on page 394 we consider patients with partial immunity, and by evaluating areas between curves we are able to decide which patients will be symptomatic and infectious (or noninfectious), as well as those who will be asymptomatic and noninfectious.

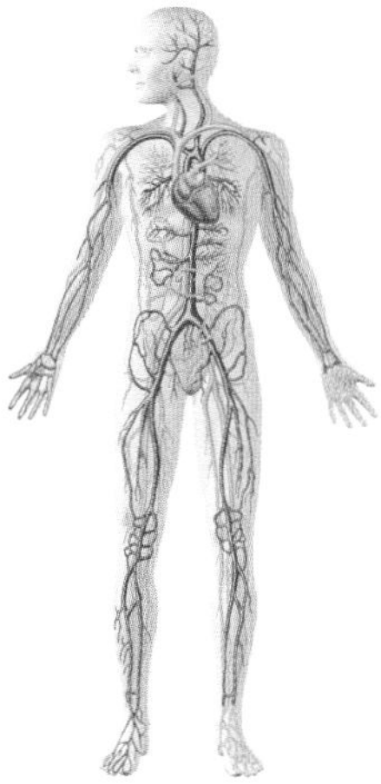
© leonello calvetti / Shutterstock.com

Blood Flow

The heart pumps blood through a series of interconnected vessels in your body. Several medical problems involve abnormal blood pressure and flow. Can we predict blood pressure and flow as a function of various physiological characteristics?

In Example 3.3.9 and Exercises 3.3.49 and 3.5.92 we use Poiseuille's law of laminar flow to calculate the rate at which the velocity of blood flow in arteries changes with respect to the distance from the center of the artery and with respect to time. In Exercise 6.3.10 we show how blood pressure depends on the radius of an artery. In the section *Cerebral Blood Flow* on page 390 we explain the Kety-Schmidt method, which is a diagnostic technique for measuring cerebral blood flow using inhaled nitrous oxide as a tracer. This method depends on knowing the area between two curves representing the concentration of nitrous oxide as blood enters the brain and the concentration as blood leaves the brain in the jugular vein. (See Example 6.1.4 and Exercises 6.1.21–22.)

© Vlad61 / Shutterstock.com

■ Conservation Biology

Human impacts arising from natural resource extraction and pollution are having devastating effects on many ecosystems. It is crucial that we be able to forecast these effects in order to better manage our impact on the environment. Mathematics is playing a central role in this endeavor.

Exercise 3.1.41 shows how derivatives can be used to study thermal pollution, while Exercises 3.5.91 and 3.8.43 use derivatives to determine the effect of habitat fragmentation on population dynamics. The project on page 239, as well as Example 4.4.5 and Exercises 4.4.21 and 4.5.21, use derivatives to explore the effect of harvesting on population sustainability. The project on page 298 then extends these ideas with an introduction to game theory. In Exercises 7.4.32–34 and Section 10.3 we use differential equations to model the effects of habitat destruction and pollution, while in Example 8.5.1 and Exercise 8.5.22 techniques from matrix algebra are used to model the conservation biology of right whales and spotted owls, respectively. The stability of coral reef ecosystems is explored using differential equations in Exercise 10.4.34.

Probability, Statistics, and Biology

The mathematical tools of probability and statistics (both of which rely on calculus) are also fundamental to many areas of modern biology. Many biological processes—like species extinctions, the inheritance of genetic diseases, and the likelihood of success of medical procedures—involve aspects of chance that can be understood only with the use of probability theory. Furthermore, the statistical analysis of data forms the basis of all of science, including biology, and the tools of statistics are rooted in calculus and probability theory. Although this book is not the place for a thorough treatment of statistics, you will be introduced to some of the central concepts of the subject in Chapters 11 and 13.

© Rena Schild / Shutterstock.com

Performance-enhancing Drugs

Erythropoietin (EPO) is a hormone that stimulates red blood cell production. Synthetic variants of EPO are sometimes used by athletes in an attempt to increase aerobic capacity during competition. How effective is EPO at increasing performance?

In Exercise 11.1.19 we summarize data for the performance of athletes both before and after they have been given EPO, using various summary statistics. In Exercises 11.3.7 and 11.3.18 we then explore these data graphically. After learning some probability theory in Chapter 12, we can then begin to analyze the effects of EPO more rigorously using statistical techniques. Examples 13.3.2, 13.3.3, and 13.3.6 illustrate how we can use these techniques to test the hypothesis that EPO alters athletic performance.

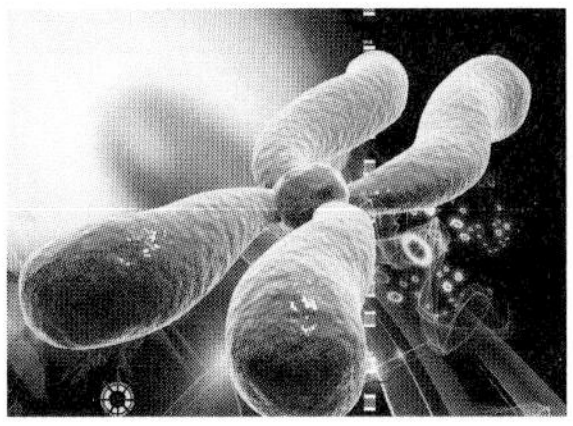

DNA Supercoiling

When DNA is packaged into chromosomes, it is often coiled and twisted to make it more compact. This is called supercoiling. Some of these coils are very dynamic, repeatedly forming and disappearing at different locations throughout the genome. What causes this process?

One hypothesis is that the coils form and disappear randomly over time, as a result of chance twisting and untwisting of the DNA. To explore whether this hypothesis provides a reasonable explanation, we need to determine the pattern of supercoiling that it would cause. In Chapter 12 we introduce the necessary ideas of probability theory to model this process. The project after Section 12.4 then uses these ideas to model the random twisting and untwisting of supercoils. You will see that the available supercoiling data match the model predictions remarkably well.

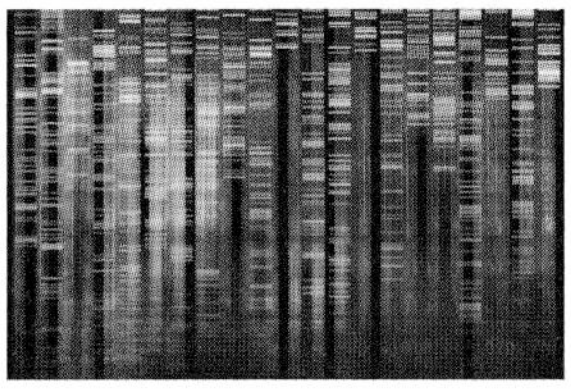

Huntington's Disease

Huntington's disease is a genetic disorder causing neurodegeneration and eventual death. Symptoms typically appear in a person's thirties and death occurs around 20 years after the onset of symptoms. What causes the variability in the age of onset, and how likely are you to inherit this disease if one of your parents has it?

In Exercise 11.1.14 we summarize data for the age of onset, and Exercises 11.2.15 and 11.2.29 explore the data graphically. Exercises 13.1.14 and 13.1.23 then use so-called "normal curves" to estimate the fraction of cases having different ages of onset. In Exercises 13.2.7 and 13.3.7 we use confidence intervals and hypothesis testing, respectively, to better understand the mean age of onset. Exercises 11.3.14 and 11.3.20 use statistical techniques to explore how the age of onset is related to different DNA sequences, and Examples 12.3.3 and 12.3.9 illustrate how probability theory can be used to predict the likelihood of a child inheriting the disease from its parents.

Case Studies in Mathematical Modeling

A **mathematical model** is a mathematical description (often by means of a function or an equation) of a real-world phenomenon, such as the size of a population, the speed of a falling object, the frequency of a particular gene, the concentration of an antibiotic in a patient, or the life expectancy of a person at birth. The purpose of the model is to understand the phenomenon and perhaps to make predictions about future behavior.

Figure 1 illustrates the process of mathematical modeling. Given a real-world problem, the first task is to formulate a mathematical model by identifying and naming the relevant quantities and making assumptions that simplify the phenomenon enough to make it mathematically tractable. We use our knowledge of the biological situation and our mathematical skills to obtain equations that relate the quantities. In situations where there is no physical law to guide us, we may need to collect data (either from a library or the Internet or by conducting our own experiments) and examine the data to discern patterns.

The second stage is to apply the mathematics that we know (such as the calculus that will be developed throughout this book) to the mathematical model that we have formulated in order to derive mathematical conclusions. Then, in the third stage, we take those mathematical conclusions and interpret them as information about the original biological phenomenon by way of offering explanations or making predictions. The final step is to test our predictions by checking them against new real data. If the predictions don't compare well with reality, we need to refine our model or to formulate a new model and start the cycle again.

A mathematical model is never a completely accurate representation of a physical situation—it is an idealization. Picasso once said that "art is a lie that makes us realize truth." The same could be said about mathematical models. A good model simplifies reality enough to permit mathematical calculations, but is nevertheless realistic enough to teach us something important about the real world. Because models are simplifications, however, it is always important to keep their limitations in mind. In the end, Mother Nature has the final say.

Throughout this book we will explore a variety of different mathematical models from the life sciences. In each case we provide a brief description of the real-world problem as well as a brief mention of the real-world predictions that result from the mathematical analysis. Nevertheless, the main body of this text is designed to teach important mathematical concepts and techniques and therefore its focus is primarily on the center portion of Figure 1.

To better illustrate the entirety of the modeling process, however, we also provide a pair of *case studies in mathematical modeling.* Each case study is an extended, self-contained example of mathematical modeling from the scientific literature. In the following pages the real-world problem at the center of each case study is introduced as motivation for learning the mathematics in this book. Then, throughout subsequent chapters, these case studies are periodically revisited as we develop our mathematical skills further. In doing so, we illustrate how these mathematical skills help to address real-world problems. Additional case studies can be found on the website www.stewartcalculus.com.

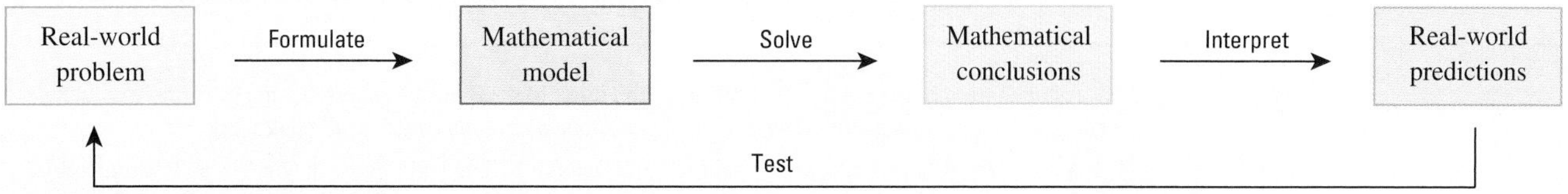

FIGURE 1 The modeling process

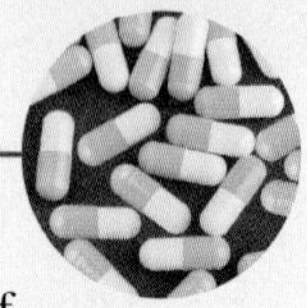

Antibiotics are often prescribed to patients who have bacterial infections. When a single dose of antibiotic is taken, its concentration at the site of infection initially increases very rapidly before slowly decaying back to zero as the antibiotic is metabolized.[1] The curve shown in Figure 1 illustrates this pattern and is referred to as the *antibiotic concentration profile*.

The clinical effectiveness of an antibiotic is determined not only by its concentration profile but also by the effect that any given concentration has on the growth rate of the bacteria population. This effect is characterized by a *dose response relationship*, which is a graph of the growth rate of the bacteria population as a function of antibiotic concentration. Bacteria typically grow well under low antibiotic concentrations, but their growth rate becomes negative (that is, their population declines) if the antibiotic concentration is high enough. Figure 2 shows an example of a dose response relationship.[2]

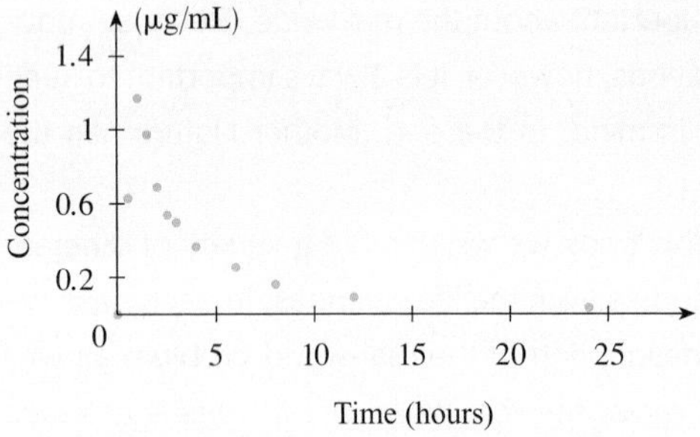

FIGURE 1
Antibiotic concentration profile in plasma of a healthy human volunteer after receiving 500 mg of ciprofloxacin

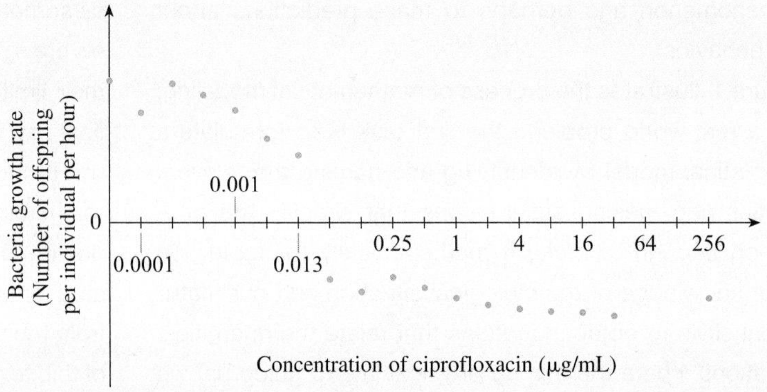

FIGURE 2
Dose response relationship for ciprofloxacin with the bacteria *E. coli*

Together, the antibiotic concentration profile and the dose response relationship determine how the bacteria population size changes over time. When the antibiotic is first administered, the concentration at the site of infection will be high and therefore the growth rate of the bacteria population will be negative (the population will decline). As the antibiotic concentration decays, the growth rate of the bacteria population eventually changes from negative to positive and the bacteria population size then rebounds. The plot of the bacteria population size as a function of time after the antibiotic is given is called the *kill curve*. An example is shown in Figure 3.

To determine how much antibiotic should be used to treat an infection, clinical researchers measure kill curves for different antibiotic doses. Figure 4 presents a family of such curves: Notice that as the dose of antibiotic increases, the bacteria population tends to decline to lower levels and to take longer to rebound.

When developing new antibiotics, clinical researchers summarize kill curves like those in Figure 4 into a simpler form to see more clearly the relationship between the

1. Adapted from S. Imre et al., "Validation of an HPLC Method for the Determination of Ciprofloxacin in Human Plasma," *Journal of Pharmaceutical and Biomedical Analysis* 33 (2003): 125–30.

2. Adapted from A. Firsov et al., "Parameters of Bacterial Killing and Regrowth Kinetics and Antimicrobial Effect Examined in Terms of Area under the Concentration-Time Curve Relationships: Action of Ciprofloxacin against *Escherichia coli* in an In Vitro Dynamic Model," *Antimicrobial Agents and Chemotherapy* 41 (1997): 1281–87.

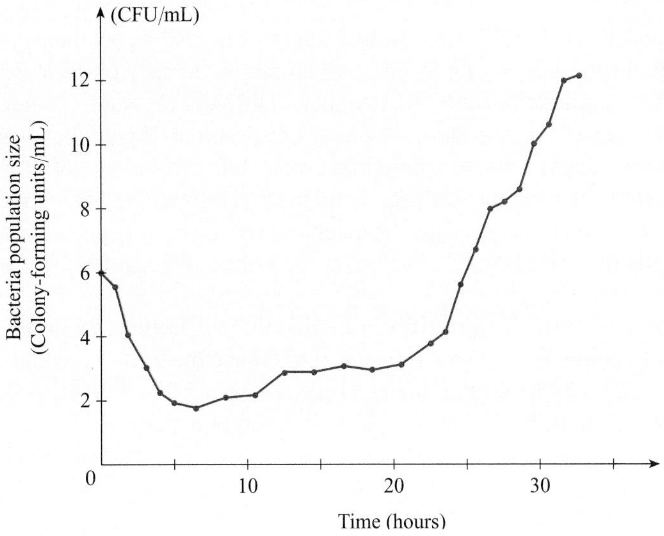

FIGURE 3
The kill curve of ciprofloxacin for *E. coli* when measured in a growth chamber. A dose corresponding to a concentration of 0.6 μg/mL was given at $t = 0$.

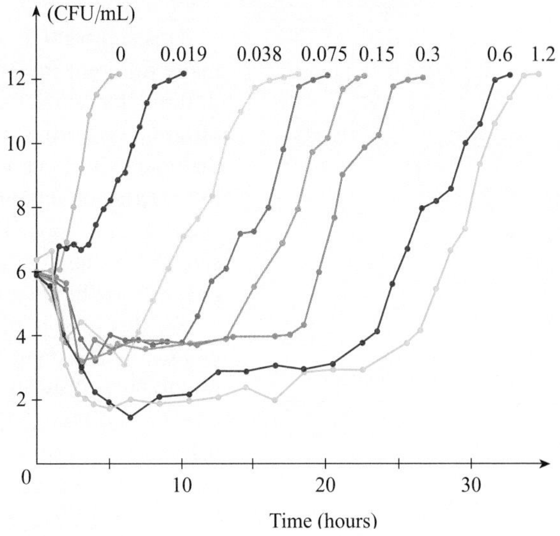

FIGURE 4
The kill curves of ciprofloxacin for *E. coli* when measured in a growth chamber. The concentration of ciprofloxin at $t = 0$ is indicated above each curve (in μg/mL).

magnitude of antibiotic treatment and its effectiveness. This is done by obtaining both a measure of the magnitude of antibiotic treatment, from the antibiotic concentration profile underlying each kill curve, and a measure of the killing effectiveness, from the kill curve itself. These measures are then plotted on a graph of killing effectiveness against the magnitude of antibiotic treatment.

As an example, Figure 5 plots the magnitude of the drop in population size before the rebound occurs (a measure of killing effectiveness) against the peak antibiotic concentration (a measure of the magnitude of antibiotic treatment). Each of the eight colored points corresponds to the associated kill curve in Figure 4. (Peak concentration is measured in dimensionless units, as will be explained in Case Study 1a.) The points indicate that, overall, as the peak concentration increases, the magnitude of the drop in population size increases as well. This relationship can then be used by the researchers to choose an antibiotic dose that gives the peak concentration required to kill the bacterial infection.

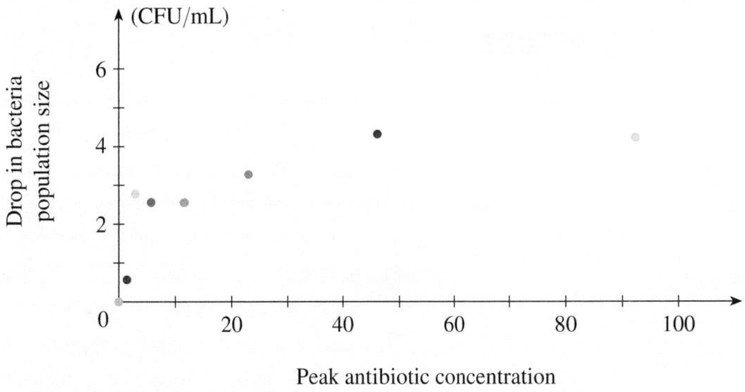

FIGURE 5

This approach for choosing a suitable antibiotic dose may seem sensible, but there are many different measures for the killing effectiveness of an antibiotic, as well as many different measures for the magnitude of antibiotic treatment. Different measures capture different properties of the bacteria–antibiotic interaction. For example, Figure 4 shows that many different antibiotic doses produce approximately the same magnitude of drop in bacteria population despite the fact that the doses result in large differences in the time necessary for population rebound to occur. Thus the magnitude of the drop in population size before rebound occurs does not completely capture the killing effectiveness of the different antibiotic doses.

For this reason, researchers typically quantify antibiotic killing effectiveness in several ways. The three most common are (1) the time taken to reduce the bacteria population to 90% of its initial value, (2) the drop in population size before rebound occurs, as was used in Figure 5, and (3) a measure that combines the drop in population size and the duration of time that the population size remains small (because effective treatment not only produces a large drop in bacteria population but maintains the population at a low level for a long period of time).

Similarly, there are many measures for the magnitude of antibiotic treatment. The most commonly used measures include (1) peak antibiotic concentration, as was used in Figure 5, (2) duration of time for which the antibiotic concentration is high enough to cause negative bacteria growth, and (3) a measure that combines both peak concentration and duration of time that the concentration remains high.

The conclusions clinical researchers obtain about suitable antibiotic doses can differ depending on which measures are used. For example, Figure 6 shows the relationship between the time taken to reduce the bacteria population to 90% of its initial value plotted against the same measure of peak antibiotic concentration as was used in Figure 5 for the kill curves shown in Figure 4. Unlike Figure 5, Figure 6 shows no consistent relationship between effectiveness (as measured by the speed of the population decline) and strength of treatment.

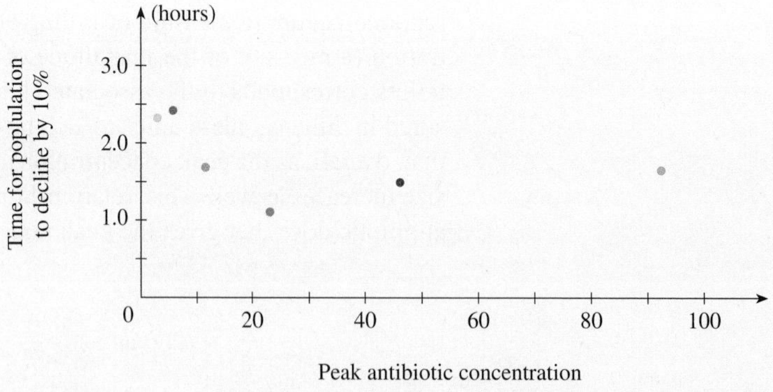

FIGURE 6

To use appropriate measures to formulate effective antibiotic doses, we therefore need to understand what determines the shape of the relationships between measures, and when and why these relationships will differ depending on the measures used. This is where mathematical modeling can play an important role: By modeling the biological processes involved, we can better understand what drives the different patterns, and we can then use models to make predictions about what we expect to observe in other situations. Making such predictions is the goal of this case study.

The order in which mathematical tools are used by researchers is not always the same as the order in which they are best learned. For example, when analyzing the problem in this case study, researchers would first use techniques from Chapter 3 and then Chapter 6 to model the dynamics of the drug and bacteria and to quantity the strength of treatment and effectiveness of killing. They would then analyze these models using the techniques of Chapters 1 and 2.

For our learning objectives, however, this case study will be developed in the opposite order: In Case Study 1a we will use a given model for the effect of antibiotics on bacteria growth to draw conclusions about the differences in the relationships shown in Figures 5 and 6. In Case Study 1b we will begin to fill in the gaps by deriving the model used in Case Study 1a. In Case Study 1c we will continue to fill in gaps from Case Study 1a by deriving different measures for the magnitude of antibiotic treatment. We will also show how a process called dose fractionation can be used to alter various aspects of these measures. Finally, in Case Study 1d we will use the model derived in Case Study 1b to make new predictions about the effectiveness of antibiotics and compare these predictions to data.

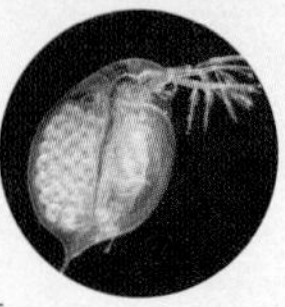

By definition, a parasite has an antagonistic relationship with the host it infects. For this reason we might expect the host to evolve strategies that resist infection, and the parasite to evolve strategies that subvert this host resistance. The end result might be a never-ending coevolutionary cycle between host and parasite, with neither party gaining the upper hand. Indeed, we might expect the ability of the parasite to infect the host to remain relatively unchanged over time despite the fact that both host and parasite are engaged in cycles of evolutionary conflict beneath this seemingly calm surface.

This is an intriguing idea, but how might it be examined scientifically? Ideally we would like to hold the parasite fixed in time and see if its ability to infect the host declines as the host evolves resistance. Alternatively, we might hold the host fixed in time and see if the parasite's ability to infect the host increases as it evolves ways to subvert the host's current defenses.

Another possibility would be to challenge the host with parasites from its evolutionary past. In this case we might expect the host to have the upper hand, since it will have evolved resistance to these ancestral parasites. Similarly, if we could challenge the host with parasites from its evolutionary future, then we might expect the parasite to have the upper hand, since it will have evolved a means of subverting the current host defenses.

Exactly this sort of "time-travel" experiment has been done using a bacterium as the host and a parasite called a *bacteriophage*.[1] To do so, researchers let the host and parasite coevolve together for several generations. During this time, they periodically took samples of both the host and the parasite and placed the samples in a freezer. After several generations they had a frozen archive of the entire temporal sequence of hosts and parasites. The power of their approach is that the host and parasite could then be resuscitated from this frozen state. This allowed the researchers to resuscitate hosts from one point in time in the sequence and then challenge them with resuscitated parasites from their past, present, and future.

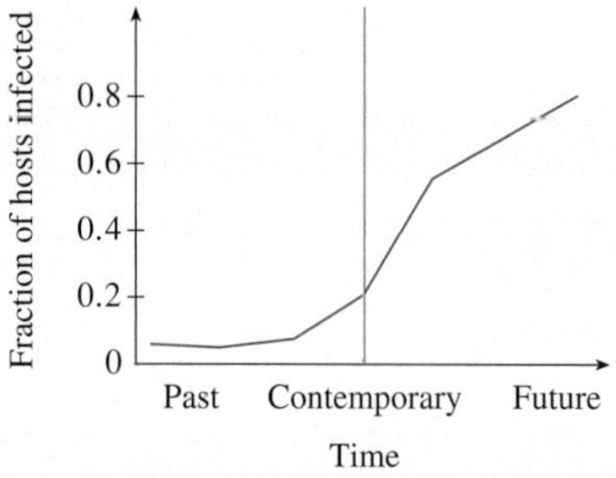

FIGURE 1
Horizontal axis is the time from which the parasite was taken, relative to the host's point in time.

The results of one such experiment are shown in Figure 1. The data show that hosts are indeed better able to resist parasites from their past, but are much more susceptible to infection by those from their future.

This is a compelling experiment but, by its very nature, it was conducted in a highly artificial setting. It would be interesting to somehow explore this idea in a natural host–parasite system. Incredibly, researchers have done exactly that with a species of freshwater crustacean and its parasite.[2]

Daphnia are freshwater crustacea that live in many lakes. They are parasitized by many different microbes, including a species of bacteria called *Pasteuria ramosa*. These two organisms have presumably been coevolving in lakes for many years, and the question is whether or not they too have been undergoing cycles of evolutionary conflict.

Occasionally, both the host and the parasite produce dormant offspring (called propagules) that sink to the bottom of the lake. As a time passes, sediment containing these propagules accumulates at the bottom of the lake. Over many years this sediment builds up, providing a historical record of the host and parasite (see Figure 2). A sediment core can then be taken from the bottom of the lake, giving an archive of the temporal sequence of hosts and parasites over evolutionary time (see Figure 3). And again, as with the first experiment, these propagules can be resuscitated and infection experiments conducted.

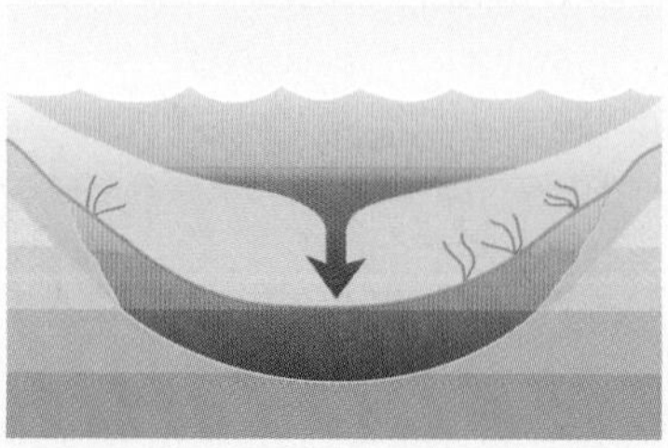

FIGURE 2
Sedimentation

1. A. Buckling et al. 2002, "Antagonistic Coevolution between a Bacterium and a Bacteriophage." *Proceedings of the Royal Society: Series B* 269 (2002): 931–36.

2. E. Decaestecker et al., "Host-Parasite 'Red Queen' Dynamics Archived in Pond Sediment." *Nature* 450 (2007): 870–73.

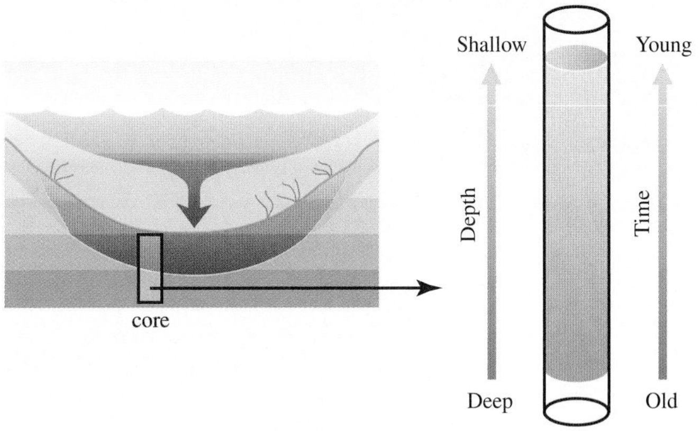

FIGURE 3

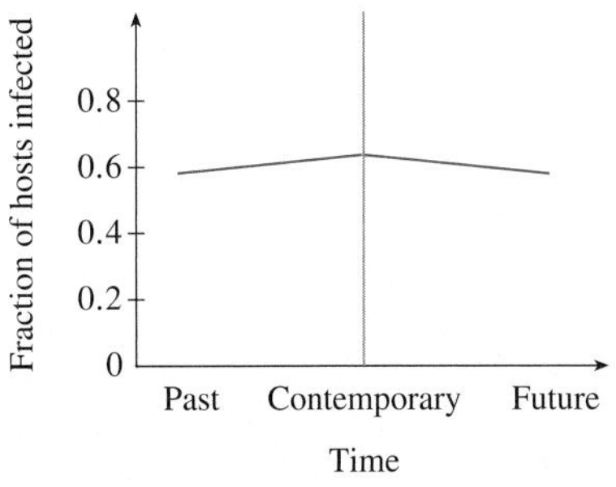

FIGURE 4

Horizontal axis is the time from which the parasite was taken, relative to the host's point in time.

Source: Adapted from S. Gandon et al., "Host-Parasite Coevolution and Patterns of Adaptation across Time and Space," *Journal of Evolutionary Biology* 21 (2008): 1861–66.

The results of the second experiment are shown in Figure 4: The pattern is quite different from that in Figure 1, with hosts being able to resist parasites from their past and their future, more than those taken from a contemporary point in time.

How can we understand these different patterns? Is it possible that this *Daphnia*–parasite system is also undergoing the same dynamic as the bacteriophage system, but that the different pattern seen in this experiment is simply due to differences in conditions? More generally, what pattern would we expect to see in the *Daphnia* experiment under different conditions if such coevolutionary conflict is actually occurring? To answer these questions we need a more quantitative approach. This is where mathematical modeling comes into play.

Models begin by simplifying reality (recall that a model is "a lie that makes us realize truth"). Thus, let's begin by supposing that there are only two possible host genotypes (A and a) and two possible parasite genotypes (B and b). Suppose that parasites of type B can infect only hosts of type A, while parasites of type b can infect only hosts of type a. Although we know reality is likely more complicated than this, these simplifying assumptions capture the essential features of an antagonistic interaction between a host and its parasite.

Under these assumptions we might expect parasites of type B to flourish when hosts of type A are common. But this will then give an advantage to hosts of type a, since they are resistant to type B parasites. As a result, type a hosts will then increase in frequency. Eventually, however, this will favor the spread of type b parasites, which then sets the stage for the return of type A hosts. At this point we might expect the cycle to repeat.

In this case study you will construct and analyze a model of this process. As is common in modeling, the order in which different mathematical tools are used by scientists is not always the same as the order in which they are best learned. For example, when scientists worked on this question they first used techniques from Chapter 7 and then Chapter 10 to formulate the model. They then used techniques from Chapter 6 and then Chapter 2 to draw important biological conclusions.[3] To fit with our learning objectives, however, this case study is developed the other way around. Following Chapter 2, in Case Study 2a, we will use given functions to draw biological conclusions about host–parasite coevolution. Following Chapter 6, in Case Study 2b, we will then begin to fill in the gaps by deriving these functions from the output of a model. Following Chapter 7, in Case Study 2c, we will then formulate this model explicitly, and following Chapter 10, in Case Study 2d, we will derive the output of the model that is used in Case Study 2b.

3. S. Gandon et al., "Host–Parasite Coevolution and Patterns of Adaptation across Time and Space," *Journal of Evolutionary Biology* 21 (2008): 1861–66.

Functions and Sequences

1

Often a graph is the best way to represent a function because it conveys so much information at a glance. The electrocardiograms shown are graphs that exhibit electrical activity in various parts of the heart (See Figure 1 on page 2.) They enable a cardiologist to view the heart from different angles and thereby diagnose possible problems.

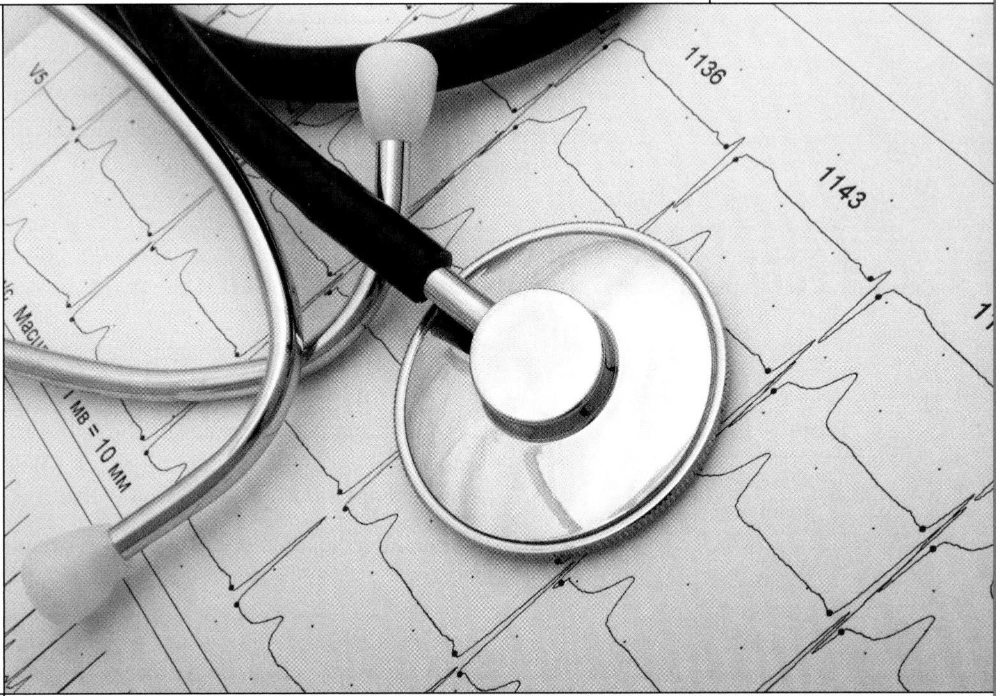

THE FUNDAMENTAL OBJECTS THAT WE deal with in calculus are functions. This chapter prepares the way for calculus by discussing the basic ideas concerning functions, their graphs, and ways of transforming and combining them. We stress that a function can be represented in different ways: by an equation, in a table, by a graph, or in words. We look at the main types of functions that occur in calculus and describe the process of using these functions as mathematical models in biology. A special type of function, namely a sequence, is often used in modeling biological phenomena. In particular, we study recursive sequences, also called difference equations, because they are useful in describing cell division, insect populations, and other biological processes.

1.1 Four Ways to Represent a Function

Functions arise whenever one quantity depends on another. Consider the following four situations.

Table 1

Year	Population (millions)
1900	1650
1910	1750
1920	1860
1930	2070
1940	2300
1950	2560
1960	3040
1970	3710
1980	4450
1990	5280
2000	6080
2010	6870

A. The area A of a circle depends on the radius r of the circle. The rule that connects r and A is given by the equation $A = \pi r^2$. With each positive number r there is associated one value of A, and we say that A is a *function* of r.

B. The human population of the world P depends on the time t. Table 1 gives estimates of the world population $P(t)$ at time t, for certain years. For instance,

$$P(1950) \approx 2{,}560{,}000{,}000$$

But for each value of the time t there is a corresponding value of P, and we say that P is a function of t.

C. The cost C of mailing an envelope depends on its weight w. Although there is no simple formula that connects w and C, the post office has a rule for determining C when w is known.

D. Figure 1 shows a graph called an electrocardiogram (ECG), or rhythm strip, one of 12 produced by an electrocardiograph. It measures the electric potential V (measured in millivolts) as a function of time in a certain direction (toward the positive electrode of a lead) corresponding to a particular part of the heart. For a given value of the time t, the graph provides a corresponding value of V.

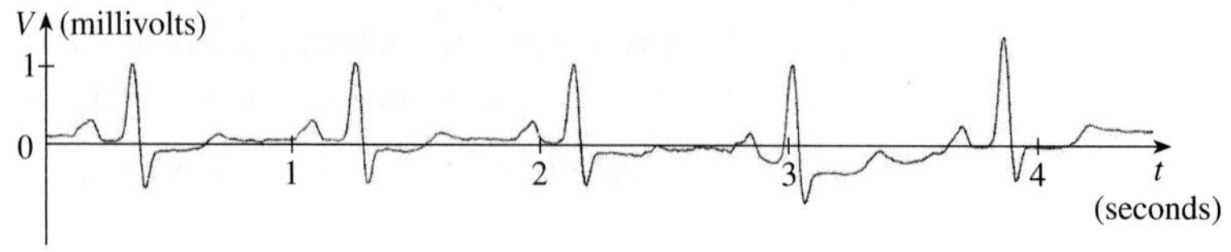

FIGURE 1
Electrocardiogram
Source: Courtesy of Dr. Brian Gilbert

Each of these examples describes a rule whereby, given a number (r, t, w, or t), another number (A, P, C, or V) is assigned. In each case we say that the second number is a function of the first number.

Definition A **function** f is a rule that assigns to each element x in a set D exactly one element, called $f(x)$, in a set E.

We usually consider functions for which the sets D and E are sets of real numbers. The set D is called the **domain** of the function. The number $f(x)$ is the **value of f at x** and is read "f of x." The **range** of f is the set of all possible values of $f(x)$ as x varies throughout the domain. A symbol that represents an arbitrary number in the *domain* of a function f is called an **independent variable**. A symbol that represents a number in the *range* of f is called a **dependent variable**. In Example A, for instance, r is the independent variable and A is the dependent variable.

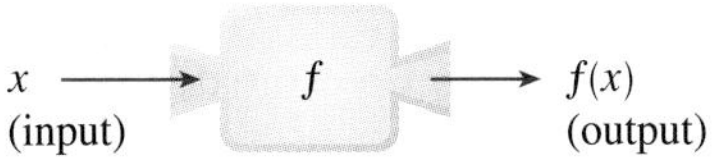

FIGURE 2
Machine diagram for a function f

It's helpful to think of a function as a **machine** (see Figure 2). If x is in the domain of the function f, then when x enters the machine, it's accepted as an input and the machine produces an output $f(x)$ according to the rule of the function. Thus we can think of the domain as the set of all possible inputs and the range as the set of all possible outputs.

The preprogrammed functions in a calculator are good examples of a function as a machine. For example, the square root key on your calculator computes such a function. You press the key labeled $\sqrt{\ }$ (or $\sqrt{x}$) and enter the input x. If $x < 0$, then x is not in the domain of this function; that is, x is not an acceptable input, and the calculator will indicate an error. If $x \geqslant 0$, then an *approximation* to $\sqrt{x}$ will appear in the display. Thus the $\sqrt{x}$ key on your calculator is not quite the same as the exact mathematical function f defined by $f(x) = \sqrt{x}$.

Another way to picture a function is by an **arrow diagram** as in Figure 3. Each arrow connects an element of D to an element of E. The arrow indicates that $f(x)$ is associated with x, $f(a)$ is associated with a, and so on.

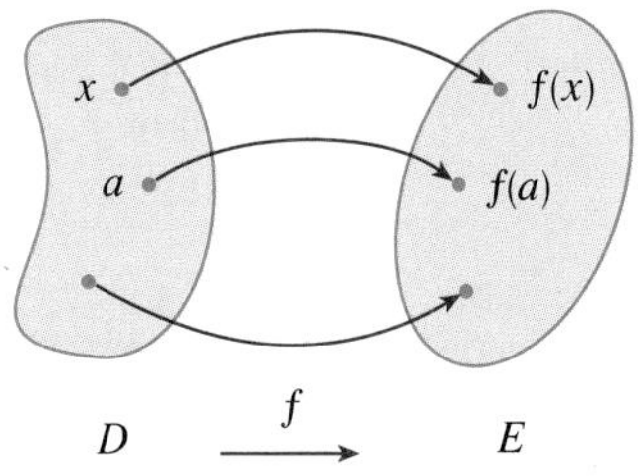

FIGURE 3
Arrow diagram for f

The most common method for visualizing a function is its graph. If f is a function with domain D, then its **graph** is the set of ordered pairs

$$\{(x, f(x)) \mid x \in D\}$$

(Notice that these are input-output pairs.) In other words, the graph of f consists of all points (x, y) in the coordinate plane such that $y = f(x)$ and x is in the domain of f.

The graph of a function f gives us a useful picture of the behavior of a function. Since the y-coordinate of any point (x, y) on the graph is $y = f(x)$, we can read the value of $f(x)$ from the graph as being the height of the graph above the point x (see Figure 4). The graph of f also allows us to picture the domain of f on the x-axis and its range on the y-axis as in Figure 5.

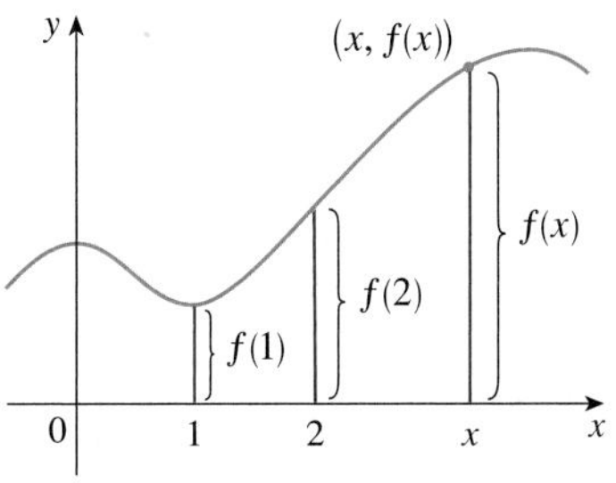

FIGURE 4

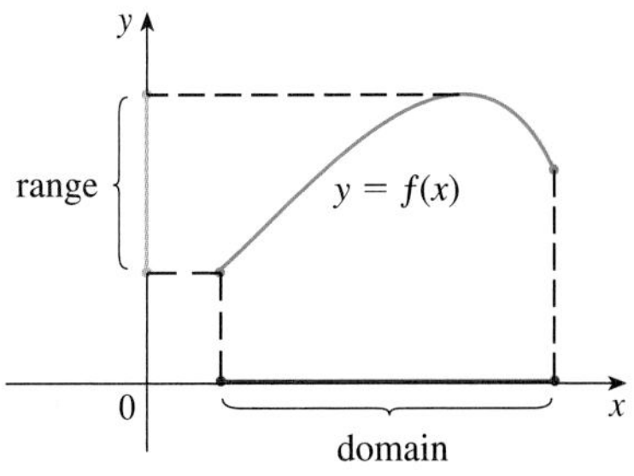

FIGURE 5

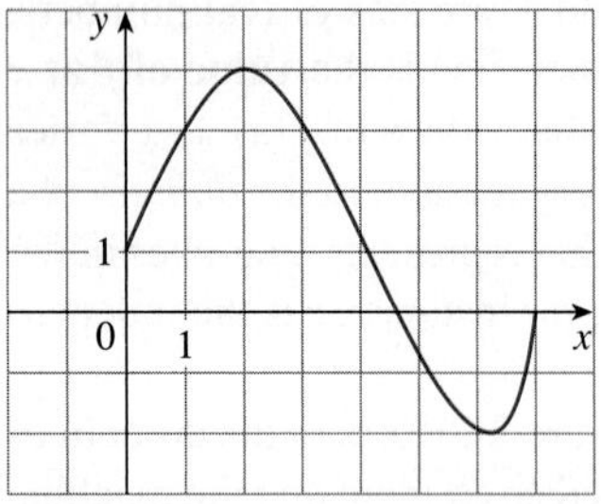

FIGURE 6

EXAMPLE 1 | The graph of a function f is shown in Figure 6.
(a) Find the values of $f(1)$ and $f(5)$.
(b) What are the domain and range of f?

SOLUTION
(a) We see from Figure 6 that the point $(1, 3)$ lies on the graph of f, so the value of f at 1 is $f(1) = 3$. (In other words, the point on the graph that lies above $x = 1$ is 3 units above the x-axis.)

When $x = 5$, the graph lies about 0.7 units below the x-axis, so we estimate that $f(5) \approx -0.7$.

(b) We see that $f(x)$ is defined when $0 \leqslant x \leqslant 7$, so the domain of f is the closed interval $[0, 7]$. Notice that f takes on all values from -2 to 4, so the range of f is

$$\{y \mid -2 \leqslant y \leqslant 4\} = [-2, 4]$$

■

The notation for intervals is given in Appendix A.

EXAMPLE 2 | Sketch the graph and find the domain and range of each function.
(a) $f(x) = 2x - 1$ (b) $g(x) = x^2$

SOLUTION
(a) The equation of the graph is $y = 2x - 1$, and we recognize this as being the equation of a line with slope 2 and y-intercept -1. (Recall the slope-intercept form of the equation of a line: $y = mx + b$. See Appendix B.) This enables us to sketch a portion of the graph of f in Figure 7. The expression $2x - 1$ is defined for all real numbers, so the domain of f is the set of all real numbers, which we denote by $\mathbb{R}$. The graph shows that the range is also $\mathbb{R}$.

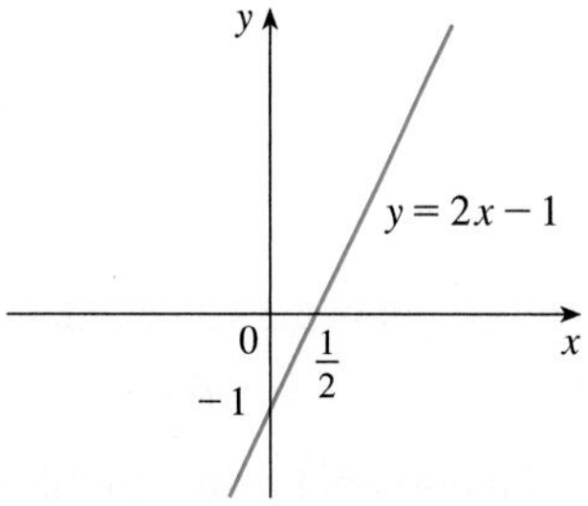

FIGURE 7

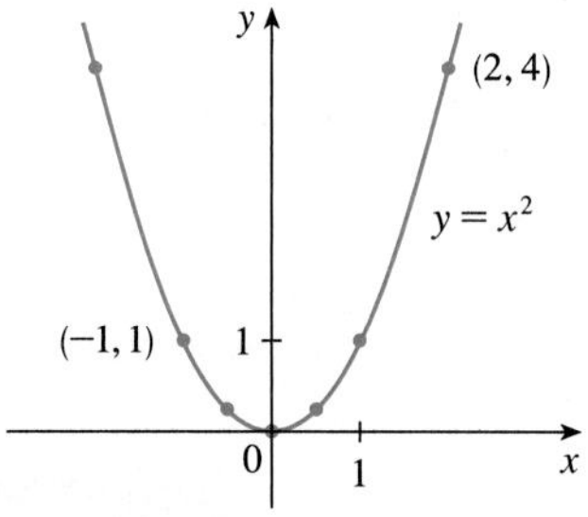

FIGURE 8

(b) Since $g(2) = 2^2 = 4$ and $g(-1) = (-1)^2 = 1$, we could plot the points $(2, 4)$ and $(-1, 1)$, together with a few other points on the graph, and join them to produce the graph (Figure 8). The equation of the graph is $y = x^2$, which represents a parabola (see Appendix B). The domain of g is $\mathbb{R}$. The range of g consists of all values of $g(x)$, that is, all numbers of the form x^2. But $x^2 \geqslant 0$ for all numbers x and any positive number y is a square. So the range of g is $\{y \mid y \geqslant 0\} = [0, \infty)$. This can also be seen from Figure 8. ■

EXAMPLE 3 | **Antihypertension medication** Figure 9 shows the effect of nifedipine tablets (antihypertension medication) on the heart rate $H(t)$ of a patient as a function of time.
(a) Estimate the heart rate after two hours.
(b) During what time period is the heart rate less than 65 beats/min?

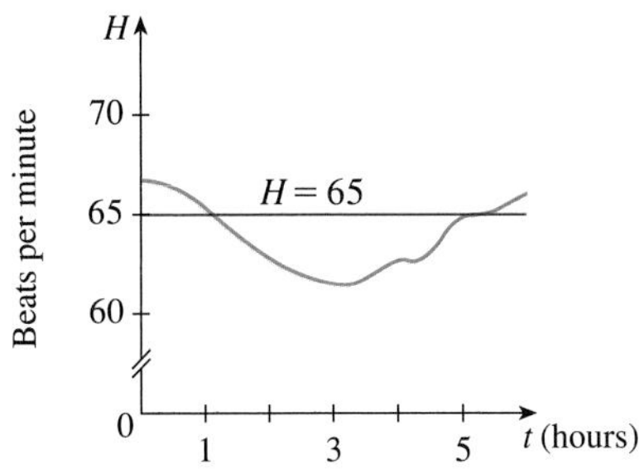

FIGURE 9

Source: Adapted from M. Brown et al., "Formulation of Long-Acting Nifedipine Tablets Influences the Heart Rate and Sympathetic Nervous System Response in Hypertensive Patients," *British Journal of Clinical Pharmacology* 65 (2008): 646–52.

SOLUTION

(a) If $H(t)$ is the rate at time t, we estimate from the graph in Figure 9 that

$$H(2) \approx 62.5 \text{ beats/min}$$

(b) Notice that the curve lies below the line $H = 65$ for $1 \leqslant t \leqslant 5$. In other words, the heart rate is less than 65 beats/min from 1 hour to 5 hours after the tablet is administered. ■

EXAMPLE 4 | If $f(x) = 2x^2 - 5x + 1$ and $h \neq 0$, evaluate $\dfrac{f(a + h) - f(a)}{h}$.

SOLUTION We first evaluate $f(a + h)$ by replacing x by $a + h$ in the expression for $f(x)$:

$$\begin{aligned} f(a + h) &= 2(a + h)^2 - 5(a + h) + 1 \\ &= 2(a^2 + 2ah + h^2) - 5(a + h) + 1 \\ &= 2a^2 + 4ah + 2h^2 - 5a - 5h + 1 \end{aligned}$$

Then we substitute into the given expression and simplify:

$$\begin{aligned} \frac{f(a + h) - f(a)}{h} &= \frac{(2a^2 + 4ah + 2h^2 - 5a - 5h + 1) - (2a^2 - 5a + 1)}{h} \\ &= \frac{2a^2 + 4ah + 2h^2 - 5a - 5h + 1 - 2a^2 + 5a - 1}{h} \\ &= \frac{4ah + 2h^2 - 5h}{h} = 4a + 2h - 5 \end{aligned}$$

■

The expression

$$\frac{f(a + h) - f(a)}{h}$$

in Example 4 is called a **difference quotient** and occurs frequently in calculus. As we will see in Chapter 2, it represents the average rate of change of $f(x)$ between $x = a$ and $x = a + h$.

■ Representations of Functions

There are four possible ways to represent a function:

- verbally (by a description in words)
- numerically (by a table of values)
- visually (by a graph)
- algebraically (by an explicit formula)

If a single function can be represented in all four ways, it's often useful to go from one representation to another to gain additional insight into the function. (In Example 2, for instance, we started with algebraic formulas and then obtained the graphs.) But certain functions are described more naturally by one method than by another. With this in mind, let's reexamine the four situations that we considered at the beginning of this section.

A. The most useful representation of the area of a circle as a function of its radius is probably the algebraic formula $A(r) = \pi r^2$, though it is possible to compile a table of values or to sketch a graph (half a parabola). Because a circle has to have a positive radius, the domain is $\{r \mid r > 0\} = (0, \infty)$, and the range is also $(0, \infty)$.

t (years since 1990)	Population (millions)
0	1650
10	1750
20	1860
30	2070
40	2300
50	2560
60	3040
70	3710
80	4450
90	5280
100	6080
110	6870

B. We are given a description of the function in words: $P(t)$ is the human population of the world at time t. Let's measure t so that $t = 0$ corresponds to the year 1900. The table of values of world population provides a convenient representation of this function. If we plot these values, we get the graph (called a *scatter plot*) in Figure 10. It too is a useful representation; the graph allows us to absorb all the data at once. What about a formula? Of course, it's impossible to devise an explicit formula that gives the exact human population $P(t)$ at any time t. But it is possible to find an expression for a function that *approximates* $P(t)$. In fact, using methods explained in Section 1.2, we obtain the approximation

$$P(t) \approx f(t) = (1.43653 \times 10^9) \cdot (1.01395)^t$$

Figure 11 shows that it is a reasonably good "fit." The function f is called a *mathematical model* for population growth. In other words, it is a function with an explicit formula that approximates the behavior of our given function. We will see, however, that the ideas of calculus can be applied to a table of values; an explicit formula is not necessary.

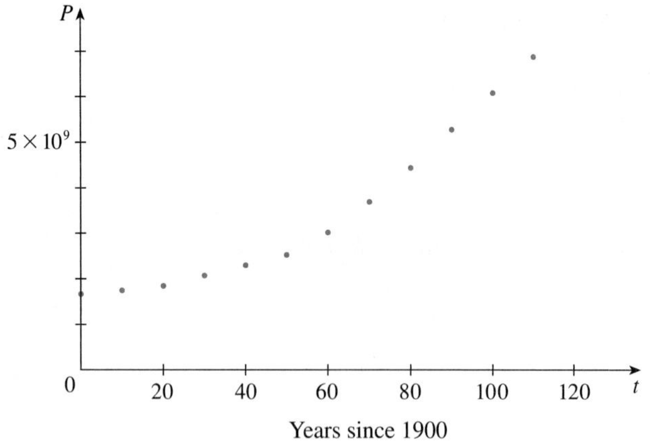

FIGURE 10

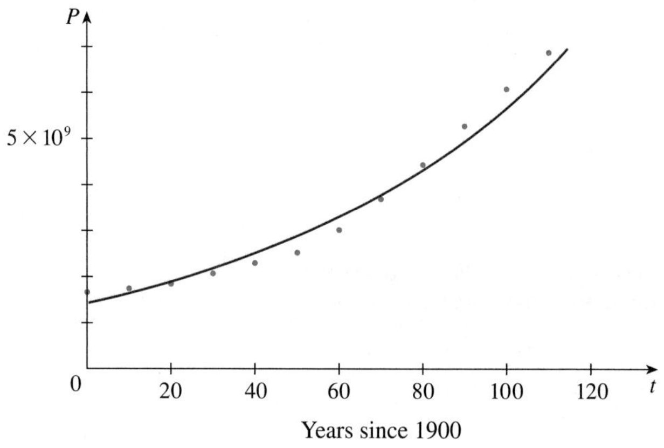

FIGURE 11

A function defined by a table of values is called a *tabular* function.

The function P is typical of the functions that arise whenever we attempt to apply calculus to the real world. We start with a verbal description of a function. Then we might be able to construct a table of values of the function, perhaps from instrument readings in a scientific experiment. Even though we don't have complete knowledge of the values of the function, we will see throughout the book that it is still possible to perform the operations of calculus on such a function.

w (ounces)	$C(w)$ (dollars)
$0 < w \leqslant 1$	0.92
$1 < w \leqslant 2$	1.12
$2 < w \leqslant 3$	1.32
$3 < w \leqslant 4$	1.52
$4 < w \leqslant 5$	1.72
⋮	⋮

C. Again the function is described in words: Let $C(w)$ be the cost of mailing a large envelope with weight w. The rule that the US Postal Service used as of 2014 is as follows: The cost is 92 cents for up to 1 oz, plus 20 cents for each additional ounce (or less) up to 13 oz. The table of values shown in the margin is the most convenient representation for this function, though it is possible to sketch a graph (see Example 11).

D. The graph shown in Figure 1 is the most natural representation of the voltage function $V(t)$ that reflects the electrical activity of the heart. It's true that a table of values could be compiled, and it is even possible to devise an approximate formula. But everything a doctor needs to know—amplitudes and patterns—can be seen easily from the graph. (The same is true for the patterns seen in polygraphs for lie-detection and seismographs for analysis of earthquakes.) The waves represent the depolarization and repolarization of the atria and ventricles of the heart. They enable a cardiologist to see whether the patient has irregular heart rhythms and help diagnose different types of heart disease.

In the next example we sketch the graph of a function that is defined verbally.

EXAMPLE 5 | When you turn on a hot-water faucet, the temperature T of the water depends on how long the water has been running. Draw a rough graph of T as a function of the time t that has elapsed since the faucet was turned on.

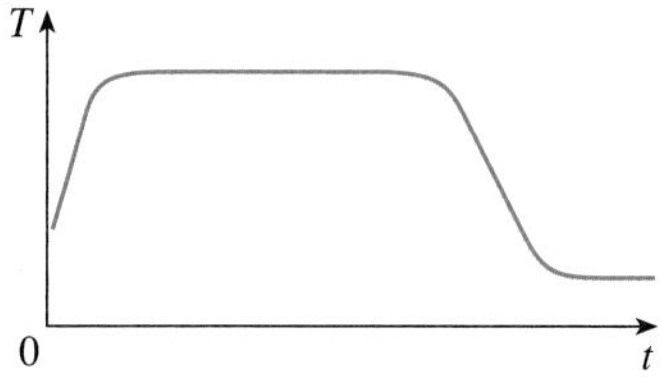

FIGURE 12

SOLUTION The initial temperature of the running water is close to room temperature because the water has been sitting in the pipes. When the water from the hot-water tank starts flowing from the faucet, T increases quickly. In the next phase, T is constant at the temperature of the heated water in the tank. When the tank is drained, T decreases to the temperature of the water supply. This enables us to make the rough sketch of T as a function of t in Figure 12. ■

EXAMPLE 6 | **BB** Bone mass A human femur (thighbone) is essentially a hollow tube filled with yellow marrow (see Figure 13). If the outer radius is r and the inner radius is r_{in}, an important quantity characterizing such bones is

$$k = \frac{r_{\text{in}}}{r}$$

The density of bone is approximately 1.8 g/cm^3 and that of marrow is about 1 g/cm^3. For a femur with length L, express its mass as a function of k.

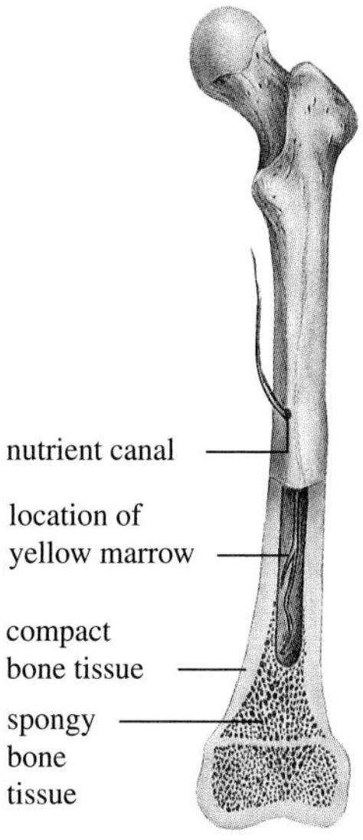

FIGURE 13
Structure of a human femur

Source: From Starr. *Biology*, 8E © 2011 Brooks/Cole, a part of Cengage Learning, Inc. Reproduced by permission. www.cengage.com/permissions

SOLUTION The mass of the tubular bone is obtained by subtracting the mass of the inner tube from the mass of the outer tube:

$$1.8\pi r^2 L - 1.8\pi r_{\text{in}}^2 L = 1.8\pi r^2 L - 1.8\pi (rk)^2 L$$

Similarly, the mass of the marrow is

$$1 \times (\pi r_{\text{in}}^2 L) = \pi (rk)^2 L$$

So the total mass as a function of k is

$$\begin{aligned} m(k) &= 1.8\pi r^2 L - 1.8\pi (rk)^2 L + \pi (rk)^2 L \\ &= \pi r^2 L(1.8 - 0.8k^2) \end{aligned}$$

■

EXAMPLE 7 | Find the domain of each function.

(a) $f(x) = \sqrt{x + 2}$ (b) $g(x) = \dfrac{1}{x^2 - x}$

SOLUTION

(a) Because the square root of a negative number is not defined (as a real number), the domain of f consists of all values of x such that $x + 2 \geqslant 0$. This is equivalent to $x \geqslant -2$, so the domain is the interval $[-2, \infty)$.

(b) Since

$$g(x) = \frac{1}{x^2 - x} = \frac{1}{x(x - 1)}$$

and division by 0 is not allowed, we see that $g(x)$ is not defined when $x = 0$ or $x = 1$.

Domain Convention
If a function is given by a formula and the domain is not stated explicitly, the convention is that the domain is the set of all numbers for which the formula makes sense and defines a real number.

Thus the domain of g is

$$\{x \mid x \neq 0, x \neq 1\}$$

which could also be written in interval notation as

$$(-\infty, 0) \cup (0, 1) \cup (1, \infty)$$ ■

The graph of a function is a curve in the xy-plane. But the question arises: Which curves in the xy-plane are graphs of functions? This is answered by the following test.

The Vertical Line Test A curve in the xy-plane is the graph of a function of x if and only if no vertical line intersects the curve more than once.

The reason for the truth of the Vertical Line Test can be seen in Figure 14. If each vertical line $x = a$ intersects a curve only once, at (a, b), then exactly one function value is defined by $f(a) = b$. But if a line $x = a$ intersects the curve twice, at (a, b) and (a, c), then the curve can't represent a function because a function can't assign two different values to a.

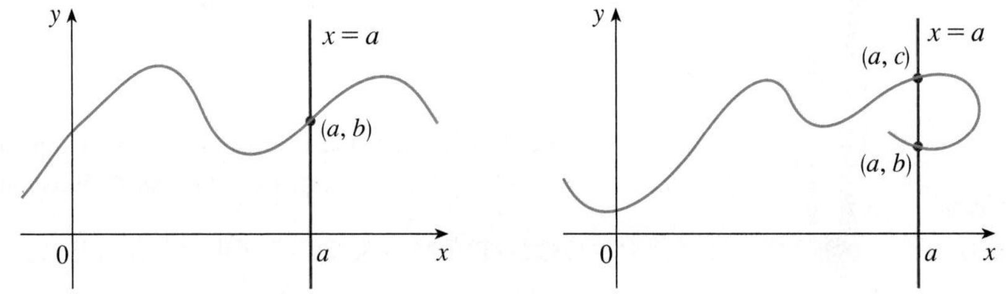

FIGURE 14

For example, the parabola $x = y^2 - 2$ shown in Figure 15(a) is not the graph of a function of x because, as you can see, there are vertical lines that intersect the parabola twice. The parabola, however, does contain the graphs of *two* functions of x. Notice that the equation $x = y^2 - 2$ implies $y^2 = x + 2$, so $y = \pm\sqrt{x + 2}$. Thus the upper and lower halves of the parabola are the graphs of the functions $f(x) = \sqrt{x + 2}$ [from Example 7(a)] and $g(x) = -\sqrt{x + 2}$. [See Figures 15(b) and (c).] We observe that if we reverse the roles of x and y, then the equation $x = h(y) = y^2 - 2$ *does* define x as a function of y (with y as the independent variable and x as the dependent variable) and the parabola now appears as the graph of the function h.

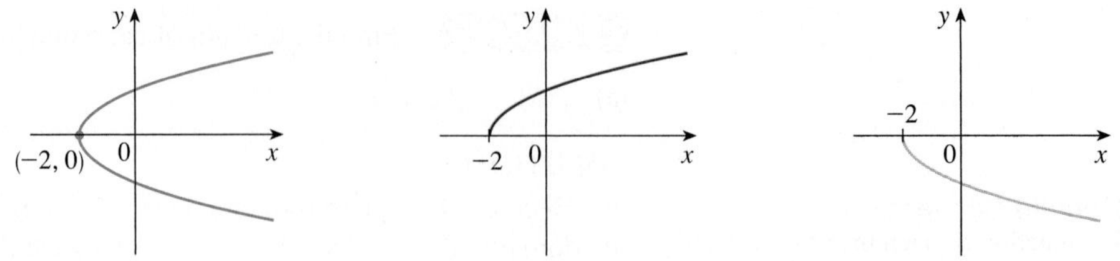

FIGURE 15 (a) $x = y^2 - 2$ (b) $y = \sqrt{x + 2}$ (c) $y = -\sqrt{x + 2}$

■ Piecewise Defined Functions

The functions in the following four examples are defined by different formulas in different parts of their domains. Such functions are called **piecewise defined functions**.

EXAMPLE 8 | A function f is defined by

$$f(x) = \begin{cases} 1 - x & \text{if } x \leqslant -1 \\ x^2 & \text{if } x > -1 \end{cases}$$

Evaluate $f(-2)$, $f(-1)$, and $f(0)$ and sketch the graph.

SOLUTION Remember that a function is a rule. For this particular function the rule is the following: First look at the value of the input x. If it happens that $x \leqslant -1$, then the value of $f(x)$ is $1 - x$. On the other hand, if $x > -1$, then the value of $f(x)$ is x^2.

Since $-2 \leqslant -1$, we have $f(-2) = 1 - (-2) = 3$.

Since $-1 \leqslant -1$, we have $f(-1) = 1 - (-1) = 2$.

Since $0 > -1$, we have $f(0) = 0^2 = 0$.

How do we draw the graph of f? We observe that if $x \leqslant -1$, then $f(x) = 1 - x$, so the part of the graph of f that lies to the left of the vertical line $x = -1$ must coincide with the line $y = 1 - x$, which has slope -1 and y-intercept 1. If $x > -1$, then $f(x) = x^2$, so the part of the graph of f that lies to the right of the line $x = -1$ must coincide with the graph of $y = x^2$, which is a parabola. This enables us to sketch the graph in Figure 16. The solid dot indicates that the point $(-1, 2)$ is included on the graph; the open dot indicates that the point $(-1, 1)$ is excluded from the graph. ■

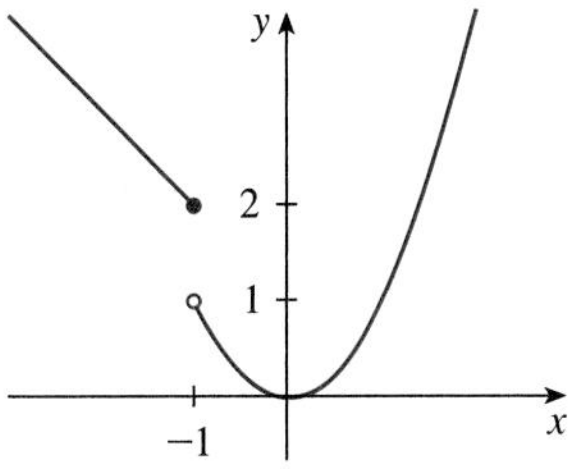

FIGURE 16

The next example of a piecewise defined function is the absolute value function. Recall that the **absolute value** of a number a, denoted by $|a|$, is the distance from a to 0 on the real number line. Distances are always positive or 0, so we have

For a more extensive review of absolute values, see Appendix A.

$$|a| \geqslant 0 \qquad \text{for every number } a$$

For example,

$$|3| = 3 \qquad |-3| = 3 \qquad |0| = 0 \qquad |\sqrt{2} - 1| = \sqrt{2} - 1 \qquad |3 - \pi| = \pi - 3$$

In general, we have

$$\boxed{\begin{aligned} |a| &= a & \text{if } a \geqslant 0 \\ |a| &= -a & \text{if } a < 0 \end{aligned}}$$

(Remember that if a is negative, then $-a$ is positive.)

EXAMPLE 9 | Sketch the graph of the absolute value function $f(x) = |x|$.

SOLUTION From the preceding discussion we know that

$$|x| = \begin{cases} x & \text{if } x \geqslant 0 \\ -x & \text{if } x < 0 \end{cases}$$

Using the same method as in Example 8, we see that the graph of f coincides with the line $y = x$ to the right of the y-axis and coincides with the line $y = -x$ to the left of the y-axis (see Figure 17). ■

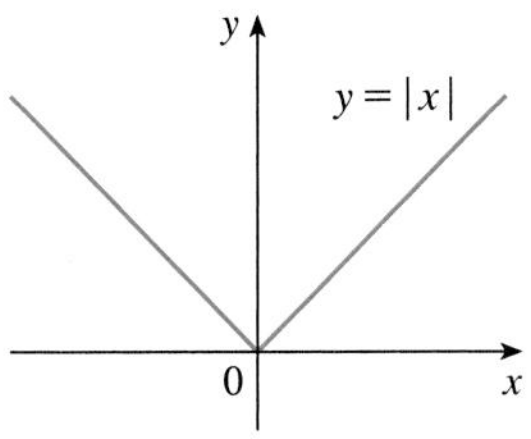

FIGURE 17

EXAMPLE 10 | BB **Metabolic power in walking and running** Suppose you are walking slowly but then increase your pace and start running more and more quickly to catch a bus. When you start running, your gait (manner of movement) changes. Figure 18 shows a graph of metabolic power consumed by men walking and running (calculated from measurements of oxygen consumption) as a function of speed. Notice that it is a piecewise defined function and the second piece starts when you begin to run.

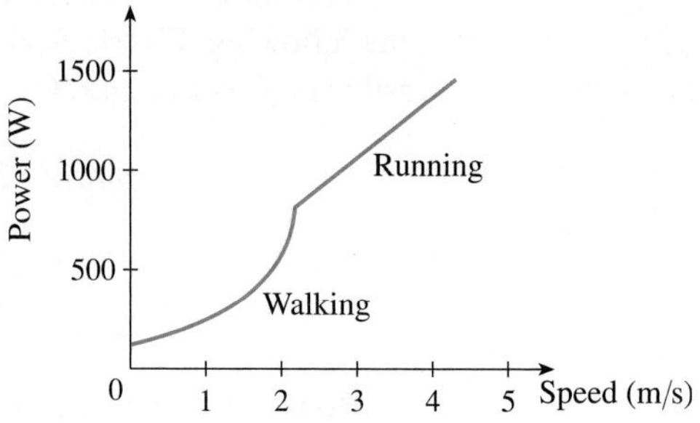

FIGURE 18
Metabolic power is a piecewise defined function of speed
Source: Adapted from R. Alexander, *Optima for Animals*, 2nd ed. (Princeton, NJ: Princeton University Press, 1996), 53.

EXAMPLE 11 | In Example C at the beginning of this section we considered the cost $C(w)$ of mailing a large envelope with weight w. In effect, this is a piecewise defined function because, from the table of values on page 6, we have

$$C(w) = \begin{cases} 0.92 & \text{if } 0 < w \leqslant 1 \\ 1.12 & \text{if } 1 < w \leqslant 2 \\ 1.32 & \text{if } 2 < w \leqslant 3 \\ 1.52 & \text{if } 3 < w \leqslant 4 \\ \vdots & \end{cases}$$

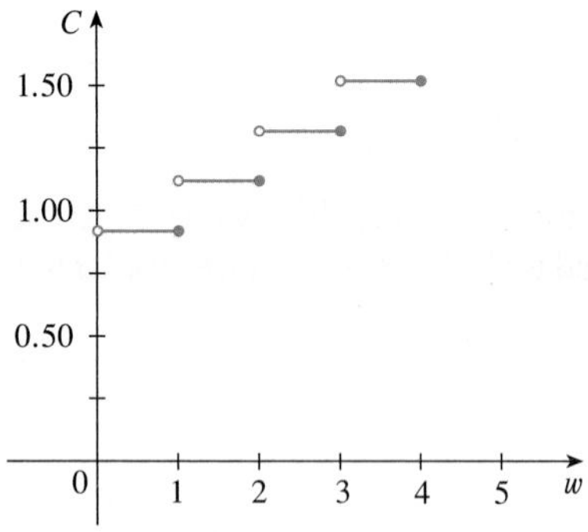

FIGURE 19

The graph is shown in Figure 19. You can see why functions similar to this one are called **step functions**—they jump from one value to the next. Such functions will be studied in Chapter 2.

■ Symmetry

If a function f satisfies $f(-x) = f(x)$ for every number x in its domain, then f is called an **even function**. For instance, the function $f(x) = x^2$ is even because

$$f(-x) = (-x)^2 = x^2 = f(x)$$

The geometric significance of an even function is that its graph is symmetric with respect to the y-axis (see Figure 20). This means that if we have plotted the graph of f for $x \geqslant 0$, we obtain the entire graph simply by reflecting this portion about the y-axis.

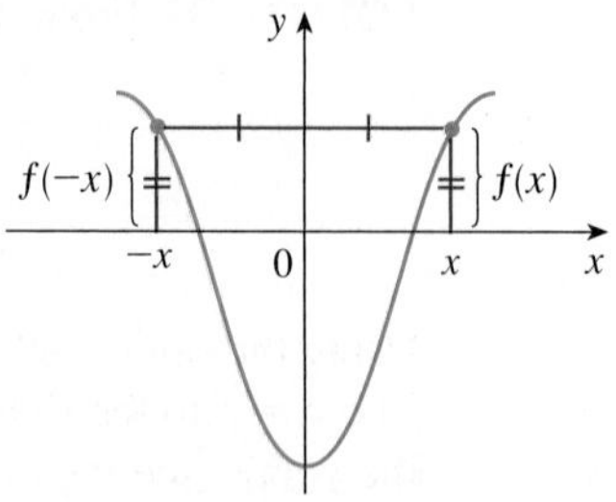

FIGURE 20
An even function

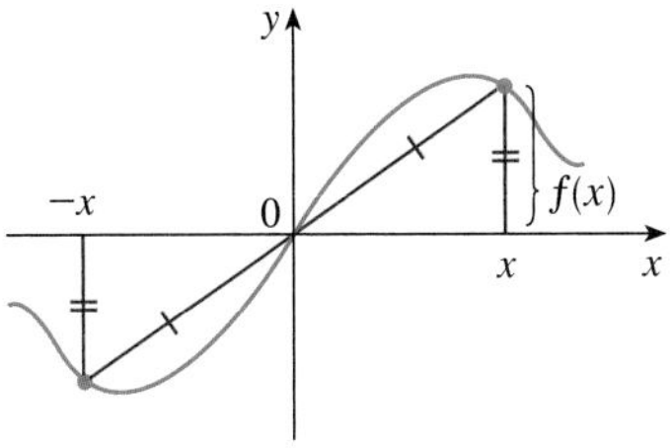

FIGURE 21
An odd function

If f satisfies $f(-x) = -f(x)$ for every number x in its domain, then f is called an **odd function**. For example, the function $f(x) = x^3$ is odd because

$$f(-x) = (-x)^3 = -x^3 = -f(x)$$

The graph of an odd function is symmetric about the origin (see Figure 21). If we already have the graph of f for $x \geqslant 0$, we can obtain the entire graph by rotating this portion through 180° about the origin.

EXAMPLE 12 | Determine whether each of the following functions is even, odd, or neither even nor odd.

(a) $f(x) = x^5 + x$ (b) $g(x) = 1 - x^4$ (c) $h(x) = 2x - x^2$

SOLUTION

(a)
$$\begin{aligned} f(-x) &= (-x)^5 + (-x) = (-1)^5x^5 + (-x) \\ &= -x^5 - x = -(x^5 + x) \\ &= -f(x) \end{aligned}$$

Therefore f is an odd function.

(b)
$$g(-x) = 1 - (-x)^4 = 1 - x^4 = g(x)$$

So g is even.

(c)
$$h(-x) = 2(-x) - (-x)^2 = -2x - x^2$$

Since $h(-x) \neq h(x)$ and $h(-x) \neq -h(x)$, we conclude that h is neither even nor odd. ■

The graphs of the functions in Example 12 are shown in Figure 22. Notice that the graph of h is symmetric neither about the y-axis nor about the origin.

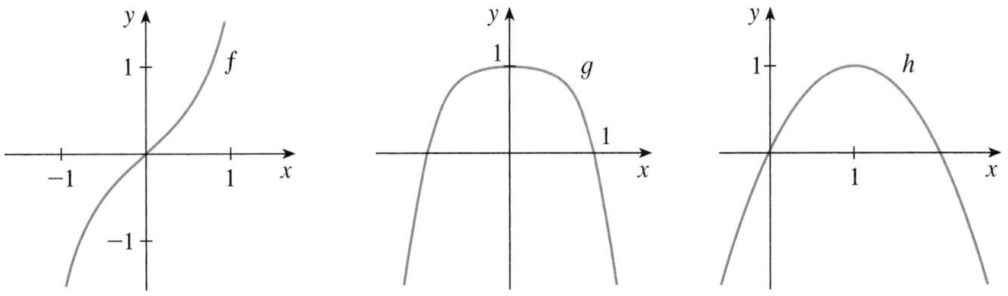

FIGURE 22 (a) Odd function (b) Even function (c) Neither even nor odd

■ Periodic Functions

Many phenomena in the life sciences display a recurring type of behavior: from breathing, to the beating of the heart, to the cycling of female reproductive hormones, to seasonal migration of butterflies. Such phenomena are referred to as *periodic*. To describe such processes mathematically we need functions that display this behavior.

Definition A function f is called **periodic** if there is a positive constant T such that $f(x + T) = f(x)$ for all values of x in the domain of f. The smallest value of T for which this is true is called the **period** of f.

The electrocardiogram shown in Figure 1 on page 2 is an example of an approximately periodic function. The period of the function V appears to be about 0.9 seconds: $V(t + 0.9) \approx V(t)$. The trigonometric functions are also periodic and are discussed in the next section.

EXAMPLE 13 | BB **Malarial fever** Figure 23 shows a typical temperature chart for a fever in humans induced by a species of malaria called *P. vivax*. Notice that the temperature approximately satisfies

$$T(t + 48) = T(t)$$

so the temperature function has a period of about 48 hours.

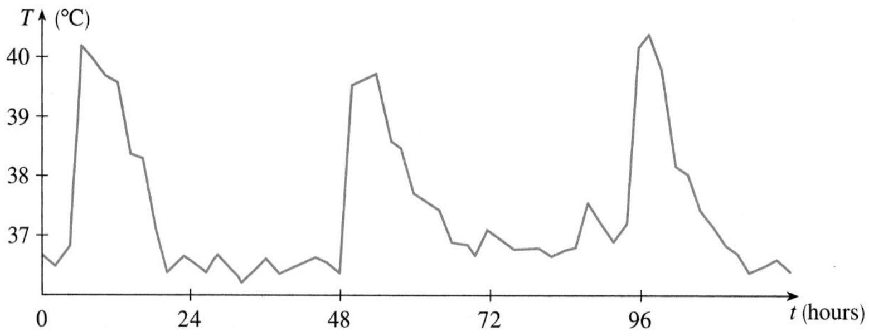

FIGURE 23
Temperature chart for *P. vivax* infection

Source: Adapted from L. Bruce-Chwatt, *Essential Malariology* (New York: Wiley, 1985).

■ Increasing and Decreasing Functions

The graph shown in Figure 24 rises from A to B, falls from B to C, and rises again from C to D. The function f is said to be increasing on the interval $[a, b]$, decreasing on $[b, c]$, and increasing again on $[c, d]$. Notice that if x_1 and x_2 are any two numbers between a and b with $x_1 < x_2$, then $f(x_1) < f(x_2)$. We use this as the defining property of an increasing function.

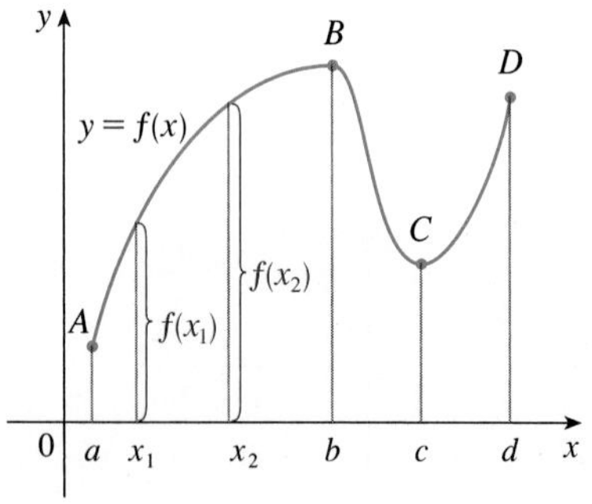

FIGURE 24

Definition A function f is called **increasing** on an interval I if

$$f(x_1) < f(x_2) \qquad \text{whenever } x_1 < x_2 \text{ in } I$$

It is called **decreasing** on I if

$$f(x_1) > f(x_2) \qquad \text{whenever } x_1 < x_2 \text{ in } I$$

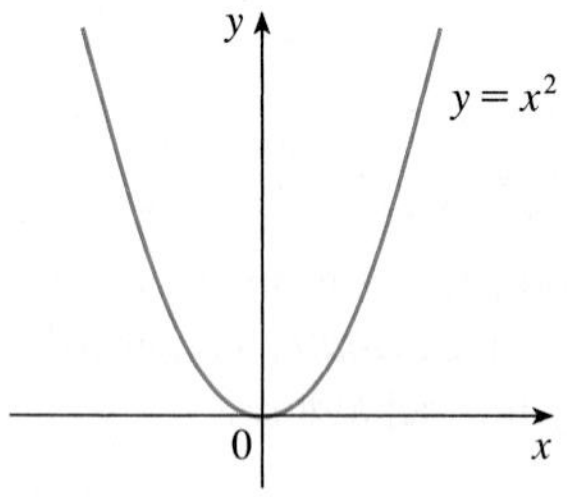

FIGURE 25

In the definition of an increasing function it is important to realize that the inequality $f(x_1) < f(x_2)$ must be satisfied for *every* pair of numbers x_1 and x_2 in I with $x_1 < x_2$.

You can see from Figure 25 that the function $f(x) = x^2$ is decreasing on the interval $(-\infty, 0]$ and increasing on the interval $[0, \infty)$.

EXERCISES 1.1

1. If $f(x) = x + \sqrt{2 - x}$ and $g(u) = u + \sqrt{2 - u}$, is it true that $f = g$?

2. If

$$f(x) = \frac{x^2 - x}{x - 1} \qquad \text{and} \qquad g(x) = x$$

is it true that $f = g$?

3. The graph of a function f is given.
(a) State the value of $f(1)$.
(b) Estimate the value of $f(-1)$.
(c) For what values of x is $f(x) = 1$?
(d) Estimate the value of x such that $f(x) = 0$.
(e) State the domain and range of f.
(f) On what interval is f increasing?

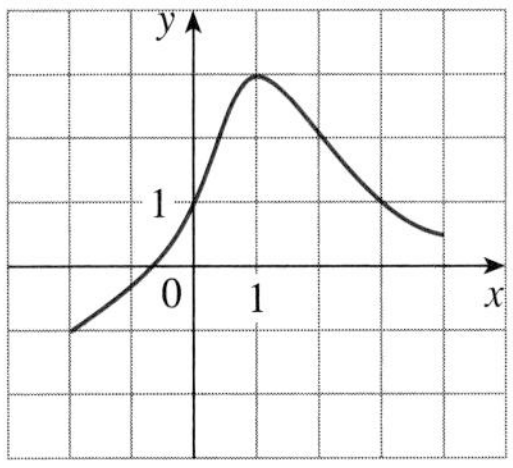

4. The graphs of f and g are given.
(a) State the values of $f(-4)$ and $g(3)$.
(b) For what values of x is $f(x) = g(x)$?
(c) Estimate the solution of the equation $f(x) = -1$.
(d) On what interval is f decreasing?
(e) State the domain and range of f.
(f) State the domain and range of g.

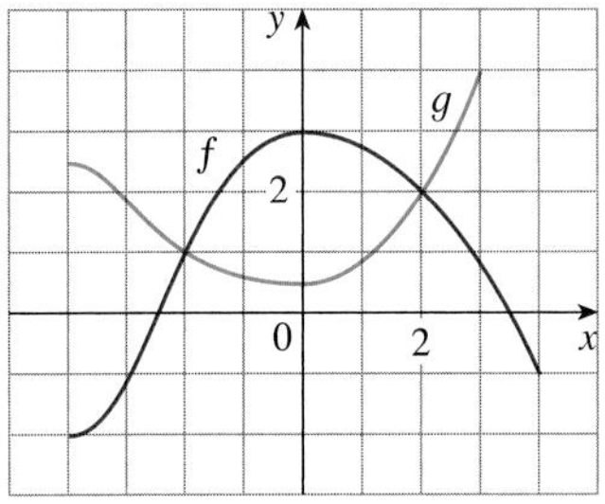

5–8 Determine whether the curve is the graph of a function of x. If it is, state the domain and range of the function.

5.

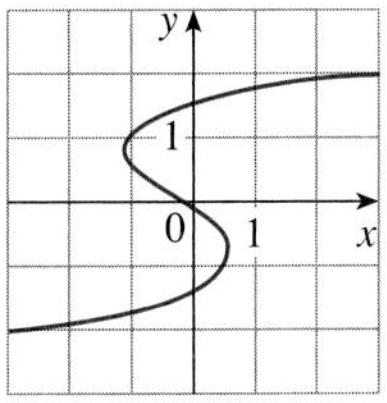

6.

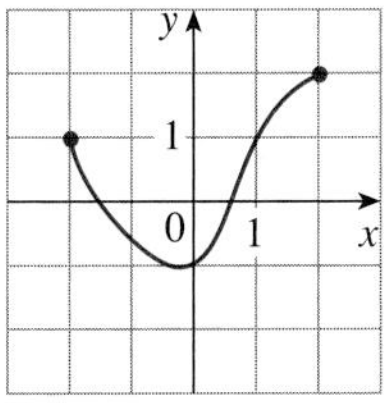

7.

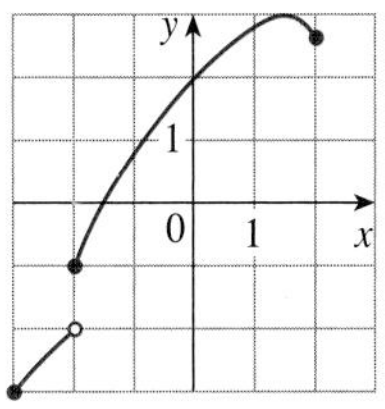

8.

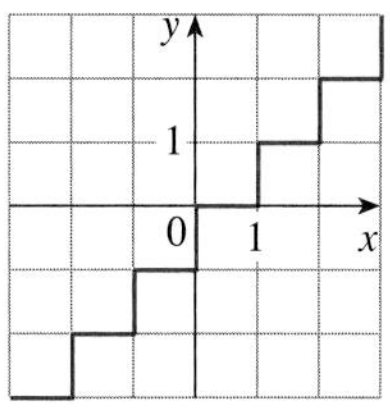

9. Global temperature Shown is a graph of the global average temperature T during the 20th century.
(a) What was the global average temperature in 1950?
(b) In what year was the average temperature 14.2°?
(c) When was the temperature smallest? Largest?
(d) Estimate the range of T.

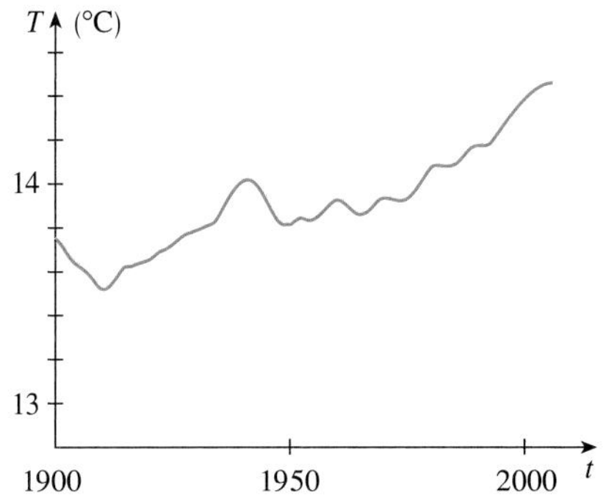

Source: Adapted from *Globe and Mail* [Toronto] 5 Dec. 2009. Print.

10. Tree ring width Trees grow faster and form wider rings in warm years and grow more slowly and form narrower rings in cooler years. The figure shows ring widths of a Siberian pine from 1500 to 2000.
(a) What is the range of the ring width function?
(b) What does the graph tend to say about the temperature of the earth? Does the graph reflect the volcanic eruptions of the mid-19th century?

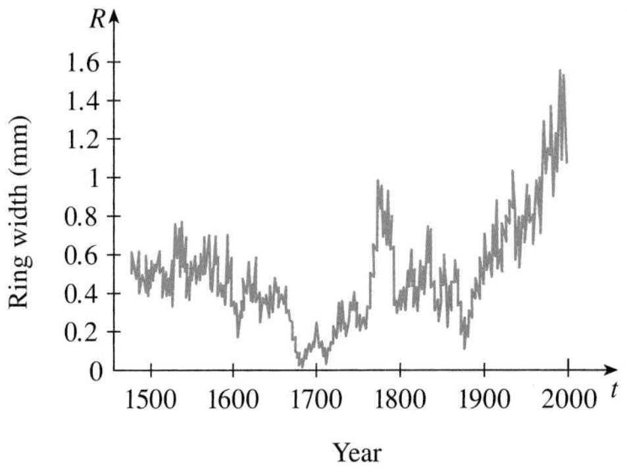

Source: Adapted from G. Jacoby et al., "Mongolian Tree Rings and 20th-Century Warming," *Science* 273 (1996): 771–73.

11. Esophageal pH A healthy esophagus has a pH of about 7.0. When acid reflux occurs, stomach acid (which has pH ranging from 1.0 to 3.0) flows backward from the stomach into the esophagus. When the pH of the esophagus is less than 4.0, the episode is called "clinical acid reflux" and can cause ulcers and damage the lining of the esophagus. The graph shows esophageal pH for a sleeping patient with acid reflux. During what time interval is the patient considered to have an episode of clinical acid reflux?

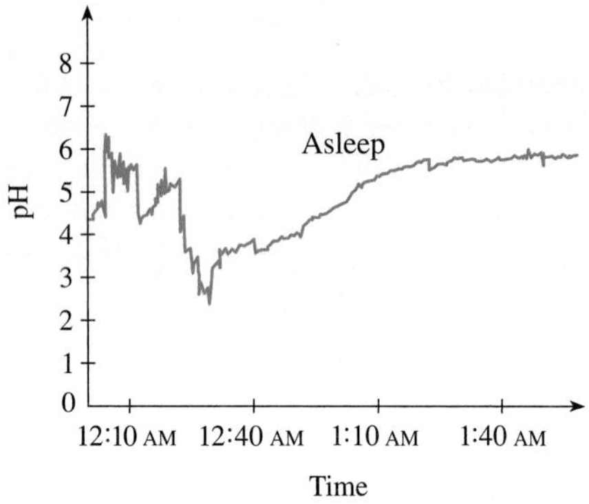

Source: Adapted from T. Demeester et al., "Patterns of Gastroesophageal Reflux in Health and Disease," *Annals of Surgery* 184 (1976): 459–70.

12. Tadpole weights The figure shows the average body weights of tadpoles raised in different densities. The function f shows body weights when the density is 10 tadpoles/L. For functions g and h the densities are 80 and 160 tadpoles/L, respectively. What do these graphs tell you about the effect of crowding?

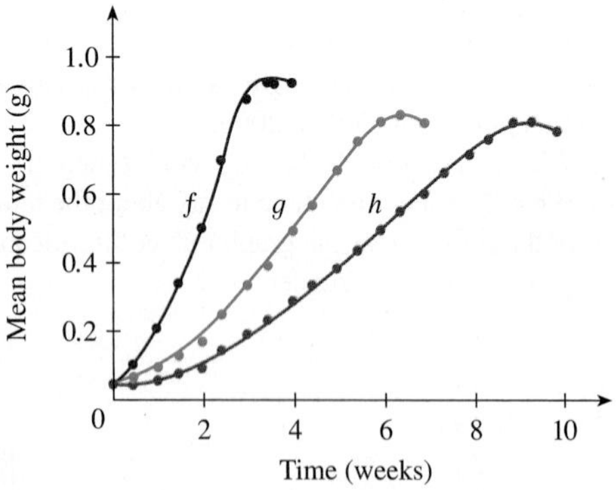

Source: Adapted from P. Russell et al., *Biology: The Dynamic Science* (Belmont, CA: Cengage Learning, 2011), 1156.

13. Species richness Tropical regions receive more rainfall and intense sunlight and have longer growing seasons than regions farther from the equator. As a result, they enjoy greater species richness, that is, greater numbers of species. The graph shows how species richness varies with latitude for ants.

(a) How many species would you expect to find at 30°S? At 20°N?

(b) If you found about 100 ant species at a certain location, at roughly what latitude would you be?

(c) What symmetry property does this function possess?

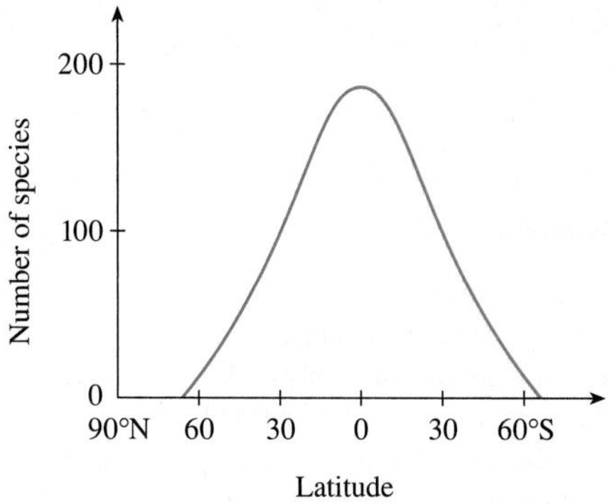

Source: Adapted from P. Russell et al., *Biology: The Dynamic Science* (Belmont, CA: Cengage Learning, 2011), 1190.

14. In this section we discussed examples of ordinary, everyday functions: Population is a function of time, postage cost is a function of weight, water temperature is a function of time. Give three other examples of functions from everyday life that are described verbally. What can you say about the domain and range of each of your functions? If possible, sketch a rough graph of each function.

15. The graph shown gives the weight of a certain person as a function of age. Describe in words how this person's weight varies over time. What do you think happened when this person was 30 years old?

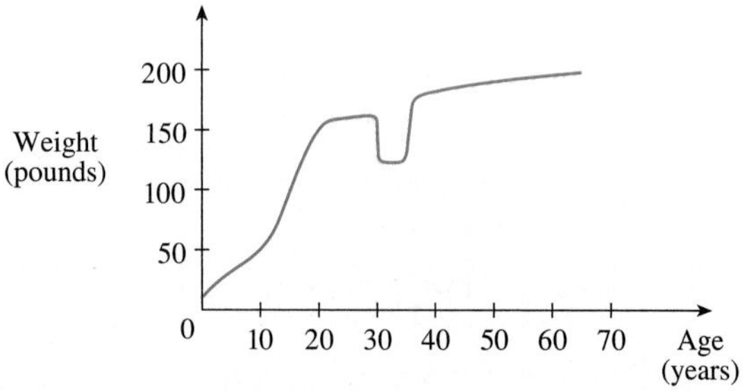

16. Ground reaction force in walking The graph shows the horizontal force exerted by the ground on a person during walking. Positive values are forces in the forward direction and negative values are forces in the backward direction. Give an explanation for the shape of the graph of the force function, including the points where it crosses the axis.

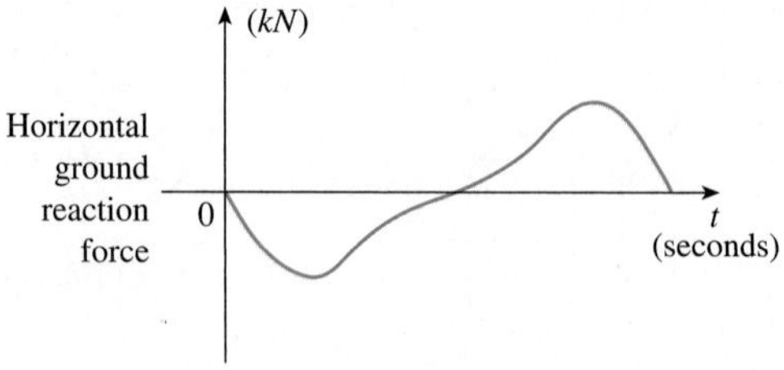

17. You put some ice cubes in a glass, fill the glass with cold water, and then let the glass sit on a table. Describe how the temperature of the water changes as time passes. Then sketch a rough graph of the temperature of the water as a function of the elapsed time.

18. Three runners compete in a 100-meter race. The graph depicts the distance run as a function of time for each runner. Describe in words what the graph tells you about this race. Who won the race? Did each runner finish the race?

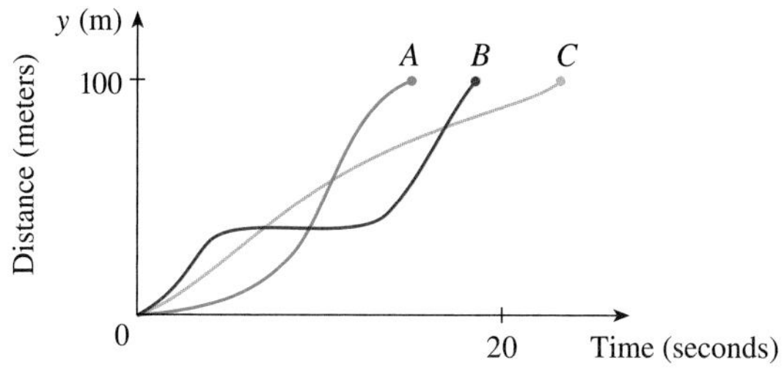

19. Bacteria count Shown is a typical graph of the number N of bacteria grown in a batch culture as a function of time t. Describe what you think is happening during each of the four phases.

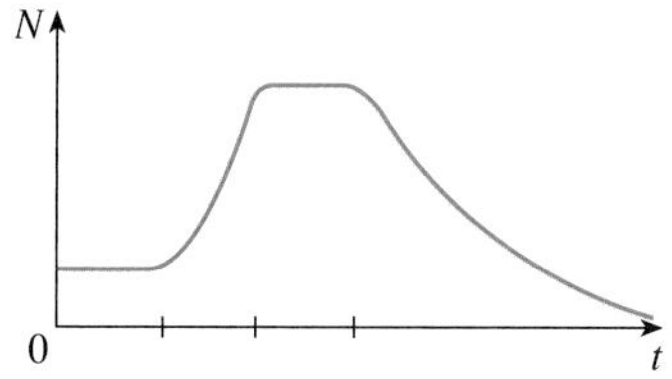

20. Sketch a rough graph of the number of hours of daylight as a function of the time of year.

21. Sketch a rough graph of the outdoor temperature as a function of time during a typical spring day.

22. You place a frozen pie in an oven and bake it for an hour. Then you take it out and let it cool before eating it. Describe how the temperature of the pie changes as time passes. Then sketch a rough graph of the temperature of the pie as a function of time.

23. Sketch the graph of the amount of a particular brand of coffee sold by a store as a function of the price of the coffee.

24. Sketch a rough graph of the market value of a new car as a function of time for a period of 20 years. Assume the car is well maintained.

25. Bird count The table shows the number of house finches, in thousands, observed in the Christmas bird count in California.

Year	1980	1985	1990	1995	2000	2005	2010
Count	74	92	88	107	70	61	78

(a) Use the data to sketch a rough graph of the bird count as a function of time.

(b) Use your graph to estimate the count in 1997.

26. Blood alcohol concentration Researchers measured the blood alcohol concentration (BAC) of eight adult male subjects after rapid consumption of 30 mL of ethanol (corresponding to two standard alcoholic drinks). The table shows the data they obtained by averaging the BAC (in mg/mL) of the eight men.

t (hours)	0.0	0.2	0.5	0.75	1.0	1.25	1.5
BAC	0	0.25	0.41	0.40	0.33	0.29	0.24

t (hours)	1.75	2.0	2.25	2.5	3.0	3.5	4.0
BAC	0.22	0.18	0.15	0.12	0.07	0.03	0.01

(a) Use the readings to sketch the graph of the BAC as a function of t.

(b) Use your graph to describe how the concentration of alcohol varies with time.

Source: Adapted from P. Wilkinson et al., "Pharmacokinetics of Ethanol after Oral Administration in the Fasting State," *Journal of Pharmacokinetics and Biopharmaceutics* 5 (1977): 207–24.

27. If $f(x) = 3x^2 - x + 2$, find $f(2)$, $f(-2)$, $f(a)$, $f(-a)$, $f(a + 1)$, $2f(a)$, $f(2a)$, $f(a^2)$, $[f(a)]^2$, and $f(a + h)$.

28. A spherical balloon with radius r inches has volume $V(r) = \frac{4}{3}\pi r^3$. Find a function that represents the amount of air required to inflate the balloon from a radius of r inches to a radius of $r + 1$ inches.

29–32 Evaluate the difference quotient for the given function. Simplify your answer.

29. $f(x) = 4 + 3x - x^2, \quad \dfrac{f(3 + h) - f(3)}{h}$

30. $f(x) = x^3, \quad \dfrac{f(a + h) - f(a)}{h}$

31. $f(x) = \dfrac{1}{x}, \quad \dfrac{f(x) - f(a)}{x - a}$

32. $f(x) = \dfrac{x + 3}{x + 1}, \quad \dfrac{f(x) - f(1)}{x - 1}$

33–39 Find the domain of the function.

33. $f(x) = \dfrac{x + 4}{x^2 - 9}$

34. $f(x) = \dfrac{2x^3 - 5}{x^2 + x - 6}$

35. $f(t) = \sqrt[3]{2t - 1}$

36. $g(t) = \sqrt{3 - t} - \sqrt{2 + t}$

37. $h(x) = \dfrac{1}{\sqrt[4]{x^2 - 5x}}$

38. $f(u) = \dfrac{u + 1}{1 + \dfrac{1}{u + 1}}$

39. $F(p) = \sqrt{2 - \sqrt{p}}$

40. Find the domain and range and sketch the graph of the function $h(x) = \sqrt{4 - x^2}$.

41–52 Find the domain and sketch the graph of the function.

41. $f(x) = 2 - 0.4x$

42. $F(x) = x^2 - 2x + 1$

43. $f(t) = 2t + t^2$

44. $H(t) = \dfrac{4 - t^2}{2 - t}$

45. $g(x) = \sqrt{x - 5}$

46. $F(x) = |2x + 1|$

47. $G(x) = \dfrac{3x + |x|}{x}$

48. $g(x) = |x| - x$

49. $f(x) = \begin{cases} x + 2 & \text{if } x < 0 \\ 1 - x & \text{if } x \geqslant 0 \end{cases}$

50. $f(x) = \begin{cases} 3 - \frac{1}{2}x & \text{if } x \leqslant 2 \\ 2x - 5 & \text{if } x > 2 \end{cases}$

51. $f(x) = \begin{cases} x + 2 & \text{if } x \leqslant -1 \\ x^2 & \text{if } x > -1 \end{cases}$

52. $f(x) = \begin{cases} x + 9 & \text{if } x < -3 \\ -2x & \text{if } |x| \leqslant 3 \\ -6 & \text{if } x > 3 \end{cases}$

53–57 Find a formula for the described function and state its domain.

53. A rectangle has perimeter 20 m. Express the area of the rectangle as a function of the length of one of its sides.

54. A rectangle has area 16 m^2. Express the perimeter of the rectangle as a function of the length of one of its sides.

55. Express the area of an equilateral triangle as a function of the length of a side.

56. Express the surface area of a cube as a function of its volume.

57. An open rectangular box with volume 2 m^3 has a square base. Express the surface area of the box as a function of the length of a side of the base.

58. A cell phone plan has a basic charge of \$35 a month. The plan includes 400 free minutes and charges 10 cents for each additional minute of usage. Write the monthly cost C as a function of the number x of minutes used and graph C as a function of x for $0 \leqslant x \leqslant 600$.

59. A hotel chain charges \$75 each night for the first two nights and \$50 for each additional night's stay. Express the total cost T as a function of the number of nights x that a guest stays.

60. The function in Example 11 is called a *step function* because its graph looks like stairs. Give two other examples of step functions that arise in everyday life.

61. Temperature chart The figure shows the temperature of a patient infected with the malaria species *P. malariae*. Estimate the period of the temperature function.

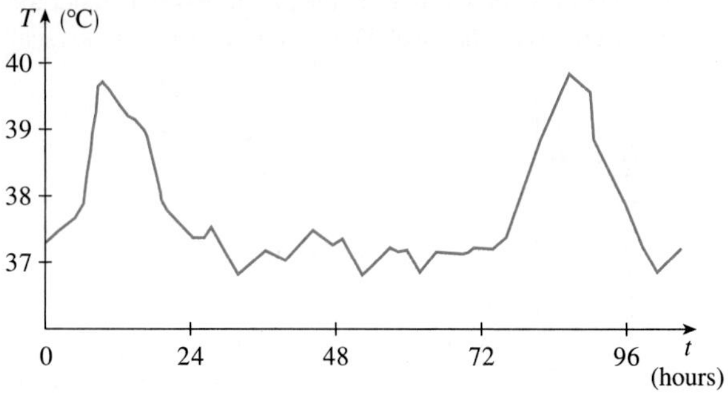

Source: Adapted from L. Bruce-Chwatt, *Essential Malariology* (New York: Wiley, 1985).

62. Malarial fever A temperature chart is shown for a patient with a fever induced by the malaria species *P. falciparum*. What do you think is happening?

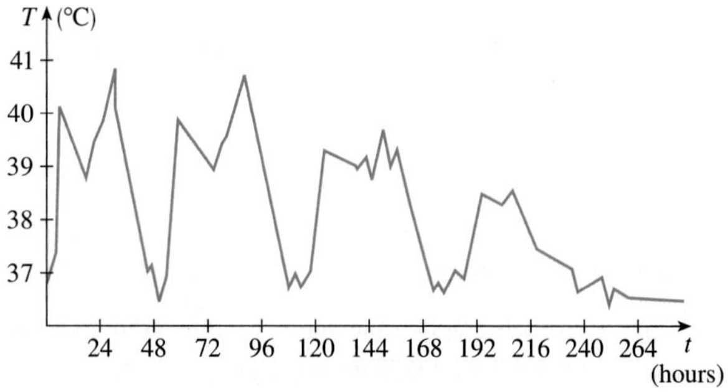

Source: Adapted from L. Bruce-Chwatt, *Essential Malariology* (New York: Wiley, 1985).

63–64 Graphs of f and g are shown. Decide whether each function is even, odd, or neither. Explain your reasoning.

63.

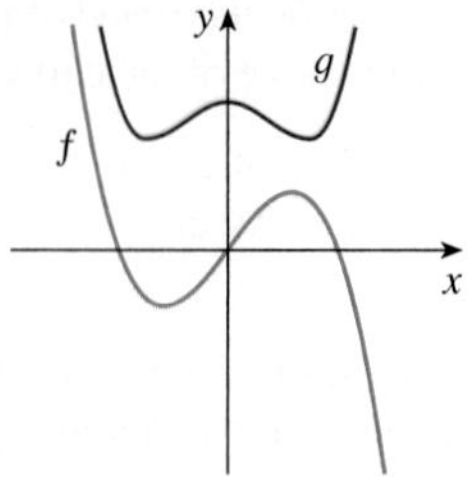

64.

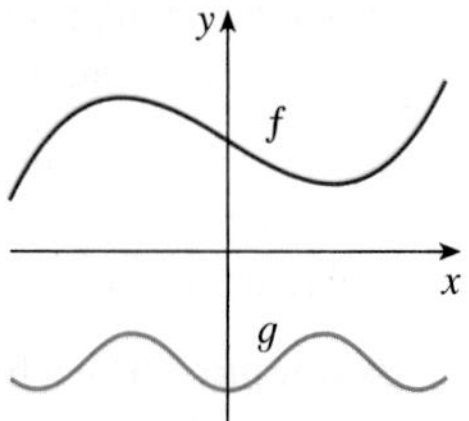

65. (a) If the point (5, 3) is on the graph of an even function, what other point must also be on the graph?
(b) If the point (5, 3) is on the graph of an odd function, what other point must also be on the graph?

66. A function f has domain $[-5, 5]$ and a portion of its graph is shown.
(a) Complete the graph of f if it is known that f is even.
(b) Complete the graph of f if it is known that f is odd.

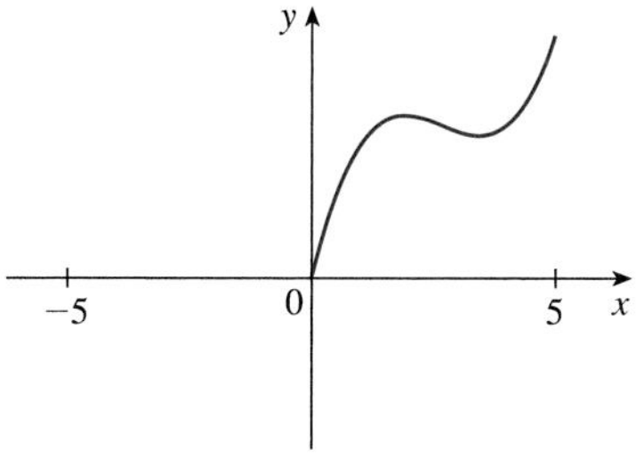

67–72 Determine whether f is even, odd, or neither. If you have a graphing calculator, use it to check your answer visually.

67. $f(x) = \dfrac{x}{x^2 + 1}$

68. $f(x) = \dfrac{x^2}{x^4 + 1}$

69. $f(x) = \dfrac{x}{x + 1}$

70. $f(x) = x|x|$

71. $f(x) = 1 + 3x^2 - x^4$

72. $f(x) = 1 + 3x^3 - x^5$

73. If f and g are both even functions, is $f + g$ even? If f and g are both odd functions, is $f + g$ odd? What if f is even and g is odd? Justify your answers.

74. If f and g are both even functions, is the product fg even? If f and g are both odd functions, is fg odd? What if f is even and g is odd? Justify your answers.

1.2 | A Catalog of Essential Functions

In *Case Studies in Mathematical Modeling* (page xli), we discussed the idea of a mathematical model and the process of mathematical modeling. There are many different types of functions that can be used to model relationships observed in the real world. In this section we discuss the behavior and graphs of these functions and give examples of situations appropriately modeled by such functions.

■ Linear Models

The coordinate geometry of lines is reviewed in Appendix B.

When we say that y is a **linear function** of x, we mean that the graph of the function is a line, so we can use the slope-intercept form of the equation of a line to write a formula for the function as

$$y = f(x) = mx + b$$

where m is the slope of the line and b is the y-intercept.

A characteristic feature of linear functions is that they grow at a constant rate. For instance, Figure 1 shows a graph of the linear function $f(x) = 3x - 2$ and a table of sample values. Notice that whenever x increases by 0.1, the value of $f(x)$ increases by 0.3. So $f(x)$ increases three times as fast as x. Thus the slope of the graph $y = 3x - 2$, namely 3, can be interpreted as the rate of change of y with respect to x.

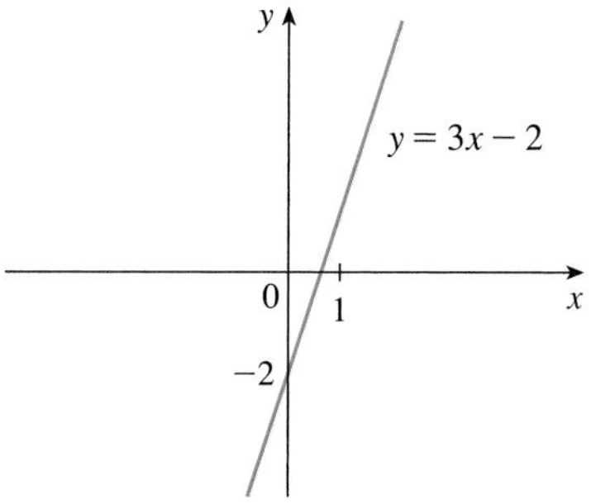

x	$f(x) = 3x - 2$
1.0	1.0
1.1	1.3
1.2	1.6
1.3	1.9
1.4	2.2
1.5	2.5

FIGURE 1

A special case of a linear function occurs when we talk about *direct variation*. If the quantities x and y are related by an equation $y = kx$ for some constant $k \neq 0$, we say that **y varies directly as x**, or **y is proportional to x**. The constant k is called the **constant of proportionality**. Equivalently, we can write $f(x) = kx$, where f is a linear function whose graph has slope k and y-intercept 0.

EXAMPLE 1

(a) As dry air moves upward, it expands and cools. If the ground temperature is 20°C and the temperature at a height of 1 km is 10°C, express the temperature T (in °C) as a function of the height h (in kilometers), assuming that a linear model is appropriate.
(b) Draw the graph of the function in part (a). What does the slope represent?
(c) What is the temperature at a height of 2.5 km?

SOLUTION

(a) Because we are assuming that T is a linear function of h, we can write

$$T = mh + b$$

We are given that $T = 20$ when $h = 0$, so

$$20 = m \cdot 0 + b = b$$

In other words, the y-intercept is $b = 20$.

We are also given that $T = 10$ when $h = 1$, so

$$10 = m \cdot 1 + 20$$

The slope of the line is therefore $m = 10 - 20 = -10$ and the required linear function is

$$T = -10h + 20$$

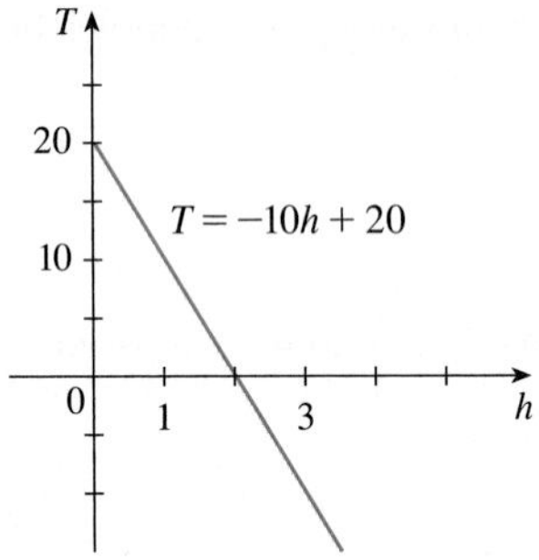

FIGURE 2

(b) The graph is sketched in Figure 2. The slope is $m = -10°\text{C/km}$, and this represents the rate of change of temperature with respect to height.
(c) At a height of $h = 2.5$ km, the temperature is

$$T = -10(2.5) + 20 = -5°\text{C}$$ ■

If there is no physical law or principle to help us formulate a model, we construct an **empirical model**, which is based entirely on collected data. We seek a curve that "fits" the data in the sense that it captures the basic trend of the data points.

EXAMPLE 2 | BB **Carbon dioxide in the atmosphere** Table 1 lists the average carbon dioxide level in the atmosphere, measured in parts per million at Mauna Loa Observatory from 1980 to 2012. Use the data in Table 1 to find a model for the carbon dioxide level.

SOLUTION We use the data in Table 1 to make the scatter plot in Figure 3, where t represents time (in years) and C represents the CO_2 level (in parts per million, ppm).

Table 1

Year	CO_2 level (in ppm)	Year	CO_2 level (in ppm)
1980	338.7	1998	366.5
1982	341.2	2000	369.4
1984	344.4	2002	373.2
1986	347.2	2004	377.5
1988	351.5	2006	381.9
1990	354.2	2008	385.6
1992	356.3	2010	389.9
1994	358.6	2012	393.8
1996	362.4		

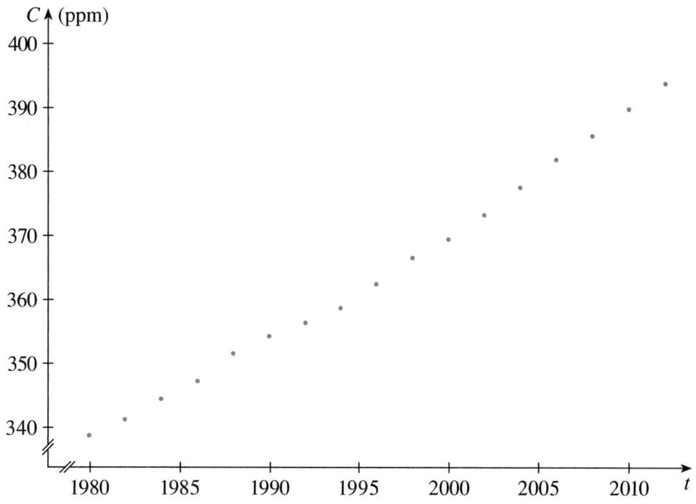

FIGURE 3 Scatter plot for the average CO_2 level

Notice that the data points appear to lie close to a straight line, so it's natural to choose a linear model in this case. But there are many possible lines that approximate these data points, so which one should we use? One possibility is the line that passes through the first and last data points. The slope of this line is

$$\frac{393.8 - 338.7}{2012 - 1980} = \frac{55.1}{32} = 1.721875 \approx 1.722$$

We write its equation as

$$C - 338.7 = 1.722(t - 1980)$$

or

$$C = 1.722t - 3070.86 \qquad (1)$$

Equation 1 gives one possible linear model for the carbon dioxide level; it is graphed in Figure 4.

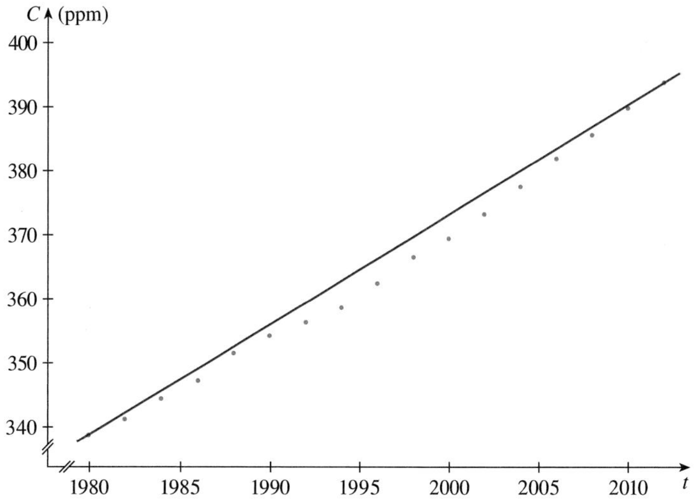

FIGURE 4 Linear model through first and last data points

A computer or graphing calculator finds the regression line by the method of **least squares**, which is to minimize the sum of the squares of the vertical distances between the data points and the line. The details are explained in Section 11.3.

Notice that our model gives values higher than most of the actual CO_2 levels. A better linear model is obtained by a procedure from statistics called *linear regression*. If we use a graphing calculator, we enter the data from Table 1 into the data editor and choose the linear regression command. (With Maple we use the fit[leastsquare] command in the stats package; with Mathematica we use the Fit command.) The machine gives the slope and y-intercept of the regression line as

$$m = 1.71262 \qquad b = -3054.14$$

So our least squares model for the CO_2 level is

$$C = 1.71262t - 3054.14 \tag{2}$$

In Figure 5 we graph the regression line as well as the data points. Comparing with Figure 4, we see that it gives a better fit than our previous linear model.

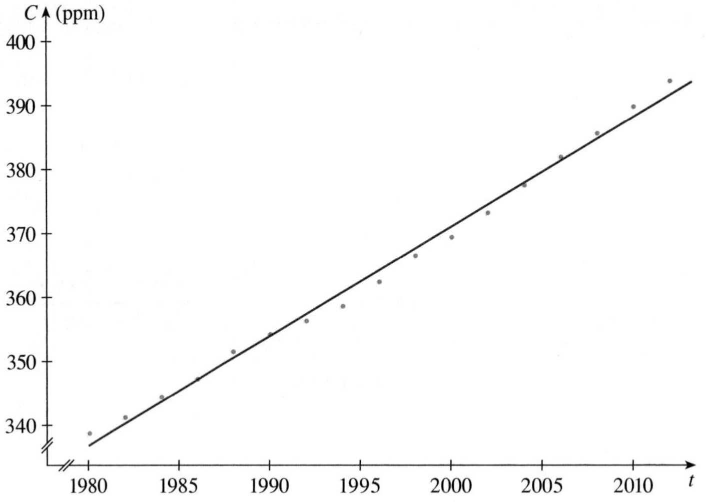

FIGURE 5
The regression line

EXAMPLE 3 | Interpolating and extrapolating the CO_2 level Use the linear model given by Equation 2 to estimate the average CO_2 level for 1987 and to predict the level for the year 2020. According to this model, when will the CO_2 level exceed 420 parts per million?

SOLUTION Using Equation 2 with $t = 1987$, we estimate that the average CO_2 level in 1987 was

$$C(1987) = (1.71262)(1987) - 3054.14 \approx 348.84$$

This is an example of *interpolation* because we have estimated a value *between* observed values. (In fact, the Mauna Loa Observatory reported that the average CO_2 level in 1987 was 348.93 ppm, so our estimate is remarkably accurate.)

With $t = 2020$, we get

$$C(2020) = (1.71262)(2020) - 3054.14 \approx 405.35$$

So we predict that the average CO_2 level in the year 2020 will be 405.4 ppm. This is an example of *extrapolation* because we have predicted a value *outside* the time frame of observations. Consequently, we are far less certain about the accuracy of our prediction.

Using Equation 2, we see that the CO_2 level exceeds 420 ppm when

$$1.71262t - 3054.14 > 420$$

Solving this inequality, we get

$$t > \frac{3474.14}{1.71262} \approx 2028.55$$

We therefore predict that the CO_2 level will exceed 420 ppm by the year 2029. This prediction is risky because it involves a time quite remote from our observations. In fact, we see from Figure 5 that the trend has been for CO_2 levels to increase rather more rapidly in recent years, so the level might exceed 420 ppm well before 2029. ■

■ Polynomials

A function P is called a **polynomial** if

$$P(x) = a_n x^n + a_{n-1} x^{n-1} + \cdots + a_2 x^2 + a_1 x + a_0$$

where n is a nonnegative integer and the numbers $a_0, a_1, a_2, \ldots, a_n$ are constants called the **coefficients** of the polynomial. The domain of any polynomial is $\mathbb{R} = (-\infty, \infty)$. If the leading coefficient $a_n \neq 0$, then the **degree** of the polynomial is n. For example, the function

$$P(x) = 2x^6 - x^4 + \tfrac{2}{5}x^3 + \sqrt{2}$$

is a polynomial of degree 6.

A polynomial of degree 1 is of the form $P(x) = mx + b$ and so it is a linear function. A polynomial of degree 2 is of the form $P(x) = ax^2 + bx + c$ and is called a **quadratic function**. Its graph is always a parabola obtained by shifting the parabola $y = ax^2$, as we will see in the next section. The parabola opens upward if $a > 0$ and downward if $a < 0$. (See Figure 6.)

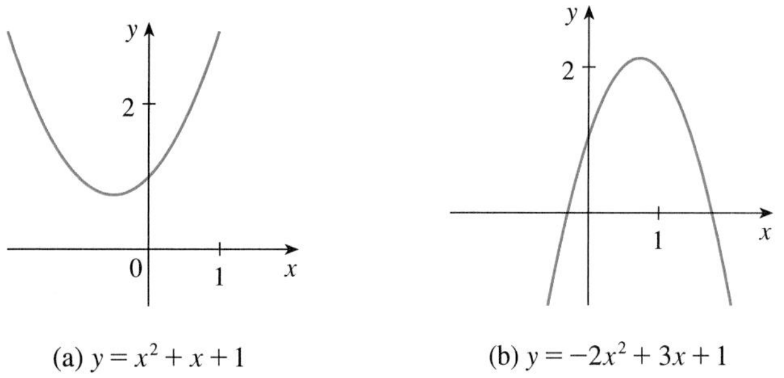

FIGURE 6
The graphs of quadratic functions are parabolas.

A polynomial of degree 3 is of the form

$$P(x) = ax^3 + bx^2 + cx + d \qquad a \neq 0$$

and is called a **cubic function**. Figure 7 shows the graph of a cubic function in part (a) and graphs of polynomials of degrees 4 and 5 in parts (b) and (c). We will see later why the graphs have these shapes.

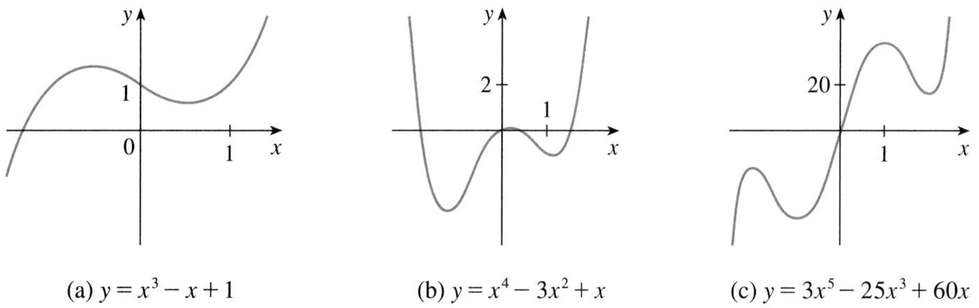

FIGURE 7

Polynomials are commonly used to model various quantities that occur in biology. Figure 8 shows a quadratic model of the vertical trajectory of zebra finches. (Digitized points representing the position of the bird's eye were used in fitting the curve.) Such birds use "flap-bounding." This means that they flap their wings rapidly to gain dynamic energy and then fold their wings into their body for a period of time and act as a projectile.

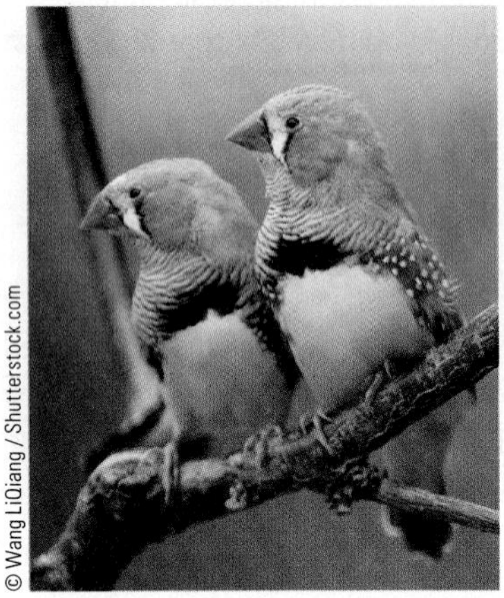

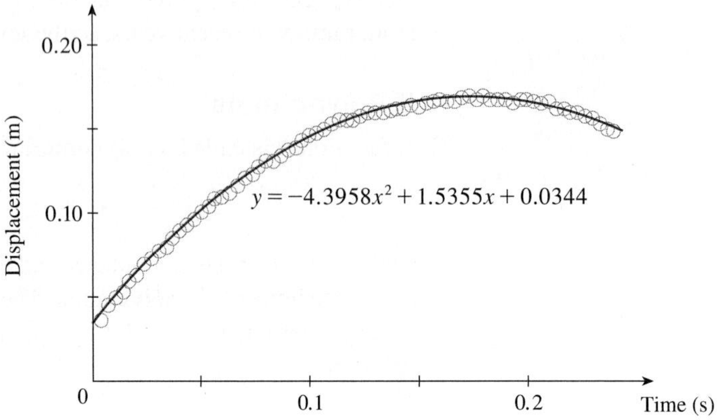

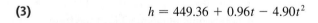

FIGURE 8 Zebra finch trajectory

Source: Adapted from B. Tobalske et al., "Kinematics of Flap-Bounding Flight in the Zebra Finch Over a Wide Range of Speeds," *Journal of Experimental Biology* 202 (1999): 1725–39.

In the following example we use a quadratic function to model the fall of a ball.

Table 2

Time (seconds)	Height (meters)
0	450
1	445
2	431
3	408
4	375
5	332
6	279
7	216
8	143
9	61

EXAMPLE 4 | A ball is dropped from the upper observation deck of the CN Tower, 450 m above the ground, and its height h above the ground is recorded at 1-second intervals in Table 2. Find a model to fit the data and use the model to predict the time at which the ball hits the ground.

SOLUTION We draw a scatter plot of the data in Figure 9 and observe that a linear model is inappropriate. But it looks as if the data points might lie on a parabola, so we try a quadratic model instead. Using a graphing calculator or computer algebra system (which uses the least squares method), we obtain the following quadratic model:

(3) $$h = 449.36 + 0.96t - 4.90t^2$$

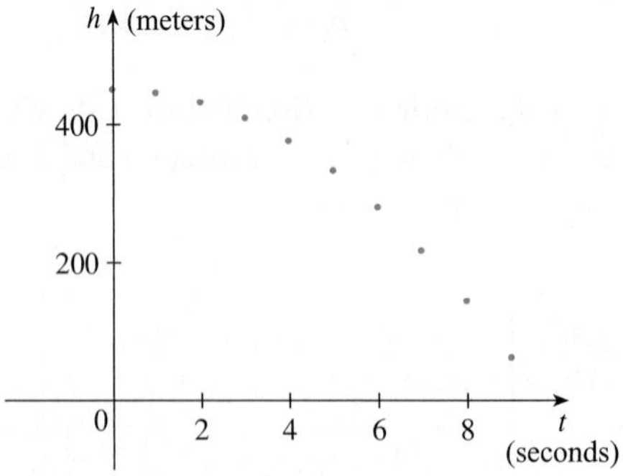

FIGURE 9
Scatter plot for a falling ball

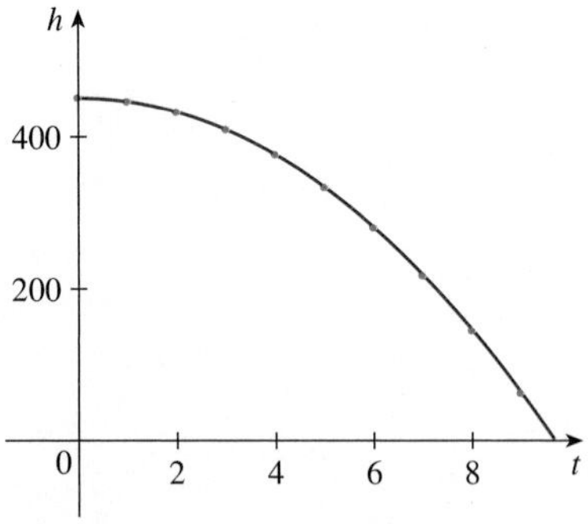

FIGURE 10
Quadratic model for a falling ball

In Figure 10 we plot the graph of Equation 3 together with the data points and see that the quadratic model gives a very good fit.

The ball hits the ground when $h = 0$, so we solve the quadratic equation

$$-4.90t^2 + 0.96t + 449.36 = 0$$

The quadratic formula gives

$$t = \frac{-0.96 \pm \sqrt{(0.96)^2 - 4(-4.90)(449.36)}}{2(-4.90)}$$

The positive root is $t \approx 9.67$, so we predict that the ball will hit the ground after about 9.7 seconds. ■

If a scatter plot of data has a single peak, then it may be appropriate to use a quadratic polynomial as a model (as in Figure 8). But the more fluctuation the data exhibit, the higher the degree of the polynomial needed to model the data. In particular, marine biologists sometimes use cubic polynomials to model the length of fish as a function of age in order to track fish populations. (See Exercise 27.) Then the model can be used to estimate the age of fish whose length has been measured.

■ Power Functions

A function of the form $f(x) = x^p$, where p is a constant, is called a **power function**. We consider several cases.

(i) $p = n$, where n is a positive integer

The graphs of $f(x) = x^n$ for $n = 1, 2, 3, 4$, and 5 are shown in Figure 11. (These are polynomials with only one term.) We already know the shape of the graphs of $y = x$ (a line through the origin with slope 1) and $y = x^2$ [a parabola, see Example 1.1.2(b)].

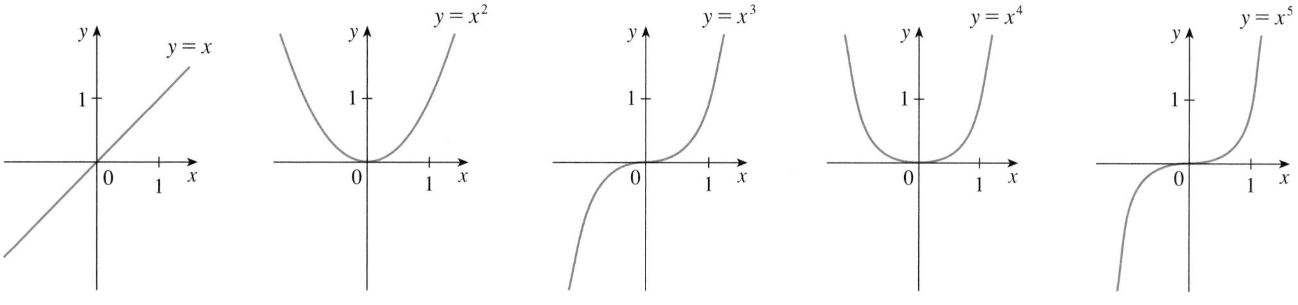

FIGURE 11 Graphs of $f(x) = x^n$ for $n = 1, 2, 3, 4, 5$

The general shape of the graph of $f(x) = x^n$ depends on whether n is even or odd. If n is even, then $f(x) = x^n$ is an even function and its graph is similar to the parabola $y = x^2$. If n is odd, then $f(x) = x^n$ is an odd function and its graph is similar to that of $y = x^3$. Notice from Figure 12 (on page 24), however, that as n increases, the graph of $y = x^n$ becomes flatter near 0 and steeper when $|x| \geqslant 1$. (If x is small, then x^2 is smaller, x^3 is even smaller, x^4 is smaller still, and so on.)

A **family of functions** is a collection of functions whose equations are related. Figure 12 shows two families of power functions, one with even powers and one with odd powers.

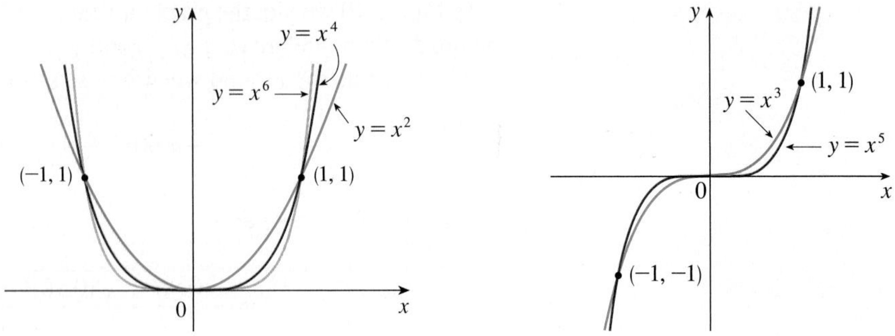

FIGURE 12

(ii) $p = 1/n$, where n is a positive integer

The function $f(x) = x^{1/n} = \sqrt[n]{x}$ is a **root function**. For $n = 2$ it is the square root function $f(x) = \sqrt{x}$, whose domain is $[0, \infty)$ and whose graph is the upper half of the parabola $x = y^2$. [See Figure 13(a).] For other even values of n, the graph of $y = \sqrt[n]{x}$ is similar to that of $y = \sqrt{x}$. For $n = 3$ we have the cube root function $f(x) = \sqrt[3]{x}$ whose domain is $\mathbb{R}$ (recall that every real number has a cube root) and whose graph is shown in Figure 13(b). The graph of $y = \sqrt[n]{x}$ for n odd ($n > 3$) is similar to that of $y = \sqrt[3]{x}$.

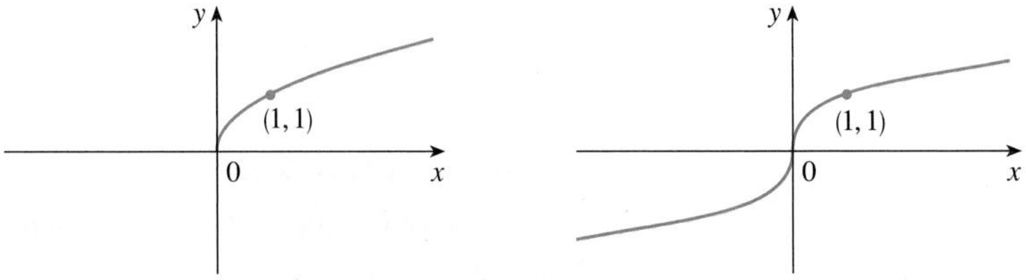

(a) $f(x) = \sqrt{x}$

(b) $f(x) = \sqrt[3]{x}$

FIGURE 13
Graphs of root functions

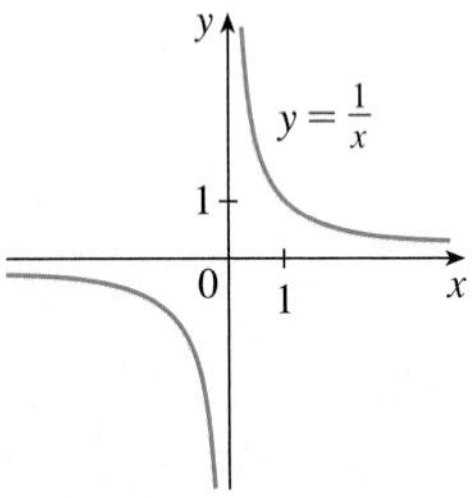

FIGURE 14
The reciprocal function

(iii) $p = -1$

The graph of the **reciprocal function** $f(x) = x^{-1} = 1/x$ is shown in Figure 14. Its graph has the equation $y = 1/x$, or $xy = 1$, and is a hyperbola with the coordinate axes as its asymptotes. This function arises in many areas of the life sciences; one such area is described in the following example.

EXAMPLE 5 | BB Anesthesiology[1] Anesthesiologists often put patients on ventilators during surgery to maintain a steady state concentration C of carbon dioxide in the lungs. If P is the rate of production of CO_2 by the body (measured in mg/min) and V is the ventilation rate (measured as lung volume exchanged per minute, mL/min), then at steady state the production of CO_2 exactly balances removal by ventilation:

$$P \frac{\text{mg}}{\text{min}} = \left(C \frac{\text{mg}}{\text{mL}}\right)\left(V \frac{\text{mL}}{\text{min}}\right)$$

Thus the steady state concentration of CO_2 is inversely proportional to the ventilation rate:

$$C = \frac{P}{V}$$

where P is a constant. The graph of C as a function of V is shown in Figure 15 and has the same general shape as the right half of Figure 14. ■

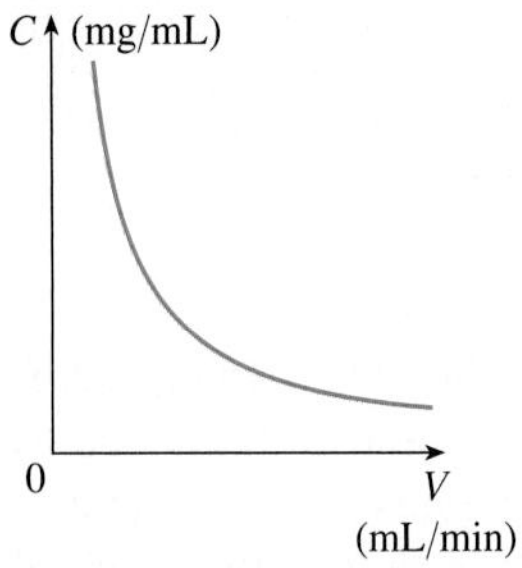

FIGURE 15
Concentration of CO_2 as a function of ventilation rate

1. Adapted from S. Cruickshank, *Mathematics and Statistics in Anaesthesia* (New York: Oxford University Press, USA, 1998).

EXAMPLE 6 | BB **Species richness in bat caves** It makes sense that the larger the area of a region, the larger the number of species that inhabit the region. Many ecologists have modeled the species–area relation with a power function and, in particular, the number of species S of bats living in caves in central Mexico has been related to the surface area A of the caves by the equation $S = 0.14A^{0.64}$. (In Example 1.5.14 this model will be derived from collected data.)

(a) The cave called Misión Imposible near Puebla, Mexico, has a surface area of $A = 60\text{ m}^2$. How many species of bats would you expect to find in that cave?

(b) If you discover that four species of bats live in a cave, estimate the area of the cave.

SOLUTION A graph of the power function model is shown in Figure 16.

(a) According to the model $S = 0.14A^{0.64}$, the expected number of species in a cave with surface area $A = 60\text{ m}^2$ is

$$S = 0.14(60)^{0.64} \approx 1.92$$

So we would expect there to be two species of bats in this cave.

(b) For a cave with four species of bats we have

$$S = 0.14A^{0.64} = 4 \quad \Rightarrow \quad A^{0.64} = \frac{4}{0.14}$$

So

$$A = \left(\frac{4}{0.14}\right)^{1/0.64} \approx 188$$

We predict that a cave with four species of bats would have a surface area of about 190 m^2. ■

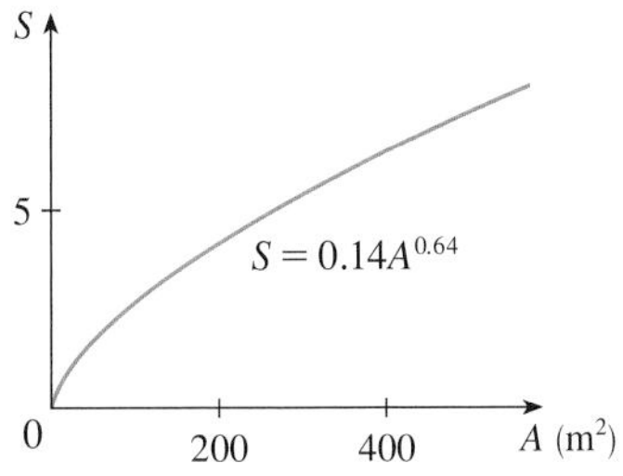

FIGURE 16

The number of different bat species in a cave is related to the size of the cave by a power function.

Source: Derived from A. Brunet et al., "The Species–Area Relationship in Bat Assemblages of Tropical Caves," *Journal of Mammalogy* 82 (2001): 1114–22.

Power functions are also used to model other species–area relationships (Exercise 25), the weight of a bird as a function of wingspan (Exercise 24), illumination as a function of distance from a light source (Exercise 23), and the period of revolution of a planet as a function of its distance from the sun (Exercise 26).

■ Rational Functions

A **rational function** f is a ratio of two polynomials:

$$f(x) = \frac{P(x)}{Q(x)}$$

where P and Q are polynomials. The domain consists of all values of x such that $Q(x) \neq 0$. A simple example of a rational function is the function $f(x) = 1/x$, whose domain is $\{x \mid x \neq 0\}$; this is the reciprocal function graphed in Figure 14. The function

$$f(p) = \frac{p^2 - 2p}{p^2 - 2}$$

arises in models for the spread of drug resistance (see the project on page 78) and is a rational function with domain $\{p \mid p \neq \pm\sqrt{2}\}$. Its complete graph is shown in Figure 17, though when we use the model we will restrict this domain.

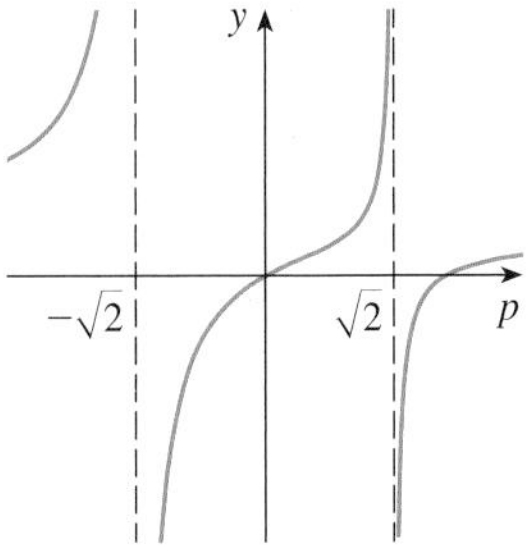

FIGURE 17

■ Algebraic Functions

A function f is called an **algebraic function** if it can be constructed using algebraic operations (such as addition, subtraction, multiplication, division, and taking roots) starting with polynomials. Any rational function is automatically an algebraic function. Here are two more examples:

$$f(x) = \sqrt{x^2 + 1} \qquad g(x) = \frac{x^4 - 16x^2}{x + \sqrt{x}} + (x - 2)\sqrt[3]{x + 1}$$

When we sketch algebraic functions in Chapter 4, we will see that their graphs can assume a variety of shapes. Figure 18 illustrates some of the possibilities.

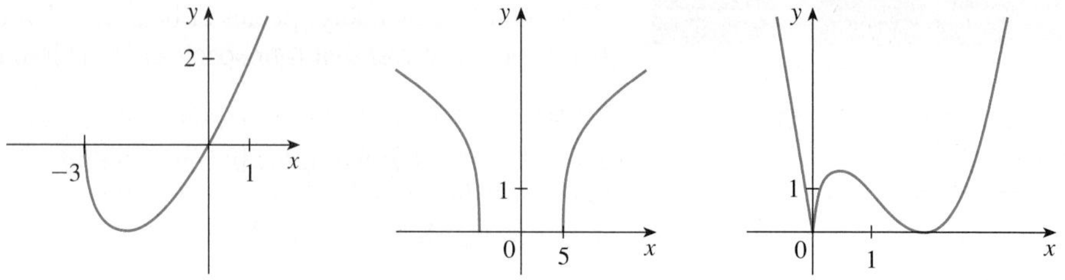

FIGURE 18 (a) $f(x) = x\sqrt{x+3}$ (b) $g(x) = \sqrt[4]{x^2 - 25}$ (c) $h(x) = x^{2/3}(x-2)^2$

An example of an algebraic function occurs in the theory of relativity. The mass of a particle with velocity v is

$$m = f(v) = \frac{m_0}{\sqrt{1 - v^2/c^2}}$$

where m_0 is the rest mass of the particle and $c = 3.0 \times 10^5$ km/s is the speed of light in a vacuum.

■ Trigonometric Functions

The Reference Pages are located at the front of the book.

Trigonometry and the trigonometric functions are reviewed on Reference Page 2 and also in Appendix C. In calculus the convention is that radian measure is always used (except when otherwise indicated). For example, when we use the function $f(x) = \sin x$, it is understood that $\sin x$ means the sine of the angle whose radian measure is x. Thus the graphs of the sine and cosine functions are as shown in Figure 19.

Curves with this general shape are sometimes called *sinusoidal*.

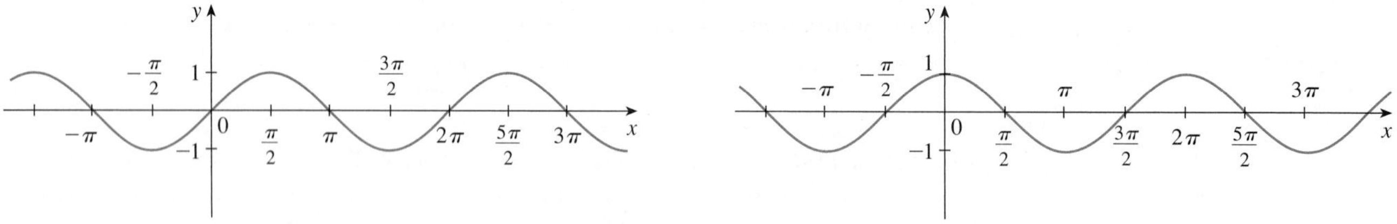

(a) $f(x) = \sin x$ (b) $g(x) = \cos x$

FIGURE 19

Notice that for both the sine and cosine functions the domain is $(-\infty, \infty)$ and the range is the closed interval $[-1, 1]$. Thus, for all values of x, we have

$$-1 \leqslant \sin x \leqslant 1 \qquad -1 \leqslant \cos x \leqslant 1$$

or, in terms of absolute values,

$$|\sin x| \leqslant 1 \qquad |\cos x| \leqslant 1$$

Also, the zeros of the sine function occur at the integer multiples of π; that is,

$$\sin x = 0 \qquad \text{when} \qquad x = n\pi \quad n \text{ an integer}$$

The sine and cosine functions are periodic functions and have period 2π; that is, for all values of x,

$$\sin(x + 2\pi) = \sin x \qquad\qquad \cos(x + 2\pi) = \cos x$$

Although the sine and cosine functions are simple periodic functions, they can be manipulated and combined in ways described in Section 1.3 to model a wide variety of periodic phenomena. For instance, in Example 1.3.4 we will see that a reasonable model for the number of hours of daylight in Philadelphia t days after January 1 is given by the function

$$L(t) = 12 + 2.8 \sin\left[\frac{2\pi}{365}(t - 80)\right]$$

The tangent function is related to the sine and cosine functions by the equation

$$\tan x = \frac{\sin x}{\cos x}$$

and its graph is shown in Figure 20. It is undefined whenever $\cos x = 0$, that is, when $x = \pm\pi/2, \pm3\pi/2, \ldots$. Its range is $(-\infty, \infty)$. Notice that the tangent function has period π:

$$\tan(x + \pi) = \tan x \qquad \text{for all } x$$

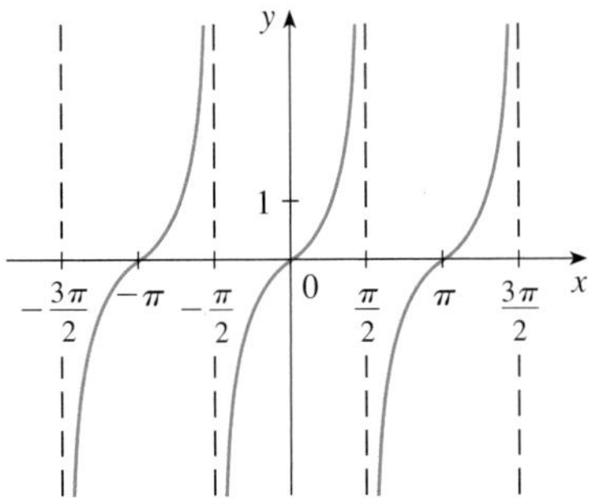

FIGURE 20
$y = \tan x$

The remaining three trigonometric functions (cosecant, secant, and cotangent) are the reciprocals of the sine, cosine, and tangent functions. Their graphs are shown in Appendix C.

■ Exponential Functions

The **exponential functions** are the functions of the form $f(x) = b^x$, where the base b is a positive constant. The graphs of $y = 2^x$ and $y = (0.5)^x$ are shown in Figure 21. In both cases the domain is $(-\infty, \infty)$ and the range is $(0, \infty)$.

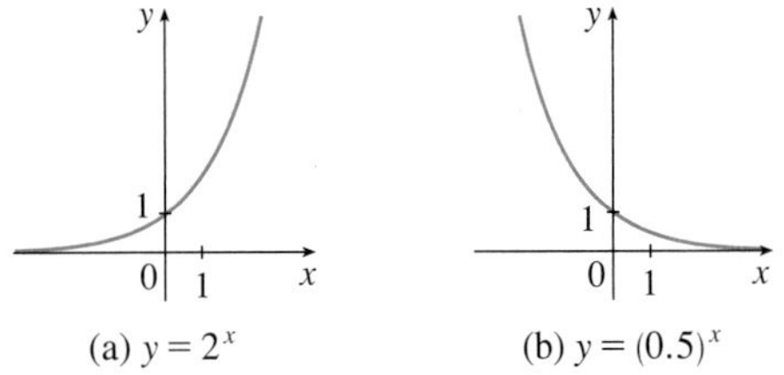

FIGURE 21

Exponential functions will be studied in detail in Section 1.4, and we will see that they are useful for modeling many natural phenomena, such as population growth (if $b > 1$) and radioactive decay (if $b < 1$).

■ Logarithmic Functions

The logarithmic functions $f(x) = \log_b x$, where the base b is a positive constant, are the inverse functions of the exponential functions. They will be studied in Section 1.5.

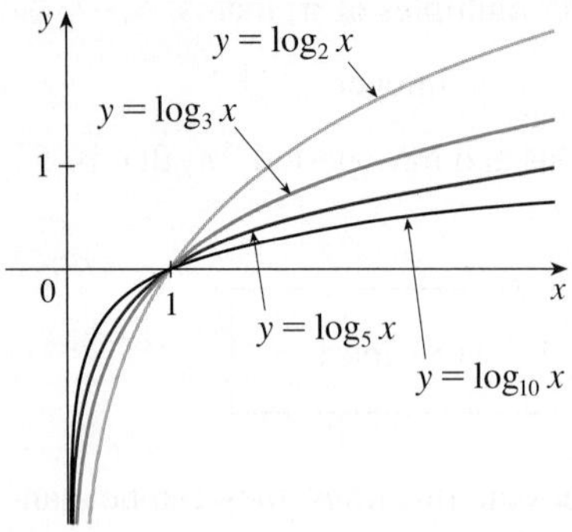

FIGURE 22

Figure 22 shows the graphs of four logarithmic functions with various bases. In each case the domain is $(0, \infty)$, the range is $(-\infty, \infty)$, and the function increases slowly when $x > 1$.

EXAMPLE 7 | Classify the following functions as one of the types of functions that we have discussed.

(a) $f(x) = 5^x$ (b) $g(x) = x^5$

(c) $h(x) = \dfrac{1 + x}{1 - \sqrt{x}}$ (d) $u(t) = 1 - t + 5t^4$

SOLUTION

(a) $f(x) = 5^x$ is an exponential function. (The x is the exponent.)

(b) $g(x) = x^5$ is a power function. (The x is the base.) We could also consider it to be a polynomial of degree 5.

(c) $h(x) = \dfrac{1 + x}{1 - \sqrt{x}}$ is an algebraic function.

(d) $u(t) = 1 - t + 5t^4$ is a polynomial of degree 4. ■

EXERCISES 1.2

1–2 Classify each function as a power function, root function, polynomial (state its degree), rational function, algebraic function, trigonometric function, exponential function, or logarithmic function.

1. (a) $f(x) = \log_2 x$ (b) $g(x) = \sqrt[4]{x}$

(c) $h(x) = \dfrac{2x^3}{1 - x^2}$ (d) $u(t) = 1 - 1.1t + 2.54t^2$

(e) $v(t) = 5^t$ (f) $w(\theta) = \sin\theta \cos^2\theta$

2. (a) $y = \pi^x$ (b) $y = x^\pi$

(c) $y = x^2(2 - x^3)$ (d) $y = \tan t - \cos t$

(e) $y = \dfrac{s}{1 + s}$ (f) $y = \dfrac{\sqrt{x^3 - 1}}{1 + \sqrt[3]{x}}$

3–4 Match each equation with its graph. Explain your choices. (Don't use a computer or graphing calculator.)

3. (a) $y = x^2$ (b) $y = x^5$ (c) $y = x^8$

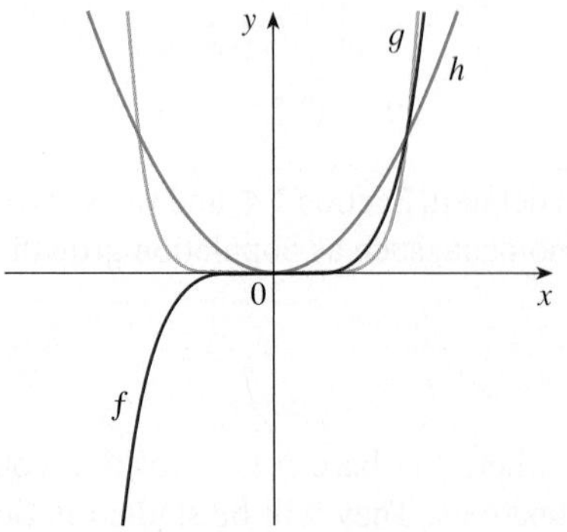

4. (a) $y = 3x$ (b) $y = 3^x$

(c) $y = x^3$ (d) $y = \sqrt[3]{x}$

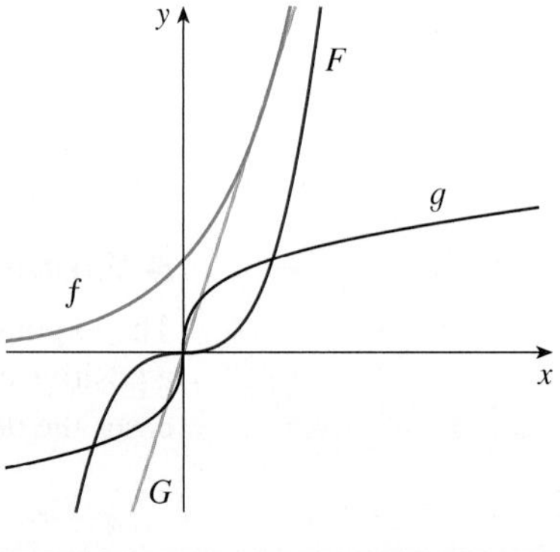

5. (a) Find an equation for the family of linear functions with slope 2 and sketch several members of the family.
(b) Find an equation for the family of linear functions such that $f(2) = 1$ and sketch several members of the family.
(c) Which function belongs to both families?

6. What do all members of the family of linear functions $f(x) = 1 + m(x + 3)$ have in common? Sketch several members of the family.

7. What do all members of the family of linear functions $f(x) = c - x$ have in common? Sketch several members of the family.

8. Find expressions for the quadratic functions whose graphs are shown.

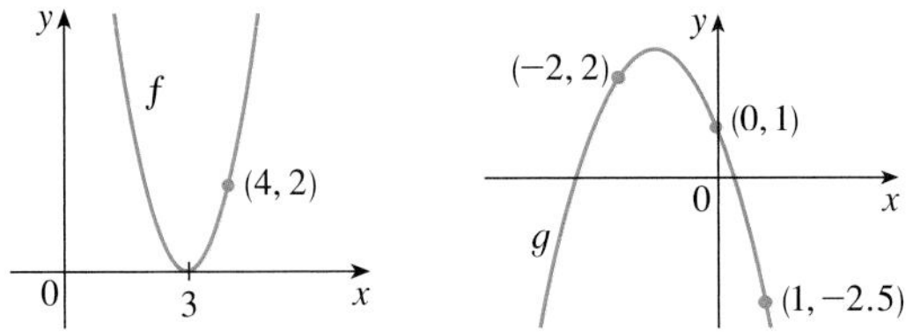

9. Find an expression for a cubic function f if $f(1) = 6$ and $f(-1) = f(0) = f(2) = 0$.

10. Climate change Recent studies indicate that the average surface temperature of the earth has been rising steadily. Some scientists have modeled the temperature by the linear function $T = 0.02t + 8.50$, where T is temperature in °C and t represents years since 1900.

(a) What do the slope and T-intercept represent?

(b) Use the equation to predict the average global surface temperature in 2100.

11. Drug dosage If the recommended adult dosage for a drug is D (in mg), then to determine the appropriate dosage c for a child of age a, pharmacists use the equation $c = 0.0417D(a + 1)$. Suppose the dosage for an adult is 200 mg.

(a) Find the slope of the graph of c. What does it represent?

(b) What is the dosage for a newborn?

12. At the surface of the ocean, the water pressure is the same as the air pressure above the water, 15 lb/in^2. Below the surface, the water pressure increases by 4.34 lb/in^2 for every 10 ft of descent.

(a) Express the water pressure as a function of the depth below the ocean surface.

(b) At what depth is the pressure 100 lb/in^2?

13. The relationship between the Fahrenheit (F) and Celsius (C) temperature scales is given by the linear function $F = \frac{9}{5}C + 32$.

(a) Sketch a graph of this function.

(b) What is the slope of the graph and what does it represent? What is the F-intercept and what does it represent?

14. Absorbing cerebrospinal fluid Cerebrospinal fluid is continually produced and reabsorbed by the body at a rate that depends on its current volume. A medical researcher finds that absorption occurs at a rate of 0.35 mL/min when the volume of fluid is 150 mL and at a rate of 0.14 mL/min when the volume is 50 mL.

(a) Suppose the absorption rate A is a linear function of the volume V. Sketch a graph of $A(V)$.

(b) What is the slope of the graph and what does it represent?

(c) What is the A-intercept of the graph and what does it represent?

15. Biologists have noticed that the **chirping rate of crickets** of a certain species is related to temperature, and the relationship appears to be very nearly linear. A cricket produces 113 chirps per minute at 70°F and 173 chirps per minute at 80°F.

(a) Find a linear equation that models the temperature T as a function of the number of chirps per minute N.

(b) What is the slope of the graph? What does it represent?

(c) If the crickets are chirping at 150 chirps per minute, estimate the temperature.

16. The monthly cost of driving a car depends on the number of miles driven. Lynn found that in May it cost her \$380 to drive 480 mi and in June it cost her \$460 to drive 800 mi.

(a) Express the monthly cost C as a function of the distance driven d, assuming that a linear relationship gives a suitable model.

(b) Use part (a) to predict the cost of driving 1500 miles per month.

(c) Draw the graph of the linear function. What does the slope represent?

(d) What does the C-intercept represent?

(e) Why does a linear function give a suitable model in this situation?

17–18 For each scatter plot, decide what type of function you might choose as a model for the data. Explain your choices.

17. (a)

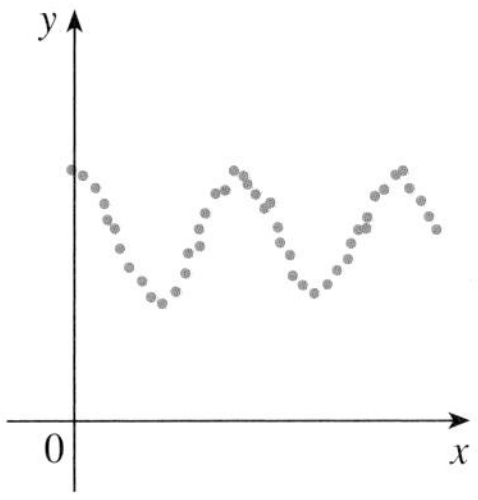

(b)

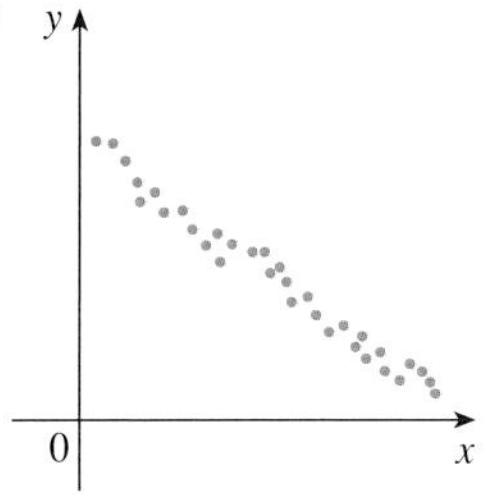

18. (a)

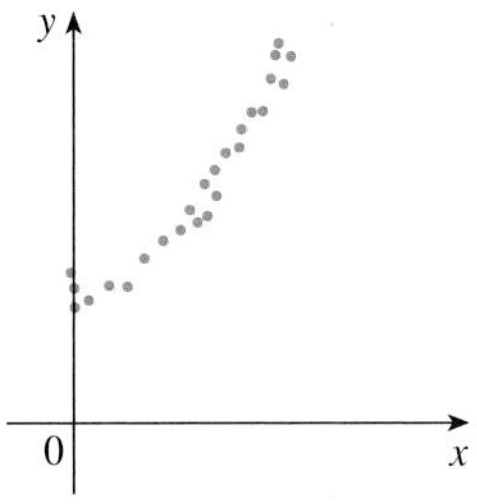

(b)

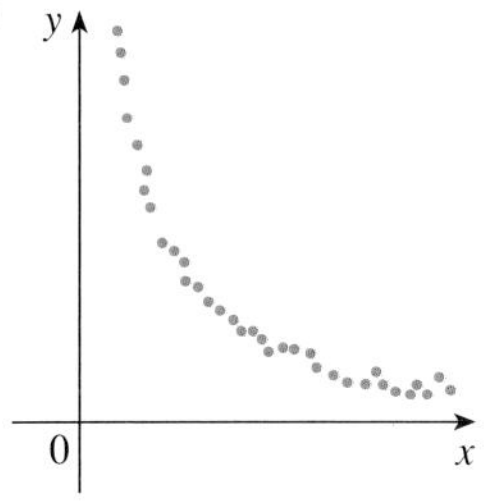

19. Peptic ulcer rate The table on page 30 shows (lifetime) peptic ulcer rates (per 100 population) for various family incomes as reported by the National Health Interview Survey.

(a) Make a scatter plot of these data and decide whether a linear model is appropriate.

(b) Find and graph a linear model using the first and last data points.

(c) Find and graph the least squares regression line.

(d) Use the linear model in part (c) to estimate the ulcer rate for an income of \$25,000.
(e) According to the model, how likely is someone with an income of \$80,000 to suffer from peptic ulcers?
(f) Do you think it would be reasonable to apply the model to someone with an income of \$200,000?

Income	Ulcer rate (per 100 population)
\$4,000	14.1
\$6,000	13.0
\$8,000	13.4
\$12,000	12.5
\$16,000	12.0
\$20,000	12.4
\$30,000	10.5
\$45,000	9.4
\$60,000	8.2

20. Cricket chirping rate In Exercise 15 we modeled temperature as a linear function of the chirping rate of crickets from limited data. Here we use more extensive data in the following table to construct a linear model.

Temperature (°F)	Chirping rate (chirps/min)	Temperature (°F)	Chirping rate (chirps/min)
50	20	75	140
55	46	80	173
60	79	85	198
65	91	90	211
70	113		

(a) Make a scatter plot of the data.
(b) Find and graph the regression line.
(c) Use the linear model in part (b) to estimate the chirping rate at 100°F.

21. Femur length Anthropologists use a linear model that relates human femur (thighbone) length to height. The model allows an anthropologist to determine the height of an individual when only a partial skeleton (including the femur) is found. Here we find the model by analyzing the data on femur length and height for the eight males given in the following table.

Femur length (cm)	Height (cm)	Femur length (cm)	Height (cm)
50.1	178.5	44.5	168.3
48.3	173.6	42.7	165.0
45.2	164.8	39.5	155.4
44.7	163.7	38.0	155.8

(a) Make a scatter plot of the data.
(b) Find and graph the regression line that models the data.
(c) An anthropologist finds a human femur of length 53 cm. How tall was the person?

22. Asbestos and lung tumors When laboratory rats are exposed to asbestos fibers, some of them develop lung tumors. The table lists the results of several experiments by different scientists.
(a) Find the regression line for the data.
(b) Make a scatter plot and graph the regression line. Does the regression line appear to be a suitable model for the data?
(c) What does the y-intercept of the regression line represent?

Asbestos exposure (fibers/mL)	Percent of mice that develop lung tumors	Asbestos exposure (fibers/mL)	Percent of mice that develop lung tumors
50	2	1600	42
400	6	1800	37
500	5	2000	38
900	10	3000	50
1100	26		

23. Many physical quantities are connected by *inverse square laws*, that is, by power functions of the form $f(x) = kx^{-2}$. In particular, the illumination of an object by a light source is inversely proportional to the square of the distance from the source. Suppose that after dark you are in a room with just one lamp and you are trying to read a book. The light is too dim and so you move halfway to the lamp. How much brighter is the light?

24. Wingspan and weight The weight W (in pounds) of a bird (that can fly) has been related to the wingspan L (in inches) of the bird by the power function $L = 30.6W^{0.3952}$. (In Exercise 1.5.66 this model will be derived from data.)
(a) The bald eagle has a wingspan of about 90 inches. Use the model to estimate the weight of the eagle.
(b) An ostrich weighs about 300 pounds. Use the model to estimate what the wingspan of an ostrich should be in order for it to fly.
(c) The wingspan of an ostrich is about 72 inches. Use your answer to part (b) to explain why ostriches can't fly.

25. Species–area relation for reptiles The table shows the number N of species of reptiles and amphibians inhabiting Caribbean islands and the area A of the island in square miles.
(a) Use a power function to model N as a function of A.
(b) The Caribbean island of Dominica has area 291 mi^2. How many species of reptiles and amphibians would you expect to find on Dominica?

Island	A	N
Saba	4	5
Monserrat	40	9
Puerto Rico	3,459	40
Jamaica	4,411	39
Hispaniola	29,418	84
Cuba	44,218	76

26. The table shows the mean (average) distances d of the planets from the sun (taking the unit of measurement to be the distance from the earth to the sun) and their periods T (time of revolution in years).

Planet	d	T
Mercury	0.387	0.241
Venus	0.723	0.615
Earth	1.000	1.000
Mars	1.523	1.881
Jupiter	5.203	11.861
Saturn	9.541	29.457
Uranus	19.190	84.008
Neptune	30.086	164.784

(a) Fit a power model to the data.

(b) Kepler's Third Law of Planetary Motion states that

"The square of the period of revolution of a planet is proportional to the cube of its mean distance from the sun."

Does your model corroborate Kepler's Third Law?

27. Fish growth The table gives the lengths of rock bass caught at different ages, as determined by their otoliths (ear bones in their heads). Scientists have proposed a cubic polynomial to model the data.

BB

Photo by Karna McKinney, AFSC, NOAA Fisheries

(a) Use a cubic polynomial to model the data. Graph the polynomial together with a scatter plot of the data.

(b) Use your model to estimate the length of a 5-year-old rock bass.

(c) A fisherman catches a rock bass that is 20 inches long. Use your model to estimate its age.

Age (years)	Length (inches)	Age (years)	Length (inches)
1	4.8	9	18.2
2	8.8	9	17.1
2	8.0	10	18.8
3	7.9	10	19.5
4	11.9	11	18.9
5	14.4	12	21.7
6	14.1	12	21.9
6	15.8	13	23.8
7	15.6	14	26.9
8	17.8	14	25.1

1.3 New Functions from Old Functions

In this section we start with the basic functions we discussed in Section 1.2 and obtain new functions by shifting, stretching, and reflecting their graphs. We also show how to combine pairs of functions by the standard arithmetic operations and by composition.

■ Transformations of Functions

By applying certain transformations to the graph of a given function we can obtain the graphs of related functions. This will give us the ability to sketch the graphs of many functions quickly by hand. It will also enable us to write equations for given graphs.

Let's first consider **translations**. If c is a positive number, then the graph of $y = f(x) + c$ is just the graph of $y = f(x)$ shifted upward a distance of c units (because each y-coordinate is increased by the same number c). Likewise, if $g(x) = f(x - c)$, where $c > 0$, then the value of g at x is the same as the value of f at $x - c$ (c units to the left of x). Therefore the graph of $y = f(x - c)$ is just the graph of $y = f(x)$ shifted c units to the right (see Figure 1).

> **Vertical and Horizontal Shifts** Suppose $c > 0$. To obtain the graph of
>
> $y = f(x) + c$, shift the graph of $y = f(x)$ a distance c units upward
>
> $y = f(x) - c$, shift the graph of $y = f(x)$ a distance c units downward
>
> $y = f(x - c)$, shift the graph of $y = f(x)$ a distance c units to the right
>
> $y = f(x + c)$, shift the graph of $y = f(x)$ a distance c units to the left

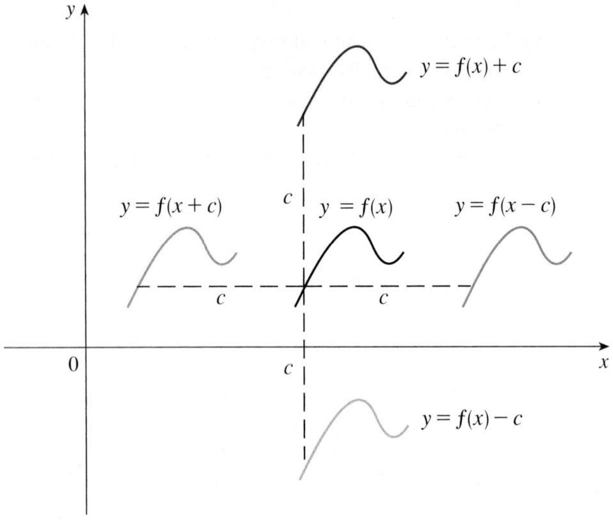

FIGURE 1 Translating the graph of f

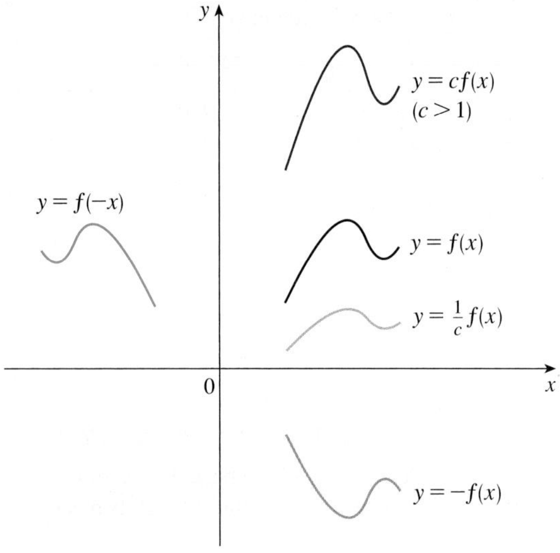

FIGURE 2 Stretching and reflecting the graph f

Now let's consider the **stretching** and **reflecting** transformations. If $c > 1$, then the graph of $y = cf(x)$ is the graph of $y = f(x)$ stretched by a factor of c in the vertical direction (because each y-coordinate is multiplied by the same number c). The graph of $y = -f(x)$ is the graph of $y = f(x)$ reflected about the x-axis because the point (x, y) is replaced by the point $(x, -y)$. (See Figure 2 and the following chart, where the results of other stretching, shrinking, and reflecting transformations are also given.)

> **Vertical and Horizontal Stretching and Reflecting** Suppose $c > 1$. To obtain the graph of
>
> $y = cf(x)$, stretch the graph of $y = f(x)$ vertically by a factor of c
>
> $y = (1/c)f(x)$, shrink the graph of $y = f(x)$ vertically by a factor of c
>
> $y = f(cx)$, shrink the graph of $y = f(x)$ horizontally by a factor of c
>
> $y = f(x/c)$, stretch the graph of $y = f(x)$ horizontally by a factor of c
>
> $y = -f(x)$, reflect the graph of $y = f(x)$ about the x-axis
>
> $y = f(-x)$, reflect the graph of $y = f(x)$ about the y-axis

Figure 3 illustrates these stretching transformations when applied to the cosine function with $c = 2$. For instance, in order to get the graph of $y = 2\cos x$ we multiply the y-coordinate of each point on the graph of $y = \cos x$ by 2. This means that the graph of $y = \cos x$ gets stretched vertically by a factor of 2.

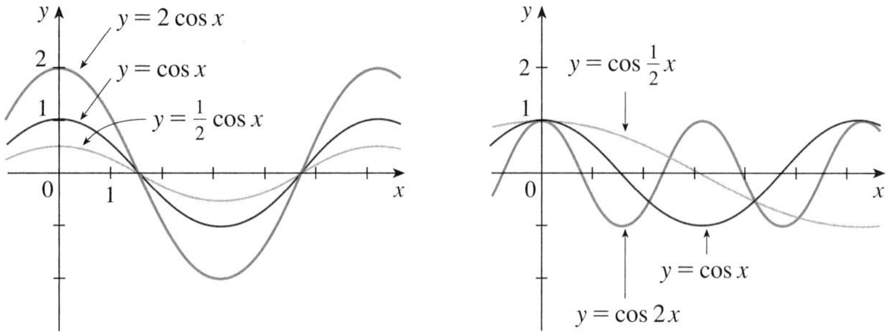

FIGURE 3

EXAMPLE 1 | Given the graph of $y = \sqrt{x}$, use transformations to graph $y = \sqrt{x} - 2$, $y = \sqrt{x - 2}$, $y = -\sqrt{x}$, $y = 2\sqrt{x}$, and $y = \sqrt{-x}$.

SOLUTION The graph of the square root function $y = \sqrt{x}$, obtained from Figure 1.2.13(a), is shown in Figure 4(a). In the other parts of the figure we sketch $y = \sqrt{x} - 2$ by shifting 2 units downward, $y = \sqrt{x - 2}$ by shifting 2 units to the right, $y = -\sqrt{x}$ by reflecting about the x-axis, $y = 2\sqrt{x}$ by stretching vertically by a factor of 2, and $y = \sqrt{-x}$ by reflecting about the y-axis.

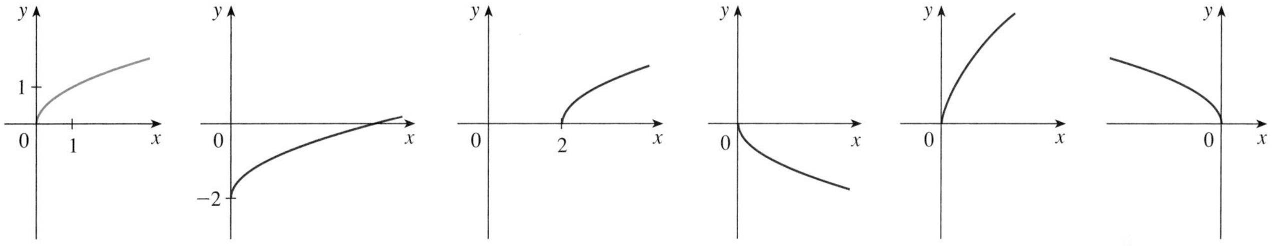

(a) $y = \sqrt{x}$ (b) $y = \sqrt{x} - 2$ (c) $y = \sqrt{x - 2}$ (d) $y = -\sqrt{x}$ (e) $y = 2\sqrt{x}$ (f) $y = \sqrt{-x}$

FIGURE 4

■

EXAMPLE 2 | Sketch the graph of the function $f(x) = x^2 + 6x + 10$.

SOLUTION Completing the square, we write the equation of the graph as

$$y = x^2 + 6x + 10 = (x + 3)^2 + 1$$

This means we obtain the desired graph by starting with the parabola $y = x^2$ and shifting 3 units to the left and then 1 unit upward (see Figure 5).

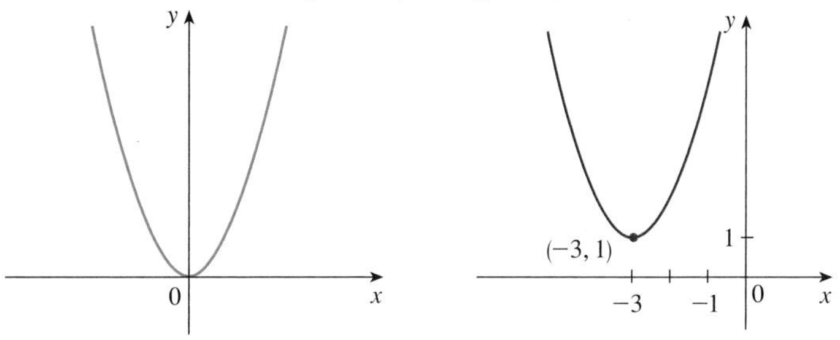

FIGURE 5

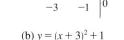

(a) $y = x^2$ (b) $y = (x + 3)^2 + 1$

■

EXAMPLE 3 | Sketch the graphs of the following functions.

(a) $y = \sin 2x$ (b) $y = 1 - \sin x$

SOLUTION

(a) We obtain the graph of $y = \sin 2x$ from that of $y = \sin x$ by compressing horizontally by a factor of 2. (See Figures 6 and 7.) Thus, whereas the period of $y = \sin x$ is 2π, the period of $y = \sin 2x$ is $2\pi/2 = \pi$.

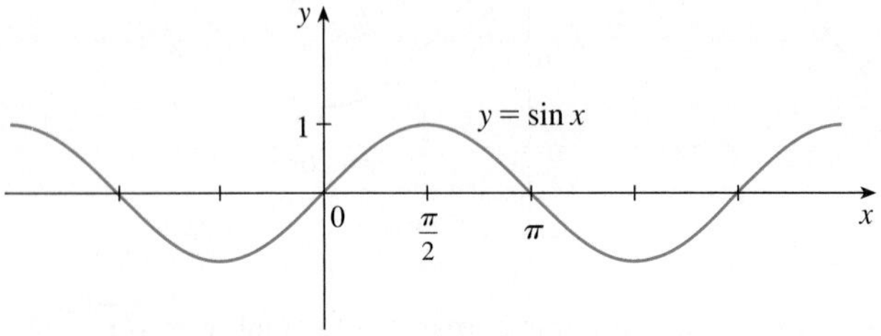

FIGURE 6

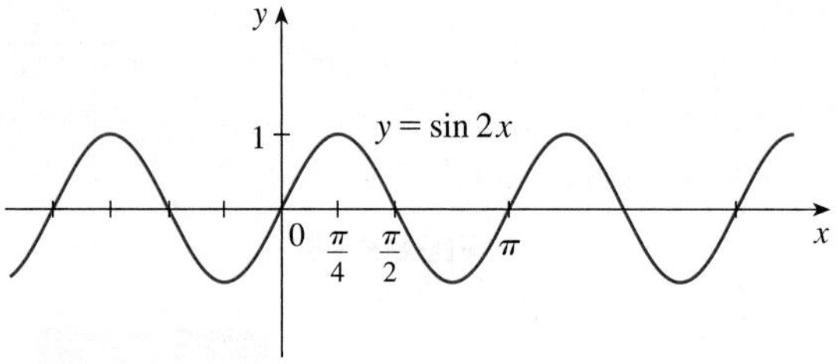

FIGURE 7

(b) To obtain the graph of $y = 1 - \sin x$, we again start with $y = \sin x$. We reflect about the x-axis to get the graph of $y = -\sin x$ and then we shift 1 unit upward to get $y = 1 - \sin x$. (See Figure 8.)

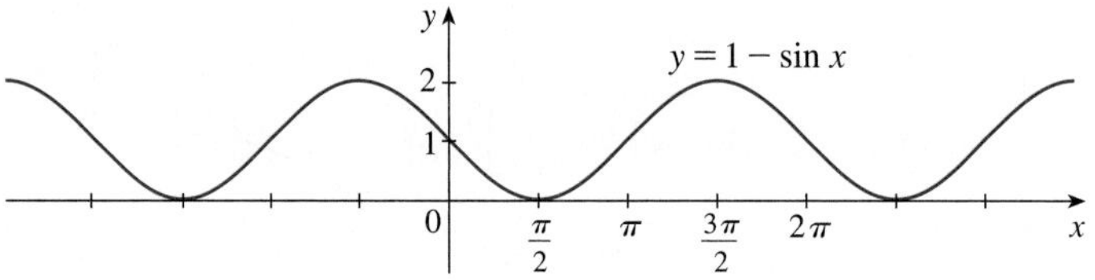

FIGURE 8

■

EXAMPLE 4 | **Hours of daylight** Figure 9 shows graphs of the number of hours of daylight as functions of the time of the year at several latitudes. Given that Philadelphia is located at approximately 40°N latitude, find a function that models the length of daylight at Philadelphia.

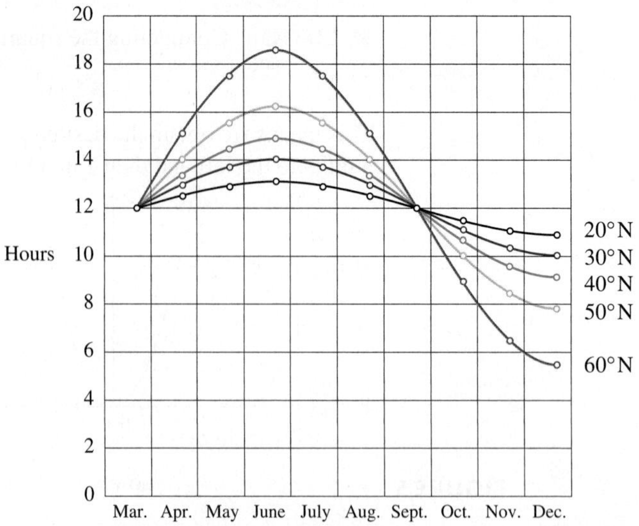

FIGURE 9

Graph of the length of daylight from March 21 through December 21 at various latitudes

Source: Adapted from L. Harrison, *Daylight, Twilight, Darkness and Time* (New York: Silver, Burdett, 1935), 40.

SOLUTION Notice that each curve resembles a shifted and stretched sine function. By looking at the blue curve we see that, at the latitude of Philadelphia, daylight lasts about 14.8 hours on June 21 and 9.2 hours on December 21, so the amplitude of the curve (the factor by which we have to stretch the sine curve vertically) is $\frac{1}{2}(14.8 - 9.2) = 2.8$.

By what factor do we need to stretch the sine curve horizontally if we measure the time t in days? Because there are about 365 days in a year, the period of our model should be 365. But the period of $y = \sin t$ is 2π, so the horizontal stretching factor is $2\pi/365$.

We also notice that the curve begins its cycle on March 21, the 80th day of the year, so we have to shift the curve 80 units to the right. In addition, we shift it 12 units upward. Therefore we model the length of daylight in Philadelphia on the tth day of the year by the function

$$L(t) = 12 + 2.8 \sin\left[\frac{2\pi}{365}(t - 80)\right]$$ ■

■ Combinations of Functions

Two functions f and g can be combined to form new functions $f + g$, $f - g$, fg, and f/g in a manner similar to the way we add, subtract, multiply, and divide real numbers. The sum and difference functions are defined by

$$(f + g)(x) = f(x) + g(x) \qquad (f - g)(x) = f(x) - g(x)$$

If the domain of f is A and the domain of g is B, then the domain of $f + g$ is the intersection $A \cap B$ because both $f(x)$ and $g(x)$ have to be defined. For example, the domain of $f(x) = \sqrt{x}$ is $A = [0, \infty)$ and the domain of $g(x) = \sqrt{2 - x}$ is $B = (-\infty, 2]$, so the domain of $(f + g)(x) = \sqrt{x} + \sqrt{2 - x}$ is $A \cap B = [0, 2]$.

Similarly, the product and quotient functions are defined by

$$(fg)(x) = f(x)g(x) \qquad \left(\frac{f}{g}\right)(x) = \frac{f(x)}{g(x)}$$

The domain of fg is $A \cap B$, but we can't divide by 0 and so the domain of f/g is $\{x \in A \cap B \mid g(x) \neq 0\}$. For instance, if $f(x) = x^2$ and $g(x) = x - 1$, then the domain of the rational function $(f/g)(x) = x^2/(x - 1)$ is $\{x \mid x \neq 1\}$, or $(-\infty, 1) \cup (1, \infty)$.

There is another way of combining two functions to obtain a new function. For example, suppose that $y = f(u) = \sqrt{u}$ and $u = g(x) = x^2 + 1$. Since y is a function of u and u is, in turn, a function of x, it follows that y is ultimately a function of x. We compute this by substitution:

$$y = f(u) = f(g(x)) = f(x^2 + 1) = \sqrt{x^2 + 1}$$

The procedure is called *composition* because the new function is *composed* of the two given functions f and g.

In general, given any two functions f and g, we start with a number x in the domain of g and calculate $g(x)$. If this number $g(x)$ is in the domain of f, then we can calculate the value of $f(g(x))$. Notice that the output of one function is used as the input to the next function. The result is a new function $h(x) = f(g(x))$ obtained by substituting g into f. It is called the *composition* (or *composite*) of f and g and is denoted by $f \circ g$ ("f circle g").

Definition Given two functions f and g, the **composite function** $f \circ g$ (also called the **composition** of f and g) is defined by

$$(f \circ g)(x) = f(g(x))$$

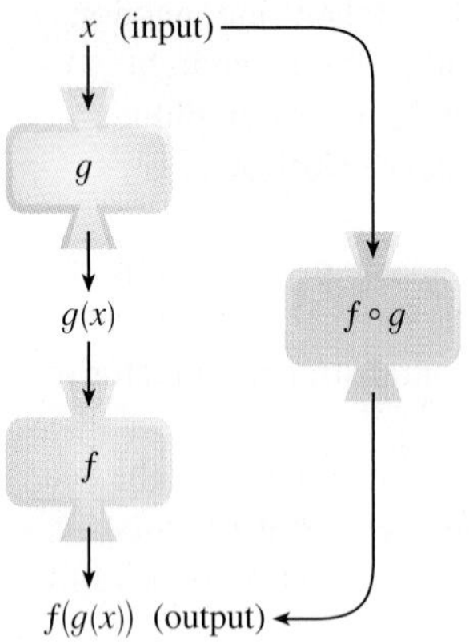

FIGURE 10
The $f \circ g$ machine is composed of the g machine (first) and then the f machine.

The domain of $f \circ g$ is the set of all x in the domain of g such that $g(x)$ is in the domain of f. In other words, $(f \circ g)(x)$ is defined whenever both $g(x)$ and $f(g(x))$ are defined. Figure 10 shows how to picture $f \circ g$ in terms of machines.

EXAMPLE 5 | If $f(x) = x^2$ and $g(x) = x - 3$, find the composite functions $f \circ g$ and $g \circ f$.

SOLUTION We have

$$(f \circ g)(x) = f(g(x)) = f(x - 3) = (x - 3)^2$$

$$(g \circ f)(x) = g(f(x)) = g(x^2) = x^2 - 3 \qquad \blacksquare$$

NOTE You can see from Example 5 that, in general, $f \circ g \neq g \circ f$. Remember, the notation $f \circ g$ means that the function g is applied first and then f is applied second. In Example 5, $f \circ g$ is the function that *first* subtracts 3 and *then* squares; $g \circ f$ is the function that *first* squares and *then* subtracts 3.

EXAMPLE 6 | If $f(x) = \sqrt{x}$ and $g(x) = \sqrt{2 - x}$, find each function and its domain.
(a) $f \circ g$ (b) $g \circ f$ (c) $f \circ f$ (d) $g \circ g$

SOLUTION

(a)
$$(f \circ g)(x) = f(g(x)) = f(\sqrt{2 - x}) = \sqrt{\sqrt{2 - x}} = \sqrt[4]{2 - x}$$

The domain of $f \circ g$ is $\{x \mid 2 - x \geq 0\} = \{x \mid x \leq 2\} = (-\infty, 2]$.

(b)
$$(g \circ f)(x) = g(f(x)) = g(\sqrt{x}) = \sqrt{2 - \sqrt{x}}$$

If $0 \leq a \leq b$, then $a^2 \leq b^2$.

For $\sqrt{x}$ to be defined we must have $x \geq 0$. For $\sqrt{2 - \sqrt{x}}$ to be defined we must have $2 - \sqrt{x} \geq 0$, that is, $\sqrt{x} \leq 2$, or $x \leq 4$. Thus we have $0 \leq x \leq 4$, so the domain of $g \circ f$ is the closed interval $[0, 4]$.

(c)
$$(f \circ f)(x) = f(f(x)) = f(\sqrt{x}) = \sqrt{\sqrt{x}} = \sqrt[4]{x}$$

The domain of $f \circ f$ is $[0, \infty)$.

(d)
$$(g \circ g)(x) = g(g(x)) = g(\sqrt{2 - x}) = \sqrt{2 - \sqrt{2 - x}}$$

This expression is defined when both $2 - x \geq 0$ and $2 - \sqrt{2 - x} \geq 0$. The first inequality means $x \leq 2$, and the second is equivalent to $\sqrt{2 - x} \leq 2$, or $2 - x \leq 4$, or $x \geq -2$. Thus $-2 \leq x \leq 2$, so the domain of $g \circ g$ is the closed interval $[-2, 2]$. $\blacksquare$

It is possible to take the composition of three or more functions. For instance, the composite function $f \circ g \circ h$ is found by first applying h, then g, and then f as follows:

$$(f \circ g \circ h)(x) = f(g(h(x)))$$

EXAMPLE 7 | BB **Antibiotic dosage** Antibiotics are used to treat bacterial sinus infections. If a dosage of x mg is taken orally, suppose that the amount absorbed into the bloodstream through the stomach is $h(x) = 8x/(x + 8)$ mg. If x mg enters the bloodstream, suppose that the amount surviving filtration by the liver is $g(x) = \frac{1}{4}x$. Finally, if x mg survives filtration by the liver, suppose that $f(x) = x - 1$ mg is absorbed into the sinus cavity, provided that $x > 1$ [otherwise $f(x) = 0$].

(a) Use composition of functions to derive the function that relates oral dosage to the amount of drug that reaches the sinus cavity.
(b) Suppose the antibiotic is instead administered by injection. Derive the function that relates dosage to the amount of drug that reaches the sinus cavity.

SOLUTON
(a) The amount of antibiotic after filtration by the liver is

$$g(h(x)) = \tfrac{1}{4}h(x) = \frac{1}{4}\left(\frac{8x}{x+8}\right) = \frac{2x}{x+8}$$

Now $$\frac{2x}{x+8} > 1 \iff 2x > x + 8 \iff x > 8$$

So if $x > 8$, the amount of the drug that reaches the sinus cavity is

$$f(g(h(x))) = f\left(\frac{2x}{x+8}\right) = \frac{2x}{x+8} - 1 = \frac{2x - (x+8)}{x+8} = \frac{x-8}{x+8}$$

Otherwise $f(g(h(x))) = 0$. So we can write the amount that reaches the sinus cavity as

$$f(g(h(x))) = \begin{cases} \dfrac{x-8}{x+8} & \text{if } x > 8 \\ 0 & \text{if } x \leqslant 8 \end{cases}$$

(b) If the drug is administered by injection, then the amount reaching the sinus cavity is

$$f(g(x)) = \begin{cases} \frac{1}{4}x - 1 & \text{if } x > 4 \\ 0 & \text{if } x \leqslant 4 \end{cases}$$

Figure 11 displays the graphs of the functions f, g, and h, as well as the composite functions showing how the amount of antibiotic reaching the sinus cavity depends on the amount administered in both cases. ■

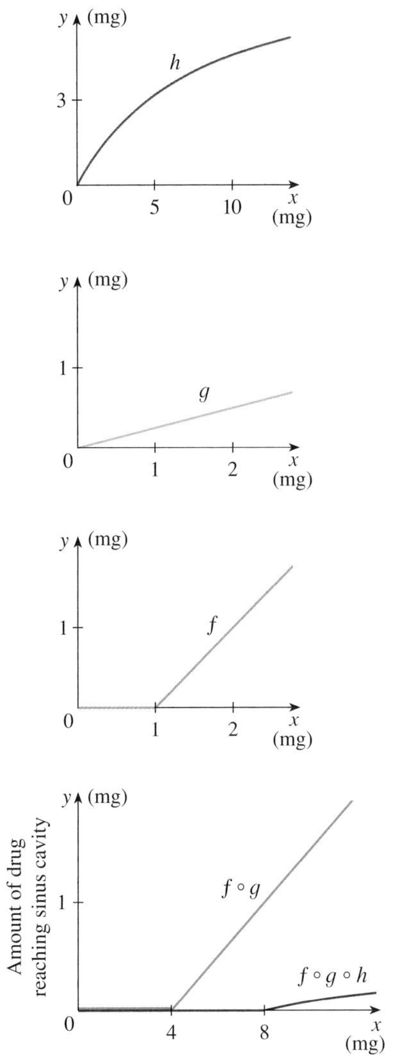

FIGURE 11

So far we have used composition to build complicated functions from simpler ones. But in calculus it is often useful to be able to *decompose* a complicated function into simpler ones, as in the following example.

EXAMPLE 8 | Given $F(x) = \cos^2(x + 9)$, find functions f, g, and h such that $F = f \circ g \circ h$.

SOLUTION Since $F(x) = [\cos(x + 9)]^2$, the formula for F says: First add 9, then take the cosine of the result, and finally square. So we let

$$h(x) = x + 9 \qquad g(x) = \cos x \qquad f(x) = x^2$$

Then

$$(f \circ g \circ h)(x) = f(g(h(x))) = f(g(x + 9)) = f(\cos(x + 9))$$

$$= [\cos(x + 9)]^2 = F(x)$$ ■

EXERCISES 1.3

1. Suppose the graph of f is given. Write equations for the graphs that are obtained from the graph of f as follows.
(a) Shift 3 units upward. (b) Shift 3 units downward.
(c) Shift 3 units to the right. (d) Shift 3 units to the left.
(e) Reflect about the x-axis. (f) Reflect about the y-axis.
(g) Stretch vertically by a factor of 3.
(h) Shrink vertically by a factor of 3.

2. Explain how each graph is obtained from the graph of $y = f(x)$.
(a) $y = f(x) + 8$ (b) $y = f(x + 8)$
(c) $y = 8f(x)$ (d) $y = f(8x)$
(e) $y = -f(x) - 1$ (f) $y = 8f(\frac{1}{8}x)$

3. The graph of $y = f(x)$ is given. Match each equation with its graph and give reasons for your choices.
(a) $y = f(x - 4)$ (b) $y = f(x) + 3$
(c) $y = \frac{1}{3}f(x)$ (d) $y = -f(x + 4)$
(e) $y = 2f(x + 6)$

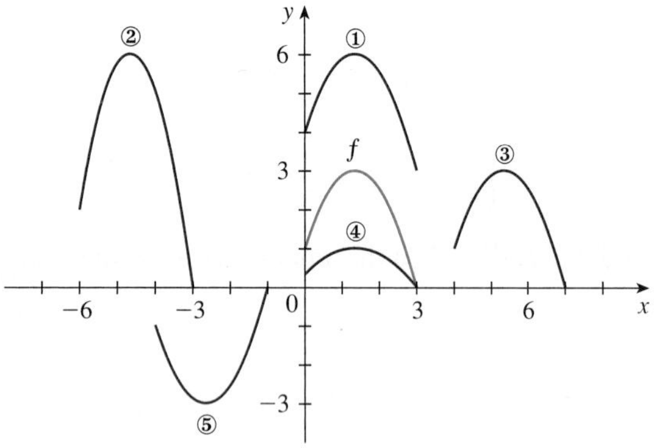

4. The graph of f is given. Draw the graphs of the following functions.
(a) $y = f(x) - 2$ (b) $y = f(x - 2)$
(c) $y = -2f(x)$ (d) $y = f(\frac{1}{3}x) + 1$

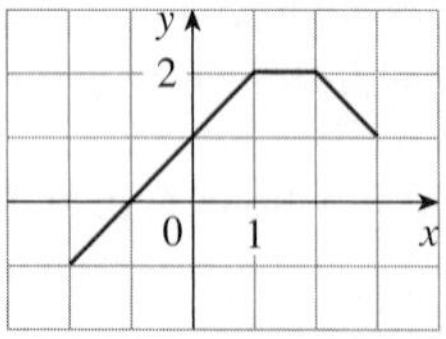

5. The graph of f is given. Use it to graph the following functions.
(a) $y = f(2x)$ (b) $y = f(\frac{1}{2}x)$
(c) $y = f(-x)$ (d) $y = -f(-x)$

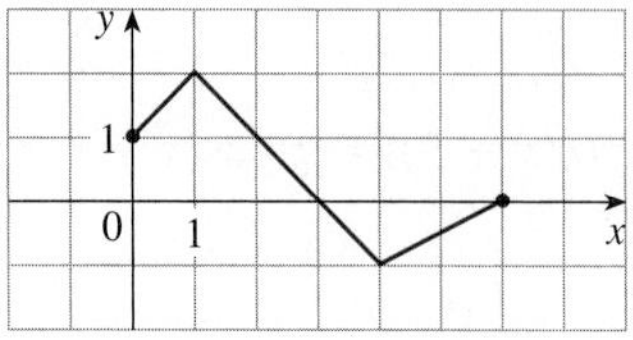

6. (a) How is the graph of $y = 2 \sin x$ related to the graph of $y = \sin x$? Use your answer and Figure 6 to sketch the graph of $y = 2 \sin x$.
(b) How is the graph of $y = 1 + \sqrt{x}$ related to the graph of $y = \sqrt{x}$? Use your answer and Figure 4(a) to sketch the graph of $y = 1 + \sqrt{x}$.

7–20 Graph the function by hand, not by plotting points, but by starting with the graph of one of the standard functions given in Section 1.2, and then applying the appropriate transformations.

7. $y = \dfrac{1}{x + 2}$

8. $y = (x - 1)^3$

9. $y = -\sqrt[3]{x}$

10. $y = x^2 + 6x + 4$

11. $y = \sqrt{x - 2} - 1$

12. $y = 4 \sin 3x$

13. $y = \sin(\frac{1}{2}x)$

14. $y = \dfrac{2}{x} - 2$

15. $y = -x^3$

16. $y = 1 - 2\sqrt{x + 3}$

17. $y = \frac{1}{2}(1 - \cos x)$

18. $y = |x| - 2$

19. $y = 1 - 2x - x^2$

20. $y = \dfrac{1}{4} \tan\left(x - \dfrac{\pi}{4}\right)$

---

21. The city of New Orleans is located at latitude 30°N. Use Figure 9 to find a function that models the number of hours of daylight at New Orleans as a function of the time of year. To check the accuracy of your model, use the fact that on March 31 the sun rises at 5:51 AM and sets at 6:18 PM in New Orleans.

22. A variable star is one whose brightness alternately increases and decreases. For the most visible variable star, Delta Cephei, the time between periods of maximum brightness is 5.4 days, the average brightness (or magnitude) of the star is 4.0, and its brightness varies by ± 0.35 magnitude. Find a function that models the brightness of Delta Cephei as a function of time.

23. Some of the highest tides in the world occur in the Bay of Fundy on the Atlantic Coast of Canada. At Hopewell Cape the water depth at low tide is about 2.0 m and at high tide it is about 12.0 m. The natural period of oscillation is about 12 hours and on June 30, 2009, high tide occurred at 6:45 AM. Find a function involving the cosine function that models the water depth $D(t)$ (in meters) as a function of time t (in hours after midnight) on that day.

24. Volume of air in lungs In a normal respiratory cycle the volume of air that moves into and out of the lungs is about 500 mL. The reserve and residual volumes of air that remain in the lungs occupy about 2000 mL and a single respiratory cycle for an average human takes about 4 seconds. Find a model for the total volume of air $V(t)$ in the lungs as a function of time.

25. **Gene frequency** The frequency of a certain gene in a parasite population undergoes sinusoidal cycles as a result of coevolution with its host. It reaches a maximum frequency of 80% and a minimum of 20%, with a complete cycle taking three years. Find a function that describes the gene frequency dynamics over time (measured in years) assuming that it starts at a frequency of 50% at time $t = 0$. See also Case Study 2 on page xlvi.

26. **Cyclic neutropenia** is a blood disorder in humans characterized by periodic fluctuations in the density of a certain kind of blood cell called *neutrophils*. The density of neutrophils reaches highs of around 2000 cells/μL of blood and lows near zero. The period of fluctuations is approximately three weeks. Model the temporal dynamics of neutrophils in days, assuming that the density is at its highest on day 0.

27–28 Find (a) $f + g$, (b) $f - g$, (c) fg, and (d) f/g and state their domains.

27. $f(x) = x^3 + 2x^2, \quad g(x) = 3x^2 - 1$

28. $f(x) = \sqrt{3 - x}, \quad g(x) = \sqrt{x^2 - 1}$

29–34 Find the functions (a) $f \circ g$, (b) $g \circ f$, (c) $f \circ f$, and (d) $g \circ g$ and their domains.

29. $f(x) = x^2 - 1, \quad g(x) = 2x + 1$

30. $f(x) = x - 2, \quad g(x) = x^2 + 3x + 4$

31. $f(x) = 1 - 3x, \quad g(x) = \cos x$

32. $f(x) = \sqrt{x}, \quad g(x) = \sqrt[3]{1 - x}$

33. $f(x) = x + \dfrac{1}{x}, \quad g(x) = \dfrac{x + 1}{x + 2}$

34. $f(x) = \dfrac{x}{1 + x}, \quad g(x) = \sin 2x$

35–38 Find $f \circ g \circ h$.

35. $f(x) = 3x - 2, \quad g(x) = \sin x, \quad h(x) = x^2$

36. $f(x) = |x - 4|, \quad g(x) = 2^x, \quad h(x) = \sqrt{x}$

37. $f(x) = \sqrt{x - 3}, \quad g(x) = x^2, \quad h(x) = x^3 + 2$

38. $f(x) = \tan x, \quad g(x) = \dfrac{x}{x - 1}, \quad h(x) = \sqrt[3]{x}$

39–44 Express the function in the form $f \circ g$.

39. $F(x) = (2x + x^2)^4$

40. $F(x) = \cos^2 x$

41. $F(x) = \dfrac{\sqrt[3]{x}}{1 + \sqrt[3]{x}}$

42. $G(x) = \sqrt[3]{\dfrac{x}{1 + x}}$

43. $v(t) = \sec(t^2)\tan(t^2)$

44. $u(t) = \dfrac{\tan t}{1 + \tan t}$

45–47 Express the function in the form $f \circ g \circ h$.

45. $R(x) = \sqrt{\sqrt{x} - 1}$

46. $H(x) = \sqrt[8]{2 + |x|}$

47. $H(x) = \sec^4(\sqrt{x})$

---

48. Use the table to evaluate each expression.

(a) $f(g(1))$ (b) $g(f(1))$ (c) $f(f(1))$

(d) $g(g(1))$ (e) $(g \circ f)(3)$ (f) $(f \circ g)(6)$

x	1	2	3	4	5	6
$f(x)$	3	1	4	2	2	5
$g(x)$	6	3	2	1	2	3

49. Use the given graphs of f and g to evaluate each expression, or explain why it is undefined.

(a) $f(g(2))$ (b) $g(f(0))$ (c) $(f \circ g)(0)$

(d) $(g \circ f)(6)$ (e) $(g \circ g)(-2)$ (f) $(f \circ f)(4)$

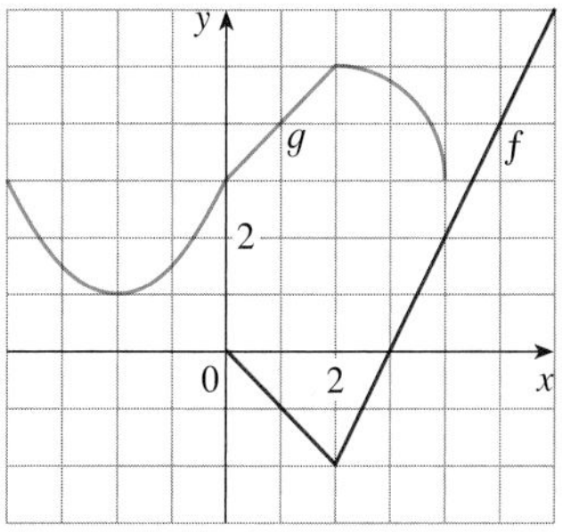

50. Use the given graphs of f and g to estimate the value of $f(g(x))$ for $x = -5, -4, -3, \ldots, 5$. Use these estimates to sketch a rough graph of $f \circ g$.

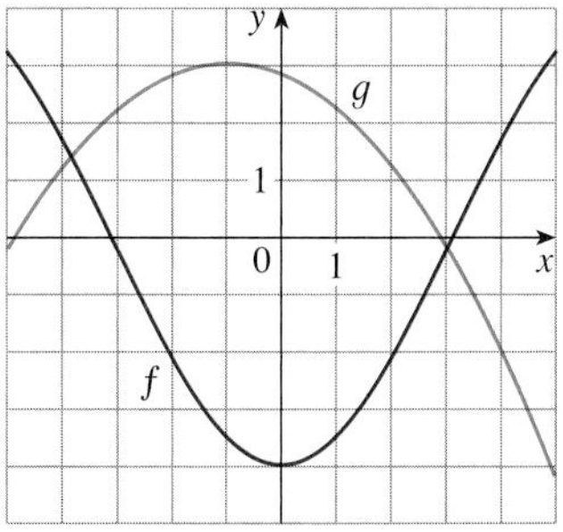

51. A stone is dropped into a lake, creating a circular ripple that travels outward at a speed of 60 cm/s.

(a) Express the radius r of this circle as a function of the time t (in seconds).

(b) If A is the area of this circle as a function of the radius, find $A \circ r$ and interpret it.

52. A spherical balloon is being inflated and the radius of the balloon is increasing at a rate of 2 cm/s.
(a) Express the radius r of the balloon as a function of the time t (in seconds).
(b) If V is the volume of the balloon as a function of the radius, find $V \circ r$ and interpret it.

53. A ship is moving at a speed of 30 km/h parallel to a straight shoreline. The ship is 6 km from shore and it passes a lighthouse at noon.
(a) Express the distance s between the lighthouse and the ship as a function of d, the distance the ship has traveled since noon; that is, find f so that $s = f(d)$.
(b) Express d as a function of t, the time elapsed since noon; that is, find g so that $d = g(t)$.
(c) Find $f \circ g$. What does this function represent?

54. Bioavailability is a term that refers to the fraction of an antibiotic dose taken orally that is absorbed into the bloodstream. Suppose that, for a dosage of x mg, the bioavailability is $h(x) = \frac{1}{2}x$ mg. If x mg enters the bloodstream, suppose that the amount eventually absorbed into the site of an infection is given by $g(x) = 4x/(x + 4)$ mg. Finally, if x mg is absorbed into the site of an infection, suppose that the number of surviving bacteria is given by $f(x) = 3200/(8 + x^2)$, measured in colony forming units, CFU.
(a) Derive the function that relates oral dosage to the number of surviving bacteria using composition of functions.
(b) Suppose the antibiotic is instead administered by injection. Derive the function that relates dosage to the number of surviving bacteria using composition of functions.
(c) Sketch the graphs of the functions found in parts (a) and (b).

55. Matrix-digesting enzymes are sometimes produced by cancer cells to digest the tissue surrounding the tumor, allowing it to grow and spread. In solid tumors, the enzymes are produced only by the cells on the surface of the tumor. Suppose the diameter of a spherical tumor is growing at a rate of g mm/year.
(a) What is the diameter d of the tumor as a function of time?
(b) Suppose the rate of enzyme production P is proportional to the surface area of the tumor S. Find $P \circ S \circ d$ and interpret it.

56. If you invest x dollars at 4% interest compounded annually, then the amount $A(x)$ of the investment after one year is $A(x) = 1.04x$. Find $A \circ A$, $A \circ A \circ A$, and $A \circ A \circ A \circ A$. What do these compositions represent? Find a formula for the composition of n copies of A.

57. Let f and g be linear functions with equations $f(x) = m_1x + b_1$ and $g(x) = m_2x + b_2$. Is $f \circ g$ also a linear function? If so, what is the slope of its graph?

58. Suppose g is an even function and let $h = f \circ g$. Is h always an even function?

59. Suppose g is an odd function and let $h = f \circ g$. Is h always an odd function? What if f is odd? What if f is even?

■ PROJECT The Biomechanics of Human Movement

BB

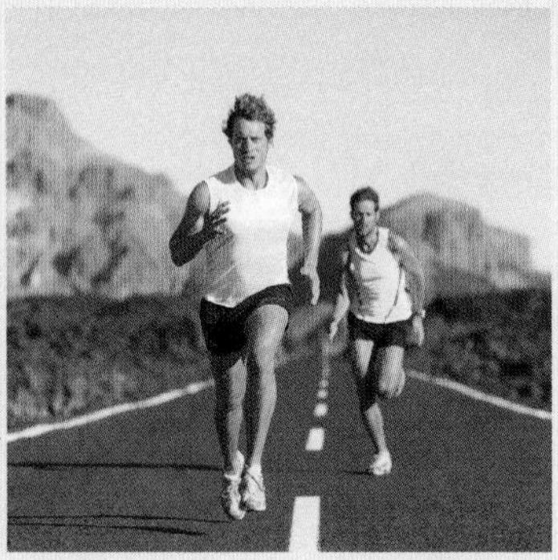

Periodic processes in biology often display much more complicated patterns than those of the sine and cosine functions. For example, the vertical force exerted on the ground during human locomotion can display a variety of patterns depending on speed. (See, for example, Figure 1.) One way to model such processes is to first construct a function that adequately models a single cycle and then to use a periodic extension of this function to represent multiple cycles. In this project you will transform the cosine function to model the vertical force exerted on the ground during running and walking using such periodic extensions.

1. (a) Consider the force exerted on the ground during one stride of the right leg during running at 3.6 m/s [Figure 1(a)]. Suppose the foot strikes the ground at $t = 0$, leaves the ground at $t = 0.25$ s, and that the maximal force exerted on the ground is 2 kN. Explain why the function $f(t) = 1 - \cos(8\pi t)$ provides a reasonable description of the force exerted over one stride.
(b) Plot the periodic extension of the function from part (a) assuming it repeats every 0.65 seconds.

2. (a) To model the force pattern of a single cycle during walking at 1.5 m/s [Figure 1(b)] we need to alter the function from Problem 1(a) so that it is wider.

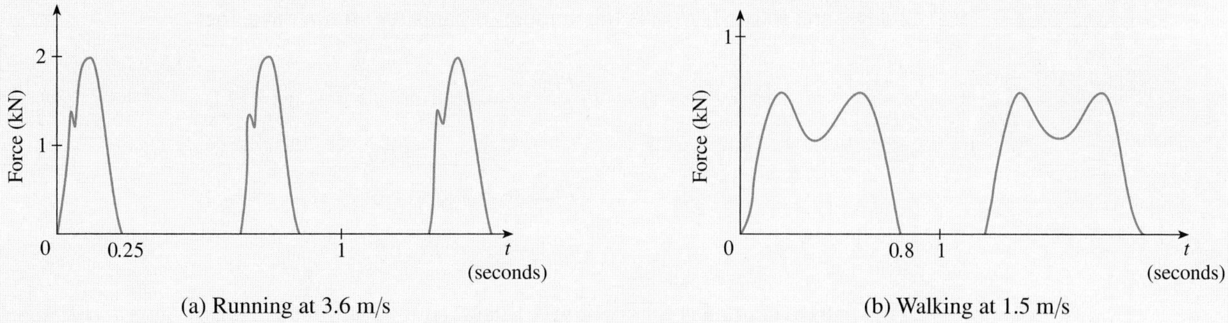

FIGURE 1

Source: Adapted from R. Alexander, "Walking and Running," *American Scientist* 72 (1984): 348–54.

Do so, assuming that the foot strikes the ground at $t = 0$ and leaves the ground at $t = 0.8$ s.

(b) The function in part (a) has the correct width but we need to create a dip in the function midway through the stride at $t = 0.4$ s. One way to do this is to add the function in part (a) to a similar function that oscillates twice as fast (that is, has half the period) and then scale the height of the resulting function appropriately. Use this approach to obtain a function with the desired shape.

(c) Plot the periodic extension of the function obtained in part (b) assuming it repeats every 1.2 seconds.

3. A variation of the function in Problem 2(b) is given by

$$g(t) = 1 - \frac{\cos(2.5\pi t) + q\cos(5\pi t)}{1 + q}$$

where q is a constant.

(a) Plot the graph of g for $q = 0.2$, 0.8, and 1.8. What role does q play?

(b) Compare the graphs of g in part (a) with the graph of the function in Problem 2(b).

1.4 Exponential Functions

We often hear people saying that something is "growing exponentially." What does that mean, exactly? We answer that question in this section by looking at examples of exponential growth and decay as modeled by exponential functions.

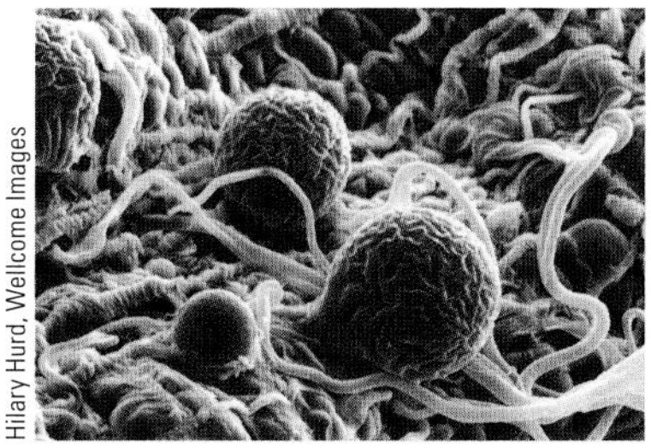

■ The Growth of Malarial Parasites

Malaria kills more than a million people every year. To understand the mechanisms that regulate malarial growth, controlled experiments have been done on mice. Individual cells of the malarial species *Plasmodium chabaudi* reproduce synchronously (at the same time) every 24 hours. The parasites develop in red blood cells for a period of 24 hours and then they all burst at the same time, quickly reinvade new blood cells, and start the process again. Each infected blood cell produces eight new parasites when it bursts. So a single parasite at time 0 produces 8 parasites after 1 day, $8 \times 8 = 64$ parasites after

2 days, and so on. If $P(n)$ is the number of parasites after n days, then

$$P(0) = 1$$

$$P(1) = 8$$

$$P(2) = 8 \times P(1) = 8^2$$

$$P(3) = 8 \times P(2) = 8^3$$

This pattern continues for six or seven days, so for that time period we have

$$P(n) = 8^n$$

The rapid growth of this function is demonstrated by the table of values and the resulting scatter plot in Figure 1.

Day n	$P(n)$
0	1
1	8
2	64
3	512
4	4,096
5	32,768

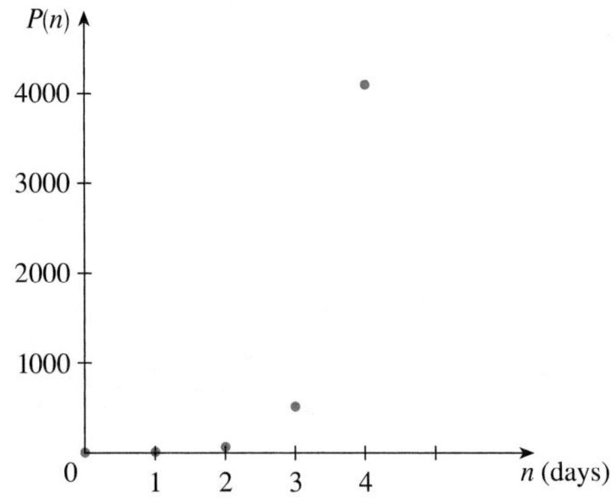

FIGURE 1

■ Exponential Functions

The function $P(n) = 8^n$ is called an *exponential function* because the variable n is the exponent. Likewise, $f(x) = 5^x$ is an exponential function because x is the exponent. The exponential function $f(x) = 2^x$ should not be confused with the power function $g(x) = x^2$, in which the variable is the base.

In general, an **exponential function** is a function of the form

$$f(x) = b^x$$

where b is a positive constant. Let's recall what this means.

If $x = n$, a positive integer, then

$$b^n = \underbrace{b \cdot b \cdot \cdots \cdot b}_{n \text{ factors}}$$

If $x = 0$, then $b^0 = 1$, and if $x = -n$, where n is a positive integer, then

$$b^{-n} = \frac{1}{b^n}$$

If x is a rational number, $x = p/q$, where p and q are integers and $q > 0$, then

$$b^x = b^{p/q} = \sqrt[q]{b^p} = \left(\sqrt[q]{b}\right)^p$$

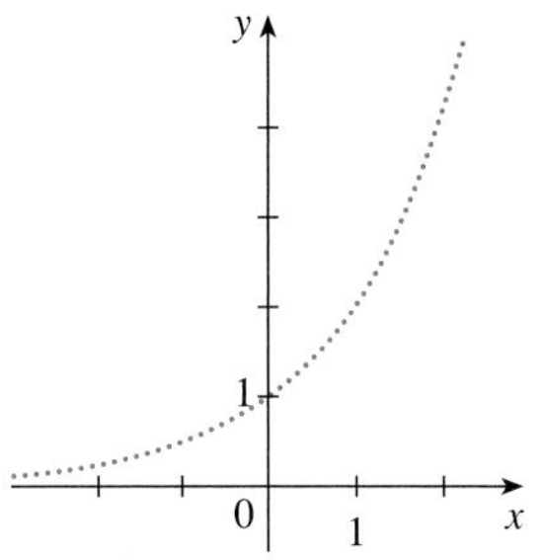

FIGURE 2
Representation of $y = 2^x$, x rational

But what is the meaning of b^x if x is an irrational number? For instance, what is meant by $2^{\sqrt{3}}$ or 5^{π}?

To help us answer this question, we first look at the graph of the function $y = 2^x$, where x is rational. A representation of this graph is shown in Figure 2. We want to enlarge the domain of $y = 2^x$ to include both rational and irrational numbers.

There are holes in the graph in Figure 2 corresponding to irrational values of x. We want to fill in the holes by defining $f(x) = 2^x$, where $x \in \mathbb{R}$, so that f is an increasing function. In particular, since the irrational number $\sqrt{3}$ satisfies

$$1.7 < \sqrt{3} < 1.8$$

we must have

$$2^{1.7} < 2^{\sqrt{3}} < 2^{1.8}$$

and we know what $2^{1.7}$ and $2^{1.8}$ mean because 1.7 and 1.8 are rational numbers. Similarly, if we use better approximations for $\sqrt{3}$, we obtain better approximations for $2^{\sqrt{3}}$:

$$\begin{aligned}
1.73 < \sqrt{3} < 1.74 \quad &\Rightarrow \quad 2^{1.73} < 2^{\sqrt{3}} < 2^{1.74} \\
1.732 < \sqrt{3} < 1.733 \quad &\Rightarrow \quad 2^{1.732} < 2^{\sqrt{3}} < 2^{1.733} \\
1.7320 < \sqrt{3} < 1.7321 \quad &\Rightarrow \quad 2^{1.7320} < 2^{\sqrt{3}} < 2^{1.7321} \\
1.73205 < \sqrt{3} < 1.73206 \quad &\Rightarrow \quad 2^{1.73205} < 2^{\sqrt{3}} < 2^{1.73206} \\
\vdots \qquad\quad \vdots \qquad & \qquad\quad \vdots \qquad\qquad \vdots
\end{aligned}$$

A proof of this fact is given in J. Marsden and A. Weinstein, *Calculus Unlimited* (Menlo Park, CA: Benjamin/Cummings, 1981).

It can be shown that there is exactly one number that is greater than all of the numbers

$$2^{1.7}, \quad 2^{1.73}, \quad 2^{1.732}, \quad 2^{1.7320}, \quad 2^{1.73205}, \quad \ldots$$

and less than all of the numbers

$$2^{1.8}, \quad 2^{1.74}, \quad 2^{1.733}, \quad 2^{1.7321}, \quad 2^{1.73206}, \quad \ldots$$

We define $2^{\sqrt{3}}$ to be this number. Using the preceding approximation process, we can compute it correct to six decimal places:

$$2^{\sqrt{3}} \approx 3.321997$$

Similarly, we can define 2^x (or b^x, if $b > 0$) where x is any irrational number. Figure 3 shows how all the holes in Figure 2 have been filled to complete the graph of the function $f(x) = 2^x$, $x \in \mathbb{R}$.

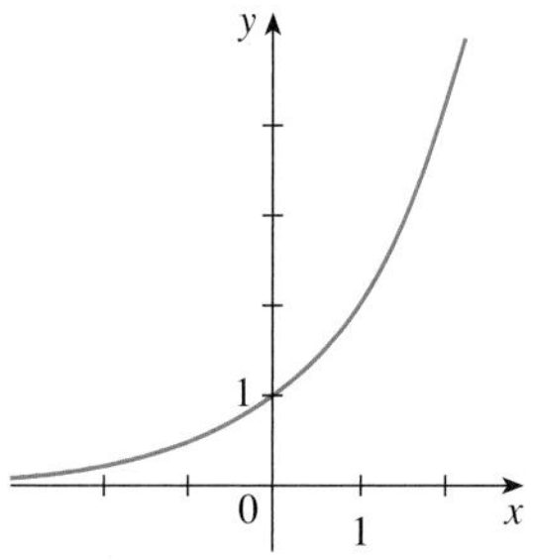

FIGURE 3
$y = 2^x$, x real

The graphs of members of the family of functions $y = b^x$ are shown in Figure 4 for various values of the base b. Notice that all of these graphs pass through the same point $(0, 1)$ because $b^0 = 1$ for $b \neq 0$. Notice also that as the base b gets larger, the exponential function grows more rapidly (for $x > 0$).

If $0 < b < 1$, then b^x approaches 0 as x becomes large. If $b > 1$, then b^x approaches 0 as x decreases through negative values. In both cases the x-axis is a horizontal asymptote. These matters are discussed in Section 2.2.

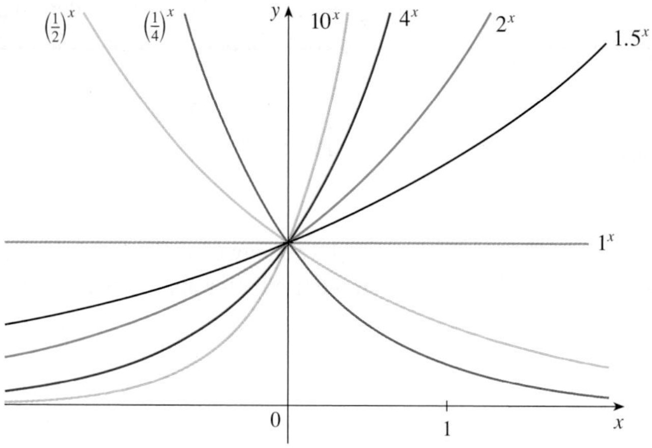

FIGURE 4

You can see from Figure 4 that there are basically three kinds of exponential functions $y = b^x$. If $0 < b < 1$, the exponential function decreases; if $b = 1$, it is a constant; and if $b > 1$, it increases. These three cases are illustrated in Figure 5. Observe that if $b \neq 1$, then the exponential function $y = b^x$ has domain $\mathbb{R}$ and range $(0, \infty)$. Notice also that, since $(1/b)^x = 1/b^x = b^{-x}$, the graph of $y = (1/b)^x$ is just the reflection of the graph of $y = b^x$ about the y-axis.

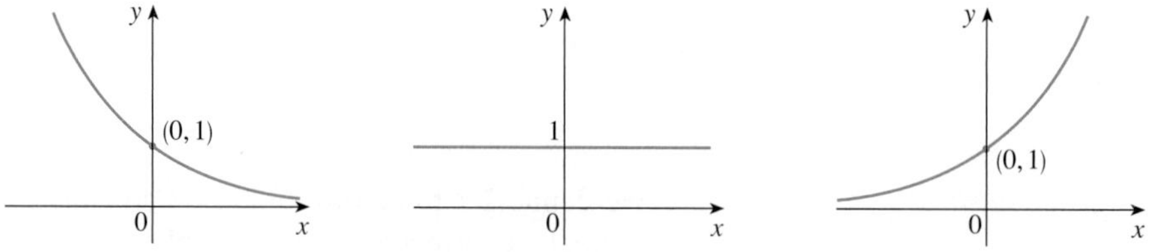

FIGURE 5 (a) $y = b^x$, $0 < b < 1$ (b) $y = 1^x$ (c) $y = b^x$, $b > 1$

One reason for the importance of the exponential function lies in the following properties. If x and y are rational numbers, then these laws are well known from elementary algebra. It can be proved that they remain true for arbitrary real numbers x and y.

www.stewartcalculus.com
For review and practice using the Laws of Exponents, click on *Review of Algebra.*

Laws of Exponents If a and b are positive numbers and x and y are any real numbers, then

1. $b^{x+y} = b^x b^y$ **2.** $b^{x-y} = \dfrac{b^x}{b^y}$ **3.** $(b^x)^y = b^{xy}$ **4.** $(ab)^x = a^x b^x$

EXAMPLE 1 | Sketch the graph of the function $y = 3 - 2^x$ and determine its domain and range.

For a review of reflecting and shifting graphs, see Section 1.3.

SOLUTION First we reflect the graph of $y = 2^x$ [shown in Figures 3 and 6(a)] about the x-axis to get the graph of $y = -2^x$ in Figure 6(b). Then we shift the graph of $y = -2^x$ upward 3 units to obtain the graph of $y = 3 - 2^x$ in Figure 6(c). The domain is $\mathbb{R}$ and the range is $(-\infty, 3)$.

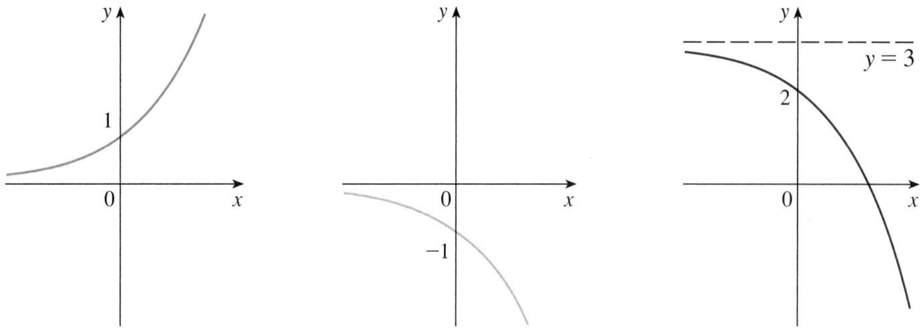

FIGURE 6 (a) $y = 2^x$ (b) $y = -2^x$ (c) $y = 3 - 2^x$ ■

EXAMPLE 2 | Use a graphing device to compare the exponential function $f(x) = 2^x$ and the power function $g(x) = x^2$. Which function grows more quickly when x is large?

SOLUTION Figure 7 shows both functions graphed in the viewing rectangle $[-2, 6]$ by $[0, 40]$. We see that the graphs intersect three times, but for $x > 4$ the graph of $f(x) = 2^x$ stays above the graph of $g(x) = x^2$. Figure 8 gives a more global view and shows that for large values of x, the exponential function $y = 2^x$ grows far more rapidly than the power function $y = x^2$.

Example 2 shows that $y = 2^x$ increases more quickly than $y = x^2$. To demonstrate just how quickly $f(x) = 2^x$ increases, let's perform the following thought experiment. Suppose we start with a piece of paper a thousandth of an inch thick and we fold it in half 50 times. Each time we fold the paper in half, the thickness of the paper doubles, so the thickness of the resulting paper would be $2^{50}/1000$ inches. How thick do you think that is? It works out to be more than 17 million miles!

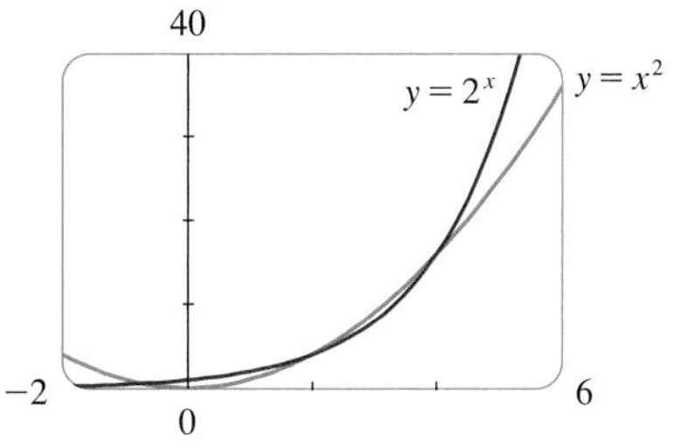

FIGURE 7

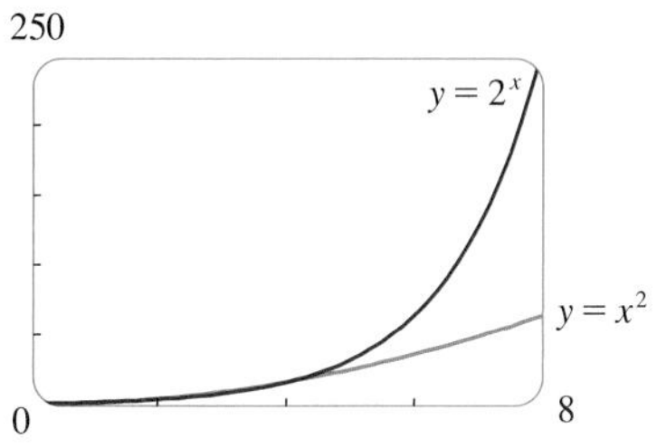

FIGURE 8 ■

■ Exponential Growth

We have seen how the exponential function $P(n) = 8^n$ models the initial growth of the malarial parasite *P. chabaudi*. In the next example we use an exponential function to model the human population in the 20th century.

EXAMPLE 3 | BB World population growth Table 1 shows data for the population of the world from 1900 to 2010. Use an exponential function to model the data.

SOLUTION Figure 9 shows the scatter plot corresponding to the data in Table 1.

Table 1

t (years after 1900)	Population (millions)	t (years after 1900)	Population (millions)
0	1650	60	3040
10	1750	70	3710
20	1860	80	4450
30	2070	90	5280
40	2300	100	6080
50	2560	110	6870

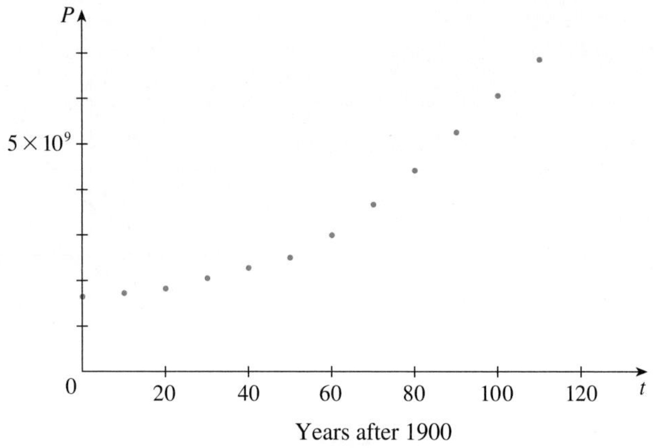

FIGURE 9 Scatter plot for world population

The pattern of the data points in Figure 9 suggests exponential growth, so we use a graphing calculator with exponential regression capability to apply the method of least squares and obtain the exponential model

$$P = (1436.53) \cdot (1.01395)^t$$

where $t = 0$ corresponds to 1900. Figure 10 shows the graph of this exponential function together with the original data points. We see that the exponential curve fits the data reasonably well. The period of relatively slow population growth is explained by the two world wars and the Great Depression of the 1930s.

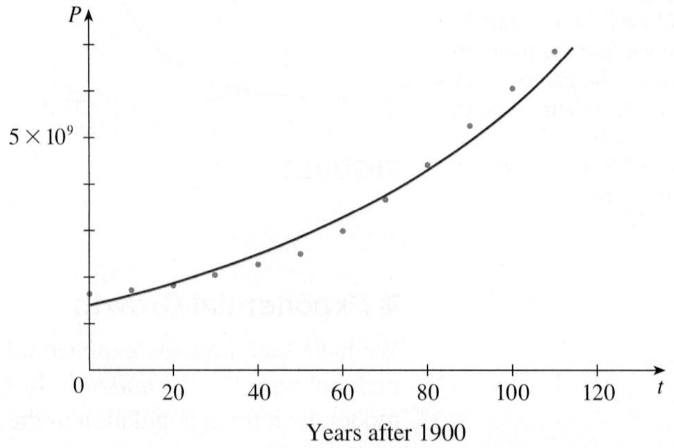

FIGURE 10 Exponential model for population growth

■

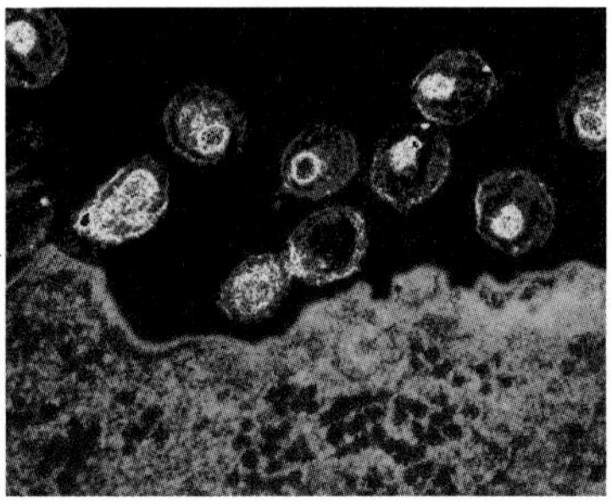
Scott Camazine / Alamy

■ HIV Density and Exponential Decay

In 1995 a paper appeared detailing the effect of the protease inhibitor ABT-538 on the human immunodeficiency virus HIV-1.[1] Table 2 shows values of the plasma viral load $V(t)$ of patient 303, measured in RNA copies per mL, t days after ABT-538 treatment was begun. The corresponding scatter plot is shown in Figure 11.

Table 2

t (days)	$V(t)$
1	76.0
4	53.0
8	18.0
11	9.4
15	5.2
22	3.6

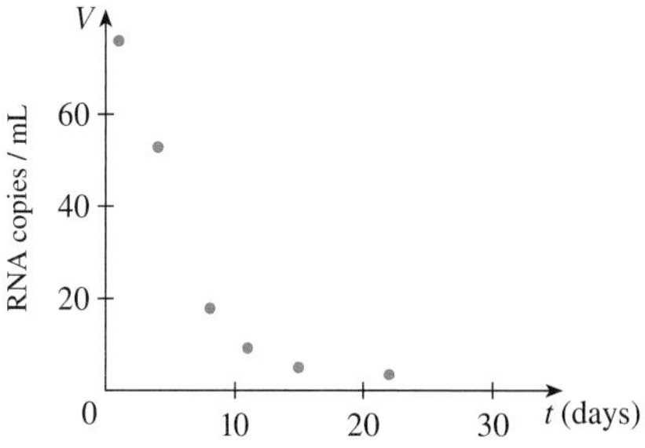

FIGURE 11 Plasma viral load in patient 303

The rather dramatic decline of the viral load that we see in Figure 11 reminds us of the graphs of the exponential function $y = b^x$ in Figures 4 and 5(a) for the case where the base b is less than 1. So let's model the function $V(t)$ by an exponential function. Using a graphing calculator or computer to fit the data in Table 2 with an exponential function of the form $y = a \cdot b^t$, we obtain the model

$$V = 96.39785 \cdot (0.818656)^t$$

In Figure 12 we graph this exponential function with the data points and see that the model represents the viral load reasonably well for the first month of treatment.

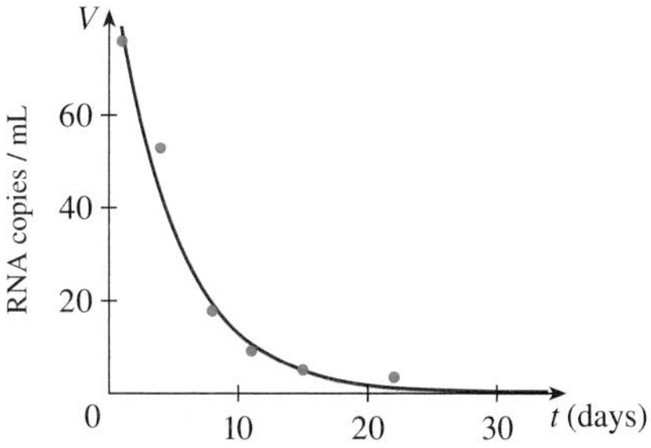

FIGURE 12
Exponential model for viral load

We could use the graph in Figure 12 to estimate the **half-life** of V, that is, the time required for the viral load to be reduced to half its initial value. (See Exercise 33.) In the next example we are given the half-life of a radioactive element and asked to find the mass of a sample at any time.

EXAMPLE 4 | The half-life of strontium-90, ^{90}Sr, is 25 years. This means that half of any given quantity of ^{90}Sr will disintegrate in 25 years.

(a) If a sample of ^{90}Sr has a mass of 24 mg, find an expression for the mass $m(t)$ that remains after t years.

(b) Find the mass remaining after 40 years, correct to the nearest milligram.

1. D. Ho et al., "Rapid Turnover of Plasma Virions and CD4 Lymphocytes in HIV-1 Infection," *Nature* 373 (1995): 123–26.

(c) Use a graphing device to graph $m(t)$ and use the graph to estimate the time required for the mass to be reduced to 5 mg.

SOLUTION

(a) The mass is initially 24 mg and is halved during each 25-year period, so

$$\begin{aligned} m(0) &= 24 \\ m(25) &= \frac{1}{2}(24) \\ m(50) &= \frac{1}{2} \cdot \frac{1}{2}(24) = \frac{1}{2^2}(24) \\ m(75) &= \frac{1}{2} \cdot \frac{1}{2^2}(24) = \frac{1}{2^3}(24) \\ m(100) &= \frac{1}{2} \cdot \frac{1}{2^3}(24) = \frac{1}{2^4}(24) \end{aligned}$$

From this pattern, it appears that the mass remaining after t years is

$$m(t) = \frac{1}{2^{t/25}}(24) = 24 \cdot 2^{-t/25} = 24 \cdot (2^{-1/25})^t$$

This is an exponential function with base $b = 2^{-1/25} = 1/2^{1/25}$.

(b) The mass that remains after 40 years is

$$m(40) = 24 \cdot 2^{-40/25} \approx 7.9 \text{ mg}$$

(c) We use a graphing calculator or computer to graph the function $m(t) = 24 \cdot 2^{-t/25}$ in Figure 13. We also graph the line $m = 5$ and use the cursor to estimate that $m(t) = 5$ when $t \approx 57$. So the mass of the sample will be reduced to 5 mg after about 57 years. ■

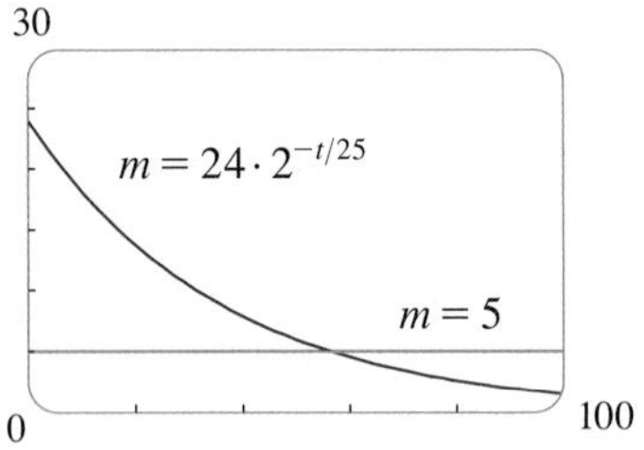

FIGURE 13

■ The Number e

Of all possible bases for an exponential function, there is one that is most convenient for the purposes of calculus. The choice of a base b is influenced by the way the graph of $y = b^x$ crosses the y-axis. Figures 14 and 15 show the tangent lines to the graphs of $y = 2^x$ and $y = 3^x$ at the point (0, 1). (Tangent lines will be defined precisely in Section 3.1. For present purposes, you can think of the tangent line to an exponential graph at a point as the line that touches the graph only at that point.) If we measure the slopes of these tangent lines at (0, 1), we find that $m \approx 0.7$ for $y = 2^x$ and $m \approx 1.1$ for $y = 3^x$.

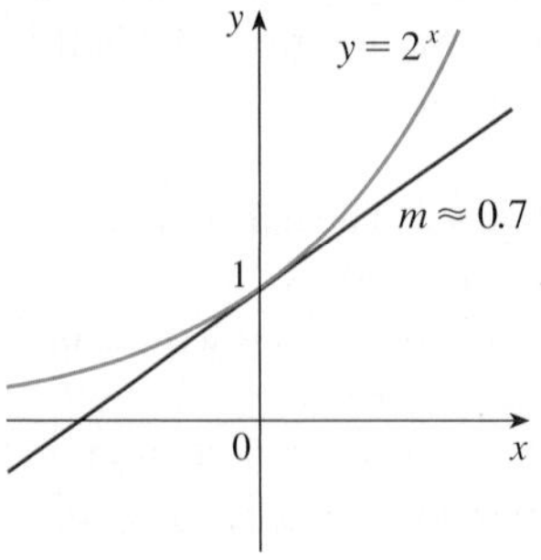

FIGURE 14

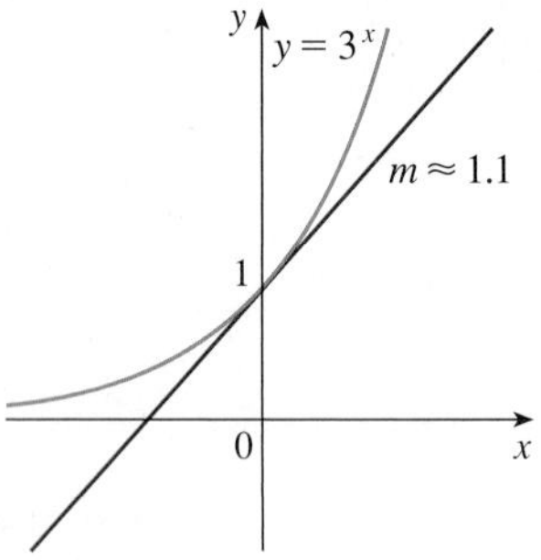

FIGURE 15

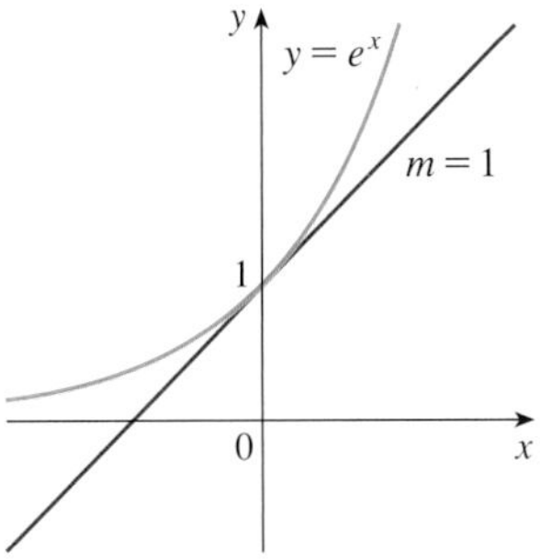

FIGURE 16
The natural exponential function crosses the y-axis with a slope of 1.

TEC Module 1.4 enables you to graph exponential functions with various bases and their tangent lines in order to estimate more closely the value of b for which the tangent has slope 1.

It turns out, as we will see in Chapter 3, that some of the formulas of calculus will be greatly simplified if we choose the base b so that the slope of the tangent line to $y = b^x$ at (0, 1) is *exactly* 1. (See Figure 16.) In fact, there *is* such a number and it is denoted by the letter e. (This notation was chosen by the Swiss mathematician Leonhard Euler in 1727, probably because it is the first letter of the word *exponential*.) In view of Figures 14 and 15, it comes as no surprise that the number e lies between 2 and 3 and the graph of $y = e^x$ lies between the graphs of $y = 2^x$ and $y = 3^x$. (See Figure 17.) In Chapter 3 we will see that the value of e, correct to five decimal places, is

$$e \approx 2.71828$$

We call the function $f(x) = e^x$ the **natural exponential function**.

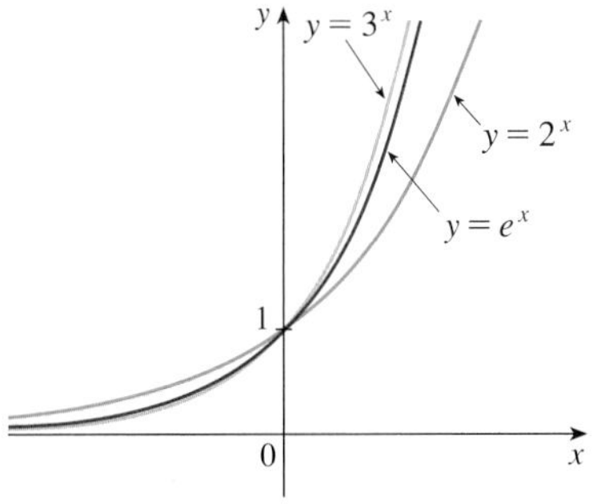

FIGURE 17

EXAMPLE 5 | Graph the function $y = \frac{1}{2}e^{-x} - 1$ and state the domain and range.

SOLUTION We start with the graph of $y = e^x$ from Figures 16 and 18(a) and reflect about the y-axis to get the graph of $y = e^{-x}$ in Figure 18(b). (Notice that the graph crosses the y-axis with a slope of -1.) Then we compress the graph vertically by a factor of 2 to obtain the graph of $y = \frac{1}{2}e^{-x}$ in Figure 18(c). Finally, we shift the graph downward one unit to get the desired graph in Figure 18(d). The domain is $\mathbb{R}$ and the range is $(-1, \infty)$.

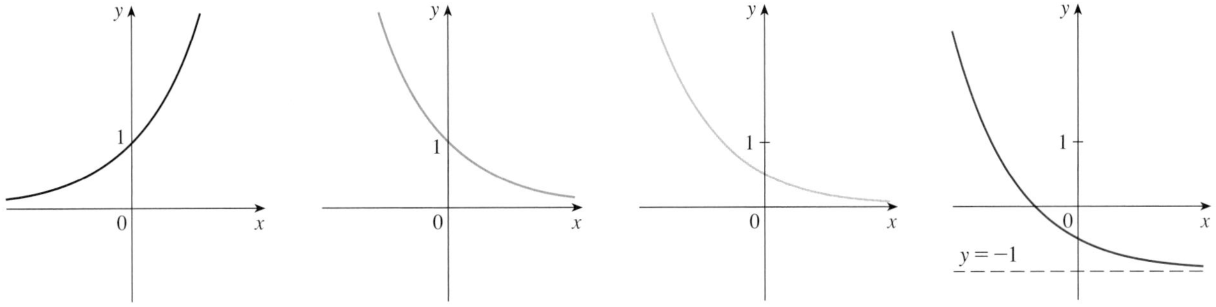

(a) $y = e^x$ (b) $y = e^{-x}$ (c) $y = \frac{1}{2}e^{-x}$ (d) $y = \frac{1}{2}e^{-x} - 1$

FIGURE 18

How far to the right do you think we would have to go for the height of the graph of $y = e^x$ to exceed a million? The next example demonstrates the rapid growth of this function by providing an answer that might surprise you.

EXAMPLE 6 | Use a graphing device to find the values of x for which $e^x > 1{,}000{,}000$.

SOLUTION In Figure 19 we graph both the function $y = e^x$ and the horizontal line $y = 1{,}000{,}000$. We see that these curves intersect when $x \approx 13.8$. Thus $e^x > 10^6$ when $x > 13.8$. It is perhaps surprising that the values of the exponential function have already surpassed a million when x is only 14. ■

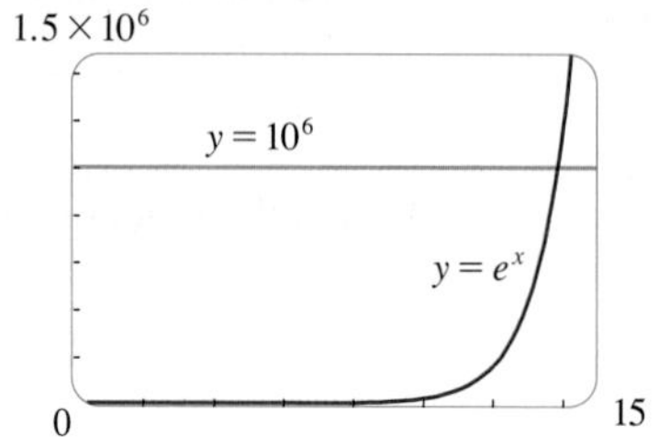

FIGURE 19

EXERCISES 1.4

1–4 Use the Law of Exponents to rewrite and simplify the expression.

1. (a) $\dfrac{4^{-3}}{2^{-8}}$ (b) $\dfrac{1}{\sqrt[3]{x^4}}$

2. (a) $8^{4/3}$ (b) $x(3x^2)^3$

3. (a) $b^8(2b)^4$ (b) $\dfrac{(6y^3)^4}{2y^5}$

4. (a) $\dfrac{x^{2n} \cdot x^{3n-1}}{x^{n+2}}$ (b) $\dfrac{\sqrt{a\sqrt{b}}}{\sqrt[3]{ab}}$

5. (a) Write an equation that defines the exponential function with base $b > 0$.
(b) What is the domain of this function?
(c) If $b \neq 1$, what is the range of this function?
(d) Sketch the general shape of the graph of the exponential function for each of the following cases.
(i) $b > 1$ (ii) $b = 1$ (iii) $0 < b < 1$

6. (a) How is the number e defined?
(b) What is an approximate value for e?
(c) What is the natural exponential function?

7–10 Graph the given functions on a common screen. How are these graphs related?

7. $y = 2^x$, $y = e^x$, $y = 5^x$, $y = 20^x$

8. $y = e^x$, $y = e^{-x}$, $y = 8^x$, $y = 8^{-x}$

9. $y = 3^x$, $y = 10^x$, $y = \left(\frac{1}{3}\right)^x$, $y = \left(\frac{1}{10}\right)^x$

10. $y = 0.9^x$, $y = 0.6^x$, $y = 0.3^x$, $y = 0.1^x$

11–16 Make a rough sketch of the graph of the function. Do not use a calculator. Just use the graphs given in Figures 4 and 17 and, if necessary, the transformations of Section 1.3.

11. $y = 10^{x+2}$ **12.** $y = (0.5)^x - 2$

13. $y = -2^{-x}$ **14.** $y = e^{|x|}$

15. $y = 1 - \frac{1}{2}e^{-x}$ **16.** $y = 2(1 - e^x)$

17. Starting with the graph of $y = e^x$, write the equation of the graph that results from
(a) shifting 2 units downward
(b) shifting 2 units to the right
(c) reflecting about the x-axis
(d) reflecting about the y-axis
(e) reflecting about the x-axis and then about the y-axis

18. Starting with the graph of $y = e^x$, find the equation of the graph that results from
(a) reflecting about the line $y = 4$
(b) reflecting about the line $x = 2$

19–20 Find the domain of each function.

19. (a) $f(x) = \dfrac{1 - e^{x^2}}{1 - e^{1-x^2}}$ (b) $f(x) = \dfrac{1 + x}{e^{\cos x}}$

20. (a) $g(t) = \sin(e^{-t})$ (b) $g(t) = \sqrt{1 - 2^t}$

21–22 Find the exponential function $f(x) = Cb^x$ whose graph is given.

21.

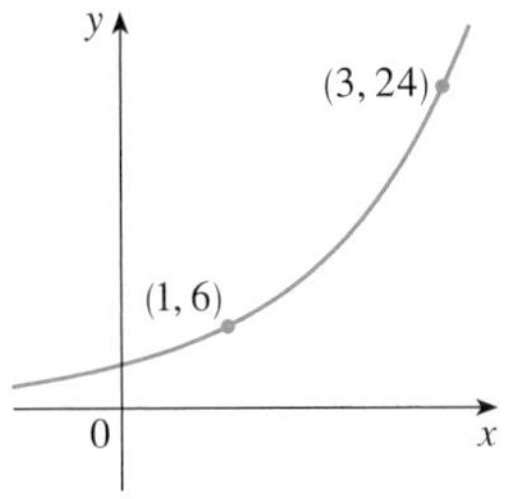

22.

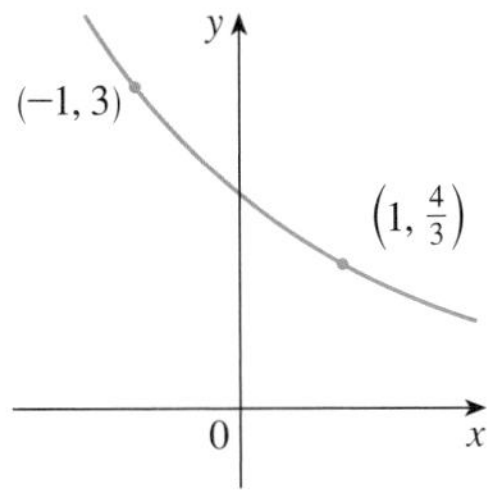

23. If $f(x) = 5^x$, show that

$$\frac{f(x+h) - f(x)}{h} = 5^x\left(\frac{5^h - 1}{h}\right)$$

24. Suppose you are offered a job that lasts one month. Which of the following methods of payment do you prefer?
I. One million dollars at the end of the month.
II. One cent on the first day of the month, two cents on the second day, four cents on the third day, and, in general, 2^{n-1} cents on the nth day.

25. Suppose the graphs of $f(x) = x^2$ and $g(x) = 2^x$ are drawn on a coordinate grid where the unit of measurement is 1 inch. Show that, at a distance 2 ft to the right of the origin, the height of the graph of f is 48 ft but the height of the graph of g is about 265 mi.

26. Compare the functions $f(x) = x^5$ and $g(x) = 5^x$ by graphing both functions in several viewing rectangles. Find all points of intersection of the graphs correct to one decimal place. Which function grows more rapidly when x is large?

27. Compare the functions $f(x) = x^{10}$ and $g(x) = e^x$ by graphing both f and g in several viewing rectangles. When does the graph of g finally surpass the graph of f?

28. Use a graph to estimate the values of x such that $e^x > 1{,}000{,}000{,}000$.

29. ***Giardia lamblia* growth** A researcher is trying to determine the doubling time of a population of the bacterium *Giardia lamblia*. He starts a culture in a nutrient solution and estimates the bacteria count every four hours. His data are shown in the table.

Time (hours)	0	4	8	12	16	20	24
Bacteria count (CFU/mL)	37	47	63	78	105	130	173

(a) Make a scatter plot of the data.
(b) Use a graphing calculator to find an exponential curve $f(t) = a \cdot b^t$ that models the bacteria population t hours later.
(c) Graph the model from part (b) together with the scatter plot in part (a). Use the TRACE feature to determine how long it takes for the bacteria count to double.

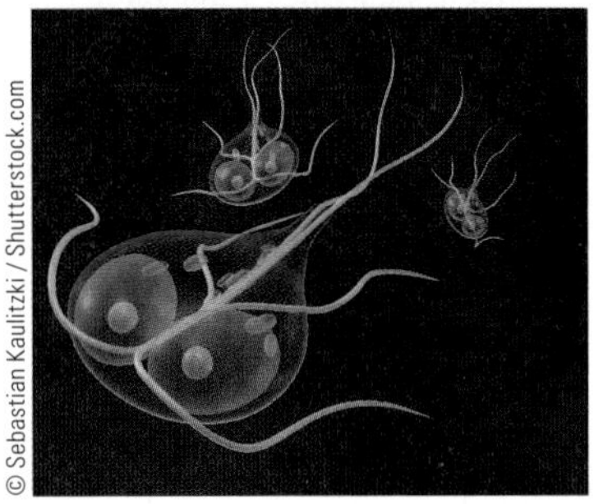

G. lamblia

30. A **bacteria culture** starts with 500 bacteria and doubles in size every half hour.
(a) How many bacteria are there after 3 hours?
(b) How many bacteria are there after t hours?
(c) How many bacteria are there after 40 minutes?
(d) Graph the population function and estimate the time for the population to reach 100,000.

31. The half-life of bismuth-210, ^{210}Bi, is 5 days.
(a) If a sample has a mass of 200 mg, find the amount remaining after 15 days.
(b) Find the amount remaining after t days.
(c) Estimate the amount remaining after 3 weeks.
(d) Use a graph to estimate the time required for the mass to be reduced to 1 mg.

32. An isotope of sodium, ^{24}Na, has a half-life of 15 hours. A sample of this isotope has mass 2 g.
(a) Find the amount remaining after 60 hours.
(b) Find the amount remaining after t hours.
(c) Estimate the amount remaining after 4 days.
(d) Use a graph to estimate the time required for the mass to be reduced to 0.01 g.

33. Half-life of HIV Use the graph of V in Figure 12 to estimate the half-life of the viral load of patient 303 during the first month of treatment.

34. Blood alcohol concentration After alcohol is fully absorbed into the body, it is metabolized with a half-life of about 1.5 hours. Suppose you have had three alcoholic drinks and an hour later, at midnight, your blood alcohol concentration (BAC) is 0.6 mg/mL.
(a) Find an exponential decay model for your BAC t hours after midnight.
(b) Graph your BAC and use the graph to determine when you can drive home if the legal limit is 0.08 mg/mL.

Source: Adapted from P. Wilkinson et al., "Pharmacokinetics of Ethanol after Oral Administration in the Fasting State," *Journal of Pharmacokinetics and Biopharmaceutics* 5 (1977): 207–24.

35. World population Use a calculator with exponential regression capability to model the population of the world with the data from 1950 to 2010 in Table 1 on page 46. Use the model to estimate the population in 1993 and to predict the population in the year 2020.

36. US population The table gives the population of the United States, in millions, for the years 1900–2010. Use a calculator with exponential regression capability to model the US population since 1900. Use the model to estimate the population in 1925 and to predict the population in the year 2020.

Year	Population	Year	Population
1900	76	1960	179
1910	92	1970	203
1920	106	1980	227
1930	123	1990	250
1940	131	2000	281
1950	150	2010	310

37. If you graph the function

$$f(x) = \frac{1 - e^{1/x}}{1 + e^{1/x}}$$

you'll see that f appears to be an odd function. Prove it.

38. Graph several members of the family of functions

$$f(x) = \frac{1}{1 + ae^{bx}}$$

where $a > 0$. How does the graph change when b changes? How does it change when a changes?

1.5 Logarithms; Semilog and Log-Log Plots

Inverse Functions

Table 1 gives data from an experiment in which a bacteria culture was started with 100 bacteria in a limited nutrient medium. The size of the bacteria population was recorded at hourly intervals. The number of bacteria N is a function of the time t: $N = f(t)$.

Suppose, however, that the biologist changes her point of view and becomes interested in the time required for the population to reach various levels. In other words, she is thinking of t as a function of N. This function is called the *inverse function* of f, denoted by f^{-1}, and read "f inverse." Thus $t = f^{-1}(N)$ is the time required for the population level to reach N. The values of f^{-1} can be found by reading Table 1 from right to left or by consulting Table 2. For instance, $f^{-1}(550) = 6$ because $f(6) = 550$.

Table 1 N as a function of t

t (hours)	$N = f(t)$ = population at time t
0	100
1	168
2	259
3	358
4	445
5	509
6	550
7	573
8	586

Table 2 t as a function of N

N	$t = f^{-1}(N)$ = time to reach N bacteria
100	0
168	1
259	2
358	3
445	4
509	5
550	6
573	7
586	8

Not all functions possess inverses. Let's compare the functions f and g whose arrow diagrams are shown in Figure 1. Note that f never takes on the same value twice (any two inputs in A have different outputs), whereas g does take on the same value twice (both 2 and 3 have the same output, 4). In symbols,

$$g(2) = g(3)$$

but $$f(x_1) \neq f(x_2) \qquad \text{whenever } x_1 \neq x_2$$

Functions that share this property with f are called *one-to-one functions.*

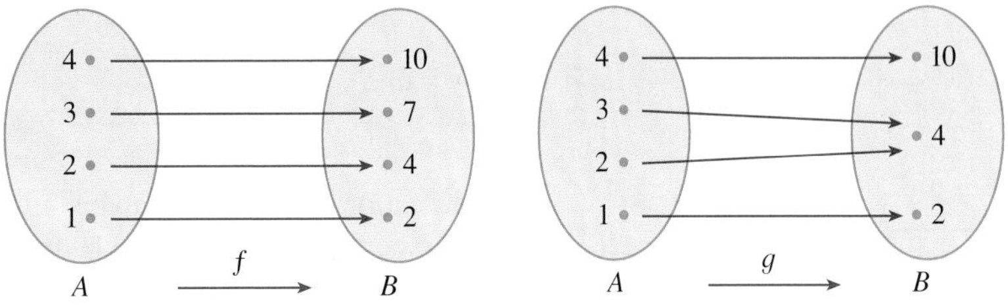

FIGURE 1
f is one-to-one; g is not.

In the language of inputs and outputs, this definition says that f is one-to-one if each output corresponds to only one input.

> **(1) Definition** A function f is called a **one-to-one function** if it never takes on the same value twice; that is,
>
> $$f(x_1) \neq f(x_2) \qquad \text{whenever } x_1 \neq x_2$$

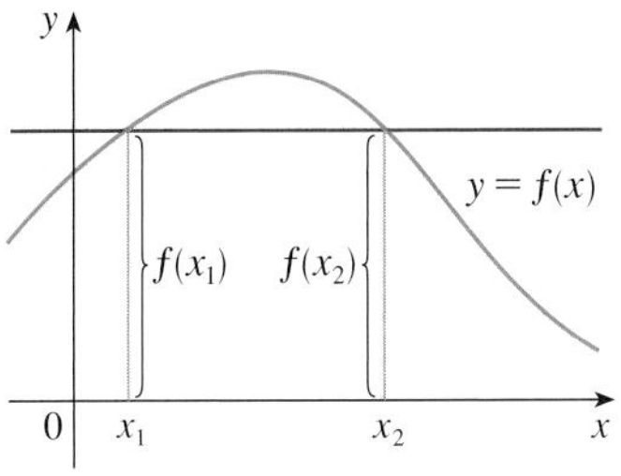

FIGURE 2
The function is not one-to-one because $f(x_1) = f(x_2)$.

If a horizontal line intersects the graph of f in more than one point, then we see from Figure 2 that there are numbers x_1 and x_2 such that $f(x_1) = f(x_2)$. This means that f is not one-to-one. Therefore we have the following geometric method for determining whether a function is one-to-one.

> **Horizontal Line Test** A function is one-to-one if and only if no horizontal line intersects its graph more than once.

EXAMPLE 1 | Is the function $f(x) = x^3$ one-to-one?

SOLUTION 1 If $x_1 \neq x_2$, then $x_1^3 \neq x_2^3$ (two different numbers can't have the same cube). Therefore, by Definition 1, $f(x) = x^3$ is one-to-one.

SOLUTION 2 From Figure 3 we see that no horizontal line intersects the graph of $f(x) = x^3$ more than once. Therefore, by the Horizontal Line Test, f is one-to-one. ■

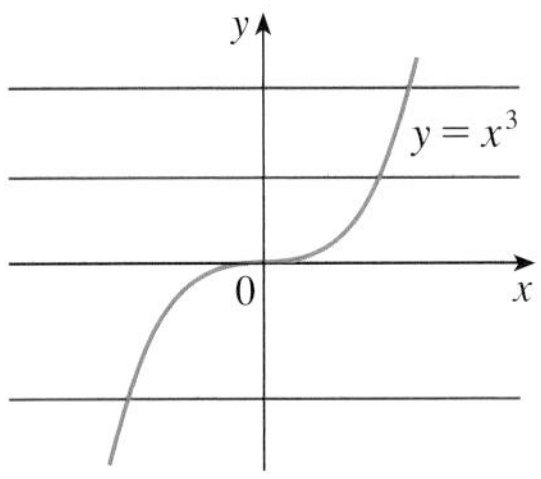

FIGURE 3
$f(x) = x^3$ is one-to-one.

EXAMPLE 2 | Is the function $g(x) = x^2$ one-to-one?

SOLUTION 1 This function is not one-to-one because, for instance,

$$g(1) = 1 = g(-1)$$

and so 1 and -1 have the same output.

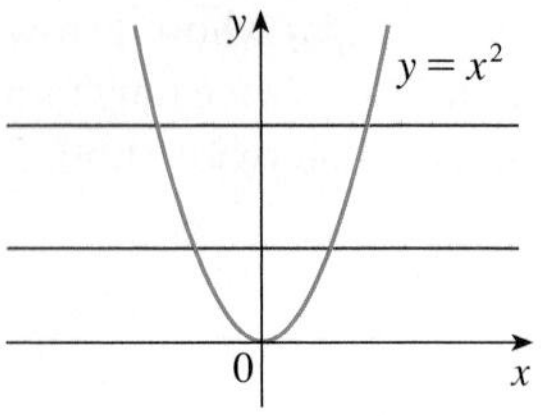

FIGURE 4
$g(x) = x^2$ is not one-to-one.

SOLUTION 2 From Figure 4 we see that there are horizontal lines that intersect the graph of g more than once. Therefore, by the Horizontal Line Test, g is not one-to-one. ■

One-to-one functions are important because they are precisely the functions that possess inverse functions according to the following definition.

(2) Definition Let f be a one-to-one function with domain A and range B. Then its **inverse function** f^{-1} has domain B and range A and is defined by

$$f^{-1}(y) = x \iff f(x) = y$$

for any y in B.

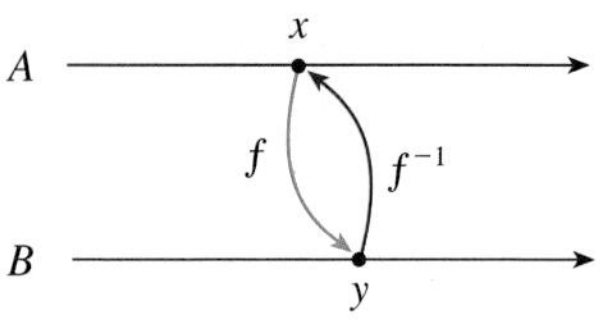

FIGURE 5

This definition says that if f maps x into y, then f^{-1} maps y back into x. (If f were not one-to-one, then f^{-1} would not be uniquely defined.) The arrow diagram in Figure 5 indicates that f^{-1} reverses the effect of f. Note that

$$\text{domain of } f^{-1} = \text{range of } f$$

$$\text{range of } f^{-1} = \text{domain of } f$$

For example, the inverse function of $f(x) = x^3$ is $f^{-1}(x) = x^{1/3}$ because if $y = x^3$, then

$$f^{-1}(y) = f^{-1}(x^3) = (x^3)^{1/3} = x$$

CAUTION Do not mistake the -1 in f^{-1} for an exponent. Thus

$$f^{-1}(x) \quad \text{does } \textit{not} \text{ mean} \quad \frac{1}{f(x)}$$

The reciprocal $1/f(x)$ could, however, be written as $[f(x)]^{-1}$.

EXAMPLE 3 | If f is one-to-one and $f(1) = 5$, $f(3) = 7$, and $f(8) = -10$, find $f^{-1}(7)$, $f^{-1}(5)$, and $f^{-1}(-10)$.

SOLUTION From the definition of f^{-1} we have

$$f^{-1}(7) = 3 \quad \text{because} \quad f(3) = 7$$

$$f^{-1}(5) = 1 \quad \text{because} \quad f(1) = 5$$

$$f^{-1}(-10) = 8 \quad \text{because} \quad f(8) = -10$$

The diagram in Figure 6 makes it clear how f^{-1} reverses the effect of f in this case.

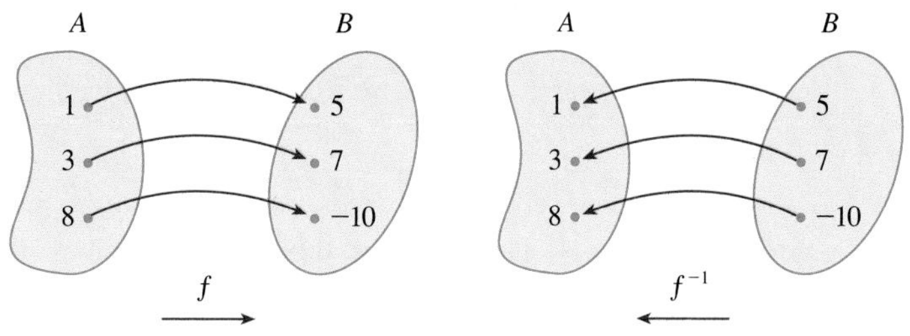

FIGURE 6
The inverse function reverses inputs and outputs.

The letter x is traditionally used as the independent variable, so when we concentrate on f^{-1} rather than on f, we usually reverse the roles of x and y in Definition 2 and write

(3)
$$f^{-1}(x) = y \iff f(y) = x$$

By substituting for y in Definition 2 and substituting for x in (3), we get the following **cancellation equations**:

(4)
$$f^{-1}(f(x)) = x \quad \text{for every } x \text{ in } A$$
$$f(f^{-1}(x)) = x \quad \text{for every } x \text{ in } B$$

The first cancellation equation says that if we start with x, apply f, and then apply f^{-1}, we arrive back at x, where we started (see the machine diagram in Figure 7). Thus f^{-1} undoes what f does. The second equation says that f undoes what f^{-1} does.

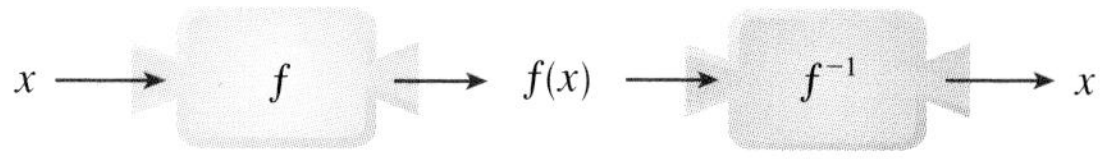

FIGURE 7

For example, if $f(x) = x^3$, then $f^{-1}(x) = x^{1/3}$ and so the cancellation equations become

$$f^{-1}(f(x)) = (x^3)^{1/3} = x$$

$$f(f^{-1}(x)) = (x^{1/3})^3 = x$$

These equations simply say that the cube function and the cube root function cancel each other when applied in succession.

Now let's see how to compute inverse functions. If we have a function $y = f(x)$ and are able to solve this equation for x in terms of y, then according to Definition 2 we must have $x = f^{-1}(y)$. If we want to call the independent variable x, we then interchange x and y and arrive at the equation $y = f^{-1}(x)$.

(5) How to Find the Inverse Function of a One-to-One Function f

STEP 1 Write $y = f(x)$.

STEP 2 Solve this equation for x in terms of y (if possible).

STEP 3 To express f^{-1} as a function of x, interchange x and y.
The resulting equation is $y = f^{-1}(x)$.

EXAMPLE 4 | Find the inverse function of $f(x) = x^3 + 2$.

SOLUTION According to (5) we first write

$$y = x^3 + 2$$

Then we solve this equation for x:

$$x^3 = y - 2$$

$$x = \sqrt[3]{y - 2}$$

Finally, we interchange x and y:

$$y = \sqrt[3]{x - 2}$$

Therefore the inverse function is $f^{-1}(x) = \sqrt[3]{x - 2}$. ■

In Example 4, notice how f^{-1} reverses the effect of f. The function f is the rule "Cube, then add 2"; f^{-1} is the rule "Subtract 2, then take the cube root."

The principle of interchanging x and y to find the inverse function also gives us the method for obtaining the graph of f^{-1} from the graph of f. Since $f(a) = b$ if and only if $f^{-1}(b) = a$, the point (a, b) is on the graph of f if and only if the point (b, a) is on the graph of f^{-1}. But we get the point (b, a) from (a, b) by reflecting about the line $y = x$. (See Figure 8.)

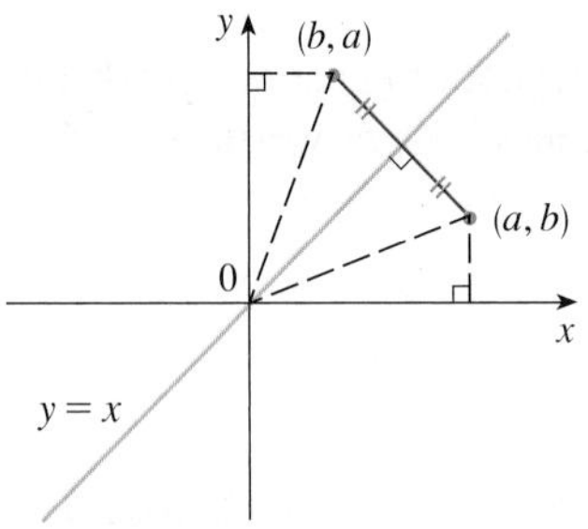

FIGURE 8

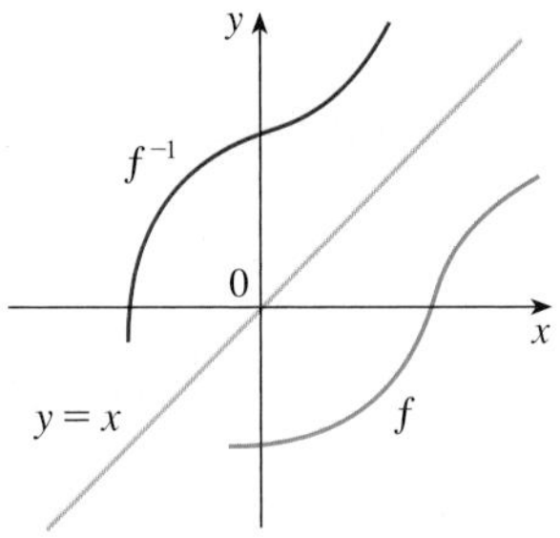

FIGURE 9

Therefore, as illustrated by Figure 9:

> The graph of f^{-1} is obtained by reflecting the graph of f about the line $y = x$.

Ranges of R_0 values for some common diseases

Disease	R_0
Measles	12–18
Diphtheria	6–7
Smallpox	5–7
Polio	5–7
Mumps	4–7
HIV	2–5

EXAMPLE 5 | **Reproduction number** One of the main quantities that epidemiologists try to measure for infectious diseases is the so-called *basic reproduction number*, denoted by R_0. Biologically, this is the expected number of new infections that an infected individual will produce when introduced into a completely susceptible population. The probability of an outbreak occurring (as opposed to the disease dying out by chance) is then modeled by the equation

$$P = 1 - \frac{1}{R_0} \qquad \text{where } R_0 \geq 1$$

Find the inverse function of P, interpret it, and graph it.

SOLUTION To find the inverse function we solve the given equation for R_0:

$$\frac{1}{R_0} = 1 - P \qquad\qquad R_0 = \frac{1}{1 - P}$$

This equation expresses R_0 as a function of P. It gives the value of R_0 that is required to obtain outbreak probability P.

In Exercise 32 this model is modified to take vaccinations into account.

Notice that we did not interchange the variables P and R_0 because they need to retain their meanings. Figure 10 shows the graphs of each of these variables as a function of the other variable. Observe that each of these graphs is a reflection of the other about the diagonal line $P = R_0$.

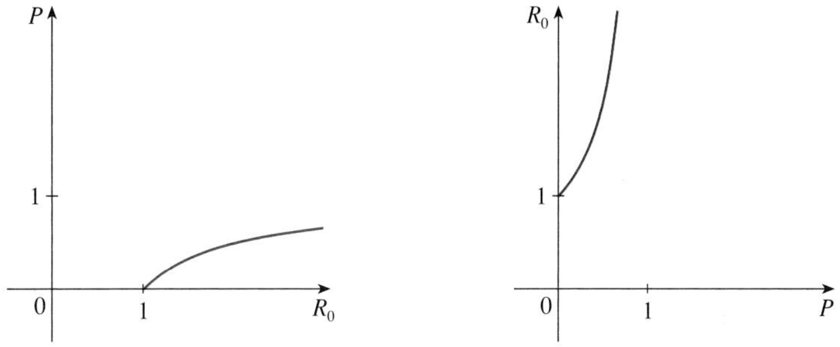

(a) P as a function of R_0 $\quad$ $P = 1 - \dfrac{1}{R_0}, \quad R_0 \geqslant 1$

(b) R_0 as a function of P $\quad$ $R_0 = \dfrac{1}{1 - P}, \quad 0 \leqslant P < 1$

FIGURE 10

■

■ Logarithmic Functions

If $b > 0$ and $b \neq 1$, the exponential function $f(x) = b^x$ is either increasing or decreasing and so it is one-to-one by the Horizontal Line Test. It therefore has an inverse function f^{-1}, which is called the **logarithmic function with base *b*** and is denoted by $\log_b$. If we use the formulation of an inverse function given by (3),

$$f^{-1}(x) = y \iff f(y) = x$$

then we have

(6)
$$\boxed{\log_b x = y \iff b^y = x}$$

Thus, if $x > 0$, then $\log_b x$ is the exponent to which the base b must be raised to give x. For example, $\log_{10} 0.001 = -3$ because $10^{-3} = 0.001$.

The cancellation equations (4), when applied to the functions $f(x) = b^x$ and $f^{-1}(x) = \log_b x$, become

(7)
$$\boxed{\begin{aligned} \log_b(b^x) &= x \quad \text{for every } x \in \mathbb{R} \\ b^{\log_b x} &= x \quad \text{for every } x > 0 \end{aligned}}$$

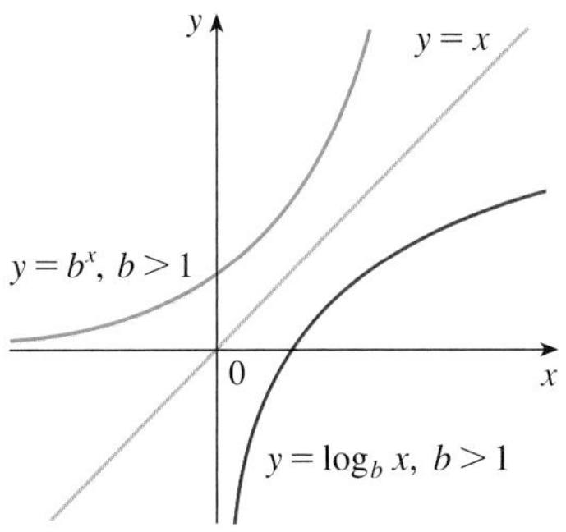

FIGURE 11

The logarithmic function $\log_b$ has domain $(0, \infty)$ and range $\mathbb{R}$. Its graph is the reflection of the graph of $y = b^x$ about the line $y = x$.

Figure 11 shows the case where $b > 1$. (The most important logarithmic functions have base $b > 1$.) The fact that $y = b^x$ is a very rapidly increasing function for $x > 0$ is reflected in the fact that $y = \log_b x$ is a very slowly increasing function for $x > 1$.

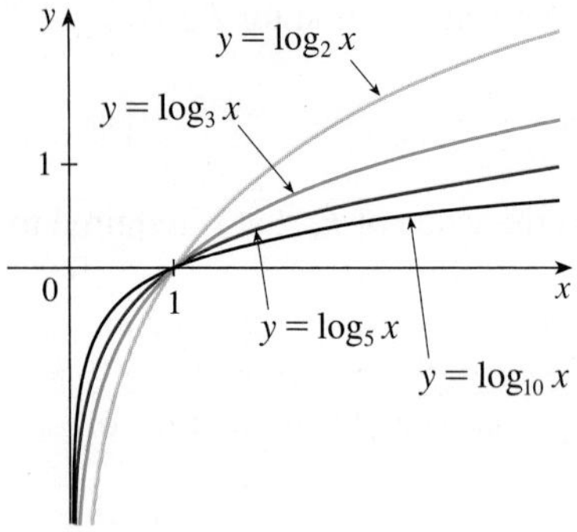

FIGURE 12

Figure 12 shows the graphs of $y = \log_b x$ with various values of the base $b > 1$. Since $\log_b 1 = 0$, the graphs of all logarithmic functions pass through the point $(1, 0)$.

The following properties of logarithmic functions follow from the corresponding properties of exponential functions given in Section 1.4.

> **Laws of Logarithms** If x and y are positive numbers, then
>
> **1.** $\log_b(xy) = \log_b x + \log_b y$
>
> **2.** $\log_b\left(\dfrac{x}{y}\right) = \log_b x - \log_b y$
>
> **3.** $\log_b(x^r) = r\log_b x$ (where r is any real number)

EXAMPLE 6 | Use the laws of logarithms to evaluate $\log_2 80 - \log_2 5$.

SOLUTION Using Law 2, we have

$$\log_2 80 - \log_2 5 = \log_2\left(\frac{80}{5}\right) = \log_2 16 = 4$$

because $2^4 = 16$. ■

■ Natural Logarithms

Notation for Logarithms
Most textbooks in calculus and the sciences, as well as calculators, use the notation $\ln x$ for the natural logarithm and $\log x$ for the "common logarithm," $\log_{10} x$. In the more advanced mathematical and scientific literature and in computer languages, however, the notation $\log x$ usually denotes the natural logarithm.

Of all possible bases b for logarithms, we will see in Chapter 3 that the most convenient choice of a base is the number e, which was defined in Section 1.4. The logarithm with base e is called the **natural logarithm** and has a special notation:

$$\log_e x = \ln x$$

If we put $b = e$ and replace $\log_e$ with "ln" in (6) and (7), then the defining properties of the natural logarithm function become

(8)
$$\ln x = y \iff e^y = x$$

(9)
$$\ln(e^x) = x \quad \text{for every } x \in \mathbb{R}$$
$$e^{\ln x} = x \quad \text{for every } x > 0$$

In particular, if we set $x = 1$, we get

$$\ln e = 1$$

EXAMPLE 7 | Find x if $\ln x = 5$.

SOLUTION 1 From (8) we see that

$$\ln x = 5 \qquad \text{means} \qquad e^5 = x$$

Therefore $x = e^5$.

(If you have trouble working with the "ln" notation, just replace it by $\log_e$. Then the equation becomes $\log_e x = 5$; so, by the definition of logarithm, $e^5 = x$.)

SOLUTION 2 Start with the equation

$$\ln x = 5$$

and apply the exponential function to both sides of the equation:

$$e^{\ln x} = e^5$$

But the second cancellation equation in (9) says that $e^{\ln x} = x$. Therefore $x = e^5$. ■

EXAMPLE 8 | Solve the equation $e^{5-3x} = 10$.

SOLUTION We take natural logarithms of both sides of the equation and use (9):

$$\ln(e^{5-3x}) = \ln 10$$

$$5 - 3x = \ln 10$$

$$3x = 5 - \ln 10$$

$$x = \tfrac{1}{3}(5 - \ln 10)$$

Since the natural logarithm is found on scientific calculators, we can approximate the solution: to four decimal places, $x \approx 0.8991$. ■

EXAMPLE 9 | Express $\ln a + \frac{1}{2}\ln b$ as a single logarithm.

SOLUTION Using Laws 3 and 1 of logarithms, we have

$$\begin{aligned} \ln a + \tfrac{1}{2}\ln b &= \ln a + \ln b^{1/2} \\ &= \ln a + \ln\sqrt{b} \\ &= \ln\left(a\sqrt{b}\right) \end{aligned}$$ ■

The following formula shows that logarithms with any base can be expressed in terms of the natural logarithm.

> **(10) Change of Base Formula** For any positive number b ($b \neq 1$), we have
>
> $$\log_b x = \frac{\ln x}{\ln b}$$

PROOF Let $y = \log_b x$. Then, from (6), we have $b^y = x$. Taking natural logarithms of both sides of this equation, we get $y \ln b = \ln x$. Therefore

$$y = \frac{\ln x}{\ln b}$$ ■

Scientific calculators have a key for natural logarithms, so Formula 10 enables us to use a calculator to compute a logarithm with any base (as shown in the following example).

EXAMPLE 10 | Evaluate $\log_8 5$ correct to six decimal places.

SOLUTION Formula 10 gives

$$\log_8 5 = \frac{\ln 5}{\ln 8} \approx 0.773976$$

■

EXAMPLE 11 | In Example 1.4.4 we showed that the mass of ^{90}Sr that remains from a 24-mg sample after t years is $m = f(t) = 24 \cdot 2^{-t/25}$. Find the inverse of this function and interpret it.

SOLUTION We need to solve the equation $m = f(t) = 24 \cdot 2^{-t/25}$ for t. We start by isolating the exponential and taking natural logarithms of both sides:

$$2^{-t/25} = \frac{m}{24}$$

$$\ln(2^{-t/25}) = \ln\left(\frac{m}{24}\right)$$

$$-\frac{t}{25}\ln 2 = \ln m - \ln 24$$

$$t = -\frac{25}{\ln 2}(\ln m - \ln 24) = \frac{25}{\ln 2}(\ln 24 - \ln m)$$

So the inverse function is

$$f^{-1}(m) = \frac{25}{\ln 2}(\ln 24 - \ln m)$$

This function gives the time required for the mass to decay to m milligrams. In particular, the time required for the mass to be reduced to 5 mg is

$$t = f^{-1}(5) = \frac{25}{\ln 2}(\ln 24 - \ln 5) \approx 56.58 \text{ years}$$

This answer agrees with the graphical estimate that we made in Example 1.4.4(c). ■

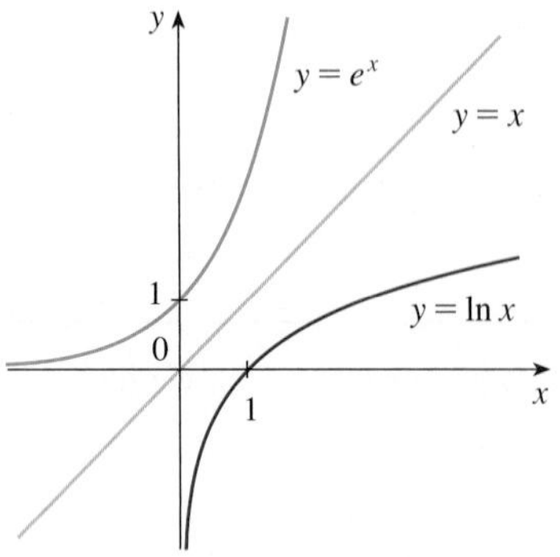

FIGURE 13
The graph of $y = \ln x$ is the reflection of the graph of $y = e^x$ about the line $y = x$.

■ Graph and Growth of the Natural Logarithm

The graphs of the exponential function $y = e^x$ and its inverse function, the natural logarithm function, are shown in Figure 13. Because the curve $y = e^x$ crosses the y-axis with a slope of 1, it follows that the reflected curve $y = \ln x$ crosses the x-axis with a slope of 1.

In common with all other logarithmic functions with base greater than 1, the natural logarithm is an increasing function defined on $(0, \infty)$ and the y-axis is a vertical asymptote. (This means that the values of $\ln x$ become very large negative as x approaches 0.)

EXAMPLE 12 | Sketch the graph of the function $y = \ln(x - 2) - 1$.

SOLUTION We start with the graph of $y = \ln x$ as given in Figure 13. Using the transformations of Section 1.3, we shift it 2 units to the right to get the graph of $y = \ln(x - 2)$ and then we shift it 1 unit downward to get the graph of $y = \ln(x - 2) - 1$. (See Figure 14.)

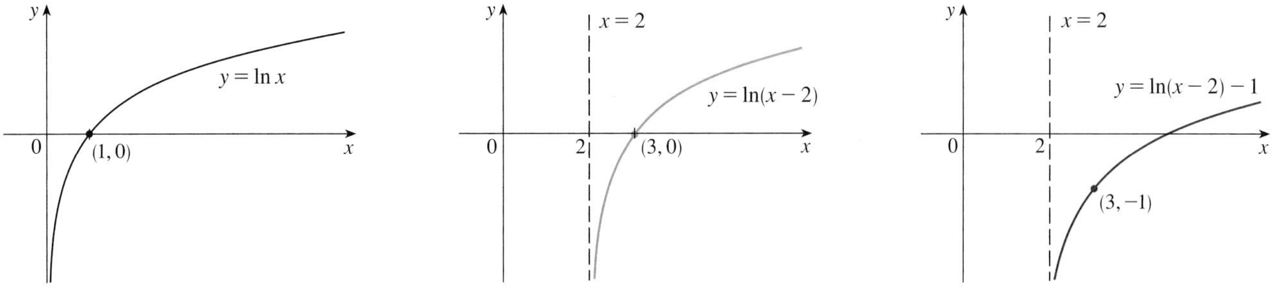

FIGURE 14

Although $\ln x$ is an increasing function, it grows *very* slowly when $x > 1$. In fact, $\ln x$ grows more slowly than any positive power of x. To illustrate this fact, we compare approximate values of the functions $y = \ln x$ and $y = x^{1/2} = \sqrt{x}$ in the following table and we graph them in Figures 15 and 16. You can see that initially the graphs of $y = \sqrt{x}$ and $y = \ln x$ grow at comparable rates, but eventually the root function far surpasses the logarithm.

x	1	2	5	10	50	100	500	1000	10,000	100,000
$\ln x$	0	0.69	1.61	2.30	3.91	4.6	6.2	6.9	9.2	11.5
$\sqrt{x}$	1	1.41	2.24	3.16	7.07	10.0	22.4	31.6	100	316
$\frac{\ln x}{\sqrt{x}}$	0	0.49	0.72	0.73	0.55	0.46	0.28	0.22	0.09	0.04

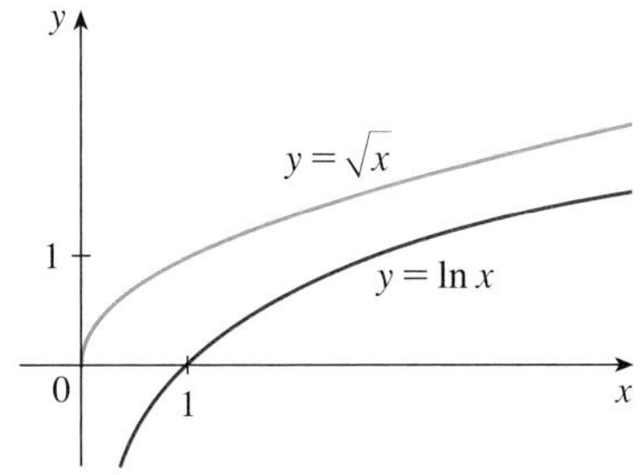

FIGURE 15

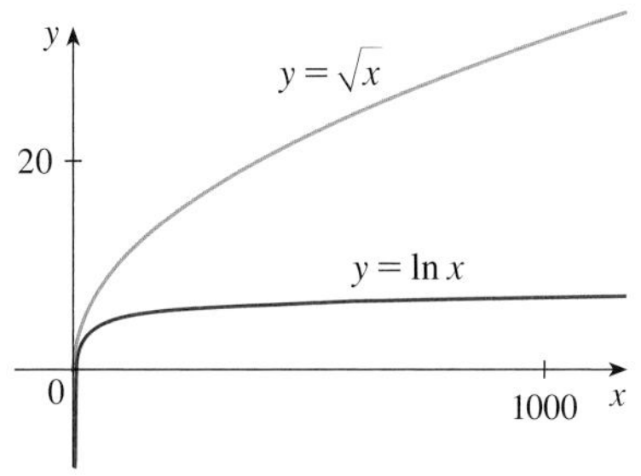

FIGURE 16

Semilog Plots

We've seen that the exponential function $y = b^x$ $(b > 1)$ increases so rapidly that it's sometimes difficult to represent data points conveniently on a single plot. (See Figure 1.4.1.) On the other hand, we have just seen that their inverse functions, the logarithmic functions, increase very slowly. For that reason, logarithmic scales are often used when real-world quantities involve a huge disparity in size: the pH scale for the acidity of a solution, the decibel scale for loudness, the Richter scale for the magnitude of an earthquake. In such cases, the equidistant marks on a **logarithmic scale** represent consecutive powers of 10. (See Figure 17.)

The marks on the "logarithmic ruler" are the logarithms of the numbers they represent.

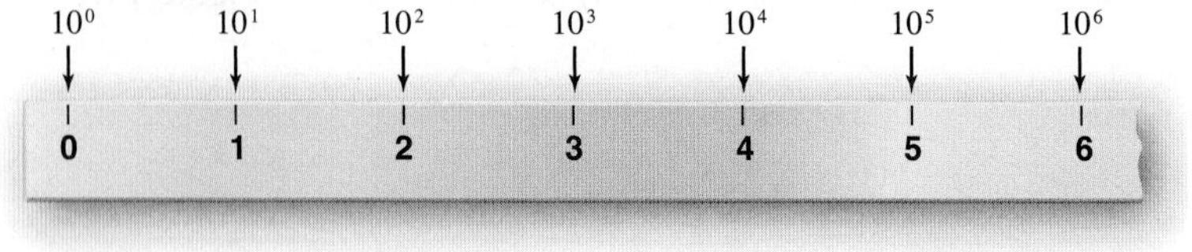

FIGURE 17

Vast differences in size occur in biology too. Figure 18 shows that, in comparing lengths, a logarithmic scale provides more manageable numbers.

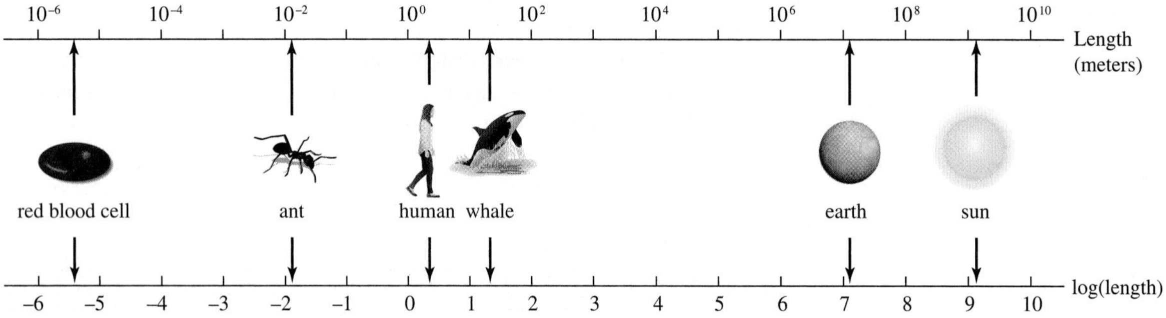

FIGURE 18

In biology it's common to use a **semilog plot** to see whether data points are appropriately modeled by an exponential function. This means that instead of plotting the points (x, y), we plot the points $(x, \log y)$. (Logarithms to the base 10 are always used, so $\log = \log_{10}$.) In other words, we use a logarithmic scale on the vertical axis.

If we start with an exponential function of the form $y = a \cdot b^x$ and take logarithms of both sides, we get

$$\log y = \log(a \cdot b^x) = \log a + \log b^x$$

(11) $$\log y = \log a + x \log b$$

If we let $Y = \log y$, $M = \log b$, and $B = \log a$, then Equation 11 becomes

$$Y = B + Mx$$

which is the equation of a line with slope M and Y-intercept B.

So if we obtain experimental data that we suspect might possibly be exponential, then we could graph a semilog scatter plot and see if it is approximately linear. If so, we could then obtain an exponential model for our original data.

EXAMPLE 13 | Viral load In Section 1.4 we presented data on the viral load $V(t)$ of patient 303 after t days of treatment with ABT-538. In the following table we calculate $\log V(t)$ and in Figure 19 we show the corresponding semilog plot.

t	$V(t)$	$\log V(t)$
1	76	1.9
4	53	1.7
8	18	1.3
11	9.4	0.97
15	5.2	0.72
22	3.6	0.56
29	2.8	0.45

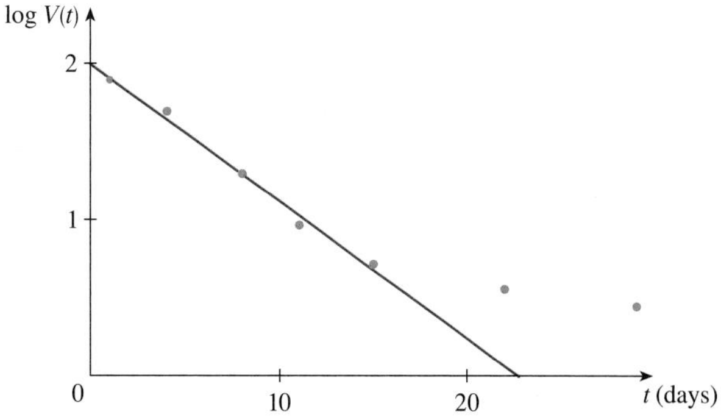

FIGURE 19

We see that the first five data points in Figure 19 lie very nearly on a straight line and, using linear regression, we get the equation

$$\log V(t) = 2.006 - 0.088t$$

for $1 \leqslant t \leqslant 15$. Then, applying the exponential function with base 10 to both sides of this equation, we obtain an equation for the viral load:

$$V(t) = 101 \cdot (0.817)^t$$

This is quite close to the exponential model we got in Section 1.4 using six data points. ■

■ Log-Log Plots

If we use logarithmic scales on both the horizontal and vertical axes, the resulting graph is called a **log-log plot**. It is used when we suspect that a power function might be a good model for our data. If we start with a power function $y = Cx^p$ and take logarithms of both sides, we get

$$\log y = \log(Cx^p) = \log C + \log x^p$$

$$\log y = \log C + p \log x \qquad \textbf{(12)}$$

Let $Y = \log y$, $A = \log C$, and $X = \log x$. Then Equation 12 becomes

$$Y = A + pX$$

We recognize that Y is a linear function of X, so the points $(\log x, \log y)$ lie on a straight line.

EXAMPLE 14 | BB **Species richness in bat caves** Table 3 on page 64 gives the areas of several caves in central Mexico and the number of bat species that live in each cave.[1]

(a) Make a scatter plot and a log-log plot of the data.
(b) Is a power model appropriate? If so, find an expression for it.
(c) The cave called El Sapo near Puebla, Mexico, has a surface area of $A = 205\ \text{m}^2$. Use the model to estimate the number of bat species you would expect to find in that cave.

1. A. Brunet et al., "The Species–Area Relationship in Bat Assemblages of Tropical Caves," *Journal of Mammalogy* 82 (2001): 1114–22.

Table 3 Species–Area Data

Cave	Area (m^2)	Number of Species
La Escondida	18	1
El Escorpion	19	1
El Tigre	58	1
Misión Imposible	60	2
San Martin	128	5
El Arenal	187	4
La Ciudad	344	6
Virgen	511	7

SOLUTION

(a) Let A denote the surface area of a cave and S the number of bat species in the cave. From the scatter plot in Figure 20 we see that the data are neither linear nor exponential. So we calculate the logarithms of the data and create the log-log plot in Figure 21.

$\log A$	$\log S$
1.26	0
1.28	0
1.76	0
1.78	0.30
2.11	0.70
2.27	0.60
2.54	0.78
2.71	0.85

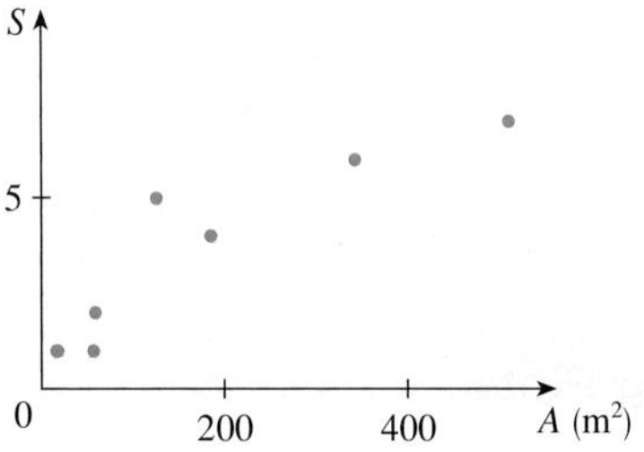

FIGURE 20

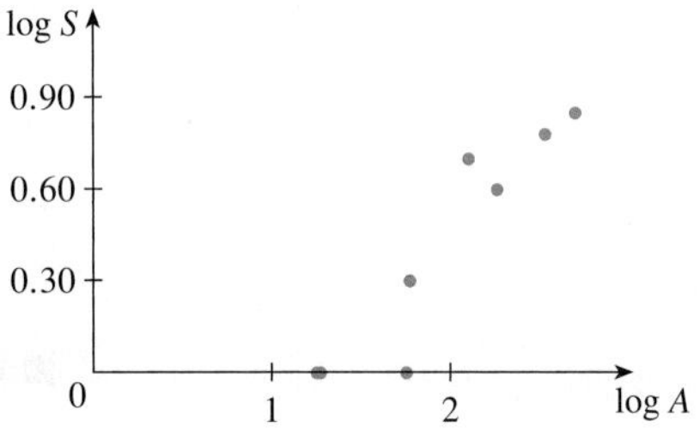

FIGURE 21

(b) It appears that $\log S$ is approximately a linear function of $\log A$. With a graphing calculator or computer, we get the linear model

$$\log S = 0.64 \log A - 0.86$$

Then we apply the exponential function with base 10 to both sides of this equation:

$$S = 10^{0.64 \log A - 0.86} = 10^{\log A^{0.64}} 10^{-0.86}$$

$$S = 0.14A^{0.64}$$

(Alternatively, after verifying from Figure 21 that a power model is appropriate, we could have used a calculator to calculate this power model directly from the original data.) Figure 22 shows that the power model is a reasonable one.

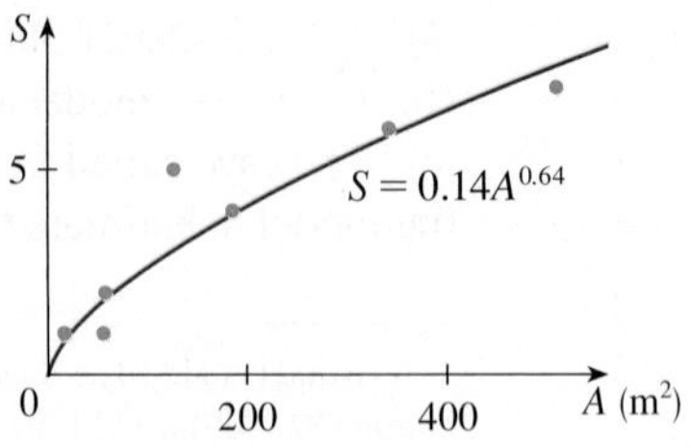

FIGURE 22

(c) Using the model from part (b) with $A = 205$, we get

$$S \approx 0.14 \cdot 205^{0.64} \approx 4.22$$

The El Sapo cave actually does have four species of bats.

So we would expect to find about four bat species in the El Sapo cave. ■

Summary: Linear, Exponential, or Power Model?
To determine whether a linear, exponential, or power model is appropriate, we make a scatter plot, a semilog plot, and a log-log plot.

- If the **scatter plot** of the data lies approximately on a line, then a linear model is appropriate.
- If the **semilog plot** of the data lies approximately on a line, then an exponential model is appropriate.
- If the **log-log plot** of the data lies approximately on a line, then a power model is appropriate.

EXERCISES 1.5

1. (a) What is a one-to-one function?
(b) How can you tell from the graph of a function whether it is one-to-one?

2. (a) Suppose f is a one-to-one function with domain A and range B. How is the inverse function f^{-1} defined? What is the domain of f^{-1}? What is the range of f^{-1}?
(b) If you are given a formula for f, how do you find a formula for f^{-1}?
(c) If you are given the graph of f, how do you find the graph of f^{-1}?

3–14 A function is given by a table of values, a graph, a formula, or a verbal description. Determine whether it is one-to-one.

3.

x	1	2	3	4	5	6
$f(x)$	1.5	2.0	3.6	5.3	2.8	2.0

4.

x	1	2	3	4	5	6
$f(x)$	1.0	1.9	2.8	3.5	3.1	2.9

5.
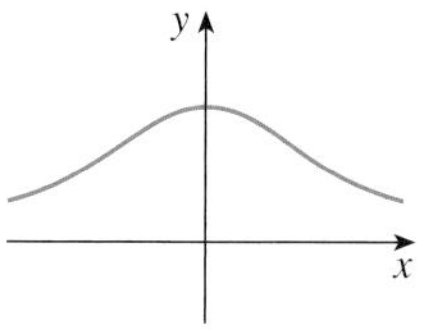

6.
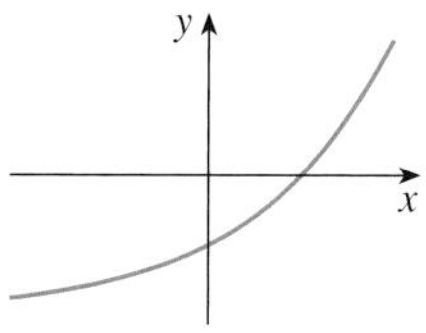

7.
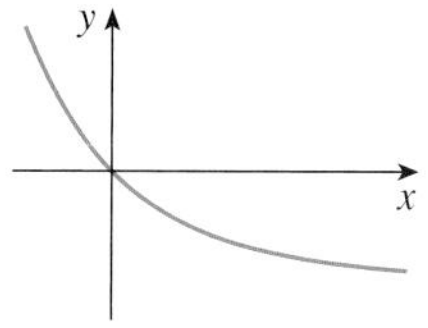

8.
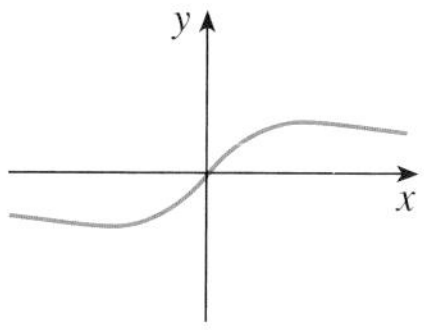

9. $f(x) = x^2 - 2x$

10. $f(x) = 10 - 3x$

11. $g(x) = 1/x$

12. $g(x) = \cos x$

13. $f(t)$ is the height of a football t seconds after kickoff.

14. $f(t)$ is your height at age t.

15. Assume that f is a one-to-one function.
(a) If $f(6) = 17$, what is $f^{-1}(17)$?
(b) If $f^{-1}(3) = 2$, what is $f(2)$?

16. If $f(x) = x^5 + x^3 + x$, find $f^{-1}(3)$ and $f(f^{-1}(2))$.

17. If $g(x) = 3 + x + e^x$, find $g^{-1}(4)$.

18. The graph of f is given.
(a) Why is f one-to-one?
(b) What are the domain and range of f^{-1}?
(c) What is the value of $f^{-1}(2)$?
(d) Estimate the value of $f^{-1}(0)$.

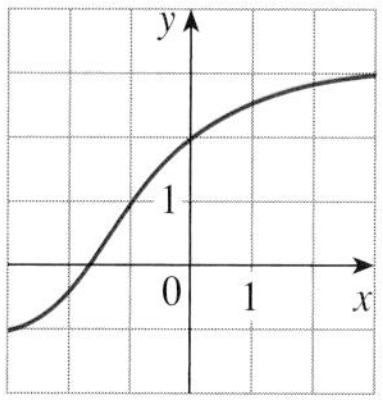

19. The formula $C = \frac{5}{9}(F - 32)$, where $F \geqslant -459.67$, expresses the Celsius temperature C as a function of the Fahrenheit temperature F. Find a formula for the inverse function and interpret it. What is the domain of the inverse function?

20. In the theory of relativity, the mass of a particle with speed v is

$$m = f(v) = \frac{m_0}{\sqrt{1 - v^2/c^2}}$$

where m_0 is the rest mass of the particle and c is the speed of light in a vacuum. Find the inverse function of f and explain its meaning.

21–26 Find a formula for the inverse of the function.

21. $f(x) = 1 + \sqrt{2 + 3x}$

22. $f(x) = \dfrac{4x - 1}{2x + 3}$

23. $f(x) = e^{2x-1}$

24. $y = x^2 - x, \quad x \geqslant \frac{1}{2}$

25. $y = \ln(x + 3)$

26. $y = \dfrac{e^x}{1 + 2e^x}$

27–28 Find an explicit formula for f^{-1} and use it to graph f^{-1}, f, and the line $y = x$ on the same screen. To check your work, see whether the graphs of f and f^{-1} are reflections about the line.

27. $f(x) = x^4 + 1, \quad x \geqslant 0$

28. $f(x) = 2 - e^x$

29–30 Use the given graph of f to sketch the graph of f^{-1}.

29.

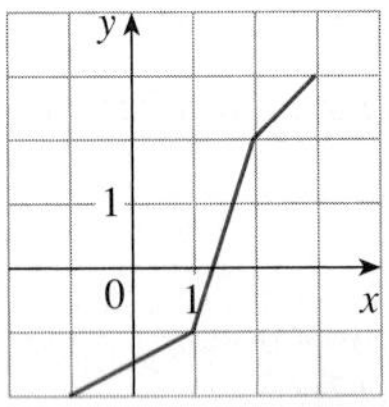

30.

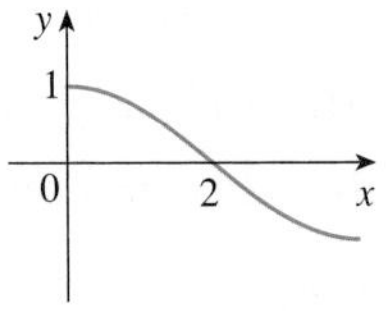

31. Let $f(x) = \sqrt{1 - x^2}, \ 0 \leqslant x \leqslant 1$.
(a) Find f^{-1}. How is it related to f?
(b) Identify the graph of f and explain your answer to part (a).

32. Vaccination coverage Suppose we modify the function in Example 5 by introducing vaccination to control the probability of an outbreak of the disease. We want to know the fraction of the population that we have to vaccinate to achieve a target outbreak probability. If v is the vaccination fraction, then the outbreak probability as a function of v is

$$P = 1 - \frac{1}{R_0(1 - v)}$$

Find the inverse of this function to obtain the vaccination coverage needed for any given target outbreak probability. What do you notice about the inverse function in relation to the original function?

33. (a) How is the logarithmic function $y = \log_b x$ defined?
(b) What is the domain of this function?
(c) What is the range of this function?
(d) Sketch the general shape of the graph of the function $y = \log_b x$ if $b > 1$.

34. (a) What is the natural logarithm?
(b) What is the common logarithm?
(c) Sketch the graphs of the natural logarithm function and the natural exponential function with a common set of axes.

35–38 Find the exact value of each expression.

35. (a) $\log_5 125$ (b) $\log_3\left(\frac{1}{27}\right)$

36. (a) $\ln(1/e)$ (b) $\log_{10}\sqrt{10}$

37. (a) $\log_2 6 - \log_2 15 + \log_2 20$
(b) $\log_3 100 - \log_3 18 - \log_3 50$

38. (a) $e^{-2\ln 5}$ (b) $\ln\left(\ln e^{e^{10}}\right)$

39–41 Express the given quantity as a single logarithm.

39. $\ln 5 + 5 \ln 3$

40. $\ln(a + b) + \ln(a - b) - 2 \ln c$

41. $\frac{1}{3}\ln(x + 2)^3 + \frac{1}{2}[\ln x - \ln(x^2 + 3x + 2)^2]$

42. Use Formula 10 to evaluate each logarithm correct to six decimal places.
(a) $\log_{12} 10$ (b) $\log_2 8.4$

43. Suppose that the graph of $y = \log_2 x$ is drawn on a coordinate grid where the unit of measurement is an inch. How many miles to the right of the origin do we have to move before the height of the curve reaches 3 ft?

44. Compare the functions $f(x) = x^{0.1}$ and $g(x) = \ln x$ by graphing both f and g in several viewing rectangles. When does the graph of f finally surpass the graph of g?

45–46 Make a rough sketch of the graph of each function. Do not use a calculator. Just use the graphs given in Figures 12 and 13 and, if necessary, the transformations of Section 1.3.

45. (a) $y = \log_{10}(x + 5)$ (b) $y = -\ln x$

46. (a) $y = \ln(-x)$ (b) $y = \ln|x|$

47–50 Solve each equation for x.

47. (a) $e^{7-4x} = 6$ (b) $\ln(3x - 10) = 2$

48. (a) $\ln(x^2 - 1) = 3$ (b) $e^{2x} - 3e^x + 2 = 0$

49. (a) $2^{x-5} = 3$ (b) $\ln x + \ln(x - 1) = 1$

50. (a) $\ln(\ln x) = 1$ (b) $e^{ax} = Ce^{bx}$, where $a \neq b$

51–52 Solve each inequality for x.

51. (a) $\ln x < 0$ (b) $e^x > 5$

52. (a) $1 < e^{3x-1} < 2$ (b) $1 - 2 \ln x < 3$

53. Dialysis time Hemodialysis is a process by which a machine is used to filter urea and other waste products from an individual's blood when the kidneys fail. The concentration of urea in the blood is often modeled as exponential decay. If K is the mass transfer coefficient (in mL/min), $c(t)$ is the urea concentration in the blood at time t (in mg/mL) and V is the blood volume, then $c(t) = c_0 e^{-Kt/V}$ where c_0 is the initial concentration at time $t = 0$.

(a) How long should a patient be put on dialysis to reduce the blood urea concentration from an initial value of 1.65 mg/mL to 0.60 mg/mL, given that $K = 340$ mL/min and $V = 32{,}941$ mL?

(b) Derive a general formula for the dialysis time T in terms of the initial urea concentration c_0 and the target urea concentration $c(T)$.

54. Dialysis treatment adequacy

(a) The quantity Kt/V in Exercise 53 is sometimes used as a measure of dialysis treatment adequacy. What does this represent and what are its units?

(b) Another quantity sometimes used to measure dialysis treatment adequacy is the fractional reduction in urea during a dialysis session, denoted by U (that is, the ratio of the amount of urea removed during dialysis to its initial amount). What is the relationship between U and Kt/V?

55. (a) Find the domain of $f(x) = \ln(e^x - 3)$.
(b) Find f^{-1} and its domain.

56. (a) What are the values of $e^{\ln 300}$ and $\ln(e^{300})$?
(b) Use your calculator to evaluate $e^{\ln 300}$ and $\ln(e^{300})$. What do you notice? Can you explain why the calculator has trouble?

57. Bacteria population If a bacteria population starts with 500 bacteria and doubles in size every half hour, then the number of bacteria after t hours is $n = f(t) = 500 \cdot 4^t$. (See Exercise 1.4.30).

(a) Find the inverse of this function and explain its meaning.

(b) When will the population reach 10,000?

58. When a camera flash goes off, the batteries immediately begin to recharge the flash's capacitor, which stores electric charge given by

$$Q(t) = Q_0(1 - e^{-t/a})$$

(The maximum charge capacity is Q_0 and t is measured in seconds.)

(a) Find the inverse of this function and explain its meaning.

(b) How long does it take to recharge the capacitor to 90% of capacity if $a = 2$?

59–64 Data points (x, y) are given.
(a) Draw a scatter plot of the data points.
(b) Make semilog and log-log plots of the data.
(c) Is a linear, power, or exponential function appropriate for modeling these data?
(d) Find an appropriate model for the data and then graph the model together with a scatter plot of the data.

59.

x	2	4	6	8	10	12
y	0.08	0.12	0.18	0.26	0.35	0.53

60.

x	1.0	2.4	3.1	3.6	4.3	4.9
y	3.2	4.8	5.8	6.2	7.2	7.9

61.

x	0.5	1.0	1.5	2.0	2.5	3.0
y	4.10	3.71	3.39	3.2	2.78	2.53

62.

x	10	20	30	40	50	60
y	29	82	150	236	330	430

63.

x	3	4	5	6	7	8
y	11.3	20.2	32.2	45.7	62.1	80.4

64.

x	5	10	15	20	25	30
y	0.013	0.046	0.208	0.930	4.131	18.002

65. Indian population The table gives the midyear population of India (in millions) for the last half of the 20th century.

Year	Population	Year	Population
1950	370	1980	685
1960	445	1990	838
1970	554	2000	1006

(a) Make a scatter plot, semilog plot, and log-log plot for these data and comment on which type of model would be most appropriate.

(b) Obtain an exponential model for the population.
(c) Use your model to estimate the population in 2010 and compare with the actual population of 1173 million. What conclusion can you make?

66. Why is the dodo extinct? Ornithologists measured and cataloged the wingspans and weights of many different species of birds that can fly. The table shows the wingspan L for a bird of weight W.
(a) Make a scatter plot, semilog plot, and log-log plot for these data. Which type of model do you think would be best?
(b) Find an exponential model and power model for the data.
(c) Graph the models from part (b). Which is better?
(d) The dodo is a bird that has been extinct since the late 17th century. It weighed about 45 pounds and had a wingspan of about 20 inches.Use the model chosen in part (c) to estimate the wingspan of a 45-lb bird. Why couldn't a dodo fly?

Bird	W (lb)	L (in)
Turkey vulture	4.40	69
Bald eagle	6.82	84
Great horned owl	3.08	44
Cooper's hawk	1.03	28
Sandhill crane	9.02	79
Atlantic puffin	0.95	24
California condor	17.82	109
Common loon	7.04	48
Yellow warbler	0.022	8
Common grackle	0.20	16
Wood stork	5.06	63
Mallard	2.42	35

Study of a Dodo (oil on canvas), Hart, F. (19th Century)/Royal Albert Memorial Museum, Exeter, Devon, UK/The Bridgeman Art Lbrary.

The dodo (now extinct)

67. Biodiversity in a rain forest To quantify the biodiversity of trees in a tropical rain forest, biologists collected data in the Pasoh Forest Reserve of Malaysia. The table shows the number of tree species S found for a given area A in the rain forest.
(a) Make a scatter plot of the data and either a semilog plot or a log-log plot, whichever you think is more appropriate.
(b) Use your preferred plot from part (a) to find a model. Graph your model with the scatter plot.

A (m^2)	S	A (m^2)	S
3.81	3	122.07	70
7.63	3	244.14	112
15.26	12	488.28	134
30.52	13	976.56	236
61.04	31		

Source: Adapted from K. Kochummen et al., "Floristic Composition of Pasoh Forest Reserve, a Lowland Rain Forest in Peninsular Malaysia," *Journal of Tropical Forest Science* 3 (1991): 1–13.

68. Malarial parasites The table, supplied by Andrew Read, shows the results of an experiment involving malarial parasites. The time t is measured in days and N is the number of parasites per microliter of blood.
(a) Make a scatter plot and a semilog plot of the data.
(b) Find an exponential model and graph your model with the scatter plot. Is it a good fit?

t (days)	N	t (days)	N
1	228	5	372,331
2	2,357	6	2,217,441
3	12,750	7	6,748,400
4	26,661		

69. Drinking and driving In a medical study, researchers measured the mean blood alcohol concentration (BAC) of eight fasting adult male subjects (in mg/mL) after rapid consumption of 30 mL of ethanol (corresponding to two standard alcoholic drinks). The BAC peaked after half an hour and the table shows measurements starting after an hour.

t (hours)	1.0	1.25	1.5	1.75	2.0
BAC	0.33	0.29	0.24	0.22	0.18

t (hours)	2.25	2.5	3.0	3.5	4.0
BAC	0.15	0.12	0.069	0.034	0.010

(a) Make a scatter plot and a semilog plot of the data.
(b) Find an exponential model and graph your model with the scatter plot. Is it a good fit?

(c) Use your model and logarithms to determine when the BAC will be less than 0.08 mg/mL, the legal limit for driving.

Source: Adapted from P. Wilkinson et al., "Pharmacokinetics of Ethanol after Oral Administration in the Fasting State," *Journal of Pharmacokinetics and Biopharmaceutics* 5 (1977): 207–24.

70. Amplifying DNA Polymerase Chain Reaction (PCR) is a biochemical technique that allows scientists to take tiny samples of DNA and amplify them into large samples that can then be examined to determine the DNA sequence. (This is useful, for example, in forensic science.) The process works by mixing the sample with appropriate enzymes and then heating it until the DNA double helix separates into two individual strands. The enzymes then copy each strand, and once the sample is cooled the number of DNA molecules will have doubled. By repeatedly performing this heating and cooling process, the number of DNA molecules continues to double every temperature cycle (referred to as a PCR cycle).

(a) Suppose a sample containg x molecules is collected from a crime scene and is amplified by PCR. Express the number of DNA molecules as a function of the number n of PCR cycles.

(b) There is a detection threshold of T molecules below which no DNA can be seen. Derive an equation for the number of PCR cycles it will take for the DNA sample to reach the detection threshold.

(c) One way scientists determine the abundance of different DNA molecules in a sample is by measuring the difference in time it takes to reach the detection threshold for each. Sketch a graph of the number of cycles needed to reach the detection threshold as a function of the initial number of molecules. Comment on the relationship between differences in initial number of molecules and differences in the time to reach the detection threshold.

PROJECT The Coding Function of DNA

BB

Proteins are made up of long chains of molecules called amino acids. Twenty different amino acids are coded by the DNA of living organisms. The "alphabet" of DNA consists of four letters A, T, C, and G, called *bases*. These bases are grouped together along the DNA sequence into "words," called *codons*. All codons contain the same number of bases (that is, the words are always the same length) and any given codon specifies exactly one amino acid (see Figure 1). The coding of amino acids by DNA described here can be viewed as a *function* that takes an input codon and produces an output amino acid.

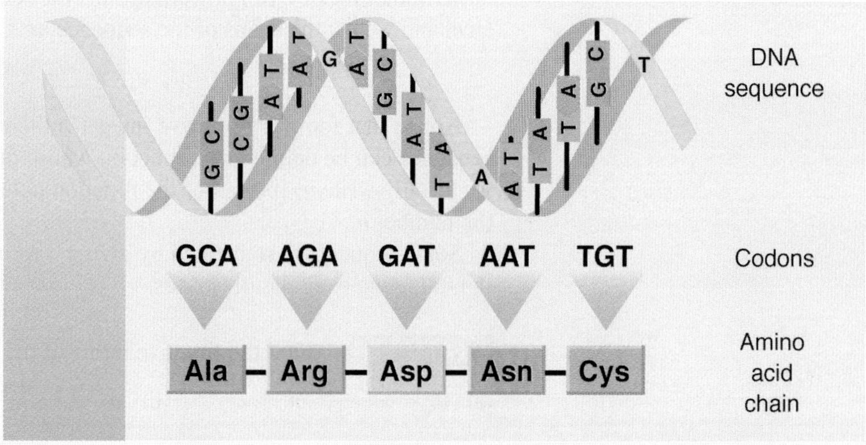

1. Suppose that all codons contained only one base. What would the domain of the coding function be? What would its biggest possible range be?

2. Suppose that all codons contained two bases. What would the domain of the coding function be? What would its biggest possible range be?

3. In fact, all codons actually contain three bases. What is the domain of the coding function in this case? What is its biggest possible range?

4. Given your answers to Problems 1–3, speculate on why codons contain three bases rather than fewer or more.

5. Explain why the coding function in which codons have three bases is not a one-to-one function.

Researchers use the fact that the coding function is not one-to-one to infer which DNA sequence variants are advantageous. In particular, because the coding function is not one-to-one, two kinds of DNA mutations can occur: those that do not alter the amino acid (called synonymous mutations) and those that do (called nonsynonymous mutations). Synonymous mutations are not expected to affect organismal functioning because they don't affect protein structure. As a result, such synonymous mutations are expected to accumulate over time, by chance, in a clocklike fashion. Nonsynonymous mutations do change the amino acid and therefore alter protein structure. If such alterations are advantageous, then we would expect these mutational changes to occur at a rate that is higher than those of the neutral, synonymous mutations. This kind of comparison is possible only because the genetic coding function isn't one-to-one; it forms the basis of nonsynonymous-to-synonymous ratio tests used in biology.

1.6 Sequences and Difference Equations

A **sequence** can be thought of as a list of numbers written in a definite order:

$$a_1, \ a_2, \ a_3, \ a_4, \ \ldots, \ a_n, \ \ldots$$

The number a_1 is called the *first term*, a_2 is the *second term*, and in general a_n is the *nth term*. Sometimes we might want to start the sequence with $n = 0$. Then a_0 is the zeroth term and we list the terms of the sequence as

$$a_0, \ a_1, \ a_2, \ a_3 \ , \ldots, \ a_n, \ \ldots$$

Notice that for every positive integer n there is a corresponding number a_n and so a sequence can be defined as a function whose domain is the set of positive integers. But we usually write a_n instead of the function notation $f(n)$ for the value of the function at the number n.

Some sequences are defined by giving a formula for the nth term a_n in terms of n, as the following example illustrates.

EXAMPLE 1 | Find the first five terms of the sequence.

(a) $a_n = \dfrac{n}{n+1}, \quad n \geqslant 1$ (b) $a_n = (-1)^{n-1}, \quad n \geqslant 1$ (c) $a_n = 8^n, \quad n \geqslant 0$

SOLUTION

(a) Putting $n = 1, 2, 3, 4, 5$, successively, in the formula for a_n we get the initial terms of the sequence:

$$\frac{1}{2}, \frac{2}{3}, \frac{3}{4}, \frac{4}{5}, \frac{5}{6}, \ldots, \frac{n}{n+1}, \ldots$$

(b) Again we start with $n = 1$, noting that $(-1)^{1-1} = (-1)^0 = 1$:

$$1, \; -1, \; 1, \; -1, \; 1, \; \ldots$$

(c) Here we start with $n = 0$. The terms are $8^0, 8^1, 8^2, 8^3, 8^4$, or

$$1, \; 8, \; 64, \; 512, \; 4096, \; \ldots$$

We've already met this sequence in describing the growth of malarial parasites in Section 1.4. ■

A sequence can be pictured by plotting its **graph**. Because a sequence is a function whose domain is the set of positive integers, its graph consists of isolated points with coordinates

$$(1, a_1) \qquad (2, a_2) \qquad (3, a_3) \qquad \ldots \qquad (n, a_n) \qquad \ldots$$

Parts (a) and (b) of Figure 1 show the graphs of the sequences in parts (a) and (b) of Example 1.

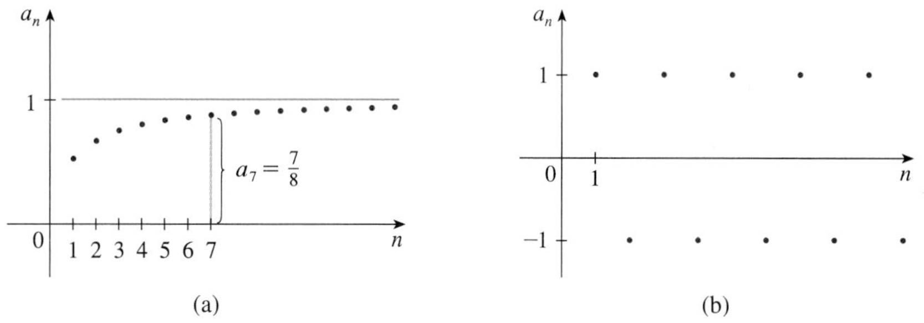

FIGURE 1

The heights of the points on the graph in Figure 1(a) appear to be approaching the number 1, whereas those in Figure 1(b) are oscillating (forever) between -1 and 1. The behavior of sequences in the long run (as n becomes large) will be discussed in Section 2.1.

If we know the first few terms of a sequence but don't have a general formula for a_n, we might be able to detect a pattern in the numbers and write a formula for the nth term. Such a formula might not be unique; we look for the simplest formula, as in the next example.

EXAMPLE 2 | Find a formula for the general term a_n of the sequence

$$\frac{3}{5}, \; -\frac{4}{25}, \; \frac{5}{125}, \; -\frac{6}{625}, \; \frac{7}{3125}, \; \ldots$$

assuming that the pattern of the first few terms continues.

SOLUTION We are given that

$$a_1 = \frac{3}{5} \qquad a_2 = -\frac{4}{25} \qquad a_3 = \frac{5}{125} \qquad a_4 = -\frac{6}{625} \qquad a_5 = \frac{7}{3125}$$

Notice that the numerators of these fractions start with 3 and increase by 1 whenever we go to next term. The second term has numerator 4, the third term has numerator 5; in general, the nth term will have numerator $n + 2$. The denominators are the powers of 5, so a_n has denominator 5^n. The signs of the terms are alternately positive and negative, so we multiply by $(-1)^{n-1}$ as in Example 1(b). Therefore

$$a_n = (-1)^{n-1}\frac{n+2}{5^n}$$ ■

■ Recursive Sequences: Difference Equations

Some sequences don't have simple defining formulas like the ones in Examples 1 and 2. The nth term of a sequence may depend on some of the terms that precede it. A sequence defined in this way is called **recursive**.

EXAMPLE 3 | Find the first five terms of the sequence defined recursively by the equations

$$a_1 = 2 \qquad a_{n+1} = \tfrac{1}{2}(a_n + 6) \qquad \text{for } n \geqslant 1$$

SOLUTION The defining formula allows us to calculate a term if we know the preceding one. We are given the first term, so we can use it to find the second term. Then we find the third term from the second one, and so on:

$$a_2 = \tfrac{1}{2}(a_1 + 6) = \tfrac{1}{2}(2 + 6) = 4$$

$$a_3 = \tfrac{1}{2}(a_2 + 6) = \tfrac{1}{2}(4 + 6) = 5$$

$$a_4 = \tfrac{1}{2}(a_3 + 6) = \tfrac{1}{2}(5 + 6) = 5.5$$

$$a_5 = \tfrac{1}{2}(a_4 + 6) = \tfrac{1}{2}(5.5 + 6) = 5.75$$

So the sequence starts like this:

$$2, \ 4, \ 5, \ 5.5, \ 5.75, \ \ldots$$ ■

EXAMPLE 4 | **Fibonacci sequence** Find the first ten terms of the recursive sequence given by

$$F_1 = 1, \qquad F_2 = 1, \qquad F_n = F_{n-1} + F_{n-2} \qquad \text{for } n \geqslant 3$$

SOLUTION We are given F_1 and F_2, so we proceed as follows:

$$F_3 = F_2 + F_1 = 1 + 1 = 2$$

$$F_4 = F_3 + F_2 = 2 + 1 = 3$$

$$F_5 = F_4 + F_3 = 3 + 2 = 5$$

Each term is the sum of the two terms that precede it, so we can easily write as many terms as we please. Here are the first ten terms:

$$1, \ 1, \ 2, \ 3, \ 5, \ 8, \ 13, \ 21, \ 34, \ 55$$ ■

FIGURE 2
The Fibonacci sequence in the branching of a tree

The sequence in Example 4 is called the **Fibonacci sequence**, named after the 13th-century Italian mathematician known as Fibonacci, who used it to solve a problem concerning the breeding of rabbits (see Exercise 23). This sequence also occurs in numerous applications in plant biology. (See Figure 2 for one such occurrence.)

A recursive sequence is also called a **difference equation**. The recursive sequence in Example 3 is called a **first-order difference equation** because a_{n+1} depends on just the preceding term a_n, whereas the Fibonacci sequence is a **second-order difference equation** because F_n depends on the two preceding terms F_{n-1} and F_{n-2}.

The general first-order difference equation is of the form

$$a_{n+1} = f(a_n)$$

where f is some function. Why is it called a *difference equation*? The word *difference* comes from the fact that such equations are often formulated in terms of the difference between one term and the next:

$$\Delta a_n = a_{n+1} - a_n$$

The equation $\Delta a_n = g(a_n)$ can be written as follows:

$$a_{n+1} - a_n = g(a_n)$$

$$a_{n+1} = a_n + g(a_n) = f(a_n)$$

where $f(x) = x + g(x)$.

BB

■ Discrete-Time Models in the Life Sciences

Difference equations are often used in biology to model cell division and insect populations, for example. In these contexts we usually replace n by t, to denote time. If we think of t as the current time, then $t + 1$ is one unit of time into the future. (For cell division, t might represent hours or days; for insect populations, it could represent days, months, or years.) We will use N_t to denote the population size, so a difference equation modeling population size has the form

$$N_{t+1} = f(N_t) \qquad t = 0, 1, 2, 3, \ldots$$

In this context we call f an **updating function** because f "updates" the population from N_t to N_{t+1}.

We have already seen an example of this in Section 1.4, where a malarial parasite produces 8 new parasites in a period of 24 hours. So

$$N_{t+1} = 8N_t \qquad N_0 = 1$$

where t is measured in days, and we saw that the solution of this difference equation is

$$N_t = 8^t$$

If an *E. coli* population starts with N_0 bacteria and its size doubles every 20 minutes, then we measure t in units of 20 minutes and write $N_{t+1} = 2N_t$. As before, we find that

$$N_t = N_0 \cdot 2^t$$

More generally, if a population of cells divides synchronously, with each cell producing R daughter cells, then the difference equation

We have discussed sequences defined by a formula and also recursive sequences. Equation 1 defines a recursive sequence and the solution given by Equation 2 is the corresponding formula.

(1) $$\boxed{N_{t+1} = RN_t}$$

relates successive generations and the solution is

(2) $$N_t = N_0 \cdot R^t$$

The number R is the number of offspring per individual and is called the **per capita growth factor**.

Similar equations arise when we consider insect populations that breed seasonally. We will take the unit of time to be the time span from one generation to the next. Then in general terms we can formulate the model

(3) $$N_{t+1} = N_t + \text{inflow} - \text{outflow}$$

If the population has a constant per capita birth rate β and constant per capita death rate μ, then the difference equation in (3) becomes

(4) $$N_{t+1} = N_t + \beta N_t - \mu N_t$$

Notice that in the case where insects die immediately after producing the next generation, we have $\mu = 1$ and Equation 4 becomes Equation 1 with $R = \beta$.

So far we have considered populations that grow under ideal conditions without limitations to growth. Let's now consider a more realistic model called the **logistic difference equation**. Let K represent the **carrying capacity**, which is the population size at which the per capita growth factor is 1. We replace the difference equation in Equation 1, $N_{t+1} = RN_t$, by the model

(5) $$N_{t+1} = \left[1 + r\left(1 - \frac{N_t}{K}\right)\right]N_t$$

where r is a positive constant. Here the per capita growth factor is

(6) $$R = 1 + r\left(1 - \frac{N_t}{K}\right)$$

whereas in (1) R is simply a constant. Notice that N_t/K is the fraction of the carrying capacity at time t and so $r(1 - N_t/K)$ is small when N_t is close to K and $r(1 - N_t/K)$ is close to its largest value r when N_t is small. Observe that R decreases linearly from $1 + r$ to 1 as N_t increases from 0 to K. This means that the logistic difference equation has a variable per capita growth factor R.

We can simplify Equation 5 by defining

$$x_t = \frac{r}{(1 + r)K}N_t$$

Then

$$x_{t+1} = \frac{r}{(1 + r)K}N_{t+1} = \frac{r}{(1 + r)K}\left[1 + r\left(1 - \frac{N_t}{K}\right)\right]N_t$$

$$= \frac{r}{(1 + r)K}\left[(1 + r)N_t - \frac{rN_t^2}{K}\right] = \frac{r}{K}N_t - \frac{r^2N_t^2}{(1 + r)K^2}$$

On the right side of this equation we use the fact that

$$N_t = \frac{(1 + r)K}{r}x_t$$

to obtain $$x_{t+1} = (1 + r)x_t - (1 + r)x_t^2 = (1 + r)x_t(1 - x_t)$$

If we now write $R_{\max} = 1 + r$, which is the largest value of R as a function of N_t in Equation 6, we obtain a simpler-looking difference equation:

(7) $$x_{t+1} = R_{max} x_t (1 - x_t)$$

Equation 7 is also called the **logistic difference equation**.

EXAMPLE 5 | BB If $x_0 = \frac{7}{8}$, graph the first ten terms of the logistic difference equation (7) for (a) $R_{max} = 1.5$ and (b) $R_{max} = 3.2$.

SOLUTION
(a) With $x_0 = \frac{7}{8}$ and $x_{t+1} = 1.5x_t(1 - x_t)$, we use a graphing calculator or computer to calculate the first ten terms approximately and then we plot them in Figure 3.

t	x_t	t	x_t
0	0.8750	6	0.3155
1	0.1641	7	0.3239
2	0.2057	8	0.3285
3	0.2451	9	0.3309
4	0.2775	10	0.3321
5	0.3008		

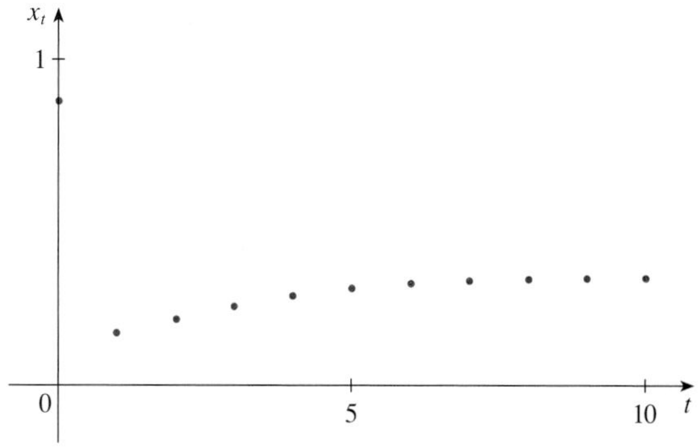

FIGURE 3

(b) With $R_{max} = 3.2$ we get the following values and plot them in Figure 4.

t	x_t	t	x_t
0	0.8750	6	0.7604
1	0.3500	7	0.5831
2	0.7280	8	0.7779
3	0.6337	9	0.5528
4	0.7428	10	0.7911
5	0.6113		

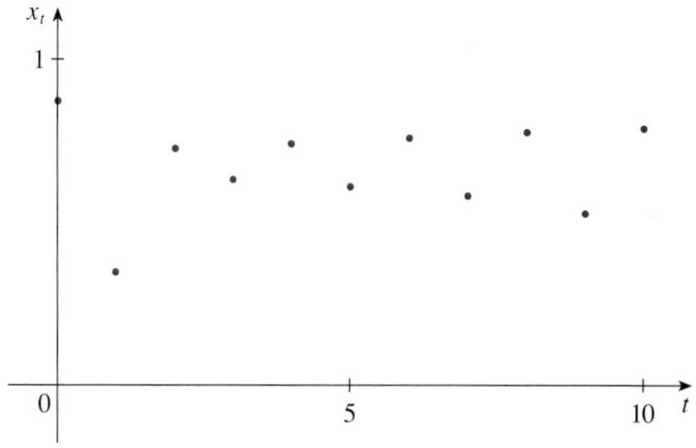

FIGURE 4 ■

Notice from Figures 3 and 4 that when we change the value of r in the logistic difference equation, and thereby change R_{max}, the sequence looks quite different. We will return to this equation in Sections 2.1 and 4.5 to explore the limiting behavior of the logistic difference equation for different values of R_{max}.

EXAMPLE 6 | BB **Maintaining cerebrospinal pressure**[1] Cerebrospinal fluid (CSF) is a clear liquid that occupies the compartment of the body containing the

1. Adapted from S. Cruickshank, *Mathematics and Statistics in Anaesthesia* (New York: Oxford University Press, USA, 1998).

brain and spinal cord (the cerebrospinal chamber). It is important to maintain an appropriate cerebrospinal pressure during medical procedures and the pressure is a function of the CSF volume. Suppose that we measure the CSF volume every five minutes during a medical procedure and that A mL of CSF is secreted into the cerebrospinal chamber every five minutes. Also, suppose that the amount of CSF reabsorbed every five minutes is proportional to its current volume.
(a) Derive a discrete-time difference equation for how the volume V of CSF changes from one measurement to the next.
(b) Ultimately we are interested in how the cerebrospinal pressure P changes from one measurement to the next. Suppose that pressure is related to volume according to the equation $P = V^2$. (This would be appropriate if pressure is nearly zero for small volumes but increases at an accelerating rate as volume increases.) Derive a discrete-time recursion for the pressure.

SOLUTION
(a) The volume at measurement $m + 1$ is the volume at measurement m plus the secreted CSF (A) minus what is reabsorbed (kV, where k is the constant of proportionality). Therefore we have

$$V_{m+1} = V_m + A - kV_m = A + (1 - k)V_m$$

(b) For any given measurement, the pressure is given by $P_m = V_m^2$. So

$$P_{m+1} = V_{m+1}^2 = [A + (1 - k)V_m]^2$$

This recursion is not yet complete because it tells us the pressure at measurement $m + 1$ as a function of the *volume* at measurement m. To use it recursively we need the recursion to give us the pressure at measurement $m + 1$ as a function of the *pressure* at measurement m. Therefore we need to write V_m in terms of P_m. Solving $P_m = V_m^2$ for V_m and keeping only the positive solution, we get $V_m = \sqrt{P_m}$. Substituting, we get

$$\begin{aligned} P_{m+1} &= \left[A + (1 - k)\sqrt{P_m}\right]^2 \\ &= A^2 + 2A(1 - k)\sqrt{P_m} + (1 - k)^2 P_m \end{aligned}$$

■

EXERCISES 1.6

1–4 List the first five terms of the sequence.

1. $a_n = \dfrac{2n}{n^2 + 1}$

2. $a_n = \dfrac{3^n}{1 + 2^n}$

3. $a_n = \dfrac{(-1)^{n-1}}{5^n}$

4. $a_n = \cos \dfrac{n\pi}{2}$

5–8 Calculate, to four decimal places, the first ten terms of the sequence and use them to plot the graph of the sequence by hand.

5. $a_n = \dfrac{3n}{1 + 6n}$

6. $a_n = 2 + \dfrac{(-1)^n}{n}$

7. $a_n = 1 + \left(-\frac{1}{2}\right)^n$

8. $a_n = 1 + \dfrac{10^n}{9^n}$

9–14 Find a formula for the general term a_n of the sequence, assuming that the pattern of the first few terms continues.

9. $1, \frac{1}{3}, \frac{1}{5}, \frac{1}{7}, \frac{1}{9}, \ldots$

10. $1, -\frac{1}{3}, \frac{1}{9}, -\frac{1}{27}, \frac{1}{81}, \ldots$

11. $-3, 2, -\frac{4}{3}, \frac{8}{9}, -\frac{16}{27}, \ldots$

12. $5, 8, 11, 14, 17, \ldots$

13. $\frac{1}{2}, -\frac{4}{3}, \frac{9}{4}, -\frac{16}{5}, \frac{25}{6}, \ldots$

14. $1, 0, -1, 0, 1, 0, -1, 0, \ldots$

15–22 Find the first six terms of the recursive sequence.

15. $a_1 = 1,\ a_{n+1} = 5a_n - 3$

16. $a_1 = 6,\ a_{n+1} = \dfrac{a_n}{n}$

17. $a_1 = 2,\ a_{n+1} = \dfrac{a_n}{1 + a_n}$

18. $a_1 = 1,\ a_{n+1} = 4 - a_n$

19. $a_1 = 1,\ a_{n+1} = \sqrt{3a_n}$

20. $a_1 = 3,\ a_{n+1} = \sqrt{3a_n}$

21. $a_1 = 2,\ a_2 = 1,\ a_{n+1} = a_n - a_{n-1}$

22. $a_1 = 1,\ a_2 = 2,\ a_{n+2} = a_{n+1} + 2a_n$

---

23. Breeding rabbits Fibonacci posed the following problem: Suppose that rabbits live forever and that every month each pair produces a new pair, which becomes reproductive at age 2 months. If we start with one newborn pair, how many pairs of rabbits will we have in the nth month? Show that the answer is F_n, the nth term of the Fibonacci sequence defined in Example 4.

24. Harvesting fish A fish farmer has 5000 catfish in his pond. The number of catfish increases by 8% per month and the farmer harvests 300 catfish per month.

(a) Show that the catfish population P_n after n months is given recursively by

$$P_n = 1.08P_{n-1} - 300 \qquad P_0 = 5000$$

(b) How many catfish are in the pond after six months?

25. Consider the difference equation

$$N_0 = 1 \qquad N_{t+1} = RN_t$$

What can you say about the solution of this equation as t becomes large in the following three cases?

(a) $R < 1$ (b) $R = 1$ (c) $R > 1$

26. (a) For a difference equation of the form $N_{t+1} = f(N_t)$, calculate the composition $(f \circ f)(N_t)$. What is the meaning of $f \circ f$ in this context?

(b) If $N_{t+1} = f(N_t)$, where f is one-to-one, what is $f^{-1}(N_{t+1})$? What is the meaning of the inverse function f^{-1} in this context?

27–31 Logistic equation For the logistic difference equation $x_{t+1} = cx_t(1 - x_t)$ and the given values of x_0 and c, calculate x_t to four decimal places for $t = 1, 2, \ldots, 10$ and graph x_t. Comment on the behavior of the sequence.

27. $x_0 = 0.5,\ c = 1.5$

28. $x_0 = 0.5,\ c = 2.5$

29. $x_0 = \frac{7}{8},\ c = 3.42$

30. $x_0 = \frac{7}{8},\ c = 3.45$

31. $x_0 = 0.5,\ c = 3.7$

---

32. Ricker equation In the logistic difference equation the factor $(1 - x_t)$ decreases linearly from 1 to 0 as x_t increases from 0 to 1. If, instead, we introduce the decreasing exponential factor e^{-x_t}, we get what is called the *Ricker difference equation*:

$$x_{t+1} = cx_te^{-x_t}$$

This model has the advantage that the factor e^{-x_t} never reaches 0.

(a) If $x_0 = 0.5$ and $c = 2.5$, calculate the first ten terms to four decimal places and graph them.

(b) Compare with Exercise 28.

33. Repeat Exercise 32(a) for $x_0 = \frac{7}{8}$ and $c = 3.42$. Compare with Exercise 29.

34. Drug concentration Suppose C_t is the concentration of a drug in the bloodstream at time t, A is the concentration of the drug that is administered at each time step, and k is the fraction of the drug metabolized in a time step.

(a) What is the recursion that models how the drug concentration changes?

(b) If the initial concentration is $C_0 = 120\ \mu g/mL$ and $A = 80\ \mu g/mL$ and $k = \frac{1}{2}$, plot some points on the graph of C_t. Is the graph similar for other values of A and k?

35. Bacteria colonies on agar plates Bacteria are often grown on agar plates and form circular colonies. The area of a colony is proportional to the number of bacteria it contains. The agar (a gelatinous substance obtained from red seaweed) is the resource that bacteria use to reproduce and so only those bacteria on the edge of the colony can produce new offspring. Therefore the population changes according to the equation $N_{t+1} = N_t + I$, where I is the input of new individuals and is proportional to the circumference of the colony, with proportionality constant R.

(a) Derive the recursion for the population size.

(b) Plot some points on the graph of N_t, assuming specific values for the proportionality constants.

36. Spherical colonies Suppose the volume of a spherical colony is proportional to the number of individuals in it and growth occurs only at the surface-resource interface of the colony. Find a difference equation that models the population.

37. Salmon and bears Pacific salmon populations have discrete breeding cycles in which they return from the ocean to streams to reproduce and then die. This occurs every one to five years, depending on the species.

(a) Suppose that each fish must first survive predation by bears while swimming upstream, and predation occurs with probability d. After swimming upstream, each fish produces b offspring before dying. The stream is then stocked with m additional newly hatched fish before all fish then swim out to sea. What is the discrete-time recursion for the population size, assuming that there is no mortality while at sea? You should count the population immediately before the upstream journey.

(b) Suppose instead that bears prey on fish only while the fish are swimming downstream. What is the discrete-time recursion for the population dynamics? (Again assume there is no mortality while at sea.)

(c) Which of the recursions obtained in parts (a) and (b) predicts the largest increase in population size from one year to the next? Justify your answer both mathematically and in terms of the underlying biology. You can assume that $0 < d < 1$ and $b > 0$.

38. Methyl groups in DNA DNA sometimes has chemical groups attached, called methyl groups, that affect gene expression. Suppose that, during each hour, first a fraction m of unmethylated locations on the DNA become methylated, and then a fraction u of methylated locations become unmethylated. Find a recursion for the fraction f of the DNA molecule that is methylated.

39. Two bacteria strains Suppose the population sizes of two strains of bacteria each grow as described by the recursions $a_{t+1} = R_a a_t$ and $b_{t+1} = R_b b_t$, respectively. The frequency of the first strain at time t is defined as $p_t = a_t/(a_t + b_t)$. Derive a recursion for p_t and show that it can be written in terms of a single constant $\alpha = R_a/R_b$.

40. Find the first 40 terms of the sequence defined by

$$a_{n+1} = \begin{cases} \frac{1}{2}a_n & \text{if } a_n \text{ is an even number} \\ 3a_n + 1 & \text{if } a_n \text{ is an odd number} \end{cases}$$

and $a_1 = 11$. Do the same if $a_1 = 25$. Make a conjecture about this type of sequence.

■ PROJECT Drug Resistance in Malaria

BB

Drug resistance in malaria is a serious problem in many parts of the world. Let's suppose that there are two different genes in the population, one that causes resistance (labeled R) and one that is sensitive to the drug (labeled S). We will construct a recursion for the frequency of the R gene by *assuming* that the entire malaria life cycle occurs synchronously.

The life cycle of malaria is quite complicated. Part of it occurs in mosquitoes and part in humans (see Figure 1). While in humans, each malaria parasite carries a single copy of the gene (either R or S) and is referred to as **haploid**. But while in the mosquito, pairs of such haploid parasites combine through a process of sexual reproduction to form **diploid** parasites. Diploid individuals carry two copies of the gene and therefore can be of three different types: RR, RS, or SS (see Figure 2).

To contruct a recursion for the frequency of the R gene, we must first choose a point in the malaria life cycle at which to census the population. Since the haploid stage of the parasite in humans is the simplest, let's choose that. We use p_t to denote the frequency of the R type in the haploid stage. (All R genes are colored yellow in Figure 2.) Our goal is to derive a recursion of the form $p_{t+1} = f(p_t)$ for some function f. We do this by dividing the entire life cycle into three steps: (1) union of pairs of haploid individuals to form diploid individuals; (2) differential survival of diploid individuals in the mosquito; and (3) the production of new haploid individuals by the surviving diploid parasites.

1. Suppose that when pairs of haploid individuals unite to form diploids in the mosquitoes, they do so randomly and independently of the gene they carry. If p_t is the frequency of the R gene in the haploids, what are the frequencies of the RR, RS, and SS types in the mosquitoes after this step occurs? (As a check, make sure the frequencies sum to 1.)

2. Suppose that the probability of survival of the three different types of diploid individuals in mosquitoes, as a result of the drug, is given by the constants W_{RR}, W_{RS}, and W_{SS}. What are the frequencies of the three different types after this differential survival?

3. After differential survival the three types produce several haploid individuals that infect humans. Suppose that each diploid individual produces a total of b haploid

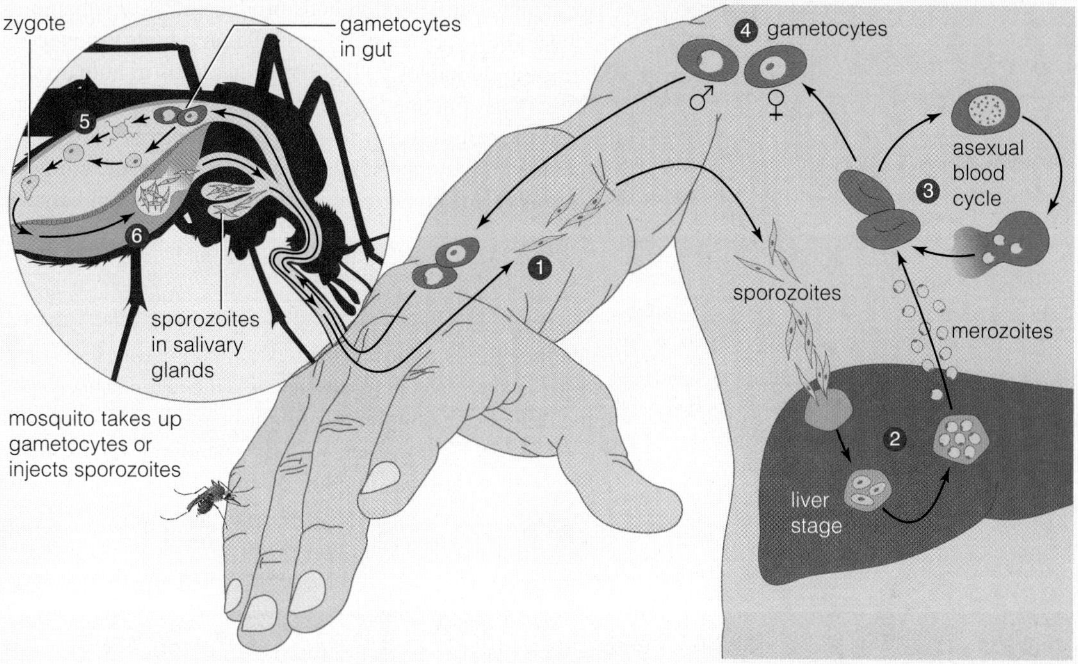

1. Infected mosquito bites a human. Sporozoites enter blood, which carries them to the liver.
2. Sporozoites reproduce asexually in liver cells, mature into merozoites. Merozoites leave the liver and infect red blood cells.
3. Merozoites reproduce asexually in some red blood cells.
4. In other red blood cells, merozoites differentiate into gametocytes.
5. A female mosquito bites and sucks blood from the infected person. Gametocytes in blood enter her gut and mature into gametes, which fuse to form zygotes.
6. Meiosis of zygotes produces cells that develop into sporozoites. The sporozoites migrate to the mosquito's salivary glands.

FIGURE 1
Life cycle of malaria

Source: From Starr. *Biology,* 8E © 2011 Brooks/Cole, a part of Cengage Learning, Inc. Reproduced by permission. www.cengage.com/permissions

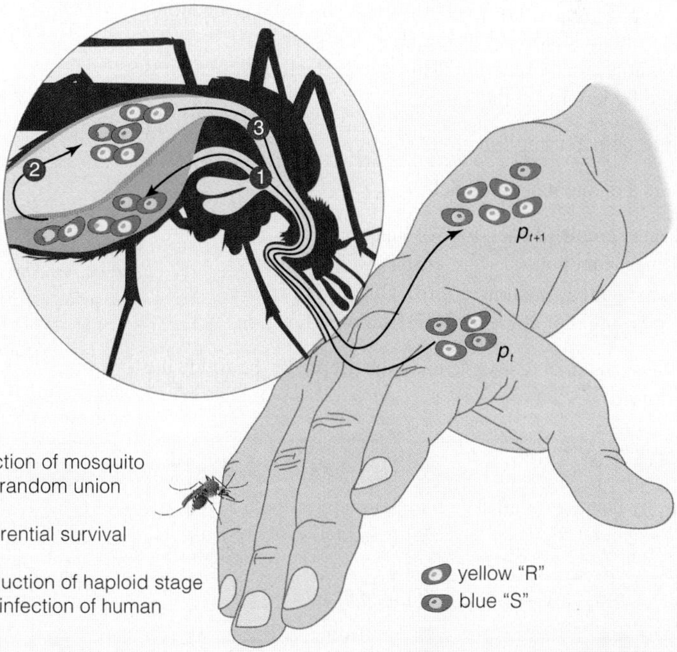

1. Infection of mosquito and random union
2. Differential survival
3. Production of haploid stage and infection of human

FIGURE 2

Source: Adapted from C. Starr et al., *Biology: Concepts and Applications*, 8th ed. (Belmont, CA: Cengage Learning, 2011), 316.

descendants, but RR individuals produce all R-type haploids, SS individuals produce all S-type haploids, and RS individuals produce a 50:50 mixture of both. What is the frequency of the R-type haploids in humans after this occurs?

We will assume the frequency doesn't change as the haploid parasites go through the rest of their life cycle in humans (as shown in Figure 1). Thus, the frequency just calculated is the value of p after the completion of the entire life cycle and so it is equal to p_{t+1}. Therefore you should arrive at the final result

$$p_{t+1} = \frac{p_t^2 W_{RR} + p_t(1 - p_t)W_{RS}}{p_t^2 W_{RR} + 2p_t(1 - p_t)W_{RS} + (1 - p_t)^2 W_{SS}}$$

4. When the drug is removed from use we sometimes see that the frequency of the drug-resistant gene decreases because the S type survives best in the absence of the drug. For example, this might be modeled by choosing $W_{RR} = \frac{1}{4}$ and $W_{RS} = W_{SS} = \frac{1}{2}$. Show that, with these choices, the expression in Problem 3 reduces to the rational function discussed in Section 1.2 on page 25.

Chapter 1 REVIEW

CONCEPT CHECK

1. (a) What is a function? What are its domain and range?
(b) What is the graph of a function?
(c) How can you tell whether a given curve is the graph of a function?

2. Discuss four ways of representing a function. Illustrate your discussion with examples.

3. (a) What is an even function? How can you tell if a function is even by looking at its graph?
(b) What is an odd function? How can you tell if a function is odd by looking at its graph?

4. What is an increasing function?

5. What is a mathematical model?

6. Give an example of each type of function.
(a) Linear function (b) Power function
(c) Exponential function (d) Quadratic function
(e) Polynomial of degree 5 (f) Rational function

7. Sketch by hand, on the same axes, the graphs of the following functions.
(a) $f(x) = x$ (b) $g(x) = x^2$
(c) $h(x) = x^3$ (d) $j(x) = x^4$

8. Draw, by hand, a rough sketch of the graph of each function.
(a) $y = \sin x$ (b) $y = \cos x$
(c) $y = \tan x$ (d) $y = e^x$
(e) $y = \ln x$ (f) $y = 1/x$
(g) $y = |x|$ (h) $y = \sqrt{x}$

9. Suppose that f has domain A and g has domain B.
(a) What is the domain of $f + g$?
(b) What is the domain of fg?
(c) What is the domain of f/g?

10. How is the composite function $f \circ g$ defined? What is its domain?

11. Suppose the graph of f is given. Write an equation for each of the graphs that are obtained from the graph of f as follows.
(a) Shift 2 units upward.
(b) Shift 2 units downward.
(c) Shift 2 units to the right.
(d) Shift 2 units to the left.
(e) Reflect about the x-axis.
(f) Reflect about the y-axis.
(g) Stretch vertically by a factor of 2.
(h) Shrink vertically by a factor of 2.
(i) Stretch horizontally by a factor of 2.
(j) Shrink horizontally by a factor of 2.

12. (a) What is a one-to-one function? How can you tell if a function is one-to-one by looking at its graph?
(b) If f is a one-to-one function, how is its inverse function f^{-1} defined? How do you obtain the graph of f^{-1} from the graph of f?

13. (a) What is a semilog plot?
(b) If a semilog plot of your data lies approximately on a line, what type of model is appropriate?

14. (a) What is a log-log plot?
(b) If a log-log plot of your data lies approximately on a line, what type of model is appropriate?

15. (a) What is a sequence?
(b) What is a recursive sequence?

16. Discrete-time models
(a) If there are N_t cells at time t and they divide according to the difference equation $N_{t+1} = RN_t$, write an expression for N_t.
(b) If a population has carrying capacity K, write the logistic difference equation for N_t.
(c) Write the logistic difference equation for

$$x_t = \frac{r}{(1 + r)K} N_t$$

Answers to the Concept Check can be found on the back endpapers.

TRUE-FALSE QUIZ

Determine whether the statement is true or false. If it is true, explain why. If it is false, explain why or give an example that disproves the statement.

1. If f is a function, then $f(s + t) = f(s) + f(t)$.
2. If $f(s) = f(t)$, then $s = t$.
3. If f is a function, then $f(3x) = 3f(x)$.
4. If $x_1 < x_2$ and f is a decreasing function, then $f(x_1) > f(x_2)$.
5. A vertical line intersects the graph of a function at most once.
6. If f and g are functions, then $f \circ g = g \circ f$.
7. If f is one-to-one, then $f^{-1}(x) = \dfrac{1}{f(x)}$.
8. You can always divide by e^x.
9. If $0 < a < b$, then $\ln a < \ln b$.
10. If $x > 0$, then $(\ln x)^6 = 6 \ln x$.
11. If $x > 0$ and $a > 1$, then $\dfrac{\ln x}{\ln a} = \ln \dfrac{x}{a}$.
12. If x is any real number, then $\sqrt{x^2} = x$.

EXERCISES

1. Let f be the function whose graph is given.
(a) Estimate the value of $f(2)$.
(b) Estimate the values of x such that $f(x) = 3$.
(c) State the domain of f.
(d) State the range of f.
(e) On what interval is f increasing?
(f) Is f one-to-one? Explain.
(g) Is f even, odd, or neither even nor odd? Explain.

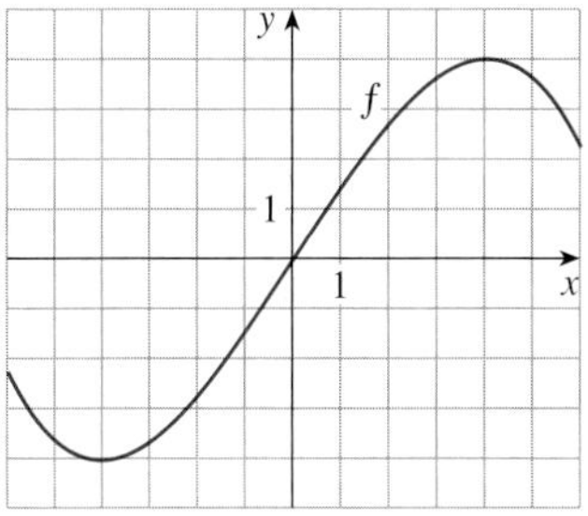

2. The graph of g is given.
(a) State the value of $g(2)$.
(b) Why is g one-to-one?
(c) Estimate the value of $g^{-1}(2)$.
(d) Estimate the domain of g^{-1}.
(e) Sketch the graph of g^{-1}.

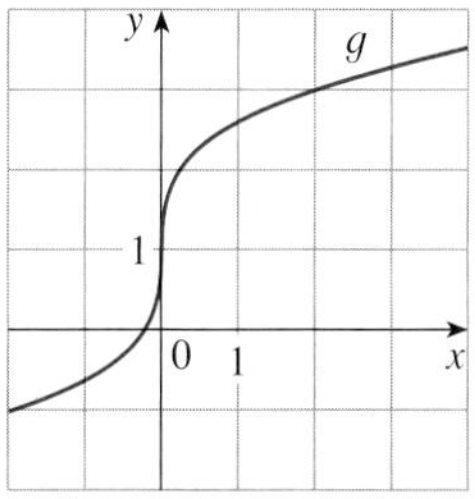

3. Sea level The figure shows how the sea level has changed over the last quarter of a million years according to data from

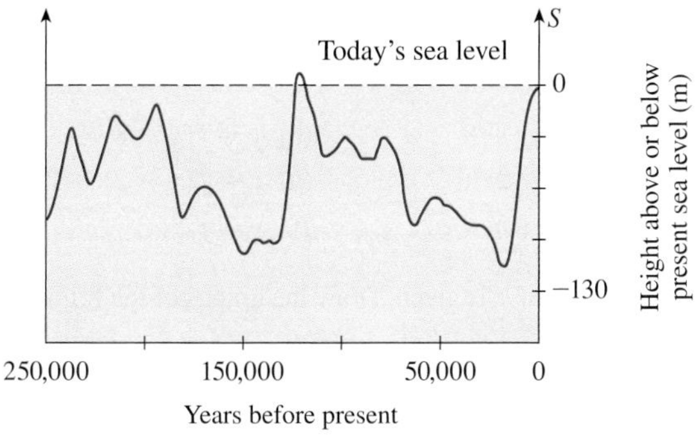

Source: Adapted from T. Garrison, *Oceanography: An Invitation to Marine Science* (Belmont, CA: Cengage Learning, 2010), 113, figure 4.15.

ocean floor cores. $S(t)$ is the sea level (in meters) relative to present sea level.
(a) What was the sea level 100,000 years ago?
(b) When was the sea level lowest? Highest?
(c) What is the range of this function?
(d) Can you account for the fluctuation of the sea level in terms of ice ages?

4. **Marine fish catch** The figure shows the worldwide commercial marine fish catch $F(t)$ in millions of tonnes (metric tons).
(a) In what year was the fish catch 70 million tonnes?
(b) What is the range of F?

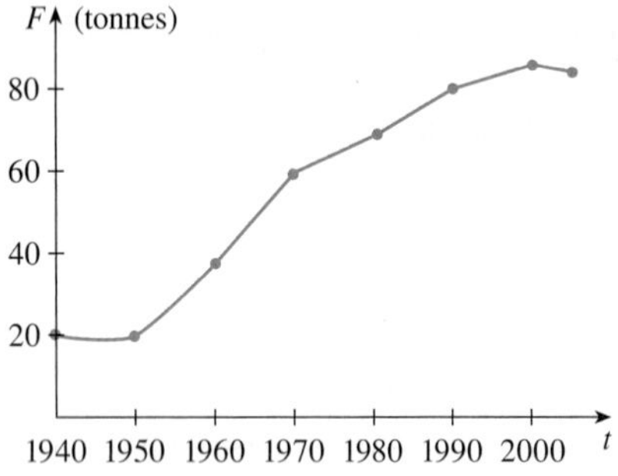

Source: Adapted from T. Garrison, *Oceanography: An Invitation to Marine Science* (Belmont, CA: Cengage Learning, 2010), 472, figure 17.13a.

5. If $f(x) = x^2 - 2x + 3$, evaluate the difference quotient

$$\frac{f(a + h) - f(a)}{h}$$

6. Sketch a rough graph of the yield of a crop as a function of the amount of fertilizer used.

7–10 Find the domain and range of the function. Write your answer in interval notation.

7. $f(x) = 2/(3x - 1)$
8. $g(x) = \sqrt{16 - x^4}$
9. $h(x) = \ln(x + 6)$
10. $F(t) = 3 + \cos 2t$

11. Suppose that the graph of f is given. Describe how the graphs of the following functions can be obtained from the graph of f.
(a) $y = f(x) + 8$
(b) $y = f(x + 8)$
(c) $y = 1 + 2f(x)$
(d) $y = f(x - 2) - 2$
(e) $y = -f(x)$
(f) $y = f^{-1}(x)$

12. The graph of f is given. Draw the graphs of the following functions.
(a) $y = f(x - 8)$
(b) $y = -f(x)$
(c) $y = 2 - f(x)$
(d) $y = \frac{1}{2}f(x) - 1$
(e) $y = f^{-1}(x)$
(f) $y = f^{-1}(x + 3)$

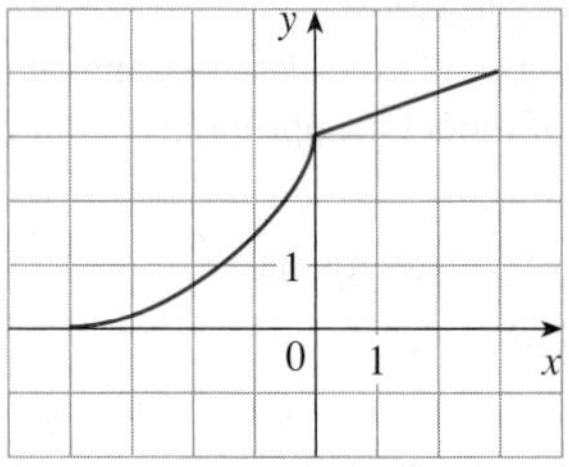

13–19 Use transformations to sketch the graph of the function.

13. $y = -\sin 2x$
14. $y = 3 \ln(x - 2)$
15. $y = \frac{1}{2}(1 + e^x)$
16. $y = 2 - \sqrt{x}$
17. $f(x) = \dfrac{1}{x + 2}$
18. $f(x) = x^2 - 2x$
19. $f(x) = \begin{cases} -x & \text{if } x < 0 \\ e^x - 1 & \text{if } x \geq 0 \end{cases}$

20. Determine whether f is even, odd, or neither even nor odd.
(a) $f(x) = 2x^5 - 3x^2 + 2$
(b) $f(x) = x^3 - x^7$
(c) $f(x) = e^{-x^2}$
(d) $f(x) = 1 + \sin x$

21. If $f(x) = \ln x$ and $g(x) = x^2 - 9$, find the functions (a) $f \circ g$, (b) $g \circ f$, (c) $f \circ f$, (d) $g \circ g$, and their domains.

22. Express the function $F(x) = 1/\sqrt{x + \sqrt{x}}$ as a composition of three functions.

BB 23. **Life expectancy** Life expectancy improved dramatically in the 20th century. The table gives the life expectancy at birth (in years) of males born in the United States. Use a scatter plot to choose an appropriate type of model. Use your model to predict the life-span of a male born in the year 2010.

Birth year	Life expectancy	Birth year	Life expectancy
1900	48.3	1960	66.6
1910	51.1	1970	67.1
1920	55.2	1980	70.0
1930	57.4	1990	71.8
1940	62.5	2000	73.0
1950	65.6		

24. A small-appliance manufacturer finds that it costs \$9000 to produce 1000 toaster ovens a week and \$12,000 to produce 1500 toaster ovens a week.
(a) Express the cost as a function of the number of toaster ovens produced, assuming that it is linear. Then sketch the graph.

(b) What is the slope of the graph and what does it represent?
(c) What is the y-intercept of the graph and what does it represent?

25. If $f(x) = 2x + \ln x$, find $f^{-1}(2)$.

26. Find the inverse function of $f(x) = \dfrac{x + 1}{2x + 1}$.

27. Find the exact value of each expression.
(a) $e^{2 \ln 3}$ (b) $\log_{10} 25 + \log_{10} 4$

28. Solve each equation for x.
(a) $e^x = 5$ (b) $\ln x = 2$
(c) $e^{e^x} = 2$

29. The half-life of palladium-100, ^{100}Pd, is four days. (So half of any given quantity of ^{100}Pd will disintegrate in four days.) The initial mass of a sample is one gram.
(a) Find the mass that remains after 16 days.
(b) Find the mass $m(t)$ that remains after t days.
(c) Find the inverse of this function and explain its meaning.
(d) When will the mass be reduced to 0.01 g?

30. Population growth The population of a certain species in a limited environment with initial population 100 and carrying capacity 1000 is

$$P(t) = \frac{100{,}000}{100 + 900e^{-t}}$$

where t is measured in years.
(a) Graph this function and estimate how long it takes for the population to reach 900.
(b) Find the inverse of this function and explain its meaning.
(c) Use the inverse function to find the time required for the population to reach 900. Compare with the result of part (a).

31. Graph members of the family of functions $f(x) = \ln(x^2 - c)$ for several values of c. How does the graph change when c changes?

32. Graph the three functions $y = x^b$, $y = b^x$, and $y = \log_b x$ on the same screen for two or three values of $b > 1$. For large values of x, which of these functions has the largest values and which has the smallest values?

33–34 Data points (x, y) are given.
(a) Draw a scatter plot of the data points.
(b) Make semilog and log-log plots of the data.
(c) Is a linear, power, or exponential function appropriate for modeling these data?
(d) Find an appropriate model for the data and then graph the model together with a scatter plot of the data.

33.

x	4	8	12	16	20	24
y	7.0	11.5	15.2	18.9	22.1	25.0

34.

x	1	3	6	10	14	16
y	7.22	4.61	2.38	0.99	0.41	0.26

35. Nigerian population The table gives the midyear population of Nigeria (in millions) from 1985 to 2010.

Year	Population	Year	Population
1985	85	2000	124
1990	97	2005	142
1995	110	2010	162

(a) Make a scatter plot, semilog plot, and log-log plot for these data and comment on which type of model would be most appropriate.
(b) Obtain an exponential model for the population.
(c) Use your model to estimate the population in 2008 and predict the population in 2020.

36–37 Find the first six terms of the sequence.

36. $a_n = \sin(n\pi/3)$

37. $a_1 = 3,\ a_{n+1} = n + 2a_n - 1$

38. Find a formula for the general term of the sequence

$$-3,\ \frac{5}{4},\ -\frac{7}{9},\ \frac{9}{16},\ -\frac{11}{25},\ \ldots$$

assuming that the pattern of the first few terms continues.

39. If $x_0 = 0.9$ and $x_{t+1} = 2.7x_t(1 - x_t)$, calculate x_t to four decimal places for $t = 1, 2, \ldots 10$ and graph x_t. Comment on the behavior of the sequence.

40. Beverton-Holt model An alternative to the logistic model for restricted population growth is the *Beverton-Holt recruitment curve*. Here the recursion model is

$$N_{t+1} = \frac{cN_t}{1 + (c - 1)N_t/K}$$

where K is the carrying capacity and c is the per capita growth factor.
(a) If $K = 50$ and $c = 1.7$, plot some points on the graph of N_t for the following values of the initial population: $N_0 = 10, 30, 70$.
(b) For the values in part (a), compare the Beverton-Holt model with the logistic model.

CASE STUDY 1a Kill Curves and Antibiotic Effectiveness

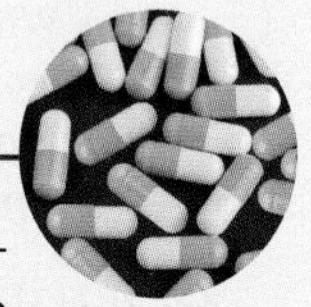

We are studying the relationship between the magnitude of antibiotic treatment and the effectiveness of the treatment. Recall that the extent of bacterial killing by an antibiotic is determined by both the *antibiotic concentration profile* and the *dose response relationship*. Figures 1 and 2 show these plots for the antibiotic ciprofloxacin when used against *E. coli*.[1]

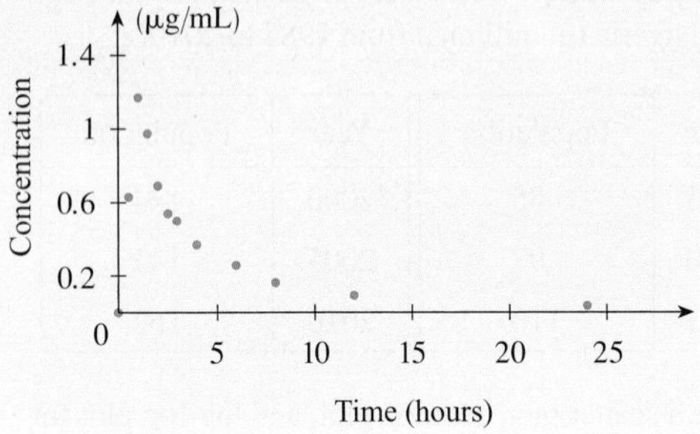

FIGURE 1
Antibiotic concentration profile in plasma of a healthy human volunteer after receiving 500 mg of ciprofloxacin

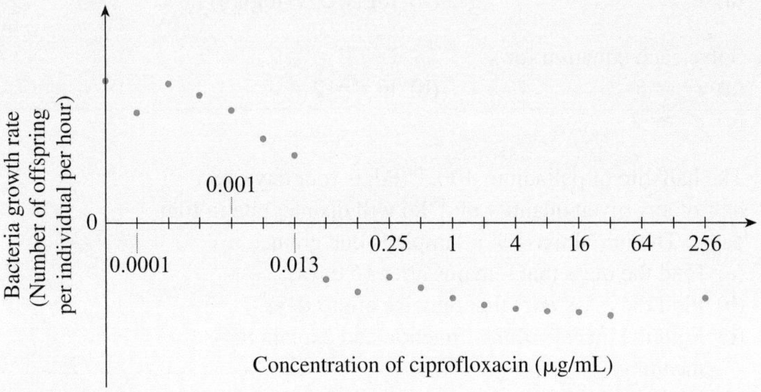

FIGURE 2
Dose response relationship for ciprofloxacin with the bacteria *E. coli*

Now, in the words of Picasso, we are viewing mathematical models as "lies that reveal truth." In other words, we don't expect our mathematical model to capture every detail of the biological system; rather, we simply want it to capture the most important features. To this end, let's describe the main patterns seen in Figures 1 and 2 mathematically.

Figure 1 shows that the antibiotic concentration increases extremely quickly, followed by a slow decay. To simplify matters let's therefore suppose that it increases instantly from zero to the peak concentration at time $t = 0$, and it then decays. As we will see in Case Study 1b, the decay can be well modeled using the exponential decay function

(1) $$c(t) = c_0 e^{-kt}$$

where c_0 is the concentration at $t = 0$ and k is a positive constant. Equation 1 is plotted in Figure 3.

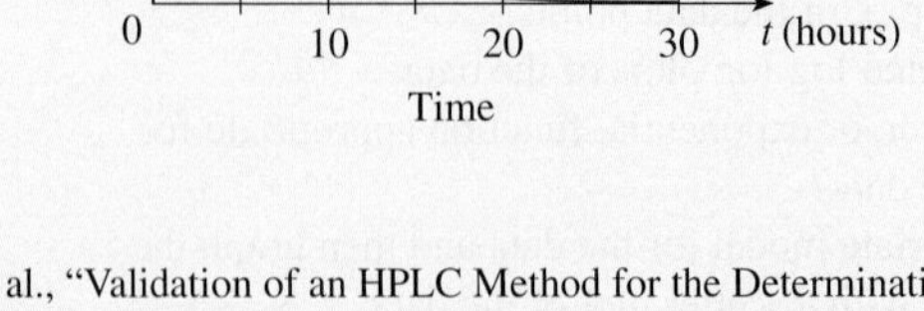

FIGURE 3
Drug concentration profile modeled by the function $c(t) = c_0 e^{-kt}$ with $c_0 = 1.2$ μg/mL and $k = 0.175$

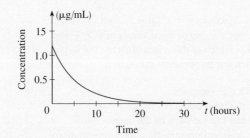

1. Adapted from S. Imre et al., "Validation of an HPLC Method for the Determination of Ciprofloxacin in Human Plasma," *Journal of Pharmaceutical and Biomedical Analysis* 33 (2003): 125–30.

In Figure 2 it looks like the dose response relationship doesn't vary much up to a concentration of around 0.013 μg/mL. It then drops suddenly to a low value and remains relatively constant as the antibiotic concentration increases further. To simplify matters, let's model the dose response relationship by the piecewise defined function

$$r(c) = \begin{cases} r_2 & \text{if } c < MIC \\ r_1 & \text{if } c \geqslant MIC \end{cases}$$

where MIC is a constant that is referred to as the *minimum inhibitory concentration* ($MIC = 0.013$ μg/mL in this case), r_1 and r_2 are constants giving the bacteria population growth rate under high and low antibiotic concentration, respectively, and $r_1 < 0$ and $r_2 > 0$. This function is plotted in Figure 4.[2]

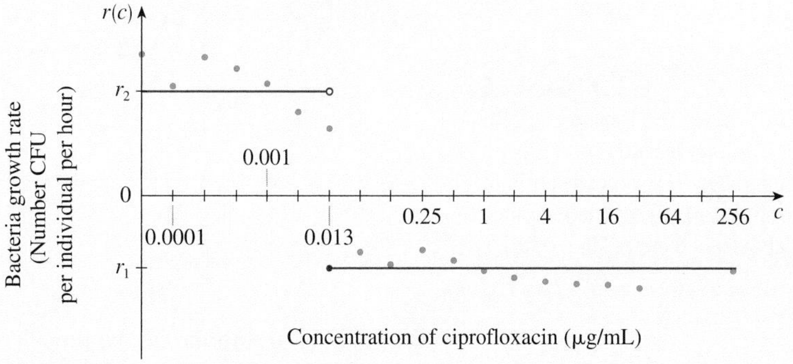

FIGURE 4
Dose response relationship modeled by the piecewise defined function $r(c)$

The functions in Figures 3 and 4 will, together, determine how the bacteria population size changes over time. At $t = 0$ the antibiotic concentration is c_0, and if c_0 is greater than $MIC = 0.013$ μg/mL, then the bacteria population size will decline. At the same time the antibiotic concentration will decay as time passes, eventually reaching a value of $MIC = 0.013$ μg/mL. At this point the growth rate of the bacteria population becomes positive.

In Case Study 1b you will show that, using the functions in Figures 3 and 4, a suitable model for the size of the bacteria population $P(t)$ (in CFU/mL) as a function of time t (in hours) is given by the piecewise defined function

(2a)
$$P(t) = \begin{cases} 6e^{t/3} & \text{if } t < 2.08 \\ 12 & \text{if } t \geqslant 2.08 \end{cases}$$

if $c_0 < 0.013$, and

(2b)
$$P(t) = \begin{cases} 6e^{-t/20} & \text{if } t < a \\ 6Ae^{t/3} & \text{if } a \leqslant t < b \\ 12 & \text{if } t \geqslant b \end{cases}$$

if $c_0 \geqslant 0.013$, where the constants a, b, and A are defined by $a = 5.7\ln(77c_0)$, $b = 6.6\ln(77c_0) + 2.08$, and $A = (77c_0)^{-2.2}$.

2. Adapted from W. Bär et al., "Rapid Method for Detection of Minimal Bactericidal Concentration of Antibiotics," *Journal of Microbiological Methods* 77 (2009): 85–89, Figure 1.

1. Plot $P(t)$ as a function of time for each of the concentrations $c_0 = 0, 0.019, 0.038, 0.075, 0.15, 0.3, 0.6, 1.2$. These are the kill curves predicted by the model. Comment on the similarities and differences between the predicted curves and those from the data in Figure 5.[3]

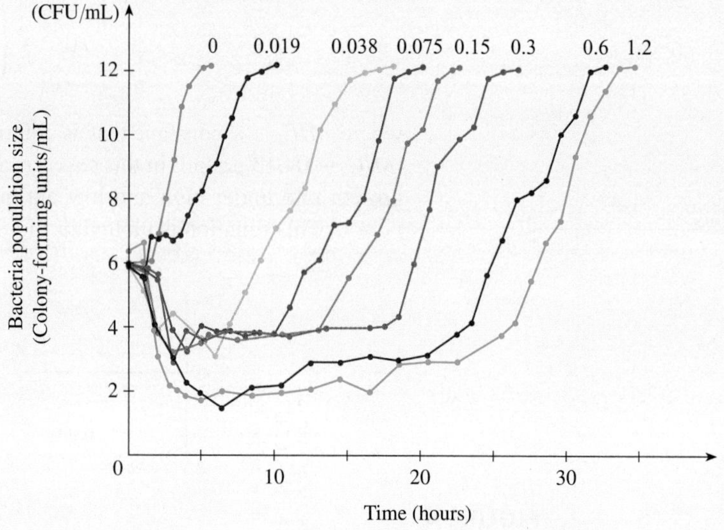

FIGURE 5
The kill curves of ciprofloxacin for *E. coli* when measured in a growth chamber. The concentration of ciprofloxacin at $t = 0$ is indicated above each curve (in μg/mL).

Our goal is to summarize the model kill curves from Problem 1 in a simpler form in order to see more clearly the relationship between the magnitude of antibiotic treatment and its predicted effectiveness.

To do this, we need to obtain a measure of the magnitude of antibiotic treatment as well as a measure of its effectiveness. We first obtain a measure of the magnitude of antibiotic treatment from the antibiotic concentration profiles that underlie each predicted kill curve. Three measures are commonly used: (1) the peak antibiotic concentration divided by *MIC*, denoted by ρ; (2) the duration of time for which the antibiotic concentration remains above *MIC*, denoted by τ; and (3) the area under the antibiotic concentration profile divided by *MIC*, denoted by α. These measures are illustrated graphically in Figure 6.

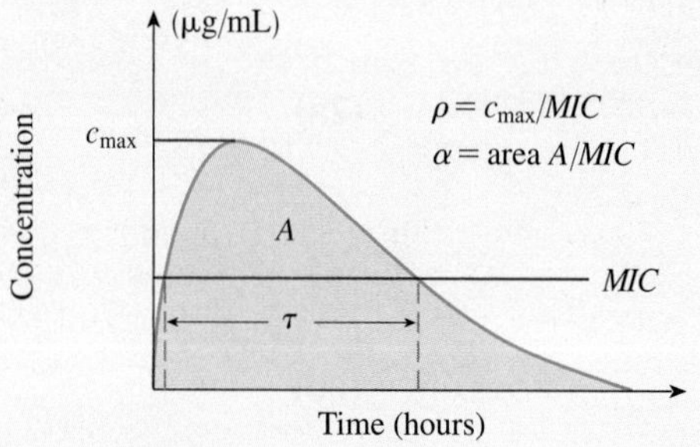

FIGURE 6
Three measures ρ, α, and τ of the magnitude of antibiotic treatment

3. Adapted from A. Firsov et al., "Parameters of Bacterial Killing and Regrowth Kinetics and Antimicrobial Effect Examined in Terms of Area under the Concentration-Time Curve Relationships: Action of Ciprofloxacin against *Escherichia coli* in an In Vitro Dynamic Model." *Antimicrobial Agents and Chemotherapy* 41 (1997): 1281–87.

2. Find expressions for ρ and τ in terms of k, c_0, and MIC, using Equation 1 for the antibiotic concentration profile.

3. In Case Study 1c you will show that $\alpha = \dfrac{1}{k}\dfrac{c_0}{MIC}$. Sketch graphs of ρ, τ, and α as functions of c_0, using the values $k = 0.175$ (1/hours) and $MIC = 0.013$ μg/mL. What are their units?

You will notice from Problem 3 that, for a given antibiotic and bacterial species (in other words, for a given value of k and MIC), all three quantities ρ, τ, and α increase with one another. For example, it is not possible to have a high value of α without also having high values of ρ and τ. Therefore, since these measures are not independent of one another, we need to consider only one of them. We will focus the remainder of our study on α, since it is the most commonly used.

Next we need to quantify the effectiveness of the antibiotic by quantifying different properties of the kill curves. Let's consider two possibilities: (i) the time taken to reduce the bacteria population size to 90% of its initial size, denoted by T, and (ii) the drop in population size before the population rebound occurs, denoted by Δ. Both measures are shown in Figure 7.

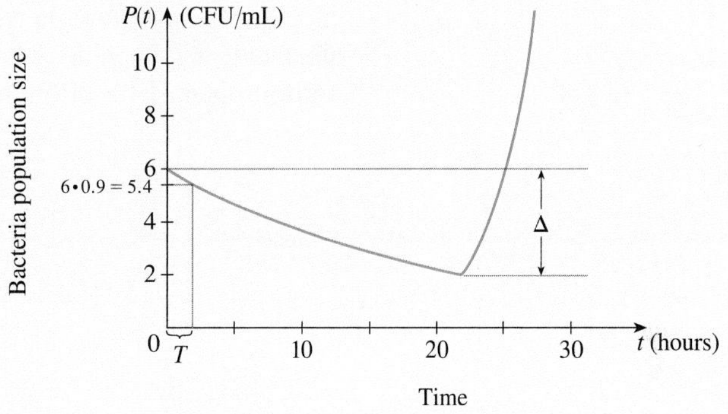

FIGURE 7
Two measures T and Δ of antibiotic effectiveness

4. Find expressions for Δ and T in the modeled populations in terms of c_0.

Our final goal is to use the results from Problem 4, along with the expression for α, to plot Δ against α and to plot T against α as well. This will give us the model's predictions for the plots in Figures 5 and 6 in Case Study 1 on page xlii.

5. Substitute the values $k = 0.175$ and $MIC = 0.013$ into the expression for α. This expression, along with the results from Problem 4, should give you functions of the form $T = f(c_0)$, $\Delta = g(c_0)$, and $\alpha = h(c_0)$ for some functions f, g, and h. [*Note:* Some of the functions might actually be independent of c_0.]

6. Using the concept of inverse functions, explain how to obtain a function that gives Δ as a function of α in terms of g and h^{-1}. Find an explicit expression for this function.

7. What is T as a function of α?

8. Plot the functions obtained in Problems 6 and 7. You should obtain the curves shown in Figures 8 and 9.

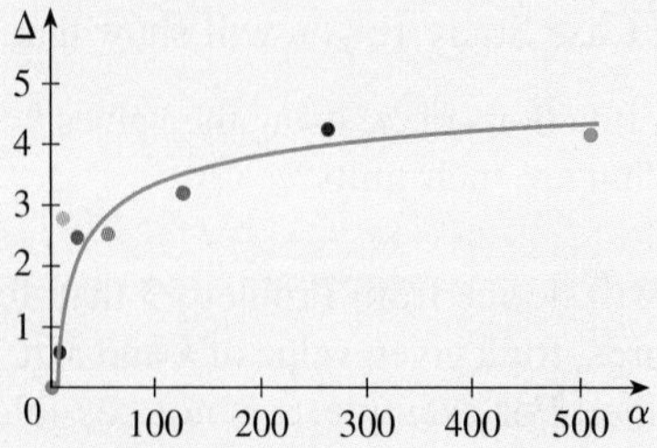

FIGURE 8
Predicted relationship between Δ and α, along with the observations obtained using the kill curve data in Figure 5

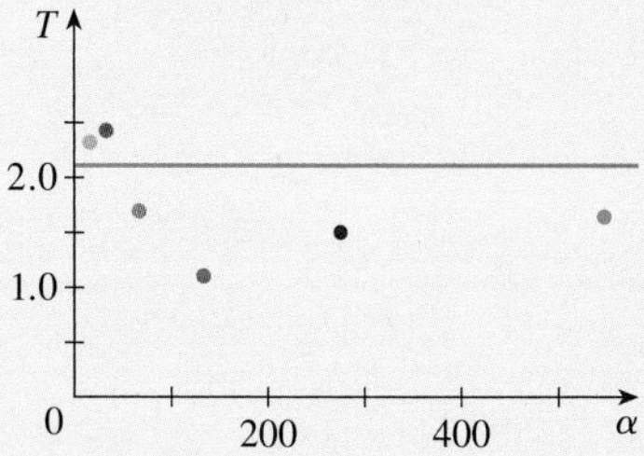

FIGURE 9
Predicted relationship between T and α, along with the observations obtained using the kill curve data in Figure 5

9. From Figures 8 and 9 you can see that this relatively simple model predicts the observed data reasonably well. In particular, T is predicted to be independent of the magnitude of antibiotic treatment, whereas Δ increases with it. Provide a biological explanation in terms of the model for why this occurs. [*Hint:* Relate the fact that T is predicted to be independent of α to the form of the kill curves from Problem 1 for different antibiotic doses.]

Limits

2

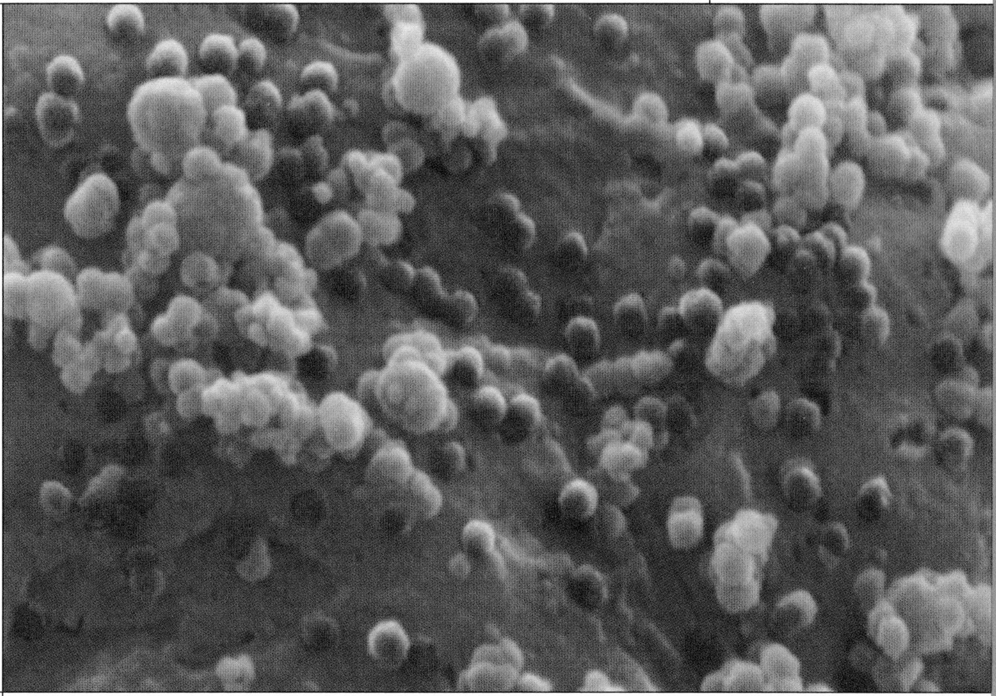

Viruses infect all living organisms and are responsible for diseases ranging from the common cold to smallpox and AIDS. In the project on page 101 you are asked to use recursive sequences to investigate the interaction among viral infection, the human immune system, and antiviral drugs.

Eye of Science / Science Source

THE IDEA OF A LIMIT is the basic concept in all of calculus. It underlies such phenomena as the long-term behavior of a population, the rate of growth of a tumor, and the area of a leaf.

2.1 Limits of Sequences

The Long-Term Behavior of a Sequence

In Section 1.6 we looked at sequences, both those given by a simple defining formula and those defined recursively by a difference equation. Here we investigate what happens to the terms a_n of a sequence in the long run. In other words, we explore what happens to a_n as n becomes large.

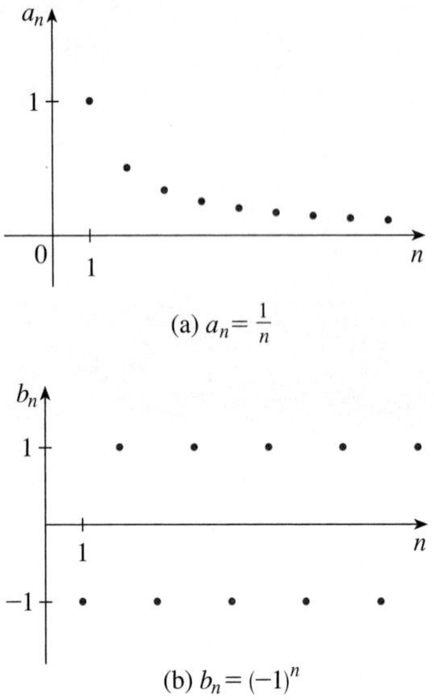

FIGURE 1

EXAMPLE 1 | What happens to the terms of the sequence when n becomes large?

(a) $a_n = \dfrac{1}{n}$ (b) $b_n = (-1)^n$

SOLUTION

(a) The first few terms of the sequence are

$$1, \quad \frac{1}{2}, \quad \frac{1}{3}, \quad \frac{1}{4}, \quad \frac{1}{5}, \quad \frac{1}{6}, \quad \ldots$$

and for larger values of n we have

$$a_{10} = 0.1, \quad a_{100} = 0.01, \quad a_{1000} = 0.001, \quad a_{1,000,000} = 0.000001, \quad \ldots$$

The larger the value of n, the smaller the value of a_n. The terms are approaching 0 as n increases. [See Figure 1(a).]

(b) Here the terms are

$$-1, \quad 1, \quad -1, \quad 1, \quad -1, \quad 1, \quad -1, \quad \ldots$$

The values of the terms alternate between 1 and -1 forever. So they don't approach any fixed number. [See Figure 1(b)]. ■

The sequences in Example 1 behave quite differently. The terms $a_n = 1/n$ approach 0 as n becomes large. (In fact we could make $1/n$ as small as we like by taking n large enough). We indicate this by saying that the sequence has *limit* 0 and by writing

$$\lim_{n\to\infty} \frac{1}{n} = 0$$

On the other hand, the sequence $b_n = (-1)^n$ does not have a limit, that is,

$$\lim_{n\to\infty} (-1)^n \quad \text{does not exist}$$

Definition of a Limit

In general we write

$$\lim_{n\to\infty} a_n = L$$

if the terms a_n approach L as n becomes large.

A more precise definition of the limit of a sequence is given in Appendix D.

(1) Definition A sequence $\{a_n\}$ has the **limit** L and we write

$$\lim_{n\to\infty} a_n = L \qquad \text{or} \qquad a_n \to L \text{ as } n \to \infty$$

if we can make the terms a_n as close to L as we like by taking n sufficiently large. If $\lim_{n\to\infty} a_n$ exists, we say the sequence **converges** (or is **convergent**). Otherwise, we say the sequence **diverges** (or is **divergent**).

Figure 2 illustrates Definition 1 by showing the graphs of two sequences that have the limit L.

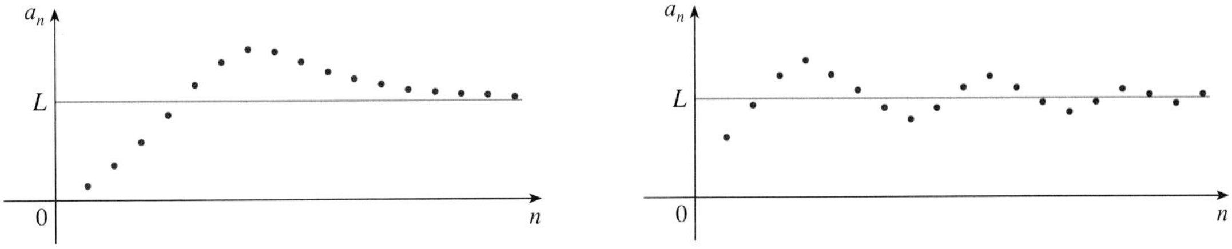

FIGURE 2
Graphs of two sequences with $\lim_{n\to\infty} a_n = L$

If a_n becomes large as n becomes large, we use the notation

$$\lim_{n\to\infty} a_n = \infty$$

In this case the sequence $\{a_n\}$ is divergent, but in a special way. We say that a_n diverges to ∞.

EXAMPLE 2 | Is the sequence $a_n = \sqrt{n}$ convergent or divergent?

SOLUTION When n is large, $\sqrt{n}$ is large, so

$$\lim_{n\to\infty} \sqrt{n} = \infty$$

and the sequence $\{\sqrt{n}\}$ is divergent. Notice that we can make $\sqrt{n}$ as big as we want by taking n big enough. For instance,

$$\sqrt{n} > 1000 \qquad \text{when} \qquad n > 1{,}000{,}000$$ ■

■ Limit Laws

The more precise version of Definition 1 in Appendix D can be used to prove the following properties of limits.

Limit Laws for Sequences

If $\{a_n\}$ and $\{b_n\}$ are convergent sequences and c is a constant, then

$$\lim_{n\to\infty} (a_n + b_n) = \lim_{n\to\infty} a_n + \lim_{n\to\infty} b_n$$

$$\lim_{n\to\infty} (a_n - b_n) = \lim_{n\to\infty} a_n - \lim_{n\to\infty} b_n$$

$$\lim_{n\to\infty} ca_n = c \lim_{n\to\infty} a_n \qquad\qquad \lim_{n\to\infty} c = c$$

$$\lim_{n\to\infty} (a_n b_n) = \lim_{n\to\infty} a_n \cdot \lim_{n\to\infty} b_n$$

$$\lim_{n\to\infty} \frac{a_n}{b_n} = \frac{\lim_{n\to\infty} a_n}{\lim_{n\to\infty} b_n} \quad \text{if } \lim_{n\to\infty} b_n \neq 0$$

$$\lim_{n\to\infty} a_n^p = \left[\lim_{n\to\infty} a_n\right]^p \quad \text{if } p > 0 \text{ and } a_n > 0$$

From the last of these laws and the fact that $\lim_{n\to\infty}(1/n) = 0$, we deduce that

(2)
$$\lim_{n\to\infty} \frac{1}{n^p} = 0 \qquad \text{for any number } p > 0$$

Combining this fact with the various limit laws, we can calculate limits of sequences as in the following example.

EXAMPLE 3 | Find $\lim_{n\to\infty} \dfrac{1 + 2n^2}{5 + 3n + 4n^2}$.

SOLUTION As n becomes large, both numerator and denominator become large, so it isn't obvious what happens to their ratio. If we divide the numerator and denominator by n^2 (the highest power of n that occurs in the denominator), then we can use the Limit Laws and take advantage of the limits we know from Equation 2:

$$\lim_{n\to\infty} \frac{1 + 2n^2}{5 + 3n + 4n^2} = \lim_{n\to\infty} \frac{\dfrac{1 + 2n^2}{n^2}}{\dfrac{5 + 3n + 4n^2}{n^2}} = \lim_{n\to\infty} \frac{\dfrac{1}{n^2} + 2}{\dfrac{5}{n^2} + \dfrac{3}{n} + 4}$$

$$= \frac{\lim_{n\to\infty} \dfrac{1}{n^2} + \lim_{n\to\infty} 2}{5 \lim_{n\to\infty} \dfrac{1}{n^2} + 3 \lim_{n\to\infty} \dfrac{1}{n} + \lim_{n\to\infty} 4}$$

$$= \frac{0 + 2}{5 \cdot 0 + 3 \cdot 0 + 4} = \frac{1}{2}$$

We have used the Limit Laws and Equation 2 with $p = 2$ and $p = 1$. ■

■ Geometric Sequences

A **geometric sequence** is a sequence of the form $b_n = ar^n$. We start with a number $b_0 = a$ and multiply repeatedly by a number r:

$$a, \quad ar, \quad ar^2, \quad ar^3, \quad ar^4, \quad \ldots$$

We saw an example of this in Section 1.4 with $a = 1$ and $r = 8$. The malarial species *P. chabaudi* reproduces every 24 hours and so a single such parasite at time $t = 0$ results in $b_n = 8^n$ parasites after n days (at least initially). From our knowledge of the exponential function, we know that the sequence $b_n = 8^n$ grows indefinitely:

$$\lim_{n\to\infty} 8^n = \infty$$

We also modeled cell division by the recursion

$$N_{t+1} = RN_t$$

whose solution is

$$N_t = N_0 R^t$$

where R is the per capita growth factor. So this is a geometric sequence with $a = N_0$ and $r = R$.

In general there are three cases for the geometric sequence $b_n = r^n$ if r is a positive number. From the graphs of the exponential functions in Figures 1.4.4 and 1.4.5 (page 44) we see that

(3)
$$\lim_{n\to\infty} r^n = \begin{cases} 0 & \text{if } 0 < r < 1 \\ 1 & \text{if } r = 1 \\ \infty & \text{if } r > 1 \end{cases}$$

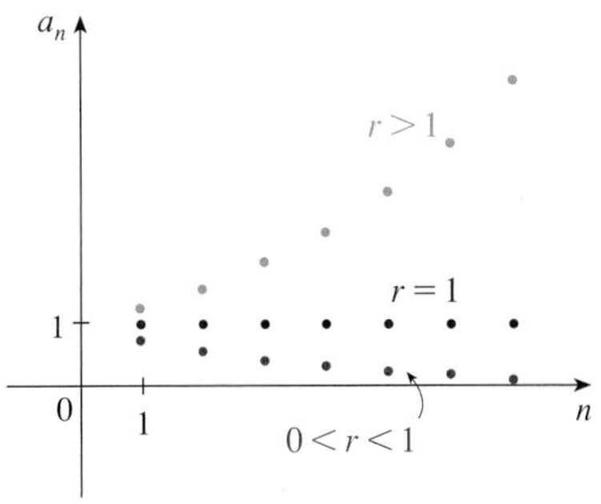

FIGURE 3
The sequence $b_n = r^n$

These three cases are illustrated in Figure 3.

EXAMPLE 4 | Calculate $\lim_{n\to\infty} \dfrac{2^n - 1}{6^n}$ if it exists.

SOLUTION Simplifying, we get

$$\lim_{n\to\infty} \frac{2^n - 1}{6^n} = \lim_{n\to\infty} \left[\frac{2^n}{6^n} - \frac{1}{6^n}\right] = \lim_{n\to\infty} \left[\left(\frac{1}{3}\right)^n - \left(\frac{1}{6}\right)^n\right]$$

$$= \lim_{n\to\infty} \left(\frac{1}{3}\right)^n - \lim_{n\to\infty} \left(\frac{1}{6}\right)^n = 0 - 0 = 0$$

where the second last equality follows from (3) with $r = \frac{1}{3}$ and $r = \frac{1}{6}$. ■

■ Recursion for Medication

Let's modify the geometric recursion $b_{n+1} = rb_n$ by adding a constant term c:

$$b_{n+1} = rb_n + c$$

This difference equation arises when a patient is given a daily medication.

EXAMPLE 5 | BB **Drug concentration** A drug is administered to a patient at the same time every day. Suppose the concentration of the drug is C_n (measured in mg/mL) after the injection on the nth day. Before the injection the next day, only 30% of the drug present on the preceding day remains in the bloodstream. If the daily dose raises the concentration by 0.2 mg/mL, find the concentration after four days.

SOLUTION Just before the daily dose of medication is administered, the concentration is reduced to 30% of the preceding day's concentration, that is, $0.3C_n$. With the new dose, the concentration is increased by 0.2 mg/mL and so

(4)
$$C_{n+1} = 0.3C_n + 0.2$$

Starting with $C_0 = 0$ and putting $n = 0, 1, 2, 3$ into this difference equation, we get

$$C_1 = 0.3C_0 + 0.2 = 0.2$$

$$C_2 = 0.3C_1 + 0.2 = 0.3(0.2) + 0.2 = 0.26$$

$$C_3 = 0.3C_2 + 0.2 = 0.3(0.26) + 0.2 = 0.278$$

$$C_4 = 0.3C_3 + 0.2 = 0.3(0.278) + 0.2 = 0.2834$$

The concentration after four days is 0.2834 mg/mL. ■

The successive daily concentrations in Example 5 appear to be increasing. Do you think they increase indefinitely? Or do they have a limit?

For now, let's *assume* there is a limiting concentration. Let's call it C, that is, $\lim_{n\to\infty} C_n = C$. As $n \to \infty$, observe that $n + 1 \to \infty$ too. So $\lim_{n\to\infty} C_{n+1} = C$ too. Therefore, taking limits in Equation 4, we have

$$\lim_{n\to\infty} C_{n+1} = \lim_{n\to\infty} (0.3C_n + 0.2) = 0.3 \lim_{n\to\infty} C_n + \lim_{n\to\infty} 0.2$$

$$C = 0.3C + 0.2$$

$$0.7C = 0.2$$

$$C = \frac{0.2}{0.7} = \frac{2}{7}$$

This shows that *if* there is a limiting concentration C, then

$$C = \frac{2}{7} \approx 0.2857 \text{ mg/mL}$$

We are going to verify that there is indeed a limiting concentration by first finding an explicit formula for C_n.

■ Geometric Series

If we add the terms of a geometric sequence, we get another sequence, this one consisting of sums:

$$s_0 = a$$

$$s_1 = a + ar$$

$$s_2 = a + ar + ar^2$$

$$s_3 = a + ar + ar^2 + ar^3$$

$$\vdots$$

$$s_n = a + ar + ar^2 + \cdots + ar^n$$

If we multiply s_n by r and align the terms with those above, we get

$$rs_n = \qquad ar + ar^2 + \cdots + ar^n + ar^{n+1}$$

If we now subtract these last two equations, most of the terms cancel in pairs:

$$s_n - rs_n = a - ar^{n+1}$$
$$s_n(1 - r) = a(1 - r^{n+1})$$

If $r \neq 1$, we can solve for s_n:

The sum of terms of a sequence is called a **series**. Formula 5 gives the sum of a finite geometric series.

(5)
$$s_n = a + ar + ar^2 + \cdots + ar^n = \frac{a(1 - r^{n+1})}{1 - r}$$

We know that the solution of the difference equation

$$b_{n+1} = rb_n \qquad b_0 = a$$

is $b_n = ar^n$. We can now use Formula 5 to find a solution of the recursion

$$b_{n+1} = c + rb_n \qquad b_0 = a$$

To see the pattern, we start by computing the first few terms:

$$b_1 = c + ra$$
$$b_2 = c + rb_1 = c + r(c + ra) = c(1 + r) + r^2a$$
$$b_3 = c + rb_2 = c + r(c + cr + r^2a) = c(1 + r + r^2) + r^3a$$
$$\vdots$$
$$b_n = c(1 + r + r^2 + \cdots r^{n-1}) + r^na$$

Using Formula 5 for the sum of a finite geometric series, we have the following formula.

(6) The solution of the difference equation $b_{n+1} = c + rb_n$, $b_0 = a$, is

$$b_n = r^na + c\left(\frac{1 - r^n}{1 - r}\right)$$

EXAMPLE 6 | **Drug concentration (continued)** What is the concentration of the drug in Example 5 after the nth dose? What is the limiting concentration?

SOLUTION The difference equation in Example 5 is of the form $C_{n+1} = rC_n + c$, where $r = 0.3$, $c = 0.2$, and $C_0 = 0$. So from Equation 6 we have

$$C_n = 0.2\left[\frac{1 - (0.3)^n}{1 - 0.3}\right] = \frac{2}{7}[1 - (0.3)^n]$$

Because $0.3 < 1$, we know from Equation 3 that $\lim_{n\to\infty} (0.3)^n = 0$. So the limiting concentration is

$$\lim_{n\to\infty} C_n = \lim_{n\to\infty} \frac{2}{7}[1 - (0.3)^n] = \frac{2}{7}(1 - 0) = \frac{2}{7} \text{ mg/mL}$$

as we had predicted. ■

What happens if we try to add all of the infinitely many terms of a geometric sequence? For instance, if $a = 1$ and $r = \frac{1}{2}$, does it make sense to talk about the value of the infinite sum

$$1 + \frac{1}{2} + \frac{1}{4} + \frac{1}{8} + \frac{1}{16} + \cdots + \frac{1}{2^n} + \cdots$$

This sum is called an **infinite series**. From Equation 5 we know that the sum of the first $n + 1$ terms is

$$s_n = 1 + \frac{1}{2} + \frac{1}{4} + \frac{1}{8} + \frac{1}{16} + \cdots + \frac{1}{2^n} = \frac{1\left(1 - \dfrac{1}{2^{n+1}}\right)}{1 - \frac{1}{2}} = 2\left(1 - \frac{1}{2^{n+1}}\right)$$

and $\lim_{n\to\infty}(1/2^{n+1}) = 0$, so $\lim_{n\to\infty} s_n = 2$. In other words, by adding enough terms of the series, we can make the sum as close as we like to 2. We say that the sum of the geometric series is 2 and we write

$$1 + \frac{1}{2} + \frac{1}{4} + \frac{1}{8} + \frac{1}{16} + \cdots + \frac{1}{2^{n-1}} + \frac{1}{2^n} + \cdots = 2$$

In general, if $-1 < r < 1$, we know that $\lim_{n\to\infty} r^n = 0$ and so, from Equation 5,

$$\lim_{n\to\infty} s_n = \lim_{n\to\infty}\left[\frac{a}{1-r} - \frac{a}{1-r}r^{n+1}\right] = \frac{a}{1-r} - \frac{a}{1-r}\lim_{n\to\infty} r^{n+1} = \frac{a}{1-r}$$

> If $-1 < r < 1$, the sum of the infinite geometric series is
>
> $$a + ar + ar^2 + \cdots + ar^n + \cdots = \frac{a}{1-r}$$

As the following example illustrates, a repeating decimal number is the sum of an infinite geometric series.

EXAMPLE 7 | Write the number $2.3\overline{17} = 2.3171717\ldots$ as a ratio of integers.

SOLUTION

$$2.3171717\ldots = 2.3 + \frac{17}{10^3} + \frac{17}{10^5} + \frac{17}{10^7} + \ldots$$

After the first term we have a geometric series with $a = 17/10^3$ and $r = 1/10^2$. Therefore

$$2.3\overline{17} = 2.3 + \frac{\dfrac{17}{10^3}}{1 - \dfrac{1}{10^2}} = 2.3 + \frac{\dfrac{17}{1000}}{\dfrac{99}{100}}$$

$$= \frac{23}{10} + \frac{17}{990} = \frac{1147}{495}$$

■

■ The Logistic Sequence in the Long Run

In Section 1.6 we looked at the logistic difference equation, which is of the form

$$x_{t+1} = cx_t(1 - x_t) \tag{7}$$

where c is a constant. Here we examine this equation again, this time exploring the limiting behavior of the terms for various values of c.

EXAMPLE 8 | BB **Logistic difference equation** Compute and plot the next 16 terms of the logistic equation (7) in the following cases. Then describe the limiting behavior of the sequence.

(a) $x_0 = 0.8,\ c = 2.8$ (b) $x_0 = 0.4,\ c = 3.4$

SOLUTION

(a) Taking $x_0 = 0.8$ and $x_{t+1} = 2.8x_t(1 - x_t)$, we use a graphing calculator or computer to compute the terms up to x_{16} to four decimal places and then we plot the graph in Figure 4.

t	x_t	t	x_t
0	0.8000	9	0.6321
1	0.4480	10	0.6511
2	0.6924	11	0.6360
3	0.5963	12	0.6482
4	0.6740	13	0.6385
5	0.6152	14	0.6463
6	0.6628	15	0.6401
7	0.6258	16	0.6450
8	0.6557		

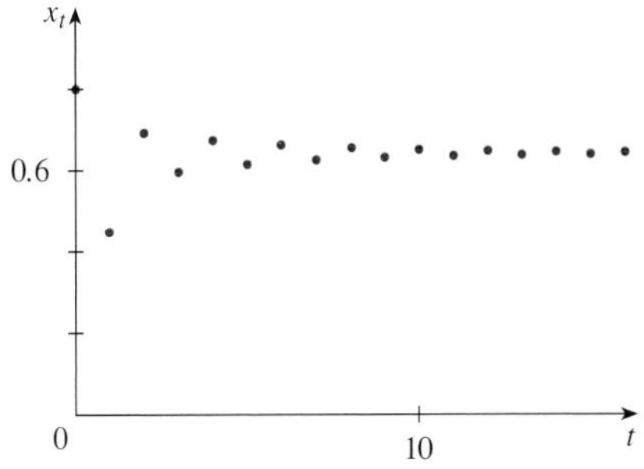

FIGURE 4

From the table of values and Figure 4 it appears that the terms of the sequence are approaching a number between 0.64 and 0.65. In fact if we assume that the limit exists and we call it L, we can evaluate it by the same technique as for the sequence in Example 5. We have both $\lim_{t\to\infty} x_t = L$ and $\lim_{t\to\infty} x_{t+1} = L$ and so, using the Limit Laws, we have

$$L = \lim_{t\to\infty} x_{t+1} = \lim_{t\to\infty} [2.8x_t(1 - x_t)] = 2.8L(1 - L)$$

$$1 = 2.8(1 - L)$$

$$1 - L = \frac{1}{2.8} \quad \Rightarrow \quad L = 1 - \frac{1}{2.8} = \frac{1.8}{2.8} = \frac{9}{14}$$

$$\lim_{n\to\infty} x_t = \frac{9}{14} \approx 0.64286$$

(b) Here $x_0 = 0.4$ and $x_{t+1} = 3.4x_t(1 - x_t)$. Again we calculate the next 16 terms and plot them in Figure 5.

t	x_t	t	x_t
0	0.4000	9	0.8385
1	0.8160	10	0.4603
2	0.5105	11	0.8446
3	0.8496	12	0.4461
4	0.4344	13	0.8401
5	0.8354	14	0.4566
6	0.4676	15	0.8436
7	0.8464	16	0.4486
8	0.4419		

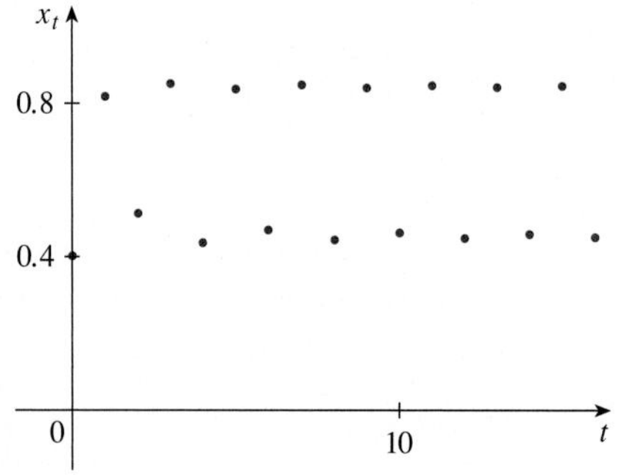

FIGURE 5

Looking at the table and Figure 5, it appears that the values of x_t do not approach any fixed number. Instead, they oscillate between values near 0.84 and values near 0.45. In other words, for $c = 3.4$,

$$\lim_{t \to \infty} x_t \quad \text{does not exist}$$ ■

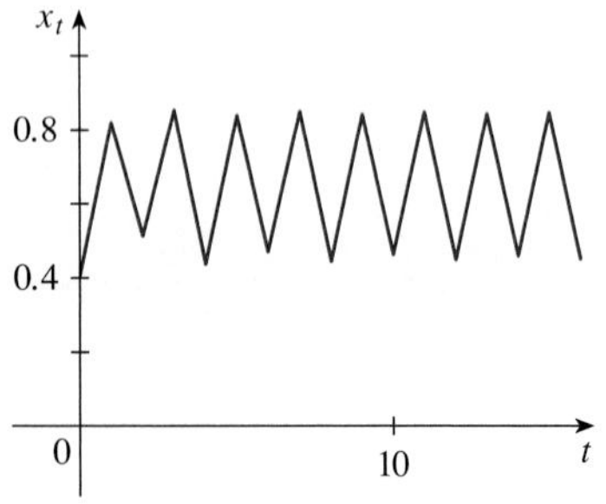

FIGURE 6

Sometimes it's easier to see what's happening in the graph of a sequence if we join consecutive points by line segments. In Figure 6 we have done this for the logistic sequence in Example 8(b) by joining the points in Figure 5. Thus Figure 6 is not truly the graph of that sequence, but it does represent the sequence in a way that might be easier to visualize.

Let's pursue that idea for other values of c. As Figure 7 shows, when we increase the value of c the behavior of the logistic sequence becomes more complex and when $c = 4$ it is quite erratic. [This behavior is part of what it means to be *chaotic*.]

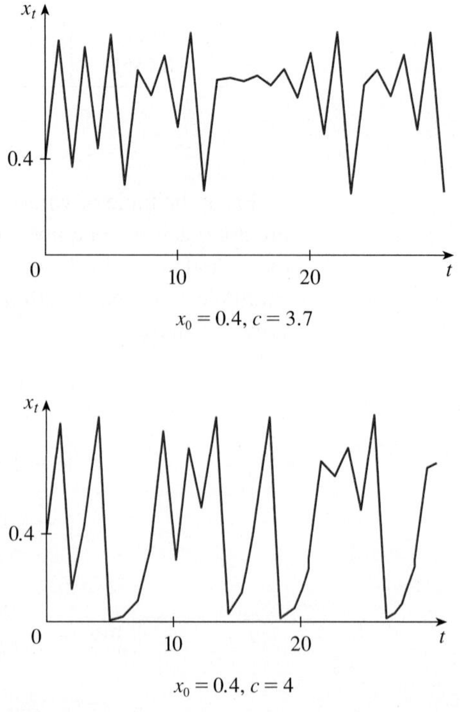

FIGURE 7
Logistic sequences

EXERCISES 2.1

1. (a) What is a sequence?
(b) What does it mean to say that $\lim_{n\to\infty} a_n = 8$?
(c) What does it mean to say that $\lim_{n\to\infty} a_n = \infty$?

2. (a) What is a convergent sequence? Give two examples.
(b) What is a divergent sequence? Give two examples.

3. World record sprint times The graph plots the sequence of the world record times for the men's 100-meter sprint every five years t. Do you think that this sequence has a nonzero limit as $t \to \infty$? What would that mean for this sporting event?

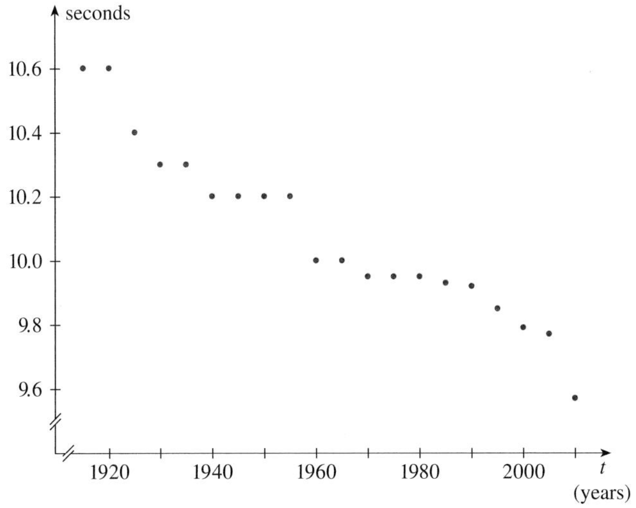

4. World record hammer throws The graph plots the sequence of the world record distances for the women's hammer throw by year t.
(a) Explain what it would mean for this sporting event if the sequence does not have a limit as $t \to \infty$.
(b) Do you think this sequence is convergent or divergent? Explain.

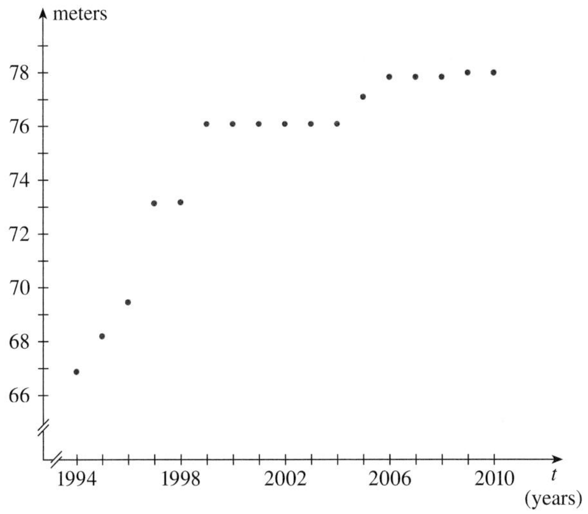

5–8 Calculate, to four decimal places, the first ten terms of the sequence and use them to plot the graph of the sequence. Does the sequence appear to have a limit? If so, calculate it. If not, explain why.

5. $a_n = \dfrac{n^2}{2n + 3n^2}$

6. $a_n = 4 - \dfrac{2}{n} + \dfrac{3}{n^2}$

7. $a_n = 3 + \left(-\frac{2}{3}\right)^n$

8. $a_n = \dfrac{n}{\sqrt{n} + 1}$

9–26 Determine whether the sequence is convergent or divergent. If it is convergent, find the limit.

9. $a_n = \dfrac{1}{3n^4}$

10. $a_n = \dfrac{5}{3^n}$

11. $a_n = \dfrac{2n^2 + n - 1}{n^2}$

12. $a_n = \dfrac{n^3 - 1}{n}$

13. $a_n = \dfrac{3 + 5n}{2 + 7n}$

14. $a_n = \dfrac{n^3 - 1}{n^3 + 1}$

15. $a_n = 1 - (0.2)^n$

16. $a_n = 2^{-n} + 6^{-n}$

17. $a_n = \dfrac{n^2}{\sqrt{n^3 + 4n}}$

18. $a_n = \sin(n\pi/2)$

19. $a_n = \cos(n\pi/2)$

20. $a_n = \dfrac{\pi^n}{3^n}$

21. $a_n = \dfrac{10^n}{1 + 9^n}$

22. $a_n = \dfrac{\sqrt[3]{n}}{\sqrt{n} + \sqrt[4]{n}}$

23. $a_n = \ln(2n^2 + 1) - \ln(n^2 + 1)$

24. $a_n = \dfrac{3^{n+2}}{5^n}$

25. $a_n = \dfrac{e^n + e^{-n}}{e^{2n} - 1}$

26. $a_n = \ln(n + 1) - \ln n$

27–34 Calculate, to four decimal places, the first eight terms of the recursive sequence. Does it appear to be convergent? If so, guess the value of the limit. Then assume the limit exists and determine its exact value.

27. $a_1 = 1, \; a_{n+1} = \frac{1}{2}a_n + 1$

28. $a_1 = 2, \; a_{n+1} = 1 - \frac{1}{3}a_n$

29. $a_1 = 2, \; a_{n+1} = 2a_n - 1$

30. $a_1 = 1, \; a_{n+1} = \sqrt{5a_n}$

31. $a_1 = 1, \; a_{n+1} = \dfrac{6}{1 + a_n}$

32. $a_1 = 3, \; a_{n+1} = 8 - a_n$

33. $a_1 = 1, \; a_{n+1} = \sqrt{2 + a_n}$

34. $a_1 = 100, \; a_{n+1} = \dfrac{1}{2}\left(a_n + \dfrac{25}{a_n}\right)$

35. Antibiotic pharmacokinetics A doctor prescribes a 100-mg antibiotic tablet to be taken every eight hours. Just before each tablet is taken, 20% of the drug present in the preceding time step remains in the body.
(a) How much of the drug is in the body just after the second tablet is taken? After the third tablet?
(b) If Q_n is the quantity of the antibiotic in the body just after the nth tablet is taken, write a difference equation that expresses Q_{n+1} in terms of Q_n.
(c) Find a formula for Q_n as a function of n.
(d) What quantity of the antibiotic remains in the body in the long run?

36. Drug pharmacokinetics A patient is injected with a drug every 12 hours. Immediately before each injection the concentration of the drug has been reduced by 90% and the new dose increases the concentration by 1.5 mg/mL.
(a) What is the concentration after three doses?
(b) If C_n is the concentration after the nth dose, write a difference equation that expresses C_{n+1} in terms of C_n.
(c) Find a formula for C_n as a function of n.
(d) What is the limiting value of the concentration?

37. Drug pharmacokinetics A patient takes 150 mg of a drug at the same time every day. Just before each tablet is taken, 5% of the drug present in the preceding time step remains in the body.
(a) What quantity of the drug is in the body after the third tablet? After the nth tablet?
(b) What quantity of the drug remains in the body in the long run?

38. Insulin injection After injection of a dose D of insulin, the concentration of insulin in a patient's system decays exponentially and so it can be written as De^{-at}, where t represents time in hours and a is a positive constant.
(a) If a dose D is injected every T hours, write an expression for the sum of the residual concentrations just before the $(n + 1)$st injection.
(b) Determine the limiting pre-injection concentration.
(c) If the concentration of insulin must always remain at or above a critical value C, determine a minimal dosage D in terms of C, a, and T.

39. Let $x = 0.99999\ldots$.
(a) Do you think that $x < 1$ or $x = 1$?
(b) Sum a geometric series to find the value of x.
(c) How many decimal representations does the number 1 have?
(d) Which numbers have more than one decimal representation?

40. A sequence is defined recursively by $a_n = (5 - n)a_{n-1}$, $a_1 = 1$. Find the sum of all the terms of the sequence.

41–46 Express the number as a ratio of integers.

41. $0.\overline{8} = 0.8888\ldots$

42. $0.\overline{46} = 0.46464646\ldots$

43. $2.\overline{516} = 2.516516516\ldots$

44. $10.1\overline{35} = 10.135353535\ldots$

45. $1.5\overline{342}$

46. $7.\overline{12345}$

47–52 Logistic equation Plot enough terms of the logistic difference equation $x_{t+1} = cx_t(1 - x_t)$ to see how the terms behave. Does the sequence appear to be convergent? If so, estimate the limit and then, assuming the limit exists, calculate its exact value. If not, describe the behavior of the terms.

47. $x_0 = 0.1, \; c = 2$

48. $x_0 = 0.8, \; c = 2.6$

49. $x_0 = 0.2, \; c = 3.2$

50. $x_0 = 0.4, \; c = 3.5$

51. $x_0 = 0.1, \; c = 3.8$

52. $x_0 = 0.6, \; c = 3.9$

53. Logistic equation: Dependence on initial values Compare plots of the first 20 terms of the logistic equation $x_{t+1} = \frac{1}{4}x_t(1 - x_t)$ for the initial values $x_0 = 0.2$ and $x_0 = 0.2001$. When the initial value changes slightly, how does the solution change?

54. Logistic equation: Dependence on initial values Repeat Exercise 53 for the equation $x_{t+1} = 4x_t(1 - x_t)$ and compare with the results of Exercise 53. [This behavior is another part of what it means to be *chaotic*.]

55–58 The **Ricker equation** $x_{t+1} = cx_te^{-x_t}$ was introduced in Exercise 1.6.32. Plot enough terms of the Ricker equation to see how the terms behave. Does the sequence appear to be convergent? If so, estimate the limit and then, assuming the limit exists, calculate its exact value. If not, describe the behavior of the terms.

55. $x_0 = 0.2, \; c = 2$

56. $x_0 = 0.4, \; c = 3$

57. $x_0 = 0.8, \; c = 10$

58. $x_0 = 0.9, \; c = 20$

59. The **Sierpinski carpet** is constructed by removing the center one-ninth of a square of side 1, then removing the centers of the eight smaller remaining squares, and so on. (The figure shows the first three steps of the construction.) Show that the sum of the areas of the removed squares is 1. This implies that the Sierpinski carpet has area 0.

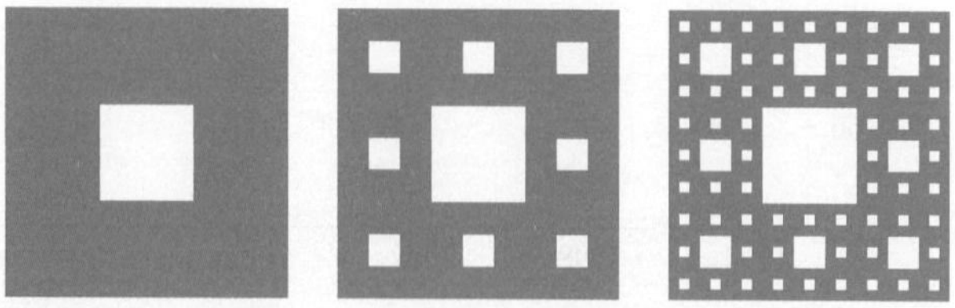

60. A right triangle ABC is given with $\angle A = \theta$ and $|AC| = b$. CD is drawn perpendicular to AB, DE is drawn perpendicular to BC, $EF \perp AB$, and this process is continued indefinitely, as shown in the figure. Find the total length of all the perpendiculars

$$|CD| + |DE| + |EF| + |FG| + \cdots$$

in terms of b and θ.

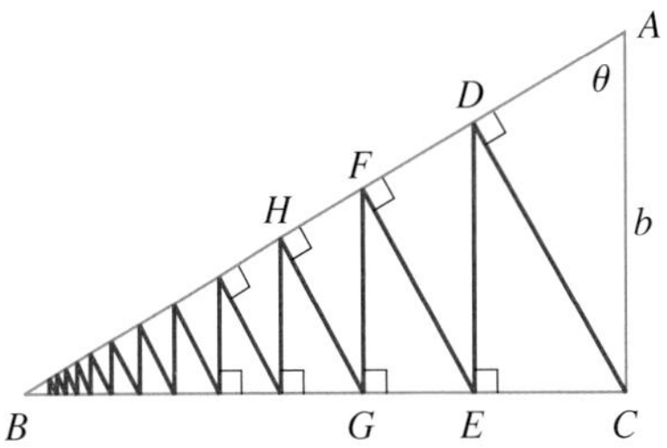

■ PROJECT Modeling the Dynamics of Viral Infections[1]

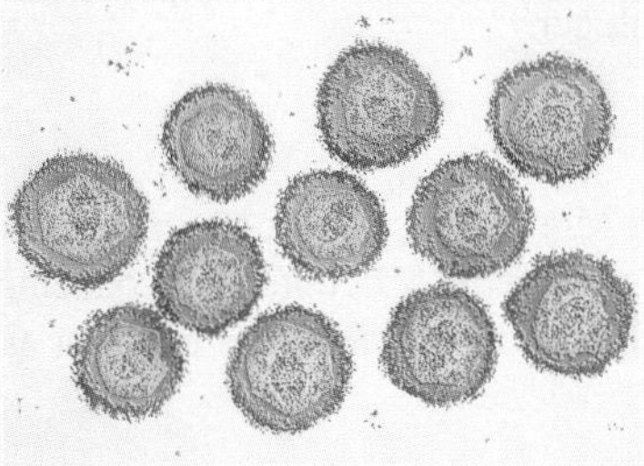

Chris Bjornberg / Science Source

A patient is infected with a virus that triples its numbers every hour. The immune system eventually kicks in and reduces the replication rate by a factor of $\frac{1}{2}$ in addition to killing 500,000 virus particles per hour, but this doesn't happen until the viral load reaches two million copies.

To combat the infection, the infected person receives hourly doses of an antiviral drug. This drug further reduces the replication rate, to a value of 1.25, and the immune system and the drug together can kill 25,000,000 copies of the virus per hour.

Let's model the phases of the infection using a discrete-time difference equation.

1. What is the recursion for the number of viral particles in the absence of treatment and before the immune response starts? What is the equation for the number of viral particles as a function of time?

2. How long does it take for the immune system to be activated after the infection starts? Derive an equation for this length of time for an arbitrary initial number of viral particles.

3. What is the recursion for the number of viral particles after the immune response has begun, but before the drug is used?

4. What is the condition for the viral population size to decrease over time solely because of the immune system? Is this possible?

5. What is the recursion for the number of viral particles in the presence of both the immune response and the drug?

6. What is the condition for the viral population size to decrease over time when the drug and immune system are both acting? Is this possible?

7. If an individual is infected by one virus, how much time do you have to start drug treatment in order to control the infection?

8. Often the outcome of an infection depends on the number of viral particles causing the infection. Derive an expression for the amount of time it takes from the initial infection for an individual to reach the critical viral load—beyond which control of the infection is impossible—as a function of the initial number of particles n_0 and an arbitrary initial replication rate R.

---

1. Adapted from F. Giordano et al., *A First Course in Mathematical Modeling* (Belmont, CA: Cengage Learning, 2014).

9. Suppose 24 hours have passed since the start of an infection with a single viral particle. How many viral particles are in the body? Can the use of the drug now control the infection? If so, how long will it take until the individual is free of the virus?

2.2 Limits of Functions at Infinity

In the preceding section we looked at the long-term behavior of a sequence $\{a_n\}$ by analyzing what happens to the terms a_n when n becomes large. Here we do something similar: We ask what happens to function values $f(x)$ when x becomes large. The only difference is that x is no longer restricted to be an integer.

■ The Monod Growth Function

In the 1930s and 1940s the French biologist Jacques Monod carried out experiments on *E. coli* bacteria feeding on a single nutrient, such as glucose. If N denotes the concentration of the nutrient, he modeled the per capita reproduction rate R of the bacteria as a function of N:

(1) $$R(N) = \frac{SN}{c + N}$$

where c is a positive constant and S is the saturation level of the nutrient. The function $R(N)$ given by Equation 1 is called the **Monod growth function**. Monod later recognized that functions of this form had been used in biochemistry to model enzyme reactions, in which case the function is called the Michaelis-Menten function.

Monod

Jacques Monod (1910–1976) was one of the fathers of molecular biology and was awarded the Nobel prize in 1965 for his "discoveries concerning genetic control of enzyme and virus synthesis." In addition to his scientific discoveries, Monod was also a philosopher of science, a musician, a political activist, and chief of staff of operations for the French Resistance in World War II.

EXAMPLE 1 | BB The Monod growth function with $S = 2$ and $c = 5$ is

$$R(N) = \frac{2N}{5 + N}$$

Evaluate it for $N = 5, 10, 50, 100, 500, 1000, 5000, 10{,}000$. Then graph the function and comment on the shape of the graph.

SOLUTION The values are shown in the table and graphs of the function are shown on the intervals $[0, 10]$, $[0, 100]$, and $[0, 1000]$ in Figure 1.

N	$R(N)$
5	1.0000
10	1.3333
50	1.8182
100	1.9048
500	1.9802
1000	1.9900
5000	1.9980
10,000	1.9990

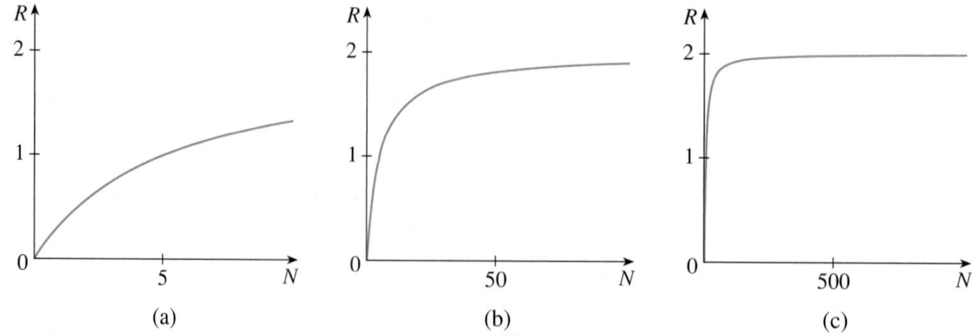

FIGURE 1

We see that $R(N)$ is an increasing function whose values are always less than 2 (the saturation level) but which approach 2 as N grows larger. Biologically, this means that the rate of reproduction of each bacterium increases with nutrient concentration, getting closer to 2 but never exceeding this value. ■

In Example 1 we saw that as N gets larger and larger the values of $R(N)$ become closer and closer to 2. In fact it seems that we can make the values of $R(N)$ as close as we like to 2 by making N sufficiently large. This situation is expressed symbolically by writing

$$\lim_{N \to \infty} R(N) = 2$$

■ Definition of a Limit at Infinity

In general, we use the notation

$$\lim_{x \to \infty} f(x) = L$$

to indicate that the values of $f(x)$ become closer and closer to L as x becomes larger and larger.

> **(2) Definition** Let f be a function defined on some interval (a, ∞). Then
>
> $$\lim_{x \to \infty} f(x) = L$$
>
> means that the values of $f(x)$ can be made arbitrarily close to L by taking x sufficiently large.

Another notation for $\lim_{x \to \infty} f(x) = L$ is

$$f(x) \to L \quad \text{as} \quad x \to \infty$$

The symbol ∞ does not represent a number. Nonetheless, the expression $\lim\limits_{x \to \infty} f(x) = L$ is often read as

"the limit of $f(x)$, as x approaches infinity, is L"

or "the limit of $f(x)$, as x becomes infinite, is L"

or "the limit of $f(x)$, as x increases without bound, is L"

The meaning of such phrases is given by Definition 2. A more precise definition is given in Appendix D.

Geometric illustrations of Definition 2 are shown in Figure 2. Notice that there are many ways for the graph of f to approach the line $y = L$ (which is called a *horizontal asymptote*) as we look to the far right of each graph.

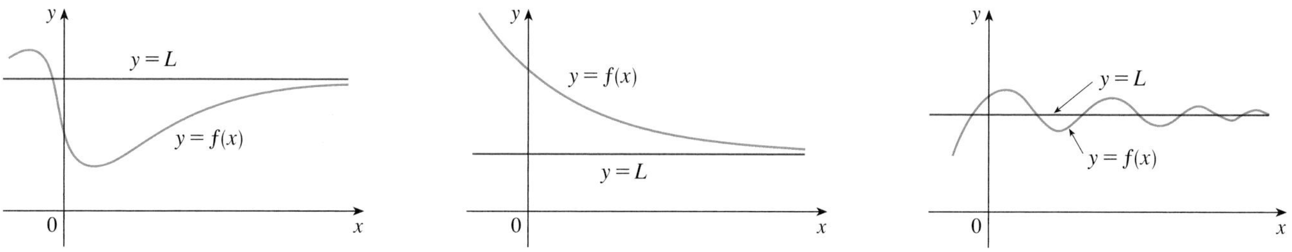

FIGURE 2 Examples illustrating $\lim\limits_{x \to \infty} f(x) = L$

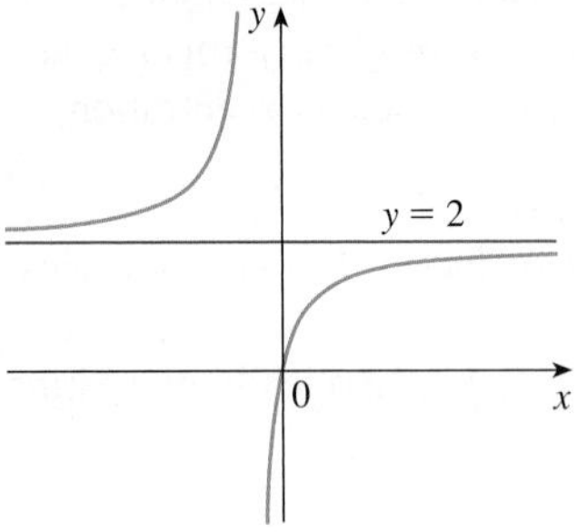

FIGURE 3

$y = \dfrac{2x}{5 + x}$

What happens if we let x decrease through negative values indefinitely? The Monod growth function has no biological meaning if N is negative, but let's consider the corresponding abstract mathematical function

$$f(x) = \frac{2x}{5 + x}$$

for numerically large negative values of x. The graph of f in Figure 3 shows that the values of $f(x)$ also approach 2 as x decreases through negative values without bound. This is expressed by writing

$$\lim_{x \to -\infty} \frac{2x}{5 + x} = 2$$

The general definition is as follows.

> **(3) Definition** Let f be a function defined on some interval $(-\infty, a)$. Then
>
> $$\lim_{x \to -\infty} f(x) = L$$
>
> means that the values of $f(x)$ can be made arbitrarily close to L by taking x sufficiently large negative.

Again, the symbol $-\infty$ does not represent a number, but the expression $\lim_{x \to -\infty} f(x) = L$ is often read as

"the limit of $f(x)$, as x approaches negative infinity, is L"

Definition 3 is illustrated in Figure 4. Notice that the graph approaches the line $y = L$ as we look to the far left of each graph.

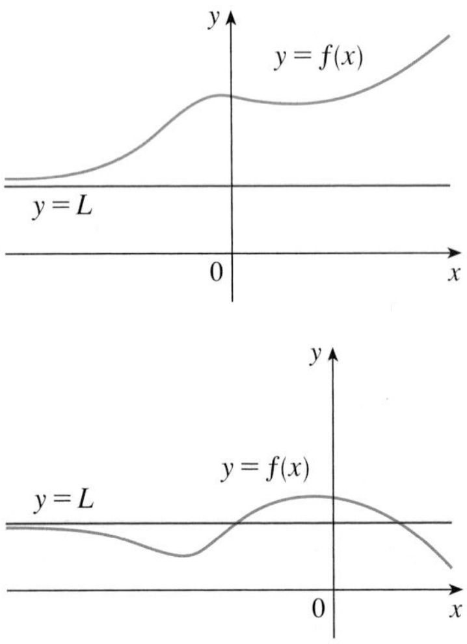

FIGURE 4
Examples illustrating $\lim_{x \to -\infty} f(x) = L$

> **(4) Definition** The line $y = L$ is called a **horizontal asymptote** of the curve $y = f(x)$ if either
>
> $$\lim_{x \to \infty} f(x) = L \qquad \text{or} \qquad \lim_{x \to -\infty} f(x) = L$$

For instance, the curves in Figures 2 and 4 have the line $y = L$ as a horizontal asymptote.

EXAMPLE 2 | What are the horizontal asymptotes of the curve shown in Figure 5?

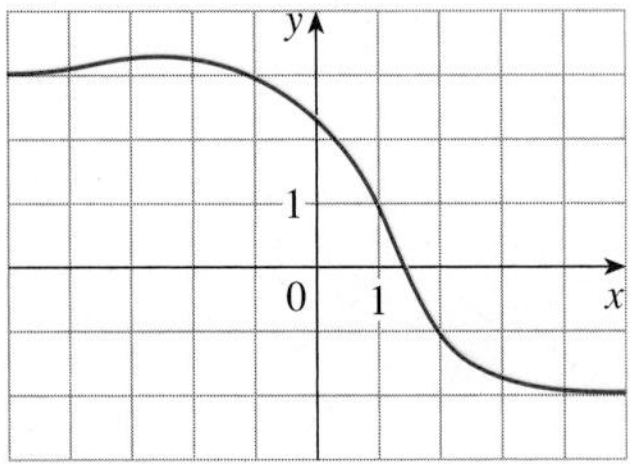

FIGURE 5

SOLUTION From the portion of the graph shown in Figure 5, it appears that $f(x)$ approaches -2 when x gets large and approaches 3 when x becomes large negative. So

$$\lim_{x \to \infty} f(x) = -2 \qquad \text{and} \qquad \lim_{x \to -\infty} f(x) = 3$$

This means that both $y = -2$ and $y = 3$ are horizontal asymptotes of the curve $y = f(x)$. ■

EXAMPLE 3 | Find $\lim_{x\to\infty} \frac{1}{x}$ and $\lim_{x\to-\infty} \frac{1}{x}$.

SOLUTION Observe that when x is large, $1/x$ is small. For instance,

$$\frac{1}{100} = 0.01 \qquad \frac{1}{10{,}000} = 0.0001 \qquad \frac{1}{1{,}000{,}000} = 0.000001$$

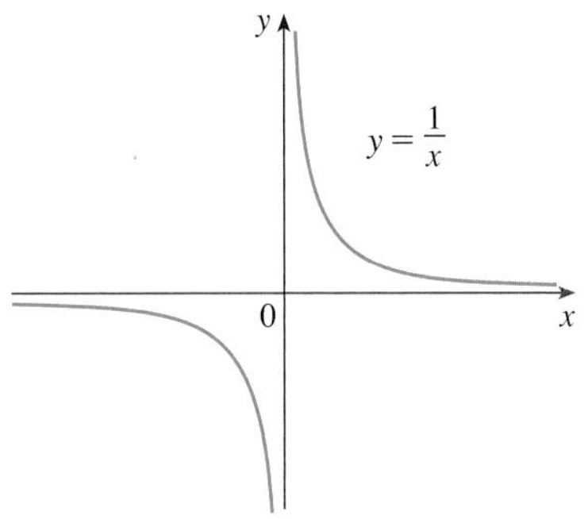

FIGURE 6
$\lim_{x\to\infty} \frac{1}{x} = 0, \quad \lim_{x\to-\infty} \frac{1}{x} = 0$

In fact, by taking x large enough, we can make $1/x$ as close to 0 as we please. Therefore, according to Definition 2, we have

$$\lim_{x\to\infty} \frac{1}{x} = 0$$

Similar reasoning shows that when x is large negative, $1/x$ is small negative, so we also have

$$\lim_{x\to-\infty} \frac{1}{x} = 0$$

It follows that the line $y = 0$ (the x-axis) is a horizontal asymptote of the curve $y = 1/x$. (See Figure 6.) ■

The limit laws for sequences that we used in Section 2.1 have their counterparts for functions. For instance, the Sum Law for functions states that

$$\lim_{x\to\infty} [f(x) + g(x)] = \lim_{x\to\infty} f(x) + \lim_{x\to\infty} g(x)$$

if these limits exist. And the following limit, which corresponds to Equation 2.1.2, is also very useful.

$$\lim_{x\to\infty} \frac{1}{x^p} = 0 \qquad \text{for any number } p > 0$$

EXAMPLE 4 | **BB** **The Monod growth function (continued)** From Example 1 it appears that the limit of the Monod function is $\lim_{N\to\infty} R(N) = 2$. Verify this using the Limit Laws.

SOLUTION Dividing the numerator and denominator by N, and using the Limit Laws, we get

$$\lim_{N\to\infty} R(N) = \lim_{N\to\infty} \frac{2N}{5+N} = \lim_{N\to\infty} \frac{2}{\dfrac{5}{N} + 1}$$

$$= \frac{2}{5 \lim_{N\to\infty} (1/N) + \lim_{N\to\infty} 1} = \frac{2}{5 \cdot 0 + 1} = 2$$

■

We conclude from Example 4 that the line $R = 2$ (corresponding to the saturation level) is a horizontal asymptote of the Monod function $R(N) = 2N/(5 + N)$. For the general Monod function given by Equation 1, we can show similarly that

$$\lim_{N\to\infty} R(N) = \lim_{N\to\infty} \frac{SN}{c + N} = S$$

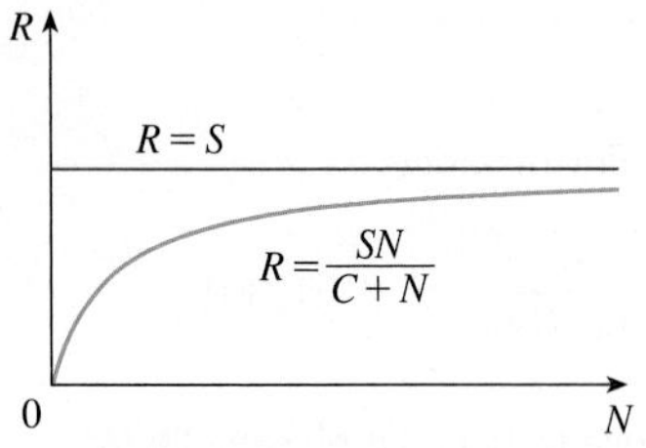

FIGURE 7

So the line $R = S$ is a horizontal asymptote (see Figure 7). Biologically, this means that no individual can have a reproductive rate greater than S, and this rate is approached asymptotically as the nutrient concentration becomes very large.

EXAMPLE 5 | Evaluate $\lim_{t\to\infty} \dfrac{8t}{1 + 4t^2}$.

SOLUTION As with any rational function, we first divide both the numerator and denominator by the highest power of t that occurs in the denominator:

$$\lim_{t\to\infty} \frac{8t}{1 + 4t^2} = \lim_{t\to\infty} \frac{\dfrac{8t}{t^2}}{\dfrac{1 + 4t^2}{t^2}} = \lim_{t\to\infty} \frac{\dfrac{8}{t}}{\dfrac{1}{t^2} + 4}$$

$$= \frac{8 \lim_{t\to\infty} \dfrac{1}{t}}{\lim_{t\to\infty} \dfrac{1}{t^2} + \lim_{t\to\infty} 4} = \frac{8 \cdot 0}{0 + 4} = 0$$

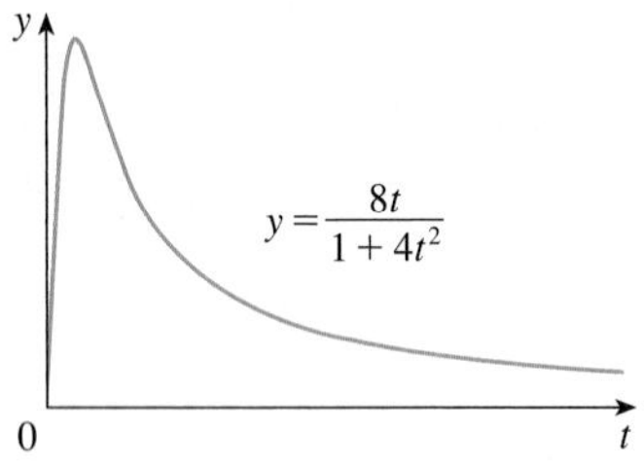

FIGURE 8

So the t-axis is a horizontal asymptote. Figure 8 shows the graph of the function $f(t) = 8t/(1 + 4t^2)$ for $t \geqslant 0$. If a biologist were looking for a function to model a quantity that starts at 0, quickly reaches a peak, and then gradually decays toward 0, then a function similar to f might be a suitable candidate. ■

EXAMPLE 6 | Compute $\lim_{x\to\infty} \left(\sqrt{x^2 + 1} - x\right)$.

SOLUTION Because both $\sqrt{x^2 + 1}$ and x are large when x is large, it's difficult to see what happens to their difference, so we use algebra to rewrite the function.

We first multiply numerator and denominator by the conjugate radical:

We can think of the given function as having a denominator of 1.

$$\lim_{x\to\infty} \left(\sqrt{x^2 + 1} - x\right) = \lim_{x\to\infty} \left(\sqrt{x^2 + 1} - x\right) \frac{\sqrt{x^2 + 1} + x}{\sqrt{x^2 + 1} + x}$$

$$= \lim_{x\to\infty} \frac{(x^2 + 1) - x^2}{\sqrt{x^2 + 1} + x} = \lim_{x\to\infty} \frac{1}{\sqrt{x^2 + 1} + x}$$

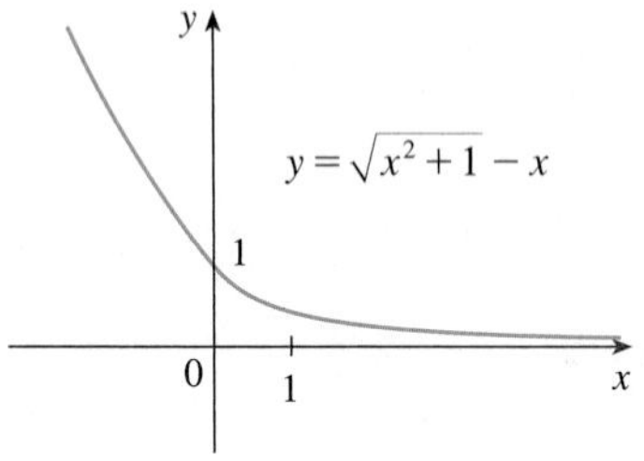

FIGURE 9

Notice that the denominator of this last expression $\left(\sqrt{x^2 + 1} + x\right)$ becomes large as $x \to \infty$ (it's bigger than x). So

$$\lim_{x\to\infty} \left(\sqrt{x^2 + 1} - x\right) = \lim_{x\to\infty} \frac{1}{\sqrt{x^2 + 1} + x} = 0$$

Figure 9 illustrates this result. ■

EXAMPLE 7 | Evaluate $\lim_{x\to\infty} \sin x$.

SOLUTION As x increases, the values of $\sin x$ oscillate between 1 and -1 infinitely often and so they don't approach any definite number. Thus $\lim_{x\to\infty} \sin x$ does not exist. ■

Limits Involving Exponential Functions

Let's recall the shapes of the graphs of the exponential functions from Figures 1.4.4 and 1.4.5. The graphs take the general shape shown in Figure 10.

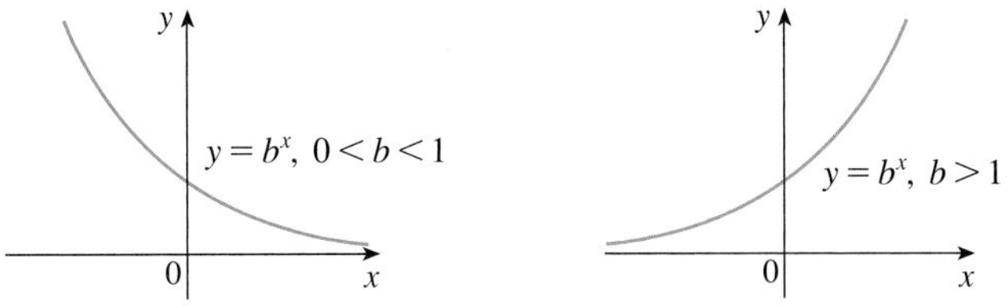

FIGURE 10

In both cases ($0 < b < 1$ and $b > 1$) the line $y = 0$ is a horizontal asymptote. Specifically:

$$\text{If } 0 < b < 1, \qquad \text{then} \qquad \lim_{x \to \infty} b^x = 0.$$

$$\text{If } b > 1, \qquad \text{then} \qquad \lim_{x \to -\infty} b^x = 0.$$

For instance, in Section 1.4 we considered the model

(5) $$V(t) = 96.39785 \cdot (0.818656)^t$$

for the viral load in a patient with HIV. Here the base of the exponential function is $b = 0.818656$, which is less than 1, and so

$$\lim_{t \to \infty} V(t) = 0$$

which is confirmed in Figure 1.4.12.

For the most important special case, $b = e$, we have the graph in Figure 11, illustrating the fact that

$$\boxed{\lim_{x \to -\infty} e^x = 0}$$

You can see that the values of e^x approach 0 very rapidly as $x \to -\infty$.

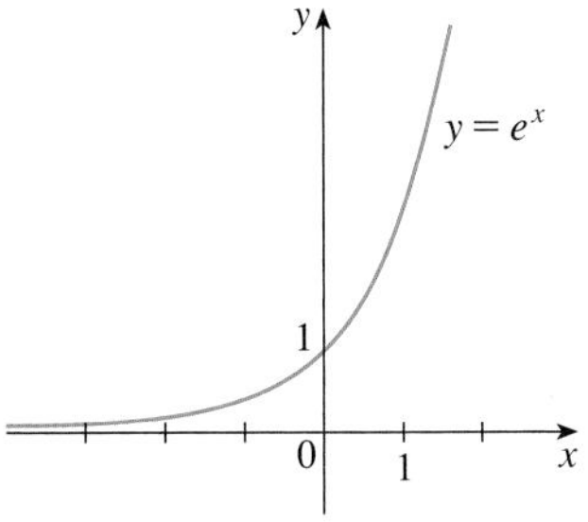

x	e^x
0	1.00000
−1	0.36788
−2	0.13534
−3	0.04979
−5	0.00674
−8	0.00034
−10	0.00005

FIGURE 11

Recall from Section 1.3 that we get the graph of $y = f(-x)$ by reflecting the graph of $y = f(x)$ about the y-axis. So the graph of the function $y = e^{-x}$ is obtained by reflecting the graph of $y = e^x$ (Figure 11) about the y-axis. This function occurs so often in biology that it's wise to be familiar with the shape of its graph and its behavior for large x:

(6) $$\boxed{\lim_{x \to \infty} e^{-x} = 0}$$

Another way to see the limit of the viral load function $V(t)$ in Equation 5 is to write $V(t)$ in terms of a power of e. In general, $b^x = (e^{\ln b})^x = e^{x \ln b}$ and so, with $b = 0.818656$ and using Equation 6, we have

$$\lim_{t \to \infty} V(t) = \lim_{t \to \infty} (96.39785 \cdot e^{-0.2001t}) = 0$$

Gause

G. F. Gause (1910–1986) was a Russian biologist whose success was a result of his training and expertise in both mathematics and experimental biology. He proposed the *competitive exclusion principle:* Two species competing for the same resources cannot coexist if one has even a slight advantage over the other.

EXAMPLE 8 | BB **Gause's logistic model** In the 1930s the biologist G. F. Gause conducted an experiment with the protozoan *Paramecium*. He modeled his data with the logistic function

$$P(t) = \frac{64}{1 + 31e^{-0.7944t}}$$

for the protozoan population after t days. Find the initial population and the limiting population.

SOLUTION The initial population was

$$P(0) = \frac{64}{1 + 31 \cdot e^0} = 2$$

The limiting population was

$$\lim_{t \to \infty} P(t) = \lim_{t \to \infty} \frac{64}{1 + 31e^{-0.7944t}} = \frac{64}{1 + 31 \lim_{t \to \infty} e^{-0.7944t}}$$

Notice that $-0.7944t \to -\infty$ as $t \to \infty$.

$$= \frac{64}{1 + 31 \cdot 0} = 64$$

Here we have used the Limit Laws as well as Equation 6. ■

■ Infinite Limits at Infinity

The notation

$$\lim_{x \to \infty} f(x) = \infty$$

is used to indicate that the values of $f(x)$ become large as x becomes large. Similar meanings are attached to the following symbols:

$$\lim_{x \to -\infty} f(x) = \infty \qquad \lim_{x \to \infty} f(x) = -\infty \qquad \lim_{x \to -\infty} f(x) = -\infty$$

EXAMPLE 9 | Find $\lim_{x \to \infty} x^3$ and $\lim_{x \to -\infty} x^3$.

SOLUTION When x becomes large, x^3 also becomes large. For instance,

$$10^3 = 1000 \qquad 100^3 = 1{,}000{,}000 \qquad 1000^3 = 1{,}000{,}000{,}000$$

In fact, we can make x^3 as big as we like by taking x large enough. Therefore we can write

$$\lim_{x \to \infty} x^3 = \infty$$

Similarly, when x is large negative, so is x^3. Thus

$$\lim_{x \to -\infty} x^3 = -\infty$$

These limit statements can also be seen from the graph of $y = x^3$ in Figure 12. ■

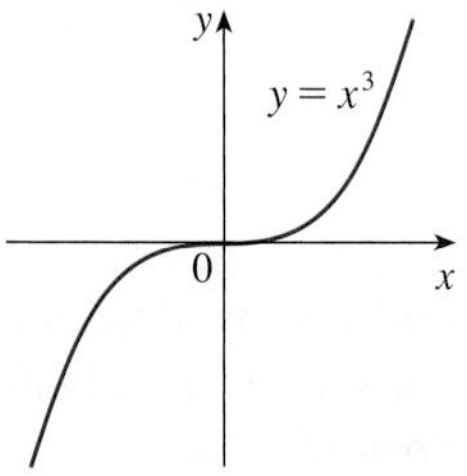

FIGURE 12
$\lim_{x \to \infty} x^3 = \infty$, $\lim_{x \to -\infty} x^3 = -\infty$

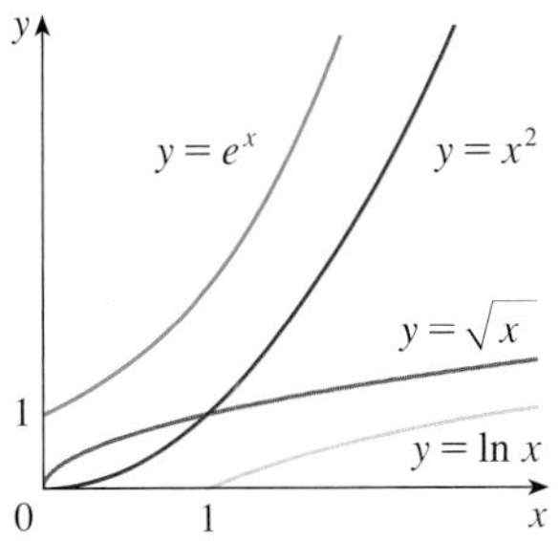

FIGURE 13

Looking at Figure 11 on page 107 we see that

$$\lim_{x \to \infty} e^x = \infty$$

It is also true that

$$\lim_{x \to \infty} x^2 = \infty \qquad \lim_{x \to \infty} \sqrt{x} = \infty \qquad \lim_{x \to \infty} \ln x = \infty$$

But as Figure 13 demonstrates, these four functions become large at different rates. In Chapter 4 we will see how to rank functions according to how quickly they grow.

EXAMPLE 10 | Find $\lim_{x \to \infty} (x^2 - x)$.

SOLUTION It would be **wrong** to write

$$\lim_{x \to \infty} (x^2 - x) = \lim_{x \to \infty} x^2 - \lim_{x \to \infty} x = \infty - \infty$$

The Limit Laws can't be applied to infinite limits because ∞ is not a number ($\infty - \infty$ can't be defined). However, we *can* write

$$\lim_{x \to \infty} (x^2 - x) = \lim_{x \to \infty} x(x - 1) = \infty$$

because both x and $x - 1$ become arbitrarily large and so their product does too. ■

EXAMPLE 11 | Find $\lim_{x \to \infty} \dfrac{x^2 + x}{3 - x}$.

SOLUTION As in Example 5, we divide the numerator and denominator by the highest power of x in the denominator, which is just x:

$$\lim_{x \to \infty} \frac{x^2 + x}{3 - x} = \lim_{x \to \infty} \frac{x + 1}{\dfrac{3}{x} - 1} = -\infty$$

because $x + 1 \to \infty$ and $3/x - 1 \to -1$ as $x \to \infty$. ■

EXERCISES 2.2

1. Explain in your own words the meaning of each of the following.
   (a) $\lim_{x \to \infty} f(x) = 5$ (b) $\lim_{x \to -\infty} f(x) = 3$

2. (a) Can the graph of $y = f(x)$ intersect a horizontal asymptote? If so, how many times? Illustrate by sketching graphs.
   (b) How many horizontal asymptotes can the graph of $y = f(x)$ have? Sketch graphs to illustrate the possibilities.

3. Guess the value of the limit

$$\lim_{x \to \infty} \frac{x^2}{2^x}$$

   by evaluating the function $f(x) = x^2/2^x$ for $x = 0, 1, 2, 3, 4, 5, 6, 7, 8, 9, 10, 20, 50$, and 100. Then use a graph of f to support your guess.

4. (a) Use a graph of

$$f(x) = \left(1 - \frac{2}{x}\right)^x$$

   to estimate the value of $\lim_{x \to \infty} f(x)$ correct to two decimal places.
   (b) Use a table of values of $f(x)$ to estimate the limit to four decimal places.

5–28 Find the limit.

5. $\lim_{x \to \infty} \dfrac{1}{2x + 3}$

6. $\lim_{x \to \infty} \dfrac{3x + 5}{x - 4}$

7. $\lim_{x\to\infty} \dfrac{3x-2}{2x+1}$

8. $\lim_{x\to\infty} \dfrac{1-x^2}{x^3-x+1}$

9. $\lim_{x\to\infty} \dfrac{1-x-x^2}{2x^2-7}$

10. $\lim_{x\to-\infty} \dfrac{4x^3+6x^2-2}{2x^3-4x+5}$

11. $\lim_{t\to-\infty} 0.6^t$

12. $\lim_{r\to\infty} \dfrac{5}{10^r}$

13. $\lim_{t\to\infty} \dfrac{\sqrt{t}+t^2}{2t-t^2}$

14. $\lim_{t\to\infty} \dfrac{t-t\sqrt{t}}{2t^{3/2}+3t-5}$

15. $\lim_{x\to\infty} \dfrac{(2x^2+1)^2}{(x-1)^2(x^2+x)}$

16. $\lim_{x\to\infty} \dfrac{x^2}{\sqrt{x^4+1}}$

17. $\lim_{x\to\infty} \left(\sqrt{9x^2+x}-3x\right)$

18. $\lim_{x\to\infty} \left(\sqrt{x^2+ax}-\sqrt{x^2+bx}\right)$

19. $\lim_{x\to\infty} \dfrac{6}{3+e^{-2x}}$

20. $\lim_{x\to\infty} \sqrt{x^2+1}$

21. $\lim_{x\to\infty} \dfrac{x^4-3x^2+x}{x^3-x+2}$

22. $\lim_{x\to\infty} (e^{-x}+2\cos 3x)$

23. $\lim_{x\to-\infty} (x^4+x^5)$

24. $\lim_{x\to-\infty} \dfrac{1+x^6}{x^4+1}$

25. $\lim_{t\to\infty} e^{-1/t^2}$

26. $\lim_{x\to\infty} \dfrac{e^{3x}-e^{-3x}}{e^{3x}+e^{-3x}}$

27. $\lim_{x\to\infty} \dfrac{1-e^x}{1+2e^x}$

28. $\lim_{x\to-\infty} [\ln(x^2)-\ln(x^2+1)]$

---

29. For the **Monod growth function** $R(N) = SN/(c+N)$, what is the significance of the constant c? [*Hint:* What is $R(c)$?]

30. The **Michaelis-Menten equation** models the rate v of an enzymatic reaction as a function of the concentration [S] of a substrate S. In the case of the enzyme chymotrypsin the equation is

$$v = \frac{0.14[\text{S}]}{0.015+[\text{S}]}$$

(a) What is the horizontal asymptote of the graph of v? What is its significance?

(b) Use a graphing calculator or computer to graph v as a function of [S].

31. Virulence and pathogen transmission The number of new infections produced by an individual infected with a pathogen such as influenza depends on the mortality rate that the pathogen causes. This pathogen-induced mortality rate is referred to as the pathogen's *virulence*. [The photo shows victims at Camp Funston of the influenza epidemic of 1918.] Extremely high levels of virulence result in very little transmission because the infected individual dies before infecting other individuals. Under certain assumptions, the number of new infections N is related to virulence v by the function

$$N(v) = \frac{8v}{1+2v+v^2}$$

where v is the mortality rate (that is, virulence) and $v \geqslant 0$. Evaluate $\lim_{v\to\infty} N(v)$ and interpret your result.

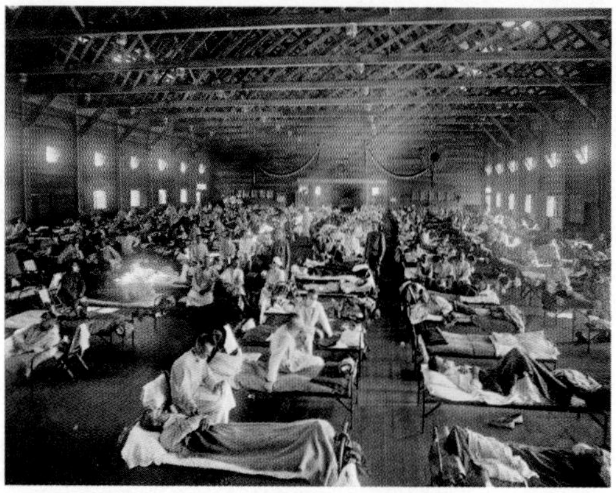

Science Source

32. The **von Bertalanffy growth function**

$$L(t) = L_\infty - (L_\infty - L_0)e^{-kt}$$

where k is a positive constant, models the length L of a fish as a function of t, the age of the fish. This model assumes that the fish has a well-defined length L_0 at birth ($t = 0$).

(a) Calculate $\lim_{t\to\infty} L(t)$. How do you interpret the answer?

(b) If $L_0 = 2$ cm and $L_\infty = 40$ cm, graph $L(t)$ for several values of k. What role does k play?

33. The **Pacific halibut fishery** has been modeled by the equation

$$B(t) = \frac{8\times 10^7}{1+3e^{-0.71t}}$$

where $B(t)$ is the biomass (the total mass of the members of the population) in kilograms at time t. What is $\lim_{t\to\infty} B(t)$? What is the significance of this limit?

34. (a) A tank contains 5000 L of pure water. Brine that contains 30 g of salt per liter of water is pumped into the tank at a rate of 25 L/min. Show that the concentration of salt after t minutes (in grams per liter) is

$$C(t) = \frac{30t}{200+t}$$

(b) What happens to the concentration as $t\to\infty$?

35. Since $\lim_{x\to\infty} e^{-x} = 0$, we should be able to make e^{-x} as small as we like by choosing x large enough. How large do we have to take x so that $e^{-x} < 0.0001$?

36. Let $f(x) = x/(x + 1)$. What is $\lim_{x\to\infty} f(x)$? How large does x have to be so that $f(x) > 0.99$?

37. The velocity $v(t)$ of a falling raindrop at time t is modeled by the equation

$$v(t) = v^*(1 - e^{-gt/v^*})$$

where g is the acceleration due to gravity and v^* is the terminal velocity of the raindrop.

(a) Find $\lim_{t\to\infty} v(t)$.

(b) For a large raindrop in moderate rainfall, a typical terminal velocity is 7.5 m/s. How long does it take for the velocity of such a raindrop to reach 99% of its terminal velocity? (Take $g = 9.8\ \text{m/s}^2$.)

2.3 | Limits of Functions at Finite Numbers

We've investigated what happens to function values $f(x)$ as x becomes large. Here we focus on what happens to $f(x)$ when x approaches a finite number. We begin by observing how such limits arise when we try to determine the speed of a falling ball. Later we'll see how similar considerations are involved in finding rates of change in biology.

■ Velocity Is a Limit

If you watch the speedometer of a car as you travel in city traffic, you see that the speed doesn't stay the same for very long; that is, the velocity of the car is not constant. We assume from watching the speedometer that the car has a definite velocity at each moment, but how is the "instantaneous" velocity defined? Let's investigate the example of a falling ball.

Steve Allen / Stockbyte / Getty Images

The CN Tower in Toronto was the tallest freestanding building in the world for 32 years.

EXAMPLE 1 | Suppose that a ball is dropped from the upper observation deck of the CN Tower in Toronto, 450 m above the ground. Find the velocity of the ball after 5 seconds.

SOLUTION Through experiments carried out four centuries ago, Galileo discovered that the distance fallen by any freely falling body is proportional to the square of the time it has been falling. (This model for free fall neglects air resistance.) If the distance fallen after t seconds is denoted by $s(t)$ and measured in meters, then Galileo's law is expressed by the equation

$$s(t) = 4.9t^2$$

The difficulty in finding the velocity after 5 s is that we are dealing with a single instant of time ($t = 5$), so no time interval is involved. However, we can approximate the desired quantity by computing the average velocity over the brief time interval of a tenth of a second from $t = 5$ to $t = 5.1$:

$$\text{average velocity} = \frac{\text{change in position}}{\text{time elapsed}}$$

$$= \frac{s(5.1) - s(5)}{0.1}$$

$$= \frac{4.9(5.1)^2 - 4.9(5)^2}{0.1} = 49.49\ \text{m/s}$$

The following table shows the results of similar calculations of the average velocity over successively smaller time periods.

Time interval	Average velocity (m/s)
$5 \leq t \leq 6$	53.9
$5 \leq t \leq 5.1$	49.49
$5 \leq t \leq 5.05$	49.245
$5 \leq t \leq 5.01$	49.049
$5 \leq t \leq 5.001$	49.0049

It appears that as we shorten the time period, the average velocity is becoming closer to 49 m/s. The **instantaneous velocity** when $t = 5$ is defined to be the limiting value of these average velocities over shorter and shorter time periods that start at $t = 5$. Thus the (instantaneous) velocity after 5 s appears to be

$$v = 49 \text{ m/s}$$

■

■ Limits: Numerical and Graphical Methods

We saw in Example 1 how limiting values arise in determining the velocity of an object. In Chapter 3 we will see how limits also occur in finding tangent lines to curves and rates of growth in biology, so we now turn our attention to limits in general and numerical and graphical methods for computing them.

Let's investigate the behavior of the function f defined by $f(x) = x^2 - x + 2$ for values of x near 2. The following table gives values of $f(x)$ for values of x close to 2 but not equal to 2.

x	$f(x)$	x	$f(x)$
1.0	2.000000	3.0	8.000000
1.5	2.750000	2.5	5.750000
1.8	3.440000	2.2	4.640000
1.9	3.710000	2.1	4.310000
1.95	3.852500	2.05	4.152500
1.99	3.970100	2.01	4.030100
1.995	3.985025	2.005	4.015025
1.999	3.997001	2.001	4.003001

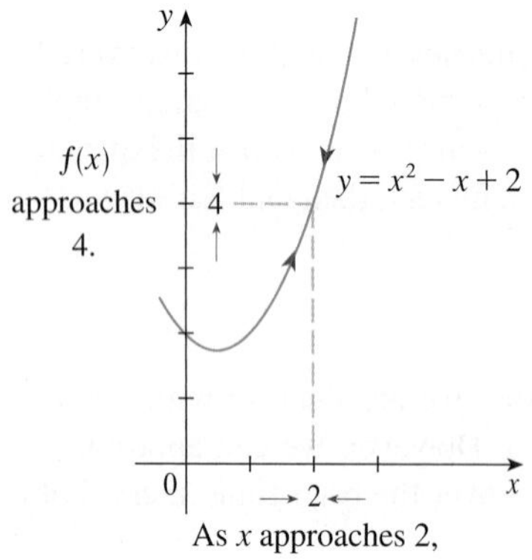

FIGURE 1

From the table and the graph of f (a parabola) shown in Figure 1 we see that when x is close to 2 (on either side of 2), $f(x)$ is close to 4. In fact, it appears that we can make the values of $f(x)$ as close as we like to 4 by taking x sufficiently close to 2. We express this by saying "the limit of the function $f(x) = x^2 - x + 2$ as x approaches 2 is equal to 4." The notation for this is

$$\lim_{x \to 2} (x^2 - x + 2) = 4$$

In general, we use the following notation.

> **(1) Definition** Suppose $f(x)$ is defined when x is near the number a. (This means that f is defined on some open interval that contains a, except possibly at a itself.) Then we write
>
> $$\lim_{x \to a} f(x) = L$$
>
> and say "the limit of $f(x)$, as x approaches a, equals L"
>
> if we can make the values of $f(x)$ arbitrarily close to L (as close to L as we like) by taking x to be sufficiently close to a (on either side of a) but not equal to a.

Roughly speaking, this says that the values of $f(x)$ approach L as x approaches a. In other words, the values of $f(x)$ tend to get closer and closer to the number L as x gets closer and closer to the number a (from either side of a) but $x \neq a$. (A more precise definition is given in Appendix D.)

An alternative notation for

$$\lim_{x \to a} f(x) = L$$

is

$$f(x) \to L \quad \text{as} \quad x \to a$$

which is usually read "$f(x)$ approaches L as x approaches a."

Notice the phrase "but $x \neq a$" in the definition of a limit. This means that in finding the limit of $f(x)$ as x approaches a, we never consider $x = a$. In fact, $f(x)$ need not even be defined when $x = a$. The only thing that matters is how f is defined *near a*.

Figure 2 shows the graphs of three functions. Note that in part (c), $f(a)$ is not defined and in part (b), $f(a) \neq L$. But in each case, regardless of what happens at a, it is true that $\lim_{x \to a} f(x) = L$.

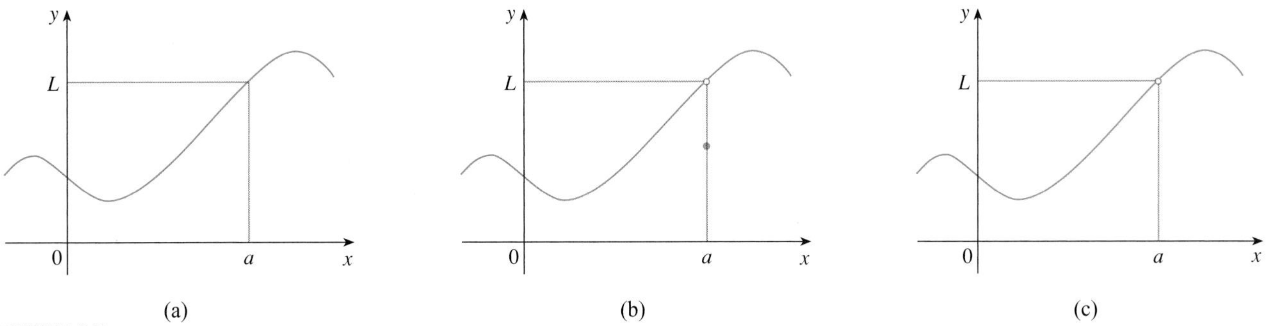

FIGURE 2
$\lim_{x \to a} f(x) = L$ in all three cases.

EXAMPLE 2 | Guess the value of $\displaystyle\lim_{x \to 1} \frac{x - 1}{x^2 - 1}$.

SOLUTION Notice that the function $f(x) = (x - 1)/(x^2 - 1)$ is not defined when $x = 1$, but that doesn't matter because the definition of $\lim_{x \to a} f(x)$ says that we consider values of x that are close to a but not equal to a.

$x < 1$	$f(x)$
0.5	0.666667
0.9	0.526316
0.99	0.502513
0.999	0.500250
0.9999	0.500025

$x > 1$	$f(x)$
1.5	0.400000
1.1	0.476190
1.01	0.497512
1.001	0.499750
1.0001	0.499975

The tables at the left give values of $f(x)$ (correct to six decimal places) for values of x that approach 1 (but are not equal to 1). On the basis of the values in the tables, we make the guess that

$$\lim_{x\to 1} \frac{x-1}{x^2-1} = 0.5$$ ■

Example 2 is illustrated by the graph of f in Figure 3. Now let's change f slightly by giving it the value 2 when $x = 1$ and calling the resulting function g:

$$g(x) = \begin{cases} \dfrac{x-1}{x^2-1} & \text{if } x \neq 1 \\ 2 & \text{if } x = 1 \end{cases}$$

This new function g still has the same limit as x approaches 1 (see Figure 4).

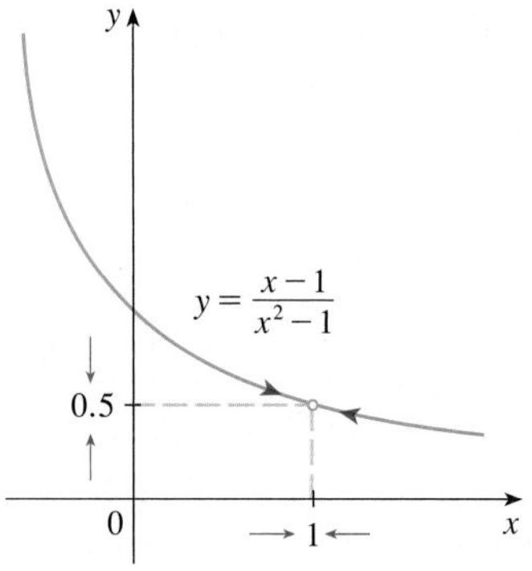

FIGURE 3

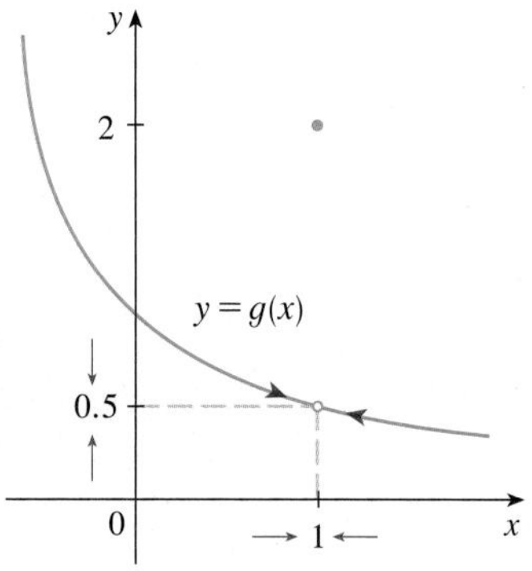

FIGURE 4

EXAMPLE 3 | Estimate the value of $\displaystyle\lim_{t\to 0} \frac{\sqrt{t^2+9}-3}{t^2}$.

SOLUTION The table lists values of the function for several values of t near 0.

t	$\dfrac{\sqrt{t^2+9}-3}{t^2}$
± 1.0	0.16228
± 0.5	0.16553
± 0.1	0.16662
± 0.05	0.16666
± 0.01	0.16667

As t approaches 0, the values of the function seem to approach 0.1666666 . . . and so we guess that

$$\lim_{t\to 0} \frac{\sqrt{t^2+9}-3}{t^2} = \frac{1}{6}$$ ■

t	$\dfrac{\sqrt{t^2+9}-3}{t^2}$
± 0.0005	0.16800
± 0.0001	0.20000
± 0.00005	0.00000
± 0.00001	0.00000

In Example 3 what would have happened if we had taken even smaller values of t? The table in the margin shows the results from one calculator; you can see that something strange seems to be happening.

If you try these calculations on your own calculator you might get different values, but eventually you will get the value 0 if you make t sufficiently small. Does this mean that the answer is really 0 instead of $\frac{1}{6}$? No, the value of the limit is $\frac{1}{6}$, as we will show in the next section. The problem is that the calculator gave false values because $\sqrt{t^2+9}$ is very close to 3 when t is small. (In fact, when t is sufficiently small, a calculator's value for $\sqrt{t^2+9}$ is 3.000. . . to as many digits as the calculator is capable of carrying.)

Something similar happens when we try to graph the function

www.stewartcalculus.com
For a further explanation of why calculators sometimes give false values, click on *Lies My Calculator and Computer Told Me.* In particular, see the section called *The Perils of Subtraction.*

$$f(t) = \frac{\sqrt{t^2+9}-3}{t^2}$$

of Example 3 on a graphing calculator or computer. Parts (a) and (b) of Figure 5 show quite accurate graphs of f, and when we use the trace mode (if available) we can estimate easily that the limit is about $\frac{1}{6}$. But if we zoom in too much, as in parts (c) and (d), then we get inaccurate graphs, again because of problems with subtraction.

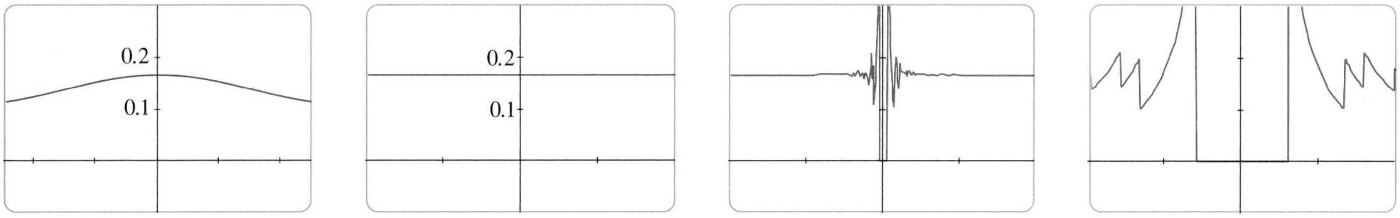

(a) $[-5, 5]$ by $[-0.1, 0.3]$ (b) $[-0.1, 0.1]$ by $[-0.1, 0.3]$ (c) $[-10^{-6}, 10^{-6}]$ by $[-0.1, 0.3]$ (d) $[-10^{-7}, 10^{-7}]$ by $[-0.1, 0.3]$

FIGURE 5

EXAMPLE 4 | Guess the value of $\displaystyle\lim_{x \to 0} \frac{\sin x}{x}$.

SOLUTION The function $f(x) = (\sin x)/x$ is not defined when $x = 0$. Using a calculator (and remembering that $\sin x$ means the sine of the angle whose *radian* measure is x), we construct a table of values correct to eight decimal places. From the table at the left and the graph in Figure 6 we guess that

$$\lim_{x \to 0} \frac{\sin x}{x} = 1$$

x	$\dfrac{\sin x}{x}$
± 1.0	0.84147098
± 0.5	0.95885108
± 0.4	0.97354586
± 0.3	0.98506736
± 0.2	0.99334665
± 0.1	0.99833417
± 0.05	0.99958339
± 0.01	0.99998333
± 0.005	0.99999583
± 0.001	0.99999983

This guess is in fact correct, as will be proved in the next section using a geometric argument.

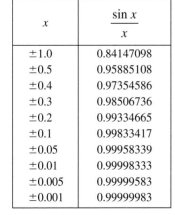

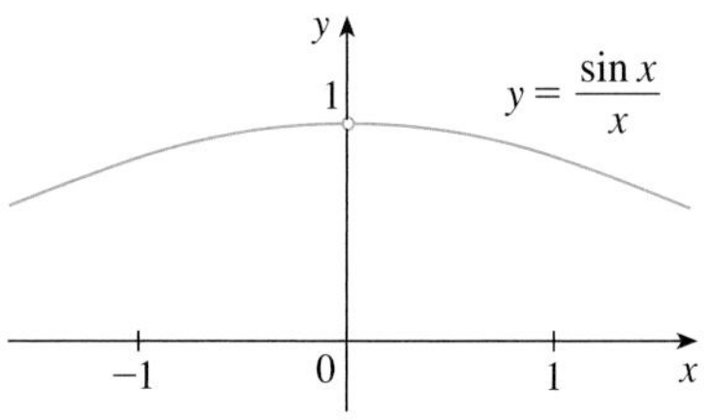

FIGURE 6

EXAMPLE 5 | Investigate $\lim_{x \to 0} \sin \frac{\pi}{x}$.

SOLUTION Again the function $f(x) = \sin(\pi/x)$ is undefined at 0. Evaluating the function for some small values of x, we get

$$f(1) = \sin \pi = 0 \qquad f\left(\tfrac{1}{2}\right) = \sin 2\pi = 0$$

$$f\left(\tfrac{1}{3}\right) = \sin 3\pi = 0 \qquad f\left(\tfrac{1}{4}\right) = \sin 4\pi = 0$$

$$f(0.1) = \sin 10\pi = 0 \qquad f(0.01) = \sin 100\pi = 0$$

Similarly, $f(0.001) = f(0.0001) = 0$. On the basis of this information we might be tempted to guess that

$$\lim_{x \to 0} \sin \frac{\pi}{x} = 0$$

but this time our guess is wrong. Note that although $f(1/n) = \sin n\pi = 0$ for any integer n, it is also true that $f(x) = 1$ for infinitely many values of x that approach 0. You can see this from the graph of f shown in Figure 7.

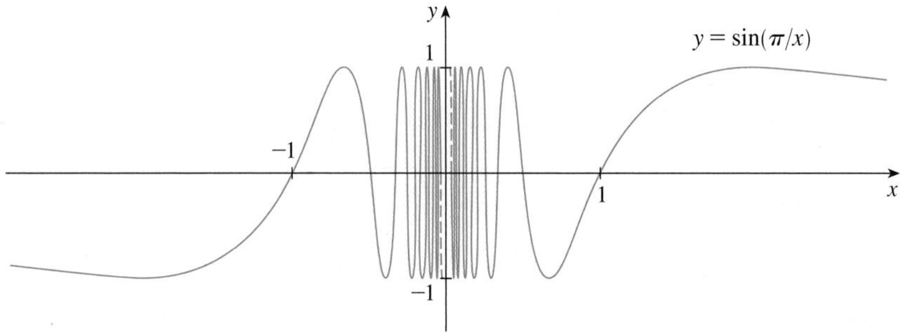

FIGURE 7

The dashed lines near the y-axis indicate that the values of $\sin(\pi/x)$ oscillate between 1 and -1 infinitely often as x approaches 0. (See Exercise 43.) Since the values of $f(x)$ do not approach a fixed number as x approaches 0,

$$\lim_{x \to 0} \sin \frac{\pi}{x} \quad \text{does not exist}$$

■

x	$x^3 + \dfrac{\cos 5x}{10{,}000}$
1	1.000028
0.5	0.124920
0.1	0.001088
0.05	0.000222
0.01	0.000101

x	$x^3 + \dfrac{\cos 5x}{10{,}000}$
0.005	0.00010009
0.001	0.00010000

EXAMPLE 6 | Find $\lim_{x \to 0} \left(x^3 + \dfrac{\cos 5x}{10{,}000} \right)$.

SOLUTION As before, we construct a table of values. From the first table in the margin it appears that

$$\lim_{x \to 0} \left(x^3 + \frac{\cos 5x}{10{,}000} \right) = 0$$

But if we persevere with smaller values of x, the second table suggests that

$$\lim_{x \to 0} \left(x^3 + \frac{\cos 5x}{10{,}000} \right) = 0.000100 = \frac{1}{10{,}000}$$

Later we will see that $\lim_{x \to 0} \cos 5x = 1$; then it follows that the limit is 0.0001. ■

Examples 5 and 6 illustrate some of the pitfalls in guessing the value of a limit. It is easy to guess the wrong value if we use inappropriate values of x, but it is difficult to know when to stop calculating values. And, as the discussion after Example 3 shows, sometimes calculators and computers give the wrong values. In the next section, however, we will develop foolproof methods for calculating limits.

EXAMPLE 7 | BB **Catastrophic population collapse** In later chapters we will explore a model for population size in which catastrophic collapse and extinction occurs if the habitat is degraded beyond a critical level. (See Examples 4.1.6 and 7.2.3 and Exercise 7.2.20.) We will see that the population size is given by the function

$$N(K) = \begin{cases} 0 & \text{if } 0 \leqslant K < a \\ K & \text{if } K \geqslant a \end{cases}$$

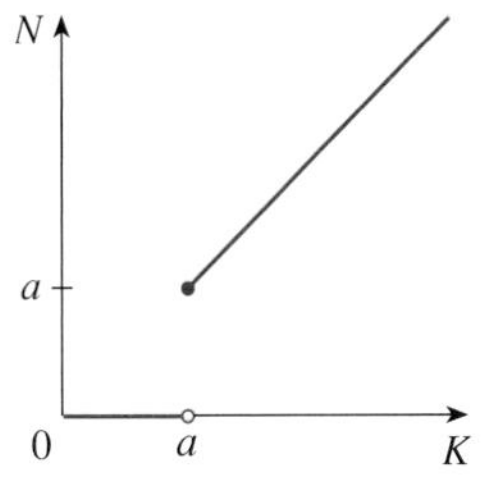

FIGURE 8

where a is a positive number and K is a nonnegative constant that measures habitat quality. (It is referred to as the carrying capacity. See Figure 8.)

As K approaches a from the left, $N(K)$ approaches 0. As K approaches a from the right, $N(K)$ approaches a. There is no single number that $N(K)$ approaches as K approaches a. Therefore $\lim_{K \to a} N(K)$ does not exist. ■

■ One-Sided Limits

We noticed in Example 7 that $N(K)$ approaches 0 as K approaches a from the left and $N(K)$ approaches a as K approaches a from the right. We indicate this situation symbolically by writing

$$\lim_{K \to a^-} N(K) = 0 \qquad \text{and} \qquad \lim_{K \to a^+} N(K) = a$$

The notation $K \to a^-$ indicates that we consider only values of K that are less than a. Likewise, $K \to a^+$ indicates that we consider only values of K that are greater than a.

(2) Definition We write

$$\lim_{x \to a^-} f(x) = L$$

and say the **left-hand limit of $f(x)$ as x approaches a** [or the **limit of $f(x)$ as x approaches a from the left**] is equal to L if we can make the values of $f(x)$ arbitrarily close to L by taking x to be sufficiently close to a and x less than a.

Notice that Definition 2 differs from Definition 1 only in that we require x to be less than a. Similarly, if we require that x be greater than a, we get "the **right-hand limit of $f(x)$ as x approaches a** is equal to L" and we write

$$\lim_{x \to a^+} f(x) = L$$

Thus the notation $x \to a^+$ means that we consider only $x > a$. These definitions are illustrated in Figure 9.

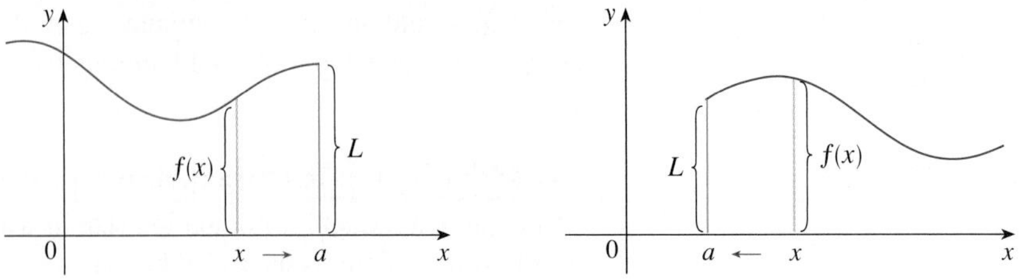

FIGURE 9

(a) $\lim_{x \to a^-} f(x) = L$ (b) $\lim_{x \to a^+} f(x) = L$

By comparing Definition 1 with the definitions of one-sided limits, we see that the following is true.

> **(3)** $\lim_{x \to a} f(x) = L$ if and only if $\lim_{x \to a^-} f(x) = L$ and $\lim_{x \to a^+} f(x) = L$

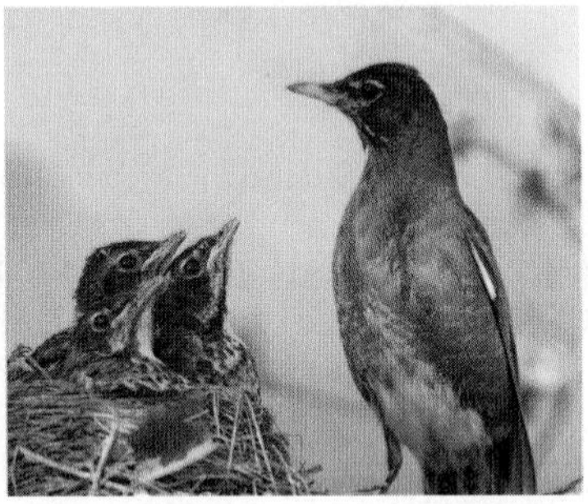

EXAMPLE 8 | **Bird population** American robins breed in early spring and pairs produce a clutch of three to five eggs, laid one day apart. Only about 25% of chicks survive their first year. The graph in Figure 10 shows the population size $P(t)$ of a flock of robins for a 7-day period in spring, where t is the number of days starting at noon on day 0.

(a) What happened at $t = 2$? At $t = 3$? Between $t = 4$ and $t = 5$?

(b) What are the values of the following limits (if they exist)?

$$\lim_{t \to 2^-} P(t) \qquad \lim_{t \to 2^+} P(t) \qquad \lim_{t \to 2} P(t) \qquad \lim_{t \to 6} P(t)$$

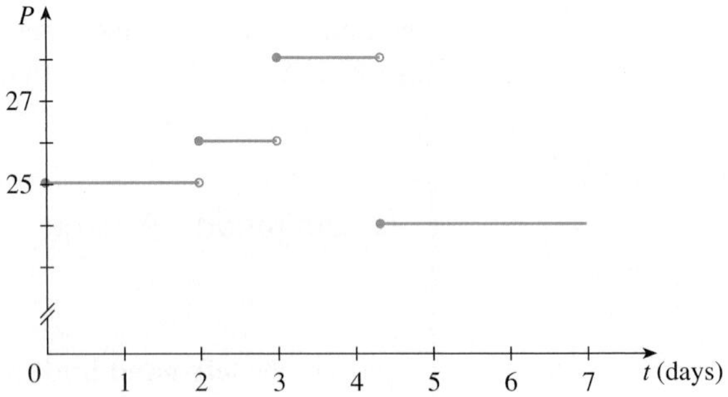

FIGURE 10

SOLUTION

(a) We see from Figure 10 that at $t = 2$ the robin population increased by one, so the most likely explanation is that a chick hatched. At $t = 3$ two more chicks hatched. Between $t = 4$ and $t = 5$ the robin population decreased by 4, so four of the robins died, perhaps because of a predator.

(b) We read the following values from the graph in Figure 10:

$$\lim_{t \to 2^-} P(t) = 25 \qquad \lim_{t \to 2^+} P(t) = 26$$

Because these limits are different, we conclude that

$$\lim_{t\to 2} P(t) \quad \text{does not exist}$$

On the other hand,

$$\lim_{t\to 6^-} P(t) = 24 = \lim_{t\to 6^+} P(t)$$

so

$$\lim_{t\to 6} P(t) = 24$$

Infinite Limits

x	$\dfrac{1}{x^2}$
± 1	1
± 0.5	4
± 0.2	25
± 0.1	100
± 0.05	400
± 0.01	10,000
± 0.001	1,000,000

EXAMPLE 9 | Find $\lim_{x\to 0} \dfrac{1}{x^2}$ if it exists.

SOLUTION As x becomes close to 0, x^2 also becomes close to 0, and $1/x^2$ becomes very large. (See the table in the margin.) In fact, it appears from the graph of the function $f(x) = 1/x^2$ shown in Figure 11 that the values of $f(x)$ can be made arbitrarily large by taking x close enough to 0. Thus the values of $f(x)$ do not approach a number, so $\lim_{x\to 0} (1/x^2)$ does not exist.

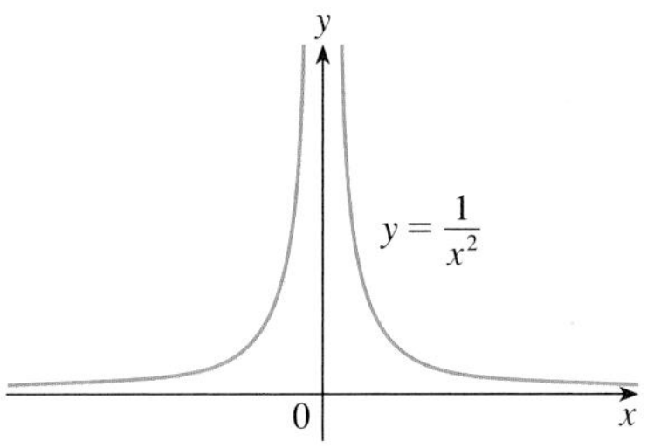

FIGURE 11

To indicate the kind of behavior exhibited in Example 9, we use the notation

$$\lim_{x\to 0} \frac{1}{x^2} = \infty$$

This does not mean that we are regarding ∞ as a number. Nor does it mean that the limit exists. It simply expresses the particular way in which the limit does not exist: $1/x^2$ can be made as large as we like by taking x close enough to 0.

In general, we write symbolically

$$\lim_{x\to a} f(x) = \infty$$

to indicate that the values of $f(x)$ tend to become larger and larger (or "increase without bound") as x becomes closer and closer to a.

> **(4) Definition** Let f be a function defined on both sides of a, except possibly at a itself. Then
>
> $$\lim_{x\to a} f(x) = \infty$$
>
> means that the values of $f(x)$ can be made arbitrarily large (as large as we please) by taking x sufficiently close to a but not equal to a.

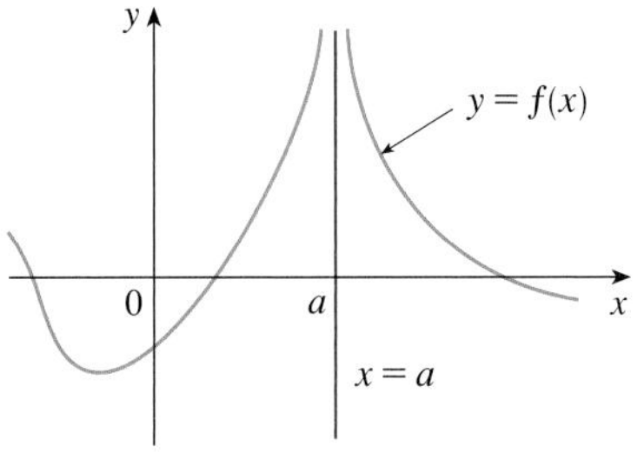

FIGURE 12
$\lim_{x\to a} f(x) = \infty$

Another notation for $\lim_{x\to a} f(x) = \infty$ is

$$f(x) \to \infty \qquad \text{as} \qquad x \to a$$

Again, the symbol ∞ is not a number, but the expression $\lim_{x\to a} f(x) = \infty$ is often read as

"the limit of $f(x)$, as x approaches a, is infinity"

or "$f(x)$ becomes infinite as x approaches a"

or "$f(x)$ increases without bound as x approaches a"

This definition is illustrated graphically in Figure 12.

When we say a number is "large negative," we mean that it is negative but its magnitude (absolute value) is large.

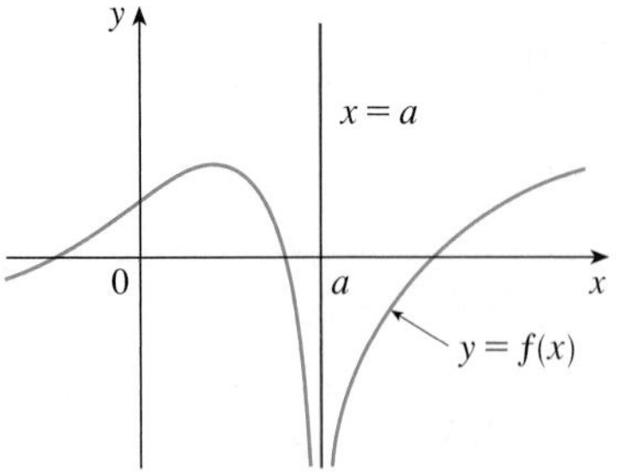

FIGURE 13
$\lim_{x \to a} f(x) = -\infty$

A similar sort of limit, for functions that become large negative as x gets close to a, is defined in Definition 5 and is illustrated in Figure 13.

> **(5) Definition** Let f be defined on both sides of a, except possibly at a itself. Then
>
> $$\lim_{x \to a} f(x) = -\infty$$
>
> means that the values of $f(x)$ can be made arbitrarily large negative by taking x sufficiently close to a but not equal to a.

The symbol $\lim_{x \to a} f(x) = -\infty$ can be read as "the limit of $f(x)$, as x approaches a, is negative infinity" or "$f(x)$ decreases without bound as x approaches a." As an example we have

$$\lim_{x \to 0} \left(-\frac{1}{x^2} \right) = -\infty$$

Similar definitions can be given for the one-sided infinite limits

$$\lim_{x \to a^-} f(x) = \infty \qquad \lim_{x \to a^+} f(x) = \infty$$

$$\lim_{x \to a^-} f(x) = -\infty \qquad \lim_{x \to a^+} f(x) = -\infty$$

remembering that $x \to a^-$ means that we consider only values of x that are less than a, and similarly $x \to a^+$ means that we consider only $x > a$. Illustrations of these four cases are given in Figure 14.

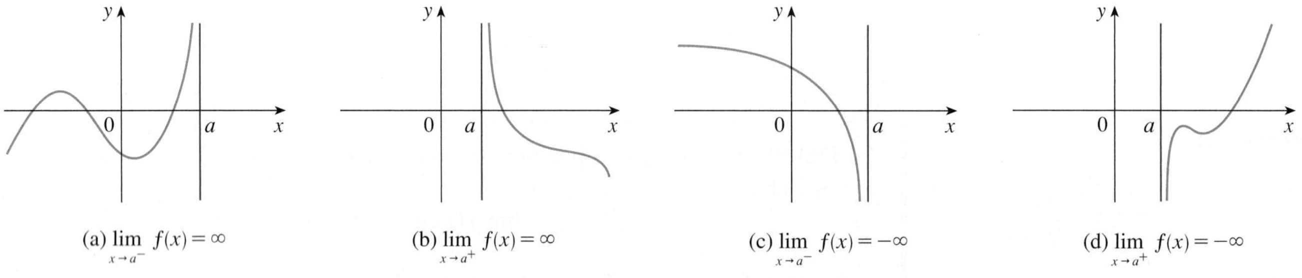

(a) $\lim_{x \to a^-} f(x) = \infty$ (b) $\lim_{x \to a^+} f(x) = \infty$ (c) $\lim_{x \to a^-} f(x) = -\infty$ (d) $\lim_{x \to a^+} f(x) = -\infty$

FIGURE 14

> **(6) Definition** The line $x = a$ is called a **vertical asymptote** of the curve $y = f(x)$ if at least one of the following statements is true:
>
> $$\lim_{x \to a} f(x) = \infty \qquad \lim_{x \to a^-} f(x) = \infty \qquad \lim_{x \to a^+} f(x) = \infty$$
>
> $$\lim_{x \to a} f(x) = -\infty \qquad \lim_{x \to a^-} f(x) = -\infty \qquad \lim_{x \to a^+} f(x) = -\infty$$

For instance, the y-axis is a vertical asymptote of the curve $y = 1/x^2$ because $\lim_{x \to 0} (1/x^2) = \infty$. In Figure 14 the line $x = a$ is a vertical asymptote in each of the four cases shown. In general, knowledge of vertical asymptotes is very useful in sketching graphs.

EXAMPLE 10 | **BB** Anesthesiology In Example 1.2.5 we discussed the use by anesthesiologists of ventilators during surgery. We saw that

$$C = \frac{P}{V}$$

where C is the steady state concentration of CO_2 in the lungs and V is the ventilation rate. As $V \to 0^+$, $C \to \infty$ when P is constant and so

$$\lim_{V \to 0^+} \frac{P}{V} = \infty$$

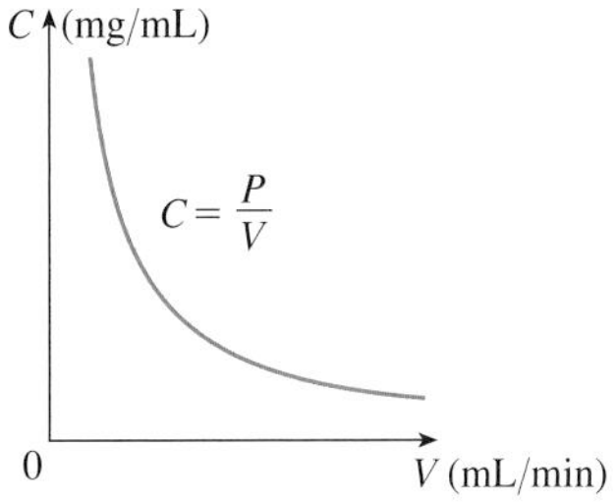

FIGURE 15

Therefore in Figure 15 the C-axis is a vertical asymptote of the graph of C. This means that as the ventilation rate becomes very low, the concentration of carbon dioxide in the lungs becomes very high. (*Note:* This does not mean that, in reality, we expect the CO_2 level in the lungs to become arbitrarily high as the ventilation rate approaches 0. Remember, this is simply a *model* of how CO_2 concentration depends on ventilation rate. It is therefore a simplification of reality. Mathematical models often fail to be accurate descriptions of reality under extreme conditions, and this is an example.) ■

EXAMPLE 11 | Find $\lim_{x \to 3^+} \frac{2x}{x - 3}$ and $\lim_{x \to 3^-} \frac{2x}{x - 3}$.

SOLUTION If x is close to 3 but larger than 3, then the denominator $x - 3$ is a small positive number and $2x$ is close to 6. So the quotient $2x/(x - 3)$ is a large *positive* number. Thus, intuitively, we see that

$$\lim_{x \to 3^+} \frac{2x}{x - 3} = \infty$$

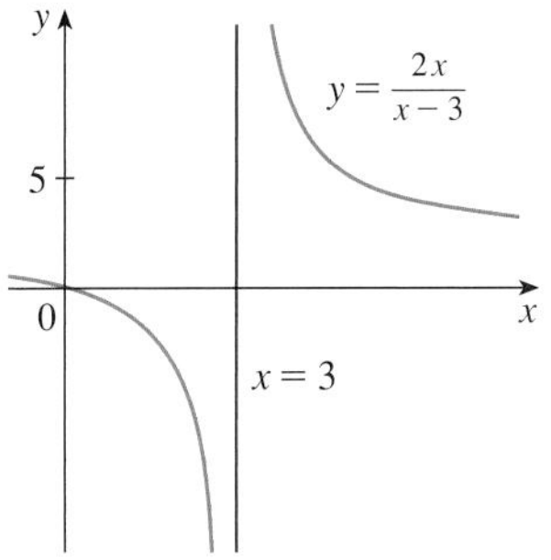

FIGURE 16

Likewise, if x is close to 3 but smaller than 3, then $x - 3$ is a small negative number but $2x$ is still a positive number (close to 6). So $2x/(x - 3)$ is a numerically large *negative* number. Thus

$$\lim_{x \to 3^-} \frac{2x}{x - 3} = -\infty$$

The graph of the curve $y = 2x/(x - 3)$ is given in Figure 16. The line $x = 3$ is a vertical asymptote. ■

EXAMPLE 12 | Find the vertical asymptotes of $f(x) = \tan x$.

SOLUTION Because

$$\tan x = \frac{\sin x}{\cos x}$$

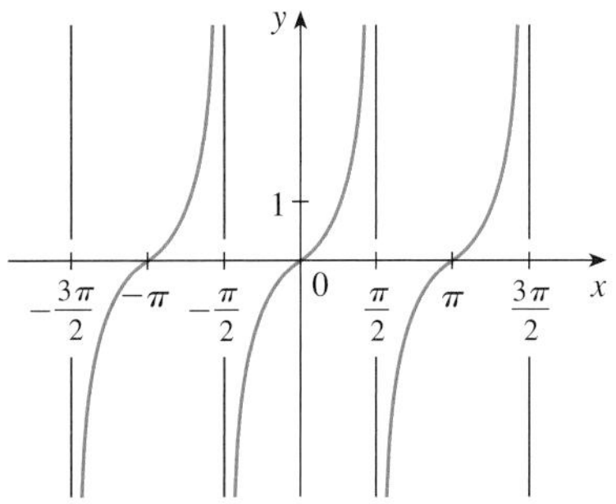

FIGURE 17
$y = \tan x$

there are potential vertical asymptotes where $\cos x = 0$. In fact, since $\cos x \to 0^+$ as $x \to (\pi/2)^-$ and $\cos x \to 0^-$ as $x \to (\pi/2)^+$, whereas $\sin x$ is positive when x is near $\pi/2$, we have

$$\lim_{x \to (\pi/2)^-} \tan x = \infty \qquad \text{and} \qquad \lim_{x \to (\pi/2)^+} \tan x = -\infty$$

This shows that the line $x = \pi/2$ is a vertical asymptote. Similar reasoning shows that the lines $x = (2n + 1)\pi/2$, where n is an integer, are all vertical asymptotes of $f(x) = \tan x$. The graph in Figure 17 confirms this. ■

Another example of a function whose graph has a vertical asymptote is the natural logarithmic function $y = \ln x$. From Figure 18 we see that

$$\lim_{x \to 0^+} \ln x = -\infty$$

and so the line $x = 0$ (the y-axis) is a vertical asymptote. In fact, the same is true for $y = \log_b x$ provided that $b > 1$. (See Figures 11 and 12 in Section 1.5.)

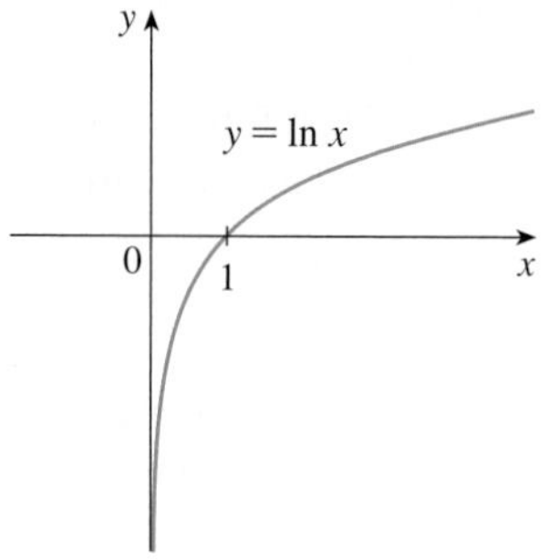

FIGURE 18
The y-axis is a vertical asymptote of the natural logarithmic function.

EXERCISES 2.3

1. Explain in your own words what is meant by the equation

$$\lim_{x \to 2} f(x) = 5$$

Is it possible for this statement to be true and yet $f(2) = 3$? Explain.

2. Explain what it means to say that

$$\lim_{x \to 1^-} f(x) = 3 \qquad \text{and} \qquad \lim_{x \to 1^+} f(x) = 7$$

In this situation is it possible that $\lim_{x \to 1} f(x)$ exists? Explain.

3. Explain the meaning of each of the following.
(a) $\lim_{x \to -3} f(x) = \infty$ (b) $\lim_{x \to 4^+} f(x) = -\infty$

4. Use the given graph of f to state the value of each quantity, if it exists. If it does not exist, explain why.
(a) $\lim_{x \to 2^-} f(x)$ (b) $\lim_{x \to 2^+} f(x)$ (c) $\lim_{x \to 2} f(x)$
(d) $f(2)$ (e) $\lim_{x \to 4} f(x)$ (f) $f(4)$

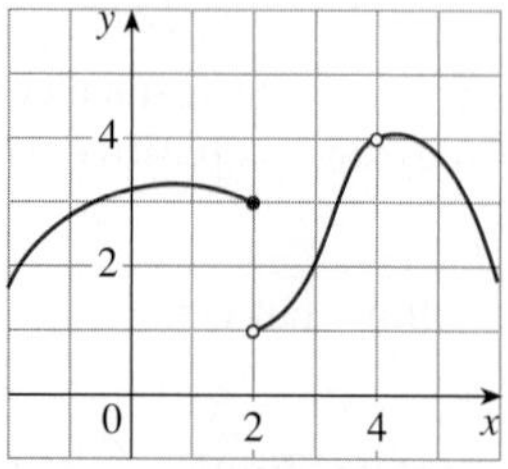

5. For the function f whose graph is given, state the value of each quantity, if it exists. If it does not exist, explain why.
(a) $\lim_{x \to 1} f(x)$ (b) $\lim_{x \to 3^-} f(x)$ (c) $\lim_{x \to 3^+} f(x)$
(d) $\lim_{x \to 3} f(x)$ (e) $f(3)$

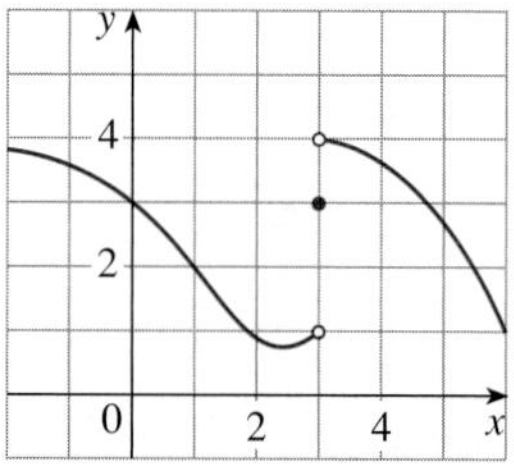

6. For the function h whose graph is given, state the value of each quantity, if it exists. If it does not exist, explain why.
(a) $\lim_{x \to -3^-} h(x)$ (b) $\lim_{x \to -3^+} h(x)$ (c) $\lim_{x \to -3} h(x)$
(d) $h(-3)$ (e) $\lim_{x \to 0^-} h(x)$ (f) $\lim_{x \to 0^+} h(x)$
(g) $\lim_{x \to 0} h(x)$ (h) $h(0)$ (i) $\lim_{x \to 2} h(x)$
(j) $h(2)$ (k) $\lim_{x \to 5^+} h(x)$ (l) $\lim_{x \to 5^-} h(x)$

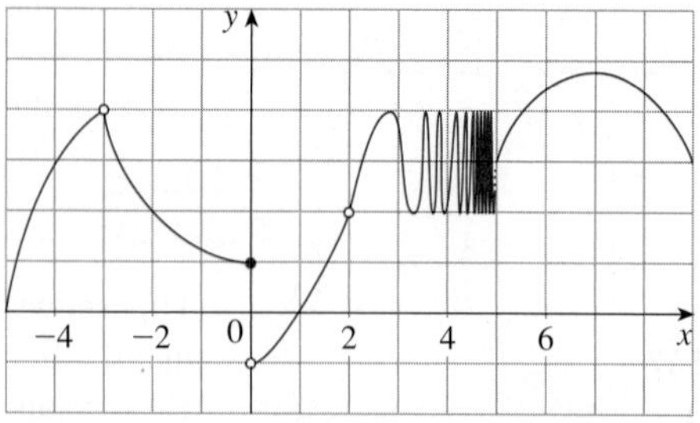

7. The **population of a village** is $P(t)$, t days after June 1. Use the graph of P to state the value of each limit, if it exists. If it doesn't exist, explain why.

(a) $\lim_{t\to 2^-} P(t)$ (b) $\lim_{t\to 2^+} P(t)$ (c) $\lim_{t\to 2} P(t)$

(d) $\lim_{t\to 4^-} P(t)$ (e) $\lim_{t\to 4^+} P(t)$ (f) $\lim_{t\to 4} P(t)$

(g) $\lim_{x\to 5} P(t)$

(h) What do you think happened on June 3rd and 5th?

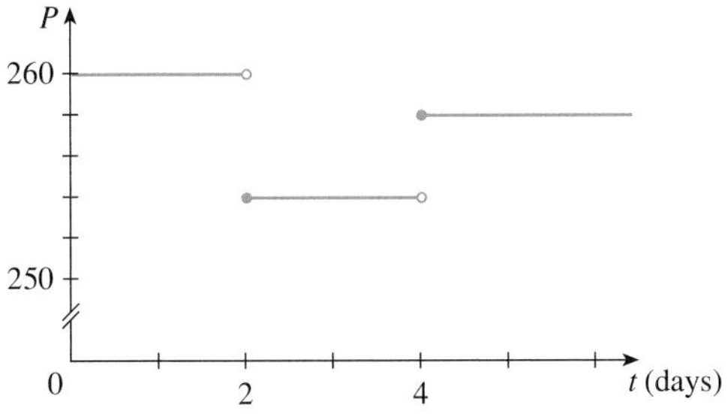

8. For the function R whose graph is shown, state the following.

(a) $\lim_{x\to 2} R(x)$ (b) $\lim_{x\to 5} R(x)$

(c) $\lim_{x\to -3^-} R(x)$ (d) $\lim_{x\to -3^+} R(x)$

(e) The equations of the vertical asymptotes.

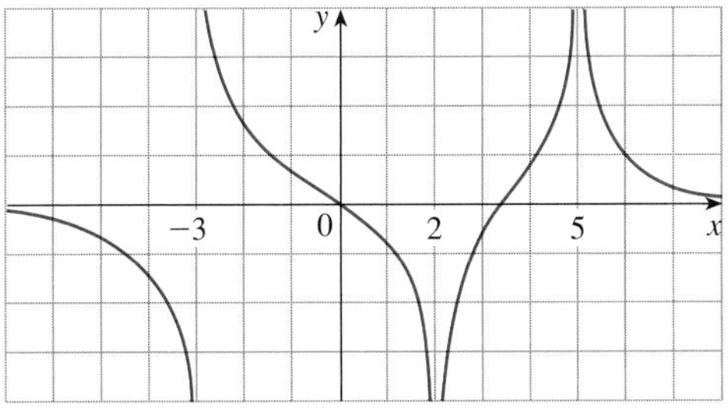

9. For the function g whose graph is given, state the following.

(a) $\lim_{x\to 0} g(x)$ (b) $\lim_{x\to 2^-} g(x)$

(c) $\lim_{x\to 2^+} g(x)$ (d) $\lim_{x\to \infty} g(x)$

(e) $\lim_{x\to -\infty} g(x)$

(f) The equations of the asymptotes

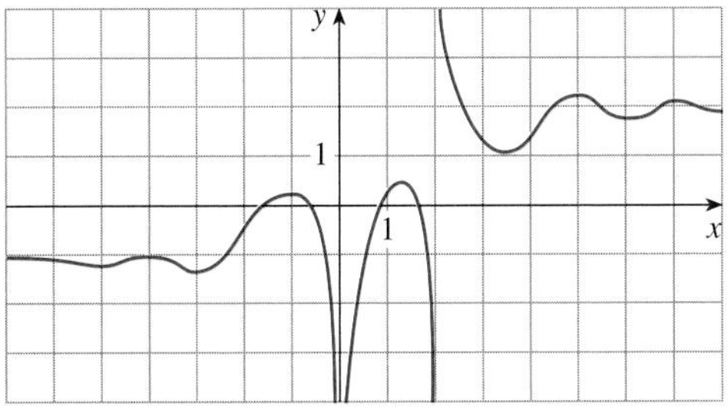

10. Drug injections A patient receives a 150-mg injection of a drug every 4 hours. The graph shows the amount $f(t)$ of the drug in the bloodstream after t hours. Find

$$\lim_{x\to 12^-} f(t) \quad \text{and} \quad \lim_{x\to 12^+} f(t)$$

and explain the significance of these one-sided limits.

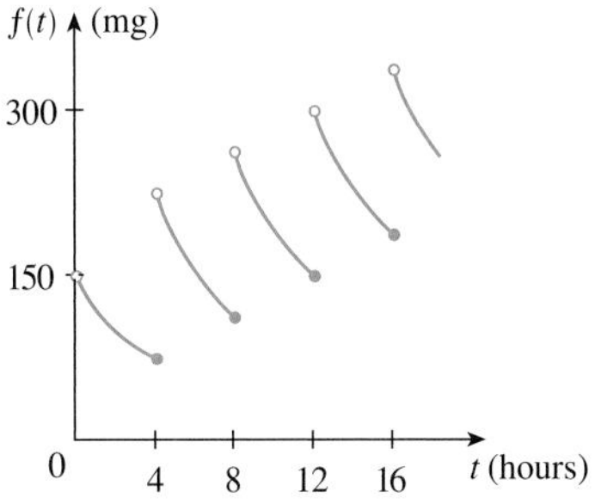

11–18 Sketch the graph of an example of a function f that satisfies all of the given conditions.

11. $\lim_{x\to 3^+} f(x) = 4$, $\lim_{x\to 3^-} f(x) = 2$, $\lim_{x\to -2} f(x) = 2$, $f(3) = 3$, $f(-2) = 1$

12. $\lim_{x\to 0^-} f(x) = 2$, $\lim_{x\to 0^+} f(x) = 0$, $\lim_{x\to 4^-} f(x) = 3$, $\lim_{x\to 4^+} f(x) = 0$, $f(0) = 2$, $f(4) = 1$

13. $\lim_{x\to 0} f(x) = -\infty$, $\lim_{x\to -\infty} f(x) = 5$, $\lim_{x\to \infty} f(x) = -5$

14. $\lim_{x\to 2} f(x) = \infty$, $\lim_{x\to -2^+} f(x) = \infty$, $\lim_{x\to -2^-} f(x) = -\infty$, $\lim_{x\to -\infty} f(x) = 0$, $\lim_{x\to \infty} f(x) = 0$, $f(0) = 0$

15. $\lim_{x\to 2} f(x) = -\infty$ $\lim_{x\to \infty} f(x) = \infty$, $\lim_{x\to -\infty} f(x) = 0$, $\lim_{x\to 0^+} f(x) = \infty$, $\lim_{x\to 0^-} f(x) = -\infty$

16. $\lim_{x\to \infty} f(x) = 3$, $\lim_{x\to 2^-} f(x) = \infty$, $\lim_{x\to 2^+} f(x) = -\infty$, f is odd

17. $f(0) = 3$, $\lim_{x\to 0^-} f(x) = 4$, $\lim_{x\to 0^+} f(x) = 2$, $\lim_{x\to -\infty} f(x) = -\infty$, $\lim_{x\to 4^-} f(x) = -\infty$, $\lim_{x\to 4^+} f(x) = \infty$, $\lim_{x\to \infty} f(x) = 3$

18. $\lim_{x\to \infty} f(x) = -\infty$, $\lim_{x\to \infty} f(x) = 2$, $f(0) = 0$, f is even

19–22 Guess the value of the limit (if it exists) by evaluating the function at the given numbers (correct to six decimal places).

19. $\displaystyle \lim_{x\to 2} \frac{x^2 - 2x}{x^2 - x - x}$,

$x = 2.5, 2.1, 2.05, 2.01, 2.005, 2.001,$

$1.9, 1.95, 1.99, 1.995, 1.999$

20. $\displaystyle\lim_{x\to-1}\frac{x^2-2x}{x^2-x-2}$,
$x = 0, -0.5, -0.9, -0.95, -0.99, -0.999,$
$-2, -1.5, -1.1, -1.01, -1.001$

21. $\displaystyle\lim_{t\to 0}\frac{e^{5t}-1}{t}$, $\quad t = \pm0.5, \pm0.1, \pm0.01, \pm0.001, \pm0.0001$

22. $\displaystyle\lim_{h\to 0}\frac{(2+h)^5-32}{h}$,
$h = \pm0.5, \pm0.1, \pm0.01, \pm0.001, \pm0.0001$

23–26 Use a table of values to estimate the value of the limit. If you have a graphing device, use it to confirm your result graphically.

23. $\displaystyle\lim_{x\to 0}\frac{\sqrt{x+4}-2}{x}$

24. $\displaystyle\lim_{x\to 0}\frac{\tan 3x}{\tan 5x}$

25. $\displaystyle\lim_{x\to 1}\frac{x^6-1}{x^{10}-1}$

26. $\displaystyle\lim_{x\to 0}\frac{9^x-5^x}{x}$

27. (a) By graphing the function $f(x) = (\cos 2x - \cos x)/x^2$ and zooming in toward the point where the graph crosses the y-axis, estimate the value of $\lim_{x\to 0} f(x)$.
(b) Check your answer in part (a) by evaluating $f(x)$ for values of x that approach 0.

28. (a) Estimate the value of

$$\lim_{x\to 0}\frac{\sin x}{\sin \pi x}$$

by graphing the function $f(x) = (\sin x)/(\sin \pi x)$. State your answer correct to two decimal places.
(b) Check your answer in part (a) by evaluating $f(x)$ for values of x that approach 0.

29–37 Determine the infinite limit.

29. $\displaystyle\lim_{x\to-3^+}\frac{x+2}{x+3}$

30. $\displaystyle\lim_{x\to-3^-}\frac{x+2}{x+3}$

31. $\displaystyle\lim_{x\to 1}\frac{2-x}{(x-1)^2}$

32. $\displaystyle\lim_{x\to 5^-}\frac{e^x}{(x-5)^3}$

33. $\displaystyle\lim_{x\to 3^+}\ln(x^2-9)$

34. $\displaystyle\lim_{x\to\pi^-}\cot x$

35. $\displaystyle\lim_{x\to 2\pi^-}x\csc x$

36. $\displaystyle\lim_{x\to 2^-}\frac{x^2-2x}{x^2-4x+4}$

37. $\displaystyle\lim_{x\to 2^+}\frac{x^2-2x-8}{x^2-5x+6}$

38. (a) Find the vertical asymptotes of the function

$$y = \frac{x^2+1}{3x-2x^2}$$

(b) Confirm your answer to part (a) by graphing the function.

39. Determine $\displaystyle\lim_{x\to 1^-}\frac{1}{x^3-1}$ and $\displaystyle\lim_{x\to 1^+}\frac{1}{x^3-1}$
(a) by evaluating $f(x) = 1/(x^3-1)$ for values of x that approach 1 from the left and from the right,
(b) by reasoning as in Example 11, and
(c) from a graph of f.

40. (a) Graph the function $f(x) = e^x + \ln|x-4|$ for $0 \leqslant x \leqslant 5$. Do you think the graph is an accurate representation of f?
(b) How would you get a graph that represents f better?

41. (a) Estimate the value of the limit $\lim_{x\to 0}(1+x)^{1/x}$ to five decimal places. Does this number look familiar?
(b) Illustrate part (a) by graphing the function $y = (1+x)^{1/x}$.

42. (a) Evaluate the function $f(x) = x^2 - (2^x/1000)$ for $x = 1, 0.8, 0.6, 0.4, 0.2, 0.1,$ and 0.05, and guess the value of

$$\lim_{x\to 0}\left(x^2 - \frac{2^x}{1000}\right)$$

(b) Evaluate $f(x)$ for $x = 0.04, 0.02, 0.01, 0.005, 0.003,$ and 0.001. Guess again.

43. Graph the function $f(x) = \sin(\pi/x)$ of Example 5 in the viewing rectangle $[-1, 1]$ by $[-1, 1]$. Then zoom in toward the origin several times. Comment on the behavior of this function.

44. In the theory of relativity, the mass of a particle with velocity v is

$$m = \frac{m_0}{\sqrt{1 - v^2/c^2}}$$

where m_0 is the mass of the particle at rest and c is the speed of light. What happens as $v \to c^-$?

2.4 Limits: Algebraic Methods

In Section 2.3 we used calculators and graphs to guess the values of limits, but we saw that such methods don't always lead to the correct answer. In this section we use the following properties of limits, called the *Limit Laws*, to calculate limits.

The Limit Laws

The following five properties of limits are similar to the ones we have seen before, but some of the other properties of limits apply only when the variable x approaches a finite number.

Limit Laws for Functions

Suppose that c is a constant and the limits

$$\lim_{x \to a} f(x) \qquad \text{and} \qquad \lim_{x \to a} g(x)$$

exist. Then

1. $\lim_{x \to a} [f(x) + g(x)] = \lim_{x \to a} f(x) + \lim_{x \to a} g(x)$

2. $\lim_{x \to a} [f(x) - g(x)] = \lim_{x \to a} f(x) - \lim_{x \to a} g(x)$

3. $\lim_{x \to a} [cf(x)] = c \lim_{x \to a} f(x)$

4. $\lim_{x \to a} [f(x)\,g(x)] = \lim_{x \to a} f(x) \cdot \lim_{x \to a} g(x)$

5. $\lim_{x \to a} \dfrac{f(x)}{g(x)} = \dfrac{\lim_{x \to a} f(x)}{\lim_{x \to a} g(x)} \quad \text{if } \lim_{x \to a} g(x) \neq 0$

These laws can be stated verbally as follows:

Sum Law 1. The limit of a sum is the sum of the limits.

Difference Law 2. The limit of a difference is the difference of the limits.

Constant Multiple Law 3. The limit of a constant times a function is the constant times the limit of the function.

Product Law 4. The limit of a product is the product of the limits

Quotient Law 5. The limit of a quotient is the quotient of the limits (provided that the limit of the denominator is not 0).

It is easy to believe that these properties are true. For instance, if $f(x)$ is close to L and $g(x)$ is close to M, it is reasonable to conclude that $f(x) + g(x)$ is close to $L + M$. This gives us an intuitive basis for believing that Law 1 is true. All of these laws can be proved using the precise definition of a limit. In Appendix E we give the proof of Law 1.

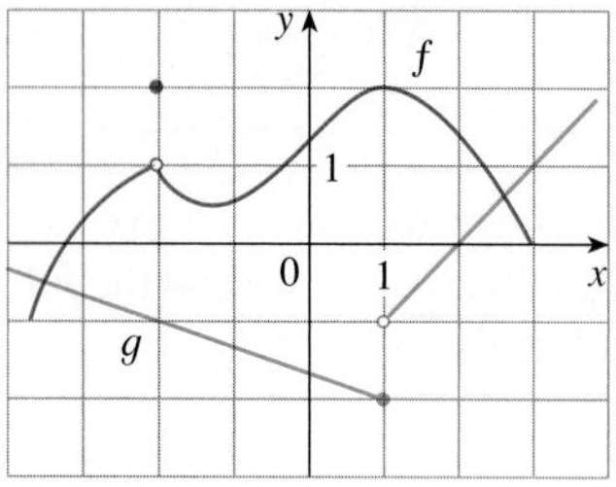

FIGURE 1

EXAMPLE 1 | Use the Limit Laws and the graphs of f and g in Figure 1 to evaluate the following limits, if they exist.

(a) $\lim_{x \to -2} [f(x) + 5g(x)]$ (b) $\lim_{x \to 1} [f(x)g(x)]$ (c) $\lim_{x \to 2} \dfrac{f(x)}{g(x)}$

SOLUTION

(a) From the graphs of f and g we see that

$$\lim_{x \to -2} f(x) = 1 \quad \text{and} \quad \lim_{x \to -2} g(x) = -1$$

Therefore we have

$$\begin{aligned} \lim_{x \to -2} [f(x) + 5g(x)] &= \lim_{x \to -2} f(x) + \lim_{x \to -2} [5g(x)] && \text{(by Law 1)} \\ &= \lim_{x \to -2} f(x) + 5 \lim_{x \to -2} g(x) && \text{(by Law 3)} \\ &= 1 + 5(-1) = -4 \end{aligned}$$

(b) We see that $\lim_{x \to 1} f(x) = 2$. But $\lim_{x \to 1} g(x)$ does not exist because the left and right limits are different:

$$\lim_{x \to 1^-} g(x) = -2 \qquad \lim_{x \to 1^+} g(x) = -1$$

So we can't use Law 4 for the desired limit. But we can use Law 4 for the one-sided limits:

$$\lim_{x \to 1^-} [f(x)g(x)] = 2 \cdot (-2) = -4 \qquad \lim_{x \to 1^+} [f(x)g(x)] = 2 \cdot (-1) = -2$$

The left and right limits aren't equal, so $\lim_{x \to 1} [f(x)g(x)]$ does not exist.

(c) The graphs show that

$$\lim_{x \to 2} f(x) \approx 1.4 \quad \text{and} \quad \lim_{x \to 2} g(x) = 0$$

Because the limit of the denominator is 0, we can't use Law 5. The given limit does not exist because the denominator approaches 0 while the numerator approaches a nonzero number. ■

If we use the Product Law repeatedly with $g(x) = f(x)$, we obtain the following law.

Power Law

6. $\lim_{x \to a} [f(x)]^n = \left[\lim_{x \to a} f(x)\right]^n$ where n is a positive integer

In applying these six limit laws, we need to use two special limits:

7. $\lim_{x \to a} c = c$ **8.** $\lim_{x \to a} x = a$

These limits are obvious from an intuitive point of view (state them in words or draw graphs of $y = c$ and $y = x$).

If we now put $f(x) = x$ in Law 6 and use Law 8, we get another useful special limit.

> **9.** $\lim_{x \to a} x^n = a^n$ where n is a positive integer

A similar limit holds for roots as follows.

> **10.** $\lim_{x \to a} \sqrt[n]{x} = \sqrt[n]{a}$ where n is a positive integer
>
> (If n is even, we assume that $a > 0$.)

More generally, we have the following law.

Root Law

> **11.** $\lim_{x \to a} \sqrt[n]{f(x)} = \sqrt[n]{\lim_{x \to a} f(x)}$ where n is a positive integer
>
> $\left[\text{If } n \text{ is even, we assume that } \lim_{x \to a} f(x) > 0\right]$.

EXAMPLE 2 | Evaluate the following limits and justify each step.

(a) $\lim_{x \to 5} (2x^2 - 3x + 4)$ (b) $\lim_{x \to -2} \dfrac{x^3 + 2x^2 - 1}{5 - 3x}$

SOLUTION

(a)

$$\begin{aligned} \lim_{x \to 5} (2x^2 - 3x + 4) &= \lim_{x \to 5} (2x^2) - \lim_{x \to 5} (3x) + \lim_{x \to 5} 4 && \text{(by Laws 2 and 1)} \\ &= 2 \lim_{x \to 5} x^2 - 3 \lim_{x \to 5} x + \lim_{x \to 5} 4 && \text{(by 3)} \\ &= 2(5^2) - 3(5) + 4 && \text{(by 9, 8, and 7)} \\ &= 39 \end{aligned}$$

(b) We start by using Law 5, but its use is fully justified only at the final stage when we see that the limits of the numerator and denominator exist and the limit of the denominator is not 0.

$$\begin{aligned} \lim_{x \to -2} \frac{x^3 + 2x^2 - 1}{5 - 3x} &= \frac{\lim_{x \to -2} (x^3 + 2x^2 - 1)}{\lim_{x \to -2} (5 - 3x)} && \text{(by Law 5)} \\ &= \frac{\lim_{x \to -2} x^3 + 2 \lim_{x \to -2} x^2 - \lim_{x \to -2} 1}{\lim_{x \to -2} 5 - 3 \lim_{x \to -2} x} && \text{(by 1, 2, and 3)} \\ &= \frac{(-2)^3 + 2(-2)^2 - 1}{5 - 3(-2)} && \text{(by 9, 8, and 7)} \\ &= -\frac{1}{11} \end{aligned}$$

■

Newton and Limits

Isaac Newton was born on Christmas Day in 1642, the year of Galileo's death. When he entered Cambridge University in 1661 Newton didn't know much mathematics, but he learned quickly by reading Euclid and Descartes and by attending the lectures of Isaac Barrow. Cambridge was closed because of the plague in 1665 and 1666, and Newton returned home to reflect on what he had learned. Those two years were amazingly productive for at that time he made four of his major discoveries: (1) his representation of functions as sums of infinite series, including the binomial theorem; (2) his work on differential and integral calculus; (3) his laws of motion and law of universal gravitation; and (4) his prism experiments on the nature of light and color. Because of a fear of controversy and criticism, he was reluctant to publish his discoveries and it wasn't until 1687, at the urging of the astronomer Halley, that Newton published *Principia Mathematica*. In this work, the greatest scientific treatise ever written, Newton set forth his version of calculus and used it to investigate mechanics, fluid dynamics, and wave motion, and to explain the motion of planets and comets.

The beginnings of calculus are found in the calculations of areas and volumes by ancient Greek scholars such as Eudoxus and Archimedes. Although aspects of the idea of a limit are implicit in their "method of exhaustion," Eudoxus and Archimedes never explicitly formulated the concept of a limit. Likewise, mathematicians such as Cavalieri, Fermat, and Barrow, the immediate precursors of Newton in the development of calculus, did not actually use limits. It was Isaac Newton who was the first to talk explicitly about limits. He explained that the main idea behind limits is that quantities "approach nearer than by any given difference." Newton stated that the limit was the basic concept in calculus, but it was left to later mathematicians like Cauchy to clarify his ideas about limits.

NOTE If we let $f(x) = 2x^2 - 3x + 4$, then $f(5) = 39$. In other words, we would have gotten the correct answer in Example 2(a) by substituting 5 for x. Similarly, direct substitution provides the correct answer in part (b). The functions in Example 2 are a polynomial and a rational function, respectively, and similar use of the Limit Laws proves that direct substitution always works for such functions (see Exercise 45). We state this fact as follows.

> **Direct Substitution Property** If f is a polynomial or a rational function and a is in the domain of f, then
>
> $$\lim_{x \to a} f(x) = f(a)$$

Functions with the Direct Substitution Property are called *continuous at a* and will be studied in Section 2.5. However, not all limits can be evaluated by direct substitution, as the following examples show.

EXAMPLE 3 | Find $\displaystyle\lim_{x \to 1} \frac{x^2 - 1}{x - 1}$.

SOLUTION Let $f(x) = (x^2 - 1)/(x - 1)$. We can't find the limit by substituting $x = 1$ because $f(1)$ isn't defined. Nor can we apply the Quotient Law, because the limit of the denominator is 0. Instead, we need to do some preliminary algebra. We factor the numerator as a difference of squares:

$$\frac{x^2 - 1}{x - 1} = \frac{(x - 1)(x + 1)}{x - 1}$$

The numerator and denominator have a common factor of $x - 1$. When we take the limit as x approaches 1, we have $x \neq 1$ and so $x - 1 \neq 0$. Therefore we can cancel the common factor and compute the limit as follows:

$$\lim_{x \to 1} \frac{x^2 - 1}{x - 1} = \lim_{x \to 1} \frac{(x - 1)(x + 1)}{x - 1} = \lim_{x \to 1} (x + 1) = 1 + 1 = 2$$ ■

NOTE In Example 3 we were able to compute the limit by replacing the given function $f(x) = (x^2 - 1)/(x - 1)$ by a simpler function, $g(x) = x + 1$, with the same limit. This is valid because $f(x) = g(x)$ except when $x = 1$, and in computing a limit as x approaches 1 we don't consider what happens when x is actually *equal* to 1. In general, we have the following useful fact.

> If $f(x) = g(x)$ when $x \neq a$, then $\displaystyle\lim_{x \to a} f(x) = \lim_{x \to a} g(x)$, provided the limits exist.

EXAMPLE 4 | Find $\displaystyle\lim_{x \to 1} g(x)$ where

$$g(x) = \begin{cases} x + 1 & \text{if } x \neq 1 \\ \pi & \text{if } x = 1 \end{cases}$$

SOLUTION Here g is defined at $x = 1$ and $g(1) = \pi$, but the value of a limit as x approaches 1 does not depend on the value of the function at 1. Since $g(x) = x + 1$ for

$x \neq 1$, we have

$$\lim_{x \to 1} g(x) = \lim_{x \to 1} (x + 1) = 2$$ ■

Note that the values of the functions in Examples 3 and 4 are identical except when $x = 1$ (see Figure 2) and so they have the same limit as x approaches 1.

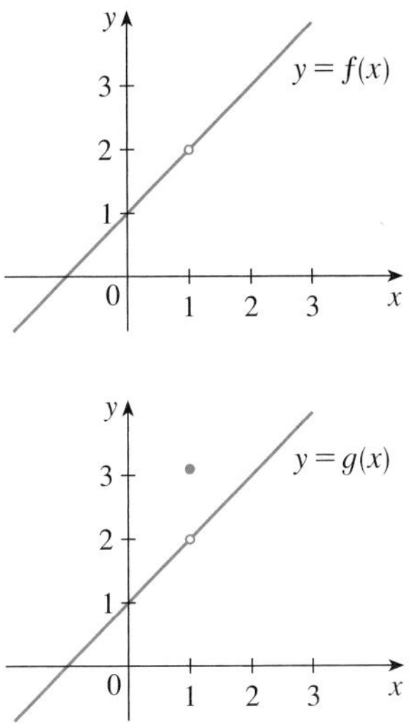

FIGURE 2
The graphs of the functions f (from Example 3) and g (from Example 4)

EXAMPLE 5 | Evaluate $\displaystyle\lim_{h \to 0} \frac{(3 + h)^2 - 9}{h}$.

SOLUTION If we define

$$F(h) = \frac{(3 + h)^2 - 9}{h}$$

then, as in Example 3, we can't compute $\lim_{h \to 0} F(h)$ by letting $h = 0$ since $F(0)$ is undefined. But if we simplify $F(h)$ algebraically, we find that

$$F(h) = \frac{(9 + 6h + h^2) - 9}{h} = \frac{6h + h^2}{h} = 6 + h$$

(Recall that we consider only $h \neq 0$ when letting h approach 0.) Thus

$$\lim_{h \to 0} \frac{(3 + h)^2 - 9}{h} = \lim_{h \to 0} (6 + h) = 6$$ ■

EXAMPLE 6 | Find $\displaystyle\lim_{t \to 0} \frac{\sqrt{t^2 + 9} - 3}{t^2}$.

SOLUTION We can't apply the Quotient Law immediately, since the limit of the denominator is 0. Here the preliminary algebra consists of rationalizing the numerator:

$$\begin{aligned}
\lim_{t \to 0} \frac{\sqrt{t^2 + 9} - 3}{t^2} &= \lim_{t \to 0} \frac{\sqrt{t^2 + 9} - 3}{t^2} \cdot \frac{\sqrt{t^2 + 9} + 3}{\sqrt{t^2 + 9} + 3} \\
&= \lim_{t \to 0} \frac{(t^2 + 9) - 9}{t^2\left(\sqrt{t^2 + 9} + 3\right)} \\
&= \lim_{t \to 0} \frac{t^2}{t^2\left(\sqrt{t^2 + 9} + 3\right)} \\
&= \lim_{t \to 0} \frac{1}{\sqrt{t^2 + 9} + 3} \\
&= \frac{1}{\sqrt{\lim_{t \to 0} (t^2 + 9)} + 3} \\
&= \frac{1}{3 + 3} = \frac{1}{6}
\end{aligned}$$

This calculation confirms the guess that we made in Example 2.3.3. ■

EXAMPLE 7 | **HIV prevalence and incidence** The number of people with HIV in New York in the early 1980s is well described by the function $P(t) = 80 + 16t + 121t^2$, where t measures the years since 1982. (The graph of P is shown in Figure 3.) Epidemiologists call $P(t)$ the *prevalence* of HIV at time t. The *incidence* of HIV is the number of new infections per unit time over a specified time interval.

(a) What was the HIV incidence during the 12 months of 1983?
(b) What was the HIV incidence during the first month of 1983?
(c) Calculate the limiting value of the incidence as the time interval shrinks to zero.

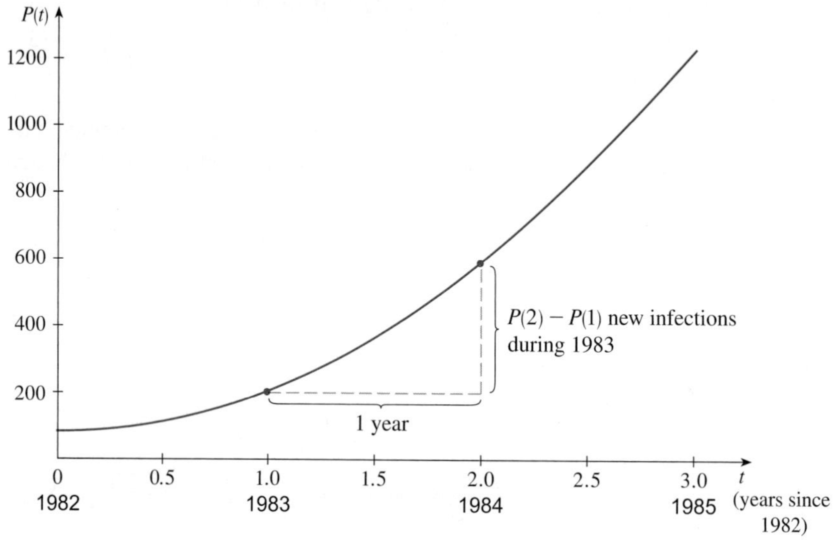

FIGURE 3

SOLUTION The number of infections in 1983 was

$$P(1) = 80 + 16(1) + 121(1^2) = 217$$

(a) The HIV incidence during the 12 months of 1983 was

$$\frac{P(2) - P(1)}{1} = \frac{80 + 16(2) + 121(2^2) - 217}{1} = 379 \text{ infections/year}$$

(b) The HIV incidence during the first month of 1983 was

$$\frac{P\left(1\frac{1}{12}\right) - P(1)}{\frac{1}{12}} = \frac{\frac{34{,}465}{144} - 217}{\frac{1}{12}} = \frac{3217}{12} \approx 268 \text{ infections/year}$$

(c) The limiting value of the incidence is

$$\lim_{h\to 0} \frac{P(1+h) - P(1)}{h} = \lim_{h\to 0} \frac{80 + 16(1+h) + 121(1+h)^2 - 217}{h}$$

$$= \lim_{h\to 0} \frac{217 + 258h + 121h^2 - 217}{h}$$

$$= \lim_{h\to 0} (258 + 121h) = 258 \text{ infections/year}$$ ■

Additional Properties of Limits

Some limits are best calculated by first finding the left- and right-hand limits. The following theorem is a reminder of what we discovered in Section 2.3. It says that a two-sided limit exists if and only if both of the one-sided limits exist and are equal.

(1) Theorem $\lim_{x \to a} f(x) = L$ if and only if $\lim_{x \to a^-} f(x) = L = \lim_{x \to a^+} f(x)$

When computing one-sided limits, we use the fact that the Limit Laws also hold for one-sided limits.

The result of Example 8 looks plausible from Figure 4.

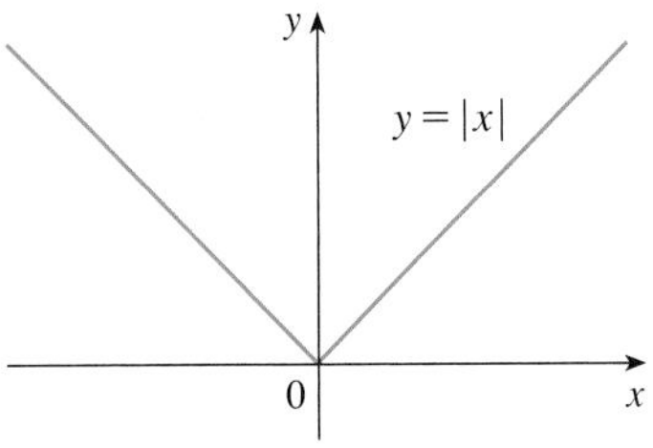

FIGURE 4

EXAMPLE 8 | Show that $\lim_{x \to 0} |x| = 0$.

SOLUTION Recall that

$$|x| = \begin{cases} x & \text{if } x \geqslant 0 \\ -x & \text{if } x < 0 \end{cases}$$

Since $|x| = x$ for $x > 0$, we have

$$\lim_{x \to 0^+} |x| = \lim_{x \to 0^+} x = 0$$

For $x < 0$ we have $|x| = -x$ and so

$$\lim_{x \to 0^-} |x| = \lim_{x \to 0^-} (-x) = 0$$

Therefore, by Theorem 1,

$$\lim_{x \to 0} |x| = 0$$

■

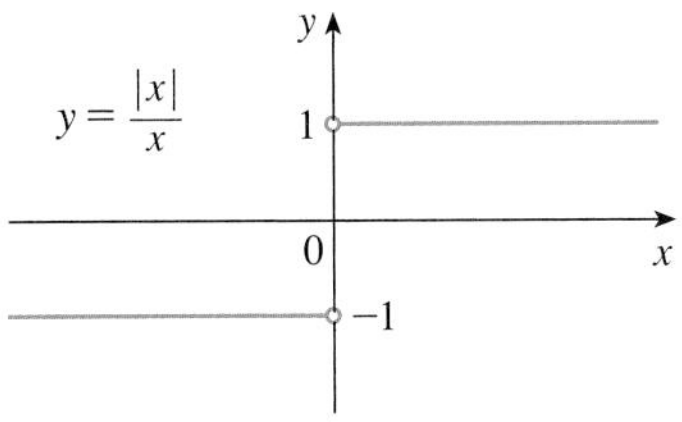

FIGURE 5

EXAMPLE 9 | Prove that $\lim_{x \to 0} \frac{|x|}{x}$ does not exist.

SOLUTION

$$\lim_{x \to 0^+} \frac{|x|}{x} = \lim_{x \to 0^+} \frac{x}{x} = \lim_{x \to 0^+} 1 = 1$$

$$\lim_{x \to 0^-} \frac{|x|}{x} = \lim_{x \to 0^-} \frac{-x}{x} = \lim_{x \to 0^-} (-1) = -1$$

Since the right- and left-hand limits are different, it follows from Theorem 1 that $\lim_{x \to 0} |x|/x$ does not exist. The graph of the function $f(x) = |x|/x$ is shown in Figure 5 and supports the one-sided limits that we found. ■

The next two theorems give two additional properties of limits. Both can be proved using the precise definition in Appendix D.

(2) Theorem If $f(x) \leqslant g(x)$ when x is near a (except possibly at a) and the limits of f and g both exist as x approaches a, then

$$\lim_{x \to a} f(x) \leqslant \lim_{x \to a} g(x)$$

(3) The Squeeze Theorem If $f(x) \leqslant g(x) \leqslant h(x)$ when x is near a (except possibly at a) and

$$\lim_{x \to a} f(x) = \lim_{x \to a} h(x) = L$$

then
$$\lim_{x \to a} g(x) = L$$

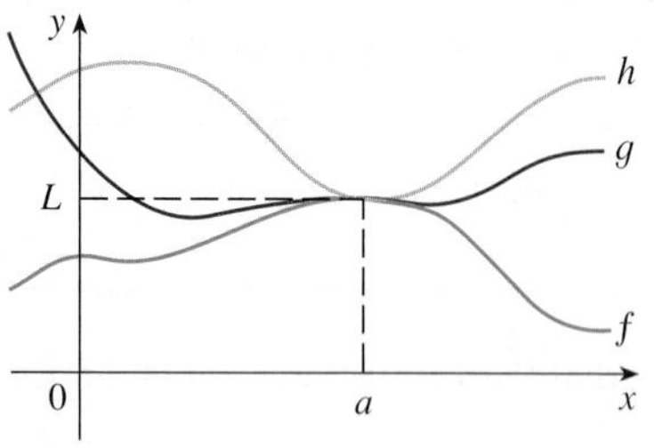

FIGURE 6

The Squeeze Theorem, which is sometimes called the Sandwich Theorem or the Pinching Theorem, is illustrated by Figure 6. It says that if $g(x)$ is squeezed between $f(x)$ and $h(x)$ near a, and if f and h have the same limit L at a, then g is forced to have the same limit L at a.

EXAMPLE 10 | Show that $\lim_{x \to 0} x^2 \sin \dfrac{1}{x} = 0$.

SOLUTION First note that we **cannot** use

$$\lim_{x \to 0} x^2 \sin \frac{1}{x} = \lim_{x \to 0} x^2 \cdot \lim_{x \to 0} \sin \frac{1}{x}$$

because $\lim_{x \to 0} \sin(1/x)$ does not exist (see Example 2.3.5).

Instead we apply the Squeeze Theorem, and so we need to find a function f smaller than $g(x) = x^2 \sin(1/x)$ and a function h bigger than g such that both $f(x)$ and $h(x)$ approach 0 as x approaches 0. To do this we use our knowledge of the sine function. Because the sine of any number lies between -1 and 1, we can write

(4)
$$-1 \leqslant \sin \frac{1}{x} \leqslant 1$$

Any inequality remains true when multiplied by a positive number. We know that $x^2 \geqslant 0$ for all x and so, multiplying each side of the inequalities in (4) by x^2, we get

$$-x^2 \leqslant x^2 \sin \frac{1}{x} \leqslant x^2$$

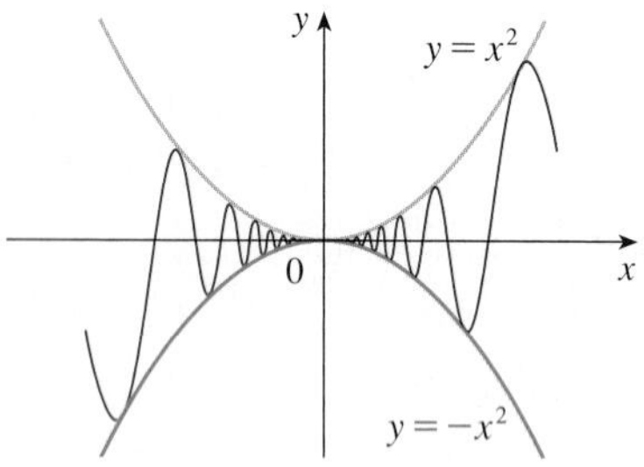

FIGURE 7
$y = x^2 \sin(1/x)$

as illustrated by Figure 7. We know that

$$\lim_{x \to 0} x^2 = 0 \qquad \text{and} \qquad \lim_{x \to 0} (-x^2) = 0$$

Taking $f(x) = -x^2$, $g(x) = x^2 \sin(1/x)$, and $h(x) = x^2$ in the Squeeze Theorem, we obtain

$$\lim_{x \to 0} x^2 \sin \frac{1}{x} = 0$$

Limits of Trigonometric Functions

We've seen that polynomials and rational functions satisfy the Direct Substitution Property. We will now show that the sine and cosine functions also satisfy that property. We know from the definitions of $\sin \theta$ and $\cos \theta$ that the coordinates of the point P in

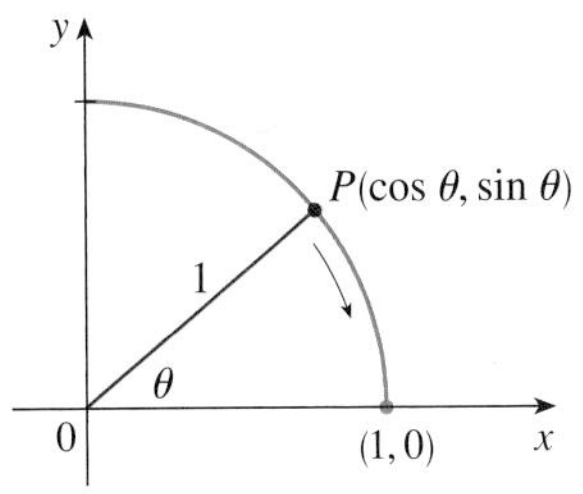

FIGURE 8

Another way to establish the limits in (5) is to use the inequality $\sin\theta < \theta$ (for $\theta > 0$), which is proved after Example 11.

Figure 8 are $(\cos\theta, \sin\theta)$. As $\theta \to 0$, we see that P approaches the point $(1, 0)$ and so $\cos\theta \to 1$ and $\sin\theta \to 0$. Thus

(5) $$\lim_{\theta\to 0} \cos\theta = 1 \qquad \lim_{\theta\to 0} \sin\theta = 0$$

Since $\cos 0 = 1$ and $\sin 0 = 0$, the equations in (5) assert that the cosine and sine functions satisfy the Direct Substitution Property at 0. The addition formulas for cosine and sine can then be used to deduce that these functions satisfy the Direct Substitution Property everywhere (see Exercises 47 and 48). In other words, for any real number a,

$$\lim_{\theta\to a} \sin\theta = \sin a \qquad \lim_{\theta\to a} \cos\theta = \cos a$$

This enables us to evaluate certain limits quite simply, as the next example shows.

EXAMPLE 11 | Evaluate $\lim_{x\to\pi} x\cos x$.

SOLUTION Using Limit Law 4 and the fact that the cosine function satisfies the Direct Substitution Property, we get

$$\lim_{x\to\pi} x\cos x = \left(\lim_{x\to\pi} x\right)\left(\lim_{x\to\pi} \cos x\right)$$
$$= \pi \cdot \cos\pi = -\pi$$

■

In Example 2.3.4 we made the guess, on the basis of numerical and graphical evidence, that

(6) $$\lim_{\theta\to 0} \frac{\sin\theta}{\theta} = 1$$

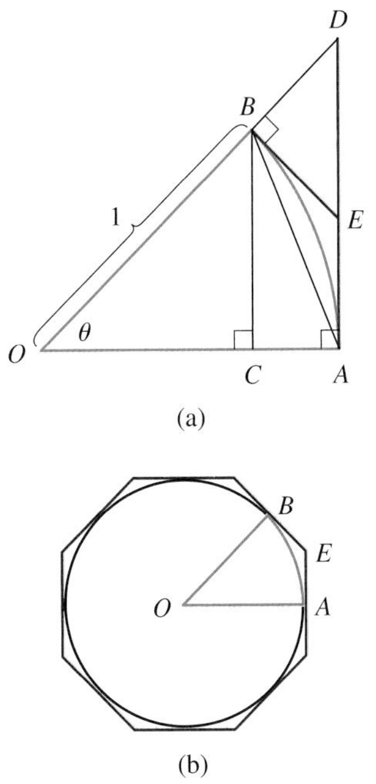

FIGURE 9

We can prove Equation 6 with help from the Squeeze Theorem. Assume first that θ lies between 0 and $\pi/2$. Figure 9(a) shows a sector of a circle with center O, central angle θ, and radius 1. BC is drawn perpendicular to OA. By the definition of radian measure, we have arc $AB = \theta$. Also $|BC| = |OB|\sin\theta = \sin\theta$. From the diagram we see that

$$|BC| < |AB| < \text{arc } AB$$

Therefore $$\sin\theta < \theta \qquad \text{so} \qquad \frac{\sin\theta}{\theta} < 1$$

Let the tangent lines at A and B intersect at E. You can see from Figure 9(b) that the circumference of a circle is smaller than the length of a circumscribed polygon, and so arc $AB < |AE| + |EB|$. Thus

$$\theta = \text{arc } AB < |AE| + |EB| < |AE| + |ED|$$
$$= |AD| = |OA|\tan\theta = \tan\theta$$

Therefore we have

$$\theta < \frac{\sin\theta}{\cos\theta} \qquad \text{and so} \qquad \cos\theta < \frac{\sin\theta}{\theta} < 1$$

We know that $\lim_{\theta \to 0} 1 = 1$ and $\lim_{\theta \to 0} \cos \theta = 1$, so by the Squeeze Theorem, we have

$$\lim_{\theta \to 0^+} \frac{\sin \theta}{\theta} = 1$$

But the function $(\sin \theta)/\theta$ is an even function, so its right and left limits must be equal. Hence, we have

$$\lim_{\theta \to 0} \frac{\sin \theta}{\theta} = 1$$

so we have proved Equation 6.

EXAMPLE 12 | Find $\lim_{x \to 0} \dfrac{\sin 7x}{4x}$.

SOLUTION In order to apply Equation 6, we first rewrite the function by multiplying and dividing by 7:

Note that $\sin 7x \neq 7 \sin x$.

$$\frac{\sin 7x}{4x} = \frac{7}{4}\left(\frac{\sin 7x}{7x}\right)$$

Notice that as $x \to 0$, we have $7x \to 0$, and so, by Equation 6 with $\theta = 7x$,

$$\lim_{x \to 0} \frac{\sin 7x}{7x} = \lim_{7x \to 0} \frac{\sin(7x)}{7x} = 1$$

Thus

$$\lim_{x \to 0} \frac{\sin 7x}{4x} = \lim_{x \to 0} \frac{7}{4}\left(\frac{\sin 7x}{7x}\right)$$

$$= \frac{7}{4} \lim_{x \to 0} \frac{\sin 7x}{7x} = \frac{7}{4} \cdot 1 = \frac{7}{4}$$

■

EXAMPLE 13 | Evaluate $\lim_{\theta \to 0} \dfrac{\cos \theta - 1}{\theta}$.

SOLUTION

We multiply numerator and denominator by $\cos \theta + 1$ in order to put the function in a form in which we can use the limits we know.

$$\lim_{\theta \to 0} \frac{\cos \theta - 1}{\theta} = \lim_{\theta \to 0}\left(\frac{\cos \theta - 1}{\theta} \cdot \frac{\cos \theta + 1}{\cos \theta + 1}\right) = \lim_{\theta \to 0} \frac{\cos^2\theta - 1}{\theta(\cos \theta + 1)}$$

$$= \lim_{\theta \to 0} \frac{-\sin^2\theta}{\theta(\cos \theta + 1)} = -\lim_{\theta \to 0}\left(\frac{\sin \theta}{\theta} \cdot \frac{\sin \theta}{\cos \theta + 1}\right)$$

$$= -\lim_{\theta \to 0} \frac{\sin \theta}{\theta} \cdot \lim_{\theta \to 0} \frac{\sin \theta}{\cos \theta + 1}$$

$$= -1 \cdot \left(\frac{0}{1 + 1}\right) = 0 \qquad \text{(by Equation 6)}$$

■

EXERCISES 2.4

1. Given that

$$\lim_{x\to 2} f(x) = 4 \qquad \lim_{x\to 2} g(x) = -2 \qquad \lim_{x\to 2} h(x) = 0$$

find the limits that exist. If the limit does not exist, explain why.

(a) $\lim_{x\to 2} [f(x) + 5g(x)]$ (b) $\lim_{x\to 2} [g(x)]^3$

(c) $\lim_{x\to 2} \sqrt{f(x)}$ (d) $\lim_{x\to 2} \dfrac{3f(x)}{g(x)}$

(e) $\lim_{x\to 2} \dfrac{g(x)}{h(x)}$ (f) $\lim_{x\to 2} \dfrac{g(x)h(x)}{f(x)}$

2. The graphs of f and g are given. Use them to evaluate each limit, if it exists. If the limit does not exist, explain why.

(a) $\lim_{x\to 2} [f(x) + g(x)]$ (b) $\lim_{x\to 1} [f(x) + g(x)]$

(c) $\lim_{x\to 0} [f(x)g(x)]$ (d) $\lim_{x\to -1} \dfrac{f(x)}{g(x)}$

(e) $\lim_{x\to 2} [x^3 f(x)]$ (f) $\lim_{x\to 1} \sqrt{3 + f(x)}$

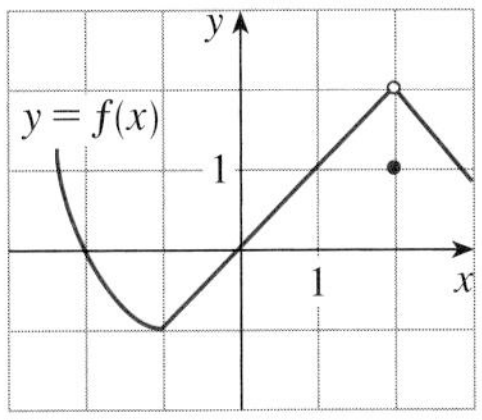

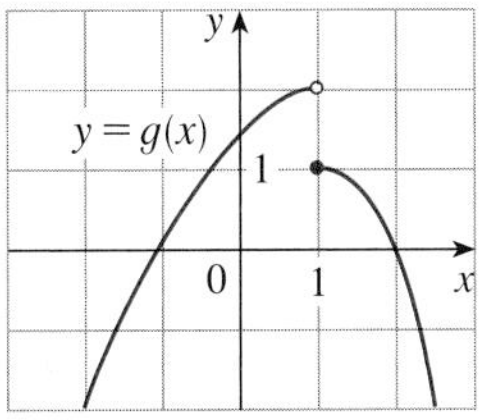

3–7 Evaluate the limit and justify each step by indicating the appropriate Limit Law(s).

3. $\lim_{x\to -2} (3x^4 + 2x^2 - x + 1)$ **4.** $\lim_{t\to -1} (t^2 + 1)^3(t + 3)^5$

5. $\lim_{x\to 2} \sqrt{\dfrac{2x^2 + 1}{3x - 2}}$ **6.** $\lim_{x\to 0} \dfrac{\cos^4 x}{5 + 2x^3}$

7. $\lim_{\theta\to \pi/2} \theta \sin\theta$

8. (a) What is wrong with the following equation?

$$\frac{x^2 + x - 6}{x - 2} = x + 3$$

(b) In view of part (a), explain why the equation

$$\lim_{x\to 2} \frac{x^2 + x - 6}{x - 2} = \lim_{x\to 2} (x + 3)$$

is correct.

9–24 Evaluate the limit, if it exists.

9. $\lim_{x\to 5} \dfrac{x^2 - 6x + 5}{x - 5}$ **10.** $\lim_{x\to 4} \dfrac{x^2 - 4x}{x^2 - 3x - 4}$

11. $\lim_{x\to 5} \dfrac{x^2 - 5x + 6}{x - 5}$ **12.** $\lim_{x\to -1} \dfrac{2x^2 + 3x + 1}{x^2 - 2x - 3}$

13. $\lim_{t\to -3} \dfrac{t^2 - 9}{2t^2 + 7t + 3}$ **14.** $\lim_{x\to -1} \dfrac{x^2 - 4x}{x^2 - 3x - 4}$

15. $\lim_{h\to 0} \dfrac{(4 + h)^2 - 16}{h}$ **16.** $\lim_{h\to 0} \dfrac{(2 + h)^3 - 8}{h}$

17. $\lim_{x\to -2} \dfrac{x + 2}{x^3 + 8}$ **18.** $\lim_{h\to 0} \dfrac{\sqrt{1 + h} - 1}{h}$

19. $\lim_{x\to -4} \dfrac{\dfrac{1}{4} + \dfrac{1}{x}}{4 + x}$ **20.** $\lim_{x\to -1} \dfrac{x^2 + 2x + 1}{x^4 - 1}$

21. $\lim_{x\to 16} \dfrac{4 - \sqrt{x}}{16x - x^2}$ **22.** $\lim_{t\to 0} \left(\dfrac{1}{t} - \dfrac{1}{t^2 + t}\right)$

23. $\lim_{t\to 0} \left(\dfrac{1}{t\sqrt{1 + t}} - \dfrac{1}{t}\right)$ **24.** $\lim_{x\to -4} \dfrac{\sqrt{x^2 + 9} - 5}{x + 4}$

25. (a) Estimate the value of

$$\lim_{x\to 0} \frac{x}{\sqrt{1 + 3x} - 1}$$

by graphing the function $f(x) = x/(\sqrt{1 + 3x} - 1)$.

(b) Make a table of values of $f(x)$ for x close to 0 and guess the value of the limit.

(c) Use the Limit Laws to prove that your guess is correct.

26. (a) Use a graph of

$$f(x) = \frac{\sqrt{3 + x} - \sqrt{3}}{x}$$

to estimate the value of $\lim_{x\to 0} f(x)$ to two decimal places.

(b) Use a table of values of $f(x)$ to estimate the limit to four decimal places.

(c) Use the Limit Laws to find the exact value of the limit.

27. Use the Squeeze Theorem to show that $\lim_{x\to 0} (x^2 \cos 20\pi x) = 0$. Illustrate by graphing the functions $f(x) = -x^2$, $g(x) = x^2 \cos 20\pi x$, and $h(x) = x^2$ on the same screen.

28. Use the Squeeze Theorem to show that

$$\lim_{x\to 0} \sqrt{x^3 + x^2}\, \sin\frac{\pi}{x} = 0$$

Illustrate by graphing the functions f, g, and h (in the notation of the Squeeze Theorem) on the same screen.

29. If $4x - 9 \leqslant f(x) \leqslant x^2 - 4x + 7$ for $x \geqslant 0$, find $\lim_{x\to 4} f(x)$.

30. If $2x \leqslant g(x) \leqslant x^4 - x^2 + 2$ for all x, evaluate $\lim_{x \to 1} g(x)$.

31. Prove that $\lim_{x \to 0} x^4 \cos \dfrac{2}{x} = 0$.

32. Gene regulation Genes produce molecules called mRNA that go on to produce proteins. High concentrations of protein inhibit the production of mRNA, leading to stable gene regulation. This process has been modeled (see Section 10.3) to show that the concentration of mRNA over time is given by the equation

$$m(t) = \tfrac{1}{2}e^{-t}(\sin t - \cos t) + \tfrac{1}{2}$$

(a) Evaluate $\lim_{t \to 0} m(t)$ and interpret your result.
(b) Use the Squeeze Theorem to evaluate $\lim_{t \to \infty} m(t)$ and interpret your result.

33–36 Find the limit, if it exists. If the limit does not exist, explain why.

33. $\lim_{x \to 3} \left(2x + |x - 3|\right)$

34. $\lim_{x \to -6} \dfrac{2x + 12}{|x + 6|}$

35. $\lim_{x \to 0^-} \left(\dfrac{1}{x} - \dfrac{1}{|x|}\right)$

36. $\lim_{x \to -2} \dfrac{2 - |x|}{2 + x}$

---

37. Let $g(x) = \dfrac{x^2 + x - 6}{|x - 2|}$.

(a) Find
 (i) $\lim_{x \to 2^+} g(x)$ (ii) $\lim_{x \to 2^-} g(x)$
(b) Does $\lim_{x \to 2} g(x)$ exist?
(c) Sketch the graph of g.

38. Let

$$f(x) = \begin{cases} x^2 + 1 & \text{if } x < 1 \\ (x - 2)^2 & \text{if } x \geqslant 1 \end{cases}$$

(a) Find $\lim_{x \to 1^-} f(x)$ and $\lim_{x \to 1^+} f(x)$.
(b) Does $\lim_{x \to 1} f(x)$ exist?
(c) Sketch the graph of f.

39–44 Find the limit.

39. $\lim_{x \to 0} \dfrac{\sin 3x}{x}$

40. $\lim_{x \to 0} \dfrac{\sin 4x}{\sin 6x}$

41. $\lim_{t \to 0} \dfrac{\tan 6t}{\sin 2t}$

42. $\lim_{t \to 0} \dfrac{\sin^2 3t}{t^2}$

43. $\lim_{\theta \to 0} \dfrac{\sin \theta}{\theta + \tan \theta}$

44. $\lim_{x \to 0} x \cot x$

---

45. (a) If p is a polynomial, show that $\lim_{x \to a} p(x) = p(a)$.
(b) If r is a rational function, use part (a) to show that $\lim_{x \to a} r(x) = r(a)$ for every number a in the domain of r.

46. In the theory of relativity, the Lorentz contraction formula

$$L = L_0\sqrt{1 - v^2/c^2}$$

expresses the length L of an object as a function of its velocity v with respect to an observer, where L_0 is the length of the object at rest and c is the speed of light. Find $\lim_{v \to c^-} L$ and interpret the result. Why is a left-hand limit necessary?

47. To prove that sine has the Direct Substitution Property we need to show that $\lim_{x \to a} \sin x = \sin a$ for every real number a. If we let $h = x - a$, then $x = a + h$ and $x \to a \iff h \to 0$. So an equivalent statement is that

$$\lim_{h \to 0} \sin(a + h) = \sin a$$

Use (5) to show that this is true.

48. Prove that cosine has the Direct Substitution Property.

49. If $\lim_{x \to 1} \dfrac{f(x) - 8}{x - 1} = 10$, find $\lim_{x \to 1} f(x)$.

50. If $\lim_{x \to 0} \dfrac{f(x)}{x^2} = 5$, find the following limits.

(a) $\lim_{x \to 0} f(x)$ (b) $\lim_{x \to 0} \dfrac{f(x)}{x}$

51. Is there a number a such that

$$\lim_{x \to -2} \frac{3x^2 + ax + a + 3}{x^2 + x - 2}$$

exists? If so, find the value of a and the value of the limit.

52. The figure shows a fixed circle C_1 with equation $(x - 1)^2 + y^2 = 1$ and a shrinking circle C_2 with radius r and center the origin. P is the point $(0, r)$, Q is the upper point of intersection of the two circles, and R is the point of intersection of the line PQ and the x-axis. What happens to R as C_2 shrinks, that is, as $r \to 0^+$?

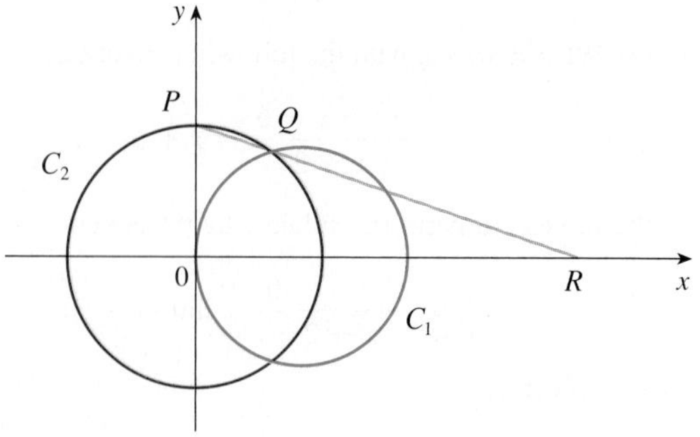

2.5 Continuity

We noticed in Section 2.4 that the limit of a function as x approaches a can often be found simply by calculating the value of the function at a. Functions with this property are called *continuous at a*. We will see that the mathematical definition of continuity corresponds closely with the meaning of the word *continuity* in everyday language. (A continuous process is one that takes place gradually, without interruption or abrupt change.)

Definition of a Continuous Function

(1) Definition A function f is **continuous at a number *a*** if

$$\lim_{x \to a} f(x) = f(a)$$

As illustrated in Figure 1, if f is continuous, then the points $(x, f(x))$ on the graph of f approach the point $(a, f(a))$ on the graph. So there is no gap in the curve.

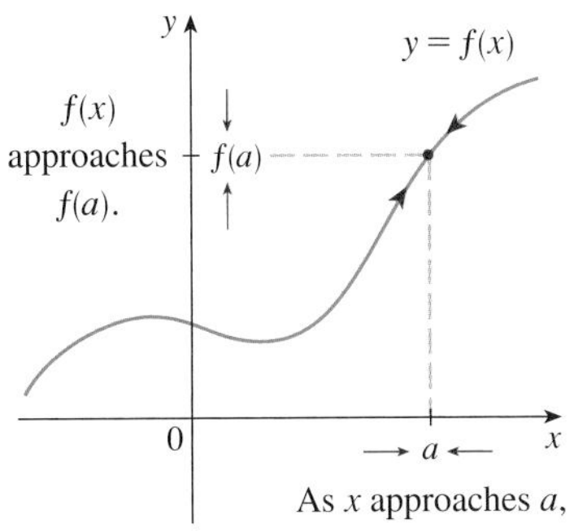

FIGURE 1

Notice that Definition 1 implicitly requires three things if f is continuous at a:

1. $f(a)$ is defined (that is, a is in the domain of f)
2. $\lim_{x \to a} f(x)$ exists
3. $\lim_{x \to a} f(x) = f(a)$

The definition says that f is continuous at a if $f(x)$ approaches $f(a)$ as x approaches a. Thus a continuous function f has the property that a small change in x produces only a small change in $f(x)$. In fact, the change in $f(x)$ can be kept as small as we please by keeping the change in x sufficiently small.

If f is defined near a (in other words, f is defined on an open interval containing a, except perhaps at a), we say that f is **discontinuous at *a*** (or f has a **discontinuity** at a) if f is not continuous at a.

Geometrically, you can think of a function that is continuous at every number in an interval as a function whose graph has no break in it. The graph can be drawn without removing your pen from the paper.

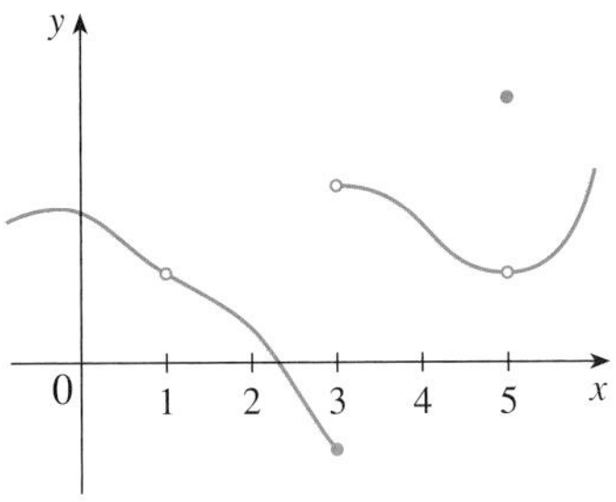

FIGURE 2

EXAMPLE 1 | Figure 2 shows the graph of a function f. At which numbers is f discontinuous? Why?

SOLUTION It looks as if there is a discontinuity when $a = 1$ because the graph has a break there. The official reason that f is discontinuous at 1 is that $f(1)$ is not defined.

The graph also has a break when $a = 3$, but the reason for the discontinuity is different. Here, $f(3)$ is defined, but $\lim_{x \to 3} f(x)$ does not exist (because the left and right limits are different). So f is discontinuous at 3.

What about $a = 5$? Here, $f(5)$ is defined and $\lim_{x \to 5} f(x)$ exists (because the left and right limits are the same). But

$$\lim_{x \to 5} f(x) \neq f(5)$$

So f is discontinuous at 5. ■

EXAMPLE 2 | Show that the function $f(x) = x^3 + 3x^2$ is continuous at every real number a.

SOLUTION Using the Limit Laws, we have

$$\lim_{x \to a} f(x) = \lim_{x \to a} (x^3 + 3x^2) = \lim_{x \to a} x^3 + 3 \lim_{x \to a} x^2$$
$$= a^3 + 3a^2 = f(a)$$

Therefore f is continuous at any number a. ■

EXAMPLE 3 | Where are each of the following functions discontinuous?

(a) $f(x) = \dfrac{x^2 - x - 2}{x - 2}$

(b) $f(x) = \begin{cases} \dfrac{1}{x^2} & \text{if } x \neq 0 \\ 1 & \text{if } x = 0 \end{cases}$

(c) $f(x) = \begin{cases} \dfrac{x^2 - x - 2}{x - 2} & \text{if } x \neq 2 \\ 1 & \text{if } x = 2 \end{cases}$

(d) $H(t) = \begin{cases} 0 & \text{if } t < 0 \\ 1 & \text{if } t \geqslant 0 \end{cases}$

SOLUTION

(a) Notice that $f(2)$ is not defined, so f is discontinuous at 2. Later we'll see why f is continuous at all other numbers.

(b) Here $f(0) = 1$ is defined but

$$\lim_{x \to 0} f(x) = \lim_{x \to 0} \frac{1}{x^2}$$

does not exist. (See Example 2.3.9.) So f is discontinuous at 0.

(c) Here $f(2) = 1$ is defined and

$$\lim_{x \to 2} f(x) = \lim_{x \to 2} \frac{x^2 - x - 2}{x - 2} = \lim_{x \to 2} \frac{(x - 2)(x + 1)}{x - 2} = \lim_{x \to 2} (x + 1) = 3$$

exists. But

$$\lim_{x \to 2} f(x) \neq f(2)$$

so f is not continuous at 2.

(d) H is called the Heaviside function after the electrical engineer Oliver Heaviside (1853–1925) and can be used to describe an electric current that is turned on at time $t = 0$. Because

$$\lim_{t \to 0^-} H(t) = 0 \qquad \text{and} \qquad \lim_{t \to 0^+} H(t) = 1$$

we know that $\lim_{t \to 0} H(t)$ does not exist. So H is discontinuous at 0. ■

Figure 3 shows the graphs of the functions in Example 3. In each case the graph can't be drawn without lifting the pen from the paper because a hole or break or jump occurs in the graph. The kind of discontinuity illustrated in parts (a) and (c) is called **removable** because we could remove the discontinuity by redefining f at just the single number 2.

[The function $g(x) = x + 1$ is continuous.] The discontinuity in part (b) is called an **infinite discontinuity**. The discontinuity in part (d) is called a **jump discontinuity** because the function "jumps" from one value to another.

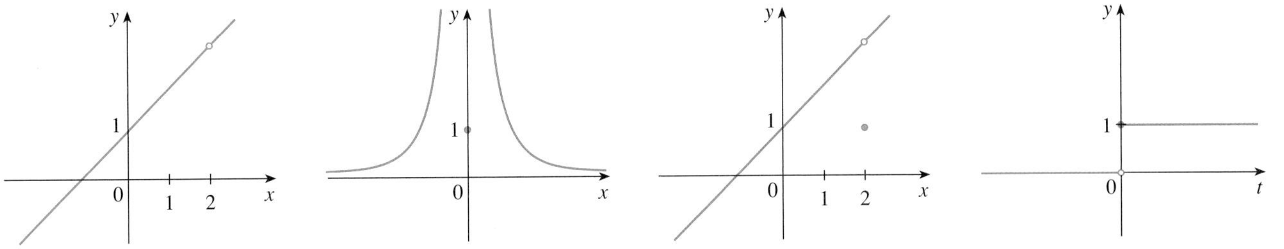

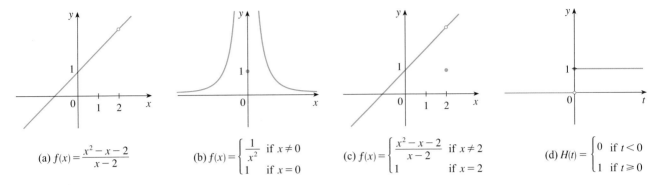

(a) $f(x) = \dfrac{x^2 - x - 2}{x - 2}$

(b) $f(x) = \begin{cases} \dfrac{1}{x^2} & \text{if } x \neq 0 \\ 1 & \text{if } x = 0 \end{cases}$

(c) $f(x) = \begin{cases} \dfrac{x^2 - x - 2}{x - 2} & \text{if } x \neq 2 \\ 1 & \text{if } x = 2 \end{cases}$

(d) $H(t) = \begin{cases} 0 & \text{if } t < 0 \\ 1 & \text{if } t \geqslant 0 \end{cases}$

FIGURE 3
Graphs of the functions in Example 3

Physical phenomena are usually continuous. For instance, the displacement or velocity of a vehicle varies continuously with time, as does a person's height. But discontinuities do occur in biology. Figure 4 shows the graph of the robin population from Example 2.3.8. The population function has a jump discontinuity whenever births or deaths occur, that is, at $t = 2$, 3, and 4.2.

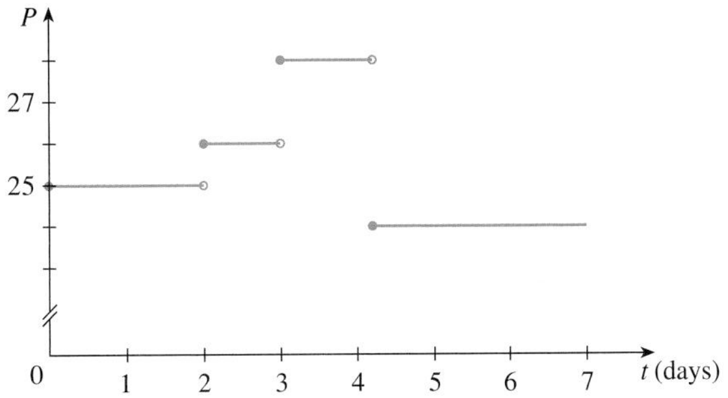

FIGURE 4

EXAMPLE 4 | BB **Population harvesting and collapse** Intensive harvesting of a population, such as occurs for some fish species, can cause population extinction. How this extinction occurs, however, depends on the nature of the harvesting. For example, two of the models for harvesting in the project on page 438 give the following equations for the population size N (measured in thousands) as a function of harvesting effort h, where $h \geqslant 0$:

Model 1 $$N(h) = \begin{cases} 2(1 - h) & \text{if } h \leqslant 1 \\ 0 & \text{if } h > 1 \end{cases}$$

Model 2 $$N(h) = \begin{cases} \frac{1}{2}\left(1 + \sqrt{9 - 8h}\right) & \text{if } h \leqslant \frac{9}{8} \\ 0 & \text{if } h > \frac{9}{8} \end{cases}$$

Plot the population size for both models and comment on their continuity properties at the point where extinction occurs. What is the biological significance of the difference in continuity between the two models?

SOLUTION The graphs of both models are shown in Figure 5. In model 1 the function is continuous and a small increase in harvesting always causes a small decrease in population size. Thus, if the harvesting pressure is only increased by small amounts, we would be able to "see extinction coming." But in model 2 a small increase in harvesting causes a discontinuous collapse in population size at $h = \frac{9}{8}$. In such cases a small increase in harvesting can cause the population to go extinct with no warning.

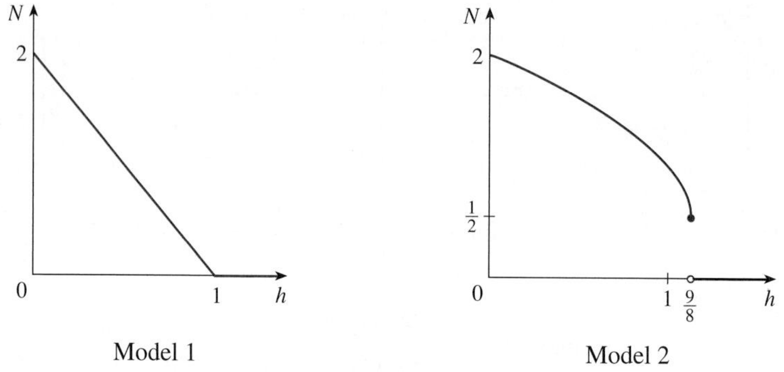

FIGURE 5

(2) Definition A function f is **continuous from the right at a number a** if

$$\lim_{x \to a^+} f(x) = f(a)$$

and f is **continuous from the left at a** if

$$\lim_{x \to a^-} f(x) = f(a)$$

Notice from Figure 4 that $P(t)$ is continuous from the right, but not from the left, at each of the jump discontinuities.

(3) Definition A function f is **continuous on an interval** if it is continuous at every number in the interval. (If f is defined only on one side of an endpoint of the interval, we understand *continuous* at the endpoint to mean *continuous from the right* or *continuous from the left.*)

Instead of using Definitions 1, 2, and 3 to verify the continuity of a function as we did in Example 2, it is often convenient to use the next theorem, which shows how to build complicated continuous functions from simple ones.

(4) Theorem If f and g are continuous at a and c is a constant, then the following functions are also continuous at a:

1. $f + g$ **2.** $f - g$ **3.** cf

4. fg **5.** $\dfrac{f}{g}$ if $g(a) \neq 0$

PROOF Each of the five parts of this theorem follows from the corresponding Limit Law in Section 2.4. For instance, we give the proof of part 1. Since f and g are continuous at a, we have

$$\lim_{x \to a} f(x) = f(a) \qquad \text{and} \qquad \lim_{x \to a} g(x) = g(a)$$

Therefore

$$\begin{aligned} \lim_{x \to a} (f + g)(x) &= \lim_{x \to a} [f(x) + g(x)] \\ &= \lim_{x \to a} f(x) + \lim_{x \to a} g(x) \qquad \text{(by Law 1)} \\ &= f(a) + g(a) \\ &= (f + g)(a) \end{aligned}$$

This shows that $f + g$ is continuous at a. ■

It follows from Theorem 4 and Definition 3 that if f and g are continuous on an interval, then so are the functions $f + g$, $f - g$, cf, fg, and (if g is never 0) f/g.

■ Which Functions Are Continuous?

It will be useful in our future work to know which functions are continuous and which are not. To that end let's begin to compile a list of functions that are known to be continuous. The following theorem was stated in Section 2.4 as the Direct Substitution Property.

(5) Theorem

(a) Any polynomial is continuous everywhere; that is, it is continuous on $\mathbb{R} = (-\infty, \infty)$.

(b) Any rational function is continuous wherever it is defined; that is, it is continuous on its domain.

PROOF

(a) A polynomial is a function of the form

$$P(x) = c_n x^n + c_{n-1} x^{n-1} + \cdots + c_1 x + c_0$$

where $c_0, c_1, \ldots, c_n$ are constants. We know that

$$\lim_{x \to a} c_0 = c_0 \qquad \text{(by Law 7)}$$

and

$$\lim_{x \to a} x^m = a^m \qquad m = 1, 2, \ldots, n \qquad \text{(by Law 9)}$$

This equation is precisely the statement that the function $f(x) = x^m$ is a continuous function. Thus, by part 3 of Theorem 4, the function $g(x) = cx^m$ is continuous. Since P is a sum of functions of this form and a constant function, it follows from part 1 of Theorem 4 that P is continuous.

(b) A rational function is a function of the form

$$f(x) = \frac{P(x)}{Q(x)}$$

where P and Q are polynomials. The domain of f is $D = \{x \in \mathbb{R} \mid Q(x) \neq 0\}$. We know from part (a) that P and Q are continuous everywhere. Thus, by part 5 of Theorem 4, f is continuous at every number in D. ■

As an illustration of Theorem 5, observe that the volume of a sphere varies continuously with its radius because the formula $V(r) = \frac{4}{3}\pi r^3$ shows that V is a polynomial function of r. Likewise, if a ball is thrown vertically into the air with a velocity of 50 ft/s, then the height of the ball in feet t seconds later is given by the formula $h = 50t - 16t^2$. Again this is a polynomial function, so the height is a continuous function of the elapsed time.

Knowledge of which functions are continuous enables us to evaluate some limits very quickly, as the following example shows. Compare it with Example 2.4.2(b).

EXAMPLE 5 | Find $\displaystyle\lim_{x \to -2} \frac{x^3 + 2x^2 - 1}{5 - 3x}$.

SOLUTION The function

$$f(x) = \frac{x^3 + 2x^2 - 1}{5 - 3x}$$

is rational, so by Theorem 5 it is continuous on its domain, which is $\{x \mid x \neq \frac{5}{3}\}$. Therefore

$$\lim_{x \to -2} \frac{x^3 + 2x^2 - 1}{5 - 3x} = \lim_{x \to -2} f(x) = f(-2)$$

$$= \frac{(-2)^3 + 2(-2)^2 - 1}{5 - 3(-2)} = -\frac{1}{11}$$ ■

It turns out that most of the familiar functions are continuous at every number in their domains. For instance, Limit Law 10 (page 127) is exactly the statement that root functions are continuous.

From the appearance of the graphs of the sine and cosine functions (Figure 1.2.19), we would certainly guess that they are continuous. In fact we showed in Section 2.4 that they both satisfy the Direct Substitution Property, so indeed they are continuous everywhere.

It follows from part 5 of Theorem 4 that

$$\tan x = \frac{\sin x}{\cos x}$$

is continuous except where $\cos x = 0$. This happens when x is an odd integer multiple of $\pi/2$, so $y = \tan x$ has infinite discontinuities when $x = \pm\pi/2, \pm 3\pi/2, \pm 5\pi/2$, and so on (see Figure 6).

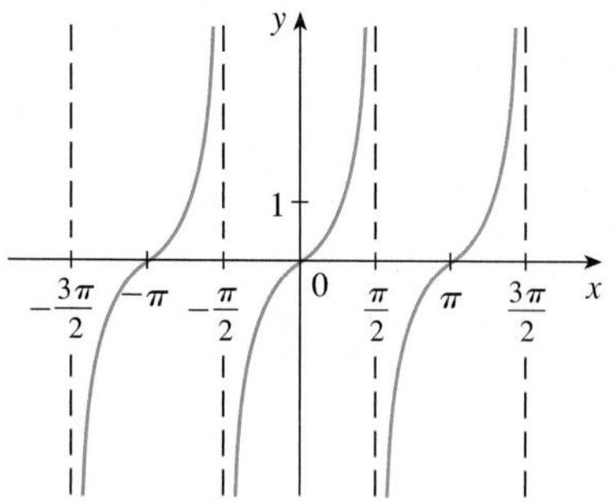

FIGURE 6
$y = \tan x$

In Section 1.4 we defined the exponential function $y = b^x$ so as to fill in the holes in the graph of $y = b^x$ where x is rational. In other words, the very definition of $y = b^x$ makes it a continuous function on $\mathbb{R}$. The inverse function of any continuous one-to-one function is also continuous. (The graph of f^{-1} is obtained by reflecting the graph of f

about the line $y = x$. So if the graph of f has no break in it, neither does the graph of f^{-1}.) Therefore the function $y = \log_b x$ is continuous on $(0, \infty)$ because it is the inverse function of $y = b^x$.

(6) Theorem The following types of functions are continuous at every number in their domains:

polynomials	rational functions
power and root functions	trigonometric functions
exponential functions	logarithmic functions

EXAMPLE 6 | Where is the function $f(x) = \dfrac{\ln x + e^x}{x^2 - 1}$ continuous?

SOLUTION We know from Theorem 6 that the function $y = \ln x$ is continuous for $x > 0$ and $y = e^x$ is continuous on $\mathbb{R}$. Thus, by part 1 of Theorem 4, $y = \ln x + e^x$ is continuous on $(0, \infty)$. The denominator, $y = x^2 - 1$, is a polynomial, so it is continuous everywhere. Therefore, by part 5 of Theorem 4, f is continuous at all positive numbers x except where $x^2 - 1 = 0$. So f is continuous on the intervals $(0, 1)$ and $(1, \infty)$. ■

EXAMPLE 7 | Evaluate $\displaystyle\lim_{x \to \pi} \frac{\sin x}{2 + \cos x}$.

SOLUTION Theorem 6 tells us that $y = \sin x$ is continuous. The function in the denominator, $y = 2 + \cos x$, is the sum of two continuous functions and is therefore continuous. Notice that this function is never 0 because $\cos x \geqslant -1$ for all x and so $2 + \cos x > 0$ everywhere. Thus the ratio

$$f(x) = \frac{\sin x}{2 + \cos x}$$

is continuous everywhere. Hence, by the definition of a continuous function,

$$\lim_{x \to \pi} \frac{\sin x}{2 + \cos x} = \lim_{x \to \pi} f(x) = f(\pi) = \frac{\sin \pi}{2 + \cos \pi} = \frac{0}{2 - 1} = 0$$ ■

Another way of combining continuous functions f and g to get a new continuous function is to form the composite function $f \circ g$. This fact is a consequence of the following theorem.

This theorem says that a limit symbol can be moved through a function symbol if the function is continuous and the limit exists. In other words, the order of these two symbols can be reversed.

(7) Theorem If f is continuous at b and $\displaystyle\lim_{x \to a} g(x) = b$, then $\displaystyle\lim_{x \to a} f(g(x)) = f(b)$. In other words,

$$\lim_{x \to a} f(g(x)) = f\left(\lim_{x \to a} g(x)\right)$$

Intuitively, Theorem 7 is reasonable because if x is close to a, then $g(x)$ is close to b, and since f is continuous at b, if $g(x)$ is close to b, then $f(g(x))$ is close to $f(b)$.

(8) Theorem If g is continuous at a and f is continuous at $g(a)$, then the composite function $f \circ g$ given by $(f \circ g)(x) = f(g(x))$ is continuous at a.

This theorem is often expressed informally by saying "a continuous function of a continuous function is a continuous function."

PROOF Since g is continuous at a, we have

$$\lim_{x \to a} g(x) = g(a)$$

Since f is continuous at $b = g(a)$, we can apply Theorem 7 to obtain

$$\lim_{x \to a} f(g(x)) = f(g(a))$$

which is precisely the statement that the function $h(x) = f(g(x))$ is continuous at a; that is, $f \circ g$ is continuous at a. ■

EXAMPLE 8 | Where are the following functions continuous?

(a) $h(x) = \sin(x^2)$ (b) $F(x) = \ln(1 + \cos x)$

SOLUTION

(a) We have $h(x) = f(g(x))$, where

$$g(x) = x^2 \qquad \text{and} \qquad f(x) = \sin x$$

Now g is continuous on $\mathbb{R}$ since it is a polynomial, and f is also continuous everywhere. Thus $h = f \circ g$ is continuous on $\mathbb{R}$ by Theorem 8.

(b) We know from Theorem 6 that $f(x) = \ln x$ is continuous and $g(x) = 1 + \cos x$ is continuous (because both $y = 1$ and $y = \cos x$ are continuous). Therefore, by Theorem 8, $F(x) = f(g(x))$ is continuous wherever it is defined. Now $\ln(1 + \cos x)$ is defined when $1 + \cos x > 0$. So it is undefined when $\cos x = -1$, and this happens when $x = \pm\pi, \pm 3\pi, \ldots$. Thus F has discontinuities when x is an odd multiple of π and is continuous on the intervals between these values (see Figure 7). ■

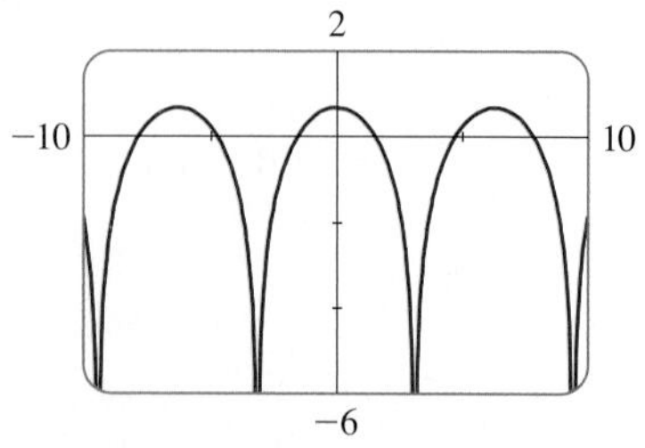

FIGURE 7
$y = \ln(1 + \cos x)$

An important property of continuous functions is expressed by the following theorem, whose proof is found in more advanced books on calculus.

(9) The Intermediate Value Theorem Suppose that f is continuous on the closed interval $[a, b]$ and let N be any number between $f(a)$ and $f(b)$, where $f(a) \neq f(b)$. Then there exists a number c in (a, b) such that $f(c) = N$.

The Intermediate Value Theorem states that a continuous function takes on every intermediate value between the function values $f(a)$ and $f(b)$. It is illustrated by Figure 8. Note that the value N can be taken on once [as in part (a)] or more than once [as in part (b)].

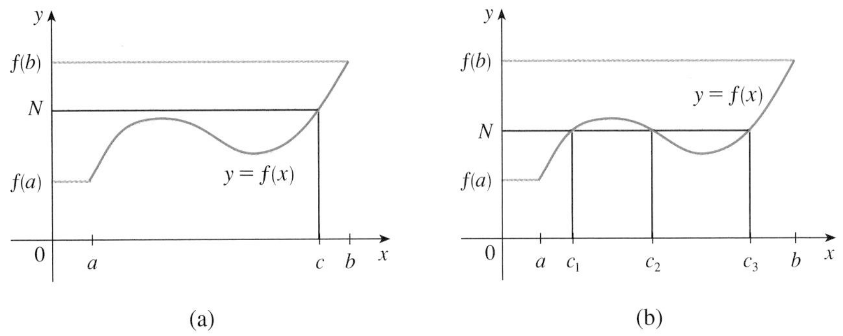

FIGURE 8

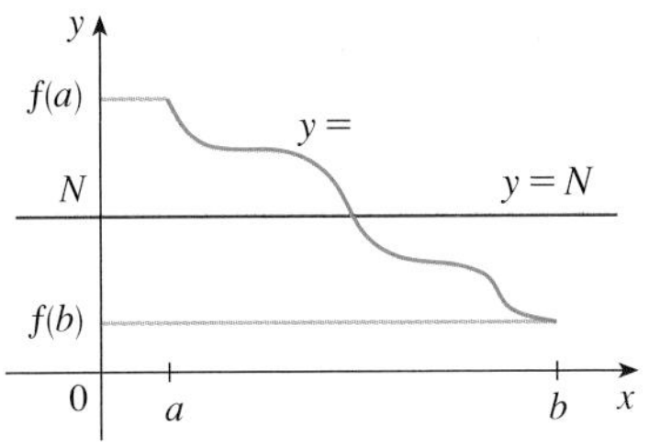

FIGURE 9

If we think of a continuous function as a function whose graph has no hole or break, then it is easy to believe that the Intermediate Value Theorem is true. In geometric terms it says that if any horizontal line $y = N$ is given between $y = f(a)$ and $y = f(b)$ as in Figure 9, then the graph of f can't jump over the line. It must intersect $y = N$ somewhere.

EXAMPLE 9 | **BB** Population harvesting and collapse (continued) Model 1 in Example 4 for the size of a population as a function of harvesting rate is continuous for all harvesting rates, whereas model 2 has a jump discontinuity at the harvesting rate $h = \frac{9}{8}$. Provide a biological interpretation of the Intermediate Value Theorem in the context of these two models.

SOLUTION Since model 1 is continuous for all values of h, there is a harvesting rate that results in any population size between 2 and 0 (extinction). The discontinuity in model 2, however, means that there are some population sizes between 2 and 0 that cannot be obtained no matter what harvesting rate is used (for example, there is no harvesting rate h that results in a population size of $N = \frac{1}{4}$). This can have important conservation implications. For instance, suppose conservation biologists decide not to impose harvesting regulations unless the population has declined to a critical value of $N = \frac{1}{4}$. If model 1 is appropriate, then such regulations can prevent extinction. But if model 2 is appropriate, then the critical population size can never be reached no matter what the harvesting rate. Instead, the population will undergo irreversible collapse as harvesting rates increase before the population ever declines to the critical level. ■

One use of the Intermediate Value Theorem is in locating roots of equations, as shown in the following example.

EXAMPLE 10 | Show that there is a root of the equation

$$4x^3 - 6x^2 + 3x - 2 = 0$$

between 1 and 2.

SOLUTION Let $f(x) = 4x^3 - 6x^2 + 3x - 2$. We are looking for a solution of the given equation, that is, a number c between 1 and 2 such that $f(c) = 0$. Therefore we take $a = 1$, $b = 2$, and $N = 0$ in Theorem 9. We have

$$f(1) = 4 - 6 + 3 - 2 = -1 < 0$$

and

$$f(2) = 32 - 24 + 6 - 2 = 12 > 0$$

Thus $f(1) < 0 < f(2)$; that is, $N = 0$ is a number between $f(1)$ and $f(2)$. Now f is continuous since it is a polynomial, so the Intermediate Value Theorem says there is a number c between 1 and 2 such that $f(c) = 0$. In other words, the equation $4x^3 - 6x^2 + 3x - 2 = 0$ has at least one root c in the interval $(1, 2)$.

In fact, we can locate a root more precisely by using the Intermediate Value Theorem again. Since

$$f(1.2) = -0.128 < 0 \qquad \text{and} \qquad f(1.3) = 0.548 > 0$$

a root must lie between 1.2 and 1.3. A calculator gives, by trial and error,

$$f(1.22) = -0.007008 < 0 \qquad \text{and} \qquad f(1.23) = 0.056068 > 0$$

so a root lies in the interval $(1.22, 1.23)$. ■

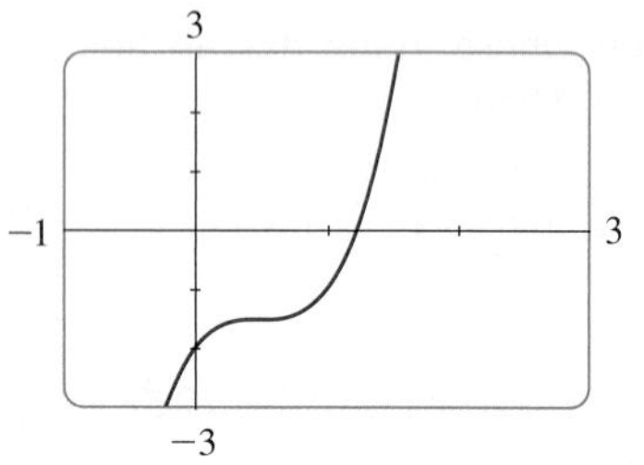

FIGURE 10

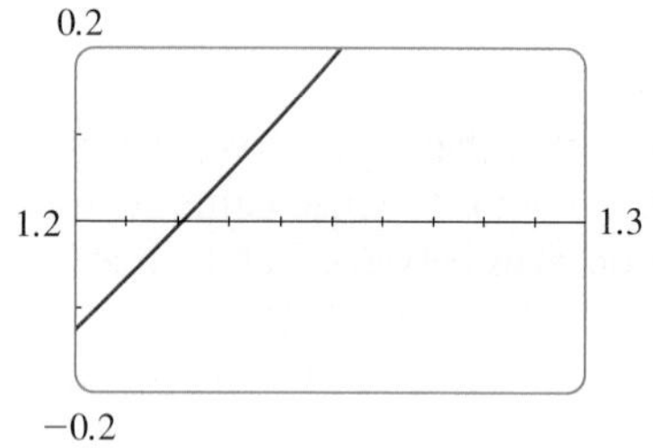

FIGURE 11

We can use a graphing calculator or computer to illustrate the use of the Intermediate Value Theorem in Example 10. Figure 10 shows the graph of f in the viewing rectangle $[-1, 3]$ by $[-3, 3]$ and you can see that the graph crosses the x-axis between 1 and 2. Figure 11 shows the result of zooming in to the viewing rectangle $[1.2, 1.3]$ by $[-0.2, 0.2]$.

In fact, the Intermediate Value Theorem plays a role in the very way these graphing devices work. A computer calculates a finite number of points on the graph and turns on the pixels that contain these calculated points. It assumes that the function is continuous and takes on all the intermediate values between two consecutive points. The computer therefore connects the pixels by turning on the intermediate pixels.

■ Approximating Discontinuous Functions by Continuous Ones

In calculus it's desirable to work with continuous functions, but the functions that arise in biology are often discontinuous. For instance, if $n = P(t)$ is the number of individuals in an animal or plant population at time t, then P is discontinuous whenever a birth or death occurs, as we saw in Figure 4. For a large animal or plant population, however, we can often replace the population function by a continuous function that approximates P. Figure 12 shows the graph of a population function, which is actually a step function, together with the graph of a continuous approximating function.

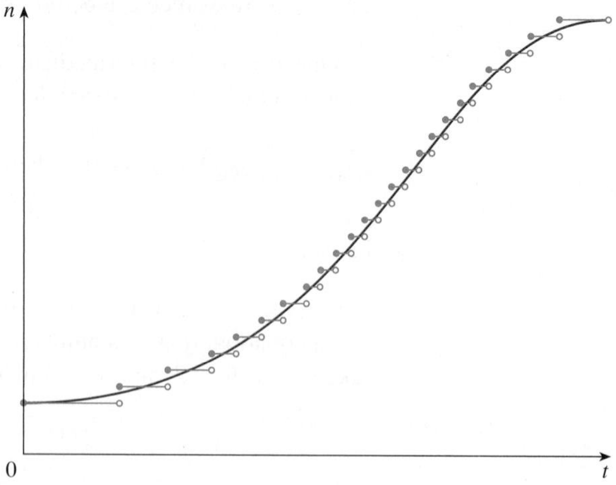

FIGURE 12

With the large numbers involved in the population of the world or HIV replication within a patient, for example, it is reasonable to model the population function with a continuous function. Indeed, in Section 1.4 we obtained the exponential model

$$P(t) \approx (1436.53) \cdot (1.01395)^t$$

for the world population after 1900 (corresponding to $t = 0$). In general, the exponential functions

$$f(t) = n_0 b^t \qquad \text{and} \qquad g(t) = n_0 e^{kt}$$

where b and k are constants, are continuous functions that are often used to model population growth, at least in the initial stages.

EXERCISES 2.5

1. Write an equation that expresses the fact that a function f is continuous at the number 4.

2. If f is continuous on $(-\infty, \infty)$, what can you say about its graph?

3. (a) From the graph of f, state the numbers at which f is discontinuous and explain why.
(b) For each of the numbers stated in part (a), determine whether f is continuous from the right, or from the left, or neither.

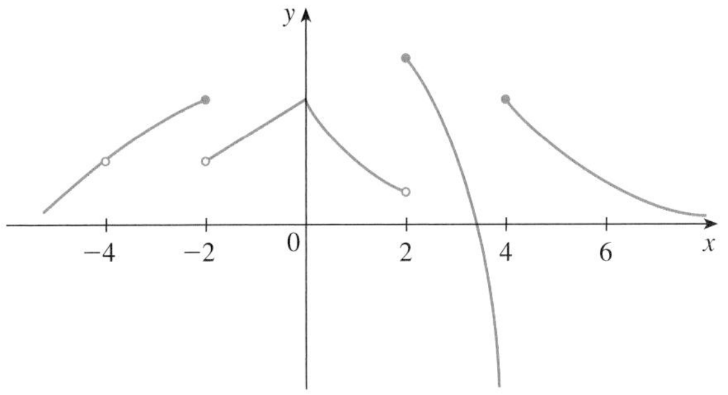

4. From the graph of g, state the intervals on which g is continuous.

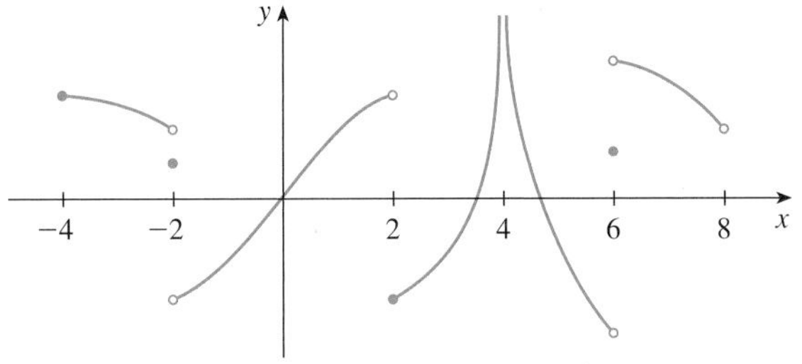

5–8 Sketch the graph of a function f that is continuous except for the stated discontinuity.

5. Discontinuous, but continuous from the right, at 2

6. Discontinuities at -1 and 4, but continuous from the left at -1 and from the right at 4

7. Removable discontinuity at 3, jump discontinuity at 5

8. Neither left nor right continuous at -2, continuous only from the left at 2

9. Drug concentration A patient is injected with a drug every 12 hours. The graph shows the concentration $C(t)$ of the drug in the bloodstream after t hours.
(a) At what values of t does C have discontinuities?
(b) What type of discontinuity does C have?

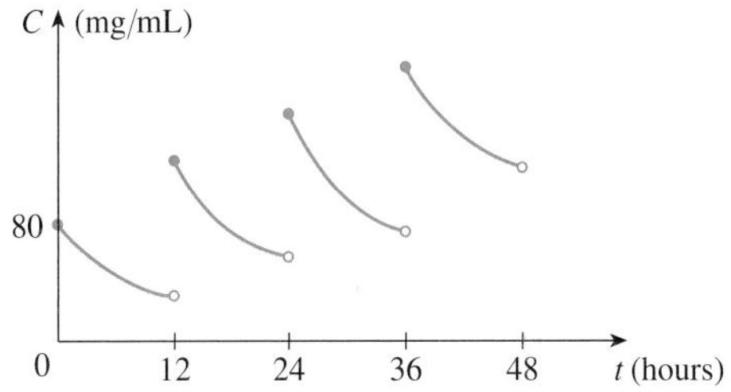

10. Squirrel population The graph of a population $P(t)$ of squirrels is shown. Identify the discontinuities of P and comment on when and why they occur.

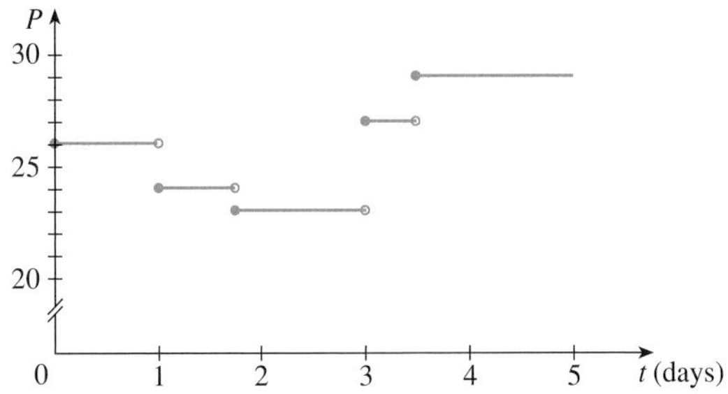

11. A parking lot charges \$3 for the first hour (or part of an hour) and \$2 for each succeeding hour (or part), up to a daily maximum of \$10.
(a) Sketch a graph of the cost of parking at this lot as a function of the time parked there.
(b) Discuss the discontinuities of this function and their significance to someone who parks in the lot.

12. Explain why each function is continuous or discontinuous.
(a) The temperature in New York City as a function of time
(b) The population of New York City as a function of time
(c) The temperature at a specific time as a function of the distance due west from New York City
(d) The altitude above sea level as a function of the distance due west from New York City
(e) The cost of a taxi ride as a function of the distance traveled

13. If f and g are continuous functions with $f(3) = 5$ and $\lim_{x\to 3}[2f(x) - g(x)] = 4$, find $g(3)$.

14–15 Use the definition of continuity and the properties of limits to show that the function is continuous at the given number a.

14. $f(x) = 3x^4 - 5x + \sqrt[3]{x^2 + 4}, \quad a = 2$

15. $f(x) = (x + 2x^3)^4, \quad a = -1$

16. Use the definition of continuity and the properties of limits to show that the following function is continuous on the interval $(2, \infty)$.

$$f(x) = \frac{2x + 3}{x - 2}$$

17–20 Explain why the function is discontinuous at the given number a. Sketch the graph of the function.

17. $f(x) = \begin{cases} e^x & \text{if } x < 0 \\ x^2 & \text{if } x \geqslant 0 \end{cases} \qquad a = 0$

18. $f(x) = \begin{cases} \dfrac{x^2 - x}{x^2 - 1} & \text{if } x \neq 1 \\ 1 & \text{if } x = 1 \end{cases} \qquad a = 1$

19. $f(x) = \begin{cases} \cos x & \text{if } x < 0 \\ 0 & \text{if } x = 0 \\ 1 - x^2 & \text{if } x > 0 \end{cases} \qquad a = 0$

20. $f(x) = \begin{cases} \dfrac{2x^2 - 5x - 3}{x - 3} & \text{if } x \neq 3 \\ 6 & \text{if } x = 3 \end{cases} \qquad a = 3$

21–26 Explain, using Theorems 4, 5, 6, and 8, why the function is continuous at every number in its domain. State the domain.

21. $R(x) = x^2 + \sqrt{2x - 1}$

22. $G(x) = \sqrt[3]{x}\,(1 + x^3)$

23. $L(t) = e^{-5t} \cos 2\pi t$

24. $h(x) = \dfrac{\sin x}{x + 1}$

25. $G(t) = \ln(t^4 - 1)$

26. $F(x) = \sin(\cos(\sin x))$

27–28 Locate the discontinuities of the function and illustrate by graphing.

27. $y = \dfrac{1}{1 + e^{1/x}}$

28. $y = \ln(\tan^2 x)$

29–32 Use continuity to evaluate the limit.

29. $\displaystyle\lim_{x\to 4} \frac{5 + \sqrt{x}}{\sqrt{5 + x}}$

30. $\displaystyle\lim_{x\to \pi} \sin(x + \sin x)$

31. $\displaystyle\lim_{x\to 1} e^{x^2 - x}$

32. Drug resistance As we have previously noted (page 25), if p is the current frequency of the resistance gene in a model for the spread of drug resistance, then the frequency in the next generation is

$$\frac{p^2 - 2p}{p^2 - 2}$$

What is the limit of this function as $p \to \frac{1}{2}$?

33–34 Show that f is continuous on $(-\infty, \infty)$.

33. $f(x) = \begin{cases} x^2 & \text{if } x < 1 \\ \sqrt{x} & \text{if } x \geqslant 1 \end{cases}$

34. $f(x) = \begin{cases} \sin x & \text{if } x < \pi/4 \\ \cos x & \text{if } x \geqslant \pi/4 \end{cases}$

35. Find the numbers at which the function

$$f(x) = \begin{cases} x + 2 & \text{if } x < 0 \\ e^x & \text{if } 0 \leqslant x \leqslant 1 \\ 2 - x & \text{if } x > 1 \end{cases}$$

is discontinuous. At which of these points is f continuous from the right, from the left, or neither? Sketch the graph of f.

36. The gravitational force exerted by the planet Earth on a unit mass at a distance r from the center of the planet is

$$F(r) = \begin{cases} \dfrac{GMr}{R^3} & \text{if } r < R \\ \dfrac{GM}{r^2} & \text{if } r \geqslant R \end{cases}$$

where M is the mass of Earth, R is its radius, and G is the gravitational constant. Is F a continuous function of r?

37. For what value of the constant c is the function f continuous on $(-\infty, \infty)$?

$$f(x) = \begin{cases} cx^2 + 2x & \text{if } x < 2 \\ x^3 - cx & \text{if } x \geqslant 2 \end{cases}$$

38. Suppose that a function f is continuous on $[0, 1]$ except at 0.25 and that $f(0) = 1$ and $f(1) = 3$. Let $N = 2$. Sketch two possible graphs of f, one showing that f might not satisfy the conclusion of the Intermediate Value Theorem and one showing that f might still satisfy the conclusion of the Intermediate Value Theorem (even though it doesn't satisfy the hypothesis).

39. If $f(x) = x^2 + 10 \sin x$, show that there is a number c such that $f(c) = 1000$.

40. Suppose f is continuous on $[1, 5]$ and the only solutions of the equation $f(x) = 6$ are $x = 1$ and $x = 4$. If $f(2) = 8$, explain why $f(3) > 6$.

41–44 Use the Intermediate Value Theorem to show that there is a solution of the given equation in the specified interval.

41. $x^4 + x - 3 = 0, \quad (1, 2)$

42. $\sqrt[3]{x} = 1 - x, \quad (0, 1)$

43. $e^x = 3 - 2x, \quad (0, 1)$

44. $\sin x = x^2 - x, \quad (1, 2)$

45–46 (a) Prove that the equation has at least one real root. (b) Use your calculator to find an interval of length 0.01 that contains a root.

45. $\cos x = x^3$

46. $\ln x = 3 - 2x$

47–48 (a) Prove that the equation has at least one real solution. (b) Use your graphing device to find the solution correct to three decimal places.

47. $100e^{-x/100} = 0.01x^2$

48. $\sqrt{x - 5} = \dfrac{1}{x + 3}$

49. Is there a number that is exactly 1 more than its cube?

50. Show that the function

$$f(x) = \begin{cases} x^4 \sin(1/x) & \text{if } x \neq 0 \\ 0 & \text{if } x = 0 \end{cases}$$

is continuous on $(-\infty, \infty)$.

51. A Tibetan monk leaves the monastery at 7:00 AM and takes his usual path to the top of the mountain, arriving at 7:00 PM. The following morning, he starts at 7:00 AM at the top and takes the same path back, arriving at the monastery at 7:00 PM. Use the Intermediate Value Theorem to show that there is a point on the path that the monk will cross at exactly the same time of day on both days.

Chapter 2 REVIEW

CONCEPT CHECK

1. (a) What is a convergent sequence?
(b) What does $\lim_{n\to\infty} a_n = 3$ mean?

2. What is $\lim\limits_{n\to\infty} r^n$ in the following three cases?
(a) $0 < r < 1$ (b) $r = 1$ (c) $r > 1$

3. (a) What is the sum of the finite geometric series $a + ar + ar^2 + \cdots + ar^n$?
(b) If $-1 < r < 1$, what is the sum of the infinite geometric series $a + ar + ar^2 + \cdots + ar^n + \cdots$?

4. Explain what each of the following means and illustrate with a sketch.
(a) $\lim\limits_{x\to a} f(x) = L$
(b) $\lim\limits_{x\to a^+} f(x) = L$
(c) $\lim\limits_{x\to a^-} f(x) = L$
(d) $\lim\limits_{x\to a} f(x) = \infty$
(e) $\lim\limits_{x\to\infty} f(x) = L$

5. State the following Limit Laws for functions.
(a) Sum Law
(b) Difference Law
(c) Constant Multiple Law
(d) Product Law
(e) Quotient Law
(f) Power Law
(g) Root Law

6. What does the Squeeze Theorem say?

7. (a) What does it mean to say that the line $x = a$ is a vertical asymptote of the curve $y = f(x)$? Draw curves to illustrate the various possibilities.
(b) What does it mean to say that the line $y = L$ is a horizontal asymptote of the curve $y = f(x)$? Draw curves to illustrate the various possibilities.

8. Which of the following curves have vertical asymptotes? Which have horizontal asymptotes?
(a) $y = x^4$
(b) $y = \sin x$
(c) $y = \tan x$
(d) $y = e^x$
(e) $y = \ln x$
(f) $y = 1/x$
(g) $y = \sqrt{x}$

9. (a) What does it mean for f to be continuous at a?
(b) What does it mean for f to be continuous on the interval $(-\infty, \infty)$? What can you say about the graph of such a function?

10. What does the Intermediate Value Theorem say?

Answers to the Concept Check can be found on the back endpapers.

TRUE-FALSE QUIZ

Determine whether the statement is true or false. If it is true, explain why. If it is false, explain why or give an example that disproves the statement.

1. If $\lim_{n\to\infty} a_n = L$, then $\lim_{n\to\infty} a_{2n+1} = L$.

2. $0.99999\ldots = 1$

3. $\displaystyle\lim_{x\to 4}\left(\frac{2x}{x-4} - \frac{8}{x-4}\right) = \lim_{x\to 4}\frac{2x}{x-4} - \lim_{x\to 4}\frac{8}{x-4}$

4. $\displaystyle\lim_{x\to 1}\frac{x^2+6x-7}{x^2+5x-6} = \frac{\lim_{x\to 1}(x^2+6x-7)}{\lim_{x\to 1}(x^2+5x-6)}$

5. $\displaystyle\lim_{x\to 1}\frac{x-3}{x^2+2x-4} = \frac{\lim_{x\to 1}(x-3)}{\lim_{x\to 1}(x^2+2x-4)}$

6. If $\lim_{x\to 5} f(x) = 2$ and $\lim_{x\to 5} g(x) = 0$, then $\lim_{x\to 5}[f(x)/g(x)]$ does not exist.

7. If $\lim_{x\to 5} f(x) = 0$ and $\lim_{x\to 5} g(x) = 0$, then $\lim_{x\to 5}[f(x)/g(x)]$ does not exist.

8. If $\lim_{x\to 6}[f(x)\,g(x)]$ exists, then the limit must be $f(6)\,g(6)$.

9. If p is a polynomial, then $\lim_{x\to b} p(x) = p(b)$.

10. If $\lim_{x\to 0} f(x) = \infty$ and $\lim_{x\to 0} g(x) = \infty$, then $\lim_{x\to 0}[f(x) - g(x)] = 0$.

11. A function can have two different horizontal asymptotes.

12. If f has domain $[0, \infty)$ and has no horizontal asymptote, then $\lim_{x\to\infty} f(x) = \infty$ or $\lim_{x\to\infty} f(x) = -\infty$.

13. If the line $x = 1$ is a vertical asymptote of $y = f(x)$, then f is not defined at 1.

14. If $f(1) > 0$ and $f(3) < 0$, then there exists a number c between 1 and 3 such that $f(c) = 0$.

15. If f is continuous at 5 and $f(5) = 2$ and $f(4) = 3$, then $\lim_{x\to 2} f(4x^2 - 11) = 2$.

16. If f is continuous on $[-1, 1]$ and $f(-1) = 4$ and $f(1) = 3$, then there exists a number r such that $|r| < 1$ and $f(r) = \pi$.

EXERCISES

1–4 Determine whether the sequence is convergent or divergent. If it is convergent, find its limit.

1. $a_n = \dfrac{2+n^3}{1+2n^3}$

2. $a_n = \dfrac{9^{n+1}}{10^n}$

3. $a_n = \dfrac{n^3}{1+n^2}$

4. $a_n = (-2)^n$

5. Calculate the first eight terms of the sequence defined by $a_1 = 1$, $a_{n+1} = \frac{1}{3}a_n + 3$. Does it appear to be convergent? Assuming the limit exists, find its exact value.

6. **Drug concentration** A patient is injected with a drug at the same time every day. Before each injection, the concentration of the drug has dropped to 20% of its original value and the new dose raises the concentration by 0.25 mg/mL.
   (a) What is the concentration after four doses?
   (b) If C_n is the concentration after n doses, write a difference equation that expresses C_{n+1} in terms of C_n.
   (c) Solve the difference equation to find a formula for C_n.
   (d) Find the limiting value of the concentration.

7. Express the repeating decimal 1.2345345345 . . . as a fraction.

8. **Logistic equation** Plot enough terms of the logistic difference equation to see the behavior of the terms. If the sequence appears to be convergent, estimate its limit and then, assuming the limit exists, find its exact value. If not, describe how the terms behave.
   (a) $x_{t+1} = 2.5x_t(1 - x_t)$, $x_0 = 0.5$
   (b) $x_{t+1} = 3.3x_t(1 - x_t)$, $x_0 = 0.4$

9. The graph of f is given.
   (a) Find each limit, or explain why it does not exist.
   (i) $\lim_{x\to 2^+} f(x)$ (ii) $\lim_{x\to -3^+} f(x)$
   (iii) $\lim_{x\to -3} f(x)$ (iv) $\lim_{x\to 4} f(x)$
   (v) $\lim_{x\to 0} f(x)$ (vi) $\lim_{x\to 2^-} f(x)$
   (vii) $\lim_{x\to\infty} f(x)$ (viii) $\lim_{x\to -\infty} f(x)$
   (b) State the equations of the horizontal asymptotes.
   (c) State the equations of the vertical asymptotes.
   (d) At what numbers is f discontinuous? Explain.

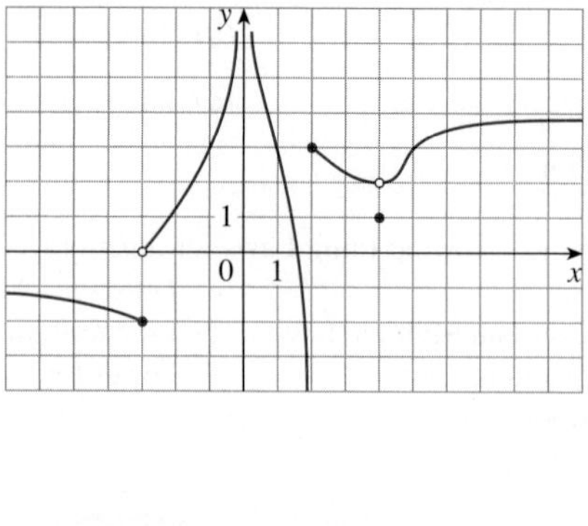

10. Sketch the graph of an example of a function f that satisfies all of the following conditions:

$\lim_{x\to-\infty} f(x) = -2, \quad \lim_{x\to\infty} f(x) = 0, \quad \lim_{x\to-3} f(x) = \infty,$

$\lim_{x\to3^-} f(x) = -\infty, \quad \lim_{x\to3^+} f(x) = 2,$

f is continuous from the right at 3

11–28 Find the limit.

11. $\lim_{x\to\infty} \dfrac{1-x}{2+5x}$

12. $\lim_{t\to\infty} 3^{-2t}$

13. $\lim_{x\to1} e^{x^3-x}$

14. $\lim_{x\to3} \dfrac{x^2-9}{x^2+2x-3}$

15. $\lim_{x\to-3} \dfrac{x^2-9}{x^2+2x-3}$

16. $\lim_{x\to1^+} \dfrac{x^2-9}{x^2+2x-3}$

17. $\lim_{h\to0} \dfrac{(h-1)^3+1}{h}$

18. $\lim_{t\to2} \dfrac{t^2-4}{t^3-8}$

19. $\lim_{r\to9} \dfrac{\sqrt{r}}{(r-9)^4}$

20. $\lim_{v\to4^+} \dfrac{4-v}{|4-v|}$

21. $\lim_{u\to1} \dfrac{u^4-1}{u^3+5u^2-6u}$

22. $\lim_{x\to3} \dfrac{\sqrt{x+6}-x}{x^3-3x^2}$

23. $\lim_{x\to\pi^-} \ln(\sin x)$

24. $\lim_{x\to-\infty} \dfrac{1-2x^2-x^4}{5+x-3x^4}$

25. $\lim_{x\to\infty} \dfrac{\sqrt{x^2-9}}{2x-6}$

26. $\lim_{x\to\infty} e^{x-x^2}$

27. $\lim_{x\to\infty} \left(\sqrt{x^2+4x+1}-x\right)$

28. $\lim_{x\to1} \left(\dfrac{1}{x-1}+\dfrac{1}{x^2-3x+2}\right)$

29. The **Michaelis-Menten equation** for the rate v of the enzymatic reaction of the concentration [S] of a substrate S, in the case of the enzyme pepsin, is

$$v = \frac{0.50[\text{S}]}{3.0 \times 10^{-4} + [\text{S}]}$$

What is $\lim_{[\text{S}]\to\infty} v$? What is the meaning of the limit in this context?

30. Prove that $\lim_{x\to0} x^2\cos(1/x^2) = 0$.

31. Let

$$f(x) = \begin{cases} \sqrt{-x} & \text{if } x < 0 \\ 3-x & \text{if } 0 \leqslant x < 3 \\ (x-3)^2 & \text{if } x > 3 \end{cases}$$

(a) Evaluate each limit, if it exists.

(i) $\lim_{x\to0^+} f(x)$ (ii) $\lim_{x\to0^-} f(x)$ (iii) $\lim_{x\to0} f(x)$

(iv) $\lim_{x\to3^-} f(x)$ (v) $\lim_{x\to3^+} f(x)$ (vi) $\lim_{x\to3} f(x)$

(b) Where is f discontinuous?

(c) Sketch the graph of f.

32. Show that each function is continuous on its domain. State the domain.

(a) $g(x) = \dfrac{\sqrt{x^2-9}}{x^2-2}$ (b) $h(x) = xe^{\sin x}$

33–34 Use the Intermediate Value Theorem to show that there is a root of the equation in the given interval.

33. $2x^3+x^2+2=0, \quad (-2,-1)$

34. $e^{-x^2}=x, \quad (0,1)$

CASE STUDY 2a Hosts, Parasites, and Time-Travel

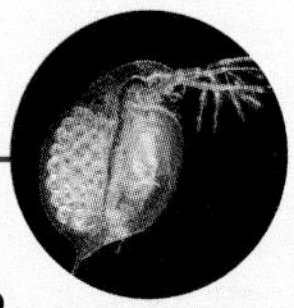

We are studying a model for the interaction between *Daphnia* and its parasite. Recall that there are two possible host genotypes (A and a) and two possible parasite genotypes (B and b). Parasites of type B can infect only hosts of type A, while parasites of type b can infect only hosts of type a. Here we will take equations that will be obtained in Case Studies 2b and 2d to explore the biological predictions that can be obtained from them.

In Case Study 2d we will derive the functions

(1a) $$q(t) = \tfrac{1}{2} + M_q\cos(ct - \phi_q)$$

(1b) $$p(t) = \tfrac{1}{2} + M_p\cos(ct - \phi_p)$$

where $q(t)$ is the predicted frequency of host genotype A at time t and $p(t)$ is the predicted frequency of parasite genotype B at time t. In these equations ϕ_q, ϕ_p, and c are positive constants, while M_q, M_p are positive constants that are strictly less than $\frac{1}{2}$.

1. Describe, in words, how the genotype frequencies of the host and parasite change over time. Provide an explanation, in biological terms, for these dynamics.
2. How do the constants M_q and M_p affect the pattern of genotype frequencies over time?
3. The constant c is determined by the consequences of infection, in terms of reproduction, for both the host and the parasite. A large difference in reproductive success between infected versus uninfected hosts makes c large. Likewise, a large difference in reproductive success between parasites that are unable to infect a host versus those that are able to infect a host also makes c large. How does c affect the pattern of genotype frequencies over time as predicted by Equations 1? Provide an explanation for this in biological terms.
4. The constants ϕ_q and ϕ_p are referred to as the *phase* of $q(t)$ and $p(t)$, respectively. How do these constants affect the pattern of genotype frequencies over time?
5. As you will see in Equation 3, the difference $\phi^* = \phi_p - \phi_q$ turns out to be important in the coevolution of the host and parasite. What does this difference represent mathematically? Explain why this quantity is a measure of the extent to which the frequency of the parasite genotype lags behind the frequency of the host genotype.

Equations 1 give the genotype frequencies as functions of time. In the *Daphnia*-parasite system described in Case Study 2 on page xlvi, these equations can also be interpreted as giving the dynamics of genotype frequencies as a function of depth in the sediment core shown in Figure 1.

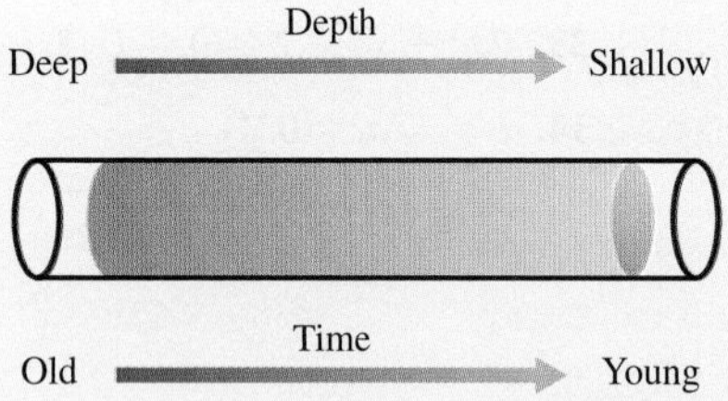

FIGURE 1

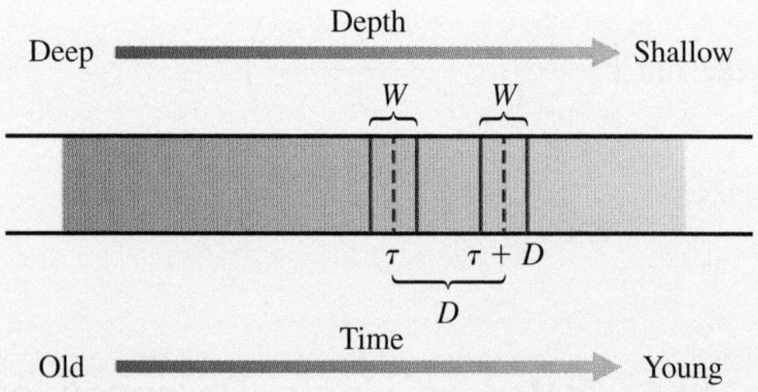

FIGURE 2

In the experiment described in Case Study 2, researchers chose a fixed depth τ and extracted a layer of sediment of width W centered around this depth (see Figure 2). The contents of this layer were mixed completely, and then hosts and parasites were extracted at random from the mixture. Researchers also took deeper and shallower layers (which represent the past and the future for hosts located in the layer at τ) and again completely mixed each layer. The center of these layers was a distance D from the center of the focal layer at τ, with $D < 0$ corresponding to a deeper layer and $D > 0$ a shallower layer (see Figure 2). The researchers then challenged hosts from the layer at τ with parasites from their past (that is, from the layer with $D < 0$), present (the layer at τ), and future (the layer with $D > 0$). For each challenge experiment the fraction of hosts becoming infected was measured.

In Case Study 2b we will show that, when a layer of sediment at location τ with width W is mixed completely, the frequency of type A hosts in this mixture is predicted to be

(2a) $$q_{\text{ave}}(\tau) = \tfrac{1}{2} + M_q \cos(c\tau - \phi_q)\frac{2\sin\left(\frac{1}{2}cW\right)}{cW}$$

Likewise, we will show that, when a layer of sediment at location τ with width W is mixed completely, the frequency of type B parasites in this mixture is predicted to be

(2b) $$p_{\text{ave}}(\tau) = \tfrac{1}{2} + M_p \cos(c\tau - \phi_p)\frac{2\sin\left(\frac{1}{2}cW\right)}{cW}$$

6. The functions (2a) and (2b) are similar to (1a) and (1b) except that the second terms are multiplied by the quantity $2\sin\left(\frac{1}{2}cW\right)/cW$. Describe how the frequency of the genotypes within a mixed layer depends on the width W of this layer. In particular, what happens as the width of the layer becomes very small (that is, when $W \to 0$)? What happens as the width becomes very large (that is, $W \to \infty$)? Provide a biological interpretation for your answers.

In the experiment introduced in Case Study 2, hosts from depth τ were challenged with parasites from depth $\tau + D$. This was repeated for different depths τ, and the overall fraction of hosts infected was measured. In Case Study 2b we will show that the predicted fraction of hosts infected from such an experiment is

(3) $$F(D) = \tfrac{1}{2} + M_p M_q \cos(cD - \phi^*)\frac{4\sin^2\left(\frac{1}{2}cW\right)}{c^2W^2}$$

where $\phi^* = \phi_p - \phi_q$ and $D < 0$ corresponds to parasites from a host's past and $D > 0$ to parasites from a host's future.

7. Sketch the graph of $F(D)$ when $\phi^* = 0$. Be as accurate as possible, showing where the maxima and minima occur as well as where the graph crosses the vertical axis. Construct similar sketches when ϕ^* is small and positive as well as when ϕ^* is small and negative. These plots depict the predicted fraction of infected hosts in the experiment as a function of the relative point in time from which the parasite was taken.

8. Suppose that cD is relatively small, meaning that the layers used in the challenge experiments are close to one another. Use your results from Problem 7 to explain how it is possible to obtain the experimental data like those shown in Figure 3.

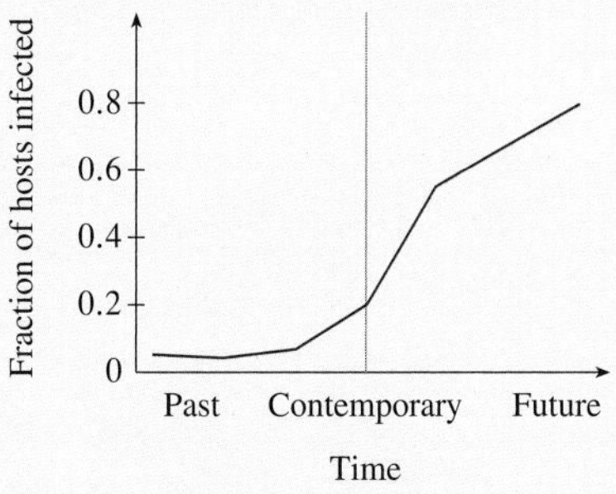

FIGURE 3
Horizontal axis is the time from which the parasite was taken, relative to the host's point in time.

Source: Adapted from S. Gandon et al., "Host-Parasite Coevolution and Patterns of Adaptation across Time and Space," *Journal of Evolutionary Biology* 21 (2008): 1861–66.

In particular, what is true about the value of ϕ^* in this case? Provide a biological interpretation of your answer.

9. Again suppose that cD is relatively small. Use your results from Problem 7 to explain how it is possible to obtain data like those shown in Figure 4. What is true about the value of ϕ^* in this case? Provide a biological interpretation of your answer.

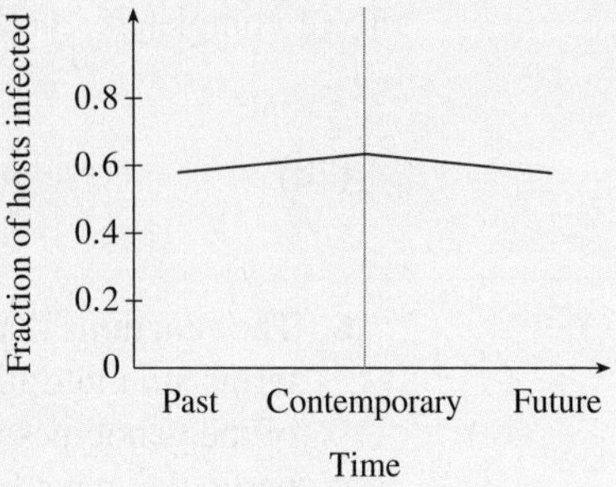

FIGURE 4
Horizontal axis is the time from which the parasite was taken, relative to the host's point in time.

Source: Adapted from S. Gandon et al., "Host-Parasite Coevolution and Patterns of Adaptation across Time and Space," *Journal of Evolutionary Biology* 21 (2008): 1861–66.

Derivatives

3

The maximum sustainable swimming speed S of salmon depends on the water temperature T. Exercise 42 in Section 3.1 asks you to analyze how S varies as T changes by estimating the derivative of S with respect to T.

IN THIS CHAPTER WE STUDY a special type of limit, called a derivative. Derivatives arise when we want to find a rate of growth, a velocity, the slope of a tangent line, or any instantaneous rate of change.

3.1 Derivatives and Rates of Change

The problem of finding a rate of change at a given instant and the problem of finding the tangent line to a curve at a given point involve finding the same type of limit, which we call a *derivative*.

■ Measuring the Rate of Increase of Blood Alcohol Concentration

Biomedical scientists have studied the chemical and physiological changes in the body that result from alcohol consumption. The reaction in the human body occurs in two stages: a fairly rapid process of absorption and a more gradual one of metabolism. To predict the effect of alcohol consumption, one needs to know the *rate* at which alcohol is absorbed and metabolized.

Medical researchers measured the blood alcohol concentration (BAC) of eight fasting adult male subjects after rapid consumption of 15 mL of ethanol (corresponding to one alcoholic drink).[1] The data they obtained were modeled by the concentration function

(1) $$C(t) = 0.0225te^{-0.0467t}$$

where t is measured in minutes after consumption and C is measured in mg/mL. The graph of C is shown in Figure 1.

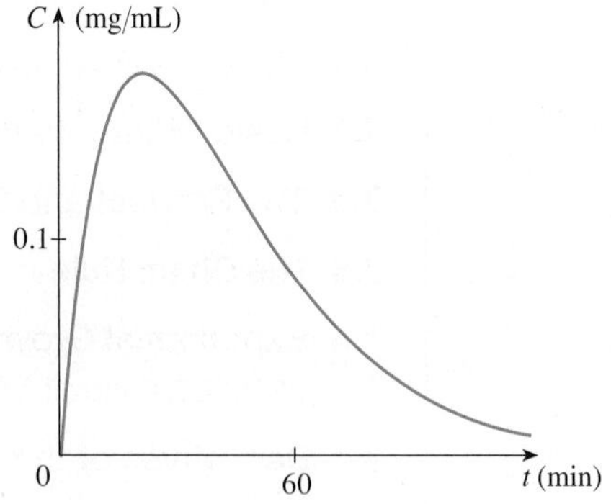

FIGURE 1

EXAMPLE 1 | Blood alcohol concentration How quickly is the BAC (given by Equation 1) increasing after 10 minutes?

SOLUTION We are asked to find the rate of change of C with respect to t when $t = 10$. The difficulty is that we are dealing with a single instant of time ($t = 10$ min) and so no time interval is involved. However, we can approximate the desired quantity

1. P. Wilkinson et al., "Pharmacokinetics of Ethanol after Oral Administration in the Fasting State," *Journal of Pharmacokinetics and Biopharmaceutics* 5 (1977): 207–24.

by calculating the *average* rate of change of C with respect to t in the time interval from $t = 10$ to $t = 11$:

$$\text{average rate of change} = \frac{\text{change in } C}{\text{change in } t} = \frac{C(11) - C(10)}{11 - 10}$$

$$\approx \frac{0.148073 - 0.141048}{1} = 0.007025 \text{ (mg/mL)/min}$$

The following table shows the results of similar calculations of the average rates of change [in (mg/mL)/min] over successively smaller time periods.

Time interval	Average rate of change	Time interval	Average rate of change
$10 \leqslant t \leqslant 11$	0.00703	$9 \leqslant t \leqslant 10$	0.00804
$10 \leqslant t \leqslant 10.5$	0.00727	$9.5 \leqslant t \leqslant 10$	0.00777
$10 \leqslant t \leqslant 10.1$	0.00747	$9.9 \leqslant t \leqslant 10$	0.00757
$10 \leqslant t \leqslant 10.01$	0.00751	$9.99 \leqslant t \leqslant 10$	0.00752

It appears that as we shorten the time period, the average rate of change is becoming closer and closer to a number between 0.00752 and 0.00753 (mg/mL)/min. The *instantaneous rate of change* at $t = 10$ is defined to be the limiting value of these average rates of change over shorter and shorter time periods that start or end at $t = 10$. So we estimate that the BAC increased at a rate of about 0.0075 (mg/mL)/min. ■

We will return to the story about blood alcohol concentration in Example 6.

■ Tangent Lines

The word *tangent* is derived from the Latin word *tangens*, which means "touching." Thus a tangent to a curve is a line that touches the curve. In other words, a tangent line should have the same direction as the curve at the point of contact. How can this idea be made precise?

For a circle we could simply follow Euclid and say that a tangent is a line that intersects the circle once and only once, as in Figure 2(a). For more complicated curves this definition is inadequate. Figure 2(b) shows two lines l and t passing through a point P on a curve C. The line l intersects C only once, but it certainly does not look like what we think of as a tangent. The line t, on the other hand, looks like a tangent but it intersects C twice.

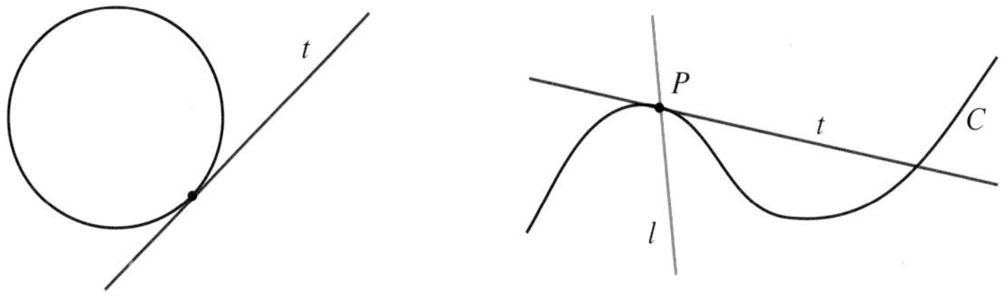

FIGURE 2 (a) (b)

To be specific, let's look at the problem of trying to find a tangent line t to the parabola $y = x^2$ in the following example.

EXAMPLE 2 | Find an equation of the tangent line to the parabola $y = x^2$ at the point $P(1, 1)$.

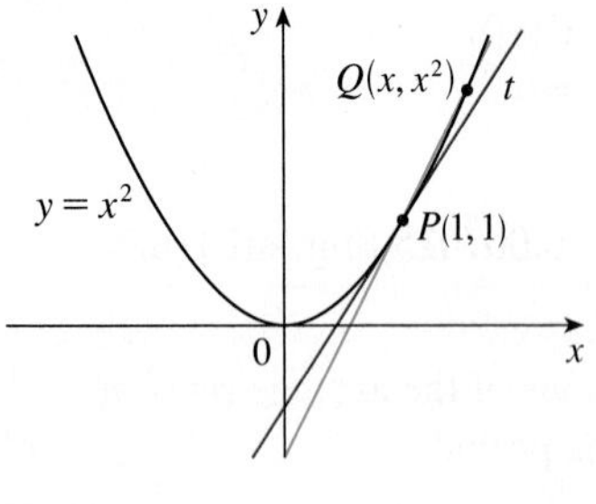

FIGURE 3

SOLUTION We will be able to find an equation of the tangent line t as soon as we know its slope m. The difficulty is that we know only one point, P, on t, whereas we need two points to compute the slope. But observe that we can compute an approximation to m by choosing a nearby point $Q(x, x^2)$ on the parabola (as in Figure 3) and computing the slope m_{PQ} of the secant line PQ. [A **secant line**, from the Latin word *secans*, meaning cutting, is a line that cuts (intersects) a curve more than once.]

We choose $x \neq 1$ so that $Q \neq P$. Then

$$m_{PQ} = \frac{x^2 - 1}{x - 1}$$

What happens as x approaches 1? From Figure 4 we see that Q approaches P along the parabola and the secant lines PQ rotate about P and approach the tangent line t.

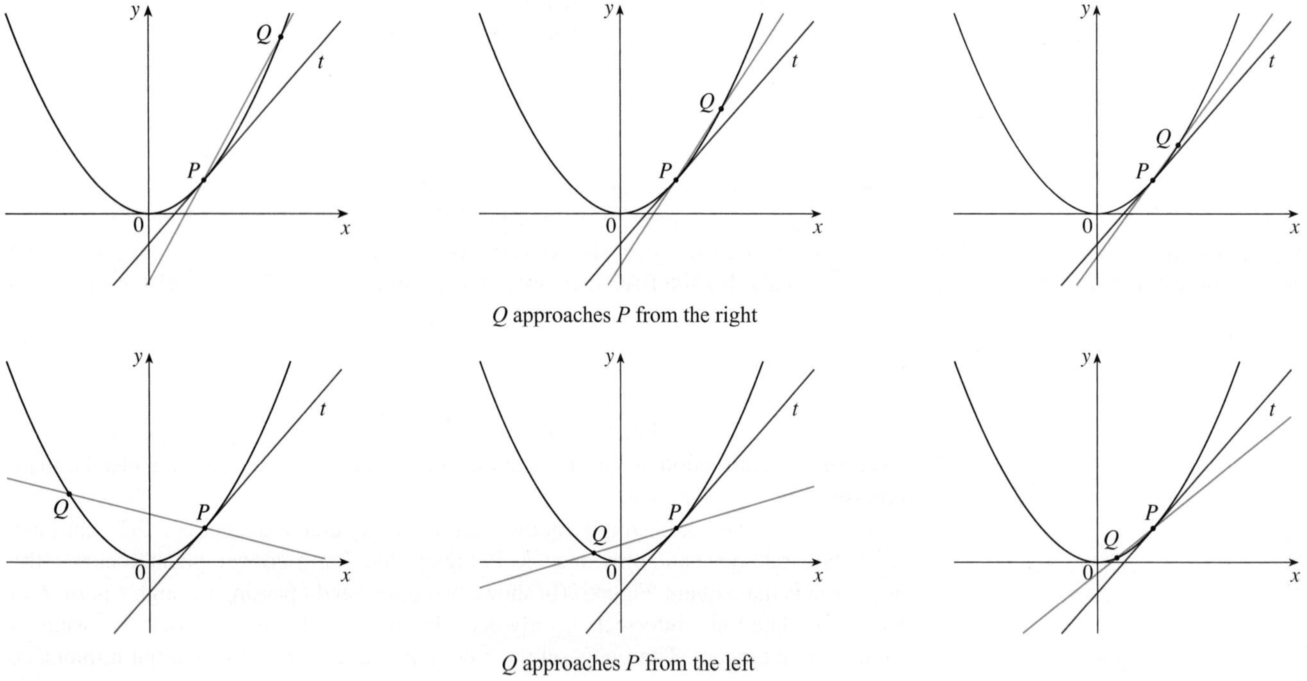

Q approaches P from the right

Q approaches P from the left

FIGURE 4

It appears that the slope m of the tangent line is the limit of the slopes of the secant lines as x approaches 1:

TEC In Visual 3.1A you can see how the process in Figure 4 works for additional functions.

$$m = \lim_{x \to 1} \frac{x^2 - 1}{x - 1} = \lim_{x \to 1} \frac{(x - 1)(x + 1)}{x - 1}$$

$$= \lim_{x \to 1} (x + 1) = 1 + 1 = 2$$

Point-slope form for a line through the point (x_1, y_1) with slope m:

$$y - y_1 = m(x - x_1)$$

Using the point-slope form of the equation of a line, we find that an equation of the tangent line at $(1, 1)$ is

$$y - 1 = 2(x - 1) \qquad \text{or} \qquad y = 2x - 1$$

■

We sometimes refer to the slope of the tangent line to a curve at a point as the **slope of the curve** at the point. The idea is that if we zoom in far enough toward the point, the curve looks almost like a straight line. Figure 5 illustrates this procedure for the curve $y = x^2$ in Example 2. The more we zoom in, the more the parabola looks like a line. In other words, the curve becomes almost indistinguishable from its tangent line.

TEC Visual 3.1B shows an animation of Figure 5.

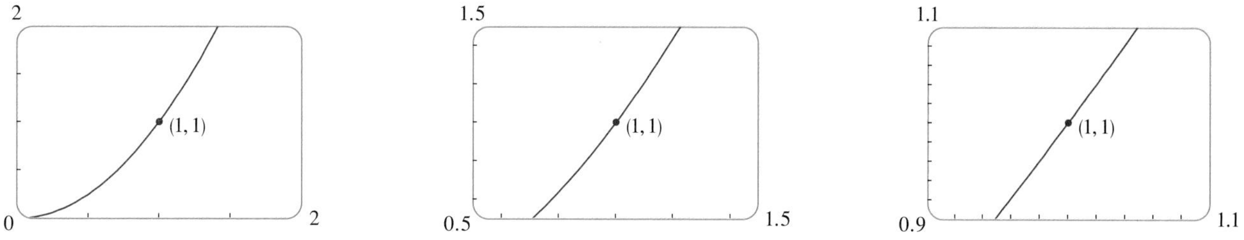

FIGURE 5 Zooming in toward the point (1, 1) on the parabola $y = x^2$

In general, if a curve C has equation $y = f(x)$ and we want to find the tangent line to C at the point $P(a, f(a))$, then we consider a nearby point $Q(x, f(x))$, where $x \neq a$, and compute the slope of the secant line PQ:

$$m_{PQ} = \frac{f(x) - f(a)}{x - a}$$

Then we let Q approach P along the curve C by letting x approach a. If m_{PQ} approaches a number m, then we define the *tangent* t to be the line through P with slope m. (This amounts to saying that the tangent line is the limiting position of the secant line PQ as Q approaches P. See Figure 6.)

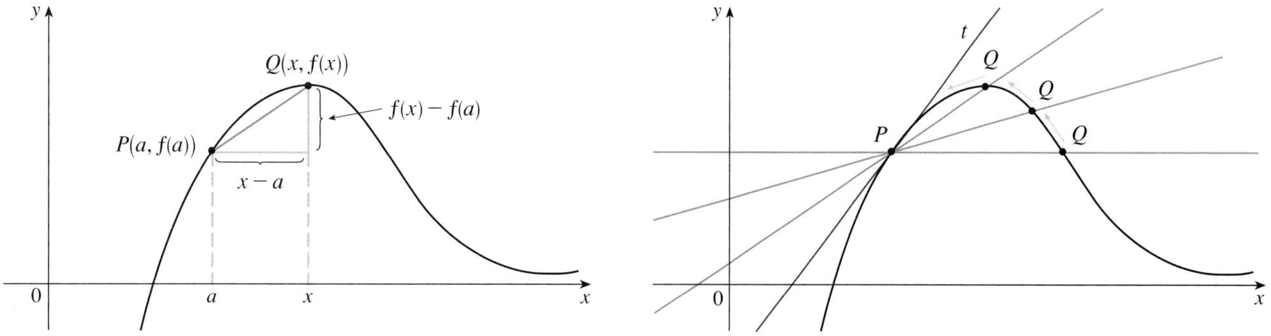

FIGURE 6

(2) Definition The **tangent line** to the curve $y = f(x)$ at the point $P(a, f(a))$ is the line through P with slope

$$m = \lim_{x \to a} \frac{f(x) - f(a)}{x - a}$$

provided that this limit exists.

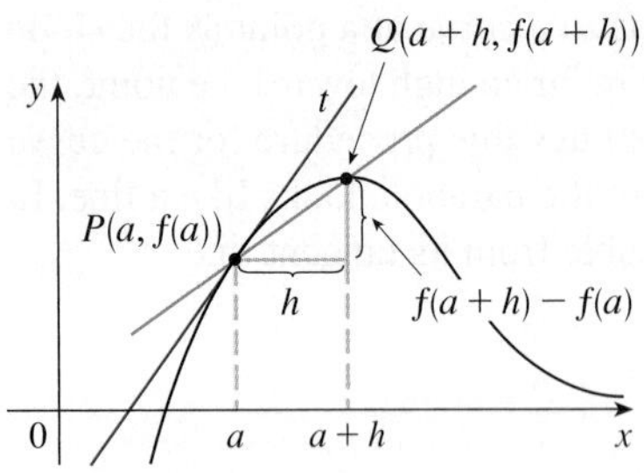

FIGURE 7

There is another expression for the slope of a tangent line that is sometimes easier to use. If $h = x - a$, then $x = a + h$ and so the slope of the secant line PQ is

$$m_{PQ} = \frac{f(a+h) - f(a)}{h}$$

(See Figure 7 where the case $h > 0$ is illustrated and Q is to the right of P. If it happened that $h < 0$, however, Q would be to the left of P.) Notice that as x approaches a, h approaches 0 (because $h = x - a$) and so the expression for the slope of the tangent line in Definition 2 becomes

(3)
$$m = \lim_{h \to 0} \frac{f(a+h) - f(a)}{h}$$

EXAMPLE 3 | Find an equation of the tangent line to the hyperbola $y = 3/x$ at the point (3, 1).

SOLUTION Let $f(x) = 3/x$. Then, by Equation 3, the slope of the tangent at (3, 1) is

$$m = \lim_{h \to 0} \frac{f(3+h) - f(3)}{h} = \lim_{h \to 0} \frac{\dfrac{3}{3+h} - 1}{h} = \lim_{h \to 0} \frac{\dfrac{3 - (3+h)}{3+h}}{h}$$

$$= \lim_{h \to 0} \frac{-h}{h(3+h)} = \lim_{h \to 0} -\frac{1}{3+h} = -\frac{1}{3}$$

Therefore an equation of the tangent at the point (3, 1) is

$$y - 1 = -\tfrac{1}{3}(x - 3)$$

which simplifies to $\quad y = 2 - \tfrac{1}{3}x \quad$ or $\quad x + 3y - 6 = 0$

The hyperbola and its tangent are shown in Figure 8. ■

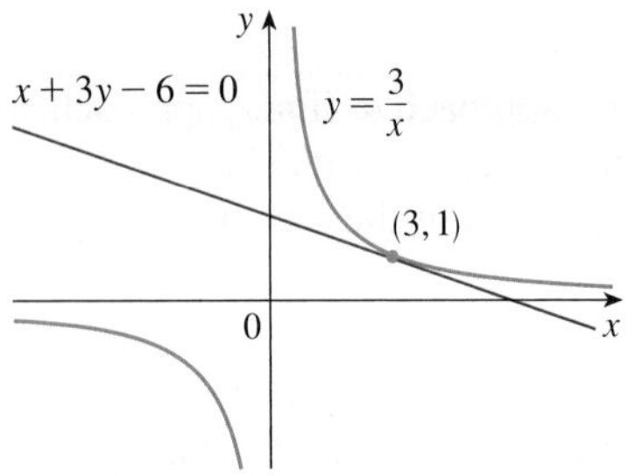

FIGURE 8

■ Derivatives

We have seen that the same type of limit arises in finding the slope of a tangent line (Equation 3) or a rate of change (Example 1). In fact, limits of the form

$$\lim_{h \to 0} \frac{f(a+h) - f(a)}{h}$$

arise whenever we calculate a rate of change in any of the sciences, such as a rate of growth in biology or a rate of reaction in chemistry. Since this type of limit occurs so widely, it is given a special name and notation.

$f'(a)$ is read "f prime of a."

(4) Definition The **derivative of a function f at a number a**, denoted by $f'(a)$, is

$$f'(a) = \lim_{h \to 0} \frac{f(a+h) - f(a)}{h}$$

if this limit exists.

If we write $x = a + h$, then we have $h = x - a$ and h approaches 0 if and only if x approaches a. Therefore an equivalent way of stating the definition of the derivative, as we saw in finding tangent lines, is

(5)
$$f'(a) = \lim_{x \to a} \frac{f(x) - f(a)}{x - a}$$

EXAMPLE 4 | Find the derivative of the function $f(x) = x^2 - 8x + 9$ at the number a.

SOLUTION From Definition 4 we have

$$\begin{aligned} f'(a) &= \lim_{h \to 0} \frac{f(a+h) - f(a)}{h} \\ &= \lim_{h \to 0} \frac{[(a+h)^2 - 8(a+h) + 9] - [a^2 - 8a + 9]}{h} \\ &= \lim_{h \to 0} \frac{a^2 + 2ah + h^2 - 8a - 8h + 9 - a^2 + 8a - 9}{h} \\ &= \lim_{h \to 0} \frac{2ah + h^2 - 8h}{h} = \lim_{h \to 0} (2a + h - 8) \\ &= 2a - 8 \end{aligned}$$

■

We defined the tangent line to the curve $y = f(x)$ at the point $P(a, f(a))$ to be the line that passes through P and has slope m given by Equation 2 or 3. Since, by Definition 4, this is the same as the derivative $f'(a)$, we can now say the following.

> The tangent line to $y = f(x)$ at $(a, f(a))$ is the line through $(a, f(a))$ whose slope is equal to $f'(a)$, the derivative of f at a.

If we use the point-slope form of the equation of a line, we can write an equation of the tangent line to the curve $y = f(x)$ at the point $(a, f(a))$:

$$y - f(a) = f'(a)(x - a)$$

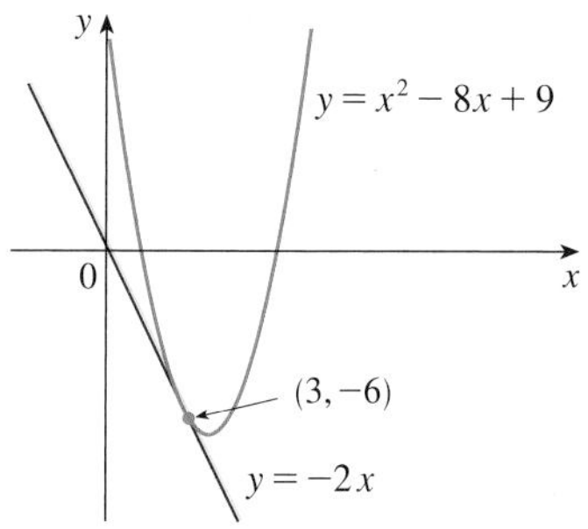

FIGURE 9

EXAMPLE 5 | Find an equation of the tangent line to the parabola $y = x^2 - 8x + 9$ at the point $(3, -6)$.

SOLUTION From Example 4 we know that the derivative of $f(x) = x^2 - 8x + 9$ at the number a is $f'(a) = 2a - 8$. Therefore the slope of the tangent line at $(3, -6)$ is $f'(3) = 2(3) - 8 = -2$. Thus an equation of the tangent line, shown in Figure 9, is

$$y - (-6) = (-2)(x - 3) \qquad \text{or} \qquad y = -2x$$

■

■ Rates of Change

We have already seen one example of a rate of change in Example 1. In general, suppose y is a quantity that depends on another quantity x. Thus y is a function of x and we write $y = f(x)$. If x changes from x_1 to x_2, then the change in x (also called the **increment** of x) is

$$\Delta x = x_2 - x_1$$

and the corresponding change in y is

$$\Delta y = f(x_2) - f(x_1)$$

The difference quotient

$$\frac{\Delta y}{\Delta x} = \frac{f(x_2) - f(x_1)}{x_2 - x_1}$$

is called the **average rate of change of *y* with respect to *x*** over the interval $[x_1, x_2]$ and can be interpreted as the slope of the secant line PQ in Figure 10.

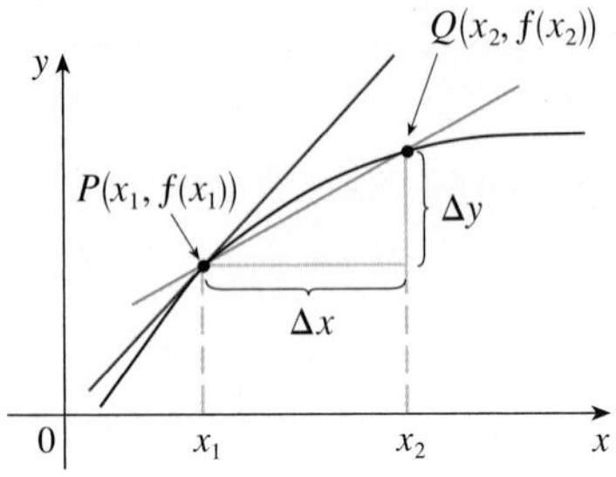

FIGURE 10
average rate of change $= m_{PQ}$
instantaneous rate of change $=$ slope of tangent at P

As we did in Example 1, we now consider the average rate of change over smaller and smaller intervals by letting x_2 approach x_1 and therefore letting Δx approach 0. The limit of these average rates of change is called the **(instantaneous) rate of change of *y* with respect to *x*** at $x = x_1$, which is interpreted as the slope of the tangent to the curve $y = f(x)$ at $P(x_1, f(x_1))$:

(6) $$\text{instantaneous rate of change} = \lim_{\Delta x \to 0} \frac{\Delta y}{\Delta x} = \lim_{x_2 \to x_1} \frac{f(x_2) - f(x_1)}{x_2 - x_1}$$

We recognize this limit as being the derivative $f'(x_1)$.

We know that one interpretation of the derivative $f'(a)$ is as the slope of the tangent line to the curve $y = f(x)$ when $x = a$. We now have a second interpretation:

> The derivative $f'(a)$ is the instantaneous rate of change of $y = f(x)$ with respect to x when $x = a$.

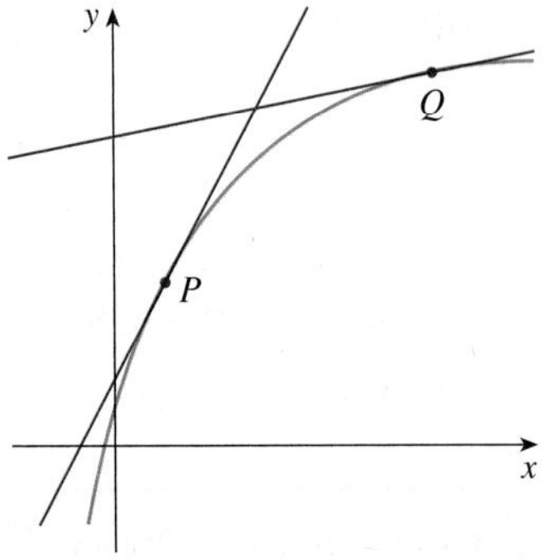

FIGURE 11
The y-values are changing rapidly at P and slowly at Q.

The connection with the first interpretation is that if we sketch the curve $y = f(x)$, then the instantaneous rate of change is the slope of the tangent to this curve at the point where $x = a$. This means that when the derivative is large (and therefore the curve is steep, as at the point P in Figure 11), the y-values change rapidly. When the derivative is small, the curve is relatively flat (as at point Q) and the y-values change slowly.

EXAMPLE 6 | **Blood alcohol concentration (continued)** Draw the tangent line to the BAC curve in Example 1 at $t = 10$ and interpret its slope.

SOLUTION In Example 1 we estimated that the rate of increase of the blood alcohol concentration when $t = 10$ is about 0.0075 (mg/mL)/min. The equation of the curve (Equation 1) is

$$C(t) = 0.0225te^{-0.0467t}$$

which gives $C(10) \approx 0.14105$. So, using the point-slope equation of a line, we get that an approximate equation of the tangent line at $t = 10$ is

$$C - 0.14105 = 0.0075(t - 10)$$

or

$$C = 0.06605 + 0.0075t$$

The concentration curve and its tangent line are graphed in Figure 12 and the slope of the tangent line is the rate of increase of BAC when $t = 10$.

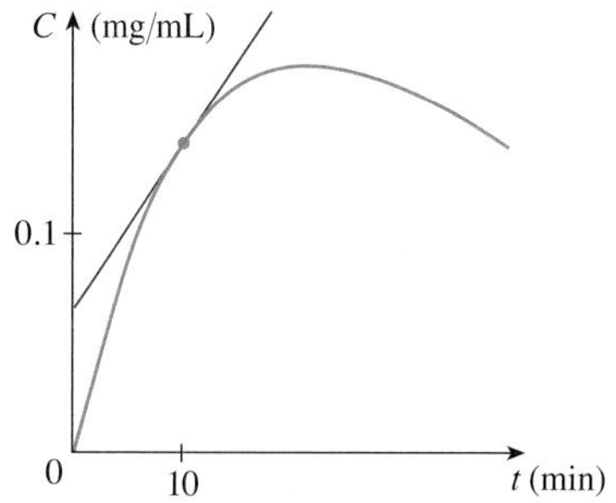

FIGURE 12

■

t	N
1	228
2	2,357
3	12,750
4	26,661
5	372,331
6	2,217,441

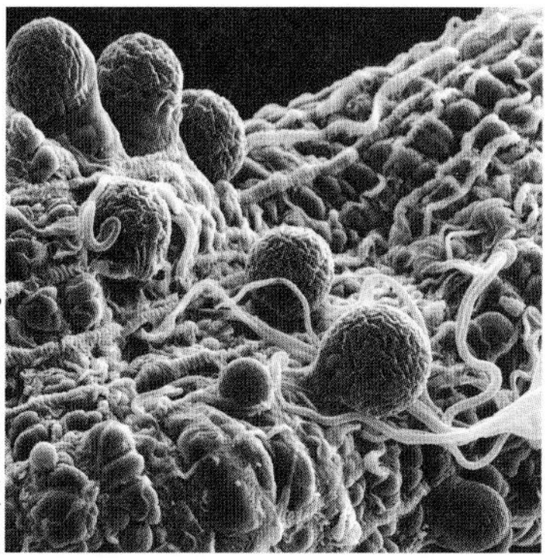

Hilary Hurd, Wellcome Images

EXAMPLE 7 | BB **Malarial parasites** The table at the left, supplied by Andrew Read, shows experimental data involving malarial parasites. The time t is measured in days and N is the number of parasites per microliter of blood.

(a) Find the average rates of change of N with respect to t over the intervals [1, 3], [2, 3], [3, 4], and [3, 5].

(b) Interpret and estimate the value of the derivative $N'(3)$.

SOLUTION

(a) The average rate of change over [1, 3] is

$$\frac{N(3) - N(1)}{3 - 1} = \frac{12{,}750 - 228}{2} = 6261 \text{ (parasites/}\mu\text{L)/day}$$

Similar calculations give the average rates of change in the following table:

Interval	Rate of change
[1, 3]	6,261
[2, 3]	10,393
[3, 4]	13,911
[3, 5]	179,791

(b) The derivative $N'(3)$ means the rate of change of N with respect to t when $t = 3$ days. According to Equation 5,

$$N'(3) = \lim_{t \to 3} \frac{N(t) - N(3)}{t - 3}$$

Although the function $N(t)$ is not a smooth function, it can be approximated by a smooth one as in Figure 2.5.12. In that sense it is meaningful to talk about the derivative $N'(3)$.

The difference quotients in this expression (for various values of t) are just the rates of change in the table in part (a). So $N'(3)$ lies somewhere 10,393 and 13,911 (parasites/μL)/day. We estimate that the rate of increase of the parasite population on day 3 was approximately the average of these two numbers, namely

$$N'(3) \approx 12{,}152 \text{ (parasites/}\mu\text{L)/day}$$

■

A familiar example of a rate of change is **velocity**. In Example 2.3.1 we found the instantaneous velocity of a ball dropped from the CN Tower as the limit of average velocities over shorter and shorter time periods. More generally, if $s = f(t)$ is the position function of a particle that moves along a straight line, then $f'(a)$ is the rate of change of the displacement s with respect to the time t. In other words, *$f'(a)$ is the velocity of the particle at time $t = a$.* The *speed* of the particle is the absolute value of the velocity, that is, $|f'(a)|$.

Let's revisit the example of the falling ball.

EXAMPLE 8 | Suppose that a ball is dropped from the upper observation deck of the CN Tower, 450 m above the ground.
(a) What is the velocity of the ball after 5 seconds?
(b) How fast is the ball traveling when it hits the ground?

SOLUTION We will need to find the velocity both when $t = 5$ and when the ball hits the ground, so it's efficient to start by finding the velocity at a general time t. Using the equation of motion $s = f(t) = 4.9t^2$, we have

$$v(t) = f'(t) = \lim_{h \to 0} \frac{f(t+h) - f(t)}{h} = \lim_{h \to 0} \frac{4.9(t+h)^2 - 4.9t^2}{h}$$

$$= \lim_{h \to 0} \frac{4.9(t^2 + 2th + h^2 - t^2)}{h} = \lim_{h \to 0} \frac{4.9(2th + h^2)}{h}$$

$$= \lim_{h \to 0} 4.9(2t + h) = 9.8t$$

(a) The velocity after 5 seconds is $v(5) = (9.8)(5) = 49$ m/s.

(b) Since the observation deck is 450 m above the ground, the ball will hit the ground at the time t_1 when $s(t_1) = 450$, that is,

$$4.9t_1^2 = 450$$

This gives

$$t_1^2 = \frac{450}{4.9} \quad \text{and} \quad t_1 = \sqrt{\frac{450}{4.9}} \approx 9.6 \text{ s}$$

The velocity of the ball as it hits the ground is therefore

$$v(t_1) = 9.8t_1 = 9.8\sqrt{\frac{450}{4.9}} \approx 94 \text{ m/s}$$ ■

EXAMPLE 9 | **HIV prevalence and incidence** In Example 2.4.7 we saw that the *prevalence* of a disease P is the number of cases as a function of time t. The *incidence* of the disease is the number of new infections per unit time over a specified time interval. We calculated the incidence of HIV in New York during the early 1980s over shorter and shorter time intervals. In the limiting case, as the interval shrinks to zero, we obtained the incidence of HIV at a point in time. Thus the incidence of HIV at a particular time can be viewed as the derivative of its prevalence function at that time. (Technically, this is true only if prevalence is changing solely due to the occurrence of new infections, as was the case for HIV in the early 1980s, since incidence is defined as the rate of generation of new infections. More generally, prevalence can change as a result of individuals dying or recovering from disease, as occurred later for HIV). ■

EXERCISES 3.1

1. A curve has equation $y = f(x)$.
(a) Write an expression for the slope of the secant line through the points $P(3, f(3))$ and $Q(x, f(x))$.
(b) Write an expression for the slope of the tangent line at P.

2. Graph the curve $y = e^x$ in the viewing rectangles $[-1, 1]$ by $[0, 2]$, $[-0.5, 0.5]$ by $[0.5, 1.5]$, and $[-0.1, 0.1]$ by $[0.9, 1.1]$. What do you notice about the curve as you zoom in toward the point $(0, 1)$?

3. (a) Find the slope of the tangent line to the parabola $y = 4x - x^2$ at the point $(1, 3)$
(i) using Definition 2 (ii) using Equation 3
(b) Find an equation of the tangent line in part (a).
(c) Graph the parabola and the tangent line. As a check on your work, zoom in toward the point $(1, 3)$ until the parabola and the tangent line are indistinguishable.

4. (a) Find the slope of the tangent line to the curve $y = x - x^3$ at the point $(1, 0)$
(i) using Definition 2 (ii) using Equation 3
(b) Find an equation of the tangent line in part (a).
(c) Graph the curve and the tangent line in successively smaller viewing rectangles centered at $(1, 0)$ until the curve and the line appear to coincide.

5–8 Find an equation of the tangent line to the curve at the given point.

5. $y = 4x - 3x^2$, $(2, -4)$

6. $y = x^3 - 3x + 1$, $(2, 3)$

7. $y = \sqrt{x}$, $(1, 1)$

8. $y = \dfrac{2x + 1}{x + 2}$, $(1, 1)$

---

9. (a) Find the slope of the tangent to the curve $y = 3 + 4x^2 - 2x^3$ at the point where $x = a$.
(b) Find equations of the tangent lines at the points $(1, 5)$ and $(2, 3)$.
(c) Graph the curve and both tangents on a common screen.

10. (a) Find the slope of the tangent to the curve $y = 1/\sqrt{x}$ at the point where $x = a$.
(b) Find equations of the tangent lines at the points $(1, 1)$ and $\left(4, \frac{1}{2}\right)$.
(c) Graph the curve and both tangents on a common screen.

11. For the function g whose graph is given, arrange the following numbers in increasing order and explain your reasoning:

$$0 \qquad g'(-2) \qquad g'(0) \qquad g'(2) \qquad g'(4)$$

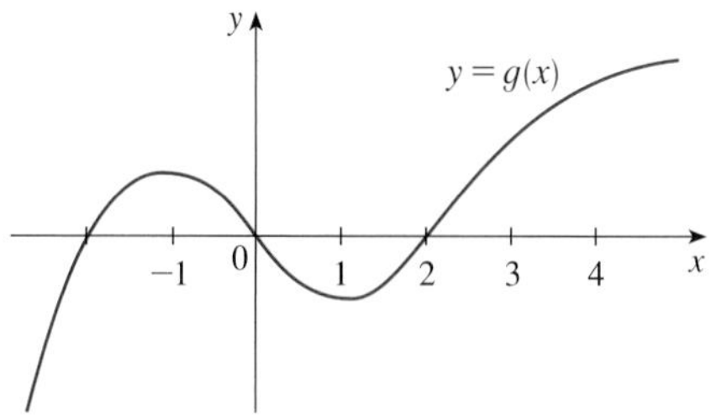

12. Find an equation of the tangent line to the graph of $y = g(x)$ at $x = 5$ if $g(5) = -3$ and $g'(5) = 4$.

13. If an equation of the tangent line to the curve $y = f(x)$ at the point where $a = 2$ is $y = 4x - 5$, find $f(2)$ and $f'(2)$.

14. If the tangent line to $y = f(x)$ at $(4, 3)$ passes through the point $(0, 2)$, find $f(4)$ and $f'(4)$.

15. Sketch the graph of a function f for which $f(0) = 0$, $f'(0) = 3$, $f'(1) = 0$, and $f'(2) = -1$.

16. Sketch the graph of a function g for which $g(0) = g(2) = g(4) = 0$, $g'(1) = g'(3) = 0$, $g'(0) = g'(4) = 1$, $g'(2) = -1$, $\lim_{x \to \infty} g(x) = \infty$, and $\lim_{x \to -\infty} g(x) = -\infty$.

17. If $f(x) = 3x^2 - x^3$, find $f'(1)$ and use it to find an equation of the tangent line to the curve $y = 3x^2 - x^3$ at the point $(1, 2)$.

18. If $g(x) = x^4 - 2$, find $g'(1)$ and use it to find an equation of the tangent line to the curve $y = x^4 - 2$ at the point $(1, -1)$.

19. (a) If $F(x) = 5x/(1 + x^2)$, find $F'(2)$ and use it to find an equation of the tangent line to the curve $y = 5x/(1 + x^2)$ at the point $(2, 2)$.
(b) Illustrate part (a) by graphing the curve and the tangent line on the same screen.

20. (a) If $G(x) = 4x^2 - x^3$, find $G'(a)$ and use it to find equations of the tangent lines to the curve $y = 4x^2 - x^3$ at the points $(2, 8)$ and $(3, 9)$.
(b) Illustrate part (a) by graphing the curve and the tangent lines on the same screen.

21–25 Find $f'(a)$.

21. $f(x) = 3x^2 - 4x + 1$

22. $f(t) = 2t^3 + t$

23. $f(t) = \dfrac{2t + 1}{t + 3}$

24. $f(x) = x^{-2}$

25. $f(x) = \sqrt{1 - 2x}$

---

26. Shown are graphs of the position functions of two runners, A and B, who run a 100-meter race and finish in a tie.
(a) Describe and compare how the runners run the race.
(b) At what time is the distance between the runners the greatest?
(c) At what time do they have the same velocity?

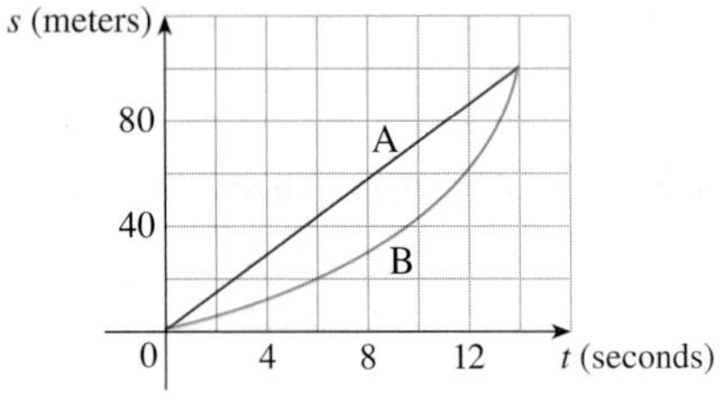

27. If a ball is thrown into the air with a velocity of 40 ft/s, its height (in feet) after t seconds is given by $y = 40t - 16t^2$. Find the velocity when $t = 2$.

28. If a rock is thrown upward on the planet Mars with a velocity of 10 m/s, its height (in meters) after t seconds is given by $H = 10t - 1.86t^2$.
(a) Find the velocity of the rock after one second.
(b) Find the velocity of the rock when $t = a$.
(c) When will the rock hit the surface?
(d) With what velocity will the rock hit the surface?

29. The displacement (in meters) of a particle moving in a straight line is given by the equation of motion $s = 1/t^2$, where t is measured in seconds. Find the velocity of the particle at times $t = a$, $t = 1$, $t = 2$, and $t = 3$.

30. Invasive species The Argentine ant is an invasive species in North America.

The graph shows the cumulative number of counties in the United States that have been invaded by this species over time.

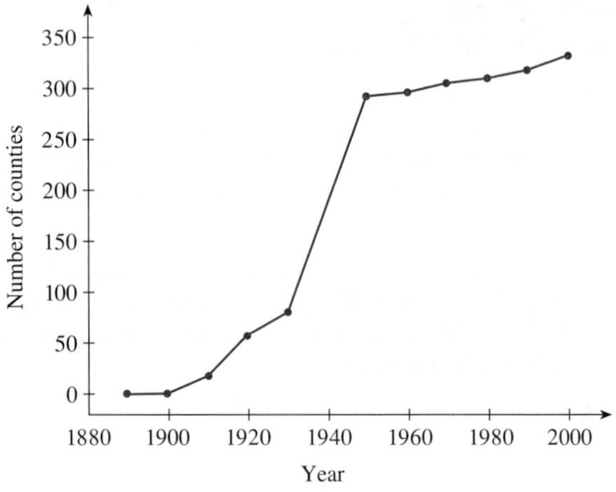

(a) Estimate the average rate of invasion between 1890 and 1920, between 1920 and 1960, and between 1960 and 2000.
(b) Estimate the instantaneous rate of invasion in 1940.

31. Population growth The table gives the US midyear population, in millions, from 1990 to 2010.

t	1990	1995	2000	2005	2010
$P(t)$	249.6	266.3	282.2	295.8	308.3

(a) Find the average rate of population increase
(i) from 1990 to 2000 (ii) from 1995 to 2000
(iii) from 2000 to 2005 (iv) from 2000 to 2010
(b) If $P(t)$ is the population at time t, estimate and interpret the value of the derivative $P'(2000)$.

32. Viral load The table shows values of the viral load $V(t)$ in HIV patient 303, measured in RNA copies/mL, t days after ABT-538 treatment was begun.

t	4	8	11	15	22
$V(t)$	53	18	9.4	5.2	3.6

(a) Find the average rate of change of V with respect to t over each time interval:
(i) [4, 11] (ii) [8, 11]
(iii) [11, 15] (iv) [11, 22]
What are the units?
(b) Estimate and interpret the value of the derivative $V'(11)$.

Source: Adapted from D. Ho et al., "Rapid Turnover of Plasma Virions and CD4 Lymphocytes in HIV-1 Infection," *Nature* 373 (1995): 123–26.

33. Blood alcohol concentration Researchers measured the average blood alcohol concentration $C(t)$ of eight men starting one hour after consumption of 30 mL of ethanol (corresponding to two alcoholic drinks):

t (hours)	1.0	1.5	2.0	2.5	3.0
$C(t)$ (mg/mL)	0.33	0.24	0.18	0.12	0.07

(a) Find the average rate of change of C with respect to t over each time interval:
(i) [1.0, 2.0] (ii) [1.5, 2.0]
(iii) [2.0, 2.5] (iv) [2.0, 3.0]
What are the units?

(b) Estimate and interpret the value of the derivative $C'(2)$.

Source: Adapted from P. Wilkinson et al., "Pharmacokinetics of Ethanol after Oral Administration in the Fasting State," *Journal of Pharmacokinetics and Biopharmaceutics* 5 (1977): 207–24.

34. Let $D(t)$ be the US national debt at time t. The table gives approximate values of the function by providing end of year estimates, in billions of dollars, from 1990 to 2010. Interpret and estimate the value of $D'(2000)$.

t	1990	1995	2000	2005	2010
$D(t)$	3233	4974	5662	8170	14,025

Source: US Dept. of the Treasury

35. Let $T(t)$ be the temperature (in °F) in Seattle t hours after midnight on May 7, 2012. The table shows values of this function recorded every two hours. What is the meaning of $T'(12)$? Estimate its value.

t	4	6	8	10	12	14	16
T	48	46	51	57	62	68	71

36. A roast turkey is taken from an oven when its temperature has reached 185°F and is placed on a table in a room where the temperature is 75°F. The graph shows how the temperature of the turkey decreases and eventually approaches room temperature. By measuring the slope of the tangent, estimate the rate of change of the temperature after an hour.

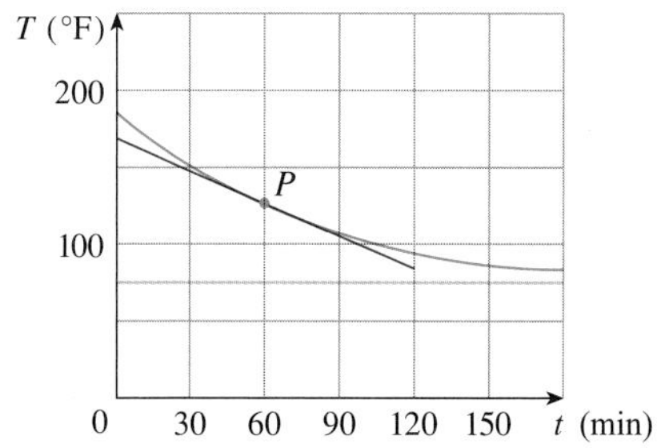

37. A warm can of soda is placed in a cold refrigerator. Sketch the graph of the temperature of the soda as a function of time. Is the initial rate of change of temperature greater or less than the rate of change after an hour?

38. Bacteria population The number of bacteria after t hours in a controlled laboratory experiment is $n = f(t)$.

(a) What is the meaning of the derivative $f'(5)$? What are its units?

(b) Suppose there is an unlimited amount of space and nutrients for the bacteria. Which do you think is larger, $f'(5)$ or $f'(10)$? If the supply of nutrients is limited, would that affect your conclusion? Explain.

39. The cost of producing x ounces of gold from a new gold mine is $C = f(x)$ dollars.

(a) What is the meaning of the derivative $f'(x)$? What are its units?

(b) What does the statement $f'(800) = 17$ mean?

(c) Do you think the values of $f'(x)$ will increase or decrease in the short term? What about the long term? Explain.

40. The quantity (in pounds) of a gourmet ground coffee that is sold by a coffee company at a price of p dollars per pound is $Q = f(p)$.

(a) What is the meaning of the derivative $f'(8)$? What are its units?

(b) Is $f'(8)$ positive or negative? Explain.

41. Oxygen solubility The quantity of oxygen that can dissolve in water depends on the temperature of the water. (So thermal pollution influences the oxygen content of water.) The graph shows how oxygen solubility S varies as a function of the water temperature T.

(a) What is the meaning of the derivative $S'(T)$? What are its units?

(b) Estimate the value of $S'(16)$ and interpret it.

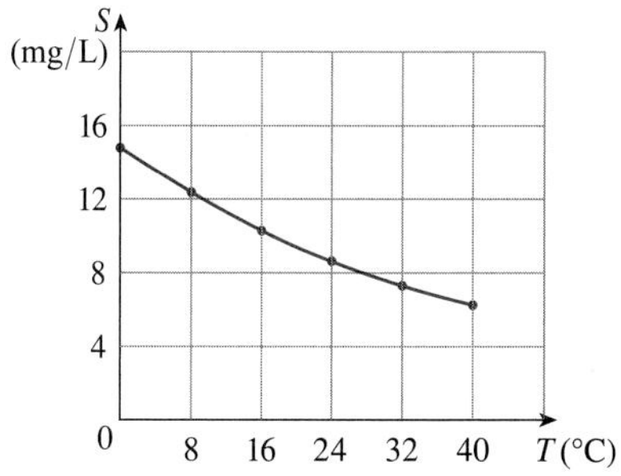

Source: Adapted from C. Kupchella et al., *Environmental Science: Living Within the System of Nature*, 2d ed. (Boston: Allyn and Bacon, 1989).

42. Swimming speed of salmon The graph at the right shows the influence of the temperature T on the maximum sustainable swimming speed S of Coho salmon.

(a) What is the meaning of the derivative $S'(T)$? What are its units?

(b) Estimate the values of $S'(15)$ and $S'(25)$ and interpret them.

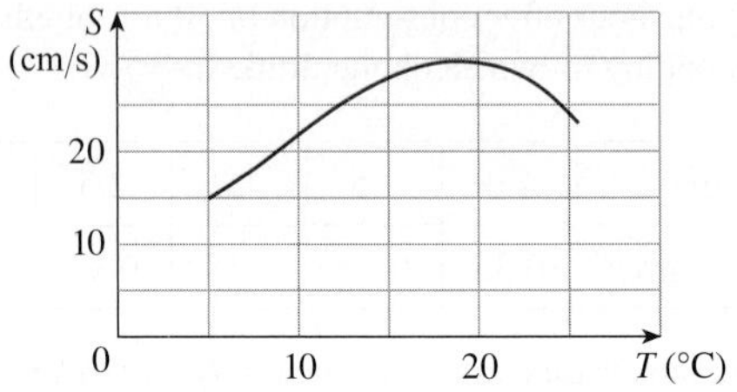

3.2 The Derivative as a Function

In the preceding section we considered the derivative of a function f at a fixed number a:

(1) $$f'(a) = \lim_{h \to 0} \frac{f(a+h) - f(a)}{h}$$

Here we change our point of view and let the number a vary. If we replace a in Equation 1 by a variable x, we obtain

(2) $$f'(x) = \lim_{h \to 0} \frac{f(x+h) - f(x)}{h}$$

Given any number x for which this limit exists, we assign to x the number $f'(x)$. So we can regard f' as a new function, called the **derivative of f** and defined by Equation 2. We know that the value of f' at x, $f'(x)$, can be interpreted geometrically as the slope of the tangent line to the graph of f at the point $(x, f(x))$.

The function f' is called the derivative of f because it has been "derived" from f by the limiting operation in Equation 2. The domain of f' is the set $\{x \mid f'(x) \text{ exists}\}$ and may be smaller than the domain of f.

■ Graphing a Derivative from a Function's Graph

EXAMPLE 1 | The graph of a function f is given in Figure 1. Use it to sketch the graph of the derivative f'.

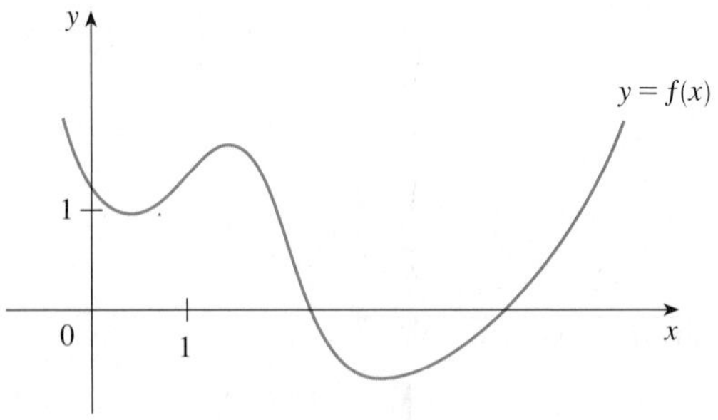

FIGURE 1

SOLUTION We can estimate the value of the derivative at any value of x by drawing the tangent at the point $(x, f(x))$ and estimating its slope. For instance, for $x = 5$ we draw the tangent at P in Figure 2(a) and estimate its slope to be about $\frac{3}{2}$, so $f'(5) \approx 1.5$. This allows us to plot the point $P'(5, 1.5)$ on the graph of f' directly beneath P. Repeating this procedure at several points, we get the graph shown in Figure 2(b). Notice that the tangents at A, B, and C are horizontal, so the derivative is 0 there and the graph of f' crosses the x-axis at the points A', B', and C', directly beneath A, B, and C. Between A and B the tangents have positive slope, so $f'(x)$ is positive there. But between B and C the tangents have negative slope, so $f'(x)$ is negative there.

TEC Visual 3.2 shows an animation of Figure 2 for several functions.

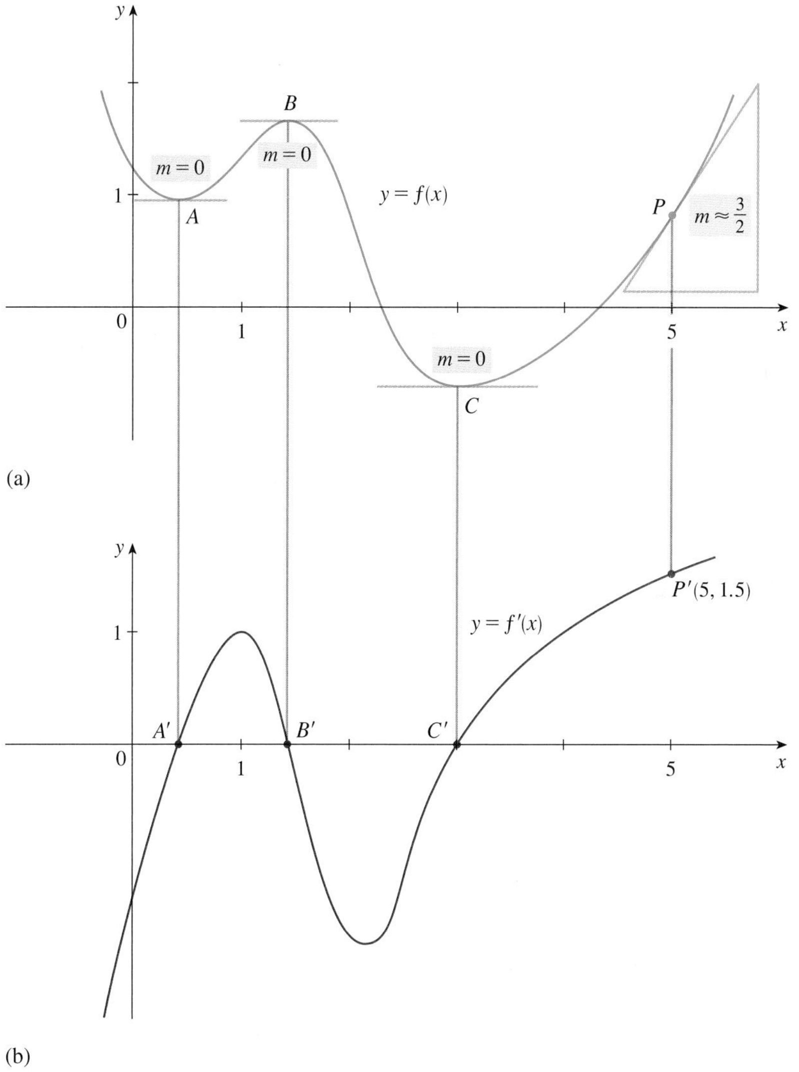

FIGURE 2

EXAMPLE 2 | HIV prevalence and incidence In epidemiology the *prevalence* of a disease, $P(t)$, is the number of individuals currently infected with the disease at time t. The *incidence* of a disease is the rate at which new individuals are contracting the disease. During the initial spread of HIV in the United States, prevalence increased, changing only as a result of new infections. Consequently, $P'(t) \approx$ incidence from 1977–1990 (see Figure 3). Sketch a graph of the incidence of HIV over this time period.

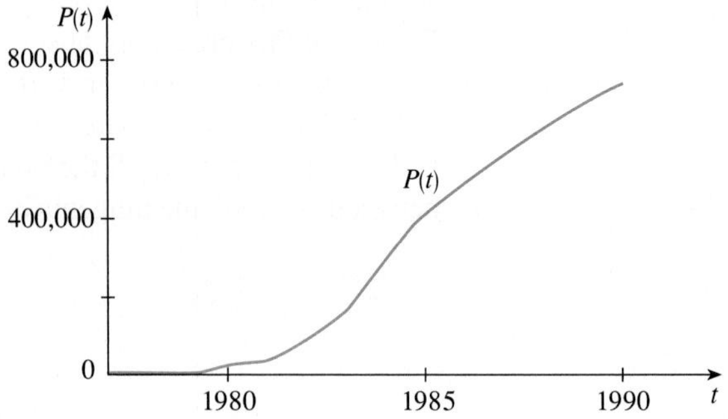

FIGURE 3
HIV prevalence in the United States 1977–1990

Source: R. Song et al., "Estimation of HIV Incidence in the United States," *Journal of the American Medical Association* 300 (2008): 520–29.

SOLUTION We take the incidence over time to be the function $P'(t)$. From the graph of $P(t)$ we see that the slope $P'(t)$ starts off near zero prior to 1980 (the rate of people becoming infected was small) and then increases. We see abrupt increases in the slope around 1981 and 1983, and an abrupt decrease around 1985, after which it remains relatively constant (see Figure 3). As a result, the incidence curve increases quickly in 1981 and again in 1983, and then decreases quickly in 1985 as shown in Figure 4. After 1985 it remains relatively flat, meaning that during this period a constant number of new infections per unit time occurred.

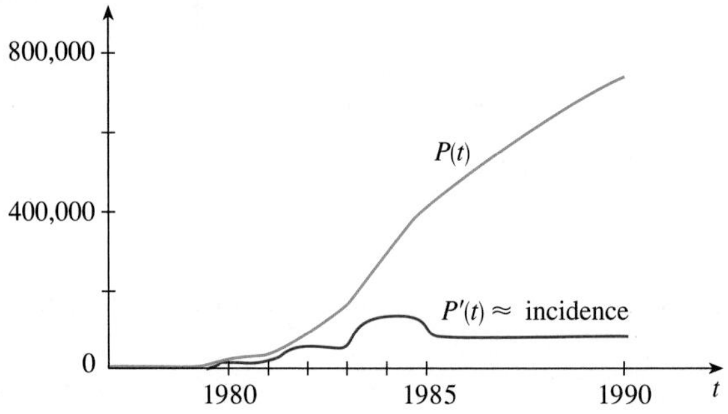

FIGURE 4
HIV prevalence and incidence in the United States 1977–1990

Source: R. Song et al., "Estimation of HIV Incidence in the United States," *Journal of the American Medical Association* 300 (2008): 520–29.

■

■ Finding a Derivative from a Function's Formula

EXAMPLE 3

(a) If $f(x) = x^3 - x$, find a formula for $f'(x)$.
(b) Illustrate by comparing the graphs of f and f'.

SOLUTION

(a) When using Equation 2 to compute a derivative, we must remember that the variable is h and that x is temporarily regarded as a constant during the calculation of the limit.

$$f'(x) = \lim_{h \to 0} \frac{f(x+h) - f(x)}{h} = \lim_{h \to 0} \frac{[(x+h)^3 - (x+h)] - [x^3 - x]}{h}$$

$$= \lim_{h \to 0} \frac{x^3 + 3x^2h + 3xh^2 + h^3 - x - h - x^3 + x}{h}$$

$$= \lim_{h \to 0} \frac{3x^2h + 3xh^2 + h^3 - h}{h} = \lim_{h \to 0} (3x^2 + 3xh + h^2 - 1) = 3x^2 - 1$$

(b) We use a graphing device to graph f and f' in Figure 5. Notice that $f'(x) = 0$ when f has horizontal tangents and $f'(x)$ is positive when the tangents have positive slope. So these graphs serve as a check on our work in part (a).

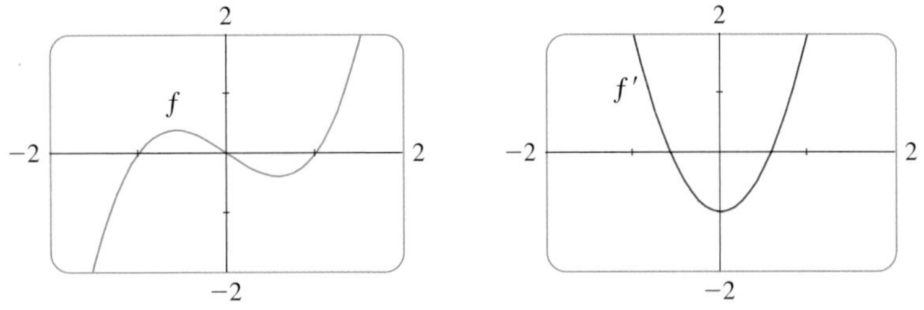

FIGURE 5

EXAMPLE 4 | If $f(x) = \sqrt{x}$, find the derivative of f. State the domain of f'.

SOLUTION

Here we rationalize the numerator.

$$f'(x) = \lim_{h \to 0} \frac{f(x+h) - f(x)}{h} = \lim_{h \to 0} \frac{\sqrt{x+h} - \sqrt{x}}{h}$$

$$= \lim_{h \to 0} \left(\frac{\sqrt{x+h} - \sqrt{x}}{h} \cdot \frac{\sqrt{x+h} + \sqrt{x}}{\sqrt{x+h} + \sqrt{x}} \right)$$

$$= \lim_{h \to 0} \frac{(x+h) - x}{h\left(\sqrt{x+h} + \sqrt{x}\right)} = \lim_{h \to 0} \frac{1}{\sqrt{x+h} + \sqrt{x}}$$

$$= \frac{1}{\sqrt{x} + \sqrt{x}} = \frac{1}{2\sqrt{x}}$$

We see that $f'(x)$ exists if $x > 0$, so the domain of f' is $(0, \infty)$. This is smaller than the domain of f, which is $[0, \infty)$.

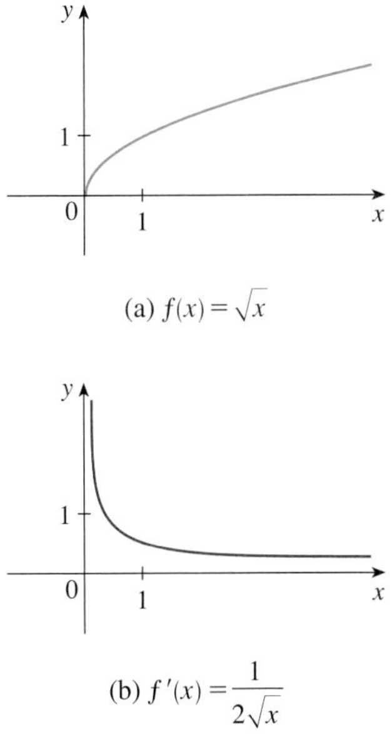

(a) $f(x) = \sqrt{x}$

(b) $f'(x) = \dfrac{1}{2\sqrt{x}}$

FIGURE 6

Let's check to see that the result of Example 4 is reasonable by looking at the graphs of f and f' in Figure 6. When x is close to 0, $\sqrt{x}$ is also close to 0, so $f'(x) = 1/\left(2\sqrt{x}\right)$ is very large and this corresponds to the steep tangent lines near $(0, 0)$ in Figure 6(a) and the large values of $f'(x)$ just to the right of 0 in Figure 6(b). When x is large, $f'(x)$ is very small and this corresponds to the flatter tangent lines at the far right of the graph of f and the horizontal asymptote of the graph of f'.

EXAMPLE 5 | Find f' if $f(x) = \dfrac{1 - x}{2 + x}$.

SOLUTION

$$\frac{\dfrac{a}{b} - \dfrac{c}{d}}{e} = \frac{ad - bc}{bd} \cdot \frac{1}{e}$$

$$f'(x) = \lim_{h \to 0} \frac{f(x+h) - f(x)}{h} = \lim_{h \to 0} \frac{\dfrac{1 - (x+h)}{2 + (x+h)} - \dfrac{1 - x}{2 + x}}{h}$$

$$= \lim_{h \to 0} \frac{(1 - x - h)(2 + x) - (1 - x)(2 + x + h)}{h(2 + x + h)(2 + x)}$$

$$= \lim_{h \to 0} \frac{(2 - x - 2h - x^2 - xh) - (2 - x + h - x^2 - xh)}{h(2 + x + h)(2 + x)}$$

$$= \lim_{h \to 0} \frac{-3h}{h(2 + x + h)(2 + x)} = \lim_{h \to 0} \frac{-3}{(2 + x + h)(2 + x)} = -\frac{3}{(2 + x)^2}$$

Differentiability

If we use the traditional notation $y = f(x)$ to indicate that the independent variable is x and the dependent variable is y, then some common alternative notations for the derivative are as follows:

$$f'(x) = y' = \frac{dy}{dx} = \frac{df}{dx} = \frac{d}{dx} f(x) = Df(x) = D_x f(x)$$

The symbols D and d/dx are called **differentiation operators** because they indicate the operation of **differentiation**, which is the process of calculating a derivative.

The symbol dy/dx, which was introduced by Leibniz, should not be regarded as a ratio (for the time being); it is simply a synonym for $f'(x)$. Nonetheless, it is a very useful and suggestive notation, especially when used in conjunction with increment notation. Referring to Equation 3.1.6, we can rewrite the definition of derivative in Leibniz notation in the form

$$\frac{dy}{dx} = \lim_{\Delta x \to 0} \frac{\Delta y}{\Delta x}$$

If we want to indicate the value of a derivative dy/dx in Leibniz notation at a specific number a, we use the notation

$$\left.\frac{dy}{dx}\right|_{x=a} \qquad \text{or} \qquad \left.\frac{dy}{dx}\right]_{x=a}$$

which is a synonym for $f'(a)$.

Leibniz

Gottfried Wilhelm Leibniz was born in Leipzig in 1646 and studied law, theology, philosophy, and mathematics at the university there, graduating with a bachelor's degree at age 17. After earning his doctorate in law at age 20, Leibniz entered the diplomatic service and spent most of his life traveling to the capitals of Europe on political missions. In particular, he worked to avert a French military threat against Germany and attempted to reconcile the Catholic and Protestant churches.

His serious study of mathematics did not begin until 1672 while he was on a diplomatic mission in Paris. There he built a calculating machine and met scientists, like Huygens, who directed his attention to the latest developments in mathematics and science. Leibniz sought to develop a symbolic logic and system of notation that would simplify logical reasoning. In particular, the version of calculus that he published in 1684 established the notation and the rules for finding derivatives that we use today.

Unfortunately, a dreadful priority dispute arose in the 1690s between the followers of Newton and those of Leibniz as to who had invented calculus first. Leibniz was even accused of plagiarism by members of the Royal Society in England. The truth is that each man invented calculus independently. Newton arrived at his version of calculus first but, because of his fear of controversy, did not publish it immediately. So Leibniz's 1684 account of calculus was the first to be published.

(3) Definition A function f is **differentiable at a** if $f'(a)$ exists. It is **differentiable on an open interval** (a, b) [or (a, ∞) or $(-\infty, a)$ or $(-\infty, \infty)$] if it is differentiable at every number in the interval.

EXAMPLE 6 | Where is the function $f(x) = |x|$ differentiable?

SOLUTION If $x > 0$, then $|x| = x$ and we can choose h small enough that $x + h > 0$ and hence $|x + h| = x + h$. Therefore, for $x > 0$, we have

$$f'(x) = \lim_{h \to 0} \frac{|x + h| - |x|}{h} = \lim_{h \to 0} \frac{(x + h) - x}{h}$$

$$= \lim_{h \to 0} \frac{h}{h} = \lim_{h \to 0} 1 = 1$$

and so f is differentiable for any $x > 0$.

Similarly, for $x < 0$ we have $|x| = -x$ and h can be chosen small enough that $x + h < 0$ and so $|x + h| = -(x + h)$. Therefore, for $x < 0$,

$$f'(x) = \lim_{h \to 0} \frac{|x + h| - |x|}{h} = \lim_{h \to 0} \frac{-(x + h) - (-x)}{h}$$

$$= \lim_{h \to 0} \frac{-h}{h} = \lim_{h \to 0} (-1) = -1$$

and so f is differentiable for any $x < 0$.

For $x = 0$ we have to investigate

$$f'(0) = \lim_{h \to 0} \frac{f(0 + h) - f(0)}{h}$$

$$= \lim_{h \to 0} \frac{|0 + h| - |0|}{h} \qquad \text{(if it exists)}$$

Let's compute the left and right limits separately:

$$\lim_{h \to 0^+} \frac{|0 + h| - |0|}{h} = \lim_{h \to 0^+} \frac{|h|}{h} = \lim_{h \to 0^+} \frac{h}{h} = \lim_{h \to 0^+} 1 = 1$$

and $$\lim_{h \to 0^-} \frac{|0 + h| - |0|}{h} = \lim_{h \to 0^-} \frac{|h|}{h} = \lim_{h \to 0^-} \frac{-h}{h} = \lim_{h \to 0^-} (-1) = -1$$

Since these limits are different, $f'(0)$ does not exist. Thus f is differentiable at all x except 0.

A formula for f' is given by

$$f'(x) = \begin{cases} 1 & \text{if } x > 0 \\ -1 & \text{if } x < 0 \end{cases}$$

and its graph is shown in Figure 7(b). The fact that $f'(0)$ does not exist is reflected geometrically in the fact that the curve $y = |x|$ does not have a tangent line at (0, 0). [See Figure 7(a).] ■

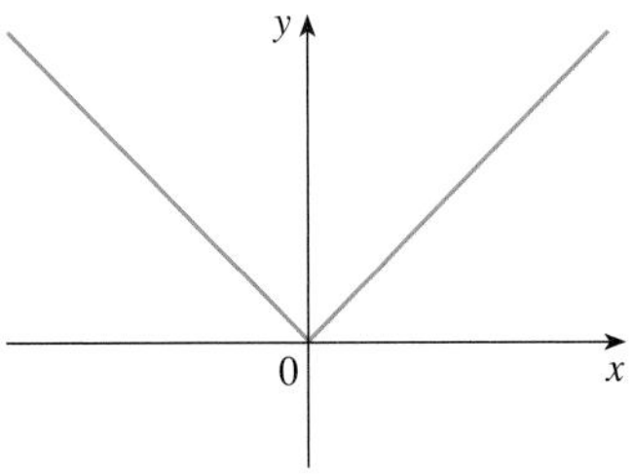

(a) $y = f(x) = |x|$

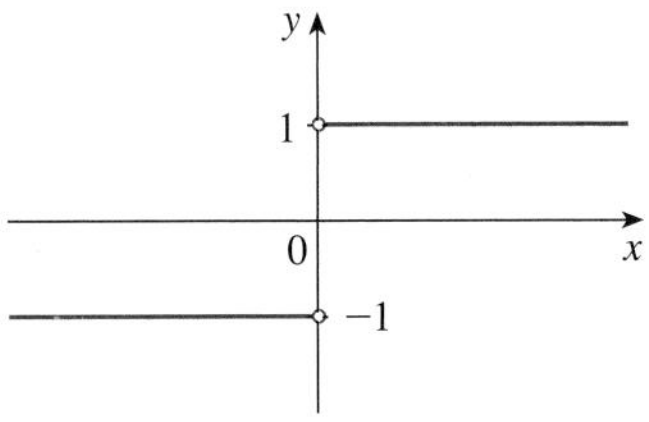

(b) $y = f'(x)$

FIGURE 7

Both continuity and differentiability are desirable properties for a function to have. The following theorem shows how these properties are related.

> **(4) Theorem** If f is differentiable at a, then f is continuous at a.

PROOF To prove that f is continuous at a, we have to show that $\lim_{x \to a} f(x) = f(a)$. We do this by first showing that the difference $f(x) - f(a)$ approaches 0.

The given information is that f is differentiable at a, that is,

$$f'(a) = \lim_{x \to a} \frac{f(x) - f(a)}{x - a}$$

exists (see Equation 3.1.5). To connect the given and the unknown, we divide and multiply $f(x) - f(a)$ by $x - a$ (which we can do when $x \neq a$):

$$f(x) - f(a) = \frac{f(x) - f(a)}{x - a}(x - a)$$

Thus, using the Product Law and (3.1.5), we can write

$$\lim_{x \to a} [f(x) - f(a)] = \lim_{x \to a} \frac{f(x) - f(a)}{x - a}(x - a)$$

$$= \lim_{x \to a} \frac{f(x) - f(a)}{x - a} \cdot \lim_{x \to a} (x - a)$$

$$= f'(a) \cdot 0 = 0$$

To use what we have just proved, we start with $f(x)$ and add and subtract $f(a)$:

$$\begin{aligned}\lim_{x \to a} f(x) &= \lim_{x \to a} [f(a) + (f(x) - f(a))] \\ &= \lim_{x \to a} f(a) + \lim_{x \to a} [f(x) - f(a)] \\ &= f(a) + 0 = f(a)\end{aligned}$$

Therefore f is continuous at a. ■

NOTE The converse of Theorem 4 is false; that is, there are functions that are continuous but not differentiable. For instance, the function $f(x) = |x|$ is continuous at 0 because

$$\lim_{x \to 0} f(x) = \lim_{x \to 0} |x| = 0 = f(0)$$

(See Example 2.4.8.) But in Example 6 we showed that f is not differentiable at 0.

We saw that the function $y = |x|$ in Example 6 is not differentiable at 0 and Figure 7(a) shows that its graph changes direction abruptly when $x = 0$. In general, if the graph of a function f has a "corner" or "kink" in it, then the graph of f has no tangent at this point and f is not differentiable there. [In trying to compute $f'(a)$, we find that the left and right limits are different.]

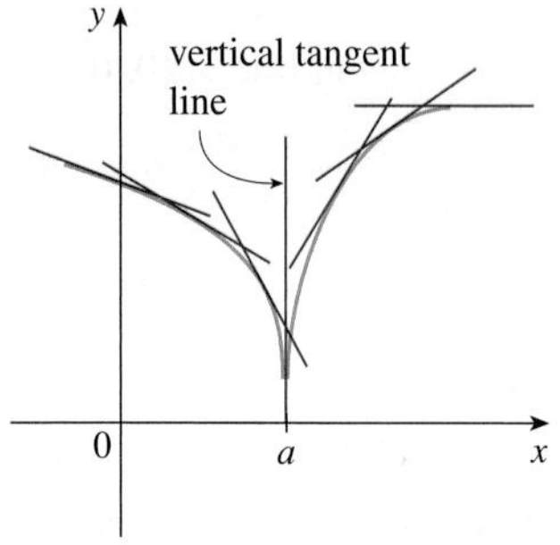

FIGURE 8

Theorem 4 gives another way for a function not to have a derivative. It says that if f is not continuous at a, then f is not differentiable at a. So at any discontinuity (for instance, a jump discontinuity) f fails to be differentiable.

A third possibility is that the curve has a **vertical tangent line** when $x = a$; that is, f is continuous at a and

$$\lim_{x \to a} |f'(x)| = \infty$$

This means that the tangent lines become steeper and steeper as $x \to a$. Figure 8 shows one way that this can happen; Figure 9(c) shows another. Figure 9 illustrates the three possibilities that we have discussed.

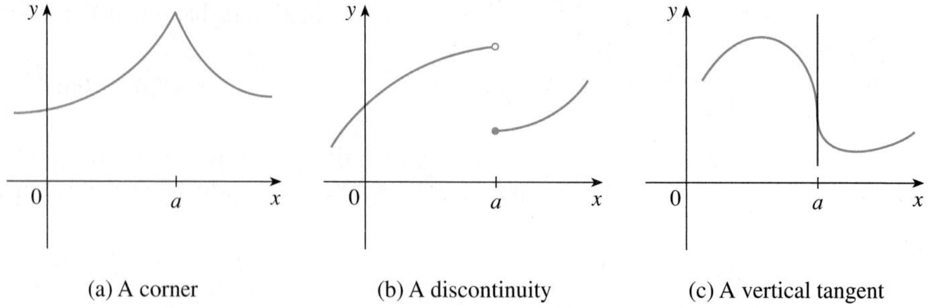

FIGURE 9
Three ways for f not to be differentiable at a

A graphing calculator or computer provides another way of looking at differentiability. If f is differentiable at a, then when we zoom in toward the point $(a, f(a))$ the graph straightens out and appears more and more like a line. (See Figure 10. We saw a specific example of this in Figure 3.1.5.) But no matter how much we zoom in toward a point like the ones in Figures 8 and 9(a), we can't eliminate the sharp point or corner (see Figure 11).

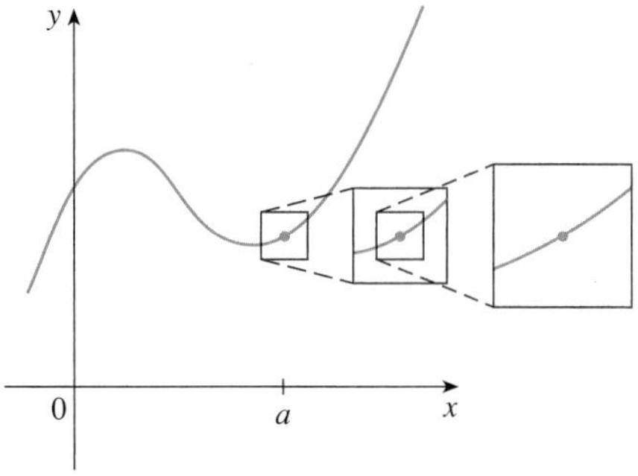

FIGURE 10
f is differentiable at a.

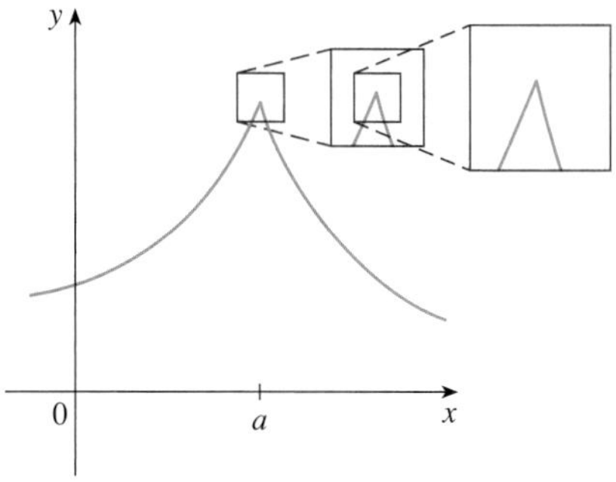

FIGURE 11
f is not differentiable at a.

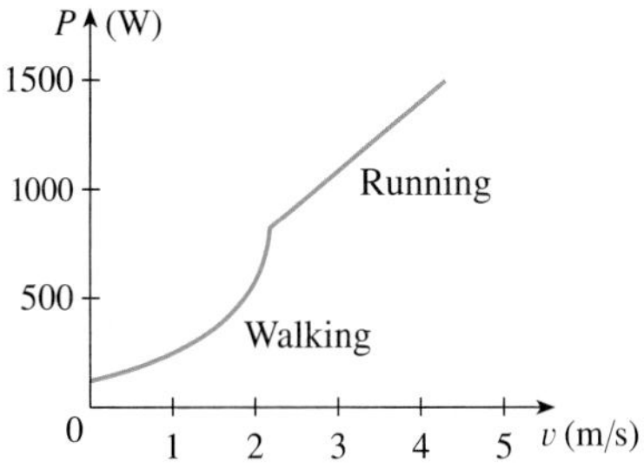

FIGURE 12

EXAMPLE 7 | **BB** Metabolic power in walking and running Figure 12 shows a graph of the metabolic power $P(v)$ consumed by humans who walk and then run at speed v.
(a) Is P a differentiable function of v?
(b) Sketch the graph of $P'(v)$.

SOLUTION
(a) We see from the graph of P in Figure 12 that P is not differentiable at speed $v \approx 2.2$ m/s because the graph has a corner there. But P is differentiable at all other speeds.
(b) Using the method of Example 1, we measure slopes at a few points (as in Figure 13) and plot the resulting points in Figure 14.

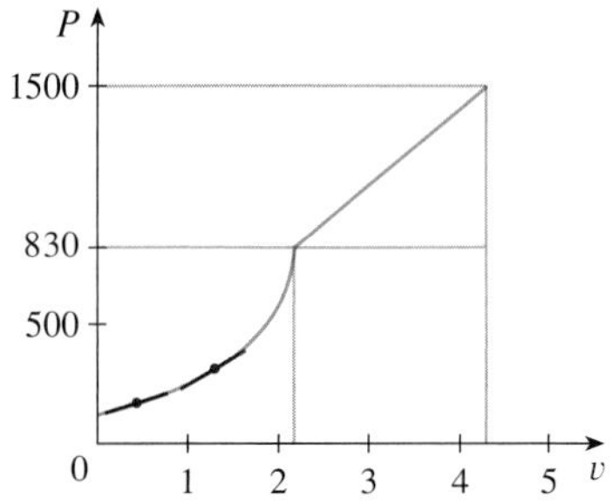

FIGURE 13

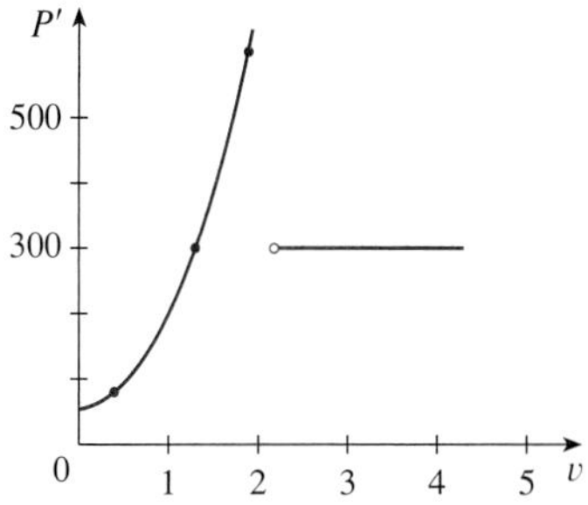

FIGURE 14

Notice, however, that unlike in Example 1, the axes have unequal scales. For instance, for $2.2 \leqslant v \leqslant 4.4$, the slope is approximately

$$\frac{1500 - 830}{4.4 - 2.2} = \frac{670}{2.2} \approx 305 \text{ W/(m/s)}$$

The discontinuity in the graph of P' at $v \approx 2.2$ reflects the fact that $P'(v)$ does not exist there. ■

■ Higher Derivatives

If f is a differentiable function, then its derivative f' is also a function, so f' may have a derivative of its own, denoted by $(f')' = f''$. This new function f'' is called the **second**

derivative of f because it is the derivative of the derivative of f. Using Leibniz notation, we write the second derivative of $y = f(x)$ as

$$\frac{d}{dx}\left(\frac{dy}{dx}\right) = \frac{d^2y}{dx^2}$$

EXAMPLE 8 | If $f(x) = x^3 - x$, find and interpret $f''(x)$.

SOLUTION In Example 3 we found that the first derivative is $f'(x) = 3x^2 - 1$. So the second derivative is

$$f''(x) = (f')'(x) = \lim_{h \to 0} \frac{f'(x+h) - f'(x)}{h}$$

$$= \lim_{h \to 0} \frac{[3(x+h)^2 - 1] - [3x^2 - 1]}{h}$$

$$= \lim_{h \to 0} \frac{3x^2 + 6xh + 3h^2 - 1 - 3x^2 + 1}{h}$$

$$= \lim_{h \to 0} (6x + 3h) = 6x$$

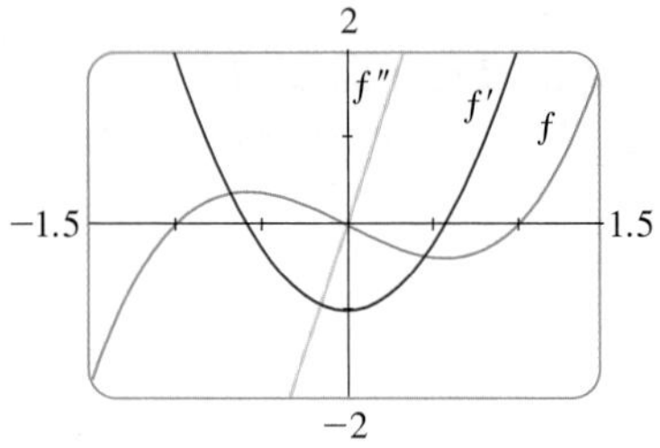

FIGURE 15

The graphs of f, f', and f'' are shown in Figure 15.

We can interpret $f''(x)$ as the slope of the curve $y = f'(x)$ at the point $(x, f'(x))$. In other words, it is the rate of change of the slope of the original curve $y = f(x)$.

Notice from Figure 15 that $f''(x)$ is negative when $y = f'(x)$ has negative slope and positive when $y = f'(x)$ has positive slope. So the graphs serve as a check on our calculations. ■

TEC In Module 3.2 you can see how changing the coefficients of a polynomial f affects the appearance of the graphs of f, f', and f''.

In general, we can interpret a second derivative as a rate of change of a rate of change. The most familiar example of this is *acceleration*, which we define as follows.

If $s = s(t)$ is the position function of an object that moves in a straight line, we know that its first derivative represents the velocity $v(t)$ of the object as a function of time:

$$v(t) = s'(t) = \frac{ds}{dt}$$

The instantaneous rate of change of velocity with respect to time is called the **acceleration** $a(t)$ of the object. Thus the acceleration function is the derivative of the velocity function and is therefore the second derivative of the position function:

$$a(t) = v'(t) = s''(t)$$

or, in Leibniz notation,

$$a = \frac{dv}{dt} = \frac{d^2s}{dt^2}$$

The **third derivative** f''' is the derivative of the second derivative: $f''' = (f'')'$. So $f'''(x)$ can be interpreted as the slope of the curve $y = f''(x)$ or as the rate of change of $f''(x)$. If $y = f(x)$, then alternative notations for the third derivative are

$$y''' = f'''(x) = \frac{d}{dx}\left(\frac{d^2y}{dx^2}\right) = \frac{d^3y}{dx^3}$$

The process can be continued. The fourth derivative f'''' is usually denoted by $f^{(4)}$. In general, the nth derivative of f is denoted by $f^{(n)}$ and is obtained from f by differentiating n times. If $y = f(x)$, we write

$$y^{(n)} = f^{(n)}(x) = \frac{d^n y}{dx^n}$$

EXAMPLE 9 | If $f(x) = x^3 - x$, find $f'''(x)$ and $f^{(4)}(x)$.

SOLUTION In Example 8 we found that $f''(x) = 6x$. The graph of the second derivative has equation $y = 6x$ and so it is a straight line with slope 6. Since the derivative $f'''(x)$ is the slope of $f''(x)$, we have

$$f'''(x) = 6$$

for all values of x. So f''' is a constant function and its graph is a horizontal line. Therefore, for all values of x,

$$f^{(4)}(x) = 0$$

■

■ What a Derivative Tells Us about a Function

Because $f'(x)$ represents the slope of the curve $y = f(x)$ at the point $(x, f(x))$, it tells us the direction in which the curve proceeds at each point. So it is reasonable to expect that information about $f'(x)$ will provide us with information about $f(x)$.

In particular, to see how the derivative of f can tell us where a function is increasing or decreasing, look at Figure 16. (Increasing functions and decreasing functions were defined in Section 1.1.) Between A and B and between C and D, the tangent lines have positive slope and so $f'(x) > 0$. Between B and C, the tangent lines have negative slope and so $f'(x) < 0$. Thus it appears that f increases when $f'(x)$ is positive and decreases when $f'(x)$ is negative.

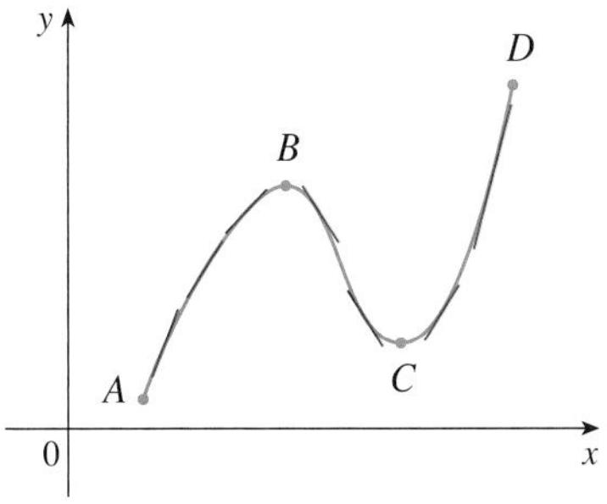

FIGURE 16

It turns out, as we will see in Chapter 4, that what we observed for the function graphed in Figure 16 is always true. We state the general result as follows.

> If $f'(x) > 0$ on an interval, then f is increasing on that interval.
>
> If $f'(x) < 0$ on an interval, then f is decreasing on that interval.

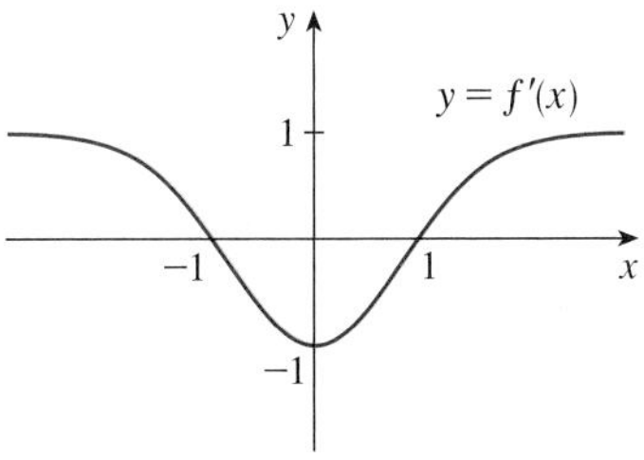

FIGURE 17

EXAMPLE 10

(a) If it is known that the graph of the derivative f' of a function is as shown in Figure 17, what can we say about f?

(b) If it is known that $f(0) = 0$, sketch a possible graph of f.

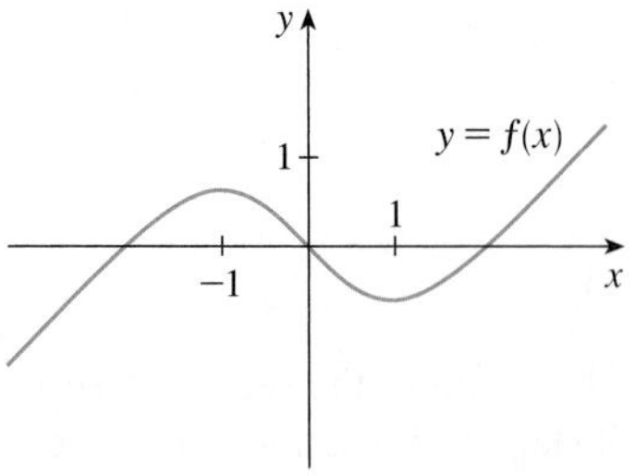

FIGURE 18

SOLUTION

(a) We observe from Figure 17 that $f'(x)$ is negative when $-1 < x < 1$, so the original function f must be decreasing on the interval $(-1, 1)$. Similarly, $f'(x)$ is positive for $x < -1$ and for $x > 1$, so f is increasing on the intervals $(-\infty, -1)$ and $(1, \infty)$. Also note that, since $f'(-1) = 0$ and $f'(1) = 0$, the graph of f has horizontal tangents when $x = \pm 1$.

(b) We use the information from part (a), and the fact that the graph passes through the origin, to sketch a possible graph of f in Figure 18. Notice that $f'(0) = -1$, so we have drawn the curve $y = f(x)$ passing through the origin with a slope of -1. Notice also that $f'(x) \to 1$ as $x \to \pm\infty$ (from Figure 17). So the slope of the curve $y = f(x)$ approaches 1 as x becomes large (positive or negative). That is why we have drawn the graph of f in Figure 18 progressively straighter as $x \to \pm\infty$. ■

EXERCISES 3.2

1–2 Use the given graph to estimate the value of each derivative. Then sketch the graph of f'.

1. (a) $f'(-3)$ (b) $f'(-2)$
(c) $f'(-1)$ (d) $f'(0)$
(e) $f'(1)$ (f) $f'(2)$
(g) $f'(3)$

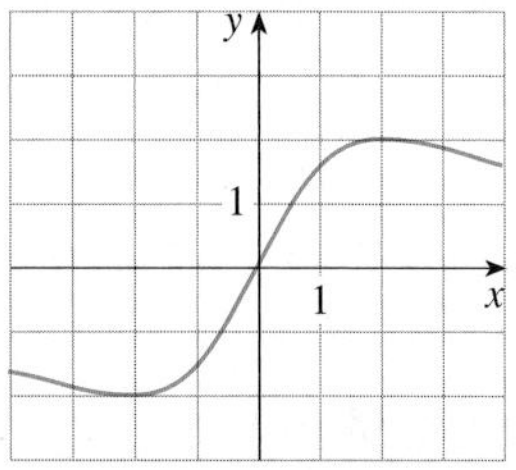

2. (a) $f'(0)$ (b) $f'(1)$
(c) $f'(2)$ (d) $f'(3)$
(e) $f'(4)$ (f) $f'(5)$
(g) $f'(6)$ (h) $f'(7)$

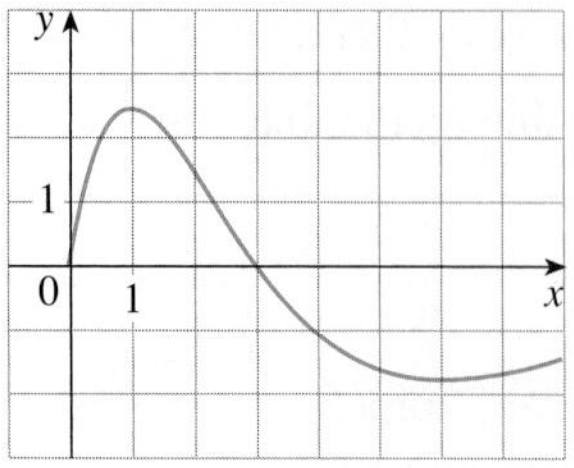

3. Match the graph of each function in (a)–(d) with the graph of its derivative in I–IV. Give reasons for your choices.

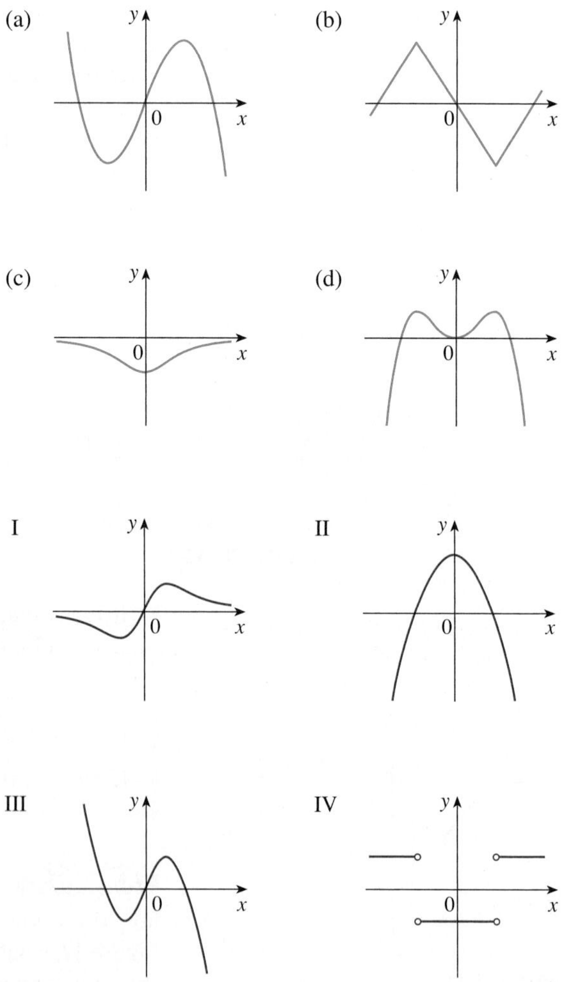

4–11 Trace or copy the graph of the given function f. (Assume that the axes have equal scales.) Then use the method of Example 1 to sketch the graph of f' below it.

4.

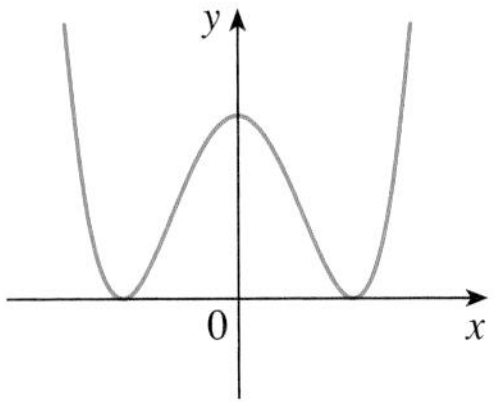

5.

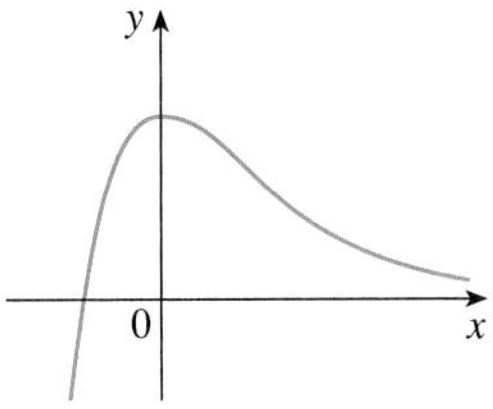

6.

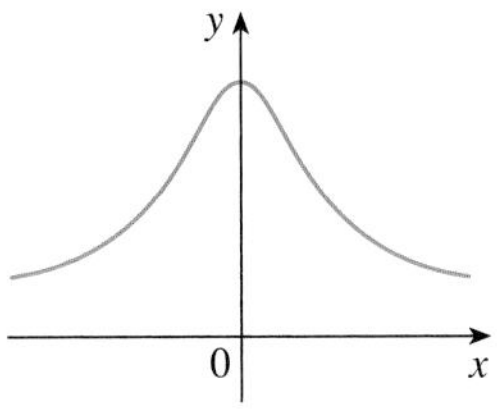

7.

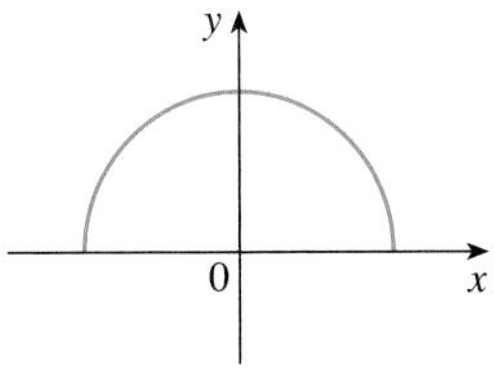

8.

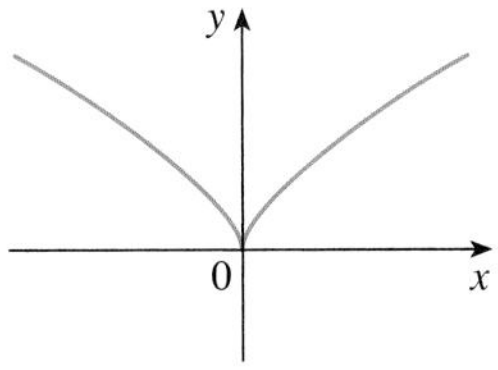

9.

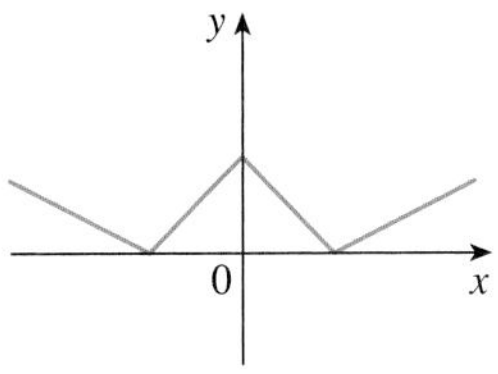

10.

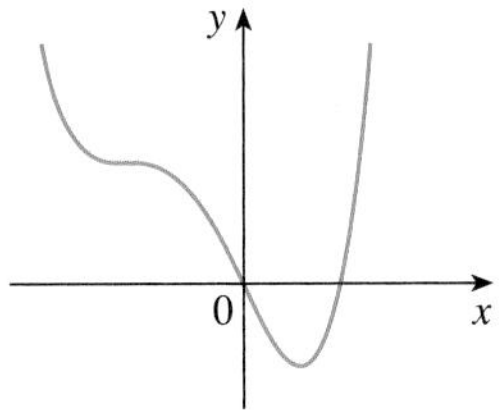

11.

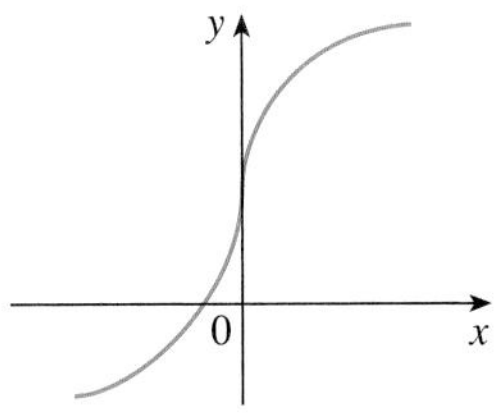

12. Yeast population Shown is the graph of the population function $P(t)$ for yeast cells in a laboratory culture. Use the method of Example 1 to graph the derivative $P'(t)$. What does the graph of P' tell us about the yeast population?

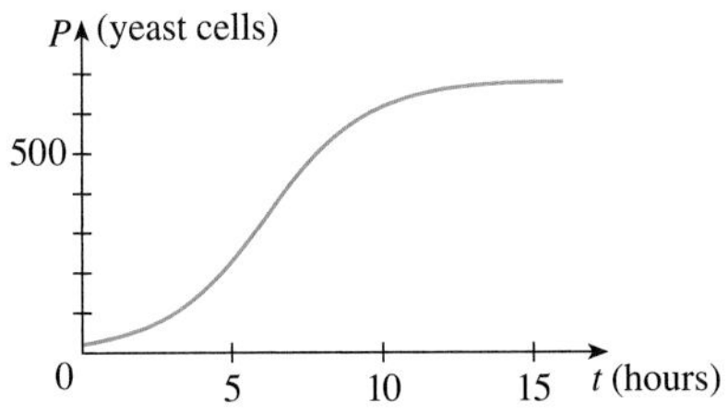

13. Tadpole weights The graph shows the average body weight W as a function of time for tadpoles raised in a density of 80 tadpoles/L.

(a) What is the meaning of the derivative $W'(t)$?

(b) Sketch the graph of $W'(t)$.

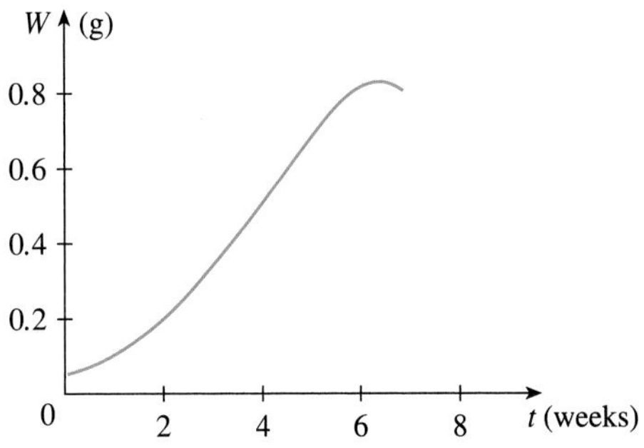

14. Ground reaction force in walking The graph shows the horizontal force $F(t)$ exerted by the ground on a person who is walking.

(a) What is the meaning of the derivative $F'(t)$?

(b) Sketch the graph of $F'(t)$.

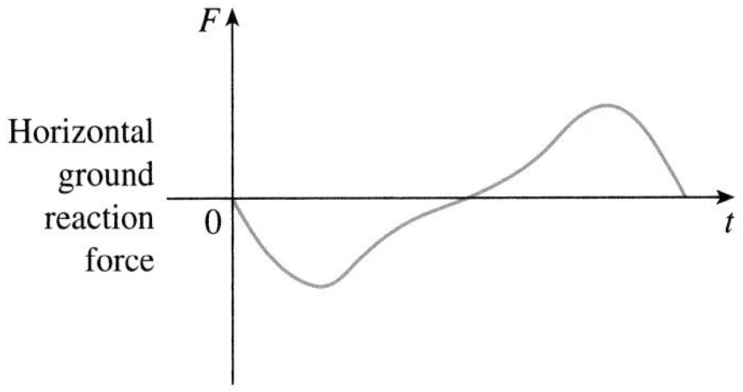

15. Marriage age The graph shows how the average age of first marriage of Japanese men varied in the last half of the 20th century. Sketch the graph of the derivative function $M'(t)$. During which years was the derivative negative?

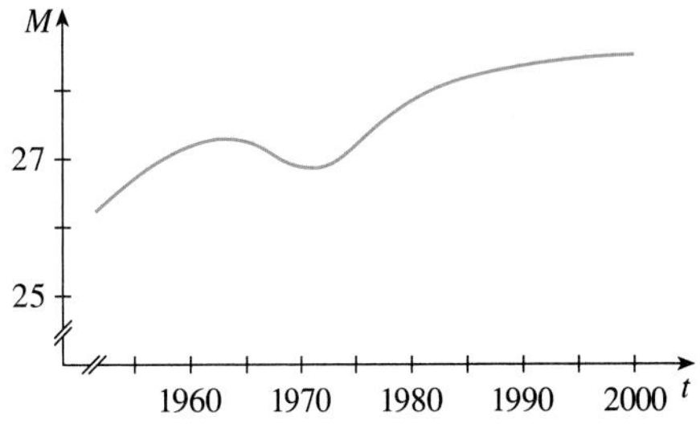

16–18 Make a careful sketch of the graph of f and below it sketch the graph of f' in the same manner as in Exercises 4–11. Can you guess a formula for $f'(x)$ from its graph?

16. $f(x) = \sin x$

17. $f(x) = e^x$

18. $f(x) = \ln x$

19. Let $f(x) = x^2$.
(a) Estimate the values of $f'(0)$, $f'(\frac{1}{2})$, $f'(1)$, and $f'(2)$ by using a graphing device to zoom in on the graph of f.
(b) Use symmetry to deduce the values of $f'(-\frac{1}{2})$, $f'(-1)$, and $f'(-2)$.
(c) Use the results from parts (a) and (b) to guess a formula for $f'(x)$.
(d) Use the definition of a derivative to prove that your guess in part (c) is correct.

20. Let $f(x) = x^3$.
(a) Estimate the values of $f'(0)$, $f'(\frac{1}{2})$, $f'(1)$, $f'(2)$, and $f'(3)$ by using a graphing device to zoom in on the graph of f.
(b) Use symmetry to deduce the values of $f'(-\frac{1}{2})$, $f'(-1)$, $f'(-2)$, and $f'(-3)$.
(c) Use the values from parts (a) and (b) to graph f'.
(d) Guess a formula for $f'(x)$.
(e) Use the definition of a derivative to prove that your guess in part (d) is correct.

21–31 Find the derivative of the function using the definition of a derivative. State the domain of the function and the domain of its derivative.

21. $f(x) = \frac{1}{2}x - \frac{1}{3}$

22. $f(x) = mx + b$

23. $f(t) = 5t - 9t^2$

24. $f(x) = 1.5x^2 - x + 3.7$

25. $f(x) = x^2 - 2x^3$

26. $f(x) = x + \sqrt{x}$

27. $g(x) = \sqrt{1 + 2x}$

28. $f(x) = \dfrac{x^2 - 1}{2x - 3}$

29. $G(t) = \dfrac{4t}{t + 1}$

30. $g(t) = \dfrac{1}{\sqrt{t}}$

31. $f(x) = x^4$

32–34
(a) Use the definition of the derivative to calculate f'.
(b) Check to see that your answer is reasonable by comparing the graphs of f and f'.

32. $f(x) = x + 1/x$

33. $f(x) = x^4 + 2x$

34. $f(t) = t^2 - \sqrt{t}$

35. Malarial parasites An experiment measured the number of malarial parasites $N(t)$ per microliter of blood, where t is measured in days. The results of the experiment are shown in the table.

t	N	t	N
1	228	5	372,331
2	2357	6	2,217,441
3	12,750	7	6,748,400
4	26,661		

(a) What is the meaning of $N'(t)$? What are its units?
(b) Construct a table of estimated values for $N'(t)$.
[See Example 3.1.7 for $N'(3)$.]

36. Blood alcohol concentration Researchers measured the blood alcohol concentration $C(t)$ of eight adult male subjects after rapid consumption of 30 mL of ethanol (corresponding to two standard alcoholic drinks). The table shows the data they obtained by averaging the BAC (in mg/mL) of the eight men.

t (hours)	0.0	0.2	0.5	0.75	1.0	1.25
$C(t)$	0	0.25	0.41	0.40	0.33	0.29

t (hours)	1.5	1.75	2.0	2.25	2.5	3.0
$C(t)$	0.24	0.22	0.18	0.15	0.12	0.07

(a) What is the meaning of $C'(t)$?
(b) Make a table of estimated values for $C'(t)$.

Source: Adapted from P. Wilkinson et al., "Pharmacokinetics of Ethanol after Oral Administration in the Fasting State," *Journal of Pharmacokinetics and Biopharmaceutics*, 5 (1977): 207–24.

37–40 The graph of f is given. State, with reasons, the numbers at which f is not differentiable.

37.

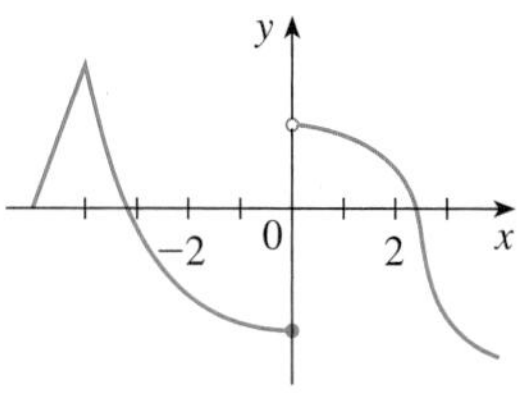

38.

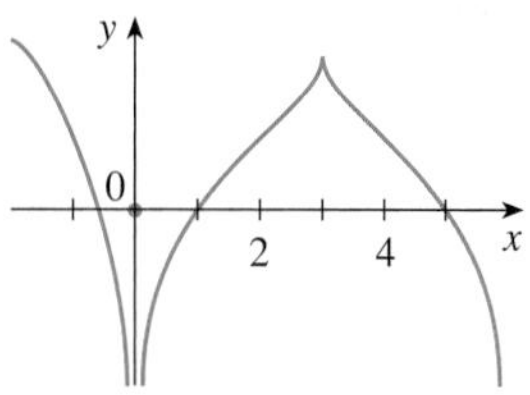

39.

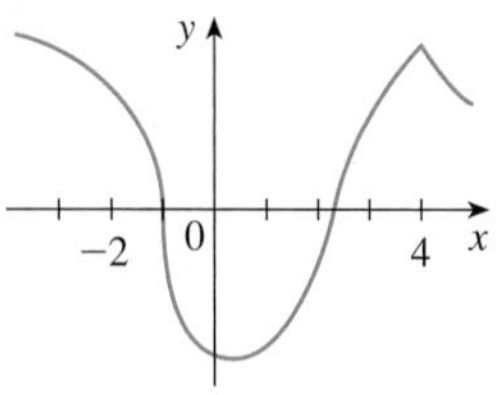

40.

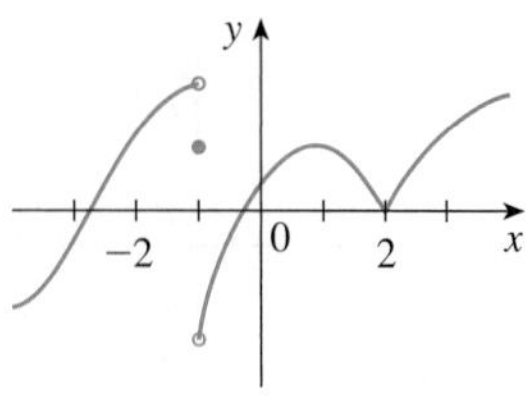

41. Graph the function $f(x) = x + \sqrt{|x|}$. Zoom in repeatedly, first toward the point $(-1, 0)$ and then toward the origin. What is different about the behavior of f in the vicinity of these two points? What do you conclude about the differentiability of f?

42. Zoom in toward the points $(1, 0)$, $(0, 1)$, and $(-1, 0)$ on the graph of the function $g(x) = (x^2 - 1)^{2/3}$. What do you notice? Account for what you see in terms of the differentiability of g.

43. The figure shows the graphs of f, f', and f''. Identify each curve, and explain your choices.

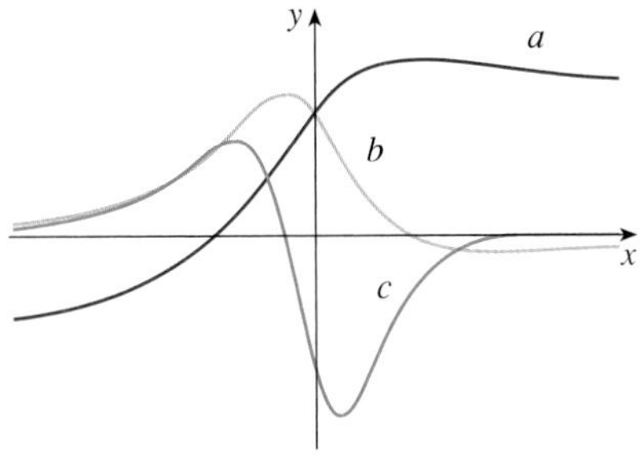

44. The figure shows graphs of f, f', f'', and f'''. Identify each curve, and explain your choices.

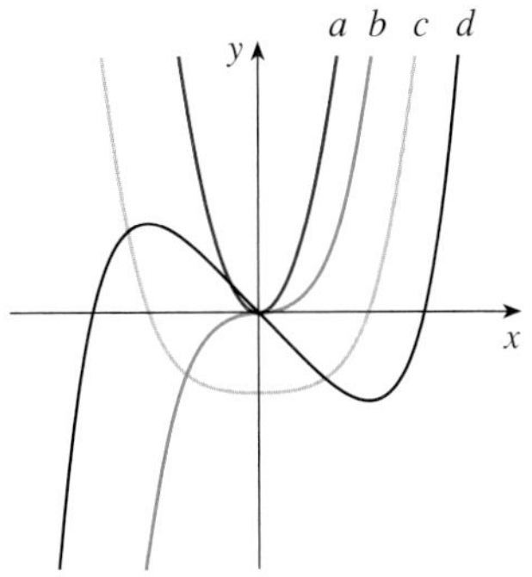

45–46 Use the definition of a derivative to find $f'(x)$ and $f''(x)$. Then graph f, f', and f'' on a common screen and check to see if your answers are reasonable.

45. $f(x) = 3x^2 + 2x + 1$ **46.** $f(x) = x^3 - 3x$

47–48 The graph of the *derivative* f' of a function f is shown.

(a) On what intervals is f increasing?

(b) If it is known that $f(0) = 0$, sketch a possible graph of f.

47.

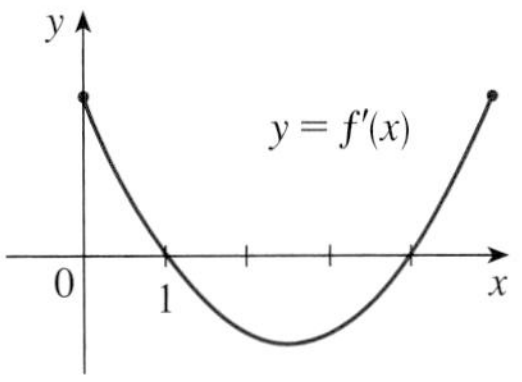

48.

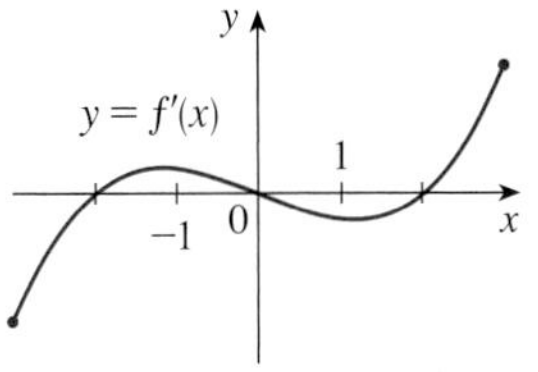

49. Recall that a function f is called *even* if $f(-x) = f(x)$ for all x in its domain and *odd* if $f(-x) = -f(x)$ for all such x. Prove each of the following.

(a) The derivative of an even function is an odd function.

(b) The derivative of an odd function is an even function.

50. When you turn on a hot-water faucet, the temperature T of the water depends on how long the water has been running.

(a) Sketch a possible graph of T as a function of the time t that has elapsed since the faucet was turned on.

(b) Describe how the rate of change of T with respect to t varies as t increases.

(c) Sketch a graph of the derivative of T.

3.3 Basic Differentiation Formulas

If it were always necessary to compute derivatives directly from the definition, as we did in the preceding section, such computations would be tedious and the evaluation of some limits would require ingenuity. Fortunately, several rules have been developed for finding derivatives without having to use the definition directly. These formulas greatly simplify the task of differentiation.

In this section we learn how to differentiate constant functions, power functions, polynomials, exponential functions, and the sine and cosine functions.

Let's start with the simplest of all functions, the constant function $f(x) = c$. The graph of this function is the horizontal line $y = c$, which has slope 0, so we must have $f'(x) = 0$. (See Figure 1.) A formal proof, from the definition of a derivative, is also easy:

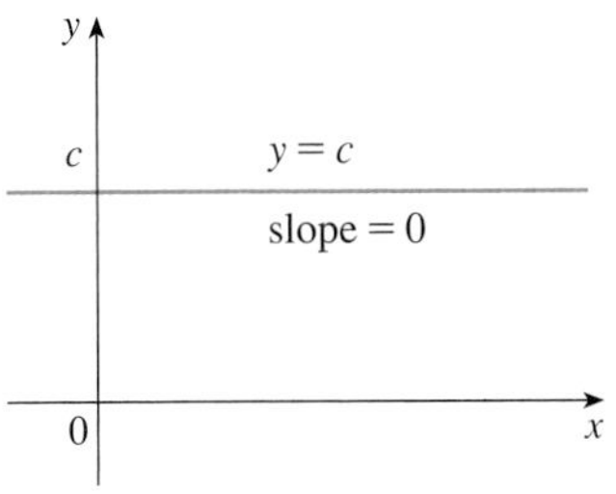

FIGURE 1
The graph of $f(x) = c$ is the line $y = c$, so $f'(x) = 0$.

$$f'(x) = \lim_{h \to 0} \frac{f(x + h) - f(x)}{h} = \lim_{h \to 0} \frac{c - c}{h} = \lim_{h \to 0} 0 = 0$$

In Leibniz notation, we write this rule as follows.

Derivative of a Constant Function

$$\frac{d}{dx}(c) = 0$$

Power Functions

We next look at the functions $f(x) = x^n$, where n is a positive integer. If $n = 1$, the graph of $f(x) = x$ is the line $y = x$, which has slope 1. (See Figure 2.) So

(1) $$\frac{d}{dx}(x) = 1$$

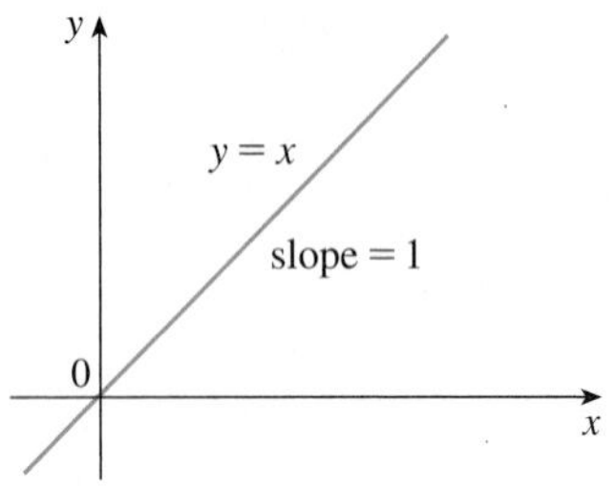

FIGURE 2
The graph of $f(x) = x$ is the line $y = x$, so $f'(x) = 1$.

(You can also verify Equation 1 from the definition of a derivative.) We have already investigated the cases $n = 2$ and $n = 3$. In fact, in Section 3.2 (Exercises 19 and 20) we found that

(2) $$\frac{d}{dx}(x^2) = 2x \qquad \frac{d}{dx}(x^3) = 3x^2$$

For $n = 4$ we find the derivative of $f(x) = x^4$ as follows:

$$\begin{aligned} f'(x) &= \lim_{h\to 0}\frac{f(x+h) - f(x)}{h} = \lim_{h\to 0}\frac{(x+h)^4 - x^4}{h} \\ &= \lim_{h\to 0}\frac{x^4 + 4x^3h + 6x^2h^2 + 4xh^3 + h^4 - x^4}{h} \\ &= \lim_{h\to 0}\frac{4x^3h + 6x^2h^2 + 4xh^3 + h^4}{h} \\ &= \lim_{h\to 0}(4x^3 + 6x^2h + 4xh^2 + h^3) = 4x^3 \end{aligned}$$

Thus

(3) $$\frac{d}{dx}(x^4) = 4x^3$$

Comparing the equations in (1), (2), and (3), we see a pattern emerging. It seems to be a reasonable guess that, when n is a positive integer, $(d/dx)(x^n) = nx^{n-1}$. This turns out to be true.

The Power Rule If n is a positive integer, then

$$\frac{d}{dx}(x^n) = nx^{n-1}$$

PROOF If $f(x) = x^n$, then

$$f'(x) = \lim_{h\to 0} \frac{f(x+h) - f(x)}{h} = \lim_{h\to 0} \frac{(x+h)^n - x^n}{h}$$

The Binomial Theorem is given on Reference Page 1.

In finding the derivative of x^4 we had to expand $(x + h)^4$. Here we need to expand $(x + h)^n$ and we use the Binomial Theorem to do so:

$$f'(x) = \lim_{h\to 0} \frac{\left[x^n + nx^{n-1}h + \dfrac{n(n-1)}{2}x^{n-2}h^2 + \cdots + nxh^{n-1} + h^n\right] - x^n}{h}$$

$$= \lim_{h\to 0} \frac{nx^{n-1}h + \dfrac{n(n-1)}{2}x^{n-2}h^2 + \cdots + nxh^{n-1} + h^n}{h}$$

$$= \lim_{h\to 0} \left[nx^{n-1} + \frac{n(n-1)}{2}x^{n-2}h + \cdots + nxh^{n-2} + h^{n-1}\right]$$

$$= nx^{n-1}$$

because every term except the first has h as a factor and therefore approaches 0. ■

We illustrate the Power Rule using various notations in Example 1.

EXAMPLE 1

(a) If $f(x) = x^6$, then $f'(x) = 6x^5$.

(b) If $y = x^{1000}$, then $y' = 1000x^{999}$.

(c) If $y = t^4$, then $\dfrac{dy}{dt} = 4t^3$.

(d) $\dfrac{d}{dr}(r^3) = 3r^2$ ■

What about power functions with negative integer exponents? In Exercise 69 we ask you to verify from the definition of a derivative that

$$\frac{d}{dx}\left(\frac{1}{x}\right) = -\frac{1}{x^2}$$

We can rewrite this equation as

$$\frac{d}{dx}(x^{-1}) = (-1)x^{-2}$$

and so the Power Rule is true when $n = -1$. In fact, we will show in the next section [Exercise 3.4.63(c)] that it holds for all negative integers.

What if the exponent is a fraction? In Example 3.2.4 we found that

$$\frac{d}{dx}\sqrt{x} = \frac{1}{2\sqrt{x}}$$

which can be written as

$$\frac{d}{dx}(x^{1/2}) = \tfrac{1}{2}x^{-1/2}$$

This shows that the Power Rule is true even when $n = \frac{1}{2}$. In fact, we will show in Section 3.7 that it is true for all real numbers n.

> **The Power Rule (General Version)** If n is any real number, then
>
> $$\frac{d}{dx}(x^n) = nx^{n-1}$$

Figure 3 shows the function y in Example 2(b) and its derivative y'. Notice that y is not differentiable at 0 (y' is not defined there). Observe that y' is positive when y increases and is negative when y decreases.

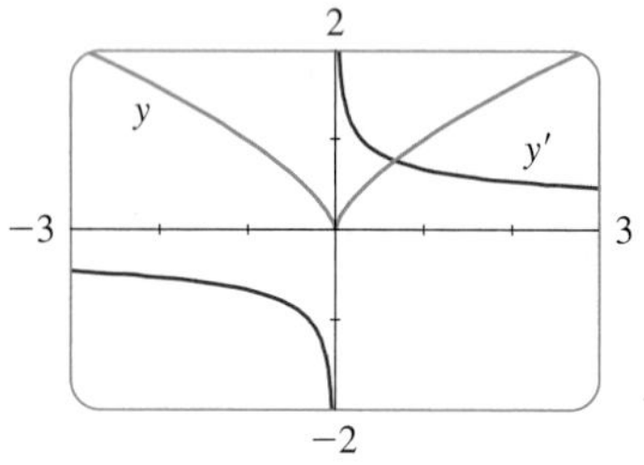

FIGURE 3
$y = \sqrt[3]{x^2}$

EXAMPLE 2 | Differentiate

(a) $f(x) = \dfrac{1}{x^2}$ (b) $y = \sqrt[3]{x^2}$

SOLUTION In each case we rewrite the function as a power of x.

(a) Since $f(x) = x^{-2}$, we use the Power Rule with $n = -2$:

$$f'(x) = \frac{d}{dx}(x^{-2}) = -2x^{-2-1} = -2x^{-3} = -\frac{2}{x^3}$$

(b) $$\frac{dy}{dx} = \frac{d}{dx}\left(\sqrt[3]{x^2}\right) = \frac{d}{dx}(x^{2/3}) = \tfrac{2}{3}x^{(2/3)-1} = \tfrac{2}{3}x^{-1/3}$$ ■

The Power Rule enables us to find tangent lines without having to resort to the definition of a derivative. It also enables us to find *normal lines*. The **normal line** to a curve C at a point P is the line through P that is perpendicular to the tangent line at P. (In the study of optics, one needs to consider the angle between a light ray and the normal line to a lens.)

EXAMPLE 3 | Find equations of the tangent line and normal line to the curve $y = x\sqrt{x}$ at the point (1, 1). Illustrate by graphing the curve and these lines.

SOLUTION The derivative of $f(x) = x\sqrt{x} = xx^{1/2} = x^{3/2}$ is

$$f'(x) = \tfrac{3}{2}x^{(3/2)-1} = \tfrac{3}{2}x^{1/2} = \tfrac{3}{2}\sqrt{x}$$

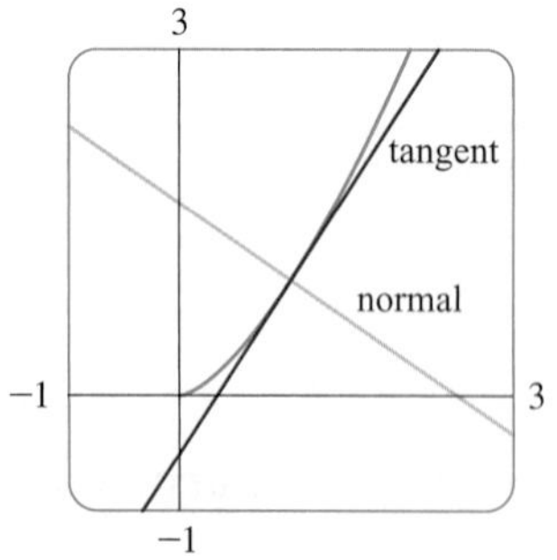

FIGURE 4
$y = x\sqrt{x}$

So the slope of the tangent line at (1, 1) is $f'(1) = \frac{3}{2}$. Therefore an equation of the tangent line is

$$y - 1 = \tfrac{3}{2}(x - 1) \quad \text{or} \quad y = \tfrac{3}{2}x - \tfrac{1}{2}$$

The normal line is perpendicular to the tangent line, so its slope is the negative reciprocal of $\frac{3}{2}$, that is, $-\frac{2}{3}$. Thus an equation of the normal line is

$$y - 1 = -\tfrac{2}{3}(x - 1) \quad \text{or} \quad y = -\tfrac{2}{3}x + \tfrac{5}{3}$$

We graph the curve and its tangent line and normal line in Figure 4. ■

■ New Derivatives from Old

When new functions are formed from old functions by addition, subtraction, or multiplication by a constant, their derivatives can be calculated in terms of derivatives of the old functions. In particular, the following formula says that *the derivative of a constant times a function is the constant times the derivative of the function.*

Geometric Interpretation of the Constant Multiple Rule

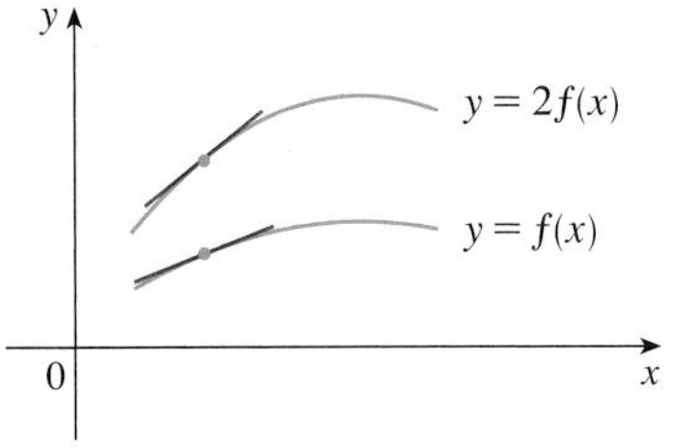

Multiplying by $c = 2$ stretches the graph vertically by a factor of 2. All the rises have been doubled but the runs stay the same. So the slopes are doubled, too.

The Constant Multiple Rule If c is a constant and f is a differentiable function, then

$$\frac{d}{dx}[cf(x)] = c\frac{d}{dx}f(x)$$

PROOF Let $g(x) = cf(x)$. Then

$$\begin{aligned} g'(x) &= \lim_{h\to 0}\frac{g(x+h)-g(x)}{h} = \lim_{h\to 0}\frac{cf(x+h)-cf(x)}{h} \\ &= \lim_{h\to 0} c\left[\frac{f(x+h)-f(x)}{h}\right] \\ &= c\lim_{h\to 0}\frac{f(x+h)-f(x)}{h} \qquad \text{(by Limit Law 3)} \\ &= cf'(x) \end{aligned}$$

■

EXAMPLE 4

(a) $\frac{d}{dx}(3x^4) = 3\frac{d}{dx}(x^4) = 3(4x^3) = 12x^3$

(b) $\frac{d}{dx}(-x) = \frac{d}{dx}[(-1)x] = (-1)\frac{d}{dx}(x) = -1(1) = -1$ ■

EXAMPLE 5 | **BB** **Anesthesiology** As explained in Example 1.2.5, when ventilators are used during surgery the steady state concentration C of CO_2 in the lungs is $C = P/V$, where P is the rate of production of CO_2 by the body and V is the ventilation rate. If P is constant, find dC/dV and interpret it.

SOLUTION

Because P is constant, we can use the Constant Multiple Rule as follows:

$$\begin{aligned} \frac{dC}{dV} &= P\cdot\frac{d}{dV}\left(\frac{1}{V}\right) = P\cdot\frac{d}{dV}(V^{-1}) \\ &= P(-V^{-2}) = -\frac{P}{V^2} \end{aligned}$$

This is the rate of change of the concentration with respect to the ventilation rate. Notice the minus sign in the expression for dC/dV: the concentration decreases as the ventilation rate increases. Notice also that, because of the V^2 in the denominator, this rate of change is close to 0 when V is large. ■

The next rule tells us that *the derivative of a sum of functions is the sum of the derivatives.*

Using prime notation, we can write the Sum Rule as

$$(f+g)' = f' + g'$$

The Sum Rule If f and g are both differentiable, then

$$\frac{d}{dx}[f(x)+g(x)] = \frac{d}{dx}f(x) + \frac{d}{dx}g(x)$$

PROOF Let $F(x) = f(x) + g(x)$. Then

$$\begin{aligned} F'(x) &= \lim_{h \to 0} \frac{F(x+h) - F(x)}{h} \\ &= \lim_{h \to 0} \frac{[f(x+h) + g(x+h)] - [f(x) + g(x)]}{h} \\ &= \lim_{h \to 0} \left[\frac{f(x+h) - f(x)}{h} + \frac{g(x+h) - g(x)}{h} \right] \\ &= \lim_{h \to 0} \frac{f(x+h) - f(x)}{h} + \lim_{h \to 0} \frac{g(x+h) - g(x)}{h} \qquad \text{(by Limit Law 1)} \\ &= f'(x) + g'(x) \end{aligned}$$

■

The Sum Rule can be extended to the sum of any number of functions. For instance, using this theorem twice, we get

$$(f + g + h)' = [(f + g) + h]' = (f + g)' + h' = f' + g' + h'$$

By writing $f - g$ as $f + (-1)g$ and applying the Sum Rule and the Constant Multiple Rule, we get the following formula.

The Difference Rule If f and g are both differentiable, then

$$\frac{d}{dx}[f(x) - g(x)] = \frac{d}{dx}f(x) - \frac{d}{dx}g(x)$$

The Constant Multiple Rule, the Sum Rule, and the Difference Rule can be combined with the Power Rule to differentiate any polynomial, as the following examples demonstrate.

EXAMPLE 6

$$\begin{aligned} &\frac{d}{dx}(x^8 + 12x^5 - 4x^4 + 10x^3 - 6x + 5) \\ &\quad = \frac{d}{dx}(x^8) + 12\frac{d}{dx}(x^5) - 4\frac{d}{dx}(x^4) + 10\frac{d}{dx}(x^3) - 6\frac{d}{dx}(x) + \frac{d}{dx}(5) \\ &\quad = 8x^7 + 12(5x^4) - 4(4x^3) + 10(3x^2) - 6(1) + 0 \\ &\quad = 8x^7 + 60x^4 - 16x^3 + 30x^2 - 6 \end{aligned}$$

■

EXAMPLE 7 | Find the points on the curve $y = x^4 - 6x^2 + 4$ where the tangent line is horizontal.

SOLUTION Horizontal tangents occur where the derivative is zero. We have

$$\begin{aligned} \frac{dy}{dx} &= \frac{d}{dx}(x^4) - 6\frac{d}{dx}(x^2) + \frac{d}{dx}(4) \\ &= 4x^3 - 12x + 0 = 4x(x^2 - 3) \end{aligned}$$

Thus $dy/dx = 0$ if $x = 0$ or $x^2 - 3 = 0$, that is, $x = \pm\sqrt{3}$. So the given curve has horizontal tangents when $x = 0$, $\sqrt{3}$, and $-\sqrt{3}$. The corresponding points are $(0, 4)$, $(\sqrt{3}, -5)$, and $(-\sqrt{3}, -5)$. (See Figure 5.) ■

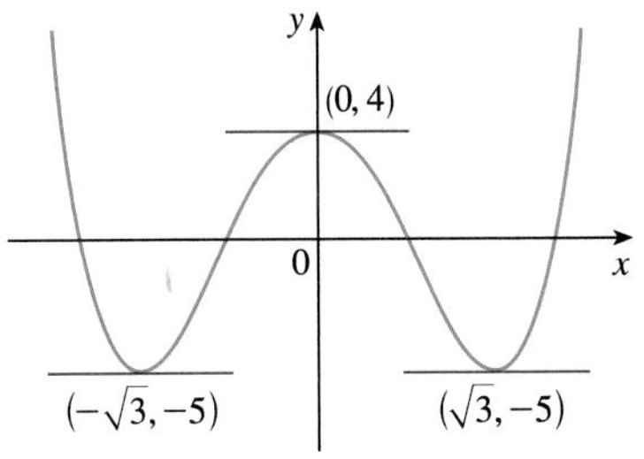

FIGURE 5
The curve $y = x^4 - 6x^2 + 4$ and its horizontal tangents

EXAMPLE 8 | The position function of a particle is $s = 2t^3 - 5t^2 + 3t + 4$, where s is measured in centimeters and t in seconds. Find the acceleration as a function of time. What is the acceleration after 2 seconds?

SOLUTION The velocity and acceleration are

$$v(t) = \frac{ds}{dt} = 6t^2 - 10t + 3$$

$$a(t) = \frac{dv}{dt} = 12t - 10$$

The acceleration after 2 s is $a(2) = 14 \text{ cm/s}^2$. ■

EXAMPLE 9 | **Blood flow** When we consider the flow of blood through a blood vessel, such as a vein or artery, we can model the shape of the blood vessel by a cylindrical tube with radius R and length l as illustrated in Figure 6.

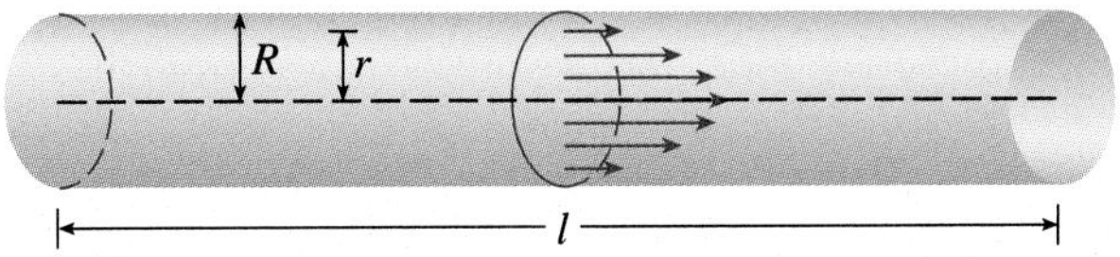

FIGURE 6
Blood flow in an artery

Because of friction at the walls of the tube, the velocity v of the blood is greatest along the central axis of the tube and decreases as the distance r from the axis increases until v becomes 0 at the wall. The relationship between v and r is given by the **law of laminar flow** discovered by the French physician Jean-Louis-Marie Poiseuille in 1840. This law states that

$$v = \frac{P}{4\eta l}(R^2 - r^2) \tag{4}$$

where η is the viscosity of the blood and P is the pressure difference between the ends of the tube. If P and l are constant, then v is a function of r with domain $[0, R]$.

For more detailed information, see W. Nichols and M. O'Rourke (eds.), *McDonald's Blood Flow in Arteries: Theoretical, Experimental, and Clinical Principles,* 5th ed. (New York, 2005).

The average rate of change of the velocity as we move from $r = r_1$ outward to $r = r_2$ is given by

$$\frac{\Delta v}{\Delta r} = \frac{v(r_2) - v(r_1)}{r_2 - r_1}$$

and if we let $\Delta r \to 0$, we obtain the **velocity gradient**, that is, the instantaneous rate of change of velocity with respect to r:

$$\text{velocity gradient} = \lim_{\Delta r \to 0} \frac{\Delta v}{\Delta r} = \frac{dv}{dr}$$

Using Equation 4, we obtain

$$\frac{dv}{dr} = \frac{P}{4\eta l}(0 - 2r) = -\frac{Pr}{2\eta l}$$

For one of the smaller human arteries we can take $\eta = 0.027$, $R = 0.008$ cm, $l = 2$ cm, and $P = 4000$ dynes/cm^2, which gives

$$v = \frac{4000}{4(0.027)2}(0.000064 - r^2)$$
$$\approx 1.85 \times 10^4(6.4 \times 10^{-5} - r^2)$$

At $r = 0.002$ cm the blood is flowing at a speed of

$$v(0.002) \approx 1.85 \times 10^4(64 \times 10^{-6} - 4 \times 10^{-6})$$
$$= 1.11 \text{ cm/s}$$

and the velocity gradient at that point is

$$\left.\frac{dv}{dr}\right|_{r=0.002} = -\frac{4000(0.002)}{2(0.027)2} \approx -74 \text{ (cm/s)/cm}$$

To get a feeling for what this statement means, let's change our units from centimeters to micrometers (1 cm = 10,000 μm). Then the radius of the artery is 80 μm. The velocity at the central axis is 11,850 μm/s, which decreases to 11,110 μm/s at a distance of $r = 20$ μm. The fact that $dv/dr = -74$ (μm/s)/μm means that, when $r = 20$ μm, the velocity is decreasing at a rate of about 74 μm/s for each micrometer that we proceed away from the center. ■

■ Exponential Functions

Let's try to compute the derivative of the exponential function $f(x) = b^x$ using the definition of a derivative:

$$f'(x) = \lim_{h \to 0} \frac{f(x+h) - f(x)}{h} = \lim_{h \to 0} \frac{b^{x+h} - b^x}{h}$$
$$= \lim_{h \to 0} \frac{b^x b^h - b^x}{h} = \lim_{h \to 0} \frac{b^x(b^h - 1)}{h}$$

The factor b^x doesn't depend on h, so we can take it in front of the limit:

$$f'(x) = b^x \lim_{h \to 0} \frac{b^h - 1}{h}$$

Notice that the limit is the value of the derivative of f at 0, that is,

$$\lim_{h \to 0} \frac{b^h - 1}{h} = f'(0)$$

Therefore we have shown that if the exponential function $f(x) = b^x$ is differentiable at 0, then it is differentiable everywhere and

(5) $$f'(x) = f'(0)b^x$$

This equation says that *the rate of change of any exponential function is proportional to the function itself.* (The slope is proportional to the height.)

h	$\dfrac{2^h - 1}{h}$	$\dfrac{3^h - 1}{h}$
0.1	0.7177	1.1612
0.01	0.6956	1.1047
0.001	0.6934	1.0992
0.0001	0.6932	1.0987

Numerical evidence for the existence of $f'(0)$ is given in the table at the left for the cases $b = 2$ and $b = 3$. (Values are stated correct to four decimal places.) It appears that the limits exist and

$$\text{for } b = 2, \quad f'(0) = \lim_{h \to 0} \frac{2^h - 1}{h} \approx 0.69$$

$$\text{for } b = 3, \quad f'(0) = \lim_{h \to 0} \frac{3^h - 1}{h} \approx 1.10$$

In fact, it can be proved that these limits exist and, correct to six decimal places, the values are

$$\frac{d}{dx}(2^x)\bigg|_{x=0} \approx 0.693147 \qquad \frac{d}{dx}(3^x)\bigg|_{x=0} \approx 1.098612$$

Thus, from Equation 5, we have

$$\textbf{(6)} \qquad \frac{d}{dx}(2^x) \approx (0.69)2^x \qquad \frac{d}{dx}(3^x) \approx (1.10)3^x$$

Of all possible choices for the base b in Equation 5, the simplest differentiation formula occurs when $f'(0) = 1$. In view of the estimates of $f'(0)$ for $b = 2$ and $b = 3$, it seems reasonable that there is a number b between 2 and 3 for which $f'(0) = 1$. It is traditional to denote this value by the letter e. (In fact, that is how we introduced e in Section 1.4.) Thus we have the following definition.

In Exercise 1 we will see that e lies between 2.7 and 2.8. Later we will be able to show that, correct to five decimal places,

$$e \approx 2.71828$$

Definition of the Number e

$$e \text{ is the number such that} \quad \lim_{h \to 0} \frac{e^h - 1}{h} = 1$$

Geometrically, this means that of all the possible exponential functions $y = b^x$, the function $f(x) = e^x$ is the one whose tangent line at (0, 1) has a slope $f'(0)$ that is exactly 1. (See Figures 7 and 8.)

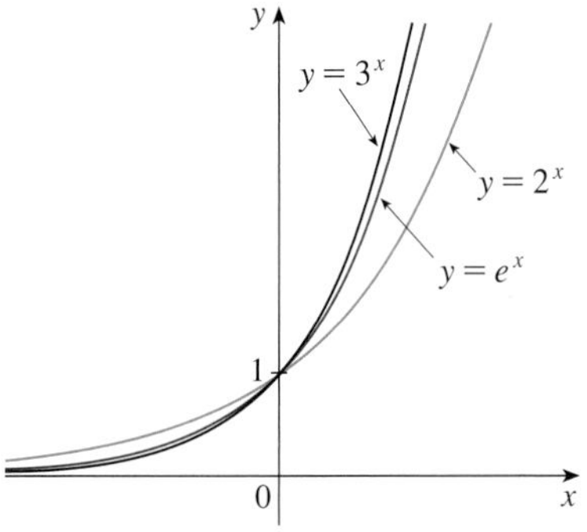

FIGURE 7

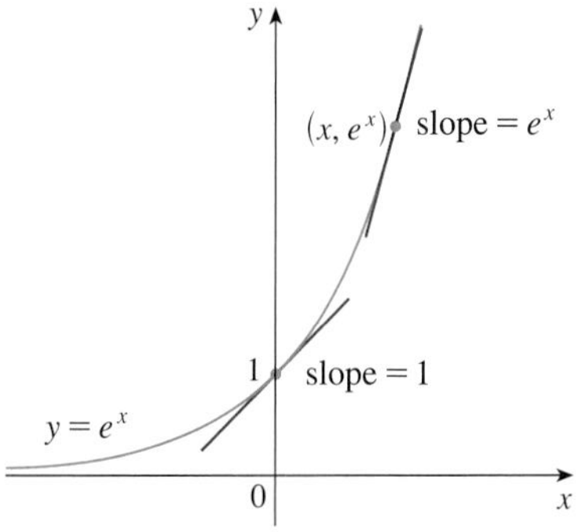

FIGURE 8

If we put $b = e$ and, therefore, $f'(0) = 1$ in Equation 5, it becomes the following important differentiation formula.

TEC Visual 3.3 uses the slope-a-scope to illustrate this formula.

Derivative of the Natural Exponential Function

$$\frac{d}{dx}(e^x) = e^x$$

Thus the exponential function $f(x) = e^x$ has the property that it is its own derivative. The geometrical significance of this fact is that the slope of a tangent line to the curve $y = e^x$ is equal to the y-coordinate of the point (see Figure 8).

EXAMPLE 10 | If $f(x) = e^x - x$, find f' and f''. Compare the graphs of f and f'.

SOLUTION Using the Difference Rule, we have

$$f'(x) = \frac{d}{dx}(e^x - x) = \frac{d}{dx}(e^x) - \frac{d}{dx}(x) = e^x - 1$$

In Section 3.2 we defined the second derivative as the derivative of f', so

$$f''(x) = \frac{d}{dx}(e^x - 1) = \frac{d}{dx}(e^x) - \frac{d}{dx}(1) = e^x$$

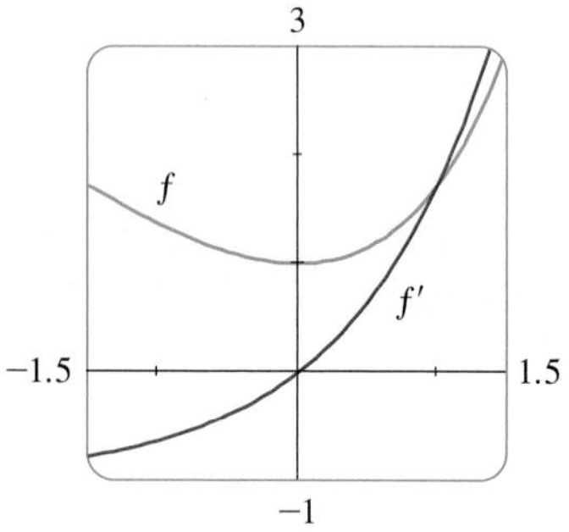

FIGURE 9

The function f and its derivative f' are graphed in Figure 9. Notice that f has a horizontal tangent when $x = 0$; this corresponds to the fact that $f'(0) = 0$. Notice also that, for $x > 0$, $f'(x)$ is positive and f is increasing. When $x < 0$, $f'(x)$ is negative and f is decreasing. ■

EXAMPLE 11 | At what point on the curve $y = e^x$ is the tangent line parallel to the line $y = 2x$?

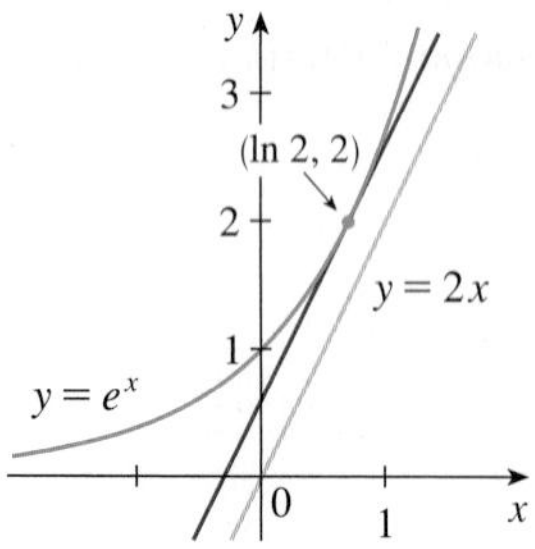

FIGURE 10

SOLUTION Since $y = e^x$, we have $y' = e^x$. Let the x-coordinate of the point in question be a. Then the slope of the tangent line at that point is e^a. This tangent line will be parallel to the line $y = 2x$ if it has the same slope, that is, 2. Equating slopes, we get

$$e^a = 2 \quad \Rightarrow \quad a = \ln 2$$

Therefore the required point is $(a, e^a) = (\ln 2, 2)$. (See Figure 10.) ■

■ Sine and Cosine Functions

If we sketch the graph of the function $f(x) = \sin x$ and use the interpretation of $f'(x)$ as the slope of the tangent to the sine curve in order to sketch the graph of f' (see Exercise 3.2.16), then it looks as if the graph of f' may be the same as the cosine curve (see Figure 11).

A review of the trigonometric functions is given in Appendix C.

TEC Visual 3.3 shows an animation of Figure 11.

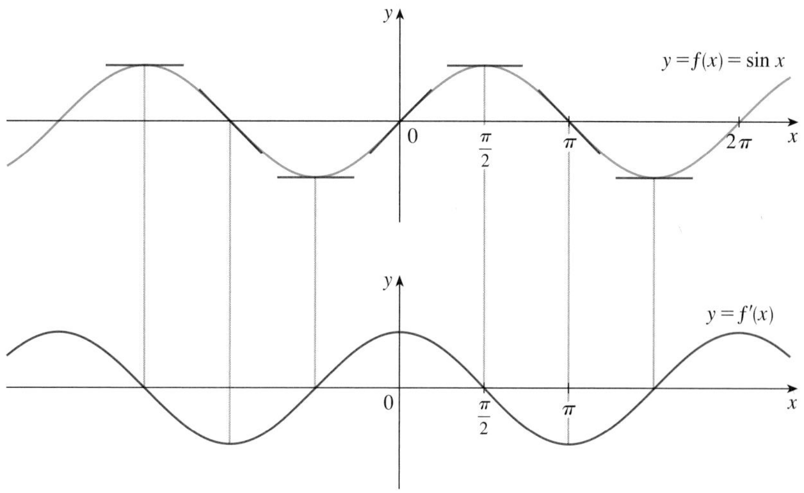

FIGURE 11

To prove that this is true we need to use two limits from Section 2.4 (see Equation 6 and Example 13 in that section):

$$\lim_{\theta \to 0} \frac{\sin \theta}{\theta} = 1 \qquad \lim_{\theta \to 0} \frac{\cos \theta - 1}{\theta} = 0$$

(7) $$\frac{d}{dx}(\sin x) = \cos x$$

PROOF If $f(x) = \sin x$, then

We have used the addition formula for sine. See Appendix C.

Note that we regard x as a constant when computing a limit as $h \to 0$, so $\sin x$ and $\cos x$ are also constants.

$$f'(x) = \lim_{h \to 0} \frac{f(x+h) - f(x)}{h} = \lim_{h \to 0} \frac{\sin(x+h) - \sin x}{h}$$

$$= \lim_{h \to 0} \frac{\sin x \cos h + \cos x \sin h - \sin x}{h}$$

$$= \lim_{h \to 0} \left[\frac{\sin x \cos h - \sin x}{h} + \frac{\cos x \sin h}{h} \right]$$

$$= \lim_{h \to 0} \left[\sin x \left(\frac{\cos h - 1}{h} \right) + \cos x \left(\frac{\sin h}{h} \right) \right]$$

$$= \lim_{h \to 0} \sin x \cdot \lim_{h \to 0} \frac{\cos h - 1}{h} + \lim_{h \to 0} \cos x \cdot \lim_{h \to 0} \frac{\sin h}{h}$$

$$= (\sin x) \cdot 0 + (\cos x) \cdot 1 = \cos x$$ ■

Using the same methods as in the proof of Formula 7, one can prove (see Exercise 70) that

(8) $$\frac{d}{dx}(\cos x) = -\sin x$$

EXAMPLE 12 | Differentiate $y = 3 \sin \theta + 4 \cos \theta$.

SOLUTION

$$\frac{dy}{d\theta} = 3 \frac{d}{d\theta}(\sin \theta) + 4 \frac{d}{d\theta}(\cos \theta) = 3 \cos \theta - 4 \sin \theta$$ ■

EXAMPLE 13 | Find the 27th derivative of $\cos x$.

SOLUTION The first few derivatives of $f(x) = \cos x$ are as follows:

$$f'(x) = -\sin x$$
$$f''(x) = -\cos x$$
$$f'''(x) = \sin x$$
$$f^{(4)}(x) = \cos x$$
$$f^{(5)}(x) = -\sin x$$

Looking for a pattern, we see that the successive derivatives occur in a cycle of length 4 and, in particular, $f^{(n)}(x) = \cos x$ whenever n is a multiple of 4. Therefore

$$f^{(24)}(x) = \cos x$$

and, differentiating three more times, we have

$$f^{(27)}(x) = \sin x$$ ■

EXERCISES 3.3

1. (a) How is the number e defined?
(b) Use a calculator to estimate the values of the limits

$$\lim_{h \to 0} \frac{2.7^h - 1}{h} \quad \text{and} \quad \lim_{h \to 0} \frac{2.8^h - 1}{h}$$

correct to two decimal places. What can you conclude about the value of e?

2. (a) Sketch, by hand, the graph of the function $f(x) = e^x$, paying particular attention to how the graph crosses the y-axis. What fact allows you to do this?
(b) What types of functions are $f(x) = e^x$ and $g(x) = x^e$? Compare the differentiation formulas for f and g.
(c) Which of the two functions in part (b) grows more rapidly when x is large?

3–32 Differentiate the function.

3. $f(x) = 186.5$

4. $f(x) = \sqrt{30}$

5. $f(x) = 5x - 1$

6. $F(x) = \frac{3}{4}x^8$

7. $f(x) = x^3 - 4x + 6$

8. $f(t) = \frac{1}{2}t^6 - 3t^4 + t$

9. $f(x) = x - 3 \sin x$

10. $y = \sin t + \pi \cos t$

11. $f(t) = \frac{1}{4}(t^4 + 8)$

12. $h(x) = (x - 2)(2x + 3)$

13. $A(s) = -\dfrac{12}{s^5}$

14. $B(y) = cy^{-6}$

15. $g(t) = 2t^{-3/4}$

16. $h(t) = \sqrt[4]{t} - 4e^t$

17. $y = 3e^x + \dfrac{4}{\sqrt[3]{x}}$

18. $y = \sqrt{x}\,(x - 1)$

19. $F(x) = (\frac{1}{2}x)^5$

20. $f(x) = \dfrac{x^2 - 3x + 1}{x^2}$

21. $y = \dfrac{x^2 + 4x + 3}{\sqrt{x}}$

22. $g(u) = \sqrt{2}\,u + \sqrt{3u}$

23. $y = 4\pi^2$

24. $L(\theta) = \dfrac{\sin \theta}{2} + \dfrac{c}{\theta}$

25. $g(y) = \dfrac{A}{y^{10}} + B \cos y$

26. $h(N) = rN\left(1 - \dfrac{N}{K}\right)$

27. $f(x) = k(a - x)(b - x)$

28. $F(v) = ae^v + \dfrac{b}{v} + \dfrac{c}{v^2}$

29. $u = \sqrt[5]{t} + 4\sqrt{t^5}$

30. $v = \left(\sqrt{x} + \dfrac{1}{\sqrt[3]{x}}\right)^2$

31. $G(y) = \dfrac{A}{y^{10}} + Be^y$

32. $y = e^{x+1} + 1$

33–34 Find an equation of the tangent line to the curve at the given point.

33. $y = \sqrt[4]{x}$, $(1, 1)$

34. $y = x^4 + 2x^2 - x$, $(1, 2)$

35–38 Find equations of the tangent line and normal line to the curve at the given point.

35. $y = 6 \cos x$, $(\pi/3, 3)$

36. $y = x^2 - x^4$, $(1, 0)$

37. $y = x^4 + 2e^x$, $(0, 2)$

38. $y = (1 + 2x)^2$, $(1, 9)$

39–40 Find an equation of the tangent line to the curve at the given point. Illustrate by graphing the curve and the tangent line on the same screen.

39. $y = x + \sqrt{x}$, $(1, 2)$

40. $y = 3x^2 - x^3$, $(1, 2)$

41–42 Find $f'(x)$. Compare the graphs of f and f' and use them to explain why your answer is reasonable.

41. $f(x) = 3x^{15} - 5x^3 + 3$

42. $f(x) = x + \dfrac{1}{x}$

43–46 Find the first and second derivatives of the function.

43. $f(x) = x^4 - 3x^3 + 16x$

44. $G(r) = \sqrt{r} + \sqrt[3]{r}$

45. $g(t) = 2 \cos t - 3 \sin t$

46. $h(t) = \sqrt{t} + 5 \sin t$

47. Fish growth Biologists have proposed a cubic polynomial to model the length L of rock bass at age A:

$$L = 0.0155A^3 - 0.372A^2 + 3.95A + 1.21$$

where L is measured in inches and A in years. (See Exercise 1.2.27.) Calculate

$$\left.\frac{dL}{dA}\right|_{A=12}$$

and interpret your answer.

48. Rain forest biodiversity The number of tree species S in a given area A in the Pasoh Forest Reserve in Malaysia has been modeled by the power function

$$S(A) = 0.882A^{0.842}$$

where A is measured in square meters. Find $S'(100)$ and interpret your answer.

(*Source:* Adapted from K. Kochummen et al., "Floristic Composition of Pasoh Forest Reserve, a Lowland Rain Forest in Peninsular Malaysia," *Journal of Tropical Forest Science* 3 (1991): 1–13.

49. Blood flow Refer to the law of laminar flow given in Example 9. Consider a blood vessel with radius 0.01 cm, length 3 cm, pressure difference 3000 dynes/cm^2, and viscosity $\eta = 0.027$.

(a) Find the velocity of the blood along the centerline $r = 0$, at radius $r = 0.005$ cm, and at the wall $r = R = 0.01$ cm.

(b) Find the velocity gradient at $r = 0$, $r = 0.005$, and $r = 0.01$.

(c) Where is the velocity the greatest? Where is the velocity changing most?

50. Invasive species often display a wave of advance as they colonize new areas. Mathematical models based on random dispersal and reproduction have demonstrated that the speed with which such waves move is given by the expression $2\sqrt{Dr}$, where r is the reproductive rate of individuals and D is a constant quantifying dispersal. Calculate the derivative of the wave speed with respect to the reproductive rate r and explain its meaning.

51. The position function of a particle is given by $s = t^3 - 4.5t^2 - 7t$, $t \geqslant 0$.

(a) Find the velocity and acceleration of the particle.

(b) When does the particle reach a velocity of 5 m/s?

(c) When is the acceleration 0? What is the significance of this value of t?

52. If a ball is given a push so that it has an initial velocity of 5 m/s rolling down a certain inclined plane, then the distance it has rolled after t seconds is $s = 5t + 3t^2$.

(a) Find the velocity after 2 s.

(b) How long does it take for the velocity to reach 35 m/s?

53. (a) A company makes computer chips from square wafers of silicon. It wants to keep the side length of a wafer very close to 15 mm and it wants to know how the area $A(x)$ of a wafer changes when the side length x changes. Find $A'(15)$ and explain its meaning in this situation.

(b) Show that the rate of change of the area of a square with respect to its side length is half its perimeter. Try to explain geometrically why this is true by drawing a square whose side length x is increased by an amount Δx. How can you approximate the resulting change in area ΔA if Δx is small?

54. (a) Sodium chlorate crystals are easy to grow in the shape of cubes by allowing a solution of water and sodium chlorate to evaporate slowly. If V is the volume of such a cube with side length x, calculate dV/dx when $x = 3$ mm and explain its meaning.

(b) Show that the rate of change of the volume of a cube with respect to its edge length is equal to half the surface area of the cube. Explain geometrically why this result is true by arguing by analogy with Exercise 53(b).

55. (a) Find the average rate of change of the area of a circle with respect to its radius r as r changes from

(i) 2 to 3 (ii) 2 to 2.5 (iii) 2 to 2.1

(b) Find the instantaneous rate of change when $r = 2$.

(c) Show that the rate of change of the area of a circle with respect to its radius (at any r) is equal to the circumference of the circle. Try to explain geometrically why this is true by drawing a circle whose radius is increased by an amount Δr. How can you approximate the resulting change in area ΔA if Δr is small?

56. (a) **Cell growth** The volume of a growing spherical cell is $V = \frac{4}{3}\pi r^3$, where the radius r is measured in micrometers ($1\ \mu\text{m} = 10^{-6}$ m). Find the average rate of change of V with respect to r when r changes from
(i) 5 to 8 μm (ii) 5 to 6 μm (iii) 5 to 5.1 μm
(b) Find the instantaneous rate of change of V with respect to r when $r = 5$ μm.
(c) Show that the rate of change of the volume of a cell with respect to its radius is equal to its surface area ($S = 4\pi r^2$). Explain geometrically why this result is true. Argue by analogy with Exercise 55(c).

57. Find $\dfrac{d^{99}}{dx^{99}}(\sin x)$.

58. Find the nth derivative of each function by calculating the first few derivatives and observing the pattern that occurs.
(a) $f(x) = x^n$ (b) $f(x) = 1/x$

59. For what values of x does the graph of $f(x) = x + 2\sin x$ have a horizontal tangent?

60. For what values of x does the graph of $f(x) = x^3 + 3x^2 + x + 3$ have a horizontal tangent?

61. Show that the curve $y = 6x^3 + 5x - 3$ has no tangent line with slope 4.

62. Find an equation of the tangent line to the curve $y = x\sqrt{x}$ that is parallel to the line $y = 1 + 3x$.

63. Find equations of both lines that are tangent to the curve $y = 1 + x^3$ and parallel to the line $12x - y = 1$.

64. At what point on the curve $y = 1 + 2e^x - 3x$ is the tangent line parallel to the line $3x - y = 5$? Illustrate by graphing the curve and both lines.

65. Find an equation of the normal line to the parabola $y = x^2 - 5x + 4$ that is parallel to the line $x - 3y = 5$.

66. Where does the normal line to the parabola $y = x - x^2$ at the point $(1, 0)$ intersect the parabola a second time? Illustrate with a sketch.

67. Draw a diagram to show that there are two tangent lines to the parabola $y = x^2$ that pass through the point $(0, -4)$. Find the coordinates of the points where these tangent lines intersect the parabola.

68. (a) Find equations of both lines through the point $(2, -3)$ that are tangent to the parabola $y = x^2 + x$.
(b) Show that there is no line through the point $(2, 7)$ that is tangent to the parabola. Then draw a diagram to see why.

69. Use the definition of a derivative to show that if $f(x) = 1/x$, then $f'(x) = -1/x^2$. (This proves the Power Rule for the case $n = -1$.)

70. Prove, using the definition of derivative, that if $f(x) = \cos x$, then $f'(x) = -\sin x$.

71. Find the parabola with equation $y = ax^2 + bx$ whose tangent line at $(1, 1)$ has equation $y = 3x - 2$.

72. Find the value of c such that the line $y = \frac{3}{2}x + 6$ is tangent to the curve $y = c\sqrt{x}$.

73. For what values of a and b is the line $2x + y = b$ tangent to the parabola $y = ax^2$ when $x = 2$?

74. A tangent line is drawn to the hyperbola $xy = c$ at a point P.
(a) Show that the midpoint of the line segment cut from this tangent line by the coordinate axes is P.
(b) Show that the triangle formed by the tangent line and the coordinate axes always has the same area, no matter where P is located on the hyperbola.

75. Evaluate $\displaystyle\lim_{x \to 1} \frac{x^{1000} - 1}{x - 1}$.

76. Draw a diagram showing two perpendicular lines that intersect on the y-axis and are both tangent to the parabola $y = x^2$. Where do these lines intersect?

3.4 The Product and Quotient Rules

The formulas of this section enable us to differentiate new functions formed from old functions by multiplication or division.

The Product Rule

By analogy with the Sum and Difference Rules, one might be tempted to guess, as Leibniz did three centuries ago, that the derivative of a product is the product of the derivatives. We can see, however, that this guess is wrong by looking at a particular example. Let $f(x) = x$ and $g(x) = x^2$. Then the Power Rule gives $f'(x) = 1$ and $g'(x) = 2x$. But

$(fg)(x) = x^3$, so $(fg)'(x) = 3x^2$. Thus $(fg)' \neq f'g'$. The correct formula was discovered by Leibniz (soon after his false start) and is called the Product Rule.

Before stating the Product Rule, let's see how we might discover it. We start by assuming that $u = f(x)$ and $v = g(x)$ are both positive differentiable functions. Then we can interpret the product uv as an area of a rectangle (see Figure 1). If x changes by an amount Δx, then the corresponding changes in u and v are

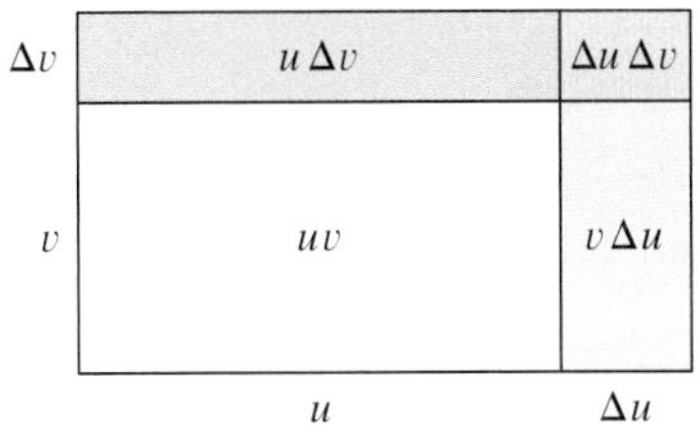

FIGURE 1
The geometry of the Product Rule

$$\Delta u = f(x + \Delta x) - f(x) \qquad \Delta v = g(x + \Delta x) - g(x)$$

and the new value of the product, $(u + \Delta u)(v + \Delta v)$, can be interpreted as the area of the large rectangle in Figure 1 (provided that Δu and Δv happen to be positive).

The change in the area of the rectangle is

(1)
$$\Delta(uv) = (u + \Delta u)(v + \Delta v) - uv = u\,\Delta v + v\,\Delta u + \Delta u\,\Delta v$$
$$= \text{the sum of the three shaded areas}$$

If we divide by Δx, we get

$$\frac{\Delta(uv)}{\Delta x} = u\frac{\Delta v}{\Delta x} + v\frac{\Delta u}{\Delta x} + \Delta u\frac{\Delta v}{\Delta x}$$

Recall that in Leibniz notation the definition of a derivative can be written as

$$\frac{dy}{dx} = \lim_{\Delta x \to 0}\frac{\Delta y}{\Delta x}$$

If we now let $\Delta x \to 0$, we get the derivative of uv:

$$\frac{d}{dx}(uv) = \lim_{\Delta x \to 0}\frac{\Delta(uv)}{\Delta x} = \lim_{\Delta x \to 0}\left(u\frac{\Delta v}{\Delta x} + v\frac{\Delta u}{\Delta x} + \Delta u\frac{\Delta v}{\Delta x}\right)$$
$$= u\lim_{\Delta x \to 0}\frac{\Delta v}{\Delta x} + v\lim_{\Delta x \to 0}\frac{\Delta u}{\Delta x} + \left(\lim_{\Delta x \to 0}\Delta u\right)\left(\lim_{\Delta x \to 0}\frac{\Delta v}{\Delta x}\right)$$
$$= u\frac{dv}{dx} + v\frac{du}{dx} + 0 \cdot \frac{dv}{dx}$$

(2)
$$\frac{d}{dx}(uv) = u\frac{dv}{dx} + v\frac{du}{dx}$$

(Notice that $\Delta u \to 0$ as $\Delta x \to 0$ since f is differentiable and therefore continuous.)

Although we started by assuming (for the geometric interpretation) that all the quantities are positive, we notice that Equation 1 is always true. (The algebra is valid whether u, v, Δu, and Δv are positive or negative.) So we have proved Equation 2, known as the Product Rule, for all differentiable functions u and v.

In prime notation:

$$(fg)' = fg' + gf'$$

The Product Rule If f and g are both differentiable, then

$$\frac{d}{dx}[f(x)g(x)] = f(x)\frac{d}{dx}[g(x)] + g(x)\frac{d}{dx}[f(x)]$$

In words, the Product Rule says that *the derivative of a product of two functions is the first function times the derivative of the second function plus the second function times the derivative of the first function.*

Figure 2 shows the graphs of the function of Example 1 and its derivative. Notice that $y' = 0$ whenever y has a horizontal tangent.

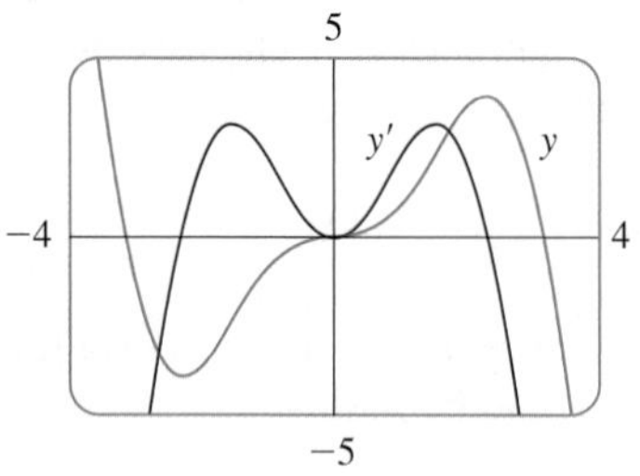

FIGURE 2

In Example 2, a and b are constants. It is customary in mathematics to use letters near the beginning of the alphabet to represent constants and letters near the end of the alphabet to represent variables.

EXAMPLE 1 | Differentiate $y = x^2 \sin x$.

SOLUTION Using the Product Rule, we have

$$\frac{dy}{dx} = x^2 \frac{d}{dx}(\sin x) + \sin x \frac{d}{dx}(x^2)$$
$$= x^2 \cos x + 2x \sin x$$

■

EXAMPLE 2 | Differentiate the function $f(t) = \sqrt{t}\,(a + bt)$.

SOLUTION 1 Using the Product Rule, we have

$$f'(t) = \sqrt{t}\,\frac{d}{dt}(a + bt) + (a + bt)\frac{d}{dt}\left(\sqrt{t}\right)$$
$$= \sqrt{t} \cdot b + (a + bt) \cdot \tfrac{1}{2}t^{-1/2}$$
$$= b\sqrt{t} + \frac{a + bt}{2\sqrt{t}} = \frac{a + 3bt}{2\sqrt{t}}$$

SOLUTION 2 If we first use the laws of exponents to rewrite $f(t)$, then we can proceed directly without using the Product Rule.

$$f(t) = a\sqrt{t} + bt\sqrt{t} = at^{1/2} + bt^{3/2}$$
$$f'(t) = \tfrac{1}{2}at^{-1/2} + \tfrac{3}{2}bt^{1/2}$$

which is equivalent to the answer given in Solution 1. ■

Example 2 shows that it is sometimes easier to simplify a product of functions before differentiating than to use the Product Rule. In Example 1, however, the Product Rule is the only possible method.

EXAMPLE 3 | If $h(x) = xg(x)$ and it is known that $g(3) = 5$ and $g'(3) = 2$, find $h'(3)$.

SOLUTION Applying the Product Rule, we get

$$h'(x) = \frac{d}{dx}[xg(x)]$$
$$= x\frac{d}{dx}[g(x)] + g(x)\frac{d}{dx}[x]$$
$$= xg'(x) + g(x)$$

Therefore

$$h'(3) = 3g'(3) + g(3) = 3 \cdot 2 + 5 = 11$$

■

■ The Quotient Rule

We find a rule for differentiating the quotient of two differentiable functions $u = f(x)$ and $v = g(x)$ in much the same way that we found the Product Rule. If x, u, and v change by amounts Δx, Δu, and Δv, then the corresponding change in the quotient u/v is

$$\Delta\left(\frac{u}{v}\right) = \frac{u + \Delta u}{v + \Delta v} - \frac{u}{v} = \frac{(u + \Delta u)v - u(v + \Delta v)}{v(v + \Delta v)}$$

$$= \frac{v\Delta u - u\Delta v}{v(v + \Delta v)}$$

so

$$\frac{d}{dx}\left(\frac{u}{v}\right) = \lim_{\Delta x \to 0} \frac{\Delta(u/v)}{\Delta x} = \lim_{\Delta x \to 0} \frac{v\dfrac{\Delta u}{\Delta x} - u\dfrac{\Delta v}{\Delta x}}{v(v + \Delta v)}$$

As $\Delta x \to 0$, $\Delta v \to 0$ also, because $v = g(x)$ is differentiable and therefore continuous. Thus, using the Limit Laws, we get

$$\frac{d}{dx}\left(\frac{u}{v}\right) = \frac{v \displaystyle\lim_{\Delta x \to 0} \frac{\Delta u}{\Delta x} - u \lim_{\Delta x \to 0} \frac{\Delta v}{\Delta x}}{v \displaystyle\lim_{\Delta x \to 0} (v + \Delta v)} = \frac{v\dfrac{du}{dx} - u\dfrac{dv}{dx}}{v^2}$$

The Quotient Rule If f and g are differentiable, then

$$\frac{d}{dx}\left[\frac{f(x)}{g(x)}\right] = \frac{g(x)\dfrac{d}{dx}[f(x)] - f(x)\dfrac{d}{dx}[g(x)]}{[g(x)]^2}$$

In prime notation:

$$\left(\frac{f}{g}\right)' = \frac{gf' - fg'}{g^2}$$

In words, the Quotient Rule says that the *derivative of a quotient is the denominator times the derivative of the numerator minus the numerator times the derivative of the denominator, all divided by the square of the denominator.*

The Quotient Rule and the other differentiation formulas enable us to compute the derivative of any rational function, as the next example illustrates.

We can use a graphing device to check that the answer to Example 4 is plausible. Figure 3 shows the graphs of the function of Example 4 and its derivative. Notice that when y grows rapidly (near -2), y' is large. And when y grows slowly, y' is near 0.

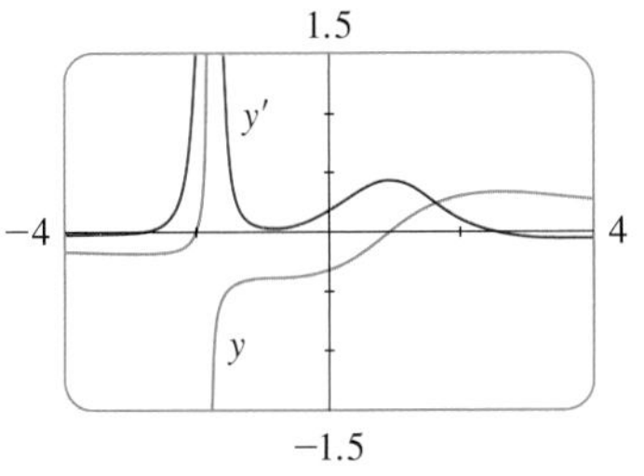

FIGURE 3

EXAMPLE 4 | Let $y = \dfrac{x^2 + x - 2}{x^3 + 6}$. Then

$$y' = \frac{(x^3 + 6)\dfrac{d}{dx}(x^2 + x - 2) - (x^2 + x - 2)\dfrac{d}{dx}(x^3 + 6)}{(x^3 + 6)^2}$$

$$= \frac{(x^3 + 6)(2x + 1) - (x^2 + x - 2)(3x^2)}{(x^3 + 6)^2}$$

$$= \frac{(2x^4 + x^3 + 12x + 6) - (3x^4 + 3x^3 - 6x^2)}{(x^3 + 6)^2}$$

$$= \frac{-x^4 - 2x^3 + 6x^2 + 12x + 6}{(x^3 + 6)^2}$$

■

EXAMPLE 5 | BB **The Monod growth function** Monod modeled the per capita growth rate R of *Escherichia coli* bacteria by the function

$$R(N) = \frac{SN}{c + N}$$

where N is the concentration of the nutrient, S is its saturation level, and c is a positive constant. Calculate dR/dN and interpret it.

SOLUTION Using the Quotient Rule, we have

$$\frac{dR}{dN} = \frac{(c + N)\dfrac{d}{dN}(SN) - SN\dfrac{d}{dN}(c + N)}{(c + N)^2}$$

$$= \frac{(c + N)S - SN \cdot 1}{(c + N)^2} = \frac{cS}{(c + N)^2}$$

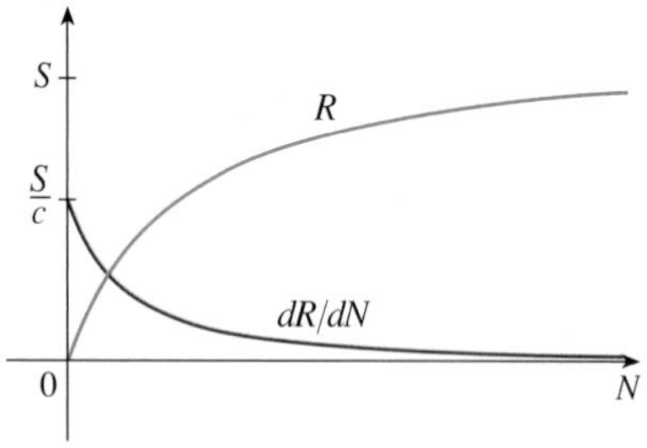

FIGURE 4

The derivative dR/dN is the rate of change of the growth rate R with respect to the concentration of the nutrient. From the expression for dR/dN we see that it is always positive, which means that R is an increasing function of N. But as the concentration becomes larger, dR/dN becomes smaller because of the denominator $(c + N)^2$. So

$$\frac{dR}{dN} \to 0 \quad \text{as} \quad N \to \infty$$

and this means that the graph of R becomes flatter as N gets larger. (See Figure 4.) ■

EXAMPLE 6 | Find an equation of the tangent line to the curve $y = e^x/(1 + x^2)$ at the point $(1, \frac{1}{2}e)$.

SOLUTION According to the Quotient Rule, we have

$$\frac{dy}{dx} = \frac{(1 + x^2)\dfrac{d}{dx}(e^x) - e^x\dfrac{d}{dx}(1 + x^2)}{(1 + x^2)^2}$$

$$= \frac{(1 + x^2)e^x - e^x(2x)}{(1 + x^2)^2} = \frac{e^x(1 - 2x + x^2)}{(1 + x^2)^2}$$

$$= \frac{e^x(1 - x)^2}{(1 + x^2)^2}$$

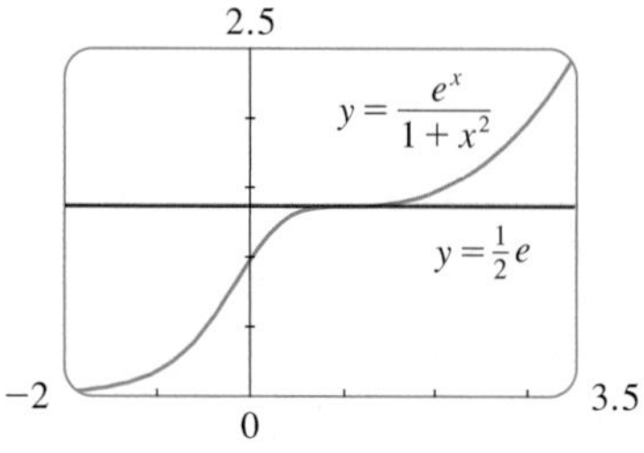

FIGURE 5

So the slope of the tangent line at $(1, \frac{1}{2}e)$ is

$$\left.\frac{dy}{dx}\right|_{x=1} = 0$$

This means that the tangent line at $(1, \frac{1}{2}e)$ is horizontal and its equation is $y = \frac{1}{2}e$. [See Figure 5. Notice that the function is increasing and crosses its tangent line at $(1, \frac{1}{2}e)$.] ■

NOTE Don't use the Quotient Rule *every* time you see a quotient. Sometimes it's easier to rewrite a quotient first to put it in a form that is simpler for the purpose of differentiation. For instance, although it is possible to differentiate the function

$$F(x) = \frac{3x^2 + 2\sqrt{x}}{x}$$

using the Quotient Rule, it is much easier to perform the division first and write the function as

$$F(x) = 3x + 2x^{-1/2}$$

before differentiating.

We summarize the differentiation formulas we have learned so far as follows.

Table of Differentiation Formulas

$\frac{d}{dx}(c) = 0$	$\frac{d}{dx}(x^n) = nx^{n-1}$	$\frac{d}{dx}(e^x) = e^x$
$(cf)' = cf'$	$(f + g)' = f' + g'$	$(f - g)' = f' - g'$
$(fg)' = fg' + gf'$	$\left(\frac{f}{g}\right)' = \frac{gf' - fg'}{g^2}$	

Trigonometric Functions

Knowing the derivatives of the sine and cosine functions, we can use the Quotient Rule to find the derivative of the tangent function:

$$\begin{aligned}\frac{d}{dx}(\tan x) &= \frac{d}{dx}\left(\frac{\sin x}{\cos x}\right)\\ &= \frac{\cos x \frac{d}{dx}(\sin x) - \sin x \frac{d}{dx}(\cos x)}{\cos^2 x}\\ &= \frac{\cos x \cdot \cos x - \sin x\,(-\sin x)}{\cos^2 x}\\ &= \frac{\cos^2 x + \sin^2 x}{\cos^2 x}\\ &= \frac{1}{\cos^2 x} = \sec^2 x\end{aligned}$$

$$\boxed{\frac{d}{dx}(\tan x) = \sec^2 x}$$

The derivatives of the remaining trigonometric functions, csc, sec, and cot, can also be found easily using the Quotient Rule (see Exercises 45–47). We collect all the differentiation formulas for trigonometric functions in the following table. Remember that they are valid only when x is measured in radians.

When you memorize this table, it is helpful to notice that the minus signs go with the derivatives of the "cofunctions," that is, cosine, cosecant, and cotangent.

Derivatives of Trigonometric Functions

$$\frac{d}{dx}(\sin x) = \cos x \qquad \frac{d}{dx}(\csc x) = -\csc x \cot x$$

$$\frac{d}{dx}(\cos x) = -\sin x \qquad \frac{d}{dx}(\sec x) = \sec x \tan x$$

$$\frac{d}{dx}(\tan x) = \sec^2 x \qquad \frac{d}{dx}(\cot x) = -\csc^2 x$$

EXAMPLE 7 | Differentiate $f(x) = \dfrac{\sec x}{1 + \tan x}$. For what values of x does the graph of f have a horizontal tangent?

SOLUTION The Quotient Rule gives

$$f'(x) = \frac{(1 + \tan x)\dfrac{d}{dx}(\sec x) - \sec x \dfrac{d}{dx}(1 + \tan x)}{(1 + \tan x)^2}$$

$$= \frac{(1 + \tan x)\sec x \tan x - \sec x \cdot \sec^2 x}{(1 + \tan x)^2}$$

$$= \frac{\sec x\,(\tan x + \tan^2 x - \sec^2 x)}{(1 + \tan x)^2}$$

$$= \frac{\sec x\,(\tan x - 1)}{(1 + \tan x)^2}$$

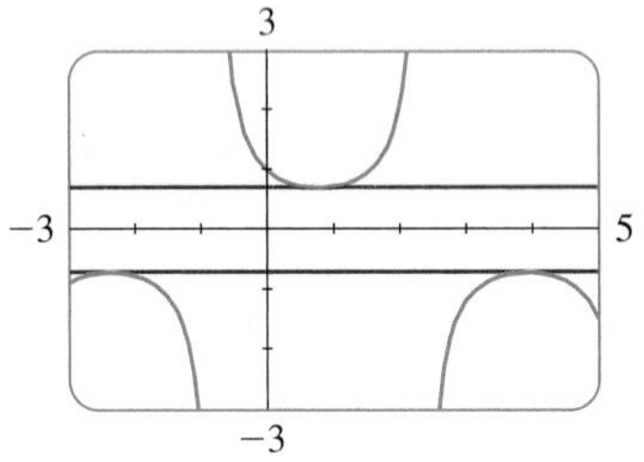

FIGURE 6
The horizontal tangents in Example 7

In simplifying the answer we have used the identity $\tan^2 x + 1 = \sec^2 x$.

Since $\sec x$ is never 0, we see that $f'(x) = 0$ when $\tan x = 1$, and this occurs when $x = n\pi + \pi/4$, where n is an integer (see Figure 6). ■

EXERCISES 3.4

1. Find the derivative of $f(x) = (1 + 2x^2)(x - x^2)$ in two ways: by using the Product Rule and by performing the multiplication first. Do your answers agree?

2. Find the derivative of the function

$$F(x) = \frac{x^4 - 5x^3 + \sqrt{x}}{x^2}$$

in two ways: by using the Quotient Rule and by simplifying first. Show that your answers are equivalent. Which method do you prefer?

3–34 Differentiate.

3. $f(x) = (x^3 + 2x)e^x$

4. $g(x) = \sqrt{x}\, e^x$

5. $g(t) = t^3 \cos t$

6. $h(t) = e^t \sin t$

7. $F(y) = \left(\dfrac{1}{y^2} - \dfrac{3}{y^4}\right)(y + 5y^3)$

8. $R(t) = (t + e^t)(3 - \sqrt{t})$

9. $f(x) = \sin x + \frac{1}{2}\cot x$

10. $y = 2\csc x + 5\cos x$

11. $h(\theta) = \theta \csc \theta - \cot \theta$

12. $y = u(a \cos u + b \cot u)$

13. $y = \dfrac{e^x}{x^2}$

14. $y = \dfrac{e^x}{1 + x}$

15. $g(x) = \dfrac{3x - 1}{2x + 1}$

16. $f(t) = \dfrac{2t}{4 + t^2}$

17. $y = \dfrac{x^3}{1 - x^2}$

18. $y = \dfrac{x + 1}{x^3 + x - 2}$

19. $y = \dfrac{t^2 + 2}{t^4 - 3t^2 + 1}$

20. $y = \dfrac{t}{(t - 1)^2}$

21. $y = (r^2 - 2r)e^r$

22. $y = \dfrac{1}{s + ke^s}$

23. $f(\theta) = \dfrac{\sec\theta}{1 + \sec\theta}$

24. $y = \dfrac{1 + \sin x}{x + \cos x}$

25. $y = \dfrac{\sin x}{x^2}$

26. $y = \dfrac{1 - \sec x}{\tan x}$

27. $y = \dfrac{v^3 - 2v\sqrt{v}}{v}$

28. $z = w^{3/2}(w + ce^w)$

29. $f(t) = \dfrac{2t}{2 + \sqrt{t}}$

30. $g(t) = \dfrac{t - \sqrt{t}}{t^{1/3}}$

31. $f(x) = \dfrac{A}{B + Ce^x}$

32. $f(x) = \dfrac{1 - xe^x}{x + e^x}$

33. $f(x) = \dfrac{x}{x + \dfrac{c}{x}}$

34. $f(x) = \dfrac{ax + b}{cx + d}$

35–36 Find an equation of the tangent line to the given curve at the specified point.

35. $y = \dfrac{2x}{x + 1}$, $(1, 1)$

36. $y = e^x \cos x$, $(0, 1)$

37–38 Find equations of the tangent line and normal line to the given curve at the specified point.

37. $y = 2xe^x$, $(0, 0)$

38. $y = \dfrac{\sqrt{x}}{x + 1}$, $(4, 0.4)$

39–40 Find $f'(x)$ and $f''(x)$.

39. $f(x) = x^4 e^x$

40. $f(x) = \dfrac{x}{x^2 - 1}$

---

41. If $H(\theta) = \theta \sin\theta$, find $H'(\theta)$ and $H''(\theta)$.

42. If $f(t) = \csc t$, find $f''(\pi/6)$.

43. If $f(x) = xe^x$, find the nth derivative, $f^{(n)}(x)$.

44. If $g(x) = x/e^x$, find $g^{(n)}(x)$.

45. Prove that $\dfrac{d}{dx}(\csc x) = -\csc x \cot x$.

46. Prove that $\dfrac{d}{dx}(\sec x) = \sec x \tan x$.

47. Prove that $\dfrac{d}{dx}(\cot x) = -\csc^2 x$.

48. Suppose that $f(\pi/3) = 4$ and $f'(\pi/3) = -2$, and let $g(x) = f(x)\sin x$ and $h(x) = (\cos x)/f(x)$. Find
(a) $g'(\pi/3)$ (b) $h'(\pi/3)$

49. Suppose that $f(5) = 1$, $f'(5) = 6$, $g(5) = -3$, and $g'(5) = 2$. Find the following values.
(a) $(fg)'(5)$ (b) $(f/g)'(5)$ (c) $(g/f)'(5)$

50. Suppose that $f(2) = -3$, $g(2) = 4$, $f'(2) = -2$, and $g'(2) = 7$. Find $h'(2)$.
(a) $h(x) = 5f(x) - 4g(x)$ (b) $h(x) = f(x)g(x)$
(c) $h(x) = \dfrac{f(x)}{g(x)}$ (d) $h(x) = \dfrac{g(x)}{1 + f(x)}$

51. If f and g are the functions whose graphs are shown, let $u(x) = f(x)g(x)$ and $v(x) = f(x)/g(x)$.
(a) Find $u'(1)$. (b) Find $v'(5)$.

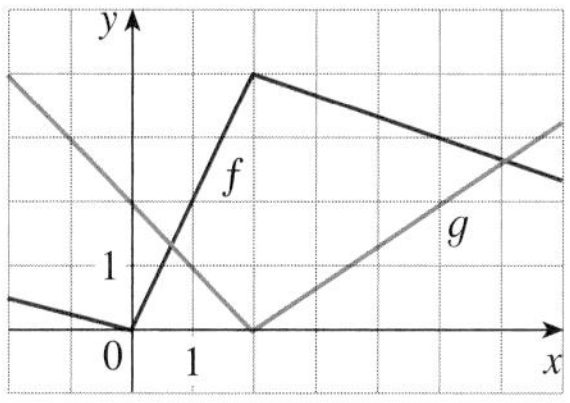

52. Let $P(x) = F(x)G(x)$ and $Q(x) = F(x)/G(x)$, where F and G are the functions whose graphs are shown.
(a) Find $P'(2)$. (b) Find $Q'(7)$.

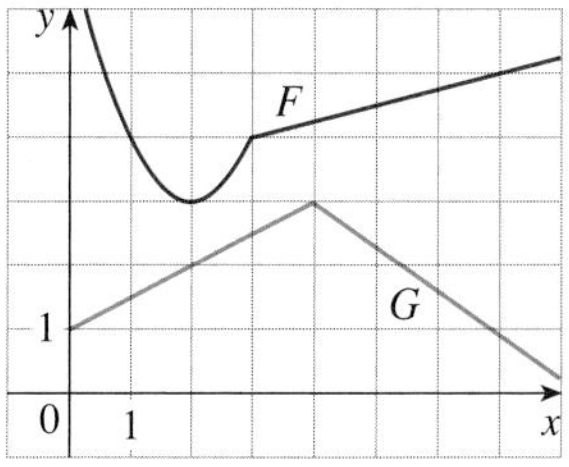

53. If g is a differentiable function, find an expression for the derivative of each of the following functions.
(a) $y = xg(x)$ (b) $y = \dfrac{x}{g(x)}$ (c) $y = \dfrac{g(x)}{x}$

54. Insecticide resistance If the frequency of a gene for insecticide resistance is p, then its frequency in the next generation is given by the expression

$$\frac{p(1 + s)}{1 + sp}$$

where s is the reproductive advantage that this gene has over the wild type in the presence of the insecticide. Determine the rate at which the gene frequency in the next generation changes as s changes.

55. The **Michaelis-Menten equation** for the enzyme chymotrypsin is

$$v = \frac{0.14[\mathrm{S}]}{0.015 + [\mathrm{S}]}$$

where v is the rate of an enzymatic reaction and [S] is the concentration of a substrate S. Calculate $dv/d[\mathrm{S}]$ and interpret it.

56. The **biomass** $B(t)$ of a fish population is the total mass of the members of the population at time t. It is the product of the number of individuals $N(t)$ in the population and the average mass $M(t)$ of a fish at time t. In the case of guppies, breeding occurs continually. Suppose that at time $t = 4$ weeks the population is 820 guppies and is growing at a rate of 50 guppies per week, while the average mass is 1.2 g and is increasing at a rate of 0.14 g/week. At what rate is the biomass increasing when $t = 4$?

57. The gas law for an ideal gas at absolute temperature T (in kelvins), pressure P (in atmospheres), and volume V (in liters) is $PV = nRT$, where n is the number of moles of the gas and $R = 0.0821$ is the gas constant. Suppose that, at a certain instant, $P = 8.0$ atm and is increasing at a rate of 0.10 atm/min and $V = 10$ L and is decreasing at a rate of 0.15 L/min. Find the rate of change of T with respect to time at that instant if $n = 10$ mol.

58. Sensitivity of the eye to brightness If R denotes the reaction of the body to some stimulus of strength x, the *sensitivity* S is defined to be the rate of change of the reaction with respect to x. A particular example is that when the brightness x of a light source is increased, the eye reacts by decreasing the area R of the pupil. The experimental formula

$$R = \frac{40 + 24x^{0.4}}{1 + 4x^{0.4}}$$

has been used to model the dependence of R on x when R is measured in square millimeters and x is measured in appropriate units of brightness.

(a) Find the sensitivity.

(b) Illustrate part (a) by graphing both R and S as functions of x. Comment on the values of R and S at low levels of brightness. Is this what you would expect?

59. How many tangent lines to the curve $y = x/(x + 1)$ pass through the point $(1, 2)$? At which points do these tangent lines touch the curve?

60. Find equations of the tangent lines to the curve

$$y = \frac{x - 1}{x + 1}$$

that are parallel to the line $x - 2y = 2$.

61. (a) Use the Product Rule twice to prove that if f, g, and h are differentiable, then $(fgh)' = f'gh + fg'h + fgh'$.

(b) Taking $f = g = h$ in part (a), show that

$$\frac{d}{dx}[f(x)]^3 = 3[f(x)]^2 f'(x)$$

(c) Use part (b) to differentiate $y = e^{3x}$.

62. (a) If $F(x) = f(x)\,g(x)$, where f and g have derivatives of all orders, show that $F'' = f''g + 2f'g' + fg''$.

(b) Find similar formulas for F''' and $F^{(4)}$.

(c) Guess a formula for $F^{(n)}$.

63. (a) If g is differentiable, the **Reciprocal Rule** says that

$$\frac{d}{dx}\left[\frac{1}{g(x)}\right] = -\frac{g'(x)}{[g(x)]^2}$$

Use the Quotient Rule to prove the Reciprocal Rule.

(b) Use the Reciprocal Rule to differentiate the function $y = 1/(x^4 + x^2 + 1)$.

(c) Use the Reciprocal Rule to verify that the Power Rule is valid for negative integers, that is,

$$\frac{d}{dx}(x^{-n}) = -nx^{-n-1}$$

for all positive integers n.

3.5 The Chain Rule

Suppose you are asked to differentiate the function

$$F(x) = \sqrt{x^2 + 1}$$

The differentiation formulas you learned in the previous sections of this chapter do not enable you to calculate $F'(x)$.

See Section 1.3 for a review of composite functions.

Observe that F is a composite function. In fact, if we let $y = f(u) = \sqrt{u}$ and let $u = g(x) = x^2 + 1$, then we can write $y = F(x) = f(g(x))$, that is, $F = f \circ g$. We know how to differentiate both f and g, so it would be useful to have a rule that tells us how to find the derivative of $F = f \circ g$ in terms of the derivatives of f and g.

It turns out that the derivative of the composite function $f \circ g$ is the product of the derivatives of f and g. This fact is one of the most important of the differentiation rules and is called the *Chain Rule*. It seems plausible if we interpret derivatives as rates of change. Regard du/dx as the rate of change of u with respect to x, dy/du as the rate of change of y with respect to u, and dy/dx as the rate of change of y with respect to x. If u changes twice as fast as x and y changes three times as fast as u, then it seems reasonable that y changes six times as fast as x, and so we expect that

$$\frac{dy}{dx} = \frac{dy}{du}\frac{du}{dx}$$

The Chain Rule If g is differentiable at x and f is differentiable at $g(x)$, then the composite function $F = f \circ g$ defined by $F(x) = f(g(x))$ is differentiable at x and F' is given by the product

$$F'(x) = f'(g(x)) \cdot g'(x)$$

In Leibniz notation, if $y = f(u)$ and $u = g(x)$ are both differentiable functions, then

$$\frac{dy}{dx} = \frac{dy}{du}\frac{du}{dx}$$

COMMENTS ON THE PROOF OF THE CHAIN RULE Let Δu be the change in u corresponding to a change of Δx in x, that is,

$$\Delta u = g(x + \Delta x) - g(x)$$

Then the corresponding change in y is

$$\Delta y = f(u + \Delta u) - f(u)$$

It is tempting to write

James Gregory

The first person to formulate the Chain Rule was the Scottish mathematician James Gregory (1638–1675), who also designed the first practical reflecting telescope. Gregory discovered the basic ideas of calculus at about the same time as Newton. He became the first Professor of Mathematics at the University of St. Andrews and later held the same position at the University of Edinburgh. But one year after accepting that position he died at the age of 36.

$$\begin{aligned}\frac{dy}{dx} &= \lim_{\Delta x \to 0} \frac{\Delta y}{\Delta x} \\ &= \lim_{\Delta x \to 0} \frac{\Delta y}{\Delta u} \cdot \frac{\Delta u}{\Delta x} \qquad \textbf{(1)} \\ &= \lim_{\Delta x \to 0} \frac{\Delta y}{\Delta u} \cdot \lim_{\Delta x \to 0} \frac{\Delta u}{\Delta x} \\ &= \lim_{\Delta u \to 0} \frac{\Delta y}{\Delta u} \cdot \lim_{\Delta x \to 0} \frac{\Delta u}{\Delta x} \qquad \text{(Note that } \Delta u \to 0 \text{ as } \Delta x \to 0 \text{ since } g \text{ is continuous.)} \\ &= \frac{dy}{du}\frac{du}{dx}\end{aligned}$$

The only flaw in this reasoning is that in (1) it might happen that $\Delta u = 0$ (even when $\Delta x \neq 0$) and, of course, we can't divide by 0. Nonetheless, this reasoning does at least suggest that the Chain Rule is true. A full proof of the Chain Rule is given at the end of this section. ■

The Chain Rule can be written either in the prime notation

(2) $$(f \circ g)'(x) = f'(g(x)) \cdot g'(x)$$

or, if $y = f(u)$ and $u = g(x)$, in Leibniz notation:

(3) $$\frac{dy}{dx} = \frac{dy}{du}\frac{du}{dx}$$

Equation 3 is easy to remember because if dy/du and du/dx were quotients, then we could cancel du. Remember, however, that du has not been defined and du/dx should not be thought of as an actual quotient.

EXAMPLE 1 | Find $F'(x)$ if $F(x) = \sqrt{x^2 + 1}$.

SOLUTION 1 (using Equation 2): At the beginning of this section we expressed F as $F(x) = (f \circ g)(x) = f(g(x))$ where $f(u) = \sqrt{u}$ and $g(x) = x^2 + 1$. Since

$$f'(u) = \tfrac{1}{2}u^{-1/2} = \frac{1}{2\sqrt{u}} \qquad \text{and} \qquad g'(x) = 2x$$

we have

$$\begin{aligned} F'(x) &= f'(g(x)) \cdot g'(x) \\ &= \frac{1}{2\sqrt{x^2+1}} \cdot 2x = \frac{x}{\sqrt{x^2+1}} \end{aligned}$$

SOLUTION 2 (using Equation 3): If we let $u = x^2 + 1$ and $y = \sqrt{u}$, then

$$F'(x) = \frac{dy}{du}\frac{du}{dx} = \frac{1}{2\sqrt{u}}(2x) = \frac{1}{2\sqrt{x^2+1}}(2x) = \frac{x}{\sqrt{x^2+1}}$$ ■

When using Formula 3 we should bear in mind that dy/dx refers to the derivative of y when y is considered as a function of x (called the *derivative of y with respect to x*), whereas dy/du refers to the derivative of y when considered as a function of u (the derivative of y with respect to u). For instance, in Example 1, y can be considered as a function of x $\left(y = \sqrt{x^2+1}\right)$ and also as a function of u $\left(y = \sqrt{u}\right)$. Note that

$$\frac{dy}{dx} = F'(x) = \frac{x}{\sqrt{x^2+1}} \qquad \text{whereas} \qquad \frac{dy}{du} = f'(u) = \frac{1}{2\sqrt{u}}$$

NOTE In using the Chain Rule we work from the outside to the inside. Formula 2 says that *we differentiate the outer function f [at the inner function $g(x)$] and then we multiply by the derivative of the inner function.*

$$\frac{d}{dx} \underbrace{f}_{\substack{\text{outer}\\\text{function}}} \underbrace{(g(x))}_{\substack{\text{evaluated}\\\text{at inner}\\\text{function}}} = \underbrace{f'}_{\substack{\text{derivative}\\\text{of outer}\\\text{function}}} \underbrace{(g(x))}_{\substack{\text{evaluated}\\\text{at inner}\\\text{function}}} \cdot \underbrace{g'(x)}_{\substack{\text{derivative}\\\text{of inner}\\\text{function}}}$$

EXAMPLE 2 | Differentiate (a) $y = \sin(x^2)$ and (b) $y = \sin^2 x$.

SOLUTION

(a) If $y = \sin(x^2)$, then the outer function is the sine function and the inner function is the squaring function, so the Chain Rule gives

$$\begin{aligned} \frac{dy}{dx} = \frac{d}{dx} \underbrace{\sin}_{\substack{\text{outer}\\\text{function}}} \underbrace{(x^2)}_{\substack{\text{evaluated}\\\text{at inner}\\\text{function}}} &= \underbrace{\cos}_{\substack{\text{derivative}\\\text{of outer}\\\text{function}}} \underbrace{(x^2)}_{\substack{\text{evaluated}\\\text{at inner}\\\text{function}}} \cdot \underbrace{2x}_{\substack{\text{derivative}\\\text{of inner}\\\text{function}}} \\ &= 2x\cos(x^2) \end{aligned}$$

(b) Note that $\sin^2 x = (\sin x)^2$. Here the outer function is the squaring function and the inner function is the sine function. So

$$\frac{dy}{dx} = \frac{d}{dx}\underbrace{(\sin x)^2}_{\text{inner function}} = \underbrace{2}_{\substack{\text{derivative}\\\text{of outer}\\\text{function}}} \cdot \underbrace{(\sin x)}_{\substack{\text{evaluated}\\\text{at inner}\\\text{function}}} \cdot \underbrace{\cos x}_{\substack{\text{derivative}\\\text{of inner}\\\text{function}}}$$

See Reference Page 2 or Appendix C.

The answer can be left as $2 \sin x \cos x$ or written as $\sin 2x$ (by a trigonometric identity known as the double-angle formula). ■

■ Combining the Chain Rule with Other Rules

In Example 2(a) we combined the Chain Rule with the rule for differentiating the sine function. In general, if $y = \sin u$, where u is a differentiable function of x, then, by the Chain Rule,

$$\frac{dy}{dx} = \frac{dy}{du}\frac{du}{dx} = \cos u \frac{du}{dx}$$

Thus

$$\frac{d}{dx}(\sin u) = \cos u \frac{du}{dx}$$

In a similar fashion, all of the differentiation formulas for trigonometric functions can be combined with the Chain Rule.

Let's make explicit the special case of the Chain Rule where the outer function f is a power function. If $y = [g(x)]^n$, then we can write $y = f(u) = u^n$ where $u = g(x)$. By using the Chain Rule and then the Power Rule, we get

$$\frac{dy}{dx} = \frac{dy}{du}\frac{du}{dx} = nu^{n-1}\frac{du}{dx} = n[g(x)]^{n-1}g'(x)$$

(4) The Power Rule Combined with the Chain Rule If n is any real number and $u = g(x)$ is differentiable, then

$$\frac{d}{dx}(u^n) = nu^{n-1}\frac{du}{dx}$$

Alternatively,

$$\frac{d}{dx}[g(x)]^n = n[g(x)]^{n-1} \cdot g'(x)$$

Notice that the derivative in Example 1 could be calculated by taking $n = \frac{1}{2}$ in Rule 4.

EXAMPLE 3 | Differentiate $y = (x^3 - 1)^{100}$.

SOLUTION Taking $u = g(x) = x^3 - 1$ and $n = 100$ in (4), we have

$$\frac{dy}{dx} = \frac{d}{dx}(x^3 - 1)^{100} = 100(x^3 - 1)^{99}\frac{d}{dx}(x^3 - 1)$$

$$= 100(x^3 - 1)^{99} \cdot 3x^2 = 300x^2(x^3 - 1)^{99}$$ ■

EXAMPLE 4 | Find $f'(x)$ if $f(x) = \dfrac{1}{\sqrt[3]{x^2 + x + 1}}$.

SOLUTION First rewrite f: $\quad f(x) = (x^2 + x + 1)^{-1/3}$

Thus

$$f'(x) = -\tfrac{1}{3}(x^2 + x + 1)^{-4/3} \frac{d}{dx}(x^2 + x + 1)$$

$$= -\tfrac{1}{3}(x^2 + x + 1)^{-4/3}(2x + 1)$$

■

EXAMPLE 5 | Find the derivative of the function

$$g(t) = \left(\frac{t - 2}{2t + 1}\right)^9$$

SOLUTION Combining the Power Rule, Chain Rule, and Quotient Rule, we get

$$g'(t) = 9\left(\frac{t - 2}{2t + 1}\right)^8 \frac{d}{dt}\left(\frac{t - 2}{2t + 1}\right)$$

$$= 9\left(\frac{t - 2}{2t + 1}\right)^8 \frac{(2t + 1) \cdot 1 - 2(t - 2)}{(2t + 1)^2} = \frac{45(t - 2)^8}{(2t + 1)^{10}}$$

■

EXAMPLE 6 | Differentiate $y = (2x + 1)^5(x^3 - x + 1)^4$.

The graphs of the functions y and y' in Example 6 are shown in Figure 1. Notice that y' is large when y increases rapidly and $y' = 0$ when y has a horizontal tangent. So our answer appears to be reasonable.

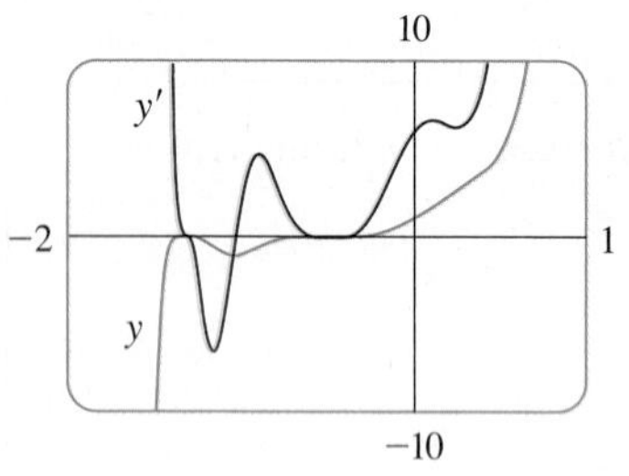

FIGURE 1

SOLUTION In this example we must use the Product Rule before using the Chain Rule:

$$\frac{dy}{dx} = (2x + 1)^5 \frac{d}{dx}(x^3 - x + 1)^4 + (x^3 - x + 1)^4 \frac{d}{dx}(2x + 1)^5$$

$$= (2x + 1)^5 \cdot 4(x^3 - x + 1)^3 \frac{d}{dx}(x^3 - x + 1)$$

$$+ (x^3 - x + 1)^4 \cdot 5(2x + 1)^4 \frac{d}{dx}(2x + 1)$$

$$= 4(2x + 1)^5(x^3 - x + 1)^3(3x^2 - 1) + 5(x^3 - x + 1)^4(2x + 1)^4 \cdot 2$$

Noticing that each term has the common factor $2(2x + 1)^4(x^3 - x + 1)^3$, we could factor it out and write the answer as

$$\frac{dy}{dx} = 2(2x + 1)^4(x^3 - x + 1)^3(17x^3 + 6x^2 - 9x + 3)$$

■

EXAMPLE 7 | Differentiate $y = e^{\sin x}$.

SOLUTION Here the inner function is $g(x) = \sin x$ and the outer function is the exponential function $f(x) = e^x$. So, by the Chain Rule,

$$\frac{dy}{dx} = \frac{d}{dx}(e^{\sin x}) = e^{\sin x}\frac{d}{dx}(\sin x) = e^{\sin x}\cos x$$

■

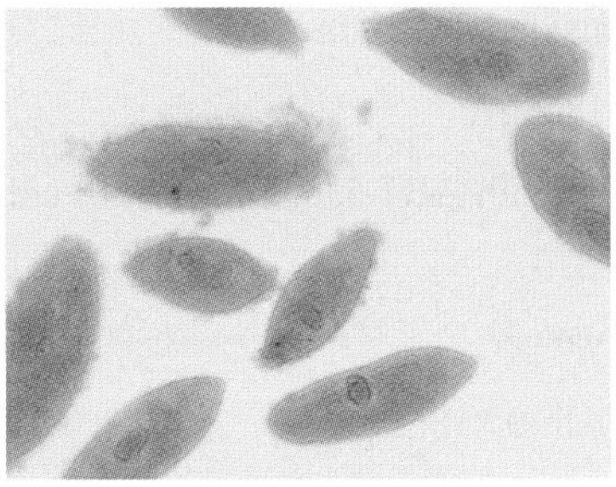

EXAMPLE 8 | BB **Gause's logistic model** In an experiment with the protozoan *Paramecium*, the biologist G. F. Gause modeled the protozoan population size with the logistic function

$$P(t) = \frac{61}{1 + 31e^{-0.7944t}}$$

where t is measured in days. According to this model, how fast was the population growing after 8 days?

SOLUTION We use the Reciprocal Rule as in Exercise 3.4.63 (or the Quotient Rule) and the Chain Rule to differentiate P:

$$P'(t) = -\frac{61}{(1 + 31e^{-0.7944t})^2} \cdot \frac{d}{dt}(1 + 31e^{-0.7944t})$$

$$= -\frac{61 \cdot 31 \cdot (-0.7944)e^{-0.7944t}}{(1 + 31e^{-0.7944t})^2} \approx \frac{1502e^{-0.7944t}}{(1 + 31e^{-0.7944t})^2}$$

When $t = 8$, we have

$$P'(8) = \frac{1502e^{-8(0.7944)}}{(1 + 31e^{-8(0.7944)})^2} \approx 2.35$$

So at that time the population was increasing at a rate of about two per day. ■

Exponential Functions with Arbitrary Bases

We can use the Chain Rule to differentiate an exponential function with any base $b > 0$. Recall from Section 1.5 that $b = e^{\ln b}$. So

$$b^x = (e^{\ln b})^x = e^{(\ln b)x}$$

and the Chain Rule gives

$$\frac{d}{dx}(b^x) = \frac{d}{dx}(e^{(\ln b)x}) = e^{(\ln b)x}\frac{d}{dx}(\ln b)x$$

$$= e^{(\ln b)x} \cdot \ln b = b^x \ln b$$

because $\ln b$ is a constant. So we have the formula

Don't confuse Formula 5 (where x is the exponent) with the Power Rule (where x is the base):

$$\frac{d}{dx}(x^n) = nx^{n-1}$$

(5)
$$\boxed{\frac{d}{dx}(b^x) = b^x \ln b}$$

In particular, if $b = 2$, we get

(6)
$$\frac{d}{dx}(2^x) = 2^x \ln 2$$

In Section 3.3 we gave the estimate

$$\frac{d}{dx}(2^x) \approx (0.69)2^x$$

This is consistent with the exact formula (6) because $\ln 2 \approx 0.693147$.

EXAMPLE 9 | Viral load In Section 1.4 we considered the function

$$V(t) = 96.39785 \cdot (0.818656)^t$$

for the viral load in a patient with HIV after treatment with ABT-538. Find $V'(10)$ and interpret it.

SOLUTION Using Formula 5 with $b = 0.818656$, we get

$$V'(t) = 96.39785 \cdot (0.818656)^t \ln(0.818656)$$

$$\approx -19.288 \cdot (0.818656)^t$$

So

$$V'(10) \approx -19.288 \cdot (0.818656)^{10} \approx -2.6$$

This means that after 10 days the viral load was decreasing at a rate of about 2.6 RNA copies/mL per day. ■

Longer Chains

The reason for the name "Chain Rule" becomes clear when we make a longer chain by adding another link. Suppose that $y = f(u)$, $u = g(x)$, and $x = h(t)$, where f, g, and h are differentiable functions. Then, to compute the derivative of y with respect to t, we use the Chain Rule twice:

$$\frac{dy}{dt} = \frac{dy}{dx}\frac{dx}{dt} = \frac{dy}{du}\frac{du}{dx}\frac{dx}{dt}$$

EXAMPLE 10 | If $f(x) = \sin(\cos(\tan x))$, then

$$f'(x) = \cos(\cos(\tan x))\frac{d}{dx}\cos(\tan x)$$

$$= \cos(\cos(\tan x))[-\sin(\tan x)]\frac{d}{dx}(\tan x)$$

$$= -\cos(\cos(\tan x))\sin(\tan x)\sec^2 x$$

Notice that we used the Chain Rule twice. ■

EXAMPLE 11 | Differentiate $y = e^{\sec 3\theta}$.

SOLUTION The outer function is the exponential function, the middle function is the secant function, and the inner function is the tripling function. So we have

$$\frac{dy}{d\theta} = e^{\sec 3\theta}\frac{d}{d\theta}(\sec 3\theta)$$

$$= e^{\sec 3\theta}\sec 3\theta \tan 3\theta \frac{d}{d\theta}(3\theta)$$

$$= 3e^{\sec 3\theta}\sec 3\theta \tan 3\theta$$

■

Implicit Differentiation

The functions that we have met so far can be described by expressing one variable explicitly in terms of another variable—for example,

$$y = \sqrt{x^3 + 1} \qquad \text{or} \qquad y = x \sin x$$

or, in general, $y = f(x)$. Some functions, however, are defined implicitly by a relation between x and y such as

(7) $$x^3 + y^3 = 6xy$$

It's not easy to solve Equation 7 for y explicitly as a function of x by hand. (A computer algebra system has no trouble, but the expressions it obtains are very complicated.) Nonetheless, (7) is the equation of a curve called the **folium of Descartes** shown in Figure 2 and it implicitly defines y as several functions of x. The graphs of three such functions are shown in Figure 3. When we say that f is a function defined implicitly by Equation 2, we mean that the equation

$$x^3 + [f(x)]^3 = 6xf(x)$$

is true for all values of x in the domain of f.

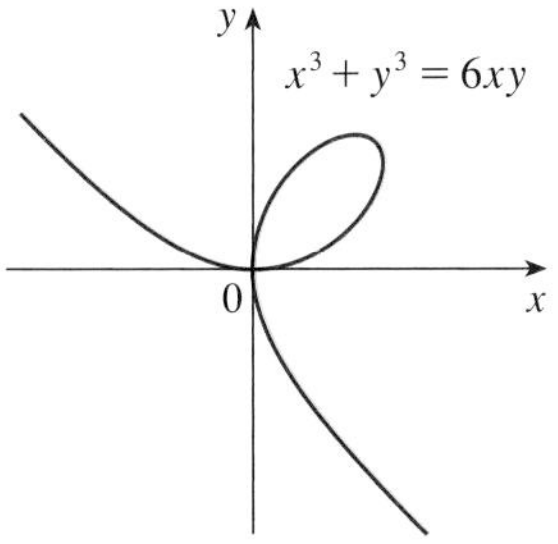

FIGURE 2
The folium of Descartes

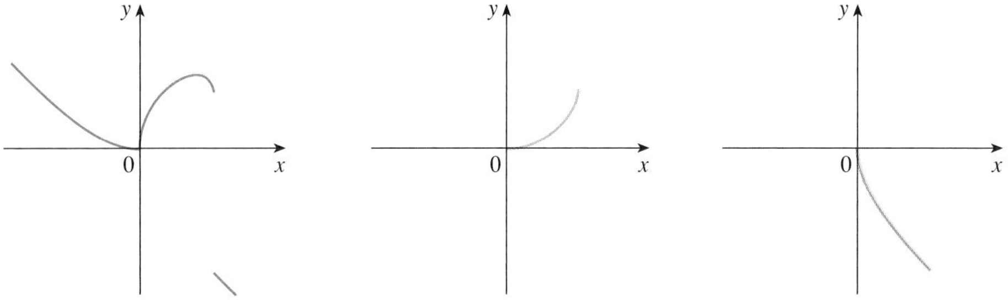

FIGURE 3
Graphs of three functions defined by the folium of Descartes

Fortunately, we don't need to solve an equation for y in terms of x in order to find the derivative of y. Instead we can use the method of **implicit differentiation**. This consists of differentiating both sides of the equation with respect to x and then solving the resulting equation for y'.

EXAMPLE 12
(a) Find y' if $x^3 + y^3 = 6xy$.
(b) Find the tangent to the folium of Descartes $x^3 + y^3 = 6xy$ at the point (3, 3).

SOLUTION
(a) Differentiating both sides of $x^3 + y^3 = 6xy$ with respect to x, we have

$$\frac{d}{dx}(x^3 + y^3) = \frac{d}{dx}(6xy)$$

Remembering that y is a function of x, and using the Chain Rule on the term y^3 and the Product Rule on the term $6xy$, we get

$$3x^2 + 3y^2y' = 6xy' + 6y$$

or $$x^2 + y^2y' = 2xy' + 2y$$

We now solve for y':

$$y^2y' - 2xy' = 2y - x^2$$

$$(y^2 - 2x)y' = 2y - x^2$$

$$y' = \frac{2y - x^2}{y^2 - 2x}$$

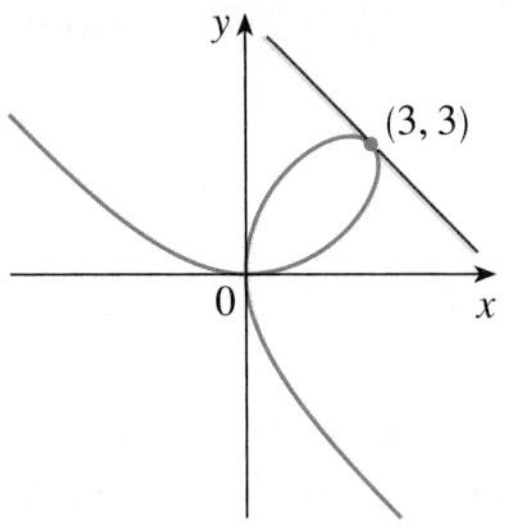

FIGURE 4

(b) When $x = y = 3$,

$$y' = \frac{2 \cdot 3 - 3^2}{3^2 - 2 \cdot 3} = -1$$

and a glance at Figure 4 confirms that this is a reasonable value for the slope at (3, 3). So an equation of the tangent to the folium at (3, 3) is

$$y - 3 = -1(x - 3) \qquad \text{or} \qquad x + y = 6$$ ■

EXAMPLE 13 | BB **Infectious disease outbreak size** Mathematical models have been used to predict the fraction of a population that will be infected by a disease under different conditions. The Kermack-McKendrick model (see Exercise 7.6.23) leads to the following equation:

$$\rho e^{-qA} = 1 - A$$

where A is the fraction of the population ultimately infected, q is a measure of disease transmissibility, and ρ is a measure of the fraction of the population that is initially susceptible to infection. How does the outbreak size A change with an increase in transmissibility q?

SOLUTION The given equation can't be solved for A, so we use implicit differentiation to find dA/dq:

$$\frac{d}{dq}(\rho e^{-qA}) = \frac{d}{dq}(1 - A)$$

Keeping in mind that A is a function of q, we have

$$\rho e^{-qA} \cdot \frac{d}{dq}(-qA) = -\frac{dA}{dq}$$

Now we use the Product Rule:

$$\rho e^{-qA}\left(-q\frac{dA}{dq} - A \cdot 1\right) = -\frac{dA}{dq}$$

Solving for dA/dq, we get

$$(-\rho q e^{-qA} + 1)\frac{dA}{dq} = A\rho e^{-qA}$$

$$\frac{dA}{dq} = \frac{A\rho e^{-qA}}{1 - \rho q e^{-qA}} = \frac{A\rho}{e^{qA} - \rho q}$$

This gives the rate of increase of the outbreak size as the transmissibility changes. ■

■ Related Rates

In a related rates problem the idea is to compute the rate of change of one quantity in terms of the rate of change of another quantity. The procedure is to find an equation that relates the two quantities and then use the Chain Rule to differentiate both sides with respect to time.

© Lightspring / Shutterstock.com

EXAMPLE 14 | **BB** Growth of a tumor When the diameter of a spherical tumor is 16 mm it is growing at a rate of 0.4 mm a day. How fast is the volume of the tumor changing at that time?

SOLUTION We model the shape of the tumor by a sphere. If r is the radius of the sphere, then its volume is

$$V = \tfrac{4}{3}\pi r^3$$

In order to use the given information, we differentiate each side of this equation with respect to t (time). To differentiate the right side we need to use the Chain Rule:

$$\frac{dV}{dt} = \frac{dV}{dr} \cdot \frac{dr}{dt} = 4\pi r^2 \frac{dr}{dt}$$

At the time in question, the radius is $r = \frac{1}{2}(16) = 8$ mm and it is increasing at a rate of

$$\frac{dr}{dt} = \tfrac{1}{2}(0.4) = 0.2 \text{ mm/day}$$

So the rate of increase of the volume of the tumor is

$$\frac{dV}{dt} = 4\pi(8)^2 \cdot 0.2 = 51.2\pi \approx 161 \text{ mm}^3\text{/day}$$ ■

■ How To Prove the Chain Rule

Recall that if $y = f(x)$ and x changes from a to $a + \Delta x$, we define the increment of y as

$$\Delta y = f(a + \Delta x) - f(a)$$

According to the definition of a derivative, we have

$$\lim_{\Delta x \to 0} \frac{\Delta y}{\Delta x} = f'(a)$$

So if we denote by ε the difference between the difference quotient and the derivative, we obtain

$$\lim_{\Delta x \to 0} \varepsilon = \lim_{\Delta x \to 0} \left(\frac{\Delta y}{\Delta x} - f'(a) \right) = f'(a) - f'(a) = 0$$

But $$\varepsilon = \frac{\Delta y}{\Delta x} - f'(a) \quad \Rightarrow \quad \Delta y = f'(a)\,\Delta x + \varepsilon\,\Delta x$$

If we define ε to be 0 when $\Delta x = 0$, then ε becomes a continuous function of Δx. Thus, for a differentiable function f, we can write

(8) $$\Delta y = f'(a)\,\Delta x + \varepsilon\,\Delta x \qquad \text{where} \quad \varepsilon \to 0 \text{ as } \Delta x \to 0$$

and ε is a continuous function of Δx. This property of differentiable functions is what enables us to prove the Chain Rule.

PROOF OF THE CHAIN RULE Suppose $u = g(x)$ is differentiable at a and $y = f(u)$ is differentiable at $b = g(a)$. If Δx is an increment in x and Δu and Δy are the corresponding increments in u and y, then we can use Equation 8 to write

(9) $$\Delta u = g'(a)\,\Delta x + \varepsilon_1\,\Delta x = [g'(a) + \varepsilon_1]\,\Delta x$$

where $\varepsilon_1 \to 0$ as $\Delta x \to 0$. Similarly

$$\textbf{(10)} \qquad \Delta y = f'(b)\,\Delta u + \varepsilon_2\,\Delta u = [f'(b) + \varepsilon_2]\,\Delta u$$

where $\varepsilon_2 \to 0$ as $\Delta u \to 0$. If we now substitute the expression for Δu from Equation 9 into Equation 10, we get

$$\Delta y = [f'(b) + \varepsilon_2][g'(a) + \varepsilon_1]\,\Delta x$$

so

$$\frac{\Delta y}{\Delta x} = [f'(b) + \varepsilon_2][g'(a) + \varepsilon_1]$$

As $\Delta x \to 0$, Equation 9 shows that $\Delta u \to 0$. So both $\varepsilon_1 \to 0$ and $\varepsilon_2 \to 0$ as $\Delta x \to 0$. Therefore

$$\frac{dy}{dx} = \lim_{\Delta x \to 0} \frac{\Delta y}{\Delta x} = \lim_{\Delta x \to 0} [f'(b) + \varepsilon_2][g'(a) + \varepsilon_1]$$

$$= f'(b)\,g'(a) = f'(g(a))\,g'(a)$$

This proves the Chain Rule. ■

EXERCISES 3.5

1–6 Write the composite function in the form $f(g(x))$. [Identify the inner function $u = g(x)$ and the outer function $y = f(u)$.] Then find the derivative dy/dx.

1. $y = \sqrt[3]{1 + 4x}$

2. $y = (2x^3 + 5)^4$

3. $y = \tan \pi x$

4. $y = \sin(\cot x)$

5. $y = e^{\sqrt{x}}$

6. $y = \sqrt{2 - e^x}$

7–36 Find the derivative of the function.

7. $F(x) = (x^4 + 3x^2 - 2)^5$

8. $F(x) = (4x - x^2)^{100}$

9. $F(x) = \sqrt{1 - 2x}$

10. $f(x) = (1 + x^4)^{2/3}$

11. $f(z) = \dfrac{1}{z^2 + 1}$

12. $f(t) = \sqrt[3]{1 + \tan t}$

13. $y = \cos(a^3 + x^3)$

14. $y = a^3 + \cos^3 x$

15. $h(t) = t^3 - 3^t$

16. $y = 3\cot(n\theta)$

17. $y = xe^{-kx}$

18. $y = e^{-2t}\cos 4t$

19. $y = (2x - 5)^4(8x^2 - 5)^{-3}$

20. $h(t) = (t^4 - 1)^3(t^3 + 1)^4$

21. $y = e^{x \cos x}$

22. $y = 10^{1-x^2}$

23. $y = \left(\dfrac{x^2 + 1}{x^2 - 1}\right)^3$

24. $G(y) = \left(\dfrac{y^2}{y + 1}\right)^5$

25. $y = \sec^2 x + \tan^2 x$

26. $y = \dfrac{e^u - e^{-u}}{e^u + e^{-u}}$

27. $y = \dfrac{r}{\sqrt{r^2 + 1}}$

28. $y = e^{k \tan \sqrt{x}}$

29. $y = \sin(\tan 2x)$

30. $f(t) = \sqrt{\dfrac{t}{t^2 + 4}}$

31. $y = 2^{\sin \pi x}$

32. $y = \sin(\sin(\sin x))$

33. $y = \cot^2(\sin \theta)$

34. $y = \sqrt{x + \sqrt{x + \sqrt{x}}}$

35. $y = \cos\sqrt{\sin(\tan \pi x)}$

36. $y = 2^{3^{x^2}}$

37–40 Find y' and y''.

37. $y = \cos(x^2)$

38. $y = \cos^2 x$

39. $y = e^{\alpha x}\sin \beta x$

40. $y = e^{e^x}$

41–44 Find an equation of the tangent line to the curve at the given point.

41. $y = (1 + 2x)^{10}$, $(0, 1)$

42. $y = \sqrt{1 + x^3}$, $(2, 3)$

43. $y = \sin(\sin x)$, $(\pi, 0)$

44. $y = \sin x + \sin^2 x$, $(0, 0)$

45. If $F(x) = f(g(x))$, where $f(-2) = 8$, $f'(-2) = 4$, $f'(5) = 3$, $g(5) = -2$, and $g'(5) = 6$, find $F'(5)$.

46. If $h(x) = \sqrt{4 + 3f(x)}$, where $f(1) = 7$ and $f'(1) = 4$, find $h'(1)$.

47. A table of values for f, g, f', and g' is given.

x	$f(x)$	$g(x)$	$f'(x)$	$g'(x)$
1	3	2	4	6
2	1	8	5	7
3	7	2	7	9

(a) If $h(x) = f(g(x))$, find $h'(1)$.
(b) If $H(x) = g(f(x))$, find $H'(1)$.

48. Let f and g be the functions in Exercise 47.
(a) If $F(x) = f(f(x))$, find $F'(2)$.
(b) If $G(x) = g(g(x))$, find $G'(3)$.

49. Suppose f is differentiable on $\mathbb{R}$. Let $F(x) = f(e^x)$ and $G(x) = e^{f(x)}$. Find expressions for (a) $F'(x)$ and (b) $G'(x)$.

50. Suppose f is differentiable on $\mathbb{R}$ and α is a real number. Let $F(x) = f(x^\alpha)$ and $G(x) = [f(x)]^\alpha$. Find expressions for (a) $F'(x)$ and (b) $G'(x)$.

51. Let $r(x) = f(g(h(x)))$, where $h(1) = 2$, $g(2) = 3$, $h'(1) = 4$, $g'(2) = 5$, and $f'(3) = 6$. Find $r'(1)$.

52. If g is a twice differentiable function and $f(x) = xg(x^2)$, find f'' in terms of g, g', and g''.

53. Find the 50th derivative of $y = \cos 2x$.

54. Find the 1000th derivative of $f(x) = xe^{-x}$.

55. The displacement of a particle on a vibrating string is given by the equation

$$s(t) = 10 + \tfrac{1}{4}\sin(10\pi t)$$

where s is measured in centimeters and t in seconds. Find the velocity and acceleration of the particle after t seconds.

56. Oral antibiotics In Example 1.3.7 we studied a model for antibiotic use in sinus infections. If x is the amount of antibiotic taken orally (in mg), then the function $h(x)$ gives the amount entering the bloodstream through the stomach. If x mg reaches the bloodstream, then $g(x)$ gives the amount that survives filtration by the liver. Finally, if x mg survives filtration by the liver, then $f(x)$ is absorbed into the sinus cavity. Thus, for a given dose x, the amount making it to the sinus cavity is $A(x) = f(g(h(x)))$. Suppose that a dose of 500 mg is given, $h(500) = 8$, $g(8) = 2$, $f(2) = 1.5$, and $h'(500) = 2.5$, $g'(8) = \tfrac{1}{4}$, and $f'(2) = 1$. Calculate $A'(x)$ and interpret your answer.

57. Gene regulation Genes produce molecules called mRNA that go on to produce proteins. High concentrations of protein inhibit the production of mRNA, leading to stable gene regulation. This process has been modeled (see Section 10.3) to show that the concentration of mRNA over time is given by the equation

$$m(t) = \tfrac{1}{2}e^{-t}(\sin t - \cos t) + \tfrac{1}{2}$$

What is the rate of change of mRNA concentration as a function of time?

58. World population growth In Example 1.4.3 we modeled the world population from 1900 to 2010 with the exponential function

$$P(t) = (1436.53) \cdot (1.01395)^t$$

where $t = 0$ corresponds to the year 1900 and $P(t)$ is measured in millions. According to this model, what was the rate of increase of world population in 1920? In 1950? In 2000?

59. Blood alcohol concentration In Section 3.1 we discussed an experiment in which the average BAC of eight male subjects was measured after consumption of 15 mL of ethanol (corresponding to one alcoholic drink). The resulting data were modeled by the concentration function

$$C(t) = 0.0225te^{-0.0467t}$$

where t is measured in minutes after consumption and C is measured in mg/mL.
(a) How rapidly was the BAC increasing after 10 minutes?
(b) How rapidly was it decreasing half an hour later?

60. Logistic growth in Japan The midyear population in Japan from 1960 to 2010 has been modeled by the function

$$P(t) = 94{,}000 + \frac{32{,}658.5}{1 + 12.75e^{-0.1706t}}$$

where t is measured in years since 1960 and $P(t)$ is measured in thousands. According to this model, how quickly was the Japanese population growing in 1970? In 1990?

61. Under certain circumstances a rumor spreads according to the equation

$$p(t) = \frac{1}{1 + ae^{-kt}}$$

where $p(t)$ is the proportion of the population that has heard the rumor at time t and a and k are positive constants.
(a) Find $\lim_{t\to\infty} p(t)$.
(b) Find the rate of spread of the rumor.
(c) Graph p for the case $a = 10$, $k = 0.5$ with t measured in hours. Use the graph to estimate how long it will take for 80% of the population to hear the rumor.

62. In Example 1.3.4 we arrived at a model for the length of daylight (in hours) in Philadelphia on the tth day of the year:

$$L(t) = 12 + 2.8\sin\left[\frac{2\pi}{365}(t - 80)\right]$$

Use this model to compare how the number of hours of daylight is increasing in Philadelphia on March 21 and May 21.

63–64
(a) Find y' by implicit differentiation.
(b) Solve the equation explicitly for y and differentiate to get y' in terms of x.
(c) Check that your solutions to parts (a) and (b) are consistent by substituting the expression for y into your solution for part (a).

63. $xy + 2x + 3x^2 = 4$ **64.** $\cos x + \sqrt{y} = 5$

65–76 Find dy/dx by implicit differentiation.

65. $x^3 + y^3 = 1$ **66.** $2\sqrt{x} + \sqrt{y} = 3$

67. $x^2 + xy - y^2 = 4$ **68.** $2x^3 + x^2y - xy^3 = 2$

69. $x^4(x + y) = y^2(3x - y)$ **70.** $y^5 + x^2y^3 = 1 + ye^{x^2}$

71. $4 \cos x \sin y = 1$ **72.** $1 + x = \sin(xy^2)$

73. $e^{x/y} = x - y$ **74.** $\tan(x - y) = \dfrac{y}{1 + x^2}$

75. $e^y \cos x = 1 + \sin(xy)$ **76.** $\sin x + \cos y = \sin x \cos y$

77–80 Use implicit differentiation to find an equation of the tangent line to the curve at the given point.

77. $x^2 + xy + y^2 = 3$, $(1, 1)$ (ellipse)

78. $x^2 + 2xy - y^2 + x = 2$, $(1, 2)$ (hyperbola)

79. $x^2 + y^2 = (2x^2 + 2y^2 - x)^2$, $\left(0, \frac{1}{2}\right)$ (cardioid)

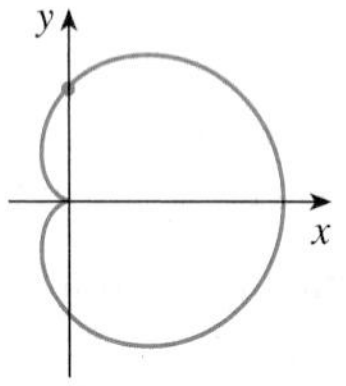

80. $x^{2/3} + y^{2/3} = 4$, $\left(-3\sqrt{3}, 1\right)$ (astroid)

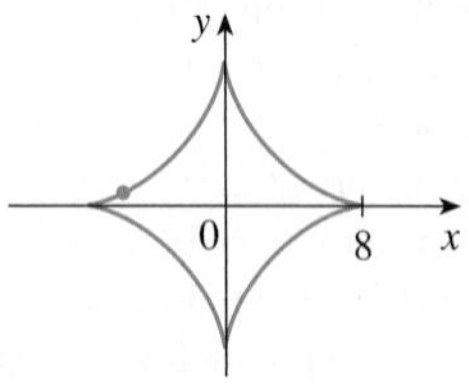

81. Infectious disease outbreak size In Example 13 we used the equation

$$\rho e^{-qA} = 1 - A$$

to determine the rate of increase of the outbreak size A with respect to the transmissibility q. Use this same equation to find the rate of change of A with respect to ρ, the fraction of the population that is initially susceptible to infection.

82. The **logistic difference equation with migration** is of the form

$$N_{t+1} = N_t + N_t(1 - N_t) + m$$

where N_t is the population at time t and m is the migration rate. Suppose that as $t \to \infty$ the population size approaches a limiting value N.
(a) What equation does N satisfy?
(b) Use implicit differentiation to find the rate of change of N with respect to m.
(c) Find an explicit expression for N as a function of m, differentiate it, and compare with your answer to part (b).

83. If V is the volume of a cube with edge length x and the cube expands as time passes, find dV/dt in terms of dx/dt.

84. (a) If A is the area of a circle with radius r and the circle expands as time passes, find dA/dt in terms of dr/dt.
(b) Suppose oil spills from a ruptured tanker and spreads in a circular pattern. If the radius of the oil spill increases at a constant rate of 1 m/s, how fast is the area of the spill increasing when the radius is 30 m?

85. Each side of a square is increasing at a rate of 6 cm/s. At what rate is the area of the square increasing when the area of the square is 16 cm^2?

86. The length of a rectangle is increasing at a rate of 8 cm/s and its width is increasing at a rate of 3 cm/s. When the length is 20 cm and the width is 10 cm, how fast is the area of the rectangle increasing?

87. Boyle's Law states that when a sample of gas is compressed at a constant temperature, the pressure P and volume V satisfy the equation $PV = C$, where C is a constant. Suppose that at a certain instant the volume is 600 cm^3, the pressure is 150 kPa, and the pressure is increasing at a rate of 20 kPa/min. At what rate is the volume decreasing at this instant?

88. When air expands adiabatically (without gaining or losing heat), its pressure P and volume V are related by the equation $PV^{1.4} = C$, where C is a constant. Suppose that at a certain instant the volume is 400 cm^3 and the pressure is 80 kPa and is decreasing at a rate of 10 kPa/min. At what rate is the volume increasing at this instant?

89. Bone mass In Example 1.1.6 we found an expression for the mass m of a human femur of length L in terms of the outer radius r, the inner radius r_{in}, and their ratio $k = r_{\text{in}}/r$. More generally, if the bone density is ρ, measured in g/cm^3, then bone mass is given by the equation

$$m = \pi r^2 L[\rho - (\rho - 1)k^2]$$

It may happen that both ρ and k change with age, t.
(a) If ρ changes during aging, find an expression for the rate of change of m with respect to t.
(b) If k changes during aging, find an expression for the rate of change of m with respect to t.

90. The **von Bertalanffy growth function**

$$L(t) = L_\infty - (L_\infty - L_0)e^{-kt}$$

where k is a positive constant, models the length L of a fish as a function of t, the age of the fish. Here L_0 is the length at birth and L_∞ is the final length. Suppose that the mass of a fish of length L is given by $M = aL^3$, where a is a positive constant. Calculate the rate of change of mass with respect to age.

91. Habitat fragmentation and species conservation The size of a class-structured population is modeled in Section 8.8. In certain situations the long-term growth rate of a population is given by $r = \frac{1}{2}(1 + \sqrt{1 + 8s})$, where s is the annual survival probability of juveniles. Suppose this survival probability is related to habitat area a by a function $s(a)$. Determine an expression for the rate of change of growth rate with respect to a change in habitat area.

92. Blood flow In Example 3.3.9 we discussed Poiseuille's law of laminar flow:

$$v = \frac{P}{4\eta l}(R^2 - r^2)$$

where v is the blood velocity at a distance r from the center of a blood vessel (a vein or artery) in the shape of a tube with radius R and length l, P is the pressure difference between the ends of the tube, and η is the viscosity of the blood. In very cold weather, blood vessels in the hands and feet contract. Suppose that a blood vessel with $l = 1$ cm, $P = 1500$ dynes/cm^2, $\eta = 0.027$, and $R = 0.01$ cm contracts so that $R'(t) = -0.0005$ cm/min at a particular moment. Calculate dv/dt, the rate of change of the blood flow, at the center of the blood vessel at that time.

93. Angiotensin-converting enzyme (ACE) inhibitors are a type of blood pressure medication that reduces blood pressure by dilating blood vessels. Suppose that the radius R of a blood vessel is related to the dosage x of the medication by the function $R(x)$. One version of Poiseuille's Law gives the relationship between the blood pressure P and the radius as $P = 4\eta lv/R^2$, where v is the blood velocity at the center of the vessel, η is the viscosity of the blood, and l is the length of the blood vessel. Determine the rate of change of blood pressure with respect to dosage.

94. Brain size in fish Brain weight B as a function of body weight W in fish has been modeled by the power function $B = 0.007W^{2/3}$, where B and W are measured in grams. A model for body weight as a function of body length L (measured in centimeters) is $W = 0.12L^{2.53}$. If, over 10 million years, the average length of a certain species of fish evolved from 15 cm to 20 cm at a constant rate, how fast was this species' brain weight increasing when the average length was 18 cm?

95. Use the Chain Rule to show that if θ is measured in degrees, then

$$\frac{d}{d\theta}(\sin\theta) = \frac{\pi}{180}\cos\theta$$

(This gives one reason for the convention that radian measure is always used when dealing with trigonometric functions in calculus: The differentiation formulas would not be as simple if we used degree measure.)

96. If $y = f(u)$ and $u = g(x)$, where f and g are twice differentiable functions, show that

$$\frac{d^2y}{dx^2} = \frac{d^2y}{du^2}\left(\frac{du}{dx}\right)^2 + \frac{dy}{du}\frac{d^2u}{dx^2}$$

3.6 | Exponential Growth and Decay

In many natural phenomena, quantities grow or decay at a rate proportional to their size. For instance, if $y = f(t)$ is the number of individuals in a population of animals or bacteria at time t, then it seems reasonable to expect that the rate of growth $f'(t)$ is proportional to the population $f(t)$; that is, $f'(t) = kf(t)$ for some constant k. Why is this reasonable? Suppose we have a population (of bacteria, for instance) with size $P = 1000$ and at a certain time it is growing at a rate of $P' = 300$ bacteria per hour. Now let's take another 1000 bacteria of the same type and put them with the first population. Each half of the new population was growing at a rate of 300 bacteria per hour. We would expect the total population of 2000 to increase at a rate of 600 bacteria per hour initially (provided there's enough room and nutrition). So if we double the size, we double the growth rate. In general, it seems reasonable that the growth rate should be proportional to the size.

Indeed, under ideal conditions (unlimited environment, adequate nutrition, immunity to disease) the mathematical model given by the equation $f'(t) = kf(t)$ predicts what actually happens fairly accurately. Another example occurs in nuclear physics where the mass of a radioactive substance decays at a rate proportional to the mass. In chemistry, the rate of a unimolecular first-order reaction is proportional to the concentration of the substance.

In general, if $y(t)$ is the value of a quantity y at time t and if the rate of change of y with respect to t is proportional to its size $y(t)$ at any time, then

(1)
$$\boxed{\frac{dy}{dt} = ky}$$

where k is a constant. Equation 1 is sometimes called the **law of natural growth** (if $k > 0$) or the **law of natural decay** (if $k < 0$). It is called a **differential equation** because it involves an unknown function y and its derivative dy/dt.

It's not hard to think of a solution of Equation 1. This equation asks us to find a function whose derivative is a constant multiple of itself. We have met such functions in this chapter. Any exponential function of the form $y(t) = Ce^{kt}$, where C is a constant, satisfies

$$y'(t) = C(ke^{kt}) = k(Ce^{kt}) = ky(t)$$

We will see in Section 7.4 that *any* function that satisfies $dy/dt = ky$ must be of the form $y = Ce^{kt}$. To see the significance of the constant C, we observe that

$$y(0) = Ce^{k \cdot 0} = C$$

Therefore C is the initial value of the function.

> **(2) Theorem** The only solutions of the differential equation $dy/dt = ky$ are the exponential functions
>
> $$y(t) = y(0)e^{kt}$$

■ Population Growth

What is the significance of the proportionality constant k? In the context of population growth, where $P(t)$ is the size of a population at time t, we can write

(3)
$$\frac{dP}{dt} = kP \qquad \text{or} \qquad \frac{1}{P}\frac{dP}{dt} = k$$

The quantity

$$\frac{1}{P}\frac{dP}{dt}$$

is the growth rate divided by the population size; it is called the **relative growth rate** or **per capita growth rate**. According to (3), instead of saying "the growth rate is proportional to population size" we could say "the relative growth rate is constant." Then (2) says that a population with constant relative growth rate must grow exponentially. Notice that the relative growth rate k appears as the coefficient of t in the exponential function Ce^{kt}. For instance, if

$$\frac{dP}{dt} = 0.02P$$

and t is measured in years, then the relative growth rate is $k = 0.02$ and the population grows at a relative rate of 2% per year. If the population at time 0 is P_0, then the expression for the population is

$$P(t) = P_0 e^{0.02t}$$

EXAMPLE 1 | Use the fact that the world population was 2560 million in 1950 and 3040 million in 1960 to model the population of the world in the second half of the 20th century. (Assume that the growth rate is proportional to the population size.) What is the relative growth rate? Use the model to estimate the population in 1993. At what rate was the population increasing in 1993? According to the model, what will the population be in the year 2020?

SOLUTION We measure the time t in years and let $t = 0$ in the year 1950. We measure the population $P(t)$ in millions of people. Then $P(0) = 2560$ and $P(10) = 3040$. Since we are assuming that $dP/dt = kP$, Theorem 2 gives

$$P(t) = P(0)e^{kt} = 2560e^{kt}$$

$$P(10) = 2560e^{10k} = 3040$$

$$k = \frac{1}{10} \ln \frac{3040}{2560} \approx 0.017185$$

The relative growth rate is about 1.7% per year and the model is

$$P(t) = 2560e^{0.017185t}$$

We estimate that the world population in 1993 was

$$P(43) = 2560e^{0.017185(43)} \approx 5360 \text{ million}$$

At that time the rate of increase was

$$P'(43) = kP(43) \approx 92 \text{ million people per year}$$

The model predicts that the population in 2020 will be

$$P(70) = 2560e^{0.017185(70)} \approx 8524 \text{ million}$$

The graph in Figure 1 shows that the model is fairly accurate to the end of the 20th century (the dots represent the actual population), so the estimate for 1993 is quite reliable. But the prediction for 2020 is riskier.

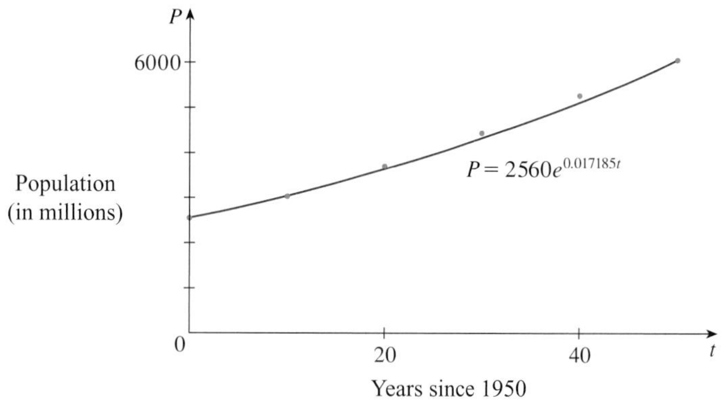

FIGURE 1
A model for world population growth in the second half of the 20th century

■

Radioactive Decay

Radioactive substances decay by spontaneously emitting radiation. If $m(t)$ is the mass remaining from an initial mass m_0 of the substance after time t, then the relative decay rate

$$-\frac{1}{m}\frac{dm}{dt}$$

has been found experimentally to be constant. (Since dm/dt is negative, the relative decay rate is positive.) It follows that

$$\frac{dm}{dt} = km$$

where k is a negative constant. In other words, radioactive substances decay at a rate proportional to the remaining mass. This means that we can use (2) to show that the mass decays exponentially:

$$m(t) = m_0 e^{kt}$$

Physicists express the rate of decay in terms of **half-life**, the time required for half of any given quantity to decay.

EXAMPLE 2 | The half-life of radium-226 is 1590 years.
(a) A sample of radium-226 has a mass of 100 mg. Find a formula for the mass of the sample that remains after t years.
(b) Find the mass after 1000 years correct to the nearest milligram.
(c) When will the mass be reduced to 30 mg?

SOLUTION
(a) Let $m(t)$ be the mass of radium-226 (in milligrams) that remains after t years. Then $dm/dt = km$ and $m(0) = 100$, so (2) gives

$$m(t) = m(0)e^{kt} = 100e^{kt}$$

In order to determine the value of k, we use the fact that $m(1590) = \frac{1}{2}(100)$. Thus

$$100e^{1590k} = 50 \qquad \text{so} \qquad e^{1590k} = \tfrac{1}{2}$$

and

$$1590k = \ln \tfrac{1}{2} = -\ln 2$$

$$k = -\frac{\ln 2}{1590}$$

Therefore

$$m(t) = 100e^{-(\ln 2)t/1590}$$

We could use the fact that $e^{\ln 2} = 2$ to write the expression for $m(t)$ in the alternative form

$$m(t) = 100 \times 2^{-t/1590}$$

(b) The mass after 1000 years is

$$m(1000) = 100e^{-(\ln 2)1000/1590} \approx 65 \text{ mg}$$

(c) We want to find the value of t such that $m(t) = 30$, that is,

$$100e^{-(\ln 2)t/1590} = 30 \qquad \text{or} \qquad e^{-(\ln 2)t/1590} = 0.3$$

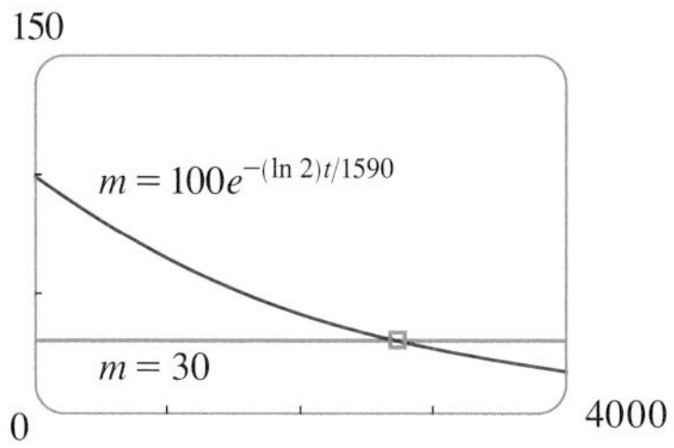

FIGURE 2

We solve this equation for t by taking the natural logarithm of both sides:

$$-\frac{\ln 2}{1590}t = \ln 0.3$$

Thus $$t = -1590\frac{\ln 0.3}{\ln 2} \approx 2762 \text{ years}$$ ■

As a check on our work in Example 2, we use a calculator to draw the graph of $m(t)$ in Figure 2 together with the horizontal line $m = 30$. These curves intersect when $t \approx 2800$, and this agrees with the answer to part (c).

■ Newton's Law of Cooling

Newton's Law of Cooling states that the rate of cooling of an object is proportional to the temperature difference between the object and its surroundings, provided that this difference is not too large. (This law also applies to warming.) If we let $T(t)$ be the temperature of the object at time t and T_s be the temperature of the surroundings, then we can formulate Newton's Law of Cooling as a differential equation:

$$\frac{dT}{dt} = k(T - T_s)$$

where k is a constant. This equation is not quite the same as Equation 1, so we make the change of variable $y(t) = T(t) - T_s$. Because T_s is constant, we have $y'(t) = T'(t)$ and so the equation becomes

$$\frac{dy}{dt} = ky$$

We can then use (2) to find an expression for y, from which we can find T.

EXAMPLE 3 | A bottle of soda pop at room temperature (72°F) is placed in a refrigerator where the temperature is 44°F. After half an hour the soda pop has cooled to 61°F.
(a) What is the temperature of the soda pop after another half hour?
(b) How long does it take for the soda pop to cool to 50°F?

SOLUTION

(a) Let $T(t)$ be the temperature of the soda after t minutes. The surrounding temperature is $T_s = 44$°F, so Newton's Law of Cooling states that

$$\frac{dT}{dt} = k(T - 44)$$

If we let $y = T - 44$, then $y(0) = T(0) - 44 = 72 - 44 = 28$, so y satisfies

$$\frac{dy}{dt} = ky \qquad y(0) = 28$$

and by (2) we have

$$y(t) = y(0)e^{kt} = 28e^{kt}$$

We are given that $T(30) = 61$, so $y(30) = 61 - 44 = 17$ and

$$28e^{30k} = 17 \qquad e^{30k} = \tfrac{17}{28}$$

Taking logarithms, we have

$$k = \frac{\ln\left(\frac{17}{28}\right)}{30} \approx -0.01663$$

Thus

$$y(t) = 28e^{-0.01663t}$$
$$T(t) = 44 + 28e^{-0.01663t}$$
$$T(60) = 44 + 28e^{-0.01663(60)} \approx 54.3$$

So after another half hour the pop has cooled to about 54°F.

(b) We have $T(t) = 50$ when

$$44 + 28e^{-0.01663t} = 50$$
$$e^{-0.01663t} = \frac{6}{28}$$
$$t = \frac{\ln\left(\frac{6}{28}\right)}{-0.01663} \approx 92.6$$

The pop cools to 50°F after about 1 hour 33 minutes. ■

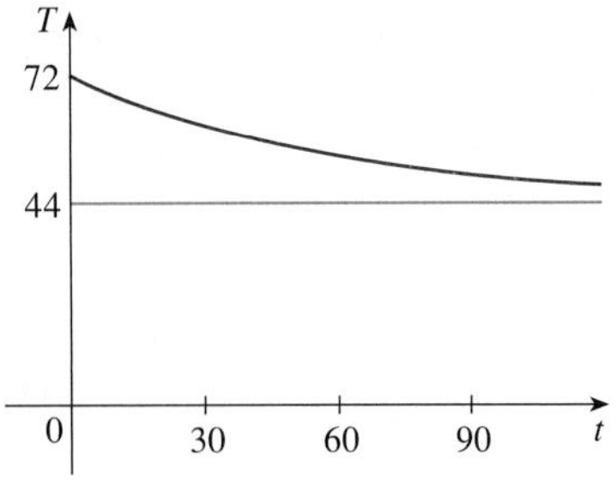

FIGURE 3

Notice that in Example 3, we have

$$\lim_{t\to\infty} T(t) = \lim_{t\to\infty} (44 + 28e^{-0.01663t}) = 44 + 28 \cdot 0 = 44$$

which is to be expected. The graph of the temperature function is shown in Figure 3.

EXERCISES 3.6

1. **Protozoan population** A population of protozoa develops with a constant relative growth rate of 0.7944 per member per day. On day zero the population consists of two members. Find the population size after six days.

2. ***E. coli* population** A common inhabitant of human intestines is the bacterium *Escherichia coli*. A cell of this bacterium in a nutrient-broth medium divides into two cells every 20 minutes. The initial population of a culture is 60 cells.
   (a) Find the relative growth rate.
   (b) Find an expression for the number of cells after t hours.
   (c) Find the number of cells after 8 hours.
   (d) Find the rate of growth after 8 hours.
   (e) When will the population reach 20,000 cells?

3. **Bacteria population** A bacteria culture initially contains 100 cells and grows at a rate proportional to its size. After an hour the population has increased to 420.
   (a) Find an expression for the number of bacteria after t hours.
   (b) Find the number of bacteria after 3 hours.
   (c) Find the rate of growth after 3 hours.
   (d) When will the population reach 10,000?

4. **Bacteria population** A bacteria culture grows with constant relative growth rate. The bacteria count was 400 after 2 hours and 25,600 after 6 hours.
   (a) What is the relative growth rate? Express your answer as a percentage.
   (b) What was the intitial size of the culture?
   (c) Find an expression for the number of bacteria after t hours.
   (d) Find the number of cells after 4.5 hours.
   (e) Find the rate of growth after 4.5 hours.
   (f) When will the population reach 50,000?

5. **World population** The table gives estimates of the world population, in millions, from 1750 to 2000.

Year	Population	Year	Population
1750	790	1900	1650
1800	980	1950	2560
1850	1260	2000	6080

   (a) Use the exponential model and the population figures for 1750 and 1800 to predict the world population in 1900 and 1950. Compare with the actual figures.

(b) Use the exponential model and the population figures for 1850 and 1900 to predict the world population in 1950. Compare with the actual population.

(c) Use the exponential model and the population figures for 1900 and 1950 to predict the world population in 2000. Compare with the actual population and try to explain the discrepancy.

6. Indonesian population The table gives the population of Indonesia, in millions, for the second half of the 20th century.

Year	Population
1950	83
1960	100
1970	122
1980	150
1990	182
2000	214

(a) Assuming the population grows at a rate proportional to its size, use the census figures for 1950 and 1960 to predict the population in 1980. Compare with the actual figure.

(b) Use the census figures for 1960 and 1980 to predict the population in 2000. Compare with the actual population.

(c) Use the census figures for 1980 and 2000 to predict the population in 2010 and compare with the actual population of 243 million.

(d) Use the model in part (c) to predict the population in 2020. Do you think the prediction will be too high or too low? Why?

7. The half-life of cesium-137 is 30 years. Suppose we have a 100-mg sample.

(a) Find the mass that remains after t years.

(b) How much of the sample remains after 100 years?

(c) What is the rate of decay after 100 years?

(d) After how long will only 1 mg remain?

8. Strontium-90 has a half-life of 28 days.

(a) A sample has a mass of 50 mg initially. Find a formula for the mass remaining after t days.

(b) Find the mass remaining after 40 days.

(c) What is the rate of decay after 40 days?

(d) How long does it take the sample to decay to a mass of 2 mg?

(e) Sketch the graph of the mass function.

9. A sample of tritium-3 decayed to 94.5% of its original amount after a year.

(a) What is the half-life of tritium-3?

(b) How long would it take the sample to decay to 20% of its original amount?

10. Radiometric dating Scientists can determine the age of ancient objects by the method of *radiometric dating*. The bombardment of the upper atmosphere by cosmic rays converts nitrogen to a radioactive isotope of carbon, ^{14}C, with a half-life of about 5730 years. Vegetation absorbs carbon dioxide through the atmosphere and animal life assimilates ^{14}C through food chains. When a plant or animal dies, it stops replacing its carbon and the amount of ^{14}C begins to decrease through radioactive decay. Therefore the level of radioactivity must also decay exponentially.

A discovery revealed a parchment fragment that had about 74% as much ^{14}C radioactivity as does plant material on the earth today. Estimate the age of the parchment.

11. Dating dinosaurs Dinosaur fossils are too old to be reliably dated using carbon-14, which has a half-life of about 5730 years. (See Exercise 10.) Suppose we had a 68-million-year-old dinosaur fossil. What fraction of the living dinosaur's ^{14}C would be remaining today? Suppose the minimum detectable amount is 0.1%. What is the maximum age of a fossil that we could date using ^{14}C?

12. Dating dinosaurs with potassium Dinosaur fossils are often dated by using an element other than carbon, such as potassium-40, that has a longer half-life (in this case, approximately 1.25 billion years). Suppose the minimum detectable amount is 0.1% and a dinosaur is dated with ^{40}K to be 68 million years old. Is such a dating possible? In other words, what is the maximum age of a fossil that we could date using ^{40}K?

13. A roast turkey is taken from an oven when its temperature has reached 185°F and is placed on a table in a room where the temperature is 75°F.

(a) If the temperature of the turkey is 150°F after half an hour, what is the temperature after 45 minutes?

(b) When will the turkey have cooled to 100°F?

14. In a murder investigation, the temperature of the corpse was 32.5°C at 1:30 PM and 30.3°C an hour later. Normal body temperature is 37.0°C and the temperature of the surroundings was 20.0°C. When did the murder take place?

15. When a cold drink is taken from a refrigerator, its temperature is 5°C. After 25 minutes in a 20°C room its temperature has increased to 10°C.

(a) What is the temperature of the drink after 50 minutes?

(b) When will its temperature be 15°C?

16. A freshly brewed cup of coffee has temperature 95°C in a 20°C room. When its temperature is 70°C, it is cooling at a rate of 1°C per minute. When does this occur?

17. The rate of change of atmospheric pressure P with respect to altitude h is proportional to P, provided that the temperature is constant. At 15°C the pressure is 101.3 kPa at sea level and 87.14 kPa at $h = 1000$ m.

(a) What is the pressure at an altitude of 3000 m?

(b) What is the pressure at the top of Mount McKinley, at an altitude of 6187 m?

■ PROJECT Controlling Red Blood Cell Loss During Surgery

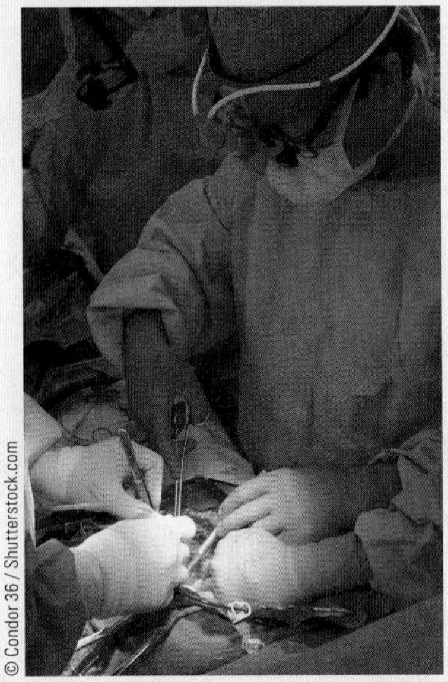

A typical volume of blood in the human body is about 5 L. A certain percentage of that volume (called the *hematocrit*) consists of red blood cells (RBCs); typically the hematocrit is about 45% in males. Suppose that a surgery takes four hours and a male patient bleeds 2 L of blood. During surgery the patient's blood volume is maintained at 5 L by injection of saline solution, which mixes quickly with the blood but dilutes it so that the hematocrit decreases as time passes.

1. Assuming that the rate of RBC loss is proportional to the concentration of RBCs, determine the patient's concentration of RBCs by the end of the operation.

2. A procedure called *acute normovolemic hemodilution* (ANH) has been developed to minimize RBC loss during surgery. In this procedure blood is extracted from the patient before the operation and replaced with saline solution. This dilutes the patient's blood, resulting in fewer RBCs being lost during the bleeding. The extracted blood is then returned to the patient after surgery. Only a certain amount of blood can be extracted, however, because the RBC concentration can never be allowed to drop below 25% during surgery. What is the maximum amount of blood that can be extracted in the ANH procedure for the surgery described in this project?

3. What is the RBC loss without the ANH procedure? What is the loss if the procedure is carried out with the volume calculated in Problem 2?

3.7 | Derivatives of the Logarithmic and Inverse Tangent Functions

In this section we use implicit differentiation to find the derivatives of the logarithmic functions $y = \log_b x$ and, in particular, the natural logarithmic function $y = \ln x$, as well as the inverse tangent function $y = \tan^{-1}x$.

■ Differentiating Logarithmic Functions

It can be proved that logarithmic functions are differentiable. This is certainly plausible from their graphs (see Figure 1.5.12). We now differentiate the general logarithmic function $y = \log_b x$ by using the fact that it is the inverse function of the exponential function with base b.

(1) $$\frac{d}{dx}(\log_b x) = \frac{1}{x \ln b}$$

PROOF Let $y = \log_b x$. Then

$$b^y = x$$

Formula 3.5.5 says that

$$\frac{d}{dx}(b^x) = b^x \ln b$$

Differentiating this equation implicitly with respect to x, using Formula 3.5.5, we get

$$b^y(\ln b)\frac{dy}{dx} = 1$$

and so

$$\frac{dy}{dx} = \frac{1}{b^y \ln b} = \frac{1}{x \ln b}$$ ■

If we put $b = e$ in Formula 1, then the factor $\ln b$ on the right side becomes $\ln e = 1$ and we get the formula for the derivative of the natural logarithmic function $\log_e x = \ln x$:

(2) $$\frac{d}{dx}(\ln x) = \frac{1}{x}$$

By comparing Formulas 1 and 2, we see one of the main reasons that natural logarithms (logarithms with base e) are used in calculus: The differentiation formula is simplest when $b = e$ because $\ln e = 1$.

EXAMPLE 1 | Differentiate $y = \ln(x^3 + 1)$.

SOLUTION To use the Chain Rule, we let $u = x^3 + 1$. Then $y = \ln u$, so

$$\frac{dy}{dx} = \frac{dy}{du}\frac{du}{dx} = \frac{1}{u}\frac{du}{dx}$$

$$= \frac{1}{x^3 + 1}(3x^2) = \frac{3x^2}{x^3 + 1}$$ ■

In general, if we combine Formula 2 with the Chain Rule as in Example 1, we get

(3) $$\frac{d}{dx}(\ln u) = \frac{1}{u}\frac{du}{dx} \quad \text{or} \quad \frac{d}{dx}[\ln g(x)] = \frac{g'(x)}{g(x)}$$

EXAMPLE 2 | Find $\frac{d}{dx}\ln(\sin x)$.

SOLUTION Using (3), we have

$$\frac{d}{dx}\ln(\sin x) = \frac{1}{\sin x}\frac{d}{dx}(\sin x) = \frac{1}{\sin x}\cos x = \cot x$$ ■

EXAMPLE 3 | Differentiate $f(x) = \sqrt{\ln x}$.

SOLUTION This time the logarithm is the inner function, so the Chain Rule gives

$$f'(x) = \tfrac{1}{2}(\ln x)^{-1/2}\frac{d}{dx}(\ln x) = \frac{1}{2\sqrt{\ln x}}\cdot\frac{1}{x} = \frac{1}{2x\sqrt{\ln x}}$$ ■

EXAMPLE 4 | Differentiate $f(x) = \log_{10}(2 + \sin x)$.

SOLUTION Using Formula 1 with $b = 10$, we have

$$\begin{aligned} f'(x) &= \frac{d}{dx}\log_{10}(2 + \sin x) \\ &= \frac{1}{(2 + \sin x)\ln 10}\frac{d}{dx}(2 + \sin x) \\ &= \frac{\cos x}{(2 + \sin x)\ln 10} \end{aligned}$$ ■

EXAMPLE 5 | Find $\dfrac{d}{dx}\ln\left(\dfrac{x+1}{\sqrt{x-2}}\right)$.

SOLUTION 1

$$\begin{aligned} \frac{d}{dx}\ln\left(\frac{x+1}{\sqrt{x-2}}\right) &= \frac{1}{\dfrac{x+1}{\sqrt{x-2}}}\frac{d}{dx}\left(\frac{x+1}{\sqrt{x-2}}\right) \\ &= \frac{\sqrt{x-2}}{x+1}\,\frac{\sqrt{x-2}\cdot 1 - (x+1)\left(\frac{1}{2}\right)(x-2)^{-1/2}}{x-2} \\ &= \frac{x - 2 - \frac{1}{2}(x+1)}{(x+1)(x-2)} \\ &= \frac{x-5}{2(x+1)(x-2)} \end{aligned}$$

Figure 1 shows the graph of the function f of Example 5 together with the graph of its derivative. It gives a visual check on our calculation. Notice that $f'(x)$ is large negative when f is rapidly decreasing.

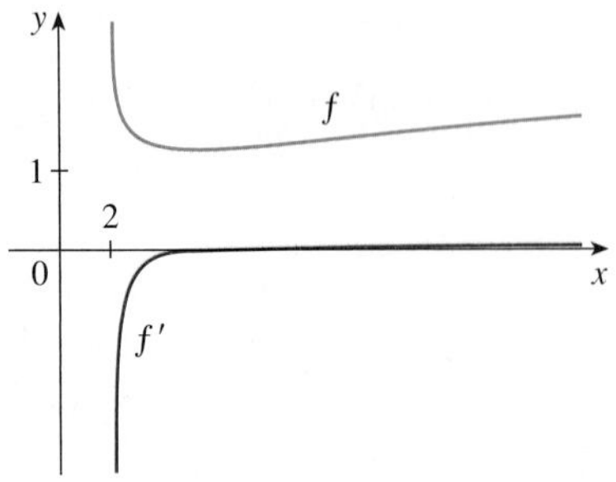

FIGURE 1

SOLUTION 2 If we first simplify the given function using the laws of logarithms, then the differentiation becomes easier:

$$\begin{aligned} \frac{d}{dx}\ln\left(\frac{x+1}{\sqrt{x-2}}\right) &= \frac{d}{dx}\left[\ln(x+1) - \tfrac{1}{2}\ln(x-2)\right] \\ &= \frac{1}{x+1} - \frac{1}{2}\left(\frac{1}{x-2}\right) \end{aligned}$$

(This answer can be left as written, but if we used a common denominator we would see that it gives the same answer as in Solution 1.) ■

Figure 2 shows the graph of the function $f(x) = \ln|x|$ in Example 6 and its derivative $f'(x) = 1/x$. Notice that when x is small, the graph of $y = \ln|x|$ is steep and so $f'(x)$ is large (positive or negative).

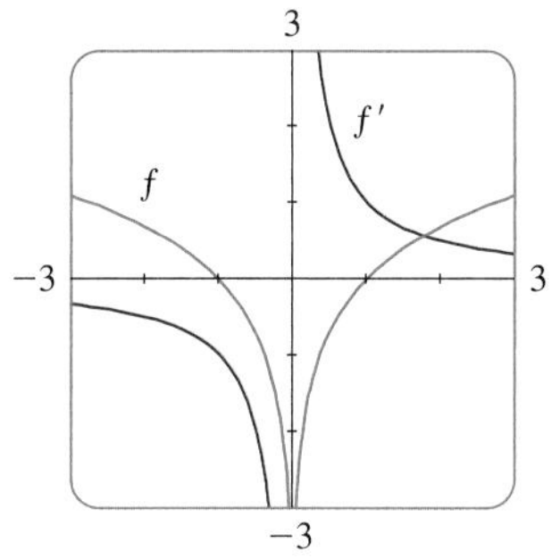

FIGURE 2

EXAMPLE 6 | Find $f'(x)$ if $f(x) = \ln|x|$.

SOLUTION Since

$$f(x) = \begin{cases} \ln x & \text{if } x > 0 \\ \ln(-x) & \text{if } x < 0 \end{cases}$$

it follows that

$$f'(x) = \begin{cases} \dfrac{1}{x} & \text{if } x > 0 \\ \dfrac{1}{-x}(-1) = \dfrac{1}{x} & \text{if } x < 0 \end{cases}$$

Thus $f'(x) = 1/x$ for all $x \neq 0$. ∎

The result of Example 6 is worth remembering:

(4) $$\frac{d}{dx} \ln|x| = \frac{1}{x}$$

■ Logarithmic Differentiation

The calculation of derivatives of complicated functions involving products, quotients, or powers can often be simplified by taking logarithms. The method used in the following example is called **logarithmic differentiation**.

EXAMPLE 7 | Differentiate $y = \dfrac{x^{3/4}\sqrt{x^2+1}}{(3x+2)^5}$.

SOLUTION We take logarithms of both sides of the equation and use the Laws of Logarithms to simplify:

$$\ln y = \tfrac{3}{4}\ln x + \tfrac{1}{2}\ln(x^2+1) - 5\ln(3x+2)$$

Differentiating implicitly with respect to x gives

$$\frac{1}{y}\frac{dy}{dx} = \frac{3}{4}\cdot\frac{1}{x} + \frac{1}{2}\cdot\frac{2x}{x^2+1} - 5\cdot\frac{3}{3x+2}$$

Solving for dy/dx, we get

$$\frac{dy}{dx} = y\left(\frac{3}{4x} + \frac{x}{x^2+1} - \frac{15}{3x+2}\right)$$

If we hadn't used logarithmic differentiation in Example 7, we would have had to use both the Quotient Rule and the Product Rule. The resulting calculation would have been horrendous.

Because we have an explicit expression for y, we can substitute and write

$$\frac{dy}{dx} = \frac{x^{3/4}\sqrt{x^2+1}}{(3x+2)^5}\left(\frac{3}{4x} + \frac{x}{x^2+1} - \frac{15}{3x+2}\right)$$ ∎

Steps in Logarithmic Differentiation

1. Take natural logarithms of both sides of an equation $y = f(x)$ and use the Laws of Logarithms to simplify.
2. Differentiate implicitly with respect to x.
3. Solve the resulting equation for y'.

If $f(x) < 0$ for some values of x, then $\ln f(x)$ is not defined, but we can write $|y| = |f(x)|$ and use Equation 4. We illustrate this procedure by proving the general version of the Power Rule, as promised in Section 3.3.

The Power Rule If n is any real number and $f(x) = x^n$, then

$$f'(x) = nx^{n-1}$$

If $x = 0$, we can show that $f'(0) = 0$ for $n > 1$ directly from the definition of a derivative.

PROOF Let $y = x^n$ and use logarithmic differentiation:

$$\ln|y| = \ln|x|^n = n\ln|x| \qquad x \neq 0$$

Therefore
$$\frac{y'}{y} = \frac{n}{x}$$

Hence
$$y' = n\frac{y}{x} = n\frac{x^n}{x} = nx^{n-1}$$ ■

To differentiate a function of the form $y = [f(x)]^{g(x)}$, where both the base and the exponent are functions, logarithmic differentiation can be used as in the following example.

EXAMPLE 8 | Differentiate $y = x^{\sqrt{x}}$.

SOLUTION 1 Because both the base and the exponent are variable, we use logarithmic differentiation:

$$\ln y = \ln x^{\sqrt{x}} = \sqrt{x}\ln x$$

$$\frac{y'}{y} = \sqrt{x}\cdot\frac{1}{x} + (\ln x)\frac{1}{2\sqrt{x}}$$

$$y' = y\left(\frac{1}{\sqrt{x}} + \frac{\ln x}{2\sqrt{x}}\right) = x^{\sqrt{x}}\left(\frac{2 + \ln x}{2\sqrt{x}}\right)$$

Figure 3 illustrates Example 8 by showing the graphs of $f(x) = x^{\sqrt{x}}$ and its derivative.

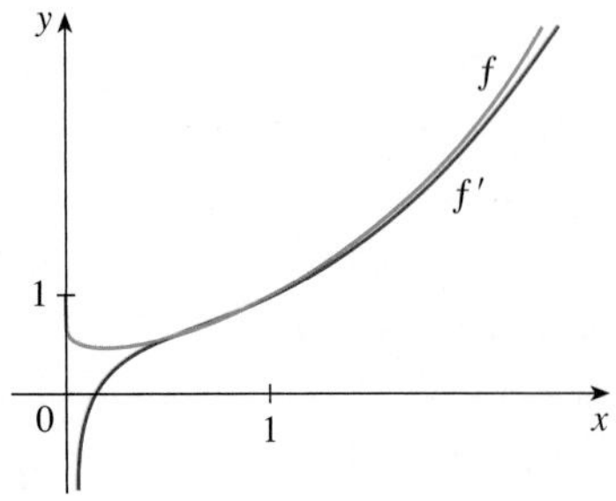

FIGURE 3

SOLUTION 2 Another method is to write $x^{\sqrt{x}} = (e^{\ln x})^{\sqrt{x}}$:

$$\frac{d}{dx}\left(x^{\sqrt{x}}\right) = \frac{d}{dx}\left(e^{\sqrt{x}\ln x}\right) = e^{\sqrt{x}\ln x}\frac{d}{dx}\left(\sqrt{x}\ln x\right)$$

$$= x^{\sqrt{x}}\left(\frac{2 + \ln x}{2\sqrt{x}}\right) \qquad \text{(as in Solution 1)}$$ ■

The Number e as a Limit

We have shown that if $f(x) = \ln x$, then $f'(x) = 1/x$. Thus $f'(1) = 1$. We now use this fact to express the number e as a limit.

From the definition of a derivative as a limit, we have

$$\begin{aligned} f'(1) &= \lim_{h \to 0} \frac{f(1+h) - f(1)}{h} = \lim_{x \to 0} \frac{f(1+x) - f(1)}{x} \\ &= \lim_{x \to 0} \frac{\ln(1+x) - \ln 1}{x} = \lim_{x \to 0} \frac{1}{x} \ln(1+x) \\ &= \lim_{x \to 0} \ln(1+x)^{1/x} \end{aligned}$$

Because $f'(1) = 1$, we have

$$\lim_{x \to 0} \ln(1+x)^{1/x} = 1$$

Then, by Theorem 2.5.7 and the continuity of the exponential function, we have

$$e = e^1 = e^{\lim_{x \to 0} \ln(1+x)^{1/x}} = \lim_{x \to 0} e^{\ln(1+x)^{1/x}} = \lim_{x \to 0} (1+x)^{1/x}$$

(5) $$\boxed{e = \lim_{x \to 0} (1+x)^{1/x}}$$

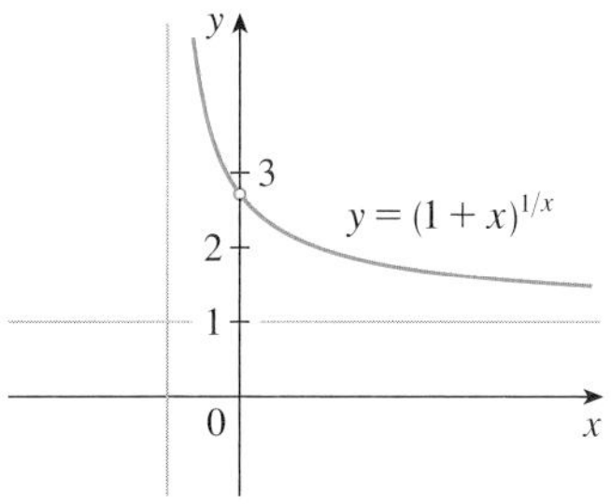

FIGURE 4

x	$(1+x)^{1/x}$
0.1	2.59374246
0.01	2.70481383
0.001	2.71692393
0.0001	2.71814593
0.00001	2.71826824
0.000001	2.71828047
0.0000001	2.71828169
0.00000001	2.71828181

Formula 5 is illustrated by the graph of the function $y = (1+x)^{1/x}$ in Figure 4 and a table of values for small values of x. This illustrates the fact that, correct to seven decimal places,

$$e \approx 2.7182818$$

If we put $n = 1/x$ in Formula 5, then $n \to \infty$ as $x \to 0^+$ and so an alternative expression for e is

(6) $$\boxed{e = \lim_{n \to \infty} \left(1 + \frac{1}{n}\right)^n}$$

Differentiating the Inverse Tangent Function

Recall from Section 1.5 that the only functions that have inverse functions are one-to-one functions. The tangent function, however, is not one-to-one and so it doesn't have an inverse function. But we can make it one-to-one by restricting its domain to the interval $(-\pi/2, \pi/2)$. Thus the **inverse tangent function** is defined as the inverse of the function $f(x) = \tan x$, $-\pi/2 < x < \pi/2$, as shown in Figure 5. It is denoted by $\tan^{-1}$ or arctan.

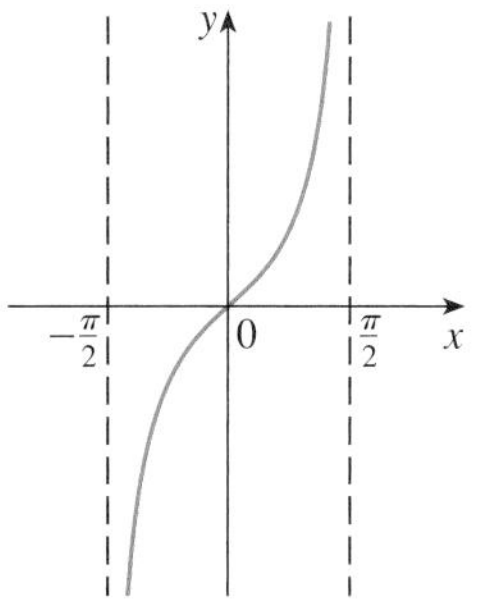

FIGURE 5
$y = \tan x$, $-\frac{\pi}{2} < x < \frac{\pi}{2}$

$$\boxed{\tan^{-1} x = y \iff \tan y = x \quad \text{and} \quad -\frac{\pi}{2} < y < \frac{\pi}{2}}$$

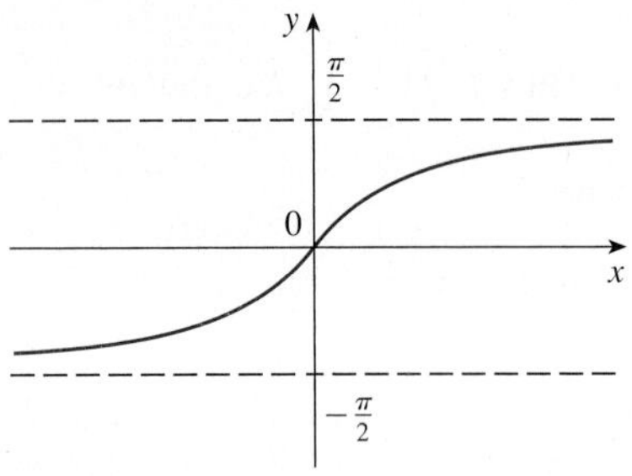

FIGURE 6
$y = \tan^{-1}x = \arctan x$

The inverse tangent function, $\tan^{-1} = \arctan$, has domain $\mathbb{R}$ and range $(-\pi/2, \pi/2)$. Its graph is shown in Figure 6.

We know that

$$\lim_{x \to (\pi/2)^-} \tan x = \infty \qquad \text{and} \qquad \lim_{x \to -(\pi/2)^+} \tan x = -\infty$$

and so the lines $x = \pm\pi/2$ are vertical asymptotes of the graph of tan. Since the graph of $\tan^{-1}$ is obtained by reflecting the graph of the restricted tangent function about the line $y = x$, it follows that the lines $y = \pi/2$ and $y = -\pi/2$ are horizontal asymptotes of the graph of $\tan^{-1}$. This fact is expressed by the following limits:

(7)
$$\lim_{x \to \infty} \tan^{-1}x = \frac{\pi}{2} \qquad \lim_{x \to -\infty} \tan^{-1}x = -\frac{\pi}{2}$$

We can use implicit differentiation to derive the formula for the derivative of the arctangent function. If $y = \tan^{-1}x$, then $\tan y = x$. Differentiating this latter equation implicitly with respect to x, we have

$$\sec^2 y \frac{dy}{dx} = 1$$

Of all the inverse trigonometric functions, the most useful for our purposes is the inverse tangent function, as we will see in Section 5.8.

and so
$$\frac{dy}{dx} = \frac{1}{\sec^2 y} = \frac{1}{1 + \tan^2 y} = \frac{1}{1 + x^2}$$

(8)
$$\frac{d}{dx}(\tan^{-1}x) = \frac{1}{1 + x^2}$$

EXAMPLE 9 | Differentiate:

(a) $y = \dfrac{1}{\tan^{-1}x}$ (b) $f(x) = x \arctan \sqrt{x}$

SOLUTION

(a)
$$\frac{dy}{dx} = \frac{d}{dx}(\tan^{-1}x)^{-1} = -(\tan^{-1}x)^{-2}\frac{d}{dx}(\tan^{-1}x)$$
$$= -\frac{1}{(\tan^{-1}x)^2(1 + x^2)}$$

(b) Using the Product Rule and the Chain Rule, we have

$$f'(x) = x\frac{1}{1 + (\sqrt{x})^2}\left(\tfrac{1}{2}x^{-1/2}\right) + \arctan\sqrt{x}$$
$$= \frac{\sqrt{x}}{2(1 + x)} + \arctan\sqrt{x}$$ ■

EXERCISES 3.7

1. Explain why the natural logarithmic function $y = \ln x$ is used much more frequently in calculus than the other logarithmic functions $y = \log_b x$.

2–20 Differentiate the function.

2. $f(x) = x \ln x - x$

3. $f(x) = \sin(\ln x)$

4. $f(x) = \ln(\sin^2 x)$

5. $f(x) = \log_2(1 - 3x)$

6. $f(x) = \log_5(xe^x)$

7. $f(x) = \sqrt[5]{\ln x}$

8. $f(x) = \ln \sqrt[5]{x}$

9. $f(x) = \sin x \ln(5x)$

10. $f(t) = \dfrac{1 + \ln t}{1 - \ln t}$

11. $F(t) = \ln \dfrac{(2t + 1)^3}{(3t - 1)^4}$

12. $h(x) = \ln\left(x + \sqrt{x^2 - 1}\right)$

13. $g(x) = \ln\left(x\sqrt{x^2 - 1}\right)$

14. $F(y) = y \ln(1 + e^y)$

15. $y = \ln|2 - x - 5x^2|$

16. $H(z) = \ln\sqrt{\dfrac{a^2 - z^2}{a^2 + z^2}}$

17. $y = \ln(e^{-x} + xe^{-x})$

18. $y = [\ln(1 + e^x)]^2$

19. $y = 2x \log_{10} \sqrt{x}$

20. $y = \log_2(e^{-x} \cos \pi x)$

21–22 Find y' and y''.

21. $y = x^2 \ln(2x)$

22. $y = \dfrac{\ln x}{x^2}$

23–24 Differentiate f and find the domain of f.

23. $f(x) = \dfrac{x}{1 - \ln(x - 1)}$

24. $f(x) = \ln \ln \ln x$

25–26 Find an equation of the tangent line to the curve at the given point.

25. $y = \ln(x^2 - 3x + 1)$, $(3, 0)$

26. $y = x^2 \ln x$, $(1, 0)$

27. If $f(x) = \dfrac{\ln x}{x^2}$, find $f'(1)$.

28. Find equations of the tangent lines to the curve $y = (\ln x)/x$ at the points $(1, 0)$ and $(e, 1/e)$. Illustrate by graphing the curve and its tangent lines.

29. **Dialysis** The project on page 458 models the removal of urea from the bloodstream via dialysis. Given that the initial urea concentration, measured in mg/mL, is c (where $c > 1$), the duration of dialysis required for certain conditions is given by the equation

$$t = \ln\left(\frac{3c + \sqrt{9c^2 - 8c}}{2}\right)$$

Calculate the derivative of t with respect to c and interpret it.

30. **Genetic drift** A population of fruit flies contains two genetically determined kinds of individuals: white-eyed flies and red-eyed flies. Suppose that a scientist maintains the population at constant size N by randomly choosing N juvenile flies after reproduction to form the next generation. Eventually, because of the random sampling in each generation, by chance the population will contain only a single type of fly. This is called *genetic drift*. Suppose that the initial fraction of the population that are white-eyed is p_0. An equation for the average number of generations required before all flies are white-eyed (given that this occurs instead of all flies being red-eyed) is

$$g = -2N\frac{1 - p_0}{p_0}\ln(1 - p_0)$$

Calculate the derivative of g with repect to p_0 and explain its meaning.

31. **Carbon dating** If N is the measured amount of ^{14}C in a fossil organism and N_0 is the amount in living organisms, then the estimated age of the fossil is given by the equation

$$a = \frac{5370}{\ln 2}\ln\left(\frac{N_0}{N}\right)$$

Calculate da/dN and interpret it.

32. Let $f(x) = \log_b(3x^2 - 2)$. For what value of b is $f'(1) = 3$?

33–41 Use logarithmic differentiation to find the derivative of the function.

33. $y = (2x + 1)^5(x^4 - 3)^6$

34. $y = \sqrt{x}\, e^{x^2}(x^2 + 1)^{10}$

35. $y = \dfrac{\sin^2 x \tan^4 x}{(x^2 + 1)^2}$

36. $y = \sqrt[4]{\dfrac{x^2 + 1}{x^2 - 1}}$

37. $y = x^x$

38. $y = x^{\cos x}$

39. $y = (\cos x)^x$

40. $y = \sqrt{x}^{\,x}$

41. $y = (\tan x)^{1/x}$

42. Predator-prey dynamics In Chapter 7 we study a model for the population sizes of a predator and its prey species. If $u(t)$ and $v(t)$ denote the prey and predator population sizes at time t, an equation relating the two is

$$ve^{-v}u^{\alpha}e^{-\alpha u} = c$$

where c and α are positive constants. Use logarithmic differentiation to obtain an equation relating the relative (per capita) rate of change of predator (that is, v'/v) to that of prey (that is, u'/u).

43–48 Find the derivative of the function. Simplify where possible.

43. $y = (\tan^{-1}x)^2$

44. $y = \tan^{-1}(x^2)$

45. $y = \arctan(\cos\theta)$

46. $f(x) = x\ln(\arctan x)$

47. $y = \tan^{-1}\left(x - \sqrt{1 + x^2}\right)$

48. $y = \arctan\sqrt{\dfrac{1 - x}{1 + x}}$

49–50 Find the limit.

49. $\lim_{x\to\infty}\arctan(e^x)$

50. $\lim_{x\to 0^+}\tan^{-1}(\ln x)$

51. Find y' if $y = \ln(x^2 + y^2)$.

52. Find y' if $x^y = y^x$.

53. Find a formula for $f^{(n)}(x)$ if $f(x) = \ln(x - 1)$.

54. Find $\dfrac{d^9}{dx^9}(x^8\ln x)$.

55. Use the definition of derivative to prove that

$$\lim_{x\to 0}\frac{\ln(1 + x)}{x} = 1$$

56. Show that $\lim_{n\to\infty}\left(1 + \dfrac{x}{n}\right)^n = e^x$ for any $x > 0$.

3.8 Linear Approximations and Taylor Polynomials

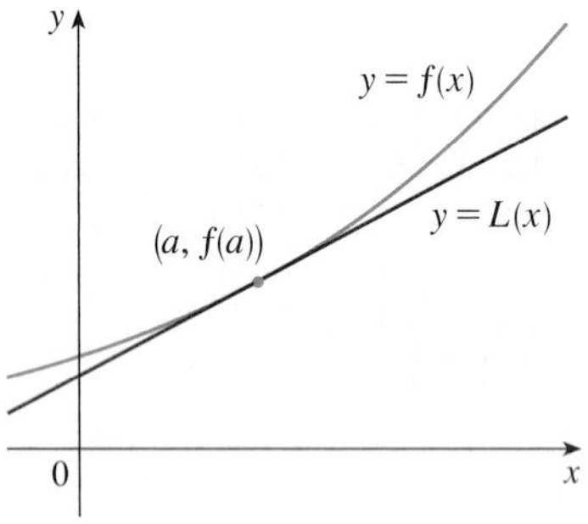

FIGURE 1

■ Tangent Line Approximations

We have seen that a curve lies very close to its tangent line near the point of tangency. In fact, by zooming in toward a point on the graph of a differentiable function, we noticed that the graph looks more and more like its tangent line. (See Figure 3.1.5.) This observation is the basis for a method of finding approximate values of functions.

The idea is that it might be easy to calculate a value $f(a)$ of a function, but difficult (or even impossible) to compute nearby values of f. So we settle for the easily computed values of the linear function L whose graph is the tangent line of f at $(a, f(a))$. (See Figure 1.)

In other words, we use the tangent line at $(a, f(a))$ as an approximation to the curve $y = f(x)$ when x is near a. An equation of this tangent line is

$$y = f(a) + f'(a)(x - a)$$

and the approximation

(1) $$f(x) \approx f(a) + f'(a)(x - a)$$

is called the **linear approximation** or **tangent line approximation** of f at a. The linear function whose graph is this tangent line, that is,

(2) $$L(x) = f(a) + f'(a)(x - a)$$

is called the **linearization** of f at a.

EXAMPLE 1 | Find the linearization of the function $f(x) = \sqrt{x + 3}$ at $a = 1$ and use it to approximate the numbers $\sqrt{3.98}$ and $\sqrt{4.05}$. Are these approximations overestimates or underestimates?

SOLUTION The derivative of $f(x) = (x + 3)^{1/2}$ is

$$f'(x) = \tfrac{1}{2}(x + 3)^{-1/2} = \frac{1}{2\sqrt{x + 3}}$$

and so we have $f(1) = 2$ and $f'(1) = \frac{1}{4}$. Putting these values into Equation 2, we see that the linearization is

$$L(x) = f(1) + f'(1)(x - 1) = 2 + \tfrac{1}{4}(x - 1) = \frac{7}{4} + \frac{x}{4}$$

The corresponding linear approximation (1) is

$$\sqrt{x + 3} \approx \frac{7}{4} + \frac{x}{4} \qquad \text{(when } x \text{ is near 1)}$$

In particular, we have

$$\sqrt{3.98} \approx \tfrac{7}{4} + \tfrac{0.98}{4} = 1.995 \qquad \text{and} \qquad \sqrt{4.05} \approx \tfrac{7}{4} + \tfrac{1.05}{4} = 2.0125$$

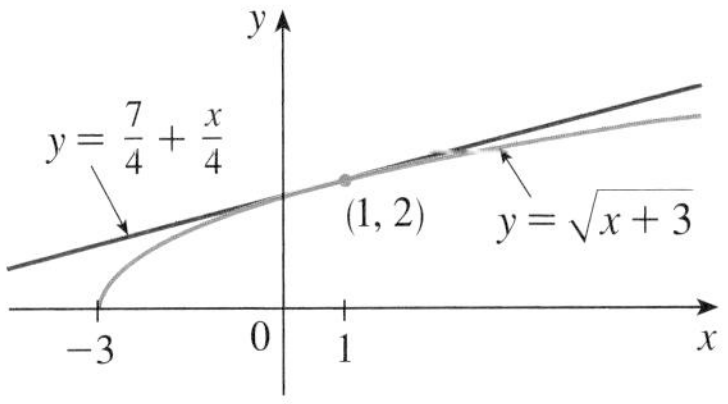

FIGURE 2

The linear approximation is illustrated in Figure 2. We see that, indeed, the tangent line approximation is a good approximation to the given function when x is near 1. We also see that our approximations are overestimates because the tangent line lies above the curve.

Of course, a calculator could give us approximations for $\sqrt{3.98}$ and $\sqrt{4.05}$, but the linear approximation gives an approximation *over an entire interval.* ■

In the following table we compare the estimates from the linear approximation in Example 1 with the true values. Notice from this table, and also from Figure 2, that the tangent line approximation gives good estimates when x is close to 1 but the accuracy of the approximation deteriorates when x is farther away from 1.

	x	From $L(x)$	Actual value
$\sqrt{3.9}$	0.9	1.975	1.97484176 . . .
$\sqrt{3.98}$	0.98	1.995	1.99499373 . . .
$\sqrt{4}$	1	2	2.00000000 . . .
$\sqrt{4.05}$	1.05	2.0125	2.01246117 . . .
$\sqrt{4.1}$	1.1	2.025	2.02484567 . . .
$\sqrt{5}$	2	2.25	2.23606797 . . .
$\sqrt{6}$	3	2.5	2.44948974 . . .

How good is the approximation that we obtained in Example 1? The next example shows that by using a graphing calculator or computer we can determine an interval throughout which a linear approximation provides a specified accuracy.

EXAMPLE 2 | For what values of x is the linear approximation

$$\sqrt{x+3} \approx \frac{7}{4} + \frac{x}{4}$$

accurate to within 0.5? What about accuracy to within 0.1?

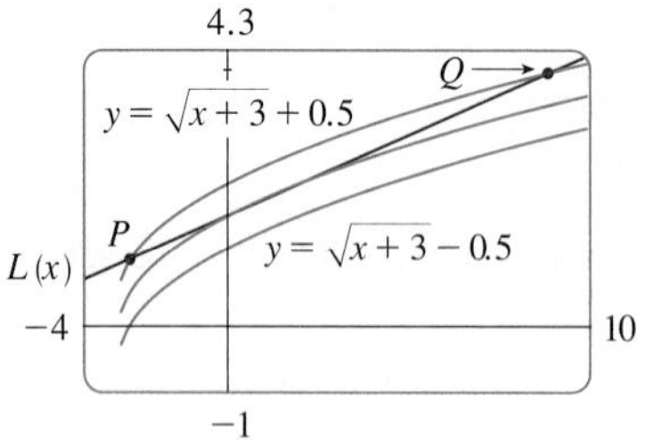

FIGURE 3

SOLUTION Accuracy to within 0.5 means that the functions should differ by less than 0.5:

$$\left| \sqrt{x+3} - \left(\frac{7}{4} + \frac{x}{4} \right) \right| < 0.5$$

Equivalently, we could write

$$\sqrt{x+3} - 0.5 < \frac{7}{4} + \frac{x}{4} < \sqrt{x+3} + 0.5$$

This says that the linear approximation should lie between the curves obtained by shifting the curve $y = \sqrt{x+3}$ upward and downward by an amount 0.5. Figure 3 shows the tangent line $y = (7 + x)/4$ intersecting the upper curve $y = \sqrt{x+3} + 0.5$ at P and Q. Zooming in and using the cursor, we estimate that the x-coordinate of P is about -2.66 and the x-coordinate of Q is about 8.66. Thus we see from the graph that the approximation

$$\sqrt{x+3} \approx \frac{7}{4} + \frac{x}{4}$$

is accurate to within 0.5 when $-2.6 < x < 8.6$. (We have rounded to a narrower interval to be safe.)

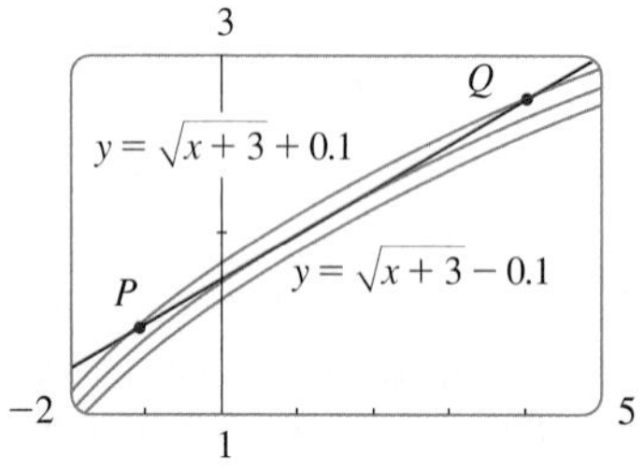

FIGURE 4

Similarly, from Figure 4 we see that the approximation is accurate to within 0.1 when $-1.1 < x < 3.9$. ■

EXAMPLE 3 | **Population growth** If $N(t)$ represents a population size at time t and the rate of growth as a function of N is $f(N)$, what is the linear approximation of the growth rate at $N = 0$?

SOLUTION The growth rate is

$$\frac{dN}{dt} = f(N)$$

The linearization of $f(N)$ at $N = 0$ is

$$L(N) = f(0) + f'(0)N$$

We can assume that $f(0) = 0$ because when the population has size $N = 0$, its growth rate will be zero. So $L(N) = f'(0)N$. If we let the initial growth rate be $f'(0) = r$, then the linear approximation is

$$\frac{dN}{dt} \approx rN$$

This means that for small population sizes, the population grows approximately exponentially. (Recall from Theorem 3.6.2 that the only solutions of the equation $dN/dt = rN$ are exponential functions.) ■

EXAMPLE 4 | Find the linear approximation of the sine function at 0.

SOLUTION If we let $f(x) = \sin x$, then $f'(x) = \cos x$ and so the linearization at 0 is

$$L(x) = f(0) + f'(0)(x - 0) = \sin 0 + (\cos 0)(x) = x$$

So the linear approximation at 0 is

$$\sin x \approx x$$

This approximation is used in optics when x is small: the results of calculations made with this linear approximation became the basic theoretical tool used to design lenses. ■

Newton's Method

How would you solve an equation like $\cos x = x$? Aside from linear and quadratic equations, most equations don't have simple formulas for their roots. Many calculators have numerical rootfinders that enable us to find approximate roots of equations, though they need to be used with care.

How do those numerical rootfinders work? They use a variety of methods, but most of them make some use of **Newton's method**, also called the **Newton-Raphson method**. We will explain how this method works, partly to show what happens inside a calculator or computer, and partly as an application of the idea of linear approximation.

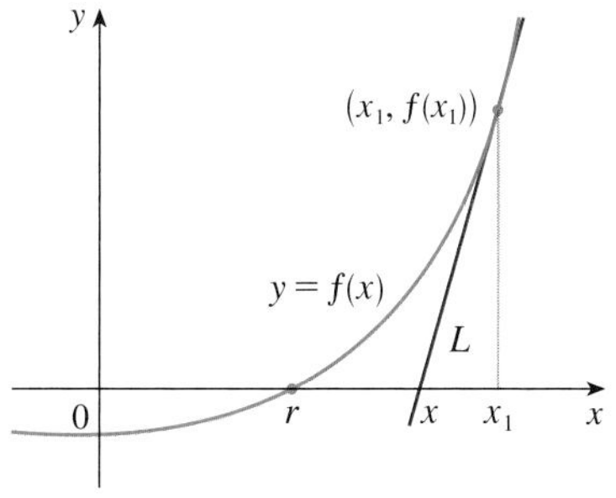

FIGURE 5

The geometry behind Newton's method is shown in Figure 5, where the root that we are trying to find is labeled r. We start with a first approximation x_1, which is obtained by guessing, or from a rough sketch of the graph of f, or from a computer-generated graph of f. Consider the tangent line L to the curve $y = f(x)$ at the point $(x_1, f(x_1))$ and look at the x-intercept of L, labeled x_2. The idea behind Newton's method is that the tangent line is close to the curve and so its x-intercept, x_2, is close to the x-intercept of the curve (namely, the root r that we are seeking). Because the tangent is a line, we can easily find its x-intercept.

To find a formula for x_2 in terms of x_1 we use the fact that the slope of L is $f'(x_1)$, so its equation is

$$y - f(x_1) = f'(x_1)(x - x_1)$$

Since the x-intercept of L is x_2, we set $y = 0$ and obtain

$$0 - f(x_1) = f'(x_1)(x_2 - x_1)$$

If $f'(x_1) \neq 0$, we can solve this equation for x_2:

$$x_2 = x_1 - \frac{f(x_1)}{f'(x_1)}$$

We use x_2 as a second approximation to r.

Next we repeat this procedure with x_1 replaced by the second approximation x_2, using the tangent line at $(x_2, f(x_2))$. This gives a third approximation:

$$x_3 = x_2 - \frac{f(x_2)}{f'(x_2)}$$

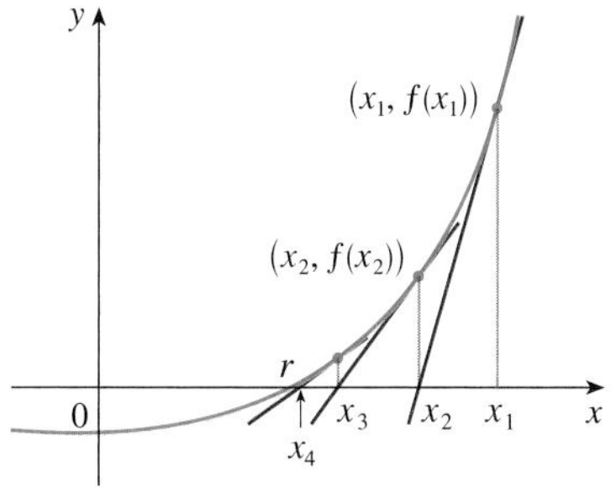

FIGURE 6

If we keep repeating this process, we obtain a sequence of approximations $x_1, x_2, x_3, x_4, \ldots$ as shown in Figure 6. In general, if the nth approximation is x_n and $f'(x_n) \neq 0$, then the next approximation is given by

(3) $$\boxed{x_{n+1} = x_n - \frac{f(x_n)}{f'(x_n)}}$$

If the numbers x_n approach the desired root r as n becomes large, that is,

$$\lim_{n \to \infty} x_n = r$$

Limits of sequences were defined in Section 2.1.

then we use them as approximations to r. Although the sequence of successive approximations converges to the desired root for functions of the type illustrated in Figure 6, in certain circumstances the sequence may not converge. For example, consider the situation shown in Figure 7. You can see that x_2 is a worse approximation than x_1. This is likely to be the case when $f'(x_1)$ is close to 0. It might even happen that an approximation (such as x_3 in Figure 7) falls outside the domain of f. Then Newton's method fails and a better initial approximation x_1 should be chosen. See Exercises 33–34 for specific examples in which Newton's method works very slowly or does not work at all.

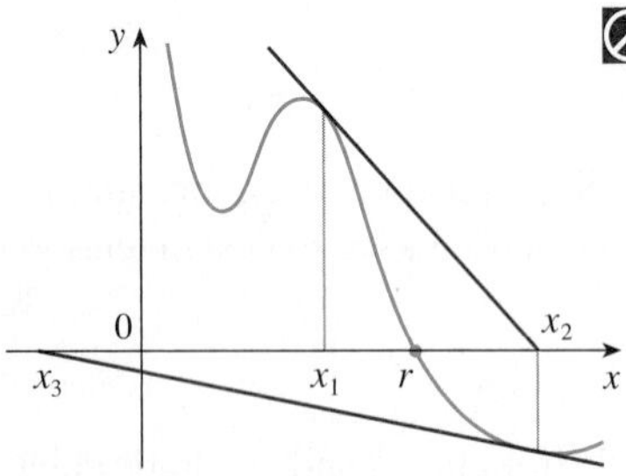

FIGURE 7

EXAMPLE 5 | Starting with $x_1 = 2$, find the third approximation x_3 to the root of the equation $x^3 - 2x - 5 = 0$.

SOLUTION We apply Newton's method with

$$f(x) = x^3 - 2x - 5 \qquad \text{and} \qquad f'(x) = 3x^2 - 2$$

TEC In Module 3.8 you can investigate how Newton's Method works for several functions and what happens when you change x_1.

Newton himself used this equation to illustrate his method and he chose $x_1 = 2$ after some experimentation because $f(1) = -6$, $f(2) = -1$, and $f(3) = 16$. Equation 3 becomes

$$x_{n+1} = x_n - \frac{x_n^3 - 2x_n - 5}{3x_n^2 - 2}$$

With $n = 1$ we have

$$x_2 = x_1 - \frac{x_1^3 - 2x_1 - 5}{3x_1^2 - 2}$$

$$= 2 - \frac{2^3 - 2(2) - 5}{3(2)^2 - 2} = 2.1$$

Figure 8 shows the geometry behind the first step in Newton's method in Example 5. Since $f'(2) = 10$, the tangent line to $y = x^3 - 2x - 5$ at $(2, -1)$ has equation $y = 10x - 21$ and so its x-intercept is $x_2 = 2.1$.

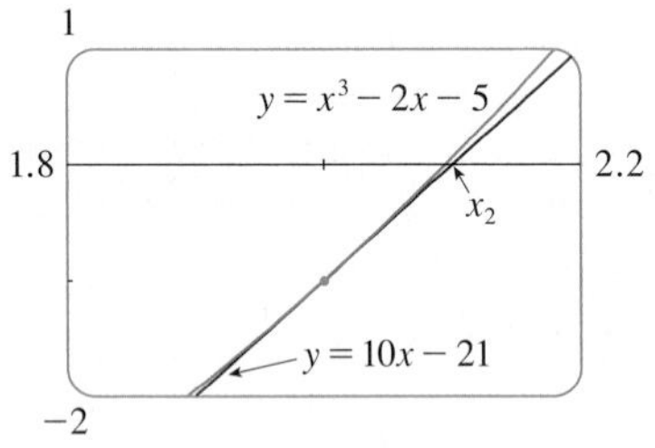

FIGURE 8

Then with $n = 2$ we obtain

$$x_3 = x_2 - \frac{x_2^3 - 2x_2 - 5}{3x_2^2 - 2} = 2.1 - \frac{(2.1)^3 - 2(2.1) - 5}{3(2.1)^2 - 2} \approx 2.0946$$

It turns out that this third approximation $x_3 \approx 2.0946$ is accurate to four decimal places. ■

Suppose that we want to achieve a given accuracy, say to eight decimal places, using Newton's method. How do we know when to stop? The rule of thumb that is generally used is that we can stop when successive approximations x_n and x_{n+1} agree to eight decimal places.

Notice that the procedure in going from n to $n + 1$ is the same for all values of n. (It is a recursive sequence, as defined in Section 1.6.) This means that Newton's method is particularly convenient for use with a programmable calculator or a computer.

EXAMPLE 6 | Find, correct to six decimal places, the root of the equation $\cos x = x$.

SOLUTION We first rewrite the equation in standard form:

$$\cos x - x = 0$$

Therefore we let $f(x) = \cos x - x$. Then $f'(x) = -\sin x - 1$, so Formula 3 becomes

$$x_{n+1} = x_n - \frac{\cos x_n - x_n}{-\sin x_n - 1} = x_n + \frac{\cos x_n - x_n}{\sin x_n + 1}$$

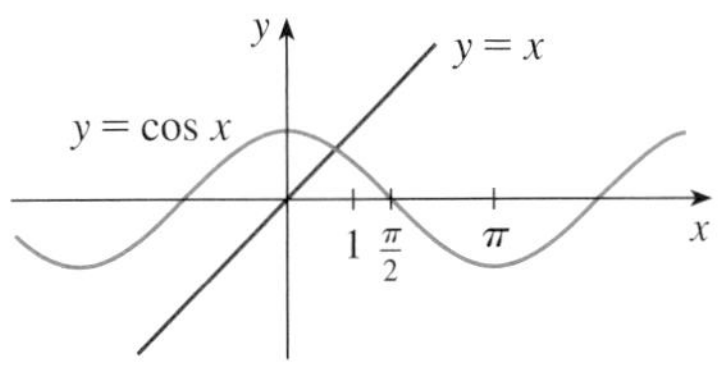

FIGURE 9

In order to guess a suitable value for x_1 we sketch the graphs of $y = \cos x$ and $y = x$ in Figure 9. It appears that they intersect at a point whose x-coordinate is somewhat less than 1, so let's take $x_1 = 1$ as a convenient first approximation. Then, remembering to put our calculator in radian mode, we get

$$x_2 \approx 0.75036387$$
$$x_3 \approx 0.73911289$$
$$x_4 \approx 0.73908513$$
$$x_5 \approx 0.73908513$$

Since x_4 and x_5 agree to six decimal places (eight, in fact), we conclude that the root of the equation, correct to six decimal places, is 0.739085. ■

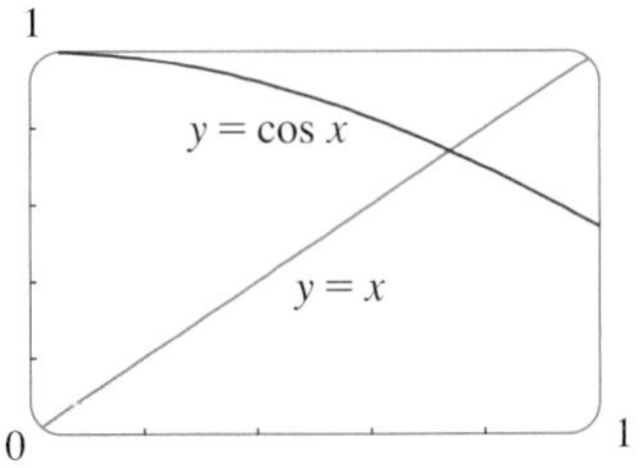

FIGURE 10

Instead of using the rough sketch in Figure 9 to get a starting approximation for Newton's method in Example 6, we could have used the more accurate graph that a calculator or computer provides. Figure 10 suggests that we use $x_1 = 0.75$ as the initial approximation. Then Newton's method gives

$$x_2 \approx 0.73911114 \qquad x_3 \approx 0.73908513 \qquad x_4 \approx 0.73908513$$

and so we obtain the same answer as before, but with one fewer step.

■ Taylor Polynomials

The tangent line approximation $L(x)$ is the best first-degree (linear) approximation to $f(x)$ near $x = a$ because $f(x)$ and $L(x)$ have the same rate of change (derivative) at a. For a better approximation than a linear one, let's try a second-degree (quadratic) approximation $P(x)$. In other words, we approximate a curve by a parabola instead of by a straight line. To make sure that the approximation is a good one, we stipulate the following:

(i) $P(a) = f(a)$ (P and f should have the same value at a.)

(ii) $P'(a) = f'(a)$ (P and f should have the same rate of change at a.)

(iii) $P''(a) = f''(a)$ (The slopes of P and f should change at the same rate at a.)

Let's write the polynomial P in the form

$$P(x) = A + B(x - a) + C(x - a)^2$$

Then

$$P'(x) = B + 2C(x - a) \qquad \text{and} \qquad P''(x) = 2C$$

Applying the conditions (i), (ii), and (iii), we get

$$P(a) = f(a) \quad \Rightarrow \quad A = f(a)$$

$$P'(a) = f'(a) \quad \Rightarrow \quad B = f'(a)$$

$$P''(a) = f''(a) \quad \Rightarrow \quad 2C = f''(a) \quad \Rightarrow \quad C = \tfrac{1}{2}f''(a)$$

So the quadratic function that satisfies the three conditions is

(4)
$$P(x) = f(a) + f'(a)(x - a) + \tfrac{1}{2}f''(a)(x - a)^2$$

This function is called the **second-degree Taylor polynomial of f centered at a** and is usually denoted by $T_2(x)$.

EXAMPLE 7 | Find the second-degree Taylor polynomial $T_2(x)$ centered at $a = 0$ for the function $f(x) = \cos x$. Illustrate by graphing T_2, f, and the linearization $L(x) = 1$.

SOLUTION Since $f(x) = \cos x$, $f'(x) = -\sin x$, and $f''(x) = -\cos x$, the second-degree Taylor polynomial centered at 0 is

$$T_2(x) = f(0) + f'(0)x + \tfrac{1}{2}f''(0)x^2$$
$$= 1 + 0 + \tfrac{1}{2}(-1)x^2 = 1 - \tfrac{1}{2}x^2$$

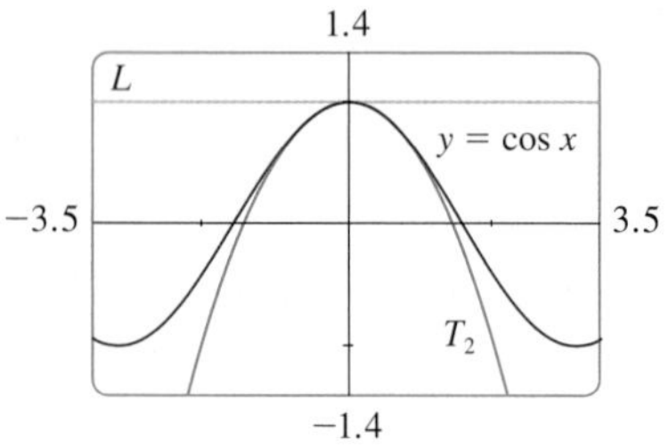

FIGURE 11

Figure 11 shows a graph of the cosine function together with its linear approximation $L(x) = 1$ and its quadratic approximation $T_2(x) = 1 - \frac{1}{2}x^2$ near 0. You can see that the quadratic approximation is much better than the linear one. ■

Instead of being satisfied with a linear or quadratic approximation to $f(x)$ near a, let's try to find better approximations with higher-degree polynomials. We look for an nth-degree polynomial

(5) $$T_n(x) = c_0 + c_1(x - a) + c_2(x - a)^2 + c_3(x - a)^3 + \cdots + c_n(x - a)^n$$

such that T_n and its first n derivatives have the same values at $x = a$ as f and its first n derivatives. By differentiating repeatedly and setting $x = a$, you are asked to show in Exercise 44 that these conditions are satisfied if $c_0 = f(a)$, $c_1 = f'(a)$, $c_2 = \frac{1}{2}f''(a)$, and in general

$$c_k = \frac{f^{(k)}(a)}{k!}$$

where $k! = 1 \cdot 2 \cdot 3 \cdot 4 \cdot \cdots \cdot k$ is called k factorial. The resulting polynomial

$$T_n(x) = f(a) + f'(a)(x - a) + \frac{f''(a)}{2!}(x - a)^2 + \cdots + \frac{f^{(n)}(a)}{n!}(x - a)^n$$

is called the **nth-degree Taylor polynomial of f centered at a**.

EXAMPLE 8 | Find the first three Taylor polynomials T_1, T_2, and T_3 for the function $f(x) = \ln x$ centered at $a = 1$.

SOLUTION We start by calculating the first three derivatives at $a = 1$:

$$f(x) = \ln x \qquad f'(x) = \frac{1}{x} \qquad f''(x) = -\frac{1}{x^2} \qquad f'''(x) = \frac{2}{x^3}$$

$$f(1) = 0 \qquad f'(1) = 1 \qquad f''(1) = -1 \qquad f'''(1) = 2$$

Then

$$T_1(x) = f(1) + f'(1)(x - 1) = x - 1$$

$$T_2(x) = T_1(x) + \frac{f''(1)}{2!}(x - 1)^2 = x - 1 - \tfrac{1}{2}(x - 1)^2$$

$$T_3(x) = T_2(x) + \frac{f'''(1)}{3!}(x - 1)^3 = x - 1 - \tfrac{1}{2}(x - 1)^2 + \tfrac{1}{3}(x - 1)^3$$

Figure 12 shows the graphs of these Taylor polynomials. Notice that these polynomial approximations are better when x is close to 1 and that each successive approximation is better than the preceding ones.

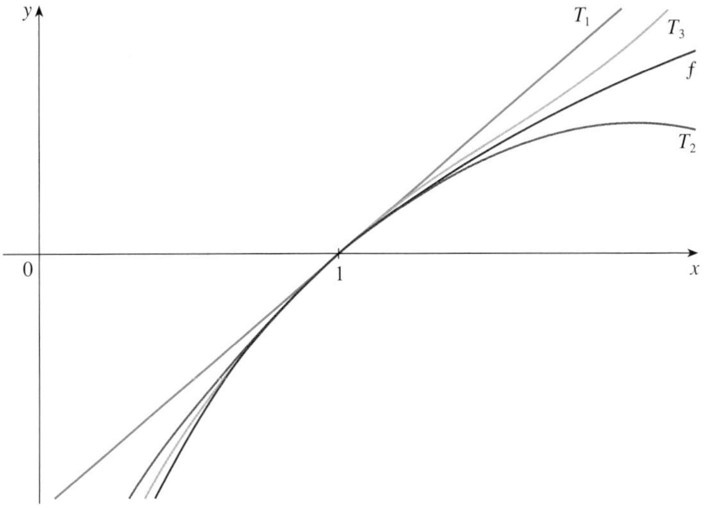

FIGURE 12

■

EXERCISES 3.8

1–4 Find the linearization $L(x)$ of the function at a.

1. $f(x) = x^4 + 3x^2, \quad a = -1$ **2.** $f(x) = \ln x, \quad a = 1$

3. $f(x) = \cos x, \quad a = \pi/2$ **4.** $f(x) = x^{3/4}, \quad a = 16$

5. Find the linear approximation of the function $f(x) = \sqrt{1 - x}$ at $a = 0$ and use it to approximate the numbers $\sqrt{0.9}$ and $\sqrt{0.99}$. Illustrate by graphing f and the tangent line.

6. Find the linear approximation of the function $g(x) = \sqrt[3]{1 + x}$ at $a = 0$ and use it to approximate the numbers $\sqrt[3]{0.95}$ and $\sqrt[3]{1.1}$. Illustrate by graphing g and the tangent line.

7–10 Verify the given linear approximation at $a = 0$. Then determine the values of x for which the linear approximation is accurate to within 0.1.

7. $\sqrt[3]{1 - x} \approx 1 - \frac{1}{3}x$ **8.** $\tan x \approx x$

9. $1/(1 + 2x)^4 \approx 1 - 8x$ **10.** $e^x \approx 1 + x$

11–12 Use a linear approximation to estimate the given number.

11. $(2.001)^5$ **12.** $e^{-0.015}$

13–14 Explain, in terms of linear approximations, why the approximation is reasonable.

13. $\ln 1.05 \approx 0.05$ **14.** $(1.01)^6 \approx 1.06$

15. Insecticide resistance If the frequency of a gene for insecticide resistance is p (a constant), then its frequency in the next generation is given by the expression

$$f = \frac{p(1 + s)}{1 + sp}$$

where s is the reproductive advantage this gene has over the wild type in the presence of the insecticide. Often the selective advantage s is very small. Approximate the frequency in the next generation with a linear approximation, given that s is small.

16. Relative change in blood velocity Suppose $y = f(x)$ and x and y change by amounts Δx and Δy. A way of expressing a linear approximation is to write $\Delta y \approx f'(x)\,\Delta x$. The *relative change* in y is $\Delta y/y$.

A special case of Poiseuille's law of laminar flow (see Example 3.3.9) is that at the central axis of a blood vessel the velocity of the blood is related to the radius R of the vessel by an equation of the form $v = cR^2$. If the radius changes, how is the relative change in the blood velocity related to the relative change in the radius? If the radius is increased by 10%, what happens to the velocity?

17. Relative change in blood flow Another law of Poiseuille says that when blood flows along a blood vessel, the flux F (the volume of blood per unit time that flows past a given point) is proportional to the fourth power of the radius R of the blood vessel:

$$F = kR^4$$

(We will show why this is true in Section 6.3.) A partially clogged artery can be expanded by an operation called angioplasty, in which a balloon-tipped catheter is inflated inside the artery in order to widen it and restore the normal blood flow.

Show that the relative change in F is about four times the relative change in R. How will a 5% increase in the radius affect the flow of blood?

18. Volume and surface area of a tumor The diameter of a tumor was measured to be 19 mm. If the diameter increases by 1 mm, use linear approximations to estimate the relative changes in the volume $\left(V = \frac{4}{3}\pi r^3\right)$ and surface area $(S = 4\pi r^2)$.

19. The figure shows the graph of a function f. Suppose that Newton's method is used to approximate the root r of the equation $f(x) = 0$ with initial approximation $x_1 = 1$.

(a) Draw the tangent lines that are used to find x_2 and x_3, and estimate the numerical values of x_2 and x_3.

(b) Would $x_1 = 5$ be a better first approximation? Explain.

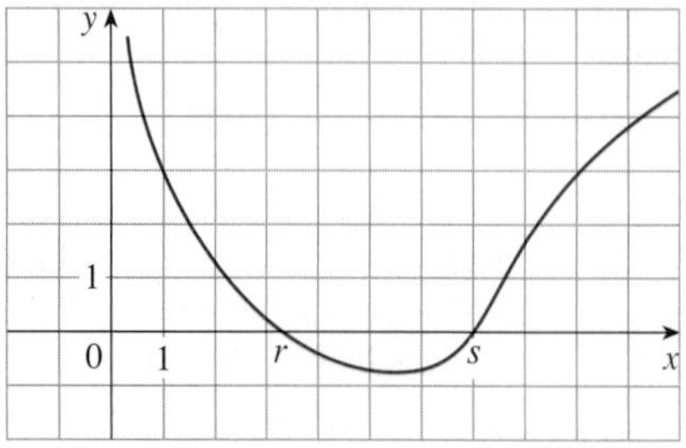

20. Follow the instructions for Exercise 19(a) but use $x_1 = 9$ as the starting approximation for finding the root s.

21–22 Use Newton's method with the specified initial approximation x_1 to find x_3, the third approximation to the root of the given equation. (Give your answer to four decimal places.)

21. $x^3 + 2x - 4 = 0, \quad x_1 = 1$

22. $\frac{1}{3}x^3 + \frac{1}{2}x^2 + 3 = 0, \quad x_1 = -3$

23–26 Use Newton's method to find all roots of the equation correct to six decimal places.

23. $x^4 = 1 + x$

24. $e^x = 3 - 2x$

25. $(x - 2)^2 = \ln x$

26. $\dfrac{1}{x} = 1 + x^3$

27–31 Use Newton's method to find all the roots of the equation correct to eight decimal places. Start by drawing a graph to find initial approximations.

27. $x^6 - x^5 - 6x^4 - x^2 + x + 10 = 0$

28. $x^2(4 - x^2) = \dfrac{4}{x^2 + 1}$

29. $x^2\sqrt{2 - x - x^2} = 1$

30. $3\sin(x^2) = 2x$

31. $4e^{-x^2}\sin x = x^2 - x + 1$

32. Infectious disease outbreak size If 99% of a population is initially uninfected and each initial infected person generates, on average, two new infections, then, according to the model we considered in Example 3.5.13,

$$0.99e^{-2A} = 1 - A$$

where A is the fraction of the population infected at the end of an outbreak. Use Newton's method to obtain an approximation (accurate to two decimal places) for the percentage of the population that is eventually infected.

33. Explain why Newton's method doesn't work for finding the root of the equation $x^3 - 3x + 6 = 0$ if the initial approximation is chosen to be $x_1 = 1$.

34. (a) Use Newton's method with $x_1 = 1$ to find the root of the equation $x^3 - x = 1$ correct to six decimal places.

(b) Solve the equation in part (a) using $x_1 = 0.6$ as the initial approximation.

(c) Solve the equation in part (a) using $x_1 = 0.57$. (You definitely need a programmable calculator for this part.)

(d) Graph $f(x) = x^3 - x - 1$ and its tangent lines at $x_1 = 1$, 0.6, and 0.57 to explain why Newton's method is so sensitive to the value of the initial approximation.

35–38 Find the Taylor polynomial of degree n centered at the number a.

35. $f(x) = e^x, \quad n = 3, \quad a = 0$

36. $f(x) = \sin \pi x, \quad n = 3, \quad a = 0$

37. $f(x) = 1/x, \quad n = 4, \quad a = 1$

38. $f(x) = \sqrt{x}, \quad n = 2, \quad a = 4$

39. Find the quadratic approximation to $f(x) = \sqrt{x + 3}$ near $a = 1$. Graph f, the quadratic approximation, and the linear approximation from Example 1 on a common screen. What do you conclude?

40. Determine the values of x for which the quadratic approximation $f(x) \approx T_2(x)$ in Example 7 is accurate to within 0.1. [*Hint:* Graph $y = T_2(x)$, $y = \cos x - 0.1$, and $y = \cos x + 0.1$ on a common screen.]

41. Find the first five Taylor polynomials for $f(x) = \sin x$ centered at 0. Graph them on the interval $[-4, 4]$ and comment on how well they approximate f.

42. Find the 8th-degree Taylor polynomial centered at $a = 0$ for the function $f(x) = \cos x$. Graph f together with the Taylor polynomials T_2, T_4, T_6, T_8 in the viewing rectangle $[-5, 5]$ by $[-1.4, 1.4]$ and comment on how well they approximate f.

43. Habitat fragmentation and species conservation The size of a class-structured population is modeled in Section 8.8. In certain situations the long-term per capita growth rate of the population is given by

$$r = \tfrac{1}{2}\left(1 + \sqrt{1 + 8s}\right)$$

where s is the annual survival probability of juveniles.

(a) Approximate the growth rate with a Taylor polynomial of degree one (linear approximation) centered at 0.

(b) Approximate the growth rate with a Taylor polynomial of degree two centered at 0.

44. Show that if a polynomial T_n of the form given in Equation 5 has the same value at a and the same derivatives at $x = a$ as a function f, then its coefficients are given by the formula

$$c_k = \frac{f^{(k)}(a)}{k!}$$

PROJECT Harvesting Renewable Resources

In Exercise 1.6.32 we considered the Ricker difference equation

$$x_{t+1} = cx_t e^{-x_t}$$

where x_t is the size of a population at time t and c is the per capita reproductive output when the population size is small. (We assume that $c > 1$, which means that individuals more than replace themselves when the population size is small.)

1. Suppose that as $t \to \infty$ the population size approaches a limiting value x. Express x in terms of the growth factor c. You will find two solutions; focus only on the one that is strictly positive.

2. Now let's consider harvesting. Assume that h individuals are harvested in each time step. Then our model becomes

$$x_{t+1} = cx_t e^{-x_t} - h$$

What equation does the limiting population size x satisfy?

3. Even though the equation you found in Problem 2 can't be solved explicitly for x, you can use implicit differentiation to find an expression for the derivative of x with respect to h. Do so.

4. Our aim is to find an expression for the limiting population size x in terms of the harvest rate h. As a first approximation, find the linearization of the function $x(h)$ at $h = 0$. [Note that $x(0)$ is known from Problem 1.]

5. Find the second-order Taylor polynomial approximation for $x(h)$ at $h = 0$.

Chapter 3 REVIEW

CONCEPT CHECK

1. Write an expression for the slope of the tangent line to the curve $y = f(x)$ at the point $(a, f(a))$.

2. Define the derivative $f'(a)$. Discuss two ways of interpreting this number.

3. If $y = f(x)$ and x changes from x_1 to x_2, write expressions for the following.
(a) The average rate of change of y with respect to x over the interval $[x_1, x_2]$
(b) The instantaneous rate of change of y with respect to x at $x = x_1$

4. Define the second derivative of f. If $f(t)$ is the position function of a particle, how can you interpret the second derivative?

5. (a) What does it mean for f to be differentiable at a?
(b) What is the relation between the differentiability and continuity of a function?
(c) Sketch the graph of a function that is continuous but not differentiable at $a = 2$.

6. Describe several ways in which a function can fail to be differentiable. Illustrate with sketches.

7. State each differentiation rule both in symbols and in words.
(a) The Power Rule (b) The Constant Multiple Rule
(c) The Sum Rule (d) The Difference Rule
(e) The Product Rule (f) The Quotient Rule
(g) The Chain Rule

8. State the derivative of each function.
(a) $y = x^n$ (b) $y = e^x$ (c) $y = b^x$
(d) $y = \ln x$ (e) $y = \log_b x$ (f) $y = \sin x$
(g) $y = \cos x$ (h) $y = \tan x$ (i) $y = \csc x$
(j) $y = \sec x$ (k) $y = \cot x$ (l) $y = \tan^{-1}x$

9. (a) How is the number e defined?
(b) Express e as a limit.
(c) Why is the natural exponential function $y = e^x$ used more often in calculus than the other exponential functions $y = b^x$?
(d) Why is the natural logarithmic function $y = \ln x$ used more often in calculus than the other logarithmic functions $y = \log_b x$?

10. (a) Explain how implicit differentiation works. When should you use it?
(b) Explain how logarithmic differentiation works. When should you use it?

11. Write an expression for the linearization of f at a.

12. Write an expression for the nth-degree Taylor polynomial of f centered at a.

Answers to the Concept Check can be found on the back endpapers.

TRUE-FALSE QUIZ

Determine whether the statement is true or false. If it is true, explain why. If it is false, explain why or give an example that disproves the statement.

1. If f is continuous at a, then f is differentiable at a.

2. If $f'(r)$ exists, then $\lim_{x \to r} f(x) = f(r)$.

3. If f and g are differentiable, then
$$\frac{d}{dx}[f(x) + g(x)] = f'(x) + g'(x)$$

4. If f and g are differentiable, then
$$\frac{d}{dx}[f(x)g(x)] = f'(x)g'(x)$$

5. If f and g are differentiable, then
$$\frac{d}{dx}[f(g(x))] = f'(g(x))g'(x)$$

6. If f is differentiable, then $\dfrac{d}{dx}\sqrt{f(x)} = \dfrac{f'(x)}{2\sqrt{f(x)}}$.

7. If f is differentiable, then $\dfrac{d}{dx}f(\sqrt{x}) = \dfrac{f'(x)}{2\sqrt{x}}$.

8. If $y = e^2$, then $y' = 2e$.

9. $\dfrac{d}{dx}(10^x) = x10^{x-1}$

10. $\dfrac{d}{dx}(\ln 10) = \dfrac{1}{10}$

11. $\dfrac{d}{dx}(\tan^2 x) = \dfrac{d}{dx}(\sec^2 x)$

12. $\dfrac{d^2y}{dx^2} = \left(\dfrac{dy}{dx}\right)^2$

13. If $g(x) = x^5$, then $\lim\limits_{x \to 2} \dfrac{g(x) - g(2)}{x - 2} = 80$

14. An equation of the tangent line to the parabola $y = x^2$ at $(-2, 4)$ is $y - 4 = 2x(x + 2)$.

EXERCISES

1. For the function f whose graph is shown, arrange the following numbers in increasing order:

$$0 \quad 1 \quad f'(2) \quad f'(3) \quad f'(5) \quad f''(5)$$

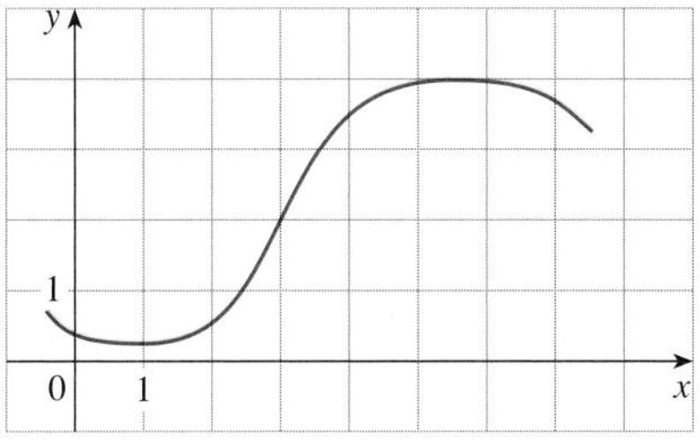

2. Life expectancy The table shows how the life expectancy $L(t)$ in Bangladesh has changed from 1990 to 2010.

t	1990	1995	2000	2005	2010
$L(t)$	56	61	65	68	71

(a) Calculate the average rate of change of the life expectancy $L(t)$ with respect to time over the following time intervals.
(i) [1990, 2000] (ii) [1995, 2000]
(iii) [2000, 2010] (iv) [2000, 2005]
(b) Estimate the value of $L'(2000)$.

3. The total cost of repaying a student loan at an interest rate of $r\%$ per year is $C = f(r)$.
(a) What is the meaning of the derivative $f'(r)$? What are its units?
(b) What does the statement $f'(10) = 1200$ mean?
(c) Is $f'(r)$ always positive or does it change sign?

4. (a) Use the definition of a derivative to find $f'(2)$, where $f(x) = x^3 - 2x$.
(b) Find an equation of the tangent line to the curve $y = x^3 - 2x$ at the point (2, 4).
(c) Illustrate part (b) by graphing the curve and the tangent line on the same screen.

5–7 Trace or copy the graph of the function. Then sketch a graph of its derivative directly beneath.

5.

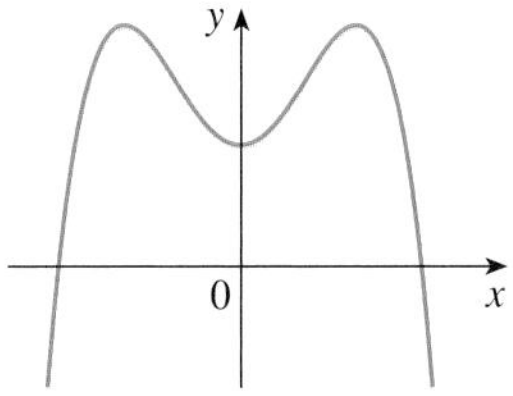

6.

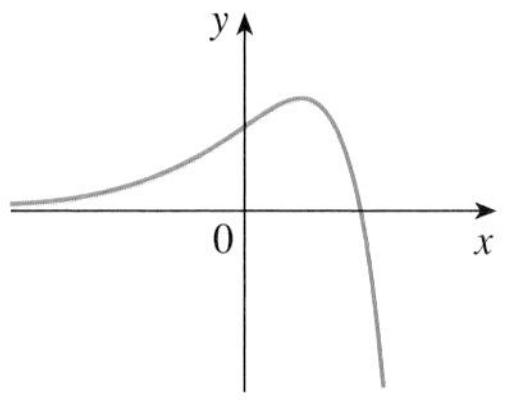

7.

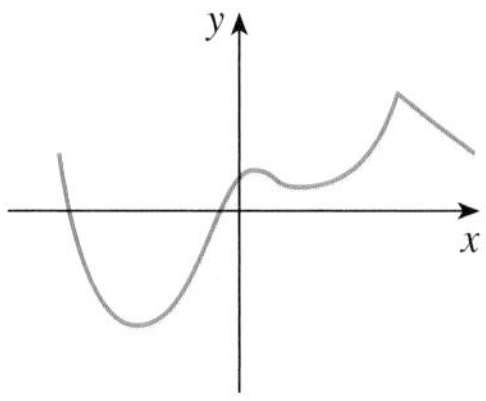

8. Bacteria count Shown is a typical graph of the number N of bacteria grown in a bacteria culture as a function of time t.
(a) What is the meaning of the derivative $N'(t)$?
(b) Sketch the graph of $N'(t)$.

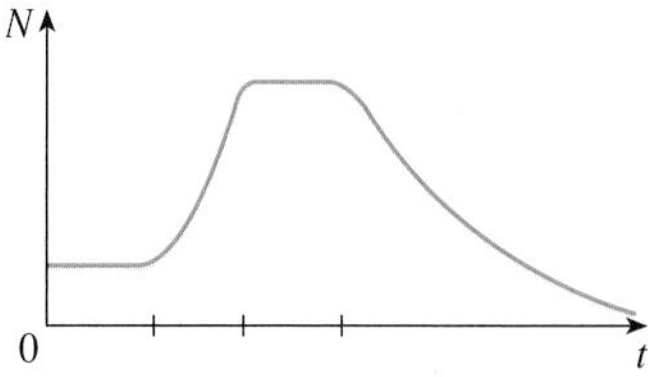

9. Antihypertension medication The figure shows the heart rate $H(t)$ after a patient has taken nifedipine tablets.
(a) What is the meaning of the derivative $H'(t)$?
(b) Sketch the graph of $H'(t)$.

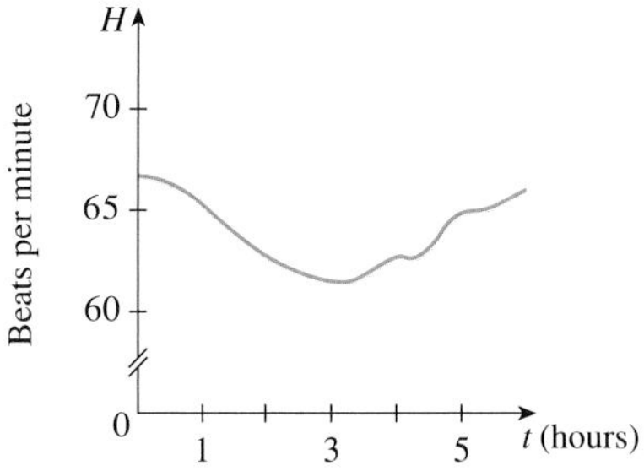

10. (a) Find the asymptotes of the graph of $f(x) = \dfrac{4 - x}{3 + x}$ and use them to sketch the graph.
(b) Use your graph from part (a) to sketch the graph of f'.
(c) Use the definition of a derivative to find $f'(x)$.
(d) Use a calculator to graph f' and compare with your sketch in part (b).

11. (a) If $f(x) = \sqrt{3 - 5x}$, use the definition of a derivative to find $f'(x)$.
(b) Find the domains of f and f'.
(c) Graph f and f' on a common screen. Compare the graphs to see whether your answer to part (a) is reasonable.

12. The figure shows the graphs of f, f', and f''. Identify each curve, and explain your choices.

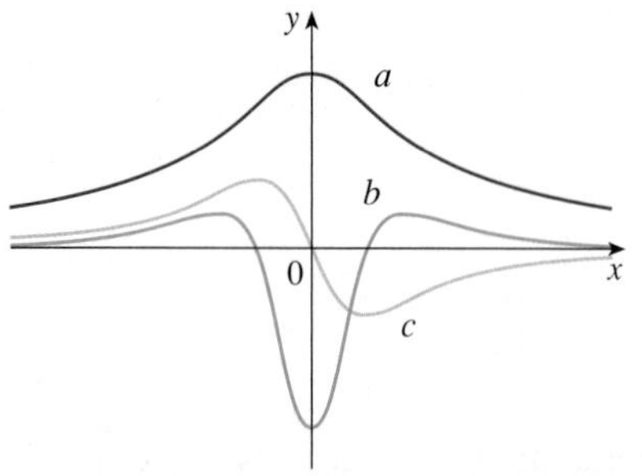

13. The graph of f is shown. State, with reasons, the numbers at which f is not differentiable.

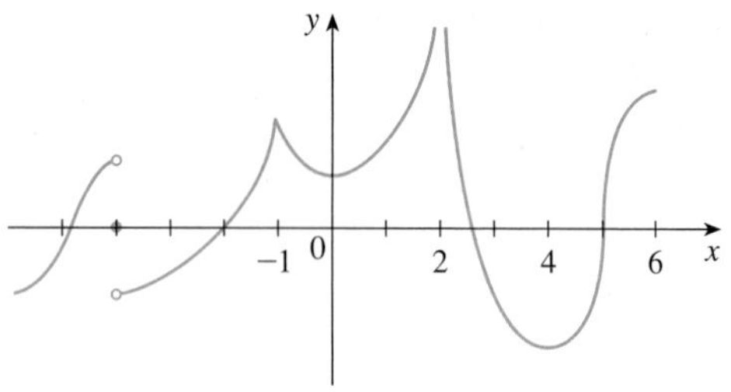

14. The **total fertility rate** at time t, denoted by $F(t)$, is an estimate of the average number of children born to each woman (assuming that current birth rates remain constant). The graph of the total fertility rate in the United States shows the fluctuations from 1940 to 2010.

(a) Estimate the values of $F'(1950)$, $F'(1965)$, and $F'(1987)$.

(b) What are the meanings of these derivatives?

(c) Can you suggest reasons for the values of these derivatives?

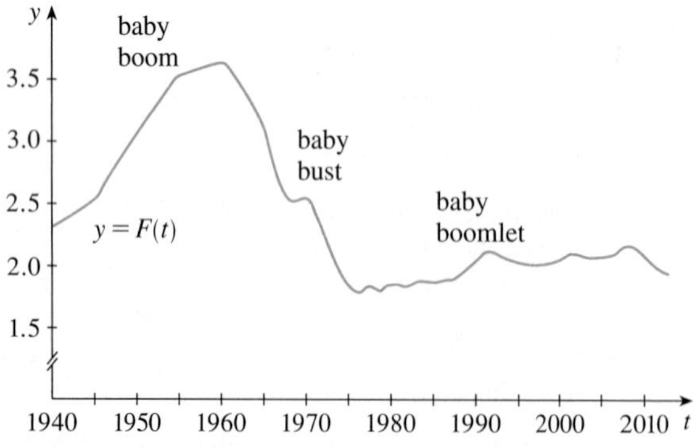

15–50 Calculate y'.

15. $y = (x^4 - 3x^2 + 5)^3$

16. $y = \cos(\tan x)$

17. $y = \sqrt{x} + \dfrac{1}{\sqrt[3]{x^4}}$

18. $y = \dfrac{3x - 2}{\sqrt{2x + 1}}$

19. $y = 2x\sqrt{x^2 + 1}$

20. $y = \dfrac{e^x}{1 + x^2}$

21. $y = e^{\sin 2\theta}$

22. $y = e^{-t}(t^2 - 2t + 2)$

23. $y = \dfrac{t}{1 - t^2}$

24. $y = e^{mx}\cos nx$

25. $y = \dfrac{e^{1/x}}{x^2}$

26. $y = \left(\dfrac{u - 1}{u^2 + u + 1}\right)^4$

27. $xy^4 + x^2y = x + 3y$

28. $y = \ln(\csc 5x)$

29. $y = \dfrac{\sec 2\theta}{1 + \tan 2\theta}$

30. $x^2 \cos y + \sin 2y = xy$

31. $y = e^{cx}(c \sin x - \cos x)$

32. $y = \ln(x^2 e^x)$

33. $y = \log_5(1 + 2x)$

34. $y = (\ln x)^{\cos x}$

35. $\sin(xy) = x^2 - y$

36. $y = \sqrt{t \ln(t^4)}$

37. $y = 3^{x \ln x}$

38. $xe^y = y - 1$

39. $y = \ln \sin x - \frac{1}{2}\sin^2 x$

40. $y = \dfrac{(x^2 + 1)^4}{(2x + 1)^3(3x - 1)^5}$

41. $y = x \tan^{-1}(4x)$

42. $y = e^{\cos x} + \cos(e^x)$

43. $y = \ln|\sec 5x + \tan 5x|$

44. $y = 10^{\tan \pi\theta}$

45. $y = \tan^2(\sin \theta)$

46. $y = \ln\left|\dfrac{x^2 - 4}{2x + 5}\right|$

47. $y = \sin\left(\tan\sqrt{1 + x^3}\right)$

48. $y = \arctan\left(\arcsin\sqrt{x}\right)$

49. $y = \cos\left(e^{\sqrt{\tan 3x}}\right)$

50. $y = \sin^2\left(\cos\sqrt{\sin \pi x}\right)$

---

51. If $f(t) = \sqrt{4t + 1}$, find $f''(2)$.

52. If $g(\theta) = \theta \sin\theta$, find $g''(\pi/6)$.

53. If $f(x) = 2^x$, find $f^{(n)}(x)$.

54. Find $f^{(n)}(x)$ if $f(x) = 1/(2 - x)$.

55–56 Find an equation of the tangent to the curve at the given point.

55. $y = 4\sin^2 x$, $(\pi/6, 1)$

56. $y = \dfrac{x^2 - 1}{x^2 + 1}$, $(0, -1)$

57–58 Find equations of the tangent line and normal line to the curve at the given point.

57. $y = (2 + x)e^{-x}$, $(0, 2)$

58. $x^2 + 4xy + y^2 = 13$, $(2, 1)$

---

59. (a) If $f(x) = x\sqrt{5 - x}$, find $f'(x)$.

(b) Find equations of the tangent lines to the curve $y = x\sqrt{5 - x}$ at the points $(1, 2)$ and $(4, 4)$.

(c) Illustrate part (b) by graphing the curve and tangent lines on the same screen.

(d) Check to see that your answer to part (a) is reasonable by comparing the graphs of f and f'.

60. (a) Graph the function $f(x) = x - 2 \sin x$ in the viewing rectangle $[0, 8]$ by $[-2, 8]$.
(b) On which interval is the average rate of change larger: $[1, 2]$ or $[2, 3]$?
(c) At which value of x is the instantaneous rate of change larger: $x = 2$ or $x = 5$?
(d) Check your visual estimates in part (c) by computing $f'(x)$ and comparing the numerical values of $f'(2)$ and $f'(5)$.

61. Suppose that $h(x) = f(x)g(x)$ and $F(x) = f(g(x))$, where $f(2) = 3$, $g(2) = 5$, $g'(2) = 4$, $f'(2) = -2$, and $f'(5) = 11$. Find (a) $h'(2)$ and (b) $F'(2)$.

62. If f and g are the functions whose graphs are shown, let $P(x) = f(x)g(x)$, $Q(x) = f(x)/g(x)$, and $C(x) = f(g(x))$. Find (a) $P'(2)$, (b) $Q'(2)$, and (c) $C'(2)$.

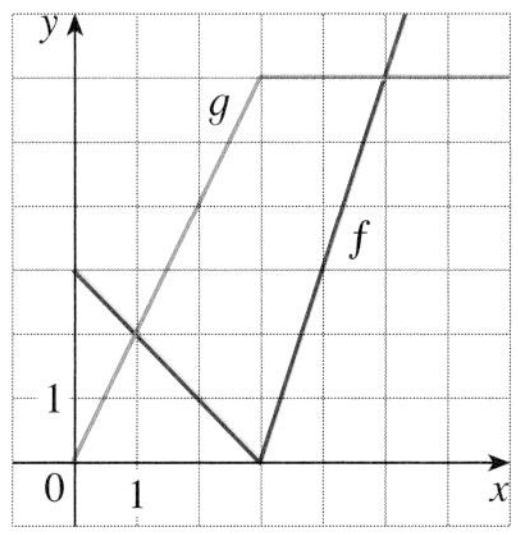

63–70 Find f' in terms of g'.

63. $f(x) = x^2 g(x)$

64. $f(x) = g(x^2)$

65. $f(x) = [g(x)]^2$

66. $f(x) = g(g(x))$

67. $f(x) = g(e^x)$

68. $f(x) = e^{g(x)}$

69. $f(x) = \ln|g(x)|$

70. $f(x) = g(\ln x)$

---

71. At what point on the curve $y = [\ln(x + 4)]^2$ is the tangent horizontal?

72. (a) Find an equation of the tangent to the curve $y = e^x$ that is parallel to the line $x - 4y = 1$.
(b) Find an equation of the tangent to the curve $y = e^x$ that passes through the origin.

73. Find the points on the ellipse $x^2 + 2y^2 = 1$ where the tangent line has slope 1.

74. Find a parabola $y = ax^2 + bx + c$ that passes through the point $(1, 4)$ and whose tangent lines at $x = -1$ and $x = 5$ have slopes 6 and -2, respectively.

75. An equation of motion of the form $s = Ae^{-ct}\cos(\omega t + \delta)$ represents damped oscillation of an object. Find the velocity and acceleration of the object.

76. A particle moves along a horizontal line so that its coordinate at time t is $x = \sqrt{b^2 + c^2t^2}$, $t \geqslant 0$, where b and c are positive constants.
(a) Find the velocity and acceleration functions.
(b) Show that the particle always moves in the positive direction.

77. The volume of a right circular cone is $V = \frac{1}{3}\pi r^2 h$, where r is the radius of the base and h is the height.
(a) Find the rate of change of the volume with respect to the height if the radius is constant.
(b) Find the rate of change of the volume with respect to the radius if the height is constant.

78. The **Michaelis-Menten equation** for the enzyme pepsin is

$$v = \frac{0.50[S]}{3.0 \times 10^{-4} + [S]}$$

where v is the rate of an enzymatic reaction and $[S]$ is the concentration of a substrate S. Calculate $dv/d[S]$ and interpret it.

79. Health care expenditures The US health care expenditures for 1970–2008 have been modeled by the function

$$E(t) = 101.35e^{0.088128t}$$

where t is the number of years elapsed since 1970 and E is measured in billions of dollars. According to this model, at what rate were health care expenditures increasing in 1980? In 2000?

80. Drug concentration The function $C(t) = K(e^{-at} - e^{-bt})$, where a, b, and K are positive constants and $b > a$, is used to model the concentration at time t of a drug injected into the bloodstream.
(a) Show that $\lim_{t \to \infty} C(t) = 0$.
(b) Find $C'(t)$, the rate of change of drug concentration in the blood.
(c) When is this rate equal to 0?

81. Bacteria growth A bacteria culture contains 200 cells initially and grows at a rate proportional to its size. After half an hour the population has increased to 360 cells.
(a) Find the number of bacteria after t hours.
(b) Find the number of bacteria after 4 hours.
(c) Find the rate of growth after 4 hours.
(d) When will the population reach 10,000?

82. Cobalt-60 has a half-life of 5.24 years.
(a) Find the mass that remains from a 100-mg sample after 20 years.
(b) How long would it take for the mass to decay to 1 mg?

83. Drug elimination Let $C(t)$ be the concentration of a drug in the bloodstream. As the body eliminates the drug, $C(t)$ decreases at a rate that is proportional to the amount of the drug that is present at the time. Thus $C'(t) = -kC(t)$, where k is a positive number called the *elimination constant* of the drug.

(a) If C_0 is the concentration at time $t = 0$, find the concentration at time t.

(b) If the body eliminates half the drug in 30 hours, how long does it take to eliminate 90% of the drug?

84. A cup of hot chocolate has temperature 80°C in a room kept at 20°C. After half an hour the hot chocolate cools to 60°C.

(a) What is the temperature of the chocolate after another half hour?

(b) When will the chocolate have cooled to 40°C?

85. The volume of a cube is increasing at a rate of 10 cm^3/min. How fast is the surface area increasing when the length of an edge is 30 cm?

86. Yeast population The number of yeast cells in a laboratory culture increases rapidly initially but levels off eventually. The population is modeled by the function

$$n = f(t) = \frac{a}{1 + be^{-0.7t}}$$

where t is measured in hours. At time $t = 0$ the population is 20 cells and is increasing at a rate of 12 cells/hour. Find the values of a and b. According to this model, what happens to the yeast population in the long run?

87. Use Newton's method to find the root of the equation $x^5 - x^4 + 3x^2 - 3x - 2 = 0$ in the interval $[1, 2]$ correct to six decimal places.

88. Use Newton's method to find all roots of the equation $\sin x = x^2 - 3x + 1$ correct to six decimal places.

89. (a) Find the linearization of $f(x) = \sqrt[3]{1 + 3x}$ at $a = 0$. State the corresponding linear approximation and use it to give an approximate value for $\sqrt[3]{1.03}$.

(b) Determine the values of x for which the linear approximation given in part (a) is accurate to within 0.1.

90. (a) Find the first three Taylor polynomials for $f(x) = 4(x - 2)^{-2}$ centered at 0.

(b) Graph f and the Taylor polynomials from part (a) on the interval $[-1, 1]$ and comment on how well the polynomials approximate f.

91. Dialysis The project on page 458 models the removal of urea from the bloodstream via dialysis. In certain situations the duration of dialysis required, given that the initial urea concentration is c, where $c > 1$, is given by the equation

$$t = \ln\left(\frac{3c + \sqrt{9c^2 - 8c}}{2}\right)$$

(a) Use a linear approximation to estimate the time required if the initial concentration is near $c = 1$.

(b) Use a second-order Taylor polynomial to give a more accurate approximation.

92. Infectious disease outbreak size We have worked with the model

$$pe^{-qA} = 1 - A$$

where A is the fraction of the population infected, q is a measure of disease transmissibility, and p is the fraction of the population that is initially susceptible to infection.

(a) Use implicit differentiation to find the linear approximation of A as a function of q at $q = 0$.

(b) Find the second-order Taylor polynomial approximation for $A(q)$ at $q = 0$.

93. Express the limit

$$\lim_{\theta \to \pi/3} \frac{\cos\theta - 0.5}{\theta - \pi/3}$$

as a derivative and thus evaluate it.

94. Find points P and Q on the parabola $y = 1 - x^2$ so that the triangle ABC formed by the x-axis and the tangent lines at P and Q is an equilateral triangle.

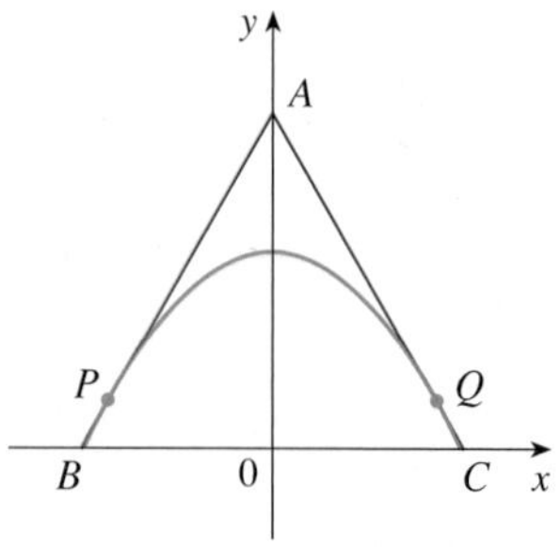

CASE STUDY 1b Kill Curves and Antibiotic Effectiveness

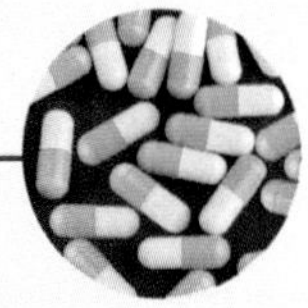

We are studying the relationship between the magnitude of antibiotic treatment and the effectiveness of the treatment. Recall that the extent of bacterial killing by an antibiotic is determined by both the *antibiotic concentration profile* and the *dose response relationship*. Figure 1 shows the antibiotic concentration profile for ciprofloxacin.[1] Our first goal here is to choose an appropriate mathematical description of this profile.

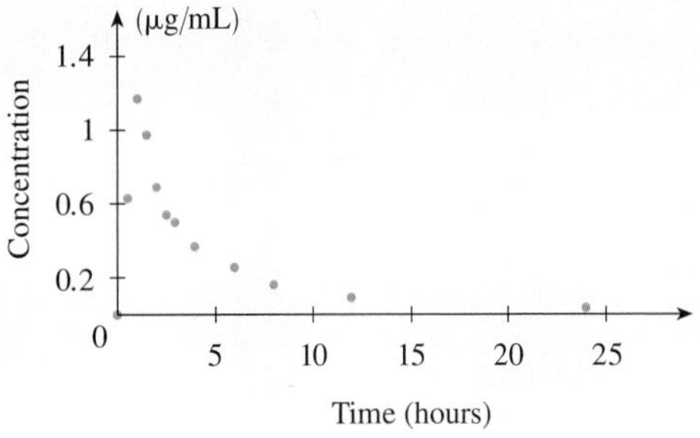

FIGURE 1
Antibiotic concentration profile in plasma of a healthy human volunteer after receiving 500 mg of ciprofloxacin

In Case Study 1a we modeled the initial increase in concentration as occurring instantly. We then need to determine how to model the decay in concentration. From Figure 1 it looks as though the rate of decay (that is, the slope of the relationship between concentration and time) is smaller for lower concentrations. We also know that the rate of decay must be zero when the concentration is zero. Therefore, as a simple model, let's suppose that the rate of decay of concentration is proportional to the current concentration; that is,

(1)
$$\frac{dc}{dt} = -kc$$

for some positive constant k. Here c is measured in $\mu g/mL$ and t is measured in hours.

1. If the concentration at $t = 0$ is c_0, verify that the concentration function $c(t) = c_0 e^{-kt}$ satisfies Equation 1. Suppose that the antibiotic ciprofloxacin has a half-life of 4 hours. What is the value of k?

Next we wish to model the bacteria population dynamics. When a bacteria population is small, it grows at a rate that is proportional to its size because each bacterium produces a constant number of offspring per unit time. A simple model for the growth of the bacteria population size P when small is therefore

(2)
$$\frac{dP}{dt} = rP$$

where r is a constant called the *per capita growth rate* (it is the rate of offspring production by each individual bacterium). As a result, if the bacteria population starts at size P_0, its predicted size at time t is $P(t) = P_0 e^{rt}$.

1. Adapted from Imre, S. et al., "Validation of an HPLC Method for the Determination of Ciprofloxacin in Human Plasma," *Journal of Pharmaceutical and Biomedical Analysis* 33 (2003): 125–30.

As the population grows, resources become depleted. Eventually the bacteria population reaches a size at which it no longer changes. For the data in Figure 2[2] it looks as though the maximum population size is around 12 CFU/mL. A simple model is therefore that the population grows according to Equation 2 if $P < 12$ and it remains constant at $P = 12$ if the value of P predicted from the model in Equation 2 is ever greater than or equal to 12.

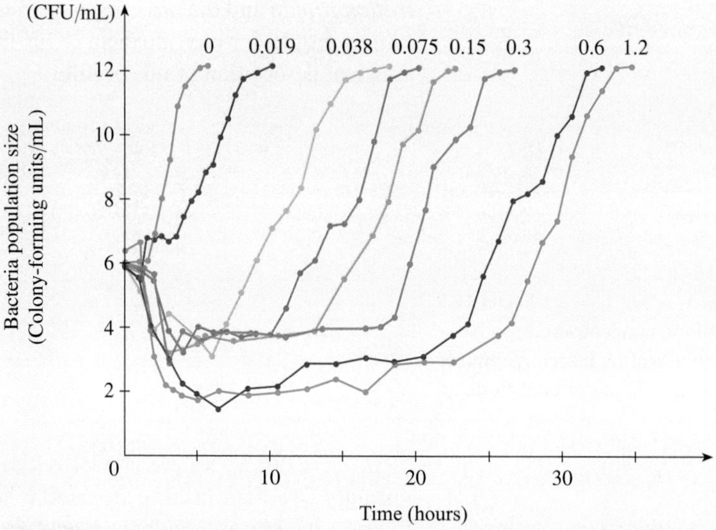

FIGURE 2
The kill curves of ciprofloxacin for *E. coli* when measured in a growth chamber. The concentration of ciprofloxacin at $t = 0$ is indicated above each curve (in μg/mL).

Our final step is to connect the model for bacteria population growth to the model for the antibiotic concentration profile. The connection between the two is given by the dose response relationship. Recall that in Case Study 1a we modeled the dose response relationship with the piecewise defined function

$$r(c) = \begin{cases} r_2 & \text{if } c < MIC \\ r_1 & \text{if } c \geqslant MIC \end{cases}$$

where $r(c)$ is the per capita growth rate of the bacteria population and MIC is a constant referred to as the *minimum inhibitory concentration* ($MIC = 0.013$ μg/mL in this case). The constants r_1 and r_2 give the per capita growth rate under high and low antibiotic concentrations, respectively, with $r_1 < 0$ and $r_2 > 0$ (Figure 3).

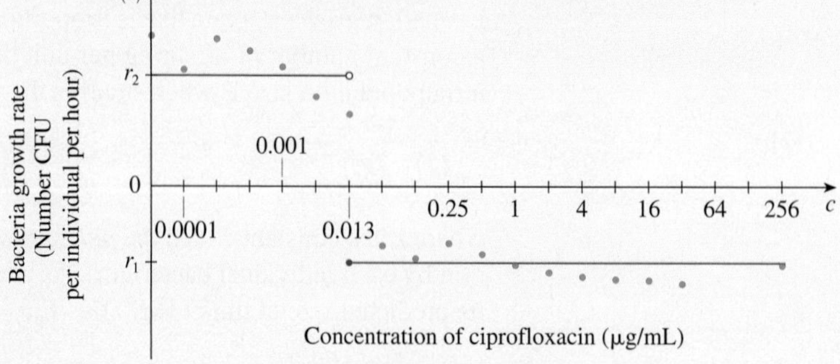

FIGURE 3
Dose response relationship modeled by the piecewise defined function $r(c)$

Source: Adapted from W. Bär et al., "Rapid Method for Detection of Minimal Bactericidal Concentration of Antibiotics," *Journal of Microbiological Methods* 77 (2009): 85–89, Figure 1.

2. Adapted from A. Firsov et al., "Parameters of Bacterial Killing and Regrowth Kinetics and Antimicrobial Effect Examined in Terms of Area Under the Concentration–Time Curve Relationships: Action of Ciprofloxacin against *Escherichia coli* in an In Vitro Dynamic Model," *Antimicrobial Agents and Chemotherapy* 41 (1997): 1281–87.

Suppose the bacteria population starts at $t = 0$ at a size of 6 CFU/mL. Suppose also that $MIC = 0.013$, $k = 0.175$, $r_1 = -\frac{1}{20}$, and $r_2 = \frac{1}{3}$.

2. Using the form of the solution to Equation 2, show that the bacteria population size at time t is given by the function

(3a)
$$P(t) = \begin{cases} 6e^{t/3} & \text{if } t < 2.08 \\ 12 & \text{if } t \geqslant 2.08 \end{cases}$$

if $c_0 < MIC$, where $MIC = 0.013$. On the other hand, if $c_0 > MIC$, show that

(3b)
$$P(t) = \begin{cases} 6e^{-t/20} & \text{if } t < a \\ 6Ae^{t/3} & \text{if } a \leqslant t < b \\ 12 & \text{if } t \geqslant b \end{cases}$$

where the constants a, b, and A are defined by $a = 5.7 \ln(77c_0)$, $b = 6.6 \ln(77c_0) + 2.08$, and $A = (77c_0)^{-2.2}$.

Equations 3 are the predicted kill curves explored in Case Study 1a.

3. Using the form of the solution to Equation 2, show that for arbitrary MIC, k, r_1, and r_2 the bacteria population size at time t is given by

(4a)
$$P(t) = \begin{cases} 6e^{r_2 t} & \text{if } t < t_2 \\ 12 & \text{if } t \geqslant t_2 \end{cases}$$

if $c_0 < MIC$, and

(4b)
$$P(t) = \begin{cases} 6e^{r_1 t} & \text{if } t < t_1 \\ 6e^{r_1 t_1} e^{r_2 (t - t_1)} & \text{if } t_1 \leqslant t < \left(1 - \dfrac{r_1}{r_2}\right) t_1 + t_2 \\ 12 & \text{if } \left(1 - \dfrac{r_1}{r_2}\right) t_1 + t_2 \leqslant t \end{cases}$$

if $c_0 \geqslant MIC$, where $t_1 = \dfrac{1}{k} \ln\left(\dfrac{c_0}{MIC}\right)$ and $t_2 = \dfrac{\ln 2}{r_2}$.

Applications of Derivatives

4

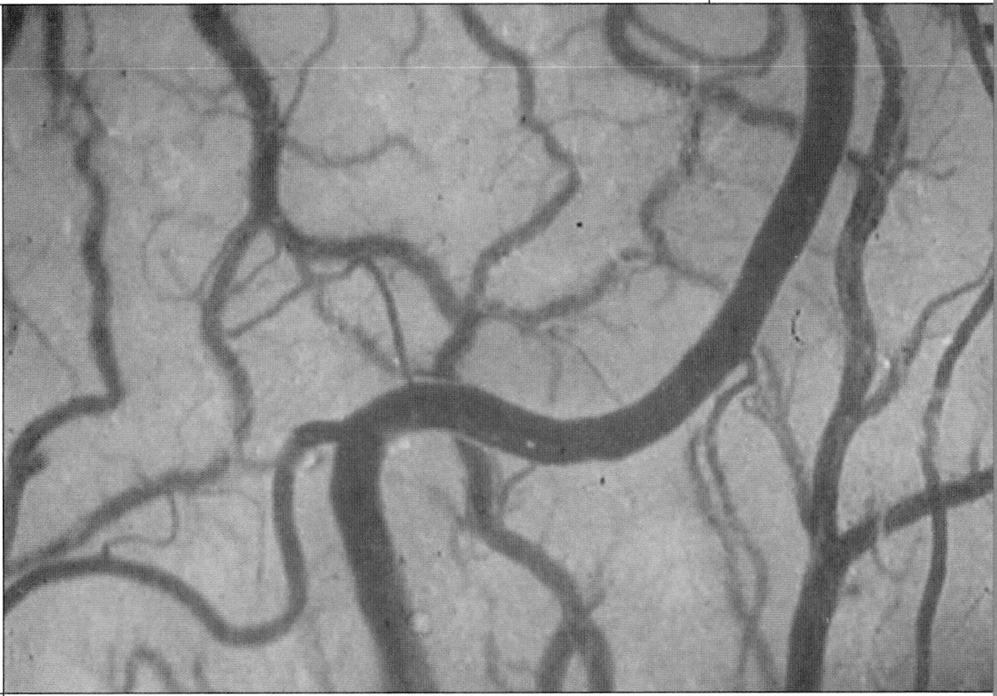

In the blood vascular system, blood vessels divide into smaller vessels at certain angles and these angles affect the resistance of the blood and therefore the energy expended by the heart in pumping blood. In Example 6 in Section 4.4 we determine the vascular branching angle that minimizes this energy.

Kage-Mikrofotografie / Agefotostock

WE HAVE ALREADY INVESTIGATED SOME of the applications of derivatives, but now that we know the differentiation rules we are in a better position to pursue the applications of differentiation in greater depth. Here we learn how derivatives affect the shape of a graph of a function and, in particular, how they help us locate maximum and minimum values of functions. In addition, we use derivatives to provide insight into the long-term behavior of discrete-time models in the life sciences.

4.1 Maximum and Minimum Values

Some of the most important applications of differential calculus are *optimization problems*, in which we are required to find the optimal (best) way of doing something. Here are examples of such problems that we will solve in this chapter:

- What is the radius of a contracted windpipe that expels air most rapidly during a cough?
- At what angle should blood vessels branch so as to minimize the energy expended by the heart in pumping blood?

These problems can be reduced to finding the maximum or minimum values of a function. Let's first explain exactly what we mean by maximum and minimum values.

■ Absolute and Local Extreme Values

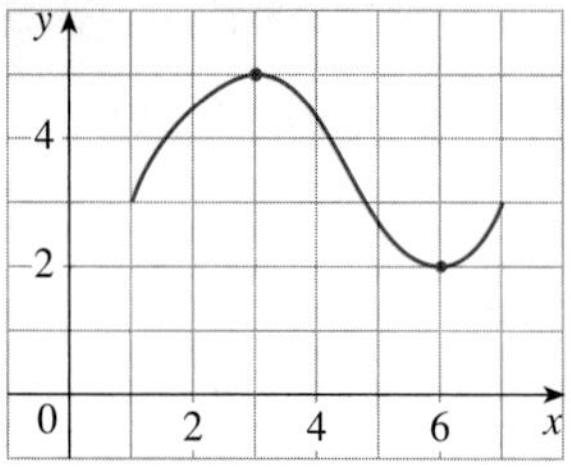

FIGURE 1

We see that the highest point on the graph of the function f shown in Figure 1 is the point $(3, 5)$. In other words, the largest value of f is $f(3) = 5$. Likewise, the smallest value is $f(6) = 2$. We say that $f(3) = 5$ is the *absolute maximum* of f and $f(6) = 2$ is the *absolute minimum*. In general, we use the following definition.

(1) Definition Let c be a number in the domain D of a function f. Then $f(c)$ is the

- **absolute maximum** value of f on D if $f(c) \geqslant f(x)$ for all x in D.
- **absolute minimum** value of f on D if $f(c) \leqslant f(x)$ for all x in D.

An absolute maximum or minimum is sometimes called a **global** maximum or minimum. The maximum and minimum values of f are called **extreme values** of f.

Figure 2 shows the graph of a function f with absolute maximum at d and absolute minimum at a. Note that $(d, f(d))$ is the highest point on the graph and $(a, f(a))$ is the lowest point. In Figure 2, if we consider only values of x near b [for instance, if we restrict our attention to the interval (a, c)], then $f(b)$ is the largest of those values of $f(x)$ and is called a *local maximum value* of f. Likewise, $f(c)$ is called a *local minimum value* of f because $f(c) \leqslant f(x)$ for x near c [in the interval (b, d), for instance]. The function f also has a local minimum at e. In general, we have the following definition.

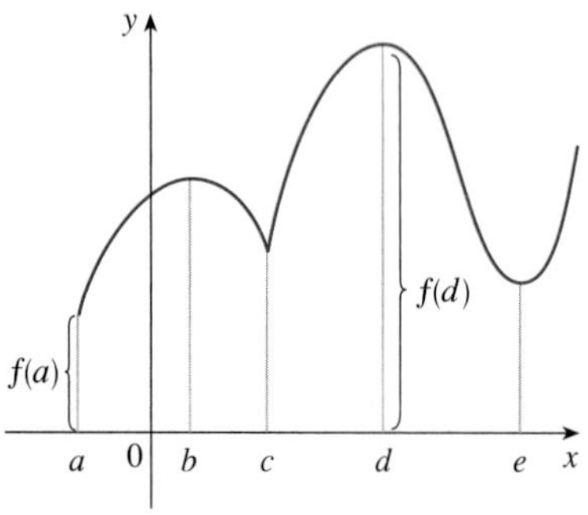

FIGURE 2
Abs min $f(a)$, abs max $f(d)$
loc min $f(c)$, $f(e)$, loc max $f(b)$, $f(d)$

(2) Definition The number $f(c)$ is a

- **local maximum** value of f if $f(c) \geqslant f(x)$ when x is near c.
- **local minimum** value of f if $f(c) \leqslant f(x)$ when x is near c.

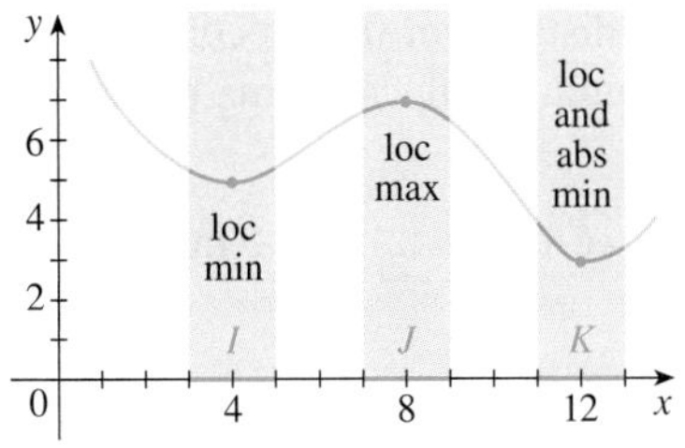

FIGURE 3

In Definition 2 (and elsewhere), if we say that something is true **near** c, we mean that it is true on some open interval containing c. For instance, in Figure 3 we see that $f(4) = 5$ is a local minimum because it's the smallest value of f on the interval I. It's not the absolute minimum because $f(x)$ takes smaller values when x is near 12 (in the interval K, for instance). In fact, $f(12) = 3$ is both a local minimum and the absolute minimum. Similarly, $f(8) = 7$ is a local maximum, but not the absolute maximum because f takes larger values near 1.

EXAMPLE 1 | The function $f(x) = \cos x$ takes on its (local and absolute) maximum value of 1 infinitely many times, since $\cos 2n\pi = 1$ for any integer n and $-1 \leqslant \cos x \leqslant 1$ for all x. Likewise, $\cos(2n + 1)\pi = -1$ is its minimum value, where n is any integer. ■

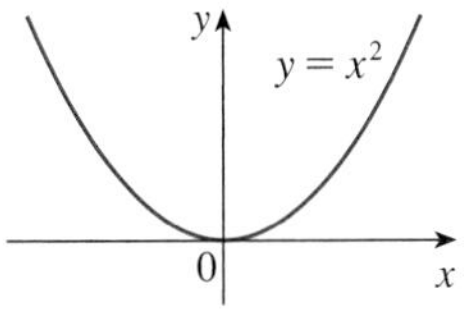

FIGURE 4
Minimum value 0, no maximum

EXAMPLE 2 | If $f(x) = x^2$, then $f(x) \geqslant f(0)$ because $x^2 \geqslant 0$ for all x. Therefore $f(0) = 0$ is the absolute (and local) minimum value of f. This corresponds to the fact that the origin is the lowest point on the parabola $y = x^2$. (See Figure 4.) However, there is no highest point on the parabola and so this function has no maximum value. ■

EXAMPLE 3 | From the graph of the function $f(x) = x^3$, shown in Figure 5, we see that this function has neither an absolute maximum value nor an absolute minimum value. In fact, it has no local extreme values either.

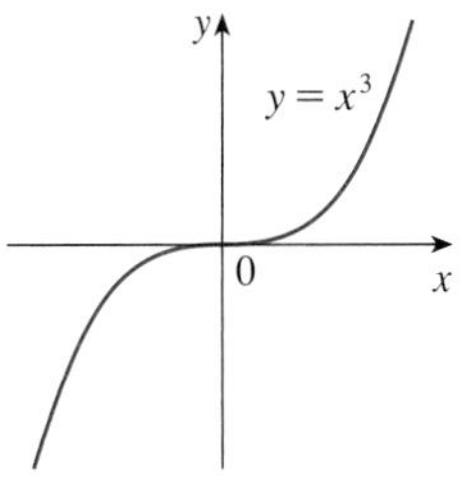

FIGURE 5
No minimum, no maximum

■

EXAMPLE 4 | BB **Electrocardiogram** Figure 6 shows a rhythm strip from an ECG. It is a graph of the electric potential function $f(t)$ (measured in millivolts) as a function of time in a certain direction corresponding to a particular part of the heart. The points P, Q, R, S, and T on the graph are labeled with the notation that is standard for cardiologists. We see that the function f has local maxima at the points P, R, and T, with an absolute maximum at R, and local minima at Q and S, with an absolute minimum at S.

Cardiologists use the relative locations of the extreme points P, Q, R, S, and T to diagnose heart problems.

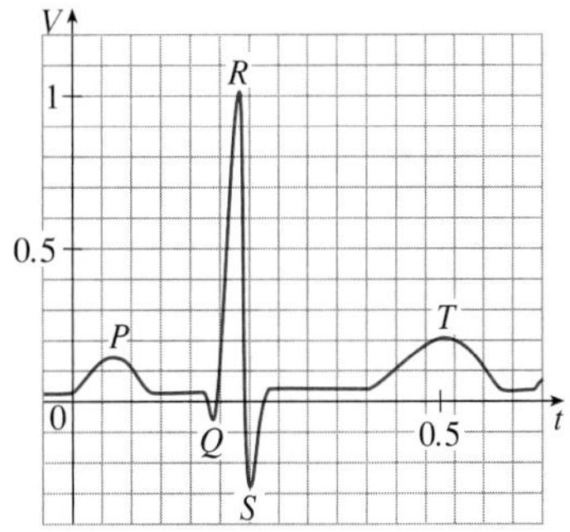

FIGURE 6

Each square in the grid corresponds to a horizontal distance of 0.04 seconds and a vertical distance of 0.1 mV. So the maximum and minimum values in this particular ECG are as follows.

Local maximum values:	At P, $f(0.05) \approx 0.15$ mV; at R, $f(0.22) \approx 1.0$ mV; at T, $f(0.5) \approx 0.2$ mV
Absolute maximum value:	At R, $f(0.22) \approx 1.0$ mV
Local minimum values:	At Q, $f(0.19) \approx -0.06$ mV; at S, $f(0.24) \approx -0.3$ mV
Absolute minimum value:	At S, $f(0.24) \approx -0.3$ mV

■

We have seen that some functions have extreme values, whereas others do not. The following theorem gives conditions under which a function is guaranteed to possess extreme values.

(3) The Extreme Value Theorem If f is continuous on a closed interval $[a, b]$, then f attains an absolute maximum value $f(c)$ and an absolute minimum value $f(d)$ at some numbers c and d in $[a, b]$.

The Extreme Value Theorem is illustrated in Figure 7. Note that an extreme value can be taken on more than once. Although the Extreme Value Theorem is intuitively very plausible, it is difficult to prove and so we omit the proof.

FIGURE 7

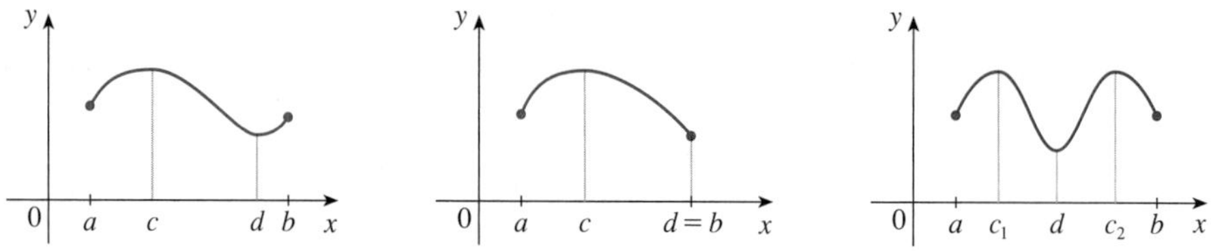

Figures 8 and 9 show that a function need not possess extreme values if either hypothesis (continuity or closed interval) is omitted from the Extreme Value Theorem.

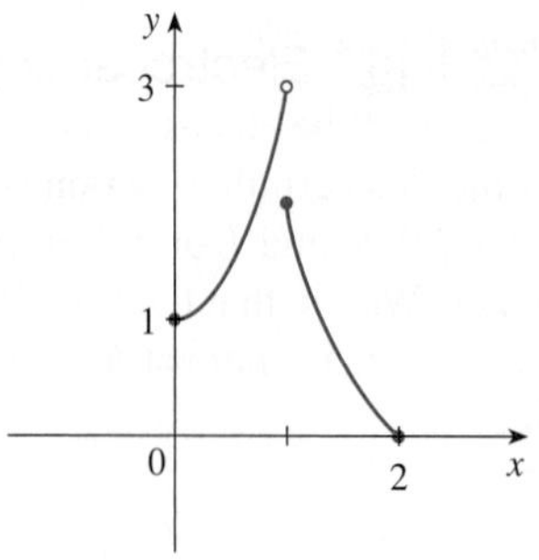

FIGURE 8
This function has a minimum value $f(2) = 0$, but no maximum value.

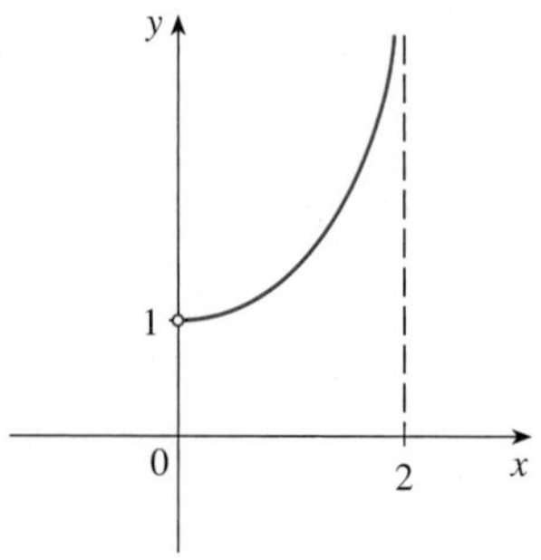

FIGURE 9
This continuous function g has no maximum or minimum.

The function f whose graph is shown in Figure 8 is defined on the closed interval $[0, 2]$ but has no maximum value. [Notice that the range of f is $[0, 3)$. The function takes on values arbitrarily close to 3, but never actually attains the value 3.] This does not contradict the Extreme Value Theorem because f is not continuous. [Nonetheless, a discontinuous function *could* have maximum and minimum values. See Exercise 13(c).]

The function g shown in Figure 9 is continuous on the open interval $(0, 2)$ but has neither a maximum nor a minimum value. [The range of g is $(1, \infty)$. The function takes on arbitrarily large values.] This does not contradict the Extreme Value Theorem because the interval $(0, 2)$ is not closed.

■ Fermat's Theorem

The Extreme Value Theorem says that a continuous function on a closed interval has a maximum value and a minimum value, but it does not tell us how to find these extreme values. We start by looking for local extreme values.

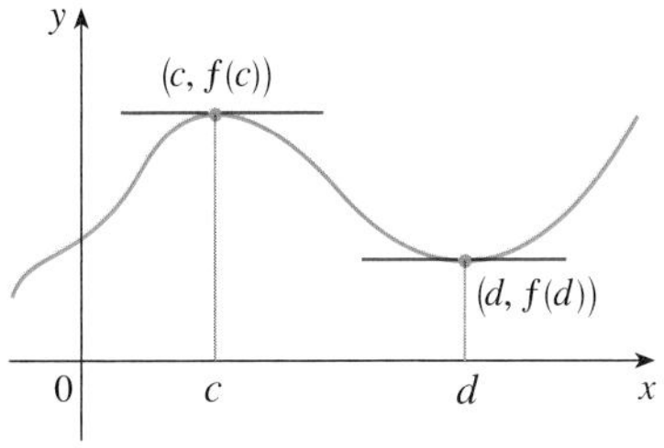

FIGURE 10

Figure 10 shows the graph of a function f with a local maximum at c and a local minimum at d. It appears that at the maximum and minimum points the tangent lines are horizontal and therefore each has slope 0. We know that the derivative is the slope of the tangent line, so it appears that $f'(c) = 0$ and $f'(d) = 0$. The following theorem says that this is always true for differentiable functions.

(4) Fermat's Theorem If f has a local maximum or minimum at c, and if $f'(c)$ exists, then $f'(c) = 0$.

Our intuition suggests that Fermat's Theorem is true. A rigorous proof, using the definition of a derivative, is given in Appendix E.

Although Fermat's Theorem is very useful, we have to guard against reading too much into it. If $f(x) = x^3$, then $f'(x) = 3x^2$, so $f'(0) = 0$. But f has no maximum or minimum at 0, as you can see from its graph in Figure 11. The fact that $f'(0) = 0$ simply means that the curve $y = x^3$ has a horizontal tangent at $(0, 0)$. Instead of having a maximum or minimum at $(0, 0)$, the curve crosses its horizontal tangent there.

Thus, when $f'(c) = 0$, f doesn't necessarily have a maximum or minimum at c. (In other words, the converse of Fermat's Theorem is false in general.)

Fermat

Fermat's Theorem is named after Pierre Fermat (1601–1665), a French lawyer who took up mathematics as a hobby. Despite his amateur status, Fermat was one of the two inventors of analytic geometry (Descartes was the other). His methods for finding tangents to curves and maximum and minimum values (before the invention of limits and derivatives) made him a forerunner of Newton in the creation of differential calculus.

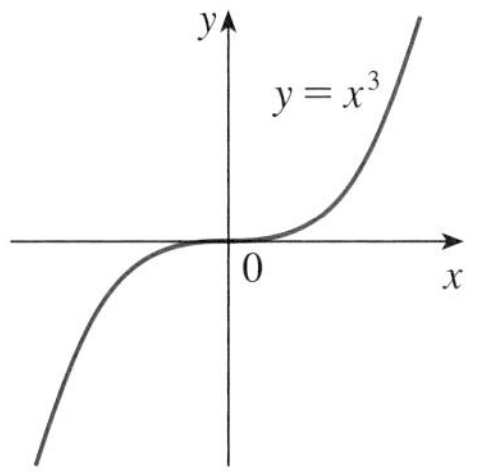

FIGURE 11
If $f(x) = x^3$, then $f'(0) = 0$ but f has no maximum or minimum.

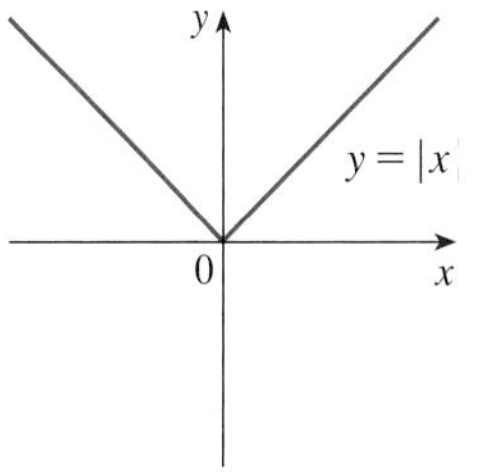

FIGURE 12
If $f(x) = |x|$, then $f(0) = 0$ is a minimum value, but $f'(0)$ does not exist.

We should bear in mind that there may be an extreme value where $f'(c)$ does not exist. For instance, the function $f(x) = |x|$ has its (local and absolute) minimum value at 0 (see Figure 12), but that value cannot be found by setting $f'(x) = 0$ because, as was shown in Example 3.2.6, $f'(0)$ does not exist.

Fermat's Theorem does suggest that we should at least *start* looking for extreme values of f at the numbers c where $f'(c) = 0$ or where $f'(c)$ does not exist. Such numbers are given a special name.

(5) Definition A **critical number** of a function f is a number c in the domain of f such that either $f'(c) = 0$ or $f'(c)$ does not exist.

Figure 13 shows a graph of the function f in Example 5. It supports our answer because there is a horizontal tangent when $x = 1.5$ and a vertical tangent when $x = 0$.

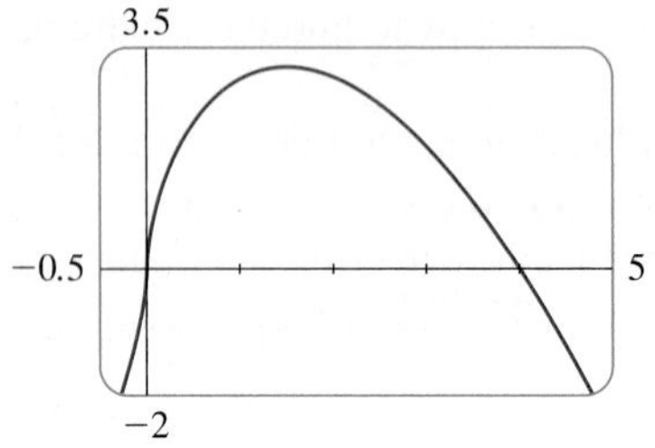

FIGURE 13

EXAMPLE 5 | Find the critical numbers of $f(x) = x^{3/5}(4 - x)$.

SOLUTION The Product Rule gives

$$f'(x) = x^{3/5}(-1) + (4 - x)\left(\tfrac{3}{5}x^{-2/5}\right) = -x^{3/5} + \frac{3(4 - x)}{5x^{2/5}}$$

$$= \frac{-5x + 3(4 - x)}{5x^{2/5}} = \frac{12 - 8x}{5x^{2/5}}$$

[The same result could be obtained by first writing $f(x) = 4x^{3/5} - x^{8/5}$.] Therefore $f'(x) = 0$ if $12 - 8x = 0$, that is, $x = \frac{3}{2}$, and $f'(x)$ does not exist when $x = 0$. Thus the critical numbers are $\frac{3}{2}$ and 0. ■

In terms of critical numbers, Fermat's Theorem can be rephrased as follows (compare Definition 5 with Theorem 4):

(6) If f has a local maximum or minimum at c, then c is a critical number of f.

The Closed Interval Method

To find an absolute maximum or minimum of a continuous function on a closed interval, we note that either it is local [in which case it occurs at a critical number by (6)] or it occurs at an endpoint of the interval. Thus the following three-step procedure always works.

The Closed Interval Method To find the *absolute* maximum and minimum values of a continuous function f on a closed interval $[a, b]$:

1. Find the values of f at the critical numbers of f in (a, b).
2. Find the values of f at the endpoints of the interval.
3. The largest of the values from Steps 1 and 2 is the absolute maximum value; the smallest of these values is the absolute minimum value.

The phenomenon whereby a population declines to extinction below a critical population size is referred to as an *Allee effect* after the American ecologist Warder Clyde Allee (1885–1955).

EXAMPLE 6 | BB **The Allee effect** One of the models for the growth rate of a population of size N at time t reflects the fact that some populations decline to extinction unless they stay above a critical value. A particular case of this model is expressed by the growth rate

$$f(N) = N(N - 3)(8 - N)$$

where N is measured in hundreds of individuals. [Notice that $f(N)$ is negative when $0 < N < 3$.] Find the absolute maximum and minimum values of the growth rate function

$$f(N) = N(N - 3)(8 - N) \qquad 0 \leqslant N \leqslant 9$$

SOLUTION Because f is continuous on the interval $[0, 9]$, we can use the Closed Interval Method:

$$f(N) = N(N - 3)(8 - N) = -N^3 + 11N^2 - 24N$$

$$f'(N) = -3N^2 + 22N - 24 = -(3N - 4)(N - 6)$$

Since $f'(N)$ exists for all N, the only critical numbers of f occur when $f'(N) = 0$, that is, $N = \frac{4}{3}$ or $N = 6$. Notice that each of these critical numbers lies in the interval $(0, 9)$. The values of f at these critical numbers are

$$f\left(\tfrac{4}{3}\right) = -\tfrac{400}{27} \qquad f(6) = 36$$

The values of f at the endpoints of the interval are

$$f(0) = 0 \qquad f(9) = -54$$

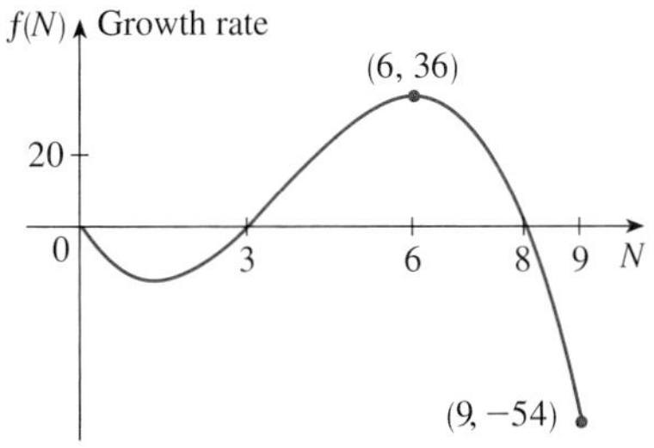

FIGURE 14

Comparing these four numbers, we see that the absolute maximum value is $f(6) = 36$ and the absolute minimum value is $f(9) = -54$.

So the population increases fastest when $N = 6$ (the population is 600) and the absolute maximum value is $f(6) = 36$, which means that the maximum rate of increase is 3600 individuals per year. The population decreases most rapidly on the given interval when $N = 9$ and the absolute minimum value is $f(9) = -54$. This means that the maximum rate of decrease is 5400 individuals per year.

Note that in this example the absolute minimum occurs at an endpoint, whereas the absolute maximum occurs at a critical number. The graph of f is sketched in Figure 14. ■

EXAMPLE 7 | Blood alcohol concentration In Section 3.1 we used the function

$$C(t) = 0.0225te^{-0.0467t}$$

to model the average blood alcohol concentration (BAC) of a group of eight male subjects after rapid consumption of 15 mL of ethanol (corresponding to one alcoholic drink), where t is measured in minutes after consumption and C is measured in mg/mL. Find the maximum value of the BAC during the first hour.

SOLUTION We begin by differentiating the concentration function using the Product Rule:

$$C'(t) = 0.0225t(-0.0467)e^{-0.0467t} + 0.0225e^{-0.0467t}$$
$$= 0.0225e^{-0.0467t}(-0.0467t + 1)$$

The critical number occurs when $C'(t) = 0$, that is,

$$0.0467t = 1 \qquad \Rightarrow \qquad t = \frac{1}{0.0467} \approx 21.4$$

The value of C at this critical number is about

$$C(21.4) \approx 0.177$$

and the values of C at the endpoints of the interval $[0, 60]$ are

$$C(0) = 0 \qquad C(60) \approx 0.0819$$

Comparing the values of C at the critical number and at the endpoints, we see that the maximum value of the BAC in the first hour was about 0.177 mg/mL and this occurred about 21 minutes after consumption. (See the graph of C in Figure 15.) Notice that the maximum value of 0.177 mg/mL was well above the legal driving limit of 0.08 mg/mL and occurred after just one drink. ■

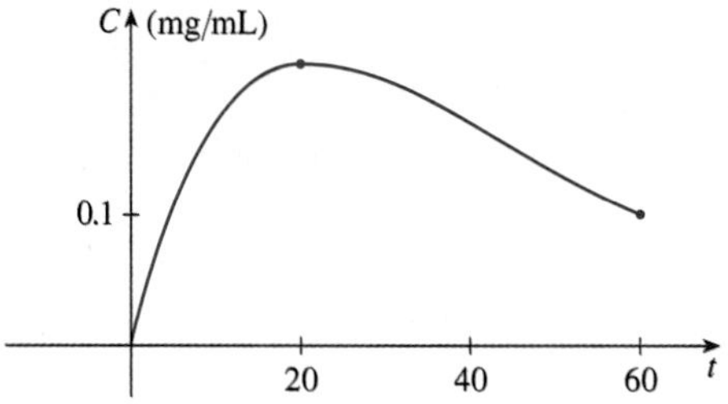

FIGURE 15

EXERCISES 4.1

1. Explain the difference between an absolute minimum and a local minimum.

2. Suppose f is a continuous function defined on a closed interval $[a, b]$.
(a) What theorem guarantees the existence of an absolute maximum value and an absolute minimum value for f?
(b) What steps would you take to find those maximum and minimum values?

3–4 For each of the numbers a, b, c, d, r, and s, state whether the function whose graph is shown has an absolute maximum or minimum, a local maximum or minimum, or neither a maximum nor a minimum.

3.

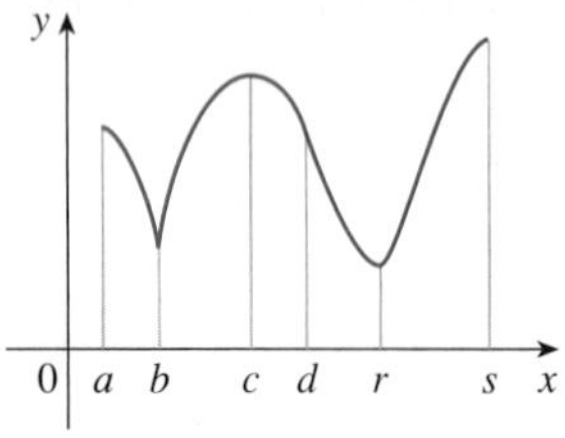

4.

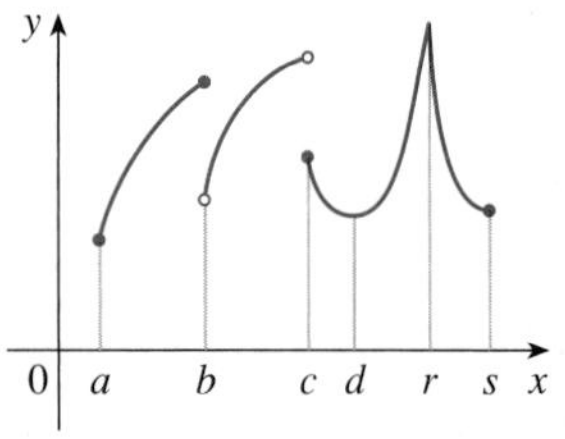

5. Electrocardiogram A cardiologist looking at the rhythm strip shown might suspect *right atrial hypertrophy* because of the relatively tall peaked wave at P (compare with Figure 6). State the local and absolute maximum and minimum values of the electric potential function $f(t)$.

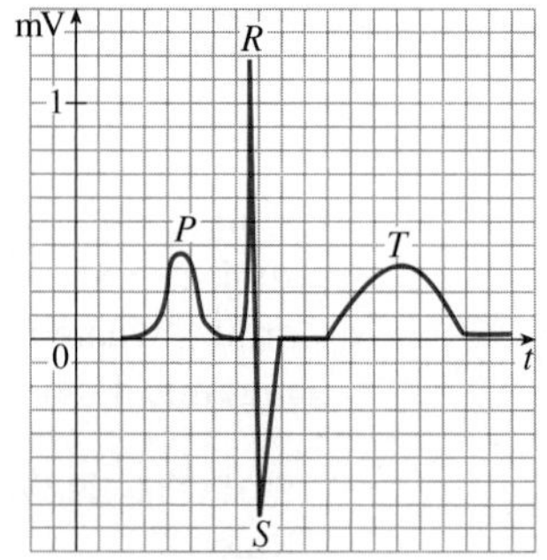

6. Electrocardiogram A cardiologist looking at this rhythm strip might suspect *infarction* because of the elevation of the graph near S and T (compare with Figure 6). State the local and absolute maximum and minimum values of the electric potential function $f(t)$.

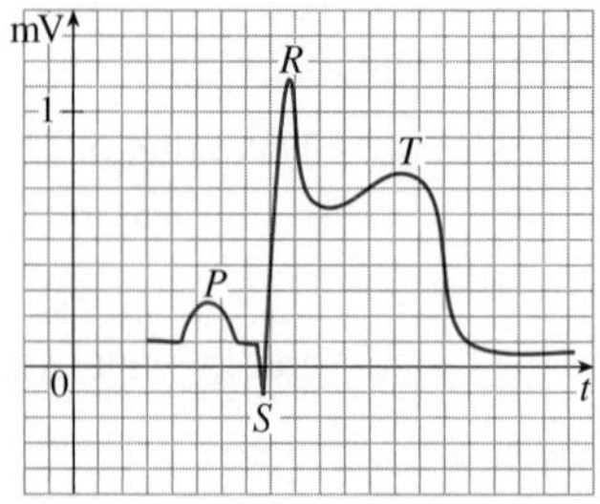

7. In the **influenza pandemic** of 1918–1919 about 40 million people died worldwide. A study in 2007 assessed the nonpharmaceutical interventions used in 43 US cities to combat the infection, including isolation, quarantines, school closures, and public gathering cancellations. The graph shows the weekly excess death rate $D(t)$ for New York City. State the local and absolute maximum and minimum values of D over the given time period and estimate when they occurred.

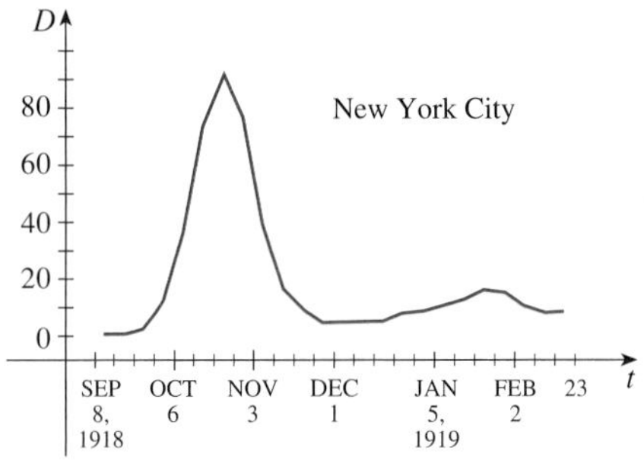

Source: H. Markel et al., "Nonpharmaceutical Interventions Implemented by US Cities During the 1918–1919 Influenza Pandemic," *J. Amer. Med. Assn.* 298 (2007): 644–54.

8. Influenza pandemic The study cited in Exercise 7 also included the corresponding graph for Denver shown here.
(a) State the corresponding local and absolute maximum and minimum values for Denver.
(b) Compare the graphs for New York and Denver. How do you think the strategies differed in the two cities?

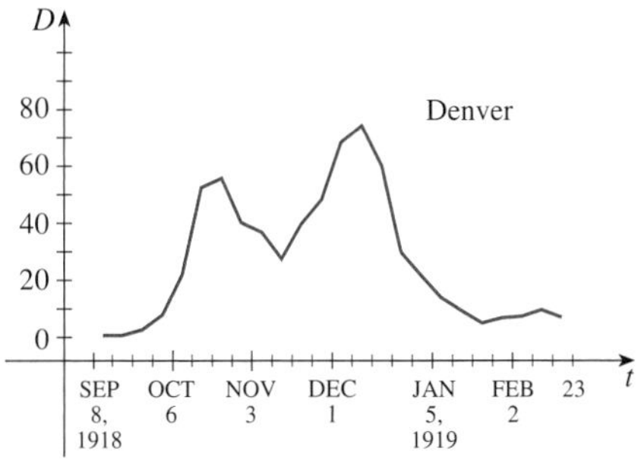

9–12 Sketch the graph of a function f that is continuous on $[1, 5]$ and has the given properties.

9. Absolute minimum at 2, absolute maximum at 3, local minimum at 4

10. Absolute minimum at 1, absolute maximum at 5, local maximum at 2, local minimum at 4

11. Absolute maximum at 5, absolute minimum at 2, local maximum at 3, local minima at 2 and 4

12. f has no local maximum or minimum, but 2 and 4 are critical numbers

13. (a) Sketch the graph of a function that has a local maximum at 2 and is differentiable at 2.
(b) Sketch the graph of a function that has a local maximum at 2 and is continuous but not differentiable at 2.
(c) Sketch the graph of a function that has a local maximum at 2 and is not continuous at 2.

14. (a) Sketch the graph of a function on $[-1, 2]$ that has an absolute maximum but no local maximum.
(b) Sketch the graph of a function on $[-1, 2]$ that has a local maximum but no absolute maximum.

15. (a) Sketch the graph of a function on $[-1, 2]$ that has an absolute maximum but no absolute minimum.
(b) Sketch the graph of a function on $[-1, 2]$ that is discontinuous but has both an absolute maximum and an absolute minimum.

16. (a) Sketch the graph of a function that has two local maxima, one local minimum, and no absolute minimum.
(b) Sketch the graph of a function that has three local minima, two local maxima, and seven critical numbers.

17–24 Sketch the graph of f by hand and use your sketch to find the absolute and local maximum and minimum values of f. (Use the graphs and transformations of Sections 1.2 and 1.3.)

17. $f(x) = \frac{1}{2}(3x - 1), \quad x \leqslant 3$

18. $f(x) = 2 - \frac{1}{3}x, \quad x \geqslant -2$

19. $f(x) = x^2, \quad 0 < x < 2$

20. $f(x) = e^x$

21. $f(x) = \ln x, \quad 0 < x \leqslant 2$

22. $f(t) = \cos t, \quad -3\pi/2 \leqslant t \leqslant 3\pi/2$

23. $f(x) = 1 - \sqrt{x}$

24. $f(x) = \begin{cases} 4 - x^2 & \text{if } -2 \leqslant x < 0 \\ 2x - 1 & \text{if } 0 \leqslant x \leqslant 2 \end{cases}$

25–40 Find the critical numbers of the function.

25. $f(x) = 4 + \frac{1}{3}x - \frac{1}{2}x^2$

26. $f(x) = x^3 + 6x^2 - 15x$

27. $f(x) = x^3 + 3x^2 - 24x$

28. $f(x) = x^3 + x^2 + x$

29. $s(t) = 3t^4 + 4t^3 - 6t^2$

30. $g(t) = |3t - 4|$

31. $g(y) = \dfrac{y - 1}{y^2 - y + 1}$

32. $h(p) = \dfrac{p - 1}{p^2 + 4}$

33. $h(t) = t^{3/4} - 2t^{1/4}$

34. $g(x) = x^{1/3} - x^{-2/3}$

35. $F(x) = x^{4/5}(x - 4)^2$

36. $g(\theta) = 4\theta - \tan\theta$

37. $f(\theta) = 2\cos\theta + \sin^2\theta$

38. $h(t) = 3t - \arcsin t$

39. $f(x) = x^2 e^{-3x}$

40. $f(x) = x^{-2}\ln x$

41–54 Find the absolute maximum and absolute minimum values of f on the given interval.

41. $f(x) = 12 + 4x - x^2$, $[0, 5]$

42. $f(x) = 5 + 54x - 2x^3$, $[0, 4]$

43. $f(x) = 2x^3 - 3x^2 - 12x + 1$, $[-2, 3]$

44. $f(x) = x^3 - 6x^2 + 9x + 2$, $[-1, 4]$

45. $f(x) = x^4 - 2x^2 + 3$, $[-2, 3]$

46. $f(x) = (x^2 - 1)^3$, $[-1, 2]$

47. $f(t) = t\sqrt{4 - t^2}$, $[-1, 2]$

48. $f(x) = \dfrac{x^2 - 4}{x^2 + 4}$, $[-4, 4]$

49. $f(x) = xe^{-x^2/8}$, $[-1, 4]$

50. $f(x) = x - \ln x$, $\left[\frac{1}{2}, 2\right]$

51. $f(x) = \ln(x^2 + x + 1)$, $[-1, 1]$

52. $f(x) = x - 2\tan^{-1}x$, $[0, 4]$

53. $f(t) = 2\cos t + \sin 2t$, $[0, \pi/2]$

54. $f(t) = t + \cot(t/2)$, $[\pi/4, 7\pi/4]$

---

55. If a and b are positive numbers, find the maximum value of $f(x) = x^a(1 - x)^b$, $0 \leqslant x \leqslant 1$.

56. Antibiotic pharmacokinetics After an antibiotic tablet is taken, the concentration of the antibiotic in the bloodstream is modeled by the function

$$C(t) = 8(e^{-0.4t} - e^{-0.6t})$$

where the time t is measured in hours and C is measured in $\mu g/mL$. What is the maximum concentration of the antibiotic during the first 12 hours?

57. Disease virulence The Kermack-McKendrick model for infectious disease transmission (see Exercise 7.6.23) can be used to predict the population size P as a function of the disease's *virulence* (that is, the extent to which the disease kills people). The population size P is large when virulence v is low and it is also large when virulence is high because the disease kills people so fast that very few people get infected. For a specific choice of constants, the population size is

$$P(v) = \frac{10 + v + v^2}{1 + v} \qquad 0 \leqslant v \leqslant 9$$

Find the smallest and largest population sizes and the virulence values for which they occur.

58. The **Maynard Smith and Slatkin model** for population growth is a discrete-time model of the form

$$n_{t+1} = \frac{\lambda n_t}{1 + \alpha n_t^k}$$

For the constants $\lambda = 2$, $\alpha = 0.25$, and $k = 2$, the model is $n_{t+1} = f(n_t)$, where the updating function is

$$f(n) = \frac{2n}{1 + 0.25n^2}$$

Find the largest value of f and interpret it. [*Hint:* Consider $\lim_{n\to\infty} f(n)$.]

59. Coughing When a foreign object lodged in the trachea (windpipe) forces a person to cough, the diaphragm thrusts upward causing an increase in pressure in the lungs. This is accompanied by a contraction of the trachea, making a narrower channel for the expelled air to flow through. For a given amount of air to escape in a fixed time, it must move faster through the narrower channel than the wider one. The greater the velocity of the airstream, the greater the force on the foreign object. X rays show that the radius of the circular tracheal tube contracts to about two-thirds of its normal radius during a cough. According to a mathematical model of coughing, the velocity v of the airstream is related to the radius r of the trachea by the equation

$$v(r) = k(r_0 - r)r^2 \qquad \tfrac{1}{2}r_0 \leqslant r \leqslant r_0$$

where k is a constant and r_0 is the normal radius of the trachea. The restriction on r is due to the fact that the tracheal wall stiffens under pressure and a contraction greater than $\frac{1}{2}r_0$ is prevented (otherwise the person would suffocate).

(a) Determine the value of r in the interval $\left[\frac{1}{2}r_0, r_0\right]$ at which v has an absolute maximum. How does this compare with experimental evidence?

(b) What is the absolute maximum value of v on the interval?

(c) Sketch the graph of v on the interval $[0, r_0]$.

60. On May 7, 1992, the space shuttle *Endeavour* was launched on mission STS-49, the purpose of which was to install a new perigee kick motor in an Intelsat communications satellite. The table gives the velocity data for the shuttle between liftoff and the jettisoning of the solid rocket boosters.

Event	Time (s)	Velocity (ft/s)
Launch	0	0
Begin roll maneuver	10	185
End roll maneuver	15	319
Throttle to 89%	20	447
Throttle to 67%	32	742
Throttle to 104%	59	1325
Maximum dynamic pressure	62	1445
Solid rocket booster separation	125	4151

(a) Use a graphing calculator or computer to find the cubic polynomial that best models the velocity of the shuttle for the time interval $t \in [0, 125]$. Then graph this polynomial.

(b) Find a model for the acceleration of the shuttle and use it to estimate the maximum and minimum values of the acceleration during the first 125 seconds.

61. Between 0°C and 30°C the volume V (in cubic centimeters) of 1 kg of water at a temperature T is given approximately by the formula

$$V = 999.87 - 0.06426T + 0.0085043T^2 - 0.0000679T^3$$

Find the temperature at which water has its maximum density.

62. A cubic function is a polynomial of degree 3; that is, it has the form $f(x) = ax^3 + bx^2 + cx + d$, where $a \neq 0$.

(a) Show that a cubic function can have two, one, or no critical number(s). Give examples and sketches to illustrate the three possibilities.

(b) How many local extreme values can a cubic function have?

■ PROJECT The Calculus of Rainbows

Rainbows are created when raindrops scatter sunlight. They have fascinated humankind since ancient times and have inspired attempts at scientific explanation since the time of Aristotle. In this project we use the ideas of Descartes and Newton to explain the shape, location, and colors of rainbows.

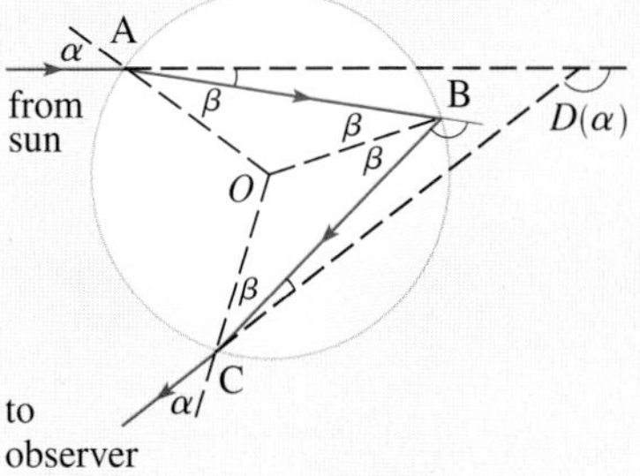

Formation of the primary rainbow

1. The figure shows a ray of sunlight entering a spherical raindrop at A. Some of the light is reflected, but the line AB shows the path of the part that enters the drop. Notice that the light is refracted toward the normal line AO and in fact Snell's Law says that $\sin \alpha = k \sin \beta$, where α is the angle of incidence, β is the angle of refraction, and $k \approx \frac{4}{3}$ is the index of refraction for water. At B some of the light passes through the drop and is refracted into the air, but the line BC shows the part that is reflected. (The angle of incidence equals the angle of reflection.) When the ray reaches C, part of it is reflected, but for the time being we are more interested in the part that leaves the raindrop at C. (Notice that it is refracted away from the normal line.) The *angle of deviation* $D(\alpha)$ is the amount of clockwise rotation that the ray has undergone during this three-stage process. Thus

$$D(\alpha) = (\alpha - \beta) + (\pi - 2\beta) + (\alpha - \beta) = \pi + 2\alpha - 4\beta$$

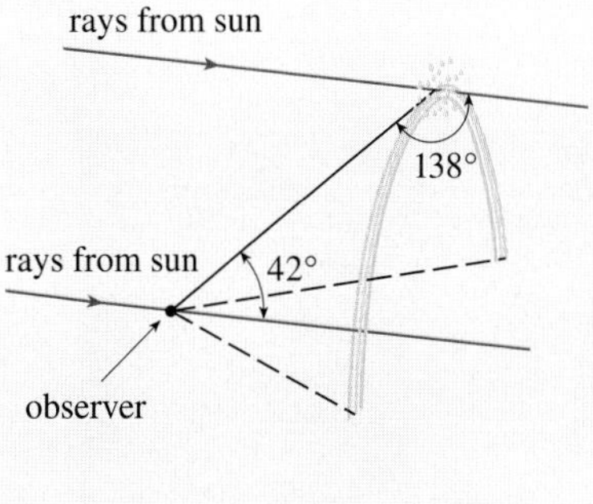

Show that the minimum value of the deviation is $D(\alpha) \approx 138°$ and occurs when $\alpha \approx 59.4°$.

The significance of the minimum deviation is that when $\alpha \approx 59.4°$ we have $D'(\alpha) \approx 0$, so $\Delta D/\Delta\alpha \approx 0$. This means that many rays with $\alpha \approx 59.4°$ become deviated by approximately the same amount. It is the *concentration* of rays coming from near the direction of minimum deviation that creates the brightness of the primary rainbow. The figure at the left shows that the angle of elevation from the observer up to the highest point on the rainbow is $180° - 138° = 42°$. (This angle is called the *rainbow angle*.)

2. Problem 1 explains the location of the primary rainbow, but how do we explain the colors? Sunlight comprises a range of wavelengths, from the red range through orange, yellow, green, blue, indigo, and violet. As Newton discovered in his prism experiments of 1666, the index of refraction is different for each color. (The effect is called *dispersion*.) For red light the refractive index is $k \approx 1.3318$ whereas for violet light it is $k \approx 1.3435$. By repeating the calculation of Problem 1 for these values of k, show that the rainbow angle is about 42.3° for the red bow and 40.6° for the violet bow. So the rainbow really consists of seven individual bows corresponding to the seven colors.

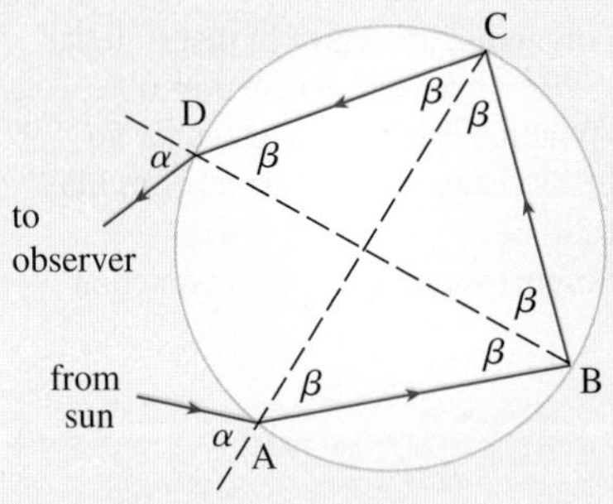

Formation of the secondary rainbow

3. Perhaps you have seen a fainter secondary rainbow above the primary bow. That results from the part of a ray that enters a raindrop and is refracted at A, reflected twice (at B and C), and refracted as it leaves the drop at D (see the figure at the left). This time the deviation angle $D(\alpha)$ is the total amount of counterclockwise rotation that the ray undergoes in this four-stage process. Show that

$$D(\alpha) = 2\alpha - 6\beta + 2\pi$$

and $D(\alpha)$ has a minimum value when

$$\cos\alpha = \sqrt{\frac{k^2 - 1}{8}}$$

Taking $k = \frac{4}{3}$, show that the minimum deviation is about 129° and so the rainbow angle for the secondary rainbow is about 51°, as shown in the following figure.

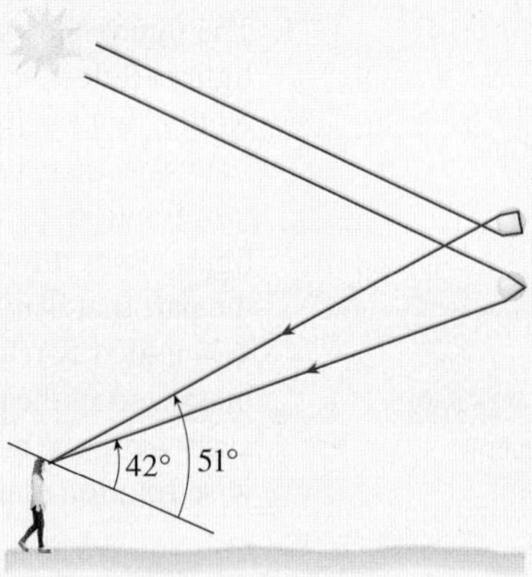

4. Show that the colors in the secondary rainbow appear in the opposite order from those in the primary rainbow.

4.2 How Derivatives Affect the Shape of a Graph

Many of the applications of calculus depend on our ability to deduce facts about a function f from information concerning its derivatives. At the end of Section 3.2 we discussed one instance of this principle by conjecturing that if f has a positive derivative, then it is an increasing function. Here we prove that fact and also see how the second derivative of a function influences the shape of its graph.

Lagrange and the Mean Value Theorem

The Mean Value Theorem was first formulated by Joseph-Louis Lagrange (1736–1813), born in Italy of a French father and an Italian mother. He was a child prodigy and became a professor in Turin at the tender age of 19. Lagrange made great contributions to number theory, theory of functions, theory of equations, and analytical and celestial mechanics. In particular, he applied calculus to the analysis of the stability of the solar system. At the invitation of Frederick the Great, he succeeded Euler at the Berlin Academy and, when Frederick died, Lagrange accepted King Louis XVI's invitation to Paris, where he was given apartments in the Louvre and became a professor at the Ecole Polytechnique. Despite all the trappings of luxury and fame, he was a kind and quiet man, living only for science.

The Mean Value Theorem

We start with a fact, known as the Mean Value Theorem, that will be useful not only for present purposes but also for explaining why some of the other basic results of calculus are true.

The Mean Value Theorem If f is a differentiable function on the interval $[a, b]$, then there exists a number c between a and b such that

(1) $$f'(c) = \frac{f(b) - f(a)}{b - a}$$

or, equivalently,

(2) $$f(b) - f(a) = f'(c)(b - a)$$

We can see that this theorem is reasonable by interpreting it geometrically. Figures 1 and 2 show the points $A(a, f(a))$ and $B(b, f(b))$ on the graphs of two differentiable functions.

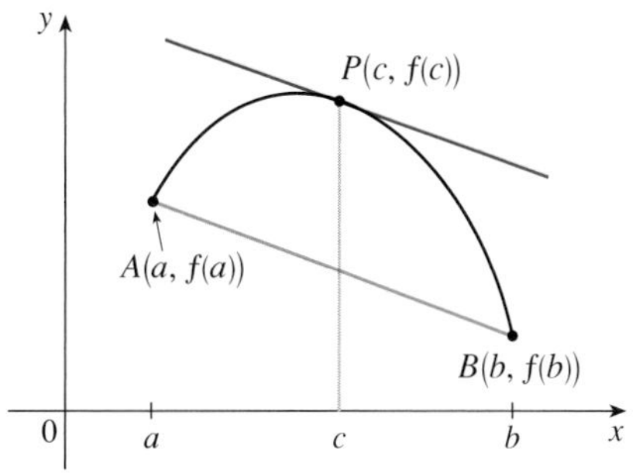

FIGURE 1

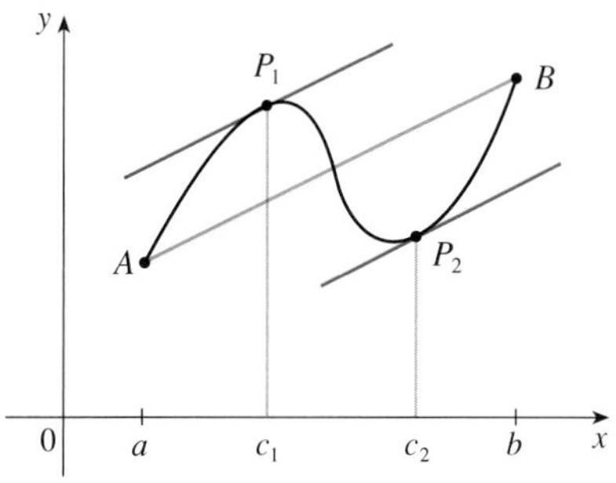

FIGURE 2

The slope of the secant line AB is

$$m_{AB} = \frac{f(b) - f(a)}{b - a}$$

which is the same expression as on the right side of Equation 1. Since $f'(c)$ is the slope of the tangent line at the point $(c, f(c))$, the Mean Value Theorem, in the form given by

Equation 1, says that there is at least one point $P(c, f(c))$ on the graph where the slope of the tangent line is the same as the slope of the secant line AB. In other words, there is a point P where the tangent line is parallel to the secant line AB. It seems clear that there is one such point P in Figure 1 and two such points P_1 and P_2 in Figure 2. (Imagine a line parallel to AB, starting far away and moving parallel to itself until it touches the graph for the first time.)

Because our intuition tells us that the Mean Value Theorem is true, we take it as the starting point for the development of the main facts of calculus. (When calculus is developed from first principles, however, the Mean Value Theorem is proved as a consequence of the axioms that define the real number system.)

EXAMPLE 1 | If an object moves in a straight line with position function $s = f(t)$, then the average velocity between $t = a$ and $t = b$ is

$$\frac{f(b) - f(a)}{b - a}$$

and the velocity at $t = c$ is $f'(c)$. Thus the Mean Value Theorem (in the form of Equation 1) tells us that at some time $t = c$ between a and b the instantaneous velocity $f'(c)$ is equal to that average velocity. For instance, if a car traveled 180 km in 2 hours, then the speedometer must have read 90 km/h at least once. ■

EXAMPLE 2 | BB **Compensatory growth** Experiments have been conducted in which individuals are deprived of food for a period of time during development and then placed back on a normal diet (see Figure 3). These experimental subjects display a period of reduced growth during the food deprivation, followed by a period of *compensatory growth* in which they catch up in size to individuals on a normal diet.
(a) Prove that there is always a time when an individual on a normal diet is growing at a rate equal to its average growth rate over the development period.
(b) Prove that there is always a time when an individual on a food-deprived diet is growing at a rate equal to its average growth rate over the development period.
(c) Given an example of each type of individual, show that there is a time after the point t_1 when food deprivation starts when the growth rate is nevertheless the same for both of them.

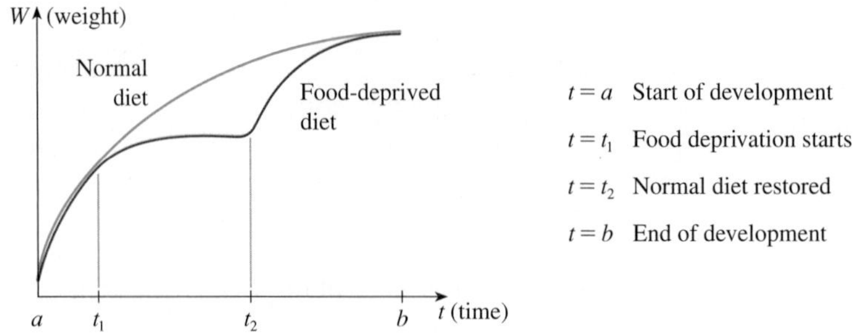

FIGURE 3

SOLUTION

(a) Let $W(t)$ be the weight of a specific individual on a normal diet at time t. Then W is a continuous function and is even differentiable. Let $t = a$ and $t = b$ be times at the beginning and end of the development period. By the Mean Value Theorem, there is a number c such that

$$W'(c) = \frac{W(b) - W(a)}{b - a}$$

This equation says that at time $t = c$ the instantaneous rate of growth is equal to the average rate of growth.

(b) Exactly the same argument as in part (a) applies here. Because of the shape of the graph in this case, the time c will probably be different from the time in part (a).

(c) Let W_1 and W_2 be the weight functions for the normal and food-deprived individuals. Then the difference function $D(t) = W_1(t) - W_2(t)$ is differentiable and $D(t_1) = W_1(t_1) - W_2(t_1) = 0$ because the two individuals start with the same weight. Similarly, $D(b) = 0$. By the Mean Value Theorem there is a time c such that

$$D'(c) = \frac{D(b) - D(t_1)}{b - t_1} = \frac{0 - 0}{b - t_1} = 0$$

But $D'(c) = W_1'(c) - W_2'(c)$ and so $W_1'(c) = W_2'(c)$. So at time $t = c$ the two individuals were growing at the same rate. ■

The main significance of the Mean Value Theorem is that it enables us to obtain information about a function from information about its derivative. Our immediate use of this principle is to prove the basic facts concerning increasing and decreasing functions.

■ Increasing and Decreasing Functions

In Section 1.1 we defined increasing functions and decreasing functions and in Section 3.2 we observed from graphs that a function with a positive derivative is increasing. We now deduce this fact from the Mean Value Theorem.

Let's abbreviate the name of this test to the I/D Test.

Increasing/Decreasing Test

(a) If $f'(x) > 0$ on an interval, then f is increasing on that interval.

(b) If $f'(x) < 0$ on an interval, then f is decreasing on that interval.

PROOF

(a) Let x_1 and x_2 be any two numbers in the interval with $x_1 < x_2$. According to the definition of an increasing function (page 12), we have to show that $f(x_1) < f(x_2)$.

Because we are given that $f'(x) > 0$, we know that f is differentiable on $[x_1, x_2]$. So, by the Mean Value Theorem, there is a number c between x_1 and x_2 such that

$$f(x_2) - f(x_1) = f'(c)(x_2 - x_1) \qquad (3)$$

Now $f'(c) > 0$ by assumption and $x_2 - x_1 > 0$ because $x_1 < x_2$. Thus the right side of Equation 3 is positive, and so

$$f(x_2) - f(x_1) > 0 \qquad \text{or} \qquad f(x_1) < f(x_2)$$

This shows that f is increasing.

Part (b) is proved similarly. ■

EXAMPLE 3 | Find where the function $f(x) = 3x^4 - 4x^3 - 12x^2 + 5$ is increasing and where it is decreasing.

SOLUTION First we calculate the derivative of f:

$$f'(x) = 12x^3 - 12x^2 - 24x = 12x(x - 2)(x + 1)$$

To use the I/D Test we have to know where $f'(x) > 0$ and where $f'(x) < 0$. This depends on the signs of the three factors of $f'(x)$, namely, $12x$, $x - 2$, and $x + 1$. We divide the real line into intervals whose endpoints are the critical numbers -1, 0, and 2 and arrange our work in a chart. A plus sign indicates that the given expression is positive, and a minus sign indicates that it is negative. The last column of the chart gives the conclusion based on the I/D Test. For instance, $f'(x) < 0$ for $0 < x < 2$, so f is decreasing on $(0, 2)$. (It would also be true to say that f is decreasing on the closed interval $[0, 2]$.)

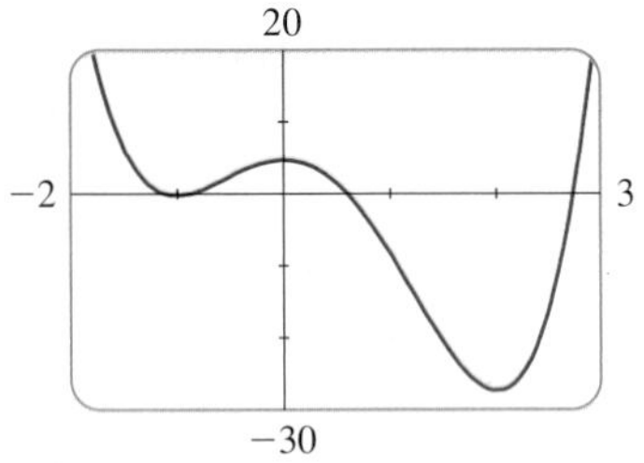

FIGURE 4

Interval	$12x$	$x - 2$	$x + 1$	$f'(x)$	f
$x < -1$	$-$	$-$	$-$	$-$	decreasing on $(-\infty, -1)$
$-1 < x < 0$	$-$	$-$	$+$	$+$	increasing on $(-1, 0)$
$0 < x < 2$	$+$	$-$	$+$	$-$	decreasing on $(0, 2)$
$x > 2$	$+$	$+$	$+$	$+$	increasing on $(2, \infty)$

The graph of f shown in Figure 4 confirms the information in the chart. ■

Recall from Section 4.1 that if f has a local maximum or minimum at c, then c must be a critical number of f (by Fermat's Theorem), but not every critical number gives rise to a maximum or a minimum. We therefore need a test that will tell us whether or not f has a local maximum or minimum at a critical number.

You can see from Figure 4 that $f(0) = 5$ is a local maximum value of f because f increases on $(-1, 0)$ and decreases on $(0, 2)$. Or, in terms of derivatives, $f'(x) > 0$ for $-1 < x < 0$ and $f'(x) < 0$ for $0 < x < 2$. In other words, the sign of $f'(x)$ changes from positive to negative at 0. This observation is the basis of the following test.

The First Derivative Test Suppose that c is a critical number of a continuous function f.

(a) If f' changes from positive to negative at c, then f has a local maximum at c.

(b) If f' changes from negative to positive at c, then f has a local minimum at c.

(c) If f' does not change sign at c (for example, if f' is positive on both sides of c or negative on both sides), then f has no local maximum or minimum at c.

The First Derivative Test is a consequence of the I/D Test. In part (a), for instance, since the sign of $f'(x)$ changes from positive to negative at c, f is increasing to the left of c and decreasing to the right of c. It follows that f has a local maximum at c.

It is easy to remember the First Derivative Test by visualizing diagrams such as those in Figure 5.

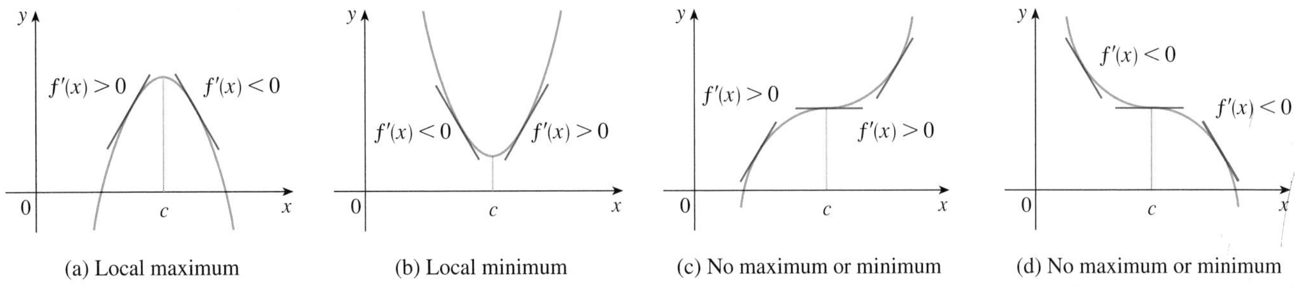

FIGURE 5

EXAMPLE 4 | Find the local minimum and maximum values of the function f in Example 3.

SOLUTION From the chart in the solution to Example 3 we see that $f'(x)$ changes from negative to positive at -1, so $f(-1) = 0$ is a local minimum value by the First Derivative Test. Similarly, f' changes from negative to positive at 2, so $f(2) = -27$ is also a local minimum value. As previously noted, $f(0) = 5$ is a local maximum value because $f'(x)$ changes from positive to negative at 0. ■

■ Concavity

Let's see how the sign of $f''(x)$ affects the appearance of the graph of f. Since $f'' = (f')'$, we know that if $f''(x)$ is positive, then f' is an increasing function. This says that the slopes of the tangent lines of the curve $y = f(x)$ increase from left to right. Figure 6 shows the graph of such a function. The slope of this curve becomes progressively larger as x increases and we observe that, as a consequence, the curve bends upward. [It can be proved that this is equivalent to saying that the graph of f lies above all of its tangent lines.] Such a curve is called **concave upward**. In Figure 7, however, $f''(x)$ is negative, which means that f' is decreasing. Thus the slopes of f decrease from left to right and the curve bends downward. This curve is called **concave downward**. We summarize our discussion as follows.

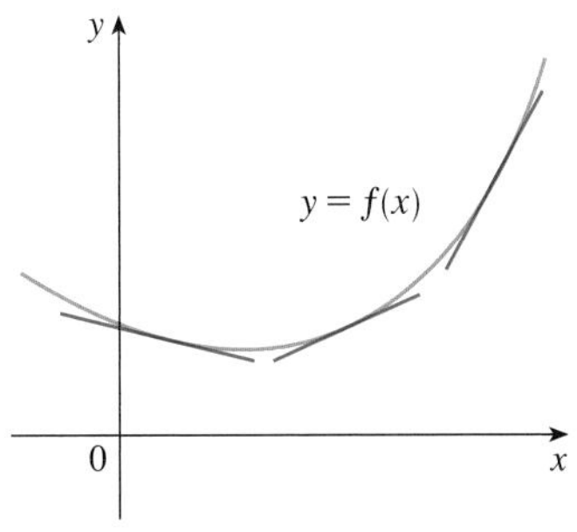

FIGURE 6
Since $f''(x) > 0$, the slopes increase and f is concave upward.

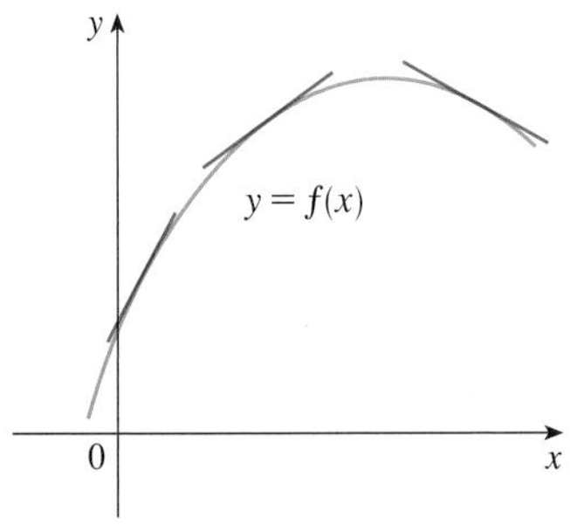

FIGURE 7
Since $f''(x) < 0$, the slopes decrease and f is concave downward.

Concavity Test

(a) If $f''(x) > 0$ for all x in I, then the graph of f is concave upward on I.

(b) If $f''(x) < 0$ for all x in I, then the graph of f is concave downward on I.

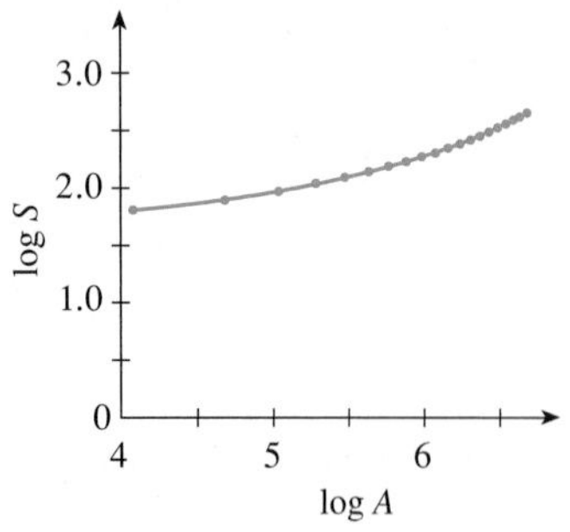

FIGURE 8

EXAMPLE 5 | BB Species–area relationship In Example 1.5.14 we explored the notion of species–area relationships by plotting the number of species S of bats against the areas A of their caves in central Mexico. We found that a log-log plot ($\log S$ as a function of $\log A$) was approximately linear and so a power model (for S as a function of A) was appropriate. We also found that a log-log plot was linear for the number of tree species in a given area of a rain forest in Malaysia (see Exercise 1.5.67). In both cases this means that, on a log scale, the rate of increase in number of species with an increase in area is the same no matter what the area is.

A recent study, however, has demonstrated that log-log plots are no longer linear when we consider areas that are substantially larger.[1] Figure 8 (from that study) shows a plot of $\log S$ versus $\log A$ for amphibians in Africa. You can see that $\log S$ is not a linear function of $\log A$ because its graph is concave upward. This means that, on a log scale, the rate of increase in the number of amphibian species with an increase in area is larger for big areas than for small ones. Put another way, the species–area relationship "accelerates" as the area increases. ■

EXAMPLE 6 | BB Risk aversion in junco foraging[2] Different junco habitats yield different amounts of seeds, and individuals can choose which habitat to feed in. The amount of energy reward E obtained from feeding in different habitats increases with the seed abundance s in the habitat but it does so at a decelerating rate (see Figure 9).

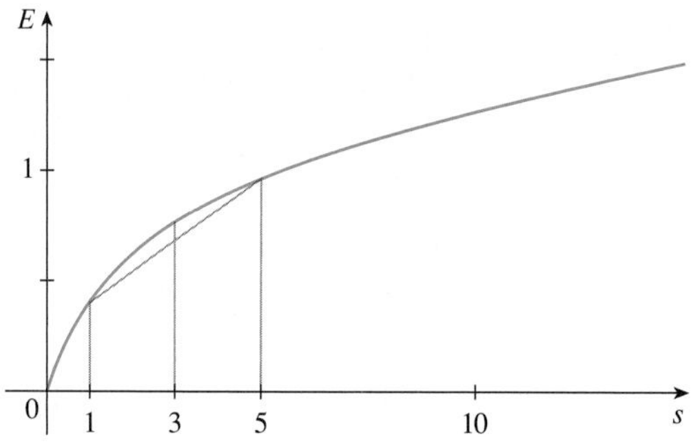

FIGURE 9

Suppose a bird can choose to feed exclusively in a habitat with $s = 3$ seeds per unit area or it can divide its time equally between two habitats with 1 and 5 seeds per unit area, respectively. For both choices the bird experiences an average of 3 seeds per unit area. Which choice provides the greatest energy reward?

SOLUTION The function $E(s)$ graphed in Figure 9 gives the energy reward as a function of seed density. Because the graph of E is concave downward, it lies below its tangent lines and above its secant lines. If we draw the secant line from $(1, E(1))$ to $(5, E(5))$, it will lie *below* the curve. The height of the secant line when $s = 3$ is the average of the heights when $s = 1$ and $s = 5$. Therefore

$$\frac{E(1) + E(5)}{2} < E(3)$$

1. D. Storch et al., "Universal Species–Area and Endemics–Area Relationships at Continental Scales," *Nature* 488 (2012): 78–83.
2. T. Caraco et al., "An Empirical Demonstration of Risk-Sensitive Foraging Preferences," *Animal Behavior* 28 (1980): 820–30.

This means that the junco gets more energy reward in a habitat with 3 seeds per unit area than it does by splitting its time between habitats with 1 and 5 seeds per unit area. ■

Figure 10 shows the graph of a function that is concave upward (abbreviated CU) on the intervals (b, c), (d, e), and (e, p) and concave downward (CD) on the intervals (a, b), (c, d), and (p, q).

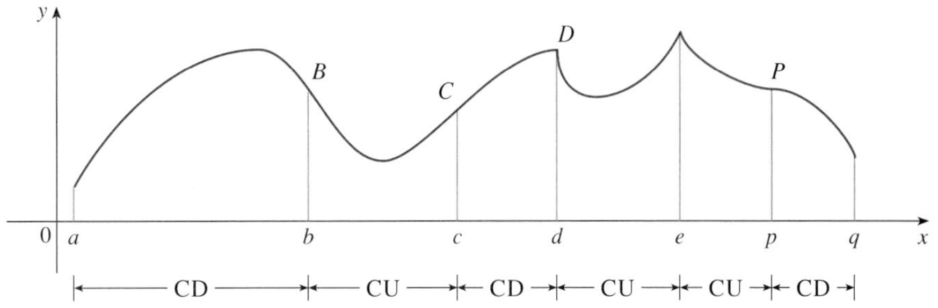

FIGURE 10

Notice in Figure 10 that the curve changes its direction of concavity when $x = b$, c, d, and p. The corresponding points on the curve (B, C, D, and P) are called *inflection points*. In general, a point P on a curve $y = f(x)$ is called an **inflection point** if f is continuous there and the curve changes from concave upward to concave downward or from concave downward to concave upward at P.

In view of the Concavity Test, there is a point of inflection at any point where the second derivative changes sign.

EXAMPLE 7 | Sketch a possible graph of a function f that satisfies the following conditions:

(i) $f'(x) > 0$ on $(-\infty, 1)$, $f'(x) < 0$ on $(1, \infty)$

(ii) $f''(x) > 0$ on $(-\infty, -2)$ and $(2, \infty)$, $f''(x) < 0$ on $(-2, 2)$

(iii) $\lim_{x \to -\infty} f(x) = -2$, $\lim_{x \to \infty} f(x) = 0$

SOLUTION Condition (i) tells us that f is increasing on $(-\infty, 1)$ and decreasing on $(1, \infty)$. Condition (ii) says that f is concave upward on $(-\infty, -2)$ and $(2, \infty)$, and concave downward on $(-2, 2)$. From condition (iii) we know that the graph of f has two horizontal asymptotes: $y = -2$ and $y = 0$.

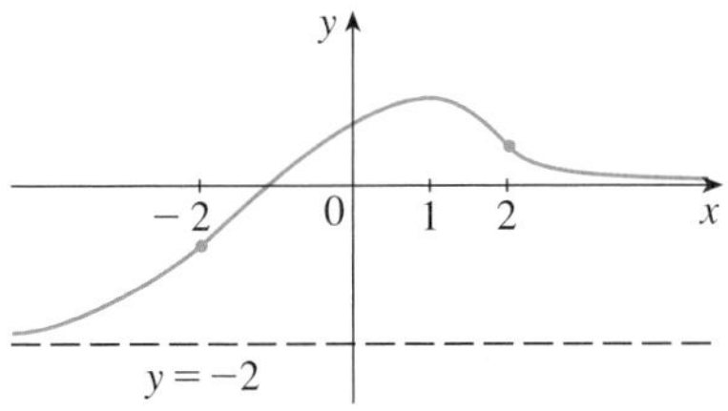

FIGURE 11

We first draw the horizontal asymptote $y = -2$ as a dashed line (see Figure 11). We then draw the graph of f approaching this asymptote at the far left, increasing to its maximum point at $x = 1$ and decreasing toward the x-axis as at the far right. We also make sure that the graph has inflection points when $x = -2$ and 2. Notice that we made the curve bend upward for $x < -2$ and $x > 2$, and bend downward when x is between -2 and 2. ■

A consequence of the Concavity Test is the following test for maximum and minimum values.

The Second Derivative Test Suppose f'' is continuous near c.

(a) If $f'(c) = 0$ and $f''(c) > 0$, then f has a local minimum at c.

(b) If $f'(c) = 0$ and $f''(c) < 0$, then f has a local maximum at c.

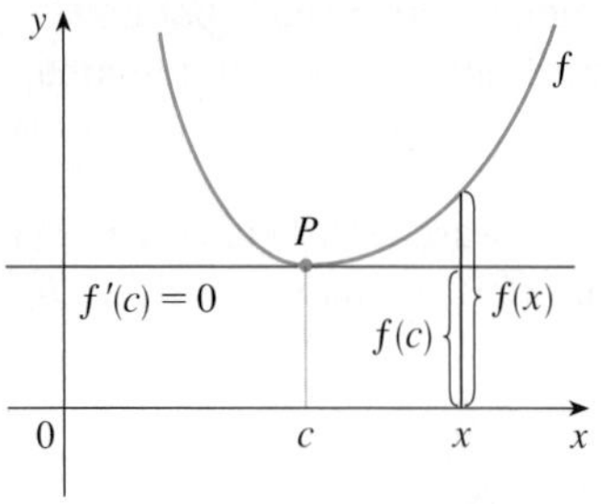

FIGURE 12
$f''(x) > 0$, f is concave upward

For instance, part (a) is true because $f''(x) > 0$ near c and so f is concave upward near c. This means that the graph of f lies *above* its horizontal tangent at c and so f has a local minimum at c. (See Figure 12.)

EXAMPLE 8 | Discuss the curve $y = x^4 - 4x^3$ with respect to concavity, points of inflection, and local maxima and minima. Use this information to sketch the curve.

SOLUTION If $f(x) = x^4 - 4x^3$, then

$$f'(x) = 4x^3 - 12x^2 = 4x^2(x - 3)$$

$$f''(x) = 12x^2 - 24x = 12x(x - 2)$$

To find the critical numbers we set $f'(x) = 0$ and obtain $x = 0$ and $x = 3$. To use the Second Derivative Test we evaluate f'' at these critical numbers:

$$f''(0) = 0 \qquad f''(3) = 36 > 0$$

Since $f'(3) = 0$ and $f''(3) > 0$, $f(3) = -27$ is a local minimum. Since $f''(0) = 0$, the Second Derivative Test gives no information about the critical number 0. But since $f'(x) < 0$ for $x < 0$ and also for $0 < x < 3$, the First Derivative Test tells us that f does not have a local maximum or minimum at 0.

Since $f''(x) = 0$ when $x = 0$ or 2, we divide the real line into intervals with these numbers as endpoints and complete the following chart.

Interval	$f''(x) = 12x(x - 2)$	Concavity
$(-\infty, 0)$	+	upward
$(0, 2)$	−	downward
$(2, \infty)$	+	upward

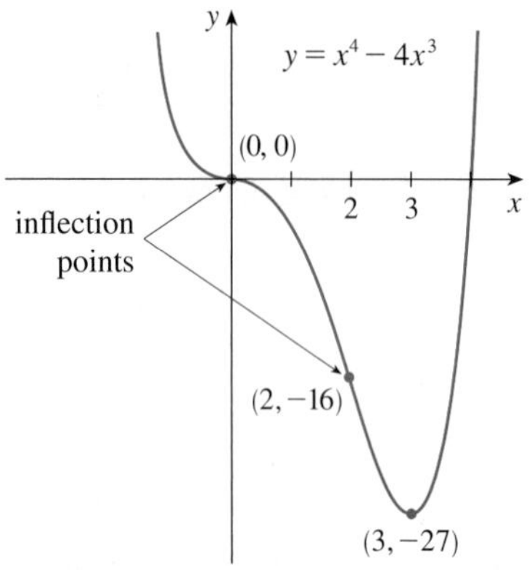

FIGURE 13

The point (0, 0) is an inflection point since the curve changes from concave upward to concave downward there. Also $(2, -16)$ is an inflection point since the curve changes from concave downward to concave upward there.

Using the local minimum, the intervals of concavity, and the inflection points, we sketch the curve in Figure 13. ■

NOTE The Second Derivative Test is inconclusive when $f''(c) = 0$. In other words, at such a point there might be a maximum, there might be a minimum, or there might be neither (as in Example 8). This test also fails when $f''(c)$ does not exist. In such cases the First Derivative Test must be used. In fact, even when both tests apply, the First Derivative Test is often the easier one to use.

EXAMPLE 9 | Sketch the graph of the function $f(x) = x^{2/3}(6 - x)^{1/3}$.

SOLUTION Calculation of the first two derivatives gives

Use the differentiation rules to check these calculations.

$$f'(x) = \frac{4 - x}{x^{1/3}(6 - x)^{2/3}} \qquad f''(x) = \frac{-8}{x^{4/3}(6 - x)^{5/3}}$$

Since $f'(x) = 0$ when $x = 4$ and $f'(x)$ does not exist when $x = 0$ or $x = 6$, the critical numbers are 0, 4, and 6.

Interval	$4 - x$	$x^{1/3}$	$(6 - x)^{2/3}$	$f'(x)$	f
$x < 0$	$+$	$-$	$+$	$-$	decreasing on $(-\infty, 0)$
$0 < x < 4$	$+$	$+$	$+$	$+$	increasing on $(0, 4)$
$4 < x < 6$	$-$	$+$	$+$	$-$	decreasing on $(4, 6)$
$x > 6$	$-$	$+$	$+$	$-$	decreasing on $(6, \infty)$

To find the local extreme values we use the First Derivative Test. Because f' changes from negative to positive at 0, $f(0) = 0$ is a local minimum. Since f' changes from positive to negative at 4, $f(4) = 2^{5/3}$ is a local maximum. The sign of f' does not change at 6, so there is no minimum or maximum there. (The Second Derivative Test could be used at 4 but not at 0 or 6 because f'' does not exist at either of these numbers.)

Looking at the expression for $f''(x)$ and noting that $x^{4/3} \geqslant 0$ for all x, we have $f''(x) < 0$ for $x < 0$ and for $0 < x < 6$ and $f''(x) > 0$ for $x > 6$. So f is concave downward on $(-\infty, 0)$ and $(0, 6)$ and concave upward on $(6, \infty)$, and the only inflection point is $(6, 0)$. The graph is sketched in Figure 14. Note that the curve has vertical tangents at $(0, 0)$ and $(6, 0)$ because $|f'(x)| \to \infty$ as $x \to 0$ and as $x \to 6$. ■

Try reproducing the graph in Figure 14 with a graphing calculator or computer. Some machines produce the complete graph, some produce only the portion to the right of the y-axis, and some produce only the portion between $x = 0$ and $x = 6$. An equivalent expression that gives the correct graph is

$$y = (x^2)^{1/3} \cdot \frac{6 - x}{|6 - x|} |6 - x|^{1/3}$$

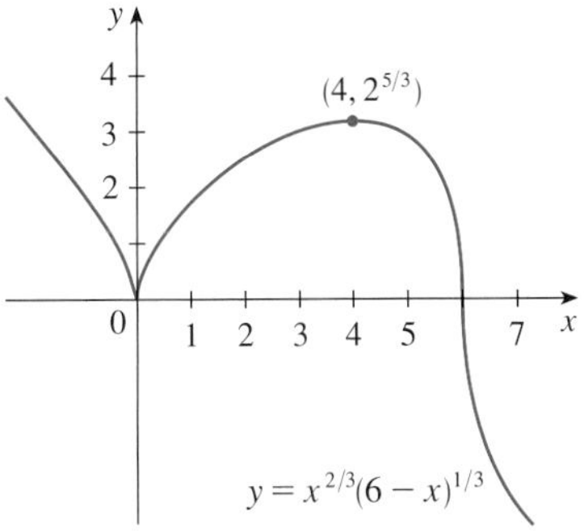

FIGURE 14

EXAMPLE 10 | Bee population A population of honeybees raised in an apiary started with 50 bees at time $t = 0$ and was modeled by the function

$$P(t) = \frac{75{,}200}{1 + 1503e^{-0.5932t}}$$

where t is the time in weeks, $0 \leqslant t \leqslant 25$. Use a graph to estimate the time at which the bee population was growing fastest. Then use derivatives to give a more accurate estimate.

SOLUTION The population grows fastest when the population curve $y = P(t)$ has the steepest tangent line. From the graph of P in Figure 15, we estimate that the steepest tangent occurs when $t \approx 12$, so the bee population was growing most rapidly after about 12 weeks.

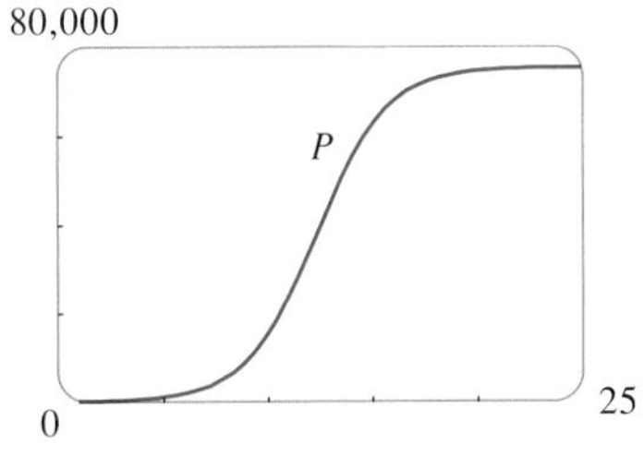

FIGURE 15

For a better estimate we calculate the derivative $P'(t)$, which is the rate of increase of the bee population:

$$P'(t) = \frac{67{,}046{,}785.92e^{-0.5932t}}{(1 + 1503e^{-0.5932t})^2}$$

We graph P' in Figure 16 and observe that P' has its maximum value when $t \approx 12.3$.

To get a still better estimate we note that f' has its maximum value when f' changes from increasing to decreasing. This happens when f changes from concave upward to concave downward, that is, when f has an inflection point. So we ask a CAS to compute the second derivative:

$$P''(t) \approx \frac{119{,}555{,}093{,}144e^{-1.1864t}}{(1 + 1503e^{-0.5932t})^3} - \frac{39{,}772{,}153e^{-0.5932t}}{(1 + 1503e^{-0.5932t})^2}$$

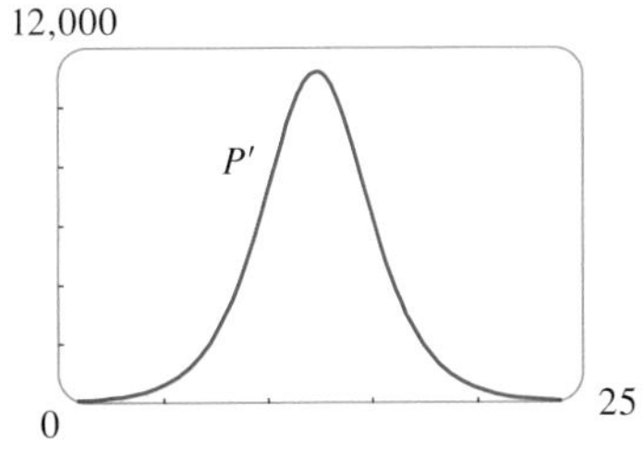

FIGURE 16

We could plot this function to see where it changes from positive to negative, but instead let's have the CAS solve the equation $P''(t) = 0$. It gives the answer $t \approx 12.3318$. ■

Graphing with Technology

When we use technology to graph a curve, our strategy is different from the one we've been using until now. Here we *start* with a graph produced by a graphing calculator or computer and then we refine it. We use calculus to make sure that we reveal all the important aspects of the curve. And with the use of graphing devices we can tackle curves that would be far too complicated to consider without technology.

EXAMPLE 11 | Graph the polynomial $f(x) = 2x^6 + 3x^5 + 3x^3 - 2x^2$. Use the graphs of f' and f'' to estimate all maximum and minimum points and intervals of concavity.

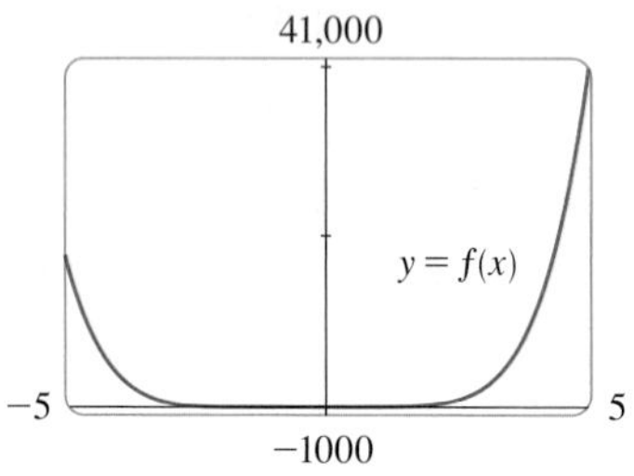

FIGURE 17

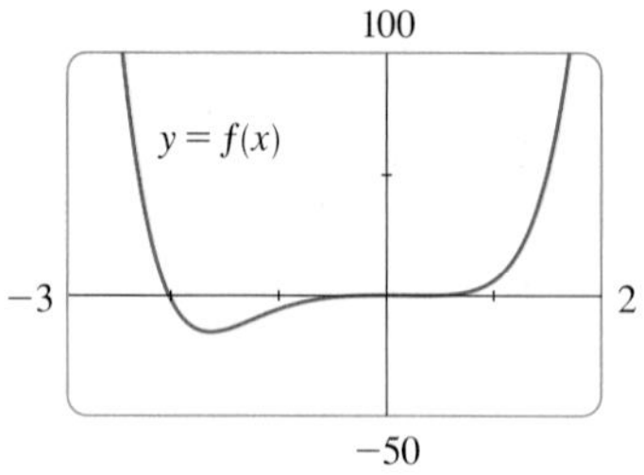

FIGURE 18

SOLUTION If we specify a domain but not a range, many graphing devices will deduce a suitable range from the values computed. Figure 17 shows the plot from one such device if we specify that $-5 \leqslant x \leqslant 5$. Although this viewing rectangle is useful for showing that the asymptotic behavior (or end behavior) is the same as for $y = 2x^6$, it is obviously hiding some finer detail. So we change to the viewing rectangle $[-3, 2]$ by $[-50, 100]$ shown in Figure 18.

From this graph it appears that there is an absolute minimum value of about -15.33 when $x \approx -1.62$ (by using the cursor) and f is decreasing on $(-\infty, -1.62)$ and increasing on $(-1.62, \infty)$. Also there appears to be a horizontal tangent at the origin and inflection points when $x = 0$ and when x is somewhere between -2 and -1.

Now let's try to confirm these impressions using calculus. We differentiate and get

$$f'(x) = 12x^5 + 15x^4 + 9x^2 - 4x$$

$$f''(x) = 60x^4 + 60x^3 + 18x - 4$$

When we graph f' in Figure 19 we see that $f'(x)$ changes from negative to positive when $x \approx -1.62$; this confirms (by the First Derivative Test) the minimum value that we found earlier. But, perhaps to our surprise, we also notice that $f'(x)$ changes from positive to negative when $x = 0$ and from negative to positive when $x \approx 0.35$. This means that f has a local maximum at 0 and a local minimum when $x \approx 0.35$, but these were hidden in Figure 18. Indeed, if we now zoom in toward the origin in Figure 20, we see what we missed before: a local maximum value of 0 when $x = 0$ and a local minimum value of about -0.1 when $x \approx 0.35$.

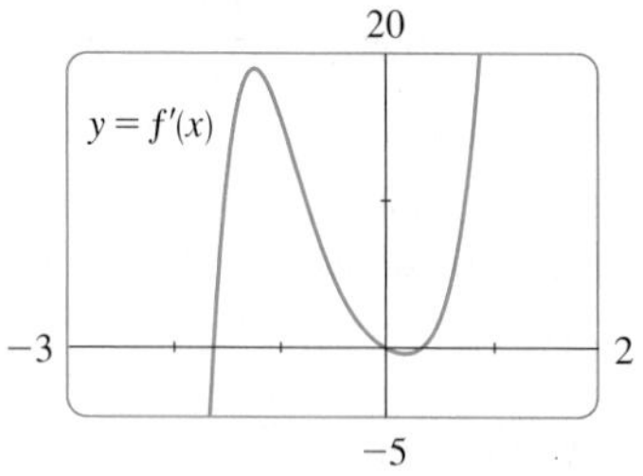

FIGURE 19

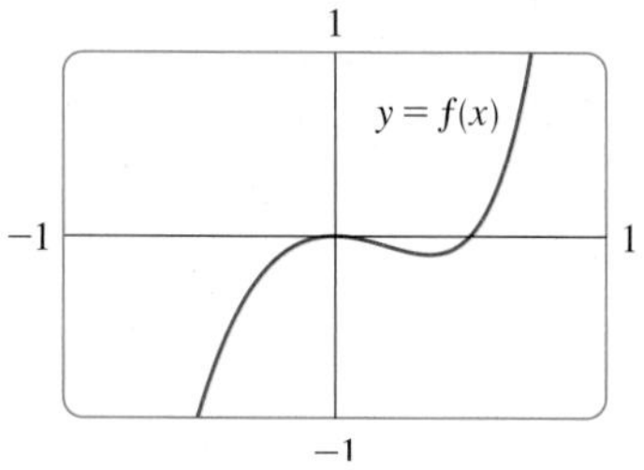

FIGURE 20

What about concavity and inflection points? From Figures 18 and 20 there appear to be inflection points when x is a little to the left of -1 and when x is a little to the right of 0. But it's difficult to determine inflection points from the graph of f, so we graph the second derivative f'' in Figure 21. We see that f'' changes from positive to negative when $x \approx -1.23$ and from negative to positive when $x \approx 0.19$. So, correct to two decimal places, f is concave upward on $(-\infty, -1.23)$ and $(0.19, \infty)$ and concave downward on $(-1.23, 0.19)$. The inflection points are $(-1.23, -10.18)$ and $(0.19, -0.05)$.

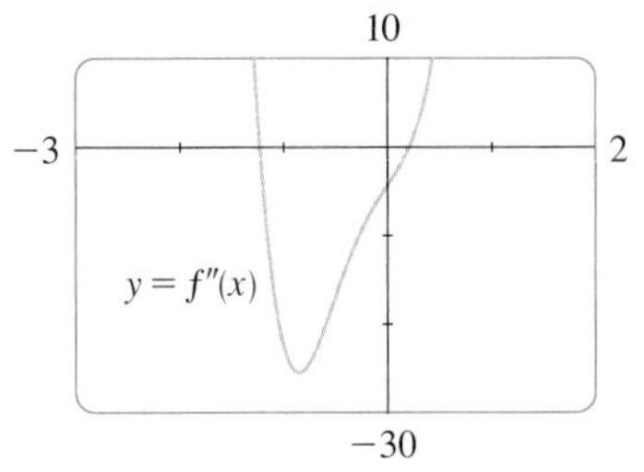

FIGURE 21

We have discovered that no single graph reveals all the important features of this polynomial. But Figures 18 and 20, when taken together, do provide an accurate picture. ■

EXERCISES 4.2

1. Use the graph of f to estimate the values of c that satisfy the conclusion of the Mean Value Theorem for the interval $[0, 8]$.

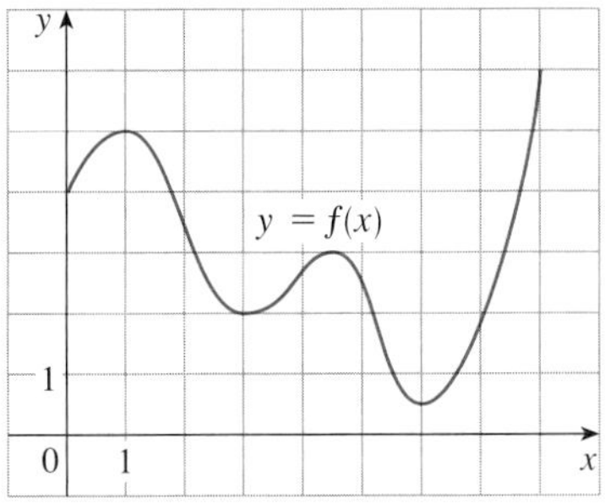

2. Foraging Many animals forage on resources that are distributed in discrete patches. For example, bumblebees visit many flowers, foraging on nectar from each. The amount of nectar $N(t)$ consumed from any flower increases with the amount of time spent at that flower, but with diminishing returns, as illustrated.

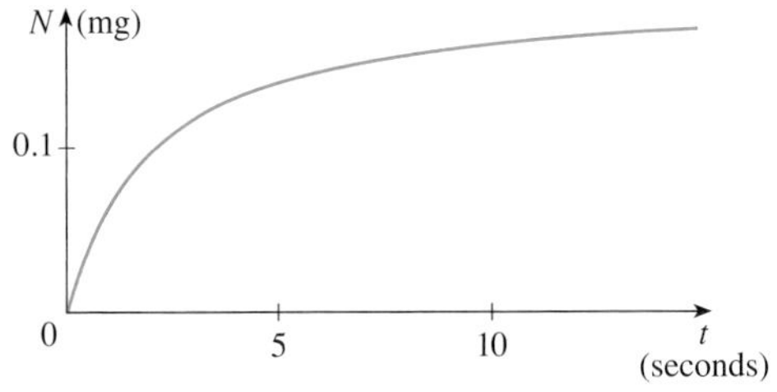

(a) What does this mean about the first and second derivatives of N?

(b) What is the average rate at which nectar is consumed over the first 10 seconds?

(c) The Mean Value Theorem tells us that there exists a time at which the instantaneous rate of nectar consumption is equal to the average found in part (b). Illustrate this idea graphically and estimate the time at which this occurs.

3. Suppose that $3 \leqslant f'(x) \leqslant 5$ for all values of x. Show that $18 \leqslant f(8) - f(2) \leqslant 30$.

4–5 Use the given graph of f to find the following.

(a) The open intervals on which f is increasing.

(b) The open intervals on which f is decreasing.

(c) The open intervals on which f is concave upward.

(d) The open intervals on which f is concave downward.

(e) The coordinates of the points of inflection.

4.

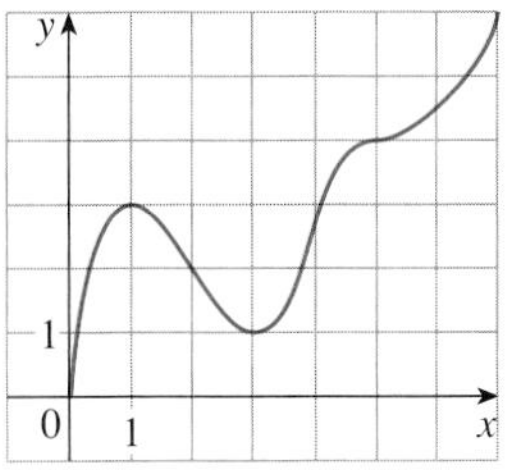

5.

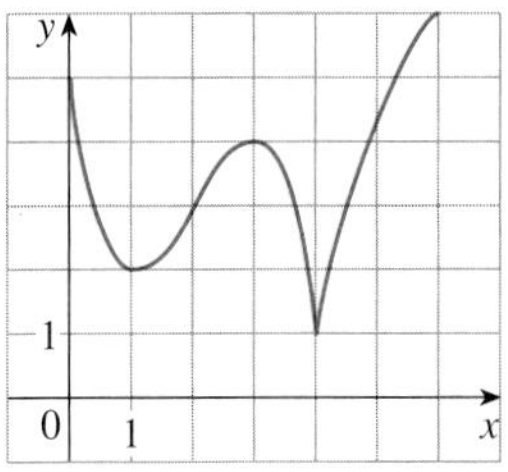

6. Suppose you are given a formula for a function f.

(a) How do you determine where f is increasing or decreasing?

(b) How do you determine where the graph of f is concave upward or concave downward?

(c) How do you locate inflection points?

7. (a) State the First Derivative Test.
(b) State the Second Derivative Test. Under what circumstances is it inconclusive? What do you do if it fails?

8. The graph of the first derivative f' of a function f is shown.
(a) On what intervals is f increasing? Explain.
(b) At what values of x does f have a local maximum or minimum? Explain.
(c) On what intervals is f concave upward or concave downward? Explain.
(d) What are the x-coordinates of the inflection points of f? Why?

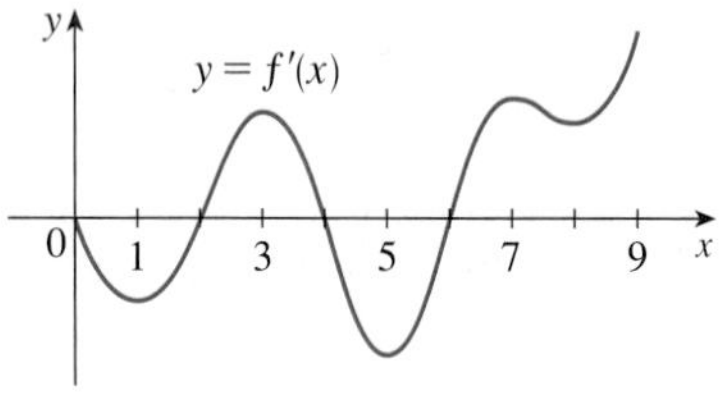

9. In each part state the x-coordinates of the inflection points of f. Give reasons for your answers.
(a) The curve is the graph of f.
(b) The curve is the graph of f'.
(c) The curve is the graph of f''.

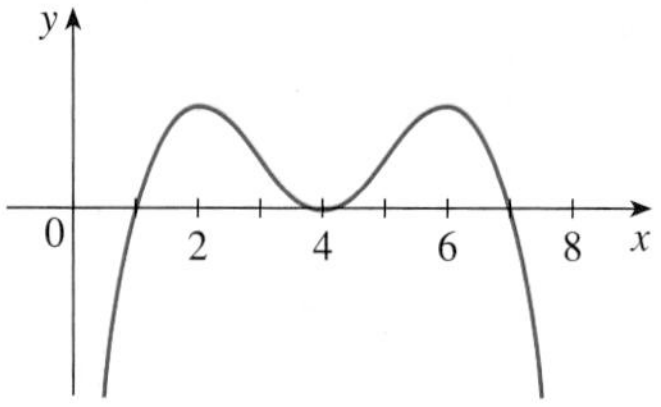

10. HIV prevalence The table gives the number of HIV-infected men in San Francisco from 1982 to 1991.

Year	Number of infections	Year	Number of infections
1982	80	1987	3500
1983	300	1988	4500
1984	700	1989	6000
1985	1500	1990	7200
1986	2500	1991	9000

(a) If $H(t)$ is the number of infected men at time t, plot the values of $H(t)$. What does the direction of concavity appear to be? Provide a biological interpretation.
(b) Use the table to construct a table of estimated values for $H'(t)$.
(c) Use the table of values of $H'(t)$ in part (b) to construct a table of values for $H''(t)$. Do the values corroborate your answer to part (a)?

11–20
(a) Find the intervals on which f is increasing or decreasing.
(b) Find the local maximum and minimum values of f.
(c) Find the intervals of concavity and the inflection points.

11. $f(x) = 2x^3 + 3x^2 - 36x$

12. $f(x) = 4x^3 + 3x^2 - 6x + 1$

13. $f(x) = x^4 - 2x^2 + 3$

14. $f(x) = \dfrac{x^2}{x^2 + 3}$

15. $f(x) = \sin x + \cos x, \quad 0 \leqslant x \leqslant 2\pi$

16. $f(x) = \cos^2 x - 2 \sin x, \quad 0 \leqslant x \leqslant 2\pi$

17. $f(x) = e^{2x} + e^{-x}$

18. $f(x) = x^2 \ln x$

19. $f(x) = (\ln x)/\sqrt{x}$

20. $f(x) = \sqrt{x}\, e^{-x}$

21–22 Find the local maximum and minimum values of f using both the First and Second Derivative Tests. Which method do you prefer?

21. $f(x) = x + \sqrt{1 - x}$

22. $f(x) = \dfrac{x}{x^2 + 4}$

23. Suppose f'' is continuous on $(-\infty, \infty)$.
(a) If $f'(2) = 0$ and $f''(2) = -5$, what can you say about f?
(b) If $f'(6) = 0$ and $f''(6) = 0$, what can you say about f?

24. (a) Find the critical numbers of $f(x) = x^4(x - 1)^3$.
(b) What does the Second Derivative Test tell you about the behavior of f at these critical numbers?
(c) What does the First Derivative Test tell you?

25–36
(a) Find the intervals of increase or decrease.
(b) Find the local maximum and minimum values.
(c) Find the intervals of concavity and the inflection points.
(d) Use the information from parts (a)–(c) to sketch the graph. Check your work with a graphing device if you have one.

25. $f(x) = 2x^3 - 3x^2 - 12x$

26. $f(x) = 2 + 3x - x^3$

27. $f(x) = 2 + 2x^2 - x^4$

28. $g(x) = 200 + 8x^3 + x^4$

29. $h(x) = (x + 1)^5 - 5x - 2$

30. $h(x) = x^5 - 2x^3 + x$

31. $A(x) = x\sqrt{x + 3}$

32. $B(x) = 3x^{2/3} - x$

33. $C(x) = x^{1/3}(x + 4)$

34. $f(x) = \ln(x^4 + 27)$

35. $f(\theta) = 2 \cos \theta + \cos^2\theta, \quad 0 \leqslant \theta \leqslant 2\pi$

36. $f(t) = t + \cos t, \quad -2\pi \leqslant t \leqslant 2\pi$

37–44

(a) Find the vertical and horizontal asymptotes.
(b) Find the intervals of increase or decrease.
(c) Find the local maximum and minimum values.
(d) Find the intervals of concavity and the inflection points.
(e) Use the information from parts (a)–(d) to sketch the graph of f.

37. $f(x) = \dfrac{x^2}{x^2 - 1}$

38. $f(x) = \dfrac{x^2}{(x-2)^2}$

39. $f(x) = \sqrt{x^2 + 1} - x$

40. $f(x) = x \tan x, \quad -\pi/2 < x < \pi/2$

41. $f(x) = \ln(1 - \ln x)$

42. $f(x) = \dfrac{e^x}{1 + e^x}$

43. $f(x) = e^{-1/(x+1)}$

44. $f(x) = e^{\arctan x}$

---

45. Suppose the derivative of a function f is $f'(x) = (x+1)^2(x-3)^5(x-6)^4$. On what interval is f increasing?

46. Use the methods of this section to sketch the curve $y = x^3 - 3a^2x + 2a^3$, where a is a positive constant. What do the members of this family of curves have in common? How do they differ from each other?

47. Let $f(t)$ be the temperature at time t where you live and suppose that at time $t = 3$ you feel uncomfortably hot. How do you feel about the given data in each case?
(a) $f'(3) = 2, \quad f''(3) = 4$
(b) $f'(3) = 2, \quad f''(3) = -4$
(c) $f'(3) = -2, \quad f''(3) = 4$
(d) $f'(3) = -2, \quad f''(3) = -4$

48. Suppose $f(3) = 2$, $f'(3) = \frac{1}{2}$, and $f'(x) > 0$ and $f''(x) < 0$ for all x.
(a) Sketch a possible graph for f.
(b) How many solutions does the equation $f(x) = 0$ have? Why?
(c) Is it possible that $f'(2) = \frac{1}{3}$? Why?

49. Coffee is being poured into the mug shown in the figure at a constant rate (measured in volume per unit time). Sketch a rough graph of the depth of the coffee in the mug as a function of time. Account for the shape of the graph in terms of concavity. What is the significance of the inflection point?

50. Antibiotic pharmacokinetics Suppose that antibiotics are injected into a patient to treat a sinus infection. The antibiotics circulate in the blood, slowly diffusing into the sinus cavity while simultaneously being filtered out of the blood by the liver. In Chapter 10 we will derive a model for the concentration of the antibiotic in the sinus cavity as a function of time since the injection:

$$c(t) = \frac{e^{-\alpha t} - e^{-\beta t}}{\beta - \alpha} \quad \text{where } \beta > \alpha > 0$$

(a) At what time does c have its maximum value?
(b) At what time does the inflection point occur? What is the significance of the inflection point for the concentration function?
(c) Sketch the graph of c.

51. A **drug-loading curve** describes the level of medication in the bloodstream after a drug is administered. A *surge function* $S(t) = At^pe^{-kt}$ is often used to model the loading curve, reflecting an initial surge in the drug level and then a more gradual decline. If, for a particular drug, $A = 0.01$, $p = 4$, $k = 0.07$, and t is measured in minutes, estimate the times corresponding to the inflection points and explain their significance. If you have a graphing device, use it to graph the drug response curve.

52. Mutation accumulation When a population is subjected to a mutagen, the fraction of the population that contains at least one mutation increases with the duration of the exposure. A commonly used equation describing this fraction is $f(t) = 1 - e^{-\mu t}$, where μ is the mutation rate and is positive. Suppose we have two populations, A and B. Population A is subjected to the mutagen for 3 min whereas, with population B, half of the individuals are subjected to the mutagen for 2 min and the other half for 4 min. Which population will have the largest fraction of mutants? Explain your answer using derivatives.

53. A **dose response curve** in pharmacology is a plot of the effectiveness R of a drug as a function of the drug concentration c. Such curves typically increase with an S-shape, a simple mathematical model being

$$R(c) = \frac{c^2}{3 + c^2}$$

(a) At what drug concentration does the inflection point occur?
(b) Suppose we have two different treatment protocols, one where the concentration is held steady at 2 and another in which the concentration varies through time, spending equal amounts of time at 1.5 and 2.5. Which protocol would have the greater response?

54. The family of bell-shaped curves

$$y = \frac{1}{\sigma\sqrt{2\pi}} e^{-(x-\mu)^2/(2\sigma^2)}$$

occurs in probability and statistics, where it is called the *normal density function*. The constant μ is called the *mean* and the positive constant σ is called the *standard deviation*. For simplicity, let's scale the function so as to remove the factor $1/(\sigma\sqrt{2\pi})$ and let's analyze the special case where $\mu = 0$. So we study the function

$$f(x) = e^{-x^2/(2\sigma^2)}$$

(a) Find the asymptote, maximum value, and inflection points of f.
(b) What role does σ play in the shape of the curve?
(c) Illustrate by graphing four members of this family on the same screen.

55. In the theory of relativity, the mass of a particle is

$$m = \frac{m_0}{\sqrt{1 - v^2/c^2}}$$

where m_0 is the rest mass of the particle, m is the mass when the particle moves with speed v relative to the observer, and c is the speed of light. Sketch the graph of m as a function of v.

56. In the theory of relativity, the energy of a particle is

$$E = \sqrt{m_0^2c^4 + h^2c^2/\lambda^2}$$

where m_0 is the rest mass of the particle, λ is its wave length, and h is Planck's constant. Sketch the graph of E as a function of λ. What does the graph say about the energy?

57. Find a cubic function $f(x) = ax^3 + bx^2 + cx + d$ that has a local maximum value of 3 at $x = -2$ and a local minimum value of 0 at $x = 1$.

58. For what values of the numbers a and b does the function

$$f(x) = axe^{bx^2}$$

have the maximum value $f(2) = 1$?

59–62 Produce graphs of f that reveal all the important aspects of the curve. In particular, you should use graphs of f' and f'' to estimate the intervals of increase and decrease, extreme values, intervals of concavity, and inflection points.

59. $f(x) = 4x^4 - 32x^3 + 89x^2 - 95x + 29$

60. $f(x) = x^6 - 15x^5 + 75x^4 - 125x^3 - x$

61. $f(x) = x^2 - 4x + 7\cos x, \quad -4 \leqslant x \leqslant 4$

62. $f(x) = \tan x + 5\cos x$

63. Growth rate A 20-year-old university student weighs 138 lb and had a birth weight of 6 lb. Prove that at some point in her life she was growing at a rate of 6.6 pounds per year.

64. Antibiotic concentration Suppose an antibiotic is administered orally. It is first absorbed into the bloodstream, from which it passes into the sinus cavity. It is also metabolized from both sites. The concentrations $C_1(t)$ in the blood and $C_2(t)$ in the sinus cavity are shown. Prove that there is a time when the concentration in each site is increasing at the same rate. Prove that there is also a time when the concentration in each site is decreasing at the same rate.

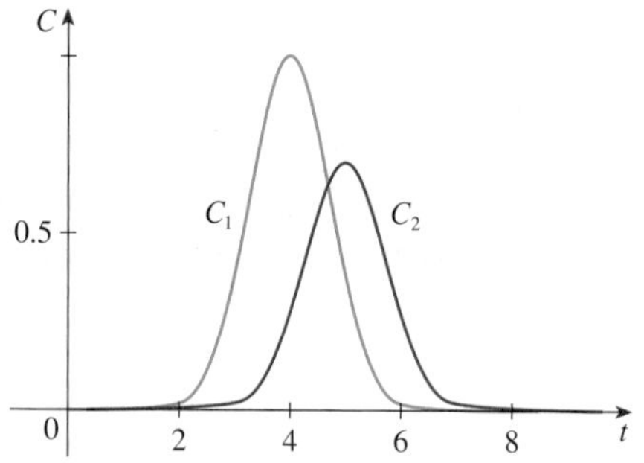

65. Show that a cubic function (a third-degree polynomial) always has exactly one point of inflection. If its graph has three x-intercepts x_1, x_2, and x_3, show that the x-coordinate of the inflection point is $(x_1 + x_2 + x_3)/3$.

66. For what values of c does the polynomial $P(x) = x^4 + cx^3 + x^2$ have two inflection points? One inflection point? None? Illustrate by graphing P for several values of c. How does the graph change as c decreases?

4.3 | L'Hospital's Rule: Comparing Rates of Growth

Indeterminate Quotients

Suppose we are trying to analyze the behavior of the function

$$F(x) = \frac{\ln x}{x - 1}$$

Although F is not defined when $x = 1$, we need to know how F behaves *near* 1. In particular, we would like to know the value of the limit

(1) $$\lim_{x \to 1} \frac{\ln x}{x - 1}$$

In computing this limit we can't apply Law 5 of limits (the limit of a quotient is the quotient of the limits, see Section 2.4) because the limit of the denominator is 0. In fact, although the limit in (1) exists, its value is not obvious because both numerator and denominator approach 0 and $\frac{0}{0}$ is not defined.

In general, if we have a limit of the form

$$\lim_{x \to a} \frac{f(x)}{g(x)}$$

where both $f(x) \to 0$ and $g(x) \to 0$ as $x \to a$, then this limit may or may not exist and is called an **indeterminate form of type $\frac{0}{0}$**. We met some limits of this type in Chapter 2. For rational functions, we can cancel common factors:

$$\lim_{x \to 1} \frac{x^2 - x}{x^2 - 1} = \lim_{x \to 1} \frac{x(x - 1)}{(x + 1)(x - 1)} = \lim_{x \to 1} \frac{x}{x + 1} = \frac{1}{2}$$

We used a geometric argument to show that

$$\lim_{x \to 0} \frac{\sin x}{x} = 1$$

But these methods do not work for limits such as (1), so in this section we introduce a systematic method, known as *l'Hospital's Rule,* for the evaluation of indeterminate forms.

Another situation in which a limit is not obvious occurs when we look for a horizontal asymptote of F and need to evaluate the limit

(2) $$\lim_{x \to \infty} \frac{\ln x}{x - 1}$$

It isn't obvious how to evaluate this limit because both numerator and denominator become large as $x \to \infty$. There is a struggle between numerator and denominator. If the numerator wins, the limit will be ∞; if the denominator wins, the answer will be 0. Or there may be some compromise, in which case the answer will be some finite positive number.

In general, if we have a limit of the form

$$\lim_{x \to a} \frac{f(x)}{g(x)}$$

where both $f(x) \to \infty$ (or $-\infty$) and $g(x) \to \infty$ (or $-\infty$), then the limit may or may not exist and is called an **indeterminate form of type ∞/∞**. We saw in Section 2.2 that this type of limit can be evaluated for certain functions, including rational functions, by dividing numerator and denominator by the highest power of x that occurs in the denominator. For instance,

$$\lim_{x \to \infty} \frac{x^2 - 1}{2x^2 + 1} = \lim_{x \to \infty} \frac{1 - \dfrac{1}{x^2}}{2 + \dfrac{1}{x^2}} = \frac{1 - 0}{2 + 0} = \frac{1}{2}$$

L'Hospital

L'Hospital's Rule is named after a French nobleman, the Marquis de l'Hospital (1661–1704), but was discovered by a Swiss mathematician, John Bernoulli (1667–1748). You might sometimes see l'Hospital spelled as l'Hôpital, but he spelled his own name l'Hospital, as was common in the 17th century. See Exercise 57 for the example that the Marquis used to illustrate his rule.

This method does not work for limits such as (2), but l'Hospital's Rule also applies to this type of indeterminate form.

L'Hospital's Rule Suppose f and g are differentiable and $g'(x) \neq 0$ near a (except possibly at a). Suppose that

$$\lim_{x \to a} f(x) = 0 \quad \text{and} \quad \lim_{x \to a} g(x) = 0$$

or that

$$\lim_{x \to a} f(x) = \pm\infty \quad \text{and} \quad \lim_{x \to a} g(x) = \pm\infty$$

(In other words, we have an indeterminate form of type $\frac{0}{0}$ or ∞/∞.) Then

$$\lim_{x \to a} \frac{f(x)}{g(x)} = \lim_{x \to a} \frac{f'(x)}{g'(x)}$$

if the limit on the right side exists (or is ∞ or $-\infty$).

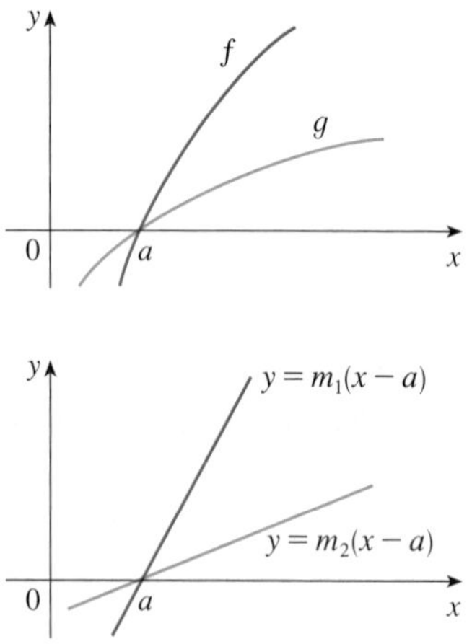

FIGURE 1

Figure 1 suggests visually why l'Hospital's Rule might be true. The first graph shows two differentiable functions f and g, each of which approaches 0 as $x \to a$. If we were to zoom in toward the point $(a, 0)$, the graphs would start to look almost linear. But if the functions actually were linear, as in the second graph, then their ratio would be

$$\frac{m_1(x - a)}{m_2(x - a)} = \frac{m_1}{m_2}$$

which is the ratio of their derivatives. This suggests that

$$\lim_{x \to a} \frac{f(x)}{g(x)} = \lim_{x \to a} \frac{f'(x)}{g'(x)}$$

NOTE 1 L'Hospital's Rule says that the limit of a quotient of functions is equal to the limit of the quotient of their derivatives, provided that the given conditions are satisfied. It is especially important to verify the conditions regarding the limits of f and g before using l'Hospital's Rule.

NOTE 2 L'Hospital's Rule is also valid for one-sided limits and for limits at infinity or negative infinity; that is, "$x \to a$" can be replaced by any of the symbols $x \to a^+$, $x \to a^-$, $x \to \infty$, or $x \to -\infty$.

NOTE 3 For the special case in which $f(a) = g(a) = 0$, f' and g' are continuous, and $g'(a) \neq 0$, it is easy to see why l'Hospital's Rule is true. In fact, using the alternative form of the definition of a derivative, we have

$$\lim_{x \to a} \frac{f'(x)}{g'(x)} = \frac{f'(a)}{g'(a)} = \frac{\displaystyle\lim_{x \to a} \frac{f(x) - f(a)}{x - a}}{\displaystyle\lim_{x \to a} \frac{g(x) - g(a)}{x - a}} = \lim_{x \to a} \frac{\dfrac{f(x) - f(a)}{x - a}}{\dfrac{g(x) - g(a)}{x - a}}$$

$$= \lim_{x \to a} \frac{f(x) - f(a)}{g(x) - g(a)} = \lim_{x \to a} \frac{f(x)}{g(x)}$$

The general version of l'Hospital's Rule is more difficult; its proof can be found in more advanced books.

EXAMPLE 1 | Find $\displaystyle\lim_{x \to 1} \frac{\ln x}{x - 1}$.

SOLUTION Since

$$\lim_{x \to 1} \ln x = \ln 1 = 0 \quad \text{and} \quad \lim_{x \to 1} (x - 1) = 0$$

we can apply l'Hospital's Rule:

Notice that when using l'Hospital's Rule we differentiate the numerator and denominator *separately*. We do *not* use the Quotient Rule.

$$\lim_{x \to 1} \frac{\ln x}{x - 1} = \lim_{x \to 1} \frac{\dfrac{d}{dx}(\ln x)}{\dfrac{d}{dx}(x - 1)} = \lim_{x \to 1} \frac{1/x}{1}$$

$$= \lim_{x \to 1} \frac{1}{x} = 1$$

EXAMPLE 2 | Calculate $\displaystyle\lim_{x \to \infty} \frac{e^x}{x^2}$.

SOLUTION We have $\lim_{x \to \infty} e^x = \infty$ and $\lim_{x \to \infty} x^2 = \infty$, so l'Hospital's Rule gives

$$\lim_{x \to \infty} \frac{e^x}{x^2} = \lim_{x \to \infty} \frac{\dfrac{d}{dx}(e^x)}{\dfrac{d}{dx}(x^2)} = \lim_{x \to \infty} \frac{e^x}{2x}$$

Since $e^x \to \infty$ and $2x \to \infty$ as $x \to \infty$, the limit on the right side is also indeterminate, but a second application of l'Hospital's Rule gives

$$\lim_{x \to \infty} \frac{e^x}{x^2} = \lim_{x \to \infty} \frac{e^x}{2x} = \lim_{x \to \infty} \frac{e^x}{2} = \infty$$

The graph of the function of Example 2 is shown in Figure 2. We have noticed previously that exponential functions grow far more rapidly than power functions, so the result of Example 2 is not unexpected. See also Exercise 55.

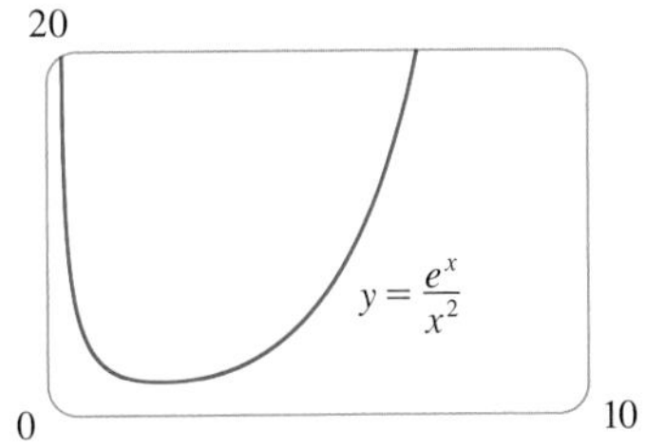

FIGURE 2

EXAMPLE 3 | Calculate $\displaystyle\lim_{x \to \infty} \frac{\ln x}{\sqrt{x}}$.

SOLUTION Since $\ln x \to \infty$ and $\sqrt{x} \to \infty$ as $x \to \infty$, l'Hospital's Rule applies:

$$\lim_{x \to \infty} \frac{\ln x}{\sqrt{x}} = \lim_{x \to \infty} \frac{1/x}{\frac{1}{2}x^{-1/2}}$$

Notice that the limit on the right side is now indeterminate of type $\frac{0}{0}$. But instead of applying l'Hospital's Rule a second time as we did in Example 2, we simplify the expression and see that a second application is unnecessary:

$$\lim_{x \to \infty} \frac{\ln x}{\sqrt{x}} = \lim_{x \to \infty} \frac{1/x}{\frac{1}{2}x^{-1/2}} = \lim_{x \to \infty} \frac{2}{\sqrt{x}} = 0$$

The graph of the function of Example 3 is shown in Figure 3. We have discussed previously the slow growth of logarithms, so it isn't surprising that this ratio approaches 0 as $x \to \infty$. See also Exercise 56.

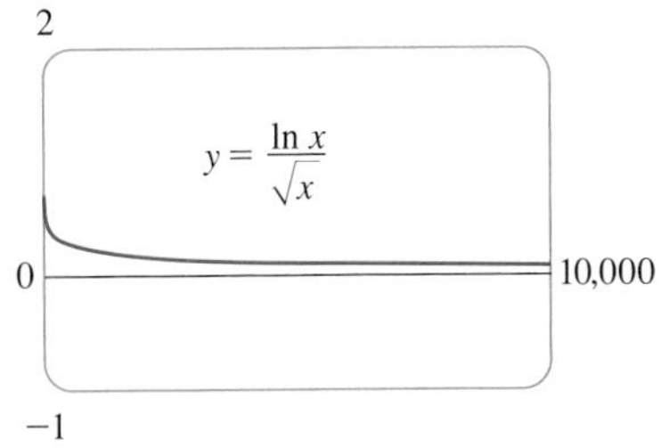

FIGURE 3

EXAMPLE 4 | **Glucose administration** In Exercise 7.4.43 you are asked to derive an equation for the concentration of glucose in the bloodstream during intravenous injection. For a specific choice of constants, the concentration one minute after injection begins is

$$C = \frac{r}{k}(1 - e^{-k})$$

where r is the injection rate and k is the rate at which glucose is metabolized from the blood. Different patients metabolize glucose at different rates, meaning they have different values of k. You can verify that C is a decreasing function of k and therefore, by taking the limit as $k \to 0$, we obtain an upper bound for the predicted concentration of any patient. Find this limit.

SOLUTION Because $k \to 0$ and $1 - e^{-k} \to 0$, we can apply l'Hospital's Rule:

$$\lim_{k \to 0} C = \lim_{k \to 0} r \cdot \frac{1 - e^{-k}}{k} = r \lim_{k \to 0} \frac{e^{-k}}{1}$$
$$= r \cdot 1 = r$$

Thus the upper bound for the concentration is r, the same as the injection rate. ■

EXAMPLE 5 | Find $\displaystyle\lim_{x \to \pi^-} \frac{\sin x}{1 - \cos x}$.

SOLUTION If we blindly attempted to use l'Hospital's Rule, we would get

$$\lim_{x \to \pi^-} \frac{\sin x}{1 - \cos x} = \lim_{x \to \pi^-} \frac{\cos x}{\sin x} = -\infty$$

⊘ This is **wrong!** Although the numerator $\sin x \to 0$ as $x \to \pi^-$, notice that the denominator $(1 - \cos x)$ does not approach 0, so l'Hospital's Rule can't be applied here.

The required limit is, in fact, easy to find because the function is continuous at π and the denominator is nonzero there:

$$\lim_{x \to \pi^-} \frac{\sin x}{1 - \cos x} = \frac{\sin \pi}{1 - \cos \pi} = \frac{0}{1 - (-1)} = 0$$ ■

Example 5 shows what can go wrong if you use l'Hospital's Rule without thinking. Other limits *can* be found using l'Hospital's Rule but are more easily found by other methods. (See Examples 2.4.3, 2.4.5, and 2.2.5 and the discussion at the beginning of this section.) So when evaluating any limit, you should consider other methods before using l'Hospital's Rule.

■ Which Functions Grow Fastest?

L'Hospital's Rule enables us to compare the rates of growth of functions. Suppose we have two functions $f(x)$ and $g(x)$ that both become large as x becomes large, that is,

$$\lim_{x \to \infty} f(x) = \infty \qquad \text{and} \qquad \lim_{x \to \infty} g(x) = \infty$$

We say that $f(x)$ approaches infinity **more quickly** than $g(x)$ if

$$\lim_{x \to \infty} \frac{f(x)}{g(x)} = \infty$$

and that $f(x)$ approaches infinity **more slowly** than $g(x)$ if

$$\lim_{x \to \infty} \frac{f(x)}{g(x)} = 0$$

For example, we used l'Hospital's Rule in Example 2 to show that

$$\lim_{x \to \infty} \frac{e^x}{x^2} = \infty$$

and so the exponential function $y = e^x$ grows more quickly than $y = x^2$. In fact $y = e^x$

grows more quickly than all the power functions $y = x^n$ (see Exercise 55). This fact is illustrated in Figure 4, where you can see that $y = x^3$ exceeds $y = e^x$ some of the time but after about $x = 4.5$ the exponential function overtakes the other functions.

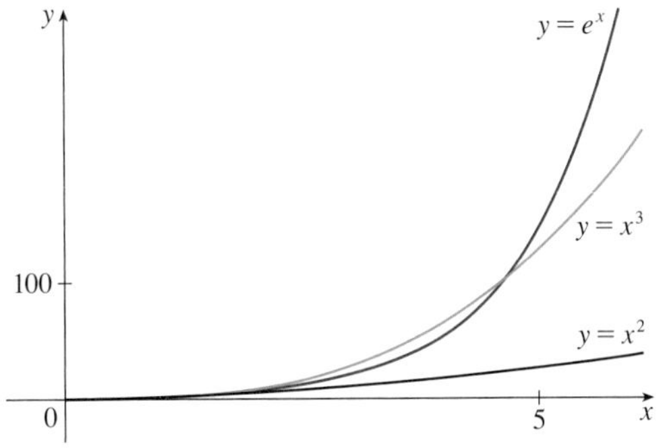

FIGURE 4

EXAMPLE 6 | Rank the following functions in order of how quickly they approach infinity as $x \to \infty$:

$$y = \ln x \qquad y = x \qquad y = e^{0.1x} \qquad y = \sqrt{x}$$

SOLUTION If we try to decide the ranking by plotting the four functions as in Figure 5(a), we get a misleading picture: it looks as if $y = x$ is the winner. But l'Hospital's Rule tells us that can't be true:

$$\lim_{x \to \infty} \frac{x}{e^{0.1x}} = \lim_{x \to \infty} \frac{1}{0.1e^{0.1x}} = 0$$

so $y = x$ grows more slowly than $y = e^{0.1x}$.

We know from Example 3 that $y = \ln x$ grows more slowly than $y = \sqrt{x}$. Also, $y = \sqrt{x}$ grows more slowly than $y = x$ because

$$\frac{\sqrt{x}}{x} = \frac{1}{\sqrt{x}} \to 0 \qquad \text{as } x \to \infty$$

So the ranking, from fastest to slowest, is as follows:

$$y = e^{0.1x} \qquad y = x \qquad y = \sqrt{x} \qquad y = \ln x$$

This ranking is illustrated in Figure 5(b). ■

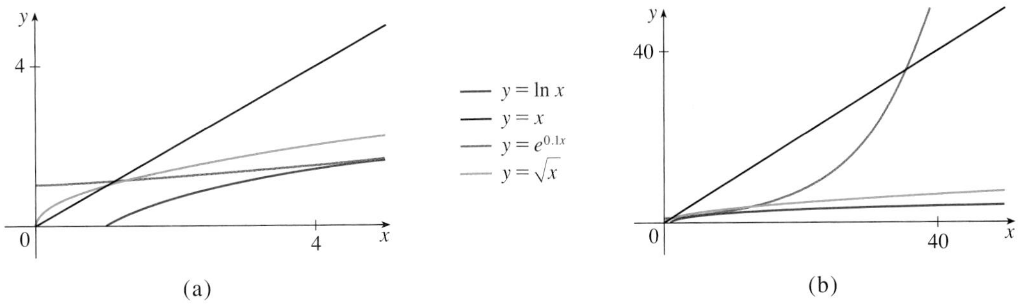

FIGURE 5

EXAMPLE 7 | Analyze the limiting behavior of the function

$$f(x) = \frac{x + x^4}{1 + x^2 + e^x}$$

by considering just the dominant terms in the numerator and denominator.

SOLUTION The dominant term in the numerator is x^4 because it becomes large more quickly than x. We know from Example 2 that e^x is the dominant term in the denominator. If we retain just the two dominant terms, we get the simpler function

$$g(x) = \frac{x^4}{e^x}$$

The exponential function grows faster than any power of x and so $\lim_{x\to\infty} g(x) = 0$. Therefore we expect that $\lim_{x\to\infty} f(x) = 0$ too. This could be verified directly with four applications of l'Hospital's Rule, but the important thing is to recognize that our intuition about orders of magnitude leads us to the correct answer. In Figure 6 we see that the functions f and g have practically identical behavior for large values of x. ■

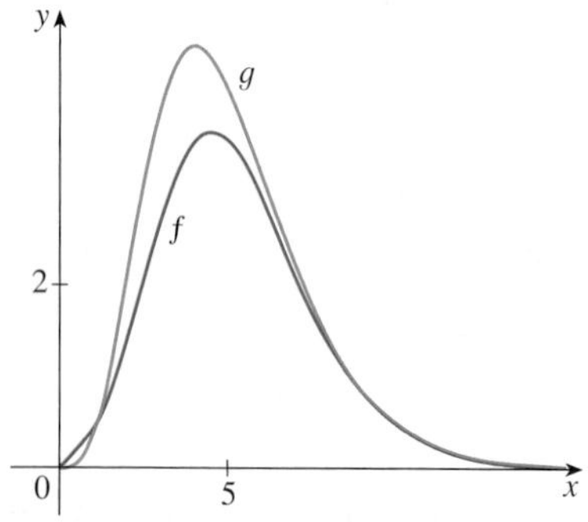

FIGURE 6

■ Indeterminate Products

If $\lim_{x\to a} f(x) = 0$ and $\lim_{x\to a} g(x) = \infty$ (or $-\infty$), then it isn't clear what the value of $\lim_{x\to a} f(x)\,g(x)$, if any, will be. There is a struggle between f and g. If f wins, the limit will be 0; if g wins, the answer will be ∞ (or $-\infty$). Or there may be a compromise where the answer is a finite nonzero number. This kind of limit is called an **indeterminate form of type 0 · ∞**. We can deal with it by writing the product fg as a quotient:

$$fg = \frac{f}{1/g} \qquad \text{or} \qquad fg = \frac{g}{1/f}$$

This converts the given limit into an indeterminate form of type $\frac{0}{0}$ or ∞/∞ so that we can use l'Hospital's Rule.

EXAMPLE 8 | Evaluate $\lim_{x\to 0^+} x \ln x$. Use the knowledge of this limit, together with information from derivatives, to sketch the curve $y = x \ln x$.

SOLUTION The given limit is indeterminate because, as $x \to 0^+$, the first factor (x) approaches 0 while the second factor ($\ln x$) approaches $-\infty$. Writing $x = 1/(1/x)$, we have $1/x \to \infty$ as $x \to 0^+$, so l'Hospital's Rule gives

$$\lim_{x\to 0^+} x \ln x = \lim_{x\to 0^+} \frac{\ln x}{1/x} = \lim_{x\to 0^+} \frac{1/x}{-1/x^2} = \lim_{x\to 0^+} (-x) = 0$$

If $f(x) = x \ln x$, then

$$f'(x) = x \cdot \frac{1}{x} + \ln x = 1 + \ln x$$

so $f'(x) = 0$ when $\ln x = -1$, which means that $x = e^{-1}$. In fact, $f'(x) > 0$ when $x > e^{-1}$ and $f'(x) < 0$ when $x < e^{-1}$, so f is increasing on $(1/e, \infty)$ and decreasing on $(0, 1/e)$. Thus, by the First Derivative Test, $f(1/e) = -1/e$ is a local (and absolute) minimum. Also, $f''(x) = 1/x > 0$, so f is concave upward on $(0, \infty)$. We use this information, together with the crucial knowledge that $\lim_{x\to 0^+} f(x) = 0$, to sketch the curve in Figure 7. ■

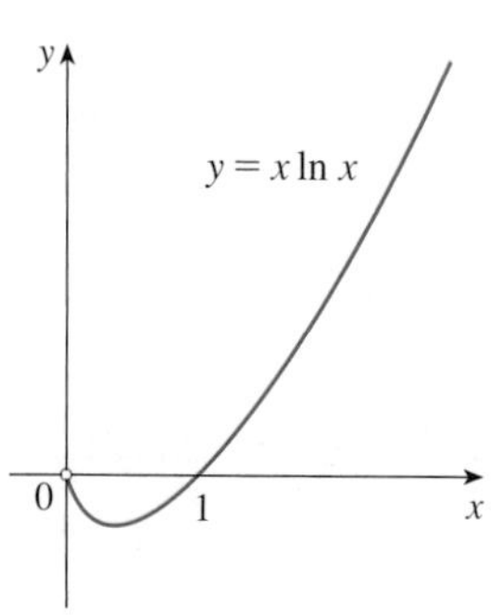

FIGURE 7

NOTE In solving Example 8 another possible option would have been to write

$$\lim_{x \to 0^+} x \ln x = \lim_{x \to 0^+} \frac{x}{1/\ln x}$$

This gives an indeterminate form of the type $\frac{0}{0}$, but if we apply l'Hospital's Rule we get a more complicated expression than the one we started with. In general, when we rewrite an indeterminate product, we try to choose the option that leads to the simpler limit.

EXAMPLE 9 | **Blood alcohol concentration** In Sections 3.1 and 4.1 we modeled the blood alcohol concentration (BAC) after rapid consumption of one alcoholic drink by the function

$$C(t) = 0.0225te^{-0.0467t}$$

where t is measured in minutes after consumption and C is measured in mg/mL. Calculate $\lim_{t \to \infty} C(t)$.

SOLUTION The limit is an indeterminate product because $0.0225t \to \infty$ and $e^{-0.0467t} \to 0$ as $t \to \infty$. If we write $C(t)$ as a quotient instead of a product and then use l'Hospital's Rule, we get

$$\lim_{t \to \infty} C(t) = \lim_{t \to \infty} 0.0225te^{-0.0467t} = \lim_{t \to \infty} \frac{0.0225t}{e^{0.0467t}}$$

$$= \lim_{t \to \infty} \frac{0.0225}{0.0467e^{0.0467t}} = 0$$

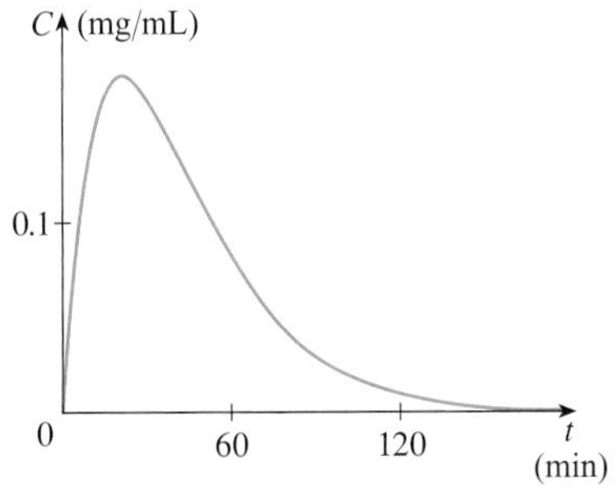

FIGURE 8

because the denominator approaches infinity as $t \to \infty$. This result is not surprising; it is to be expected that the BAC eventually approaches 0 as time passes. Figure 8 illustrates the result. ■

■ Indeterminate Differences

If $\lim_{x \to a} f(x) = \infty$ and $\lim_{x \to a} g(x) = \infty$, then the limit

$$\lim_{x \to a} [f(x) - g(x)]$$

is called an **indeterminate form of type $\infty - \infty$**. Again there is a contest between f and g. Will the answer be ∞ (f wins) or will it be $-\infty$ (g wins) or will they compromise on a finite number? To find out, we try to convert the difference into a quotient (for instance, by using a common denominator, or rationalization, or factoring out a common factor) so that we have an indeterminate form of type $\frac{0}{0}$ or ∞/∞.

EXAMPLE 10 | Compute $\lim_{x \to (\pi/2)^-} (\sec x - \tan x)$.

SOLUTION First notice that $\sec x \to \infty$ and $\tan x \to \infty$ as $x \to (\pi/2)^-$, so the limit is indeterminate. Here we use a common denominator:

$$\lim_{x \to (\pi/2)^-} (\sec x - \tan x) = \lim_{x \to (\pi/2)^-} \left(\frac{1}{\cos x} - \frac{\sin x}{\cos x} \right)$$

$$= \lim_{x \to (\pi/2)^-} \frac{1 - \sin x}{\cos x} = \lim_{x \to (\pi/2)^-} \frac{-\cos x}{-\sin x} = 0$$

Note that the use of l'Hospital's Rule is justified because $1 - \sin x \to 0$ and $\cos x \to 0$ as $x \to (\pi/2)^-$. ■

EXAMPLE 11 | Calculate $\lim_{x\to\infty} (e^x - x)$.

SOLUTION This is an indeterminate difference because both e^x and x approach infinity. We would certainly expect the answer to be infinity because $e^x \to \infty$ much faster than x. But we can verify this by factoring out x:

$$e^x - x = x\left(\frac{e^x}{x} - 1\right)$$

The term $e^x/x \to \infty$ as $x \to \infty$ by l'Hospital's Rule and so we now have a product in which both factors grow large:

$$\lim_{x\to\infty} (e^x - x) = \lim_{x\to\infty} \left[x\left(\frac{e^x}{x} - 1\right)\right] = \infty$$

■

EXERCISES 4.3

1–34 Find the limit. Use l'Hospital's Rule where appropriate. If there is a more elementary method, consider using it. If l'Hospital's Rule doesn't apply, explain why.

1. $\lim_{x\to 1} \frac{x^2 - 1}{x^2 - x}$

2. $\lim_{x\to 1} \frac{x^a - 1}{x^b - 1}$

3. $\lim_{x\to(\pi/2)^+} \frac{\cos x}{1 - \sin x}$

4. $\lim_{x\to 0} \frac{\sin 4x}{\tan 5x}$

5. $\lim_{t\to 0} \frac{e^t - 1}{t^3}$

6. $\lim_{t\to 0} \frac{e^{3t} - 1}{t}$

7. $\lim_{x\to\infty} \frac{\ln x}{\sqrt{x}}$

8. $\lim_{\theta\to\pi/2} \frac{1 - \sin\theta}{\csc\theta}$

9. $\lim_{x\to 0^+} \frac{\ln x}{x}$

10. $\lim_{x\to\infty} \frac{(\ln x)^2}{x}$

11. $\lim_{x\to 0} \frac{\sqrt{1 + 2x} - \sqrt{1 - 4x}}{x}$

12. $\lim_{x\to 1} \frac{\ln x}{\sin \pi x}$

13. $\lim_{t\to 0} \frac{5^t - 3^t}{t}$

14. $\lim_{u\to\infty} \frac{e^{u/10}}{u^3}$

15. $\lim_{x\to 0} \frac{e^x - 1 - x}{x^2}$

16. $\lim_{x\to 0} \frac{\cos mx - \cos nx}{x^2}$

17. $\lim_{x\to 1} \frac{1 - x + \ln x}{1 + \cos \pi x}$

18. $\lim_{x\to 0} \frac{x}{\tan^{-1}(4x)}$

19. $\lim_{x\to 1} \frac{x^a - ax + a - 1}{(x - 1)^2}$

20. $\lim_{x\to 0} \frac{e^x - e^{-x} - 2x}{x - \sin x}$

21. $\lim_{x\to 0} \frac{\cos x - 1 + \frac{1}{2}x^2}{x^4}$

22. $\lim_{x\to a^+} \frac{\cos x \ln(x - a)}{\ln(e^x - e^a)}$

23. $\lim_{x\to\infty} x \sin(\pi/x)$

24. $\lim_{x\to-\infty} x^2 e^x$

25. $\lim_{x\to 0} \cot 2x \sin 6x$

26. $\lim_{x\to 0^+} \sin x \ln x$

27. $\lim_{x\to\infty} x^3 e^{-x^2}$

28. $\lim_{x\to\infty} x \tan(1/x)$

29. $\lim_{x\to 1} \left(\frac{x}{x - 1} - \frac{1}{\ln x}\right)$

30. $\lim_{x\to 0} (\csc x - \cot x)$

31. $\lim_{x\to\infty} \left(\sqrt{x^2 + x} - x\right)$

32. $\lim_{x\to 0^+} \left(\cot x - \frac{1}{x}\right)$

33. $\lim_{x\to\infty} (x - \ln x)$

34. $\lim_{x\to\infty} (xe^{1/x} - x)$

35–38 Use l'Hospital's Rule to help find the asymptotes of f. Then use them, together with information from f' and f'', to sketch the graph of f. Check your work with a graphing device.

35. $f(x) = xe^{-x}$

36. $f(x) = e^x/x$

37. $f(x) = (\ln x)/x$

38. $f(x) = xe^{-x^2}$

39–42 Rank the functions in order of how quickly they grow as $x \to \infty$.

39. $y = x^5$, $y = \ln(x^{10})$, $y = e^{2x}$, $y = e^{3x}$

40. $y = 2^x$, $y = 3^x$, $y = e^{x/2}$, $y = e^{x/3}$

41. $y = (\ln x)^2$, $y = (\ln x)^3$, $y = \sqrt{x}$, $y = \sqrt[3]{x}$

42. $y = x + e^{-x}$, $y = 10 \ln x$, $y = 5\sqrt{x}$, $y = x\sqrt{x}$

43–44 Guess the value of the limit by considering the dominant terms in the numerator and denominator. Then use l'Hospital's Rule to confirm your guess.

43. $\lim_{x\to\infty} \dfrac{e^{-2x} + x + e^{0.1x}}{x^3 - x^2}$ **44.** $\lim_{x\to\infty} \dfrac{x^2 - x + \ln x}{x + 2^x}$

45–46 Rank the functions in order of how quickly they approach 0 as $x \to \infty$.

45. $y = \dfrac{1}{x}$, $y = \dfrac{1}{x^2}$, $y = e^{-x}$, $y = x^{-1/2}$

46. $y = e^{-x}$, $y = xe^{-x}$, $y = e^{-x^2}$, $y = xe^{-x^2}$

47–48 What happens if you try to use l'Hospital's Rule to find the limit? Evaluate the limit using another method.

47. $\lim_{x\to\infty} \dfrac{x}{\sqrt{x^2 + 1}}$ **48.** $\lim_{x\to(\pi/2)^-} \dfrac{\sec x}{\tan x}$

49. Models of population growth have the general form $dN/dt = f(N)$, where $f(N)$ is a function such that $f(0) = 0$ and $f(N)$ is positive for some positive values of N. The per capita growth rate is defined to be the population growth rate divided by the population size.

(a) What is the per capita growth rate for the model $dN/dt = rN$?

(b) What is the per capita growth rate for the model $dN/dt = rN(1 - N/K)$ when N is small (that is, when $N \to 0$)?

(c) Express the per capita growth rate when $N \to 0$ for the general model $dN/dt = f(N)$ in terms of the function f and/or its derivative.

50. Foraging In Exercise 4.2.2 we let $N(t)$ be the amount of nectar foraged from a flower by a bumblebee in t seconds.

(a) What is the average rate of nectar consumption over a period of t seconds?

(b) Find the average rate of nectar consumption for very short foraging visits by using l'Hospital's Rule to calculate the limit of the answer to part (a) as $t \to 0$.

51. Stiles-Crawford effect Light enters the eye through the pupil and strikes the retina, where photoreceptor cells sense light and color. W. Stanley Stiles and B. H. Crawford studied the phenomenon in which measured brightness decreases as light enters farther from the center of the pupil. (See the figure.) They detailed their findings of this phenomenon, known as the *Stiles-Crawford effect of the first kind*, in a paper published in 1933. In particular, they observed that the amount of luminance sensed was not proportional to the area of the pupil. The percentage P of the luminance entering a pupil at a distance of r mm from the center that is sensed at the retina can be described by

$$P = \frac{100(1 - 10^{-\rho r^2})}{\rho r^2 \ln 10}$$

where ρ is an experimentally determined constant, typically about 0.05.

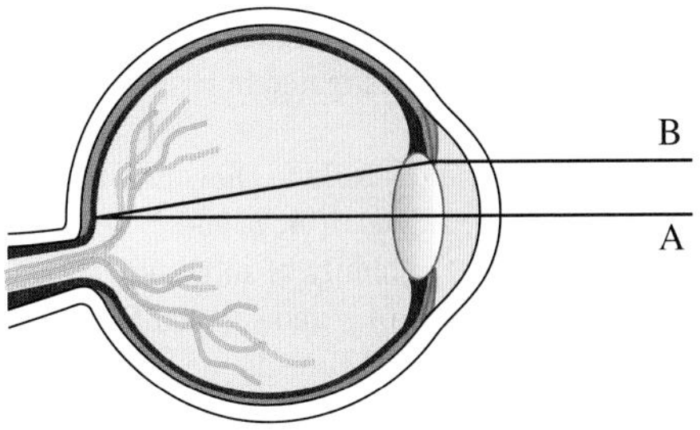

A light beam A that enters through the center of the pupil measures brighter than a beam B entering near the edge of the pupil.

(a) If light enters a pupil at a distance of 3 mm from the center, what is the percentage of luminance that is sensed at the retina?

(b) Compute the percentage of luminance sensed if light enters the pupil at a distance of 2 mm from the center. Does it make sense that it is larger than the answer to part (a)?

(c) Compute $\lim_{r\to 0} P$. Is the result what you would expect?

Source: Adapted from W. Stiles et al., "The Luminous Efficiency of Rays Entering the Eye Pupil at Different Points." *Proceedings of the Royal Society of London. Series B*, 112 (1933): 428–50.

52. Antibiotic concentration for large patients Suppose an antibiotic is administered at a constant rate through intravenous supply to a patient and is metabolized. It can be shown using the type of mixing models discussed in Exercises 7.4.45–48 that the concentration of antibiotic after one unit of time is

$$c(V) = c_0 e^{-1/V} + \theta V(1 - e^{-1/V})$$

where c_0 is the initial concentration, θ is the rate of supply, and V is the volume of the patient's blood. What is the predicted value for large patients (that is, for large values of V)?

53. Drug pharmacokinetics So-called two-compartment models are often used to describe drug pharmacokinetics, with the blood being one compartment and the internal organs being the other (see Section 10.3). If the rate of flow from the blood to the organs is α and the rate of metabolism from the organs is β, then under certain conditions the concentration of drug in the organ at time t is given by

$$\frac{e^{-\alpha t} - e^{-\beta t}}{\beta - \alpha}$$

What is the predicted concentration at time t if the values of α and β are very close to one another?

54. Drug pharmacokinetics The level of medication in the bloodstream after a drug is administered is often modeled by a function of the form

$$S(t) = At^p e^{-kt}$$

where A, p, and k are positive constants. This is called a *surge function* because its graph shows an initial surge in the drug level followed by a more gradual decline. (Particular cases have been investigated in Example 9 and Exercise 4.2.51.)

(a) Use l'Hospital's Rule to show that $\lim_{t\to\infty} S(t) = 0$ for all positive values of A, p, and k.

(b) Investigate the family of surge functions for $A = 1$ and positive values of p and k. What features do these curves have in common? How do they differ from one another? In particular, what happens to the maximum and minimum points and inflection points as p and k change? Illustrate by graphing several members of the family.

55. Prove that

$$\lim_{x\to\infty} \frac{e^x}{x^n} = \infty$$

for any positive integer n. This shows that the exponential function approaches infinity faster than any power of x.

56. Prove that

$$\lim_{x\to\infty} \frac{\ln x}{x^p} = 0$$

for any number $p > 0$. This shows that the logarithmic function approaches infinity more slowly than any power of x.

57. The first appearance in print of l'Hospital's Rule was in the book *Analyse des Infiniment Petits* published by the Marquis de l'Hospital in 1696. This was the first calculus *textbook* ever published and the example that the Marquis used in that book to illustrate his rule was to find the limit of the function

$$y = \frac{\sqrt{2a^3x - x^4} - a\sqrt[3]{aax}}{a - \sqrt[4]{ax^3}}$$

as x approaches a, where $a > 0$. (At that time it was common to write aa instead of a^2.) Solve this problem.

58. If f' is continuous, $f(2) = 0$, and $f'(2) = 7$, evaluate

$$\lim_{x\to 0} \frac{f(2 + 3x) + f(2 + 5x)}{x}$$

■ PROJECT Mutation-Selection Balance in Genetic Diseases

Several human diseases—such as Tay-Sachs disease, phenylketonuria, neurofibromatosis, and Huntington's disease—occur if an individual carries a mutated copy of a specific gene. Because the carriers of such deleterious genes are less likely to survive and reproduce, the frequency of such genes in the population is usually quite low.

In population-genetic terms the frequency of such disease-causing genes reflects a balance between their spontaneous appearance through mutation and selection against them through reduced survival. Moreover, some diseases (like Tay-Sachs and phenylketonuria) are *recessive*, meaning that only individuals carrying two copies of the mutated gene are affected, while others (like neurofibromatosis and Huntington's disease) are *dominant*, meaning that individuals carrying even a single copy will be affected.

Let's model the mutation-selection balance for an arbitrary disease to predict the frequency of the mutation in the population. We will use A to denote the disease-causing variant and a to denote the normal version of the gene. For simplicity we will assume that generations are discrete in time and we will model the dynamics of the disease-causing gene using a difference equation.

Suppose the probability of surviving one time step for an aa individual is standardized to 1, and that for an AA individual is $1 - s$, where s is a constant and $0 < s < 1$. We can incorporate different levels of dominance by supposing that the probability of survival of Aa individuals is $1 - sh$, where h is a constant and $0 \leqslant h \leqslant 1$. When $h = 1$ the disease is dominant since Aa individuals are just as affected by the disease as AA individuals. When $h = 0$ the disease is recessive because only AA individuals are affected. The case $h = \frac{1}{2}$ is called *codominant*, meaning that the survival of Aa individuals lies exactly halfway between those of AA and aa individuals.

To describe how selection will change the frequency of the A gene, we can follow a derivation similar to that used in the project on page 78. Using p_t for the frequency of A at time t, we obtain the following expression for the frequency of A after selection has occurred:

$$\frac{p_t^2(1 - s) + p_t(1 - p_t)(1 - sh)}{p_t^2(1 - s) + 2p_t(1 - p_t)(1 - sh) + (1 - p_t)^2}$$

1. Suppose that, in a single time step, first selection occurs as described by this expression and then a fraction μ of the a variants mutate to the disease-causing variant A. Obtain an expression for p_{t+1} in terms of p_t.

2. At mutation-selection balance, the frequency of A will no longer change, meaning that $p_{t+1} = p_t$. Using $\hat{p}$ to denote this equilibrium frequency, show that $\hat{p}$ satisfies a cubic equation. Further, show that this cubic equation can be reduced to a quadratic equation if we are interested only in solutions satisfying $\hat{p} \neq 1$.

3. Show that one root of the quadratic equation obtained in Problem 2 is

$$\hat{p} = \frac{sh(1 + \mu) - \sqrt{[sh(1 + \mu)]^2 - 4s\mu(2h - 1)}}{2s(2h - 1)}$$

4. Assume that $2\mu < s$. What is the predicted frequency of genes that cause recessive diseases (that is, the frequency as $h \to 0$)?

5. Assume that $2\mu < s$. Use l'Hospital's Rule to determine the predicted frequency of genes that cause codominant diseases (that is, the frequency as $h \to \frac{1}{2}$).

4.4 Optimization Problems

There are many situations in the life sciences in which it is desirable to find an optimal outcome: maximizing the yield of an agricultural crop by controlling the nitrogen level of the soil, minimizing the energy required for fish to swim or birds to fly during migration, determining the dosage of a drug for the best result. In this section we use the methods of this chapter to determine such optimal outcomes.

In solving these problems the greatest challenge is often to convert the word problem into a mathematical optimization problem by setting up the function that is to be maximized or minimized. The following steps may be useful.

Steps in Solving Optimization Problems

1. **Understand the Problem** The first step is to read the problem carefully until it is clearly understood. Ask yourself: What is the unknown? What are the given quantities? What are the given conditions?

2. **Draw a Diagram** In most problems it is useful to draw a diagram and identify the given and required quantities on the diagram.

3. **Introduce Notation** Assign a symbol to the quantity that is to be maximized or minimized (let's call it Q for now). Also select symbols ($a, b, c, \ldots, x, y$) for other unknown quantities and label the diagram with these symbols. It may help to use initials as suggestive symbols—for example, A for area, h for height, t for time.

4. Express Q in terms of some of the other symbols from Step 3.
5. If Q has been expressed as a function of more than one variable in Step 4, use the given information to find relationships (in the form of equations) among these variables. Then use these equations to eliminate all but one of the variables in the expression for Q. Thus Q will be expressed as a function of *one* variable x, say, $Q = f(x)$. Write the domain of this function.
6. Use the methods of Sections 4.1 and 4.2 to find the *absolute* maximum or minimum value of f. In particular, if the domain of f is a closed interval, then the Closed Interval Method in Section 4.1 can be used.

EXAMPLE 1 | A farmer has 2400 ft of fencing and wants to fence off a rectangular field that borders a river with a straight bank. He needs no fence along the river. What are the dimensions of the field that has the largest area?

SOLUTION In order to get a feeling for what is happening in this problem, let's experiment with some special cases. Figure 1 (not to scale) shows three possible ways of laying out the 2400 ft of fencing.

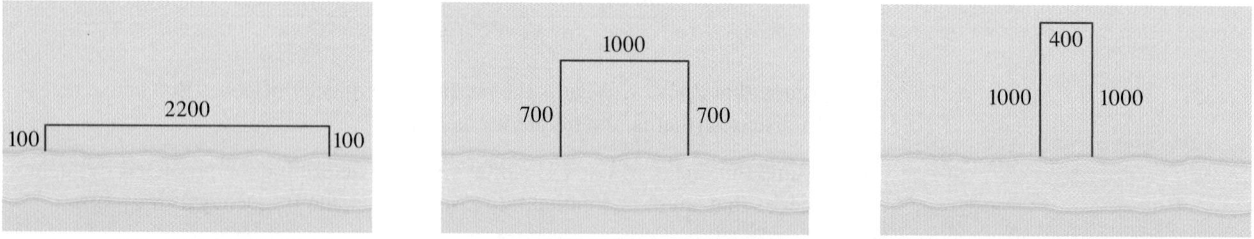

Area = 100 · 2200 = 220,000 ft^2 Area = 700 · 1000 = 700,000 ft^2 Area = 1000 · 400 = 400,000 ft^2

FIGURE 1

We see that when we try shallow, wide fields or deep, narrow fields, we get relatively small areas. It seems plausible that there is some intermediate configuration that produces the largest area.

Figure 2 illustrates the general case. We wish to maximize the area A of the rectangle. Let x and y be the depth and width of the rectangle (in feet). Then we express A in terms of x and y:

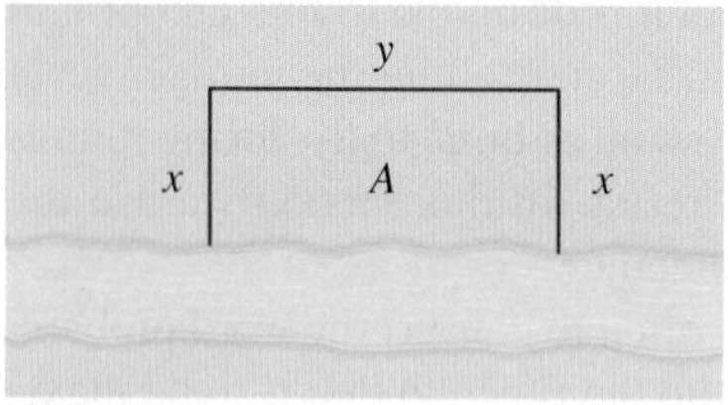

FIGURE 2

$$A = xy$$

We want to express A as a function of just one variable, so we eliminate y by expressing it in terms of x. To do this we use the given information that the total length of the fencing is 2400 ft. Thus

$$2x + y = 2400$$

From this equation we have $y = 2400 - 2x$, which gives

$$A = x(2400 - 2x) = 2400x - 2x^2$$

Note that $x \geqslant 0$ and $x \leqslant 1200$ (otherwise $A < 0$). So the function that we wish to maximize is

$$A(x) = 2400x - 2x^2 \qquad 0 \leqslant x \leqslant 1200$$

The derivative is $A'(x) = 2400 - 4x$, so to find the critical numbers we solve the equation

$$2400 - 4x = 0$$

which gives $x = 600$. The maximum value of A must occur either at this critical number or at an endpoint of the interval. Since $A(0) = 0$, $A(600) = 720{,}000$, and $A(1200) = 0$, the Closed Interval Method gives the maximum value as $A(600) = 720{,}000$.

[Alternatively, we could have observed that $A''(x) = -4 < 0$ for all x, so A is always concave downward and the local maximum at $x = 600$ must be an absolute maximum.]

Thus the rectangular field should be 600 ft deep and 1200 ft wide. ■

EXAMPLE 2 | BB **Nectar foraging by bumblebees** Many animals forage on resources that are distributed in discrete patches. For example, bumblebees visit many flowers, foraging on nectar from each. The amount of nectar $N(t)$ consumed from any flower increases with the amount of time spent at that flower, but with diminishing returns (see Figure 3). Suppose this function is given by

$$N(t) = \frac{0.3t}{t + 2}$$

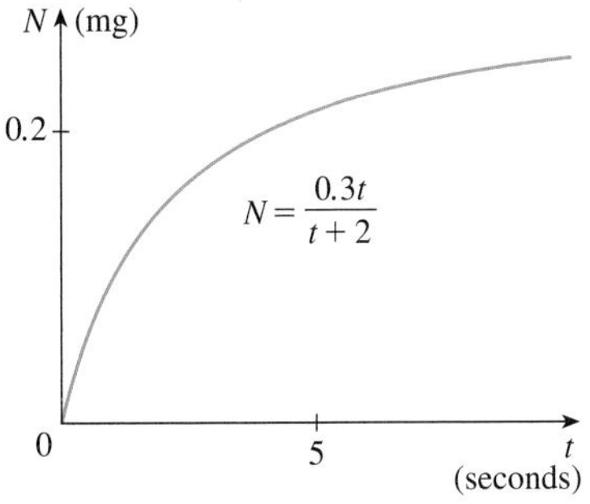

FIGURE 3

where t is measured in seconds and N in milligrams. Suppose also that the time it takes a bee to travel from one flower to the next is 4 seconds.

(a) If a bee spends t seconds at each flower, find an equation for the average amount of nectar consumed per second, from the beginning of a visit to a flower until the beginning of the visit to the next flower.

(b) Suppose bumblebees forage on a given flower for an amount of time that maximizes the average rate of energy gain obtained in part (a). What is this optimal foraging time?

SOLUTION

(a) The total time to complete one cycle of foraging and traveling is $t + 4$. So the average amount of nectar consumed per second over one cycle is

$$f(t) = \frac{N(t)}{t + 4} = \frac{0.3t}{(t + 2)(t + 4)} = \frac{0.3t}{t^2 + 6t + 8}$$

(b) The derivative of f is

$$f'(t) = \frac{(t^2 + 6t + 8)(0.3) - (0.3t)(2t + 6)}{(t^2 + 6t + 8)^2} = \frac{-0.3t^2 + 2.4}{(t^2 + 6t + 8)^2}$$

Then $f'(t) = 0$ when $t^2 = 2.4/0.3 = 8$. But $t \geqslant 0$, so the only critical number is $t = \sqrt{8} = 2\sqrt{2}$.

Since the domain of f is $[0, \infty)$, we can't use the argument of Example 1 concerning endpoints. But we observe that $f'(t) > 0$ for $0 \leqslant t < 2\sqrt{2}$ and $f'(t) < 0$ for $t > 2\sqrt{2}$, so f is increasing for *all* t to the left of the critical number and decreasing for *all* t to the right. So the optimal foraging time is

$$t = 2\sqrt{2} \approx 2.83 \text{ seconds}$$

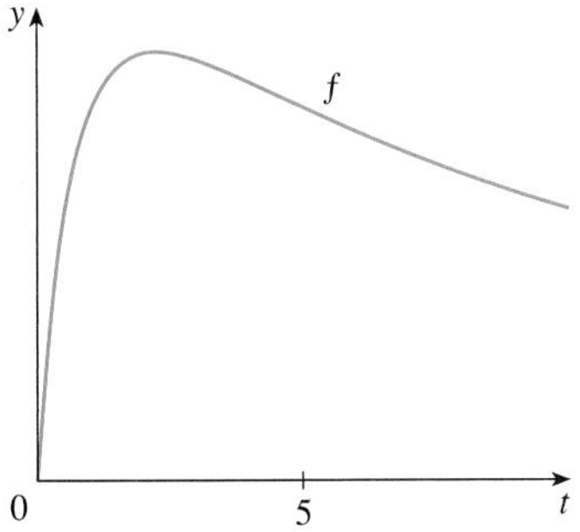

FIGURE 4

This result is illustrated by the graph of f in Figure 4. ■

NOTE 1 The argument used in Example 2 to justify the absolute maximum is a variant of the First Derivative Test (which applies only to *local* maximum or minimum values) and is stated here for future reference.

TEC Module 4.4 takes you through six additional optimization problems, including animations of the physical situations.

> **First Derivative Test for Absolute Extreme Values** Suppose that c is a critical number of a continuous function f defined on an interval.
>
> (a) If $f'(x) > 0$ for all $x < c$ and $f'(x) < 0$ for all $x > c$, then $f(c)$ is the absolute maximum value of f.
>
> (b) If $f'(x) < 0$ for all $x < c$ and $f'(x) > 0$ for all $x > c$, then $f(c)$ is the absolute minimum value of f.

NOTE 2 In Example 4.1.7 we found the maximum value of the blood alcohol concentration during the first hour by using the Closed Interval Method. But if we use the First Derivative Test for Absolute Extreme Values, we can see that it is the absolute maximum. The formula

$$C'(t) = 0.0225e^{-0.0467t}(1 - 0.0467t)$$

shows that $C'(t) > 0$ for all $t < 1/0.0467$ and $C'(t) < 0$ for all $t > 1/0.0467$. It follows that the absolute maximum of the BAC is $C(1/0.0467) \approx 0.177$ mg/mL.

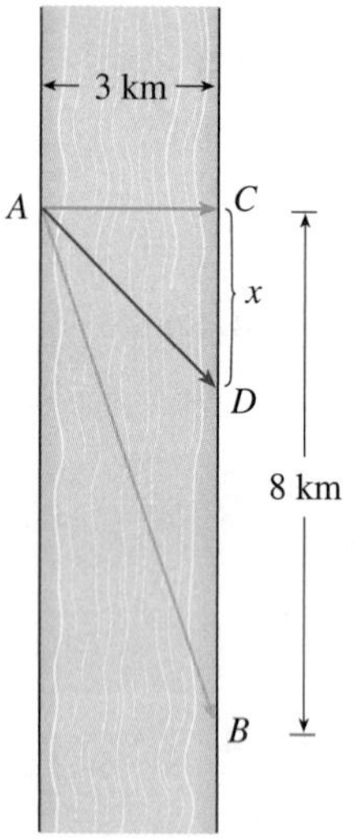

FIGURE 5

EXAMPLE 3 | A man launches his boat from point A on a bank of a straight water channel, 3 km wide, and wants to reach point B, 8 km south on the opposite bank, as quickly as possible (see Figure 5). He could row his boat directly across the channel to point C and then run to B, or he could row directly to B, or he could row to some point D between C and B and then run to B. If he can row 6 km/h and run 8 km/h, where should he land to reach B as soon as possible? (We assume that the water in the channel is not moving.)

SOLUTION If we let x be the distance from C to D, then the running distance is $|DB| = 8 - x$ and the Pythagorean Theorem gives the rowing distance as $|AD| = \sqrt{x^2 + 9}$. We use the equation

$$\text{time} = \frac{\text{distance}}{\text{rate}}$$

Then the rowing time is $\sqrt{x^2 + 9}/6$ and the running time is $(8 - x)/8$, so the total time T as a function of x is

$$T(x) = \frac{\sqrt{x^2 + 9}}{6} + \frac{8 - x}{8}$$

The domain of this function T is $[0, 8]$. Notice that if $x = 0$, he rows to C and if $x = 8$, he rows directly to B. The derivative of T is

$$T'(x) = \frac{x}{6\sqrt{x^2 + 9}} - \frac{1}{8}$$

Thus, using the fact that $x \geqslant 0$, we have

$$T'(x) = 0 \iff \frac{x}{6\sqrt{x^2+9}} = \frac{1}{8} \iff 4x = 3\sqrt{x^2+9}$$

$$\iff 16x^2 = 9(x^2+9) \iff 7x^2 = 81$$

$$\iff x = \frac{9}{\sqrt{7}}$$

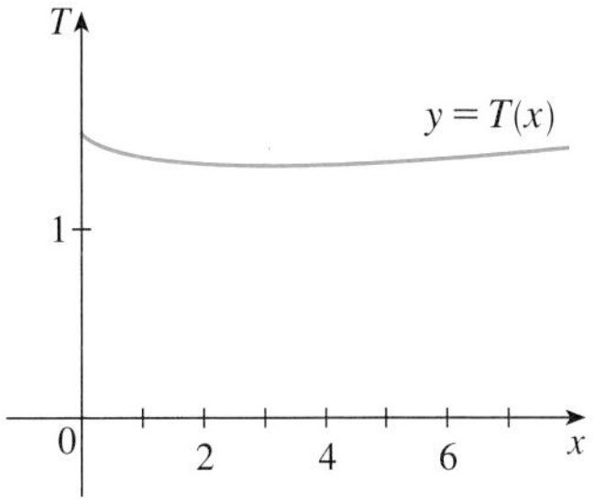

FIGURE 6

The only critical number is $x = 9/\sqrt{7}$. To see whether the minimum occurs at this critical number or at an endpoint of the domain [0, 8], we evaluate T at all three points:

$$T(0) = 1.5 \qquad T\left(\frac{9}{\sqrt{7}}\right) = 1 + \frac{\sqrt{7}}{8} \approx 1.33 \qquad T(8) = \frac{\sqrt{73}}{6} \approx 1.42$$

Since the smallest of these values of T occurs when $x = 9/\sqrt{7}$, the absolute minimum value of T must occur there. Figure 6 illustrates this calculation by showing the graph of T.

Thus the man should land the boat at a point $9/\sqrt{7}$ km ($\approx$3.4 km) south from his starting point. ■

© Brian Lasenby / Shutterstock.com

EXAMPLE 4 | **BB** Aquatic birds[1] forage underwater and periodically return to the surface to replenish their oxygen stores. Oxygen stores increase with the amount of time spent on the surface but in a diminishing way, according to the model

$$O(t) = \frac{20t}{5+t}$$

where t is the amount of time spent at the surface (in seconds). Suppose the round-trip travel time to and from the underwater foraging area is T seconds and oxygen is depleted at a constant rate of r mL/s while a bird is underwater. Furthermore, suppose the bird forages until it has just enough oxygen to return to the surface.

(a) If Q is the fraction of a single dive cycle that the bird spends foraging, find an equation for Q as a function of the surface time t.

(b) If $T = 2$ seconds and $r = 1$ mL/s, find the surface time t that maximizes the fraction of time spent foraging.

SOLUTION

(a) Let f be the time spent foraging during a single cycle. Because the surface time is t seconds and the total travel time for a single cycle is T seconds, the fraction of a cycle spent foraging is

$$Q = \frac{f}{t + T + f}$$

This expression for Q involves the constant T and the two variables t and f. We want to express Q in terms of just one variable and so we eliminate f by expressing it in terms

---

1. L. Halsey et al., "Optimal Diving Behaviour and Respiratory Gas Exchange in Birds," *Respiratory Physiology and Neurobiology* 154 (2006): 268–83.

of t. To do this we use the information given—that all of the oxygen $O(t)$ obtained at the surface is used for foraging and travel. In other words, $O(t) = rT + rf$. Solving this equation for f gives

$$f = \frac{O(t) - rT}{r}$$

Therefore

$$Q(t) = \frac{\dfrac{O(t) - rT}{r}}{t + T + \dfrac{O(t) - rT}{r}} = \frac{O(t) - rT}{O(t) + rt}$$

$$= \frac{\dfrac{20t}{5 + t} - rT}{\dfrac{20t}{5 + t} + rt} = \frac{(20 - rT)t - 5rT}{20t + 5rt + rt^2}$$

(b) With $r = 1$ and $T = 2$ we have

$$Q(t) = \frac{18t - 10}{t^2 + 25t}$$

and so the derivative of Q is

$$Q'(t) = \frac{(t^2 + 25t)(18) - (18t - 10)(2t + 25)}{(t^2 + 25t)^2} = -\frac{2(9t^2 - 10t - 125)}{(t^2 + 25t)^2}$$

Solving the equation $Q'(t) = 0$ with the quadratic formula, and retaining only the positive root, we get

$$t = \tfrac{5}{9}(1 + \sqrt{46}) \approx 4.32 \text{ seconds}$$

This value of t gives the largest fraction of foraging time because $Q'(t)$ is positive to the left and negative to the right of $t = \frac{5}{9}(1 + \sqrt{46})$. ■

EXAMPLE 5 | BB **Sustainable harvesting**[2] For many natural fish populations, the net number of new recruits to the population in a given year can be modeled as a function of the existing population size N by an equation of the form

$$R(N) = rN\left(1 - \frac{N}{K}\right)$$

where r and K are positive constants. (K is called the carrying capacity.) The population will increase if the net number of recruits $R(N)$ is positive and it will decrease if $R(N)$ is negative. Thus, because $R(N)$ is positive when $0 < N < K$ and $R(N)$ is negative when $N > K$, we expect the population to stabilize at a constant size of $N = K$.

If the population is subject to harvesting, N will begin to change and once the population has stabilized again, the number of fish harvested each year, which we denote by H, must equal the net recruitment for that year; that is, $R(N) = H$.

(a) Suppose $H = hN$, where h is a measure of the "fishing effort" expended. What is the population size once it has stabilized?

2. C. Clark, "The Economics of Overexploitation," *Science* 181 (1973): 630–34.

(b) Express the total harvest rate H as a function of h once the population has stabilized, and determine the fishing effort that results in the largest possible harvest rate.
(c) What is the population size once it has stabilized if this fishing effort is used, and what is the total harvest?

SOLUTION

(a) Since $H = hN$, once the population has stabilized we have

$$hN = H = rN\left(1 - \frac{N}{K}\right)$$

and so

$$h = r\left(1 - \frac{N}{K}\right) \qquad \Rightarrow \qquad \frac{rN}{K} = r - h$$

Solving this equation for N, we get

$$N = \frac{K}{r}(r - h) = K\left(1 - \frac{h}{r}\right)$$

This is the population size once it has become stable.

(b) With the expression for N from part (a), we have

$$H = hN = K\left(h - \frac{h^2}{r}\right)$$

and so

$$\frac{dH}{dh} = K\left(1 - \frac{2h}{r}\right) = 0 \quad \text{when } h = \frac{r}{2}$$

Also, dH/dh is positive when $h < \frac{1}{2}r$ and negative when $h > \frac{1}{2}r$. So the largest possible total harvest rate occurs when $h = \frac{1}{2}r$.

(c) If this optimal fishing effort is used, then the ultimate population size is

$$N = K\left(1 - \frac{\frac{1}{2}r}{r}\right) = K\left(1 - \tfrac{1}{2}\right) = \frac{K}{2}$$

and the total harvest is

$$H = hN = \frac{r}{2} \cdot \frac{K}{2} = \frac{rK}{4}$$

■

EXAMPLE 6 | Branching blood vessels The blood vascular system consists of blood vessels (arteries, arterioles, capillaries, and veins) that convey blood from the heart to the organs and back to the heart. This system should work so as to minimize the energy expended by the heart in pumping the blood. In particular, this energy is reduced when the resistance of the blood is lowered. One of Poiseuille's Laws gives the resistance R of the blood as

$$R = C\frac{L}{r^4}$$

where L is the length of the blood vessel, r is the radius, and C is a positive constant determined by the viscosity of the blood. (Poiseuille established this law experimentally, but it also follows from the results of Section 6.3.) Figure 7 shows a main blood vessel with radius r_1 branching at an angle θ into a smaller vessel with radius r_2.

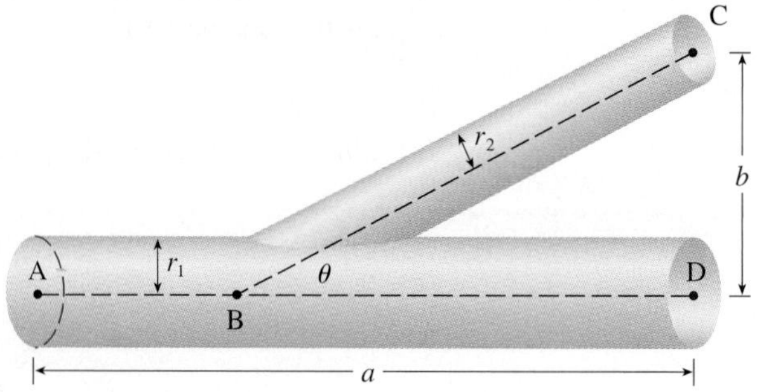

FIGURE 7
Vascular branching

(a) Express the total resistance R of the blood along the path ABC as a function of the branching angle θ.
(b) Show that R is minimized when

$$\cos\theta = \frac{r_2^4}{r_1^4}$$

(c) Find the optimal branching angle (correct to the nearest degree) when the radius of the smaller blood vessel is two-thirds the radius of the larger vessel.

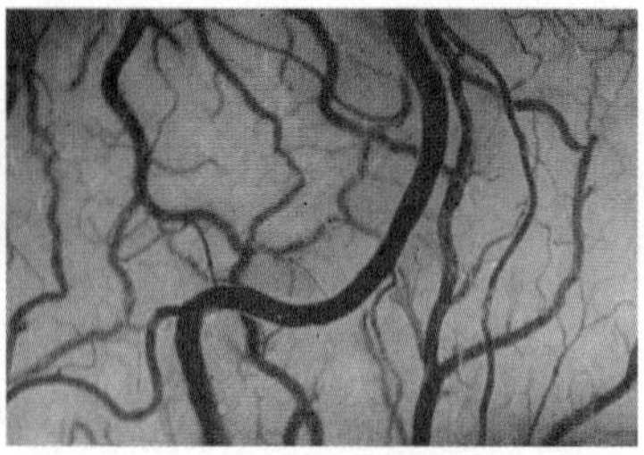

SOLUTION

(a) From Figure 7 we see that

$$\sin\theta = \frac{b}{|BC|} \qquad \Rightarrow \qquad |BC| = b\csc\theta$$

and

$$\cos\theta = \frac{|BD|}{|BC|} \qquad \Rightarrow \qquad |BC| = (a - |AB|)\sec\theta$$

Equating the expressions for $|BC|$, we have

$$b\csc\theta = (a - |AB|)\sec\theta \qquad \Rightarrow \qquad b\cot\theta = a - |AB|$$

so

$$|AB| = a - b\cot\theta$$

Then, from Poiseuille's Law, the total resistance is

$$R(\theta) = C\frac{|AB|}{r_1^4} + C\frac{|BC|}{r_2^4} = C\left(\frac{a - b\cot\theta}{r_1^4} + \frac{b\csc\theta}{r_2^4}\right)$$

(b) Differentiating the expression for $R(\theta)$ in part (a), we get

$$R'(\theta) = C\left(\frac{b\csc^2\theta}{r_1^4} - \frac{b\csc\theta\cot\theta}{r_2^4}\right) = bC\csc\theta\left(\frac{\csc\theta}{r_1^4} - \frac{\cot\theta}{r_2^4}\right)$$

So

$$R'(\theta) = 0 \qquad \Longleftrightarrow \qquad \frac{\csc\theta}{r_1^4} = \frac{\cot\theta}{r_2^4} \qquad \Longleftrightarrow \qquad \frac{r_2^4}{r_1^4} = \frac{\cot\theta}{\csc\theta} = \cos\theta$$

Now

$$R'(\theta) > 0 \iff \frac{\csc\theta}{r_1^4} > \frac{\cot\theta}{r_2^4} \iff \cos\theta < \frac{r_2^4}{r_1^4}$$

and

$$R'(\theta) < 0 \quad \text{when} \quad \cos\theta > \frac{r_2^4}{r_1^4}$$

Noting that $\cos\theta$ is a decreasing function of θ for $0 < \theta < \pi$, we conclude from the First Derivative Test that the resistance has an absolute minimum value when

$$\cos\theta = \frac{r_2^4}{r_1^4}$$

(c) When $r_2 = \frac{2}{3}r_1$ we have $\cos\theta = \left(\frac{2}{3}\right)^4$ and so

$$\theta = \cos^{-1}\left(\tfrac{2}{3}\right)^4 \approx 79^\circ$$

■

EXERCISES 4.4

1. Consider the following problem: Find two numbers whose sum is 23 and whose product is a maximum.

(a) Make a table of values, like the following one, so that the sum of the numbers in the first two columns is always 23. On the basis of the evidence in your table, estimate the answer to the problem.

First number	Second number	Product
1	22	22
2	21	42
3	20	60
⋮	⋮	⋮

(b) Use calculus to solve the problem and compare with your answer to part (a).

2. Find two numbers whose difference is 100 and whose product is a minimum.

3. Find two positive numbers whose product is 100 and whose sum is a minimum.

4. The sum of two positive numbers is 16. What is the smallest possible value of the sum of their squares?

5. Find the dimensions of a rectangle with perimeter 100 m whose area is as large as possible.

6. Photosynthesis The rate (in mg carbon/m^3/h) at which photosynthesis takes place for a species of phytoplankton is modeled by the function

$$P = \frac{100I}{I^2 + I + 4}$$

where I is the light intensity (measured in thousands of foot-candles). For what light intensity is P a maximum?

7. Crop yield A model used for the yield Y of an agricultural crop as a function of the nitrogen level N in the soil (measured in appropriate units) is

$$Y = \frac{kN}{1 + N^2}$$

where k is a positive constant. What nitrogen level gives the best yield?

8. The **measles pathogenesis** function

$$f(t) = -t(t - 21)(t + 1)$$

is used in Section 5.1 to model the development of the disease, where t is measured in days and $f(t)$ represents the number of infected cells per milliliter of plasma. What is the peak infection time for the measles virus?

9. Consider the following problem: A farmer with 750 ft of fencing wants to enclose a rectangular area and then divide it into four pens with fencing parallel to one side of the rectangle. What is the largest possible total area of the four pens?

(a) Draw several diagrams illustrating the situation, some with shallow, wide pens and some with deep, narrow pens. Find the total areas of these configurations. Does it appear that there is a maximum area? If so, estimate it.

(b) Draw a diagram illustrating the general situation. Introduce notation and label the diagram with your symbols.

(c) Write an expression for the total area.

(d) Use the given information to write an equation that relates the variables.

(e) Use part (d) to write the total area as a function of one variable.
(f) Finish solving the problem and compare the answer with your estimate in part (a).

10. Consider the following problem: A box with an open top is to be constructed from a square piece of cardboard, 3 ft wide, by cutting out a square from each of the four corners and bending up the sides. Find the largest volume that such a box can have.
(a) Draw several diagrams to illustrate the situation, some short boxes with large bases and some tall boxes with small bases. Find the volumes of several such boxes. Does it appear that there is a maximum volume? If so, estimate it.
(b) Draw a diagram illustrating the general situation. Introduce notation and label the diagram with your symbols.
(c) Write an expression for the volume.
(d) Use the given information to write an equation that relates the variables.
(e) Use part (d) to write the volume as a function of one variable.
(f) Finish solving the problem and compare the answer with your estimate in part (a).

11. If 1200 cm^2 of material is available to make a box with a square base and an open top, find the largest possible volume of the box.

12. A box with a square base and open top must have a volume of 32,000 cm^3. Find the dimensions of the box that minimize the amount of material used.

13. (a) Show that of all the rectangles with a given area, the one with smallest perimeter is a square.
(b) Show that of all the rectangles with a given perimeter, the one with greatest area is a square.

14. A rectangular storage container with an open top is to have a volume of 10 m^3. The length of its base is twice the width. Material for the base costs \$10 per square meter. Material for the sides costs \$6 per square meter. Find the cost of materials for the cheapest such container.

15. Find the point on the line $y = 2x + 3$ that is closest to the origin.

16. Find the point on the curve $y = \sqrt{x}$ that is closest to the point $(3, 0)$.

17. Age and size at maturity Most organisms grow for a period of time before maturing reproductively. For many species of insects and fish, the later the age a at maturity, the larger the individual will be, and this translates into a greater reproductive output. At the same time, however, the probability of surviving to maturity decreases as the age of maturity increases. These contrasting effects can be combined into a single measure of reproductive success in different ways. Suppose μ is a constant representing mortality rate. Find the optimal age at maturity for the following models.

(a) $r = \dfrac{\ln(ae^{-\mu a})}{a}$ (b) $R = ae^{-\mu a}$

Source: Adapted from D. Roff, *The Evolution of Life Histories: Theory and Analysis* (New York: Chapman and Hall, 1992).

18. Enzootic stability Suppose the rate at which people of age a get infected with a pathogen is given by $\lambda e^{-\lambda a}$, where λ is a positive constant. Not all infections develop into disease. Suppose that, of those individuals of age a that get infected, a fraction pa develop the disease (where p is a number chosen so that $0 < pa < 1$ over the ages of interest). At what age is the rate of disease development the highest?

Source: Adapted from P. Coleman et al., "Endemic Stability—A Veterinary Idea Applied to Public Health," *The Lancet* 357 (2001): 1284–86.

19. If $C(x)$ is the cost of producing x units of a commodity, then the **average cost** per unit is $c(x) = C(x)/x$. The **marginal cost** is the rate of change of the cost with respect to the number of items produced, that is, the derivative $C'(x)$.
(a) Show that if the average cost is a minimum, then the marginal cost equals the average cost.
(b) If $C(x) = 16{,}000 + 200x + 4x^{3/2}$, in dollars, find (i) the cost, average cost, and marginal cost at a production level of 1000 units; (ii) the production level that will minimize the average cost; and (iii) the minimum average cost.

20. If $R(x)$ is the revenue that a company receives when it sells x units of a product, then the **marginal revenue function** is the derivative $R'(x)$. The profit function is

$$P(x) = R(x) - C(x)$$

where C is the cost function from Exercise 19.
(a) Show that if the profit $P(x)$ is a maximum, then the marginal revenue equals the marginal cost.
(b) If $C(x) = 16{,}000 + 500x - 1.6x^2 + 0.004x^3$ is the cost function and $R(x) = 1700x - 7x^2$ is the revenue function, find the production level that will maximize profit.

21. Sustainable harvesting Example 5 was based on the assumption that we want to maximize the total harvest size H, but instead we might want to maximize profit.
(a) Suppose the selling price of a unit of harvest is p dollars and the cost per unit harvested is C dollars. (Assume $p > C$.) Show that the fishing effort that maximizes profit is the same as the effort that maximizes the harvest size H.
(b) Suppose the selling price of a unit of harvest is p dollars and the unit cost is inversely proportional to the fishing effort h (that is, $C = \alpha/h$). Assume $\alpha < rp$. What is the fishing effort that maximizes profit? What is the limiting population size, assuming this fishing effort is used?

(c) Explain how and why your answer to part (b) differs from that of part (a).

22. The **von Bertalanffy model** for the growth of an individual fish assumes that an individual acquires energy at a rate that is proportional to its length squared, but that it uses energy for metabolism at a rate proportional to its length cubed. The underlying idea is that energy acquisition is proportional to a fish's surface area, while energy utilization for metabolism is proportional to its mass (and thus its volume). Suppose the net energy available for growth at any time is the difference between acquisition and utilization through metabolism. What is the length at which growth will be fastest, as a function of the constants of proportionality?

23. Nectar foraging by bumblebees Suppose that, instead of the specific nectar function in Example 2, we have an arbitrary function N with $N(0) = 0$, $N(t) \geqslant 0$, $N'(t) > 0$, $N''(t) < 0$, and arbitrary travel time T.

(a) Interpret the conditions on the function N.

(b) Show that the optimal foraging time t satisfies the equation

$$N'(t) = \frac{N(t)}{t + T}$$

(c) Show that, for any foraging time t satisfying the equation in part (b), the second derivative condition for a maximum value of the foraging function f in Example 2 is satisfied.

24. Aquatic birds Suppose that, instead of the specific oxygen function in Example 4, we have an arbitrary function O with $O(0) = 0$, $O(t) \geqslant 0$, $O'(t) > 0$, $O''(t) < 0$, and arbitrary travel time T.

(a) Interpret the conditions on the function O.

(b) Show that the optimal surface time t satisfies the equation

$$O'(t) = \frac{O(t) - rT}{t + T}$$

(c) Show that, for any surface time t satisfying the equation in part (b), the second derivative condition for a maximum value of the foraging function Q in Example 4 is satisfied.

25. Let v_1 be the velocity of light in air and v_2 the velocity of light in water. According to Fermat's Principle, a ray of light will travel from a point A in the air to a point B in the water by a path ACB that minimizes the time taken. Show that

$$\frac{\sin\theta_1}{\sin\theta_2} = \frac{v_1}{v_2}$$

where θ_1 (the angle of incidence) and θ_2 (the angle of refraction) are as shown. This equation is known as Snell's Law.

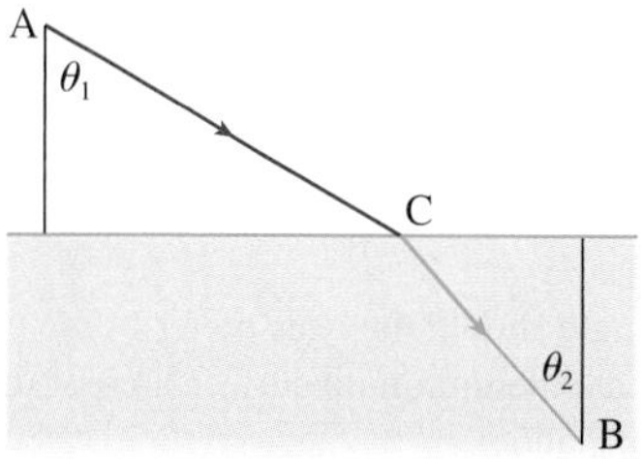

26. The graph shows the fuel consumption c of a car (measured in gallons per hour) as a function of the speed v of the car. At very low speeds the engine runs inefficiently, so initially c decreases as the speed increases. But at high speeds the fuel consumption increases. You can see that $c(v)$ is minimized for this car when $v \approx 30$ mi/h. However, for fuel efficiency, what must be minimized is not the consumption in gallons per hour but rather the fuel consumption in gallons *per mile*. Let's call this consumption G. Using the graph, estimate the speed at which G has its minimum value.

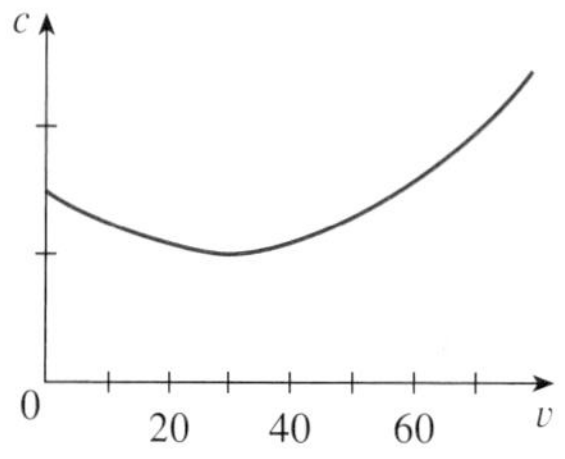

27. Beehives In a beehive, each cell is a regular hexagonal prism, open at one end with a trihedral angle at the other end as in the figure.

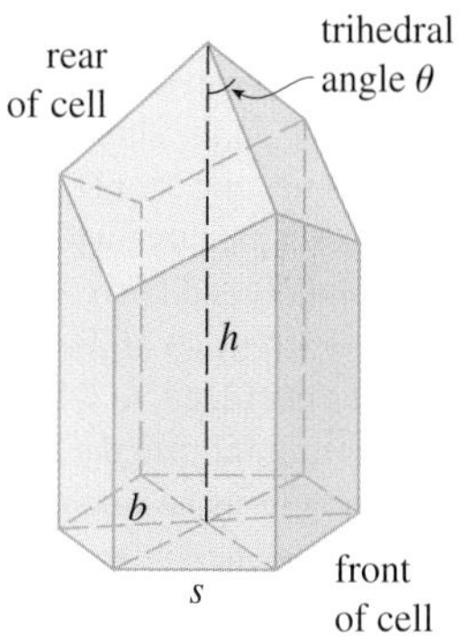

It is believed that bees form their cells in such a way as to minimize the surface area for a given side length and height, thus using the least amount of wax in cell construction. Examination of these cells has shown that the measure of the apex angle θ is amazingly consistent. Based on the geometry of the cell, it can be shown that the surface area S is given by

$$S = 6sh - \tfrac{3}{2}s^2 \cot\theta + \left(3s^2\sqrt{3}/2\right)\csc\theta$$

where s, the length of the sides of the hexagon, and h, the height, are constants.

(a) Calculate $dS/d\theta$.
(b) What angle should the bees prefer?
(c) Determine the minimum surface area of the cell (in terms of s and h).

Note: Actual measurements of the angle θ in beehives have been made, and the measures of these angles seldom differ from the calculated value by more than 2°.

28. Swimming speed of fish For a fish swimming at a speed v relative to the water, the energy expenditure per unit time is proportional to v^3. It is believed that migrating fish try to minimize the total energy required to swim a fixed distance. If the fish are swimming against a current u ($u < v$), then the time required to swim a distance L is $L/(v - u)$ and the total energy E required to swim the distance is given by

$$E(v) = av^3 \cdot \frac{L}{v - u}$$

where a is the proportionality constant.

(a) Determine the value of v that minimizes E.
(b) Sketch the graph of E.

Note: This result has been verified experimentally; migrating fish swim against a current at a speed 50% greater than the current speed.

29. Bird flight paths Ornithologists have determined that some species of birds tend to avoid flights over large bodies of water during daylight hours. It is believed that more energy is required to fly over water than over land because air generally rises over land and falls over water during the day. A bird with these tendencies is released from an island that is 5 km from the nearest point B on a straight shoreline, flies to a point C on the shoreline, and then flies along the shoreline to its nesting area D. Assume that the bird instinctively chooses a path that will minimize its energy expenditure. Points B and D are 13 km apart.

(a) In general, if it takes 1.4 times as much energy to fly over water as it does over land, to what point C should the bird fly in order to minimize the total energy expended in returning to its nesting area?
(b) Let W and L denote the energy (in joules) per kilometer flown over water and land, respectively. What would a large value of the ratio W/L mean in terms of the bird's flight? What would a small value mean? Determine the ratio W/L corresponding to the minimum expenditure of energy.
(c) What should the value of W/L be in order for the bird to fly directly to its nesting area D? What should the value of W/L be for the bird to fly to B and then along the shoreline to D?
(d) If the ornithologists observe that birds of a certain species reach the shore at a point 4 km from B, how many times more energy does it take a bird to fly over water than over land?

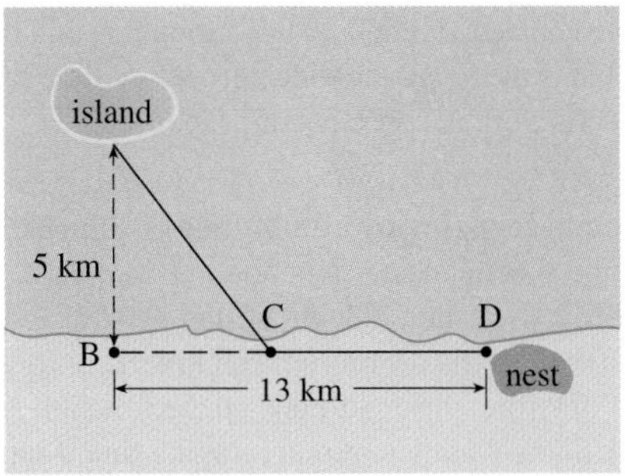

30. Crows and whelks Crows on the west coast of Canada feed on whelks by carrying them to heights of about 5 m and dropping them onto rocks (several times if necessary) to break open their shells. Two of the questions raised by the author of a study of this phenomenon were "Do crows drop whelks from the best height for breaking?" and "How energetically profitable is dropping of whelks?" The author constructed poles and dropped whelks from various heights. A model based on the study's data for the number of times a whelk needs to be dropped from a height h to be broken is

$$n(h) = \frac{h + 14.8}{h - 1.2}$$

where h is measured in meters. The energy expended by a crow in this activity is proportional to the height h and to the number $n(h)$:

$$E = khn(h) = \frac{kh(h + 14.8)}{h - 1.2}$$

(a) What value of h minimizes the energy expended by the crows?
(b) How does your answer to part (a) compare with the observed average dropping height of 5.3 m that is actually used by crows? Does the model support the existence of an optimal foraging strategy?

Source: Adapted from R. Zach, "Decision-making and Optimal Foraging in Northwestern Crows," *Behaviour* 68 (1979): 106–17.

PROJECT Flapping and Gliding

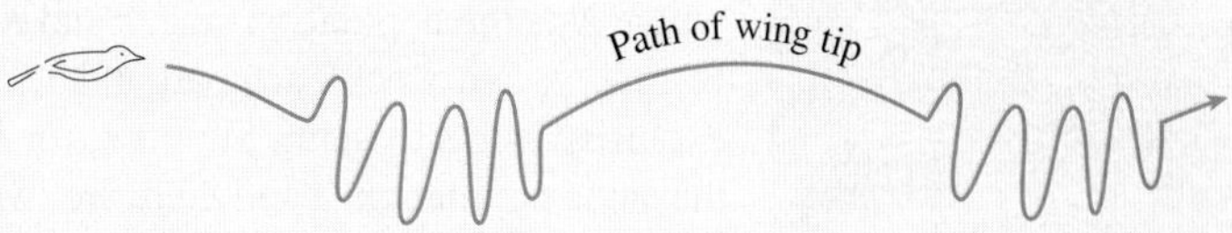

FIGURE 1

Small birds like finches alternate between flapping their wings and keeping them folded while gliding (see Figure 1). In this project we analyze this phenomenon and try to determine how frequently a bird should flap its wings. Some of the principles are the same as for fixed-wing aircraft and so we begin by considering how required power and energy depend on the speed of airplanes.[1]

1. The power needed to propel an airplane forward at velocity v is

$$P = Av^3 + \frac{BL^2}{v}$$

where A and B are positive constants specific to the particular aircraft and L is the lift, the upward force supporting the weight of the plane. Find the speed that minimizes the required power.

2. The speed found in Problem 1 minimizes power but a faster speed might use less fuel. The energy needed to propel the airplane a unit distance is $E = P/v$. At what speed is energy minimized?

3. How much faster is the speed for minimum energy than the speed for minimum power?

4. When applying the equation of Problem 1 to bird flight, we split the term Av^3 into two parts: $A_b v^3$ for the bird's body and $A_w v^3$ for its wings. Let x be the fraction of flying time spent in flapping mode. If m is the bird's mass and all the lift occurs during flapping, then the lift is mg/x, where g is the acceleration due to gravity, and so the power needed during flapping is

$$P_{\text{flap}} = (A_b + A_w)v^3 + \frac{B(mg/x)^2}{v}$$

The power while wings are folded is $P_{\text{fold}} = A_b v^3$. Show that the average power over an entire flight cycle is

$$\overline{P} = xP_{\text{flap}} + (1 - x)P_{\text{fold}} = A_b v^3 + xA_w v^3 + \frac{Bm^2g^2}{xv}$$

5. For what value of x is the average power a minimum? What can you conclude if the bird flies slowly? What can you conclude if the bird flies increasingly faster?

6. The average energy over a cycle is $\overline{E} = \overline{P}/v$. What value of x minimizes $\overline{E}$?

7. Compare your answers to Problems 5 and 6. What do you notice?

---

1. Adapted from R. Alexander, *Optima for Animals* (Princeton, NJ: Princeton University Press, 1996).

PROJECT The Tragedy of the Commons: An Introduction to Game Theory

In Example 4.4.5 we explored sustainable fish harvesting. We assumed that a single company is exploiting the resource and found that the steady-state population size in the presence of harvesting satisfied the equation

$$rN\left(1 - \frac{N}{K}\right) = hN$$

where N is the population size, r and K are positive constants, and h is the fishing effort. In reality, fish stocks are part of the "Commons," meaning that no single person has exclusive rights to them. Suppose, for example, that a second company begins to exploit the same population. Then there are two fishing efforts, h_1 and h_2, one for each company. Once the population size has stabilized, the equation

$$rN\left(1 - \frac{N}{K}\right) = h_1N + h_2N$$

must hold, where h_1N and h_2N are the total harvests for companies 1 and 2, respectively. Suppose you run company 1 and before company 2 arrives you are using the optimal h calculated in Example 4.4.5, that is, $h_1 = \frac{1}{2}r$.

1. When company 2 arrives, it needs to decide upon a fishing effort h_2. What value of h_2 maximizes its harvest once the population has reached a steady state, assuming that you continue using h_1?

2. Once your competitor is using their rate obtained in Problem 1, your harvesting rate will no longer be optimal for you. What is your new optimal rate h_1^*, given that your competitor continues to use the rate found in Problem 1?

3. More generally, determine your optimal fishing effort as a function of the rate your competitor uses and your competitor's optimal fishing effort as a function of the rate you use. These are referred to as the "best response" fishing efforts.

4. The harvesting problem can be viewed as a game played between the two companies, where the payoff to each depends on both of their choices of fishing effort. An area of mathematics called **game theory** has been developed to analyze such problems. A key concept in game theory is that of a **Nash equilibrium**, which is a pair of values h_1^* and h_2^* that simultaneously satisfy both best response functions. At a Nash equilibrium each party is doing the best that it can, given the choice of its competitor. What is the Nash equilibrium pair of fishing efforts?

5. What is the total population size at the Nash equilibrium, and what are the total harvests of you and your competitor?

6. Demonstrate that both you and your competitor could have a higher total harvest than that attained at the Nash equilibrium if you could agree to cooperate and to split the harvest that you were obtaining before your competitor showed up.

7. Problem 6 shows that both you and your competitor are worse off at the Nash equilibrium than you would be if you agreed to cooperate. Show that, in terms of population size, the fish population is also worse off. This is the "Tragedy of the Commons."

Nash

John F. Nash, Jr. (1928–2015) was an American mathematician best known for his work in game theory. He developed the idea now known as a Nash equilibrium in his 28-page doctoral thesis in 1950. In 1994 he was awarded the Nobel Prize in Economics for this work. Nash also made several other foundational contributions to advanced mathematics, despite suffering from schizophrenia. His extraordinary life is chronicled in the book *A Beautiful Mind* by Sylvia Nasar and in a Hollywood movie of the same name.

4.5 Recursions: Equilibria and Stability

In Section 1.6 we looked at recursive sequences, which we also called difference equations or discrete-time models. These are defined by a recursion of the form

$$a_{n+1} = f(a_n) \qquad \text{or} \qquad x_{t+1} = f(x_t) \qquad \text{or} \qquad N_{t+1} = f(N_t)$$

where f is the updating function, N_t is the number of individuals in a population at time t, and N_{t+1} is the population one unit of time into the future. Then in Section 2.1 we investigated the long-term behavior of such recursions. In particular, we saw that some recursive sequences approach a limiting value as t becomes large:

$$\lim_{t\to\infty} x_t = L$$

Here we assume that the updating function f that defines the recursion is a differentiable function and learn that the values of its derivative play a role in determining the limiting behavior of the sequence.

Equilibria

(1) Definition An **equilibrium** of a recursive sequence $x_{t+1} = f(x_t)$ is a number $\hat{x}$ that is left unchanged by the function f, that is,

$$f(\hat{x}) = \hat{x}$$

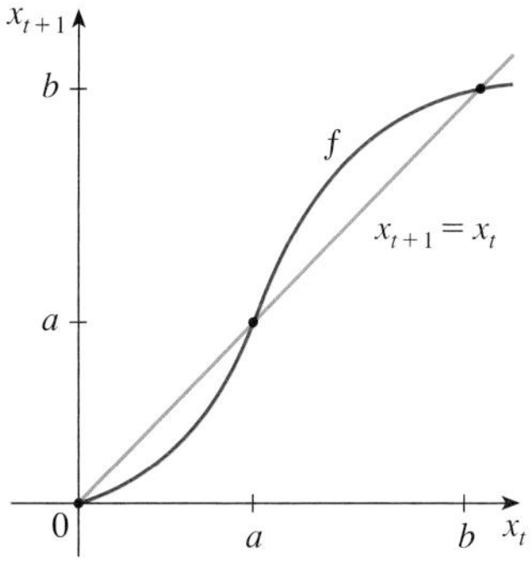

FIGURE 1
The recursion $x_{t+1} = f(x_t)$ has three equilibria

It's helpful to think of an equilibrium as a point on a number line. An equilibrium is sometimes called a **fixed point** because f leaves the point $\hat{x}$ fixed. Notice that if $\hat{x}$ is an equilibrium and if, for instance, $x_6 = \hat{x}$, then

$$x_7 = f(x_6) = f(\hat{x}) = \hat{x}$$

and, similarly, all of the following terms in the sequence are also equal to $\hat{x}$.

To find the equilibria algebraically, we solve the equation $f(x) = x$, if possible. To locate them geometrically we graph the curves $y = f(x)$ and $y = x$ (the diagonal line) and see where they intersect. Because the recursion is $x_{t+1} = f(x_t)$, when we graph f we label the horizontal and vertical axes x_t and x_{t+1}, as in Figure 1. For that particular recursion we see that there are three points of intersection and therefore three equilibria, 0, a, and b.

(2) Definition An equilibrium is called **stable** if solutions that begin close to the equilibrium approach that equilibrium. It is called **unstable** if solutions that start close to the equilibrium move away from it.

So when we say that $\hat{x}$ is a stable equilibrium of the recursion $x_{t+1} = f(x_t)$ we mean that if x_t is a solution of the recursion and x_0 is sufficiently close to $\hat{x}$, then $x_t \to \hat{x}$ as $t \to \infty$.

EXAMPLE 1 | Determine the equilibrium of the difference equation $N_{t+1} = RN_t$, where $R > 0$, and classify it as stable or unstable.

SOLUTION The equilibrium $\hat{N}$ satisfies the equation $\hat{N} = R\hat{N}$. The only solution of $(R - 1)\hat{N} = 0$ is $\hat{N} = 0$, unless $R = 1$. We know that the solution of the recursion $N_{t+1} = RN_t$ is $N_t = N_0 \cdot R^t$. There are three cases:

- If $0 < R < 1$, then $N_t = N_0 \cdot R^t \to 0$ as $t \to \infty$, so $N_t \to \hat{N} = 0$. Therefore the equilibrium $\hat{N} = 0$ is stable in this case.
- If $R > 1$, then $N_t = N_0 \cdot R^t \to \infty$ as $t \to \infty$, and so the equilibrium $\hat{N} = 0$ is unstable in this case.
- If $R = 1$, then $N_t = N_0$ for all t. This case is called *neutral*. ■

■ Cobwebbing

There is a graphical method for finding equilibria and determining whether they are stable or unstable. It is called **cobwebbing** and is illustrated in Figure 2. We start with an initial value x_0 on the horizontal axis and locate $x_1 = f(x_0)$ as the distance from the point x_0 up to the point (x_0, x_1) on the graph of f. Then we draw the horizontal line segment from (x_0, x_1) to the point (x_1, x_1) on the diagonal line. The point x_1 lies directly beneath (x_1, x_1) on the horizontal axis.

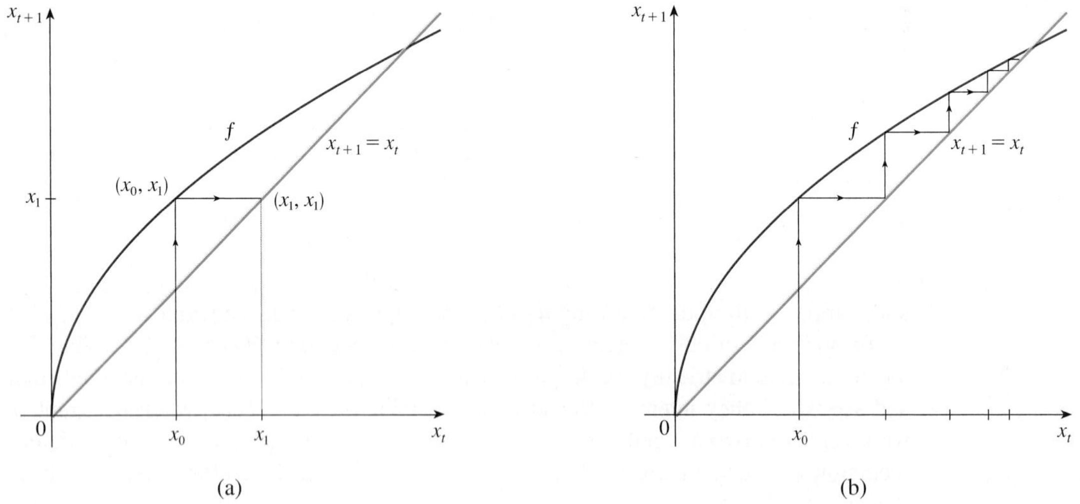

FIGURE 2
Cobwebbing

In Figure 2(b) we repeat this procedure to obtain x_2 from x_1, drawing a vertical line segment from (x_1, x_1) to (x_1, x_2) on the graph of f and then a horizontal line segment over to the diagonal. Continuing in this manner we create a zigzag path that reflects off the diagonal line and shows how the successive terms in the sequence can be obtained geometrically.

EXAMPLE 2 | Use cobwebbing to determine whether the equilibria $\hat{x} = 0$, $\hat{x} = a$, and $\hat{x} = b$ in Figure 1 are stable or unstable.

SOLUTION Figure 3 is a larger version of Figure 1. We experiment with different initial values and use cobwebbing to visualize the values of x_t. We notice that

$$\text{if } a < x_0 < b, \text{ then } \lim_{t \to \infty} x_t = b$$

$$\text{but} \quad \text{if } 0 < x_0 < a, \text{ then } \lim_{t \to \infty} x_t = 0$$

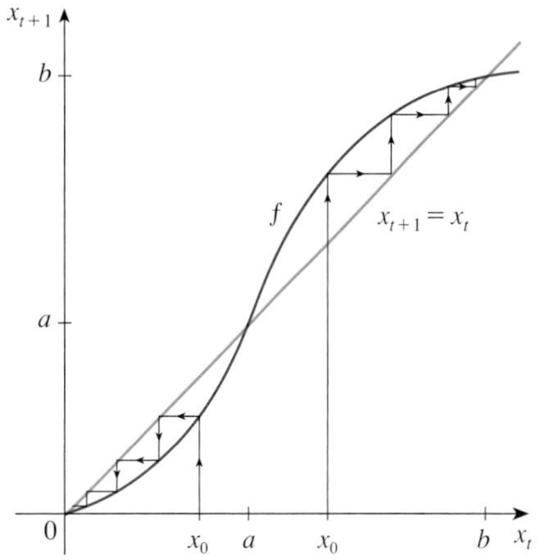

FIGURE 3
Cobwebbing with stable and unstable equilibria

Solutions that start close to b approach b, so $\hat{x} = b$ is a stable equilibrium. Likewise, solutions that start close to 0 approach 0, so $\hat{x} = 0$ is also a stable equilibrium. But solutions that start close to a (on either side of a) move away from a. So $\hat{x} = a$ is an unstable equilibrium. ■

So far we have used cobwebbing only with increasing functions f. Figure 4 shows what happens when f decreases. We apply cobwebbing with initial value x_0 to a difference equation $x_{t+1} = f(x_t)$ with decreasing f. Instead of the zigzag paths in Figures 2 and 3, you can see that we get spiral paths and the values of x_t oscillate around the equilibrium $\hat{x}$. In Figure 4(a), $x_t \to \hat{x}$ as $t \to \infty$, so $\hat{x}$ is stable. In Figure 4(b), however, the values of x_t move away from $\hat{x}$, so $\hat{x}$ is unstable.

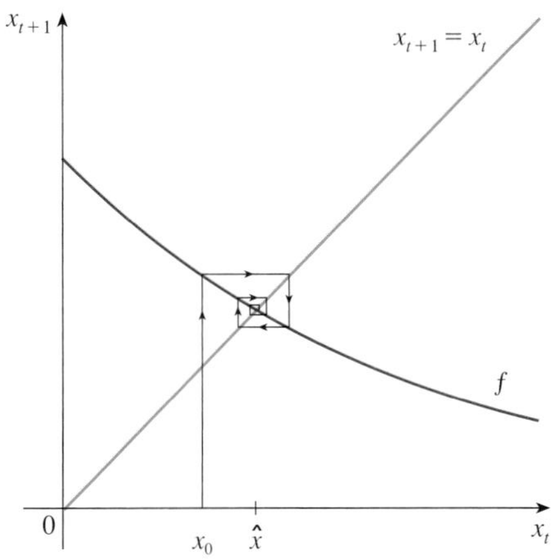

(a) Stable spiral

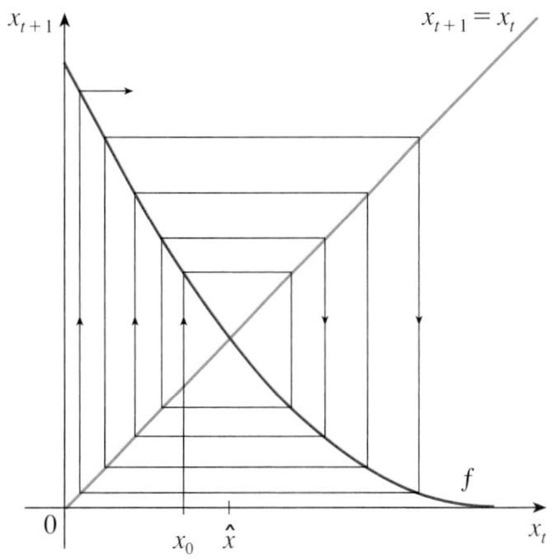

(b) Unstable spiral

FIGURE 4 Spiral cobwebbing

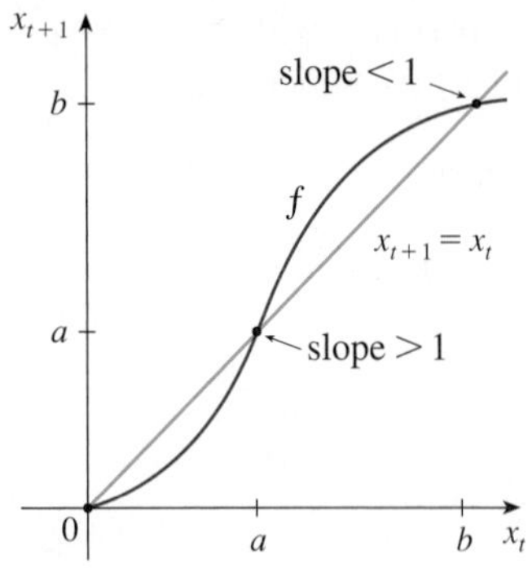

FIGURE 5

■ Stability Criterion

An equilibrium occurs when the graph of f crosses the diagonal line, which has slope 1. Figure 5 shows the increasing function f from Figure 3 and we see that at the stable equilibrium $\hat{x} = b$ the curve crosses the diagonal from above to below, so $f'(b) < 1$. At the unstable equilibrium $\hat{x} = a$ the curve crosses the diagonal from below to above, so $f'(a) > 1$.

If f is decreasing, we see from diagrams like Figure 4 that stable spirals occur when $-1 < f'(\hat{x}) < 0$ and unstable spirals occur for steeper curves, that is, $f'(\hat{x}) < -1$.

To summarize, our intuition tells us that equilibria are stable when $-1 < f'(\hat{x}) < 1$ and unstable when $f'(\hat{x}) > 1$ or $f'(\hat{x}) < -1$. So the following theorem appears plausible. A proof, using the Mean Value Theorem, appears in Appendix E.

> **(3) The Stability Criterion for Recursive Sequences** Suppose that $\hat{x}$ is an equilibrium of the recursive sequence $x_{t+1} = f(x_t)$, where f' is continuous. If $|f'(\hat{x})| < 1$, the equilibrium is stable. If $|f'(\hat{x})| > 1$, the equilibrium is unstable.

Let's revisit some of the difference equations we studied in Section 2.1 and see how the Stability Criterion applies to those equations.

EXAMPLE 3 | BB **Drug concentration** In Example 2.1.5 we considered the difference equation

$$C_{n+1} = 0.3C_n + 0.2$$

where C_n is the concentration of a drug in the bloodstream of a patient after injection on the nth day, 30% of the drug remains in the bloodstream the next day, and the daily dose raises the concentration by 0.2 mg/mL.

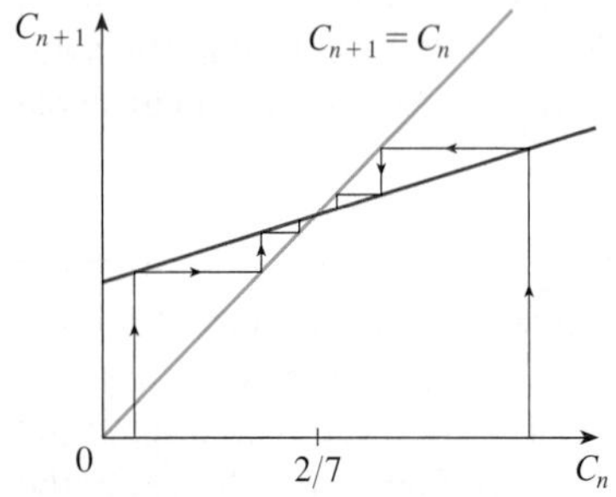

FIGURE 6

Here the recursion is of the form $C_{n+1} = f(C_n)$, where $f(x) = 0.3x + 0.2$. The equilibrium concentration is $\hat{C}$, where $0.3\hat{C} + 0.2 = \hat{C}$. Solving this equation gives $\hat{C} = \frac{2}{7}$. The derivative of f is $f'(\hat{C}) = 0.3$, which is less than 1, so the equilibrium is stable, as illustrated by the cobwebbing in Figure 6. In fact, in Section 2.1 we calculated that

$$\lim_{n \to \infty} C_n = \frac{2}{7}$$

■

EXAMPLE 4 | BB **Logistic difference equation** In Example 2.1.8 we examined the long-term behavior of the terms defined by the logistic difference equation

$$x_{t+1} = cx_t(1 - x_t)$$

for different positive values of c. Use the Stability Criterion to explain that behavior.

SOLUTION We can write the logistic equation as $x_{t+1} = f(x_t)$, where

$$f(x) = cx(1 - x)$$

We first find the equilibria by solving the equation $f(x) = x$:

$$cx(1 - x) = x \iff x = 0 \quad \text{or} \quad c(1 - x) = 1$$

So one equilibrium is $\hat{x} = 0$. To find the other one, note that

$$c - cx = 1 \iff c - 1 = cx \iff x = \frac{c - 1}{c} = 1 - \frac{1}{c}$$

So the other equilibrium is

$$\hat{x} = 1 - \frac{1}{c}$$

The derivative of $f(x) = c(x - x^2)$ is $f'(x) = c(1 - 2x)$. For the first equilibrium, $\hat{x} = 0$, we have $f'(0) = c$, so the Stability Criterion tells us that $\hat{x} = 0$ is stable if $0 < c < 1$ and unstable if $c > 1$.

For the second equilibrium, $\hat{x} = 1 - 1/c$, we have

$$f'\left(1 - \frac{1}{c}\right) = c\left[1 - 2\left(1 - \frac{1}{c}\right)\right] = c\left(\frac{2}{c} - 1\right) = 2 - c$$

The Stability Criterion says that this equilibrium is stable if $|2 - c| < 1$. But

$$|2 - c| < 1 \iff -1 < 2 - c < 1 \iff 1 < c < 3$$

We also note that $f'(\hat{x})$ is negative when $2 - c < 0$, that is, $c > 2$, so oscillation occurs when $c > 2$. Let's compile all this information in the following chart:

	$\hat{x} = 0$	$\hat{x} = 1 - \frac{1}{c}$
$0 < c < 1$	stable	
$1 < c < 2$	unstable	stable
$2 < c < 3$	unstable	stable (oscillation)
$c > 3$	unstable	unstable (oscillation)

In Exercises 17–20 you are asked to illustrate the four cases in the chart in Example 4 both by cobwebbing and by graphing the recursive sequence.

Referring to the chart, we find an explanation for what we noticed in Example 2.1.8. When $c < 3$, one of the equilibria is stable and so the terms converge to that number. But when $c > 3$ both equilibria are unstable and so the terms have nowhere to go; they don't approach any fixed number. ■

EXAMPLE 5 | BB **Ricker equation** W. E. Ricker introduced the discrete-time model

$$x_{t+1} = cx_t e^{-x_t} \qquad c > 0$$

in the context of modeling fishery populations. Find the equilibria and determine the values of c for which they are stable.

SOLUTION The Ricker equation is $x_{t+1} = f(x_t)$, where

$$f(x) = cxe^{-x}$$

To find the equilibria we solve the equation $f(x) = x$:

$$cxe^{-x} = x \iff x = 0 \text{ or } ce^{-x} = 1$$

One equilibrium is $\hat{x} = 0$. The other satisfies

$$ce^{-x} = 1 \iff c = e^x \iff x = \ln c$$

The second equilibrium is $\hat{x} = \ln c$.

We use the Product Rule to differentiate f:

$$f'(x) = cx(-e^{-x}) + ce^{-x} = c(1 - x)e^{-x}$$

For $\hat{x} = 0$ we have $f'(0) = c$, so it is stable if $0 < c < 1$ and unstable if $c > 1$. For $\hat{x} = \ln c$ we get

$$f'(\ln c) = c(1 - \ln c)e^{-\ln c} = c(1 - \ln c) \cdot \frac{1}{c} = 1 - \ln c$$

Therefore

$$|f'(\hat{x})| < 1 \iff |1 - \ln c| < 1 \iff -1 < 1 - \ln c < 1$$

Now

$$1 - \ln c < 1 \iff \ln c > 0 \iff c > 1$$

and

$$-1 < 1 - \ln c \iff \ln c < 2 \iff c < e^2$$

By the Stability Criterion, $\hat{x} = \ln c$ is stable when

$$1 < c < e^2$$

When $0 < c < 1$ or $c > e^2$, $\hat{x} = \ln c$ is unstable.

We also note that oscillation occurs when $f'(\hat{x}) < 0$, so

$$1 - \ln c < 0 \Rightarrow \ln c > 1 \Rightarrow c > e$$

Figure 7 illustrates cobwebbing for the Ricker equation for three values of c.

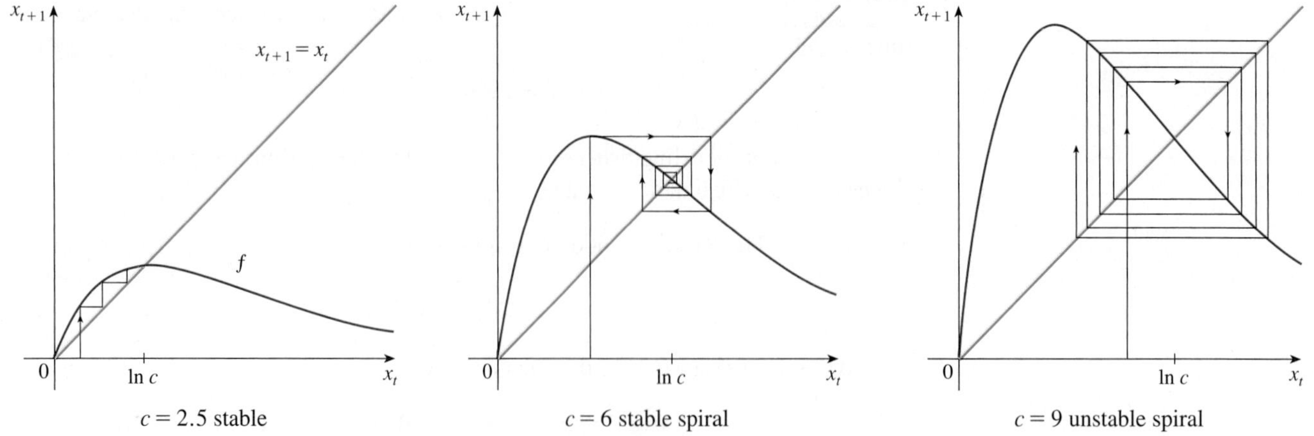

FIGURE 7

■

EXERCISES 4.5

1–4 The graph of the function f for a recursive sequence $x_{t+1} = f(x_t)$ is shown. Estimate the equilibria and classify them as stable or unstable. Confirm your answer by cobwebbing.

1.

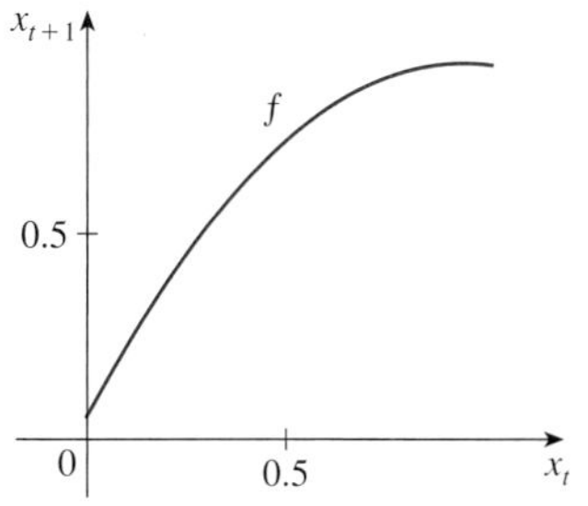

2.

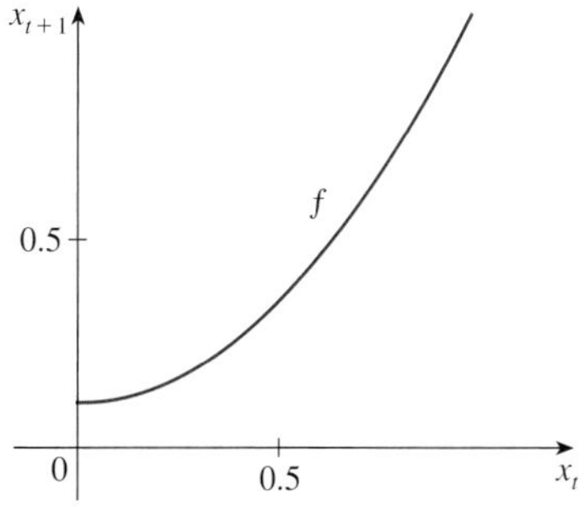

3.

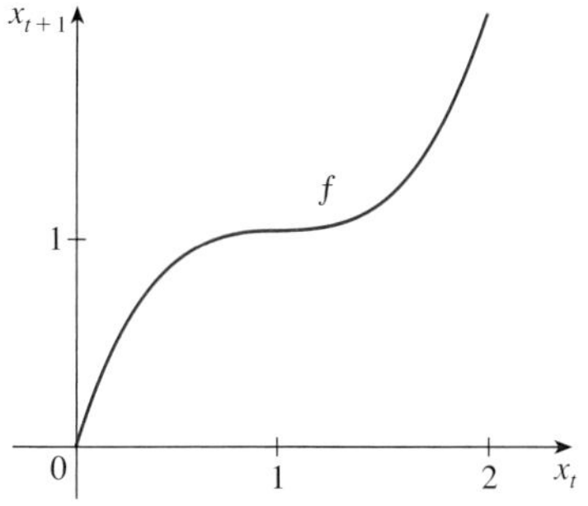

4.

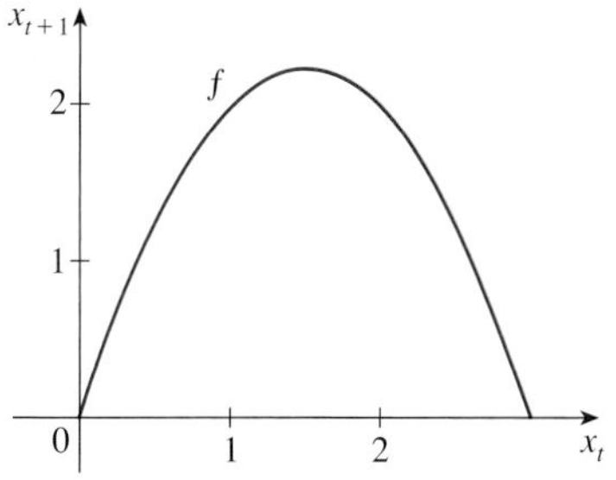

5–10 Find the equilibria of the difference equation and classify them as stable or unstable.

5. $x_{t+1} = \frac{1}{2}x_t^2$

6. $x_{t+1} = 1 - x_t^2$

7. $x_{t+1} = \dfrac{x_t}{0.2 + x_t}$

8. $x_{t+1} = \dfrac{3x_t}{1 + x_t}$

9. $x_{t+1} = 10x_t e^{-2x_t}$

10. $x_{t+1} = x_t^3 - 3x_t^2 + 3x_t$

11–12 Find the equilibria of the difference equation and classify them as stable or unstable. Use cobwebbing to find $\lim_{t\to\infty} x_t$ for the given initial values.

11. $x_{t+1} = \dfrac{4x_t^2}{x_t^2 + 3}$, $x_0 = 0.5$, $x_0 = 2$

12. $x_{t+1} = \dfrac{7x_t^2}{x_t^2 + 10}$, $x_0 = 1$, $x_0 = 3$

13–14 Find the equilibria of the difference equation. Determine the values of c for which each equilibrium is stable.

13. $x_{t+1} = \dfrac{cx_t}{1 + x_t}$

14. $x_{t+1} = \dfrac{x_t}{c + x_t}$

15. Drug pharmacokinetics A patient takes 200 mg of a drug at the same time every day. Just before each tablet is taken, 10% of the drug remains in the body.
(a) If Q_n is the quantity of the drug in the body just after the nth tablet is taken, write a difference equation expressing Q_{n+1} in terms of Q_n.
(b) Find the equilibria of the equation in part (a).
(c) Draw a cobwebbing diagram for the equation.

16. Drug pharmacokinetics A patient is injected with a drug every 8 hours. Immediately before each injection the concentration of the drug has been reduced by 40% and the new dose increases the concentration by 1.2 mg/mL.
(a) If Q_n is the concentration of the drug in the body just after the nth injection is given, write a difference equation expressing Q_{n+1} in terms of Q_n.
(b) Find the equilibria of the equation in part (a).
(c) Draw a cobwebbing diagram for the equation.

17–20 Logistic difference equation Illustrate the results of Example 4 for the logistic difference equation by cobwebbing and by graphing the first ten terms of the sequence for the given values of c and x_0.

17. $c = 0.8$, $x_0 = 0.6$

18. $c = 1.8$, $x_0 = 0.1$

19. $c = 2.7$, $x_0 = 0.1$

20. $c = 3.6$, $x_0 = 0.4$

21. Sustainable harvesting In Example 4.4.5 we looked at a model of sustainable harvesting, which can be formulated as a discrete-time model:

$$N_{t+1} = N_t + rN_t\left(1 - \frac{N_t}{K}\right) - hN_t$$

Find the equilibria and determine when each is stable.

22. Heart excitation A simple model for the time x_t it takes for an electrical impulse in the heart to travel through the atrioventricular node of the heart is

$$x_{t+1} = \frac{375}{x_t - 90} + 100 \qquad x_t > 90$$

(a) Find the relevant equilibrium and determine when it is stable.

(b) Draw a cobwebbing diagram.

Source: Adapted from D. Kaplan et al., *Understanding Nonlinear Dynamics* (New York: Springer-Verlag, 1995).

23. Species discovery curves A common assumption is that the rate of discovery of new species is proportional to the fraction of currently undiscovered species. If d_t is the fraction of species discovered by time t, a recursion equation describing this process is

$$d_{t+1} = d_t + a(1 - d_t)$$

where a is a constant representing the discovery rate and satisfies $0 < a < 1$. Find the equilibria and determine the stability.

24. Drug resistance in malaria In the project on page 78 we developed the following recursion equation for the spread of a gene for drug resistance in malaria:

$$p_{t+1} = \frac{p_t^2 W_{\text{RR}} + p_t(1 - p_t)W_{\text{RS}}}{p_t^2 W_{\text{RR}} + 2p_t(1 - p_t)W_{\text{RS}} + (1 - p_t)^2 W_{\text{SS}}}$$

where W_{RR}, W_{RS}, and W_{SS} are constants representing the probability of survival of the three genotypes. In fact this model applies to the evolutionary dynamics of any gene in a population of diploid individuals.

(a) Find the equilibria of the model in terms of the constants.

(b) Suppose that $W_{\text{RR}} = \frac{3}{4}$, $W_{\text{RS}} = \frac{1}{2}$, and $W_{\text{SS}} = \frac{1}{4}$. Determine the stability of each equilibrium (provided it lies between 0 and 1). Plot the cobwebbing diagram and interpret your results.

(c) Suppose that $W_{\text{RR}} = \frac{1}{2}$, $W_{\text{RS}} = \frac{3}{4}$, and $W_{\text{SS}} = \frac{1}{4}$. Determine the stability of each equilibrium. Plot the cobwebbing diagram and interpret your results.

25. Blood cell production A simple model of blood cell production is given by

$$R_{t+1} = R_t(1 - d) + F(R_t)$$

where d is the fraction of red blood cells that die from one day to the next and $F(x)$ is a function specifying the number of new cells produced in a day, given that the current number is x. Find the equilibria and determine the stability in each case.

(a) $F(x) = \theta(K - x)$, where θ and K are positive constants

(b) $F(x) = \dfrac{ax}{b + x^2}$, where a and b are positive constants and $a > bd$

Source: Adapted from N. Mideo et al., "Understanding and Predicting Strain-Specific Patterns of Pathogenesis in the Rodent Malaria *Plasmodium chabaudi*," *The American Naturalist* 172 (2008): E214–E328.

4.6 Antiderivatives

Suppose you know the rate at which a bacteria population is increasing and want to know the size of the population at some future time. Or suppose you know the rate of decrease of your blood alcohol concentration and want to know your BAC an hour from now. In each case, the problem is to find a function F whose derivative is a known function f. If such a function F exists, it is called an *antiderivative* of f.

> **Definition** A function F is called an **antiderivative** of f on an interval I if $F'(x) = f(x)$ for all x in I.

For instance, let $f(x) = x^2$. It isn't difficult to discover an antiderivative of f if we keep the Power Rule in mind. In fact, if $F(x) = \frac{1}{3}x^3$, then $F'(x) = x^2 = f(x)$. But the function $G(x) = \frac{1}{3}x^3 + 100$ also satisfies $G'(x) = x^2$. Therefore both F and G are antiderivatives of f. Indeed, any function of the form $H(x) = \frac{1}{3}x^3 + C$, where C is a constant, is an antiderivative of f. The following theorem says that f has no other antiderivative. A proof of Theorem 1, using the Mean Value Theorem, is outlined in Exercise 46.

> **(1) Theorem** If F is an antiderivative of f on an interval I, then the most general antiderivative of f on I is
>
> $$F(x) + C$$
>
> where C is an arbitrary constant.

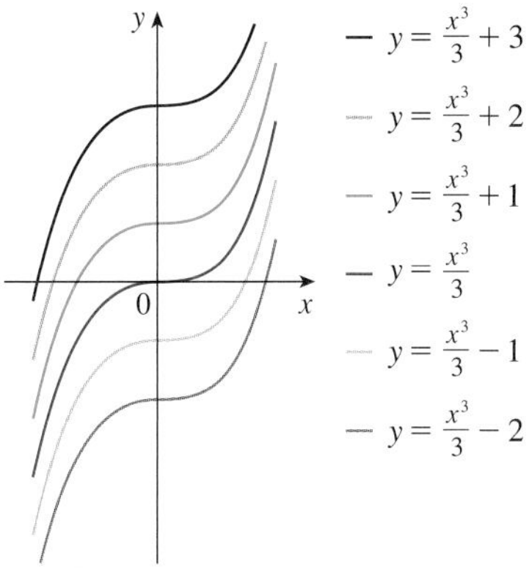

FIGURE 1
Members of the family of antiderivatives of $f(x) = x^2$

Going back to the function $f(x) = x^2$, we see that the general antiderivative of f is $\frac{1}{3}x^3 + C$. By assigning specific values to the constant C, we obtain a family of functions whose graphs are vertical translates of one another (see Figure 1). This makes sense because each curve must have the same slope at any given value of x.

EXAMPLE 1 | Find the most general antiderivative of each of the following functions.

(a) $f(x) = \sin x$ (b) $f(x) = 1/x$ (c) $f(x) = x^n$, $n \neq -1$

SOLUTION

(a) If $F(x) = -\cos x$, then $F'(x) = \sin x$, so an antiderivative of $\sin x$ is $-\cos x$. By Theorem 1, the most general antiderivative is $G(x) = -\cos x + C$.

(b) Recall from Section 3.7 that

$$\frac{d}{dx}(\ln x) = \frac{1}{x}$$

So on the interval $(0, \infty)$ the general antiderivative of $1/x$ is $\ln x + C$. We also learned that

$$\frac{d}{dx}(\ln |x|) = \frac{1}{x}$$

for all $x \neq 0$. Theorem 1 then tells us that the general antiderivative of $f(x) = 1/x$ is $\ln |x| + C$ on any interval that doesn't contain 0. In particular, this is true on each of the intervals $(-\infty, 0)$ and $(0, \infty)$. So the general antiderivative of f is

$$F(x) = \begin{cases} \ln x + C_1 & \text{if } x > 0 \\ \ln(-x) + C_2 & \text{if } x < 0 \end{cases}$$

(c) We use the Power Rule to discover an antiderivative of x^n. In fact, if $n \neq -1$, then

$$\frac{d}{dx}\left(\frac{x^{n+1}}{n+1}\right) = \frac{(n+1)x^n}{n+1} = x^n$$

Thus the general antiderivative of $f(x) = x^n$ is

$$F(x) = \frac{x^{n+1}}{n+1} + C$$

This is valid for $n \geqslant 0$ since then $f(x) = x^n$ is defined on an interval. If n is negative (but $n \neq -1$), it is valid on any interval that doesn't contain 0. ■

As in Example 1, every differentiation formula, when read from right to left, gives rise to an antidifferentiation formula. In Table 2 we list some particular antiderivatives. Each formula in the table is true because the derivative of the function in the right column appears in the left column. In particular, the first formula says that the antiderivative of a constant times a function is the constant times the antiderivative of the function. The second formula says that the antiderivative of a sum is the sum of the antiderivatives. (We use the notation $F' = f$, $G' = g$.)

(2) Table of Antidifferentiation Formulas

Function	Particular antiderivative	Function	Particular antiderivative
$cf(x)$	$cF(x)$	$\cos x$	$\sin x$
$f(x) + g(x)$	$F(x) + G(x)$	$\sin x$	$-\cos x$
$x^n \ (n \neq -1)$	$\dfrac{x^{n+1}}{n+1}$	$\sec^2 x$	$\tan x$
$1/x$	$\ln \lvert x \rvert$	$\sec x \tan x$	$\sec x$
e^x	e^x	$\dfrac{1}{1+x^2}$	$\tan^{-1} x$
e^{cx}	$\dfrac{1}{c}e^{cx}$		

To obtain the most general antiderivative from the particular ones in Table 2, we have to add a constant (or constants), as in Example 1.

EXAMPLE 2 | Find all functions g such that

$$g'(x) = 4 \sin x + \frac{2x^5 - \sqrt{x}}{x}$$

SOLUTION We first rewrite the given function as follows:

$$g'(x) = 4 \sin x + \frac{2x^5}{x} - \frac{\sqrt{x}}{x} = 4 \sin x + 2x^4 - \frac{1}{\sqrt{x}}$$

Thus we want to find an antiderivative of

$$g'(x) = 4 \sin x + 2x^4 - x^{-1/2}$$

Using the formulas in Table 2 together with Theorem 1, we obtain

$$g(x) = 4(-\cos x) + 2\frac{x^5}{5} - \frac{x^{1/2}}{\frac{1}{2}} + C$$

$$= -4 \cos x + \tfrac{2}{5}x^5 - 2\sqrt{x} + C$$

In applications of calculus it is very common to have a situation as in Example 2, where it is required to find a function, given knowledge about its derivatives. An equation that involves the derivatives of a function is called a **differential equation**. These will be studied in some detail in Chapter 7, but for the present we can solve some elementary differential equations. The general solution of a differential equation involves an arbitrary constant (or constants) as in Example 2. However, there may be some extra conditions given that will determine the constants and therefore uniquely specify the solution.

A differential equation of the form

$$\frac{dy}{dt} = f(t)$$

is called a **pure-time differential equation** because the right side of the equation does not depend on y; it depends only on t (time). The solution will be a family of antiderivatives of f. The initial value of the solution may be specified by an **initial condition** of the form $y = y_0$ when $t = t_0$. Then the problem of finding a solution of the differential equation that also satisfies the initial condition is called an **initial-value problem**:

$$\frac{dy}{dt} = f(t) \qquad y = y_0 \text{ when } t = t_0$$

Figure 2 shows the graphs of the function f' in Example 3 and its antiderivative f. Notice that $f'(x) > 0$, so f is always increasing. Also notice that when f' has a maximum or minimum, f appears to have an inflection point. So the graph serves as a check on our calculation.

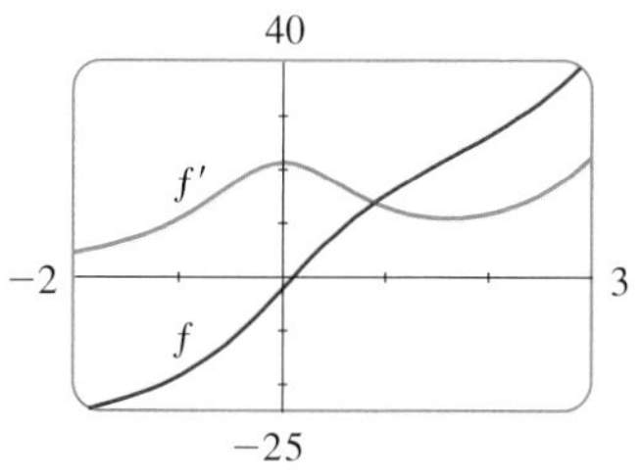

FIGURE 2

EXAMPLE 3 | Find f if $f'(x) = e^x + 20(1 + x^2)^{-1}$ and $f(0) = -2$.

SOLUTION The general antiderivative of

$$f'(x) = e^x + \frac{20}{1 + x^2}$$

is

$$f(x) = e^x + 20 \tan^{-1}x + C$$

To determine C we use the fact that $f(0) = -2$:

$$f(0) = e^0 + 20 \tan^{-1} 0 + C = -2$$

Thus we have $C = -2 - 1 = -3$, so the particular solution is

$$f(x) = e^x + 20 \tan^{-1}x - 3$$

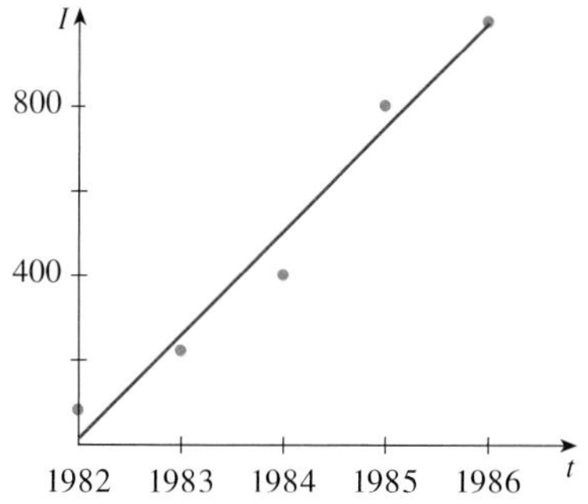

FIGURE 3

EXAMPLE 4 | HIV incidence and prevalence The rate I at which people were becoming infected with HIV (termed the incidence) in New York in the early 1980s is plotted in Figure 3. We can see from the figure that the data are well approximated by the linear function $I(t) = 16 + 242t$, where t measures the number of years since 1982. Suppose there were 80 infections at year $t = 0$. What is the number of infections expected in 1990 (termed the prevalence)?

SOLUTION Let $P(t)$ be the prevalence in year t, that is, the number of infections. We are given that

$$\frac{dP}{dt} = I(t) = 16 + 242t \qquad P(0) = 80$$

This is an initial-value problem for a pure-time differential equation. The general solution is given by the antiderivative of dP/dt:

$$P(t) = 16t + 121t^2 + C$$

Then $P(0) = C$, but we are given that $P(0) = 80$, so $C = 80$. The solution is

$$P(t) = 80 + 16t + 121t^2$$

The projected number of infections in 1990 is

$$P(8) = 80 + 16 \cdot 8 + 121 \cdot 8^2 = 7952$$

The actual number was estimated to be 7200. ■

EXAMPLE 5 | Find f if $f''(x) = 12x^2 + 6x - 4$, $f(0) = 4$, and $f(1) = 1$.

SOLUTION The general antiderivative of $f''(x) = 12x^2 + 6x - 4$ is

$$f'(x) = 12\frac{x^3}{3} + 6\frac{x^2}{2} - 4x + C = 4x^3 + 3x^2 - 4x + C$$

Using the antidifferentiation rules once more, we find that

$$f(x) = 4\frac{x^4}{4} + 3\frac{x^3}{3} - 4\frac{x^2}{2} + Cx + D = x^4 + x^3 - 2x^2 + Cx + D$$

To determine C and D we use the given conditions that $f(0) = 4$ and $f(1) = 1$. Since $f(0) = 0 + D = 4$, we have $D = 4$. Since

$$f(1) = 1 + 1 - 2 + C + 4 = 1$$

we have $C = -3$. Therefore the required function is

$$f(x) = x^4 + x^3 - 2x^2 - 3x + 4$$ ■

Antidifferentiation is particularly useful in analyzing the motion of an object moving in a straight line. Recall that if the object has position function $s = f(t)$, then the velocity function is $v(t) = s'(t)$. This means that the position function is an antiderivative of the velocity function. Likewise, the acceleration function is $a(t) = v'(t)$, so the velocity function is an antiderivative of the acceleration. If the acceleration and the initial values $s(0)$ and $v(0)$ are known, then the position function can be found by antidifferentiating twice.

EXAMPLE 6 | A particle moves in a straight line and has acceleration given by $a(t) = 6t + 4$. Its initial velocity is $v(0) = -6$ cm/s and its initial displacement is $s(0) = 9$ cm. Find its position function $s(t)$.

SOLUTION Since $v'(t) = a(t) = 6t + 4$, antidifferentiation gives

$$v(t) = 6\frac{t^2}{2} + 4t + C = 3t^2 + 4t + C$$

Note that $v(0) = C$. But we are given that $v(0) = -6$, so $C = -6$ and

$$v(t) = 3t^2 + 4t - 6$$

Since $v(t) = s'(t)$, s is the antiderivative of v:

$$s(t) = 3\frac{t^3}{3} + 4\frac{t^2}{2} - 6t + D = t^3 + 2t^2 - 6t + D$$

This gives $s(0) = D$. We are given that $s(0) = 9$, so $D = 9$ and the required position function is

$$s(t) = t^3 + 2t^2 - 6t + 9$$ ■

EXERCISES 4.6

1–20 Find the most general antiderivative of the function. (Check your answer by differentiation.)

1. $f(x) = \frac{1}{2} + \frac{3}{4}x^2 - \frac{4}{5}x^3$

2. $f(x) = 8x^9 - 3x^6 + 12x^3$

3. $f(x) = (x + 1)(2x - 1)$

4. $f(x) = x(2 - x)^2$

5. $f(x) = 5x^{1/4} - 7x^{3/4}$

6. $f(x) = 2x + 3x^{1.7}$

7. $f(x) = 6\sqrt{x} - \sqrt[6]{x}$

8. $f(x) = \sqrt[4]{x^3} + \sqrt[3]{x^4}$

9. $f(x) = \sqrt{2}$

10. $f(x) = e^2$

11. $c(t) = \dfrac{3}{t^2},\ t > 0$

12. $h(m) = \dfrac{2}{\sqrt{m}}$

13. $g(\theta) = \cos\theta - 5\sin\theta$

14. $f(x) = 2\sqrt{x} + 6\cos x$

15. $v(s) = 4s + 3e^s$

16. $u(r) = e^{-2r}$

17. $f(u) = \dfrac{u^4 + 3\sqrt{u}}{u^2}$

18. $f(x) = 3e^x + 7\sec^2 x$

19. $f(t) = \dfrac{t^4 - t^2 + 1}{t^2}$

20. $f(x) = \dfrac{1 + x - x^2}{x}$

21–28 Solve the initial-value problem.

21. $\dfrac{dy}{dt} = t^2 + 1,\ \ t \geqslant 0,\ \ y = 6 \text{ when } t = 0$

22. $\dfrac{dy}{dt} = 1 + \dfrac{2}{t},\ \ t > 0,\ \ y = 5 \text{ when } t = 1$

23. $\dfrac{dP}{dt} = 2e^{3t},\ \ t \geqslant 0,\ \ P(0) = 1$

24. $\dfrac{dm}{dt} = 100e^{-0.4t},\ \ t \geqslant 0,\ \ m(0) = 50$

25. $\dfrac{dr}{d\theta} = \cos\theta + \sec\theta\tan\theta,\ \ 0 < \theta < \pi/2,\ \ r(\pi/3) = 4$

26. $\dfrac{dy}{dx} = x^2 + 1 + \dfrac{1}{x^2 + 1},\ \ y = 0 \text{ when } x = 1$

27. $\dfrac{du}{dt} = \sqrt{t} + \dfrac{2}{\sqrt{t}},\ \ t > 0,\ \ u(1) = 5$

28. $\dfrac{dv}{dt} = e^{-t}(1 + e^{2t}),\ \ t \geqslant 0,\ \ v(0) = 3$

29–40 Find f.

29. $f''(x) = 6x + 12x^2$

30. $f''(x) = 2 + x^3 + x^6$

31. $f''(x) = \frac{2}{3}x^{2/3}$

32. $f''(x) = 6x + \sin x$

33. $f'(x) = 1 - 6x,\ \ f(0) = 8$

34. $f'(x) = 8x^3 + 12x + 3,\ \ f(1) = 6$

35. $f'(x) = \sqrt{x}(6 + 5x),\ \ f(1) = 10$

36. $f'(x) = 2x - 3/x^4,\ \ x > 0,\ \ f(1) = 3$

37. $f''(\theta) = \sin\theta + \cos\theta,\ \ f(0) = 3,\ \ f'(0) = 4$

38. $f''(x) = 8x^3 + 5,\ \ f(1) = 0,\ \ f'(1) = 8$

39. $f''(x) = 2 - 12x,\ \ f(0) = 9,\ \ f(2) = 15$

40. $f''(t) = 2e^t + 3\sin t,\ \ f(0) = 0,\ \ f(\pi) = 0$

41. Bacteria culture A culture of the bacterium *Rhodobacter sphaeroides* initially has 25 bacteria and t hours later increases at a rate of $3.4657e^{0.1386t}$ bacteria per hour. Find the population size after four hours.

42. A sample of cesium-37 with an initial mass of 75 mg decays t years later at a rate of $1.7325e^{-0.0231t}$ mg/year. Find the mass of the sample after 20 years.

43. A particle moves along a straight line with velocity function $v(t) = \sin t - \cos t$ and its initial displacement is $s(0) = 0$ m. Find its position function $s(t)$.

44. A particle moves with acceleration function $a(t) = 5 + 4t - 2t^2$. Its initial velocity is $v(0) = 3$ m/s and its initial displacement is $s(0) = 10$ m. Find its position after t seconds.

45. A stone is dropped from the upper observation deck (the Space Deck) of the CN Tower, 450 m above the ground.

(a) Find the distance of the stone above ground level at time t. Use the fact that acceleration due to gravity is $g \approx 9.8\ \text{m/s}^2$.

(b) How long does it take the stone to reach the ground?

(c) With what velocity does it strike the ground?

46. To prove Theorem 1, let F and G be any two antiderivatives of f on I and let $H = G - F$.

(a) If x_1 and x_2 are any two numbers in I with $x_1 < x_2$, apply the Mean Value Theorem on the interval $[x_1, x_2]$ to show that $H(x_1) = H(x_2)$. Why does this show that H is a constant function?

(b) Deduce Theorem 1 from the result of part (a).

47. The graph of f' is shown in the figure. Sketch the graph of f if f is continuous on $[0, 3]$ and $f(0) = -1$.

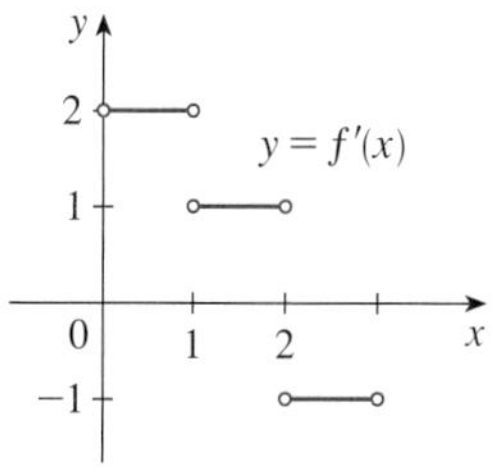

48. Find a function f such that $f'(x) = x^3$ and the line $x + y = 0$ is tangent to the graph of f.

Chapter 4 REVIEW

CONCEPT CHECK

1. Explain the difference between an absolute maximum and a local maximum. Illustrate with a sketch.

2. (a) What does the Extreme Value Theorem say?
   (b) Explain how the Closed Interval Method works.

3. (a) State Fermat's Theorem.
   (b) Define a critical number of f.

4. State the Mean Value Theorem and give a geometric interpretation.

5. (a) State the Increasing/Decreasing Test.
   (b) What does it mean to say that f is concave upward on an interval I?
   (c) State the Concavity Test.
   (d) What are inflection points? How do you find them?

6. (a) State the First Derivative Test.
   (b) State the Second Derivative Test.
   (c) What are the relative advantages and disadvantages of these tests?

7. (a) What does l'Hospital's Rule say?
   (b) How can you use l'Hospital's Rule if you have a product $f(x)\,g(x)$ where $f(x) \to 0$ and $g(x) \to \infty$ as $x \to a$?
   (c) How can you use l'Hospital's Rule if you have a difference $f(x) - g(x)$ where $f(x) \to \infty$ and $g(x) \to \infty$ as $x \to a$?

8. (a) What is an equilibrium of the recursive sequence $x_{t+1} = f(x_t)$?
   (b) What is a stable equilibrium? An unstable equilibrium?
   (c) State the Stability Criterion.

9. (a) What is an antiderivative of a function f?
   (b) Suppose F_1 and F_2 are both antiderivatives of f on an interval I. How are F_1 and F_2 related?

Answers to the Concept Check can be found on the back endpapers.

TRUE-FALSE QUIZ

Determine whether the statement is true or false. If it is true, explain why. If it is false, explain why or give an example that disproves the statement.

1. If $f'(c) = 0$, then f has a local maximum or minimum at c.

2. If f has an absolute minimum value at c, then $f'(c) = 0$.

3. If f is continuous on (a, b), then f attains an absolute maximum value $f(c)$ and an absolute minimum value $f(d)$ at some numbers c and d in (a, b).

4. If f is differentiable and $f(-1) = f(1)$, then there is a number c such that $|c| < 1$ and $f'(c) = 0$.

5. If $f'(x) < 0$ for $1 < x < 6$, then f is decreasing on $(1, 6)$.

6. If $f''(2) = 0$, then $(2, f(2))$ is an inflection point of the curve $y = f(x)$.

7. If $f'(x) = g'(x)$ for $0 < x < 1$, then $f(x) = g(x)$ for $0 < x < 1$.

8. There exists a function f such that $f(1) = -2$, $f(3) = 0$, and $f'(x) > 1$ for all x.

9. There exists a function f such that $f(x) > 0$, $f'(x) < 0$, and $f''(x) > 0$ for all x.

10. There exists a function f such that $f(x) < 0$, $f'(x) < 0$, and $f''(x) > 0$ for all x.

11. If f and g are increasing on an interval I, then $f + g$ is increasing on I.

12. If f and g are increasing on an interval I, then $f - g$ is increasing on I.

13. If f and g are increasing on an interval I, then fg is increasing on I.

14. If f and g are positive increasing functions on an interval I, then fg is increasing on I.

15. If f is increasing and $f(x) > 0$ on I, then $g(x) = 1/f(x)$ is decreasing on I.

16. If f is even, then f' is even.

17. If f is periodic, then f' is periodic.

18. $\displaystyle\lim_{x \to 0} \frac{x}{e^x} = 1$

EXERCISES

1–6 Find the local and absolute extreme values of the function on the given interval.

1. $f(x) = x^3 - 6x^2 + 9x + 1, \quad [2, 4]$
2. $f(x) = x\sqrt{1 - x}, \quad [-1, 1]$
3. $f(x) = \dfrac{3x - 4}{x^2 + 1}, \quad [-2, 2]$
4. $f(x) = (x^2 + 2x)^3, \quad [-2, 1]$
5. $f(x) = x + \sin 2x, \quad [0, \pi]$
6. $f(x) = (\ln x)/x^2, \quad [1, 3]$

7–14
(a) Find the vertical and horizontal asymptotes, if any.
(b) Find the intervals of increase or decrease.
(c) Find the local maximum and minimum values.
(d) Find the intervals of concavity and the inflection points.
(e) Use the information from parts (a)–(d) to sketch the graph of f. Check your work with a graphing device.

7. $f(x) = 2 - 2x - x^3$
8. $f(x) = x^4 + 4x^3$
9. $f(x) = x + \sqrt{1 - x}$
10. $f(x) = \dfrac{1}{1 - x^2}$
11. $y = \sin^2 x - 2\cos x$
12. $y = e^{2x - x^2}$
13. $y = e^x + e^{-3x}$
14. $y = \ln(x^2 - 1)$

15. **Antibiotic pharmacokinetics** A model for the concentration of an antibiotic drug in the bloodstream t hours after the administration of the drug is
$$C(t) = 2.5(e^{-0.3t} - e^{-0.7t})$$
where C is measured in μg/mL.
(a) At what time does the concentration have its maximum value? What is the maximum value?
(b) At what time does the inflection point occur? What is the significance of the inflection point?

16. **Drug pharmacokinetics** Another model for the concentration of a drug in the bloodstream is
$$C(t) = 0.5t^2 e^{-0.6t}$$
where t is measured in hours and C is measured in μg/mL.
(a) At what time does the concentration have its largest value? What is the largest value?
(b) How many inflection points are there? At what times do they occur? What is the significance of each inflection point?
(c) Compare the graphs of this concentration function and the one in Exercise 15. How are the graphs similar? How are they different?

17. **Population bound** Suppose that an initial population size is 300 individuals and the population grows at a rate of at most 120 individuals per week. What can you say about the population size after five weeks?

18. Sketch the graph of a function that satisfies the following conditions:
$f(0) = 0$, $\quad f$ is continuous and even,
$f'(x) = 2x$ if $0 < x < 1$, $\quad f'(x) = -1$ if $1 < x < 3$,
$f'(x) = 1$ if $x > 3$

19–25 Evaluate the limit.

19. $\displaystyle\lim_{x \to 0} \frac{\tan \pi x}{\ln(1 + x)}$
20. $\displaystyle\lim_{x \to 0} \frac{1 - \cos x}{x^2 + x}$
21. $\displaystyle\lim_{x \to 0} \frac{e^{4x} - 1 - 4x}{x^2}$
22. $\displaystyle\lim_{x \to \infty} \frac{e^{4x} - 1 - 4x}{x^2}$
23. $\displaystyle\lim_{x \to \infty} x^3 e^{-x}$
24. $\displaystyle\lim_{x \to 0^+} x^2 \ln x$
25. $\displaystyle\lim_{x \to 1^+} \left(\frac{x}{x - 1} - \frac{1}{\ln x} \right)$

26. Rank the functions in order of how quickly they grow as $x \to \infty$.
$$y = \sqrt[4]{x} \qquad y = \ln(10x) \qquad y = 10^x \qquad y = \sqrt{1 + e^x}$$

27. Find two positive integers such that the sum of the first number and four times the second number is 1000 and the product of the numbers is as large as possible.

28. Find the point on the hyperbola $xy = 8$ that is closest to the point $(3, 0)$.

29. The velocity of a wave of length L in deep water is
$$v = K\sqrt{\frac{L}{C} + \frac{C}{L}}$$
where K and C are known positive constants. What is the length of the wave that gives the minimum velocity?

30. The **Ricker model** for population growth is a discrete-time model of the form
$$n_{t+1} = cn_t e^{-\lambda n_t}$$
For the constants $c = 2$ and $\lambda = 3$, the model is $n_{t+1} = f(n_t)$, where the updating function is
$$f(n) = 2ne^{-3n}$$
Find the largest value of f and interpret it.

31. **Drug resistance evolution** A simple model for the spread of drug resistance is given by $\Delta p = p(1 - p)s$, where s is a measure of the reproductive advantage of the drug resistance gene in the presence of drugs, p is the

frequency of the drug resistance gene in the population, and Δp is the change in the frequency of the drug resistance gene in the population after one year. Notice that the amount of change in the frequency Δp differs depending on the gene's current frequency p. What current frequency makes the rate of evolution Δp the largest?

32. The **thermic effect of food** (TEF) is the increase in metabolic rate after a meal. Researchers used the functions

$$f(t) = 175.9te^{-t/1.3} \qquad g(t) = 113.6te^{-t/1.85}$$

to model the TEF (measured in kJ/h) for a lean person and an obese person, respectively.

(a) Find the maximum value of the TEF for both individuals.

(b) Graph the TEF functions for both individuals. Describe how the graphs are similar and how they differ.

Source: Adapted from G. Reed et al., "Measuring the Thermic Effects of Food," *American Journal of Clinical Nutrition* 63 (1996): 164–69.

33–34 Find the equilibria of the difference equation and classify them as stable or unstable.

33. $x_{t+1} = \dfrac{4x_t}{1 + 5x_t}$ **34.** $x_{t+1} = 5x_t e^{-4x_t}$

35. Find the equilibria of the difference equation

$$x_{t+1} = \frac{6x_t^2}{x_t^2 + 8}$$

and classify them as stable or unstable. Use cobwebbing to evaluate $\lim_{t\to\infty} x_t$ for $x_0 = 1$ and $x_0 = 3$.

36. Let $f(x) = 1.07x + \sin x$, $0 \leqslant x \leqslant 11$. How many equilibria does the recursion $x_{t+1} = f(x_t)$ have? Estimate their values and explain why they are stable or unstable.

37–40 Find the most general antiderivative of the function.

37. $f(x) = \sin x + \sec x \tan x, \quad 0 \leqslant x < \pi/2$

38. $g(t) = (1 + t)/\sqrt{t}$

39. $q(t) = 2 + (t + 1)(t^2 - 1)$

40. $w(\theta) = 2\theta - 3\cos\theta$

41–42 Solve the initial-value problem.

41. $\dfrac{dy}{dt} = 1 - e^{\pi t}, \quad y = 0 \text{ when } t = 0$

42. $\dfrac{dr}{dt} = \dfrac{4}{1 + t^2}, \quad r(1) = 2$

43–44 Find $f(x)$.

43. $f''(x) = 1 - 6x + 48x^2, \quad f(0) = 1, \quad f'(0) = 2$

44. $f''(x) = 2x^3 + 3x^2 - 4x + 5, \quad f(0) = 2, \quad f(1) = 0$

45. A particle moves in a straight line with acceleration $a(t) = \sin t + 3\cos t$, initial displacement $s(0) = 0$, and initial velocity $v(0) = 2$. Find its position function $s(t)$.

46. Sketch the graph of a continuous, even function f such that $f(0) = 0$, $f'(x) = 2x$ if $0 < x < 1$, $f'(x) = -1$ if $1 < x < 3$, and $f'(x) = 1$ if $x > 3$.

47. If a rectangle has its base on the x-axis and two vertices on the curve $y = e^{-x^2}$, show that the rectangle has the largest possible area when the two vertices are at the points of inflection of the curve.

Integrals

5

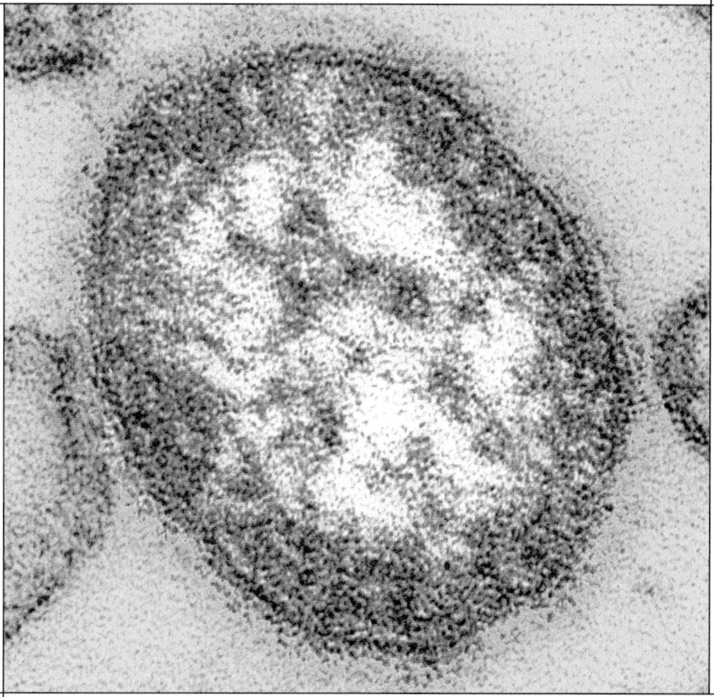

Shown is a transmission electron micrograph of a measles virus. In Sections 5.1 and 5.3 we will see how an integral can be used to represent the total amount of infection needed to develop symptoms of measles.

Scott Camazine / Alamy

IN CHAPTER 3 WE USED the problems of finding tangent lines and rates of increase to introduce the derivative, which is the central idea in differential calculus. In much the same way, this chapter starts with the area and distance problems and uses them to formulate the idea of a definite integral, which is the basic concept of integral calculus. We will see in this chapter and the next how to use integrals to solve problems concerning disease development, population dynamics, biological control, blood flow, and cardiac output, among others.

There is a connection between integral calculus and differential calculus. The Fundamental Theorem of Calculus relates the integral to the derivative, and we will see in this chapter that it greatly simplifies the solution of many problems.

5.1 Areas, Distances, and Pathogenesis

In this section we discover that in trying to find the area under a curve or the distance traveled by a car, we end up with the same special type of limit.

Why would a biologist be interested in calculating an area? A botanist might want to know the area of a leaf and compare it with the leaf's area at other stages of its development. An ecologist might want to know the area of a lake and compare it with the area in previous years. An oncologist might want to know the area of a tumor and compare it with the areas at prior times to see how quickly it is growing. But there are also indirect ways in which areas are of interest. We will see at the end of this section, for example, that the area beneath part of the pathogenesis curve for a measles infection represents the amount of infection needed to develop symptoms of the disease.

■ The Area Problem

We begin by attempting to solve the *area problem*: Find the area of the region S that lies under the curve $y = f(x)$ from a to b. This means that S, illustrated in Figure 1, is bounded by the graph of a continuous function f [where $f(x) \geq 0$], the vertical lines $x = a$ and $x = b$, and the x-axis.

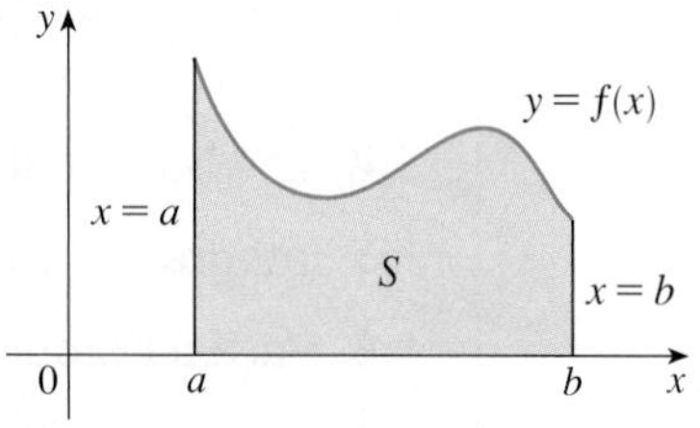

FIGURE 1

$S = \{(x, y) \mid a \leq x \leq b, 0 \leq y \leq f(x)\}$

In trying to solve the area problem we have to ask ourselves: What is the meaning of the word *area*? This question is easy to answer for regions with straight sides. For a rectangle, the area is defined as the product of the length and the width. The area of a triangle is half the base times the height. The area of a polygon is found by dividing it into triangles (as in Figure 2) and adding the areas of the triangles.

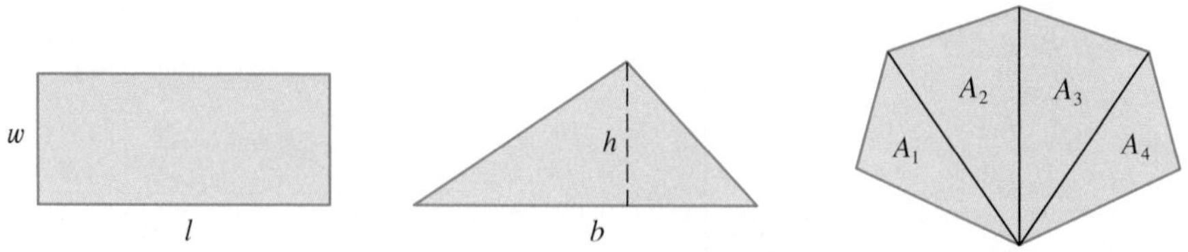

FIGURE 2

$A = lw$ $\quad$ $A = \frac{1}{2}bh$ $\quad$ $A = A_1 + A_2 + A_3 + A_4$

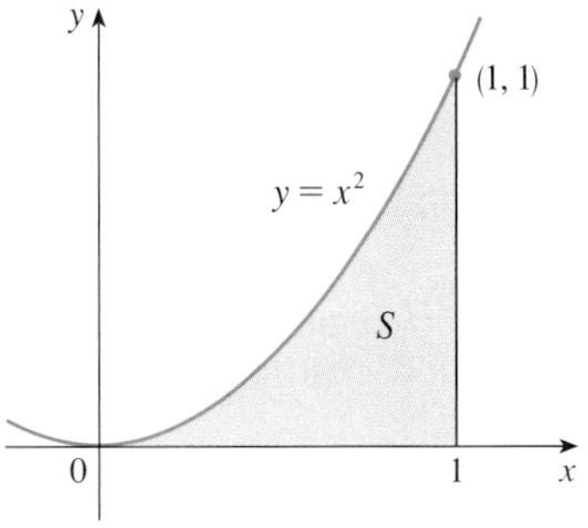

FIGURE 3

However, it isn't so easy to find the area of a region with curved sides. We all have an intuitive idea of what the area of a region is. But part of the area problem is to make this intuitive idea precise by giving an exact definition of area.

Recall that in defining a tangent we first approximated the slope of the tangent line by slopes of secant lines and then we took the limit of these approximations. We pursue a similar idea for areas. We first approximate the region S by rectangles and then we take the limit of the areas of these rectangles as we increase the number of rectangles. The following example illustrates the procedure.

EXAMPLE 1 | Use rectangles to estimate the area under the parabola $y = x^2$ from 0 to 1 (the parabolic region S illustrated in Figure 3).

SOLUTION We first notice that the area of S must be somewhere between 0 and 1 because S is contained in a square with side length 1, but we can certainly do better than that. Suppose we divide S into four strips S_1, S_2, S_3, and S_4 by drawing the vertical lines $x = \frac{1}{4}$, $x = \frac{1}{2}$, and $x = \frac{3}{4}$ as in Figure 4(a).

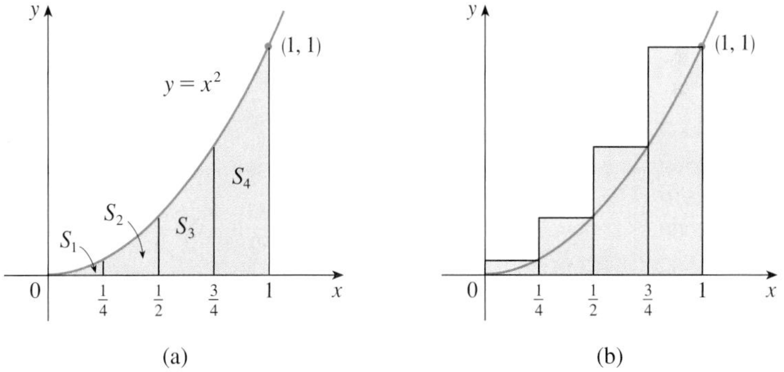

FIGURE 4

We can approximate each strip by a rectangle that has the same base as the strip and whose height is the same as the right edge of the strip [see Figure 4(b)]. In other words, the heights of these rectangles are the values of the function $f(x) = x^2$ at the right endpoints of the subintervals $\left[0, \frac{1}{4}\right]$, $\left[\frac{1}{4}, \frac{1}{2}\right]$, $\left[\frac{1}{2}, \frac{3}{4}\right]$, and $\left[\frac{3}{4}, 1\right]$.

Each rectangle has width $\frac{1}{4}$ and the heights are $\left(\frac{1}{4}\right)^2$, $\left(\frac{1}{2}\right)^2$, $\left(\frac{3}{4}\right)^2$, and 1^2. If we let R_4 be the sum of the areas of these approximating rectangles, we get

$$R_4 = \frac{1}{4} \cdot \left(\frac{1}{4}\right)^2 + \frac{1}{4} \cdot \left(\frac{1}{2}\right)^2 + \frac{1}{4} \cdot \left(\frac{3}{4}\right)^2 + \frac{1}{4} \cdot 1^2 = \frac{15}{32} = 0.46875$$

From Figure 4(b) we see that the area A of S is less than R_4, so

$$A < 0.46875$$

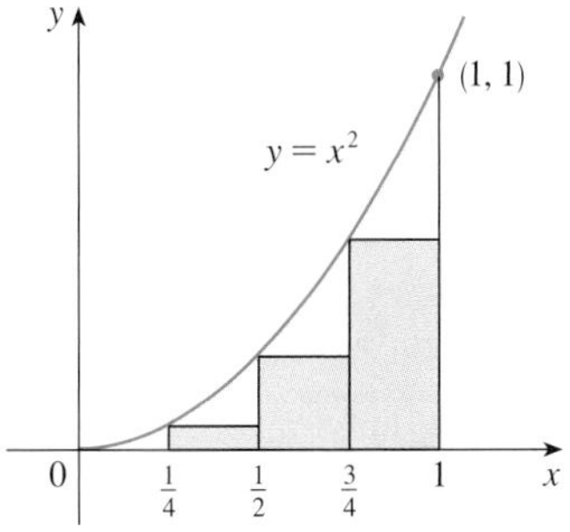

FIGURE 5

Instead of using the rectangles in Figure 4(b) we could use the smaller rectangles in Figure 5 whose heights are the values of f at the *left* endpoints of the subintervals. (The leftmost rectangle has collapsed because its height is 0.) The sum of the areas of these approximating rectangles is

$$L_4 = \frac{1}{4} \cdot 0^2 + \frac{1}{4} \cdot \left(\frac{1}{4}\right)^2 + \frac{1}{4} \cdot \left(\frac{1}{2}\right)^2 + \frac{1}{4} \cdot \left(\frac{3}{4}\right)^2 = \frac{7}{32} = 0.21875$$

We see that the area of S is larger than L_4, so we have lower and upper estimates for A:

$$0.21875 < A < 0.46875$$

We can repeat this procedure with a larger number of strips. Figure 6 shows what happens when we divide the region S into eight strips of equal width.

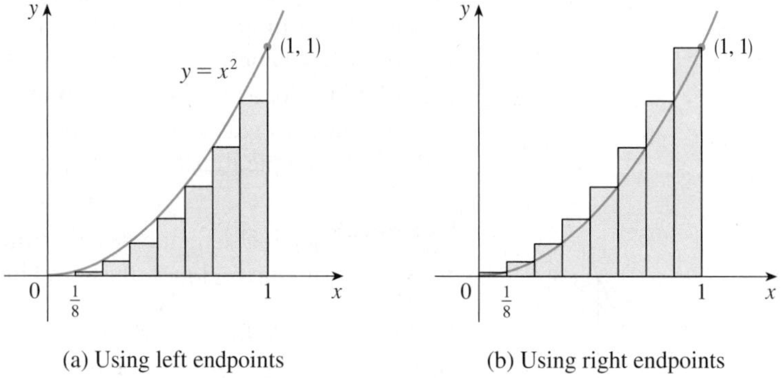

FIGURE 6
Approximating S with eight rectangles

By computing the sum of the areas of the smaller rectangles (L_8) and the sum of the areas of the larger rectangles (R_8), we obtain better lower and upper estimates for A:

$$0.2734375 < A < 0.3984375$$

So one possible answer to the question is to say that the true area of S lies somewhere between 0.2734375 and 0.3984375.

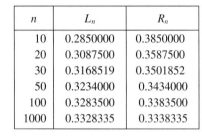

n	L_n	R_n
10	0.2850000	0.3850000
20	0.3087500	0.3587500
30	0.3168519	0.3501852
50	0.3234000	0.3434000
100	0.3283500	0.3383500
1000	0.3328335	0.3338335

We could obtain better estimates by increasing the number of strips. The table at the left shows the results of similar calculations (with a computer) using n rectangles whose heights are found with left endpoints (L_n) or right endpoints (R_n). In particular, we see by using 50 strips that the area lies between 0.3234 and 0.3434. With 1000 strips we narrow it down even more: A lies between 0.3328335 and 0.3338335. A good estimate is obtained by averaging these numbers: $A \approx 0.3333335$. ■

From the values in the table in Example 1, it looks as if R_n is approaching $\frac{1}{3}$ as n increases. We confirm this in the next example.

EXAMPLE 2 | For the region S in Example 1, show that the sum of the areas of the upper approximating rectangles approaches $\frac{1}{3}$, that is,

$$\lim_{n \to \infty} R_n = \tfrac{1}{3}$$

SOLUTION R_n is the sum of the areas of the n rectangles in Figure 7. Each rectangle has width $1/n$ and the heights are the values of the function $f(x) = x^2$ at the points $1/n, 2/n, 3/n, \ldots, n/n$; that is, the heights are $(1/n)^2, (2/n)^2, (3/n)^2, \ldots, (n/n)^2$. Thus

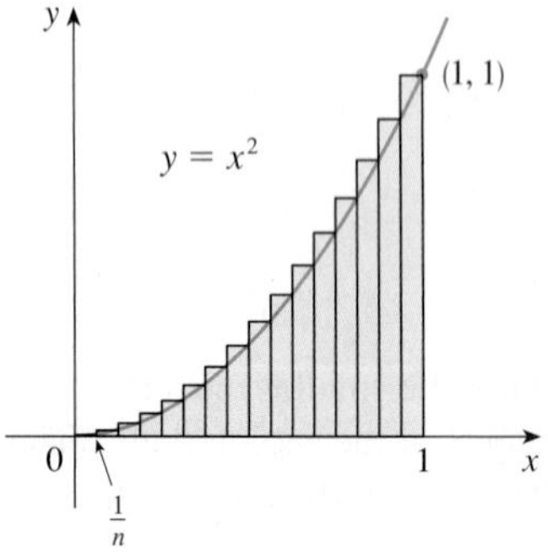

FIGURE 7

$$\begin{aligned} R_n &= \frac{1}{n}\left(\frac{1}{n}\right)^2 + \frac{1}{n}\left(\frac{2}{n}\right)^2 + \frac{1}{n}\left(\frac{3}{n}\right)^2 + \cdots + \frac{1}{n}\left(\frac{n}{n}\right)^2 \\ &= \frac{1}{n} \cdot \frac{1}{n^2}(1^2 + 2^2 + 3^2 + \cdots + n^2) \\ &= \frac{1}{n^3}(1^2 + 2^2 + 3^2 + \cdots + n^2) \end{aligned}$$

Here we need the formula for the sum of the squares of the first n positive integers:

(1) $$1^2 + 2^2 + 3^2 + \cdots + n^2 = \frac{n(n+1)(2n+1)}{6}$$

Perhaps you have seen this formula before. It is proved in Example 5 in Appendix F.

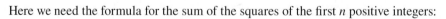

Putting Formula 1 into our expression for R_n, we get

$$R_n = \frac{1}{n^3} \cdot \frac{n(n+1)(2n+1)}{6} = \frac{(n+1)(2n+1)}{6n^2}$$

Limits of sequences were introduced in Section 2.1.

Now we compute the limit of the sequence R_n:

$$\lim_{n\to\infty} R_n = \lim_{n\to\infty} \frac{(n+1)(2n+1)}{6n^2}$$
$$= \lim_{n\to\infty} \frac{1}{6}\left(\frac{n+1}{n}\right)\left(\frac{2n+1}{n}\right)$$
$$= \lim_{n\to\infty} \frac{1}{6}\left(1+\frac{1}{n}\right)\left(2+\frac{1}{n}\right)$$
$$= \frac{1}{6} \cdot 1 \cdot 2 = \frac{1}{3}$$

It can be shown that the lower approximating sums also approach $\frac{1}{3}$, that is,

$$\lim_{n\to\infty} L_n = \tfrac{1}{3}$$

TEC In Visual 5.1 you can create pictures like those in Figures 8 and 9 for other values of n.

From Figures 8 and 9 it appears that, as n increases, both L_n and R_n become better and better approximations to the area of S. Therefore we *define* the area A to be the limit of the sums of the areas of the approximating rectangles, that is,

$$A = \lim_{n\to\infty} R_n = \lim_{n\to\infty} L_n = \tfrac{1}{3}$$

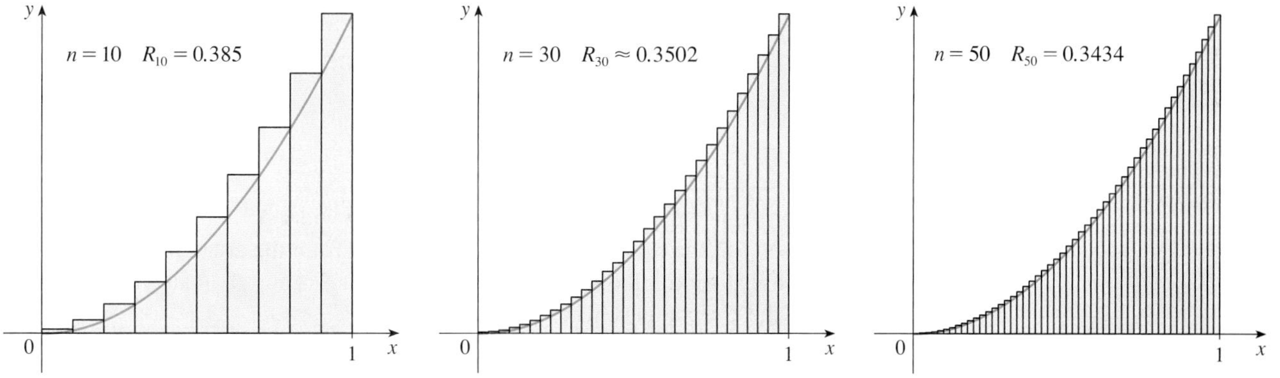

FIGURE 8

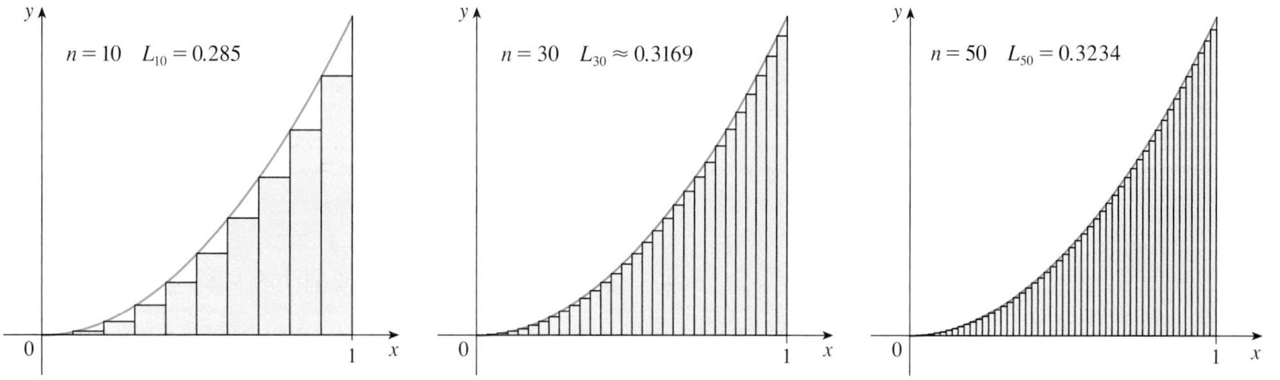

FIGURE 9 The area is the number that is smaller than all upper sums and larger than all lower sums.

Let's apply the idea of Examples 1 and 2 to the more general region S of Figure 1. We start by subdividing S into n strips $S_1, S_2, \ldots, S_n$ of equal width as in Figure 10.

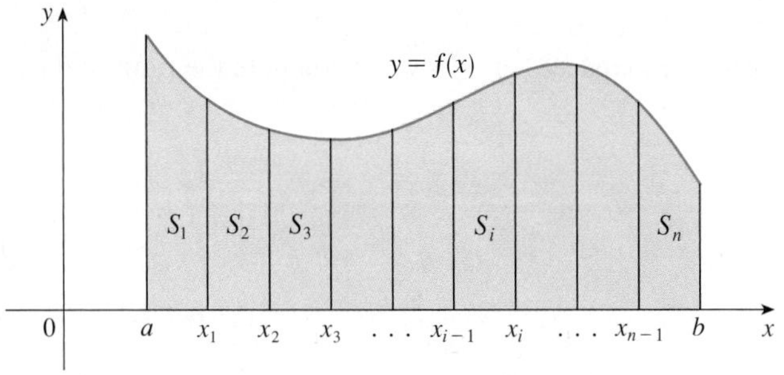

FIGURE 10

The width of the interval $[a, b]$ is $b - a$, so the width of each of the n strips is

$$\Delta x = \frac{b - a}{n}$$

These strips divide the interval $[a, b]$ into n subintervals

$$[x_0, x_1], \quad [x_1, x_2], \quad [x_2, x_3], \quad \ldots, \quad [x_{n-1}, x_n]$$

where $x_0 = a$ and $x_n = b$. The right endpoints of the subintervals are

$$x_1 = a + \Delta x,$$

$$x_2 = a + 2\,\Delta x,$$

$$x_3 = a + 3\,\Delta x,$$

$$\vdots$$

Let's approximate the ith strip S_i by a rectangle with width Δx and height $f(x_i)$, which is the value of f at the right endpoint (see Figure 11). Then the area of the ith rectangle is $f(x_i)\,\Delta x$. What we think of intuitively as the area of S is approximated by the sum of the areas of these rectangles, which is

$$R_n = f(x_1)\,\Delta x + f(x_2)\,\Delta x + \cdots + f(x_n)\,\Delta x$$

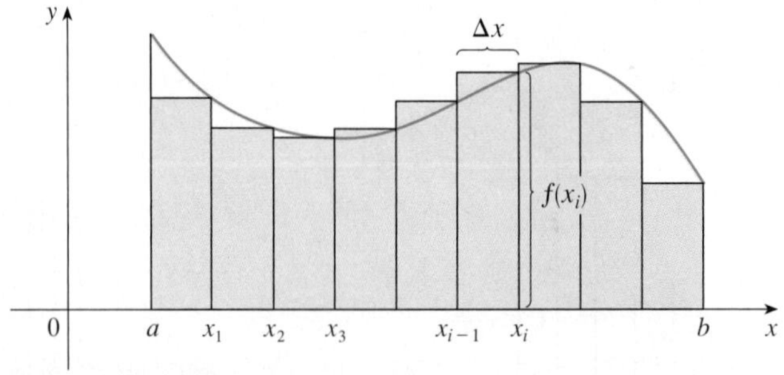

FIGURE 11

Figure 12 shows this approximation for $n = 2, 4, 8$, and 12. Notice that this approximation appears to become better and better as the number of strips increases, that is, as $n \to \infty$. Therefore we define the area A of the region S in the following way.

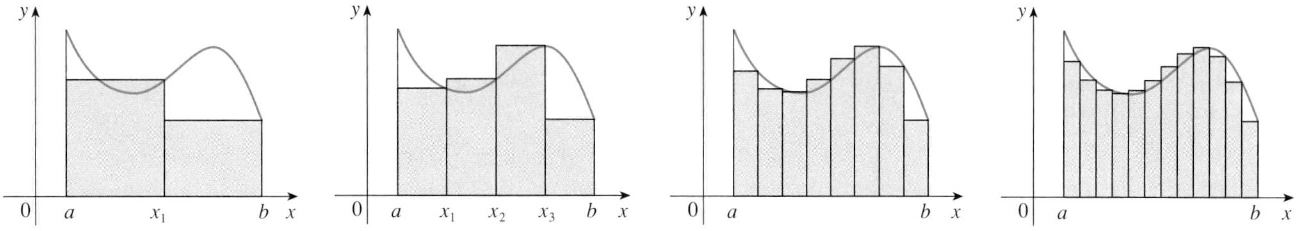

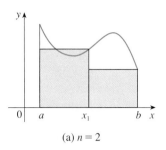

(a) $n = 2$ (b) $n = 4$ (c) $n = 8$ (d) $n = 12$

FIGURE 12

> **(2) Definition** The **area** A of the region S that lies under the graph of the continuous function f is the limit of the sum of the areas of approximating rectangles:
>
> $$A = \lim_{n \to \infty} R_n = \lim_{n \to \infty} [f(x_1)\,\Delta x + f(x_2)\,\Delta x + \cdots + f(x_n)\,\Delta x]$$

It can be proved that the limit in Definition 2 always exists, since we are assuming that f is continuous. It can also be shown that we get the same value if we use left endpoints:

(3) $$A = \lim_{n \to \infty} L_n = \lim_{n \to \infty} [f(x_0)\,\Delta x + f(x_1)\,\Delta x + \cdots + f(x_{n-1})\,\Delta x]$$

In fact, instead of using left endpoints or right endpoints, we could take the height of the ith rectangle to be the value of f at any number x_i^* in the ith subinterval $[x_{i-1}, x_i]$. We call the numbers $x_1^*, x_2^*, \ldots, x_n^*$ the **sample points**. Figure 13 shows approximating rectangles when the sample points are not chosen to be endpoints. So a more general expression for the area of S is

(4) $$A = \lim_{n \to \infty} [f(x_1^*)\,\Delta x + f(x_2^*)\,\Delta x + \cdots + f(x_n^*)\,\Delta x]$$

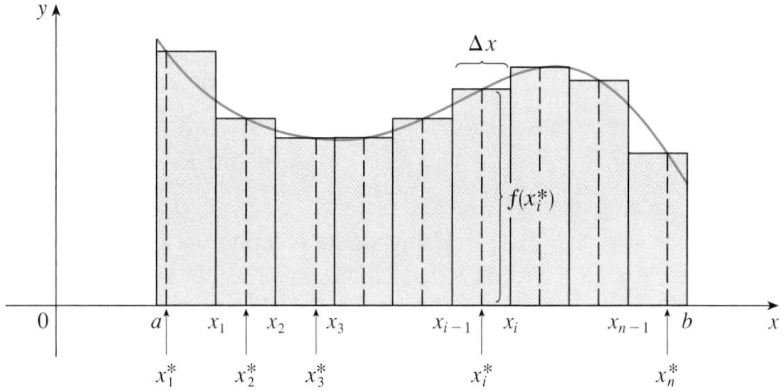

FIGURE 13

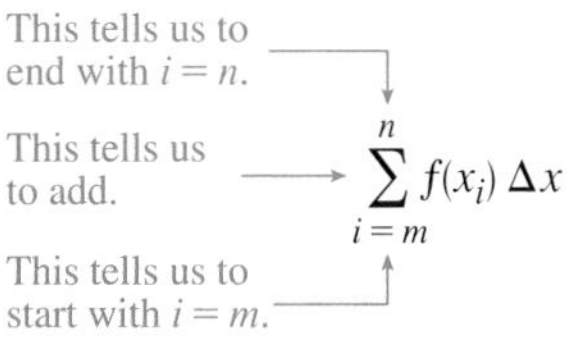

We often use **sigma notation** to write sums with many terms more compactly. For instance,

$$\sum_{i=1}^{n} f(x_i)\,\Delta x = f(x_1)\,\Delta x + f(x_2)\,\Delta x + \cdots + f(x_n)\,\Delta x$$

So the expressions for area in Equations 2, 3, and 4 can be written as follows:

If you need practice with sigma notation, look at the examples and try some of the exercises in Appendix F.

$$A = \lim_{n \to \infty} \sum_{i=1}^{n} f(x_i)\,\Delta x$$

$$A = \lim_{n \to \infty} \sum_{i=1}^{n} f(x_{i-1})\,\Delta x$$

$$A = \lim_{n \to \infty} \sum_{i=1}^{n} f(x_i^*)\,\Delta x$$

We can also rewrite Formula 1 in the following way:

$$\sum_{i=1}^{n} i^2 = \frac{n(n+1)(2n+1)}{6}$$

EXAMPLE 3 | Let A be the area of the region that lies under the graph of $f(x) = e^{-x}$ between $x = 0$ and $x = 2$.

(a) Using right endpoints, find an expression for A as a limit. Do not evaluate the limit.

(b) Estimate the area by taking the sample points to be midpoints and using four subintervals and then ten subintervals.

SOLUTION

(a) Since $a = 0$ and $b = 2$, the width of a subinterval is

$$\Delta x = \frac{2 - 0}{n} = \frac{2}{n}$$

So $x_1 = 2/n$, $x_2 = 4/n$, $x_3 = 6/n$, $x_i = 2i/n$, and $x_n = 2n/n$. The sum of the areas of the approximating rectangles is

$$\begin{aligned} R_n &= f(x_1)\,\Delta x + f(x_2)\,\Delta x + \cdots + f(x_n)\,\Delta x \\ &= e^{-x_1}\,\Delta x + e^{-x_2}\,\Delta x + \cdots + e^{-x_n}\,\Delta x \\ &= e^{-2/n}\left(\frac{2}{n}\right) + e^{-4/n}\left(\frac{2}{n}\right) + \cdots + e^{-2n/n}\left(\frac{2}{n}\right) \end{aligned}$$

According to Definition 2, the area is

$$A = \lim_{n \to \infty} R_n = \lim_{n \to \infty} \frac{2}{n}\left(e^{-2/n} + e^{-4/n} + e^{-6/n} + \cdots + e^{-2n/n}\right)$$

Using sigma notation we could write

$$A = \lim_{n \to \infty} \frac{2}{n} \sum_{i=1}^{n} e^{-2i/n}$$

It is difficult to evaluate this limit directly by hand, though it isn't hard with the help of a computer algebra system. In Section 5.3 we will be able to find A more easily using a different method.

(b) With $n = 4$ the subintervals of equal width $\Delta x = 0.5$ are $[0, 0.5]$, $[0.5, 1]$, $[1, 1.5]$, and $[1.5, 2]$. The midpoints of these subintervals are $x_1^* = 0.25$, $x_2^* = 0.75$, $x_3^* = 1.25$, and $x_4^* = 1.75$, and the sum of the areas of the four approximating rectangles (see

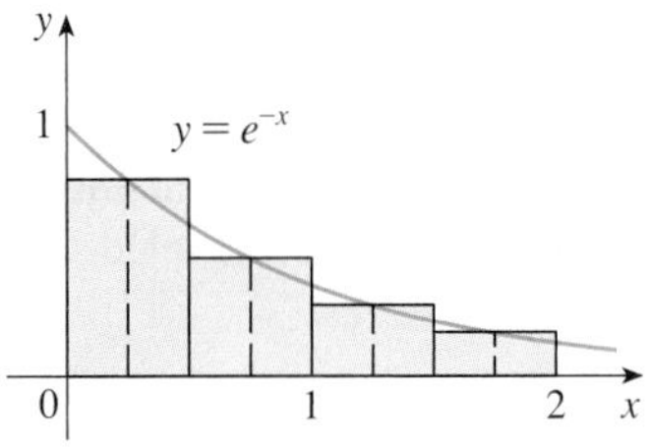

FIGURE 14

Figure 14) is

$$M_4 = \sum_{i=1}^{4} f(x_i^*)\,\Delta x$$
$$= f(0.25)\,\Delta x + f(0.75)\,\Delta x + f(1.25)\,\Delta x + f(1.75)\,\Delta x$$
$$= e^{-0.25}(0.5) + e^{-0.75}(0.5) + e^{-1.25}(0.5) + e^{-1.75}(0.5)$$
$$= \tfrac{1}{2}(e^{-0.25} + e^{-0.75} + e^{-1.25} + e^{-1.75}) \approx 0.8557$$

So an estimate for the area is

$$A \approx 0.8557$$

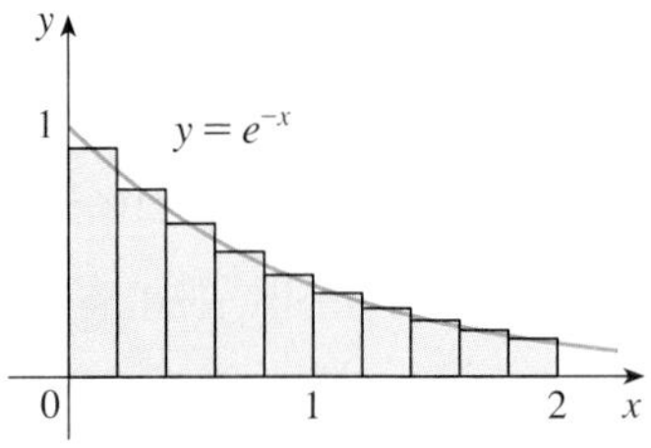

FIGURE 15

With $n = 10$ the subintervals are $[0, 0.2], [0.2, 0.4], \ldots, [1.8, 2]$ and the midpoints are $x_1^* = 0.1, x_2^* = 0.3, x_3^* = 0.5, \ldots, x_{10}^* = 1.9$. Thus

$$A \approx M_{10} = f(0.1)\,\Delta x + f(0.3)\,\Delta x + f(0.5)\,\Delta x + \cdots + f(1.9)\,\Delta x$$
$$= 0.2(e^{-0.1} + e^{-0.3} + e^{-0.5} + \cdots + e^{-1.9}) \approx 0.8632$$

From Figure 15 it appears that this estimate is better than the estimate with $n = 4$. ■

■ The Distance Problem

Now let's consider the *distance problem*: Find the distance traveled by an object during a certain time period if the velocity of the object is known at all times. (In a sense this is the inverse problem of the velocity problem that we discussed in Section 2.3.) If the velocity remains constant, then the distance problem is easy to solve by means of the formula

$$\text{distance} = \text{velocity} \times \text{time}$$

But if the velocity varies, it's not so easy to find the distance traveled. We investigate the problem in the following example.

EXAMPLE 4 | Suppose the odometer on our car is broken and we want to estimate the distance driven over a 30-second time interval. We take speedometer readings every five seconds and record them in the following table:

Time (s)	0	5	10	15	20	25	30
Velocity (mi/h)	17	21	24	29	32	31	28

In order to have the time and the velocity in consistent units, let's convert the velocity readings to feet per second (1 mi/h = 5280/3600 ft/s):

Time (s)	0	5	10	15	20	25	30
Velocity (ft/s)	25	31	35	43	47	45	41

During the first five seconds the velocity doesn't change very much, so we can estimate the distance traveled during that time by assuming that the velocity is constant. If we take the velocity during that time interval to be the initial velocity (25 ft/s), then we obtain the approximate distance traveled during the first five seconds:

$$25 \text{ ft/s} \times 5 \text{ s} = 125 \text{ ft}$$

Similarly, during the second time interval the velocity is approximately constant and we take it to be the velocity when $t = 5$ s. So our estimate for the distance traveled from $t = 5$ s to $t = 10$ s is

$$31 \text{ ft/s} \times 5 \text{ s} = 155 \text{ ft}$$

If we add similar estimates for the other time intervals, we obtain an estimate for the total distance traveled:

$$(25 \times 5) + (31 \times 5) + (35 \times 5) + (43 \times 5) + (47 \times 5) + (45 \times 5) = 1130 \text{ ft}$$

We could just as well have used the velocity at the *end* of each time period instead of the velocity at the beginning as our assumed constant velocity. Then our estimate becomes

$$(31 \times 5) + (35 \times 5) + (43 \times 5) + (47 \times 5) + (45 \times 5) + (41 \times 5) = 1210 \text{ ft}$$

If we had wanted a more accurate estimate, we could have taken velocity readings every two seconds, or even every second. ■

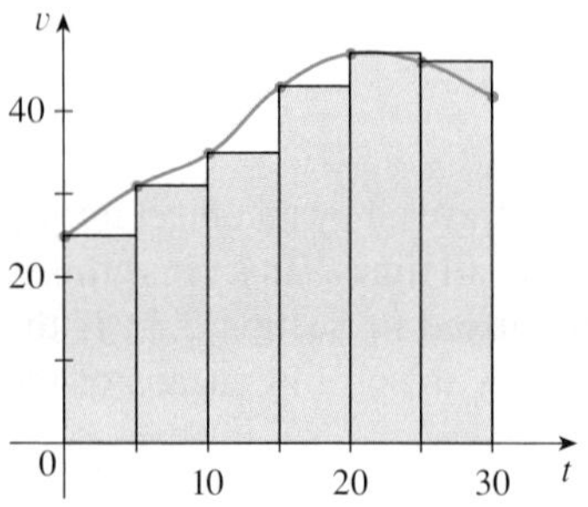

FIGURE 16

Perhaps the calculations in Example 4 remind you of the sums we used earlier to estimate areas. The similarity is explained when we sketch a graph of the velocity function of the car in Figure 16 and draw rectangles whose heights are the initial velocities for each time interval. The area of the first rectangle is $25 \times 5 = 125$, which is also our estimate for the distance traveled in the first five seconds. In fact, the area of each rectangle can be interpreted as a distance because the height represents velocity and the width represents time. The sum of the areas of the rectangles in Figure 16 is $L_6 = 1130$, which is our initial estimate for the total distance traveled.

In general, suppose an object moves with velocity $v = f(t)$, where $a \leq t \leq b$ and $f(t) \geq 0$ (so the object always moves in the positive direction). We take velocity readings at times $t_0 (= a), t_1, t_2, \ldots, t_n (= b)$ so that the velocity is approximately constant on each subinterval. If these times are equally spaced, then the time between consecutive readings is $\Delta t = (b - a)/n$. During the first time interval the velocity is approximately $f(t_0)$ and so the distance traveled is approximately $f(t_0)\,\Delta t$. Similarly, the distance traveled during the second time interval is about $f(t_1)\,\Delta t$ and the total distance traveled during the time interval $[a, b]$ is approximately

$$f(t_0)\,\Delta t + f(t_1)\,\Delta t + \cdots + f(t_{n-1})\,\Delta t = \sum_{i=1}^{n} f(t_{i-1})\,\Delta t$$

If we use the velocity at right endpoints instead of left endpoints, our estimate for the total distance becomes

$$f(t_1)\,\Delta t + f(t_2)\,\Delta t + \cdots + f(t_n)\,\Delta t = \sum_{i=1}^{n} f(t_i)\,\Delta t$$

The more frequently we measure the velocity, the more accurate our estimates become, so it seems plausible that the *exact* distance d traveled is the *limit* of such expressions:

(5) $$d = \lim_{n \to \infty} \sum_{i=1}^{n} f(t_{i-1})\,\Delta t = \lim_{n \to \infty} \sum_{i=1}^{n} f(t_i)\,\Delta t$$

We will see in Section 5.3 that this is indeed true.

Because Equation 5 has the same form as our expressions for area in Equations 2 and 3, it follows that the distance traveled is equal to the area under the graph of the velocity function.

■ Pathogenesis

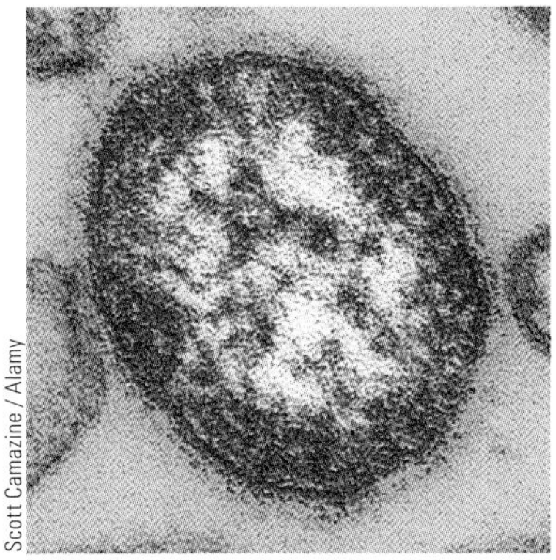

Measles is a highly contagious infection of the respiratory system and is caused by a virus. Despite the fact that more than 80% of the world's population is vaccinated for it, measles remains the fifth leading cause of death worldwide.

In general, the term *pathogenesis* refers to the way a disease originates and develops over time. In the case of measles, the virus enters through the respiratory tract and replicates there before spreading into the bloodstream and then the skin. In a person with no immunity to measles the characteristic rash usually appears about 12 days after infection. The virus reaches a peak density in the blood at about 14 days. The virus level then decreases fairly rapidly over the next few days as a result of the immune response. This progression is reflected in the pathogenesis curve in Figure 17. Notice that the vertical axis is measured in units of number of infected cells per mL of blood plasma.

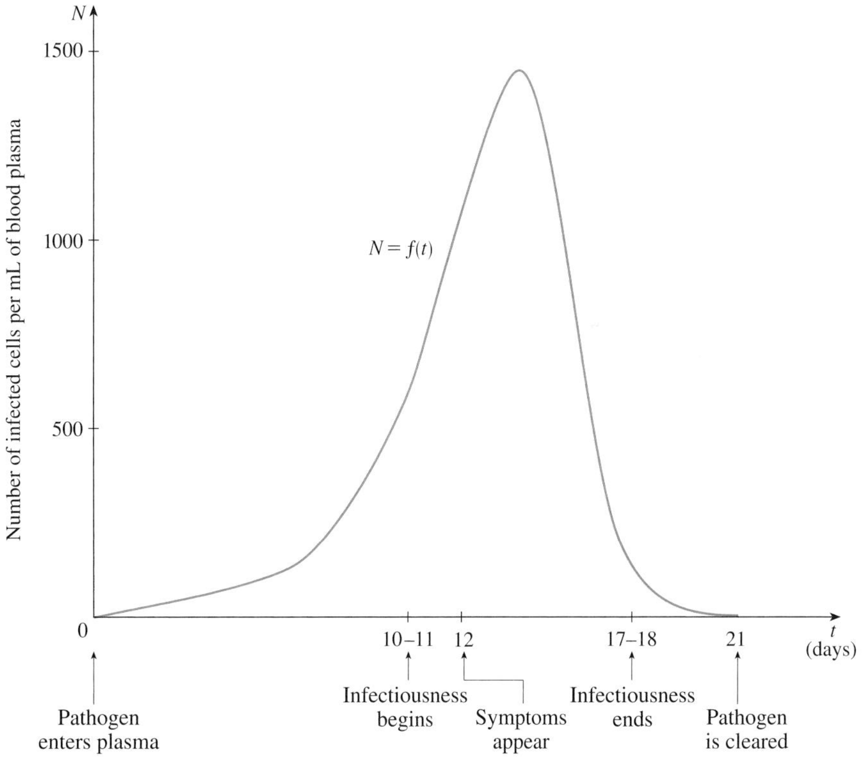

FIGURE 17
Measles pathogenesis curve
Source: J. M. Heffernan et al., "An In-Host Model of Acute Infection: Measles as a Case Study," *Theoretical Population Biology* 73 (2008): 134–47.

Let's denote by f the measles pathogenesis function in Figure 17. Therefore $f(t)$ gives the number of infected cells per mL of plasma on day t. Measles symptoms are thought to develop only after the immune system has been exposed to a threshold "amount of infection." The amount of infection is determined by both the number of infected cells per mL and by the duration over which these cells are exposed to the immune system. If the density of infected cells were constant during infection, then the total

amount of infection would be measured as

$$\text{amount of infection} = \text{density of infected cells} \times \text{time}$$

with the units being (number of cells per mL) $\times$ days. Of course the density is not constant, but we can break the duration of infection into shorter time intervals over which the density changes very little. If each of these shorter time intervals has width Δt, we could add the areas $f(t_i)\,\Delta t$ of the rectangles in Figure 18 and get an approximation to the amount of infection over the first 12 days. Then we take the limit as $\Delta t \to 0$ and the number of rectangles becomes large. By an argument like the one that led to Equation 5, we conclude that the amount of infection needed to stimulate the appearance of symptoms is as follows.

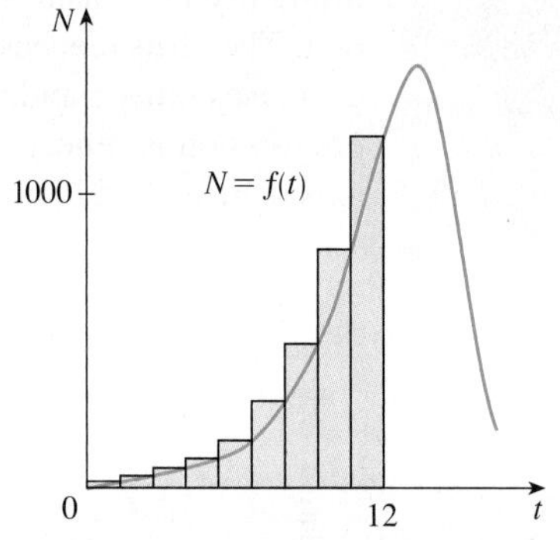

FIGURE 18

> The area under the pathogenesis curve $N = f(t)$ from $t = 0$ to $t = 12$ (shaded in Figure 19) is equal to the total amount of infection needed to develop symptoms.

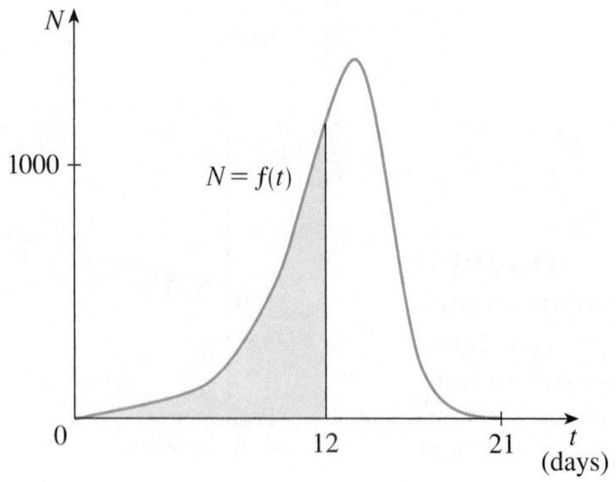

FIGURE 19
Area under pathogenesis curve up to 12 days is the amount of infection needed for symptoms.

The measles pathogenesis curve has been modeled by the polynomial

$$f(t) = -t(t - 21)(t + 1)$$

In Exercise 9 you are asked to use this model to estimate the area under the curve $N = f(t)$ and therefore the total amount of infection needed for the patient to be symptomatic.

EXERCISES 5.1

1. (a) By reading values from the given graph of f, use four rectangles to find a lower estimate and an upper estimate for the area under the given graph of f from $x = 0$ to $x = 8$. In each case sketch the rectangles that you use.

(b) Find new estimates using eight rectangles in each case.

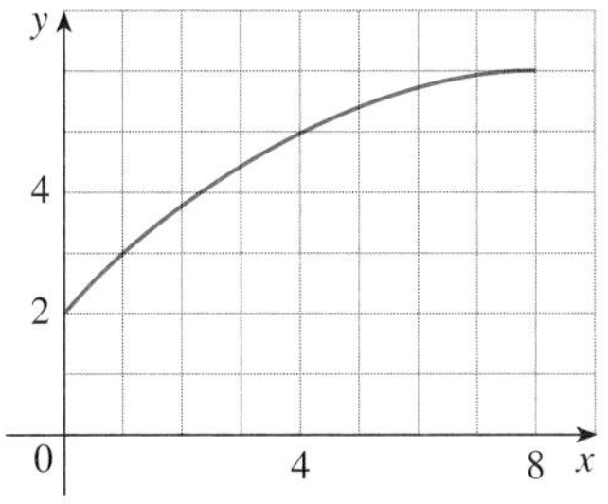

2. (a) Use six rectangles to find estimates of each type for the area under the given graph of f from $x = 0$ to $x = 12$.

(i) L_6 (sample points are left endpoints)

(ii) R_6 (sample points are right endpoints)

(iii) M_6 (sample points are midpoints)

(b) Is L_6 an underestimate or overestimate of the true area?

(c) Is R_6 an underestimate or overestimate of the true area?

(d) Which of the numbers L_6, R_6, or M_6 gives the best estimate? Explain.

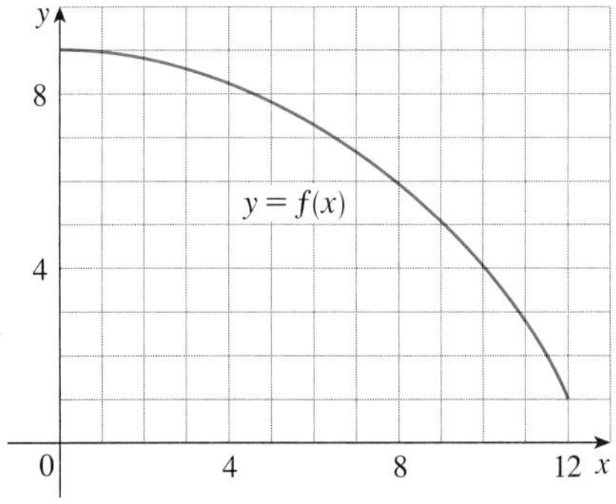

3. (a) Estimate the area under the graph of $f(x) = \cos x$ from $x = 0$ to $x = \pi/2$ using four approximating rectangles and right endpoints. Sketch the graph and the rectangles. Is your estimate an underestimate or an overestimate?

(b) Repeat part (a) using left endpoints.

4. (a) Estimate the area under the graph of $f(x) = \sqrt{x}$ from $x = 0$ to $x = 4$ using four approximating rectangles and right endpoints. Sketch the graph and the rectangles. Is your estimate an underestimate or an overestimate?

(b) Repeat part (a) using left endpoints.

5. (a) Estimate the area under the graph of $f(x) = 1 + x^2$ from $x = -1$ to $x = 2$ using three rectangles and right endpoints. Then improve your estimate by using six rectangles. Sketch the curve and the approximating rectangles.

(b) Repeat part (a) using left endpoints.

(c) Repeat part (a) using midpoints.

(d) From your sketches in parts (a)–(c), which appears to be the best estimate?

6. (a) Graph the function

$$f(x) = x - 2 \ln x \qquad 1 \leqslant x \leqslant 5$$

(b) Estimate the area under the graph of f using four approximating rectangles and taking the sample points to be (i) right endpoints and (ii) midpoints. In each case sketch the curve and the rectangles.

(c) Improve your estimates in part (b) by using eight rectangles.

7. The speed of a runner increased steadily during the first three seconds of a race. Her speed at half-second intervals is given in the table. Find lower and upper estimates for the distance that she traveled during these three seconds.

t (s)	0	0.5	1.0	1.5	2.0	2.5	3.0
v (ft/s)	0	6.2	10.8	14.9	18.1	19.4	20.2

8. Speedometer readings for a motorcycle at 12-second intervals are given in the table.

(a) Estimate the distance traveled by the motorcycle during this time period using the velocities at the beginning of the time intervals.

(b) Give another estimate using the velocities at the end of the time periods.

(c) Are your estimates in parts (a) and (b) upper and lower estimates? Explain.

t (s)	0	12	24	36	48	60
v (ft/s)	30	28	25	22	24	27

9. Measles pathogenesis The function

$$f(t) = -t(t - 21)(t + 1)$$

can be used to model the measles pathogenesis curve in Figures 17 and 19. Suppose symptoms appear after 12 days. Use six subintervals and their midpoints to estimate the total amount of infection needed to develop symptoms.

10. Measles pathogenesis If a patient has had previous exposure to measles, the immune system responds more quickly. This results in a suppression of the level of virus in the plasma during an infection. Suppose that such previous exposure causes the viral density in the plasma to be $\frac{3}{5}$ of that in a patient with no previous exposure. This means that the level of virus in the plasma is given by $\frac{3}{5}f(t)$, where $f(t) = -t(t - 21)(t + 1)$ as in Exercise 9.

(a) Use subintervals of width 2 days and their midpoints to estimate the total amount of infection at $t = 12$ days.

(b) If 7848 cells per mL $\times$ days is the total amount of infection required to develop symptoms, use subintervals of width 2 days and their midpoints to estimate the day when symptoms will appear.

11. SARS incidence The table shows the number of people per day who died from SARS in Singapore at two-week intervals beginning on March 1, 2003.

Date	Deaths per day	Date	Deaths per day
March 1	0.0079	April 26	0.5620
March 15	0.0638	May 10	0.4630
March 29	0.1944	May 24	0.2897
April 12	0.4435		

Estimate the number of people who died of SARS in Singapore between March 1 and May 24, 2003, using both left endpoints and right endpoints.

Source: Adapted from A. Gumel et al., "Modelling Strategies for Controlling SARS Outbreaks," *Proceedings of the Royal Society of London: Series B* 271 (2004): 2223–32.

12. Niche overlap The extent to which species compete for resources is often measured by the *niche overlap*. If the horizontal axis represents a continuum of different resource types (for example, seed sizes for certain bird species), then a plot of the degree of preference for these resources is called a *species' niche*. The degree of overlap of two species' niches is then a measure of the extent to which they compete for resources. The niche overlap for a species is the fraction of the area under its preference curve that is also under the other species' curve. The niches displayed in the figure are given by

$$n_1(x) = (x - 1)(3 - x) \qquad 1 \leqslant x \leqslant 3$$

$$n_2(x) = (x - 2)(4 - x) \qquad 2 \leqslant x \leqslant 4$$

Estimate the niche overlap for species 1 using midpoints. (Choose the number of subintervals yourself.)

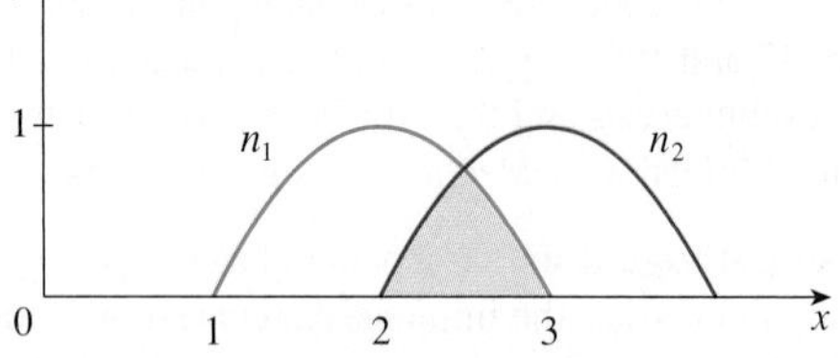

13. The velocity graph of a braking car is shown. Use it to estimate the distance traveled by the car while the brakes are applied.

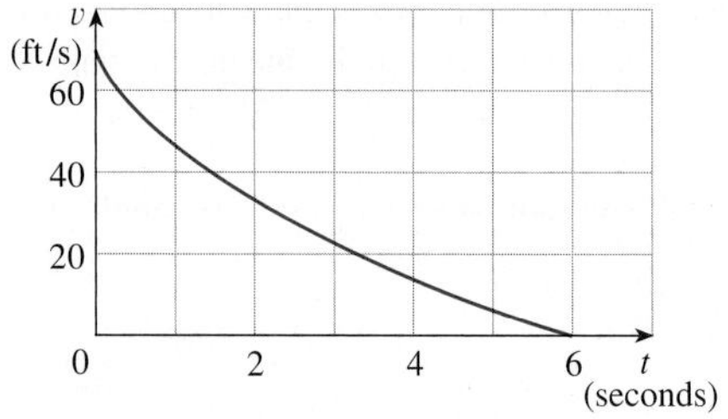

14. The velocity graph of a car accelerating from rest to a speed of 120 km/h over a period of 30 seconds is shown. Estimate the distance traveled during this period.

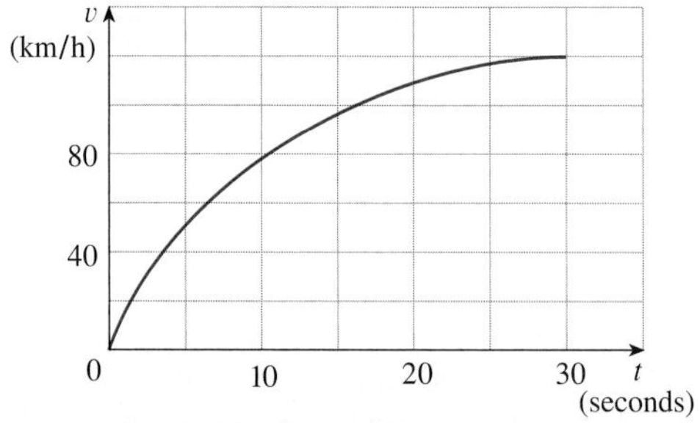

15–17 Use Definition 2 to find an expression for the area under the graph of f as a limit. Do not evaluate the limit.

15. $f(x) = \dfrac{2x}{x^2 + 1}, \quad 1 \leqslant x \leqslant 3$

16. $f(x) = x^2 + \sqrt{1 + 2x}, \quad 4 \leqslant x \leqslant 7$

17. $f(x) = x \cos x, \quad 0 \leqslant x \leqslant \pi/2$

---

18. (a) Use Definition 2 to find an expression for the area under the curve $y = x^3$ from 0 to 1 as a limit.

(b) The following formula for the sum of the cubes of the first n integers is proved in Appendix E. Use it to evaluate the limit in part (a).

$$1^3 + 2^3 + 3^3 + \cdots + n^3 = \left[\frac{n(n + 1)}{2}\right]^2$$

19. Let A be the area under the graph of an increasing continuous function f from a to b, and let L_n and R_n be the approximations to A with n subintervals using left and right endpoints, respectively.

(a) How are A, L_n, and R_n related?

(b) Show that

$$R_n - L_n = \frac{b - a}{n}[f(b) - f(a)]$$

(c) Deduce that

$$R_n - A < \frac{b-a}{n}[f(b) - f(a)]$$

20. If A is the area under the curve $y = e^x$ from 1 to 3, use Exercise 19 to find a value of n such that $R_n - A < 0.0001$.

CAS **21.** (a) Express the area under the curve $y = x^5$ from 0 to 2 as a limit.

(b) Use a computer algebra system to find the sum in your expression from part (a).

(c) Evaluate the limit in part (a).

CAS **22.** Find the exact area of the region under the graph of $y = e^{-x}$ from 0 to 2 by using a computer algebra system to evaluate the sum and then the limit in Example 3(a). Compare your answer with the estimate obtained in Example 3(b).

CAS **23.** Find the exact area under the cosine curve $y = \cos x$ from $x = 0$ to $x = b$, where $0 \leqslant b \leqslant \pi/2$. (Use a computer algebra system both to evaluate the sum and compute the limit.) In particular, what is the area if $b = \pi/2$?

24. (a) Let A_n be the area of a polygon with n equal sides inscribed in a circle with radius r. By dividing the polygon into n congruent triangles with central angle $2\pi/n$, show that

$$A_n = \tfrac{1}{2}nr^2 \sin\left(\frac{2\pi}{n}\right)$$

(b) Show that $\lim_{n\to\infty} A_n = \pi r^2$. [*Hint:* Use Equation 2.4.6 on page 133.]

5.2 The Definite Integral

We saw in Section 5.1 that a limit of the form

(1)
$$\lim_{n\to\infty} \sum_{i=1}^{n} f(x_i^*)\,\Delta x = \lim_{n\to\infty} [f(x_1^*)\,\Delta x + f(x_2^*)\,\Delta x + \cdots + f(x_n^*)\,\Delta x]$$

arises when we compute an area. We also saw that it arises when we try to find the distance traveled by an object or the total amount of infection needed to show measles symptoms. It turns out that this same type of limit occurs in a wide variety of situations even when f is not necessarily a positive function. In Chapter 6 we will see that limits of the form (1) also arise in finding volumes of tumors, cardiac output, blood flow, and many other quantities. We therefore give this type of limit a special name and notation.

(2) Definition of a Definite Integral If f is a function defined for $a \leqslant x \leqslant b$, we divide the interval $[a, b]$ into n subintervals of equal width $\Delta x = (b - a)/n$. We let $x_0\ (= a), x_1, x_2, \ldots, x_n\ (= b)$ be the endpoints of these subintervals and we let $x_1^*, x_2^*, \ldots, x_n^*$ be any **sample points** in these subintervals, so x_i^* lies in the ith subinterval $[x_{i-1}, x_i]$. Then the **definite integral of f from a to b** is

$$\int_a^b f(x)\,dx = \lim_{n\to\infty} \sum_{i=1}^{n} f(x_i^*)\,\Delta x$$

provided that this limit exists. If it does exist, we say that f is **integrable** on $[a, b]$.

A precise definition of this type of limit is given in Appendix D.

NOTE 1 The symbol $\int$ was introduced by Leibniz and is called an **integral sign**. It is an elongated S and was chosen because an integral is a limit of sums. In the notation $\int_a^b f(x)\,dx$, $f(x)$ is called the **integrand** and a and b are called the **limits of integration**; a is the **lower limit** and b is the **upper limit**. For now, the symbol dx has no meaning by itself; $\int_a^b f(x)\,dx$ is all one symbol. The dx simply indicates that the independent variable is x. The procedure of calculating an integral is called **integration**.

NOTE 2 The definite integral $\int_a^b f(x)\,dx$ is a number; it does not depend on x. In fact, we could use any letter in place of x without changing the value of the integral:

$$\int_a^b f(x)\,dx = \int_a^b f(t)\,dt = \int_a^b f(r)\,dr$$

Riemann

Bernhard Riemann received his Ph.D. under the direction of the legendary Gauss at the University of Göttingen and remained there to teach. Gauss, who was not in the habit of praising other mathematicians, spoke of Riemann's "creative, active, truly mathematical mind and gloriously fertile originality." The definition (2) of an integral that we use is due to Riemann. He also made major contributions to the theory of functions of a complex variable, mathematical physics, number theory, and the foundations of geometry. Riemann's broad concept of space and geometry turned out to be the right setting, 50 years later, for Einstein's general relativity theory. Riemann's health was poor throughout his life, and he died of tuberculosis at the age of 39.

NOTE 3 The sum

$$\sum_{i=1}^{n} f(x_i^*)\,\Delta x$$

that occurs in Definition 2 is called a **Riemann sum** after the German mathematician Bernhard Riemann (1826–1866). So Definition 2 says that the definite integral of an integrable function can be approximated to within any desired degree of accuracy by a Riemann sum.

We know that if f happens to be positive, then the Riemann sum can be interpreted as a sum of areas of approximating rectangles (see Figure 1). By comparing Definition 2 with the definition of area in Section 5.1, we see that the definite integral $\int_a^b f(x)\,dx$ can be interpreted as the area under the curve $y = f(x)$ from a to b. (See Figure 2.)

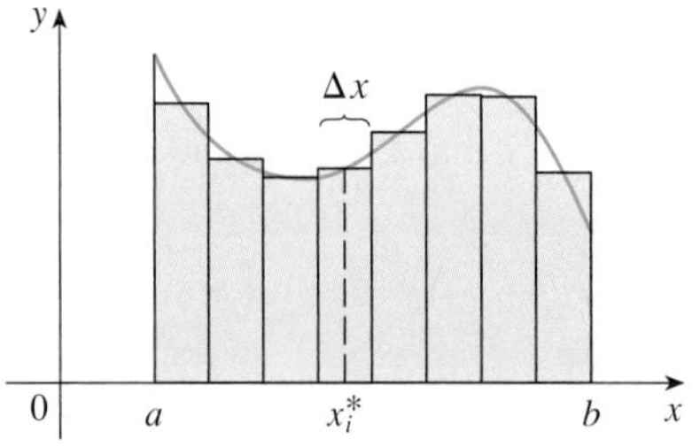

FIGURE 1
If $f(x) \geqslant 0$, the Riemann sum $\sum f(x_i^*)\,\Delta x$ is the sum of areas of rectangles.

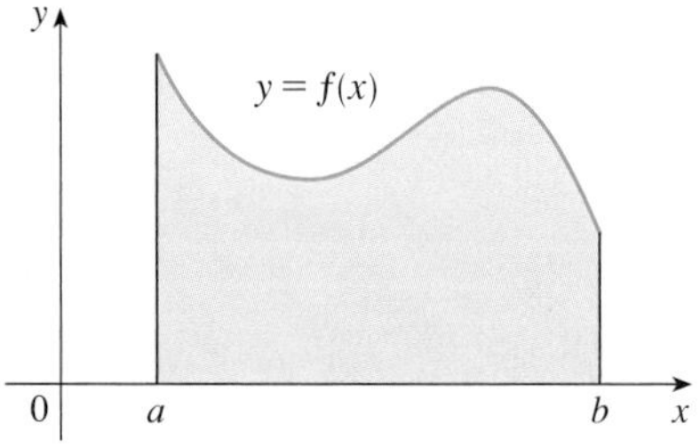

FIGURE 2
If $f(x) \geqslant 0$, the integral $\int_a^b f(x)\,dx$ is the area under the curve $y = f(x)$ from a to b.

If f takes on both positive and negative values, as in Figure 3, then the Riemann sum is the sum of the areas of the rectangles that lie above the x-axis and the *negatives* of the areas of the rectangles that lie below the x-axis (the areas of the blue rectangles *minus* the areas of the gold rectangles). When we take the limit of such Riemann sums, we get the situation illustrated in Figure 4. A definite integral can be interpreted as a **net area**, that is, a difference of areas:

$$\int_a^b f(x)\,dx = A_1 - A_2$$

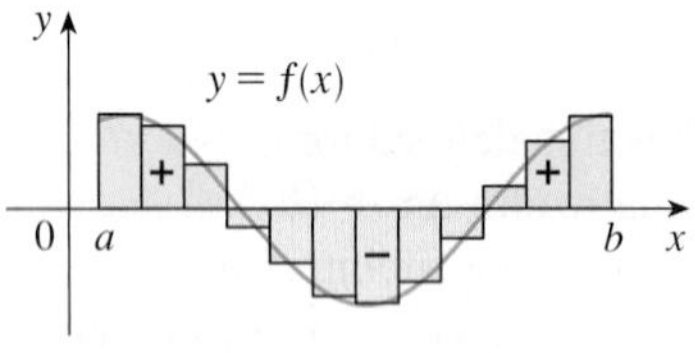

FIGURE 3
$\sum f(x_i^*)\,\Delta x$ is an approximation to the net area.

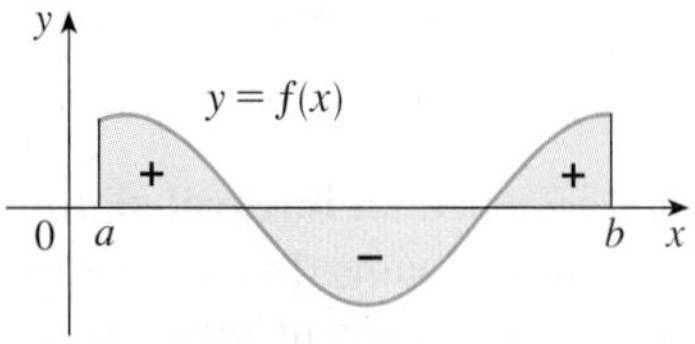

FIGURE 4
$\int_a^b f(x)\,dx$ is the net area.

where A_1 is the area of the region above the x-axis and below the graph of f, and A_2 is the area of the region below the x-axis and above the graph of f.

NOTE 4 Although we have defined $\int_a^b f(x)\,dx$ by dividing $[a, b]$ into subintervals of equal width, there are situations in which it is advantageous to work with subintervals of unequal width. For instance, if in a biological experiment data are collected at times that are not equally spaced, then we can still estimate the area under a curve. If the subinterval widths are $\Delta x_1, \Delta x_2, \ldots, \Delta x_n$, we have to ensure that all these widths approach 0 in the limiting process. This happens if the largest width, $\max \Delta x_i$, approaches 0. So in this case the definition of a definite integral becomes

$$\int_a^b f(x)\,dx = \lim_{\max \Delta x_i \to 0} \sum_{i=1}^{n} f(x_i^*)\,\Delta x_i$$

NOTE 5 We have defined the definite integral for an integrable function, but not all functions are integrable. The following theorem shows that the most commonly occurring functions are in fact integrable. It is proved in more advanced courses.

> **(3) Theorem** If f is continuous on $[a, b]$, or if f has only a finite number of jump discontinuities, then f is integrable on $[a, b]$; that is, the definite integral $\int_a^b f(x)\,dx$ exists.

If f is integrable on $[a, b]$, then the limit in Definition 2 exists and gives the same value no matter how we choose the sample points x_i^*. To simplify the calculation of the integral we often take the sample points to be right endpoints. Then $x_i^* = x_i$ and the definition of an integral simplifies as follows.

> **(4) Theorem** If f is integrable on $[a, b]$, then
>
> $$\int_a^b f(x)\,dx = \lim_{n\to\infty} \sum_{i=1}^{n} f(x_i)\,\Delta x$$
>
> where $\quad \Delta x = \dfrac{b - a}{n} \quad$ and $\quad x_i = a + i\,\Delta x$

EXAMPLE 1 | Express

$$\lim_{n\to\infty} \sum_{i=1}^{n} (x_i^3 + x_i \sin x_i)\,\Delta x$$

as an integral on the interval $[0, \pi]$.

SOLUTION Comparing the given limit with the limit in Theorem 4, we see that they will be identical if we choose $f(x) = x^3 + x \sin x$. We are given that $a = 0$ and $b = \pi$. Therefore, by Theorem 4, we have

$$\lim_{n\to\infty} \sum_{i=1}^{n} (x_i^3 + x_i \sin x_i)\,\Delta x = \int_0^\pi (x^3 + x \sin x)\,dx$$ ■

Later, when we apply the definite integral in biological contexts, it will be important to recognize limits of sums as integrals, as we did in Example 1. When Leibniz chose the notation for an integral, he chose the ingredients as reminders of the limiting process. In general, when we write

$$\lim_{n\to\infty} \sum_{i=1}^{n} f(x_i^*)\,\Delta x = \int_a^b f(x)\,dx$$

we replace $\lim \Sigma$ by $\int$, x_i^* by x, and Δx by dx.

■ Calculating Integrals

When we use a limit to evaluate a definite integral, we need to know how to work with sums. The following three equations give formulas for sums of powers of positive integers. Equation 5 may be familiar to you from a course in algebra. Equations 6 and 7 were discussed in Section 5.1 and are proved in Appendix F.

(5) $$\sum_{i=1}^{n} i = \frac{n(n+1)}{2}$$

(6) $$\sum_{i=1}^{n} i^2 = \frac{n(n+1)(2n+1)}{6}$$

(7) $$\sum_{i=1}^{n} i^3 = \left[\frac{n(n+1)}{2}\right]^2$$

The remaining formulas are simple rules for working with sigma notation:

Formulas 8–11 are proved by writing out each side in expanded form. The left side of Equation 9 is

$$ca_1 + ca_2 + \cdots + ca_n$$

The right side is

$$c(a_1 + a_2 + \cdots + a_n)$$

These are equal by the distributive property. The other formulas are discussed in Appendix F.

(8) $$\sum_{i=1}^{n} c = nc$$

(9) $$\sum_{i=1}^{n} ca_i = c\sum_{i=1}^{n} a_i$$

(10) $$\sum_{i=1}^{n} (a_i + b_i) = \sum_{i=1}^{n} a_i + \sum_{i=1}^{n} b_i$$

(11) $$\sum_{i=1}^{n} (a_i - b_i) = \sum_{i=1}^{n} a_i - \sum_{i=1}^{n} b_i$$

EXAMPLE 2

(a) Evaluate the Riemann sum for $f(x) = x^3 - 6x$, taking the sample points to be right endpoints and $a = 0$, $b = 3$, and $n = 6$.

(b) Evaluate $\int_0^3 (x^3 - 6x)\,dx$.

SOLUTION

(a) With $n = 6$ the interval width is

$$\Delta x = \frac{b-a}{n} = \frac{3-0}{6} = \frac{1}{2}$$

and the right endpoints are $x_1 = 0.5$, $x_2 = 1.0$, $x_3 = 1.5$, $x_4 = 2.0$, $x_5 = 2.5$, and $x_6 = 3.0$. So the Riemann sum is

$$\begin{aligned} R_6 &= \sum_{i=1}^{6} f(x_i)\,\Delta x \\ &= f(0.5)\,\Delta x + f(1.0)\,\Delta x + f(1.5)\,\Delta x + f(2.0)\,\Delta x + f(2.5)\,\Delta x + f(3.0)\,\Delta x \\ &= \tfrac{1}{2}(-2.875 - 5 - 5.625 - 4 + 0.625 + 9) \\ &= -3.9375 \end{aligned}$$

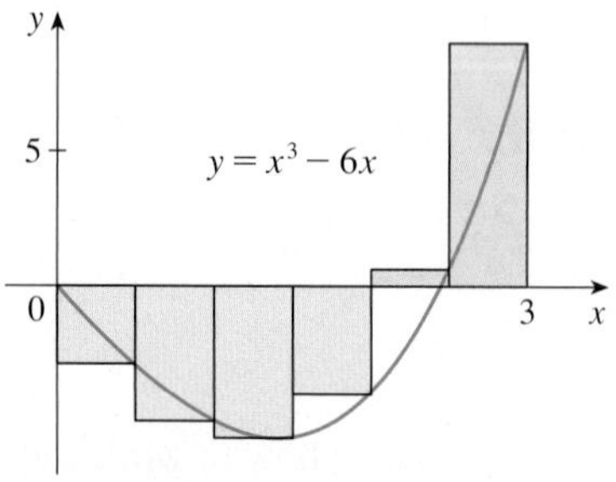

FIGURE 5

Notice that f is not a positive function and so the Riemann sum does not represent a sum of areas of rectangles. But it does represent the sum of the areas of the blue rectangles (above the x-axis) minus the sum of the areas of the gold rectangles (below the x-axis) in Figure 5.

(b) With n subintervals we have

$$\Delta x = \frac{b - a}{n} = \frac{3}{n}$$

Thus $x_0 = 0$, $x_1 = 3/n$, $x_2 = 6/n$, $x_3 = 9/n$, and, in general, $x_i = 3i/n$. Since we are using right endpoints, we can use Theorem 4:

In the sum, n is a constant (unlike i), so we can move $3/n$ in front of the Σ sign.

$$\int_0^3 (x^3 - 6x)\,dx = \lim_{n\to\infty} \sum_{i=1}^{n} f(x_i)\,\Delta x = \lim_{n\to\infty} \sum_{i=1}^{n} f\left(\frac{3i}{n}\right)\frac{3}{n}$$

$$= \lim_{n\to\infty} \frac{3}{n} \sum_{i=1}^{n} \left[\left(\frac{3i}{n}\right)^3 - 6\left(\frac{3i}{n}\right)\right] \qquad \text{(Equation 9 with } c = 3/n\text{)}$$

$$= \lim_{n\to\infty} \frac{3}{n} \sum_{i=1}^{n} \left[\frac{27}{n^3} i^3 - \frac{18}{n} i\right]$$

$$= \lim_{n\to\infty} \left[\frac{81}{n^4} \sum_{i=1}^{n} i^3 - \frac{54}{n^2} \sum_{i=1}^{n} i\right] \qquad \text{(Equations 11 and 9)}$$

$$= \lim_{n\to\infty} \left\{\frac{81}{n^4}\left[\frac{n(n+1)}{2}\right]^2 - \frac{54}{n^2}\,\frac{n(n+1)}{2}\right\} \qquad \text{(Equations 7 and 5)}$$

$$= \lim_{n\to\infty} \left[\frac{81}{4}\left(1 + \frac{1}{n}\right)^2 - 27\left(1 + \frac{1}{n}\right)\right]$$

$$= \frac{81}{4} - 27 = -\frac{27}{4} = -6.75$$

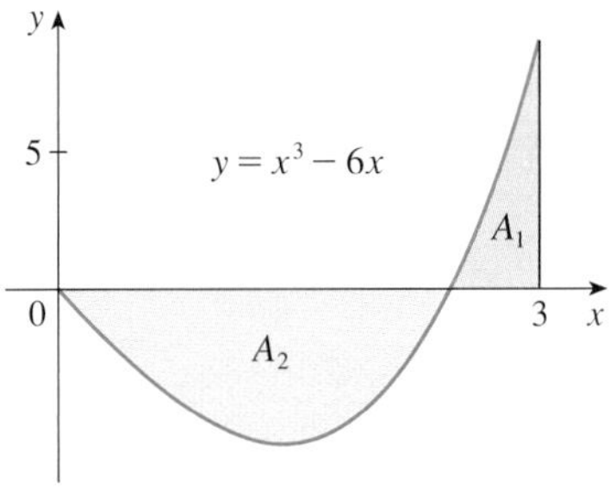

FIGURE 6
$\int_0^3 (x^3 - 6x)\,dx = A_1 - A_2 = -6.75$

This integral can't be interpreted as an area because f takes on both positive and negative values. But it can be interpreted as the difference of areas $A_1 - A_2$, where A_1 and A_2 are shown in Figure 6.

Figure 7 illustrates the calculation by showing the positive and negative terms in the right Riemann sum R_n for $n = 40$. The values in the table show the Riemann sums approaching the exact value of the integral, -6.75, as $n \to \infty$.

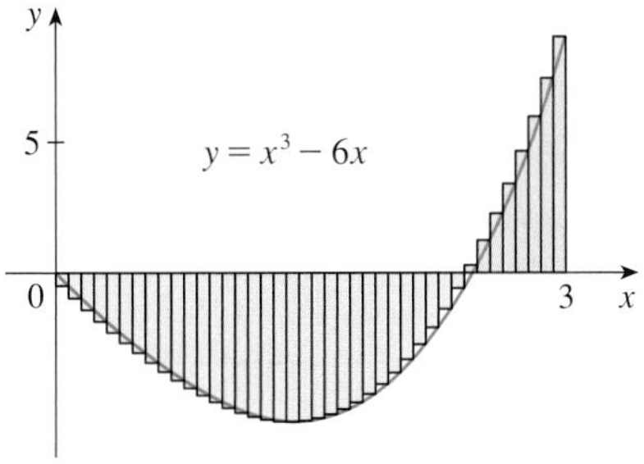

FIGURE 7
$R_{40} \approx -6.3998$

n	R_n
40	−6.3998
100	−6.6130
500	−6.7229
1000	−6.7365
5000	−6.7473

■

A much simpler method for evaluating the integral in Example 2 will be given in Section 5.3 after we have proved the Evaluation Theorem.

EXAMPLE 3 | Evaluate the following integrals by interpreting each in terms of areas.

(a) $\int_0^1 \sqrt{1 - x^2}\, dx$ (b) $\int_0^3 (x - 1)\, dx$

SOLUTION

(a) Since $f(x) = \sqrt{1 - x^2} \geqslant 0$, we can interpret this integral as the area under the curve $y = \sqrt{1 - x^2}$ from 0 to 1. But, since $y^2 = 1 - x^2$, we get $x^2 + y^2 = 1$, which shows that the graph of f is the quarter-circle with radius 1 in Figure 8. Therefore

$$\int_0^1 \sqrt{1 - x^2}\, dx = \tfrac{1}{4}\pi(1)^2 = \frac{\pi}{4}$$

(b) The graph of $y = x - 1$ is the line with slope 1 shown in Figure 9. We compute the integral as the difference of the areas of the two triangles:

$$\int_0^3 (x - 1)\, dx = A_1 - A_2 = \tfrac{1}{2}(2 \cdot 2) - \tfrac{1}{2}(1 \cdot 1) = 1.5$$

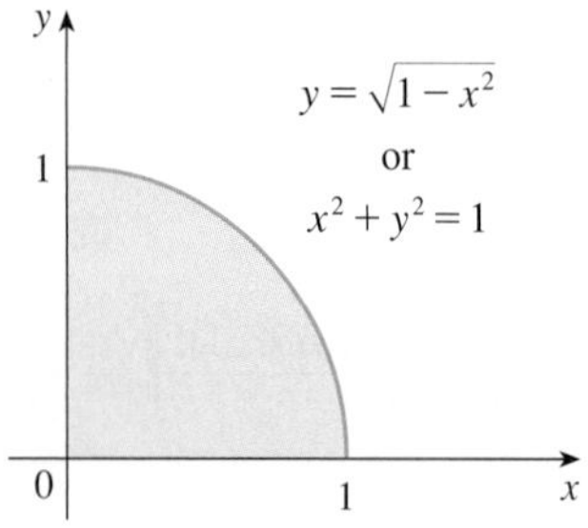

FIGURE 8

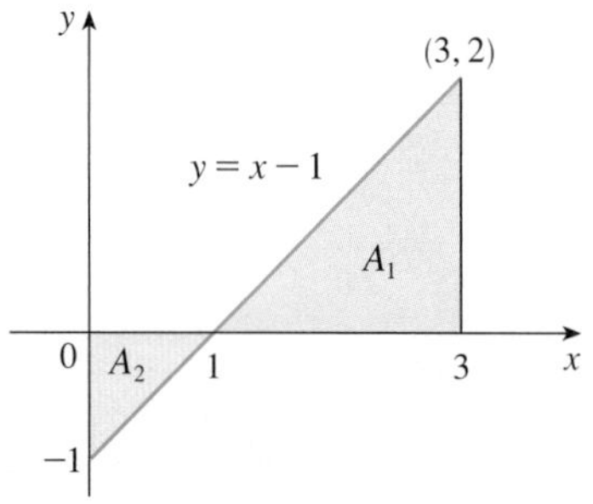

FIGURE 9

The Midpoint Rule

We often choose the sample point x_i^* to be the right endpoint of the ith subinterval because it is convenient for computing the limit. But if the purpose is to find an *approximation* to an integral, it is usually better to choose x_i^* to be the midpoint of the interval, which we denote by $\bar{x}_i$. Any Riemann sum is an approximation to an integral, but if we use midpoints we get the following approximation.

TEC Module 5.2 shows how the Midpoint Rule estimates improve as n increases.

Midpoint Rule

$$\int_a^b f(x)\, dx \approx \sum_{i=1}^{n} f(\bar{x}_i)\, \Delta x = \Delta x \,[f(\bar{x}_1) + \cdots + f(\bar{x}_n)]$$

where $\Delta x = \dfrac{b - a}{n}$

and $\bar{x}_i = \tfrac{1}{2}(x_{i-1} + x_i) = \text{midpoint of } [x_{i-1}, x_i]$

EXAMPLE 4 | Use the Midpoint Rule with $n = 5$ to approximate $\displaystyle\int_1^2 \frac{1}{x}\, dx$.

SOLUTION The endpoints of the five subintervals are 1, 1.2, 1.4, 1.6, 1.8, and 2.0, so the midpoints are 1.1, 1.3, 1.5, 1.7, and 1.9. The width of the subintervals is

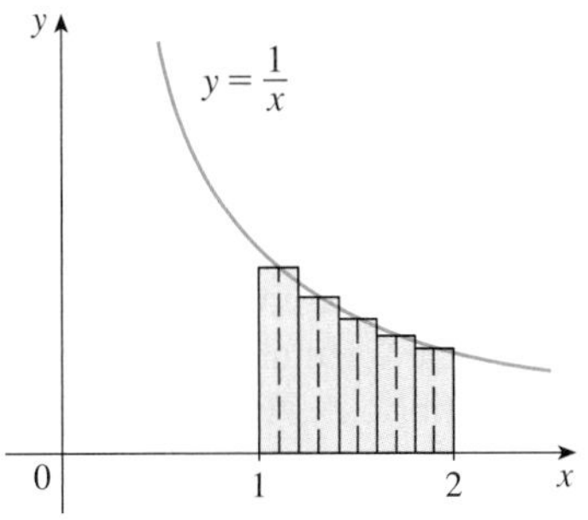

FIGURE 10

$\Delta x = (2 - 1)/5 = \frac{1}{5}$, so the Midpoint Rule gives

$$\int_1^2 \frac{1}{x}\,dx \approx \Delta x\,[f(1.1) + f(1.3) + f(1.5) + f(1.7) + f(1.9)]$$

$$= \frac{1}{5}\left(\frac{1}{1.1} + \frac{1}{1.3} + \frac{1}{1.5} + \frac{1}{1.7} + \frac{1}{1.9}\right)$$

$$\approx 0.691908$$

Since $f(x) = 1/x > 0$ for $1 \leqslant x \leqslant 2$, the integral represents an area, and the approximation given by the Midpoint Rule is the sum of the areas of the rectangles shown in Figure 10. ■

If we apply the Midpoint Rule to the integral in Example 2, we get the picture in Figure 11. The approximation $M_{40} \approx -6.7563$ is much closer to the true value -6.75 than the right endpoint approximation, $R_{40} \approx -6.3998$, shown in Figure 7.

TEC In Visual 5.2 you can compare left, right, and midpoint approximations to the integral in Example 2 for different values of n.

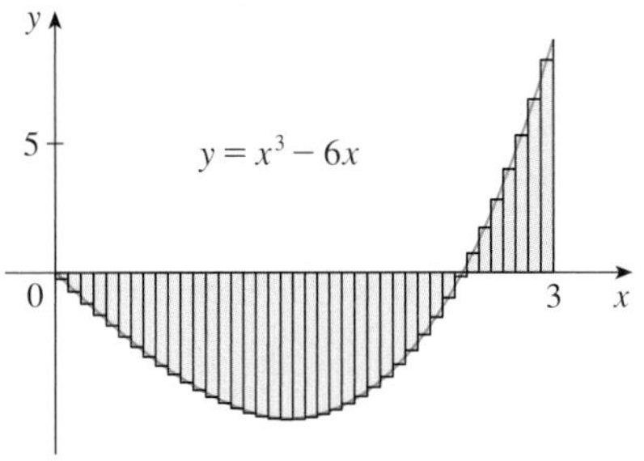

FIGURE 11
$M_{40} \approx -6.7563$

EXAMPLE 5 | Aspirin pharmacokinetics In a study[1] of the effects of low-dose aspirin, 10 young males were given a single 80-mg dose of ASA (acetylsalicylic acid) on three separate days. Blood samples were obtained for 24 hours after each dose and peak plasma ASA levels of about 1 μg/mL were reached within 30 minutes. After the results for the 10 volunteers were averaged, the concentration of ASA in the bloodstream was modeled by the function

$$C(t) = 32t^2e^{-4.2t}$$

where t is measured in hours and C is measured in μg/mL. The graph of C is shown in Figure 12. Use the Midpoint Rule with 10 subintervals to estimate the value of the integral $\int_0^2 C(t)\,dt$. What are the units?

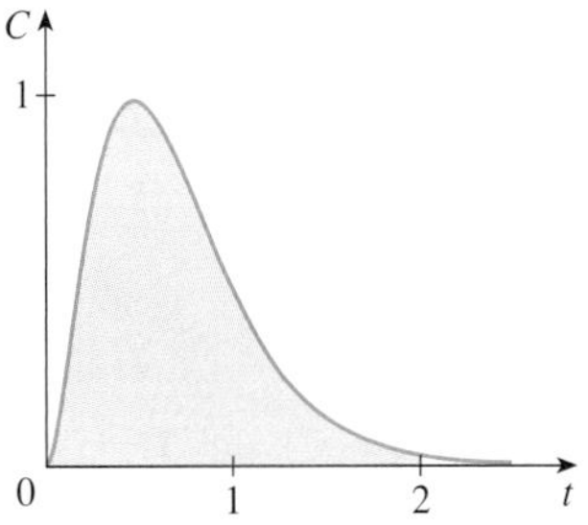

FIGURE 12
Area under the ASA concentration function

SOLUTION If we divide the interval [0, 2] into 10 subintervals, then the midpoints are 0.1, 0.3, 0.5, . . . , 1.7, 1.9 and the width of each subinterval is $\Delta t = 0.2$. Using the Midpoint Rule, we have

$$\int_0^2 C(t)\,dt \approx \Delta t[f(0.1) + f(0.3) + f(0.5) + \cdots + f(1.9)]$$

$$\approx 0.2[0.2103 + 0.8169 + 0.9797 + 0.8289 + 0.5916$$
$$+ 0.3815 + 0.2300 + 0.1322 + 0.0733 + 0.0395]$$
$$= 0.8568$$

1. I. H. Benedek et al., "Variability in the Pharmacokinetics and Pharmacodynamics of Low Dose Aspirin in Healthy Male Volunteers," *Journal of Clinical Pharmacology* 35 (1995): 1181–86.

The units for C are micrograms per milliliter (μg/mL) and the units for t are hours, so the units for the integral are micrograms per milliliter times hours:

$$\int_0^2 C(t)\,dt \approx 0.8568\ (\mu\text{g/mL})\cdot\text{h}$$ ■

How would we interpret biologically the integral we computed in Example 5? In the biological literature it is denoted simply by AUC (area under the curve), but what does that mean in this context? For a drug to work, it needs to be "available" to interact with the target tissue (or target pathogen if the drug is an antibiotic). Availability can be increased by increasing the concentration or by increasing the time the drug lingers before it is cleared through metabolism. AUC is a combined measure of these, and therefore is a composite measure of the overall "availability."

■ Properties of the Definite Integral

When we defined the definite integral $\int_a^b f(x)\,dx$, we implicitly assumed that $a < b$. But the definition as a limit of Riemann sums makes sense even if $a > b$. Notice that if we reverse a and b, then Δx changes from $(b - a)/n$ to $(a - b)/n$. Therefore

$$\int_b^a f(x)\,dx = -\int_a^b f(x)\,dx$$

If $a = b$, then $\Delta x = 0$ and so

$$\int_a^a f(x)\,dx = 0$$

We now develop some basic properties of integrals that will help us to evaluate integrals in a simple manner. We assume that f and g are continuous functions.

Properties of the Integral

1. $\int_a^b c\,dx = c(b - a)$, where c is any constant

2. $\int_a^b [f(x) + g(x)]\,dx = \int_a^b f(x)\,dx + \int_a^b g(x)\,dx$

3. $\int_a^b cf(x)\,dx = c\int_a^b f(x)\,dx$, where c is any constant

4. $\int_a^b [f(x) - g(x)]\,dx = \int_a^b f(x)\,dx - \int_a^b g(x)\,dx$

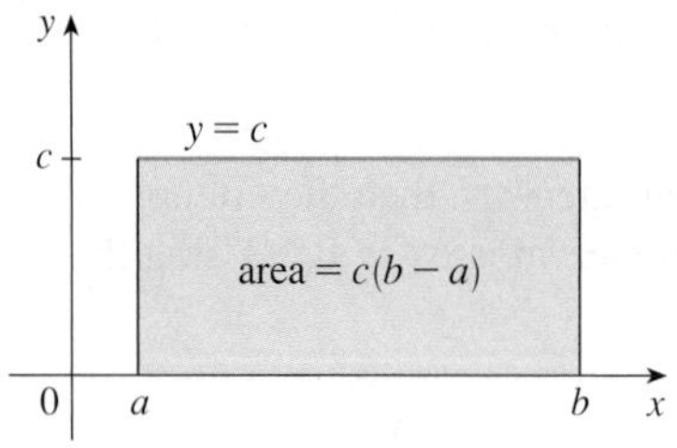

FIGURE 13

$\int_a^b c\,dx = c(b - a)$

Property 1 says that the integral of a constant function $f(x) = c$ is the constant times the length of the interval. If $c > 0$ and $a < b$, this is to be expected because $c(b - a)$ is the area of the shaded rectangle in Figure 13.

Property 2 says that the integral of a sum is the sum of the integrals. For positive functions it says that the area under $f + g$ is the area under f plus the area under g. Figure 14 helps us understand why this is true: In view of how graphical addition works, the corresponding vertical line segments have equal height.

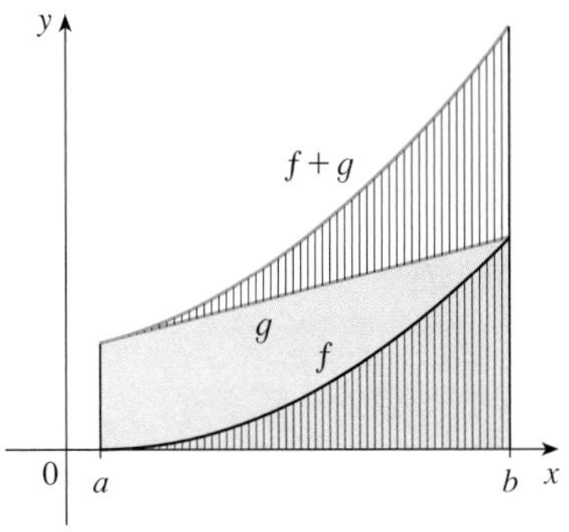

FIGURE 14

$\int_a^b [f(x)+g(x)]\,dx = \int_a^b f(x)\,dx + \int_a^b g(x)\,dx$

Property 3 seems intuitively reasonable because we know that multiplying a function by a positive number c stretches or shrinks its graph vertically by a factor of c. So it stretches or shrinks each approximating rectangle by a factor c and therefore it has the effect of multiplying the area by c.

In general, Property 2 follows from Theorem 4 and the fact that the limit of a sum is the sum of the limits:

$$\begin{aligned}\int_a^b [f(x) + g(x)]\,dx &= \lim_{n\to\infty} \sum_{i=1}^{n} [f(x_i) + g(x_i)]\,\Delta x \\ &= \lim_{n\to\infty} \left[\sum_{i=1}^{n} f(x_i)\,\Delta x + \sum_{i=1}^{n} g(x_i)\,\Delta x\right] \\ &= \lim_{n\to\infty} \sum_{i=1}^{n} f(x_i)\,\Delta x + \lim_{n\to\infty} \sum_{i=1}^{n} g(x_i)\,\Delta x \\ &= \int_a^b f(x)\,dx + \int_a^b g(x)\,dx\end{aligned}$$

Property 3 can be proved in a similar manner and says that the integral of a constant times a function is the constant times the integral of the function. In other words, a constant (but *only* a constant) can be taken in front of an integral sign. Property 4 is proved by writing $f - g = f + (-g)$ and using Properties 2 and 3 with $c = -1$.

EXAMPLE 6 | Use the properties of integrals to evaluate $\int_0^1 (4 + 3x^2)\,dx$.

SOLUTION Using Properties 2 and 3 of integrals, we have

$$\int_0^1 (4 + 3x^2)\,dx = \int_0^1 4\,dx + \int_0^1 3x^2\,dx = \int_0^1 4\,dx + 3\int_0^1 x^2\,dx$$

We know from Property 1 that

$$\int_0^1 4\,dx = 4(1 - 0) = 4$$

and we found in Example 5.1.2 that $\int_0^1 x^2\,dx = \frac{1}{3}$. So

$$\begin{aligned}\int_0^1 (4 + 3x^2)\,dx &= \int_0^1 4\,dx + 3\int_0^1 x^2\,dx \\ &= 4 + 3 \cdot \tfrac{1}{3} = 5\end{aligned}$$

■

The next property tells us how to combine integrals of the same function over adjacent intervals:

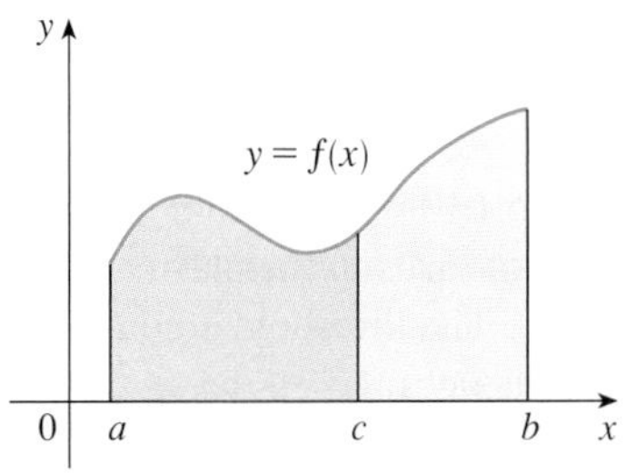

FIGURE 15

5.	$\int_a^c f(x)\,dx + \int_c^b f(x)\,dx = \int_a^b f(x)\,dx$

This is not easy to prove in general, but for the case where $f(x) \geqslant 0$ and $a < c < b$ Property 5 can be seen from the geometric interpretation in Figure 15: The area under $y = f(x)$ from a to c plus the area from c to b is equal to the total area from a to b.

EXAMPLE 7 | If it is known that $\int_0^{10} f(x)\,dx = 17$ and $\int_0^{8} f(x)\,dx = 12$, find $\int_8^{10} f(x)\,dx$.

SOLUTION By Property 5, we have

$$\int_0^8 f(x)\,dx + \int_8^{10} f(x)\,dx = \int_0^{10} f(x)\,dx$$

so
$$\int_8^{10} f(x)\,dx = \int_0^{10} f(x)\,dx - \int_0^8 f(x)\,dx = 17 - 12 = 5$$
■

Properties 1–5 are true whether $a < b$, $a = b$, or $a > b$. The following properties, in which we compare sizes of functions and sizes of integrals, are true only if $a \leqslant b$. Again we assume that the functions are continuous.

Comparison Properties of the Integral

6. If $f(x) \geqslant 0$ for $a \leqslant x \leqslant b$, then $\int_a^b f(x)\,dx \geqslant 0$.

7. If $f(x) \geqslant g(x)$ for $a \leqslant x \leqslant b$, then $\int_a^b f(x)\,dx \geqslant \int_a^b g(x)\,dx$.

8. If $m \leqslant f(x) \leqslant M$ for $a \leqslant x \leqslant b$, then

$$m(b - a) \leqslant \int_a^b f(x)\,dx \leqslant M(b - a)$$

Because we are assuming that f and g are continuous, the inequalities $\geqslant$ and $\leqslant$ can be replaced by the strict inequalities $>$ and $<$.

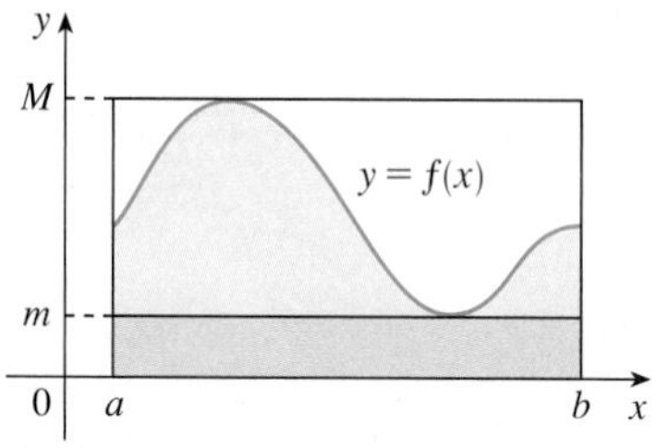

FIGURE 16

If $f(x) \geqslant 0$, then $\int_a^b f(x)\,dx$ represents the area under the graph of f, so the geometric interpretation of Property 6 is simply that areas are positive. (It also follows directly from the definition because all the quantities involved are positive.) Property 7 says that a bigger function has a bigger integral. It follows from Properties 6 and 4 because $f - g \geqslant 0$.

Property 8 is illustrated by Figure 16 for the case where $f(x) \geqslant 0$. If f is continuous we could take m and M to be the absolute minimum and maximum values of f on the interval $[a, b]$. In this case Property 8 says that the area under the graph of f is greater than the area of the rectangle with height m and less than the area of the rectangle with height M.

PROOF OF PROPERTY 8 Since $m \leqslant f(x) \leqslant M$, Property 7 gives

$$\int_a^b m\,dx \leqslant \int_a^b f(x)\,dx \leqslant \int_a^b M\,dx$$

Using Property 1 to evaluate the integrals on the left and right sides, we obtain

$$m(b - a) \leqslant \int_a^b f(x)\,dx \leqslant M(b - a)$$
■

EXAMPLE 8 | **Measles pathogenesis** In Exercise 5.1.10 we were told that if a patient has had previous exposure to measles, then the level of virus in the plasma during an infection is suppressed (due to stronger immunity). The threshold amount of infection required to develop symptoms is 7848 infected cells per mL $\times$ days. Suppose you and your friend Bob are both infected with measles at the same time but Bob has stronger immunity than you (so that your pathogenesis curve is always at least as high as his). Use Property 7 to show that, if Bob starts to display symptoms on day T, then you must necessarily also display symptoms by this day.

SOLUTION Let's use $f(t)$ and $g(t)$ to be the pathogenesis curves for you and Bob. Because Bob has stronger immunity than you, we know that $f(t) \geqslant g(t)$ at all times (your pathogenesis curve is always at least as high as his). Now if Bob starts to display symptoms on day T, then $\int_0^T g(t)\,dt = 7848$. Furthermore, after T days the level of infection experienced by you will be $\int_0^T f(t)\,dt$. Using Property 7, we see that

$$\int_0^T f(t)\,dt \geqslant \int_0^T g(t)\,dt$$

Therefore we have that $\int_0^T f(t)\,dt \geqslant 7848$, meaning that by day T you will also have reached the threshold amount of infection required to display symptoms. ■

As the next example shows, Property 8 is useful when all we want is a rough estimate of the size of an integral without going to the bother of using the Midpoint Rule.

EXAMPLE 9 | Use Property 8 to estimate $\int_0^1 e^{-x^2}\,dx$.

SOLUTION Because $f(x) = e^{-x^2}$ is a decreasing function on $[0, 1]$, its absolute maximum value is $M = f(0) = 1$ and its absolute minimum value is $m = f(1) = e^{-1}$. Thus, by Property 8,

$$e^{-1}(1 - 0) \leqslant \int_0^1 e^{-x^2}\,dx \leqslant 1(1 - 0)$$

or

$$e^{-1} \leqslant \int_0^1 e^{-x^2}\,dx \leqslant 1$$

Since $e^{-1} \approx 0.3679$, we can write

$$0.367 \leqslant \int_0^1 e^{-x^2}\,dx \leqslant 1$$ ■

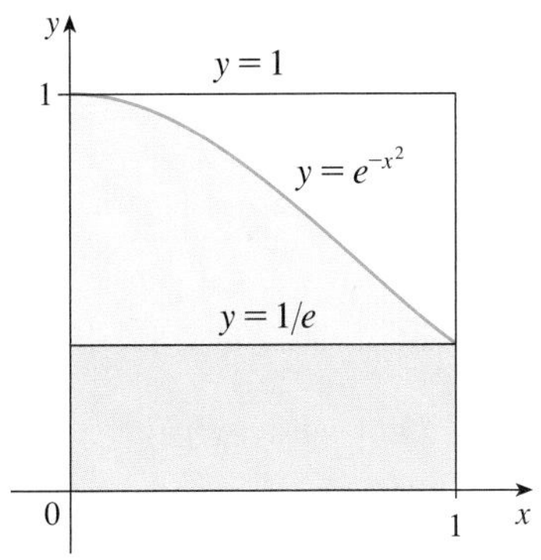

FIGURE 17

The result of Example 9 is illustrated in Figure 17. The integral is greater than the area of the lower rectangle and less than the area of the square.

EXERCISES 5.2

1. Evaluate the Riemann sum for $f(x) = 3 - \frac{1}{2}x$, $2 \leqslant x \leqslant 14$, with six subintervals, taking the sample points to be left endpoints. Explain, with the aid of a diagram, what the Riemann sum represents.

2. If $f(x) = x^2 - 2x$, $0 \leqslant x \leqslant 3$, evaluate the Riemann sum with $n = 6$, taking the sample points to be right endpoints. What does the Riemann sum represent? Illustrate with a diagram.

3. If $f(x) = e^x - 2$, $0 \leqslant x \leqslant 2$, find the Riemann sum with $n = 4$ correct to six decimal places, taking the sample points to be midpoints. What does the Riemann sum represent? Illustrate with a diagram.

4. (a) Find the Riemann sum for $f(x) = \sin x$, $0 \leqslant x \leqslant 3\pi/2$, with six terms, taking the sample points to be right endpoints. (Give your answer correct to six decimal places.) Explain what the Riemann sum represents with the aid of a sketch.
   (b) Repeat part (a) with midpoints as the sample points.

5. The graph of a function f is given. Estimate $\int_0^8 f(x)\,dx$ using four subintervals with (a) right endpoints, (b) left endpoints, and (c) midpoints.

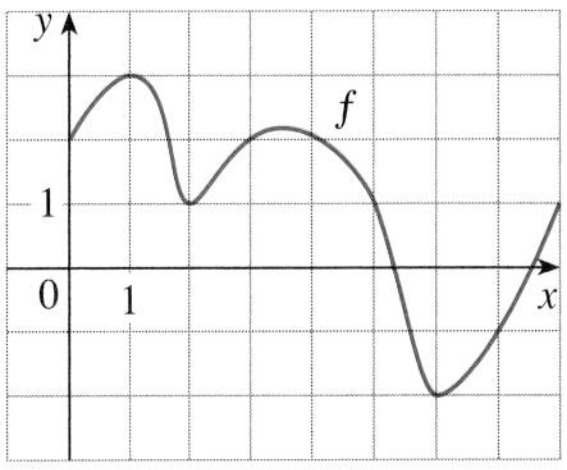

6. The graph of g is shown. Estimate $\int_{-3}^{3} g(x)\,dx$ with six subintervals using (a) right endpoints, (b) left endpoints, and (c) midpoints.

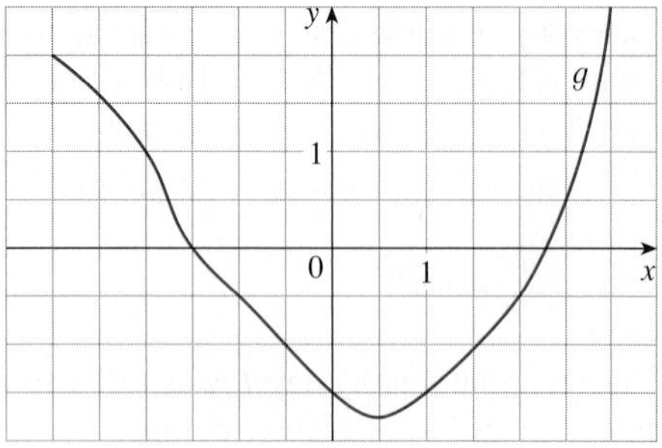

7. A table of values of an increasing function f is shown. Use the table to find lower and upper estimates for $\int_{10}^{30} f(x)\,dx$.

x	10	14	18	22	26	30
$f(x)$	−12	−6	−2	1	3	8

8. The table gives the values of a function obtained from an experiment. Use them to estimate $\int_{3}^{9} f(x)\,dx$ using three equal subintervals with (a) right endpoints, (b) left endpoints, and (c) midpoints. If the function is known to be an increasing function, can you say whether your estimates are less than or greater than the exact value of the integral?

x	3	4	5	6	7	8	9
$f(x)$	−3.4	−2.1	−0.6	0.3	0.9	1.4	1.8

9–12 Use the Midpoint Rule with the given value of n to approximate the integral. Round the answer to four decimal places.

9. $\int_{2}^{10} \sqrt{x^3 + 1}\,dx, \quad n = 4$

10. $\int_{0}^{\pi/2} \cos^4 x\,dx, \quad n = 4$

11. $\int_{0}^{1} \sin(x^2)\,dx, \quad n = 5$

12. $\int_{1}^{5} x^2 e^{-x}\,dx, \quad n = 4$

13. Drug pharmacokinetics During testing of a new drug, researchers measured the plasma drug concentration of each test subject at 10-minute intervals. The average concentrations $C(t)$ are shown in the table, where t is measured in minutes and C is measured in $\mu g/mL$. Use the Midpoint Rule to estimate the integral $\int_{0}^{100} C(t)\,dt$. State the units.

t	0	10	20	30	40	50
$C(t)$	0	1.3	1.8	2.2	2.4	2.5

t	60	70	80	90	100
$C(t)$	2.4	2.3	2.0	1.6	1.1

14. Salicylic acid pharmacokinetics In the study cited in Example 5, the metabolite salicylic acid (SA) was rapidly formed and peak SA levels of about 4.2 $\mu g/mL$ were reached after an hour. The concentration of SA was modeled by the function

$$C(t) = 11.4te^{-t}$$

where t is measured in hours and C is measured in $\mu g/mL$. Use the Midpoint Rule with eight subintervals to estimate the integral $\int_{0}^{4} C(t)\,dt$. State the units.

15–18 Express the limit as a definite integral on the given interval.

15. $\lim_{n\to\infty} \sum_{i=1}^{n} x_i \ln(1 + x_i^2)\,\Delta x, \quad [2, 6]$

16. $\lim_{n\to\infty} \sum_{i=1}^{n} \frac{\cos x_i}{x_i}\,\Delta x, \quad [\pi, 2\pi]$

17. $\lim_{n\to\infty} \sum_{i=1}^{n} \sqrt{2x_i^* + (x_i^*)^2}\,\Delta x, \quad [1, 8]$

18. $\lim_{n\to\infty} \sum_{i=1}^{n} [4 - 3(x_i^*)^2 + 6(x_i^*)^5]\,\Delta x, \quad [0, 2]$

19–23 Use the form of the definition of the integral given in Theorem 4 to evaluate the integral.

19. $\int_{-1}^{5} (1 + 3x)\,dx$

20. $\int_{1}^{4} (x^2 + 2x - 5)\,dx$

21. $\int_{0}^{2} (2 - x^2)\,dx$

22. $\int_{0}^{5} (1 + 2x^3)\,dx$

23. $\int_{1}^{2} x^3\,dx$

24. (a) Find an approximation to the integral $\int_{0}^{4} (x^2 - 3x)\,dx$ using a Riemann sum with right endpoints and $n = 8$.
(b) Draw a diagram like Figure 3 to illustrate the approximation in part (a).
(c) Use Theorem 4 to evaluate $\int_{0}^{4} (x^2 - 3x)\,dx$.
(d) Interpret the integral in part (c) as a difference of areas and illustrate with a diagram like Figure 4.

25–26 Express the integral as a limit of Riemann sums. Do not evaluate the limit.

25. $\int_{2}^{6} \frac{x}{1 + x^5}\,dx$

26. $\int_{1}^{10} (x - 4\ln x)\,dx$

27. The graph of f is shown. Evaluate each integral by interpreting it in terms of areas.

(a) $\int_{0}^{2} f(x)\,dx$ (b) $\int_{0}^{5} f(x)\,dx$

(c) $\int_{5}^{7} f(x)\,dx$ (d) $\int_{0}^{9} f(x)\,dx$

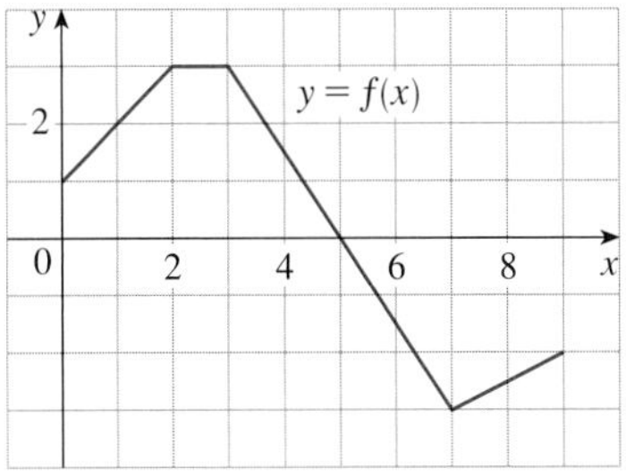

28. The graph of g consists of two straight lines and a semicircle. Use it to evaluate each integral.

(a) $\int_0^2 g(x)\,dx$ (b) $\int_2^6 g(x)\,dx$ (c) $\int_0^7 g(x)\,dx$

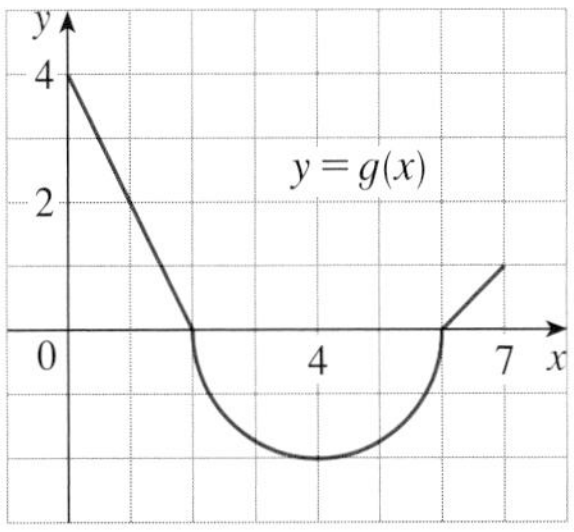

29–34 Evaluate the integral by interpreting it in terms of areas.

29. $\int_0^3 \left(\frac{1}{2}x - 1\right) dx$

30. $\int_{-2}^2 \sqrt{4 - x^2}\,dx$

31. $\int_{-3}^0 \left(1 + \sqrt{9 - x^2}\right) dx$

32. $\int_{-1}^3 (3 - 2x)\,dx$

33. $\int_{-1}^2 |\,x\,|\,dx$

34. $\int_0^{10} |\,x - 5\,|\,dx$

35. Evaluate $\int_\pi^\pi \sin^2 x \cos^4 x\,dx$.

36. Given that $\int_0^1 3x\sqrt{x^2 + 4}\,dx = 5\sqrt{5} - 8$, what is $\int_1^0 3u\sqrt{u^2 + 4}\,du$?

37. Write as a single integral in the form $\int_a^b f(x)\,dx$:

$$\int_{-2}^2 f(x)\,dx + \int_2^5 f(x)\,dx - \int_{-2}^{-1} f(x)\,dx$$

38. If $\int_1^5 f(x)\,dx = 12$ and $\int_4^5 f(x)\,dx = 3.6$, find $\int_1^4 f(x)\,dx$.

39. If $\int_0^9 f(x)\,dx = 37$ and $\int_0^9 g(x)\,dx = 16$, find $\int_0^9 [2f(x) + 3g(x)]\,dx$.

40. Find $\int_0^5 f(x)\,dx$ if

$$f(x) = \begin{cases} 3 & \text{for } x < 3 \\ x & \text{for } x \geqslant 3 \end{cases}$$

41. For the function f whose graph is shown, list the following quantities in increasing order, from smallest to largest, and explain your reasoning.

(A) $\int_0^8 f(x)\,dx$ (B) $\int_0^3 f(x)\,dx$ (C) $\int_3^8 f(x)\,dx$

(D) $\int_4^8 f(x)\,dx$ (E) $f'(1)$

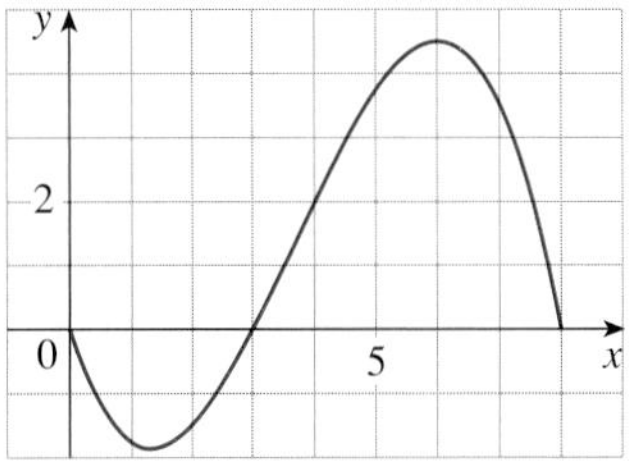

42. If $F(x) = \int_2^x f(t)\,dt$, where f is the function whose graph is given, which of the following values is largest?

(A) $F(0)$ (B) $F(1)$ (C) $F(2)$

(D) $F(3)$ (E) $F(4)$

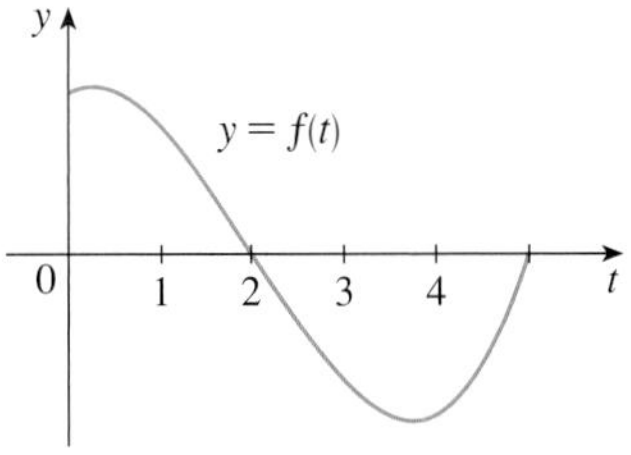

43. Each of the regions A, B, and C bounded by the graph of f and the x-axis has area 3. Find the value of

$$\int_{-4}^2 [f(x) + 2x + 5]\,dx$$

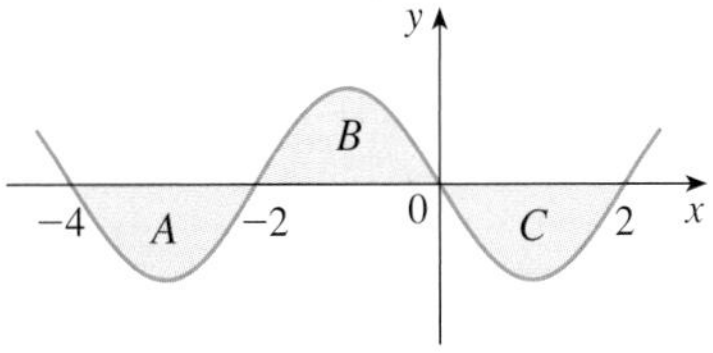

44. Suppose f has absolute minimum value m and absolute maximum value M. Between what two values must $\int_0^2 f(x)\,dx$ lie? Which property of integrals allows you to make your conclusion?

45. Use the properties of integrals to verify that

$$2 \leqslant \int_{-1}^1 \sqrt{1 + x^2}\,dx \leqslant 2\sqrt{2}$$

46. Use Property 8 to estimate the value of the integral

$$\int_0^2 \frac{1}{1 + x^2}\,dx$$

47–48 Express the limit as a definite integral.

47. $\lim_{n \to \infty} \sum_{i=1}^{n} \frac{i^4}{n^5}$ [*Hint:* Consider $f(x) = x^4$.]

48. $\lim_{n \to \infty} \frac{1}{n} \sum_{i=1}^{n} \frac{1}{1 + (i/n)^2}$

5.3 The Fundamental Theorem of Calculus

The Fundamental Theorem of Calculus is appropriately named because it establishes a connection between the two branches of calculus: differential calculus and integral calculus. Differential calculus arose from the tangent problem, whereas integral calculus arose from a seemingly unrelated problem, the area problem. Newton's mentor at Cambridge, Isaac Barrow (1630–1677), discovered that these two problems are actually closely related. In fact, he realized that differentiation and integration are inverse processes. The Fundamental Theorem of Calculus gives the precise inverse relationship between the derivative and the integral. It was Newton and Leibniz who exploited this relationship and used it to develop calculus into a systematic mathematical method.

■ Evaluating Definite Integrals

In Section 5.2 we computed integrals from the definition as a limit of Riemann sums and we saw that this procedure is sometimes long and difficult. Newton and Leibniz realized that they could calculate $\int_a^b f(x)\,dx$ if they happened to know an antiderivative F of f. Their discovery, called the Evaluation Theorem, is part of the Fundamental Theorem of Calculus, which is discussed later in this section.

Evaluation Theorem If f is continuous on the interval $[a, b]$, then

$$\int_a^b f(x)\,dx = F(b) - F(a)$$

where F is any antiderivative of f, that is, $F' = f$.

This theorem states that if we know an antiderivative F of f, then we can evaluate $\int_a^b f(x)\,dx$ simply by subtracting the values of F at the endpoints of the interval $[a, b]$. It is very surprising that $\int_a^b f(x)\,dx$, which was defined by a complicated procedure involving all of the values of $f(x)$ for $a \leqslant x \leqslant b$, can be found by knowing the values of $F(x)$ at only two points, a and b.

For instance, we know from Section 4.6 that an antiderivative of the function $f(x) = x^2$ is $F(x) = \frac{1}{3}x^3$, so the Evaluation Theorem tells us that

$$\int_0^1 x^2\,dx = F(1) - F(0) = \tfrac{1}{3} \cdot 1^3 - \tfrac{1}{3} \cdot 0^3 = \tfrac{1}{3}$$

Comparing this method with the calculation in Example 5.1.2, where we found the area under the parabola $y = x^2$ from 0 to 1 by computing a limit of sums, we see that the Evaluation Theorem provides us with a simple and powerful method.

Although the Evaluation Theorem may be surprising at first glance, it becomes plausible if we interpret it in physical terms. If $v(t)$ is the velocity of an object and $s(t)$ is its position at time t, then $v(t) = s'(t)$, so s is an antiderivative of v. In Section 5.1 we considered an object that always moves in the positive direction and made the guess that the area under the velocity curve is equal to the distance traveled. In symbols:

$$\int_a^b v(t)\,dt = s(b) - s(a)$$

That is exactly what the Evaluation Theorem says in this context.

PROOF OF THE EVALUATION THEOREM We divide the interval $[a, b]$ into n subintervals with endpoints $x_0\ (= a), x_1, x_2, \ldots, x_n\ (= b)$ and with length $\Delta x = (b - a)/n$. Let F be any antiderivative of f. By subtracting and adding like terms, we can express the total difference in the F values as the sum of the differences over the subintervals:

$$\begin{aligned} F(b) - F(a) &= F(x_n) - F(x_0) \\ &= F(x_n) - F(x_{n-1}) + F(x_{n-1}) - F(x_{n-2}) + \cdots + F(x_2) - F(x_1) + F(x_1) - F(x_0) \\ &= \sum_{i=1}^{n} [F(x_i) - F(x_{i-1})] \end{aligned}$$

The Mean Value Theorem was discussed in Section 4.2.

Now F is continuous (because it's differentiable) and so we can apply the Mean Value Theorem to F on each subinterval $[x_{i-1}, x_i]$. Thus there exists a number x_i^* between x_{i-1} and x_i such that

$$F(x_i) - F(x_{i-1}) = F'(x_i^*)(x_i - x_{i-1}) = f(x_i^*)\,\Delta x$$

Therefore $$F(b) - F(a) = \sum_{i=1}^{n} f(x_i^*)\,\Delta x$$

Now we take the limit of each side of this equation as $n \to \infty$. The left side is a constant and the right side is a Riemann sum for the function f, so

$$F(b) - F(a) = \lim_{n\to\infty} \sum_{i=1}^{n} f(x_i^*)\,\Delta x = \int_a^b f(x)\,dx$$ ■

When applying the Evaluation Theorem we use the notation

$$F(x)\Big]_a^b = F(b) - F(a)$$

and so we can write

$$\int_a^b f(x)\,dx = F(x)\Big]_a^b \qquad \text{where} \qquad F' = f$$

Other common notations are $F(x)\big|_a^b$ and $[F(x)]_a^b$.

EXAMPLE 1 | Evaluate $\int_1^3 e^x\,dx$.

SOLUTION An antiderivative of $f(x) = e^x$ is $F(x) = e^x$, so we use the Evaluation Theorem as follows:

In applying the Evaluation Theorem we use a particular antiderivative F of f. It is not necessary to use the most general antiderivative $(e^x + C)$.

$$\int_1^3 e^x\,dx = e^x\Big]_1^3 = e^3 - e$$ ■

You can see from Example 1 that it is quite easy to calculate $\int_1^3 e^x\,dx$ with the Evaluation Theorem. Without the Evaluation Theorem it would be very difficult to calculate the integral. In fact if we use Theorem 5.2.4 with $f(x) = e^x$, $a = 1$, and $b = 3$, we get a challenging limit:

$$\int_1^3 e^x\,dx = \lim_{n\to\infty} \sum_{i=1}^{n} f(x_i)\,\Delta x = \lim_{n\to\infty} \frac{2}{n} \sum_{i=1}^{n} e^{1+2i/n}$$

This limit can be evaluated but it isn't easy to do so. The method of Example 1 is *much* easier.

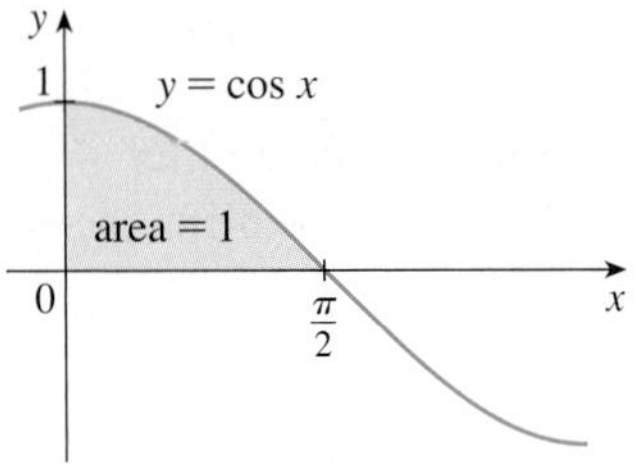

FIGURE 1

EXAMPLE 2 | Find the area under the cosine curve from 0 to b, where $0 \leqslant b \leqslant \pi/2$.

SOLUTION Since an antiderivative of $f(x) = \cos x$ is $F(x) = \sin x$, we have

$$A = \int_0^b \cos x \, dx = \sin x \Big]_0^b = \sin b - \sin 0 = \sin b$$

In particular, taking $b = \pi/2$, we have proved that the area under the cosine curve from 0 to $\pi/2$ is $\sin(\pi/2) = 1$. (See Figure 1.) ■

When the French mathematician Gilles de Roberval first found the area under the sine and cosine curves in 1635, this was a very challenging problem that required a great deal of ingenuity. If we didn't have the benefit of the Evaluation Theorem, we would have to compute a difficult limit of sums using obscure trigonometric identities (or a computer algebra system as in Exercise 5.1.23). It was even more difficult for Roberval because the apparatus of limits had not been invented in 1635. But in the 1660s and 1670s, when the Evaluation Theorem was discovered by Newton and Leibniz, such problems became very easy, as you can see from Example 2.

■ Indefinite Integrals

We need a convenient notation for antiderivatives that makes them easy to work with. Because of the relation given by the Evaluation Theorem between antiderivatives and integrals, the notation $\int f(x)\,dx$ is traditionally used for an antiderivative of f and is called an **indefinite integral**. Thus

$$\int f(x)\,dx = F(x) \qquad \text{means} \qquad F'(x) = f(x)$$

You should distinguish carefully between definite and indefinite integrals. A definite integral $\int_a^b f(x)\,dx$ is a *number*, whereas an indefinite integral $\int f(x)\,dx$ is a *function* (or family of functions). The connection between them is given by the Evaluation Theorem: If f is continuous on $[a, b]$, then

$$\int_a^b f(x)\,dx = \int f(x)\,dx \Big]_a^b$$

Recall from Section 4.6 that if F is an antiderivative of f on an interval I, then the most general antiderivative of f on I is $F(x) + C$, where C is an arbitrary constant. For instance, the formula

$$\int \frac{1}{x}\,dx = \ln|x| + C$$

is valid (on any interval that doesn't contain 0) because $(d/dx)\ln|x| = 1/x$. So an indefinite integral $\int f(x)\,dx$ can represent either a particular antiderivative of f or an entire *family* of antiderivatives (one for each value of the constant C).

The effectiveness of the Evaluation Theorem depends on having a supply of antiderivatives of functions. We therefore restate the Table of Antidifferentiation Formulas from Section 4.6, together with a few others, in the notation of indefinite integrals. Any formula can be verified by differentiating the function on the right side and obtaining the integrand. For instance,

$$\int \sec^2 x\,dx = \tan x + C \qquad \text{because} \qquad \frac{d}{dx}(\tan x + C) = \sec^2 x$$

(1) Table of Indefinite Integrals

$$\int [f(x) + g(x)]\,dx = \int f(x)\,dx + \int g(x)\,dx \qquad \int cf(x)\,dx = c\int f(x)\,dx$$

$$\int x^n\,dx = \frac{x^{n+1}}{n+1} + C \quad (n \neq -1) \qquad \int \frac{1}{x}\,dx = \ln|x| + C$$

$$\int e^x\,dx = e^x + C \qquad \int e^{kx}\,dx = \frac{1}{k}e^{kx} + C$$

$$\int a^x\,dx = \frac{a^x}{\ln a} + C \qquad \int \sin x\,dx = -\cos x + C$$

$$\int \cos x\,dx = \sin x + C \qquad \int \sec^2 x\,dx = \tan x + C$$

$$\int \csc^2 x\,dx = -\cot x + C \qquad \int \sec x \tan x\,dx = \sec x + C$$

$$\int \csc x \cot x\,dx = -\csc x + C \qquad \int \frac{1}{x^2+1}\,dx = \tan^{-1}x + C$$

EXAMPLE 3 | Find the general indefinite integral

$$\int (10x^4 - 2\sec^2 x)\,dx$$

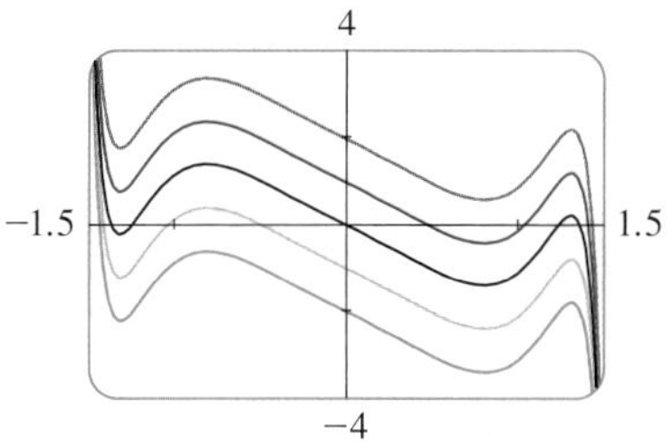

FIGURE 2

The indefinite integral in Example 3 is graphed in Figure 2 for several values of C. Here the value of C is the y-intercept.

SOLUTION Using our convention and Table 1 and properties of integrals, we have

$$\int (10x^4 - 2\sec^2 x)\,dx = 10\int x^4\,dx - 2\int \sec^2 x\,dx$$

$$= 10\frac{x^5}{5} - 2\tan x + C$$

$$= 2x^5 - 2\tan x + C$$

You should check this answer by differentiating it. ■

EXAMPLE 4 | Evaluate $\int_0^3 (x^3 - 6x)\,dx$.

SOLUTION Using the Evaluation theorem and Table 1, we have

$$\int_0^3 (x^3 - 6x)\,dx = \frac{x^4}{4} - 6\frac{x^2}{2}\Bigg]_0^3$$

$$= \left(\tfrac{1}{4}\cdot 3^4 - 3\cdot 3^2\right) - \left(\tfrac{1}{4}\cdot 0^4 - 3\cdot 0^2\right)$$

$$= \tfrac{81}{4} - 27 - 0 + 0 = -6.75$$

Compare this calculation with Example 5.2.2(b). ■

EXAMPLE 5 | Find $\int_0^2 \left(2x^3 - 6x + \frac{3}{x^2+1}\right) dx$ and interpret the result in terms of areas.

SOLUTION The Evaluation Theorem gives

$$\int_0^2 \left(2x^3 - 6x + \frac{3}{x^2+1}\right) dx = 2\frac{x^4}{4} - 6\frac{x^2}{2} + 3\tan^{-1}x \Big]_0^2$$

$$= \tfrac{1}{2}x^4 - 3x^2 + 3\tan^{-1}x \Big]_0^2$$

$$= \tfrac{1}{2}(2^4) - 3(2^2) + 3\tan^{-1}2 - 0$$

$$= -4 + 3\tan^{-1}2$$

This is the exact value of the integral. If a decimal approximation is desired, we can use a calculator to approximate $\tan^{-1}2$. Doing so, we get

$$\int_0^2 \left(2x^3 - 6x + \frac{3}{x^2+1}\right) dx \approx -0.67855$$

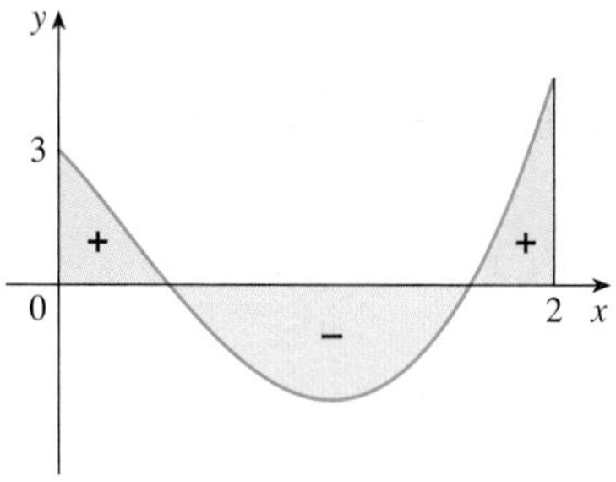

FIGURE 3

Figure 3 shows the graph of the integrand. We know from Section 5.2 that the value of the integral can be interpreted as a net area: the sum of the areas labeled with a plus sign minus the area labeled with a minus sign. ■

EXAMPLE 6 | Evaluate $\int_1^9 \frac{2t^2 + t^2\sqrt{t} - 1}{t^2}\, dt$.

SOLUTION First we need to write the integrand in a simpler form by carrying out the division:

$$\int_1^9 \frac{2t^2 + t^2\sqrt{t} - 1}{t^2}\, dt = \int_1^9 (2 + t^{1/2} - t^{-2})\, dt$$

$$= 2t + \frac{t^{3/2}}{\frac{3}{2}} - \frac{t^{-1}}{-1}\Big]_1^9 = 2t + \tfrac{2}{3}t^{3/2} + \frac{1}{t}\Big]_1^9$$

$$= \left(2\cdot 9 + \tfrac{2}{3}\cdot 9^{3/2} + \tfrac{1}{9}\right) - \left(2\cdot 1 + \tfrac{2}{3}\cdot 1^{3/2} + \tfrac{1}{1}\right)$$

$$= 18 + 18 + \tfrac{1}{9} - 2 - \tfrac{2}{3} - 1 = 32\tfrac{4}{9}$$ ■

■ The Net Change Theorem

The Evaluation Theorem says that if f is continuous on $[a, b]$, then

$$\int_a^b f(x)\, dx = F(b) - F(a)$$

where F is any antiderivative of f. This means that $F' = f$, so the equation can be rewritten as

$$\int_a^b F'(x)\, dx = F(b) - F(a)$$

We know that $F'(x)$ represents the rate of change of $y = F(x)$ with respect to x and $F(b) - F(a)$ is the change in y when x changes from a to b. [Note that y could, for instance, increase, then decrease, then increase again. Although y might change in both directions, $F(b) - F(a)$ represents the net change in y.] So we can reformulate the Evaluation Theorem in words as follows.

Net Change Theorem The integral of a rate of change is the net change:

$$\int_a^b F'(x)\,dx = F(b) - F(a)$$

EXAMPLE 7 | **Integrating rate of growth** If $N(t)$ is the size of a population at time t, explain the biological meaning of

$$\int_{t_1}^{t_2} \frac{dN}{dt}\,dt$$

SOLUTION The derivative dN/dt is the rate of growth of the population. According to the Net Change Theorem, we have

$$\int_{t_1}^{t_2} \frac{dN}{dt}\,dt = N(t_2) - N(t_1)$$

This is the net change in population during the time period from t_1 to t_2. The population increases when births happen and decreases when deaths occur. The net change takes into account both births and deaths. ■

EXAMPLE 8 | If an object moves along a straight line with position function $s(t)$, then its velocity is $v(t) = s'(t)$, so the Net Change Theorem says, in this context, that

$$\int_{t_1}^{t_2} v(t)\,dt = s(t_2) - s(t_1)$$

This is the net change of position, or *displacement*, of the particle during the time period from t_1 to t_2. In Section 5.1 we guessed that this was true for the case where the object moves in the positive direction, but now we have proved that it is always true. ■

■ The Fundamental Theorem

The first part of the Fundamental Theorem deals with functions defined by an equation of the form

(2) $$g(x) = \int_a^x f(t)\,dt$$

where f is a continuous function on $[a, b]$ and x varies between a and b. Observe that g depends only on x, which appears as the variable upper limit in the integral. If x is a fixed number, then the integral $\int_a^x f(t)\,dt$ is a definite number. If we then let x vary, the number $\int_a^x f(t)\,dt$ also varies and defines a function of x denoted by $g(x)$.

If f happens to be a positive function, then $g(x)$ can be interpreted as the area under the graph of f from a to x, where x can vary from a to b. (Think of g as the "area so far" function; see Figure 4.)

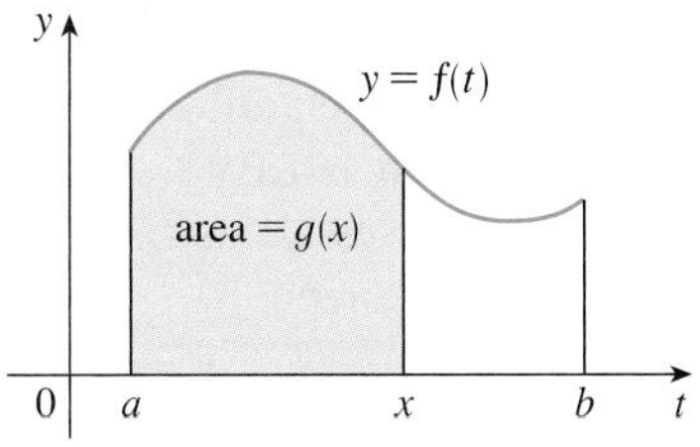

FIGURE 4

EXAMPLE 9 | If $g(x) = \int_a^x f(t)\,dt$, where $a = 1$ and $f(t) = t^2$, find a formula for $g(x)$ and calculate $g'(x)$.

SOLUTION In this case we can compute $g(x)$ explicitly using the Evaluation Theorem:

$$g(x) = \int_1^x t^2\,dt = \frac{t^3}{3}\bigg]_1^x = \frac{x^3 - 1}{3}$$

Then
$$g'(x) = \frac{d}{dx}\left(\tfrac{1}{3}x^3 - \tfrac{1}{3}\right) = x^2$$ ■

For the function in Example 9, notice that $g'(x) = x^2$, that is, $g' = f$. In other words, if g is defined as the integral of f by Equation 2, then g turns out to be an antiderivative of f, at least in this case. To see why this might be generally true we consider any continuous function f with $f(x) \geqslant 0$. Then $g(x) = \int_a^x f(t)\,dt$ can be interpreted as the area under the graph of f from a to x, as in Figure 4.

In order to compute $g'(x)$ from the definition of a derivative we first observe that, for $h > 0$, $g(x + h) - g(x)$ is obtained by subtracting areas, so it is the area under the graph of f from x to $x + h$ (the blue area in Figure 5). For small h you can see from the figure that this area is approximately equal to the area of the rectangle with height $f(x)$ and width h:

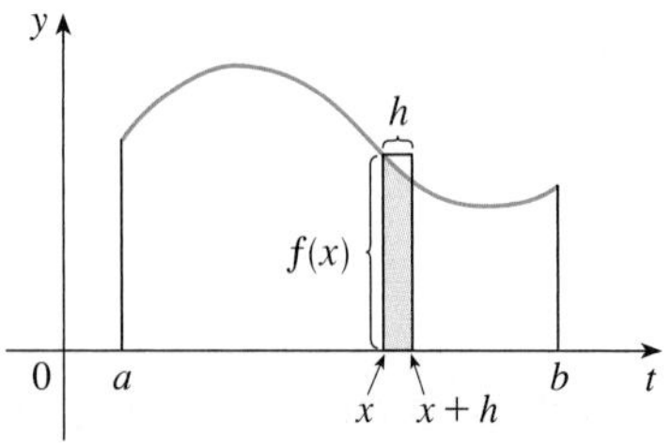

FIGURE 5

$$g(x + h) - g(x) \approx hf(x)$$

so
$$\frac{g(x + h) - g(x)}{h} \approx f(x)$$

Intuitively, we therefore expect that

$$g'(x) = \lim_{h \to 0} \frac{g(x + h) - g(x)}{h} = f(x)$$

The fact that this is true, even when f is not necessarily positive, is the first part of the Fundamental Theorem of Calculus.

We abbreviate the name of this theorem as FTC1. In words, it says that the derivative of a definite integral with respect to its upper limit is the integrand evaluated at the upper limit.

The Fundamental Theorem of Calculus, Part 1 If f is continuous on $[a, b]$, then the function g defined by

$$g(x) = \int_a^x f(t)\,dt \qquad a \leqslant x \leqslant b$$

is an antiderivative of f, that is, $g'(x) = f(x)$ for $a < x < b$.

Using Leibniz notation for derivatives, we can write this theorem as

$$\frac{d}{dx}\int_a^x f(t)\,dt = f(x)$$

when f is continuous. Roughly speaking, this equation says that if we first integrate f and then differentiate the result, we get back to the original function f.

TEC Module 5.3 provides visual evidence for FTC1.

It is easy to prove the Fundamental Theorem if we make the assumption that f possesses an antiderivative F. (This is certainly plausible.) Then, by the Evaluation Theorem,

$$\int_a^x f(t)\,dt = F(x) - F(a)$$

for any x between a and b. Therefore

$$\frac{d}{dx}\int_a^x f(t)\,dt = \frac{d}{dx}[F(x) - F(a)] = F'(x) = f(x)$$

as required.

The amount of infection was defined on page 326.

EXAMPLE 10 | **Measles pathogenesis** In Section 5.1 we saw that the amount of infection exposed to the immune system by day t of a measles infection is

$$A(t) = \int_0^t f(s)\,ds$$

where $f(s) = -s(s - 21)(s + 1)$. What is the rate of change of the total amount of infection $A(t)$ at time t?

SOLUTION Since f is continuous, Part 1 of the Fundamental Theorem of Calculus implies that the rate of change of $A(t)$ at time t is

$$A'(t) = \frac{d}{dt}\int_0^t f(s)\,ds = f(t) = -t(t - 21)(t + 1)$$

■

EXAMPLE 11 | Find $\displaystyle\frac{d}{dx}\int_1^{x^4} \sec t\,dt$.

SOLUTION Here we have to be careful to use the Chain Rule in conjunction with Part 1 of the Fundamental Theorem. Let $u = x^4$. Then

$$\begin{aligned}\frac{d}{dx}\int_1^{x^4} \sec t\,dt &= \frac{d}{dx}\int_1^{u} \sec t\,dt \\ &= \frac{d}{du}\left[\int_1^u \sec t\,dt\right]\frac{du}{dx} && \text{(by the Chain Rule)} \\ &= \sec u\,\frac{du}{dx} && \text{(by FTC1)} \\ &= \sec(x^4)\cdot 4x^3\end{aligned}$$

■

■ Differentiation and Integration as Inverse Processes

We now bring together the two parts of the Fundamental Theorem. We regard Part 1 as fundamental because it relates integration and differentiation. But the Evaluation Theorem also relates integrals and derivatives, so we rename it as Part 2 of the Fundamental Theorem.

The Fundamental Theorem of Calculus Suppose f is continuous on $[a, b]$.

1. If $g(x) = \int_a^x f(t)\,dt$, then $g'(x) = f(x)$.

2. $\int_a^b f(x)\,dx = F(b) - F(a)$, where F is any antiderivative of f, that is, $F' = f$.

We noted that Part 1 can be rewritten as

$$\frac{d}{dx}\int_a^x f(t)\,dt = f(x)$$

which says that if f is integrated and then the result is differentiated, we arrive back at the original function f. And we reformulated Part 2 as the Net Change Theorem:

$$\int_a^b F'(x)\,dx = F(b) - F(a)$$

This version says that if we take a function F, first differentiate it, and then integrate the result, we arrive back at the original function F, but in the form $F(b) - F(a)$. Taken together, the two parts of the Fundamental Theorem of Calculus say that differentiation and integration are inverse processes. Each undoes what the other does.

The Fundamental Theorem of Calculus is unquestionably the most important theorem in calculus and, indeed, it ranks as one of the great accomplishments of the human mind. Before it was discovered, from the time of Eudoxus and Archimedes to the time of Galileo and Fermat, problems of finding areas, volumes, and lengths of curves were so difficult that only a genius could meet the challenge. But now, armed with the systematic method that Newton and Leibniz fashioned out of the Fundamental Theorem, we will see in the chapters to come that these challenging problems are accessible to all of us.

EXERCISES 5.3

1–28 Evaluate the integral.

1. $\int_{-2}^{3} (x^2 - 3)\,dx$

2. $\int_{1}^{2} x^{-2}\,dx$

3. $\int_{0}^{2} \left(x^4 - \frac{3}{4}x^2 + \frac{2}{3}x - 1\right) dx$

4. $\int_{0}^{1} \left(1 + \frac{1}{2}u^4 - \frac{2}{5}u^9\right) du$

5. $\int_{0}^{1} x^{4/5}\,dx$

6. $\int_{1}^{8} \sqrt[3]{x}\,dx$

7. $\int_{-1}^{0} (2x - e^x)\,dx$

8. $\int_{-5}^{5} e\,dx$

9. $\int_{1}^{2} (1 + 2y)^2\,dy$

10. $\int_{0}^{2} (y - 1)(2y + 1)\,dy$

11. $\int_{1}^{9} \frac{x - 1}{\sqrt{x}}\,dx$

12. $\int_{-1}^{1} t(1 - t)^2\,dt$

13. $\int_{0}^{1} x\left(\sqrt[3]{x} + \sqrt[4]{x}\right) dx$

14. $\int_{0}^{\pi/4} \sec\theta \tan\theta\,d\theta$

15. $\int_{0}^{\pi/4} \sec^2 t\,dt$

16. $\int_{1}^{18} \sqrt{\frac{3}{z}}\,dz$

17. $\int_{1}^{9} \frac{1}{2x}\,dx$

18. $\int_{0}^{5} (2e^x + 4\cos x)\,dx$

19. $\int_{0}^{1} (x^e + e^x)\,dx$

20. $\int_{0}^{1} 10^x\,dx$

21. $\int_{-1}^{1} e^{u+1}\,du$

22. $\int_{0}^{1} \frac{4}{t^2 + 1}\,dt$

23. $\int_{1}^{2} \frac{v^3 + 3v^6}{v^4}\,dv$

24. $\int_{0}^{\pi/3} \frac{\sin\theta + \sin\theta\tan^2\theta}{\sec^2\theta}\,d\theta$

25. $\int_{0}^{\pi/4} \frac{1 + \cos^2\theta}{\cos^2\theta}\,d\theta$

26. $\int_{1}^{2} \frac{(x - 1)^3}{x^2}\,dx$

27. $\int_{0}^{1/\sqrt{3}} \frac{t^2 - 1}{t^4 - 1}\,dt$

28. $\int_{0}^{2} |2x - 1|\,dx$

29–30 What is wrong with the equation?

29. $\int_{-1}^{3} \frac{1}{x^2}\,dx = \left.\frac{x^{-1}}{-1}\right]_{-1}^{3} = -\frac{4}{3}$

30. $\int_{0}^{\pi} \sec^2 x\,dx = \tan x\Big]_{0}^{\pi} = 0$

31–32 Use a graph to give a rough estimate of the area of the region that lies beneath the given curve. Then find the exact area.

31. $y = \sin x,\ 0 \leqslant x \leqslant \pi$

32. $y = \sec^2 x,\ 0 \leqslant x \leqslant \pi/3$

33–34 Evaluate the integral and interpret it as a difference of areas. Illustrate with a sketch.

33. $\int_{-1}^{2} x^3\,dx$

34. $\int_{-\pi/2}^{2\pi} \cos x\,dx$

35–36 Verify by differentiation that the formula is correct.

35. $\int \cos^3 x\,dx = \sin x - \frac{1}{3}\sin^3 x + C$

36. $\int x\cos x\,dx = x\sin x + \cos x + C$

37–38 Find the general indefinite integral. Illustrate by graphing several members of the family on the same screen.

37. $\int \left(\cos x + \frac{1}{2}x\right) dx$ **38.** $\int (e^x - 2x^2)\, dx$

39–44 Find the general indefinite integral.

39. $\int (1 - t)(2 + t^2)\, dt$ **40.** $\int v(v^2 + 2)^2\, dv$

41. $\int (1 + \tan^2\alpha)\, d\alpha$ **42.** $\int \sec t\,(\sec t + \tan t)\, dt$

43. $\int \dfrac{\sin x}{1 - \sin^2 x}\, dx$ **44.** $\int \dfrac{\sin 2x}{\sin x}\, dx$

45. Measles pathogenesis The function

$$f(t) = -t(t - 21)(t + 1)$$

has been used to model the measles virus concentration in an infected individual. The area under the graph of f represents the total amount of infection. We saw in Section 5.1 that at $t = 12$ days this total amount of infection reaches the threshold beyond which symptoms appear. Use the Evaluation Theorem to calculate this threshold value.

46. If $V'(t)$ is the rate at which water flows into a reservoir at time t, what does the integral

$$\int_{t_1}^{t_2} V'(t)\, dt$$

represent?

47. Growth rate If $w'(t)$ is the rate of growth of a child in pounds per year, what does $\int_5^{10} w'(t)\, dt$ represent?

48. Age-structured populations Suppose the number of individuals of age a is given by the function $N(a)$ (number of individuals per age a). What does the integral $\int_0^{15} N(a)\, da$ represent?

49. Sea urchins Integration is sometimes used when censusing a population. For example, suppose the density of sea urchins at different points x along a coastline is given by the function $f(x)$ individuals per meter, where x is the distance (in meters) along the coast from the start of the species' range. What does the integral $\int_a^b f(x)\, dx$ represent?

© Vilainecrevette / Shutterstock.com

50. Bacteria growth A bacteria colony increases in size at a rate of $4.0553e^{1.8t}$ bacteria per hour. If the initial population is 46 bacteria, find the population four hours later.

51. In a chemical reaction, the rate of reaction is the derivative of the concentration $[\mathrm{C}](t)$ of the product of the reaction. What does

$$\int_{t_1}^{t_2} \frac{d[\mathrm{C}]}{dt}\, dt$$

represent?

52. A **honeybee population** starts with 100 bees and increases at a rate of $n'(t)$ bees per week. What does the expression $100 + \int_0^{15} n'(t)\, dt$ represent?

53. If oil leaks from a tank at a rate of $r(t)$ gallons per minute at time t, what does $\int_0^{120} r(t)\, dt$ represent?

54. Suppose that a volcano is erupting and readings of the rate $r(t)$ at which solid materials are spewed into the atmosphere are given in the table. The time t is measured in seconds and the units for $r(t)$ are tonnes (metric tons) per second.

t	0	1	2	3	4	5	6
$r(t)$	2	10	24	36	46	54	60

(a) Give upper and lower estimates for the total quantity $Q(6)$ of erupted materials after 6 seconds.
(b) Use the Midpoint Rule to estimate $Q(6)$.

55. Water flows from the bottom of a storage tank at a rate of $r(t) = 200 - 4t$ liters per minute, where $0 \leq t \leq 50$. Find the amount of water that flows from the tank during the first 10 minutes.

56. Water flows into and out of a storage tank. A graph of the rate of change $r(t)$ of the volume of water in the tank, in liters per day, is shown. If the amount of water in the tank at time $t = 0$ is 25,000 L, use the Midpoint Rule to estimate the amount of water in the tank four days later.

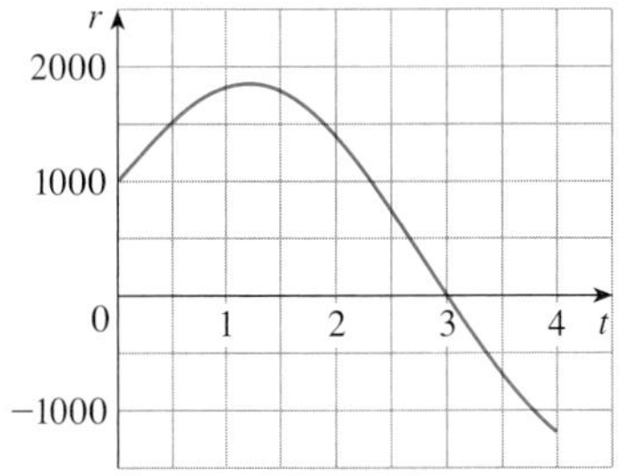

57. Von Bertalanffy growth Many fish grow in a way that is described by the von Bertalanffy growth equation. For a fish that starts life with a length of 1 cm and has a maximum length of 30 cm, this equation predicts that the growth rate is $29e^{-a}$ cm/year, where a is the age of the fish. How long will the fish be after 5 years?

58. Niche overlap The extent to which species compete for resources is often measured by the *niche overlap*. If the horizontal axis represents a continuum of different resource types (for example, seed sizes for certain bird species), then a plot of the degree of preference for these resources is called a *species' niche*. The degree of overlap of two species' niches is then a measure of the extent to which they compete for resources. The niche overlap for a species is the fraction of the area under its preference curve that is also under the other species' curve. The niches displayed in the figure are given by

$$n_1(x) = (x - 1)(3 - x) \qquad 1 \leqslant x \leqslant 3$$

$$n_2(x) = (x - 2)(4 - x) \qquad 2 \leqslant x \leqslant 4$$

Use the Evaluation Theorem to calculate the niche overlap for species 1.

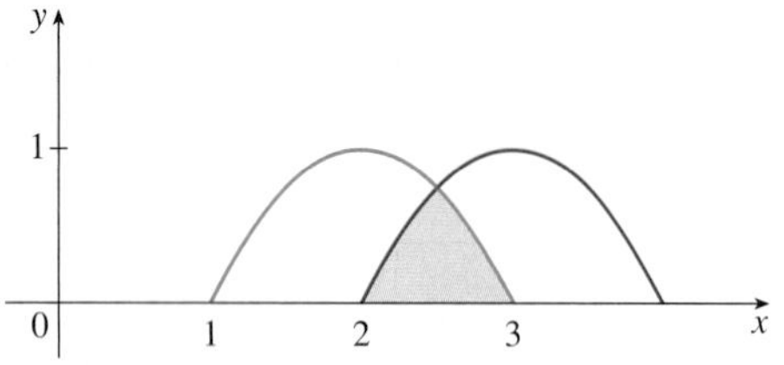

59. Medical imaging devices like CT scans work by passing an X-ray beam through part of the body and measuring how much the intensity of the beam attenuates (in other words, is reduced). The amount of attenuation depends on the density and composition of the tissue.

(a) If $A(x)$ is the attenuation rate at position x in the tissue and L is the thickness of the tissue through which the beam passes, what is the total attenuation of the beam during this procedure?

(b) Suppose the maximum and minimum attenuation rates in the tissue are β and α, respectively. Find upper and lower bounds for the total attenuation of the beam as it passes through the tissue. Justify your answer using the properties of integrals in Section 5.2.

Source: Adapted from T. Feeman, *The Mathematics of Medical Imaging* (New York: Springer Science + Business Media, 2011).

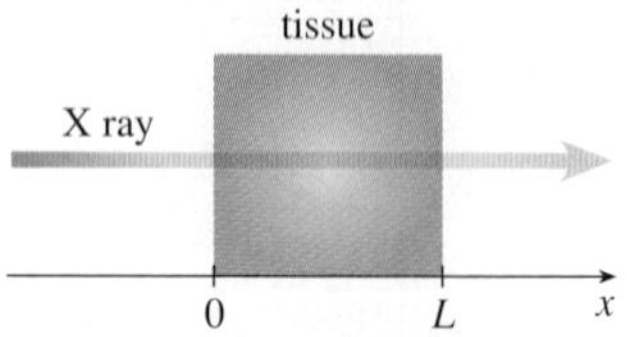

60. Incidence and prevalence The incidence $i(t)$ of an infectious disease at time t is the rate at which new infections are occurring at that time. The prevalence $P(t)$ at time t is the total number of infected individuals at that time. Let's suppose that $P(0) = 0$.

(a) Express the total number of new infections between times $t = 0$ and $t = a$ as a definite integral.

(b) Suppose that all individuals either die or recover from infection, and that D is the total number that have done so between times $t = 0$ and $t = a$. Express D in terms of $P(a)$ and your result from part (a).

(c) Let $d(t)$ be the rate at which people are dying or recovering from infection at time t. What is the relationship between D and $d(t)$?

61. Photosynthesis The *rate of primary production* refers to the rate of conversion of inorganic carbon to organic carbon via photosynthesis. It is measured as a mass of carbon fixed per unit biomass, per unit time. The rate of primary production depends on light intensity, measured as the flux of photons (that is, number of photons per unit area per unit time). One model for this relationship is

$$P(I) = P_{\max}(1 - e^{-aI})$$

where P is the rate of primary production as a function of light intensity I. Suppose that light intensity changes with time according to the equation $I(t) = kt$, where k is a constant.

(a) What is the rate of primary production as a function of time?

(b) What is the total amount of primary production over the first five units of time?

(c) What is the total amount of primary production over the first t units of time?

(d) What is the rate of change of total primary production at time t?

Source: Adapted from A. Jassby et al., "Mathematical Formulation of the Relationship between Photosynthesis and Light for Phytoplankton," *Limnology and Oceanography* 21 (1976): 540–7.

62. Photosynthesis Much of the earth's photosynthesis occurs in the oceans. The rate of primary production (as discussed in Exercise 61) depends on light intensity, measured as the flux of photons (that is, number of photons per unit area per unit time). For monochromatic light, intensity decreases with water depth according to Beer's Law, which states that $I(x) = e^{-kx}$, where x is water depth. A simple model for the relationship between rate of photosynthesis and light intensity is $P(I) = aI$, where a is a constant and P is measured as a mass of carbon fixed per volume of water, per unit time.

(a) What is the rate of photosynthesis as a function of water depth?

(b) What is the total rate of photosynthesis of a water column that is one unit in surface area and four units deep?

(c) What is the total rate of photosynthesis of a water column that is one unit in surface area and x units deep?

(d) What is the rate of change of the total photosynthesis with respect to the depth x?

Source: Adapted from A. Jassby et al., "Mathematical Formulation of the Relationship between Photosynthesis and Light for Phytoplankton," *Limnology and Oceanography* 21 (1976): 540–7.

63. (a) Show that $1 \leqslant \sqrt{1 + x^3} \leqslant 1 + x^3$ for $x \geqslant 0$.

(b) Show that $1 \leqslant \int_0^1 \sqrt{1 + x^3}\, dx \leqslant 1.25$.

64. (a) Show that $\cos(x^2) \geqslant \cos x$ for $0 \leqslant x \leqslant 1$.
(b) Deduce that $\int_0^{\pi/6} \cos(x^2)\, dx \geqslant \frac{1}{2}$.

65. Let $g(x) = \int_0^x f(t)\, dt$, where f is the function whose graph is shown.

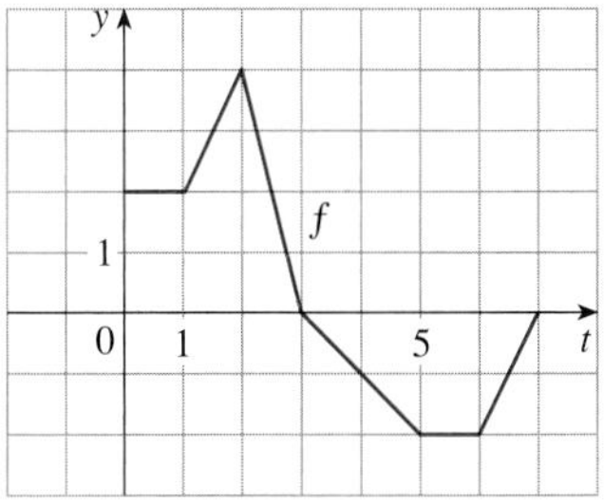

(a) Evaluate $g(0)$, $g(1)$, $g(2)$, $g(3)$, and $g(6)$.
(b) On what interval is g increasing?
(c) Where does g have a maximum value?
(d) Sketch a rough graph of g.

66. Let $g(x) = \int_0^x f(t)\, dt$, where f is the function whose graph is shown.
(a) Evaluate $g(x)$ for $x = 0, 1, 2, 3, 4, 5$, and 6.
(b) Estimate $g(7)$.
(c) Where does g have a maximum value? Where does it have a minimum value?
(d) Sketch a rough graph of g.

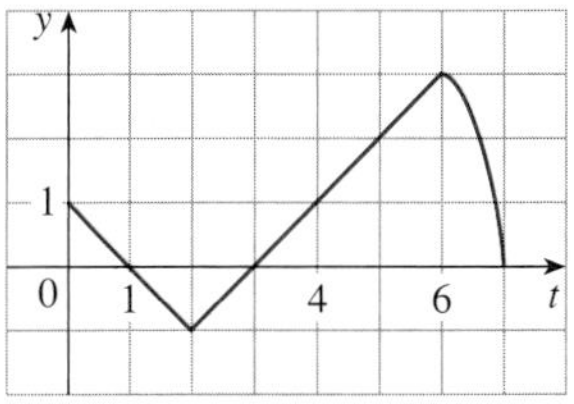

67–68 Sketch the area represented by $g(x)$. Then find $g'(x)$ in two ways: (a) by using Part 1 of the Fundamental Theorem and (b) by evaluating the integral using Part 2 and then differentiating.

67. $g(x) = \int_0^x (1 + t^2)\, dt$

68. $g(x) = \int_0^x (1 + \sqrt{t})\, dt$

69–78 Use Part 1 of the Fundamental Theorem of Calculus to find the derivative of the function.

69. $g(x) = \int_1^x \frac{1}{t^3 + 1}\, dt$

70. $g(x) = \int_3^x e^{t^2 - t}\, dt$

71. $g(y) = \int_2^y t^2 \sin t\, dt$

72. $g(r) = \int_0^r \sqrt{x^2 + 4}\, dx$

73. $F(x) = \int_x^0 \sqrt{1 + \sec t}\, dt$

$$\left[\textit{Hint: } \int_x^0 \sqrt{1 + \sec t}\, dt = -\int_0^x \sqrt{1 + \sec t}\, dt \right]$$

74. $G(x) = \int_x^1 \cos \sqrt{t}\, dt$

75. $h(x) = \int_2^{1/x} \arctan t\, dt$

76. $h(x) = \int_0^{x^2} \sqrt{1 + r^3}\, dr$

77. $y = \int_0^{\tan x} \sqrt{t + \sqrt{t}}\, dt$

78. $y = \int_{e^x}^0 \sin^3 t\, dt$

79. If $f(1) = 12$, f' is continuous, and $\int_1^4 f'(x)\, dx = 17$, what is the value of $f(4)$?

80. The *error function*

$$\operatorname{erf}(x) = \frac{2}{\sqrt{\pi}} \int_0^x e^{-t^2}\, dt$$

is used in probability, statistics, and engineering.
(a) Show that $\int_a^b e^{-t^2}\, dt = \frac{1}{2}\sqrt{\pi}\,[\operatorname{erf}(b) - \operatorname{erf}(a)]$.
(b) Show that the function $y = e^{x^2}\operatorname{erf}(x)$ satisfies the differential equation $y' = 2xy + 2/\sqrt{\pi}$.

81. Suppose h is a function such that $h(1) = -2$, $h'(1) = 2$, $h''(1) = 3$, $h(2) = 6$, $h'(2) = 5$, $h''(2) = 13$, and h'' is continuous everywhere. Evaluate $\int_1^2 h''(u)\, du$.

82. The area labeled B is three times the area labeled A. Express b in terms of a.

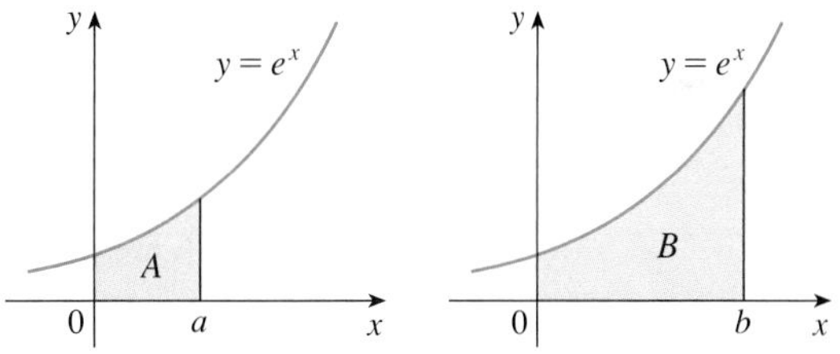

83–84 Evaluate the limit by first recognizing the sum as a Riemann sum for a function defined on $[0, 1]$.

83. $\lim_{n \to \infty} \sum_{i=1}^{n} \frac{i^3}{n^4}$

84. $\lim_{n \to \infty} \frac{1}{n} \left(\sqrt{\frac{1}{n}} + \sqrt{\frac{2}{n}} + \sqrt{\frac{3}{n}} + \cdots + \sqrt{\frac{n}{n}} \right)$

85. Find a function f and a number a such that

$$6 + \int_a^x \frac{f(t)}{t^2}\, dt = 2\sqrt{x} \qquad \text{for all } x > 0$$

■ PROJECT The Outbreak Size of an Infectious Disease BB

In Sections 7.6 and 10.4 we will analyze the Kermack-McKendrick model for infectious disease dynamics. If $S(t)$ is the number of people susceptible to infection at time t and $I(t)$ is the number of people already infected at this time, then a simplified version of the model specifies the derivatives of S and I with respect to time as

(1) $$\frac{dS}{dt} = -\beta SI$$

(2) $$\frac{dI}{dt} = \beta SI - \mu I$$

where β and μ are positive constants and the numbers of susceptible and infected people at time $t = 0$ are $S(0)$ and $I(0)$, respectively.

1. Use Equation 1 to express the quantity $-\beta\int_0^T S(t)I(t)\,dt$ in terms of the function $S(t)$.

2. Use Equation 1 to express the quantity $-\beta\int_0^T I(t)\,dt$ in terms of the function $S(t)$.

3. Use Equation 2 to express the quantity $\beta\int_0^T S(t)I(t)\,dt - \mu\int_0^T I(t)\,dt$ in terms of the function $I(t)$.

4. Use the results from Problems 1–3 to obtain a single equation that $S(0)$, $S(T)$, $I(T)$, and $I(0)$ must satisfy and that does not involve integrals.

5. It can be shown that $\lim_{T\to\infty} I(T) = 0$ and that $\lim_{T\to\infty} S(T) = S_\infty$, where S_∞ is a positive constant. What equation do you get if you let $T \to \infty$ in your answer to Problem 4?

6. Suppose that the number of people initially infected, $I(0)$, is negligibly small and define $q = \beta S(0)/\mu$ and $A = 1 - S_\infty/S(0)$. Here q is a measure of the transmissibility of the infection and A is the fraction of the original susceptible population $S(0)$ that ultimately gets infected (that is, the outbreak size). Show that your answer to Problem 5 can be written as

(3) $$e^{-qA} = 1 - A$$

Equation 3 is a special case of an equation seen previously in many exercises and examples. For instance, see Example 3.5.13, Exercises 3.5.81, 3.8.32, 3.Review.92, and Example 9.4.6.

5.4 The Substitution Rule

■ Substitution in Indefinite Integrals

Because of the Fundamental Theorem, it's important to be able to find antiderivatives. But our antidifferentiation formulas don't tell us how to evaluate integrals such as

(1) $$\int 2x\sqrt{1 + x^2}\,dx$$

Our strategy is to simplify the integral by introducing a new variable. Suppose that we let u be the quantity under the root sign in (1): $u = 1 + x^2$. Then $du/dx = 2x$. Up until now we have not interpreted du/dx as a ratio of two quantities du and dx, but in this context it is useful to do so. Here we regard the symbols du and dx as separate entities called *differentials* and we write the equation $du/dx = 2x$ as

$$du = 2x\,dx$$

Then formally, without justifying our calculation, we could write

(2)
$$\int 2x\sqrt{1+x^2}\,dx = \int \sqrt{1+x^2}\,2x\,dx = \int \sqrt{u}\,du$$
$$= \tfrac{2}{3}u^{3/2} + C = \tfrac{2}{3}(1+x^2)^{3/2} + C$$

But now we can check that we have the correct answer by using the Chain Rule to differentiate the final function of Equation 2:

$$\frac{d}{dx}\left[\tfrac{2}{3}(1+x^2)^{3/2} + C\right] = \tfrac{2}{3}\cdot\tfrac{3}{2}(1+x^2)^{1/2}\cdot 2x = 2x\sqrt{1+x^2}$$

In general, this method works whenever we have an integral that we can write in the form $\int f(g(x))\,g'(x)\,dx$. Observe that if $F' = f$, then

(3)
$$\int F'(g(x))\,g'(x)\,dx = F(g(x)) + C$$

because, by the Chain Rule,

$$\frac{d}{dx}[F(g(x))] = F'(g(x))\,g'(x)$$

If we make the "change of variable" or "substitution" $u = g(x)$, then from Equation 3 we have

$$\int F'(g(x))\,g'(x)\,dx = F(g(x)) + C = F(u) + C = \int F'(u)\,du$$

or, writing $F' = f$, we get

$$\int f(g(x))\,g'(x)\,dx = \int f(u)\,du$$

Thus we have proved the following rule.

(4) The Substitution Rule If $u = g(x)$ is a differentiable function whose range is an interval I and f is continuous on I, then

$$\int f(g(x))\,g'(x)\,dx = \int f(u)\,du$$

Notice that the Substitution Rule for integration was proved using the Chain Rule for differentiation. Notice also that if $u = g(x)$, then $du/dx = g'(x)$, which we write in terms of **differentials**: $du = g'(x)\,dx$. This is probably the best way to remember the Substitution Rule.

EXAMPLE 1 | Find $\int x^3 \cos(x^4 + 2)\,dx$.

SOLUTION We make the substitution $u = x^4 + 2$ because its differential is $du = 4x^3\,dx$, which, apart from the constant factor 4, occurs in the integral. Thus, using $x^3\,dx = \frac{1}{4}\,du$ and the Substitution Rule, we have

$$\begin{aligned}\int x^3\cos(x^4+2)\,dx &= \int \cos u \cdot \tfrac{1}{4}\,du = \tfrac{1}{4}\int \cos u\,du \\ &= \tfrac{1}{4}\sin u + C \\ &= \tfrac{1}{4}\sin(x^4+2) + C\end{aligned}$$

Check the answer by differentiating it.

Notice that at the final stage we had to return to the original variable x. ■

The idea behind the Substitution Rule is to replace a relatively complicated integral by a simpler integral. This is accomplished by changing from the original variable x to a new variable u that is a function of x. Thus in Example 1 we replaced the integral $\int x^3\cos(x^4+2)\,dx$ by the simpler integral $\frac{1}{4}\int \cos u\,du$.

The main challenge in using the Substitution Rule is to think of an appropriate substitution. You should try to choose u to be some function in the integrand whose differential also occurs (except for a constant factor). This was the case in Example 1. If that is not possible, try choosing u to be some complicated part of the integrand (perhaps the inner function in a composite function). Finding the right substitution is a bit of an art. It's not unusual to guess wrong; if your first guess doesn't work, try another substitution.

EXAMPLE 2 | Evaluate $\int \sqrt{2x+1}\,dx$.

SOLUTION 1 Let $u = 2x + 1$. Then $du = 2\,dx$, so $dx = \frac{1}{2}\,du$. Thus the Substitution Rule gives

$$\begin{aligned}\int \sqrt{2x+1}\,dx &= \int \sqrt{u}\cdot \tfrac{1}{2}\,du = \tfrac{1}{2}\int u^{1/2}\,du \\ &= \frac{1}{2}\cdot\frac{u^{3/2}}{3/2} + C = \tfrac{1}{3}u^{3/2} + C \\ &= \tfrac{1}{3}(2x+1)^{3/2} + C\end{aligned}$$

SOLUTION 2 Another possible substitution is $u = \sqrt{2x+1}$. Then

$$du = \frac{dx}{\sqrt{2x+1}} \qquad \text{so} \qquad dx = \sqrt{2x+1}\,du = u\,du$$

(Or observe that $u^2 = 2x + 1$, so $2u\,du = 2\,dx$.) Therefore

$$\begin{aligned}\int \sqrt{2x+1}\,dx &= \int u \cdot u\,du = \int u^2\,du \\ &= \frac{u^3}{3} + C = \tfrac{1}{3}(2x+1)^{3/2} + C\end{aligned}$$ ■

EXAMPLE 3 | Find $\displaystyle\int \frac{x}{\sqrt{1-4x^2}}\,dx$.

SOLUTION Let $u = 1 - 4x^2$. Then $du = -8x\,dx$, so $x\,dx = -\frac{1}{8}\,du$ and

$$\begin{aligned}\int \frac{x}{\sqrt{1-4x^2}}\,dx &= -\tfrac{1}{8}\int \frac{1}{\sqrt{u}}\,du = -\tfrac{1}{8}\int u^{-1/2}\,du \\ &= -\tfrac{1}{8}(2\sqrt{u}) + C = -\tfrac{1}{4}\sqrt{1-4x^2} + C\end{aligned}$$ ■

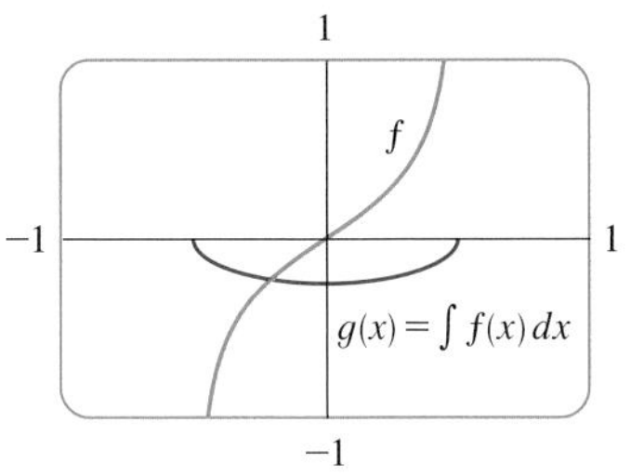

FIGURE 1

$f(x) = \dfrac{x}{\sqrt{1-4x^2}}$

$g(x) = \int f(x)\,dx = -\frac{1}{4}\sqrt{1-4x^2}$

The answer to Example 3 could be checked by differentiation, but instead let's check it with a graph. In Figure 1 we have used a computer to graph both the integrand $f(x) = x/\sqrt{1-4x^2}$ and its indefinite integral $g(x) = -\frac{1}{4}\sqrt{1-4x^2}$ (we take the case $C = 0$). Notice that $g(x)$ decreases when $f(x)$ is negative, increases when $f(x)$ is positive, and has its minimum value when $f(x) = 0$. So it seems reasonable, from the graphical evidence, that g is an antiderivative of f.

EXAMPLE 4 | Calculate $\int e^{5x}\,dx$.

SOLUTION If we let $u = 5x$, then $du = 5\,dx$, so $dx = \frac{1}{5}\,du$. Therefore

$$\int e^{5x}\,dx = \frac{1}{5}\int e^u\,du = \frac{1}{5}e^u + C = \frac{1}{5}e^{5x} + C$$ ■

NOTE With some experience, you might be able to evaluate integrals like those in Examples 1–4 without going to the trouble of making an explicit substitution. By recognizing the pattern in Equation 3, where the integrand on the left side is the product of the derivative of an outer function and the derivative of the inner function, we could work Example 1 as follows:

$$\int x^3\cos(x^4+2)\,dx = \int \cos(x^4+2)\cdot x^3\,dx = \frac{1}{4}\int \cos(x^4+2)\cdot(4x^3)\,dx$$

$$= \frac{1}{4}\int \cos(x^4+2)\cdot\frac{d}{dx}(x^4+2)\,dx = \frac{1}{4}\sin(x^4+2) + C$$

Similarly, the solution to Example 4 could be written like this:

$$\int e^{5x}\,dx = \frac{1}{5}\int 5e^{5x}\,dx = \frac{1}{5}\int \frac{d}{dx}(e^{5x})\,dx = \frac{1}{5}e^{5x} + C$$

The following example, however, is more complicated and so an explicit substitution is advisable.

EXAMPLE 5 | Calculate $\int \tan x\,dx$.

SOLUTION First we write tangent in terms of sine and cosine:

$$\int \tan x\,dx = \int \frac{\sin x}{\cos x}\,dx$$

This suggests that we should substitute $u = \cos x$, since then $du = -\sin x\,dx$ and so $\sin x\,dx = -du$:

$$\int \tan x\,dx = \int \frac{\sin x}{\cos x}\,dx = -\int \frac{1}{u}\,du$$

$$= -\ln|u| + C = -\ln|\cos x| + C$$ ■

Since $-\ln|\cos x| = \ln(|\cos x|^{-1}) = \ln(1/|\cos x|) = \ln|\sec x|$, the result of Example 5 can also be written as

$$\int \tan x\,dx = \ln|\sec x| + C$$

Substitution in Definite Integrals

When evaluating a *definite* integral by substitution, two methods are possible. One method is to evaluate the indefinite integral first and then use the Evaluation Theorem. For instance, using the result of Example 2, we have

$$\int_0^4 \sqrt{2x+1}\,dx = \int \sqrt{2x+1}\,dx\Big]_0^4$$
$$= \tfrac{1}{3}(2x+1)^{3/2}\Big]_0^4 = \tfrac{1}{3}(9)^{3/2} - \tfrac{1}{3}(1)^{3/2}$$
$$= \tfrac{1}{3}(27-1) = \tfrac{26}{3}$$

Another method, which is usually preferable, is to change the limits of integration when the variable is changed.

This rule says that when using a substitution in a definite integral, we must put everything in terms of the new variable u, not only x and dx but also the limits of integration. The new limits of integration are the values of u that correspond to $x = a$ and $x = b$.

(5) The Substitution Rule for Definite Integrals If g' is continuous on $[a, b]$ and f is continuous on the range of $u = g(x)$, then

$$\int_a^b f(g(x))\,g'(x)\,dx = \int_{g(a)}^{g(b)} f(u)\,du$$

PROOF Let F be an antiderivative of f. Then, by (3), $F(g(x))$ is an antiderivative of $f(g(x))\,g'(x)$, so by the Evaluation Theorem, we have

$$\int_a^b f(g(x))\,g'(x)\,dx = F(g(x))\Big]_a^b = F(g(b)) - F(g(a))$$

But, applying the Evaluation Theorem a second time, we also have

$$\int_{g(a)}^{g(b)} f(u)\,du = F(u)\Big]_{g(a)}^{g(b)} = F(g(b)) - F(g(a))$$ ■

EXAMPLE 6 | Evaluate $\int_0^4 \sqrt{2x+1}\,dx$ using (5).

SOLUTION Using the substitution from Solution 1 of Example 2, we have $u = 2x + 1$ and $dx = \frac{1}{2}\,du$. To find the new limits of integration we note that

$$\text{when } x = 0,\ u = 2(0) + 1 = 1 \qquad \text{and} \qquad \text{when } x = 4,\ u = 2(4) + 1 = 9$$

Therefore
$$\int_0^4 \sqrt{2x+1}\,dx = \int_1^9 \tfrac{1}{2}\sqrt{u}\,du = \tfrac{1}{2}\cdot\tfrac{2}{3}u^{3/2}\Big]_1^9$$
$$= \tfrac{1}{3}(9^{3/2} - 1^{3/2}) = \tfrac{26}{3}$$

Observe that when using (5) we do *not* return to the variable x after integrating. We simply evaluate the expression in u between the appropriate values of u. ■

The integral given in Example 7 is an abbreviation for

$$\int_1^2 \frac{1}{(3-5x)^2}\,dx$$

EXAMPLE 7 | Evaluate $\displaystyle\int_1^2 \frac{dx}{(3-5x)^2}$.

SOLUTION Let $u = 3 - 5x$. Then $du = -5\,dx$, so $dx = -\frac{1}{5}\,du$. When $x = 1$,

$u = -2$ and when $x = 2$, $u = -7$. Thus

$$\int_1^2 \frac{dx}{(3-5x)^2} = -\frac{1}{5}\int_{-2}^{-7} \frac{du}{u^2}$$

$$= -\frac{1}{5}\left[-\frac{1}{u}\right]_{-2}^{-7} = \frac{1}{5u}\bigg]_{-2}^{-7}$$

$$= \frac{1}{5}\left(-\frac{1}{7}+\frac{1}{2}\right) = \frac{1}{14}$$

■

EXAMPLE 8 | Metabolism A model for the basal metabolism rate, in kcal/h, of a young man is

$$R(t) = 85 - 0.18\cos(\pi t/12)$$

where t is the time in hours measured from 5:00 AM. What is the total basal metabolism of this man, $\int_0^{24} R(t)\,dt$, over a 24-hour time period?

SOLUTION We make the substitution $u = \pi t/12$. Then $du = (\pi/12)\,dt$ and $u = 2\pi$ when $t = 24$. So

$$\int_0^{24} R(t)\,dt = \int_0^{24}\left[85 - 0.18\cos\left(\frac{\pi t}{12}\right)\right]dt = \int_0^{2\pi}(8.5 - 0.18\cos u)\cdot\frac{12}{\pi}\,du$$

$$= \frac{12}{\pi}\Big[85u - 0.18\sin u\Big]_0^{2\pi} = \frac{12}{\pi}(85\cdot 2\pi - 0) = 2040$$

The total metabolism for the 24-hour period was 2040 kcal. ■

Symmetry

The next theorem uses the Substitution Rule for Definite Integrals (5) to simplify the calculation of integrals of functions that possess symmetry properties.

> **(6) Integrals of Symmetric Functions** Suppose f is continuous on $[-a, a]$.
>
> (a) If f is even $[f(-x) = f(x)]$, then $\int_{-a}^{a} f(x)\,dx = 2\int_0^a f(x)\,dx$.
>
> (b) If f is odd $[f(-x) = -f(x)]$, then $\int_{-a}^{a} f(x)\,dx = 0$.

PROOF We split the integral in two:

$$\textbf{(7)}\quad \int_{-a}^{a} f(x)\,dx = \int_{-a}^{0} f(x)\,dx + \int_0^a f(x)\,dx = -\int_0^{-a} f(x)\,dx + \int_0^a f(x)\,dx$$

In the first integral on the far right side we make the substitution $u = -x$. Then $du = -dx$ and when $x = -a$, $u = a$. Therefore

$$-\int_0^{-a} f(x)\,dx = -\int_0^a f(-u)(-du) = \int_0^a f(-u)\,du$$

and so Equation 7 becomes

(8) $$\int_{-a}^{a} f(x)\,dx = \int_{0}^{a} f(-u)\,du + \int_{0}^{a} f(x)\,dx$$

(a) If f is even, then $f(-u) = f(u)$ so Equation 8 gives

$$\int_{-a}^{a} f(x)\,dx = \int_{0}^{a} f(u)\,du + \int_{0}^{a} f(x)\,dx = 2\int_{0}^{a} f(x)\,dx$$

(b) If f is odd, then $f(-u) = -f(u)$ and so Equation 8 gives

$$\int_{-a}^{a} f(x)\,dx = -\int_{0}^{a} f(u)\,du + \int_{0}^{a} f(x)\,dx = 0$$ ■

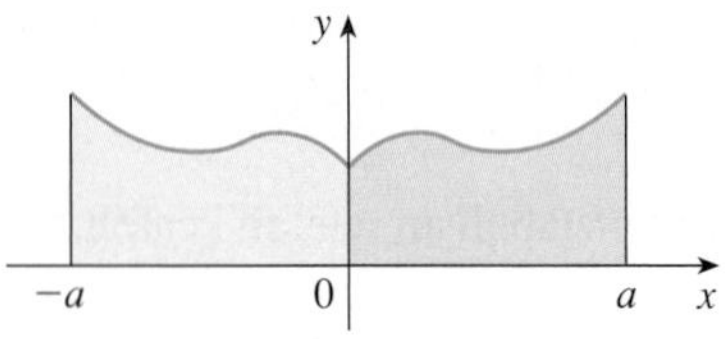

(a) f even, $\int_{-a}^{a} f(x)\,dx = 2\int_{0}^{a} f(x)\,dx$

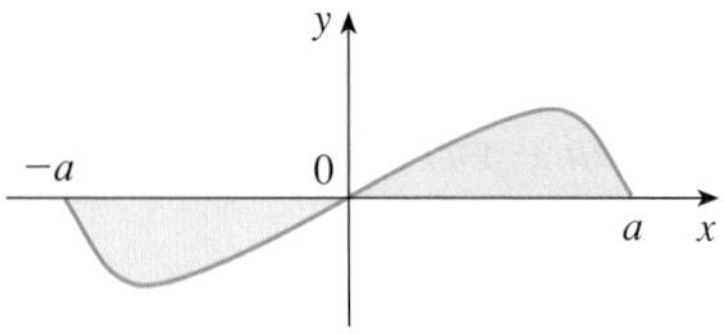

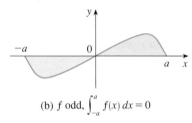

(b) f odd, $\int_{-a}^{a} f(x)\,dx = 0$

FIGURE 2

Theorem 6 is illustrated by Figure 2. For the case where f is positive and even, part (a) says that the area under $y = f(x)$ from $-a$ to a is twice the area from 0 to a because of symmetry. Recall that an integral $\int_a^b f(x)\,dx$ can be expressed as the area above the x-axis and below $y = f(x)$ minus the area below the axis and above the curve. Thus part (b) says the integral is 0 because the areas cancel.

EXAMPLE 9 | Since $f(x) = x^6 + 1$ satisfies $f(-x) = f(x)$, it is even and so

$$\int_{-2}^{2} (x^6 + 1)\,dx = 2\int_{0}^{2} (x^6 + 1)\,dx$$

$$= 2\left[\tfrac{1}{7}x^7 + x\right]_0^2 = 2\left(\tfrac{128}{7} + 2\right) = \tfrac{284}{7}$$ ■

EXAMPLE 10 | Since $f(x) = (\tan x)/(1 + x^2 + x^4)$ satisfies $f(-x) = -f(x)$, it is odd and so

$$\int_{-1}^{1} \frac{\tan x}{1 + x^2 + x^4}\,dx = 0$$ ■

EXERCISES 5.4

1–6 Evaluate the integral by making the given substitution.

1. $\int e^{-x}\,dx, \quad u = -x$

2. $\int x^3(2 + x^4)^5\,dx, \quad u = 2 + x^4$

3. $\int x^2\sqrt{x^3 + 1}\,dx, \quad u = x^3 + 1$

4. $\int \dfrac{dt}{(1 - 6t)^4}, \quad u = 1 - 6t$

5. $\int \cos^3\theta\,\sin\theta\,d\theta, \quad u = \cos\theta$

6. $\int \dfrac{\sec^2(1/x)}{x^2}\,dx, \quad u = 1/x$

7–36 Evaluate the indefinite integral.

7. $\int x\sin(x^2)\,dx$

8. $\int x^2(x^3 + 5)^9\,dx$

9. $\int (3x - 2)^{20}\,dx$

10. $\int (3t + 2)^{2.4}\,dt$

11. $\int \sin \pi t\,dt$

12. $\int e^x\cos(e^x)\,dx$

13. $\int \dfrac{(\ln x)^2}{x}\,dx$

14. $\int \dfrac{x}{(x^2 + 1)^2}\,dx$

15. $\int \dfrac{dx}{5 - 3x}$

16. $\int \dfrac{\sin\sqrt{x}}{\sqrt{x}}\,dx$

17. $\int \dfrac{a + bx^2}{\sqrt{3ax + bx^3}}\,dx$

18. $\int \dfrac{z^2}{z^3 + 1}\,dz$

19. $\int e^x\sqrt{1 + e^x}\,dx$

20. $\int \sec 2\theta\,\tan 2\theta\,d\theta$

21. $\int \dfrac{\cos x}{\sin^2 x}\,dx$

22. $\int \dfrac{\tan^{-1}x}{1 + x^2}\,dx$

23. $\int (x^2 + 1)(x^3 + 3x)^4\,dx$

24. $\int \dfrac{\sin(\ln x)}{x}\,dx$

25. $\int \sqrt{\cot x}\,\csc^2 x\,dx$

26. $\int \dfrac{\cos(\pi/x)}{x^2}\,dx$

27. $\int e^{2r}\sin(e^{2r})\,dr$

28. $\int \dfrac{dt}{\cos^2 t\sqrt{1+\tan t}}$

29. $\int \sec^3 x\tan x\,dx$

30. $\int x^2\sqrt{2+x}\,dx$

31. $\int x(2x+5)^8\,dx$

32. $\int \dfrac{e^x}{e^x+1}\,dx$

33. $\int \dfrac{\sin 2x}{1+\cos^2 x}\,dx$

34. $\int \dfrac{\sin x}{1+\cos^2 x}\,dx$

35. $\int \dfrac{1+x}{1+x^2}\,dx$

36. $\int \dfrac{x}{1+x^4}\,dx$

37–51 Evaluate the definite integral.

37. $\int_0^1 \cos(\pi t/2)\,dt$

38. $\int_0^1 (3t-1)^{50}\,dt$

39. $\int_0^1 \sqrt[3]{1+7x}\,dx$

40. $\int_0^{\sqrt{\pi}} x\cos(x^2)\,dx$

41. $\int_0^1 x^2(1+2x^3)^5\,dx$

42. $\int_{1/6}^{1/2} \csc\pi t\,\cot\pi t\,dt$

43. $\int_1^4 \dfrac{e^{\sqrt{x}}}{\sqrt{x}}\,dx$

44. $\int_0^{\pi/2} \cos x\sin(\sin x)\,dx$

45. $\int_{-\pi/4}^{\pi/4} (x^3 + x^4\tan x)\,dx$

46. $\int_{-\pi/2}^{\pi/2} \dfrac{x^2\sin x}{1+x^6}\,dx$

47. $\int_1^2 x\sqrt{x-1}\,dx$

48. $\int_0^a x\sqrt{a^2-x^2}\,dx$

49. $\int_0^1 \dfrac{e^z+1}{e^z+z}\,dz$

50. $\int_0^{T/2} \sin(2\pi t/T - \alpha)\,dt$

51. $\int_0^1 \dfrac{dx}{(1+\sqrt{x})^4}$

---

52. Verify that $f(x) = \sin\sqrt[3]{x}$ is an odd function and use that fact to show that

$$0 \leqslant \int_{-2}^{3} \sin\sqrt[3]{x}\,dx \leqslant 1$$

53. Evaluate $\int_{-2}^{2} (x+3)\sqrt{4-x^2}\,dx$ by writing it as a sum of two integrals and interpreting one of those integrals in terms of an area.

54. Evaluate $\int_0^1 x\sqrt{1-x^4}\,dx$ by making a substitution and interpreting the resulting integral in terms of an area.

55. Which of the following areas are equal? Why?

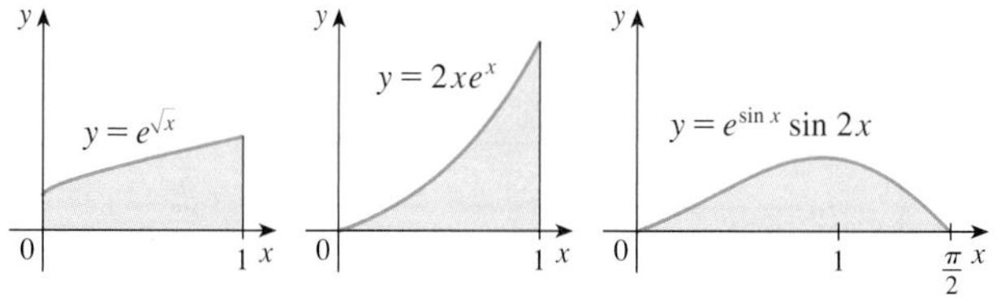

56. A **bacteria population** starts with 400 bacteria and grows at a rate of $r(t) = (450.268)e^{1.12567t}$ bacteria per hour. How many bacteria will there be after three hours?

57. An oil storage tank ruptures at time $t = 0$ and oil leaks from the tank at a rate of $r(t) = 100e^{-0.01t}$ liters per minute. How much oil leaks out during the first hour?

58. Fish biomass The rate of growth of a fish population was modeled by the equation

$$G(t) = \frac{60{,}000e^{-0.6t}}{(1+5e^{-0.6t})^2}$$

where t is measured in years since 2000 and G in kilograms per year. If the biomass was 25,000 kg in the year 2000, what is the predicted biomass for the year 2020?

59. Breathing is cyclic and a full respiratory cycle from the beginning of inhalation to the end of exhalation takes about 5 s. The maximum rate of air flow into the lungs is about 0.5 L/s. This explains, in part, why the function $f(t) = \frac{1}{2}\sin(2\pi t/5)$ has often been used to model the rate of air flow into the lungs. Use this model to find the volume of inhaled air in the lungs at time t.

60. Dialysis treatment removes urea and other waste products from a patient's blood by diverting some of the bloodflow externally through a machine called a dialyzer. The rate at which urea is removed from the blood (in mg/min) is often well described by the equation

$$c(t) = \frac{K}{V}c_0 e^{-Kt/V}$$

where K is the rate of flow of blood through the dialyzer (in mL/min), V is the volume of the patient's blood (in mL), and c_0 is the amount of urea in the blood (in mg) at time $t = 0$. Evaluate the integral $\int_0^{30} c(t)\,dt$ and interpret it.

61. Gompertz tumor growth In Chapter 7 we will explore a model for tumor growth in which the growth rate is given by

$$g(t) = 2^{1-e^{-t}}e^{-t}\ln 2\ \text{mm}^3/\text{month}$$

By how much is the volume of the tumor predicted to increase over the first year?

62. Photosynthesis The *rate of primary production* refers to the rate of conversion of inorganic carbon to organic carbon via photosynthesis. It is measured as a mass of carbon fixed per unit biomass, per unit time. A common model for this relationship is

$$P(I) = \frac{aI}{\sqrt{1+bI^2}}$$

where P is the rate of primary production as a function of light intensity I, and a and b are positive constants. Suppose the light intensity changes with time according to the equation $I(t) = kt$, where k is a positive constant.

(a) What is the rate of primary production as a function of time?

(b) What is the total amount of primary production over the first five units of time?

Source: Adapted from A. Jassby et al., "Mathematical Formulation of the Relationship between Photosynthesis and Light for Phytoplankton," *Limnology and Oceanography* 21 (1976): 540–7.

63. Growing degree days The rate of development of many plant species depends on the temperature of the environment in such a way that maturity is reached only after a certain number of "degree-days." Suppose that temperature T as a function of time t is given by

$$T(t) = 15\left(1 + \sin\frac{2\pi t}{60}\right) \qquad 0 \leqslant t \leqslant 60$$

where t is measured in days. If maturity is reached on day $t = 20$, what is the number of degree-days required? [*Hint:* What are the units for $\int_0^{20} T(t)\,dt$?]

64. If f is continuous and $\int_0^9 f(x)\,dx = 4$, find $\int_0^3 xf(x^2)\,dx$.

65. If f is continuous and $\int_0^4 f(x)\,dx = 10$, find $\int_0^2 f(2x)\,dx$.

66. If f is continuous on $\mathbb{R}$, prove that

$$\int_a^b f(x + c)\,dx = \int_{a+c}^{b+c} f(x)\,dx$$

For the case where $f(x) \geqslant 0$, draw a diagram to interpret this equation geometrically as an equality of areas.

67. If a and b are positive numbers, show that

$$\int_0^1 x^a(1 - x)^b\,dx = \int_0^1 x^b(1 - x)^a\,dx$$

5.5 Integration by Parts

Every differentiation rule has a corresponding integration rule. For instance, the Substitution Rule for integration corresponds to the Chain Rule for differentiation. The rule that corresponds to the Product Rule for differentiation is called the rule for *integration by parts.*

Indefinite Integrals

The Product Rule states that if f and g are differentiable functions, then

$$\frac{d}{dx}[f(x)g(x)] = f(x)g'(x) + g(x)f'(x)$$

In the notation for indefinite integrals this equation becomes

$$\int [f(x)g'(x) + g(x)f'(x)]\,dx = f(x)g(x)$$

or

$$\int f(x)g'(x)\,dx + \int g(x)f'(x)\,dx = f(x)g(x)$$

We can rearrange this equation as

(1)
$$\boxed{\int f(x)g'(x)\,dx = f(x)g(x) - \int g(x)f'(x)\,dx}$$

Formula 1 is called the **formula for integration by parts**. It is perhaps easier to remember in the following notation. Let $u = f(x)$ and $v = g(x)$. Then the differentials are $du = f'(x)\,dx$ and $dv = g'(x)\,dx$, so, by the Substitution Rule, the formula for integration by parts becomes

(2)
$$\boxed{\int u\,dv = uv - \int v\,du}$$

EXAMPLE 1 | Find $\int x \sin x \, dx$.

SOLUTION USING FORMULA 1 Suppose we choose $f(x) = x$ and $g'(x) = \sin x$. Then $f'(x) = 1$ and $g(x) = -\cos x$. (For g we can choose *any* antiderivative of g'.) Thus, using Formula 1, we have

$$\begin{aligned}\int x \sin x \, dx &= f(x)g(x) - \int g(x)f'(x)\, dx \\ &= x(-\cos x) - \int (-\cos x)\, dx \\ &= -x \cos x + \int \cos x \, dx \\ &= -x \cos x + \sin x + C\end{aligned}$$

It's wise to check the answer by differentiating it. If we do so, we get $x \sin x$, as expected.

It is helpful to use the pattern:

$u = \square \qquad dv = \square$

$du = \square \qquad v = \square$

SOLUTION USING FORMULA 2 Let

$$u = x \qquad dv = \sin x \, dx$$

Then

$$du = dx \qquad v = -\cos x$$

and so

$$\begin{aligned}\int x \sin x \, dx = \int \overset{u}{\overbrace{x}} \, \overset{dv}{\overbrace{\sin x \, dx}} &= \overset{u}{\overbrace{x}} \, \overset{v}{\overbrace{(-\cos x)}} - \int \overset{v}{\overbrace{(-\cos x)}} \, \overset{du}{\overbrace{dx}} \\ &= -x \cos x + \int \cos x \, dx \\ &= -x \cos x + \sin x + C\end{aligned}$$

■

NOTE Our aim in using integration by parts is to obtain a simpler integral than the one we started with. Thus in Example 1 we started with $\int x \sin x \, dx$ and expressed it in terms of the simpler integral $\int \cos x \, dx$. If we had instead chosen $u = \sin x$ and $dv = x \, dx$, then $du = \cos x \, dx$ and $v = x^2/2$, so integration by parts gives

$$\int x \sin x \, dx = (\sin x)\frac{x^2}{2} - \frac{1}{2}\int x^2 \cos x \, dx$$

Although this is true, $\int x^2 \cos x \, dx$ is a more difficult integral than the one we started with. In general, when deciding on a choice for u and dv, we usually try to choose $u = f(x)$ to be a function that becomes simpler when differentiated (or at least not more complicated) as long as $dv = g'(x)\, dx$ can be readily integrated to give v.

EXAMPLE 2 | Evaluate $\int \ln x \, dx$.

SOLUTION Here we don't have much choice for u and dv. Let

$$u = \ln x \qquad dv = dx$$

Then

$$du = \frac{1}{x}\, dx \qquad v = x$$

Integrating by parts, we get

$$\int \ln x\,dx = x \ln x - \int x\,\frac{dx}{x}$$

It's customary to write $\int 1\,dx$ as $\int dx$.

$$= x \ln x - \int dx$$

Check the answer by differentiating it.

$$= x \ln x - x + C$$

Integration by parts is effective in this example because the derivative of the function $f(x) = \ln x$ is simpler than f. ■

EXAMPLE 3 | Find $\int t^2 e^t\,dt$.

SOLUTION Notice that t^2 becomes simpler when differentiated (whereas e^t is unchanged when differentiated or integrated), so we choose

$$u = t^2 \qquad dv = e^t\,dt$$

Then $$du = 2t\,dt \qquad v = e^t$$

Integration by parts gives

(3) $$\int t^2 e^t\,dt = t^2 e^t - 2\int te^t\,dt$$

The integral that we obtained, $\int te^t\,dt$, is simpler than the original integral but is still not obvious. Therefore we use integration by parts a second time, this time with $u = t$ and $dv = e^t\,dt$. Then $du = dt$, $v = e^t$, and

$$\int te^t\,dt = te^t - \int e^t\,dt$$

$$= te^t - e^t + C$$

Putting this in Equation 3, we get

$$\int t^2 e^t\,dt = t^2 e^t - 2\int te^t\,dt$$

$$= t^2 e^t - 2(te^t - e^t + C)$$

$$= t^2 e^t - 2te^t + 2e^t + C_1 \qquad \text{where } C_1 = -2C$$ ■

EXAMPLE 4 | Evaluate $\int e^x \sin x\,dx$.

An easier method, using complex numbers, is given in Exercise 50 in Appendix G.

SOLUTION Neither e^x nor $\sin x$ becomes simpler when differentiated, but we try choosing $u = e^x$ and $dv = \sin x\,dx$ anyway. Then $du = e^x\,dx$ and $v = -\cos x$, so integration by parts gives

(4) $$\int e^x \sin x\,dx = -e^x \cos x + \int e^x \cos x\,dx$$

The integral that we have obtained, $\int e^x \cos x\,dx$, is no simpler than the original one, but at least it's no more difficult. Having had success in the preceding example integrating by parts twice, we persevere and integrate by parts again. This time we use $u = e^x$ and $dv = \cos x\,dx$. Then $du = e^x\,dx$, $v = \sin x$, and

(5) $$\int e^x \cos x\,dx = e^x \sin x - \int e^x \sin x\,dx$$

Figure 1 illustrates Example 4 by showing the graphs of $f(x) = e^x \sin x$ and $F(x) = \frac{1}{2}e^x(\sin x - \cos x)$. As a visual check on our work, notice that $f(x) = 0$ when F has a maximum or minimum.

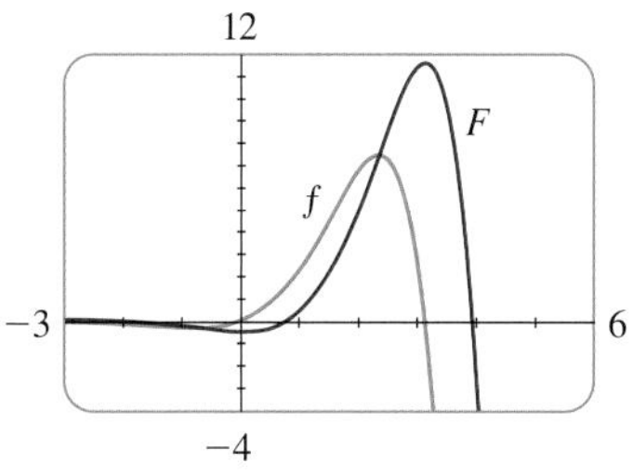

FIGURE 1

At first glance, it appears as if we have accomplished nothing because we have arrived at $\int e^x \sin x\,dx$, which is where we started. However, if we put the expression for $\int e^x \cos x\,dx$ from Equation 5 into Equation 4 we get

$$\int e^x \sin x\,dx = -e^x \cos x + e^x \sin x - \int e^x \sin x\,dx$$

This can be regarded as an equation to be solved for the unknown integral. Adding $\int e^x \sin x\,dx$ to both sides, we obtain

$$2\int e^x \sin x\,dx = -e^x \cos x + e^x \sin x$$

Dividing by 2 and adding the constant of integration, we get

$$\int e^x \sin x\,dx = \tfrac{1}{2}e^x(\sin x - \cos x) + C$$ ■

■ Definite Integrals

If we combine the formula for integration by parts with the Evaluation Theorem, we can evaluate definite integrals by parts. Evaluating both sides of Formula 1 between a and b, assuming f' and g' are continuous, and using the Evaluation Theorem, we obtain

(6)
$$\int_a^b f(x)g'(x)\,dx = f(x)g(x)\Big]_a^b - \int_a^b g(x)f'(x)\,dx$$

EXAMPLE 5 | Aspirin pharmacokinetics In Example 5.2.5 we used data from a paper[1] and modeled the average concentration of low-dose aspirin in the bloodstream of 10 volunteers by the function

$$C(t) = 32t^2e^{-4.2t}$$

where t is measured in hours and C is measured in μg/mL. There we used the Midpoint Rule to estimate $\int_0^2 C(t)\,dt$ and interpreted the integral in terms of the availability of the drug. Here we use integration by parts to evaluate $\int_0^2 C(t)\,dt$.

SOLUTION Notice that t^2 becomes simpler when differentiated (whereas $e^{-4.2t}$ doesn't). So we choose

$$u = 32t^2 \qquad dv = e^{-4.2t}\,dt$$

Then

$$du = 64t\,dt \qquad v = -\frac{1}{4.2}e^{-4.2t}$$

Formula 6 gives

(7)
$$\int_0^2 32t^2e^{-4.2t}\,dt = -\frac{32}{4.2}t^2e^{-4.2t}\Big]_0^2 - \int_0^2\left(-\frac{64}{4.2}\right)te^{-4.2t}\,dt$$

$$= -\frac{32}{4.2}(4e^{-8.4}) + \frac{64}{4.2}\int_0^2 te^{-4.2t}\,dt$$

1. I. H. Benedek et al., "Variability in the Pharmacokinetics and Pharmacodynamics of Low Dose Aspirin in Healthy Male Volunteers," *Journal of Clinical Pharmacology* 35 (1995): 1181–86.

The integral that we obtained, $\int_0^2 te^{-4.2t}\,dt$, is simpler than the original integral but is still not obvious. Therefore we use integration by parts a second time, this time with $u = t$ and $dv = e^{-4.2t}\,dt$. Then

$$du = dt \qquad \text{and} \qquad v = -\frac{1}{4.2}e^{-4.2t}$$

so

$$\int_0^2 te^{-4.2t}\,dt = -\frac{t}{4.2}e^{-4.2t}\Big]_0^2 + \frac{1}{4.2}\int_0^2 e^{-4.2t}\,dt$$

$$= -\frac{2}{4.2}e^{-8.4} + \frac{1}{4.2}\left[\frac{e^{-4.2t}}{-4.2}\right]_0^2 = -\frac{2}{4.2}e^{-8.4} - \frac{e^{-8.4}-1}{(4.2)^2}$$

Putting this in Equation 7, we get

$$\int_0^2 C(t)\,dt = \frac{1}{4.2}\left[-128e^{-8.4} + 64\left(-\frac{2}{4.2}e^{-8.4} - \frac{e^{-8.4}-1}{(4.2)^2}\right)\right]$$

Comparing the answer in Example 5 with the estimate of 0.8568 that we got in Example 5.2.5, we see that the Midpoint Rule gave a reasonably accurate estimate.

Approximating this expression to four decimal places, we get

$$\int_0^2 C(t)\,dt \approx 0.8552\ (\mu\text{g/mL}) \cdot \text{h}$$ ■

EXAMPLE 6 | Calculate $\int_0^1 \tan^{-1}x\,dx$.

SOLUTION Let

$$u = \tan^{-1}x \qquad dv = dx$$

Then

$$du = \frac{dx}{1+x^2} \qquad v = x$$

So Formula 6 gives

$$\int_0^1 \tan^{-1}x\,dx = x\tan^{-1}x\Big]_0^1 - \int_0^1 \frac{x}{1+x^2}\,dx$$

$$= 1\cdot\tan^{-1}1 - 0\cdot\tan^{-1}0 - \int_0^1 \frac{x}{1+x^2}\,dx$$

$$= \frac{\pi}{4} - \int_0^1 \frac{x}{1+x^2}\,dx$$

Since $\tan^{-1}x \geqslant 0$ for $x \geqslant 0$, the integral in Example 6 can be interpreted as the area of the region shown in Figure 2.

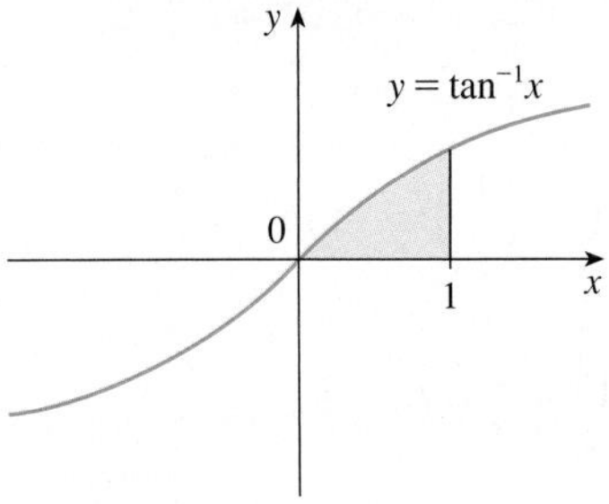

FIGURE 2

To evaluate this integral we use the substitution $t = 1 + x^2$ (since u has another meaning in this example). Then $dt = 2x\,dx$, so $x\,dx = \frac{1}{2}\,dt$. When $x = 0$, $t = 1$; when $x = 1$, $t = 2$; so

$$\int_0^1 \frac{x}{1+x^2}\,dx = \frac{1}{2}\int_1^2 \frac{dt}{t} = \frac{1}{2}\ln|t|\Big]_1^2$$

$$= \frac{1}{2}(\ln 2 - \ln 1) = \frac{1}{2}\ln 2$$

Therefore

$$\int_0^1 \tan^{-1}x\,dx = \frac{\pi}{4} - \int_0^1 \frac{x}{1+x^2}\,dx = \frac{\pi}{4} - \frac{\ln 2}{2}$$ ■

EXERCISES 5.5

1–2 Evaluate the integral using integration by parts with the indicated choices of u and dv.

1. $\int x^2 \ln x\, dx$; $u = \ln x$, $dv = x^2\, dx$

2. $\int \theta \cos\theta\, d\theta$; $u = \theta$, $dv = \cos\theta\, d\theta$

3–20 Evaluate the integral.

3. $\int x \cos 5x\, dx$

4. $\int xe^{-x}\, dx$

5. $\int re^{r/2}\, dr$

6. $\int t \sin 2t\, dt$

7. $\int x^2 \sin \pi x\, dx$

8. $\int x^2 \cos mx\, dx$

9. $\int \ln \sqrt[3]{x}\, dx$

10. $\int p^5 \ln p\, dp$

11. $\int e^{2\theta} \sin 3\theta\, d\theta$

12. $\int e^{-\theta} \cos 2\theta\, d\theta$

13. $\int_0^{\pi} t \sin 3t\, dt$

14. $\int_0^1 (x^2 + 1)e^{-x}\, dx$

15. $\int_1^2 \frac{\ln x}{x^2}\, dx$

16. $\int_4^9 \frac{\ln y}{\sqrt{y}}\, dy$

17. $\int_0^1 \frac{y}{e^{2y}}\, dy$

18. $\int_1^{\sqrt{3}} \arctan(1/x)\, dx$

19. $\int_1^2 (\ln x)^2\, dx$

20. $\int_0^1 \frac{r^3}{\sqrt{4 + r^2}}\, dr$

21–26 First make a substitution and then use integration by parts to evaluate the integral.

21. $\int \cos \sqrt{x}\, dx$

22. $\int t^3 e^{-t^2}\, dt$

23. $\int_{\sqrt{\pi/2}}^{\sqrt{\pi}} \theta^3 \cos(\theta^2)\, d\theta$

24. $\int_0^{\pi} e^{\cos t} \sin 2t\, dt$

25. $\int x \ln(1 + x)\, dx$

26. $\int \sin(\ln x)\, dx$

27. (a) If $n \geqslant 2$ is an integer, show that

$$\int \sin^n x\, dx = -\frac{1}{n} \cos x \sin^{n-1} x + \frac{n-1}{n} \int \sin^{n-2} x\, dx$$

This is called a *reduction formula* because the exponent n has been *reduced* to $n - 1$ and $n - 2$.

(b) Use the reduction formula in part (a) to show that

$$\int \sin^2 x\, dx = \frac{x}{2} - \frac{\sin 2x}{4} + C$$

(c) Use parts (a) and (b) to evaluate $\int \sin^4 x\, dx$.

28. (a) Prove the reduction formula

$$\int \cos^n x\, dx = \frac{1}{n} \cos^{n-1} x \sin x + \frac{n-1}{n} \int \cos^{n-2} x\, dx$$

(b) Use part (a) to evaluate $\int \cos^2 x\, dx$.

(c) Use parts (a) and (b) to evaluate $\int \cos^4 x\, dx$.

29–30 Use integration by parts to prove the reduction formula.

29. $\int (\ln x)^n\, dx = x(\ln x)^n - n \int (\ln x)^{n-1}\, dx$

30. $\int x^n e^x\, dx = x^n e^x - n \int x^{n-1} e^x\, dx$

31. Use Exercise 29 to find $\int (\ln x)^3\, dx$.

32. Use Exercise 30 to find $\int x^4 e^x\, dx$.

33. Salicylic acid pharmacokinetics In the article cited in Example 5 the authors also studied the formation and concentration of salicylic acid in the bloodstream of 10 volunteers. A model for the concentration is

$$C(t) = 11.4te^{-t}$$

where t is measured in hours and C in μg/mL. Calculate $\int_0^4 C(t)\, dt$ and include the units in your answer.

34. Rumen microbial ecosystem The rumen is the first chamber in the stomach of ruminants such as cattle, sheep, and deer. Fermentation reactions by symbiotic organisms begin digesting plant matter in the rumen. If μ is the fraction of matter entering or leaving the rumen per unit time in a model for continuous fermentation, then the integral

$$\int_0^1 \mu e^{-\mu t}(1 - t)\, dt$$

is the fraction of soluble material passing from the rumen in the first hour without being fermented. Evaluate this integral.

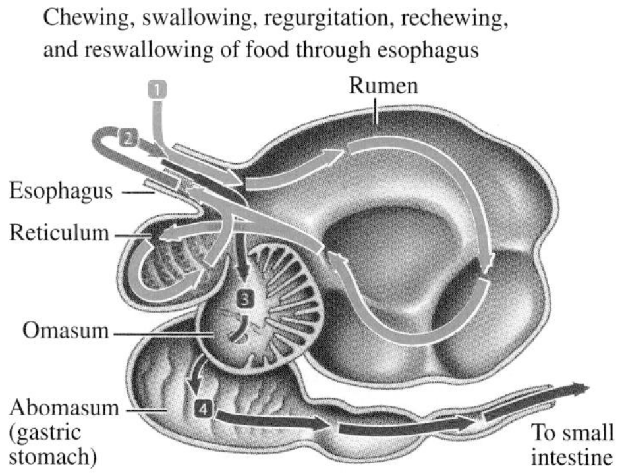

Source: Adapted from R. E. Hungate, "The Rumen Microbial Ecosystem," *Annual Review of Ecology and Systematics* 6 (1975): 39–66.

35. Gene regulation In Section 10.3 a model of gene regulation is analyzed and it is shown that the concentration of protein in a cell as a function of time is given by the equation

$$p(t) = \tfrac{1}{2} - \tfrac{1}{2}e^{-t}(\sin t + \cos t)$$

The bioavailability of this protein is defined as the integral of this concentration over time. What is the bioavailability of the protein over the first unit of time?

36. Insect metamorphosis The rate of development of many insects increases gradually with temperature up to a maximum and then rapidly falls to zero. This can be approximated by the function $T - T^k$, where k is a positive integer and T is a standardized measure of temperature such that $0 \leqslant T \leqslant 1$ and $T = 0$ results in no development, whereas $T = 1$ is a lethal temperature. Suppose the daily temperature oscillates according to the equation $T(t) = a\cos^2 t$, where π units of time equals one day, and a is the factor between 0 and 1 by which the temperature oscillates over the course of a day. Use Exercise 28 to determine the cumulative amount of development that happens over the first week for the case where $k = 3$.

37. Suppose that $f(1) = 2$, $f(4) = 7$, $f'(1) = 5$, $f'(4) = 3$, and f'' is continuous. Find the value of $\int_1^4 xf''(x)\,dx$.

38. If $f(0) = g(0) = 0$ and f'' and g'' are continuous, show that

$$\int_0^a f(x)g''(x)\,dx = f(a)g'(a) - f'(a)g(a) + \int_0^a f''(x)g(x)\,dx$$

5.6 Partial Fractions

In Chapter 7 we will study the logistic model for population growth and in solving the logistic differential equation we will need to be able to integrate the function

$$F(N) = \frac{K}{N(K - N)}$$

where K is a positive constant. The integration techniques we have learned so far don't enable us to integrate this function, but in this section we learn how to do so. (See Example 2.)

The idea is to express a rational function as a sum of simpler functions, called *partial fractions*, that we already know how to integrate. To illustrate the method, observe that by taking the fractions $2/(x - 1)$ and $1/(x + 2)$ to a common denominator we obtain

$$\frac{2}{x - 1} - \frac{1}{x + 2} = \frac{2(x + 2) - (x - 1)}{(x - 1)(x + 2)} = \frac{x + 5}{x^2 + x - 2}$$

If we now reverse the procedure, we see how to integrate the function on the right side of this equation:

$$\int \frac{x + 5}{x^2 + x - 2}\,dx = \int \left(\frac{2}{x - 1} - \frac{1}{x + 2}\right) dx$$

$$= 2\ln|x - 1| - \ln|x + 2| + C$$

More generally, let's consider a rational function

$$f(x) = \frac{P(x)}{Q(x)}$$

where P and Q are polynomials. It's possible to express f as a sum of simpler fractions provided that the degree of P is less than the degree of Q. If that's not the case, then we must take the preliminary step of dividing Q into P (by long division) until a remainder $R(x)$ is obtained whose degree is less than the degree of Q. As the following example illustrates, sometimes this preliminary step is all that is required.

EXAMPLE 1 | Find $\displaystyle\int \frac{x^3 + x}{x - 1}\,dx$.

SOLUTION Since the degree of the numerator is greater than the degree of the

$$\begin{array}{r} x^2 + x + 2 \\ x - 1\,\overline{)\,x^3 \qquad + x} \\ \underline{x^3 - x^2} \\ x^2 + x \\ \underline{x^2 - x} \\ 2x \\ \underline{2x - 2} \\ 2 \end{array}$$

denominator, we first perform the long division. This enables us to write

$$\int \frac{x^3 + x}{x - 1}\,dx = \int \left(x^2 + x + 2 + \frac{2}{x - 1} \right) dx$$

$$= \frac{x^3}{3} + \frac{x^2}{2} + 2x + 2\ln|x - 1| + C \qquad \blacksquare$$

The next step is to factor the denominator $Q(x)$ as far as possible. We concentrate on the case where $Q(x)$ is a product of distinct linear factors of the form $ax + b$. (Other cases are explored in the exercises.)

Then we express the function $R(x)/Q(x)$ as a sum of **partial fractions** of the form

$$\frac{A}{ax + b}$$

Each factor $ax + b$ of $Q(x)$ has a corresponding partial fraction. The constants in the numerators can be determined as in the following examples.

EXAMPLE 2 | BB Logistic model If K is a constant, evaluate the integral

$$\int \frac{K}{N(K - N)}$$

SOLUTION Because there are two linear factors, we write

$$\textbf{(1)} \qquad \frac{K}{N(K - N)} = \frac{A}{N} + \frac{B}{K - N}$$

where A and B are constants. To determine the values of A and B, we multiply both sides of Equation 1 by the product of denominators, $N(K - N)$:

$$K = A(K - N) + BN$$

$$K = (B - A)N + AK$$

Comparing the left and right sides, we see that this equation is satisfied if $B - A = 0$ and $A = 1$. So the solution is $A = B = 1$ and we have

$$\frac{K}{N(K - N)} = \frac{1}{N} + \frac{1}{K - N}$$

This new form of the function allows us to integrate more easily:

Notice that in integrating the second term we made the mental substitution $u = K - N$ and so $du = -dN$.

$$\int \frac{K}{N(K - N)}\,dN = \int \left(\frac{1}{N} + \frac{1}{K - N} \right) dN$$

$$= \ln|N| - \ln|K - N| + C \qquad \blacksquare$$

EXAMPLE 3 | Evaluate $\displaystyle\int \frac{x^2 + 2x - 1}{2x^3 + 3x^2 - 2x}\,dx$.

SOLUTION Since the degree of the numerator is less than the degree of the denominator, we don't need to divide. We factor the denominator as

$$2x^3 + 3x^2 - 2x = x(2x^2 + 3x - 2) = x(2x - 1)(x + 2)$$

Since the denominator has three distinct linear factors, the partial fraction decomposition of the integrand has the form

$$\frac{x^2 + 2x - 1}{x(2x-1)(x+2)} = \frac{A}{x} + \frac{B}{2x-1} + \frac{C}{x+2} \tag{2}$$

Another method for finding A, B, and C is given in the note after this example.

To determine the values of A, B, and C, we multiply both sides of this equation by the product of the denominators, $x(2x - 1)(x + 2)$, obtaining

$$x^2 + 2x - 1 = A(2x-1)(x+2) + Bx(x+2) + Cx(2x-1) \tag{3}$$

Expanding the right side of Equation 3 and writing it in the standard form for polynomials, we get

$$x^2 + 2x - 1 = (2A + B + 2C)x^2 + (3A + 2B - C)x - 2A \tag{4}$$

The polynomials in Equation 4 are identical, so their coefficients must be equal. The coefficient of x^2 on the right side, $2A + B + 2C$, must equal the coefficient of x^2 on the left side—namely, 1. Likewise, the coefficients of x are equal and the constant terms are equal. This gives the following system of equations for A, B, and C:

$$\begin{aligned} 2A + B + 2C &= 1 \\ 3A + 2B - C &= 2 \\ -2A &= -1 \end{aligned}$$

Solving, we get $A = \frac{1}{2}$, $B = \frac{1}{5}$, and $C = -\frac{1}{10}$, and so

We could check our work by taking the terms to a common denominator and adding them.

$$\int \frac{x^2 + 2x - 1}{2x^3 + 3x^2 - 2x}\,dx = \int \left[\frac{1}{2}\frac{1}{x} + \frac{1}{5}\frac{1}{2x-1} - \frac{1}{10}\frac{1}{x+2}\right] dx$$

$$= \tfrac{1}{2}\ln|x| + \tfrac{1}{10}\ln|2x-1| - \tfrac{1}{10}\ln|x+2| + K$$

Figure 1 shows the graphs of the integrand in Example 3 and its indefinite integral (with $K = 0$). Which is which?

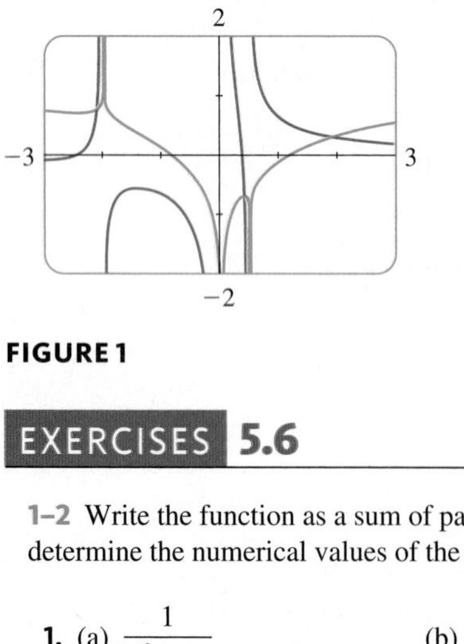

FIGURE 1

In integrating the middle term we have made the mental substitution $u = 2x - 1$, which gives $du = 2\,dx$ and $dx = \frac{1}{2}\,du$. ■

NOTE We can use an alternative method to find the coefficients A, B, and C in Example 3. Equation 3 is an identity; it is true for every value of x. Let's choose values of x that simplify the equation. If we put $x = 0$ in Equation 3, then the second and third terms on the right side vanish and the equation then becomes $-2A = -1$, or $A = \frac{1}{2}$. Likewise, $x = \frac{1}{2}$ gives $5B/4 = \frac{1}{4}$ and $x = -2$ gives $10C = -1$, so $B = \frac{1}{5}$ and $C = -\frac{1}{10}$. (You may object that Equation 2 is not valid for $x = 0, \frac{1}{2}$, or -2, so why should Equation 3 be valid for those values? In fact, Equation 4 is true for all values of x, even $x = 0, \frac{1}{2}$, and -2. See Exercise 23 for the reason.)

EXERCISES 5.6

1–2 Write the function as a sum of partial fractions. Do not determine the numerical values of the coefficients.

1. (a) $\dfrac{1}{x^2 - 1}$ (b) $\dfrac{2}{x^2 + x}$

2. (a) $\dfrac{x}{x^2 + x - 2}$ (b) $\dfrac{2 - x}{x^2 - 2x - 8}$

3–14 Evaluate the integral.

3. $\displaystyle\int \frac{x}{x-6}\,dx$

4. $\displaystyle\int \frac{r^2}{r+4}\,dr$

5. $\displaystyle\int \frac{x-9}{(x+5)(x-2)}\,dx$

6. $\displaystyle\int \frac{1}{(t+4)(t-1)}\,dt$

7. $\displaystyle\int_2^3 \frac{1}{x^2-1}\,dx$

8. $\displaystyle\int_0^1 \frac{x-1}{x^2+3x+2}\,dx$

9. $\displaystyle\int \frac{ax}{x^2 - bx}\,dx$

10. $\displaystyle\int \frac{1}{(x + a)(x + b)}\,dx$

11. $\displaystyle\int_0^1 \frac{2}{2x^2 + 3x + 1}\,dx$

12. $\displaystyle\int_0^1 \frac{x^3 - 4x - 10}{x^2 - x - 6}\,dx$

13. $\displaystyle\int_1^2 \frac{4y^2 - 7y - 12}{y(y + 2)(y - 3)}\,dy$

14. $\displaystyle\int \frac{x^2 + 2x - 1}{x^3 - x}\,dx$

15–18 Make a substitution to express the integrand as a rational function and then evaluate the integral.

15. $\displaystyle\int_9^{16} \frac{\sqrt{x}}{x - 4}\,dx$

16. $\displaystyle\int \frac{dx}{2\sqrt{x + 3} + x}$

17. $\displaystyle\int \frac{e^{2x}}{e^{2x} + 3e^x + 2}\,dx$

18. $\displaystyle\int \frac{\cos x}{\sin^2 x + \sin x}\,dx$

19. If a linear factor in the denominator of a rational function is repeated, there will be two corresponding partial fractions. For instance,

$$f(x) = \frac{5x^2 + 3x - 2}{x^2(x + 2)} = \frac{A}{x} + \frac{B}{x^2} + \frac{C}{x + 2}$$

Determine the values of A, B, and C and use them to evaluate $\int f(x)\,dx$.

20. Use the method of Exercise 19 to evaluate

$$\int \frac{x^2 - 5x + 16}{(2x + 1)(x - 2)^2}\,dx$$

21. If a factor of the denominator is an irreducible quadratic, such as $x^2 + 1$, the corresponding partial fraction has a linear numerator. For instance,

$$f(x) = \frac{2x^2 + x + 1}{x(x^2 + 1)} = \frac{A}{x} + \frac{Bx + C}{x^2 + 1}$$

Determine the values of A, B, and C and use them to evaluate $\int f(x)\,dx$.

22. Use the method of Exercise 21 to evaluate

$$\int \frac{3x^2 - 2x + 3}{(x - 1)(x^2 + 1)}\,dx$$

23. Suppose that F, G, and Q are polynomials and

$$\frac{F(x)}{Q(x)} = \frac{G(x)}{Q(x)}$$

for all x except when $Q(x) = 0$. Prove that $F(x) = G(x)$ for all x. [*Hint:* Use continuity.]

BB **24. Sterile insect technique** One method of slowing the growth of an insect population without using pesticides is to introduce into the population a number of sterile males that mate with fertile females but produce no offspring. (The photo shows a screw-worm fly, the first pest effectively eliminated from a region by this method.) Let P represent the number of female insects in a population and S the number of sterile males introduced each generation. Let r be the per capita rate of production of females by females, provided their chosen mate is not sterile. Then the female population is related to time t by

$$t = \int \frac{P + S}{P[(r - 1)P - S]}\,dP$$

Suppose an insect population with 10,000 females grows at a rate of $r = 1.1$ and 900 sterile males are added. Evaluate the integral to give an equation relating the female population to time. (Note that the resulting equation can't be solved explicitly for P.)

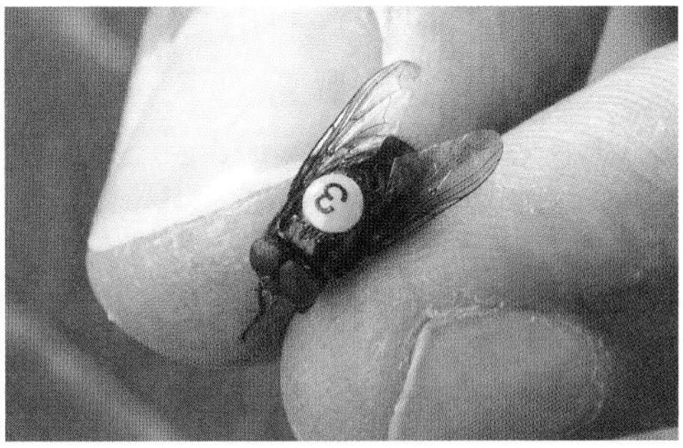

USDA

5.7 Integration Using Tables and Computer Algebra Systems

We have not given an exhaustive treatment of techniques of integration, so it's important to know that there are resources available to scientists and others who encounter a difficult integral. Here we describe how to evaluate integrals using tables and computer algebra systems. For a *definite* integral that seems intractable, one can always approximate it by using the Midpoint Rule or the numerical integration capability of a scientific calculator.

Tables of Integrals

Tables of indefinite integrals are very useful when we are confronted by an integral that is difficult to evaluate by hand and we don't have access to a computer algebra system. A relatively brief table of 120 integrals, categorized by form, is provided on the Reference Pages at the front of the book. More extensive tables are available in the *CRC Standard Mathematical Tables and Formulae*, 32nd ed. by Daniel Zwillinger (Boca Raton, FL, 2011) (709 entries) or in Gradshteyn and Ryzhik's *Table of Integrals, Series, and Products*, 7e (San Diego, 2007), which contains hundreds of pages of integrals. It should be remembered, however, that integrals do not often occur in exactly the form listed in a table. Usually we need to use the Substitution Rule or algebraic manipulation to transform a given integral into one of the forms in the table.

EXAMPLE 1 | Use the Table of Integrals to evaluate $\int_0^2 \frac{x^2 + 12}{x^2 + 4}\, dx$.

The Table of Integrals appears on Reference Pages 6–10 at the front of the book.

SOLUTION The only formula in the table that resembles our given integral is entry 17:

$$\int \frac{du}{a^2 + u^2} = \frac{1}{a} \tan^{-1} \frac{u}{a} + C$$

If we perform long division, we get

$$\frac{x^2 + 12}{x^2 + 4} = 1 + \frac{8}{x^2 + 4}$$

Now we can use Formula 17 with $a = 2$:

$$\int_0^2 \frac{x^2 + 12}{x^2 + 4}\, dx = \int_0^2 \left(1 + \frac{8}{x^2 + 4}\right) dx = x + 8 \cdot \tfrac{1}{2} \tan^{-1} \frac{x}{2}\Big]_0^2$$

$$= 2 + 4 \tan^{-1} 1 = 2 + \pi$$

■

EXAMPLE 2 | Use the Table of Integrals to find $\int x^3 \sin x\, dx$.

SOLUTION If we look in the section called *Trigonometric Forms*, we see that none of the entries explicitly includes a u^3 factor. However, we can use the reduction formula in entry 84 with $n = 3$:

$$\int x^3 \sin x\, dx = -x^3 \cos x + 3 \int x^2 \cos x\, dx$$

85. $\int u^n \cos u\, du$

$= u^n \sin u - n \int u^{n-1} \sin u\, du$

We now need to evaluate $\int x^2 \cos x\, dx$. We can use the reduction formula in entry 85 with $n = 2$, followed by entry 82:

$$\int x^2 \cos x\, dx = x^2 \sin x - 2 \int x \sin x\, dx$$

$$= x^2 \sin x - 2(\sin x - x \cos x) + K$$

Combining these calculations, we get

$$\int x^3 \sin x\, dx = -x^3 \cos x + 3x^2 \sin x + 6x \cos x - 6 \sin x + C$$

where $C = 3K$.

■

EXAMPLE 3 | Use the Table of Integrals to find $\int x\sqrt{x^2 + 2x + 4}\, dx$.

SOLUTION Since the table gives forms involving $\sqrt{a^2 + x^2}$, $\sqrt{a^2 - x^2}$, and $\sqrt{x^2 - a^2}$, but not $\sqrt{ax^2 + bx + c}$, we first complete the square:

$$x^2 + 2x + 4 = (x + 1)^2 + 3$$

If we make the substitution $u = x + 1$ (so $x = u - 1$), the integrand will involve the pattern $\sqrt{a^2 + u^2}$:

$$\int x\sqrt{x^2 + 2x + 4}\, dx = \int (u - 1)\sqrt{u^2 + 3}\, du$$

$$= \int u\sqrt{u^2 + 3}\, du - \int \sqrt{u^2 + 3}\, du$$

The first integral is evaluated using the substitution $t = u^2 + 3$:

$$\int u\sqrt{u^2 + 3}\, du = \tfrac{1}{2}\int \sqrt{t}\, dt = \tfrac{1}{2} \cdot \tfrac{2}{3}t^{3/2} = \tfrac{1}{3}(u^2 + 3)^{3/2}$$

21. $\displaystyle\int \sqrt{a^2 + u^2}\, du = \frac{u}{2}\sqrt{a^2 + u^2} + \frac{a^2}{2}\ln\left(u + \sqrt{a^2 + u^2}\right) + C$

For the second integral we use Formula 21 with $a = \sqrt{3}$:

$$\int \sqrt{u^2 + 3}\, du = \frac{u}{2}\sqrt{u^2 + 3} + \tfrac{3}{2}\ln\left(u + \sqrt{u^2 + 3}\right)$$

Thus

$$\int x\sqrt{x^2 + 2x + 4}\, dx$$

$$= \tfrac{1}{3}(x^2 + 2x + 4)^{3/2} - \frac{x + 1}{2}\sqrt{x^2 + 2x + 4} - \tfrac{3}{2}\ln\left(x + 1 + \sqrt{x^2 + 2x + 4}\right) + C$$

■

■ Computer Algebra Systems

We have seen that the use of tables involves matching the form of the given integrand with the forms of the integrands in the tables. Computers are particularly good at matching patterns. And just as we used substitutions in conjunction with tables, a CAS can perform substitutions that transform a given integral into one that occurs in its stored formulas. So it isn't surprising that computer algebra systems excel at integration. That doesn't mean that integration by hand is an obsolete skill. We will see that a hand computation sometimes produces an indefinite integral in a form that is more convenient than a machine answer.

To begin, let's see what happens when we ask a machine to integrate the relatively simple function $y = 1/(3x - 2)$. Using the substitution $u = 3x - 2$, an easy calculation by hand gives

$$\int \frac{1}{3x - 2}\, dx = \tfrac{1}{3}\ln|3x - 2| + C$$

whereas Derive, Mathematica, and Maple all return the answer

$$\tfrac{1}{3}\ln(3x - 2)$$

The first thing to notice is that computer algebra systems omit the constant of integration. In other words, they produce a *particular* antiderivative, not the most general one. Therefore, when making use of a machine integration, we might have to add a constant.

Second, the absolute value signs are omitted in the machine answer. That is fine if our problem is concerned only with values of x greater than $\frac{2}{3}$. But if we are interested in other values of x, then we need to insert the absolute value symbol.

EXAMPLE 4 | Use a CAS to evaluate $\int x(x^2 + 5)^8\,dx$.

SOLUTION Maple and Mathematica give the same answer:

$$\tfrac{1}{18}x^{18} + \tfrac{5}{2}x^{16} + 50x^{14} + \tfrac{1750}{3}x^{12} + 4375x^{10} + 21875x^8 + \tfrac{218750}{3}x^6 + 156250x^4 + \tfrac{390625}{2}x^2$$

It's clear that both systems must have expanded $(x^2 + 5)^8$ by the Binomial Theorem and then integrated each term.

If we integrate by hand instead, using the substitution $u = x^2 + 5$, we get

Derive and the TI-89 and TI-92 also give this answer.

$$\int x(x^2 + 5)^8\,dx = \tfrac{1}{18}(x^2 + 5)^9 + C$$

For most purposes, this is a more convenient form of the answer. ■

EXAMPLE 5 | Use a CAS to find $\int \sin^5 x \cos^2 x\,dx$.

SOLUTION Derive and Maple report the answer

$$-\tfrac{1}{7}\sin^4 x\cos^3 x - \tfrac{4}{35}\sin^2 x\cos^3 x - \tfrac{8}{105}\cos^3 x$$

whereas Mathematica produces

$$-\tfrac{5}{64}\cos x - \tfrac{1}{192}\cos 3x + \tfrac{3}{320}\cos 5x - \tfrac{1}{448}\cos 7x$$

We suspect that there are trigonometric identities that we could use to show that these answers are equivalent. Indeed, if we ask Derive, Maple, and Mathematica to simplify their expressions using trigonometric identities, they ultimately produce the same form of the answer:

$$\int \sin^5 x\cos^2 x\,dx = -\tfrac{1}{3}\cos^3 x + \tfrac{2}{5}\cos^5 x - \tfrac{1}{7}\cos^7 x$$ ■

■ Can We Integrate All Continuous Functions?

The question arises: Will our basic integration formulas, together with the Substitution Rule, integration by parts, tables of integrals, and computer algebra systems, enable us to find the integral of every continuous function? In particular, can we use these techniques to evaluate $\int e^{x^2}\,dx$? The answer is No, at least not in terms of the functions that we are familiar with.

Most of the functions that we have been dealing with in this book are called **elementary functions**. These are the polynomials, rational functions, power functions (x^a), exponential functions (a^x), logarithmic functions, trigonometric and inverse trigonometric functions, and all functions that can be obtained from these by the five operations of addition, subtraction, multiplication, division, and composition. For instance, the function

$$f(x) = \sqrt{\frac{x^2 - 1}{x^3 + 2x - 1}} + \ln(\cos x) - xe^{\sin 2x}$$

is an elementary function.

If f is an elementary function, then f' is an elementary function but $\int f(x)\,dx$ need not be an elementary function. Consider $f(x) = e^{x^2}$. Since f is continuous, its integral exists, and if we define the function F by

$$F(x) = \int_0^x e^{t^2}\,dt$$

then we know from Part 1 of the Fundamental Theorem of Calculus that

$$F'(x) = e^{x^2}$$

Thus $f(x) = e^{x^2}$ has an antiderivative F, but it has been proved that F is not an elementary function. This means that no matter how hard we try, we will never succeed in evaluating $\int e^{x^2}\,dx$ in terms of the functions we know. The same can be said of the following integrals:

$$\int \frac{e^x}{x}\,dx \qquad \int \sin(x^2)\,dx \qquad \int \cos(e^x)\,dx$$

$$\int \sqrt{x^3+1}\,dx \qquad \int \frac{1}{\ln x}\,dx \qquad \int \frac{\sin x}{x}\,dx$$

In fact, the majority of elementary functions don't have elementary antiderivatives.

EXERCISES 5.7

1–18 Use the Table of Integrals on Reference Pages 6–10 to evaluate the integral.

1. $\int \tan^3(\pi x)\,dx$

2. $\int e^{2\theta}\sin 3\theta\,d\theta$

3. $\int \frac{dx}{x^2\sqrt{4x^2+9}}$

4. $\int_2^3 \frac{1}{x^2\sqrt{4x^2-7}}\,dx$

5. $\int e^{2x}\arctan(e^x)\,dx$

6. $\int \frac{\sqrt{2y^2-3}}{y^2}\,dy$

7. $\int_0^\pi x^3\sin x\,dx$

8. $\int \frac{dx}{2x^3-3x^2}$

9. $\int \frac{\tan^3(1/z)}{z^2}\,dz$

10. $\int x\sin(x^2)\cos(3x^2)\,dx$

11. $\int \sin^2 x\cos x\ln(\sin x)\,dx$

12. $\int \frac{\sin 2\theta}{\sqrt{5-\sin\theta}}\,d\theta$

13. $\int \frac{e^x}{3-e^{2x}}\,dx$

14. $\int_0^1 x^4 e^{-x}\,dx$

15. $\int \frac{x^4\,dx}{\sqrt{x^{10}-2}}$

16. $\int \frac{\sec^2\theta\tan^2\theta}{\sqrt{9-\tan^2\theta}}\,d\theta$

17. $\int \frac{\sqrt{4+(\ln x)^2}}{x}\,dx$

18. $\int e^t\sin(\alpha t-3)\,dt$

19. Verify Formula 53 in the Table of Integrals (a) by differentiation and (b) by using the substitution $t = a + bu$.

20. Verify Formula 31 (a) by differentiation and (b) by substituting $u = a\sin\theta$.

CAS **21–27** Use a computer algebra system to evaluate the integral. Compare the answer with the result of using tables. If the answers are not the same, show that they are equivalent.

21. $\int \sec^4 x\,dx$

22. $\int x^2(1+x^3)^4\,dx$

23. $\int x\sqrt{1+2x}\,dx$

24. $\int \frac{dx}{e^x(3e^x+2)}$

25. $\int \tan^5 x\,dx$

26. $\int \sin^4 x\,dx$

27. $\int \frac{1}{\sqrt{1+\sqrt[3]{x}}}\,dx$

CAS **28.** Computer algebra systems sometimes need a helping hand from human beings. Try to evaluate

$$\int (1+\ln x)\sqrt{1+(x\ln x)^2}\,dx$$

with a computer algebra system. If it doesn't return an answer, make a substitution that changes the integral into one that the CAS *can* evaluate.

5.8 Improper Integrals

In defining a definite integral $\int_a^b f(x)\,dx$ we dealt with a function f defined on a finite interval $[a, b]$. In this section we extend the concept of a definite integral to the case where the interval is infinite. In this case the integral is called an *improper* integral. One of the most important applications of this idea, probability distributions, will be studied in Chapter 12.

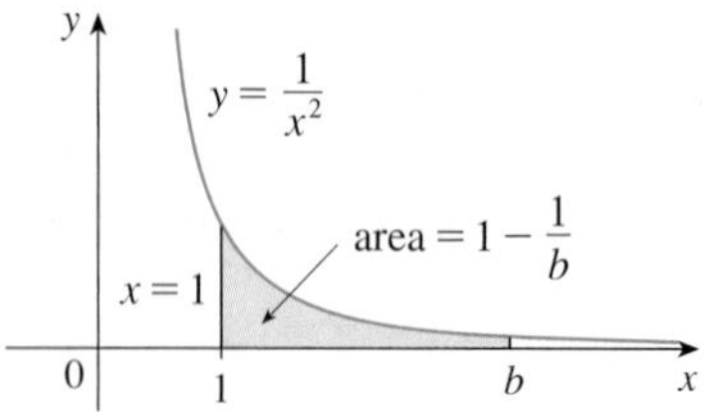

FIGURE 1

Consider the infinite region S that lies under the curve $y = 1/x^2$, above the x-axis, and to the right of the line $x = 1$. You might think that, since S is infinite in extent, its area must be infinite, but let's take a closer look. The area of the part of S that lies to the left of the line $x = b$ (shaded in Figure 1) is

$$A(b) = \int_1^b \frac{1}{x^2}\,dx = -\frac{1}{x}\bigg]_1^b = 1 - \frac{1}{b}$$

Notice that $A(b) < 1$ no matter how large b is chosen.

We also observe that

$$\lim_{b\to\infty} A(b) = \lim_{b\to\infty}\left(1 - \frac{1}{b}\right) = 1$$

The area of the shaded region approaches 1 as $b \to \infty$ (see Figure 2), so we say that the area of the infinite region S is equal to 1 and we write

$$\int_1^\infty \frac{1}{x^2}\,dx = \lim_{b\to\infty}\int_1^b \frac{1}{x^2}\,dx = 1$$

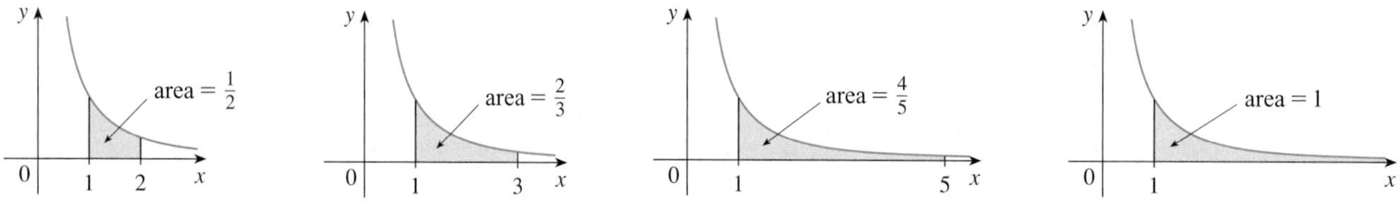

FIGURE 2

Using this example as a guide, we define the integral of f (not necessarily a positive function) over an infinite interval as the limit of integrals over finite intervals.

> **(1) Definition of an Improper Integral** If $\int_a^b f(x)\,dx$ exists for every number $b \geq a$, then
>
> $$\int_a^\infty f(x)\,dx = \lim_{b\to\infty}\int_a^b f(x)\,dx$$
>
> provided this limit exists (as a finite number). An improper integral is called **convergent** if the corresponding limit exists and **divergent** if the limit does not exist.

The improper integral in Definition 1 can be interpreted as an area provided that f is a positive function. For instance, if $f(x) \geq 0$ and the integral $\int_a^\infty f(x)\,dx$ is convergent, then we define the area of the region $S = \{(x, y) \mid x \geq a, 0 \leq y \leq f(x)\}$ in Figure 3 to be

$$A(S) = \int_a^\infty f(x)\,dx$$

This is appropriate because $\int_a^\infty f(x)\,dx$ is the limit as $b \to \infty$ of the area under the graph of f from a to b.

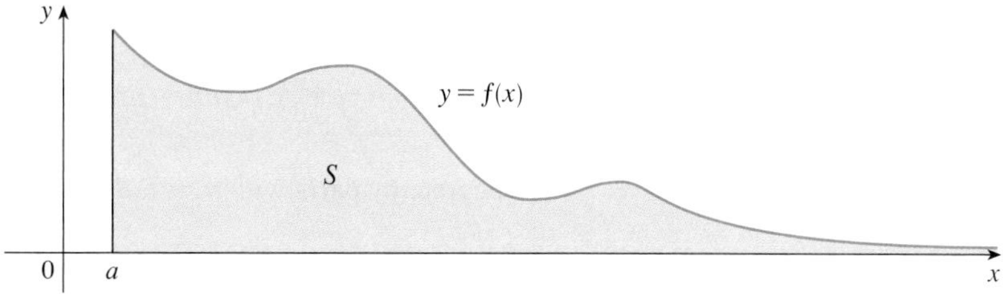

FIGURE 3

EXAMPLE 1 | Determine whether the integral $\int_1^\infty (1/x)\,dx$ is convergent or divergent.

SOLUTION According to Definition 1, we have

$$\int_1^\infty \frac{1}{x}\,dx = \lim_{b\to\infty}\int_1^b \frac{1}{x}\,dx = \lim_{b\to\infty} \ln|x|\Big]_1^b$$

$$= \lim_{b\to\infty}(\ln b - \ln 1) = \lim_{b\to\infty}\ln b = \infty$$

The limit does not exist as a finite number and so the improper integral $\int_1^\infty (1/x)\,dx$ is divergent. ■

Let's compare the result of Example 1 with the example given at the beginning of this section:

$$\int_1^\infty \frac{1}{x^2}\,dx \text{ converges} \qquad \int_1^\infty \frac{1}{x}\,dx \text{ diverges}$$

Geometrically, this says that although the curves $y = 1/x^2$ and $y = 1/x$ look very similar for $x > 0$, the region under $y = 1/x^2$ to the right of $x = 1$ (the shaded region in Figure 4) has finite area whereas the corresponding region under $y = 1/x$ (in Figure 5) has infinite area. Note that both $1/x^2$ and $1/x$ approach 0 as $x \to \infty$ but $1/x^2$ approaches 0 faster than $1/x$. The values of $1/x$ don't decrease fast enough for its integral to have a finite value.

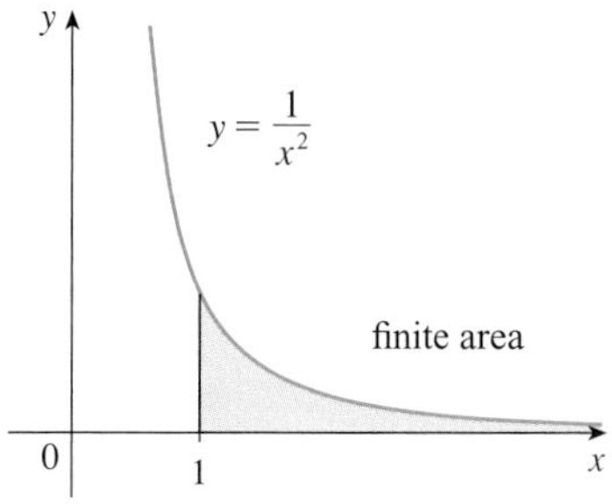

FIGURE 4

$\int_1^\infty (1/x^2)\,dx$ converges.

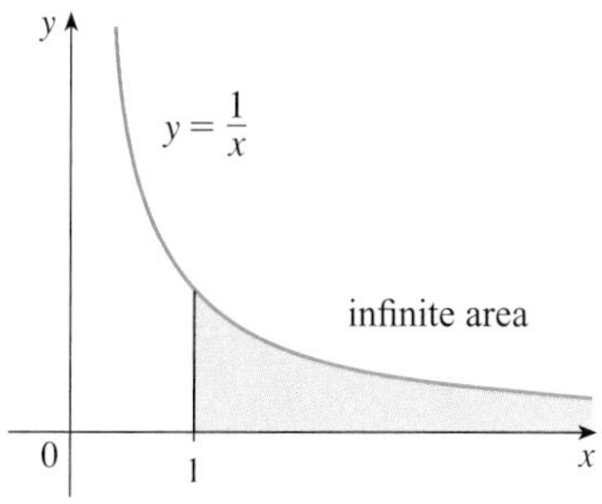

FIGURE 5

$\int_1^\infty (1/x)\,dx$ diverges.

EXAMPLE 2 | **Salicylic acid pharmacokinetics** In Exercises 5.2.14 and 5.5.33 we used the function $C(t) = 11.4te^{-t}$ to model the concentration of SA in the bloodstream of volunteers, where t is measured in hours and C is measured in $\mu g/mL$. Calculate $\int_0^\infty C(t)\,dt$ and interpret it.

SOLUTION The definition of an improper integral says that

$$\int_0^\infty C(t)\,dt = \lim_{b\to\infty}\int_0^b C(t)\,dt = 11.4\lim_{b\to\infty}\int_0^b te^{-t}\,dt$$

We integrate by parts with $u = t$, $dv = e^{-t}\,dt$, so $du = dt$ and $v = -e^{-t}$:

$$\int_0^b te^{-t}\,dt = -te^{-t}\Big]_0^b - \int_0^b (-e^{-t})\,dt$$

$$= -be^{-b} - 0 - e^{-t}\Big]_0^b = -be^{-b} - e^{-b} + 1$$

TEC In Module 5.8 you can investigate visually and numerically whether several improper integrals are convergent or divergent.

We know that $e^{-b} \to 0$ as $b \to \infty$, and by l'Hospital's Rule we have

$$\lim_{b\to\infty} be^{-b} = \lim_{b\to\infty}\frac{b}{e^b} = \lim_{b\to\infty}\frac{1}{e^b}$$

$$= \lim_{b\to\infty} e^{-b} = 0$$

Therefore

$$\int_0^\infty C(t)\,dt = 11.4\lim_{b\to\infty}(-be^{-b} - e^{-b} + 1)$$

$$= 11.4(-0 - 0 + 1) = 11.4\ (\mu g/mL)\cdot h$$

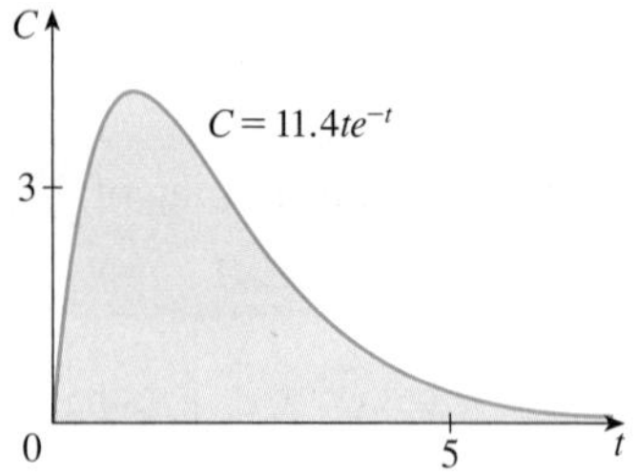

FIGURE 6
$\int_0^\infty C(t)\,dt = 11.4\ (\mu g/mL)\cdot h$

We have previously interpreted $\int_0^1 C(t)\,dt$ in terms of the "availability" of the drug during the first hour, with units of concentration times time. Similarly, $\int_0^4 C(t)\,dt$ measures the availability over the first four hours. The improper integral $\int_0^\infty C(t)\,dt$ (the total area under the concentration curve in Figure 6) measures the availability for all time. In other words, it measures the long-term availability of SA. ■

We next define an improper integral over an interval that extends infinitely far in the negative direction.

> **(2) Definition** If $\int_a^b f(x)\,dx$ exists for every number $a \leqslant b$, then
>
> $$\int_{-\infty}^b f(x)\,dx = \lim_{a\to-\infty}\int_a^b f(x)\,dx$$
>
> provided this limit exists (as a finite number).

EXAMPLE 3 | Evaluate $\int_{-\infty}^0 e^x\,dx$, if it is convergent.

SOLUTION Using Definition 2, we have

$$\int_{-\infty}^0 e^x\,dx = \lim_{a\to-\infty}\int_a^0 e^x\,dx = \lim_{a\to-\infty}\Big[e^x\Big]_a^0$$

$$= \lim_{a\to-\infty}(1 - e^a) = 1$$

because $\lim_{a\to-\infty} e^a = 0$. ■

Finally, we define an integral over the entire real line by splitting the real line into separate parts.

(3) Definition If both $\int_c^\infty f(x)\,dx$ and $\int_{-\infty}^c f(x)\,dx$ are convergent, then we define

$$\int_{-\infty}^{\infty} f(x)\,dx = \int_{-\infty}^{c} f(x)\,dx + \int_{c}^{\infty} f(x)\,dx$$

where c is any real number. (See Exercise 30).

EXAMPLE 4 | Evaluate $\displaystyle\int_{-\infty}^{\infty} \frac{1}{1+x^2}\,dx$.

SOLUTION It's convenient to choose $c = 0$ in Definition 3:

$$\int_{-\infty}^{\infty} \frac{1}{1+x^2}\,dx = \int_{-\infty}^{0} \frac{1}{1+x^2}\,dx + \int_{0}^{\infty} \frac{1}{1+x^2}\,dx$$

We must now evaluate the integrals on the right side separately:

$$\begin{aligned}\int_{0}^{\infty} \frac{1}{1+x^2}\,dx &= \lim_{b\to\infty} \int_0^b \frac{dx}{1+x^2} = \lim_{b\to\infty} \tan^{-1}x\Big]_0^b \\ &= \lim_{b\to\infty} (\tan^{-1}b - \tan^{-1}0) = \lim_{b\to\infty} \tan^{-1}b = \frac{\pi}{2}\end{aligned}$$

$$\begin{aligned}\int_{-\infty}^{0} \frac{1}{1+x^2}\,dx &= \lim_{a\to-\infty} \int_a^0 \frac{dx}{1+x^2} = \lim_{a\to-\infty} \tan^{-1}x\Big]_a^0 \\ &= \lim_{a\to-\infty} (\tan^{-1}0 - \tan^{-1}a) \\ &= 0 - \left(-\frac{\pi}{2}\right) = \frac{\pi}{2}\end{aligned}$$

Since both of these integrals are convergent, the given integral is convergent and

$$\int_{-\infty}^{\infty} \frac{1}{1+x^2}\,dx = \frac{\pi}{2} + \frac{\pi}{2} = \pi$$

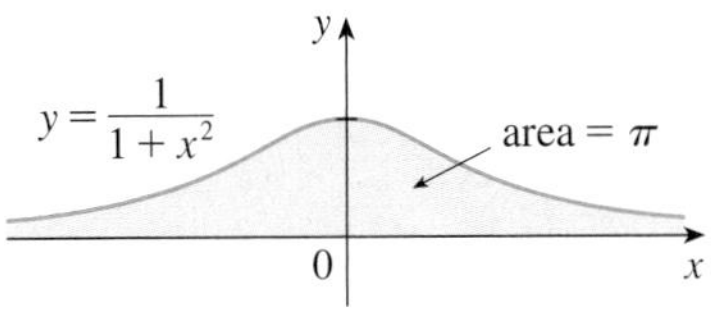

FIGURE 7

Since $1/(1 + x^2) > 0$, the given improper integral can be interpreted as the area of the infinite region that lies under the curve $y = 1/(1 + x^2)$ and above the x-axis (see Figure 7). ■

We will study continuous probability distributions in Chapter 12.

Improper integrals are used in various areas of biology, a few of which are illustrated in Example 2 and Exercises 25–28. In the study of continuous probability distributions many improper integrals arise. Probability density functions f have the property that

$$\int_{-\infty}^{\infty} f(x)\,dx = 1$$

and a basic fact for normal distributions is that

$$\int_{-\infty}^{\infty} e^{-x^2}\,dx = \sqrt{\pi}$$

(The proof of this fact requires methods that are beyond the scope of this book.)

Improper integrals also occur with great frequency in the mathematics required for medical imaging. In particular, functions defined in terms of improper integrals have become the basic tools needed to develop the theory of the CT scan, which creates images from X-ray data.

EXERCISES 5.8

1. Find the area under the curve $y = 1/x^3$ from $x = 1$ to $x = b$ and evaluate it for $b = 10$, 100, and 1000. Then find the total area under this curve for $x \geq 1$.

2. (a) Graph the functions $f(x) = 1/x^{1.1}$ and $g(x) = 1/x^{0.9}$ in the viewing rectangles $[0, 10]$ by $[0, 1]$ and $[0, 100]$ by $[0, 1]$.
(b) Find the areas under the graphs of f and g from $x = 1$ to $x = b$ and evaluate for $b = 10$, 100, 10^4, 10^6, 10^{10}, and 10^{20}.
(c) Find the total area under each curve for $x \geq 1$, if it exists.

3–22 Determine whether each integral is convergent or divergent. Evaluate those that are convergent.

3. $\displaystyle\int_3^\infty \frac{1}{(x-2)^{3/2}}\,dx$

4. $\displaystyle\int_0^\infty \frac{1}{\sqrt[4]{1+x}}\,dx$

5. $\displaystyle\int_{-\infty}^{-1} \frac{1}{\sqrt{2-w}}\,dw$

6. $\displaystyle\int_0^\infty \frac{x}{(x^2+2)^2}\,dx$

7. $\displaystyle\int_4^\infty e^{-y/2}\,dy$

8. $\displaystyle\int_{-\infty}^{-1} e^{-2t}\,dt$

9. $\displaystyle\int_{2\pi}^\infty \sin\theta\,d\theta$

10. $\displaystyle\int_{-\infty}^\infty (y^3 - 3y^2)\,dy$

11. $\displaystyle\int_{-\infty}^\infty xe^{-x^2}\,dx$

12. $\displaystyle\int_1^\infty \frac{e^{-\sqrt{x}}}{\sqrt{x}}\,dx$

13. $\displaystyle\int_1^\infty \frac{x+1}{x^2+2x}\,dx$

14. $\displaystyle\int_{-\infty}^\infty \cos\pi t\,dt$

15. $\displaystyle\int_0^\infty se^{-5s}\,ds$

16. $\displaystyle\int_{-\infty}^6 re^{r/3}\,dr$

17. $\displaystyle\int_1^\infty \frac{\ln x}{x}\,dx$

18. $\displaystyle\int_{-\infty}^\infty x^3e^{-x^4}\,dx$

19. $\displaystyle\int_{-\infty}^\infty \frac{x^2}{9+x^6}\,dx$

20. $\displaystyle\int_1^\infty \frac{\ln x}{x^3}\,dx$

21. $\displaystyle\int_e^\infty \frac{1}{x(\ln x)^3}\,dx$

22. $\displaystyle\int_0^\infty \frac{e^x}{e^{2x}+3}\,dx$

23–24 Sketch the region and find its area (if the area is finite).

23. $S = \{(x, y) \mid x \leq 1,\ 0 \leq y \leq e^x\}$

24. $S = \{(x, y) \mid x \geq -2,\ 0 \leq y \leq e^{-x/2}\}$

25. Drug pharmacokinetics The plasma drug concentration of a new drug was modeled by the function $C(t) = 23te^{-2t}$, where t is measured in hours and C in μg/mL.
(a) What is the maximum drug concentration and when did it occur?
(b) Calculate $\int_0^\infty C(t)\,dt$ and explain its significance.

BB **26. Spread of drug use** In a study of the spread of illicit drug use from an enthusiastic user to a population of N users, the authors model the number of expected new users by the equation

$$\gamma = \int_0^\infty \frac{cN(1 - e^{-kt})}{k}\,e^{-\lambda t}\,dt$$

where c, k, and λ are positive constants. Evaluate this integral to express γ in terms of c, N, k, and λ.

Source: Adapted from F. C. Hoppensteadt et al., "Threshold Analysis of a Drug Use Epidemic Model," *Mathematical Biosciences* 53 (1981): 79–87.

27. Photosynthesis Much of the earth's photosynthesis occurs in the oceans. The rate of primary production depends on light intensity, measured as the flux of photons (that is, number of photons per unit area per unit time). For monochromatic light, intensity decreases with water depth according to Beer's Law, which states that $I(x) = e^{-kx}$, where x is water depth. A simple model for the relationship between rate of photosynthesis and light intensity is $P(I) = aI$, where a is a constant and P is measured as a mass of carbon fixed per volume of water, per unit time. Calculate $\int_0^\infty P(I(x))\,dx$ and interpret it.

Source: Adapted from A. Jassby et al., "Mathematical Formulation of the Relationship between Photosynthesis and Light for Phytoplankton," *Limnology and Oceanography* 21 (1976): 540–7.

28. Dialysis treatment removes urea and other waste products from a patient's blood by diverting some of the bloodflow externally through a machine called a dialyzer. The rate at which urea is removed from the blood (in mg/min) is often well described by the equation

$$c(t) = \frac{K}{V}c_0e^{-Kt/V}$$

where K is the rate of flow of blood through the dialyzer (in mL/min), V is the volume of the patient's blood (in mL), and c_0 is the amount of urea in the blood (in mg) at time $t = 0$. Evaluate the integral $\int_0^\infty c(t)\,dt$ and interpret it.

29. A manufacturer of lightbulbs wants to produce bulbs that last about 700 hours but, of course, some bulbs burn out faster than others. Let $F(t)$ be the fraction of the company's bulbs that burn out before t hours, so $F(t)$ always lies between 0 and 1.
(a) Make a rough sketch of what you think the graph of F might look like.
(b) What is the meaning of the derivative $r(t) = F'(t)$?
(c) What is the value of $\int_0^\infty r(t)\,dt$? Why?

30. If $\int_{-\infty}^\infty f(x)\,dx$ is convergent and a and b are real numbers, show that

$$\int_{-\infty}^a f(x)\,dx + \int_a^\infty f(x)\,dx = \int_{-\infty}^b f(x)\,dx + \int_b^\infty f(x)\,dx$$

31. (a) Show that $\int_{-\infty}^\infty x\,dx$ is divergent.
(b) Show that

$$\lim_{t\to\infty} \int_{-t}^t x\,dx = 0$$

This shows that we can't define

$$\int_{-\infty}^\infty f(x)\,dx = \lim_{t\to\infty} \int_{-t}^t f(x)\,dx$$

32. For what values of p is the integral

$$\int_1^\infty \frac{1}{x^p}\,dx$$

convergent? Evaluate the integral for those values of p.

33–35 Evaluate the integral, given that

$$\int_0^\infty e^{-x^2}\,dx = \tfrac{1}{2}\sqrt{\pi}$$

33. $\int_0^\infty e^{-x^2/2}\,dx$

34. $\int_0^\infty x^2 e^{-x^2}\,dx$

35. $\int_0^\infty \sqrt{x}\,e^{-x}\,dx$

Chapter 5 REVIEW

CONCEPT CHECK

1. (a) Write an expression for a Riemann sum of a function f. Explain the meaning of the notation that you use.
(b) If $f(x) \geqslant 0$, what is the geometric interpretation of a Riemann sum? Illustrate with a diagram.
(c) If $f(x)$ takes on both positive and negative values, what is the geometric interpretation of a Riemann sum? Illustrate with a diagram.

2. (a) Write the definition of the definite integral of a continuous function from a to b.
(b) What is the geometric interpretation of $\int_a^b f(x)\,dx$ if $f(x) \geqslant 0$?
(c) What is the geometric interpretation of $\int_a^b f(x)\,dx$ if $f(x)$ takes on both positive and negative values? Illustrate with a diagram.

3. State the Midpoint Rule.

4. (a) State the Evaluation Theorem.
(b) State the Net Change Theorem.

5. If $r(t)$ is the rate of growth of a population at time t, where t is measured in months, what does $\int_6^{10} r(t)\,dt$ represent?

6. (a) Explain the meaning of the indefinite integral $\int f(x)\,dx$.
(b) What is the connection between the definite integral $\int_a^b f(x)\,dx$ and the indefinite integral $\int f(x)\,dx$?

7. State both parts of the Fundamental Theorem of Calculus.

8. (a) State the Substitution Rule. In practice, how do you use it?
(b) State the rule for integration by parts. In practice, how do you use it?

9. Define the following improper integrals.
(a) $\int_a^\infty f(x)\,dx$ (b) $\int_{-\infty}^b f(x)\,dx$ (c) $\int_{-\infty}^\infty f(x)\,dx$

10. Explain exactly what is meant by the statement that "differentiation and integration are inverse processes."

Answers to the Concept Check can be found on the back endpapers.

TRUE-FALSE QUIZ

Determine whether the statement is true or false. If it is true, explain why. If it is false, explain why or give an example that disproves the statement.

1. If f and g are continuous on $[a, b]$, then

$$\int_a^b [f(x) + g(x)]\,dx = \int_a^b f(x)\,dx + \int_a^b g(x)\,dx$$

2. If f and g are continuous on $[a, b]$, then

$$\int_a^b [f(x)g(x)]\,dx = \left(\int_a^b f(x)\,dx\right)\left(\int_a^b g(x)\,dx\right)$$

3. If f is continuous on $[a, b]$, then

$$\int_a^b 5f(x)\,dx = 5\int_a^b f(x)\,dx$$

4. If f is continuous on $[a, b]$, then

$$\int_a^b xf(x)\,dx = x\int_a^b f(x)\,dx$$

5. If f is continuous on $[a, b]$ and $f(x) \geqslant 0$, then

$$\int_a^b \sqrt{f(x)}\,dx = \sqrt{\int_a^b f(x)\,dx}$$

6. If f' is continuous on $[1, 3]$, then $\int_1^3 f'(v)\,dv = f(3) - f(1)$.

7. If f and g are continuous and $f(x) \geqslant g(x)$ for $a \leqslant x \leqslant b$, then

$$\int_a^b f(x)\,dx \geqslant \int_a^b g(x)\,dx$$

8. If f and g are differentiable and $f(x) \geqslant g(x)$ for $a < x < b$, then $f'(x) \geqslant g'(x)$ for $a < x < b$.

9. $\displaystyle\int_{-1}^{1}\left(x^5 - 6x^9 + \frac{\sin x}{(1 + x^4)^2}\right)dx = 0$

10. $\displaystyle\int_{-5}^{5}(ax^2 + bx + c)\,dx = 2\int_0^5 (ax^2 + c)\,dx$

11. $\int_0^2 (x - x^3)\,dx$ represents the area under the curve $y = x - x^3$ from 0 to 2.

12. All continuous functions have antiderivatives.

13. All continuous functions have derivatives.

14. If f is continuous, then $\int_{-\infty}^{\infty} f(x)\,dx = \lim_{t\to\infty}\int_{-t}^{t} f(x)\,dx$.

15. If f is a continuous, decreasing function on $[1, \infty)$ and $\lim_{t\to\infty} f(x) = 0$, then $\int_1^\infty f(x)\,dx$ is convergent.

16. If $\int_a^\infty f(x)\,dx$ and $\int_a^\infty g(x)\,dx$ are both convergent, then $\int_a^\infty [f(x) + g(x)]\,dx$ is convergent.

17. If $\int_a^\infty f(x)\,dx$ and $\int_a^\infty g(x)\,dx$ are both divergent, then $\int_a^\infty [f(x) + g(x)]\,dx$ is divergent.

18. If f is continuous on $[a, b]$, then

$$\frac{d}{dx}\left(\int_a^b f(x)\,dx\right) = f(x)$$

EXERCISES

1. Use the given graph of f to find the Riemann sum with six subintervals. Take the sample points to be (a) left endpoints and (b) midpoints. In each case draw a diagram and explain what the Riemann sum represents.

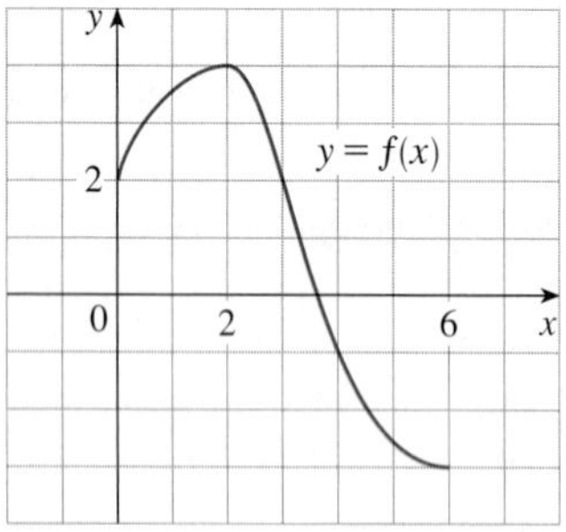

2. (a) Evaluate the Riemann sum for

$$f(x) = x^2 - x \qquad 0 \leqslant x \leqslant 2$$

with four subintervals, taking the sample points to be right endpoints. Explain, with the aid of a diagram, what the Riemann sum represents.

(b) Use the definition of a definite integral (with right endpoints) to calculate the value of the integral

$$\int_0^2 (x^2 - x)\,dx$$

(c) Use the Evaluation Theorem to check your answer to part (b).

(d) Draw a diagram to explain the geometric meaning of the integral in part (b).

3. Evaluate

$$\int_0^1 \left(x + \sqrt{1 - x^2}\right)dx$$

by interpreting it in terms of areas.

4. Express

$$\lim_{n\to\infty}\sum_{i=1}^{n} \sin x_i\,\Delta x$$

as a definite integral on the interval $[0, \pi]$ and then evaluate the integral.

5. If $\int_0^6 f(x)\,dx = 10$ and $\int_0^4 f(x)\,dx = 7$, find $\int_4^6 f(x)\,dx$.

6. (a) Write $\int_0^3 e^{-x/2}\,dx$ as a limit of Riemann sums, taking the sample points to be right endpoints.

(b) Use the Midpoint Rule with six subintervals to estimate the value of the integral in part (a). State your answer correct to three decimal places.

(c) Use the Fundamental Theorem to evaluate $\int_0^3 e^{-x/2}\,dx$. Round your answer to three decimal places and compare with your estimate in part (b).

7. The following figure shows the graphs of f, f', and $\int_0^x f(t)\,dt$. Identify each graph, and explain your choices.

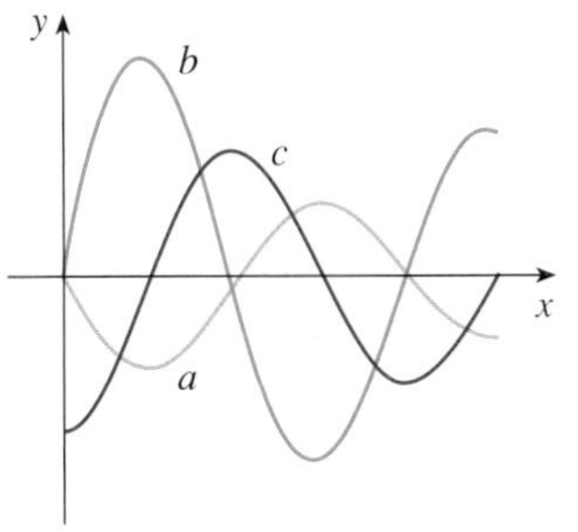

8. Evaluate:

(a) $\displaystyle\int_0^1 \frac{d}{dx}\left(e^{\arctan x}\right) dx$ (b) $\displaystyle\frac{d}{dx}\int_0^1 e^{\arctan x}\,dx$

(c) $\displaystyle\frac{d}{dx}\int_0^x e^{\arctan t}\,dt$

9–32 Evaluate the integral.

9. $\displaystyle\int_1^2 (8x^3 + 3x^2)\,dx$ **10.** $\displaystyle\int_0^T (x^4 - 8x + 7)\,dx$

11. $\displaystyle\int_0^1 (1 - x^9)\,dx$ **12.** $\displaystyle\int_0^1 (1 - x)^9\,dx$

13. $\displaystyle\int \left(\frac{1 - x}{x}\right)^2 dx$ **14.** $\displaystyle\int_0^1 (\sqrt[4]{u} + 1)^2\,du$

15. $\displaystyle\int_0^1 \frac{x}{x^2 + 1}\,dx$ **16.** $\displaystyle\int \frac{\csc^2 x}{1 + \cot x}\,dx$

17. $\displaystyle\int_0^1 v^2 \cos(v^3)\,dv$ **18.** $\displaystyle\int_0^1 \sin(3\pi t)\,dt$

19. $\displaystyle\int_0^1 e^{\pi t}\,dt$ **20.** $\displaystyle\int_1^2 \frac{1}{2 - 3x}\,dx$

21. $\displaystyle\int \frac{x + 2}{\sqrt{x^2 + 4x}}\,dx$ **22.** $\displaystyle\int_1^2 x^3 \ln x\,dx$

23. $\displaystyle\int_0^5 \frac{x}{x + 10}\,dx$ **24.** $\displaystyle\int_0^5 ye^{-0.6y}\,dy$

25. $\displaystyle\int_{-\pi/4}^{\pi/4} \frac{t^4 \tan t}{2 + \cos t}\,dt$ **26.** $\displaystyle\int_1^4 \frac{dt}{(2t + 1)^3}$

27. $\displaystyle\int_1^4 x^{3/2} \ln x\,dx$ **28.** $\displaystyle\int \sin x \cos(\cos x)\,dx$

29. $\displaystyle\int e^{\sqrt[3]{x}}\,dx$ **30.** $\displaystyle\int \tan^{-1} x\,dx$

31. $\displaystyle\int \frac{\sec\theta\,\tan\theta}{1 + \sec\theta}\,d\theta$ **32.** $\displaystyle\int_0^1 \frac{e^x}{1 + e^{2x}}\,dx$

33. Use a graph to give a rough estimate of the area of the region that lies under the curve $y = x\sqrt{x}$, $0 \leqslant x \leqslant 4$. Then find the exact area.

34–35 Find the derivative of the function.

34. $\displaystyle F(x) = \int_0^x \frac{t^2}{1 + t^3}\,dt$

35. $\displaystyle g(x) = \int_1^{\sin x} \frac{1 - t^2}{1 + t^4}\,dt$

36–38 Use the Table of Integrals on the Reference Pages to evaluate the integral.

36. $\displaystyle\int \csc^5 t\,dt$ **37.** $\displaystyle\int e^x\sqrt{1 - e^{2x}}\,dx$

38. $\displaystyle\int \frac{\cot x}{\sqrt{1 + 2\sin x}}\,dx$

39. Use Property 8 of integrals (page 338) to estimate the value of

$$\int_1^3 \sqrt{x^2 + 3}\,dx$$

40. Use the properties of integrals to verify that

$$0 \leqslant \int_0^1 x^4 \cos x\,dx \leqslant 0.2$$

41–43 Evaluate the integral or show that it is divergent.

41. $\displaystyle\int_1^\infty \frac{1}{(2x + 1)^3}\,dx$ **42.** $\displaystyle\int_0^\infty \frac{\ln x}{x^4}\,dx$

43. $\displaystyle\int_{-\infty}^0 e^{-2x}\,dx$

44. The speedometer reading v on a car was observed at one-minute intervals and recorded in the chart. Use the Midpoint Rule to estimate the distance traveled by the car.

t (min)	v (mi/h)	t (min)	v (mi/h)
0	40	6	56
1	42	7	57
2	45	8	57
3	49	9	55
4	52	10	56
5	54		

45. Let $r(t)$ be the rate at which the world's oil is consumed, where t is measured in years starting at $t = 0$ on January 1, 2000, and $r(t)$ is measured in barrels per year. What does $\int_0^{15} r(t)\,dt$ represent?

46. A **population of honeybees** increased at a rate of $r(t)$ bees per week, where the graph of r is as shown. Use the Midpoint Rule with six subintervals to estimate the increase in the bee population during the first 24 weeks.

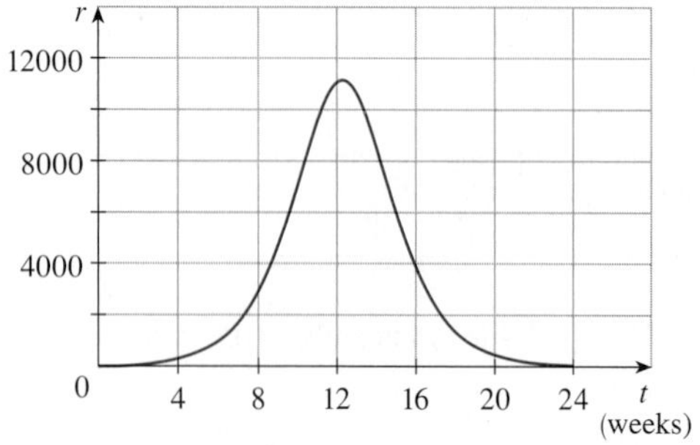

47. An oil leak from a well is causing pollution at a rate of $r(t) = 90e^{-0.12t}$ gallons per month. If the leak is never fixed, what is the total amount of oil that will be spilled?

48. Antibiotic pharmacokinetics An antibiotic tablet is taken and t hours later the concentration in the bloodstream is

$$C(t) = 3(e^{-0.8t} - e^{-1.2t})$$

where C is measured in μg/mL.

(a) What is the maximum concentration of the antibiotic and when does it occur?

(b) Calculate $\int_0^2 C(t)\,dt$ and interpret your result.

(c) Calculate $\int_0^\infty C(t)\,dt$ and explain its meaning.

49. Population dynamics Suppose that the birth and death rates in a population change through time according to the functions $b(t)$ and $d(t)$. The net rate of change is defined as $r(t) = b(t) - d(t)$.

(a) Find an expression for the net change in population size between times $t = a$ and $t = b$ in terms of $r(t)$.

(b) Use Property 4 of integrals (page 336) to show that your answer to part (a) can also be expressed in terms of the total number of births and the total number of deaths over this period.

50. Angiotensin-converting enzyme inhibitors are medications that reduce blood pressure by dilating blood vessels. The rate of change of blood pressure with respect to dosage is given by the equation

$$P'(d) = -\frac{8\eta l v R'(d)}{R(d)^3}$$

where v is blood velocity, η is blood viscosity, l is the length of the blood vessel, and $R(d)$ is the radius of the vessel as a function of the dose d. Use a substitution to integrate $P'(d)$ and show that you obtain Poiseuille's Law:

$$P(d) = \frac{4\eta l v}{R(d)^2}$$

51. Environmental pollutants In Section 10.3 a model for the transport of environmental pollutants between three lakes is analyzed. It is shown that, for certain parameter values, the concentration of pollutant in one of the lakes as a function of time is given by an equation of the form

$$x(t) = k - ke^{-at}\cos bt$$

The environmental impact of the pollutant is a function of both its concentration and the duration of time that it persists. The integral of the concentration over time is a summary measure of this impact. Calculate this measure of impact over the first unit of time.

52. Niche overlap The extent to which species compete for resources is often measured by the *niche overlap*. If the horizontal axis represents a continuum of different resource types (for example, seed sizes for certain bird species), then a plot of the degree of preference for these resources is called a *species' niche*. The degree of overlap of two species' niches is then a measure of the extent to which they compete for resources. The niche overlap for a species is the fraction of the area under its preference curve that is also under the other species' curve. Many species' niches are best modeled by a function that has a peak at some intermediate resource type and decreases to 0 asymptotically. The niches displayed in the figure are given by

$$n_1(x) = e^{-|x-8|} \qquad n_2(x) = e^{-|x-12|}$$

for species 1 and 2, respectively. Calculate the niche overlap for species 1.

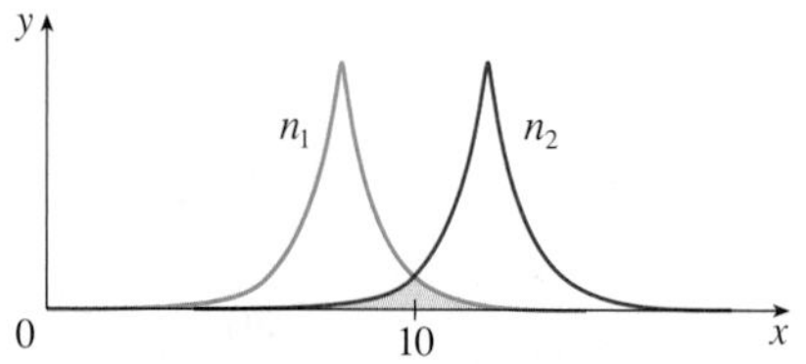

53. If f is a continuous function such that

$$\int_0^x f(t)\,dt = xe^{2x} + \int_0^x e^{-t} f(t)\,dt$$

for all x, find an explicit formula for $f(x)$.

54. Find a function f and a value of the constant a such that

$$2\int_a^x f(t)\,dt = 2\sin x - 1$$

55. If f' is continuous on $[a, b]$, show that

$$2\int_a^b f(x) f'(x)\,dx = [f(b)]^2 - [f(a)]^2$$

56. If f' is continuous on $[0, \infty)$ and $\lim_{x\to\infty} f(x) = 0$, show that

$$\int_0^\infty f'(x)\,dx = -f(0)$$

CASE STUDY 1c Kill Curves and Antibiotic Effectiveness

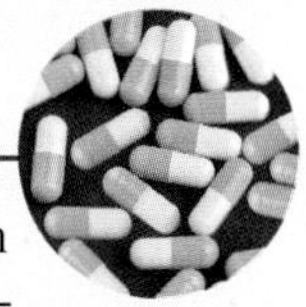

Recall that in this case study we are exploring the relationship between the magnitude of antibiotic treatment and the effectiveness of the treatment. One of the most important components of our analysis is the antibiotic concentration profile, which is a plot of the antibiotic concentration as a function of time.

In the simple model of Case Studies 1a and 1b we modeled a single dose of antibiotic using the equation

(1) $$\frac{dc}{dt} = -kc$$

for some positive constant k. From this we saw that the concentration as a function of time is

(2) $$c(t) = c_0e^{-kt}$$

where c_0 is the concentration at $t = 0$. (See Figure 1.)

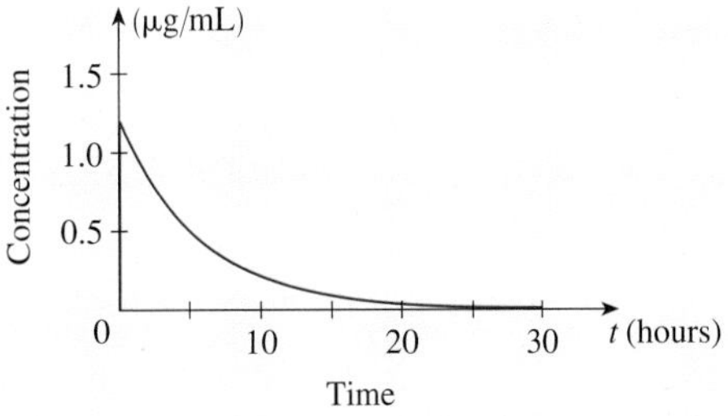

FIGURE 1
Antibiotic concentration profile modeled by the function $c(t) = c_0e^{-kt}$ with $c_0 = 1.2$ μg/mL and $k = 0.175$

Three of the most common measures of the magnitude of antibiotic treatment are (1) the peak antibiotic concentration divided by *MIC*, denoted by ρ; (2) the duration of time for which the antibiotic concentration remains above *MIC*, denoted by τ; and (3) the area under the antibiotic concentration profile divided by *MIC*, denoted by α. These are shown in Figure 2. In Case Study 1a you derived expressions for the first two.

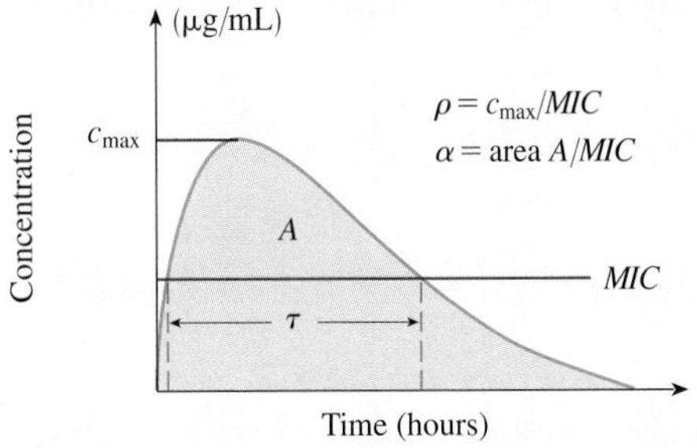

FIGURE 2
Three measures of the magnitude of antibiotic treatment

1. Find an expression for α in terms of k, c_0, and *MIC*, using Equation 2. The area under the concentration profile should be calculated from $t = 0$ to ∞.

We saw in Case Study 1a that, for a given antibiotic and bacteria species (in other words, for a given value of k and *MIC*), all three quantities ρ, τ, and α increase with one another. In other words, it is not possible to have a high value of α without also having high values of ρ and τ. Here you will show that we can break this dependency if we divide the total amount of antibiotic given c_0 into multiple smaller doses. This is referred to as *dose fractionation*.

In the simplest case, suppose that instead of giving a total amount of c_0 μg/mL of antibiotic at $t = 0$, we instead give $c_0/2$ at $t = 0$ and another dose of $c_0/2$ at time $t = \hat{t}$. The time $\hat{t}$ is called the interdose interval. Furthermore, suppose that at each dose the concentration instantly increases by $c_0/2$, and otherwise it decays according to Equation 1.

2. Find an equation for the concentration as a function of time. Figure 3 plots this function for a specific choice of constants, along with the concentration profile when a single dose of c_0 μg/mL is given at $t = 0$.

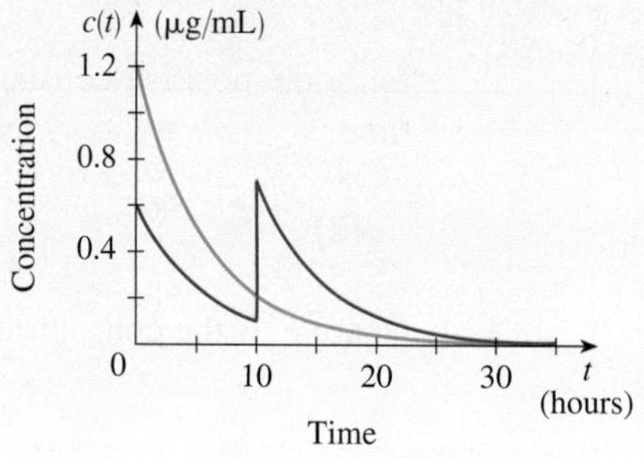

FIGURE 3
Red curve is the concentration profile modeled by the function from Problem 2 with $c_0 = 1.2$ μg/mL, $\hat{t} = 10$, and $k = 0.175$. Blue curve is the concentration profile when all the antibiotic is given at $t = 0$.

3. Use your answer to Problem 2 to find an expression for ρ, the peak concentration divided by MIC.

4. Use your answer to Problem 2 to show that α is the same under dose fractionation as it is for the single dose case in Problem 1.

5. Using your answers from Problems 3 and 4, explain how it is possible to use dose fractionation to increase ρ without also increasing α.

One of the reasons different drug doses and interdose intervals are used for different infections is to achieve different values of α, ρ, and τ.

Applications of Integrals

6

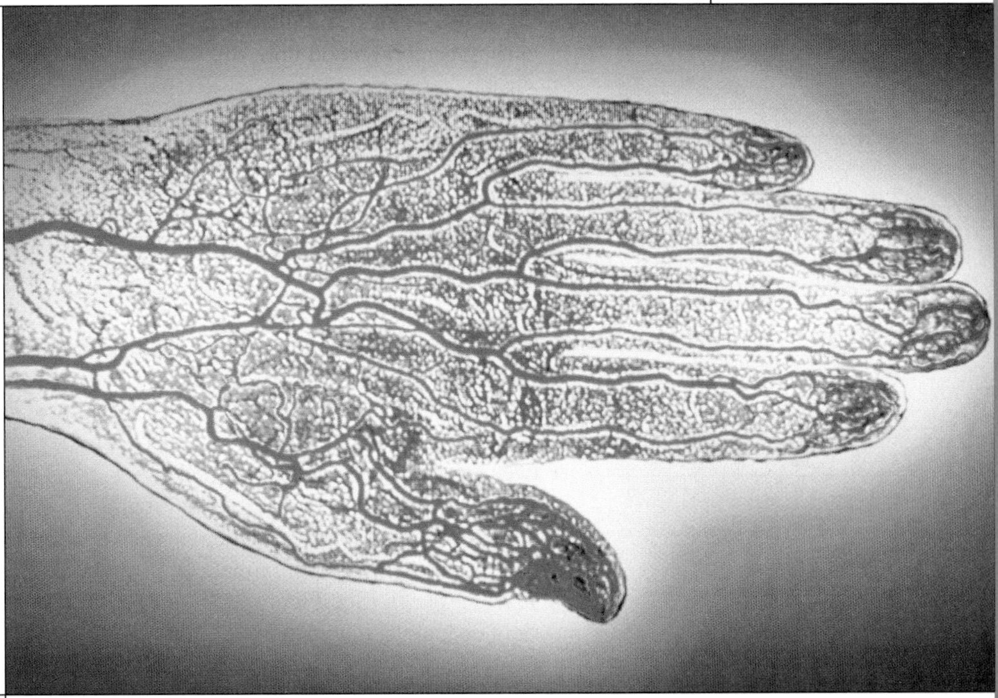

Arteries of the human hand are shown in a colorized X ray. In Section 6.3 we use an integral to calculate the flux in an artery (the volume of blood that passes a cross-section of an artery per unit time).

GJLP / CNRI / SPL / Science Source

WE HAVE ALREADY SEEN some of the applications of integrals in Chapter 5: drug pharmacokinetics, measles pathogenesis, bacterial growth, basal metabolism, breathing cycles, dialysis treatment, tumor growth, photosynthesis, medical imaging, bioavailability, the sterile insect technique, the rumen microbial ecosystem, and the spread of drug use.

In this chapter we explore some additional applications of integration: areas between curves, cerebral blood flow, disease progresssion, average values, survival and renewal, cardiac output, and volumes.

6.1 Areas Between Curves

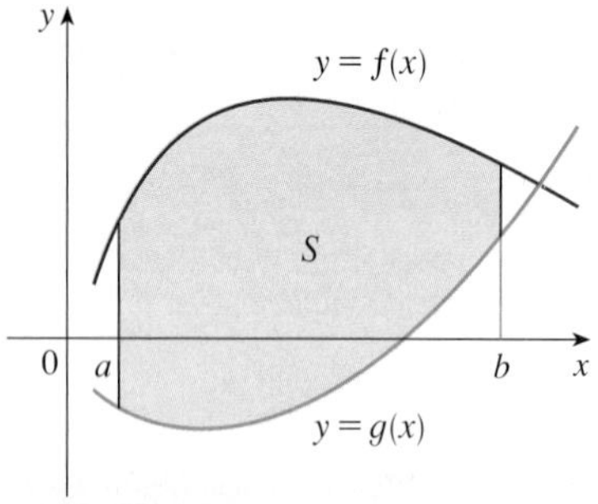

FIGURE 1
$S = \{(x, y) \mid a \leq x \leq b, g(x) \leq y \leq f(x)\}$

In Chapter 5 we defined and calculated areas of regions that lie under the graphs of functions. Here we use integrals to find areas of regions that lie between the graphs of two functions and then we show how this idea occurs in cerebral blood flow. As well, in the project after this section we see how the study of measles pathogenesis involves areas between curves.

Consider the region S that lies between two curves $y = f(x)$ and $y = g(x)$ and between the vertical lines $x = a$ and $x = b$, where f and g are continuous functions and $f(x) \geq g(x)$ for all x in $[a, b]$. (See Figure 1.)

Just as we did for areas under curves in Section 5.1, we divide S into n strips of equal width and then we approximate the ith strip by a rectangle with base Δx and height $f(x_i^*) - g(x_i^*)$. (See Figure 2. If we like, we could take all of the sample points to be right endpoints, in which case $x_i^* = x_i$.) The Riemann sum

$$\sum_{i=1}^{n} [f(x_i^*) - g(x_i^*)]\,\Delta x$$

is therefore an approximation to what we intuitively think of as the area of S.

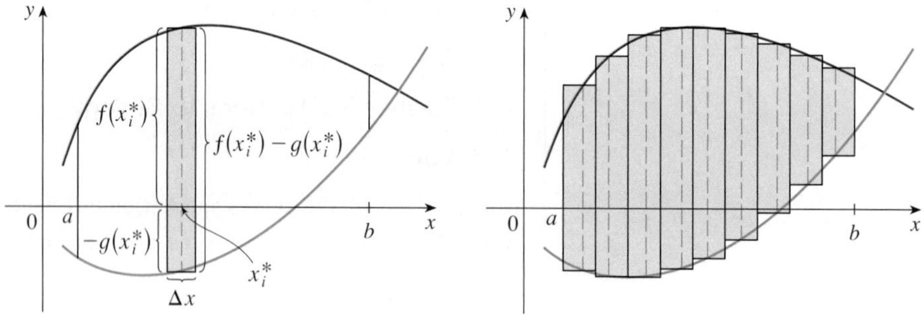

FIGURE 2 (a) Typical rectangle (b) Approximating rectangles

This approximation appears to become better and better as $n \to \infty$. Therefore we define the **area** A of the region S as the limiting value of the sum of the areas of these approximating rectangles.

(1)
$$A = \lim_{n \to \infty} \sum_{i=1}^{n} [f(x_i^*) - g(x_i^*)]\,\Delta x$$

We recognize the limit in (1) as the definite integral of $f - g$. Therefore we have the following formula for area.

> **(2)** The area A of the region bounded by the curves $y = f(x)$, $y = g(x)$, and the lines $x = a$, $x = b$, where f and g are continuous and $f(x) \geqslant g(x)$ for all x in $[a, b]$, is
>
> $$A = \int_a^b [f(x) - g(x)]\, dx$$

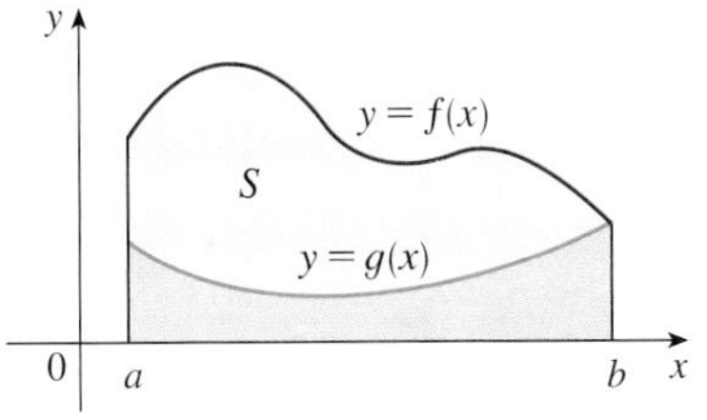

FIGURE 3

$A = \int_a^b f(x)\,dx - \int_a^b g(x)\,dx$

Notice that in the special case where $g(x) = 0$, S is the region under the graph of f and our general definition of area (1) reduces to our previous definition (Definition 5.1.2).

In the case where both f and g are positive, you can see from Figure 3 why (2) is true:

$$A = [\text{area under } y = f(x)] - [\text{area under } y = g(x)]$$
$$= \int_a^b f(x)\, dx - \int_a^b g(x)\, dx = \int_a^b [f(x) - g(x)]\, dx$$

EXAMPLE 1 | Find the area of the region bounded above by $y = e^x$, bounded below by $y = x$, and bounded on the sides by $x = 0$ and $x = 1$.

SOLUTION The region is shown in Figure 4. The upper boundary curve is $y = e^x$ and the lower boundary curve is $y = x$. So we use the area formula (2) with $f(x) = e^x$, $g(x) = x$, $a = 0$, and $b = 1$:

$$A = \int_0^1 (e^x - x)\, dx = e^x - \tfrac{1}{2}x^2 \Big]_0^1$$
$$= e - \tfrac{1}{2} - 1 = e - 1.5$$

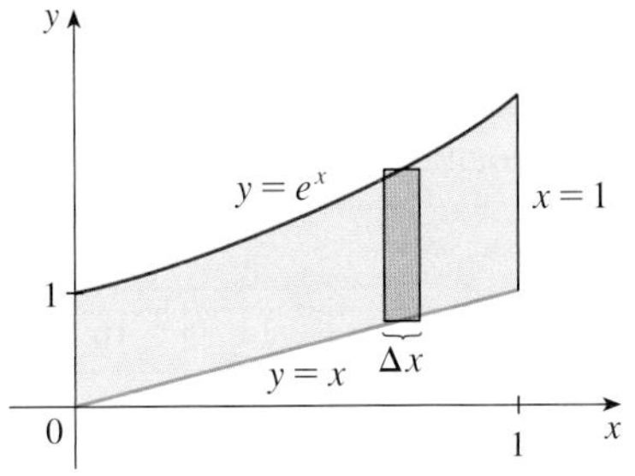

FIGURE 4

In Figure 4 we drew a typical approximating rectangle with width Δx as a reminder of the procedure by which the area is defined in (1). In general, when we set up an integral for an area, it's helpful to sketch the region to identify the top curve y_T, the bottom curve y_B, and a typical approximating rectangle as in Figure 5. Then the area of a typical rectangle is $(y_T - y_B)\,\Delta x$ and the equation

$$A = \lim_{n \to \infty} \sum_{i=1}^{n} (y_T - y_B)\,\Delta x = \int_a^b (y_T - y_B)\, dx$$

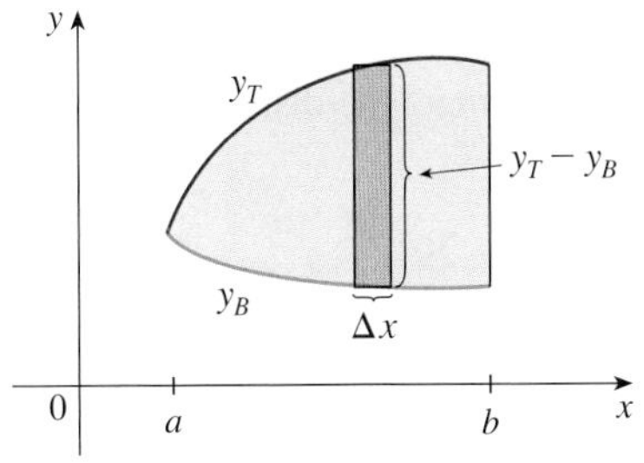

FIGURE 5

summarizes the procedure of adding (in a limiting sense) the areas of all the typical rectangles.

Notice that in Figure 5 the left-hand boundary reduces to a point, whereas in Figure 3 the right-hand boundary reduces to a point. In the next example both of the side boundaries reduce to a point, so the first step is to find a and b.

EXAMPLE 2 | Find the area of the region enclosed by the parabolas $y = x^2$ and $y = 2x - x^2$.

SOLUTION We first find the points of intersection of the parabolas by solving their equations simultaneously. This gives $x^2 = 2x - x^2$, or $2x^2 - 2x = 0$. Thus $2x(x - 1) = 0$, so $x = 0$ or 1. The points of intersection are $(0, 0)$ and $(1, 1)$.

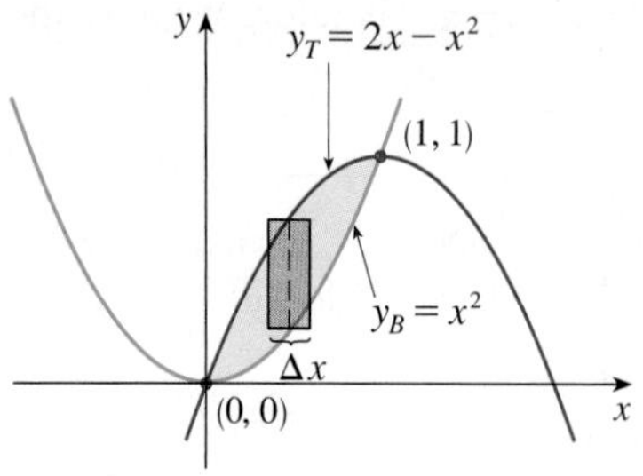

FIGURE 6

We see from Figure 6 that the top and bottom boundaries are

$$y_T = 2x - x^2 \quad \text{and} \quad y_B = x^2$$

The area of a typical rectangle is

$$(y_T - y_B)\,\Delta x = (2x - x^2 - x^2)\,\Delta x = (2x - 2x^2)\,\Delta x$$

and the region lies between $x = 0$ and $x = 1$. So the total area is

$$A = \int_0^1 (2x - 2x^2)\,dx = 2\int_0^1 (x - x^2)\,dx$$

$$= 2\left[\frac{x^2}{2} - \frac{x^3}{3}\right]_0^1 = 2\left(\frac{1}{2} - \frac{1}{3}\right) = \frac{1}{3}$$

■

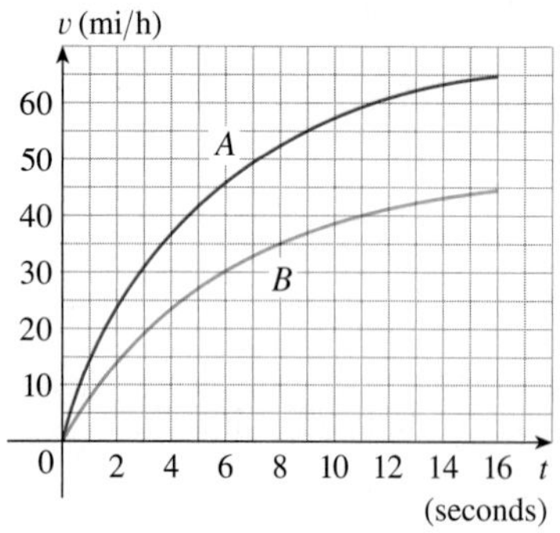

FIGURE 7

EXAMPLE 3 | **Interpreting the area between velocity curves** Figure 7 shows velocity curves for two cars, A and B, that start side by side and move along the same road. What does the area between the curves represent? Use the Midpoint Rule to estimate it.

SOLUTION We know from Section 5.3 that the area under the velocity curve A represents the distance traveled by car A during the first 16 seconds. Similarly, the area under curve B is the distance traveled by car B during that time period. So the area between these curves, which is the difference of the areas under the curves, is the distance between the cars after 16 seconds. We read the velocities from the graph and convert them to feet per second ($1 \text{ mi/h} = \frac{5280}{3600} \text{ ft/s}$).

t	0	2	4	6	8	10	12	14	16
v_A	0	34	54	67	76	84	89	92	95
v_B	0	21	34	44	51	56	60	63	65
$v_A - v_B$	0	13	20	23	25	28	29	29	30

We use the Midpoint Rule with $n = 4$ intervals, so that $\Delta t = 4$. The midpoints of the intervals are $\bar{t}_1 = 2$, $\bar{t}_2 = 6$, $\bar{t}_3 = 10$, and $\bar{t}_4 = 14$. We estimate the distance between the cars after 16 seconds as follows:

$$\int_0^{16} (v_A - v_B)\,dt \approx \Delta t\,[13 + 23 + 28 + 29]$$

$$= 4(93) = 372 \text{ ft}$$

■

■ Cerebral Blood Flow

In a paper[1] published in 1948, Seymour Kety and Carl Schmidt described a method for measuring cerebral blood flow in which the patient inhales a mixture of gases including a tracer of 15% nitrous oxide. Let $A(t)$ be the arterial concentration of N_2O measured as blood enters the brain and $V(t)$ the venous concentration of N_2O in blood flowing out of the brain in the jugular vein. Figure 8 shows typical graphs of $A(t)$ and $V(t)$, which are measured in units of mL of N_2O per mL of blood. Although $A(t) > V(t)$, you can see

1. S. Kety et al., "The Nitrous Oxide Method for the Quantitative Determination of Cerebral Blood Flow in Man: Theory, Procedure and Normal Values," *Journal of Clinical Investigation* 27 (1948): 476–83.

that after about 10 minutes $A(t)$ and $V(t)$ are almost the same because the brain is becoming saturated with nitrous oxide.

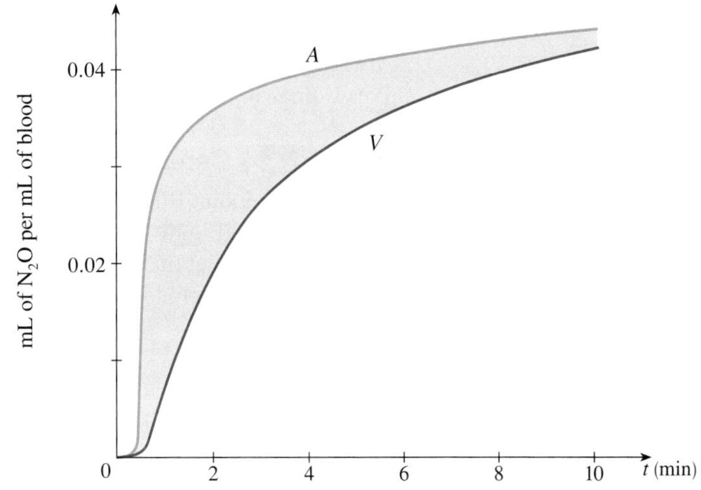

FIGURE 8
Arterial and venous concentrations of N_2O

The area between the curves A and V, shaded in Figure 8, plays a key role in calculating the **cerebral blood flow** F, which we measure in mL/min. We first consider $\int_0^{10} A(t)\,dt$. If we divide the interval $[0, 10]$ into subintervals of equal length Δt, then the volume of N_2O that flows past a point in the artery during such a subinterval from $t = t_{i-1}$ to $t = t_i$ is approximately

$$\text{(concentration of N}_2\text{O in blood)} \cdot \text{(volume of blood)} = A(t_i)(F\,\Delta t)$$

Assuming that F remains constant, we see that the total volume of N_2O that enters the brain during the first 10 minutes is approximately

$$\sum_{i=1}^{n} A(t_i)\,F\,\Delta t = F \sum_{i=1}^{n} A(t_i)\,\Delta t$$

If we now let $n \to \infty$, we get the total quantity of N_2O brought to the brain during the first 10 minutes:

$$F \int_0^{10} A(t)\,dt$$

A similar argument shows that the quantity of N_2O that leaves the brain during this time period is

$$F \int_0^{10} V(t)\,dt$$

The difference of these quantities

$$F \int_0^{10} [A(t) - V(t)]\,dt$$

is therefore the quantity of N_2O that is taken up by the whole brain during the 10 minutes of inhalation. Let's call this quantity $Q_B(10)$. Then

$$Q_B(10) = F \int_0^{10} [A(t) - V(t)]\,dt$$

and therefore

(3)
$$F = \frac{Q_B(10)}{\int_0^{10} [A(t) - V(t)]\, dt}$$

It turns out that $Q_B(10)$ can be found by other methods, so if the area between the curves is known, Equation 3 can then be used to calculate the cerebral blood flow.

EXAMPLE 4 | Cerebral blood flow

(a) Use the Midpoint Rule with five subintervals to estimate the area between the curves A and V in Figure 8.
(b) If it is known that the amount of N_2O absorbed by the brain is $Q_B(10) = 60$ mL, determine the cerebral blood flow.

SOLUTION

(a) We divide the interval $[0, 10]$ into five subintervals, whose midpoints are $t = 1, 3, 5, 7$, and 9. Then we use the graphs in Figure 8 to estimate the values of $A(t)$ and $V(t)$ at these midpoints and calculate their differences:

t	$A(t)$	$V(t)$	$A(t) - V(t)$
1	0.029	0.007	0.022
3	0.038	0.027	0.011
5	0.040	0.033	0.007
7	0.042	0.037	0.005
9	0.043	0.041	0.002

Using the Midpoint Rule with $\Delta t = 2$ min, we get the following estimate:

$$\int_0^{10} [A(t) - V(t)]\, dt \approx 2[0.022 + 0.011 + 0.007 + 0.005 + 0.002] = 0.094$$

So the area between the curves A and V is approximately 0.094 (mL/mL) · min.
(b) With $Q_B(10) = 60$ mL and our result from part (a), we use Equation 3 to calculate the cerebral blood flow:

$$F = \frac{Q_B(10)}{\int_0^{10} [A(t) - V(t)]\, dt} \approx \frac{60}{0.094} \approx 640 \text{ mL/min}$$

■

EXERCISES 6.1

1–2 Find the area of the shaded region.

1.

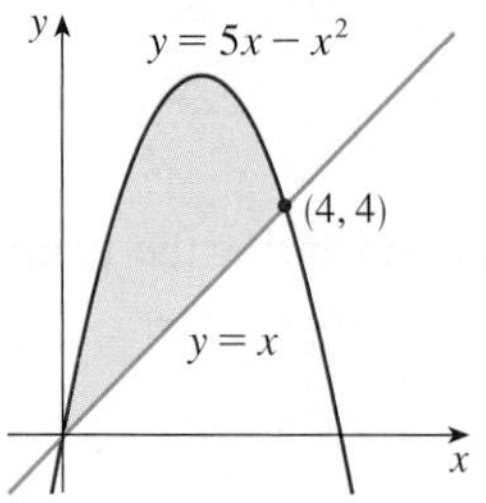

2.

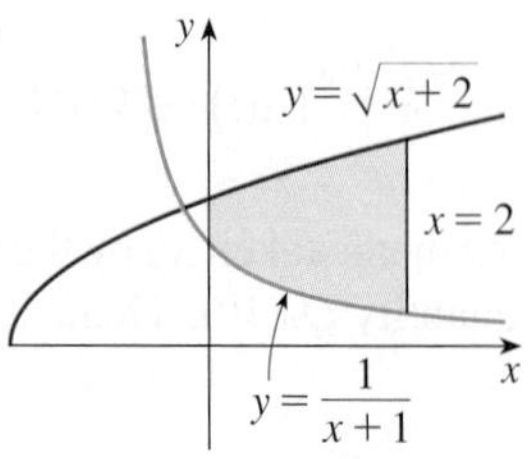

3–6 Sketch the region enclosed by the given curves. Draw a typical approximating rectangle and label its height and width. Then find the area of the region.

3. $y = e^x, \quad y = x^2 - 1, \quad x = -1, \quad x = 1$

4. $y = \ln x, \quad xy = 4, \quad x = 1, \quad x = 3$

5. $y = x^2, \quad y^2 = x$

6. $y = x^2 - 2x, \quad y = x + 4$

7–10 Sketch the region enclosed by the given curves and find its area.

7. $y = 12 - x^2, \quad y = x^2 - 6$

8. $y = x^2, \quad y = 4x - x^2$

9. $y = e^x, \quad y = xe^x, \quad x = 0$

10. $y = \cos x, \quad y = 2 - \cos x, \quad 0 \leqslant x \leqslant 2\pi$

11–12 Use a graph to find approximate x-coordinates of the points of intersection of the given curves. Then find (approximately) the area of the region bounded by the curves.

11. $y = x\sin(x^2), \quad y = x^4$

12. $y = x\cos x, \quad y = x^{10}$

---

13. Sketch the region that lies between the curves $y = \cos x$ and $y = \sin 2x$ and between $x = 0$ and $x = \pi/2$. Notice that the region consists of two separate parts. Find the area of this region.

14. Sketch the curves $y = \cos x$ and $y = 1 - \cos x$, $0 \leqslant x \leqslant \pi$, and observe that the region between them consists of two separate parts. Find the area of this region.

15. Sometimes it's easier to find an area by regarding x as a function of y instead of y as a function of x. To illustrate this idea, let S be the region enclosed by the line $y = x - 1$ and the parabola $y^2 = 2x + 6$.

(a) By sketching S, observe that if you want to integrate with respect to x you have to split S into two parts with different boundary curves.

(b) If you integrate with respect to y, observe that there is a left boundary curve and a right boundary curve.

(c) Find the area of S using the method of either part (a) or part (b).

16. Find the area of the region enclosed by the curves $y = x$ and $4x + y^2 = 12$.

17. A **laurel leaf** is shown. Estimate its area using the Midpoint Rule with six subintervals.

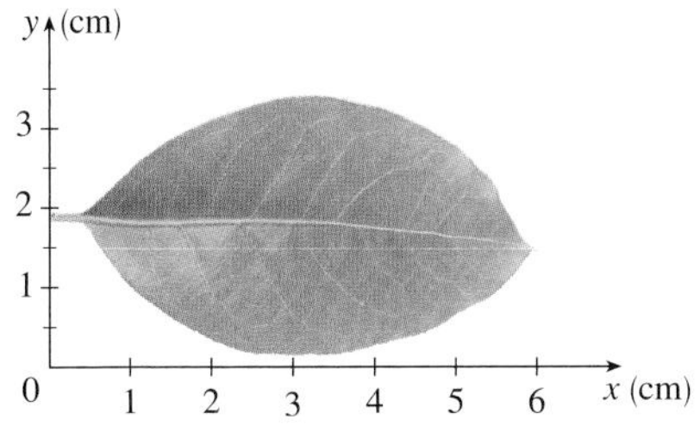

© Vasilius / Shutterstock.com

18. A **cicada wing** is shown. Estimate its area using the Midpoint Rule with six subintervals.

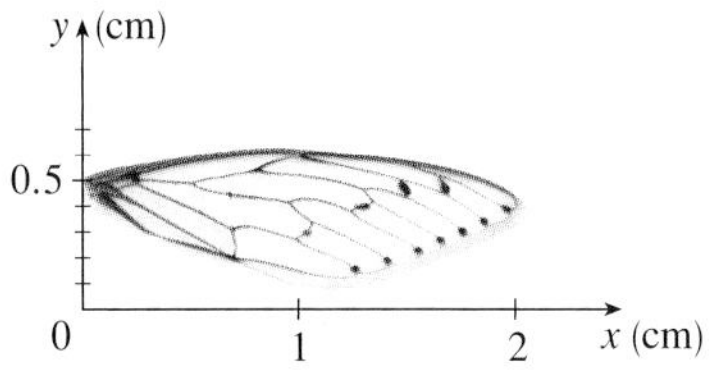

© Tropper2000 / Shutterstock.com

19. A cross-section of an airplane wing is shown. Measurements of the thickness of the wing, in centimeters, at 20-centimeter intervals are 5.8, 20.3, 26.7, 29.0, 27.6, 27.3, 23.8, 20.5, 15.1, 8.7, and 2.8. Use the Midpoint Rule to estimate the area of the wing's cross-section.

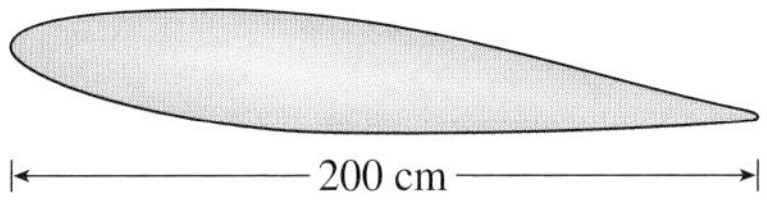

20. The widths (in meters) of a kidney-shaped swimming pool were measured at 2-meter intervals as indicated in the figure. Use the Midpoint Rule to estimate the area of the pool.

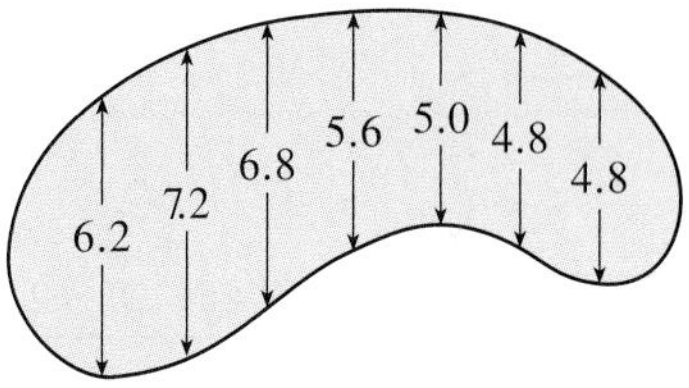

21. Cerebral blood flow The table shows measurements of $A(t)$, the concentration of N_2O flowing into a patient's brain, and $V(t)$, the concentration of N_2O flowing out of the brain, where t is measured in minutes and $A(t)$ and $V(t)$ are measured in mL of N_2O per mL of blood.

t	$A(t)$	$V(t)$
1	0.031	0.008
3	0.041	0.029
5	0.042	0.035
7	0.044	0.042
9	0.045	0.044

(a) Use the Midpoint Rule to estimate $\int_0^{10} [A(t) - V(t)]\,dt$.

(b) If the volume of N_2O absorbed by the brain in the first 10 minutes is 64 mL, calculate the cerebral blood flow.

22. Cerebral blood flow Models for the arterial and venous concentration functions in Figure 8 are given by

$$A(t) = \frac{0.05t^2}{t^2 + 1} \qquad V(t) = \frac{0.05t^2}{t^2 + 7}$$

(a) Find the area between the graphs of A and V for $0 \leqslant t \leqslant 10$.
(b) If the volume of N_2O absorbed by the brain in the first 10 minutes is 60 mL, determine the cerebral blood flow.

23. Racing cars driven by Chris and Kelly are side by side at the start of a race. The table shows the velocities of each car (in miles per hour) during the first 10 seconds of the race. Use the Midpoint Rule to estimate how much farther Kelly travels than Chris does during the first 10 seconds.

t	v_C	v_K	t	v_C	v_K
0	0	0	6	69	80
1	20	22	7	75	86
2	32	37	8	81	93
3	46	52	9	86	98
4	54	61	10	90	102
5	62	71			

24. Two cars, A and B, start side by side and accelerate from rest. The figure shows the graphs of their velocity functions.
(a) Which car is ahead after one minute? Explain.
(b) What is the meaning of the area of the shaded region?
(c) Which car is ahead after two minutes? Explain.
(d) Estimate the time at which the cars are again side by side.

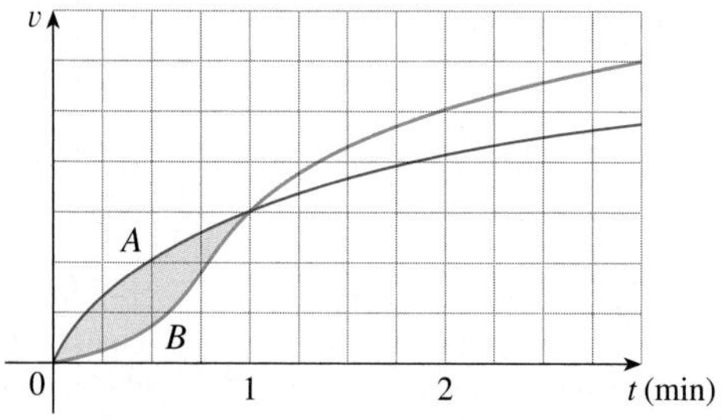

25. Birth and death rates If the birth rate of a population is $b(t) = 2200e^{0.024t}$ people per year and the death rate is $d(t) = 1460e^{0.018t}$ people per year, find the area between these curves for $0 \leqslant t \leqslant 10$. What does this area represent?

26. Find the number a such that the line $x = a$ bisects the area under the curve $y = 1/x^2$, $1 \leqslant x \leqslant 4$.

27. Find the values of c such that the area of the region bounded by the parabolas $y = x^2 - c^2$ and $y = c^2 - x^2$ is 576.

28. Find the area of the region bounded by the parabola $y = x^2$, the tangent line to this parabola at $(1, 1)$, and the x-axis.

PROJECT Disease Progression and Immunity

BB

In Section 5.1 we considered the progression of measles in a patient with no immunity and graphed the measles pathogenesis curve $N = f(t)$ (page 325), which we modeled with the function

$$f(t) = -t(t - 21)(t + 1)$$

We saw that the total amount of infection by time t (measured by the area under the pathogenesis curve up to time t) played an important role in determining whether individuals develop symptoms. In particular, symptoms appear only after the total amount of infection exceeds 7848 (cells/mL) $\times$ days, which occurs at day 12 of the infection for individuals with no immunity, that is,

$$\int_0^{12} f(t)\,dt = 7848 \text{ (cells/mL)} \times \text{days}$$

1. Plot the curves $cf(t)$ for $c = 0.9, 0.85, 0.8, 0.6$, and 0.4. These resemble curves for patients that have increasing levels of immunity against the virus at the time of the infection.

2. Some of the patients in Problem 1 will develop symptoms and some will not. Find the areas under the curve from $t = 0$ to $t = 21$ for each value of c and compare them with the value 7848 that is needed to display symptoms. Which patients will become symptomatic at some point during their infection?

The term *infectiousness* refers to the extent to which the disease is transmitted between individuals. For patients without immunity, we saw in Section 5.1 that infectiousness begins around day $t_1 = 10$ and ends around day $t_2 = 18$. Infectiousness begins on day 10 because the concentration of virus in the plasma after 10 days [that is, the value $f(10)$] is the threshold concentration required before any transmission can occur. Further, infectiousness ends on day 18 because the immune system manages to prevent further transmission from this point onward.

3. Plot the points $P_1 = (t_1, f(t_1))$ and $P_2 = (t_2, f(t_2))$ on the graph of f. These points show the values of f at the beginning and end of the infectious period. Draw a line between the points. What is the slope of this line? Find an equation of this line.

4. Given that $L = f(t_1)$ is the threshold concentration of the virus required for transmission to begin, plot the point P_3 on the curve $N = 0.9f(t)$ where $0.9f(t) = L$. The value of t satisfying this equation is the time at which infectiousness begins for a patient with $c = 0.9$. It has been shown that the time at which infectiousness ends for such patients can again be determined by drawing a line through P_3 with the same slope as that in Problem 3 and then determining the time t_4 at which it intersects the curve $N = 0.9f(t)$. Draw this line on the graph and determine t_4.

5. Repeat Problem 4 for $cf(t)$ with $c = 0.85$, 0.8, 0.6, and 0.4. Note that some patients may not have a point corresponding to P_3.

6. Find the area between the graph of f and the line in Problem 3. This area represents the total level of infectiousness of an infected person. (See Figure 1.)

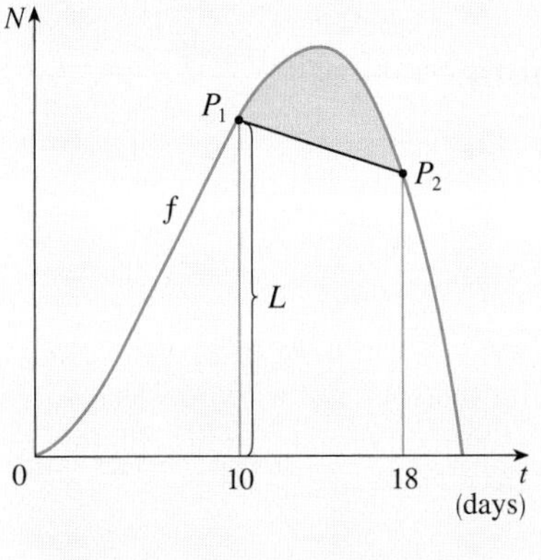

FIGURE 1

7. Find the areas enclosed by the curves $N = cf(t)$ and their corresponding intersecting lines for $c = 0.9$, 0.85, 0.8, 0.6, and 0.4. (Note that this will not be feasible for patients that have no infectious period.) Compare these areas to the one found in Problem 6.)

8. Which patients are
(a) symptomatic and infectious?
(b) symptomatic and noninfectious?
(c) asymptomatic and noninfectious?

PROJECT The Gini Index

How is it possible to measure the distribution of income among the inhabitants of a given country? One such measure is the *Gini index*, named after the Italian economist Corrado Gini, who first devised the index in 1912.

We first rank all households in a country by income and then we compute the percentage of households whose income is at most a given percentage of the country's total income. We define a **Lorenz curve** $y = L(x)$ on the interval $[0, 1]$ by plotting the point $(a/100, b/100)$ on the curve if the bottom $a\%$ of households receive at most $b\%$ of the

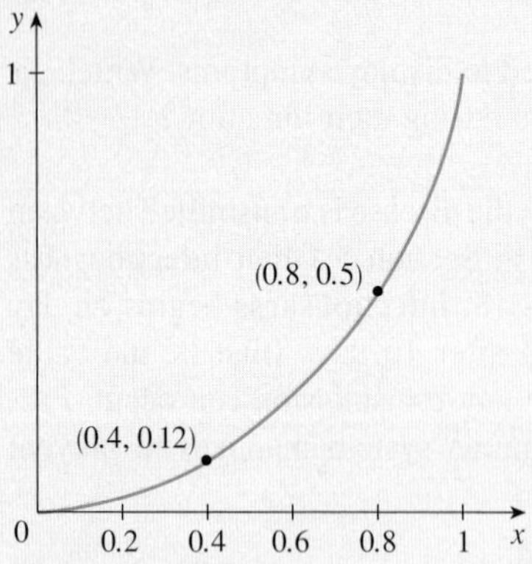

FIGURE 1
Lorenz curve for the United States in 2010

total income. For instance, in Figure 1 the point (0.4, 0.12) is on the Lorenz curve for the United States in 2010 because the poorest 40% of the population received just 12% of the total income. Likewise, the bottom 80% of the population received 50% of the total income, so the point (0.8, 0.5) lies on the Lorenz curve. (The Lorenz curve is named after the American economist Max Lorenz.)

Figure 2 shows some typical Lorenz curves. They all pass through the points (0, 0) and (1, 1) and are concave upward. In the extreme case $L(x) = x$, society is perfectly egalitarian: The poorest a% of the population receives a% of the total income and so everybody receives the same income. The area between a Lorenz curve $y = L(x)$ and the line $y = x$ measures how much the income distribution differs from absolute equality. The **Gini index** (sometimes called the **Gini coefficient** or the **coefficient of inequality**) is the area between the Lorenz curve and the line $y = x$ (shaded in Figure 3) divided by the area under $y = x$.

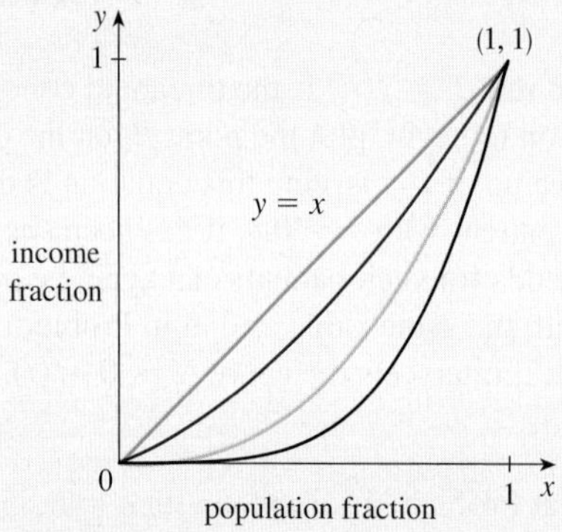

FIGURE 2

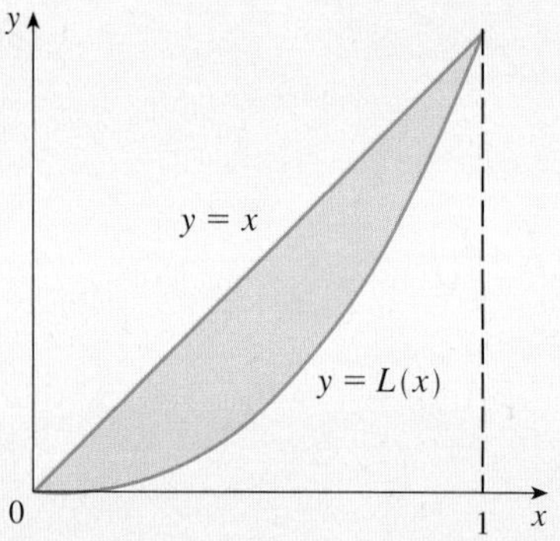

FIGURE 3

1. (a) Show that the Gini index G is twice the area between the Lorenz curve and the line $y = x$, that is,

$$G = 2\int_0^1 [x - L(x)]\,dx$$

(b) What is the value of G for a perfectly egalitarian society (everybody has the same income)? What is the value of G for a perfectly totalitarian society (a single person receives all the income)?

2. The following table (derived from data supplied by the US Census Bureau) shows values of the Lorenz function for income distribution in the United States for the year 2010.

x	0.0	0.2	0.4	0.6	0.8	1.0
$L(x)$	0.000	0.034	0.120	0.266	0.498	1.000

(a) What percentage of the total US income was received by the richest 20% of the population in 2010?

(b) Use a calculator or computer to fit a quadratic function to the data in the table. Graph the data points and the quadratic function. Is the quadratic model a reasonable fit?

(c) Use the quadratic model for the Lorenz function to estimate the Gini index for the United States in 2010.

3. The following table gives values for the Lorenz function in the years 1970, 1980, 1990, and 2000. Use the method of Problem 2 to estimate the Gini index for the United States for those years and compare with your answer to Problem 2(c). Do you notice a trend?

x	0.0	0.2	0.4	0.6	0.8	1.0
1970	0.000	0.041	0.149	0.323	0.568	1.000
1980	0.000	0.042	0.144	0.312	0.559	1.000
1990	0.000	0.038	0.134	0.293	0.530	1.000
2000	0.000	0.036	0.125	0.273	0.503	1.000

CAS **4.** A power model often provides a more accurate fit than a quadratic model for a Lorenz function. If you have a computer with Maple or Mathematica, fit a power function ($y = ax^k$) to the data in Problem 2 and use it to estimate the Gini index for the United States in 2010. Compare with your answer to parts (b) and (c) of Problem 2.

6.2 Average Values

It is easy to calculate the average value of finitely many numbers $y_1, y_2, \ldots, y_n$:

$$y_{\text{ave}} = \frac{y_1 + y_2 + \cdots + y_n}{n}$$

But how do we compute the average temperature during a day if infinitely many temperature readings are possible? Figure 1 shows the graph of a temperature function $T(t)$, where t is measured in hours and T in °C, and a guess at the average temperature, T_{ave}.

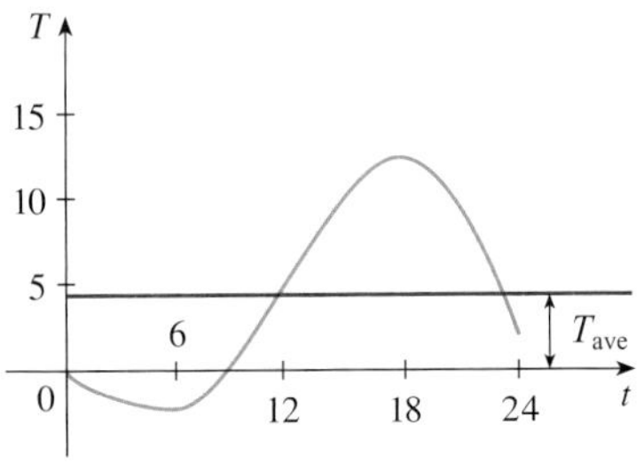

FIGURE 1

In general, let's try to compute the average value of a function $y = f(x)$, $a \leqslant x \leqslant b$. We start by dividing the interval $[a, b]$ into n equal subintervals, each with length $\Delta x = (b - a)/n$. Then we choose points $x_1^*, \ldots, x_n^*$ in successive subintervals and calculate the average of the numbers $f(x_1^*), \ldots, f(x_n^*)$:

$$\frac{f(x_1^*) + \cdots + f(x_n^*)}{n}$$

(For example, if f represents a temperature function and $n = 24$, this means that we take temperature readings every hour and then average them.) Since $\Delta x = (b - a)/n$, we can write $n = (b - a)/\Delta x$ and the average value becomes

$$\frac{f(x_1^*) + \cdots + f(x_n^*)}{\dfrac{b-a}{\Delta x}} = \frac{1}{b-a}\left[f(x_1^*)\,\Delta x + \cdots + f(x_n^*)\,\Delta x\right]$$

$$= \frac{1}{b-a}\sum_{i=1}^{n} f(x_i^*)\,\Delta x$$

If we let n increase, we would be computing the average value of a large number of closely spaced values. (For example, we would be averaging temperature readings taken every minute or even every second.) The limiting value is

$$\lim_{n\to\infty} \frac{1}{b-a} \sum_{i=1}^{n} f(x_i^*)\,\Delta x = \frac{1}{b-a} \int_a^b f(x)\,dx$$

by the definition of a definite integral.

Therefore we define the **average value of f** on the interval $[a, b]$ as

For a positive function, we can think of this definition as saying

$$\frac{\text{area}}{\text{width}} = \text{average height}$$

$$\boxed{f_{\text{ave}} = \frac{1}{b-a} \int_a^b f(x)\,dx}$$

EXAMPLE 1 | Find the average value of the function $f(x) = 1 + x^2$ on the interval $[-1, 2]$.

SOLUTION With $a = -1$ and $b = 2$ we have

$$f_{\text{ave}} = \frac{1}{b-a} \int_a^b f(x)\,dx = \frac{1}{2-(-1)} \int_{-1}^{2} (1 + x^2)\,dx$$

$$= \frac{1}{3}\left[x + \frac{x^3}{3}\right]_{-1}^{2} = 2$$

■

EXAMPLE 2 | **World population** In Chapter 1 we modeled the size of the human population of the world with the exponential function

$$P(t) = (1.43653 \times 10^9) \cdot (1.01395)^t$$

where t is measured in years and $t = 0$ corresponds to the year 1900. What was the average population in the 20th century?

SOLUTION The average population for $0 \leqslant t \leqslant 100$ was

$$P_{\text{ave}} = \frac{1}{100} \int_0^{100} (1.43653 \times 10^9) \cdot (1.01395)^t\,dt$$

$$= \frac{1}{100} \cdot (1.43653 \times 10^9) \cdot \left[\frac{(1.01395)^t}{\ln(1.01395)}\right]_{t=0}^{t=100}$$

$$= 1.43653 \times 10^7 \cdot \frac{(1.01395)^{100} - 1}{\ln(1.01395)}$$

$$\approx 310 \times 10^7 = 3.1 \times 10^9$$

So the average world population in the 20th century was about 3.1 billion. ■

If $T(t)$ is the temperature at time t, we might wonder if there is a specific time when the temperature is the same as the average temperature. For the temperature function graphed in Figure 1, we see that there are two such times—just before noon and just before midnight. In general, is there a number c at which the value of a function f is exactly equal to the average value of the function, that is, $f(c) = f_{\text{ave}}$? The following theorem says that this is true for continuous functions.

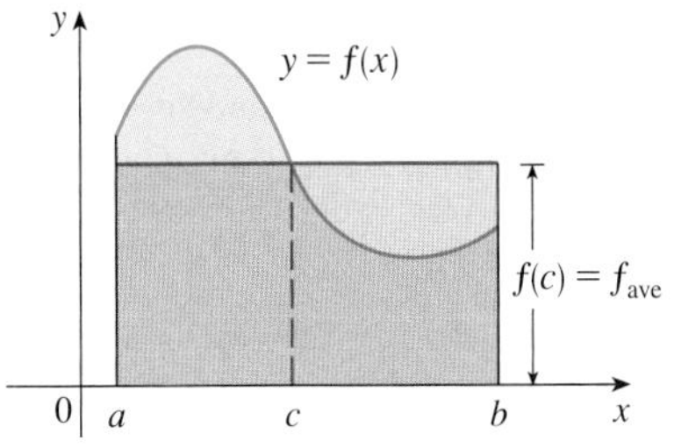

FIGURE 2

You can always chop off the top of a (two-dimensional) mountain at a certain height and use it to fill in the valleys so that the mountain becomes completely flat.

The Mean Value Theorem for Integrals If f is continuous on $[a, b]$, then there exists a number c in $[a, b]$ such that

$$f(c) = f_{ave} = \frac{1}{b-a}\int_a^b f(x)\,dx$$

that is,

$$\int_a^b f(x)\,dx = f(c)(b-a)$$

The Mean Value Theorem for Integrals is a consequence of the Mean Value Theorem for derivatives and the Fundamental Theorem of Calculus. The proof is outlined in Exercise 23.

The geometric interpretation of the Mean Value Theorem for Integrals is that, for *positive* functions f, there is a number c such that the rectangle with base $[a, b]$ and height $f(c)$ has the same area as the region under the graph of f from a to b. (See Figure 2 and the more picturesque interpretation in the margin note.)

EXAMPLE 3 | Since $f(x) = 1 + x^2$ is continuous on the interval $[-1, 2]$, the Mean Value Theorem for Integrals says there is a number c in $[-1, 2]$ such that

$$\int_{-1}^{2} (1 + x^2)\,dx = f(c)[2 - (-1)]$$

In this particular case we can find c explicitly. From Example 1 we know that $f_{ave} = 2$, so the value of c satisfies

$$f(c) = f_{ave} = 2$$

Therefore $\quad 1 + c^2 = 2 \quad$ so $\quad c^2 = 1$

So in this case there happen to be two numbers $c = \pm 1$ in the interval $[-1, 2]$ that work in the Mean Value Theorem for Integrals. ■

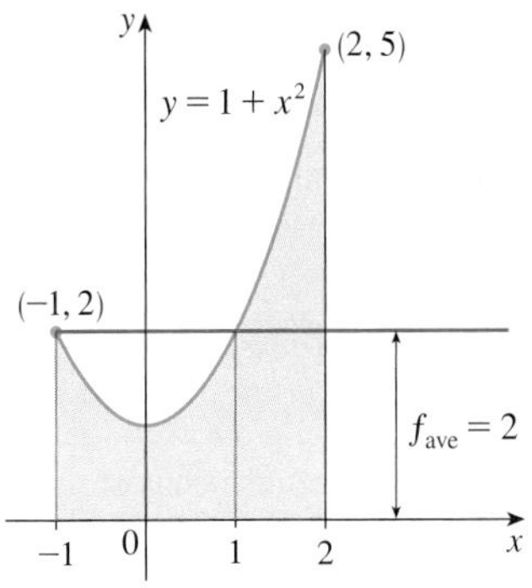

FIGURE 3

Examples 1 and 3 are illustrated by Figure 3.

EXERCISES 6.2

1–6 Find the average value of the function on the given interval.

1. $f(x) = 4x - x^2, \quad [0, 4]$
2. $f(x) = \sin 4x, \quad [-\pi, \pi]$
3. $g(x) = \sqrt[3]{x}, \quad [1, 8]$
4. $f(\theta) = \sec^2(\theta/2), \quad [0, \pi/2]$
5. $h(x) = \cos^4 x \sin x, \quad [0, \pi]$
6. $h(u) = (3 - 2u)^{-1}, \quad [-1, 1]$

7–10
(a) Find the average value of f on the given interval.
(b) Find c such that $f_{ave} = f(c)$.
(c) Sketch the graph of f and a rectangle whose area is the same as the area under the graph of f.

7. $f(x) = (x - 3)^2, \quad [2, 5]$
8. $f(x) = \ln x, \quad [1, 3]$
9. $f(x) = 2\sin x - \sin 2x, \quad [0, \pi]$
10. $f(x) = 2x/(1 + x^2)^2, \quad [0, 2]$

11. Find the average value of f on $[0, 8]$.

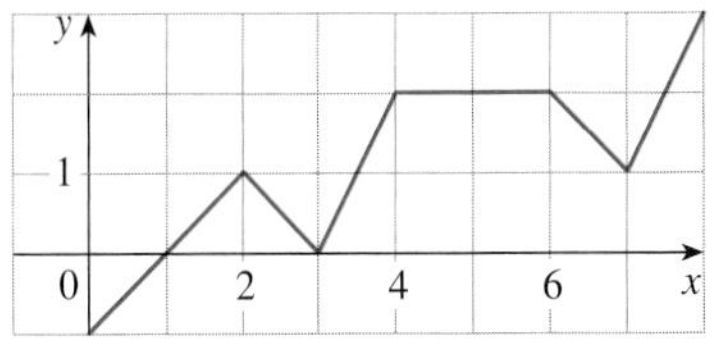

12. The velocity graph of an accelerating car is shown.
(a) Use the Midpoint rule to estimate the average velocity of the car during the first 12 seconds.
(b) At what time was the instantaneous velocity equal to the average velocity?

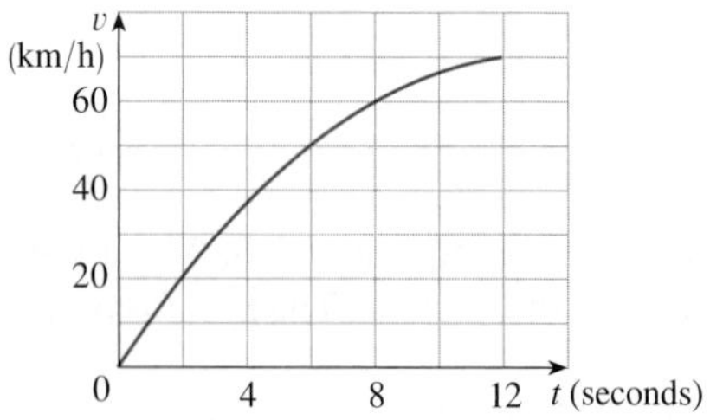

13. In a certain city the temperature (in °F) t hours after 9 AM was modeled by the function

$$T(t) = 50 + 14 \sin \frac{\pi t}{12}$$

Find the average temperature during the period from 9 AM to 9 PM.

14. If a cup of coffee has temperature 95°C in a room where the temperature is 20°C, then, according to Newton's Law of Cooling, the temperature of the coffee after t minutes is $T(t) = 20 + 75e^{-t/50}$. What is the average temperature of the coffee during the first half hour?

15. The **population of Indonesia** from 1950 to 2000 has been modeled with the function

$$P(t) = 83e^{0.018t}$$

where P is measured in millions and t is measured in years with $t = 0$ in the year 1950. What was the average population of Indonesia in the second half of the 20th century?

16. Blood alcohol concentration In Section 3.1 we modeled the BAC of male adult subjects after rapid consumption of 15 mL of ethanol (corresponding to one alcoholic drink) by the concentration function

$$C(t) = 0.0225te^{-0.0467t}$$

where t is measured in minutes after consumption and $C(t)$ is measured in mg/mL. What was the average BAC during the first hour?

17. Measles pathogenesis In Section 5.1 we modeled the infection level of the measles virus in a patient by the function

$$f(t) = -t(t - 21)(t + 1)$$

where t is measured in days and $f(t)$ is measured in the number of infected cells per mL of blood plasma. Over the course of the 21-day infection, what is the average level of infection?

18. Length of a fish For a fish that starts life with a length of 1 cm and has a maximum length of 30 cm, the von Bertalanffy growth model predicts that the growth rate is $29e^{-a}$ cm/year. What is the average length of the fish over its first five years?

19. Breathing is cyclic and a full respiratory cycle from the beginning of inhalation to the end of exhalation takes about 5 s. The maximum rate of air flow into the lungs is about 0.5 L/s. This explains, in part, why the function

$$f(t) = \frac{1}{2} \sin \frac{2\pi t}{5}$$

has often been used to model the rate of air flow into the lungs. If inhalation occurs during the interval $0 \leqslant t \leqslant 2.5$, what is the average rate of air flow during inhalation?

20. Blood flow The velocity v of blood that flows in a blood vessel with radius R and length l at a distance r from the central axis is

$$v(r) = \frac{P}{4\eta l}(R^2 - r^2)$$

where P is the pressure difference between the ends of the vessel and η is the viscosity of the blood (see Example 3.3.9). Find the average velocity (with respect to r) over the interval $0 \leqslant r \leqslant R$. Compare the average velocity with the maximum velocity.

21. If f is continuous and $\int_1^3 f(x)\,dx = 8$, show that f takes on the value 4 at least once on the interval $[1, 3]$.

22. Find the numbers b such that the average value of $f(x) = 2 + 6x - 3x^2$ on the interval $[0, b]$ is equal to 3.

23. Prove the Mean Value Theorem for Integrals by applying the Mean Value Theorem for derivatives (see Section 4.2) to the function $F(x) = \int_a^x f(t)\,dt$.

6.3 Further Applications to Biology

In Chapter 5 and in Sections 6.1 and 6.2 we presented several applications of integration to biology. In this section we consider additional examples of how integrals are used in biological settings: survival and renewal of populations, blood flow in veins and arteries, and cardiac output. Other applications are explored in the exercises.

Survival and Renewal

A population may be continually adding members while some of the existing members die. If we can model how these changes occur with suitable functions, we can predict the population size at any point in the future.

Suppose we start with an initial population P_0 and new members are added at the rate $R(t)$, where t is the number of years from now. We call R a **renewal function**. In addition, the proportion of the population that survives at least t years from now is given by a **survival function** S. [So if $S(5) = 0.8$, 80% of the current population remains after 5 years.]

To predict the population in T years, we first note that $S(t) \cdot P_0$ members of the current population survive. To account for the newly added members, we divide the time interval $[0, T]$ into n subintervals, each of length $\Delta t = T/n$, and let t_i be the right endpoint of the ith subinterval. During this time interval, approximately $R(t_i)\,\Delta t$ members are added, and the proportion of them that survive until time T is given by $S(T - t_i)$. Thus the remaining members of those added during this time interval is

$$(\text{proportion surviving})(\text{number of members}) = S(T - t_i)\,R(t_i)\,\Delta t$$

Then the total number of new members to the population who survive after T years is approximately

$$\sum_{i=1}^{n} S(T - t_i)\,R(t_i)\,\Delta t$$

If we let $n \to \infty$, this Riemann sum approaches the integral

$$\int_0^T S(T - t)\,R(t)\,dt$$

By adding this integral to the number of initial members who survived, we get the total population after T years.

Equation 1 is also valid if t represents any other unit of time, such as weeks or months.

A population begins with P_0 members and members are added at a rate given by the renewal function $R(t)$, where t is measured in years. The proportion of the population that remains after t years is given by the survival function $S(t)$. Then the population T years from now is given by

(1) $$P(T) = S(T) \cdot P_0 + \int_0^T S(T - t)R(t)\,dt$$

EXAMPLE 1 | Predicting a future population There are currently 5600 trout in a lake and the trout are reproducing at the rate $R(t) = 720e^{0.1t}$ fish/year. However, pollution is killing many of the trout; the proportion that survive after t years is given by $S(t) = e^{-0.2t}$. How many trout will there be in the lake in 10 years?

SOLUTION We have $P_0 = 5600$ and $T = 10$, so by Formula 1, the population in ten years is

$$\begin{aligned} P(10) &= S(10) \cdot 5600 + \int_0^{10} S(10 - t)R(t)\,dt \\ &= 5600e^{-0.2(10)} + \int_0^{10} e^{-0.2(10-t)} \cdot 720e^{0.1t}\,dt \\ &= 5600e^{-2} + 720\int_0^{10} e^{0.3t-2}\,dt \end{aligned}$$

Writing $e^{0.3t-2}$ as $e^{0.3t}e^{-2}$ gives

$$\begin{aligned} P(10) &= 5600e^{-2} + 720e^{-2}\int_0^{10} e^{0.3t}\,dt \\ &= 5600e^{-2} + 720e^{-2}\left[\frac{e^{0.3t}}{0.3}\right]_0^{10} \\ &= 5600e^{-2} + \frac{720}{0.3}e^{-2}(e^3 - e^0) \\ &= 5600e^{-2} + 2400(e - e^{-2}) \approx 6956.95 \end{aligned}$$

Thus we predict that there will be about 6960 trout in the lake ten years from now. ■

Although we presented Formula 1 in the context of populations, it applies to other settings as well, such as administering a drug over time as the body works to eliminate the drug. The exercises investigate additional applications.

Blood Flow

In Example 3.3.9 we discussed the law of laminar flow:

$$v(r) = \frac{P}{4\eta l}(R^2 - r^2)$$

which gives the velocity v of blood that flows along a blood vessel with radius R and length l at a distance r from the central axis, where P is the pressure difference between the ends of the vessel and η is the viscosity of the blood (see Figure 1).

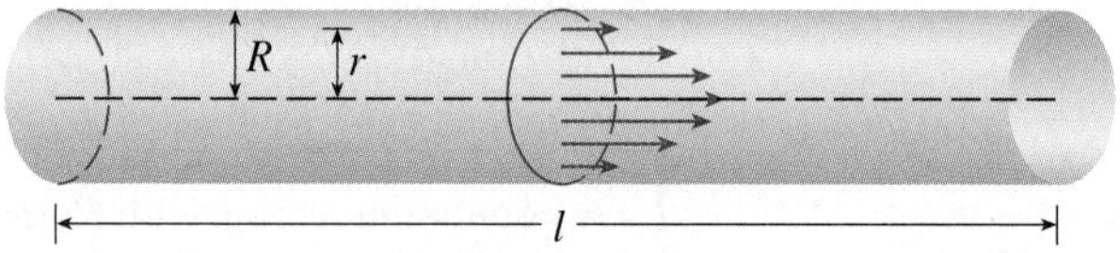

FIGURE 1
Blood flow in an artery

In order to compute the rate of blood flow, or *flux* (volume per unit time), we consider smaller, equally spaced radii $r_1, r_2, \ldots$. The approximate area of the ring with inner radius r_{i-1} and outer radius r_i is

$$2\pi r_i\,\Delta r \qquad \text{where} \quad \Delta r = r_i - r_{i-1}$$

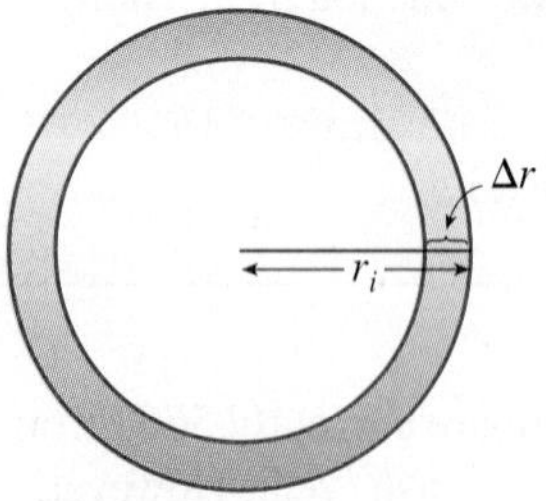

FIGURE 2

(See Figure 2.) If Δr is small, then the velocity is almost constant throughout this ring and can be approximated by $v(r_i)$. Thus the volume of blood per unit time that flows across the ring is approximately

$$(2\pi r_i\,\Delta r)\,v(r_i) = 2\pi r_i\,v(r_i)\,\Delta r$$

and the total volume of blood that flows across a cross-section per unit time is about

$$\sum_{i=1}^{n} 2\pi r_i\,v(r_i)\,\Delta r$$

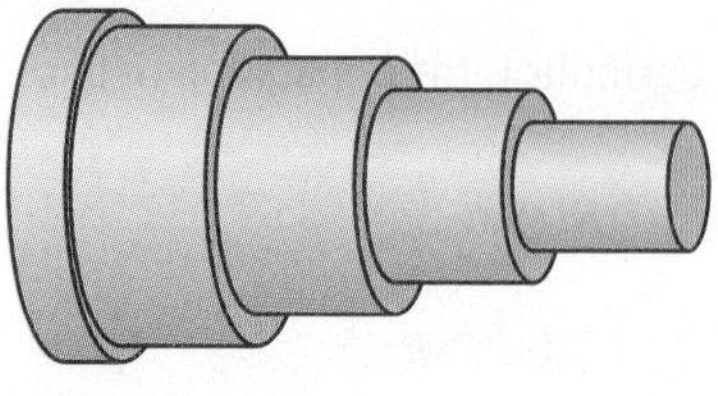

FIGURE 3

This approximation is illustrated in Figure 3. Notice that the velocity (and hence the volume per unit time) increases toward the center of the blood vessel. The approximation gets better as n increases. When we take the limit as $n \to \infty$ we get an integral that gives the exact value of the **flux** (or *discharge*), which is the volume of blood that passes a cross-section per unit time:

$$
\begin{aligned}
F &= \int_0^R 2\pi r v(r)\, dr \\
&= \int_0^R 2\pi r \frac{P}{4\eta l}(R^2 - r^2)\, dr \\
&= \frac{\pi P}{2\eta l}\int_0^R (R^2 r - r^3)\, dr = \frac{\pi P}{2\eta l}\left[R^2\frac{r^2}{2} - \frac{r^4}{4}\right]_{r=0}^{r=R} \\
&= \frac{\pi P}{2\eta l}\left[\frac{R^4}{2} - \frac{R^4}{4}\right] = \frac{\pi P R^4}{8\eta l}
\end{aligned}
$$

The resulting equation

(2) $$F = \frac{\pi P R^4}{8\eta l}$$

is called **Poiseuille's Law**; it shows that the flux is proportional to the fourth power of the radius of the blood vessel.

In Exercise 10 you are asked to investigate the effect on blood pressure if the radius of an artery is reduced to three-fourths of its normal value.

■ Cardiac Output

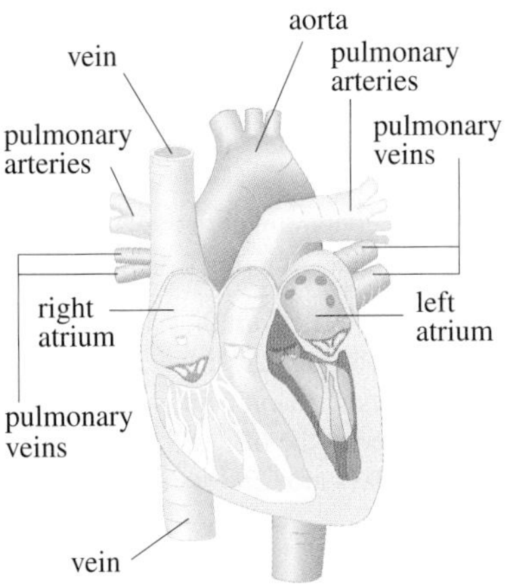

FIGURE 4

Figure 4 shows the human heart and associated blood vessels. Blood returns from the body through the veins, enters the right atrium of the heart, and is pumped to the lungs through the pulmonary arteries for oxygenation. It then flows back into the left atrium through the pulmonary veins and then out to the rest of the body through the aorta. The **cardiac output** of the heart is the volume of blood pumped by the heart per unit time, that is, the rate of flow into the aorta.

The *dye dilution method* is used to measure the cardiac output. Dye is injected into the right atrium and flows through the heart into the aorta. A probe inserted into the aorta measures the concentration of the dye leaving the heart at equally spaced times over a time interval $[0, T]$ until the dye has cleared. Let $c(t)$ be the concentration of the dye at time t. If we divide $[0, T]$ into subintervals of equal length Δt, then the amount of dye that flows past the measuring point during the subinterval from $t = t_{i-1}$ to $t = t_i$ is approximately

$$\text{(concentration of dye in blood)(volume of blood)} = c(t_i)(F\,\Delta t)$$

where F is the rate of flow that we are trying to determine. Thus the total amount of dye is approximately

$$c(t_1)F\,\Delta t + c(t_2)F\,\Delta t + \cdots + c(t_n)F\,\Delta t$$

and, letting $n \to \infty$, we find that the amount of dye is

$$A = \int_0^T c(t)F\, dt = F\int_0^T c(t)\, dt$$

Thus the cardiac output is given by

(3) $$F = \frac{A}{\displaystyle\int_0^T c(t)\, dt}$$

where the amount of dye A is known and the integral can be approximated from the concentration readings.

t	$c(t)$	t	$c(t)$
0	0	6	6.1
1	0.4	7	4.0
2	2.8	8	2.3
3	6.5	9	1.1
4	9.8	10	0
5	8.9		

EXAMPLE 2 | **Cardiac output** A 5-mg dose (called a bolus) of dye is injected into a right atrium. The concentration of the dye (in milligrams per liter) is measured in the aorta at one-second intervals as shown in the chart. Estimate the cardiac output.

SOLUTION Here $A = 5$ and $T = 10$. We can use the Midpoint Rule with $n = 5$ subdivisions to approximate the integral of the concentration. Then $\Delta t = 2$ and

$$\int_0^{10} c(t)\,dt \approx [c(1) + c(3) + c(5) + c(7) + c(9)]\,\Delta t$$

$$= [0.4 + 6.5 + 8.9 + 4.0 + 1.1](2)$$

$$= 41.8$$

Thus Formula 3 gives the cardiac output to be

$$F = \frac{A}{\int_0^{10} c(t)\,dt} \approx \frac{5}{41.8} \approx 0.12 \text{ L/s} = 7.2 \text{ L/min}$$ ■

EXERCISES 6.3

1. **Animal survival and renewal** An animal population currently has 7400 members and is reproducing at the rate $R(t) = 2240 + 60t$ members/year. The proportion of members that survive after t years is given by $S(t) = 1/(t + 1)$.
   (a) How many of the original members survive four years?
   (b) How many new members are added during the next four years?
   (c) Explain why the animal population four years from now is not the same as the sum of your answers from parts (a) and (b).

2. **City population** A city currently has 36,000 residents and is adding new residents steadily at the rate of 1600 per year. If the proportion of residents that remain after t years is given by $S(t) = 1/(t + 1)$, what is the population of the city seven years from now?

3. **Insect survival and renewal** A population of insects currently numbers 22,500 and is increasing at a rate of $R(t) = 1225e^{0.14t}$ insects/week. If the survival function for the insects is $S(t) = e^{-0.2t}$, where t is measured in weeks, how many insects are there after 12 weeks?

4. **Animal survival and renewal** There are currently 3800 birds of a particular species in a national park and their number is increasing at a rate of $R(t) = 525e^{0.05t}$ birds/year. If the proportion of birds that survive t years is given by $S(t) = e^{-0.1t}$, what do you predict the bird population will be 10 years from now?

5. **Drug concentration** A drug is administered intravenously to a patient at the rate of 12 mg/h. The patient's body eliminates the drug over time so that after t hours the proportion that remains is $e^{-0.25t}$. If the patient currently has 50 mg of the drug in her bloodstream, how much of the drug is present eight hours from now?

6. **Drug concentration** A patient receives a drug at a constant rate of 30 mg/h. The drug is eliminated from the bloodstream over time so that the fraction $e^{-0.2t}$ remains after t hours. The patient currently has 80 mg of the drug present in the bloodstream. How much will be present in 24 hours?

7. **Water pollution** A contaminant is leaking into a lake at a rate of

$$R(t) = 1600e^{0.06t} \text{ gallons/h}$$

Enzymes have been added to the lake that neutralize the contaminant over time so that after t hours the fraction of the contaminant that remains is $S(t) = e^{-0.32t}$. If there are currently 10,000 gallons of the contaminant in the lake, how many gallons are present in the lake 18 hours from now?

8. **Insect survival and renewal** Sterile fruit flies are used in an experiment where the proportion that survive at least t days is given by $e^{-0.15t}$. If the experiment begins with 200 fruit flies, and flies are added at the rate of 5 per hour, how many flies are present 14 days after the start of the experiment?

9. **Blood flow** Use Poiseuille's Law to calculate the rate of flow in a small human artery where we can take $\eta = 0.027 \text{ dyn} \cdot \text{s/cm}^2$, $R = 0.008$ cm, $l = 2$ cm, and $P = 4000 \text{ dyn/cm}^2$.

10. **Blood flow** High blood pressure results from constriction of the arteries. To maintain a normal flow rate (flux), the heart has to pump harder, thus increasing the blood pressure. Use Poiseuille's Law to show that if R_0 and P_0 are normal values of the radius and pressure in an artery and the constricted values are R and P, then for the flux to remain constant, P and R are related by the equation

$$\frac{P}{P_0} = \left(\frac{R_0}{R}\right)^4$$

Deduce that if the radius of an artery is reduced to three-fourths of its former value, then the pressure is more than tripled.

11. Cardiac output The dye dilution method is used to measure cardiac output with 6 mg of dye. The dye concentrations, in mg/L, are modeled by $c(t) = 20te^{-0.6t}$, $0 \leqslant t \leqslant 10$, where t is measured in seconds. Find the cardiac output. [*Hint*: Integration by parts is required.]

12. Cardiac output After an 8-mg injection of dye, the readings of dye concentration, in mg/L, at two-second intervals are as shown in the table. Use the Midpoint Rule to estimate the cardiac output.

t	$c(t)$	t	$c(t)$
0	0	12	3.9
2	2.4	14	2.3
4	5.1	16	1.6
6	7.8	18	0.7
8	7.6	20	0
10	5.4		

13. Cardiac output The graph of the concentration function $c(t)$ is shown after a 7-mg injection of dye into a heart. Use the Midpoint Rule to estimate the cardiac output.

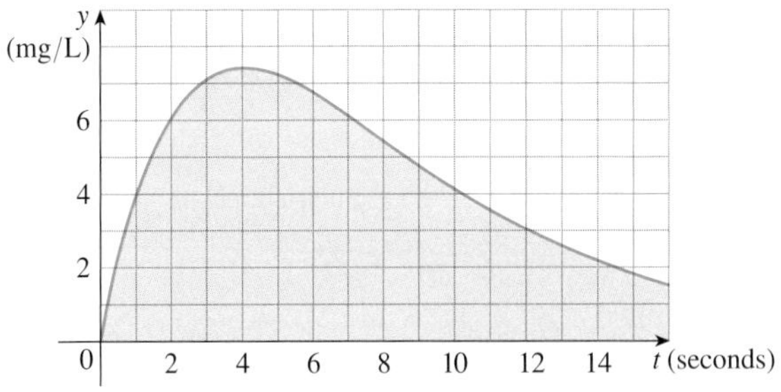

14. Drug administration A patient is continually receiving a drug. If the drug is eliminated from the body over time so that the fraction that remains after t hours is $e^{-0.4t}$, at what constant rate should the drug be administered to maintain a steady level of the drug in the bloodstream?

6.4 Volumes

In trying to find the volume of a solid we face the same type of problem as in finding areas. We have an intuitive idea of what volume means, but we must make this idea precise by using calculus to give an exact definition of volume.

We start with a simple type of solid called a **cylinder** (or, more precisely, a *right cylinder*). As illustrated in Figure 1(a), a cylinder is bounded by a plane region B_1, called the base, and a congruent region B_2 in a parallel plane. The cylinder consists of all points on line segments that are perpendicular to the base and join B_1 to B_2. If the area of the base is A and the height of the cylinder (the distance from B_1 to B_2) is h, then the volume V of the cylinder is defined as

$$V = Ah$$

In particular, if the base is a circle with radius r, then the cylinder is a circular cylinder with volume $V = \pi r^2 h$ [see Figure 1(b)], and if the base is a rectangle with length l and width w, then the cylinder is a rectangular box (also called a *rectangular parallelepiped*) with volume $V = lwh$ [see Figure 1(c)].

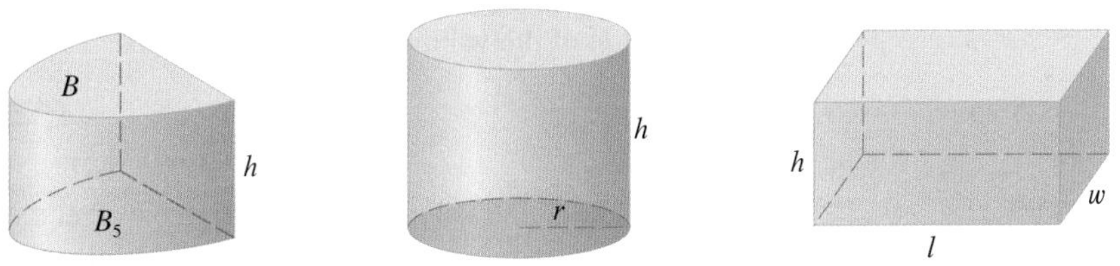

(a) Cylinder $V = Ah$ (b) Circular cylinder $V = \pi r^2 h$ (c) Rectangular box $V = lwh$

FIGURE 1

For a solid S that isn't a cylinder we first "cut" S into pieces and approximate each piece by a cylinder. We estimate the volume of S by adding the volumes of the cylinders. We arrive at the exact volume of S through a limiting process in which the number of pieces becomes large.

We start by intersecting S with a plane and obtaining a plane region that is called a **cross-section** of S. Let $A(x)$ be the area of the cross-section of S in a plane P_x perpendicular to the x-axis and passing through the point x, where $a \leqslant x \leqslant b$. (See Figure 2. Think of slicing S with a knife through x and computing the area of this slice.) The cross-sectional area $A(x)$ will vary as x increases from a to b.

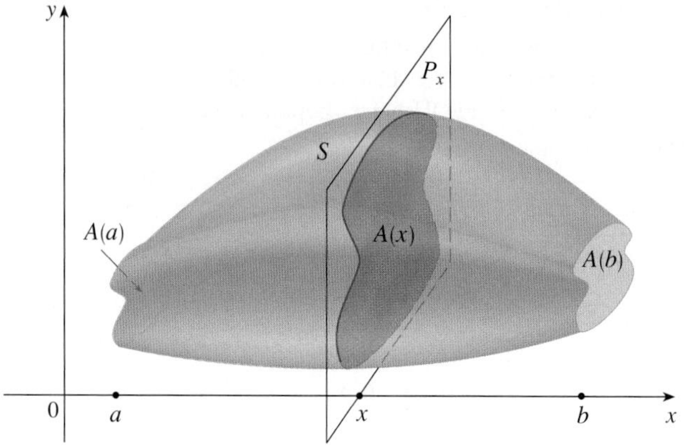

FIGURE 2

Let's divide S into n "slabs" of equal width Δx by using the planes $P_{x_1}, P_{x_2}, \ldots$ to slice the solid. (Think of slicing a loaf of bread.) If we choose sample points x_i^* in $[x_{i-1}, x_i]$, we can approximate the ith slab S_i (the part of S that lies between the planes $P_{x_{i-1}}$ and P_{x_i}) by a cylinder with base area $A(x_i^*)$ and "height" Δx. (See Figure 3.)

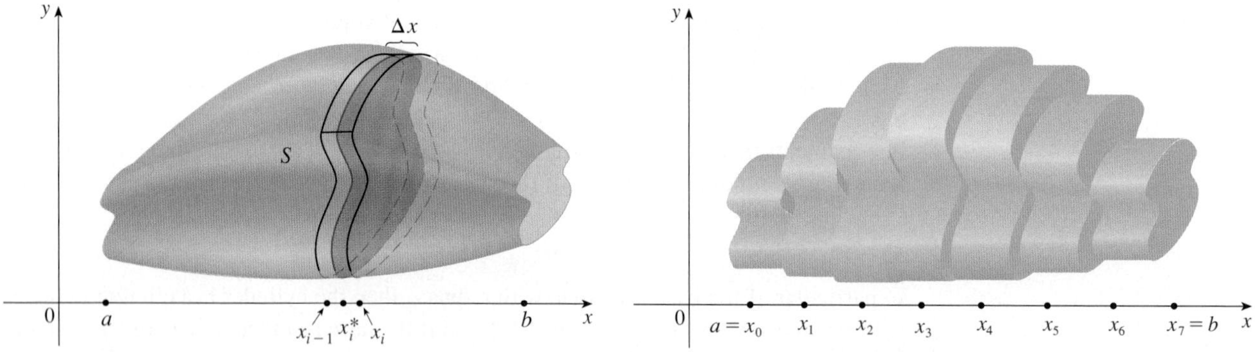

FIGURE 3

The volume of this cylinder is $A(x_i^*)\,\Delta x$, so an approximation to our intuitive conception of the volume of the ith slab S_i is

$$V(S_i) \approx A(x_i^*)\,\Delta x$$

Adding the volumes of these slabs, we get an approximation to the total volume (that is, what we think of intuitively as the volume):

$$V \approx \sum_{i=1}^{n} A(x_i^*)\,\Delta x$$

This approximation appears to become better and better as $n \to \infty$. (Think of the slices as becoming thinner and thinner.) Therefore we *define* the volume as the limit of these sums as $n \to \infty$. But we recognize the limit of Riemann sums as a definite integral and so we have the following definition.

It can be proved that this definition is independent of how S is situated with respect to the x-axis. In other words, no matter how we slice S with parallel planes, we always get the same answer for V.

Definition of Volume Let S be a solid that lies between $x = a$ and $x = b$. If the cross-sectional area of S in the plane P_x, through x and perpendicular to the x-axis, is $A(x)$, where A is a continuous function, then the **volume** of S is

$$V = \lim_{n \to \infty} \sum_{i=1}^{n} A(x_i^*)\,\Delta x = \int_a^b A(x)\,dx$$

When we use the volume formula $V = \int_a^b A(x)\,dx$, it is important to remember that $A(x)$ is the area of a moving cross-section obtained by slicing through x perpendicular to the x-axis.

Notice that, for a cylinder, the cross-sectional area is constant: $A(x) = A$ for all x. So our definition of volume gives $V = \int_a^b A\,dx = A(b - a)$; this agrees with the formula $V = Ah$.

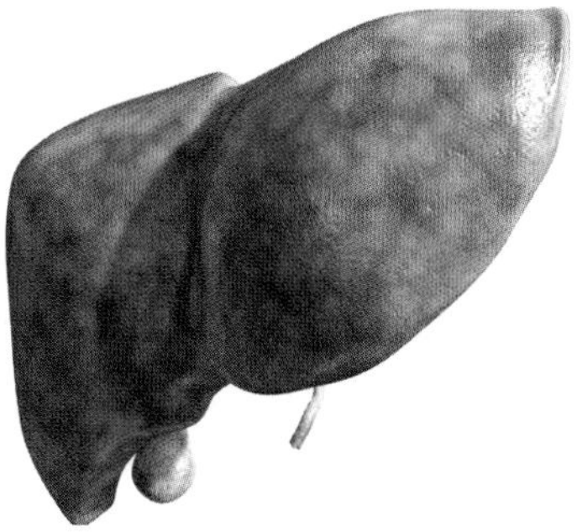

MedicalRF.com / Getty Images

EXAMPLE 1 | Volume of a liver A CAT scan produces equally spaced cross-sectional views of a human organ that provide information about the organ otherwise obtained only by surgery. For example, such measurements of liver volume can be related to diseases such as cirrhosis.[1] Suppose that a CAT scan of a human liver shows cross-sections spaced 2 cm apart. The liver is 20 cm long and the cross-sectional areas, in square centimeters, are 0, 39, 63, 128, 117, 106, 94, 79, 58, 18, and 0. Use the Midpoint Rule to estimate the volume of the liver.

SOLUTION In using the Midpoint Rule we will use $n = 5$ subintervals; their midpoints are 2, 6, 10, 14, and 18. If $A(x)$ is the area of the cross-section of the liver at a distance of x centimeters from one end, then the volume is

$$\begin{aligned} V &= \int_0^{20} A(x)\,dx \\ &= \frac{20 - 0}{5}[A(2) + A(6) + A(10) + A(14) + A(18)] \\ &= 4[39 + 128 + 106 + 79 + 18] \\ &= 4 \cdot 370 = 1480 \end{aligned}$$

The volume of the liver is approximately 1480 cm^3. ■

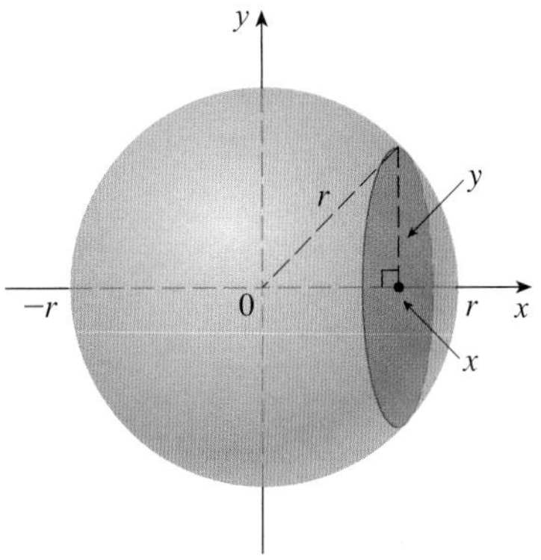

FIGURE 4

EXAMPLE 2 | Show that the volume of a sphere of radius r is $V = \frac{4}{3}\pi r^3$.

SOLUTION If we place the sphere so that its center is at the origin (see Figure 4), then the plane P_x intersects the sphere in a circle whose radius (from the Pythagorean Theorem) is $y = \sqrt{r^2 - x^2}$. So the cross-sectional area is

$$A(x) = \pi y^2 = \pi(r^2 - x^2)$$

1. J.-Y. Zhu et al., "Measurement of Liver Volume and Its Clinical Significance in Cirrhotic Portal Hypertensive Patients," *World Journal of Gastroenterology* 5 (1999): 525–26.

Using the definition of volume with $a = -r$ and $b = r$, we have

$$V = \int_{-r}^{r} A(x)\,dx = \int_{-r}^{r} \pi(r^2 - x^2)\,dx$$

$$= 2\pi \int_{0}^{r} (r^2 - x^2)\,dx \qquad \text{(The integrand is even.)}$$

$$= 2\pi \left[r^2x - \frac{x^3}{3} \right]_0^r = 2\pi \left(r^3 - \frac{r^3}{3} \right)$$

$$= \tfrac{4}{3}\pi r^3$$

■

Figure 5 illustrates the definition of volume when the solid is a sphere with radius $r = 1$. From the result of Example 2, we know that the volume of the sphere is $\frac{4}{3}\pi$, which is approximately 4.18879. Here the slabs are circular cylinders, or *disks*, and the three parts of Figure 5 show the geometric interpretations of the Riemann sums

$$\sum_{i=1}^{n} A(\overline{x}_i)\,\Delta x = \sum_{i=1}^{n} \pi(1^2 - \overline{x}_i^2)\,\Delta x$$

when $n = 5$, 10, and 20 if we choose the sample points x_i^* to be the midpoints $\overline{x}_i$. Notice that as we increase the number of approximating cylinders, the corresponding Riemann sums become closer to the true volume.

TEC Visual 6.4A shows an animation of Figure 5.

(a) Using 5 disks, $V \approx 4.2726$ (b) Using 10 disks, $V \approx 4.2097$ (c) Using 20 disks, $V \approx 4.1940$

FIGURE 5 Approximating the volume of a sphere with radius 1

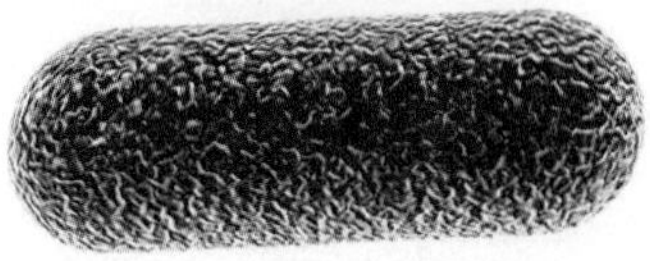

You have probably seen the formula $V = \frac{4}{3}\pi r^3$ for the volume of a sphere before. But calculus is required to prove it, so this may be the first time you have have seen it proved.

This formula is useful in biology because the shape of a tumor is often modeled by a sphere. (See, for instance, Example 3.5.14.) And the shape of a bacterium can often be modeled as a circular cylinder capped by two hemispheres as shown in the margin.

EXAMPLE 3 | Find the volume of the solid obtained by rotating about the x-axis the region under the curve $y = \sqrt{x}$ from 0 to 1. Illustrate the definition of volume by sketching a typical approximating cylinder.

SOLUTION The region is shown in Figure 6(a). If we rotate about the x-axis, we get the solid shown in Figure 6(b). When we slice through the point x, we get a disk with radius $\sqrt{x}$. The area of this cross-section is

$$A(x) = \pi\left(\sqrt{x}\right)^2 = \pi x$$

and the volume of the approximating cylinder (a disk with thickness Δx) is

$$A(x)\,\Delta x = \pi x\,\Delta x$$

The solid lies between $x = 0$ and $x = 1$, so its volume is

$$V = \int_0^1 A(x)\,dx = \int_0^1 \pi x\,dx = \pi \frac{x^2}{2}\bigg]_0^1 = \frac{\pi}{2}$$ ■

Did we get a reasonable answer in Example 3? As a check on our work, let's replace the given region by a square with base [0, 1] and height 1. If we rotate this square, we get a cylinder with radius 1, height 1, and volume $\pi \cdot 1^2 \cdot 1 = \pi$. We computed that the given solid has half this volume. That seems about right.

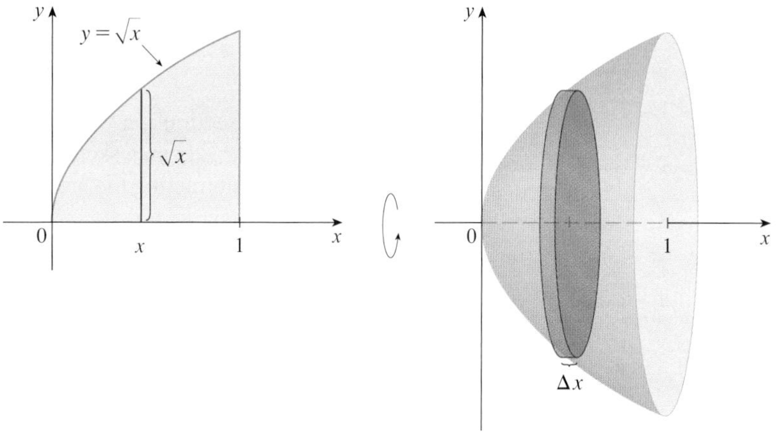

FIGURE 6 (a) (b)

EXAMPLE 4 | The region enclosed by the curves $y = x$ and $y = x^2$ is rotated about the x-axis. Find the volume of the resulting solid.

SOLUTION The curves $y = x$ and $y = x^2$ intersect at the points (0, 0) and (1, 1). The region between them, the solid of rotation, and a cross-section perpendicular to the x-axis are shown in Figure 7. A cross-section in the plane P_x has the shape of a *washer* (an annular ring) with inner radius x^2 and outer radius x, so we find the cross-sectional area by subtracting the area of the inner circle from the area of the outer circle:

$$A(x) = \pi x^2 - \pi(x^2)^2 = \pi(x^2 - x^4)$$

TEC Visual 6.4B shows how solids of revolution are formed.

Therefore we have

$$V = \int_0^1 A(x)\,dx = \int_0^1 \pi(x^2 - x^4)\,dx = \pi\left[\frac{x^3}{3} - \frac{x^5}{5}\right]_0^1 = \frac{2\pi}{15}$$

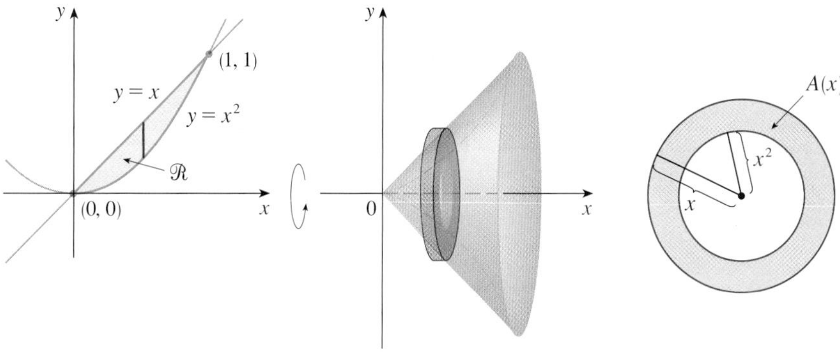

FIGURE 7 (a) (b) (c) ■

The solids in Examples 2–4 are all called **solids of revolution** because they are obtained by revolving a region about a line. In general, we calculate the volume of a solid of revolution by using the basic defining formula

$$V = \int_a^b A(x)\,dx$$

and we find the cross-sectional area $A(x)$ in one of the following ways:

- If the cross-section is a disk (as in Examples 2 and 3), we find the radius of the disk (in terms of x) and use

$$A = \pi(\text{radius})^2$$

- If the cross-section is a washer (as in Example 4), we find the inner radius r_{in} and outer radius r_{out} from a sketch (as in Figures 7 and 8) and compute the area of the washer by subtracting the area of the inner disk from the area of the outer disk:

$$A = \pi(\text{outer radius})^2 - \pi(\text{inner radius})^2$$

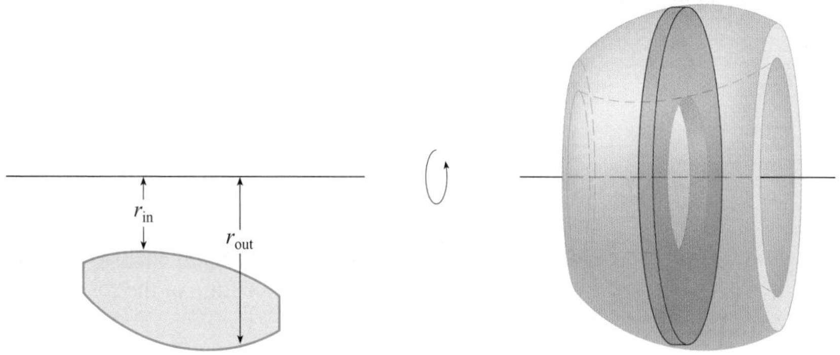

FIGURE 8

We now find the volume of a solid that is *not* a solid of revolution.

TEC Visual 6.4C shows how the solid in Figure 9 is generated.

EXAMPLE 5 | Figure 9 shows a solid with a circular base of radius 1. Parallel cross-sections perpendicular to the base are equilateral triangles. Find the volume of the solid.

SOLUTION Let's take the circle to be $x^2 + y^2 = 1$. The solid, its base, and a typical cross-section at a distance x from the origin are shown in Figure 10.

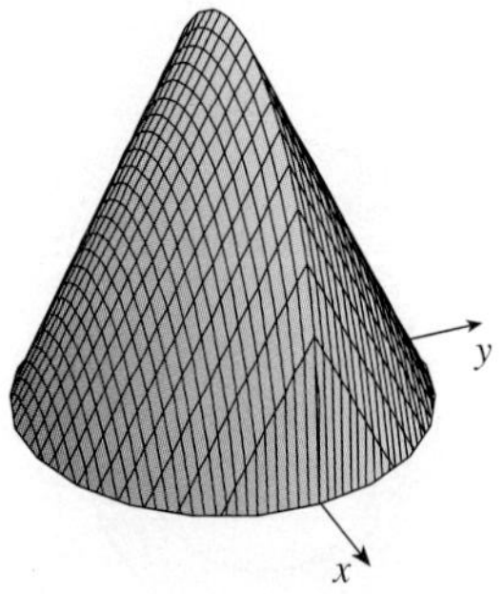

FIGURE 9
Computer-generated picture of the solid in Example 5

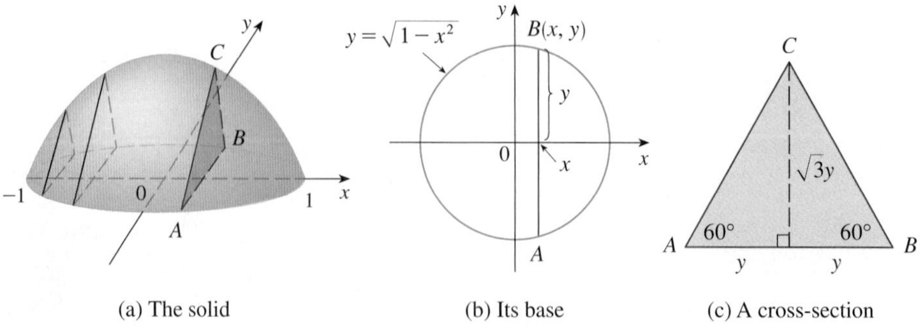

(a) The solid (b) Its base (c) A cross-section

FIGURE 10

Since B lies on the circle, we have $y = \sqrt{1 - x^2}$ and so the base of the triangle ABC is $|AB| = 2\sqrt{1 - x^2}$. Since the triangle is equilateral, we see from Figure 10(c) that its height is $\sqrt{3}\,y = \sqrt{3}\sqrt{1 - x^2}$. The cross-sectional area is therefore

$$A(x) = \tfrac{1}{2} \cdot 2\sqrt{1 - x^2} \cdot \sqrt{3}\sqrt{1 - x^2} = \sqrt{3}\,(1 - x^2)$$

and the volume of the solid is

$$V = \int_{-1}^{1} A(x)\,dx = \int_{-1}^{1} \sqrt{3}\,(1 - x^2)\,dx$$

$$= 2\int_{0}^{1} \sqrt{3}\,(1 - x^2)\,dx = 2\sqrt{3}\left[x - \frac{x^3}{3}\right]_0^1 = \frac{4\sqrt{3}}{3} \qquad \blacksquare$$

EXERCISES 6.4

1–6 Find the volume of the solid obtained by rotating the region bounded by the given curves about the x-axis. Sketch the region, the solid, and a typical disk or washer.

1. $y = 2 - \frac{1}{2}x,\ \ y = 0,\ \ x = 1,\ \ x = 2$

2. $y = 1 - x^2,\ \ y = 0$

3. $y = \sqrt{x - 1},\ \ y = 0,\ \ x = 5$

4. $y = \sqrt{25 - x^2},\ \ y = 0,\ \ x = 2,\ \ x = 4$

5. $y = x^3,\ \ y = x,\ \ x \geqslant 0$

6. $y = \frac{1}{4}x^2,\ \ y = 5 - x^2$

7–8 Here we rotate about the y-axis instead of the x-axis. Find the volume of the solid obtained by rotating the region bounded by the given curves about the y-axis. Sketch the region, the solid, and a typical disk.

7. $x = 2\sqrt{y},\ \ x = 0,\ \ y = 9$

8. $y = \ln x,\ \ y = 1,\ \ y = 2,\ \ x = 0$

9. Volume of a pancreas A CAT scan of a human pancreas shows cross-sections spaced 1 cm apart. The pancreas is 12 cm long and the cross-sectional areas, in square centimeters, are 0, 7.7, 15.2, 18.0, 10.3, 10.8, 9.7, 8.7, 7.7, 5.5, 4.0, 2.7, and 0. Use the Midpoint Rule to estimate the volume of the pancreas.

10. A log 10 m long is cut at 1-meter intervals and its cross-sectional areas A (at a distance x from the end of the log) are listed in the table. Use the Midpoint Rule with $n = 5$ to estimate the volume of the log.

x (m)	A (m^2)	x (m)	A (m^2)
0	0.68	6	0.53
1	0.65	7	0.55
2	0.64	8	0.52
3	0.61	9	0.50
4	0.58	10	0.48
5	0.59		

11. (a) If the region shown in the figure is rotated about the x-axis to form a solid, use the Midpoint Rule with $n = 4$ to estimate the volume of the solid.

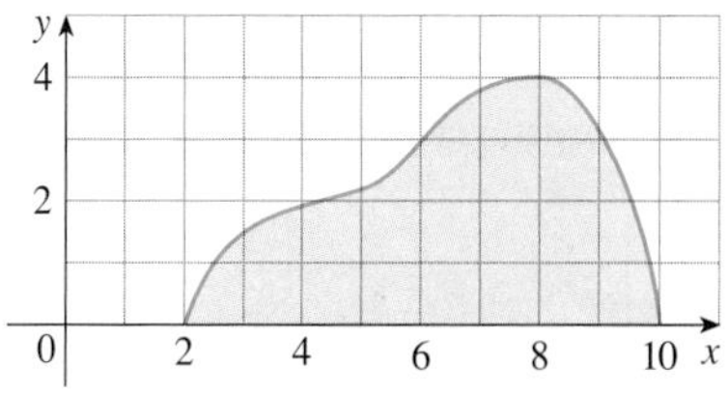

(b) Estimate the volume if the region is rotated about the y-axis. Again use the Midpoint Rule with $n = 4$.

12. Volume of a bird's egg

CAS (a) A model for the shape of a bird's egg is obtained by rotating about the x-axis the region under the graph of

$$f(x) = (ax^3 + bx^2 + cx + d)\sqrt{1 - x^2}$$

Use a CAS to find the volume of such an egg.

(b) For a red-throated loon, $a = -0.06$, $b = 0.04$, $c = 0.1$, and $d = 0.54$. Graph f and find the volume of an egg of this species.

13–15 Find the volume of the described solid S.

13. S is a right circular cone with height h and base radius r

14. The base of S is a circular disk with radius r. Parallel cross-sections perpendicular to the base are squares.

15. The base of S is an elliptical region with boundary curve $9x^2 + 4y^2 = 36$. Cross-sections perpendicular to the x-axis are isosceles right triangles with hypotenuse in the base.

16. The base of S is a circular disk with radius r. Parallel cross-sections perpendicular to the base are isosceles triangles with height h and unequal side in the base.
(a) Set up an integral for the volume of S.
(b) By interpreting the integral as an area, find the volume of S.

17. Find the volume common to two spheres, each with radius r, if the center of each sphere lies on the surface of the other sphere.

18. Find the volume common to two circular cylinders, each with radius r, if the axes of the cylinders intersect at right angles.

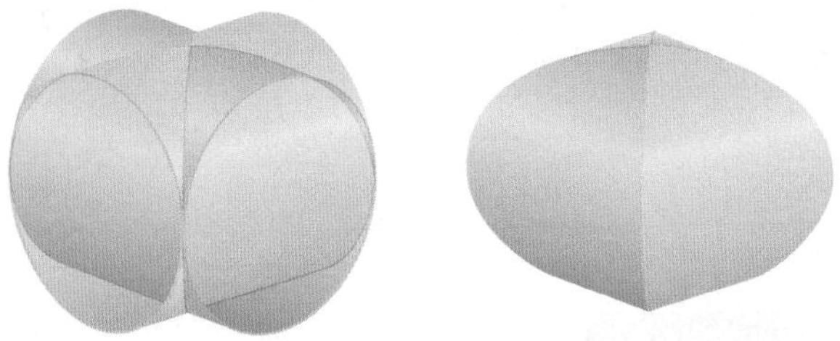

Chapter 6 REVIEW

CONCEPT CHECK

1. Draw two typical curves $y = f(x)$ and $y = g(x)$, where $f(x) \geqslant g(x)$ for $a \leqslant x \leqslant b$. Show how to approximate the area between these curves by a Riemann sum and sketch the corresponding approximating rectangles. Then write an expression for the exact area.

2. Suppose that Sue runs faster than Kathy throughout a 1500-meter race. What is the physical meaning of the area between their velocity curves for the first minute of the race?

3. (a) What is the average value of a function f on an interval $[a, b]$?
(b) What does the Mean Value Theorem for Integrals say? What is its geometric interpretation?

4. If we have survival and renewal functions for a population, how do we predict the size of the population T years from now?

5. (a) What is the cardiac output of the heart?
(b) Explain how the cardiac output can be measured by the dye dilution method.

6. (a) Suppose S is a solid with known cross-sectional areas. Explain how to approximate the volume of S by a Riemann sum. Then write an expression for the exact volume.
(b) If S is a solid of revolution, how do you find the cross-sectional areas?

Answers to the Concept Check can be found on the back endpapers.

EXERCISES

1–4 Find the area of the region bounded by the given curves.

1. $y = x^2, \quad y = 4x - x^2$

2. $y = 1/x, \quad y = x^2, \quad y = 0, \quad x = e$

3. $y = 1 - 2x^2, \quad y = |x|$

4. $x + y = 0, \quad x = y^2 + 3y$

5. **MRI brain scan** Shown is a cross-section of a human brain obtained with an MRI. Use the Midpoint Rule to estimate the area of the cross-section.

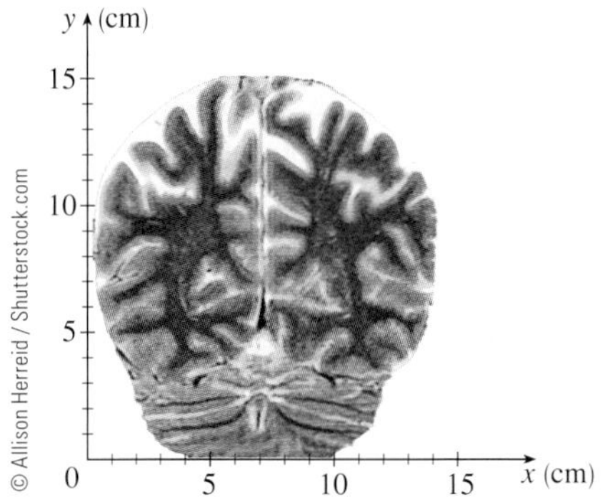

6. **Birth and death rates** The birth rate of a population is $b(t) = 1240e^{0.0197t}$ people per year and the death rate is $d(t) = 682e^{0.008t}$ people per year. Find the area between these curves for $0 \leqslant t \leqslant 20$. What does this area represent?

7. Find the average value of the function $f(t) = t \sin(t^2)$ on the interval $[0, 10]$.

8. Find the average value of the function $f(x) = x^2\sqrt{1 + x^3}$ on the interval $[0, 2]$.

9. **Antibiotic pharmacokinetics** When an antibiotic tablet is taken, the concentration of the antibiotic in the bloodstream is modeled by the function

$$C(t) = 8(e^{-0.4t} - e^{-0.6t})$$

where the time t is measured in hours and C is measured in μg/mL. What is the average concentration of the antibiotic during the first two hours?

10. **Salicylic acid pharmacokinetics** In a study of the effects of aspirin, salicylic acid was formed and its concentration was modeled by the function

$$C(t) = 11.4te^{-t}$$

where the time t is measured in hours and C is measured in μg/mL. What is the average concentration of the salicylic acid during the first four hours?

11. **Survival and renewal** Suppose a city's population is currently 75,000 and the renewal function is

$$R(t) = 3200e^{0.05t}$$

If the survival function is $S(t) = e^{-0.1t}$, predict the population in 10 years.

12. **Animal survival and renewal** The fish population in a lake is currently 3400 and is increasing at a rate of $R(t) = 650e^{0.04t}$ fish per month. If the proportion of fish that remain after t months is $S(t) = e^{-0.09t}$, how many fish will be in the lake in three years?

13. **Cardiac output** After a 6-mg injection of dye into a heart, the readings of dye concentration, in mg/L, at two-second intervals are as shown in the table. Use the Midpoint Rule to estimate the cardiac output.

t	$C(t)$	t	$C(t)$
0	0	14	4.7
2	1.9	16	3.3
4	3.3	18	2.1
6	5.1	20	1.1
8	7.6	22	0.5
10	7.1	24	0
12	5.8		

14. Find the volume of the solid obtained by rotating about the x-axis the region bounded by the curves $y = e^{-2x}$, $y = 1 + x$, and $x = 1$.

15. Let $\mathcal{R}$ be the region bounded by the curves $y = \tan(x^2)$, $x = 1$, and $y = 0$. Use the Midpoint Rule with $n = 4$ to estimate the following quantities.
(a) The area of $\mathcal{R}$
(b) The volume obtained by rotating $\mathcal{R}$ about the x-axis

16. Let $\mathcal{R}$ be the region in the first quadrant bounded by the curves $y = x^3$ and $y = 2x - x^2$. Calculate the following quantities.
(a) The area of $\mathcal{R}$
(b) The volume obtained by rotating $\mathcal{R}$ about the x-axis

17. Find the volumes of the solids obtained by rotating the region bounded by the curves $y = x$ and $y = x^2$ about the following lines.
(a) The x-axis (b) The y-axis

18. Let $\mathcal{R}$ be the region bounded by the curves $y = 1 - x^2$ and $y = x^6 - x + 1$. Estimate the following quantities.
(a) The x-coordinates of the points of intersection of the curves
(b) The area of $\mathcal{R}$
(c) The volume generated when $\mathcal{R}$ is rotated about the x-axis

19. The base of a solid is a circular disk with radius 3. Find the volume of the solid if parallel cross-sections perpendicular to the base are isosceles right triangles with hypotenuse lying along the base.

20. The height of a monument is 20 m. A horizontal cross-section at a distance x meters from the top is an equilateral triangle with side $\frac{1}{4}x$ meters. Find the volume of the monument.

CASE STUDY 1d Kill Curves and Antibiotic Effectiveness

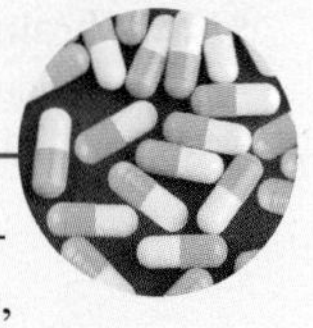

In this case study we have explored the relationship between the magnitude of antibiotic treatment and the effectiveness of the treatment. To do so, in Case Study 1b we showed that a suitable model for the size of the bacteria population $P(t)$ (in CFU/mL) as a function of time t (in hours) is given by the piecewise defined function

(1a)
$$P(t) = \begin{cases} 6e^{t/3} & \text{if } t < 2.08 \\ 12 & \text{if } t \geqslant 2.08 \end{cases}$$

Recall that *MIC* is a constant referred to as the minimum inhibitory concentration of the antibiotic and c_0 is the concentration at time $t = 0$.

if $c_0 < MIC$, where $MIC = 0.013\ \mu\text{g/mL}$. On the other hand, if $c_0 > MIC$

(1b)
$$P(t) = \begin{cases} 6e^{-t/20} & \text{if } t < a \\ 6Ae^{t/3} & \text{if } a \leqslant t < b \\ 12 & \text{if } t \geqslant b \end{cases}$$

where the constants a, b, and A are defined by $a = 5.7 \ln(77c_0)$, $b = 6.6 \ln(77c_0) + 2.08$, and $A = (77c_0)^{-2.2}$.

In Case Study 1a we used these so-called kill curves to plot the relationship between α (a measure of the magnitude of antibiotic treatment) and two different measures of the killing effectiveness, denoted by Δ and T. The quantity Δ is the drop in population size before the population rebound occurs, and T is the time taken to reduce the bacteria population size to 90% of its initial size. (Refer to Figure 1.)

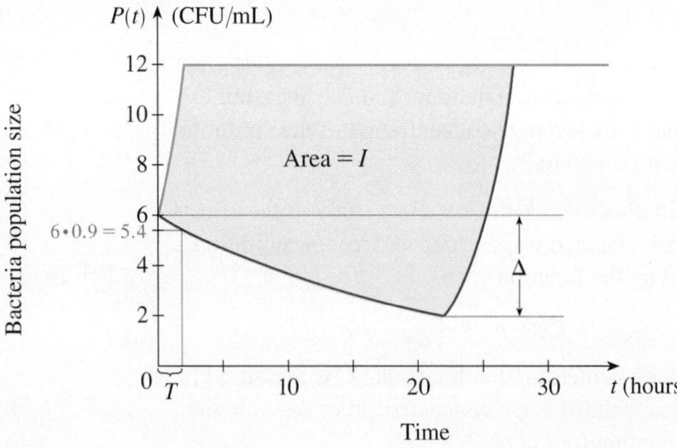

FIGURE 1
Three measures of killing effectiveness. The blue curve is bacteria population size in the absence of antibiotic. The red curve is bacteria population size in the presence of antibiotic.

Now that we have a model that works reasonably well, we can use it to make predictions about other patterns, and then compare these with available data. As an example, another measure of the killing effectiveness of an antibiotic is the area I between the population size curve in the absence of antibiotic, and the kill curve in the presence of the antibiotic as shown in Figure 1. In many cases this measure might be preferable because it incorporates both the drop in bacteria population size, and the length of time for which this reduced population size is maintained. Let's see what our model predicts about the relationship between I and the magnitude of antibiotic treatment α.

1. Suppose that $a > 2.08$ [that is, $5.7 \ln(77c_0) > 2.08$]. Find an expression for I in the modeled populations in terms of c_0. You should assume that $c_0 > MIC$.

2. The result from Problem 1 should give you a function of the form $I = g(c_0)$ for some function g. Substitute the values $k = 0.175$ and $MIC = 0.013$ into the expression for α obtained in Case Study 1c. This will give $\alpha = h(c_0)$ for some function h.

3. Using the concept of a function's inverse, explain how to obtain an expression giving I as a function of α in terms of g and h^{-1}. Find an explicit expression for this function.

4. Plot the function obtained in Problem 3.

The experimental kill curves shown in Figure 2 have also been used to quantify the relationship between I and α.[1] In other words, the values of I and α have been calculated for each experimental kill curve in Figure 2. If we overlay these data points on the plot from Problem 4, we obtain Figure 3. You can see that, again, our model predicts the observed data reasonably well.

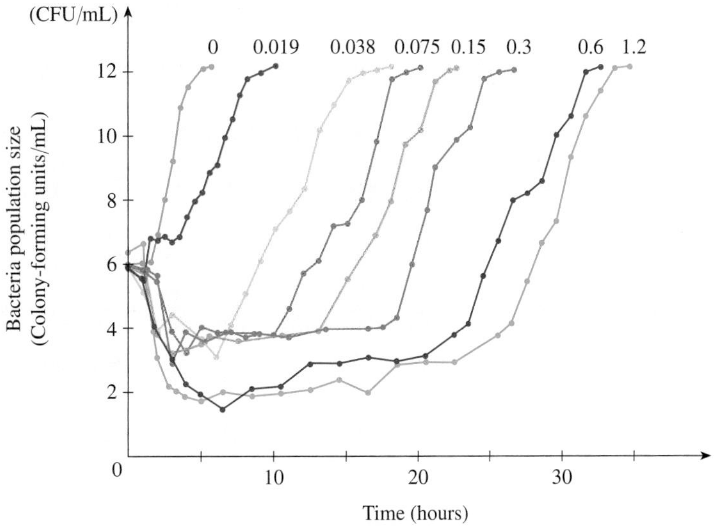

FIGURE 2
The kill curves of ciprofloxacin for *E. coli* when measured in a growth chamber. The concentration of ciprofloxacin at $t = 0$ is indicated above each curve (in μg/mL).

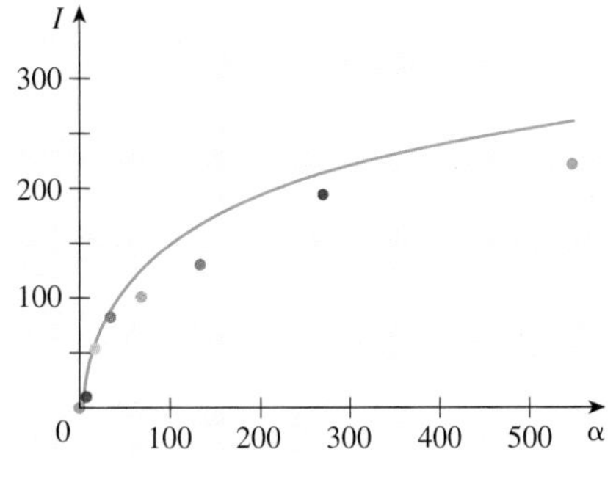

FIGURE 3
Predicted relationship between I and α, along with experimental observations obtained using the kill curve data in Figure 2.

1. Adapted from A. Firsov et al., "Parameters of Bacterial Killing and Regrowth Kinetics and Antimicrobial Effect in Terms of Area Under the Concentration-Time Curve Relationships: Action of Ciprofloxacin against *Escherichia coli* in an In Vitro Dynamic Model." *Antimicrobial Agents and Chemotherapy* 41 (1997): 1281.

CASE STUDY 2b Hosts, Parasites, and Time-Travel

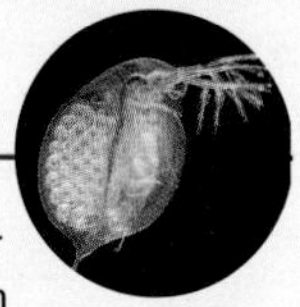

In Case Study 2c you will derive a model for the dynamics of the genotypes of *Daphnia* and its parasite. Recall that we are modeling a situation involving two possible host genotypes (A and a) and two possible parasite genotypes (B and b). Parasites of type B can infect only hosts of type A, while parasites of type b can infect only hosts of type a. You will then derive an explicit solution of a simplified version of the model in Case Study 2d. This will give the frequency of host genotype A and parasite genotype B as functions of time. These functions are

$$q(t) = \tfrac{1}{2} + M_q \cos(ct - \phi_q)$$

(1)

$$p(t) = \tfrac{1}{2} + M_p \cos(ct - \phi_p)$$

where $q(t)$ is the predicted frequency of host genotype A at time t and $p(t)$ is the predicted frequency of the parasite genotype B at time t. In these equations ϕ_q, ϕ_p, and c are positive constants, and M_q and M_p are positive constants that are strictly less than $\frac{1}{2}$ (the biological significance of these constants is explored in Case Study 2a).

In this part of the case study you will use Equations 1 to make predictions from the model that can be compared with data from the experiments.

Recall that, in the experiment, a host from a fixed layer of the sediment core was challenged with infection by a parasite from either the same layer, a layer above the fixed layer (that is, from its future), or a layer below it (that is, from its past). We can view different depths in the sediment core as representing different points of time in the history of the *Daphnia*-parasite interaction (see Figure 1). In this way Equations 1 can equally be viewed as specifying the frequency of the host and parasite genotypes as functions of location in the sediment core. Increasing values of t correspond to shallower points in the core as shown in Figure 2.

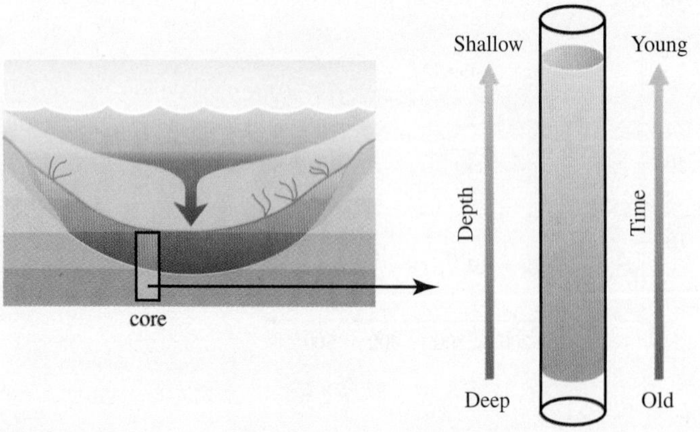

FIGURE 1

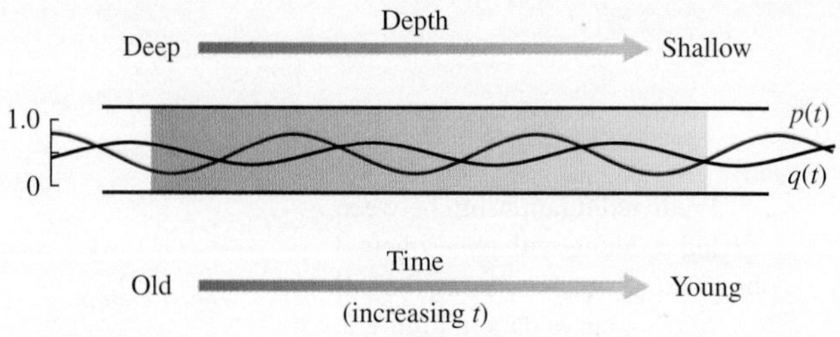

FIGURE 2

In the experiment introduced in Case Study 2 on page xlvii, researchers chose a fixed depth τ and extracted a layer of sediment of width W centered around this depth. This layer is shown in Figure 3.

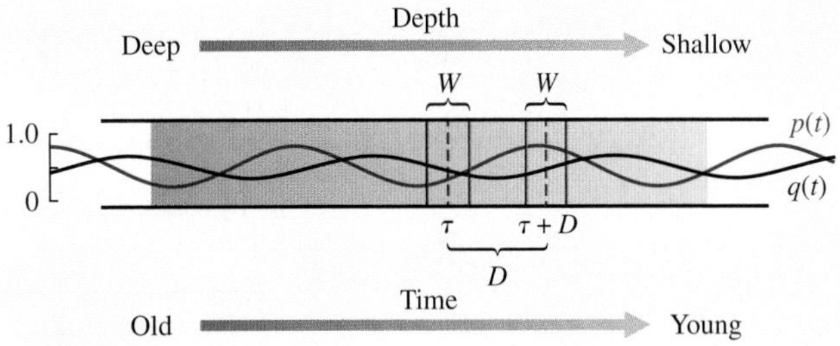

FIGURE 3

After the contents of this layer were completely mixed, hosts and parasites were extracted at random from the mixture. Researchers also took deeper and shallower layers (that represent the past and the future for hosts located in the layer at τ) and completely mixed each. The center of these layers was a distance D from the center of the focal layer (see Figure 3). They then challenged hosts from the layer at τ with parasites from their past (that is, from the layer with $D < 0$), present (the layer at τ), and future (the layer with $D > 0$). For each challenge experiment the fraction of hosts becoming infected was measured.

We can use our model to predict the fraction of hosts infected. To do so, we first need to know the predicted frequency of hosts of type A in the layer at τ as well as the frequency of the parasites of type B in the layer at $\tau + D$.

1. Consider a focal layer at location τ with width W as shown in Figure 4.

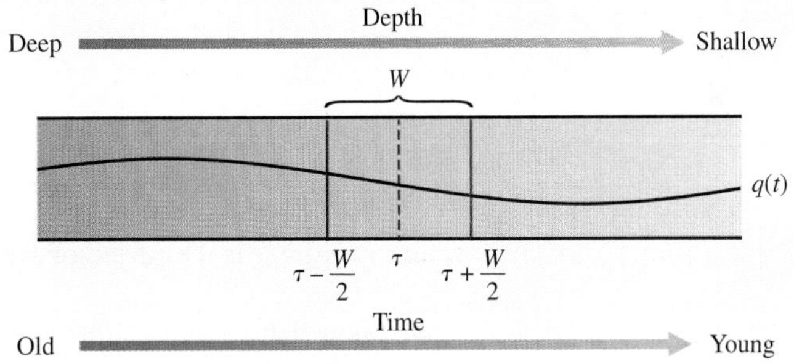

FIGURE 4

The frequency of host type A will vary across the depth of this layer as specified by the function $q(t)$. Show that, when this layer is completely mixed, the frequency of A in the mixture is given by

$$q_{\text{ave}}(\tau) = \frac{1}{2} + M_q \cos(c\tau - \phi_q)\,\frac{2\sin\left(\frac{1}{2}cW\right)}{cW}$$

Hint: You might want to use the trigonometric identity

$$\sin(x + y) - \sin(x - y) = 2\cos x \sin y$$

2. Consider another layer at a location a distance D from τ, again with thickness W, as shown in Figure 5. The frequency of parasite type B will vary across depth in this layer, as specified by the function $p(t)$. Show that, when this layer is completely mixed, the frequency of B in the mixture is given by

$$p_{\text{ave}}(\tau + D) = \frac{1}{2} + M_p \cos(c(\tau + D) - \phi_p)\,\frac{2\sin\left(\frac{1}{2}cW\right)}{cW}$$

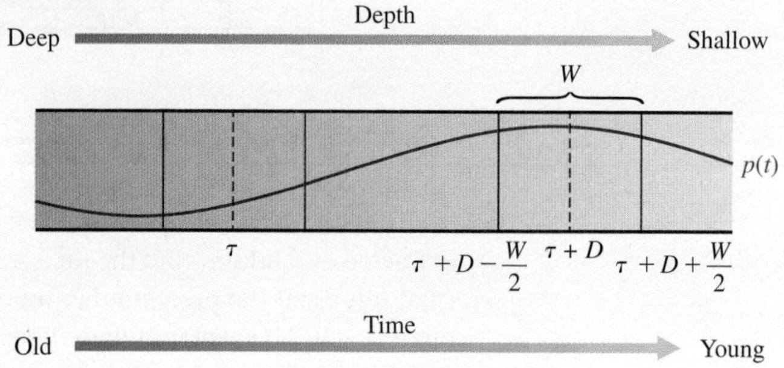

FIGURE 5

3. Suppose that hosts from the layer at τ are challenged with parasites from the layer at $\tau + D$. Use the facts that only B parasites can infect A hosts and only b parasites can infect a hosts to explain why the fraction of hosts infected in this challenge experiment is predicted to be

$$I(\tau) = p_{\text{ave}}(\tau + D)q_{\text{ave}}(\tau) + [1 - p_{\text{ave}}(\tau + D)][1 - q_{\text{ave}}(\tau)]$$

The final step is to recognize that the experiment was actually conducted with several different, randomly chosen depths τ. Therefore we need to average $I(\tau)$ in Problem 3 over the possible depths. Because $I(\tau)$ is periodic, we need only average over one period of its cycle. Its average is therefore

$$F = \frac{1}{T}\int_0^T I(\tau)\,d\tau$$

where $T = 2\pi/c$ is the period of $I(\tau)$.

4. Show that

$$F(D) = \tfrac{1}{2} + M_p M_q \cos(cD - \phi^*)\,\frac{4\sin^2\left(\frac{1}{2}cW\right)}{c^2W^2} \qquad \textbf{(2)}$$

where $\phi^* = \phi_p - \phi_q$.

Hint: You might want to use the trigonometric identity

$$\cos x \cos y = \tfrac{1}{2}(\cos(x + y) + \cos(x - y))$$

Equation 2 was used in Case Study 2a to predict the experimental outcome expected under different conditions.

7 Differential Equations

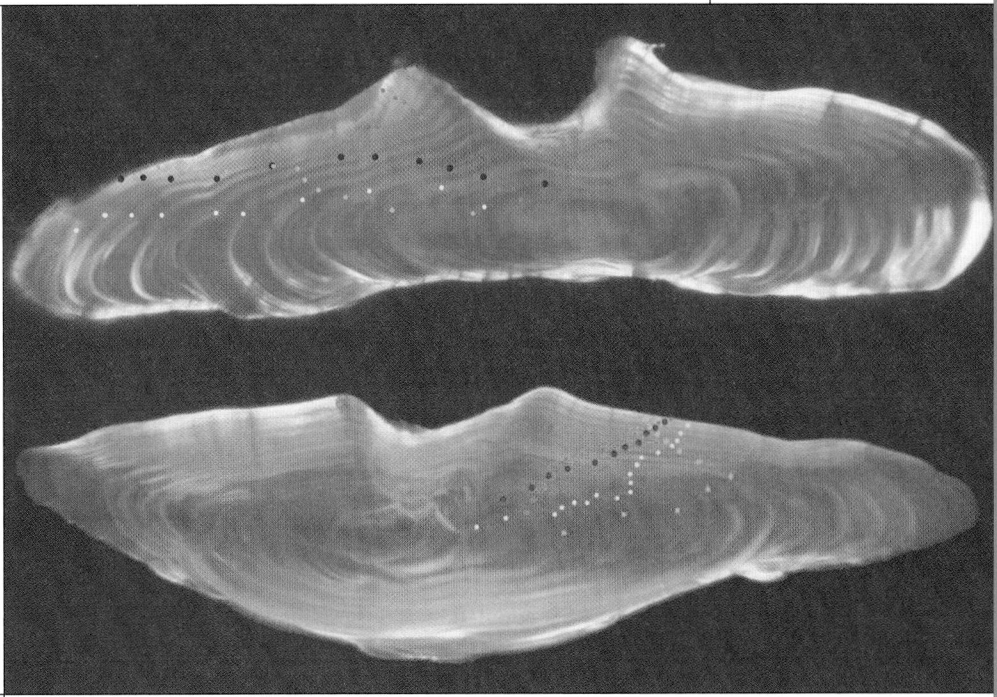

Shown are otoliths from Atlantic redfish—they were used to estimate fish age when fitting the von Bertalanffy differential equation in Example 7.4.2.

Dr. Cristoph Stransky / Thuenen Institute of Sea Fisheries

ONE OF THE MOST IMPORTANT applications of calculus is to differential equations. A wide variety of biological processes can be modeled using differential equations, and such equations have provided enormous insight into our understanding of the dynamics of living organisms—how individuals and populations change over time.

7.1 Modeling with Differential Equations

Many biological processes occur continuously through time. Examples include the change in concentration of a drug in the bloodstream of a patient, or the growth in mass of individual organisms. Even the population dynamics of many species, from size of bacteria colonies to the size of the human population, are sometimes best modeled by assuming the quantity of interest (population size, in this case) changes continuously through time. (For example, see page 146.) As we will see in this chapter, differential equations provide a convenient and natural way to construct such models.

Models of Population Growth

A **differential equation** is an equation that contains an unknown function and one or more of its derivatives. Such equations arise in a variety of situations but one of the most common is in models of population growth.

Consider the growth of a population of yeast. Yeast are single-celled organisms used for a variety of purposes, including alcohol production and baking. Researchers collected the data in Table 1 from a yeast population grown in liquid culture, measuring the population size (in number of individuals per mL of culture) at different points in time (in hours).[1] Figure 1 is a scatter plot of these data.

Table 1

Time (h)	Pop. size ($\times 10^6$/mL)	Time (h)	Pop. size ($\times 10^6$/mL)
0	0.200	19	209
1	0.330	20	190
2	0.500	21	210
3	1.10	22	200
4	1.40	23	215
5	3.10	24	220
6	3.50	25	200
7	9.00	26	180
8	10.0	27	213
9	25.4	28	210
10	27.0	29	210
11	55.0	30	220
12	76.0	31	213
13	115	32	200
14	160	33	211
15	162	34	200
16	190	35	208
17	193	36	230
18	190		

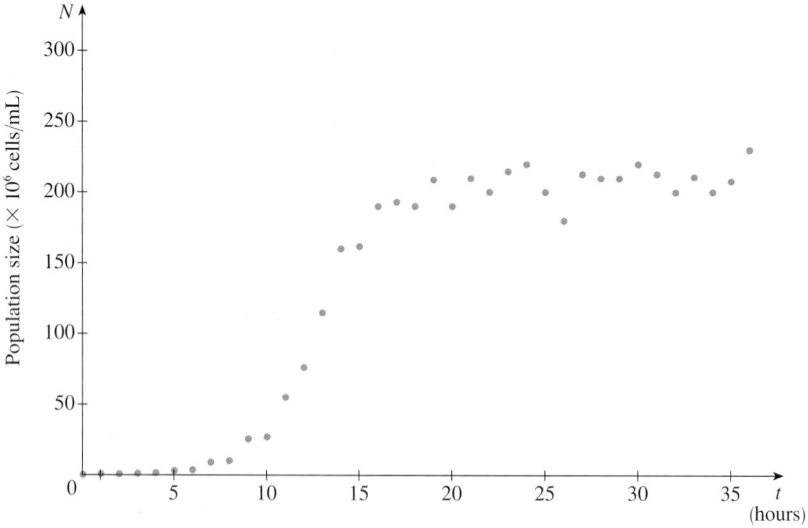

FIGURE 1 A scatter plot of the data in Table 1

1. B. K. Mable et al., "Masking and Purging Mutations following EMS Treatment in Haploid, Diploid, and Tetraploid Yeast (*Saccharomyces cerevisiae*)," *Genetical Research* 77 (2001): 9–26.

Although the population size in Table 1 was measured at one-hour intervals, the yeast themselves are replicating in a way that is nearly continuous in time. In other words, no matter how small we make the interval of time between successive measurements, some reproduction and death will likely have occurred.

BB How can we model such processes? Let's start simply and assume that each individual yeast cell produces offspring at a constant rate β. Thus the total rate of offspring production (that is, the total birth rate) at time t is $\beta N(t)$, where $N(t)$ is the number of yeast cells present at time t. Likewise, suppose the total loss rate of yeast cells through death at time t is $\mu N(t)$, where μ is a constant death rate per individual cell.

With the preceding assumptions, we see that the rate of change of the number of yeast cells at time t is the total birth rate minus the total death rate, $\beta N(t) - \mu N(t)$. And since the rate of change of $N(t)$, the number of yeast cells, can also be written as $dN(t)/dt$, we can write

$$\textbf{(1)} \qquad \frac{dN(t)}{dt} = \beta N(t) - \mu N(t)$$

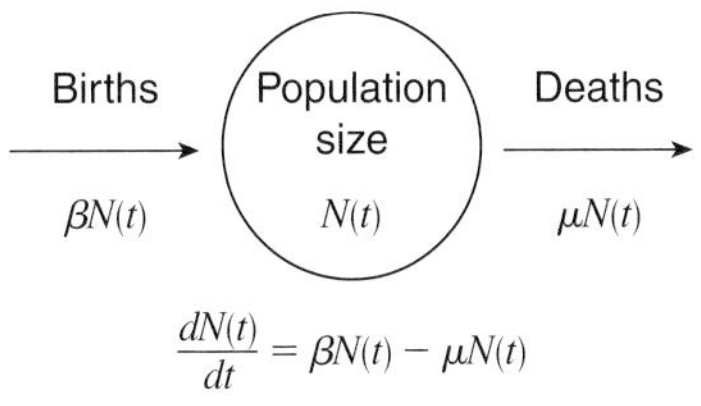

FIGURE 2

(See Figure 2.) Now if we define the constant r as

$$\textbf{(2)} \qquad r = \beta - \mu$$

then Equation 1 can be written more simply as

$$\textbf{(3)} \qquad \frac{dN(t)}{dt} = rN(t)$$

Equation 3 can be derived directly by simply *assuming* that the yeast population grows at a rate proportional to its size. The rate of growth of the population is the derivative dN/dt, and therefore we obtain Equation 3, where r is a constant of proportionality. See Equation 3.6.1, where it was called the *law of natural growth*.

The quantity r in Equation 2 is called the **per capita growth rate**. It is the rate of growth of the population *per individual* in the population. Since dN/dt is the rate of growth of the population, the rate of growth *per individual* is dN/dt divided by $N(t)$. From Equation 3, we get

$$\frac{dN(t)}{dt}\frac{1}{N(t)} = r$$

showing that r is indeed the per capita growth rate.

Equation 3 involves the unknown function $N(t)$ along with its first derivative and is therefore a differential equation. The population size N is the *dependent variable* and time t is the *independent variable*. This differential equation tells us that the rate of change of the population size of yeast at any time is proportional to the size of the population at that time. Put another way, the rate of reproduction of each individual in the population (that is, the *per capita* rate of reproduction) is constant and equal to r.

The model given by Equation 3 is one of the simplest models for population growth. Let's see how well it predicts the data in Table 1. First notice that if $r > 0$, then from Equation 3

$$\frac{dN(t)}{dt} = rN(t) > 0$$

Biologically, if the per capita growth rate is positive (meaning that the birth rate β is larger than the death rate μ), then the yeast population will increase. On the other hand, if $r < 0$ (the birth rate β is smaller than the death rate μ), then from Equation 3

$$\frac{dN(t)}{dt} = rN(t) < 0$$

and the yeast population will decrease.

To make more progress, we would like to obtain an explicit function $N(t)$ that tells us exactly what the population size will be at any time. Such a function $N(t)$ is called a *solution* of the differential equation. It is a function that, when substituted into both sides of the differential equation, produces an equality.

Equation 3 tells us that $N(t)$ is a function whose derivative is equal to the function itself, multiplied by a constant, r. As we have seen in Chapter 3, exponential functions have exactly this property. In fact we can see that the function $N(t) = Ce^{rt}$ satisfies the differential equation. In particular, substituting this choice of $N(t)$ into Equation 3, we obtain

$$N'(t) = C(re^{rt}) = r(Ce^{rt}) = rN(t)$$

demonstrating that $N(t) = Ce^{rt}$ does, in fact, satisfy the differential equation. (We will see in Section 7.4 that there is no other solution.) Here C is an arbitrary constant. We can obtain a biological interpretation of this constant by setting $t = 0$: This gives $N(0) = Ce^{r(0)} = C$, revealing that C is the population size at $t = 0$. Figure 3 shows examples of the solution curves for different values of C when $r > 0$.

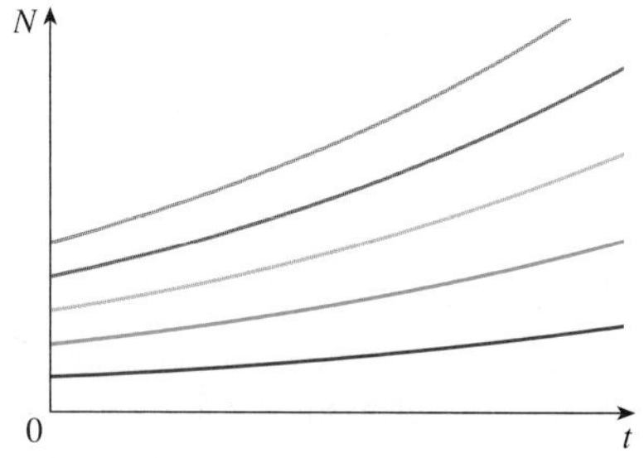

FIGURE 3
The family of solutions $N(t) = Ce^{rt}$ with $r > 0$, $t \geqslant 0$, and different values of C

We can already see from Figure 3 that Equation 3 does not capture all of the features of the data in Figure 1. For example, it appears to predict continued population growth. To obtain a more satisfying comparison, however, we should choose appropriate values for the constants C and r.

From the data in Table 1 we see that $N(0) = 0.200$ and therefore $C = 0.200$. One way to obtain a suitable value for r is to consider the factor by which the population of yeast grew over some fixed period of time. For example, in the first hour the yeast population grew by a factor of

$$\frac{0.330}{0.200} = 1.65$$

On the other hand, according to the model, the factor by which this population is predicted to have grown is

$$\frac{N(1)}{N(0)} = \frac{Ce^{r \cdot 1}}{Ce^{r \cdot 0}} = e^r$$

Therefore a reasonable choice for r would be the value for which $e^r = 1.65$. Solving this equation for r gives $r = \ln 1.65 \approx 0.5$. Thus, our final model is

$$N(t) = 0.2e^{0.5t}$$

Figures 4(a) and 4(b) both plot this equation along with the data from Table 1, but on two different intervals of time. The model provides remarkably accurate predictions over the first 13 or so hours, as shown in Figure 4(a), but its predictions are extremely inaccurate for later time points in the data [see Figure 4(b)].

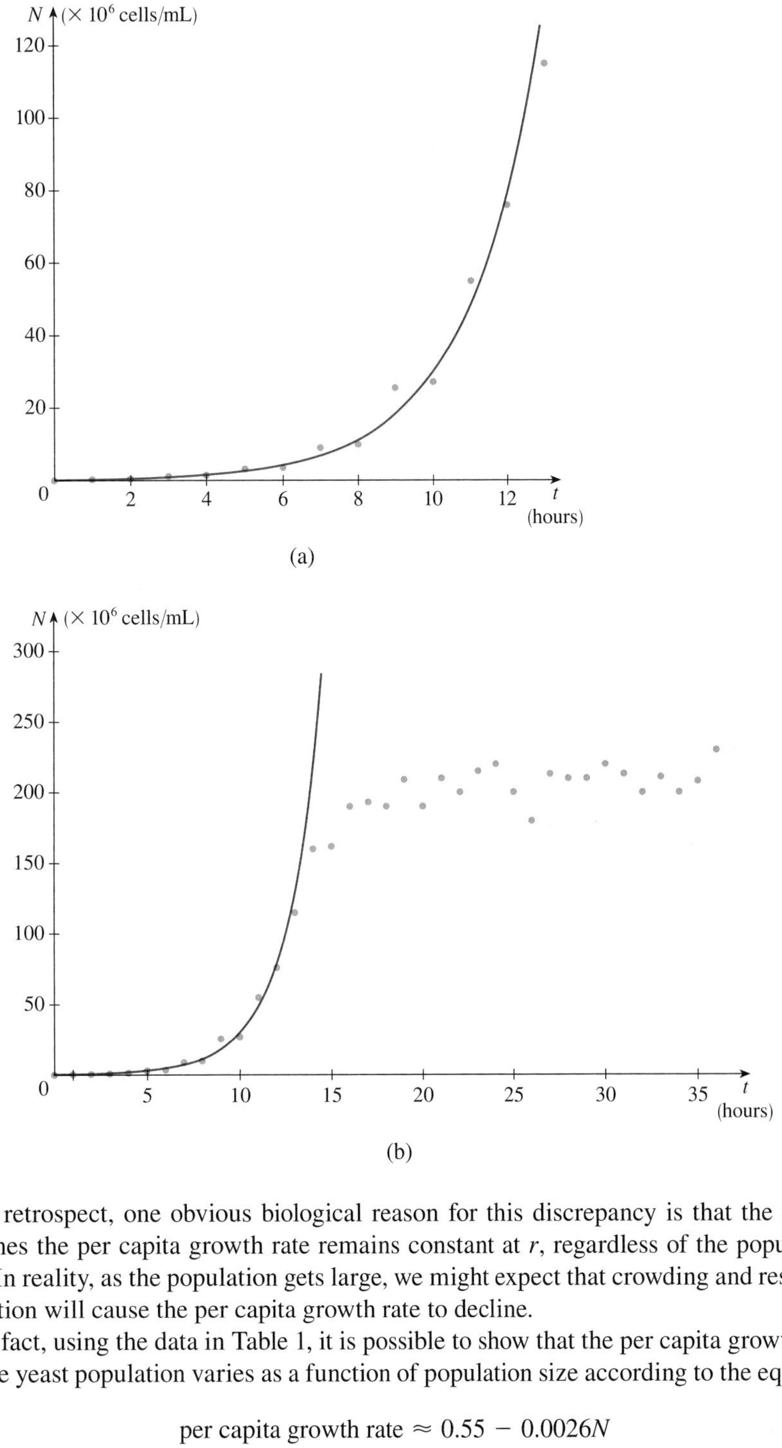

FIGURE 4

In retrospect, one obvious biological reason for this discrepancy is that the model assumes the per capita growth rate remains constant at r, regardless of the population size. In reality, as the population gets large, we might expect that crowding and resource depletion will cause the per capita growth rate to decline.

BB In fact, using the data in Table 1, it is possible to show that the per capita growth rate for the yeast population varies as a function of population size according to the equation

See Exercise 11.3.25.

$$\text{per capita growth rate} \approx 0.55 - 0.0026N$$

In other words,

$$\frac{dN(t)}{dt}\frac{1}{N(t)} \approx 0.55 - 0.0026N(t)$$

Thus a better differential equation for modeling the yeast population is

$$\frac{dN}{dt} = (0.55 - 0.0026N)N$$

We will learn how to analyze differential equations of this form in later sections. For now we simply note that these techniques can be used to show that the solution is

$$N(t) = \frac{42e^{0.55t}}{209.8 + 0.2e^{0.55t}}$$

(See Exercise 18.) This function is plotted in Figure 5 along with the data from Table 1. We see that this model provides quite accurate predictions over the entire time period of the experiment.

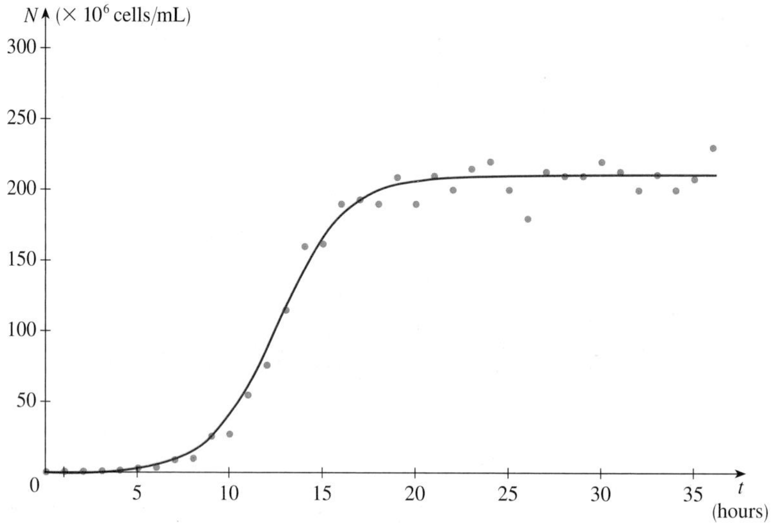

FIGURE 5

Our revised yeast model is a specific example of a more general model for population growth called the *logistic differential equation.* Suppose that the per capita growth rate of a population decreases linearly as the population size increases, from a value of r when $N = 0$ to a value of 0 when $N = K$. The positive constant K is referred to as the *carrying capacity*; it is the population size at which crowding and resource depletion cause the per capita growth rate to be zero. In Exercise 16 you are asked to show that this results in the differential equation

The logistic growth equation was first proposed by Dutch mathematical biologist Pierre-François Verhulst in the 1840s as a model for world population growth.

(4)
$$\frac{dN}{dt} = r\left(1 - \frac{N}{K}\right)N$$

Equation 4 is called the **logistic differential equation**, or more simply the **logistic equation**. In Exercise 17 you are asked to show that, for the yeast model, $r = 0.55$ and $K \approx 210$.

We can obtain some qualitative features of the solutions of Equation 4 by inspection. We first observe that the constant functions $N(t) = 0$ and $N(t) = K$ are solutions because, in either case, the left side of Equation 4 is then zero (the derivative of a constant is zero), and the right side is zero as well. Such constant solutions are called *equilibrium solutions.* (A formal definition of an equilibrium solution will be given in Section 7.2.)

If the initial population $N(0)$ lies between 0 and K, then the right side of Equation 4 is positive, so $N'(t) > 0$ and the population increases (assuming $r > 0$). But if the population exceeds the carrying capacity ($N > K$), then $1 - N/K$ is negative, so $N'(t) < 0$ and the population decreases. In either case, if the population approaches the carrying capacity ($N \to K$), then $N'(t) \to 0$, which means the population levels off.

■ Classifying Differential Equations

Differential equations involve an unknown function and its derivatives. The **order** of the differential equation is the order of the highest derivative appearing in the equation. For example, $y'(t) + 2y(t) = 3$ is a first-order differential equation, whereas $5y''(t) - y'(t) = y(t)$ is a second-order differential equation. The **solution** of a differential equation is a function that, when substituted into the equation, produces an equality. For example, we can verify that the function $y(t) = e^t - 2$ is a solution of the differential equation $dy/dt = 2 + y(t)$ as follows: Substituting the function into the left side of this differential equation gives

$$\frac{dy}{dt} = \frac{d}{dt}(e^t - 2) = e^t$$

and substituting it into the right side gives

$$2 + (e^t - 2) = e^t$$

The right and left sides evaluate to the same expression, demonstrating that the function $y(t) = e^t - 2$ is indeed a solution.

Typically, there are several solutions to a differential equation. In many problems we need to find the particular solution that satisfies an additional condition of the form $y(t_0) = y_0$. This is called an **initial condition**. The problem of finding a solution of the differential equation that also satisfies an initial condition is called an **initial-value problem**. Graphically, when we impose an initial condition, we look at the family of solution curves and pick the one that passes through the point (t_0, y_0). For an example involving the logistic equation, see Figure 6. This corresponds to measuring the state of a system at time t_0 and using the solution of the initial-value problem to predict the future behavior of the system.

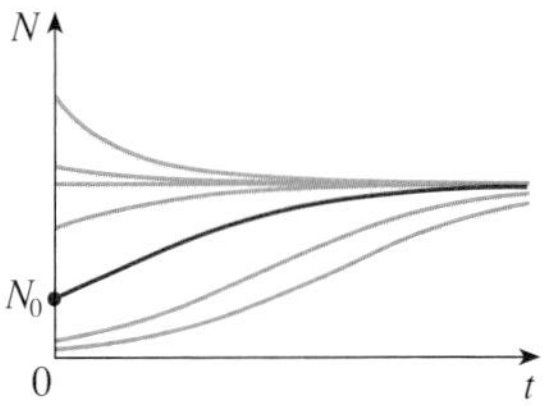

FIGURE 6
The family of solutions of the logistic equation. The solution curve satisfying the initial condition $N(0) = N_0$ is shown in red.

Verifying a solution is relatively easy, but obtaining a solution in the first place may not be. The difficulty of this task—and indeed whether or not it is even possible—is determined by the type of the differential equation. We consider three types of first-order differential equations: pure time, autonomous, and nonautonomous differential equations.

Pure-Time Differential Equations

Pure-time differential equations involve the derivative of the function but not the function itself. For example, if the rate of change of population size y depends on time only, this results in a differential equation of the form

$$\frac{dy}{dt} = f(t)$$

We have already studied this type of equation in the context of antidifferentiation (in Section 4.6) and integration (in Chapter 5). We can obtain the solution $y(t)$ by calculating the antiderivative of $f(t)$. Although we refer to such equations as *pure-time differential equations*, the independent variable need not be time.

EXAMPLE 1 | **Spatial species distributions** As we move up a stream from its mouth toward its source, suppose that the population size n of a species of insect at a fixed point in time changes over space according to

$$\frac{dn}{dx} = 1 - 2e^{-x}$$

where $0 \leq x \leq 10$ is the spatial location (in km) between the mouth ($x = 0$ km) and a dam ($x = 10$ km). (This situation is described in Figure 7.) Suppose the population size at the dam is $n(10) = 20$. Obtain an expression for the population size as a function of distance from the mouth.

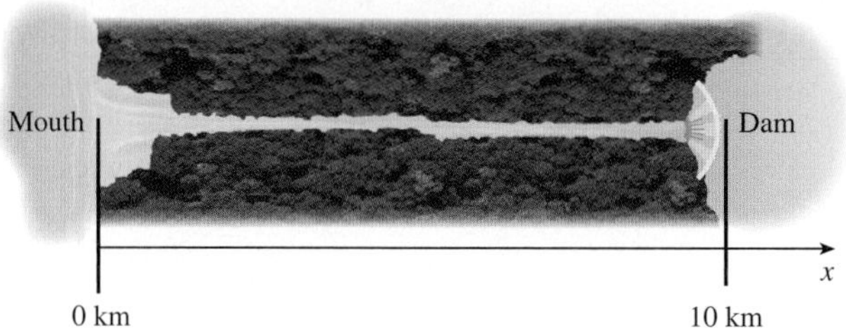

FIGURE 7
Population density along a stream

SOLUTION We first seek a function $n(x)$ that satisfies the given differential equation. This function can be obtained by integrating both sides of the differential equation with respect to x:

$$\frac{dn}{dx} = 1 - 2e^{-x}$$

$$\int \frac{dn}{dx}\,dx = \int (1 - 2e^{-x})\,dx$$

$$n(x) = x + 2e^{-x} + C$$

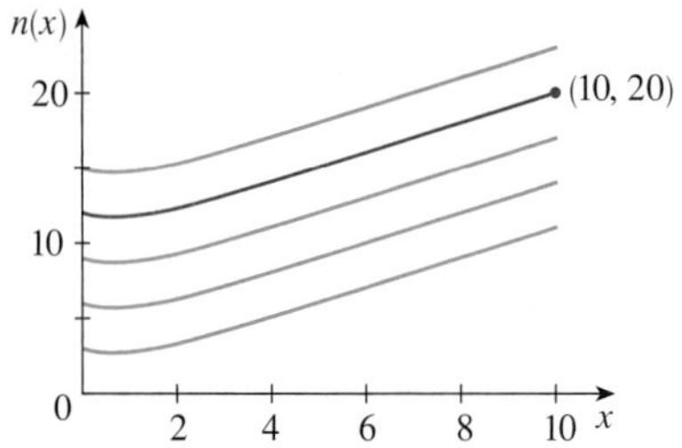

FIGURE 8
The family of solutions giving population size along the stream. The solution curve satisfying the initial condition $n(10) = 20$ is shown in red.

The function $n(x) = x + 2e^{-x} + C$ is a family of solutions. We now need to choose the specific function from this family that satisfies the condition $n(10) = 20$. Substituting $x = 10$ into $n(x)$ gives

$$n(10) = 10 + 2e^{-10} + C = 20$$

This tells us that we must choose $C = 10 - 2e^{-10}$. Therefore the population size as a function of x is $n(x) = x + 2e^{-x} + 10 - 2e^{-10}$. See Figure 8. ■

Autonomous Differential Equations

Autonomous differential equations arise when the equation involves both the derivative of the function and the function itself, but when there is no explicit dependence on the independent variable. Such equations have the general form

$$\frac{dy}{dt} = g(y)$$

where y is the unknown function of the independent variable t. Equations 3 and 4 are examples of autonomous differential equations.

EXAMPLE 2 | BB **Modeling intravenous drug delivery** Often the rate at which the body metabolizes a drug is proportional to the current concentration of the drug. In other words, if $y(t)$ is the concentration of a drug in the bloodstream at time t

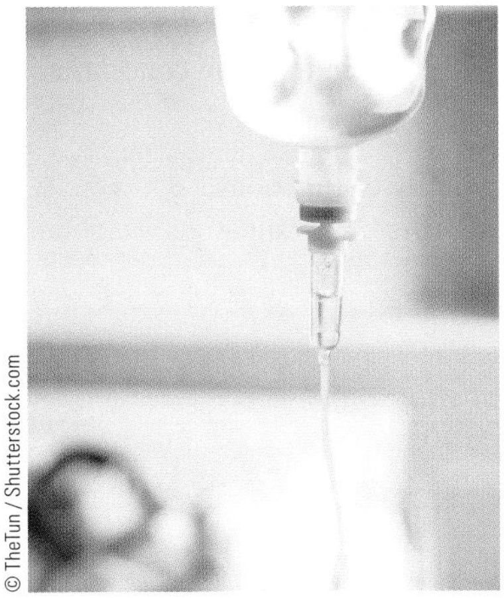

(measured in mg/mL), then

$$\text{outflow of drug through metabolism} = ky$$

where k is a positive constant of proportionality (with units 1/hour).

For drugs administered through a constant intravenous supply, the concentration in the bloodstream will also be replenished at a rate that is determined by the drug concentration in the supply:

$$\text{inflow of drug through IV supply} = A$$

where A is a positive constant with units mg/(mL hour). The total rate of change of concentration resulting from both processes (that is, dy/dt) is therefore

$$\frac{dy}{dt} = \text{inflow} - \text{outflow}$$

or

$$\frac{dy}{dt} = A - ky \qquad (5)$$

Equation 5 is an autonomous differential equation because it involves the dependent variable y but not the independent variable t. Figure 9 shows a family of solutions to differential equation (5), and suggests that the drug concentration is predicted to approach a limiting value of A/k at time passes, regardless of the initial concentration. ■

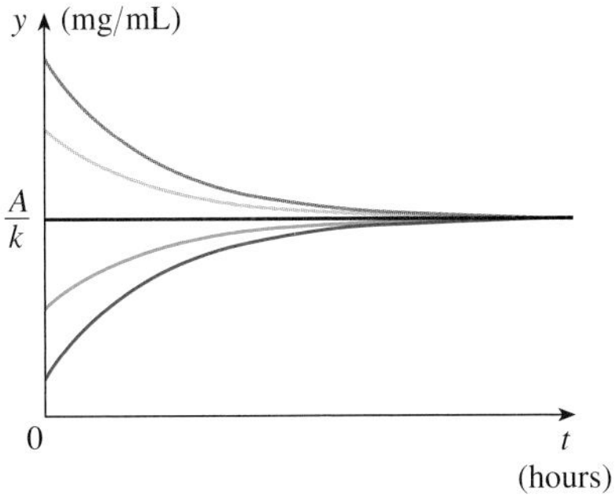

FIGURE 9
The family of solutions of Equation 5

Nonautonomous Differential Equations

Nonautonomous differential equations are a combination of pure-time and autonomous differential equations. They arise when the equation involves the function and its derivative, and the independent variable appears explicitly as well.

EXAMPLE 3 | Administering drugs A drug is administered to a patient intravenously at a time-varying rate of $A(t) = 1 + \sin t$ mg/(mL hour), and is metabolized at a rate of $y(t)$ mg/(mL hour), where $y(t)$ is the concentration at time t (in units of mg/mL). Thus y obeys the differential equation

$$\frac{dy}{dt} = 1 + \sin t - y \qquad (6)$$

Verify that the family of functions $y(t) = Ce^{-t} + \frac{1}{2}(2 - \cos t + \sin t)$ satisfies the differential equation.

SOLUTION Substituting $y(t)$ into the left side of the differential equation (6) gives $-Ce^{-t} + \frac{1}{2}(\sin t + \cos t)$. Substituting it into the right side gives

$$\begin{aligned} 1 + \sin t - y &= 1 + \sin t - Ce^{-t} - \tfrac{1}{2}(2 - \cos t + \sin t) \\ &= -Ce^{-t} + \tfrac{1}{2}(\sin t + \cos t) \end{aligned}$$

Since both quantities are the same, this family of functions y satisfies the differential equation. ■

EXAMPLE 4 | Administering drugs (continued) What is the drug concentration as a function of time for the model in Example 3 if the initial drug concentration at $t = 0$ is zero?

SOLUTION We seek the specific member from the family of functions, $y(t) = Ce^{-t} + \frac{1}{2}(2 - \cos t + \sin t)$, that also satisfies $y(0) = 0$. Evaluating, we obtain

$$y(0) = Ce^{-0} + \tfrac{1}{2}(2 - \cos 0 + \sin 0) = C + \tfrac{1}{2} = 0$$

and therefore $C = -\frac{1}{2}$. Thus the solution to the *initial-value problem* is $y(t) = \frac{1}{2}(2 - e^{-t} - \cos t + \sin t)$, as shown in Figure 10.

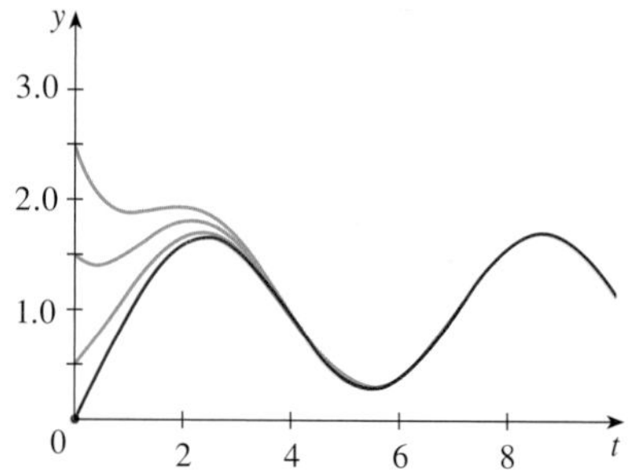

FIGURE 10
The family of solutions giving drug concentration $y(t)$. The solution curve satisfying the initial condition $y(0) = 0$ is shown in red.

■

EXERCISES 7.1

1. Show that $y = \frac{2}{3}e^x + e^{-2x}$ is a solution of the differential equation $y' + 2y = 2e^x$. Is this differential equation pure-time, autonomous, or nonautonomous?

2. Verify that $y = -t \cos t - t$ is a solution of the initial-value problem

$$t\frac{dy}{dt} = y + t^2 \sin t \qquad y(\pi) = 0$$

Is this differential equation pure-time, autonomous, or nonautonomous?

3. Show that $y = e^{-at} \cos t$ is a solution of the differential equation $y' = -e^{-at}(a \cos t + \sin t)$. Is this differential equation pure-time, autonomous, or nonautonomous?

4. (a) Show that every member of the family of functions $y = (\ln x + C)/x$ is a solution of the differential equation $x^2y' + xy = 1$.
(b) Illustrate part (a) by graphing several members of the family of solutions on a common screen.
(c) Find a solution of the differential equation that satisfies the initial condition $y(1) = 2$.
(d) Find a solution of the differential equation that satisfies the initial condition $y(2) = 1$.

5. (a) What can you say about a solution of the equation $y' = -y^2$ just by looking at the differential equation?
(b) Verify that all members of the family $y = 1/(x + C)$ are solutions of the equation in part (a).
(c) Can you think of a solution of the differential equation $y' = -y^2$ that is not a member of the family in part (b)?
(d) Find a solution of the initial-value problem

$$y' = -y^2 \qquad y(0) = 0.5$$

6. (a) What can you say about the graph of a solution of the equation $y' = xy^3$ when x is close to 0? What if x is large?
(b) Verify that all members of the family $y = (c - x^2)^{-1/2}$ are solutions of the differential equation $y' = xy^3$.
(c) Graph several members of the family of solutions on a common screen. Do the graphs confirm what you predicted in part (a)?
(d) Find a solution of the initial-value problem

$$y' = xy^3 \qquad y(0) = 2$$

7. Logistic growth A population is modeled by the differential equation

$$\frac{dN}{dt} = 1.2N\left(1 - \frac{N}{4200}\right)$$

where $N(t)$ is the number of individuals at time t (measured in days).
(a) For what values of N is the population increasing?
(b) For what values of N is the population decreasing?
(c) What are the equilibrium solutions?

8. The Fitzhugh-Nagumo model for the electrical impulse in a neuron states that, in the absence of relaxation effects, the electrical potential in a neuron $v(t)$ obeys the differential equation

$$\frac{dv}{dt} = -v[v^2 - (1 + a)v + a]$$

where a is a constant and $0 < a < 1$.

(a) For what values of v is v unchanging (that is, $dv/dt = 0$)?
(b) For what values of v is v increasing?
(c) For what values of v is v decreasing?

9. Explain why the functions with the given graphs *can't* be solutions of the differential equation

$$\frac{dy}{dt} = e^t(y - 1)^2$$

(a)

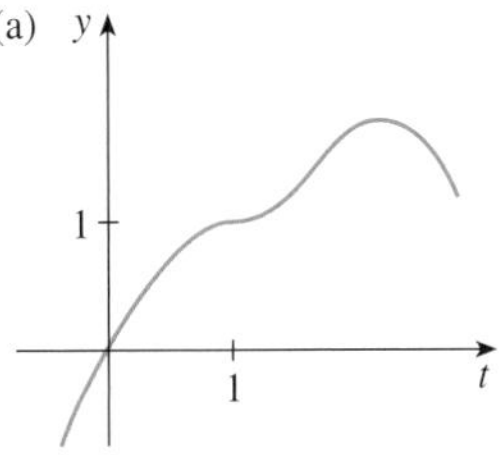

(b)

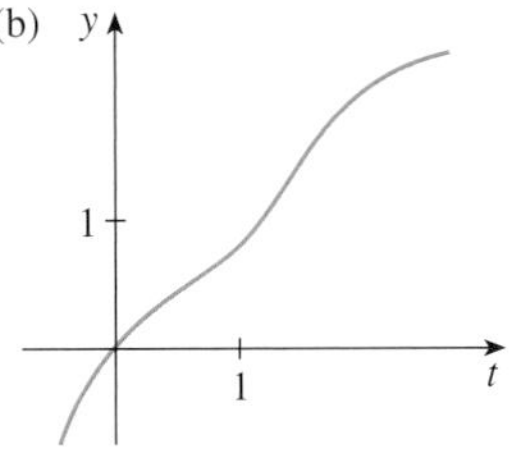

10. The function with the given graph is a solution of one of the following differential equations. Decide which is the correct equation and justify your answer.

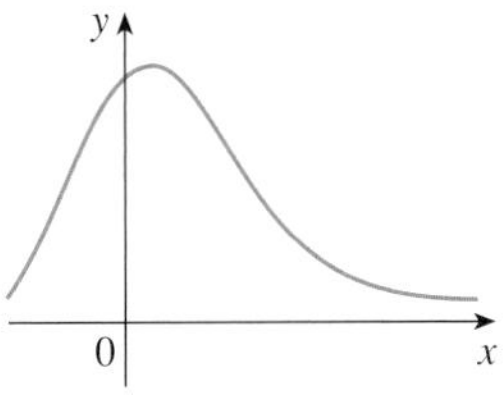

A. $y' = 1 + xy$ **B.** $y' = -2xy$ **C.** $y' = 1 - 2xy$

11. Match the differential equations with the solution graphs labeled I–IV. Give reasons for your choices.

(a) $y' = 1 + x^2 + y^2$ (b) $y' = xe^{-x^2-y^2}$

(c) $y' = \dfrac{1}{1 + e^{x^2+y^2}}$ (d) $y' = \sin(xy)\cos(xy)$

I

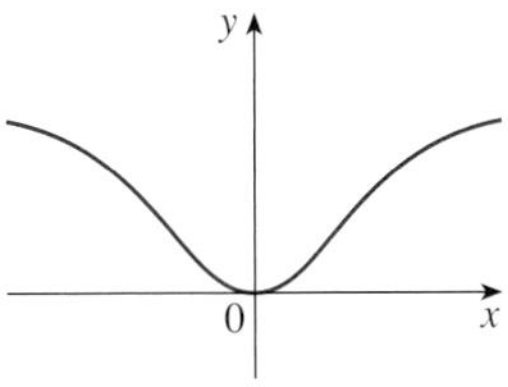

II

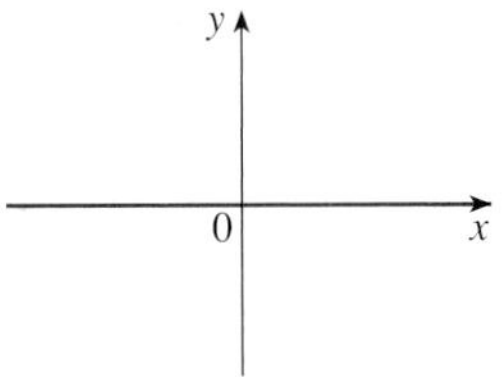

III

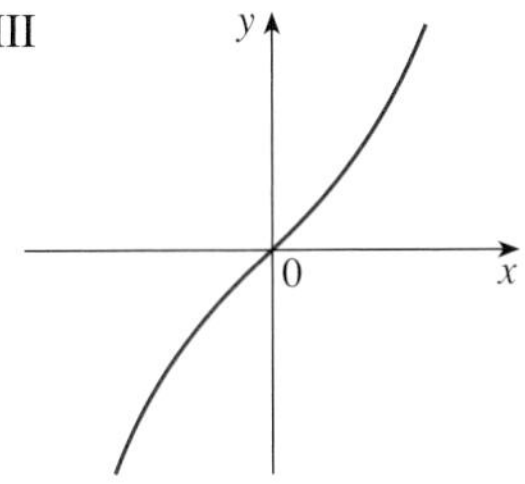

IV

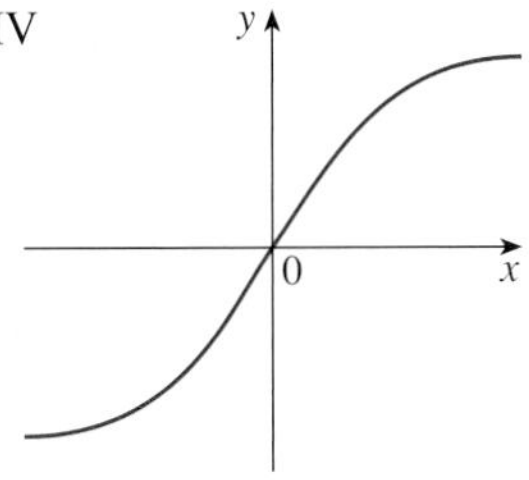

12. Von Bertalanffy's equation states that the rate of growth in length of an individual fish is proportional to the difference between the current length L and the asymptotic length L_∞ (in cm).

(a) Write a differential equation that expresses this idea.
(b) Make a rough sketch of the graph of a solution to a typical initial-value problem for this differential equation.

13–15 Drug dissolution Differential equations have been used extensively in the study of drug dissolution for patients given oral medications. The three simplest equations used are the zero-order kinetic equation, the Noyes-Whitney equation, and the Weibull equation. All assume that the initial concentration is zero but make different assumptions about how the concentration increases over time during the dissolution of the medication.

13. The **zero-order kinetic equation** states that the rate of change in the concentration of drug c (in mg/mL) during dissolution is governed by the differential equation

$$\frac{dc}{dt} = k$$

where k is a positive constant. Is this differential equation pure-time, autonomous, or nonautonomous? State in words what this differential equation says about how drug dissolution occurs. What is the solution of this differential equation with the initial condition $c(0) = 0$?

14. The **Noyes-Whitney equation** for the dynamics of the drug concentration is

$$\frac{dc}{dt} = k(c_s - c)$$

where k and c_s are positive constants. Is this differential equation pure-time, autonomous, or nonautonomous? State in words what this differential equation says about how drug dissolution occurs. Verify that $c = c_s(1 - e^{-kt})$ is the solution to this equation for the initial condition $c(0) = 0$.

15. The **Weibull equation** for the dynamics of the drug concentration is

$$\frac{dc}{dt} = \frac{k}{t^b}(c_s - c)$$

where k, c_s, and b are positive constants and $b < 1$. Notice that this differential equation is undefined when $t = 0$. Is this differential equation pure-time, autonomous, or nonautonomous? State in words what this differential equation says about how drug dissolution occurs. Verify that

$$c = c_s\left(1 - e^{-\alpha t^{1-b}}\right)$$

is a solution for $t \neq 0$, where $\alpha = k/(1 - b)$.

---

16. The logistic differential equation Suppose that the per capita growth rate of a population of size N declines linearly from a value of r when $N = 0$ to a value of 0 when $N = K$.

Show that the differential equation for N is

$$\frac{dN}{dt} = r\left(1 - \frac{N}{K}\right)N$$

17. Modeling yeast populations Use the fact that the per capita growth rate of the yeast population in Table 1 is $0.55 - 0.0026N$ to show that, in terms of the logistic equation (4), $r = 0.55$ and $K \approx 210$.

18. Modeling yeast populations (cont.) Verify that

$$N(t) = \frac{42e^{0.55t}}{209.8 + 0.2e^{0.55t}}$$

is an approximate solution of the differential equation

$$\frac{dN}{dt} = (0.55 - 0.0026N)N$$

PROJECT Chaotic Blowflies and the Dynamics of Populations

BB

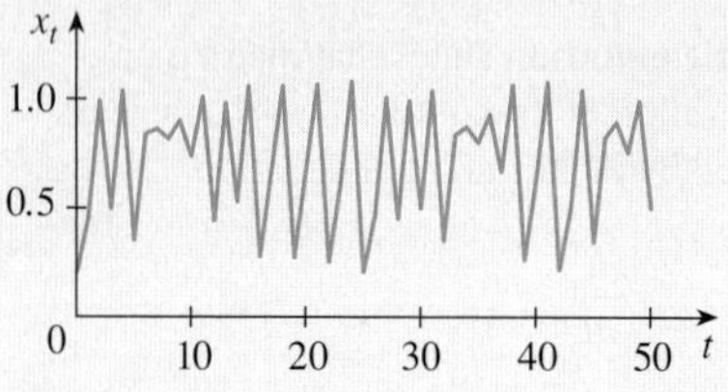

FIGURE 1
x is plotted against time with $x_0 = 0.1$ and $R_{max} = 3.89$.

In Section 1.6 we explored the dynamics of the logistic difference equation. After some simplification, the population size in successive times steps was given by the recursion

(1) $$x_{t+1} = R_{max}x_t(1 - x_t)$$

where R_{max} is a positive constant. See Equation 1.6.7. For large enough values of R_{max} the recursion exhibits very complicated behavior, as shown in Figure 1. In fact, Equation 1 is famous for being one of the simplest recursions that exhibits chaotic dynamics.[1]

The plots for the logistic differential equation that we have seen in Section 7.1 do not exhibit this type of complicated behavior. Here we explore why. To do so, we will derive the logistic differential equation from the logistic difference equation.

1. In Section 1.6 we obtained Equation 1 by starting with the equation

$$N_{t+1} = [1 + r(1 - N_t/K)]\,N_t$$

(See Equation 1.6.5.) If the time interval is of length h instead, where $h < 1$, then this equation becomes

$$N_{t+h} = [1 + rh(1 - N_t/K)]\,N_t$$

Use this result to derive a differential equation for N by writing an expression for $\dfrac{N_{t+h} - N_t}{h}$ and then letting $h \to 0$.

2. Show that with the change of variables $y = N/K$ the differential equation from Problem 1 can be written as $dy/dt = ry(1 - y)$.

3. In Section 7.4 we will learn how to solve differential equations like the one in Problem 2. If $y(0) = y_0$, the solution is $y(t) = y_0/[e^{-rt} + y_0(1 - e^{-rt})]$. Sketch this solution for different choices of y_0 and r. This solution can never exhibit the sort of behavior of Equation 1 that is displayed in Figure 1. Explain why from a biological standpoint.

A blowfly

The reason for the complicated dynamics in Figure 1 is the existence of a time-lag between the current population size x_t and its effects on population regulation. This allows the population to overshoot its carrying capacity. Once an overshoot occurs, a dramatic population decline will ensue. The resulting low population size then sets the stage for a large population rebound and another overshoot of the carrying capacity. Some

1. R. May, "Simple Mathematical Models with Very Complicated Dynamics," *Nature* 261 (1976): 459–67.

insect species, like blowflies, are believed to be subject to such dynamics, as shown in Figure 2. The exploration of chaos in nature is a fascinating area of research.[2]

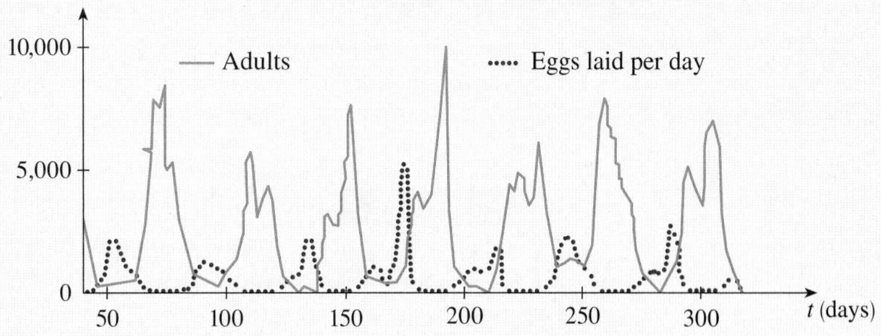

FIGURE 2
Irregular fluctuations in a lab population of blowflies

Source: Adapted from W. Gurney et al., "Nicholson's Blowflies Revisited," *Nature* 287 (1980): 17–21.

2. A. Hastings et al., "Chaos in Ecology: Is Mother Nature a Strange Attractor?" *Annual Review of Ecology, Evolution, and Systematics* 24 (1993): 1–33.

7.2 Phase Plots, Equilibria, and Stability

We now examine some techniques for analyzing the behavior of differential equations. Here we focus on a graphical technique known as a *phase plot.*

Phase Plots

Phase plots provide a way to visualize the dynamics of autonomous differential equations, to locate their equilibria, and to determine the stability properties of these equilibria. Consider the autonomous differential equation

$$\frac{dy}{dt} = g(y)$$

To construct a phase plot we graph the right side of the differential equation, $g(y)$, as a function of the dependent variable y. Where this plot lies above the horizontal axis, $y'(t) > 0$ and so y is increasing. Where it lies below the horizontal axis, $y'(t) < 0$ and so y is decreasing. Points where the plot crosses the axis correspond to values of the variable at which $y'(t) = 0$. We can use these considerations to place arrows on the horizontal axis indicating the direction of change in y, as shown in Figure 1.

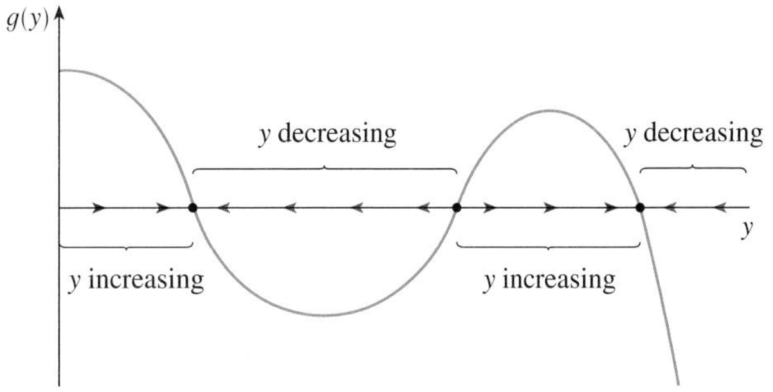

FIGURE 1
A typical phase plot

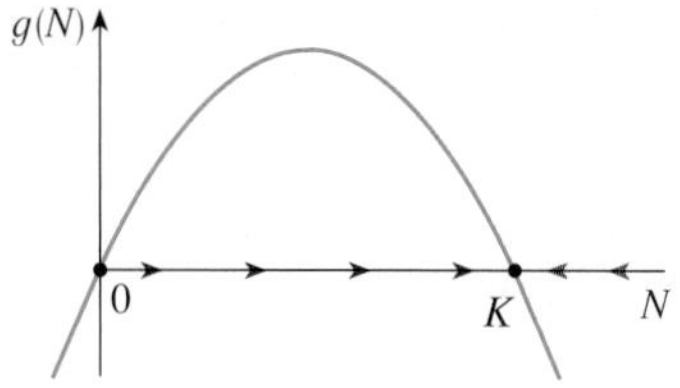

FIGURE 2
Phase plot for logistic model

EXAMPLE 1 | **The logistic equation** Construct a phase plot for the logistic growth model $dN/dt = r(1 - N/K)N$ assuming that $r > 0$.

SOLUTION We need to plot $g(N) = r(1 - N/K)N$ as a function of N. This is a parabola opening downward, crossing the horizontal axis at $N = 0$ and $N = K$. In Figure 2 we can see that N increases when taking on values between 0 and K and decreases when taking on values greater than K. ■

EXAMPLE 2 | **Modeling intravenous drug delivery** In Example 7.1.2 we developed a model for the dynamics of the concentration of a drug in the bloodstream. If A is a positive constant representing the rate of drug delivery through an intravenous supply and k is a positive constant related to the rate of metabolism of the drug, we obtained

$$\frac{dy}{dt} = A - ky$$

where $y(t)$ is the concentration of drug in the bloodstream at time t (in mg/mL). Construct a phase plot for this model.

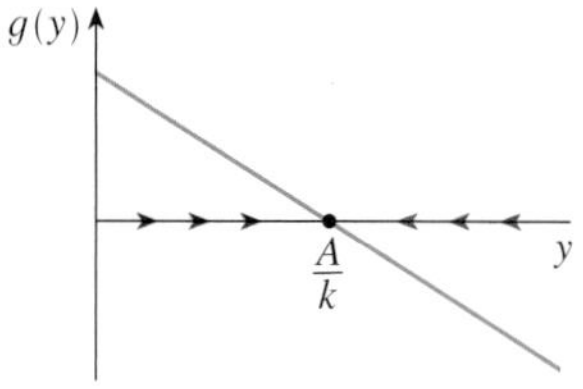

FIGURE 3
Phase plot for Example 2

SOLUTION We plot $g(y) = A - ky$ as a function of y. This is a straight line with slope $-k$, intersecting the horizontal axis at $y = A/k$. We see that y increases when it is less than A/k and decreases when greater than A/k. (See Figure 3.) This agrees with the plot of y as a function of time in Figure 7.1.9. ■

EXAMPLE 3 | BB **The Allee effect** Some populations decline to extinction once their size is less than a critical value. For example, if the population size is too small, then individuals might have difficulty finding mates for reproduction. This is referred to as an *Allee effect* after the American ecologist Warder Clyde Allee (1885–1955). A simple extension of the logistic model that incorporates this effect is given by

$$\frac{dN}{dt} = r(N - a)\left(1 - \frac{N}{K}\right)N$$

where $0 < a < K$. Construct a phase plot assuming that $r > 0$.

SOLUTION We plot $g(N) = r(N - a)(1 - N/K)N$ as a function of N. This is a cubic polynomial whose graph crosses the horizontal axis at $N = 0$, $N = a$, and $N = K$. The graph lies below the horizontal axis for values of N between 0 and a and for values of $N > K$. [See Figure 4(a).] Therefore N will approach 0 if it starts between 0 and a, whereas it will approach K if it starts anywhere greater than a. Figure 4(b) displays data for the phase plot of an experimental microbial population.

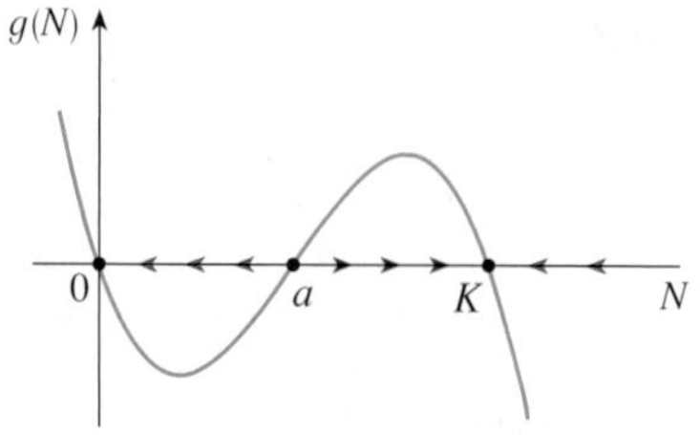

(a) Phase plot for an Allee effect

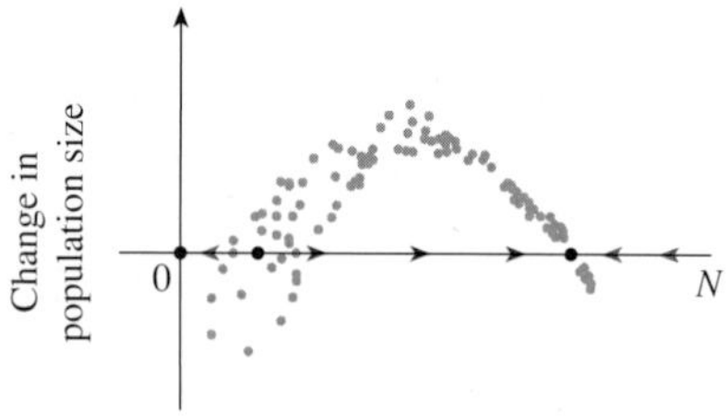

(b) Data for the phase plot of a microbial population

FIGURE 4

Source: Part (b) adapted from L. Dai et al., "Generic Indicators for Loss of Resilience before a Tipping Point Leading to Population Collapse," *Science* 336 (2012): 1175–77.

■

Equilibria and Stability

Phase plots provide information about how the dependent variable changes, as well as values of the variable at which no change occurs. Such values are referred to as *equilibria.*

> **Definition** Consider the autonomous differential equation
>
> **(1)** $$\frac{dy}{dt} = g(y)$$
>
> An **equilibrium** solution is a constant value of y (denoted $\hat{y}$) such that $dy/dt = 0$ when $y = \hat{y}$.

Equilibria are found by determining values of $\hat{y}$ that satisfy $g(\hat{y}) = 0$. They correspond to places where the phase plot crosses the horizontal axis.

EXAMPLE 4 | The logistic equation (continued) Show that $\hat{N} = 0$ and $\hat{N} = K$ are equilibria of the logistic growth model from Example 1.

SOLUTION Substituting $\hat{N} = 0$ and $\hat{N} = K$ into the equation $g(N) = r(1 - N/K)N$ gives $g(\hat{N}) = 0$ in both cases. In the yeast data of Section 7.1, $\hat{N} = K$ corresponds to the steady number of yeast cells reached as the experiment progressed. ($\hat{N} = 0$ corresponds to the absence of yeast.) ■

EXAMPLE 5 | Modeling intravenous drug delivery (continued) Find all equilibria of the model in Example 2.

SOLUTION We seek values of $\hat{y}$ that make $g(\hat{y}) = 0$; that is, values of $\hat{y}$ that satisfy the equation $A - k\hat{y} = 0$. The only solution is $\hat{y} = A/k$. Again this agrees with the plot of y as a function of time t in Figure 7.1.9. ■

In addition to providing a way to visualize equilibria, phase plots provide information about their stability properties.

> **Definition** An equilibrium $\hat{y}$ of differential equation (1) is **locally stable** if y approaches the value $\hat{y}$ as $t \to \infty$ for all initial values of y sufficiently close to $\hat{y}$.

An equilibrium that is not stable is referred to as **unstable**. In Example 1, the arrows on the horizontal axis of Figure 2 reveal that $\hat{N} = 0$ is an unstable equilibrium whereas $\hat{N} = K$ is a locally stable equilibrium. In Example 2, the arrows in Figure 3 reveal that $\hat{y} = A/k$ is a locally stable equilibrium.

EXAMPLE 6 | BB The Allee effect (continued) Find all equilibria for the model of an Allee effect in Example 3 and determine their stability properties from the phase plot in Figure 4(a).

SOLUTION We need to find the values of $\hat{N}$ that satisfy the equation

$$r(\hat{N} - a)(1 - \hat{N}/K)\hat{N} = 0$$

We can see that these are $\hat{N} = 0$, $\hat{N} = a$, and $\hat{N} = K$. These are the points at which the phase plot in Figure 4(a) crosses the horizontal axis. From the arrows on the figure we can also see that both $\hat{N} = 0$ and $\hat{N} = K$ are locally stable whereas $\hat{N} = a$ is unstable—no matter how close we start N to the value a, it always moves farther away from a as time passes. ■

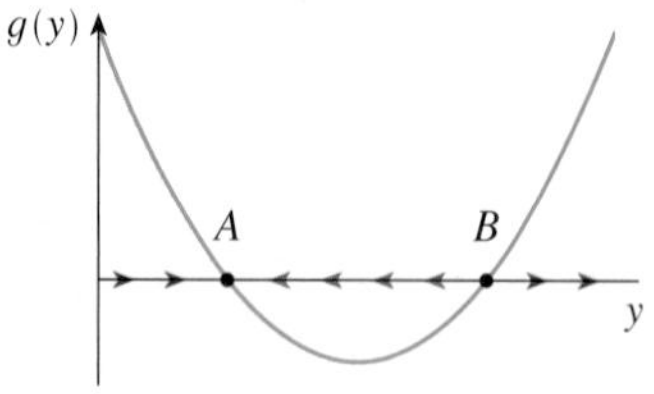

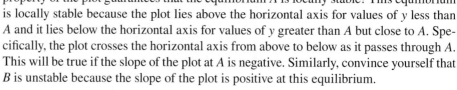

FIGURE 5
A phase plot

We have used phase plots to determine stability graphically, but these plots also provide a guide for using calculus to do so. Let's examine the phase plot in Figure 5. What property of the plot guarantees that the equilibrium A is locally stable? This equilibrium is locally stable because the plot lies above the horizontal axis for values of y less than A and it lies below the horizontal axis for values of y greater than A but close to A. Specifically, the plot crosses the horizontal axis from above to below as it passes through A. This will be true if the slope of the plot at A is negative. Similarly, convince yourself that B is unstable because the slope of the plot is positive at this equilibrium.

Local Stability Criterion Suppose that $\hat{y}$ is an equilibrium of the differential equation

$$\frac{dy}{dt} = g(y)$$

Then $\hat{y}$ is *locally stable* if $g'(\hat{y}) < 0$, and $\hat{y}$ is *unstable* if $g'(\hat{y}) > 0$. If $g'(\hat{y}) = 0$, then the analysis is inconclusive.

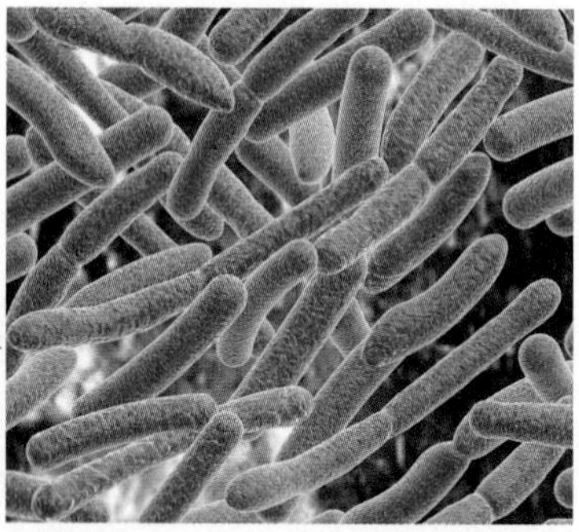

FIGURE 6
Cross-feeding bacteria

EXAMPLE 7 | Population genetics Two bacterial strains sometimes feed on chemicals excreted by one another: strain A feeds on chemicals produced by strain B, and vice versa. This phenomenon is referred to as *cross-feeding*. Suppose that two strains of bacteria are engaged in cross-feeding (strain 1 and strain 2). Exercise 17 asks you to show that, for a relatively simple model of cross-feeding, the frequency $p(t)$ of the strain 1 bacteria is governed by the differential equation

$$\frac{dp}{dt} = p(1 - p)[\alpha(1 - p) - \beta p]$$

where α and β are positive constants. Suppose that $\alpha = 1$ and $\beta = 2$. Then the differential equation simplifies to

$$\frac{dp}{dt} = p(1 - p)[(1 - p) - 2p] = p(1 - p)(1 - 3p)$$

(a) Find all equilibria.
(b) Determine the stability properties of each equilibrium found in part (a).

SOLUTION

(a) Equilibria are values of $\hat{p}$ satisfying the equation

$$\hat{p}(1 - \hat{p})(1 - 3\hat{p}) = 0$$

This gives $\hat{p} = 0$, $\hat{p} = 1$, and $\hat{p} = \frac{1}{3}$.
(b) We first need to calculate the derivative of $g(p) = p(1 - p)(1 - 3p)$ with respect to p. After some simplification we obtain

$$g'(p) = 1 - 8p + 9p^2$$

We then need to evaluate this at each of the equilibria.

- $\hat{p} = 0$: $g'(0) = 1 - 8(0) + 9(0)^2 = 1$

  which is positive, meaning that $\hat{p} = 0$ is unstable.

- $\hat{p} = 1$: $g'(1) = 1 - 8(1) + 9(1)^2 = 2$

  which is positive, meaning that $\hat{p} = 1$ is unstable.

- $\hat{p} = \frac{1}{3}$: $g'\left(\frac{1}{3}\right) = 1 - 8\left(\frac{1}{3}\right) + 9\left(\frac{1}{3}\right)^2 = -\frac{2}{3}$

  which is negative, meaning that $\hat{p} = \frac{1}{3}$ is locally stable.

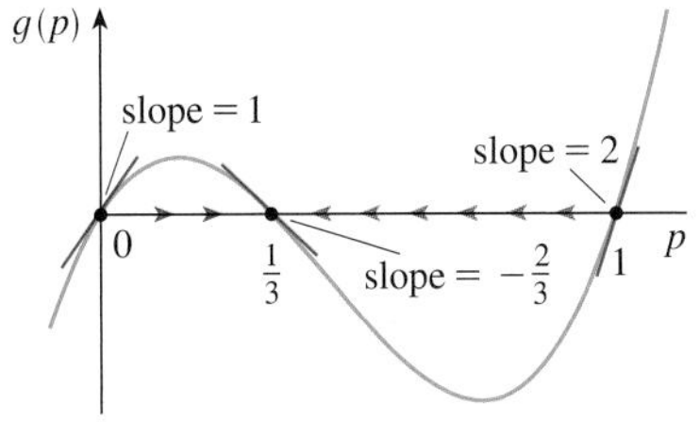

FIGURE 7
Phase plot for cross-feeding model

These results suggest that, over time, we expect the frequency of strain 1 bacteria in the population to approach $\hat{p} = \frac{1}{3}$ as indicated by the arrows in Figure 7. ■

■ A Mathematical Derivation of the Local Stability Criterion

We obtained the local stability criterion from graphical considerations, but we can also derive it more rigorously. Consider the autonomous differential equation

$$\frac{dy}{dt} = g(y)$$

Suppose that $\hat{y}$ is an equilibrium [that is, $g(\hat{y}) = 0$]. To determine if this equilibrium is locally stable, we need to determine whether y will approach the value $\hat{y}$ provided that we start the variable sufficiently close to this value.

Let's consider starting the value of y a small distance ε from $\hat{y}$. In other words, $\varepsilon(t) = y(t) - \hat{y}$. If $\hat{y}$ is locally stable, then the magnitude of $\varepsilon(t)$ must decrease as time passes.

We can derive a differential equation that governs the dynamics of $\varepsilon(t)$ by differentiating $\varepsilon(t)$ with respect to t and using the differential equation for y:

$$\frac{d\varepsilon}{dt} = \frac{d}{dt}[y(t) - \hat{y}] = \frac{dy}{dt} = g(y) = g(\varepsilon + \hat{y})$$

where we have obtained the final equality by using the fact that $\varepsilon(t) = y(t) - \hat{y}$. Now, provided that we start the value of y near $\hat{y}$, ε will be small and we can therefore approximate the function $g(\varepsilon + \hat{y})$ using a linear approximation near the value $\varepsilon = 0$ (see Section 3.8). In other words, we can write

$$g(\varepsilon + \hat{y}) \approx g(\hat{y}) + g'(\hat{y}) \cdot \varepsilon$$

provided that ε is small. Furthermore, because $\hat{y}$ is an equilibrium we know that $g(\hat{y}) = 0$, and therefore the differential equation for ε simplifies to

$$\frac{d\varepsilon}{dt} \approx g'(\hat{y}) \cdot \varepsilon$$

At this point we note that the quantity $g'(\hat{y})$ is a constant (that is, it is not time-varying). As a result, ε approximately obeys a differential equation of the form $d\varepsilon/dt = k\varepsilon$, where $k = g'(\hat{y})$ is a constant. This suggests that $\varepsilon(t)$ grows over time (meaning equilibrium $\hat{y}$ is unstable) if $g'(\hat{y}) > 0$. And $\varepsilon(t)$ decays over time (meaning equilibrium $\hat{y}$ is locally stable) if $g'(\hat{y}) < 0$.

EXERCISES 7.2

1–3 Consider the differential equation $dy/dt = g(y)$.

1. For each graph, determine whether the equilibria (i) and (ii) are locally stable.

(a)
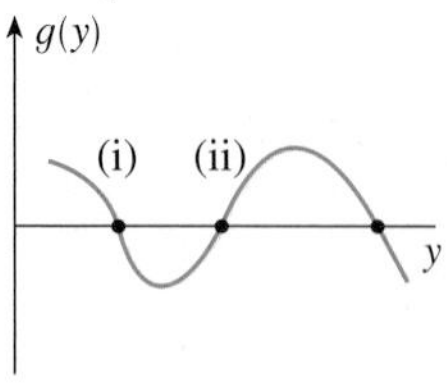

(b)
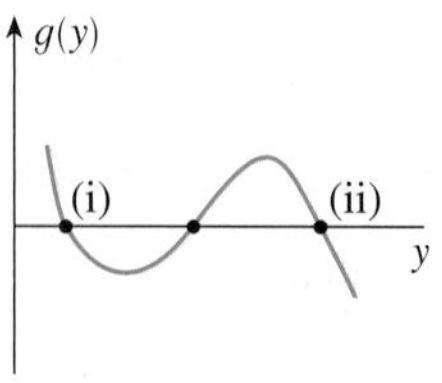

2. For each graph, determine whether the equilibria (i) and (ii) are unstable.

(a)
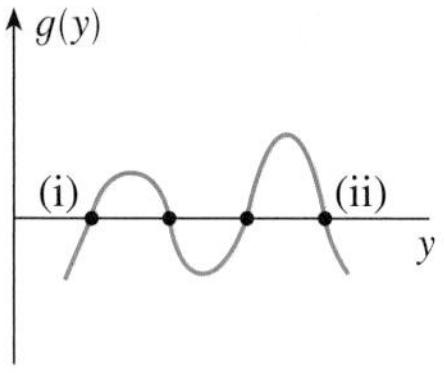

(b)
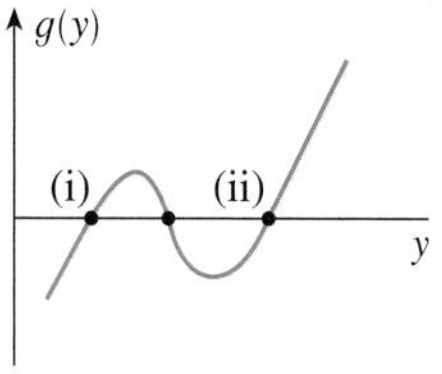

3. Complete the phase plot for each graph by locating the equilibria and indicating the direction in which y changes on the horizontal axis.

(a)
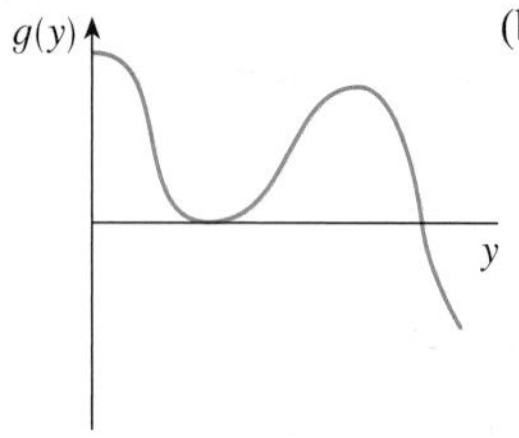

(b)
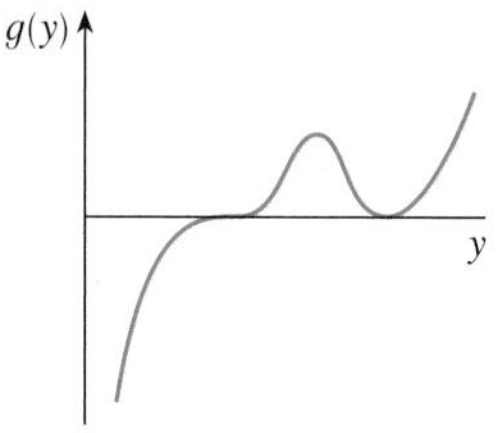

(c)
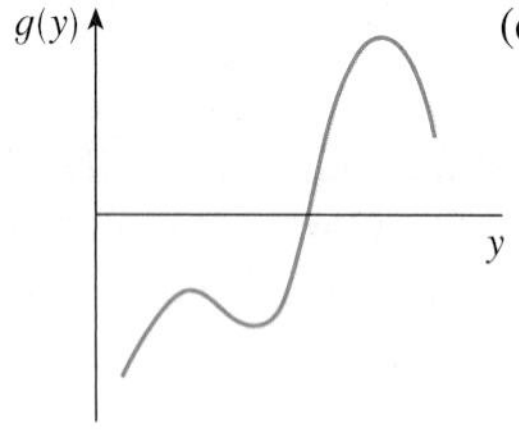

(d)
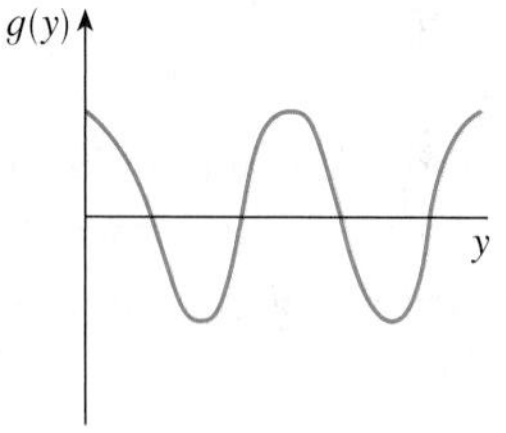

4–6 Find all equilibria of the autonomous differential equation and construct the phase plot.

4. (a) $y' = y^2 - 2$

(b) $y' = \dfrac{y - 3}{y + 9}, \quad y \geqslant 0$

(c) $y' = y(3 - y)(25 - y^2)$

5. (a) $y' = y + 2 \ln y, \quad y > 0$

(b) $y' = y^3 - a, \quad a \geqslant 0$

(c) $y' = \dfrac{5}{2 + y}, \quad y \geqslant 0$

6. (a) $y' = \dfrac{y^2 - a}{y + 1}, \quad 0 \leqslant a < 1, y > -1$

(b) $y' = y(a - y)^2(b - y), \quad b > a \geqslant 0$

(c) $y' = \dfrac{a}{b + y} - 1, \quad a > b > 0, y \geqslant 0$

7–8 Find all equilibria and use the local stability criterion to determine if each is locally stable or unstable. Then construct the phase plot.

7. (a) $y' = 5 - 3y$ (b) $y' = 2y - 3y^2$

8. (a) $y' = 5y(2e^{-y} - 1)$
(b) $y' = y^{5/3} - 2y \quad$ (assume $y \geqslant 0$)

9. Find the equilibria for the differential equation and determine the values of a for which each equilibrium is locally stable. Assume $a \neq 0$.
(a) $y' = 1 + ay$
(b) $y' = 1 - e^{-ay}$
(c) $y' = ae^y \cos y, \quad 0 < y < \pi$
(d) $y' = y(a - y)$

10. The Allee effect For the model of population dynamics from Example 3, use the local stability criterion to verify that $\hat{N} = 0$ and $\hat{N} = K$ are locally stable whereas $\hat{N} = a$ is unstable.

11. Suppose that the population dynamics of a species obeys a modified version of the logistic differential equation having the following form:

$$\frac{dN}{dt} = r\left(1 - \frac{N}{K}\right)^2 N$$

where $r \neq 0$ and $K > 0$.
(a) Show that $\hat{N} = 0$ and $\hat{N} = K$ are equilibria.
(b) For which values of r is the equilibrium $\hat{N} = 0$ unstable?
(c) Apply the local stability criterion to the equilibrium $\hat{N} = K$. What do you think your answer means about the stability of this equilibrium? (*Note:* This is an example in which the local stability criterion is inconclusive.)
(d) Construct two phase plots, one for the case where $r > 0$ and the other for $r < 0$, and determine the stability of $\hat{N} = K$ in each case. Does the answer match your reasoning in part (c)?

12. Infectious disease dynamics The spread of an infectious disease, such as influenza, is often modeled using the following autonomous differential equation:

$$\frac{dI}{dt} = \beta I(N - I) - \mu I$$

where I is the number of infected people, N is the total size of the population being modeled, β is a constant determining the rate of transmission, and μ is the rate at which people recover from infection.

(a) Suppose $\beta = 0.01$, $N = 1000$, and $\mu = 2$. Find all equilibria.

(b) For the equilibria in part (a), determine whether each is stable or unstable.

(c) Leaving the constants unspecified, what are the equilibria of the model in terms of these constants?

(d) The epidemiological quantity R_0 that was introduced in Example 1.5.5 is calculated by rearranging the condition for $\hat{I} = 0$ to be unstable, giving an inequality of the form $R_0 > 1$. Show that R_0 can be written as $R_0 = \beta N/\mu$.

13. Harvesting of renewable resources Suppose a population grows according to the logistic equation but is subject to a constant per capita harvest rate of h. If $N(t)$ is the population size at time t, the population dynamics are

$$\frac{dN}{dt} = r\left(1 - \frac{N}{K}\right)N - hN$$

Different values of h result in different equilibrium population sizes; if h is large enough, we might expect extinction.

(a) Suppose $r = 2$ and $K = 1000$. Find all equilibria. [*Hint:* One will be a function of h.]

(b) For the nonzero equilibrium in part (a), what is the critical value of h greater than which the population will go extinct?

(c) Determine the values of h that make the nonzero equilibrium in part (a) locally stable. Assume $h \neq 2$.

14. Harvesting of renewable resources Suppose a population grows according to the logistic equation but is subject to a constant total harvest rate of H. If $N(t)$ is the population size at time t, the population dynamics are

$$\frac{dN}{dt} = r\left(1 - \frac{N}{K}\right)N - H$$

Different values of H result in different equilibrium population sizes; if H is large enough, we might expect extinction.

(a) Suppose $r = 2$, $K = 1000$, and $H = 100$. Find all equilibria.

(b) Determine whether each of the equilibria in part (a) is locally stable or unstable. Is the population predicted to go extinct?

15. Levins' metapopulation model Many species are made up of several small subpopulations that occasionally go extinct but that are subsequently recolonized. The entire collection of subpopulations is referred to as a metapopulation. One way to model this phenomenon is to keep track of only the fraction of subpopulations that are currently not extinct. Suppose $p(t)$ is the fraction of subpopulations that are not extinct at time t. The Levins model states that $p(t)$ obeys the differential equation

$$\frac{dp}{dt} = cp(1 - p) - mp$$

where c and m are positive constants reflecting the colonization and extinction rates, respectively. (See also Review Exercise 15 on page 482.) Assume $c \neq m$.

(a) What are the equilibria of this model in terms of the constants?

(b) What is the condition on the constants for the nonzero equilibrium in part (a) to lie between zero and one? Interpret this condition.

(c) What are the conditions on the constants for the nonzero equilibrium in part (a) to be locally stable? Assume $m \neq c$.

16. Bacteria population genetics Suppose there are two bacterial strains 1 and 2, each undergoing growth according to the differential equations

$$\frac{dN_1}{dt} = r_1 N_1 \quad \text{and} \quad \frac{dN_2}{dt} = r_2 N_2$$

respectively, where $r_1 \neq r_2$. Define

$$p(t) = \frac{N_1(t)}{N_1(t) + N_2(t)}$$

to be the frequency of strain 1 at time t.

(a) Differentiate p using its definition to show that p obeys the differential equation

$$\frac{dp}{dt} = sp(1 - p)$$

where $s = r_1 - r_2$.

(b) What are the equilibria of the differential equation of part (a)? For what values of s is each locally stable?

17. Bacterial cross-feeding The differential equation from Exercise 16 can be extended to model the effects of bacterial cross-feeding. Suppose that the growth rate of strain 1 (r_1) is zero when the frequency of strain 2 is zero and that it increases linearly to a maximum value of α when the frequency of strain 2 is 1. Likewise, suppose that the growth rate of strain 2 (r_2) is zero when the frequency of strain 1 is zero and that it increases linearly to a maximum value of β when the frequency of strain 1 is one. Show that these assumptions, combined with the differential equation from Exercise 16 (a), result in the cross-feeding model of Example 7.

18. Bacterial cross-feeding (cont.) The differential equation derived in Exercise 17 (and from Example 7) is

$$\frac{dp}{dt} = p(1 - p)[\alpha(1 - p) - \beta p]$$

where α and β are both positive constants.

(a) Find all equilibria.

(b) Determine the values of α and β for which each equilibrium in part (a) is locally stable.

19. Mutation-selection balance The population-genetic differential equation in Exercise 16 assumes there is no mutation. Suppose that strain 1 bacteria mutate to strain 2 at a per capita rate of μ, but otherwise their dynamics are exactly as given in Exercise 16.
(a) What is the resulting differential equation for p?
(b) Determine the equilibria of the differential equation from part (a).
(c) Determine the constant values under which each equilibrium in part (b) is locally stable. Assume $s \neq \mu$.

20. Catastrophic population collapse The graph depicting the rate of change of a population as a function of population size is given by the green curve in the figure, with N measured in thousands of individuals. Suppose the population size starts between 1000 and 2000 individuals.
(a) What is the predicted population size in the long term?
(b) Now suppose that habitat degradation begins and that K is a constant quantifying habitat quality (sometimes referred to as the *carrying capacity*). $K = 2$ represents a pristine habitat and corresponds to the green curve. As K decreases, the curve changes continuously, from green to blue to purple to red. Purple corresponds to $K = 1$. Sketch a curve of the long-term population size as a function of K, depicting the main qualitative features. You can assume that the curve is linear except at points of discontinuity.

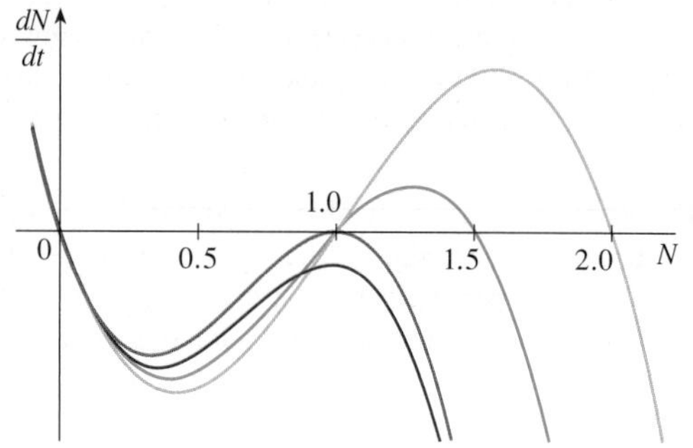

■ PROJECT Catastrophic Population Collapse: An Introduction to Bifurcation Theory

BB

Overexploitation of fish stocks by commercial fisheries can cause catastrophic population collapse.

Biological populations exhibit a wide range of dynamics, from stable population sizes, to oscillatory dynamics, to extinction. As conditions change—for example, through global warming, pollution, or increased harvesting pressure—populations can switch from one of these dynamical regimes to another. When this occurs, a *bifurcation* is said to have happened. **Bifurcation theory** provides a set of tools from the study of differential equations that we can use to study such phenomena.

Consider the following model for the dynamics of a population of size N (measured as number of individuals $\times 10^4$) over time (in months) that is subject to harvesting:

(1) $$\frac{dN}{dt} = \left(1 - \frac{N}{2}\right)N - hN$$

The population grows according to a logistic equation in the absence of harvesting and h is a constant per capita harvest rate (see also Exercise 7.2.13).

1. Find all equilibria and determine the values of h for which each is stable or unstable.

A **bifurcation plot** is a plot of all equilibria of a differential equation as a function of a constant of interest (called the bifurcation parameter). Solid curves are used to represent locally stable equilibria and dashed curves represent unstable equilibria.

2. Plot the equilibria from Problem 1 as a function of h for $0 \leqslant h \leqslant 2$, using solid and dashed curves as described. In other words, construct the bifurcation plot.

You should obtain the plot in Figure 1.

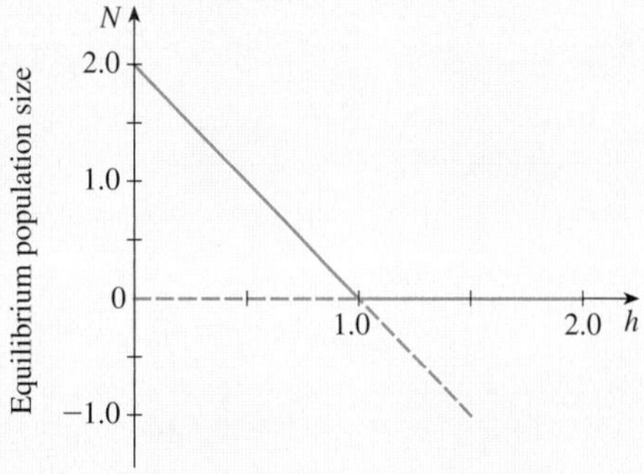

FIGURE 1
Bifurcation plot for Equation 1

3. Describe in words what happens to the predicted population size as h increases continuously from $h = 0$ to $h = 2$. (You can assume that the initial population size is positive.) Equation 1 exhibits what is called a *transcritical bifurcation*. As the bifurcation parameter is increased, two equilibria move continuously, eventually merging and exchanging stability properties before once again diverging.

Now consider the following alternative model:

(2) $$\frac{dN}{dt} = \left(1 - \frac{N}{2}\right)N - h\frac{N}{1 + N}$$

The only difference between (1) and (2) is in the form of the loss rate through harvesting.

4. Describe, in words, the form of the loss rate through harvesting as a function of population size in Equation 2. Provide an explanation for why this form is a reasonable way to model harvesting.

5. Verify that the bifurcation plot for Equation 2 is as given in Figure 2.

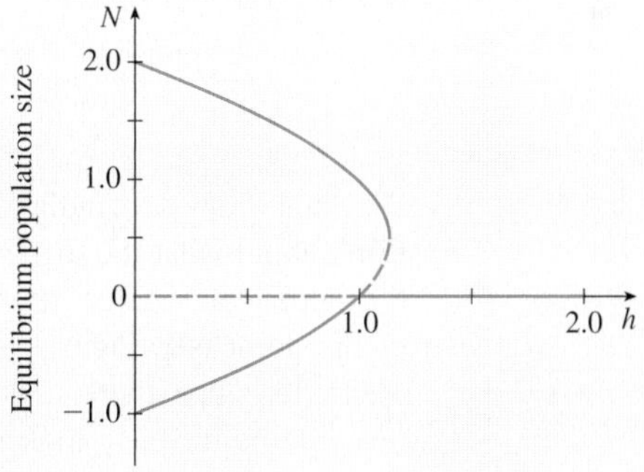

FIGURE 2
Bifurcation plot for Equation 2

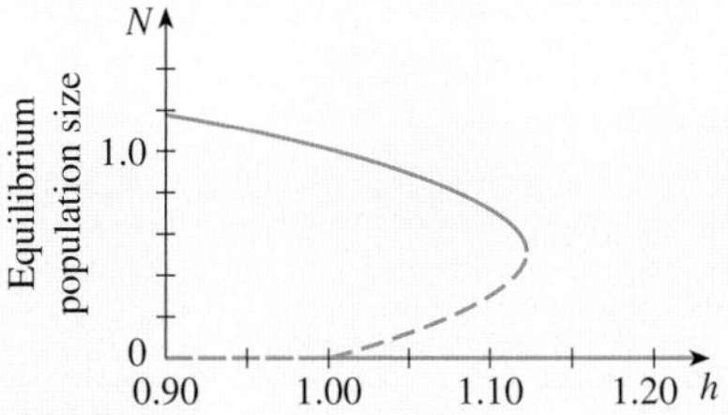

FIGURE 3
Bifurcation plot for Equation 2

6. Describe in words what happens to the predicted population size as h increases continuously from $h = 0$ to $h = 2$. (You can assume that the initial population size is positive.)

Equation 2 exhibits a *saddle-node bifurcation*: As the bifurcation parameter is increased, two equilibria move continuously, eventually merging and annihilating one another. This can result in an abrupt or discontinuous change in model predictions as the bifurcation parameter changes.

Figure 3 zooms in on the bifurcation plot from Figure 2. You can see that, for values of h between $h = 1$ and $h = 1.125$, there are two locally stable equilibria separated by an unstable equilibrium. This type of bifurcation has been observed in real populations by experimentally manipulating the loss rate (see Figure 4).[1]

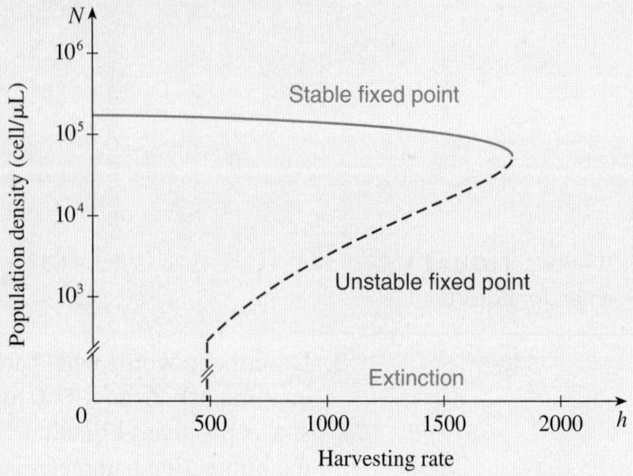

FIGURE 4
An experimentally measured bifurcation plot for a microbial population

1. Adapted from L. Dai et al., "Generic Indicators for Loss of Resilience before a Tipping Point Leading to Population Collapse," *Science* 336 (2012): 1175–77.

7.3 Direction Fields and Euler's Method

In this section we develop another graphical technique known as a direction field. This will allow us to gain more information about the shapes of solution curves. It will also lead us to Euler's method for solving differential equations numerically.

■ Direction Fields

Suppose we are asked to sketch the graph of the solution of the initial-value problem

$$y' = t + y \qquad y(0) = 1$$

We don't know a formula for the solution, so how can we possibly sketch its graph? Let's think about what the differential equation means. The equation $y' = t + y$ tells us that the slope at any point (t, y) on the graph is equal to the sum of the t- and y-coordinates of the point (see Figure 1). In particular, because the curve passes through the point $(0, 1)$, its slope there must be $0 + 1 = 1$. So a small portion of the solution curve near the point $(0, 1)$ looks like a short line segment through $(0, 1)$ with slope 1. (See Figure 2.)

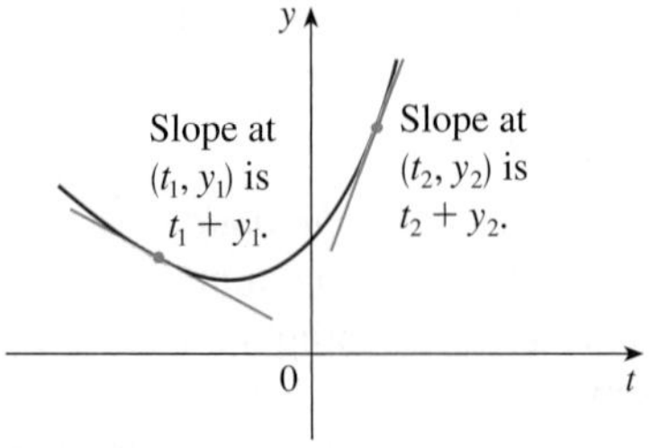

FIGURE 1
A solution of $y' = t + y$

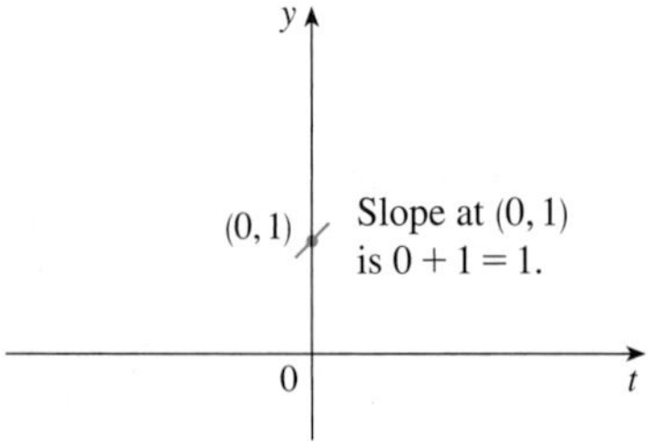

FIGURE 2
Beginning of the solution curve through (0, 1)

As a guide to sketching the rest of the curve, let's draw short line segments at a number of points (t, y) with slope $t + y$. The result, called a *direction field,* is shown in Figure 3. For instance, the line segment at the point $(1, 2)$ has slope $1 + 2 = 3$. The direction field allows us to visualize the general shape of the solution curves by indicating the direction in which the curves proceed at each point.

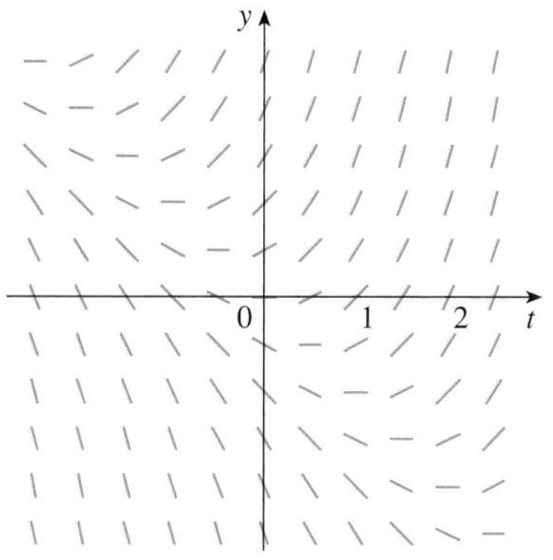

FIGURE 3
Direction field for $y' = t + y$

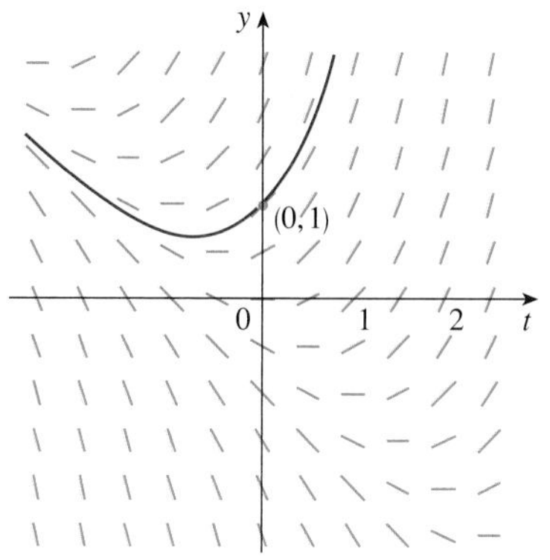

FIGURE 4
The solution curve through $(0, 1)$

Now we can sketch the solution curve through the point $(0, 1)$ by following the direction field, as shown in Figure 4. We can visualize $y(t)$ as being a curve that snakes its way through the direction field, always being parallel to nearby line segements. Such a curve is referred to as a **solution curve**.

In general, suppose we have the first-order differential equation

$$y' = F(t, y)$$

where $F(t, y)$ is some function in t and y. The differential equation says that the slope of a solution curve at a point (t, y) on the curve is $F(t, y)$. If we draw short line segments with slope $F(t, y)$ at several points (t, y), the result is called a **direction field** (or **slope field**). These line segments indicate the direction in which a solution curve is heading.

EXAMPLE 1

(a) Sketch the direction field for the differential equation $\dfrac{dy}{dx} = x^2 + y^2 - 1$.

(b) Use part (a) to sketch the solution curve that passes through the origin.

SOLUTION

(a) We start by computing the slope at several points, as given in the following table:

$$y' = x^2 + y^2 - 1$$

y \ x	−3	−2	−1	0	1	2	3
2	⋯	⋮	⋮	⋮	⋮	⋮	⋯
1	⋯	4	1	0	1	4	⋯
0	⋯	3	0	−1	0	3	⋯
−1	⋯	4	1	0	1	4	⋯
−2	⋯	⋮	⋮	⋮	⋮	⋮	⋯

Now we draw short line segments with these slopes at the indicated points. In Figure 5 we have used a computer to draw line segments at these and several other points.

TEC Module 7.3A shows direction fields and solution curves for a variety of differential equations.

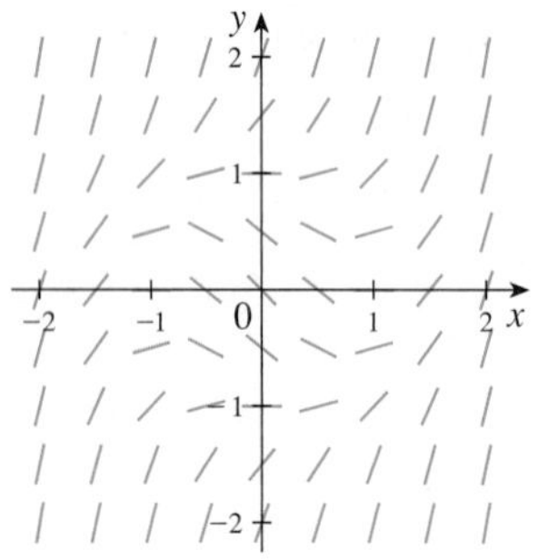

FIGURE 5

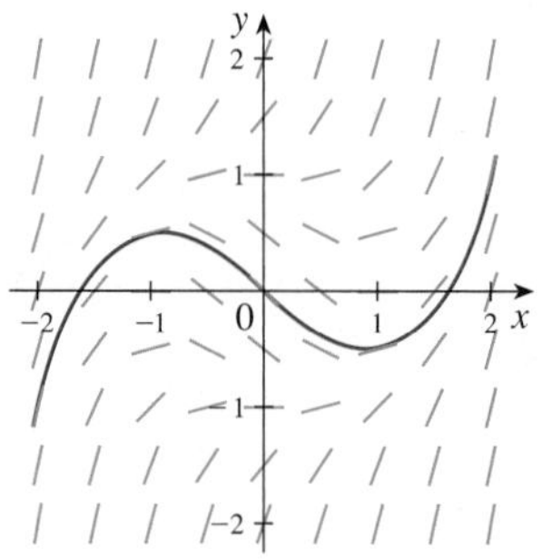

FIGURE 6

(b) We start at the origin and move to the right in the direction of the line segment (which has slope -1). We continue to draw the solution curve so that it moves parallel to the nearby line segments. The resulting solution curve is shown in Figure 6. Returning to the origin, we draw the solution curve to the left as well. ■

The more line segments we draw in a direction field, the clearer the solution curves appear. It's tedious to compute slopes and draw line segments for a large number of points by hand, but computers are well suited for this task. Figure 7 shows an even more detailed computer-drawn direction field for the differential equation in Example 1. It enables us to draw, with reasonable accuracy, the solution curves shown in Figure 8 with y-intercepts -2, -1, 0, 1, and 2.

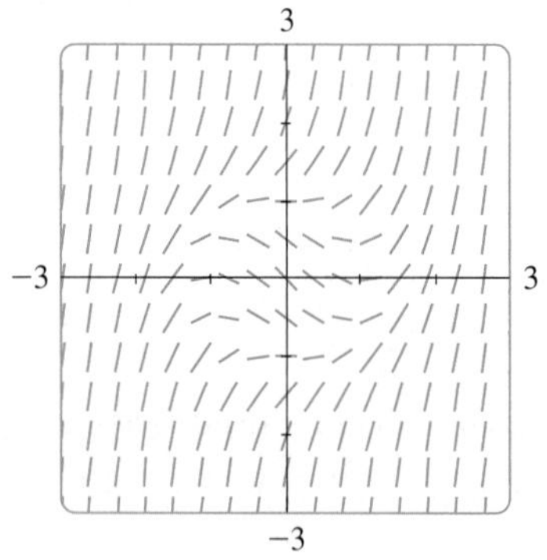

FIGURE 7

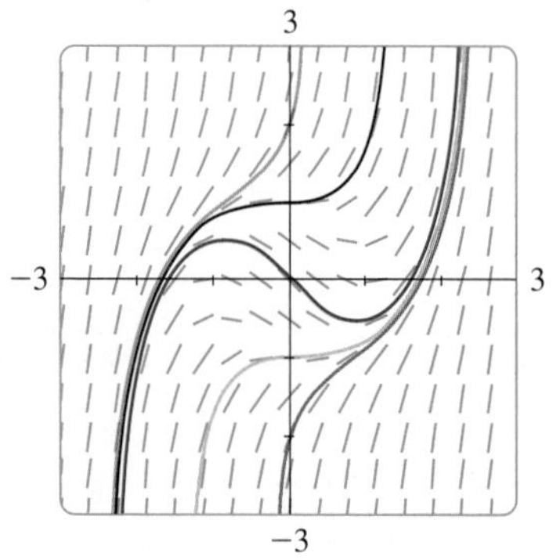

FIGURE 8

EXAMPLE 2 | Population genetics (continued) The model presented in Example 7.2.7 reduces to

$$\frac{dp}{dt} = p(1 - p)(2 - 5p)$$

when $\alpha = 2$ and $\beta = 3$.

(a) Draw the direction field for $0 \leqslant t \leqslant 8$ and $0 \leqslant p \leqslant 1$.
(b) Identify all equilibria on the plot.
(c) Sketch the solution curve that starts at 0.8 when $t = 0$; that is, $p(0) = 0.8$.
(d) What happens to the solution curve plotted in part (c) as $t \to \infty$?

SOLUTION

(a) We start by computing the slope at several points in the following table:

$$p' = p(1 - p)(2 - 5p)$$

p										
1	0	0	0	0	0	0	0	0	0	
0.8	−0.32	−0.32	−0.32	−0.32	−0.32	−0.32	−0.32	−0.32	−0.32	
0.6	−0.24	−0.24	−0.24	−0.24	−0.24	−0.24	−0.24	−0.24	−0.24	
0.4	0	0	0	0	0	0	0	0	0	
0.2	0.16	0.16	0.16	0.16	0.16	0.16	0.16	0.16	0.16	
0	0	0	0	0	0	0	0	0	0	
	0	1	2	3	4	5	6	7	8	t

Notice that all the columns are identical. This is because the independent variable t does not occur on the right side of the equation for p' (it is *autonomous*). As a result, the slopes corresponding to two different points with the same p-coordinate must be equal. Thus, if we know one solution to an autonomous differential equation, then we can obtain infinitely many others just by shifting the graph of the known solution to the right or left, as shown in Figure 9.

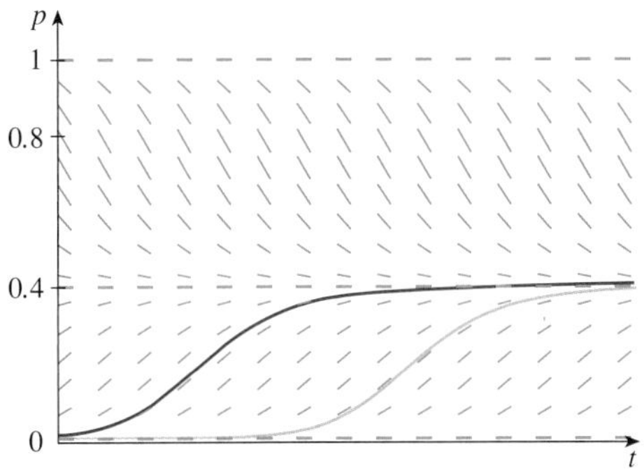

FIGURE 9

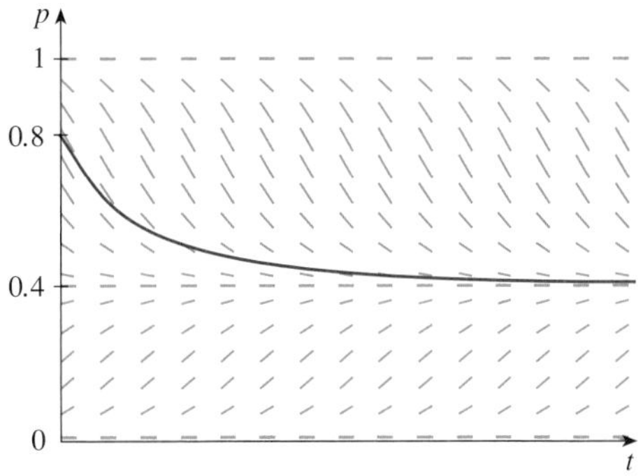

FIGURE 10

(b) The equilibria are values of the variable at which the direction field is horizontal. From Figure 10 we can see that equilibria occur at $\hat{p} = 0$, $\hat{p} = 0.4$, and $\hat{p} = 1$.

(c) The solution curve corresponding to $p(0) = 0.8$ is obtained by starting at this point and drawing a curve that moves in the direction of the line segment at this point and then makes its way through the direction field, always remaining parallel to nearby line segments (see Figure 10).

(d) From the solution curve plotted in part (c) it appears that $p(t) \to 0.4$ as $t \to \infty$. ■

■ Euler's Method

The basic idea behind direction fields can be used to find numerical approximations to solutions of differential equations. We illustrate the method on the initial-value problem that we used to introduce direction fields:

$$y' = t + y \qquad y(0) = 1$$

The differential equation tells us that $y'(0) = 0 + 1 = 1$, so the solution curve has slope

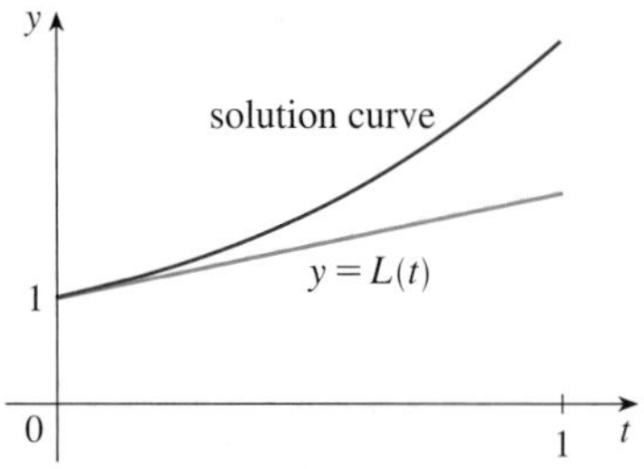

FIGURE 11
First Euler approximation

1 at the point (0, 1). As a first approximation to the solution we could use the linear approximation $L(t) = t + 1$. In other words, we could use the tangent line at (0, 1) as a rough approximation to the solution curve (see Figure 11).

Euler's idea was to improve on this approximation by proceeding only a short distance along this tangent line and then making a midcourse correction by changing direction as indicated by the direction field. Figure 12 shows what happens if we start out along the tangent line but stop when $t = 0.5$. (This horizontal distance traveled is called the *step size.*) Since $L(0.5) = 1.5$, we have $y(0.5) \approx 1.5$ and we take (0.5, 1.5) as the starting point for a new line segment. The differential equation tells us that $y'(0.5) = 0.5 + 1.5 = 2$, so we use the linear function

$$y = 1.5 + 2(t - 0.5) = 2t + 0.5$$

as an approximation to the solution for $t > 0.5$ (the green segment in Figure 12). If we decrease the step size from 0.5 to 0.25, we get the better Euler approximation shown in Figure 13.

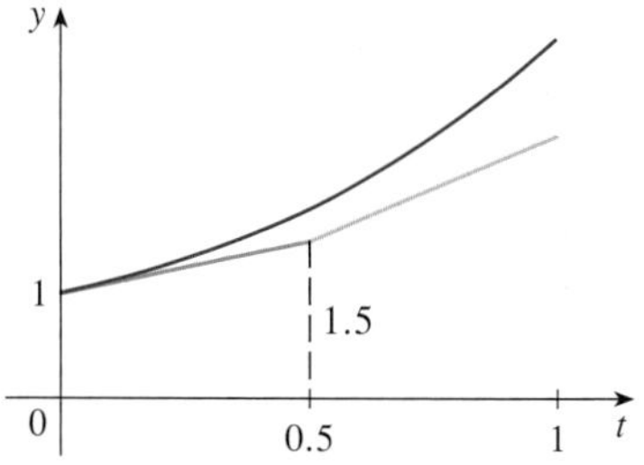

FIGURE 12
Euler approximation with step size 0.5

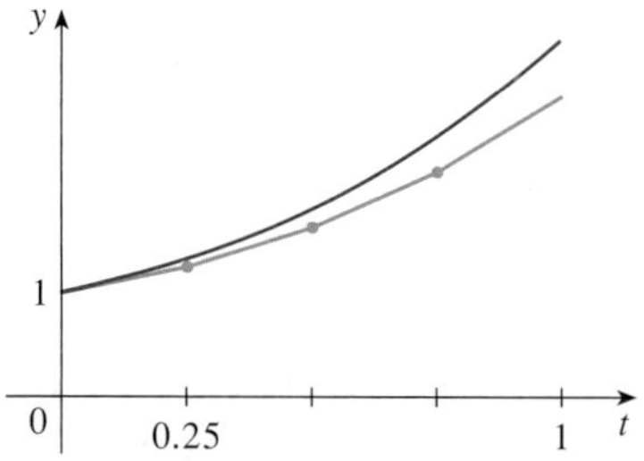

FIGURE 13
Euler approximation with step size 0.25

Euler

Leonhard Euler (1707–1783) was the leading mathematician of the mid-18th century and the most prolific mathematician of all time. He was born in Switzerland but spent most of his career at the academies of science supported by Catherine the Great in St. Petersburg and Frederick the Great in Berlin. The collected works of Euler (pronounced *Oiler*) fill about 100 large volumes. As the French physicist Arago said, "Euler calculated without apparent effort, as men breathe or as eagles sustain themselves in the air." Euler's calculations and writings were not diminished by raising 13 children or being totally blind for the last 17 years of his life. In fact, when blind, he dictated his discoveries to his helpers from his prodigious memory and imagination. His treatises on calculus and most other mathematical subjects became the standard for mathematics instruction, and the equation $e^{i\pi} + 1 = 0$ that he discovered brings together the five most famous numbers in all of mathematics.

In general, Euler's method says to start at the point given by the initial value and proceed in the direction indicated by the direction field. Stop after a short time, look at the slope at the new location, and proceed in that direction. Keep stopping and changing direction according to the direction field. Euler's method does not produce the exact solution to an initial-value problem—it gives approximations. But by decreasing the step size (and therefore increasing the number of midcourse corrections), we obtain successively better approximations to the exact solution. (Compare Figures 11, 12, and 13.)

For the general first-order initial-value problem $y' = F(t, y)$, $y(t_0) = y_0$, our aim is to find approximate values for the solution at equally spaced numbers t_0, $t_1 = t_0 + h$, $t_2 = t_1 + h$, . . . , where h is the step size. The differential equation tells us that the slope at (t_0, y_0) is $y' = F(t_0, y_0)$, so Figure 14 shows that the approximate value of the solution when $t = t_1$ is

$$y_1 = y_0 + hF(t_0, y_0)$$

Similarly, $$y_2 = y_1 + hF(t_1, y_1)$$

In general, $$y_{n+1} = y_n + hF(t_n, y_r)$$

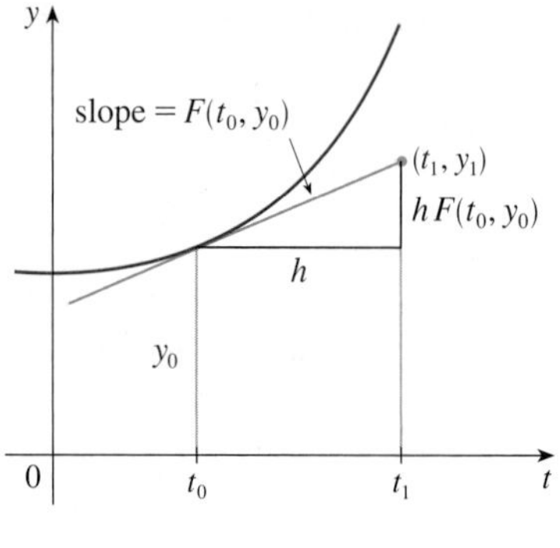

FIGURE 14

Euler's Method Approximate values for the solution of the initial-value problem $y' = F(t, y)$, $y(t_0) = y_0$, with step size h, at $t_{n+1} = t_n + h$, are

$$y_{n+1} = y_n + hF(t_n, y_n) \qquad n = 1, 2, 3, \ldots$$

Euler's method amounts to approximating the differential equation with a difference equation (see Section 1.6); the smaller the time step, the better the approximation.

EXAMPLE 3 | Use Euler's method with step size 0.1 to construct a table of approximate values for the solution of the initial-value problem

$$\frac{dy}{dx} = x + y \qquad y(0) = 1$$

SOLUTION We are given that $h = 0.1$, $x_0 = 0$, $y_0 = 1$, and $F(x, y) = x + y$. So we have

$$y_1 = y_0 + hF(x_0, y_0) = 1 + 0.1(0 + 1) = 1.1$$

$$y_2 = y_1 + hF(x_1, y_1) = 1.1 + 0.1(0.1 + 1.1) = 1.22$$

$$y_3 = y_2 + hF(x_2, y_2) = 1.22 + 0.1(0.2 + 1.22) = 1.362$$

This means that if $y(x)$ is the exact solution, then $y(0.3) \approx 1.362$.

Proceeding with similar calculations, we get the values in the table:

TEC Module 7.3B shows how Euler's method works numerically and visually for a variety of differential equations and step sizes.

n	x_n	y_n	n	x_n	y_n
1	0.1	1.100000	6	0.6	1.943122
2	0.2	1.220000	7	0.7	2.197434
3	0.3	1.362000	8	0.8	2.487178
4	0.4	1.528200	9	0.9	2.815895
5	0.5	1.721020	10	1.0	3.187485

■

Computer software packages that produce numerical approximations to solutions of differential equations use methods that are refinements of Euler's method. Although Euler's method is simple and not as accurate, it is the basic idea on which the more accurate methods are based.

For a more accurate table of values in Example 3 we could decrease the step size. But for a large number of small steps the amount of computation is considerable and so we need to program a calculator or computer to carry out these calculations. The following table shows the results of applying Euler's method with decreasing step size to the initial-value problem of Example 3.

Step size	Euler estimate of $y(0.5)$	Euler estimate of $y(1)$
0.500	1.500000	2.500000
0.250	1.625000	2.882813
0.100	1.721020	3.187485
0.050	1.757789	3.306595
0.020	1.781212	3.383176
0.010	1.789264	3.409628
0.005	1.793337	3.423034
0.001	1.796619	3.433848

Notice that the Euler estimates in the table seem to be approaching limits, namely, the true values of $y(0.5)$ and $y(1)$. Figure 15 shows graphs of the Euler approximations with step sizes 0.5, 0.25, 0.1, 0.05, 0.02, 0.01, and 0.005. They are approaching the exact solution curve as the step size h approaches 0.

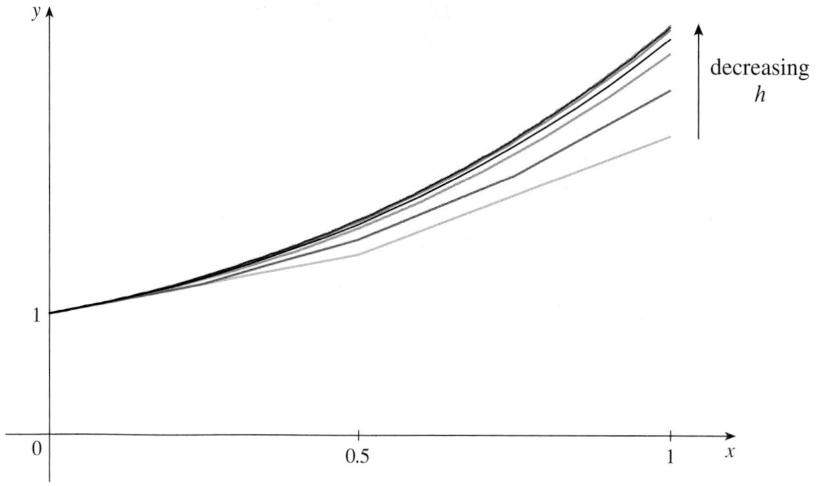

FIGURE 15
Euler approximations approaching the exact solution

EXAMPLE 4 | Administering drugs Estimate the drug concentration (in mg/mL) in the bloodstream after $\frac{1}{2}$ hour, assuming the concentration changes according to the differential equation

$$\frac{dy}{dt} = 1 + \sin t - y \qquad y(0) = 2.5$$

where t is measured in hours.

SOLUTION We use Euler's method with $F(t, y) = 1 + \sin t - y$, $t_0 = 0$, $y_0 = 2.5$, and a step size $h = 0.1$ (which corresponds to six minutes):

$$y_1 = 2.5 + 0.1(1 + \sin 0 - 2.5) = 2.35$$

$$y_2 = 2.35 + 0.1(1 + \sin 0.1 - 2.35) \approx 2.22$$

$$y_3 = 2.22 + 0.1(1 + \sin 0.2 - 2.22) \approx 2.12$$

$$y_4 = 2.12 + 0.1(1 + \sin 0.3 - 2.12) \approx 2.04$$

$$y_5 = 2.04 + 0.1(1 + \sin 0.4 - 2.04) \approx 1.97$$

So the concentration after $\frac{1}{2}$ hour is

$$y(0.5) \approx 1.97 \text{ mg/mL}$$

■

EXERCISES 7.3

1. A direction field for the differential equation $y' = x \cos \pi y$ is shown.

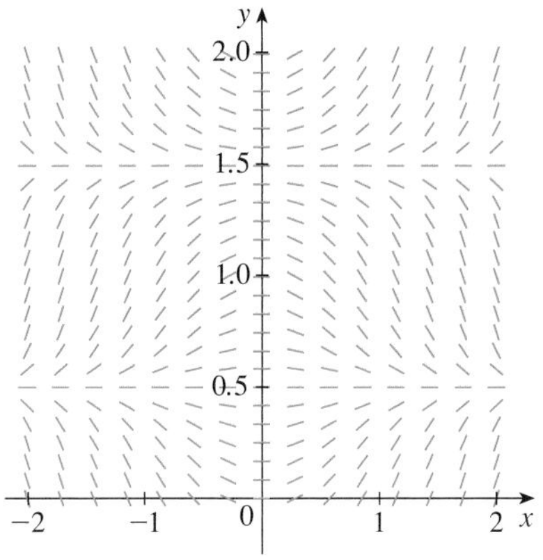

(a) Sketch the graphs of the solutions that satisfy the given initial conditions.

(i) $y(0) = 0$ (ii) $y(0) = 0.5$

(iii) $y(0) = 1$ (iv) $y(0) = 1.6$

(b) Find all the equilibrium solutions.

2. A direction field for the differential equation $y' = \tan\left(\frac{1}{2}\pi y\right)$ is shown.

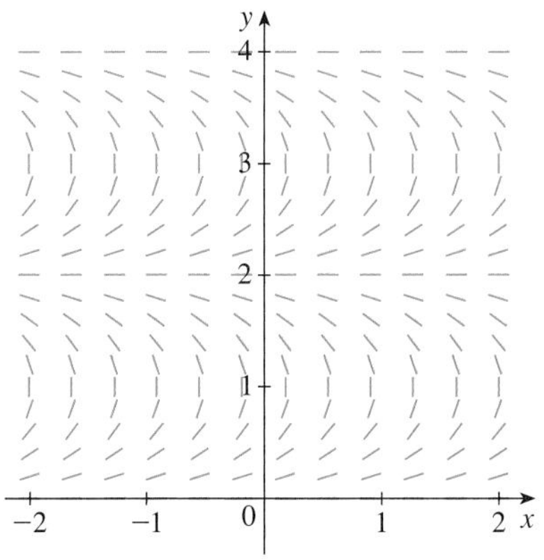

(a) Sketch the graphs of the solutions that satisfy the given initial conditions.

(i) $y(0) = 1$ (ii) $y(0) = 0.2$

(iii) $y(0) = 2$ (iv) $y(1) = 3$

(b) Find all the equilibrium solutions.

3–6 Match the differential equation with its direction field (labeled I–IV). Give reasons for your answer.

3. $y' = 2 - y$

4. $y' = x(2 - y)$

5. $y' = x + y - 1$

6. $y' = \sin x \sin y$

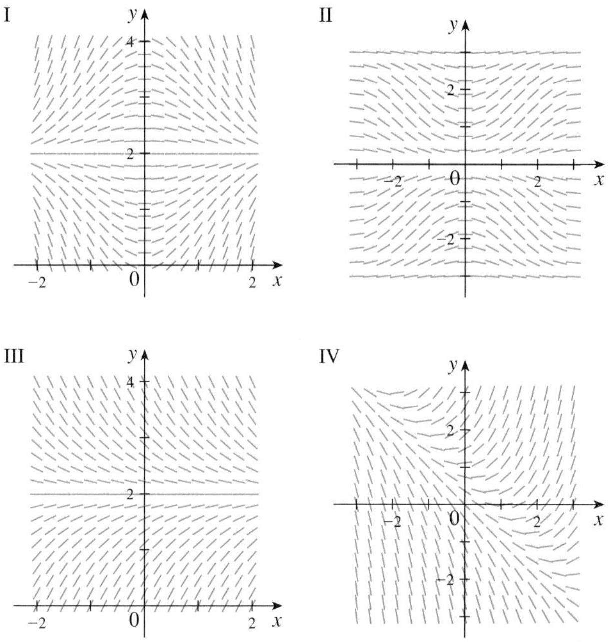

7. Use the direction field labeled II (above) to sketch the graphs of the solutions that satisfy the given initial conditions.

(a) $y(0) = 1$ (b) $y(0) = 2$ (c) $y(0) = -1$

8. Use the direction field labeled IV (above) to sketch the graphs of the solutions that satisfy the given initial conditions.

(a) $y(0) = -1$ (b) $y(0) = 0$ (c) $y(0) = 1$

9–10 Sketch a direction field for the differential equation. Then use it to sketch three solution curves.

9. $y' = \frac{1}{2}y$

10. $y' = x - y + 1$

11–14 Sketch the direction field of the differential equation. Then use it to sketch a solution curve that passes through the given point.

11. $y' = y - 2x$, $(1, 0)$

12. $y' = xy - x^2$, $(0, 1)$

13. $y' = y + xy$, $(0, 1)$

14. $y' = x + y^2$, $(0, 0)$

CAS **15–16** Use a computer algebra system to draw a direction field for the given differential equation. Get a printout and sketch on it the solution curve that passes through (0, 1). Then use the CAS to draw the solution curve and compare it with your sketch.

15. $y' = x^2 \sin y$

16. $y' = x(y^2 - 4)$

17. Use a computer algebra system to draw a direction field for the differential equation $y' = y^3 - 4y$. Get a printout and sketch on it solutions that satisfy the initial condition $y(0) = c$ for various values of c. For what values of c does $\lim_{t\to\infty} y(t)$ exist? What are the possible values for this limit?

18. Make a rough sketch of a direction field for the autonomous differential equation $y' = g(y)$, where the graph of g is as shown. How does the limiting behavior of solutions depend on the value of $y(0)$?

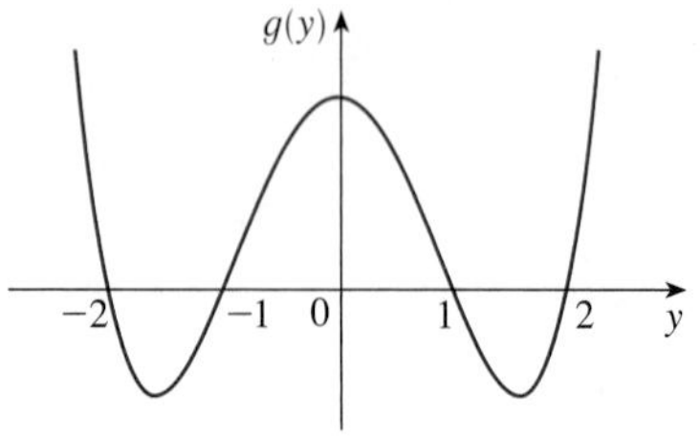

19. (a) Use Euler's method with each of the following step sizes to estimate the value of $y(0.4)$, where y is the solution of the initial-value problem $y' = y$, $y(0) = 1$.
(i) $h = 0.4$ (ii) $h = 0.2$ (iii) $h = 0.1$

(b) We know that the exact solution of the initial-value problem in part (a) is $y = e^x$. Draw, as accurately as you can, the graph of $y = e^x$, $0 \leqslant x \leqslant 0.4$, together with the Euler approximations using the step sizes in part (a). (Your sketches should resemble Figures 12, 13, and 15.) Use your sketches to decide whether your estimates in part (a) are underestimates or overestimates.

(c) The error in Euler's method is the difference between the exact value and the approximate value. Find the errors made in part (a) in using Euler's method to estimate the true value of $y(0.4)$, namely, $e^{0.4}$. What happens to the error each time the step size is halved?

20–21 Modeling yeast populations In Section 7.1 we introduced the following differential equation to describe the dynamics of an experimental yeast population:

$$\frac{dN}{dt} = (0.55 - 0.0026N)N$$

where $N(t)$ is the population size (in millions of individuals per mL) at time t (in hours).

20. Sketch the direction field of the differential equation for values of N between 0 and 250.

21. Suppose $N(0) = 0.2$. Use Euler's method with a step size of $h = 0.5$ to estimate the population size after four hours. Compare your result to the data in Table 7.1.1.

---

22. A direction field for a differential equation is shown in the figure. Draw, with a ruler, the graphs of the Euler approximations to the solution curve that passes through the origin. Use step sizes $h = 1$ and $h = 0.5$. Will the Euler estimates be underestimates or overestimates? Explain.

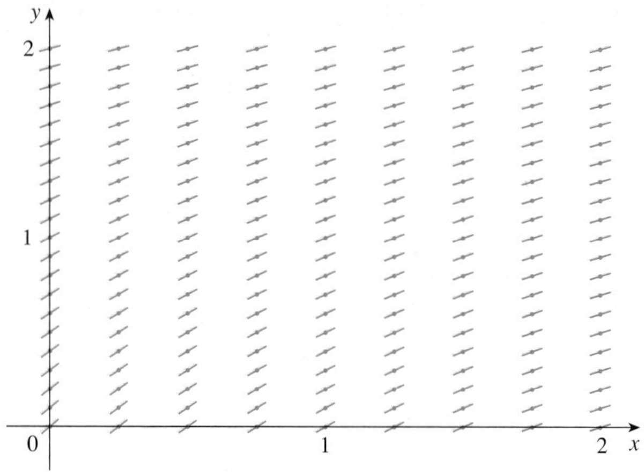

23. Use Euler's method with step size 0.5 to compute the approximate y-values y_1, y_2, y_3, and y_4 of the solution of the initial-value problem $y' = y - 2x$, $y(1) = 0$.

24. Use Euler's method with step size 0.2 to estimate $y(1)$, where $y(x)$ is the solution of the initial-value problem $y' = xy - x^2$, $y(0) = 1$.

25. Use Euler's method with step size 0.1 to estimate $y(0.5)$, where $y(x)$ is the solution of the initial-value problem $y' = y + xy$, $y(0) = 1$.

26. (a) Use Euler's method with step size 0.2 to estimate $y(0.4)$, where $y(x)$ is the solution of the initial-value problem $y' = x + y^2$, $y(0) = 0$.
(b) Repeat part (a) with step size 0.1.

27. (a) Program a calculator or computer to use Euler's method to compute $y(1)$, where $y(x)$ is the solution of the initial-value problem

$$\frac{dy}{dx} + 3x^2y = 6x^2 \qquad y(0) = 3$$

for each of the given step sizes.
(i) $h = 1$ (ii) $h = 0.1$
(iii) $h = 0.01$ (iv) $h = 0.001$

(b) Verify that $y = 2 + e^{-x^3}$ is the exact solution of the differential equation.

(c) Find the errors in using Euler's method to compute $y(1)$ with the step sizes in part (a). What happens to the error when the step size is divided by 10?

28. (a) Program your computer algebra system, using Euler's method with step size 0.01, to calculate $y(2)$, where y is the solution of the initial-value problem

$$y' = x^3 - y^3 \qquad y(0) = 1$$

(b) Check your work by using the CAS to draw the solution curve.

7.4 Separable Equations

We have looked at first-order differential equations from a geometric point of view (phase plots and direction fields) and from a numerical point of view (Euler's method). What about the symbolic point of view? It would be helpful to have an explicit formula for a solution of a differential equation. Although this is not always possible, in this section we examine a commonly encountered type of differential equation that *can* be solved explicitly.

A **separable equation** is a first-order differential equation in which the expression for dy/dt can be factored as a function of t times a function of y. In other words, it can be written in the form

$$\frac{dy}{dt} = f(t)\, g(y)$$

The name separable comes from the fact that the expression on the right side can be separated into a function of t and a function of y. Equivalently, if $g(y) \neq 0$, we could write

$$\textbf{(1)} \qquad \frac{dy}{dt} = \frac{f(t)}{h(y)}$$

where $h(y) = 1/g(y)$. To solve this equation we rewrite it in the differential form

$$h(y)\, dy = f(t)\, dt$$

The technique for solving separable differential equations was first used by James Bernoulli (in 1690) in solving a problem about pendulums and by Leibniz (in a letter to Huygens in 1691). John Bernoulli explained the general method in a paper published in 1694.

so that all y's are on one side of the equation and all t's are on the other side. Then we integrate both sides of the equation:

$$\textbf{(2)} \qquad \int h(y)\, dy = \int f(t)\, dt$$

Equation 2 defines y implicitly as a function of t. In some cases we may be able to solve for y in terms of t.

We can verify that Equation 2 is indeed a solution using the Chain Rule: If h and f satisfy (2), then

$$\frac{d}{dt}\left(\int h(y)\, dy\right) = \frac{d}{dt}\left(\int f(t)\, dt\right)$$

so

$$\frac{d}{dy}\left(\int h(y)\, dy\right)\frac{dy}{dt} = f(t)$$

and

$$h(y)\frac{dy}{dt} = f(t)$$

Thus Equation 1 is satisfied.

One of the simplest applications of the technique of separation of variables is to the differential equation for exponential growth introduced in Sections 3.6 and 7.1. In particular, if $y(t)$ is the value of some quantity at time t and if the rate of change of y with respect to t is proportional to its size $y(t)$, then

$$\frac{dy}{dt} = ky$$

where k is a constant. If $y \neq 0$ we can write this equation in terms of differentials and integrate both sides as follows:

$$\int \frac{dy}{y} = \int k\,dt$$

$$\ln|y| = kt + C$$

$$|y| = e^{kt+C} = e^C e^{kt}$$

$$y = Ae^{kt}$$

The absolute value can be cleared by noting that we can write

$$y = e^C e^{kt} \quad \text{if } y > 0$$

$$y = -e^C e^{kt} \quad \text{if } y < 0$$

Therefore $y = Ae^{kt}$, where $A = \pm e^C$.

where A $(= \pm e^C)$ is an arbitrary constant. This is the solution presented in Sections 3.6 and 7.1. If $y = 0$ we cannot divide the differential equation by y. However, we can readily verify that, in this case, $y = 0$ is also a solution. Therefore the constant A in the solution $y = Ae^{kt}$ can also be 0. This corresponds to an equilibrium solution.

EXAMPLE 1

(a) Solve the differential equation $\dfrac{dy}{dx} = \dfrac{x^2}{y^2}$.

(b) Find the solution of this equation that satisfies the initial condition $y(0) = 2$.

SOLUTION

(a) We write the equation in terms of differentials and integrate both sides:

$$y^2\,dy = x^2\,dx$$

$$\int y^2\,dy = \int x^2\,dx$$

$$\tfrac{1}{3}y^3 = \tfrac{1}{3}x^3 + C$$

where C is an arbitrary constant. (We could have used a constant C_1 on the left side and another constant C_2 on the right side. But then we could combine these constants by writing $C = C_2 - C_1$.)

Solving for y, we get

$$y = \sqrt[3]{x^3 + 3C}$$

We could leave the solution like this or we could write it in the form

$$y = \sqrt[3]{x^3 + K}$$

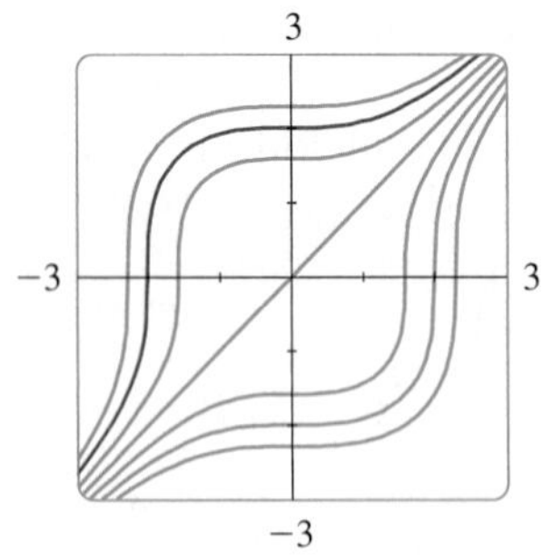

FIGURE 1
Graphs of several members of the family of solutions of the differential equation in Example 1. The solution of the initial-value problem in part (b) is shown in red.

where $K = 3C$. (Since C is an arbitrary constant, so is K.) Figure 1 plots this family of solutions.

(b) If we put $x = 0$ in the general solution in part (a), we get $y(0) = \sqrt[3]{K}$. To satisfy the initial condition $y(0) = 2$, we must have $\sqrt[3]{K} = 2$ and so $K = 8$. Thus the solution of the initial-value problem is

$$y = \sqrt[3]{x^3 + 8}$$

■

EXAMPLE 2 | **The von Bertalanffy growth equation** A commonly used differential equation for the growth, in length, of an individual fish is

$$\frac{dL}{da} = k(L_\infty - L)$$

where $L(a)$ is length (in cm) at age a (in years), L_∞ is the asymptotic length, and k is a positive constant whose units are 1/year.
(a) Find a family of solutions for length as a function of age.
(b) Find the solution that has an initial length of $L(0) = 2$.

Von Bertalanffy
Ludwig von Bertalanffy (1901–1972) was an Austrian-born biologist who first published this differential equation for individual growth in 1934. It captures the idea that the rate of growth in length is proportional to the difference between current length and asymptotic length.

SOLUTION

(a) Assuming $L \neq L_\infty$, we can write the equation in differential form as

$$\frac{dL}{L_\infty - L} = k\,da$$

Now integrate to obtain

$$\int \frac{dL}{L_\infty - L} = \int k\,da$$

or

$$-\ln|L_\infty - L| = ka + C_1$$

Now we can solve for L:

$$|L_\infty - L| = e^{-ka}e^{-C_1}$$

or

$$L = L_\infty - Ce^{-ka}$$

where $C = \pm e^{-C_1}$ is an arbitrary constant. An example of this solution with particular constant values is shown in Figure 2.

If $L = L_\infty$, we cannot divide the differential equation by $L - L_\infty$, but we can verify that $L = L_\infty$ is itself another solution. Thus the constant C in the preceding solution can be 0 as well, and this again corresponds to an equilibrium solution.

(b) Setting $a = 0$ in the family of solutions from part (a) gives $L(0) = L_\infty - C$. To satisfy the initial condition $L(0) = 2$, we therefore require that $L_\infty - C = 2$, or $C = L_\infty - 2$. The desired solution is thus $L = L_\infty - (L_\infty - 2)e^{-ka}$ or

$$L = L_\infty(1 - e^{-ka}) + 2e^{-ka}$$

From this we can see why L_∞ is called the asymptotic length. As $a \to \infty$, $L \to L_\infty$. ■

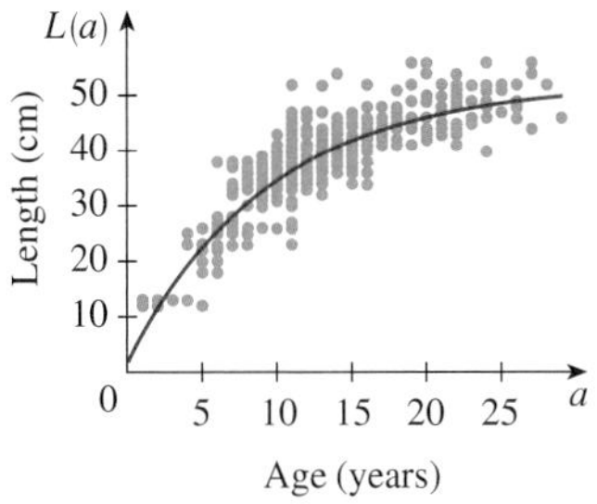

FIGURE 2
Age-length relationship for Atlantic redfish along with the solution to the von Bertalanffy equation with constants specific to redfish.

Source: Adapted from C. Stransky et al., "Age Determination and Growth of Atlantic Redfish (*Sebastes marinus* and *S. mentella*): Bias and Precision of Age Readers and Otolith Preparation Methods," *ICES Journal of Marine Science* 62 (2005): 655–70.

EXAMPLE 3 | Allometric growth During growth, the claw of fiddler crabs increases in length at a *per unit rate* that is 1.57 times larger than that of its overall body width. In other words, if L and B denote claw length and body width, respectively (in mm), then

$$\frac{dL}{dt}\frac{1}{L} = 1.57\,\frac{dB}{dt}\frac{1}{B}$$

(See Figure 3.) Find an equation that specifies claw length as a function of body width at any point during growth.

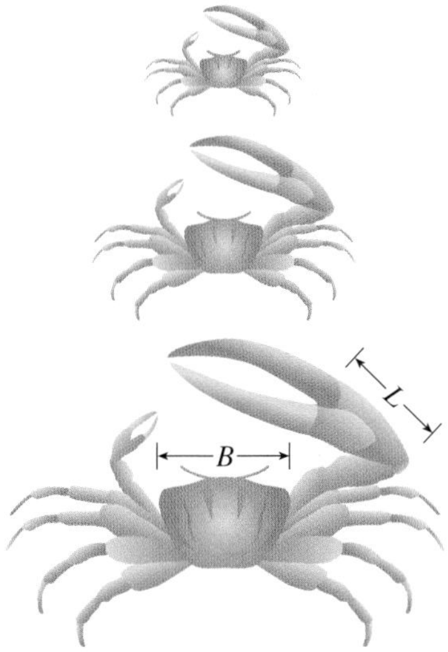

FIGURE 3
Three crabs of different sizes, along with their claw lengths

SOLUTION Multiplying both sides by dt gives

$$\frac{dL}{L} = 1.57\,\frac{dB}{B}$$

Now integrate both sides to get

$$\ln L = 1.57 \ln B + C$$

On a log-log plot this is a straight line. Figure 4 plots data displaying this relationship. We can also rearrange our solution to the differential equation to obtain a power function for allometric scaling like those in Sections 1.2 and 1.5. Defining $k = e^C$, we obtain

$$L = kB^{1.57}$$

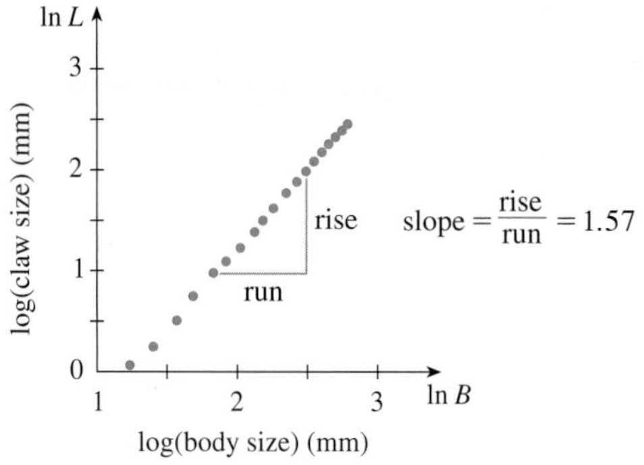

FIGURE 4
Data for the relationship between claw length and body width on a log-log plot

EXAMPLE 4 | Population dynamics Suppose the *per capita growth rate* of a population decreases as the population size n increases, in a way that is described by the expression $1/(1 + n)$. The differential equation for n is therefore

$$\frac{1}{n}\frac{dn}{dt} = \frac{1}{1 + n}$$

Solve this differential equation.

SOLUTION Writing the equation in differential form gives

$$\int \frac{1 + n}{n}\, dn = \int dt$$

(3) $$\ln n + n = t + C$$

where C is an arbitrary constant. Equation 3 gives the family of solutions implicitly. In this case it's impossible to solve the equation to express n explicitly as a function of t. ■

The Gompertz differential equation assumes that the *per volume growth rate* of the tumor declines as the tumor volume gets larger according to the expression $a(\ln b - \ln V)$. Notice that the tumor growth rate is zero when $V = b$, where b represents the asymptotic tumor volume.

EXAMPLE 5 | Gompertz model of tumor growth The Gompertz differential equation models the growth of a tumor in volume V (in mm^3) and is given by

$$\frac{dV}{dt} = a(\ln b - \ln V)V$$

where a and b are positive constants.

(a) Find a family of solutions for tumor volume as a function of time.
(b) Find the solution that has an initial tumor volume of $V(0) = 1 \text{ mm}^3$.

SOLUTION First note that $\ln b - \ln V = \ln(b/V)$. Therefore, assuming $V \neq 0$ and $V \neq b$, we can write the equation in differential form and integrate as

$$\int \frac{dV}{V[\ln(b/V)]} = \int a\, dt$$

We can then integrate the left side using the substitution $u = \ln(b/V)$. We get

$$-\ln|\ln(b/V)| = at + C_1$$

Now we can solve for V by exponentiating both sides twice:

$$\ln(b/V) = Ce^{-at}$$

and then

$$b/V = e^{Ce^{-at}}$$

or

$$V = be^{-Ce^{-at}}$$

where $C = \pm e^{-C_1}$ is an arbitrary constant.

On the other hand, we can verify that $V = b$ is also an (equilibrium) solution.

(b) Setting $t = 0$ in the family of solutions from part (a) gives $V(0) = be^{-C}$. To satisfy the initial condition $V(0) = 1$, we therefore require that $1 = be^{-C}$ or $0 = \ln b - C$. Therefore the desired solution is $V = be^{-(\ln b)e^{-at}}$ or

$$V = b(e^{-\ln b})^{e^{-at}} \quad \Rightarrow \quad V = b\left(\frac{1}{b}\right)^{e^{-at}}$$

Figure 5 shows model predictions and data for three different sets of constant values with initial condition $V(0) = 35$. ■

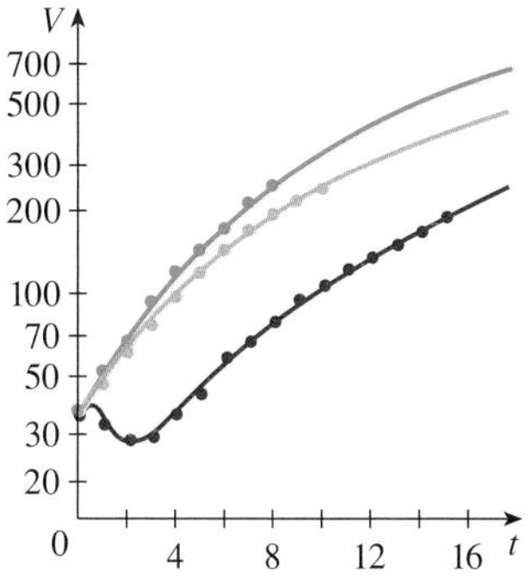

FIGURE 5
The solution of the Gompertz model fitted to tumor data.

Source: Adapted from D. Miklavčič et al., "Mathematical Modelling of Tumor Growth in Mice Following Electrotherapy and Bleomycin Treatment," *Mathematics and Computers in Simulation* 39 (1995): 597–602.

EXAMPLE 6 | The logistic equation Find the solution to the following initial-value problem involving the logistic equation:

$$\frac{dN}{dt} = r\left(1 - \frac{N}{K}\right)N \qquad N(0) = N_0$$

SOLUTION Assuming $N \neq 0$ and $N \neq K$, we can write the equation in differential form and integrate as

(4)
$$\int \frac{dN}{(1 - N/K)N} = \int r\,dt$$

To evaluate the integral on the left side, we write

$$\frac{1}{(1 - N/K)N} = \frac{K}{N(K - N)}$$

Using partial fractions (see Section 5.6), we get

$$\frac{K}{N(K - N)} = \frac{1}{N} + \frac{1}{K - N}$$

This enables us to rewrite Equation 4:

$$\int \left(\frac{1}{N} + \frac{1}{K - N} \right) dN = \int r\, dt$$

$$\ln|N| - \ln|K - N| = rt + C$$

$$\ln \left| \frac{K - N}{N} \right| = -rt - C$$

$$\left| \frac{K - N}{N} \right| = e^{-rt-C} = e^{-C}e^{-rt}$$

(5) $$\frac{K - N}{N} = Ae^{-rt}$$

where $A = \pm e^{-C}$. Solving Equation 5 for N, we get

$$\frac{K}{N} - 1 = Ae^{-rt} \quad \Rightarrow \quad \frac{N}{K} = \frac{1}{1 + Ae^{-rt}}$$

so $$N = \frac{K}{1 + Ae^{-rt}}$$

We find the value of A by putting $t = 0$ in Equation 5. If $t = 0$, then $N = N_0$ (the initial population), so

$$\frac{K - N_0}{N_0} = Ae^0 = A$$

Thus the solution to the logistic equation is

$$N(t) = \frac{K}{1 + Ae^{-rt}} \qquad \text{where } A = \frac{K - N_0}{N_0}$$

On the other hand, if $N = 0$, then we can verify that this is also an (equilibrium) solution. Likewise, $N = K$ is an equilibrium solution.

We can now return to the model of yeast growth from page 424. As mentioned in Section 7.1, the model output in Figure 7.1.5 comes from the logistic growth equation with constant values $N_0 = 0.2$, $K = 210$, and $r = 0.55$. Substituting these values into the solution that we just obtained gives (after some rearrangement)

$$N(t) = \frac{42e^{0.55t}}{209.8 + 0.2e^{0.55t}}$$

This is exactly the solution presented on page 424. ■

EXERCISES 7.4

1–10 Solve the differential equation.

1. $\dfrac{dy}{dx} = xy^2$

2. $\dfrac{dy}{dx} = xe^{-y}$

3. $(x^2 + 1)y' = xy$

4. $(y^2 + xy^2)y' = 1$

5. $(y + \sin y)y' = x + x^3$

6. $\dfrac{du}{dr} = \dfrac{1 + \sqrt{r}}{1 + \sqrt{u}}$

7. $\dfrac{dy}{dt} = \dfrac{te^t}{y\sqrt{1 + y^2}}$

8. $\dfrac{dy}{d\theta} = \dfrac{e^y \sin^2\theta}{y \sec\theta}$

9. $\dfrac{du}{dt} = 2 + 2u + t + tu$

10. $\dfrac{dz}{dt} + e^{t+z} = 0$

11–18 Find the solution of the differential equation that satisfies the given initial condition.

11. $\dfrac{dy}{dx} = \dfrac{x}{y}, \quad y(0) = -3$

12. $\dfrac{dy}{dx} = \dfrac{\ln x}{xy}, \quad y(1) = 2$

13. $\dfrac{du}{dt} = \dfrac{2t + \sec^2 t}{2u}, \quad u(0) = -5$

14. $y' = \dfrac{xy \sin x}{y + 1}, \quad y(0) = 1$

15. $x \ln x = y\left(1 + \sqrt{3 + y^2}\right)y', \quad y(1) = 1$

16. $\dfrac{dP}{dt} = \sqrt{Pt}, \quad P(1) = 2$

17. $y' \tan x = a + y, \; y(\pi/3) = a, \quad 0 < x < \pi/2$

18. $\dfrac{dL}{dt} = kL^2 \ln t, \; L(1) = -1$

19. Find an equation of the curve that passes through the point $(0, 1)$ and whose slope at (x, y) is xy.

20. Find the function f such that $f'(x) = f(x)[1 - f(x)]$ and $f(0) = \frac{1}{2}$.

21. Solve the differential equation $y' = x + y$ by making the change of variable $u = x + y$.

22. Solve the differential equation $xy' = y + xe^{y/x}$ by making the change of variable $v = y/x$.

23. (a) Solve the differential equation $y' = 2x\sqrt{1 - y^2}$.
(b) Solve the initial-value problem $y' = 2x\sqrt{1 - y^2}$, $y(0) = 0$, and graph the solution.
(c) Does the initial-value problem $y' = 2x\sqrt{1 - y^2}$, $y(0) = 2$, have a solution? Explain.

24. Solve the equation $e^{-y}y' + \cos x = 0$ and graph several members of the family of solutions. How does the solution curve change as the constant C varies?

CAS **25.** Solve the initial-value problem $y' = (\sin x)/\sin y$, $y(0) = \pi/2$, and graph the solution (if your CAS does implicit plots).

CAS **26.** Solve the equation $y' = x\sqrt{x^2 + 1}/(ye^y)$ and graph several members of the family of solutions (if your CAS does implicit plots). How does the solution curve change as the constant C varies?

CAS **27–28**
(a) Use a computer algebra system to draw a direction field for the differential equation. Get a printout and use it to sketch some solution curves without solving the differential equation.
(b) Solve the differential equation.
(c) Use the CAS to draw several members of the family of solutions obtained in part (b). Compare with the curves from part (a).

27. $y' = y^2$

28. $y' = xy$

29–31 An **integral equation** is an equation that contains an unknown function $y(x)$ and an integral that involves $y(x)$. Solve the given integral equation. [*Hint:* Use an initial condition obtained from the integral equation.]

29. $y(x) = 2 + \displaystyle\int_2^x [t - ty(t)]\, dt$

30. $y(x) = 2 + \displaystyle\int_1^x \frac{dt}{ty(t)}, \quad x > 0$

31. $y(x) = 4 + \displaystyle\int_0^x 2t\sqrt{y(t)}\, dt$

32–34 Seasonality and habitat destruction The per capita growth rate of many species varies temporally for a variety of reasons, including seasonality and habitat destruction. Suppose $n(t)$ represents the population size at time t, where n is measured in individuals and t is measured in years. Solve the differential equation for habitat destruction and describe the predicted population dynamics.

32. $n' = e^{-t}n \qquad n(0) = n_0$

Here the per capita growth rate declines over time, but always remains positive. It is modeled by the function e^{-t}.

33. $n' = (e^{-t} - 1)n \qquad n(0) = n_0$

Here the per capita growth rate declines over time, starting at zero and becoming negative. It is modeled by the function $e^{-t} - 1$.

34. $n' = (r - at)n \qquad n(0) = n_0$

Here the per capita growth rate declines over time, going

from positive to negative. It is modeled by the function $r - at$, where r and a are positive constants.

---

35. Noyes-Whitney drug dissolution Solve the initial-value problem in Exercise 7.1.14 for the Noyes-Whitney drug dissolution equation.

36. Weibull drug dissolution Solve the Weibull drug dissolution equation given in Exercise 7.1.15.

37–38 Bacteria colony growth In Exercises 1.6.35–36, we obtained difference equations for the growth of circular and spherical colonies of bacteria. These equations are based on the idea that nutrients for growth are available only at the colony–environment interface. Continous-time versions of these equations are presented here, where k is a positive constant and n is the number of bacteria (in thousands). Solve each differential equation to find the size of the colony as a function of time. Assume $n(0) = 1$.

37. $\dfrac{dn}{dt} = kn^{1/2}$ (circular colony)

38. $\dfrac{dn}{dt} = kn^{2/3}$ (spherical colony)

---

39. Tumor growth The Gompertz equation in Example 5 is not the only possibility for modeling tumor growth. Suppose that a tumor can be modeled as a spherical collection of cells and it acquires resources for growth only through its surface area (like the spherical bacterial colony in Exercise 38). All cells in a tumor are also subject to a constant per capita death rate. The dynamics of tumor mass M (in grams) might therefore be modeled as

$$\frac{dM}{dt} = kM^{2/3} - \mu M$$

where μ and k are positive constants. The first term represents tumor growth via nutrients entering through the surface. The second term represents a constant per capita death rate.

(a) Assuming that $k = 1$ and $M(0) = 1$, find M as a function of t.

(b) What happens to the tumor mass as $t \to \infty$?

(c) Assuming tumor mass is proportional to its volume, the diameter of the tumor is related to its mass as $D = aM^{1/3}$, where $a > 0$. Derive a differential equation for D and show that it has the form of the von Bertalanffy equation in Example 2.

40. In an elementary chemical reaction, single molecules of two reactants A and B form a molecule of the product C: $A + B \to C$. The law of mass action states that the rate of reaction is proportional to the product of the concentrations of A and B:

$$\frac{d[C]}{dt} = k[A][B]$$

Thus, if the initial concentrations are $[A] = a$ moles/L and $[B] = b$ moles/L and we write $x = [C]$, then we have

$$\frac{dx}{dt} = k(a - x)(b - x)$$

CAS (a) Assuming that $a \neq b$, find x as a function of t. Use the fact that the initial concentration of C is 0.

(b) Find $x(t)$ assuming that $a = b$. How does this expression for $x(t)$ simplify if it is known that $[C] = \frac{1}{2}a$ after 20 seconds?

41. Population genetics Exercise 7.2.16 derives the following equation from population genetics that specifies the evolutionary dynamics of the frequency of a bacterial strain of interest:

$$\frac{dp}{dt} = sp(1 - p) \qquad p(0) = p_0$$

where s is a constant. Find the solution, $p(t)$.

42. Mutation-selection balance The equation of Exercise 41 can be extended to account for a deleterious mutation that destroys the bacterial strain of interest. The differential equation becomes

$$\frac{dp}{dt} = sp(1 - p) - \mu p \qquad p(0) = p_0$$

where μ is the mutation rate and $\mu > 0$ (see Exercise 7.2.19). Solve this initial-value problem for $s \neq \mu$.

43. Glucose administration A glucose solution is administered intravenously to the bloodstream at a constant rate r. As the glucose is added, it is converted into other substances and removed from the bloodstream at a rate that is proportional to the concentration at that time. Thus a model for the concentration $C(t)$ (in mg/mL) of the glucose solution in the bloodstream is

$$\frac{dC}{dt} = r - kC$$

where k is a positive constant.

(a) Suppose that the concentration at time $t = 0$ is C_0. Determine the concentration at any time t by solving the initial-value problem.

(b) Assuming that $C_0 < r/k$, find $\lim_{t\to\infty} C(t)$ and interpret your answer.

44. mRNA transcription The intermediate molecule mRNA arises in the decoding of DNA: it is produced by a process called transcription and it eventually decays. Suppose that the rate of transcription is changing exponentially according to the expression e^{bt}, where b is a positive constant and mRNA has a constant per capita decay rate of k. The number of mRNA transcript molecules T thus changes as

$$\frac{dT}{dt} = e^{bt} - kT$$

Although the form of this equation is similar to that from Exercise 43, the first term on the right side is now time-varying. As a result, the differential equation is no longer separable; however, the equation can be solved using the change of variables $y(t) = e^{kt}T(t)$. Solve the differential equation using this technique.

45–48 Mixing problems Mixing problems arise in many areas of science. They typically involve a tank of fixed capacity filled with a well-mixed solution of some substance (such as salt). Solution of a given concentration enters the tank at a fixed rate and the mixture, thoroughly stirred, leaves at a fixed rate. We will focus on examples where the inflow and outflow rates are the same, so that the volume of solution in the tank remains constant. If $y(t)$ denotes the amount of substance in the tank at time t, then $y' = (\text{rate in}) - (\text{rate out})$.

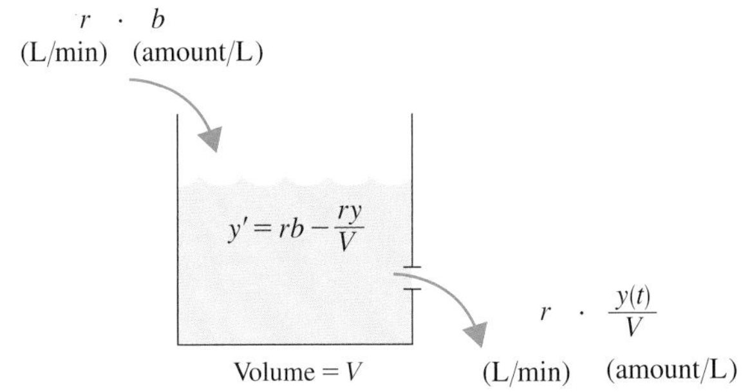

45. A tank contains 1000 L of brine with 15 kg of dissolved salt. Pure water enters the tank at a rate of 10 L/min. The solution is kept thoroughly mixed and drains from the tank at the same rate. How much salt is in the tank (a) after t minutes? (b) After 20 minutes?

46. Dialysis treatment removes urea and other waste products from a patient's blood by diverting some of the blood flow externally through a machine called a dialyzer. Suppose that a patient's blood volume is V mL and blood is diverted through the dialyzer at a rate of K mL/min. At the start of treatment the patient's blood contains $c(0) = c_0$ mg/mL of urea.

(a) Formulate the process of dialysis as an initial-value problem.

(b) What is the concentration of urea in the patient's blood after t minutes of dialysis? Compare your answer to Exercise 1.5.53.

47. A vat with 500 gallons of beer contains 4% alcohol (by volume). Beer with 6% alcohol is pumped into the vat at a rate of 5 gal/min and the mixture is pumped out at the same rate. What is the percentage of alcohol after an hour?

48. Lung ventilation A patient is placed on a ventilator to remove CO_2 from the lungs. Suppose that the rate of ventilation is 100 mL/s, with the percentage of CO_2 (by volume) in the inflow being zero. Suppose also that air is absorbed by the lungs at a rate of 10 mL/s and gas consisting of 100% CO_2 is excreted back into the lungs at the same rate. The volume of a typical pair of lungs is around 4000 mL. If the patient starts ventilation with 20% of lung volume being CO_2, what volume of CO_2 will remain in the lungs after 30 minutes?

49. When a raindrop falls, it increases in size and so its mass at time t is a function of t, namely, $m(t)$. The rate of growth of the mass is $km(t)$ for some positive constant k. When we apply Newton's Law of Motion to the raindrop, we get $(mv)' = gm$, where v is the velocity of the raindrop (directed downward) and g is the acceleration due to gravity. The terminal velocity of the raindrop is $\lim_{t\to\infty} v(t)$. Find an expression for the terminal velocity in terms of g and k.

50. Homeostasis refers to a state in which the nutrient content of a consumer is independent of the nutrient content of its food. In the absence of homeostasis, a model proposed by Sterner and Elser is given by

$$\frac{dy}{dx} = \frac{1}{\theta}\frac{y}{x}$$

where x and y represent the nutrient content of the food and the consumer, respectively, and θ is a constant with $\theta \geqslant 1$.

(a) Solve the differential equation.

(b) What happens when $\theta = 1$? What happens when $\theta \to \infty$?

Source: Adapted from R. Sterner et al., *Ecological Stoichiometry: The Biology of Elements from Molecules to the Biosphere* (Princeton, NJ: Princeton University Press, 2002).

51. Tissue culture Let $A(t)$ be the area of a tissue culture at time t and let M be the final area of the tissue when growth is complete. Most cell divisions occur on the periphery of the tissue and the number of cells on the periphery is proportional to $\sqrt{A(t)}$. So a reasonable model for the growth of tissue is obtained by assuming that the rate of growth of the area is jointly proportional to $\sqrt{A(t)}$ and $M - A(t)$.

(a) Formulate a differential equation and use it to show that the tissue grows fastest when $A(t) = \frac{1}{3}M$.

CAS (b) Solve the differential equation to find an expression for $A(t)$. Use a computer algebra system to perform the integration.

52. According to Newton's Law of Universal Gravitation, the gravitational force on an object of mass m that has been projected vertically upward from the earth's surface is

$$F = \frac{mgR^2}{(x + R)^2}$$

where $x = x(t)$ is the object's distance above the surface at time t, R is the earth's radius, and g is the acceleration due to gravity. Also, by Newton's Second Law, $F = ma = m\,(dv/dt)$ and so

$$m\frac{dv}{dt} = -\frac{mgR^2}{(x + R)^2}$$

(a) Suppose a rocket is fired vertically upward with an initial velocity v_0. Let h be the maximum height above the surface reached by the object. Show that

$$v_0 = \sqrt{\frac{2gRh}{R + h}}$$

[*Hint:* By the Chain Rule, $m\,(dv/dt) = mv\,(dv/dx)$.]

(b) Calculate $v_e = \lim_{h\to\infty} v_0$. This limit is called the *escape velocity* for the earth.
(c) Use $R = 3960$ mi and $g = 32$ ft/s^2 to calculate v_e in feet per second and in miles per second.

53. Species–area relationship The number of species found on an island typically increases with the area of the island. Suppose that this relationship is such that the rate of increase with island area is always proportional to the density of species (that is, number of species per unit area) with a proportionality constant between 0 and 1. Find the function that describes the species-area relationship. Compare your answer to Example 1.5.14.

■ PROJECT Why Does Urea Concentration Rebound after Dialysis?

A patient undergoes dialysis treatment to remove urea from the bloodstream when the kidneys are not functioning properly. Blood is diverted from the patient through a machine that filters out the urea. In many patients, once a dialysis session ends there is a relatively rapid rebound in the concentration of urea in the blood—too rapid to be accounted for by the production of new urea (see Figure 1).

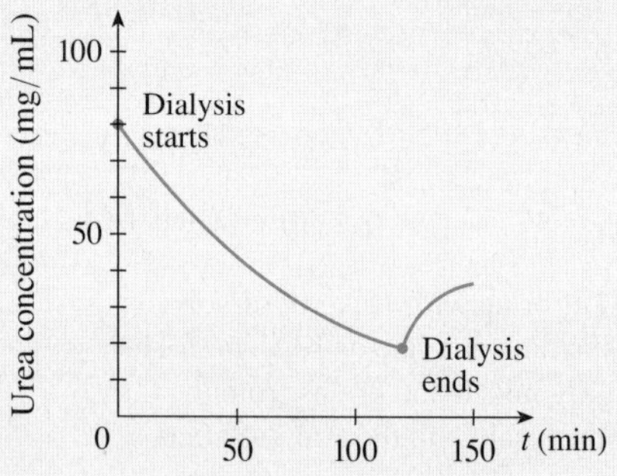

FIGURE 1
Urea rebound after dialysis

One explanation for this rebound is that urea also exists in other parts of the body, and there is continual movement of urea from these other areas into the bloodstream. Modeling this movement results in a so-called "two-compartment" model, as shown in Figure 2.

In Exercise 7.4.46 we saw that a common, one-compartment model for dialysis is

$$\frac{dc}{dt} = -\frac{K}{V}c$$

where K and V are positive constants and c is the concentration of urea in the blood (in mg/mL). To construct a two-compartment model we need to describe the dynamics using two variables, c for the concentration in the blood and p for the concentration in the inaccessible pool (both measured in mg/mL). A model for this process is

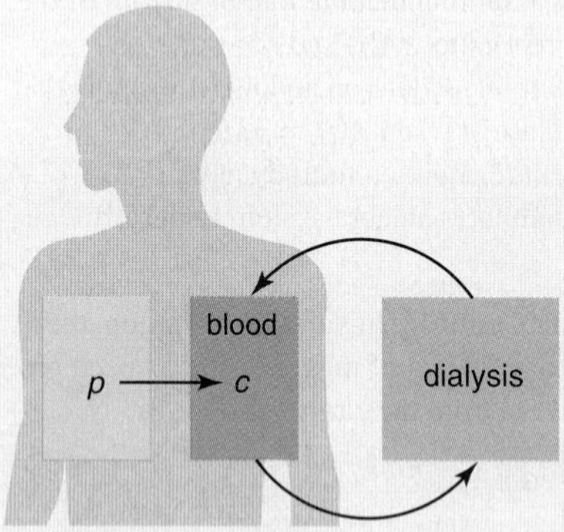

FIGURE 2
A schematic diagram of the two-compartment model

$$\text{(1)} \qquad \frac{dc}{dt} = -\frac{K}{V}c + ap \qquad \frac{dp}{dt} = -ap$$

where K, V, and a are positive constants.

1. Explain the terms in Equations 1 and the assumptions that underlie them.

2. The dynamics of c depend on both the concentration in the blood c and in the inaccessible pool p. However, the dynamics of p depend only on p and so we can solve the differential equation for p independently of the differential equation for c. What is this solution, assuming that the initial concentration of urea in the pool is c_0?

3. Use the solution you obtained in Problem 2, along with the equation for the dynamics of c in Equations 1, to write a single nonautomonous differential equation for the dynamics of c.

Nonautonomous differential equations of the form $y'(t) = f(t) - by(t)$ can be solved using the change of variables $x(t) = e^{bt}y(t)$. This is closely related to a technique for solving differential equations called "integrating factors."

4. Suppose that the initial concentration of urea in the blood is also c_0. Solve the initial-value problem for the concentration of urea in the blood as a function of time during dialysis using the change of variables described in the margin. Plot the concentration over time, assuming that $c_0 = 80$, $a = 0.015$, and $K/V = 0.03$.

5. Suppose dialysis treatment ends after 110 minutes. If we use $c^*(t)$ and $p^*(t)$ to denote the concentration of urea in the blood and in the inaccesible pool, t units of time after treatment has stopped, what is the initial-value problem for $c^*(t)$ and $p^*(t)$?

6. Solve the initial-value problem in Problem 5. Plot the solution along with that from Problem 4 on the same graph, using the constant values from Problem 4. What is the limiting value of the urea concentration in the blood after it has fully rebounded? Explain biologically why this limiting value occurs.

If there were flow of urea from the bloodstream back into the inaccessible pool, how do you think this would complicate the analysis of the model?

7.5 Systems of Differential Equations

The differential equations we have explored so far involve the dynamics of a single variable: population size, gene frequency, body size, and so on. Nevertheless, many biological phenomena—from the dynamics of interacting species to the dynamics of electrical impulses in a neuron—are described by multiple variables. This leads to models involving **coupled systems of differential equations**, meaning that the dynamics of each variable depend on the values of all variables in the system. This coupling means that we can't solve one equation and then the other; we have to solve all equations simultaneously.

We begin our exploration of such models in this section and the next. A more complete treatment of coupled systems requires some background in multivariable calculus and is presented in Chapter 10. A brief introduction is included in this chapter, however, because it offers the flavor of the ideas and applications without requiring any further preparation. Here we focus on graphical techniques for *systems of two autonomous differential equations*. These techniques involve generalizing the ideas from Section 7.2 from plotting the dynamics along a *phase line* to plotting the dynamics in a *phase plane*. In order to do so, we first briefly consider the idea of parametric curves.

■ Parametric Curves

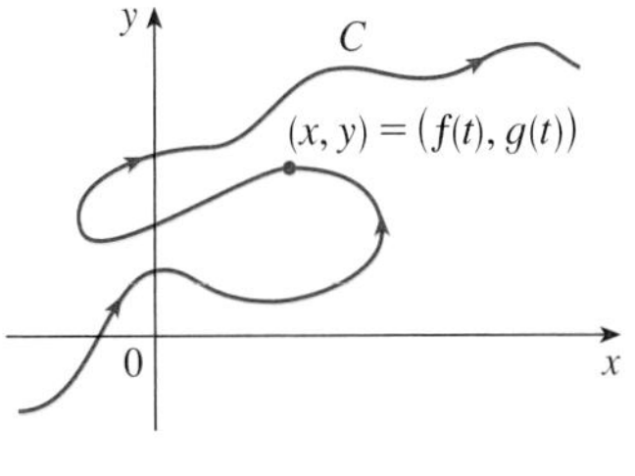

FIGURE 1

Imagine that a particle moves along the curve C shown in Figure 1. It is impossible to describe C by an equation of the form $y = f(x)$ because C fails the Vertical Line Test. But the x- and y-coordinates of the particle are functions of time and so we can write $x = f(t)$ and $y = g(t)$. Such a pair of equations is often a convenient way of describing a curve and gives rise to the following definition.

Suppose that x and y are both given as functions of a third variable t (called a **parameter**) by the equations

$$x = f(t) \qquad y = g(t)$$

(called **parametric equations**). Each value of t determines a point (x, y) that we can plot in a coordinate plane. As t varies, the point $(x, y) = (f(t), g(t))$ varies and traces out a curve C, which we call a **parametric curve**. The parameter t does not necessarily represent time and, in fact, we could use a letter other than t for the parameter. But in the applications of parametric curves to systems of differential equations the parameter will be time and therefore we can interpret $(x, y) = (f(t), g(t))$ as the position of a particle at time t.

EXAMPLE 1 | Sketch and identify the curve defined by the parametric equations

$$x = t^2 - 2t \qquad y = t + 1$$

SOLUTION Each value of t gives a point on the curve, as shown in the table. For instance, if $t = 0$, then $x = 0$, $y = 1$ and so the corresponding point is $(0, 1)$. In Figure 2 we plot the points (x, y) determined by several values of the parameter t and we join them to produce a curve.

t	x	y
-2	8	-1
-1	3	0
0	0	1
1	-1	2
2	0	3
3	3	4
4	8	5

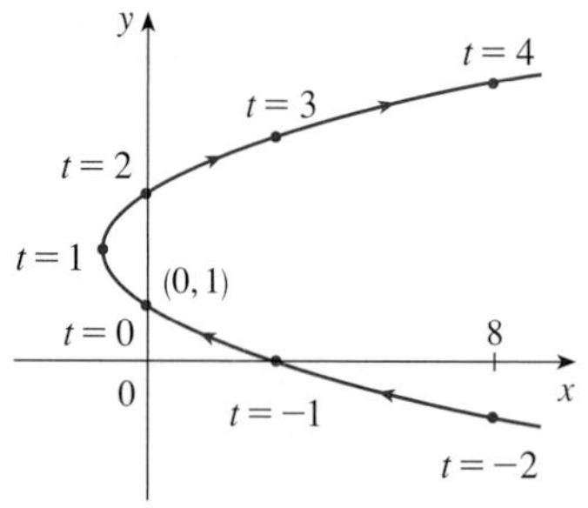

FIGURE 2

A particle whose position is given by the parametric equations moves along the curve in the direction of the arrows as t increases. Notice that the consecutive points marked on the curve appear at equal time intervals but not at equal distances. That is because the particle slows down and then speeds up as t increases.

It appears from Figure 2 that the curve traced out by the particle may be a parabola. This can be confirmed by eliminating the parameter t as follows. We obtain $t = y - 1$ from the second equation and substitute into the first equation. This gives

$$x = t^2 - 2t = (y - 1)^2 - 2(y - 1) = y^2 - 4y + 3$$

and so the curve represented by the given parametric equations is the parabola $x = y^2 - 4y + 3$. ■

This equation in x and y describes *where* the particle has been, but it doesn't tell us *when* the particle was at a particular point. The parametric equations have an advantage—they tell us *when* the particle was at a point. They also indicate the *direction* of the motion.

The notion of a parametric curve is very useful in the study of systems of differential equations.

■ Systems of Two Autonomous Differential Equations

Consider a situation in which one species, called the *prey*, has an ample food supply and the second species, called the *predator*, feeds on the prey. Examples of prey and predators include rabbits and wolves in an isolated forest, small fish and sharks, aphids and ladybugs, and bacteria and amoebas. Our model will have two dependent variables, and both are functions of time. We let $R(t)$ be the number of prey (using R for rabbits) and $W(t)$ be the number of predators (with W for wolves) at time t.

In the absence of predators, the ample food supply would support exponential growth of the prey, that is,

$$\frac{dR}{dt} = rR \qquad \text{where } r \text{ is a positive constant}$$

In the absence of prey, we assume that the predator population would decline through mortality at a rate proportional to itself, that is,

$$\frac{dW}{dt} = -kW \qquad \text{where } k \text{ is a positive constant}$$

With both species present, however, we assume that the principal cause of death among the prey is being eaten by a predator, and the birth rate of the predators depends on their available food supply, namely, the prey. We also assume that the two species encounter each other at a rate that is proportional to both populations and is therefore proportional to the product RW. This is referred to as the **principle of mass action**: the rate of encounter of two entities is proportional to the densities of each. A system of two *coupled* differential equations that incorporates these assumptions is as follows:

W represents the predator.
R represents the prey.

(1) $$\frac{dR}{dt} = rR - aRW \qquad \frac{dW}{dt} = -kW + bRW$$

where k, r, a, and b are positive constants. Notice that the term $-aRW$ decreases the growth rate of the prey and the term bRW increases the growth rate of the predators.

The equations in (1) are known as the **predator-prey equations**, or the **Lotka-Volterra equations**. A **solution** of this system of equations is a pair of functions $R(t)$ and $W(t)$ that describe the populations of prey and predator as functions of time. Therefore a solution can be represented as a parametric curve $(x, y) = (R(t), W(t))$ in the plane. The graphical techniques developed here and in the next section are formulated in terms of such parametric curves.

The Lotka-Volterra equations were proposed as a model to explain the variations in the shark and food-fish populations in the Adriatic Sea by the Italian mathematician Vito Volterra (1860–1940).

EXAMPLE 2 | Lotka-Volterra equations Suppose that populations of rabbits and wolves are described by the Lotka-Volterra equations (1) with $r = 0.08$, $a = 0.001$, $k = 0.02$, and $b = 0.00002$. The time t is measured in months.

(a) Use the system of differential equations to find an expression for dW/dR.

(b) Draw a direction field for the resulting differential equation in the RW-plane. Then use that direction field to sketch some parametric curves representing solutions of Equations 1.

(c) Suppose that, at some point in time, there are 1000 rabbits and 40 wolves. Draw the corresponding parametric curve representing the solution for these initial conditions. Use it to describe the changes in both population levels.

(d) Use part (c) to make sketches of R and W as functions of t.

SOLUTION

(a) We use the Chain Rule to eliminate t:

$$\frac{dW}{dt} = \frac{dW}{dR}\frac{dR}{dt}$$

so

$$\frac{dW}{dR} = \frac{\frac{dW}{dt}}{\frac{dR}{dt}} = \frac{-0.02W + 0.00002RW}{0.08R - 0.001RW}$$

(b) If we think of W as a function of R, we have the differential equation

$$\frac{dW}{dR} = \frac{-0.02W + 0.00002RW}{0.08R - 0.001RW}$$

We draw the direction field for this differential equation in Figure 3. This direction field is always tangent to the parametric curves representing solutions to Equations 1, as shown in Figure 4. If we move along a curve, we observe how the relationship between R and W changes as time passes. Although it is not obvious from the direction field, it can be shown that the curves are closed in the sense that if we travel along a curve, we always return to our starting point.

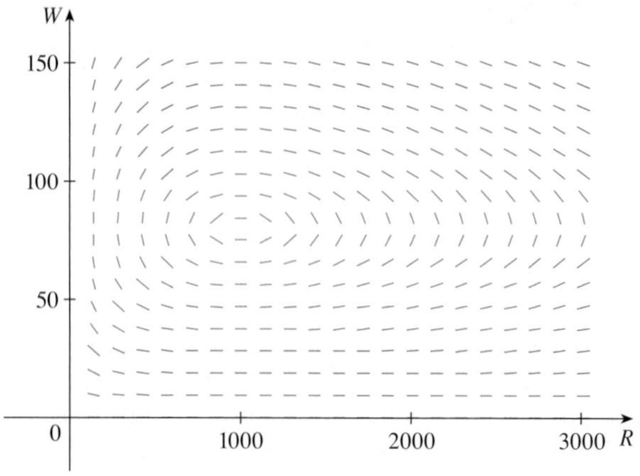

FIGURE 3 Direction field for the predator-prey system

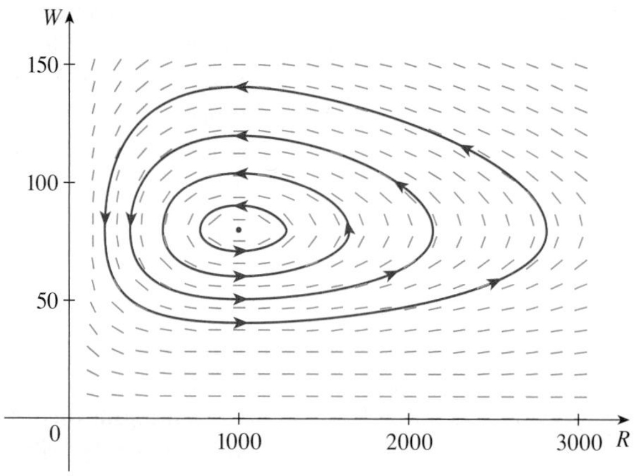

FIGURE 4 Phase portrait of the system

When we represent solutions of a system of differential equations as parametric curves in Figure 4, we refer to the RW-plane as the **phase plane**. This plane is the two-dimensional counterpart of the phase plots considered in Section 7.2. There, when we had a single variable, we plotted arrows on a line corresponding to the variable. Movement could be in either direction along the line and the arrows indicated the direction of this movement. Now, with two variables, we plot arrows in the plane corresponding to the two variables. Movement can now be in any direction in the plane and again the arrows indicate the direction of this movement. The parametric curves in the phase plane are called **phase trajectories**, and so a phase trajectory is a path traced out by solutions (R, W) as time goes by. A **phase portrait** consists of typical phase trajectories, as shown in Figure 4.

(c) Starting with 1000 rabbits and 40 wolves corresponds to drawing the parametric curve through the point $P_0(1000, 40)$. Figure 5 shows this phase trajectory with the direction field removed. Starting at the point P_0 at time $t = 0$ and letting t increase, do we move clockwise or counterclockwise around the phase trajectory? If we put $R = 1000$ and $W = 40$ in the first differential equation, we get

$$\frac{dR}{dt} = 0.08(1000) - 0.001(1000)(40) = 80 - 40 = 40$$

Since $dR/dt > 0$, we conclude that R is increasing at P_0 and so we move counterclockwise around the phase trajectory.

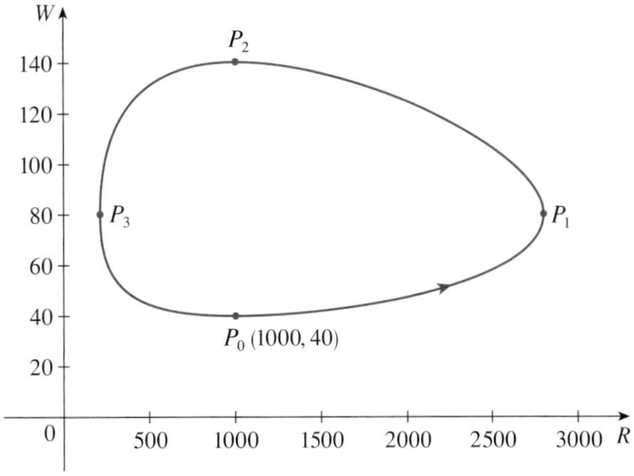

FIGURE 5
Phase trajectory through (1000, 40)

We see that at P_0 there aren't enough wolves to maintain a balance between the populations, so the rabbit population increases. That results in more wolves and eventually there are so many wolves that the rabbits have a hard time avoiding them. So the number of rabbits begins to decline (at P_1, where we estimate that R reaches its maximum population of about 2800). This means that at some later time the wolf population starts to fall (at P_2, where $R = 1000$ and $W \approx 140$). But this benefits the rabbits, so their population later starts to increase (at P_3, where $W = 80$ and $R \approx 210$). As a consequence, the wolf population eventually starts to increase as well. This happens when the populations return to their initial values of $R = 1000$ and $W = 40$, and the entire cycle begins again.

(d) From the description in part (c) of how the rabbit and wolf populations rise and fall, we can sketch the graphs of $R(t)$ and $W(t)$. Suppose the points P_1, P_2, and P_3 in Figure 5 are reached at times t_1, t_2, and t_3. Then we can sketch graphs of R and W as in Figure 6.

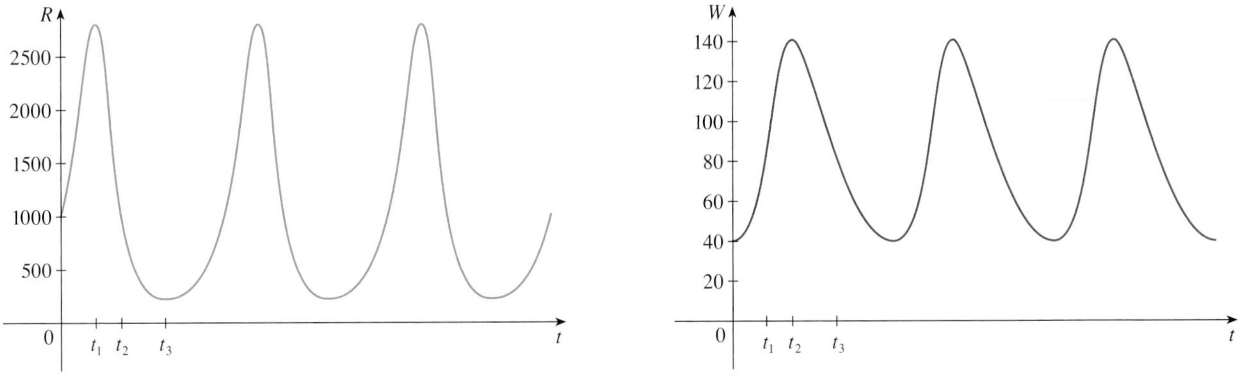

FIGURE 6 Graphs of the rabbit and wolf populations as functions of time

TEC In Module 7.5 you can change the coefficients in the Lotka-Volterra equations and observe the resulting changes in the phase trajectory and graphs of the rabbit and wolf populations.

To make the graphs easier to compare, we draw the graphs on the same axes but with different scales for R and W, as in Figure 7. Notice that the rabbits reach their maximum population size about a quarter of a cycle before the wolves.

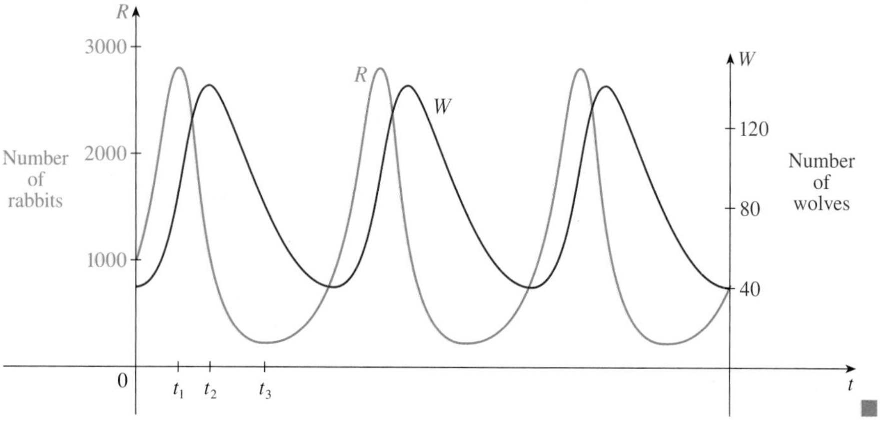

FIGURE 7
Comparison of the rabbit and wolf populations

An important part of the modeling process is to interpret our mathematical conclusions as real-world predictions and to test the predictions against real data. The Hudson's Bay Company, which started trading in animal furs in Canada in 1670, has kept records that date back to the 1840s. Figure 8 shows graphs of the number of pelts of the snowshoe hare and its predator, the Canada lynx, traded by the company over a 90-year period. You can see that the coupled oscillations in the hare and lynx populations predicted by the Lotka-Volterra model do actually occur and the period of these cycles is roughly 10 years.

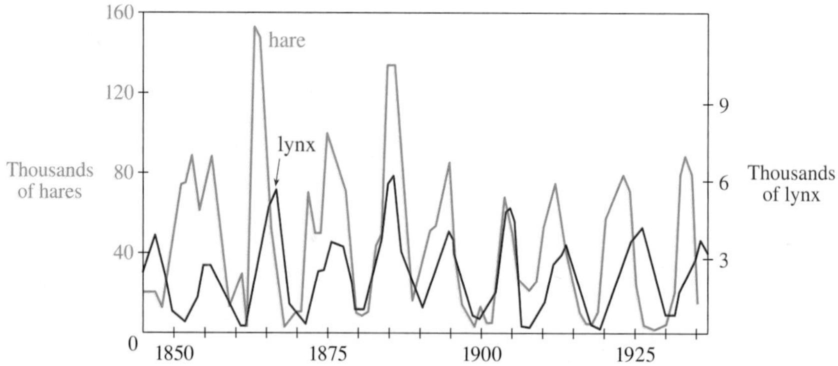

FIGURE 8
Relative abundance of hare and lynx from Hudson's Bay Company records

Although the relatively simple Lotka-Volterra model has had some success in explaining and predicting coupled populations, more sophisticated models have also been proposed. One way to modify the Lotka-Volterra equations is to assume that, in the absence of predators, the prey grow according to a logistic model with carrying capacity M. Then the Lotka-Volterra equations (1) are replaced by the system of differential equations

$$\frac{dR}{dt} = rR\left(1 - \frac{R}{M}\right) - aRW \qquad \frac{dW}{dt} = -kW + bRW$$

This model is investigated in Exercises 23 and 24.

Models have also been proposed to describe and predict population levels of two or more species that compete for the same resources or cooperate for mutual benefit. Such models are explored in Exercises 14 and 15.

EXERCISES 7.5

1–4 Sketch the curve by using the parametric equations to plot points. Indicate with an arrow the direction in which the curve is traced as t increases.

1. $x = t^2 + t, \quad y = t^2 - t, \quad -2 \leq t \leq 2$

2. $x = t^2, \quad y = t^3 - 4t, \quad -3 \leq t \leq 3$

3. $x = \cos^2 t, \quad y = 1 - \sin t, \quad 0 \leq t \leq \pi/2$

4. $x = e^{-t} + t, \quad y = e^t - t, \quad -2 \leq t \leq 2$

5–8

(a) Sketch the curve by using the parametric equations to plot points. Indicate with an arrow the direction in which the curve is traced as t increases.

(b) Eliminate the parameter to find a Cartesian equation of the curve.

5. $x = 3t - 5, \quad y = 2t + 1$

6. $x = 1 + 3t, \quad y = 2 - t^2$

7. $x = \sqrt{t}, \quad y = 1 - t$

8. $x = t^2, \quad y = t^3$

9–12 Describe the motion of a particle with position (x, y) as t varies in the given interval.

9. $x = 3 + 2\cos t, \quad y = 1 + 2\sin t, \quad \pi/2 \leq t \leq 3\pi/2$

10. $x = 2\sin t, \quad y = 4 + \cos t, \quad 0 \leq t \leq 3\pi/2$

11. $x = 5\sin t, \quad y = 2\cos t, \quad -\pi \leq t \leq 5\pi$

12. $x = \sin t, \quad y = \cos^2 t, \quad -2\pi \leq t \leq 2\pi$

13. Predator-prey equations For each predator-prey system, determine which of the variables, x or y, represents the prey population and which represents the predator population. Is the growth of the prey restricted just by the predators or by other factors as well? Do the predators feed only on the prey or do they have additional food sources? Explain.

(a) $$\frac{dx}{dt} = -0.05x + 0.0001xy$$
$$\frac{dy}{dt} = 0.1y - 0.005xy$$

(b) $$\frac{dx}{dt} = 0.2x - 0.0002x^2 - 0.006xy$$
$$\frac{dy}{dt} = -0.015y + 0.00008xy$$

14. Competition and cooperation Each system of differential equations is a model for two species that either compete for the same resources or cooperate for mutual benefit (flowering plants and insect pollinators, for instance). Decide whether each system describes competition or cooperation and explain why it is a reasonable model. (Ask yourself what effect an increase in one species has on the growth rate of the other.)

(a) $$\frac{dx}{dt} = 0.12x - 0.0006x^2 + 0.00001xy$$
$$\frac{dy}{dt} = 0.08x + 0.00004xy$$

(b) $$\frac{dx}{dt} = 0.15x - 0.0002x^2 - 0.0006xy$$
$$\frac{dy}{dt} = 0.2y - 0.00008y^2 - 0.0002xy$$

15. Cooperation, competition, or predation? The system of differential equations

$$\frac{dx}{dt} = 0.5x - 0.004x^2 - 0.001xy$$
$$\frac{dy}{dt} = 0.4y - 0.001y^2 - 0.002xy$$

is a model for the populations of two species. Does the model describe cooperation, or competition, or a predator-prey relationship?

16. A food web Lynx eat snowshoe hares, and snowshoe hares eat woody plants, such as willows. Suppose that, in the absence of hares, the willow population will grow exponentially and the lynx population will decay exponentially. In the absence of lynx and willow, the hare population will decay exponentially. If $L(t)$, $H(t)$, and $W(t)$ represent the populations of these three species at time t, write a system of differential equations as a model for their dynamics. If the constants in your equation are all positive, explain why you have used plus or minus signs.

17–18 Rabbits and foxes A phase trajectory is shown for populations of rabbits (R) and foxes (F).

(a) Describe how each population changes as time goes by.

(b) Use your description to make a rough sketch of the graphs of R and F as functions of time.

17.

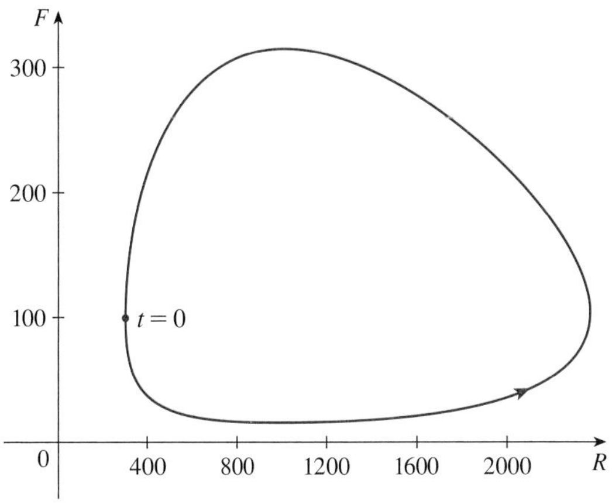

18.

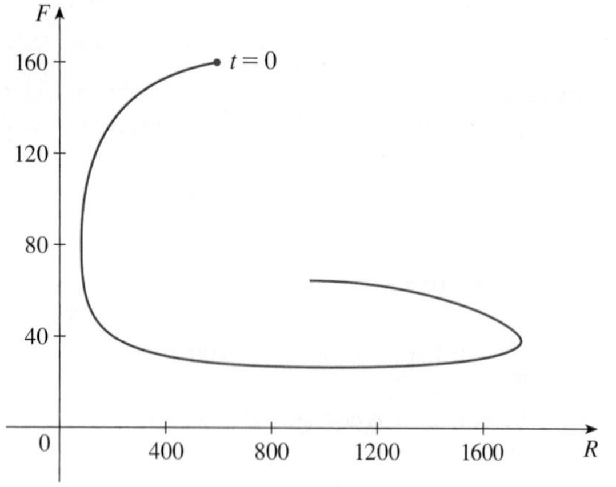

19–20 Graphs of populations of two species are shown. Use them to sketch the corresponding phase trajectory.

19.

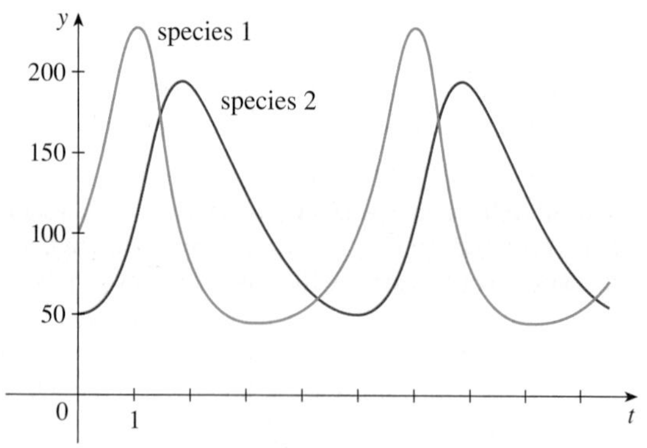

20.

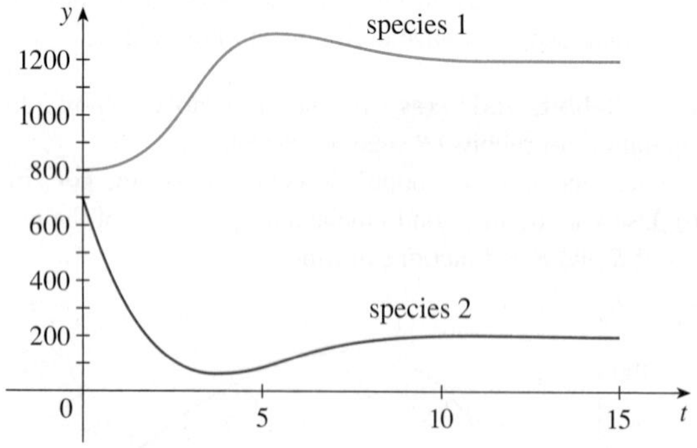

21. Lotka-Volterra equations In Example 1(a) we showed that parametric curves describing the rabbit and wolf populations in the phase plane satisfy the differential equation

$$\frac{dW}{dR} = \frac{-0.02W + 0.00002RW}{0.08R - 0.001RW}$$

By solving this separable differential equation, show that

$$\frac{R^{0.02}W^{0.08}}{e^{0.00002R}e^{0.001W}} = C$$

where C is a constant.

It is impossible to solve this equation for W as an explicit function of R (or vice versa). If you have a computer algebra system that graphs implicitly defined curves, use this equation and your CAS to draw the solution curve that passes through the point (1000, 40) and compare with Figure 5.

22. Aphid-ladybug dynamics Populations of aphids and ladybugs are modeled by the equations

$$\frac{dA}{dt} = 2A - 0.01AL$$

$$\frac{dL}{dt} = -0.5L + 0.0001AL$$

(a) Find an expression for dL/dA.

(b) The direction field for the differential equation in part (b) is shown. Use it to sketch a phase portrait. What do the phase trajectories have in common?

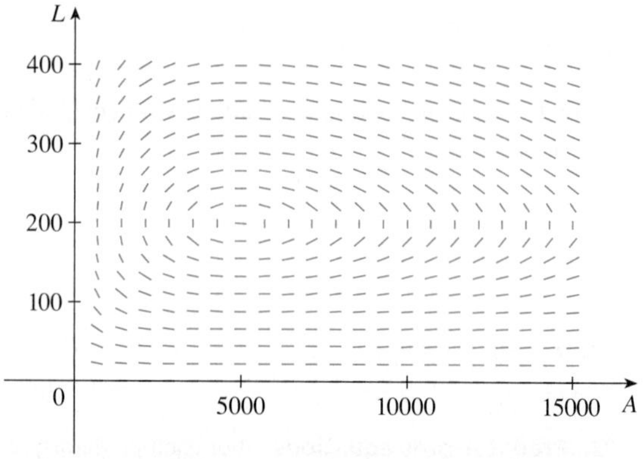

(c) Suppose that at time $t = 0$ there are 1000 aphids and 200 ladybugs. Draw the corresponding phase trajectory and use it to describe how both populations change.

(d) Use part (c) to make rough sketches of the aphid and ladybug populations as functions of t. How are these two graphs related?

23. Modified predator-prey dynamics In Example 1 we used Lotka-Volterra equations to model populations of rabbits and wolves. Let's modify those equations as follows:

$$\frac{dR}{dt} = 0.08R(1 - 0.0002R) - 0.001RW$$

$$\frac{dW}{dt} = -0.02W + 0.00002RW$$

(a) According to these equations, what happens to the rabbit population in the absence of wolves?

(b) The figure shows the phase trajectory that starts at the point (1000, 40). Describe what eventually happens to the rabbit and wolf populations.

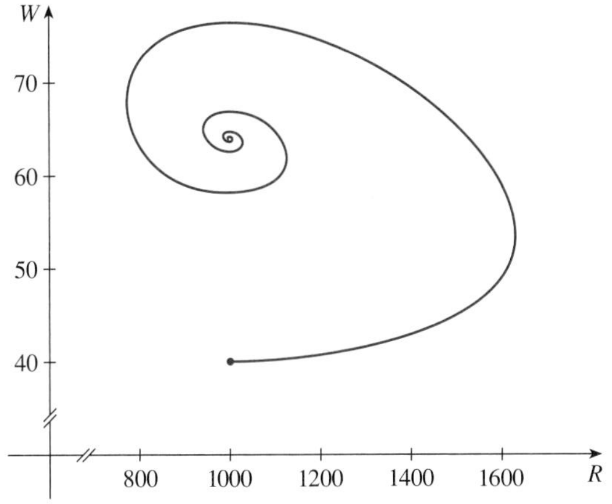

(c) Sketch graphs of the rabbit and wolf populations as functions of time.

24. Modified aphid-ladybug dynamics In Exercise 22 we modeled populations of aphids and ladybugs with a Lotka-Volterra system. Suppose we modify those equations as follows:

$$\frac{dA}{dt} = 2A(1 - 0.0001A) - 0.01AL$$

$$\frac{dL}{dt} = -0.5L + 0.0001AL$$

(a) In the absence of ladybugs, what does the model predict about the aphids?

(b) Find an expression for dL/dA.

CAS (c) Use a computer algebra system to draw a direction field for the differential equation in part (b). Then use the direction field to sketch a phase portrait. What do the phase trajectories have in common?

(d) Suppose that at time $t = 0$ there are 1000 aphids and 200 ladybugs. Draw the corresponding phase trajectory and use it to describe how both populations change.

(e) Use part (d) to make rough sketches of the aphid and ladybug populations as functions of t. How are these two graphs related?

PROJECT The Flight Path of Hunting Raptors

Many raptors, such as falcons and hawks, circle in on their prey while hunting rather than flying directly toward them (see Figure 1). One reason for this behavior is that they must aim one eye directly at the prey for maximum visual acuity and, because of the position of their eyes, this requires that they keep their direction of flight at a constant angle to the prey. Can we predict the flight path that a raptor will take toward its prey by describing this behavior mathematically?

FIGURE 1

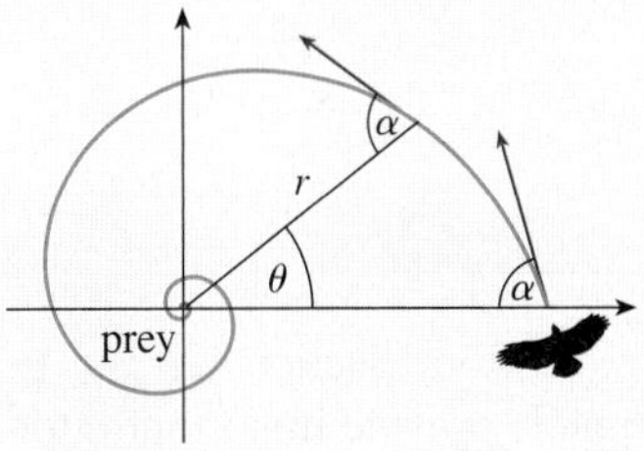

FIGURE 2

To simplify matters, let's consider the flight path in the horizontal plane only, under the assumption that the bird maintains its flight path at a constant angle α to the prey, where $0 < \alpha < \pi/2$ (see Figure 2). We need to describe the position of the bird in the plane at any time t. We could use the Cartesian coordinate system to do this, but the spiral structure of the flight path suggests a simpler approach: specify the distance r between the bird and its prey, along with the angle of rotation θ that identifies the location of the bird. Such coordinates are called polar coordinates.

The bird will circle the prey repeatedly as it closes in, meaning that the distance to the prey will decrease as the angle of rotation θ increases. To describe this process we will therefore derive a differential equation for the distance to the prey r as a function of the angle of rotation θ.

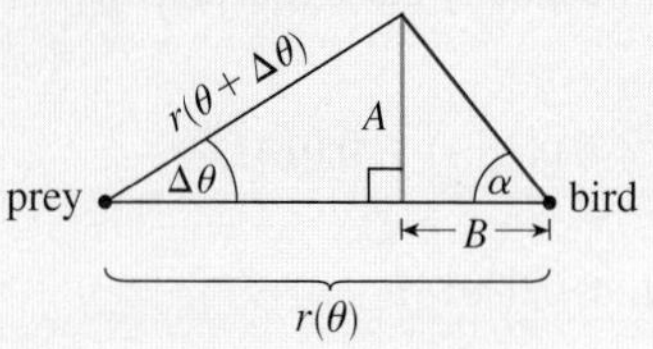

FIGURE 3

1. In a small interval of rotation $\Delta\theta$ we can approximate the movement of the bird as a straight line (the red line in Figure 3). The initial distance to the prey is $r(\theta)$ and after the small change in θ it is $r(\theta + \Delta\theta)$. The known angles are also labeled in Figure 3. Our goal is to express $r(\theta + \Delta\theta)$ as a function of $r(\theta)$. To do this we first calculate some intermediate quantities. Show that the following relationships hold:

$$r(\theta + \Delta\theta) = \frac{r(\theta) - B}{\cos \Delta\theta} \qquad B = \frac{\sin \Delta\theta}{\tan \alpha} r(\theta + \Delta\theta)$$

2. Use the relationships given in Problem 1 to obtain a formula for $r(\theta + \Delta\theta)$ in terms of $r(\theta)$, $\cos \Delta\theta$, $\sin \Delta\theta$, and $\tan \alpha$.

3. From your answer to Problem 2, show that the differential equation for r as a function of θ is given by

$$\frac{dr}{d\theta} = -r \cot \alpha$$

Hint: First obtain an expression for $\dfrac{r(\theta + \Delta\theta) - r(\theta)}{\Delta\theta}$ and then take the limit as $\Delta\theta \to 0$.

4. Suppose that we choose the initial angle of rotation to be zero when the prey is first spotted and the prey is a distance $r(0) = r_0$ from the bird. Solve the initial-value problem corresponding to the differential equation from Problem 3 with this initial condition. What is the distance to the prey as a function of the angle of rotation?

5. By what factor does the distance to the prey get reduced every time the bird circles the prey? Explain your answer.

6. Express the curve found in Problem 4 in Cartesian coordinates by specifying it as a parametric curve in x and y.

FIGURE 4

The curve describing the flight path of a hunting raptor found in Problem 4 is known as a **logarithmic spiral**. Its defining feature is that the tangent to the spiral at any point is always at a constant angle from a radial line joining the center of the spiral to this point. This means that the local geometry of the curve remains fixed and therefore it retains its shape no matter how big the spiral becomes. Perhaps because of this feature of maintaining a constant shape, logarithmic spirals are quite common in living organisms. One of the most famous examples is the spiral shell of a nautilus shown in Figure 4.

7.6 Phase Plane Analysis

The preceding section illustrated how the the phase plots from Section 7.2 can be extended to systems of two differential equations. These plots provide important information about the dynamics of the system but they can be tedious to construct in the absence of a computer. In this section we develop some tools to obtain a very general qualitative understanding of the dynamics of *systems of two autonomous differential equations*. The technique—referred to as phase plane analysis—involves identifying equilibria of the equations and then determining the qualitative nature of the dynamics around these equilibria.

Equilibria

Recall the predator-prey equations from Section 7.5 for the population sizes of rabbits and wolves:

$$\textbf{(1)} \qquad \frac{dR}{dt} = rR - aRW \qquad \frac{dW}{dt} = -kW + bRW$$

The corresponding phase plane is shown in Figure 1. An *equilibrium* of this system of differential equations is a constant population size of rabbits $\hat{R}$ and of wolves $\hat{W}$ at which no further change in either occurs. This requires that both $dR/dt = 0$ and $dW/dt = 0$. Using Equations 1, this gives two equations in two unknowns: $rR - aRW = 0$ and $-kW + bRW = 0$.

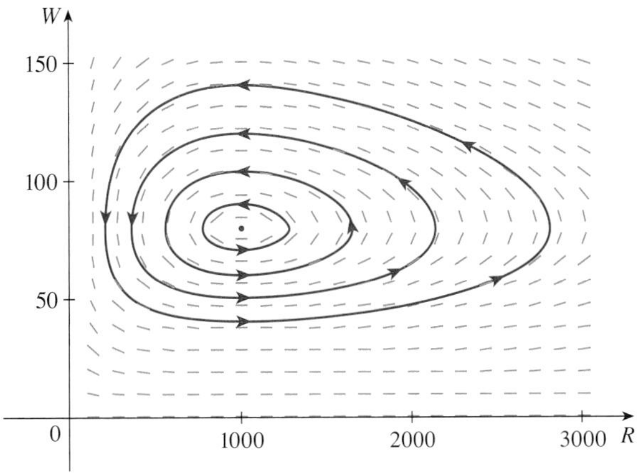

FIGURE 1
The predator-prey phase plot when $r = 0.08$, $a = 0.001$, $k = 0.02$, and $b = 0.00002$

Definition Consider the autonomous system of differential equations

$$\textbf{(2)} \qquad \frac{dx}{dt} = f(x, y) \qquad \frac{dy}{dt} = g(x, y)$$

An **equilibrium** is a pair of values $(\hat{x}, \hat{y})$ such that both $dx/dt = 0$ and $dy/dt = 0$ when $x = \hat{x}$ and $y = \hat{y}$. This gives a pair of equations $f(\hat{x}, \hat{y}) = 0$ and $g(\hat{x}, \hat{y}) = 0$ that define the values of $\hat{x}$ and $\hat{y}$.

We can connect the pair of equations defining the equilibria of the predator-prey model to the phase plane in Figure 1. The equation $rR - aRW = 0$ must hold if the population size of rabbits is to remain constant; it can be factored to give $R(r - aW) = 0$. Therefore the population size of rabbits will remain constant if either $R = 0$ or $W = r/a$. The first of these equations corresponds to absence of rabbits altogether. The second corresponds to the population size of wolves at which the birth rate of rabbits is exactly balanced by their death rate through predation. The lines defined by these equations are called the *R-nullclines* and are plotted on the phase plane shown in Figure 2(a).

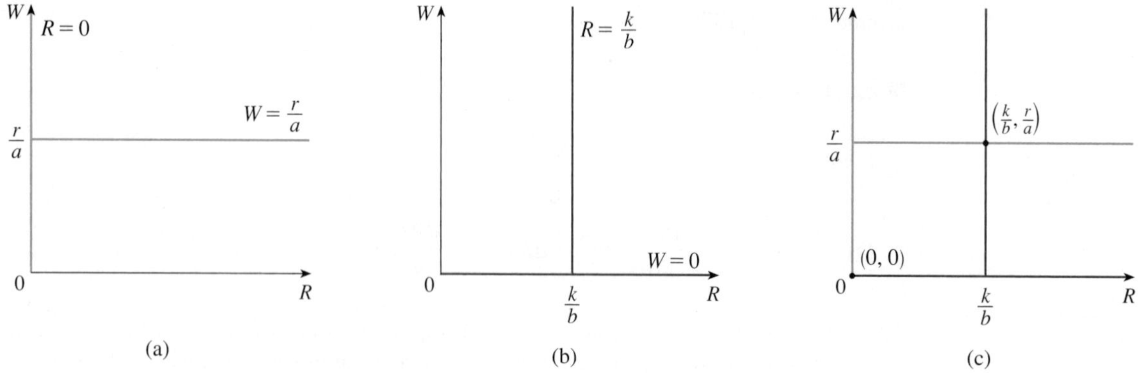

FIGURE 2
R-nullclines are blue, *W*-nullclines are red, and equilibria are black dots.

The second equation, $-kW + bRW = 0$, must hold if the population size of wolves is to remain constant; it can be factored and solved to give $W = 0$ and $R = k/b$. The first of these equations corresponds to absence of wolves. The second corresponds to the population size of rabbits at which the birth rate of wolves (through consumption of rabbits) is exactly balanced by their death rate. The lines defined by these equations are the *W-nullclines* and are plotted on the phase plane shown in Figure 2(b). Figure 2(c) shows the *R*- and *W*-nullclines plotted together.

> **Definition** The ***x*-nullclines** of differential equations (2) are the curves in the *xy*-plane that satisfy the equation $f(x, y) = 0$. Along these curves, $dx/dt = 0$. The ***y*-nullclines** of differential equations (2) are the curves in the *xy*-plane that satisfy the equation $g(x, y) = 0$. Along these curves, $dy/dt = 0$.

Whenever a trajectory in the phase plane crosses a nullcline, it must do so either horizontally or vertically, depending on the nullcline in question. This is because movement in either the vertical or horizontal direction is zero on a nullcline since either $dy/dt = 0$ or $dx/dt = 0$.

Nullclines of the differential equations provide a graphical way to visualize the equilibria of differential equations. For example, in the predator-prey model, an equilibrium requires that *both* the predator and the prey population sizes remain constant through time. Geometrically, an equilibrium will therefore occur at any point where an *R*-nullcline intersects a *W*-nullcline. It will be precisely at such intersection points that both variables remain constant through time [see Figure 2(c)].

> **Finding Equilibria Graphically** For differential equations (2) any point at which an *x*-nullcline intersects a *y*-nullcline is an equilibrium.

In addition to this visualization of equilibrium points and nullclines, we can sometimes derive expressions for the equilibria algebraically. The predator-prey model has two equilibria: (i) $\hat{R} = 0$, $\hat{W} = 0$ and (ii) $\hat{R} = k/b$, $\hat{W} = r/a$.

EXAMPLE 1 | **Lotka-Volterra competition equations** The differential equation for logistic population growth (Equation 7.1.4) can be extended to model competitive interactions between two species. Let's use $N_1(t)$ and $N_2(t)$ to denote the population size of species 1 and 2 at time t. Suppose that the per capita growth rate of each species decreases linearly with the population size of each species. Specifically, the per capita growth rate of species 1 is

$$r\left(1 - \frac{N_1 + \alpha N_2}{K_1}\right)$$

where α, r, and K_1 are positive constants. Likewise, the per capita growth rate of species 2 is

$$r\left(1 - \frac{N_2 + \beta N_1}{K_2}\right)$$

where β and K_2 are positive constants. (Compare these per capita growth rates to the per capita growth rate in Equation 7.1.4.) This gives the system

$$\textbf{(3)} \qquad \frac{dN_1}{dt} = r\left(1 - \frac{N_1 + \alpha N_2}{K_1}\right)N_1 \qquad \frac{dN_2}{dt} = r\left(1 - \frac{N_2 + \beta N_1}{K_2}\right)N_2$$

(a) Suppose $r = 1$, $K_1 = 1000$, $K_2 = 600$, $\alpha = 2$, and $\beta = 1$. Find the N_1- and N_2-nullclines and plot them on the phase plane. Indicate all equilibria.
(b) Calculate the equilibria algebraically.
(c) Suppose instead that $\beta = 0$, but all other constants have the same values as in part (a). Calculate the equilibria algebraically.

SOLUTION

(a) The N_1-nullclines satisfy $dN_1/dt = 0$ or

$$\left(1 - \frac{N_1 + 2N_2}{1000}\right)N_1 = 0$$

Therefore the N_1-nullclines are $N_1 = 0$ and $N_1 + 2N_2 = 1000$. The second equation can be rewritten as $N_2 = 500 - \frac{1}{2}N_1$. It is plotted in Figure 3(a). The N_2-nullclines satisfy $dN_2/dt = 0$ or

$$\left(1 - \frac{N_2 + N_1}{600}\right)N_2 = 0$$

The N_2-nullclines are therefore $N_2 = 0$ and $N_1 + N_2 = 600$. The second equation can be rewritten as $N_2 = 600 - N_1$ and is plotted in Figure 3(b). The nullclines are plotted together, along with the equilibria, in Figure 3(c).

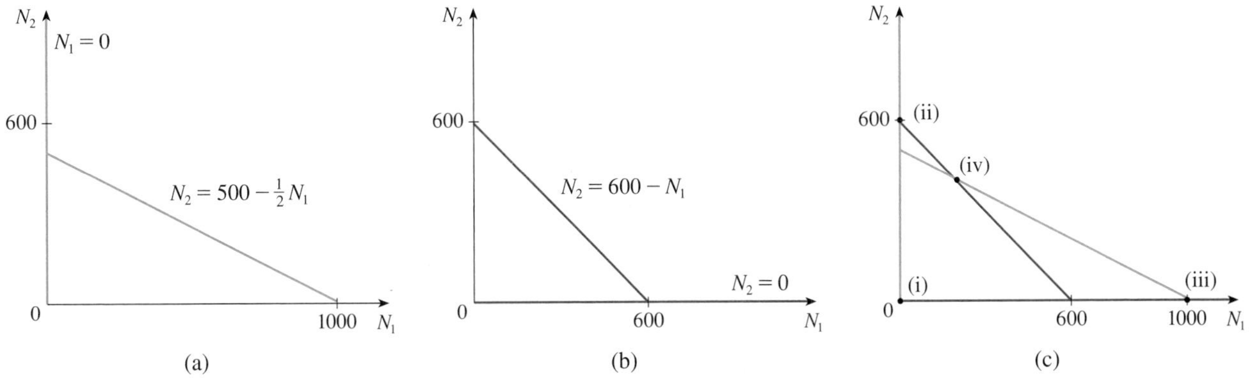

FIGURE 3 N_1-nullclines are blue, N_2-nullclines are red, and equilibria are black dots.

(b) Equilibria are pairs of values $(\hat{N}_1, \hat{N}_2)$ that simultaneously satisfy the pair of equations

$$\left(1 - \frac{\hat{N}_1 + 2\hat{N}_2}{1000}\right)\hat{N}_1 = 0 \qquad \text{and} \qquad \left(1 - \frac{\hat{N}_2 + \hat{N}_1}{600}\right)\hat{N}_2 = 0$$

We can calculate the equilibria by solving the first equation for $\hat{N}_1$, substituting the result into the second equation, and then solving it for $\hat{N}_2$. There are two solutions to the first equation: $\hat{N}_1 = 0$ and $\hat{N}_1 = 1000 - 2\hat{N}_2$. We consider each of these in turn.

Substituting $\hat{N}_1 = 0$ into the second equation gives

$$\left(1 - \frac{\hat{N}_2}{600}\right)\hat{N}_2 = 0$$

Solving for $\hat{N}_2$ gives $\hat{N}_2 = 0$ and $\hat{N}_2 = 600$. Therefore two equilibria are (i) $\hat{N}_1 = 0$, $\hat{N}_2 = 0$ and (ii) $\hat{N}_1 = 0$, $\hat{N}_2 = 600$ [see Figure 3(c)].

Substituting $\hat{N}_1 = 1000 - 2\hat{N}_2$ into the second equation gives

$$\left(1 - \frac{\hat{N}_2 + (1000 - 2\hat{N}_2)}{600}\right)\hat{N}_2 = 0$$

Solving for $\hat{N}_2$ gives $\hat{N}_2 = 0$ and $\hat{N}_2 = 400$. In the first case we then have an $\hat{N}_1$ value of $\hat{N}_1 = 1000 - 2 \cdot 0 = 1000$, and in the second case we have an $\hat{N}_1$ value of $\hat{N}_1 = 1000 - 2 \cdot 400 = 200$. Therefore a third equilibrium is (iii) $\hat{N}_1 = 1000$, $\hat{N}_2 = 0$ and a fourth is (iv) $\hat{N}_1 = 200$, $\hat{N}_2 = 400$ as shown in Figure 3(c).

(c) With $\beta = 0$, the equilibria are now pairs of values, $(\hat{N}_1, \hat{N}_2)$ that simultaneously satisfy the equations

$$\left(1 - \frac{\hat{N}_1 + 2\hat{N}_2}{1000}\right)\hat{N}_1 = 0 \qquad \text{and} \qquad \left(1 - \frac{\hat{N}_2}{600}\right)\hat{N}_2 = 0$$

The second equation no longer involves $\hat{N}_1$ and therefore we can solve it immediately for $\hat{N}_2$. We obtain $\hat{N}_2 = 0$ and $\hat{N}_2 = 600$.

Substituting $\hat{N}_2 = 0$ into the first equation gives $[1 - \hat{N}_1/1000]\hat{N}_1 = 0$. Solving this for $\hat{N}_1$ gives $\hat{N}_1 = 0$ and $\hat{N}_1 = 1000$. Therefore two equilibria are (i) $\hat{N}_1 = 0$, $\hat{N}_2 = 0$ and (ii) $\hat{N}_1 = 1000$, $\hat{N}_2 = 0$.

If instead we substitute $\hat{N}_2 = 600$ into the first equation, we get

$$\left(1 - \frac{\hat{N}_1 + 1200}{1000}\right)\hat{N}_1 = 0$$

Solving this for $\hat{N}_1$ gives $\hat{N}_1 = 0$ and $\hat{N}_1 = -200$. Therefore two additional equilibria are (iii) $\hat{N}_1 = 0$, $\hat{N}_2 = 600$ and (iv) $\hat{N}_1 = -200$, $\hat{N}_2 = 600$.

Notice that equilibrium (iv) involves a negative value of $\hat{N}_1$. From a mathematical standpoint this is a perfectly fine equilibrium, but from a biological standpoint it is not of interest because it would correspond to a negative population size. Equilibria that are biologically relevant [(i), (ii), and (iii) in this example] are referred to as **biologically feasible**. ■

■ Qualitative Dynamics in the Phase Plane

Let's return to Figure 2 from the predator-prey model of Equations 1. In Figure 2(a) we plotted the R-nullclines. These are curves in the plane along which $dR/dt = 0$.

Therefore these curves separate the plane into regions within which either $dR/dt > 0$ or $dR/dt < 0$. We can determine which of these two situations applies in each region. Consider the region above the line $W = r/a$ in Figure 2(a). This corresponds to large values of W, in which case Equations 1 give

$$\frac{dR}{dt} = rR - aRW \approx -aRW < 0$$

for large enough values of W. Therefore $dR/dt < 0$ in this region. Conversely, if W is close to zero, we have

$$\frac{dR}{dt} = rR - aRW \approx rR > 0$$

We can therefore indicate whether R is increasing or decreasing in each of these regions of the phase plane with a single arrow as in Figure 4(a).

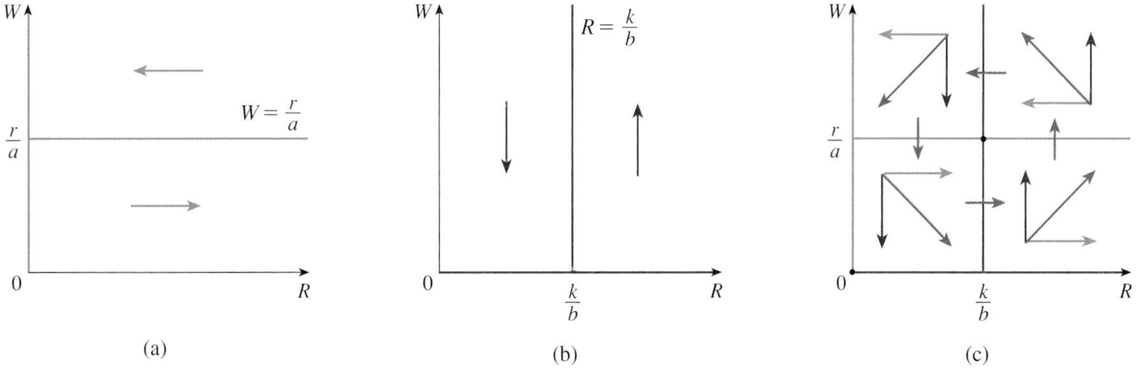

FIGURE 4
Purple arrows indicate direction of motion in part (c).

We can follow the same procedure for the W-nullclines in Figure 2(b). To the right of the line $R = k/b$ the value of R will be very large. From Equations 1 we have

$$dW/dt = -kW + bRW \approx bRW > 0$$

Conversely, if R is close to zero then

$$dW/dt = -kW + bRW \approx -kW < 0$$

This gives the direction arrows for W in Figure 4(b). Putting these two plots together then gives the overall direction of movement by the purple arrows in the phase plane shown in Figure 4(c). This provides a very general, qualitative picture of the dynamics without having to plot a large number of direction arrows. In this case we see that spiraling trajectories in the phase plane are expected.

EXAMPLE 2 | Lotka-Volterra competition equations (continued)

(a) Determine the qualitative dynamics in the phase plane for the Lotka-Volterra competition equations of Example 1.
(b) Plot the variables as a function of time.

SOLUTION

(a) We begin by first considering the N_1-nullclines shown in Figure 3(a). Above the nullcline the value of N_2 will be very large. From Equations 3 we have

$$\frac{dN_1}{dt} = \left(1 - \frac{N_1 + 2N_2}{1000}\right)N_1 \approx -\frac{2N_2}{1000}N_1 < 0$$

for large enough N_2. But as we move closer to the origin, N_1 and N_2 will be very small. From Equations 3 we have

$$\frac{dN_1}{dt} = \left(1 - \frac{N_1 + 2N_2}{1000}\right)N_1 \approx N_1 > 0$$

[See Figure 5(a).]

For the N_2-nullclines, a similar argument shows that to the right of the N_2-nullcline, $dN_2/dt < 0$. Likewise, as we move close to the origin, $dN_2/dt > 0$ [see Figure 5(b)]. Putting together these two plots gives the qualitative dynamics shown in Figure 5(c). This reveals that equilibria (ii) and (iii) are both locally stable (if we start near either of these equilibria, we will move toward the equilibrium). And we can see that equilibria (i) and (iv) are unstable (if we start near either of these, we will move away). Thus, the two species do not coexist. One will competitively exclude the other, and the initial conditions determine which species "wins." Exercise 27 explores this model for other parameter values.

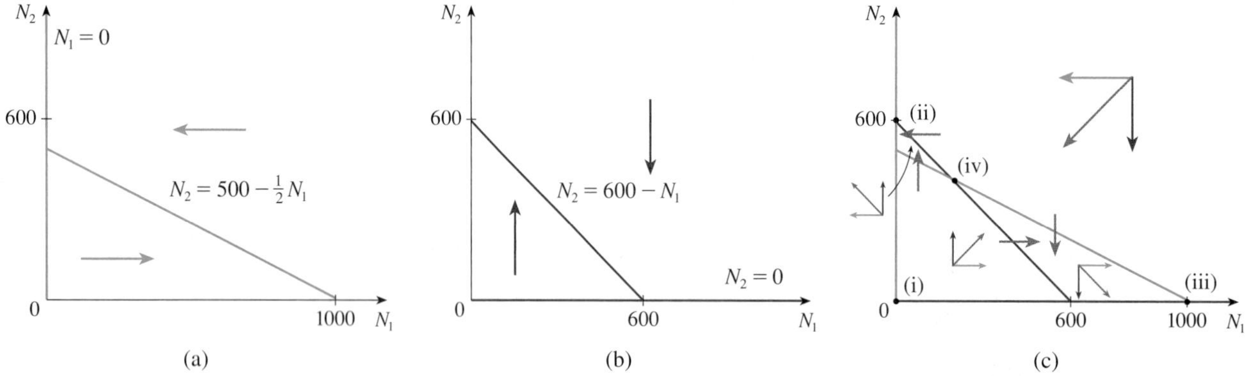

FIGURE 5
Purple arrows indicate direction of motion in part (c).

(b) In Figure 6 we have used a CAS to plot the variables against time for two sets of initial conditions. In part (a) N_2 initially increases but then decays to zero while N_1 continually increases. In part (b) the opposite occurs. These correspond to the variables moving toward equilibria (iii) and (ii) in Figure 5(c), respectively.

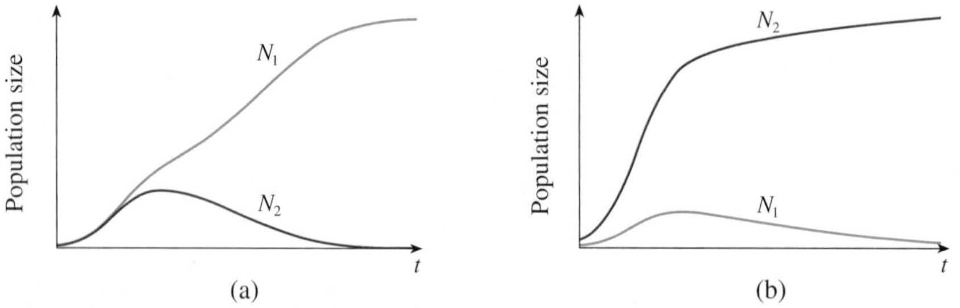

FIGURE 6

■

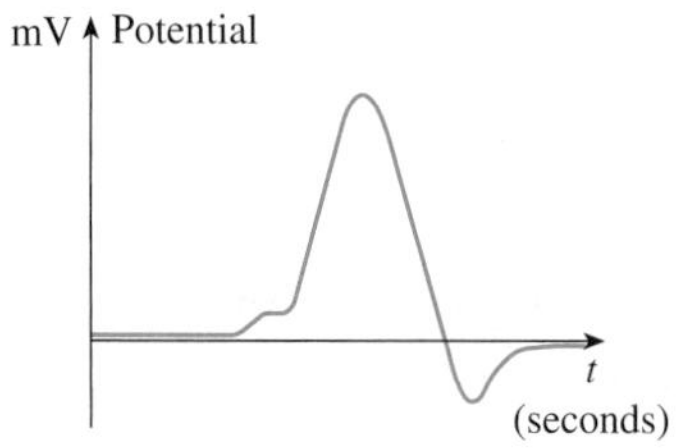

FIGURE 7
Action potential

EXAMPLE 3 | **Fitzhugh-Nagumo equations for a neuron** Neurons carry electrical impulses throughout the body and display what is called an *all-or-nothing* response. Low levels of electrical stimulation have little effect on the neuron. If the electrical stimulus rises above a certain threshold intensity, however, a large electrical impulse called an action potential is generated and travels along the length of the neuron (see Figure 7). Increasing the level of electrical stimulation further does not seem to alter the impulse.

The electrical potential of a neuron has a resting value (which we take to be zero) and a threshold value a above which an impulse is triggered. When an impulse is triggered it grows in potential to a maximum possible value that we take to be 1. Using $v(t)$ for the neuron's potential (in mV) at time t (in seconds), its initial dynamics are therefore similar to those of population dynamics with an Allee effect (recall Example 7.2.3); namely, $dv/dt = (v - a)(1 - v)v$, where a is a constant satisfying $0 < a < 1$. The potential eventually returns to its resting state, however, as the permeability of the neuron's cell wall changes and allows exchange of charged molecules (ions of potassium and sodium). This behavior can be modeled by simply appending a loss term to the equation for the dynamics of v. Assuming the rate of loss is proportional to the magnitude of ion exchange, and using w to denote this magnitude of ion exchange, the system is therefore

$$\textbf{(4)} \qquad \frac{dv}{dt} = (v - a)(1 - v)v - w \qquad \frac{dw}{dt} = bv - cw$$

Both v and w can be positive or negative. Notice that the dampening effect of ion exchange w increases in proportion to the potential at rate b, and it is also subject to a constant per unit decay rate of c.

(a) Suppose $a = 0.2$, $b = 0.01$, and $c = 0.04$. Identify the equilibria in the phase plane and determine the qualitative dynamics.

(b) Plot the potential $v(t)$ as a function of time.

In the 1950s Alan Lloyd Hodgkin and Andrew Fielding Huxley developed a model for neuron activity based on a system of four differential equations. This work, coupled with their study of squid neurons, earned them the 1963 Nobel Prize in Medicine and Physiology. The Fitzhugh-Nagumo equations are a simplified, two-dimensional version of their model.

SOLUTION

(a) The v-nullclines satisfy $dv/dt = 0$, that is, $w = (v - 0.2)(1 - v)v$, a cubic equation whose graph crosses the horizontal axis at $v = 0$, $v = 0.2$, and $v = 1$ [see Figure 8(a)]. Rather than substituting extreme values of the variables as we did in Examples 1 and 2, here we use an alternative method to determine the direction of movement in the phase plane.

From Equations 4 we can see that $dv/dt > 0$ whenever $w < (v - 0.2)(1 - v)v$. Since $(v - 0.2)(1 - v)v$ defines the cubic polynomial in Figure 8(a), v will therefore be increasing below the curve and decreasing above it.

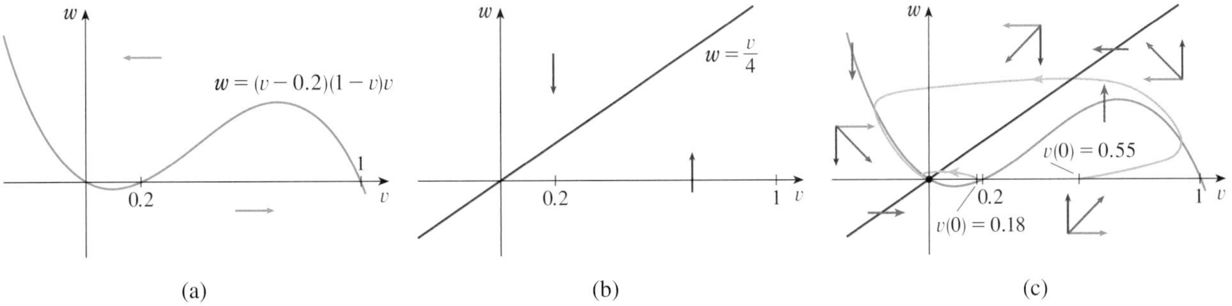

FIGURE 8
Purple arrows indicate direction of motion. Green curves are solution curves.

The w-nullcline satisfies $0.01v - 0.04w = 0$, or $w = v/4$ as in Figure 8(b). From Equations 4 we can see that $dw/dt > 0$ whenever $0.01v > 0.04w$. We can rewrite this inequality as $w < v/4$, and since $w = v/4$ defines the line plotted in Figure 8(b), w will be increasing below the line and decreasing above it.

The nullclines are plotted together, along with the single equilibrium at the origin, in Figure 8(c): this shows that the system exhibits an oscillatory behavior around the equilibrium. But the qualitative nature of these oscillatory dynamics depends on the initial conditions. Using a CAS, we have plotted two solution curves, both with $w(0) = 0$. If the initial potential $v(0)$ is below the threshold of 0.2, the neuron potential immediately decays towards zero. If the initial potential is above 0.2, then the potential initially grows in magnitude before eventually decaying back to the equilibrium state.

(b) The potential for each set of initial conditions in Figure 8(c) is plotted against time in Figure 9 using a CAS. Figure 9(b) resembles an actual action potential when the initial stimulus of the neuron is above the threshold value.

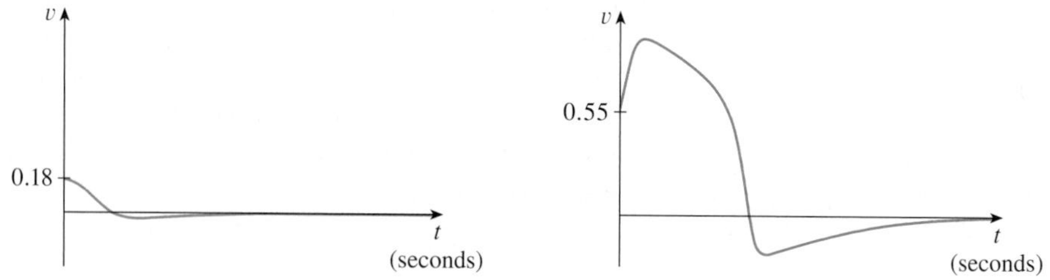

FIGURE 9 (a) $v(0) = 0.18$, $w(0) = 0$ (b) $v(0) = 0.55$, $w(0) = 0$ ■

As a final remark, we note that a phase plane analysis can sometimes give us information about the stability of equilibria. In Example 2 the analysis gave us conclusive information about the stability properties of all the equilibria. In Example 3, however, the phase plane analysis itself is insufficient to allow us to reach any conclusion (although the equilibrium is, in fact, locally stable). We can determine the qualitative tendency to oscillate around the equilibrium point, but we cannot determine whether these oscillations converge towards the equilibrium or move away from it. The same is true for the predator-prey model in Figure 4. In Section 10.4 we will derive mathematical criteria that distinguish between these possibilities.

EXERCISES 7.6

1–6 In each phase plane the x-nullclines are blue and the y-nullclines are red. Use the information given to indicate the direction of movement in the phase plane and label all equilibria. For each equilibrium, determine if it is locally stable or unstable, or if the information is inconclusive.

1. The variable x is increasing in the region below the curved nullcline and decreasing elsewhere. The variable y is decreasing between $x = 0$ and $x = a$ and increasing elsewhere.

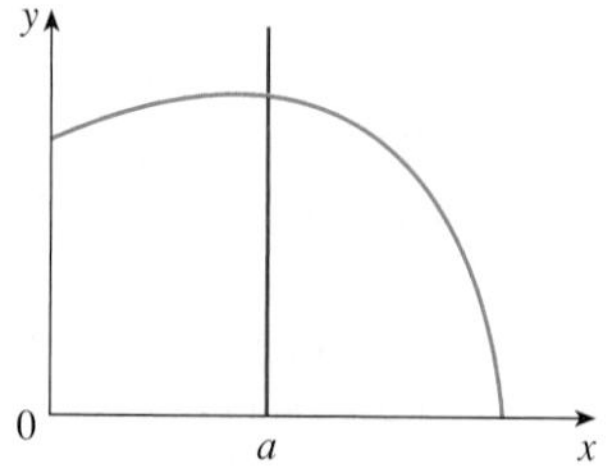

2. The variable x is increasing between zero and its nullcline and decreasing elsewhere. The variable y is increasing below its nullcline and decreasing above it.

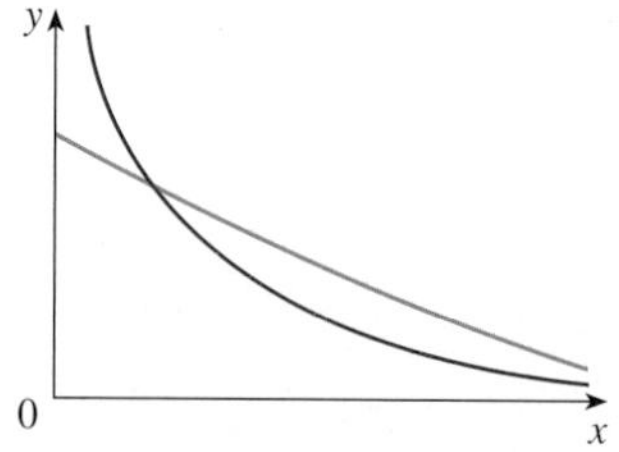

3. The variable x is increasing above its nullcline and decreasing below it. The variable y is decreasing above its nullcline and increasing below it.

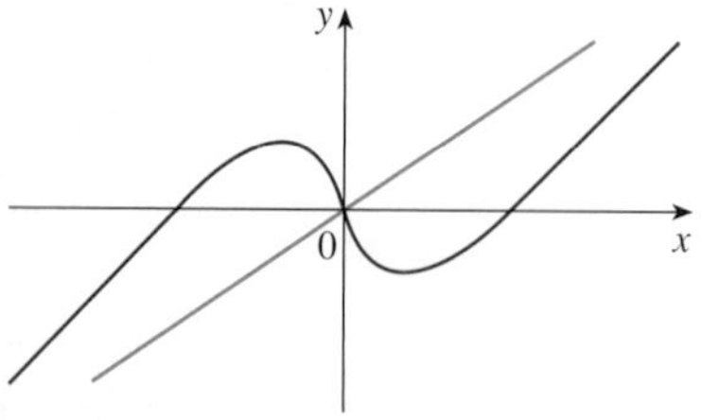

4. The variable x is increasing below its nullcline and decreasing above it. The variable y is decreasing above its nullcline and increasing below it.

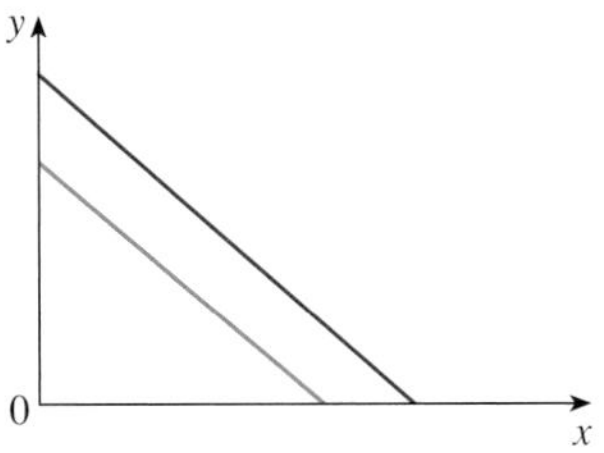

5. The variable x is increasing to the left of its nullcline and decreasing to the right of it. The variable y is decreasing above its nullcline and increasing below it.

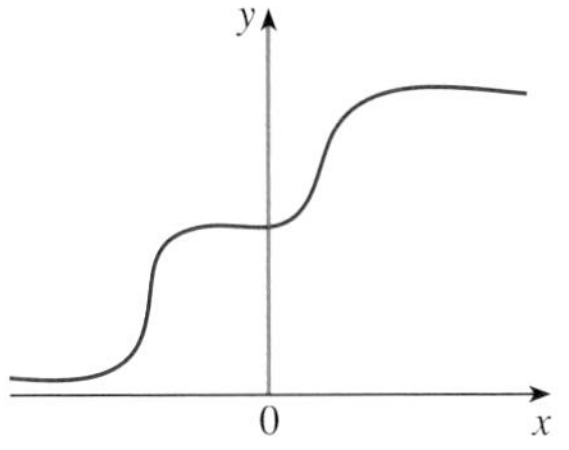

6. The variable x is increasing in quadrants I and III and decreasing in quadrants II and IV. The variable y is increasing everywhere.

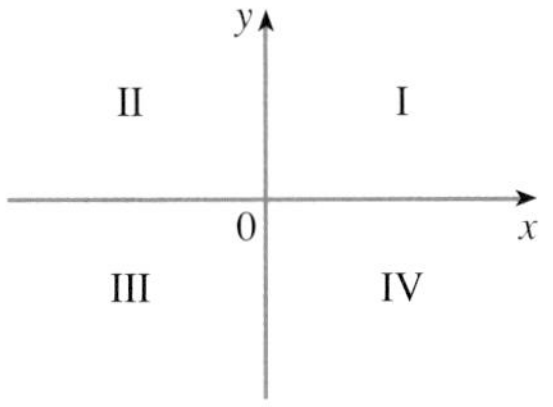

7–15 A system of differential equations is given.
(a) Construct the phase plane, plotting all nullclines, labeling all equilibria, and indicating the direction of motion.
(b) Obtain an expression for each equilibrium.

7. $x' = x(3 - x - y), \quad y' = y(2 - x - y), \quad x, y \geqslant 0$

8. $p' = p(1 - p - q), \quad q' = q(2 - 3p - q), \quad p, q \geqslant 0$

9. $n' = n(1 - 2m), \quad m' = m(2 - 2n - m), \quad n, m \geqslant 0$

10. $x' = x(2 - x), \quad y' = y(3 - y)$

11. $p' = -p^2 + q - 1, \quad q' = q(2 - p - q)$

12. $p' = 2q - 1, \quad q' = q^2 - q - p$

13. $x' = 5 - 2x - xy, \quad y' = xy - y, \quad x, y \geqslant 0$

14. $z' = z^3 - 4z^2 + 3z - 2w, \quad w' = z - w - 1$

15. $x' = -(x - 2)\ln(xy), \quad y' = e^x(x - y), \quad x, y > 0$

16–20 A system of differential equations is given.
(a) Use a phase plane analysis to determine the values of the constant a for which the sole equilibrium of the differential equations is locally stable.
(b) Obtain an expression for each equilibrium (it may be a function of the constant a).

16. $x' = a(x - 3), \quad y' = 5 - y, \quad a \neq 0$

17. $x' = y - ax, \quad y' = x - y, \quad a > 0, \quad a \neq 1$

18. $x' = a(x - a), \quad y' = 4 - y - x, \quad a \neq 0$

19. $x' = ay^2 - x + 1, \quad y' = 2(1 - y)$

20. $x' = -(y - 1) - a(x - 1),$
$y' = -(y - 1) - \dfrac{1}{a}(x - 1), \quad a \neq -1, 0, 1, \quad x, y > 0$

21. Hooke's Law states that the force F exerted by a spring on a mass is proportional to the displacement from its resting position.

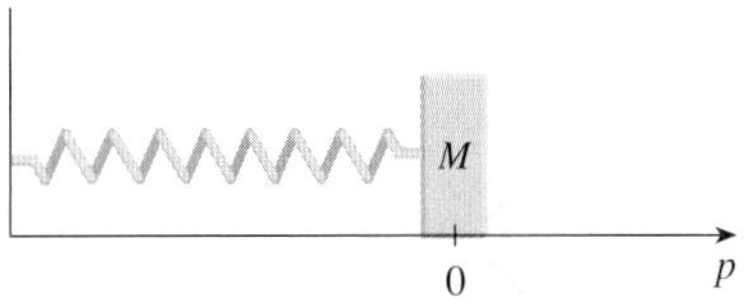

From the figure we have $F = -kp$ for some positive constant k, where p is position. Also, Newton's Second Law tells us that $F = ma$, where m is the mass of the object and a is its acceleration. Further, since we know that acceleration is the second derivative of position with respect to time, we obtain the differential equation

$$m\frac{d^2p}{dt^2} = -kp$$

This is a **second-order differential equation** because it involves the second derivative of the unknown function, p.
(a) Define a new variable $q = dp/dt$. Show that the second-order differential equation can then be written as the following system of two first-order differential equations

$$\frac{dp}{dt} = q \qquad \frac{dq}{dt} = -\frac{kp}{m}$$

(b) Construct the phase plane for the equations in part (a), including nullclines, the equilibrium, and the direction of movement.
(c) What does the phase plane analysis from part (b) tell you about the position of the mass over time?
(d) What does the phase plane analysis from part (b) tell you about the velocity of the mass over time?

22. The **van der Pol equation** is a second-order differential equation describing oscillatory dynamics in a variable x:

$$\frac{dx^2}{dt^2} - \mu(1 - x^2)\frac{dx}{dt} + x = 0$$

where μ is a positive constant. This equation was first obtained by an electrical engineer named Balthasar van der Pol, but has since been used as a model for a variety of phenomena, including sustained oscillatory dynamics of neural impulses.

(a) Convert the Van der Pol equation into a system of two first-order differential equations by defining the new variable y as

$$y = x - \frac{x^3}{3} - \frac{dx/dt}{\mu}$$

(b) Construct the phase plane for the equations obtained in part (a), including nullclines, the equilibrium, and the direction of movement.

(c) What does the phase plane analysis from part (b) tell you about the dynamics of x?

23. The **Kermack-McKendrick equations** are first-order differential equations describing an infectious disease outbreak. Using S and I to denote the number of susceptible and infected people in a population, the equations are

$$S' = -\beta SI \qquad I' = \beta SI - \mu I$$

where β and μ are positive constants representing the transmission rate and rate of recovery, respectively.

(a) Provide a biological explanation for each term of the equations.

(b) Suppose $\beta = 1$ and $\mu = 5$. Construct the phase plane including all nullclines, equilibria, and arrows indicating the direction of movement in the plane.

(c) Construct the phase plane for arbitrary values of β and μ, including all nullclines, equilibria, and arrows indicating direction of movement in the plane.

24. The **Kermack-McKendrick equations** from Exercise 23 can be extended to model persistent diseases rather than single outbreaks by including an inflow of susceptible individuals and their natural death. This gives the differential equations

$$S' = \theta - \gamma S - \beta SI \qquad I' = \beta SI - \mu I$$

where θ and γ are positive constants representing inflow and mortality of susceptible individuals, respectively.

(a) Suppose $\beta = 1$, $\mu = 1$, $\gamma = 1$, and $\theta = 10$. Construct the phase plane including all nullclines, equilibria, and arrows indicating the direction of movement in the plane.

(b) Suppose $\beta = \frac{1}{15}$, $\mu = 1$, $\gamma = 1$, and $\theta = 10$. Construct the phase plane including all nullclines, equilibria, and arrows indicating the direction of movement in the plane.

(c) What is the difference in the predicted dynamics between part (a) and part (b)?

25. The **Michaelis-Menten equations** describe a biochemical reaction in which an enzyme E and substrate S bind to form a complex C. This complex can then either dissociate back into its original components or undergo a reaction in which a product P is produced along with the free enzyme:

$$\text{E} + \text{S} \leftrightarrows \text{C} \rightarrow \text{E} + \text{P}$$

This can be expressed by the differential equations

$$\frac{dx}{dt} = -k_f xyM + k_r(1 - y)M$$

$$\frac{dy}{dt} = -k_f xyM + k_r(1 - y)M + k_{\text{cat}}(1 - y)M$$

$$\frac{dz}{dt} = k_{\text{cat}}(1 - y)M$$

where M is the total number of enzymes (both free and bound), x and z are the numbers of substrate and product molecules, y is the fraction of the enzyme pool that is free, and the k_i's are positive constants.

(a) Explain all the terms in this system of differential equations.

(b) Although this is a system of three differential equations, its dynamics can be understood by constructing a phase plane for the variables x and y alone. Explain why.

(c) Construct the phase plane mentioned in part (b), including all nullclines and equilibria, and indicate the direction of movement in the plane.

26. Metastasis of malignant tumors Metastasis is the process by which cancer cells spread throughout the body and initiate tumors in various organs. This sometimes happens via the bloodstream, where cancer cells become lodged in capillaries of organs and then move across the capillary wall into the organ. Using C to denote the number of cells lodged in a capillary and I for the number that have invaded the organ, we can model this as

$$C' = -\alpha C - \beta C \qquad I' = \alpha C - \delta I + \rho I$$

where all constants are positive, α is the rate of movement across the capillary wall, β is the rate of dislodgment from the capillary, δ is the rate at which cancer cells in the organ die, and ρ is their growth rate.

(a) Suppose $\rho < \delta$. Construct the phase plane, including all nullclines, equilibria, and arrows indicating the direction of movement in the plane.

(b) Suppose $\rho > \delta$. Construct the phase plane, including all nullclines, equilibria, and arrows indicating the direction of movement in the plane.

(c) What is the difference in the predicted dynamics between part (a) and part (b)?

Source: Adapted from D. Kaplan et al., *Understanding Nonlinear Dynamics* (New York: Springer-Verlag, 1995).

27. Lotka-Volterra competition equations For each case, derive the equations for all nullclines of the Lotka-Volterra model in Example 1 and use them to construct the phase plane, including all nullclines, equilibria, and arrows indicating the direction of movement. (Assume that all constants are positive.)

(a) $K_1 > \alpha K_2$ and $K_2 < \beta K_1$
(b) $K_1 < \alpha K_2$ and $K_2 > \beta K_1$
(c) $K_1 < \alpha K_2$ and $K_2 < \beta K_1$
(d) $K_1 > \alpha K_2$ and $K_2 > \beta K_1$

28–30 Consumer-resource models often have the following general form

$$R' = f(R) - g(R, C) \qquad C' = \varepsilon g(R, C) - h(C)$$

where $f(R)$ is a function describing the rate of replenishment of the resource, $g(R, C)$ describes the rate of consumption of the resource, and $h(C)$ is the rate of loss of the consumer. The constant ε is the conversion efficiency of resources into consumers and lies between zero and one. Construct the phase plane, including all nullclines, equilibria, and arrows indicating the direction of movement in the plane. Describe how consumer and resource abundance are predicted to change over time.

28. A **chemostat** is an experimental consumer-resource system. If the resource is not self-reproducing, then it can be modeled by choosing $f(R) = \theta$, $g(R, C) = bRC$, and $h(C) = \mu C$, where all constants are positive.

29. A model for self-reproducing resources is obtained by choosing $f(R) = rR$, $g(R, C) = bRC$, and $h(C) = \mu C$, where all constants are positive.

30. A model for self-reproducing resources with limited growth is obtained by choosing $f(R) = rR(1 - R/K)$, $g(R, C) = bRC$, and $h(C) = \mu C$. Assume all constants are positive and $K > \mu/(\varepsilon b)$.

---

31. Hemodialysis is a process by which a machine is used to filter urea and other waste products from a patient's blood if their kidneys fail. The concentration of a patient's urea during dialysis is sometimes modeled by supposing there are two compartments within the patient—the blood, which is directly filtered by the dialysis machine, and another compartment that cannot be directly filtered but that is connected to the blood. A system of two differential equations describing this is

$$\frac{dc}{dt} = -\frac{K}{V}c + ap - bc \qquad \frac{dp}{dt} = -ap + bc$$

where c and p are the urea concentrations in the blood (in mg/mL) and in the inaccessible pool, respectively, and all constants are positive.

(a) Explain each term of the system of differential equations.

(b) Construct the phase plane, including all nullclines, equilibria, and arrows indicating the direction of movement in the plane. What happens to the urea concentration as $t \to \infty$?

32. Fitzhugh-Nagumo equations Consider the following alternative form of the Fitzhugh-Nagumo equations from Example 3:

$$\frac{dv}{dt} = (v - a)(1 - v)v - w \qquad \frac{dw}{dt} = \varepsilon(v - w)$$

where $\varepsilon > 0$ and $0 < a < 1$. Construct the phase plane, including all nullclines, equilibria, and arrows indicating the direction of movement in the plane.

33. The **Rosenzweig-MacArthur model** is a consumer-resource model similar to that from Exercise 30, but with a different consumption function. A simplified version is

$$R' = R(K - R) - \frac{R}{a + R}C \qquad C' = \frac{R}{a + R}C - bC$$

Suppose that all constants are positive and that $K > ab/(1 - b) > 0$. Construct the phase plane, including all nullclines, equilibria, and arrows indicating the direction of movement in the plane.

■ PROJECT Determining the Critical Vaccination Coverage

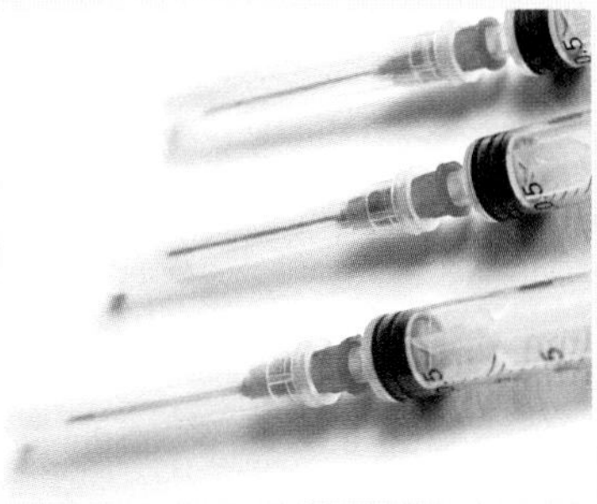

Vaccines are preventative medications that are administered before an individual becomes infected by a pathogen. Either vaccinated people do not become infected or they are able to clear the infection more quickly. Let's suppose a vaccine shortens the duration of the infection by causing the vaccinated individual's immune system to clear the infection more quickly. We will model the dynamics of infected people only, using the following pair of autonomous differential equations:

$$\frac{dN}{dt} = N\beta S(1 - p) + V\beta S(1 - p) - (m + c)N$$

$$\frac{dV}{dt} = N\beta Sp + V\beta Sp - (m + c\gamma)V$$

where N and V are the numbers of nonvaccinated and vaccinated individuals, respectively, S is the number of susceptible individuals, and p is the fraction of these that are vaccinated (with $1 - p$ remaining unvaccinated). For simplicity, these equations assume that the total number of susceptible people remains approximately constant at S over the time frame of interest, and that all infected individuals suffer a constant per capita mortality rate of m regardless of vaccination status. The constant β quantifies the transmissibility of the disease and is positive. Infected individuals are removed through their clearing of the infection via an immune response, with per capita rate c for unvaccinated individuals and $c\gamma$ for vaccinated individuals (with $\gamma > 1$, which reflects the heightened clearance rate caused by the vaccine).

Our task is to determine the value of p needed to ensure that the epidemic will die out, written as a function of the vaccine effectiveness γ.

1. Explain all the terms in the differential equations.

2. Let's assume that $\beta S > m + c$. What does this imply about the disease dynamics if nobody is vaccinated (that is, if $p = 0$)? Let's also assume that $\beta S < m + c\gamma$. What does this imply about the disease dynamics if everybody is vaccinated (that is, if $p = 1$)?

3. Assuming $\beta S > m + c$ and $\beta S < m + c\gamma$, construct the phase plane and nullclines for the case where the vaccination coverage is very low (that is, p is very small). Indicate the sole equilibrium.

4. Determine whether the equilibrium found in Problem 3 is stable or not using the phase plane.

5. As you increase the vaccination coverage p from near zero to near one, the nullclines move. Determine the direction in which each nullcline moves as p increases.

6. From the phase plane in Problem 3 and your answer to Problem 5 you should be able to determine that there is a critical value of p greater than which the number of infected individuals always decreases to zero. Illustrate a phase plane diagram for the case where p is just less than this critical value, and another for the case where p is just greater than this critical value.

7. Use your result from Problem 6 to obtain a mathematical inequality that must be satisfied by p for the vaccination coverage to be adequate to prevent the epidemic. Your inequality should involve the vaccine effectiveness γ.

Chapter 7 REVIEW

CONCEPT CHECK

1. (a) What is a differential equation?
   (b) What is the order of a differential equation?
   (c) What is an initial condition?
   (d) What are the differences between pure-time, autonomous, and nonautonomous differential equations?

2. What can you say about the solutions of the equation $y' = x^2 + y^2$ just by looking at the differential equation?

3. What is a phase plot for the differential equation $y' = g(y)$?

4. What is a direction field for the differential equation $y' = F(x, y)$?

5. Explain how Euler's method works.

6. What is a separable differential equation? How do you solve it?

7. (a) Write the logistic equation.
   (b) Under what circumstances is this an appropriate model for population growth?

8. (a) Write Lotka-Volterra equations to model populations of sharks S and their food F.
   (b) What do these equations say about each population in the absence of the other?

9. What is a nullcline?

10. (a) Write Lotka-Volterra competition equations for two competing fish species, x and y.
    (b) What would the nullclines have to look like for species x to always outcompete species y?

Answers to the Concept Check can be found on the back endpapers.

TRUE-FALSE QUIZ

Determine whether the statement is true or false. If it is true, explain why. If it is false, explain why or give an example that disproves the statement.

1. All solutions of the differential equation $y' = -1 - y^4$ are decreasing functions.

2. The function $f(x) = (\ln x)/x$ is a solution of the differential equation $x^2y' + xy = 1$.

3. Consider the differential equation $y' = g(y)$ where $g(y)$ is a differentiable function of y. It is not possible for y to exhibit oscillatory behavior.

4. The equation $y' = x + y$ is separable.

5. The equation $y' = 3y - 2x + 6xy - 1$ is separable.

6. If y is the solution of the initial-value problem

$$\frac{dy}{dt} = 2y\left(1 - \frac{y}{5}\right) \qquad y(0) = 1$$

then $\lim_{t\to\infty} y = 5$.

EXERCISES

1–4 A differential equation is given.

(a) Determine all equilibria as a function of the constant a.

(b) Construct a phase plot and use it to determine the stability of the equilibria found in part (a) for three different values of the constant: (i) $a < 0$, (ii) $a = 0$, and (iii) $a > 0$.

(c) Use the local stability criterion to verify your answers to part (b).

1. $x' = ax - x^2$

2. $x' = a - x^2$

3. $x' = ax - x^3$

4. $x' = ax + x^3$

5. (a) A direction field for the differential equation $y' = y(y - 2)(y - 4)$ is shown. Sketch the graphs of the solutions that satisfy the given initial conditions.

   (i) $y(0) = -0.3$ (ii) $y(0) = 1$

   (iii) $y(0) = 3$ (iv) $y(0) = 4.3$

   (b) If the initial condition is $y(0) = c$, for what values of c is $\lim_{t\to\infty} y(t)$ finite? What are the equilibrium solutions?

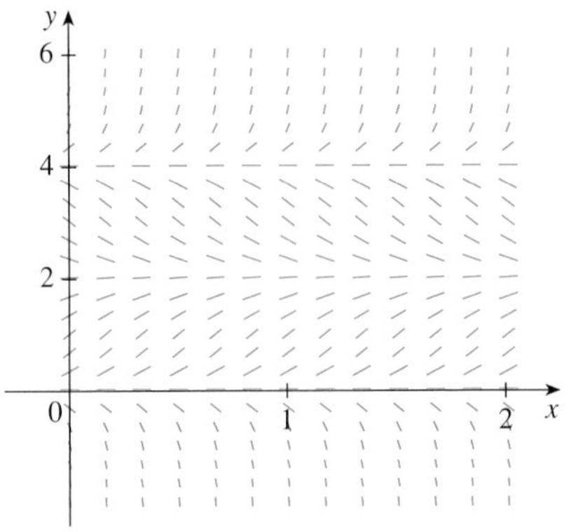

6. (a) Sketch a direction field for the differential equation $y' = x/y$. Then use it to sketch the four solutions that satisfy the initial conditions $y(0) = 1$, $y(0) = -1$, $y(2) = 1$, and $y(-2) = 1$.

   (b) Check your work in part (a) by solving the differential equation explicitly. What type of curve is each solution curve?

7. (a) A direction field for the differential equation $y' = x^2 - y^2$ is shown. Sketch the solution of the initial-value problem

$$y' = x^2 - y^2 \qquad y(0) = 1$$

   Use your graph to estimate the value of $y(0.3)$.

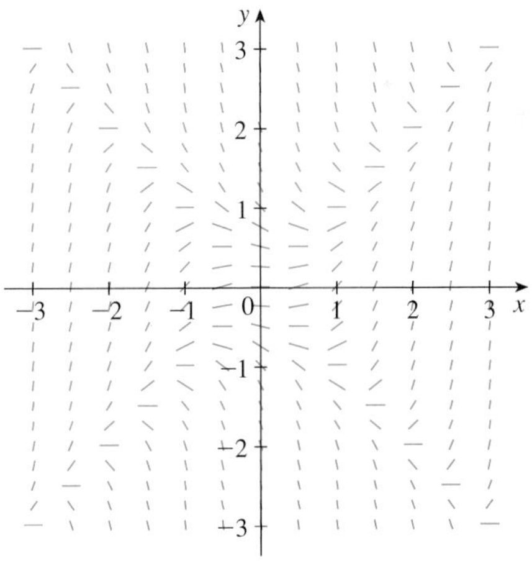

   (b) Use Euler's method with step size 0.1 to estimate $y(0.3)$, where $y(x)$ is the solution of the initial-value problem in part (a). Compare with your estimate from part (a).

   (c) On what lines are the centers of the horizontal line segments of the direction field in part (a) located? What happens when a solution curve crosses these lines?

8. (a) Use Euler's method with step size 0.2 to estimate $y(0.4)$, where $y(x)$ is the solution of the initial-value problem

$$y' = 2xy^2 \qquad y(0) = 1$$

   (b) Repeat part (a) with step size 0.1.

(c) Find the exact solution of the differential equation and compare the value at 0.4 with the approximations in parts (a) and (b).

9–10 Solve the differential equation.

9. $2ye^{y^2}y' = 2x + 3\sqrt{x}$

10. $\dfrac{dx}{dt} = 1 - t + x - tx$

11–12 Solve the initial-value problem.

11. $\dfrac{dr}{dt} + 2tr = r, \quad r(0) = 5$

12. $(1 + \cos x)y' = (1 + e^{-y})\sin x, \quad y(0) = 0$

13. Seasonality and population dynamics The per capita growth rate of a population varies seasonally. The population dynamics are modeled as

$$n' = \cos\left(\frac{2\pi t}{365}\right)n \qquad n(0) = n_0$$

where $n(t)$ is the population size at time t (measured in days). Determine the population size at time t.

14. Seasonality and population dynamics The per capita growth rate of a population varies seasonally and habitat destuction is also occurring. This is modeled as

$$n' = r\left(\cos\left[\frac{2\pi t}{365}\right] - at\right)n \qquad n(0) = n_0$$

where $n(t)$ is the population size at time t (measured in days) and r and a are positive constants. Determine the population size at time t.

15. Levins' metapopulation model from Exercise 7.2.15 describes a population consisting of patches that can be either occupied or vacant. Occupied patches create more occupied patches by sending individuals to unoccupied patches. If the frequency of occupied patches is p, we would therefore expect that the rate at which new patches become occupied is proportional to both p and $1 - p$. Patches also become unoccupied through mortality at a constant rate. The differential equation for p is

$$\frac{dp}{dt} = cp(1 - p) - mp \qquad p(0) = p_0$$

where c and m are constants.

(a) Find the solution to this initial-value problem.

(b) Under what conditions on the constants will the frequency of occupied patches go to zero as $t \to \infty$?

16. The Brentano-Stevens Law in psychology models the way that a subject reacts to a stimulus. It states that if R represents the reaction to an amount S of stimulus, then the relative rates of increase are proportional:

$$\frac{1}{R}\frac{dR}{dt} = \frac{k}{S}\frac{dS}{dt}$$

where k is a positive constant. Find R as a function of S.

17. Lung preoxygenation Some medical procedures require a patient's airway to be temporarily blocked, preventing the inspiration of oxygen. The duration of time over which such procedures can be performed safely may be increased by replacing a large fraction of the air in the patient's lungs with oxygen prior to the procedure. Suppose the lung volume is 3 L and well-mixed air in the lungs is replaced with pure oxygen at a rate of 10 mL/s.

(a) What is the amount of oxygen in the lungs as a function of time if they initially contain 20% oxygen?

(b) Use your answer from part (a) to determine how long the process of oxygenation should be run to result in an 80% oxygen content in the lungs.

18. A tank contains 100 L of pure water. Brine that contains 0.1 kg of salt per liter enters the tank at a rate of 10 L/min. The solution is kept thoroughly mixed and drains from the tank at the same rate. How much salt is in the tank after six minutes?

19. Hormone transport In lung physiology, the transport of a substance across a capillary wall has been modeled by the differential equation

$$\frac{dh}{dt} = -\frac{R}{V}\left(\frac{h}{k + h}\right)$$

where h is the hormone concentration in the bloodstream (in mg/mL), t is time (in seconds), R is the maximum transport rate, V is the volume of the capillary, and k is a positive constant that measures the affinity between the hormones and the enzymes that assist the process. Solve this differential equation to find a relationship between h and t.

20. Predator-prey dynamics Populations of birds and insects are modeled by the equations

$$\frac{dx}{dt} = 0.4x - 0.002xy$$

$$\frac{dy}{dt} = -0.2y + 0.000008xy$$

(a) Which of the variables, x or y, represents the bird population, and which represents the insect population? Explain.

(b) Find the equilibrium solutions and explain their significance.

(c) Find an expression for dy/dx.

(d) The direction field for the differential equation in part (c) is shown. Use it to sketch the phase trajectory corresponding to initial populations of 100 birds and 40,000 insects. Then use the phase trajectory to describe how both populations change.

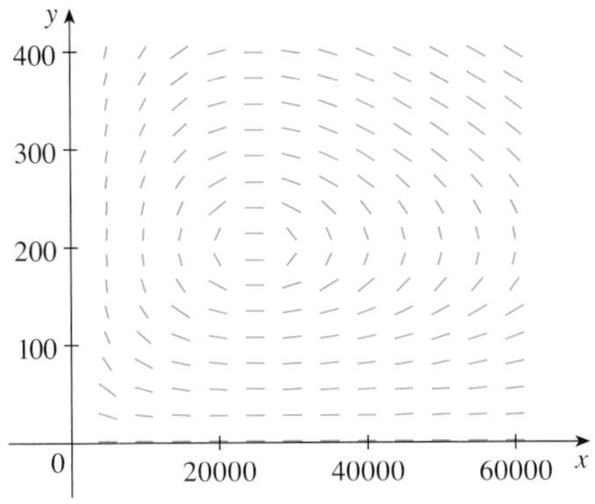

(e) Use part (d) to make rough sketches of the bird and insect populations as functions of time. How are these two graphs related?

21. Suppose the model of Exercise 20 is replaced by the equations

$$\frac{dx}{dt} = 0.4x(1 - 0.000005x) - 0.002xy$$

$$\frac{dy}{dt} = -0.2y + 0.000008xy$$

(a) According to these equations, what happens to the insect population in the absence of birds?
(b) Find the equilibrium solutions and explain their significance.
(c) The figure shows the phase trajectory that starts with 100 birds and 40,000 insects. Describe what eventually happens to the bird and insect populations.

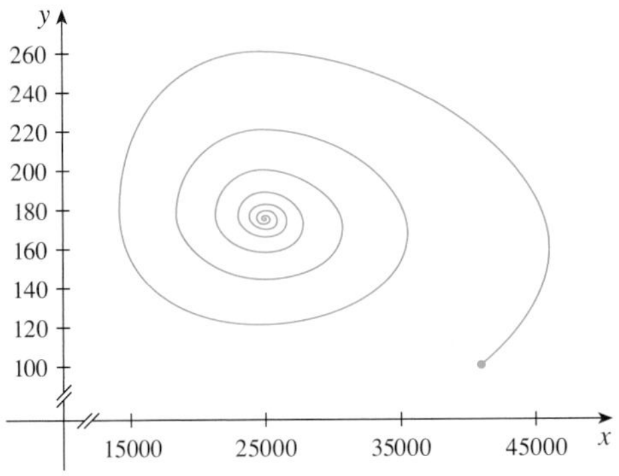

(d) Sketch graphs of the bird and insect populations as functions of time.

22. Cancer progression The development of many cancers, such as colorectal cancer, proceeds through a series of precancerous stages. Suppose there are $n - 1$ precancerous stages before cancer develops at stage n. A simple system of differential equations modeling this is

$$x_0' = -u_0 x_0$$

$$x_i' = u_{i-1}x_{i-1} - u_i x_i$$

$$x_n' = u_{n-1}x_{n-1}$$

where x_i is the fraction of the population in state i, the u_i's are constants, and $i = 1, \ldots, n - 1$.
(a) Suppose $n = 2$. What is the system of three differential equations?
(b) Suppose $u_0 = 1$ and $u_1 = 1$. Construct the phase plane for the variables x_0 and x_1 alone, including all nullclines, equilibria, and arrows indicating the direction of movement in the plane.
(c) From your answer to part (a), obtain a differential equation for dx_1/dx_0 for the parametric curves describing the trajectories in the x_0-x_1 plane. Use this equation to plot the vector field. Then sketch several solutions curves for x_1 as a function of x_0, assuming that $x_0(0) = a$ and $x_1(0) = 1 - a$, where $0 < a < 1$.
(d) Your answer to part (c) will produce a family of functions in the constant a. Provide a biological interpretation for these parametric curves.

23. Competition-colonization models The metapopulation model from Exercise 15 can be extended to include two species, where one is a superior competitor. The equations are

$$\frac{dp_1}{dt} = c_1 p_1(1 - p_1) - m_1 p_1$$

$$\frac{dp_2}{dt} = c_2 p_2(1 - p_1 - p_2) - m_2 p_2 - c_1 p_1 p_2$$

where p_1 and p_2 are the fractions of patches occupied by species 1 and 2, respectively. These equations model a process in which any patch has at most one species, and where species 2 patches can be "taken over" by species 1, but not vice versa.
(a) Explain how the terms in the equations reflect the assumption that species 1 is the superior competitor.
(b) Suppose that $m_1 = m_2 = 3$, $c_1 = 5$, and $c_2 = 30$. Construct the phase plane, including all nullclines, equilibria, and arrows indicating the direction of movement in the plane. Show that, despite species 1 being a better competitor, the two species are predicted to coexist.

24. Habitat destruction The model of Exercise 23 can be extended to include the effects of habitat destruction. Suppose that only a fraction h of the patches are habitable

$(0 < h < 1)$. The equations become

$$\frac{dp_1}{dt} = c_1p_1(h - p_1) - m_1p_1$$

$$\frac{dp_2}{dt} = c_2p_2(h - p_1 - p_2) - m_2p_2 - c_1p_1p_2$$

Suppose that $m_1 = m_2 = 3$, $c_1 = 5$, and $c_2 = 30$.

(a) Construct the phase plane, including all nullclines, equilibria, and arrows indicating the direction of movement in the plane when $\frac{3}{5} < h < 1$.

(b) Construct the phase plane, including all nullclines, equilibria, and arrows indicating the direction of movement in the plane when $\frac{1}{10} < h < \frac{3}{5}$.

(c) Construct the phase plane, including all nullclines, equilibria, and arrows indicating the direction of movement in the plane when $0 < h < \frac{1}{10}$.

(d) From your results to parts (a), (b), and (c), determine how habitat destruction is expected to affect the coexistence of the two species.

Source: Adapted from S. Nee et al., "Dynamics of Metapopulations: Habitat Destruction and Competitive Coexistence," *Journal of Animal Ecology* 61 (1992): 37–40.

25. Cell cycle dynamics The process of cell division is periodic, with repeated growth and division phases as the cell population multiplies. It has been suggested that the division phase is triggered by high concentrations of a molecule called MPF (maturation promoting factor). The production of this factor is stimulated by another molecule called cyclin, and MPF eventually inhibits its own production. Using M and C to denote the concentrations of these two biomolecules (in mg/mL), a simple model for their interaction is

$$\frac{dM}{dt} = \alpha C + \beta CM^2 - \frac{\gamma M}{1 + M}$$

$$\frac{dC}{dt} = \delta - M$$

(a) Suppose that $\alpha = 2$, $\beta = 1$, $\gamma = 10$, and $\delta = 1$. Construct the phase plane, including all nullclines, equilibria, and arrows indicating the direction of movement in the plane.

(b) From your answer to part (a), what is the qualitative nature of the dynamics of M predicted by this model? What does this predict about the dynamics of cell division?

(c) For any equilibrium found in part (a), specify whether it is locally stable, unstable, or if the information is inconclusive.

Source: Adapted from R. Norel et al., "A Model for the Adjustment of the Mitotic Clock by Cyclin and MPF Levels," *Science* 251 (1991): 1076–78.

CASE STUDY 2c Hosts, Parasites, and Time-Travel

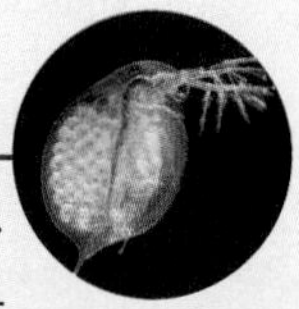

In this part of the case study you will formulate a mathematical model for the antagonistic interactions between *Daphnia* and its parasite using differential equations. Let's suppose that there are two possible host genotypes (A and a) and two possible parasite genotypes (B and b). Parasites of type B can infect only hosts of type A, while parasites of type b can infect only hosts of type a (see Table 1). We will derive a set of two coupled differential equations that model the dynamics of the frequency of A in the host population and B in the parasite population.

Table 1

The outcome of challenges between different host and parasite genotypes.

	Host A	Host a
Parasite B	Infection occurs	Infection does not occur
Parasite b	Infection does not occur	Infection occurs

A common differential equation used in biology to model the frequency dynamics of a particular genotype is

BB **(1)**
$$\frac{df}{dt} = f(1 - f)(r_1 - r_2)$$

where f is the frequency of type 1, and r_1 and r_2 are the per capita reproduction rates of the two types. For example, see Exercise 7.2.16. We will use an equation of this form for both the host and the parasite populations.

Suppose the per capita reproduction rate of uninfected hosts is r_q and that for infected hosts is $r_q - s_q$. The constant s_q is assumed to satisfy the inequality $0 < s_q < r_q$ and represents the reduction in reproductive output of a host due to infection. Similarly, the per capita reproduction rate of a parasite that is able to infect a host is r_p and that for one unable to infect a host is $r_p - s_p$ (the parasite can reproduce in the absence of the host, but it does so less well). The constant s_p is assumed to satisfy the inequality $0 < s_p < r_p$ and represents the reduction in reproductive output of a parasite if it is unable to infect a host.

Let's use q to denote the frequency of type A individuals in the host population and p to denote the frequency of type B individuals in the parasite population. Suppose that host–parasite encounters occur at random with respect to genotype.

1. With random encounters, the average per capita reproduction rate for hosts of a given type is $r_B p + r_b(1 - p)$, where r_B and r_b are the reproduction rates of the host when encountering a type B or type b parasite, respectively. Show that the average per capita reproduction rates of hosts of type A and a are therefore

type A: $r_q - ps_q$

type a: $r_q - (1 - p)s_q$

2. With random encounters, the average per capita reproduction rate for parasites of a given type is $r_A q + r_a(1 - q)$, where r_A and r_a are the reproduction rates of the parasite when encountering a type A or type a host, respectively. Show that the average per capita reproduction rates of parasites of type B and b are therefore

type B: $r_p - (1 - q)s_p$

type b: $r_p - qs_p$

3. Suppose both q and p satisfy differential equations of the form given in Equation 1. Show that q and p therefore satisfy

$$\frac{dq}{dt} = s_q q(1 - q)(1 - 2p)$$

$$\frac{dp}{dt} = s_p p(1 - p)(2q - 1)$$

4. Construct the phase plane including all nullclines, equilibria, and arrows indication the direction of movement in the plane.

5. Explain, qualitatively, how the frequencies of the two parasite genotypes are predicted to change over time. Similarly, explain how the frequencies of the two host genotypes are predicted to change over time.

Vectors and Matrix Models

8

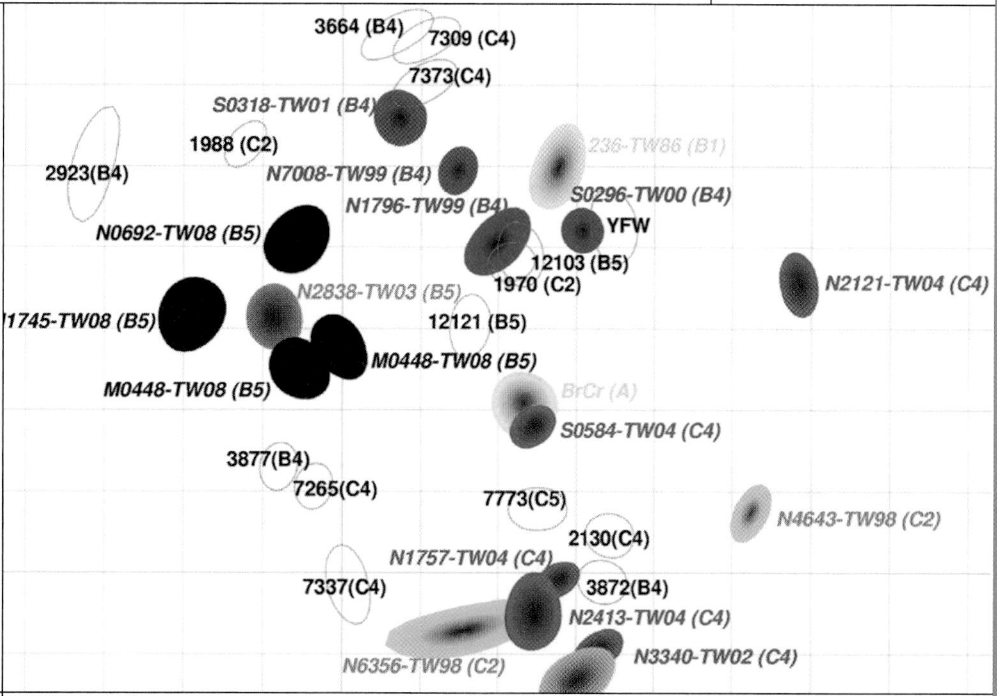

Shown is an antigenic map of isolates of human enterovirus. Similar maps are constructed in Examples 8.1.3, 8.1.6, and 8.1.8.

Source: S.-W. Huang et al., "Reemergence of Enterovirus 71 in 2008 in Taiwan: Dynamics of Genetic and Antigenic Evolution from 1998 to 2008," *Journal of Clinical Microbiology* 47 (2009): 3653–62.

IN THIS CHAPTER WE INTRODUCE coordinate systems and vectors for three-dimensional space and higher. This will lead us into important ideas in linear algebra and the powerful techniques of matrix models. It also sets the stage for a more complete analysis of systems of differential equations and the study of functions of multiple variables.

8.1 Coordinate Systems

To locate a point in a plane, two numbers are necessary. We know that any point in the plane can be represented as an ordered pair (a, b) of real numbers, where a is the x-coordinate and b is the y-coordinate. For this reason, a plane is called two-dimensional.

Three-Dimensional Space

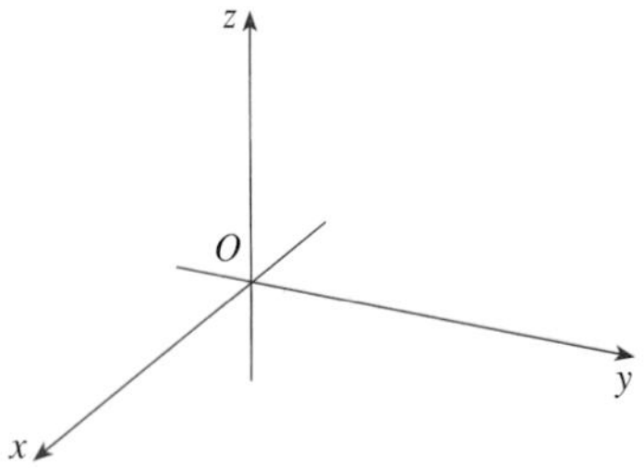

FIGURE 1
Coordinate axes

To locate a point in space, three numbers are required. We represent any point in space by an ordered triple (a, b, c) of real numbers. In order to do so we first choose a fixed point O (the origin) and three directed lines through O that are perpendicular to each other, called the **coordinate axes** and labeled the x-axis, y-axis, and z-axis. Usually we think of the x- and y-axes as being horizontal and the z-axis as being vertical, and we draw the orientation of the axes as in Figure 1. The direction of the z-axis is determined by the **right-hand rule** as illustrated in Figure 2: If you curl the fingers of your right hand around the z-axis in the direction of a 90° counterclockwise rotation from the positive x-axis to the positive y-axis, then your thumb points in the positive direction of the z-axis.

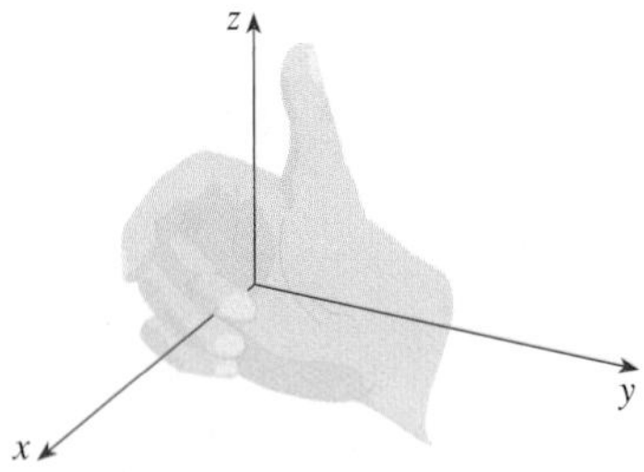

FIGURE 2
Right-hand rule

The three coordinate axes determine the three **coordinate planes** illustrated in Figure 3(a). The xy-plane is the plane that contains the x- and y-axes; the yz-plane contains the y- and z-axes; the xz-plane contains the x- and z-axes. These three coordinate planes divide space into eight parts, called **octants**. The **first octant**, in the foreground, is determined by the positive axes.

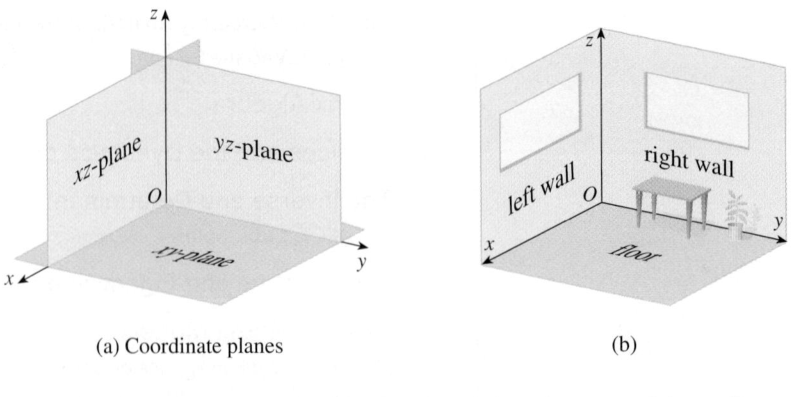

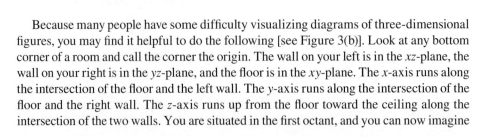

FIGURE 3 (a) Coordinate planes (b)

Because many people have some difficulty visualizing diagrams of three-dimensional figures, you may find it helpful to do the following [see Figure 3(b)]. Look at any bottom corner of a room and call the corner the origin. The wall on your left is in the xz-plane, the wall on your right is in the yz-plane, and the floor is in the xy-plane. The x-axis runs along the intersection of the floor and the left wall. The y-axis runs along the intersection of the floor and the right wall. The z-axis runs up from the floor toward the ceiling along the intersection of the two walls. You are situated in the first octant, and you can now imagine

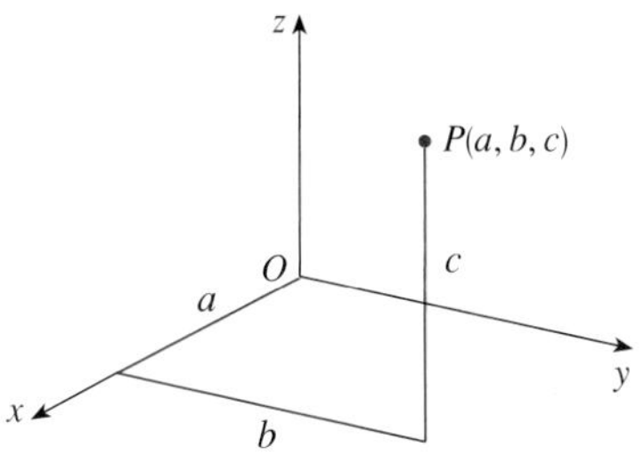

FIGURE 4

seven other rooms situated in the other seven octants (three on the same floor and four on the floor below), all connected by the common corner point O.

Now if P is any point in space, let a be the (directed) distance from the yz-plane to P, let b be the distance from the xz-plane to P, and let c be the distance from the xy-plane to P. We represent the point P by the ordered triple (a, b, c) of real numbers and we call a, b, and c the **coordinates** of P; a is the x-coordinate, b is the y-coordinate, and c is the z-coordinate. Thus, to locate the point (a, b, c), we can start at the origin O and move a units along the x-axis, then b units parallel to the y-axis, and then c units parallel to the z-axis as in Figure 4.

The point $P(a, b, c)$ determines a rectangular box as in Figure 5. If we drop a perpendicular from P to the xy-plane, we get a point Q with coordinates $(a, b, 0)$ called the **projection** of P onto the xy-plane. Similarly, $R(0, b, c)$ and $S(a, 0, c)$ are the projections of P onto the yz-plane and xz-plane, respectively.

As numerical illustrations, the points $(-4, 3, -5)$ and $(3, -2, -6)$ are plotted in Figure 6.

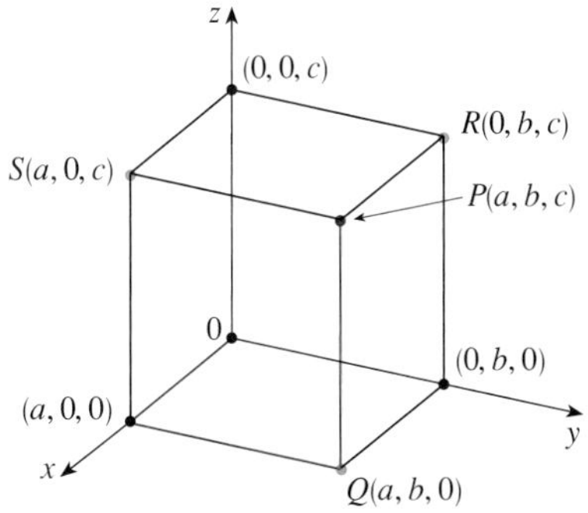

FIGURE 5

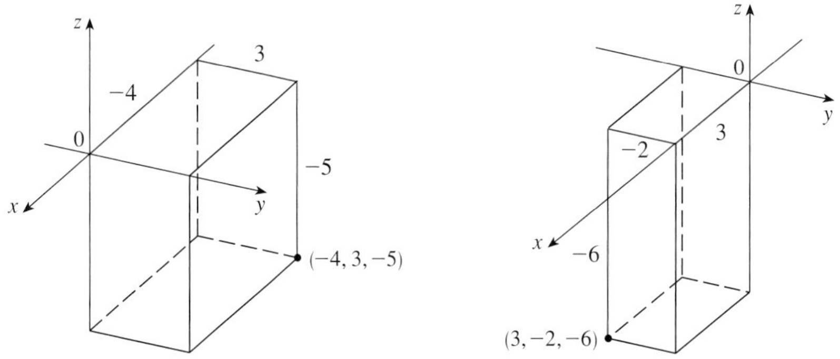

FIGURE 6

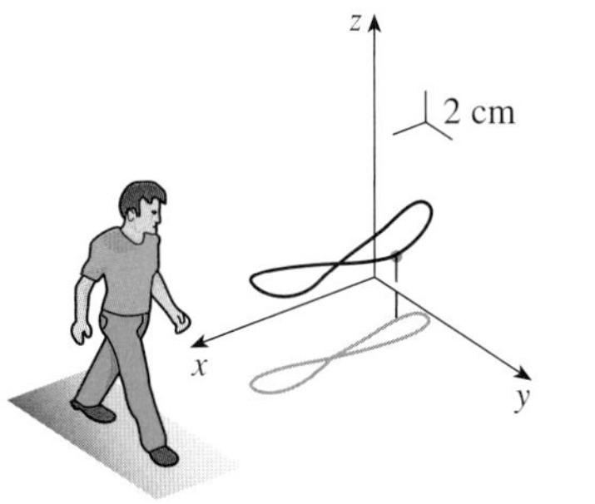

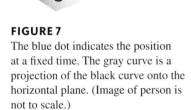

FIGURE 7
The blue dot indicates the position at a fixed time. The gray curve is a projection of the black curve onto the horizontal plane. (Image of person is not to scale.)

EXAMPLE 1 | Human biomechanics The position of the center of a human walking on a treadmill at any point in time is described by its x, y, and z coordinates (see Figure 7). Over time the point moves and traces out a curve in space. Which directions of movement are revealed by the projection of this point onto the horizontal coordinate plane? Explain why the curve projected onto this plane has the shape shown in Figure 7.

SOLUTION The projection onto the horizontal coordinate plane shows motion from side to side as well as from front to back. It does not reveal any motion in the vertical direction. The projected curve shown in the figure resembles a "figure eight" that is elongated in the direction of side-to-side movement (the x-direction). People sway from side to side with each step while walking, causing the movement in the x-direction. At the same time, each step that the person takes moves him forward slightly, and the turning of the treadmill then moves him backward again. Together these movements trace out a figure eight in the horizontal plane. In this example, the magnitude of the side-to-side movement is larger than the magnitude of the movement front to back, which causes the projected curve to appear elongated in the x-direction. ■

The Cartesian product $\mathbb{R} \times \mathbb{R} \times \mathbb{R} = \{(x, y, z) \mid x, y, z \in \mathbb{R}\}$ is the set of all ordered triples of real numbers and is denoted by $\mathbb{R}^3$. We have given a one-to-one correspondence between points P in space and ordered triples (a, b, c) in $\mathbb{R}^3$. It is called a **three-dimensional rectangular coordinate system**. Notice that, in terms of coordinates, the first octant can be described as the set of points whose coordinates are all positive.

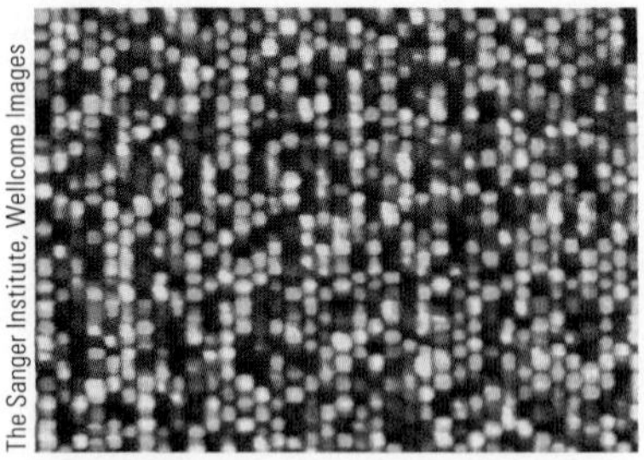

FIGURE 8
Expression level of each gene is measured as color intensity of a dot on a microarray, as shown here. The entire microarray displays the genome expression profile. See also the Project on page 513.

EXAMPLE 2 | BB **Genome expression profiles** Due to advances in biotechnology, researchers can now quantify gene expression across the entire genome of organisms in response to various experimental perturbations. The set of the levels of expression of a collection of genes is called a **genome expression profile** (see Figure 8). Most perturbations result in some genes being upregulated (positive expression) and others being suppressed (negative expression). Consider three hypothetical genes, A, B, and C, and suppose two different experiments are conducted that result in the following dimensionless expression profiles: Experiment 1 (A, B, C) $= (1, 1, -1.5)$ and Experiment 2 (A, B, C) $= (1.5, -1, 0.2)$. Plot these data points in $\mathbb{R}^3$. Interpret their projections on each of the three coordinate planes.

SOLUTION Using the x-, y-, and z-coordinates for the expression levels of genes A, B, and C, respectively, we obtain Figure 9(a) for Experiment 1 and Figure 9(b) for Experiment 2. The projections of each data point on the three different coordinate planes reveal the various possible two-gene expression profiles.

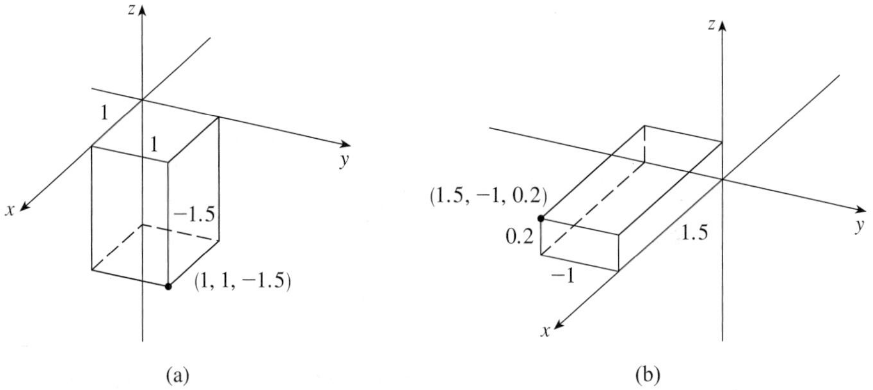

FIGURE 9

EXAMPLE 3 | BB **Antigenic cartography** The extent to which viruses react with the immune system is sometimes measured by testing their ability to bind with a large panel of different immune molecules called antisera. Suppose the binding ability of four influenza strains with three different kinds of antiserum molecules are given by the dimensionless numbers in the following table.

	Antiserum 1	Antiserum 2	Antiserum 3
Strain 1	2.06	1.92	2.96
Strain 2	1.79	1.91	2.44
Strain 3	2.68	3.53	3.31
Strain 4	2.39	4.05	4.46

Plot the data for the four strains in three-dimensional space. This space is referred to as **antigenic space** and construction of such an antigenicity plot is referred to as **antigenic cartography**.

SOLUTION Using the x-, y-, and z-coordinates for the binding ability to antisera 1, 2, and 3, we obtain Figure 10. The binding ability to any pair of antisera is given by the projection of these four red points onto the various coordinate planes.

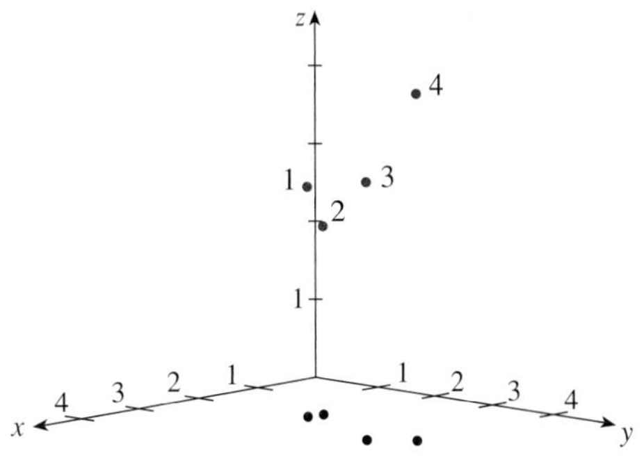

FIGURE 10
Four strains plotted in three-dimensional antigenic space (red dots), as well as the projection onto the two-dimensional antigenic space of antisera 1 and 2 (black dots).

In two-dimensional analytic geometry, the graph of an equation involving x and y is a curve in $\mathbb{R}^2$. In three-dimensional analytic geometry, an equation in x, y, and z represents a *surface* in $\mathbb{R}^3$.

EXAMPLE 4 | What surfaces in $\mathbb{R}^3$ are represented by the following equations?
(a) $z = 3$ (b) $y = 5$

SOLUTION

(a) The equation $z = 3$ represents the set $\{(x, y, z) \mid z = 3\}$, which is the set of all points in $\mathbb{R}^3$ whose z-coordinate is 3. This is the horizontal plane that is parallel to the xy-plane and three units above it as in Figure 11(a).

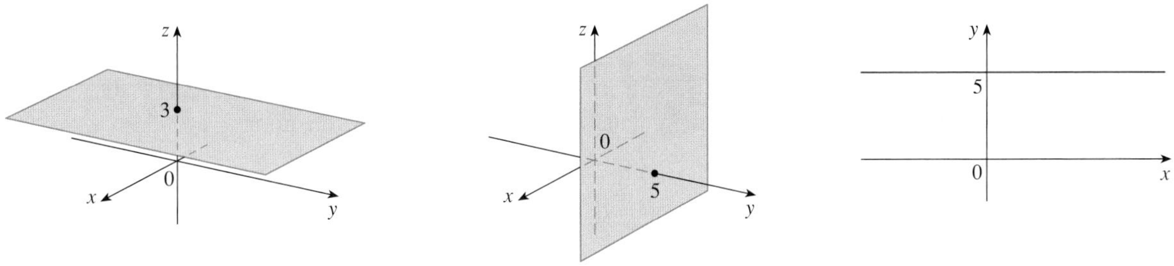

(a) $z = 3$, a plane in $\mathbb{R}^3$ (b) $y = 5$, a plane in $\mathbb{R}^3$ (c) $y = 5$, a line in $\mathbb{R}^2$

FIGURE 11

(b) The equation $y = 5$ represents the set of all points in $\mathbb{R}^3$ whose y-coordinate is 5. This is the vertical plane that is parallel to the xz-plane and five units to the right of it as in Figure 11(b).

Note that when an equation is given, we must understand from the context whether it represents a curve in $\mathbb{R}^2$ or a surface in $\mathbb{R}^3$. For example, $y = 5$ represents a plane in $\mathbb{R}^3$, but of course $y = 5$ can also represent a line in $\mathbb{R}^2$ if we are dealing with two-dimensional analytic geometry. See Figure 11(b) and (c). ■

The familiar formula for the distance between two points in a plane is easily extended to the following three-dimensional formula.

Distance Formula in Three Dimensions The distance $|P_1P_2|$ between the points $P_1(x_1, y_1, z_1)$ and $P_2(x_2, y_2, z_2)$ is

$$|P_1P_2| = \sqrt{(x_2 - x_1)^2 + (y_2 - y_1)^2 + (z_2 - z_1)^2}$$

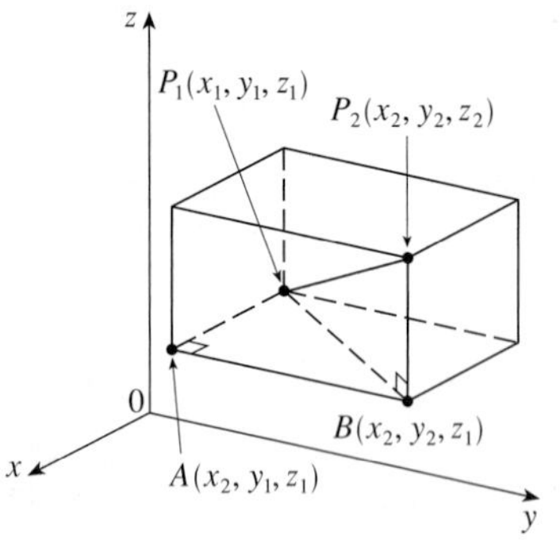

FIGURE 12

To see why this formula is true, we construct a rectangular box as in Figure 12, where P_1 and P_2 are opposite vertices and the faces of the box are parallel to the coordinate planes. If $A(x_2, y_1, z_1)$ and $B(x_2, y_2, z_1)$ are the vertices of the box indicated in the figure, then

$$|P_1A| = |x_2 - x_1| \qquad |AB| = |y_2 - y_1| \qquad |BP_2| = |z_2 - z_1|$$

Because triangles P_1BP_2 and P_1AB are both right-angled, two applications of the Pythagorean Theorem give

$$|P_1P_2|^2 = |P_1B|^2 + |BP_2|^2$$

and

$$|P_1B|^2 = |P_1A|^2 + |AB|^2$$

Combining these equations, we get

$$\begin{aligned} |P_1P_2|^2 &= |P_1A|^2 + |AB|^2 + |BP_2|^2 \\ &= |x_2 - x_1|^2 + |y_2 - y_1|^2 + |z_2 - z_1|^2 \\ &= (x_2 - x_1)^2 + (y_2 - y_1)^2 + (z_2 - z_1)^2 \end{aligned}$$

Therefore

$$|P_1P_2| = \sqrt{(x_2 - x_1)^2 + (y_2 - y_1)^2 + (z_2 - z_1)^2}$$

EXAMPLE 5 | The distance from the point $P(2, -1, 7)$ to the point $Q(1, -3, 5)$ is

$$|PQ| = \sqrt{(1 - 2)^2 + (-3 + 1)^2 + (5 - 7)^2} = \sqrt{1 + 4 + 4} = 3$$ ■

EXAMPLE 6 | BB **Antigenic cartography (continued)** What is the antigenic distance between influenza strain 2 and strain 4 in Example 3?

SOLUTION The distance between strain 2 and strain 4 is the distance between point $P(1.79, 1.91, 2.44)$ and point $Q(2.39, 4.05, 4.46)$. This is

$$|PQ| = \sqrt{(2.39 - 1.79)^2 + (4.05 - 1.91)^2 + (4.46 - 2.44)^2} \approx 3$$ ■

EXAMPLE 7 | Find an equation of a sphere with radius r and center $C(h, k, l)$.

SOLUTION By definition, a sphere is the set of all points $P(x, y, z)$ whose distance from C is r. (See Figure 13.) Thus P is on the sphere if and only if $|PC| = r$. Squaring both sides, we have $|PC|^2 = r^2$ or

$$(x - h)^2 + (y - k)^2 + (z - l)^2 = r^2$$ ■

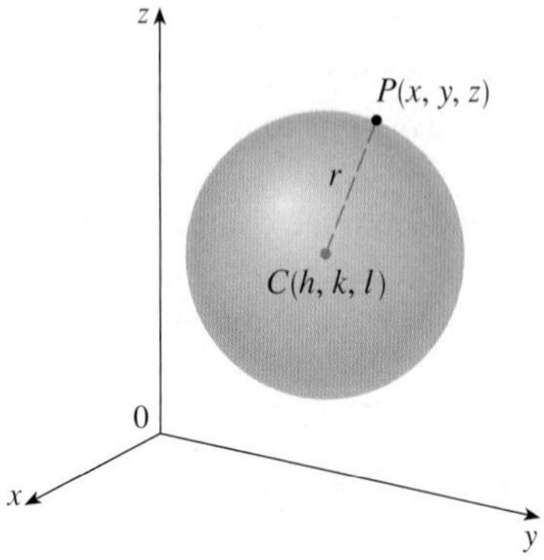

FIGURE 13

The result of Example 7 is worth remembering.

Equation of a Sphere An equation of a sphere with center $C(h, k, l)$ and radius r is

$$(x - h)^2 + (y - k)^2 + (z - l)^2 = r^2$$

In particular, if the center is the origin O, then an equation of the sphere is

$$x^2 + y^2 + z^2 = r^2$$

Higher-Dimensional Space

Just as we extended two-dimensional space to three-dimensional space, we can go further and generalize to n-dimensional space. Although we cannot visualize such spaces, we can still work with them mathematically. This is extremely important in the life sciences because the systems and objects that we seek to describe mathematically are often characterized by several variables. The Cartesian product $\mathbb{R} \times \mathbb{R} \times \cdots \times \mathbb{R}$, where the product involves n copies of $\mathbb{R}$, is denoted $\mathbb{R}^n$ and is defined as $\mathbb{R}^n = \{(x_1, x_2, \ldots, x_n) \mid x_1, x_2, \ldots, x_n \in \mathbb{R}\}$. This is the set of all ordered n-tuples of real numbers.

Distance Formula in n Dimensions The distance $|P_1P_2|$ between the points $P_1(a_1, \ldots, a_n)$ and $P_2(b_1, \ldots, b_n)$ is

$$|P_1P_2| = \sqrt{(b_1 - a_1)^2 + \cdots + (b_n - a_n)^2}$$

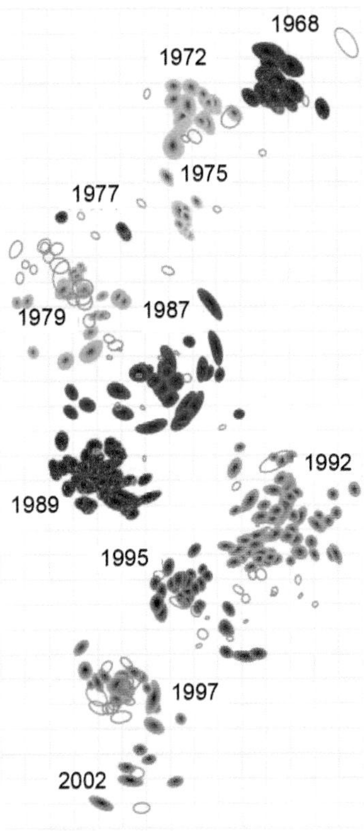

FIGURE 14

EXAMPLE 8 | BB **Antigenic cartography (continued)** Antigenic cartography has been carried out for 273 human influenza viruses collected over time from 1968 through to 2002.[1] Each virus was tested for its binding ability to 79 different antisera, producing a 79-dimensional antigenic space. The result is a plot of 273 points in $\mathbb{R}^{79}$. We cannot visualize such a high-dimensional space but, just as in three-dimensional space, we can project the points onto lower-dimensional surfaces. Although the way in which this is done can be complicated, Figure 14 gives an example of these data projected onto a two-dimensional plane. Despite the virus data existing in 79-dimensional antigenic space, we get a remarkably good visualization of the viruses, and how they have changed antigenically over a 34-year period, by projecting the points onto a plane representing a two-dimensional antigenic space. ■

EXERCISES 8.1

1. Suppose you start at the origin, move along the x-axis a distance of 4 units in the positive direction, and then move downward a distance of 3 units. What are the coordinates of your position?

2. Sketch the points $(0, 5, 2)$, $(4, 0, -1)$, $(2, 4, 6)$, and $(1, -1, 2)$ on a single set of coordinate axes.

3. Which of the points $P(6, 2, 3)$, $Q(-5, -1, 4)$, and $R(0, 3, 8)$ is closest to the xz-plane? Which point lies in the yz-plane?

4. What are the projections of the point $(2, 3, 5)$ on the xy-, yz-, and xz-planes? Draw a rectangular box with the origin and $(2, 3, 5)$ as opposite vertices and with its faces parallel to the coordinate planes. Label all vertices of the box. Find the length of the diagonal of the box.

5. Describe and sketch the surface in $\mathbb{R}^3$ represented by the equation $x + y = 2$.

6. (a) What does the equation $x = 4$ represent in $\mathbb{R}^2$? What does it represent in $\mathbb{R}^3$? Illustrate with sketches.
   (b) What does the equation $y = 3$ represent in $\mathbb{R}^3$? What does $z = 5$ represent? What does the pair of equations $y = 3$, $z = 5$ represent? In other words, describe the set of points (x, y, z) such that $y = 3$ and $z = 5$. Illustrate with a sketch.

7. Find the lengths of the sides of the triangle PQR. Is it a right triangle? Is it an isosceles triangle?
   (a) $P(3, -2, -3)$, $Q(7, 0, 1)$, $R(1, 2, 1)$
   (b) $P(2, -1, 0)$, $Q(4, 1, 1)$, $R(4, -5, 4)$

8. Find the distance from $(3, 7, -5)$ to each of the following.
   (a) The xy-plane (b) The yz-plane
   (c) The xz-plane (d) The x-axis
   (e) The y-axis (f) The z-axis

1. D. Smith et al., "Mapping the Antigenic and Genetic Evolution of Influenza Virus," *Science* 305 (2004): 371–76.

9. Determine whether the points lie on a straight line.
(a) $A(2, 4, 2)$, $B(3, 7, -2)$, $C(1, 3, 3)$
(b) $D(0, -5, 5)$, $E(1, -2, 4)$, $F(3, 4, 2)$

10. Find an equation of the sphere with center $(2, -6, 4)$ and radius 5. Describe its intersection with each of the coordinate planes.

11. Find an equation of the sphere that passes through the point $(4, 3, -1)$ and has center $(3, 8, 1)$.

12. Find an equation of the sphere that passes through the origin and whose center is $(1, 2, 3)$.

13–16 Show that the equation represents a sphere, and find its center and radius.

13. $x^2 + y^2 + z^2 - 6x + 4y - 2z = 11$

14. $x^2 + y^2 + z^2 + 8x - 6y + 2z + 17 = 0$

15. $2x^2 + 2y^2 + 2z^2 = 8x - 24z + 1$

16. $3x^2 + 3y^2 + 3z^2 = 10 + 6y + 12z$

17. (a) Prove that the midpoint of the line segment from $P_1(x_1, y_1, z_1)$ to $P_2(x_2, y_2, z_2)$ is

$$\left(\frac{x_1 + x_2}{2}, \frac{y_1 + y_2}{2}, \frac{z_1 + z_2}{2}\right)$$

(b) A median of a triangle is a line segment joining a vertex to the midpoint of the opposite side. Find the lengths of the medians of the triangle with vertices $A(1, 2, 3)$, $B(-2, 0, 5)$, and $C(4, 1, 5)$.

18. Find an equation of a sphere if one of its diameters has endpoints $(2, 1, 4)$ and $(4, 3, 10)$.

19. Find equations of the spheres with center $(2, -3, 6)$ that touch (a) the xy-plane, (b) the yz-plane, (c) the xz-plane.

20. Find an equation of the largest sphere with center $(5, 4, 9)$ that is contained in the first octant.

21–30 Describe in words the region of $\mathbb{R}^3$ represented by the equations or inequalities.

21. $x = 5$

22. $y = -2$

23. $y < 8$

24. $x \geqslant -3$

25. $0 \leqslant z \leqslant 6$

26. $z^2 = 1$

27. $x^2 + y^2 = 4, \quad z = -1$

28. $y^2 + z^2 = 16$

29. $x^2 + y^2 + z^2 \leqslant 3$

30. $x = z$

31–34 Write inequalities to describe the region.

31. The region between the yz-plane and the vertical plane $x = 5$

32. The solid cylinder that lies on or below the plane $z = 8$ and on or above the disk in the xy-plane with center the origin and radius 2

33. The region consisting of all points between (but not on) the spheres of radius r and R centered at the origin, where $r < R$

34. The solid upper hemisphere of the sphere of radius 2 centered at the origin

35. The figure shows a line L_1 in space and a second line L_2, which is the projection of L_1 onto the xy-plane. (In other words, the points on L_2 are directly beneath, or above, the points on L_1.)
(a) Find the coordinates of the point P on the line L_1.
(b) Locate on the diagram the points A, B, and C, where the line L_1 intersects the xy-plane, the yz-plane, and the xz-plane, respectively.

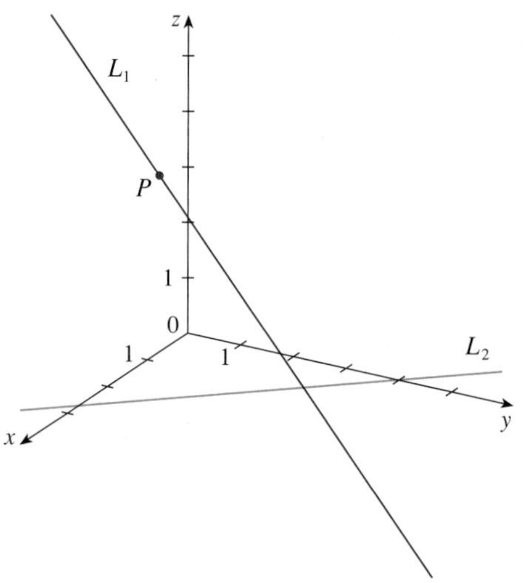

36. Darwin's finches have been used to study how differences in bird morphology are related to differences in diet. Morphological measurements (in mm) of three species are given in the table for three traits.

Species	Wing length	Tarsus length	Beak length
G. difficilis	64	18.1	9.6
G. fuliginosa	62.1	17.9	8.6
G. scandens	73.1	21.1	14.5

The proportion of time spent feeding on different types of food for these three species is given in the following table.

Species	Seeds	Pollen	Other
G. difficilis	0.67	0.23	0.1
G. fuliginosa	0.7	0.28	0.02
G. scandens	0.14	0	0.86

(a) Thinking of the morphology of each species as a point in $\mathbb{R}^3$, calculate the morphological distance between each pair of species.
(b) Thinking of the diet of each species as a point in $\mathbb{R}^3$, calculate the diet distance between each pair of species.

(c) Do species that are morphologically most similar also tend to have the most similar diets?

Source: Adapted from D. Schluter et al., "Ecological Correlates of Morphological Evolution in a Darwin's Finch, *Geospiza difficilis*," *Evolution* 38 (1984): 856–69.

37. Human biomechanics The trajectory of the center of a human walking on a treadmill is shown in the figure. Here x denotes the lateral position, y the position forward or backward, and z the vertical position. The coordinates at time $t = 0$ are (1.2, 0, 1.5) and at $t = 2$ are (1, 2, 0.5), where distances are measured in cm.

(a) What is magnitude of the the net lateral distance traveled over the first two seconds?

(b) What is the magnitude of the net distance traveled in the vertical direction over the first two seconds?

(c) What is the net distance traveled through three-dimensional space over the first two seconds?

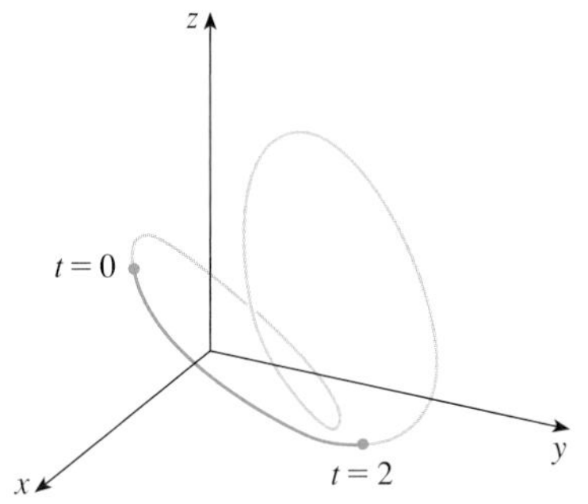

Source: Adapted from L. Tesio et al., "The 3D Trajectory of the Body Centre of Mass during Adult Human Walking: Evidence for a Speed-Curvature Power Law," *Journal of Biomechanics* 44 (2011): 732–40.

38. Vaccine design Most vaccines protect only against pathogens that fall within a certain region of antigenic space. Suppose that a vaccine protects against any strain falling within a sphere of radius 2 centered at the point (2, 1, 0) in antigenic space. For each strain, determine whether this vaccine will be effective.

(a) A strain located at (0, 0, 0) in antigenic space

(b) A strain located at (1, 0, 3) in antigenic space

(c) A strain located at (1, 0, 1) in antigenic space

(d) A strain located at (1/4, 2, 1) in antigenic space

39. Antigenic evolution and vaccination Antigenic data like those in Figure 14 can be summarized by taking the points from each year and drawing the smallest possible circle that encompasses these data. This results in a temporal sequence of circles in antigenic space, one for each year. Similarly, if the antigenic data are in three dimensions, spheres can be drawn around the data for each year, as shown in the figure. If the circles (or spheres) from years x and $x + 1$ overlap, then the amount of antigenic change between these years is relatively small. In such cases we might expect a single vaccine to work for both years. If the circles or spheres do not overlap, then we might need different vaccines for each year.

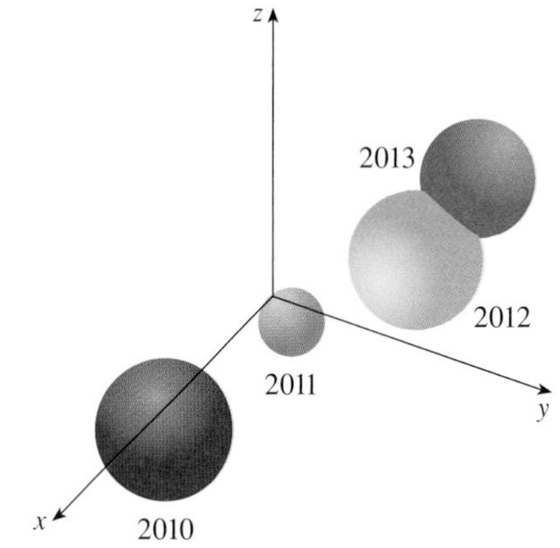

(a) Suppose the antigenic data are two-dimensional, and the circles for two successive years are given by the equations $(x - 2)^2 + (y - 3)^2 = 1$ and $(x - 3)^2 + (y - 2)^2 = \frac{1}{4}$. Would a single vaccine work for both years?

(b) Suppose the antigenic data are three-dimensional, and the spheres for two successive years are given by the equations

$$(x - 2)^2 + (y - 3)^2 + (z - 1)^2 = 1$$

and $$(x - 3)^2 + (y - 2)^2 + z^2 = \frac{1}{4}$$

Would a single vaccine work for both years?

(c) Notice that the x- and y-coordinates of the centers of the circles in part (a) are the same as the x- and y-coordinates of the centers of the spheres in part (b), and the radii are the same as well. What is the relationship between the plot of the circles in part (a) and the plot of the spheres in part (b)?

40. Describe and sketch a solid with the following properties: When illuminated by rays parallel to the z-axis, its shadow is a circular disk. If the rays are parallel to the y-axis, its shadow is a square. If the rays are parallel to the x-axis, its shadow is an isosceles triangle.

8.2 Vectors

The term **vector** is used by scientists to indicate a quantity (such as displacement or velocity or force) that has both magnitude and direction. A vector is often represented by an arrow or a directed line segment. The length of the arrow represents the magnitude of the vector and the arrow points in the direction of the vector. We denote a vector by printing a letter in lowercase boldface (**v**) or by putting an arrow above the letter ($\vec{v}$).

For instance, suppose a particle moves along a line segment from point A to point B. The corresponding **displacement vector v**, shown in Figure 1, has **initial point** A (the tail) and **terminal point** B (the tip) and we indicate this by writing $\mathbf{v} = \overrightarrow{AB}$. Notice that the vector $\mathbf{u} = \overrightarrow{CD}$ has the same length and the same direction as **v** even though it is in a different position. We say that **u** and **v** are **equivalent** (or **equal**) and we write $\mathbf{u} = \mathbf{v}$. The **zero vector**, denoted by **0**, has length 0. It is the only vector with no specific direction.

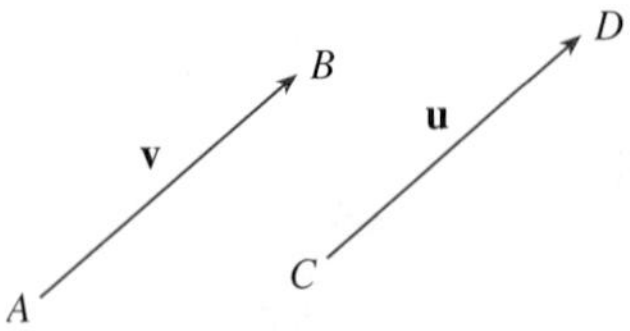

FIGURE 1
Equivalent vectors

■ Combining Vectors

Suppose a particle moves from A to B, so its displacement vector is $\overrightarrow{AB}$. Then the particle changes direction and moves from B to C, with displacement vector $\overrightarrow{BC}$ as in Figure 2. The combined effect of these displacements is that the particle has moved from A to C. The resulting displacement vector $\overrightarrow{AC}$ is called the *sum* of $\overrightarrow{AB}$ and $\overrightarrow{BC}$ and we write

$$\overrightarrow{AC} = \overrightarrow{AB} + \overrightarrow{BC}$$

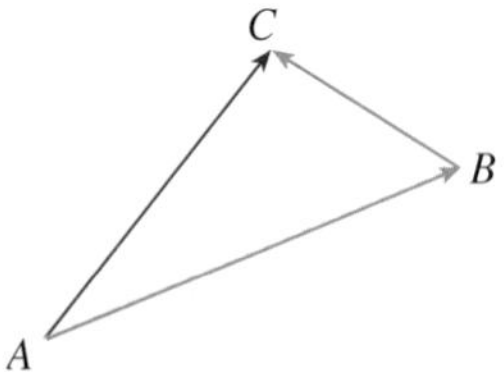

FIGURE 2

In general, if we start with vectors **u** and **v**, we first move **v** so that its tail coincides with the tip of **u** and define the sum of **u** and **v** as follows.

> **Definition of Vector Addition** If **u** and **v** are vectors positioned so the initial point of **v** is at the terminal point of **u**, then the sum $\mathbf{u} + \mathbf{v}$ is the vector from the initial point of **u** to the terminal point of **v**.

The definition of vector addition is illustrated in Figure 3. You can see why this definition is sometimes called the **Triangle Law**.

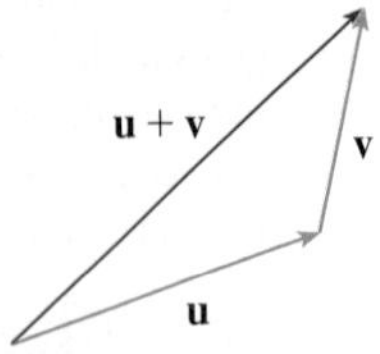

FIGURE 3
The Triangle Law

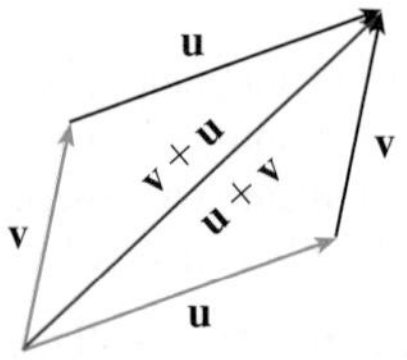

FIGURE 4
The Parallelogram Law

In Figure 4 we start with the same vectors **u** and **v** as in Figure 3 and draw another copy of **v** with the same initial point as **u**. Completing the parallelogram, we see that $\mathbf{u} + \mathbf{v} = \mathbf{v} + \mathbf{u}$. This also gives another way to construct the sum: If we place **u** and **v** so they start at the same point, then $\mathbf{u} + \mathbf{v}$ lies along the diagonal of the parallelogram with **u** and **v** as sides. (This is called the **Parallelogram Law**.)

EXAMPLE 1 | BB **Antigenic cartography** The evolution of influenza in antigenic space across years can be viewed as a vector, since it is characterized by both its magnitude and direction of change (see also Example 8.1.8). Figure 5 shows clusters of influenza virus in two-dimensional antigenic space from several years. Illustrate how the vector of antigenic change between the centers of the viral clusters in 1989 and 1995 is the sum of the corresponding vectors from 1989–1992 and 1992–1995. Use both the Triangle and the Parallelogram Law.

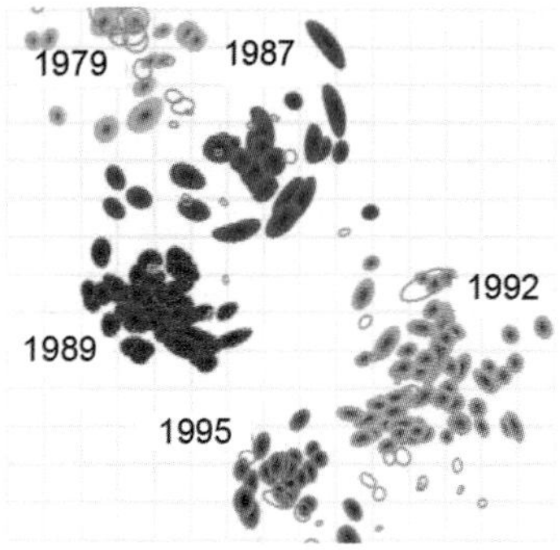

FIGURE 5

SOLUTION The vector corresponding to the antigenic change from 1989 to 1995 is labeled as **c** in Figure 6. For the Triangle Law, first we draw a vector from the 1989 cluster to the 1992 cluster (labeled **a**). Then we draw a vector from the 1992 cluster to the 1995 cluster (labeled **b**). We see that $\mathbf{a} + \mathbf{b} = \mathbf{c}$. Figure 6(a) illustrates the Triangle Law.

For the Parallelogram Law, we again draw a vector from the 1989 cluster to the 1992 cluster. Then we draw a vector from the 1992 cluster to the 1995 cluster and translate it so that it starts where **a** starts. The vector **c** then lies on the diagonal of the corresponding parallelogram, as shown in Figure 6(b).

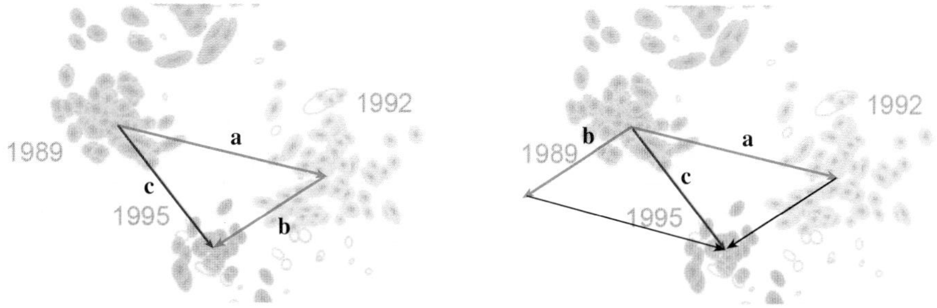

FIGURE 6 (a) The Triangle Law (b) The Parallelogram Law ■

It is possible to multiply a vector by a real number c. (In this context we call the real number c a **scalar** to distinguish it from a vector.) For instance, we want $2\mathbf{v}$ to be the same vector as $\mathbf{v} + \mathbf{v}$, which has the same direction as $\mathbf{v}$ but is twice as long. In general, we multiply a vector by a scalar as follows.

Definition of Scalar Multiplication If c is a scalar and $\mathbf{v}$ is a vector, then the **scalar multiple** $c\mathbf{v}$ is the vector whose length is $|c|$ times the length of $\mathbf{v}$ and whose direction is the same as $\mathbf{v}$ if $c > 0$ and is opposite to $\mathbf{v}$ if $c < 0$. If $c = 0$ or $\mathbf{v} = \mathbf{0}$, then $c\mathbf{v} = \mathbf{0}$.

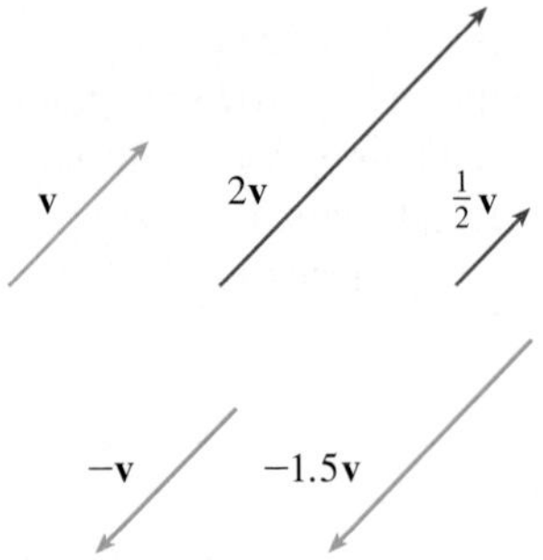

FIGURE 7
Scalar multiples of **v**

This definition is illustrated in Figure 7. We see that real numbers work like scaling factors here; that's why we call them scalars. Notice that two nonzero vectors are **parallel** if they are scalar multiples of one another. In particular, the vector $-\mathbf{v} = (-1)\mathbf{v}$ has the same length as **v** but points in the opposite direction. We call it the **negative** of **v**.

By the **difference** $\mathbf{u} - \mathbf{v}$ of two vectors we mean

$$\mathbf{u} - \mathbf{v} = \mathbf{u} + (-\mathbf{v})$$

So we can construct $\mathbf{u} - \mathbf{v}$ by first drawing the negative of **v**, $-\mathbf{v}$, and then adding it to **u** by the Parallelogram Law as in Figure 8(a). Alternatively, since $\mathbf{v} + (\mathbf{u} - \mathbf{v}) = \mathbf{u}$, the vector $\mathbf{u} - \mathbf{v}$, when added to **v**, gives **u**. So we could construct $\mathbf{u} - \mathbf{v}$ as in Figure 8(b) by means of the Triangle Law.

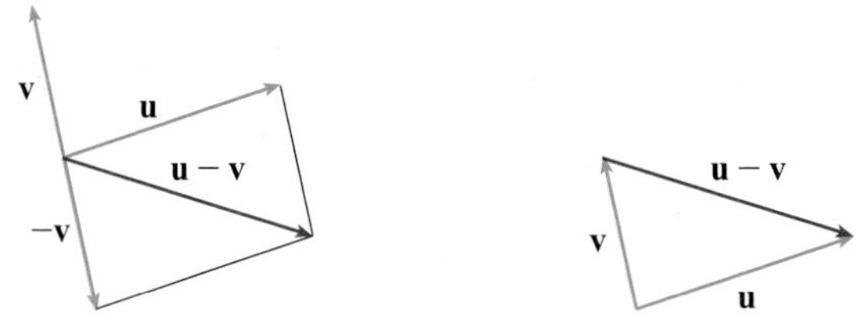

FIGURE 8
Drawing $\mathbf{u} - \mathbf{v}$

(a) (b)

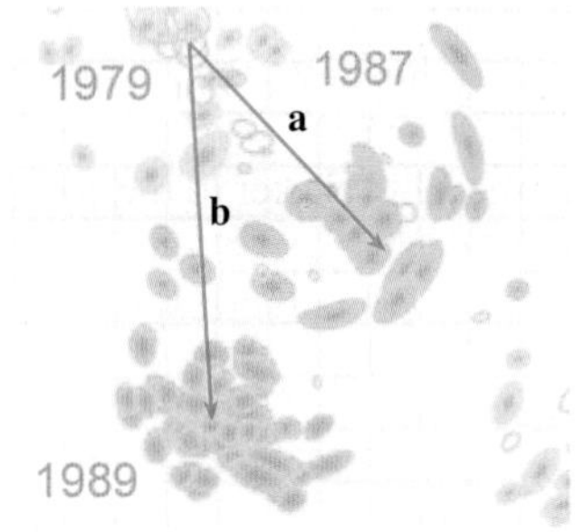

FIGURE 9

EXAMPLE 2 | BB Antigenic cartography (continued) Using the vectors **a** and **b** in Figure 9, illustrate the vector of antigenic change $\mathbf{b} - \mathbf{a}$ with both the Triangle and the Parallelogram Laws.

SOLUTION For the Triangle Law we seek the vector that, when added to **a** gives **b**. This is obtained as the vector from the terminal point of **a** to the terminal point of **b**, as shown in Figure 10(a).

For the Parallelogram Law we first construct the vector $-\mathbf{a}$. Then we add it to **b** using the Parallelogram Law as in Figure 10(b). Although the vectors constructed in Figures 10(a) and 10(b) are in different positions, they are equivalent vectors because they represent the same direction and magnitude of antigenic change.

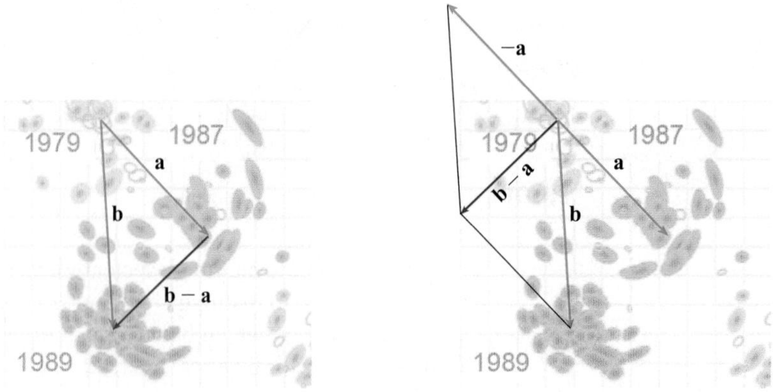

FIGURE 10

(a) The Triangle Law (b) The Parallelogram Law ■

■ Components

For some purposes it's best to introduce a coordinate system and treat vectors algebraically. If we place the initial point of a vector **a** at the origin of a rectangular coordinate system, then the terminal point of **a** has coordinates of the form (a_1, a_2) or (a_1, a_2, a_3), depending on whether our coordinate system is two- or three-dimensional (see Figure 11).

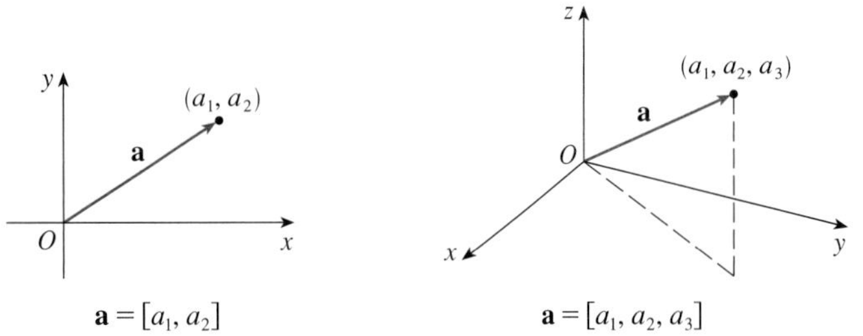

FIGURE 11

These coordinates are called the **components** of **a** and we write

$$\mathbf{a} = [a_1, a_2] \qquad \text{or} \qquad \mathbf{a} = [a_1, a_2, a_3]$$

The component form of a vector is sometimes written using angled brackets rather than square brackets. So, for instance, vector $\mathbf{a} = [a_1, a_2, a_3]$ can be equivalently written as $\mathbf{a} = \langle a_1, a_2, a_3 \rangle$.

We use the notation $[a_1, a_2]$ for the ordered pair that refers to a vector so as not to confuse it with the ordered pair (a_1, a_2) that refers to a point in the plane.

For instance, the vectors shown in Figure 12 are all equivalent to the vector $\overrightarrow{OP} = [3, 2]$ whose terminal point is $P(3, 2)$. What they have in common is that the terminal point is reached from the initial point by a displacement of three units to the right and two upward. We can think of all these geometric vectors as **representations** of the algebraic vector $\mathbf{a} = [3, 2]$. The particular representation $\overrightarrow{OP}$ from the origin to the point $P(3, 2)$ is called the **position vector** of the point P.

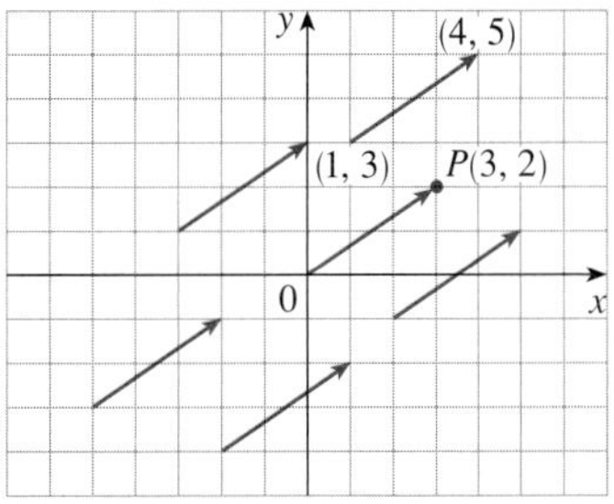

FIGURE 12
Representations of the vector $\mathbf{a} = [3, 2]$

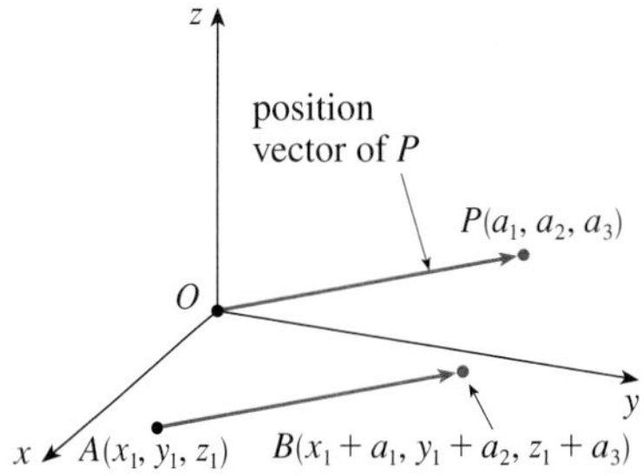

FIGURE 13
Representations of $\mathbf{a} = [a_1, a_2, a_3]$

In three dimensions, the vector $\mathbf{a} = \overrightarrow{OP} = [a_1, a_2, a_3]$ is the **position vector** of the point $P(a_1, a_2, a_3)$. (See Figure 13.) Let's consider any other representation $\overrightarrow{AB}$ of **a**, where the initial point is $A(x_1, y_1, z_1)$ and the terminal point is $B(x_2, y_2, z_2)$. Then we must have $x_2 = x_1 + a_1$, $y_2 = y_1 + a_2$, and $z_2 = z_1 + a_3$ and so $a_1 = x_2 - x_1$, $a_2 = y_2 - y_1$, and $a_3 = z_2 - z_1$. Thus we have the following result.

> **(1)** Given the points $A(x_1, y_1, z_1)$ and $B(x_2, y_2, z_2)$, the vector **a** with representation $\overrightarrow{AB}$ is
>
> $$\mathbf{a} = [x_2 - x_1, y_2 - y_1, z_2 - z_1]$$

EXAMPLE 3 | BB **Antigenic cartography (continued)** Given the coordinate system overlaid on the two-dimensional antigenic space in Figure 14, find the components of the vector represented by the directed line segment from the cluster A in 1987 to cluster B in 1989.

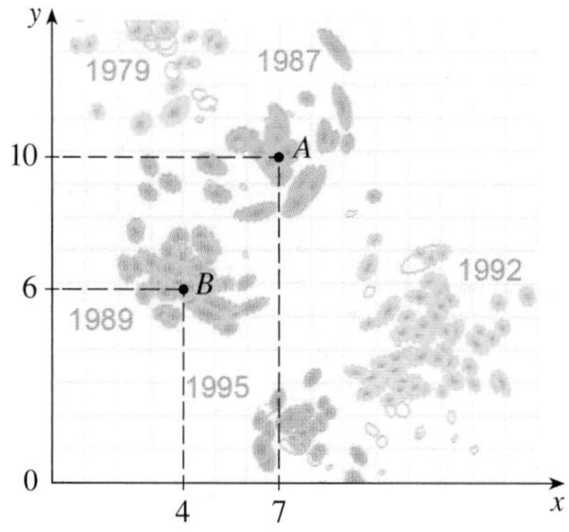

FIGURE 14

SOLUTION By result (1), the vector corresponding to $\overrightarrow{AB}$ is

$$\mathbf{a} = [4 - 7, 6 - 10] = [-3, -4]$$

The **magnitude** or **length** of a vector $\mathbf{a}$ is the length of any of its representations and is denoted by the symbol $|\mathbf{a}|$ or $\|\mathbf{a}\|$. By using the distance formula to compute its length, we obtain the following formulas.

The length of the two-dimensional vector $\mathbf{a} = [a_1, a_2]$ is

$$|\mathbf{a}| = \sqrt{a_1^2 + a_2^2}$$

The length of the three-dimensional vector $\mathbf{a} = [a_1, a_2, a_3]$ is

$$|\mathbf{a}| = \sqrt{a_1^2 + a_2^2 + a_3^2}$$

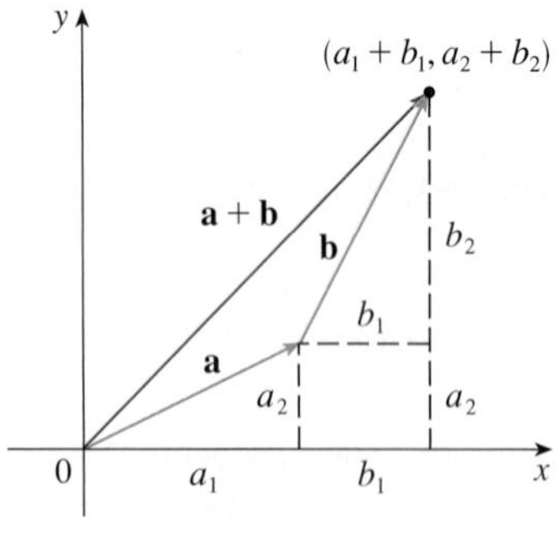

FIGURE 15

How do we add vectors algebraically? Figure 15 shows that if $\mathbf{a} = [a_1, a_2]$ and $\mathbf{b} = [b_1, b_2]$, then the sum is $\mathbf{a} + \mathbf{b} = [a_1 + b_1, a_2 + b_2]$, at least for the case where the components are positive. In other words, *to add algebraic vectors we add their components.* Similarly, *to subtract vectors we subtract components.* From the similar triangles in Figure 16 we see that the components of $c\mathbf{a}$ are ca_1 and ca_2. So *to multiply a vector by a scalar we multiply each component by that scalar.*

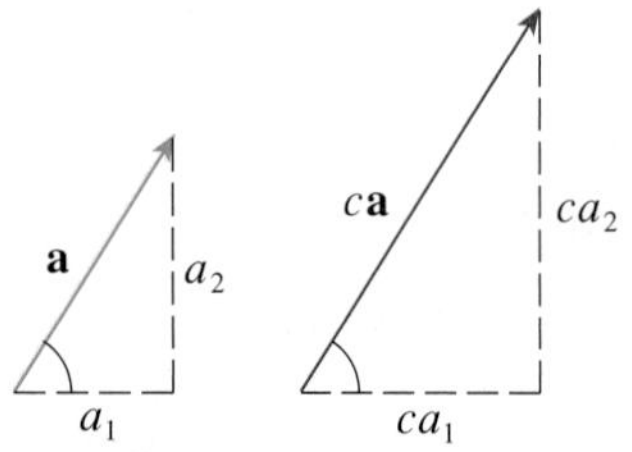

FIGURE 16

If $\mathbf{a} = [a_1, a_2]$ and $\mathbf{b} = [b_1, b_2]$, then

$$\mathbf{a} + \mathbf{b} = [a_1 + b_1, a_2 + b_2] \qquad \mathbf{a} - \mathbf{b} = [a_1 - b_1, a_2 - b_2]$$

$$c\mathbf{a} = [ca_1, ca_2]$$

Similarly, for three-dimensional vectors,

$$[a_1, a_2, a_3] + [b_1, b_2, b_3] = [a_1 + b_1, a_2 + b_2, a_3 + b_3]$$

$$[a_1, a_2, a_3] - [b_1, b_2, b_3] = [a_1 - b_1, a_2 - b_2, a_3 - b_3]$$

$$c[a_1, a_2, a_3] = [ca_1, ca_2, ca_3]$$

A common alternative notation for vectors in three dimensions makes use of unit vectors in the x-, y-, and z-directions. By convention these three special vectors are denoted by **i**, **j**, and **k**, respectively. Exercises 8.2.30–32 show how any three-dimensional vector can be written in terms of **i**, **j**, and **k**.

A **unit vector** is a vector whose length is 1. In general, if $\mathbf{a} \neq \mathbf{0}$, then the unit vector that has the same direction as **a** is

$$\text{(2)} \qquad \mathbf{u} = \frac{1}{|\mathbf{a}|}\mathbf{a} = \frac{\mathbf{a}}{|\mathbf{a}|}$$

EXAMPLE 4 | Find the unit vector in the direction of the vector $\mathbf{a} = [2, -1, -2]$.

SOLUTION The given vector has length

$$|\mathbf{a}| = \sqrt{2^2 + (-1)^2 + (-2)^2} = \sqrt{9} = 3$$

so, by Equation 2, the unit vector with the same direction is

$$\tfrac{1}{3}[2, -1, -2] = \left[\tfrac{2}{3}, -\tfrac{1}{3}, -\tfrac{2}{3}\right]$$

■

EXAMPLE 5 | If $\mathbf{a} = [4, 0, 3]$ and $\mathbf{b} = [-2, 1, 5]$, find $|\mathbf{a}|$ and the vectors $\mathbf{a} + \mathbf{b}$, $\mathbf{a} - \mathbf{b}$, $3\mathbf{b}$, and $2\mathbf{a} + 5\mathbf{b}$.

SOLUTION

$$|\mathbf{a}| = \sqrt{4^2 + 0^2 + 3^2} = \sqrt{25} = 5$$

$$\mathbf{a} + \mathbf{b} = [4, 0, 3] + [-2, 1, 5]$$

$$= [4 + (-2), 0 + 1, 3 + 5] = [2, 1, 8]$$

$$\mathbf{a} - \mathbf{b} = [4, 0, 3] - [-2, 1, 5]$$

$$= [4 - (-2), 0 - 1, 3 - 5] = [6, -1, -2]$$

$$3\mathbf{b} = 3[-2, 1, 5] = [3(-2), 3(1), 3(5)] = [-6, 3, 15]$$

$$2\mathbf{a} + 5\mathbf{b} = 2[4, 0, 3] + 5[-2, 1, 5]$$

$$= [8, 0, 6] + [-10, 5, 25] = [-2, 5, 31]$$

■

FIGURE 17

EXAMPLE 6 | Biomechanics A force is represented by a vector because it has a magnitude (in newtons) and a direction. If more than one force is acting on an object, the **resultant force** is the vector sum of these forces. Consider the horizontal and vertical forces exerted by an athlete at the start of the 100-meter sprint as shown in Figure 17.

(a) What is the resultant force vector exerted by the athlete?

(b) What is the total magnitude of the force exerted by the athlete?

SOLUTION

(a) The force vector in the horizontal direction is $[180, 0]$ and in the vertical direction, $[0, 490]$. The resultant force vector is therefore $[180, 0] + [0, 490] = [180, 490]$.

(b) The total magnitude of the force is the length of the resultant force vector. This is $\sqrt{180^2 + 490^2} \approx 522$ newtons. ■

We denote by V_2 the set of all two-dimensional vectors and by V_3 the set of all three-dimensional vectors. More generally, V_n is the set of all n-dimensional vectors where an n-dimensional vector is an ordered n-tuple:

$$\mathbf{a} = [a_1, a_2, \ldots, a_n]$$

and $a_1, a_2, \ldots, a_n$ are real numbers called the components of $\mathbf{a}$. Addition and scalar multiplication are defined in terms of components just as for the cases $n = 2$ and $n = 3$. Likewise, the length of a vector from V_n is calculated by using the distance formula on page 493.

Vectors in n dimensions are used to list various quantities in an organized way. For instance, the components of a six-dimensional vector

$$\mathbf{p} = [p_1, p_2, p_3, p_4, p_5, p_6]$$

might represent the expression levels of six different genes.

Properties of Vectors If $\mathbf{a}$, $\mathbf{b}$, and $\mathbf{c}$ are vectors in V_n and c and d are scalars, then

1. $\mathbf{a} + \mathbf{b} = \mathbf{b} + \mathbf{a}$	2. $\mathbf{a} + (\mathbf{b} + \mathbf{c}) = (\mathbf{a} + \mathbf{b}) + \mathbf{c}$
3. $\mathbf{a} + \mathbf{0} = \mathbf{a}$	4. $\mathbf{a} + (-\mathbf{a}) = \mathbf{0}$
5. $c(\mathbf{a} + \mathbf{b}) = c\mathbf{a} + c\mathbf{b}$	6. $(c + d)\mathbf{a} = c\mathbf{a} + d\mathbf{a}$
7. $(cd)\mathbf{a} = c(d\mathbf{a})$	8. $1\mathbf{a} = \mathbf{a}$

These eight properties of vectors can be readily verified either geometrically or algebraically. For instance, Property 1 can be seen from Figure 4 (it's equivalent to the Parallelogram Law) or as follows for the case $n = 2$:

$$\mathbf{a} + \mathbf{b} = [a_1, a_2] + [b_1, b_2] = [a_1 + b_1, a_2 + b_2]$$
$$= [b_1 + a_1, b_2 + a_2] = [b_1, b_2] + [a_1, a_2]$$
$$= \mathbf{b} + \mathbf{a}$$

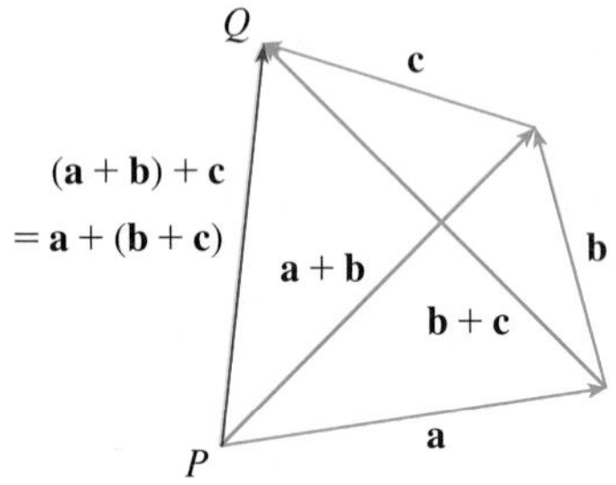

FIGURE 18

We can see why Property 2 (the associative law) is true by looking at Figure 18 and applying the Triangle Law several times: The vector $\overrightarrow{PQ}$ is obtained either by first constructing $\mathbf{a} + \mathbf{b}$ and then adding $\mathbf{c}$ or by adding $\mathbf{a}$ to the vector $\mathbf{b} + \mathbf{c}$.

EXERCISES 8.2

Note: Vector notations using square brackets [1, 1] and angle brackets ⟨1, 1⟩ are equivalent. You may see either notation in your online homework.

1. Are the following quantities vectors or scalars? Explain.
(a) The cost of a theater ticket
(b) The current in a river
(c) The initial flight path from Houston to Dallas
(d) The population of the world

2. What is the relationship between the point (4, 7) and the vector [4, 7]? Illustrate with a sketch.

3. Name all the equal vectors in the parallelogram shown.

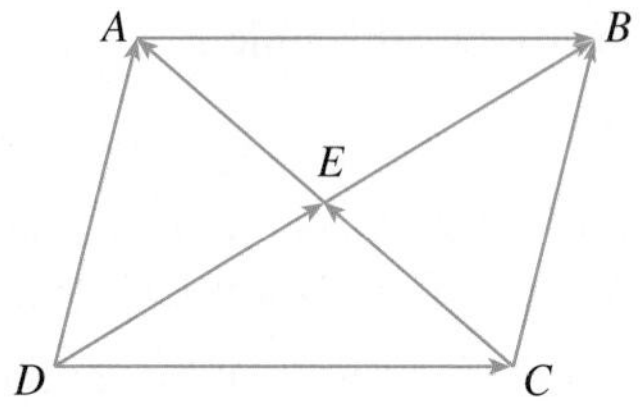

4. Write each combination of vectors as a single vector.
(a) $\overrightarrow{PQ} + \overrightarrow{QR}$ (b) $\overrightarrow{RP} + \overrightarrow{PS}$
(c) $\overrightarrow{QS} - \overrightarrow{PS}$ (d) $\overrightarrow{RS} + \overrightarrow{SP} + \overrightarrow{PQ}$

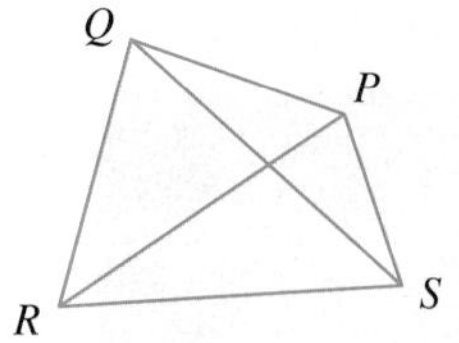

5. Copy the vectors in the figure and use them to draw the following vectors.
(a) $\mathbf{u} + \mathbf{v}$ (b) $\mathbf{u} - \mathbf{v}$
(c) $\mathbf{v} + \mathbf{w}$ (d) $\mathbf{w} + \mathbf{v} + \mathbf{u}$

6. Copy the vectors in the figure and use them to draw the following vectors.

(a) $\mathbf{a} + \mathbf{b}$ (b) $\mathbf{a} - \mathbf{b}$
(c) $\frac{1}{2}\mathbf{a}$ (d) $-3\mathbf{b}$
(e) $\mathbf{a} + 2\mathbf{b}$ (f) $2\mathbf{b} - \mathbf{a}$

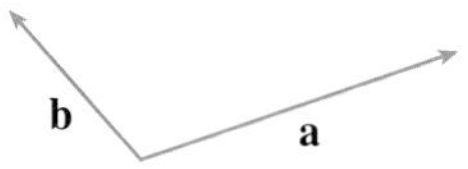

7–10 Find a vector $\mathbf{a}$ with representation given by the directed line segment $\overrightarrow{AB}$. Draw $\overrightarrow{AB}$ and the equivalent representation starting at the origin.

7. $A(-1, 3)$, $B(2, 2)$ **8.** $A(2, 1)$, $B(0, 6)$

9. $A(0, 3, 1)$, $B(2, 3, -1)$ **10.** $A(4, 0, -2)$, $B(4, 2, 1)$

11–14 Find the sum of the given vectors and illustrate geometrically.

11. $[-1, 4]$, $[6, -2]$ **12.** $[-2, -1]$, $[5, 7]$

13. $[0, 1, 2]$, $[0, 0, -3]$ **14.** $[-1, 0, 2]$, $[0, 4, 0]$

15–18 Find $\mathbf{a} + \mathbf{b}$, $2\mathbf{a} + 3\mathbf{b}$, $|\mathbf{a}|$, and $|\mathbf{a} - \mathbf{b}|$.

15. $\mathbf{a} = [5, -12]$, $\mathbf{b} = [-3, -6]$

16. $\mathbf{a} = [4, 1]$, $\mathbf{b} = [1, -2]$

17. $\mathbf{a} = [1, 2, -3]$, $\mathbf{b} = [-2, -1, 5]$

18. $\mathbf{a} = [2, -4, 4]$, $\mathbf{b} = [0, 2, -1]$

19–21 Find a unit vector that has the same direction as the given vector.

19. $[-3, 7]$ **20.** $[-4, 2, 4]$

21. $[8, -1, 4]$

22. Find a vector that has the same direction as $[-2, 4, 2]$ but has length 6.

23. If $\mathbf{v}$ lies in the first quadrant and makes an angle $\pi/3$ with the positive x-axis and $|\mathbf{v}| = 4$, find $\mathbf{v}$ in component form.

24. If a child pulls a sled through the snow on a level path with a force of 50 N exerted at an angle of 38° above the horizontal, find the horizontal and vertical components of the force.

25. A quarterback throws a football with angle of elevation 40° and speed 60 ft/s. Find the horizontal and vertical components of the velocity vector.

26–27 Find the magnitude of the resultant force and the angle it makes with the positive x-axis.

26.

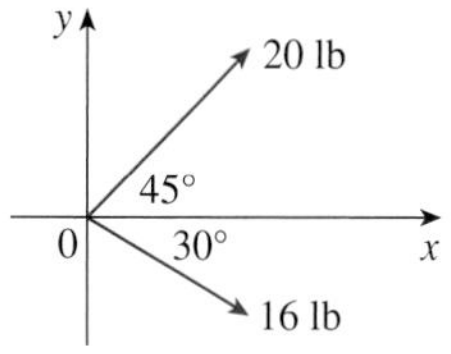

27.

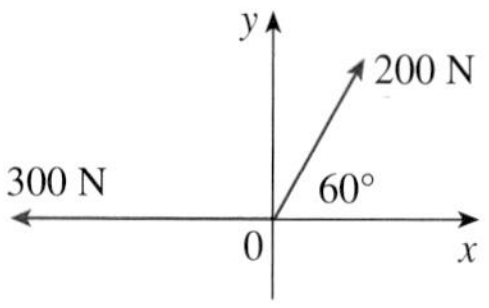

28. The magnitude of a velocity vector is called *speed*. Suppose that a wind is blowing from the direction N45°W at a speed of 50 km/h. (This means that the direction from which the wind blows is 45° west of the northerly direction.) A pilot is steering a plane in the direction N60°E at an airspeed (speed in still air) of 250 km/h. The *true course*, or *track*, of the plane is the direction of the resultant of the velocity vectors of the plane and the wind. The *ground speed* of the plane is the magnitude of the resultant. Find the true course and the ground speed of the plane.

29. A woman walks due west on the deck of a ship at 3 mi/h. The ship is moving north at a speed of 22 mi/h. Find the speed and direction of the woman relative to the surface of the water.

30–32 The unit vectors in V_3 that coincide with the coordinate axes are called the **standard basis vectors** and are denoted by $\mathbf{i} = [1, 0, 0]$, $\mathbf{j} = [0, 1, 0]$, and $\mathbf{k} = [0, 0, 1]$. (See the figure.)

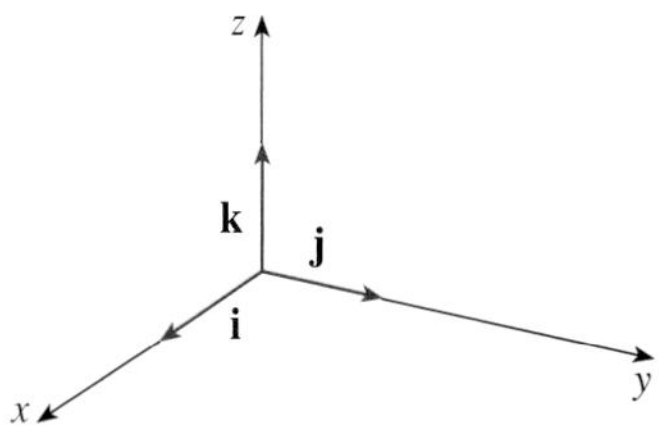

If $\mathbf{a} = [a_1, a_2, a_3]$, we can write

$$\mathbf{a} = [a_1, 0, 0] + [0, a_2, 0] + [0, 0, a_3]$$
$$= a_1[1, 0, 0] + a_2[0, 1, 0] + a_3[0, 0, 1]$$

Therefore any vector $\mathbf{a}$ in V_3 can be expressed in terms of the standard basis vectors as $\mathbf{a} = a_1\mathbf{i} + a_2\mathbf{j} + a_3\mathbf{k}$.

30. Express the following vectors in terms of the standard basis vectors.

(a) $[-1, 4]$ (b) $[5, 7]$
(c) $[-2, 1, 2]$ (d) $[-1, 0, 2]$

31. If $\mathbf{a} = \mathbf{i} + 2\,\mathbf{j} - 3\,\mathbf{k}$ and $\mathbf{b} = 4\,\mathbf{i} + 7\,\mathbf{k}$, evaluate the following in terms of the standard basis vectors.
(a) $\mathbf{a} + \mathbf{b}$ (b) $\mathbf{a} - \mathbf{b}$
(c) $2\,\mathbf{a} + 3\,\mathbf{b}$ (d) $5\,\mathbf{a} - 7\,\mathbf{b}$

32. Find the unit vector that points in the same direction as the given vector and express it in terms of the standard basis vectors.
(a) $\mathbf{i} + \mathbf{j}$ (b) $\mathbf{i} + \mathbf{j} + \mathbf{k}$
(c) $2\,\mathbf{i} - \mathbf{k}$ (d) $4\,\mathbf{i} + 6\,\mathbf{j} - \mathbf{k}$

---

33. Find the unit vectors that are parallel to the tangent line to the parabola $y = x^2$ at the point $(2, 4)$.

34. (a) Find the unit vectors that are parallel to the tangent line to the curve $y = 2 \sin x$ at the point $(\pi/6, 1)$.
(b) Find the unit vectors that are perpendicular to the tangent line.
(c) Sketch the curve $y = 2 \sin x$ and the vectors in parts (a) and (b), all starting at $(\pi/6, 1)$.

35. (a) Draw the vectors $\mathbf{a} = [3, 2]$, $\mathbf{b} = [2, -1]$, and $\mathbf{c} = [7, 1]$.
(b) Show, by means of a sketch, that there are scalars s and t such that $\mathbf{c} = s\mathbf{a} + t\mathbf{b}$.
(c) Use the sketch to estimate the values of s and t.
(d) Find the exact values of s and t.

36. Suppose that $\mathbf{a}$ and $\mathbf{b}$ are nonzero vectors that are not parallel and $\mathbf{c}$ is any vector in the plane determined by $\mathbf{a}$ and $\mathbf{b}$. Give a geometric argument to show that $\mathbf{c}$ can be written as $\mathbf{c} = s\mathbf{a} + t\mathbf{b}$ for suitable scalars s and t. Then give an argument using components.

37. Suppose $\mathbf{a}$ is a three-dimensional unit vector in the first octant that starts at the origin and makes angles of $60°$ and $72°$ with the positive x- and y-axes, respectively. Express $\mathbf{a}$ in terms of its components.

38. Biomechanics Two sprinters of equal mass leave the starting blocks with the following horizontal and vertical force vectors:

Runner	Horizontal	Vertical
Runner 1	150 N	300 N
Runner 2	200 N	250 N

Newton's Second Law states that force is equal to mass times acceleration ($\mathbf{F} = m\mathbf{a}$). Which runner has the greater acceleration out of the blocks?

39. Vectorcardiography As the heart beats it generates differences in electrical potential (that is, voltage) across the body. Cardiologists view the voltage at any point in time during a heartbeat as a vector. During ventricular contraction in healthy individuals this vector points downward and to the left of the patient (see the figure). Abnormalities in either the magnitude or direction can be used to diagnose cardiac problems.

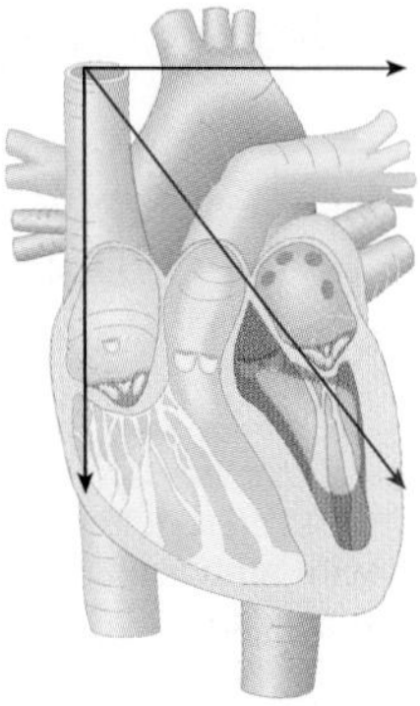

We can think of the voltage vector as being the resultant vector of a vertical and a horizontal component of voltage (see the figure). Conductive abnormalities in different parts of the heart can reduce the voltage in either of these directions, thereby altering the direction and magnitude of the resultant voltage vector.
(a) A *left anterior hemiblock* reverses the direction of the vertical component of voltage and reduces the magnitude of the horizontal component. Describe how such a condition could be diagnosed based on the resultant voltage vector.
(b) A *left posterior hemiblock* reverses the direction of the horizontal component of voltage. Describe how such a condition could be diagnosed based on the resultant voltage vector.

40. Suppose a vector $\mathbf{a}$ makes angles α, β, and γ with the positive x-, y-, and z-axes, respectively. Find the components of $\mathbf{a}$ and show that $\cos^2\alpha + \cos^2\beta + \cos^2\gamma = 1$. (The numbers $\cos\alpha$, $\cos\beta$, and $\cos\gamma$ are called the *direction cosines* of $\mathbf{a}$.)

41. If $\mathbf{r} = [x, y, z]$ and $\mathbf{r}_0 = [x_0, y_0, z_0]$, describe the set of all points (x, y, z) such that $|\mathbf{r} - \mathbf{r}_0| = 1$.

42. If $\mathbf{r} = [x, y]$, $\mathbf{r}_1 = [x_1, y_1]$, and $\mathbf{r}_2 = [x_2, y_2]$, describe the set of all points (x, y) such that $|\mathbf{r} - \mathbf{r}_1| + |\mathbf{r} - \mathbf{r}_2| = k$, where $k > |\mathbf{r}_1 - \mathbf{r}_2|$.

43. Figure 18 gives a geometric demonstration of Property 2 of vectors. Use components to give an algebraic proof of this fact for the case $n = 2$.

44. Prove Property 5 of vectors algebraically for the case $n = 3$. Then use similar triangles to give a geometric proof.

45. Use vectors to prove that the line joining the midpoints of two sides of a triangle is parallel to the third side and half its length.

46. Antigenic cartography The Triangle Inequality for vectors (see Exercise 8.3.48) is

$$|\mathbf{a} + \mathbf{b}| \leqslant |\mathbf{a}| + |\mathbf{b}|$$

Suppose that $\mathbf{a}$ denotes the vector of antigenic change in influenza from year 2012 to 2013, and $\mathbf{b}$ denotes the vector of antigenic change from 2013 to 2014. Explain what the Triangle Inequality means in terms of this antigenic change.

8.3 The Dot Product

So far we have added two vectors and multiplied a vector by a scalar. The question arises: Is it possible to multiply two vectors so that their product is a useful quantity? One such product is the dot product.

(1) Definition If $\mathbf{a} = [a_1, a_2, a_3]$ and $\mathbf{b} = [b_1, b_2, b_3]$, then the **dot product** of $\mathbf{a}$ and $\mathbf{b}$ is the number $\mathbf{a} \cdot \mathbf{b}$ given by

$$\mathbf{a} \cdot \mathbf{b} = a_1b_1 + a_2b_2 + a_3b_3$$

Thus, to find the dot product of **a** and **b**, we multiply corresponding components and add. The result is not a vector. It is a real number, that is, a scalar. For this reason, the dot product is sometimes called the **scalar product**. Although Definition 1 is given for three-dimensional vectors, the dot product of two-dimensional vectors is defined in a similar fashion:

$$[a_1, a_2] \cdot [b_1, b_2] = a_1b_1 + a_2b_2$$

Likewise, for n-dimensional vectors we have:

$$[a_1, \ldots, a_n] \cdot [b_1, \ldots, b_n] = a_1b_1 + \cdots + a_nb_n$$

EXAMPLE 1

$$[2, 4] \cdot [3, -1] = 2(3) + 4(-1) = 2$$

$$[-1, 7, 4] \cdot \left[6, 2, -\tfrac{1}{2}\right] = (-1)(6) + 7(2) + 4\left(-\tfrac{1}{2}\right) = 6$$

$$[1, 2, -3] \cdot [0, 2, -1] = 1(0) + 2(2) + (-3)(-1) = 7$$

■

The dot product obeys many of the laws that hold for ordinary products of real numbers.

(2) Properties of the Dot Product If **a**, **b**, and **c** are vectors in V_3 and c is a scalar, then

1. $\mathbf{a} \cdot \mathbf{a} = |\mathbf{a}|^2$

2. $\mathbf{a} \cdot \mathbf{b} = \mathbf{b} \cdot \mathbf{a}$

3. $\mathbf{a} \cdot (\mathbf{b} + \mathbf{c}) = \mathbf{a} \cdot \mathbf{b} + \mathbf{a} \cdot \mathbf{c}$

4. $(c\mathbf{a}) \cdot \mathbf{b} = c(\mathbf{a} \cdot \mathbf{b}) = \mathbf{a} \cdot (c\mathbf{b})$

5. $\mathbf{0} \cdot \mathbf{a} = 0$

These properties are easily proved using Definition 1. For instance, here are the proofs of Properties 1 and 3:

1. $\mathbf{a} \cdot \mathbf{a} = a_1^2 + a_2^2 + a_3^2 = |\mathbf{a}|^2$

3.
$$\begin{aligned} \mathbf{a} \cdot (\mathbf{b} + \mathbf{c}) &= [a_1, a_2, a_3] \cdot [b_1 + c_1, b_2 + c_2, b_3 + c_3] \\ &= a_1(b_1 + c_1) + a_2(b_2 + c_2) + a_3(b_3 + c_3) \\ &= a_1b_1 + a_1c_1 + a_2b_2 + a_2c_2 + a_3b_3 + a_3c_3 \\ &= (a_1b_1 + a_2b_2 + a_3b_3) + (a_1c_1 + a_2c_2 + a_3c_3) \\ &= \mathbf{a} \cdot \mathbf{b} + \mathbf{a} \cdot \mathbf{c} \end{aligned}$$

The proofs of the remaining properties are left as exercises. ■

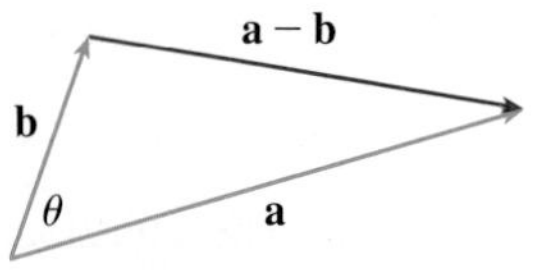

FIGURE 1

The dot product $\mathbf{a} \cdot \mathbf{b}$ can be given a geometric interpretation in terms of the angle θ between $\mathbf{a}$ and $\mathbf{b}$. This angle is defined to be the angle between the representations of $\mathbf{a}$ and $\mathbf{b}$ that start at the origin, where $0 \leqslant \theta \leqslant \pi$ (see Figure 1). From the Law of Cosines for the two vectors we have:

(3) $$|\mathbf{a} - \mathbf{b}|^2 = |\mathbf{a}|^2 + |\mathbf{b}|^2 - 2|\mathbf{a}||\mathbf{b}| \cos \theta$$

Using Properties 1, 2, and 3 of the dot product, we can rewrite the left side of this equation as follows:

$$\begin{aligned} |\mathbf{a} - \mathbf{b}|^2 &= (\mathbf{a} - \mathbf{b}) \cdot (\mathbf{a} - \mathbf{b}) \\ &= \mathbf{a} \cdot \mathbf{a} - \mathbf{a} \cdot \mathbf{b} - \mathbf{b} \cdot \mathbf{a} + \mathbf{b} \cdot \mathbf{b} \\ &= |\mathbf{a}|^2 - 2\mathbf{a} \cdot \mathbf{b} + |\mathbf{b}|^2 \end{aligned}$$

Therefore Equation 3 gives

$$|\mathbf{a}|^2 - 2\mathbf{a} \cdot \mathbf{b} + |\mathbf{b}|^2 = |\mathbf{a}|^2 + |\mathbf{b}|^2 - 2|\mathbf{a}||\mathbf{b}| \cos \theta$$

$$-2\mathbf{a} \cdot \mathbf{b} = -2|\mathbf{a}||\mathbf{b}| \cos \theta$$

or

> **(4) An Alternative Formula for the Dot Product**
>
> $$\mathbf{a} \cdot \mathbf{b} = |\mathbf{a}||\mathbf{b}| \cos \theta$$
>
> where θ is the angle between $\mathbf{a}$ and $\mathbf{b}$ ($0 \leqslant \theta \leqslant \pi$). ($\theta$ is the smaller angle between the two vectors when drawn from the same initial point.)

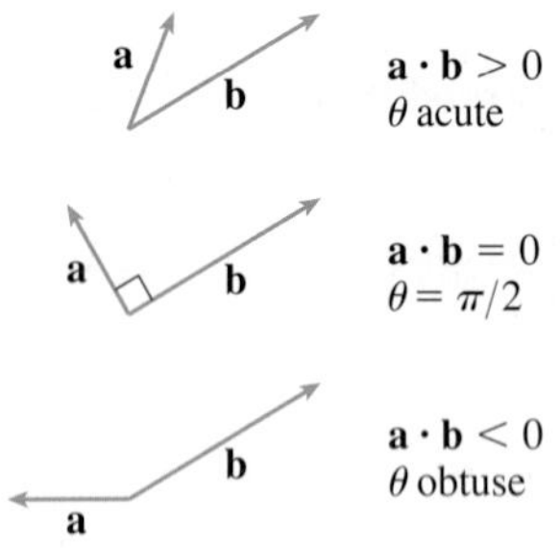

FIGURE 2

TEC Visual 8.3A shows an animation of Figure 2.

Notice that, if the dot product of two nonzero vectors is zero, then $\cos \theta = 0$. Thus $\theta = \pi/2$ and the two vectors are perpendicular (sometimes called **orthogonal**; see Figure 2). Also, because $\cos \theta > 0$ if $0 \leqslant \theta < \pi/2$ and $\cos \theta < 0$ if $\pi/2 < \theta \leqslant \pi$, we see that $\mathbf{a} \cdot \mathbf{b}$ is positive for $\theta < \pi/2$ and negative for $\theta > \pi/2$. We can therefore think of $\mathbf{a} \cdot \mathbf{b}$ as measuring the extent to which $\mathbf{a}$ and $\mathbf{b}$ point in the same direction. The dot product $\mathbf{a} \cdot \mathbf{b}$ is positive if $\mathbf{a}$ and $\mathbf{b}$ point in the same general direction, 0 if they are perpendicular, and negative if they point in generally opposite directions (see Figure 2). In the extreme case where $\mathbf{a}$ and $\mathbf{b}$ point in exactly the same direction we have $\theta = 0$, so $\cos \theta = 1$ and

$$\mathbf{a} \cdot \mathbf{b} = |\mathbf{a}||\mathbf{b}|$$

If $\mathbf{a}$ and $\mathbf{b}$ point in exactly opposite directions, then $\theta = \pi$ and so $\cos \theta = -1$ and $\mathbf{a} \cdot \mathbf{b} = -|\mathbf{a}||\mathbf{b}|$.

EXAMPLE 2 | Show that $[2, 2, -1]$ is perpendicular to $[5, -4, 2]$.

SOLUTION Since

$$[2, 2, -1] \cdot [5, -4, 2] = 2(5) + 2(-4) + (-1)(2) = 0$$

these vectors are perpendicular. ■

EXAMPLE 3 | Find the angle between the vectors $\mathbf{a} = [2, 2, -1]$ and $\mathbf{b} = [5, -3, 2]$.

SOLUTION Let θ be the required angle. Since

$$|\mathbf{a}| = \sqrt{2^2 + 2^2 + (-1)^2} - 3 \qquad \text{and} \qquad |\mathbf{b}| = \sqrt{5^2 + (-3)^2 + 2^2} = \sqrt{38}$$

and since

$$\mathbf{a} \cdot \mathbf{b} = 2(5) + 2(-3) + (-1)(2) = 2$$

we obtain

$$\cos\theta = \frac{\mathbf{a} \cdot \mathbf{b}}{|\mathbf{a}||\mathbf{b}|} = \frac{2}{3\sqrt{38}}$$

So the angle between **a** and **b** is

$$\theta = \cos^{-1}\left(\frac{2}{3\sqrt{38}}\right) \approx 1.46 \quad (\text{or } 84°)$$ ■

We now have techniques for comparing vectors both in terms of their magnitudes (in Section 8.2) and in terms of their directions (the dot product). Both can provide important information in a variety of biological contexts.

EXAMPLE 4 | BB Using the dot product for biological discovery[1]

Recall from Example 8.1.2 that a genome expression profile gives the level of expression of each of a collection of genes. Thus, it can be represented as an n-dimensional vector, where n is the number of genes assayed. For some cells a large number of such expression profiles have been characterized in response to different, known biochemical compounds. These can then be used to discover the mode of action of new biochemical compounds by quantifying the genome expression profile induced by the new compound and determining which known profile it most closely resembles. The new compound likely affects cell function through a mechanism similar to that of the best-matching, known biochemical compound.

Consider expression profiles **a** and **b** that are induced by two known biochemical compounds. Suppose that four genes are assayed and the expression profiles are given by the vectors $\mathbf{a} = [2, 5, 0, 1]$ and $\mathbf{b} = [1, 2, 4, 3]$, where the expression levels are dimensionless numbers. Suppose further that a new biochemical compound induces an expression profile given by the vector $\mathbf{n} = [1, 0, 5, 2]$.

(a) Which known expression profile vector points in a direction most similar to that of **n**: profile **a** or profile **b**?

(b) What is the angle between profile **n** and the profile identified in part (a)?

SOLUTION The magnitudes of the profiles are given by

$$|\mathbf{a}| = \sqrt{2^2 + 5^2 + 0^2 + 1^2} = \sqrt{30}$$

$$|\mathbf{b}| = \sqrt{1^2 + 2^2 + 4^2 + 3^2} = \sqrt{30}$$

$$|\mathbf{n}| = \sqrt{1^2 + 0^2 + 5^2 + 2^2} = \sqrt{30}$$

---

1. F. Kuruvilla et al., "Vector Algebra in the Analysis of Genome-Wide Expression Data," *Genome Biology* 3 (2002): research0011.1–11.

Because all vectors have the same magnitude, $\sqrt{30}$, the dot product of **n** with either **a** or **b** will have a maximum value of 30 if they point in the same direction and a minimum value of -30 if they point in opposite directions.

(a) Calculating the dot products gives

$$\mathbf{n} \cdot \mathbf{a} = [1, 0, 5, 2] \cdot [2, 5, 0, 1] = 4$$

$$\mathbf{n} \cdot \mathbf{b} = [1, 0, 5, 2] \cdot [1, 2, 4, 3] = 27$$

Therefore **n** points in a direction most similar to **b**. We conclude that the new compound likely affects the cell through a mechanism similar to that of the known biochemical compound whose profile is **b**.

(b) The angle between **n** and **b** is given by $\cos^{-1}(27/30) \approx 0.45$ (or 25.8°). ■

The dot product also enables us to write equations of planes. A plane is determined by a point $P_0(x_0, y_0, z_0)$ in the plane and a vector $\mathbf{n} = [a, b, c]$ that is orthogonal to the plane. This orthogonal vector **n** is called a **normal vector**. If $P(x, y, z)$ is any point in the plane, then the vector $[x - x_0, y - y_0, z - z_0]$ is perpendicular to **n** and so

$$[a, b, c] \cdot [x - x_0, y - y_0, z - z_0] = 0$$

This gives us the following equation for the plane.

> **(5)** An equation of the plane that passes through the point $P_0(x_0, y_0, z_0)$ and is perpendicular to the vector $[a, b, c]$ is
>
> $$a(x - x_0) + b(y - y_0) + c(z - z_0) = 0$$

EXAMPLE 5 | Find an equation of the plane through the point $(2, 4, -1)$ with normal vector $\mathbf{n} = [2, 3, 4]$.

SOLUTION Putting $a = 2$, $b = 3$, $c = 4$, $x_0 = 2$, $y_0 = 4$, and $z_0 = -1$ in Equation 5, we see that an equation of the plane is

$$2(x - 2) + 3(y - 4) + 4(z + 1) = 0$$

or

$$2x + 3y + 4z = 12$$

■

■ Projections

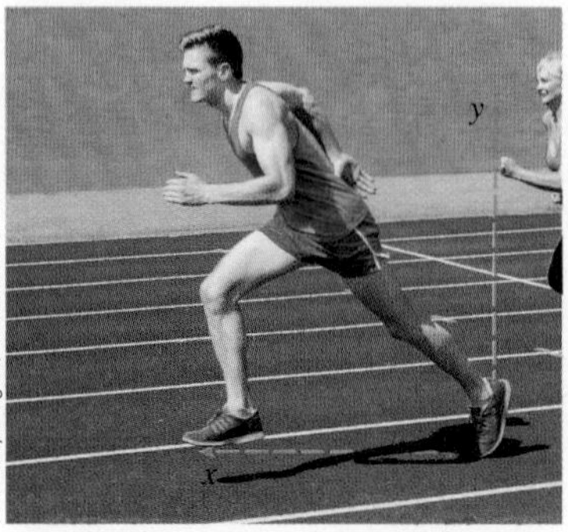

FIGURE 3

In addition to providing a means of comparing the direction of vectors, the dot product is useful for other analyses. Consider the force exerted by the foot of a runner as he pushes off the ground. The force propels him both forward and upward as shown in Figure 3. This force is a vector because it has both magnitude and direction. But what component of this force acts in the forward direction? Put another way, can we construct a horizontal vector that corresponds to the same forward force? From Figure 3 we can see that such a vector will have an x-component equal to the x-component of the original force vector.

More generally, Figure 4 shows representations $\overrightarrow{PQ}$ and $\overrightarrow{PR}$ of two arbitrary vectors **a** and **b** with the same initial point P. If S is the base of the perpendicular from R to the line containing $\overrightarrow{PQ}$, then the vector with representation $\overrightarrow{PS}$ is called the **vector projection** of **b** onto **a** and is denoted by $\text{proj}_{\mathbf{a}}\, \mathbf{b}$. (You can think of it as a shadow of **b**).

The scalar projection of **b** onto **a** (also called the **component of b along a**) is defined to be the signed magnitude of the vector projection, which is the number $|\mathbf{b}| \cos \theta$, where

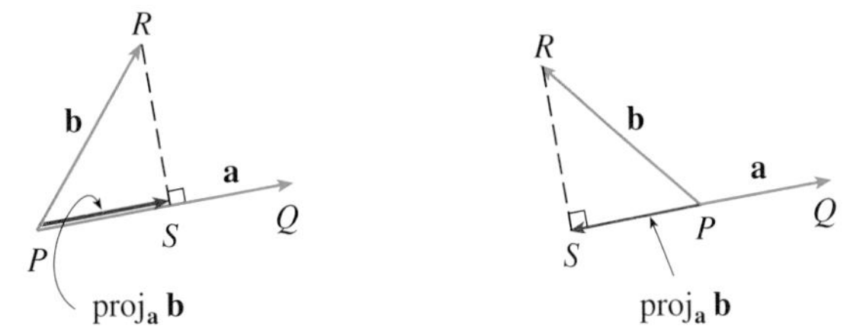

FIGURE 4
Vector projections

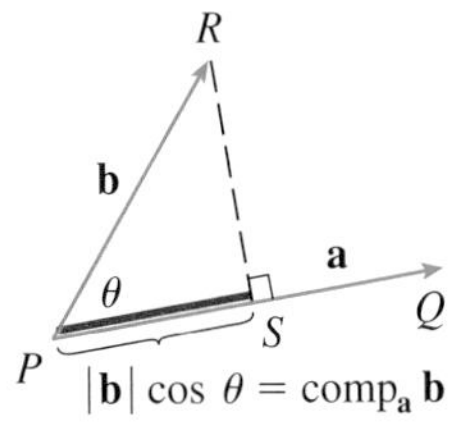

FIGURE 5
Scalar projection

θ is the angle between **a** and **b**. (See Figure 5.) This is denoted by $\text{comp}_\mathbf{a}\,\mathbf{b}$. Observe that it is negative if $\pi/2 < \theta \leqslant \pi$.

The equation

$$\mathbf{a}\cdot\mathbf{b} = |\mathbf{a}||\mathbf{b}|\cos\theta = |\mathbf{a}|(|\mathbf{b}|\cos\theta)$$

shows that the dot product of **a** and **b** can be interpreted as the length of **a** times the scalar projection of **b** onto **a**. Since

$$|\mathbf{b}|\cos\theta = \frac{\mathbf{a}\cdot\mathbf{b}}{|\mathbf{a}|} = \frac{\mathbf{a}}{|\mathbf{a}|}\cdot\mathbf{b}$$

the component of **b** along **a** can be computed by taking the dot product of **b** with the unit vector in the direction of **a**. We summarize these ideas as follows.

Scalar projection of **b** onto **a**: $\text{comp}_\mathbf{a}\,\mathbf{b} = \dfrac{\mathbf{a}\cdot\mathbf{b}}{|\mathbf{a}|}$

Vector projection of **b** onto **a**: $\text{proj}_\mathbf{a}\,\mathbf{b} = \left(\dfrac{\mathbf{a}\cdot\mathbf{b}}{|\mathbf{a}|}\right)\dfrac{\mathbf{a}}{|\mathbf{a}|} = \dfrac{\mathbf{a}\cdot\mathbf{b}}{|\mathbf{a}|^2}\,\mathbf{a}$

EXAMPLE 6 | Find the scalar projection and vector projection of $\mathbf{b} = [1, 1, 2]$ onto $\mathbf{a} = [-2, 3, 1]$.

SOLUTION Since $|\mathbf{a}| = \sqrt{(-2)^2 + 3^2 + 1^2} = \sqrt{14}$, the scalar projection of **b** onto **a** is

$$\text{comp}_\mathbf{a}\,\mathbf{b} = \frac{\mathbf{a}\cdot\mathbf{b}}{|\mathbf{a}|} = \frac{(-2)(1) + 3(1) + 1(2)}{\sqrt{14}} = \frac{3}{\sqrt{14}}$$

The vector projection is this scalar projection times the unit vector in the direction of **a**:

$$\text{proj}_\mathbf{a}\,\mathbf{b} = \frac{3}{\sqrt{14}}\frac{\mathbf{a}}{|\mathbf{a}|} = \frac{3}{14}\mathbf{a} = \left[-\frac{3}{7}, \frac{9}{14}, \frac{3}{14}\right]$$ ■

EXAMPLE 7 | BB **Vectorcardiography and Einthoven's triangle** As the heart beats, it generates differences in electrical potential (that is, voltage) across the body. Cardiologists view the electrical potential generated by the heart at any point in time during a heartbeat as a vector. This vector points in the direction of the greatest potential, and its magnitude represents the voltage in this direction. Abnormalities in either the magnitude or direction can be used to diagnose cardiac problems.

In Chapter 1 we saw an example of measurements of the voltage between two points on a patient's body. Most electrocardiographs actually take readings from among at

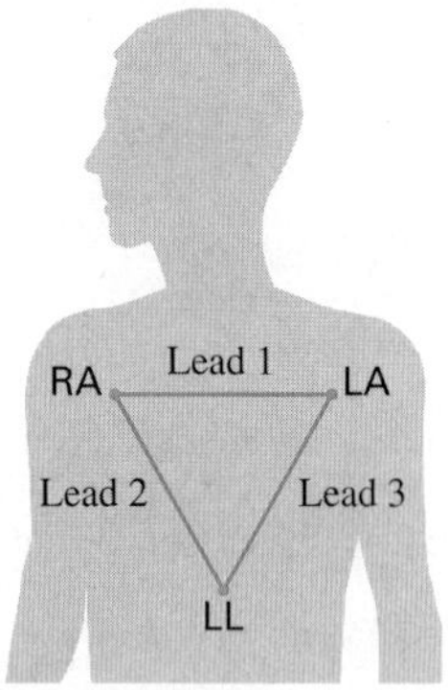

FIGURE 6
Einthoven's triangle

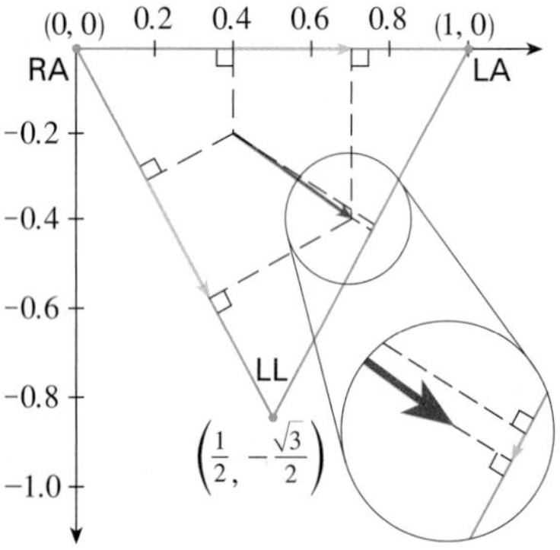

FIGURE 7
Red arrow is the heart voltage vector. Green arrows are the vector projections of the heart voltage vector onto the sides of Einthoven's triangle. The signed magnitude of each vector projection (that is, it's scalar projection) is measured by the corresponding lead.

least three points on the body—the right arm (RA), the left arm (LA), and the left leg (LL). These points make up a triangle known as Einthoven's triangle, as shown in Figure 6. The voltage between all pairs of vertices of this triangle is measured continuously, with each side of the triangle being referred to as a *lead*. For example, the measured voltage between RA and LA at any point in time gives the voltage in the direction of a vector corresponding to the top edge of the triangle at that time (labeled lead 1 in Figure 6).

To capture this idea mathematically, Einthoven's triangle is typically viewed as an equilateral triangle with unit sides, and with vertices having the coordinates $(0, 0)$, $(1, 0)$, and $(1/2, -\sqrt{3}/2)$ as displayed in Figure 7. At any point in time, each of the three leads measures the voltage in the direction of the vector corresponding to that side of the triangle.

Now consider a heart voltage vector at some point during a heartbeat. We can't measure this vector directly, but we can use the measurements from the three leads to infer its magnitude and direction. In particular, the measurement from lead 1 represents the scalar projection of the heart voltage vector onto the edge of the triangle between RA and LA (see Figure 7). Likewise, the measurement from lead 2 represents the scalar projection of the heart voltage vector onto the edge of the triangle between RA and LL, and so on. Thus the voltages measured in the three leads at any time gives us information about the magnitude and direction of the heart voltage vector at that time.

Suppose that heart voltage vector at some time during the heartbeat points from location $(0.4, -0.2)$ to location $(0.7, -0.4)$ in the coordinate system containing Einthoven's triangle shown in Figure 7. Assume voltage is measured in mV.

(a) What is the voltage of the heart vector (that is, its magnitude) at this time?

(b) What will be the electrocardiograph reading for each of the three leads at this time?

SOLUTION

(a) From result 8.2.1, a representation of the heart voltage vector **h** is calculated as

$$\mathbf{h} = [0.7 - 0.4, -0.4 - (-0.2)] = [0.3, -0.2]$$

The magnitude of the heart vector is therefore

$$|\mathbf{h}| = \sqrt{0.3^2 + (-0.2)^2} \approx 0.36 \text{ mV}$$

(b) The three leads correspond to the three sides of Einthoven's triangle in Figure 7. From result 8.2.1, a vector representation of each side is

$$\mathbf{l}_1 = [1 - 0, 0 - 0] = [1, 0] \quad \text{for RA to LA}$$

$$\mathbf{l}_2 = \left[\frac{1}{2} - 0, -\frac{\sqrt{3}}{2} - 0\right] = \left[\frac{1}{2}, -\frac{\sqrt{3}}{2}\right] \quad \text{for RA to LL}$$

$$\mathbf{l}_3 = \left[\frac{1}{2} - 1, -\frac{\sqrt{3}}{2} - 0\right] = \left[-\frac{1}{2}, -\frac{\sqrt{3}}{2}\right] \quad \text{for LA to LL}$$

Notice that Einthoven's triangle is defined to have sides of unit length. We verify this as follows:

$$|\mathbf{l}_1| = \sqrt{1^2 + 0^2} = 1 \qquad |\mathbf{l}_2| = \sqrt{\left(\tfrac{1}{2}\right)^2 + \left(-\sqrt{3}/2\right)^2} = 1$$

$$|\mathbf{l}_3| = \sqrt{\left(-\tfrac{1}{2}\right)^2 + \left(-\sqrt{3}/2\right)^2} = 1$$

The electrocardiograph reading from each lead is the scalar projection of the heart voltage vector on that lead. Therefore the reading in lead 1 will be $\mathbf{l}_1 \cdot \mathbf{h}/|\mathbf{l}_1| = 0.3$. The reading in lead 2 will be $\mathbf{l}_2 \cdot \mathbf{h}/|\mathbf{l}_2| \approx 0.32$. Finally, the reading in lead 3 will be $\mathbf{l}_3 \cdot \mathbf{h}/|\mathbf{l}_3| \approx 0.02$. As expected from Figure 7, real readings from leads 1 and 2 are normally relatively large, while the reading from lead 3 is normally nearly zero because the heart voltage vector is usually almost orthogonal to this lead. ■

EXERCISES 8.3

Note: Vector notations using square brackets [1, 1] and angle brackets ⟨1, 1⟩ are equivalent. You may see either notation in your online homework.

1. Which of the following expressions are meaningful? Which are meaningless? Explain.
(a) $(\mathbf{a} \cdot \mathbf{b}) \cdot \mathbf{c}$ (b) $(\mathbf{a} \cdot \mathbf{b})\mathbf{c}$
(c) $|\mathbf{a}|(\mathbf{b} \cdot \mathbf{c})$ (d) $\mathbf{a} \cdot (\mathbf{b} + \mathbf{c})$
(e) $\mathbf{a} \cdot \mathbf{b} + \mathbf{c}$ (f) $|\mathbf{a}| \cdot (\mathbf{b} + \mathbf{c})$

2–10 Find $\mathbf{a} \cdot \mathbf{b}$. For Exercises 9–10, refer to the notation introduced in Exercises 8.2.30–32.

2. $|\mathbf{a}| = 3$, $|\mathbf{b}| = \sqrt{6}$, the angle between $\mathbf{a}$ and $\mathbf{b}$ is 45°

3. $|\mathbf{a}| = 6$, $|\mathbf{b}| = 5$, the angle between $\mathbf{a}$ and $\mathbf{b}$ is $2\pi/3$

4. $\mathbf{a} = [-2, 3]$, $\mathbf{b} = [0.7, 1.2]$

5. $\mathbf{a} = \left[-2, \frac{1}{3}\right]$, $\mathbf{b} = [-5, 12]$

6. $\mathbf{a} = [6, -2, 3]$, $\mathbf{b} = [2, 5, -1]$

7. $\mathbf{a} = \left[4, 1, \frac{1}{4}\right]$, $\mathbf{b} = [6, -3, -8]$

8. $\mathbf{a} = [p, -p, 2p]$, $\mathbf{b} = [2q, q, -q]$

9. $\mathbf{a} = 2\mathbf{i} + \mathbf{j}$, $\mathbf{b} = \mathbf{i} - \mathbf{j} + \mathbf{k}$

10. $\mathbf{a} = 3\mathbf{i} + 2\mathbf{j} - \mathbf{k}$, $\mathbf{b} = 4\mathbf{i} + 5\mathbf{k}$

11–12 If $\mathbf{u}$ is a unit vector, find $\mathbf{u} \cdot \mathbf{v}$ and $\mathbf{u} \cdot \mathbf{w}$.

11.

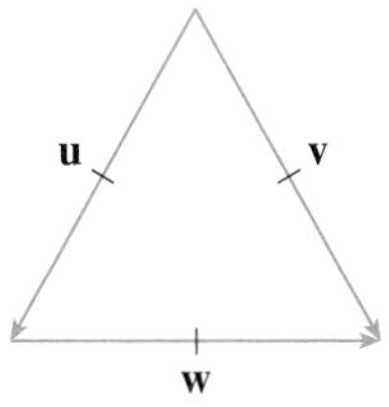

12.

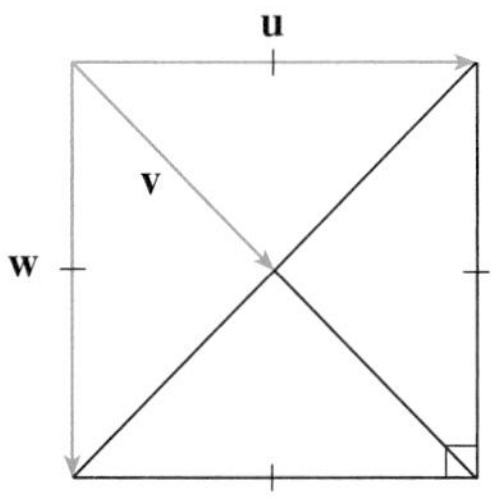

13. Exercises 8.2.30–32 introduced the standard basis vectors $\mathbf{i} = [1, 0, 0]$, $\mathbf{j} = [0, 1, 0]$, and $\mathbf{k} = [0, 0, 1]$.
(a) Show that $\mathbf{i} \cdot \mathbf{j} = \mathbf{j} \cdot \mathbf{k} = \mathbf{k} \cdot \mathbf{i} = 0$.
(b) Show that $\mathbf{i} \cdot \mathbf{i} = \mathbf{j} \cdot \mathbf{j} = \mathbf{k} \cdot \mathbf{k} = 1$.

14. Population dynamics Suppose a population of fish contains a females that are 1 year old, b females between 2 and 4 years old, and c females more than 4 years old. During the summer each one-year-old female produces 15 offspring, each female between 2 and 4 years old produces 35 offspring, and each female more than 4 years old produces 20 offspring. If $\mathbf{a} = [a, b, c]$ and $\mathbf{p} = [15, 35, 20]$, what is the meaning of the dot product $\mathbf{a} \cdot \mathbf{p}$?

15–18 Find the angle between the vectors. (First find an exact expression and then approximate to the nearest degree.)

15. $\mathbf{a} = [-8, 6]$, $\mathbf{b} = \left[\sqrt{7}, 3\right]$

16. $\mathbf{a} = \left[\sqrt{3}, 1\right]$, $\mathbf{b} = [0, 5]$

17. $\mathbf{a} = [0, 1, 1]$, $\mathbf{b} = [1, 2, -3]$

18. $\mathbf{a} = [1, 2, -2]$, $\mathbf{b} = [4, 0, -3]$

19–20 Find, correct to the nearest degree, the three angles of the triangle with the given vertices.

19. $A(1, 0)$, $B(3, 6)$, $C(-1, 4)$

20. $D(0, 1, 1)$, $E(-2, 4, 3)$, $F(1, 2, -1)$

21–22 Determine whether the given vectors are orthogonal, parallel, or neither.

21. (a) $\mathbf{a} = [-5, 3, 7]$, $\mathbf{b} = [6, -8, 2]$
(b) $\mathbf{a} = [4, 6]$, $\mathbf{b} = [-3, 2]$
(c) $\mathbf{a} = [-1, 2, 5]$, $\mathbf{b} = [3, 4, -1]$
(d) $\mathbf{a} = [2, 6, -4]$, $\mathbf{b} = [-3, -9, 6]$

22. (a) $\mathbf{u} = [-3, 9, 6]$, $\mathbf{v} = [4, -12, -8]$
(b) $\mathbf{u} = [1, -1, 2]$, $\mathbf{v} = [2, -1, 1]$
(c) $\mathbf{u} = [a, b, c]$, $\mathbf{v} = [-b, a, 0]$

23. Use vectors to decide whether the triangle with vertices $P(1, -3, -2)$, $Q(2, 0, -4)$, and $R(6, -2, -5)$ is right-angled.

24. For what values of b are the vectors $[-6, b, 2]$ and $[b, b^2, b]$ orthogonal?

25. Find a unit vector that is orthogonal to both [1, 1, 0] and [1, 0, 1].

26. Find two unit vectors that make an angle of 60° with $\mathbf{v} = [3, 4]$.

27. Find an equation of the plane that passes through the origin and is perpendicular to the vector $[1, -2, 5]$.

28. Find an equation of the plane through the point $\left(-1, \frac{1}{2}, 3\right)$ with normal vector [1, 4, 1].

29–32 Find the scalar and vector projections of $\mathbf{b}$ onto $\mathbf{a}$.

29. $\mathbf{a} = [3, -4]$, $\mathbf{b} = [5, 0]$

30. $\mathbf{a} = [1, 2]$, $\mathbf{b} = [-4, 1]$

31. $\mathbf{a} = [2, -1, 4]$, $\mathbf{b} = \left[0, 1, \frac{1}{2}\right]$

32. $\mathbf{a} = [1, 1, 1]$, $\mathbf{b} = [1, -1, 1]$

---

33. Show that the vector $\text{orth}_{\mathbf{a}}\mathbf{b} = \mathbf{b} - \text{proj}_{\mathbf{a}}\mathbf{b}$ is orthogonal to $\mathbf{a}$. (It is called an **orthogonal projection** of $\mathbf{b}$.)

34. For the vectors in Exercise 30, find $\text{orth}_{\mathbf{a}}\mathbf{b}$ and illustrate by drawing the vectors $\mathbf{a}$, $\mathbf{b}$, $\text{proj}_{\mathbf{a}}\mathbf{b}$, and $\text{orth}_{\mathbf{a}}\mathbf{b}$.

35. If $\mathbf{a} = [3, 0, -1]$, find a vector $\mathbf{b}$ such that $\text{comp}_{\mathbf{a}}\mathbf{b} = 2$.

36. Suppose that $\mathbf{a}$ and $\mathbf{b}$ are nonzero vectors.
(a) Under what circumstances is $\text{comp}_{\mathbf{a}}\mathbf{b} = \text{comp}_{\mathbf{b}}\mathbf{a}$?
(b) Under what circumstances is $\text{proj}_{\mathbf{a}}\mathbf{b} = \text{proj}_{\mathbf{b}}\mathbf{a}$?

37. Antigenic evolution Influenza viruses from North America and Asia are sampled in two successive years and plotted in two-dimensional antigenic space. The coordinates of each are (i) North American viruses: (2, 1) for 2013 and (4, 3) for 2014, and (ii) Asian viruses: (4, 18) for 2013 and (5, 17) for 2014.
(a) How does the magnitude of antigenic change compare between the two regions?
(b) How does the direction of antigenic change compare between the two regions?

38. Community ecology Community ecologists study the factors that determine the abundance of different species. The abundance of three algae species (in mg/mL) is quantified in two different lakes before and after an unusually hot summer. The coordinates of these four samples in $\mathbb{R}^3$ are (i) Lake A: (97, 84, 43) and (100, 80, 50) before and after summer, respectively, and (ii) Lake B: (23, 59, 22) and (20, 63, 15) before and after summer, respectively.
(a) How does the magnitude of the change in algal community compare between the two lakes?
(b) How does the direction of the change in algal community compare between the two lakes?

39. Genome expression profiles and drug design Researchers are trying to design a drug that alters the expression level of three genes in a specific way. The desired change in expression, written as a vector in V_3, is $[3, 9, -5]$, where all measurements are dimensionless. Suppose two potential drugs are being studied, and the expression profiles that they induce are (A) $[2, 4, 1]$ and (B) $[-2, 3, -5]$.
(a) What is the magnitude of the desired change in expression?
(b) Using scalar projections, determine the fraction of this desired change that is induced by each drug.
(c) Which drug is closest to having the desired effect?

40. Vectorcardiography In each of the following questions the heart voltage vector of a patient with different pathologies is given. Determine the electrocardiograph reading that would result in each of the three leads of Einthoven's triangle, and draw a sketch like that in Figure 7. Recall from Example 7 that vector representations of the sides of Einthoven's triangle are

$$\mathbf{l}_1 = [1, 0] \qquad \mathbf{l}_2 = [1/2, -\sqrt{3}/2]$$
$$\mathbf{l}_3 = [-1/2, -\sqrt{3}/2]$$

(a) Left anterior hemiblock: $\mathbf{h} = [0.3, 0.2]$
(b) Left posterior hemiblock: $\mathbf{h} = [-0.3, -0.2]$
(c) Apical ischemia: $\mathbf{h} = [-0.3, 0.2]$
(d) Chronic obstructive pulmonary disease: $\mathbf{h} = [0.1, -0.0667]$

41. Find the angle between a diagonal of a cube and one of its edges.

42. Find the angle between a diagonal of a cube and a diagonal of one of its faces.

43. If $\mathbf{r} = [x, y, z]$, $\mathbf{a} = [a_1, a_2, a_3]$, and $\mathbf{b} = [b_1, b_2, b_3]$, show that the vector equation $(\mathbf{r} - \mathbf{a}) \cdot (\mathbf{r} - \mathbf{b}) = 0$ represents a sphere, and find its center and radius.

44. If $\mathbf{c} = |\mathbf{a}|\mathbf{b} + |\mathbf{b}|\mathbf{a}$, where $\mathbf{a}$, $\mathbf{b}$, and $\mathbf{c}$ are all nonzero vectors, show that $\mathbf{c}$ bisects the angle between $\mathbf{a}$ and $\mathbf{b}$.

45. Prove Property 4 of the dot product. Use either the definition of a dot product (considering the cases $c > 0$, $c = 0$, and $c < 0$ separately) or the component form.

46. Suppose that all sides of a quadrilateral are equal in length and opposite sides are parallel. Use vector methods to show that the diagonals are perpendicular.

47. Prove the Cauchy-Schwarz Inequality:

$$|\mathbf{a} \cdot \mathbf{b}| \leqslant |\mathbf{a}||\mathbf{b}|$$

48. The Triangle Inequality for vectors is

$$|\mathbf{a} + \mathbf{b}| \leqslant |\mathbf{a}| + |\mathbf{b}|$$

Use the Cauchy-Schwarz Inequality from Exercise 47 to prove the Triangle Inequality. [*Hint:* Use the fact that $|\mathbf{a} + \mathbf{b}|^2 = (\mathbf{a} + \mathbf{b}) \cdot (\mathbf{a} + \mathbf{b})$ and use Property 3 of the dot product.]

49. The Parallelogram Law states that

$$|\mathbf{a} + \mathbf{b}|^2 + |\mathbf{a} - \mathbf{b}|^2 = 2|\mathbf{a}|^2 + 2|\mathbf{b}|^2$$

(a) Give a geometric interpretation of the Parallelogram Law.
(b) Prove the Parallelogram Law. (See the hint in Exercise 48.)

50. Show that if $\mathbf{u} + \mathbf{v}$ and $\mathbf{u} - \mathbf{v}$ are orthogonal, then the vectors $\mathbf{u}$ and $\mathbf{v}$ must have the same length.

■ PROJECT Microarray Analysis of Genome Expression

BB

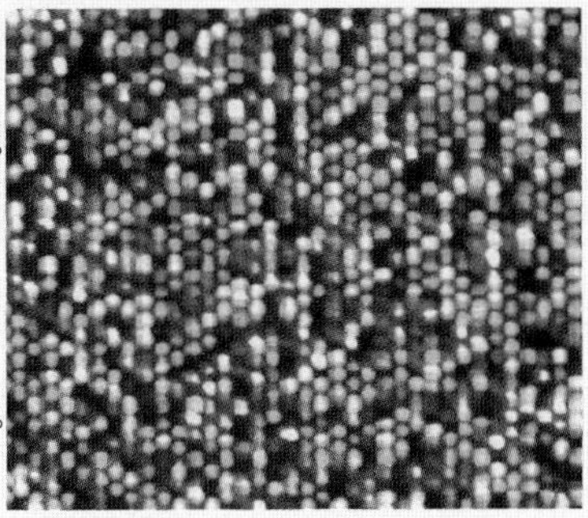

The Sanger Institute, Wellcome Images

FIGURE 1
A microarray

Microarrays provide a technique for quantifying how the expression of different genes in the genome changes in response to changes in conditions. Suppose we are interested in how a drug affects genome expression. Cells are maintained under a reference, or normal, condition as well as in the presence of the drug. All gene products generated under the reference condition are labeled with a green dye, and all gene products generated in the presence of the drug are labeled with a red dye. The next step is to take all known genes individually and adhere copies of each to different locations on a glass plate. The dyed extracts from cells are then washed over the plate, and each of the dyed gene products binds with its corresponding gene.

The result is a glass plate containing thousands of colored dots, one for each gene. The dots vary in color from green to yellow to red. A dot is green if the corresponding gene tends to be expressed less in the presence of the drug than in the reference condition because most of the extract binding to this gene on the plate will have come from reference cells. A dot is yellow if the corresponding gene is expressed equally in both cell types, and it is red if the gene tends to be expressed more in the presence of the drug than in the reference condition. The entire collection of colored dots on the plate is called a microarray. An example is shown in Figure 1.

To analyze microarray data the color intensity for each gene is translated into a numerical score in the interval $(-\infty, \infty)$, with negative values indicating green and positive values indicating red. For simplicity, let's suppose there are five genes on the microarray and we have four different potential drugs, giving the following data:

	Gene 1	Gene 2	Gene 3	Gene 4	Gene 5
Drug A	1	3	5	6	9
Drug B	2	6	9	12	19
Drug C	−1	6	7	0	3
Drug D	5	2	5	7	12

Suppose our goal is to determine the similarity of drugs A, B, C, and D in terms of the patterns of genome expression that they induce. To do so, we need a measure of similarity of any two profiles.

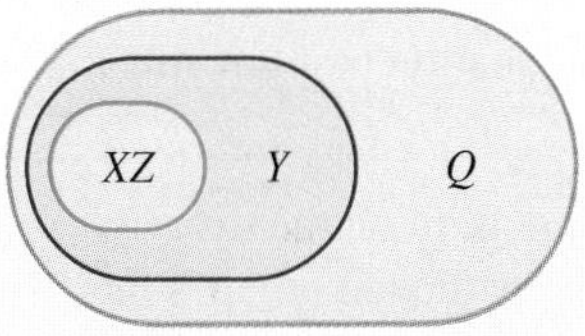

FIGURE 2
An example of hierarchical clustering. Drugs X and Z are most similar, followed by drug Y, and then drug Q.

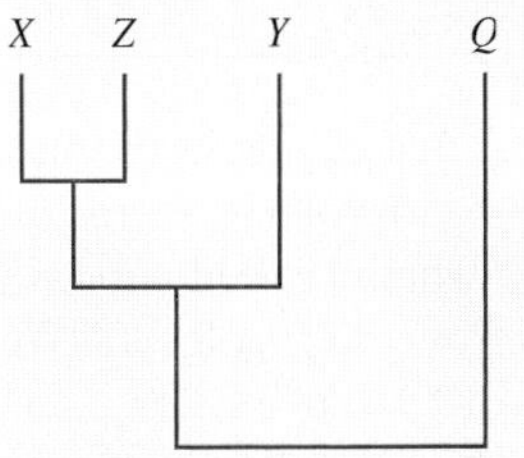

FIGURE 3
An example of a dendogram corresponding to the clusters in Figure 2

1. Treating the expression profile for each drug as a point in $\mathbb{R}^5$, calculate the distances between all pairs of profiles.

2. The next step is to group the drugs hierarchically according to their similarity. To do so, first take the two drugs that are closest and draw a circle around them. Then, take the drug that is next closest to either of these first two and draw a curve that encloses all three (see Figure 2). Proceed until all drugs are included, giving a diagram depicting hierarchical clusters of decreasing similarity as shown in Figure 2. It is also possible to depict the drug similarities using a *dendogram*, where the length of the path in the diagram between two drugs reflects the similarity between the drugs (see Figure 3). Construct both.

3. The other commonly used measure of similarity between profiles involves the dot product. Treating the expression profile for each drug as a vector in V_5, calculate the cosine of the angle between each of the pairs of profiles as a measure of similarity. Repeat the clustering analysis in Problem 2 for this measure.

Problems 1 and 3 provide different measures of the similarity between pairs of profiles, and both are commonly used in microarray analysis. The similarity measure in Problem 1 is sometimes referred to as the Euclidean distance and the similarity measure in Problem 3 is sometimes referred to as the Pearson correlation coefficient.

■ PROJECT Vaccine Escape

BB

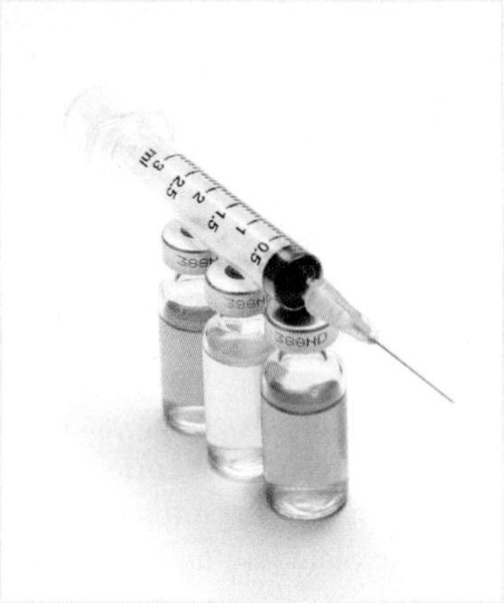

Vaccine escape refers to the loss of effectiveness of a vaccine due to the antigenic evolution of the pathogen. Suppose that a vaccine protects against all pathogens within a circle of radius 5 centered at the origin in two-dimensional antigenic space. Further, suppose that the vector **v** giving the instantaneous velocity of evolution in antigenic space is always of fixed magnitude k and angle θ from a line connecting the pathogen to the origin, with $0 < \theta < \pi/2$. (See Figure 1.)

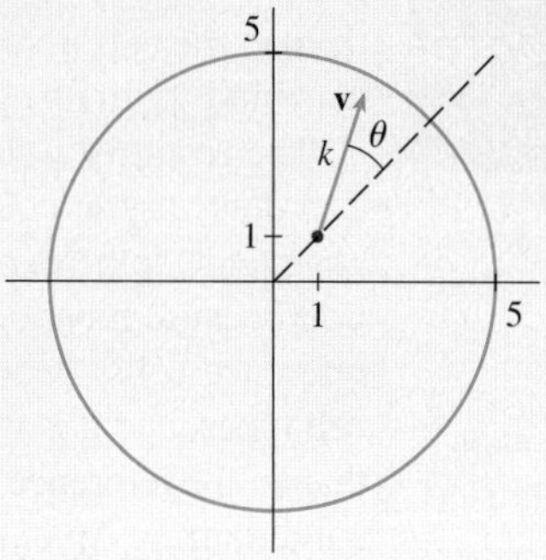

FIGURE 1

1. Suppose the pathogen is currently at location $(1, 1)$ in antigenic space. Using a vector projection, determine the vector that specifies the velocity at which the pathogen is evolving directly away from the origin.

2. Suppose the pathogen is currently at location $(1, 1)$ in antigenic space. Using a scalar projection, determine the speed at which the pathogen is evolving directly away from the origin.

3. In general, suppose the pathogen is at location (x, y). Using a scalar projection, determine the speed at which the pathogen is evolving directly away from the origin.

4. Use your result from Problem 3 to write a differential equation that governs the distance of the pathogen from the origin.

5. Integrate the differential equation from Problem 4 and determine the time at which vaccine escape occurs (that is, the time when the pathogen escapes the vaccine's circle of protection), assuming the distance to the origin at time $t = 0$ is $\sqrt{2}$.

8.4 Matrix Algebra

Many of the examples from Sections 8.2 and 8.3 involve vectors that change over time; for example, antigenic evolution of the influenza virus or the heart voltage vector. In Section 8.5 we will consider how to model such changes using what is called a matrix model. Before doing so, however, we first introduce some ideas from matrix algebra. We will return to biological applications in Section 8.5.

■ Matrix Notation

A **matrix** is a rectangular array of numbers. We use uppercase symbols to denote matrices. The **size** of a matrix is defined by the number of rows and columns it contains. For example, an $m \times n$ matrix has m rows and n columns. If A is an $m \times n$ matrix, then the ijth entry of A is the entry in the ith row and the jth column. We denote this entry by a_{ij}. For example, the matrix

(1)
$$A = \begin{bmatrix} 0 & 7 & 1 \\ 2 & 9 & 2 \end{bmatrix}$$

is a 2×3 matrix, and $a_{12} = 7$, $a_{21} = 2$, $a_{23} = 2$, and so forth.

A **square** matrix is one in which the number of rows is the same as the number of columns. We say that a square matrix is $n \times n$, or has size n, if it has n rows and columns.

The **transpose** of a matrix is obtained by interchanging its rows and columns and is denoted by a superscript T. For example, the transpose of the matrix in Equation 1, denoted by A^T, is

(2)
$$A^T = \begin{bmatrix} 0 & 2 \\ 7 & 9 \\ 1 & 2 \end{bmatrix}$$

If A is an $m \times n$ matrix, then A^T is an $n \times m$ matrix.

Vectors are often treated using the notation of matrices by introducing a distinction between those that are written as columns and those that are written as rows. All vectors we have seen so far have been written as **row vectors** in that their components are listed as a row. For example, we have been writing the vector with components x and y as $[x, y]$. Another possibility is to write this vector by placing its components in a column, giving the **column vector** $\begin{bmatrix} x \\ y \end{bmatrix}$. The row and column forms of a vector quantify the same thing but they are each used in different contexts in matrix algebra. We can view the row form of a vector as a $1 \times n$ matrix and the column form as an $n \times 1$ matrix. Furthermore, if $\mathbf{v}$ denotes a vector written in row form, then $\mathbf{v}^T$ is the same vector written in column form.

■ Matrix Addition and Scalar Multiplication

Only matrices of the same size can be added. If A and B are both $m \times n$ matrices with entries a_{ij} and b_{ij}, then $A + B$ is a new $m \times n$ matrix whose entries are $a_{ij} + b_{ij}$. Thus the matrix sum $A + B$ is calculated by adding the corresponding entries of each matrix.

EXAMPLE 1 | Evaluate the following sums, if possible.

(a) $M + N$, where $M = \begin{bmatrix} 2 & x & 9 \\ 4 & 5 & 6 \end{bmatrix}$ and $N = \begin{bmatrix} 92 & 6 & 2 \\ 15 & 3 & 1 \end{bmatrix}$

(b) $X + Y$, where $X = \begin{bmatrix} 5 & 3 \\ 7 & 13 \end{bmatrix}$ and $Y = \begin{bmatrix} 3 & 9 & 21 \\ 5 & 7 & 6 \end{bmatrix}$

SOLUTION

(a) Both matrices are 2×3 and therefore can be added. Adding the corresponding entries gives $M + N = \begin{bmatrix} 94 & x + 6 & 11 \\ 19 & 8 & 7 \end{bmatrix}$.

(b) Matrix X is 2×2 while Y is 2×3. Since they are not the same size, they cannot be added. ■

Scalar multiplication with matrices works just as it does with vectors. If A is an $m \times n$ matrix with entries a_{ij} and c is a scalar, then the product cA is an $m \times n$ matrix with entries ca_{ij}. In other words, the product cA is calculated by multiplying each entry of A by c.

Matrix subtraction can be defined through a combination of scalar multiplication and matrix addition. If A and B are both $m \times n$ matrices with entries a_{ij} and b_{ij}, then $A - B$ is calculated by first multiplying B by -1 and then adding this to A. Thus the difference $A - B$ is calculated by subtracting entry b_{ij} from a_{ij}. Again notice that matrix subtraction can be performed only with matrices of the same size.

Finally, two matrices A and B are said to be **equal** if $A - B = 0$ where 0 is an $m \times n$ matrix of zeros.

Properties of Matrix Addition If A, B, and C are $m \times n$ matrices and a and b are scalars, then

1. $A + B = B + A$ **2.** $A + (B + C) = (A + B) + C$

3. $A + 0 = A$ **4.** $A + (-A) = 0$

5. $a(A + B) = aA + aB$ **6.** $(a + b)A = aA + bA$

Properties 1–6 can be verified using the definitions of matrix addition and multiplication by a scalar.

■ Matrix Multiplication

Some matrices can also be multiplied with one another. Matrix multiplication can be viewed as an extension of the dot product of vectors. If we wish to calculate the matrix product AB, we view the matrix A as a collection of row vectors and B as a collection of column vectors. The ijth entry of the resulting product is then the dot product of the ith row of A with the jth column of B. (See Figure 1.) For example, if A is a 2×3 matrix and B is a 3×2 matrix, we have

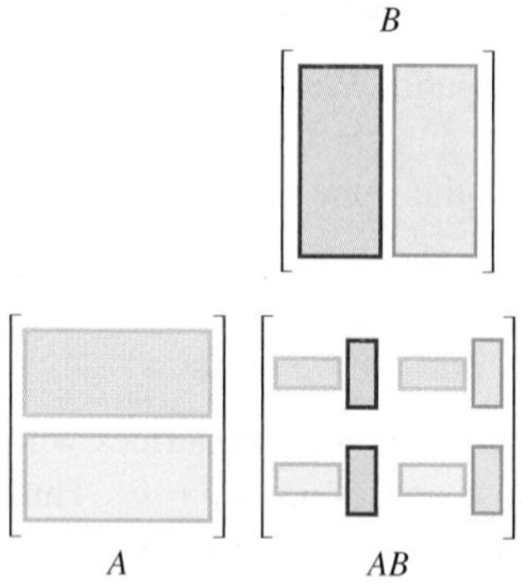

FIGURE 1

$$\begin{bmatrix} a_{11} & a_{12} & a_{13} \\ a_{21} & a_{22} & a_{23} \end{bmatrix} \begin{bmatrix} b_{11} & b_{12} \\ b_{21} & b_{22} \\ b_{31} & b_{32} \end{bmatrix} = \begin{bmatrix} a_{11}b_{11} + a_{12}b_{21} + a_{13}b_{31} & a_{11}b_{12} + a_{12}b_{22} + a_{13}b_{32} \\ a_{21}b_{11} + a_{22}b_{21} + a_{23}b_{31} & a_{21}b_{12} + a_{22}b_{22} + a_{23}b_{32} \end{bmatrix}$$

Note that the resulting matrix is 2×2. More generally, if A is an $m \times n$ matrix and B is an $n \times p$ matrix, then $C = AB$ is an $m \times p$ matrix whose ijth entry is found by taking the dot product of the ith row of A with the jth column of B, as shown in Figure 2.

$$\underset{A}{\begin{bmatrix} - & \mathbf{r}_1 & - \\ - & \mathbf{r}_2 & - \\ & \vdots & \\ - & \mathbf{r}_m & - \end{bmatrix}} \underset{B}{\begin{bmatrix} | & | & & | \\ \mathbf{c}_1 & \mathbf{c}_2 & \cdots & \mathbf{c}_p \\ | & | & & | \end{bmatrix}} = \underset{AB}{\begin{bmatrix} \mathbf{r}_1 \cdot \mathbf{c}_1 & \mathbf{r}_1 \cdot \mathbf{c}_2 & \cdots & \mathbf{r}_1 \cdot \mathbf{c}_p \\ \mathbf{r}_2 \cdot \mathbf{c}_1 & \mathbf{r}_2 \cdot \mathbf{c}_2 & \cdots & \mathbf{r}_2 \cdot \mathbf{c}_p \\ \vdots & \vdots & \ddots & \vdots \\ \mathbf{r}_m \cdot \mathbf{c}_1 & \mathbf{r}_m \cdot \mathbf{c}_2 & \cdots & \mathbf{r}_m \cdot \mathbf{c}_p \end{bmatrix}}$$

FIGURE 2
The ith row of A is indicated by $\mathbf{r}_i$.
The jth column of B is indicated by $\mathbf{c}_j$.

Matrix multiplication is not defined for matrices where the number of columns of the first matrix is different from the number of rows of the second matrix. A simple way to determine if a given matrix multiplication is defined is to write the size of the first matrix,

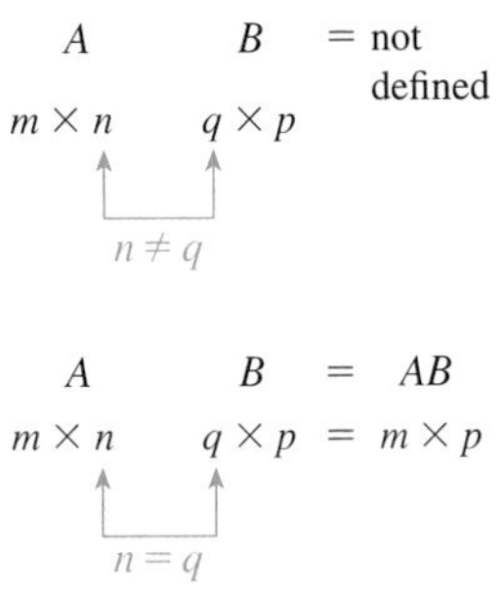

FIGURE 3

followed by the size of the second matrix (see Figure 3). If the two "inner" numbers are not the same, then the matrix multiplication is not defined. If they are the same, then the resulting matrix has size given by the two "outer" numbers as illustrated in Figure 3.

Matrix multiplication is summarized by the following rule.

> **Matrix Multiplication** If A is an $m \times n$ matrix and B is an $n \times p$ matrix, then their product $C = AB$ is an $m \times p$ matrix whose entries are given by
>
> $$c_{ij} = \sum_{k=1}^{n} a_{ik}b_{kj} = a_{i1}b_{1j} + a_{i2}b_{2j} + \cdots + a_{in}b_{nj}$$
>
> for $1 \leqslant i \leqslant m$ and $1 \leqslant j \leqslant p$.

EXAMPLE 2 | Determine each matrix product if it is defined.

$$A = \begin{bmatrix} 2 & 7 \\ 9 & -3 \end{bmatrix} \qquad B = \begin{bmatrix} 3 & -7 & 2 \\ 1 & 5 & 9 \end{bmatrix} \qquad C = \begin{bmatrix} 5 & -6 \\ 8 & 2 \end{bmatrix}$$

(a) AB (b) BA (c) AC (d) CA

SOLUTION

(a) Matrices A and B have sizes 2×2 and 2×3, respectively. From Figure 3, since $n = q = 2$, matrix multiplication is therefore defined and the resulting matrix is 2×3. Performing the calculation, we obtain

$$\begin{bmatrix} 2 & 7 \\ 9 & -3 \end{bmatrix}\begin{bmatrix} 3 & -7 & 2 \\ 1 & 5 & 9 \end{bmatrix} = \begin{bmatrix} 6 + 7 & -14 + 35 & 4 + 63 \\ 27 - 3 & -63 - 15 & 18 - 27 \end{bmatrix} = \begin{bmatrix} 13 & 21 & 67 \\ 24 & -78 & -9 \end{bmatrix}$$

(b) Matrices B and A have sizes 2×3 and 2×2, respectively. From Figure 3, since $n = 3$ and $q = 2$, $n \neq q$ and the product is not defined.

(c) Using the rule from Figure 3, we have $n = q = 2$. The resulting matrix is 2×2. Performing the calculation, we obtain

$$\begin{bmatrix} 2 & 7 \\ 9 & -3 \end{bmatrix}\begin{bmatrix} 5 & -6 \\ 8 & 2 \end{bmatrix} = \begin{bmatrix} 10 + 56 & -12 + 14 \\ 45 - 24 & -54 - 6 \end{bmatrix} = \begin{bmatrix} 66 & 2 \\ 21 & -60 \end{bmatrix}$$

(d) Using the rule from Figure 3, we have $n = q = 2$. The resulting matrix is 2×2 and, performing the calculation, we obtain

$$\begin{bmatrix} 5 & -6 \\ 8 & 2 \end{bmatrix}\begin{bmatrix} 2 & 7 \\ 9 & -3 \end{bmatrix}\begin{bmatrix} 10 - 54 & 35 + 18 \\ 16 + 18 & 56 - 6 \end{bmatrix} = \begin{bmatrix} -44 & 53 \\ 34 & 50 \end{bmatrix}$$

■

Multiplication of two quantities α and β is said to be **commutative** if $\alpha\beta = \beta\alpha$. Parts (c) and (d) of Example 2 illustrate the important fact that matrix multiplication is not, in general, commutative ($AC \neq CA$ in this example).

An $n \times n$ matrix D is called **diagonal** if all off-diagonal entries are zero; that is, $d_{ij} = 0$ for all $i \neq j$. Multiplication of a diagonal matrix with itself is especially easy.

EXAMPLE 3 | For an arbitrary 2 × 2 diagonal matrix D with entries d_{ii}, calculate the matrix DD.

SOLUTION Calculating the matrix product, we obtain

$$DD = \begin{bmatrix} d_{11} & 0 \\ 0 & d_{22} \end{bmatrix}\begin{bmatrix} d_{11} & 0 \\ 0 & d_{22} \end{bmatrix} = \begin{bmatrix} d_{11}^2 & 0 \\ 0 & d_{22}^2 \end{bmatrix}$$

Thus DD is a diagonal matrix with entries d_{ii}^2. ■

An alternative notation for the matrix product MM is M^2. Likewise, M^k represents the matrix M multiplied by itself k times. It is referred to as the kth power of M. Exercise 7 shows that the simple pattern for the second power of a diagonal matrix illustrated in Example 3 also holds for larger matrices as well as for higher matrix powers.

A diagonal matrix is called an **identity** matrix if all the entries on the diagonal are 1. Identity matrices are usually denoted by I and play the same role in matrix multiplication that the number 1 plays in regular multiplication.

EXAMPLE 4 | For an arbitrary 2 × 2 matrix A show that $AI = IA = A$.

SOLUTION Performing the required matrix products we obtain

$$AI = \begin{bmatrix} a_{11} & a_{12} \\ a_{21} & a_{22} \end{bmatrix}\begin{bmatrix} 1 & 0 \\ 0 & 1 \end{bmatrix} = \begin{bmatrix} a_{11} & a_{12} \\ a_{21} & a_{22} \end{bmatrix} = A$$

and

$$IA = \begin{bmatrix} 1 & 0 \\ 0 & 1 \end{bmatrix}\begin{bmatrix} a_{11} & a_{12} \\ a_{21} & a_{22} \end{bmatrix} = \begin{bmatrix} a_{11} & a_{12} \\ a_{21} & a_{22} \end{bmatrix} = A$$ ■

We finish by summarizing some important properties of matrix multiplication. These can be verified using the definitions presented in this section (see Exercise 11).

Properties of Matrix Multiplication Suppose A, B, and C are matrices and a and b are scalars. Provided the required matrix multiplications are defined, then

1. $A(BC) = (AB)C$
2. $(aA)(bB) = abAB$
3. $A(B + C) = AB + AC$
4. $(B + C)A = BA + CA$
5. $IA = A, AI = A$
6. $0A = 0, A0 = 0$

Note: Matrix multiplication is *not*, in general, commutative; that is, $AB \neq BA$.

EXERCISES 8.4

1–2 Suppose A, B, and C are 2 × 2 matrices, E, F, and G are 3 × 3 matrices, H and K are 2 × 3 matrices, and L and M are 3 × 2 matrices. For each of the following, if the operation is defined, specify the size of the matrix that results.

1. (a) $A + 4C$ (b) $\frac{1}{2}K + L$
(c) $5K + 3H$ (d) $0G + 3(E + F)$
(e) $3A + 6(B + M)$ (f) $12M - L$
(g) $F + G - 2C$
(h) $\alpha F + \beta G$, where α and β are scalars

2. (a) $AB + C$ (b) $3GF$
(c) $CK + B$ (d) $CK + H$
(e) EMC (f) GLH
(g) HLG (h) $2EL + 5MB$

3. Using the given matrices, calculate the quantities, if they are defined.

$$A = \begin{bmatrix} 2 & 5 \\ 1 & 7 \end{bmatrix} \quad B = \begin{bmatrix} 7 & x \\ a & 5 \end{bmatrix} \quad C = \begin{bmatrix} 9 & 2 \\ 7 & 10 \end{bmatrix}$$

$$E = \begin{bmatrix} 0 & 3 & 1 \\ 7 & 6 & 0 \\ 9 & 13 & 5 \end{bmatrix} \quad F = \begin{bmatrix} x & 2 & 9 \\ 6 & y & 13 \\ 0 & 1 & 0 \end{bmatrix}$$

$$G = \begin{bmatrix} 13 & 2 & 0 \\ 5 & 9 & 12 \\ 7 & 0 & 1 \end{bmatrix}$$

$$H = \begin{bmatrix} 3 & 1 & 7 \\ 15 & 0 & 2 \end{bmatrix} \quad K = \begin{bmatrix} 1 & 0 & 1 \\ 0 & y & 1 \end{bmatrix}$$

(a) $A - 3C$ (b) $3F + G - E$
(c) $5K + 9H$ (d) $H - 12G$
(e) $5B - A$ (f) $B - 9K$
(g) $3F - F$ (h) $H - K + 2H$

4. Using the given matrices, calculate the quantities, if they are defined.

$$A = \begin{bmatrix} 7 & 0 \\ 1 & 9 \end{bmatrix} \quad B = \begin{bmatrix} 0 & 2 \\ 13 & 6 \end{bmatrix} \quad C = \begin{bmatrix} 15 & 0 \\ 0 & 2 \end{bmatrix}$$

$$E = \begin{bmatrix} x & 5 & 3 \\ 0 & 9 & 1 \\ 6 & 4 & 0 \end{bmatrix} \quad F = \begin{bmatrix} y & 0 & y \\ 0 & 1 & 0 \\ 0 & 1 & 1 \end{bmatrix}$$

$$G = \begin{bmatrix} 1 & 2 & 3 \\ 5 & 4 & 6 \\ 9 & 7 & 8 \end{bmatrix}$$

$$H = \begin{bmatrix} 2 & 1 & 3 \\ 3 & 7 & 6 \end{bmatrix} \quad K = \begin{bmatrix} 9 & 2 \\ 1 & 5 \\ 7 & 8 \end{bmatrix}$$

(a) $A - 2BC$ (b) EG
(c) $FK - 2K$ (d) HEA
(e) $3BC - HK$ (f) $7K - H$
(g) $AHKB$ (h) $BKHA$

5. Find the transpose of each matrix.

(a) $\begin{bmatrix} 3X \\ 1 \\ 2 \end{bmatrix}$ (b) $\begin{bmatrix} 3 & 3 & 9 \end{bmatrix}$

(c) $\begin{bmatrix} 2 & 1 & 7 \\ 8 & 3 & 6 \end{bmatrix}$ (d) $\begin{bmatrix} 2 & 1 \\ 1 & 3 \end{bmatrix}$

6–10 The notation A^k means the matrix A multiplied with itself k times.

6. (a) For the 2×2 identity matrix I, show that $I^2 = I$.
(b) For the $n \times n$ identity matrix I, show that $I^2 = I$.
(c) What do you think the entries of I^k are?

7. (a) Show that, for all $n \times n$ diagonal matrices D having entries d_{ii}, the matrix D^2 is also diagonal with entries d_{ii}^2.
(b) For an arbitrary $n \times n$ diagonal matrix D, what do you think the entries of D^k are?

8. Show that, in general, if A is a 2×2 matrix, then

$$\begin{bmatrix} a_{11} & a_{12} \\ a_{21} & a_{22} \end{bmatrix}^2 \neq \begin{bmatrix} a_{11}^2 & a_{12}^2 \\ a_{21}^2 & a_{22}^2 \end{bmatrix}$$

9. Consider the matrix $C = \begin{bmatrix} 0 & 1 \\ 1 & 0 \end{bmatrix}$.

(a) Calculate C^2, C^3, C^4, and C^5.
(b) What do you think C^k is?

10. Find an example of a nonzero 2×2 matrix whose square is the zero matrix.

---

11. For arbitrary 2×2 matrices A, B, and C, verify the following properties.
(a) $A(BC) = (AB)C$
(b) $A(B + C) = AB + AC$
(c) $(B + C)A = BA + CA$

12. Suppose that A and B are arbitrary 2×2 matrices. Is the quantity given always equal to $(A + B)^2$?
(a) $A^2 + 2AB + B^2$
(b) $(B + A)^2$
(c) $A(A + B) + B(A + B)$
(d) $A^2 + AB + BA + B^2$

13. Suppose that A is a 2×2 matrix that commutes with all possible 2×2 matrices. Show that, necessarily, $a_{11} = a_{22}$ and $a_{12} = a_{21} = 0$. *Hint:* Try using

$$X = \begin{bmatrix} 1 & 0 \\ 0 & 0 \end{bmatrix} \quad \text{and} \quad Y = \begin{bmatrix} 0 & 1 \\ 0 & 0 \end{bmatrix}$$

as two particular 2×2 matrices with which A must commute.

14. (a) For any 2×2 matrices A and B show that

$$(A + B)^T = A^T + B^T$$

(b) For any $n \times n$ matrices A and B show that

$$(A + B)^T = A^T + B^T$$

15. (a) For any 2×2 matrices A and B show that

$$(AB)^T = B^T A^T$$

(b) For any $n \times n$ matrices A and B show that

$$(AB)^T = B^T A^T$$

16. A matrix A is called **symmetric** if $A^T = A$. Verify, for all 3×3 matrices B, that $B + B^T$ is symmetric.

17. Suppose a matrix A satisfies the equation $A + A^T = 0$. What must be true about the entries of A? Such matrices are called **skew-symmetric**.

18. An **upper triangular matrix** is a square matrix with all zeros below its diagonal. A **lower triangular matrix** is a square matrix with all zeros above its diagonal.
(a) For all 3×3 upper triangular matrices U, verify that U^2 is also upper triangular.
(b) For all 3×3 lower triangular matrices L, verify that L^2 is also lower triangular.

8.5 Matrices and the Dynamics of Vectors

In this section we show how matrices can be used to model changes in a vector from one point in time to the next. Such models are called matrix models.

■ Systems of Difference Equations: Matrix Models

In Sections 1.6 and 2.1 we studied the following difference equation for population size:

(1) $$n_{t+1} = Rn_t$$

Note: We have changed notation slightly from Sections 1.6 and 2.1—we are now using a lowercase n to denote population size.

Equation 1 has an equilibrium at $\hat{n} = 0$. For other values of n we saw that n grows in magnitude if $R > 1$ and decays to 0 if $0 < R < 1$. In the special case where $R = 1$, any value of n is an equilibrium because Equation 1 then becomes $n_{t+1} = n_t$.

Suppose we want to model the population size when there are adults and juveniles. Each adult produces two juveniles in one time step, and half the juveniles survive to become adults. Finally, $\frac{1}{3}$ of adults survive from one time step to the next. Using j_t and a_t to denote the number of juveniles and adults at time t, we obtain the *system of difference equations*

(2) $$j_{t+1} = 2a_t$$
$$a_{t+1} = \tfrac{1}{2}j_t + \tfrac{1}{3}a_t$$

We can view the state of the population at any time t as a vector $\begin{bmatrix} j_t \\ a_t \end{bmatrix}$ (see Figure 1). We have written this as a column vector in anticipation of using the techniques of matrix algebra from Section 8.4 to simplify the notation. In particular, let's view this vector as a 2×1 matrix and consider calculating the following matrix product using the rules of matrix multiplication from Section 8.4:

$$\begin{bmatrix} 0 & 2 \\ \frac{1}{2} & \frac{1}{3} \end{bmatrix}\begin{bmatrix} j_t \\ a_t \end{bmatrix}$$

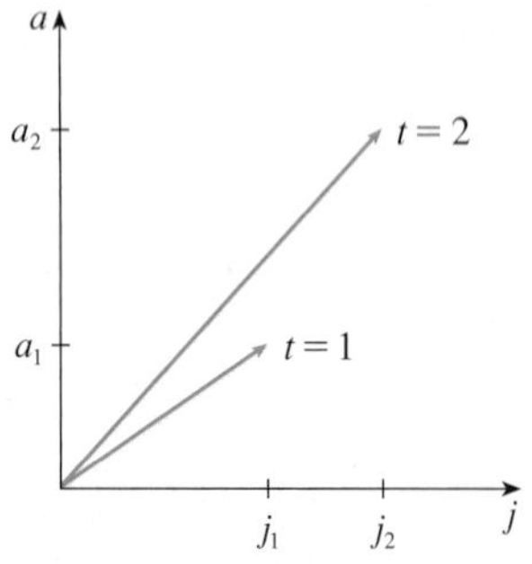

FIGURE 1

This is the product of a 2×2 matrix with a 2×1 matrix, and so matrix multiplication is defined. From the rule in Figure 8.4.3, we anticipate the result to be a 2×1 matrix (that is, a column vector).

Performing the multiplication, we obtain

$$\begin{bmatrix} 0 & 2 \\ \frac{1}{2} & \frac{1}{3} \end{bmatrix}\begin{bmatrix} j_t \\ a_t \end{bmatrix} = \begin{bmatrix} 0 \cdot j_t + 2a_t \\ \frac{1}{2}j_t + \frac{1}{3}a_t \end{bmatrix} = \begin{bmatrix} 2a_t \\ \frac{1}{2}j_t + \frac{1}{3}a_t \end{bmatrix}$$

From Equations 2 we see that the resulting vector can be written as $\begin{bmatrix} j_{t+1} \\ a_{t+1} \end{bmatrix}$, and therefore the system of difference equations in (2) can be written in matrix notation as

$$\begin{bmatrix} j_{t+1} \\ a_{t+1} \end{bmatrix} = \begin{bmatrix} 0 & 2 \\ \frac{1}{2} & \frac{1}{3} \end{bmatrix}\begin{bmatrix} j_t \\ a_t \end{bmatrix}$$

As a final step we define Q to be the matrix $\begin{bmatrix} 0 & 2 \\ \frac{1}{2} & \frac{1}{3} \end{bmatrix}$ and $\mathbf{n}_t$ to be the column vector $\begin{bmatrix} j_t \\ a_t \end{bmatrix}$. We then have

(3) $$\mathbf{n}_{t+1} = Q\mathbf{n}_t$$

The parallel between the single-variable recursion in Equation 1 for n_t and Equation 3 for the vector $\mathbf{n}_t$ is now apparent. The change in the vector $\begin{bmatrix} j_t \\ a_t \end{bmatrix}$ is described by the product of the matrix of constants Q with this vector, just as the change in n_t in Equation 1 is described by the product of the constant R with n_t. In the vector setting the growth factor R in Equation 1 is replaced by the matrix Q in Equation 3. Models such as Equation 3 are called **matrix models**.

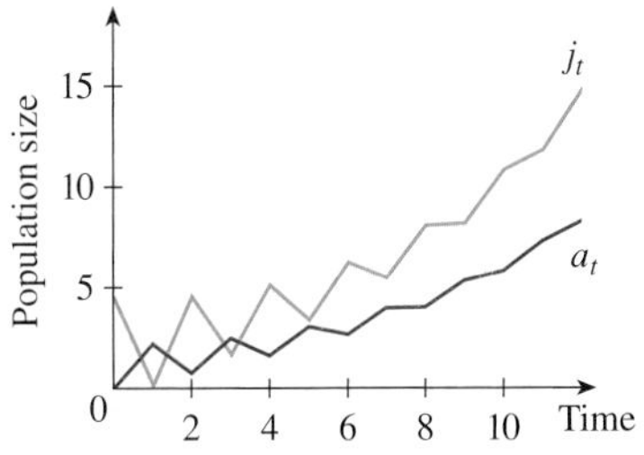

FIGURE 2

As with Equation 1, how the vector $\mathbf{n}_t$ changes over time depends on the initial condition and on the matrix of constants Q. And as with Equation 1, we can recursively apply Equation 3 to an initial vector to generate numerical solutions. Figure 2 illustrates an example of this for Equations 2. Both j_t and a_t initially oscillate, but the population size of both juveniles and adults appears to grow in the long run. In later sections of this chapter we will learn how to determine the behavior of such matrix models mathematically. As with Equation 1, the long-term behavior of the model depends on a measure of the magnitude of the "growth factor" Q.

HIROYA MINAKUCHI / MINDEN PICTURES / National Geographic Creative

FIGURE 3
A right whale

EXAMPLE 1 | BB **Conservation biology of right whales** Matrix models are often used to predict the population size of endangered species such as right whales. Right whales have important stage structure to their life cycle. A simplified model considers three stages: (i) newborn calves, (ii) mature but nonreproducing adults, and (iii) reproductive adults. Let's use c_t, a_t, and r_t, respectively, to denote the number of individuals in each of these stages. Suppose that newborn calves mature into adults with probability 0.072 each time step (otherwise they die). Nonreproducing adults remain that way with probability 0.8 or begin reproducing with probability 0.19 (otherwise they die). Reproducing individuals produce, on average, 0.3 calves each time step and revert to the nonreproducing adult class with probability 0.88 (otherwise they die). Construct the matrix model for this population.

SOLUTION First we summarize the model using what is called a *matrix diagram* in Figure 4.

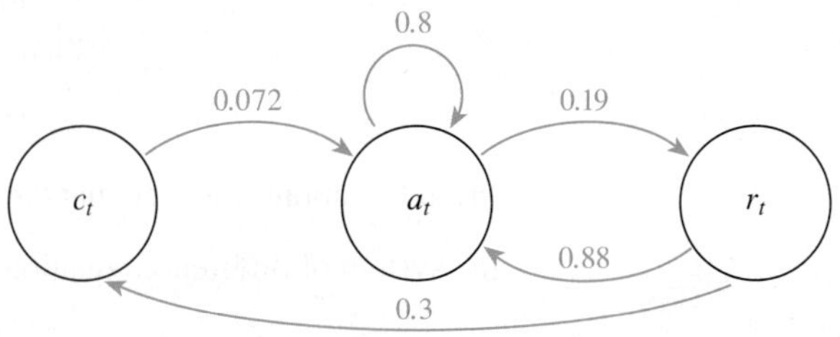

FIGURE 4
Each circle represents a life stage. An arrow is drawn from circle j to i if stage j contributes to the number of individuals in stage i in the next time step. The numbers next to the arrows indicate the *per capita* contribution.

We proceed by deriving a difference equation for each of the three variables in turn. From Figure 4 the number of calves at time $t + 1$ is the per capita number produced by each reproductive individual, 0.3, multiplied by the number of these individuals at time t. Therefore, $c_{t+1} = 0.3r_t$. The number of adults at time $t + 1$ is the number of calves at time t multiplied by the per capita probability that they mature, 0.072, plus the number of adults at time t multiplied by the per capita probability that they remain as nonreproducing adults, 0.8, plus the number of reproductive individuals at time t multiplied by the per capita probability that they revert to the nonreproducing adult stage, 0.88. This gives

$$a_{t+1} = 0.072c_t + 0.8a_t + 0.88r_t$$

Finally, the number of reproductive individuals at time $t + 1$ is the number of adults at time t multiplied by the per capita probability that they move into the reproductive class, 0.19, giving $r_{t+1} = 0.19a_t$. You can verify that organizing this using matrix notation gives

$$\begin{bmatrix} c_{t+1} \\ a_{t+1} \\ r_{t+1} \end{bmatrix} = \begin{bmatrix} 0 & 0 & 0.3 \\ 0.072 & 0.8 & 0.88 \\ 0 & 0.19 & 0 \end{bmatrix} \begin{bmatrix} c_t \\ a_t \\ r_t \end{bmatrix}$$ ■

Example 1 introduces the idea of drawing a **matrix diagram** to organize the construction of a matrix model. In general, such diagrams consists of a single circle for each variable of the model, and an arrow from circle j to i whenever type j contributes to the amount of type i in the next time step (see Figure 5). The corresponding matrix can then be written directly from the diagram by noting that the entry in the ith row and jth column of the matrix corresponds to the arrow from circle j to i.

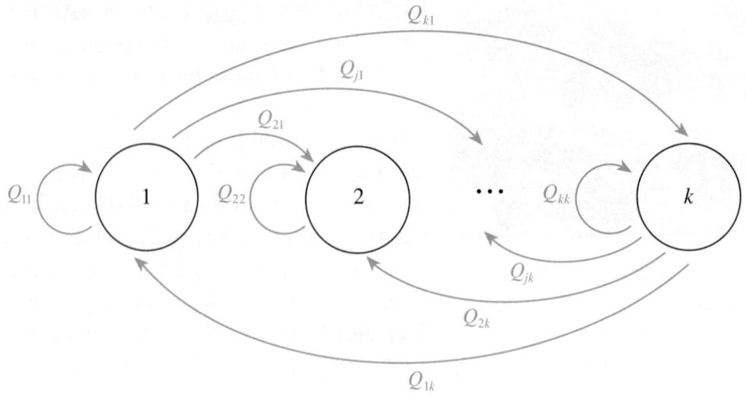

FIGURE 5
A matrix diagram

EXAMPLE 2 | **Breast cancer** Consider a population of patients at risk of developing breast cancer. Each individual's health status can be in one of three states: (1) healthy, (2) early-stage disease, or (3) late-stage disease. Each year 0.2% of healthy females develop early-stage disease, with the rest remaining healthy. Likewise, 45% of women with early-stage disease recover through treatment, 45% develop late-stage disease, and 10% remain in the early-stage category. All individuals with late-stage disease remain in that state.

(a) Using h_t, e_t, and l_t to denote the number of individuals in each category, construct the matrix diagram for this example.

(b) Construct the matrix model corresponding to the diagram from part (a).

SOLUTION

(a) Each of the three states is denoted by a circle in Figure 6, and we draw arrows corresponding to the possible transitions in health status. Finally, we label each arrow with its appropriate value.

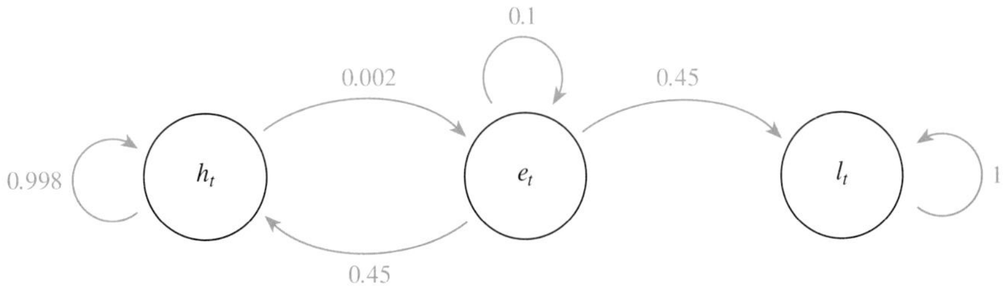

FIGURE 6

(b) We find the entry in the ith row and jth column of the matrix by looking for an arrow from circle j to circle i in Figure 6, giving

$$\begin{bmatrix} h_{t+1} \\ e_{t+1} \\ l_{t+1} \end{bmatrix} = \begin{bmatrix} 0.998 & 0.45 & 0 \\ 0.002 & 0.1 & 0 \\ 0 & 0.45 & 1 \end{bmatrix} \begin{bmatrix} h_t \\ e_t \\ l_t \end{bmatrix}$$

■

■ Leslie Matrices

Matrix models are often used to model the size of age-structured populations. Consider a simplified model for a Coho salmon population. Individuals hatch from eggs in rivers on the West Coast of North America and migrate to the ocean. At around three years of age they migrate back to the river, reproduce, and die. Let's use $n_{1,t}$, $n_{2,t}$, and $n_{3,t}$ to denote the number of individuals in each age class at time t, where time is measured in years. Figure 7 gives the matrix diagram for the model, along with estimated values for the constants.

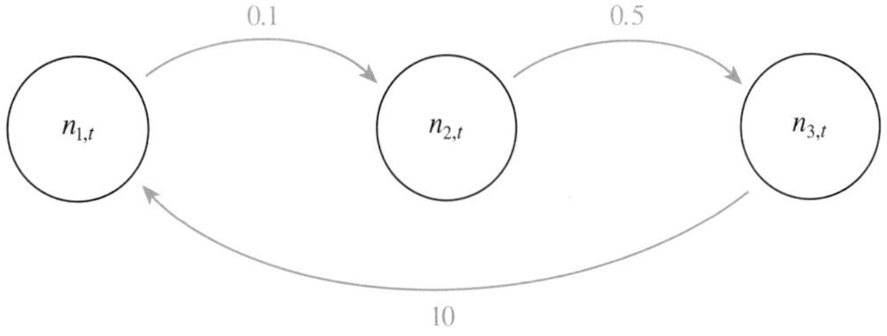

FIGURE 7

The resulting matrix model is

$$\begin{bmatrix} n_{1,t+1} \\ n_{2,t+1} \\ n_{3,t+1} \end{bmatrix} = \begin{bmatrix} 0 & 0 & 10 \\ 0.1 & 0 & 0 \\ 0 & 0.5 & 0 \end{bmatrix} \begin{bmatrix} n_{1,t} \\ n_{2,t} \\ n_{3,t} \end{bmatrix}$$

Figure 8 shows the number of individuals in each age class over time. The graph looks rather complicated, but the population size appears to decay to zero in the long run.

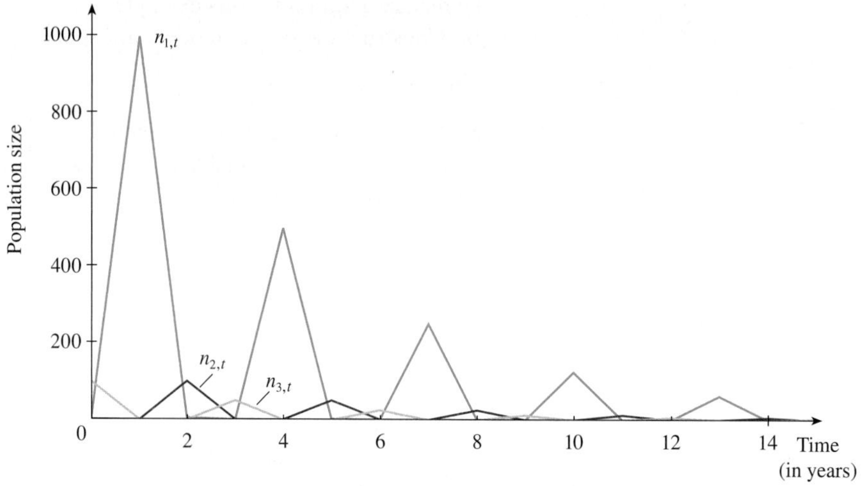

FIGURE 8

Matrices for age-structured models like that of the Coho salmon always have a special structure. They are called **Leslie matrices** after Patrick H. Leslie and are often denoted by the symbol L. To see this structure better, let's examine the general case of k age classes. Using b_i for the number of offspring produced by an age i individual and s_i for the probability that an age i individual survives to become an age $i + 1$ individual, we have the matrix diagram shown in Figure 9.

Leslie

Patrick H. Leslie was a mathematical biologist who worked at Oxford University and pioneered the use of matrix models in population biology in the 1940s.

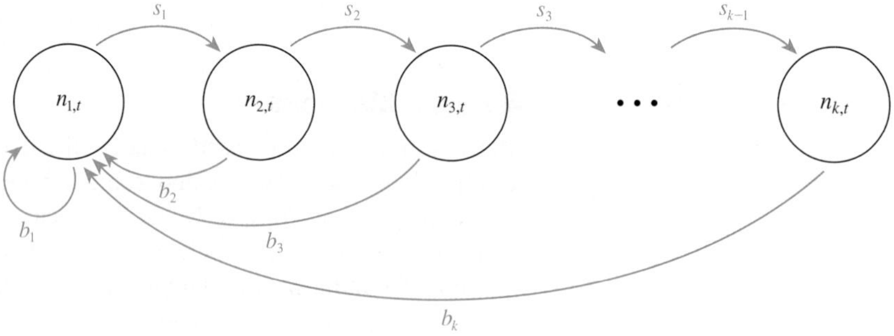

FIGURE 9

Algebraically, we have

$$\begin{aligned} n_{1,t+1} &= b_1 n_{1,t} + b_2 n_{2,t} + \cdots + b_k n_{k,t} \\ n_{2,t+1} &= s_1 n_{1,t} \\ n_{3,t+1} &= s_2 n_{2,t} \\ &\vdots \\ n_{k,t+1} &= s_{k-1} n_{k-1,t} \end{aligned}$$

or, in matrix notation,

$$\begin{bmatrix} n_{1,t+1} \\ n_{2,t+1} \\ n_{3,t+1} \\ \vdots \\ n_{k,t+1} \end{bmatrix} = \begin{bmatrix} b_1 & b_2 & b_3 & \cdots & b_k \\ s_1 & 0 & 0 & \cdots & 0 \\ 0 & s_2 & 0 & \cdots & 0 \\ \vdots & \vdots & \vdots & \ddots & \vdots \\ 0 & 0 & 0 & s_{k-1} & 0 \end{bmatrix} \begin{bmatrix} n_{1,t} \\ n_{2,t} \\ n_{3,t} \\ \vdots \\ n_{k,t} \end{bmatrix}$$

Again we have assumed that a single time step corresponds to the time it takes to move from one age class to the next. As a result, all surviving individuals must progress from one age class to the next each time step, giving the matrix a special form. The age-specific birth rates make up the first row of the matrix, and the age-specific survival probabilities lie along the *subdiagonal* of the matrix (that is, the diagonal lying immediately below the diagonal that runs from the top left to the bottom right corner of the matrix).

■ Summary

Matrix models are a special kind of dynamical system that govern how vectors change from one time step to the next. They consist of a column vector of variables, $\mathbf{n}_t$, that changes from one time step to the next, along with a matrix of constants, A, that determines how the vector changes. They take the form

$$\mathbf{n}_{t+1} = A\mathbf{n}_t$$

For any given initial vector we can recursively apply the equation to obtain numerical solutions. As we will see in Section 8.8, we can also analyze such equations quite extensively from a mathematical perspective and thereby gain a complete understanding of how the vector changes through time.

EXERCISES 8.5

1. Construct the corresponding matrix for each of the following matrix diagrams by placing an X as the ijth entry of the matrix if the entry is nonzero, and a 0 otherwise.

(a)

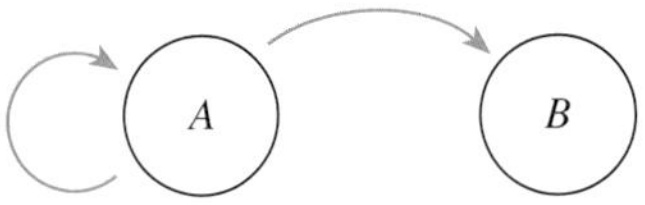

(b)

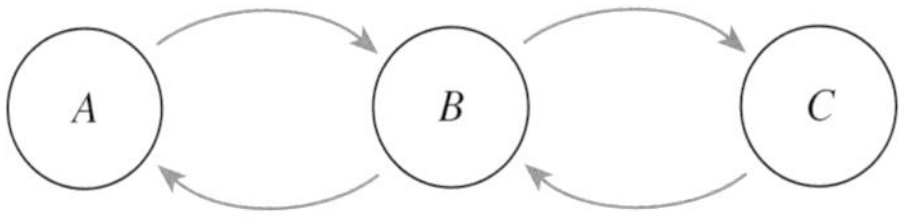

(c)

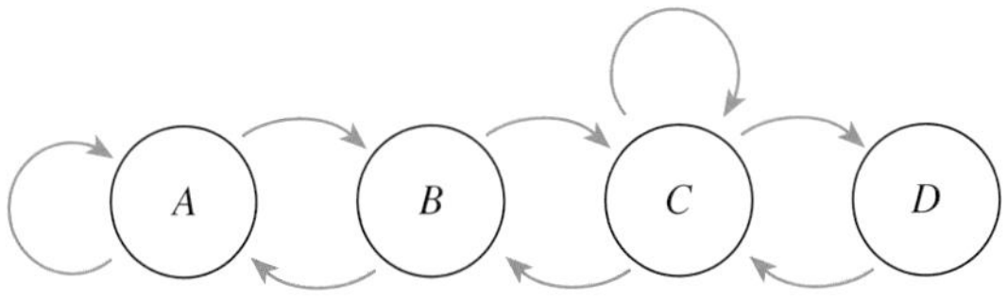

(d)

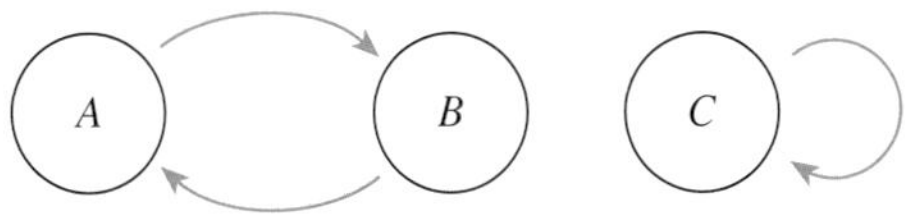

(e)

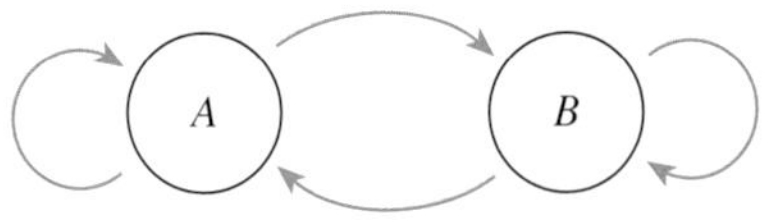

2. For each of the following matrices an X in entry ij represents the fact that type j contributes to type i in the next time step. Construct the corresponding matrix diagram for each.

(a) $\begin{bmatrix} X & X \\ 0 & X \end{bmatrix}$

(b) $\begin{bmatrix} X & 0 & X \\ 0 & 0 & X \\ X & 0 & 0 \end{bmatrix}$

(c) $\begin{bmatrix} 0 & X & 0 & X \\ X & 0 & X & 0 \\ X & X & 0 & 0 \\ 0 & 0 & X & X \end{bmatrix}$

(d) $\begin{bmatrix} X & 0 & 0 \\ 0 & X & 0 \\ 0 & 0 & X \end{bmatrix}$

(e) $\begin{bmatrix} 0 & X \\ X & 0 \end{bmatrix}$

3–4 Construct the corresponding matrix for each of the following matrix diagrams.

3. (a)

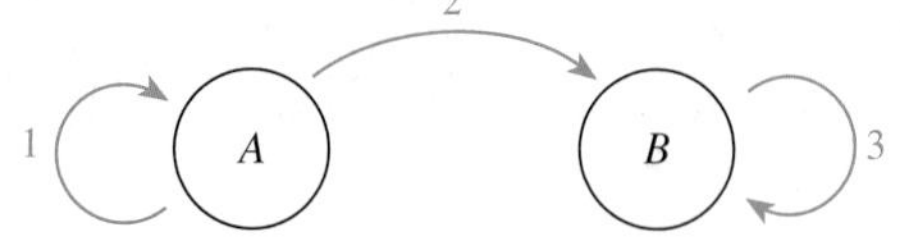

(b)

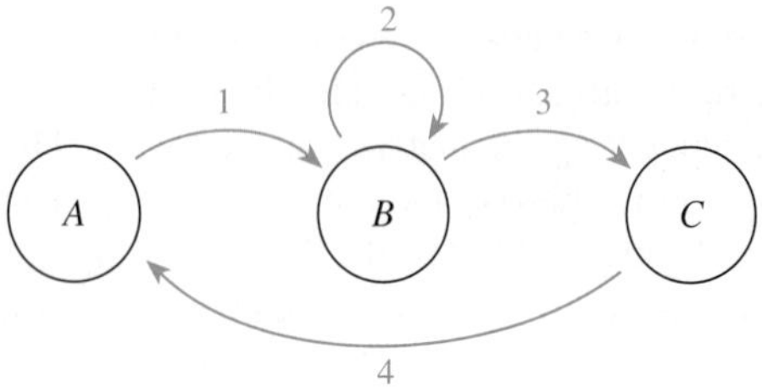

(c)

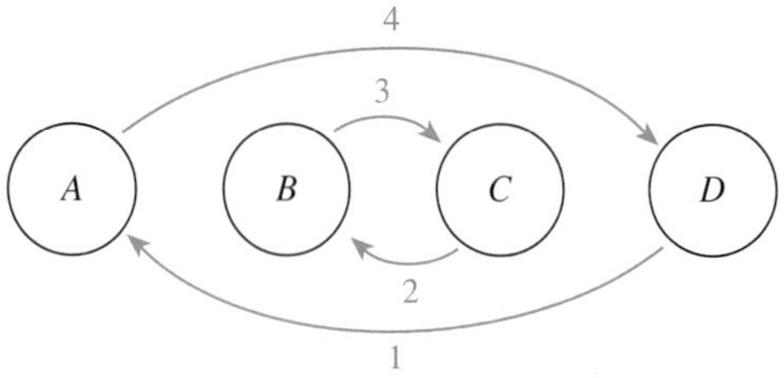

(d)

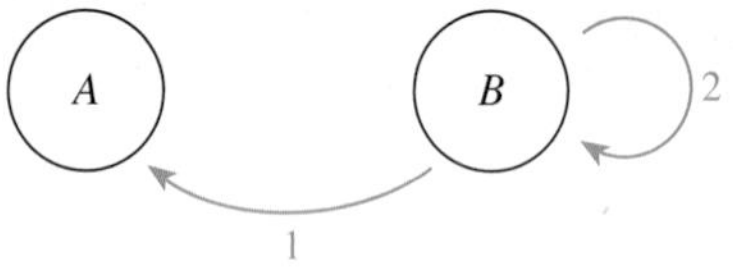

4. (a)

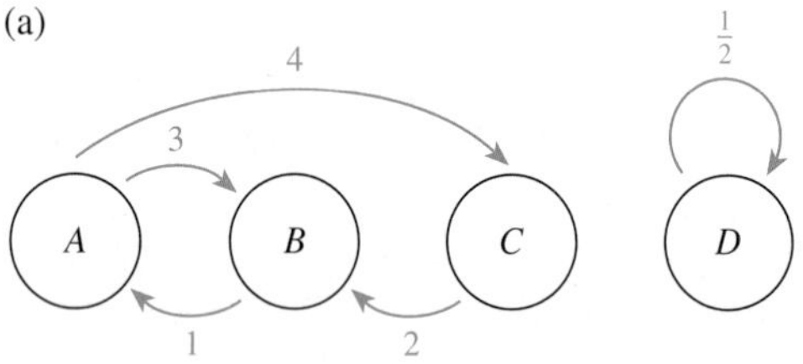

(b)

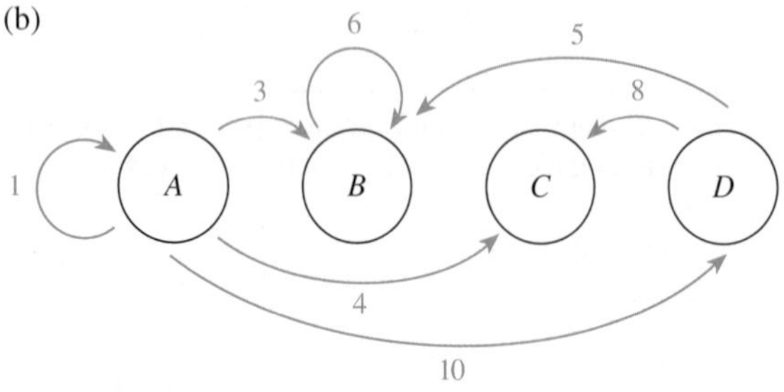

(c)

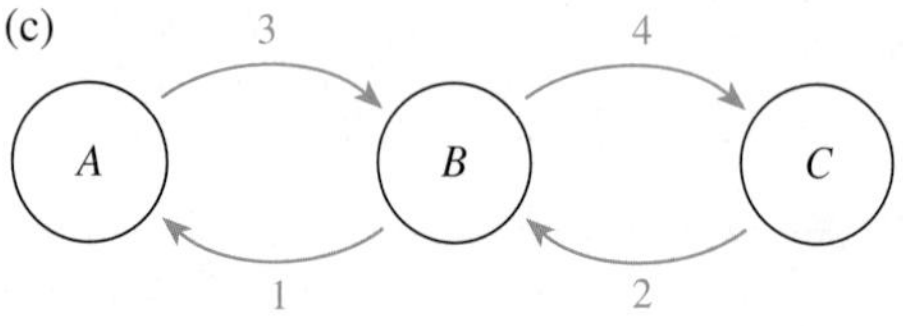

(d)

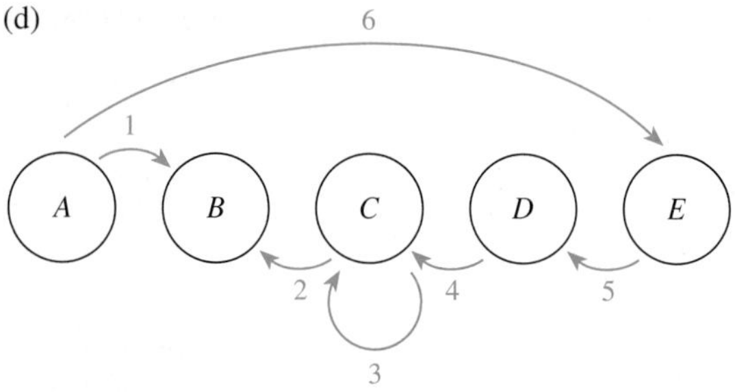

5–6 Each of the following matrices represents a system of difference equations. Construct the corresponding matrix diagram.

5. (a) $\begin{bmatrix} 1 & 6 \\ 5 & 0 \end{bmatrix}$ (b) $\begin{bmatrix} 2 & 0 & 3 \\ 5 & 1 & 2 \\ 7 & 9 & 0 \end{bmatrix}$

(c) $\begin{bmatrix} 1 & 1 \\ 1 & 1 \end{bmatrix}$ (d) $\begin{bmatrix} 7 & 6 & 0 \\ 0 & 2 & 3 \\ 0 & 1 & 0 \end{bmatrix}$

6. (a) $\begin{bmatrix} 3 & 0 & 7 & 9 & 11 \\ 0.1 & 0.1 & 0.8 & 0 & 0 \\ 9 & 0 & 1 & 1 & 1 \\ 6 & 5 & 4 & 3 & 2 \\ 0 & 0 & 0 & 0 & 0 \end{bmatrix}$

(b) $\begin{bmatrix} 0 & 1 & 3 & 0 \\ 3 & 1 & 3 & 1 \\ 0 & 0 & 9 & 0 \\ 7 & 0 & 0 & 0 \end{bmatrix}$

(c) $\begin{bmatrix} 1 & 0 & 0 \\ 0.25 & 0.5 & 0.25 \\ 0 & 0 & 1 \end{bmatrix}$

(d) $\begin{bmatrix} 1 & 0 & 1 & 0 \\ 0 & 1 & 0 & 1 \\ 1 & 0 & 1 & 0 \\ 0 & 1 & 0 & 1 \end{bmatrix}$

7. Suppose the matrix representing a system of difference equations contains a row of zeros. What does this imply about the system?

8. Suppose the matrix representing a system of difference equations contains a column of zeros. What does this imply about the system?

9–15 Construct the matrix diagram for each of the following models.

9. Dispersal A population of dragonflies occupies two ponds. Each day 20% of the individuals in pond A move to pond B, and 30% of the individuals in pond B move to pond

A. Using A_t and B_t to denote the number of individuals in each pond at time t, we have

$$A_{t+1} = 0.8A_t + 0.3B_t$$
$$B_{t+1} = 0.2A_t + 0.7B_t$$

10. **Mutation** Each generation 5% of individuals carrying allele X mutate to carry allele Y. Using X_t and Y_t to denote the number of individuals carrying each allele at time t, we have

$$X_{t+1} = 0.95X_t$$
$$Y_{t+1} = 0.05X_t + Y_t$$

11. **Cancer progression** Consider a population of women at risk of developing breast cancer. Each individual's health status can be in one of three states: (1) healthy, (2) early-stage disease, or (3) late-stage disease. Each year 0.2% of healthy females develop early-stage disease, with the rest remaining healthy. Likewise, 65% of women with early-stage disease recover through treatment and 35% develop late-stage disease. Finally, 10% of individuals with late-stage disease revert to early stage disease while the rest remain in late stage. Using h_t, e_t, and l_t to denote the number of individuals in each state at time t, we have

$$h_{t+1} = 0.998h_t + 0.65e_t$$
$$e_{t+1} = 0.002h_t + 0.1l_t$$
$$l_{t+1} = 0.35e_t + 0.9l_t$$

12. **Yellow perch** Yellow perch is a species of freshwater fish. Suppose a population of interest lives to a maximum age of 4 years. Only 5% of age 1 individuals survive to age 2, 20% of age 2 individuals survive to age 3, and 75% of age 3 individuals survive to age 4. Both age 3 and age 4 individuals reproduce, producing 100 and 150 age 1 offspring, respectively. Using $n_{i,t}$ for the number of individuals of age i at time t, we have

$$n_{1,t+1} = 100n_{3,t} + 150n_{4,t}$$
$$n_{2,t+1} = 0.05n_{1,t}$$
$$n_{3,t+1} = 0.2n_{2,t}$$
$$n_{4,t+1} = 0.75n_{3,t}$$

13. **Drug diffusion** A drug taken orally eventually diffuses into the bloodstream through the stomach. Let s_t be the amount of drug in the stomach at time t and b_t the amount that has been absorbed into the bloodstream. Each time step, 50% of the drug is absorbed from the stomach into the bloodstream and 80% of the drug in the bloodstream is metabolized. We have

$$s_{t+1} = 0.5s_t$$
$$b_{t+1} = 0.5s_t + 0.2b_t$$

14. **Breeding systems** Cooperatively breeding bird species are those in which some individuals forgo their own reproduction to help others reproduce. Therefore the population consists of three kinds of individuals: juveniles, helpers, and reproductives. Using j_t, h_t, and r_t to denote the number of each kind at time t, a model for the population size of each is

$$j_{t+1} = 2r_t$$
$$h_{t+1} = 0.9j_t + 0.8h_t + 0.1r_t$$
$$r_{t+1} = 0.1j_t + 0.2h_t + 0.9r_t$$

15. **Ecological succession** is a process by which a biological community progresses between different states. Suppose the community can be either primarily grassland, primarily shrub, or primarily forest, and this can change from year to year. Using g_t, s_t, and f_t to denote the probability that the community is in each of the three possible states at time t, we have

$$g_{t+1} = 0.9g_t + 0.001s_t$$
$$s_{t+1} = 0.1g_t + 0.799s_t + 0.001f_t$$
$$f_{t+1} = 0.2s_t + 0.999f_t$$

16–19 Write the system of difference equations and construct the corresponding matrix that describes this system.

16. **Vectorcardiography** Suppose the components X and Y of the heart's voltage vector change from one beat to the next according to the diagram

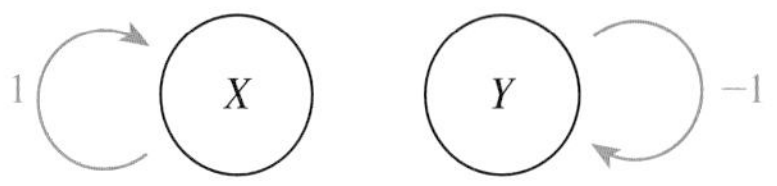

17. **Antigenic evolution** Suppose influenza viruses are plotted in two-dimensional antigenic space, and that the x- and y-coordinates change from one year to the next according to the diagram

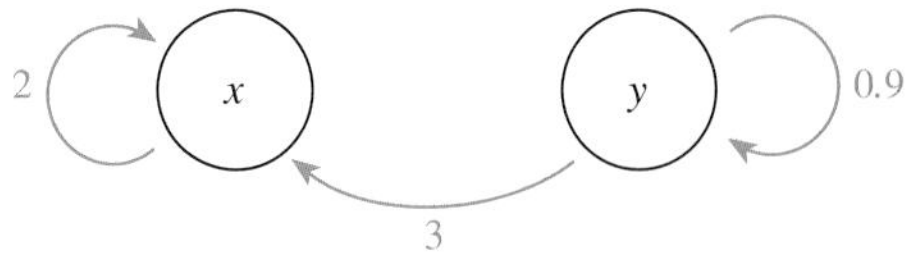

18. **Genome expression profiles** A genome expression profile can be characterized as a vector. Consider a species of frog and suppose that global warming causes the components W, X, Y, and Z of its genome expression vector to change from one year to the next according to the diagram

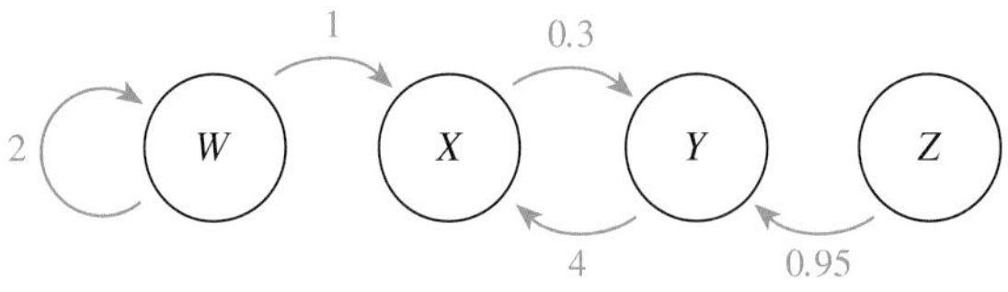

19. **Genetics of inbreeding** Plants are sometimes bred with themselves to generate homozygous individuals (that is,

individuals that carry two identical copies of a gene of interest). Suppose the numbers of plants that are AA, Aa, and aa in generation t are denoted by d_t, h_t, and r_t, respectively. These variables then change from one generation to the next according to the diagram

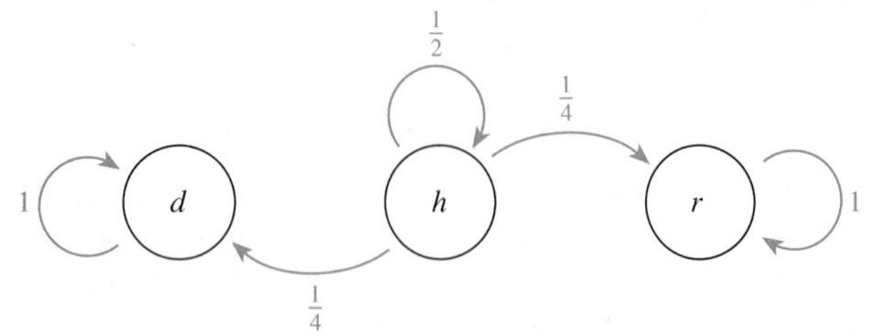

20–22 Write the system of equations and construct the corresponding matrix model for each of the following exercises.

20. DNA methylation DNA methylation is a process by which methyl chemical groups are attached to the DNA. Thus each gene in the genome can be methylated or unmethylated. Suppose that 80% of methylated genes remain methylated and 50% of unmethylated genes remain unmethylated each generation.

21. Mutation Suppose that one of three different alleles is present in each individual in a population. In each generation the following happens: 5% of individuals carrying allele X mutate to carry allele Y, 3% mutate to allele Z, and the rest remain unchanged; 0.1% of individuals carrying allele Y mutate to carry allele Z and the rest remain unchanged; 90% of individuals carrying allele Z mutate to carry allele X and the rest remain unchanged.

22. Spotted owls were at the center of a debate between conservation biologists and the logging industry in the 1990s. Part of resolving the debate involved constructing a matrix model for their population. The owl population in year t can be divided into juveniles, subadults, and adults. Adults produce, on average, 0.33 juveniles each year. Approximately 60% of juveniles survive to be subadults the next year, while 71% of subadults survive to become adults. Approximately 94% of the adult population survives from one year to the next.

© C. M. Corcoran / Shutterstock.com

8.6 The Inverse and Determinant of a Matrix

This section introduces the ideas of matrix inverses and determinants and illustrates how they can be used to solve systems of equations. The inverse and determinant are defined only for square matrices. Our focus in the main text is on the mathematical details, with applications explored in the exercises and subsequent sections.

■ The Inverse of a Matrix

In Section 8.4 the identity matrix I was introduced and we noted that I plays the role of the number 1 in matrix multiplication. In particular, $AI = A$ and $IA = A$ just as $a1 = a$ and $1a = a$ in scalar multiplication. In the scalar case we can rearrange the equation to obtain $a^{-1}a = 1$, where $a^{-1} = 1/a$ provided that $a \neq 0$. In other words, a^{-1} is the quantity that, when multiplied with a, gives 1. And if $a = 0$, then no such quantity exists.

Can something similar be defined for matrices? Is there a matrix that, when multiplied with A, gives the identity matrix I? We will see that, just as with the scalar case, sometimes there is and sometimes there isn't. The existence of a matrix B such that $BA = I$ requires that the matrix A, in some sense, not "act like zero." One of the goals of this section is to make this idea precise. When such a matrix exists, it is called the *inverse* of A.

EXAMPLE 1 | Show that B is the inverse of A, where

$$A = \begin{bmatrix} 1 & 2 \\ 7 & 5 \end{bmatrix} \qquad B = \begin{bmatrix} -\frac{5}{9} & \frac{2}{9} \\ \frac{7}{9} & -\frac{1}{9} \end{bmatrix}$$

SOLUTION Calculating the matrix product gives

$$BA = \begin{bmatrix} -\frac{5}{9} & \frac{2}{9} \\ \frac{7}{9} & -\frac{1}{9} \end{bmatrix} \begin{bmatrix} 1 & 2 \\ 7 & 5 \end{bmatrix} = \begin{bmatrix} -\frac{5}{9}(1) + \frac{2}{9}(7) & -\frac{5}{9}(2) + \frac{2}{9}(5) \\ \frac{7}{9}(1) - \frac{1}{9}(7) & \frac{7}{9}(2) - \frac{1}{9}(5) \end{bmatrix} = \begin{bmatrix} 1 & 0 \\ 0 & 1 \end{bmatrix} = I$$ ■

> **Definition** Suppose that A is an $n \times n$ matrix. If there exists an $n \times n$ matrix B such that
>
> $$AB = BA = I$$
>
> then B is called the **inverse** of A and is denoted by A^{-1}.

It is possible to show that, when the inverse of a matrix exists, it is unique (see Exercise 8). It is also possible to show that, if there exists a matrix B such that $AB = I$, then necessarily $BA = I$ as well (and vice versa). Therefore we need check only one order of multiplication when finding an inverse. If A has an inverse, then we say that A is **invertible** or **nonsingular**. Otherwise A is called **singular**.

The following example illustrates how to calculate a matrix inverse in a 2×2 case, provided that the inverse exists.

EXAMPLE 2 | Derive the inverse of the matrix A from Example 1 by using the definition of an inverse.

SOLUTION From the definition, if a matrix $B = \begin{bmatrix} b_{11} & b_{12} \\ b_{21} & b_{22} \end{bmatrix}$ is the inverse of $A = \begin{bmatrix} 1 & 2 \\ 7 & 5 \end{bmatrix}$, then $AB = BA = I$. We need focus on only one of these orders of mulitplication. Choosing $AB = I$, we have

$$\begin{bmatrix} 1 & 2 \\ 7 & 5 \end{bmatrix} \begin{bmatrix} b_{11} & b_{12} \\ b_{21} & b_{22} \end{bmatrix} = \begin{bmatrix} 1 & 0 \\ 0 & 1 \end{bmatrix}$$

or

$$\begin{bmatrix} b_{11} + 2b_{21} & b_{12} + 2b_{22} \\ 7b_{11} + 5b_{21} & 7b_{12} + 5b_{22} \end{bmatrix} = \begin{bmatrix} 1 & 0 \\ 0 & 1 \end{bmatrix}$$

Looking at this equation entry by entry, we see that it represents four equations in four unknowns. The four equations can be split into two pairs of equations, each with two unknowns:

$$\begin{aligned} b_{11} + 2b_{21} &= 1 \\ 7b_{11} + 5b_{21} &= 0 \end{aligned} \qquad \text{and} \qquad \begin{aligned} b_{12} + 2b_{22} &= 0 \\ 7b_{12} + 5b_{22} &= 1 \end{aligned}$$

We can now solve each pair by substitution. If we focus on the first pair, we see that the second equation gives $b_{21} = -7b_{11}/5$. Substituting this into the first equation of this pair then gives $b_{11} + 2(-7b_{11}/5) = 1$, whose solution is $b_{11} = -5/9$. This can then be back-substituted into $b_{21} = -7b_{11}/5$ to give $b_{21} = 7/9$.

Similarly, the first equation of the second pair gives $b_{12} = -2b_{22}$. Substituting this into the second equation of this pair then gives $7(-2b_{22}) + 5b_{22} = 1$, whose solution is $b_{22} = -1/9$. This can then be back-substituted into $b_{12} = -2b_{22}$ to give $b_{12} = 2/9$.

Putting these results together in the matrix B gives

$$B = \begin{bmatrix} -\frac{5}{9} & \frac{2}{9} \\ \frac{7}{9} & -\frac{1}{9} \end{bmatrix}$$ ■

The approach used in Example 2 leads to the following formula for the inverse of a 2×2 matrix (see Exercise 4):

The Inverse of a 2 x 2 Matrix Suppose

$$A = \begin{bmatrix} a_{11} & a_{12} \\ a_{21} & a_{22} \end{bmatrix}$$

and $a_{11}a_{22} - a_{12}a_{21} \neq 0$. Then A is invertible and

$$A^{-1} = \frac{1}{a_{11}a_{22} - a_{12}a_{21}} \begin{bmatrix} a_{22} & -a_{12} \\ -a_{21} & a_{11} \end{bmatrix}$$

If $a_{11}a_{22} - a_{12}a_{21} = 0$, then A is not invertible (that is, A is singular).

EXAMPLE 3 | If possible, find the inverse.

(a) $M = \begin{bmatrix} 7 & 9 \\ 5 & 6 \end{bmatrix}$ (b) $N = \begin{bmatrix} 14 & 6 \\ 7 & 3 \end{bmatrix}$ (c) $J = \begin{bmatrix} 2 & 3 \\ -1 & -2 \end{bmatrix}$

SOLUTION From the formula for the inverse of a 2×2 matrix, an inverse will exist if and only if $a_{11}a_{22} - a_{12}a_{21} \neq 0$.

(a) $m_{11}m_{22} - m_{12}m_{21} = (7)(6) - (9)(5) = 42 - 45 = -3$. Therefore an inverse exists. Using the formula, we obtain

$$M^{-1} = \frac{1}{-3}\begin{bmatrix} 6 & -9 \\ -5 & 7 \end{bmatrix} = \begin{bmatrix} -2 & 3 \\ \frac{5}{3} & -\frac{7}{3} \end{bmatrix}$$

(b) $n_{11}n_{22} - n_{12}n_{21} = (14)(3) - (6)(7) = 42 - 42 = 0$. Therefore an inverse does not exist.

(c) $j_{11}j_{22} - j_{12}j_{21} = (2)(-2) - (3)(-1) = -4 + 3 = -1$. Therefore an inverse exists. Using the formula, we obtain

$$J^{-1} = \frac{1}{-1}\begin{bmatrix} -2 & -3 \\ 1 & 2 \end{bmatrix} = \begin{bmatrix} 2 & 3 \\ -1 & -2 \end{bmatrix}$$

Notice that this last example has the interesting property that $J^{-1} = J$. In other words, J is its own inverse. ■

The following properties of matrix inverses are often useful. These are proved in Exercises 6, 7, and 8.

Properties of Matrix Inverses Suppose A and B are both invertible $n \times n$ matrices. Then

1. $(A^{-1})^{-1} = A$
2. $(AB)^{-1} = B^{-1}A^{-1}$
3. A^{-1} is unique.

■ The Determinant of a Matrix

The formula for the inverse of a 2×2 matrix allows us to determine when such a matrix is invertible. But how can we tell this for matrices of other sizes? As already mentioned, a matrix A is invertible if, in some sense, it does not "act like zero." We now make this idea precise.

To any $n \times n$ matrix A we can assign a scalar quantity called its *determinant*, denoted by $\det A$. If we view scalars as 1×1 matrices, then the definition of the determinant for matrices of sizes $n = 1$ through $n = 3$ is as follows.

The Determinant Suppose A is an $n \times n$ matrix.

1. If $n = 1$, then $\det A = a_{11}$.

2. If $n = 2$, then $\det A = a_{11}a_{22} - a_{12}a_{21}$.

3. If $n = 3$, then $\det A = a_{11}a_{22}a_{33} + a_{12}a_{23}a_{31} + a_{13}a_{21}a_{32} - a_{13}a_{22}a_{31} - a_{11}a_{23}a_{32} - a_{12}a_{21}a_{33}$.

There is an algorithmic procedure for calculating the determinant of larger matrices but it becomes impractical to do so by hand for matrices larger than 3×3. There is also a simple graphical device for remembering the formula for the determinant of a 3×3 matrix. To begin, duplicate the first two columns to the right of the matrix. You can then compute the determinant by multiplying the entries on six diagonals, adding the rightward products and subtracting the leftward products.

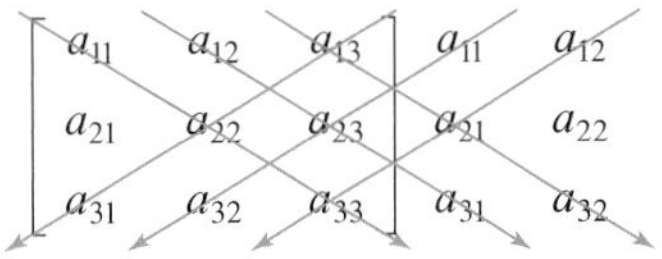

Given the determinant of a matrix we have the following theorem:

(1) Theorem If A is an $n \times n$ matrix, then A is invertible if and only if $\det A \neq 0$.

Notice that the quantity in the denominator of the formula for the inverse of a 2×2 matrix is its determinant.

EXAMPLE 4 | Which of the following matrices are invertible?

(a) $A = \begin{bmatrix} 2 \end{bmatrix}$ (b) $B = \begin{bmatrix} 2 & 3 \\ 6 & 9 \end{bmatrix}$ (c) $C = \begin{bmatrix} 5 & 3 \\ 2 & 1 \end{bmatrix}$ (d) $D = \begin{bmatrix} 10 & 7 & 3 \\ 13 & 5 & 8 \\ 6 & -1 & 7 \end{bmatrix}$

SOLUTION

(a) The matrix A is 1×1, with $\det A = 2$. Therefore it is invertible.

(b) The matrix B is 2×2, with $\det B = (2)(9) - (3)(6) = 0$. Therefore it is not invertible.

(c) The matrix C is 2×2, with $\det C = (5)(1) - (3)(2) = -1$. Therefore it is invertible.

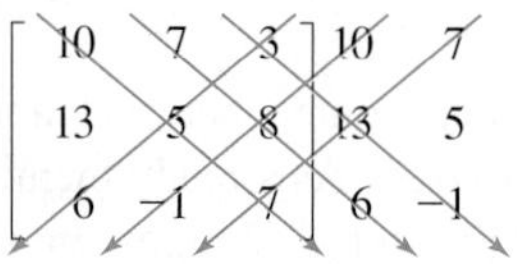

(d) The matrix D is 3×3, and

$$\begin{aligned}\det D &= (10)(5)(7) + (7)(8)(6) + (3)(13)(-1) - (3)(5)(6) - (10)(8)(-1) - (7)(13)(7)\\ &= 350 + 336 + (-39) - 90 - (-80) - 637\\ &= 0\end{aligned}$$

Therefore D is not invertible. ■

■ Solving Systems of Linear Equations

One reason that matrix inverses (and therefore determinants) are useful is for solving linear systems of n equations with n unknowns. Consider, for example, the following pair of equations with two unknowns:

$$3x_1 - 2x_2 = -4$$
$$7x_1 + x_2 = 19$$

Writing these equations in matrix notation gives

(2) $$A\mathbf{x} = \mathbf{b}$$

where $A = \begin{bmatrix} 3 & -2 \\ 7 & 1 \end{bmatrix}$ is a matrix of coefficients, $\mathbf{x} = \begin{bmatrix} x_1 \\ x_2 \end{bmatrix}$ and $\mathbf{b} = \begin{bmatrix} -4 \\ 19 \end{bmatrix}$. Such a system of equations in which $\mathbf{b} \neq \mathbf{0}$ is called an **inhomogeneous** system. In general, there are three possibilities for such a system: (i) there is a unique nonzero solution for $\mathbf{x}$, (ii) there are infinitely many solutions for $\mathbf{x}$, or (iii) there is no solution for $\mathbf{x}$.

To solve Equation 2 let's first compare its form to the scalar equation.

(3) $$ax = b$$

Provided that $a \neq 0$ we can solve Equation 3 by multiplying both sides by $a^{-1} = 1/a$ to get $a^{-1}ax = a^{-1}b$, or $x = a^{-1}b$. In a similar fashion, provided that $\det A \neq 0$ the matrix A will be invertible and we can multiply both sides of Equation 2 by A^{-1} to obtain

$$A^{-1}A\mathbf{x} = A^{-1}\mathbf{b}$$

or

$$\mathbf{x} = A^{-1}\mathbf{b}$$

Using the formula for the inverse of a 2×2 matrix, we obtain

$$\mathbf{x} = \frac{1}{(3)(1) - (-2)(7)}\begin{bmatrix} 1 & 2 \\ -7 & 3 \end{bmatrix}\begin{bmatrix} -4 \\ 19 \end{bmatrix} = \begin{bmatrix} \frac{1}{17} & \frac{2}{17} \\ -\frac{7}{17} & \frac{3}{17} \end{bmatrix}\begin{bmatrix} -4 \\ 19 \end{bmatrix} = \begin{bmatrix} \dfrac{-4 + 38}{17} \\ \dfrac{28 + 57}{17} \end{bmatrix} = \begin{bmatrix} 2 \\ 5 \end{bmatrix}$$

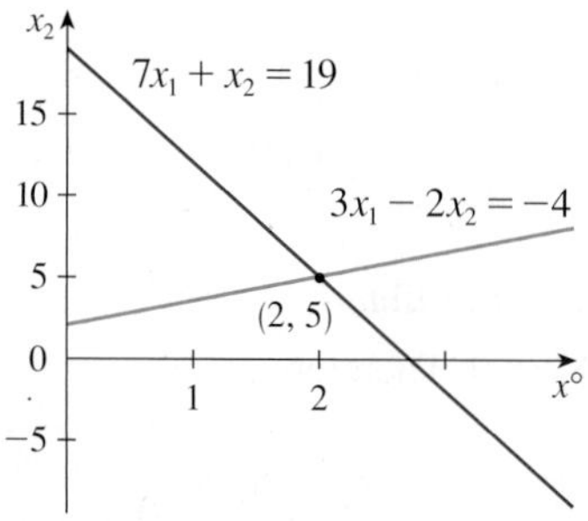

FIGURE 1

Equation 2 thus has a unique solution given by $x_1 = 2$, $x_2 = 5$. Graphically, the two equations represent straight lines in the x_1x_2-plane and their intersection point is the unique solution illustrated in Figure 1. More generally, we have the following theorem.

> **(4) Theorem** Suppose A is an $n \times n$ matrix, $\mathbf{x}$ is an $n \times 1$ vector of unknowns, and $\mathbf{b}$ is an $n \times 1$ vector of constants. If A is invertible, then the inhomogeneous system of equations $A\mathbf{x} = \mathbf{b}$ has a unique solution given by $\mathbf{x} = A^{-1}\mathbf{b}$.

If A in Theorem 4 is not invertible, then the system of equations can have either infinitely many solutions or no solution.

EXAMPLE 5 | Using $\mathbf{b} = \begin{bmatrix} 2 \\ 1 \end{bmatrix}$, obtain the solution to the inhomogeneous system of equations for each of the following matrices.

(a) $A = \begin{bmatrix} 1 & -1 \\ 2 & 1 \end{bmatrix}$ (b) $B = \begin{bmatrix} 1 & -1 \\ -1 & 1 \end{bmatrix}$ (c) $C = \begin{bmatrix} 1 & -1 \\ \frac{1}{2} & -\frac{1}{2} \end{bmatrix}$

SOLUTION

(a) First we calculate $\det A = 1 - (-2) = 3$ and therefore the matrix A is invertible. Theorem 4 then tells us that the inhomogeneous equation $A\mathbf{x} = \mathbf{b}$ has a unique solution given by $\mathbf{x} = A^{-1}\mathbf{b}$. Using the formula of the inverse of a 2×2 matrix, we obtain

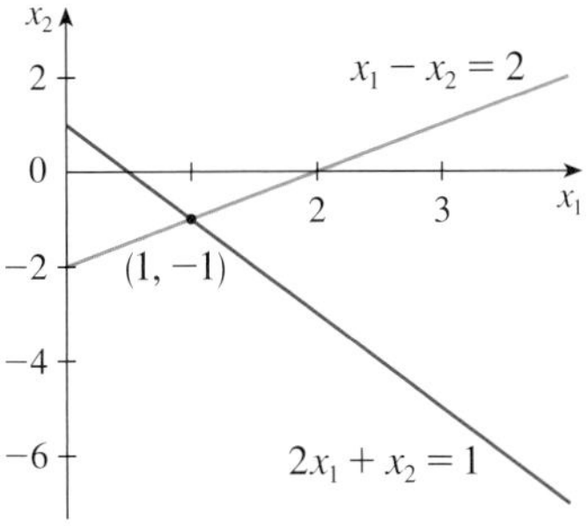

FIGURE 2

$$\mathbf{x} = \frac{1}{3}\begin{bmatrix} 1 & 1 \\ -2 & 1 \end{bmatrix}\begin{bmatrix} 2 \\ 1 \end{bmatrix} = \begin{bmatrix} \frac{1}{3} & \frac{1}{3} \\ -\frac{2}{3} & \frac{1}{3} \end{bmatrix}\begin{bmatrix} 2 \\ 1 \end{bmatrix} = \begin{bmatrix} \dfrac{2+1}{3} \\ \dfrac{-4+1}{3} \end{bmatrix} = \begin{bmatrix} 1 \\ -1 \end{bmatrix}$$

The solution corresponds to the intersection point of the lines defined by $x_1 - x_2 = 2$ and $2x_1 + x_2 = 1$, as shown in Figure 2.

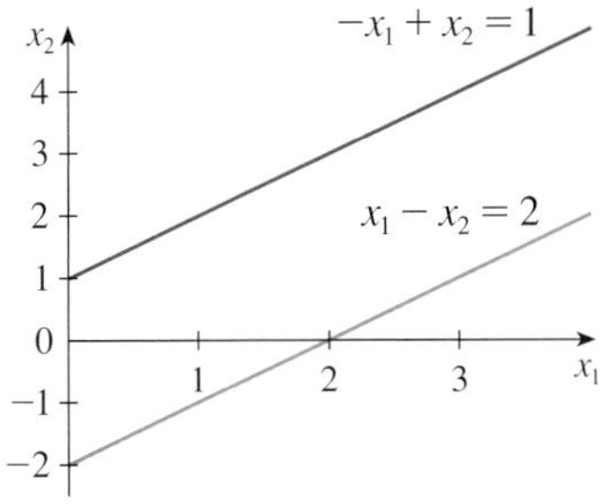

FIGURE 3

(b) Calculating the determinant of B gives $\det B = 1 - 1 = 0$. Therefore the matrix B is not invertible. The inhomogeneous equation $B\mathbf{x} = \mathbf{b}$ might have an infinite number of solutions or no solution. To determine which is the case, we need to investigate the equations in more depth. Carrying out the matrix multiplication gives the two equations $x_1 - x_2 = 2$ and $-x_1 + x_2 = 1$. Solving the second equation for x_2 gives $x_2 = 1 + x_1$, and substituting this into the first equation gives $x_1 - (1 + x_1) = 2$. This simplifies to $-1 = 2$. Because there is no choice of x_1 that will make this true, there is no solution. Graphically, the two equations represent parallel straight lines in the x_1x_2-plane and the lack of a solution corresponds to their not having an intersection point (see Figure 3).

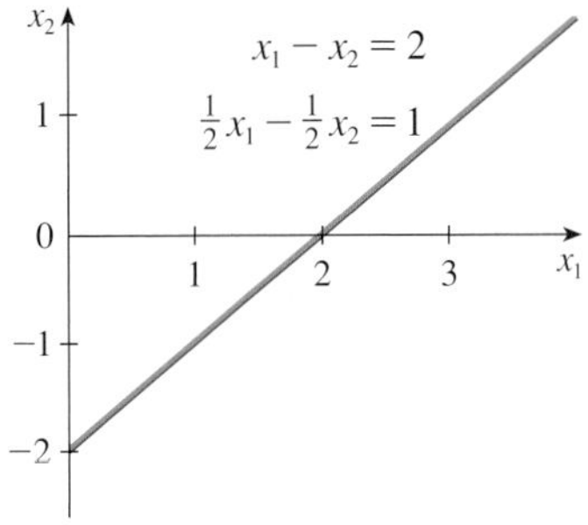

FIGURE 4

(c) Calculating the determinant of C gives $\det C = -\frac{1}{2} - \left(-\frac{1}{2}\right) = 0$. Therefore the matrix C is not invertible and the inhomogeneous equation $C\mathbf{x} = \mathbf{b}$ might have an infinite number of solutions or no solution. Carrying out the matrix multiplication gives the two equations $x_1 - x_2 = 2$ and $\frac{1}{2}x_1 - \frac{1}{2}x_2 = 1$. Solving the second equation for x_1 gives $x_1 = 2 + x_2$, and substituting this into the first equation gives $(2 + x_2) - x_2 = 2$. This simplifies to $2 = 2$. This equation holds true no matter what value of x_2 is chosen and therefore there are an infinite number of solutions. Graphically, the two equations represent the same straight line in the x_1x_2-plane, and therefore there are an infinite number of points of intersection as illustrated in Figure 4. ■

The solution to a system of linear equations is simplified considerably in the special case where $\mathbf{b} = \mathbf{0}$. This results in an equation of the form $A\mathbf{x} = \mathbf{0}$, which is called a **homogeneous** system. Clearly $\mathbf{x} = \mathbf{0}$ is always a solution and it is referred to as the **trivial solution**. There are therefore now only two possibilties: (i) the trivial solution is the unique solution, or (ii) there are an infinite number of nontrivial solutions. The following theorem tells us when each of these outcomes occurs.

(5) Theorem Suppose A is an $n \times n$ matrix, $\mathbf{x}$ is an $n \times 1$ vector of unknowns, and $\mathbf{0}$ is an $n \times 1$ vector of zeros. If A is invertible, then the homogeneous system of equations $A\mathbf{x} = \mathbf{0}$ has a unique solution given by the trivial solution $\mathbf{x} = \mathbf{0}$. If A is not invertible, then there are infinitely many nontrivial solutions.

EXERCISES 8.6

1. Determine if matrices A and B are inverses of one another.

(a) $A = \begin{bmatrix} 1 & 5 \\ 2 & 7 \end{bmatrix}$ $B = \begin{bmatrix} -\frac{7}{3} & \frac{5}{3} \\ \frac{2}{3} & -\frac{1}{3} \end{bmatrix}$

(b) $A = \begin{bmatrix} 1 & 0 & 3 \\ 2 & 7 & 9 \\ 0 & 2 & 1 \end{bmatrix}$ $B = \begin{bmatrix} -11 & 6 & -21 \\ -2 & 1 & -3 \\ 4 & -2 & 7 \end{bmatrix}$

(c) $A = \begin{bmatrix} 0 & 1 \\ 3 & 2 \end{bmatrix}$ $B = \begin{bmatrix} 2 & 3 \\ 1 & 0 \end{bmatrix}$

(d) $A = \begin{bmatrix} 0 & 1 \\ 1 & 1 \end{bmatrix}$ $B = \begin{bmatrix} -1 & 1 \\ 1 & 0 \end{bmatrix}$

(e) $A = \begin{bmatrix} 1 & 2 & 3 \\ 0 & 1 & 7 \\ 0 & 2 & 1 \end{bmatrix}$ $B = \begin{bmatrix} 1 & 0 & 1 \\ 0 & 5 & 3 \\ 7 & 0 & 1 \end{bmatrix}$

(f) $A = \begin{bmatrix} 9 & 0 \\ 2 & 3 \end{bmatrix}$ $B = \begin{bmatrix} \frac{1}{9} & 0 \\ -\frac{2}{27} & \frac{1}{3} \end{bmatrix}$

2. Find the inverse of the matrix.

(a) $\begin{bmatrix} 5 & 0 \\ 9 & 6 \end{bmatrix}$ (b) $\begin{bmatrix} 1 & 3 \\ 1 & 0 \end{bmatrix}$

(c) $\begin{bmatrix} 9 & 2 \\ 7 & 5 \end{bmatrix}$ (d) $\begin{bmatrix} 1 & 1 \\ 2 & 3 \end{bmatrix}$

(e) $\begin{bmatrix} 3x & y \\ 2x & y \end{bmatrix}$ (f) $\begin{bmatrix} x^2 & 2x \\ x^3 & x \end{bmatrix}$

3. Suppose that D is an $n \times n$ diagonal matrix with entries d_{ii}. Show that D^{-1} is an $n \times n$ diagonal matrix with entries $1/d_{ii}$.

4. Suppose A is an nonsingular 2×2 matrix. Derive the formula for its inverse, namely,

$$A^{-1} = \frac{1}{a_{11}a_{22} - a_{12}a_{21}} \begin{bmatrix} a_{22} & -a_{12} \\ -a_{21} & a_{11} \end{bmatrix}$$

5. Find all 2×2 matrices A such that $\det A = 1$ and $A = A^{-1}$.

6. If A is nonsingular, then $(A^{-1})^{-1} = A$.
(a) Verify this theorem for 2×2 matrices.
(b) Prove the theorem for any $n \times n$ matrix.

7. Suppose that A and B are nonsingular matrices. Then AB is also nonsingular. Furthermore, a theorem from linear algebra then states that $(AB)^{-1} = B^{-1}A^{-1}$.
(a) Verify this theorem for 2×2 matrices.
(b) Prove the theorem for any $n \times n$ matrix.

8. Suppose that A is nonsingular. Prove that its inverse A^{-1} is unique. [*Hint:* Show that if there exists two matrices, B and C, such that $BA = I$ and $AC = I$, then necessarily $B = C$.]

9. If A is nonsingular, then $(A^T)^{-1} = (A^{-1})^T$.
(a) Verify this theorem for 2×2 matrices.
(b) Prove the theorem for any $n \times n$ matrix. [*Hint:* You might wish to use the property from Exercise 8.4.15(b).]

10. Use the determinant to decide whether each matrix is singular or nonsingular.

(a) $\begin{bmatrix} 3 & 2 \\ 6 & 4 \end{bmatrix}$

(b) $\begin{bmatrix} a & b \\ 7a & 9b \end{bmatrix}$ $a \neq 0,\ b \neq 0$

(c) $\begin{bmatrix} 1 & 0 & 5 \\ 4 & 2 & 6 \\ 3 & 2 & 1 \end{bmatrix}$

(d) $\begin{bmatrix} 9 & 1 & 0 \\ 1 & 0 & 1 \\ -3 & 2 & 0 \end{bmatrix}$

(e) $\begin{bmatrix} x & x + x^2 \\ 3x & 0 \end{bmatrix}$ $x \neq 0,\ x \neq -1$

(f) $\begin{bmatrix} y^3 & 4y^3 \\ 3x^2 & 12x^2 \end{bmatrix}$

11. For which values of a is the matrix singular?

$$\begin{bmatrix} a & 4 \\ 2 & 8 \end{bmatrix}$$

12. For which values of a and b is the matrix singular?

$$\begin{bmatrix} a & b \\ 7 & 9 \end{bmatrix}$$

13. For which values of a is the matrix singular?

$$\begin{bmatrix} 1 & 1 & 1 \\ 1 & 2 & a \\ 1 & 4 & a^2 \end{bmatrix}$$

14. If A and B are $n \times n$ matrices, then $\det(AB) = \det A \det B$. Verify this theorem for 2×2 matrices.

15. Using the theorem from Exercise 14, what can you say about the determinant of the matrix power A^k, where k is a positive integer, if the matrix A is nonsingular?

16. If A is an $n \times n$ matrix, then $\det A = \det A^T$. Verify this theorem for 2×2 matrices.

17. If D is an $n \times n$ diagonal matrix with entries d_{ii}, then $\det D = d_{11}d_{22} \cdots d_{nn}$. Verify this theorem for 2×2 matrices.

18. Write each system of linear equations in the form $A\mathbf{x} = \mathbf{b}$ by identifying the matrix A and the vectors $\mathbf{x}$ and $\mathbf{b}$.

(a) $3x_1 - x_2 = 9$
$2x_1 + 9x_2 = -10$

(b) $2x_1 = 10$
$3x_1 - x_2 = 14$

(c) $x_1 + x_2 - x_3 = 0$
$2x_1 - 2x_2 + 9x_3 = 5$
$x_1 + 9x_3 = 1$

(d) $2x_1 - x_2 + 9x_3 = 12$
$x_1 - \frac{1}{2}x_2 + x_3 = 1$
$x_2 = 2$

19. For which values of k does the system of linear equations have zero, one, or an infinite number of solutions? [*Note:* Not all three possibilities need occur.]

$$3x_1 + x_2 = 2$$
$$kx_1 + 2x_2 = 4$$

20. For which values of k does the system of linear equations have zero, one, or an infinite number of solutions? [*Note:* Not all three possibilities need occur.]

$$2x_1 - x_2 = 3$$
$$4x_1 - 2x_2 = k$$

21. Consider the linear system of equations

$$a_{11}x_1 + a_{12}x_2 = b_1$$
$$a_{21}x_1 + a_{22}x_2 = b_2$$

Suppose that the matrix $A = \begin{bmatrix} a_{11} & a_{12} \\ a_{21} & a_{22} \end{bmatrix}$ is nonsingular. Derive expressions for the solutions x_1 and x_2.

22. Suppose that $\mathbf{x} = \mathbf{p}$ and $\mathbf{x} = \mathbf{q}$ are both solutions to the inhomogeneous system of equations $A\mathbf{x} = \mathbf{b}$. Show that $\mathbf{z} = \alpha\mathbf{p} + (1 - \alpha)\mathbf{q}$ is then also a solution, where α is a scalar.

23. Suppose that $\mathbf{x} = \mathbf{p}$ and $\mathbf{x} = \mathbf{q}$ are both solutions to the homogeneous system of equations $A\mathbf{x} = \mathbf{0}$. Show that $\mathbf{z} = \alpha\mathbf{p} + \beta\mathbf{q}$ is then also a solution, where α and β are scalars.

24–29 Find all solutions to the system of linear equations.

24. $x_1 + 2x_2 = 0$
$2x_1 + 5x_2 = 1$

25. $x_1 - 3x_2 = 5$
$x_1 + x_2 = 2$

26. $x_1 = 1$
$2x_1 + 3x_2 = 6$

27. $3x_1 + 6x_2 = 2$
$9x_1 + 12x_2 = 1$

28. $5x_1 + x_2 = 0$
$25x_1 + 5x_2 = 0$

29. $8x_1 + 4x_2 = 4$
$4x_1 + 2x_2 = 2$

---

30. Vectorcardiography Suppose that the voltage vector $\mathbf{v}_t$ of the heart changes from one beat to the next according to the equation

$$\mathbf{v}_{t+1} = \begin{bmatrix} 1 & 0 \\ 0 & -1 \end{bmatrix} \mathbf{v}_t$$

If the voltage vector during the current heartbeat is $\begin{bmatrix} 0.3 \\ -0.2 \end{bmatrix}$, what was the voltage vector in the previous heartbeat?

31. Antigenic evolution Suppose the vector $\mathbf{x}_t$ characterizing the antigenic state of an influenza virus population changes from one season to the next according to the equation

$$\mathbf{x}_{t+1} = \begin{bmatrix} 2 & 3 \\ 0 & 0.9 \end{bmatrix} \mathbf{x}_t$$

If the vector in the current season is $\begin{bmatrix} 8 \\ 1.8 \end{bmatrix}$, what was the vector in the previous influenza season?

32. Find a second-degree polynomial (that is, an equation of the form $y = a + bx + cx^2$) that goes through the three points $(0, 1)$, $(1, 0)$, and $(-1, 0)$. Is there more than one possibility?

33. Leslie matrices Consider the following model for the population size $\mathbf{n}_t$ of an age-structured population with two age classes:

$$\mathbf{n}_{t+1} = \begin{bmatrix} b & 2 \\ \frac{1}{2} & 0 \end{bmatrix} \mathbf{n}_t$$

An equilibrium is a value of the vector for which no change occurs (that is, $\mathbf{n}_{t+1} = \mathbf{n}_t$). Denoting such values by $\hat{\mathbf{n}}$, they must therefore satisfy the equation

$$\hat{\mathbf{n}} = \begin{bmatrix} b & 2 \\ \frac{1}{2} & 0 \end{bmatrix} \hat{\mathbf{n}}$$

(a) Suppose that $b \neq 0$. Find all possible equilibrium values.

(b) Suppose that $b = 0$. Find all possible equilibrium values.

34. Mutation Suppose that, as a result of mutation, the number of individuals in a population carrying allele A and allele B changes in a way described by the equation

$$\mathbf{y}_{t+1} = \begin{bmatrix} 0.95 & 0 \\ 0.05 & 1 \end{bmatrix} \mathbf{y}_t$$

where $\mathbf{y}_t$ is the vector whose components are the numbers of individuals carrying each allele at time t. An equilibrium is a value of the vector for which no change occurs (that is, $\mathbf{y}_{t+1} = \mathbf{y}_t$). Denoting such values by $\hat{\mathbf{y}}$, they must therefore satisfy the equation

$$\hat{\mathbf{y}} = \begin{bmatrix} 0.95 & 0 \\ 0.05 & 1 \end{bmatrix} \hat{\mathbf{y}}$$

Find all possible equilibrium values.

35. Vectorcardiography Suppose that the voltage vector $\mathbf{v}_t$ of the heart changes from one beat to the next according to the equation

$$\mathbf{v}_{t+1} = \begin{bmatrix} 1 & 0 \\ 0 & -1 \end{bmatrix} \mathbf{v}_t$$

An equilibrium is a value of the vector for which no change occurs (that is, $\mathbf{v}_{t+1} = \mathbf{v}_t$). Denoting such values by $\hat{\mathbf{v}}$, they must therefore satisfy the equation

$$\hat{\mathbf{v}} = \begin{bmatrix} 1 & 0 \\ 0 & -1 \end{bmatrix} \hat{\mathbf{v}}$$

Find all possible equilibrium values.

36. Resource allocation Each day an organism has 100 J of energy to divide between growth and reproduction. Each millimeter of growth costs 3 J and each egg produced costs 5 J. Denote the amount of growth per day by x_1 and the number of eggs produced per day by x_2.

(a) Suppose that the organism divides its energy so that for every millimeter of growth that occurs each day, it produces (on average) two eggs. What is the amount of growth and number of eggs produced on each day?

(b) Suppose that the organism divides its energy each day so that the total energy spent on eggs is twice that spent on growth. What is the amount of growth and number of eggs produced on each day?

PROJECT Cubic Splines

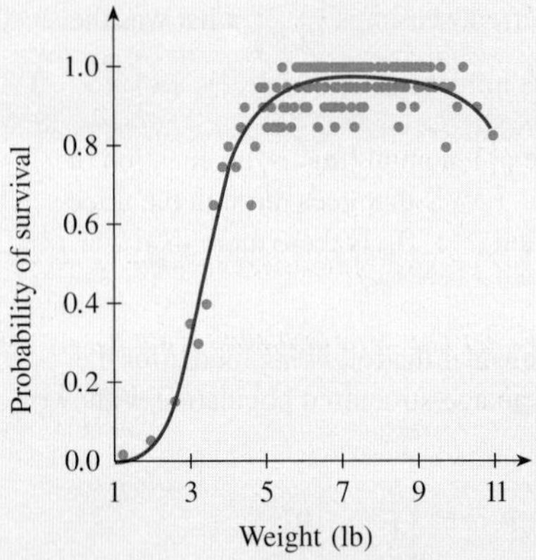

FIGURE 1

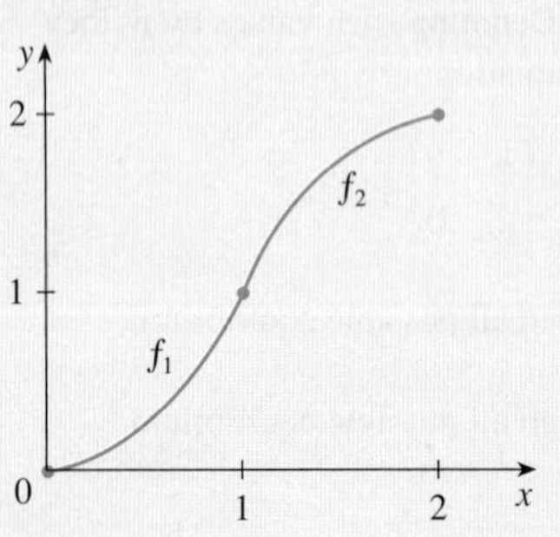

FIGURE 2

In many situations it is desirable to construct a curve that goes through specified points with certain slope and curvature properties at these points. For example, if we are interested in the relationship between birth weight and infant survival, as illustrated in Figure 1, we might wish to fit a relatively flexible curve to data.[1]

One way to do this is to construct a curve that is made up of cubic polynomials, "pieced together" at specific points. Such a curve is called a *cubic spline*. We consider how to fit curves to data in more detail in Chapter 11, but for the present purposes we will simply construct a cubic spline that satisfies certain conditions.

Suppose we wish to construct a cubic spline that is made up of two segments and satisfies the following properties: (i) the spline goes through the points (0, 0), (1, 1) and (2, 2); (ii) the two segments are joined together at the point (1, 1); (iii) the slope and second derivative of the spline are both zero at the points (0, 0) and (2, 2); and (iv) the spline has continuous first and second derivatives everywhere between 0 and 2. Such a curve is shown in Figure 2.

Each segment of the cubic spline consists of a cubic polynomial of the form $f_i = a_i + b_i x + c_i x^2 + d_i x^3$.

1. Consider the segment f_1 that joins the points (0, 0) and (1, 1). Obtain a system of four linear equations in four unknowns that characterizes the coefficients a_1, b_1, c_1, and d_1, and write it in matrix form.

2. Solve the system of equations from Problem 1.

3. Consider the segment f_2 that joins the points (1, 1) and (2, 2). Obtain a system of four linear equations in four unknowns that characterizes the coefficients a_2, b_2, c_2, and d_2, and write it in matrix form.

CAS **4.** Solve the system of equations from Problem 3.

1. D. Schluter, "Estimating the Form of Natural Selection on a Quantitative Trait," *Evolution* 42 (1988): 849–61.

For Problem 2 you should be able to solve the system of equations by inspection. For Problem 4, however, to solve the system of equations using matrix inverses you will need to calculate the inverse of a 4×4 matrix by using a computer algebra system.

8.7 Eigenvectors and Eigenvalues

The mathematical background of Section 8.6 can now be used to return to our study of matrix models. In this section we introduce eigenvectors and eigenvalues. In Section 8.8 we then show how they are fundamental to understanding the behavior of matrix models.

■ Characterizing How Matrix Multiplication Changes Vectors

In Section 8.5 we studied models for the dynamics of vectors having the form

$$\mathbf{n}_{t+1} = A\mathbf{n}_t$$

where $\mathbf{n}_t$ is a vector of variables and A is a square matrix. Our goal now is to characterize in general what such multiplication of a vector by a matrix does to the vector.

As an example, suppose that

$$A = \begin{bmatrix} 1 & \frac{1}{2} \\ 1 & \frac{3}{2} \end{bmatrix}$$

Figure 1 shows how this matrix changes the vector from one time step to the next when $\mathbf{n}_0 = \begin{bmatrix} -3 \\ 4 \end{bmatrix}$. The changes in the vector look quite complicated. For example, the direction of the vector keeps changing and sometimes the vector is compressed from one step to the next (for example, between time 0 and time 1) and sometimes it is stretched (for example, between time 2 and time 3).

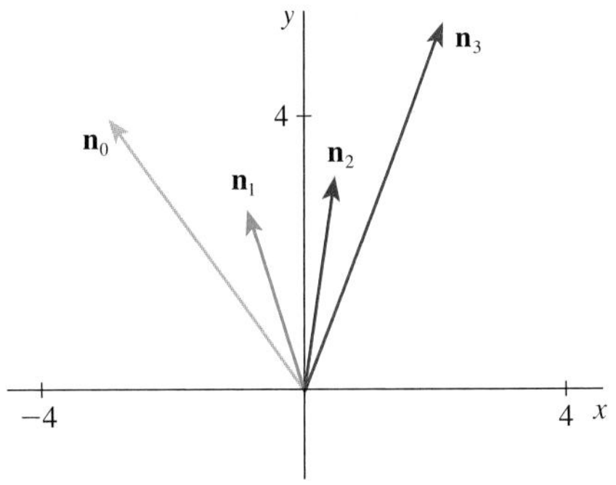

FIGURE 1

If you were to experiment with different initial vectors $\mathbf{n}_0$, you would get different patterns. However, you might happen across an initial vector that results in a particularly simple result. For example, suppose $\mathbf{n}_0 = \begin{bmatrix} 1 \\ 2 \end{bmatrix}$. In this case the vector remains on its initial axis, and it is simply stretched by a factor of 2 each time step, as shown in Figure 2(a). Likewise, if $\mathbf{n}_0 = \begin{bmatrix} -1 \\ 1 \end{bmatrix}$ then again the vector remains on its initial axis, but now it is compressed by a factor of $\frac{1}{2}$ each time step, as shown in Figure 2(b). Thus, for some special initial vectors, the change in the vector is relatively easy to describe.

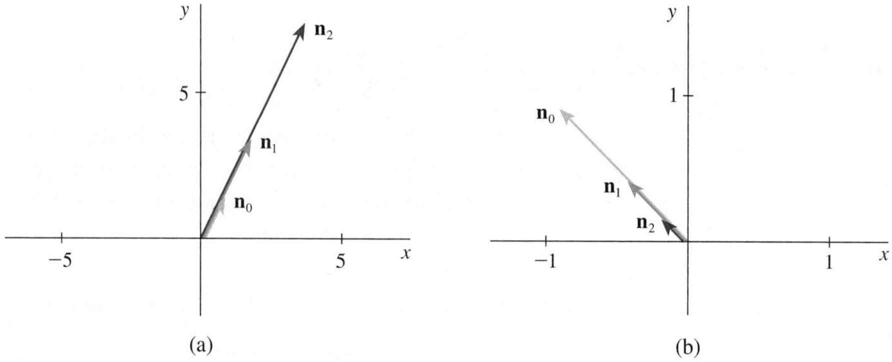

FIGURE 2

As we will see in Section 8.8, such special vectors play a central role in understanding the behavior of matrix models. These vectors are called *eigenvectors* of the matrix A, and the factor by which each vector is stretched or shrunk is called its corresponding *eigenvalue.* In the remainder of this section we will learn how to calculate eigenvectors and eigenvalues. We will see that a matrix of size n has exactly n eigenvalues, although our focus will be primarily on $n = 2$. We also focus on matrices whose eigenvalues are distinct (that is, none of them have the same value). This is a sufficient condition for each eigenvalue to be associated with its own distinct eigenvector. Matrices for which this is not true are called **defective** and are studied in courses on linear algebra.

Eigenvectors and Eigenvalues

An eigenvector of a matrix is a vector that, when multiplied by the matrix, is simply changed in length.

(1) Definition Suppose that A is an $n \times n$ matrix. A *nonzero* vector $\mathbf{v}$ that satisfies the equation

$$A\mathbf{v} = \lambda\mathbf{v}$$

is called an **eigenvector** of the matrix A. The scalar λ is the **eigenvalue** associated with this eigenvector.

Definition 1 provides a precise description of the idea that eigenvectors always remain on their initial axis when multiplied by a matrix. If $\mathbf{v}$ is an eigenvector of A, then when we multiply $\mathbf{v}$ by A we obtain the vector $\mathbf{v}$ multiplied by a scalar λ.

How do we find eigenvectors and eigenvalues? As the following calculations reveal, it is easiest to first calculate the eigenvalues of a matrix and then calculate their associated eigenvectors.

Consider the matrix $A = \begin{bmatrix} 1 & \frac{1}{2} \\ 1 & \frac{3}{2} \end{bmatrix}$ from our earlier example. From Definition (1) we have

$$A\mathbf{v} = \lambda\mathbf{v}$$

or, equivalently,

$$A\mathbf{v} - \lambda\mathbf{v} = \mathbf{0}$$

where $\mathbf{0}$ is the zero vector. To proceed further, we factor out the vector $\mathbf{v}$. To do so, we must first multiply λ by the identity matrix I in order for the elements of the equation to maintain compatible sizes:

(2) $$(A - \lambda I)\mathbf{v} = \mathbf{0}$$

Notice that, although λ is a scalar, $A - \lambda I$ is a 2×2 matrix.

We now seek a vector $\mathbf{v}$ that satisfies Equation 2. Recalling Theorem 8.6.5, we see that, if the matrix $A - \lambda I$ is invertible, then the only solution is the trivial solution $\mathbf{v} = \mathbf{0}$. Therefore, to have a nonzero eigenvector $\mathbf{v}$, we require that $A - \lambda I$ be singular. From Theorem 8.6.1, this requires that

$$\det(A - \lambda I) = 0$$

Our considerations have allowed us to remove $\mathbf{v}$ from the equation, and therefore we now have an equation that determines the eigenvalues λ. Calculating the matrix $A - \lambda I$, we get

$$A - \lambda I = \begin{bmatrix} 1 - \lambda & \frac{1}{2} \\ 1 & \frac{3}{2} - \lambda \end{bmatrix}$$

and, using the definition of the determinant of a 2×2 matrix, we have

$$\begin{aligned} \det\begin{bmatrix} 1 - \lambda & \frac{1}{2} \\ 1 & \frac{3}{2} - \lambda \end{bmatrix} &= (1 - \lambda)\left(\tfrac{3}{2} - \lambda\right) - \tfrac{1}{2} \cdot 1 \\ &= \tfrac{3}{2} - \lambda - \tfrac{3}{2}\lambda + \lambda^2 - \tfrac{1}{2} \\ &= \lambda^2 - \tfrac{5}{2}\lambda + 1 \end{aligned}$$

Setting this result equal to zero and multiplying the equation by 2 then gives $2\lambda^2 - 5\lambda + 2 = 0$. This can be factored to produce $(\lambda - 2)(2\lambda - 1) = 0$, the solutions of which are $\lambda = 2$ and $\lambda = \frac{1}{2}$. These are the eigenvalues of A.

We can now calculate the eigenvector associated with each of the eigenvalues $\lambda = 2$ and $\lambda = \frac{1}{2}$. First let's find the eigenvector associated with $\lambda = 2$. We seek a vector $\mathbf{v}$ such that, when $\lambda = 2$ is substituted into the left side of Equation 2, we get the zero vector. In other words,

$$\begin{bmatrix} 1 - \lambda & \frac{1}{2} \\ 1 & \frac{3}{2} - \lambda \end{bmatrix}\begin{bmatrix} v_1 \\ v_2 \end{bmatrix} = \begin{bmatrix} 1 - 2 & \frac{1}{2} \\ 1 & \frac{3}{2} - 2 \end{bmatrix}\begin{bmatrix} v_1 \\ v_2 \end{bmatrix} = \begin{bmatrix} 0 \\ 0 \end{bmatrix}$$

This gives the following pair of equations in two unknowns:

$$-v_1 + \tfrac{1}{2}v_2 = 0$$

$$v_1 - \tfrac{1}{2}v_2 = 0$$

These two equations are redundant because both specify that $2v_1 = v_2$. As a result, there are infinitely many solutions—we are free to choose either v_1 or v_2 arbitrarily, and the other is then determined by this choice.

To make our calculations simple, it is usually best to work with whole numbers. For example, we might choose $v_1 = 1$, in which case we then have $v_2 = 2$. The eigenvector associated with eigenvalue $\lambda = 2$ is then $\mathbf{v} = \begin{bmatrix} 1 \\ 2 \end{bmatrix}$. If we had made a different choice for v_1, we would have ended up with a different vector $\mathbf{v}$. For example, if we choose $v_1 = 2$, we would then have $v_2 = 4$ and thus $\mathbf{v} = \begin{bmatrix} 2 \\ 4 \end{bmatrix}$. Regardless of our choice, however, all the resulting vectors are eigenvectors and all lie on the same line because they all have the form $\mathbf{v} = \begin{bmatrix} a \\ 2a \end{bmatrix}$ or $\mathbf{v} = a\begin{bmatrix} 1 \\ 2 \end{bmatrix}$ for some scalar a as shown in Figure 3.

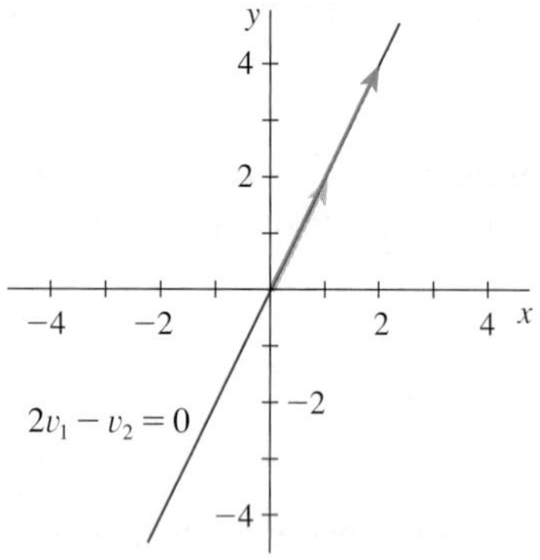

FIGURE 3
All eigenvectors associated with the eigenvalue $\lambda = 2$ lie on the red line.

We now calculate the eigenvector associated with $\lambda = \frac{1}{2}$. We obtain

$$\begin{bmatrix} 1-\lambda & \frac{1}{2} \\ 1 & \frac{3}{2}-\lambda \end{bmatrix}\begin{bmatrix} v_1 \\ v_2 \end{bmatrix} = \begin{bmatrix} 1-\frac{1}{2} & \frac{1}{2} \\ 1 & \frac{3}{2}-\frac{1}{2} \end{bmatrix}\begin{bmatrix} v_1 \\ v_2 \end{bmatrix} = \begin{bmatrix} 0 \\ 0 \end{bmatrix}$$

which gives the pair of equations

$$\begin{aligned} \tfrac{1}{2}v_1 + \tfrac{1}{2}v_2 &= 0 \\ v_1 + v_2 &= 0 \end{aligned}$$

Again, these equations are redundant, both specifying that $v_1 = -v_2$. Choosing $v_2 = 1$, we then have $v_1 = -1$ and therefore $\mathbf{v} = \begin{bmatrix} -1 \\ 1 \end{bmatrix}$. Again the choice $v_2 = 1$ is arbitrary, but all choices result in vectors lying on the same line because they all have the form $\mathbf{v} = a\begin{bmatrix} -1 \\ 1 \end{bmatrix}$ for some scalar a. (See Figure 4.)

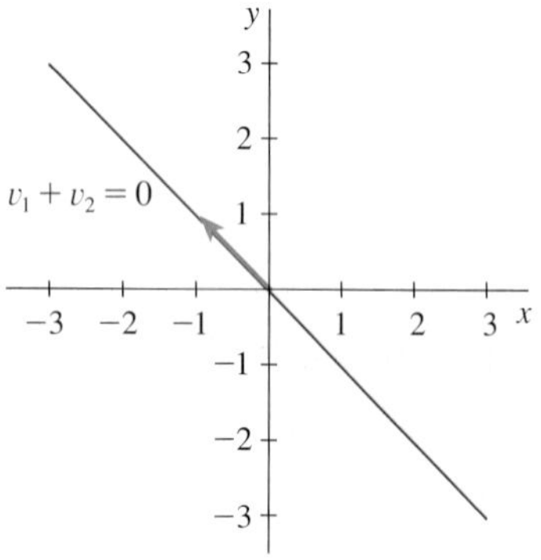

FIGURE 4
All eigenvectors associated with the eigenvalue $\lambda = \frac{1}{2}$ lie on the red line.

Note that the eigenvector-eigenvalue pairs we just obtained are exactly those identified in Figure 2.

EXAMPLE 1 | Consider the matrix $A = \begin{bmatrix} 2 & 2 \\ 1 & 3 \end{bmatrix}$.

(a) Find its eigenvalues.
(b) Find the eigenvectors associated with the eigenvalues from part (a).

SOLUTION

(a) Using the equation $\det(A - \lambda I) = 0$, we get

$$\det\begin{bmatrix} 2-\lambda & 2 \\ 1 & 3-\lambda \end{bmatrix} = 0$$

or

$$(2-\lambda)(3-\lambda) - 2 = \lambda^2 - 5\lambda + 4 = (\lambda - 1)(\lambda - 4) = 0$$

The eigenvalues are therefore $\lambda = 1$ and $\lambda = 4$.

(b) Beginning with eigenvalue $\lambda = 1$, we can use Equation 2 to obtain

$$\begin{bmatrix} 2-\lambda & 2 \\ 1 & 3-\lambda \end{bmatrix}\begin{bmatrix} v_1 \\ v_2 \end{bmatrix} = \begin{bmatrix} 2-1 & 2 \\ 1 & 3-1 \end{bmatrix}\begin{bmatrix} v_1 \\ v_2 \end{bmatrix} = \begin{bmatrix} 0 \\ 0 \end{bmatrix}$$

This is the pair of equations

$$\begin{aligned} v_1 + 2v_2 &= 0 \\ v_1 + 2v_2 &= 0 \end{aligned}$$

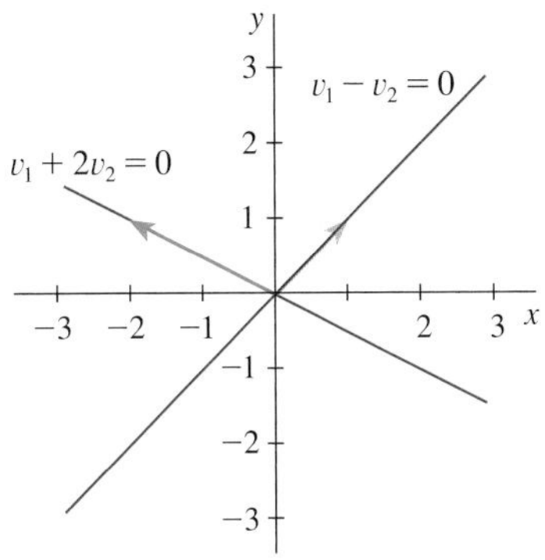

FIGURE 5
The eigenvectors for both eigenvalues, along with red lines indicating all scalar multiples of these eigenvectors.

These are identical, both specifying that $v_1 = -2v_2$. We choose (arbitrarily) $v_2 = 1$, giving $v_1 = -2$ and thus $\mathbf{v} = \begin{bmatrix} -2 \\ 1 \end{bmatrix}$.

For the eigenvalue $\lambda = 4$, we use Equation 2 to obtain

$$\begin{bmatrix} 2-\lambda & 2 \\ 1 & 3-\lambda \end{bmatrix}\begin{bmatrix} v_1 \\ v_2 \end{bmatrix} = \begin{bmatrix} 2-4 & 2 \\ 1 & 3-4 \end{bmatrix}\begin{bmatrix} v_1 \\ v_2 \end{bmatrix} = \begin{bmatrix} 0 \\ 0 \end{bmatrix}$$

or, equivalently,

$$\begin{aligned} -2v_1 + 2v_2 &= 0 \\ v_1 - v_2 &= 0 \end{aligned}$$

Both equations specify that $v_1 = v_2$ and we choose (arbitrarily) $v_1 = 1$, giving $v_2 = 1$ and thus $\mathbf{v} = \begin{bmatrix} 1 \\ 1 \end{bmatrix}$. (See Figure 5.) ■

EXAMPLE 2 | Consider the matrix $A = \begin{bmatrix} 2 & 1 \\ 1 & \frac{1}{2} \end{bmatrix}$.

(a) Find its eigenvalues.
(b) Find the eigenvectors associated with the eigenvalues from part (a).

SOLUTION

(a) Using the equation $\det(A - \lambda I) = 0$, we get

$$\det\begin{bmatrix} 2-\lambda & 1 \\ 1 & \frac{1}{2}-\lambda \end{bmatrix} = 0$$

or

$$(2-\lambda)\left(\tfrac{1}{2}-\lambda\right) - 1 = \lambda\left(\lambda - \tfrac{5}{2}\right) = 0$$

The eigenvalues are therefore $\lambda = 0$ and $\lambda = \frac{5}{2}$.

(b) Beginning with eigenvalue $\lambda = 0$, we have

$$\begin{bmatrix} 2-\lambda & 1 \\ 1 & \frac{1}{2}-\lambda \end{bmatrix}\begin{bmatrix} v_1 \\ v_2 \end{bmatrix} = \begin{bmatrix} 2-0 & 1 \\ 1 & \frac{1}{2}-0 \end{bmatrix}\begin{bmatrix} v_1 \\ v_2 \end{bmatrix} = \begin{bmatrix} 0 \\ 0 \end{bmatrix}$$

or, equivalently,

$$\begin{aligned} 2v_1 + v_2 &= 0 \\ v_1 + \tfrac{1}{2}v_2 &= 0 \end{aligned}$$

Both equations specify that $2v_1 = -v_2$. We choose $v_1 = 1$, giving the eigenvector $\mathbf{v} = \begin{bmatrix} 1 \\ -2 \end{bmatrix}$ associated with the eigenvalue $\lambda = 0$.

For the eigenvalue $\lambda = \frac{5}{2}$, we have

$$\begin{bmatrix} 2-\lambda & 1 \\ 1 & \frac{1}{2}-\lambda \end{bmatrix}\begin{bmatrix} v_1 \\ v_2 \end{bmatrix} = \begin{bmatrix} 2-\frac{5}{2} & 1 \\ 1 & \frac{1}{2}-\frac{5}{2} \end{bmatrix}\begin{bmatrix} v_1 \\ v_2 \end{bmatrix} = \begin{bmatrix} 0 \\ 0 \end{bmatrix}$$

or, equivalently,

$$\begin{aligned} -\tfrac{1}{2}v_1 + v_2 &= 0 \\ v_1 - 2v_2 &= 0 \end{aligned}$$

These equations both specify that $v_1 = 2v_2$. We choose $v_2 = 1$, giving the eigenvector $\mathbf{v} = \begin{bmatrix} 2 \\ 1 \end{bmatrix}$ associated with the eigenvalue $\lambda = \frac{5}{2}$. (See Figure 6).

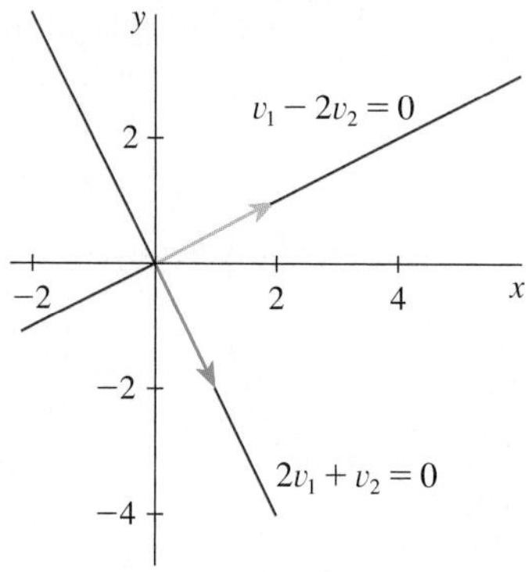

FIGURE 6
The eigenvectors for both eigenvalues, along with red lines indicating all scalar multiples of these eigenvectors.

This example illustrates that a matrix can have an eigenvalue that is zero. In Example 8.8.2, we will see what happens geometrically when vectors are multiplied by such matrices. ■

It is possible to derive an explicit formula for the eigenvalues of an arbitrary 2×2 matrix. Consider the matrix

$$A = \begin{bmatrix} a_{11} & a_{12} \\ a_{21} & a_{22} \end{bmatrix}$$

Again, using the equation $\det(A - \lambda I) = 0$, we get

$$\det \begin{bmatrix} a_{11} - \lambda & a_{12} \\ a_{21} & a_{22} - \lambda \end{bmatrix} = 0$$

or

$$(a_{11} - \lambda)(a_{22} - \lambda) - a_{12}a_{21} = \lambda^2 - (a_{11} + a_{22})\lambda + (a_{11}a_{22} - a_{12}a_{21}) = 0$$

The last equation is a quadratic polynomial in λ, and it can therefore be solved using the quadratic formula (or perhaps by factoring).

In the preceding calculation we saw that the equation $\det(A - \lambda I) = 0$ is a second-degree polynomial in λ. Although we won't prove it, for a square matrix A of size n, the equation $\det(A - \lambda I) = 0$ that determines its eigenvalues is an nth-degree polynomial in λ. This polynomial is referred to as the **characteristic polynomial** of the matrix A. Our focus on matrices that have distinct eigenvalues is therefore a focus on matrices whose characteristic polynomial has no repeated roots.

Given that the eigenvalues of a matrix are the roots of a polynomial, we also must expect that eigenvalues are sometimes complex numbers. In general, if we define the quantity $i = \sqrt{-1}$, the eigenvalues of matrices whose entries are real numbers always come in complex conjugate pairs, having the form $\lambda = a + bi$ and $\lambda = a - bi$ for some real numbers a and b. (See Exercise 8.8.27.)

Appendix G provides a review of complex numbers.

EXAMPLE 3 | Consider the matrix $A = \begin{bmatrix} \sqrt{3} & -1 \\ 1 & \sqrt{3} \end{bmatrix}$.

(a) Find its eigenvalues.
(b) Find the eigenvectors associated with the eigenvalues from part (a).

(c) Consider an arbitrary initial vector $\mathbf{n}_0 = \begin{bmatrix} x_0 \\ y_0 \end{bmatrix}$. How does multiplication of $\mathbf{n}_0$ by A affect the length of this vector?

(d) Use the dot product to determine how multiplication by A affects the direction of the vector.

(e) Describe, overall, what multiplication by the matrix A does to vectors.

SOLUTION

(a) Calculating the characteristic polynomial and solving it using the quadratic formula gives the eigenvalues:

$$\begin{aligned}\lambda &= \tfrac{1}{2}\left[2\sqrt{3} \pm \sqrt{(-2\sqrt{3})^2 - 4(4)}\right] \\ &= \tfrac{1}{2}\left(2\sqrt{3} \pm \sqrt{12 - 16}\right) \\ &= \sqrt{3} \pm i\end{aligned}$$

(b) Beginning with eigenvalue $\lambda = \sqrt{3} + i$, we have

$$\begin{bmatrix} \sqrt{3} - \lambda & -1 \\ 1 & \sqrt{3} - \lambda \end{bmatrix}\begin{bmatrix} v_1 \\ v_2 \end{bmatrix} = \begin{bmatrix} -i & -1 \\ 1 & -i \end{bmatrix}\begin{bmatrix} v_1 \\ v_2 \end{bmatrix} = \begin{bmatrix} 0 \\ 0 \end{bmatrix}$$

Both of these equations specify that $v_2 = -iv_1$. We choose $v_1 = 1$, giving $\mathbf{v} = \begin{bmatrix} 1 \\ -i \end{bmatrix}$.

For eigenvalue $\lambda = \sqrt{3} - i$, we have

$$\begin{bmatrix} \sqrt{3} - \lambda & -1 \\ 1 & \sqrt{3} - \lambda \end{bmatrix}\begin{bmatrix} v_1 \\ v_2 \end{bmatrix} = \begin{bmatrix} i & -1 \\ 1 & i \end{bmatrix}\begin{bmatrix} v_1 \\ v_2 \end{bmatrix} = \begin{bmatrix} 0 \\ 0 \end{bmatrix}$$

These equations both specify that $v_2 = iv_1$. We choose $v_1 = 1$, giving $\mathbf{v} = \begin{bmatrix} 1 \\ i \end{bmatrix}$. Notice that, when the eigenvalues are complex, their corresponding eigenvectors are also complex.

(c) Carrying out the matrix multiplication with the initial vector $\mathbf{n}_0 = \begin{bmatrix} x_0 \\ y_0 \end{bmatrix}$ gives

$$\mathbf{n}_1 = \begin{bmatrix} \sqrt{3} & -1 \\ 1 & \sqrt{3} \end{bmatrix}\begin{bmatrix} x_0 \\ y_0 \end{bmatrix} = \begin{bmatrix} \sqrt{3}x_0 - y_0 \\ x_0 + \sqrt{3}y_0 \end{bmatrix}$$

The length of the initial vector is $|\mathbf{n}_0| = \sqrt{x_0^2 + y_0^2}$, and the length of the vector $\mathbf{n}_1$ is

$$\begin{aligned}|\mathbf{n}_1| &= \sqrt{(\sqrt{3}x_0 - y_0)^2 + (x_0 + \sqrt{3}y_0)^2} \\ &= \sqrt{3x_0^2 - 2\sqrt{3}x_0y_0 + y_0^2 + x_0^2 + 2\sqrt{3}x_0y_0 + 3y_0^2} \\ &= 2\sqrt{x_0^2 + y_0^2}\end{aligned}$$

Therefore multiplication by A increases the length of the vector by a factor of 2.

(d) From the definition of the dot product, we have

$$\cos\theta = \frac{\mathbf{n}_0 \cdot \mathbf{n}_1}{|\mathbf{n}_0||\mathbf{n}_1|}$$

where θ is the angle between $\mathbf{n}_0$ and $\mathbf{n}_1$. Substituting the vectors $\mathbf{n}_0$ and $\mathbf{n}_1$ gives

Here we are writing $\mathbf{n}_0$ and $\mathbf{n}_1$ as row vectors.

$$\cos\theta = \frac{[x_0, y_0] \cdot \left[\sqrt{3}x_0 - y_0, x_0 + \sqrt{3}y_0\right]}{2\sqrt{x_0^2 + y_0^2}\sqrt{x_0^2 + y_0^2}}$$

$$= \frac{\sqrt{3}x_0^2 - x_0y_0 + x_0y_0 + \sqrt{3}y_0^2}{2(x_0^2 + y_0^2)}$$

$$= \frac{\sqrt{3}}{2}$$

Solving for θ shows that the angle between $\mathbf{n}_0$ and $\mathbf{n}_1$ is $\theta = \pi/6$. Thus multiplication by A rotates the vector by 30 degrees.

(e) From parts (c) and (d) and from Figure 7, we see that each multiplication by A rotates the vector 30 degrees in the counterclockwise direction and stretches its length by a factor of 2.

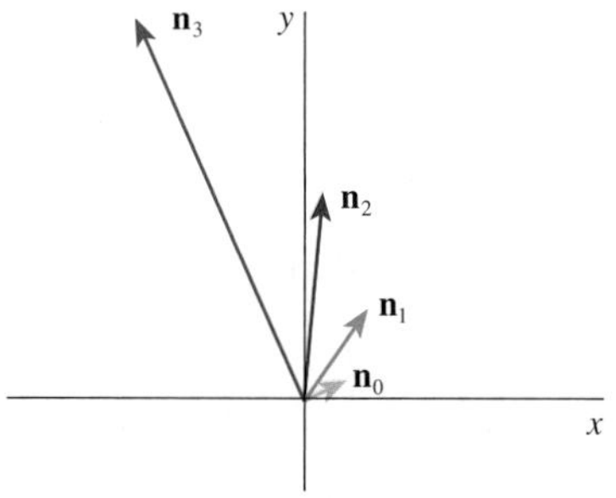

FIGURE 7

Example 3 suggests that when the eigenvalues are complex, there are no longer vectors in the x_1x_2-plane that remain on a fixed axis when multiplied by the matrix. (There are still such vectors, but they do not lie in the x_1x_2-plane because they are complex.) Instead, it appears that all vectors in the x_1x_2-plane get rotated through a fixed angle. As the next section will show, this is true in general when eigenvalues are complex.

EXERCISES 8.7

1–4 Multiplication by a matrix has an interesting geometric interpretation. Consider the letter L in the plane, made up of the two vectors [0, 2] and [1, 0]:

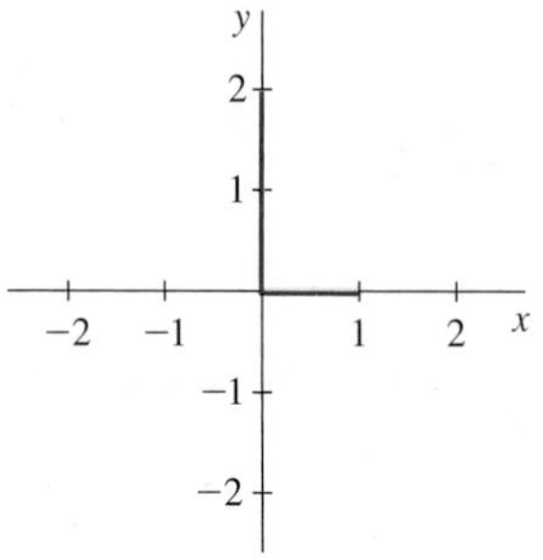

Describe how multiplication by the given matrix changes L.

1. The following matrices are sometimes called *reflections*.

(a) $\begin{bmatrix} 1 & 0 \\ 0 & -1 \end{bmatrix}$ (b) $\begin{bmatrix} -1 & 0 \\ 0 & 1 \end{bmatrix}$

(c) $\begin{bmatrix} -1 & 0 \\ 0 & -1 \end{bmatrix}$ (d) $\begin{bmatrix} 0 & -1 \\ -1 & 0 \end{bmatrix}$

2. The following matrices are sometimes called *contractions* or *expansions*.

(a) $\begin{bmatrix} 2 & 0 \\ 0 & 1 \end{bmatrix}$ (b) $\begin{bmatrix} 1 & 0 \\ 0 & \frac{1}{3} \end{bmatrix}$

(c) $\begin{bmatrix} \frac{1}{5} & 0 \\ 0 & 2 \end{bmatrix}$ (d) $\begin{bmatrix} \frac{1}{2} & 0 \\ 0 & \frac{1}{2} \end{bmatrix}$

3. The following matrices are sometimes called *shears*.

(a) $\begin{bmatrix} 1 & 1 \\ 0 & 1 \end{bmatrix}$ (b) $\begin{bmatrix} 1 & 0 \\ -1 & 1 \end{bmatrix}$

(c) $\begin{bmatrix} 1 & -2 \\ 0 & 1 \end{bmatrix}$ (d) $\begin{bmatrix} 1 & 0 \\ 1 & 1 \end{bmatrix}$

4. The following matrices are sometimes called *rotations*.

(a) $\begin{bmatrix} \frac{\sqrt{3}}{2} & \frac{1}{2} \\ -\frac{1}{2} & \frac{\sqrt{3}}{2} \end{bmatrix}$ (b) $\begin{bmatrix} \frac{1}{2} & -\frac{\sqrt{3}}{2} \\ \frac{\sqrt{3}}{2} & \frac{1}{2} \end{bmatrix}$

---

5. Consider a unit square in the first quadrant.

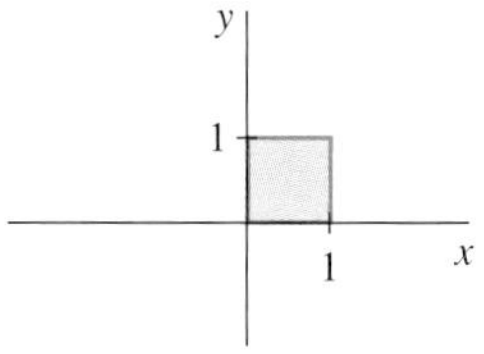

Each point in the square can be viewed as the tip of its position vector. What is the matrix that results in each of the following images when applied to these vectors?

(a)

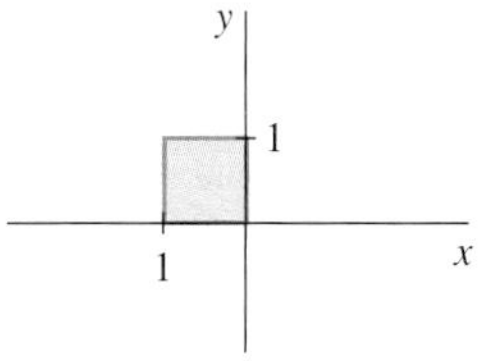

(b)

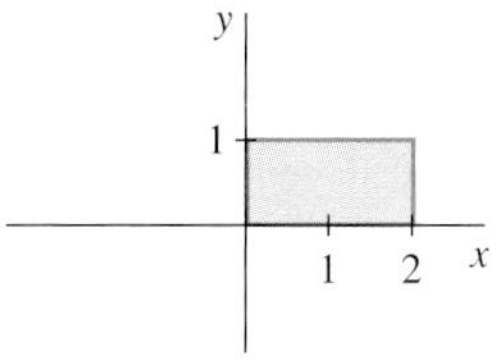

(c)

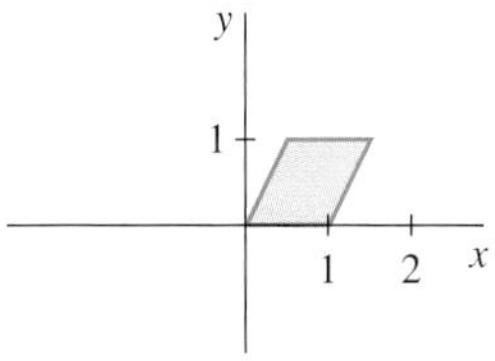

(d)

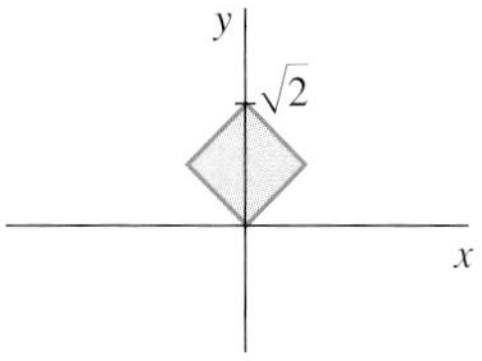

6. Use the dot product to show that the matrix

$$\begin{bmatrix} \cos\theta & -\sin\theta \\ \sin\theta & \cos\theta \end{bmatrix}$$

rotates vectors through an angle θ. Choose a particular value of θ and convince yourself that the direction of rotation is counterclockwise.

7. Vectorcardiography Suppose the voltage vector of a healthy heart is given by $[0.3, -0.2]$. Cardiac abnormalities result in changes to this vector. Each abnormality can be viewed as a change of a healthy heart voltage vector into a pathological voltage vector through matrix multiplication. For each of the following heart pathologies, find a matrix that produces the pathological voltage vector from a healthy voltage vector.

(a) Left anterior hemiblock: $\mathbf{h} = [0.3, 0.2]$

(b) Left posterior hemiblock: $\mathbf{h} = [-0.3, -0.2]$

(c) Apical ischemia: $\mathbf{h} = [-0.3, 0.2]$

(d) Chronic obstructive pulmonary disease: $\mathbf{h} = [0.1, -0.0667]$

8. Morphometrics D'arcy Thompson was a Scottish scientist who pioneered the use of mathematics for describing differences in morphology between species. For example, he demonstrated how the shape of the fish species *Argyropelecus offers* can be related to the shape of another species *Sternoptyx diaphana* through matrix multiplication:

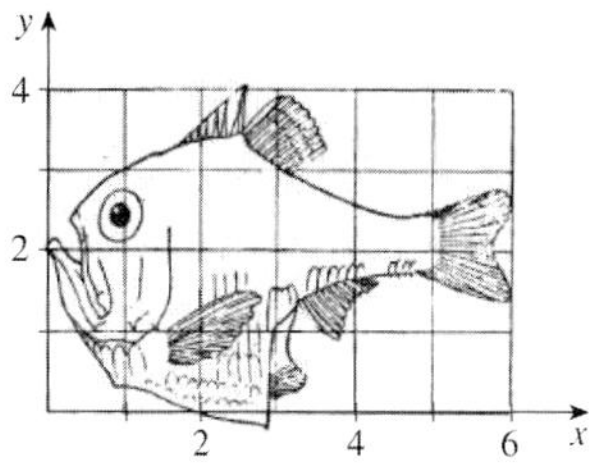

Argyropelecus offers

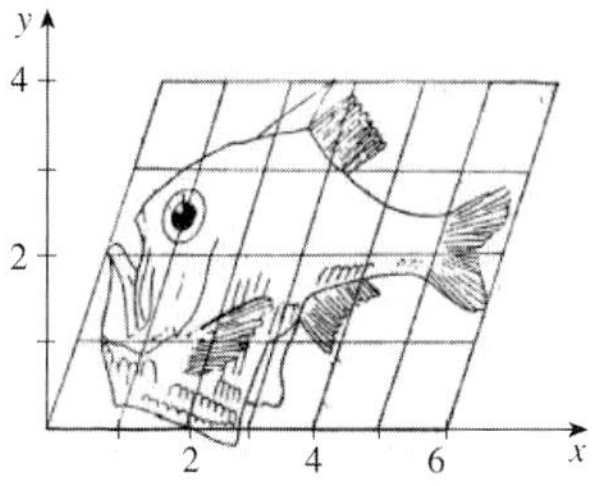

Sternoptyx diaphana

Source: Title: ON GROWTH AND FORM ABRIDGED EDITION authored by D'Arcy W. Thompson, edited by John Tyler Bonner; Copyright © 1961 Cambridge University Press. Reprinted with the permission of Cambridge University Press.

What is the form of the matrix that describes this change?

9. Determine whether or not the given scalar k is an eigenvalue of the matrix A.

(a) $A = \begin{bmatrix} 1 & 2 \\ 2 & 1 \end{bmatrix}$ $k = 3$

(b) $A = \begin{bmatrix} 0 & 2 & 1 \\ 2 & 1 & 0 \\ 0 & 2 & 1 \end{bmatrix}$ $k = 0$

(c) $A = \begin{bmatrix} 5 & 2 \\ 0 & 1 \end{bmatrix}$ $k = 2$

(d) $A = \begin{bmatrix} 1 & -1 \\ 1 & 1 \end{bmatrix}$ $k = 1 - i$

(e) $A = \begin{bmatrix} 1 & 2 & 1 \\ 0 & 2 & 0 \\ 2 & 1 & 0 \end{bmatrix}$ $k = 0$

(f) $A = \begin{bmatrix} 1 & a \\ 1 & 1 \end{bmatrix}$ $k = 1 + \sqrt{a}$

10. Determine whether or not $\mathbf{x}$ is an eigenvector of A. If it is, determine its associated eigenvalue.

(a) $A = \begin{bmatrix} 3 & -1 \\ 2 & 0 \end{bmatrix}$ $\quad \mathbf{x} = \begin{bmatrix} 1 \\ 2 \end{bmatrix}$

(b) $A = \begin{bmatrix} -3 & -1 & 5 \\ -2 & 1 & 2 \\ -2 & -1 & 4 \end{bmatrix}$ $\quad \mathbf{x} = \begin{bmatrix} 2 \\ 1 \\ 1 \end{bmatrix}$

(c) $A = \begin{bmatrix} 1 & -1 \\ -2 & 0 \end{bmatrix}$ $\quad \mathbf{x} = \begin{bmatrix} 1 \\ 0 \end{bmatrix}$

(d) $A = \begin{bmatrix} 1 & 2 \\ -2 & 1 \end{bmatrix}$ $\quad \mathbf{x} = \begin{bmatrix} 2 \\ 2i \end{bmatrix}$

(e) $A = \begin{bmatrix} 2 + 3a & -2 - 2a \\ 3 + 3a & -3 - 2a \end{bmatrix}$ $\quad \mathbf{x} = \begin{bmatrix} 2 \\ 3 \end{bmatrix}$

(f) $A = \begin{bmatrix} -9 & 4 & 6 \\ -6 & 3 & 4 \\ -9 & 4 & 6 \end{bmatrix}$ $\quad \mathbf{x} = \begin{bmatrix} 2 \\ 0 \\ 3 \end{bmatrix}$

11. Find the eigenvalues of each matrix.

(a) $\begin{bmatrix} 2 & 0 \\ 3 & 0 \end{bmatrix}$ (b) $\begin{bmatrix} 5 & -4 \\ 6 & -5 \end{bmatrix}$

(c) $\begin{bmatrix} 3 & -1 \\ 0 & 2 \end{bmatrix}$ (d) $\begin{bmatrix} 0 & 2 \\ -\frac{1}{2} & 0 \end{bmatrix}$

(e) $\begin{bmatrix} 6 & -4 & -4 \\ 0 & 0 & 0 \\ 6 & -4 & -4 \end{bmatrix}$ (f) $\begin{bmatrix} -1 & 4 & 2 \\ 0 & 3 & 0 \\ -3 & 4 & 4 \end{bmatrix}$

12. Find an eigenvector associated with the given eigenvalue of A.

(a) $A = \begin{bmatrix} 9 & 0 \\ 2 & 3 \end{bmatrix}$ $\quad \lambda = 9$

(b) $A = \begin{bmatrix} 1 & 5 \\ 2 & 7 \end{bmatrix}$ $\quad \lambda = 4 + \sqrt{19}$

(c) $A = \begin{bmatrix} 1 & 0 & 3 \\ 2 & 0 & 0 \\ 0 & 2 & 1 \end{bmatrix}$ $\quad \lambda = 3$

(d) $A = \begin{bmatrix} 0 & 1 \\ 3 & 2 \end{bmatrix}$ $\quad \lambda = -1$

(e) $A = \begin{bmatrix} 0 & 1 \\ 1 & 1 \end{bmatrix}$ $\quad \lambda = \dfrac{1 + \sqrt{5}}{2}$

(f) $A = \begin{bmatrix} 1 & 2 & 3 \\ 0 & 1 & 7 \\ 0 & 2 & 1 \end{bmatrix}$ $\quad \lambda = 1 + \sqrt{14}$

13. Find the eigenvalues and eigenvectors of the matrix.

(a) $\begin{bmatrix} 1 & 0 \\ 0 & -1 \end{bmatrix}$ (b) $\begin{bmatrix} 1 & 2 \\ 2 & 1 \end{bmatrix}$

(c) $\begin{bmatrix} 1 & -2 \\ 2 & 1 \end{bmatrix}$ (d) $\begin{bmatrix} 2 & 7 \\ 0 & 5 \end{bmatrix}$

(e) $\begin{bmatrix} 1 & 2 \\ 3 & -3 \end{bmatrix}$ (f) $\begin{bmatrix} 2 & 6 \\ 5 & 0 \end{bmatrix}$

(g) $\begin{bmatrix} 1 & 0 & 1 \\ 2 & 1 & 0 \\ 3 & 0 & 1 \end{bmatrix}$ (h) $\begin{bmatrix} 1 & 2 & 3 \\ 0 & 1 & 7 \\ 0 & 2 & 1 \end{bmatrix}$

14. Derive a general formula for the eigenvalues of the 2×2 matrix

$$\begin{bmatrix} a & b \\ c & d \end{bmatrix}$$

15. The case of matrices with *repeated eigenvalues* is treated in courses on linear algebra. As an example, try calculating the eigenvalues and eigenvectors of the following matrices. Comment on anything unusual that occurs.

(a) $\begin{bmatrix} 1 & 0 \\ 0 & 1 \end{bmatrix}$ (b) $\begin{bmatrix} 1 & 1 \\ 0 & 1 \end{bmatrix}$ (c) $\begin{bmatrix} 1 & 0 & 0 \\ 0 & 1 & 0 \\ 0 & 0 & 2 \end{bmatrix}$

16. In general, the eigenvalues of a diagonal matrix are given by the entries on the diagonal. Verify this for 2×2 and 3×3 matrices.

17. In general, the eigenvalues of an upper triangular matrix are given by the entries on the diagonal. The same is true for a lower triangular matrix. Verify this for 2×2 and 3×3 matrices.

18. The *trace* of a matrix is given by the sum of its diagonal entries. In general, the trace of a matrix is equal to the sum of its eigenvalues, and its determinant is equal to the product of its eigenvalues. Verify this for 2×2 matrices.

19. Suppose that $A^2 = 0$ for some matrix A. Prove that the only possible eigenvalues of A are then 0.

20. Suppose that an eigenvalue of matrix A is zero. Prove that A must therefore be singular.

21. Show that the characteristic polynomials of A and A^T are the same.

22. Suppose that A is a nonsingular matrix with an eigenvalue λ. Show that $1/\lambda$ is then an eigenvalue of A^{-1}.

23. Suppose that λ is an eigenvalue of A. Show that 2λ is then an eigenvalue of $2A$.

24. Suppose that λ is an eigenvalue of A. Show that λ^2 is then an eigenvalue of A^2.

25. Suppose that $\mathbf{v}$ is an eigenvector of matrix A with eigenvalue λ_A, and it is also an eigenvector of matrix B with eigenvalue λ_B.

(a) Show that $\mathbf{v}$ is an eigenvector of $A + B$ and find its associated eigenvalue.

(b) Show that $\mathbf{v}$ is an eigenvector of AB and find its associated eigenvalue.

26. Vectorcardiography The voltage vector of a heart with a certain pathology changes from one beat to the next as described by the matrix

$$\begin{bmatrix} 1 & 0 \\ 0 & -1 \end{bmatrix}$$

Find its eigenvalues and associated eigenvectors.

27. The **antigenic evolution** of a virus in one season is described by the matrix

$$\begin{bmatrix} 2 & 3 \\ 0 & \frac{9}{10} \end{bmatrix}$$

Find its eigenvalues and associated eigenvectors.

28. The change in **population size** of a species with juvenile and adult individuals is described by the matrix

$$\begin{bmatrix} 0 & 2 \\ \frac{1}{2} & \frac{1}{3} \end{bmatrix}$$

Find its eigenvalues and associated eigenvectors.

29. The **Leslie matrix** for an age-structured population is given by

$$\begin{bmatrix} b & 2 \\ \frac{1}{2} & 0 \end{bmatrix}$$

Find its eigenvalues and associated eigenvectors.

CAS **30. Cancer progression** Example 8.5.2 introduced a model for the progression of cancer that involved the matrix

$$\begin{bmatrix} 0.998 & 0.45 & 0 \\ 0.002 & 0.1 & 0 \\ 0 & 0.45 & 1 \end{bmatrix}$$

Find its eigenvalues and associated eigenvectors.

31. The **Leslie matrix** for an age-structured population is given by

$$\begin{bmatrix} 1 & 2 & 4 \\ \frac{1}{2} & 0 & 0 \\ 0 & \frac{1}{3} & 0 \end{bmatrix}$$

Find its characteristic polynomial.

32. The genetics of inbreeding A model for the genetics of inbreeding was introduced in Exercise 8.5.19 and can be described by the matrix

$$\begin{bmatrix} 1 & \frac{1}{4} & 0 \\ 0 & \frac{1}{2} & 0 \\ 0 & \frac{1}{4} & 1 \end{bmatrix}$$

Try calculating its eigenvalues and associated eigenvectors. You will find that one of the eigenvalues is repeated.

8.8 Iterated Matrix Models

Matrix models are used in the life sciences to model how vectors change over multiple time steps. Section 8.5 introduced several examples. One way to determine the long-term behavior of such vectors over time is to iterate the matrix multiplication repeatedly. Given an initial vector $\mathbf{n}_0$, we calculate successive vectors $\mathbf{n}_1, \mathbf{n}_2, \ldots$ using the recursion

(1) $$\mathbf{n}_{t+1} = A\mathbf{n}_t$$

We used this idea in Chapters 1 and 2 to explore the scalar recursion equation $n_{t+1} = Rn_t$. As in the scalar case, however, this becomes cumbersome for large values of t. A better alternative is to find the solution to the recursion.

■ Solving Matrix Models

To solve the recursion equation (1), let's begin by repeatedly iterating it, starting with initial vector $\mathbf{n}_0$. For $\mathbf{n}_1$ we obtain

$$\mathbf{n}_1 = A\mathbf{n}_0$$

Likewise, for $\mathbf{n}_2$ we get

$$\mathbf{n}_2 = A\mathbf{n}_1 = A(A\mathbf{n}_0) = A^2\mathbf{n}_0$$

where A^2 is A multiplied with itself. Continuing, we see the general pattern

(2) $$\mathbf{n}_t = A^t\mathbf{n}_0$$

where A^t is A multiplied with itself t times.

Equation 2 is the solution to recursion equation (1), just as $n_t = R^t n_0$ is the solution to the scalar recursion equation $n_{t+1} = Rn_t$. To verify this we substitute $A^t\mathbf{n}_0$ into the left side of Equation 1:

$$A^{t+1}\mathbf{n}_0 = AA^t\mathbf{n}_0 = A\mathbf{n}_t$$

This is identical to the right side of Equation 1, demonstrating that it is a solution.

Unlike the scalar case, however, the matrix multiplication required to calculate A^t becomes very tedious, particularly if there are unspecified constants in A (try calculating A^3 for an arbitrary 2×2 matrix). Furthermore, such calculations do little to help us understand the long-term behavior of $\mathbf{n}_t$. This is where the eigenvalues and eigenvectors of A enter the picture. We can use them to rewrite A in a more useful way.

Recall the definition of the eigenvectors and eigenvalues of A:

(3) $$A\mathbf{v}_i = \lambda_i\mathbf{v}_i$$

Note that we have now used the subscript i on $\mathbf{v}$ and λ to reflect the fact that there are n such pairs for an $n \times n$ matrix (ignoring *defective* matrices; see Section 8.7). For example, in the 2×2 case we have two eigenvectors, which we will denote as column vectors by $\begin{bmatrix} | \\ \mathbf{v}_1 \\ | \end{bmatrix}$ and $\begin{bmatrix} | \\ \mathbf{v}_2 \\ | \end{bmatrix}$. These eigenvectors are associated with eigenvalues λ_1 and λ_2, respectively. We can write Equation 3 simultaneously for both pairs as

$$A\begin{bmatrix} | & | \\ \mathbf{v}_1 & \mathbf{v}_2 \\ | & | \end{bmatrix} = \begin{bmatrix} | & | \\ \mathbf{v}_1 & \mathbf{v}_2 \\ | & | \end{bmatrix}\begin{bmatrix} \lambda_1 & 0 \\ 0 & \lambda_2 \end{bmatrix}$$

where $\begin{bmatrix} | & | \\ \mathbf{v}_1 & \mathbf{v}_2 \\ | & | \end{bmatrix}$ is a 2×2 matrix whose columns are the eigenvectors $\begin{bmatrix} | \\ \mathbf{v}_1 \\ | \end{bmatrix}$ and $\begin{bmatrix} | \\ \mathbf{v}_2 \\ | \end{bmatrix}$.

More generally, for an $n \times n$ matrix A we can define an $n \times n$ matrix P whose columns are the n eigenvectors of A, and an $n \times n$ diagonal matrix D whose entries along the diagonal are the corresponding eigenvalues. We can then write

$$AP = PD$$

Our goal now is to use this equation to rewrite A in a different form, using its eigenvectors and eigenvalues. A theorem from linear algebra states that if A is not defective, then P is invertible. Therefore, multiplying both sides of the equation on the right by P^{-1} gives

(4) $$A = PDP^{-1}$$

How does the expression for A in Equation 4 help us evaluate $A^t\mathbf{n}_0$ in Equation 2? Let's iterate Equation 1 again, this time using our new way of expressing A. For $\mathbf{n}_1$ we obtain

$$\mathbf{n}_1 = A\mathbf{n}_0 = PDP^{-1}\mathbf{n}_0$$

For $\mathbf{n}_2$ we get

$$\begin{aligned}\mathbf{n}_2 &= A\mathbf{n}_1\\ &= (PDP^{-1})\mathbf{n}_1\\ &= (PDP^{-1})(PDP^{-1})\mathbf{n}_0\\ &= (PDP^{-1}PDP^{-1})\mathbf{n}_0\end{aligned}$$

Recalling that $P^{-1}P = I$, this last equation simplifies as

$$\begin{aligned}\mathbf{n}_2 &= PDIDP^{-1}\mathbf{n}_0\\ &= PDDP^{-1}\mathbf{n}_0\\ &= PD^2P^{-1}\mathbf{n}_0\end{aligned}$$

where D^2 is D multiplied with itself. Proceeding further, we see the general pattern:

(5) $$\mathbf{n}_t = PD^tP^{-1}\mathbf{n}_0$$

Although Equation 5 looks more complicated than Equation 2, it is much easier to work with. In particular, as we saw in Example 8.4.3 and Exercise 8.4.7, for any diagonal matrix D, the matrix D^k is simply a diagonal matrix with each of the entries of D raised to the power of k. Therefore, once we have calculated the eigenvalues and eigenvectors of A, calculating the value of $\mathbf{n}_t$ is straightforward for any t of interest.

Not only does Equation 5 often provide a computational advantage over Equation 2, it can also be used to better understand the structure of the solution. Let's return to the case of $n = 2$. Writing Equation 5 more explicitly, we have

(6) $$\mathbf{n}_t = \begin{bmatrix} | & | \\ \mathbf{v}_1 & \mathbf{v}_2 \\ | & | \end{bmatrix}\begin{bmatrix} \lambda_1^t & 0 \\ 0 & \lambda_2^t \end{bmatrix}P^{-1}\mathbf{n}_0$$

If we define two new constants, c_1 and c_2, as

$$\begin{bmatrix} c_1 \\ c_2 \end{bmatrix} = P^{-1}\mathbf{n}_0$$

then Equation 6 can be written as

(7) $$\begin{aligned}\mathbf{n}_t &= \begin{bmatrix} | & | \\ \mathbf{v}_1 & \mathbf{v}_2 \\ | & | \end{bmatrix}\begin{bmatrix} \lambda_1^t & 0 \\ 0 & \lambda_2^t \end{bmatrix}\begin{bmatrix} c_1 \\ c_2 \end{bmatrix}\\ &= \begin{bmatrix} | & | \\ \mathbf{v}_1 & \mathbf{v}_2 \\ | & | \end{bmatrix}\begin{bmatrix} c_1\lambda_1^t \\ c_2\lambda_2^t \end{bmatrix}\\ &= c_1\begin{bmatrix} | \\ \mathbf{v}_1 \\ | \end{bmatrix}\lambda_1^t + c_2\begin{bmatrix} | \\ \mathbf{v}_2 \\ | \end{bmatrix}\lambda_2^t\end{aligned}$$

Equation 7 shows that the solution $\mathbf{n}_t$ can be viewed as the sum of two parts, each of which corresponds to one of the eigenvector-eigenvalue pairs of the matrix A. More generally, for an $n \times n$ matrix we have

$$\textbf{(8)} \qquad \mathbf{n}_t = c_1 \begin{bmatrix} | \\ \mathbf{v}_1 \\ | \end{bmatrix} \lambda_1^t + c_2 \begin{bmatrix} | \\ \mathbf{v}_2 \\ | \end{bmatrix} \lambda_2^t + \cdots + c_n \begin{bmatrix} | \\ \mathbf{v}_n \\ | \end{bmatrix} \lambda_n^t$$

EXAMPLE 1 | Consider the matrix $A = \begin{bmatrix} 1 & \frac{1}{2} \\ 1 & \frac{3}{2} \end{bmatrix}$. On page 539 we found the eigenvalues to be $\lambda_1 = 2$ and $\lambda_2 = \frac{1}{2}$, and their corresponding eigenvectors to be $\mathbf{v}_1 = \begin{bmatrix} 1 \\ 2 \end{bmatrix}$ and $\mathbf{v}_2 = \begin{bmatrix} -1 \\ 1 \end{bmatrix}$.

(a) Express the solution to the recursion $\mathbf{n}_{t+1} = A\mathbf{n}_t$ in terms of these eigenvalues and eigenvectors, assuming $\mathbf{n}_0 = \begin{bmatrix} 2 \\ 1 \end{bmatrix}$.

(b) What happens to the vector $\mathbf{n}_t$ as $t \to \infty$?

SOLUTION

(a) To evaluate Equation 7 explicitly we need to calculate the constants c_1 and c_2. These are determined from P^{-1} and the vector $\mathbf{n}_0$. To calculate P^{-1} we note that $P = \begin{bmatrix} 1 & -1 \\ 2 & 1 \end{bmatrix}$ and therefore, using the formula for the inverse of a 2×2 matrix on page 530, we get $P^{-1} = \begin{bmatrix} \frac{1}{3} & \frac{1}{3} \\ -\frac{2}{3} & \frac{1}{3} \end{bmatrix}$. The constants c_1 and c_2 are calculated as $\begin{bmatrix} c_1 \\ c_2 \end{bmatrix} = P^{-1}\mathbf{n}_0 = \begin{bmatrix} 1 \\ -1 \end{bmatrix}$. We obtain

$$\mathbf{n}_t = (1)\begin{bmatrix} 1 \\ 2 \end{bmatrix} 2^t + (-1)\begin{bmatrix} -1 \\ 1 \end{bmatrix}\left(\frac{1}{2}\right)^t = \begin{bmatrix} 1 \\ 2 \end{bmatrix} 2^t + \begin{bmatrix} 1 \\ -1 \end{bmatrix}\left(\frac{1}{2}\right)^t$$

(b) As $t \to \infty$, the part of the solution corresponding to eigenvector $\mathbf{v}_2$ decays to zero because $\lim_{t\to\infty} \left(\frac{1}{2}\right)^t = 0$. At the same time, the part of the solution corresponding to eigenvector $\mathbf{v}_1$ grows by a factor of 2 each time step. As a result, the vector $\mathbf{n}_t$ ultimately increases in magnitude and approaches the direction of the first eigenvector, $\mathbf{v}_1 = \begin{bmatrix} 1 \\ 2 \end{bmatrix}$, as shown in Figure 1. ■

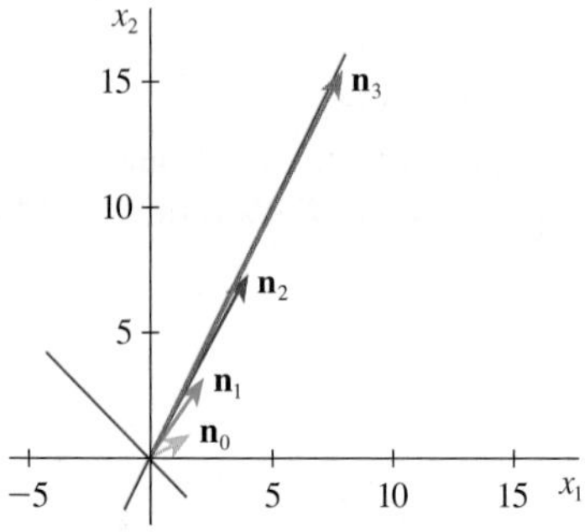

FIGURE 1
The red lines indicate all scalar multiples of the eigenvectors. The vector $\mathbf{n}_1$ was obtained by multiplying $\mathbf{n}_0$ by A. The vector $\mathbf{n}_2$ was obtained by multiplying $\mathbf{n}_1$ by A, and so on.

EXAMPLE 2 | In Example 8.7.2 (on page 541) we found the eigenvectors and eigenvalues of matrix $A = \begin{bmatrix} 2 & 1 \\ 1 & \frac{1}{2} \end{bmatrix}$ to be $\lambda_1 = 0$ and $\lambda_2 = \frac{5}{2}$, and $\mathbf{v}_1 = \begin{bmatrix} 1 \\ -2 \end{bmatrix}$ and $\mathbf{v}_2 = \begin{bmatrix} 2 \\ 1 \end{bmatrix}$.

(a) Express the solution to the recursion $\mathbf{n}_{t+1} = A\mathbf{n}_t$ in terms of these eigenvalues and eigenvectors, assuming $\mathbf{n}_0 = \begin{bmatrix} a \\ b \end{bmatrix}$.

(b) What happens to the vector $\mathbf{n}_t$ as $t \to \infty$?

SOLUTION

(a) We have $P = \begin{bmatrix} 1 & 2 \\ -2 & 1 \end{bmatrix}$ and therefore $P^{-1} = \begin{bmatrix} \frac{1}{5} & -\frac{2}{5} \\ \frac{2}{5} & \frac{1}{5} \end{bmatrix}$. Calculating $\begin{bmatrix} c_1 \\ c_2 \end{bmatrix} = P^{-1}\mathbf{n}_0$

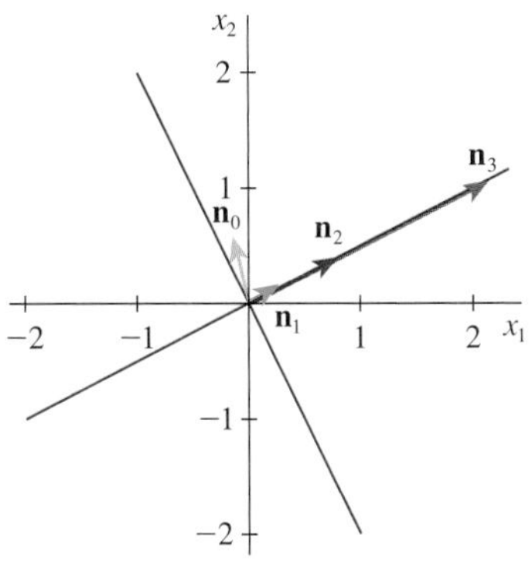

FIGURE 2
The red lines indicate all scalar multiples of the eigenvectors. The vector $\mathbf{n}_1$ was obtained by multiplying $\mathbf{n}_0$ by A. The vector $\mathbf{n}_2$ was obtained by multiplying $\mathbf{n}_1$ by A, and so on.

gives the constants $c_1 = (a - 2b)/5$ and $c_2 = (2a + b)/5$. The solution is therefore

$$\mathbf{n}_t = \frac{a - 2b}{5}\begin{bmatrix} 1 \\ -2 \end{bmatrix} 0^t + \frac{2a + b}{5}\begin{bmatrix} 2 \\ 1 \end{bmatrix}\left(\frac{5}{2}\right)^t = \frac{2a + b}{5}\begin{bmatrix} 2 \\ 1 \end{bmatrix}\left(\frac{5}{2}\right)^t$$

Note that this solution is restricted to values of t satisfying $t \geqslant 1$ because 0^t is undefined when $t = 0$. The exclusion of the case $t = 0$ in the solution of matrix models is typical whenever the matrix has a zero eigenvalue.

(b) Because $\lambda_1 = 0$, the part of the solution corresponding to its eigenvector drops out of the solution. As a result, although the initial vector $\mathbf{n}_0$ can lie anywhere in the plane, $\mathbf{n}_1$ always lies on the line containing the eigenvector $\mathbf{v}_2$. The solution $\mathbf{n}_t$ lies on this axis for all subsequent times, growing in magnitude by a factor of $\frac{5}{2}$ each time step. This is illustrated in Figure 2 for the specific initial vector $\mathbf{n}_0 = [-0.15, 0.65]$. ■

EXAMPLE 3 | BB **Class-structured population dynamics** On page 520 we introduced matrix models with an example of a population of juveniles and adults that changes in size through time according to the recursion $\mathbf{n}_{t+1} = Q\mathbf{n}_t$, where

$$Q = \begin{bmatrix} 0 & 2 \\ \frac{1}{2} & \frac{1}{3} \end{bmatrix}$$

(a) Use the eigenvectors and eigenvalues of Q to find an equation that gives the number of juveniles and adults at time t for an arbitrary initial condition

$$\mathbf{n}_0 = \begin{bmatrix} j_0 \\ a_0 \end{bmatrix}$$

(b) How do the numbers of juveniles and adults change each time step once t is very large?

SOLUTION

(a) The eigenvalues of Q can be calculated as $\lambda_1 = \left(1 + \sqrt{37}\right)/6 \approx 1.18$ and $\lambda_2 = \left(1 - \sqrt{37}\right)/6 \approx -0.85$. After some calculation, the corresponding eigenvectors are found to be

$$\mathbf{v}_1 = \begin{bmatrix} -1 + \sqrt{37} \\ 3 \end{bmatrix} \approx \begin{bmatrix} 5.1 \\ 3 \end{bmatrix} \qquad \mathbf{v}_2 = \begin{bmatrix} -1 - \sqrt{37} \\ 3 \end{bmatrix} \approx \begin{bmatrix} -7.1 \\ 3 \end{bmatrix}$$

Therefore we have

$$P = \begin{bmatrix} -1 + \sqrt{37} & -1 - \sqrt{37} \\ 3 & 3 \end{bmatrix} \quad \text{and} \quad P^{-1} = \begin{bmatrix} \dfrac{1}{2\sqrt{37}} & \dfrac{1 + \sqrt{37}}{6\sqrt{37}} \\ -\dfrac{1}{2\sqrt{37}} & \dfrac{-1 + \sqrt{37}}{6\sqrt{37}} \end{bmatrix}$$

where P^{-1} follows from the definition on page 530.

Finally, calculating $\begin{bmatrix} c_1 \\ c_2 \end{bmatrix} = P^{-1}\mathbf{n}_0$ gives the constants

$$c_1 = \frac{j_0}{2\sqrt{37}} + a_0\left(\frac{1 + \sqrt{37}}{6\sqrt{37}}\right) \approx 0.08j_0 + 0.19a_0$$

$$c_2 = \frac{-j_0}{2\sqrt{37}} + a_0\left(\frac{-1 + \sqrt{37}}{6\sqrt{37}}\right) \approx -0.08j_0 + 0.14a_0$$

The solution, in decimal form, is therefore

$$\mathbf{n}_t = (0.08j_0 + 0.19a_0)\begin{bmatrix} 5.1 \\ 3 \end{bmatrix}(1.18)^t + (-0.08j_0 + 0.14a_0)\begin{bmatrix} -7.1 \\ 3 \end{bmatrix}(-0.85)^t$$

(b) As t increases, the part of the solution corresponding to the first eigenvector grows by a factor $\lambda_1 \approx 1.18$ each time step. The part of the solution corresponding to the second eigenvector alternates in sign and decays in magnitude each time step, asymptotically going to zero. Therefore, for any choice of initial vector $\mathbf{n}_0$, once t gets large the solution vector $\mathbf{n}_t$ will grow in magnitude by a factor ≈ 1.18 each time step and asymptotically approach the line defined by the eigenvector

$$\mathbf{v}_1 = \begin{bmatrix} -1 + \sqrt{37} \\ 3 \end{bmatrix} \approx \begin{bmatrix} 5.1 \\ 3 \end{bmatrix}$$

Figure 3 illustrates the specific case where $\mathbf{n}_0 = [0.15, 0.65]$ (measured in thousands of individuals).

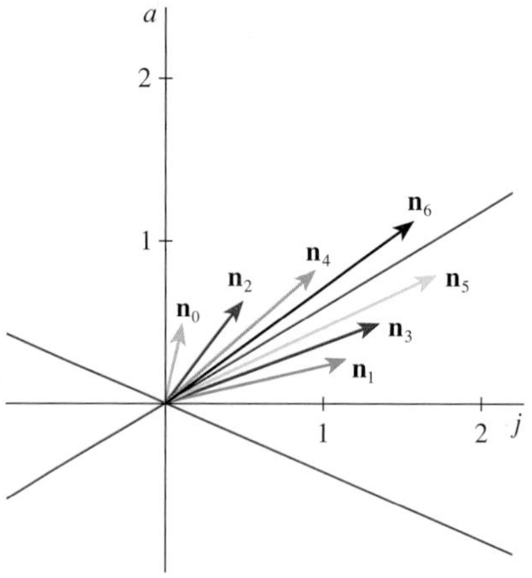

FIGURE 3
The red lines indicate all scalar multiples of the eigenvectors. The vector $\mathbf{n}_1$ was obtained by multiplying $\mathbf{n}_0$ by A. The vector $\mathbf{n}_2$ was obtained by multiplying $\mathbf{n}_1$ by A, and so on.

Thus, once t is large, both juvenile and adult subpopulations grow by a factor 1.18 each time step [see Figure 4(a)]. Their relative abundances approach a constant ratio given by the ratio of the components of $\mathbf{v}_1$. This is illustrated in Figure 4(b).

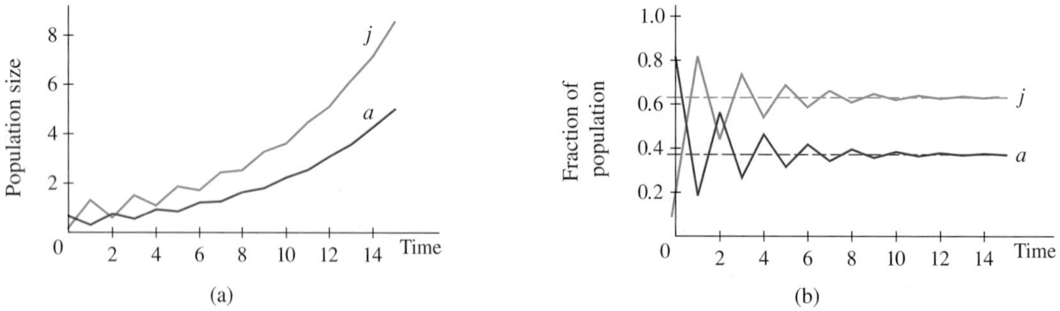

FIGURE 4 Plots of the components of vectors in Figure 3 as functions of time. Dashed lines in part (b) indicate the components of the vector $k\mathbf{v}_1$, where $k = 1/(5.1 + 3) = 1/8.1$. This scaling of $\mathbf{v}_1$ was chosen so that the components of the resulting vector add to 1. ■

■ Solutions with Complex Eigenvalues

In Section 8.7 we saw that the eigenvector-eigenvalue pairs of some matrices involve complex numbers. How do we interpret Equation 8 in this case? An important result from complex analysis called De Moivre's Theorem provides the key.

Complex numbers and De Moivre's Theorem are discussed in Appendix G.

De Moivre's Theorem provides a relationship between powers of complex numbers and the trigonometric functions sine and cosine. If $a + bi$ is a complex number (where a and b are real numbers) and k is a positive integer, then

$$(a + bi)^k = r^k(\cos k\theta + i \sin k\theta)$$

where $r = \sqrt{a^2 + b^2}$ is called the modulus of $a + bi$ and θ is called its argument. Specifically, $\theta = \tan^{-1}(b/a)$.

To illustrate the consequences of De Moivre's Theorem, we return to Example 8.7.3. There we studied the matrix $A = \begin{bmatrix} \sqrt{3} & -1 \\ 1 & \sqrt{3} \end{bmatrix}$ and found its eigenvalues to be $\lambda_1 = \sqrt{3} + i$ and $\lambda_2 = \sqrt{3} - i$, with associated eigenvectors $\mathbf{v}_1 = \begin{bmatrix} 1 \\ -i \end{bmatrix}$ and $\mathbf{v}_2 = \begin{bmatrix} 1 \\ i \end{bmatrix}$. Substituting these into Equation 7 gives the solution

$$\begin{aligned} \mathbf{n}_t &= c_1 \begin{bmatrix} 1 \\ -i \end{bmatrix} (\sqrt{3} + i)^t + c_2 \begin{bmatrix} 1 \\ i \end{bmatrix} (\sqrt{3} - i)^t \\ &= c_1 \begin{bmatrix} 1 \\ -i \end{bmatrix} 2^t\left(\cos\left(\tfrac{\pi}{6}t\right) + i \sin\left(\tfrac{\pi}{6}t\right)\right) + c_2 \begin{bmatrix} 1 \\ i \end{bmatrix} 2^t\left(\cos\left(\tfrac{\pi}{6}t\right) - i \sin\left(\tfrac{\pi}{6}t\right)\right) \end{aligned}$$

where we have used the facts that

$$r = \sqrt{(\sqrt{3})^2 + 1^2} = 2 \qquad \theta = \tan^{-1}\left(\frac{1}{\sqrt{3}}\right) = \frac{\pi}{6} \qquad \theta = \tan^{-1}\left(-\frac{1}{\sqrt{3}}\right) = -\frac{\pi}{6}$$

as well as the fact that sine is an odd function.

The final step is to calculate the constants c_1 and c_2. First, we have $P = \begin{bmatrix} 1 & 1 \\ -i & i \end{bmatrix}$ and therefore $P^{-1} = \begin{bmatrix} \frac{1}{2} & \frac{i}{2} \\ \frac{1}{2} & -\frac{i}{2} \end{bmatrix}$. The constants c_1 and c_2 are then

$$\begin{bmatrix} c_1 \\ c_2 \end{bmatrix} = P^{-1}\mathbf{n}_0 = \begin{bmatrix} (x_0 + y_0 i)/2 \\ (x_0 - y_0 i)/2 \end{bmatrix}$$

Substituting these into the solution gives

$$\begin{aligned} \mathbf{n}_t &= \frac{x_0 + y_0 i}{2} \begin{bmatrix} 1 \\ -i \end{bmatrix} 2^t\left(\cos\left(\tfrac{\pi}{6}t\right) + i \sin\left(\tfrac{\pi}{6}t\right)\right) + \frac{x_0 - y_0 i}{2} \begin{bmatrix} 1 \\ i \end{bmatrix} 2^t\left(\cos\left(\tfrac{\pi}{6}t\right) - i \sin\left(\tfrac{\pi}{6}t\right)\right) \\ &= \frac{1}{2} \begin{bmatrix} x_0 + y_0 i \\ -x_0 i + y_0 \end{bmatrix} 2^t\left(\cos\left(\tfrac{\pi}{6}t\right) + i \sin\left(\tfrac{\pi}{6}t\right)\right) + \frac{1}{2} \begin{bmatrix} x_0 - y_0 i \\ x_0 i + y_0 \end{bmatrix} 2^t\left(\cos\left(\tfrac{\pi}{6}t\right) - i \sin\left(\tfrac{\pi}{6}t\right)\right) \\ &= 2^t \begin{bmatrix} x_0 \cos\left(\tfrac{\pi}{6}t\right) - y_0 \sin\left(\tfrac{\pi}{6}t\right) \\ x_0 \sin\left(\tfrac{\pi}{6}t\right) + y_0 \cos\left(\tfrac{\pi}{6}t\right) \end{bmatrix} = 2^t \begin{bmatrix} \cos\left(\tfrac{\pi}{6}t\right) & -\sin\left(\tfrac{\pi}{6}t\right) \\ \sin\left(\tfrac{\pi}{6}t\right) & \cos\left(\tfrac{\pi}{6}t\right) \end{bmatrix} \begin{bmatrix} x_0 \\ y_0 \end{bmatrix} \end{aligned}$$

The second-last equality follows after considerable simplification, in which all of the imaginary parts of the solution cancel. As conjectured in Example 8.7.3, we can now see, using the result of Exercise 8.7.6, that $\mathbf{n}_t$ gets stretched by a factor of 2 each time step and rotated through an angle of $\pi/6$ radians. Figure 5 illustrates a specific example where $\mathbf{n}_0 = [0.35, -0.25]$. The components of $\mathbf{n}_t$ from Figure 5 are plotted as functions of time in Figure 6.

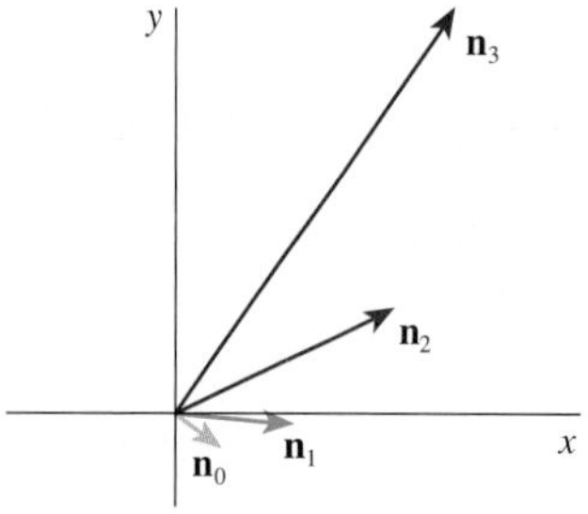

FIGURE 5
The vector $\mathbf{n}_1$ was obtained by multiplying $\mathbf{n}_0$ by A. The vector $\mathbf{n}_2$ was obtained by multiplying $\mathbf{n}_1$ by A, and so on.

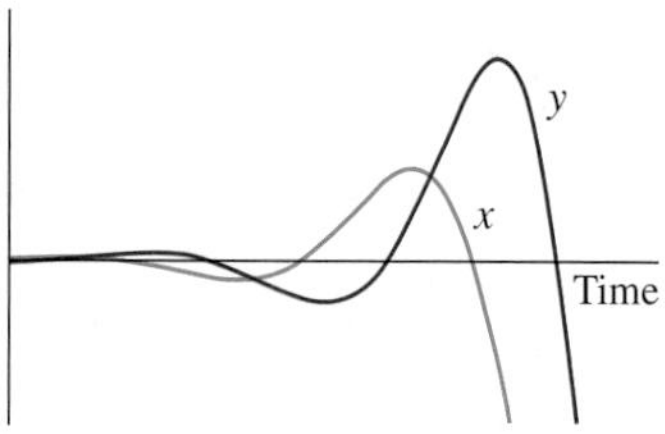

FIGURE 6
The components of $\mathbf{n}_t$ as functions of time

The cancellation of imaginary parts in the solution of the preceding example occurs in general when the eigenvalues of A are complex. The solution in such cases always consists of the oscillatory functions sine and cosine, and the solution grows or decays in magnitude depending on the modulus of the eigenvalues (see Exercise 29).

■ Perron-Frobenius Theory

In Examples 1–3 the long-term behavior of the solution (that is, the behavior of $\mathbf{n}_t$ as $t \to \infty$) was determined by just one of the eigenvector-eigenvalue pairs of the matrix A. The reason can be seen from Equation 8: Suppose that the eigenvalues in Equation 8 are ordered in terms of their absolute value (or modulus, if they are complex) from the largest λ_1 to the smallest λ_n, with $|\lambda_1| > |\lambda_i|$ for all $i \neq 1$ (that is, λ_1 is strictly larger in modulus than all other eigenvalues). Dividing both sides of Equation 8 by λ_1^t gives

$$\frac{\mathbf{n}_t}{\lambda_1^t} = c_1 \begin{bmatrix} | \\ \mathbf{v}_1 \\ | \end{bmatrix} + c_2 \begin{bmatrix} | \\ \mathbf{v}_2 \\ | \end{bmatrix} \frac{\lambda_2^t}{\lambda_1^t} + \cdots + c_n \begin{bmatrix} | \\ \mathbf{v}_n \\ | \end{bmatrix} \frac{\lambda_n^t}{\lambda_1^t}$$

Now taking the limit $t \to \infty$, we have

$$\lim_{t \to \infty} \frac{\mathbf{n}_t}{\lambda_1^t} = c_1 \begin{bmatrix} | \\ \mathbf{v}_1 \\ | \end{bmatrix}$$

This shows that, in the long term, as $t \to \infty$ the components of $\mathbf{n}_t$ become proportional to the components of the eigenvector $\mathbf{v}_1$. Thus the solution vector asymptotically approaches the line containing $\mathbf{v}_1$, regardless of the initial vector $\mathbf{n}_0$. This is a powerful piece of information: In the long run, all components of the vector $\mathbf{n}_t$ change by a factor λ_1 each time step, with the relative sizes of its components being proportional to the components of $\mathbf{v}_1$.

The preceding considerations are true only when λ_1 is strictly larger in magnitude than the other eigenvalues, and when the components of the associated eigenvector are positive. The power of matrix models in the life sciences stems, in large part, from the fact that many biological processes correspond to matrices whose eigenvectors and eigenvalues satisfy these conditions. The following definition and theorem specify precisely the kinds of matrices that have these properties.

Definition An $n \times n$ matrix A whose entries are nonnegative is called **primitive** if there exists a positive integer k such that all the entries of A^k are positive.

With this definition, we can now state the following theorem.

Perron-Frobenius Theorem Suppose that A is an $n \times n$ matrix whose entries are all nonnegative. If A is primitive, then all of the following are true:

1. There exists an eigenvalue of A, call it λ_1, that is real and positive.
2. $|\lambda_1| > |\lambda_i|$ for all $i \neq 1$ (that is, λ_1 is greater in magnitude than all other eigenvalues).
3. The components of the eigenvector associated with λ_1 are all positive.

EXAMPLE 4 | Consider the recursion $\mathbf{n}_{t+1} = A\mathbf{n}_t$, where $A = \begin{bmatrix} 2 & 1 \\ 1 & 2 \end{bmatrix}$.
(a) What happens to the magnitude of $\mathbf{n}_t$ as $t \to \infty$?
(b) What happens to the direction of $\mathbf{n}_t$ as $t \to \infty$?

SOLUTION Since A is primitive, we can apply the Perron-Frobenius Theorem. One eigenvalue is real and positive, larger in modulus than the other, and has an eigenvector with positive components. In this case we can calculate them explicitly, obtaining $\lambda_1 = 3$ and $\lambda_2 = 1$, along with the eigenvectors $\mathbf{v}_1 = \begin{bmatrix} 1 \\ 1 \end{bmatrix}$ and $\mathbf{v}_2 = \begin{bmatrix} -1 \\ 1 \end{bmatrix}$.

(a) As $t \to \infty$ the vector $\mathbf{n}_t$ eventually grows in magnitude by a factor of $\lambda_1 = 3$ each time step, regardless of the initial vector $\mathbf{n}_0$.

(b) As $t \to \infty$ the vector $\mathbf{n}_t$ eventually points in the direction of $\mathbf{v}_1 = \begin{bmatrix} 1 \\ 1 \end{bmatrix}$, regardless of the initial vector $\mathbf{n}_0$. Figure 7 illustrates this behavior when $\mathbf{n}_0 = [-0.35, 0.45]$. ■

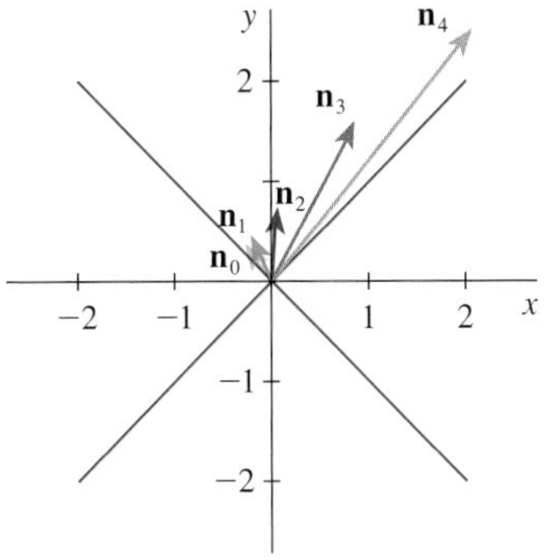

FIGURE 7
The red lines indicate all scalar multiples of the eigenvectors. Each vector was obtained from the preceding one through multiplication by A.

EXAMPLE 5 | BB **Age structure and the Leslie matrix** Consider an age-structured population whose size is governed by the equation

$$\begin{bmatrix} n_{1,t+1} \\ n_{2,t+1} \\ n_{3,t+1} \end{bmatrix} = \begin{bmatrix} 2 & 2 & 2 \\ \frac{1}{10} & 0 & 0 \\ 0 & \frac{1}{2} & 0 \end{bmatrix} \begin{bmatrix} n_{1,t} \\ n_{2,t} \\ n_{3,t} \end{bmatrix}$$

What is the long-term fate of the population?

SOLUTION Denoting the Leslie matrix in the equation by L, we begin by exploring the first few powers of L. We find that L^3 has entries that are all positive, meaning that L is primitive. Therefore we can apply the Perron-Frobenius Theorem and so there is a positive, real eigenvalue that is larger in magnitude than all others. In the long term the population will grow or decay depending on whether this eigenvalue is larger or smaller than 1.

Calculating the characteristic polynomial of L gives $f(\lambda) = \lambda^3 - 2\lambda^2 - \frac{1}{5}\lambda - \frac{1}{10}$. This polynomial does not factor and therefore its roots are difficult to find. But the Perron-Frobenius Theorem tells us that the largest root of the equation $f(\lambda) = 0$ is positive. Therefore, since $\lim_{\lambda \to \infty} f(\lambda) = \infty$, as the value of λ increases, the graph of the function $f(\lambda)$ crosses the horizontal axis for the last time in one of the two ways illustrated in Figure 8.

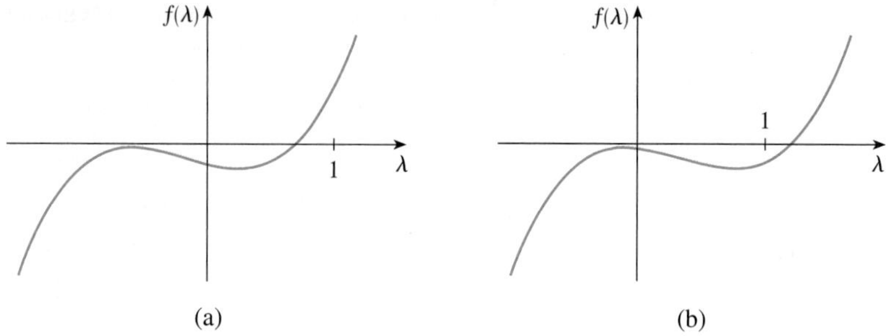

FIGURE 8

Furthermore, by noting that $f(1) = -13/10 < 0$, we can see that Figure 8(a) cannot be correct because it requires $f(1) > 0$. Indeed, because $f(1) = -13/10$ and $\lim_{\lambda \to \infty} f(\lambda) = \infty$, the Intermediate Value Theorem tells us that the largest root in the Perron-Frobenius Theorem must be larger than 1. Consequently, in the long term the population will grow. The Perron-Frobenius Theorem also tells us that the fraction of the population made up of each age class stabilizes to constant values that are proportional to the components of the eigenvector associated with λ_1. Figure 9 illustrates the case where $n_{1,0} = 0.1$, $n_{2,0} = 1$, and $n_{3,0} = 1$.

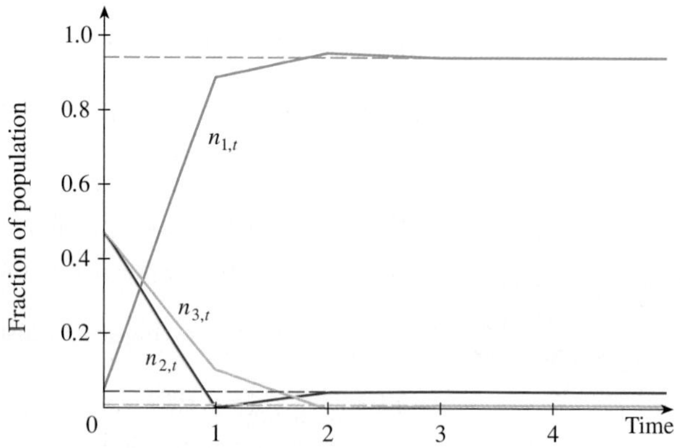

FIGURE 9
Dashed lines indicate the components of the eigenvector associated with the largest eigenvalue.

■

EXERCISES 8.8

1–8 Show that $A = PDP^{-1}$, where P is a matrix whose columns are the eigenvectors of A, and D is a diagonal matrix with the corresponding eigenvalues.

1. $A = \begin{bmatrix} 2 & 0 \\ 0 & 1 \end{bmatrix}$

2. $A = \begin{bmatrix} 1 & 1 \\ 0 & 3 \end{bmatrix}$

3. $A = \begin{bmatrix} 1 & 2 \\ 2 & 1 \end{bmatrix}$

4. $A = \begin{bmatrix} 1 & -1 \\ 1 & 1 \end{bmatrix}$

5. $A = \begin{bmatrix} 1 & 2 \\ -3 & 3 \end{bmatrix}$

6. $A = \begin{bmatrix} 0 & a \\ 0 & b \end{bmatrix}$ with $b \neq 0$

CAS **7.** $A = \begin{bmatrix} 0 & 0 & 1 \\ 0 & 1 & 0 \\ 0 & 0 & 2 \end{bmatrix}$

CAS **8.** $A = \begin{bmatrix} 1 & 0 & 1 \\ 2 & 1 & 0 \\ 3 & 0 & 1 \end{bmatrix}$

9–14 Show that $A^2 = PD^2P^{-1}$, where P is a matrix whose columns are the eigenvectors of A, and D is a diagonal matrix with the corresponding eigenvalues.

9. $A = \begin{bmatrix} 1 & 0 \\ 0 & 4 \end{bmatrix}$

10. $A = \begin{bmatrix} 1 & 4 \\ 0 & 4 \end{bmatrix}$

11. $A = \begin{bmatrix} 2 & 2 \\ 0 & 1 \end{bmatrix}$

12. $A = \begin{bmatrix} 1 & -1 \\ 2 & 1 \end{bmatrix}$

13. $A = \begin{bmatrix} a & 0 \\ 0 & b \end{bmatrix}$ with $a \neq b$

14. $A = \begin{bmatrix} a & b \\ 0 & c \end{bmatrix}$ with $a \neq c$

15–20 Express the solution to the recursion $\mathbf{n}_{t+1} = A\mathbf{n}_t$ in terms of the eigenvectors and eigenvalues of A, assuming that $\mathbf{n}_0 = \begin{bmatrix} 1 \\ 1 \end{bmatrix}$.

15. $A = \begin{bmatrix} 2 & 0 \\ 0 & 1 \end{bmatrix}$

16. $A = \begin{bmatrix} 0 & 1 \\ 1 & 0 \end{bmatrix}$

17. $A = \begin{bmatrix} 1 & 1 \\ 0 & 2 \end{bmatrix}$

18. $A = \begin{bmatrix} 1 & 1 \\ 1 & 1 \end{bmatrix}$

19. $A = \begin{bmatrix} 1 & a \\ 0 & 2 \end{bmatrix}$ with $a \neq 0$

20. $A = \begin{bmatrix} 1 & a \\ 0 & b \end{bmatrix}$ with $a \neq 0$ and $b \neq 1$

21–25 Express the solution to the recursion $\mathbf{n}_{t+1} = A\mathbf{n}_t$ in terms of the eigenvectors and eigenvalues of A, assuming arbitrary initial conditions.

21. $A = \begin{bmatrix} a & 0 \\ 0 & b \end{bmatrix}$ with $a \neq b$

22. $A = \begin{bmatrix} 0 & 1 \\ 1 & 0 \end{bmatrix}$

23. $A = \begin{bmatrix} 2 & 1 \\ 0 & 1 \end{bmatrix}$

24. $A = \begin{bmatrix} a & 1 \\ 0 & 1 \end{bmatrix}$ with $a \neq 1$

25. $A = \begin{bmatrix} a & b \\ b & a \end{bmatrix}$ with $b \neq 0$

26–29 Like scalars, vectors with complex components can be broken into real and imaginary parts. For example, $\mathbf{v} = \mathbf{a} + \mathbf{b}i$, where $\mathbf{a}$ and $\mathbf{b}$ are the real and imaginary parts of the vector, respectively. The complex conjugate of $\mathbf{v}$ is $\overline{\mathbf{v}} = \mathbf{a} - \mathbf{b}i$. The analogous notation holds for matrices. The following rules from complex analysis, familiar from the scalar case, carry over to vectors and matrices: $\overline{r\mathbf{x}} = \overline{r}\,\overline{\mathbf{x}}$, $\overline{B\mathbf{x}} = \overline{B}\,\overline{\mathbf{x}}$, $\overline{BC} = \overline{B}\,\overline{C}$, and $\overline{rB} = \overline{r}\,\overline{B}$, where r is a scalar, $\mathbf{x}$ is a vector, and B and C are matrices. (*Note:* Complex numbers are discussed in Appendix G.)

26. Suppose that A is a matrix with real entries. Show that if λ is a complex eigenvalue of A, then its associated eigenvector will also be complex.

27. Suppose that A is a matrix with real entries. If λ is a complex eigenvalue of A with associated eigenvector $\mathbf{x}$, show that $\overline{\lambda}$ is also an eigenvalue of A with associated eigenvector $\overline{\mathbf{x}}$ (that is, eigenvalues and eigenvectors of real matrices always come in complex conjugate pairs).

28. Suppose that $T = \begin{bmatrix} a & -b \\ b & a \end{bmatrix}$ with $a \neq 0$ and $b \neq 0$.

(a) Show that the eigenvalues of T are $\lambda = a \pm bi$.

(b) Show that T can be written as

$$T = r\begin{bmatrix} \cos\theta & -\sin\theta \\ \sin\theta & \cos\theta \end{bmatrix}$$

where $r = \sqrt{a^2 + b^2}$ is the modulus of the eigenvalues and θ is the argument. This shows that the matrix T represents a rotation that also scales all vectors by a factor r. (*Note:* Recall Exercise 8.7.6.)

29. Suppose that A is any 2×2 matrix with complex eigenvalues $\lambda = a \pm bi$. We can write $A = PDP^{-1}$, where P is a matrix whose columns contain the (complex) eigenvectors of A, and D is a diagonal matrix with their associated (complex) eigenvalues.

(a) Show that $D = R^{-1}TR$, where

$$R = \begin{bmatrix} i & -i \\ 1 & 1 \end{bmatrix} \quad \text{and} \quad T = \begin{bmatrix} a & -b \\ b & a \end{bmatrix}$$

(b) Part (a) shows that A can be written as $A = PR^{-1}TRP^{-1}$. Show that $S = PR^{-1}$ is a 2×2 matrix whose columns are the real and imaginary components of the eigenvectors of A.

(c) Part (b) shows that $A = STS^{-1}$, where S is a 2×2 matrix whose columns are the real and imaginary components of the eigenvectors of A. In light of Exercise 28, show that we can then write A as

$$A = rS\begin{bmatrix} \cos\theta & -\sin\theta \\ \sin\theta & \cos\theta \end{bmatrix}S^{-1}$$

where $r = \sqrt{a^2 + b^2}$ is the modulus of the eigenvalues and θ is the argument.

(d) Show that

$$A^t = r^t S \begin{bmatrix} \cos\theta t & -\sin\theta t \\ \sin\theta t & \cos\theta t \end{bmatrix} S^{-1}$$

This proves that all 2×2 matrices with complex eigenvalues rotate vectors by an angle θ and stretch them by a factor r each time step.

30–34 Express the solution to the recursion $\mathbf{n}_{t+1} = A\mathbf{n}_t$ in terms of the eigenvectors and eigenvalues of A. Use De Moivre's Theorem to simplify the solution if appropriate.

30. $A = \begin{bmatrix} 0 & -1 \\ 1 & 0 \end{bmatrix}$ with $\mathbf{n}_0 = \begin{bmatrix} 1 \\ 1 \end{bmatrix}$

31. $A = \begin{bmatrix} 1 & -1 \\ 1 & 1 \end{bmatrix}$ with $\mathbf{n}_0 = \begin{bmatrix} 1 \\ 1 \end{bmatrix}$

32. $A = \begin{bmatrix} 1 & -1 \\ 2 & 1 \end{bmatrix}$ with $\mathbf{n}_0 = \begin{bmatrix} 1 \\ 0 \end{bmatrix}$

33. $A = \begin{bmatrix} 1 & 5 \\ -1 & 1 \end{bmatrix}$ with $\mathbf{n}_0 = \begin{bmatrix} 0 \\ 1 \end{bmatrix}$

34. $A = \begin{bmatrix} 1 & -a \\ a & 1 \end{bmatrix}$ with $\mathbf{n}_0 = \begin{bmatrix} 1 \\ 0 \end{bmatrix}$ where a is a real number

35. Vectorcardiography In Exercise 8.7.26 we modeled the heart voltage vector with the recursion $\mathbf{v}_{t+1} = A\mathbf{v}_t$, where $A = \begin{bmatrix} 1 & 0 \\ 0 & -1 \end{bmatrix}$.

(a) Can you apply the Perron-Frobenius Theorem to this model?

(b) Using the initial condition $\mathbf{v}_0 = \begin{bmatrix} 0.3 \\ -0.2 \end{bmatrix}$, express the solution to the recursion in terms of the eigenvectors and eigenvalues of A.

(c) Describe the long-term behavior of $\mathbf{v}_t$.

36. Antigenic evolution In Exercise 8.7.27 we modeled antigenic evolution with the recursion $\mathbf{x}_{t+1} = A\mathbf{x}_t$, where

$$A = \begin{bmatrix} 2 & 3 \\ 0 & \frac{9}{10} \end{bmatrix}$$

(a) Using the initial condition $\mathbf{x}_0 = \begin{bmatrix} 1 \\ 1 \end{bmatrix}$, express the solution to the recursion in terms of the eigenvectors and eigenvalues of A.

(b) Describe the long-term behavior of $\mathbf{x}_t$.

37. Mutation In Exercise 8.6.34 we modeled mutation with a recursion of the form $\mathbf{y}_{t+1} = A\mathbf{y}_t$. Suppose

$$A = \begin{bmatrix} \frac{9}{10} & \frac{1}{5} \\ \frac{1}{10} & \frac{4}{5} \end{bmatrix}$$

(a) Calculate the eigenvalues of A.

(b) In light of the Perron-Frobenius Theorem, what is the long-term behavior of $\mathbf{y}_t$?

(c) Using the initial condition $\mathbf{y}_0 = \begin{bmatrix} 1 \\ 0 \end{bmatrix}$, express the solution to the recursion in terms of the eigenvectors and eigenvalues of A.

(d) Verify your answer to part (b) using the solution obtained in part (c).

38. Leslie matrices In Exercise 8.7.29 we modeled an age-structured population using the recursion $\mathbf{n}_{t+1} = L\mathbf{n}_t$, where

$$L = \begin{bmatrix} b & 2 \\ \frac{1}{2} & 0 \end{bmatrix} \qquad b > 0$$

(a) Verify that the Perron-Frobenius Theorem can be applied to this model.

(b) Calculate the eigenvalues of L.

(c) Given your answers to parts (a) and (b), what is the long-term behavior of $\mathbf{n}_t$ as a function of b?

(d) Using the initial condition $\mathbf{n}_0 = \begin{bmatrix} 1 \\ 0 \end{bmatrix}$, express the solution to the recursion in terms of the eigenvectors and eigenvalues of L.

(e) Verify your answer to part (c) using the solution you obtained in part (d).

39. Methylation In Exercise 8.5.20 we modeled DNA methylation with the recursion $\mathbf{x}_{t+1} = A\mathbf{x}_t$, where

$$A = \begin{bmatrix} \frac{8}{10} & \frac{1}{2} \\ \frac{2}{10} & \frac{1}{2} \end{bmatrix}$$

(a) Calculate the eigenvalues of A.

(b) The Perron-Frobenius Theorem can clearly be applied to the matrix A. Given your answer to part (a), what is the long-term behavior of $\mathbf{x}_t$?

(c) Using the initial condition $\mathbf{x}_0 = \begin{bmatrix} a \\ 1-a \end{bmatrix}$, express the solution to the recursion in terms of the eigenvectors and eigenvalues of A.

(d) Verify your answer to part (b) using the solution you obtained in part (c).

CAS **40. Breast cancer** In Example 8.5.2 a model for the dynamics of breast cancer was given by the recursion $\mathbf{n}_{t+1} = A\mathbf{n}_t$, where

$$A = \begin{bmatrix} 0.998 & 0.45 & 0 \\ 0.002 & 0.1 & 0 \\ 0 & 0.45 & 1 \end{bmatrix}$$

(a) Using the initial condition $\mathbf{n}_0 = \begin{bmatrix} 1 \\ 0 \\ 0 \end{bmatrix}$, express the solution to the recursion in terms of the eigenvectors and eigenvalues of A.

(b) Describe the long-term behavior of $\mathbf{n}_t$.

41. Fibonacci numbers The nth Fibonacci number is generated by the recursion equation $F_n = F_{n-1} + F_{n-2}$ with $F_0 = F_1 = 1$. Define the two new variables $x_n = F_n$ and $y_n = F_{n-1}$.

(a) Show that the vector $\mathbf{z}_n = \begin{bmatrix} x_n \\ y_n \end{bmatrix}$ obeys the recursion

$$\mathbf{z}_n = A\mathbf{z}_{n-1}, \text{ where } A = \begin{bmatrix} 1 & 1 \\ 1 & 0 \end{bmatrix}.$$

(b) Find the eigenvalues of A.

(c) Can the Perron-Frobenius Theorem be applied to the matrix A? If so, what is the long-term behavior of the vector $\mathbf{z}_n$?

(d) Find a formula for the nth Fibonacci number.

PROJECT The Emergence of Geometric Order in Proliferating Cells

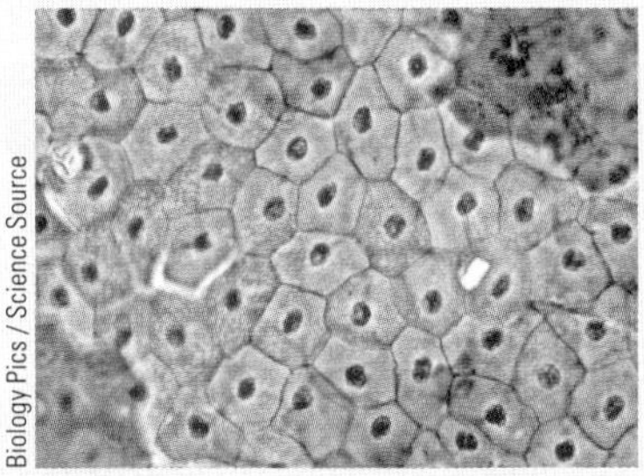

Biology Pics / Science Source

FIGURE 1

When some cells divide they form thin sheets in which the majority of cells are each adjacent to six other cells. This gives the sheet a hexagonal pattern as shown in Figure 1. In this project we will see how a mathematical model[1] predicts the emergence of this pattern simply by the way cells divide.

Each cell in a sheet can be viewed as a polygon having a certain number of edges and vertices (see Figure 2). We assume that each cell has a minimum of four edges and that, when a cell division occurs (that is, when a cell splits into two daughter cells), the line of cell division always connects two edges, subdividing each edge (see Figure 3). The cells divide asynchronously and we take a single time step to be the time after which all cells in the sheet have divided once. Let c_t, v_t, and e_t denote the total number of cells, vertices, and edges after t cell divisions. Since each cell divides into two in a single time step, we have $c_{t+1} = 2c_t$.

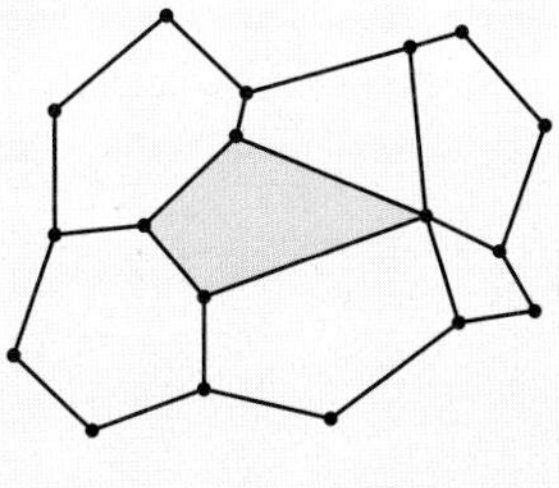

FIGURE 2
The shaded cell has four vertices and edges.

1. Explain why the total number of vertices in the cell sheet obeys the recursion $v_{t+1} = v_t + 2c_t$, and why the total number of edges obeys the recursion $e_{t+1} = e_t + 3c_t$. (Figure 3 might be helpful for this.)

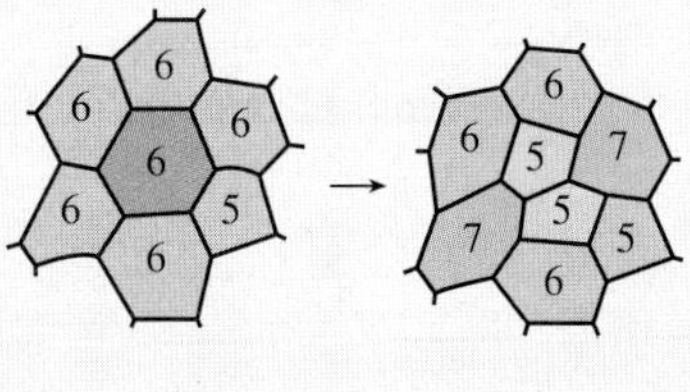

FIGURE 3
Numerals in each cell indicate the number of edges.

1. M. Gibson et al., "The Emergence of Geometric Order in Proliferating Metazoan Epithelia," *Nature* 442 (2006): 1038–41.

2. Write the system of recursion equations for c_t, v_t, and e_t in matrix notation.

3. What are the eigenvalues of the matrix from Problem 2? Show that

$$\begin{bmatrix} 1 \\ 2 \\ 3 \end{bmatrix}, \begin{bmatrix} 0 \\ 1 \\ 0 \end{bmatrix}, \text{ and } \begin{bmatrix} 0 \\ 0 \\ 1 \end{bmatrix}$$ are eigenvectors corresponding to these eigenvalues.

4. Write the solution to the recursion from Problem 2 in terms of its eigenvectors and eigenvalues for an arbitrary initial condition.

5. If we ignore complications arising from the fact that cells on the boundary are not completely surrounded by other cells, then the average number of sides per cell at time t is $s_t = 2e_t/c_t$. Why?

6. Show that, in the long term (that is, when $t \to \infty$), the average number of sides per cell s_t approaches six.

Chapter 8 REVIEW

CONCEPT CHECK

1. What is the difference between a vector and a scalar?

2. How do you add two vectors geometrically? How do you add them algebraically?

3. If $\mathbf{a}$ is a vector and c is a scalar, how is $c\mathbf{a}$ related to $\mathbf{a}$ geometrically? How do you find $c\mathbf{a}$ algebraically?

4. How do you find the vector from one point to another algebraically?

5. How do you find the dot product $\mathbf{a} \cdot \mathbf{b}$ of two vectors if you know their lengths and the angle between them? What if you know their components?

6. How are dot products useful?

7. Write expressions for the scalar and vector projections of $\mathbf{b}$ onto $\mathbf{a}$. Illustrate with diagrams.

8. What is the equation of a sphere?

9. (a) How do you tell if two vectors are parallel?
(b) How do you tell if two vectors are perpendicular?

10. What is a symmetric matrix?

11. If a matrix A rotates vectors counterclockwise by θ degrees, what does the matrix A^{-1} do?

12. If a 2×2 matrix has complex eigenvalues, what does this matrix do to vectors upon multiplication?

13. Suppose A is a matrix and k is a positive integer. What does the notation A^k mean?

14. What is the relationship between the inverse of a matrix and the determinant of a matrix?

15. Explain what eigenvalues and eigenvectors are.

16. Why does a 2×2 matrix have two eigenvalues?

17. Suppose a 2×2 matrix with real entries has complex eigenvalues. Why must the eigenvalues be complex conjugates?

18. Why is it sometimes useful to write a matrix A in the form $A = PDP^{-1}$, where D is a diagonal matrix?

19. What does the Perron-Frobenius Theorem say, and why is it useful?

Answers to the Concept Check can be found on the back endpapers.

TRUE-FALSE QUIZ

Determine whether the statement is true or false. If it is true, explain why. If it is false, explain why or give an example that disproves the statement.

1. For any vectors $\mathbf{u}$ and $\mathbf{v}$ in V_3, $\mathbf{u} \cdot \mathbf{v} = \mathbf{v} \cdot \mathbf{u}$.

2. For any scalar c and vectors $\mathbf{u}$ and $\mathbf{v}$, $c(\mathbf{u} + \mathbf{v}) = c\mathbf{u} + c\mathbf{v}$.

3. If $\mathbf{u} \cdot \mathbf{v} = 0$, then $\mathbf{u} = \mathbf{0}$ or $\mathbf{v} = \mathbf{0}$.

4. If $\mathbf{u} = [u_1, u_2]$ and $\mathbf{v} = [v_1, v_2]$, then $\mathbf{u} \cdot \mathbf{v} = [u_1v_1, u_2v_2]$.

5. The set of points $\{(x, y, z) \mid x^2 + y^2 = 1\}$ is a circle.

6. The dot product of parallel vectors is zero.

7. For any square matrices of the same size, $AB = BA$.

8. For any square matrices of the same size, $A(B + C) = AB + AC$.

9. The determinant of the identity matrix is 1.

10. If $A\mathbf{x} = a\mathbf{x}$ for a square matrix A, vector $\mathbf{x}$, and scalar a, where $\mathbf{x} \neq \mathbf{0}$, then a is an eigenvalue of A.

11. If $\mathbf{y}$ is an eigenvector of A, then $A\mathbf{y} = k\mathbf{y}$ for some scalar k.

12. If a square matrix A is singular, it has a zero eigenvalue.

13. The eigenvalues of any square matrix lie on its diagonal.

14. If a 2×2 matrix A has complex eigenvalues, then it stretches and rotates vectors.

15. All square matrices A with distinct eigenvalues can be written as $A = PDP^{-1}$.

16. The inhomogeneous system of equations $A\mathbf{x} = \mathbf{b}$, where $\mathbf{b} \neq \mathbf{0}$, has either the trivial solution or an infinite number of solutions.

17. The homogeneous system of equations $A\mathbf{x} = \mathbf{0}$ has either the trivial solution or an infinite number of solutions.

18. If $\mathbf{u}$ and $\mathbf{v}$ are in V_3, then $|\mathbf{u} \cdot \mathbf{v}| \leqslant |\mathbf{u}||\mathbf{v}|$.

EXERCISES

1. (a) Find an equation of the sphere that passes through the point $(6, -2, 3)$ and has center $(-1, 2, 1)$.
(b) Find the curve in which this sphere intersects the yz-plane.
(c) Find the center and radius of the sphere

$$x^2 + y^2 + z^2 - 8x + 2y + 6z + 1 = 0$$

2. Copy the vectors in the figure and use them to draw each of the following vectors.
(a) $\mathbf{a} + \mathbf{b}$ (b) $\mathbf{a} - \mathbf{b}$
(c) $-\frac{1}{2}\mathbf{a}$ (d) $2\mathbf{a} + \mathbf{b}$

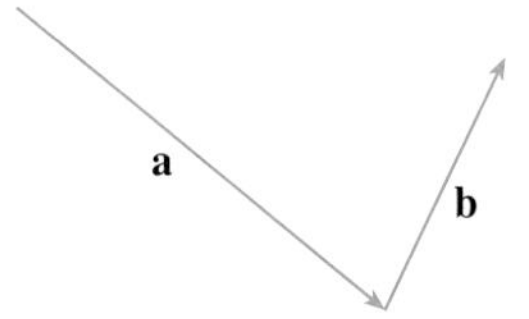

3. A hypersphere with center at point P and radius r in $\mathbb{R}^n$ is the set of all points that are a distance r from P. Give the equation of a hypersphere.

4. Calculate the given quantity if

$$\mathbf{a} = [1, 1, -2]$$

$$\mathbf{b} = [3, -2, 1]$$

$$\mathbf{c} = [0, 1, -5]$$

(a) $2\mathbf{a} + 3\mathbf{b}$ (b) $|\mathbf{b}|$
(c) $\mathbf{a} \cdot \mathbf{b}$ (d) $\text{comp}_{\mathbf{a}}\mathbf{b}$
(e) $\text{proj}_{\mathbf{a}}\mathbf{b}$
(f) The angle between $\mathbf{a}$ and $\mathbf{b}$ (correct to the nearest degree)

5. Find the values of x such that the vectors $[3, 2, x]$ and $[2x, 4, x]$ are orthogonal.

6. Find two unit vectors that are orthogonal to both $[0, 1, 2]$ and $[1, -2, 3]$.

7. Find the acute angle between the lines.
(a) $2x - y = 3$, $3x + y = 7$
(b) $x + 2y = 7$, $5x - y = 2$

8. Find the acute angle between two diagonals of a cube.

9. Use a scalar projection to show that the distance from a point $P_1(x_1, y_1)$ to the line $ax + by + c = 0$ is

$$\frac{|ax_1 + by_1 + c|}{\sqrt{a^2 + b^2}}$$

Use this formula to find the distance from the point $(-2, 3)$ to the line $3x - 4y + 5 = 0$.

10. Viral identification Viruses from the same geographic location tend to cluster together in antigenic space. Suppose all viruses from Asia fall within a sphere of radius 1 centered at $(2, 3, 1)$ in antigenic space, viruses from North America fall within a sphere of radius 1 centered at $(-3, 2, 0)$, and those from Europe fall within a sphere of radius 1 centered at $(1, 0, 1)$. (All numbers are dimensionless quantities.) A patient is infected with a virus located at

(2.5, 3, 1.75) in antigenic space. In which geographic location was this patient infected?

11. Genome expression divergence The genome expression profiles of two closely related species are given by the points $(-1, 3)$ and $(1, 1)$ in dimensionless quantities.
(a) Calculate the difference in expression profile as measured by distance in "expression space."
(b) Suppose the points in expression space are treated as the tips of position vectors. Calculate the difference in expression profile as measured by the angle between the expression vectors.

12. Influenza Each month all individuals in a city are characterized as being either susceptible to influenza infection, currently infected, or resistant to infection. Suppose that susceptible individuals can become infected, infected individuals can either become susceptible again or resistant, and resistant individuals remain resistant. Draw a matrix diagram for this situation.

13. Construct the matrix model for the following matrix diagram.

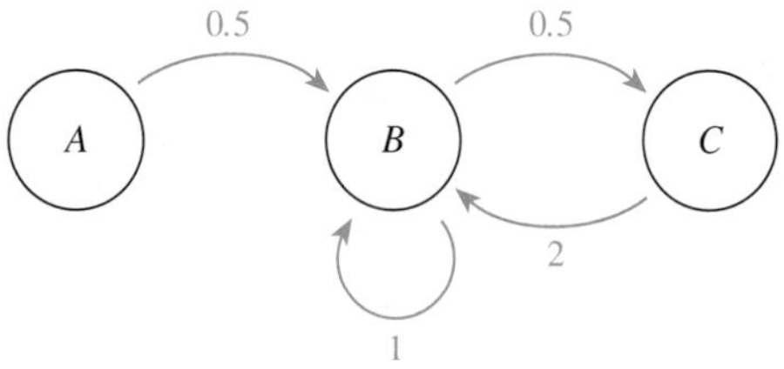

14. Leslie matrix modeling Suppose that a species lives for a maximum of three years and individuals of all age classes produce three new offspring each year. Further, suppose the annual survival rates are 10% at age 1 and 30% at age 2.
(a) Construct the matrix diagram for this situation.
(b) Construct the matrix model from the diagram in part (a).

15. Calculate the given quantity, if it is defined, using the given matrices:

$$A = \begin{bmatrix} 7 & 0 \\ 1 & 9 \end{bmatrix} \qquad B = \begin{bmatrix} 1 & 4 \\ 1 & 0 \\ 0 & 9 \end{bmatrix}$$

$$C = \begin{bmatrix} 1 & 2 \\ 9 & 6 \end{bmatrix} \qquad D = \begin{bmatrix} 15 & 0 & 7 \\ 0 & 5 & 2 \end{bmatrix}$$

(a) $A - D$ (b) $2C + A$
(c) $A + 2B + C$ (d) $D - B$
(e) $B^T + D$ (f) $2C - A^{-1}$

16. Calculate the given quantity, if it is defined, using the given matrices:

$$A = \begin{bmatrix} 1 & 0 \\ 1 & 9 \end{bmatrix} \qquad B = \begin{bmatrix} 1 & 1 \\ 1 & 9 \\ 0 & 9 \end{bmatrix}$$

$$C = \begin{bmatrix} 1 & 2 \\ 9 & 2 \end{bmatrix} \qquad D = \begin{bmatrix} 1 & 1 & 7 \\ 0 & 3 & 2 \end{bmatrix}$$

(a) AD (b) $C + C^T$
(c) $A(B + C)$ (d) $(DB)^{-1}$
(e) $(C - 2A)B$ (f) $ABCD$

17. For all matrices $\begin{bmatrix} a_{11} & a_{12} \\ a_{21} & a_{22} \end{bmatrix}$ where all components a_{ij} are nonzero, show that if $A = A^{-1}$, then $\det A = -1$ and $a_{11} = -a_{22}$.

18. If A is nonsingular, then $\det A^{-1} = 1/(\det A)$. Verify this theorem for 2×2 matrices.

19. If I is the $n \times n$ identity matrix, then $\det I = 1$. Verify this theorem for 2×2 matrices.

20. Stage-structured population Suppose a population contains juveniles and adults, and the vector $\mathbf{n}_t$ of the numbers of each of these changes from one year to the next according to the equation

$$\mathbf{n}_{t+1} = \begin{bmatrix} 0 & 2 \\ \frac{1}{2} & \frac{1}{3} \end{bmatrix} \mathbf{n}_t$$

If the number of juveniles and adults in the current year is given by the vector $\begin{bmatrix} 30 \\ 8 \end{bmatrix}$, find the numbers of juveniles and adults in the previous year.

21–24 Solve the system of equations.

21. $3x - y = 2$
$x + 7y = 4$

22. $2x + y = 2$
$4x - y = 4$

23. $7x + y = 0$
$14x + 2y = 0$

24. $x - 2y = 2$
$3x - 6y = 4$

25. Leslie matrix Consider the following model for an age-structured population with three age classes:

$$\mathbf{n}_{t+1} = \begin{bmatrix} 1 & 2 & 4 \\ \frac{1}{2} & 0 & 0 \\ 0 & \frac{1}{3} & 0 \end{bmatrix} \mathbf{n}_t$$

An equilibrium is a value of the vector for which no change occurs (that is, $\mathbf{n}_{t+1} = \mathbf{n}_t$). Denoting this value by $\hat{\mathbf{n}}$, it must therefore satisfy the equation

$$\hat{\mathbf{n}} = \begin{bmatrix} 1 & 2 & 4 \\ \frac{1}{2} & 0 & 0 \\ 0 & \frac{1}{3} & 0 \end{bmatrix} \hat{\mathbf{n}}$$

Find all equilibria and explain what each represents biologically.

26. CAT scans Computed Axial Tomography uses narrow X-ray beams directed though the body at several angles to construct an image of internal organs and structures. The density of the material through which the beam passes affects the extent to which it is diminished before exiting the body. In most cases the part of the body to be scanned is overlaid with a grid, and X-ray beams are directed through the elements of this grid at different angles. Measurements of the exiting beams can then be used to calculate the density of each element of the grid.

(a) Consider a simple grid with three elements, and three X-ray beams directed through them at different angles, as shown in the figure. The unknown densities of the material in each element are x_1, x_2 and x_3.

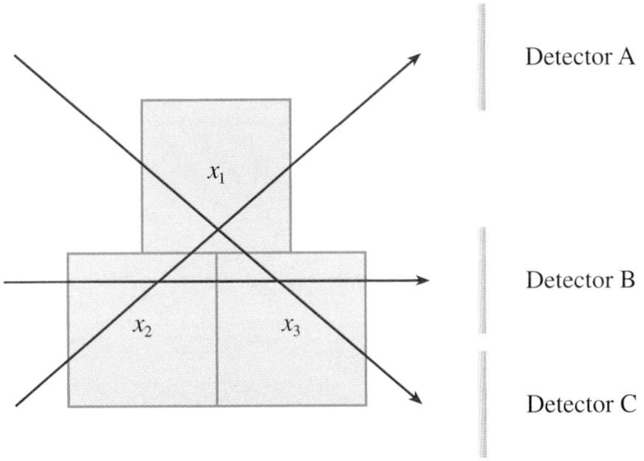

Suppose that, if a beam travels through elements i and j, its intensity upon exiting is $e^{-(x_i+x_j)}$. If the measured intensities of the beams at detectors A, B, and C are a, b, and c, respectively, construct a system of equations for the unknown densities.

(b) Using $a = 0.6$, $b = 0.5$, and $c = 0.7$, determine the density of each element. In CAT scans these densities are then used to color the pixels of the image.

27–30 Find the eigenvectors and eigenvalues for the matrix.

27. $A = \begin{bmatrix} 3 & 1 \\ 1 & 3 \end{bmatrix}$

28. $A = \begin{bmatrix} 2 & -2 \\ -4 & 4 \end{bmatrix}$

29. $A = \begin{bmatrix} 2 & 1 \\ -1 & 2 \end{bmatrix}$

30. $A = \begin{bmatrix} a & -1 \\ 1 & a \end{bmatrix}$

31. Suppose A is a 2×2 matrix whose columns sum to one. Show that $\lambda = 1$ is an eigenvalue of A.

32–35 Express the solution to the recursion $\mathbf{n}_{t+1} = A\mathbf{n}_t$ in terms of the eigenvectors and eigenvalues of A. Use De Moivre's Theorem to simplify the solution, if appropriate.

32. $A = \begin{bmatrix} 1 & 1 \\ 2 & 1 \end{bmatrix}$ and $\mathbf{n}_0 = \begin{bmatrix} 1 \\ 0 \end{bmatrix}$

33. $A = \begin{bmatrix} 2 & 3 \\ -4 & -6 \end{bmatrix}$ and $\mathbf{n}_0 = \begin{bmatrix} 1 \\ 0 \end{bmatrix}$

34. $A = \begin{bmatrix} 2 & -1 \\ 1 & 1 \end{bmatrix}$ and $\mathbf{n}_0 = \begin{bmatrix} 1 \\ 0 \end{bmatrix}$

35. $A = \begin{bmatrix} a & -1 \\ 1 & a \end{bmatrix}$ and $\mathbf{n}_0 = \begin{bmatrix} 1 \\ 1 \end{bmatrix}$, where a is a real number

36. Dispersal In Exercise 8.5.9 we modeled the dispersal of dragonflies using the recursion $\mathbf{x}_{t+1} = A\mathbf{x}_t$, where

$$A = \begin{bmatrix} 0.8 & 0.3 \\ 0.2 & 0.7 \end{bmatrix}$$

(a) Verify that the Perron-Frobenius Theorem can be applied to this model.
(b) Calculate the eigenvalues of A.
(c) Given your answers to parts (a) and (b), what is the long-term behavior of $\mathbf{x}_t$?
(d) Using the initial condition $\mathbf{x}_0 = \begin{bmatrix} 1 \\ 1 \end{bmatrix}$, express the solution to the recursion in terms of the eigenvectors and eigenvalues of A.
(e) Verify your answer to part (c) using the solution you obtained in part (d).

37. Succession In Exercise 8.5.15 we modeled ecological succession using the recursion $\mathbf{y}_{t+1} = A\mathbf{y}_t$, where

$$A = \begin{bmatrix} 0.9 & 0.001 & 0 \\ 0.1 & 0.799 & 0.001 \\ 0 & 0.2 & 0.999 \end{bmatrix}$$

Can the Perron-Frobenius Theorem be applied to this model?

38. Inbreeding In Exercise 8.5.19 we modeled inbreeding using the recursion $\mathbf{x}_{t+1} = A\mathbf{x}_t$, where

$$A = \begin{bmatrix} 1 & \frac{1}{4} & 0 \\ 0 & \frac{1}{2} & 0 \\ 0 & \frac{1}{4} & 1 \end{bmatrix}$$

(a) Calculate the eigenvalues of A. (You will find a repeated eigenvalue.)

(b) Verify that the eigenvectors of A are

$$\mathbf{v}_1 = \begin{bmatrix} 1 \\ 0 \\ 0 \end{bmatrix} \quad \mathbf{v}_2 = \begin{bmatrix} 0 \\ 0 \\ 1 \end{bmatrix} \quad \mathbf{v}_3 = \begin{bmatrix} 1 \\ -2 \\ 1 \end{bmatrix}$$

CAS (c) Using the initial condition $\mathbf{x}_0 = \begin{bmatrix} 0 \\ 1 \\ 0 \end{bmatrix}$, express the solution to the recursion in terms of the eigenvectors and eigenvalues of A.

39. A sequence is given recursively by the equation

$$a_n = \frac{(a_{n-1} + a_{n-2})}{2}$$

Define the two new variables $x_n = a_n$ and $y_n = a_{n-1}$.

(a) Find a matrix A such that the vector $\mathbf{z}_n = \begin{bmatrix} x_n \\ y_n \end{bmatrix}$ obeys the recursion $\mathbf{z}_n = A\mathbf{z}_{n-1}$.

(b) Verify that the Perron-Frobenius Theorem can be applied.

(c) Find the eigenvalues of A.

(d) In light of the Perron-Frobenius Theorem, what is the long-term behavior of the vector $\mathbf{z}_n$?

(e) Evaluate $\lim_{n\to\infty} a_n$.

Multivariable Calculus

9

The energy needed by a lizard (such as this iguana) to walk or run depends on both its weight and its speed. So the energy is a function of two variables, which we investigate in Exercise 47 in Section 9.2.

BIOLOGICAL QUANTITIES OFTEN DEPEND ON two or more variables. In this chapter we extend the basic ideas of differential calculus to functions of several variables.

9.1 | Functions of Several Variables

In this section we study functions of two or more variables from four points of view:

- verbally (by a description in words)
- numerically (by a table of values)
- algebraically (by an explicit formula)
- visually (by a graph or level curves)

Functions of Two Variables

The temperature T at a point on the surface of the earth at any given time depends on the longitude x and latitude y of the point. We can think of T as being a function of the two variables x and y, or as a function of the pair (x, y). We indicate this functional dependence by writing $T = f(x, y)$.

The resistance R of blood flowing through an artery depends on the radius r and length L of the artery. In fact, one of Poiseuille's laws states that $R = CL/r^4$, where C is a positive constant determined by the viscosity of the blood. We say that R is a function of r and L, and we write

$$R(r, L) = C\frac{L}{r^4}$$

> **Definition** A **function f of two variables** is a rule that assigns to each ordered pair of real numbers (x, y) in a set D a unique real number denoted by $f(x, y)$. The set D is the **domain** of f and its **range** is the set of values that f takes on, that is, $\{f(x, y) \mid (x, y) \in D\}$.

We often write $z = f(x, y)$ to make explicit the value taken on by f at the general point (x, y). The variables x and y are **independent variables** and z is the **dependent variable**. [Compare this with the notation $y = f(x)$ for functions of a single variable.]

The domain is a subset of $\mathbb{R}^2$, the xy-plane. We can think of the domain as the set of all possible inputs and the range as the set of all possible outputs. If a function f is given by a formula and no domain is specified, then the domain of f is understood to be the set of all pairs (x, y) for which the given expression is a well-defined real number.

EXAMPLE 1 | Wind-chill index In regions with severe winter weather, the *wind-chill index* is often used to describe the apparent severity of the cold. This index W is a subjective temperature that depends on the actual temperature T and the wind speed v. So W is a function of T and v, and we can write $W = f(T, v)$. Table 1 records values of W compiled by the National Weather Service of the US and the Meteorological Service of Canada.

The Wind-Chill Index
A new wind-chill index was introduced in November of 2001 and is more accurate than the old index for measuring how cold it feels when it's windy. The new index is based on a model of how fast a human face loses heat. It was developed through clinical trials in which volunteers were exposed to a variety of temperatures and wind speeds in a refrigerated wind tunnel.

Table 1 Wind-chill index as a function of air temperature and wind speed

Wind speed (km/h); rows: Actual temperature (°C)

T \ v	5	10	15	20	25	30	40	50	60	70	80
5	4	3	2	1	1	0	−1	−1	−2	−2	−3
0	−2	−3	−4	−5	−6	−6	−7	−8	−9	−9	−10
−5	−7	−9	−11	−12	−12	−13	−14	−15	−16	−16	−17
−10	−13	−15	−17	−18	−19	−20	−21	−22	−23	−23	−24
−15	−19	−21	−23	−24	−25	−26	−27	−29	−30	−30	−31
−20	−24	−27	−29	−30	−32	−33	−34	−35	−36	−37	−38
−25	−30	−33	−35	−37	−38	−39	−41	−42	−43	−44	−45
−30	−36	−39	−41	−43	−44	−46	−48	−49	−50	−51	−52
−35	−41	−45	−48	−49	−51	−52	−54	−56	−57	−58	−60
−40	−47	−51	−54	−56	−57	−59	−61	−63	−64	−65	−67

For instance, the table shows that if the temperature is $-5°C$ and the wind speed is 50 km/h, then subjectively it would feel as cold as a temperature of about $-15°C$ with no wind. So

$$f(-5, 50) = -15°C$$ ■

EXAMPLE 2 | The **body mass index** (BMI) of a person is defined by the equation

$$B(m, h) = \frac{m}{h^2}$$

where m is the person's mass (in kilograms) and h is the height (in meters). So B is a function of the two variables m and h. A rough guideline is that a person is underweight if the BMI is less than 18.5; optimal if the BMI lies between 18.5 and 25; overweight if the BMI lies between 25 and 30; and obese if the BMI exceeds 30. If someone weighs 64 kg and is 168 cm high, what is the BMI?

SOLUTION With $m = 64$ kg and $h = 1.68$ m, the BMI is

$$B(64, 1.68) = \frac{64}{(1.68)^2} \approx 22.7 \text{ kg/m}^2$$

Because

$$18.5 \leqslant B(64, 1.68) \leqslant 25$$

this person is considered to have optimal weight. ■

EXAMPLE 3 | For each of the following functions, evaluate $f(3, 2)$ and find and sketch the domain.

(a) $f(x, y) = \dfrac{\sqrt{x + y + 1}}{x - 1}$ (b) $f(x, y) = x \ln(y^2 - x)$

SOLUTION

(a)
$$f(3, 2) = \frac{\sqrt{3 + 2 + 1}}{3 - 1} = \frac{\sqrt{6}}{2}$$

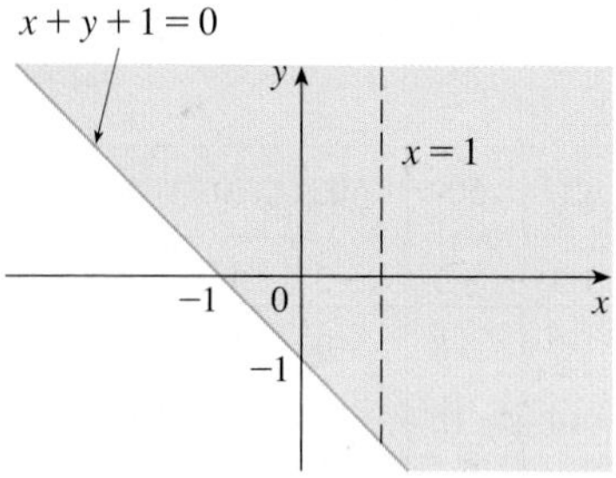

FIGURE 1

Domain of $f(x, y) = \dfrac{\sqrt{x+y+1}}{x-1}$

The expression for f makes sense if the denominator is not 0 and the quantity under the square root sign is nonnegative. So the domain of f is

$$D = \{(x, y) \mid x + y + 1 \geqslant 0,\ x \neq 1\}$$

The inequality $x + y + 1 \geqslant 0$, or $y \geqslant -x - 1$, describes the points that lie on or above the line $y = -x - 1$, while $x \neq 1$ means that the points on the line $x = 1$ must be excluded from the domain (see Figure 1).

(b) $$f(3, 2) = 3 \ln(2^2 - 3) = 3 \ln 1 = 0$$

Since $\ln(y^2 - x)$ is defined only when $y^2 - x > 0$, that is, $x < y^2$, the domain of f is $D = \{(x, y) \mid x < y^2\}$. This is the set of points to the left of the parabola $x = y^2$. (See Figure 2.)

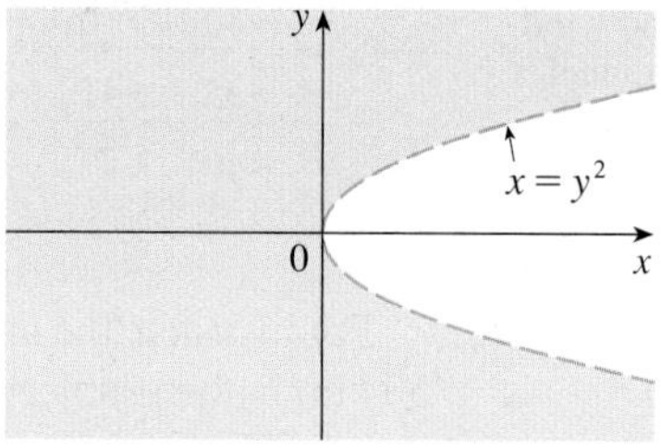

FIGURE 2

Domain of $f(x, y) = x \ln(y^2 - x)$

■

EXAMPLE 4 | Find the domain and range of $g(x, y) = \sqrt{9 - x^2 - y^2}$.

SOLUTION The domain of g is

$$D = \{(x, y) \mid 9 - x^2 - y^2 \geqslant 0\} = \{(x, y) \mid x^2 + y^2 \leqslant 9\}$$

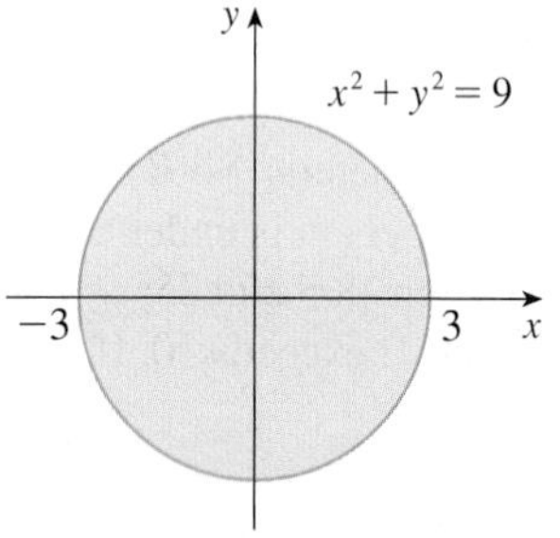

FIGURE 3

Domain of $g(x, y) = \sqrt{9 - x^2 - y^2}$

which is the disk with center (0, 0) and radius 3. (See Figure 3.) The range of g is

$$\left\{z \mid z = \sqrt{9 - x^2 - y^2},\ (x, y) \in D\right\}$$

Since z is a positive square root, $z \geqslant 0$. Also, because $9 - x^2 - y^2 \leqslant 9$, we have

$$\sqrt{9 - x^2 - y^2} \leqslant 3$$

So the range is

$$\{z \mid 0 \leqslant z \leqslant 3\} = [0, 3]$$

■

■ Graphs

One way of visualizing the behavior of a function of two variables is to consider its graph.

> **Definition** If f is a function of two variables with domain D, then the **graph** of f is the set of all points (x, y, z) in $\mathbb{R}^3$ such that $z = f(x, y)$ and (x, y) is in D.

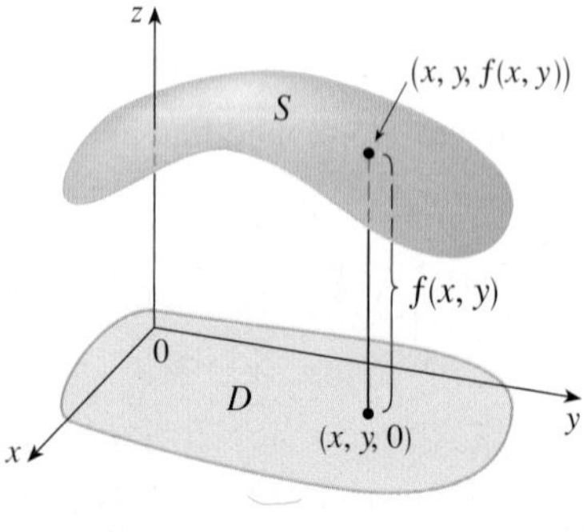

FIGURE 4

Just as the graph of a function f of one variable is a curve C with equation $y = f(x)$, so the graph of a function f of two variables is a surface S with equation $z = f(x, y)$. We can visualize the graph S of f as lying directly above or below its domain D in the xy-plane (see Figure 4).

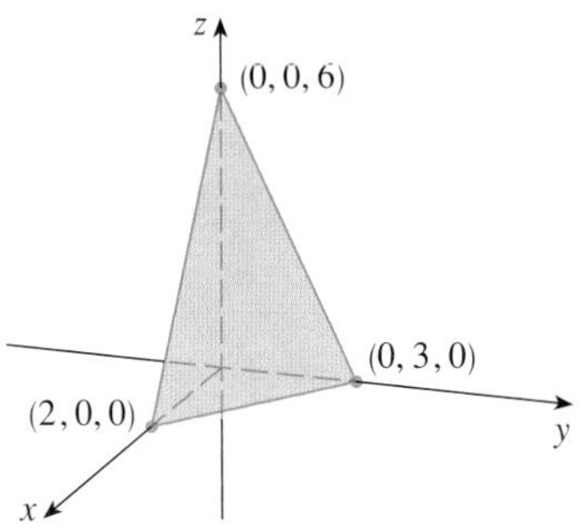

FIGURE 5

EXAMPLE 5 | Sketch the graph of the function $f(x, y) = 6 - 3x - 2y$.

SOLUTION The graph of f has the equation $z = 6 - 3x - 2y$, or $3x + 2y + z = 6$, which represents a plane (see Section 8.3). To graph the plane we first find the intercepts. Putting $y = z = 0$ in the equation, we get $x = 2$ as the x-intercept. Similarly, the y-intercept is 3 and the z-intercept is 6. This helps us sketch the portion of the graph that lies in the first octant in Figure 5. ■

The function in Example 5 is a special case of the function

$$f(x, y) = ax + by + c$$

which is called a **linear function**. The graph of such a function has the equation

$$z = ax + by + c \qquad \text{or} \qquad ax + by - z + c = 0$$

so it is a plane. In much the same way that linear functions of one variable are important in single-variable calculus, we will see that linear functions of two variables play a central role in multivariable calculus.

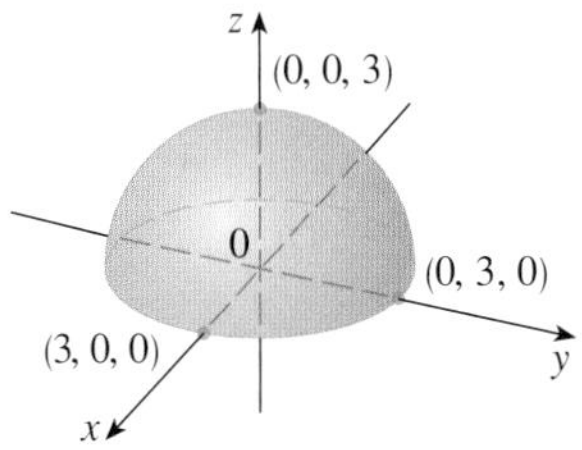

FIGURE 6

EXAMPLE 6 | Sketch the graph of $g(x, y) = \sqrt{9 - x^2 - y^2}$.

SOLUTION The graph has equation $z = \sqrt{9 - x^2 - y^2}$. We square both sides of this equation to obtain $z^2 = 9 - x^2 - y^2$, or $x^2 + y^2 + z^2 = 9$, which we recognize as an equation of the sphere with center the origin and radius 3. (See Section 8.1.) But, since $z \geqslant 0$, the graph of g is just the top half of this sphere (see Figure 6). ■

NOTE An entire sphere can't be represented by a single function of x and y. As we saw in Example 6, the upper hemisphere of the sphere $x^2 + y^2 + z^2 = 9$ is represented by the function $g(x, y) = \sqrt{9 - x^2 - y^2}$. The lower hemisphere is represented by the function $h(x, y) = -\sqrt{9 - x^2 - y^2}$.

EXAMPLE 7 | Find the domain and range and sketch the graph of $h(x, y) = x^2 + y^2$.

SOLUTION Notice that $h(x, y)$ is defined for all possible ordered pairs of real numbers (x, y), so the domain is $\mathbb{R}^2$, the entire xy-plane. The range of h is the set $[0, \infty)$ of all nonnegative real numbers. [Notice that $x^2 \geqslant 0$ and $y^2 \geqslant 0$, so $h(x, y) \geqslant 0$ for all x and y.]

The graph of h has the equation $z = x^2 + y^2$. If we put $x = 0$, we get $z = y^2$, so the yz-plane intersects the surface in a parabola. If we put $x = k$ (a constant), we get $z = y^2 + k^2$. This means that if we slice the graph with any plane parallel to the yz-plane, we obtain a parabola that opens upward. (These curves that we get by slicing a surface with a plane parallel to one of the coordinate planes are called **traces**.) Similarly, if $y = k$, the trace is $z = x^2 + k^2$, which is again a parabola that opens upward. If we put $z = k$, we get the horizontal traces $x^2 + y^2 = k$, which we recognize as a family of circles. Knowing the shapes of the traces, we can sketch the graph of f in Figure 7. Because of the parabolic traces, the surface $z = x^2 + y^2$ is called a **paraboloid**. ■

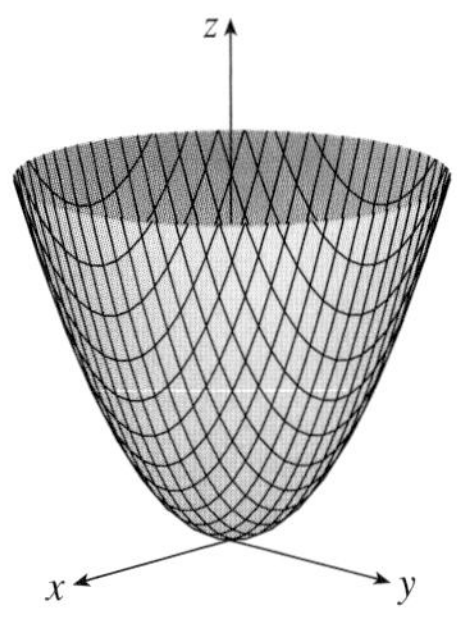

FIGURE 7

Computer programs are readily available for graphing functions of two variables. In most such programs, traces in the vertical planes $x = k$ and $y = k$ are drawn for equally spaced values of k and parts of the graph are eliminated using hidden line removal.

Figure 8 shows computer-generated graphs of several functions. Notice that we get an especially good picture of a function when rotation is used to give views from different vantage points. In parts (a) and (b) the graph of f is very flat and close to the xy-plane except near the origin; this is because $e^{-x^2-y^2}$ is very small when x or y is large.

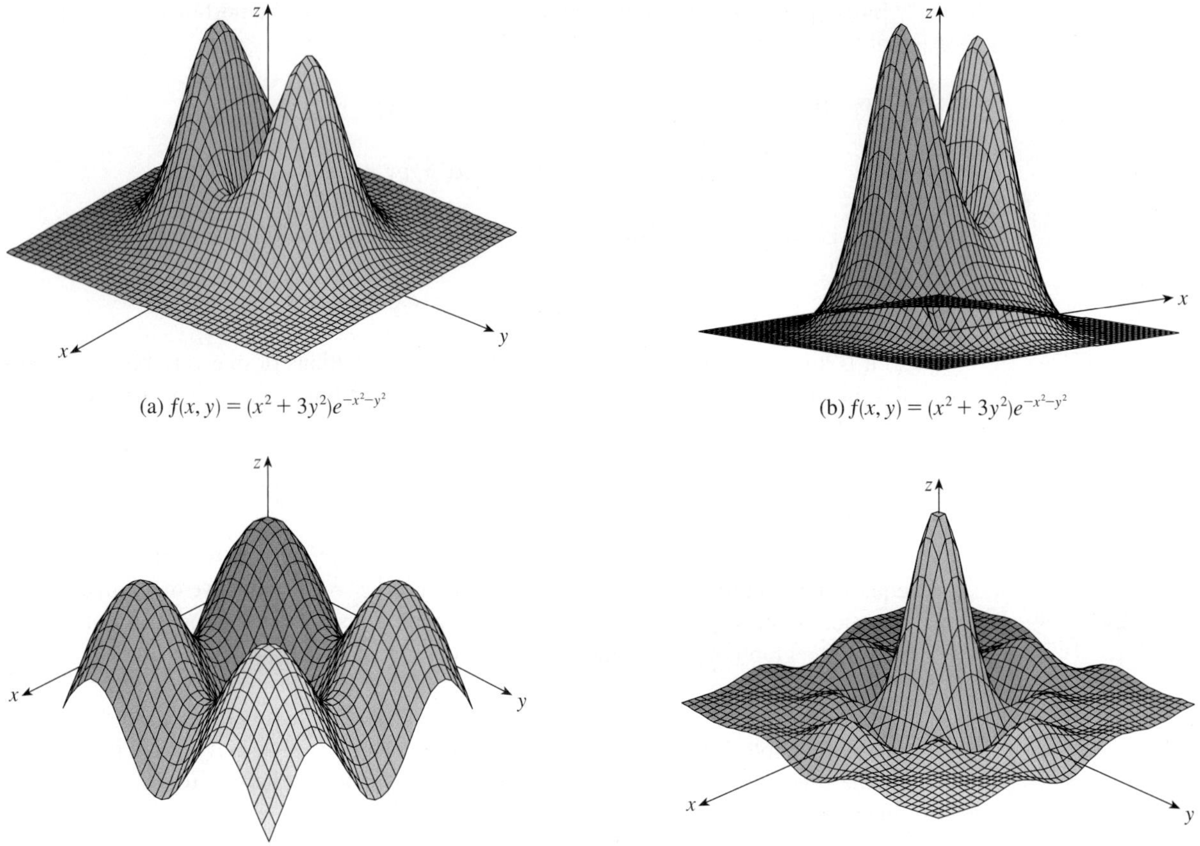

(a) $f(x, y) = (x^2 + 3y^2)e^{-x^2-y^2}$

(b) $f(x, y) = (x^2 + 3y^2)e^{-x^2-y^2}$

(c) $f(x, y) = \sin x + \sin y$

(d) $f(x, y) = \dfrac{\sin x \sin y}{xy}$

FIGURE 8

■ Level Curves

Another method for visualizing functions, borrowed from mapmakers, is a contour map on which points of constant elevation are joined to form *contour lines*, or *level curves*.

> **Definition** The **level curves** of a function f of two variables are the curves with equations $f(x, y) = k$, where k is a constant (in the range of f).

A level curve $f(x, y) = k$ is the set of all points in the domain of f at which f takes on a given value k. In other words, it shows where the graph of f has height k.

You can see from Figure 9 the relation between level curves and horizontal traces. The level curves $f(x, y) = k$ are just the traces of the graph of f in the horizontal plane $z = k$ projected down to the xy-plane. So if you draw the level curves of a function and visualize them being lifted up to the surface at the indicated height, then you can mentally piece together a picture of the graph. The surface is steep where the level curves are close together. It is somewhat flatter where they are farther apart.

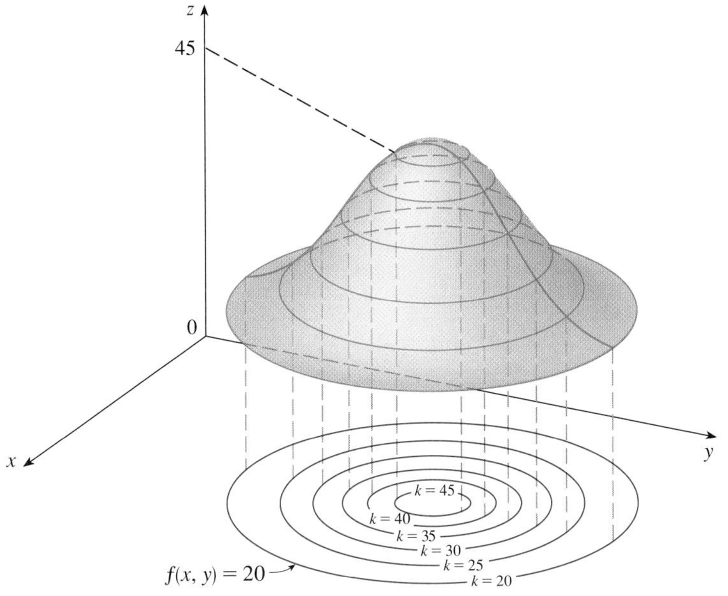

FIGURE 9

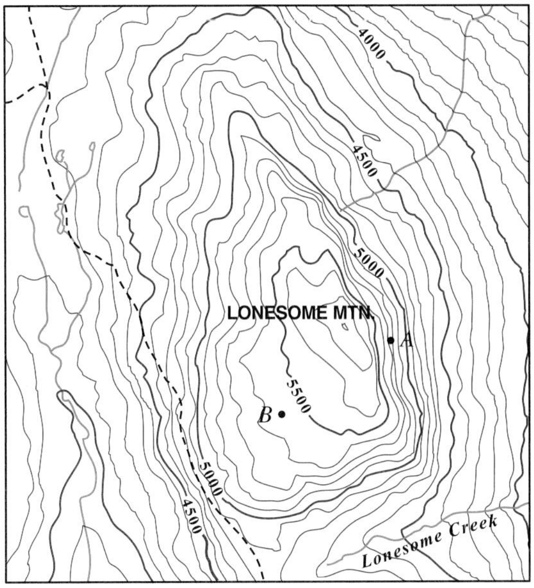

FIGURE 10

TEC Visual 9.1A animates Figure 9 by showing level curves being lifted up to graphs of functions.

One common example of level curves occurs in topographic maps of mountainous regions, such as the map in Figure 10. The level curves are curves of constant elevation above sea level. If you walk along one of these contour lines, you neither ascend nor descend. Another common example is the temperature at locations (x, y) with longitude x and latitude y. Here the level curves are called **isothermals** and join locations with the same temperature. Figure 11 shows a weather map of the world indicating the average July temperatures. The isothermals are the curves that separate the colored bands.

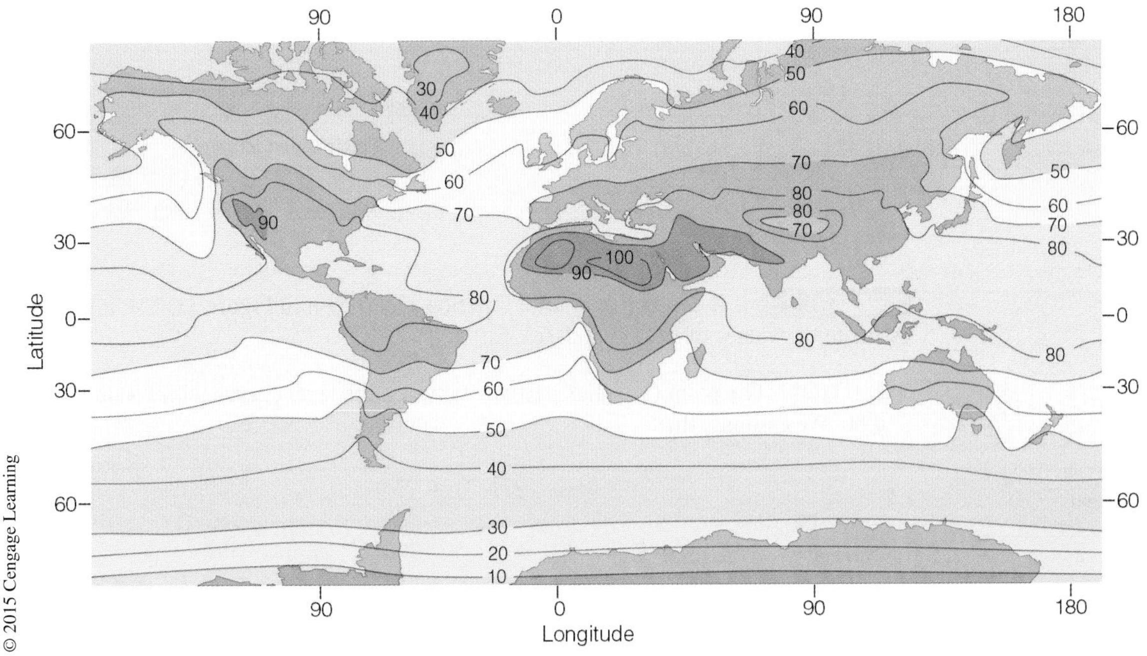

FIGURE 11 Average air temperature near sea level in July (°F)

In weather maps of atmospheric pressure at a given time as a function of longitude and latitude, the level curves are called **isobars** and join locations with the same pressure (see Exercise 28). Surface winds tend to flow from areas of high pressure across the isobars toward areas of low pressure, and are strongest where the isobars are tightly packed.

A contour map of worldwide precipitation is shown in Figure 12. Here the level curves are not labeled but they separate the colored regions and the amount of precipitation in each region is indicated in the color key.

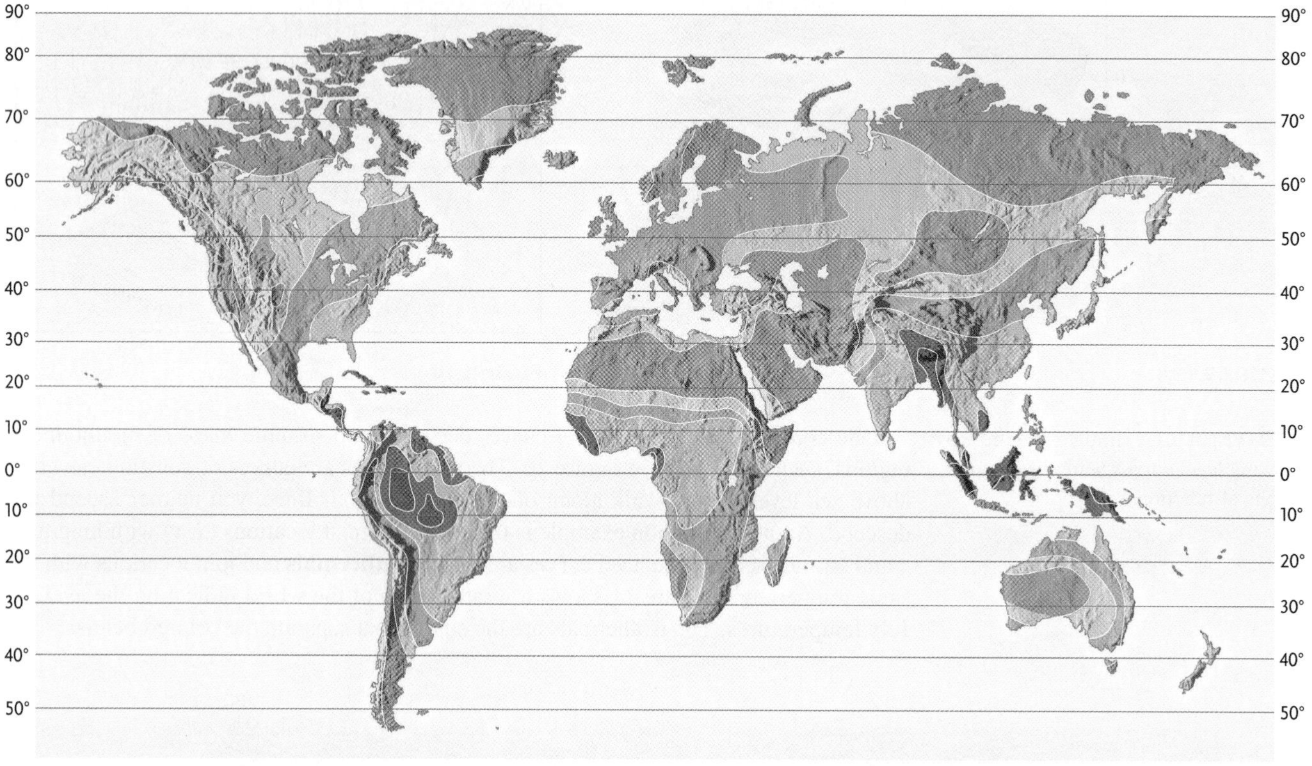

FIGURE 12 Precipitation

From Russell/Hertz/McMillan, *Biology*, 2E. © 2012 Cengage Learning.

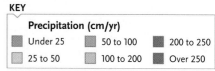

KEY

Precipitation (cm/yr)

Under 25 | 50 to 100 | 200 to 250
25 to 50 | 100 to 200 | Over 250

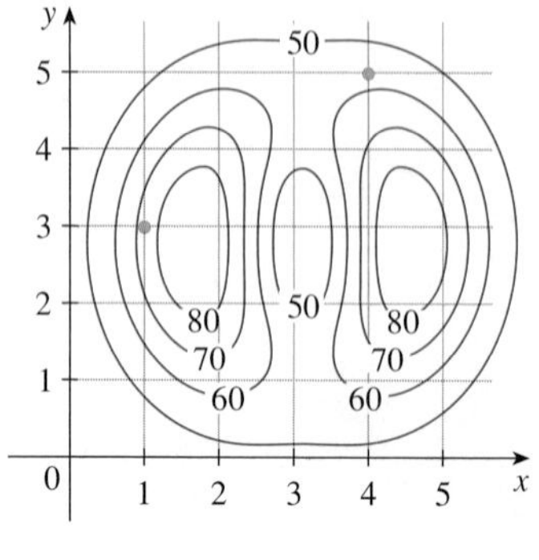

FIGURE 13

EXAMPLE 8 | A contour map for a function f is shown in Figure 13. Use it to estimate the values of $f(1, 3)$ and $f(4, 5)$.

SOLUTION The point $(1, 3)$ lies partway between the level curves with z-values 70 and 80. We estimate that

$$f(1, 3) \approx 73$$

Similarly, we estimate that

$$f(4, 5) \approx 56$$

■

EXAMPLE 9 | Sketch the level curves of the function $f(x, y) = 6 - 3x - 2y$ for the values $k = -6, 0, 6, 12$.

SOLUTION The level curves are

$$6 - 3x - 2y = k \qquad \text{or} \qquad 3x + 2y + (k - 6) = 0$$

This is a family of lines with slope $-\frac{3}{2}$. The four particular level curves with $k = -6, 0, 6,$ and 12 are $3x + 2y - 12 = 0$, $3x + 2y - 6 = 0$, $3x + 2y = 0$, and $3x + 2y + 6 = 0$. They are sketched in Figure 14. The level curves are equally spaced parallel lines because the graph of f is a plane (see Figure 5). ■

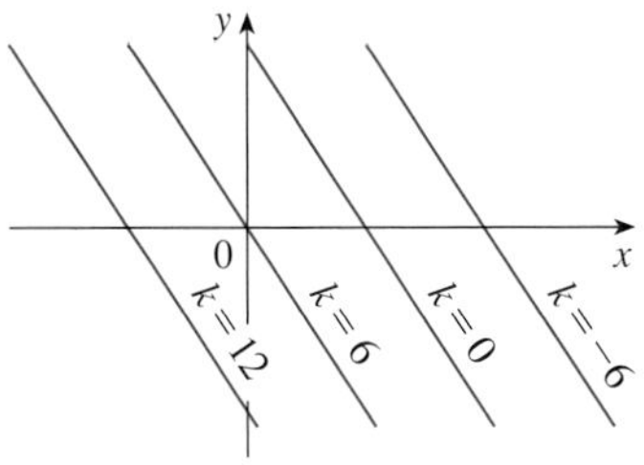

FIGURE 14
Contour map of $f(x, y) = 6 - 3x - 2y$

EXAMPLE 10 | Sketch the level curves of the function

$$g(x, y) = \sqrt{9 - x^2 - y^2} \qquad \text{for} \quad k = 0, 1, 2, 3$$

SOLUTION The level curves are

$$\sqrt{9 - x^2 - y^2} = k \qquad \text{or} \qquad x^2 + y^2 = 9 - k^2$$

This is a family of concentric circles with center $(0, 0)$ and radius $\sqrt{9 - k^2}$. The cases $k = 0, 1, 2, 3$ are shown in Figure 15. Try to visualize these level curves lifted up to form a surface and compare with the graph of g (a hemisphere) in Figure 6. (See TEC Visual 9.1A.)

TEC Visual 9.1B demonstrates the connection between surfaces and their contour maps.

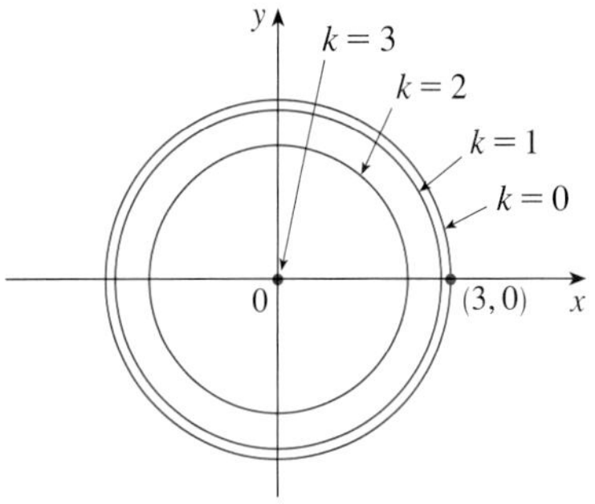

FIGURE 15
Contour map of $g(x, y) = \sqrt{9 - x^2 - y^2}$

■

EXAMPLE 11 | **Body mass index (continued)** In Example 2 we discussed the body mass index function

$$B(m, h) = \frac{m}{h^2}$$

Sketch some level curves for this function.

SOLUTION The level curves have the equations

$$\frac{m}{h^2} = k \qquad \text{or} \qquad m = kh^2$$

for various values of k. These are parabolas in the hm-plane but, since $h > 0$ and

$m > 0$, the level curves are the portions that lie in the first quadrant. In Figure 16 these curves are labeled with the value of the BMI. For instance, the curve labeled 20 shows all values of the height h and mass m that result in a BMI of 20.

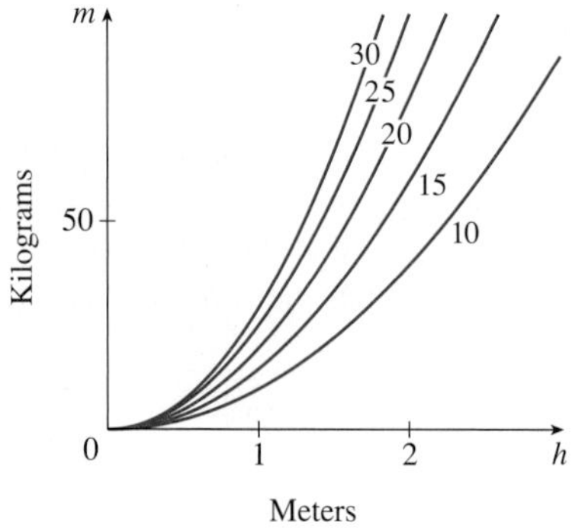

FIGURE 16
Level curves for body mass index

■

EXAMPLE 12 | **BB** **Infectious disease control** A quantity of central importance in the spread of infectious diseases is the basic reproduction number R_0 (see Example 5 on page 56). This gives the average number of new infections that each infected individual produces when introduced into a completely susceptible population. Models for the spread of SARS have been constructed to determine the effect of vaccination and quarantine on R_0.[1] In the simplest case, we have

$$R_0(d, v) = 5(1 - v)\frac{d}{1 + d}$$

where v is the fraction of the population that is vaccinated and d is the average number of days that individuals remain in the population while infectious. (Quarantine reduces this number.)

Figure 17(a) shows a graph of $R_0(d, v)$ drawn by a computer algebra system. Notice how the reproduction number approaches 0 as v approaches 1 (the entire population is vaccinated). From the rotated graph in Figure 17(b) we see how R_0 approaches 5 as d increases and v approaches 0. Thus, qualitatively, the spread of SARS in the population is reduced when the vaccination fraction is high ($v \to 1$) and when quarantine levels are high ($d \to 0$). In the coming sections we will explore the effects of these two interventions more quantitatively. The level curves of R_0 are shown in Figure 17(c).

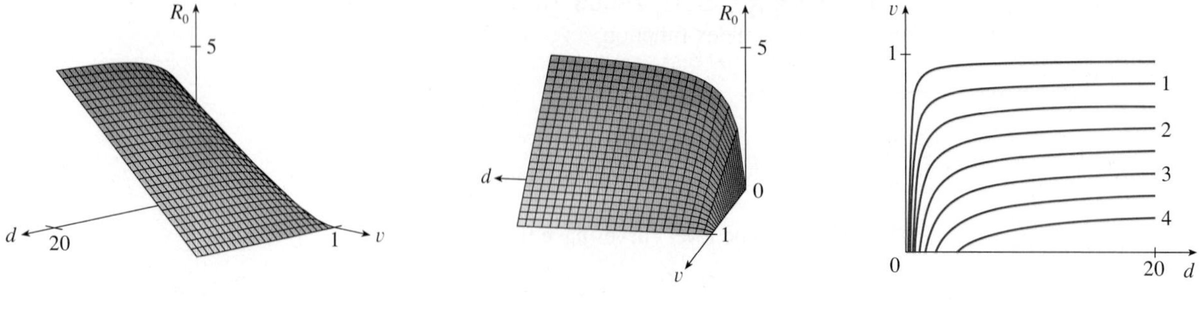

(a) (b) (c)

■

FIGURE 17
Visualizing the basic reproduction number

1. A. Gumel et al., "Modelling Strategies for Controlling SARS Outbreaks," *Proceedings of the Royal Society: Series B* 271 (2004): 2223–32.

Figure 18 shows some computer-generated level curves together with the corresponding computer-generated graphs. Notice that the level curves in part (c) crowd together near the origin. That corresponds to the fact that the graph in part (d) is very steep near the origin.

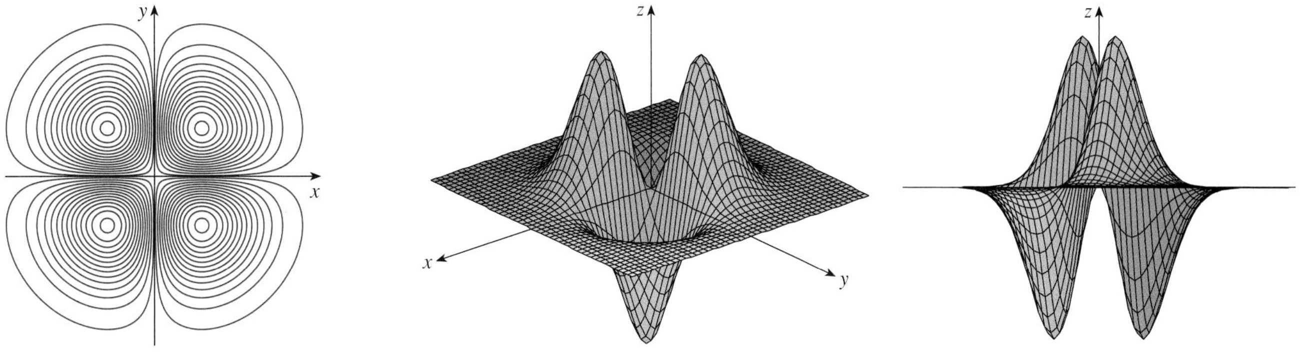

(a) Level curves of $f(x, y) = -xye^{-x^2-y^2}$

(b) Two views of $f(x, y) = -xye^{-x^2-y^2}$

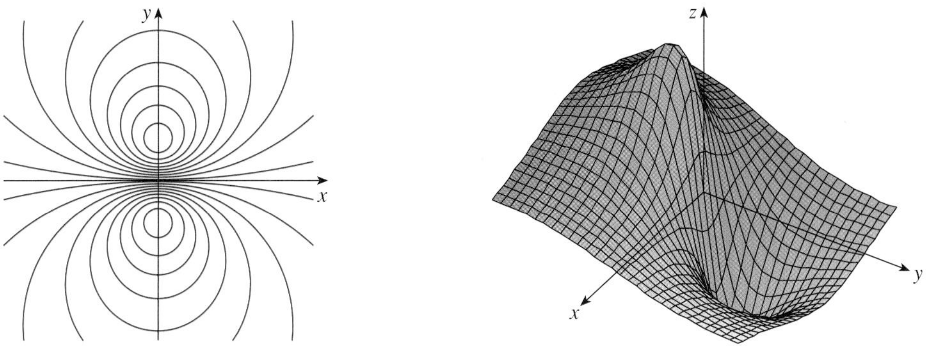

(c) Level curves of $f(x, y) = \dfrac{-3y}{x^2 + y^2 + 1}$

(d) $f(x, y) = \dfrac{-3y}{x^2 + y^2 + 1}$

FIGURE 18

■ Functions of Three Variables

A **function of three variables**, f, is a rule that assigns to each ordered triple (x, y, z) in a domain $D \subset \mathbb{R}^3$ a unique real number denoted by $f(x, y, z)$. For instance, the temperature T at a point on the surface of the earth depends on the longitude x and latitude y of the point and on the time t, so we could write $T = f(x, y, t)$.

EXAMPLE 13 | When migrating fish swim at a speed v relative to the water, the energy expenditure per unit time is proportional to v^3. Suppose the fish swim for a distance d against a current that has speed u, where $u < v$. Write the required energy E as a function of the three variables u, v, and d.

SOLUTION When swimming against the current, the ground speed of the fish is $v - u$. If T is the time required to travel the ground distance d, then $d = (v - u)T$, so

$$T = \frac{d}{v - u}$$

The energy per unit time is av^3, where a is the proportionality constant. So the total energy is

$$E = av^3 \cdot T = av^3 \cdot \frac{d}{v - u}$$

Thus the energy as a function of u, v, and d is

$$E(u, v, d) = \frac{av^3 d}{v - u}$$ ■

EXAMPLE 14 | Find the domain of f if

$$f(x, y, z) = \ln(z - y) + xy \sin z$$

SOLUTION The expression for $f(x, y, z)$ is defined as long as $z - y > 0$, so the domain of f is

$$D = \{(x, y, z) \in \mathbb{R}^3 \mid z > y\}$$

This is a **half-space** consisting of all points that lie above the plane $z = y$. ■

Functions of any number of variables can be considered. A **function of n variables** is a rule that assigns a number $z = f(x_1, x_2, \ldots, x_n)$ to an n-tuple $(x_1, x_2, \ldots, x_n)$ of real numbers. We denote by $\mathbb{R}^n$ the set of all such n-tuples. For example, if a company uses n different ingredients in making a food product, c_i is the cost per unit of the ith ingredient, and x_i units of the ith ingredient are used, then the total cost C of the ingredients is a function of the n variables $x_1, x_2, \ldots, x_n$:

$$C = f(x_1, x_2, \ldots, x_n) = c_1x_1 + c_2x_2 + \cdots + c_nx_n$$

■ Limits and Continuity

For functions of a single variable, recall from Section 2.3 the meaning of a limit: $\lim_{x \to a} f(x) = L$ means that we can make the values of $f(x)$ as close to L as we like by taking x to be sufficiently close to a, but not equal to a. For functions of two variables we use the notation

$$\lim_{(x, y) \to (a, b)} f(x, y) = L$$

to indicate that the values of $f(x, y)$ approach the number L as the point (x, y) approaches the point (a, b) along any path that stays within the domain of f.

A more precise definition of the limit of a function of two variables is given in Appendix D.

(1) Definition We write

$$\lim_{(x, y) \to (a, b)} f(x, y) = L$$

and we say that the **limit of $f(x, y)$ as (x, y) approaches (a, b)** is L if we can make the values of $f(x, y)$ as close to L as we like by taking the point (x, y) sufficiently close to the point (a, b), but not equal to (a, b).

For functions of a single variable, when we let x approach a, there are only two possible directions of approach, from the left or from the right. We recall from Chapter 2 that if $\lim_{x \to a^-} f(x) \neq \lim_{x \to a^+} f(x)$, then $\lim_{x \to a} f(x)$ does not exist.

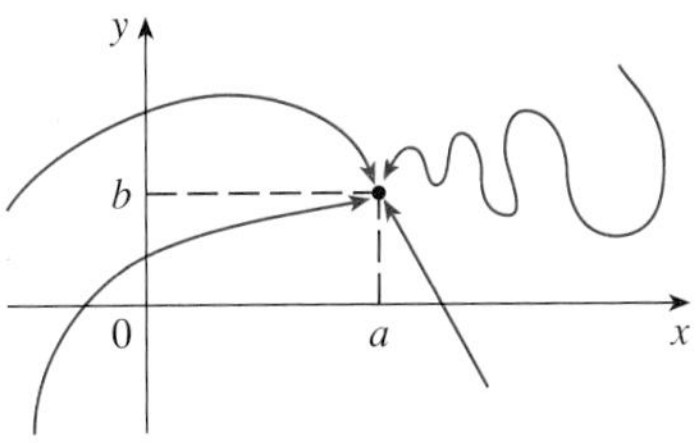

FIGURE 19

For functions of two variables the situation is not as simple because we can let (x, y) approach (a, b) from an infinite number of directions in any manner whatsoever (see Figure 19) as long as (x, y) stays within the domain of f.

Definition 1 says that the distance between $f(x, y)$ and L can be made arbitrarily small by making the distance from (x, y) to (a, b) sufficiently small (but not 0). The definition refers only to the *distance* between (x, y) and (a, b). It does not refer to the direction of approach. Therefore, if the limit exists, then $f(x, y)$ must approach the same limit no matter how (x, y) approaches (a, b). Thus, if we can find two different paths of approach along which the function $f(x, y)$ has different limits, then it follows that $\lim_{(x, y)\to(a, b)} f(x, y)$ does not exist.

> If $f(x, y) \to L_1$ as $(x, y) \to (a, b)$ along a path C_1 and $f(x, y) \to L_2$ as $(x, y) \to (a, b)$ along a path C_2, where $L_1 \neq L_2$, then $\lim_{(x, y)\to(a, b)} f(x, y)$ does not exist.

EXAMPLE 15 | If $f(x, y) = xy/(x^2 + y^2)$, does $\displaystyle\lim_{(x, y)\to(0, 0)} f(x, y)$ exist?

SOLUTION First let's approach $(0, 0)$ along the x-axis. If $y = 0$, then $f(x, 0) = 0/x^2 = 0$. Therefore

$$f(x, y) \to 0 \qquad \text{as} \qquad (x, y) \to (0, 0) \text{ along the } x\text{-axis}$$

If $x = 0$, then $f(0, y) = 0/y^2 = 0$, so

$$f(x, y) \to 0 \qquad \text{as} \qquad (x, y) \to (0, 0) \text{ along the } y\text{-axis}$$

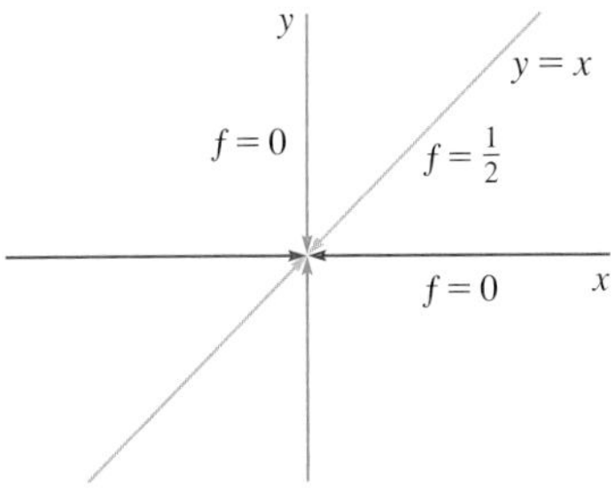

FIGURE 20

Although we have obtained identical limits along the axes, that does not show that the given limit is 0. Let's now approach $(0, 0)$ along another line, say $y = x$. For all $x \neq 0$,

$$f(x, x) = \frac{x^2}{x^2 + x^2} = \frac{1}{2}$$

Therefore $\qquad f(x, y) \to \frac{1}{2} \qquad \text{as} \qquad (x, y) \to (0, 0) \text{ along } y = x$

(See Figure 20.) Since we have obtained different limits along different paths, the given limit does not exist. ■

Figure 21 sheds some light on Example 15. The ridge that occurs above the line $y = x$ corresponds to the fact that $f(x, y) = \frac{1}{2}$ for all points (x, y) on that line except the origin.

TEC In Visual 9.1C a rotating line on the surface in Figure 21 shows different limits at the origin from different directions.

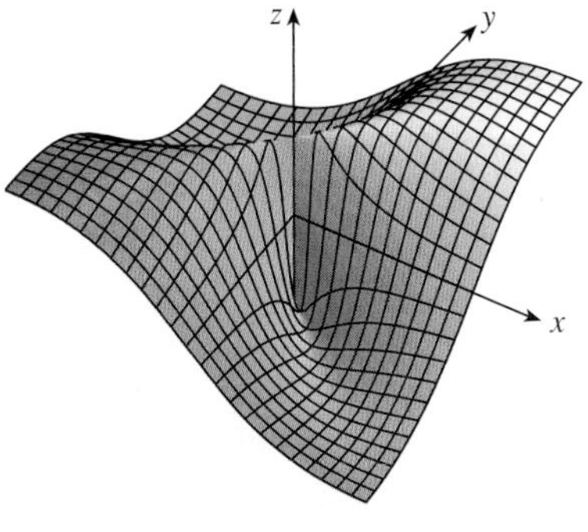

FIGURE 21

$$f(x, y) = \frac{xy}{x^2 + y^2}$$

Just as for functions of one variable, the calculation of limits for functions of two variables can be greatly simplified by the use of properties of limits. The Limit Laws listed in Section 2.4 can be extended to functions of two variables: The limit of a sum is the sum of the limits, the limit of a product is the product of the limits, and so on. In particular, the following equations are true.

(2) $$\lim_{(x,y)\to(a,b)} x = a \qquad \lim_{(x,y)\to(a,b)} y = b \qquad \lim_{(x,y)\to(a,b)} c = c$$

The Squeeze Theorem also holds.

EXAMPLE 16 | Find $\displaystyle\lim_{(x,y)\to(0,0)} \frac{3x^2y}{x^2+y^2}$ if it exists.

SOLUTION If we let $(x, y) \to (0, 0)$ along any line through the origin, we find that $f(x, y) \to 0$. This doesn't prove that the given limit is 0, but let's look at the distance from $f(x, y)$ to 0:

$$\left|\frac{3x^2y}{x^2+y^2} - 0\right| = \left|\frac{3x^2y}{x^2+y^2}\right| = \frac{3x^2|y|}{x^2+y^2}$$

Notice that $x^2 \leqslant x^2 + y^2$ because $y^2 \geqslant 0$. So

$$\frac{x^2}{x^2+y^2} \leqslant 1$$

Thus $$0 \leqslant \frac{3x^2|y|}{x^2+y^2} \leqslant 3|y|$$

Now we use the Squeeze Theorem. Since

$$\lim_{(x,y)\to(0,0)} 0 = 0 \qquad \text{and} \qquad \lim_{(x,y)\to(0,0)} 3|y| = 0 \qquad \text{[by (2)]}$$

we conclude that $$\lim_{(x,y)\to(0,0)} \frac{3x^2y}{x^2+y^2} = 0$$ ■

Recall that evaluating limits of *continuous* functions of a single variable is easy. It can be accomplished by direct substitution because the defining property of a continuous function is $\lim_{x\to a} f(x) = f(a)$. Continuous functions of two variables are also defined by the direct substitution property.

(3) Definition A function f of two variables is called **continuous at** (a, b) if

$$\lim_{(x,y)\to(a,b)} f(x, y) = f(a, b)$$

We say f is **continuous on** D if f is continuous at every point (a, b) in D.

The intuitive meaning of continuity is that if the point (x, y) changes by a small amount, then the value of $f(x, y)$ changes by a small amount. This means that a surface that is the graph of a continuous function has no hole or break.

Using the properties of limits, you can see that sums, differences, products, and quotients of continuous functions are continuous on their domains. Let's use this fact to give examples of continuous functions.

A **polynomial function of two variables** (or polynomial, for short) is a sum of terms of the form $cx^m y^n$, where c is a constant and m and n are nonnegative integers. A **rational function** is a ratio of polynomials. For instance,

$$f(x, y) = x^4 + 5x^3y^2 + 6xy^4 - 7y + 6$$

is a polynomial, whereas

$$g(x, y) = \frac{2xy + 1}{x^2 + y^2}$$

is a rational function.

The limits in (2) show that the functions $f(x, y) = x$, $g(x, y) = y$, and $h(x, y) = c$ are continuous. Since any polynomial can be built up out of the simple functions f, g, and h by multiplication and addition, it follows that *all polynomials are continuous on* $\mathbb{R}^2$. Likewise, any rational function is continuous on its domain because it is a quotient of continuous functions.

EXAMPLE 17 | Evaluate $\displaystyle\lim_{(x, y)\to(1, 2)} (x^2y^3 - x^3y^2 + 3x + 2y)$.

SOLUTION Since $f(x, y) = x^2y^3 - x^3y^2 + 3x + 2y$ is a polynomial, it is continuous everywhere, so we can find the limit by direct substitution:

$$\lim_{(x, y)\to(1, 2)} (x^2y^3 - x^3y^2 + 3x + 2y) = 1^2 \cdot 2^3 - 1^3 \cdot 2^2 + 3 \cdot 1 + 2 \cdot 2 = 11 \quad ■$$

Figure 22 shows the graph of the continuous function in Example 18.

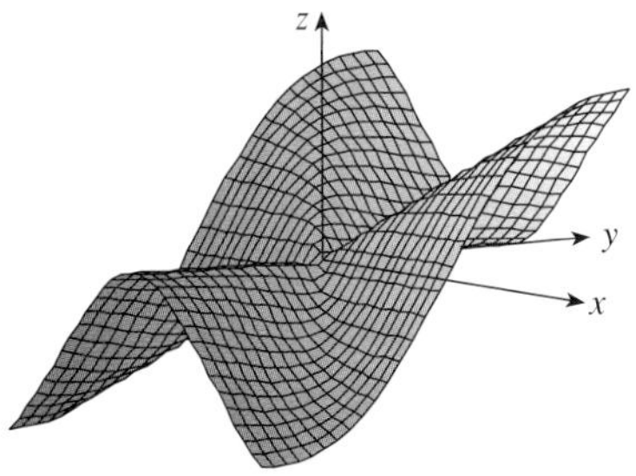

FIGURE 22

EXAMPLE 18 | Let

$$f(x, y) = \begin{cases} \dfrac{3x^2y}{x^2 + y^2} & \text{if } (x, y) \neq (0, 0) \\ 0 & \text{if } (x, y) = (0, 0) \end{cases}$$

We know f is continuous for $(x, y) \neq (0, 0)$ since it is equal to a rational function there. Also, from Example 16, we have

$$\lim_{(x, y)\to(0, 0)} f(x, y) = \lim_{(x, y)\to(0, 0)} \frac{3x^2y}{x^2 + y^2} = 0 = f(0, 0)$$

Therefore f is continuous at $(0, 0)$, and so it is continuous on $\mathbb{R}^2$. ■

Just as for functions of one variable, composition is another way of combining two continuous functions to get a third. In fact, it can be shown that if f is a continuous function of two variables and g is a continuous function of a single variable that is defined on the range of f, then the composite function $h = g \circ f$ defined by $h(x, y) = g(f(x, y))$ is also a continuous function.

EXAMPLE 19 | Where is the function $h(x, y) = \arctan(y/x)$ continuous?

SOLUTION The function $f(x, y) = y/x$ is a rational function and therefore continuous except on the line $x = 0$. The function $g(t) = \arctan t$ is continuous everywhere. So the composite function

$$g(f(x, y)) = \arctan(y/x) = h(x, y)$$

is continuous except where $x = 0$. Figure 23 shows the break in the graph of h above the y-axis.

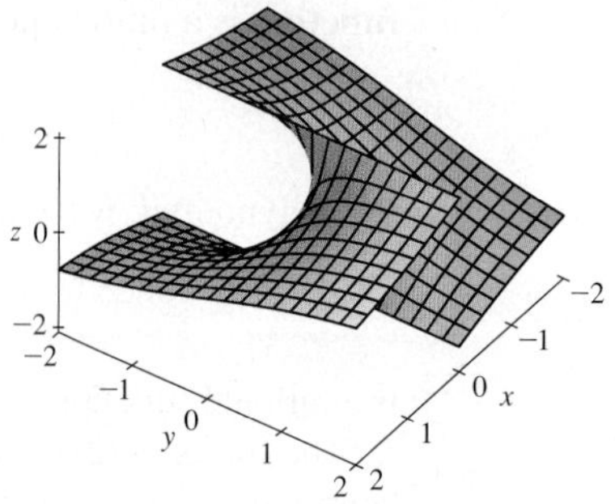

FIGURE 23
The function $h(x, y) = \arctan(y/x)$ is discontinuous where $x = 0$.

Everything that we have done in this section can be extended to functions of three or more variables. The notation

$$\lim_{(x, y, z)\to(a, b, c)} f(x, y, z) = L$$

means that the values of $f(x, y, z)$ approach the number L as the point (x, y, z) approaches the point (a, b, c) along any path in the domain of f. The function f is **continuous** at (a, b, c) if

$$\lim_{(x, y, z)\to(a, b, c)} f(x, y, z) = f(a, b, c)$$

For instance, the function

$$f(x, y, z) = \frac{1}{x^2 + y^2 + z^2 - 1}$$

is a rational function of three variables and so is continuous at every point in $\mathbb{R}^3$ except where $x^2 + y^2 + z^2 = 1$. In other words, it is discontinuous on the sphere with center the origin and radius 1.

EXERCISES 9.1

1. **Wind chill** In Example 1 we considered the function $W = f(T, v)$, where W is the wind-chill index, T is the actual temperature, and v is the wind speed. A numerical representation is given in Table 1.
   (a) What is the value of $f(-15, 40)$? What is its meaning?
   (b) Describe in words the meaning of the question "For what value of v is $f(-20, v) = -30$?" Then answer the question.
   (c) Describe in words the meaning of the question "For what value of T is $f(T, 20) = -49$?" Then answer the question.
   (d) What is the meaning of the function $W = f(-5, v)$? Describe the behavior of this function.
   (e) What is the meaning of the function $W = f(T, 50)$? Describe the behavior of this function.

2. The **temperature-humidity index** I (or humidex, for short) is the perceived air temperature when the actual temperature is T and the relative humidity is h, so we can write $I = f(T, h)$. The following table of values of I is an excerpt from a table compiled by the National Oceanic & Atmospheric Administration.

Table 2 Apparent temperature as a function of temperature and humidity

Relative humidity (%)

Actual temperature (°F) T \ h	20	30	40	50	60	70
80	77	78	79	81	82	83
85	82	84	86	88	90	93
90	87	90	93	96	100	106
95	93	96	101	107	114	124
100	99	104	110	120	132	144

   (a) What is the value of $f(95, 70)$? What is its meaning?
   (b) For what value of h is $f(90, h) = 100$?
   (c) For what value of T is $f(T, 50) = 88$?
   (d) What are the meanings of the functions $I = f(80, h)$ and $I = f(100, h)$? Compare the behavior of these two functions of h.

3. **Body surface area** A model for the surface area of a human body is given by the function

$$S = f(w, h) = 0.1091w^{0.425}h^{0.725}$$

where w is the weight (in pounds), h is the height (in inches), and S is measured in square feet.
(a) Find $f(160, 70)$ and interpret it.
(b) What is your own surface area?

4. The **wind-chill index** W discussed in Example 1 has been modeled by the following function:

$$W(T, v) = 13.12 + 0.6215T - 11.37v^{0.16} + 0.3965Tv^{0.16}$$

Check to see how closely this model agrees with the values in Table 1 for a few values of T and v.

5. A manufacturer has modeled its yearly production function P (the monetary value of its entire production) as a so-called Cobb-Douglas function

$$P(L, K) = 1.47L^{0.65}K^{0.35}$$

where L is the number of labor hours (in thousands) and K is the invested capital (in millions of dollars).
(a) Find $P(120, 20)$ and interpret it.
(b) If both the amount of labor and the amount of capital are doubled, verify that the production is also doubled.

6. A company makes two kinds of chocolate bars: plain, and with almonds. Fixed production costs are \$10,000 and it costs \$1.10 to make a plain chocolate bar and \$1.25 to make one with almonds.
(a) Express the cost of making x plain bars and y bars with almonds as a function of two variables $C = f(x, y)$.
(b) Find $f(2000, 1000)$ and interpret it.
(c) What is the domain of f?

7. **Snake reversals and stripes** In a study of the survivorship of juvenile garter snakes, a researcher arrived at the model

$$F = 4.2 + 0.008R + 0.102S + 0.017R^2 - 0.034S^2 - 0.268RS$$

where F is a measure of the fitness of the snake, R measures the number of reversals of direction during flight from a predator, and S measures the degree of stripedness in the color pattern of the snake. Which is likelier to survive longer, a snake with $R = 3$ and $S = 1$ or one with $R = 1$ and $S = 3$?

Source: Adapted from E. Brodie III, "Correlational Selection for Color Pattern and Antipredator Behavior in the Garter Snake *Thamnophis Ordinoides*," *Evolution* 46 (1992): 1284–98.

8. **Blood flow** The shape of a blood vessel (a vein or artery) can be modeled by a cylindrical tube with radius R and length L. The velocity v of the blood is modeled by Poiseuille's law of laminar flow, which expresses v as a function of five variables:

$$v = f(P, \eta, L, R, r) = \frac{P}{4\eta L}(R^2 - r^2)$$

where η is the viscosity of the blood, P is the pressure difference between the ends of the tube (in dynes/cm^2), r is the distance from the central axis of the tube, and r, R, and L are measured in centimeters.
(a) Evaluate $f(4000, 0.027, 2, 0.008, 0.002)$ and interpret it. (These values are typical for some of the smaller human arteries.)
(b) Where in the artery is the flow the greatest? Where is it least?

9. Let $g(x, y) = \cos(x + 2y)$.
(a) Evaluate $g(2, -1)$.
(b) Find the domain of g.
(c) Find the range of g.

10. Let $F(x, y) = 1 + \sqrt{4 - y^2}$.
(a) Evaluate $F(3, 1)$.
(b) Find and sketch the domain of F.
(c) Find the range of F.

11. Let $f(x, y, z) = \sqrt{x} + \sqrt{y} + \sqrt{z} + \ln(4 - x^2 - y^2 - z^2)$.
(a) Evaluate $f(1, 1, 1)$.
(b) Find and describe the domain of f.

12. Let $g(x, y, z) = x^3y^2z\sqrt{10 - x - y - z}$.
(a) Evaluate $g(1, 2, 3)$.
(b) Find and describe the domain of g.

13–20 Find and sketch the domain of the function.

13. $f(x, y) = \sqrt{2x - y}$

14. $f(x, y) = \sqrt{xy}$

15. $f(x, y) = \sqrt{1 - x^2} - \sqrt{1 - y^2}$

16. $f(x, y) = \ln(x^2 + y^2 - 2)$

17. $f(x, y) = \dfrac{\sqrt{y - x^2}}{1 - x^2}$

18. $f(x, y) = \sqrt{y} + \sqrt{25 - x^2 - y^2}$

19. $f(x, y, z) = \sqrt{1 - x^2 - y^2 - z^2}$

20. $f(x, y, z) = \sqrt{z} + \ln(1 - x^2 - y^2)$

21–25 Sketch the graph of the function.

21. $f(x, y) = 3$

22. $f(x, y) = y$

23. $f(x, y) = 10 - 4x - 5y$

24. $f(x, y) = 1 + 2x^2 + 2y^2$

25. $f(x, y) = y^2 + 1$

26. Match the function with its graph (labeled I–VI). Give reasons for your choices.

(a) $f(x, y) = |x| + |y|$ (b) $f(x, y) = |xy|$

(c) $f(x, y) = \dfrac{1}{1 + x^2 + y^2}$ (d) $f(x, y) = (x^2 - y^2)^2$

(e) $f(x, y) = (x - y)^2$ (f) $f(x, y) = \sin(|x| + |y|)$

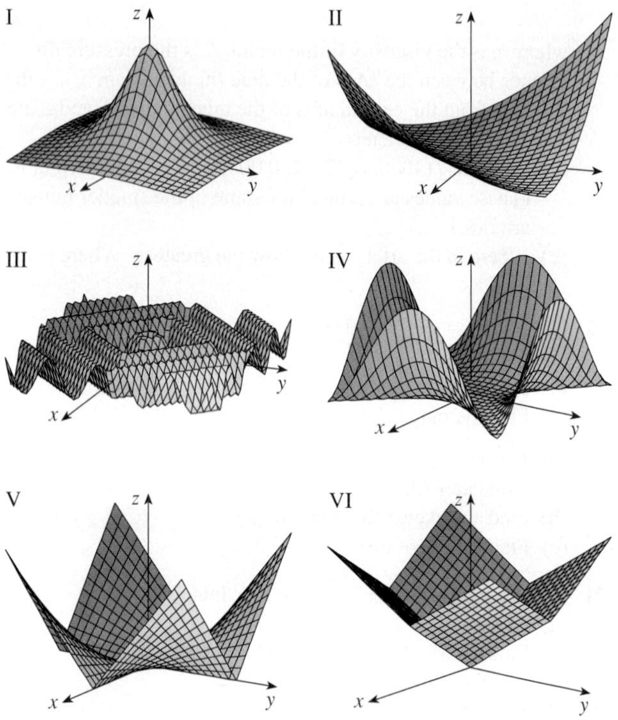

27. A contour map for a function f is shown. Use it to estimate the values of $f(-3, 3)$ and $f(3, -2)$. What can you say about the shape of the graph?

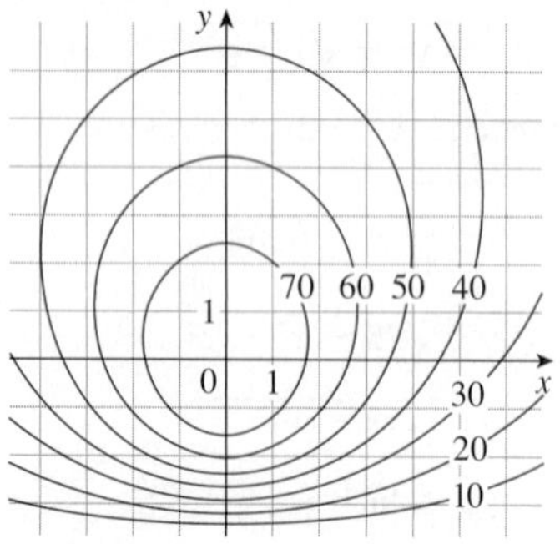

28. Shown is a contour map of atmospheric pressure in North America on August 12, 2008. On the level curves (called isobars) the pressure is indicated in millibars (mb).

(a) Estimate the pressure at C (Chicago), N (Nashville), S (San Francisco), and V (Vancouver).

(b) At which of these locations were the winds strongest?

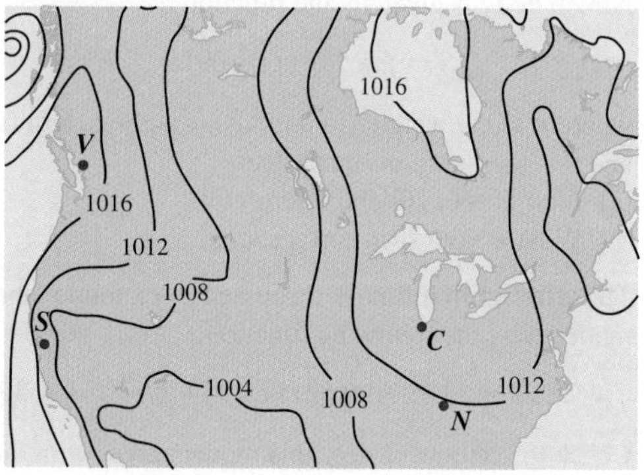

29. Level curves (isothermals) are shown for the water temperature (in °C) in Long Lake (Minnesota) in 1998 as a function of depth and time of year. Estimate the temperature in the lake on June 9 (day 160) at a depth of 10 m and on June 29 (day 180) at a depth of 5 m.

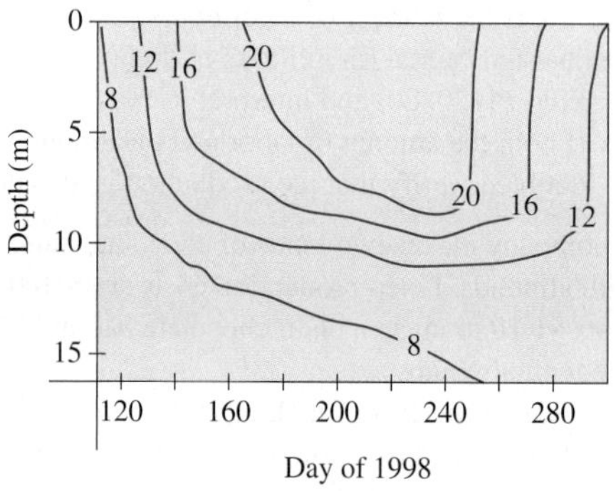

30. Two contour maps are shown. One is for a function f whose graph is a cone. The other is for a function g whose graph is a paraboloid. Which is which, and why?

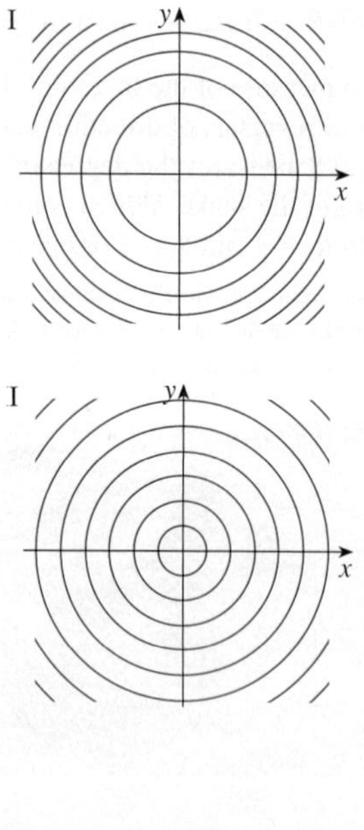

31. Locate the points A and B on the map of Lonesome Mountain (see Figure 10). How would you describe the terrain near A? Near B?

32. Make a rough sketch of a contour map for the function whose graph is shown.

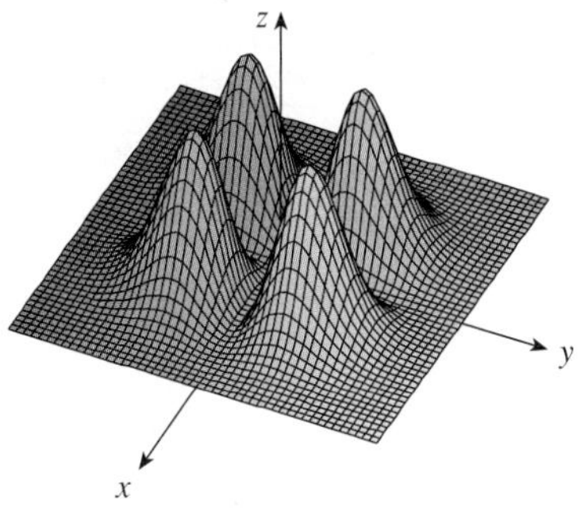

33–38 Draw a contour map of the function showing several level curves.

33. $f(x, y) = 2x - y$

34. $f(x, y) = y - x^2$

35. $f(x, y) = xy$

36. $f(x, y) = \sqrt{36 - x^2 - y^2}$

37. $f(x, y) = ye^x$

38. $f(x, y) = \dfrac{y}{x + 1}$

39. Body mass index was discussed in Examples 2 and 11. Draw the level curves $B(m, h) = 18.5$, $B(m, h) = 25$, $B(m, h) = 30$, and $B(m, h) = 40$. Then shade the region corresponding to optimal BMI. Does someone who weighs 62 kg and is 152 cm tall fall into this category?

40. Body mass index Draw the level curve of the body mass index function corresponding to someone who is 200 cm tall and weighs 80 kg. Find the weights and heights of two other people with that same level curve.

41–46 Match the function (a) with its graph (labeled A–F on page 584) and (b) with its contour map (labeled I–VI). Give reasons for your choices.

41. $z = \sin(xy)$

42. $z = e^x \cos y$

43. $z = \sin(x - y)$

44. $z = \sin x - \sin y$

45. $z = (1 - x^2)(1 - y^2)$

46. $z = \dfrac{x - y}{1 + x^2 + y^2}$

CAS **47. Snake reversals and stripes** Use computer software to graph the fitness function F in Exercise 7 as well as a contour map. How would you describe the shape of the surface?

CAS **48.** Use computer software to graph the function

$$f(x, y) = xy^2 - x^3$$

Why do you think the surface is called a monkey saddle?

49–58 Find the limit, if it exists, or show that the limit does not exist.

49. $\displaystyle\lim_{(x, y)\to(1, 2)} (5x^3 - x^2y^2)$

50. $\displaystyle\lim_{(x, y)\to(1, -1)} e^{-xy}\cos(x + y)$

51. $\displaystyle\lim_{(x, y)\to(2, 1)} \frac{4 - xy}{x^2 + 3y^2}$

52. $\displaystyle\lim_{(x, y, z)\to(\pi, 0, 1/3)} e^{y^2}\tan(xz)$

53. $\displaystyle\lim_{(x, y)\to(0, 0)} \frac{y^4}{x^4 + 3y^4}$

54. $\displaystyle\lim_{(x, y)\to(0, 0)} \frac{x^2 + \sin^2 y}{2x^2 + y^2}$

55. $\displaystyle\lim_{(x, y)\to(0, 0)} \frac{xy\cos y}{3x^2 + y^2}$

56. $\displaystyle\lim_{(x, y)\to(0, 0)} \frac{6x^3y}{2x^4 + y^4}$

57. $\displaystyle\lim_{(x, y)\to(0, 0)} \frac{xy}{\sqrt{x^2 + y^2}}$

58. $\displaystyle\lim_{(x, y)\to(0, 0)} \frac{x^2\sin^2 y}{x^2 + 2y^2}$

59–66 Determine the set of points at which the function is continuous.

59. $F(x, y) = \arctan\left(x + \sqrt{y}\right)$

60. $F(x, y) = \cos\sqrt{1 + x - y}$

61. $G(x, y) = \ln(x^2 + y^2 - 4)$

62. $H(x, y) = \dfrac{e^x + e^y}{e^{xy} - 1}$

63. $f(x, y, z) = \dfrac{\sqrt{y}}{x^2 - y^2 + z^2}$

64. $f(x, y, z) = \sqrt{x + y + z}$

65. $f(x, y) = \begin{cases} \dfrac{x^2y^3}{2x^2 + y^2} & \text{if } (x, y) \neq (0, 0) \\ 1 & \text{if } (x, y) = (0, 0) \end{cases}$

66. $f(x, y) = \begin{cases} \dfrac{xy}{x^2 + xy + y^2} & \text{if } (x, y) \neq (0, 0) \\ 0 & \text{if } (x, y) = (0, 0) \end{cases}$

Graphs and Contour Maps for Exercises 41–46

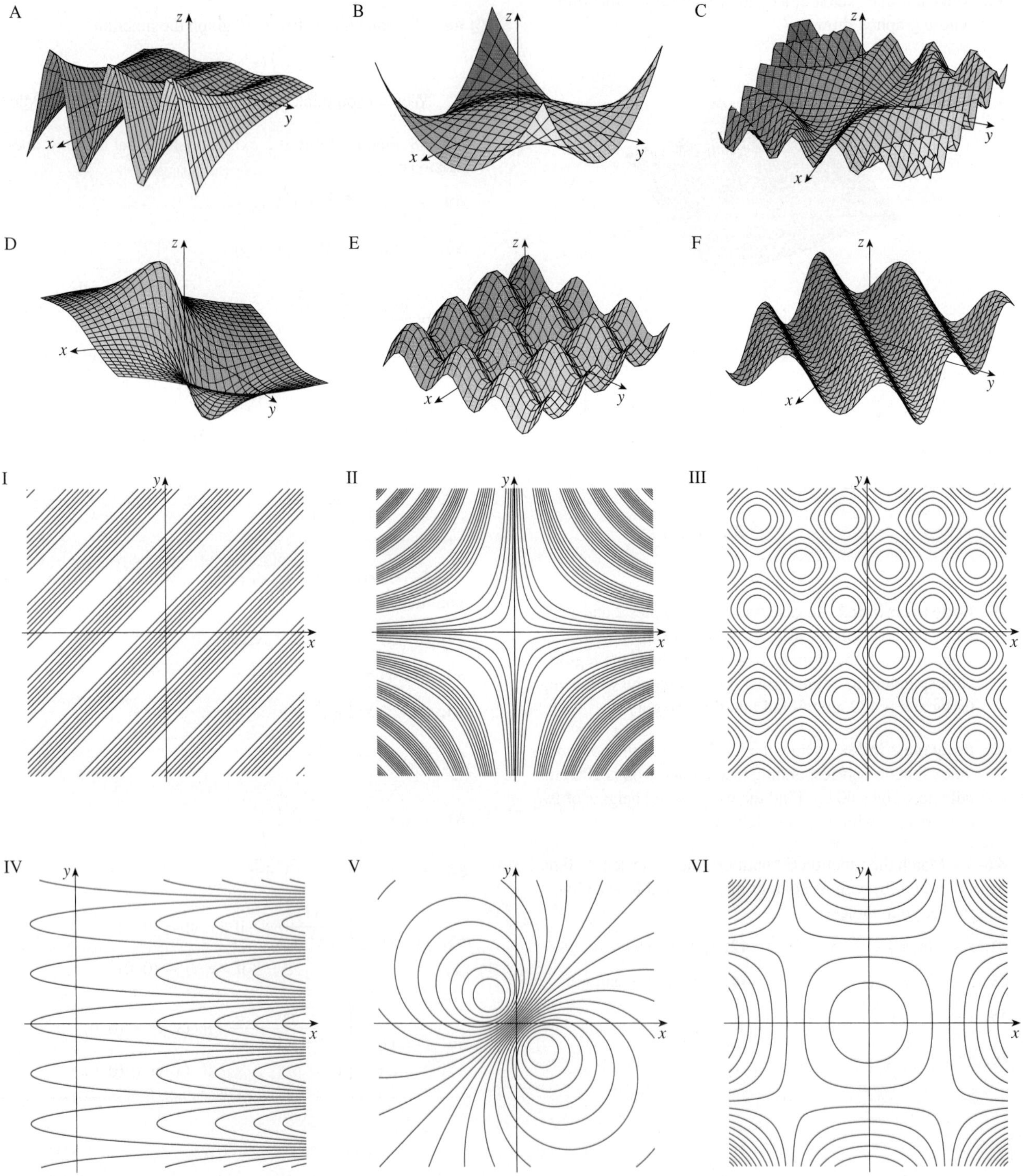

9.2 Partial Derivatives

On a hot day, extreme humidity makes us think the temperature is higher than it really is, whereas in very dry air we perceive the temperature to be lower than the thermometer indicates. The National Weather Service has devised the *heat index* (also called the temperature-humidity index, or humidex, in some countries) to describe the combined effects of temperature and humidity. The heat index I is the perceived air temperature when the actual temperature is T and the relative humidity is H. So I is a function of T and H and we can write $I = f(T, H)$. The following table of values of I is an excerpt from a table compiled by the National Weather Service.

Table 1 Heat index I as a function of temperature and humidity

Relative humidity (%)

Actual temperature (°F)	T \ H	50	55	60	65	70	75	80	85	90
	90	96	98	100	103	106	109	112	115	119
	92	100	103	105	108	112	115	119	123	128
	94	104	107	111	114	118	122	127	132	137
	96	109	113	116	121	125	130	135	141	146
	98	114	118	123	127	133	138	144	150	157
	100	119	124	129	135	141	147	154	161	168

If we concentrate on the highlighted column of the table, which corresponds to a relative humidity of $H = 70\%$, we are considering the heat index as a function of the single variable T for a fixed value of H. Let's write $g(T) = f(T, 70)$. Then $g(T)$ describes how the heat index I increases as the actual temperature T increases when the relative humidity is 70%. The derivative of g when $T = 96°\text{F}$ is the rate of change of I with respect to T when $T = 96°\text{F}$:

$$g'(96) = \lim_{h \to 0} \frac{g(96 + h) - g(96)}{h} = \lim_{h \to 0} \frac{f(96 + h, 70) - f(96, 70)}{h}$$

We can approximate $g'(96)$ using the values in Table 1 by taking $h = 2$ and -2:

$$g'(96) \approx \frac{g(98) - g(96)}{2} = \frac{f(98, 70) - f(96, 70)}{2} = \frac{133 - 125}{2} = 4$$

$$g'(96) \approx \frac{g(94) - g(96)}{-2} = \frac{f(94, 70) - f(96, 70)}{-2} = \frac{118 - 125}{-2} = 3.5$$

Averaging these values, we can say that the derivative $g'(96)$ is approximately 3.75. This means that, when the actual temperature is 96°F and the relative humidity is 70%, the apparent temperature (heat index) rises by about 3.75°F for every degree that the actual temperature rises!

Now let's look at the highlighted row in Table 1, which corresponds to a fixed temperature of $T = 96°\text{F}$. The numbers in this row are values of the function $G(H) = f(96, H)$, which describes how the heat index increases as the relative humidity H increases when the actual temperature is $T = 96°\text{F}$. The derivative of this function when $H = 70\%$ is the rate of change of I with respect to H when $H = 70\%$:

$$G'(70) = \lim_{h \to 0} \frac{G(70 + h) - G(70)}{h} = \lim_{h \to 0} \frac{f(96, 70 + h) - f(96, 70)}{h}$$

By taking $h = 5$ and -5, we approximate $G'(70)$ using the tabular values:

$$G'(70) \approx \frac{G(75) - G(70)}{5} = \frac{f(96, 75) - f(96, 70)}{5} = \frac{130 - 125}{5} = 1$$

$$G'(70) \approx \frac{G(65) - G(70)}{-5} = \frac{f(96, 65) - f(96, 70)}{-5} = \frac{121 - 125}{-5} = 0.8$$

By averaging these values we get the estimate $G'(70) \approx 0.9$. This says that, when the temperature is 96°F and the relative humidity is 70%, the heat index rises about 0.9°F for every percent that the relative humidity rises.

In general, if f is a function of two variables x and y, suppose we let only x vary while keeping y fixed, say $y = b$, where b is a constant. Then we are really considering a function of a single variable x, namely, $g(x) = f(x, b)$. If g has a derivative at a, then we call it the **partial derivative of f with respect to x at (a, b)** and denote it by $f_x(a, b)$. Thus

(1) $$f_x(a, b) = g'(a) \qquad \text{where} \qquad g(x) = f(x, b)$$

By the definition of a derivative, we have

$$g'(a) = \lim_{h \to 0} \frac{g(a + h) - g(a)}{h}$$

and so Equation 1 becomes

(2) $$f_x(a, b) = \lim_{h \to 0} \frac{f(a + h, b) - f(a, b)}{h}$$

Similarly, the **partial derivative of f with respect to y at (a, b)**, denoted by $f_y(a, b)$, is obtained by keeping x fixed ($x = a$) and finding the ordinary derivative at b of the function $G(y) = f(a, y)$:

(3) $$f_y(a, b) = \lim_{h \to 0} \frac{f(a, b + h) - f(a, b)}{h}$$

With this notation for partial derivatives, we can write the rates of change of the heat index I with respect to the actual temperature T and relative humidity H when $T = 96$°F and $H = 70\%$ as follows:

$$f_T(96, 70) \approx 3.75 \qquad f_H(96, 70) \approx 0.9$$

If we now let the point (a, b) vary in Equations 2 and 3, f_x and f_y become functions of two variables.

(4) If f is a function of two variables, its **partial derivatives** are the functions f_x and f_y defined by

$$f_x(x, y) = \lim_{h \to 0} \frac{f(x + h, y) - f(x, y)}{h}$$

$$f_y(x, y) = \lim_{h \to 0} \frac{f(x, y + h) - f(x, y)}{h}$$

An alternative notation for partial derivatives is similar to Leibniz notation but uses the symbol ∂ instead of d. If $z = f(x, y)$, we write

$$f_x(x, y) = f_x = \frac{\partial f}{\partial x} = \frac{\partial}{\partial x} f(x, y) = \frac{\partial z}{\partial x}$$

$$f_y(x, y) = f_y = \frac{\partial f}{\partial y} = \frac{\partial}{\partial y} f(x, y) = \frac{\partial z}{\partial y}$$

To compute partial derivatives, all we have to do is remember from Equation 1 that the partial derivative with respect to x is just the *ordinary* derivative of the function g of a single variable that we get by keeping y fixed. Thus we have the following rule.

Rule for Finding Partial Derivatives of $z = f(x, y)$

1. To find f_x, regard y as a constant and differentiate $f(x, y)$ with respect to x.

2. To find f_y, regard x as a constant and differentiate $f(x, y)$ with respect to y.

EXAMPLE 1 | If $f(x, y) = x^3 + x^2y^3 - 2y^2$, find $f_x(2, 1)$ and $f_y(2, 1)$.

SOLUTION Holding y constant and differentiating with respect to x, we get

$$f_x(x, y) = 3x^2 + 2xy^3$$

and so

$$f_x(2, 1) = 3 \cdot 2^2 + 2 \cdot 2 \cdot 1^3 = 16$$

Holding x constant and differentiating with respect to y, we get

$$f_y(x, y) = 3x^2y^2 - 4y$$

$$f_y(2, 1) = 3 \cdot 2^2 \cdot 1^2 - 4 \cdot 1 = 8$$

■

■ Interpretations of Partial Derivatives

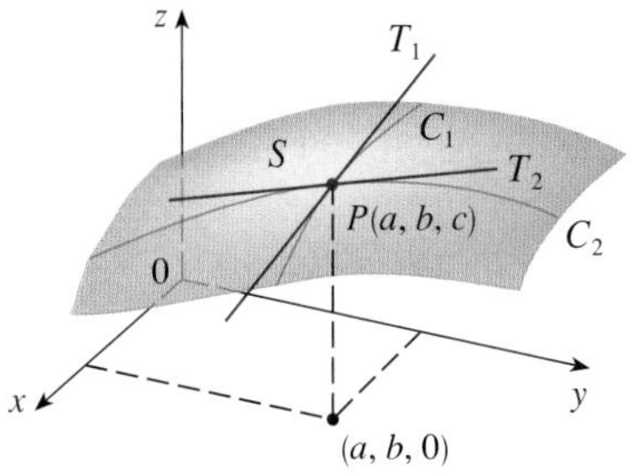

FIGURE 1
The partial derivatives of f at (a, b) are the slopes of the tangents to C_1 and C_2.

To give a geometric interpretation of partial derivatives, we recall that the equation $z = f(x, y)$ represents a surface S (the graph of f). If $f(a, b) = c$, then the point $P(a, b, c)$ lies on S. By fixing $y = b$, we are restricting our attention to the curve C_1 in which the vertical plane $y = b$ intersects S. (In other words, C_1 is the trace of S in the plane $y = b$.) Likewise, the vertical plane $x = a$ intersects S in a curve C_2. Both of the curves C_1 and C_2 pass through the point P. (See Figure 1.)

Notice that the curve C_1 is the graph of the function $g(x) = f(x, b)$, so the slope of its tangent T_1 at P is $g'(a) = f_x(a, b)$. The curve C_2 is the graph of the function $G(y) = f(a, y)$, so the slope of its tangent T_2 at P is $G'(b) = f_y(a, b)$.

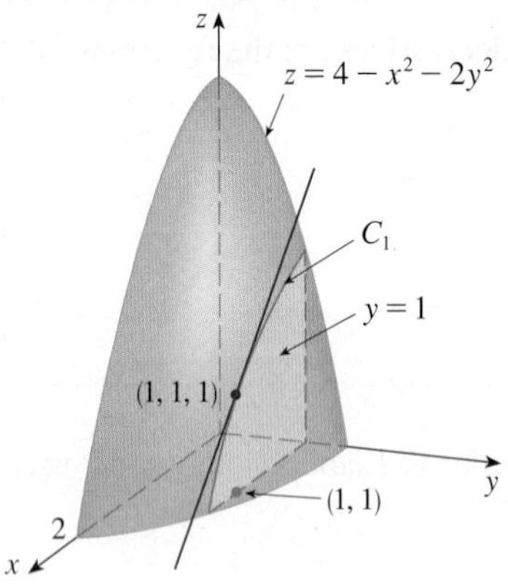

FIGURE 2

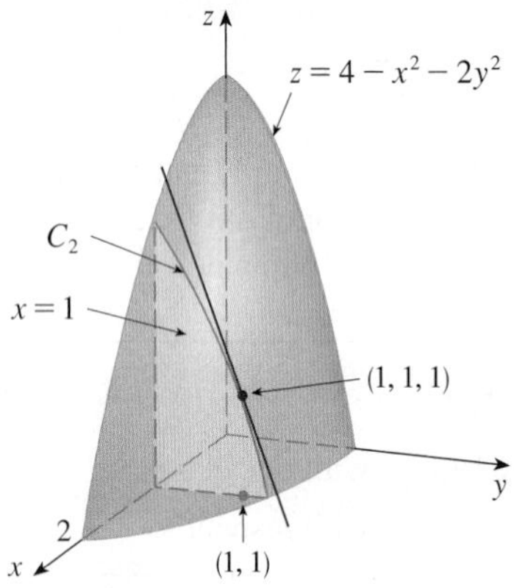

FIGURE 3

Thus the partial derivatives $f_x(a, b)$ and $f_y(a, b)$ can be interpreted geometrically as the slopes of the tangent lines at $P(a, b, c)$ to the traces C_1 and C_2 of S in the planes $y = b$ and $x = a$.

EXAMPLE 2 | If $f(x, y) = 4 - x^2 - 2y^2$, find $f_x(1, 1)$ and $f_y(1, 1)$ and interpret these numbers as slopes.

SOLUTION We have

$$f_x(x, y) = -2x \qquad f_y(x, y) = -4y$$

$$f_x(1, 1) = -2 \qquad f_y(1, 1) = -4$$

The graph of f is the paraboloid $z = 4 - x^2 - 2y^2$ and the vertical plane $y = 1$ intersects it in the parabola $z = 2 - x^2$, $y = 1$. (As in the preceding discussion, we label it C_1 in Figure 2.) The slope of the tangent line to this parabola at the point $(1, 1, 1)$ is $f_x(1, 1) = -2$. Similarly, the curve C_2 in which the plane $x = 1$ intersects the paraboloid is the parabola $z = 3 - 2y^2$, $x = 1$, and the slope of the tangent line at $(1, 1, 1)$ is $f_y(1, 1) = -4$. (See Figure 3.) ■

As we have seen in the case of the heat index function, partial derivatives can also be interpreted as *rates of change*. If $z = f(x, y)$, then $\partial z/\partial x$ represents the rate of change of z with respect to x when y is fixed. Similarly, $\partial z/\partial y$ represents the rate of change of z with respect to y when x is fixed.

EXAMPLE 3 | Body mass index In Example 9.1.2 we defined the body mass index of a person as

$$B(m, h) = \frac{m}{h^2}$$

Calculate the partial derivatives of B for a man with $m = 64$ kg and $h = 1.68$ m and interpret them.

SOLUTION Regarding h as a constant, we see that the partial derivative with respect to m is

$$\frac{\partial B}{\partial m}(m, h) = \frac{\partial}{\partial m}\left(\frac{m}{h^2}\right) = \frac{1}{h^2}$$

so

$$\frac{\partial B}{\partial m}(64, 1.68) = \frac{1}{(1.68)^2} \approx 0.35\ (\text{kg/m}^2)/\text{kg}$$

This is the rate at which his BMI increases with respect to his weight when he weighs 64 kg and his height is 1.68 m. So if his weight increases by a small amount, a kilogram for instance, and his height remains unchanged, then his BMI will increase by about 0.35 kg/m^2.

Now we regard m as a constant. The partial derivative with respect to h is

$$\frac{\partial B}{\partial h}(m, h) = \frac{\partial}{\partial h}\left(\frac{m}{h^2}\right) = m\left(-\frac{2}{h^3}\right) = -\frac{2m}{h^3}$$

so

$$\frac{\partial B}{\partial h}(64, 1.68) = -\frac{2 \cdot 64}{(1.68)^3} \approx -27\ (\text{kg/m}^2)/\text{m}$$

This is the rate at which his BMI increases with respect to his height when he weighs 64 kg and his height is 1.68 m. So if his weight stays unchanged while his height increases by a small amount, say 1 cm, then his BMI will *decrease* by about $27(0.01) = 0.27\ \text{kg/m}^2$. ■

EXAMPLE 4 | If $f(x, y) = \sin\left(\dfrac{x}{1 + y}\right)$, calculate $\dfrac{\partial f}{\partial x}$ and $\dfrac{\partial f}{\partial y}$.

SOLUTION Using the Chain Rule for functions of one variable, we have

$$\frac{\partial f}{\partial x} = \cos\left(\frac{x}{1 + y}\right) \cdot \frac{\partial}{\partial x}\left(\frac{x}{1 + y}\right) = \cos\left(\frac{x}{1 + y}\right) \cdot \frac{1}{1 + y}$$

$$\frac{\partial f}{\partial y} = \cos\left(\frac{x}{1 + y}\right) \cdot \frac{\partial}{\partial y}\left(\frac{x}{1 + y}\right) = -\cos\left(\frac{x}{1 + y}\right) \cdot \frac{x}{(1 + y)^2}$$ ■

EXAMPLE 5 | BB **Infectious disease control** In Example 9.1.12 we explored a model for the effect of vaccination and quarantine on the spread of SARS. We used the equation

$$R_0(d, v) = 5(1 - v)\frac{d}{1 + d}$$

where R_0 is the average number of new infections that each infected individual produces, v is the fraction of the population that is vaccinated, and d is the average number of days an individual remains in the population while infectious. Suppose that an outbreak has just begun and there is currently no vaccination ($v = 0$). Also suppose that there is currently no quarantine, and infectious individuals therefore circulate in the population for an average of four days. Evaluate R_0, $\partial R_0/\partial d$, and $\partial R_0/\partial v$ at $d = 4$ and $v = 0$, and provide a biological interpretation of the results.

SOLUTION The value $R_0(4, 0) = 5 \cdot \frac{4}{5} = 4$ means that each infected individual is currently causing, on average, four new infections.

The partial derivative of R_0 with respect to d is

$$\frac{\partial R_0}{\partial d}(d, v) = 5(1 - v) \cdot \frac{(1 + d) \cdot 1 - d \cdot 1}{(1 + d)^2} = \frac{5(1 - v)}{(1 + d)^2}$$

so

$$\frac{\partial R_0}{\partial d}(4, 0) = \frac{5}{5^2} = \frac{1}{5}$$

This means that, if we were to introduce a small amount of quarantine (which would cause d to decrease slightly and thus reduce the spread of SARS), the value of R_0 would decline at a rate of $\frac{1}{5}$ new infections per day of decrease in d.

The partial derivative of R_0 with respect to v is

$$\frac{\partial R_0}{\partial v}(d, v) = -\frac{5d}{1 + d} \qquad \text{so} \qquad \frac{\partial R_0}{\partial v}(4, 0) = -\frac{5 \cdot 4}{5} = -4$$

This value means that, if we were to introduce a small amount of vaccination (which would increase v slightly and thus reduce the spread of SARS), the value of R_0 would decline at a rate of four new infections per unit increase in v. Thus the spread of SARS is more sensitive to vaccination than to quarantine. ■

Some computer algebra systems can plot surfaces defined by implicit equations in three variables. Figure 4 shows such a plot of the surface defined by the equation in Example 6.

FIGURE 4

EXAMPLE 6 | Find $\partial z/\partial x$ and $\partial z/\partial y$ if z is defined implicitly as a function of x and y by the equation

$$x^3 + y^3 + z^3 + 6xyz = 1$$

SOLUTION To find $\partial z/\partial x$, we differentiate implicitly with respect to x, being careful to treat y as a constant:

$$3x^2 + 3z^2 \frac{\partial z}{\partial x} + 6yz + 6xy \frac{\partial z}{\partial x} = 0$$

Solving this equation for $\partial z/\partial x$, we obtain

$$\frac{\partial z}{\partial x} = -\frac{x^2 + 2yz}{z^2 + 2xy}$$

Similarly, implicit differentiation with respect to y gives

$$\frac{\partial z}{\partial y} = -\frac{y^2 + 2xz}{z^2 + 2xy}$$ ■

■ Functions of More Than Two Variables

Partial derivatives can also be defined for functions of three or more variables. For example, if f is a function of three variables x, y, and z, then its partial derivative with respect to x is defined as

$$f_x(x, y, z) = \lim_{h \to 0} \frac{f(x + h, y, z) - f(x, y, z)}{h}$$

and it is found by regarding y and z as constants and differentiating $f(x, y, z)$ with respect to x. If $w = f(x, y, z)$, then $f_x = \partial w/\partial x$ can be interpreted as the rate of change of w with respect to x when y and z are held fixed. But we can't interpret it geometrically because the graph of f lies in four-dimensional space.

In general, if u is a function of n variables, $u = f(x_1, x_2, \ldots, x_n)$, its partial derivative with respect to the ith variable x_i is

$$\frac{\partial u}{\partial x_i} = \lim_{h \to 0} \frac{f(x_1, \ldots, x_{i-1}, x_i + h, x_{i+1}, \ldots, x_n) - f(x_1, \ldots, x_i, \ldots, x_n)}{h}$$

and we also write

$$\frac{\partial u}{\partial x_i} = \frac{\partial f}{\partial x_i} = f_{x_i} = D_i f$$

EXAMPLE 7 | Find f_x, f_y, and f_z if $f(x, y, z) = e^{xy} \ln z$.

SOLUTION Holding y and z constant and differentiating with respect to x, we have

$$f_x = ye^{xy} \ln z$$

Similarly, $$f_y = xe^{xy} \ln z \qquad \text{and} \qquad f_z = \frac{e^{xy}}{z}$$ ■

EXAMPLE 8 | **Migrating fish** In Example 9.1.13 we derived an expression for the energy expended by migrating fish:

$$E(u, v, d) = \frac{av^3 d}{v - u}$$

where a is a positive constant, v is the speed of the fish relative to the water, u is the speed of the current, and d is the distance the fish swim upstream. Find the partial derivatives of E with repect to u, v, and d. What are their biological meanings?

SOLUTION Keeping v and d constant and differentiating with respect to u, we get

$$\frac{\partial E}{\partial u} = (av^3d)\left[-\frac{1}{(v-u)^2}(-1)\right] = \frac{av^3d}{(v-u)^2}$$

This represents the rate of change in energy expended that occurs from an increase in current speed while holding swimming speed and distance constant. It is positive, meaning that the energy expended increases as current speed increases.

Now we keep u and d constant and differentiate with respect to v using the Quotient Rule:

$$\frac{\partial E}{\partial v} = \frac{(v-u)(3av^2d) - (av^3d)\cdot 1}{(v-u)^2} = \frac{2av^3d - 3auv^2d}{(v-u)^2} = \frac{av^2d(2v-3u)}{(v-u)^2}$$

This represents the rate of change in energy expended that occurs from an increase in swimming speed while holding current speed and distance constant. This quantity can be positive (when $2v > 3u$) or negative (when $2v < 3u$). Thus the energy expended can increase or decrease as swimming speed increases, depending on the situation.

Finally we keep u and v constant and differentiate with respect to d:

$$\frac{\partial E}{\partial d} = \frac{av^3}{v-u}$$

This represents the rate of change in energy expended that occurs from an increase in distance while holding swimming speed and current speed constant. It is positive, meaning that the energy expended increases as the distance increases. ■

Higher Derivatives

If f is a function of two variables, then its partial derivatives f_x and f_y are also functions of two variables, so we can consider their partial derivatives $(f_x)_x$, $(f_x)_y$, $(f_y)_x$, and $(f_y)_y$, which are called the **second partial derivatives** of f. If $z = f(x, y)$, we use the following notation:

$$(f_x)_x = f_{xx} = \frac{\partial}{\partial x}\left(\frac{\partial f}{\partial x}\right) = \frac{\partial^2 f}{\partial x^2} = \frac{\partial^2 z}{\partial x^2}$$

$$(f_x)_y = f_{xy} = \frac{\partial}{\partial y}\left(\frac{\partial f}{\partial x}\right) = \frac{\partial^2 f}{\partial y\,\partial x} = \frac{\partial^2 z}{\partial y\,\partial x}$$

$$(f_y)_x = f_{yx} = \frac{\partial}{\partial x}\left(\frac{\partial f}{\partial y}\right) = \frac{\partial^2 f}{\partial x\,\partial y} = \frac{\partial^2 z}{\partial x\,\partial y}$$

$$(f_y)_y = f_{yy} = \frac{\partial}{\partial y}\left(\frac{\partial f}{\partial y}\right) = \frac{\partial^2 f}{\partial y^2} = \frac{\partial^2 z}{\partial y^2}$$

Thus the notation f_{xy} (or $\partial^2 f/\partial y\,\partial x$) means that we first differentiate with respect to x and then with respect to y, whereas in computing f_{yx} the order is reversed.

EXAMPLE 9 | Find the second partial derivatives of

$$f(x, y) = x^3 + x^2y^3 - 2y^2$$

SOLUTION In Example 1 we found that

$$f_x(x, y) = 3x^2 + 2xy^3 \qquad f_y(x, y) = 3x^2y^2 - 4y$$

Therefore

$$f_{xx} = \frac{\partial}{\partial x}(3x^2 + 2xy^3) = 6x + 2y^3 \qquad f_{xy} = \frac{\partial}{\partial y}(3x^2 + 2xy^3) = 6xy^2$$

$$f_{yx} = \frac{\partial}{\partial x}(3x^2y^2 - 4y) = 6xy^2 \qquad f_{yy} = \frac{\partial}{\partial y}(3x^2y^2 - 4y) = 6x^2y - 4$$ ■

Clairaut

Alexis Clairaut was a child prodigy in mathematics: he read l'Hospital's textbook on calculus when he was ten and presented a paper on geometry to the French Academy of Sciences when he was 13. At the age of 18, Clairaut published *Recherches sur les courbes à double courbure,* which was the first systematic treatise on three-dimensional analytic geometry and included the calculus of space curves.

Notice that $f_{xy} = f_{yx}$ in Example 9. This is not just a coincidence. It turns out that the mixed partial derivatives f_{xy} and f_{yx} are equal for most functions that one meets in practice. The following theorem, which was discovered by the French mathematician Alexis Clairaut (1713–1765), gives conditions under which we can assert that $f_{xy} = f_{yx}$. The proof is given in Appendix E.

Clairaut's Theorem Suppose f is defined on a disk D that contains the point (a, b). If the functions f_{xy} and f_{yx} are both continuous on D, then

$$f_{xy}(a, b) = f_{yx}(a, b)$$

Partial derivatives of order 3 or higher can also be defined. For instance,

$$f_{xyy} = (f_{xy})_y = \frac{\partial}{\partial y}\left(\frac{\partial^2 f}{\partial y\, \partial x}\right) = \frac{\partial^3 f}{\partial y^2\, \partial x}$$

and using Clairaut's Theorem it can be shown that $f_{xyy} = f_{yxy} = f_{yyx}$ if these functions are continuous.

EXAMPLE 10 | Calculate f_{xxyz} if $f(x, y, z) = \sin(3x + yz)$.

SOLUTION

$$f_x = 3\cos(3x + yz)$$

$$f_{xx} = -9\sin(3x + yz)$$

$$f_{xxy} = -9z\cos(3x + yz)$$

$$f_{xxyz} = -9\cos(3x + yz) + 9yz\sin(3x + yz)$$ ■

■ Partial Differential Equations

Partial derivatives occur in *partial differential equations* that express certain physical and biological laws. For instance, the partial differential equation

$$\frac{\partial^2 u}{\partial t^2} = a^2 \frac{\partial^2 u}{\partial x^2}$$

is called the **wave equation** and describes the motion of a waveform, which could be an ocean wave, a sound wave, a light wave, or a wave traveling along a vibrating string. For instance, if $u(x, t)$ represents the displacement of a vibrating violin or guitar string at time t and at a distance x from one end of the string (as in Figure 5), then $u(x, t)$ satisfies the wave equation. Here the constant a depends on the density of the string and on the tension in the string. The wave equation also arises in the description of such biological phenomena as the spread of colonies of bacteria or viruses and the propagation of nerve signals in neurons.

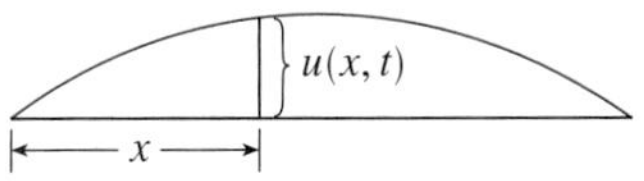

FIGURE 5

EXAMPLE 11 | Verify that the function $u(x, t) = \sin(x - at)$ satisfies the wave equation.

SOLUTION

$$u_x = \cos(x - at) \qquad u_t = -a\cos(x - at)$$

$$u_{xx} = -\sin(x - at) \qquad u_{tt} = -a^2\sin(x - at) = a^2 u_{xx}$$

So u satisfies the wave equation. ■

Another partial differential equation that arises frequently in biology is the **diffusion equation**:

$$\frac{\partial c}{\partial t} = D\frac{\partial^2 c}{\partial x^2}$$

where D is a positive constant called the *diffusion constant.* This equation describes several biological processes. The function $c(x, t)$ could represent, for example, the concentration of a pollutant at time t at a distance x from the source of the pollution. The diffusion equation can also describe the invasion of alien species into a new habitat, or the diffusion of heat through a solid, or the movement of organisms along chemical gradients.

In Exercise 70 you are asked to verify that the function

$$c(x, t) = \frac{1}{\sqrt{4\pi Dt}}e^{-x^2/(4Dt)}$$

is a solution of the diffusion equation.

EXERCISES 9.2

1. The temperature T (in °C) at a location in the Northern Hemisphere depends on the longitude x, latitude y, and time t, so we can write $T = f(x, y, t)$. Let's measure time in hours from the beginning of January.
   (a) What are the meanings of the partial derivatives $\partial T/\partial x$, $\partial T/\partial y$, and $\partial T/\partial t$?
   (b) Honolulu has longitude 158°W and latitude 21°N. Suppose that at 9:00 AM on January 1 the wind is blowing hot air to the northeast, so the air to the west and south is warm and the air to the north and east is cooler. Would you expect $f_x(158, 21, 9)$, $f_y(158, 21, 9)$, and $f_t(158, 21, 9)$ to be positive or negative? Explain.

2. At the beginning of this section we discussed the function $I = f(T, H)$, where I is the heat index, T is the temperature, and H is the relative humidity. Use Table 1 to estimate $f_T(92, 60)$ and $f_H(92, 60)$. What are the practical interpretations of these values?

3. The **wind-chill index** W is the perceived temperature when the actual temperature is T and the wind speed is v, so we can write $W = f(T, v)$. The following table of values is an excerpt from Table 9.1.1.

Wind speed (km/h)

Actual temperature (°C): T \ v	20	30	40	50	60	70
−10	−18	−20	−21	−22	−23	−23
−15	−24	−26	−27	−29	−30	−30
−20	−30	−33	−34	−35	−36	−37
−25	−37	−39	−41	−42	−43	−44

   (a) Estimate the values of $f_T(-15, 30)$ and $f_v(-15, 30)$. What are the practical interpretations of these values?

(b) In general, what can you say about the signs of $\partial W/\partial T$ and $\partial W/\partial v$?

(c) What appears to be the value of the following limit?

$$\lim_{v\to\infty} \frac{\partial W}{\partial v}$$

4. The wave heights h in the open sea depend on the speed v of the wind and the length of time t that the wind has been blowing at that speed. Values of the function $h = f(v, t)$ are recorded in feet in the following table.

Duration (hours)

Wind speed (knots)

v \ t	5	10	15	20	30	40	50
10	2	2	2	2	2	2	2
15	4	4	5	5	5	5	5
20	5	7	8	8	9	9	9
30	9	13	16	17	18	19	19
40	14	21	25	28	31	33	33
50	19	29	36	40	45	48	50
60	24	37	47	54	62	67	69

(a) What are the meanings of the partial derivatives $\partial h/\partial v$ and $\partial h/\partial t$?

(b) Estimate the values of $f_v(40, 15)$ and $f_t(40, 15)$. What are the practical interpretations of these values?

(c) What appears to be the value of the following limit?

$$\lim_{t\to\infty} \frac{\partial h}{\partial t}$$

5–6 Determine the signs of the partial derivatives for the function f whose graph is shown.

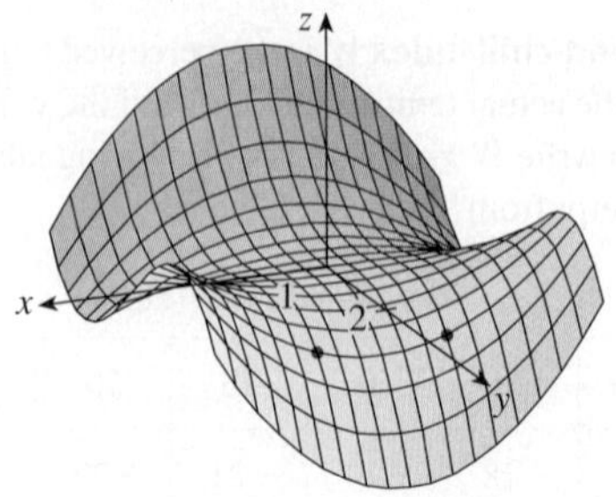

5. (a) $f_x(1, 2)$ (b) $f_y(1, 2)$

6. (a) $f_x(-1, 2)$ (b) $f_y(-1, 2)$

7. If $f(x, y) = 16 - 4x^2 - y^2$, find $f_x(1, 2)$ and $f_y(1, 2)$ and interpret these numbers as slopes. Illustrate with either hand-drawn sketches or computer plots.

8. A contour map is given for a function f. Use it to estimate $f_x(2, 1)$ and $f_y(2, 1)$.

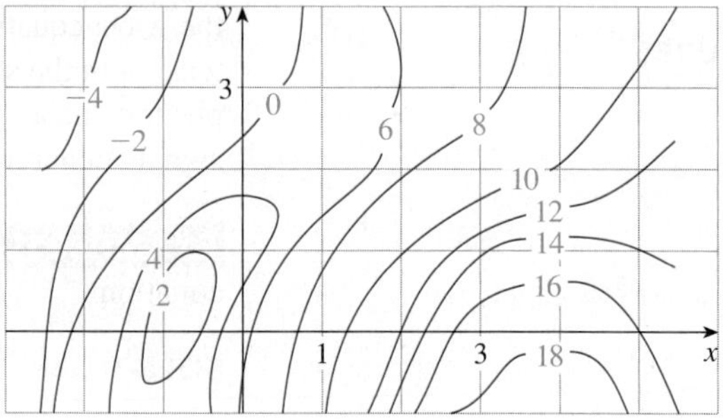

9–32 Find the first partial derivatives of the function.

9. $f(x, y) = y^5 - 3xy$

10. $f(x, y) = x^4y^3 + 8x^2y$

11. $f(x, t) = e^{-t}\cos \pi x$

12. $f(x, t) = \sqrt{x}\,\ln t$

13. $z = (2x + 3y)^{10}$

14. $z = \tan xy$

15. $f(x, y) = \dfrac{x - y}{x + y}$

16. $f(x, y) = x^y$

17. $w = \sin\alpha\cos\beta$

18. $w = \dfrac{e^v}{u + v^2}$

19. $f(r, s) = r\ln(r^2 + s^2)$

20. $f(x, t) = \arctan(x\sqrt{t})$

21. $u = te^{w/t}$

22. $f(x, y) = \int_y^x \cos(t^2)\,dt$

23. $f(x, y, z) = xz - 5x^2y^3z^4$

24. $f(x, y, z) = x\sin(y - z)$

25. $w = \ln(x + 2y + 3z)$

26. $w = ze^{xyz}$

27. $u = xe^{-t}\sin\theta$

28. $u = x^{y/z}$

29. $f(x, y, z, t) = xyz^2\tan(yt)$

30. $f(x, y, z, t) = \dfrac{xy^2}{t + 2z}$

31. $u = \sqrt{x_1^2 + x_2^2 + \cdots + x_n^2}$

32. $u = \sin(x_1 + 2x_2 + \cdots + nx_n)$

33–36 Find the indicated partial derivative.

33. $f(x, y) = \ln\left(x + \sqrt{x^2 + y^2}\right)$; $f_x(3, 4)$

34. $f(x, y) = \arctan(y/x)$; $f_x(2, 3)$

35. $f(x, y, z) = \dfrac{y}{x + y + z}$; $f_y(2, 1, -1)$

36. $f(x, y, z) = \sqrt{\sin^2 x + \sin^2 y + \sin^2 z}$; $f_z(0, 0, \pi/4)$

37–40 Use implicit differentiation to find $\partial z/\partial x$ and $\partial z/\partial y$.

37. $x^2 + y^2 + z^2 = 3xyz$

38. $yz = \ln(x + z)$

39. $x - z = \arctan(yz)$

40. $\sin(xyz) = x + 2y + 3z$

41. Body surface area A model for the surface area of a human body is given by the function

$$S = f(w, h) = 0.1091w^{0.425}h^{0.725}$$

where w is the weight (in pounds), h is the height (in inches), and S is measured in square feet. Calculate and interpret the partial derivatives.

(a) $\dfrac{\partial S}{\partial w}(160, 70)$ (b) $\dfrac{\partial S}{\partial h}(160, 70)$

42. The **wind-chill index** is modeled by the function

$$W = 13.12 + 0.6215T - 11.37v^{0.16} + 0.3965Tv^{0.16}$$

where T is the temperature (in °C) and v is the wind speed (in km/h). When $T = -15$ °C and $v = 30$ km/h, by how much would you expect the apparent temperature W to drop if the actual temperature decreases by 1°C? What if the wind speed increases by 1 km/h?

43. Blood flow One of Poiseuille's laws states that the resistance of blood flowing through an artery is

$$R = C\frac{L}{r^4}$$

where L and r are the length and radius of the artery and C is a positive constant determined by the viscosity of the blood. Calculate $\partial R/\partial L$ and $\partial R/\partial r$ and interpret them.

44. Antibiotic concentration If an antibiotic is administered to a patient at a constant rate through intravenous supply and is metabolized, then the concentration of antibiotic after one unit of time is

$$c(\theta, V) = c_0 e^{-1/V} + \theta V(1 - e^{-1/V})$$

where c_0 is the initial concentration, θ is the rate of supply, and V is the volume of the patient's blood. Calculate $\partial c/\partial \theta$ and $\partial c/\partial V$ and interpret them.

45. Flapping and gliding In the project on page 297 we expressed the power needed by a bird during its flapping mode as

$$P(v, x, m) = Av^3 + \frac{B(mg/x)^2}{v}$$

where A and B are constants specific to a species of bird, v is the velocity of the bird, m is the mass of the bird, g is the acceleration due to gravity, and x is the fraction of the flying time spent in flapping mode. Calculate $\partial P/\partial v$, $\partial P/\partial x$, and $\partial P/\partial m$ and interpret them.

46. Dialysis removes urea from a patient's blood by diverting some blood flow externally through a dialyzer. The rate at which urea is removed from the blood (in mg/min) is modeled by the equation

$$c(K, t, V) = \frac{K}{V}c_0 e^{-Kt/V}$$

where K is the rate of blood flow through the dialyzer (in mL/min), t is the time (in min), V is the volume of the patient's blood (in mL), and c_0 is the amount of urea in the blood (in mg) at time $t = 0$. Calculate $\partial c/\partial K$, $\partial c/\partial t$, and $\partial c/\partial V$ and interpret them.

47. Lizard energy expenditure The average energy E (in kcal) needed for a lizard to walk or run a distance of 1 km has been modeled by the equation

$$E(m, v) = 2.65m^{0.66} + \frac{3.5m^{0.75}}{v}$$

where m is the body mass of the lizard (in grams) and v is its speed (in km/h). Calculate $E_m(400, 8)$ and $E_v(400, 8)$ and interpret your answers.

Source: Adapted from C. Robbins, *Wildlife Feeding and Nutrition*, 2nd ed. (San Diego: Academic Press, 1993).

© tratong / Shutterstock.com

48. Snake reversals and stripes In a study of the survivorship of juvenile garter snakes, a researcher arrived at the model

$$F(R, S) = 4.2 + 0.008R + 0.102S + 0.017R^2 - 0.034S^2 - 0.268RS$$

where F is a measure of the fitness of the snake, R measures the number of reversals of direction during flight from a predator, and S measures the degree of stripedness in the color pattern of the snake. Calculate and interpret the partial derivatives.

(a) $\dfrac{\partial F}{\partial R}(2, 3)$ (b) $\dfrac{\partial F}{\partial S}(2, 3)$

Source: Adapted from E. Brodie III, "Correlational Selection for Color Pattern and Antipredator Behavior in the Garter Snake *Thamnophis ordinoides*," *Evolution* 46 (1992): 1284–98.

49. The *van der Waals equation* for n moles of a gas is

$$\left(P + \frac{n^2a}{V^2}\right)(V - nb) = nRT$$

where P is the pressure, V is the volume, and T is the temperature of the gas. The constant R is the universal gas constant and a and b are positive constants that are characteristic of a particular gas. Calculate $\partial T/\partial P$ and $\partial P/\partial V$.

50. (a) The gas law for a fixed mass m of an ideal gas at absolute temperature T, pressure P, and volume V is $PV = mRT$, where R is the gas constant. Show that

$$\frac{\partial P}{\partial V}\frac{\partial V}{\partial T}\frac{\partial T}{\partial P} = -1$$

(b) Show that, for an ideal gas,

$$T\frac{\partial P}{\partial T}\frac{\partial V}{\partial T} = mR$$

51–56 Find all the second partial derivatives.

51. $f(x, y) = x^3y^5 + 2x^4y$

52. $f(x, y) = \sin^2(mx + ny)$

53. $w = \sqrt{u^2 + v^2}$

54. $v = \dfrac{xy}{x - y}$

55. $z = \arctan \dfrac{x + y}{1 - xy}$

56. $v = e^{xe^y}$

57–58 Verify that the conclusion of Clairaut's Theorem holds, that is, $u_{xy} = u_{yx}$.

57. $u = xe^{xy}$

58. $u = \tan(2x + 3y)$

59–64 Find the indicated partial derivative(s).

59. $f(x, y) = 3xy^4 + x^3y^2$; f_{xxy}, f_{yyy}

60. $f(x, t) = x^2e^{-ct}$; f_{ttt}, f_{txx}

61. $f(x, y, z) = \cos(4x + 3y + 2z)$; f_{xyz}, f_{yzz}

62. $f(r, s, t) = r\ln(rs^2t^3)$; f_{rss}, f_{rst}

63. $u = e^{r\theta}\sin\theta$; $\dfrac{\partial^3 u}{\partial r^2\,\partial\theta}$

64. $u = x^ay^bz^c$; $\dfrac{\partial^6 u}{\partial x\,\partial y^2\,\partial z^3}$

65. If $f(x, y, z) = xy^2z^3 + \sec^2(x\sqrt{z})$, find f_{xzy}. [*Hint:* Which order of differentiation is easiest?]

66. If $g(x, y, z) = \sqrt{1 + xz} + \sqrt{1 - xy}$, find g_{xyz}. [*Hint:* Use a different order of differentiation for each term.]

67. Verify that the function $u = e^{-\alpha^2k^2t}\sin kx$ is a solution of the *heat conduction equation* $u_t = \alpha^2 u_{xx}$.

68. Determine whether each of the following functions is a solution of Laplace's equation $u_{xx} + u_{yy} = 0$.

(a) $u = x^2 + y^2$
(b) $u = x^2 - y^2$
(c) $u = x^3 + 3xy^2$
(d) $u = \ln\sqrt{x^2 + y^2}$
(e) $u = e^{-x}\cos y - e^{-y}\cos x$

69. Show that each of the following functions is a solution of the wave equation $u_{tt} = a^2u_{xx}$.

(a) $u = \sin(kx)\sin(akt)$
(b) $u = t/(a^2t^2 - x^2)$
(c) $u = (x - at)^6 + (x + at)^6$
(d) $u = \sin(x - at) + \ln(x + at)$

70. Diffusion equation Verify that the function

$$c(x, t) = \frac{1}{\sqrt{4\pi Dt}}e^{-x^2/(4Dt)}$$

is a solution of the diffusion equation

$$\frac{\partial c}{\partial t} = D\frac{\partial^2 c}{\partial x^2}$$

71. If $u = xe^y + ye^x$, show that

$$\frac{\partial^3 u}{\partial x^3} + \frac{\partial^3 u}{\partial y^3} = x\frac{\partial^3 u}{\partial x\,\partial y^2} + y\frac{\partial^3 u}{\partial x^2\,\partial y}$$

72. Show that the Cobb-Douglas production function $P = bL^\alpha K^\beta$ satisfies the equation

$$L\frac{\partial P}{\partial L} + K\frac{\partial P}{\partial K} = (\alpha + \beta)P$$

73. You are told that there is a function f whose partial derivatives are $f_x(x, y) = x + 4y$ and $f_y(x, y) = 3x - y$. Should you believe it?

9.3 Tangent Planes and Linear Approximations

One of the most important ideas in single-variable calculus is that as we zoom in toward a point on the graph of a differentiable function, the graph becomes indistinguishable from its tangent line and we can approximate the function by a linear function. (See Section 3.8.) Here we develop similar ideas in three dimensions. As we zoom in toward a point on a surface that is the graph of a differentiable function of two variables, the surface looks more and more like a plane (its tangent plane) and we can approximate the function by a linear function of two variables.

Tangent Planes

Suppose a surface S has equation $z = f(x, y)$, where f has continuous first partial derivatives, and let $P(x_0, y_0, z_0)$ be a point on S. As in the preceding section, let C_1 and C_2 be the curves obtained by intersecting the vertical planes $y = y_0$ and $x = x_0$ with the surface S. Then the point P lies on both C_1 and C_2. Let T_1 and T_2 be the tangent lines to the curves C_1 and C_2 at the point P. Then the **tangent plane** to the surface S at the point P is defined to be the plane that contains both tangent lines T_1 and T_2. (See Figure 1.)

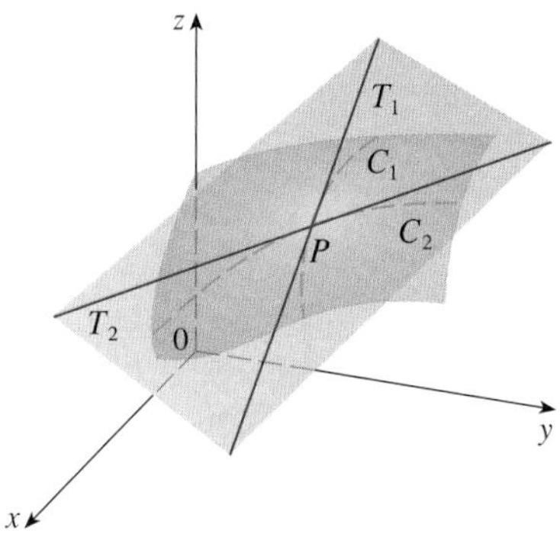

FIGURE 1
The tangent plane contains the tangent lines T_1 and T_2.

It can be proved that if C is any other curve that lies on the surface S and passes through P, then its tangent line at P also lies in the tangent plane. Therefore you can think of the tangent plane to S at P as consisting of all possible tangent lines at P to curves that lie on S and pass through P. The tangent plane at P is the plane that most closely approximates the surface S near the point P.

We know from Section 8.3 that any plane passing through the point $P(x_0, y_0, z_0)$ has an equation of the form

$$A(x - x_0) + B(y - y_0) + C(z - z_0) = 0$$

By dividing this equation by C and letting $a = -A/C$ and $b = -B/C$, we can write it in the form

(1) $$z - z_0 = a(x - x_0) + b(y - y_0)$$

If Equation 1 represents the tangent plane at P, then its intersection with the plane $y = y_0$ must be the tangent line T_1. Setting $y = y_0$ in Equation 1 gives

$$z - z_0 = a(x - x_0) \qquad y = y_0$$

and we recognize this as the equation (in point-slope form) of a line with slope a. But from Section 9.2 we know that the slope of the tangent T_1 is $f_x(x_0, y_0)$. Therefore $a = f_x(x_0, y_0)$.

Similarly, putting $x = x_0$ in Equation 1, we get $z - z_0 = b(y - y_0)$, which must represent the tangent line T_2, so $b = f_y(x_0, y_0)$.

Note the similarity between the equation of a tangent plane and the equation of a tangent line:

$$y - y_0 = f'(x_0)(x - x_0)$$

(2) Suppose f has continuous partial derivatives. An equation of the tangent plane to the surface $z = f(x, y)$ at the point $P(x_0, y_0, z_0)$ is

$$z - z_0 = f_x(x_0, y_0)(x - x_0) + f_y(x_0, y_0)(y - y_0)$$

EXAMPLE 1 | Find the tangent plane to the paraboloid $z = 2x^2 + y^2$ at the point $(1, 1, 3)$.

SOLUTION Let $f(x, y) = 2x^2 + y^2$. Then

$$f_x(x, y) = 4x \qquad f_y(x, y) = 2y$$

$$f_x(1, 1) = 4 \qquad f_y(1, 1) = 2$$

Then (2) gives the equation of the tangent plane at $(1, 1, 3)$ as

$$z - 3 = 4(x - 1) + 2(y - 1)$$

or

$$z = 4x + 2y - 3$$

■

TEC Visual 9.3 shows an animation of Figures 2 and 3.

Figure 2(a) shows the paraboloid and its tangent plane at (1, 1, 3) that we found in Example 1. In parts (b) and (c) we zoom in toward the point (1, 1, 3) by restricting the domain of the function $f(x, y) = 2x^2 + y^2$. Notice that the more we zoom in, the flatter the graph appears and the more it resembles its tangent plane.

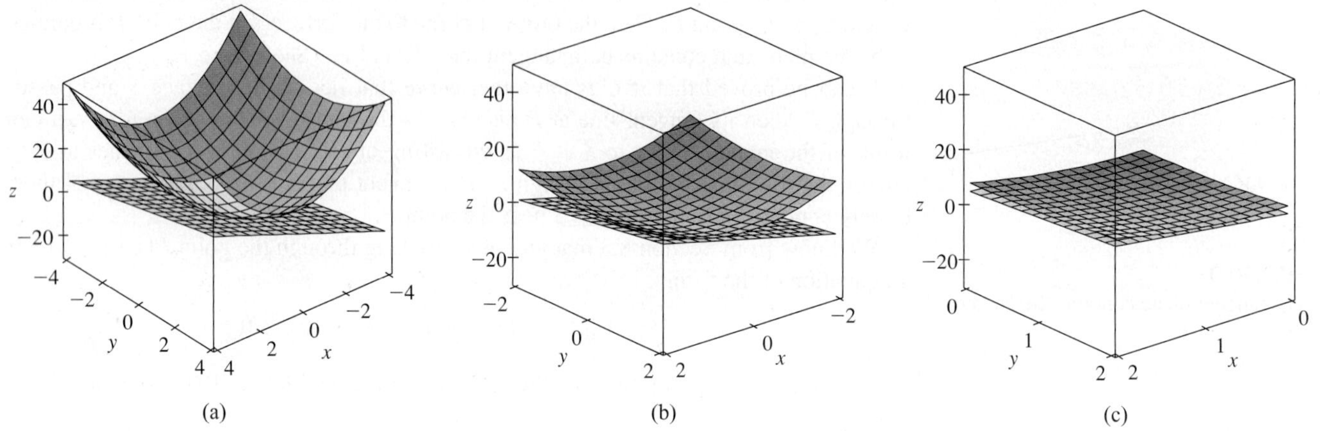

FIGURE 2 The paraboloid $z = 2x^2 + y^2$ appears to coincide with its tangent plane as we zoom in toward (1, 1, 3).

In Figure 3 we corroborate this impression by zooming in toward the point (1, 1) on a contour map of the function $f(x, y) = 2x^2 + y^2$. Notice that the more we zoom in, the more the level curves look like equally spaced parallel lines, which is characteristic of a plane.

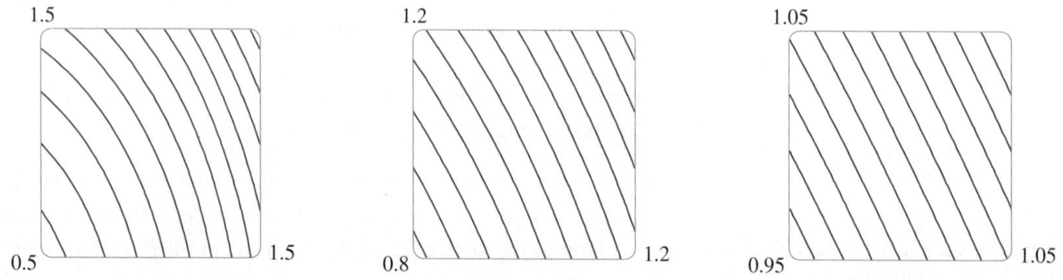

FIGURE 3
Zooming in toward (1, 1) on a contour map of $f(x, y) = 2x^2 + y^2$

Linear Approximations

In Example 1 we found that an equation of the tangent plane to the graph of the function $f(x, y) = 2x^2 + y^2$ at the point (1, 1, 3) is $z = 4x + 2y - 3$. Therefore, in view of the visual evidence in Figures 2 and 3, the linear function of two variables

$$L(x, y) = 4x + 2y - 3$$

is a good approximation to $f(x, y)$ when (x, y) is near (1, 1). The function L is called the *linearization* of f at (1, 1) and the approximation

$$f(x, y) \approx 4x + 2y - 3$$

is called the *linear approximation* or *tangent plane approximation* of f at (1, 1).

For instance, at the point (1.1, 0.95) the linear approximation gives

$$f(1.1, 0.95) \approx 4(1.1) + 2(0.95) - 3 = 3.3$$

which is quite close to the true value of $f(1.1, 0.95) = 2(1.1)^2 + (0.95)^2 = 3.3225$. But if we take a point farther away from (1, 1), such as (2, 3), we no longer get a good approximation. In fact, $L(2, 3) = 11$ whereas $f(2, 3) = 17$.

In general, we know from (2) that an equation of the tangent plane to the graph of a function f of two variables at the point $(a, b, f(a, b))$ is

$$z = f(a, b) + f_x(a, b)(x - a) + f_y(a, b)(y - b)$$

The linear function whose graph is this tangent plane, namely,

$$\textbf{(3)} \qquad L(x, y) = f(a, b) + f_x(a, b)(x - a) + f_y(a, b)(y - b)$$

is called the **linearization** of f at (a, b) and the approximation

$$\textbf{(4)} \qquad f(x, y) \approx f(a, b) + f_x(a, b)(x - a) + f_y(a, b)(y - b)$$

is called the **linear approximation** or the **tangent plane approximation** of f at (a, b).

We have defined tangent planes for surfaces $z = f(x, y)$, where f has continuous first partial derivatives. What happens if f_x and f_y are not continuous? Figure 4 pictures such a function; its equation is

$$f(x, y) = \begin{cases} \dfrac{xy}{x^2 + y^2} & \text{if } (x, y) \neq (0, 0) \\ 0 & \text{if } (x, y) = (0, 0) \end{cases}$$

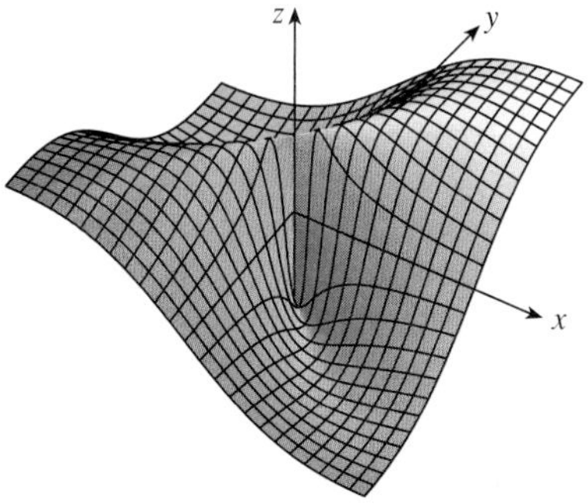

FIGURE 4

$f(x, y) = \dfrac{xy}{x^2 + y^2}$ if $(x, y) \neq (0, 0)$,
$f(0, 0) = 0$

It can be shown that its partial derivatives exist at the origin and, in fact, $f_x(0, 0) = 0$ and $f_y(0, 0) = 0$, but f_x and f_y are not continuous. The linear approximation would be $f(x, y) \approx 0$, but $f(x, y) = \frac{1}{2}$ at all points on the line $y = x$. So a function of two variables can behave badly even though both of its partial derivatives exist. To rule out such behavior, we formulate the idea of a differentiable function of two variables by requiring the distance between $f(x, y)$ and its linear approximation $L(x, y)$ to approach 0 faster than the distance from (x, y) to (a, b).

(5) Definition The function $f(x, y)$ is **differentiable** at (a, b) if

$$\lim_{(x, y) \to (a, b)} \frac{|f(x, y) - L(x, y)|}{\sqrt{(x - a)^2 + (y - b)^2}} = 0$$

Definition 5 says that a differentiable function is one for which the linear approximation (4) is a good approximation when (x, y) is near (a, b). In other words, the tangent plane approximates the graph of f well near the point of tangency.

It's sometimes hard to use Definition 5 directly to check the differentiability of a function, but the next theorem (which we will not prove) provides a convenient sufficient condition for differentiability.

(6) Theorem If the partial derivatives f_x and f_y exist near (a, b) and are continuous at (a, b), then f is differentiable at (a, b).

EXAMPLE 2 | Show that $f(x, y) = xe^{xy}$ is differentiable at $(1, 0)$ and find its linearization there. Then use it to approximate $f(1.1, -0.1)$.

SOLUTION The partial derivatives are

$$f_x(x, y) = e^{xy} + xye^{xy} \qquad f_y(x, y) = x^2e^{xy}$$

$$f_x(1, 0) = 1 \qquad f_y(1, 0) = 1$$

Figure 5 shows the graphs of the function f and its linearization L in Example 2.

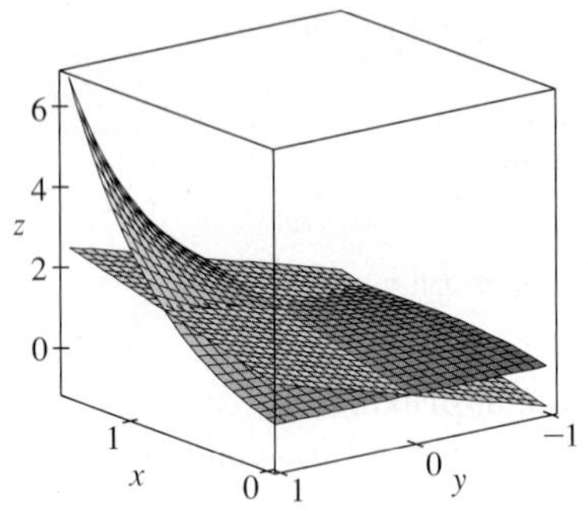

FIGURE 5

Both f_x and f_y are continuous functions, so f is differentiable by Theorem 6. The linearization is

$$L(x, y) = f(1, 0) + f_x(1, 0)(x - 1) + f_y(1, 0)(y - 0)$$
$$= 1 + 1(x - 1) + 1 \cdot y = x + y$$

The corresponding linear approximation is

$$xe^{xy} \approx x + y$$

so
$$f(1.1, -0.1) \approx 1.1 - 0.1 = 1$$

Compare this with the actual value of $f(1.1, -0.1) = 1.1e^{-0.11} \approx 0.98542$. ■

EXAMPLE 3 | **Heat index** At the beginning of Section 9.2 we discussed the heat index (perceived temperature) I as a function of the actual temperature T and the relative humidity H and gave the following table of values from the National Weather Service.

Relative humidity (%)

Actual temperature (°F)	T \ H	50	55	60	65	70	75	80	85	90
	90	96	98	100	103	106	109	112	115	119
	92	100	103	105	108	112	115	119	123	128
	94	104	107	111	114	118	122	127	132	137
	96	109	113	116	121	125	130	135	141	146
	98	114	118	123	127	133	138	144	150	157
	100	119	124	129	135	141	147	154	161	168

Find a linear approximation for the heat index $I = f(T, H)$ when T is near 96°F and H is near 70%. Use it to estimate the heat index when the temperature is 97°F and the relative humidity is 72%.

SOLUTION We read from the table that $f(96, 70) = 125$. In Section 9.2 we used the tabular values to estimate that $f_T(96, 70) \approx 3.75$ and $f_H(96, 70) \approx 0.9$. (See pages 585–86.) So the linear approximation is

$$f(T, H) \approx f(96, 70) + f_T(96, 70)(T - 96) + f_H(96, 70)(H - 70)$$
$$\approx 125 + 3.75(T - 96) + 0.9(H - 70)$$

In particular,

$$f(97, 72) \approx 125 + 3.75(1) + 0.9(2) = 130.55$$

Therefore, when $T = 97°\text{F}$ and $H = 72\%$, the heat index is

$$I \approx 131°\text{F}$$

■

EXAMPLE 4 | The **wind-chill index** $W(T, v)$ is the subjective temperature that we perceive when the actual temperature is T and the wind speed is v. (See Example 9.1.1.) If T is measured in °C and v in km/h, then a model for W is

$$W(T, v) = 13.12 + 0.6215T - 11.37v^{0.16} + 0.3965Tv^{0.16}$$

Use this equation to find the linearization of W at $(-12, 20)$.

SOLUTION The partial derivatives of W are

$$W_T(T, v) = 0.6215 + 0.3965v^{0.16}$$

$$W_v(T, v) = -1.8192v^{-0.84} + 0.06344Tv^{-0.84}$$

Evaluating W, W_T, and W_v to two decimal places when $T = -12$ and $v = 20$, we get

$$W(-12, 20) = -20.38 \qquad W_T(-12, 20) = 1.26 \qquad W_v(-12, 20) = -0.21$$

So the linearization of the wind chill function at $(-12, 20)$ is

$$L(T, v) = -20.38 + 1.26(T + 12) - 0.21(v - 20)$$

This linear function is a good approximation to the wind chill function $W(T, v)$ when T is near $-12°\text{C}$ and v is near 20 km/h. The accuracy of this approximation decreases as we move away from these values. ■

Linear approximations and differentiability can be defined in a similar manner for functions of more than two variables. A differentiable function is defined by an expression similar to the one in Definition 5. For functions of three variables the **linear approximation** is

$$f(x, y, z) \approx f(a, b, c) + f_x(a, b, c)(x - a) + f_y(a, b, c)(y - b) + f_z(a, b, c)(z - c)$$

and the linearization $L(x, y, z)$ is the right side of this expression.

EXERCISES 9.3

1–6 Find an equation of the tangent plane to the given surface at the specified point.

1. $z = 3y^2 - 2x^2 + x, \quad (2, -1, -3)$

2. $z = 3(x - 1)^2 + 2(y + 3)^2 + 7, \quad (2, -2, 12)$

3. $z = \sqrt{xy}, \quad (1, 1, 1)$

4. $z = xe^{xy}, \quad (2, 0, 2)$

5. $z = y\cos(x - y), \quad (2, 2, 2)$

6. $z = \ln(x - 2y), \quad (3, 1, 0)$

7–12 Explain why the function is differentiable at the given point. Then find the linearization of the function at that point.

7. $f(x, y) = x\sqrt{y}, \quad (1, 4)$

8. $f(x, y) = x^3y^4, \quad (1, 1)$

9. $f(x, y) = \dfrac{x}{x + y}$, $(2, 1)$

10. $f(x, y) = \sqrt{x + e^{4y}}$, $(3, 0)$

11. $f(x, y, z) = x^2y + y^2z$, $(1, 2, 3)$

12. $f(x, y, z) = e^{-xy} \cos z$, $(0, 2, 0)$

13–14 Verify the linear approximation at (0, 0).

13. $\dfrac{2x + 3}{4y + 1} \approx 3 + 2x - 12y$ **14.** $\sqrt{y + \cos^2 x} \approx 1 + \frac{1}{2}y$

---

15. Given that f is a differentiable function with $f(2, 5) = 6$, $f_x(2, 5) = 1$, and $f_y(2, 5) = -1$, use a linear approximation to estimate $f(2.2, 4.9)$.

16. Find the linear approximation of the function

$$f(x, y) = \ln(x - 3y)$$

at (7, 2) and use it to approximate $f(6.9, 2.06)$.

17. Find the linear approximation of the function

$$f(x, y, z) = \sqrt{x^2 + y^2 + z^2}$$

at (3, 2, 6) and use it to approximate the number

$$\sqrt{(3.02)^2 + (1.97)^2 + (5.99)^2}$$

18. The wave heights h in the open sea depend on the speed v of the wind and the length of time t that the wind has been blowing at that speed. Values of the function $h = f(v, t)$ are recorded in feet in the following table. Use the table to find a linear approximation to the wave height function when v is near 40 knots and t is near 20 hours. Then estimate the wave heights when the wind has been blowing for 24 hours at 43 knots.

Duration (hours)

Wind speed (knots) v \ t	5	10	15	20	30	40	50
20	5	7	8	8	9	9	9
30	9	13	16	17	18	19	19
40	14	21	25	28	31	33	33
50	19	29	36	40	45	48	50
60	24	37	47	54	62	67	69

19. Use the table in Example 3 to find a linear approximation to the heat index function when the temperature is near 94°F and the relative humidity is near 80%. Then estimate the heat index when the temperature is 95°F and the relative humidity is 78%.

20. The **wind-chill index** W is the perceived temperature when the actual temperature is T and the wind speed is v, so we can write $W = f(T, v)$. The following table of values is an excerpt from Table 9.1.1. Use the table to find a linear approximation to the wind-chill index function when T is near -15°C and v is near 50 km/h. Then estimate the wind-chill index when the temperature is -17°C and the wind speed is 55 km/hr.

Wind speed (km/h)

Actual temperature (°C) T \ v	20	30	40	50	60	70
−10	−18	−20	−21	−22	−23	−23
−15	−24	−26	−27	−29	−30	−30
−20	−30	−33	−34	−35	−36	−37
−25	−37	−39	−41	−42	−43	−44

21. Lizard energy expenditure The average energy E (in kcal) needed for a lizard to walk or run a distance of 1 km has been modeled by the equation

$$E(m, v) = 2.65m^{0.66} + \frac{3.5m^{0.75}}{v}$$

where m is the body mass of the lizard (in grams) and v is its speed (in km/h). Find the linearization of the energy function at (400, 8).

Source: Adapted from C. Robbins, *Wildlife Feeding and Nutrition*, 2nd ed. (San Diego: Academic Press, 1993).

22. Body surface area A model for the surface area of a human body is given by the function

$$S = f(w, h) = 0.1091w^{0.425}h^{0.725}$$

where w is the weight (in pounds), h is the height (in inches), and S is measured in square feet.

(a) Calculate the linearization $L(w, h)$ of this function when $w = 160$ lb and $h = 70$ in.

(b) Compare the values of $f(w, h)$ and $L(w, h)$ when $(w, h) = (162, 73)$ and when $(w, h) = (170, 80)$ Comment on the relative accuracy of the approximations.

PROJECT The Speedo LZR Racer

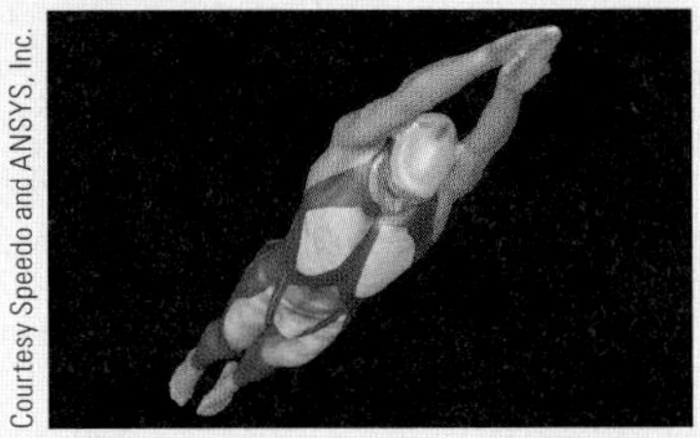

Many technological advances have occurred in sports that have contributed to increased athletic performance. One of the best known is the introduction, in 2008, of the Speedo LZR racer. It was claimed that this full-body swimsuit reduced a swimmer's drag in the water. Figure 1 shows the number of world records broken in men's and women's long-course freestyle swimming events from 1990 to 2011.[1] The dramatic increase in 2008 when the suit was introduced led people to claim that such suits are a form of technological doping. As a result all full-body suits were banned from competition starting in 2010.

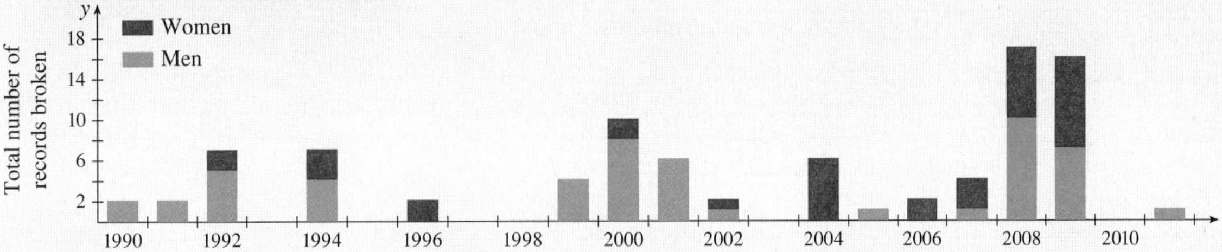

FIGURE 1 Number of world records set in long-course men's and women's freestyle swimming event 1990–2011

It might be surprising that a simple reduction in drag could have such a big effect on performance. We can gain some insight into this using a simple mathematical model.[2]

The speed v of an object being propelled through water is given by

$$v(P, C) = \left(\frac{2P}{kC}\right)^{1/3}$$

where P is the power being used to propel the object, C is the drag coefficient, and k is a positive constant. Athletes can therefore increase their swimming speeds by increasing their power or reducing their drag coefficients. But how effective is each of these?

To compare the effect of increasing power versus reducing drag, we need to somehow compare the two in common units. The most common approach is to determine the *percentage change* in speed that results from a given *percentage change* in power and in drag.

If we work with percentages as fractions, then when power is changed by a fraction x (with x corresponding to $100x$ percent), P changes from P to $P + xP$. Likewise, if the drag coefficient is changed by a fraction y, this means that it has changed from C to $C + yC$. Finally, the fractional change in speed resulting from both effects is

$$\textbf{(1)} \qquad \frac{v(P + xP, C + yC) - v(P, C)}{v(P, C)}$$

1. Expression 1 gives the fractional change in speed that results from a change x in power and a change y in drag. Show that this reduces to the function

$$f(x, y) = \left(\frac{1 + x}{1 + y}\right)^{1/3} - 1$$

Given the context, what is the domain of f?

1. L. Foster et al., "Influence of Full Body Swimsuits on Competitive Performance," *Procedia Engineering* 34 (2012): 712–17.

2. Adapted from http://plus.maths.org/content/swimming.

2. Suppose that the possible changes in power x and drag y are small. Find the linear approximation to the function $f(x, y)$. What does this approximation tell you about the effect of a small increase in power versus a small decrease in drag?

3. Calculate $f_{xx}(x, y)$ and $f_{yy}(x, y)$. Based on the signs of these derivatives, does the linear approximation in Problem 2 result in an overestimate or an underestimate for an increase in power? What about for a decrease in drag? Use your answer to explain why, for changes in power or drag that are not very small, a decrease in drag is more effective.

4. Graph the level curves of $f(x, y)$. Explain how the shapes of these curves relate to your answers to Problems 2 and 3.

9.4 The Chain Rule

Recall that the Chain Rule for functions of a single variable gives the rule for differentiating a composite function: If $y = f(x)$ and $x = g(t)$, where f and g are differentiable functions, then y is indirectly a differentiable function of t and

(1)
$$\frac{dy}{dt} = \frac{dy}{dx}\frac{dx}{dt}$$

For functions of two variables we have $z = f(x, y)$, where each of the variables x and y is a function of a variable t. This means that z is indirectly a function of t, $z = f(g(t), h(t))$, and the Chain Rule gives a formula for differentiating z as a function of t. We assume that f is differentiable (Definition 9.3.5). Recall that this is the case when f_x and f_y are continuous (Theorem 9.3.6).

(2) The Chain Rule Suppose that $z = f(x, y)$ is a differentiable function of x and y, where $x = g(t)$ and $y = h(t)$ are both differentiable functions of t. Then z is a differentiable function of t and

$$\frac{dz}{dt} = \frac{\partial z}{\partial x}\frac{dx}{dt} + \frac{\partial z}{\partial y}\frac{dy}{dt}$$

A rigorous proof of the Chain Rule is rather technical and is similar to the proof of the single-variable case at the end of Section 3.5. Instead, we give a more intuitive indication of why it is true.

Near a point (a, b) in the domain of f, we approximate the differentiable function f by its linear approximation:

$$f(x, y) \approx L(x, y) = f(a, b) + f_x(a, b)(x - a) + f_y(a, b)(y - b)$$

Let $\Delta x = x - a$, $\Delta y = y - b$, and $\Delta z = f(x, y) - f(a, b)$. Then we can write

$$\Delta z \approx \frac{\partial z}{\partial x}\Delta x + \frac{\partial z}{\partial y}\Delta y$$

A change of Δt in t produces changes of Δx in x and Δy in y which, in turn, produce a change of Δz in z. Dividing both sides of this last equation by Δt, we get

$$\frac{\Delta z}{\Delta t} \approx \frac{\partial z}{\partial x}\frac{\Delta x}{\Delta t} + \frac{\partial z}{\partial y}\frac{\Delta y}{\Delta t}$$

If we now let $\Delta t \to 0$, this approximation becomes better and we have

$$\frac{\Delta z}{\Delta t} \to \frac{dz}{dt} \qquad \frac{\Delta x}{\Delta t} \to \frac{dx}{dt} \qquad \frac{\Delta y}{\Delta t} \to \frac{dy}{dt}$$

So it seems reasonable that

$$\frac{dz}{dt} = \frac{\partial z}{\partial x}\frac{dx}{dt} + \frac{\partial z}{\partial y}\frac{dy}{dt}$$

EXAMPLE 1 | If $z = x^2y + 3xy^4$, where $x = \sin 2t$ and $y = \cos t$, find dz/dt when $t = 0$.

SOLUTION The Chain Rule gives

$$\begin{aligned}\frac{dz}{dt} &= \frac{\partial z}{\partial x}\frac{dx}{dt} + \frac{\partial z}{\partial y}\frac{dy}{dt} \\ &= (2xy + 3y^4)(2\cos 2t) + (x^2 + 12xy^3)(-\sin t)\end{aligned}$$

It's not necessary to substitute the expressions for x and y in terms of t. We simply observe that when $t = 0$, we have $x = \sin 0 = 0$ and $y = \cos 0 = 1$. Therefore

$$\left.\frac{dz}{dt}\right|_{t=0} = (0 + 3)(2\cos 0) + (0 + 0)(-\sin 0) = 6$$ ■

Strictly speaking, it's not absolutely necessary to use Theorem 2 to solve Example 1. Instead we could have substituted $x = \sin 2t$ and $y = \cos t$ into the expression for z and then used the ordinary chain rule. But in the next example it really is necessary to use Theorem 2.

© Ugo Montaldo / Shutterstock.com

EXAMPLE 2 | BB Tuna biomass depends on the availability of the small fish that they eat (such as puffer fish and trigger fish) as well as the size of the catch of the annual commercial tuna fishery (currently about four million tons per year). Use the Chain Rule to discuss whether the tuna biomass is increasing or decreasing.

SOLUTION Let T be the tuna biomass and S the biomass of the small fish that tuna eat. Let C be the size of the annual tuna catch. Then we can write $T = f(S, C)$. If t denotes time, then the Chain Rule says that

$$\text{(3)} \qquad \frac{dT}{dt} = \frac{\partial T}{\partial S}\frac{dS}{dt} + \frac{\partial T}{\partial C}\frac{dC}{dt}$$

If we suppose that the population of small fish is increasing, then dS/dt is positive. What about $\partial T/\partial S$? If S increases and C remains the same, then the tuna have more to eat and so T also increases. Thus $\partial T/\partial S > 0$. So the first term on the right side of (3) is positive.

If S remains the same and the catch increases, then T will decrease, which means that $\partial T/\partial C$ is negative. What is the sign of dC/dt? The size of the tuna catch has been increasing in recent decades but overfishing has resulted in international quotas, which are sometimes ignored. If C continues to increase, then $dC/dt > 0$ and so the second term in (3) is negative. From Equation 3 we see that dT/dt is the sum of a positive term and a negative term, so we are unable to determine whether dT/dt is positive or negative. On the other hand, if the catch decreases, then the second term is positive and so dT/dt is positive and the tuna biomass increases. ■

EXAMPLE 3 | The pressure P (in kilopascals), volume V (in liters), and temperature T (in kelvins) of a mole of an ideal gas are related by the equation $PV = 8.31T$. Find the rate at which the pressure is changing when the temperature is 300 K and increasing at a rate of 0.1 K/s and the volume is 100 L and increasing at a rate of 0.2 L/s.

SOLUTION If t represents the time elapsed in seconds, then at the given instant we have $T = 300$, $dT/dt = 0.1$, $V = 100$, $dV/dt = 0.2$. Since

$$P = 8.31\frac{T}{V}$$

the Chain Rule gives

$$\frac{dP}{dt} = \frac{\partial P}{\partial T}\frac{dT}{dt} + \frac{\partial P}{\partial V}\frac{dV}{dt} = \frac{8.31}{V}\frac{dT}{dt} - \frac{8.31T}{V^2}\frac{dV}{dt}$$

$$= \frac{8.31}{100}(0.1) - \frac{8.31(300)}{100^2}(0.2) = -0.04155$$

The pressure is decreasing at a rate of about 0.042 kPa/s. ■

For functions of three variables, where $w = f(x, y, z)$ is differentiable and x, y, and z are differentiable functions of t, the Chain Rule has an extra term:

$$\frac{dw}{dt} = \frac{\partial w}{\partial x}\frac{dx}{dt} + \frac{\partial w}{\partial y}\frac{dy}{dt} + \frac{\partial w}{\partial z}\frac{dz}{dt}$$

EXAMPLE 4 | Find dw/dt if $w = xe^{y/z}$, where $x = 3t + 2$, $y = t^2$, and $z = t^3 - 1$.

SOLUTION By the Chain Rule, we have

$$\frac{dw}{dt} = \frac{\partial w}{\partial x}\frac{dx}{dt} + \frac{\partial w}{\partial y}\frac{dy}{dt} + \frac{\partial w}{\partial z}\frac{dz}{dt}$$

$$= e^{y/z} \cdot 3 + xe^{y/z} \cdot \frac{1}{z} \cdot 2t + xe^{y/z}\left(-\frac{y}{z^2}\right) \cdot 3t^2$$

$$= e^{y/z}\left(3 + \frac{2xt}{z} - \frac{3xyt^2}{z^2}\right)$$ ■

■ Implicit Differentiation

The Chain Rule can be used to give a more complete description of the process of implicit differentiation that was introduced in Sections 3.5 and 9.2. We suppose that an equation of the form $F(x, y) = 0$ defines y implicitly as a differentiable function of x, that is, $y = f(x)$, where $F(x, f(x)) = 0$ for all x in the domain of f. If F is differentiable, we can apply the Chain Rule to differentiate both sides of the equation $F(x, y) = 0$ with respect to x. Since both x and y are functions of x, we obtain

$$\frac{\partial F}{\partial x}\frac{dx}{dx} + \frac{\partial F}{\partial y}\frac{dy}{dx} = 0$$

But $dx/dx = 1$, so if $\partial F/\partial y \neq 0$ we solve for dy/dx and obtain

(4)
$$\frac{dy}{dx} = -\frac{\dfrac{\partial F}{\partial x}}{\dfrac{\partial F}{\partial y}} = -\frac{F_x}{F_y}$$

To derive this equation we assumed that $F(x, y) = 0$ defines y implicitly as a function of x. The **Implicit Function Theorem**, proved in advanced calculus, gives conditions under which this assumption is valid: It states that if F is defined on a disk containing (a, b), where $F(a, b) = 0$, $F_y(a, b) \neq 0$, and F_x and F_y are continuous on the disk, then the equation $F(x, y) = 0$ defines y as a function of x near the point (a, b) and the derivative of this function is given by Equation 4.

EXAMPLE 5 | Find y' if $x^3 + y^3 = 6xy$.

The solution to Example 5 should be compared to the one in Example 3.5.12.

SOLUTION The given equation can be written as

$$F(x, y) = x^3 + y^3 - 6xy = 0$$

so Equation 4 gives

$$\frac{dy}{dx} = -\frac{F_x}{F_y} = -\frac{3x^2 - 6y}{3y^2 - 6x} = -\frac{x^2 - 2y}{y^2 - 2x}$$ ■

Now we suppose that z is given implicitly as a function $z = f(x, y)$ by an equation of the form $F(x, y, z) = 0$. This means that $F(x, y, f(x, y)) = 0$ for all (x, y) in the domain of f. If F and f are differentiable, then we can use the Chain Rule to differentiate the equation $F(x, y, z) = 0$ as follows:

$$\frac{\partial F}{\partial x}\frac{\partial x}{\partial x} + \frac{\partial F}{\partial y}\frac{\partial y}{\partial x} + \frac{\partial F}{\partial z}\frac{\partial z}{\partial x} = 0$$

But $$\frac{\partial}{\partial x}(x) = 1 \qquad \text{and} \qquad \frac{\partial}{\partial x}(y) = 0$$

so this equation becomes

$$\frac{\partial F}{\partial x} + \frac{\partial F}{\partial z}\frac{\partial z}{\partial x} = 0$$

If $\partial F/\partial z \neq 0$, we solve for $\partial z/\partial x$ and obtain the first formula in Equations 5. The formula for $\partial z/\partial y$ is obtained in a similar manner.

(5)
$$\frac{\partial z}{\partial x} = -\frac{\dfrac{\partial F}{\partial x}}{\dfrac{\partial F}{\partial z}} \qquad \frac{\partial z}{\partial y} = -\frac{\dfrac{\partial F}{\partial y}}{\dfrac{\partial F}{\partial z}}$$

Again, a version of the **Implicit Function Theorem** stipulates conditions under which our assumption is valid: If F is defined within a sphere containing (a, b, c), where $F(a, b, c) = 0$, $F_z(a, b, c) \neq 0$, and F_x, F_y, and F_z are continuous inside the sphere, then the equation $F(x, y, z) = 0$ defines z as a function of x and y near the point (a, b, c) and this function is differentiable, with partial derivatives given by (5).

EXAMPLE 6 | BB Infectious disease outbreak size Epidemiologists often wish to predict the fraction of the population that will ultimately be infected when a disease begins to spread. Mathematical models have been used to do so. The Kermack-McKendrick model leads to the following equation (see the project on page 354):

$$\rho e^{-qA} = 1 - A$$

where A is the fraction of the population ultimately infected, q is a measure of disease transmissibility, and ρ is a measure of the fraction of the population that is initially susceptible to infection. How does the outbreak size A change with an increase in the transmissibility q?

SOLUTION Let $F(\rho, q, A) = \rho e^{-qA} - 1 + A$. Then, from Equations 5, the rate of change of A with respect to q is

$$\frac{\partial A}{\partial q} = -\frac{F_q}{F_A} = -\frac{-\rho A e^{-qA}}{-\rho q e^{-qA} + 1} = \frac{\rho A}{e^{qA} - \rho q}$$

This is the rate of increase of the outbreak size as the transmissibility increases while ρ remains constant. ■

EXERCISES 9.4

1–6 Use the Chain Rule to find dz/dt or dw/dt.

1. $z = x^2 + y^2 + xy$, $x = \sin t$, $y = e^t$

2. $z = \cos(x + 4y)$, $x = 5t^4$, $y = 1/t$

3. $z = \sqrt{1 + x^2 + y^2}$, $x = \ln t$, $y = \cos t$

4. $z = \tan^{-1}(y/x)$, $x = e^t$, $y = 1 - e^{-t}$

5. $w = xe^{y/z}$, $x = t^2$, $y = 1 - t$, $z = 1 + 2t$

6. $w = \ln\sqrt{x^2 + y^2 + z^2}$, $x = \sin t$, $y = \cos t$, $z = \tan t$

7. Suppose $z = f(x, y)$, where $x = g(s, t)$ and $y = h(s, t)$ are differentiable functions of s and t. Use Theorem 2 to show that

$$\frac{\partial z}{\partial s} = \frac{\partial z}{\partial x}\frac{\partial x}{\partial s} + \frac{\partial z}{\partial y}\frac{\partial y}{\partial s}$$

$$\frac{\partial z}{\partial t} = \frac{\partial z}{\partial x}\frac{\partial x}{\partial t} + \frac{\partial z}{\partial y}\frac{\partial y}{\partial t}$$

8–10 Use Exercise 7 to find $\partial z/\partial s$ and $\partial z/\partial t$.

8. $z = e^x \sin y$, $x = st^2$, $y = s^2 t$

9. $z = x^2 y^3$, $x = s\cos t$, $y = s\sin t$

10. $z = e^{x+2y}$, $x = s/t$, $y = t/s$

11. If $z = f(x, y)$, where f is differentiable, and

$$x = g(t) \qquad y = h(t)$$
$$g(3) = 2 \qquad h(3) = 7$$
$$g'(3) = 5 \qquad h'(3) = -4$$
$$f_x(2, 7) = 6 \qquad f_y(2, 7) = -8$$

find dz/dt when $t = 3$.

12. Let $W(s, t) = F(u(s, t), v(s, t))$, where F, u, and v are differentiable, and

$$u(1, 0) = 2 \qquad v(1, 0) = 3$$
$$u_s(1, 0) = -2 \qquad v_s(1, 0) = 5$$
$$u_t(1, 0) = 6 \qquad v_t(1, 0) = 4$$
$$F_u(2, 3) = -1 \qquad F_v(2, 3) = 10$$

Find $W_s(1, 0)$ and $W_t(1, 0)$.

13–16 Use Equation 4 to find dy/dx.

13. $y \cos x = x^2 + y^2$

14. $y^5 + x^2y^3 = 1 + ye^{x^2}$

15. $\cos(x - y) = xe^y$

16. $\sin x + \cos y = \sin x \cos y$

17–20 Use Equations 5 to find $\partial z/\partial x$ and $\partial z/\partial y$.

17. $x^2 + y^2 + z^2 = 3xyz$

18. $xyz = \cos(x + y + z)$

19. $x - z = \arctan(yz)$

20. $yz = \ln(x + z)$

21. Gray wolves range over Canada, the northern United States, Europe, and Asia. They feed on animals such as moose, deer, elk, and caribou and smaller animals such as hare and lynx. Their competitors are principally coyotes and bears. They have few predators, aside from humans. If W denotes the size of the gray wolf population, we can write

$$W = f(F, C)$$

where F is the size of the food supply and C is the number of competitors.

(a) The variables W, F, and C are all functions of time t. Write out the Chain Rule for this situation.

(b) Are the partial derivatives $\partial W/\partial F$ and $\partial W/\partial C$ positive or negative?

(c) What happens if the food supply increases while the competition decreases?

(d) What can you say if both F and C increase?

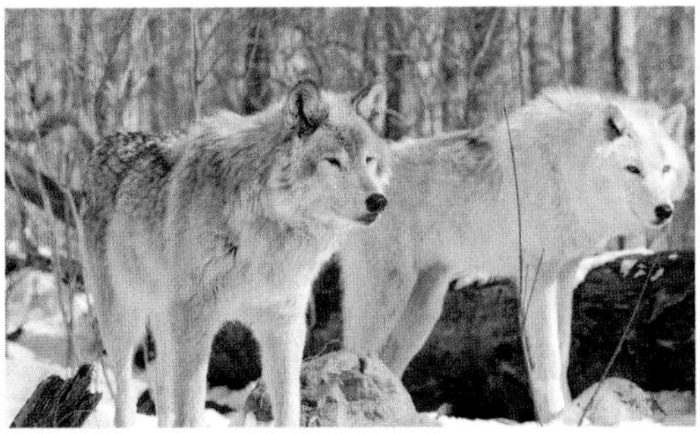

22. Wheat production W in a given year depends on the average temperature T and the annual rainfall R. Scientists estimate that the average temperature is rising at a rate of 0.15°C/year and rainfall is decreasing at a rate of 0.1 cm/year. They also estimate that, at current production levels, $\partial W/\partial T = -2$ and $\partial W/\partial R = 8$.

(a) What is the significance of the signs of these partial derivatives?

(b) Estimate the current rate of change of wheat production, dW/dt.

23. Body mass index Recall from Example 9.1.2 that the BMI of a person is

$$B(m, h) = \frac{m}{h^2}$$

where m is the person's mass (in kilograms) and h is the height (in meters). Both m and h depend on the person's age a. Use the Chain Rule to find an expression for the rate of change of the BMI with respect to age.

24. Infectious disease control In Example 9.1.12 we looked at the following model for the spread of SARS:

$$R_0(d, v) = 5(1 - v)\frac{d}{1 + d}$$

where R_0 is the basic reproduction number, v is the fraction of the population that is vaccinated, and d is the average number of days that an infected individual remains in the population before being quarantined. The quantities d and v depend on the investment I that the government makes to contain the disease. Use the Chain Rule to obtain an expression for the rate of change of R_0 with respect to investment.

25. Infectious disease outbreak size In Example 6 we used the equation $\rho e^{-qA} = 1 - A$ to model the relationship among A (the fraction of the population ultimately infected by the disease), q (a measure of disease transmissibility), and ρ (a measure of the fraction of the population that is initially susceptible to infection). How does the outbreak size A change with an increase in ρ?

26. The speed of sound traveling through ocean water with salinity 35 parts per thousand has been modeled by the equation

$$C = 1449.2 + 4.6T - 0.055T^2 + 0.00029T^3 + 0.016D$$

where C is the speed of sound (in meters per second), T is the temperature (in degrees Celsius), and D is the depth below the ocean surface (in meters). A scuba diver began a leisurely dive into the ocean water; the diver's depth and the surrounding water temperature over time are recorded in the following graphs. Estimate the rate of change (with respect to time) of the speed of sound through the ocean water experienced by the diver 20 minutes into the dive. What are the units?

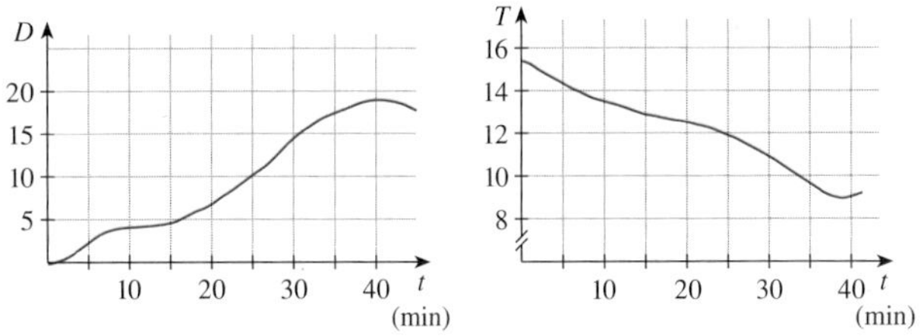

27. The pressure of 1 mole of an ideal gas is increasing at a rate of 0.05 kPa/s and the temperature is increasing at a rate of 0.15 K/s. Use the equation in Example 3 to find the rate of change of the volume when the pressure is 20 kPa and the temperature is 320 K.

28. If a sound with frequency f_s is produced by a source traveling along a line with speed v_s and an observer is traveling with speed v_o along the same line from the opposite direction toward the source, then the frequency of the sound heard by the observer is

$$f_o = \left(\frac{c + v_o}{c - v_s}\right) f_s$$

where c is the speed of sound, about 332 m/s. (This is the *Doppler effect.*) Suppose that, at a particular moment, you are in a train traveling at 34 m/s and accelerating at 1.2 m/s^2. A train is approaching you from the opposite direction on the other track at 40 m/s, accelerating at 1.4 m/s^2, and sounds its whistle, which has a frequency of 460 Hz. At that instant, what is the perceived frequency that you hear and how fast is it changing?

29. Suppose that the equation $F(x, y, z) = 0$ implicitly defines each of the three variables x, y, and z as functions of the other two: $z = f(x, y)$, $y = g(x, z)$, $x = h(y, z)$. If F is differentiable and F_x, F_y, and F_z are all nonzero, show that

$$\frac{\partial z}{\partial x}\frac{\partial x}{\partial y}\frac{\partial y}{\partial z} = -1$$

30. Equation 4 is a formula for the derivative dy/dx of a function defined implicitly by an equation $F(x, y) = 0$, provided that F is differentiable and $F_y \neq 0$. Prove that if F has continuous second derivatives, then a formula for the second derivative of y is

$$\frac{d^2y}{dx^2} = -\frac{F_{xx}F_y^2 - 2F_{xy}F_xF_y + F_{yy}F_x^2}{F_y^3}$$

9.5 Directional Derivatives and the Gradient Vector

The weather map in Figure 1 shows a contour map of the temperature function $T(x, y)$ for the states of California and Nevada at 3:00 PM on a day in October. The level curves, or isothermals, join locations with the same temperature. The partial derivative T_x at a location such as Reno is the rate of change of temperature with respect to distance if we travel east from Reno; T_y is the rate of change of temperature if we travel north. But what if we want to know the rate of change of temperature when we travel southeast (toward Las Vegas), or in some other direction? In this section we introduce a type of derivative, called a *directional derivative*, that enables us to find the rate of change of a function of two variables in any direction.

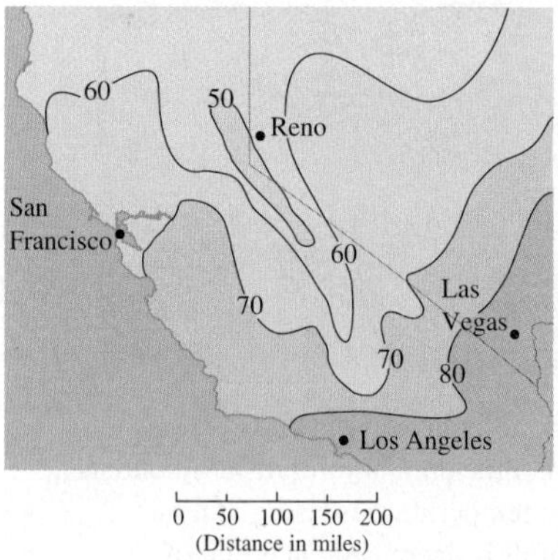

FIGURE 1

Directional Derivatives

Recall that if $z = f(x, y)$, then the partial derivatives f_x and f_y are defined as

(1)
$$f_x(x_0, y_0) = \lim_{h \to 0} \frac{f(x_0 + h, y_0) - f(x_0, y_0)}{h}$$

$$f_y(x_0, y_0) = \lim_{h \to 0} \frac{f(x_0, y_0 + h) - f(x_0, y_0)}{h}$$

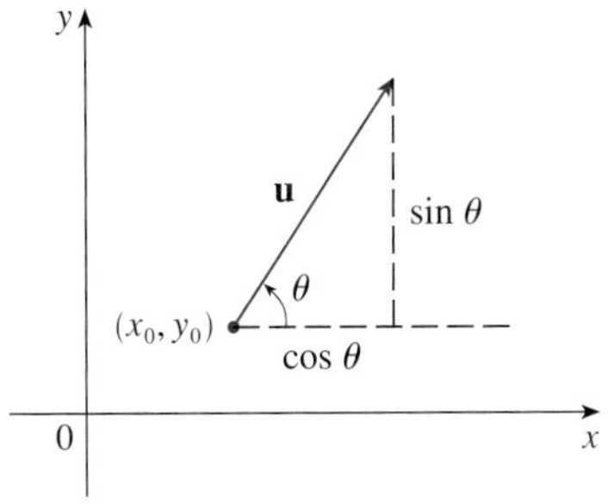

FIGURE 2
A unit vector
$\mathbf{u} = [a, b] = [\cos\theta, \sin\theta]$

and represent the rates of change of z in the x- and y-directions, that is, in the directions of the unit vectors $\mathbf{i} = [1, 0]$ and $\mathbf{j} = [0, 1]$.

Suppose that we now wish to find the rate of change of z at (x_0, y_0) in the direction of an arbitrary unit vector $\mathbf{u} = [a, b]$. (See Figure 2.) To do this we consider the surface S with the equation $z = f(x, y)$ (the graph of f) and we let $z_0 = f(x_0, y_0)$. Then the point $P(x_0, y_0, z_0)$ lies on S. The vertical plane that passes through P in the direction of $\mathbf{u}$ intersects S in a curve C. (See Figure 3.) The slope of the tangent line T to C at the point P is the rate of change of z in the direction of $\mathbf{u}$.

TEC Visual 9.5A animates Figure 3 by rotating $\mathbf{u}$ and therefore T.

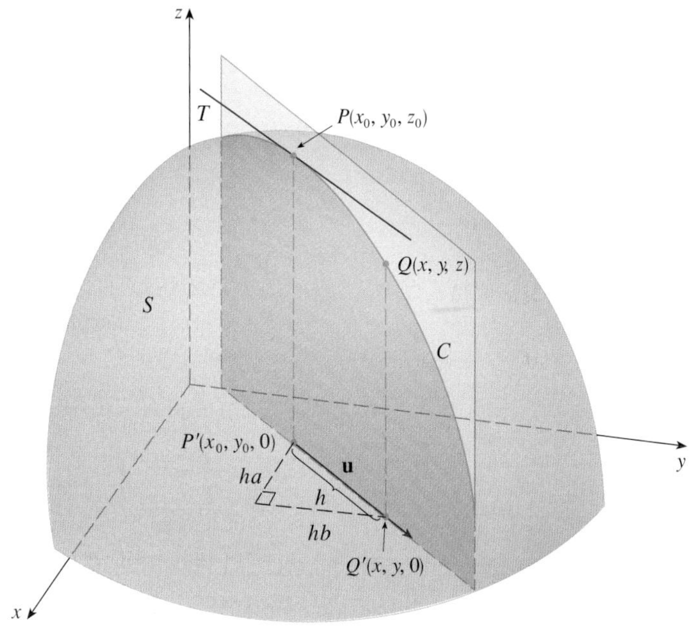

FIGURE 3

If $Q(x, y, z)$ is another point on C and P', Q' are the projections of P, Q onto the xy-plane, then the vector $\overrightarrow{P'Q'}$ is parallel to $\mathbf{u}$ and so

$$\overrightarrow{P'Q'} = h\mathbf{u} = [ha, hb]$$

for some scalar h. Therefore $x - x_0 = ha$, $y - y_0 = hb$, so $x = x_0 + ha$, $y = y_0 + hb$, and

$$\frac{\Delta z}{h} = \frac{z - z_0}{h} = \frac{f(x_0 + ha, y_0 + hb) - f(x_0, y_0)}{h}$$

If we take the limit as $h \to 0$, we obtain the rate of change of z (with respect to distance) in the direction of $\mathbf{u}$, which is called the directional derivative of f in the direction of $\mathbf{u}$.

(2) Definition The **directional derivative** of f at (x_0, y_0) in the direction of a unit vector $\mathbf{u} = [a, b]$ is

$$D_{\mathbf{u}} f(x_0, y_0) = \lim_{h \to 0} \frac{f(x_0 + ha, y_0 + hb) - f(x_0, y_0)}{h}$$

if this limit exists.

By comparing Definition 2 with Equations 1 we see that if $\mathbf{u} = \mathbf{i} = [1, 0]$, then $D_{\mathbf{i}} f = f_x$ and if $\mathbf{u} = \mathbf{j} = [0, 1]$, then $D_{\mathbf{j}} f = f_y$. In other words, the partial derivatives of f with respect to x and y are just special cases of the directional derivative.

EXAMPLE 1 | Use the weather map in Figure 1 to estimate the value of the directional derivative of the temperature function at Reno in the southeasterly direction.

SOLUTION The unit vector directed toward the southeast is $\mathbf{u} = \left[1/\sqrt{2}, -1/\sqrt{2}\right]$, but we won't need to use this expression. We start by drawing a line through Reno toward the southeast (see Figure 4).

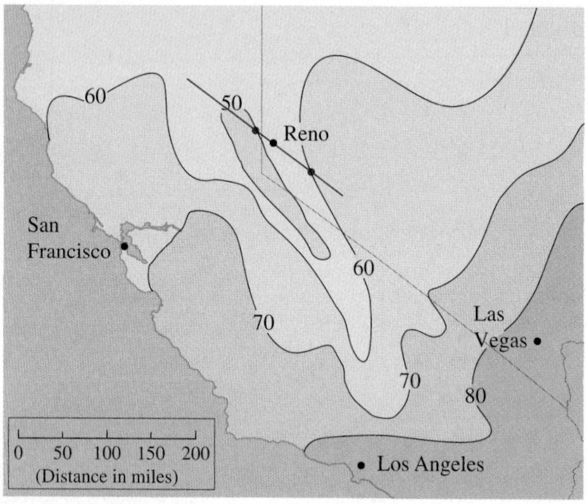

FIGURE 4

We approximate the directional derivative $D_{\mathbf{u}} T$ by the average rate of change of the temperature between the points where this line intersects the isothermals $T = 50$ and $T = 60$. The temperature at the point southeast of Reno is $T = 60°\text{F}$ and the temperature at the point northwest of Reno is $T = 50°\text{F}$. The distance between these points looks to be about 75 miles. So the rate of change of the temperature in the southeasterly direction is

$$D_{\mathbf{u}} T \approx \frac{60 - 50}{75} = \frac{10}{75} \approx 0.13°\text{F/mi}$$ ■

When we compute the directional derivative of a function defined by a formula, we generally use the following theorem.

> **(3) Theorem** If f is a differentiable function of x and y, then f has a directional derivative in the direction of any unit vector $\mathbf{u} = [a, b]$ and
>
> $$D_{\mathbf{u}} f(x, y) = f_x(x, y)\, a + f_y(x, y)\, b$$

PROOF If we define a function g of the single variable h by

$$g(h) = f(x_0 + ha, y_0 + hb)$$

then, by the definition of a derivative, we have

(4)
$$g'(0) = \lim_{h \to 0} \frac{g(h) - g(0)}{h} = \lim_{h \to 0} \frac{f(x_0 + ha, y_0 + hb) - f(x_0, y_0)}{h}$$
$$= D_{\mathbf{u}} f(x_0, y_0)$$

On the other hand, we can write $g(h) = f(x, y)$, where $x = x_0 + ha$, $y = y_0 + hb$, so the Chain Rule (Theorem 9.4.2) gives

$$g'(h) = \frac{\partial f}{\partial x}\frac{dx}{dh} + \frac{\partial f}{\partial y}\frac{dy}{dh} = f_x(x, y)\,a + f_y(x, y)\,b$$

If we now put $h = 0$, then $x = x_0$, $y = y_0$, and

(5) $$g'(0) = f_x(x_0, y_0)\,a + f_y(x_0, y_0)\,b$$

Comparing Equations 4 and 5, we see that

$$D_{\mathbf{u}} f(x_0, y_0) = f_x(x_0, y_0)\,a + f_y(x_0, y_0)\,b \qquad \blacksquare$$

If the unit vector $\mathbf{u}$ makes an angle θ with the positive x-axis (as in Figure 2), then we can write $\mathbf{u} = [\cos\theta, \sin\theta]$ and the formula in Theorem 3 becomes

(6) $$D_{\mathbf{u}} f(x, y) = f_x(x, y)\cos\theta + f_y(x, y)\sin\theta$$

The directional derivative $D_{\mathbf{u}} f(1, 2)$ in Example 2 represents the rate of change of z in the direction of $\mathbf{u}$. This is the slope of the tangent line to the curve of intersection of the surface $z = x^3 - 3xy + 4y^2$ and the vertical plane through (1, 2, 0) in the direction of $\mathbf{u}$ shown in Figure 5.

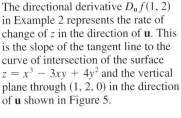

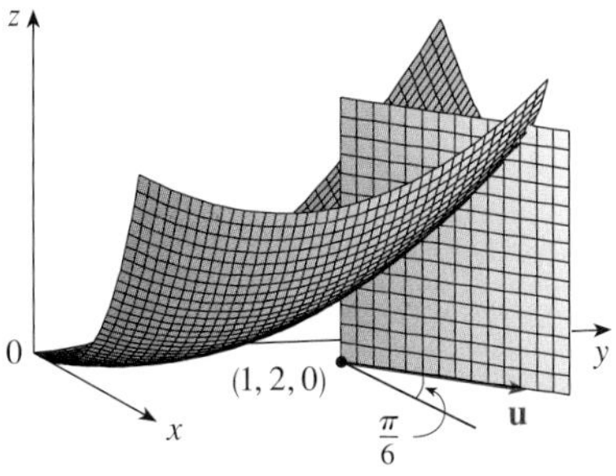

FIGURE 5

EXAMPLE 2 | Find the directional derivative $D_{\mathbf{u}} f(x, y)$ if

$$f(x, y) = x^3 - 3xy + 4y^2$$

and $\mathbf{u}$ is the unit vector given by angle $\theta = \pi/6$. What is $D_{\mathbf{u}} f(1, 2)$?

SOLUTION Formula 6 gives

$$\begin{aligned} D_{\mathbf{u}} f(x, y) &= f_x(x, y)\cos\frac{\pi}{6} + f_y(x, y)\sin\frac{\pi}{6} \\ &= (3x^2 - 3y)\frac{\sqrt{3}}{2} + (-3x + 8y)\frac{1}{2} \\ &= \tfrac{1}{2}\left[3\sqrt{3}\,x^2 - 3x + \left(8 - 3\sqrt{3}\right)y\right] \end{aligned}$$

Therefore

$$D_{\mathbf{u}} f(1, 2) = \tfrac{1}{2}\left[3\sqrt{3}(1)^2 - 3(1) + \left(8 - 3\sqrt{3}\right)(2)\right] = \frac{13 - 3\sqrt{3}}{2} \qquad \blacksquare$$

The Gradient Vector

Notice from Theorem 3 that the directional derivative of a differentiable function can be written as the dot product of two vectors:

(7) $$\begin{aligned} D_{\mathbf{u}} f(x, y) &= f_x(x, y)\,a + f_y(x, y)\,b \\ &= [f_x(x, y), f_y(x, y)] \cdot [a, b] \\ &= [f_x(x, y), f_y(x, y)] \cdot \mathbf{u} \end{aligned}$$

The first vector in this dot product occurs not only in computing directional derivatives but in many other contexts as well. So we give it a special name (the *gradient* of f) and a special notation (**grad** f or ∇f, which is read "del f").

(8) Definition If f is a function of two variables x and y, then the **gradient** of f is the vector function ∇f defined by

$$\nabla f(x, y) = [f_x(x, y), f_y(x, y)]$$

EXAMPLE 3 | If $f(x, y) = \sin x + e^{xy}$, then

$$\nabla f(x, y) = [f_x, f_y] = [\cos x + ye^{xy}, xe^{xy}]$$

and $$\nabla f(0, 1) = [2, 0]$$ ■

With this notation for the gradient vector, we can rewrite the expression (7) for the directional derivative of a differentiable function as

(9) $$D_{\mathbf{u}} f(x, y) = \nabla f(x, y) \cdot \mathbf{u}$$

This expresses the directional derivative in the direction of **u** as the scalar projection of the gradient vector onto **u**.

The gradient vector $\nabla f(2, -1)$ in Example 4 is shown in Figure 6 with initial point $(2, -1)$. Also shown is the vector **v** that gives the direction of the directional derivative. Both of these vectors are superimposed on a contour plot of the graph of f.

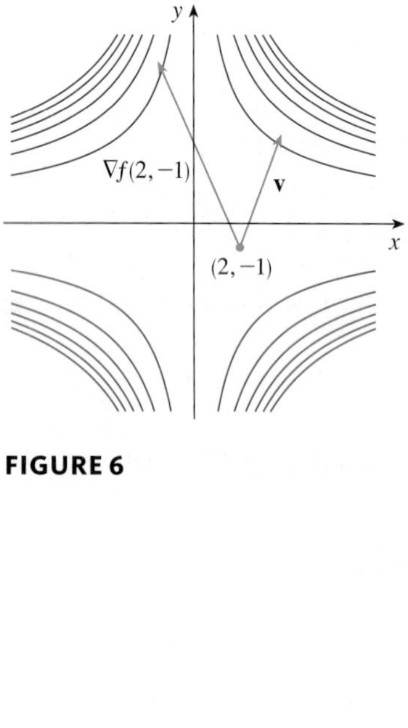

FIGURE 6

EXAMPLE 4 | Find the directional derivative of the function $f(x, y) = x^2y^3 - 4y$ at the point $(2, -1)$ in the direction of the vector $\mathbf{v} = [2, 5]$.

SOLUTION We first compute the gradient vector at $(2, -1)$:

$$\nabla f(x, y) = [2xy^3, 3x^2y^2 - 4]$$

$$\nabla f(2, -1) = [-4, 8]$$

Note that **v** is not a unit vector, but since $|\mathbf{v}| = \sqrt{29}$, the unit vector in the direction of **v** is

$$\mathbf{u} = \frac{\mathbf{v}}{|\mathbf{v}|} = \left[\frac{2}{\sqrt{29}}, \frac{5}{\sqrt{29}}\right]$$

Therefore, by Equation 9, we have

$$D_{\mathbf{u}} f(2, -1) = \nabla f(2, -1) \cdot \mathbf{u} = [-4, 8] \cdot \left[\frac{2}{\sqrt{29}}, \frac{5}{\sqrt{29}}\right]$$

$$= \frac{-4 \cdot 2 + 8 \cdot 5}{\sqrt{29}} = \frac{32}{\sqrt{29}}$$ ■

EXAMPLE 5 | BB **Snake reversals and stripes** In a study of the survivorship of juvenile garter snakes, a researcher[1] arrived at the model

$$F = 4.2 + 0.008R + 0.102S + 0.017R^2 - 0.034S^2 - 0.268RS$$

where F is a measure of the fitness of the snake, R measures the number of reversals of direction during flight from a predator, and S measures the degree of stripedness in the

1. E. Brodie III, "Correlational Selection for Color Pattern and Antipredator Behavior in the Garter Snake *Thamnophis ordinoides*," *Evolution* 46 (1992): 1284–98.

color pattern of the snake. We have previously considered how F changes as R or S changes by computing the partial derivatives with respect to R and S (Exercise 9.2.48). How does F change when $R = 3$ and $S = 2$ if the phenotype changes so that R and S increase by equal amounts?

SOLUTION The gradient vector of F is

$$\nabla F(R, S) = [F_R, F_S] = [0.008 + 0.034R - 0.268S, 0.102 - 0.068S - 0.268R]$$

and when $R = 3$ and $S = 2$ this becomes

$$\nabla F(3, 2) = [-0.426, -0.838]$$

We want the derivative in the direction halfway between $\mathbf{i}$ and $\mathbf{j}$, that is, in the direction $\mathbf{v} = [1, 1]$. The unit vector in this direction is $\mathbf{u} = (1/\sqrt{2})[1, 1]$, so the directional derivative is

$$D_{\mathbf{u}}F(3, 2) = \nabla F(3, 2) \cdot \mathbf{u} = [-0.426, -0.838] \cdot \left[\frac{1}{\sqrt{2}}, \frac{1}{\sqrt{2}}\right]$$

$$= -\frac{1.264}{\sqrt{2}} \approx -0.894$$

This means that the fitness function at (3, 2) decreases at a rate of 0.894 units when R and S increase by equal amounts. ■

■ Maximizing the Directional Derivative

Suppose we have a function f of two or three variables and we consider all possible directional derivatives of f at a given point. These give the rates of change of f in all possible directions. We can then ask the questions: In which of these directions does f change fastest and what is the maximum rate of change? The answers are provided by the following theorem.

TEC Visual 9.5B provides visual confirmation of Theorem 10.

(10) Theorem If f is a differentiable function and (a, b) is in the domain of f, then the maximum value of the directional derivative $D_{\mathbf{u}}f(a, b)$ is $|\nabla f(a, b)|$ and it occurs when $\mathbf{u}$ has the same direction as the gradient vector $\nabla f(a, b)$.

PROOF From Equation 9 we have

$$D_{\mathbf{u}}f = \nabla f \cdot \mathbf{u} = |\nabla f||\mathbf{u}|\cos\theta = |\nabla f|\cos\theta$$

where θ is the angle between ∇f and $\mathbf{u}$. The maximum value of $\cos\theta$ is 1 and this occurs when $\theta = 0$. Therefore the maximum value of $D_{\mathbf{u}}f$ is $|\nabla f|$ and it occurs when $\theta = 0$, that is, when $\mathbf{u}$ has the same direction as ∇f. ■

EXAMPLE 6

(a) If $f(x, y) = xe^y$, find the rate of change of f at the point $P(2, 0)$ in the direction from P to $Q(\frac{1}{2}, 2)$.

(b) In what direction does f have the maximum rate of change? What is this maximum rate of change?

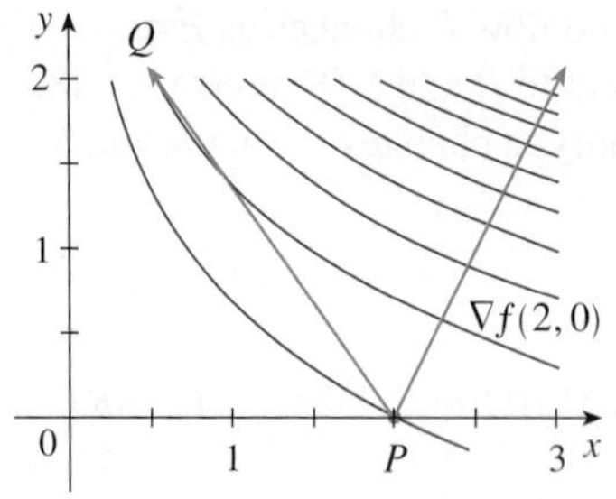

FIGURE 7

At (2, 0) the function in Example 6 increases fastest in the direction of the gradient vector $\nabla f(2, 0) = [1, 2]$. Notice from Figure 7 that this vector appears to be perpendicular to the level curve through (2, 0). Figure 8 shows the graph of f and the gradient vector.

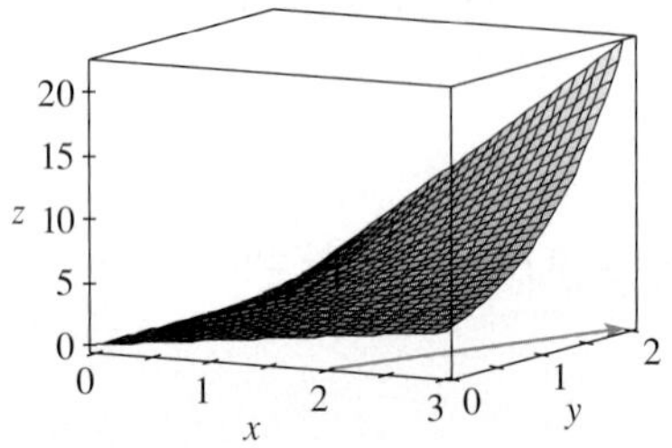

FIGURE 8

SOLUTION

(a) We first compute the gradient vector:

$$\nabla f(x, y) = [f_x, f_y] = [e^y, xe^y]$$

$$\nabla f(2, 0) = [1, 2]$$

The unit vector in the direction of $\overrightarrow{PQ} = [-1.5, 2]$ is $\mathbf{u} = \left[-\frac{3}{5}, \frac{4}{5}\right]$, so the rate of change of f in the direction from P to Q is

$$D_{\mathbf{u}} f(2, 0) = \nabla f(2, 0) \cdot \mathbf{u} = [1, 2] \cdot \left[-\tfrac{3}{5}, \tfrac{4}{5}\right]$$

$$= 1\left(-\tfrac{3}{5}\right) + 2\left(\tfrac{4}{5}\right) = 1$$

(b) According to Theorem 10, f increases fastest in the direction of the gradient vector $\nabla f(2, 0) = [1, 2]$. The maximum rate of change is

$$|\nabla f(2, 0)| = |[1, 2]| = \sqrt{5}$$ ■

Another geometric aspect of the gradient vector is illustrated in Figure 9:

> The gradient vector $\nabla f(a, b)$ is perpendicular to the level curve $f(x, y) = k$ that passes through the point (a, b).

Although we won't prove this fact, it is intuitively plausible because the values of f remain constant as we move along the level curve and so it seems reasonable that the values of f should change most quickly in the perpendicular direction, which we know is the direction of $\nabla f(a, b)$ by Theorem 10.

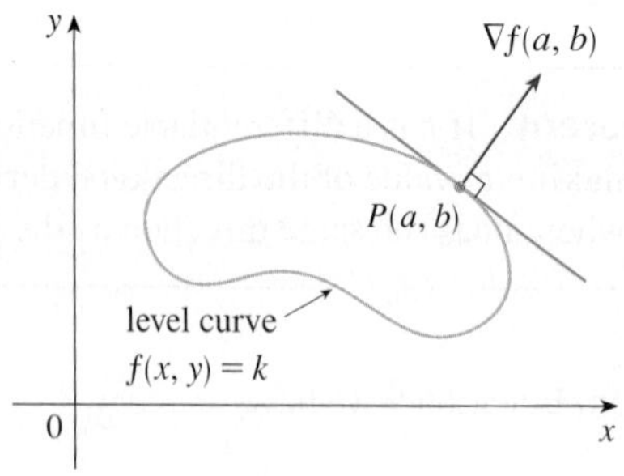

FIGURE 9

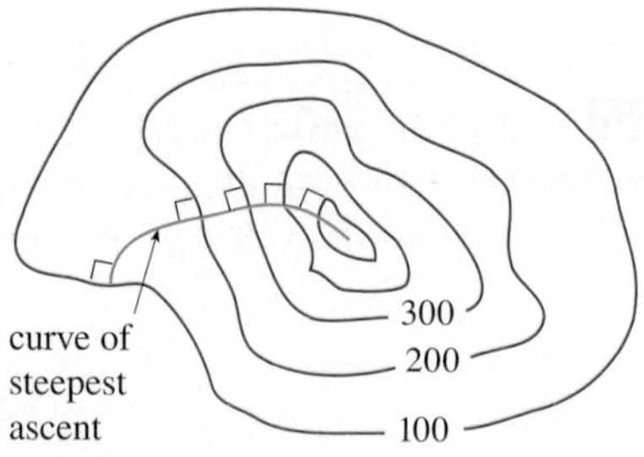

FIGURE 10

If we consider a topographical map of a hill and let $f(x, y)$ represent the height above sea level at a point with coordinates (x, y), then a curve of steepest ascent can be drawn as in Figure 10 by making it perpendicular to all of the contour lines. This phenomenon can also be noticed in Figure 9.1.10, where Lonesome Creek follows a curve of steepest descent.

EXERCISES 9.5

1. Level curves for barometric pressure (in millibars) are shown for 6:00 AM on November 10, 1998. A deep low with pressure 972 mb is moving over northeast Iowa. The distance along the red line from K (Kearney, Nebraska) to S (Sioux City, Iowa) is 300 km. Estimate the value of the directional derivative of the pressure function at Kearney in the direction of Sioux City. What are the units of the directional derivative?

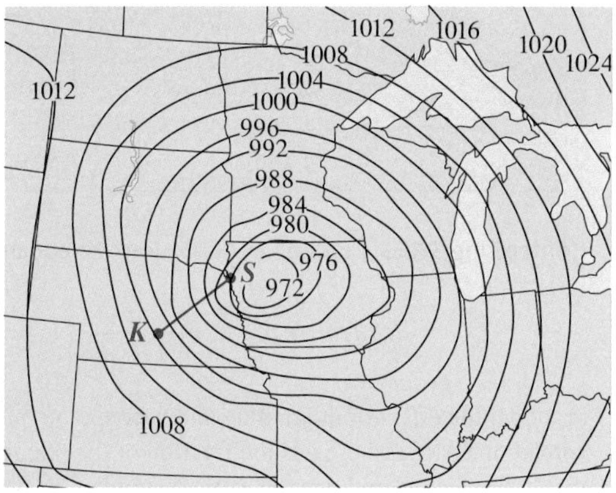

2. The contour map shows the average maximum temperature for November 2004 (in °C). Estimate the value of the directional derivative of this temperature function at Dubbo, New South Wales, in the direction of Sydney. What are the units?

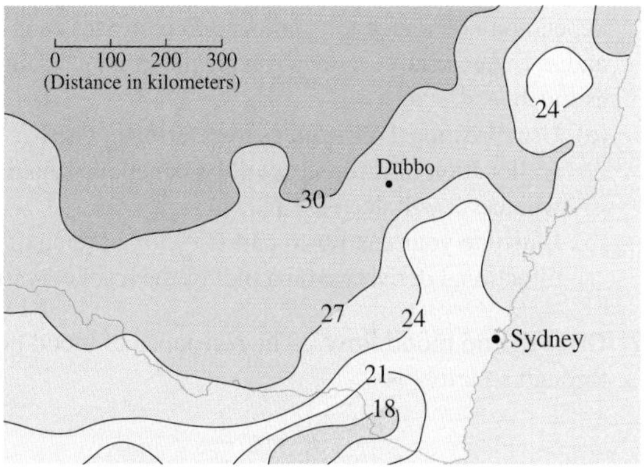

3. A table of values for the wind-chill index $W = f(T, v)$ is given in Exercise 9.2.3. Use the table to estimate the value of $D_{\mathbf{u}} f(-20, 30)$, where $\mathbf{u} = \left[1/\sqrt{2}, 1/\sqrt{2}\right]$.

4–6 Find the directional derivative of f at the given point in the direction indicated by the angle θ.

4. $f(x, y) = x^2y^3 - y^4$, $(2, 1)$, $\theta = \pi/4$

5. $f(x, y) = ye^{-x}$, $(0, 4)$, $\theta = 2\pi/3$

6. $f(x, y) = x \sin(xy)$, $(2, 0)$, $\theta = \pi/3$

7–10
(a) Find the gradient of f.
(b) Evaluate the gradient at the point P.
(c) Find the rate of change of f at P in the direction of the vector $\mathbf{u}$.

7. $f(x, y) = 5xy^2 - 4x^3y$, $P(1, 2)$, $\mathbf{u} = \left[\frac{5}{13}, \frac{12}{13}\right]$

8. $f(x, y) = y \ln x$, $P(1, -3)$, $\mathbf{u} = \left[-\frac{4}{5}, \frac{3}{5}\right]$

9. $f(x, y) = \sin(2x + 3y)$, $P(-6, 4)$, $\mathbf{u} = \left[\frac{1}{2}\sqrt{3}, -\frac{1}{2}\right]$

10. $f(x, y) = y^2/x$, $P(1, 2)$, $\mathbf{u} = \left[\frac{2}{3}, \frac{1}{3}\sqrt{5}\right]$

11–15 Find the directional derivative of the function at the given point in the direction of the vector $\mathbf{v}$.

11. $f(x, y) = 1 + 2x\sqrt{y}$, $(3, 4)$, $\mathbf{v} = [4, -3]$

12. $f(x, y) = \ln(x^2 + y^2)$, $(2, 1)$, $\mathbf{v} = [-1, 2]$

13. $g(p, q) = p^4 - p^2q^3$, $(2, 1)$, $\mathbf{v} = [1, 3]$

14. $g(r, s) = \tan^{-1}(rs)$, $(1, 2)$, $\mathbf{v} = [5, 10]$

15. $V(u, t) = e^{-ut}$, $(0, 3)$, $\mathbf{v} = [2, -1]$

16. Use the figure to estimate $D_{\mathbf{u}} f(2, 2)$.

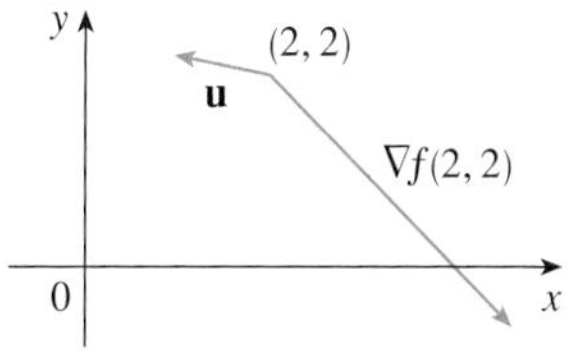

17. Find the directional derivative of $f(x, y) = \sqrt{xy}$ at $P(2, 8)$ in the direction of $Q(5, 4)$.

18. Find the directional derivative of $g(r, \theta) = e^{-r} \sin \theta$ at $P(0, \pi/3)$ in the direction of $Q(1, \pi/2)$.

19–22 Find the maximum rate of change of f at the given point and the direction in which it occurs.

19. $f(x, y) = y^2/x$, $(2, 4)$

20. $g(x, y) = x^2 - xy + y^2$, $(2, 1)$

21. $f(x, y) = \sin(xy)$, $(1, 0)$

22. $f(p, q) = qe^{-p} + pe^{-q}$, $(0, 0)$

23. (a) Show that a differentiable function f decreases most rapidly at (a, b) in the direction opposite to the gradient vector, that is, in the direction of $-\nabla f(a, b)$.

(b) Use the result of part (a) to find the direction in which the function $f(x, y) = x^4y - x^2y^3$ decreases fastest at the point $(2, -3)$.

24. Snake reversals and stripes In Example 5 we calculated $\nabla F(3, 2)$, the gradient vector of the snake fitness function F when $R = 3$ and $S = 2$. In which direction $\mathbf{u}$ is the directional derivative of F at $(3, 2)$ a maximum? Calculate and interpret the maximum value.

25. Find all points at which the direction of fastest change of the function $f(x, y) = x^2 + y^2 - 2x - 4y$ is $[1, 1]$.

26. Sketch the gradient vector $\nabla f(4, 6)$ for the function f whose level curves are shown. Explain how you chose the direction and length of this vector.

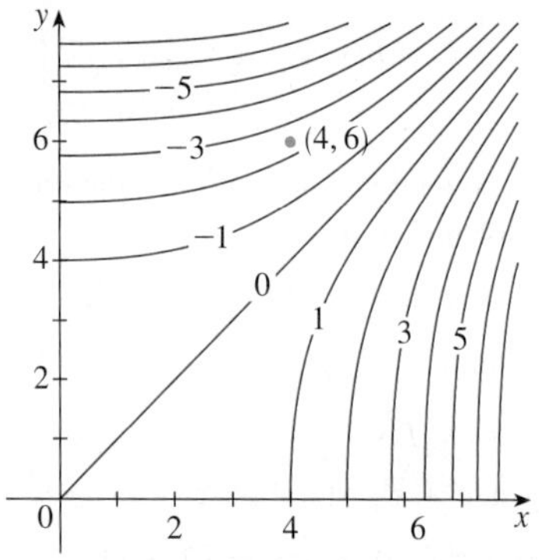

27. Near a buoy, the depth of a lake at the point with coordinates (x, y) is $z = 200 + 0.02x^2 - 0.001y^3$, where x, y, and z are measured in meters. A fisherman in a small boat starts at the point $(80, 60)$ and moves toward the buoy, which is located at $(0, 0)$. Is the water under the boat getting deeper or shallower when he departs? Explain.

28. Suppose you are climbing a hill whose shape is given by the equation $z = 1000 - 0.005x^2 - 0.01y^2$, where x, y, and z are measured in meters, and you are standing at a point with coordinates $(60, 40, 966)$. The positive x-axis points east and the positive y-axis points north.

(a) If you walk due south, will you start to ascend or descend? At what rate?

(b) If you walk northwest, will you start to ascend or descend? At what rate?

(c) In which direction is the slope largest? What is the rate of ascent in that direction? At what angle above the horizontal does the path in that direction begin?

29. Chemotaxis is the phenomenon in which organisms move toward or away from certain chemicals. Suppose that a bacterium finds food by always moving toward the highest concentration of glucose. If it starts at the point $(1, 2)$ in the xy-plane and the concentration of glucose, in certain units, is

$$f(x, y) = \frac{1}{1 + x^2 + y^2}$$

find the direction in which the bacterium will move initially.

30. Shown is a topographic map of Blue River Pine Provincial Park in British Columbia. Draw curves of steepest descent from point A (descending to Mud Lake) and from point B.

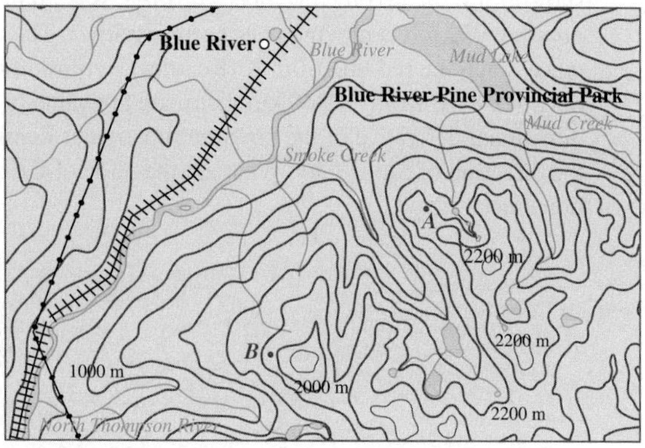

31. Controlling SARS Example 9.1.12 used the equation

$$R_0(d, v) = 5(1 - v)\frac{d}{1 + d}$$

to model the effect of quarantine and vaccination on the spread of SARS, where v is the fraction of the population that is vaccinated and d is the average number of days an infectious individual remains in the population (recall that R_0 is the average number of new infections generated by each infected individual). Suppose that $d = 2$, $v = 0.1$, and public health authorities are considering a reallocation of resources that will increase v (which decreases R_0) but that comes at the cost of also increasing d (which increases R_0). The net effect of this reallocation on R_0 will therefore depend on how much the reallocation increases each of v and d. Suppose any reallocation results in d increasing twice as much as v.

(a) Use directional derivatives to determine whether a small reallocation of resources will be beneficial (that is, whether it will decrease R_0).

(b) Illustrate your answer to part (a) with a sketch of the directional derivative on a plot of the level curves of R_0.

32. Growth and blood flow The resistance of blood flowing through an artery is

$$R = C\frac{L}{r^4}$$

where L and r are the length and radius of the artery and C is a positive constant. Both L and r increase during growth. Suppose $r = 0.1$ mm, $L = 1$ mm, and $C = 1$.

(a) Suppose the length increases 10 mm for every mm increase in radius during growth. Use a directional derivative to determine the rate at which the resistance of blood flow changes with respect to a unit of growth in the rL-plane.

(b) Use a directional derivative to determine how much faster the length of the artery can change relative to that of its radius before the rate of change of resistance with respect to growth will be positive.

(c) Illustrate your answers to parts (a) and (b) with a sketch of the directional derivatives on a plot of the level curves of R.

33. If $f(x, y) = xy$, find the gradient vector $\nabla f(3, 2)$ and use it to find the tangent line to the level curve $f(x, y) = 6$ at the point $(3, 2)$. Sketch the level curve, the tangent line, and the gradient vector.

34. If $g(x, y) = x^2 + y^2 - 4x$, find the gradient vector $\nabla g(1, 2)$ and use it to find the tangent line to the level curve $g(x, y) = 1$ at the point $(1, 2)$. Sketch the level curve, the tangent line, and the gradient vector.

35. The second directional derivative of $f(x, y)$ is

$$D_{\mathbf{u}}^2 f(x, y) = D_{\mathbf{u}}[D_{\mathbf{u}} f(x, y)]$$

If $f(x, y) = x^3 + 5x^2y + y^3$ and $\mathbf{u} = \left[\frac{3}{5}, \frac{4}{5}\right]$, calculate $D_{\mathbf{u}}^2 f(2, 1)$.

36. (a) If $\mathbf{u} = [a, b]$ is a unit vector and f has continuous second partial derivatives, show that

$$D_{\mathbf{u}}^2 f = f_{xx}a^2 + 2f_{xy}ab + f_{yy}b^2$$

(b) Find the second directional derivative of $f(x, y) = xe^{2y}$ in the direction of $\mathbf{v} = [4, 6]$.

9.6 Maximum and Minimum Values

As we saw in Chapter 4, one of the main uses of ordinary derivatives is in finding maximum and minimum values (extreme values). In this section we see how to use partial derivatives to locate maxima and minima of functions of two variables. In particular, in Example 6 we use Brodie's model to determine whether the survivorship function for snakes has a maximum value.

Look at the hills and valleys in the graph of f shown in Figure 1. There are two points (a, b) where f has a *local maximum*, that is, where $f(a, b)$ is larger than nearby values of $f(x, y)$. The larger of these two values is the *absolute maximum*. Likewise, f has two *local minima*, where $f(a, b)$ is smaller than nearby values. The smaller of these two values is the *absolute minimum*.

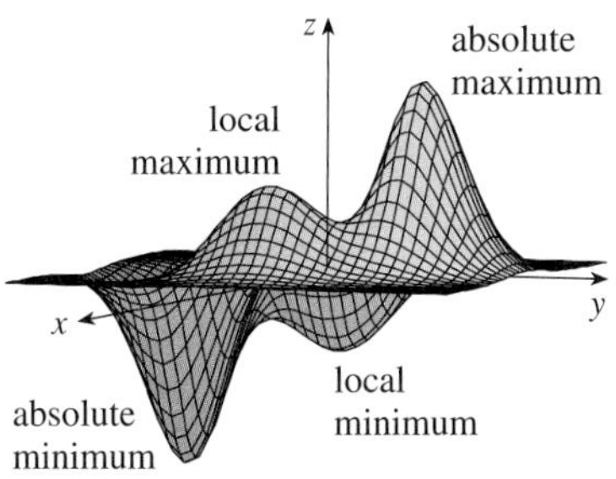

FIGURE 1

(1) Definition A function of two variables has a **local maximum** at (a, b) if $f(x, y) \leqslant f(a, b)$ when (x, y) is near (a, b). [This means that $f(x, y) \leqslant f(a, b)$ for all points (x, y) in some disk with center (a, b).] The number $f(a, b)$ is called a **local maximum value**. If $f(x, y) \geqslant f(a, b)$ when (x, y) is near (a, b), then f has a **local minimum** at (a, b) and $f(a, b)$ is a **local minimum value**.

If the inequalities in Definition 1 hold for all points (x, y) in the domain of f, then f has an **absolute maximum** (or **absolute minimum**) at (a, b). An absolute maximum or minimum is sometimes called a **global** maximum or minimum.

Notice that the conclusion of Theorem 2 can be stated in the notation of gradient vectors as $\nabla f(a, b) = \mathbf{0}$.

(2) Fermat's Theorem for Functions of Two Variables If f has a local maximum or minimum at (a, b) and the first-order partial derivatives of f exist there, then $f_x(a, b) = 0$ and $f_y(a, b) = 0$.

PROOF Let $g(x) = f(x, b)$. If f has a local maximum (or minimum) at (a, b), then g has a local maximum (or minimum) at a, so $g'(a) = 0$ by Fermat's Theorem for functions of one variable (see Theorem 4.1.4). But $g'(a) = f_x(a, b)$ (see Equation 9.2.1) and so $f_x(a, b) = 0$. Similarly, by applying Fermat's Theorem to the function $G(y) = f(a, y)$, we obtain $f_y(a, b) = 0$. ■

If we put $f_x(a, b) = 0$ and $f_y(a, b) = 0$ in the equation of a tangent plane (Equation 9.3.2), we get $z = z_0$. Thus the geometric interpretation of Theorem 2 is that if the graph of f has a tangent plane at a local maximum or minimum, then the tangent plane must be horizontal.

A point (a, b) is called a **critical point** (or *stationary point*) of f if $f_x(a, b) = 0$ and $f_y(a, b) = 0$, or if one of these partial derivatives does not exist. Theorem 2 says that if f has a local maximum or minimum at (a, b), then (a, b) is a critical point of f. However, as in single-variable calculus, not all critical points give rise to maxima or minima. At a critical point, a function could have a local maximum or a local minimum or neither.

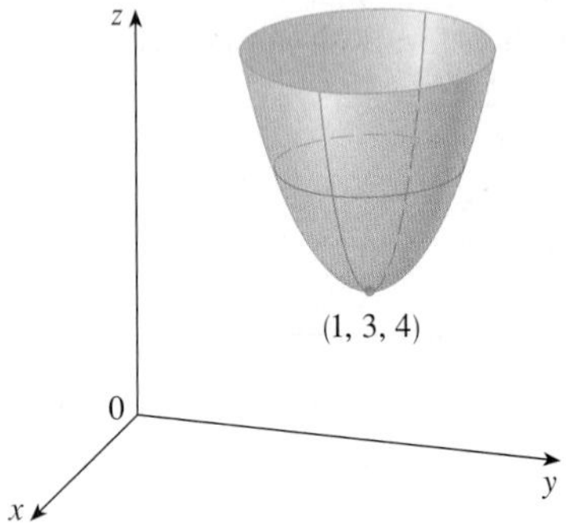

FIGURE 2
$z = x^2 + y^2 - 2x - 6y + 14$

EXAMPLE 1 | Let $f(x, y) = x^2 + y^2 - 2x - 6y + 14$. Then

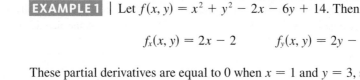

$$f_x(x, y) = 2x - 2 \qquad f_y(x, y) = 2y - 6$$

These partial derivatives are equal to 0 when $x = 1$ and $y = 3$, so the only critical point is $(1, 3)$. By completing the square, we find that

$$f(x, y) = 4 + (x - 1)^2 + (y - 3)^2$$

Since $(x - 1)^2 \geqslant 0$ and $(y - 3)^2 \geqslant 0$, we have $f(x, y) \geqslant 4$ for all values of x and y. Therefore $f(1, 3) = 4$ is a local minimum, and in fact it is the absolute minimum of f. This can be confirmed geometrically from the graph of f, which is the paraboloid with vertex $(1, 3, 4)$ shown in Figure 2. ■

EXAMPLE 2 | Find the extreme values of $f(x, y) = y^2 - x^2$.

SOLUTION Since $f_x = -2x$ and $f_y = 2y$, the only critical point is $(0, 0)$. Notice that for points on the x-axis we have $y = 0$, so $f(x, y) = -x^2 < 0$ (if $x \neq 0$). However, for points on the y-axis we have $x = 0$, so $f(x, y) = y^2 > 0$ (if $y \neq 0$). Thus every disk with center $(0, 0)$ contains points where f takes positive values as well as points where f takes negative values. Therefore $f(0, 0) = 0$ can't be an extreme value for f, so f has no extreme value. ■

Example 2 illustrates the fact that a function need not have a maximum or minimum value at a critical point. Figure 3 shows how this is possible. The graph of f is the hyperbolic paraboloid $z = y^2 - x^2$, which has a horizontal tangent plane ($z = 0$) at the origin. You can see that $f(0, 0) = 0$ is a maximum in the direction of the x-axis but a minimum in the direction of the y-axis. Near the origin the graph has the shape of a saddle and so $(0, 0)$ is called a *saddle point* of f.

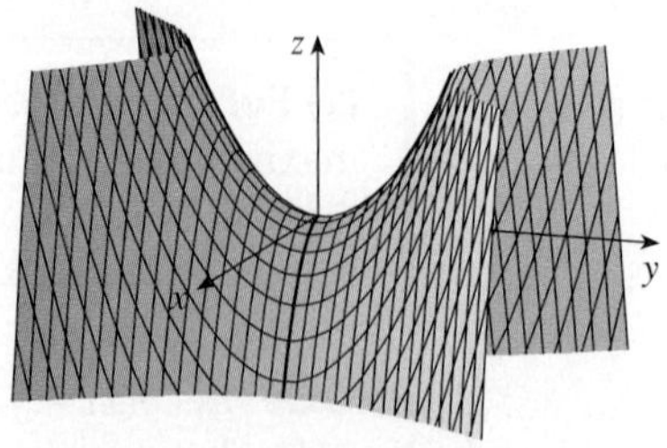

FIGURE 3
$z = y^2 - x^2$

We need to be able to determine whether or not a function has an extreme value at a critical point. The following test, which is proved in Appendix E, is analogous to the Second Derivative Test for functions of one variable.

(3) Second Derivatives Test Suppose the second partial derivatives of f are continuous on a disk with center (a, b), and suppose that $f_x(a, b) = 0$ and $f_y(a, b) = 0$ [that is, (a, b) is a critical point of f]. Let

$$D = D(a, b) = f_{xx}(a, b) f_{yy}(a, b) - [f_{xy}(a, b)]^2$$

(a) If $D > 0$ and $f_{xx}(a, b) > 0$, then $f(a, b)$ is a local minimum.
(b) If $D > 0$ and $f_{xx}(a, b) < 0$, then $f(a, b)$ is a local maximum.
(c) If $D < 0$, then $f(a, b)$ is not a local maximum or minimum.

NOTE 1 In case (c) the point (a, b) is called a **saddle point** of f and the graph of f crosses its tangent plane at (a, b).

NOTE 2 If $D = 0$, the test gives no information: f could have a local maximum or local minimum at (a, b), or (a, b) could be a saddle point of f.

NOTE 3 To remember the formula for D, it's helpful to write it as a determinant:

$$D = \begin{vmatrix} f_{xx} & f_{xy} \\ f_{yx} & f_{yy} \end{vmatrix} = f_{xx} f_{yy} - (f_{xy})^2$$

EXAMPLE 3 | Find the local maximum and minimum values and saddle points of $f(x, y) = x^4 + y^4 - 4xy + 1$.

SOLUTION We first locate the critical points:

$$f_x = 4x^3 - 4y \qquad f_y = 4y^3 - 4x$$

Setting these partial derivatives equal to 0, we obtain the equations

$$x^3 - y = 0 \qquad \text{and} \qquad y^3 - x = 0$$

To solve these equations we substitute $y = x^3$ from the first equation into the second one. This gives

$$0 = x^9 - x = x(x^8 - 1) = x(x^4 - 1)(x^4 + 1) = x(x^2 - 1)(x^2 + 1)(x^4 + 1)$$

so there are three real roots: $x = 0, 1, -1$. The three critical points are $(0, 0)$, $(1, 1)$, and $(-1, -1)$.

Next we calculate the second partial derivatives and $D(x, y)$:

$$f_{xx} = 12x^2 \qquad f_{xy} = -4 \qquad f_{yy} = 12y^2$$

$$D(x, y) = f_{xx} f_{yy} - (f_{xy})^2 = 144x^2y^2 - 16$$

Since $D(0, 0) = -16 < 0$, it follows from case (c) of the Second Derivatives Test that the origin is a saddle point; that is, f has no local maximum or minimum at $(0, 0)$. Since $D(1, 1) = 128 > 0$ and $f_{xx}(1, 1) = 12 > 0$, we see from case (a) of the test that $f(1, 1) = -1$ is a local minimum. Similarly, we have $D(-1, -1) = 128 > 0$ and $f_{xx}(-1, -1) = 12 > 0$, so $f(-1, -1) = -1$ is also a local minimum.

The graph of f is shown in Figure 4. ■

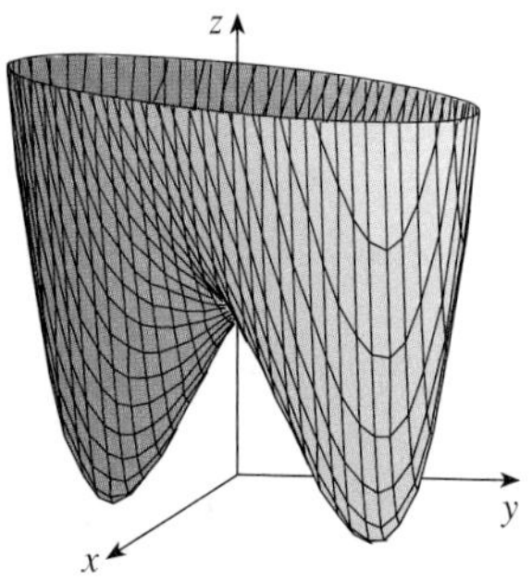

FIGURE 4
$z = x^4 + y^4 - 4xy + 1$

A contour map of the function f in Example 3 is shown in Figure 5. The level curves near $(1, 1)$ and $(-1, -1)$ are oval in shape and indicate that as we move away from $(1, 1)$ or $(-1, -1)$ in any direction the values of f are increasing. The level curves near $(0, 0)$, on the other hand, resemble hyperbolas. They reveal that as we move away from the origin (where the value of f is 1), the values of f decrease in some directions but increase in other directions. Thus the contour map suggests the presence of the minima and saddle point that we found in Example 3.

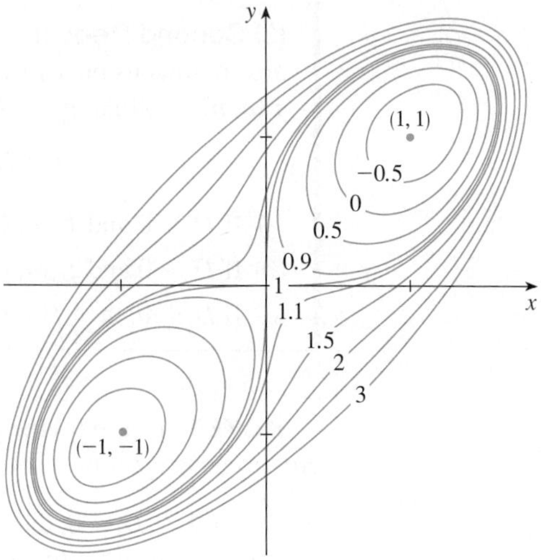

FIGURE 5

EXAMPLE 4 | Find the shortest distance from the point $(1, 0, -2)$ to the plane $x + 2y + z = 4$.

SOLUTION The distance from any point (x, y, z) to the point $(1, 0, -2)$ is

$$d = \sqrt{(x - 1)^2 + y^2 + (z + 2)^2}$$

TEC In Module 9.6 you can use contour maps to estimate the locations of critical points.

but if (x, y, z) lies on the plane $x + 2y + z = 4$, then $z = 4 - x - 2y$ and so we have $d = \sqrt{(x - 1)^2 + y^2 + (6 - x - 2y)^2}$. We can minimize d by minimizing the simpler expression

$$d^2 = f(x, y) = (x - 1)^2 + y^2 + (6 - x - 2y)^2$$

By solving the equations

$$f_x = 2(x - 1) - 2(6 - x - 2y) = 4x + 4y - 14 = 0$$

$$f_y = 2y - 4(6 - x - 2y) = 4x + 10y - 24 = 0$$

we find that the only critical point is $\left(\frac{11}{6}, \frac{5}{3}\right)$. Since $f_{xx} = 4$, $f_{xy} = 4$, and $f_{yy} = 10$, we have $D(x, y) = f_{xx} f_{yy} - (f_{xy})^2 = 24 > 0$ and $f_{xx} > 0$, so by the Second Derivatives Test f has a local minimum at $\left(\frac{11}{6}, \frac{5}{3}\right)$. Intuitively, we can see that this local minimum is actually an absolute minimum because there must be a point on the given plane that is closest to $(1, 0, -2)$. If $x = \frac{11}{6}$ and $y = \frac{5}{3}$, then

$$d = \sqrt{(x - 1)^2 + y^2 + (6 - x - 2y)^2} = \sqrt{\left(\tfrac{5}{6}\right)^2 + \left(\tfrac{5}{3}\right)^2 + \left(\tfrac{5}{6}\right)^2} = \tfrac{5}{6}\sqrt{6}$$

The shortest distance from $(1, 0, -2)$ to the plane $x + 2y + z = 4$ is $\frac{5}{6}\sqrt{6}$. ■

EXAMPLE 5 | A rectangular box without a lid is to be made from 12 m^2 of cardboard. Find the maximum volume of such a box.

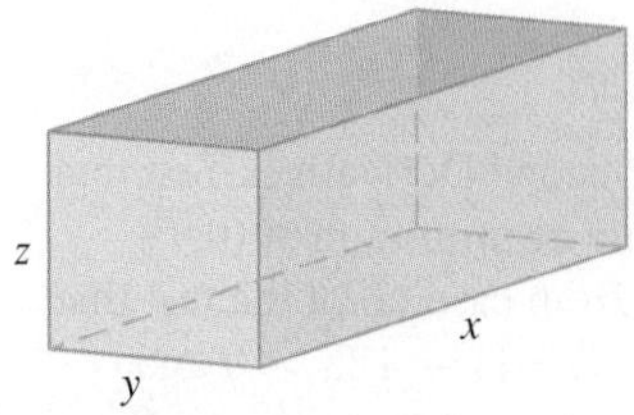

FIGURE 6

SOLUTION Let the length, width, and height of the box (in meters) be x, y, and z, as shown in Figure 6. Then the volume of the box is

$$V = xyz$$

We can express V as a function of just two variables x and y by using the fact that the area of the four sides and the bottom of the box is

$$2xz + 2yz + xy = 12$$

Solving this equation for z, we get $z = (12 - xy)/[2(x + y)]$, so the expression for V becomes

$$V = xy\frac{12 - xy}{2(x + y)} = \frac{12xy - x^2y^2}{2(x + y)}$$

We compute the partial derivatives:

$$\frac{\partial V}{\partial x} = \frac{y^2(12 - 2xy - x^2)}{2(x + y)^2} \qquad \frac{\partial V}{\partial y} = \frac{x^2(12 - 2xy - y^2)}{2(x + y)^2}$$

If V is a maximum, then $\partial V/\partial x = \partial V/\partial y = 0$, but $x = 0$ or $y = 0$ gives $V = 0$, so we must solve the equations

$$12 - 2xy - x^2 = 0 \qquad 12 - 2xy - y^2 = 0$$

These imply that $x^2 = y^2$ and so $x = y$. (Note that x and y must both be positive in this problem.) If we put $x = y$ in either equation, we get $12 - 3x^2 = 0$, which gives $x = 2$, $y = 2$, and $z = (12 - 2 \cdot 2)/[2(2 + 2)] = 1$.

We could use the Second Derivatives Test to show that this gives a local maximum of V, or we could simply argue from the physical nature of this problem that there must be an absolute maximum volume, which has to occur at a critical point of V, so it must occur when $x = 2$, $y = 2$, $z = 1$. Then $V = 2 \cdot 2 \cdot 1 = 4$, so the maximum volume of the box is 4 m^3. ■

EXAMPLE 6 | **BB** Snake reversals and stripes In Example 9.5.5 we investigated Brodie's model for the survivorship of young garter snakes:

$$F = 4.2 + 0.008R + 0.102S + 0.017R^2 - 0.034S^2 - 0.268RS$$

where F is a measure of the fitness of the snake, R measures the number of reversals of direction during flight from a predator, and S measures the degree of stripedness in the color pattern of the snake. Is there a maximum value of the fitness?

SOLUTION The partial derivatives of F are

$$F_R = 0.008 + 0.034R - 0.268S \qquad F_S = 0.102 - 0.068S - 0.268R$$

To determine the critical point we need to solve the system of linear equations

$$0.034R - 0.268S = -0.008$$

$$0.268R + 0.068S = 0.102$$

Solving this system, we get $R \approx 0.361$, $S \approx 0.076$. The second-order derivatives are

$$F_{RR} = 0.034 \qquad F_{RS} = -0.268 \qquad F_{SS} = -0.068$$

So $$D = F_{RR}F_{SS} - F_{RS}^2 = -0.074136$$

Because D is negative, the Second Derivatives Test tells us that the fitness function has no maximum or minimum value. This fact is illustrated by the graph and contour map of F in Figure 7. The function F has a saddle point at approximately $P(0.361, 0.076)$.

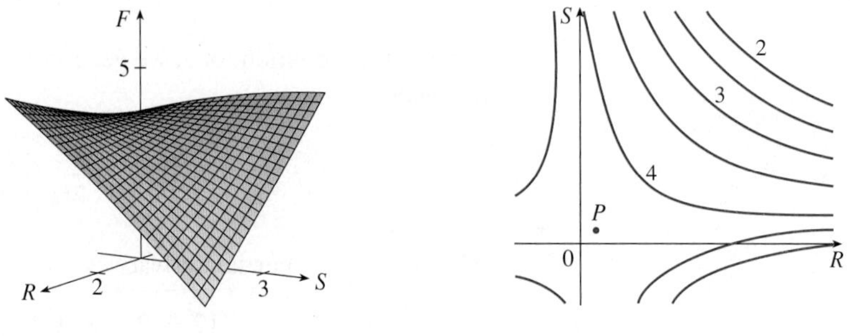

(a) (b)

FIGURE 7
Fitness function for garter snakes

Absolute Maximum and Minimum Values

For a function f of one variable, the Extreme Value Theorem says that if f is continuous on a closed interval $[a, b]$, then f has an absolute minimum value and an absolute maximum value. According to the Closed Interval Method in Section 4.1, we found these by evaluating f not only at the critical numbers but also at the endpoints a and b.

There is a similar situation for functions of two variables. Just as a closed interval contains its endpoints, a **closed set** in $\mathbb{R}^2$ is one that contains all its boundary points. [A boundary point of D is a point (a, b) such that every disk with center (a, b) contains points in D and also points not in D.] For instance, the disk

$$D = \{(x, y) \mid x^2 + y^2 \leq 1\}$$

which consists of all points on and inside the circle $x^2 + y^2 = 1$, is a closed set because it contains all of its boundary points (which are the points on the circle $x^2 + y^2 = 1$). But if even one point on the boundary curve were omitted, the set would not be closed. (See Figure 8.)

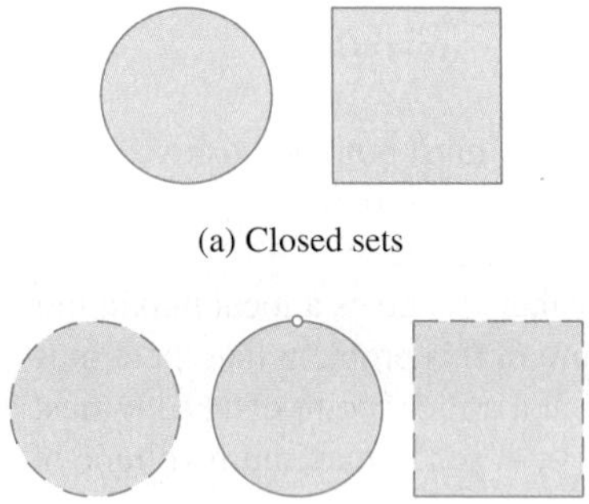

(a) Closed sets

(b) Sets that are not closed

FIGURE 8

A **bounded set** in $\mathbb{R}^2$ is one that is contained within some disk. In other words, it is finite in extent. Then, in terms of closed and bounded sets, we can state the following counterpart of the Extreme Value Theorem in two dimensions.

(4) Extreme Value Theorem for Functions of Two Variables If f is continuous on a closed, bounded set D in $\mathbb{R}^2$, then f attains an absolute maximum value $f(x_1, y_1)$ and an absolute minimum value $f(x_2, y_2)$ at some points (x_1, y_1) and (x_2, y_2) in D.

To find the extreme values guaranteed by Theorem 4, we note that, by Theorem 2, if f has an extreme value at (x_1, y_1), then (x_1, y_1) is either a critical point of f or a boundary point of D. Thus we have the following extension of the Closed Interval Method.

(5) To find the absolute maximum and minimum values of a continuous function f on a closed, bounded set D:

1. Find the values of f at the critical points of f in D.
2. Find the extreme values of f on the boundary of D.
3. The largest of the values from steps 1 and 2 is the absolute maximum value; the smallest of these values is the absolute minimum value.

EXAMPLE 7 | Find the absolute maximum and minimum values of the function $f(x, y) = x^2 - 2xy + 2y$ on the rectangle $D = \{(x, y) \mid 0 \leqslant x \leqslant 3, 0 \leqslant y \leqslant 2\}$.

SOLUTION Since f is a polynomial, it is continuous on the closed, bounded rectangle D, so Theorem 4 tells us there is both an absolute maximum and an absolute minimum. According to step 1 in (5), we first find the critical points. These occur when

$$f_x = 2x - 2y = 0 \qquad f_y = -2x + 2 = 0$$

so the only critical point is $(1, 1)$, and the value of f there is $f(1, 1) = 1$.

In step 2 we look at the values of f on the boundary of D, which consists of the four line segments L_1, L_2, L_3, L_4 shown in Figure 9. On L_1 we have $y = 0$ and

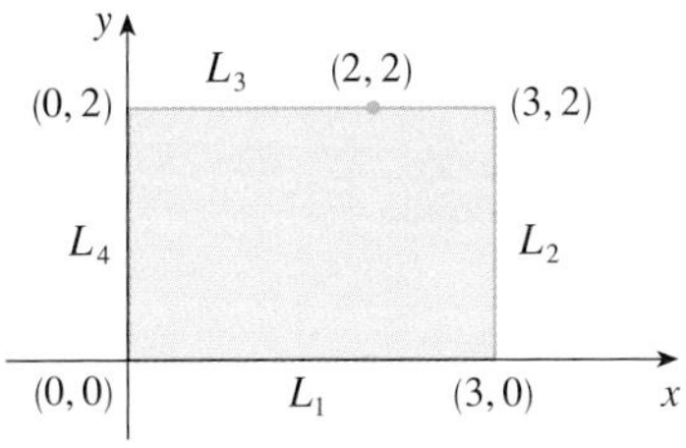

FIGURE 9

$$f(x, 0) = x^2 \qquad 0 \leqslant x \leqslant 3$$

This is an increasing function of x, so its minimum value is $f(0, 0) = 0$ and its maximum value is $f(3, 0) = 9$. On L_2 we have $x = 3$ and

$$f(3, y) = 9 - 4y \qquad 0 \leqslant y \leqslant 2$$

This is a decreasing function of y, so its maximum value is $f(3, 0) = 9$ and its minimum value is $f(3, 2) = 1$. On L_3 we have $y = 2$ and

$$f(x, 2) = x^2 - 4x + 4 \qquad 0 \leqslant x \leqslant 3$$

By the methods of Chapter 4, or simply by observing that $f(x, 2) = (x - 2)^2$, we see that the minimum value of this function is $f(2, 2) = 0$ and the maximum value is $f(0, 2) = 4$. Finally, on L_4 we have $x = 0$ and

$$f(0, y) = 2y \qquad 0 \leqslant y \leqslant 2$$

with maximum value $f(0, 2) = 4$ and minimum value $f(0, 0) = 0$. Thus, on the boundary, the minimum value of f is 0 and the maximum is 9.

In step 3 we compare these values with the value $f(1, 1) = 1$ at the critical point and conclude that the absolute maximum value of f on D is $f(3, 0) = 9$ and the absolute minimum value is $f(0, 0) = f(2, 2) = 0$. Figure 10 shows the graph of f.

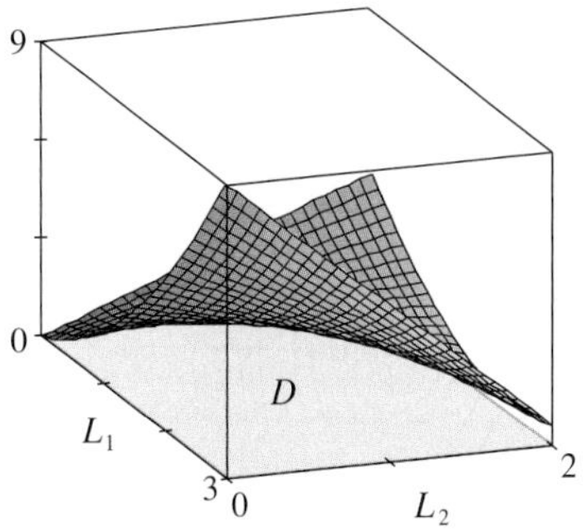

FIGURE 10
$f(x, y) = x^2 - 2xy + 2y$

■

EXERCISES 9.6

1. Suppose (1, 1) is a critical point of a function f with continuous second derivatives. In each case, what can you say about f?

(a) $f_{xx}(1, 1) = 4, \quad f_{xy}(1, 1) = 1, \quad f_{yy}(1, 1) = 2$
(b) $f_{xx}(1, 1) = 4, \quad f_{xy}(1, 1) = 3, \quad f_{yy}(1, 1) = 2$

2. Suppose (0, 2) is a critical point of a function g with continuous second derivatives. In each case, what can you say about g?

(a) $g_{xx}(0, 2) = -1, \quad g_{xy}(0, 2) = 6, \quad g_{yy}(0, 2) = 1$
(b) $g_{xx}(0, 2) = -1, \quad g_{xy}(0, 2) = 2, \quad g_{yy}(0, 2) = -8$
(c) $g_{xx}(0, 2) = 4, \quad g_{xy}(0, 2) = 6, \quad g_{yy}(0, 2) = 9$

3–4 Use the level curves in the figure to predict the location of the critical points of f and whether f has a saddle point or a local maximum or minimum at each critical point. Explain your reasoning. Then use the Second Derivatives Test to confirm your predictions.

3. $f(x, y) = 4 + x^3 + y^3 - 3xy$

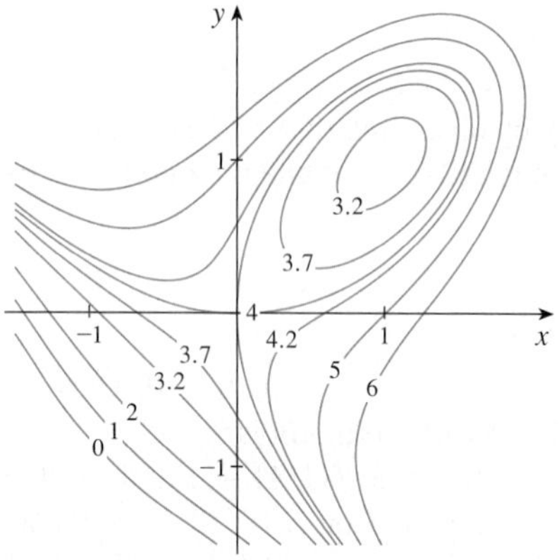

4. $f(x, y) = 3x - x^3 - 2y^2 + y^4$

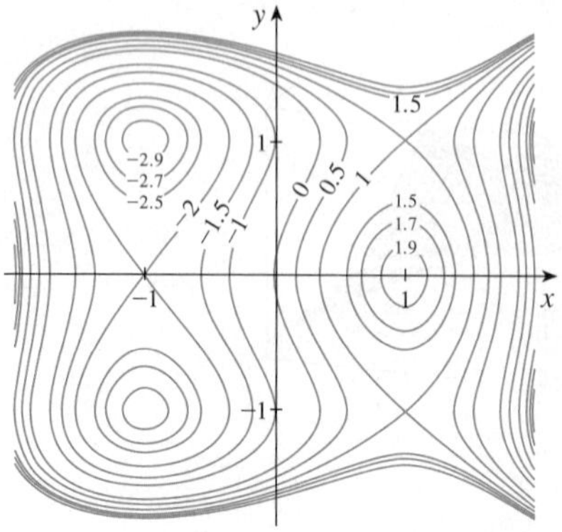

5–16 Find the local maximum and minimum values and saddle point(s) of the function. If you have three-dimensional graphing software, graph the function with a domain and viewpoint that reveal all the important aspects of the function.

5. $f(x, y) = x^2 + xy + y^2 + y$

6. $f(x, y) = x^3y + 12x^2 - 8y$

7. $f(x, y) = x^4 + y^4 - 4xy + 2$

8. $f(x, y) = xe^{-2x^2-2y^2}$

9. $f(x, y) = x^3 - 12xy + 8y^3$

10. $f(x, y) = xy + \dfrac{1}{x} + \dfrac{1}{y}$

11. $f(x, y) = e^x \cos y$

12. $f(x, y) = y \cos x$

13. $f(x, y) = (x^2 + y^2)e^{y^2-x^2}$

14. $f(x, y) = e^y(y^2 - x^2)$

15. $f(x, y) = y^2 - 2y \cos x, \quad -1 \leqslant x \leqslant 7$

16. $f(x, y) = \sin x \sin y, \quad -\pi < x < \pi, \quad -\pi < y < \pi$

17–20 Find the absolute maximum and minimum values of f on the set D.

17. $f(x, y) = 1 + 4x - 5y$, D is the closed triangular region with vertices (0, 0), (2, 0), and (3, 0)

18. $f(x, y) = 3 + xy - x - 2y$, D is the closed triangular region with vertices (1, 0), (5, 0), and (1, 4)

19. $f(x, y) = x^2 + y^2 + x^2y + 4$,
$D = \{(x, y) \mid |x| \leqslant 1, |y| \leqslant 1\}$

20. $f(x, y) = 4x + 6y - x^2 - y^2$,
$D = \{(x, y) \mid 0 \leqslant x \leqslant 4, 0 \leqslant y \leqslant 5\}$

21. For functions of one variable it is impossible for a continuous function to have two local maxima and no local minimum. But for functions of two variables such functions exist. Show that the function

$$f(x, y) = -(x^2 - 1)^2 - (x^2y - x - 1)^2$$

has only two critical points, but has local maxima at both of them. Then use a graphing device to produce a graph with a carefully chosen domain and viewpoint to see how this is possible.

22. If a function of one variable is continuous on an interval and has only one critical number, then a local maximum has to be an absolute maximum. But this is not true for functions of two variables. Show that the function

$$f(x, y) = 3xe^y - x^3 - e^{3y}$$

has exactly one critical point, and that f has a local maximum there that is not an absolute maximum. Then use a graphing device to produce a graph with a carefully chosen domain and viewpoint to see how this is possible.

23. Find the shortest distance from the point $(2, 1, -1)$ to the plane $x + y - z = 1$.

24. Find the point on the plane $x - y + z = 4$ that is closest to the point $(1, 2, 3)$.

25. Find the points on the cone $z^2 = x^2 + y^2$ that are closest to the point $(4, 2, 0)$.

26. Find the points on the surface $y^2 = 9 + xz$ that are closest to the origin.

27. Find three positive numbers whose sum is 100 and whose product is a maximum.

28. Find three positive numbers whose sum is 12 and the sum of whose squares is as small as possible.

29. Find the maximum volume of a rectangular box that is inscribed in a sphere of radius r.

30. Find the dimensions of the box with volume 1000 cm^3 that has minimal surface area.

31. Find the volume of the largest rectangular box in the first octant with three faces in the coordinate planes and one vertex in the plane $x + 2y + 3z = 6$.

32. Find the dimensions of the rectangular box with largest volume if the total surface area is given as 64 cm^2.

33. Find the dimensions of a rectangular box of maximum volume such that the sum of the lengths of its 12 edges is a constant c.

34. The base of an aquarium with given volume V is made of slate and the sides are made of glass. If slate costs five times as much (per unit area) as glass, find the dimensions of the aquarium that minimize the cost of the materials.

35. A cardboard box without a lid is to have a volume of $32{,}000 \text{ cm}^3$. Find the dimensions that minimize the amount of cardboard used.

36. **Crop yield** A model for the yield Y of an agricultural crop as a function of the nitrogen level N and phosphorus level P in the soil (measured in appropriate units) is

$$Y(N, P) = kNPe^{-N-P}$$

where k is a positive constant. What levels of nitrogen and phosphorus result in the best yield?

37. The **Shannon index** (sometimes called the Shannon-Wiener index or Shannon-Weaver index) is a measure of diversity in an ecosystem. For the case of three species, it is defined as

$$H = -p_1 \ln p_1 - p_2 \ln p_2 - p_3 \ln p_3$$

where p_i is the proportion of the ecosystem made up of species i.

(a) Express H as a function of two variables using the fact that $p_1 + p_2 + p_3 = 1$.

(b) What is the domain of H?

(c) Find the maximum value of H. For what values of p_1, p_2, p_3 does it occur?

38. A rectangular building is being designed to minimize heat loss. The east and west walls lose heat at a rate of 10 units/m^2 per day, the north and south walls at a rate of 8 units/m^2 per day, the floor at a rate of 1 unit/m^2 per day, and the roof at a rate of 5 units/m^2 per day. Each wall must be at least 30 m long, the height must be at least 4 m, and the volume must be exactly 4000 m^3.

(a) Find and sketch the domain of the heat loss as a function of the lengths of the sides.

(b) Find the dimensions that minimize heat loss. (Check both the critical points and the points on the boundary of the domain.)

(c) Could you design a building with even less heat loss if the restrictions on the lengths of the walls were removed?

39. **Infectious disease control** In Example 9.1.12 we considered a model for the spread of SARS:

$$R_0(d, v) = 5(1 - v)\frac{d}{1 + d}$$

where R_0 is the basic reproduction number, v is the fraction of the population that is vaccinated, and d is the average number of days that an infected individual remains in the population before being quarantined. If we restrict our attention to the case where $0 \leqslant d \leqslant 20$, what is the maximum value of R_0?

40. **Hardy-Weinberg Law** Three alleles (alternative versions of a gene) A, B, and O determine the four blood types A (AA or AO), B (BB or BO), O (OO), and AB. The Hardy-Weinberg Law states that the proportion of individuals in a population who carry two different alleles is

$$P = 2pq + 2pr + 2rq$$

where p, q, and r are the proportions of A, B, and O in the population. Use the fact that $p + q + r = 1$ to show that P is at most $\frac{2}{3}$.

Chapter 9 REVIEW

CONCEPT CHECK

1. (a) What is a function of two variables?
   (b) Describe two methods for visualizing a function of two variables.
2. What does

$$\lim_{(x, y) \to (a, b)} f(x, y) = L$$

mean? How can you show that such a limit does not exist?

3. (a) What does it mean to say that f is continuous at (a, b)?
   (b) If f is continuous on $\mathbb{R}^2$, what can you say about its graph?
4. (a) Write expressions for the partial derivatives $f_x(a, b)$ and $f_y(a, b)$ as limits.
   (b) How do you interpret $f_x(a, b)$ and $f_y(a, b)$ geometrically? How do you interpret them as rates of change?
   (c) If $f(x, y)$ is given by a formula, how do you calculate f_x and f_y?
5. What does Clairaut's Theorem say?
6. How do you find an equation for the tangent plane to a surface $z = f(x, y)$?
7. Define the linearization of f at (a, b). What is the corresponding linear approximation? What is the geometric interpretation of the linear approximation?
8. (a) What does it mean to say that f is differentiable at (a, b)?
   (b) How do you usually verify that f is differentiable?
9. State the Chain Rule for the case where $z = f(x, y)$ and x and y are functions of a variable t.
10. If z is defined implicitly as a function of x and y by an equation of the form $F(x, y, z) = 0$, how do you find $\partial z/\partial x$ and $\partial z/\partial y$?
11. (a) Write an expression as a limit for the directional derivative of f at (x_0, y_0) in the direction of a unit vector $\mathbf{u} = [a, b]$. How do you interpret it as a rate? How do you interpret it geometrically?
    (b) If f is differentiable, write an expression for $D_{\mathbf{u}} f(x_0, y_0)$ in terms of f_x and f_y.
12. (a) Define the gradient vector ∇f for a function f of two variables.
    (b) Express $D_{\mathbf{u}} f$ in terms of ∇f.
    (c) Explain the geometric significance of the gradient.
13. What do the following statements mean?
    (a) f has a local maximum at (a, b).
    (b) f has an absolute maximum at (a, b).
    (c) f has a local minimum at (a, b).
    (d) f has an absolute minimum at (a, b).
    (e) f has a saddle point at (a, b).
14. (a) If f has a local maximum at (a, b), what can you say about its partial derivatives at (a, b)?
    (b) What is a critical point of f?
15. State the Second Derivatives Test.
16. (a) What is a closed set in $\mathbb{R}^2$? What is a bounded set?
    (b) State the Extreme Value Theorem for functions of two variables.
    (c) How do you find the values that the Extreme Value Theorem guarantees?

Answers to the Concept Check can be found on the back endpapers.

TRUE-FALSE QUIZ

Determine whether the statement is true or false. If it is true, explain why. If it is false, explain why or give an example that disproves the statement.

1. $f_y(a, b) = \lim_{y \to b} \dfrac{f(a, y) - f(a, b)}{y - b}$
2. There exists a function f with continuous second-order partial derivatives such that $f_x(x, y) = x + y^2$ and $f_y(x, y) = x - y^2$.
3. $f_{xy} = \dfrac{\partial^2 f}{\partial x \, \partial y}$
4. $D_{\mathbf{u}} f(x, y, z) = f_z(x, y, z)$, where $\mathbf{u} = [0, 0, 1]$
5. If $f(x, y) \to L$ as $(x, y) \to (a, b)$ along every straight line through (a, b), then $\lim_{(x, y) \to (a, b)} f(x, y) = L$.
6. If $f_x(a, b)$ and $f_y(a, b)$ both exist, then f is differentiable at (a, b).
7. If f has a local minimum at (a, b) and f is differentiable at (a, b), then $\nabla f(a, b) = \mathbf{0}$.
8. If f is a function, then

$$\lim_{(x, y) \to (2, 5)} f(x, y) = f(2, 5)$$

9. If $f(x, y) = \ln y$, then $\nabla f(x, y) = 1/y$.

10. If (2, 1) is a critical point of f and

$$f_{xx}(2, 1)\, f_{yy}(2, 1) < [f_{xy}(2, 1)]^2$$

then f has a saddle point at (2, 1).

11. If $f(x, y) = \sin x + \sin y$, then $-\sqrt{2} \leqslant D_{\mathbf{u}} f(x, y) \leqslant \sqrt{2}$.

12. If $f(x, y)$ has two local maxima, then f must have a local minimum.

EXERCISES

1–2 Find and sketch the domain of the function.

1. $f(x, y) = \ln(x + y + 1)$

2. $f(x, y) = \sqrt{4 - x^2 - y^2} + \sqrt{1 - x^2}$

3–4 Sketch the graph of the function.

3. $f(x, y) = 1 - y^2$

4. $f(x, y) = 1 - \frac{1}{5}x - \frac{1}{3}y$

5–6 Sketch several level curves of the function.

5. $f(x, y) = \ln(1 + x^2 + y^2)$ **6.** $f(x, y) = e^x + y$

7–8 Evaluate the limit or show that it does not exist.

7. $\displaystyle\lim_{(x, y)\to(1, 1)} \frac{2xy}{x^2 + 2y^2}$ **8.** $\displaystyle\lim_{(x, y)\to(0, 0)} \frac{2xy}{x^2 + 2y^2}$

9. A metal plate is situated in the xy-plane and occupies the rectangle $0 \leqslant x \leqslant 10$, $0 \leqslant y \leqslant 8$, where x and y are measured in meters. The temperature at the point (x, y) in the plate is $T(x, y)$, where T is measured in degrees Celsius. Temperatures at equally spaced points were measured and recorded in the table.

(a) Estimate the values of the partial derivatives $T_x(6, 4)$ and $T_y(6, 4)$. What are the units?

(b) Estimate the value of $D_{\mathbf{u}} T(6, 4)$, where $\mathbf{u} = \left[1/\sqrt{2}, 1/\sqrt{2}\right]$. Interpret your result.

(c) Estimate the value of $T_{xy}(6, 4)$.

x \ y	0	2	4	6	8
0	30	38	45	51	55
2	52	56	60	62	61
4	78	74	72	68	66
6	98	87	80	75	71
8	96	90	86	80	75
10	92	92	91	87	78

10. Find a linear approximation to the temperature function $T(x, y)$ in Exercise 9 near the point (6, 4). Then use it to estimate the temperature at the point (5, 3.8).

11–15 Find the first partial derivatives.

11. $f(x, y) = \sqrt{2x + y^2}$ **12.** $u = e^{-r} \sin 2\theta$

13. $g(u, v) = u \tan^{-1} v$ **14.** $w = \dfrac{x}{y - z}$

15. $T(p, q, r) = p \ln(q + e^r)$

16. Predators handling prey C. S. Holling proposed the following model for the number of prey P_e eaten by a predator during a fixed time period:

$$P_e(N, T_h) = \frac{aN}{1 + aT_hN}$$

where a is a positive constant called the attack rate, N is the prey density, and T_h is the handling time. (Handling refers to catching the prey, moving it, eating it, and digesting it.) Calculate the partial derivatives $\partial P_e/\partial N$ and $\partial P_e/\partial T_h$ and interpret them. Are they positive or negative? What do you conclude? Are your conclusions reasonable?

Source: Adapted from C. Holling, "Some Characteristics of Simple Types of Predation and Parasitism," *The Canadian Entomologist* 91 (1959): 385–98.

17. Tadpole predation In an experiment, the probability that a tadpole with mass m (in grams) will escape capture by predators is modeled as

$$P(m, T) = \frac{1}{1 + e^{3.222 - 31.669m + 0.083T}}$$

where T is the temperature of the water (in degrees Celsius).

(a) Calculate $\partial P/\partial m$ and $\partial P/\partial T$ when $m = 0.2$ g and $T = 20°C$.

(b) Will increasing the water temperature improve or worsen the tadpoles' chance of escape?

(c) As the tadpoles grow bigger, do they have a better chance of escape?

Source: Adapted from M. Anderson et al., "The Direct and Indirect Effects of Temperature on a Predator-Prey Relationship," *Canadian Journal of Zoology* 79 (2001): 1834–41.

18. The speed of sound traveling through ocean water is a function of temperature, salinity, and pressure. It has been modeled by the function

$$C = 1449.2 + 4.6T - 0.055T^2 + 0.00029T^3 + (1.34 - 0.01T)(S - 35) + 0.016D$$

where C is the speed of sound (in meters per second), T is the temperature (in degrees Celsius), S is the salinity (the concentration of salts in parts per thousand, which means the number of grams of dissolved solids per 1000 g of water), and D is the depth below the ocean surface (in meters). Compute $\partial C/\partial T$, $\partial C/\partial S$, and $\partial C/\partial D$ when $T = 10°\text{C}$, $S = 35$ parts per thousand, and $D = 100$ m. Explain the physical significance of these partial derivatives.

19–22 Find all second partial derivatives of f.

19. $f(x, y) = 4x^3 - xy^2$ **20.** $z = xe^{-2y}$

21. $f(x, y, z) = x^k y^l z^m$ **22.** $v = r\cos(s + 2t)$

---

23. If $z = xy + xe^{y/x}$, show that $x\dfrac{\partial z}{\partial x} + y\dfrac{\partial z}{\partial y} = xy + z$.

24. If $z = \sin(x + \sin t)$, show that

$$\frac{\partial z}{\partial x}\frac{\partial^2 z}{\partial x\,\partial t} = \frac{\partial z}{\partial t}\frac{\partial^2 z}{\partial x^2}$$

25–28 Find an equation of the tangent plane to the given surface at the specified point.

25. $z = 4x^2 - y^2 + 2y$, $(-1, 2, 4)$

26. $z = 9x^2 + y^2 + 6x - 3y + 5$, $(1, 2, 18)$

27. $z = \sqrt{4 - x^2 - 2y^2}$, $(1, -1, 1)$

28. $z = y\ln x$, $(1, 4, 0)$

---

29. Explain why the function $f(x, y) = xy + y^3$ is differentiable. Then find the linearization of f at the point $(2, 1)$.

30. Find the linear approximation of the function $f(x, y, z) = x^3\sqrt{y^2 + z^2}$ at the point $(2, 3, 4)$ and use it to estimate the number $(1.98)^3\sqrt{(3.01)^2 + (3.97)^2}$.

31. If $u = x^2y^3 + z^4$, where $x = p + 3p^2$, $y = pe^p$, and $z = p\sin p$, use the Chain Rule to find du/dp.

32. Body surface area A model for the surface area of a human body is given by the function

$$S = f(w, h) = 0.1091w^{0.425}h^{0.725}$$

where w is the weight (in pounds), h is the height (in inches), and S is measured in square feet. Both w and h depend on the age a of the person. Use the Chain Rule to find an expression for the rate of change of S with respect to a.

33. If $z = f(u, v)$, where $u = xy$, $v = y/x$, and f has continuous second partial derivatives, show that

$$x^2\frac{\partial^2 z}{\partial x^2} - y^2\frac{\partial^2 z}{\partial y^2} = -4uv\frac{\partial^2 z}{\partial u\,\partial v} + 2v\frac{\partial z}{\partial v}$$

34. If $\cos(xyz) = 1 + x^2y^2 + z^2$, find $\dfrac{\partial z}{\partial x}$ and $\dfrac{\partial z}{\partial y}$.

35. Find the gradient of the function $g(p, q) = pq^2e^{-pq}$.

36. (a) When is the directional derivative of f a maximum?
(b) When is it a minimum?
(c) When is it 0?
(d) When is it half of its maximum value?

37. Find the directional derivative of $f(x, y) = x^2e^{-y}$ at the point $(-2, 0)$ in the direction toward the point $(2, -3)$.

38. Find the maximum rate of change of $f(x, y) = x^2y + \sqrt{y}$ at the point $(2, 1)$. In which direction does it occur?

39. The contour map shows wind speed in knots during Hurricane Andrew on August 24, 1992. Use it to estimate the value of the directional derivative of the wind speed at Homestead, Florida, in the direction of the eye of the hurricane.

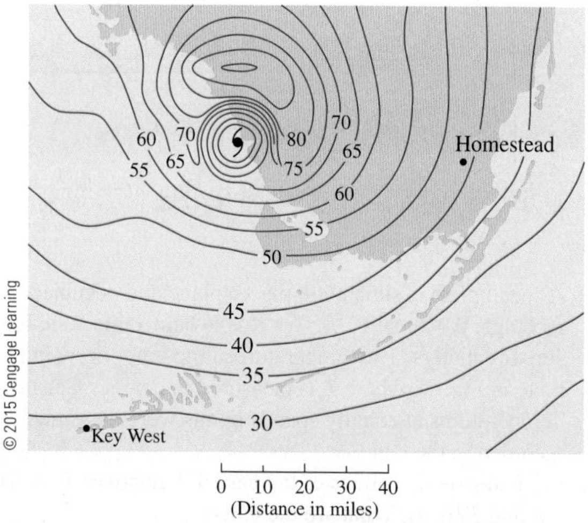

40–42 Find the local maximum and minimum values and saddle points of the function. If you have three-dimensional graphing software, graph the function with a domain and viewpoint that reveal all the important aspects of the function.

40. $f(x, y) = x^3 - 6xy + 8y^3$

41. $f(x, y) = 3xy - x^2y - xy^2$ **42.** $f(x, y) = (x^2 + y)e^{y/2}$

43–44 Find the absolute maximum and minimum values of f on the set D.

43. $f(x, y) = 4xy^2 - x^2y^2 - xy^3$; D is the closed triangular region in the xy-plane with vertices $(0, 0)$, $(0, 6)$, and $(6, 0)$

44. $f(x, y) = e^{-x^2-y^2}(x^2 + 2y^2)$; D is the disk $x^2 + y^2 \leq 4$

Systems of Linear Differential Equations

10

Jellyfish, like those shown here, move by contracting part of their body to create a high-pressure water jet. We model this phenomenon in Exercises 10.1.34 and 10.3.1.

IN SECTIONS 7.5 AND 7.6 we introduced systems of differential equations. In this chapter we deal with a special kind of system called a system of linear differential equations. Such systems deserve special attention because they arise in a number of areas in the life sciences and because it is possible to develop a complete understanding of their behavior. As we will see in Section 10.4, they also form the foundation for analyzing systems of nonlinear differential equations.

10.1 Qualitative Analysis of Linear Systems

In this section we introduce systems of linear differential equations and give some examples of the kinds of behavior they can exhibit. Our focus is on two-dimensional systems.

■ Terminology

A two-dimensional **system of linear differential equations** has the form

$$\frac{dx_1}{dt} = a_{11}(t)x_1 + a_{12}(t)x_2 + g_1(t)$$

$$\frac{dx_2}{dt} = a_{21}(t)x_1 + a_{22}(t)x_2 + g_2(t)$$

where $x_1(t)$ and $x_2(t)$ are unknown functions, $a_{ij}(t)$ are coefficients, and $g_i(t)$ are functions of time (sometimes called forcing, or input, functions). In matrix notation the system can be written

(1) $$\frac{d\mathbf{x}}{dt} = A(t)\mathbf{x} + \mathbf{g}(t)$$

where $\mathbf{x}(t)$ is a column vector whose components are the unknown functions $x_1(t)$ and $x_2(t)$, $d\mathbf{x}/dt$ is a column vector whose components are dx_1/dt and dx_2/dt, $A(t)$ is a matrix whose entries are the coefficients, and $\mathbf{g}(t)$ is a column vector whose components are the forcing functions.

Using the terminology introduced in Section 7.1, we say that Equation 1 is a system of linear *first-order nonautonomous* differential equations. If A and $\mathbf{g}$ are independent of time, then Equation 1 is a system of linear first-order *autonomous* differential equations.

In this chapter we will study only autonomous differential equations. In fact, most of our considerations will also assume $\mathbf{g} = \mathbf{0}$. Such systems are called **homogeneous**. As shown in Exercise 23, any autonomous system of equations for which $\mathbf{g} \neq \mathbf{0}$ (which is called a **nonhomogeneous** system) can be reduced to a homogeneous system in which $\mathbf{g} = \mathbf{0}$ through a change of variables. Thus we consider systems of the form

$$\frac{dx_1}{dt} = a_{11}x_1 + a_{12}x_2$$

$$\frac{dx_2}{dt} = a_{21}x_1 + a_{22}x_2$$

or, in matrix notation,

(2) $$\frac{d\mathbf{x}}{dt} = A\mathbf{x}$$

EXAMPLE 1 | **Radioimmunotherapy** is a cancer treatment in which radioactive atoms are attached to tumor-specific antibody molecules and then injected into the bloodstream. The antibody molecules then attach only to tumor cells, where they deliver the cell-killing radioactivity. Mathematical models have been used to optimize this treatment. Let's use $x_1(t)$ and $x_2(t)$ to denote the amount of antibody (in μg) in the blood and the tumor, respectively, at time t (in minutes after the start of treatment). If a denotes the per unit rate of clearance from the blood, b the per unit rate of movement from the blood into the tumor, and c the per unit rate of clearance from the tumor, a simple model is[1]

$$\frac{dx_1}{dt} = -ax_1 - bx_1$$

$$\frac{dx_2}{dt} = bx_1 - cx_2$$

What is the matrix A for this model when it is written in the form of Equation 2?

SOLUTION Using $\mathbf{x}$ to denote the vector with components x_1 and x_2, we can write this system as Equation 2 with

$$A = \begin{bmatrix} -(a + b) & 0 \\ b & -c \end{bmatrix}$$ ■

EXAMPLE 2 | **Metapopulation dynamics** A population of deer mice is split into two patches through habitat fragmentation. The population in patch A reproduces at a per capita rate of 2, while that in patch B reproduces at a per capita rate of -1 (where time is measured in years). The reproductive rate in patch 2 is negative because the mortality rate in this patch is larger than the birth rate. The per capita movement rate from patch A to B is 3, and from patch B to A is 2. Write a system of two linear differential equations that describes the two patches.

SOLUTION Let's use $x_A(t)$ and $x_B(t)$ to denote the number of individuals in patches A and B, respectively. Individuals in patch A reproduce at a per capita rate of 2, and they leave the patch at a per capita rate of 3. At the same time, individuals migrate into the patch from patch B at a per capita rate of 2. Therefore,

$$\begin{aligned} \frac{dx_A}{dt} &= 2x_A - 3x_A + 2x_B \\ &= -x_A + 2x_B \end{aligned}$$

Similarly, individuals in patch B reproduce at a per capita rate of -1, they leave the patch at per capita rate 2, and individuals from patch A enter the patch at per capita rate 3. We obtain

$$\begin{aligned} \frac{dx_B}{dt} &= -x_B - 2x_B + 3x_A \\ &= 3x_A - 3x_B \end{aligned}$$

Using $\mathbf{x}$ to denote the vector with components x_A and x_B, we can write this system in matrix notation in the form of Equation 2 with

$$A = \begin{bmatrix} -1 & 2 \\ 3 & -3 \end{bmatrix}$$ ■

1. A. Flynn et al., "Effectiveness of Radiolabelled Antibodies for Radio-Immunotherapy in a Colorectal Xenograft Model: A Comparative Study using the Linear-Quadratic Formulation," *International Journal of Radiation Biology* 77 (2001): 507–17.

We can employ some of the techniques of phase-plane analysis introduced in Section 7.6 to systems of linear differential equations. This includes plotting the nullclines, finding equilibria, and assessing the stability of any equilibria.

We recall the definition of an equilibrium given in 7.6.2, but we make it specific to systems of two linear differential equations.

(3) Definition An **equilibrium** of the system of differential equations given by (2) is a pair of values $(\hat{x}_1, \hat{x}_2)$ such that both $dx_1/dt = 0$ and $dx_2/dt = 0$ when $x_1 = \hat{x}_1$ and $x_2 = \hat{x}_2$. Equivalently, from Equation 2, the vector of equilibrium values $\hat{\mathbf{x}}$ satisfies the equation $A\hat{\mathbf{x}} = \mathbf{0}$.

Theorem 8.6.5 tells us that the equation $A\hat{\mathbf{x}} = \mathbf{0}$ always has the solution $\hat{\mathbf{x}} = \mathbf{0}$. Thus the origin $(\hat{x}_1, \hat{x}_2) = (0, 0)$ is always an equilibrium of Equation 2. This theorem also says that, in the event that $\det A \neq 0$, the origin is the only equilibrium. This is referred to as the **generic** case. The case where $\det A = 0$ is referred to as **nongeneric**. In most of this chapter we will concentrate on the generic case because it is most common in life science applications.

Throughout this chapter we use a single color (purple) for the nullclines of both variables.

Recall also the definition of a nullcline from Section 7.6. In the context of Equation 2, we have the following definition.

Definition The x_1-nullcline of differential equation (2) is the set of points in the x_1x_2-plane satisfying the equation $dx_1/dt = 0$. From Equation 2, this is the line defined by the equation $a_{11}x_1 + a_{12}x_2 = 0$. The x_2-nullcline of differential equation (2) is the set of points in the x_1x_2-plane satisfying the equation $dx_2/dt = 0$. From Equation 2, this is the line defined by the equation $a_{21}x_1 + a_{22}x_2 = 0$.

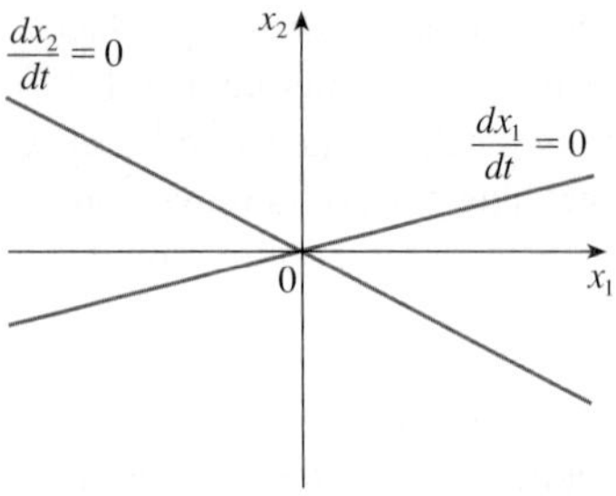

FIGURE 1

Thus a system of two linear differential equations has two nullclines, one for each variable. Both nullclines are straight lines and they intersect at the origin because both $dx_1/dt = 0$ and $dx_2/dt = 0$ at the point $(x_1, x_2) = (0, 0)$. (See Figure 1.) This is consistent with our earlier observation that the origin is an equilibrium.

EXAMPLE 3 | Metapopulation dynamics (continued) Determine the equilibrium of the model in Example 2 and interpret it biologically.

SOLUTION From the definition on page 531 for the determinant of a 2×2 matrix, we calculate $\det A$ to be

$\det A = a_{11}a_{22} - a_{12}a_{21}$

$$\det A = (-1)(-3) - (2)(3) = -3$$

Since $\det A \neq 0$, the only equilibrium is the origin $\hat{\mathbf{x}} = \mathbf{0}$. This represents the situation in which the subpopulations in patches A and B are both extinct. ■

EXAMPLE 4 | Radioimmunotherapy (continued) Suppose that $a + b \neq 0$ and $c \neq 0$ in the model of Example 1. Provide a biological interpretation of these assumptions and determine the equilibrium of the model.

SOLUTION The assumptions mean that there is a nonzero loss rate of antibody from both the bloodstream $(a + b \neq 0)$ and from the tumor $(c \neq 0)$. We calculate $\det A$ to be

$$\det A = -(a + b)(-c) - (b)(0) = c(a + b)$$

This is nonzero given our assumptions about the constants. Therefore the only equilibrium is the origin, $\hat{\mathbf{x}} = \mathbf{0}$. This represents the situation in which there is no antibody in the bloodstream or in the tumor. ■

In Section 7.2 we introduced the concept of *local stability* in the context of single-variable differential equations. An analogous concept applies to systems of differential equations.

Definition An equilibrium $\hat{\mathbf{x}}$ of differential equation (2) is **locally stable** if $\mathbf{x}$ approaches the value $\hat{\mathbf{x}}$ as $t \to \infty$ for all initial values of $\mathbf{x}$ sufficiently close to $\hat{\mathbf{x}}$.

For systems of differential equations, stability of an equilibrium requires that all components of the vector $\mathbf{x}$ approach their equilibrium values as $t \to \infty$. As we will see in Section 10.2, for *generic* systems, if $\hat{\mathbf{x}}$ is locally stable, then it is stable for all initial conditions (not just those sufficiently close to $\hat{\mathbf{x}}$). This is referred to as **global stability**. Consequently, in this chapter we will use the terms *locally stable* and *stable* interchangeably. An equilibrium that is not stable is called **unstable**.

The remainder of this section presents three qualitatively different kinds of behavior that the system of differential equations (2) can exhibit. These are classified in terms of the nature of the equilibrium $\hat{\mathbf{x}} = \mathbf{0}$ and are called saddles, nodes, and spirals. Their qualitative features are introduced here and they are defined more precisely in Section 10.2.

■ Saddles

The equilibrium at the origin, $\hat{\mathbf{x}} = \mathbf{0}$, is called a *saddle* if, roughly speaking, the motion in the x_1x_2-plane is toward the origin in some directions and away from it in others. As an example, consider Equation 2 with the matrix of coefficients

$$A = \begin{bmatrix} 0 & 1 \\ 2 & -1 \end{bmatrix}$$

Figure 2(a) displays the nullclines along with several solution curves in the x_1x_2-plane. By definition, all solution curves that cross the x_1-nullcline do so vertically because x_1 is not changing on this nullcline. Likewise, solution curves that cross the x_2-nullcline do so horizontally. Figure 2(b) plots the components of the red solution curve in Figure 2(a) against time.

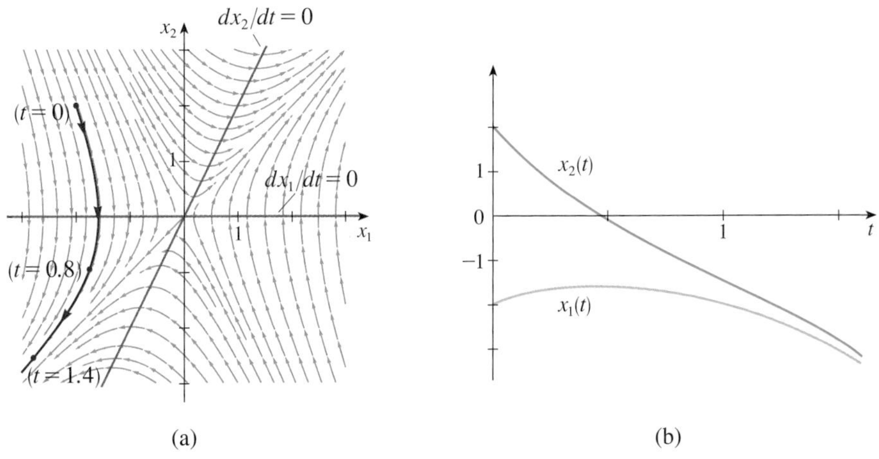

FIGURE 2
Part (a) plots the phase plane, with nullclines in purple and solution curves in blue. The red curve is a particular solution whose components are plotted against time in part (b).

When the equilibrium at the origin is a saddle, it is always unstable because there is always some initial condition for which the system moves away from the origin.

Nodes

The equilibrium at the origin, $\hat{\mathbf{x}} = \mathbf{0}$, is called a *node* if, roughly speaking, the motion in the phase plane is either toward the origin from all directions or away from the origin in all directions. As two examples, consider Equation 2 with the coefficients given by the matrices

$$B = \begin{bmatrix} -3 & -1 \\ -2 & -2 \end{bmatrix} \quad \text{or} \quad C = \begin{bmatrix} 3 & 1 \\ 2 & 2 \end{bmatrix}$$

Figure 3(a) displays the nullclines for matrix B, along with several solution curves of the system in the x_1x_2-plane. Part (b) plots the components of the red solution curve in part (a) against time. Figure 4 displays the corresponding information for matrix C. Notice in Figure 3(a) and Figure 4(a) that solution curves cross the nullclines either vertically or horizontally, depending on whether the nullcline is for x_1 or x_2, respectively.

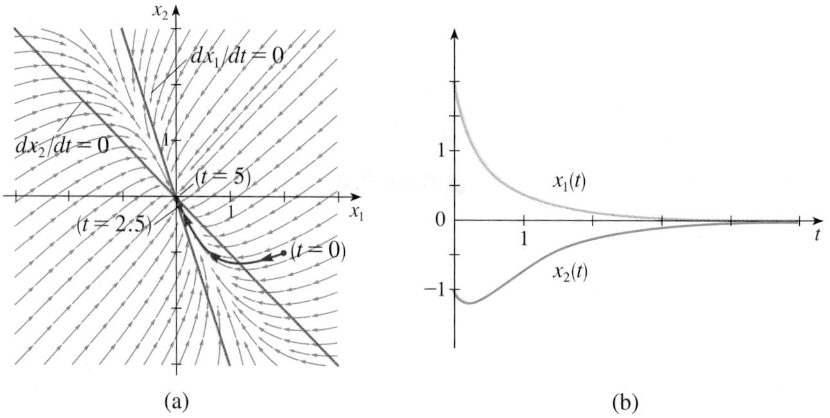

FIGURE 3
Part (a) plots the phase plane for matrix B with nullclines in purple and solution curves in blue. The red curve is a particular solution whose components are plotted against time in part (b).

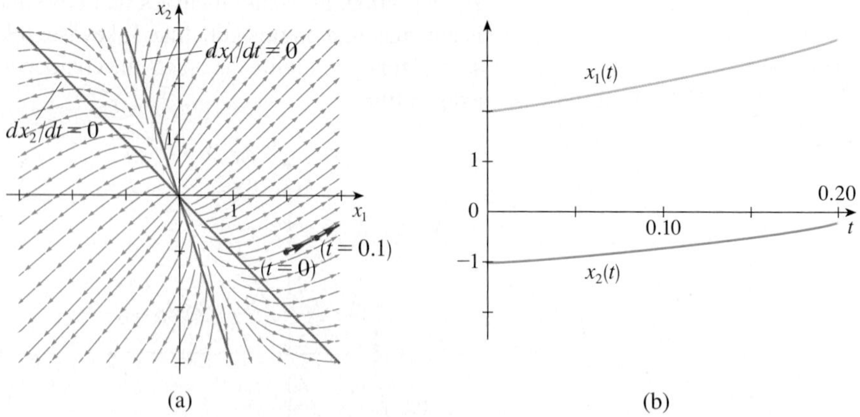

FIGURE 4
Part (a) plots the phase plane for matrix C with nullclines in purple and solution curves in blue. The red curve is a particular solution whose components are plotted against time in part (b).

The equilibrium at the origin can be a stable node, as in Figure 3(a), or an unstable node, as in Figure 4(a), depending on whether movement is toward or away from this equilibrium.

Spirals

The equilibrium at the origin, $\hat{\mathbf{x}} = \mathbf{0}$, is called a *spiral* if the motion in the phase plane is either toward the origin or away from the origin, and occurs in a spiraling fashion. As two examples, consider Equation 2 with the coefficients given by the matrices

$$D = \begin{bmatrix} 1 & -1 \\ 1 & 1 \end{bmatrix} \quad \text{or} \quad E = \begin{bmatrix} -1 & 1 \\ -1 & -1 \end{bmatrix}$$

Figure 5 displays the nullclines for matrix D, along with several solution curves of the system in the x_1x_2-plane. Part (b) plots the components of the red solution curve against time. Figure 6 displays the corresponding information for matrix E. Again you should make note of the orientation of solution curves where they cross a nullcline.

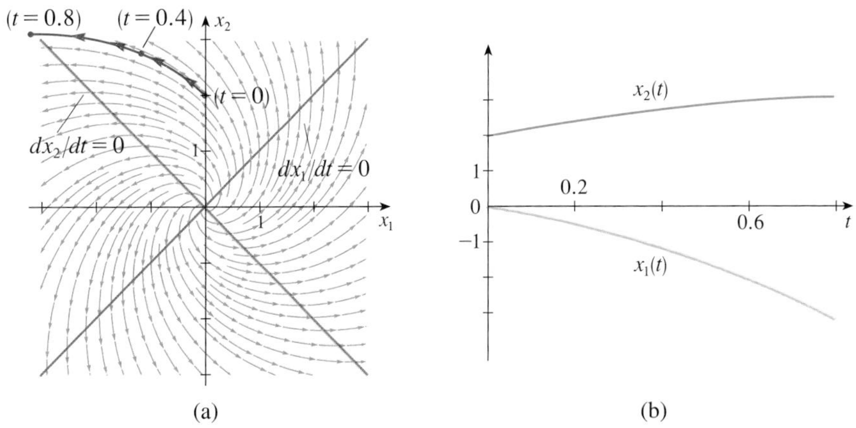

FIGURE 5
Part (a) plots the phase plane for matrix D with nullclines in purple and solution curves in blue. The red curve is a particular solution whose components are plotted against time in part (b).

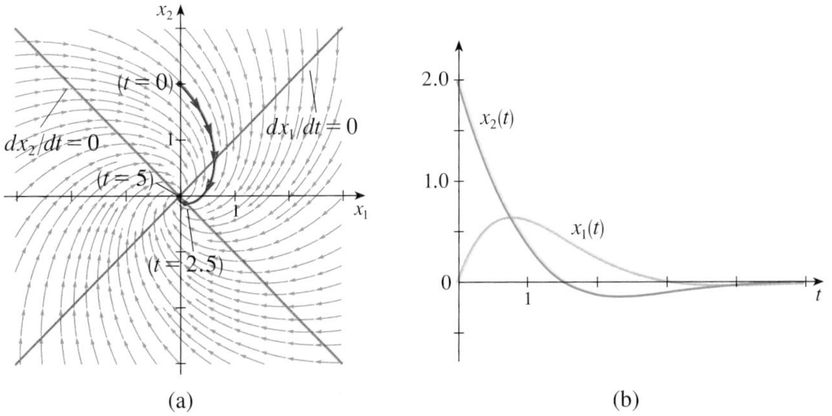

FIGURE 6
Part (a) plots the phase plane for matrix E with nullclines in purple and solution curves in blue. The red curve is a particular solution whose components are plotted against time in part (b).

The equilibrium at the origin can be an unstable spiral, as in Figure 5(a), or a stable spiral, as in Figure 6(a), depending on whether movement is away from or toward this equilibrium.

It turns out that, except for one special case called a *center*, the three qualitative types of behavior (saddle, nodes, and spirals) are the only possibilities for generic linear systems of differential equations whose matrix of coefficients is not defective. This will be demonstrated in the next section by deriving an explicit formula for the solutions of such systems.

EXERCISES 10.1

1–6 Specify whether each system is autonomous or nonautonomous, and whether it is linear or nonlinear. If it is linear, specify whether it is homogeneous or nonhomogeneous.

1. $dx/dt = x - y, \quad dy/dt = -3ty + x$

2. $dy/dx = 2y, \quad dz/dx = x - z + 3$

3. $dy/dt = 3yz - 2z, \quad dz/dt = 2z + 5y$

4. $dy/dx = 3y - 2, \quad dz/dx = 7z + y$

5. $dx/dz = 3x - 2y, \quad dy/dz = 2z + 3y$

6. $dx/dt = xy - y, \quad dy/dt = 4tx - xy$

7–14 Write each system of linear differential equations in matrix notation.

7. $dx/dt = 5x - 3y, \quad dy/dt = 2y - x$

8. $dx/dt = x - 2, \quad dy/dt = 2y + 3x - 1$

9. $dx/dt = 3ty - 7, \quad dy/dt = 2x - 3y$

10. $dx/dt = 5y, \quad dy/dt = 2x - y$

11. $dx/dt = 2x - 5, \quad dy/dt = 3x + 7y$

12. $dx/dt = 2x - y\sin t, \quad dy/dt = y - x$

13. $dx/dt = x + 4y - 3t, \quad dy/dt = y - x$

14. $dx/dt = y - 2x\sqrt{t} + 7, \quad dy/dt = 3x + 2$

15–22 Given the system of differential equations $d\mathbf{x}/dt = A\mathbf{x}$, construct the phase plane, including the nullclines. Does the equilibrium look like a saddle, a node, or a spiral?

15. $A = \begin{bmatrix} -3 & 1 \\ 2 & -2 \end{bmatrix}$ **16.** $A = \begin{bmatrix} -2 & 1 \\ -1 & -1 \end{bmatrix}$

17. $A = \begin{bmatrix} 1 & 2 \\ -2 & 1 \end{bmatrix}$ **18.** $A = \begin{bmatrix} 1 & 2 \\ 2 & -1 \end{bmatrix}$

19. $A = \begin{bmatrix} -1 & 2 \\ -3 & 0 \end{bmatrix}$ **20.** $A = \begin{bmatrix} 1 & 1 \\ 0 & 1 \end{bmatrix}$

21. $A = \begin{bmatrix} 2 & -1 \\ -1 & 2 \end{bmatrix}$ **22.** $A = \begin{bmatrix} 0 & 1 \\ 1 & 0 \end{bmatrix}$

23. Consider the system of linear differential equations $d\mathbf{x}/dt = A\mathbf{x} + \mathbf{g}$, where $\mathbf{g}$ is a vector of constants. Suppose that A is nonsingular.
(a) What is the equilibrium of this system of equations?
(b) Using $\hat{\mathbf{x}}$ to denote the equilibrium found in part (a), define a new vector of variables $\mathbf{y} = \mathbf{x} - \hat{\mathbf{x}}$. What do the components of $\mathbf{y}$ represent?
(c) Show that $\mathbf{y}$ satisfies the differential equation $d\mathbf{y}/dt = A\mathbf{y}$. This demonstrates how we can reduce a nonhomogeneous system of linear differential equations to a system that is homogenous by using a change of variables.

24. Consider the system of linear differential equations

$$\frac{d\mathbf{x}}{dt} = \begin{bmatrix} -2 & -1 \\ 2 & 1 \end{bmatrix}\mathbf{x}$$

The system is *nongeneric*, that is, the determinant of the matrix of coefficients is zero.
(a) There are an infinite number of equilibria, all lying on a line in the phase plane. What is the equation of this line?
(b) Construct the phase plane for this system.

25. Consider an autonomous homogeneous system of linear differential equations with coefficient matrix

$$A = \begin{bmatrix} a & b \\ c & d \end{bmatrix}$$

Suppose that $\det A = 0$. Show that there are an infinite number of equilibria.

26. Consider the following homogeneous system of three linear differential equations:

$$dx/dt = 3x + 2y - z$$
$$dy/dt = x - y - z$$
$$dz/dt = y + 3z$$

Suppose that $x + y = 5$ at all times. Show that this system can be reduced to two nonhomogenous linear differential equations given by

$$dx/dt = x - z + 10$$
$$dz/dt = -x + 3z + 5$$

27. Consider the following homogeneous system of four linear differential equations:

$$dw/dt = 2x + y - z$$
$$dx/dt = 3x + z$$
$$dy/dt = -y + 2z$$
$$dz/dt = 3x - 5y$$

Suppose that $x + z = 2$ and $y + w = 3$ at all times. Show that this system can be reduced to two nonhomogeneous linear differential equations given by

$$dw/dt = 3x - w + 1$$
$$dx/dt = 2x + 2$$

28. Consider any homogeneous, autonomous system of three linear differential equations for which the variables satisfy $ax_1 + bx_2 + cx_3 = d$, where a, b, c, and d are constants, not all of which are zero. Show that the system can be reduced to a nonhomogeneous system of two linear differential equations.

29. Second-order linear differential equations take the form

$$y''(t) + p(t)\,y'(t) + q(t)\,y(t) = g(t)$$

where p, q, and g are continuous functions of t. Suppose we have initial conditions $y(0) = a$ and $y'(0) = b$. Show that this equation can be rewritten as a system of two first-order linear differential equations having the form

$$\frac{d\mathbf{x}}{dt} = \begin{bmatrix} 0 & 1 \\ -q(t) & -p(t) \end{bmatrix}\mathbf{x} + \begin{bmatrix} 0 \\ g(t) \end{bmatrix}$$

with $$\mathbf{x}(0) = \begin{bmatrix} a \\ b \end{bmatrix}$$

where $x_1(t) = y(t)$ and $x_2(t) = y'(t)$.

30. Metapopulation dynamics Example 2 presents a model for a population of deer mice that is split into two patches through habitat fragmentation. The model is

$$\frac{dx_A}{dt} = -x_A + 2x_B \qquad \frac{dx_B}{dt} = 3x_A - 3x_B$$

where x_A and x_B are the population sizes in patches A and B, respectively.
(a) Construct the phase plane, including the nullclines.
(b) Describe what happens to the population in each patch as $t \to \infty$ if both start with nonzero sizes.

31. Gene regulation Genes produce molecules called mRNA that then produce proteins. High levels of protein can inhibit the production of mRNA, resulting in a feedback that regulates gene expression. Using m and p to denote the amounts of mRNA and protein in a cell ($\times\ 10^2$ copies/cell), a simple model of gene regulation is

$$dm/dt = 1 - p - m$$
$$dp/dt = m - p$$

Construct the phase plane, including the nullclines. [*Hint:* This system is nonhomogeneous.]

32. Prostate cancer During treatment, tumor cells in the prostate can become resistant through a variety of biochemical mechanisms. Some of these are reversible—the cells revert to being sensitive once treatment stops—and some are not. Using x_1, x_2, and x_3 to denote the fraction of cells that are sensitive, temporarily resistant, and permanently resistant, respectively, a simple model for their dynamics during treatment is

$$dx_1/dt = -ax_1 - cx_1 + bx_2$$
$$dx_2/dt = ax_1 - bx_2 - dx_2$$
$$dx_3/dt = cx_1 + dx_2$$

Use the fact that $x_1 + x_2 + x_3 = 1$ to reduce this to a nonhomogeneous system of two linear differential equations for x_1 and x_3.

Source: Adapted from Y. Hirata et al., "Development of a Mathematical Model that Predicts the Outcome of Hormone Therapy for Prostate Cancer," *Journal of Theoretical Biology* 264 (2010): 517–27.

33. Radioimmunotherapy Example 1 presents a model of radioimmunotherapy. The model is

$$\frac{dx_1}{dt} = -ax_1 - bx_1 \qquad \frac{dx_2}{dt} = bx_1 - cx_2$$

where x_1 and x_2 denote the amount of antibody (in μg) in the bloodstream and tumor, respectively, at time t (in minutes), and all constants are positive.
(a) Construct the phase plane, including the nullclines, to determine the qualitative behavior of the system.
(b) Describe what happens to the amount of antibody in each part of the body as $t \to \infty$.

Source: Adapted from A. Flynn et al., "Effectiveness of Radiolabelled Antibodies for Radio-Immunotherapy in a Colorectal Xenograft Model: A Comparative Study using the Linear-Quadratic Formulation," *International Journal of Radiation Biology* 77 (2001): 507–17.

34. Jellyfish locomotion Jellyfish move by contracting an elastic part of their body, called a bell, that creates a high-pressure jet of water. When the contractive force stops, the bell then springs back to its natural shape. Jellyfish locomotion has been modeled using a second-order linear differential equation having the form

$$mx''(t) + bx'(t) + kx(t) = 0$$

where $x(t)$ is the displacement of the bell at time t, m is the mass of the bell (in grams), b is a measure of the friction between the bell and the water (in units of N/m·s), and k is a measure of the stiffness of the bell (in units of N/m). Suppose that $m = 100$ g, $b = 0.1$ N/m·s, and $k = 1$ N/m.
(a) Define the new variables $z_1(t) = x(t)$ and $z_2(t) = x'(t)$, and show that the model can be expressed as a system of two first-order linear differential equations.
(b) Construct the phase plane, including the nullclines, for the equations from part (a).

Source: Adapted from M. DeMont et al., "Mechanics of Jet Propulsion in the Hydromedusan Jellyfish, *Polyorchis Penicillatus:* III. A Natural Resonating Bell—The Presence and Importance of a Resonant Phenomenon in the Locomotor Structure," *Journal of Experimental Biology* 134 (1988): 347–61.

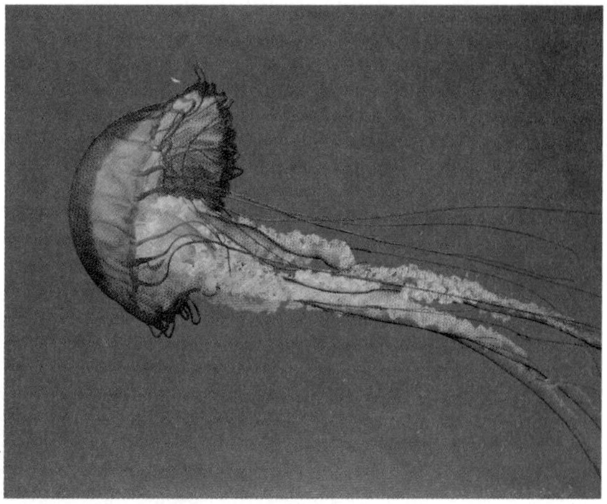

10.2 Solving Systems of Linear Differential Equations

Most systems of linear differential equations that arise in the life sciences take the form of an *initial-value problem*. Suppose A is a 2×2 matrix of coefficients and $\mathbf{x}$ is a two-dimensional vector whose components are variables.

(1) Definition The system of linear differential equations

$$\frac{d\mathbf{x}}{dt} = A\mathbf{x}$$

together with an initial condition $\mathbf{x}(t_0) = \mathbf{x}_0$ is called an **initial-value problem**.

The initial condition $\mathbf{x}(t_0) = \mathbf{x}_0$ specifies the values that each of the component variables in $\mathbf{x}(t)$ must take at $t = t_0$.

A solution to the initial-value problem (1) is a vector $\mathbf{x}(t)$ whose two components are each functions of t and that together satisfy the differential equation and initial condition. More specifically, we say that a function $\boldsymbol{\varphi}(t)$ is a solution of the differential equation in (1) if, when substituted into the left and right sides of the differential equation, it produces an equality. Further, if $\boldsymbol{\varphi}(t_0) = \mathbf{x}_0$ as well, then we say $\boldsymbol{\varphi}(t)$ solves the initial-value problem (1).

Just as there were many solutions to the differential equations we examined in Chapter 7, there are typically many solutions to a system of linear differential equations. Only one of these, however, will satisfy the initial condition. The following result can be proved.

(2) Theorem: Existence and Uniqueness of Solutions The initial-value problem (1) has one and only one solution $\mathbf{x}(t)$, and this solution is defined for all $t \in \mathbb{R}$.

Theorem 2 tells us two things: First, it tells us that there will always be a solution to the initial-value problem. Second, it tells us that there will be only one solution. Therefore, if we can manage to find a solution, no matter how we do it, then we are guaranteed that it will be the only one.

How can we find a solution to the initial-value problem? Consider a simple example:

(3)
$$\frac{dx_1}{dt} = -x_1 \qquad \frac{dx_2}{dt} = -3x_2$$

with initial condition $x_1(0) = 2$ and $x_2(0) = 5$. In matrix form this is

(4)
$$\frac{d\mathbf{x}}{dt} = A\mathbf{x}$$

where
$$A = \begin{bmatrix} -1 & 0 \\ 0 & -3 \end{bmatrix} \quad \text{and} \quad \mathbf{x}(0) = \begin{bmatrix} 2 \\ 5 \end{bmatrix}$$

In Equations 3 the differential equation for x_1 does not involve x_2, and vice versa. Their solutions are therefore readily found to be

$$x_1(t) = c_1e^{-t} \qquad x_2(t) = c_2e^{-3t}$$

where c_1 and c_2 are constants to be determined from the initial conditions (for example, see Section 3.6). Writing these solutions in vector form, we have

$$\textbf{(5)} \qquad \mathbf{x}(t) = \begin{bmatrix} x_1(t) \\ x_2(t) \end{bmatrix} = \begin{bmatrix} c_1 e^{-t} \\ 0 \end{bmatrix} + \begin{bmatrix} 0 \\ c_2 e^{-3t} \end{bmatrix} = c_1 e^{-t} \begin{bmatrix} 1 \\ 0 \end{bmatrix} + c_2 e^{-3t} \begin{bmatrix} 0 \\ 1 \end{bmatrix}$$

Solution (5) is composed of the sum of two parts, each being an exponential function multiplied by a vector. Defining

$$\mathbf{x}_1(t) = e^{-t} \begin{bmatrix} 1 \\ 0 \end{bmatrix} \quad \text{and} \quad \mathbf{x}_2(t) = e^{-3t} \begin{bmatrix} 0 \\ 1 \end{bmatrix}$$

we can write solution (5) as

$$\mathbf{x}(t) = c_1 \mathbf{x}_1(t) + c_2 \mathbf{x}_2(t)$$

In fact, $\mathbf{x}_1(t)$ and $\mathbf{x}_2(t)$ are each, individually, solutions to the system of differential equations (4). To see this, substitute $\mathbf{x}_1(t)$ into the left and right sides of Equation 4. The left side gives

$$\frac{d\mathbf{x}_1}{dt} = -e^{-t} \begin{bmatrix} 1 \\ 0 \end{bmatrix}$$

and the right side gives

$$Ae^{-t} \begin{bmatrix} 1 \\ 0 \end{bmatrix} = e^{-t} \begin{bmatrix} -1 & 0 \\ 0 & -3 \end{bmatrix} \begin{bmatrix} 1 \\ 0 \end{bmatrix} = e^{-t} \begin{bmatrix} -1 \\ 0 \end{bmatrix} = -e^{-t} \begin{bmatrix} 1 \\ 0 \end{bmatrix}$$

The two sides are identical, which means that $\mathbf{x}_1(t)$ is a solution. It is plotted in the phase plane in Figure 1(a). Likewise, substituting $\mathbf{x}_2(t)$ into the left side results in

$$\frac{d\mathbf{x}_2}{dt} = -3e^{-3t} \begin{bmatrix} 0 \\ 1 \end{bmatrix}$$

and the right side gives

$$Ae^{-3t} \begin{bmatrix} 0 \\ 1 \end{bmatrix} = e^{-3t} \begin{bmatrix} -1 & 0 \\ 0 & -3 \end{bmatrix} \begin{bmatrix} 0 \\ 1 \end{bmatrix} = e^{-3t} \begin{bmatrix} 0 \\ -3 \end{bmatrix} = -3e^{-3t} \begin{bmatrix} 0 \\ 1 \end{bmatrix}$$

Again these sides are identical and so $\mathbf{x}_2(t)$ is a solution as well. It is plotted in Figure 1(b).

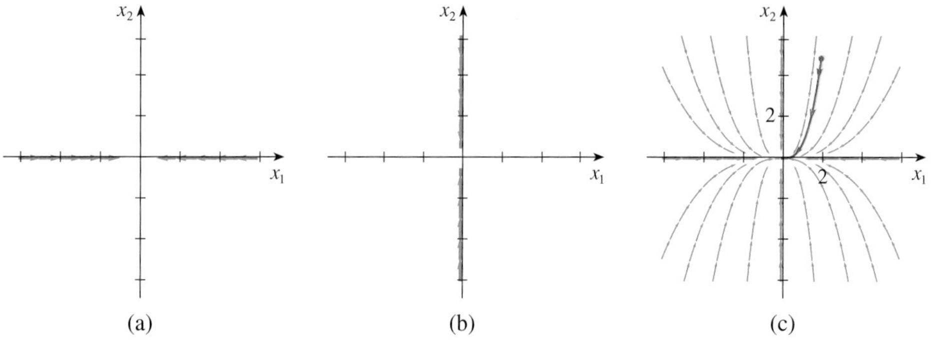

FIGURE 1 Blue curves are solutions to the system of differential equations. Part (a) shows solutions $c_1\mathbf{x}_1(t)$ for different values of c_1. Part (b) shows solutions $c_2\mathbf{x}_2(t)$ for different values of c_2. Part (c) shows solutions $c_1\mathbf{x}_1(t) + c_2\mathbf{x}_2(t)$ for different values of c_1 and c_2. The red curve is the solution to the initial-value problem.

Equation 5 is a family of solutions to the system of differential equations (4). Theorem 2 tells us, however, that there is a unique solution to this system of equations once we take the initial condition into account. If solution (5) is to satisfy the initial condition, then we require that

$$\mathbf{x}(0) = c_1\begin{bmatrix}1\\0\end{bmatrix} + c_2\begin{bmatrix}0\\1\end{bmatrix} = \begin{bmatrix}2\\5\end{bmatrix}$$

Therefore we must choose $c_1 = 2$ and $c_2 = 5$, and the unique solution to the initial-value problem (4) is

$$\mathbf{x}(t) = 2e^{-t}\begin{bmatrix}1\\0\end{bmatrix} + 5e^{-3t}\begin{bmatrix}0\\1\end{bmatrix}$$

This solution is plotted in the phase plane in Figure 1(c).

■ The General Solution

The preceding calculations illustrate some general principles that apply to the initial-value problem (1): First, as we will see, solutions to the differential equation in (1) are always functions of the form $e^{\lambda t}\mathbf{v}$, where λ is a scalar and $\mathbf{v}$ is a vector. Second, the sum of any two solutions to the differential equation is itself a solution. This latter fact is summarized as follows. (See also Exercise 9.)

> **(6) The Superposition Principle** Suppose $\mathbf{x}_1(t)$ and $\mathbf{x}_2(t)$ are solutions of the system of differential equations in (1). Then any function of the form $\mathbf{x}(t) = c_1\mathbf{x}_1(t) + c_2\mathbf{x}_2(t)$, where c_1 and c_2 are scalar quantities, is also a solution.

Typically, a family of solutions to the differential equation in (1) is obtained by finding the values of λ and $\mathbf{v}$ for which $e^{\lambda t}\mathbf{v}$ is a solution, and then combining these solutions using the Superposition Principle (6). The unique solution to the initial-value problem is then obtained by choosing appropriate constants c_1 and c_2.

Let's put these ideas into practice for the initial-value problem (1) with

$$A = \begin{bmatrix}a_{11} & a_{12}\\a_{21} & a_{22}\end{bmatrix} \qquad \text{and} \qquad \mathbf{x}(0) = \begin{bmatrix}x_1(0)\\x_2(0)\end{bmatrix}$$

We begin by looking for solutions having the form $e^{\lambda t}\mathbf{v}$ for some λ and $\mathbf{v}$. Substituting this function into the differential equation in (1) gives

$$\frac{d}{dt}(e^{\lambda t}\mathbf{v}) = Ae^{\lambda t}\mathbf{v}$$

or

$$\lambda e^{\lambda t}\mathbf{v} = Ae^{\lambda t}\mathbf{v}$$

Because $e^{\lambda t}$ is never zero, we can divide both sides by $e^{\lambda t}$, giving

(7) $$\lambda\mathbf{v} = A\mathbf{v}$$

From Definition 8.7.2 we see that the values of $\mathbf{v}$ and λ that satisfy Equation 7 are the eigenvector-eigenvalue pairs of the matrix A. This shows that a function of the form $e^{\lambda t}\mathbf{v}$ is a solution to the differential equation in (1) provided that $\mathbf{v}$ and λ are an eigenvector-eigenvalue pair of A.

Our focus in the remainder of this chapter will be on coefficient matrices that are not defective and therefore that have distinct eigenvalues (advanced courses in differential equations treat the general case). Recall from page 538 that such matrices are guaranteed to have distinct eigenvector-eigenvalue pairs. Denoting these by $\mathbf{v}_i$ and λ_i, we can then use the Superposition Principle (6) to obtain the family of solutions

$$\mathbf{x}(t) = c_1 e^{\lambda_1 t}\begin{bmatrix} | \\ \mathbf{v}_1 \\ | \end{bmatrix} + c_2 e^{\lambda_2 t}\begin{bmatrix} | \\ \mathbf{v}_2 \\ | \end{bmatrix} \tag{8}$$

where $\begin{bmatrix} | \\ \mathbf{v}_i \\ | \end{bmatrix}$ is the ith eigenvector. Equation 8 is called the **general solution** of the differential equation in (1). It is valid for any two-dimensional system whose coefficient matrix has distinct eigenvalues (even the nongeneric case where $\det A = 0$).

Finally, we can use the initial condition to find the values of c_1 and c_2 that provide the unique solution to the initial-value problem guaranteed by Theorem 2. We have

$$\mathbf{x}(0) = c_1\begin{bmatrix} | \\ \mathbf{v}_1 \\ | \end{bmatrix} + c_2\begin{bmatrix} | \\ \mathbf{v}_2 \\ | \end{bmatrix} = \begin{bmatrix} x_1(0) \\ x_2(0) \end{bmatrix}$$

or

$$\begin{bmatrix} | & | \\ \mathbf{v}_1 & \mathbf{v}_2 \\ | & | \end{bmatrix}\begin{bmatrix} c_1 \\ c_2 \end{bmatrix} = \begin{bmatrix} x_1(0) \\ x_2(0) \end{bmatrix} \tag{9}$$

Since the eigenvalues of A are distinct, a theorem from linear algebra says that the matrix whose columns are the associated eigenvectors is nonsingular (see Exercise 10). Therefore Theorem 8.6.4 tells us that Equation 9 has a unique solution for c_1 and c_2.

■ Nullclines versus Eigenvectors

The general solution (8) shows that the eigenvectors of the matrix A in the initial-value problem (1) play an important role in the behavior of the variable $\mathbf{x}(t)$. Section 10.1 revealed that the nullclines of the differential equation in (1) also play an important role. What is the relationship between the two?

Recall that the nullclines for two-variable systems of linear differential equations are straight lines in the plane along which the rate of change of one of the variables is zero (see Section 10.1). This means that the motion in the plane is either vertical or horizontal along nullclines [see Figure 2(a)].

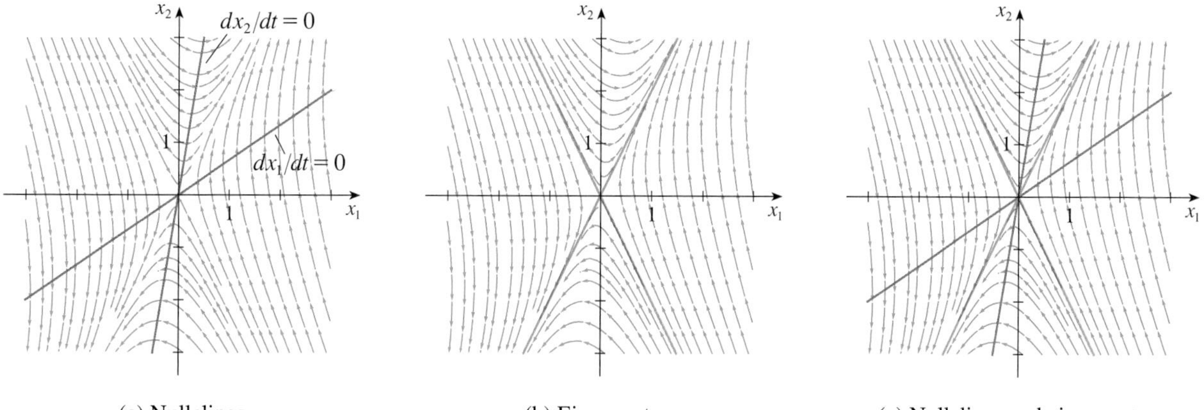

(a) Nullclines (b) Eigenvectors (c) Nullclines and eigenvectors

FIGURE 2 Purple lines are nullclines. Orange lines indicate all scalar multiples of the eigenvectors. Blue curves are solutions.

The eigenvectors of the matrix A appear in the general solution (8) in two terms, each having the form

$$ce^{\lambda t}\mathbf{v} \tag{10}$$

and each of these two terms is itself a solution to differential equation (1). What is the geometric interpretation of (10)? The factor c is a scalar, as is the factor $e^{\lambda t}$. Therefore $ce^{\lambda t}$ is a scalar quantity and it varies in magnitude with t. The remaining factor $\mathbf{v}$ is the eigenvector associated with λ. Thus (10) is a scalar multiple of the vector $\mathbf{v}$, where $ce^{\lambda t}$ is the scalar. Section 8.2 demonstrated that, geometrically, scalar multiplication simply scales the length of a vector. Therefore the plot of (10) in the phase plane traces out a straight line, moving in the direction of eigenvector $\mathbf{v}$ as t increases.

Exercise 39 demonstrates that solution curves in the phase plane cannot cross. Since (10) defines a solution curve that is a scalar multiple of the eigenvector $\mathbf{v}$, the lines corresponding to all scalar multiples of the eigenvectors therefore constrain the direction of the solution curves in the phase plane [see Figure 2(b)]. Figure 2(c) shows how, together, the nullclines and the eigenvectors dictate the shape of the solution curves in the phase plane.

We now return to the three qualitative kinds of behavior documented in Section 10.1: saddles, nodes, and spirals. Throughout the remainder of this section we assume that $\det A \neq 0$ and therefore the origin is the only equilibrium. We also assume that the eigenvalues of A are distinct.

■ Saddles

The origin is called a **saddle** if the eigenvalues of the coefficient matrix are real and have opposite signs. The form of solution (8) immediately shows that the origin is unstable in this case because one of the exponential terms will always increase without bound. As a result, there is always some initial condition for which $\mathbf{x}(t)$ does not approach the origin as $t \to \infty$.

The saddle explored on page 635 used the matrix

$$A = \begin{bmatrix} 0 & 1 \\ 2 & -1 \end{bmatrix}$$

The eigenvalues of A are $\lambda_1 = -2$ and $\lambda_2 = 1$. The associated eigenvectors are

$$\mathbf{v}_1 = \begin{bmatrix} -1 \\ 2 \end{bmatrix} \quad \text{and} \quad \mathbf{v}_2 = \begin{bmatrix} 1 \\ 1 \end{bmatrix}$$

The general solution (8) is therefore

$$\mathbf{x}(t) = c_1 e^{-2t}\begin{bmatrix} -1 \\ 2 \end{bmatrix} + c_2 e^{t}\begin{bmatrix} 1 \\ 1 \end{bmatrix}$$

To obtain a unique solution we need to specify an initial condition. Suppose that $\mathbf{x}(0) = \begin{bmatrix} -2 \\ 2 \end{bmatrix}$. Equation 9 then becomes

$$\begin{bmatrix} -1 & 1 \\ 2 & 1 \end{bmatrix}\begin{bmatrix} c_1 \\ c_2 \end{bmatrix} = \begin{bmatrix} -2 \\ 2 \end{bmatrix}$$

The solution to this equation is $c_1 = \frac{4}{3}$ and $c_2 = -\frac{2}{3}$. Figure 3 displays several solution curves, along with the nullclines, the eigenvectors, and the solution to the initial-value problem.

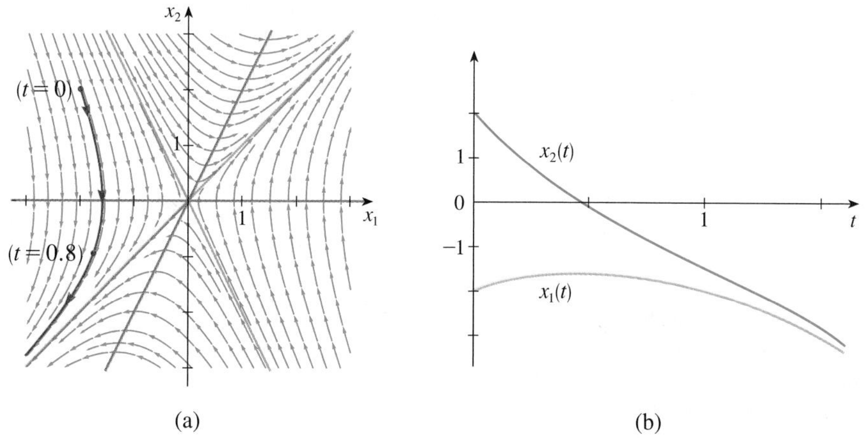

FIGURE 3
Part (a) plots the phase plane, with nullclines in purple, all scalar multiples of the eigenvectors in orange, and solution curves in blue. The red curve is the solution to the initial-value problem, and its components are plotted against time in part (b).

■ Nodes

The origin is called a **node** if the eigenvalues of A are real and have the same sign. The form of solution (8) reveals that the origin is unstable if the eigenvalues are positive (an *unstable node*) and stable if the eigenvalues are negative (a *stable node*).

For the nodes explored on page 636 the coefficient matrices were

$$B = \begin{bmatrix} -3 & -1 \\ -2 & -2 \end{bmatrix} \quad \text{and} \quad C = \begin{bmatrix} 3 & 1 \\ 2 & 2 \end{bmatrix}$$

The eigenvalues of B are $\lambda_1 = -1$ and $\lambda_2 = -4$, with eigenvectors

$$\mathbf{v}_1 = \begin{bmatrix} -1 \\ 2 \end{bmatrix} \quad \text{and} \quad \mathbf{v}_2 = \begin{bmatrix} 1 \\ 1 \end{bmatrix}$$

The eigenvalues of C are $\lambda_1 = 1$ and $\lambda_2 = 4$, with eigenvectors again being

$$\mathbf{v}_1 = \begin{bmatrix} -1 \\ 2 \end{bmatrix} \quad \text{and} \quad \mathbf{v}_2 = \begin{bmatrix} 1 \\ 1 \end{bmatrix}$$

The general solutions are therefore

$$\mathbf{x}(t) = c_1e^{-t}\begin{bmatrix} -1 \\ 2 \end{bmatrix} + c_2e^{-4t}\begin{bmatrix} 1 \\ 1 \end{bmatrix} \qquad \text{for matrix } B$$

and

$$\mathbf{x}(t) = c_1e^{t}\begin{bmatrix} -1 \\ 2 \end{bmatrix} + c_2e^{4t}\begin{bmatrix} 1 \\ 1 \end{bmatrix} \qquad \text{for matrix } C$$

To obtain unique solutions in each case we need to specify an initial condition. Suppose that, in both cases,

$$\mathbf{x}(0) = \begin{bmatrix} 2 \\ -1 \end{bmatrix}$$

Equation 9 then becomes

$$\begin{bmatrix} -1 & 1 \\ 2 & 1 \end{bmatrix}\begin{bmatrix} c_1 \\ c_2 \end{bmatrix} = \begin{bmatrix} 2 \\ -1 \end{bmatrix}$$

in both cases. The solution to this equation is $c_1 = -1$ and $c_2 = 1$. Figures 4 and 5 display several solution curves, along with the nullclines, the eigenvectors, and the solutions to each of these initial-value problems.

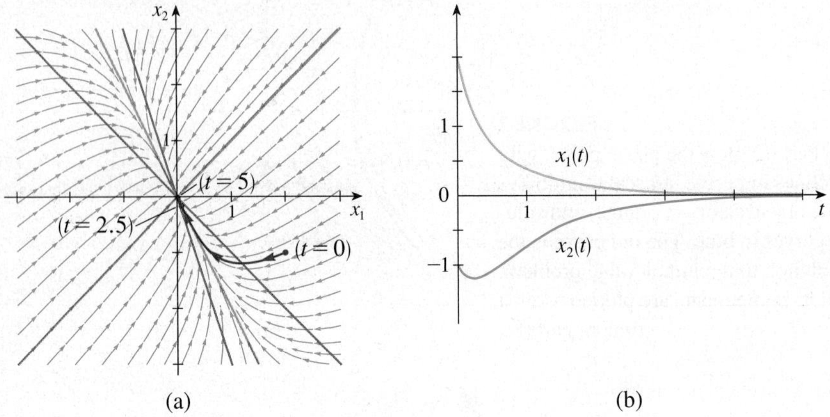

FIGURE 4
Part (a) plots the phase plane for matrix B, with nullclines in purple, all scalar multiples of the eigenvectors in orange, and solution curves in blue. The red curve is the solution to the initial-value problem, and its components are plotted against time in part (b).

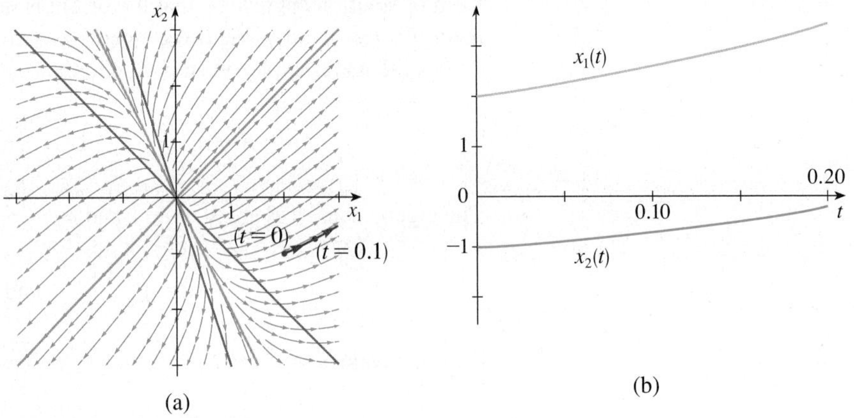

FIGURE 5
Part (a) plots the phase plane for matrix C, with nullclines in purple, all scalar multiples of the eigenvectors in orange, and solution curves in blue. The red curve is the solution to the initial-value problem, and its components are plotted against time in part (b).

■ Spirals

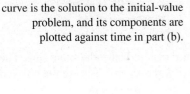

See Appendix G for a review of complex numbers.

The remaining possibility is that the eigenvalues of the coefficient matrix are complex conjugates, taking the form $\lambda = a \pm bi$ for some real numbers a and b. In this case the origin is a spiral.

To see how complex eigenvalues lead to such behavior, we first note that the eigenvectors associated with complex conjugate eigenvalues are themselves complex conjugates (see Exercises 8.8.26 and 8.8.27). Therefore the eigenvectors associated with the eigenvalues $\lambda = a \pm bi$ can be written $\mathbf{v} = \mathbf{u} \pm i\mathbf{w}$, where $\mathbf{u}$ and $\mathbf{w}$ are real-valued vectors. We take each eignvector-eigenvalue pair in turn.

To evaluate the first term of solution (8) we substitute $\lambda = a + ib$ and $\mathbf{v} = \mathbf{u} + i\mathbf{w}$ into $e^{\lambda t}\mathbf{v}$ to obtain

(11) $$e^{(a+bi)t}(\mathbf{u} + i\mathbf{w})$$

Euler's formula

$$e^{iy} = \cos y + i \sin y$$

is discussed in Appendix G.

Euler's formula allows us to simply this further. If $x + iy$ is a complex number, then e^{x+iy} can be rewritten as

$$\begin{aligned} e^{x+iy} &= e^x e^{iy} \\ &= e^x(\cos y + i \sin y) \end{aligned}$$

Applying this result with $x = at$ and $y = bt$, we can write (11) as

$$\begin{aligned} e^{(a+bi)t}(\mathbf{u} + i\mathbf{w}) &= e^{at}(\cos bt + i\sin bt)(\mathbf{u} + i\mathbf{w}) \\ &= e^{at}(\mathbf{u}\cos bt - \mathbf{w}\sin bt + i\mathbf{w}\cos bt + i\mathbf{u}\sin bt) \\ &= \mathbf{g}(t) + i\mathbf{h}(t) \end{aligned}$$

where, to simplify notation, we have defined

$$\mathbf{g}(t) = e^{at}(\mathbf{u}\cos bt - \mathbf{w}\sin bt)$$

and

$$\mathbf{h}(t) = e^{at}(\mathbf{w}\cos bt + \mathbf{u}\sin bt)$$

To calculate the second term of solution (8) we can follow the same steps for the complex conjugate eigenvector-eigenvalue pair, obtaining

$$e^{(a-bi)t}(\mathbf{u} - i\mathbf{w}) = \mathbf{g}(t) - i\mathbf{h}(t)$$

Therefore solution (8) can be written as

(12) $$\mathbf{x}(t) = c_1[\mathbf{g}(t) + i\mathbf{h}(t)] + c_2[\mathbf{g}(t) - i\mathbf{h}(t)]$$

Although (12) is the general solution in the case of complex eigenvalues, this family of solutions is complex-valued and we are interested only in real-valued solutions. We can restrict our attention to a family of real-valued solutions by choosing the constants $c_1 = (k_1 - ik_2)/2$ and $c_2 = (k_1 + ik_2)/2$, where k_1 and k_2 are arbitrary, real constants. Although we have not formally justified the use of complex values of c_1 and c_2 in solution (8), this does, in fact, turn out to be valid. With these choices, the imaginary part of Equation 12 then disappears:

$$\begin{aligned} \mathbf{x}(t) &= \frac{k_1 - ik_2}{2}[\mathbf{g}(t) + i\mathbf{h}(t)] + \frac{k_1 + ik_2}{2}[\mathbf{g}(t) - i\mathbf{h}(t)] \\ &= \tfrac{1}{2}[2k_1\mathbf{g}(t) + ik_1\mathbf{h}(t) - ik_1\mathbf{h}(t) - ik_2\mathbf{g}(t) + ik_2\mathbf{g}(t) + 2k_2\mathbf{h}(t)] \\ &= k_1\mathbf{g}(t) + k_2\mathbf{h}(t) \end{aligned}$$

Therefore, from the definitions of $\mathbf{g}(t)$ and $\mathbf{h}(t)$, the family of real solutions is

(13) $$\mathbf{x}(t) = k_1e^{at}(\mathbf{u}\cos bt - \mathbf{w}\sin bt) + k_2e^{at}(\mathbf{w}\cos bt + \mathbf{u}\sin bt)$$

Equation 13 is the (real-valued) general solution when the eigenvalues are complex.

To obtain the unique solution to the initial-value problem, we need to make use of an initial condition. In particular, supposing that

$$\mathbf{x}(0) = \begin{bmatrix} x_1(0) \\ x_2(0) \end{bmatrix}$$

then, from Equation 13, we have

$$\mathbf{x}(0) = k_1\mathbf{u} + k_2\mathbf{w}$$

or

(14) $$\begin{bmatrix} | & | \\ \mathbf{u} & \mathbf{w} \\ | & | \end{bmatrix}\begin{bmatrix} k_1 \\ k_2 \end{bmatrix} = \begin{bmatrix} x_1(0) \\ x_2(0) \end{bmatrix}$$

Although we won't prove it, a matrix whose columns are **u** and **w** (that is, whose columns are the real and imaginary parts of the eigenvectors) is nonsingular. Therefore Theorem 8.6.4 says that Equation 14 has a unique solution for k_1 and k_2. This then gives the unique solution to the initial-value problem.

The form of Equation 13 also reveals that the origin will be an *unstable spiral* if the real part a of the eigenvalues is positive, and a *stable spiral* if the real part a is negative. The case where $a = 0$ is called a **center** and the solutions in this case form closed curves in the phase plane (see Exercise 30).

Let's consider the two coefficient matrices we explored in Section 10.1:

$$D = \begin{bmatrix} 1 & -1 \\ 1 & 1 \end{bmatrix} \quad \text{and} \quad E = \begin{bmatrix} -1 & 1 \\ -1 & -1 \end{bmatrix}$$

The eigenvalues of D are $\lambda = 1 \pm i$ with eigenvectors

$$\mathbf{v} = \begin{bmatrix} 0 \\ 1 \end{bmatrix} \pm i \begin{bmatrix} 1 \\ 0 \end{bmatrix}$$

Therefore

$$a = 1, \quad b = 1, \quad \mathbf{u} = \begin{bmatrix} 0 \\ 1 \end{bmatrix}, \quad \text{and} \quad \mathbf{w} = \begin{bmatrix} 1 \\ 0 \end{bmatrix}$$

Equation 13 then simplifies to

$$\begin{aligned} \mathbf{x}(t) &= k_1 e^t \left(\begin{bmatrix} 0 \\ 1 \end{bmatrix} \cos t - \begin{bmatrix} 1 \\ 0 \end{bmatrix} \sin t \right) + k_2 e^t \left(\begin{bmatrix} 1 \\ 0 \end{bmatrix} \cos t + \begin{bmatrix} 0 \\ 1 \end{bmatrix} \sin t \right) \\ &= e^t \begin{bmatrix} -k_1 \sin t + k_2 \cos t \\ k_1 \cos t + k_2 \sin t \end{bmatrix} \end{aligned}$$

The eigenvalues of E are $\lambda = -1 \pm i$, with eigenvectors

$$\mathbf{v} = \begin{bmatrix} 0 \\ 1 \end{bmatrix} \pm i \begin{bmatrix} -1 \\ 0 \end{bmatrix}$$

Therefore

$$a = -1, \quad b = 1, \quad \mathbf{u} = \begin{bmatrix} 0 \\ 1 \end{bmatrix}, \quad \text{and} \quad \mathbf{w} = \begin{bmatrix} -1 \\ 0 \end{bmatrix}$$

Equation 13 in this case becomes

$$\mathbf{x}(t) = e^{-t} \begin{bmatrix} k_1 \sin t - k_2 \cos t \\ k_1 \cos t + k_2 \sin t \end{bmatrix}$$

Finally, to obtain unique solutions in each case we need to specify an initial condition. Suppose that, in both cases,

$$\mathbf{x}(0) = \begin{bmatrix} 0 \\ 2 \end{bmatrix}$$

Equation 14 then becomes

$$\begin{bmatrix} 0 & 1 \\ 1 & 0 \end{bmatrix} \begin{bmatrix} k_1 \\ k_2 \end{bmatrix} = \begin{bmatrix} 0 \\ 2 \end{bmatrix} \qquad \text{for matrix } D$$

or

$$\begin{bmatrix} 0 & -1 \\ 1 & 0 \end{bmatrix} \begin{bmatrix} k_1 \\ k_2 \end{bmatrix} = \begin{bmatrix} 0 \\ 2 \end{bmatrix} \qquad \text{for matrix } E$$

In both cases the solution is $k_1 = 2$ and $k_2 = 0$. Figures 6 and 7 display several solution curves, along with the nullclines and the solutions to each of these initial-value problems. Note that the eigenvectors cannot be plotted in the phase plane because they are complex-valued.

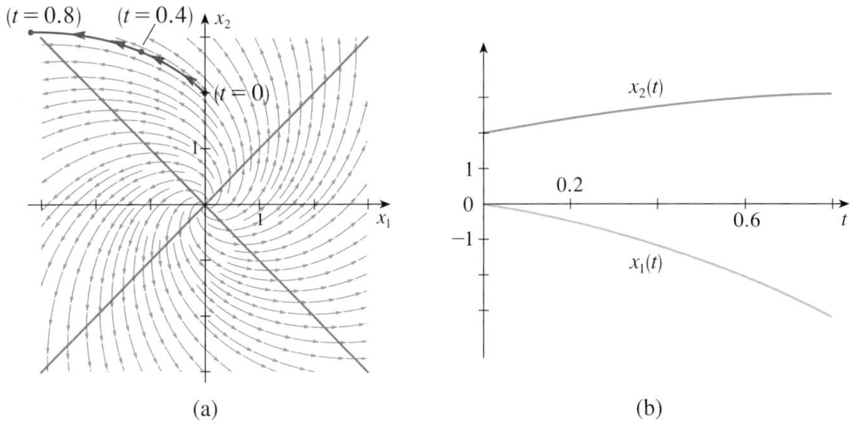

FIGURE 6
Part (a) plots the phase plane for matrix D, with nullclines in purple and solution curves in blue. The red curve is the solution to the initial-value problem, and its components are plotted against time in part (b).

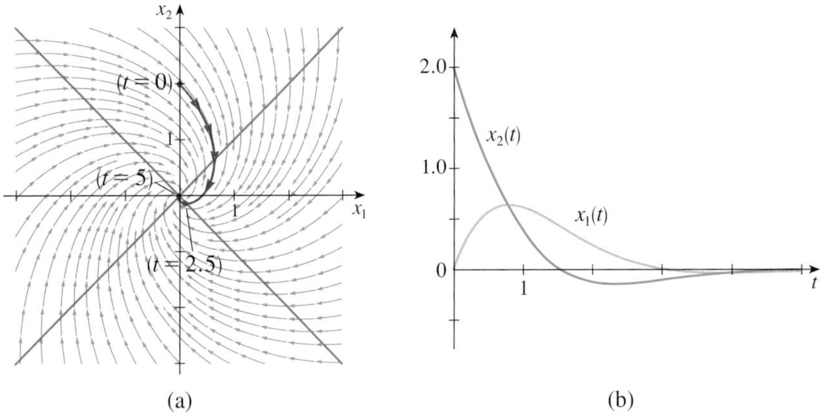

FIGURE 7
Part (a) plots the phase plane for matrix E, with nullclines in purple and solution curves in blue. The red curve is the solution to the initial-value problem, and its components are plotted against time in part (b).

Long-Term Behavior

Equation 8 for the solution to two-dimensional systems can be used to prove the following theorem about long-term behavior.

> **(15) Theorem** The origin of the initial-value problem (1) is a stable equilibrium if and only if the real parts of both eigenvalues of A are negative.

Although we have focused only on the case where the coefficient matrix has distinct eigenvalues, Theorem 15 holds in general.

The **trace** of a square matrix is the sum of its diagonal elements. Another useful result relates the determinant and the trace of A to the stability of the origin.

> **(16) Theorem** The origin of initial-value problem (1) is a stable equilibrium if and only if $\det A > 0$ and $\text{trace } A < 0$.

Theorems 15 and 16 are proved in Exercises 36 and 37. The stability properties of the equilibrium at the origin can be summarized graphically in terms of the determinant and trace of A. In particular, Exercises 37 and 38 show that the stability of the origin can be described as in Figure 8.

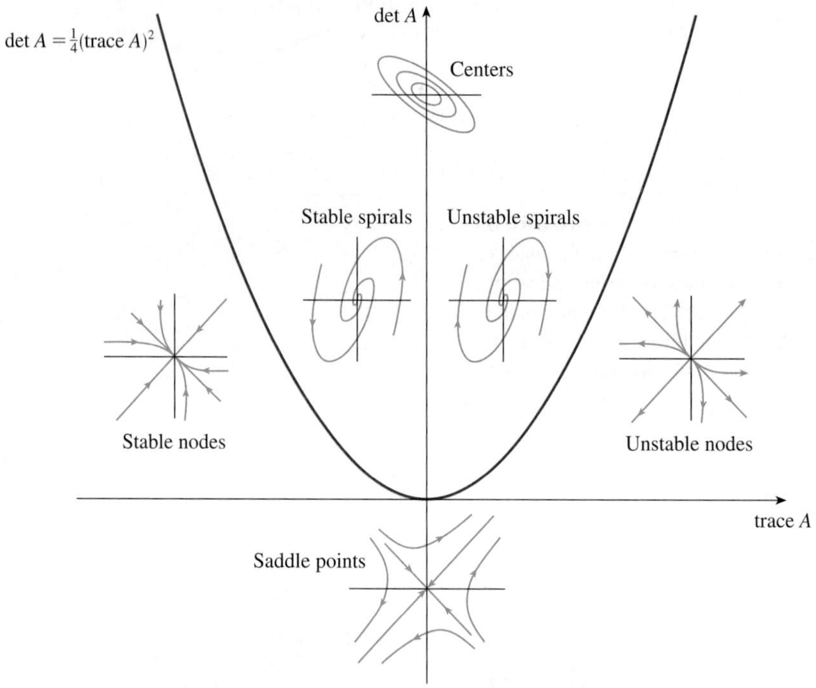

FIGURE 8
The stability properties of the equilibrium at the origin in terms of the determinant and trace of A.

EXERCISES 10.2

1–4 Show that $x_1(t)$ and $x_2(t)$ are solutions to the system of differential equations $d\mathbf{x}/dt = A\mathbf{x}$.

1. $A = \begin{bmatrix} 3 & -2 \\ 2 & -2 \end{bmatrix}$

$x_1(t) = \frac{1}{3}(4e^{2t} - e^{-t}),\quad x_2(t) = \frac{2}{3}(e^{2t} - e^{-t})$

2. $A = \begin{bmatrix} -2 & 1 \\ 1 & -2 \end{bmatrix}$

$x_1(t) = \frac{1}{2}(e^{-3t} + e^{-t}),\quad x_2(t) = \frac{1}{2}(-e^{-3t} + e^{-t})$

3. $A = \begin{bmatrix} -1 & -4 \\ 1 & -1 \end{bmatrix}$

$x_1(t) = e^{-t}\cos 2t,\quad x_2(t) = \frac{1}{2}e^{-t}\sin 2t$

4. $A = \begin{bmatrix} 1 & 2 \\ -4 & 1 \end{bmatrix}$

$x_1(t) = e^{t}\cos\left(2\sqrt{2}\,t\right),\quad x_2(t) = -\sqrt{2}\,e^{t}\sin\left(2\sqrt{2}\,t\right)$

5–8 Show that $x_1(t)$ and $x_2(t)$ are solutions to the initial-value problem $d\mathbf{x}/dt = A\mathbf{x}$ with $\mathbf{x}(0) = \mathbf{x}_0$.

5. $A = \begin{bmatrix} 3 & -2 \\ 6 & -4 \end{bmatrix} \quad \mathbf{x}_0 = \begin{bmatrix} 1 \\ 2 \end{bmatrix}$

$x_1(t) = e^{-t},\quad x_2(t) = 2e^{-t}$

6. $A = \begin{bmatrix} -2 & 1 \\ -5 & 4 \end{bmatrix} \quad \mathbf{x}_0 = \begin{bmatrix} 1 \\ 5 \end{bmatrix}$

$x_1(t) = e^{3t},\quad x_2(t) = 5e^{3t}$

7. $A = \begin{bmatrix} 1 & -1 \\ 1 & 1 \end{bmatrix} \quad \mathbf{x}_0 = \begin{bmatrix} 2 \\ 1 \end{bmatrix}$

$x_1(t) = e^{t}(2\cos t - \sin t),\quad x_2(t) = e^{t}(\cos t + 2\sin t)$

8. $A = \begin{bmatrix} -3 & 2 \\ -1 & -1 \end{bmatrix} \quad \mathbf{x}_0 = \begin{bmatrix} -2 \\ 1 \end{bmatrix}$

$x_1(t) = e^{-2t}(-2\cos t + 4\sin t),$

$x_2(t) = e^{-2t}(\cos t + 3\sin t)$

9. Prove the Superposition Principle.

10. Show that if the eigenvalues of a 2×2 matrix are real and distinct, then the matrix P whose columns are the corresponding eigenvectors is nonsingular.

11–16 Sketch several solution curves in the phase plane of the system of differential equations $d\mathbf{x}/dt = A\mathbf{x}$ using the given eigenvalues and eigenvectors of A.

11. $\lambda_1 = -1, \quad \lambda_2 = -2; \qquad \mathbf{v}_1 = \begin{bmatrix} 1 \\ 1 \end{bmatrix} \quad \mathbf{v}_2 = \begin{bmatrix} -1 \\ 1 \end{bmatrix}$

12. $\lambda_1 = 2, \quad \lambda_2 = 4; \qquad \mathbf{v}_1 = \begin{bmatrix} 1 \\ 2 \end{bmatrix} \quad \mathbf{v}_2 = \begin{bmatrix} 0 \\ 1 \end{bmatrix}$

13. $\lambda_1 = 2, \quad \lambda_2 = -2; \qquad \mathbf{v}_1 = \begin{bmatrix} 3 \\ 1 \end{bmatrix} \quad \mathbf{v}_2 = \begin{bmatrix} 1 \\ 1 \end{bmatrix}$

14. $\lambda_1 = -3, \quad \lambda_2 = -2; \qquad \mathbf{v}_1 = \begin{bmatrix} 1 \\ 0 \end{bmatrix} \quad \mathbf{v}_2 = \begin{bmatrix} 0 \\ 1 \end{bmatrix}$

15. $\lambda_1 = 5, \quad \lambda_2 = 1; \qquad \mathbf{v}_1 = \begin{bmatrix} 2 \\ 2 \end{bmatrix} \quad \mathbf{v}_2 = \begin{bmatrix} -2 \\ 1 \end{bmatrix}$

16. $\lambda_1 = 1, \quad \lambda_2 = -1; \qquad \mathbf{v}_1 = \begin{bmatrix} 3 \\ 2 \end{bmatrix} \quad \mathbf{v}_2 = \begin{bmatrix} -4 \\ 1 \end{bmatrix}$

17–28 Solve the initial value problem $d\mathbf{x}/dt = A\mathbf{x}$ with $\mathbf{x}(0) = \mathbf{x}_0$.

17. $A = \begin{bmatrix} -\frac{3}{2} & \frac{1}{2} \\ \frac{1}{2} & -\frac{3}{2} \end{bmatrix} \qquad \mathbf{x}_0 = \begin{bmatrix} 1 \\ 2 \end{bmatrix}$

18. $A = \begin{bmatrix} \frac{1}{2} & -\frac{3}{2} \\ -\frac{3}{2} & \frac{1}{2} \end{bmatrix} \qquad \mathbf{x}_0 = \begin{bmatrix} 1 \\ 2 \end{bmatrix}$

19. $A = \begin{bmatrix} 1 & 0 \\ 4 & -1 \end{bmatrix} \qquad \mathbf{x}_0 = \begin{bmatrix} 3 \\ 2 \end{bmatrix}$

20. $A = \begin{bmatrix} -1 & -2 \\ 2 & -2 \end{bmatrix} \qquad \mathbf{x}_0 = \begin{bmatrix} 1 \\ 5 \end{bmatrix}$

21. $A = \begin{bmatrix} -3 & 4 \\ -6 & 7 \end{bmatrix} \qquad \mathbf{x}_0 = \begin{bmatrix} -1 \\ 3 \end{bmatrix}$

22. $A = \begin{bmatrix} 0 & 1 \\ -6 & -5 \end{bmatrix} \qquad \mathbf{x}_0 = \begin{bmatrix} -1 \\ -2 \end{bmatrix}$

23. $A = \begin{bmatrix} -1 & 2 \\ -3 & -1 \end{bmatrix} \qquad \mathbf{x}_0 = \begin{bmatrix} 2 \\ 0 \end{bmatrix}$

24. $A = \begin{bmatrix} 3 & 0 \\ 0 & 1 \end{bmatrix} \qquad \mathbf{x}_0 = \begin{bmatrix} -2 \\ 4 \end{bmatrix}$

25. $A = \begin{bmatrix} 0 & -1 \\ -1 & 0 \end{bmatrix} \qquad \mathbf{x}_0 = \begin{bmatrix} 2 \\ 1 \end{bmatrix}$

26. $A = \begin{bmatrix} 4 & -2 \\ 3 & 1 \end{bmatrix} \qquad \mathbf{x}_0 = \begin{bmatrix} 0 \\ 1 \end{bmatrix}$

27. $A = \begin{bmatrix} 2 & -5 \\ 2 & 1 \end{bmatrix} \qquad \mathbf{x}_0 = \begin{bmatrix} 1 \\ 1 \end{bmatrix}$

28. $A = \begin{bmatrix} 3 & -4 \\ 1 & -3 \end{bmatrix} \qquad \mathbf{x}_0 = \begin{bmatrix} 2 \\ 3 \end{bmatrix}$

29. In Exercise 10.1.24 we considered the *nongeneric* system of differential equations

$$\frac{d\mathbf{x}}{dt} = \begin{bmatrix} -2 & -1 \\ 2 & 1 \end{bmatrix}\mathbf{x}$$

Theorem 2 applies to this system, and we can obtain the general solution (8) in the usual way. Do so.

30. When the eigenvalues of the coefficient matrix are complex, the origin is a spiral. If the eigenvalues are purely imaginary (that is, the real parts are zero), then the origin is called a *center*. For example, this is true of the following system:

$$\frac{d\mathbf{x}}{dt} = \begin{bmatrix} -1 & 2 \\ -3 & 1 \end{bmatrix}\mathbf{x}$$

Theorem 2 still applies to this system, and we can obtain the general solution (13) in the usual way.

(a) Find the general solution.

(b) Use the general solution in part (a) to prove that the solutions form closed curves in the phase plane.

Repeated Eigenvalues

31. Our focus has been on systems whose coefficient matrices have distinct eigenvalues. A simple example of a system with **repeated eigenvalues** is

$$\frac{d\mathbf{x}}{dt} = \begin{bmatrix} -1 & 0 \\ 0 & -1 \end{bmatrix}\mathbf{x}$$

(a) Show that $x_1(t) = c_1e^{-t}$ and $x_2(t) = c_2e^{-t}$ is a solution. The origin in this case is called a **proper node**.

(b) Try obtaining this general solution by calculating the eigenvectors and eigenvalues of the coefficient matrix. Comment on anything unusual that occurs.

32. A slightly more complicated system with **repeated eigenvalues** is

$$\frac{d\mathbf{x}}{dt} = \begin{bmatrix} -1 & 1 \\ 0 & -1 \end{bmatrix}\mathbf{x}$$

(a) Show that $x_1(t) = c_1e^{-t} + c_2te^{-t}$ and $x_2(t) = c_2e^{-t}$ is a solution. The origin in this case is called an **improper node**.

(b) Try obtaining this general solution by calculating the eigenvectors and eigenvalues of the coefficient matrix. Comment on anything unusual that occurs.

33–35 The system of differential equations $d\mathbf{x}/dt = A\mathbf{x}$ depends on a real-valued constant a. Use the eigenvalues to determine the stability properties of the equilibrium at the origin for all values of a.

33. $A = \begin{bmatrix} -1 & 1 \\ a & -1 \end{bmatrix}$

34. $A = \begin{bmatrix} -1 & a \\ 1 & 1 \end{bmatrix}$

35. $A = \begin{bmatrix} a & 1 \\ 1 & a \end{bmatrix}$

36. Use the general solution (Equation 8) and Euler's formula to prove Theorem 15 for the case where the eigenvalues of the coefficient matrix A are distinct.

37. Use Theorem 15 to prove the **trace and determinant condition for stability** given by Theorem 16. [*Hint:* Express the equations for the eigenvalues of a 2×2 matrix in terms of the trace and determinant, and then use Theorem 15].

38. Justify the summary of the qualitative behavior depicted in Figure 8. Be sure to explain where the curve defined by $\det A = \frac{1}{4}(\text{trace } A)^2$ comes from.

39. Provide an argument, based on Theorem 2, for why solution curves in the phase plane of a two-dimensional, autonomous system of linear differential equations cannot cross.

10.3 Applications

We now illustrate how the results from the preceding sections can be applied to a variety of areas in the life sciences.

Metapopulations

Many biological populations are subdivided into smaller subpopulations, each living in its own patch, with limited movement between them. The entire collection of such subpopulations is called a **metapopulation**. Consider the following model of two subpopulations, where x_A and x_B are the number of individuals in each:

$$\frac{dx_A}{dt} = r_A x_A - m_A x_A + m_B x_B$$

$$\frac{dx_B}{dt} = r_B x_B - m_B x_B + m_A x_A$$

Here r_i is the per capita growth rate of subpopulation i, and m_i is the per capita movement rate from patch i into the other patch. A specific example of the model is shown in Figure 1, where $r_A = 1$, $r_B = -2$, $m_A = 2$, and $m_B = 0$.

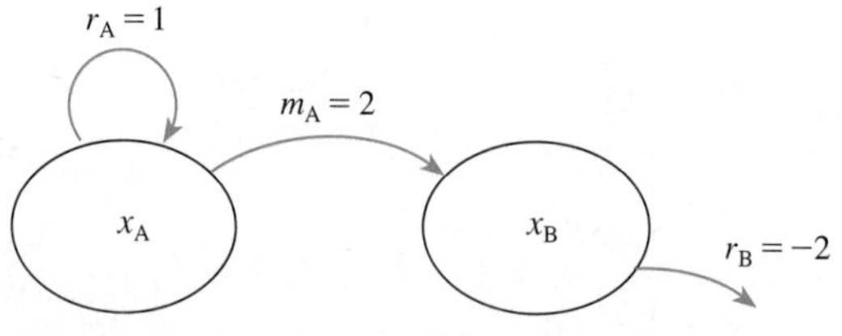

FIGURE 1

With these choices the equations become

$$\frac{dx_A}{dt} = x_A - 2x_A$$

$$\frac{dx_B}{dt} = -2x_B + 2x_A$$

Individuals in subpopulation A have a net per capita growth rate of $r_A = 1$, while those in subpopulation B die off, having a net per capita growth rate of $r_B = -2$. Individuals in subpopulation A also move to subpopulation B at per capita rate $m_A = 2$, whereas individuals in subpopulation B never move (see Figure 1). We suppose the initial size of each subpopulation is $x_A(0) = 100$ and $x_B(0) = 50$.

What is the size of each subpopulation as a function of time? Answering this question requires that we solve an initial-value problem. In matrix notation we have

(1) $$\frac{d\mathbf{x}}{dt} = A\mathbf{x}$$

with

(2) $$A = \begin{bmatrix} -1 & 0 \\ 2 & -2 \end{bmatrix} \quad \text{and} \quad \mathbf{x}(0) = \begin{bmatrix} 100 \\ 50 \end{bmatrix}$$

The eigenvalues of matrix A are readily found to be $\lambda_1 = -2$ and $\lambda_2 = -1$. (Recall from Exercise 8.7.17 that the eigenvalues of a lower triangular matrix lie on the diagonal.) Therefore, from the results of Section 10.2, the origin in this model is a stable node. The metapopulation therefore goes extinct as $t \to \infty$.

The eigenvectors are found to be

$$\mathbf{v}_1 = \begin{bmatrix} 0 \\ 1 \end{bmatrix} \quad \text{and} \quad \mathbf{v}_2 = \begin{bmatrix} 1 \\ 2 \end{bmatrix}$$

Therefore the general solution to the differential equation (1) is

(3) $$\mathbf{x}(t) = c_1 e^{-2t}\begin{bmatrix} 0 \\ 1 \end{bmatrix} + c_2 e^{-t}\begin{bmatrix} 1 \\ 2 \end{bmatrix}$$

To find the constants c_1 and c_2 we make use of the initial condition. At $t = 0$, Equation 3 is

$$\mathbf{x}(0) = c_1\begin{bmatrix} 0 \\ 1 \end{bmatrix} + c_2\begin{bmatrix} 1 \\ 2 \end{bmatrix} = \begin{bmatrix} 0 & 1 \\ 1 & 2 \end{bmatrix}\begin{bmatrix} c_1 \\ c_2 \end{bmatrix}$$

Therefore we must choose c_1 and c_2 to satisfy the equation

$$\begin{bmatrix} 0 & 1 \\ 1 & 2 \end{bmatrix}\begin{bmatrix} c_1 \\ c_2 \end{bmatrix} = \begin{bmatrix} 100 \\ 50 \end{bmatrix}$$

We obtain $c_1 = -150$ and $c_2 = 100$. The solution to the initial-value problem is therefore

(4) $$\mathbf{x}(t) = -150e^{-2t}\begin{bmatrix} 0 \\ 1 \end{bmatrix} + 100e^{-t}\begin{bmatrix} 1 \\ 2 \end{bmatrix}$$

Figure 2 shows the phase plane for this model, along with the nullclines, the eigenvectors, and the solution (4). Biologically, we are interested in only the positive quadrant of the phase plane.

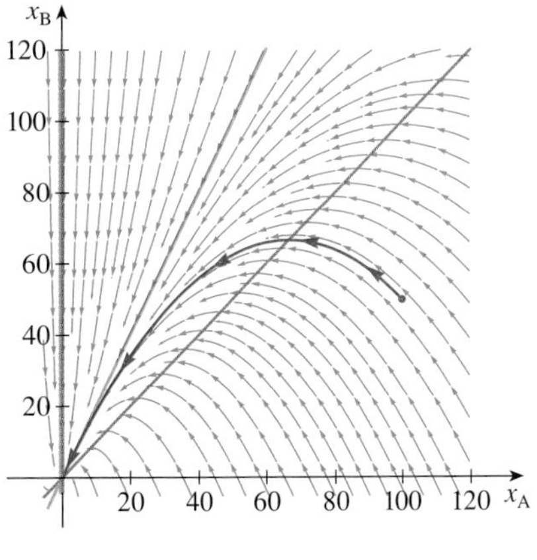

FIGURE 2
Purple lines are nullclines. Orange lines are all scalar multiples of the eigenvectors. Blue curves show a family of solutions. The red curve is the solution to the initial-value problem.

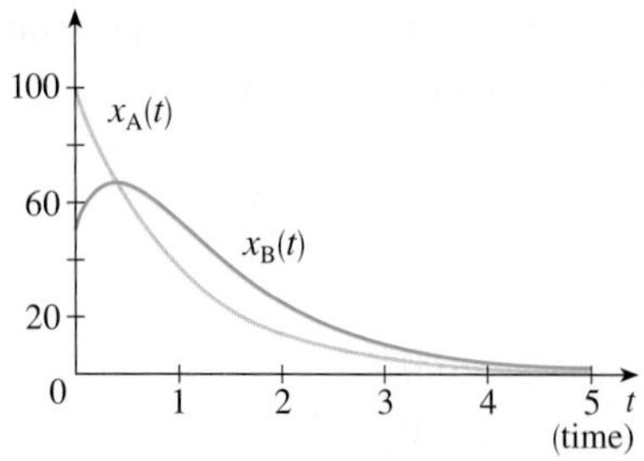

FIGURE 3

To better understand the fate of each individual subpopulation, we can look at each component of Equation 4 individually. They are

$$x_A(t) = 100e^{-t}$$

$$x_B(t) = 200e^{-t} - 150e^{-2t}$$

Figure 3 plots each component of the solution as a function of time. Both Figure 2 and Figure 3 reveal that the metapopulation ultimately goes extinct. Subpopulation A immediately declines to extinction, whereas subpopulation B initially increases in size before eventually going extinct.

■ Natural Killer Cells and Immunity

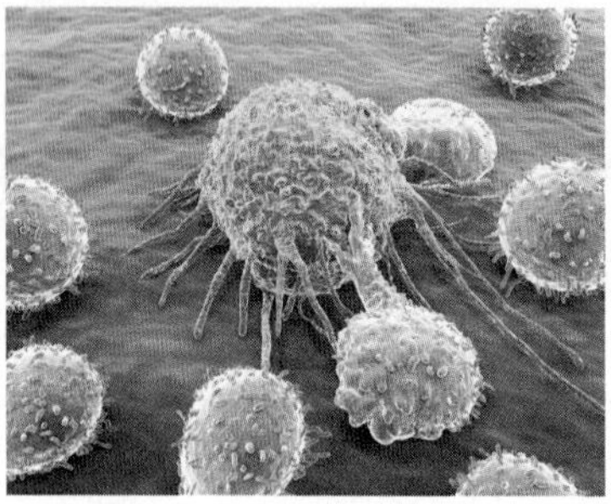

FIGURE 4
NK cells attacking a pathogen

Natural killer cells (often abbreviated NK cells) are components of the human immune system that defend the body against a variety of diseases, including cancers and pathogens (see Figure 4). NK cells are typically found in one of two states: latent or actively dividing. Suppose that latent cells die at a per capita rate d and are recruited to the dividing state at per capita rate r. Dividing cells replicate at a per capita rate k and die at a per capita rate h. All constants are positive.

These assumptions lead to the following system of differential equations describing each kind of NK cell, where $L(t)$ and $D(t)$ are the numbers of latent and dividing cells:[1]

$$\frac{dL}{dt} = -rL - dL$$

$$\frac{dD}{dt} = rL + (k - h)D$$

At $t = 0$ all NK cells are in the latent state and therefore the initial condition is $L(0) = L_0$ and $D(0) = 0$. We want to determine the number of each NK cell type as a function of time.

Estimates exist for each of the constants in this model, but these values change under different conditions. We could substitute any given constant values into the model and then solve the initial-value problem, but it would be better to have a solution in terms of arbitrary constants so that we could apply our results under different conditions without repeatedly having to solve the initial-value problem. This approach is illustrated here.

If $\mathbf{x}(t)$ denotes the vector whose components are $L(t)$ and $D(t)$, the model can be written in matrix notation as

$$\frac{d\mathbf{x}}{dt} = A\mathbf{x}$$

with

$$A = \begin{bmatrix} -(r + d) & 0 \\ r & k - h \end{bmatrix} \quad \text{and} \quad \mathbf{x}(0) = \begin{bmatrix} L_0 \\ 0 \end{bmatrix}$$

Matrix A is lower triangular and therefore its eigenvalues lie on the diagonal. They are $\lambda_1 = -(r + d)$ and $\lambda_2 = k - h$. All constants are positive and therefore $\lambda_1 < 0$, whereas λ_2 is positive if $k > h$ and negative if the reverse inequality holds. Therefore, from the results of Section 10.2, the origin is a stable node if $k < h$ and it is a saddle if $k > h$.

1. Y. Zhao et al., "Two-Compartment Model of NK Cell Proliferation: Insights from Population Response to IL-15 Stimulation," *Journal of Immunology* 188 (2012): 2981–90.

The eigenvectors are found to be

$$\mathbf{v}_1 = \begin{bmatrix} h - d - k - r \\ r \end{bmatrix} \quad \text{and} \quad \mathbf{v}_2 = \begin{bmatrix} 0 \\ 1 \end{bmatrix}$$

Therefore the general solution is

$$\textbf{(5)} \qquad \mathbf{x}(t) = c_1 e^{-(r+d)t} \begin{bmatrix} h - d - k - r \\ r \end{bmatrix} + c_2 e^{(k-h)t} \begin{bmatrix} 0 \\ 1 \end{bmatrix}$$

To obtain the unique member of this family that solves the initial-value problem, we use the initial condition. At $t = 0$, Equation 5 is

$$\mathbf{x}(0) = c_1 \begin{bmatrix} h - d - k - r \\ r \end{bmatrix} + c_2 \begin{bmatrix} 0 \\ 1 \end{bmatrix} = \begin{bmatrix} h - d - k - r & 0 \\ r & 1 \end{bmatrix} \begin{bmatrix} c_1 \\ c_2 \end{bmatrix}$$

Therefore c_1 and c_2 must satisfy the equation

$$\begin{bmatrix} h - d - k - r & 0 \\ r & 1 \end{bmatrix} \begin{bmatrix} c_1 \\ c_2 \end{bmatrix} = \begin{bmatrix} L_0 \\ 0 \end{bmatrix}$$

Solving, we obtain

$$c_1 = \frac{L_0}{h - d - k - r} \quad \text{and} \quad c_2 = \frac{-rL_0}{h - d - k - r}$$

Therefore the solution to the initial-value problem is

$$\textbf{(6)} \qquad \mathbf{x}(t) = L_0 e^{-(r+d)t} \begin{bmatrix} 1 \\ \dfrac{r}{h - d - k - r} \end{bmatrix} - L_0 e^{(k-h)t} \begin{bmatrix} 0 \\ \dfrac{r}{h - d - k - r} \end{bmatrix}$$

The phase plane for Equation 6 can now be constructed for any constants of interest. Figure 5 shows two examples, one in which the origin is a stable node and another in which it is a saddle.

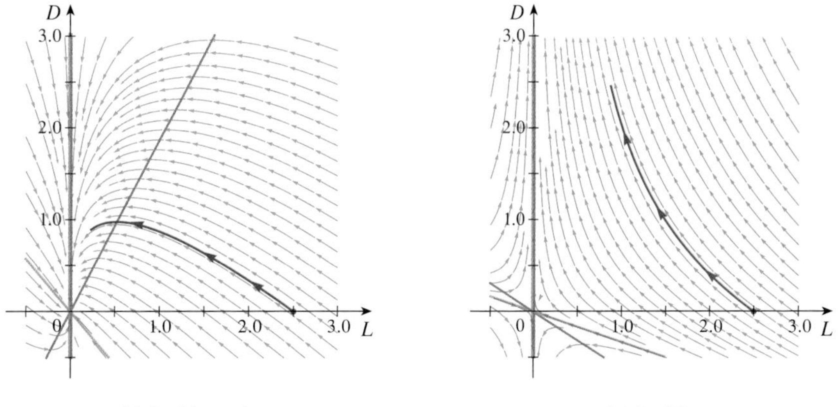

(a) Stable node

(b) Saddle

FIGURE 5
Purple lines are nullclines. Orange lines are all scalar multiples of the eigenvectors. Blue curves show a family of solutions. The red curve is the solution to the initial-value problem.

In this example it is of primary interest to examine the total number of cells $L + D$ as a function of time. Writing each component of Equation 6 individually gives

$$L(t) = L_0 e^{-(r+d)t}$$

$$D(t) = L_0 r \frac{e^{-(r+d)t} - e^{(k-h)t}}{h - d - k - r}$$

Figure 6 plots the sum $L(t) + D(t)$ for two different sets of values for the constants, along with experimental data for each.[2] In both cases the constants are such that the origin is a saddle.

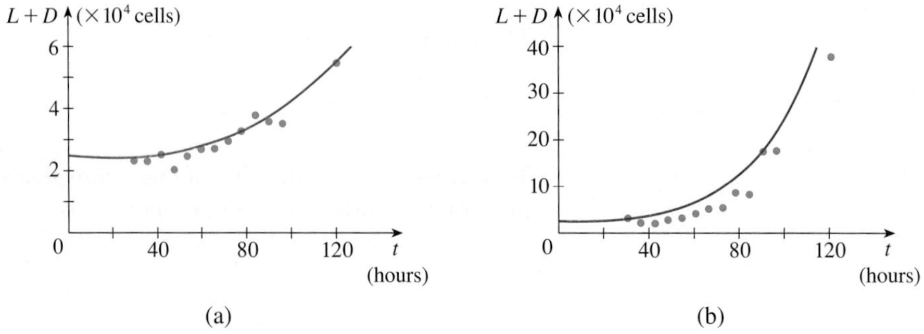

FIGURE 6

Gene Regulation

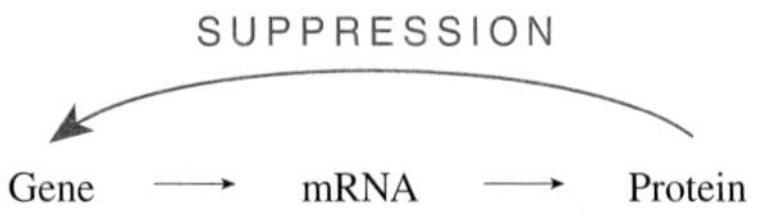

FIGURE 7

A gene produces protein by first creating an intermediate molecule called mRNA. High gene expression leads to a lot of mRNA molecules which, in turn, leads to a lot of protein. High levels of protein can suppress the production of mRNA, however, thereby regulating gene expression, as depicted in Figure 7.

Suppose $m(t)$ and $p(t)$ are the amounts of mRNA and protein in a cell (in hundreds of copies) at time t (in hours). A simple model of gene regulation is

$$\frac{dm}{dt} = f(p) - dm \qquad \frac{dp}{dt} = am - hp$$

where d is the rate at which mRNA molecules are degraded, a is the rate at which each mRNA molecule produces protein, and h is the rate at which protein molecules are degraded. The function $f(p)$ specifies the rate of production of mRNA (in hundreds of copies per hour) as a function of the amount of protein in the cell.

In this example we suppose that $f(p) = 1 - p$, meaning that the maximal rate of mRNA production is 100 copies per hour, and it decreases to 0 as the amount of protein increases to 100 copies. Further, for simplicity we use the values $d = 1$, $a = 1$, and $h = 1$. The model then reduces to

$$\frac{dm}{dt} = 1 - p - m \qquad \frac{dp}{dt} = m - p$$

or, in matrix notation,

(7)
$$\begin{bmatrix} \dfrac{dm}{dt} \\ \dfrac{dp}{dt} \end{bmatrix} = \begin{bmatrix} -1 & -1 \\ 1 & -1 \end{bmatrix} \begin{bmatrix} m \\ p \end{bmatrix} + \begin{bmatrix} 1 \\ 0 \end{bmatrix}$$

Suppose that, initially, there is no mRNA or protein; that is, $m(0) = 0$ and $p(0) = 0$.

Equation 7 is different from most models we have studied so far in that it is nonhomogeneous. As Exercise 10.1.23 shows, however, equations like (7) can be reduced to a system of homogeneous linear differential equations with a change of variables. We first find the equilibrium $(\hat{m}, \hat{p})$ of Equation 7 and then define the new variables $x_1(t) = m(t) - \hat{m}$ and $x_2(t) = p(t) - \hat{p}$.

2. Y. Zhao et al., "Two-Compartment Model of NK Cell Proliferation: Insights from Population Response to IL-15 Stimulation," *Journal of Immunology* 188 (2012): 2981–90.

Setting both $dm/dt = 0$ and $dp/dt = 0$ gives the equation

$$\begin{bmatrix} -1 & -1 \\ 1 & -1 \end{bmatrix}\begin{bmatrix} \hat{m} \\ \hat{p} \end{bmatrix} = -\begin{bmatrix} 1 \\ 0 \end{bmatrix}$$

We find that the equilibrium is $\hat{m} = \frac{1}{2}$ and $\hat{p} = \frac{1}{2}$. Therefore the new variables are $x_1(t) = m(t) - \frac{1}{2}$ and $x_2(t) = p(t) - \frac{1}{2}$. These represent the difference in the amounts from their equilibrium values.

With this change of variables we now proceed to model x_1 and x_2. Following the approach of Exercise 10.1.23, we see that the variables x_1 and x_2 satisfy the initial-value problem:

(8) $$\frac{d\mathbf{x}}{dt} = A\mathbf{x}$$

with

$$A = \begin{bmatrix} -1 & -1 \\ 1 & -1 \end{bmatrix} \quad \text{and} \quad \mathbf{x}(0) = \begin{bmatrix} 0 \\ 0 \end{bmatrix} - \begin{bmatrix} \frac{1}{2} \\ \frac{1}{2} \end{bmatrix} = \begin{bmatrix} -\frac{1}{2} \\ -\frac{1}{2} \end{bmatrix}$$

The eigenvalues of A are the complex conjugates $\lambda_1 = -1 + i$ and $\lambda_2 = -1 - i$. Therefore, from the results of Section 10.2, the origin is a stable spiral. In other words, both x_1 and x_2 approach zero as $t \to \infty$. This means that the variables m and p approach their equilibrium values as $t \to \infty$.

The eigenvectors associated with each eigenvalue are

$$\mathbf{v}_1 = \begin{bmatrix} i \\ 1 \end{bmatrix} \quad \text{and} \quad \mathbf{v}_2 = \begin{bmatrix} -i \\ 1 \end{bmatrix}$$

Using the notation in Expression 10.2.11, we have

$$\mathbf{u} = \begin{bmatrix} 0 \\ 1 \end{bmatrix} \quad \text{and} \quad \mathbf{w} = \begin{bmatrix} 1 \\ 0 \end{bmatrix}$$

Therefore, using Equation 10.2.13, we obtain the general solution

(9) $$\mathbf{x}(t) = k_1 e^{-t}(\mathbf{u}\cos t - \mathbf{w}\sin t) + k_2 e^{-t}(\mathbf{w}\cos t + \mathbf{u}\sin t)$$

To find the constants k_1 and k_2 we make use of the initial condition. At $t = 0$, Equation 9 is

$$\mathbf{x}(0) = k_1\mathbf{u} + k_2\mathbf{w} = k_1\begin{bmatrix} 0 \\ 1 \end{bmatrix} + k_2\begin{bmatrix} 1 \\ 0 \end{bmatrix} = \begin{bmatrix} 0 & 1 \\ 1 & 0 \end{bmatrix}\begin{bmatrix} k_1 \\ k_2 \end{bmatrix}$$

Therefore k_1 and k_2 must satisfy the equation

$$\begin{bmatrix} 0 & 1 \\ 1 & 0 \end{bmatrix}\begin{bmatrix} k_1 \\ k_2 \end{bmatrix} = \begin{bmatrix} -\frac{1}{2} \\ -\frac{1}{2} \end{bmatrix}$$

This equation requires that $k_1 = -\frac{1}{2}$ and $k_2 = -\frac{1}{2}$. Thus the solution to the initial-value problem for $\mathbf{x}(t)$ is

(10) $$\mathbf{x}(t) = -\tfrac{1}{2}e^{-t}(\mathbf{u}\cos t - \mathbf{w}\sin t) - \tfrac{1}{2}e^{-t}(\mathbf{w}\cos t + \mathbf{u}\sin t)$$

Substituting in the definitions of $\mathbf{u}$ and $\mathbf{w}$ gives the components of the solution (10) as

$$x_1(t) = \tfrac{1}{2}e^{-t}(\sin t - \cos t)$$

$$x_2(t) = -\tfrac{1}{2}e^{-t}(\sin t + \cos t)$$

The final step is to obtain the solution to the initial-value problem for the original variables m and p. From the definitions of x_1 and x_2 we simply need to add $\frac{1}{2}$ to each component of the solutions for x_1 and x_2, giving

$$m(t) = \tfrac{1}{2}e^{-t}(\sin t - \cos t) + \tfrac{1}{2}$$

$$p(t) = -\tfrac{1}{2}e^{-t}(\sin t + \cos t) + \tfrac{1}{2}$$

Figure 8 shows the phase plane for this model, along with the nullclines and the solution. Figure 9 plots m and p as functions of time. Figures 8 and 9 together show that the amounts of mRNA and protein in the cell oscillate slightly and both approach a nonzero equilibrium value.

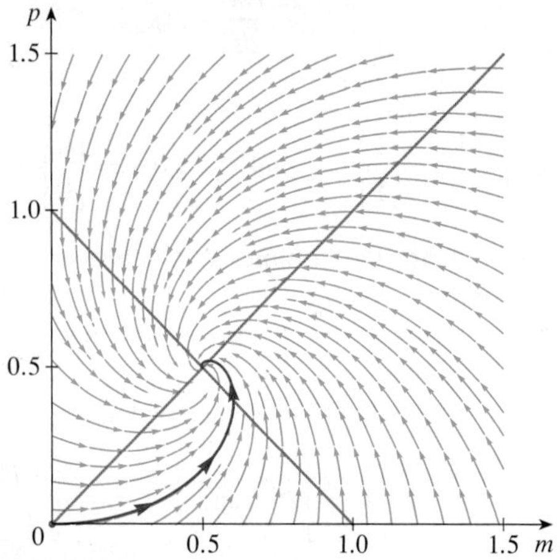

FIGURE 8
Purple lines are nullclines. Blue curves show a family of solutions. The red curve is the solution to the initial-value problem.

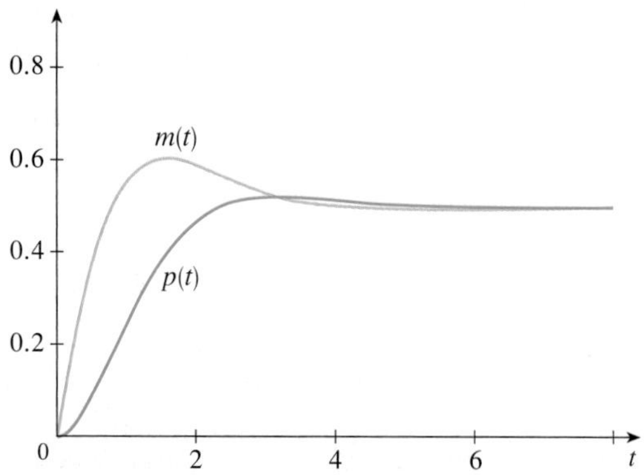

FIGURE 9
The components $m(t)$ and $p(t)$ of the red solution curve in Figure 8 are plotted against time.

■ Transport of Environmental Pollutants

This final example illustrates how we must sometimes manipulate a model before attempting to solve it. Suppose three lakes of equal volume are interconnected, as in Figure 10, with a net flow of water in the directions shown. A shipping accident releases 300,000 kg of a chemical pollutant into Lake 1. We would like to predict the amount of pollutant in each lake as a function of time.

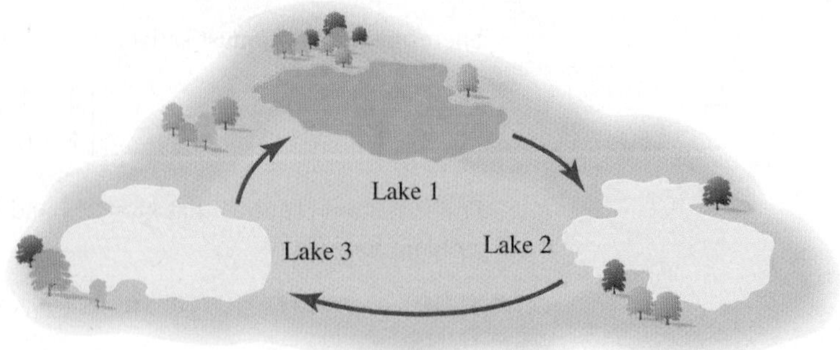

FIGURE 10

Let's construct a simplified model in which each lake is assumed to be well mixed and none of the pollutant settles out. The variables $l_1(t)$, $l_2(t)$, and $l_3(t)$ denote the mass of pollutant in each lake at time t. Each lake is viewed as a container with volume V, and

Our model is a mixing model similar to those of Exercises 13–18 in the Review Section of this chapter.

we suppose that the rate of flow of water between any pair of lakes is r, as illustrated in Figure 11.

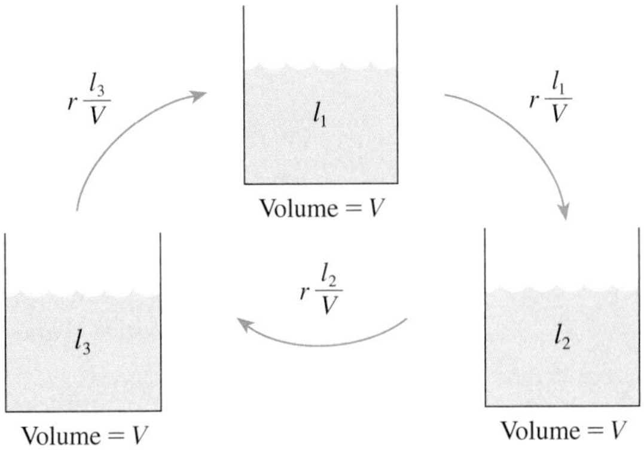

FIGURE 11

Each lake has an outflow and an inflow of pollutant. The outflow is the rate of flow of water r (in L/day) multiplied by the concentration of pollutant in the lake (in kg/L). The inflow is similarly the rate of flow r multiplied by the concentration of pollutant within the inflow (see Figure 11). This gives the following system of differential equations

$$\frac{dl_1}{dt} = -r\frac{l_1}{V} + r\frac{l_3}{V}$$
$$\frac{dl_2}{dt} = -r\frac{l_2}{V} + r\frac{l_1}{V}$$
$$\frac{dl_3}{dt} = -r\frac{l_3}{V} + r\frac{l_2}{V}$$

The initial condition is $l_1(0) = 300{,}000$ kg, $l_2(0) = 0$, and $l_3(0) = 0$.

Our model is a homogeneous system of linear differential equations, but it is three-dimensional, and our techniques so far have focused on systems of two equations. A bit of insight, however, allows us to reduce the system to two equations.

Since none of the pollutant ever settles out of the lakes, we would expect the total amount of pollutant in all three lakes to remain constant (at 300,000 kg). That is, we would expect the sum $l_1(t) + l_2(t) + l_3(t)$ to remain constant through time. To see that this is so, we differentiate this sum, obtaining

$$\begin{aligned} \frac{d}{dt}(l_1 + l_2 + l_3) &= \frac{dl_1}{dt} + \frac{dl_2}{dt} + \frac{dl_3}{dt} \\ &= -r\frac{l_1}{V} + r\frac{l_3}{V} - r\frac{l_2}{V} + r\frac{l_1}{V} - r\frac{l_3}{V} + r\frac{l_2}{V} \\ &= 0 \end{aligned}$$

This reveals that we need only two differential equations to track the pollutant. Let's track the amount in Lakes 1 and 2, with the amount in Lake 3 being $l_3(t) = 300{,}000 - l_1(t) - l_2(t)$. Making this substitution in the first two differential equations of our model then gives the system

$$\frac{dl_1}{dt} = -r\frac{l_1}{V} + r\frac{300{,}000 - l_1 - l_2}{V}$$
$$\frac{dl_2}{dt} = -r\frac{l_2}{V} + r\frac{l_1}{V}$$

where the third, redundant equation has been dropped.

Before proceeding, let's simplify the notation by defining the constant $\alpha = r/V$. In matrix notation the model is then

$$\frac{d\mathbf{l}}{dt} = \begin{bmatrix} -2\alpha & -\alpha \\ \alpha & -\alpha \end{bmatrix} \mathbf{l} + \begin{bmatrix} 300{,}000\alpha \\ 0 \end{bmatrix}$$

where $\mathbf{l}$ is the vector whose components are l_1 and l_2.

We have reduced the system of equations from three to two, but it is no longer homogeneous. Therefore the next step is to change variables as we did in the preceding example. We must first find the equilibrium values, $\hat{l}_1$ and $\hat{l}_2$, and then define the new variables $x_1(t) = l_1(t) - \hat{l}_1$ and $x_2(t) = l_2(t) - \hat{l}_2$.

Setting both $dl_1/dt = 0$ and $dl_2/dt = 0$ gives the equation

$$\begin{bmatrix} -2\alpha & -\alpha \\ \alpha & -\alpha \end{bmatrix} \begin{bmatrix} \hat{l}_1 \\ \hat{l}_2 \end{bmatrix} = -\begin{bmatrix} 300{,}000\alpha \\ 0 \end{bmatrix}$$

from which we obtain the values $\hat{l}_1 = 10^5$ and $\hat{l}_2 = 10^5$. Therefore the new variables are $x_1(t) = l_1(t) - 10^5$ and $x_2(t) = l_2(t) - 10^5$.

With this change of variables we then obtain the following initial-value problem for x_1 and x_2:

(11) $$\frac{d\mathbf{x}}{dt} = A\mathbf{x}$$

with

$$A = \begin{bmatrix} -2\alpha & -\alpha \\ \alpha & -\alpha \end{bmatrix} \quad \text{and} \quad \mathbf{x}(0) = \begin{bmatrix} 3 \times 10^5 \\ 0 \end{bmatrix} - \begin{bmatrix} 10^5 \\ 10^5 \end{bmatrix} = \begin{bmatrix} 2 \times 10^5 \\ -10^5 \end{bmatrix}$$

The eigenvalues of A are the complex conjugates

$$\lambda_1 = -\frac{3}{2}\alpha + i\frac{\sqrt{3}}{2}\alpha \qquad \lambda_2 = -\frac{3}{2}\alpha - i\frac{\sqrt{3}}{2}\alpha$$

Therefore the origin is a stable spiral because the real parts of the eigenvalues are negative. In other words, both x_1 and x_2 approach zero as $t \to \infty$. This means that the original variables l_1 and l_2 approach their equilibrium values as $t \to \infty$.

The eigenvectors associated with each eigenvalue are

$$\mathbf{v}_1 = \begin{bmatrix} -\frac{1}{2} + \frac{\sqrt{3}}{2}i \\ 1 \end{bmatrix} \quad \text{and} \quad \mathbf{v}_2 = \begin{bmatrix} -\frac{1}{2} - \frac{\sqrt{3}}{2}i \\ 1 \end{bmatrix}$$

Using the notation in Expression 10.2.11, we have

$$\mathbf{u} = \begin{bmatrix} -\frac{1}{2} \\ 1 \end{bmatrix} \quad \text{and} \quad \mathbf{w} = \begin{bmatrix} \frac{\sqrt{3}}{2} \\ 0 \end{bmatrix}$$

Therefore, using Equation 10.2.13, we obtain the general solution

(12) $$\mathbf{x}(t) = k_1 e^{-(3/2)\alpha t}\left(\mathbf{u}\cos\frac{\sqrt{3}}{2}\alpha t - \mathbf{w}\sin\frac{\sqrt{3}}{2}\alpha t\right) + k_2 e^{-(3/2)\alpha t}\left(\mathbf{w}\cos\frac{\sqrt{3}}{2}\alpha t + \mathbf{u}\sin\frac{\sqrt{3}}{2}\alpha t\right)$$

At $t = 0$, Equation 12 reduces to

$$\mathbf{x}(0) = k_1\mathbf{u} + k_2\mathbf{w} = k_1\begin{bmatrix} -\dfrac{1}{2} \\ 1 \end{bmatrix} + k_2\begin{bmatrix} \dfrac{\sqrt{3}}{2} \\ 0 \end{bmatrix} = \begin{bmatrix} -\dfrac{1}{2} & \dfrac{\sqrt{3}}{2} \\ 1 & 0 \end{bmatrix}\begin{bmatrix} k_1 \\ k_2 \end{bmatrix}$$

Therefore k_1 and k_2 must satisfy the equation

$$\begin{bmatrix} -\dfrac{1}{2} & \dfrac{\sqrt{3}}{2} \\ 1 & 0 \end{bmatrix}\begin{bmatrix} k_1 \\ k_2 \end{bmatrix} = \begin{bmatrix} 2 \times 10^5 \\ -10^5 \end{bmatrix}$$

Using Theorem 8.6.4, we therefore obtain $k_1 = -10^5$ and $k_2 = \sqrt{3} \times 10^5$.

The final step is to obtain the solution to the initial-value problem for the original variables l_1 and l_2 using (12). From the definitions of x_1 and x_2 we simply need to add 10^5 to each component of the vector given in (12). Figure 12 shows the phase plane for this model, along with the nullclines and the solution when $\alpha = 0.2$. Figure 13 plots l_1, l_2, and l_3 as functions of time. Figure 12 and Figure 13 both show that the amount of pollutant in each lake displays a slight oscillatory behavior. All lakes eventually contain the same amount of pollutant as t gets large.

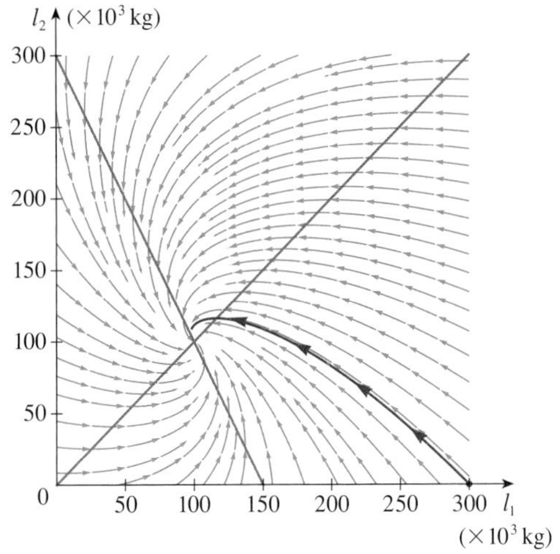

FIGURE 12
Purple lines are nullclines. Blue curves show a family of solutions. The red curve is the solution to the initial-value problem.

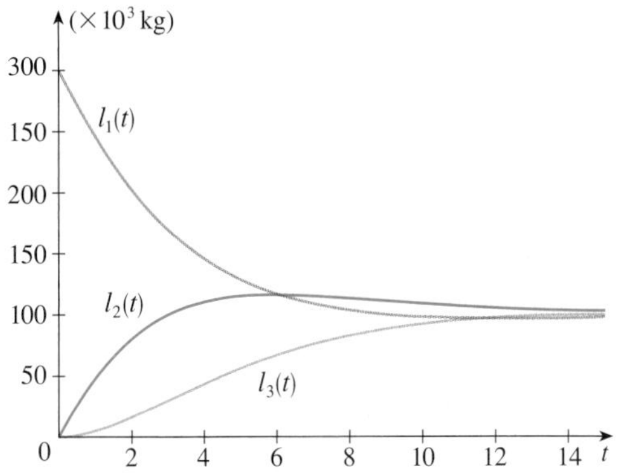

FIGURE 13
The components l_1 and l_2 of the red solution curve in Figure 12 are plotted against time, along with $l_3 = 300{,}000 - l_1 - l_2$.

EXERCISES 10.3

1. Jellyfish locomotion In Exercise 10.1.34 we introduced a model of jellyfish locomotion. It can be written as the following system of two linear differential equations

$$z_1' = z_2 \qquad z_2' = -\frac{k}{m}z_1 - \frac{b}{m}z_2$$

where z_1 is the displacement of the bell at time t and z_2 is its velocity. Find the general solution when $m = 100$ g, $b = 0.1$ N/m·s, and $k = 1$ N/m.

2. Hemodialysis is a process by which a machine is used to filter urea and other waste products from a patient's blood if the kidneys fail. The amount of urea within a patient during dialysis is sometimes modeled by supposing there are two compartments within the patient: the blood, which is directly filtered by the dialysis machine, and another compartment that cannot be directly filtered but that is connected to the blood. A system of two differential equations describing this is

$$\frac{dc}{dt} = -\frac{K}{V}c + ap - bc \qquad \frac{dp}{dt} = -ap + bc$$

where c and p are the urea concentrations in the blood and

the inaccessible pool (in mg/mL) and all constants are positive (see also Exercise 14 in the Review Section of this chapter). Suppose that $K/V = 1$, $a = b = \frac{1}{2}$, and the initial urea concentration is $c(0) = c_0$ and $p(0) = c_0$ mg/mL.
(a) Classify the equilibrium of this system.
(b) Solve this initial-value problem.

3. **Prostate cancer treatment** During the treatment of prostate cancer some tumor cells become resistant to medication through a variety of biochemical changes. Some of these changes are reversible and some are irreversible. In Exercise 10.1.32 we introduced a three-variable model for the different cell types. Under certain assumptions this can be reduced to the following two differential equations:

$$y_1' = -4y_1 \qquad y_2' = y_1 - 2y_2 + d$$

where y_1 and y_2 are the fractions of cells that are sensitive and irreversibly resistant to treatment, respectively.
(a) Use a change of variable to reduce this system to a homogeneous system of linear differential equations.
(b) Find the general solution to the system from part (a).
(c) Suppose that, at the beginning of treatment, all cells are sensitive [that is, $y_1(0) = 1$ and $y_2(0) = 0$]. What is the solution to this initial-value problem?
(d) What is the function specifying the fraction of cells that are irreversibly resistant as a function of time?
(e) How long after the start of treatment will 50% of the cells be irreversibly resistant? (Assume that $d > 1$.)

Source: Adapted from Y. Hirata et al., "Development of a Mathematical Model that Predicts the Outcome of Hormone Therapy for Prostate Cancer," *Journal of Theoretical Biology* 264 (2010): 517–27.

4. **Soil contamination** A crop is planted in soil that is contaminated with a pollutant. The pollutant gradually leaches out of the soil but is also absorbed by the growing crop. A simple model of this process is

$$\frac{ds}{dt} = -\alpha s - \beta s \qquad \frac{dc}{dt} = \beta s$$

where s and c are the amounts of pollutant in the soil and crop (in mg), respectively, and α and β are positive constants.
(a) Suppose that $s(0) = s_0$ and $c(0) = 0$. What is the solution to the initial-value problem?
(b) In the long term (that is, as $t \to \infty$), what is the amount of pollutant in the crop?

5. **Metastasis of malignant tumors** Metastasis is the process by which cancer cells spread throughout the body and initiate tumors in various organs. This sometimes happens via the bloodstream, where cancer cells become lodged in capillaries of organs and then move across the capillary wall into the organ. Using C to denote the number of cells lodged in a capillary and I for the number that have invaded the organ, we can model this as

$$C' = -\alpha C - \beta C \qquad I' = \alpha C - \delta I + \rho I$$

where all constants are positive, α is the rate of movement across the capillary wall, β is the rate of dislodgment from the capillary, δ is the rate at which cancer cells in the organ die, and ρ is their growth rate.
(a) Find the general solution.
(b) Classify the equilibrium at the origin when $\rho > \delta$ and when $\rho < \delta$.
(c) What is the solution to the initial-value problem if $C(0) = C_0$ and $I(0) = 0$?
(d) Use your result from part (c) to show that the tumor will grow in the long term if and only if $\rho > \delta$.

Source: Adapted from D. Kaplan et al., *Understanding Nonlinear Dynamics* (New York: Springer-Verlag, 1995).

6. **Radioimmunotherapy** is a cancer treatment in which radioactive atoms are attached to tumor-specific antibody molecules and then injected into the blood. The antibody molecules then attach only to tumor cells, where they then deliver the cell-killing radioactivity. The following model for this process was introduced in Example 10.1.1:

$$\frac{dx_1}{dt} = -ax_1 - bx_1 \qquad \frac{dx_2}{dt} = bx_1 - cx_2$$

where x_1 denotes the amount of antibody in the blood and x_2 the amount of antibody taken up by the tumor (both in μg). All constants are positive.
(a) Find the general solution.
(b) Suppose that $x_1(0) = x_0$ and $x_2(0) = 0$. What is the solution to this initial-value problem?

Source: Adapted from A. Flynn et al., "Effectiveness of Radiolabelled Antibodies for Radio-Immunotherapy in a Colorectal Xenograft Model: A Comparative Study Using the Linear-Quadratic Formulation," *International Journal of Radiation Biology* 77 (2001): 507–17.

7. **Cancer progression** The development of many cancers, such as colorectal cancer, proceed through a series of precancerous stages. Suppose there are $n - 1$ precancerous stages before developing into cancer at stage n. A simple system of differential equations modeling this is

$$x_0' = -u_0 x_0$$
$$x_i' = u_{i-1}x_{i-1} - u_i x_i$$
$$x_n' = u_{n-1}x_{n-1}$$

where x_i is the fraction of the population in state i, the u_i's are positive constants, and $i = 1, \ldots, n - 1$.
(a) Suppose $n = 2$. What is the system of differential equations for the three stages?
(b) Note that the variable x_2 does not appear in the equations for the rate of change of x_0 or x_1. Consequently, we can solve the two-dimensional system for x_0 and x_1 separately. Do so, assuming that $x_0(0) = k$ and $x_1(0) = 0$.
(c) Use your solution for $x_1(t)$ obtained in part (b) to write a differential equation for $x_2(t)$.
(d) Solve the differential equation from part (c), assuming $x_2(0) = 0$.

8. **Metapopulations** Consider a simple metapopulation in which subpopulation A grows at a per capita rate of $r_A = 1$

and subpopulation B declines at a per capita rate of $r_B = -1$. Suppose the per capita rate of movement between subpopulation patches is m in both directions. This gives

$$\frac{dx_A}{dt} = (1 - m)x_A + mx_B$$

$$\frac{dx_B}{dt} = -(1 + m)x_B + mx_A$$

where x_A and x_B are the numbers of individuals in patches A and B, respectively.

(a) Classify the equilibrium at the origin.
(b) Find the general solution.
(c) What is the solution to the initial-value problem if $x_A(0) = 1$ and $x_B(0) = 0$?

9. Suppose a glass of cold water is sitting in a warm room and you place a coin at room temperature R into the glass. The coin gradually cools down while, at the same time, the glass of water warms up. Newton's law of cooling suggests the following system of differential equations to describe the process

$$\frac{dw}{dt} = -k_w(w - R) \qquad \frac{dp}{dt} = -k_p(p - w)$$

where w and p are the temperatures of the water and coin (in °C), respectively, and the k's are positive constants.

(a) Explain the form of the system of differential equations and the assumptions that underlie them.
(b) Use a change of variables to obtain a homogeneous system.
(c) What is the general solution to the system you found in part (b)?
(d) What is the solution to the original initial-value problem if $w(0) = w_0$ and $p(0) = p_0$?

10. Vaccine coverage The project on page 479 explores an epidemiological model of vaccine coverage. For certain values of the constants the system of differential equations is

$$\frac{dN}{dt} = 2(1 - p)N + 2(1 - p)V - N$$

$$\frac{dV}{dt} = 2pV + 2pN - 3V$$

where N and V are the numbers of nonvaccinated and vaccinated infected people, respectively, and p is the fraction of the population that is vaccinated.

(a) Classify the equilibrium at the origin for all values of the vaccination coverage p.
(b) What is the critical vaccination coverage above which the infection dies out?
(c) What is the general solution?
(d) What is the solution to the initial-value problem where $N(0) = k$ and $V(0) = 0$?

11. Pulmonary air embolism is a type of blood clot, which can occur during surgery, whereby part of the lung is no longer able to exchange CO_2 for O_2. It is detected by a marked drop in the CO_2 concentration in exhaled air. A simple model of this process involves two lung compartments: a deep and a shallow compartment, each of which removes CO_2 from the blood.

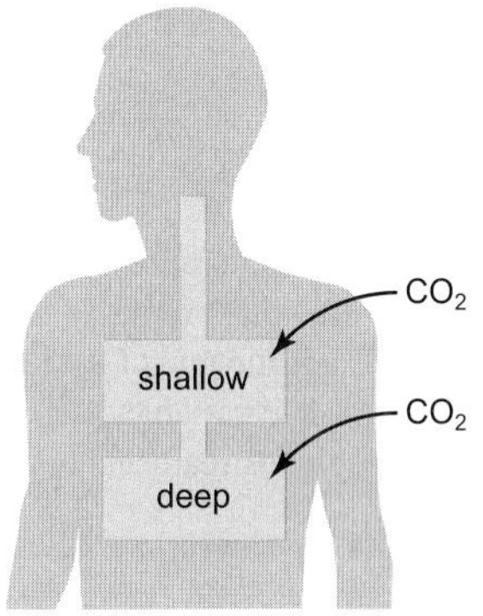

Using a for the rate of absorption of gas into each compartment from the blood, c as the concentration of CO_2 in this gas, and V as the volume of each compartment, we can model gas exchange during normal lung functioning as

$$\frac{dx_1}{dt} = ac + a\frac{x_2}{V} - 2a\frac{x_1}{V} \qquad \frac{dx_2}{dt} = ac - a\frac{x_2}{V}$$

where x_1 and x_2 are the amounts of CO_2 in the shallow and deep compartments, respectively. The concentration of CO_2 in the exhaled air at any time is given by x_1/V (see also Exercise 15 in the Review Section of this chapter).

(a) What is the equilibrium amount of CO_2 in the two compartments during normal lung functioning?
(b) If an embolism occurs in the deep lung, the equation for x_2 becomes $dx_2/dt = -ax_2/V$ because CO_2 is no longer entering this compartment. Use a change of variables to obtain a homogeneous system of differential equations for gas exchange during an embolism.
(c) The equilibrium values of x_1 and x_2 from part (a) can be used as the initial condition for the system of differential equations in part (b) to obtain an initial-value problem for the gas exchange once an embolism occurs. What is its solution?
(d) Use the solution found in part (c) to obtain a solution in term of the original variables x_1 and x_2.

Source: Adapted from S. Cruickshank, *Mathematics and Statistics in Anaesthesia.* (New York: Oxford University Press, USA, 2004).

12. Systemic lupus erythematosus is an autoimmune disease in which some immune molecules, called antibodies, target DNA instead of pathogens. This can be treated by injecting drugs that absorb the offending antibodies. The antibodies are found in both the bloodstream and in organs, and this can be modeled using a two-compartment model:

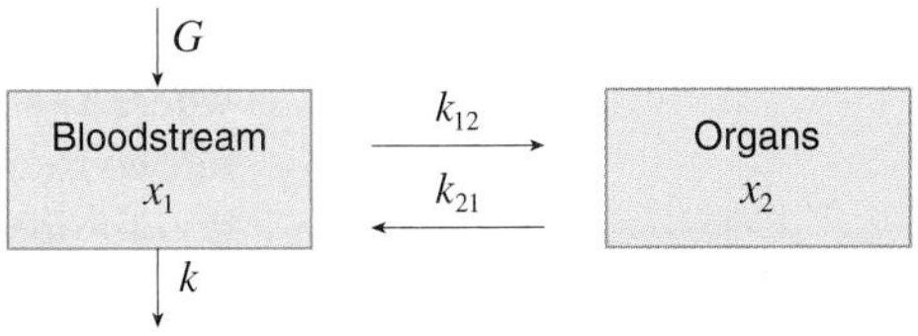

A system of differential equations describing the amount of antibody in each compartment is

$$\frac{dx_1}{dt} = G + k_{21}x_2 - k_{12}x_1 - kx_1$$

$$\frac{dx_2}{dt} = k_{12}x_1 - k_{21}x_2$$

where G is the rate of generation of antibodies, k is the rate at which the drug treatment removes antibody from the bloodstream, and k_{ij} is the rate of flow of antibody from compartment i to j. The variables x_1 and x_2 are the amounts of antibody in the bloodstream and organs, respectively, measured in μg. (See also Exercise 16 in the Review Section of this chapter.)

(a) Use a change of variables to obtain a homogeneous system of differential equations describing the situation.

(b) What is the general solution to the differential equations in part (a)?

(c) What is the general solution obtained in part (b) in terms of the original variables x_1 and x_2?

Source: Adapted from K. Suzuki et al., "Anti-DNA Antibody Kinetics Following Selective Removal by Adsorption using Dextran Sulphate Cellulose Columns in Patients with Systemic Lupus Erythematosus," *Journal of Clinical Apheresis* 11 (1996): 16–22.

■ PROJECT Pharmacokinetics of Antimicrobial Dosing

The term **pharmacokinetics** refers to the change in drug concentration within the body during treatment. Figure 1 gives an example of the serum concentration (in μg/mL) of the antimicrobial drug panobacumab in a patient during three consecutive infusions. The concentration increases nearly instantaneously during an infusion and it then declines through metabolism until the next infusion occurs.

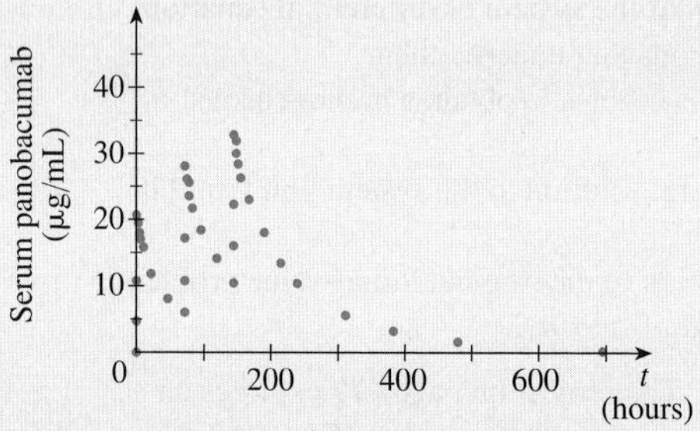

FIGURE 1

Mathematical models are routinely used to determine the drug dose and time between infusions required to achieve a desired peak serum level. A common model is a two-compartment mixing model. One compartment is the blood serum, and the other is the remainder of the body. The drug is infused directly into the serum and is metabolized from there. It also flows back and forth between the serum and the rest of the body (see Figure 2). Following the approach used in Exercises 7.4.45–48, we can obtain the following system of linear differential equations:

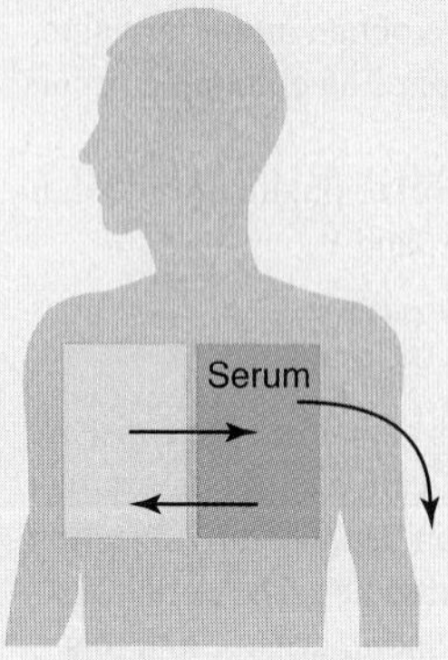

FIGURE 2

$$\frac{dx_1}{dt} = \frac{k_{21}}{V_2}x_2 - \frac{k_{12}}{V_1}x_1 - \frac{k_{10}}{V_1}x_1 \tag{1}$$

$$\frac{dx_2}{dt} = \frac{k_{12}}{V_1}x_1 - \frac{k_{21}}{V_2}x_2$$

where x_1 and x_2 are the amounts in the serum and body, respectively (in μg/mL), k_{ij} are rate constants, and V_i is the effective volume of compartment i (in mL). The initial condition for the model is $x_1(0) = d$ and $x_2(0) = 0$, where d is the drug dose that is infused. Models like (1) often provide an excellent fit to data, as shown in Figure 3.

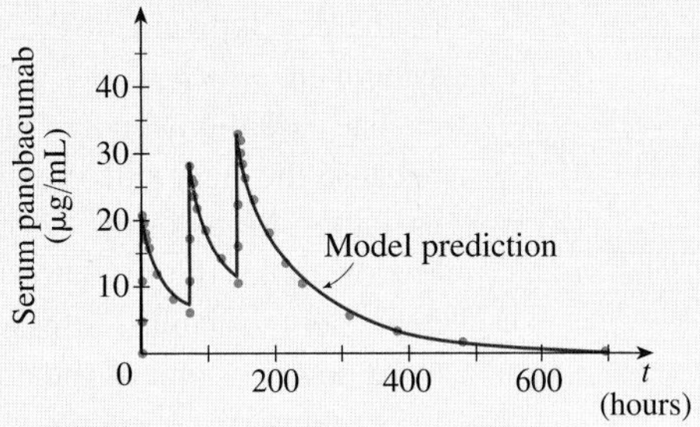

FIGURE 3

1. Explain the terms in Equations 1 and the assumptions that underlie them.

2. As an example, suppose that $k_{10}/V_1 = 2$, $k_{12}/V_1 = 1$, and $k_{21}/V_2 = 2$. Find the general solution to Equations 1.

3. Show that the general solution we obtained in Problem 2 can be written as $\varphi(t) \cdot \mathbf{c}$, where

$$\varphi(t) = \begin{bmatrix} | & | \\ \mathbf{v}_1 & \mathbf{v}_2 \\ | & | \end{bmatrix} \begin{bmatrix} e^{\lambda_1 t} & 0 \\ 0 & e^{\lambda_2 t} \end{bmatrix} \qquad \mathbf{c} = \begin{bmatrix} c_1 \\ c_2 \end{bmatrix}$$

with λ_i and $\mathbf{v}_i$ being the eigenvalues and eigenvectors of the coefficient matrix.

4. Using our result from Problem 3, show that the solution to the initial-value problem can be written as $\mathbf{x}(t) = B(t)\mathbf{x}_0$, where $B(t) = \varphi(t)\varphi^{-1}(0)$ and $\mathbf{x}_0$ is the vector of initial values (that is, its components are d and 0).

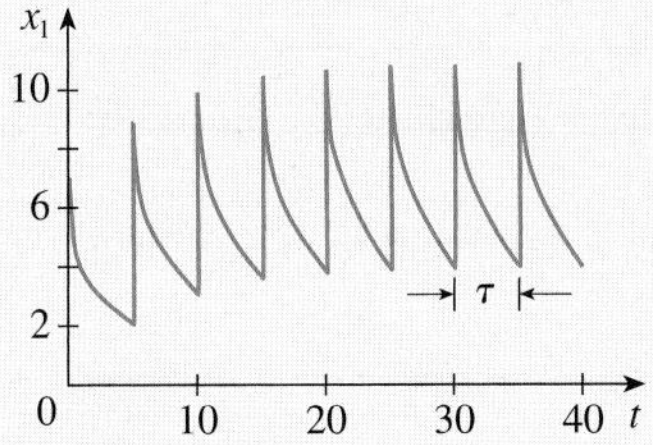

FIGURE 4

5. If drug infusions continue repeatedly at fixed intervals, the level of drug in the serum eventually displays periodic behavior, as shown in Figure 4. Suppose τ is the time between infusions. If the amount of drug in the serum and body immediately after an infusion is denoted by the vector $\mathbf{y}$, then from Problem 4 the amounts immediately before the next infusion are $B(\tau)\mathbf{y}$. Immediately after the next infusion the amounts are $B(\tau)\mathbf{y} + \mathbf{x}_0$. (Why?) Further, if the amounts in each compartment are displaying periodic behavior, then immediately after infusion they must be at the same level as they were immediately after the previous infusion; that is, $B(\tau)\mathbf{y} + \mathbf{x}_0 = \mathbf{y}$. Use this last equation to find the function relating peak serum level to the dose and infusion constants d and τ.

10.4 | Systems of Nonlinear Differential Equations

Systems of linear differential equations form the basis for analyzing the stability properties of equilibria of systems of nonlinear differential equations. The approach is illustrated in this section. As with the rest of this chapter, we focus on two-dimensional systems.

■ Linear and Nonlinear Differential Equations

In the preceding sections we studied autonomous systems of differential equations having the form

$$\textbf{(1)} \qquad \frac{dx_1}{dt} = a_{11}x_1 + a_{12}x_2 + g_1 \qquad \frac{dx_2}{dt} = a_{21}x_1 + a_{22}x_2 + g_2$$

In the development of the general techniques we saw that it was sufficient to analyze systems in which $g_1 = 0$ and $g_2 = 0$.

The equations in (1) are referred to as a system of linear differential equations because the right side of each differential equation is a linear function of the dependent variables. Many of the systems introduced in Sections 7.5 and 7.6 do not have this form. For example, the predator-prey equations on page 469 involve the product of the two dependent variables. Similarly, the Fitzhugh-Nagumo equations on page 475 involve the third power of one of the dependent variables.

Systems of autonomous differential equations that are not linear functions of the dependent variables are referred to as **systems of autonomous nonlinear differential equations**. They have the more general form

(2)
$$\frac{dx_1}{dt} = f_1(x_1, x_2)$$
$$\frac{dx_2}{dt} = f_2(x_1, x_2)$$

where f_1 and f_2 are arbitrary functions of the dependent variables. We will assume that both f_1 and f_2 have continuous first partial derivatives.

The nullclines of linear systems of differential equations are straight lines and therefore such systems typically have a single equilibrium. As we saw in Chapter 7, nonlinear systems often have several nonlinear nullclines and therefore they often have several equilibria as well. In general, it is not possible to solve systems of equations like (2) and so we focus on finding equilibria and determining their stability properties.

(3) Definition An **equilibrium** of Equations 2 is a pair of values $(\hat{x}_1, \hat{x}_2)$ such that both $dx_1/dt = 0$ and $dx_2/dt = 0$ when $x_1 = \hat{x}_1$ and $x_2 = \hat{x}_2$. The functions $x_1(t)$ and $x_2(t)$ are sometimes written in vector notation as $\mathbf{x}(t)$ and the equilibria as $\hat{\mathbf{x}}$.

Stability of equilibria in systems of nonlinear differential equations is defined in a way analogous to that for single nonlinear differential equations (see Section 7.2).

Definition An equilibrium $\hat{\mathbf{x}}$ of the system of differential equations (2) is **locally stable** if $\mathbf{x}$ approaches the value $\hat{\mathbf{x}}$ as $t \to \infty$ for all initial conditions sufficiently close to $\hat{\mathbf{x}}$.

Local Stability Analyses

Recall the Fitzhugh-Nagumo equations for the electrical potential of a neuron from page 475. Using $v(t)$ for the neuron's electrical potential at time t and $w(t)$ to denote the magnitude of ion exchange, we had

(4)
$$\frac{dv}{dt} = (v - a)(1 - v)v - w$$
$$\frac{dw}{dt} = bv - cw$$

Both v and w can be positive or negative. In the example on page 474 we used the values $a = 0.2$, $b = 0.01$, and $c = 0.04$. Figure 1 displays the nullclines, along with two solution curves in the phase plane.

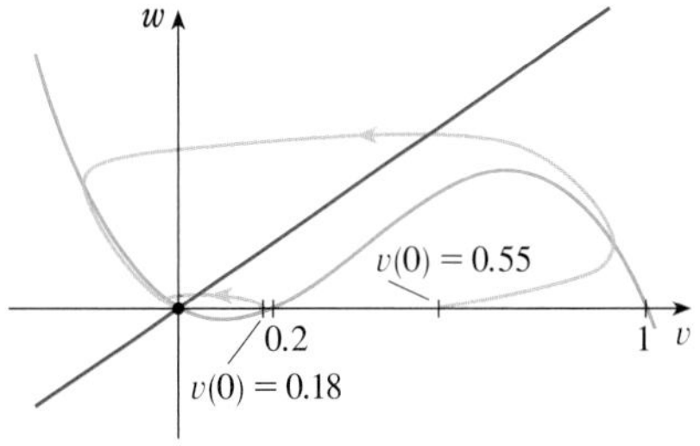

FIGURE 1
The v-nullcline is shown in blue, w-nullcline in red, and solution curves in green.

Notice that the w-nullcline is linear, with a slope of $b/c = \frac{1}{4}$, because the differential equation for w is linear. The v-nullcline is nonlinear because the differential equation for v is nonlinear—it is the graph of a cubic polynomial in v. The origin, $(v, w) = (0, 0)$, is the only equilibrium because it is the only place at which these nullclines intersect.

The two solution curves plotted on Figure 1 both ultimately move toward the origin but it is not possible to determine from the phase plane if this is also true for other initial conditions. Consequently, we need a more precise approach.

To begin, suppose the initial condition is very close to the origin. By zooming in on the origin in Figure 1 and removing the solution curves, we get Figure 2. Locally (that is, near the origin) the cubic nullcline now looks approximately linear. In fact, as we know from Section 3.8, we can approximate this cubic polynomial near the origin with a line. If we define $g(v) = (v - a)(1 - v)v$, then the cubic nullcline is defined by the equation $w = g(v)$. Now, near the origin we can approximate $g(v)$ as

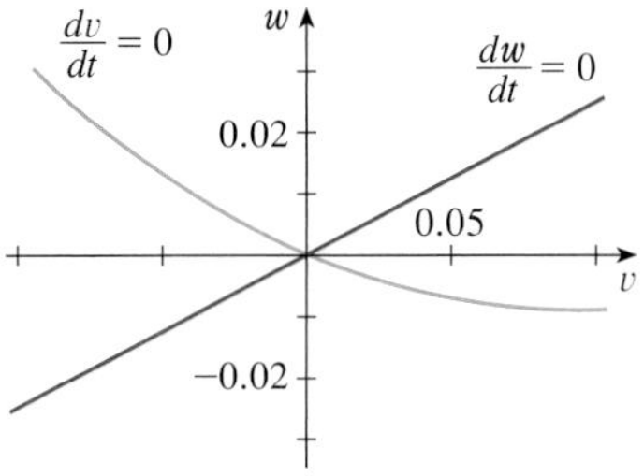

FIGURE 2

(5) $$g(v) \approx g(0) + g'(0)v = -av$$

Therefore the nullcline is approximately given by $w \approx -av$, as shown in Figure 3.

We can also use approximation (5) in Equations 4 to obtain the following system of linear differential equations:

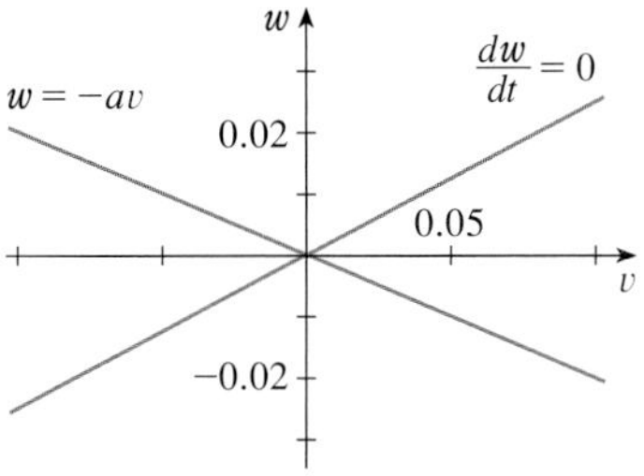

FIGURE 3

(6) $$\frac{dv}{dt} = -av - w$$
$$\frac{dw}{dt} = bv - cw$$

The lines drawn in Figure 3 are precisely the nullclines of Equations 6. Equations 6 are referred to as the linear approximation to Equations 4 near the equilibrium $(v, w) = (0, 0)$ or, equivalently, the *linearization* of Equations 4 near this point.

Using the values $a = 0.2$, $b = 0.01$, and $c = 0.04$, we can write system (6) in matrix notation as

(7) $$\frac{d\mathbf{x}}{dt} = A\mathbf{x} \qquad \text{with} \qquad A = \begin{bmatrix} -0.2 & -1 \\ 0.01 & -0.04 \end{bmatrix}$$

The eigenvalues of A are $\lambda = -0.12 \pm 0.06i$, indicating that the origin in system (7) is a stable spiral. But what does this mean in terms of the local stability of the origin for system (4)? Provided that the initial condition of system (4) is close enough to the origin, the linearization given by system (7) should be a reasonable approximation. Moreover, the origin in this linear system is a stable spiral. Therefore we might expect all solutions of system (4) to approach the origin in a spiraling fashion as well, provided they start sufficiently close to the origin (see Figure 4).

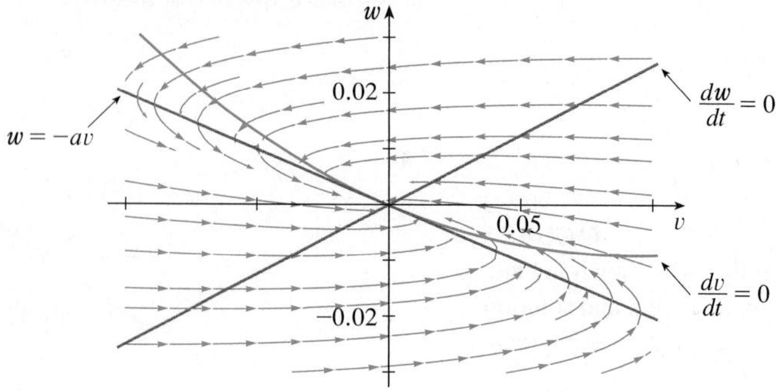

FIGURE 4
Solution curves for the linear system in Equations 7, along with its nullclines in purple. The v-nullcline of the original nonlinear model is blue.

Linearization

The preceding calculations can be extended to the following general system of autonomous nonlinear differential equations:

(8)
$$\frac{dx_1}{dt} = f_1(x_1, x_2)$$
$$\frac{dx_2}{dt} = f_2(x_1, x_2)$$

Let's consider an arbitrary equilibrium of Equations 8, denoted $(\hat{x}_1, \hat{x}_2)$. Our derivation will follow that of Section 7.2 for one-variable, nonlinear differential equations.

The equilibrium of interest need no longer be the origin and therefore we first change variables so that we can again focus attention on the origin. Defining $\varepsilon_1 = x_1 - \hat{x}_1$ and $\varepsilon_2 = x_2 - \hat{x}_2$, and noting that

$$\frac{d\varepsilon_i}{dt} = \frac{d}{dt}[x_i - \hat{x}_i] = \frac{dx_i}{dt}$$

we have, from system (8),

(9)
$$\frac{d\varepsilon_1}{dt} = f_1(\varepsilon_1 + \hat{x}_1, \varepsilon_2 + \hat{x}_2)$$
$$\frac{d\varepsilon_2}{dt} = f_2(\varepsilon_1 + \hat{x}_1, \varepsilon_2 + \hat{x}_2)$$

The variables ε_1 and ε_2 are the deviations of x_1 and x_2 from their equilibrium values, and therefore they have an equilibrium at $(\varepsilon_1, \varepsilon_2) = (0, 0)$. Furthermore, this equilibrium is locally stable if and only if the equilibrium $(\hat{x}_1, \hat{x}_2)$ of system (8) is locally stable.

The next step is to obtain the linear approximation of the right side of each differential equation in (9), near the origin $(\varepsilon_1, \varepsilon_2) = (0, 0)$. To do so, we need to use the two-variable tangent plane approximation in Equation 4 on page 599. We have

(10) $$f_1(\varepsilon_1 + \hat{x}_1, \varepsilon_2 + \hat{x}_2) \approx f_1(\hat{x}_1, \hat{x}_2) + \frac{\partial f_1(\hat{x}_1, \hat{x}_2)}{\partial x_1}\varepsilon_1 + \frac{\partial f_1(\hat{x}_1, \hat{x}_2)}{\partial x_2}\varepsilon_2$$

and

(11) $$f_2(\varepsilon_1 + \hat{x}_1, \varepsilon_2 + \hat{x}_2) \approx f_2(\hat{x}_1, \hat{x}_2) + \frac{\partial f_2(\hat{x}_1, \hat{x}_2)}{\partial x_1}\varepsilon_1 + \frac{\partial f_2(\hat{x}_1, \hat{x}_2)}{\partial x_2}\varepsilon_2$$

Equations 10 and 11 can be further simplified by noting that $f_1(\hat{x}_1, \hat{x}_2) = 0$ and $f_2(\hat{x}_1, \hat{x}_2) = 0$ because the point $(\hat{x}_1, \hat{x}_2)$ is an equilibrium of system (8). Substituting the resulting expressions into the right side of system (9) gives

$$\frac{d\varepsilon_1}{dt} \approx \frac{\partial f_1(\hat{x}_1, \hat{x}_2)}{\partial x_1}\varepsilon_1 + \frac{\partial f_1(\hat{x}_1, \hat{x}_2)}{\partial x_2}\varepsilon_2$$

$$\frac{d\varepsilon_2}{dt} \approx \frac{\partial f_2(\hat{x}_1, \hat{x}_2)}{\partial x_1}\varepsilon_1 + \frac{\partial f_2(\hat{x}_1, \hat{x}_2)}{\partial x_2}\varepsilon_2$$

or, in matrix notation,

(12)
$$\frac{d\boldsymbol{\varepsilon}}{dt} \approx \begin{bmatrix} \dfrac{\partial f_1(\hat{x}_1, \hat{x}_2)}{\partial x_1} & \dfrac{\partial f_1(\hat{x}_1, \hat{x}_2)}{\partial x_2} \\ \dfrac{\partial f_2(\hat{x}_1, \hat{x}_2)}{\partial x_1} & \dfrac{\partial f_2(\hat{x}_1, \hat{x}_2)}{\partial x_2} \end{bmatrix} \boldsymbol{\varepsilon}$$

where $\boldsymbol{\varepsilon}$ is the vector whose components are ε_1 and ε_2. Equation 12 is the **linearization** of Equations 8 near the equilibrium point $(\hat{x}_1, \hat{x}_2)$.

Finally, the key observation is that the entries of the matrix of coefficients in Equation 12 are evaluated at the equilibrium $(\hat{x}_1, \hat{x}_2)$. As a result, they do not involve the variables. Consequently, Equation 12 is an autonomous, homogeneous system of linear differential equations for $\boldsymbol{\varepsilon}(t)$. Although Equation 12 is an approximation, we might expect the eigenvalues of its coefficient matrix to determine the local stability of the equilibrium $(\varepsilon_1, \varepsilon_2) = (0, 0)$, and therefore the local stability of $(\hat{x}_1, \hat{x}_2)$. This is borne out by the following definition and theorem.

(13) Definition Consider the autonomous system of differential equations

$$\frac{dx_1}{dt} = f_1(x_1, x_2) \qquad \frac{dx_2}{dt} = f_2(x_1, x_2)$$

The matrix

$$J(x_1, x_2) = \begin{bmatrix} \dfrac{\partial f_1(x_1, x_2)}{\partial x_1} & \dfrac{\partial f_1(x_1, x_2)}{\partial x_2} \\ \dfrac{\partial f_2(x_1, x_2)}{\partial x_1} & \dfrac{\partial f_2(x_1, x_2)}{\partial x_2} \end{bmatrix}$$

is called the **Jacobian** matrix. Note that J is a function of x_1 and x_2 because its entries are functions of x_1 and x_2.

We then have the following theorem:

(14) Theorem (Local Stability) Suppose $(\hat{x}_1, \hat{x}_2)$ is an equilibrium of the system of differential equations in Definition 13. Let r be the largest eigenvalue of $J(\hat{x}_1, \hat{x}_2)$, or the largest real part of the eigenvalues if they are complex. If $r < 0$, then the equilibrium is locally stable. If $r > 0$, then the equilibrium is unstable. If $r = 0$, then the analysis is inconclusive.

Examples

EXAMPLE 1 | Lotka-Volterra competition equations In Example 7.6.1 we studied a pair of equations that describe competition between two species. These equations were obtained by extending the logistic equation for population growth in a single species. The equations are

$$\textbf{(15)}\qquad \frac{dN_1}{dt} = \left(1 - \frac{N_1 + 2N_2}{1000}\right)N_1 \qquad \frac{dN_2}{dt} = \left(1 - \frac{N_2 + N_1}{600}\right)N_2$$

where N_1 and N_2 are the population sizes of species 1 and 2. We identified four equilibria:

(a) $\hat{N}_1 = 0, \qquad \hat{N}_2 = 0$

(b) $\hat{N}_1 = 0, \qquad \hat{N}_2 = 600$

(c) $\hat{N}_1 = 1000, \qquad \hat{N}_2 = 0$

(d) $\hat{N}_1 = 200, \qquad \hat{N}_2 = 400$

Determine the local stability properties of each equilibrium.

SOLUTION The first step is to calculate the Jacobian matrix. Defining

$$f_1(N_1, N_2) = \left(1 - \frac{N_1 + 2N_2}{1000}\right)N_1 \qquad f_2(N_1, N_2) = \left(1 - \frac{N_2 + N_1}{600}\right)N_2$$

we obtain

$$J = \begin{bmatrix} \dfrac{\partial f_1(N_1, N_2)}{\partial N_1} & \dfrac{\partial f_1(N_1, N_2)}{\partial N_2} \\ \dfrac{\partial f_2(N_1, N_2)}{\partial N_1} & \dfrac{\partial f_2(N_1, N_2)}{\partial N_2} \end{bmatrix} = \begin{bmatrix} 1 - \dfrac{N_1 + N_2}{500} & -\dfrac{N_1}{500} \\ -\dfrac{N_2}{600} & 1 - \dfrac{N_1 + 2N_2}{600} \end{bmatrix}$$

(a) Evaluating the Jacobian at $\hat{N}_1 = 0$, $\hat{N}_2 = 0$ gives

$$J = \begin{bmatrix} 1 & 0 \\ 0 & 1 \end{bmatrix}$$

The eigenvalues of J are $\lambda_1 = 1$ and $\lambda_2 = 1$. Thus $r = 1$ in Theorem 14 and the equilibrium is unstable.

(b) Evaluating the Jacobian at $\hat{N}_1 = 0$, $\hat{N}_2 = 600$ gives

$$J = \begin{bmatrix} -\frac{1}{5} & 0 \\ -1 & -1 \end{bmatrix}$$

The eigenvalues of J are $\lambda_1 = -\frac{1}{5}$ and $\lambda_2 = -1$. Thus $r = -\frac{1}{5}$ in Theorem 14 and the equilibrium is locally stable.

(c) Evaluating the Jacobian at $\hat{N}_1 = 1000$, $\hat{N}_2 = 0$ gives

$$J = \begin{bmatrix} -1 & -2 \\ 0 & -\frac{2}{3} \end{bmatrix}$$

The eigenvalues of J are $\lambda_1 = -1$ and $\lambda_2 = -\frac{2}{3}$. Thus $r = -\frac{2}{3}$ in Theorem 14 and the equilibrium is locally stable.

(d) Evaluating the Jacobian at $\hat{N}_1 = 200$, $\hat{N}_2 = 400$ gives

$$J = \begin{bmatrix} -\frac{1}{5} & -\frac{2}{5} \\ -\frac{2}{3} & -\frac{2}{3} \end{bmatrix}$$

The eigenvalues of J are $\lambda_1 = -1$ and $\lambda_2 = \frac{2}{15}$. Thus $r = \frac{2}{15}$ in Theorem 14 and the equilibrium is unstable.

The preceding analysis matches the phase plane analysis in Figure 7.6.5 by showing that the only stable equilibria are those where a single species is present. This is referred to as *competitive exclusion.* In this example, the species that ultimately survives is determined by the initial size of each population. ■

The following theorem is often useful when evaluating local stability. It follows directly from a combination of Theorem 10.2.15 and Theorem 10.2.16.

(16) Theorem Suppose $(\hat{x}_1, \hat{x}_2)$ is an equilibrium of the system of differential equations in Definition 13. If

(i) $\det J(\hat{x}_1, \hat{x}_2) > 0$, and

(ii) $\operatorname{trace} J(\hat{x}_1, \hat{x}_2) < 0$

then the equilibrium is locally stable. If the inequalities in either (i), (ii), or both are reversed, then the equilibrium is unstable. (Recall that the trace of a matrix is the sum of its diagonal entries.)

EXAMPLE 2 | Kermack-McKendrick infectious disease model In Exercise 7.6.24 we studied a pair of equations that describe the spread of an infectious disease in the human population. The equations are

(17) $$\frac{dS}{dt} = \theta - \gamma S - \beta SI \qquad \frac{dI}{dt} = \beta SI - \mu I$$

where S and I are the numbers of susceptible and infected people, respectively. The constant θ is the rate of arrival of susceptible people, γ is their per capita mortality rate, β is the transmission rate of the disease from infected to susceptible people, and μ is the per capita mortality rate of infected people. All constants are positive.

Definition 3 can be used to verify that the following are both equilibria:

(a) $\hat{S} = \theta/\gamma, \hat{I} = 0$

(b) $\hat{S} = \mu/\beta, \hat{I} = (\beta\theta - \gamma\mu)/\beta\mu$

Note that the second equilibrium is of biological interest only if $\beta\theta - \gamma\mu > 0$. Assuming $\beta\theta - \gamma\mu > 0$, determine the local stability properties of both equilibria.

SOLUTION The first step is to calculate the Jacobian matrix. Defining

$$f_1(S, I) = \theta - \gamma S - \beta SI \qquad f_2(S, I) = \beta SI - \mu I$$

we obtain

$$J = \begin{bmatrix} \dfrac{\partial f_1(S, I)}{\partial S} & \dfrac{\partial f_1(S, I)}{\partial I} \\ \dfrac{\partial f_2(S, I)}{\partial S} & \dfrac{\partial f_2(S, I)}{\partial I} \end{bmatrix} = \begin{bmatrix} -\gamma - \beta I & -\beta S \\ \beta I & \beta S - \mu \end{bmatrix}$$

(a) Evaluating the Jacobian at $\hat{S} = \theta/\gamma$ and $\hat{I} = 0$ gives

$$J = \begin{bmatrix} -\gamma & -\dfrac{\beta\theta}{\gamma} \\ 0 & \dfrac{\beta\theta}{\gamma} - \mu \end{bmatrix}$$

The eigenvalues are therefore $\lambda_1 = -\gamma$ and $\lambda_2 = \beta\theta/\gamma - \mu$. The assumption that $\beta\theta - \gamma\mu > 0$ means that $\lambda_2 > 0$ and therefore this equilibrium is unstable. This illustrates that the disease will spread in the population.

(b) Evaluating the Jacobian at $\hat{S} = \mu/\beta$ and $\hat{I} = (\beta\theta - \gamma\mu)/\beta\mu$ gives

$$J = \begin{bmatrix} -\gamma - \dfrac{\beta\theta - \gamma\mu}{\mu} & -\mu \\ \dfrac{\beta\theta - \gamma\mu}{\mu} & 0 \end{bmatrix}$$

The eigenvalues of J can be readily calculated, but their signs are difficult to determine. Instead we make use of Theorem 16. We have

$$\det J = \beta\theta - \gamma\mu$$

and

$$\operatorname{trace} J = -\gamma - \frac{\beta\theta - \gamma\mu}{\mu}$$

The assumption that $\beta\theta - \gamma\mu > 0$ means that $\det J > 0$ and trace $J < 0$. Therefore, from Theorem 16, this equilibrium is locally stable. At this stable equilibrium, part of the population will be infected and the remainder will be susceptible to infection. ■

EXAMPLE 3 | Lotka-Volterra predator-prey equations In Section 7.5 we studied a pair of equations that describe a predator and its prey. The equations are

$$\textbf{(18)} \qquad \frac{dR}{dt} = rR - aRW \qquad \frac{dW}{dt} = -kW + bRW$$

where R and W are the population sizes of prey and predator, respectively. The constant r is the per capita reproductive rate of prey, k is the per capita death rate of the predator, a is the rate of consumption of prey by predators, and b is the rate at which this consumption is converted into predator offspring. We identified two equilibria of these equations in Section 7.6, one of which was $(\hat{R}, \hat{W}) = (k/b, r/a)$. Evaluate the stability of this equilibrium.

SOLUTION Defining

$$f_1(R, W) = rR - aRW \qquad f_2(R, W) = -kW + bRW$$

the Jacobian is

$$J = \begin{bmatrix} \dfrac{\partial f_1(R, W)}{\partial R} & \dfrac{\partial f_1(R, W)}{\partial W} \\ \dfrac{\partial f_2(R, W)}{\partial R} & \dfrac{\partial f_2(R, W)}{\partial W} \end{bmatrix} = \begin{bmatrix} r - aW & -aR \\ bW & -k + bR \end{bmatrix}$$

Evaluating this at the equilibrium gives

$$J = \begin{bmatrix} 0 & -\dfrac{ak}{b} \\ \dfrac{rb}{a} & 0 \end{bmatrix}$$

The eigenvalues of J are the complex conjugates $\lambda_1 = i\sqrt{rk}$ and $\lambda_2 = -i\sqrt{rk}$. The real part of these eigenvalues is 0 and therefore $r = 0$ in Theorem 14. As a result, the local stability analysis is inconclusive. Figure 7.5.4 on page 462 suggests that, in this case, the equilibrium is *neutrally stable*, that is, solution curves move neither toward the equilibrium nor away from it as $t \to \infty$. This can, in fact, be proven, and you should consider how you might do this. Thus the predator and prey populations are predicted to undergo neverending oscillations. ■

EXERCISES 10.4

1–6 Each of the nonlinear systems has an equilibrium at $(\hat{x}_1, \hat{x}_2) = (0, 0)$. Find the linearization near this point.

1. $\dfrac{dx_1}{dt} = 4x_1 - 2x_1x_2$

$\dfrac{dx_2}{dt} = -2x_2 + 8x_1x_2$

2. $\dfrac{dx_1}{dt} = 4x_1(1 - 5x_1) - 2x_1x_2$

$\dfrac{dx_2}{dt} = -2x_2 + 8x_1x_2$

3. $\dfrac{dx_1}{dt} = \sin x_1 + x_1x_2 + 3x_2^2$

$\dfrac{dx_2}{dt} = \cos x_2 - 1 + x_1(x_1 - 1) + 7x_2$

4. $\dfrac{dx_1}{dt} = 5(1 + \cos x_1) + ax_1 - bx_2 - 10$

$\dfrac{dx_2}{dt} = 3x_2 + bx_1x_2$

5. $\dfrac{dx_1}{dt} = 1 + x_1^3 - \dfrac{1 + x_1}{1 + x_2}$

$\dfrac{dx_2}{dt} = 2x_2 + x_1^2$

6. $\dfrac{dx_1}{dt} = x_2 - \dfrac{2 + ax_1}{2 + bx_2} + \cos x_2$

$\dfrac{dx_2}{dt} = \dfrac{2x_1}{1 + x_2} - ax_1$

7–12 Find all equilibria. Then find the linearization near each equilibrium.

7. $\dfrac{dx_1}{dt} = -5x_1 + x_1x_2$

$\dfrac{dx_2}{dt} = x_2 - 5x_1x_2$

8. $\dfrac{dx_1}{dt} = x_2 - 5x_1x_2$

$\dfrac{dx_2}{dt} = 2x_1 - 6x_1x_2$

9. $\dfrac{dx_1}{dt} = x_1 - 6x_2^2 + x_1x_2$

$\dfrac{dx_2}{dt} = 8x_1 + 4x_1x_2$

10. $\dfrac{dx_1}{dt} = x_1 - 2x_1^2 - 6x_1x_2$

$\dfrac{dx_2}{dt} = 2x_2 - 8x_2^2 - 2x_1x_2$

11. $\dfrac{dx_1}{dt} = e^{-x_1}(x_1 - x_2)$

$\dfrac{dx_2}{dt} = x_1 - x_2^2 + 2x_1x_2$

12. $\dfrac{dx_1}{dt} = \ln x_1 - x_2$

$\dfrac{dx_2}{dt} = x_1(1 - x_1 - x_2)$

13–18 A Jacobian matrix and two equlibria are given. Determine if each is locally stable, unstable, or if the analysis is inconclusive.

13. $J = \begin{bmatrix} (x_1 - 2)x_2 + x_1x_2 & x_1(x_1 - 2) \\ 0 & -1 + 2x_2 \end{bmatrix}$

(i) $\hat{x}_1 = 0,\ \hat{x}_2 = 2$

(ii) $\hat{x}_1 = 2,\ \hat{x}_2 = -1$

14. $J = \begin{bmatrix} -1 + 2x_1 & 0 \\ 0 & -\frac{1}{3} + 2x_2 \end{bmatrix}$

(i) $\hat{x}_1 = -1,\ \hat{x}_2 = 0$

(ii) $\hat{x}_1 = 2,\ \hat{x}_2 = \frac{1}{3}$

15. $J = \begin{bmatrix} 1 - \cos x_2 & (x_1 - 1)\sin x_2 \\ \cos x_1 & -\sin 1 \end{bmatrix}$

(i) $\hat{x}_1 = 0,\ \hat{x}_2 = 0$

(ii) $\hat{x}_1 = 1,\ \hat{x}_2 = 1$

16. $J = \begin{bmatrix} 2x_1 & -\sin x_2 \\ \cos x_1 & 0 \end{bmatrix}$

(i) $\hat{x}_1 = 1,\ \hat{x}_2 = -\pi$

(ii) $\hat{x}_1 = 1,\ \hat{x}_2 = \pi$

17. $J = \begin{bmatrix} -\dfrac{1}{2 + x_2} & -1 + \dfrac{x_1}{(2 + x_2)^2} \\ \dfrac{x_2}{(1 + x_1)^2} & -1 - \dfrac{1}{1 + x_1} \end{bmatrix}$

(i) $\hat{x}_1 = 0,\ \hat{x}_2 = 0$

(ii) $\hat{x}_1 = -2,\ \hat{x}_2 = -1 - \sqrt{3}$

18. $J = \begin{bmatrix} -\dfrac{1}{1 + x_2} & \dfrac{x_1}{(1 + x_2)^2} \\ -1 + \dfrac{x_2}{(1 + x_1)^2} & -\dfrac{1}{1 + x_1} \end{bmatrix}$

(i) $\hat{x}_1 = -2,\ \hat{x}_2 = -2$

(ii) $\hat{x}_1 = \frac{1}{2},\ \hat{x}_2 = -\frac{3}{4}$

19–23 Find all equilibria and determine their local stability properties.

19. $x' = x(3 - x - y),\quad y' = y(2 - x - y)$

20. $p' = p(1 - p - q),\quad q' = q(2 - 3p - q)$

21. $n' = n(1 - 2m),\quad m' = m(2 - 2n - m)$

22. $x' = x(2 - x),\quad y' = y(3 - y)$

23. $p' = -p^2 + q - 1,\quad q' = q(2 - p - q)$

24–25 Find all equilibria and determine their stability properties. Your answer might be a function of the constant a.

24. $x' = -xy + y + ax,\quad y' = 2y - xy$

25. $x' = ax^2 + ay - x,\quad y' = x - y,\quad a \neq 0$

26. Cell cycle In Exercise 7.Review.25 a model for the cell cycle was introduced. It modeled the concentrations of a molecule called MPF (maturation promoting factor) and another molecule called cyclin. MPF production is stimulated by cyclin, but the presence of MPF also inhibits its own production. Using M and C to denote the concentrations of these two biomolecules (in mg/mL), the model for their interaction is

$$\frac{dM}{dt} = \alpha C + \beta C M^2 - \frac{\gamma M}{1 + M}$$

$$\frac{dC}{dt} = \delta - M$$

(a) Suppose that $\alpha = 2$, $\beta = 1$, $\gamma = 10$, and $\delta = 1$. Find the only equilibrium.

(b) Calculate the Jacobian matrix.

(c) Determine the local stability properties of the equilibrium found in part (a) using the Jacobian from part (b).

(d) Describe how M and C change near the equilibrium point.

Source: Adapted from R. Norel et al., "A Model for the Adjustment of the Mitotic Clock by Cyclin and MPF Levels," *Science* 251 (1991): 1076–78.

27. Competition-colonization models In Exercise 7.Review.23 a metapopulation model for two species was introduced. The equations were

$$\frac{dp_1}{dt} = c_1p_1(1 - p_1) - m_1p_1$$

$$\frac{dp_2}{dt} = c_2p_2(1 - p_1 - p_2) - m_2p_2 - c_1p_1p_2$$

where p_i is the fraction of patches occupied by species i, and c_i and m_i are the species-specific rates of colonization and extinction of patches, respectively. These equations assume that any patch has at most one species, and species 2 patches can be taken over by species 1, but not vice versa.

(a) Suppose that $m_1 = m_2 = 3$, $c_1 = 5$, and $c_2 = 30$. Find all equilibria.

(b) Calculate the Jacobian matrix.

(c) Determine the local stability properties of each equilibrium found in part (a) using the Jacobian from part (b).

(d) Are the species predicted to be able to coexist at a stable equilibrium?

28. Gene regulation The model of gene regulation from Section 10.3 is often extended to nonlinear gene regulation by specifying a nonlinear function for how the concentration of protein in a cell affects mRNA production. One such example, called an **auto-repression** model, is

$$\frac{dm}{dt} = \frac{1}{1+p} - m \qquad \frac{dp}{dt} = m - p$$

(a) Find all equilibria.
(b) Calculate the Jacobian matrix.
(c) Only one equilibrium found in part (a) is of biological interest. Determine its local stability properties using the Jacobian from part (b).
(d) Describe how m and p change near the equilibrium point.

29–31 Consumer resource models often have the following general form

$$R' = f(R) - g(R, C) \qquad C' = \varepsilon g(R, C) - h(C)$$

where R is the number of individuals of the resource and C is the number of consumers. The function $f(R)$ gives the rate of replenishment of the resource, $g(R, C)$ describes the rate of consumption of the resource, and $h(C)$ is the rate of loss of the consumer. The constant ε, where $0 < \varepsilon < 1$, is the conversion efficiency of resources into consumers. Find all equilibria of the following examples and determine their stability properties.

29. A **chemostat** is an experimental consumer-resource system. If the resource is not self-reproducing, then it can be modeled by choosing $f(R) = \theta$, $g(R, C) = bRC$, and $h(C) = \mu C$. Suppose $\theta = 2$, $b = 1$, $\varepsilon = 1$, and $\mu = 1$.

30. A model for self-reproducing resources is obtained by choosing $f(R) = rR$, $g(R, C) = bRC$, and $h(C) = \mu C$. Suppose $r = 2$, $b = 1$, $\varepsilon = 1$, and $\mu = 1$.

31. A model for self-reproducing resources with limited growth is obtained by choosing $f(R) = rR(1 - R/K)$, $g(R, C) = bRC$, and $h(C) = \mu C$. Suppose $r = 2$, $K = 5$, $b = 1$, $\varepsilon = 1$, and $\mu = 1$.

---

32. The **Kermack-McKendrick equations** describe the outbreak of an infectious disease. Using S and I to denote the number of susceptible and infected people in a population, respectively, the equations are

$$S' = -\beta SI \qquad I' = \beta SI - \mu I$$

where β and μ are positive constants representing the transmission rate and rate of recovery.
(a) Verify that $\hat{I} = 0$, along with any value of S, is an equilibrium.
(b) Calculate the Jacobian matrix.
(c) Using your answer to part (b), determine how large S must be to guarantee that the disease will spread when rare.

33. The **Michaelis-Menten equations** describe a biochemical reaction in which an enzyme E and substrate S bind to form a complex C. This complex can then either dissociate back into its original components or undergo a reaction in which a product P is produced along with the free enzyme: $E + S \leftrightharpoons C \rightarrow E + P$. This can be expressed by the differential equations

$$\frac{dx}{dt} = -k_f xyM + k_r(1 - y)M$$
$$\frac{dy}{dt} = -k_f xyM + k_r(1 - y)M + k_{cat}(1 - y)M$$
$$\frac{dz}{dt} = k_{cat}(1 - y)M$$

where M is the total number of enzymes (both free and bound), x and z are the numbers of substrate and product molecules, y is the fraction of the enzyme pool that is free, and the k_i's are positive constants.
(a) Although this is a system of three differential equations, x and y can be analyzed separately. Explain why.
(b) Find the only equilibrium.
(c) Calculate the Jacobian matrix.
(d) Determine the local stability properties of the equilibrium.

34. Stability of Caribbean reefs Coral and macroalgae compete for space when colonizing Caribbean reefs. A modification of the model in Exercise 27 has been used to describe this process. The equations are

$$\frac{dM}{dt} = \gamma M(1 - M) - \frac{gM}{1 - C}$$
$$\frac{dC}{dt} = rC(1 - M - C) - \gamma CM - dC$$

where M is the fraction of the reef occupied by macroalgae, C is the fraction occupied by coral, r is the colonization rate of empty space by coral, d is the death rate of coral, γ is the rate of colonization by macroalgae (in both empty space and space occupied by coral), and g is a constant governing the death rate of macroalgae. Notice that the per capita death rate of macroalgae decreases as coral cover increases.
(a) Suppose that $r = 3$, $d = 1$, $\gamma = 2$, and $g = 1$. Find all equilibria. There are five, but only four of them are biologically relevant.
(b) Calculate the Jacobian matrix.
(c) Determine the local stability properties of the four relevant equilibria found in part (a).
(d) In part (c) you should find two equilibria that are locally stable. What do they represent in terms of the structure of the reef?

Source: Adapted from P. Mumby et al., "Thresholds and the Resilience of Caribbean Coral Reefs," *Nature* 450 (2007): 98–101.

35. Fitzhugh-Nagumo equations Consider the following alternative form of the Fitzhugh-Nagumo equations:

$$\frac{dv}{dt} = (v - a)(1 - v)v - w \qquad \frac{dw}{dt} = \varepsilon(v - w)$$

where $\varepsilon > 0$ and $0 < a < 1$.

(a) Verify that the origin is an equilibrium.
(b) Calculate the Jacobian matrix.
(c) Determine the local stability properties of the origin as a function of the constants.

Chapter 10 REVIEW

CONCEPT CHECK

1. What is the difference between an autonomous and a nonautonomous system of differential equations?
2. What is an equilibrium of a system of differential equations?
3. Explain the difference between local and global stability in systems of differential equations.
4. What is the difference between the solution of an initial-value problem and the general solution of a system of differential equations?
5. What does the Existence and Uniqueness Theorem tell us about homogeneous systems of linear, autonomous differential equations?
6. Explain the Superposition Principle.
7. Explain the difference between nullclines and eigenvectors in systems of linear autonomous differential equations.
8. What is the linearization of a system of nonlinear differential equations?
9. Explain what a Jacobian matrix is.
10. What do the eigenvalues of a Jacobian matrix from a system of nonlinear differential equations tell us?

Answers to the Concept Check can be found on the back endpapers.

TRUE-FALSE QUIZ

Determine whether the statement is true or false. If it is true, explain why. If it is false, explain why or give an example that disproves the statement.

1. The following system of differential equations is linear:

$$\frac{dx}{dt} = 3x + y + 3$$

$$\frac{dy}{dt} = 2xy - x + 5$$

2. The following system of differential equations is homogeneous:

$$\frac{dx_1}{dt} = a(t)x_1 + b(t)x_2$$

$$\frac{dx_2}{dt} = c(t)x_1 + d(t)x_2$$

3. The nullclines of a two-variable system of linear differential equations must be straight lines.
4. If an equilibrium of an autonomous system of linear, homogeneous differential equations is locally stable, then it is gobally stable as well.
5. A saddle equilibrium of an autonomous system of linear, homogeneous differential equations is always unstable.
6. A stable node must have two eigenvalues of the same sign.
7. Spirals have complex eigenvalues.
8. Stable equilibria of systems of linear autonomous differential equations must have eigenvalues with negative real parts.
9. Nonlinear systems of differential equations have at most one equilibrium.
10. If the Jacobian matrix for a nonlinear system of differential equations has eigenvalues with negative real parts at an equilibrium, then this equilibrium is globally stable.

EXERCISES

1–4 Is the system linear or nonlinear?

1. $p' = 2q - 1, \quad q' = q^2 - q - p$

2. $x' = 5 - 2x - y, \quad y' = 2x - y$

3. $z' = tz - 2w, \quad w' = z - w - 1$

4. $x' = -(x - 2)\ln(xy), \quad y' = e^x(x - y)$

5–8 Show that $x_1(t)$ and $x_2(t)$ are solutions to the initial-value problem $d\mathbf{x}/dt = A\mathbf{x}$ with $\mathbf{x}(0) = \mathbf{x}_0$.

5. $A = \begin{bmatrix} 1 & -2 \\ 1 & -1 \end{bmatrix} \quad \mathbf{x}_0 = \begin{bmatrix} 1 \\ 2 \end{bmatrix}$

$x_1(t) = \cos t - 3\sin t, \; x_2(t) = 2\cos t - \sin t$

6. $A = \begin{bmatrix} 0 & 1 \\ 2 & 1 \end{bmatrix} \quad \mathbf{x}_0 = \begin{bmatrix} 1 \\ 5 \end{bmatrix}$

$x_1(t) = 2e^{2t} - e^{-t}, \; x_2(t) = 4e^{2t} + e^{-t}$

7. $A = \begin{bmatrix} 1 & -1 \\ 0 & 1 \end{bmatrix} \quad \mathbf{x}_0 = \begin{bmatrix} 2 \\ 1 \end{bmatrix}$

$x_1(t) = 2e^t - te^t, \; x_2(t) = e^t$

8. $A = \begin{bmatrix} 1 & 1 \\ 1 & 1 \end{bmatrix} \quad \mathbf{x}_0 = \begin{bmatrix} 2 \\ 1 \end{bmatrix}$

$x_1(t) = \frac{1}{2} + \frac{3}{2}e^{2t}; \; x_2(t) = -\frac{1}{2} + \frac{3}{2}e^{2t}$

9–12 Solve the initial-value problem $d\mathbf{x}/dt = A\mathbf{x}$ with $\mathbf{x}(0) = \mathbf{x}_0$.

9. $A = \begin{bmatrix} 0 & 1 \\ -1 & 0 \end{bmatrix} \quad \mathbf{x}_0 = \begin{bmatrix} 1 \\ 1 \end{bmatrix}$

10. $A = \begin{bmatrix} 2 & 1 \\ 1 & 0 \end{bmatrix} \quad \mathbf{x}_0 = \begin{bmatrix} -1 \\ 1 \end{bmatrix}$

11. $A = \begin{bmatrix} 1 & 1 \\ 0 & -2 \end{bmatrix} \quad \mathbf{x}_0 = \begin{bmatrix} 1 \\ -2 \end{bmatrix}$

12. $A = \begin{bmatrix} -1 & 1 \\ -1 & 0 \end{bmatrix} \quad \mathbf{x}_0 = \begin{bmatrix} 1 \\ 0 \end{bmatrix}$

13–18 Two-compartment mixing problems are similar to the mixing problems in Exercises 7.4.45–48, except there are two connected tanks of fixed capacities filled with a well-mixed solution of some substance (for example, salt). Solution of a given concentration can enter each tank at a fixed rate, and solution also flows back and forth between the tanks. Solution can also leave each tank at a fixed rate.

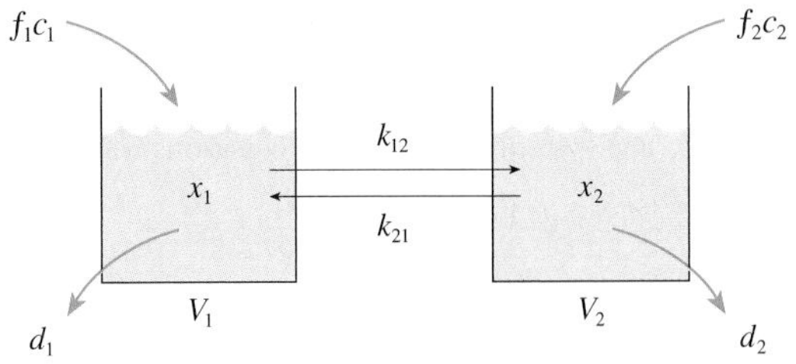

Using $x_1(t)$ and $x_2(t)$ to denote the amount of substance in each tank at time t, we obtain the following system of linear differential equations:

$$\frac{dx_1}{dt} = f_1c_1 - k_{12}\frac{x_1}{V_1} + k_{21}\frac{x_2}{V_2} - d_1\frac{x_1}{V_1}$$

$$\frac{dx_2}{dt} = f_2c_2 - k_{21}\frac{x_2}{V_2} + k_{12}\frac{x_1}{V_1} - d_2\frac{x_2}{V_2}$$

13. Two tanks containing a salt solution are connected as in the figure:

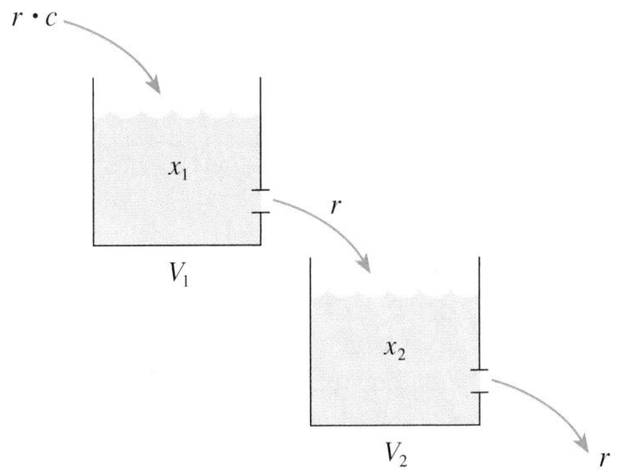

Suppose that $x_1(t)$ and $x_2(t)$ denote the amount of salt (in grams) in each tank at time t (in minutes), and $r = 8$ L/min, $c = 2$ g/L, $V_1 = 50$ L, and $V_2 = 25$ L. The initial amounts of salt in each tank are $x_1(0) = 8$ g and $x_2(0) = 0$ g.

(a) What is the initial-value problem describing this system?

(b) Solve the initial-value problem from part (a).

(c) What is the maximum amount of salt that is ever in tank 2?

14. Hemodialysis In Exercise 10.3.2 we modeled the process of hemodialysis, in which a machine filters urea from a patient's blood. The system of differential equations was

$$\frac{dc}{dt} = -\frac{K}{V}c + ap - bc$$

$$\frac{dp}{dt} = -ap + bc$$

where c and p are the urea concentrations in the blood and in other parts of the body, respectively (in mg/mL). Explain how these equations can be viewed as a special case of a general two-compartment mixing problem.

15. Pulmonary air embolism In Exercise 10.3.11 we modeled a pulmonary air embolism during surgery by viewing the lung as having two compartments: a deep and a shallow compartment, each of which removes CO_2 from the blood. The system of differential equations was

$$\frac{dx_1}{dt} = ac + a\frac{x_2}{V} - 2a\frac{x_1}{V}$$

$$\frac{dx_2}{dt} = ac - a\frac{x_2}{V}$$

where x_1 and x_2 are the amounts of CO_2 in the shallow and deep compartments, respectively, a is the rate of absorption of gas into each compartment from the blood, c is the concentration of CO_2 in this gas, and V is the volume of each compartment. Explain how these equations can be viewed as a special case of a general two-compartment mixing problem.

16. Systemic lupus erythematosus In Exercise 10.3.12 we modeled the amount of antibodies in the bloodstream and in other organs. The system of differential equations describing the amount of antibody in each compartment is

$$\frac{dx_1}{dt} = G + k_{21}x_2 - k_{12}x_1 - kx_1$$

$$\frac{dx_2}{dt} = k_{12}x_1 - k_{21}x_2$$

where x_1 and x_2 are the amounts in the blood and organs, respectively (in μg), G is the rate of generation of antibodies, k is the rate at which the drug treatment removes antibody from the bloodstream, and k_{ij} is the rate of flow of antibody from compartment i to j. Explain how these equations can be viewed as a special case of a general two-compartment mixing problem.

17. Explain how the equations in the Project on page 458 can be viewed as a special case of a general two-compartment mixing problem.

18. Explain how the equations in the Project on page 664 can be viewed as a special case of a general two-compartment mixing problem.

19–22 A Jacobian matrix and equilibrium are given. Is the equilibrium locally stable, unstable, or can you not tell?

19. $J = \begin{bmatrix} 0 & 2 \\ -1 & 2q - 1 \end{bmatrix}$ $\quad \hat{p} = -\frac{1}{4}, \ \hat{q} = \frac{1}{2}$

20. $J = \begin{bmatrix} -2 - y & -x \\ y & x - 1 \end{bmatrix}$ $\quad \hat{x} = \frac{5}{2}, \ \hat{y} = 0$

21. $J = \begin{bmatrix} 3z^2 - 8z + 3 & -2 \\ 1 & -1 \end{bmatrix}$ $\quad \hat{z} = 1, \ \hat{w} = 0$

22. $J = \begin{bmatrix} \dfrac{2 - x}{x} - \ln xy & \dfrac{2 - x}{y} \\ e^x + e^x(x - y) & -e^x \end{bmatrix}$ $\quad \hat{x} = 2, \ \hat{y} = 2$

23–26 Find all equilibria and determine their local stability properties.

23. $\dfrac{dx}{dt} = 5x - xy, \quad \dfrac{dy}{dt} = 4y - y^2 - 2xy$

24. $\dfrac{dA}{dt} = 2A - AL, \quad \dfrac{dL}{dt} = -5L + AL$

25. $\dfrac{dR}{dt} = 8R(1 - 2R) - RW, \quad \dfrac{dW}{dt} = -2W + 2RW$

26. $\dfrac{dA}{dt} = 2A(1 - A) - AL, \quad \dfrac{dL}{dt} = -5L + AL$

---

27. The **Lotka-Volterra competition equations** are

$$\frac{dN_1}{dt} = \left(1 - \frac{N_1 + \alpha N_2}{K_1}\right)N_1$$

$$\frac{dN_2}{dt} = \left(1 - \frac{N_2 + \beta N_1}{K_2}\right)N_2$$

where all constants α, β, K_1, and K_2 are postive.

(a) What are the equilibria?
(b) Calculate the Jacobian matrix.
(c) Determine the local stability properties of all biologically feasible equilibria when $K_1 > \alpha K_2$ and $K_2 < \beta K_1$.
(d) Determine the local stability properties of all biologically feasible equilibria when $K_1 < \alpha K_2$ and $K_2 > \beta K_1$.
(e) Determine the local stability properties of all biologically feasible equilibria when $K_1 < \alpha K_2$ and $K_2 < \beta K_1$.
(f) Determine the local stability properties of all biologically feasible equilibria when $K_1 > \alpha K_2$ and $K_2 > \beta K_1$.

28. The **Rosenzweig-MacArthur model** is a consumer-resource model. A simplified version is

$$R' = R(K - R) - \frac{R}{a + R}C \qquad C' = \frac{R}{a + R}C - bC$$

where R is the number of resource individuals and C is the number of consumers. Suppose that all constants are positive.

(a) What are the equilibria?
(b) Calculate the Jacobian matrix.
(c) Determine the values of the constants for which the extinction equilibrium is unstable.
(d) Determine the values of the constants for which the equilibrium with only the resource present is unstable.

29. Habitat destruction The model of Exercise 10.4.27 can be extended to include the effects of habitat destruction. Suppose that only a fraction, h, of the patches are habitable ($0 < h < 1$). The equations become

$$\frac{dp_1}{dt} = c_1p_1(h - p_1) - m_1p_1$$

$$\frac{dp_2}{dt} = c_2p_2(h - p_1 - p_2) - m_2p_2 - c_1p_1p_2$$

(a) What are the equilibria?
(b) Calculate the Jacobian matrix.
(c) Determine the values of h for which the extinction equilibrium is stable.

Source: Adapted from S. Nee et al., "Dynamics of Metapopulations: Habitat Destruction and Competitive Coexistence," *Journal of Animal Ecology* 61 (1992): 37–40.

30. Gene regulation The model of gene regulation from Section 10.3 is often extended to nonlinear gene regulation by specifying a nonlinear function for how the concentration of protein in a cell affects mRNA production. One such example, called an **auto-activation** model, is

$$\frac{dm}{dt} = \frac{2p}{1 + 2p} - m$$

$$\frac{dp}{dt} = m - p$$

(a) Find all equilibria.
(b) Calculate the Jacobian matrix.
(c) Determine the local stability properties of all equilibria.

31. Sterile insect technique Sterile insects are sometimes released as a means of controlling insect populations. Fertile insects mate with the sterile individuals and therefore fail to produce offspring. The following differential equations model this idea:

$$\frac{df}{dt} = af\frac{f}{f + s} - gf(f + s)$$

$$\frac{ds}{dt} = r - gs(f + s)$$

where $f(t)$ and $s(t)$ are the numbers of fertile and sterile individuals at time t, r is the rate of release of sterile insects, a is a positive birth rate constant, and g is a positive death rate constant.

(a) Show that, when no sterile insects are being released (that is, where $r = 0$), there is a locally stable equilibrium where $\hat{f} = a/g$ and $\hat{s} = 0$.
(b) Show that, when sterile insects are being released (that is, where $r > 0$), there is a locally stable equilibrium where $\hat{f} = 0$ and $\hat{s} = \sqrt{r/g}$.

Source: Adapted from H. Barclay et al., "Effects of Sterile Insect Releases on a Population under Predation or Parasitism," *Researches on Population Ecology* 22 (1980): 136–146.

CASE STUDY 2d Hosts, Parasites, and Time-Travel

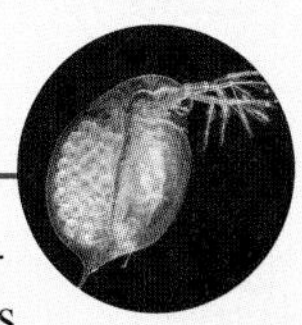

In this part of the case study we will take the formulation of the mathematical model for the antagonistic interactions between *Daphnia* and its parasite from Case Study 2c on page 484 and simplify it by linearization near one of its equilibrium points. We will then obtain an explicit solution for the frequency of the host and parasite genotypes as functions of time.

The analysis starts with equations

$$\textbf{(1)} \qquad \frac{dq}{dt} = s_q q(1 - q)(1 - 2p)$$

$$\frac{dp}{dt} = s_p p(1 - p)(2q - 1)$$

that were obtained in Case Study 2c. Recall that there are two possible host genotypes (A and a) and two possible parasite genotypes (B and b). Parasites of type B can infect only hosts of type A, while parasites of type b can infect only hosts of type a. In Equations 1, q is the frequency of the type A host and p is the frequency of the type B parasite. The constant s_q represents the reduction in reproductive output of a host due to infection, and s_p is the reduction in reproductive output of a parasite if it is unable to infect a host. In

Case Study 2c we found that $q = \frac{1}{2}$ and $p = \frac{1}{2}$ is an equilibrium of this system of differential equations.

Let's define $\varepsilon_q(t) = q(t) - \frac{1}{2}$ and $\varepsilon_p(t) = p(t) - \frac{1}{2}$ to be the deviations of q and p from these equilibrium values, respectively.

1. Linearize Equations 1 near the equilibrium $q = \frac{1}{2}$, $p = \frac{1}{2}$ to show that ε_q and ε_p satisfy the differential equations

(2)
$$\frac{d\varepsilon_q}{dt} = -\frac{s_q}{2}\varepsilon_p$$
$$\frac{d\varepsilon_p}{dt} = \frac{s_p}{2}\varepsilon_q$$

2. Show that the solution to system (2), with initial conditions $\varepsilon_q(0)$ and $\varepsilon_p(0)$, is

$$\varepsilon_q(t) = \varepsilon_q(0)\cos\left(\tfrac{1}{2}\sqrt{s_q s_p}\,t\right) - \varepsilon_p(0)\sqrt{\frac{s_q}{s_p}}\,\sin\left(\tfrac{1}{2}\sqrt{s_q s_p}\,t\right)$$

$$\varepsilon_p(t) = \varepsilon_p(0)\cos\left(\tfrac{1}{2}\sqrt{s_q s_p}\,t\right) + \varepsilon_q(0)\sqrt{\frac{s_p}{s_q}}\,\sin\left(\tfrac{1}{2}\sqrt{s_q s_p}\,t\right)$$

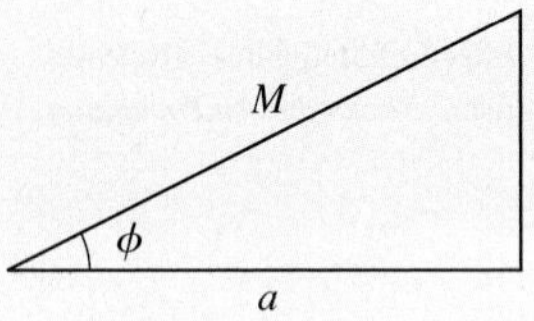

FIGURE 1

3. A useful trigonometric identity is

$$a\cos(ct) + b\sin(ct) = M\cos(ct - \phi)$$

where $M = \sqrt{a^2 + b^2}$ and ϕ is the angle between 0 and 2π whose cosine and sine satisfy the equations $\cos\phi = a/M$ and $\sin\phi = b/M$ (see Figure 1). Note that if $a > 0$ and $b > 0$ (so that we are in the first quadrant), then $\phi = \tan^{-1}(b/a)$. More generally, however, ϕ is *not* given by the principal branch of $\tan^{-1}$. Instead, if $a < 0$ (second or third quadrant), then $\phi = \pi + \tan^{-1}(b/a)$, whereas if $a < 0$ and $b < 0$ (fourth quadrant), then $\phi = 2\pi + \tan^{-1}(b/a)$. Use the identity to show that the solutions in Problem 2 can be written as

$$\varepsilon_q(t) = M_q\cos(ct - \phi_q)$$
$$\varepsilon_p(t) = M_p\cos(ct - \phi_p)$$

where $c = \frac{1}{2}\sqrt{s_q s_p}$,

$$M_q = \sqrt{\varepsilon_q(0)^2 + \varepsilon_p(0)^2\frac{s_q}{s_p}}$$

$$M_p = \sqrt{\varepsilon_p(0)^2 + \varepsilon_q(0)^2\frac{s_p}{s_q}}$$

and ϕ_q and ϕ_p are given by

$$\phi_q = \begin{cases} \tan^{-1}\left(-\dfrac{\varepsilon_p(0)\sqrt{s_q}}{\varepsilon_q(0)\sqrt{s_p}}\right) & \text{if } \varepsilon_q(0) > 0,\ \varepsilon_p(0) < 0 \\ \pi + \tan^{-1}\left(-\dfrac{\varepsilon_p(0)\sqrt{s_q}}{\varepsilon_q(0)\sqrt{s_p}}\right) & \text{if } \varepsilon_q(0) < 0 \\ 2\pi + \tan^{-1}\left(-\dfrac{\varepsilon_p(0)\sqrt{s_q}}{\varepsilon_q(0)\sqrt{s_p}}\right) & \text{if } \varepsilon_q(0) > 0,\ \varepsilon_p(0) > 0 \end{cases}$$

and

$$\phi_q = \begin{cases} \tan^{-1}\left(-\dfrac{\varepsilon_q(0)\sqrt{s_p}}{\varepsilon_p(0)\sqrt{s_q}}\right) & \text{if } \varepsilon_p(0) > 0,\ \varepsilon_q(0) > 0 \\ \pi + \tan^{-1}\left(\dfrac{\varepsilon_q(0)\sqrt{s_p}}{\varepsilon_p(0)\sqrt{s_q}}\right) & \text{if } \varepsilon_p(0) < 0 \\ 2\pi + \tan^{-1}\left(\dfrac{\varepsilon_q(0)\sqrt{s_p}}{\varepsilon_p(0)\sqrt{s_q}}\right) & \text{if } \varepsilon_p(0) > 0,\ \varepsilon_q(0) < 0 \end{cases}$$

Using the definitions of $\varepsilon_q(t)$ and $\varepsilon_p(t)$, we can see that the frequencies $q(t)$ and $p(t)$, as functions of time, are given by the equations

(3)

$$q(t) = \tfrac{1}{2} + M_q \cos(ct - \phi_q)$$

$$p(t) = \tfrac{1}{2} + M_p \cos(ct - \phi_p)$$

4. Describe, qualitatively, the predicted behavior of q and p from Equations 3.

5. How do the constants M_q and M_p affect the behavior? How do the constants ϕ_q and ϕ_p affect the behavior? How does the constant c affect the behavior?

6. Use your answers to Problem 5 to explain how the constants s_q and s_p affect the frequency of type A hosts and type B parasites over time. Can you provide a biological explanation for your answer?

The properties of Equations 3, and the predictions that can be obtained from them in terms of experimental data, are explored in Case Study 2a and 2b.

Descriptive Statistics

11

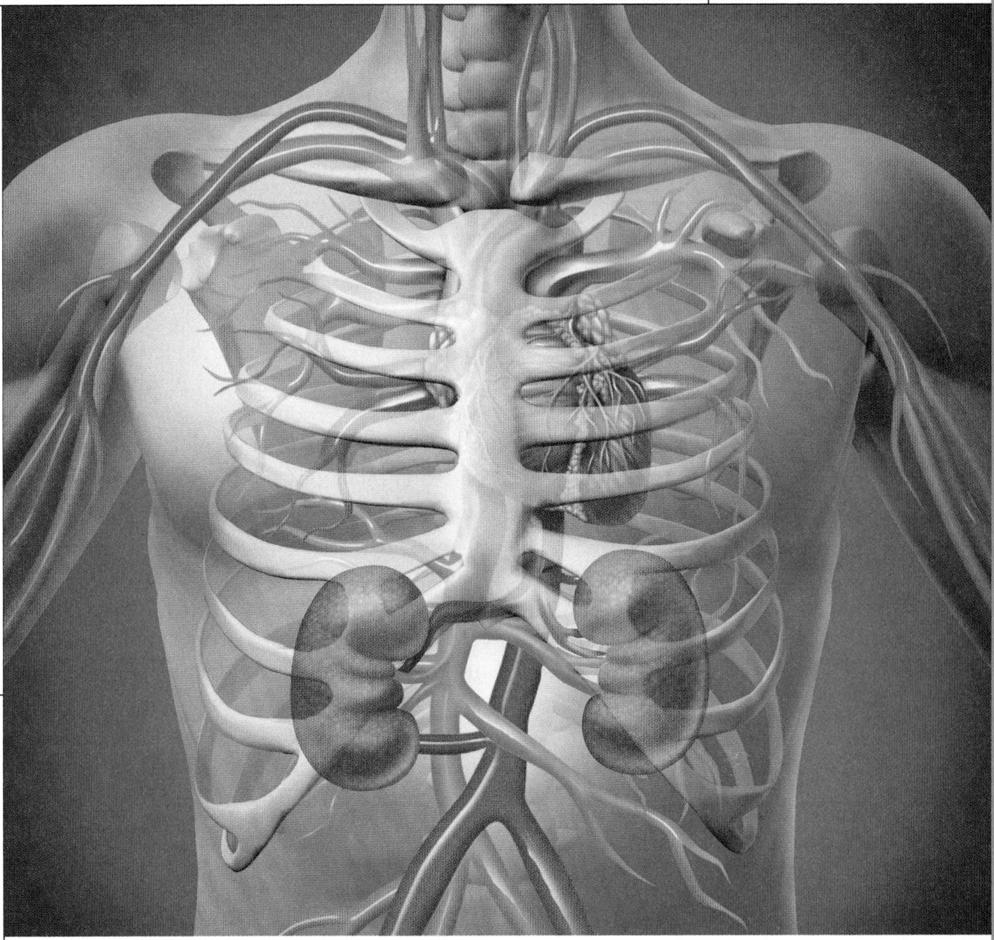

Shown is an image of the location of human kidneys. In Section 11.4 we will explore data involving the surgical removal of crystalline deposits, called kidney stones.

WE NOW BEGIN OUR STUDY of data and data analysis. In this chapter we focus on summarizing the information presented in a data set. In Chapter 13 we will use such data to make inferences about the population from which they were collected. Doing so, however, requires some ideas from probability theory, and this is the subject of Chapter 12.

We will see two important ways in which data analysis is related to the mathematics that we have learned so far. First, although we have used mathematical techniques in previous chapters to build models of various biological phenomena, we require data to inform this process. Second, we will see that the analysis of data uses many of the ideas of calculus that we have developed in previous chapters.

11.1 Numerical Descriptions of Data

© enciktat / Shutterstock.com

Athletes sometimes take performance-enhancing drugs to give them an edge in competition. One such drug is EPO, or erythropoietin. EPO is a natural hormone that stimulates red blood cell production, and synthetic variants of this hormone have been injected by athletes to increase their aerobic capacity.

Does EPO injection actually give an athlete a substantial advantage? And if so, how big is this advantage? The field of statistics provides a set of tools and techniques that can be used to answer these questions.

In statistics a *population* is a collection of individuals that we would like to draw conclusions about. In the case of EPO the population consists of all athletes who might use EPO because we wish to know, in general, if EPO use affects athletic performance. Usually we are not able to collect data from all individuals in the population and therefore we instead gather data from a subset of these individuals. This subset is referred to as a *sample*.

Table 1

Change in Aerobic Capacity (in L oxygen/min)

0.4	0.4
0.3	0.3
0.3	0.4
0.5	0.5
0.3	0.1
0.2	0.5
0.3	0.3

We will study populations and samples in detail in Section 11.4, but as an example, let's consider the aerobic performance data in Table 1 from an experiment involving a sample of 14 athletes. The athletes first had their aerobic capacity measured (as liters of oxygen consumed per minute during exercise). They were then injected with EPO for six weeks, and their aerobic capacity was measured again. The values in Table 1 give the change in aerobic capacity for each of the 14 athletes as a result of EPO injection (measured in L oxygen/min).[1]

The goal with data like those of Table 1 is to use the data to draw conclusions about the entire population of athletes. Do these data provide evidence that we can expect the aerobic performance of any athlete to change as a result of EPO injection? And if so, by how much?

The first step to answering these questions is to summarize various properties of the data set. **Descriptive statistics** provides a set of techniques for doing do, and these techniques are the topic of this chapter. For example, although a quick glance at the data in Table 1 shows that the aerobic capacity increased for all 14 individuals in the sample, can we obtain a single number that gives the typical magnitude of this change? Such measures are called **summary statistics**.

Descriptive statistics allow us to summarize the magnitude of change in aerobic performance in the sample of athletes, but how representative is this change of the population of all athletes in general? The population-level counterpart of a sample summary

1. B. Ekblom and B. Berglund, "Effect of Erythropoietin Administration on Maximal Aerobic Power," *Scandinavian Journal of Medicine & Science in Sports* 1 (1991): 88–93.

statistic is called a population *parameter*. Summary statistics can be used as *estimates* of these parameters. **Inferential statistics** provides a set of techniques (based on probability theory) that allows us to quantify how well a summary statistic estimates a population parameter. We begin an exploration of these ideas in Section 11.4, but a detailed examination is postponed until Chapter 13, after we have acquired the necessary background in probability theory.

■ Types of Variables

The things that a data set describes are called **individuals**. Individuals can be anything on which measurements can be made—for example, people, species, or proteins. The property of an individual that is measured is called a **variable**.

Categorical variables describe membership in categories. Examples include letter grades, genotypes, and presence or absence of a disease. A categorical variable is referred to as **ordinal** if the category has an inherent ordering (for example, letter grades) or **nominal** if the category has no such inherent ordering (for example, genotype or disease status).

Numerical variables describe numeric measurements that can be meaningfully ordered and for which arithmetic operations make sense. Examples include number of mutations, body mass, and life span. Numerical variables are referred to as **discrete** if they take values from a discrete (that is, unconnected) subset of the real numbers (for example, number of mutations) or **continuous** if they take values in some interval of the real numbers (for example, body mass or aerobic capacity).

More than one variable can be measured on each individual. Indeed, much of statistics is devoted to analyzing the relationships between variables. In Section 11.3 we begin studying such relationships. In the remainder of this section and in Section 11.2 we consider data in which a single variable is measured.

■ Categorical Data

Categorical data are usually summarized by listing the number of individuals in each category. These counts are also often used to calculate the fraction of the sample that falls into each category.

A pea plant

EXAMPLE 1 | Mendelian genetics In a now-classic study from 1865 that laid the foundation for our understanding of genetic inheritance, Gregor Mendel created crosses by interbreeding different types of pea plants and recorded various attributes of their offspring. In one particular type of cross he measured flower position in the offspring (either axial or terminal). He found that, of 858 offspring measured, 651 had axial flowers and 207 had terminal flowers. What type of variable is flower position?

SOLUTION There are two categories, axial and terminal, and these do not have any natural ordering. Thus flower position is a nominal categorical variable. ■

■ Numerical Data: Measures of Central Tendency

Because numerical data can display many more patterns than categorical data, many different summary statistics can be obtained for numerical variables. One way to make sense of such data is to find a typical number in the "center" of the data. Any such number is called a **measure of central tendency**. One of the most common measures of central tendency is the *average* (or the *mean*).

(1) Definition Let $x_1, x_2, \ldots, x_n$ be n data points. The **mean**, or **average**, denoted by $\bar{x}$, is the sum of the values of x divided by n:

$$\bar{x} = \frac{x_1 + x_2 + \cdots + x_n}{n}$$

© Butterfly Hunter / Shutterstock.com

EXAMPLE 2 | BB **Cuckoo bird egg size** The lengths of 10 eggs from cuckoo birds were measured, giving the following data (in mm):

19 19 20 22 22 22 23 24 25 35

Calculate the mean egg length.

SOLUTION From Definition 1, the mean egg length is

$$\bar{x} = \frac{19 + 19 + 20 + 22 + 22 + 22 + 23 + 24 + 25 + 35}{10} = 23.1 \text{ mm}$$ ■

Another measure of central tendency is the *median*, which is the middle number of an ordered list of numbers. There are as many data points that are greater than the median as there are that are less than the median.

(2) Definition Let $x_1, x_2, \ldots, x_n$ be n data points written in increasing order.
(a) If n is odd, the **median** is the middle number.
(b) If n is even, the **median** is the average of the two middle numbers.

EXAMPLE 3 | **Cuckoo bird egg size (continued)** Calculate the median egg length from the data in Example 2.

SOLUTION There are an even number of data points and therefore the median is the average of the two middle numbers:

$$\text{median} = \frac{22 + 22}{2} = 22 \text{ mm}$$ ■

The *mode* of a data set is a third measure of central tendency, but it is usually less informative than the mean or median.

(3) Definition The **mode** of a data set is the element that appears most often in the data set.

The mode of the data set in Example 2 is 22 mm.

■ Numerical Data: Measures of Spread

Measures of central tendency identify a typical value of the data. **Measures of spread** (also called **measures of dispersion**) describe the spread, or variability, of the data around a central value. For example, each of the following data sets has mean 72, but it is clear that the first set of data has more variability than the second.

Data set 1: 50 58 78 81 93

Data set 2: 72 71 72 72 73

The most important measures of spread in statistics are the *standard deviation* and *variance*.

(4) Definition Let $x_1, x_2, \ldots, x_n$ be n data points, and let $\bar{x}$ be their mean. The **standard deviation** of the data is

$$\text{s.d.} = \sqrt{\frac{1}{n}\sum_{k=1}^{n}(x_k - \bar{x})^2}$$

The **variance** is $(\text{s.d.})^2$, the square of the standard deviation.

The variance of a set of numbers, $x_1, x_2, \ldots, x_n$, can be viewed as the average of the numbers $(x_1 - \bar{x})^2, (x_2 - \bar{x})^2, \ldots, (x_n - \bar{x})^2$, where $\bar{x}$ is their mean. In other words, from Definitions 1 and 4, the variance is the average squared difference of the x_i values from their mean $\bar{x}$.

EXAMPLE 4 | BB **Antibiotic effectiveness** is often quantified by measuring the minimum concentration required to inhibit bacterial growth (called the *minimum inhibitory concentration*, or *MIC*). The *MIC* of methacycline was measured on ten bacterial colonies, giving the following data (in μg/mL):

0.12 0.04 0.03 0.42 0.05 0.10 0.13 0.12 0.09 0.11

Summarize the data using the mean and standard deviation.

SOLUTION Using Definition 1, the mean *MIC* is approximately 0.12 μg/mL. To use Definition 4 for calculating the standard deviation, we first construct a table:

x_i	$(x_i - \bar{x})$	$(x_i - \bar{x})^2$
0.12	0	0
0.04	−0.08	0.0064
0.03	−0.09	0.0081
0.42	0.30	0.0900
0.05	−0.07	0.0049
0.10	−0.02	0.0004
0.13	0.01	0.0001
0.12	0	0
0.09	−0.03	0.0009
0.11	−0.01	0.0001

We then have

$$\text{s.d.} = \sqrt{\frac{0 + 0.0064 + 0.0081 + 0.0900 + 0.0049 + 0.0004 + 0.0001 + 0 + 0.0009 + 0.0001}{10}}$$

$$\approx 0.11\ \mu\text{g/mL}$$

■

Numerical Data: The Five-Number Summary

Another simple indicator of the spread of data is the location of the **minimum** and **maximum** values. Other indicators of spread are the *quartiles*. Recall that the median divides a data set in half. The median of the lower half of the data is called the **first quartile**, Q_1. The median of the upper half of the data is called the **third quartile**, Q_3. The minimum, maximum, median, and first and third quartiles together give a good picture of the spread and center of the data.

There are two different conventions for calculating Q_1 and Q_3 for a data set that consists of an odd number of data points. The first convention includes the median data point in the calculation of both Q_1 and Q_3. The second convention excludes the median data point from both calculations. We will follow the second convention.

(5) Definition The **five-number summary** for a data set is the set of these five numbers, written in the indicated order:

Minimum Q_1 Median Q_3 Maximum

Figure 1 shows how these five numbers are related for a data set listed in order.

1 2 2 4 8 9 10 12 14 16 20 21 31

Min ↑ (1), Q_1 ↑ (between 2 and 4), Med ↑ (10), Q_3 ↑ (between 16 and 20), Max ↑ (31)

FIGURE 1
The five-number summary

For the data in Figure 1, $Q_1 = (2 + 4)/2 = 3$ and $Q_3 = (16 + 20)/2 = 18$.

The five-number summary allows us to quickly calculate two different measures of spread. A simple indicator of spread is the **range**, which is the difference between the maximum and minimum values:

$$\text{Range} = \text{Maximum} - \text{Minimum}$$

This can be compared to the spread of the middle of the data as measured by the **interquartile range** (**IQR**), which is the difference between the third and the first quartiles:

$$\text{IQR} = Q_3 - Q_1$$

For the data in Figure 1, the range is $31 - 1 = 30$ and the interquartile range is $\text{IQR} = 18 - 3 = 15$.

EXAMPLE 5 | Cuckoo bird egg size (continued) Calculate the five-number summary for the data given in Example 2.

SOLUTION The minimum and maximum are 19 mm and 35 mm, respectively, and from Example 3, the median is 22 mm. The upper and lower halves of the data contain an odd number of data points and therefore we can read Q_1 and Q_3 directly from the data: $Q_1 = 20$ mm and $Q_3 = 24$ mm. The five number summary is

$$\text{Min} = 19 \qquad Q_1 = 20 \qquad \text{Median} = 22 \qquad Q_3 = 24 \qquad \text{Max} = 35$$ ■

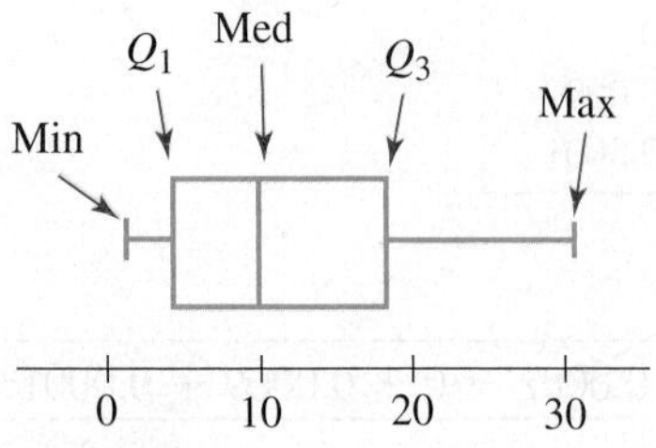

FIGURE 2

A **box plot** (also called a **box-and-whisker plot**) is a method for graphically displaying the five-number summary. The plot consists of a rectangle whose left- and right-hand sides correspond to Q_1 and Q_3, respectively. The box is divided by a vertical line segment at the location of the median. The **whiskers** are horizontal line segments that extend from both edges of the box, and there are different conventions for how these are constructed. The simplest approach extends the whiskers to the location of the maximum and minimum values. Figure 2 shows such a box plot for the data in Figure 1.

EXAMPLE 6 | **Cuckoo bird egg size (continued)** Construct a box plot for the data from Example 2.

SOLUTION Using the five-number summary for the data from Example 5, we construct the box plot as in Figure 3. (We have indicated the data points on the horizontal axis for illustrative purposes.)

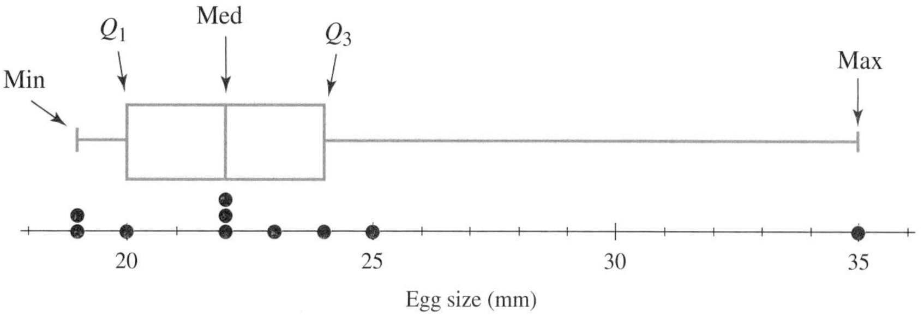

FIGURE 3

The median of the data is also called the **second quartile**, Q_2. So the three quartiles, Q_1, Q_2, and Q_3 separate the data into four parts, each with the same number of data points. The number of parts is arbitrary; statisticians also use *quintiles*, *deciles*, and *percentiles*. **Percentiles** divide the data into 100 parts, each with the same number of data points. Standardized test scores are often reported as percentiles. A score in the 99th percentile is a score in the top one-hundredth part of the scores on that test.

Outliers

If a data set includes a number that is "far out," or far away, from the rest of the data, that data point is called an **outlier**. Outliers are quite common in data from the life sciences. They can be the result of mistakes made in recording data, but they can also be legitimate data that represent natural variability.

There is no single mathematical definition of what constitutes an outlier, but a common approach is to classify a data point as an outlier if it falls more than a distance $1.5 \times \text{IQR}$ above Q_3 or more than $1.5 \times \text{IQR}$ below Q_1. These distances are illustrated in Figure 4 for the cuckoo egg data.

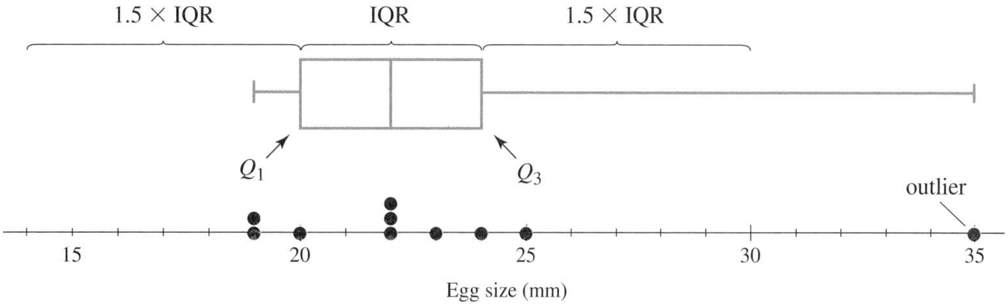

FIGURE 4 Identifying outliers using the cuckoo egg data on a box plot

The IQR for the cuckoo egg data in Example 5 is $\text{IQR} = 24 - 20 = 4$ mm. Multiplying this by 1.5 gives a value of 6 mm. Therefore, using the values of $Q_1 = 20$ mm and $Q_3 = 24$ mm from Example 5, any egg size smaller than $20 - 6 = 14$ mm or larger than $24 + 6 = 30$ mm would be considered an outlier. Figure 4 shows that the point 35 mm is the only outlier in this data set.

Another common box plot convention is to extend the whiskers to the largest and smallest data points that are not outliers. Outliers are then identified with single points. We refer to this as an **outlier box plot**. Figure 5 shows the outlier box plot for the cuckoo egg data from Example 2. Note that the left and right edges of the box still correspond to the first and third quartile of all the data points.

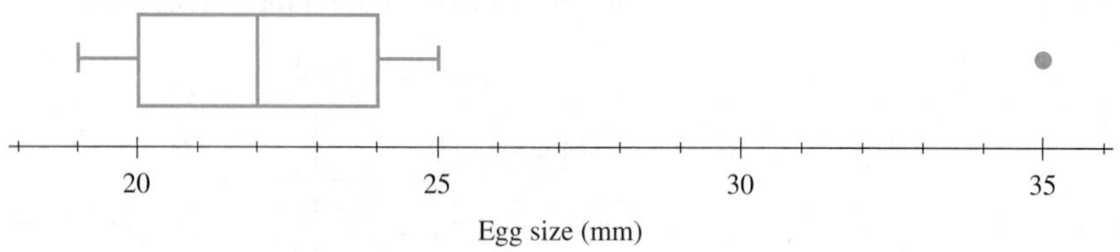

FIGURE 5
The outlier box plot for the cuckoo egg data

Outliers can have substantial effects on summary statistics and so it is important to be aware of their presence in data.

EXAMPLE 7 | Antibiotic effectiveness (continued) Closer examination of the antibiotic data in Example 4 suggests that the data point 0.42 μg/mL is an outlier. It is very different from all the other measurements. Indeed, you should verify that $Q_1 = 0.05$ μg/mL, $Q_3 = 0.12$ μg/mL, and therefore IQR $= 0.07$ μg/mL. Thus 0.42 is an outlier because it is larger than

$$Q_3 + 1.5 \times \text{IQR} = 0.12 + 1.5 \times 0.07 = 0.225$$

(a) Calculate the mean and standard deviation of methacycline *MIC* after the outlier 0.42 is removed from the data set, and compare the result to Example 4.
(b) Compare the median *MIC* for the full methacycline data set to that with the outlier removed.

SOLUTION

(a) Using Definition 1, the mean *MIC* is calculated as approximately 0.088 μg/mL as compared to 0.12 μg/mL in Example 4. Using Definition 4, the standard deviation is calculated as s.d. ≈ 0.036 μg/mL as compared to 0.11 μg/mL. Both the mean and the standard deviation in *MIC* are considerably lower with the outlier removed.

(b) Ordering the data, we have

$$0.03 \quad 0.04 \quad 0.05 \quad 0.09 \quad 0.10 \quad 0.11 \quad 0.12 \quad 0.12 \quad 0.13 \quad 0.42$$

The median is $(0.10 + 0.11)/2 = 0.105$. With the outlier removed the data set is

$$0.03 \quad 0.04 \quad 0.05 \quad 0.09 \quad 0.10 \quad 0.11 \quad 0.12 \quad 0.12 \quad 0.13$$

and the median is then the middle data point, 0.10.

Thus, while part (a) shows that the mean is strongly affected by the outlier, we see that the median is relatively unaffected. ■

Example 7 suggests that when an outlier is present in the data, the median is a better indicator of central tendency than the mean.

Whether outliers are included in calculations of summary statistics is an important issue. In general, excluding an outlier from an analysis is not good practice unless the researcher can determine that it has resulted from measurement error. In many instances, outliers reveal something that can be biologically important and that indicates need for further investigation. The graphical techniques of the next section provide a helpful way to readily identify outliers.

EXERCISES 11.1

1–6 For each question the variables may be Categorical and Ordinal, Categorical and Nominal, Numerical and Discrete, or Numerical and Continuous.

1. **Darwin's finches** live in the Galápagos Islands. A study measured the beak width of individual birds on one of the islands. What type of variable is "beak width" and what are its possible values?

2. **Antiviral medication** Researchers gave volunteers a newly developed antiviral drug and then experimentally infected the individuals with influenza. They quantified severity of infection as mild, moderate, or severe. What type of variable is "severity of infection"?

3. ***Drosophila* abdominal bristles** are hairlike structures on the abdomen of fruit flies and have been the focus of numerous genetic studies. What type of variable is "number of bristles" and what are its possible values?

4. **Coral bleaching** is a process through which coral reefs lose their color during extreme temperature events. Different coral species bleach at different temperatures, and a study quantified this temperature for 10 different species. What type of variable is "bleaching temperature" and what are its possible values?

5. **Health-care usage** A survey asked respondents about the different kinds of health-care facilities that they used in the past year. Choices included hospital emergency room, family doctor, and walk-in clinic. What type of variable is "health-care facility"?

6. **Conservation biology** The effect of logging on the presence of different bird species was quantified by determining the presence or absence of 12 specific bird species. What type of variable is "number of species present" and what are its possible values?

BB 7. **Weddell seals** Twenty individual Weddell seals from Antarctica were weighed, giving the following data (in pounds).

902	920	924	932	937
939	945	948	949	951
957	958	961	965	969
970	975	982	987	991

Calculate the mean, median, standard deviation, and IQR.

BB 8. **Monarch butterflies** migrate thousands of miles each year to overwinter in areas of Mexico. The following data give the wingspans (in centimeters) of 28 individuals captured in their overwintering grounds.

8.6	10.9	9.8	12.4	10.6	10.7	10.9
10.1	10.2	10.3	9.4	9.1	11.2	9.7
11.6	12.0	10.4	9.5	10.9	10.5	9.9
11.1	10.5	11.3	11.7	11.3	10.3	11.5

Calculate the mean, median, standard deviation, and IQR.

BB 9. **Damselflies** A study measured the number of external parasites on a sample of 20 damselflies, giving the following data.

5	0	0	54	5
12	27	24	36	5
56	43	15	42	12
62	36	34	58	23

Calculate the mean, median, standard deviation, and IQR.

Damselfly with eight ectoparasites on its abdomen

BB 10. ***Clostridium difficile*** is a bacterium that can cause potentially fatal gastrointestinal infections in hospital patients who are receiving antibiotics. Fifteen patients were tested for the presence of *C. difficile*, with each patient receiving the test five times, and the fraction of positive tests recorded. This gave the following data.

$\frac{2}{5}$	1	1	$\frac{3}{5}$	$\frac{4}{5}$
$\frac{3}{5}$	1	1	$\frac{2}{5}$	1
1	1	1	$\frac{4}{5}$	1

Calculate the mean, median, standard deviation, and IQR.

BB **11. Coral bleaching** is a process through which coral reefs lose their color during extreme temperature events. The extent of bleaching after an extreme temperature event was measured at 20 sites, giving the following data. Values are dimensionless measures of bleaching, ranging from 0 (no bleaching) to 1 (complete bleaching of all corals at the site).

0.19	0.06	0.26	0.32	0.35
0.09	0.05	0.21	0.48	0.21
0.23	0.12	0.22	0.16	0.35
0.13	0.34	0.07	0.16	0.45

Calculate the five-number summary and construct a box plot.

BB **12. Breast cancer tumors** were measured for a sample of 15 patients by estimating tumor diameter in centimeters. This gave the following data.

1.5	1.5	1.6	1.6	1.7
2.0	2.1	2.2	2.6	3.2
4.0	4.9	5.5	7.5	7.6

Calculate the five-number summary and construct a box plot.

BB **13. Antibiotic *MIC*** The minimum inhibitory concentration (*MIC*) of a new antibiotic is measured on 10 bacterial colonies, giving the following data (measured in μg/mL).

0.2	0.4	0.3	0.5	1.1
1.0	1.2	1.3	0.9	1.1

Calculate the five-number summary and construct a box plot.

BB **14. Huntington's disease** is a genetic disorder causing neurodegeneration and eventual death. The age at death of 15 patients with the disease is given in the following data.

71	70	69	68	67
66	64	62	58	57
55	53	52	48	45

Calculate the five-number summary and construct a box plot.

Source: Adapted from E. Andrew et al., "The Relationship between Trinucleotide (CAG) Repeat Length and Clinical Features of Huntington's Disease," *Nature Genetics* 4 (1993): 398–403.

15. The Suleman octuplets were born on January 26, 2009. The following data set gives the birth weight of each child, in kilograms.

1.22 1.25 1.47 1.13 0.68 1.25 1.94 1.22

Construct the outlier box plot.

BB **16. Hospital stays** The duration of stay for 20 patients at a major US hospital is given in the following data set.

6	10	5	12	5
3	41	12	6	15
20	4	3	7	13
16	18	29	23	17

Construct the outlier box plot.

BB **17. Anemia** When individuals become infected with malaria the parasite consumes red blood cells, causing the patient to become anemic. A group of 18 mice had their red blood cell concentration measured 10 days after infection, giving the following data (measured as cells $\times 10^6/\mu L$).

3.2	4.5	3.7	3.3	2.5	3.6
3.3	3.6	4.2	3.5	2.0	3.1
2.3	2.9	6.9	4.5	7.3	3.0

Construct the outlier box plot.

BB **18. Multiple sclerosis** is a serious neurological disease. The following data set gives the age of onset of multiple sclerosis for 32 individuals.

23	52	32	27	28	30	13	37
33	29	25	31	26	30	5	29
41	27	30	34	28	19	31	29
32	28	33	28	31	29	29	30

Construct the outlier box plot.

19. Performance-enhancing drugs Calculate the mean and standard deviation of the change in aerobic capacity for the data from Table 1 and construct the outlier box plot. Interpret your results in the context of how EPO affects aerobic capacity.

BB **20. HAART**, or "Highly Active Antiretroviral Therapy," is a treatment for HIV infection. The reduction in the concentration of HIV in the blood (in units of $\log_{10}$ copies/mL) was measured in 15 patients who received HAART treatment, giving the following data.

3.7	2.5	3.5	3.6	0.7
2.2	2.5	1.4	3.6	1.8
2.4	2.3	4.7	2.3	3.0

Calculate the mean and standard deviation of the reduction in HIV concentration and construct the outlier box plot. Interpret your results in the context of how HAART affects HIV concentration.

11.2 | Graphical Descriptions of Data

In the preceding section we described data using summary statistics for central tendency and spread. In this section we describe data using graphs. This allows us to see the entire data set at a glance.

■ Displaying Categorical Data

Two simple visual summaries are often used for categorical data: *bar graphs* and *pie charts*.

A **bar graph** consists of vertical bars, one bar for each category. The height of each bar is proportional to the number of individuals in that category. So the vertical axis has a numerical scale corresponding to the number (or sometimes the proportion) of individuals in each category. The labels on the horizontal axis describe the categories.

EXAMPLE 1 | **Human papillomavirus (HPV) cancers** A sample of people who developed cancer because of HPV infection was taken, and each cancer was classified according to its location in the body. These data can be displayed using a bar graph as shown in Figure 1. ■

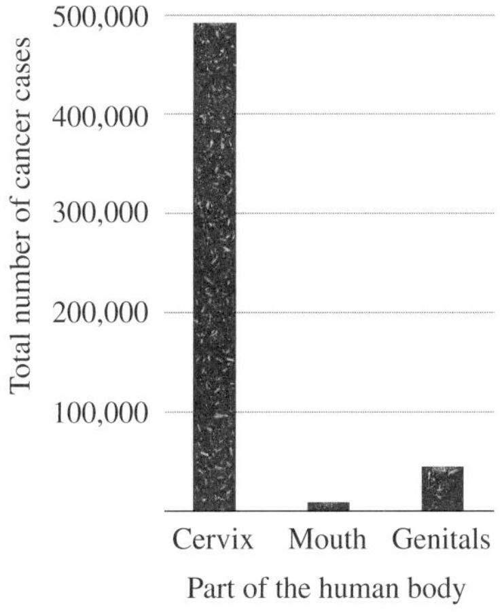

FIGURE 1

Adapted from J. Crow, "HPV: The Global Burden," *Nature* 488 (2012): S2–S3.

A **pie chart** consists of a circle divided into sectors, one sector for each category. The central angle of each sector, and therefore the area of each sector, is proportional to the number of individuals in that category. Each sector is labeled with the name of the corresponding category.

EXAMPLE 2 | **Genetic diseases** A study examined whether genes that are associated with human diseases are located in regulatory regions (that is, regions of the DNA that regulate gene expression), near regulatory regions, or are in other regions of the genome. A data set for several diseases was compiled and the results displayed as the pie chart in Figure 2.

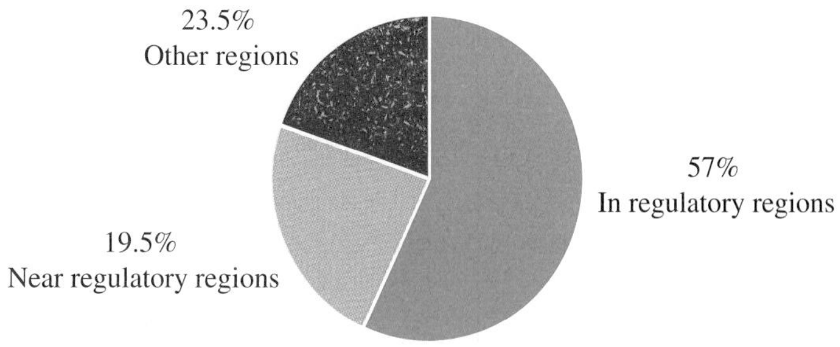

FIGURE 2

Adapted from M. Maurano et al., "Systematic Localization of Common Disease-Associated Variation in Regulatory DNA," *Science* 337 (2012): 1190–95.

■

■ Displaying Numerical Data: Histograms

To visualize numerical data, we can first convert the data into (ordinal) categorical data. For example, "age" is a numerical variable, but we can convert it into an ordinal categorical variable by placing individuals in ordered age categories (for example, 0–10 years, 11–20 years, and so on) and recording the number of individuals in each category.

In general, to put numerical data into categories we divide the range of the data into nonoverlapping adjacent intervals of equal lengths, called **bins**. We can then plot a graph called a **histogram** by first labeling the bins on the horizontal axis, and then placing a rectangle on each bin. The height of each rectangle is proportional to the number of data points in the bin.

The actual numerical value for the height of the rectangles in a histogram is determined using one of three conventions:

(i) **frequency histograms** set the height of each rectangle equal to the number of data points in the corresponding bin (beware of this somewhat confusing terminology where "frequency" is used synonymously with "number");

(ii) **relative frequency histograms** set the height of each rectangle equal to the fraction of data points in the corresponding bin; and

(iii) **density histograms** set the height of each rectangle so that its area is equal to the fraction of data points in the corresponding bin.

Regardless of the convention used, the resulting histogram has the same shape. The conventions differ only in the scaling of the vertical axis.

EXAMPLE 3 | BB Heart rates The heart rates of 40 patients in an intensive care unit are given below.

52	68	69	70	78	78	79	81	83	88
88	88	88	89	89	89	92	92	95	95
95	98	98	98	99	99	101	101	103	106
108	108	109	109	113	115	115	119	128	139

Construct a histogram for the data with bins of width 10 beats/min, using each of the three conventions. Start the first bin at 50 beats/min.

SOLUTION We first organize the data into bins of width 10, determining the number of data points falling into each as shown in the second column of Table 1. By convention the lower endpoint of each interval is closed and the upper endpoint is open. These counts can be used to construct the frequency histogram.

Table 1

Interval	Frequency	Relative Frequency	Density
$50 \leq x < 60$	1	0.025	0.0025
$60 \leq x < 70$	2	0.05	0.005
$70 \leq x < 80$	4	0.1	0.01
$80 \leq x < 90$	9	0.225	0.0225
$90 \leq x < 100$	10	0.25	0.025
$100 \leq x < 110$	8	0.2	0.02
$110 \leq x < 120$	4	0.1	0.01
$120 \leq x < 130$	1	0.025	0.0025
$130 \leq x < 140$	1	0.025	0.0025

To construct the relative frequency histogram we compute the fraction of all data points that fall within each bin. The total number of data points is 40 and so we divide the counts in column two of Table 1 by 40, giving the third column of Table 1. The density histogram specifies the area of each rectangle as being equal to the relative frequency or fraction of data points in the corresponding bin. Because the area of a rectangle is just its height multiplied by its width, the height of each rectangle in a density histogram can be calculated by dividing the relative frequency by the bin width, which is 10 beats/min in this case. This gives the fourth column in Table 1.

To construct each type of histogram we place rectangles over each interval, with heights equal to the values of the corresponding column in Table 1. All three types of histogram can be displayed in a single plot by simply changing the scale on the vertical axis, as shown in Figure 3.

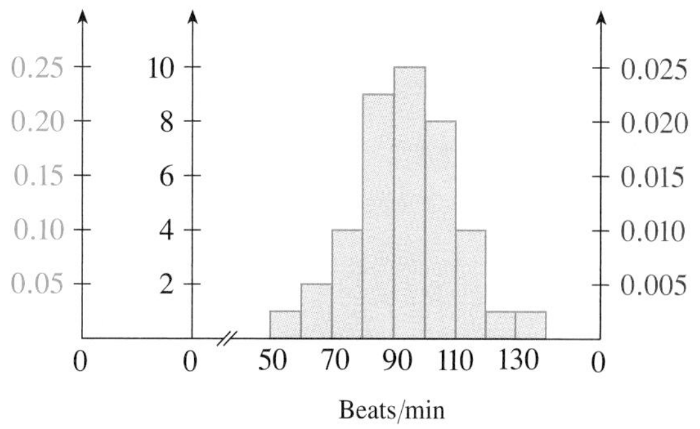

FIGURE 3
Histograms of heart rate data: frequency histogram (*black units*): relative frequency histogram (*blue units*); and density histogram (*red units*)

There is no single best approach for choosing the bin width when constructing a histogram. However, **Sturges' Rule** provides a good starting point. First we calculate a quantity k as the nearest integer to $1 + \log_2 n \approx 1 + 1.44 \ln n$, where n is the number of data points. Then, if R is the range of the data, we choose the bin width equal to R/k. This is only a starting place, however, and it is a good idea to vary the bin width to see how it affects the shape of the histogram.

Warning The shape of a histogram obtained for a data set can change slightly as the location of the first bin is changed.

Likewise, there is no single best approach for choosing where to place the first bin. Usually the lower endpoint of the interval for the first bin is chosen to make the tabulation of the data as simple as possible.

EXAMPLE 4 | BB **EPO usage by athletes** The change in aerobic capacity for 14 athletes injected with EPO is given in the following table (measured as L oxygen consumed per minute):

0.4	0.5	0.3	0.4	0.5
0.3	0.3	0.4	0.5	0.3
0.3	0.2	0.3	0.1	

Construct a density histogram for the data.

SOLUTION There are 14 data points and therefore Sturges' Rule gives $1 + 1.44 \ln 14 \approx 4.8$, which rounds to $k = 5$. The range of the data is $0.5 - 0.1 = 0.4$ and therefore we start with a bin width of $R/k = 0.4/5 = 0.08$. If we choose the lower endpoint of the first interval to be 0.1, we obtain a total of six bins, as shown in the following table.

Interval	Frequency	Density
$0.1 \leq x < 0.18$	1	0.893
$0.18 \leq x < 0.26$	1	0.893
$0.26 \leq x < 0.34$	6	5.36
$0.34 \leq x < 0.42$	3	2.68
$0.42 \leq x < 0.50$	0	0
$0.50 \leq x < 0.58$	3	2.68

As in Example 3, the density column was obtained by dividing the relative frequency by the bin width, 0.08. Note that, unlike relative frequencies, densities can be larger than 1 if the bin width is less than 1. Nevertheless, the area of all rectangles still sums to 1.

Figure 4(b) plots the resulting density histogram. Although Sturges' Rule is a good place to start for choosing bin widths, it is useful to explore choices both greater than and less than this value. Figures 4(a) and 4(c) show histograms when using one fewer and one more bin, respectively.

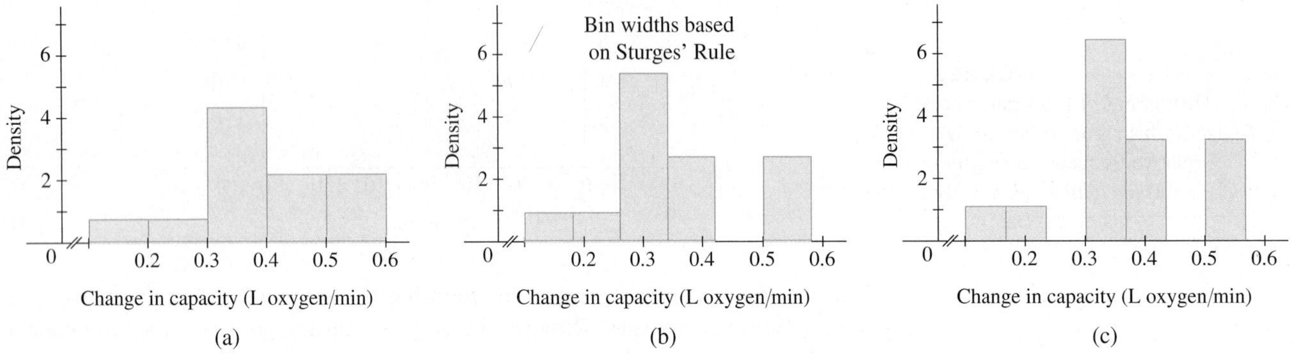

FIGURE 4

A histogram gives a visual representation of how the data are distributed across the different bins. The histogram allows us to determine whether the data are symmetric around the mean or whether the data are *skewed.* If the histogram has a long 'tail' on the right, we say that the data are **skewed to the right**. Similarly, if there is a long tail to the left, the data are **skewed to the left** (see Figure 5). The heart rate data in Figure 3 appear to be roughly symmetric, whereas the EPO data in Figure 4 appear slightly skewed to the left.

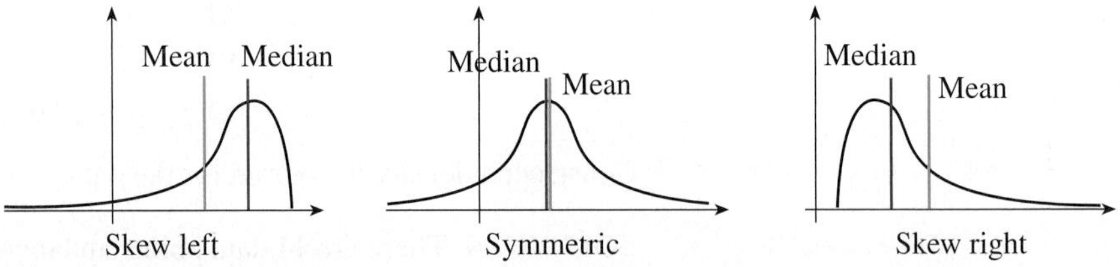

FIGURE 5
Symmetric and skewed distributions

In addition to skew, histograms also provide information about other aspects of shape. The *modality* of a histogram refers to the number of peaks. A **unimodal** histogram has a single peak located at the mode of the data set. **Bimodal** and **multimodal** histograms have two or multiple peaks, respectively. Unless there is more than one peak with the same height, however, the data still have a unique mode.

■ Interpreting Area in Histograms

Of the three different histogram conventions illustrated in Example 3, density histograms are often the most useful. This is because the area of such a histogram has a simple interpretation. Recall that the area of a bar in a density histogram is equal to the fraction of data points in the corresponding bin. Thus the area of several bars taken together is equal to the fraction of the sample that falls into any of the corresponding bins.

EXAMPLE 5 | **Heart rates (continued)** Indicate the fraction of patients whose measured heart rate lies between 70 and 100 beats per minute as an area on the histogram in Figure 3.

SOLUTION If we use a density histogram, the desired fraction is equal to the area of the three bars lying in the interval between 70 and 100, as shown in Figure 6.

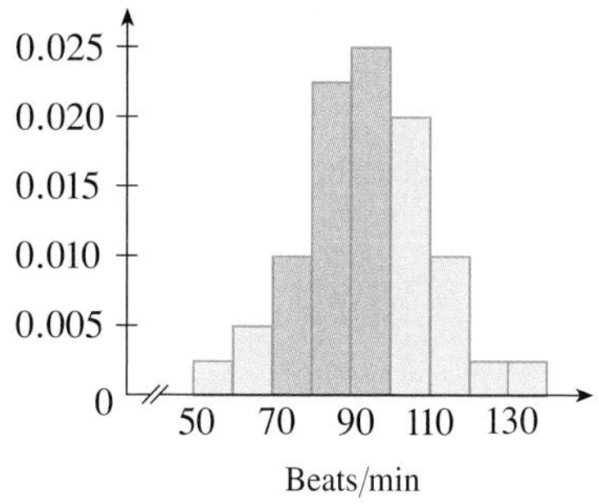

FIGURE 6

The area interpretation for histograms has a close relationship with the five-number summary and boxplots introduced in Section 11.1. Recall that the median of the data is the value of the variable that divides the data set in half. Therefore, because the area of each bar in any histogram is proportional to the number of data points in that bin, it follows that the *median* is located at the value of the variable that divides the area of a histogram in half. Likewise, the first quartile is the value of the variable below which one quarter of the area occurs, while the third quartile is the value above which one quarter of the area occurs. Figure 7 illustrates this idea using a frequency histogram.

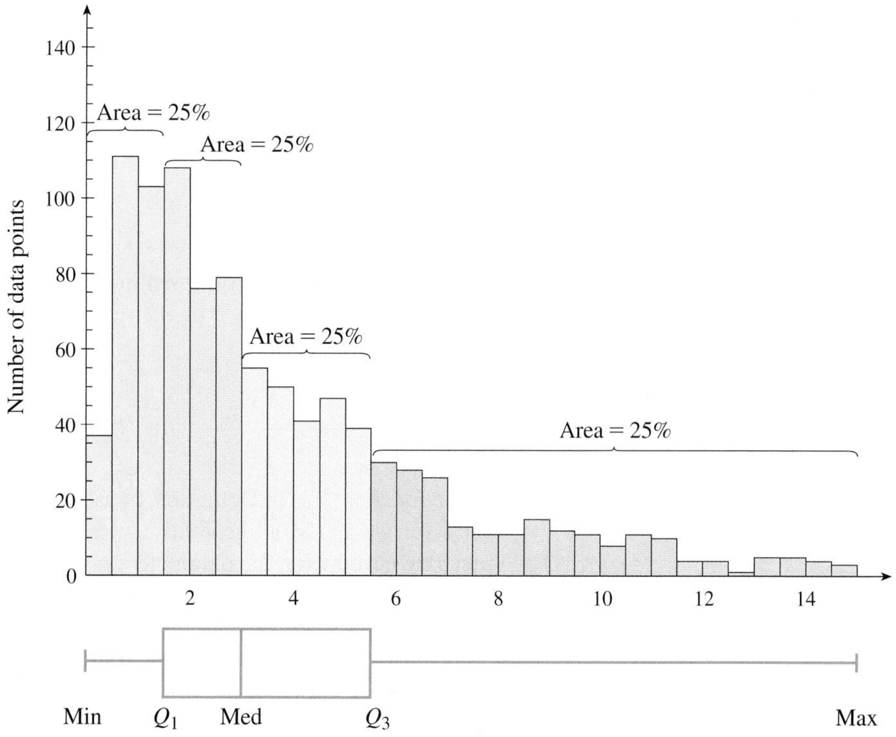

FIGURE 7
Relationship between a histogram and box plot

■ The Normal Curve

Sometimes we can approximate discrete data, such as those in histograms, with a continuous function. For large data sets even very narrow bins will often contain multiple data points. Consequently, the tops of the histogram bars will look approximately like a continuous function.

As an example, Figure 8 shows how the density histogram for heart rates might look with more data points and smaller bin widths. This idea is reminiscent of our study of integration in Chapter 5.

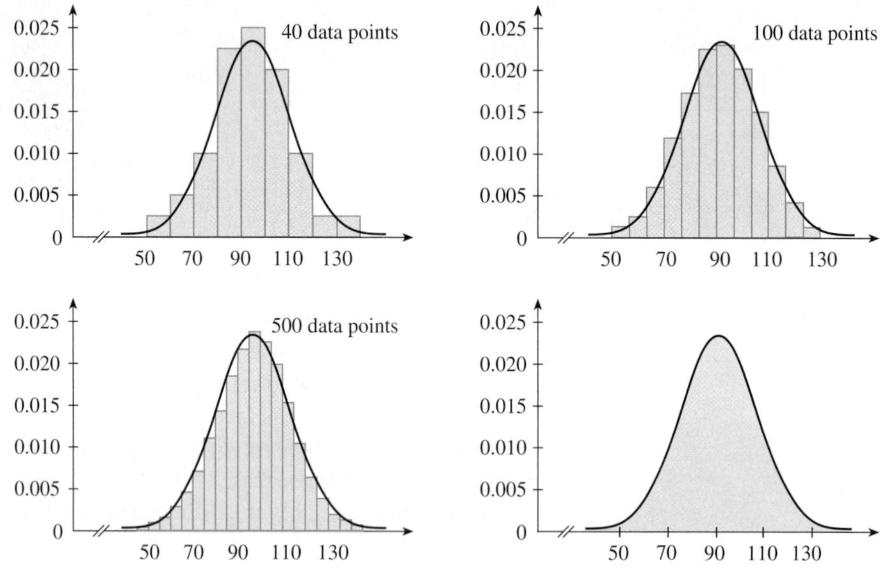

FIGURE 8
Histogram approximated by normal curve

It turns out that for many real-world variables the phenomenon illustrated in Figure 8 occurs; the histogram appears to closely resemble a continuous function as the data set gets large and the bin width gets small. In fact, for many real-world variables this limiting function, when obtained using a density histogram, is well described by what is called the *normal curve*.

> **(1) Definition** The **standard normal distribution** (also called the **standard normal curve**) is defined by the function
>
> $$f(x) = \frac{1}{\sqrt{2\pi}} e^{-x^2/2}$$

The function $f(x)$ in Definition 1 has a single, absolute, maximum at the value $x = 0$. (See Exercise 19.) More generally, we can translate the curve horizontally and alter its spread by introducing the parameters μ and σ as shown in Figure 9.

> **(2) Definition** The **normal distribution** (also called the **normal curve**) with center μ and spread σ is defined by the function
>
> $$f(x) = \frac{1}{\sigma\sqrt{2\pi}} e^{-(x-\mu)^2/(2\sigma^2)}$$

All normal curves are symmetric around their centers and have the same "bell" shape. They are sometimes referred to as bell curves. When $\mu = 0$ and $\sigma = 1$, we obtain the standard normal curve.

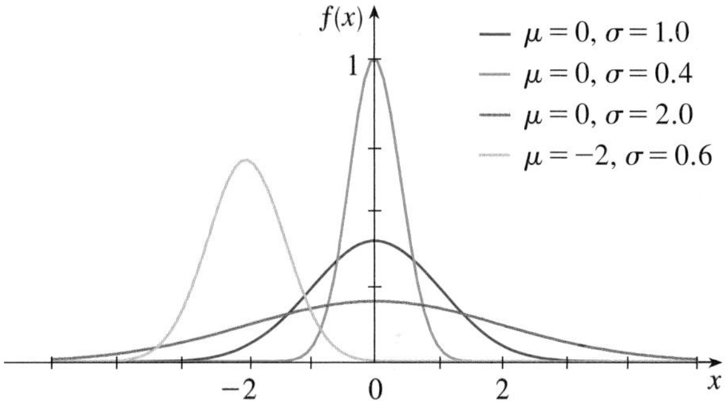

FIGURE 9

As we will see in Section 12.5, the parameter μ represents the mean, or average value, of this distribution, whereas σ represents its standard deviation. Thus, if a normal distribution is used to approximate a histogram, μ approximates the mean value of the data and σ approximates the standard deviation.

Using a normal curve to approximate data simplifies the estimation of various areas in histograms. The most important results are summarized in the following box.

(3) Areas under a Normal Curve For a normal curve with center μ and spread σ we have the following:

(a) Approximately 68.2% of the area under the curve lies between $\mu - \sigma$ and $\mu + \sigma$.

(b) Approximately 95.5% of the area under the curve lies between $\mu - 2\sigma$ and $\mu + 2\sigma$.

(b) Approximately 99.7% of the area under the curve lies between $\mu - 3\sigma$ and $\mu + 3\sigma$.

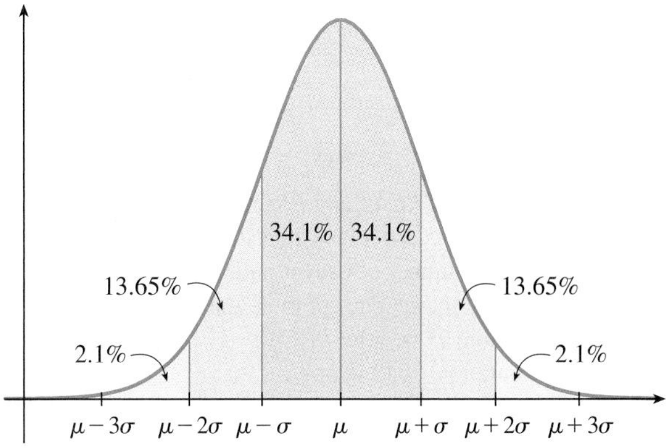

FIGURE 10

EXAMPLE 6 | A sample of IQ scores are normally distributed with a mean of 100 and a standard deviation of 15. Find the proportion of the sample with IQ scores in the given interval.

(a) Between 85 and 115

(b) At least 130

(c) At most 130

SOLUTION

(a) IQ scores between 85 and 115 lie within one standard deviation of the mean:

$$100 - 15 = 85 \qquad 100 + 15 = 115$$

Therefore approximately 68.2% of the sample has an IQ score between 85 and 115. (See Figure 11.)

(b) IQ scores between 70 and 130 lie within two standard deviations of the mean:

$$100 - 2(15) = 70 \qquad 100 + 2(15) = 130$$

Therefore approximately 95.5% of the sample has an IQ score between 70 and 130. Since the distribution is symmetric, the remaining 4.5% are split equally below 70 and above 130. Therefore, 2.25% have an IQ of at least 130. (See Figure 11.)

(c) From part (b), 2.25% of the individuals in the sample have an IQ score of at least 130. Consequently, 97.75% have a score of at most 130. (See Figure 11.)

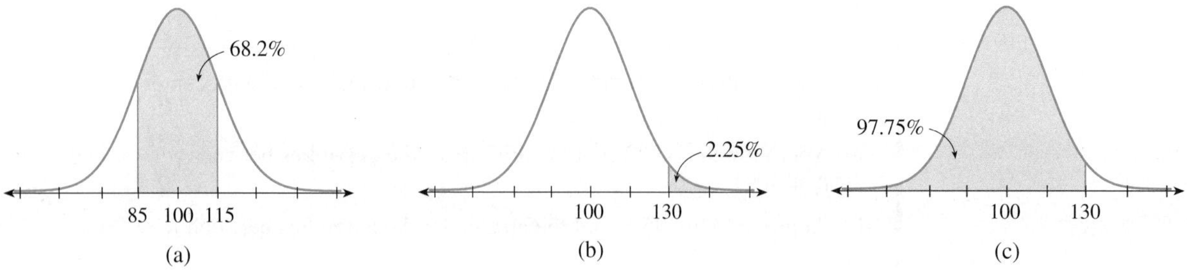

FIGURE 11

■

EXERCISES 11.2

1–4 Construct a bar graph for the data.

1. **Cancer genetics** Mutations in *BRCA1* and *BRCA2* genes increase the risk of cancer (particularly breast cancer). A study tabulated the number of cancer patients who had mutations in either of these genes or in neither. The data are: *BRCA1* mutation, 75 people; *BRCA2* mutation, 54 people; no mutation, 1042 people.

   Source: H. Risch et al., "Population BRCA1 and BRCA2 Mutation Frequencies and Cancer Penetrances: A Kin–Cohort Study in Ontario, Canada," *Journal of the National Cancer Institute* 98 (2006): 1694–1706.

2. **Species composition** The presence of different types of animals was quantified along a transect through a forest, giving the following data: hares, 65; grouse, 32; squirrels, 43; mice, 31; moose, 86; wolves, 7.

3. **Obesity** A study compiled data on the percentage of people in different countries who were obese in 2003, giving the following table.

Country	Percentage
United States	30
Mexico	24
Canada	14
Australia	22
Spain	13
France	9
Japan	3
United Kingdom	23

Source: "Health Statistics: Obesity (Most Recent) by Country," NationMaster, accessed March 11, 2015, http://www.nationmaster.com/country-info/stats/Health/Obesity.

4. **Blood types** The approximate frequencies of the four main blood types in the United States are type O, 44%; type A, 42%; type B, 11%; and type AB, 3%.

5–8 Construct a pie chart for the data.

5. Worldwide mortality rates A study categorized the total number of deaths in 1990 in certain regions of the world, giving the following numbers: communicable diseases, 17,241; noncommunicable diseases, 28,141; injuries (intentional or otherwise), 5,084.

Source: C. Murray and A. Lopez, "Mortality by Cause for Eight Regions of the World: Global Burden of Disease Study," *Lancet* 349 (1997): 1269–76.

6. HIV prevalence In 2011 the total number of people living with HIV was estimated to include 16.7 million women, 14 million men, and 3.3 million children.

7. Myopia (short-sightedness) and ambient light A well-known study examined the number of children who sleep with different levels of ambient light, in an attempt to explore whether nighttime ambient light affects vision. In a survey the researchers found that 172 children slept in darkness, 232 slept with a small night-light, and 75 slept with room-level lighting.

Source: Adapted from G. Quinn et al., "Myopia and Ambient Lighting at Night," *Nature* 399 (1999): 113–14.

8. SARS (Severe Acute Respiratory Syndrome) deaths from 2002–03 by country are given in the following table.

Canada	China	Hong Kong	Taiwan	Singapore
43	349	299	37	33

BB **9. Weddell seals** Twenty individual Weddell seals from Antarctica were weighed, giving the following data (in pounds). Construct a density histogram with bins of width 10. Indicate the fraction of seals with a weight between 910 and 930 pounds as an area in the histogram.

902	920	924	932	937
939	945	948	949	951
957	958	961	965	969
970	975	982	987	991

BB **10. Multiple sclerosis** is a serious neurological disease. The following data set gives the age of onset of multiple sclerosis for 32 individuals. Construct a density histogram with bins of width 4. Indicate the fraction of people with an age of onset less than 28 years as an area in the histogram.

23	52	32	27	28	30	13	37
33	29	25	31	26	30	5	29
41	27	30	34	28	19	31	29
32	28	33	28	31	29	29	30

BB **11. Hospital stays** The duration of stay at a major US hospital for 20 patients is given in the following data set. Construct a density histogram with bins of width 5. Indicate the fraction of stays 15 days or longer as an area in the histogram.

6	10	5	12	5
3	41	12	6	15
20	4	3	7	13
16	18	29	23	17

BB **12. Damselflies** A study measured the number of external parasites on a sample of 20 damselflies, giving the following data. Construct a density histogram with bins of width 10. Indicate the fraction of damselflies having 30 parasites or more as an area in the histogram.

5	0	0	54	5
12	27	24	36	5
56	43	15	42	12
62	36	34	58	23

BB **13. Anemia** When individuals become infected with malaria the parasite consumes red blood cells, causing the patient to become anemic. A group of 18 mice had their red blood cell concentration measured 10 days after infection, giving the following data (measured as cells $\times 10^6/\mu L$). Construct a density histogram with bins of width 0.5. Indicate the fraction of mice having a concentration less than $3 \times 10^6/\mu L$ as an area in the histogram.

3.2	4.5	3.7	3.3	2.5	3.6
3.3	3.6	4.2	3.5	2.0	3.1
2.3	2.9	6.9	4.5	7.3	3.0

BB **14. Coral bleaching** is a process through which coral reefs lose their color during extreme temperature events. The extent of bleaching after an extreme temperature event was measured at 20 sites, giving the following data. Values are dimensionless measures of bleaching, ranging from 0 (no bleaching) to 1 (complete bleaching of all corals at the site). Construct a frequency histogram.

0.19	0.06	0.26	0.32	0.35
0.09	0.05	0.21	0.48	0.21
0.23	0.12	0.22	0.16	0.35
0.13	0.34	0.07	0.16	0.45

BB **15. Huntington's disease** is a genetic disorder causing neurodegeneration and eventual death. The age at death of 15 patients with the disease is given in the following data. Construct a frequency histogram.

71	70	69	68	67
66	64	62	58	57
55	53	52	48	45

Source: Adapted from S. Andrew et al., "The Relationship between Trinucleotide (CAG) Repeat Length and Clinical Features of Huntington's Disease." *Nature Genetics* 4 (1993): 398–403.

BB **16. Monarch butterflies** migrate thousands of miles each year to overwinter in areas of Mexico. The following data give the wingspans (in centimeters) of 28 individuals captured in their overwintering grounds. Construct a frequency histogram.

8.6	10.9	9.8	12.4	10.6	10.7	10.9
10.1	10.2	10.3	9.4	9.1	11.2	9.7
11.6	12.0	10.4	9.5	10.9	10.5	9.9
11.1	10.5	11.3	11.7	11.3	10.3	11.5

BB **17. Breast cancer tumors** were measured for a sample of 15 patients by estimating tumor diameter in centimeters. This gave the following data. Construct a frequency histogram.

1.5	1.5	1.6	1.6	1.7
2.0	2.1	2.2	2.6	3.2
4.0	4.9	5.5	7.5	7.6

BB **18. HAART** (Highly Active Antiretroviral Therapy) is a treatment for HIV infection.

(a) The concentration of HIV in the blood (in units of $\log_{10}$ copies/mL) was measured in 15 patients prior to HAART treatment, giving the following data.

7.4	5.1	6.9	7.2	1.4
4.3	5.1	2.9	7.2	3.5
4.7	4.7	9.3	4.5	6.0

Construct a density histogram with bins of width 1. Indicate the fraction of patients having at least 4 $\log_{10}$ copies/mL as an area in the histogram.

(b) The concentration of HIV in the blood (in units of $\log_{10}$ copies/mL) was measured in the same 15 patients after six months of HAART treatment, giving the following data.

3.7	2.6	3.4	3.6	0.7
2.1	2.6	1.5	3.6	1.7
2.3	2.4	4.6	2.2	3.0

Construct a density histogram with bins of width 1. Indicate the fraction of patients having at least 4 $\log_{10}$ copies/mL as an area in the histogram.

(c) Interpret the difference between the histograms of part (a) and part (b) in terms of the effectiveness of HAART.

19. Show that the standard normal curve has a single critical point at $x = 0$ and that it takes on its absolute maximum value at this point.

20. Show that the general normal curve has a single critical point at $x = \mu$ and that it takes on its absolute maximum value at this point.

21–26 Use a normal curve with center and spread equal to the mean and standard deviation of the data to answer the following.

BB **21.** Estimate the fraction of the **Weddell seal** data in Exercise 9 that lies between 930 pounds and 975 pounds.

BB **22.** Estimate the fraction of the **multiple sclerosis** data in Exercise 10 that is less than 13.9 years.

BB **23.** Estimate the fraction of the **anemia** data in Exercise 13 that is greater than 6.5×10^6 cells/μL.

BB **24.** Estimate the fraction of the **coral bleaching** data in Exercise 14 that lies between 0.1 and 0.35.

BB **25.** Estimate the fraction of the **Monarch butterfly** data in Exercise 16 that lies between 9.7 cm and 12.3 cm.

BB **26.** Estimate the fraction of the **HAART** data in Exercise 18(a) that is greater than 5.3 $\log_{10}$ copies/mL.

27–30 In each question we approximate a density histogram with a continuous function. You can then calculate the corresponding area under the function using integration.

27. Hospital stays We can approximate the density histogram for hospital stay duration in Exercise 11 by the function $f(x) = \lambda e^{-\lambda x}$ where $\lambda = 1/\bar{x}$ and $\bar{x}$ is the average hospital stay. Use this approximation to estimate the fraction of stays longer than 20 days.

28. Damselflies We can approximate the density histogram for damselfly parasite count in Exercise 12 by the function $f(x) = \frac{1}{70}$ on the domain $0 \leqslant x \leqslant 70$. Use this approximation to estimate the fraction of damselflies having between 20 and 40 parasites.

29. Huntington's disease We can approximate the density histogram for Huntington's disease data in Exercise 15 by the function $f(x) = \frac{1}{30}$ on the domain $45 \leqslant x \leqslant 75$. Use this approximation to estimate the fraction of patients whose ages at death were between 45 and a, where a is an arbitrary age.

30. Breast cancer We can approximate the density histogram for the tumor size data in Exercise 17 by the function $f(x) = \lambda e^{-\lambda x}$ where $\lambda = 1/\bar{x}$ and $\bar{x}$ is the average tumor diameter. Use this approximation to estimate the fraction of tumors smaller than d, where d is an arbitrary diameter.

11.3 Relationships between Variables

Often more than one variable is measured for each individual in a sample with the objective being to determine if there are any relationships among the variables. In this section we begin to study these issues by exploring how relationships among variables in a data set can be described.

Recall that variables can be categorical or numerical. We consider three different kinds of possible relationships: relationships between two categorical variables, between a categorical variable and a numerical variable, and between two numerical variables.

■ Two Categorical Variables

Rimantadine is an antiviral drug that can be administered to prevent infection by influenza. A study[1] divided a sample of 76 children into two groups and administered rimantadine to one group and a placebo to the other. The researchers then determined whether each child was infected with influenza during the flu season.

The data set contains 76 individuals, and each individual had two categorical variables measured: (1) drug status (rimantadine or placebo), and (2) infection status (infected or not infected).

These data can most easily be summarized using a **contingency table**. In such tables the rows represent one variable and the columns another. The number of data points falling into each combination of categories, along with the row and column totals, are then entered (see Table 1). Notice that each of the 76 individuals is represented in exactly one of the four cells of the table.

Table 1 Rimantadine Effectiveness

	Infected	Not infected	Row total
Placebo	20	21	41
Rimantadine	1	34	35
Column total	21	55	76

The table illustrates how membership in one group might be contingent on membership in another. For example, Table 1 suggests that "infection status" is contingent on "drug status" since almost none of the individuals receiving rimantadine became infected (only 1 out of 35) whereas approximately half of the individuals receiving the placebo became infected (20 out of 41).

We can also depict relationships between categorical variables using bar graphs. A common approach is to use a grouped bar graph, where the bars corresponding to membership for one of the variables are grouped. Figure 1 displays the grouped bar graph for the rimantadine data.

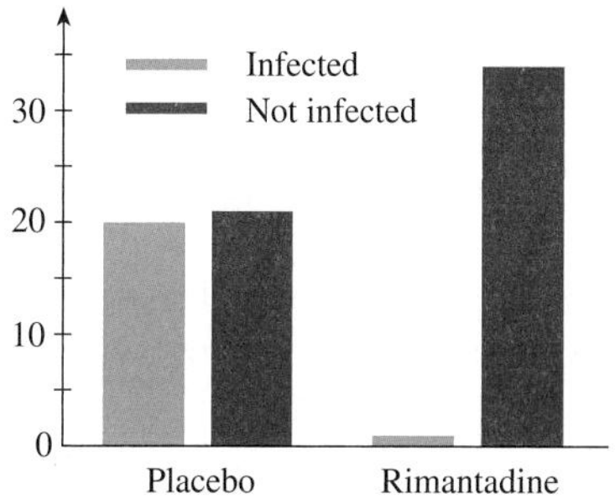

FIGURE 1

1. R. Clover et al. "Effectiveness of Rimantadine Prophylaxis of Children Within Families," *American Journal of Diseases of Children* 140 (1986): 706–9.

Do the data provide good evidence that, in the population at large, rimantadine protects against influenza? In Chapter 13 we will study how to answer this question. Doing so requires the use of *inferential statistics*.

EXAMPLE 1 | **Bat conservation biology** A study[2] looked for an association between the endangerment status of bat species and their use of caves for roosting. Researchers classified species as either "fragile" or "vulnerable," according to their endangerment, as well as whether they used caves as a main roosting site.

Eleven species used caves as their main roosting site, and nine of these were listed as fragile. Seven species did not use caves as their main roosting site, and five of these were listed as fragile. Construct a contingency table and a grouped bar graph for these data.

SOLUTION The data are compiled in the following contingency table.

	Fragile	Vulnerable	Row total
Cave use	9	2	11
No cave use	5	2	7
Column total	14	4	18

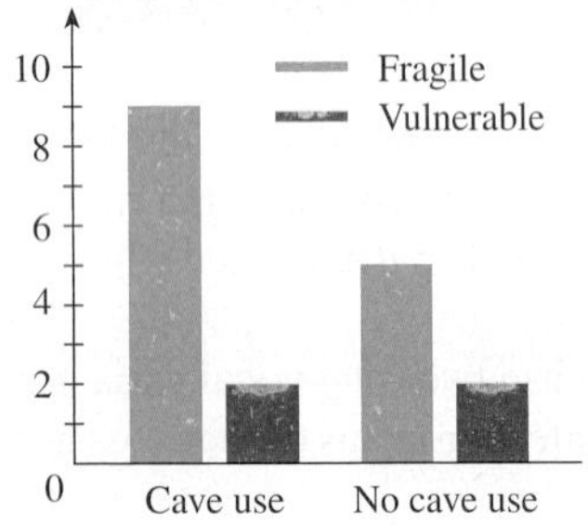

FIGURE 2

A grouped bar graph for these data is presented in Figure 2. There does not appear to be a very strong relationship between endangerment and the use of caves in these data, as seen by the fact that the fraction of vulnerable species is similar for cave-roosting and non-cave-roosting bats. ■

Categorical and Numerical Variables

When both a categorical and a numerical variable are measured, we can explore the relationship between them by summarizing the numerical variables using the techniques from Sections 11.1 and 11.2, and then comparing these summaries across the groups of the categorical variable.

EXAMPLE 2 | **Antibiotic effectiveness** The data from Example 11.1.4 were for the antibiotic methacycline. The minimum inhibitory concentration (*MIC*) of oxytetracycline was also measured on 10 bacterial colonies, giving the following data (in μg/mL):

0.32 0.22 0.35 0.17 0.14 0.23 0.15 0.57 0.55 0.63

(a) Compare the mean and standard deviation of the data for oxytetracycline with those for methacycline.
(b) Compare the two data sets using outlier box plots.

SOLUTION

(a) The mean *MIC* for oxytetracycline is

$$\frac{0.32 + 0.22 + 0.35 + 0.17 + 0.14 + 0.23 + 0.15 + 0.57 + 0.55 + 0.63}{10} \approx 0.33\ \mu\text{g/mL}$$

compared with 0.12 μg/mL for methacycline. Using Definition 11.1.4, we calculate the standard deviation as approximately 0.18 μg/mL for oxytetracycline compared with 0.11 μg/mL for methacycline. Thus both the mean and the standard deviation of the data are higher for oxytetracycline.

2. H. Arita, "Conservation Biology of the Cave Bats of Mexico," *Journal of Mammalogy* 74 (1993): 693–702.

(b) To construct an outlier box plot for each data set we first tabulate the five-number summary for each:

Oxytetracycline:	Min = 0.14	$Q_1 = 0.17$	Median = 0.275	$Q_3 = 0.55$	Max = 0.63
Methacycline:	Min = 0.03	$Q_1 = 0.05$	Median = 0.105	$Q_3 = 0.12$	Max = 0.42

We have already identified 0.42 as an outlier for methacycline in Example 11.1.7, and you should verify that there are no outliers for oxytetracycline. Therefore the outlier box plots are as shown in Figure 3.

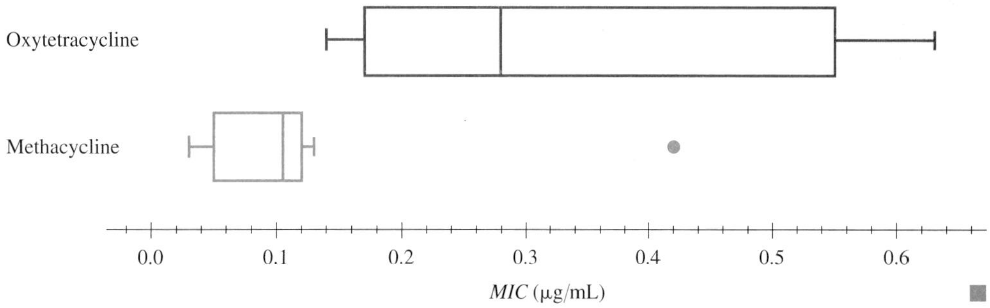

FIGURE 3

Two Numerical Variables

When two numerical variables are measured on each individual, the possible relationships between them can be quite complicated. The best way to start is by constructing a *scatter plot*.

Let's label the two variables y and x. A **scatter plot** is then a plot of y against x. Each individual in the data set is represented by a single point in the plane whose x-coordinate is its value of the x-variable, and whose y-coordinate is its value of the y-variable.

EXAMPLE 3 | BB **Smoking and mortality rates** A study in the United Kingdom examined the relationship between smoking and lung cancer. A sample of men was divided into 25 occupational groups, and for each group the number of cigarettes smoked per day was measured, as well as the likelihood of death from lung cancer. Both variables were then transformed into dimensionless measures so that $x = 100$ represents the average amount of smoking in the entire sample and $y = 100$ represents the average mortality rate in the sample.[3] The data are given in the following table.

x	y	x	y	x	y
77	84	102	88	115	128
137	116	91	104	105	115
117	123	104	129	87	79
94	128	107	86	91	85
116	155	112	96	100	120
102	101	113	144	76	60
111	118	110	139	66	51
93	113	125	113		
88	104	133	146		

Construct a scatter plot for the data and interpret it.

3. "Smoking and Cancer," The Data and Story Library, accessed March 11, 2015, http://lib.stat.cmu.edu/DASL/Stories/SmokingandCancer.html.

SOLUTION Each of the 25 data points is plotted in the scatterplot in Figure 4.

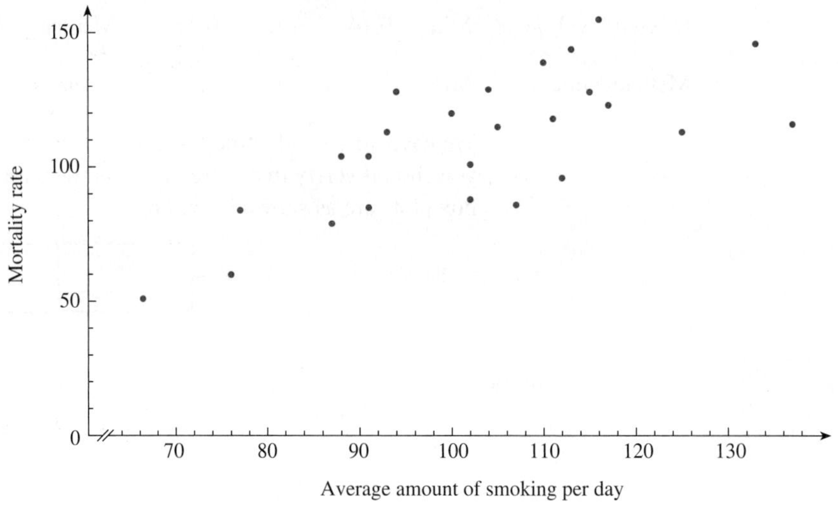

FIGURE 4

We can see that occupational groups that have higher amounts of smoking also tend to have higher mortality rates. Although suggestive, this pattern alone does not imply that smoking causes mortality because the data are *observational*. *Experimental* data are required to infer a causative relationship. These different kinds of data are discussed in Section 11.4. ■

Scatter plots help to reveal relationships between numerical variables. For example, Figure 4 shows that people in occupational groups who smoke more tend to have higher mortality rates. In general, however, relationships between numerical variables can be more complex than this. Figure 5 illustrates some possible relationships.

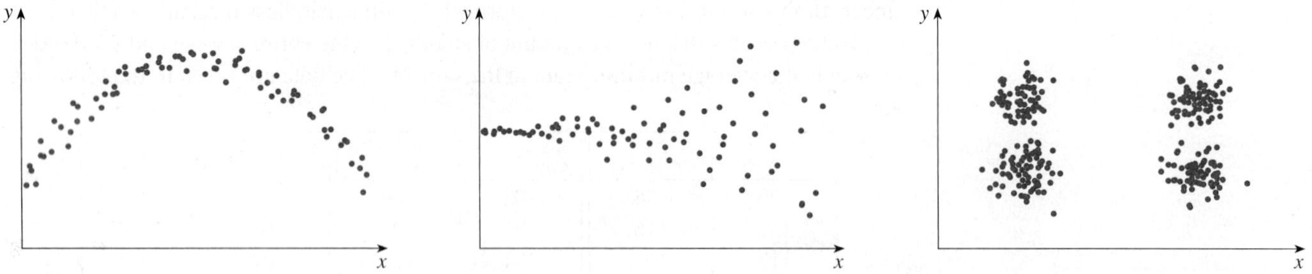

FIGURE 5
Some possible patterns in scatter plots

One of the most important kinds of relationships studied in statistics is a linear one. For instance, the data in Example 3 suggest that the relationship between smoking and mortality rate is approximately linear. One way to quantify this more formally is to draw a line through the scatter plot that, in some sense, best fits the data. But how do we find the equation of this line?

The most common approach to fitting a line to data is called the *least-squares* fit or *linear regression*. The **least-squares line** is the line that minimizes the squared vertical distance of each data point from the line, summed across all the points (see Figure 6). In other words, for each data point (x_i, y_i) we calculate the vertical distance Δy_i between the point and the line (see Figure 6), square this value, and then add these values together. The result is called the **total sum of squared deviations**. In Exercise 16 you will see why these vertical deviations are squared before being added together.

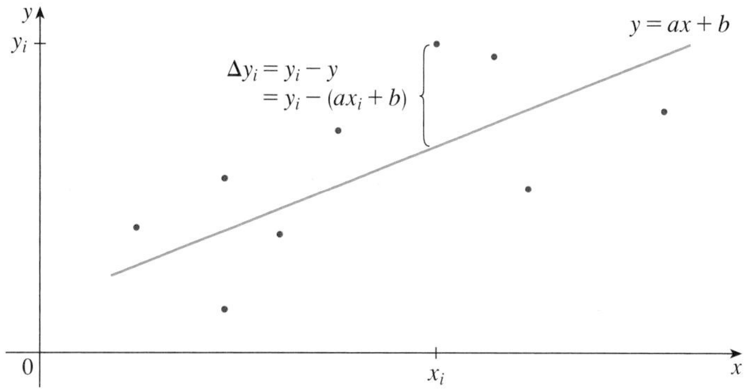

FIGURE 6
The least-squares line with the vertical distance of the data from the line, shown for one data point

If $y(x) = ax + b$ is the line to be fitted to the data, where a is its slope and b is its y-intercept, we have

$$S = [y_1 - y(x_1)]^2 + \cdots + [y_n - y(x_n)]^2$$

where n is the number of data points and S denotes the total sum of squared deviations. Using the equation for $y(x)$, we can write this as

$$S(a, b) = (y_1 - ax_1 - b)^2 + \cdots + (y_n - ax_n - b)^2$$

where we have now indicated that we are treating S as a function of a and b. Any line is completely specified by the values of a and b, and different choices will give different values of S. The goal is to choose a and b so that $S(a, b)$ is as small as possible. This is a two-variable optimization problem like those studied in Section 9.6.

To find the optimal choice of a and b, we first note that the domain of $S(a, b)$ is $\mathbb{R}^2$. We look for the minimum of this function by first determining its critical points.

Differentiating $S(a, b)$ with respect to a and setting the result equal to zero gives

$$\frac{\partial S}{\partial a} = -2x_1(y_1 - ax_1 - b) - \cdots - 2x_n(y_n - ax_n - b) = 0$$

or

$$(x_1y_1 + \cdots + x_ny_n) - a(x_1^2 + \cdots + x_n^2) - b(x_1 + \cdots + x_n) = 0$$

$$\sum_{i=1}^{n} x_iy_i - a\sum_{i=1}^{n} x_i^2 - b\sum_{i=1}^{n} x_i = 0$$

Dividing both sides of the preceding equation by n and using the definition of an average, we obtain

(1) $$\overline{xy} - a\overline{x^2} - b\overline{x} = 0$$

where $\overline{x^2}$ is the average of the square of the x-values and $\overline{xy}$ is the average of the product of the x- and y-values for each individual.

In a similar fashion, differentiating $S(a, b)$ with respect to b and setting the result equal to zero gives

$$\frac{\partial S}{\partial b} = -2(y_1 - ax_1 - b) - \cdots - 2(y_n - ax_n - b) = 0$$

or

$$(y_1 + \cdots + y_n) - a(x_1 + \cdots + x_n) - (b + \cdots + b) = 0$$

$$\sum_{i=1}^{n} y_i - a\sum_{i=1}^{n} x_i - nb = 0$$

Again dividing through by n, we obtain

(2) $$\overline{y} - a\overline{x} - b = 0$$

Equation 1 and Equation 2 can be solved simultaneously for a and b to find the critical point (most data sets will have only one critical point). It can also be shown (see Exercise 17) that this critical point is an absolute minimum of $S(a, b)$. Therefore we have the following result.

(3) Definition Let $(x_1, y_1), (x_2, y_2), \ldots, (x_n, y_n)$ be the values of two numerical variables measured on n individuals. The **least-squares line** $y = ax + b$ for the data has a and b given by

$$a = \frac{\overline{xy} - \overline{x}\,\overline{y}}{\overline{x^2} - \overline{x}^2}$$

$$b = \overline{y} - a\overline{x}$$

where $\overline{x}$ and $\overline{y}$ are the average values of x and y, $\overline{x^2}$ is the average of the square of the x values, and $\overline{xy}$ is the average of the product of the x- and y-values.

EXAMPLE 4 | BB Smoking and mortality (continued) Calculate the least-squares line for the data in Example 3 and plot it on the scatter plot.

SOLUTION The average values of x and y are readily calculated as $\overline{x} = 2572/25$ and $\overline{y} = 2725/25$. We calculate $\overline{x^2}$ and $\overline{xy}$ by first constructing a table of values for x^2 and xy, and then calculating their averages. This is shown in Table 2.

Table 2

x^2	xy	x^2	xy
5929	6468	12544	10752
18769	15892	12769	16272
13689	14391	12100	15290
8836	12032	15625	14125
13456	17980	17689	19418
10404	10302	13225	14720
12321	13098	11025	12075
8649	10509	7569	6873
7744	9152	8281	7735
10404	8976	10000	12000
8281	9464	5776	4560
10816	13416	4356	3366
11449	9202		
		$\overline{x^2} = \frac{271706}{25}$	$\overline{xy} = \frac{288068}{25}$

From Definition 3 we then obtain

$$a = \frac{\overline{xy} - \overline{x}\,\overline{y}}{\overline{x^2} - \overline{x}^2} = \frac{\frac{288068}{25} - \left(\frac{2572}{25}\right)\left(\frac{2725}{25}\right)}{\frac{271706}{25} - \left(\frac{2572}{25}\right)^2} \approx 1.0875$$

$$b = \overline{y} - a\overline{x} \approx \frac{2725}{25} - (1.0875)\frac{2572}{25} \approx -2.882$$

The least-squares line is therefore $y = 1.0875x - 2.882$. Figure 7 plots this line along with a scatter plot of the data.

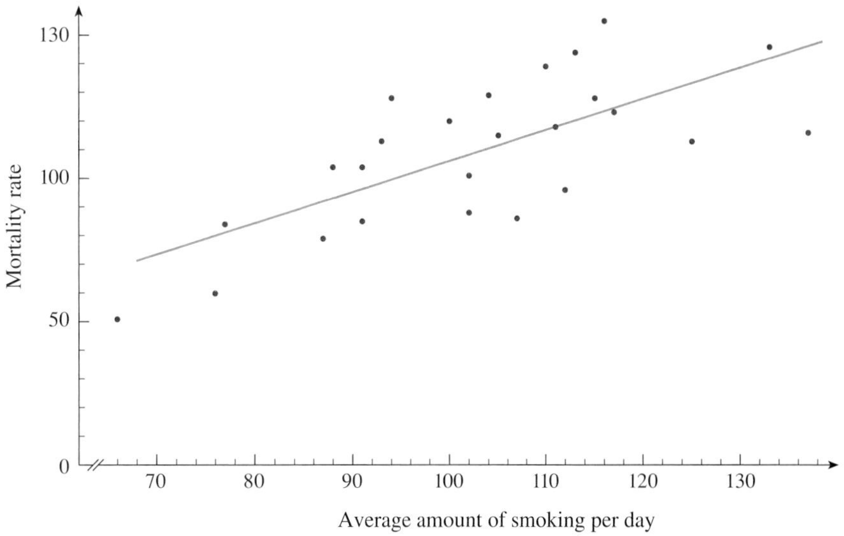

FIGURE 7

EXERCISES 11.3

1. ***Bacillus anthracis*** is the bacterial agent of anthrax. In 1881 Louis Pasteur conducted a famous experiment in which he vaccinated 25 sheep and kept an additional 25 sheep unvaccinated. He then injected live *B. anthracis* into all 50 sheep. He found that all of the unvaccinated sheep died, whereas all of the vaccinated sheep survived. Construct a contingency table. Describe any relationship that you see between the variables.

2. **Sentencing and biological mechanisms of psychopathy** A study asked several US judges to pass sentencing on a hypothetical defendant who had been convicted of a crime and was deemed to be a psychopath. Twenty-seven judges were told that psychopathy has an underlying biological basis, and 18 of these judges then cited psychopathy as a mitigating factor in reducing their sentence. Twelve judges were told that psychopathy does not have an underlying biological basis, and four of these judges then cited psychopathy as a mitigating factor in reducing their sentence. Construct a contingency table. Describe any relationship that you see between the variables.

   Source: Adapted from G. Aspinwall et al., "The Double-Edged Sword: Does Biomechanism Increase or Decrease Judges' Sentencing of Psychopathy?" *Science* 337 (2012): 846–49.

3. **Myopia (short-sightedness) and ambient light** A well-known study examined the number of children who sleep with different levels of ambient light, in an attempt to explore whether nighttime ambient light affects vision. In a survey researchers found that 9 of 172 children who slept in darkness developed myopia, 31 of 232 who slept with a small night-light developed myopia, and 48 of 75 who slept with room-level lighting developed myopia. Construct a contingency table. Do the data suggest that there is a relationship between short-sightedness and ambient light in the sample?

   Source: Adapted from G. Quinn et al., "Myopia and Ambient Lighting at Night," *Nature* 399 (1999): 113–14.

4. Antidepressants A study gave 28 patients a new antidepressant medication and 27 patients a placebo for six months. The number of patients having a depressive episode during this period was recorded. Eleven patients who received the medication had at least one depressive episode, while twenty of the patients receiving the placebo had at least one episode. Construct a contingency table. Describe any relationship observed in the data between the medication and depression.

5. An **indicator species** is one whose presence is used as an indication of the presence of other groups of species of interest. For example, suppose we wish to know whether a community is dominated by herbivores or carnivores, and we want to use the presence or absence of species X as a simple indicator of this. A study is therefore conducted and it is found that, of 12 communities that are herbivore-dominated, 9 of them have species X present. Of 15 communities that are carnivore-dominated, 4 of them have species X present. Do you think species X is a good indicator species (use a contingency table in your argument)?

6. Breast cancer A study was conducted to determine the relationship between childbearing and breast cancer. Of 4323 respondents who had cancer, 3220 of these had given birth to at least one child. Of 12,699 respondents who did not have cancer, 10,245 of these had given birth to at least one child. Do you think there is a relationship between childbearing and breast cancer (use a contingency table in your argument)?

Source: Adapted from B. MacMahon et al., "Age at First Birth and Breast Cancer Risk," *Bulletin of the World Health Organization* 43 (1970): 209–21.

7–11 Use a box plot to explore any potential differences between the two data sets in each exercise.

BB **7. EPO**, or erythropoietin, is a hormone that stimulates red blood cell production. A synthetic variant of EPO has been used by athletes as a performance-enhancing drug to increase aerobic capacity. The aerobic capacity of 14 athletes was quantified prior to injection of EPO by measuring the volume (in liters) of oxygen consumed per minute during exercise, giving the following data.

3.6	4.0	4.1	4.2	4.2
4.3	4.4	4.6	4.6	4.8
4.8	5.0	5.1	5.2	

The aerobic capacity of the same 14 athletes was quantified after injection with EPO, giving the following data.

4.0	4.5	4.4	4.6	4.7
4.6	4.7	5.0	5.1	5.1
5.1	5.2	5.4	5.3	

Does there appear to be an effect of EPO injection in the data? Explain.

Source: Adapted from B. Ekblom and B. Berglund, "Effect of Erythropoietin Administration on Maximal Aerobic Power," *Scandinavian Journal of Medicine & Science in Sports* 1 (1991): 88–93.

BB **8. HAART** (Highly Active Antiretroviral Therapy) is a treatment for HIV infection. The concentration of HIV in the blood (in units of $\log_{10}$ copies/mL) was measured in 15 patients prior to HAART treatment, giving the following data.

7.4	5.1	6.9	7.2	1.4
4.3	5.1	2.9	7.2	3.5
4.7	4.7	9.3	4.5	6.0

The concentration of HIV in the blood (in units of $\log_{10}$ copies/mL) was measured in the same 15 patients after six months of HAART treatment, giving the following data.

3.7	2.6	3.4	3.6	0.7
2.1	2.6	1.5	3.6	1.7
2.3	2.4	4.6	2.2	3.0

Does HAART appear to have an effect on HIV con-centration in the patients sampled? Explain.

BB **9. Fish morphology** The width of the mouth (in mm) of fish feeding on benthic prey (that is, prey found on the bottom of lakes) was compared with that of fish feeding on limnetic prey (that is, prey found in the water column). A sample of benthic feeders gave the following data

4.6	3.5	4.3	4.5	1.6
3.1	3.5	2.4	4.5	2.6
3.2	3.2	5.5	3.1	3.9

A sample of limnetic feeders gave

3.5	4.2	2.2	3.5	3.0
3.7	3.3	1.9	0.4	3.6
1.8	2.6	1.6	2.2	2.1

Describe any difference seen in the data between the two groups.

BB **10. Pesticides and bees** Bees are important pollinators of many agricultural crops. A study examined the impact of pesticides on the productivity of bee colonies by counting the number of worker bees produced by colonies either exposed, or not exposed, to a pesticide. For 12 colonies exposed to pesticide, the number of worker bees produced by each colony was as follows:

12	12	31	14	17
18	20	28	32	32
24	27			

For 10 colonies not exposed to pesticide, the data are

15	38	35	44	44
25	34	29	28	27

Does there appear to be any effect of pesticides in the data? Explain.

Source: Adapted from R. Gill et al., "Combined Pesticide Exposure Severely Affects Individual- and Colony-Level Traits in Bees," *Nature* 491 (2012): 105–8.

BB **11. High-density lipoprotein (HDL) cholesterol** is sometimes referred to as the "good" cholesterol. A sample of 13 female and 17 male volunteers had their HDL cholesterol levels measured (in mg/dL) giving the following data:

Females	58	60	48	55	56
	46	54	50	49	51
	53	54	53		
Males	32	52	30	42	44
	52	55	29	29	31
	43	46	41	34	51
	36	30			

Describe any difference between the two groups seen in the data.

12–15 Construct a scatter plot for the data.

BB **12. Phytoplankton growth** The optimal temperature for growth of phytoplankton taken from 16 different latitudes is given in the following table.

Latitude (°North)	Temperature (°C)	Latitude (°North)	Temperature (°C)
−70	12	5	28
−60	4	15	25
−50	22	35	21
−35	15	42	26
−30	22	50	19
−20	19	60	18
−15	30	75	10
−10	24	80	12

Describe the relationship between latitude and optimal temperature seen in the data.

Source: Adapted from M. Thomas et al., "A Global Pattern of Thermal Adaptation in Marine Phytoplankton," *Science* 338 (2012): 1085–88.

BB **13. Coral bleaching** is a process through which coral reefs lose their color during extreme temperature events. It is thought that bleaching ultimately leads to the death of corals. To explore this idea, a study was conducted to quantify the relationship between bleaching and subsequent mortality in 18 kinds of coral. The data are given in the following table.

Bleaching index	Percentage mortality	Bleaching index	Percentage mortality
0.01	58	0.2	95
0.02	0	0.2	100
0.04	24	0.26	36
0.11	0	0.27	50
0.12	0	0.34	0
0.18	8	0.35	98
0.18	41	0.36	0
0.19	78	0.47	80
0.19	22	0.47	75

Do the data show any relationship between bleaching and mortality?

Source: Adapted from T. McClanahan, "The Relationship between Bleaching and Mortality of Common Corals," *Marine Biology* 144 (2004): 1239–45.

BB **14. Huntington's disease** is a genetic disorder causing neurodegeneration and eventual death. The segment of the DNA that is associated with the disease displays a pattern of nucleotide repeats, and the length of this segment (measured as number of repeats) is thought to determine the age at death. The age at death, as well as the number of repeats, was measured for 15 patients, giving the following data.

Number of repeats	Age at death	Number of repeats	Age at death
41	71	45	58
42	70	45	57
42	69	47	55
42	68	48	53
43	67	50	52
43	66	50	48
44	64	52	45
45	62		

Does there appear to be a relationship between number of repeats and age at death? Explain.

Source: Adapted from S. Andrew, "The Relationship between Trinucleotide (CAG) Repeat Length and Clinical Features of Huntington's Disease," *Nature Genetics* 4 (1993): 398–403.

BB **15. Damselfly morphology** The mass (in mg) and wingspan (in mm) of 20 damselflies is given in the following table.

Mass	Wingspan	Mass	Wingspan
681	15.1	573	14.6
638	15.0	675	14.3
408	13.9	622	14.1
505	13.1	380	13.1
785	15.2	449	13.5
622	15.3	385	13.1
726	16.0	458	13.8
700	15.1	431	13.0
713	16.1	420	14.1
460	14.3	522	13.8

What sort of relationship between mass and wingspan is seen in the data?

16. In the derivation of Definition 3 for the least-squares line, the deviation of each data point from the line was squared.
(a) Obtain an expression for the sum of the deviations of each data point from the line (in other words, do not square the deviations).
(b) Describe how your answer to part (a) varies as a function of a and b.

17. Definition 3 for the least-squares line was derived by finding the single critical point of the function $S(a, b)$.
(a) Show that this critical point is a local minimum of $S(a, b)$.
(b) Provide an argument for the critical point being an absolute minimum of $S(a, b)$.

18–23 Calculate the least-squares line for each data set. Plot this line along with the data and interpret the result.

BB **18.** The **EPO** data from Exercise 7 give "before" and "after" measurements for each of 14 individuals. We can tabulate these data as follows.

Before	After	Before	After
3.6	4.0	4.6	5.0
4.0	4.5	4.6	5.1
4.1	4.4	4.8	5.1
4.2	4.6	4.8	5.1
4.2	4.7	5.0	5.2
4.3	4.6	5.1	5.4
4.4	4.7	5.2	5.3

BB **19.** The **HAART** data from Exercise 8 give "before" and "after" measurements for each of 15 individuals. We can tabulate these data as follows.

Before	After	Before	After
7.4	3.7	7.2	3.6
5.1	2.6	3.5	1.7
6.9	3.4	4.7	2.3
7.2	3.6	4.7	2.4
1.4	0.7	9.3	4.6
4.3	2.1	4.5	2.2
5.1	2.6	6.0	3.0
2.9	1.5		

BB **20. Huntington's disease** Use the data from Exercise 14.

BB **21. Coral bleaching** Use the data from Exercise 13.

BB **22. Breast cancer** tumors were measured for a sample of 15 patients by determining tumor diameter (in cm). These data were compared with an estimate of tumor size obtained via ultrasound, giving the following data.

Actual size	Ultrasound size	Actual size	Ultrasound size
1.5	1.4	2.6	2.5
1.5	1.2	3.2	3.5
1.6	1.8	4.0	4.2
1.6	1.5	4.9	4.4
1.7	1.7	5.5	5.5
2.0	1.9	7.5	7.2
2.1	2.0	7.6	7.9
2.2	2.2		

BB **23. Climate change and hibernation** A study examined whether climate change might be affecting the timing of emergence from hibernation for Columbian ground squirrels. Researchers measured the average emergence date (quantified as the number of days past January 1) over several years, giving the following data.

Year	Emergence date	Year	Emergence date
1992	112	2002	121
1993	115	2003	119
1994	111	2004	108
1995	116	2005	116
1996	115	2006	115
1997	111	2007	118
1998	112	2008	120
1999	112	2009	122
2000	116	2010	111
2001	116	2011	125

Source: Adapted from J. Lane et al., "Delayed Phenology and Reduced Fitness Associated with Climate Change in a Wild Hibernator," *Nature* 489 (2012): 554–57.

BB **24–25 Yeast population growth** In Section 7.1 we explored a data set for the size of a population of yeast cells (measured as number of cells $\times 10^6$/mL) at different times (in hours). The data are reproduced in the following table.

Time (h)	Pop. size ($\times 10^6$/mL)	Time (h)	Pop. size ($\times 10^6$/mL)
0	0.200	19	209
1	0.330	20	190
2	0.500	21	210
3	1.10	22	200
4	1.40	23	215
5	3.10	24	220
6	3.50	25	200
7	9.00	26	180
8	10.0	27	213
9	25.4	28	210
10	27.0	29	210
11	55.0	30	220
12	76.0	31	213
13	115	32	200
14	160	33	211
15	162	34	200
16	190	35	208
17	193	36	230
18	190		

24. In Section 7.1 we saw that a model of the form $N(t) = Ce^{rt}$, where C and r are constants, describes the data well for the first 12 hours.

(a) Define the new variable $y(t) = \ln N(t)$. Describe the graph of $y(t)$ as a function of t.

(b) The variable $y(t)$ in part (a) is called the *log transform* of population size. Calculate the log transform of the data for the first 12 hours.

(c) Calculate the least-squares line for the transformed data from part (b) and use it to determine the values of C and r. Plot the transformed data versus time for the first 12 hours and add the least-squares line to the graph.

CAS (d) Plot the untransformed data versus time for the first 12 hours and add the function $N(t)$ to the graph.

25. In Section 7.4 we saw that the following model described the data well for the entire time period:

$$N(t) = \frac{210}{1 + Ae^{-rt}}$$

where A and r are constants.

(a) Define the new variable

$$y(t) = \ln\left(\frac{210}{N(t)} - 1\right)$$

Describe the graph of $y(t)$ as a function of t.

(b) Calculate the value of $y(t)$ at each time for the first 20 hours in the data set.

(c) Calculate the least-squares line for the data from part (b) and use it to determine the values of A and r. Plot this least-squares line on the transformed data for the first 20 hours.

CAS (d) Plot the function $N(t)$ on the untransformed data for the entire time period, using the values of A and r obtained in part (c).

26. The procedure used to obtain the best-fit values of a and b in the least-squares line of Definition 3 can also be used to fit nonlinear equations to data. Suppose we wish to fit the parabola $y(x) = \alpha + \beta x + \gamma x^2$ to a data set by choosing the values of α, β, and γ that minimize the sum of squared deviations of the data from the curve.

(a) Obtain an expression for the quantity to be minimized, and call it $S(\alpha, \beta, \gamma)$.

(b) Obtain three equations in three unknowns that must be satisfied by any critical point of $S(\alpha, \beta, \gamma)$.

11.4 | Populations, Samples, and Inference

Descriptive statistics provide a means of summarizing a data set but we are usually interested in more than this. Most data are used to draw conclusions about the population of individuals from which they were collected. This is referred to as **statistical inference**—we seek to infer something about the population from a data set.

A rigorous treatment of statistical inference is possible only with the use of probability theory. In this section we introduce some of these ideas and discuss the importance of how data sets are collected.

■ Populations and Samples

In statistics a **population** is a collection of individuals for which we would like to draw conclusions. As explained in Section 11.1, usually we are not able to observe or measure all individuals in a population and therefore we instead gather data from a subset of these individuals. This subset is referred to as a **sample**. Here are some examples of populations and corresponding samples:

Population: All possible proteins in the human body.
Sample: A selection of 25 human proteins.

Population: All species of bats.
Sample: The bat species found in northern Mexico.

Population: All human newborns.
Sample: All babies born at a specific hospital.

Population: All athletes who might use EPO.
Sample: A selection of 14 athletes.

The preceding list shows that the "individuals" in a population of interest can also be groups. For instance, in the second example a species of bat is treated as an individual. Likewise, for the data on smoking and mortality in Example 11.3.3, occupational groups are treated as the individual and the population of interest is the population of all occupational groups.

All of the descriptive statistics presented in this chapter are summary measures of data for samples. The population-level counterparts of these summary statistics are referred to as **parameters**. Thus we seek to make inferences about population-level parameters based on statistics that are quantified for samples. This process is also called **estimation**, and quantities measured for a sample are referred to as **estimates** of the population parameters.

EXAMPLE 1 | **Cuckoo bird egg size** The average egg length for the sample of cuckoo bird eggs in Example 11.1.2 was 23.1 mm. The corresponding population-level parameter is the average length of all eggs in the entire cuckoo population from which this sample was taken. It would be natural to use 23.1 mm as an *estimate* for this population average. ■

EXAMPLE 2 | **EPO usage by athletes** Table 11.1.1 presented the change in aerobic capacity for 14 athletes who were injected with EPO. The average change is approximately 0.34 L oxygen/min. The corresponding population-level parameter is the average change in aerobic capacity of all athletes who might use EPO. It would be natural to use the value of 0.34 L oxygen/min as an estimate of the population parameter. ■

■ Properties of Samples

To make valid inferences about a population, we must use a sample that is representative of the population. Only then would we expect estimates from the sample to reflect the true population-level parameter value.

A **simple random sample** is a subset of individuals from a population chosen in such a way that every individual in the population is equally likely to be selected and the selection of each individual is independent. The second condition means that whether individual i is selected for the sample has no influence on the likelihood that individual j is selected, for all individuals i and j. With a random sample, each possible subset of n individuals is equally likely. In this book we will focus primarily on simple random samples.

Estimates made from samples will always differ somewhat from the population parameters. One reason for this is because of chance. **Sampling error** refers to this chance difference between the estimate and the true parameter value, simply because the sample is a subset of the population.

One way to visualize sampling error is to imagine taking repeated samples from the same population. For instance, in Example 1, suppose we selected another ten cuckoo eggs and calculated their average length. This, too, is an estimate of the population average egg length in Example 1, but by chance we would expect it to differ from the estimate of 23.1 mm obtained previously. And if we repeatedly took such samples and calculated their averages, we would obtain a distribution of the estimates. This distribution is called the *sampling distribution* and its spread is due to *sampling error.*

The spread of the sampling distribution is referred to as the **precision** of the estimate. If all of the estimates obtained from repeated samples tend to cluster closely together (so that the sampling distribution is very narrow), then we say that the estimate is very *precise* (see Figure 1). One way to increase the precision of an estimate is to increase the sample size. This seems intuitively reasonable since, the larger the fraction of the population that is included in each repeated sample, the less variable the samples will be.

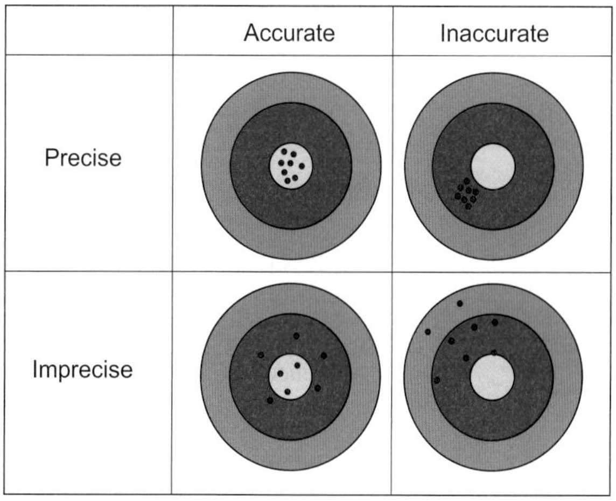

FIGURE 1
An analogy between target shooting and estimation. The bull's-eye represents the population parameter value.

Sampling error arises by chance, but other types of error can also arise, particularly if the sample is not random. **Bias** refers to the systematic over- or underestimation of a parameter. For example, suppose that when sampling cuckoo eggs, nests with large eggs are more noticeable and therefore are more likely to be included in the sample. Any estimate of egg length based on such a sample would therefore be biased, tending to *overestimate* the true egg length in the population.

We can also use the sampling distribution to visualize bias. Recall that the *spread* of the sampling distribution arises because of sampling error. The *center* of the sampling distribution (as measured by the mean or average) is also important because it will be the typical, or expected, value of the estimate obtained from any sample. An estimate is therefore called **unbiased** if the average of the sampling distribution is equal to the true population parameter value. The deviation of this average from the population parameter value quantifies the **accuracy** of the estimate, as illustrated in Figure 1. (This idea will be developed in greater depth in Section 13.1 using probability theory). ■

EXAMPLE 3 | Rimantadine In Study A, 100 people who frequently use a health clinic are recruited to take the influenza drug rimantadine and have their infection status monitored for six months. In Study B, 10 individuals are randomly selected using census data and agree to take rimantadine and have their infection status monitored for six months. In both studies the fraction of the sample who become infected with influenza is recorded, with the goal of inferring, for the population at large, the likelihood of infection while taking rimantadine.

(a) Which study would probably have the most accurate estimate?

(b) Which study would probably have the most precise estimate?

SOLUTION

(a) Accuracy refers to the extent of bias in the estimate. Study A consists of individuals who frequently use a health clinic and we might therefore be concerned that such individuals are more prone to illness than the typical person in the population at large (for example, this might be the reason why they frequently use a health clinic). If so, any estimate based on this sample will be biased toward a higher likelihood of infection. Choosing the individuals randomly guards against this problem. Therefore Study B is likely to be more accurate.

(b) Precision refers to the degree of sampling error. Study A consists of 100 individuals while Study B has only 10. Therefore we might expect that Study B has a larger sampling error. If we took repeated samples of size 10 and estimated the likelihood of infection from each, this estimate would probably vary a lot from sample to sample. On the other hand, with samples of size 100 we might expect the estimate to vary less from sample to sample. Consequently, Study A is probably more precise. ■

■ Types of Data

Our ability to draw valid inferences from data depends crucially on how the data are collected.

Anecdotal data come from a single observation, or relatively few, that have been made in an unplanned or haphazard way. In general, anecdotes provide poor data for inference because the sample on which they are based is usually small and often unrepresentative of the population. Anecdotes do, however, sometimes point to real underlying patterns that can then be more rigorously examined using other types of data.

Observational data are collected systematically from a population of interest through sampling, but for which the researcher merely acts as an observer. The researcher does not directly manipulate or alter any aspect of the individuals that are sampled. For example, the smoking and mortality-rate data in Example 11.3.3 is observational data because the researcher simply observed the data and did not influence the tendency of each occupational group to smoke.

Experimental data differ from the first two types of data in that the researcher determines the value of at least one of the variables for individuals in the sample. For instance, the rimantadine study in Section 11.3 involved experimental data because the researcher determined which individuals in the sample received rimantadine and which received the placebo instead.

Observational data are often sufficient if the primary goal is to infer *patterns* and *associations* among variables in the population, provided that the data come from a random sample. For example, the observational data on smoking and mortality from Section 11.3 might be used to infer an association between smoking and high mortality in the population as a whole.

■ Causation

Usually we are interested in more than simply making inferences about population-level patterns among variables. Rather, we often would like to know whether there is a *causal* relationship between variables. For example, although smoking and mortality rates are associated with one another, is it the case that smoking *causes* a higher mortality rate?

This is again a question of inference about the population from a sample, but the issue of causality needs to be carefully distinguished from inference in general. Although we can use observational data to infer population-level patterns, we typically require experimental data to unambiguously infer causal relationships.

As an example, suppose smoking does not increase mortality rate, but some socio-economic groups have a higher mortality rate and they smoke a lot, while others have a low level of both. In this case smoking and mortality rate will be positively associated with one another in the population because individuals have either a high mortality rate and smoke a lot or a low mortality rate and smoke very little. We could correctly infer this population-level *pattern* with randomly collected observational data. However, to show that smoking causes a high mortality rate, we would need to perform an experiment in which we manipulate the level of smoking, independent of an individual's socio-economic group. Of course, in practical terms, such experiments are unethical to perform.

In the preceding example an unknown variable (in this case socio-economic status) is the "true" cause of both smoking levels and mortality rate. Such hidden, or unintended, variables that complicate or confound our ability to infer causation are called **confounding variables**.

Causal relationships among variables can be depicted graphically using *causal diagrams*. A **causal diagram** consists of a set of nodes that represent the variables, and arrows between the nodes that depict causal relationships. An arrow from variable A to variable B indicates that changes in the value of A can, all else being equal, cause a change in the value of B. Figure 2 illustrates the two causal relationships discussed for the association between smoking and mortality rate.

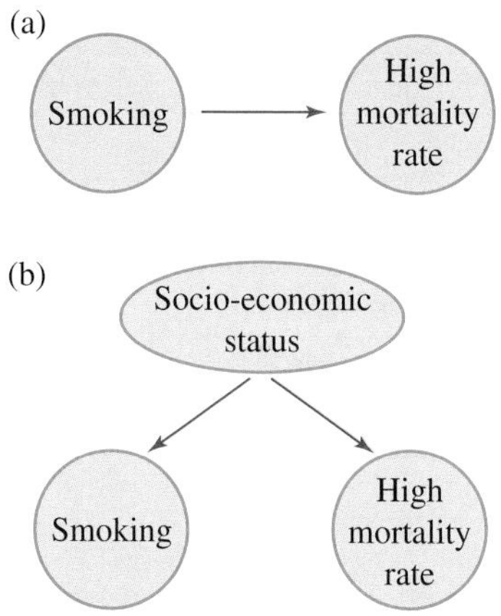

FIGURE 2

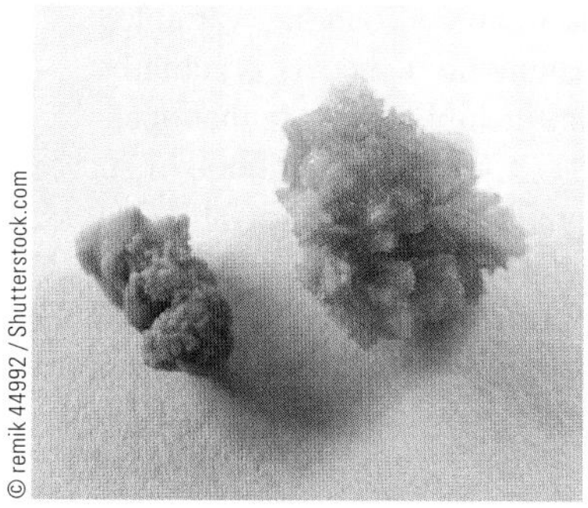

Kidney stones

EXAMPLE 4 | Keyhole surgery is a relatively noninvasive procedure for removing kidney stones. This contrasts with open surgery in which a large incision is made to remove the stones. Keyhole surgery requires a shorter recovery time and therefore we would like to know if it is as effective as open surgery. An observational study was conducted to compare the success of both types of surgery.[1] The data are presented in the following contingency table.

	Success	Failure	Row total
Open surgery	273	77	350
Keyhole surgery	289	59	348
Column total	562	136	698

We see that keyhole surgery actually has a higher likelihood of success than open surgery in these data ($289/348 \approx 83\%$ likelihood of success for keyhole surgery versus $273/350 = 78\%$ likelihood of success for open surgery). Given that the data are observational, however, we can't say for certain whether the differences in surgery *caused* the differences in success.

Can we think of any potential confounding variables that might call the causative link between surgery type and success rate into question? One possibility is the size of the kidney stone. Perhaps patients with small stones tend to be offered keyhole surgery more frequently than those with large stones, and perhaps patients with small stones also tend to have a higher success rate, regardless of the type of surgery used. Fortunately, the study also documented stone size.

1. C. Charig et al., "Comparison of Treatment of Renal Calculi by Open Surgery, Percutaneous Nephrolithotomy, and Extracorporeal Shockwave Lithotripsy," *British Medical Journal* 292 (1986): 879–82.

The following two contingency tables separate the above data into patients with stones smaller than 2 cm and those with stones larger than 2 cm.

Table 1 Stones Smaller than 2 cm

	Success	Failure	Row total
Open surgery	81	6	87
Keyhole surgery	234	35	269
Column total	315	41	356

Table 2 Stones Larger than 2 cm

	Success	Failure	Row total
Open surgery	192	71	263
Keyhole surgery	55	24	79
Column total	247	95	342

By comparing Table 1 and Table 2 we can see that keyhole surgery was indeed much more frequent for patients with small stones ($269/356 \approx 76\%$ for patients with small stones versus $79/342 \approx 23\%$ for patients with large stones).

Furthermore, stone size does appear to be an important confounding variable. In fact, the relationship between surgery type and likelihood of success is actually reversed when we control for stone size! In particular, in Table 1 we see that open surgery is actually more effective for small stones ($81/87 \approx 93\%$ likelihood of success versus $234/269 \approx 87\%$ likelihood of success for keyhole surgery). And Table 2 shows that open surgery is more effective for large stones as well ($192/263 \approx 73\%$ likelihood of success versus $55/79 \approx 70\%$ likelihood of success for keyhole surgery). This kind of paradox—that open surgery is best for each stone size group, but that keyhole surgery appears best when the data are lumped together—is refered to as *Simpson's paradox*. In this case it is due to the fact that the *confounding variable* "stone size" is an important cause of both surgery type and success rate. This is a compelling reminder that inferring causation from observational data can be a very risky business! ■

EXERCISES 11.4

1–8 Answer the following questions.

(a) What is the parameter of interest?

(b) Are the data observational or experimental?

(c) Are the data likely to have come from a random sample? If not, what sort of bias might arise in the estimates?

1. **Contaminants** Do contaminants affect the reproductive success of yellow perch fish? Ecologists used nets to collect a sample of yellow perch from a lake. They then exposed half of them (chosen randomly) to contaminants and the other half to clean water. The reproductive success of the two groups was then compared.

2. **Colon cancer** Does a new test for colon cancer actually work? Scientists used census data to randomly select a sample of individuals in the United States and each was given the test (with the outcome being positive or negative). Individuals were then followed for 10 years to see if they developed colon cancer.

3. What is the prevalence of **polio** in Nigeria? Epidemiologists randomly chose individuals who use health clinics for other reasons and tested them for polio.

4. **Population genetics** Is the frequency of a specific gene higher in bird populations near the equator compared with populations at higher latitudes? Researchers used nets to

capture birds from each location and took blood samples. DNA was then extracted from the blood in a lab to determine whether each individual carried the gene of interest.

5. **Tuberculosis (TB)** Does a new antibiotic work against TB? Medical researchers took a sample of individuals newly diagnosed with TB and gave half of them (randomly chosen) the antibiotic and the other half a placebo. All individuals were then tested for TB.

6. **Waiting times** Does the waiting time for cancer treatment after first diagnosis affect the outcome of the treatment? Cancer researchers obtained the records of all individuals diagnosed with cancer at a particular hospital and looked at the relationship between waiting time and treatment outcome.

7. Will **climate change** alter the reproductive success of plants? Botanists took the species *Arabidopsis thaliana* and maintained two populations in the lab. In one they slowly increased the temperature over a period of five years, while in the other they maintained a constant temperature. The reproductive success of the two groups was compared throughout the five years.

8. Is **schizophrenia** genetically determined? Geneticists took a random sample of patients diagnosed with schizophrenia and determined whether each individual carried a specific gene. They then did the same thing for a sample of randomly chosen, presumed healthy people and compared the two groups.

9–12 Answer the following questions.
(a) What is the parameter of interest?
(b) Which sample is likely to produce the more accurate estimate of the parameter?
(c) Which sample is likely to produce the more precise estimate of the parameter?

9. **Fish ecology** An ecologist wants to know the size distribution of salmon on their breeding grounds and considers two possible approaches: (1) she uses a mesh net to collect 500 fish and determines the fraction of these that are longer than 30 cm; (2) she uses underwater observations of 100 fish and determines the fraction of these that are longer than 30 cm.

10. **Tuberculosis (TB)** A researcher wants to know the effectiveness of a new TB drug and considers two possible approaches: (1) he takes 500 patients newly diagnosed with TB and gives half of them the new drug and half of them a placebo; (2) he takes 2000 people from an HIV clinic who have TB and gives half of them the new drug and half of them a placebo.

11. **Health care** A sociologist wants to know the fraction of people in a city who are in favor of socialized health care. He considers two possible approaches: (1) he uses census data to randomly select 2000 people and then asks their opinion; (2) he uses census data to randomly select 500 people and then asks them, as well as any housemates, for their opinion (giving a total sample of 1500 people).

12. **Heart surgery** A cardiologist wants to determine the effectiveness of a new type of heart surgery and considers two approaches: (1) she takes a sample of 12 patients whose previous type of heart surgery has failed and performs the new surgery on half of them and repeats the older type of surgery on the other half; (2) she takes a sample of 25 patients with no previous heart surgery and performs the new surgery on half of them and the older type of surgery on the other half.

13–18 For each study,
(a) draw the causal diagram that underlies the idea being explored,
(b) determine a potential confounding variable, and
(c) draw the causal diagram corresponding to your answer to part (b).

13. **Climate change** The variability of temperature is expected to increase as a result of climate change. This might have important effects on the mortality rate of different species. To explore this idea, researchers compared the mortality rates of two species of insects, one that experiences high temperature variability and the other that experiences low variability. They found that higher temperature variability was indeed associated with a higher morality rate.

14. **Myopia (short-sightedness) and ambient light** An observational study examined the number of children who sleep with different levels of ambient light, in an attempt to explore whether nighttime ambient light affects vision. They found that children who sleep with higher light levels are also more likely to develop myopia. This study is well known as one for which a confounding variable plays an important role.

    Source: Adapted from G. Quinn et al., "Myopia and Ambient Lighting at Night," *Nature* 399 (1999): 113–14.

15. **Age at first birth and breast cancer** An observational study explored the idea that having children early in life reduces the chance of breast cancer. The researchers showed that, indeed, women who have their first child at an earlier age are less likely to develop breast cancer later in life.

    Source: Adapted from B. MacMahon et al., "Age at First Birth and Breast Cancer Risk," *Bulletin of the World Health Organization* 43 (1970): 209–21.

16. **Hospital errors and death** An observational study was set up to examine whether errors made by doctors in a hospital tended to result in patient deaths. Researchers looked at all instances where an error in treatment was determined to have occurred and quantified the fraction of patients who subsequently died. They then quantified the fraction of patients who died for those not experiencing an error and compared the two groups.

17. Cell biology It is believed that a high level of DNA methylation (the attachment of methyl chemical groups to DNA) suppresses gene expression. To test this idea researchers compared the level of gene expression in cells having high levels of DNA methylation with those having low levels of methylation.

18. Bypass surgery Cardiologists wanted to compare the effectiveness of a new form of noninvasive heart surgery to traditional open-heart bypass surgery. They collected observational data on the success rates of both types of surgery from several hospitals throughout the United States and compared the success rate of each.

PROJECT The Birth Weight Paradox

Birth weight is an important predictor of infant mortality rate, with low-birth-weight children typically having a higher mortality rate. Figure 1 displays data from an observational study showing this trend.

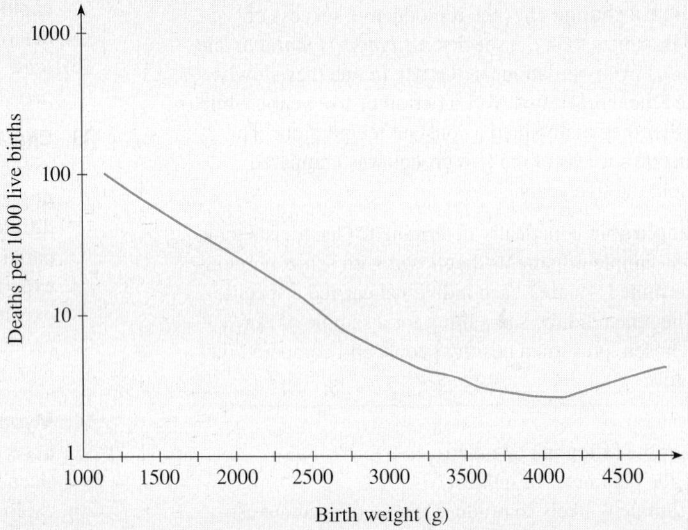

FIGURE 1
Mortality rate as a function of birth weight

Figures 1, 2, and 3 adapted from S. Hernández-Díaz, E. Schisterman, and M. Hernán, "The Birth Weight 'Paradox' Uncovered?" *American Journal of Epidemiology* 164 (2006): 1115–20.

In the same study, it was also shown that mothers who smoke during pregnancy tend to have children with a lower birth weight (see Figure 2). Together this might suggest that the children of mothers who smoke will tend to have a higher mortality rate (due to a lower birth weight).

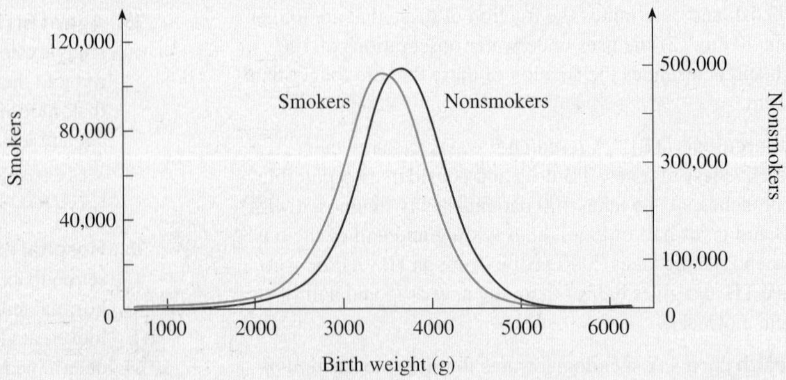

FIGURE 2
A continuous curve approximating the histogram for a sample of 4,115,494 births in the United States in 1991, separately for smokers and nonsmokers

This expectation is met for children born with relatively high birth weights but, paradoxically, the opposite is true for children born with relatively low birth weights. In other words, if we look only at those children with a relatively high birth weight, then those whose mothers smoked during pregnancy tend to have a higher mortality rate. But if we look only at those children with a relatively low birth weight, then those whose mothers smoked during pregnancy tend to have a *lower* mortality rate. Figure 3 illustrates this pattern. It would seem that, for low-birth-weight children, maternal smoking provides a survival benefit!

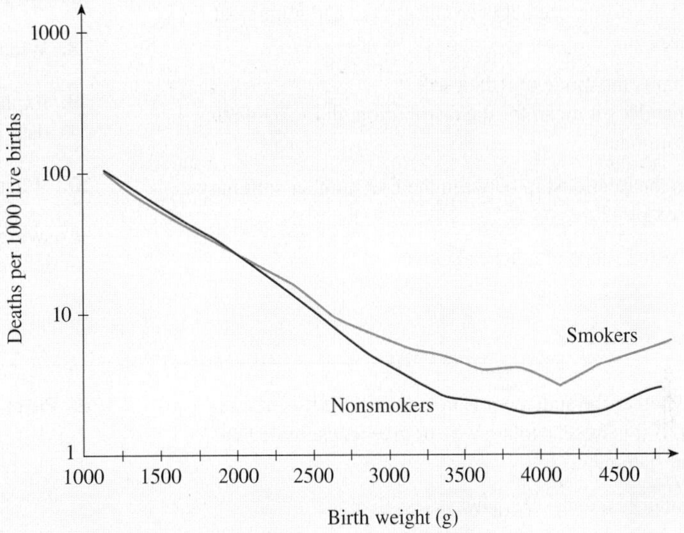

FIGURE 3
Mortality rate as a function of birth weight for smokers and nonsmokers

This peculiar pattern has been observed several times and is referred to as the "birth weight paradox." Is smoking during pregnancy actually beneficial for low-birth-weight children? Ample evidence from other sources suggests that this is not the case, so how can we understand this unexpected pattern?

The answer lies in the fact that the data are from an observational study and therefore are subject to confounding variables. We cannot reliably infer *causation* from observational data. For example, the association between low birth weight and mortality rate in Figure 1 cannot be inferred as causative from the data, nor can the association between smoking and low birth weight in Figure 2. We thus need to consider the possible causal pathways that might have led to this pattern.

1. Draw the causal diagram that corresponds to the initial expectation that smoking causes higher infant mortality rate by causing a lower birth weight.

2. Suppose that smoking causes low birth weight and also directly causes higher infant mortality rate, but there is no causal link between birth weight and mortality rate. Draw the causal diagram for this hypothesis and determine which aspects of the data this causal pathway can explain and which it cannot.

3. Suppose that smoking causes low birth weight and also directly causes higher infant mortality rate. Suppose also that lower birth weight causes higher infant mortality rate. Draw the causal diagram for this hypothesis and determine which aspects of the data this causal pathway can explain and which it cannot.

4. Suppose another (unmeasured) confounding variable (such as the number of deleterious mutations carried by a child) causes lower birth weight and also causes higher mortality rate, in addition to the causative links described in Problem 3. Explain how this could give rise to all of the patterns seen in the data.

Chapter 11 REVIEW

CONCEPT CHECK

1. How do the goals of descriptive statistics differ from those of inferential statistics?
2. What is the difference between categorical and numerical variables?
3. (a) What is the mode of a data set?
   (b) What does it mean for the distribution of a variable to be bimodal?
4. What is the relationship between the five-number summary and a box plot?
5. Describe the shape of a normal curve.
6. What is a contingency table and what is it used for?
7. What is a scatter plot and what is it used for?
8. What is a confounding variable?
9. Explain the way in which a least-squares line best fits the data in a scatter plot.
10. What is a causal diagram?

Answers to the Concept Check can be found on the back endpapers.

TRUE-FALSE QUIZ

Determine whether the statement is true or false. If it is true, explain why. If it is false, explain why or give an example that disproves the statement.

1. The median of a data set is always less than the mean.
2. The median of a data set tends to be less affected by outliers than the mean.
3. The interquartile range is always less than or equal to the range of a data set.
4. A nominal variable is a categorical variable for which there is a natural ordering.
5. The area of a histogram bar is proportional to the number of data points in the corresponding bin.
6. Pie charts are best used to display ordinal, categorical, data.
7. The term 'bias' refers to the amount of sampling error in a data set.
8. A simple random sample is one for which every individual in the population is equally likely to be selected, and all selections are independent of one another.
9. The best type of data for inferring causative relationships among variables is observational data.
10. The units of the standard deviation of the data are the same as those of the data.

EXERCISES

1–4 For each exercise
(a) calculate the mean,
(b) calculate the five-number summary, and
(c) construct a box plot.

BB **1. Pollination** The amount of pollen (in mg) removed from a flower by 10 bees is measured, giving the following data.

8	15	9	22	18
18	16	21	20	18

BB **2.** The **red blood cell count** of 15 patients is given in the table (in millions of cells/μL).

4.2	4.8	5.1	5.1	5.0
5.2	4.7	4.8	4.5	5.5
5.0	5.1	5.3	4.5	4.7

BB **3. Damselflies** The mass (in mg) of 12 damselflies is given in the table.

677	553	720	520	652
559	597	471	577	807
502	733			

BB **4.** The **mRNA** concentration for a specific gene was measured in 15 cells, giving the following data (in number of transcripts/cell).

42	21	131	51	190
165	176	98	178	77
150	201	154	173	99

5. **Mendel's peas** In a particular cross between different genotypes of pea plants, Mendel obtained the following data: 5474 plants with round seeds and 1850 plants with angular seeds. What type of data is "seed shape"? Display the data graphically.

6. The **waiting time** for lung cancer treatment in 28 patients is given in the following table. Construct a bar graph for the data.

<7 days	7–14 days	15–21 days	>21 days
4	16	2	6

7. **Cancer progression** The stage of colon cancer progression for 14 patients was classified into one of five categories, giving the following data.

Stage 0 (local cancer)	2
Stage 1 (spread to inner colon)	6
Stage 2 (spread to nearby tissues)	3
Stage 3 (spread to lymph nodes)	2
Stage 4 (spread widely)	1

Display the data graphically.

8. **Mate choice** The mate preference of female guppies has been studied extensively. Each of 10 females was allowed to choose a mate from three possibilities, giving the following data.

Orange male	Male with black spots	Plain male
6	3	1

What type of data is "mate preference"? Display the data graphically.

BB 9. The **population size** (in millions) of 10 countries in 2010 is given in the table. Construct a density histogram for population size using bins of width 200 and indicate the fraction of countries with a population size of 400 million or larger as an area on the histogram.

Country	Population (in millions)
China	1347
India	1210
United States	314
Indonesia	237
Brazil	193
Pakistan	181
Nigeria	166
Bangladesh	152
Russia	143
Japan	127

10. **HIV** The number of adults and children living with HIV (in millions) in different regions of the world is given in the table.

Region	HIV prevalence (in millions)
Sub-Saharan Africa	23.5
Middle East and North Africa	0.3
South and Southeast Asia	4
East Asia	0.83
Oceania	0.053
Latin America	1.4
Caribbean	0.23
Eastern Europe and Central Asia	1.4
Western and Central Europe	0.9
North America	1.4

Source: Adapted from "UNAIDS Report on the Global AIDS Epidemic 2012," Joint United Nations Programme on HIV/AIDS (UNAIDS), 2012, accessed March 11, 2015, http://www.unaids.org/sites/default/files/media_asset/20121120_UNAIDS_Global_Report_2012_with_annexes_en_1.pdf.

Construct a density histogram, and indicate the fraction of regions with 20 million or more people living with HIV as an area on the histogram.

11. **Fish morphology** The density histogram for mass of individuals of a certain fish species can be approximated by a normal curve with center 500 g and spread 25 g. Estimate the fraction of the data that lies between masses 475 g and 550 g.

12. **Population census** The number of moose observed along a 5-km transect was recorded for several different transects. A density histogram of the data can be approximated by a normal curve with center 58 and spread 10. Estimate the fraction of the transects for which more than 78 moose were observed.

13. **Time until antibiotic failure** The number of years elapsing between the first use of an antibiotic and its ultimate failure as a result of the evolution of resistance was estimated for several antibiotics. The density histogram of times until failure can be approximated by the function $f(x) = \lambda e^{-\lambda x}$ where $\lambda = 1/5$. Use this approximation to estimate the fraction of antibiotics that remained useful for at least 8 years.

14. **Species abundance distributions** are histograms in which the horizontal axis is the natural logarithm of abundance and the vertical axis is the density of species with that abundance. It has been argued that this histogram is well approximated by a normal curve. If the curve is centered at a log abundance of 4 and has a spread of 2, estimate how many species have an abundance less than e^2.

15–18 Construct a contingency table.

15. Prostate-specific antigen (PSA) testing is a simple but controversial test for identifying individuals who are likely to have (or develop) prostate cancer. In a sample of 490 men with prostate cancer, 75% had a positive PSA test. In a sample of 346 men without prostate cancer, 55% had a positive PSA test. Describe the relationship between PSA testing and cancer seen in the data.

16. A new **blood pressure medication** was tested by administering the medication to a random sample of people and giving another random sample a placebo. Researchers then recorded whether each person's blood pressure declined. Of 112 individuals who received the drug, 62 had a decline in blood pressure. Of 92 individuals who received the placebo, 48 had a decline in blood pressure. Describe any relationship between the variables in the data.

17. A **fecal occult blood test** is often used as an indicator of colon cancer. A sample of 13 out of 21 men had positive test results, whereas 10 out of 15 women had positive results. Describe any relationship between the variables in the data.

18. Genetic crossing Two different kinds of crosses were conducted with pea plants and the shape of the seeds in the offspring was recorded. In cross A, 3170 of the 4010 offspring had round seeds with the remainder being angular. In cross B, 2013 of the 4053 offspring had round seeds with the remainder being angular. Interpret the contingency table.

19–20 Use an outlier box plot to explore any potential differences between the two data sets.

BB **19. Crop rotation** is the practice of planting dissimilar crops sequentially in one location to avoid the buildup of pathogens and the depletion of nutrients. A researcher conducted an experiment with soybeans: she used crop rotation on 12 plots and left another 12 with no rotation. Below are the number of pounds of soybeans harvested from each plot in the final season.

Crop rotation	37	42	44	46	41	39
	29	31	37	41	42	45
No crop rotation	29	35	34	27	31	36
	41	40	37	29	31	36

BB **20. Carbohydrates** The daily carbohydrate intake (in grams) of 18 Asians and 20 Hispanics who participated in a health survey is listed below.

Asians	344	285	244	382	310	265
	272	310	291	286	288	305
	410	350	230	273	286	291
Hispanics	310	330	289	287	370	305
	314	267	273	325	347	364
	268	272	329	340	289	295
	349	278				

21–24 Construct a scatter plot for the data, including a plot of the least-squares line.

BB **21. Brain size and intelligence** A study was conducted to look at the relationship between brain size (measured as volume, in mL) and a standardized measure of intelligence. The data for 17 people are as follows.

Brain size	Intelligence	Brain size	Intelligence
965	52	1050	75
1050	59	1055	74
1060	52	1060	76
1090	53	1100	76
1090	55	1190	78
1130	61	1210	79
1060	65	1220	82
1100	67	1300	82
1055	75		

Source: Adapted from S. Witelson, H. Beresh, and D. Kigar, "Intelligence and Brain Size in 100 Postmortem Brains: Sex, Lateralization and Age Factors," *Brain* 129 (2006): 386–98.

Describe, in words, the relationship between brain size and intelligence in the data.

BB **22. Chickadee alarm calls** Chickadees perform an alarm call when they spot a predator; this call sounds like "chick-a-dee-dee-dee." The number of "dees" varies, however, and researchers wanted to know if this was related to the size of predator (measured in kg). They therefore exposed flocks of chickadees to different-sized predators and recorded the average number of "dees" in their calls, giving the following data.

Predator mass	Average number of "dees"	Predator mass	Average number of "dees"
0.07	3.95	0.72	2.80
0.08	4.08	1.40	2.45
0.12	2.75	0.99	1.33
0.19	3.03	1.40	2.45
0.35	2.27	1.08	2.56
0.45	3.16	1.08	2.06
0.72	2.19		

Source: Adapted from M. Whitlock and D. Schluter, *The Analysis of Biological Data* (Greenwood Village, CO: Roberts and Company, 2009).

Describe any relationship between the variables in the data.

BB **23. Paternal age and mutations** A study examined the number of new mutations found in children as a function of their father's age when the child was conceived, giving the following data.

Father's age	Number of mutations	Father's age	Number of mutations
16	40	25	56
17	41	27	50
19	39	29	52
18	49	30	57
23	49	31	61
24	51	34	75
24	52	36	70
24	60	37	65
25	55		

Source: Adapted from A. Kong et al., "Rate of De Novo Mutations and the Importance of Father's Age to Disease Risk." *Nature* 488 (2012): 471–75.

Do the data show any relationship between paternal age and number of mutations?

BB **24.** The winning **Olympic times** for the 100-meter sprint are given for men and for women. Plot the men's times and the women's times on the same graph.

Men's Times

Year	Time (s)	Year	Time (s)	Year	Time (s)
1900	11.00	1948	10.30	1984	9.99
1904	11.00	1952	10.40	1988	9.92
1908	10.80	1956	10.50	1992	9.96
1912	10.80	1960	10.20	1996	9.84
1920	10.80	1964	10.00	2000	9.87
1924	10.60	1968	9.95	2004	9.85
1928	10.80	1972	10.14	2008	9.69
1932	10.30	1976	10.06	2012	9.63
1936	10.30	1980	10.25		

Women's Times

Year	Time (s)	Year	Time (s)	Year	Time (s)
1928	12.20	1964	11.40	1992	10.82
1932	11.90	1968	11.08	1996	10.94
1936	11.50	1972	11.07	2000	10.75
1948	11.90	1976	11.08	2004	10.93
1952	11.50	1980	11.06	2008	10.78
1956	11.50	1984	10.97	2012	10.75
1960	11.00	1988	10.54		

Describe, in words, the relationship between the winning times and year for men and women. How does the relationship compare between men and women?

25–27 For each study,

(a) draw the causal diagram that underlies the idea being explored,

(b) determine a potential confounding variable, and

(c) draw the causal diagram corresponding to your answer to part (b).

25. Breast-feeding and infant health It is thought that children who are breast-fed develop better immune systems. To test this idea, a survey was conducted in which new mothers were asked (i) whether they breast-fed their child and (ii) how many times their child had a fever within the first year of life. It was found that the number of instances of fever was smaller for children who were breast-fed.

26. Health and depression It has been suggested that individuals with chronic illnesses are more likely to suffer from depression because the illness prevents them from participating in certain activities. A study was conducted comparing the depression levels of people with and without chronic illnesses. It was found that people with chronic illnesses did tend to have higher levels of depression.

27. Plant productivity An ecologist suggests that a particular chemical increases the productivity of plants, and as support for this idea he shows that habitats with high plant productivity also tend to have high levels of the chemical.

Probability

12

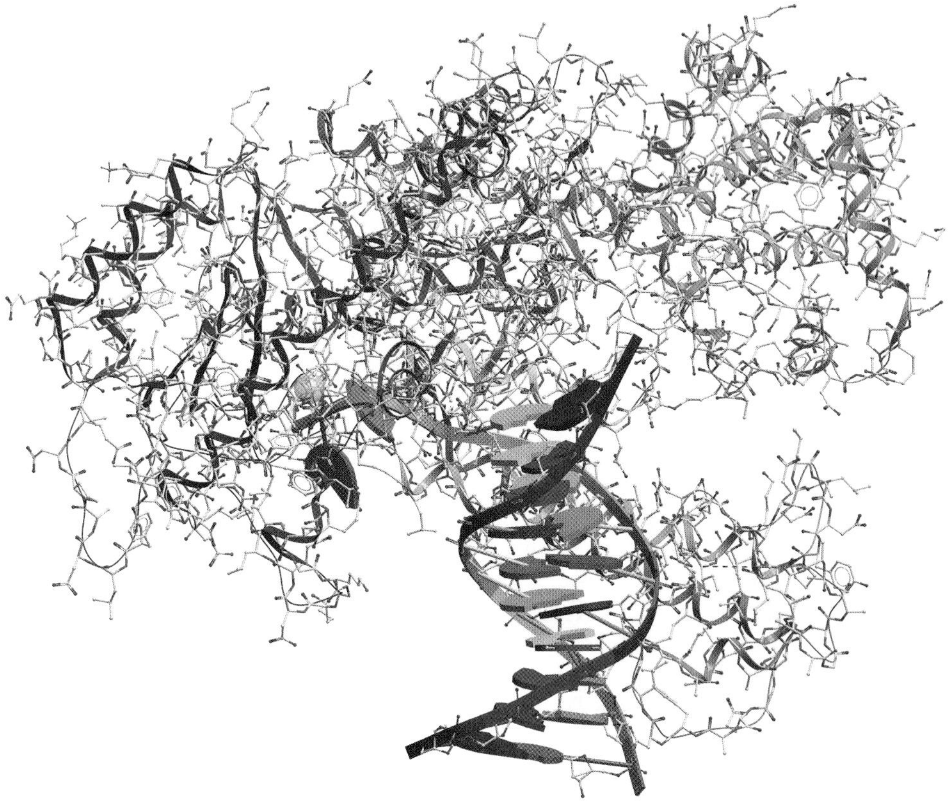

Shown is an image of replication of DNA by the enzyme DNA polymerase. In Section 12.3 we will explore an example of a genetic disease (Huntington's disease) that occurs as a result of mutations arising during DNA replication.

IN THIS CHAPTER WE BEGIN our study of probability theory. Probability theory is important to the life sciences for two reasons. First, many biological processes are inherently probabilistic. For example, we usually cannot predict with certainty the number of offspring that will be produced by a given individual or whether an individual will contract a particular disease. But we can often say something about the likelihood of the different possible outcomes. Doing so requires probability theory. Second, the techniques for inferring properties of populations from data (called *inferential statistics*; Chapter 13) are based on concepts from probability theory. We begin our study of probability with some ideas about counting.

12.1 Principles of Counting

Peptides are made of short sequences of amino acids. There are 20 main kinds of amino acids, as listed in the following table.

Alanine (Ala)	Serine (Ser)	Valine (Val)	Threonine (Thr)
Leucine (Leu)	Tyrosine (Tyr)	Isoleucine (Ile)	Asparagine (Asn)
Glycine (Gly)	Glutamine (Gln)	Cysteine (Cys)	Aspartic acid (Asp)
Phenylalanine (Phe)	Glutamic acid (Glu)	Tryptophan (Trp)	Lysine (Lys)
Methionine (Met)	Arginine (Arg)	Proline (Pro)	Histidine (His)

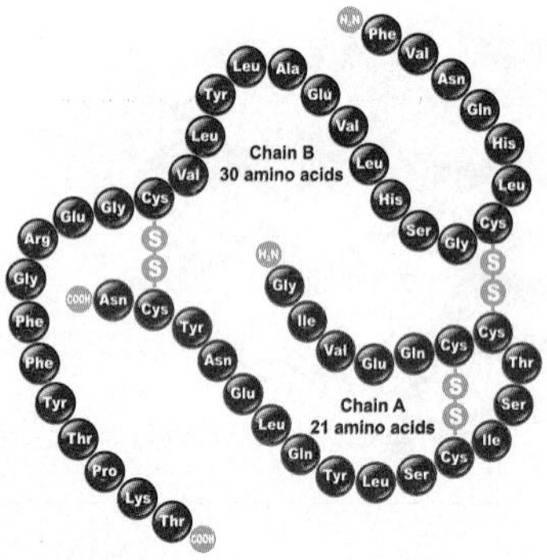

FIGURE 1
Human insulin

As an example, human *insulin* is made up of two connected peptides, one with 21 amino acids (blue chain shown in Figure 1), and the other with 30 amino acids (purple chain).

Two other important peptides in mammals are *vasopressin* and *oxytocin*. Vasopressin is involved in water absorption in the kidneys and oxytocin plays a role in reproduction. Despite their very different functions, both peptides are chains of nine amino acids and they differ from one another only in the amino acids present at position 3 and position 8 (see Figure 2).

Peptide Orientation

A peptide has a well-defined orientation determined by the nature of the unused bonds on its two ends. One end is always a so-called "N-terminus," whereas the other is a "C-terminus." By convention, peptides are drawn with the amino acid at the N-terminus end in position 1.

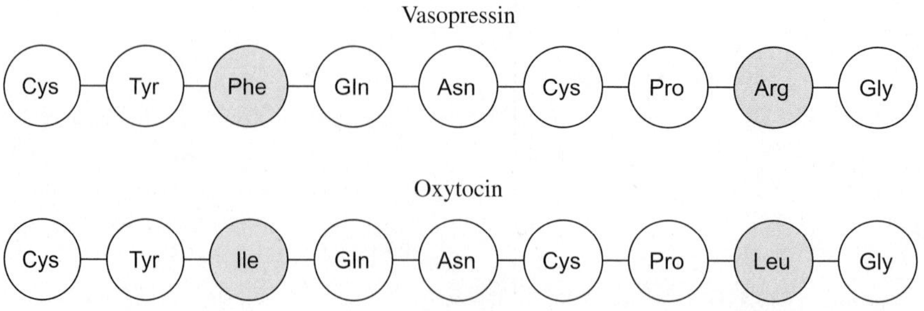

FIGURE 2 Position: 1 2 3 4 5 6 7 8 9

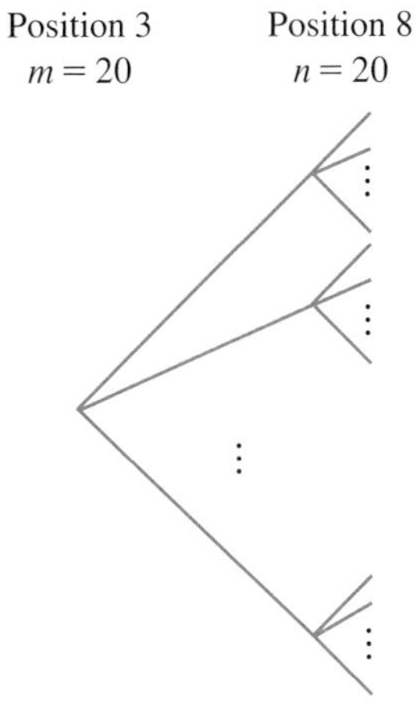

FIGURE 3

© Mr.JPEG / Shutterstock.com

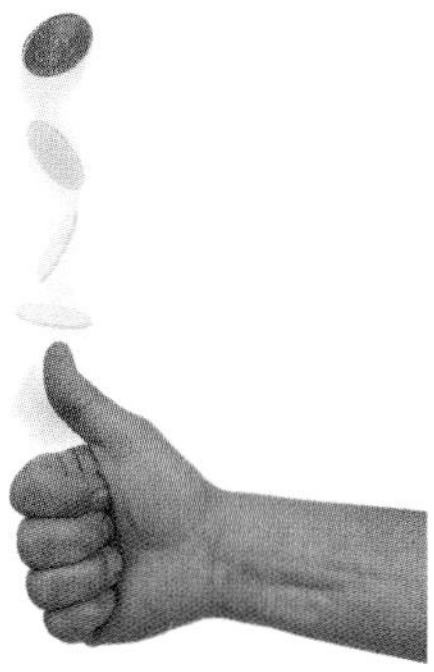

© James Steidl / Shutterstock.com

In fact, one important consequence of peptides being composed of amino acid subunits is that many different peptides can be constructed by changing relatively few amino acids. Consider the total number of nine-amino-acid peptides that can be constructed by changing only the amino acids present at positions 3 and 8.

We can calculate this number with the aid of the tree diagram in Figure 3. First, there are 20 possible amino acids at position 3. And, given any one of these choices at position 3, there are again 20 possible amino acids at position 8. This gives a total of $20 \cdot 20 = 400$ possible peptides that differ only at positions 3 and 8. Vasopressin and oxytocin are two of these, leaving 398 others.

A general principle of counting underlies this example:

> Suppose two selections can be made independently, the first having m possible choices and the second having n possible choices. Then, taken together, the total number of possible choices is
>
> $$m \cdot n$$

EXAMPLE 1 | Suppose you roll a die and flip a coin. Taken together, how many possible outcomes are there?

SOLUTION We can take the first selection to be the roll of a die and therefore $m = 6$. Similarly, we can take the second selection to be the coin toss and therefore $n = 2$. There are then $6 \cdot 2 = 12$ possible outcomes. ■

Example 1 introduces two procedures—rolling a die and flipping a coin—that we will revisit several times throughout this chapter. These procedures are important because they serve as useful models for many phenomena in the life sciences. Thinking about biological processes in these terms allows us to avoid being distracted by interesting but irrelevant biological details and instead focus on the important probabilistic features of the system.

EXAMPLE 2 | Methylation Suppose there are eight possible variants for a gene (each called an *allele*). Further, suppose each allele can be methylated or not (methylated alleles have a methyl chemical group attached to them). The expression level of a gene depends on both the allele that is present and its methylation status. Considering both allele type and methylation status, how many different possibilities are there?

SOLUTION Taking the first selection to be the choice of allele, we can view this as rolling an eight-sided die. Therefore we have $m = 8$. The second selection is the methylation pattern (yes or no), and we can view this as flipping a coin. Therefore we have $n = 2$. As a result, there are $8 \cdot 2 = 16$ possibilities. ■

The preceding examples involve only two selections, but a similar principle holds more generally.

> **(1) Fundamental Principle of Counting** Suppose k selections can be made independently, the first having n_1 possible choices, the second having n_2 possible choices, . . . , and the kth having n_k possible choices. Then, taken together, the total number of possible choices is
>
> $$n_1 \cdot n_2 \cdot \cdots \cdot n_k$$

Figure 4 illustrates the Fundamental Principle of Counting using a tree diagram with three selections.

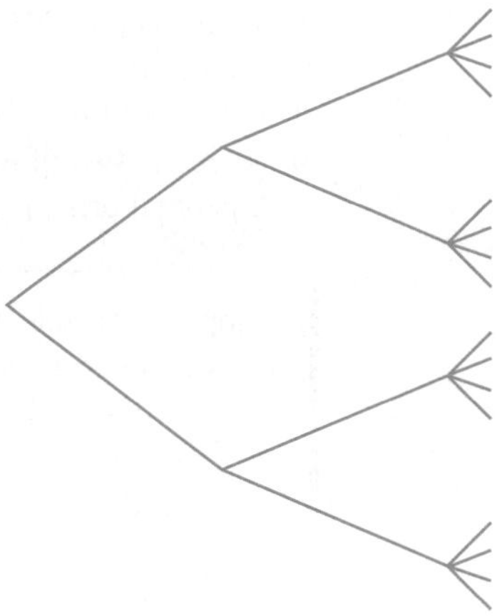

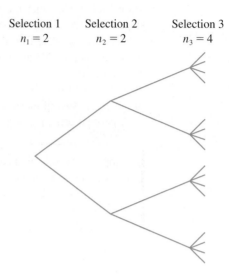

FIGURE 4
A tree diagram with three selections. Selection 1 has two possible choices, selection 2 has two possible choices, and selection 3 has four possible choices. This gives a total of $2 \cdot 2 \cdot 4 = 16$ possible choices.

EXAMPLE 3 | Nonapeptides Vasopression and oxytocin are referred to as *nonapeptides* because they each contain nine amino acids.
(a) Given that there are 20 different amino acids, how many different nonapeptides are possible?
(b) How many different nonapeptides are possible if no amino acid is ever repeated?

SOLUTION

(a) We take the choice of amino acid for positions 1 through 9 to be the selections. For each position there are 20 choices and therefore $n_1 = n_2 = \cdots = n_9 = 20$. From the Fundamental Principle of Counting the number of possibilities is therefore

$$n_1 \cdot n_2 \cdot \cdots \cdot n_9 = 20 \cdot 20 \cdot \cdots \cdot 20 = 20^9 = 512{,}000{,}000{,}000$$

(b) Again we take the choice of amino acid for positions 1 through 9 to be the selections. For position 1 there are $n_1 = 20$ choices. For position 2, if we are to avoid repeating the amino acid at position 1, there are $n_2 = 19$ remaining possibilities. Similarly, for position 3, if we are to avoid repeating the amino acids present at positions 1 and 2, there are $n_3 = 18$ remaining possibilities. Following this logic, we obtain

$$n_1 \cdot n_2 \cdot \cdots \cdot n_9 = 20 \cdot 19 \cdot 18 \cdot \cdots \cdot 12 = 60{,}949{,}324{,}800$$ ■

Example 3(b) is an instance of what is called a *permutation*.

■ Permutations

With permutations, order matters.

A **permutation** of a set of objects is an *ordered* arrangement of the objects without repetition.

Scientists studying bird populations often place a unique, color-coded band on the leg of each bird so that birds can be individually identified and followed over their lifetimes. Suppose a scientist is studying the seasonal migration of a bird population, and she is interested in the arrival of three particular individuals with band colors red, blue, and yellow. After overwintering in the tropics the birds return, one at a time. The pattern of arrival of these three individuals is an ordered arrangement of the set of three birds. Figure 5 illustrates two possible arrival orders. Each ordering of arrivals represents a particular permutation of the three individuals. How many permutations are there?

Given the small number of birds, we can easily enumerate all the possibilities. Using R, B, and Y for the colors, we have

(2) {R, B, Y} {R, Y, B} {B, R, Y} {B, Y, R} {Y, R, B} {Y, B, R}

(a)

(b)

FIGURE 5

We see that there are six permutations.

Although this approach works when there are few objects, it quickly becomes unmanageable as the number of objects increases. We can gain better insight, and generalize our solution, by systematically working through the possibilities using the Fundamental Principle of Counting.

The first bird to arrive can be any of the three individuals. Therefore $n_1 = 3$. Given a particular choice for the first bird, there are two remaining possibilities for the second arrival. Therefore $n_2 = 2$. Finally, after the first two birds have arrived, there is only one remaining choice for the third arrival, giving $n_3 = 1$. The number of permutations is thus

Recall that $n!$, read as "n factorial," is the number $n(n - 1)(n - 2) \cdots 1$.

$$n_1 \cdot n_2 \cdot n_3 = 3 \cdot 2 \cdot 1 = 3! = 6$$

More generally, using P_n to denote the number of permutations of a set of n objects, we have the following.

> **(3) Counting Permutations** The number of **permutations of a set of n objects** is
>
> $$P_n = n(n - 1)(n - 2) \cdots 1 = n!$$

A common thought experiment used as a model for permutations involves drawing balls from a jar (also called an "urn" in this context).

FIGURE 6

EXAMPLE 4 | A jar contains four colored balls: red, blue, green, and yellow (see Figure 6). Suppose we remove one ball at a time until the jar is empty. How many different sequences of colors are possible?

SOLUTION Any sequence of colors is a permutation of the set of balls. Therefore, from (3), the number of permutations (that is, the number of possible sequences of colors) is $P_4 = 4! = 24$. ■

Permutations of a set of objects need not involve all of the objects in the set. Suppose there are five color-banded migrating birds of interest. How many orderings are there for the arrival of the first two birds? Again, using the Fundamental Principle of Counting, we see that the first bird to arrive can be any of the five individuals. Therefore $n_1 = 5$. And given any particular choice for the first bird, there are then four possibilities for the second arrival. This gives $n_2 = 4$. As a result, there are $5 \cdot 4 = 20$ different orderings for the arrival of the first two birds. We say that there are 20 permutations of a set of 5 objects *when taken 2 at a time*.

More generally, using $P_{n,k}$ to denote the number of permutations of n objects when taken k at at time (with $k \leq n$), we have

(4) $$P_{n,k} = n(n - 1)(n - 2) \cdots (n - (k - 1))$$

We can rewrite Equation 4 more simply by first noting that $n!$ can be written as

$$\begin{aligned} n! &= n(n-1)(n-2)(n-3)(n-4)\cdots 1 \\ &= n(n-1)(n-2)\cdots(n-(k-1))\cdot(n-k)(n-k-1)\cdots 1 \\ &= n(n-1)(n-2)\cdots(n-(k-1))\cdot(n-k)! \\ &= P_{n,k}\cdot(n-k)! \end{aligned}$$

where the final equality makes use of Equation 4. Solving this last equation for $P_{n,k}$ gives the following result.

> **(5) Counting Permutations (continued)** The number of **permutations of a set of n objects when taken k at a time** is
>
> $$P_{n,k} = \frac{n!}{(n-k)!}$$

Notice from Formulas 5 and 3 that $P_{n,n} = P_n$ (recall that $0! = 1$). Therefore (5) is a generalization of (3).

EXAMPLE 5 | A jar contains four colored balls: red, blue, green, and yellow. Suppose we sequentially remove two balls. How many different orderings are possible?

SOLUTION Any sequence of two balls removed is a permutation of the set of four balls when taken two at a time. From (5) the number of permutations is

$$P_{4,2} = \frac{4!}{(4-2)!} = 12$$ ■

African cichlids

EXAMPLE 6 | **African cichlid mate preferences** African cichlids are freshwater fish found in the Rift Valley lakes of Mozambique and Tanzania. During the mating season each male defends a territory and females sequentially visit and assess a subset of these males before choosing a mate. If there are 100 males in the population and a female can assess only three of them before choosing a mate, how many possible assessment sequences are there?

SOLUTION The set of objects of interest is the 100 males in the population, and any assessment sequence by a female can be viewed as a permutation of this set of males taken three males at a time. Put another way, we can view the males as being 100 different-colored balls in a jar, and a female sequentially draws three of these. Therefore, from (5), the number of possible assessment sequences is

$$\begin{aligned} P_{100,3} &= \frac{100!}{(100-3)!} = \frac{(100)(99)(98)\cdot 97!}{97!} \\ &= (100)(99)(98) = 970{,}200 \end{aligned}$$ ■

■ Combinations

With combinations, order does not matter.

A **combination** of a set of objects is an *unordered* subset of the objects without repetition. With combinations, all that matters is the identity of the objects in the subset. Their order is irrelevant.

Returning to the population of three color-banded birds, suppose that only the first two birds to arrive will be successful in obtaining a nesting site. And suppose a scientist is interested in the identity of only the two successful birds. As we have seen, the number of orderings for the first two arrivals is the number of permutations of three objects taken two at a time. Enumerating these explicitly we have

(6) $\{R, B\} \quad \{R, Y\} \quad \{B, R\} \quad \{B, Y\} \quad \{Y, R\} \quad \{Y, B\}$

We see that there are six possible orderings (or permutations), as expected from Formula 5 [because $P_{3,2} = 3!/(3-2)! = 6$].

The list in (6) gives the different possible orderings, but we are interested only in the identity of the individuals. And the identity of the individuals in some of the permutations in (6) are the same. For example, the permutations {R, B} and {B, R} both contain the same individuals. Therefore they represent the same *combination*. Similarly, permutations {R, Y} and {Y, R} represent the same combination, as do permutations {B, Y} and {Y, B}.

FIGURE 7
The three combinations, represented by the three ellipses

In fact, there are 2! permutations that are equivalent for each combination because, from (3), there are 2! possible orderings of the two individuals that make up any combination. Therefore the total number of combinations can be calculated by dividing the number of permutations by 2! (that is, dividing $P_{3,2} = 6$ by 2!). There are thus three combinations, as shown in Figure 7.

More generally, suppose we have a set of n objects and we wish to know the number of combinations possible when taken k objects at a time (with $k \leq n$). The number of permutations is $P_{n,k}$ but, as in the preceding example, some of these will correspond to the same combination.

There will be $k!$ permutations that correspond to each combination because, from (3), this is the number of different orderings of the k objects that make up a combination. Therefore the number of combinations is $P_{n,k}/k!$. Using (5), along with $C_{n,k}$ to denote the number of combinations of a set of n objects when taken k at at time, we have the following.

(7) Counting Combinations The number of **combinations of a set of n objects when taken k at a time** is

$$C_{n,k} = \frac{n!}{k!(n-k)!}$$

It is also common to use the notation $\binom{n}{k}$ for $C_{n,k}$, which is read "n choose k," and this is referred to as the **binomial coefficient** (see Exercise 41). This notation should not be confused with the fraction n/k or with the vector whose components are n and k.

Drawing balls from a jar is also a common thought experiment for modeling combinations.

FIGURE 8

EXAMPLE 7 | A jar contains six colored balls: red, blue, green, yellow, orange, and purple (see Figure 8). Suppose we select four balls. How many different combinations of colors are possible?

SOLUTION We are asked for the number of combinations of six objects taken four at a time. If we set $n = 6$ and $k = 4$ in Formula 7, we obtain

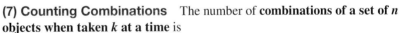

$$C_{6,4} = \binom{6}{4} = \frac{6!}{4!(6-4)!} = \frac{6 \cdot 5 \cdot \cancel{4 \cdot 3 \cdot 2 \cdot 1}}{(\cancel{4 \cdot 3 \cdot 2 \cdot 1})(2 \cdot 1)} = 15$$

Note that the number of combinations $C_{6,4} = 15$ is different from the number of permulations $P_{6,4} = 360$ because, for example, the permutation "red, blue, yellow, green" represents the same *combination* as the permutation "red, blue, green, yellow." ■

EXAMPLE 8 | HIV transmission Once an individual is infected with HIV, a very large number of different virus genotypes arise through mutation. However, during transmission between individuals, a bottleneck occurs in which only a subset of these genotypes are passed on. Suppose 150 HIV genotypes are present in an infected individual, but only 10 distinct genotypes are passed on during transmission. How many different possibilities are there for the pool of genotypes transmitted?

SOLUTION The transmission pool contains 10 distinct genotypes drawn from a set of 150. We can view this as drawing 10 balls from a jar containing 150, where each ball is distinct and we are interested only in the identity of the balls. The number of possibilities is therefore the number of combinations of 150 objects when taken 10 at a time. If we set $n = 150$ and $k = 10$ in Formula 7, we obtain

$$C_{150,10} = \binom{150}{10} = \frac{150!}{10!(150 - 10)!} = 1{,}169{,}554{,}298{,}222{,}310$$

different possibilities. This is an enormous number and it reveals that an enormous number of infection types can be generated by a single infected individual. ■

EXERCISES 12.1

1. **Experimental design** An experiment is to be conducted to examine the growth rate of juvenile coho salmon under different temperature and food conditions. To get reliable results we need to have 20 fish in each set of conditions. If there are two different kinds of food and three distinct temperatures of interest, and if we wish to conduct the experiment separately for males and females, how many fish will we need?

2. **Dengue fever** is an infectious disease of humans that is caused by a virus. There are four main viral strains, and an individual can never be infected by the same strain twice. However, a second infection can occur by any of the other strains. The severity of the second infection depends on the identity of both the strain that caused the first infection and the strain causing the second infection, as well as the order in which these were acquired. How many different levels of severity are possible for an individual acquiring a second infection?

3. Telephone numbers in North America consist of 10 digits (including the area code). How many different telephone numbers are possible, given that the first digit cannot be 0 or 1?

4. A four-member student council is elected each year and consists of one student from each of the first-, second-, third-, and fourth-year classes. Suppose that, on the ballot, there are two first-year students, two second-year students, three third-year students, and four fourth-year students. How many different elected councils are possible?

5. **Bioengineering** Bioengineers construct peptides with new functions by starting with a particular peptide and then mutating the amino acid present at different positions. They then screen the resulting mutant peptide for the desired function. Suppose the starting peptide is 10 amino acids long.
   (a) How many peptides need to be screened if the amino acids at positions 3, 5, and 7 are all mutated?
   (b) If we consider all possible mutants for the amino acid at one position only, but this can be at any of the 10 positions, how many new peptides will need to be screened?

6. A computer password is seven characters in length.
   (a) How many passwords are possible if the characters are all lowercase letters?
   (b) How many passwords are possible if the characters are either lowercase letters or the numerals 0 through 9?
   (c) How many passwords are possible if the characters are either uppercase or lowercase letters?

7. A *palindrome* is a word, such as "radar," that reads the same forward and backward. How many different five-

letter palindromes are possible if they do not need to be real words?

8. Given a set of n distinct elements, how many subsets are possible, including the empty set? [*Hint:* Each element can be either included or not included when forming a subset.]

9. A **DNA sequence** consists of a string of elements, called nucleotides, in a defined order. Suppose the DNA sequence of a virus is 800 nucleotides long. If each nucleotide can be either a G, T, C, or A, how many different sequences are possible?

10. **Experimental design** An experiment is to be conducted to examine the timing of flowering in a small plant under different temperature, water, and light conditions. Suppose there are three distinct temperatures of interest, two water levels, and four light levels. If our greenhouse can hold a total of 96 plants, and if the plants are to be spread evenly among all sets of conditions, how many individuals can we have in each set of conditions?

11. **Blood types** Certain blood types can be classified as either A, B, AB, or O, and by whether they are Rh^+ or Rh^-. How many different blood types are possible?

12. **Genotypes** Suppose there are five possible variants of a gene: A_1, A_2, A_3, A_4, and A_5. *Diploid* individuals carry two such genes. How many different diploid individuals are possible if the two variants they carry are different?

13. **Influenza reassortment** Influenza A viruses carry eight kinds of DNA segments. When an individual is infected with two different strains of the virus, each new viral particle is formed by selecting each of its eight segments from either of the two strains. How many different viral particles are possible if
(a) all eight segments differ between the two strains?
(b) only five of the segments differ between the two strains?

14. **DNA translation** When DNA is decoded into protein, the "letters" of the DNA sequence are "read" in ordered triplets. For example, the sequence ACGTACTTA is read as the three ordered triplets ACG, TAC, and TTA. Each letter can be either G, T, C, or A, and each distinct triplet can code for a different amino acid. How many different amino acids can be coded?

15. **Genetic code** Suppose that when DNA is decoded into protein, the "letters" of the DNA sequence are "read" in ordered groups of length k (where k need not equal 3). For example, if $k = 4$, then the sequence AGTACTAGACCG is read as AGTA, CTAG, and ACCG. Suppose that each group of k letters codes for a different amino acid, and each letter can be either G, T, C, or A. If there are 20 amino acids, what is the smallest value of k that allows coding for them all?

16–18 A jar contains seven differently colored balls.

16. If you remove all the balls, one at a time, how many color sequences are possible?

17. If you remove five balls, one at a time, how many color sequences are possible? How many combinations of colors are possible?

18. If you remove two balls, one at a time, how many color sequences are possible? How many combinations of colors are possible?

---

19. How many different ways are there to put four tires on a car?

20. Three cars arrive simultaneously at a four-way stop and all want to proceed straight through the intersection. How many ways are there for this to happen if they proceed one at a time?

21. Suppose five runners finish a race with no ties.
(a) In how many ways can the race be completed?
(b) In how many ways can the first, second, and third places be decided?

22. Obtain a formula for the number of permutations of n objects when taken k at a time if a particular one of these objects must always be included.

23. Obtain a formula for the number of permutations of n objects when taken k at a time if a particular one of these objects must never be included.

24. **Medical treatment** Four different antidepressant drugs are tested and their ability to reduce depression is ranked from highest to lowest. How many different rankings are possible?

25. **Dominance hierarchies** are orderings of individuals in terms of their dominance rank in social interactions. How many hierarchies are possible if there are six individuals and there are no ties?

26. **Mate choice** A mate-choice experiment is conducted in which a female cichlid is allowed to exhibit her preference for each of n different males. How many different rankings are possible?

27. **Genetic architecture** An analysis shows that four genes are all located sequentially next to one another on the same chromosome. How many arrangements of the genes on the chromosome are possible?

28–31 The formula for calculating the number of permutations must be used more than once for some problems.

28. A collection of nine black blocks with different shapes and eight white blocks with different shapes are to be lined up with alternating colors, starting with a black block. How many arrangements are possible?

29. An **operon** is a set of genes located sequentially next to one another along the DNA so that they can be controlled by a common mechanism. Suppose there are three operons, labeled A, B, and C. Operon A contains three different genes, operon B contains five different genes, and operon C contains six different genes. All genes within an operon must be located next to one another, but they can be placed in any order and still function properly. Similarly, the three operons can be placed in any order along the DNA and still function properly. How many genomic arrangements are possible?

30. Plant breeding Five male and four female plants are to be arranged in a single row for a breeding experiment. To ensure breeding, no two male plants can be next to one another. How many arrangements are possible?

31. Tropical fish Seven aquariums, each containing a different species of tropical fish, are to be arranged on a shelf. Because of aggressive interactions, species 1 and species 2 cannot be placed next to one another. How many arrangements are possible?

---

32. Suppose you toss six coins.
(a) How many ways are there to obtain four heads?
(b) How many ways are there to obtain two tails?

33. Suppose you toss five coins.
(a) How many ways are there to obtain at least three heads?
(b) How many ways are there to obtain at most two tails?

34. Random sampling Suppose a sample of 14 fish is to be taken from a population containing 100. How many different samples are possible?

35. In a raffle with 10 entries, in how many ways can three winners be selected?

36. A jogger jogs every morning to his health club, which is eight blocks east and five blocks north of his home. He always takes a route that is as short as possible, but he likes to vary it (see the figure). How many different routes can he take? [*Hint:* The route shown can be thought of as ENNEEENENEENE, where E is East and N is North.

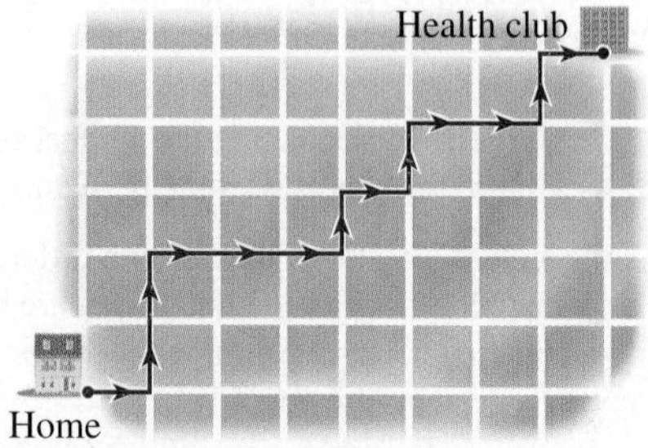

37. HIV testing Ten people are tested for HIV. How many different outcomes have six people being HIV$^+$?

38. Genetic architecture Of 10 genes studied in yeast, 7 are found to form a network of metabolic interactions. If this set of 7 genes was chosen arbitrarily from the 10 that were studied, how many different sets of 7 would be possible?

39. Mortality A small population of 20 birds on an island are marked so that each individual can be identified. After a storm, only 8 individuals remain. How many different possibilities are there for the identity of the 8 individuals?

40. Prove the identity

$$\binom{n}{k} = \binom{n}{n-k}$$

41. Prove the identity

$$(x+y)^n = \sum_{k=0}^{n} \binom{n}{k} x^k y^{n-k}$$

Hint: Use a counting argument to determine the number of ways k x's and $n - k$ y's would appear in the product

$$(x+y)(x+y)\cdots(x+y)$$

The quantity $\binom{n}{k}$ is called the *binomial coefficient* because it gives the coefficients in the expansion of the binomial $(x + y)^n$.

42–44 The formula for calculating the number of combinations must be used more than once for some problems.

42. From a group of 10 male and 10 female tennis players, 2 men and 2 women are to play each other in a men-versus-women doubles match. In how many different ways can this match be arranged?

43. A jar contains four white balls, eight black balls, and six red balls. In how many ways can two balls of each color be chosen if order does not matter?

44. Viral transmission You are riding the subway and inhale an aerosol droplet containing 10 drug-resistant and 6 drug-sensitive viral particles. Only 5 particles make their way into your lungs.
(a) How many different ways are there for these 5 particles to be made up of 2 resistant and 3 sensitive particles?
(b) How many different ways are there for at least 3 of these particles to be drug sensitive?

45–47 A statistical test, known as *Fisher's exact test*, is often used to determine if there is a relationship between two categorical variables. The test is based on the counting principles presented in this section. You will begin to explore the test here, although a complete exposition is not possible until Section 12.2 (see Exercises 12.2.55–57).

45. Consider a collection of 12 balls, 5 of which are red. Suppose that two groups are formed from these 12 balls. Group A contains 4 balls and Group B contains 8 balls.
(a) Ignoring color, how many different ways are there to form these two groups?
(b) Now suppose that 3 of the 4 balls in Group A are red (with the remaining 2 red balls being in Group B). In this case, how many different ways are there to form these two groups?

Fisher's exact test can be used to determine whether the red balls tend to be clustered in one group more than we would expect by chance.

46. Genetic markers of disease Consider a collection of 14 people, 6 of whom carry a genetic mutation. Suppose that 5 of the 14 people develop a disease and 9 do not.
(a) Ignoring the genetic status of each individual, how many ways are there to have 5 of the 14 people develop the disease?
(b) Now suppose that 4 of the 5 people who develop the disease also carry the genetic mutation (with the remaining 2 people who carry the mutation being healthy). How many different ways are there for this to occur?

Fisher's exact test can be used to determine if the occurrence of the mutation tends to be clustered in the group of people that develop the disease. If so, the gene can then be used as a marker to indicate which people are likely to develop the disease.

47. Bat conservation biology Example 11.3.1 presents data on the relationship between cave use in bat species and their endangerment status. Of the 18 bat species studied, 11 of them use caves. From the data we also see that 14 of the species are classified as being fragile and 4 are classified as being vulnerable.
(a) If we ignore cave use then how many ways are there to have 14 of the 18 species classified as fragile?
(b) In the data of Example 11.3.1 we see that 9 of the 11 species that use caves are classified as being fragile (with the remaining 2 species being classified as vulnerable). How many different ways are there for this to occur?

Fisher's exact test can be used to determine if cave-roosting bats tend to be clustered in the fragile category.

48. Human genetics Humans carry two copies of most genes. Individuals who carry two identical variants of a gene are called *homozygous*, whereas those who carry two different variants are called *heterozygous*. For example, if there are two possible variants, labeled A and a, then AA and aa individuals are homozygous, whereas Aa and aA individuals are heterozygous. Suppose two heterozygous parents have a child, and each parent gives one of its genes to the offspring.
(a) How many types of offspring are possible if all that matters is the identity of the genes?
(b) In some cases the same variant behaves differently if it is inherited from the mother versus the father. How many types of offspring are possible if we account for such effects?

49. You have five textbooks and six novels. In how many ways can three of these textbooks and two of these novels be selected and arranged on a bookshelf?

50. Mate choice (cont.) A mate-choice experiment is conducted in which a female cichlid is allowed to display her preference for each of n different males. Suppose ties are possible and therefore the result is an ordered list of groups of males, with the first group being her first choice, the next being her second choice, and so on. Let $N(n)$ denote the number of different possible outcomes. For example, $N(2) = 3$ since, with two males, male 1 can be uniquely preferred, male 2 can be uniquely preferred, or they can be equally preferred.
(a) List all possibilities when $n = 3$.
(b) Let's define $N(0) = 1$. Show that $N(n)$ obeys the recursion equation

$$N(n) = \sum_{i=1}^{n} \binom{n}{i} N(n-i)$$

[*Hint:* How many outcomes are there in which i males are equally the least preferred?]
(c) Calculate $N(3)$ and $N(4)$ using the recursion from part (b).

Source: Adapted from S. Ross, *A First Course in Probability*, 8e (Upper Saddle River, N.J.: Pearson Prentice Hall, 2010).

12.2 What Is Probability?

Probability theory is the study of uncertainty. People typically have an instinctive sense of some elements of probability theory because we very often use probabilistic language informally. Statements like "it is very likely to rain tomorrow" or "there is a low probability of an influenza pandemic this winter" are common in everyday language.

While these informal ideas of probability are familiar, they need to be made more precise if we are to use them productively in science. We begin to do so in this section.

■ Experiments, Trials, Outcomes, and Events

A central concept in probability theory is that of an *experiment*. An **experiment** is any procedure, such as flipping a coin or rolling a die, that can be repeated an indefinite number of times under identical conditions (in principle, at least) and that has a set of distinct possible **outcomes**. (*Note*: This is somewhat different from the notion of an experiment and experimental data discussed in Chapter 11).

Each repeated instance of an experiment is called a **trial**. For the purpose of studying probability, we are interested in experiments whose outcome on any given trial is uncertain.

The set of all possible outcomes of an experiment is called the **sample space** of the experiment and is usually denoted by the Greek symbol Ω (pronounced "omega"). The set containing no outcome is called the empty set and is denoted by $\varnothing$.

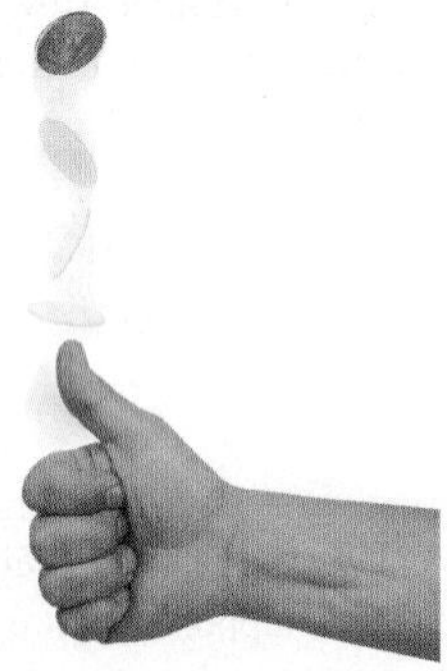

EXAMPLE 1 | What is the sample space for each experiment?

(a) Flipping a coin
(b) Rolling a six-sided die
(c) Testing two people for the presence of antibodies to HIV

SOLUTION

(a) There are two possible outcomes, namely, heads or tails. Denoting these by H and T, the sample space is $\Omega = \{H, T\}$.

(b) There are six possible outcomes, with sample space $\Omega = \{1, 2, 3, 4, 5, 6\}$.

(c) Each person can have antibodies to HIV, or not. Therefore, from the Fundamental Principle of Counting in Section 12.1, there are four possible outcomes. Using + and − to indicate the presence or absence of antibodies and listing the results for the two people in order, we can write the sample space as

$$\Omega = \{(+, +), (+, -), (-, +), (-, -)\}$$ ■

We can also consider subsets of the sample space of an experiment. For instance, two possible subsets for Example 1(b) are $\{2, 4, 6\}$, the set of outcomes that are even, and $\{4, 5, 6\}$, the set of outcomes that are greater than 3.

Any subset of the sample space is called an **event**. A **simple** event is an event consisting of a single outcome, whereas a **compound** event consists of more than one outcome.

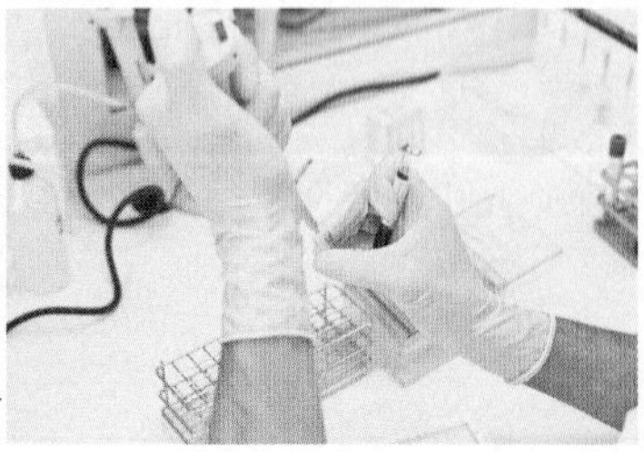

EXAMPLE 2 | **HIV antibodies** Example 1(c) gives the sample space when two people are tested for the presence of antibodies to HIV. Specify each of the following events in terms of the outcome(s) in this sample space, and state whether the event is simple or compound.

(a) Both people test negative.
(b) Exactly one person tests positive.

SOLUTION

(a) There is one outcome in which both people test negative. This event is therefore *simple* and consists of the subset $\{(-, -)\}$.

(b) There are two outcomes in which exactly one person tests positive. This event is therefore *compound* and consists of the subset $\{(+, -), (-, +)\}$. ■

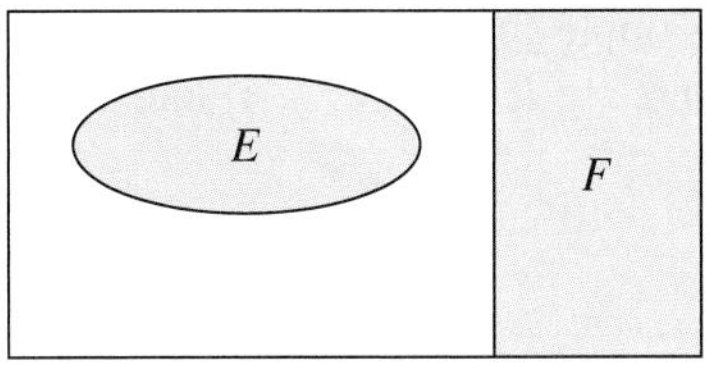

FIGURE 1
A Venn diagram, where E and F represent events

Venn diagrams provide a pictorial description of the sample space. To construct a Venn diagram, we first draw a rectangle to represent the sample space. Regions within the rectangle are then used to represent events, as shown in Figure 1.

EXAMPLE 3 | **Human genetics** Humans carry two copies of most genes. An individual who carries two identical variants of a gene is called *homozygous*. One who carries two different variants is called *heterozygous*. For example, if the two possible variants are labeled A and a, then AA and aa individuals are homozygous, whereas Aa and aA individuals are heterozygous.

Suppose two Aa individuals have a child, each giving one of their gene variants to the offspring. Draw a Venn diagram indicating each of the following events.

(a) The offspring is homozygous.
(b) The offspring is heterozygous.
(c) The offspring carries at least one A variant.

SOLUTION There are four possible outcomes in terms of the offspring's genotype: AA, Aa, aA, and aa. Depicting each of these simple events by a rectangular region and the compound events for each question by shading, we obtain the diagram in Figure 2.

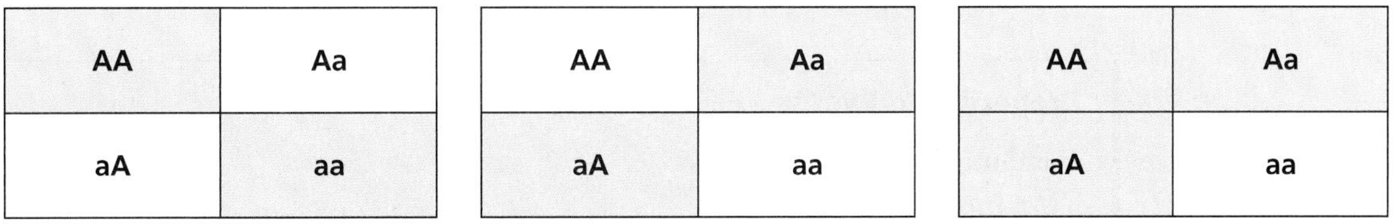

(a) (b) (c)

FIGURE 2

In biology, figures like these are referred to as *Punnett squares*, after British geneticist Reginald C. Punnett. ■

The following definitions from set theory are useful in the study of probability. They are illustrated in Figure 3.

(1) Definitions

(a) The **complement** of an event E, denoted by E^c, is the event consisting of all outcomes that are not in E.

(b) The **union** of two events E and F, denoted by $E \cup F$, is the event consisting of all outcomes that are either in E *or* F (or in both).

(c) The **intersection** of two events E and F, denoted by $E \cap F$, is the event consisting of all outcomes that are in both E *and* F.

(d) The events E and F are **mutually exclusive**, or **disjoint**, if their intersection is empty; that is $E \cap F = \varnothing$.

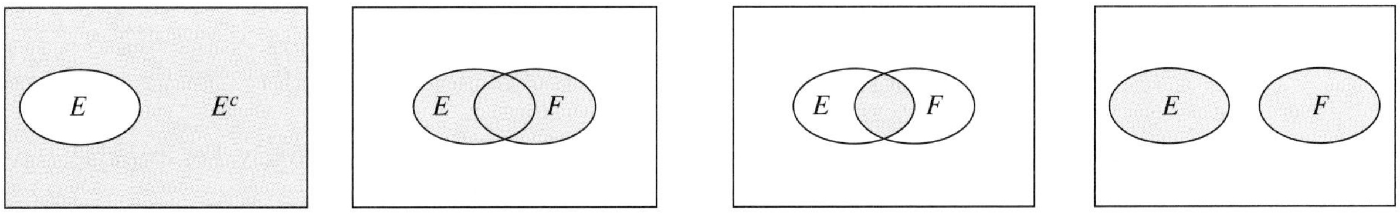

(a) Complement (b) Union (c) Intersection (d) Mutually exclusive events

FIGURE 3

EXAMPLE 4 | Example 1(b) demonstrated that the sample space for rolling a six-sided die is $\Omega = \{1, 2, 3, 4, 5, 6\}$. Two possible events are $E = \{1, 2, 3, 4\}$ and $F = \{2, 3, 6\}$.
(a) What is the event $E \cup F$?
(b) What is the event $E \cap F$?
(c) Are the events E and F^c mutually exclusive?

SOLUTION

(a) From Definition 1(b), $E \cup F$ (the union of E and F) is the event consisting of all outcomes in either E or F. Therefore $E \cup F = \{1, 2, 3, 4, 6\}$.

(b) From Definition 1(c), $E \cap F$ (the intersection of E and F) is the event consisting of all outcomes in both E and F. Therefore $E \cap F = \{2, 3\}$.

(c) From Definition 1(a), F^c (the complement of F) is the event consisting of all outcomes in the sample space Ω that are not in F, giving $F^c = \{1, 4, 5\}$. Therefore $E \cap F^c = \{1, 4\}$ and, from Definition 1(d), E and F^c are not mutually exclusive. ■

The operations of forming unions and intersections of events obey the commutative, associative, and distributive laws of ordinary algebra (see Exercises 16–18).

Properties of Events For any events E, F, and G, we have

Commutative Law:

$E \cup F = F \cup E$ $\qquad$ $E \cap F = F \cap E$

Associative Law:

$(E \cup F) \cup G = E \cup (F \cup G)$ $\qquad$ $(E \cap F) \cap G = E \cap (F \cap G)$

Distributive Law:

$(E \cup F) \cap G = (E \cap G) \cup (F \cap G)$ $\qquad$ $(E \cap F) \cup G = (E \cup G) \cap (F \cup G)$

Definitions 1(b), 1(c), and 1(d) can also be extended to multiple events. For instance, the union of a collection of events E_i, denoted $\bigcup_i E_i$, is the event consisting of the outcomes that are in any of the events E_i. Their intersection, denoted $\bigcap_i E_i$, is the event consisting of the outcomes that are in every event E_i. Finally, the events E_i are said to be **pairwise disjoint** if no two events have any outcomes in common, that is, if the intersection $E_j \cap E_k = \varnothing$ for all j and k with $j \neq k$.

■ Probability When Outcomes Are Equally Likely

Informally, the probability of event E is the fraction of outcomes in which event E occurs if the experiment was repeated a very large number of times.

Given the sample space Ω of an experiment and a set of events, we now define the term *probability*. For each event E we wish to define a number $P(E)$, referred to as the probability of event E, that represents the chance that event E will occur during a trial of the experiment. Intuitively, we think of this as the fraction of outcomes in which event E occurs if we repeated the experiment a very large number of times. Notice that $P(E)$ is a function whose domain is the set of events of interest. It is therefore sometimes referred to as a *probability function*.

We first consider experiments whose outcomes are equally likely. For example, suppose we roll a die and the event E is rolling a 2. What is $P(E)$?

We can't say what the outcome will be for any particular roll, but if we repeated this experiment many times, we would expect that roughly $\frac{1}{6}$ of the trials would result in rolling a 2. Indeed, because the six possible outcomes are equally likely, it is natural to

Frequentist versus Subjectivist Perspectives

Our focus is on analyzing probability in the context of experiments, and this is called the *frequentist* perspective of probability. However, the language of probability is often used in reference to situations that are inherently not repeatable. For example, what does it mean to say that the probability of rain today is 50%? Within the life sciences such situations commonly arise. For example, *phylogenetics* is the study of the evolutionary relationships among organisms, and probabilistic statements about these relationships can be made. But there is no obvious sense in which the history of life on earth could be repeated under identical conditions (which is required if we are to take the frequentist perspective of probabilistic statements). It is possible, however, to conduct a rigorous analysis of probability theory in such cases, where $P(E)$ is interpreted as the degree of belief that E is true. This is referred to as the *subjectivist*, or *Bayesian*, perspective.

specify the probability of any of these six outcomes to be $\frac{1}{6}$. Therefore the probability of rolling a 2 on any trial is $P(E) = \frac{1}{6}$.

In a similar way we can specify the probability of any event when all outcomes of an experiment are equally likely. For example, consider the probability of rolling an even number with a single die. There are three possible outcomes that correspond to an even roll, namely $\{2\}$, $\{4\}$, and $\{6\}$. Therefore, since three of the six possible outcomes are even, the probability of rolling an even number is $\frac{3}{6} = \frac{1}{2}$.

(2) Definition: Probabilty When Outcomes Are Equally Likely Suppose an experiment with sample space Ω has $n(\Omega)$ equally likely outcomes. If E is an event consisting of $n(E)$ outcomes, then

$$P(E) = \frac{n(E)}{n(\Omega)}$$

We interpret $P(E)$ pictorially in a Venn diagram as the fraction of the sample space that consists of the event E. By convention, the total area of the sample space is often set equal to 1. In this case $P(E)$ can be interpreted more simply as the area corresponding to event E in the Venn diagram.

We can see from Definition 2 why the counting principles of Section 12.1 are useful. To determine the probability of different events when the outcomes of an experiment are equally likely, we need to count the total number of possible outcomes as well as the number of outcomes corresponding to the event of interest.

Notice that, because any event E is a subset of the sample space Ω, we therefore have $0 \leqslant n(E) \leqslant n(\Omega)$. Thus the probability function $P(E)$ satisfies the following two conditions:

(3) $$0 \leqslant P(E) \leqslant 1$$

(4) $$P(\Omega) = 1$$

EXAMPLE 5 | Suppose you draw a single card from a standard deck of 52 playing cards. What is probability of drawing a heart?

SOLUTION The experiment consists of drawing of a single card. There are $n(\Omega) = 52$ possible outcomes and we assume that each is equally likely. The event E in question is drawing a heart, which corresponds to $n(E) = 13$ of the outcomes. Therefore, from Definition 2, the probability of drawing a heart is

$$P(E) = \frac{n(E)}{n(\Omega)} = \frac{13}{52} = \frac{1}{4}$$ ■

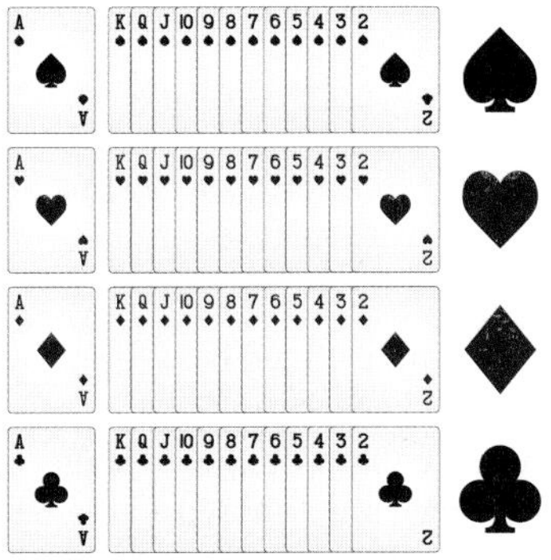

A standard deck of playing cards contains four "suits" (spades, hearts, diamonds, and clubs), with each suit containing 13 cards. Each suit has three face cards—jack, queen, and king.

EXAMPLE 6 | **Human genetics (continued)** The biochemical mechanisms of inheritance usually ensure that each of the outcomes for the offspring genotype described in Example 3 are equally likely.

(a) What is the probability that the offspring is heterozygous?

(b) What is the probability that the offspring carries at least one A variant?

SOLUTION There are $n(\Omega) = 4$ possible outcomes for the offspring genotype; namely, AA, Aa, aA, or aa.

(a) The event E is the offspring being heterozygous. The outcomes Aa and aA correspond to the offspring being heterozygous and therefore $n(E) = 2$. The probability that the offspring is heterozygous is, from Definition 2,

$$P(E) = \frac{n(E)}{n(\Omega)} = \frac{2}{4} = \frac{1}{2}$$

(b) The event E is the offspring carrying at least one A variant. The three outcomes AA, Aa, and aA correspond to the offspring carrying at least one A variant and therefore $n(E) = 3$. From Definition 2 we have

$$P(E) = \frac{n(E)}{n(\Omega)} = \frac{3}{4}$$ ■

EXAMPLE 7 | Suppose a five-card poker hand is drawn from a standard deck of playing cards. What is the probability that all five cards are spades? (Such a hand is called a *flush*.)

SOLUTION The sample space of the experiment is the set of all possible five-card hands. Therefore the number of possible outcomes is the number of combinations of 52 objects when taken 5 at a time. From Formula 12.1.7, this is

$$n(\Omega) = \binom{52}{5} = \frac{52!}{5!(52-5)!} = 2{,}598{,}960$$

The event E consists of choosing 5 spades. Since there are 13 spades in the deck, the number of outcomes with 5 spades is the number of combinations of these 13 cards when taken 5 at a time:

$$n(E) = \binom{13}{5} = \frac{13!}{5!(13-5)!} = 1287$$

Therefore, the probability of drawing five spades is

$$P(E) = \frac{n(E)}{n(\Omega)} = \frac{1287}{2{,}598{,}960} \approx 0.0005$$

This result is approximately equal to $1/2000$ and therefore, informally, we say that 1 in 2000 poker hands will result in five spades. ■

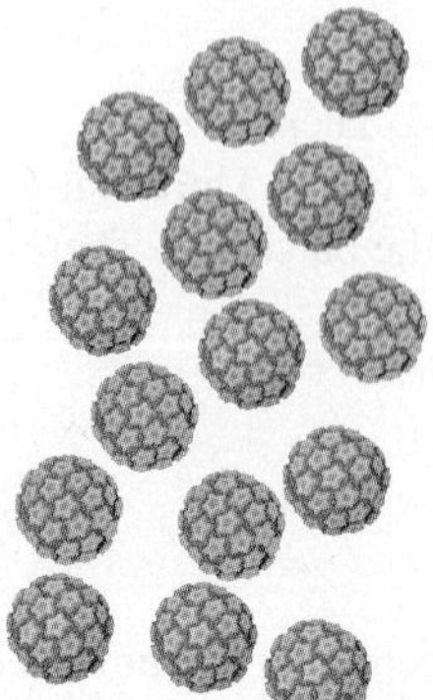

FIGURE 4
Red particles are drug resistant; purple particles are drug sensitive.

EXAMPLE 8 | **Viral transmission** Some viruses are transmitted between people via aerosol droplets. Upon inhalation, only a few of the viral particles in the droplet typically reach a person's lungs and cause infection. Suppose that a droplet contains 15 viral particles, 6 of which are drug resistant and 9 of which are drug sensitive (see Figure 4). If only 5 particles in total reach the lungs, and if each of the 15 particles is equally likely to be selected, what is the probability that an infection with only drug-resistant viruses occurs?

SOLUTION The experiment consists of the selection of 5 viral particles from the 15. The number of possible outcomes is therefore the number of combinations of

15 particles when taken 5 at a time. From Formula 12.1.7, this is

$$n(\Omega) = \binom{15}{5} = \frac{15!}{5!(15-5)!} = 3003$$

The event E consists of selecting 5 resistant particles and 0 sensitive particles. The number of ways to choose 5 resistant particles is the number of combinations of the 6 resistant particles when taken 5 at a time; that is, $\binom{6}{5} = 6$. Similarly, the number of ways to choose 0 sensitive particles is the number of combinations of the 9 sensitive particles when taken 0 at a time; that is, $\binom{9}{0} = 1$. From the Fundamental Principle of Counting in Section 12.1, the number of ways to choose 5 resistant and 0 sensitive particles is therefore

$$n(E) = 6 \cdot 1 = 6$$

Thus the probability of a completely drug-resistant infection is

$$P(E) = \frac{n(E)}{n(\Omega)} = \frac{6}{3003} \approx 0.002$$

This result is approximately equal to $1/500$ and therefore, informally, we say that 1 in 500 transmission events will result in a completely drug-resistant infection. ■

Suppose E and F are two mutually exclusive events. The number of outcomes in their union is then the sum of the number of outcomes in each event; that is, $n(E \cup F) = n(E) + n(F)$. The probability of the event $E \cup F$ is therefore

$$P(E \cup F) = \frac{n(E \cup F)}{n(\Omega)} = \frac{n(E) + N(F)}{n(\Omega)} = \frac{n(E)}{n(\Omega)} + \frac{n(F)}{n(\Omega)} = P(E) + P(F)$$

This is summarized in the following rule.

(5) Suppose all outcomes of an experiment are equally likely, and E and F are two mutually exclusive events. Then

$$P(E \cup F) = P(E) + P(F)$$

EXAMPLE 9 | Viral transmission (continued) Consider Example 8. What is the probability of an infection with a single type of virus, either resistant or sensitive?

SOLUTION We need to calculate the probability of selecting *either* (i) 5 resistant particles and 0 sensitive particles, *or* (ii) 0 resistant particles and 5 sensitive particles. Let's use E to represent event (i) and F to represent event (ii). From Definitions 1, we want to calculate the probability of the union of E and F. Furthermore, E and F are mutually exclusive events, so we can use Equation 5. We have already calculated $P(E) = 6/3003$ in Example 8.

For event F, there is one way to choose 0 resistant particles, and the number of ways to choose 5 sensitive particles is the number of combinations of the 9 sensitive particles when taken 5 at a time; that is, $\binom{9}{5} = 126$. Therefore $n(F) = 1 \cdot 126 = 126$, and we have

$$P(F) = \frac{n(F)}{n(\Omega)} = \frac{126}{3003}$$

Finally, using Equation 5, the probability of an infection with a single type of virus is

$$P(E \cup F) = P(E) + P(F) = \frac{6}{3003} + \frac{126}{3003} = \frac{132}{3003} \approx 0.04$$

This result is approximately equal to $1/25$ and therefore, informally, we say that 1 in 25 transmission events will result in an infection with a single virus type. ■

■ Probability in General

Until now we have focused on experiments having equally likely outcomes. But this is not true for many experiments of interest. For example, there are two outcomes for an HIV antibody test and they are not, in general, equally likely.

When the outcomes of an experiment are not equally likely, Definition 2 no longer applies. However, in order to ensure that the probability function $P(E)$ conforms as closely as possible to Definition 2, it is still required that $P(E)$ satisfy Conditions 3, 4, and 5. These are referred to as the *Probability Axioms*.

Kolmogorov

Andrey Kolmogorov (1903–1987) was an influential Soviet mathematician who is credited with being the first to formulate probability theory through a set of axioms. Kolmogorov also contributed to the development of mathematical biology, notably in the development of general models of ecological interactions and in the analysis of differential equations for traveling waves in the spatial spread of genes.

(6) Probability Axioms Suppose Ω is the sample space of an experiment with a finite number of outcomes (not necessarily equally likely). A function $P(E)$ is called a **probability function** if it satisfies the following conditions:

(a) $0 \leqslant P(E) \leqslant 1$ for any event E

(b) $P(\Omega) = 1$

(c) $P(E \cup F) = P(E) + P(F)$ for any two mutually exclusive events E and F

Notice that Definition 2, for the case of equally likely outcomes, is a specific example of a probability function because we have seen that it satisfies Axioms 6(a)–(c). In general, we interpret $P(E)$ in (6) pictorially in a Venn diagram as the fraction of the sample space that consists of the event E. And as in the case with equally likely outcomes, $P(E)$ can be thought of as the fraction of outcomes in which event E occurs if we were to repeat the experiment a very large number of times.

EXAMPLE 10 | BB **Annual human mortality** The following table gives estimates for the worldwide annual probability of death from different causes in the human population.[1]

Cause of death	Probability
Communicable diseases (C)	0.34
Noncommunicable diseases (N)	0.56
Other (O)	0.1

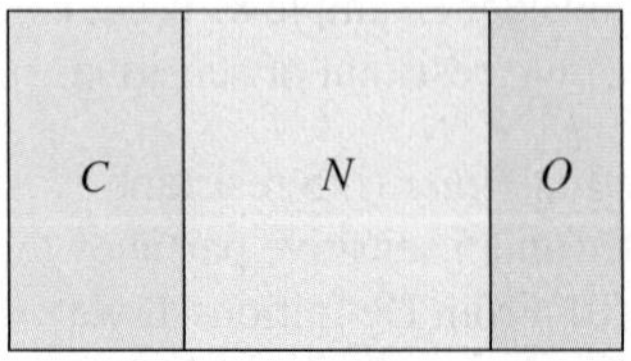

FIGURE 5
A Venn diagram for annual mortality

The experiment in question is the death of an individual and the three causes of death are the possible outcomes, as shown in the Venn diagram in Figure 5. We assume that a person cannot die of more than one cause. Interpret the Probability Axioms in the context of this example.

1. C. Murray and A. Lopez, "Mortality by Cause for Eight Regions of the World: Global Burden of Disease Study," *The Lancet* 349 (1997): 1269–76.

SOLUTION The outcomes of the experiment are not equally likely and therefore we no longer have a simple equation like Definition 2 for calculating the probability of any event. Instead we are given three values—one for each of the three outcomes—and told that they are the probabilities of these outcomes. Mathematically, this means that there is a probability function $P(E)$ for this experiment that satisfies Axioms 6, and for which $P(C) = 0.34$, $P(N) = 0.56$, and $P(O) = 0.1$.

Axiom 6(a) states the reasonable requirement that the probability of any event E is no smaller than 0 and no larger than 1. This is clearly true for the three simple events listed in the table, but this axiom requires that it be true for all compound events as well.

Axiom 6(b) states that the probability of the event consisting of the entire sample space is 1. In the context of this example, this means that the probability of dying from one of these three causes is 1.

Axiom 6(c) states, for example, that

$$P(C \cup N) = P(C) + P(N) = 0.34 + 0.56 = 0.9$$

because the events C and N are mutually exclusive. This is simply the requirement that the probability of death from a transmissible *or* a nontransmissible disease is the sum of the probabilities of each. ■

The calculation of probabilities is often greatly simplified by the following result.

(7) The Complement Rule For any event E,

$$P(E^c) = 1 - P(E)$$

PROOF Note that $E \cup E^c = \Omega$ from Definition 1(a) and therefore $P(E \cup E^c) = 1$ from Axiom 6(b). Furthermore, events E and E^c are mutually exclusive. Therefore, using Axiom 6(c), we have $1 = P(\Omega) = P(E \cup E^c) = P(E) + P(E^c)$, which can be solved for $P(E^c)$ to obtain the desired result. ■

As a consequence of Axioms 6 and Complement Rule 7, we can also derive the intuitively reasonable result, $P(\varnothing) = 0$. In other words, the probability of no outcome (sometimes called the *null event*) is 0. (See Exercise 67.)

EXAMPLE 11 | **Viral transmission (continued)** Consider Example 8. What is the probability of an infection with at least one drug-resistant viral particle?

SOLUTION Your first instinct might be to apply Definition 2 directly by counting the number of ways that an infection with at least one resistant viral particle can occur. This is certainly possible, but it is also very tedious. An easier approach is to notice that if G is the event "infection with at least one resistant particle," then G^c is the event "infection with 0 resistant particles (and 5 sensitive particles)." The probability $P(G^c)$ is much easier to calculate.

In fact, the event G^c is precisely the event F in Example 9; therefore we have $P(G^c) = P(F) = 126/3003$. Using Rule 7, we then obtain the probability of infection with at least one resistant virus:

$$P(G) = 1 - P(G^c) = 1 - \frac{126}{3003} \approx 0.96$$ ■

Example 11 illustrates a general rule of thumb: if you are asked for the probability of "at least one" outcome occurring, the complement rule often simplifies the calculation.

Probability Axiom 6(c) allows us to compute probabilities for the union of mutually exclusive events. The following rule gives an equation that applies for any two events.

(8) The Union Rule For any two events E and F,

$$P(E \cup F) = P(E) + P(F) - P(E \cap F)$$

PROOF $E \cup F$ can be viewed as the union of the mutually exclusive events E and $E^c \cap F$ (as shown in Figure 6). Furthermore, $(E^c \cap F)$ and $(E \cap F)$ are mutually exclusive and their union is

$$(E^c \cap F) \cup (E \cap F) = (E^c \cup E) \cap F = F$$ (from the distributive law on page 740)

Therefore, from Axiom 6(c),

$$P(F) = P((E^c \cap F) \cup (E \cap F)) = P(E^c \cap F) + P(E \cap F)$$

This can be rearranged to give $P(E^c \cap F) = P(F) - P(E \cap F)$. Thus we have

$$P(E \cup F) = P(E) + P(E^c \cap F) = P(E) + P(F) - P(E \cap F)$$

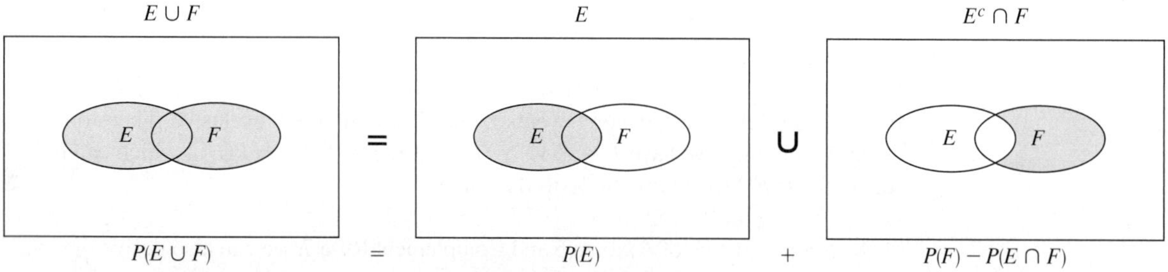

FIGURE 6 The probability corresponding to the shaded area in the third panel follows from the fact that $P(E^c \cap F) = P(F) - P(E \cap F)$. ■

EXAMPLE 12 | A card is drawn at random from a standard deck. What is the probability that the card is either a face card or a spade? (See Figure 7.)

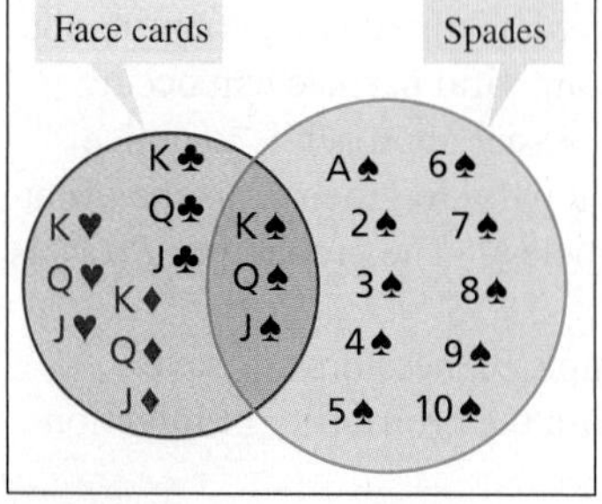

FIGURE 7

SOLUTION We use E and F to represent the events "face card" and "spade," respectively. There are 12 face cards and 13 spades in a standard deck. Therefore,

$$P(E) = \frac{12}{52} \qquad P(F) = \frac{13}{52}$$

Furthermore, since 3 cards are both face card and spade,

$$P(E \cap F) = \frac{3}{52}$$

Therefore, using Rule 8, we obtain

$$P(E \cup F) = P(E) + P(F) - P(E \cap F) = \frac{12}{52} + \frac{13}{52} - \frac{3}{52} = \frac{11}{26}$$ ■

EXERCISES 12.2

1–4 What is the sample space in each of the following experiments?

1. You take a test consisting of 20 questions, worth 1 mark each, and the outcome is your grade on the test.

2. You draw two cards from a standard deck.

3. You are assigned a seat on a plane that has 30 rows of seats, with each row having seats A, B, C, and D.

4. You roll a die and toss a coin.

---

5. An experiment consists of rolling a die. List the elements in the following sets.
(a) The sample space
(b) The event "getting an even number"
(c) The event "getting a number greater than 4"

6. An experiment consists of tossing a coin and drawing a card from a deck.
(a) How many elements does the sample space have?
(b) List the elements in the event "getting heads and an ace."
(c) List the elements in the event "getting tails and a face card."
(d) List the elements in the event "getting heads and a spade."

7. Each of three students either passes (P) or fails (F) a course.
(a) How many elements are in the sample space?
(b) List the elements of the sample space.
(c) List the outcomes in the event "exactly two of the students pass."
(d) List the outcomes in the event "at most two students fail."
(e) List the outcomes in the event "all students have the same result."

8. Genotypes In humans all individuals inherit two copies of most genes, one from their mother and one from their father. Suppose that, for a gene of interest, each copy can be in one of three states, A_1, A_2, or A_3. The identity of the pair of genes that an individual carries, including the parent from which each was inherited, is referred to as their *genotype*. Suppose the experiment of interest is the selection of a random individual from the population, and the outcome of interest is this individual's genotype for this specific gene.
(a) How many elements are in the sample space?
(b) List the elements of the sample space.
(c) List the elements in the event "the individual's two genes have the same state."
(d) List the elements in the event "the individual's two genes have different states."

9. Household infections Four students share a house and each may become infected with the flu during the winter, or not.
(a) How many elements are in the sample space?
(b) List the elements of the sample space.
(c) List the elements in the event "at least three students contract the flu."
(d) List the elements in the event "no more than one student contracts the flu."
(e) List the elements in the event "all four students contract the flu."

10–15 Evaluate each set, given that

$$A = \{2, 4, 5, 6, 8\} \qquad B = \{1, 2, 5\}$$
$$C = \{1, 3, 5, 7, 9\} \qquad \Omega = \{1, 2, 3, 4, 5, 6, 7, 8, 9\}$$

10. $A \cap B$ **11.** $A \cup C$

12. A^c **13.** $(A \cap B)^c$

14. $B^c \cap C$ **15.** $(A^c \cap B)^c$

---

16. Illustrate the commutative law for events by constructing Venn diagrams for both sides of the following equations.
(a) $E \cup F = F \cup E$
(b) $E \cap F = F \cap E$

17. Illustrate the associative law for events by constructing Venn diagrams for both sides of the following equations.
(a) $(E \cup F) \cup G = E \cup (F \cup G)$
(b) $(E \cap F) \cap G = E \cap (F \cap G)$

18. Illustrate the distributive law for events by constructing Venn diagrams for both sides of the following equations.
(a) $(E \cup F) \cap G = (E \cap G) \cup (F \cap G)$
(b) $(E \cap F) \cup G = (E \cup G) \cap (F \cup G)$

19–24 Evaluate each set, given that $A = \{1, 7, 9\}$, $B = \{2, 5, 7\}$, $C = \{1, 3, 5, 8, 9\}$, and $\Omega = \{1, 2, 3, 4, 5, 6, 7, 8, 9\}$.

19. $(A \cup B) \cup C$ **20.** $(C \cup B^c) \cup A$

21. $C \cap (A \cap B)$ **22.** $A \cap (A \cup B)$

23. $(A \cap B) \cup C$ **24.** $(A \cap B^c) \cup C$

25. **Blood types** are classified on the basis of the presence of three antigens, called the A, B, and Rh antigens. For example, type "AB Rh^-" indicates the presence of antigens A and B and the absence of antigen Rh. Type "O Rh^+" indicates the absence of antigens A and B and the presence of the Rh antigen. Draw a Venn diagram with regions representing the presence of the A, B, and Rh antigens, and make them overlap in such a way as to produce subregions for the eight different possible blood types.

26. In **Mendel's genetic experiments** many characteristics of the plants were quantified, such as their height (tall or short), flower position (axial or terminal), and seed color (colored or white). Suppose 100 plants are measured, and 35 are tall with axial flowers and colored seeds. Further, suppose a total of 25 plants have white seeds, 4 of them being short and 6 having terminal flowers. There are no short plants with terminal flowers, and the total number of short plants is 25. Construct a Venn diagram with circles illustrating the three characteristics short, terminal, and white. Indicate the number of plants in each region of the diagram.

27. **Community ecology** A survery of mammals (foxes and wolves) and birds (chickadees and blue jays) is conducted. Let the event E_1 be "foxes present," E_2 be "wolves present," E_3 be "chickadees present," and E_4 be "blue jays present. Interpret the following events

(a) $E_1 \cap E_2$ (b) $E_1^c \cup E_2$
(c) $(E_2 \cup E_3) \cap E_4$ (d) E_3^c

28–29 A die is rolled. Find the probability of the given event.

28. (a) The number showing is a six.
(b) The number showing is an even number.
(c) The number showing is greater than five.

29. (a) The number showing is a two or a three.
(b) The number showing is an odd number.
(c) The number showing is a number evenly divisible by three.

30–31 A card is drawn randomly from a standard deck. Find the probability of the given event.

30. (a) The card drawn is a king.
(b) The card drawn is a face card.
(c) The card drawn is not a face card.

31. (a) The card drawn is a heart.
(b) The card drawn is either a heart or a spade.
(c) The card drawn is a heart, a diamond, or a spade.

32–33 A coin is tossed twice. List the elements of the sample space and then find the probability of the given event.

32. (a) Not getting any heads
(b) Getting heads exactly two times

33. (a) Getting heads at least one time
(b) Getting heads exactly one time

34–35 A ball is drawn randomly from a jar that contains five red balls, two white balls, and one yellow ball. Find the probability of the given event.

34. (a) A red ball is drawn
(b) The ball drawn is not yellow
(c) A black ball is drawn

35. (a) Neither a white nor yellow ball is drawn
(b) A red, white, or yellow ball is drawn
(c) The ball that is drawn is not white

---

36. **Genome composition** The information coded in DNA consists of a sequence of letters, each of which can be an A, G, C, or T. These letters each represent one of four different biomolecules called *nucleotides*: adenine, guanine, cytosine, and thymine. Adenine and guanine are called *purines*; cytosine and thymine are called *pyrimidines*. Suppose a nucleotide is chosen at random from the following DNA sequence

ATCGATTGAGCTCTAGCG

Find the probability of the given event.
(a) A thymine is chosen.
(b) A purine is chosen.

37. A pair of dice is rolled, and the numbers showing are observed.
(a) List the sample space of this experiment.
(b) Find the probability of getting a sum of seven.
(c) Find the probability of getting a sum of nine.
(d) Find the probability that the two dice show doubles (the same number).
(e) Find the probability that the two dice show different numbers.
(f) Find the probability of getting a sum of nine or higher.

38. Two balls are picked at random from a jar that contains three red and five white balls. Find the probability of the following events.
(a) Both balls are red.
(b) Both balls are white.

39. A poker hand, consisting of five cards, is dealt from a standard deck. Find the probability that the hand contains the cards described.
(a) Five hearts
(b) Five cards of the same suit
(c) Five face cards
(d) An ace, king, queen, jack, and a ten, all of the same suit (royal flush).

40. **Mendelian inheritance** Many genetic traits are controlled by two genes, one dominant and one recessive. In Gregor Mendel's original experiments with peas, the

genes controlling the height of the plant are denoted by T (tall) and t (short). The gene T is dominant, so a plant with the genotype TT or Tt is tall, whereas one with genotype tt is short. Mendel concluded that offspring inherit one gene from each parent and that each possible combination of the two genes is equally likely. If each parent has the genotype Tt, find the probability of the following events.
(a) The offspring is tall.
(b) The offspring is short.

41. Mendelian inheritance (cont.) Refer to Exercise 40. If one parent has genotype Tt and the other has tt, find the probability of the following events.
(a) The offspring is tall.
(b) The offspring is short.

42–44 Sex ratios A couple plan to have three children. Suppose that male and female children are equally likely.

42. What is the probability that all of the children will be girls?

43. What is the probability that they will have exactly one boy?

44. What is the probability that they will have at least one boy?

---

45. In **mark-recapture studies** a certain number of individuals in a population are captured, tagged, and then released. The population is then sampled again, and the fraction of tagged individuals in the sample (that is, the fraction of individuals in the sample that have been recaptured) is used to estimate the population size. Suppose 6 moose out of a population of 20 are captured, tagged, and released. If 6 individuals from this population are then subsequently captured, what is the probability that half of them will be tagged?

46–49 Genetic diversity Scientists often collect DNA from a sample of individuals to determine the genetic diversity that exists in a population.

46. In a population of barn cats, five individuals carry gene A and six do not. DNA is taken from three individuals. What is the probability that two of them will carry gene A?

47. A population of cheetahs contains 20 individuals, 13 of which carry gene A while the rest carry gene B. DNA is taken from 5 individuals. What is the probability that all 5 will carry the same gene?

48. A population of song sparrows contains 30 individuals, 10 of which carry gene A, 12 carry gene B, and the rest carry gene C. DNA is taken from 4 individuals. What is the probability that all 4 will carry the same gene?

49. A population of moose contains 25 individuals, 10 of which carry gene A, 11 carry gene B, and the rest carry gene C. DNA is taken from 3 individuals. What is the probability that there will be at least two different genes in the sample?

---

50. Viral transmission Six viral particles of genotype A, 4 of genotype B, and 10 of genotype C are inhaled. If only three randomly chosen particles then go on to reach the lungs and cause an infection, what is the probability that the infection will be caused by a single type of particle?

51. Prove Conditions 3 and 4 using Definition 2 and the fact that $0 \leqslant n(E) \leqslant n(\Omega)$.

52. Use Definition 2 to prove the Complement Rule 7 for the case of equally likely outcomes.

53. Use Definition 2 to prove the Union Rule 8 for the case of equally likely outcomes.

54. Two dice are rolled. Let A be the event that the first die shows a 3 and B be the event that the sum of the numbers showing on the two dice is 6.
(a) Draw a Venn diagram indicating the elements of the sample space and the two events A and B, showing that they are not mutually exclusive.
(b) Find $P(A \cup B)$.

55–57 A statistical test, known as *Fisher's exact test*, is often used to determine if there is a relationship between two categorical variables. We began exploring this test in Exercises 12.1.45–47 using the counting principles presented in Section 12.1. The test starts by supposing that there is no relationship between the variables (this is referred to as the *null hypothesis*; see also Section 13.3). We then calculate the probability of obtaining data like that in the sample. If this probability is very small (for example, less than 0.05) we conclude that the null hypothesis must not be true. Rather, there must be some relationship between the variables in the population from which the sample was taken.

55. Consider a collection of 12 balls, 5 of which are red. Now imagine the 12 balls are divided into two groups, using a rule that is unknown to you. Group A contains 4 balls and Group B contains 8 balls. Suppose further that 3 of the 5 red balls are in Group A. Exercise 12.1.45 asked you to calculate the number of different ways 12 balls can be divided into groups of 4 and 8, as well as the number of ways they can be divided so that 3 of the 5 red balls end up in Group A.
(a) Suppose the groups are created randomly, without reference to ball color. What is the probability of 3 of the 5 red balls ending up in Group A?
(b) If we consider a probability of less than 0.05 as being very small, do you think the balls were randomly assigned to the two groups? Explain.

56. Genetic markers of disease Consider a collection of 14 people, 6 of whom carry a genetic mutation. Suppose that 5 of the 14 people go on to develop a specific disease and 9 do not. Further, suppose that 4 of the individuals with the disease also carry the mutation. Exercise 12.1.46 asked you to calculate the number of different ways there are for 5 of the 14 people to develop the disease, as well as the number of different ways for this to be true while also having 4 of the individuals with the disease carry the mutation.

(a) Suppose there is no relationship between carrying the mutation and developing the disease (this means that all possible outcomes for the composition of the sample are equally likely). What is the probability of obtaining the data given?

(b) If we consider a probability of less than 0.05 as being very small, do you think there is a relationship between carrying the mutation and developing the disease? Explain.

57. Bat conservation biology Example 11.3.1 presented data on the relationship between bat species' cave use and endangerment status. There were 18 bat species, and these were classified as either fragile or vulnerable, with 14 of them being fragile and 4 vulnerable. We also saw that 11 of the 18 bat species use caves, and 9 of these are in the fragile group. Exercise 12.1.47 asked you to calculate the number of different ways there are for 14 of the 18 species to be fragile, and the number of different ways for this to be true while also having 9 of the 14 fragile species use caves.

(a) Suppose there is no relationship between cave use and endangerment status (this means that all possible outcomes for the composition of the sample are equally likely). What is the probability of obtaining the data given?

(b) If we consider a probability of less than 0.05 as being very small, do you think there is a relationship between cave use and endangerment status? Explain.

58–59 A die is rolled, and the number showing is observed. Determine whether the events E and F are mutually exclusive. Then find the probability of the event $E \cup F$.

58. (a) E: The number is even.
F: The number is odd.

(b) E: The number is even.
F: The number is greater than 4.

59. (a) E: The number is greater than 3.
F: The number is less than 5.

(b) E: The number is divisible by 3.
F: The number is less than 3.

60–61 A card is drawn at random from a standard deck. Determine whether the events E and F are mutually exclusive. Then find the probability of the event $E \cup F$.

60. (a) E: The card is a face card.
F: The card is a spade.

(b) E: The card is a heart.
F: The card is a spade.

61. (a) E: The card is a club.
F: The card is a king.

(b) E: The card is an ace.
F: The card is a spade.

62. Resting heart rates Let E_1 be the event that a person has a resting heart rate no greater than 70 beats per minute, and let E_2 be the event that a person has a resting heart rate greater than 60 and no greater than 70 beats per minute. Interpret the quantity $P(E_1 \cup E_2)$.

63. HIV and tuberculosis (TB) are two infections that often occur together in parts of Africa. In some areas, around 20 percent of adults have HIV, 40 percent have TB, and 15 percent have both. Suppose a person is chosen at random from this population. What is the probability that the individual has HIV or TB?

64. Alcohol and marijuana consumption In a survey, 65 percent of respondents said they consume alcohol, 20 percent said they consume marijuana, and 5 percent said they consume both. What is the probability that a randomly chosen person from this population will be a consumer of at least one of these substances?

65. Cancer testing All patients suspected of having a particular type of cancer are tested by two separate labs. The probability that Lab A produces a positive test for any patient with cancer is 0.1, whereas that for Lab B is 0.05. If the probability that both labs simultaneously produce a positive test for any patient with cancer is 0.03, what is the probability that at least one lab will produce a positive test?

66. Color blindness is a genetic disorder that results in a lack of color vision. Suppose we randomly choose an individual from the population. Let C be the event "individual is color-blind" and M be the event "individual is male."

(a) Interpret the quantity $P(M \cup C)$.

(b) Interpret the quantity $P(M \cap C)$.

(c) It is estimated that

$$P(M \cup C) = 0.55$$

and $$P(M \cap C) = 0.05$$

If males make up 50 percent of the population, what is the probability that the randomly chosen individual is color-blind?

67. Use Axioms 6 along with Complement Rule 7 to prove that $P(\varnothing) = 0$.

68. In Section 12.4 we will need to generalize Axiom 6(c) to the case of an infinite collection of pairwise disjoint events E_i. In this case it becomes

$$P\left(\bigcup_{i=1}^{\infty} E_i\right) = \sum_{i=1}^{\infty} P(E_i)$$

Use this generalization of Axioms 6 to show that if A_i are k pairwise disjoint events, then

$$P\left(\bigcup_{i=1}^{k} A_i\right) = \sum_{i=1}^{k} P(A_i)$$

[*Hint:* Define the events $A_i = \varnothing$ for $i \in \{k + 1, k + 2, \ldots\}$ and use the fact that $P(\varnothing) = 0$.]

69. Use Axioms 6 and the result from Exercise 68 to show that, for an experiment with sample space Ω that has $n(\Omega)$ equally likely outcomes and A being an event consisting of $n(A)$ outcomes,

$$P(A) = \frac{n(A)}{n(\Omega)}$$

This shows that Definition 2 is, in fact, a consequence of the general probability Axioms 6.

12.3 Conditional Probability

Conditional probability is one of the most important ideas in probability theory. It allows us to update the probability of different events when we obtain new information, and it often provides the simplest approach to calculate many quantities of interest.

■ Conditional Probability

Example 12.2.10 presented the following table for the worldwide probability of death from different causes:

Cause of death	Probability
Communicable diseases (C)	0.34
Noncommunicable diseases (N)	0.56
Other (O)	0.1

While these values are informative, we might expect considerable variation in them, depending on where in the world the death occurs. For example, if we restrict attention to deaths that occur in economically developed countries, we might expect the probability of dying from communicable diseases to be lower than 0.34.

This is indeed the case. For example, the probability of death from communicable diseases in economically developed countries is only 0.09, whereas it is 0.41 in economically underdeveloped nations.[1]

When we restrict the sample space of an experiment by imposing some condition (for example, the condition that the death occurs in an economically developed country), the resulting probability is called a *conditional probability.*

Let's begin our study of conditional probability with a simpler example. Suppose a fair die is rolled. We know from Section 12.2 that the probability of rolling a 4, for example, is $\frac{1}{6}$. Figure 1(a) displays the Venn diagram for this outcome.

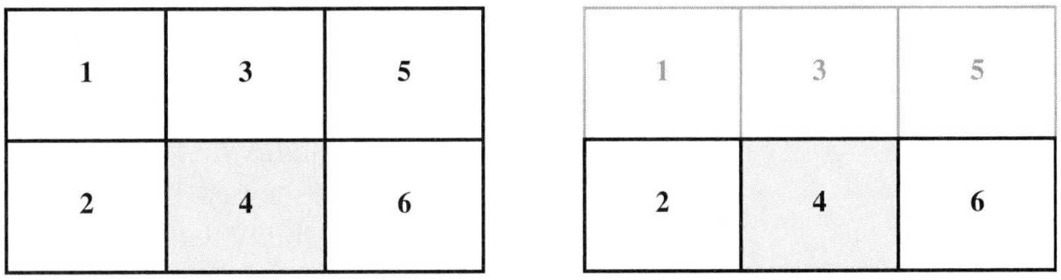

FIGURE 1 (a) Venn diagram for the probability of rolling a 4 (b) Venn diagram for the probability of rolling a 4, given the roll is even

Now suppose we restrict attention to only those outcomes that are even. What then is the probability of rolling a 4? By imposing this condition, the sample space is restricted

1. C. Murray and A. Lopez, "Mortality by Cause for Eight Regions of the World: Global Burden of Disease Study," *The Lancet* 349 (1997): 1269–76.

More formally, we can view the condition "the roll is even" as giving rise to the following revised experiment: (i) a fair die is rolled and (ii) if the roll is even, then the number showing is taken to be the outcome; otherwise, repeat (i). The possible outcomes of this revised experiment are clearly 2, 4, and 6 because, if an odd number is rolled, we discard the roll and start again (in effect acting as though nothing has happened). The sample space is therefore reduced to $\Omega = \{2, 4, 6\}$.

to the outcomes 2, 4, and 6, as shown in the Venn diagram of Figure 1(b). Further, since only one of these three outcomes in the restricted sample space corresponds to rolling a 4, the probability of rolling a 4, *given the roll is even*, is $\frac{1}{3}$.

Let's introduce some notation to describe the situation. Using E to denote the event "rolling a 4" and F to denote the event "the roll is even," we write $P(E|F) = \frac{1}{3}$. This is read as "the probability of E given F" and it represents the probability of event E occurring if we restrict the sample space to only those outcomes in which event F occurs.

This example suggests a general way to calculate $P(E|F)$. Consider two events, E and F, as shown in the Venn diagram of Figure 2(a). If we restrict the sample space to only those outcomes in which event F occurs, we obtain Figure 2(b). The probability of E given F is then represented pictorially as the fraction of the area F in which E also occurs. This fraction is given by the area of $E \cap F$ divided by the area of F in Figure 2(c).

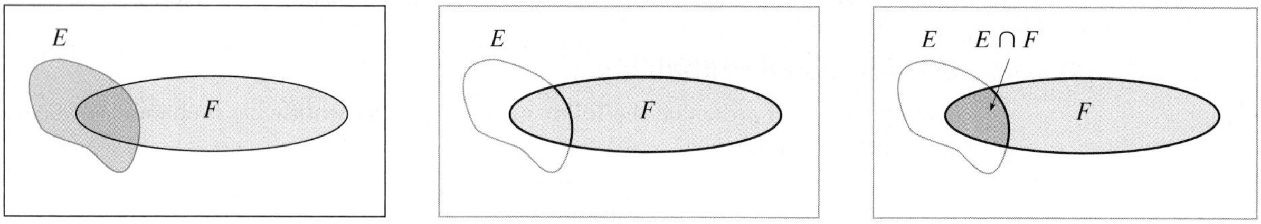

(a) Events E and F (b) Sample space restricted to event F. (c) The intersection of events E and F

FIGURE 2

If we use the fact that for any event A, we can interpret $P(A)$ as the area corresponding to this event in the Venn diagram, then the preceding considerations motivate the following definition.

> **(1) Definition** Consider any two events E and F with $P(F) > 0$. The **conditional probability of E given F** is
>
> $$P(E|F) = \frac{P(E \cap F)}{P(F)}$$

EXAMPLE 1 | Use Definition 1 to show that $P(E|F) = \frac{1}{3}$ in the die experiment.

SOLUTION The probability that a roll is both a 4 and even is $P(E \cap F) = \frac{1}{6}$. We also know that $P(F) = \frac{1}{2}$ because half of the possible outcomes are even. Therefore, using Definition 1, we have

$$P(E|F) = \frac{P(E \cap F)}{P(F)} = \frac{1/6}{1/2} = \frac{1}{3}$$

■

EXAMPLE 2 | Two dice are rolled.

(a) What is the probability that the sum of the numbers showing is greater than 8, given that the first die shows a 3?

(b) What is the probability that the sum of the numbers showing is greater than 8, given that the second die shows a 1?

SOLUTION

(a) We use E for the event that the sum is greater than 8 and F for the event that the first die shows a 3. From Figure 3(a) we have $P(E \cap F) = \frac{1}{36}$ and $P(F) = \frac{1}{6}$. Therefore, from Definition 1,

$$P(E|F) = \frac{P(E \cap F)}{P(F)} = \frac{1/36}{1/6} = \frac{1}{6}$$

(1, 1)	(1, 2)	(1, 3)	(1, 4)	(1, 5)	(1, 6)
(2, 1)	(2, 2)	(2, 3)	(2, 4)	(2, 5)	(2, 6)
(3, 1)	(3, 2)	(3, 3)	(3, 4)	(3, 5)	(3, 6)
(4, 1)	(4, 2)	(4, 3)	(4, 4)	(4, 5)	(4, 6)
(5, 1)	(5, 2)	(5, 3)	(5, 4)	(5, 5)	(5, 6)
(6, 1)	(6, 2)	(6, 3)	(6, 4)	(6, 5)	(6, 6)

(a) Dark portion of the diagram indicates sample space restricted to first die showing a 3. Blue shading indicates the event $E \cap F$.

(1, 1)	(1, 2)	(1, 3)	(1, 4)	(1, 5)	(1, 6)
(2, 1)	(2, 2)	(2, 3)	(2, 4)	(2, 5)	(2, 6)
(3, 1)	(3, 2)	(3, 3)	(3, 4)	(3, 5)	(3, 6)
(4, 1)	(4, 2)	(4, 3)	(4, 4)	(4, 5)	(4, 6)
(5, 1)	(5, 2)	(5, 3)	(5, 4)	(5, 5)	(5, 6)
(6, 1)	(6, 2)	(6, 3)	(6, 4)	(6, 5)	(6, 6)

(b) Dark portion indicates sample space restricted to the second die showing a 1.

FIGURE 3
The number on the first die is given by the row and that on the second die is given by the column.

(b) We use E for the event that the sum is greater than 8 and F for the event that the second die shows a 1. From Figure 3(b) we have $P(E \cap F) = 0$ (it is impossible for the sum to be greater than 8 if the second die shows a 1). We again have $P(F) = \frac{1}{6}$ and, therefore, from Definition 1,

$$P(E \mid F) = \frac{P(E \cap F)}{P(F)} = \frac{0}{1/6} = 0$$

Part (b) illustrates that it is possible to impose a condition (event F) for which the event E of interest can no longer occur. ■

EXAMPLE 3 | **Huntington's disease** is a genetic disorder causing neurodegeneration. Symptoms typically appear in a person's 30s and include the loss of muscle control and cognitive impairment. Affected people typically die around 20 years after the onset of symptoms. The disease is caused by a mutation in a single gene, and people can carry zero, one, or two copies of this mutation. These cases are called "homozygous normal," "heterozygous," and "homozygous mutant," respectively (see Example 12.2.3). Heterozygous and homozygous mutant individuals develop the disease. Suppose that two heterozygous individuals have a child. The possible outcomes for the offspring's genotype, along with the associated probabilities, are given in Figure 4(a). Figures 4(b) and 4(c) indicate the events "disease" and "homozygous," respectively. What is the probability that the child is homozygous, given that they have the disease?

AA (1/4)	Aa (1/4)
aA (1/4)	aa (1/4)

(a) The possible genotypes for the child.

AA (1/4)	Aa (1/4)
aA (1/4)	aa (1/4)

(b) Blue shading indicates the outcomes corresponding to the child having the disease.

AA (1/4)	Aa (1/4)
aA (1/4)	aa (1/4)

(c) Blue shading indicates the outcomes corresponding to the child being homozygous.

FIGURE 4
A is the mutant variant and a is the normal variant.

SOLUTION We use D for the event "disease" and H for the event "homozygous." We wish to calculate $P(H \mid D)$. From Figure 4 we obtain $P(H \cap D) = \frac{1}{4}$ because H and D both occur only when the child has genotype AA. Also, from Figure 4(b) we see that $P(D) = \frac{3}{4}$. Therefore, using Definition 1, we obtain

$$P(H \mid D) = \frac{P(H \cap D)}{P(D)} = \frac{1/4}{3/4} = \frac{1}{3}$$

Thus, if the offspring has the disease, then with probability $\frac{1}{3}$ it is homozygous, carrying two copies of the mutant gene. ■

EXAMPLE 4 | **BB** **Annual human mortality (continued)** Figure 5(a) shows the Venn diagram from Example 12.2.10. For any given death we might also consider the set of outcomes specifying whether the death is in an economically underdeveloped or developed country, denoted by events U and D, respectively. Estimates suggest that, for any given death, the probability of it occurring in an economically underdeveloped country is 0.78, whereas it is 0.22 in an economically developed country.[2] This gives the Venn diagram in Figure 5(b). The two diagrams are superimposed in Figure 5(c), along with the probabilities corresponding to each of the regions.

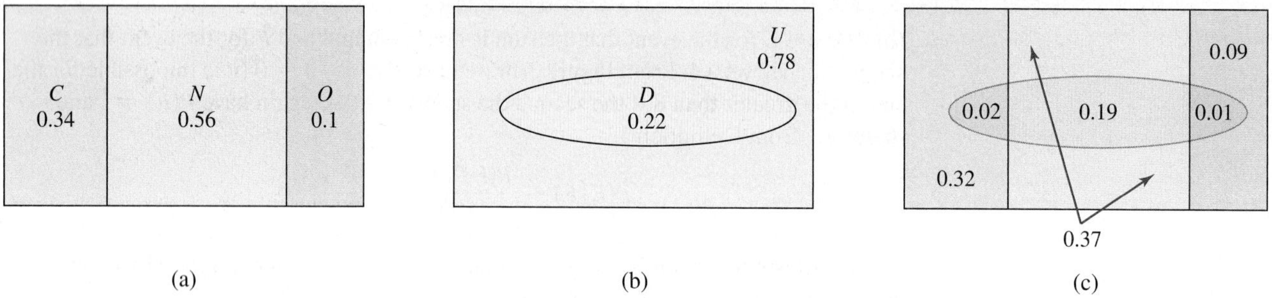

FIGURE 5

(a) Construct a Venn diagram depicting the probability of a death from communicable disease, given that the death occurs in an economically developed country.
(b) What is the probability of a death from communicable disease, given that it occurs in an economically developed country?

SOLUTION

(a) We restrict the sample space to deaths occurring in economically developed countries, which is represented by the elliptical area in Figure 6. The fraction of this area corresponding to deaths from communicable diseases then represents the probability of interest. This fraction is shaded green in Figure 6.

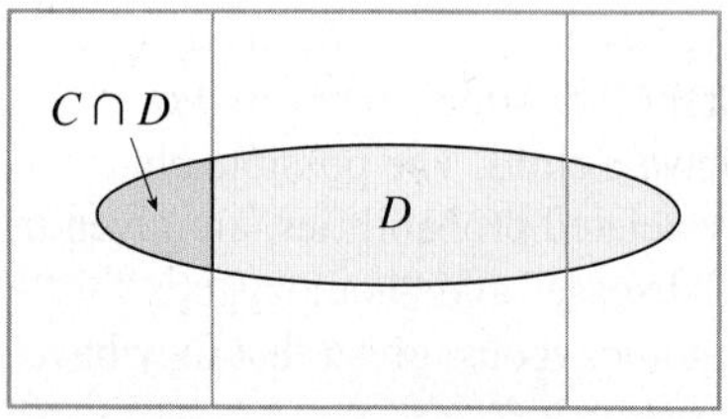

FIGURE 6

(b) The desired probability is $P(C \mid D)$. From Figure 5 we see that $P(C \cap D) = 0.02$ and $P(D) = 0.22$. Therefore, from Definition 1, we obtain

$$P(C \mid D) = \frac{P(C \cap D)}{P(D)} = \frac{0.02}{0.22} \approx 0.09$$ ■

As you might guess from visualizing the process of conditioning on an event (for example, as in Figure 2), conditional probabilities satisfy properties similar to those of ordinary probabilities given in Rules and Axioms 6–8 of Section 12.2. The following rule is one important example.

(2) The Complement Rule for Conditional Probability Consider two events E and F with $P(F) > 0$. Then

$$P(E^c \mid F) = 1 - P(E \mid F)$$

2. C. Murray et al., "Mortality by Cause for Eight Regions of the World: Global Burden of Disease Study," *The Lancet* 349 (1997): 1269–76.

PROOF From the distributive law on page 740, we can write event F as $F = (E \cap F) \cup (E^c \cap F)$. Furthermore, events $(E \cap F)$ and $(E^c \cap F)$ are mutually exclusive and therefore, from Union Rule 12.2.8, we have $P(F) = P(E \cap F) + P(E^c \cap F)$. This can be rearranged to give $P(E^c \cap F) = P(F) - P(E \cap F)$. Finally, using this in Definition 1, we obtain

$$P(E^c \mid F) = \frac{P(E^c \cap F)}{P(F)} = \frac{P(F) - P(E \cap F)}{P(F)}$$

$$= 1 - \frac{P(E \cap F)}{P(F)} = 1 - P(E \mid F)$$ ■

EXAMPLE 5 | BB **Muscular dystrophy** is a genetic disease that results in muscle degeneration and eventual death. It is caused by a mutation on the X chromosome and therefore its probability of occurrence is different in males and females. Suppose a couple with normal X chromosomes plan to have a child. Let M and F be the events that the child is a male or female, respectively, and let D be the event that the child has muscular dystrophy. Exercise 7 shows that $P(D \mid F) = \mu^2$, where μ is the probability of the muscular dystrophy mutation spontaneously occurring on any given X chromosome. What is the probability that, if the couple has a daughter, she will not have the disease?

SOLUTION We are asked to calculate the probability of no disease, given the child is a female. In terms of the given notation, this is denoted by $P(D^c \mid F)$. We are also given the quantity $P(D \mid F) = \mu^2$ and therefore, using (2), we obtain

$$P(D^c \mid F) = 1 - P(D \mid F) = 1 - \mu^2$$ ■

■ The Multiplication Rule and Independence

Definition 1 for conditional probability can be rewritten to give a useful formula for calculating the probability of events E and F both occurring (that is, the intersection of the events E and F).

(3) $$P(E \cap F) = P(E \mid F)\, P(F)$$

Equation 3 can be used for any two events E and F unless $P(F) = 0$. It is referred to as the **Multiplication Rule**. If $P(F) = 0$, then Equation 3 cannot be used because $P(E \mid F)$ is undefined. In this case we have $P(E \cap F) = 0$ because the probability of E and F both occurring must be zero.

The Multiplication Rule is simpler if the occurrence of event F tells us nothing about the probability that E will occur. In this case the two events are said to be *independent*. Mathematically, this means that

$$P(E \mid F) = P(E)$$

In words, this says that the probability of event E occurring is the same regardless of whether F occurs or not. In this case Equation 3 simplifies, suggesting the following definition.

(4) Definition Two events E and F are **independent** if

$$P(E \cap F) = P(E)\, P(F)$$

In Exercise 38 you are asked to show that you arrive at the same definition of independence if, instead, you suppose that $P(F \mid E) = P(F)$.

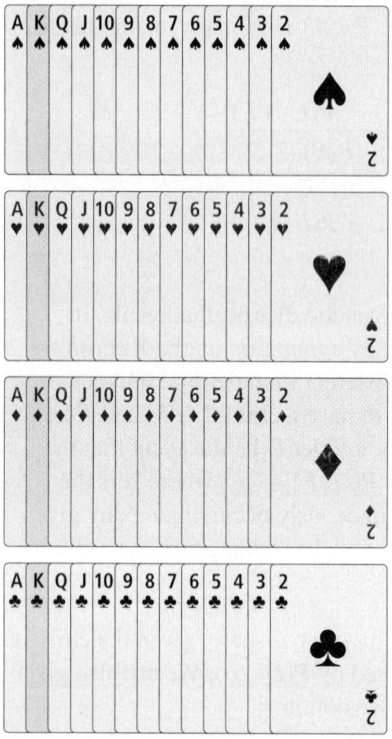

FIGURE 7

EXAMPLE 6 | Let K be the event "drawing a king" from a standard deck of cards and D be the event "drawing a diamond." Show that events K and D are independent (see Figure 7).

SOLUTION From Definition 4 we must show that $P(K \cap D) = P(K)\,P(D)$. The experiment in question is the drawing of a single card, and there are 52 possible outcomes, all of which are equally likely. From Definition 12.2.2, we therefore have

$$P(K) = \frac{n(K)}{n(\Omega)} = \frac{4}{52} = \frac{1}{13}$$

where $n(K) = 4$ is the number of outcomes corresponding to a king and $n(\Omega) = 52$ is the number of possible outcomes. Likewise,

$$P(D) = \frac{n(D)}{n(\Omega)} = \frac{13}{52} = \frac{1}{4}$$

We can also calculate $P(K \cap D)$ by noting that the event $K \cap D$ is the event "drawing the king of diamonds." Therefore

$$P(K \cap D) = \frac{n(K \cap D)}{n(\Omega)} = \frac{1}{52}$$

Because $\frac{1}{52} = \frac{1}{13} \cdot \frac{1}{4}$, it can be seen that $P(K \cap D) = P(K)\,P(D)$, meaning that K and D are independent. ■

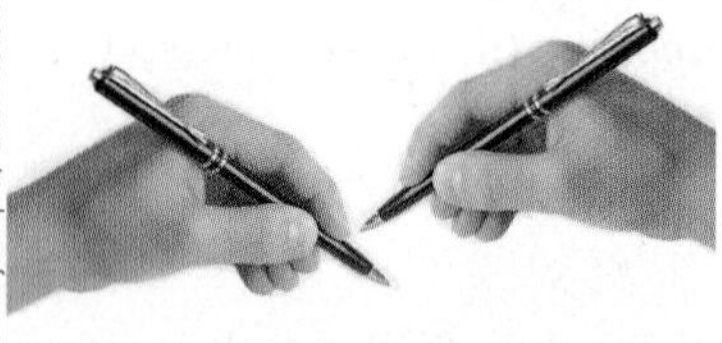

EXAMPLE 7 | Handedness and sex Approximately 10% of all people are left-handed. Approximately 50% of all people are female.

(a) If handedness and sex are independent, what is the probability that a randomly chosen person will be a left-handed female?

(b) In fact, handedness and sex are not independent, with approximately 30% of all left-handed people being female. In this case, what is the probability of a randomly chosen person being a left-handed female?

SOLUTION

(a) Using F and L to denote the events "female" and "left-handed," we are asked to calculate $P(F \cap L)$. Because the two events are independent, we have

$$P(F \cap L) = P(F)\,P(L) = 0.5 \times 0.1 = 0.05$$

(b) We are told that $P(F \mid L) = 0.3$ and asked to calculate $P(F \cap L)$. Using Equation 3, we obtain

$$P(F \cap L) = P(F \mid L)\,P(L) = 0.3 \times 0.1 = 0.03$$

which is different from part (a), revealing that F and L are not independent. In fact, most left-handed people are male. ■

Definition 4 for independence can be extended to more than two events. A set of k events is said to be **mutually independent** if the probability of the intersection of all subsets of the events is equal to the product of the probabilities of these events. For example, in the case of three events A, B, and C, this requires that all of the following equations be satisfied.

$$P(A \cap B \cap C) = P(A)\,P(B)\,P(C)$$

$$P(A \cap B) = P(A)\,P(B)$$

$$P(A \cap C) = P(A)\,P(C)$$

$$P(B \cap C) = P(B)\,P(C)$$

EXAMPLE 8 | **Sex of offspring** In humans the sexes of all children born to a mother are thought to be mutually independent. If a woman has three children and if male and female children are equally likely, what is the probability that they will all be male?

SOLUTION Using M_i to denote the event "the ith child is male," we are asked to calculate $P(M_1 \cap M_2 \cap M_3)$. The events M_i are mutually independent and therefore

$$P(M_1 \cap M_2 \cap M_3) = P(M_1)\,P(M_2)\,P(M_3) = 0.5 \times 0.5 \times 0.5 = 0.125$$ ■

■ The Law of Total Probability

Conditional probability can also be helpful when we are calculating regular, unconditional probabilities by allowing us to break the calculations into more manageable parts.

To motivate this idea, we work with an extension of Example 3. Suppose a man with Huntington's disease and a woman without the disease decide to have a child. We know the woman must be an aa individual since, if she carried any A gene, she would have the disease. However, there are two possibilities for the man. He could be an AA individual (case 1 in Figure 8) or an Aa individual (case 2 in Figure 8). Suppose that the probability of case 1 is 0.05 and the probability of case 2 is 0.95.

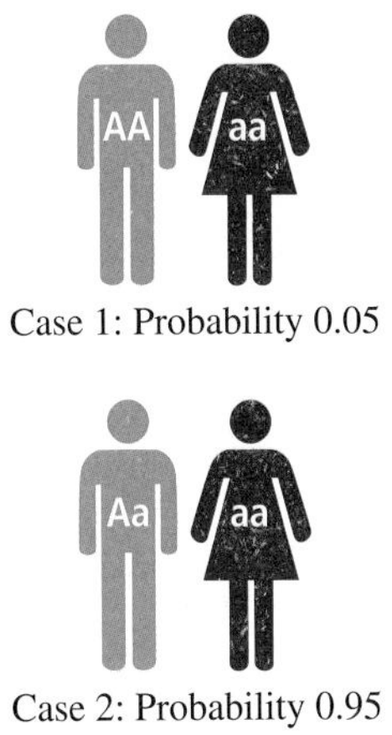

FIGURE 8

During reproduction, both parents give one of their two gene variants to the offspring with each variant being equally likely. What is the probability that their child will have Huntington's disease?

Let's first consider the cases in Figure 8 separately. If case 1 holds, then the child will always receive an A variant from the father and an a variant from the mother. The child will therefore have genotype Aa and thus will have the disease.

If case 2 holds, then the child can receive either an A or an a from the father, with each event occurring with probability $\frac{1}{2}$. And the child will again always receive an a variant from the mother. Therefore the child will be either Aa or aa, each of these events occurring with probability $\frac{1}{2}$. Only Aa children will have the disease, however, and therefore the probability of the child having Huntington's disease is $\frac{1}{2}$.

We now put these two cases together to calculate the overall probability of the child having Huntington's disease. In particular, it seems plausible that this overall probability is given by the sum of the probabilities of each case just calculated, with each being weighted by the probability of its occurrence as given in Figure 8, that is,

(5)
$$\begin{aligned} P(\text{child has Huntington's}) &= 1 \times P(\text{case } 1) + \tfrac{1}{2} \times P(\text{case } 2) \\ &= 1 \times 0.05 + \tfrac{1}{2} \times 0.95 \\ &= 0.525 \end{aligned}$$

This calculation is in fact correct and it is an example of the Law of Total Probability. It can also be visualized by using a tree diagram like those introduced in Section 12.1 (see Figure 9).

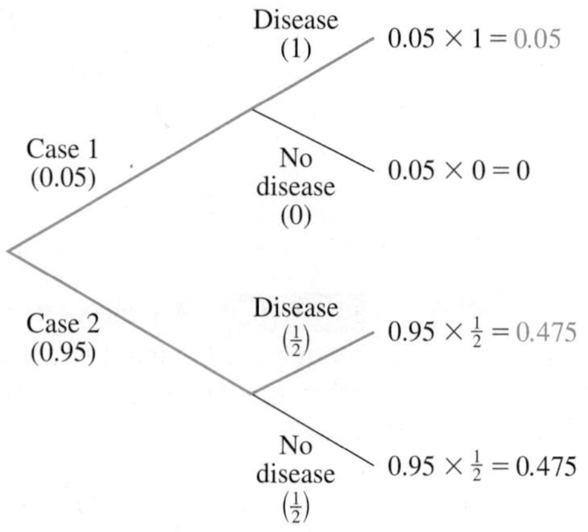

FIGURE 9
Tree diagram showing all four mutually exclusive outcomes in the Huntington's disease example. Blue lines and numbers correspond to outcomes in which the child has the disease. Therefore

$$P(\text{child has Huntington's}) = 0.05 + 0.475 = 0.525$$

Recall that a set of events $F_1, F_2, \ldots, F_k$ is called pairwise disjoint if no two events have any outcomes in common. A set of events is called **exhaustive** if their union consists of the entire sample space, that is, $F_1 \cup F_2 \cup \cdots \cup F_k = \Omega$. Therefore a set of pairwise disjoint, exhaustive events covers the entire sample space without overlapping. With this definition we then have the following result.

(6) The Law of Total Probability Suppose $F_1, F_2, \ldots, F_k$ are pairwise disjoint, exhaustive events. Then, for any other event E,

$$P(E) = P(E \mid F_1)\,P(F_1) + P(E \mid F_2)\,P(F_2) + \cdots + P(E \mid F_k)\,P(F_k)$$
$$= \sum_{i=1}^{k} P(E \mid F_i)\,P(F_i)$$

PROOF That events $F_1, F_2, \ldots, F_k$ are pairwise disjoint and exhaustive means that they partition the entire sample space into nonoverlapping pieces (see Figure 10). Event E can then be represented by the union of the pairwise disjoint events $E \cap F_i$.

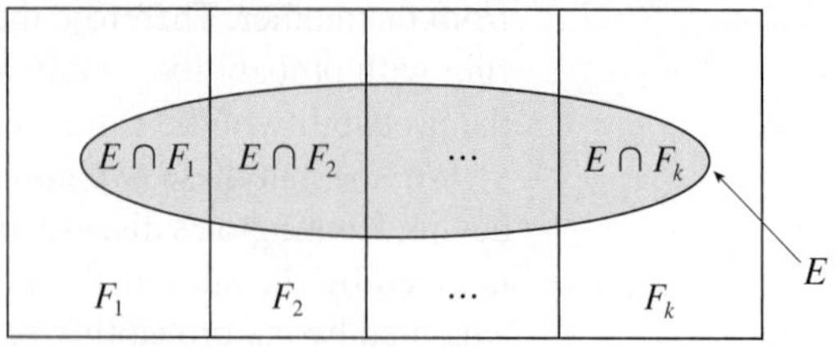

FIGURE 10
Blue ellipse represents the event E.

From Exercise 12.2.68, $P(E)$ is therefore

$$P(E) = P(E \cap F_1) + P(E \cap F_2) + \cdots + P(E \cap F_k)$$

Finally, Equation 3 shows that $P(E \cap F_i) = P(E \mid F_i)\,P(F_i)$. Therefore we obtain

$$P(E) = P(E \mid F_1)\,P(F_1) + P(E \mid F_2)\,P(F_2) + \cdots + P(E \mid F_k)\,P(F_k)$$ ∎

EXAMPLE 9 | **Huntington's disease** Interpret the calculation (5) for the probability of a child having Huntington's disease in terms of the Law of Total Probability.

SOLUTION The event E is "child has Huntington's disease" and the two cases in Figure 8 are a set of pairwise disjoint, exhaustive events F_1 and F_2. Figure 8 (on page 757) specifies that $P(F_1) = 0.05$ and $P(F_2) = 0.95$, and our preceding calculations showed that $P(E \mid F_1) = 1$ and $P(E \mid F_2) = \frac{1}{2}$. From the Law of Total Probability (6), we therefore have

$$P(E) = P(E \mid F_1)\,P(F_1) + P(E \mid F_2)\,P(F_2)$$
$$= 1 \times 0.05 + \tfrac{1}{2} \times 0.95 = 0.525$$

The term 1×0.05 corresponds to the top blue path in the tree diagram of Figure 9, and $\frac{1}{2} \times 0.95$ corresponds to the bottom blue path. ■

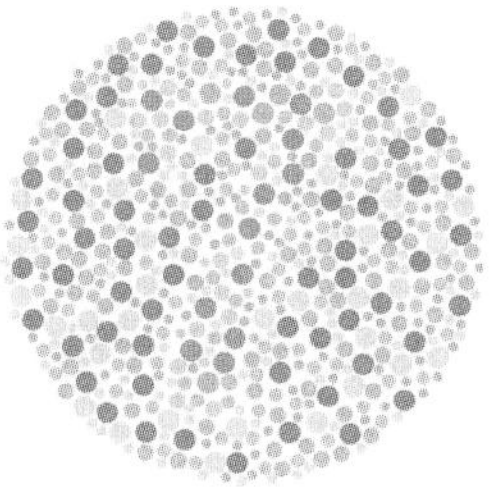

FIGURE 11
People with red-green color blindness cannot see the number 7 in the circle.

EXAMPLE 10 | **BB** **Color blindness** is caused by a genetic mutation and results in people not being able to distinguish between certain colors (see Figure 11). Color blindness is more common in males than in females, occurring in approximately 10% of males and only 1% of females. Assuming that there is an equal number of males and females in the population, what is the probability that a randomly chosen individual will be color-blind? Construct a tree diagram to illustrate the calculation.

SOLUTION We use C to denote the event "color blind" and M and F to denote the events "male" and "female." We are told that $P(C \mid M) = 0.1$ and $P(C \mid F) = 0.01$. We are also told that $P(M) = P(F)$ and, since the events M and F are disjoint and exhaustive, $P(M) + P(F) = 1$. Thus $P(M) = P(F) = 0.5$. Therefore, from (6),

$$P(C) = P(C \mid M)\,P(M) + P(C \mid F)\,P(F)$$
$$= 0.1 \times 0.5 + 0.01 \times 0.5 = 0.055$$

The corresponding tree diagram is shown in Figure 12.

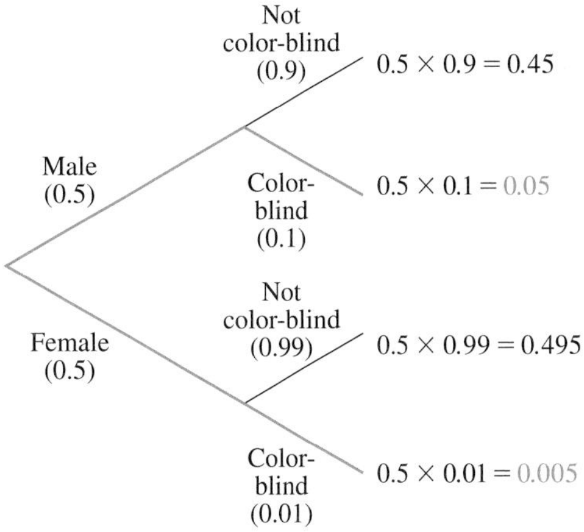

FIGURE 12
Tree diagram showing all four mutually exclusive outcomes. Blue lines and numbers correspond to outcomes where the individual is color-blind. Therefore

$$P(C) = 0.05 + 0.005$$
$$= 0.055$$

■

■ Bayes' Rule

A common mistake when first learning conditional probability is to assume that $P(A \mid B) = P(B \mid A)$ for events A and B. The following example illustrates that this is not, in general, true.

EXAMPLE 11 | In the die-rolling experiment of Example 1, we saw that the probability of rolling a 4, given that the role is even, is $\frac{1}{3}$. Formally, $P(E|F) = \frac{1}{3}$, where E is the event "roll is a 4" and F is the event "roll is even." Show that $P(F|E) \neq \frac{1}{3}$ and interpret the difference between this and $P(E|F)$ using a Venn diagram.

SOLUTION Definition 1 gives $P(F|E) = P(F \cap E)/P(E)$. Therefore we need the quantities $P(F \cap E)$ and $P(E)$. The quantity $P(F \cap E)$ is the probability that the roll is both even and a 4. Since 4 is even, this simply requires that the roll be a 4, which happens with probability $\frac{1}{6}$. $P(E)$ is also the probability that the roll is 4 and therefore it equals $\frac{1}{6}$ as well. Consequently

$$P(F|E) = \frac{1/6}{1/6} = 1$$

This is different from $\frac{1}{3}$. The Venn diagram for $P(E|F)$ in Figure 1(b) is reproduced in Figure 13(a). In it, the sample space is restricted to even rolls only, and the fraction of this restricted sample space consisting of the outcome "4" is $\frac{1}{3}$. Figure 13(b) gives the Venn diagram for $P(F|E)$. In this case the sample space is restricted to the roll "4", and the fraction of this restricted sample space consisting of the outcome "roll is even" is 1.

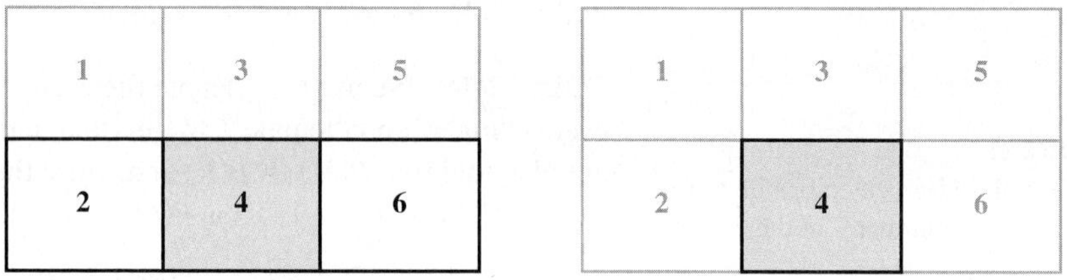

FIGURE 13

(a) Venn diagram for the probability of "rolling a 4," given "the roll is even"

(b) Venn diagram for the probability of "roll is even," given "the roll is a 4" ■

Often we know the probability of an event E conditional on an event F and we would like to reverse the conditioning. Bayes' Rule provides an equation for doing so.

> **Thomas Bayes**
>
> Thomas Bayes (1701–1761) was an English mathematician and minister who is credited with first discovering Equation 7. Despite this being his most well known accomplishment, it was not actually published until after his death. The modern statistical field of Bayesian inference is named in his honor.

(7) Bayes' Rule For any two events E and F with $P(E) > 0$,

$$P(F|E) = \frac{P(E|F)\,P(F)}{P(E)}$$

PROOF From Definition 1,

$$P(E \cap F) = P(E|F)\,P(F)$$

Likewise $$P(F \cap E) = P(F|E)\,P(E)$$

Furthermore, $P(E \cap F) = P(F \cap E)$ from the commutative law on page 740, and therefore we can combine these two equations and solve for $P(F|E)$ to obtain the desired result. ■

EXAMPLE 12 | We can use Bayes' Rule to calculate $P(F \mid E)$ in Example 11. We know that $P(E \mid F) = \frac{1}{3}$ from Example 1 and that $P(F) = \frac{1}{2}$ and $P(E) = \frac{1}{6}$. Substituting these into (7) gives

$$P(F \mid E) = \frac{P(E \mid F)\, P(F)}{P(E)} = \frac{\frac{1}{3} \times \frac{1}{2}}{\frac{1}{6}} = 1$$

■

EXAMPLE 13 | BB **Color blindness** In Example 10 we saw that the probability of a randomly chosen person being color-blind, given that the individual is male, is $P(C \mid M) = 0.1$. Suppose a randomly selected person is color-blind. What is the probability that the individual is male?

SOLUTION Using the notation of Example 10, we are asked to find $P(M \mid C)$. From Bayes' Rule we therefore need the quantities $P(C \mid M)$, $P(M)$, and $P(C)$, all of which are given or calculated in Example 10. We obtain

$$P(M \mid C) = \frac{P(C \mid M)\, P(M)}{P(C)} = \frac{0.1 \times 0.5}{0.055} \approx 0.91$$

■

One of the most important uses of Bayes' Rule in the life sciences is for disease testing. As the next example shows, when applying Bayes' Rule in these cases, we often need to use the Law of Total Probability as well.

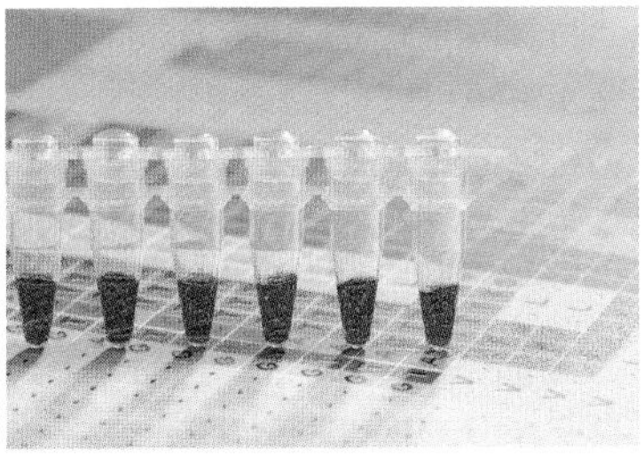

EXAMPLE 14 | **HIV testing** A test has been developed to determine if a person is infected with HIV, but such medical tests are never perfect. The probability of the test showing a positive result if the person really does have HIV is 0.99. The probability of the test showing a negative result if the person does not have HIV is 0.97. Suppose a person is tested and the result comes back positive. What is the probability that they actually have HIV? Assume the person is drawn randomly from a population in which 10% of the people are infected with HIV.

SOLUTION Let's use HIV^+ and HIV^- to denote the possible infection states and T^+ and T^- for the two possible test results. We are told that $P(T^+ \mid HIV^+) = 0.99$, $P(T^- \mid HIV^-) = 0.97$, and $P(HIV^+) = 0.1$. We are asked to calculate $P(HIV^+ \mid T^+)$. We can do so using Bayes' Rule provided we know the quantities $P(T^+ \mid HIV^+)$, $P(HIV^+)$, and $P(T^+)$. The first two of these are given. The third is not given but can be calculated using the Law of Total Probability, by conditioning on infection state:

$$P(T^+) = P(T^+ \mid HIV^+)\, P(HIV^+) + P(T^+ \mid HIV^-)\, P(HIV^-) \tag{8}$$

We are given $P(T^+ \mid HIV^+)$ and $P(HIV^+)$. Further, since HIV^+ and HIV^- are disjoint and exhaustive events, we know that $P(HIV^+) + P(HIV^-) = 1$. Therefore $P(HIV^-) = 1 - 0.1 = 0.9$. What remains is to calculate $P(T^+ \mid HIV^-)$. We are given $P(T^- \mid HIV^-)$ and therefore, using Equation 2, we obtain

$$P(T^+ \mid HIV^-) = 1 - P(T^- \mid HIV^-) = 1 - 0.97 = 0.03$$

Putting these results together in Equation 8 gives

$$P(T^+) = 0.99 \times 0.1 + 0.03 \times 0.9 = 0.126$$

We can then use Bayes' Rule to obtain

$$P(\text{HIV}^+ \mid \text{T}^+) = \frac{P(\text{T}^+ \mid \text{HIV}^+)\, P(\text{HIV}^+)}{P(\text{T}^+)} = \frac{0.99 \times 0.1}{0.126} \approx 0.79$$

This example assumes that HIV infection is relatively common in the population. The project on page 766 shows that very counterintuitive results are obtained when the disease in question is rare. ■

EXERCISES 12.3

1. A die is rolled. Find the following conditional probability.
(a) A "5" shows, given that the number showing is greater than three.
(b) A "3" shows, given that the number showing is odd.

2. A card is drawn from a standard deck. Find each of the following conditional probabilities.
(a) The card is a queen, given that it is a face card.
(b) The card is a king, given that it is a spade.
(c) The card is a spade, given that it is a king.

3. A fair coin is tossed twice. Calculate the probability.
(a) Two heads occur, given that the first toss is a head.
(b) Both tosses are the same, given that the first toss is a tail.
(c) Two heads occur, given at least one head occurs.

4. Two dice are rolled.
(a) What is the probability that the sum of the numbers is greater than eight?
(b) What is the probability that the sum of the numbers is greater than eight, given that the first die shows a 3?
(c) What is the probability that the first die shows a 3, given that the sum of the numbers is greater than eight?

5. A jar contains five black balls and seven white balls. Two balls are drawn sequentially, but the first ball is replaced before the second is drawn.
(a) What is the probability that both balls are black, given that the first one is black?
(b) What is the probability of drawing two white balls, given that at least one of the balls is white?

6–7 Sex chromosomes In humans the sex of an individual is determined by inherited sex chromosomes. All individuals carry two sex chromosomes, but males carry a so-called X and a Y chromosome (and are thus called XY), whereas females carry two X chromosomes (and are thus called XX). When a male and female produce an offspring, each gives the offspring one of its sex chromosomes, chosen with equal probability. Thus a child always inherits an X chromosome from its mother since, regardless of which of her sex chromosomes is chosen, it will always be an X. But the child inherits the X chromosome from its father with probability $\frac{1}{2}$ (in which case it will be female) or the Y chromosome from its father with probability $\frac{1}{2}$ (in which case it will be male).

6. X-linked mutations are mutations that are found on the X chromosome. For example, color blindness is caused by an X-linked mutation.
(a) Suppose that one of a woman's X chromosomes carries a mutation for color blindness, and she has a child with a man carrying a normal X chromosome. What is the probability that their child has the mutation, given that it is a female? What is the probability that their child has the mutation, given that it is a male?
(b) Suppose that a woman with normal X chromosomes has a child with a man whose X chromosome carries a mutation for color blindness. What is the probability that their child has the mutation, given that it is a female? What is the probability that their child has the mutation, given that it is a male?

7. Muscular dystrophy is a genetic disease caused by a mutation on the X chromosome. During reproduction this mutation arises spontaneously on each of the X chromosomes of the parents, independently and with probability μ. We can therefore view the production of a child as consisting of two probabilistic experiments. First, one of the two sex chromosomes is chosen with equal probability from each parent. Second, for each X chromosome that is chosen, the muscular dystrophy mutation arises independently with probability μ. All male children carrying the mutation will develop muscular dystrophy, whereas only females whose X chromosomes both carry the mutation will develop the disease.
(a) List the sample space of all possible outcomes when producing a single child, in terms of its chromosomes and their mutational states.
(b) Construct a tree diagram for the production of a single child, where the first branching represents male versus female, and the second represents their possible genotypes with respect to the mutation. Indicate the probability of each outcome and whether it results in disease.
(c) What is the probability that the child will develop the disease?
(d) What is the probability that the child will develop the disease, given that it is male?
(e) What is the probability that the child will develop the disease, given that it is female?

8–9 A jar contains five red balls numbered 1 to 5, and seven green balls numbered 1 to 7.

8. A ball is drawn at random from the jar. Find each of the following conditional probabilities.
(a) The ball is red, given that it is numbered 3.
(b) The ball is green, given that it is numbered 7.
(c) The ball is red, given that it is even-numbered.
(d) The ball has an even number, given that it is red.

9. Two balls are drawn at random from the jar. Find each of the following conditional probabilities.
(a) The second ball drawn is red, given that the first is red.
(b) The second ball drawn is red, given that the first is green.
(c) The second ball drawn is even-numbered, given that the first is odd-numbered.
(d) The second ball drawn is even-numbered, given that the first is even-numbered.

10. Jar A contains one white and two black balls, jar B contains five white and two black balls, and jar C contains four white and two black balls. One ball is to be selected from each jar. What is the probability that the ball chosen from jar B will be white, given that exactly two black balls are selected?

11. A blue die and a red die are rolled. Suppose k is a number between 1 and 12. What is the probability that the blue die shows a 1, given that the sum of the numbers on the dice is k?

12–14 Genetic crosses The two variants of the gene that determines flower position in pea plants are labeled A and T for "axial" and "terminal." A is dominant to T: only one copy of A is required to make the flowers axial, whereas individuals must be of genotype TT to have terminal flowers. In crosses between plants, each parent gives one of their two genes to the offspring with equal probability. Calculate the following probabilities.

12. The probability that an offspring of an AT × TT cross has genotype AT, given that it has axial flowers.

13. The probability that an offspring of an AT × AA cross has genotype AT, given that it has axial flowers.

14. The probability that an offspring of an AT × AT cross has genotype AT, given that it has axial flowers. (*Note:* Consider AT and TA to be identical genotypes.)

15–19 Causes of mortality Refer to the probabilities in Figure 5 to calculate the following.

15. The probability that a death is from a communicable disease, given that it occurs in an economically underdeveloped country

16. The probability that a death is from a noncommunicable disease, given that it occurs in an economically developed country

17. The probability that a death is from causes other than communicable or noncommunicable disease, given that it occurs in an economically underdeveloped country

18. The probability that a death occurs in an economically underdeveloped country, given that it occurs from a communicable disease

19. The probability that a death occurs in an economically developed country, given that it occurs from causes other than a communicable or noncommunicable disease

20. Influenza pandemics occasionally occur, and thus a new viral subtype spreads globally. Historically, three subtypes have caused pandemics: H1N1, H1N2, and H3N2 (the names refer to different molecules on the surface of the virus). Suppose that, in any given year, there are four outcomes: (i) no pandemic, with probability $96/100$, (ii) pandemic by subtype H1N1, with probability $2/100$, (iii) pandemic by subtype H1N2, with probability $1/100$, or (iv) pandemic by subtype H3N2, with probability $1/100$. Given a pandemic occurs, what is the probability that it is caused by H1N2?

21. Barred owl sex ratios Barred owls often lay two eggs, and each resulting chick has an equal probability of being male or female.
(a) What is the probability that both chicks are male, given that at least one of them is male?
(b) What is the probability that both chicks will be female, given that the first one is female?

© mlorenz / Shutterstock.com

22. Suppose events E and F are mutually exclusive and $P(F) \neq 0$. Show that $P(E \mid F) = 0$ and illustrate this result with a Venn diagram.

23. An event E is called a subset of event F if all outcomes in E are also in F. Show that in this case $P(F \mid E) = 1$ and illustrate this result with a Venn diagram. You can assume that $P(E) \neq 0$.

24. Zoonotic disease outbreaks are infectious disease outbreaks in humans that are caused by pathogens from wild or domesticated animals. Suppose two pathogen genotypes occur in chickens, and one of them is accidentally transmitted to a farmer, each being selected with probability q_1 or q_2, respectively (where $q_1 + q_2 = 1$). If type 1 is

transmitted, then an outbreak will occur with probability α_1, whereas if type 2 is transmitted, then an outbreak will occur with probability α_2.
(a) What is the overall probability of an outbreak?
(b) What is the probability of no outbreak occurring, given that type 2 is transmitted?

25. A jar contains seven black balls and three white balls. Two balls are drawn, without replacement, from the jar. Find the probability of each of the following events.
(a) The first ball drawn is black, and the second is white.
(b) The first ball drawn is black, and the second is black.

26. Two cards are drawn from a standard deck without replacement. Find the probability of each of the following events.
(a) The first is an ace and the second is a king.
(b) Both cards are aces.

27. A die is rolled twice. Let E and F be the events "the first roll shows a 6" and "the second roll shows a 6," respectively.
(a) Find the probability of showing a 6 on both rolls.
(b) Are the events E and F independent?

28. A die is rolled twice. What is the probability that it shows a 1 on the first roll and an even number on the second roll?

29–30 Tree diagrams, similar to those of Section 12.1, provide a way to visualize probability calculations for multiple experiments. Suppose there are two experiments, each with two possible outcomes: A and B for the first experiment and E and F for the second experiment. We can calculate the probability of different outcomes for the pair of experiments by tracing along the appropriate branch of the tree, multiplying the associated probabilities using the *multiplication rule*.

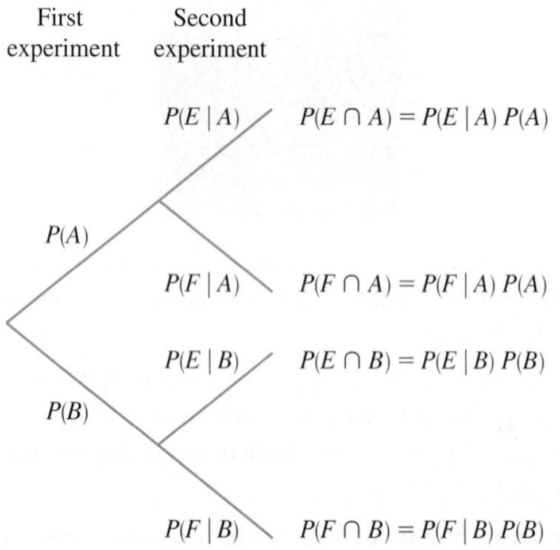

29. Jar A contains one red ball and one blue ball, whereas jar B contains two white balls and three black balls. One ball is drawn, and with probability 0.25 it comes from jar A and with probability 0.75 it comes from jar B. Use a tree diagram to calculate the probability of each of the four possible outcomes.

30. Handedness and sex A population has an equal number of males and females. The probability that a female is left-handed is $P(L \mid F) = 0.06$ and the probability that a male is left-handed is $P(L \mid M) = 0.14$. Use a tree diagram to calculate the probability of each of the four possible combinations of handedness and sex.

31–34 Consider the sample space $\{1, 2, 3, 4\}$, where $P(\{1\}) = 0.16$, $P(\{2\}) = 0.04$, $P(\{3\}) = 0.16$, and $P(\{4\}) = 0.64$. Determine whether events E and F are independent. (Recall that $E = \{i, j\}$ denotes the event that an i or a j occurs).

31. $E = \{1, 2\}$ and $F = \{3, 4\}$

32. $E = \{1, 2\}$ and $F = \{1, 2\}$

33. $E = \{1, 2\}$ and $F = \{2, 3\}$

34. $E = \{1, 2\}$ and $F = \{2, 4\}$

35. A coin is tossed and a die is rolled.
(a) Find the probability of showing a tail and an even number.
(b) Are the events "tails" and "even number" independent?

36. The tree diagrams from Exercises 29–30 simplify when the outcomes of the experiments are independent. Explain how.

37. Suppose events E and F are independent. Illustrate what this means in terms of a Venn diagram.

38. Prove that if $P(F \mid E) = P(F)$, then $P(E \mid F) = P(E)$.

39. Family planning A family plans to have two children. If the sexes of the first child and the second child are independent, and if the birth sex ratio is 50 : 50, what is the probability that the family will have a daughter and a son, irrespective of their birth order?

40. Hardy-Weinberg gene frequencies Two gene variants A and a are present in a population at frequencies p and $1 - p$, respectively. Suppose two genes are drawn randomly from the population to produce an offspring: its genotype will be either AA, Aa, or aa. If the two draws are independent, what is the probability of each genotype? (Treat genotypes Aa and aA as the same.)

41. A die is tossed. If the number showing is even, it is recorded. If the number is odd, the die is tossed again and the number recorded. What is the probability that the recorded number is a 4?

42. A fair coin is tossed. If the outcome is heads, it is recorded. If it is tails, the coin is tossed again and the outcome recorded. What is the probability that the recorded outcome is heads?

43. One card is randomly removed from a standard deck. If a second card is then removed, what is the probability that the second card is a face card?

44. A jar contains two white balls and three black balls. A fair coin is tossed and if the outcome is heads, then two balls are removed from the jar. If the outcome is tails, then one ball is removed. What is the probability that exactly one white ball is removed from the jar?

45. ***BRCA1* and *BRCA2* genes and breast cancer** *BRCA1* and *BRCA2* are two genes in humans; mutations of these genes are associated with breast cancer. Of all females that carry mutations, $\frac{1}{3}$ of them carry a mutation in the *BRCA1* gene and $\frac{2}{3}$ in the *BRCA2* gene. (Suppose that nobody carries mutations in both genes.) If the probability of developing cancer is $\frac{3}{5}$ for women who carry a *BRCA1* mutation and $\frac{1}{5}$ for women who carry a *BRCA2* mutation, what is the probability that a randomly chosen female who carries a mutation will develop breast cancer?

46. **Tuberculosis (TB) testing** Most medical tests are not completely accurate. For example, a microscopy test for TB comes out positive with probability 0.01 when the tested individual doesn't actually have TB; it comes out negative with probability 0.2 when the tested individual does have TB. Find the probability that a randomly chosen individual will test positive if
(a) the frequency of TB in the population is 2%.
(b) the frequency of TB in the population is 30%.

47. **Heart disease** A year-long US study of heart attacks in men and women is to be conducted. Census data shows that 51% of the population is female, and previous research has shown that each year approximately 0.2% of females have heart attacks whereas 0.3% of males do. What is the probability that a randomly chosen person will have a heart attack during the study?

48–50 Mendelian pea crosses The gene that determines flower position in pea plants has two variants, A and T, for "axial" and "terminal." A is dominant to T: only one copy of A is required to make the flowers axial, whereas individuals must be of genotype TT to have terminal flowers. In crosses between plants, each parent gives one of their two genes to the offspring with equal probability. In each of the following crosses, calculate the probability that an offspring plant will have axial flowers.

48. A plant with genotype AT is crossed to the offspring of an AA × AT cross.

49. A plant with terminal flowers is crossed to the offspring of an AA × AT cross.

50. Two plants with terminal flowers are crossed.

51–54 Calculate $P(F \mid E)$.

51. $P(E \mid F) = 0.3, \; P(E) = 0.4, \; P(F) = 0.1$

52. $P(E \mid F) = 0.2, \; P(E) = 0.15, \; P(F) = 0.3$

53. $P(E \mid F) = 0.4, \; P(E) = 0.4, \; P(F) = 0.25$

54. $P(E \mid F) = 0.3, \; P(E) = 0.1, \; P(F) = 0.1$

55. **Dengue-fever** viruses are carried by mosquitoes. There are several different serotypes of the dengue virus, and one study showed that, of all mosquitoes that carried serotype A, 20% also carried serotype B. If the frequency of serotypes A and B in the mosquito population are 10% and 35%, respectively, what is the probability that a randomly selected mosquito carrying serotype B also carries serotype A?

56. **Handedness and sex** Let F be the event that an offspring is female and L be the event that it is left-handed. In Example 7 we were told that if we do not know the sex of a child, then the probability that it will be left-handed is 0.1 [that is, $P(L) = 0.1$]. We were also told that 30% of all left-handed people are female [that is, $P(F \mid L) = 0.3$]. Suppose an ultrasound shows that a child to be born is a female. What is the probability that she will be left-handed? (Assume that 50 percent of newborns are females).

57. ***BRCA1*** is a gene that, when mutated, is associated with breast cancer. Suppose the probability of developing cancer is $\frac{3}{5}$ for women who carry a *BRCA1* mutation and that the frequency of the mutant gene is 1% in the population. If the probability that a randomly chosen woman will develop breast cancer is 0.1, then what is the probability that a randomly chosen breast cancer patient will carry the mutant *BRCA1* gene?

58. **Genetic screening** A genetic test is used to determine whether a patient has a specific type of cancer. The probability of the test coming back positive when cancer is not present (called a false-positive) is 0.02, whereas the probability of the test coming back positive when cancer is present is 0.9. If the frequency of cancer in the population is 15%, what is the probability that a person is cancer-free if their test is positive?

59. **Diabetes** A blood test for diabetes gives a positive result 90% of the time when an individual has diabetes and a negative result 95% of the time when he or she does not. In 2012 approximately 8.3% of the US population had diabetes. If a randomly chosen citizen tested negative, what is the probability that he or she actually has diabetes?

60. **DNA evidence** is often used in criminal trials. Suppose a murder has been committed and the perpetrator's blood is found at the crime scene. DNA from the blood is analyzed and it is found that the probability of an innocent person having DNA that matches the sample is 10^{-8}. Let M be the event "DNA match" and I be the event "innocent." Therefore we have $P(M \mid I) = 10^{-8}$. What is really of interest is the quantity $P(I \mid M)$, the probability that a person whose DNA matches is innocent. Calculate this quantity if the community has 1000 people and therefore the probability that a randomly chosen person is innocent is $P(I) = 999/1000$. You can assume that the probability of a match for the guilty person is 1 [that is, $P(M \mid I^c) = 1$].

PROJECT Testing for Rare Diseases

The application of Bayes' Rule to disease testing can produce some very counterintuitive results when the disease in question is rare. Consider HIV testing, where HIV^+ and HIV^- denote the possible infection states of an individual, and T^+ and T^- represent the two possible test results.

Two important measures of test quality are (i) $P(T^+ \mid HIV^+)$, the probability that the test correctly identifies an infected individual, and (ii) $P(T^- \mid HIV^-)$, the probability that the test correctly identifies an uninfected individual. The complements of these are $P(T^- \mid HIV^+)$, the probability that the test misidentifies an infected individual, and $P(T^+ \mid HIV^-)$, the probability that the test misidentifies an uninfected individual.

Let's simplify notation by using τ for the probability that the test correctly identifies an infected individual and ϕ for the probability that the test misidentifies an uninfected individual. Mnemonically, "tau" is associated with a "true" positive test and "phi" with a "false" positive test.

In epidemiological studies, researchers refer to the quantity

$$\frac{P(T^+ \mid HIV^+)}{P(T^+)}$$

as the *positive predictive value* of the test.

1. Bayes' Rule states that

$$P(HIV^+ \mid T^+) = \frac{P(T^+ \mid HIV^+)}{P(T^+)} P(HIV^+)$$

Using p to denote the frequency of HIV^+ individuals in the population, express $P(HIV^+ \mid T^+)$ in terms of p, τ, and ϕ.

2. Suppose the probability that the test correctly identifies an infected individual is $\tau = 0.99$, and the probability that the test misidentifies an uninfected individual is $\phi = 0.01$. Plot $P(HIV^+ \mid T^+)$ as a function of p to show that, even though the test is very effective at correctly identifying HIV^+ and HIV^- individuals, the probability of being HIV^+ if the test is positive nevertheless approaches 0 as $p \to 0$. In other words, if HIV^+ individuals are very rare, then positive test results for randomly chosen individuals are not necessarily indicative of whether they actually have the disease.

3. Using your result from Problem 1, show that $P(HIV^+ \mid T^+)$ approaches 0 as $p \to 0$ no matter how effective the test is, provided that $\phi \neq 0$. In other words, if there is any possibility of misidentifying an uninfected individual, then for very rare diseases positive test results for randomly chosen individuals are not necessarily indicative of whether they actually have the disease.

4. The result from Problem 3 is counterintuitive, but it can be understood using a Venn diagram. Construct a Venn diagram indicating the two events HIV^+ and HIV^-, under the assumption that HIV^+ individuals are rare.

5. Indicate on your diagram from Problem 4 an area representing the quantity $\tau = P(T^+ \mid HIV^+)$ under the assumption that this is close to 1 (that is, the test is very good at correctly identifying HIV^+ individuals). Do the same for the quantity $\phi = P(T^+ \mid HIV^-)$ under the assumption it is close to 0 (that is, the test is very good at correctly identifying HIV^- individuals).

6. Indicate on your diagram from Problem 5 the area representing the quantity $P(HIV^+ \mid T^+)$, and explain how this can be small despite the fact that τ is large and ϕ is small.

One way to alleviate the problem of $P(\text{HIV}^+ \mid \text{T}^+)$ being small, even for very effective tests, is to restrict the use of such tests to subpopulations of individuals for whom the disease is more common (that is, where p is not small).

12.4 Discrete Random Variables

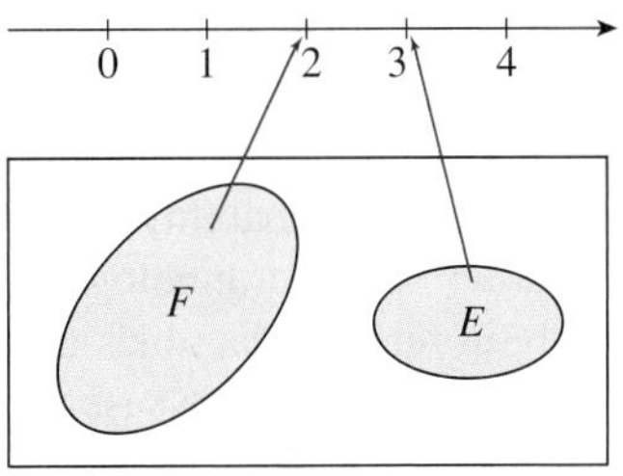

FIGURE 1

For many probabilistic experiments, we are often more interested in some numerical property of the outcomes than in the outcomes themselves. For instance, when flipping three coins, we might be more interested in the number of heads that occur than in the way that this number is obtained. When observing the process of offspring production in a particular organism, we might be more interested in the number of offspring produced than in the many probabilistic physiological events that led to this number.

Such numerical quantities are known as *random variables*. A random variable is a function whose domain is the sample space of some experiment and whose range is a subset of the real numbers (see Figure 1). Random variables are denoted by uppercase letters such as X or Y.

More formally, a discrete set of numbers is a set that can be written as a list, with one number on each row of the list. Often we will be interested in finite sets of numbers, but sometimes the set will be infinite (that is, the "list" of numbers will continue indefinitely).

> **(1) Definition** A **discrete random variable** X is a function from a sample space Ω to a *discrete* (unconnected) subset of the real numbers.

EXAMPLE 1 | Two coins are tossed. Using H and T to denote heads and tails, the sample space is

$$\Omega = \{\text{HH, HT, TH, TT}\}$$

If we use X to denote the number of heads, then X is a random variable defined by the following function (see Figure 2):

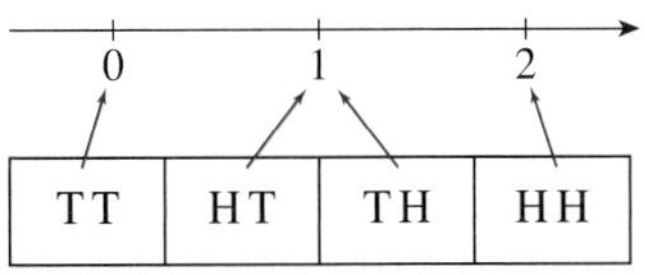

FIGURE 2

$$X(\text{HH}) = 2$$

$$X(\text{HT}) = 1$$

$$X(\text{TH}) = 1$$

$$X(\text{TT}) = 0$$

■

Once we specify a random variable X, the probability of different numerical values of X occurring is then determined by the underlying experiment. For instance, in Example 1, if the coin is fair, then all four outcomes in the sample space are equally likely. Therefore, from Definition 12.2.2, the probability that $X = 2$ is $\frac{1}{4}$, the probability that $X = 1$ is $\frac{1}{2}$, and the probability that $X = 0$ is $\frac{1}{4}$.

Although (1) gives the formal definition of a random variable, in many instances we skip the step of specifying the more detailed underlying sample space and instead focus directly on the numerical values. We view the set of possible numerical values as the set of outcomes and we assign a probability to each of these directly.

EXAMPLE 2 | A random variable Y takes on a value of 0 with probability $\frac{1}{3}$ and a value of 1 with probability $\frac{2}{3}$. Mathematically we can express this as

$$P(Y = 0) = \frac{1}{3} \qquad P(Y = 1) = \frac{2}{3}$$

■

Most discrete random variables in the life sciences represent counts of something. For example, number of species, population size, and number of mRNA molecules can each be modeled using a discrete random variable. Therefore we will restrict our attention to random variables taking on nonnegative integer values.

■ Describing Discrete Random Variables

Although we could use a Venn diagram to give a pictorial represention of a random variable, it is more common to use a *histogram* (see Section 11.2). By convention, we choose the height of each bar in the histogram so that its area is equal to the probability of the corresponding outcome (this is referred to as a density histogram in the terminology of Section 11.2). However, since we will be interested primarily in random variables that take on successive integer values, we will set the bar widths to equal 1. In this case the height of each bar is also equal to the probability of the corresponding outcome.

The histogram of a random variable X is sometimes referred to as its **probability distribution** because it illustrates how probability is distributed across the possible values of X. A random variable can also be described mathematically by its *probability density function*. If we use i to denote a particular value of the random variable X, we have the following definition.

As with sequences, we write functions of the form $p(i)$ using the subscript notation p_i.

(2) Definition The **probability density function (PDF)** of the discrete random variable X is the function p_i defined by

$$p_i = P(X = i)$$

Definition 2 says that the probability density function p_i gives the probability that X takes on the particular value i. There is a close connection between a random variable's PDF and its histogram. The value p_i specifies the area (and also the height) of the histogram bar located at the value $X = i$. Therefore the histogram of a random variable is a graph of its PDF.

Figure 3 displays the histogram of a random variable in which the probability that $X = 2$ is highlighted.

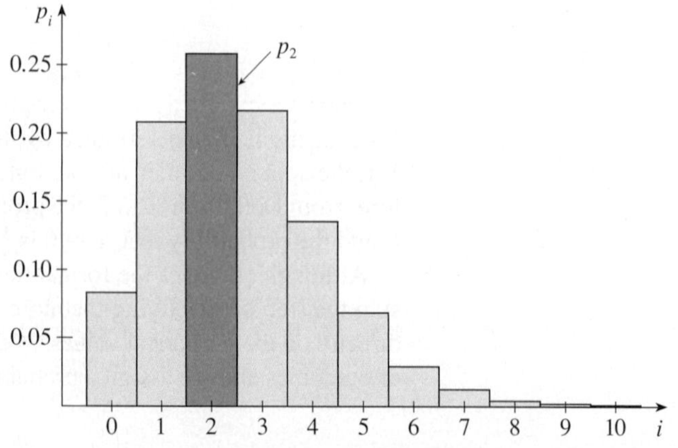

FIGURE 3
The quantity p_2 gives both the area and the height of the highlighted bar at $i = 2$.

The values p_i are probabilities and therefore they must satisfy

(3) $$p_i \geqslant 0 \qquad \text{for all } i$$

Graphically, all bars in the histogram have nonnegative height. Furthermore, because the union of all possible values of X corresponds to the entire sample space, p_i must also satisfy the condition

The summation notation in Equation 4 indicates that the sum is to be taken over all relevant values of i, from the smallest value to the largest value.

(4) $$\sum_i p_i = 1$$

Graphically, if the bars of the histogram have unit width, then the summed area of all bars of the histogram equals 1.

EXAMPLE 3 | **Offspring number** Each individual female of a particular fish species produces a number of offspring that is described by a random variable with probability density function $p_i = (1/2)^{i+1}$, where i is a nonnegative integer. Verify Condition 4 for this random variable and construct its histogram.

SOLUTION To verify Condition 4 we must sum the PDF p_i from $i = 0$ to ∞.

$$\sum_{i=0}^{\infty} p_i = \sum_{i=0}^{\infty} \left(\frac{1}{2}\right)^{i+1} = \frac{1}{2}\sum_{i=0}^{\infty} \left(\frac{1}{2}\right)^{i} = \frac{1}{2} \cdot \frac{1}{1 - \frac{1}{2}} = \frac{1}{2} \cdot 2 = 1$$

The third equality follows from the fact that the summation is a geometric series with $a = 1$ and $r = \frac{1}{2}$ (see Section 2.1). To plot the histogram for this random variable we first construct a table of values for the PDF, as shown in the margin. The histogram is displayed in Figure 4. Our calculations show that the summed area of all bars in this histogram is 1.

i	p_i
0	1/2
1	1/4
2	1/8
3	1/16
4	1/32
5	1/64
6	1/128
7	1/256
8	1/512
9	1/1024
⋮	⋮

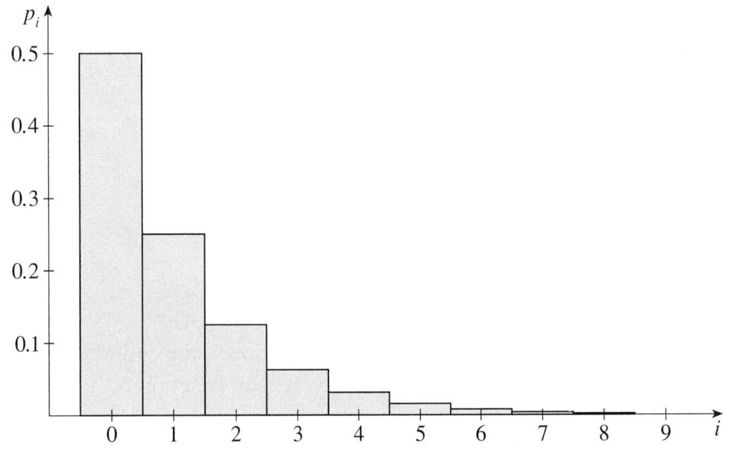

FIGURE 4
Histogram shows only up to $i = 9$.

■

Another useful description of a discrete random variable is given by its *cumulative distribution function.*

> **(5) Definition** The **cumulative distribution function (CDF)** of the discrete random variable X is the function F_i defined by
>
> $$F_i = P(X \leqslant i)$$

F_i is the probability that X takes on a value no larger than i. The CDF has a simple geometric interpretation. The area of each bar in the histogram is the probability of that value of X, and therefore F_i is the summed area of all bars of the histogram up to and including the bar at $X = i$. Figure 5 illustrates the idea by plotting the cumulative distribution function of a random variable in Figure 5(b) directly below the graph of its probability density function in Figure 5(a). The vertical coordinate of each point in part (b) is the cumulative area in the histogram of part (a) up to that value of i.

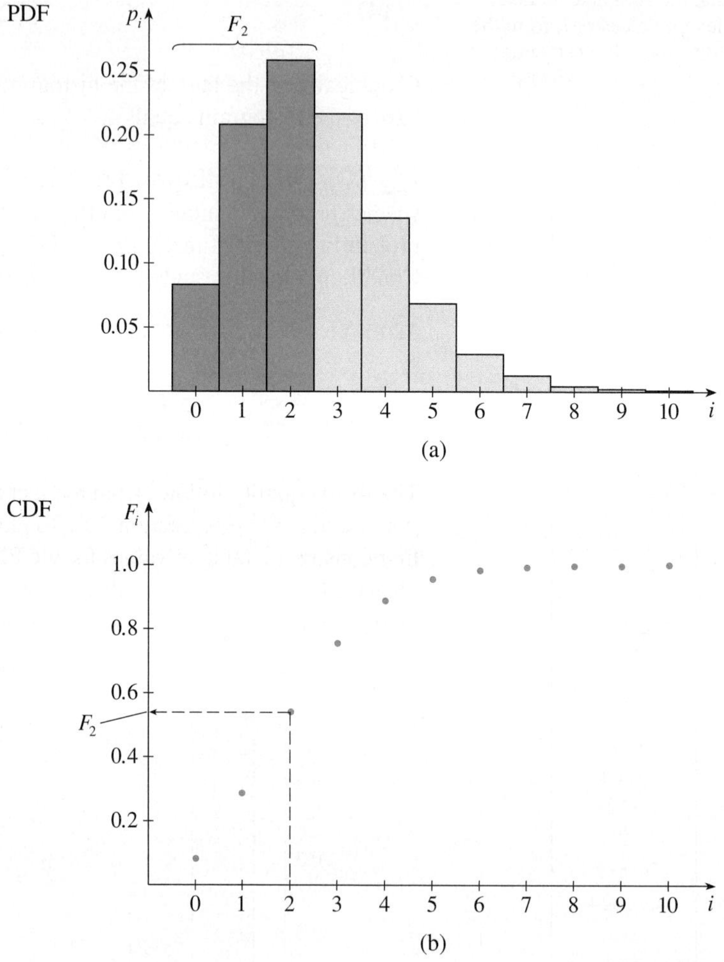

FIGURE 5

From Definition 5, the cumulative distribution function can be written in terms of the probability density function as

(6)
$$F_i = \sum_{k=0}^{i} p_k$$

EXAMPLE 4 | **Offspring number (continued)** Calculate the CDF for the probability density function $p_i = (1/2)^{i+1}$ from Example 3 and plot it below a graph of its PDF. Indicate the value F_3 on both the CDF and the PDF. What is the biological meaning of F_3?

SOLUTION From Equation 6 we obtain

$$F_i = \sum_{k=0}^{i} \left(\frac{1}{2}\right)^{k+1} = \frac{1}{2} \sum_{k=0}^{i} \left(\frac{1}{2}\right)^{k}$$

$$= \frac{1}{2}\left[2 - \left(\frac{1}{2}\right)^{i}\right] = 1 - \left(\frac{1}{2}\right)^{i+1}$$

where we have used the formula for the sum of a geometric series given in (2.1.5). The histogram of Figure 4 is reproduced in Figure 6(a), along with a plot of the CDF in Figure 6(b). The value F_3 is indicated in each plot. The quantity F_3 is the probability that an individual has 3 or fewer offspring.

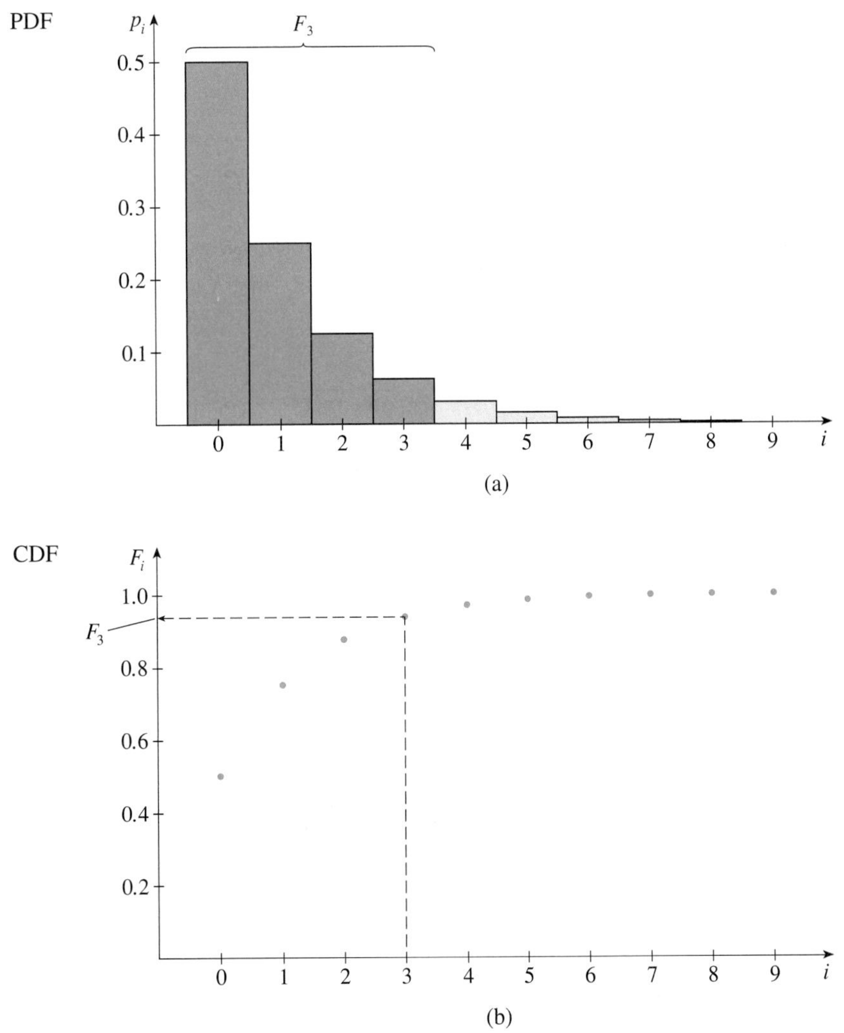

FIGURE 6

The cumulative distribution function can also be used to calculate the probability that the random variable falls *between* two values.

EXAMPLE 5 | Consider the probability density function $p_i = i/55$ for random variable X, where X can take on nonnegative integer values from 0 to 10. Figure 7 plots a graph of the PDF.
(a) Find the CDF.
(b) Use your answer from part (a) to calculate the probability that X takes on a value between 2 and 4, inclusive.

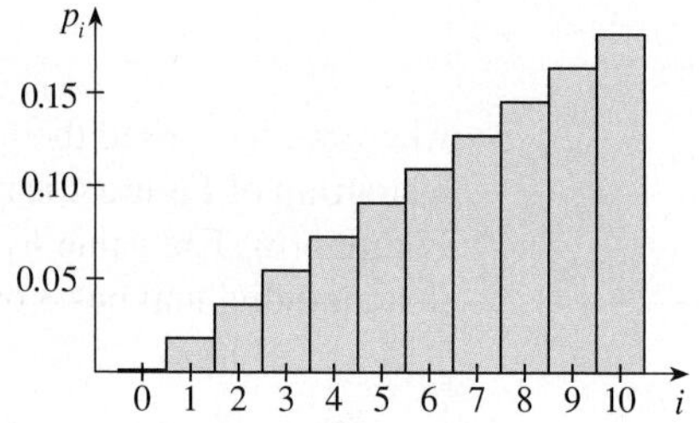

FIGURE 7

SOLUTION

(a) From Equation 6,

$$F_i = \sum_{k=0}^{i} \frac{k}{55} = \frac{1}{55} \sum_{k=0}^{i} k = \frac{1}{55} \frac{i(i+1)}{2} = \frac{i(i+1)}{110}$$

where the summation was simplified using Formula 5.2.5 on page 332.
(b) The probability that X takes a value between 2 and 4, inclusive, is

$$P(X \in \{2, 3, 4\}) = p_2 + p_3 + p_4$$

We also note that $F_4 = p_0 + p_1 + p_2 + p_3 + p_4$ and $F_1 = p_0 + p_1$. Therefore

$$P(X \in \{2, 3, 4\}) = F_4 - F_1$$

$$= \frac{4(4+1)}{110} - \frac{1(1+1)}{110} = \frac{10}{55} - \frac{1}{55} = \frac{9}{55}$$

Figure 8 illustrates the idea.

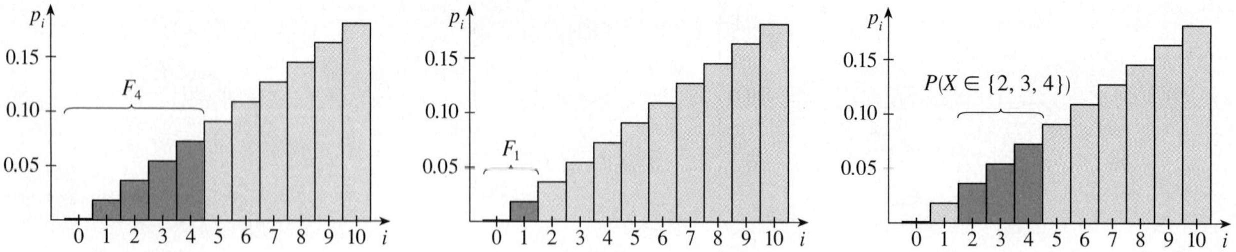

FIGURE 8

■

Equation 6 shows how the cumulative distribution function can be obtained from knowledge of the probability density function. The following example illustrates that the reverse is also possible.

EXAMPLE 6 | Illustrate how the PDF can be obtained from the cumulative distribution function graphically, using a histogram. Given the CDF $F_i = 1 - (1/2)^{i+1}$, derive the probability density function.

SOLUTION Consider a particular value of i. F_i is then the cumulative area of the histogram up to and including the bar at $X = i$ [see Figure 9(a)]. Similarly, F_{i-1} is the cumulative area of the histogram up to and including the bar at $X = i - 1$ [see Figure 9(b)]. Since p_i is the area of the bar located at $X = i$, we can obtain this area as the difference between F_i and F_{i-1} [see Figure 9(c)]; that is,

$$p_i = F_i - F_{i-1}$$

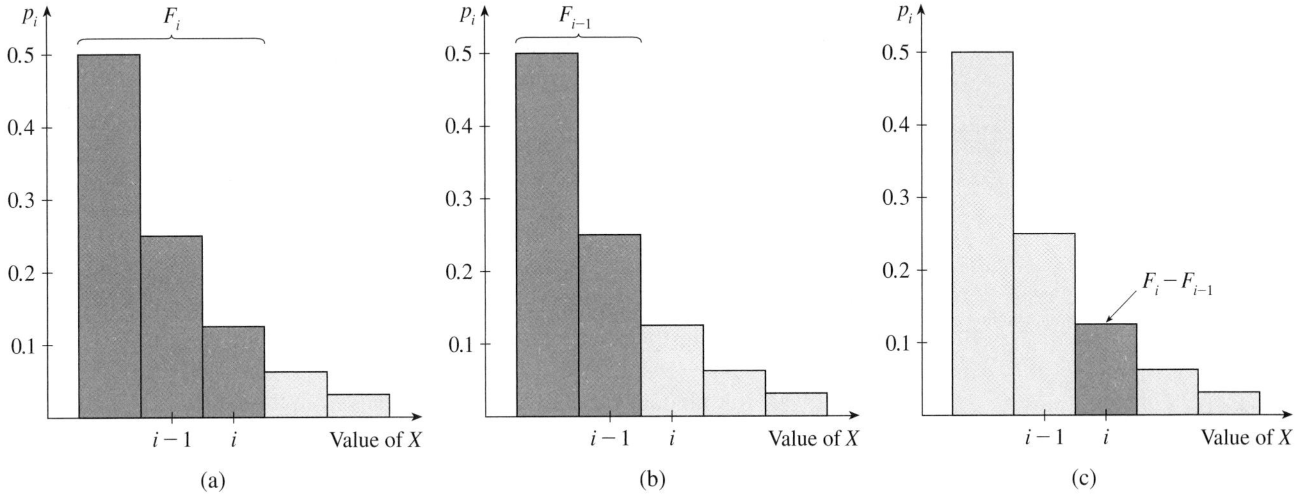

FIGURE 9

Using the given example for F_i, we have

$$\begin{aligned} p_i &= F_i - F_{i-1} \\ &= 1 - \left(\frac{1}{2}\right)^{i+1} - \left[1 - \left(\frac{1}{2}\right)^i\right] \\ &= \left(\frac{1}{2}\right)^i - \left(\frac{1}{2}\right)^{i+1} = \left(\frac{1}{2}\right)^i\left[1 - \frac{1}{2}\right] = \left(\frac{1}{2}\right)^{i+1} \end{aligned}$$ ■

■ Mean and Variance of Discrete Random Variables

In Section 11.1 we saw how to calculate the mean, or average, of a set of n numbers $x_1, \ldots, x_n$ as

$$\bar{x} = \frac{x_1 + x_2 + \cdots + x_n}{n}$$

Now suppose the number x_i is repeated k_i times. If there are m distinct values of x, then $\Sigma_{i=1}^{m} k_i = n$. The expression for the mean can then be written as

$$\begin{aligned} \bar{x} &= \frac{x_1k_1 + x_2k_2 + \cdots + x_mk_m}{n} \\ &= x_1\frac{k_1}{n} + x_2\frac{k_2}{n} + \cdots + x_m\frac{k_m}{n} \end{aligned}$$

or

(7) $$\bar{x} = x_1 p_{x_1} + x_2 p_{x_2} + \cdots + x_m p_{x_m}$$

where $p_{x_i} = k_i/n$ is the fraction of all the numbers that have value x_i.

Equation 7 motivates the following definition for the mean of a discrete random variable X.

(8) Definition The **mean** of a discrete random variable X with probability density function p_i, denoted by $E[X]$, is

$$E[X] = \sum_i i p_i$$

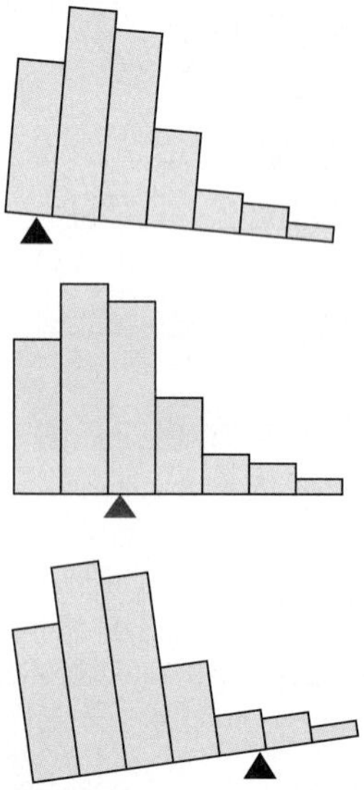

FIGURE 10
The mean is sometimes called the *center of mass* of the PDF. If we view each bar of the histogram as a solid, then the mean value of X is the value at which the histogram would exactly balance on a fulcrum.

The mean value of random variable X is also sometimes called the **expected value** of X, which is where the notation $E[X]$ comes from. It is also sometimes called the **center of mass**; Figure 10 illustrates where this terminology comes from.

EXAMPLE 7 | A random variable X takes on values from the set $\{0, 1, 2\}$ with probability density function $p_0 = \frac{1}{4}$, $p_1 = \frac{1}{2}$, and $p_2 = \frac{1}{4}$. Calculate $E[X]$.

SOLUTION From Definition 8 we obtain

$$E[X] = 0 \cdot p_0 + 1 \cdot p_1 + 2 \cdot p_2 = 0 \cdot \frac{1}{4} + 1 \cdot \frac{1}{2} + 2 \cdot \frac{1}{4} = 1$$ ■

In Definition 11.1.4 we also saw how to calculate the variance of a set of n numbers $x_1, \ldots, x_n$ as

$$\frac{(x_1 - \bar{x})^2 + (x_2 - \bar{x})^2 + \cdots + (x_n - \bar{x})^2}{n}$$

This can be viewed as the average of the numbers $(x_1 - \bar{x})^2, (x_2 - \bar{x})^2, \ldots, (x_n - \bar{x})^2$. In other words, it is the average squared difference of the x_i values from their mean $\bar{x}$. If we again suppose that the number x_i is repeated k_i times, with $\Sigma_{i=1}^{m} k_i = n$, then this expression can be written as

$$\frac{(x_1 - \bar{x})^2 k_1 + (x_2 - \bar{x})^2 k_2 + \cdots + (x_m - \bar{x})^2 k_m}{n}$$

$$= (x_1 - \bar{x})^2 \frac{k_1}{n} + (x_2 - \bar{x})^2 \frac{k_2}{n} + \cdots + (x_m - \bar{x})^2 \frac{k_m}{n}$$

or

(9) $$(x_1 - \bar{x})^2 p_{x_1} + (x_2 - \bar{x})^2 p_{x_2} + \cdots + (x_m - \bar{x})^2 p_{x_m}$$

Equation 9 motivates the following definition for the variance of a discrete random variable X.

(10) Definition The **variance** of a discrete random variable X with probability density function p_i, denoted by $\mathrm{Var}[X]$, is

$$\mathrm{Var}[X] = \sum_i (i - E[X])^2 p_i$$

A more convenient formula can be obtained by expanding the square in the summation and simplifying to get

$$\mathrm{Var}[X] = \sum_i i^2 p_i - E[X]^2$$

(See Exercise 44.) The **standard deviation** of a random variable is the square root of its variance.

The notation $E[X]^2$ denotes the square of the expected value of X. To calculate this quantity we first calculate the expected value of X and then we square the result.

EXAMPLE 8 | Calculate the variance of the random variable from Example 7.

SOLUTION From the second formula in Definition 10, we obtain

$$\begin{aligned}\mathrm{Var}[X] &= 0^2 p_0 + 1^2 p_1 + 2^2 p_2 - E[X]^2 \\ &= 0^2 \cdot \frac{1}{4} + 1^2 \cdot \frac{1}{2} + 2^2 \cdot \frac{1}{4} - 1^2 = \frac{1}{2}\end{aligned}$$

Bernoulli Random Variables

Many phenomena can be represented by a probabilistic experiment for which there are two possible outcomes. Examples include the presence or absence of a particular species, the survival or death of an individual, or a newborn child being male or female. These experiments are called **Bernoulli experiments**, and a *Bernoulli random variable* maps the sample space of such experiments to the set $\{0, 1\}$. (See Figure 11.) The outcome corresponding to 1 is arbitrarily referred to as a "success" and that corresponding to 0 as a "failure."

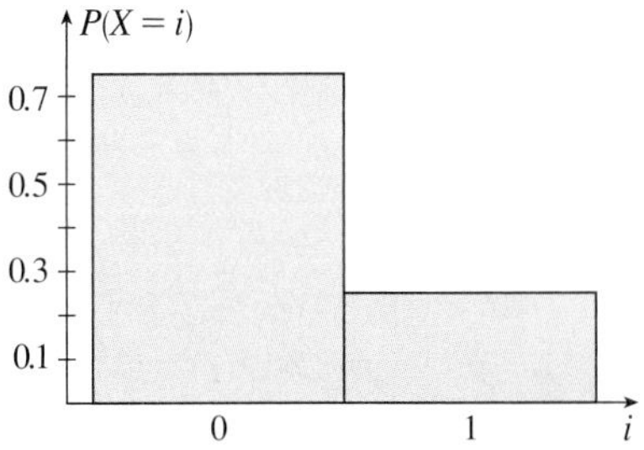

FIGURE 11
Histogram for a Bernoulli random variable

Because a Bernoulli random variable has only two possible outcomes, the convention is to use p (with no subscript) to denote the probability of success and $1 - p$ to denote the probability of failure.

(11) Definition A **Bernoulli random variable** X with parameter p takes values from the set $\{0, 1\}$, with $P(X = 1) = p$ and $P(X = 0) = 1 - p$.

EXAMPLE 9 | Calculate the mean and variance of a Bernoulli random variable.

SOLUTION From Definition 8, the mean is

$$E[X] = 0(1 - p) + 1 \cdot p = p$$

From the second formula in Definition 10, the variance is

$$\mathrm{Var}[X] = 0^2(1 - p) + 1^2 p - p^2 = p(1 - p)$$

Binomial Random Variables

A *binomial random variable* generalizes a single Bernoulli experiment to the case of n independent Bernoulli experiments. Examples include the number of species present from a set of n possible species, the number of survivors in a group of n individuals, or the number of females in a family of n children.

A simple example involves coin tosses. Tossing a single (potentially biased) coin can be represented by a Bernoulli random variable Y in which the outcomes correspond to $Y = 1$ for heads and $Y = 0$ for tails. Now imagine flipping three identical coins, each showing heads ($Y = 1$) with probability p and tails ($Y = 0$) with probability $1 - p$. Let's assume that the outcomes for each coin are mutually independent, and let's define the random variable X to be the total number of heads obtained.

What is the probability that $X = 0$? To obtain $X = 0$, all three coins must show tails. Because the event "tails" for each coin toss occurs with probability $1 - p$, and because all three coin tosses are assumed to be mutually independent, this outcome occurs with probability $(1 - p)^3$ (from the definition of mutual independence on page 756). Thus

(12) $$P(X = 0) = (1 - p)^3$$

To obtain $X = 1$, we need one coin to show heads and the other two to show tails. We can count the number of ways that this might occur by noting that this number corresponds to the number of combinations of three objects (the coins) when taken one at a time (the coin that shows heads). From Formula 12.1.7, this number of combinations is

$$C_{3,1} = \binom{3}{1} = \frac{3!}{1!(3 - 1)!} = \frac{6}{2} = 3$$

The three possibilities, along with their probabilities of occurrence, are shown in Figure 12.

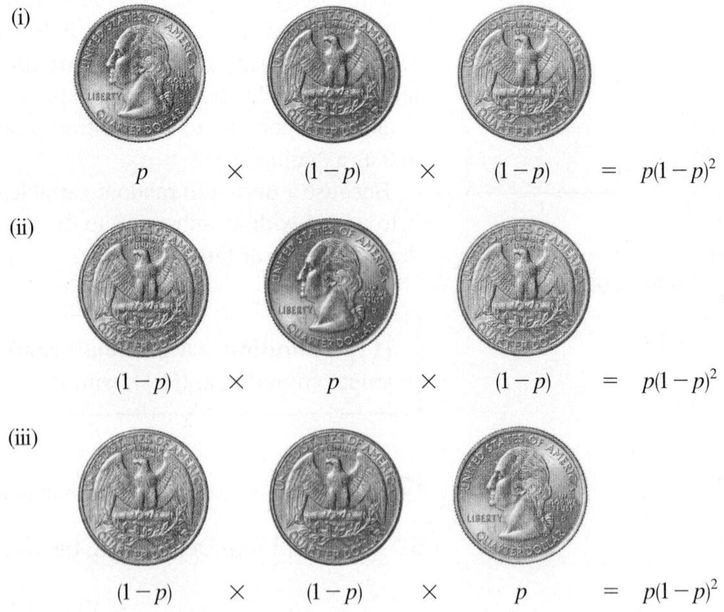

(i)
$p \times (1-p) \times (1-p) = p(1-p)^2$
(ii)
$(1-p) \times p \times (1-p) = p(1-p)^2$
(iii)
$(1-p) \times (1-p) \times p = p(1-p)^2$

FIGURE 12
The head can show on either the red, the blue, or the yellow coin. The probabilities follow from the fact that the three coin tosses are mutually independent.
Quarters: © Asaf Eliason / Shutterstock.com

Thus

(13) $$P(X = 1) = \binom{3}{1} p(1 - p)^2 = 3p(1 - p)^2$$

In a similar fashion, to obtain $X = 2$ we need two coins to show heads and one to show tails. The number of ways that this might occur corresponds to the number of combinations of three objects (the coins) when taken two at a time (the two coins showing heads). Furthermore, the probability of each of these outcomes is $p^2(1 - p)$ because each corresponds to the occurrence of two heads and one tail. Therefore

(14) $$P(X = 2) = \binom{3}{2} p^2(1 - p) = 3p^2(1 - p)$$

Finally, since there is only one way to obtain $X = 3$ (namely, all three coins show heads), from the mutual independence of the coin tosses we have

(15) $$P(X = 3) = p^3$$

Equations 12–15 specify the probability density function for X.

The preceding considerations can be extended to the case of n identical Bernoulli experiments, where X represents the total number of successes.

(16) Definition A **binomial random variable** X with parameters n and p takes values from the set $\{0, 1, \ldots, n\}$, and has probability density function

$$P(X = i) = \binom{n}{i} p^i(1 - p)^{n-i}$$

$$= \frac{n!}{i!(n - i)!} p^i(1 - p)^{n-i}$$

Figure 13 plots a histogram of the probability density function for a binomial random variable.

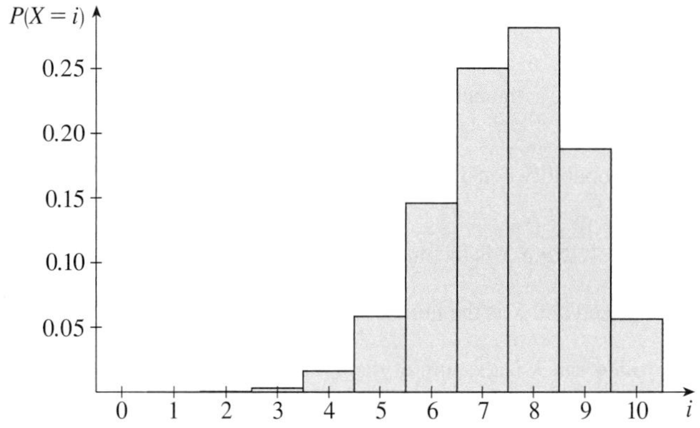

FIGURE 13
Histogram for a binomial random variable with parameters $n = 10$ and $p = 0.75$

EXAMPLE 10 | **Family sex ratio** In humans the probability that a newborn is a female is approximately 0.48. Consider a family with four children. What is the probability that three of the children will be girls if the children's sexes are mutually independent?

SOLUTION The sex of each of the four children can be modeled as a Bernoulli random variable, and these are assumed to be mutually independent. Therefore the number of girls can be modeled as a binomial random variable X with $n = 4$ and probability of success $p = 0.48$. The probability density function for X is

$$P(X = i) = \binom{4}{i} 0.48^i (1 - 0.48)^{4-i} = \frac{4!}{i!(4-i)!} 0.48^i \, 0.52^{4-i}$$

The probability that three of the children are girls is therefore

$$\begin{aligned} P(X = 3) &= \frac{4!}{3!(4-3)!} \times 0.48^3 \times 0.52^{4-3} \\ &= \frac{4 \cdot 3 \cdot 2 \cdot 1}{3 \cdot 2 \cdot 1} \times 0.48^3 \times 0.52 \\ &= 4 \times 0.48^3 \times 0.52 \\ &\approx 0.23 \end{aligned}$$

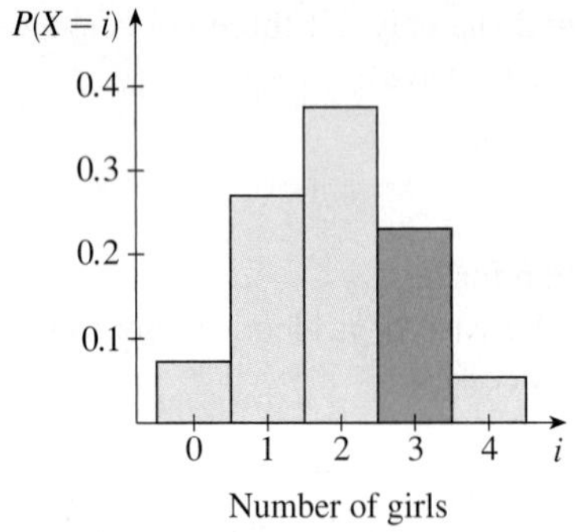

FIGURE 14

This calculation reflects the fact that there are four ways to have three girls in this family—the single boy can be born first, second, third, or fourth. Furthermore, the probability of each of these outcomes is $p^3(1 - p) = 0.48^3 \times 0.52$ because each outcome corresponds to the birth of three girls and one boy. Figure 14 displays the probability density function with the outcome $X = 3$ highlighted. ■

EXERCISES 12.4

1–2 (a) Determine the values that the random variable X can take on and specify its probability density function in a table. (b) Draw a histogram for X.

1. A jar contains five balls, numbered 1 to 5. A ball is drawn at random and X is the number on the ball.

2. A jar contains five balls numbered 1, three balls numbered 2, one ball numbered 3, and one ball numbered 4. A ball is drawn at random and X is the number on the ball.

3–7 Determine the probability density function for the random variable X.

3. Three coins are tossed and X is the number of tails.

4. Four coins are tossed and X is the number of heads.

5. Two dice are rolled and X is the sum of the numbers.

6. Two dice are rolled and X is the larger of the two numbers.

7. Two dice are rolled and X is the absolute value of the difference between the two numbers.

8–10 Determine the probability density function for the random variable X.

8. Two cards are removed from a standard deck and X is the number of spades in this sample.

9. Three cards are removed from a standard deck and X is the number of face cards in this sample.

10. Four cards are removed from a standard deck and X is the number of aces in this sample.

11–13 A jar contains seven white balls and two red balls. Determine the probability density function for the random variable X.

11. One ball is removed and X is the number of red balls in this sample.

12. Two balls are removed and X is the number of red balls in this sample.

13. Three balls are removed and X is the number of red balls in this sample.

14. A probability density function for the random variable X is given by $p_i = ki$, where k is a constant.
(a) What value must k be if X takes on integer values between 0 and 15, inclusive?
(b) What value must k be if X takes on integer values between 0 and n, inclusive?

15. A probability density function for the random variable X is given by $p_i = k(1/4)^i$, where k is a constant.
(a) What value must k be if X takes on integer values between 1 and 10, inclusive?

(b) What value must k be if X takes on integer values between 1 and n, inclusive?

16. A random variable takes on values in the set $\{0, 1, 2\}$ with probabilities $p_0 = \frac{1}{4}$, $p_1 = \frac{1}{2}$, and $p_2 = \frac{1}{4}$.
(a) Construct a histogram for the random variable.
(b) Construct a table for the CDF.
(c) Plot the CDF below the histogram from part (a).

17. A random variable takes on values in the set $\{0, 1, 2, 3\}$ with probabilities $p_0 = \frac{1}{3}$, $p_1 = \frac{1}{3}$, $p_2 = \frac{1}{4}$, $p_3 = \frac{1}{12}$.
(a) Construct a histogram for the random variable.
(b) Construct a table for the CDF.
(c) Plot the CDF below the histogram from part (a).

18. A random variable takes on integer values from 1 to 15, inclusive, and has PDF given by $p_i = \frac{1}{15}$.
(a) Construct a histogram for the random variable.
(b) Obtain an equation for the CDF.
(c) Plot the CDF below the histogram from part (a).

19. A random variable takes on integer values from 1 to 12, inclusive, and has PDF given by $p_i = i/78$.
(a) Construct a histogram for the random variable.
(b) Obtain an equation for the CDF.
(c) Plot the CDF below the histogram from part (a).

20. A random variable has CDF given by

$$F_i = \begin{cases} 0 & i = 0 \\ 0.1 & i = 1 \\ 0.2 & i = 2 \\ 0.5 & i = 3, 4 \\ 1 & i = 5 \end{cases}$$

defined for integer values of X between 0 and 5, inclusive. Construct a table for the PDF of X.

21. A random variable has CDF given by

$$F_i = \begin{cases} 0.5 & i = 0, 1, 2 \\ 1 & i = 3 \end{cases}$$

defined for integer values of X between 0 and 3, inclusive. Construct a table for the PDF of X.

22. A random variable takes on integer values between 1 and 8, inclusive, and has PDF $p_i = \frac{1}{8}$. What is the probability that it takes on a value between 3 and 6, inclusive?

23. A random variable has PDF $p_i = 2(1/3)^i$ and takes on values in the positive integers. What is the probability that it takes on a value between 10 and 20, inclusive?

24. A random variable has PDF $p_i = \frac{1}{2870} i^2$ and takes on values between 0 and 20, inclusive. What is the probability that it takes on a value between 5 and 10, inclusive? The following formula may be useful:

$$\sum_{i=1}^{n} i^2 = \frac{n(n+1)(2n+1)}{6}$$

25. A random variable has PDF $p_i = \frac{1}{3025} i^3$ and takes on values between 0 and 10, inclusive. What is the probability that it takes on a value between 2 and 6, inclusive? The following formula may be useful:

$$\sum_{i=1}^{n} i^3 = \frac{n^2(n+1)^2}{4}$$

26. Infant fever The number of times an infant has a fever during the first two years of life is a random variable X with PDF

$$p_i = \begin{cases} 0.05 & i = 0 \\ 0.31 & i = 1 \\ 0.41 & i = 2 \\ 0.11 & i = 3 \\ 0.10 & i = 4 \\ 0.02 & i = 5 \end{cases}$$

(a) Construct a table for the CDF, and interpret it.
(b) What is the probability that an infant will have at least two fevers?

27. Community ecology The number of different mammal species observed along a transect through a forest is a random variable X with CDF

$$F_i = \begin{cases} 0.01 & i = 0 \\ 0.16 & i = 1 \\ 0.36 & i = 2 \\ 0.71 & i = 3 \\ 0.96 & i = 4 \\ 1 & i = 5 \end{cases}$$

(a) What is the biological interpretation of F_3?
(b) Construct a table for the PDF, and plot the histogram.

28–36 In the games described: (a) find the expected value of your winnings and (b) find the variance of your winnings

28. You win \$2 if a coin toss shows heads or \$1 if it shows tails.

29. You win \$10 if a die roll shows a 6 or lose \$1 otherwise.

30. A card is drawn from a standard deck. You win \$100 if you draw the ace of spades or lose \$1 if you draw any other card.

31. You win \$3 if a coin toss shows heads or \$2 if it shows tails.

32. You win \$3 if a die roll shows a 6 or win \$0.50 otherwise.

33. You win \$2 if a die roll shows an even number or pay \$2 otherwise.

34. A card is drawn from a deck. You win \$104 if the card is an ace, \$26 if it is a face card, or \$13 if it is the 8 of clubs.

35. A bag contains two silver dollars and eight slugs. You pay 50 cents to reach into the bag and take a coin, which you get to keep.

36. A bag contains eight white balls and two black balls. You pick two balls at random from the bag. You win \$5 if you do not pick a black ball.

37–40 Guess the expected value of the random variable using the center-of-mass interpretation.

37.

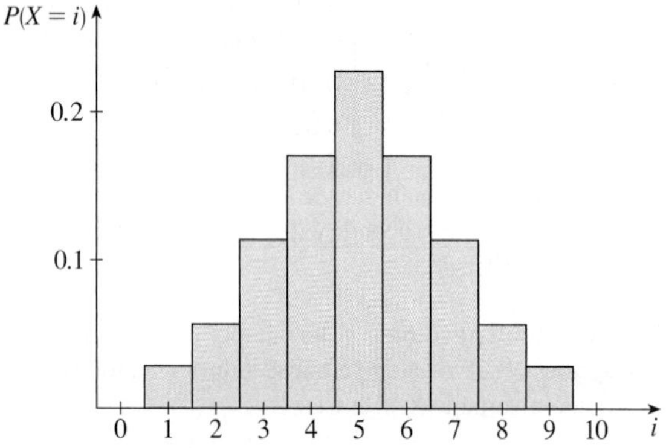

38.

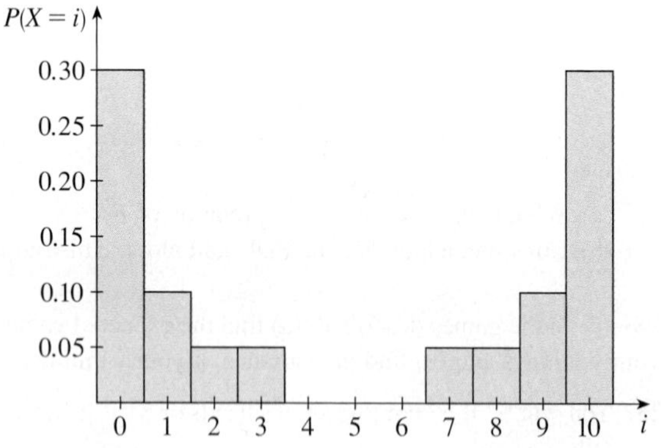

39.

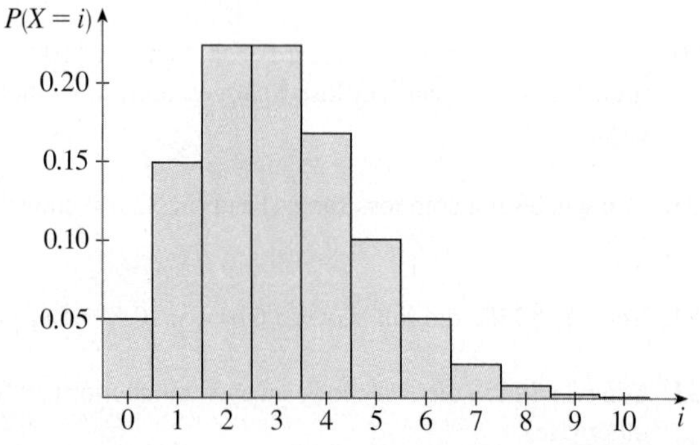

40.

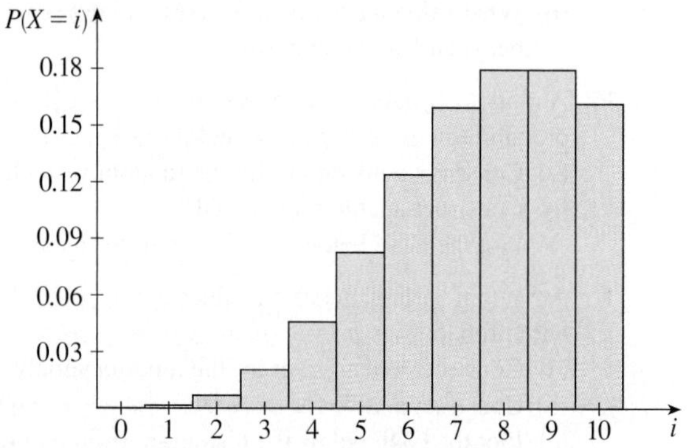

41. Clutch size refers to the number of eggs in a bird's nest. The clutch size of bald eagles is a random variable with PDF $p_1 = 0.3$, $p_2 = 0.55$, and $p_3 = 0.15$. Calculate the mean and variance of the clutch size.

42. *Drosophila* development time is a random variable with PDF $p_{13} = 0.02$, $p_{14} = 0.35$, $p_{15} = 0.35$, $p_{16} = 0.20$, and $p_{17} = 0.08$, where time is measured in days. Calculate the mean and variance of the development time.

43. Infant fever The number of times an infant has a fever during the first two years of life is a random variable X with PDF

$$p_i = \begin{cases} 0.05 & i = 0 \\ 0.31 & i = 1 \\ 0.41 & i = 2 \\ 0.11 & i = 3 \\ 0.10 & i = 4 \\ 0.02 & i = 5 \end{cases}$$

Calculate the mean and variance of X.

44. Derive the equation $\text{Var}[X] = \sum_i i^2 p_i - E[X]^2$.

45–48 A Bernoulli experiment with probability of success p is performed n times.
(a) Make a table of the probability density function.
(b) Draw a histogram for the random variable.

45. $n = 4,\ p = 0.5$

46. $n = 5,\ p = 0.4$

47. $n = 7,\ p = 0.2$

48. $n = 6,\ p = 0.9$

49. You roll a die three times. What is the probability of getting three 1's?

50. You toss a coin four times. What is the probability of getting an equal number of heads and tails?

51. An archer hits his target 80% of the time. If he shoots seven arrows, what is the probability of each event?
(a) He never hits the target.
(b) He hits the target every time.

(c) He hits the target more than once.
(d) He hits the target at least five times.

52. Five independent trials of a Bernoulli experiment with probability of success $\frac{1}{3}$ are performed. Find the probability of each event.
(a) Exactly two successes
(b) Exactly three successes
(c) No successes
(d) All successes
(e) Exactly one success
(f) Exactly one failure
(g) At least four successes
(h) At least three successes

53. Sickle-cell anemia is a hereditary blood disorder caused by a recessive gene. If both parents carry the gene but do not have the disease, then there is a 0.25 probability that an offspring will develop the disease. A couple finds through genetic testing that they both carry the gene (but do not have the disease). If they intend to have four children, find the probability of each event.
(a) At least one child develops the disease.
(b) At least three of the children develop the disease.

54. Medical testing A medical test correctly diagnoses a person having a disease with probability 0.8. Suppose that 100 people are to be tested, and 10 of these actually have the disease. Find the probability of each event.
(a) All ten cases will be detected.
(b) One case will be missed.
(c) At least two cases will be missed.

55. Rabies Health authorities estimate that 10% of the raccoons in a certain rural county are carriers of rabies. A dog is bitten by four different raccoons in this county. What is the probability that he was bitten by at least one rabies carrier?

56. Blood typing About 45% of the population of the United States and Canada have type O blood.
(a) If a random sample of 10 people is selected, what is the probability that exactly 5 have type O blood?
(b) What is the probability that at least 3 of the random sample of 10 have type O blood?

57. Handedness A psychologist needs 12 left-handed subjects for an experiment, and she interviews 15 potential subjects. About 10% of the population is left-handed.
(a) What is the probability that exactly 12 of the potential subjects are left-handed?
(b) What is the probability that 12 or more are left-handed?

58. Offpsring sex ratio Assume that for any given live human birth, the chances that the child is a boy or a girl are equally likely.
(a) What is the probability that in a family of five children a majority are boys?
(b) What is the probability that in a family of seven children a majority are girls?

59. Offpsring sex ratio The ratio of male to female births is in fact not exactly one to one. The probability that a newborn is a male is nearly 0.52. A family has ten children.
(a) What is the probability that all 10 children are boys?
(b) What is the probability that all are girls?
(c) What is the probability that 5 are girls and 5 are boys?

60. Testing drug effectiveness A certain disease causes death with probability 0.6. A new drug is tested for its effectiveness against this disease. Ten patients are given the drug, and 8 of them recover.
(a) Find the probability that 8 or more of the patients would have recovered without the drug.
(b) Does the drug appear to be effective? [Consider the drug to be effective if the probability in part (a) is 0.05 or less.]

61. Natural killer (NK) cells are a type of lymphocyte, and approximately 7% of lymphocytes are NK cells in healthy individuals. Suppose a sample of 10 lymphocytes is taken from a patient. Assuming the patient is healthy, calculate each of the following probabilities.
(a) None of the cells are NK cells.
(b) Two of the cells are NK cells.
(c) No more than three of the cells are NK cells.

62. GC content The fraction of all nucleotides in an organism's genome that are either guanine or cytosine is referred to as its GC content. Suppose the GC content is 30% and you sample 10 nucleotides.
(a) What is the probability of getting 3 nucleotides that are either guanine or cytosine?
(b) What is the probability of getting more than 8 nucleotides that are either guanine or cytosine?

63. Viral identification and GC content Influenza A viruses infect both birds and humans, but the fraction of all nucleotides in their genomes that are either guanine or cytosine (referred to as GC content) is higher in viruses found in birds than in viruses found in humans. This fact can be used to help determine the source of a given virus. Suppose the genome of bird influenza has a GC content of 50%, whereas that of human influenza has a GC content of 30%. Suppose further that 12 nucleotides are sampled from an unknown virus, and 3 of them are either guanine or cytosine.
(a) What is the probability of obtaining the sample if the virus is from birds?
(b) What is the probability of obtaining the sample if the virus is from humans?
(c) We can use the sample to estimate the GC content of the virus from which it came. Suppose the GC content of the virus, written as a decimal, is p. What is the probability of obtaining the sample as a function of p?
(d) The function obtained in part (c) is referred to as a *likelihood function*. The *maximum likelihood estimate*

for the GC content of the sampled virus is the value of p that maximizes the likelihood function. Find this value of p.

64. Maximum likelihood estimate of sex ratio Imagine that you wish to estimate the fraction of a population that is female. Suppose you take a random sample of n individuals from the population, and k of them are female. Suppose further that the population is large enough relative to n that your sample has a negligible effect on the fraction of females in the remaining population. In this case we can model the number of females in the sample as a binomial random variable with unknown parameter p.

(a) What is the probability of obtaining the observed sample as a function of p?

(b) The function obtained in part (a) is referred to as a *likelihood function*. The *maximum likelihood estimate* for p is the value of p that maximizes the likelihood function. Prove that the maximum likelihood estimate of the fraction of females in the population is $\hat{p} = k/n$, where the hat indicates that this value of p is the maximum likelihood estimate.

65. Consider a binomial random variable X with parameters n and p. Prove that the mean of X is np.

Hint: Use the definition of expected value, factor out np, and use the Binomial Theorem

$$(x + y)^n = \sum_{k=0}^{n} \binom{n}{k} x^k y^{n-k}$$

66–69 The following exercises involve hypergeometric random variables.

66. *Binomial and hypergeometric random variables*
A jar contains five black balls and three white balls.

(a) You draw a ball, record its color, and return it to the jar (this is called *sampling with replacement*). If you repeat this process three times, what is the probability that two of the sampled balls are black? [*Note:* This is given by the binomial probability density function.]

(b) You draw three balls without replacement and record their color (this is called *sampling without replacement*). What is the probability that two of the sampled balls are black? [*Note:* This is given by what is called the *hypergeometric* probability density function.]

67. Hypergeometric random variables A jar contains b black balls and w white balls, where $b + w = N$. Suppose we draw n balls *without replacement*. Show that the probability of obtaining exactly k black balls is

$$p_k = \frac{\binom{b}{k}\binom{w}{n-k}}{\binom{N}{n}}$$

68. Nesting sites A total of 100 suitable nest sites are available for a migratory bird species: 35 sites are on a cliff and 65 are in an open field. Female birds return from migration one at a time and choose a nest site. If there is no preference for one type of site over another, what is the probability that half of the first 10 birds to return will choose cliff sites? (Use the equation of Exercise 67 to answer this question.)

69. Foraging behavior Bivalves and chironomids are two types of prey eaten by certain fish species. In an experiment, 50 individuals of each type were presented to a fish, which ate 30 bivalves and 15 chironomids. What is the probability of this choice of 45 food items if the fish has no preference for one type over the other? (Use the equation of Exercise 67 to answer this question.)

---

70. A fair coin is tossed repeatedly until it shows heads. The number of tosses required for this to occur is a random variable X.

(a) What is the probability that $X = 1$?

(b) What is the probability that $X = 2$?

(c) What is the probability that $X = 3$?

(d) What is the probability that $X = k$?

71. *Geometric random variables* A Bernoulli experiment with probability of success p is repeated indefinitely until the first success occurs. The number of trials required for this to occur is called a *geometric random variable X*.

(a) Show that the PDF of a geometric random variable is $P(X = i) = p(1 - p)^{i-1}$.

(b) Use the PDF from part (a) to find the CDF.

72. Courtship displays Male ruby-throated hummingbirds exhibit dramatic courtship displays in which they fly straight upward for nearly 15 m and then dive straight downward at high speeds. The number of such displays required before a male is actually chosen as a mate can be described by a geometric random variable with PDF

$$P(X = i) = \tfrac{1}{4}\left(\tfrac{3}{4}\right)^{i-1}$$

with $i \geq 1$. What is the probability that a male will display at least six times before being chosen as a mate?

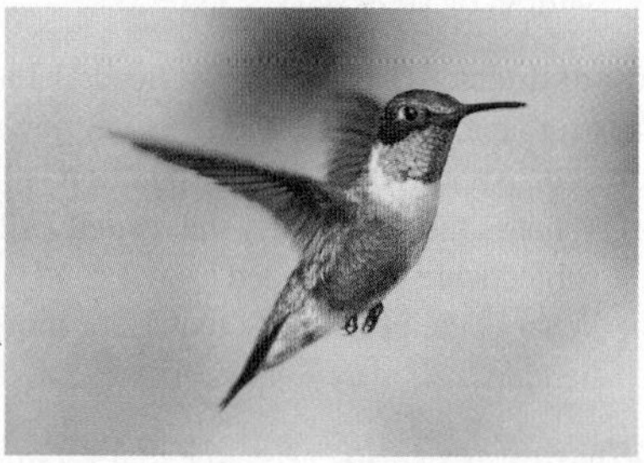

73. Extinction An endangered species is censused every year and has a constant probability of 0.01 of going extinct between each census. What is the probability that the population will go extinct in year 5?

■ PROJECT DNA Supercoiling

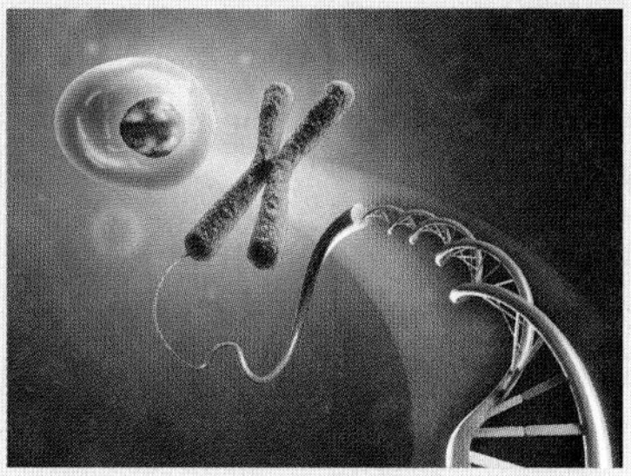

When DNA is packaged into chromosomes it is often coiled and twisted to make it more compact. This is called supercoiling. One of the simplest types of supercoils is called a *plectoneme*, which occurs when a strand of DNA is twisted into loops, as shown in Figure 1.

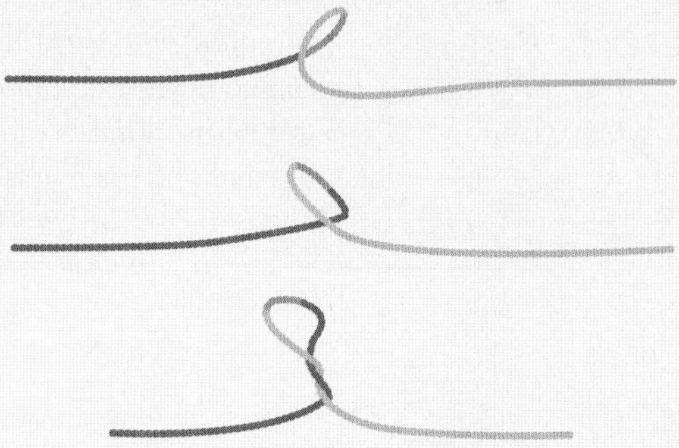

FIGURE 1

The formation of plectonemes is a dynamic process: plectonemes having different numbers of twists continually form and disappear at locations along the DNA. Probabilistic mathematical models have been used to describe this process.

We will model a single plectoneme by keeping track of the number of twists it contains. A twist can be left- or right-handed, and therefore we can describe a plectoneme mathematically by an integer, with negative values indicating the number of left-hand twists and positive values indicating the number of right-hand twists. A value of 0 corresponds to the plectoneme disappearing (see Figure 2).

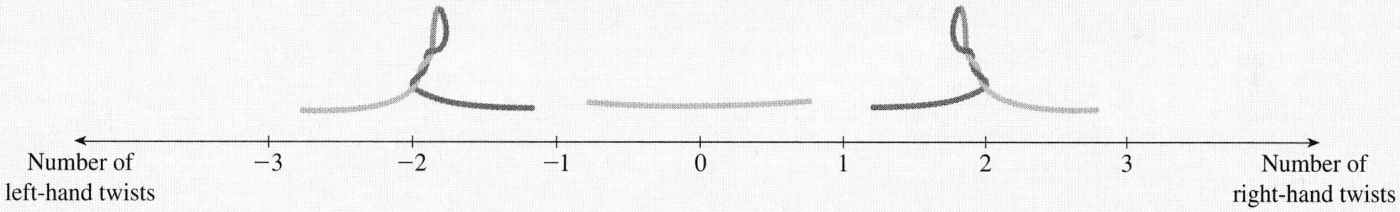

FIGURE 2

We model the change in the number of twists in discrete time. Suppose that, in each time step, a twist occurs in the positive or negative direction, each twist happening with probability $\frac{1}{2}$. Thus, as time passes, the plectoneme will twist one way or the other randomly, sometimes "passing through zero" and changing from a right-hand to a left-hand plectoneme, or vice versa.

We use X_n to denote a random variable that specifies the state of the plectoneme at time step n. As time passes, the plectoneme state is therefore described by an integer that changes every time step, going up or down one unit with equal probability. We seek to describe this process mathematically by determining the probability that the plectoneme is in state i at time step n, which we denote by $p_{i,n}$.

Suppose the plectoneme starts in state $i = 0$ (that is, there is no plectoneme present). First notice that the plectoneme can be in even-numbered states only at even-numbered time steps.

Here we are using the term *sequence* in its everyday sense, not in its mathematical sense.

1. Consider the state of the plectoneme at time step 1 . How many different sequences of left- or right-hand twisting and untwisting are possible in one time step?

2. Consider the state of the plectoneme at time step 2. How many different sequences of left- or right-hand twisting and untwisting are possible in two time steps?

3. Consider the state of the plectoneme at time step n. How many different sequences of left- or right-hand twisting and untwisting are possible in n time steps?

4. For any given time step all possible sequences are equally likely. Therefore, from Definition 12.2.2, we can calculate $p_{i,n}$ by finding the total number of sequences that end up at state i in n steps, and then divide this by the result from Problem 3. Let r be the number of steps to the right in Figure 2 (that is, in the positive direction) and l be the number of steps to the left in the figure (that is, in the negative direction). We have $n = r + l$ and, to have arrived at state i, we must also have $i = r - l$. Solve these two equations for r and l. [*Note:* If the result is not an integer, then state i cannot be reached in n steps; this will occur if n is odd and i is even, or vice versa].

5. Express the number of sequences that end up at state i in n steps as a binomial coefficient involving n and i.

6. Obtain an expression for $p_{i,n}$.

This probabilistic model is also called a *random walk*. It has been used to describe the distribution of life spans of plectonemes in DNA supercoiling, where a plectoneme's life is considered to end when the state i passes through zero. Figure 3 shows the model's predictions along with a plot of data.

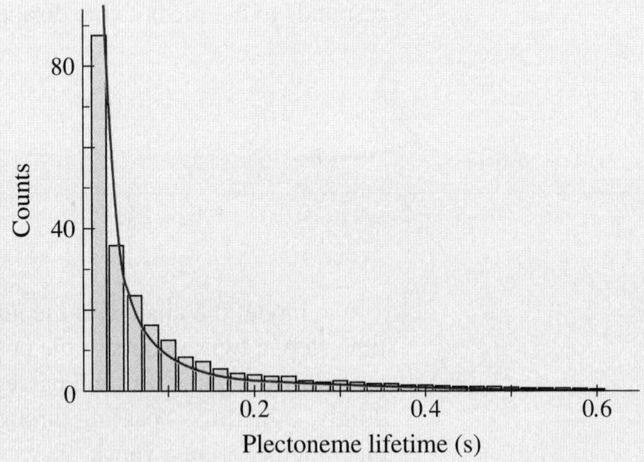

FIGURE 3
Blue bars are data and red curve is model prediction.

Sourse: M. van Loenhout et al., "Dynamics of DNA Supercoils," *Science* 338 (2012): 94–97.

■ PROJECT The Probability of an Avian Influenza Pandemic in Humans

When humans come into close contact with birds, such as waterfowl or chickens, influenza viruses can be transferred from the birds to the humans. Suppose a single person (referred to as *patient zero*) has become infected with an avian flu virus. At this point there are two possible outcomes:

(i) patient zero can infect other people who themselves infect other people, and so on, resulting in a growing pandemic; or

(ii) although patient zero might infect other people, by chance the chain of transmission ends. This results in the extinction of the avian flu virus in the human population, and no pandemic occurs.

In Example 1.5.5 we introduced the following equation for the probability P that a growing pandemic occurs:

(1) $$P = 1 - \frac{1}{R_0}$$

where R_0 is the expected number of new infections generated by a single infected individual. Here we will derive this equation.

Suppose the number of new infections caused by patient zero is a random variable X with PDF p_i. These new infections are referred to as *secondary infections*. A commonly used PDF for the number of secondary infections generated by patient zero is $p_i = r^i(1 - r)$, where r is a positive constant that is strictly smaller than one and i is a nonnegative integer. Furthermore, since R_0 is the expected number of infections generated by patient zero, we have

(2) $$E[X] = \sum_i r^i(1 - r)i = \frac{r}{1 - r} = R_0$$

1. Let Q denote the probability that extinction ultimately occurs (with $Q = 1 - P$). It can be shown that the Law of Total Probability (12.3.6) is valid even when the upper limit of summation is infinite, provided that the resulting series is defined. Using this fact, and conditioning on the number of secondary infections caused by patient zero, show that

(3) $$Q = \sum_{i=0}^{\infty} q_i r^i(1 - r)$$

where q_i is the probability of ultimate extinction, given that there are i secondary infections.

2. Suppose that each secondary infection is identical to patient zero, and each of these infections spawns their own independent chain of transmission events. Thus the outbreak stemming from each secondary infection will ultimately go extinct with probability Q. Since the chains of transmission spawned by all secondary infections are assumed to be mutually independent, it can be shown that the probability that all infections in the population go extinct is $q_i = Q^i$. Use this result in (3) to obtain an expression giving Q as a function of r.

3. Use your answer to Problem 2, along with Equation 2 and the fact that $Q = 1 - P$, to obtain Equation 1.

Equation 1 shows that the probability of a pandemic increases with the expected number of secondary infections generated R_0, but it does so in a decelerating way. For example, if on average every infected person infects two others, then $P = \frac{1}{2}$. But if on average every infected person infects four others, then $P = \frac{3}{4}$. Thus doubling R_0 does not double the probability of a pandemic. However, in the limit as $R_0 \to \infty$, the probability of a pandemic asymptotically approaches 1.

12.5 Continuous Random Variables

Suppose we are interested in the cholesterol level of a person chosen at random, or the height of an adult female chosen at random, or the lifetime of a bird chosen at random from a certain population. Such quantities are called *continuous random variables* because their values range over an interval of real numbers.

Like discrete random variables, a continuous random variable is defined formally as a function whose domain is the sample space of some experiment and whose range is a subset of the real numbers.

> **(1) Definition** A **continuous random variable** X is a function from a sample space Ω onto an *interval* of the real numbers.

As with discrete random variables, in practice this formal definition is often sidestepped by specifying continuous random variables directly, without reference to the more detailed underlying sample space.

■ Describing Continuous Random Variables

Consider the level of high-density lipoprotein (HDL) cholesterol in a randomly chosen male from the United States. Suppose we construct a histogram by creating nonoverlapping bins of equal length along the horizontal axis and then lumping together all measurements that fall within a given bin [see Figure 1(a)].

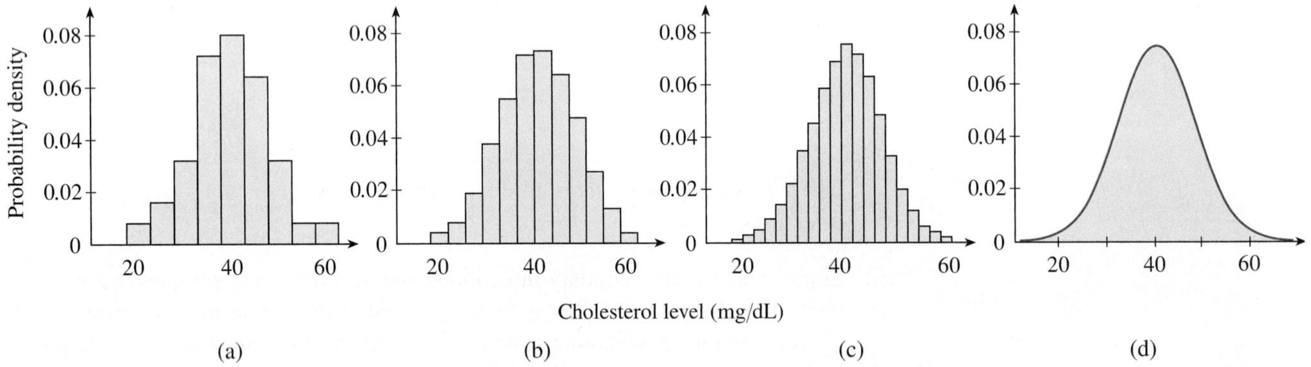

FIGURE 1

We can make the widths of the bins as narrow as we like because cholesterol level is a continuous random variable. For any bin width, however, let's choose the bar heights so that the area of a bar is always equal to the fraction of the population falling into that bin. Therefore the total area of the histogram is always equal to 1. Then, as we make the bins more and more narrow [as in Figures 1(b) and (c)], we might expect the histogram to more and more closely resemble a continuous curve as shown in Figure 1(d). This limiting process is reminiscent of our introduction to integration in Section 5.1.

All of the continuous random variables that we will consider can be described graphically by such continuous curves. Figure 2 displays a typical example. The function specifying this curve is called the *probability density function* $f(x)$ for the random variable X. By convention, the lowercase letter x is used to denote a particular value taken on by the random variable X.

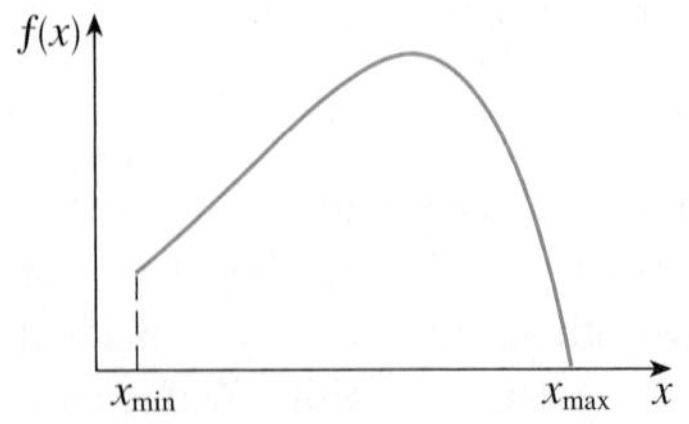

FIGURE 2

The probability density function for a continuous random variable plays a role similar to that for a discrete random variable. We can no longer talk about the probability of a single value of X occurring because there are no longer histogram bars associated with each possible value of X. But we can still visualize the probability of X falling within any given subinterval by using integration to calculate the area under the graph within this subinterval.

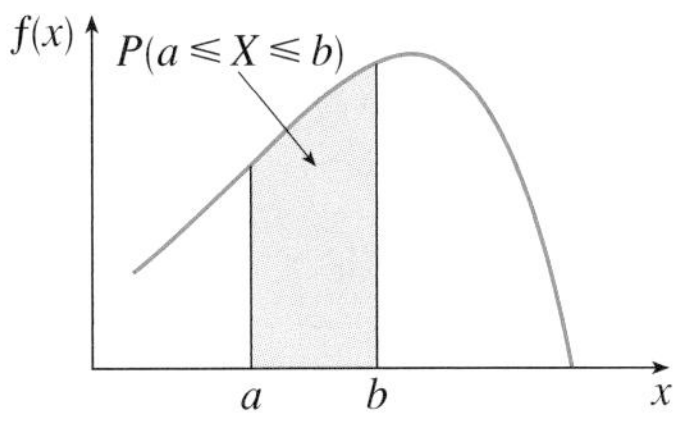

FIGURE 3

(2) Definition The **probability density function (PDF)** for the continuous random variable X is a function $f(x)$ such that

$$P(a \leqslant X \leqslant b) = \int_a^b f(x)\,dx$$

for any a and b (see Figure 3).

For the examples encountered in this book, the quantity $P(a \leqslant X \leqslant b)$ can be replaced with $P(a \leqslant X < b)$, $P(a < X \leqslant b)$, or $P(a < X < b)$ because, from Definition 2, the probability of any single value of X is 0.

In words, Definition 2 says that we obtain the probability of X falling within any subinterval of interest by calculating the definite integral of the probability density function over that subinterval.

EXAMPLE 1 | **Female height** Figure 4 shows the graph of a model for the probability density function f for a random variable X defined to be the height (in inches) of an adult female in the United States (according to data from the National Health Survey). The probability that the height of a woman chosen at random from this population is between 60 and 70 inches is equal to the area under the graph of f from 60 to 70.

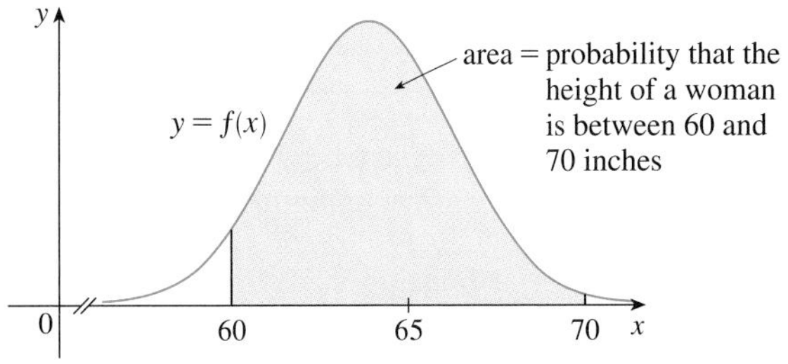

FIGURE 4
Probability density function for the height of an adult female

■

As with discrete random variables, the probability density function $f(x)$ of a continuous random variable X satisfies the condition

$$f(x) \geqslant 0 \tag{3}$$

Furthermore, since the union of all possible values of X corresponds to the entire sample space, $f(x)$ must also satisfy

$$\int_{-\infty}^{\infty} f(x)\,dx = 1 \tag{4}$$

For many continuous random variables the PDF $f(x)$ is nonzero only on some interval of the real numbers. In such cases we can restrict the limits of integration to this interval.

Thus the total area under the graph of the PDF equals 1.

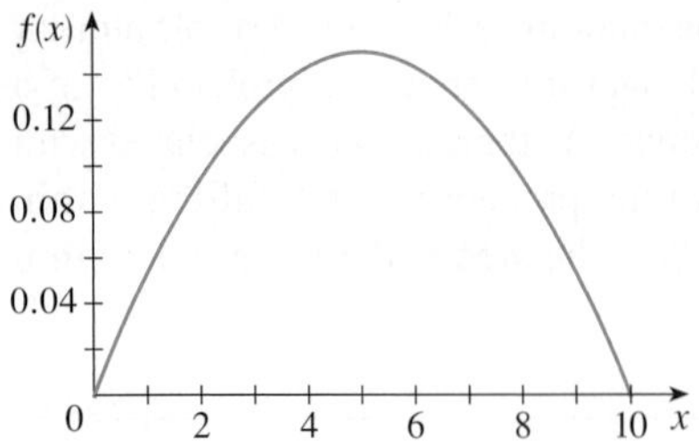

FIGURE 5

EXAMPLE 2 | Let $f(x) = 0.006x(10 - x)$ for $0 \leq x \leq 10$ and $f(x) = 0$ for all other values of x. Figure 5 displays a graph of this function.
(a) Verify that f satisfies Conditions 3 and 4 for probability density functions.
(b) Find $P(4 \leq X \leq 8)$.

SOLUTION

(a) For $0 \leq x \leq 10$ we have $0.006x(10 - x) \geq 0$, so $f(x) \geq 0$. Next we need to check that Condition 4 is satisfied:

$$\int_{-\infty}^{\infty} f(x)\,dx = \int_0^{10} 0.006x(10 - x)\,dx = 0.006\int_0^{10} (10x - x^2)\,dx$$

$$= 0.006\Big[5x^2 - \tfrac{1}{3}x^3\Big]_0^{10} = 0.006\left(500 - \tfrac{1000}{3}\right) = 1$$

Therefore f satisfies Conditions 3 and 4.

(b) The probability that X lies between 4 and 8 is illustrated in Figure 6, and is calculated as

$$P(4 \leq X \leq 8) = \int_4^8 f(x)\,dx = 0.006\int_4^8 (10x - x^2)\,dx$$

$$= 0.006\Big[5x^2 - \tfrac{1}{3}x^3\Big]_4^8 = 0.544$$

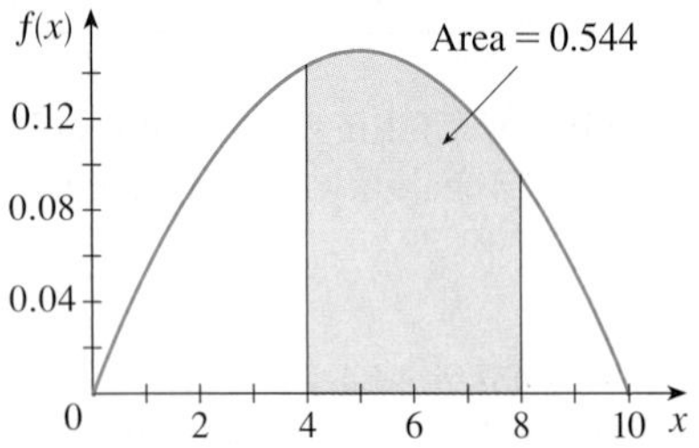

FIGURE 6

The *cumulative distribution function* for a continuous random variable is defined in an analogous way to that for a discrete random variable.

> **(5) Definition** The **cumulative distribution function (CDF)** for the continuous random variable X is the function $F(x)$ defined by
>
> $$F(x) = P(X \leq x)$$

The CDF $F(x)$ is the area under the graph of the PDF to the left of the value $X = x$ as shown in Figure 7. It is calculated from the probability density function with the integral

(6) $$F(x) = \int_{-\infty}^{x} f(s)\,ds$$

which is defined on the same domain as the probability density function $f(x)$.

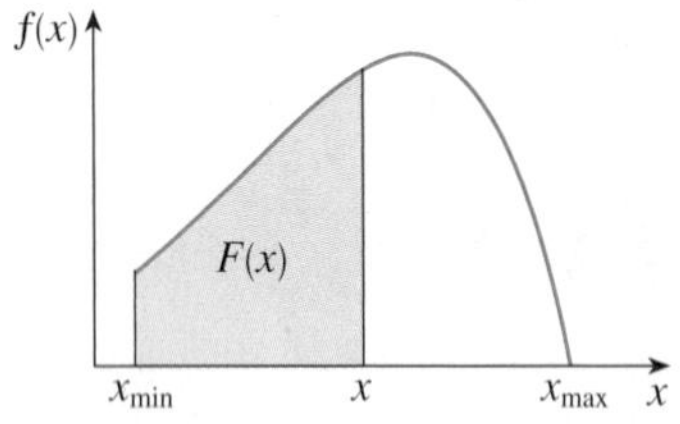

FIGURE 7

EXAMPLE 3 | Find the cumulative distribution function (CDF) for the PDF given in Example 2 and plot this CDF below the PDF to illustrate the relationship between the two.

SOLUTION The PDF in Example 2 is defined on the interval [0, 10] and the CDF is defined on the same interval. From Equation 6, we have

$$F(x) = \int_0^x f(s)\,ds = \int_0^x 0.006(10s - s^2)\,ds$$

$$= 0.006\Big[5s^2 - \tfrac{1}{3}s^3\Big]_0^x = 0.006\left(5x^2 - \tfrac{1}{3}x^3\right)$$

Figure 8 shows the relationship between the area under the graph of the PDF $f(x)$ up to a specific point x^* [see Figure 8(a)] and the value of the CDF $F(x)$ at this point [see Figure 8(b)].

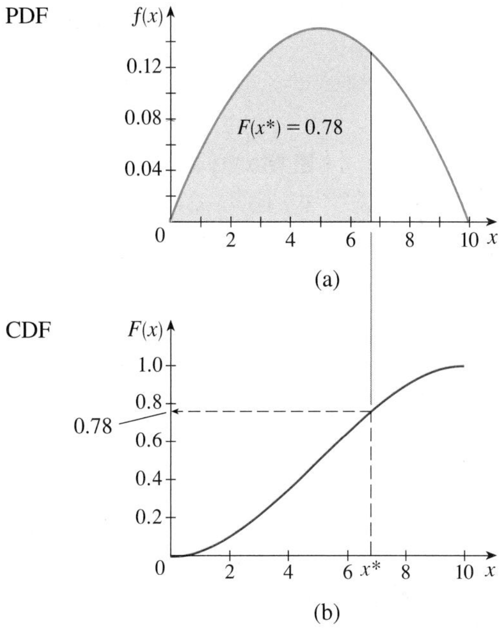

FIGURE 8

We have seen that tools from calculus (namely, integration) are central to working with continuous random variables. But the connection with calculus runs deeper than this. Note that the cumulative distribution function $F(x)$ defined in (6) is an antiderivative of the probability density function $f(x)$. In fact the Fundamental Theorem of Calculus (see Section 5.3) gives the following two important results.

1. The PDF of a random variable X can be obtained from the CDF via differentiation (FTC1):

(7) $$f(x) = F'(x)$$

2. The probability of X falling between any two values a and b can be obtained from the CDF as (FTC2):

(8) $$P(a \leqslant X \leqslant b) = \int_a^b f(x)\,dx = F(b) - F(a)$$

EXAMPLE 4 | A continuous random variable X is defined on the interval $[-\pi, \pi]$ and has cumulative distribution function

$$F(x) = \frac{1}{2\pi}(\pi + x + \sin x)$$

(a) Calculate $P(0 \leqslant X \leqslant \pi)$.
(b) What is the PDF of X?

SOLUTION

(a) From Equation 8 we have

$$\begin{aligned} P(0 \leqslant X \leqslant \pi) &= F(\pi) - F(0) \\ &= \frac{1}{2\pi}(\pi + \pi + \sin \pi) - \frac{1}{2\pi}(\pi + 0 + \sin 0) \\ &= 1 - \tfrac{1}{2} = \tfrac{1}{2} \end{aligned}$$

(b) From Equation 7 we have

$$f(x) = F'(x) = \frac{d}{dx}\left[\frac{1}{2\pi}(\pi + x + \sin x)\right] = \frac{1}{2\pi}(1 + \cos x)$$ ■

■ Mean and Variance of Continuous Random Variables

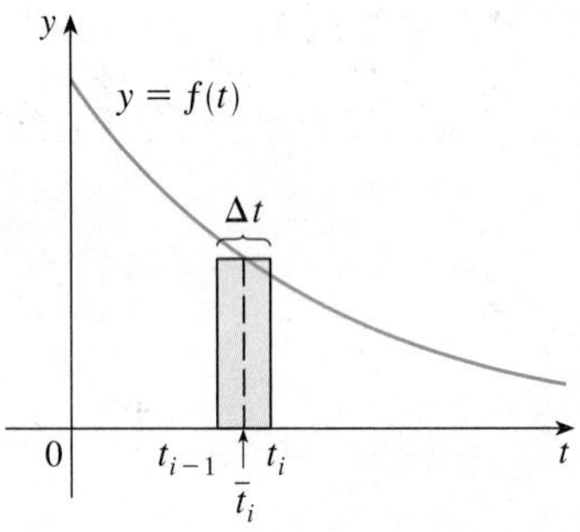

FIGURE 9

We now consider calculating the mean and variance of a continuous random variable. As an example, suppose you're waiting for a company to answer your phone call and you wonder how long, on average, customers can expect to wait. Let T be a random variable that represents the waiting time for a random customer (measured in minutes) and $f(t)$ be the corresponding probability density function. Let's suppose the waiting time is never longer than 60 minutes; therefore we can restrict attention to the interval $0 \leqslant t \leqslant 60$.

We begin by dividing the interval $[0, 60]$ into n smaller intervals of length Δt and endpoints $0, t_1, t_2, \ldots, t_n = 60$. (Think of Δt as lasting a minute, or half a minute, or 10 seconds, or even a second.) The probability that a call gets answered during the time period from t_{i-1} to t_i is the area under the curve $y = f(t)$ from t_{i-1} to t_i, which is approximately equal to $f(\bar{t}_i)\,\Delta t$. (This is the area of the approximating rectangle in Figure 9, where $\bar{t}_i$ is the midpoint of the interval.)

We can now use Definition 12.4.8, along with this approximation, to calculate the approximate average waiting time as

$$\sum_{i=1}^{n} \bar{t}_i f(\bar{t}_i)\,\Delta t$$

Again, using our knowledge of calculus, we recognize this as a Riemann sum for the function $tf(t)$. As the time interval shrinks (that is, $\Delta t \to 0$ and $n \to \infty$), this Riemann sum approaches the integral

$$\int_0^{60} tf(t)\,dt$$

This integral gives the *mean*, or *average*, waiting time.

> **(9) Definition** The **mean** of a continuous random variable X with probability density function $f(x)$ is
>
> $$E[X] = \int_{-\infty}^{\infty} x f(x)\,dx$$

As with discrete random variables, the mean value of X is also sometimes called the **expected value** of X.

EXAMPLE 5 | What is the mean value of the random variable X in Example 2?

SOLUTION From Definition 9 we have

$$E[X] = \int_{-\infty}^{\infty} x f(x)\,dx = \int_0^{10} x\,0.006x(10 - x)\,dx$$

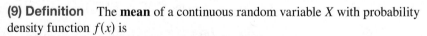

$$= 0.006\int_0^{10} (10x^2 - x^3)\,dx = 0.006\left[\tfrac{10}{3}x^3 - \tfrac{1}{4}x^4\right]_0^{10}$$

$$= 0.006\left(\frac{10{,}000}{3} - \frac{10{,}000}{4}\right) = 5$$ ■

We can also use a similar procedure to derive an equation for the variance of a continuous random variable.

(10) Definition The **variance** of a continuous random variable X with probability density function $f(x)$ is

$$\text{Var}[X] = \int_{-\infty}^{\infty} (x - E[X])^2 f(x)\,dx$$

A more convenient formula can be obtained by expanding the square in the integral and simplifying to get

$$\text{Var}[X] = \int_{-\infty}^{\infty} x^2 f(x)\,dx - E[X]^2$$

(See Exercise 52.) The **standard deviation** of a random variable is the square root of its variance.

EXAMPLE 6 | Calculate the variance of the random variable from Example 2.

SOLUTION From Definition 10,

$$\begin{aligned}\text{Var}[X] &= \int_{-\infty}^{\infty} x^2 f(x)\,dx - E[X]^2\\ &= \int_0^{10} x^2\,0.006x(10 - x)\,dx - 5^2 = 0.006\int_0^{10} (10x^3 - x^4)\,dx - 25\\ &= 0.006\Big[\tfrac{10}{4}x^4 - \tfrac{1}{5}x^5\Big]_0^{10} - 25 = 5\end{aligned}$$

Thus the mean and the variance of X are both equal to 5. ■

■ Exponential Random Variables

The waiting time for many events can be modeled by a continuous random variable whose probability density function decreases exponentially, as shown in Figure 10.

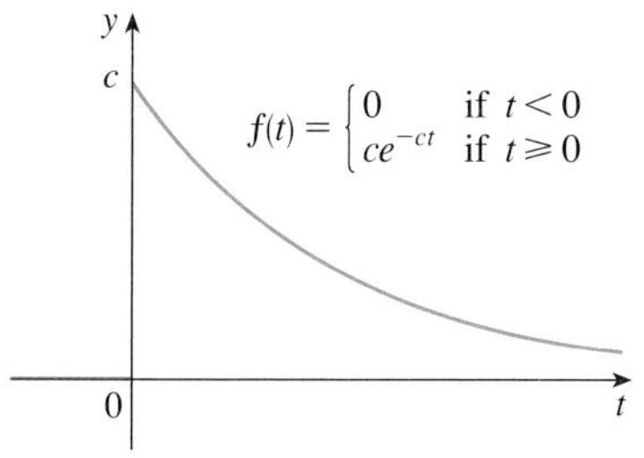

FIGURE 10
An exponential density function

EXAMPLE 7 | Find the exact form of an exponentially decreasing probability density function.

SOLUTION We are thinking of the random variable as being a waiting time (for example, the time until a particular individual dies). So instead of X, let's use T as the random variable, with time measured in years. If f is the probability density function and the individual is born at time $t = 0$, then, from Definition 2, $\int_0^2 f(t)\,dt$ represents the probability that the individual dies within the first two years and $\int_4^5 f(t)\,dt$ is the probability that the individual dies during the fifth year.

It's clear that $f(t) = 0$ for $t < 0$ (the age at death can't be negative) and, in principle, there is no upper bound on T. Thus T takes values in the interval $[0, \infty)$. We are told to use an exponentially decreasing function, that is, a function of the form $f(t) = Ae^{-ct}$, where A and c are positive constants. Thus

$$f(t) = Ae^{-ct} \qquad t \geq 0$$

We use Condition 4 to determine the value of A:

$$\begin{aligned}
1 &= \int_0^\infty f(t)\,dt \\
&= \int_0^\infty Ae^{-ct}\,dt = \lim_{x\to\infty}\int_0^x Ae^{-ct}\,dt \\
&= \lim_{x\to\infty}\left[-\frac{A}{c}e^{-ct}\right]_0^x = \lim_{x\to\infty}\frac{A}{c}(1-e^{-cx}) \\
&= \frac{A}{c}
\end{aligned}$$

Therefore $A/c = 1$ and so $A = c$. Thus, as displayed in Figure 10, every exponential probability density function has the form

$$f(t) = ce^{-ct} \qquad t \geqslant 0$$

Notice that, unlike the previous examples, the interval of values that this random variable can take on is infinite. Thus we needed to use our knowledge of improper integrals from Section 5.8 to perform the calculation. ■

(11) Definition An **exponential random variable** T with positive parameter c takes on values from the set $[0, \infty)$ and has probability density function $f(t) = ce^{-ct}$.

EXAMPLE 8 | Find the mean of the exponential random variable in Definition 11.

SOLUTION From Definition 9 we have

$$E[T] = \int_0^\infty t f(t)\,dt = \int_0^\infty t\,ce^{-ct}\,dt$$

To evaluate this integral we use integration by parts with $u = t$ and $dv = ce^{-ct}\,dt$:

$$\begin{aligned}
\int_0^\infty t\,ce^{-ct}\,dt &= \lim_{x\to\infty}\int_0^x t\,ce^{-ct}\,dt = \lim_{x\to\infty}\left(-te^{-ct}\Big]_0^x + \int_0^x e^{-ct}\,dt\right) \\
&= \lim_{x\to\infty}\left(-xe^{-cx} + \frac{1}{c} - \frac{e^{-cx}}{c}\right) = \frac{1}{c}
\end{aligned}$$

Again we have relied on our study of calculus, noting that the limit of the first term is 0 by l'Hospital's Rule (Section 4.3).

The mean is therefore $E[T] = 1/c$. If we use μ to denote the mean, then we can rewrite the exponential probability density function in Definition 11 as

$$f(t) = \mu^{-1}e^{-t/\mu}$$

■

EXAMPLE 9 | Age at death Suppose the age at death of individuals in a certain population can be modeled by an exponential random variable with a mean of 60 years.
(a) Find the probability that an individual from this population dies during the first 20 years.
(b) Find the probability that an individual from this population lives longer than 70 years.

SOLUTION

(a) We are given that the mean of the exponential random variable is $\mu = 60$ years and so, from the result of Example 8, we know that the probability density function is

$$f(t) = \frac{1}{60}e^{-t/60}$$

Thus the probability that death occurs during the first 20 years is

$$P(0 \leqslant T \leqslant 20) = \int_0^{20} f(t)\,dt = \int_0^{20} \frac{1}{60}e^{-t/60}\,dt$$

$$= -e^{-t/60}\Big]_0^{20} = 1 - e^{-20/60} \approx 0.2835$$

So about 28% of the individuals in this population die during the first 20 years.

(b) The probability that an individual lives longer than 70 years is

$$P(T > 70) = \int_{70}^{\infty} f(t)\,dt = \int_{70}^{\infty} \frac{1}{60}\,e^{-t/60}\,dt$$

$$= \lim_{x\to\infty} \int_{70}^{x} \frac{1}{60}\,e^{-t/60}\,dt = \lim_{x\to\infty} (e^{-70/60} - e^{-x/60})$$

$$\approx 0.3114$$

Therefore about 31% of the population lives longer than 70 years. ■

■ Normal Random Variables

Many important biological phenomena—such as scores on aptitude tests or heights and weights of individuals—are modeled by a *normal random variable.*

(12) Definition A **normal random variable** X with parameters μ and σ takes on values from the interval $(-\infty, \infty)$ and has probability density function

$$f(x) = \frac{1}{\sigma\sqrt{2\pi}}\,e^{-(x-\mu)^2/(2\sigma^2)}$$

The quantity σ (sigma) is the standard deviation of a normal random variable.

You can verify that the mean of this random variable is μ (see Exercise 75) and the variance is σ^2 (see Exercise 76). Figure 11 displays graphs of the PDFs for three different normal random variables. We see that the PDFs are symmetric around the mean μ, and for small values of σ the values of X are clustered about the mean, whereas for larger values of σ the values of X are more spread out (also see Section 11.2). In Chapter 13 we will explore how data can be used to estimate μ and σ.

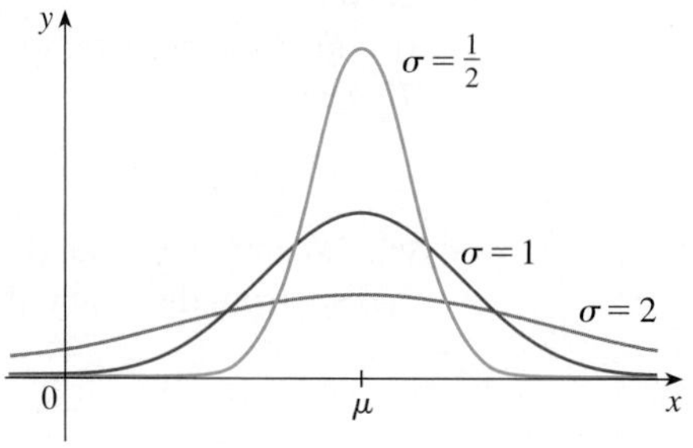

FIGURE 11
Normal distributions

The factor $1/(\sigma\sqrt{2\pi})$ is needed to make f a probability density function. In fact, it can be verified using the methods of multivariable calculus that

$$\int_{-\infty}^{\infty} \frac{1}{\sigma\sqrt{2\pi}} e^{-(x-\mu)^2/(2\sigma^2)} dx = 1$$

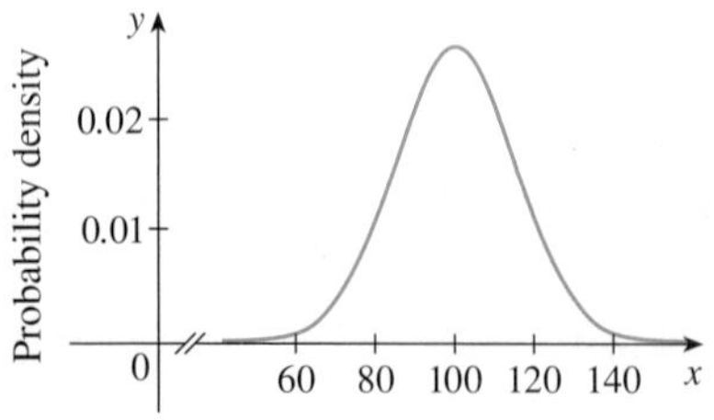

FIGURE 12
Distribution of IQ scores

EXAMPLE 10 | **Intelligence Quotient (IQ) scores** are distributed normally with mean 100 and standard deviation 15. (Figure 12 shows the corresponding probability density function.)

(a) What percentage of the population has an IQ score between 85 and 115?
(b) What percentage of the population has an IQ score above 140?

SOLUTION

(a) Since IQ scores are normally distributed, we use the probability density function given by Definition 12 with $\mu = 100$ and $\sigma = 15$:

$$P(85 \leqslant X \leqslant 115) = \int_{85}^{115} \frac{1}{15\sqrt{2\pi}} e^{-(x-100)^2/(2\cdot 15^2)} dx$$

Recall from Section 5.7 that the function $y = e^{-x^2}$ doesn't have an elementary antiderivative, so we can't evaluate the integral exactly. But we can use the numerical integration capability of a calculator or computer to estimate the integral. Doing so, we find that

$$P(85 \leqslant X \leqslant 115) \approx 0.683$$

So about 68.3% of the population has an IQ score between 85 and 115, that is, within one standard deviation of the mean (also see Section 11.2).

(b) The probability that the IQ score of a person chosen at random is more than 140 is

$$P(X > 140) = \int_{140}^{\infty} \frac{1}{15\sqrt{2\pi}} e^{-(x-100)^2/450} dx$$

To avoid an improper integral, we could approximate this integral by one from 140 to 200. (It's quite safe to say that people with an IQ score over 200 are extremely rare.) Then

$$P(X > 140) \approx \int_{140}^{200} \frac{1}{15\sqrt{2\pi}} e^{-(x-100)^2/450} dx \approx 0.0038$$

Therefore about 0.4% of the population has an IQ score over 140. ■

When we study inferential statistics in Chapter 13, we will see that it is often convenient to work with a *standard normal distribution*, whose mean is 0 and standard deviation is 1.

(13) Definition A **standard normal random variable** Z has probability density function

$$f(z) = \frac{1}{\sqrt{2\pi}} e^{-z^2/2}$$

Any normal random variable can be converted to a standard normal random variable with an appropriate change in the scale of measurement. Specifically, if X is a normal random variable with mean μ and standard deviation σ, then the random variable Z defined by

(14)
$$Z = \frac{X - \mu}{\sigma}$$

is a standard normal random variable. This is proved in Exercise 13.2.15.

Equation 14 says that any normal random variable can be converted to a standard normal random variable by measuring it as a difference from its mean, in units of its standard deviation.

EXERCISES 12.5

1. Let $f(x)$ be the probability density function for the lifetime of a manufacturer's highest quality automobile tire, where x is measured in miles. Explain the meaning of each integral.

(a) $\int_{30,000}^{40,000} f(x)\,dx$ (b) $\int_{25,000}^{\infty} f(x)\,dx$

2. Let $f(t)$ be the probability density function for the time it takes you to drive to school in the morning, where t is measured in minutes. Express the following probabilities as integrals.
(a) The probability that you drive to school in less than 15 minutes
(b) The probability that it takes you more than half an hour to drive to school

3. Age at death Let $f(t)$ be the probability density function for the age at death of *Drosophila melanogaster* fruit flies, where t is measured in days. Explain the meaning of each integral.

(a) $\int_{25}^{40} f(t)\,dt$ (b) $\int_{0}^{14} f(t)\,dt$

4. Let $f(x) = \frac{3}{64}x\sqrt{16 - x^2}$ for $0 \leqslant x \leqslant 4$ and $f(x) = 0$ for all other values of x.
(a) Verify that $f(x)$ satisfies Conditions 3 and 4.
(b) Find $P(X < 2)$.

5. Let $f(x) = xe^{-x}$ if $x \geqslant 0$ and $f(x) = 0$ if $x < 0$.
(a) Verify that $f(x)$ satisfies Conditions 3 and 4.
(b) Find $P(1 \leqslant X \leqslant 2)$.

6. Let $f(t) = 4t^3$ if $0 \leqslant t \leqslant 1$ and $f(t) = 0$ otherwise.
(a) Verify that $f(t)$ satisfies Conditions 3 and 4.
(b) Find $P(0 \leqslant T \leqslant \frac{1}{2})$.

7. Let $f(x) = c/(1 + x^2)$, where $x \in \mathbb{R}$.
(a) For what value of c does f satisfy Conditions 3 and 4?
(b) For the value of c from part (a), find $P(-1 < X < 1)$.

8. Let $f(x) = kx^2(1 - x)$ if $0 \leqslant x \leqslant 1$ and $f(x) = 0$ if $x < 0$ or $x > 1$.
(a) For what value of k does f satisfy Conditions 3 and 4?
(b) For the value of k from part (a), find $P(X \geqslant \frac{1}{2})$.

9–18 Verify that $f(x)$ is a PDF, find the CDF, and use the CDF to calculate the given probability.

9. Let $f(x) = 1$ for $0 \leqslant x \leqslant 1$ and $f(x) = 0$ otherwise. Find $P(\frac{1}{3} \leqslant X \leqslant \frac{2}{3})$.

10. Let $f(x) = \frac{1}{10}$ for $0 \leqslant x \leqslant 10$ and $f(x) = 0$ otherwise. Find $P(2 \leqslant X \leqslant 5)$.

11. Let $f(x) = x^{-2}$ for $\frac{1}{2} \leqslant x \leqslant 1$ and $f(x) = 0$ otherwise. Find $P(\frac{2}{3} \leqslant X \leqslant \frac{3}{4})$.

12. Let $f(x) = 2/(1+x)^2$ for $0 \leqslant x \leqslant 1$ and $f(x) = 0$ otherwise. Find $P(\frac{1}{2} \leqslant X \leqslant \frac{3}{4})$.

13. Let $f(t) = \frac{3}{2}\sqrt{t}$ for $0 \leqslant t \leqslant 1$ and $f(t) = 0$ otherwise. Find $P(\frac{1}{4} \leqslant T \leqslant \frac{3}{4})$.

14. Let $f(t) = e^{-t}$ for $t \geqslant 0$. Find $P(2 \leqslant T \leqslant 15)$.

15. Let $f(x) = \frac{1}{2}e^{-|x|}$ for $x \in \mathbb{R}$. Find $P(-5 \leqslant X \leqslant 10)$.

16. Let $f(x) = \ln x$ for $1 \leqslant x \leqslant e$. Find $P(2 \leqslant X \leqslant e)$.

17. Let $f(x) = \dfrac{2}{\pi(1+x^2)}$ for $x \geqslant 0$ and $f(x) = 0$ otherwise. Find $P(10 \leqslant X \leqslant 20)$.

18. Let $f(x) = \beta/x^2$ for $x \geqslant \beta$ and $f(x) = 0$ otherwise, where $\beta > 0$. Find $P(\beta + 5 \leqslant X \leqslant \beta + 10)$.

---

19. Streptococcal incubation periods The time between infection and the display of symptoms for streptococcal sore throat is a random variable whose PDF can be approximated by the function

$$f(t) = \frac{1}{15{,}676}t^2 e^{-0.05t}$$

for $0 \leqslant t \leqslant 150$ and $f(t) = 0$ otherwise, where t is measured in hours.

(a) What is the probability that an infected patient will display symptoms within the first 48 hours?

(b) What is the probability that an infected patient will not display symptoms until after 36 hours?

Source: Adapted from P. Sartwell, "The Distribution of Incubation Periods of Infectious Disease. 1949." *American Journal of Epidemiology* 141 (1995): 386–94.

20. HIV eclipse phase The length of time from the HIV virus first entering a cell to producing new viral particles is called the intracellular eclipse phase. This time can be described by a random variable with PDF $f(t) = 6.25te^{-5t/2}$ where $t \in [0, \infty)$ and time is measured in days.

(a) What is the probability that the length of the eclipse phase is less than half a day?

(b) What is the probability that the length of the eclipse phase is between 1 and 2 days?

Source: Adapted from A. Perelson et al., "HIV-1 Dynamics In Vivo: Virion Clearance Rate, Infected Cell Life-span, and Viral Generation Time." *Science* 271 (1996): 1582–86.

21. REM sleep is the phase of sleep when most active dreaming occurs. The amount of REM sleep during the first four hours of sleep can be described by a random variable T with PDF $f(t) = t/1600$ for $0 \leqslant t \leqslant 40$, $f(t) = (80 - t)/1600$ for $40 < t \leqslant 80$, and $f(t) = 0$ otherwise, where T is measured in minutes.

(a) What is the probability that the amount of REM sleep is less than 30 minutes?

(b) What is the probability that the amount of REM sleep is between 30 and 60 minutes?

22. Genome size The genome size of *Festuca pallens* plants, measured relative to the size of a standard plant, is described by a random variable X with PDF

$$f(x) = \frac{1}{\beta}e^{-(x-\mu)/\beta}e^{-(e^{-(x-\mu)/\beta})}$$

where $x \in \mathbb{R}$ and β and μ are positive parameters. This is called a *Gumbel random variable*. A study showed that $\mu \approx 1.90$ and $\beta \approx 0.05$.

(a) What is the probability that a randomly chosen plant will have a relative genome size between 1.9 and 1.95?

(b) What is the probability that a randomly chosen plant will have a relative genome size larger than 1.95?

Source: P. Šmarda et al., "Random Distribution Pattern and Non-adaptivity of Genome Size in a Highly Variable Population of *Festuca pallens*." *Annals of Botany* 100 (2007): 141–50.

23–28 Find the PDF $f(x)$ given the CDF $F(x)$.

23. $F(x) = x$ for $0 \leqslant x \leqslant 1$

24. $F(t) = \frac{1}{9}t^2$ for $0 \leqslant t \leqslant 3$

25. $F(t) = \dfrac{t^2}{1+t^2}$ for $t \geqslant 0$

26. $F(x) = \dfrac{1}{1+e^{-x}}$ for $x \in \mathbb{R}$

27. $F(t) = \dfrac{Ce^t}{1+Ce^t}$ for $t \in \mathbb{R}$, where C is a positive constant

28. $F(t) = 1 - (\beta/t)^\alpha$ for $t \geqslant \beta$, where α and β are positive constants

---

29–35 For each PDF, calculate (a) the mean and (b) the variance.

29. $f(t) = 1$ for $0 \leqslant t \leqslant 1$ and $f(t) = 0$ otherwise

30. $f(t) = \frac{1}{2}t$ for $0 \leqslant t \leqslant 2$ and $f(t) = 0$ otherwise

31. $f(x) = x^{-2}$ for $\frac{1}{2} \leqslant x \leqslant 1$ and $f(x) = 0$ otherwise

32. $f(t) = 4t^3$ for $0 \leqslant t \leqslant 1$ and $f(t) = 0$ otherwise

33. $f(t) = e^{-t}$ for $t \geqslant 0$ and $f(t) = 0$ otherwise

34. $f(x) = 2/(1+x)^2$ for $0 \leqslant x \leqslant 1$ and $f(x) = 0$ otherwise

35. $f(x) = \frac{1}{2}e^{-|x|}$ for $x \in \mathbb{R}$

---

36. A spinner from a board game randomly indicates a real number between 0 and 10. The spinner is fair in the sense that it indicates a number in a given interval with the same probability as it indicates a number in any other interval of the same length.

(a) Explain why the function

$$f(x) = \begin{cases} 0.1 & \text{if } 0 \leqslant x \leqslant 10 \\ 0 & \text{if } x < 0 \text{ or } x > 10 \end{cases}$$

is a probability density function for the spinner's values.

(b) What does your intuition tell you about the value of the mean? Check your guess by evaluating an integral.

37. A *uniform random variable* is a random variable, the graph of whose PDF is a horizontal line. Suppose X is a uniform random variable on the interval $[a, b]$.
(a) What is its probability density function $f(x)$?
(b) Calculate the mean of X.
(c) Calculate the variance of X.

38. Mating behavior It has been hypothesized that when *Drosophila* mating experiments are conducted, the time at which mating occurs after a male and female are placed together is a random variable with PDF $f(t) = \frac{1}{600}$ for $0 \leqslant t \leqslant 600$ and $f(t) = 0$ otherwise, where t is measured in seconds. Assuming this hypothesis is correct, calculate the following.
(a) The probability that mating does not occur within the first two minutes
(b) The probability that mating does occur within the first t minutes
(c) The mean time until mating occurs
(d) The variance in the time until mating occurs

39. Chromosomal recombination is a process by which two chromosomes join together and exchange DNA. The point along the DNA at which the join occurs is randomly located. Suppose X is a random variable denoting this location, measured as the distance from the centromere (in morgans), with $0 \leqslant X \leqslant 2$. In an experiment, the mean location of the join is $E[X] = 1$ and the variance is $\text{Var}[X] = \frac{1}{3}$. Are these experimental findings consistent with the hypothesis that all locations along the chromosome are equally likely to contain the join point? Explain.

40. Birth dates Suppose that humans are equally likely to be born on any day during the year. Use T to denote the random variable representing time of birth (in days), with $T = 0$ the start of the year and $T = 365$ the last day.
(a) What is the probability that a randomly chosen birth will occur during the first 60 days?
(b) What is probability that a randomly chosen birth will occur during the last 90 days?
(c) What is the mean birth date?

41. REM sleep is the phase of sleep when most active dreaming occurs. The length of REM sleep during the first four hours of sleep can be described by a random variable T with PDF $f(t) = t/1600$ for $0 \leqslant t \leqslant 40$, $f(t) = (80 - t)/1600$ for $40 < t \leqslant 80$, and $f(t) = 0$ otherwise, where T is measured in minutes.
(a) What is the mean length of REM sleep?
(b) What is the variance in the length of REM sleep?

42. HIV eclipse phase The length of time from the HIV virus first entering a cell to producing new viral particles is called the intracellular eclipse phase. This time can be described by a random variable with PDF

$$f(t) = 6.25te^{-5t/2} \qquad \text{where } t \in [0, \infty)$$

and time is measured in days.
(a) What is the mean length of the eclipse phase?
(b) What is the variance in the length of the eclipse phase?

43–48 Another measure of centrality of a probability density function is the median. The *median* is a number m such that half the area of the PDF lies to the left of m and the other half to the right. In general, the **median** of a probability density function is the number m such that

$$\int_{-\infty}^{m} f(x)\,dx = \tfrac{1}{2}$$

Calculate the median for each PDF.

43. $f(x) = 1$ for $0 \leqslant x \leqslant 1$ and $f(x) = 0$ otherwise

44. $f(x) = x^{-2}$ for $\frac{1}{2} \leqslant x \leqslant 1$ and $f(x) = 0$ otherwise

45. $f(x) = \dfrac{2}{(1 + x)^2}$ for $0 \leqslant x \leqslant 1$ and $f(x) = 0$ otherwise

46. $f(t) = e^{-t}$ for $t \geqslant 0$ and $f(t) = 0$ otherwise

47. $f(x) = \ln x$ for $1 \leqslant x \leqslant e$ and $f(x) = 0$ otherwise. You will need to use a numerical rootfinder to obtain the solution.

48. $f(t) = \frac{1}{2}t$ for $0 \leqslant t \leqslant 2$ and $f(t) = 0$ otherwise

49–51 The **mode** of a continuous random variable having density $f(x)$ is the value of x for which $f(x)$ attains its maximum. Compute the mode of X.

49. $f(x) = x^{-2}$ for $\frac{1}{2} \leqslant x \leqslant 1$ and $f(x) = 0$ otherwise

50. $f(x) = \dfrac{1}{\sigma\sqrt{2\pi}}e^{-(x-\mu)^2/(2\sigma^2)}$ for $x \in \mathbb{R}$

51. $f(x) = \frac{1}{2}e^{-|x|}$ for $x \in \mathbb{R}$

52. Derive the equation $\text{Var}[X] = \int_{-\infty}^{\infty} x^2 f(x)\,dx - E[X]^2$.

53–56 The *standard deviation* of a random variable is the square root of its variance. Calculate the standard deviation using the given PDF.

53. $f(t) = \frac{1}{2}t$ for $0 \leqslant t \leqslant 2$ and $f(t) = 0$ otherwise

54. $f(t) = 2e^{-2t}$ for $t \geqslant 0$ and $f(t) = 0$ otherwise

55. $f(x) = \dfrac{2}{(1 + x)^2}$ for $0 \leqslant x \leqslant 1$ and $f(x) = 0$ otherwise

56. $f(x) = \frac{1}{2}e^{-|x|}$ for $x \in \mathbb{R}$

57. The length of time T that a customer spends in line at a supermarket before being served is described by the exponential random variable with mean $E[T] = 10$ minutes. What is the probability that a customer will wait longer than 20 minutes?

58. A type of lightbulb is labeled as having an average lifetime of 1000 hours. It's reasonable to model the probability of failure of these bulbs by an exponential density function with mean $\mu = 1000$.
(a) What is the probability that a bulb fails within the first 200 hours?
(b) What is the probability that a bulb burns for more than 800 hours?

59. The manager of a fast-food restaurant determines that the average time that her customers wait for service is 2.5 minutes, and this waiting time is well described by an exponential random variable.
(a) Find the probability that a customer has to wait more than 4 minutes.
(b) Find the probability that a customer is served within the first 2 minutes.
(c) The manager wants to advertise that anybody who isn't served within a certain number of minutes gets a free hamburger. But she doesn't want to give away free hamburgers to more than 2% of her customers. What should the advertisement say?

60. Foraging behavior Exponential random variables also arise when measuring the distance traveled until a certain event occurs, assuming that the event occurs at a constant rate per distance. For example, if a bee is foraging and stops at flowers at a constant rate per meter, then the distance until the bee stops at a flower would be described by an exponential random variable. Suppose the bee stops on average once every 5 meters.
(a) What is the probability that a bee will travel at least 20 meters before stopping?
(b) What is the variance in distance between stops?

61. Pharmaceutical development A study of the time it takes for new drug therapies to be developed showed that the time between the development of new drugs is described by an exponential random variable T. Suppose the mean time between the development of new drugs is 1 year.
(a) What is the probability that at least 10 years will pass before a new drug is produced?
(b) What is the probability that less than 1 year will pass before a new drug is produced?
(c) What is the variance in the time between drug developments?

Source: B. Munos, "Lessons from 60 Years of Pharmaceutical Innovation." *Nature Reviews Drug Discovery* 8 (2009): 959–68.

62. Evolution of drug resistance Drugs eventually lose their effectiveness as a result of the evolution of drug resistance. A recent study showed that the time it takes for resistance to any specific drug to evolve is described by an exponential random variable with a mean time of 10 years.
(a) What is the probability that a drug will lose its effectiveness through the evolution of resistance in less than 2 years?
(b) What is the probability that a drug will remain effective for more than 15 years?

63. The coalescent process Any sample of genes from different individuals in a population can ultimately be traced backward in time to a single ancestral copy of the gene. This is modeled mathematically by imagining moving backward in time and assuming that pairs of genes randomly "coalesce" into common ancestors. If there are n genes in the current population, then there will be $n - 1$ such "coalescent events" as we move backward before we reach the single ancestral gene. A common assumption is that the length of time before the first coalescent event occurs has PDF given by

$$f(t) = \binom{n}{2} e^{-\binom{n}{2}t}$$

where $t \geqslant 0$ and $\binom{n}{2}$ is the binomial coefficient.
(a) What is the mean length of time before the first coalescent event as a function of n?
(b) What is the variance in the length of time before the first coalescent event as a function of n?
(c) What is the probability that the first coalescent event occured earlier than 20 time units ago, as a function of n?

64. Suppose that T is an exponential random variable.
(a) Use the definition of conditional probability to show that

$$P(T > t + s \mid T > s) = P(T > t)$$

for all nonnegative values of s and t.
(b) The equation in part (a) is sometimes taken to mean that exponential random variables are "memoryless." Thinking of T as the waiting time for some event to occur, provide a verbal interpretation of what this memoryless property means.

65–67 A *Laplace random variable* has PDF given by

$$f(x) = \tfrac{1}{2}ce^{-c|x|}$$

where $c > 0$. This is sometimes referred to as the "double exponential" because it is the PDF obtained by reflecting an exponential distribution in the vertical axis and rescaling to satisfy Condition 4.

65. What is the mean value of a Laplace random variable?

66. What is the variance of a Laplace random variable?

67. Dispersal The Laplace distribution is often used to model dispersal along a one-dimensional habitat, such as a coastline. Suppose $c = 1$ and x is measured in meters.
(a) What is the probability that an individual disperses more than 2 meters?
(b) What is the probability that an individual disperses less than 1 meter?

68–74 Use a calculator or computer to obtain a numerical approximation for each exercise.

68. Boxes are labeled as containing 500 g of cereal. The machine filling the boxes produces weights that are normally distributed with standard deviation 12 g.
(a) If the target weight is 500 g, what is the probability that the machine produces a box with less than 480 g of cereal?
(b) Suppose a law states that no more than 5% of a manufacturer's cereal boxes can contain less than the stated weight of 500 g. At what target weight should the manufacturer set its filling machine?

69. The speeds of vehicles on a highway with speed limit 100 km/h are normally distributed with mean 112 km/h and standard deviation 8 km/h.
(a) What is the probability that a randomly chosen vehicle is traveling at a legal speed?
(b) If police are instructed to ticket motorists driving 125 km/h or more, what percentage of motorists are targeted?

70. The "Garbage Project" at the University of Arizona reports that the amount of paper discarded by households per week is normally distributed with mean 9.4 lb and standard deviation 4.2 lb. What percentage of households throw out at least 10 lb of paper a week?

71. Human heights According to the National Health Survey, the heights of adult males in the United States are normally distributed with mean 69.0 inches and standard deviation 2.8 inches.
(a) What is the probability that an adult male chosen at random is between 65 inches and 73 inches tall?
(b) What percentage of the adult male population is more than 72 inches tall?

72. Fish length in sticklebacks is described by a normal random variable with a mean of 10 cm and a variance of 1 cm^2.
(a) What is the probability that a randomly chosen stickleback is between 8 cm and 12 cm long?
(b) What fraction of the fish population is longer than 15 cm?

73. Dispersal The normal distribution is often used to model dispersal along a one-dimensional habitat, such as a coastline. Suppose the mean dispersal distance is 0 and the variance is 2 m^2.
(a) What is the probability that an individual disperses more than 2 meters?
(b) What is the probability that an individual disperses less than 1 meter?

74. Blood pressure The diastolic blood pressure for men in a particular population is described by a normal random variable with a mean of 80 mm Hg and a variance of 30 $(\text{mm Hg})^2$.
(a) What is the probability that an individual has a blood pressure higher than 90 mm Hg?
(b) What fraction of the population has blood pressure between 75 and 85 mm Hg?

75–77 Suppose X is a normal random variable with PDF

$$f(x) = \frac{1}{\sigma\sqrt{2\pi}}\, e^{-(x-\mu)^2/(2\sigma^2)}$$

75. Calculate the mean of X.

76. Calculate the variance of X.

77. Show that the probability density function for a normally distributed random variable has inflection points at $x = \mu \pm \sigma$.

Chapter 12 REVIEW

CONCEPT CHECK

1. What is the Fundamental Principle of Counting?

2. What is the difference between a permutation and a combination?

3. Explain the relationship between an experiment, a trial, an outcome, and an event.

4. What is the difference between a simple and a compound event?

5. In your own words explain what it means for two events to be independent.

6. What is Bayes' Rule?

7. What is a random variable?

8. (a) Name two specific examples of a discrete random variable.
(b) Name two specific examples of a continuous random variable.

9. (a) What is a probability density function (PDF)?
(b) What is a cumulative distribution function (CDF)?

10. Explain the relationship between the Fundamental Theorem of Calculus, PDFs, and CDFs for continuous random variables.

Answers to the Concept Check can be found on the back endpapers.

TRUE-FALSE QUIZ

Determine whether the statement is true or false. If it is true, explain why. If it is false, explain why or give an example that disproves the statement.

1. When counting permutations, the order of the objects matters.
2. The complement of an event consists of all events in the sample space.
3. Two events E and F are disjoint if $E \cap F = \varnothing$.
4. In general, for two events E and F, $P(E \mid F) = P(F \mid E)$.
5. If E and F are two events, the Multiplication Rule can be applied to calculate $P(E \cap F)$ regardless of whether the events are independent.
6. The probability density function for a discrete random variable can be calculated from its cumulative distribution function.
7. The cumulative distribution function for a continuous random variable can be obtained by differentiating its probability density function.
8. The CDF of a continuous random variable satisfies $\lim_{x\to\infty} F(x) = 1$.
9. The mean of a continuous random variable that takes on values in the interval $[x_{\min}, x_{\max}]$ is always $(x_{\min} + x_{\max})/2$.
10. The units of the variance of a random variable are the same as those of its mean.

EXERCISES

1. A pizza restaurant offers 12 kinds of toppings. How many different pizzas are possible?

2. Explain why in any group of 677 people, at least two people must have the same pair of initials.

3. Suppose license plates contain three letters followed by three numerals. How many different plates are possible if repeats are (a) allowed? (b) Not allowed?

4. How many different batting orders are possible on a baseball team with nine players?

5. Ten people arrive at a bank and line up for service.
   (a) In how many ways can they form the line?
   (b) Suppose only six people can fit in the line. In how many ways can the line be formed?

6. **Gene expression** Cells from four tissue types are assayed for the expression (yes or no) of a particular gene. How many different patterns are possible?

7. **Tetraploid** individuals carry four copies of each gene. Suppose there are four variants of a gene: A_1, A_2, A_3, and A_4. How many different tetraploid individuals are possible if all that matters is the identity of the variants that they carry?

8. **Paired versus unpaired controls** In medical studies, researchers often assess the effect of a treatment by comparing treated individuals with control individuals. Suppose there are 6 treated individuals and we wish to choose 6 control individuals from a group of 10 people.
   (a) "Paired control individuals" are matched specifically to particular treated individuals and the pair then compared. How many possibilities are there for assigning paired controls?
   (b) "Unpaired control individuals" are not matched to specific treated individuals. Instead, the treated group is compared with the control group. How many possibilities are there for choosing the unpaired controls?

9. Prove the following identity.

$$\binom{n}{r}\binom{n-r}{k} = \binom{n}{k}\binom{n-k}{r}$$

10. Is it possible for two nonempty sets to have the same intersection and union? If so give an example.

11–14 Evaluate each, given that

$$A = \{1, 3, 5, 8\} \qquad B = \{1, 2, 5, 9\}$$
$$C = \{6\} \qquad \Omega = \{1, 2, 3, 4, 5, 6, 7, 8, 9\}$$

11. $B \cup C^c$
12. $(A \cap B) \cup C$
13. $A \cup B \cup C$
14. $A \cap B \cap C$

15. De Morgan's laws are stated below. Illustrate each by constructing Venn diagrams for both sides of the equation.
    (a) $(E \cup F)^c = E^c \cap F^c$ (b) $(E \cap F)^c = E^c \cup F^c$

16–17 A card is drawn randomly from a standard deck. Find the probability of the given event.

16. (a) The card drawn is a 3.
    (b) The card drawn is not an ace.
    (c) The card drawn is of a red suit.

17. (a) The card drawn has numerical value between 3 and 5, inclusive.
(b) The card drawn is not a 7 and not a heart.
(c) The card drawn is not a heart, not a diamond, and not a spade.

18–19 A ball is drawn randomly from a jar that contains four red balls, three white balls, and six yellow balls. Find the probability of the given event.

18. (a) A red ball is drawn.
(b) The ball drawn is not yellow.
(c) A white ball is drawn.

19. (a) Neither a white nor a yellow ball is drawn.
(b) A red, white, or yellow ball is drawn.
(c) The ball that is drawn is not white.

20. A student has locked her locker with a combination lock, showing numbers from 1 to 40, but she has forgotten the three-number combination that opens the lock. She remembers that all three numbers in the combination are different. To open the lock, she decides to try all possible combinations. If she can try 10 different combinations every minute, what is the probability that she will open the lock within one hour?

21. Five people at a restaurant each pay their portion of the bill by giving the waiter their credit cards.
(a) If the waiter gives back their cards randomly, what is the probability that none of the people get their own card?
(b) Calculate the probability in part (a) if there are n people at the restaurant.

22. Performance-enhancing drugs Three long-distance runners, A, B, and C, are tested for the performance-enhancing drug EPO. The result for each test is either positive or negative.
(a) List all elements of the sample space.
(b) List all elements corresponding to the event "two runners test positive."
(c) List all elements corresponding to the event "at least one runner tests positive."

23. Mark-recapture studies Thirty fish from a small pond are captured, marked, and released back into the pond. Researchers then capture 10 fish from this same pond. What is the probability that 4 of these will be marked if the total population size in the pond is 50?

24. *Hox* genes and developmental biology In some animals the ordering of eight *Hox genes* on the chromosome is the same as the ordering of their expression along the anterior-posterior axis of the developing animal (see the figure). If these eight genes were placed randomly on the chromosome, what is the probability that their order would match the ordering of expression shown in the figure?

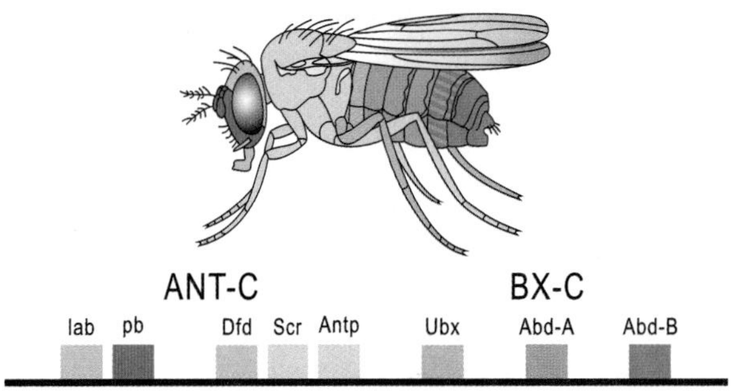

Source: Adapted from M. Whitlock and D. Schluter, *The Analysis of Biological Data* (Greenwood Village, CO: Roberts and Company, 2009).

25–26 Suppose event E is a subset of event F. We write this as $E \subset F$.

25. The difference $F - E$ consists of those elements in F but not in E.
(a) Illustrate this situation with a Venn diagram.
(b) Prove that $P(F - E) = P(F) - P(E)$.

26. Use the result from Exercise 25(b) to prove $P(E) \leqslant P(F)$.

27. Suppose that $P(E) = \frac{1}{2}$ and $P(F) = \frac{1}{2}$.
(a) What is $P(E \cap F)$ if E and F are independent?
(b) What is $P(E \mid F)$ if E and F are independent?

28. A bucket contains five black balls and seven white balls. Two balls are drawn *without* replacement.
(a) What is the probability of drawing two black balls, given that the first ball is black?
(b) What is the probability of drawing two black balls, given that at least one of the balls is black?

29. Comorbidity refers to the presence of two or more disorders. For example, in many parts of Africa people suffer from both HIV infection and tuberculosis infection. Many studies have shown that the events "being HIV$^+$" and "displaying symptoms of TB" are not independent. Explain, from a biological standpoint, what this means.

30. Sex-specific drug effectiveness Some drugs have a different effectiveness in males versus females. Suppose a drug works 80% of the time in males and 95% of the time in females. If 75% of the patients taking the drug are females, what is the probability that a randomly chosen patient will have successful treatment?

31. Prove that $P(E \mid F) < P(E)$ if $P(F \mid E) < P(F)$.

32. A random variable takes on integer values from 1 to 8, inclusive, and has PDF given by $p_i = (256/255)2^{-i}$.
(a) Construct a histogram for the random variable.
(b) Obtain an equation for the CDF.
(c) Plot the CDF below the histogram from part (a).

33. Germination rates A certain plant species has a 0.75 probability of germinating. A plot of land contains four seeds.

(a) What is the probability that at least one seed germinates?

(b) What is the probability that two or more seeds germinate?

(c) What is the probability that all four seeds germinate?

34. Pharmaceuticals A drug that is used to prevent motion sickness is found to be effective about 75% of the time. Six friends, all prone to seasickness, go on a sailing cruise, and all take the drug. Find the probability of each event.

(a) None of the friends gets seasick.

(b) All the friends get seasick.

(c) Exactly three get seasick.

(d) At least two get seasick.

35. Consider the function $f(x) = \frac{3}{2}\sqrt{x}$.

(a) Verify that $f(x)$ satisfies Conditions 12.5.3–4 for being a PDF, where $x \in [0, 1]$.

(b) Find the probability that X is larger than $\frac{1}{2}$.

(c) Find the probability that X lies between $\frac{1}{4}$ and $\frac{3}{4}$.

(d) What is the mean value of X?

(e) What is the variance of X?

36. Consider the function $f(x) = 2xe^{-x^2}$. Recall that $\int_0^\infty e^{-x^2} = \frac{1}{2}\sqrt{\pi}$.

(a) Verify that $f(x)$ satisfies Conditions 12.5.3–4 for being a PDF, where $x \geqslant 0$.

(b) What is the probability that X is larger than 3?

(c) What is the probability that X lies between 5 and 10?

(d) What is the mean value of X?

(e) What is the variance of X?

37. The length of time spent waiting in line at a certain bank is modeled by an exponential random variable with mean 8 minutes.

(a) What is the probability that a customer is served in the first 3 minutes?

(b) What is the probability that a customer has to wait more than 10 minutes?

(c) What is the median waiting time?

38. Human gestation The length of human pregnancy is normally distributed with mean 268 days and standard deviation 15 days. Use a calculator with numerical integration capability to determine the percentage of pregnancies that last between 250 days and 280 days.

39. A *Cauchy random variable* is one with probability density function

$$f(x) = \frac{1}{\pi(1 + x^2)} \qquad \text{for } x \in \mathbb{R}$$

(a) Verify that $f(x)$ satisfies Conditions 12.5.3–4.

(b) Try calculating the mean value of X. What do you notice?

Inferential Statistics

13

We use inferential statistics in Section 13.3 to determine the effectiveness of the performance-enhancing drug EPO.

IN THIS CHAPTER WE COMBINE the ideas of descriptive statistics and samples from Chapter 11 with the ideas of probability theory from Chapter 12 to explore how samples can be used to infer properties of the population from which they were taken. We do not give a comprehensive treatment of inferential statistics; instead, we focus on the introduction of some key ideas in classical statistics, including sampling distributions, confidence intervals, and hypothesis testing.

13.1 The Sampling Distribution

In Example 12.5.10 we saw that IQ scores are normally distributed with a mean of 100 and a standard deviation of 15. We used this fact to calculate the probabilities of various events.

Often we do not know the underlying probability distribution for the variable of interest. Instead, we wish to infer the properties of this distribution by analyzing a sample taken from it. This is the goal of inferential statistics. In the terminology introduced in Section 11.4, the properties of this distribution, such as its mean or standard deviation, are called *parameters*. We seek to infer these parameters from the corresponding *summary statistics*, which we can calculate from a sample.

The process of inference can be made more tractable by first narrowing our scope. We focus on normally distributed variables because many continuous variables are normally distributed. The process of inference then reduces to the task of inferring the mean and standard deviation of the distribution because a normal distribution is completely determined by these two quantities (see Definition 12.5.12).

EXAMPLE 1 | IQ scores Suppose the IQ scores for a population of interest are normally distributed with unknown mean and standard deviation. If X is a random variable representing the IQ score of a randomly chosen person, then the probability density function of X is

(1) $$f(x) = \frac{1}{\sigma\sqrt{2\pi}}\, e^{-(x-\mu)^2/(2\sigma^2)}$$

where μ and σ are unknown parameters. If we randomly sample five individuals from the population and record their IQ scores, then each score can be viewed as an outcome of a copy of the random variable X. Our goal is to use such data to infer the values of μ and σ. (See Figure 1.)

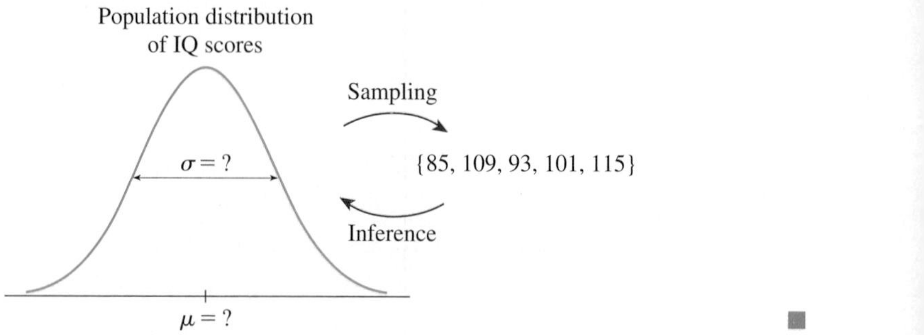

FIGURE 1

Example 1 shows that the data from a sample can be viewed as a set of outcomes of random variables. For example, the five IQ scores in Example 1 can be thought of as

The **sample mean** is the average value of the variable in a sample.

the outcomes of five random variables, where each random variable has the same probability density function $f(x)$ as given in Equation 1. Thus the *sample mean* is a summary statistic given by the average value of the outcomes of these five random variables. For this reason it will be helpful to understand some properties of the sum (and the average) of a set of random variables.

■ Sums of Random Variables

We begin by defining *independence* for random variables. Suppose X_1 and X_2 are two random variables, and consider the events $A = \{X_1 \leqslant a_1\}$ and $B = \{X_2 \leqslant a_2\}$ for some constants a_1 and a_2. From Definition 12.3.4, A and B are independent if $P(A \cap B) = P(A)\,P(B)$. This gives the following definition of independence for random variables.

Informally, X_1 and X_2 are independent if knowledge of the outcome of X_2 does not alter the probability distribution of X_1, and vice versa.

(2) Definition The random variables X_1 and X_2 are **independent** if

$$P(X_1 \leqslant a_1 \cap X_2 \leqslant a_2) = P(X_1 \leqslant a_1)\,P(X_2 \leqslant a_2)$$

for every pair of values a_1 and a_2.

Definition 2 applies only to two random variables but it can be extended to define the mutual independence of any finite set of random variables, just as we did for finite sets of events on page 756.

Another important concept is that of independent and identically distributed random variables.

(3) Definition A set of random variables $X_1, X_2, \ldots, X_n$ is said to be **independent and identically distributed (i.i.d.)** if the X_i are mutually independent and all X_i have the same probability density function.

In Section 11.4 we defined a *simple random sample:* all individuals in the population are equally likely to be sampled and the selection of all individuals is independent. In light of Definition 3, we can view the values of a variable measured on the individuals of a simple random sample as a set of i.i.d. random variables.

EXAMPLE 2 | IQ scores (continued) A sample of five IQ scores is collected by testing all members of a randomly chosen family of five. Is it reasonable to assume that the scores are i.i.d. random variables? Explain.

SOLUTION The IQ scores are probably not i.i.d. random variables. We might expect the scores of family members to be similar because of shared genes and environmental factors. Therefore, knowledge of one individual's IQ score would tell us something about the likely scores of other family members, and so the IQ scores would not be mutually independent. ■

EXAMPLE 3 | EPO, or erythropoietin, is a hormone that stimulates red blood cell production. A synthetic variant of EPO has been used by athletes as a performance-enhancing drug to increase aerobic capacity. A random sample of 14 male university students was chosen from the registration records. The athletes first had their aerobic capacity measured (as liters of oxygen consumed per minute during exercise). They were then injected with EPO for six weeks, and their aerobic capacity was measured again. The change in aerobic capacity of each individual was quantified as aerobic capacity after injection minus aerobic capacity before injection. Is it reasonable to assume that the resulting 14 values are i.i.d. random variables?

SOLUTION It is probably reasonable to assume that the changes in aerobic capacity are i.i.d. random variables. The 14 individuals represent a simple random sample and therefore their scores will have a common probability density function and be mutually independent. ■

We are mainly interested in i.i.d. normal random variables because we are assuming that the variable X in the population from which a sample is taken is a normal random variable. In this case we have the following important result.

(4) Theorem Suppose $X_1, X_2, \dots, X_n$ are n independent, identically distributed, normal random variables with mean μ and standard deviation σ. Define the random variable $Y_n = (X_1 + \cdots + X_n)/n$ to be the average of the X_i. Then Y_n is also normally distributed and has mean and standard deviation given by

$$\text{Mean of } Y_n\text{:} \quad \mu$$

$$\text{Standard deviation of } Y_n\text{:} \quad \frac{\sigma}{\sqrt{n}}$$

The proof of Theorem 4 requires multivariable integration and is therefore omitted. Theorem 4 is central to statistical inference. We can view the measurements of a normally distributed random variable in a simple random sample of n individuals as n i.i.d. normal random variables. And so, if the population mean and standard deviation of the variable are μ and σ, respectively, then the sample mean will itself be a normal random variable with mean μ and standard deviation $\sigma/\sqrt{n}$.

Beware of the different use of subscripting for X versus for Y. For example, X_5 denotes the fifth outcome in the sample, whereas Y_5 denotes the mean value in a sample of size $n = 5$.

EXAMPLE 4 | IQ scores (continued) Suppose a simple random sample of five individuals' IQ scores is taken from a population whose scores follow the normal distribution $f(x)$ given in Equation 1, with mean $\mu = 110$ and standard deviation $\sigma = 12$. The mean IQ score of this sample is itself a random variable defined by

$$Y_5 = \frac{X_1 + X_2 + X_3 + X_4 + X_5}{5}$$

where the X_i are i.i.d. random variables with common PDF $f(x)$. Figure 2 illustrates the relationship between the population PDF $f(x)$ and the sample means of 15 samples of size $n = 5$ (the red dots). Theorem 4 tells us that the sample mean Y_5 is a normal random variable with mean 110 and standard deviation $12/\sqrt{5} \approx 5.37$. Graphically, this is illustrated by the sample means in Figure 2 being relatively closely clustered around the population mean of 110. ■

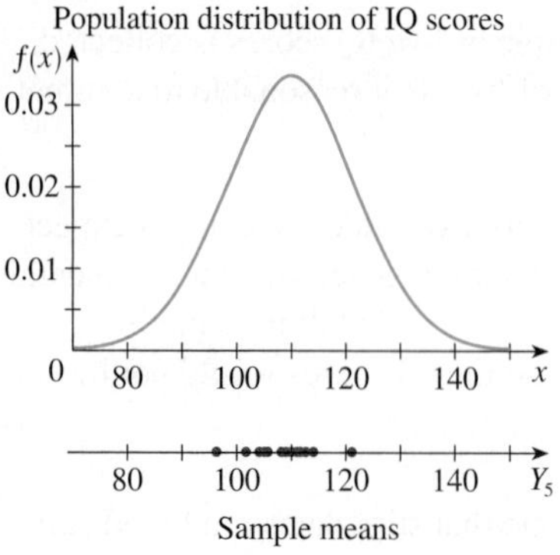

FIGURE 2

■ The Sampling Distribution of the Mean

When we take a random sample from a population and calculate a summary statistic for a variable measured on these individuals (such as a sample mean), this summary statistic is itself a random variable. The probability distribution of this random variable is called the **sampling distribution** of this statistic. The sampling distribution characterizes the distribution of summary statistic values that we expect to see if we were to repeat the process of sampling over and over again.

How is this helpful for statistical inference? Let's consider trying to infer the population mean μ of some variable. Imagine that we take a random sample from the population and use the sample mean as an estimate for the population mean μ. We know that, by chance, the sample mean will deviate from μ to some extent (for example, see Figure 2). The probability distribution for the sample mean is called the sampling distribution of the sample mean or, more concisely, the **sampling distribution of the mean.**

How far off do we expect the sample mean to be from the true population mean value μ? The sampling distribution allows us to address this question. From Theorem 4 we know that the sampling distribution of the mean is a normal distribution whose mean and standard deviation are μ and $\sigma/\sqrt{n}$, respectively. As the following examples illustrate, this theorem allows us to calculate the probability of the sample mean deviating from μ by any amount of interest.

EXAMPLE 5 | IQ scores (continued) Suppose a simple random sample of five individuals' IQ scores is taken from a population whose values follow a normal distribution with unknown mean μ and standard deviation $\sigma = 15$. What is the probability that the sample mean deviates by no more than 10 IQ points from the true population mean?

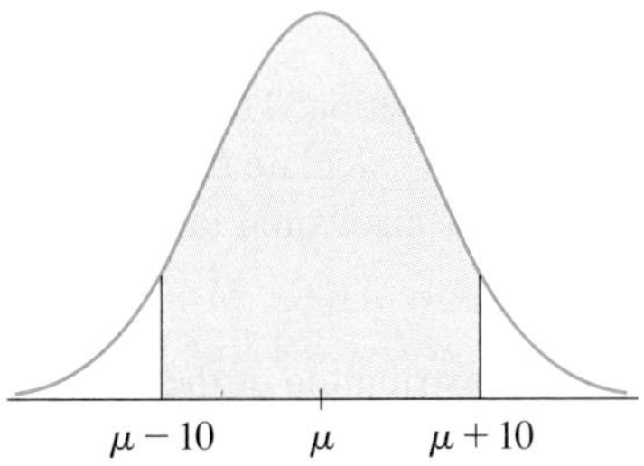

FIGURE 3

SOLUTION From Theorem 4 we know that the sample mean is a normal random variable whose mean and standard deviation are μ and $\sigma/\sqrt{5} = 15/\sqrt{5} \approx 6.71$, respectively. In other words, the sampling distribution of the mean is a normal distribution centered at a value of μ (see Figure 3), with a standard deviation of approximately 6.71. The probability that the sample mean will deviate by no more than 10 IQ points from the population mean μ is therefore the area under the curve in Figure 3 between $\mu - 10$ and $\mu + 10$. This is calculated as

$$\int_{\mu-10}^{\mu+10} \frac{1}{6.71\sqrt{2\pi}} e^{-(y-\mu)^2/(2(6.71)^2)}\, dy$$

We can simplify this integral by using the Substitution Rule. If we substitute

$$u = \frac{y - \mu}{6.71\sqrt{2}}$$

then $dy = 6.71\sqrt{2}\, du$. Also, when $y = \mu + 10$, $u = 10/(6.71\sqrt{2})$ and when $y = \mu - 10$, $u = -10/(6.71\sqrt{2})$. Therefore

$$\int_{\mu-10}^{\mu+10} \frac{1}{6.71\sqrt{2\pi}} e^{-(y-\mu)^2/(2(6.71)^2)}\, dy = \frac{1}{\sqrt{\pi}} \int_{-10/(6.71\sqrt{2})}^{10/(6.71\sqrt{2})} e^{-u^2}\, du$$

Like many of the functions discussed in Section 5.7, the antiderivative of e^{-u^2} is not an elementary function. Therefore we must use a numerical integrator to obtain an approximation. Doing so, we get

$$\frac{1}{\sqrt{\pi}} \int_{-10/(6.71\sqrt{2})}^{10/(6.71\sqrt{2})} e^{-u^2}\, du \approx 0.86$$

Thus the probability that the sample mean lies within 10 IQ points of the true population mean is approximately 0.86. Put another way, if we were to take a large number of samples, each of size $n = 5$, and calculate the mean IQ score of each sample, then roughly 86% of these sample means would fall within 10 IQ points of the true population mean IQ score. ■

EXAMPLE 6 | Butterfly wingspan The wingspan, in centimeters, of a population of Monarch butterflies is normally distributed with a mean of μ cm and a standard deviation of 5 cm. Suppose the wingspan of ten randomly selected individuals is measured and we calculate the mean of this sample. What is the probability that the sample mean will deviate by no more than 2 cm from the true population mean?

SOLUTION Again, from Theorem 4 we know that the sample mean is a normal random variable whose mean and standard deviation are μ and $5/\sqrt{10} \approx 1.58$, respectively. The probability that the sample mean will deviate by no more than 2 cm from the population mean is therefore approximately

$$\int_{\mu-2}^{\mu+2} \frac{1}{1.58\sqrt{2\pi}} e^{-(y-\mu)^2/(2(1.58)^2)}\, dy$$

Using the Substitution Rule with

$$u = \frac{y-\mu}{1.58\sqrt{2}}$$

we have

$$\int_{\mu-2}^{\mu+2} \frac{1}{1.58\sqrt{2\pi}} e^{-(y-\mu)^2/(2(1.58)^2)}\, dy = \frac{1}{\sqrt{2\pi}} \int_{-2/(1.58\sqrt{2})}^{2/(1.58\sqrt{2})} e^{-u^2} du \approx 0.79$$

Again we used a numerical integrator to obtain the final approximation. Thus, if we were to take a large number of samples, each of size $n = 10$, and calculate the mean wingspan for each sample, then roughly 79% of these sample means would fall within 2 cm of the true population mean wingspan. ■

Examples 5 and 6 illustrate how we can use the sampling distribution of the mean to determine how well a sample mean is likely to estimate the true population mean. Notice that in order to carry out these calculations, we do not need to know the population value μ. This can be shown to be true more generally (see Exercise 18).

We can now be more formal about the terms *precision* and *bias*, which were introduced in Section 11.4. Recall that the *precision* of an estimate refers to the variability of this estimate from sample to sample. An estimate is precise if it does not vary substantially from one sample to the next: its sampling distribution is relatively narrow. We reasoned intuitively in Section 11.4 that larger sample sizes should result in more precise estimates. We can now see from Theorem 4 why this is true for the sample mean of a normally distributed variable: the width of the sampling distribution for the sample mean, as measured by its standard deviation, is inversely related to the square root of the sample size, as illustrated in Figure 4.

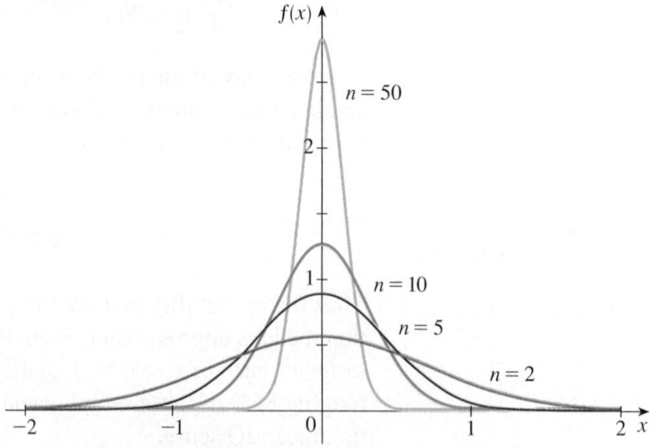

FIGURE 4
The sampling distribution of the mean for different sample sizes

Recall that *bias* refers to any systematic over- or underestimation of a population parameter. This idea can now be formalized.

(5) Definition Suppose Y is a summary statistic calculated from a random sample. We say that Y provides an **unbiased** estimate of the population parameter p if

$$E[Y] = p$$

Figure 5 provides illustrations of biased and unbiased estimates of the population mean.

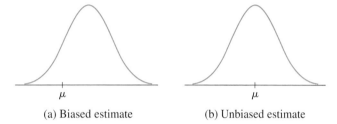

FIGURE 5
Sampling distributions for statistics that are biased and unbiased estimates for the population mean

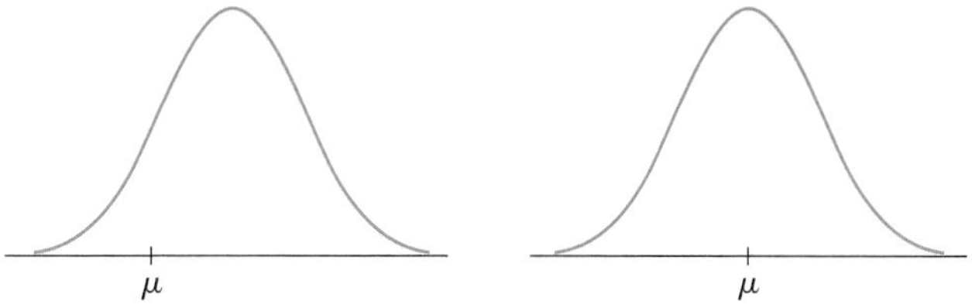

(a) Biased estimate (b) Unbiased estimate

Theorem 4 shows that, at least for normally distributed variables, the sample mean does in fact provide an unbiased estimate of the population mean because the mean value of its sampling distribution coincides with the true population mean value μ.

The Sampling Distribution of the Standard Deviation

Although our primary focus will be on estimating the population mean of continuous variables using the sampling distribution for the sample mean, in principle we can consider the sampling distribution of any summary statistic. Here we consider the standard deviation.

In Chapter 11 we gave a formula for calculating the standard deviation of a data set. Using this formula, we can ask questions about the form of the sampling distribution for standard deviation, just as we did for the sample mean. And we could also ask questions about the precision and bias of this quantity as an estimate of the true population standard deviation.

This may come as a surprise, but the standard deviation of a data set is actually a biased estimate of the true population standard deviation. In other words, the mean of the sampling distribution for the standard deviation of the data is not equal to the true population standard deviation.

Beware of the potentially confusing distinction between the *standard deviation of the data* (denoted by s.d.), defined in Definition 11.1.4, and the *sample standard deviation* (denoted by s), defined in Definition 13.1.6. The former, s.d., is simply the standard deviation of the set of numbers that make up the data set. The latter, s, is a slightly different summary statistic that provides an unbiased estimate for the population standard deviation. It is the *sample standard deviation* that is most often of interest in inferential statistics.

The most commonly used summary statistic that *does* provide an unbiased estimate of the population standard deviation is called the *sample standard deviation*, and it differs from the standard deviation of the data given in Definition 11.1.4 in that we divide by $n - 1$ rather than by n.

(6) Definition Let $x_1, x_2, \ldots, x_n$ be n data points, and let $\bar{x}$ be their mean. The **sample standard deviation** is

$$s = \sqrt{\frac{1}{n-1} \sum_{k=1}^{n} (x_k - \bar{x})^2}$$

The **sample variance** is s^2, the square of the sample standard deviation.

Whenever we wish to estimate a population standard deviation, we will use the sample standard deviation defined in (6) rather than the standard deviation of the data given in Definition 11.1.4.

EXERCISES 13.1

1. Suppose X_1 and X_2 are nonnegative, continuous, independent random variables with cumulative distribution functions $F_1(x)$ and $F_2(x)$.

(a) Write an expression in terms of the CDFs for the probability that $X_1 \leqslant 10$ *and* $X_2 \leqslant 8$.

(b) Write an expression in terms of the CDFs for the probability that $X_1 \leqslant 5$ *and* $X_2 \geqslant 15$.

(c) Write an expression in terms of the CDFs for the probability that $5 \leqslant X_1 \leqslant 10$ *and* $3 \leqslant X_2 \leqslant 8$.

2. Suppose Y_1, Y_2, and Y_3 are nonnegative, continuous, independent random variables with cumulative distribution functions $F_1(y)$, $F_2(y)$, and $F_3(y)$. Write an expression in terms of the CDFs for the probability that all three random variables are larger than or equal to 5.

3. Suppose T_1 and T_2 are independent exponential random variables with mean values of 1 and 2, respectively. What is the probability that $T_1 \geqslant 2$ *and* $T_2 \geqslant 5$?

4–7 Explain why the data set might not constitute a set of i.i.d. random variables.

4. A professor wants to infer the mathematical knowledge of all first-year university students and so gives an exam to the students of three randomly chosen first-year classes and records the scores.

5. Influenza antibodies A person's exposure to the influenza virus causes the development of antibodies against the virus. A scientist wants to infer the level of antibodies against influenza for people living in a particular city. She randomly chooses 100 households and measures the antibody level of all individuals in each of these households.

6. Suppose we wish to estimate the average annual household income in California. We record the income of all households in 50 randomly chosen neighborhoods in the state.

7. Tree heights A study is conducted to estimate the average tree height in a forest. The forest is divided into regions of 1 square km each, and the height of all trees in 35 of these regions is measured.

8–12 Use a numerical integrator to calculate the probability of the specified outcome.

8. Weddell seals The weight of newborn Weddell seals is well described by a normal random variable with unknown mean and with standard deviation 22.6 lb. The weights of 20 seals are recorded. Find the probability that the sample mean falls within 10 lb of the true population mean.

9. Multiple sclerosis The age of onset of multiple sclerosis is well described by a normal random variable with unknown mean and with standard deviation 7.6 years. The age of onset is measured for 32 individuals. Find the probability that the sample mean falls within 2 years of the true population mean.

10. Hospital stays The duration of stays at a major US hospital is well described by a normal random variable with unknown mean and with standard deviation 9.5 days. The duration of stay for 20 patients is recorded. Find the probability that the sample mean deviates by more than 5 days from the true population mean.

11. Damselflies Suppose that number of external parasites on damselflies is well described by a normal random variable with unknown mean and with standard deviation 20. A study measured the number of parasites on 20 damselflies. Find the probability that the sample mean deviates by more than 8 parasites from the true population mean.

12. Anemia The concentration of red blood cells in mice 10 days after infection with malaria is well described by a normal random variable with unknown mean and with standard deviation (1.4×10^6) cells/μL. Red blood cell concentration is measured in a sample of 18 mice. Find the probability that the sample mean deviates by more than (0.5×10^6) cells/μL from the true population mean.

13–17 Use a numerical integrator.

13. Coral bleaching is a process through which coral reefs lose their color during extreme temperature events. The extent of bleaching after an extreme temperature event can be measured using a dimensionless index, and the measurements are well described by a normal random variable with unknown mean and with standard deviation 0.12. Suppose that 14 independent studies each measure the index at 20 sites (and no two studies have any sites in common). Roughly how many of these studies would you expect to have a mean index that falls further than 0.0181 from the true population mean?

14. Huntington's disease is a genetic disorder causing neurodegeneration and eventual death. Suppose that six independent studies were conducted, and each quantified the age at death of 15 patients. If age at death is well described by a normal random variable with unknown mean and with standard deviation 8.0 years, roughly how many of these studies would you expect to have a sample mean that falls within 12 years of the true population mean?

15. Monarch butterflies The wingspan of Monarch butterflies is well described by a normal random variable with unknown mean and with standard deviation 0.9 cm. The wingspans of 28 butterflies are measured (in centimeters). In an independent follow-up study, the wingspans of another 28 butterflies are measured. What is the probability that, in both studies, the sample means deviate by more than 0.1 cm from the true population mean?

16. Breast cancer tumors were measured for a sample of 15 patients by estimating tumor diameter (in centimeters). In a second independent study, tumors were measured in another 12 patients. Suppose that diameter is well described by a normal random variable with unknown mean and with standard deviation 2.1 cm. What is the probability that neither study has a sample mean deviating by more than 0.5 cm from the true population mean?

17. Antibiotic *MIC* The minimum inhibitory concentration (*MIC*) of a new antibiotic is measured in 10 bacterial colonies. Two additional, independent studies are conducted and the *MIC* is measured in another 10 bacterial colonies in each of these studies. Suppose that *MIC* is well described by a normal random variable with unknown mean and with standard deviation 0.043 μg/mL. What is the probability that at least one of the sample means from these three studies deviates by more than 0.01 μg/mL from the true population mean?

---

18. Suppose a simple random sample of size n is taken from a population, and the variable of interest follows a normal distribution with mean μ and with standard deviation σ. Use the Substitution Rule (5.4.5) to show that the probability that the sample mean lies within z units of the true population mean is independent of μ and is given by

$$\frac{1}{\sqrt{\pi}} \int_{-z\sqrt{n}/(\sigma\sqrt{2})}^{z\sqrt{n}/(\sigma\sqrt{2})} e^{-u^2}\, du$$

19–24 Calculate the sample standard deviation.

BB **19. Weddell seals** Twenty individual Weddell seals from Antarctica were weighed, giving the following data (in pounds).

902	920	924	932	937
939	945	948	949	951
957	958	961	965	969
970	975	982	987	991

BB **20. Multiple sclerosis** is a serious neurological disease. The following data set gives the age of onset of multiple sclerosis for 32 individuals.

23	52	32	27	28	30	13	37
33	29	25	31	26	30	5	29
41	27	30	34	28	19	31	29
32	28	33	28	31	29	29	30

BB **21. Hospital stays** The duration of stay for 20 patients at a major US hospital is given in the following data set.

6	10	5	12	5
3	41	12	6	15
20	4	3	7	13
16	18	29	23	17

BB **22. Damselflies** A study measured the number of external parasites on a sample of 20 damselflies, giving the following data.

5	0	0	54	5
12	27	24	36	5
56	43	15	42	12
62	36	34	58	23

BB **23. Huntington's disease** is a genetic disorder causing neurodegeneration and eventual death. The age at death of 15 patients with the disease is given in the following data.

71	70	69	68	67
66	64	62	58	57
55	53	52	48	45

Source: Adapted from S. Andrew et al., "The Relationship between Trinucleotide (CAG) Repeat Length and Clinical Features of Huntington's Disease," *Nature Genetics* 4 (1993): 398–403.

BB **24. Monarch butterflies** migrate thousands of miles each year to overwinter in areas of Mexico. The following data give the wingspans (in centimeters) of 28 individuals captured in their overwintering grounds.

8.6	10.9	9.8	12.4	10.6	10.7	10.9
10.1	10.2	10.3	9.4	9.1	11.2	9.7
11.6	12.0	10.4	9.5	10.9	10.5	9.9
11.1	10.5	11.3	11.7	11.3	10.3	11.5

---

25. Maximum likelihood estimate of mean human height Human height is well described by a normal random variable. Suppose a simple random sample of two individuals is taken, and X_1 and X_2 are their heights.

The random variables X_1 and X_2 are i.i.d. normal random variables with common probability density function given by Equation 1. Let $F(x)$ be their common CDF.

(a) Write an expression for the probability that $X_1 \leq x_1$ *and* $X_2 \leq x_2$, in terms of the function F.

(b) The expression in part (a) is called the *joint CDF* of X_1 and X_2. If this function is denoted by $G(x_1, x_2)$, then it can be proved that the joint probability density function of X_1 and X_2, denoted by $g(x_1, x_2)$, is given by

$$g(x_1, x_2) = \frac{\partial^2 G}{\partial x_1 \, \partial x_2}$$

Calculate the joint probability density function.

(c) Suppose the heights of the two individuals are 160 cm and 170 cm. Find the value of the joint PDF for this sample, as a function of the population mean μ.

(d) The function in part (c) is referred to as a *likelihood function*. It gives the value of the joint PDF when evaluated at the sampled data points, for different values of the population mean μ. The *maximum likelihood estimate* for the population mean height is the value of μ that maximizes this likelihood function. In other words, it is the value of μ that makes the joint PDF at the sampled data as large as possible. Find this value of μ.

26. Consider a continuous variable, such as cholesterol level, that can be described by a normal random variable. Suppose a simple random sample of n individuals is taken, and let $X_1, \ldots, X_n$ be the resulting values of the variable. Thus the X_i are i.i.d. normal random variables with common probability density function given by Equation 1. Let $F(x)$ be their common CDF.

(a) Write an expression for the probability that $X_1 \leq x_1$ *and* $X_2 \leq x_2$ *and . . . and* $X_n \leq x_n$ in terms of the function F.

(b) The expression in part (a) is the joint CDF of the X_i. If this function is denoted by $G(x_1, \ldots, x_n)$, then it can be proved that the joint probability density function of the X_i, denoted by $g(x_1, \ldots, x_n)$, is given by

$$g(x_1, \ldots, x_n) = \frac{\partial^n G}{\partial x_1 \cdots \partial x_n}$$

Calculate the joint probability density function.

(c) If we substitute the sampled values of the X_i into the function found in part (b), the result can be viewed as a function of the population mean μ. This function is referred to as a *likelihood function*. It gives the value of the joint PDF when evaluated at the sampled data points $x_1, \ldots, x_n$ for different values of the population mean μ. The *maximum likelihood estimate* for the population mean is the value of μ that maximizes this likelihood function. Show that the maximum likelihood value for the population mean is equal to the sample mean.

13.2 Confidence Intervals

In this section we use the ideas developed in Section 13.1 to define an interval in such a way that we can have a certain degree of confidence that it contains the population mean μ. This is called a *confidence interval.*

Recall that Definition 12.5.13 gave the PDF of a standard normal random variable Z as

(1)
$$f(z) = \frac{1}{\sqrt{2\pi}} e^{-z^2/2}$$

As explained in Section 12.5, any normal random variable can be converted to a standard normal random variable with an appropriate change in the scale of measurement. Specifically, if X is a normal random variable with mean μ and standard deviation σ, then the random variable Z defined by

(2)
$$Z = \frac{X - \mu}{\sigma}$$

is a standard normal random variable. In Exercise 15 you are asked to show that this fact is true.

Equation 2 says that any normal random variable can be converted to a standard normal random variable by measuring it as a difference from its mean, in units of its standard deviation. For example, $Z = 2$ means that X is greater than its mean by 2 standard deviations. We can see this by setting $Z = 2$ in Equation 2 and rearranging:

$$2 = \frac{X - \mu}{\sigma}$$

$$2\sigma = X - \mu$$

$$X = \mu + 2\sigma$$

where the last two equations show that X is 2 standard deviations greater than its mean.

Throughout this section we will refer to the following summary table for the area under different portions of a standard normal curve.

(3) Areas under a Standard Normal Curve

(a) Approximately 90% of the area under the curve lies between -1.64 and 1.64.

(b) Approximately 95% of the area under the curve lies between -1.96 and 1.96.

(c) Approximately 99% of the area under the curve lies between -2.58 and 2.58.

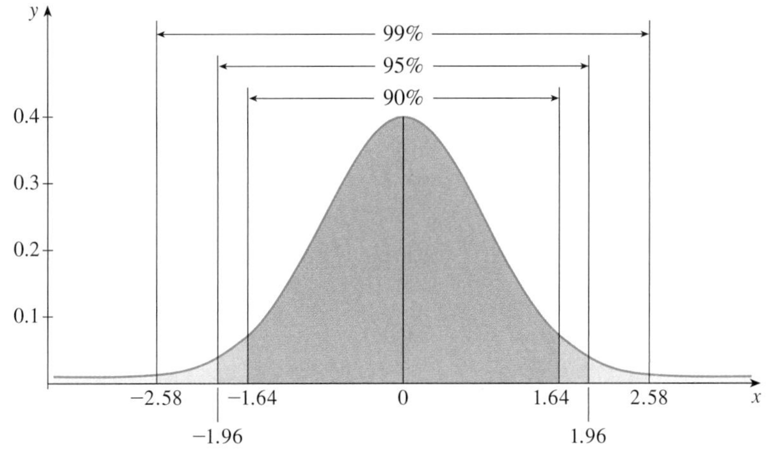

FIGURE 1

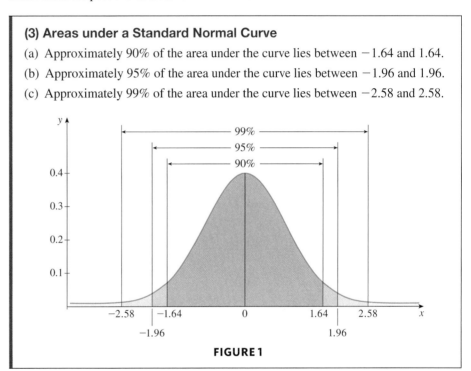

■ Interval Estimates

The results from Section 13.1 showed that the sample mean Y_n is a normal random variable with mean equal to the true population mean μ and standard deviation $\sigma/\sqrt{n}$, where σ is the population standard deviation. As a result we could calculate the probability that the sample mean falls within various distances of the true mean, provided we knew the population standard deviation.

The idea of a confidence interval turns this calculation around: we seek to determine an interval of some length that, with a specified probability, will contain the true population mean. The specified probability is called the **confidence level**.

To begin, let's define the new random variable Z_n as

(4) $$Z_n = \frac{Y_n - \mu}{\sigma/\sqrt{n}}$$

Since $\sigma/\sqrt{n}$ is the standard deviation of Y_n, from Equation 2 we can see that Z_n is a standard normal random variable. It is interpreted as the difference between the sample mean and the true population mean, measured in units of its standard deviation $\sigma/\sqrt{n}$.

Now suppose we want to construct an interval around the sample mean Y_n that we can be 95% confident will contain the population mean μ. We know from (3) that

From Box 3 we know that these equations are each approximations, but it is customary in statistics to write them as exact equalities.

$$P(-1.96 < Z < 1.96) = 0.95$$

where Z is a standard normal random variable. We also know that Z_n in (4) is a standard normal random variable and therefore

$$\text{(5)} \qquad P\left(-1.96 < \frac{Y_n - \mu}{\sigma/\sqrt{n}} < 1.96\right) = 0.95$$

But Equation 5 can be written equivalently as

$$P\left(-1.96\frac{\sigma}{\sqrt{n}} < Y_n - \mu < 1.96\frac{\sigma}{\sqrt{n}}\right) = 0.95$$

or

$$\text{(6)} \qquad P\left(-Y_n - 1.96\frac{\sigma}{\sqrt{n}} < -\mu < -Y_n + 1.96\frac{\sigma}{\sqrt{n}}\right) = 0.95$$

Recall that the direction of an inequality changes when multiplying both sides by -1.

and, by multiplying the inequality in the argument of P through by -1, we then obtain

$$\text{(7)} \qquad P\left(Y_n - 1.96\frac{\sigma}{\sqrt{n}} < \mu < Y_n + 1.96\frac{\sigma}{\sqrt{n}}\right) = 0.95$$

From (7) we see that the interval whose endpoints are

$$\text{(8)} \qquad Y_n \pm 1.96\frac{\sigma}{\sqrt{n}}$$

will contain the population mean μ with probability 0.95. The interval whose endpoints are given by (8) is called a **95% confidence interval for the mean.**

Notice that a confidence interval is defined with reference to the sample mean Y_n; therefore we think of this interval as differing from sample to sample. Sometimes the interval will contain the true population mean, and sometimes it will not. Figure 2 illustrates this idea. The confidence level tells us the percentage of such intervals that will contain the true population mean.

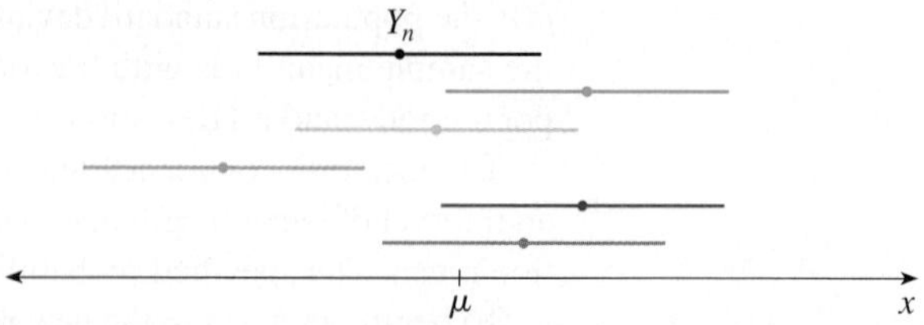

FIGURE 2
Examples of 95% confidence intervals calculated from different samples

EXAMPLE 1 | IQ scores We return to Example 13.1.1. The IQ scores of a population are normally distributed with a standard deviation of 15 IQ points and an unknown mean μ. A random sample of five individuals' IQ scores is collected:

85 109 93 101 115

Calculate the 95% confidence interval for the population mean and interpret the result.

SOLUTION The endpoints of the interval of interest are given by (8). We first calculate the sample mean, obtaining

$$Y_5 = \frac{85 + 109 + 93 + 101 + 115}{5} = 100.6$$

Substituting this, along with $n = 5$ and the given value $\sigma = 15$, into (8) gives

$$100.6 \pm 1.96 \frac{15}{\sqrt{5}}$$

Therefore the 95% confidence interval based on this sample is approximately

$$(87.5, 113.7)$$

If we constructed such confidence intervals for repeated samples of size $n = 5$, then 95% of them would contain the true population mean value μ. As Figure 3 illustrates, we can't know whether the particular interval (87.5, 113.7) does in fact contain the true mean value, since it is simply one instance of such an interval, and it may, by chance, happen to be in the unlucky 5%.

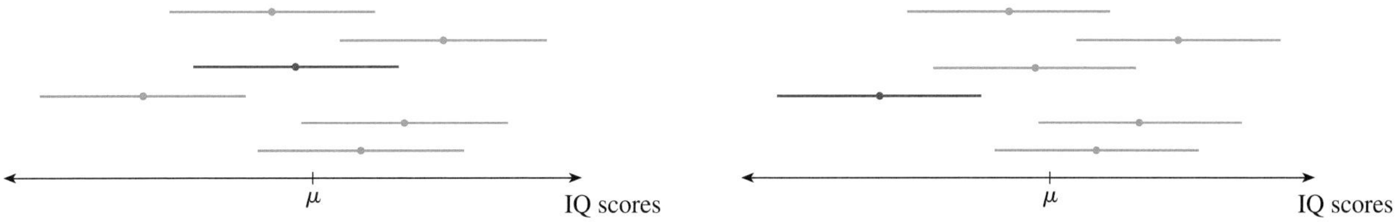

(a) Calculated interval overlaps true mean. (b) Calculated interval does not overlap true mean. ■

FIGURE 3
The red confidence interval represents the one calculated. Blue confidence intervals are hypothetical ones that would be obtained if we repeatedly took samples of size $n = 5$.

The preceding calculations can be followed for any confidence level of interest, although 95% is typically used, by convention. More generally, a C% confidence interval can be calculated using the following steps (see Exercise 17).

> **(9) Confidence Intervals When the Population Standard Deviation σ Is Known** A C% confidence interval for the mean is constructed as follows.
>
> (a) Calculate the sample mean Y_n, where n is the sample size.
>
> (b) Determine the value z_C such that $P(-z_C < Z < z_C) = \frac{C}{100}$ from the standard normal distribution.
>
> (c) Construct the interval with endpoints $Y_n \pm z_C \frac{\sigma}{\sqrt{n}}$.
>
> These steps assume that the variable of interest is normally distributed.

EXAMPLE 2 | **IQ scores (continued)** Construct the 90% and the 99% confidence intervals for the mean, using the data in Example 1.

SOLUTION For the 90% confidence interval we see that $z_{90} = 1.64$ from (3). Therefore we have the endpoints

$$100.6 \pm 1.64 \frac{15}{\sqrt{5}}$$

and the approximate confidence interval is

$$(89.6, 111.6)$$

For the 99% confidence interval we see that $z_{99} = 2.58$ from (3). Therefore we have endpoints

$$100.6 \pm 2.58 \frac{15}{\sqrt{5}}$$

and the approximate confidence interval is

$$(83.3, 117.9)$$

The 90%, 95%, and 99% confidence intervals are displayed in Figure 4. In general, for a given data set, the greater the confidence level that is specified, the larger the confidence interval will be.

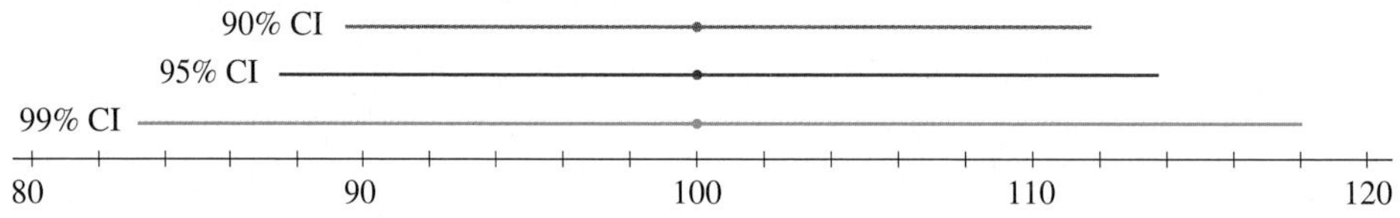

FIGURE 4
CI denotes confidence interval

■ Student's t-Distribution

Box 9 provides a set of steps for calculating the confidence interval of the mean when the population standard deviation σ is known. Unfortunately, this is almost never the case for real data. After all, if we knew the true population standard deviation, we would probably already know the true population mean as well.

To address this issue we go back to Equation 4. We can't proceed as before because we don't know the value of σ in this new scenario. A natural choice then is to substitute an *estimate* for σ into Equation 4.

The discovery of Student's t-distribution is credited to British scientist W. S. Gosset, who published the result under the pseudonym Student in 1908 while an employee of the Guinness Brewery.

From Section 13.1, we know that the sample standard deviation s given in Definition 13.1.6 is an unbiased estimate of σ. Using this estimate, we then define a new random variable, which we'll call T_{n-1}:

(10)
$$T_{n-1} = \frac{Y_n - \mu}{s/\sqrt{n}}$$

The reason for using a subscript $n - 1$ instead of n will be explained in a moment. The quantity $s/\sqrt{n}$ is sometimes referred to as the **standard error of the mean**.

It turns out that T_{n-1} is not a standard normal random variable, but its probability density function can be derived. Suppose that the sample mean Y_n is a normal random variable (which, from Theorem 13.1.4, will be true whenever the variable of interest is normally distributed). Then the PDF of the random variable (10) is called **Student's t-distribution**. As shown in Figure 5, it is symmetric about zero and has a "bell" shape, much like that of the standard normal. We will not give the explicit PDF here, but the exact form of the t-distribution depends on a quantity called the **degrees of freedom** (df), which equals $n - 1$. The t-distribution converges to the standard normal distribution as $n \to \infty$. (See Figure 5.)

The quantity $n - 1$ is called the *degrees of freedom* because once we know the sample mean, we can vary a total of only $n - 1$ of the data points when considering the potential values of the standard deviation and still have the same sample mean.

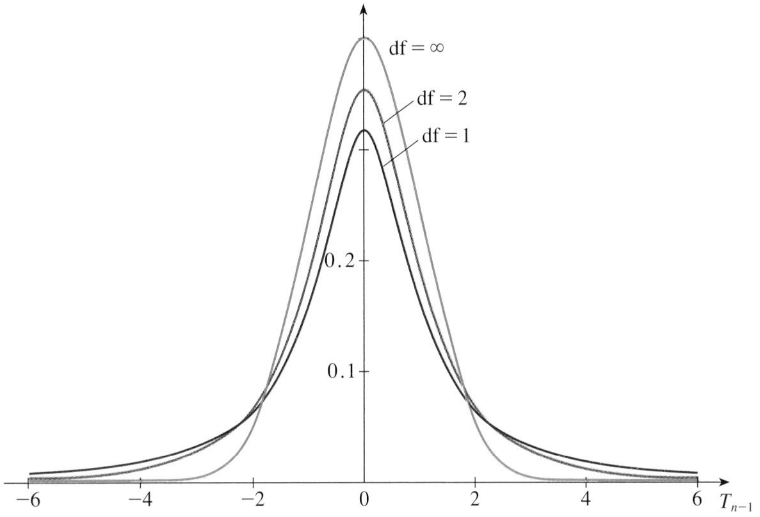

FIGURE 5
Student's t-distribution with different degrees of freedom

Given Student's t-distribution, we can now continue with steps analogous to those for Equations 5–7. Suppose $t_{95,n-1}$ is the value of the random variable T_{n-1} for which 95% of the area under the t-distribution with $n - 1$ degrees of freedom lies between $-t_{95,n-1}$ and $t_{95,n-1}$. (Such a value exists because, like the standard normal distribution, the t-distribution is symmetric about zero.) Then we have

$$P(-t_{95,n-1} < T_{n-1} < t_{95,n-1}) = 0.95$$

Substituting the expression for T_{n-1} gives

$$P\left(-t_{95,n-1} < \frac{Y_n - \mu}{s/\sqrt{n}} < t_{95,n-1}\right) = 0.95$$

We then rearrange to obtain

$$P\left(Y_n - t_{95,n-1}\frac{s}{\sqrt{n}} < \mu < Y_n + t_{95,n-1}\frac{s}{\sqrt{n}}\right) = 0.95$$

which is similar to Equation 7.

As a result, a C% confidence interval can be calculated using the following steps.

(11) Confidence Intervals When the Population Standard Deviation Is Unknown A C% confidence interval for the mean is constructed as follows.

(a) Calculate the sample mean Y_n and the sample standard deviation s, where n is the sample size.

(b) Determine the value $t_{C,n-1}$ such that $P(-t_{C,n-1} < T_{n-1} < t_{C,n-1}) = \dfrac{C}{100}$ from Student's t-distribution, with $n - 1$ degrees of freedom.

(c) Construct the interval with endpoints $Y_n \pm t_{C,n-1}\dfrac{s}{\sqrt{n}}$.

These steps assume that the variable of interest is normally distributed.

The formula for the sample standard deviation s is given in Definition 13.1.6.

To carry out step 11(b) we can, in principle, integrate the probability density function for the t-distribution. In practice, however, many calculators, spreadsheets, and computer algebra systems (CAS's) are programmed to produce these values numerically. Alternatively, we could consult a table of such values (see Appendix H, Table 1).

EXAMPLE 3 | **IQ scores (continued)** Construct the 95% confidence interval for the mean of the data in Example 1, assuming we don't know the population standard deviation.

SOLUTION Following the steps outlined in (11), we find that the sample mean is $Y_5 = 100.6$ (as in Example 1) and, using Definition 13.1.6, the sample standard deviation is

$$s = \sqrt{\frac{(85 - 100.6)^2 + (109 - 100.6)^2 + (93 - 100.6)^2 + (101 - 100.6)^2 + (115 - 100.6)^2}{5 - 1}} \approx 12$$

From Table 1 of Appendix H we read that the t-value for 95% and $n - 1 = 5 - 1 = 4$ degrees of freedom is $t_{95,4} = 2.78$. Substituting these values, along with $n = 5$, into step 11(c) gives the endpoints

$$100.6 \pm 2.78 \frac{12}{\sqrt{5}}$$

The confidence interval is therefore approximately

$$(85.7, 115.5)$$

Notice that the confidence interval is slightly larger than the interval we found in Example 1, where we knew the population standard deviation. This reflects the greater uncertainty that results from not having this information. ■

EXAMPLE 4 | BB **EPO** Table 11.1.1 on page 684 gives the following data for the change in aerobic capacity for 14 athletes after injection with EPO (measured as L oxygen/min):

0.4	0.5	0.3	0.4	0.5
0.3	0.3	0.4	0.5	0.3
0.3	0.2	0.3	0.1	

Construct the 95% confidence interval for the mean magnitude of change in performance due to EPO injection.

SOLUTION Before applying (11), we must ensure that the data come from a normally distributed population. Advanced statistics classes introduce formal tests for normality, but here we simply check the histogram of the data to see, visually, if there are any extreme deviations from normality.

As we saw in Figure 11.2.4, the histogram of the EPO data is slightly skewed, but it nevertheless looks approximately normal. Therefore, following the steps outlined in (11), the sample mean is $Y_{14} \approx 0.343$ and the sample standard deviation is $s \approx 0.116$. From Table 1 of Appendix H we read that the t-value for 95% and $n - 1 = 14 - 1 = 13$ degrees of freedom is between 2.13 and 2.23. We therefore take $t_{95,13} \approx 2.18$. Substituting these values, along with $n = 14$, into step 11(c) gives the endpoints

$$0.343 \pm 2.18 \frac{0.116}{\sqrt{14}}$$

The confidence interval therefore spans a change in performance from $\approx$0.275 to $\approx$0.411 L oxygen/min. ■

EXERCISES 13.2

1–4 Calculate the 95% confidence interval for the mean using the given population standard deviation. Assume that the data come from a normally distributed population.

BB **1. Weddell seals** Twenty individual Weddell seals from Antarctica were weighed, giving the following data (in pounds).

902	920	924	932	937
939	945	948	949	951
957	958	961	965	969
970	975	982	987	991

Suppose the population standard deviation is 22.6 lb.

BB **2. Multiple sclerosis** is a serious neurological disease. The following data set gives the age of onset of multiple sclerosis for 32 individuals.

23	52	32	27	28	30	13	37
33	29	25	31	26	30	5	29
41	27	30	34	28	19	31	29
32	28	33	28	31	29	29	30

Suppose the population standard deviation is 7.6 years.

BB **3. Hospital stays** The duration of stay for 20 patients at a major US hospital is given in the following data set.

6	10	5	12	5
3	41	12	6	15
20	4	3	7	13
16	18	29	23	17

Suppose the population standard deviation is 9.5 days.

BB **4. Anemia** When individuals become infected with malaria the parasite consumes red blood cells, causing the patient to become anemic. Red blood cell concentration was measured in a group of 18 mice, giving the following data, measured as $(\text{cells} \times 10^6)/\mu\text{L}$.

3.2	4.5	3.7	3.3	2.5	3.6
3.3	3.6	4.2	3.5	2.0	3.1
2.3	2.9	6.9	4.5	7.3	3.0

Suppose the population standard deviation is (1.4×10^6) cells/μL.

5–8 For each exercise, do the following.
(a) Construct a histogram of the data and decide if it is approximately normal.
(b) If the histogram in part (a) is approximately normal, calculate the 90% confidence interval for the population mean and provide a biological interpretation of your answer. If the histogram is not normal, calculate the 90% confidence interval for the mean as if it were normal, and then explain where in the calculation the lack of normality matters.

BB **5. Coral bleaching** is a process through which coral reefs lose their color during extreme temperature events. The extent of bleaching after an extreme temperature event was measured at 20 sites, giving the following data. Values are dimensionless measures of the extent of bleaching.

0.19	0.06	0.26	0.32	0.35
0.09	0.05	0.21	0.48	0.21
0.23	0.12	0.22	0.16	0.35
0.13	0.34	0.07	0.16	0.45

BB **6. Monarch butterflies** migrate thousands of miles each year to overwinter in areas of Mexico. The following data give the wingspans (in centimeters) of 28 individuals captured in their overwintering grounds.

8.6	10.9	9.8	12.4	10.6	10.7	10.9
10.1	10.2	10.3	9.4	9.1	11.2	9.7
11.6	12.0	10.4	9.5	10.9	10.5	9.9
11.1	10.5	11.3	11.7	11.3	10.3	11.5

BB **7. Huntington's disease** is a genetic disorder causing neurodegeneration and eventual death. The age at death of 15 patients with the disease is given in the following data.

71	70	69	68	67
66	64	62	58	57
55	53	52	48	45

Suppose the population standard deviation is 8 years.

Source: Adapted from S. Andrew et al., "The Relationship between Trinucleotide (CAG) Repeat Length and Clinical Features of Huntington's Disease," *Nature Genetics* 4 (1993): 398–403.

BB **8. Damselflies** A study measured the number of external parasites on a sample of 20 damselflies, giving the following data.

5	0	0	54	5
12	27	24	36	5
56	43	15	42	12
62	36	34	58	23

Suppose the population standard deviation is 20.

9–11 For each exercise, do the following.

(a) Construct a histogram of the data and decide if it is approximately normal.

(b) If the histogram in part (a) is approximately normal, calculate the 90% confidence interval for the population mean using the given standard deviation. Then assume that the standard deviation is unknown, repeat the calculation, and compare the results. If the histogram in part (a) is not normal, perform these calculations as if it were normal, and then explain where in the calculations the lack of normality matters.

BB **9. Breast cancer tumors** were measured for a sample of 15 patients by estimating tumor diameter (in centimeters). This gave the following data.

1.5	1.5	1.6	1.6	1.7
2.0	2.1	2.2	2.6	3.2
4.0	4.9	5.5	7.5	7.6

Suppose the population standard deviation is 2.1 cm.

10. Antibiotic *MIC* The minimum inhibitory concentration (*MIC*) of a new antibiotic is measured on 10 bacterial colonies, giving the following data (measured in $\mu g/mL$).

0.2	0.4	0.3	0.5	1.1
1.0	1.2	1.3	0.9	1.1

Suppose the population standard deviation is 0.1. $\mu g/mL$.

BB **11. HAART** (Highly Active Antiretroviral Therapy) is a treatment for HIV infection. The concentration of HIV in the blood (in units of $\log_{10}$ copies/mL) was measured in 15 patients before and after HAART treatment. The following data give the reduction in concentration.

3.7	2.5	3.5	3.6	0.7
2.2	2.5	1.4	3.6	1.8
2.4	2.3	4.7	2.3	3.0

Suppose the population standard deviation is 0.98.

12–14 Assume that the variables are normally distributed. Calculate the 95% confidence interval for each of the given groups, and plot them both on the same graph.

BB **12. Fish morphology** The width of the mouth (in mm) of fish feeding on benthic prey (that is, prey found on the bottom of lakes) was compared with that of fish feeding on limnetic prey (that is, prey found in the water column). A sample of benthic feeders gave the following data.

4.6	3.5	4.3	4.5	1.6
3.1	3.5	2.4	4.5	2.6
3.2	3.2	5.5	3.1	3.9

A sample of limnetic feeders gave this data:

3.5	4.2	2.2	3.5	3.0
3.7	3.3	1.9	0.4	3.6
1.8	2.6	1.6	2.2	2.1

BB **13. High-density lipoprotein (HDL) cholesterol** is sometimes referred to as the "good" cholesterol. A sample of 13 female and 17 male volunteers had their HDL cholesterol levels measured (in mg/dL), giving the following data:

Females	58	60	48	55	56
	46	54	50	49	51
	53	54	53		
Males	32	52	30	42	44
	52	55	29	29	31
	43	46	41	34	51
	36	30			

BB **14. Pesticides and bees** Bees are important pollinators of many agricultural crops. A study examined the impact of pesticides on the productivity of bee colonies by counting the number of worker bees produced by colonies either exposed, or not exposed, to a pesticide. For 12 colonies exposed to pesticide, the number of worker bees produced by each colony was as follows:

12	12	31	14	17
18	20	28	32	32
24	27			

For 10 colonies not exposed to pesticide, the data are

15	38	35	44	44
25	34	29	28	27

Source: Adapted from R. Gill et al., "Combined Pesticide Exposure Severely Affects Individual- and Colony-Level Traits in Bees," *Nature* 491 (2012): 105–8.

15–16 Suppose X is a normal random variable with PDF

$$f(x) = \frac{1}{\sigma\sqrt{2\pi}} e^{-(x-\mu)^2/(2\sigma^2)}$$

15. Define a new random variable

$$Z = \frac{X - \mu}{\sigma}$$

where $\sigma > 0$.

(a) Let $F_X(x)$ be the CDF of X and $F_Z(z)$ be the CDF of Z. Show that $F_Z(z) = F_X(\sigma z + \mu)$.

(b) Use the fact that the derivative of a variable's CDF is its PDF to show that Z is a standard normal random variable.

16. Define a new random variable $Y = aX + b$, where a is positive. Follow an approach similar to that of Exercise 15 to show that Y is a normal random variable with mean $a\mu + b$ and standard deviation $a\sigma$.

17. Use equations similar to Equations 5–8 to show that the $C\%$ confidence interval for a mean is given as in (9).

13.3 | Hypothesis Testing

We have used ideas from probability theory to estimate the value of population parameters such as the mean and standard deviation of a continuous variable. Sometimes, rather than estimating a parameter, we are primarily interested in using the data to determine whether the parameter is equal to a specific value. This is called *hypothesis testing*.

EXAMPLE 1 | BB **Heart rates** The heart rates (in beats/min) of 40 patients in an intensive care unit (ICU) are given:

52	68	69	70	78	78	79	81	83	88
88	88	88	89	89	89	92	92	95	95
95	98	98	98	99	99	101	101	103	106
108	108	109	109	113	115	115	119	128	139

Are heart rates from ICU patients unusual? We can formalize this question by assuming that the heart rate measurements of ICU patients come from a normal distribution with unknown mean and standard deviation. We then ask, is the mean of this distribution different from a normal resting heart rate of 72 beats/min? ■

© enciktat / Shutterstock.com

EXAMPLE 2 | BB **EPO** Example 13.2.4 gave the change in aerobic capacity of 14 athletes after injection with EPO (measured as L oxygen/min):

0.4	0.5	0.3	0.4	0.5
0.3	0.3	0.4	0.5	0.3
0.3	0.2	0.3	0.1	

Does EPO affect aerobic capacity? We can formalize this question by assuming that the measurements of the change in capacity come from a normal distribution with unknown mean and standard deviation. We then ask, is the mean of this distribution different from a value of 0? If so, then this result implies that EPO injection causes a change in aerobic capacity. ■

■ The Null and Alternative Hypotheses

Hypothesis testing begins by first specifying what are called the *null* and *alternative hypotheses.*

The **null hypothesis**, denoted by H_0, is the hypothesis that there is no effect. For instance, in Example 1 the null hypothesis is that the mean heart rate of ICU patients is 72 beats/min. This is written as

(1) $$H_0\colon\ \mu = 72$$

The **alternative hypothesis**, denoted by H_A, is the hypothesis that some effect is apparent. For instance, in Example 1, the alternative hypothesis is that the mean heart rate of ICU patients in not 72 beats/min. This is written as

(2) $$H_A\colon\ \mu \neq 72$$

In Example 1 we are interested only in whether the heart rates of ICU patients are unusual, but other alternative hypotheses might also be of interest. For example, we might wish to restrict attention to the question of whether the heart rates are *unusually high*. This would be represented by the following null and alternative hyptheses:

$$H_0\colon\ \mu \leqslant 72 \qquad H_A\colon\ \mu > 72$$

For simplicity, we will restrict our attention to null and alternative hypotheses of the form (1) and (2). Such hypotheses are referred to as *two-tailed* (for reasons that will be explained on page 824).

EXAMPLE 3 | EPO (continued) What are the null and alternative hypotheses for Example 2?

SOLUTION The null hypothesis is that the mean change in performance is zero, that is,

$$H_0\colon\ \mu = 0$$

The alternative hypothesis is that the change is nonzero, either being positive or negative, that is,

$$H_A\colon\ \mu \neq 0$$

■

■ The *t*-Statistic

We now consider the logic behind hypothesis testing. We start by assuming that the null hypothesis of no effect is true. We then ask, what is the probability of obtaining data like those that we have collected, given that the null hypothesis is true? If this probability turns out to be very small, then we should doubt our initial premise that the null hypothesis is true. In such a case we reject the null hypothesis. On the other hand, if there is a reasonably large probability of obtaining such data under the null hypothesis, then we conclude that the data are consistent with the null hypothesis.

Another way to view this procedure is to think of the null hypothesis as our "default" position. We then seek to determine whether the data are compatible with this default. If they are not, then we reject the null hypothesis. Otherwise, we conclude that the data are compatible with the null hypothesis.

Let's see how this is done mathematically when we are interested in testing hypotheses about a population mean. We will use the data from Example 1 on heart rates to illustrate the calculations.

Under the null hypothesis (1), the heart rates of ICU patients come from a normal distribution with a mean of 72 and an unknown standard deviation. Now imagine drawing a sample of 40 patients from this population and calculating the sample mean Y_{40} and the sample standard deviation s. We have seen in Equation 13.2.10 that these can then be combined into the single random variable

$$T_{39} = \frac{Y_{40} - 72}{s/\sqrt{40}} \qquad (3)$$

The quantity in (3) is referred to as the ***t*-statistic**. It is a measure of how far the sample mean lies from the hypothesized mean of 72, measured in units of the standard error of the mean, $s/\sqrt{40}$.

We expect the t-statistic to vary from sample to sample by chance. However, if the null hypothesis is true, then we expect the t-statistic to typically be relatively small because the sample mean is not expected to deviate greatly from the true population mean (see Section 13.2). On the other hand, if the t-statistic is very large (either positive or negative), then this suggests that the null hypothesis is not correct.

We can formalize this idea by noting that if the null hypothesis is true, then the probability density function of T_{39} in (3) is given by the t-distribution with 39 degrees of freedom (see Figure 1). If the value of T_{39} calculated from a data set is quite likely to occur with this distribution, then the data are compatible with the null hypothesis [see Figure 2(a)]. On the other hand, if the value of T_{39} calculated from a data set lies far into one of the "tails" of the distribution, then the data are not compatible with the null hypothesis because the observed deviation cannot be readily explained as having arisen by chance using this distribution [see Figure 2(b)].

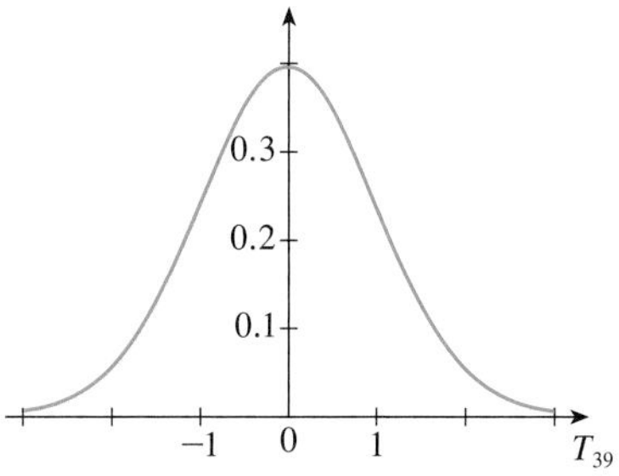

FIGURE 1
t-distribution with 39 degrees of freedom

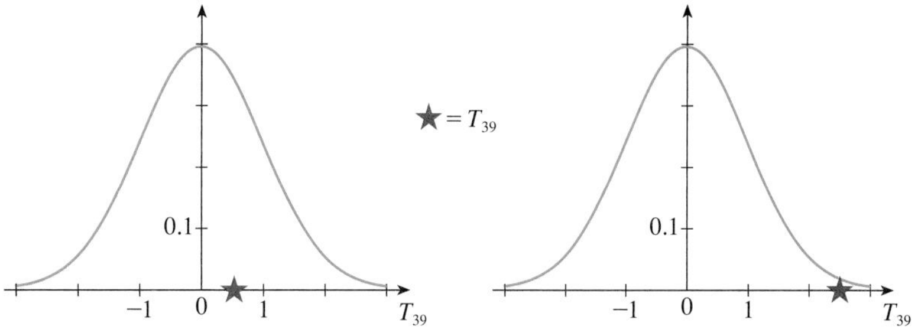

(a) Null hypothesis is not rejected

(b) Null hypothesis is rejected

FIGURE 2
t-distribution with 39 degrees of freedom

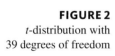

EXAMPLE 4 | BB **Heart rates (continued)** Calculate the t-statistic for the data in Example 1.

SOLUTION The sample mean is $Y_{40} = 95.3$ beats/min and the sample standard deviation is $s \approx 16.9$ beats/min. Therefore

$$T_{39} = \frac{95.3 - 72}{16.9/\sqrt{40}} \approx 8.72$$

The value of 8.72 is plotted on the graph of the PDF of T_{39} in Figure 3 on page 824. It looks as though the t-statistic for the data lies quite far into the upper tail of the distribution. Such a value is extremely unlikely to occur with this distribution, and therefore the null hypothesis is unlikely to be true.

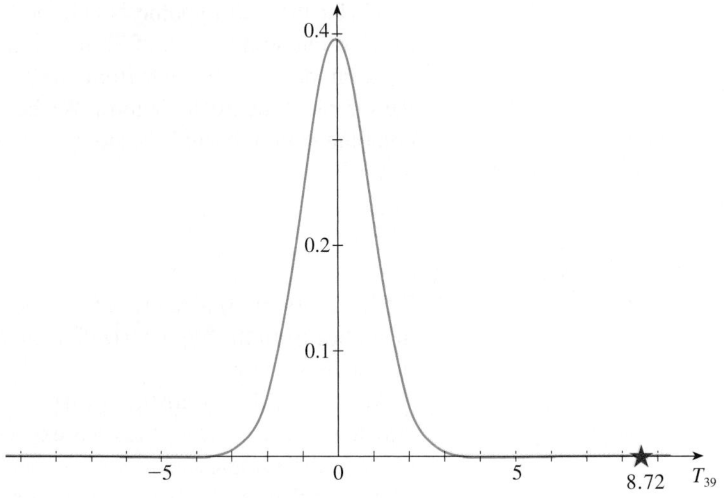

FIGURE 3
t-distribution with 39 degrees of freedom

The *P*-Value

To judge whether an observed value of the t-statistic lies far enough into the tail of the distribution to reject the null hypothesis, we need a quantitative yardstick to measure location within the distribution. This yardstick is called the P-value.

(4) Definition The ***P*-value** associated with a specific value of the t-statistic, for example, $T_{n-1} = t^*$, is the area in the tails of the Student's t-distribution with $n - 1$ degrees of freedom, beyond the values $-t^*$ and t^*. (See Figure 4.)

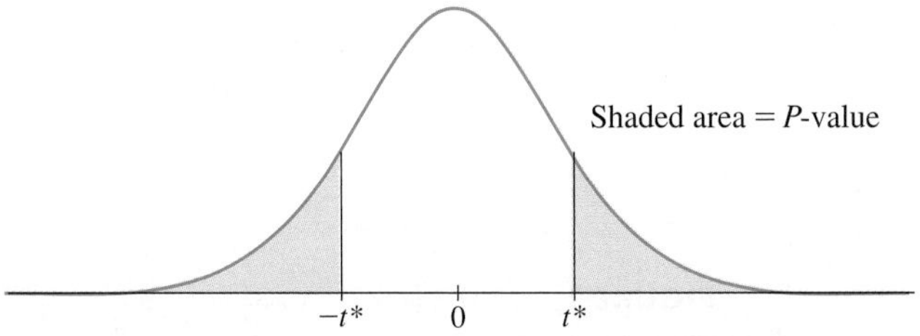

FIGURE 4

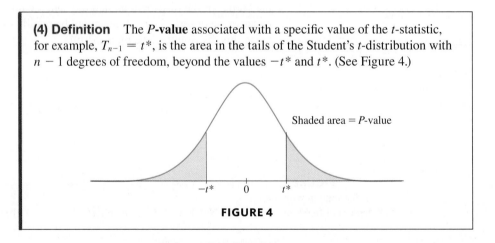

Notice that the P-value gives the summed area in *both* tails of the distribution. For this reason it is called the **two-tailed** value. It is interpreted as the probability that, if the null hypothesis were true and we took another sample from the population, then we would obtain a t-statistic value that is *at least as extreme as the observed value*. Since the P-value is two-tailed, it includes the possibility of both positive and negative extreme values. This is why the hypotheses in (1) and (2) are referred to as *two-tailed*.

EXAMPLE 5 | **Heart rates (continued)** Calculate the P-value associated with the t-statistic from Example 4.

SOLUTION Using $f_{39}(t)$ to denote the PDF of T_{39}, the P-value is given by

$$P = \int_{-\infty}^{-8.72} f_{39}(t)\,dt + \int_{8.72}^{\infty} f_{39}(t)\,dt \tag{5}$$

Without an explicit formula for the probability density function $f_{39}(t)$, we cannot evaluate the integrals in (5). However, tables of such values are often given for different values of the t-statistic (see Appendix H, Table 2). To use these, we choose the row with the correct degrees of freedom ($n - 1 = 39$ in this case) and then find the t-statistic values that bracket the *absolute value* of the calculated t-statistic. From Table 2 of Appendix H, the row with degrees of freedom closest to 39 is the row for 40 degrees of freedom. Looking across this row, we see that the largest value of the t-statistic given is 3.551, which corresponds to a P-value of 0.001. Furthermore, from Figure 4 we can see that larger values of the t-statistic correspond to smaller P-values. Now the absolute value of the t-statistic calculated from the data in Example 4 is 8.72, and since this is larger than 3.551, its corresponding P-value is less than 0.001. ■

Most scientific calculators can also calculate numerical approximations of P-values.

The P-value is a measure of the compatibility of the data with the null hypothesis. A large P-value (close to 1) indicates that the t-statistic is near the center of the distribution, meaning that the data are compatible with H_0. A small P-value (close to 0) indicates that the t-statistic is far into one of the tails of the distribution, meaning that the data are incompatible with H_0. In such a case we reject H_0.

The P-value provides an objective quantification of the extent to which the data provide evidence against the null hypothesis. For example, suppose that $P = 0.0001$ for a particular data set. This means that if the null hypothesis were true, and if we repeatedly took new samples from the population and calculated the t-statistic for each sample, then roughly only 1 in 10,000 of these t-statistic values would be as extreme as that calculated from the data set. The value obtained from the data set is therefore very unlikely under the null hypothesis. On the other hand, if $P = 0.5$ for example, then roughly half of all samples would give a value this extreme. In such a case the null hypothesis seems quite plausible. But where do we draw the line?

The P-value below which the null hypothesis is rejected is typically referred to as the **significance level**. There is no objective answer as to what this value should be. By convention, it is usually taken to be a P-value of 0.05. With this significance level, the data in Example 5 provide evidence that ICU patients have heart rates that are statistically significantly different from the normal rate of 72 beats/min.

A significance level of $\alpha = 0.05$ is typically used by convention.

■ Summary

The preceding approach to testing hypotheses about the mean of a normally distributed variable is summarized as follows.

(6) Hypothesis Testing for the Mean of a Normally Distributed Variable

(a) Specify the null and alternative hypotheses:

$$H_0\colon\ \mu = \mu_0 \qquad\qquad H_A\colon\ \mu \neq \mu_0$$

(b) Choose a significance level α.

(c) The t-statistic has the form

$$T_{n-1} = \frac{Y_n - \mu_0}{s/\sqrt{n}}$$

Calculate the value of this statistic from the data, where n is the sample size.

(d) Determine the P-value associated with the t-statistic calculated in part (c).

(i) If $P \leqslant \alpha$, then reject H_0 (data provide evidence that $\mu \neq \mu_0$).

(ii) If $P > \alpha$, then do not reject H_0 (data are compatible with the assumption that $\mu = \mu_0$).

We must check that the data come from a normal distribution before applying the steps in (6).

EXAMPLE 6 | BB EPO (continued) Test the hypotheses given in Example 3 for the EPO data from Example 2.

SOLUTION We have already checked the data for normality in Example 13.2.4. Example 3 specifies the null and alternative hypotheses, and we will use a significance level of $\alpha = 0.05$. The sample mean is $Y_{14} \approx 0.343$ L/min and the sample standard deviation is $s \approx 0.116$ L/min. Therefore

$$T_{13} = \frac{0.343 - 0}{0.116/\sqrt{14}} \approx 11.1$$

We now consult the row of Table 2 in Appendix H that corresponds to $n - 1 = 14 - 1 = 13$ degrees of freedom. The largest value of the t-statistic given is 4.221, which corresponds to a P-value of 0.001. The value calculated from the data ($T_{13} = 11.1$) is larger than 4.221, meaning that its corresponding P-value is less than 0.001. This is less than the significance level $\alpha = 0.05$ and therefore we reject the null hypothesis—the data provide evidence that EPO injection affects an athlete's aerobic capacity. ■

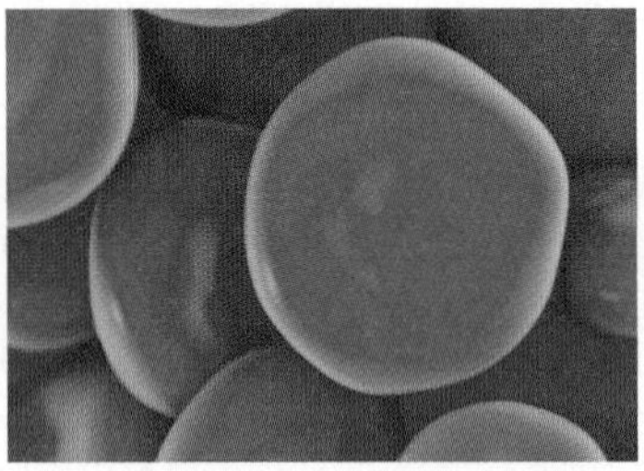

EXAMPLE 7 | BB Anemia When individuals become infected with malaria, the parasite consumes red blood cells, causing the patient to become anemic. For example, in mice the red blood cell (RBC) concentration measured 10 days after infection is (on average) (4×10^6) cells/μL. A new drug is being tested to see if it affects the level of anemia in mice. The following are data for RBC concentration on day 10 of infection from a sample of 18 mice.

4.2	4.5	3.7	4.3	4.6
3.6	3.3	4.6	4.2	3.5
4.0	3.1	4.3	4.9	6.9
4.5	7.3	3.0		

Do these data provide evidence that the drug affects RBC concentration?

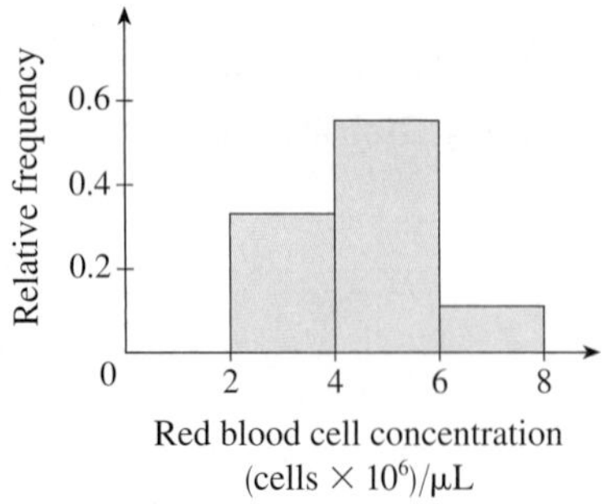

FIGURE 5

SOLUTION Before applying (6), we must first verify that the data are plausibly drawn from a normal distribution. Figure 5 plots a histogram of the data, illustrating that it is approximately normal. Therefore, following (6), we begin by specifying the null and alternative hypotheses. If the drug has no effect, then we expect the mean RBC concentration μ to be (4×10^6) cells/μL. If the drug does have an effect, then the mean concentration will differ from this value. Therefore, measuring RBC in units of $(\text{cells} \times 10^6)/\mu\text{L}$,

$$H_0\colon \mu = 4 \qquad H_A\colon \mu \neq 4$$

We use a significance level of $\alpha = 0.05$. The sample mean is calculated to be $Y_{18} \approx 4.36$ and the sample standard deviation is $s \approx 1.14$. Therefore

$$T_{17} = \frac{4.36 - 4.0}{1.14/\sqrt{18}} \approx 1.34$$

We now consult the row of Table 2 in Appendix H that corresponds to $n - 1 = 18 - 1 = 17$ degrees of freedom. The calculated value of $T_{17} = 1.34$ falls between the tabulated values of 1.333 and 1.740. The P-values corresponding to these tabulated values are 0.20 and 0.10, respectively. Therefore the P-value for the data satisfies $0.10 < P < 0.20$. (See Figure 6.) Since P is greater than the significance level $\alpha = 0.05$, we do not reject the null hypothesis—the data are consistent with the population having a mean RBC concentration of (4×10^6) cells/μL. In other words, the data do not provide evidence that the drug has any effect on anemia.

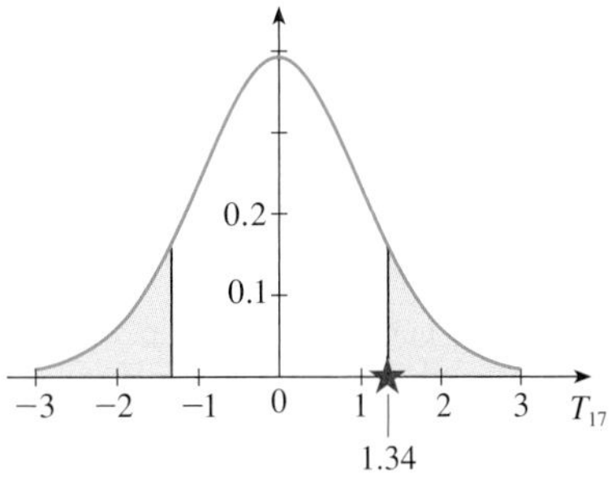

FIGURE 6
t-distribution with 17 degrees of freedom

■

Failing to reject the null hypothesis is not the same as accepting the null hypothesis. Accepting the null hypothesis suggests that the null hypothesis is true, but failure to reject the null hypothesis simply means that the data at hand are not sufficiently compelling for us to prefer the alternative hypothesis over the null hypothesis.

EXERCISES 13.3

BB **1. Weddell seals** Twenty individual Weddell seals from Antarctica were weighed, giving the following data (in pounds).

902	920	924	932	937
939	945	948	949	951
957	958	961	965	969
970	975	982	987	991

Check the data for normality. Use hypothesis testing to determine whether the data suggest that the population mean weight differs from 935 lb.

BB **2. Multiple sclerosis** is a serious neurological disease. The following data set gives the age of onset of multiple sclerosis for 32 individuals.

23	52	32	27	28	30	13	37
33	29	25	31	26	30	5	29
41	27	30	34	28	19	31	29
32	28	33	28	31	29	29	30

Check the data for normality. Use hypothesis testing to determine whether it is plausible that the mean age of onset in the population is 30 years.

BB **3. Hospital stays** The duration of stay for 20 patients at a major US hospital is given in the following data set.

6	10	5	12	5
3	41	12	6	15
20	4	3	7	13
16	18	29	23	17

Assume the data come from a normal distribution and use hypothesis testing to determine if the mean hospital stay in the population differs from 20 days. Where in your calculations have you used the assumption of normality?

BB **4. Damselflies** A study measured the number of external parasites on a sample of 20 damselflies, giving the following data.

5	0	0	54	5
12	27	24	36	5
56	43	15	42	12
62	36	34	58	23

Assume the data come from a normal distribution and use hypothesis testing to determine if the mean number of parasites on a damselfly is different from 40. Where in your calculations have you used the assumption of normality?

BB **5. Anemia** When individuals become infected with malaria the parasite consumes red blood cells, causing the patient to become anemic. Red blood cell concentration was measured in a group of 18 mice, 10 days after infection, giving the following data, in $(\text{cells} \times 10^6)/\mu\text{L}$.

3.2	4.5	3.7	3.3	2.5	3.6
3.3	3.6	4.2	3.5	2.0	3.1
2.3	2.9	6.9	4.5	7.3	3.0

Assume the data come from a normal distribution. Do the data provide evidence that the mean red blood cell concentration differs from (3.0×10^6) cells/μL at the $\alpha = 0.01$ significance level? State the null and alternative hypotheses.

BB **6. Coral bleaching** is a process through which coral reefs lose their color during extreme temperature events. The extent of bleaching after an extreme temperature event was measured at 20 sites, giving the following data. Values are dimensionless measures of the extent of bleaching.

0.19	0.06	0.26	0.32	0.35
0.09	0.05	0.21	0.48	0.21
0.23	0.12	0.22	0.16	0.35
0.13	0.34	0.07	0.16	0.45

Assume the data come from a normal distribution. Would we reject the null hypothesis that the mean bleaching index is 0.15 at the $\alpha = 0.01$ significance level?

BB **7. Huntington's disease** is a genetic disorder causing neurodegeneration and eventual death. The age at death of 15 patients with the disease is given in the following data.

71	70	69	68	67
66	64	62	58	57
55	53	52	48	45

Assume the data come from a normal distribution and use them to estimate the P-value for the data under the null hypothesis that age at death has a mean of 15 years. Provide an interpretation of your answer. Where in your calculation have you used the assumption of normality?

Source: Adapted from S. Andrew et al., "The Relationship between Trinucleotide (CAG) Repeat Length and Clinical Features of Huntington's Disease," *Nature Genetics* 4 (1993): 398–403.

BB **8. Breast cancer tumors** were measured for a sample of 15 patients by estimating tumor diameter in centimeters. This gave the following data.

1.5	1.5	1.6	1.6	1.7
2.0	2.1	2.2	2.6	3.2
4.0	4.9	5.5	7.5	7.6

Assume the data come from a normal distribution and use them to estimate the P-value for the data under the null hypothesis that the mean tumor diameter is 4.0 cm. Provide an interpretation of your answer. Where in your calculation have you used the assumption of normality?

BB **9. Monarch butterflies** migrate thousands of miles each year to overwinter in areas of Mexico. The following data give the wingspans (in centimeters) of 28 individuals captured in their overwintering grounds.

8.6	10.9	9.8	12.4	10.6	10.7	10.9
10.1	10.2	10.3	9.4	9.1	11.2	9.7
11.6	12.0	10.4	9.5	10.9	10.5	9.9
11.1	10.5	11.3	11.7	11.3	10.3	11.5

Check the data for normality. Estimate the P-value for the data under the null hypothesis that the population mean wingspan is 10 cm and interpret the result.

10. Antibiotic *MIC* The minimum inhibitory concentration (*MIC*) of a new antibiotic is measured on 10 bacterial colonies, giving the following data (measured in μg/mL).

0.2	0.4	0.3	0.5	1.1
1.0	1.2	1.3	0.9	1.1

Assume the data come from a normal distribution and use them to estimate the P-value for the data under the null hypothesis that the mean *MIC* is 0.25 μg/mL. Provide an interpretation of your answer.

BB **11. HAART** (Highly Active Antiretroviral Therapy) is a treatment for HIV infection. The concentration of HIV in the blood (in units of $\log_{10}$ copies/mL) was measured in 15 patients before and after HAART treatment. The following data give the reduction in concentration.

3.7	2.5	3.5	3.6	0.7
2.2	2.5	1.4	3.6	1.8
2.4	2.3	4.7	2.3	3.0

Assuming the data come from a normal distribution, would you take this as evidence that HAART is effective? Explain your answer.

BB **12. Fish morphology** The width of the mouth (in mm) of fish feeding on benthic prey (that is, prey found on the bottom of lakes) was measured. A sample of benthic feeders gave the following data:

4.6	3.5	4.3	4.5	1.6
3.1	3.5	2.4	4.5	2.6
3.2	3.2	5.5	3.1	3.9

Assume that the data come from a normal distribution and calculate the value of the t-statistic for the null hypothesis that the population mean mouth width is 2.64 mm. Suppose a follow-up study is conducted that measures another 15 fish. What is the probability that it will give a t-statistic that is less extreme than that for the given data, assuming the null hypothesis is true?

BB **13. High-density lipoprotein (HDL) cholesterol** is sometimes referred to as the "good" cholesterol. A sample of 13 female volunteers had their HDL cholesterol levels measured (in mg/dL), giving the following data:

58	60	48	55	56
46	54	50	49	51
53	54	53		

Assume that the data come from a normal distribution and calculate the value of the t-statistic for the null hypothesis that the population mean cholesterol level is 50 mg/dL. Suppose two follow-up studies are conducted: each measures the cholesterol in another 13 females. If the null hypothesis is true, what is the probability that both studies will give t-statistic values that are more extreme than that for the given data?

BB **14. Pesticides and bees** The number of worker bees produced by a colony after exposure to a pesticide was measured for 12 colonies, giving the following data.

12	12	31	14	17
18	20	28	32	32
24	27			

Assume that the data come from a normal distribution. It has been suggested that the mean number of worker bees produced per colony after exposure is 31.9. Are the data consistent with this view? Explain.

Source: Adapted from R. Gill et al., "Combined Pesticide Exposure Severely Affects Individual- and Colony-Level Traits in Bees," *Nature* 491 (2012): 105–8.

13.4 Contingency Table Analysis

We have explored confidence intervals and hypothesis testing for data sets that measure a single continuous variable for each individual. In this final section we consider a statistical technique for analyzing the relationship between two variables. Our focus is on categorical variables.

Hypothesis Testing with Contingency Tables

In Section 11.3 we introduced *contingency tables* as a way of depicting potential relationships between two categorical variables. Let's revisit an example from Section 11.3.

EXAMPLE 1 | Rimantadine is an antiviral drug that can be administered to prevent infection by influenza. A study[1] divided a group of 76 children into two groups and administered rimantadine to one group and a placebo to the other (this is the variable "drug status"). Researchers then determined whether each child was infected with influenza during the flu season (this is the variable "infection status"). In Section 11.3 we summarized these data in the following contingency table.

Table 1 Rimantadine Effectiveness

	Infected	Not infected	Row total
Placebo	20	21	41
Rimantadine	1	34	35
Column total	21	55	76

■

The goal with data like those in Example 1 is to assess whether there is a relationship between the variables *drug status* and *infection status*. We can phrase this question more precisely using the language of conditional probability.

Let R be the event that a child receives rimantadine and let R^c denote the complement (namely, the child receives a placebo). Then let I be the event that a child becomes infected with influenza and I^c denote the event that they do not. Our interest is in determining if the data suggest that drug status is associated with infection status. In terms of our notation, we wish to know if $P(I \mid R)$ differs from $P(I \mid R^c)$. In words, we want to know if the probability of infection given that a child receives rimantadine differs from

1. R. Clover et al. "Effectiveness of Rimantadine Prophylaxis of Children Within Families," *American Journal of Diseases of Children* 140 (1986): 706–9.

the probability of infection given that a child receives a placebo. We answer this question by using hypothesis testing.

Our first step is to specify the null and alternative hypotheses. The null hypothesis is that $P(I \mid R)$ and $P(I \mid R^c)$ are both equal to a common value q:

(1) $$H_0\colon\ P(I \mid R) = P(I \mid R^c) = q$$

The alternative hypothesis is

(2) $$H_A\colon\ P(I \mid R) \neq P(I \mid R^c)$$

As with all hypothesis testing, our next step is to determine what we would expect to observe in the data if the null hypothesis were true.

■ The Chi-squared Test Statistic

If the null hypothesis were true, then the probability of being infected with influenza would be the same (and equal to q), regardless of drug status. In such a case, what would we expect to observe?

Table 1 displays four cells, corresponding to the four drug-status and infection-status combinations. Let's use O_i to denote the observed number in each cell i. (The way in which the cells are numbered does not matter.) We note that the O_i are random variables.

Now let's consider the number of children we would expect to observe in each of the four cells if the null hypothesis were true. From Table 1 we see that a total of 41 children received the placebo. If each of these has probability q of becoming infected, then the number of children infected is a binomial random variable with $n = 41$ and probability of "success" q. The expected number of infections in the placebo group is therefore $nq = 41q$ (from Exercise 12.4.65).

Likewise, a total of 35 children received rimantadine, and the number of these that would become infected is a binomial random variable with $n = 35$ and probability of success q. The expected number of infections in the rimantadine group is therefore $nq = 35q$.

What remains is to obtain an estimate for q, assuming that the null hypothesis is true. The most natural estimate for q is simply the fraction of the total number of children in the study who became infected. In other words, $\hat{q} = 21/76 \approx 0.28$, where the "hat" ($\hat{\ }$) indicates that this is an estimate of q.

With this estimate, the expected number of infected children in the placebo group is $41 \times 0.28 \approx 11.5$ and the expected number of infected children in the rimantadine group is $35 \times 0.28 \approx 9.8$. We can also calculate the expected number of children in the remaining two cells, as shown in Table 2. We will use E_i to denote the expected number in each cell. Note that the E_i are also random variables because the estimate $\hat{q}$ is calculated from the sample.

Table 2 Observed Numbers
(With Expected Numbers Given in Parentheses)

	Infected	Not infected	Row total
Placebo	20 (11.5)	21 (29.5)	41
Rimantadine	1 (9.8)	34 (25.2)	35
Column total	21	55	76

We can combine the observed and the expected numbers into a single test statistic that gives us an overall measure of the departure of the observations from what is expected under the null hypothesis:

(3)
$$X = \sum_{i=1}^{4} \frac{(O_i - E_i)^2}{E_i}$$

The test statistic in (3) plays a role very similar to the t-statistic from Section 13.3 for testing hypotheses about a population mean. It is a measure of the squared difference of all the observations from their expectations, with each squared difference normalized by dividing by E_i. The greater the deviation of the observations from their expectations, the greater the value of X.

The test statistic in (3) is a random variable; for any given contingency table, we can calculate the value that this random variable takes on.

EXAMPLE 2 | **Rimantadine (continued)** Using the information in Table 2, calculate the value of the test statistic (3) for the data from Example 1.

SOLUTION The test statistic (3) for these data is

$$X = \frac{(20 - 11.5)^2}{11.5} + \frac{(21 - 29.5)^2}{29.5} + \frac{(1 - 9.8)^2}{9.8} + \frac{(34 - 25.2)^2}{25.2} \approx 19.7$$ ■

The test statistic (3) is a random variable, but its probability density function is rather complicated. Nevertheless, it has been shown that, provided the entries of the table are not too small, the test statistic (3) is well approximated by what is called a *chi-squared random variable* with one degree of freedom. The PDF of such a random variable is

(4)
$$f(x) = \frac{1}{\sqrt{2\pi}} x^{-1/2} e^{-x/2}$$

where $x > 0$. The graph of this PDF is displayed in Figure 1. For this reason, (3) is sometimes called the **chi-squared test statistic** and it is sometimes denoted by χ^2, which is the square of the Greek letter *chi*.

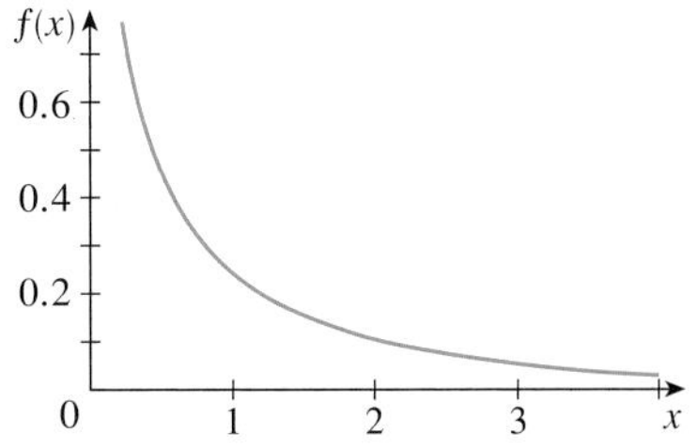

FIGURE 1
Probability density function for a chi-squared variable with 1 degree of freedom

As a rule of thumb, Equation 4 provides a good approximation for the PDF of the test statistic (3) if none of the cells in the table has an expected value less than 1 and no more than one cell has an expected value less than 5.

■ The Hypothesis Test

If the value of the test statistic (3) calculated from a data set is unusually large—that is, the observations deviate substantially from the expectations under the null hypothesis—then the null hypothesis is rejected. The chi-squared PDF gives us an approximation for how we expect X to vary from sample to sample under the assumption that the null hypothesis is true. If we obtain a value of X that is very unusual for this PDF, then presumably the null hypothesis is not true.

We decide what constitutes an unusually large value of X by using P-values, just as we did in Section 13.3.

(5) Definition The ***P*-value** associated with a specific value of the test statistic $X = x^*$ is the area in the tail of the probability density function (4) beyond this value of X. (See Figure 2.)

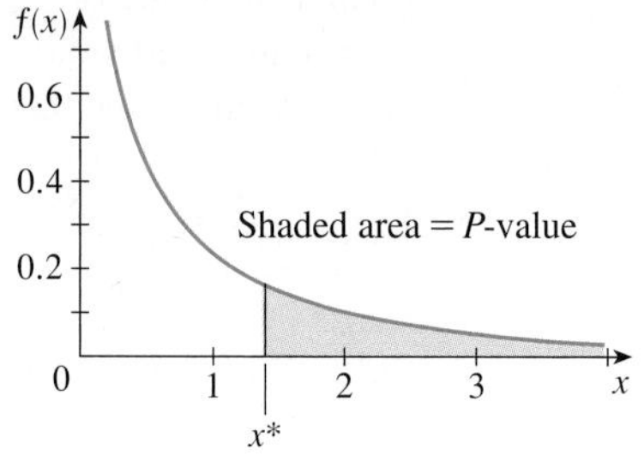

FIGURE 2

The P-value is interpreted as the probability that, if the null hypothesis were true and we took another sample from the population, then we would obtain a test statistic value that is *at least as extreme as that observed.*

EXAMPLE 3 | **Rimantadine (continued)** Calculate the P-value associated with the test statistic from Example 2.

SOLUTION We can see from Table 2 that the expected values are large enough to satisfy the conditions under which (3) is approximately a chi-squared random variable. From the PDF given by Equation 4, the P-value is therefore

$$P = \int_{19.7}^{\infty} \frac{1}{\sqrt{2\pi}} x^{-1/2} e^{-x/2}\, dx \approx 9 \times 10^{-6} \qquad (6)$$

An explicit value for the integral in (6) in terms of elementary functions does not exist, so we have used a numerical integrator to obtain an approximate solution. ■

As in Section 13.3, the P-value is a measure of the compatibility of the data with the null hypothesis. A large P-value (close to 1) indicates that the data are compatible with H_0. A small P-value (close to 0) indicates that the data are incompatible with H_0.

To test the null hypothesis, we again need to specify a significance level, which is the P-value below which the null hypothesis is rejected. For example, using the convention of a significance level of 0.05, the results of Example 3 show that null hypothesis (1) should be rejected. In other words, the data provide evidence that ramantadine does affect infection status.

■ Summary

We now summarize the preceding approach for testing hypotheses about contingency tables in which each of the two variables has two possible values.

(7) Testing For a Relationship Between Two Categorical Variables, Each with Two Possible Values

(a) Specify the null and alternative hypotheses, H_0 and H_A.

(b) Choose a significance level α.

(c) Fill in each cell of the contingency table with the observed and expected numbers, O_i and E_i, respectively. Check that the following conditions hold:

(i) None of the cells has an expected value less than 1.

(ii) No more than one cell has an expected value less than 5.

(d) The test statistic has the form

$$X = \sum_{i=1}^{4} \frac{(O_i - E_i)^2}{E_i}$$

Calculate the value of this statistic from the entries in the table.

(e) Determine the P-value associated with the test statistic calculated in part (d) by evaluating the integral

$$P = \int_X^\infty \frac{1}{\sqrt{2\pi}} x^{-1/2} e^{-x/2}\,dx$$

(i) If $P \leqslant \alpha$, then reject H_0 (the data provide evidence that the variables are associated).

(ii) If $P > \alpha$, then do not reject H_0 (the data do not provide evidence that the variables are associated).

Note that Step 7(c) can be carried out using the following equation (see Exercise 9):

(8) The Expected Number for a Cell

$$E_i = \frac{\text{(Row total)} \times \text{(Column total)}}{\text{(Table total)}}$$

The test in (7) can be extended to larger, more complex contingency tables, but we do not pursue this idea here.

EXAMPLE 4 | **Bat conservation biology** In Example 11.3.1 we explored a study[2] that looked for an association between the endangerment status of bat species and their use of caves for roosting. Researchers classified species as either "fragile" or "vulnerable," according to their endangerment status, as well as whether they used caves as a main roosting site. This gave the following contingency table.

	Fragile	Vulnerable	Row total
Cave use	9	2	11
No cave use	5	2	7
Column total	14	4	18

Do the data provide evidence that endangerment status is associated with cave use?

2. H. Arita, "Conservation Biology of the Cave Bats of Mexico," *Journal of Mammalogy* 74 (1993): 693–702.

SOLUTION For Step 7(a), we use C for the event "cave use" and C^c for the complement (namely, no cave use). And we use F for the event "fragile" and F^c for the event "vulnerable." Our interest is in determining whether the data suggest that $P(F \mid C)$ differs from $P(F \mid C^c)$. The null and alternative hypotheses are therefore

$$H_0\colon\ P(F \mid C) = P(F \mid C^c) = q \qquad H_A\colon\ P(F \mid C) \neq P(F \mid C^c)$$

For Step 7(b), we choose a significance level of $\alpha = 0.05$.

For Step 7(c), we use Equation 8. For example, the expected number of observations in the cell corresponding to C and F is

$$\frac{(\text{Row total}) \times (\text{Column total})}{(\text{Table total})} = \frac{11 \times 14}{18} \approx 8.6$$

Using similar calculations, we obtain the following table:

	Fragile	Vulnerable	Row total
Cave use	9 (8.6)	2 (2.4)	11
No cave use	5 (5.4)	2 (1.6)	7
Column total	14	4	18

We then check the expected values of the cells in the table to determine whether they satisfy the two conditions in Step 7(c). We see that none of the cells has an expected value less than 1, but two cells have an expected value less than 5. This violates the second condition in Step 7(c) and therefore we cannot expect the test statistic (3) to be well approximated by a chi-squared random variable.

In this situation we should use a different approach. For example, in Exercises 12.1.45–47 and 12.2.55–57 we developed a technique called *Fisher's exact test*, which can be used here. Exercises 12.1.47 and 12.2.57 applied this technique to the bat data and found no evidence for an association between cave use and endangerment status.

Because the expected values in the table very nearly satisfy the required conditions in Step 7(c), we might cautiously proceed, using Step 7(d), to see what happens.

For Step 7(d), we calculate the test statistic value as

$$X = \frac{(9 - 8.6)^2}{8.6} + \frac{(2 - 2.4)^2}{2.4} + \frac{(5 - 5.4)^2}{5.4} + \frac{(2 - 1.6)^2}{1.6} \approx 0.21$$

Finally, for Step 7(e), we calculate the P-value as

$$P = \int_{0.21}^{\infty} \frac{1}{\sqrt{2\pi}} x^{-1/2} e^{-x/2}\, dx \approx 0.65$$

Since $P > 0.05$, we do not reject the null hypothesis (see Figure 3). Thus, although the use of the chi-squared approximation is questionable in this example, it provides the same conclusion as Fisher's exact test. The data are consistent with the hypothesis that there is no association between a bat species' cave use and endangerment status. ■

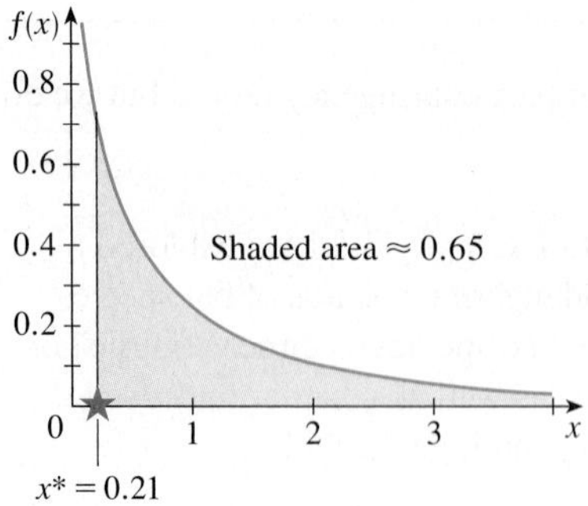

FIGURE 3

EXERCISES 13.4

1. ***Bacillus anthracis*** is the bacterial agent of anthrax. In 1881 Louis Pasteur conducted a famous experiment in which he vaccinated 25 sheep and kept an additional 25 sheep unvaccinated. He then injected live *B. anthracis* into all 50 sheep. He found that all of the unvaccinated sheep died, whereas all of the vaccinated sheep survived. Construct a contingency table for these data and test for an association between vaccination status and survival.

2. **Sentencing and biological mechanisms of psychopathy** A study asked several US judges to pass sentencing on a hypothetical defendant who had been convicted of a crime and was deemed to be a psychopath. Twenty-seven judges were told that psychopathy has an underlying biological basis, and 18 of these judges then cited psychopathy as a mitigating factor in reducing their sentence. Twelve judges were told that psychopathy does not have an underlying biological basis, and 4 of these judges then cited psychopathy as a mitigating factor in reducing their sentence. Construct a contingency table for these data. Test for an association between whether or not a judge is told psychopathy has a biological basis and whether they use this as a mitigating factor to reduce the sentence.

   Source: Adapted from G. Aspinwall et al., "The Double-Edged Sword: Does Biomechanism Increase or Decrease Judges' Sentencing of Psychopathy?" *Science* 337 (2012): 846–49.

3. **Myopia (short-sightedness) and ambient light** A well-known study examined the number of children who sleep with different levels of ambient light in an attempt to explore whether nighttime ambient light affects vision. In a survey, researchers found that 9 of 172 children who slept in darkness developed myopia, and 31 of 232 who slept with a small night-light developed myopia. Construct a contingency table and test for an association between nighttime ambient light and myopia.

   Source: Adapted from G. Quinn et al., "Myopia and Ambient Lighting at Night," *Nature* 399 (1999): 113–14.

4. **Antidepressants** A study gave 28 patients a new anti-depressant medication and 27 patients a placebo for six months. The number of patients having a depressive episode during this period was recorded. Eleven patients who received the medication had at least one depressive episode, while 20 of the patients receiving the placebo had at least one episode. Construct a contingency table for these data and test for an association between treatment type and depression.

5. An **indicator species** is one whose presence is used as an indication of the presence of other groups of species of interest. For example, suppose we wish to know whether a community is dominated by herbivores or carnivores, and we want to use the presence or absence of species X as a simple indicator of this. A study is therefore conducted and it is found that, of 12 communities that are herbivore-dominated, 9 of them have species X present. Of 15 communities that are carnivore-dominated, 4 of them have species X present. Use a contingency table to test whether species X is a good indicator species.

6. **Breast cancer** A study was conducted to determine the relationship between childbearing and breast cancer. Of 4323 respondents who had cancer, 3220 of these had given birth to at least one child. Of 12,699 respondents who did not have cancer, 10,245 of these had given birth to at least one child. Construct a contingency table for these data and test for an association between childbearing and cancer.

   Source: Adapted from B. MacMahon et al., "Age at First Birth and Breast Cancer Risk," *Bulletin of the World Health Organization* 43 (1970): 209–21.

7. **Prostate-specific antigen,** or PSA testing, is a simple but controversial test for identifying individuals who are likely to have (or develop) prostate cancer. In a sample of 490 men with prostate cancer, 75% had a positive PSA test. In a sample of 346 men without prostate cancer, 55% had a positive PSA test. Do these data support the idea that a positive test is indicative of cancer?

8. A new **blood pressure medication** was tested by administering the medication to a random sample of people and giving another random sample a placebo. Of 112 individuals who received the drug, 62 had their blood pressure decline. Of 92 individuals who received the placebo, 48 had their blood pressure decline. Do these data support the idea that the drug affects blood pressure?

9. Provide a derivation of Equation 8.

Chapter 13 REVIEW

CONCEPT CHECK

1. What are i.i.d. random variables?
2. What is the sampling distribution of a statistic?
3. Explain what is meant by the precision of an estimate with reference to its sampling distribution.
4. Explain what is meant by the bias of an estimate with reference to its sampling distribution.
5. What is the difference between the standard deviation of the data and the sample standard deviation?
6. Provide an interpretation of a 95% confidence interval.
7. What is the difference between confidence intervals calculated when the population standard deviation is known versus when it is unknown?
8. Explain what a null hypothesis and an alternative hypothesis are.
9. What is the t-statistic, and what is it used for?
10. What is the chi-squared statistic, and what is it used for?

Answers to the Concept Check can be found on the back endpapers.

TRUE-FALSE QUIZ

Determine whether the statement is true or false. If it is true, explain why. If it is false, explain why or give an example that disproves the statement.

1. If $X_1, \ldots, X_n$ are i.i.d. normal random variables, then the average of these is also a normal random variable.
2. The sampling distribution of the mean of a normally distributed variable becomes wider as the sample size increases.
3. Suppose Y is an estimate of the population mean μ and $E[Y] = \mu$. Then Y is an unbiased estimate of the mean.
4. The standard deviation of a data set is an unbiased estimate for the population standard deviation.
5. The 95% confidence interval calculated from a data set will be wider than the 90% confidence interval calculated from the same data.
6. Student's t-distribution is well approximated by a standard normal distribution when the number of degrees of freedom is large.
7. A P-value associated with a particular value of a statistic is the area under the graph of its sampling distribution that corresponds to more extreme values of the statistic.
8. The significance level for a hypothesis test is the critical P-value above which the null hypothesis is rejected.
9. Contingency table analysis is used to test for an association between two categorical variables.
10. In a hypothesis test, the alternative hypothesis is rejected if the P-value is very large.

EXERCISES

1. **Forest height** The height of 12 pine trees is measured in meters. Suppose that height is well described by a normal random variable with unknown mean and with standard deviation 15 m. Use a numerical integrator to determine the probability that the sample mean deviates by no more than 10 m from the true population mean.

2. The **body mass index (BMI)** of 10 males is measured. Suppose that BMI is well described by a normal random variable with unknown mean and with standard deviation 4. Use a numerical integrator to determine the probability that the sample mean deviates by no more than 2 from the true population mean.

3. **Lifespans** The lifespan of 14 *Drosophila* flies is measured. Suppose that lifetimes are well described by a normal random variable with unknown mean and with standard deviation 10 days. Use a numerical integrator to determine the probability that the sample mean lifespan deviates by at least 5 days from the true population mean.

4. **Damselflies** A study measured the number of external parasites on a sample of 10 damselflies. Although number of parasites is not a continuous variable, suppose that it can be approximated by a normal random variable with unknown mean and with standard deviation 5. Use a numerical integrator to determine the probability that the sample mean deviates by at least 2 parasites from the true population mean.

5. **Pollination** The amount of pollen removed from a flower (in mg) by 10 bees is measured, giving the following data.

8	15	9	22	18
18	16	21	20	18

Calculate the 95% confidence interval for the mean amount of pollen removed. Assume that the amount of pollen removed is a normally distributed random variable.

BB 6. The **red blood cell count** of 15 patients is given in the following table, in (cells $\times 10^6)/\mu$L.

4.2	4.8	5.1	5.1	5.0
5.2	4.7	4.8	4.5	5.5
5.0	5.1	5.3	4.5	4.7

Calculate the 95% confidence interval for the mean count. Assume that red blood cell count is a normally distributed random variable.

BB 7. **Damselflies** The mass (in mg) of 12 damselflies is given in the following table.

677	553	720	520	652
559	597	471	577	807
502	733			

Calculate the 95% confidence interval for the mean mass. Assume that mass is a normally distributed random variable.

BB 8. The **mRNA concentration** for a specific gene was measured in 15 cells, giving the following data (in number of transcripts/cell).

42	21	131	51	190
165	176	98	178	77
150	201	154	173	99

Calculate the 95% confidence interval for the mean number of transcripts per cell. Assume that concentration is a normally distributed random variable.

BB **9. Crop rotation** is the practice of planting dissimilar crops sequentially in one location to avoid the buildup of pathogens and the depletion of nutrients. A researcher conducted an experiment with soybeans, using crop rotation on 12 plots and leaving another 12 with no rotation. Below are the pounds of soybeans harvested from each plot in the final season.

Crop rotation	37	42	44	46	41	39
	29	31	37	41	42	45
No crop rotation	29	35	34	27	31	36
	41	40	37	29	31	36

Calculate the 95% confidence intervals for the pounds of soybeans in each data set and plot them on the same graph. You can assume that pounds of soybeans is a normally distributed random variable.

BB **10. Carbohydrates** The daily carbohydrate intakes (in grams) of 18 Asians and 20 Hispanics who participated in a health survey are listed below.

Asians	344	285	244	382	310	265
	272	310	291	286	288	305
	410	350	230	273	286	291
Hispanics	310	330	289	287	370	305
	314	267	273	325	347	364
	268	272	329	340	289	295
	349	278				

Calculate the 95% confidence intervals for the carbohydrate intake of each group and plot them on the same graph. You can assume that carbohydrate intake is a normally distributed random variable.

11–14 Use the sample to test whether the population mean differs from the given value.

11. Pollination Use the data from Exercise 5 with $\mu_0 = 15$ mg.

BB **12. Red blood cell count** Use the data from Exercise 6 with $\mu_0 = (4.7 \times 10^6)$ cells/μL.

BB **13. Damselflies** Use the data from Exercise 7 with $\mu_0 = 750$ mg.

BB **14. mRNA concentration** Use the data from Exercise 8 with $\mu_0 = 150$ transcripts/cell.

---

15. A **fecal occult blood test** is often used as an indicator of colon cancer. A sample of 13 out of 21 men had positive test results, whereas 10 out of 15 women had positive results. Construct a contingency table for these data and test for an association between test outcome and sex.

16. Genetic crossing Two kinds of crosses were conducted with pea plants and the shape of the offspring seeds were quantified. In cross A, 3170 of the 4010 offspring had round seeds, whereas the remainder of the offspring had angular seeds. In cross B, 2013 of the 4053 offspring had round seeds, whereas the remainder of the offspring had angular seeds. Construct a contingency table for these data and test for an association between seed shape and cross type.

APPENDIXES

A | Intervals, Inequalities, and Absolute Values

Certain sets of real numbers, called **intervals**, occur frequently in calculus and correspond geometrically to line segments. For example, if $a < b$, the **open interval** from a to b consists of all numbers between a and b and is denoted by the symbol (a, b). Using set-builder notation, we can write

$$(a, b) = \{x \mid a < x < b\}$$

FIGURE 1
Open interval (a, b)

Notice that the endpoints of the interval—namely, a and b—are excluded. This is indicated by the round brackets ( ) and by the open dots in Figure 1. The **closed interval** from a to b is the set

$$[a, b] = \{x \mid a \leqslant x \leqslant b\}$$

FIGURE 2
Closed interval $[a, b]$

Here the endpoints of the interval are included. This is indicated by the square brackets [ ] and by the solid dots in Figure 2. It is also possible to include only one endpoint in an interval, as shown in Table 1.

(1) Table of Intervals

Table 1 lists the nine possible types of intervals. When these intervals are discussed, it is always assumed that $a < b$.

Notation	Set description	Picture
(a, b)	$\{x \mid a < x < b\}$	
$[a, b]$	$\{x \mid a \leqslant x \leqslant b\}$	
$[a, b)$	$\{x \mid a \leqslant x < b\}$	
$(a, b]$	$\{x \mid a < x \leqslant b\}$	
(a, ∞)	$\{x \mid x > a\}$	
$[a, \infty)$	$\{x \mid x \geqslant a\}$	
$(-\infty, b)$	$\{x \mid x < b\}$	
$(-\infty, b]$	$\{x \mid x \leqslant b\}$	
$(-\infty, \infty)$	$\mathbb{R}$ (set of all real numbers)	

We also need to consider infinite intervals such as

$$(a, \infty) = \{x \mid x > a\}$$

This does not mean that ∞ ("infinity") is a number. The notation (a, ∞) stands for the set of all numbers that are greater than a, so the symbol ∞ simply indicates that the interval extends indefinitely far in the positive direction.

Inequalities

When working with inequalities, note the following rules.

Rules for Inequalities

1. If $a < b$, then $a + c < b + c$.
2. If $a < b$ and $c < d$, then $a + c < b + d$.
3. If $a < b$ and $c > 0$, then $ac < bc$.
4. If $a < b$ and $c < 0$, then $ac > bc$.
5. If $0 < a < b$, then $1/a > 1/b$.

Rule 1 says that we can add any number to both sides of an inequality, and Rule 2 says that two inequalities can be added. However, we have to be careful with multiplication. Rule 3 says that we can multiply both sides of an inequality by a *positive number,* but Rule 4 says that if we multiply both sides of an inequality by a negative number, then we reverse the direction of the inequality. For example, if we take the inequality $3 < 5$ and multiply by 2, we get $6 < 10$, but if we multiply by -2, we get $-6 > -10$. Finally, Rule 5 says that if we take reciprocals, then we reverse the direction of an inequality (provided the numbers are positive).

EXAMPLE 1 | Solve the inequality $1 + x < 7x + 5$.

SOLUTION The given inequality is satisfied by some values of x but not by others. To *solve* an inequality means to determine the set of numbers x for which the inequality is true. This is called the *solution set.*

First we subtract 1 from each side of the inequality (using Rule 1 with $c = -1$):

$$x < 7x + 4$$

Then we subtract $7x$ from both sides (Rule 1 with $c = -7x$):

$$-6x < 4$$

Now we divide both sides by -6 $\left(\text{Rule 4 with } c = -\frac{1}{6}\right)$:

$$x > -\tfrac{4}{6} = -\tfrac{2}{3}$$

These steps can all be reversed, so the solution set consists of all numbers greater than $-\frac{2}{3}$. In other words, the solution of the inequality is the interval $\left(-\frac{2}{3}, \infty\right)$. ■

EXAMPLE 2 | Solve the inequality $x^2 - 5x + 6 \leqslant 0$.

SOLUTION First we factor the left side:

$$(x - 2)(x - 3) \leqslant 0$$

We know that the corresponding equation $(x - 2)(x - 3) = 0$ has the solutions 2 and 3. The numbers 2 and 3 divide the real line into three intervals:

$$(-\infty, 2) \qquad (2, 3) \qquad (3, \infty)$$

A visual method for solving Example 2 is to use a graphing device to graph the parabola $y = x^2 - 5x + 6$ (as in Figure 3) and observe that the curve lies on or below the x-axis when $2 \leqslant x \leqslant 3$.

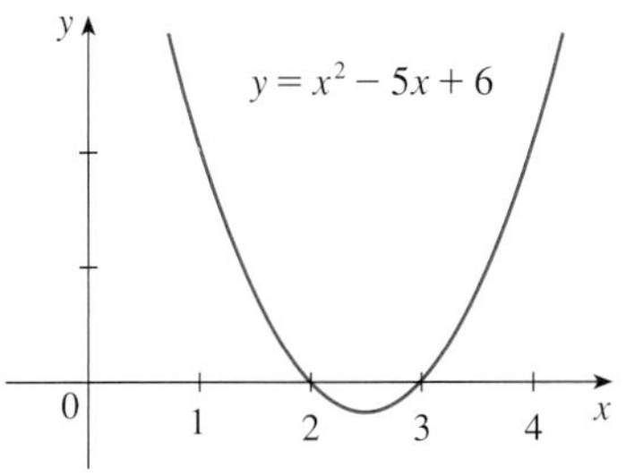

FIGURE 3

On each of these intervals we determine the signs of the factors. For instance,

$$x \in (-\infty, 2) \quad \Rightarrow \quad x < 2 \quad \Rightarrow \quad x - 2 < 0$$

Then we record these signs in the following chart:

Interval	$x - 2$	$x - 3$	$(x - 2)(x - 3)$
$x < 2$	$-$	$-$	$+$
$2 < x < 3$	$+$	$-$	$-$
$x > 3$	$+$	$+$	$+$

Another method for obtaining the information in the chart is to use *test values.* For instance, if we use the test value $x = 1$ for the interval $(-\infty, 2)$, then substitution in $x^2 - 5x + 6$ gives

$$1^2 - 5(1) + 6 = 2$$

The polynomial $x^2 - 5x + 6$ doesn't change sign inside any of the three intervals, so we conclude that it is positive on $(-\infty, 2)$.

Then we read from the chart that $(x - 2)(x - 3)$ is negative when $2 < x < 3$. Thus the solution of the inequality $(x - 2)(x - 3) \leqslant 0$ is

$$\{x \mid 2 \leqslant x \leqslant 3\} = [2, 3]$$

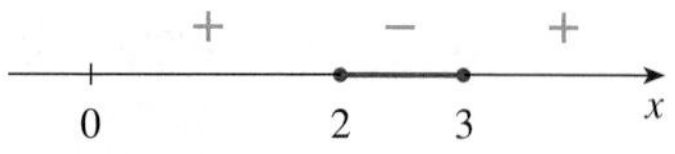

FIGURE 4

Notice that we have included the endpoints 2 and 3 because we are looking for values of x such that the product is either negative or zero. The solution is illustrated in Figure 4. ■

EXAMPLE 3 | Solve $x^3 + 3x^2 > 4x$.

SOLUTION First we take all nonzero terms to one side of the inequality sign and factor the resulting expression:

$$x^3 + 3x^2 - 4x > 0 \qquad \text{or} \qquad x(x - 1)(x + 4) > 0$$

As in Example 2 we solve the corresponding equation $x(x - 1)(x + 4) = 0$ and use the solutions $x = -4$, $x = 0$, and $x = 1$ to divide the real line into four intervals $(-\infty, -4)$, $(-4, 0)$, $(0, 1)$, and $(1, \infty)$. On each interval the product keeps a constant sign as shown in the following chart:

Interval	x	$x - 1$	$x + 4$	$x(x - 1)(x + 4)$
$x < -4$	−	−	−	−
$-4 < x < 0$	−	−	+	+
$0 < x < 1$	+	−	+	−
$x > 1$	+	+	+	+

Then we read from the chart that the solution set is

$$\{x \mid -4 < x < 0 \text{ or } x > 1\} = (-4, 0) \cup (1, \infty)$$

FIGURE 5

The solution is illustrated in Figure 5. ■

■ Absolute Value

The **absolute value** of a number a, denoted by $|a|$, is the distance from a to 0 on the real number line. Distances are always positive or 0, so we have

$$|a| \geqslant 0 \qquad \text{for every number } a$$

For example,

$$|3| = 3 \qquad |-3| = 3 \qquad |0| = 0 \qquad |\sqrt{2} - 1| = \sqrt{2} - 1 \qquad |3 - \pi| = \pi - 3$$

In general, we have

Remember that if a is negative, then $-a$ is positive.

(2)
$$|a| = a \quad \text{if } a \geqslant 0$$
$$|a| = -a \quad \text{if } a < 0$$

EXAMPLE 4 | Express $|3x - 2|$ without using the absolute-value symbol.

SOLUTION

$$|3x - 2| = \begin{cases} 3x - 2 & \text{if } 3x - 2 \geqslant 0 \\ -(3x - 2) & \text{if } 3x - 2 < 0 \end{cases}$$

$$= \begin{cases} 3x - 2 & \text{if } x \geqslant \frac{2}{3} \\ 2 - 3x & \text{if } x < \frac{2}{3} \end{cases}$$

■

Recall that the symbol $\sqrt{}$ means "the positive square root of." Thus $\sqrt{r} = s$ means $s^2 = r$ and $s \geqslant 0$. Therefore the equation $\sqrt{a^2} = a$ is not always true. It is true only when $a \geqslant 0$. If $a < 0$, then $-a > 0$, so we have $\sqrt{a^2} = -a$. In view of (2), we then have the equation

(3)
$$\boxed{\sqrt{a^2} = |a|}$$

which is true for all values of a.

Hints for the proofs of the following properties are given in the exercises.

Properties of Absolute Values Suppose a and b are any real numbers and n is an integer. Then

1. $|ab| = |a||b|$ **2.** $\left|\dfrac{a}{b}\right| = \dfrac{|a|}{|b|} \quad (b \neq 0)$ **3.** $|a^n| = |a|^n$

For solving equations or inequalities involving absolute values, it's often very helpful to use the following statements.

Suppose $a > 0$. Then

4. $|x| = a$ if and only if $x = \pm a$

5. $|x| < a$ if and only if $-a < x < a$

6. $|x| > a$ if and only if $x > a$ or $x < -a$

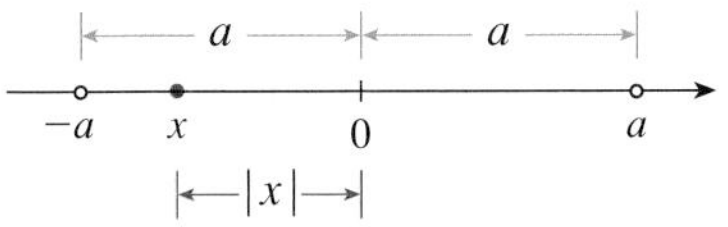

FIGURE 6

For instance, the inequality $|x| < a$ says that the distance from x to the origin is less than a, and you can see from Figure 6 that this is true if and only if x lies between $-a$ and a.

If a and b are any real numbers, then the distance between a and b is the absolute value of the difference, namely, $|a - b|$, which is also equal to $|b - a|$. (See Figure 7.)

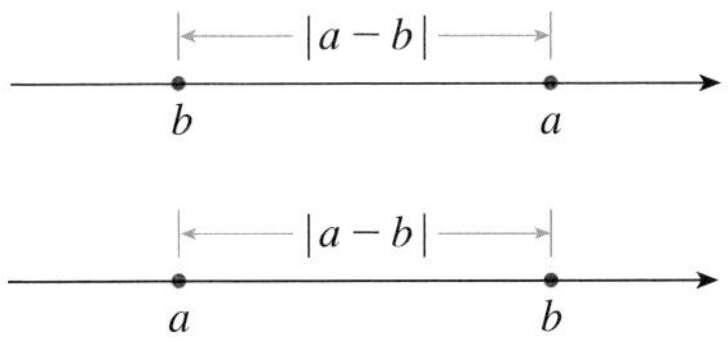

FIGURE 7
Length of a line segment $= |a - b|$

EXAMPLE 5 | Solve $|2x - 5| = 3$.

SOLUTION By Property 4 of absolute values, $|2x - 5| = 3$ is equivalent to

$$2x - 5 = 3 \quad \text{or} \quad 2x - 5 = -3$$

So $2x = 8$ or $2x = 2$. Thus $x = 4$ or $x = 1$. ■

EXAMPLE 6 | Solve $|x - 5| < 2$.

SOLUTION 1 By Property 5 of absolute values, $|x - 5| < 2$ is equivalent to

$$-2 < x - 5 < 2$$

Therefore, adding 5 to each side, we have

$$3 < x < 7$$

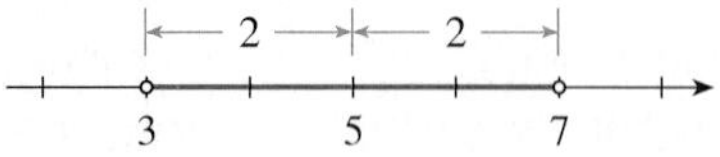

FIGURE 8

and the solution set is the open interval (3, 7).

SOLUTION 2 Geometrically, the solution set consists of all numbers x whose distance from 5 is less than 2. From Figure 8 we see that this is the interval (3, 7). ■

EXAMPLE 7 | Solve $|3x + 2| \geqslant 4$.

SOLUTION By Properties 4 and 6 of absolute values, $|3x + 2| \geqslant 4$ is equivalent to

$$3x + 2 \geqslant 4 \qquad \text{or} \qquad 3x + 2 \leqslant -4$$

In the first case $3x \geqslant 2$, which gives $x \geqslant \frac{2}{3}$. In the second case $3x \leqslant -6$, which gives $x \leqslant -2$. So the solution set is

$$\left\{x \mid x \leqslant -2 \text{ or } x \geqslant \tfrac{2}{3}\right\} = (-\infty, -2] \cup \left[\tfrac{2}{3}, \infty\right)$$ ■

EXERCISES A

1–10 Rewrite the expression without using the absolute-value symbol.

1. $|5 - 23|$

2. $|\pi - 2|$

3. $|\sqrt{5} - 5|$

4. $||-2| - |-3||$

5. $|x - 2|$ if $x < 2$

6. $|x - 2|$ if $x > 2$

7. $|x + 1|$

8. $|2x - 1|$

9. $|x^2 + 1|$

10. $|1 - 2x^2|$

11–26 Solve the inequality in terms of intervals and illustrate the solution set on the real number line.

11. $2x + 7 > 3$

12. $4 - 3x \geqslant 6$

13. $1 - x \leqslant 2$

14. $1 + 5x > 5 - 3x$

15. $0 \leqslant 1 - x < 1$

16. $1 < 3x + 4 \leqslant 16$

17. $(x - 1)(x - 2) > 0$

18. $x^2 < 2x + 8$

19. $x^2 < 3$

20. $x^2 \geqslant 5$

21. $x^3 - x^2 \leqslant 0$

22. $(x + 1)(x - 2)(x + 3) \geqslant 0$

23. $x^3 > x$

24. $x^3 + 3x < 4x^2$

25. $\dfrac{1}{x} < 4$

26. $-3 < \dfrac{1}{x} \leqslant 1$

27. The relationship between the Celsius and Fahrenheit temperature scales is given by $C = \frac{5}{9}(F - 32)$, where C is the temperature in degrees Celsius and F is the temperature in degrees Fahrenheit. What interval on the Celsius scale corresponds to the temperature range $50 \leqslant F \leqslant 95$?

28. Use the relationship between C and F given in Exercise 27 to find the interval on the Fahrenheit scale corresponding to the temperature range $20 \leqslant C \leqslant 30$.

29. As dry air moves upward, it expands and in so doing cools at a rate of about 1°C for each 100-m rise, up to about 12 km.
(a) If the ground temperature is 20°C, write a formula for the temperature at height h.
(b) What range of temperature can be expected if a plane takes off and reaches a maximum height of 5 km?

30. If a ball is thrown upward from the top of a building 128 ft high with an initial velocity of 16 ft/s, then the height h above the ground t seconds later will be

$$h = 128 + 16t - 16t^2$$

During what time interval will the ball be at least 32 ft above the ground?

31–32 Solve the equation for x.

31. $|x + 3| = |2x + 1|$

32. $|3x + 5| = 1$

33–40 Solve the inequality.

33. $|x| < 3$

34. $|x| \geqslant 3$

35. $|x - 4| < 1$

36. $|x - 6| < 0.1$

37. $|x + 5| \geqslant 2$

38. $|x + 1| \geqslant 3$

39. $|2x - 3| \leqslant 0.4$

40. $|5x - 2| < 6$

41. Solve the inequality $a(bx - c) \geqslant bc$ for x, assuming that a, b, and c are positive constants.

42. Solve the inequality $ax + b < c$ for x, assuming that a, b, and c are negative constants.

43. Prove that $|ab| = |a||b|$. [*Hint:* Use Equation 3.]

44. Show that if $0 < a < b$, then $a^2 < b^2$.

B Coordinate Geometry

The points in a plane can be identified with ordered pairs of real numbers. We start by drawing two perpendicular coordinate lines that intersect at the origin O on each line. Usually one line is horizontal with positive direction to the right and is called the x-axis; the other line is vertical with positive direction upward and is called the y-axis.

Any point P in the plane can be located by a unique ordered pair of numbers as follows: Draw lines through P perpendicular to the x- and y-axes. These lines intersect the axes in points with coordinates a and b as shown in Figure 1. Then the point P is assigned the ordered pair (a, b). The first number a is called the ***x*-coordinate** of P; the second number b is called the ***y*-coordinate** of P. We say that P is the point with coordinates (a, b), and we denote the point by the symbol $P(a, b)$. Several points are labeled with their coordinates in Figure 2.

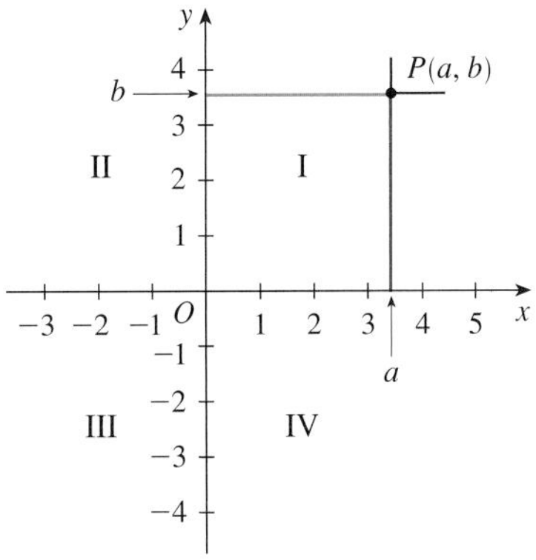

FIGURE 1

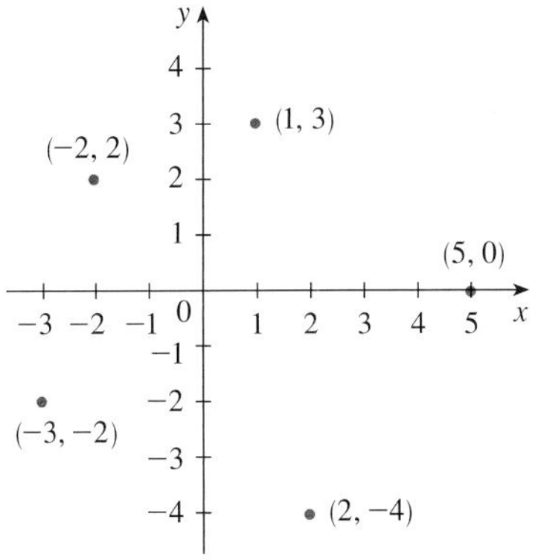

FIGURE 2

By reversing the preceding process we can start with an ordered pair (a, b) and arrive at the corresponding point P. Often we identify the point P with the ordered pair (a, b) and refer to "the point (a, b)." [Although the notation used for an open interval (a, b) is the same as the notation used for a point (a, b), you will be able to tell from the context which meaning is intended.]

This coordinate system is called the **rectangular coordinate system** or the **Cartesian coordinate system** in honor of the French mathematician René Descartes (1596–1650), even though another Frenchman, Pierre Fermat (1601–1665), invented the principles of analytic geometry at about the same time as Descartes. The plane supplied with this coordinate system is called the **coordinate plane** or the **Cartesian plane** and is denoted by $\mathbb{R}^2$.

The x- and y-axes are called the **coordinate axes** and divide the Cartesian plane into four quadrants, which are labeled I, II, III, and IV in Figure 1. Notice that the first quadrant consists of those points whose x- and y-coordinates are both positive.

EXAMPLE 1 | Describe and sketch the regions given by the following sets.

(a) $\{(x, y) \mid x \geqslant 0\}$ (b) $\{(x, y) \mid y = 1\}$ (c) $\{(x, y) \mid |y| < 1\}$

SOLUTION

(a) The points whose x-coordinates are 0 or positive lie on the y-axis or to the right of it as indicated by the shaded region in Figure 3(a).

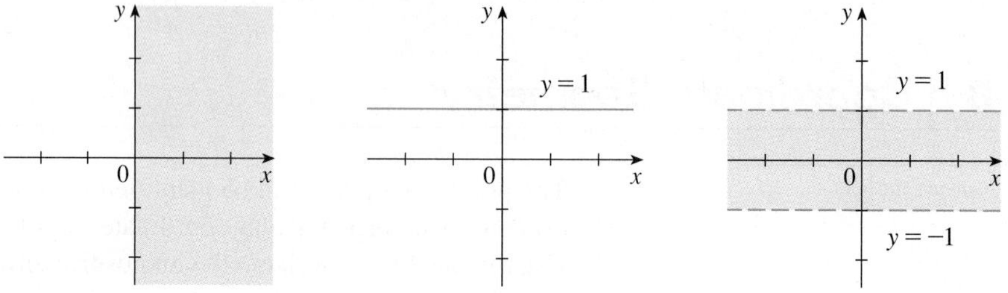

FIGURE 3 (a) $x \geqslant 0$ (b) $y = 1$ (c) $|y| < 1$

(b) The set of all points with y-coordinate 1 is a horizontal line one unit above the x-axis [see Figure 3(b)].

(c) Recall from Appendix A that

$$|y| < 1 \quad \text{if and only if} \quad -1 < y < 1$$

The given region consists of those points in the plane whose y-coordinates lie between -1 and 1. Thus the region consists of all points that lie between (but not on) the horizontal lines $y = 1$ and $y = -1$. [These lines are shown as dashed lines in Figure 3(c) to indicate that the points on these lines don't lie in the set.] ■

Recall from Appendix A that the distance between points a and b on a number line is $|a - b| = |b - a|$. Thus the distance between points $P_1(x_1, y_1)$ and $P_3(x_2, y_1)$ on a horizontal line must be $|x_2 - x_1|$ and the distance between $P_2(x_2, y_2)$ and $P_3(x_2, y_1)$ on a vertical line must be $|y_2 - y_1|$. (See Figure 4.)

To find the distance $|P_1P_2|$ between any two points $P_1(x_1, y_1)$ and $P_2(x_2, y_2)$, we note that triangle $P_1P_2P_3$ in Figure 4 is a right triangle, and so by the Pythagorean Theorem we have

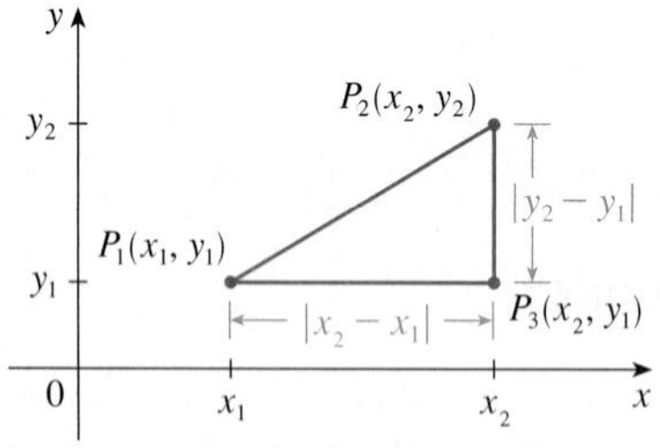

FIGURE 4

$$|P_1P_2| = \sqrt{|P_1P_3|^2 + |P_2P_3|^2} = \sqrt{|x_2 - x_1|^2 + |y_2 - y_1|^2}$$
$$= \sqrt{(x_2 - x_1)^2 + (y_2 - y_1)^2}$$

Distance Formula The distance between the points $P_1(x_1, y_1)$ and $P_2(x_2, y_2)$ is

$$|P_1P_2| = \sqrt{(x_2 - x_1)^2 + (y_2 - y_1)^2}$$

For instance, the distance between $(1, -2)$ and $(5, 3)$

$$\sqrt{(5 - 1)^2 + [3 - (-2)]^2} = \sqrt{4^2 + 5^2} = \sqrt{41}$$

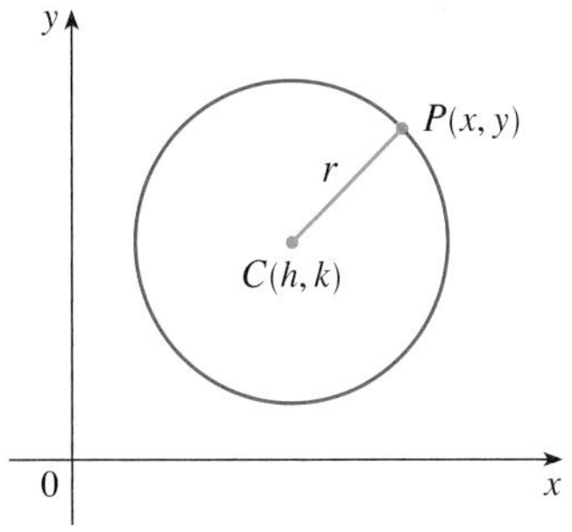

FIGURE 5

Circles

An **equation of a curve** is an equation satisfied by the coordinates of the points on the curve and by no other points. Let's use the distance formula to find the equation of a circle with radius r and center (h, k). By definition, the circle is the set of all points $P(x, y)$ whose distance from the center $C(h, k)$ is r. (See Figure 5.) Thus P is on the circle if and only if $|PC| = r$. From the distance formula, we have

$$\sqrt{(x - h)^2 + (y - k)^2} = r$$

or equivalently, squaring both sides, we get

$$(x - h)^2 + (y - k)^2 = r^2$$

This is the desired equation.

> **Equation of a Circle** An equation of the circle with center (h, k) and radius r is
>
> $$(x - h)^2 + (y - k)^2 = r^2$$
>
> In particular, if the center is the origin $(0, 0)$, the equation is
>
> $$x^2 + y^2 = r^2$$

For instance an equation of the circle with radius 3 and center $(2, -5)$ is

$$(x - 2)^2 + (y + 5)^2 = 9$$

EXAMPLE 2 | Sketch the graph of the equation $x^2 + y^2 + 2x - 6y + 7 = 0$ by first showing that it represents a circle and then finding its center and radius.

SOLUTION We first group the x-terms and y-terms as follows:

$$(x^2 + 2x) + (y^2 - 6y) = -7$$

Then we complete the square within each grouping, adding the appropriate constants (the squares of half the coefficients of x and y) to both sides of the equation:

$$(x^2 + 2x + 1) + (y^2 - 6y + 9) = -7 + 1 + 9$$

or

$$(x + 1)^2 + (y - 3)^2 = 3$$

Comparing this equation with the standard equation of a circle, we see that $h = -1$, $k = 3$, and $r = \sqrt{3}$, so the given equation represents a circle with center $(-1, 3)$ and radius $\sqrt{3}$. It is sketched in Figure 6.

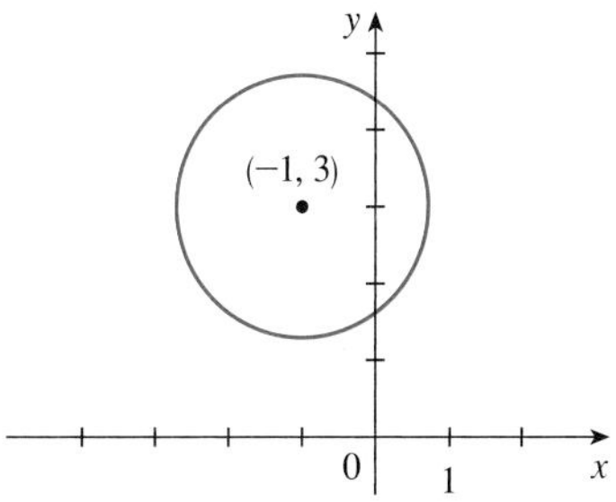

FIGURE 6
$x^2 + y^2 + 2x - 6y + 7 = 0$

■

Lines

To find the equation of a line L we use its *slope,* which is a measure of the steepness of the line.

> **Definition** The **slope** of a nonvertical line that passes through the points $P_1(x_1, y_1)$ and $P_2(x_2, y_2)$ is
>
> $$m = \frac{\Delta y}{\Delta x} = \frac{y_2 - y_1}{x_2 - x_1}$$
>
> The slope of a vertical line is not defined.

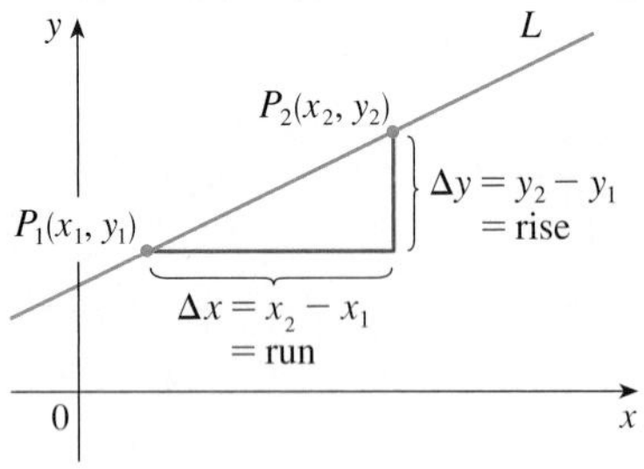

FIGURE 7

Thus the slope of a line is the ratio of the change in y, Δy, to the change in x, Δx. (See Figure 7.) The slope is therefore the rate of change of y with respect to x. The fact that the line is straight means that the rate of change is constant.

Figure 8 shows several lines labeled with their slopes. Notice that lines with positive slope slant upward to the right, whereas lines with negative slope slant downward to the right. Notice also that the steepest lines are the ones for which the absolute value of the slope is largest, and a horizontal line has slope 0.

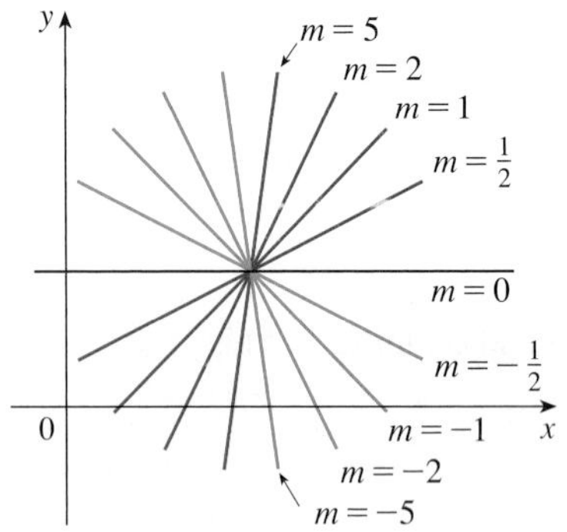

FIGURE 8

Now let's find an equation of the line that passes through a given point $P_1(x_1, y_1)$ and has slope m. A point $P(x, y)$ with $x \neq x_1$ lies on this line if and only if the slope of the line through P_1 and P is equal to m; that is,

$$\frac{y - y_1}{x - x_1} = m$$

This equation can be rewritten in the form

$$y - y_1 = m(x - x_1)$$

and we observe that this equation is also satisfied when $x = x_1$ and $y = y_1$. Therefore it is an equation of the given line.

> **Point-Slope Form of the Equation of a Line** An equation of the line passing through the point $P_1(x_1, y_1)$ and having slope m is
>
> $$y - y_1 = m(x - x_1)$$

EXAMPLE 3 | Find an equation of the line through the points $(-1, 2)$ and $(3, -4)$.

SOLUTION The slope of the line is

$$m = \frac{-4 - 2}{3 - (-1)} = -\frac{3}{2}$$

Using the point-slope form with $x_1 = -1$ and $y_1 = 2$, we obtain

$$y - 2 = -\tfrac{3}{2}(x + 1)$$

which simplifies to

$$3x + 2y = 1$$

■

Suppose a nonvertical line has slope m and y-intercept b. (See Figure 9.) This means it intersects the y-axis at the point $(0, b)$, so the point-slope form of the equation of the line, with $x_1 = 0$ and $y_1 = b$, becomes

$$y - b = m(x - 0)$$

This simplifies as follows.

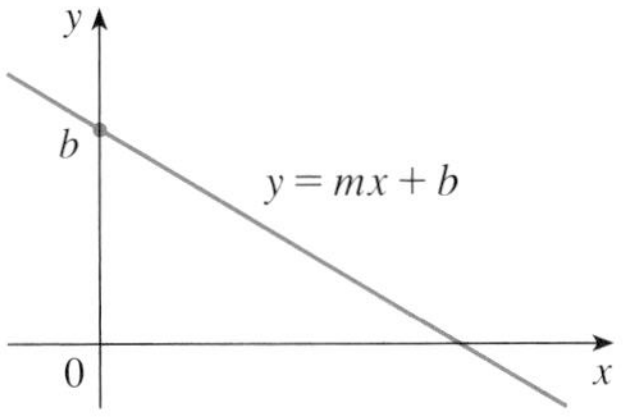

FIGURE 9

> **Slope-Intercept Form of the Equation of a Line** An equation of the line with slope m and y-intercept b is
>
> $$y = mx + b$$

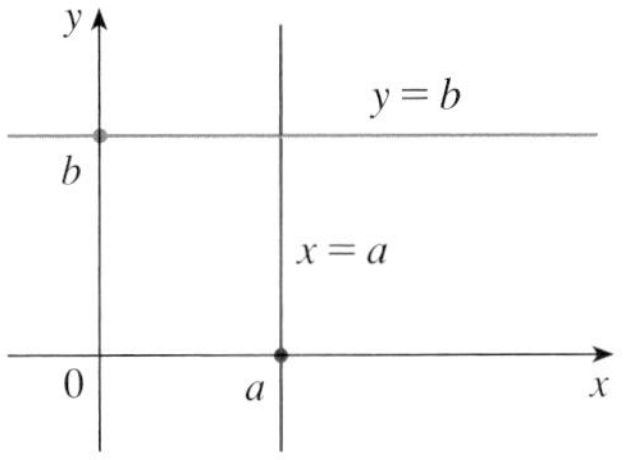

FIGURE 10

In particular, if a line is horizontal, its slope is $m = 0$, so its equation is $y = b$, where b is the y-intercept (see Figure 10). A vertical line does not have a slope, but we can write its equation as $x = a$, where a is the x-intercept, because the x-coordinate of every point on the line is a.

EXAMPLE 4 | Graph the inequality $x + 2y > 5$.

SOLUTION We are asked to sketch the graph of the set $\{(x, y) \mid x + 2y > 5\}$ and we begin by solving the inequality for y:

$$x + 2y > 5$$
$$2y > -x + 5$$
$$y > -\tfrac{1}{2}x + \tfrac{5}{2}$$

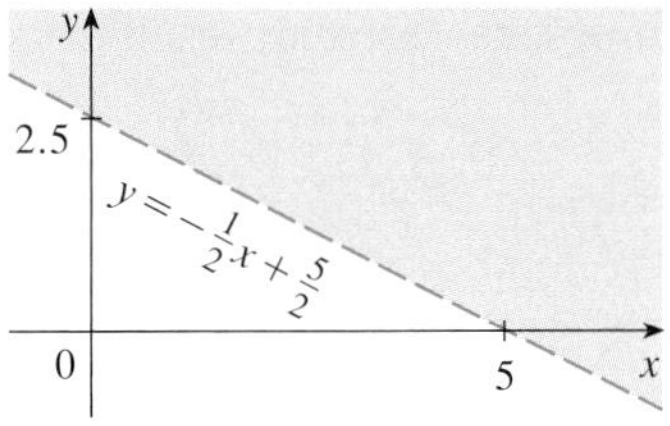

FIGURE 11

Compare this inequality with the equation $y = -\frac{1}{2}x + \frac{5}{2}$, which represents a line with slope $-\frac{1}{2}$ and y-intercept $\frac{5}{2}$. We see that the given graph consists of points whose y-coordinates are *larger* than those on the line $y = -\frac{1}{2}x + \frac{5}{2}$. Thus the graph is the region that lies *above* the line, as illustrated in Figure 11. ■

■ Parallel and Perpendicular Lines

Slopes can be used to show that lines are parallel or perpendicular. The following facts are proved, for instance, in *Precalculus: Mathematics for Calculus,* Seventh Edition, by Stewart, Redlin, and Watson (Boston: Cengage Learning, 2016).

> **Parallel and Perpendicular Lines**
>
> **1.** Two nonvertical lines are parallel if and only if they have the same slope.
>
> **2.** Two lines with slopes m_1 and m_2 are perpendicular if and only if $m_1 m_2 = -1$; that is, their slopes are negative reciprocals:
>
> $$m_2 = -\frac{1}{m_1}$$

EXAMPLE 5 | Find an equation of the line through the point $(5, 2)$ that is parallel to the line $4x + 6y + 5 = 0$.

SOLUTION The given line can be written in the form

$$y = -\tfrac{2}{3}x - \tfrac{5}{6}$$

which is in slope-intercept form with $m = -\frac{2}{3}$. Parallel lines have the same slope, so the required line has slope $-\frac{2}{3}$ and its equation in point-slope form is

$$y - 2 = -\tfrac{2}{3}(x - 5)$$

We can write this equation as $2x + 3y = 16$. ■

EXAMPLE 6 | Show that the lines $2x + 3y = 1$ and $6x - 4y - 1 = 0$ are perpendicular.

SOLUTION The equations can be written as

$$y = -\tfrac{2}{3}x + \tfrac{1}{3} \qquad \text{and} \qquad y = \tfrac{3}{2}x - \tfrac{1}{4}$$

from which we see that the slopes are

$$m_1 = -\tfrac{2}{3} \qquad \text{and} \qquad m_2 = \tfrac{3}{2}$$

Since $m_1 m_2 = -1$, the lines are perpendicular. ■

■ Conic Sections

Here we review the geometric definitions of parabolas, ellipses, and hyperbolas and their standard equations. They are called **conic sections**, or **conics**, because they result from intersecting a cone with a plane as shown in Figure 12.

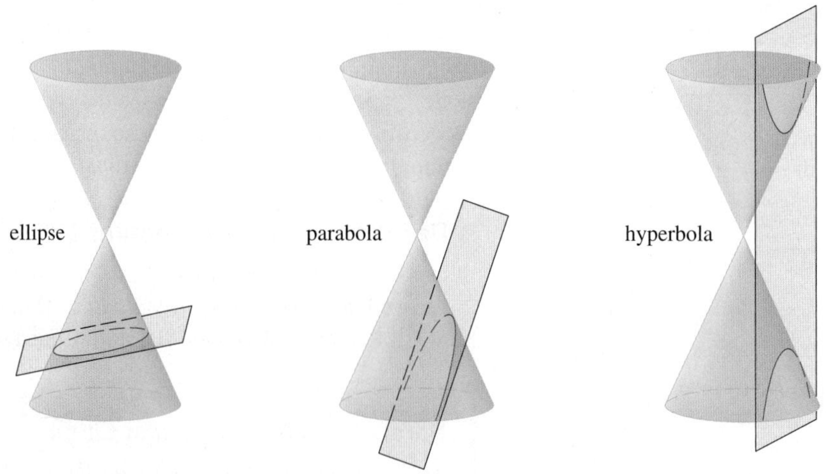

FIGURE 12
Conics

■ Parabolas

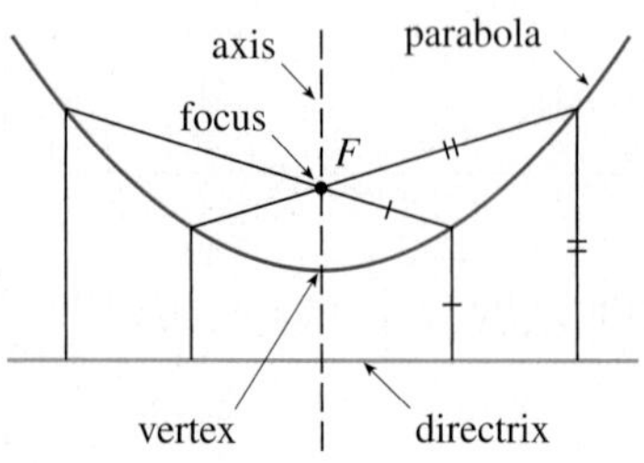

FIGURE 13

A **parabola** is the set of points in a plane that are equidistant from a fixed point F (called the **focus**) and a fixed line (called the **directrix**). This definition is illustrated by Figure 13. Notice that the point halfway between the focus and the directrix lies on the parabola; it is called the **vertex**. The line through the focus perpendicular to the directrix is called the **axis** of the parabola.

In the 16th century Galileo showed that the path of a projectile that is shot into the air at an angle to the ground is a parabola. Since then, parabolic shapes have been used in designing automobile headlights, reflecting telescopes, and suspension bridges.

We obtain a particularly simple equation for a parabola if we place its vertex at the origin O and its directrix parallel to the x-axis as in Figure 14. If the focus is the point

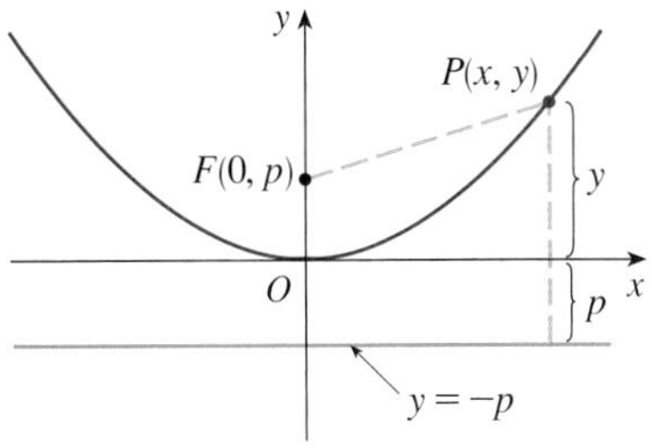

FIGURE 14

(0, p), then the directrix has the equation $y = -p$ and the parabola has the equation

$$x^2 = 4py$$

(See Exercise 47.)

If we write $a = 1/(4p)$, then the equation of a parabola becomes

$$\boxed{y = ax^2}$$

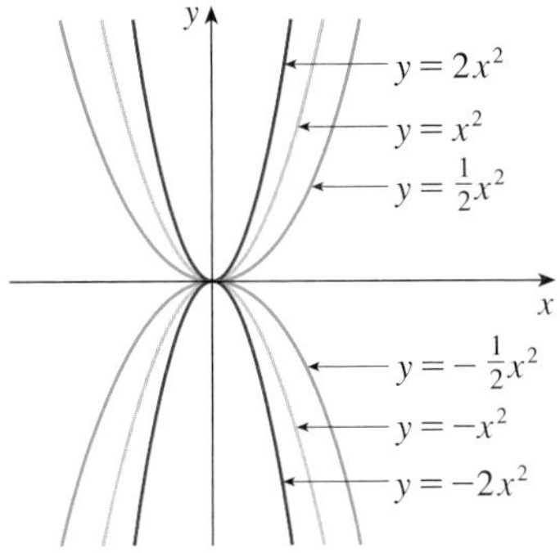

FIGURE 15

Figure 15 shows the graphs of several parabolas with equations of the form $y = ax^2$ for various values of the number a. We see that the parabola $y = ax^2$ opens upward if $a > 0$ and downward if $a < 0$ (as in Figure 16). The graph is symmetric with respect to the y-axis because its equation is unchanged when x is replaced by $-x$. This corresponds to the fact that the function $f(x) = ax^2$ is an even function.

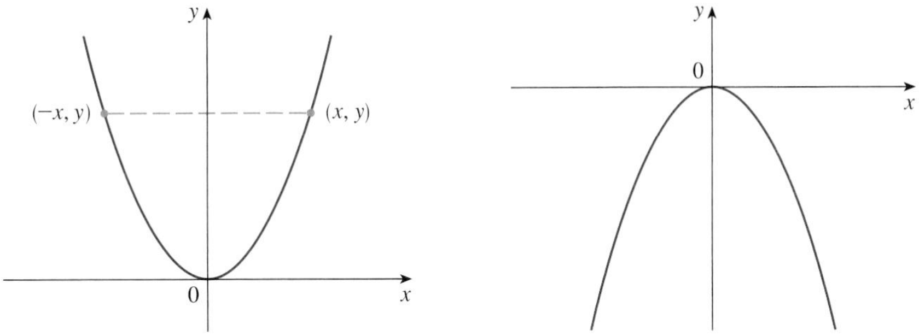

FIGURE 16 (a) $y = ax^2$, $a > 0$ (b) $y = ax^2$, $a < 0$

If we interchange x and y in the equation $y = ax^2$, the result is $x = ay^2$, which also represents a parabola. (Interchanging x and y amounts to reflecting about the diagonal line $y = x$.) The parabola $x = ay^2$ opens to the right if $a > 0$ and to the left if $a < 0$. (See Figure 17.) This time the parabola is symmetric with respect to the x-axis because the equation is unchanged when y is replaced by $-y$.

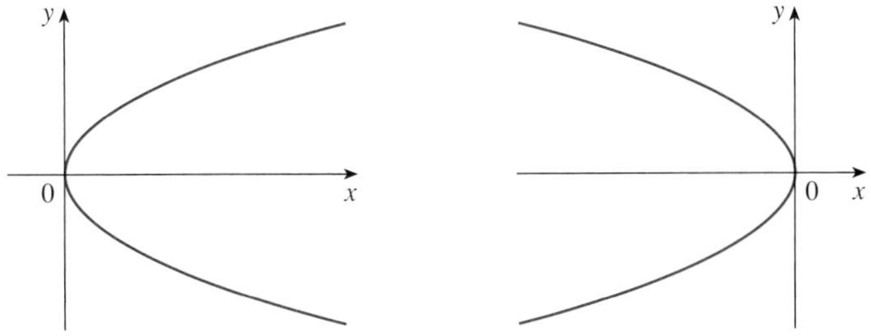

FIGURE 17 (a) $x = ay^2$, $a > 0$ (b) $x = ay^2$, $a < 0$

EXAMPLE 7 | Sketch the region bounded by the parabola $x = 1 - y^2$ and the line $x + y + 1 = 0$.

SOLUTION First we find the points of intersection by solving the two equations. Substituting $x = -y - 1$ into the equation $x = 1 - y^2$, we get $-y - 1 = 1 - y^2$, which gives

$$0 = y^2 - y - 2 = (y - 2)(y + 1)$$

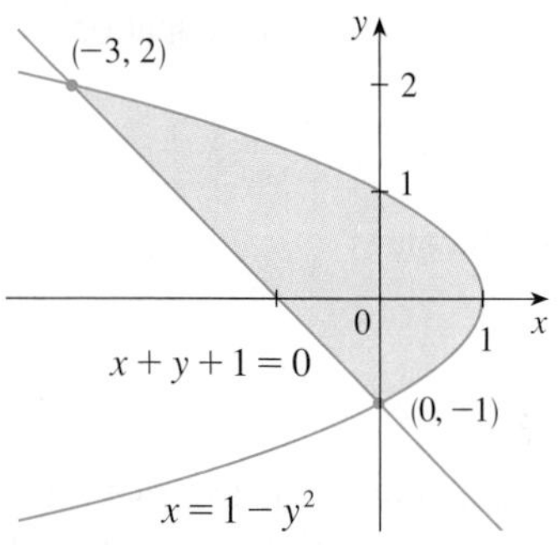

FIGURE 18

so $y = 2$ or -1. Thus the points of intersection are $(-3, 2)$ and $(0, -1)$ and we draw the line $x + y + 1 = 0$ passing through these points.

To sketch the parabola $x = 1 - y^2$ we start with the parabola $x = -y^2$ in Figure 17(b) and shift one unit to the right. We also make sure it passes through the points $(-3, 2)$ and $(0, -1)$. The region bounded by $x = 1 - y^2$ and $x + y + 1 = 0$ means the finite region whose boundaries are these curves. It is sketched in Figure 18. ■

Ellipses

An **ellipse** is the set of points in a plane the sum of whose distances from two fixed points F_1 and F_2 is a constant (see Figure 19). These two fixed points are called the **foci** (plural of **focus**). One of Kepler's laws is that the orbits of the planets in the solar system are ellipses with the sun at one focus.

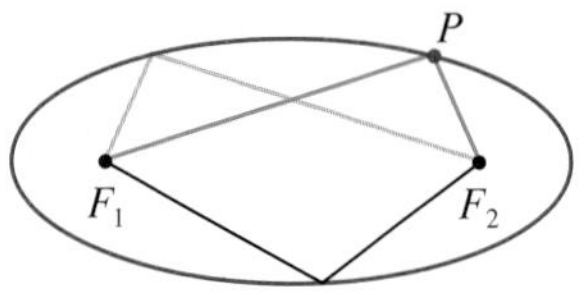

FIGURE 19

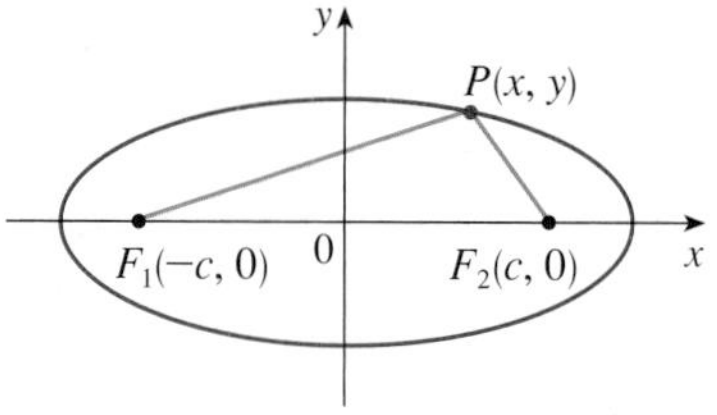

FIGURE 20

In order to obtain the simplest equation for an ellipse, we place the foci on the x-axis at the points $(-c, 0)$ and $(c, 0)$ as in Figure 20 so that the origin is halfway between the foci. If we let the sum of the distances from a point on the ellipse to the foci be $2a$, then we can write an equation of the ellipse as

(1) $$\frac{x^2}{a^2} + \frac{y^2}{b^2} = 1$$

where $c^2 = a^2 - b^2$. (See Exercise 49 and Figure 21.) Notice that the x-intercepts are $\pm a$, the y-intercepts are $\pm b$, the foci are $(\pm c, 0)$, and the ellipse is symmetric with respect to both axes. If the foci of an ellipse are located on the y-axis at $(0, \pm c)$, then we can find its equation by interchanging x and y in (1).

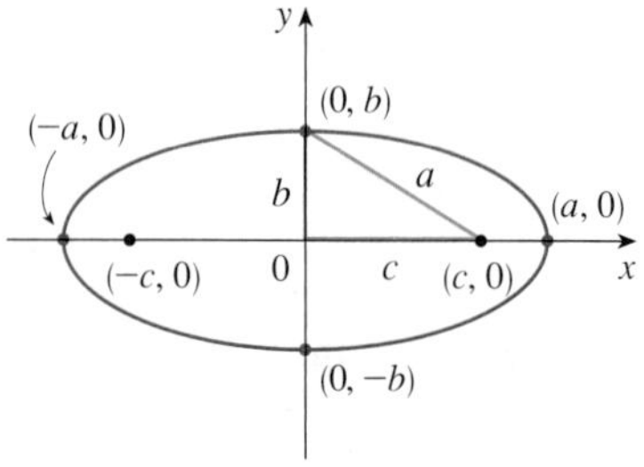

FIGURE 21

$\frac{x^2}{a^2} + \frac{y^2}{b^2} = 1, a \geqslant b$

EXAMPLE 8 | Sketch the graph of $9x^2 + 16y^2 = 144$ and locate the foci.

SOLUTION Divide both sides of the equation by 144:

$$\frac{x^2}{16} + \frac{y^2}{9} = 1$$

The equation is now in the standard form for an ellipse, so we have $a^2 = 16$, $b^2 = 9$, $a = 4$, and $b = 3$. The x-intercepts are ± 4 and the y-intercepts are ± 3. Also, $c^2 = a^2 - b^2 = 7$, so $c = \sqrt{7}$ and the foci are $(\pm\sqrt{7}, 0)$. The graph is sketched in Figure 22. ■

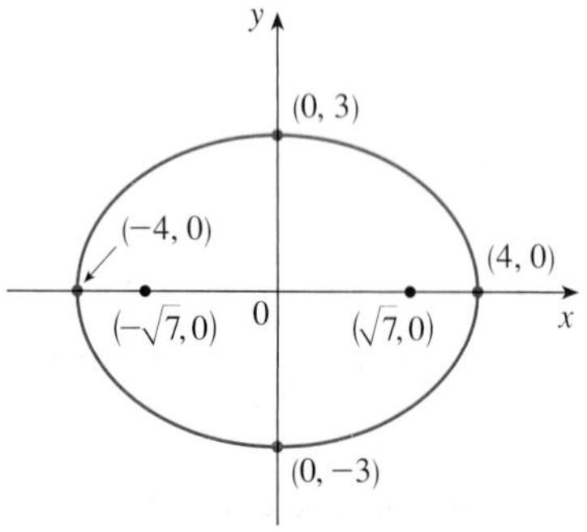

FIGURE 22

$9x^2 + 16y^2 = 144$

Like parabolas, ellipses have an interesting reflection property that has practical consequences. If a source of light or sound is placed at one focus of a surface with elliptical cross-sections, then all the light or sound is reflected off the surface to the other focus. This principle is used in *lithotripsy*, a treatment for kidney stones. A reflector with elliptical cross-section is placed in such a way that the kidney stone is at one focus. High-intensity sound waves generated at the other focus are reflected to the stone and destroy it without damaging surrounding tissue. The patient is spared the trauma of surgery and recovers within a few days.

Hyperbolas

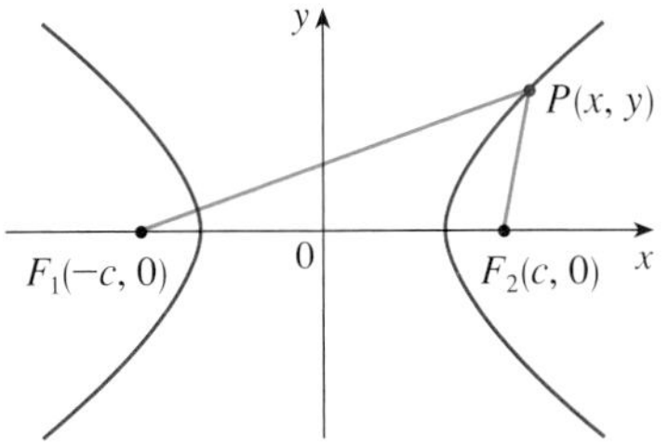

FIGURE 23
P is on the hyperbola when $|PF_1| - |PF_2| = \pm 2a$.

A **hyperbola** is the set of all points in a plane the difference of whose distances from two fixed points F_1 and F_2 (the foci) is a constant. This definition is illustrated in Figure 23.

Notice that the definition of a hyperbola is similar to that of an ellipse; the only change is that the sum of distances has become a difference of distances. It is left as Exercise 51 to show that when the foci are on the x-axis at $(\pm c, 0)$ and the difference of distances is $|PF_1| - |PF_2| = \pm 2a$, then the equation of the hyperbola is

(2)
$$\frac{x^2}{a^2} - \frac{y^2}{b^2} = 1$$

where $c^2 = a^2 + b^2$. Notice that the x-intercepts are again $\pm a$. But if we put $x = 0$ in Equation 2 we get $y^2 = -b^2$, which is impossible, so there is no y-intercept. The hyperbola is symmetric with respect to both axes.

To analyze the hyperbola further, we look at Equation 2 and obtain

$$\frac{x^2}{a^2} = 1 + \frac{y^2}{b^2} \geqslant 1$$

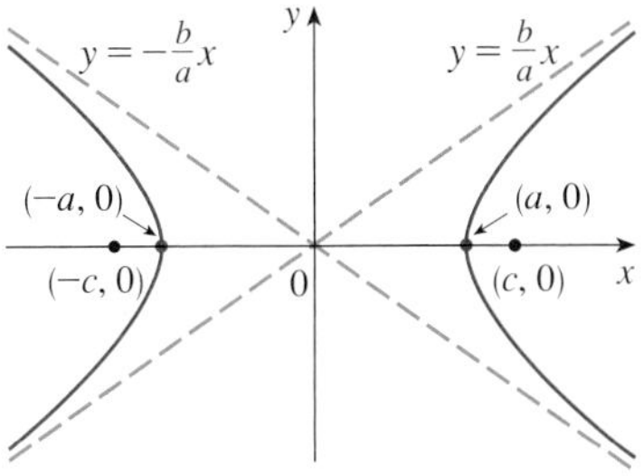

FIGURE 24
$\dfrac{x^2}{a^2} - \dfrac{y^2}{b^2} = 1$

This shows that $x^2 \geqslant a^2$, so $|x| = \sqrt{x^2} \geqslant a$. Therefore we have $x \geqslant a$ or $x \leqslant -a$. This means that the hyperbola consists of two parts, called its *branches.*

When we draw a hyperbola it is useful to first draw its *asymptotes*, which are the lines $y = (b/a)x$ and $y = -(b/a)x$ shown in Figure 24. Both branches of the hyperbola approach the asymptotes; that is, they come arbitrarily close to the asymptotes. If the foci of a hyperbola are on the y-axis, we find its equation by reversing the roles of x and y.

EXAMPLE 9 | Find the foci and asymptotes of the hyperbola $9x^2 - 16y^2 = 144$ and sketch its graph.

SOLUTION If we divide both sides of the equation by 144, it becomes

$$\frac{x^2}{16} - \frac{y^2}{9} = 1$$

which is of the form given in (2) with $a = 4$ and $b = 3$. Since $c^2 = 16 + 9 = 25$, the foci are $(\pm 5, 0)$. The asymptotes are the lines $y = \frac{3}{4}x$ and $y = -\frac{3}{4}x$. The graph is shown in Figure 25.

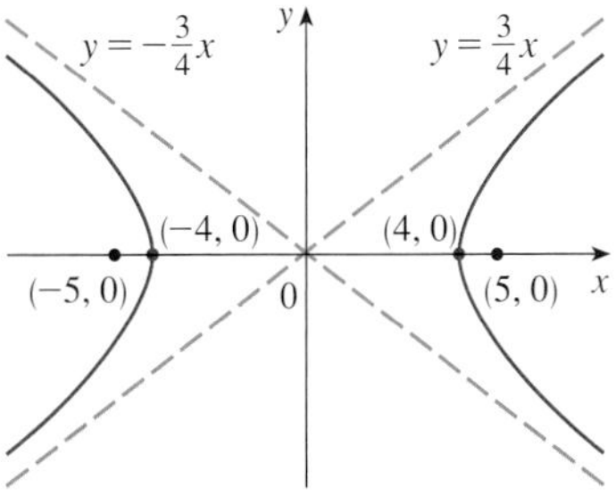

FIGURE 25
$9x^2 - 16y^2 = 144$

■

EXERCISES B

1–2 Find the distance between the points.

1. $(1, 1)$, $(4, 5)$

2. $(1, -3)$, $(5, 7)$

3–4 Find the slope of the line through P and Q.

3. $P(-3, 3)$, $Q(-1, -6)$

4. $P(-1, -4)$, $Q(6, 0)$

5. Show that the points $(-2, 9)$, $(4, 6)$, $(1, 0)$, and $(-5, 3)$ are the vertices of a square.

6. (a) Show that the points $A(-1, 3)$, $B(3, 11)$, and $C(5, 15)$ are collinear (lie on the same line) by showing that $|AB| + |BC| = |AC|$.
(b) Use slopes to show that A, B, and C are collinear.

7–10 Sketch the graph of the equation.

7. $x = 3$

8. $y = -2$

9. $xy = 0$

10. $|y| = 1$

11–24 Find an equation of the line that satisfies the given conditions.

11. Through $(2, -3)$, slope 6

12. Through $(-3, -5)$, slope $-\frac{7}{2}$

13. Through $(2, 1)$ and $(1, 6)$

14. Through $(-1, -2)$ and $(4, 3)$

15. Slope 3, y-intercept -2

16. Slope $\frac{2}{5}$, y-intercept 4

17. x-intercept 1, y-intercept -3

18. x-intercept -8, y-intercept 6

19. Through $(4, 5)$, parallel to the x-axis

20. Through $(4, 5)$, parallel to the y-axis

21. Through $(1, -6)$, parallel to the line $x + 2y = 6$

22. y-intercept 6, parallel to the line $2x + 3y + 4 = 0$

23. Through $(-1, -2)$, perpendicular to the line $2x + 5y + 8 = 0$

24. Through $(\frac{1}{2}, -\frac{2}{3})$, perpendicular to the line $4x - 8y = 1$

25–28 Find the slope and y-intercept of the line and draw its graph.

25. $x + 3y = 0$

26. $2x - 3y + 6 = 0$

27. $3x - 4y = 12$

28. $4x + 5y = 10$

29–36 Sketch the region in the xy-plane.

29. $\{(x, y) \mid x < 0\}$

30. $\{(x, y) \mid x \geq 1 \text{ and } y < 3\}$

31. $\{(x, y) \mid |x| \leq 2\}$

32. $\{(x, y) \mid |x| < 3 \text{ and } |y| < 2\}$

33. $\{(x, y) \mid 0 \leq y \leq 4 \text{ and } x \leq 2\}$

34. $\{(x, y) \mid y > 2x - 1\}$

35. $\{(x, y) \mid 1 + x \leq y \leq 1 - 2x\}$

36. $\{(x, y) \mid -x \leq y < \frac{1}{2}(x + 3)\}$

37–38 Find an equation of a circle that satisfies the given conditions.

37. Center $(3, -1)$, radius 5

38. Center $(-1, 5)$, passes through $(-4, -6)$

39–40 Show that the equation represents a circle and find the center and radius.

39. $x^2 + y^2 - 4x + 10y + 13 = 0$

40. $x^2 + y^2 + 6y + 2 = 0$

41. Show that the lines $2x - y = 4$ and $6x - 2y = 10$ are not parallel and find their point of intersection.

42. Show that the lines $3x - 5y + 19 = 0$ and $10x + 6y - 50 = 0$ are perpendicular and find their point of intersection.

43. Show that the midpoint of the line segment from $P_1(x_1, y_1)$ to $P_2(x_2, y_2)$ is

$$\left(\frac{x_1 + x_2}{2}, \frac{y_1 + y_2}{2}\right)$$

44. Find the midpoint of the line segment joining the points $(1, 3)$ and $(7, 15)$.

45. Find an equation of the perpendicular bisector of the line segment joining the points $A(1, 4)$ and $B(7, -2)$.

46. (a) Show that if the x- and y-intercepts of a line are nonzero numbers a and b, then the equation of the line can be put in the form

$$\frac{x}{a} + \frac{y}{b} = 1$$

This equation is called the **two-intercept form** of an equation of a line.

(b) Use part (a) to find an equation of the line whose x-intercept is 6 and whose y-intercept is -8.

47. Suppose that $P(x, y)$ is any point on the parabola with focus $(0, p)$ and directrix $y = -p$. (See Figure 14.) Use the definition of a parabola to show that $x^2 = 4py$.

48. Find the focus and directrix of the parabola $y = x^2$. Illustrate with a diagram.

49. Suppose an ellipse has foci $(\pm c, 0)$ and the sum of the distances from any point $P(x, y)$ on the ellipse to the foci is $2a$. Show that the coordinates of P satisfy Equation 1.

50. Find the foci of the ellipse $x^2 + 4y^2 = 4$ and sketch its graph.

51. Use the definition of a hyperbola to derive Equation 2 for a hyperbola with foci $(\pm c, 0)$.

52. (a) Find the foci and asymptotes of the hyperbola $x^2 - y^2 = 1$ and sketch its graph.
(b) Sketch the graph of $y^2 - x^2 = 1$.

53–54 Sketch the region bounded by the curves.

53. $x + 4y = 8$ and $x = 2y^2 - 8$

54. $y = 4 - x^2$ and $x - 2y = 2$

C Trigonometry

Here we review the aspects of trigonometry that are used in calculus: radian measure, trigonometric functions, trigonometric identities, and inverse trigonometric functions.

■ Angles

Angles can be measured in degrees or in radians (abbreviated as rad). The angle given by a complete revolution contains 360°, which is the same as 2π rad. Therefore

(1)
$$\boxed{\pi \text{ rad} = 180°}$$

and

(2)
$$1 \text{ rad} = \left(\frac{180}{\pi}\right)^{\circ} \approx 57.3° \qquad 1° = \frac{\pi}{180} \text{ rad} \approx 0.017 \text{ rad}$$

EXAMPLE 1

(a) Find the radian measure of 60°. (b) Express $5\pi/4$ rad in degrees.

SOLUTION

(a) From Equation 1 or 2 we see that to convert from degrees to radians we multiply by $\pi/180$. Therefore

$$60° = 60\left(\frac{\pi}{180}\right) = \frac{\pi}{3} \text{ rad}$$

(b) To convert from radians to degrees we multiply by $180/\pi$. Thus

$$\frac{5\pi}{4} \text{ rad} = \frac{5\pi}{4}\left(\frac{180}{\pi}\right) = 225°$$

■

In calculus we use radians to measure angles except when otherwise indicated. The following table gives the correspondence between degree and radian measures of some common angles.

Degrees	0°	30°	45°	60°	90°	120°	135°	150°	180°	270°	360°
Radians	0	$\frac{\pi}{6}$	$\frac{\pi}{4}$	$\frac{\pi}{3}$	$\frac{\pi}{2}$	$\frac{2\pi}{3}$	$\frac{3\pi}{4}$	$\frac{5\pi}{6}$	π	$\frac{3\pi}{2}$	2π

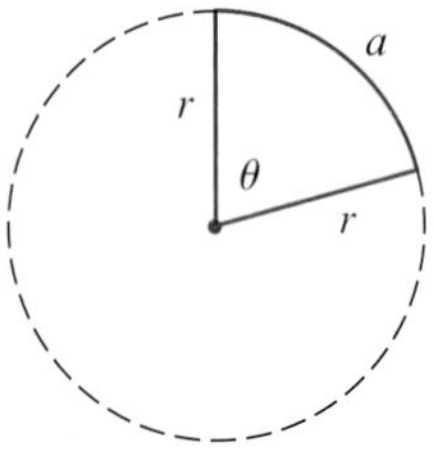

FIGURE 1

Figure 1 shows a sector of a circle with central angle θ and radius r subtending an arc with length a. Since the length of the arc is proportional to the size of the angle, and since the entire circle has circumference $2\pi r$ and central angle 2π, we have

$$\frac{\theta}{2\pi} = \frac{a}{2\pi r}$$

Solving this equation for θ and for a, we obtain

(3)
$$\boxed{\theta = \frac{a}{r}} \qquad \boxed{a = r\theta}$$

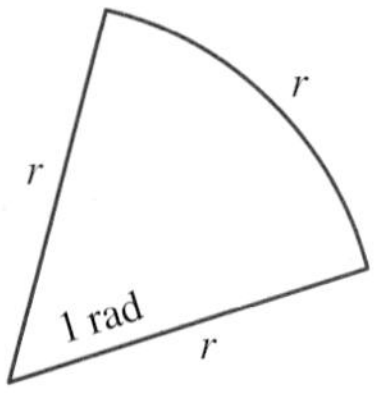

FIGURE 2

Remember that Equations 3 are valid only when θ is measured in radians.

In particular, putting $a = r$ in Equation 3, we see that an angle of 1 rad is the angle subtended at the center of a circle by an arc equal in length to the radius of the circle (see Figure 2).

EXAMPLE 2

(a) If the radius of a circle is 5 cm, what angle is subtended by an arc of 6 cm?
(b) If a circle has radius 3 cm, what is the length of an arc subtended by a central angle of $3\pi/8$ rad?

SOLUTION

(a) Using Equation 3 with $a = 6$ and $r = 5$, we see that the angle is

$$\theta = \tfrac{6}{5} = 1.2 \text{ rad}$$

(b) With $r = 3$ cm and $\theta = 3\pi/8$ rad, the arc length is

$$a = r\theta = 3\left(\frac{3\pi}{8}\right) = \frac{9\pi}{8} \text{ cm}$$ ■

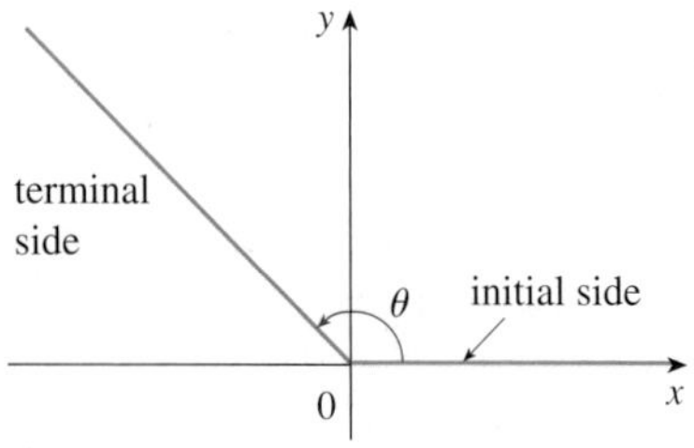

FIGURE 3 $\theta \geq 0$

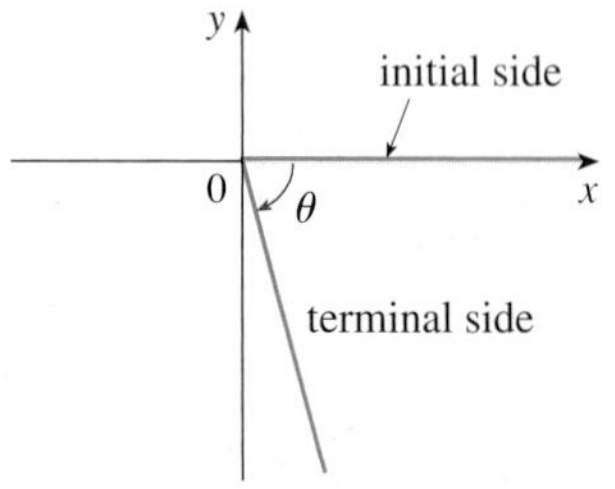

FIGURE 4 $\theta < 0$

The **standard position** of an angle occurs when we place its vertex at the origin of a coordinate system and its initial side on the positive x-axis as in Figure 3. A **positive** angle is obtained by rotating the initial side counterclockwise until it coincides with the terminal side. Likewise, **negative** angles are obtained by clockwise rotation as in Figure 4.

Figure 5 shows several examples of angles in standard position. Notice that different angles can have the same terminal side. For instance, the angles $3\pi/4$, $-5\pi/4$, and

$11\pi/4$ have the same initial and terminal sides because

$$\frac{3\pi}{4} - 2\pi = -\frac{5\pi}{4} \qquad \frac{3\pi}{4} + 2\pi = \frac{11\pi}{4}$$

and 2π rad represents a complete revolution.

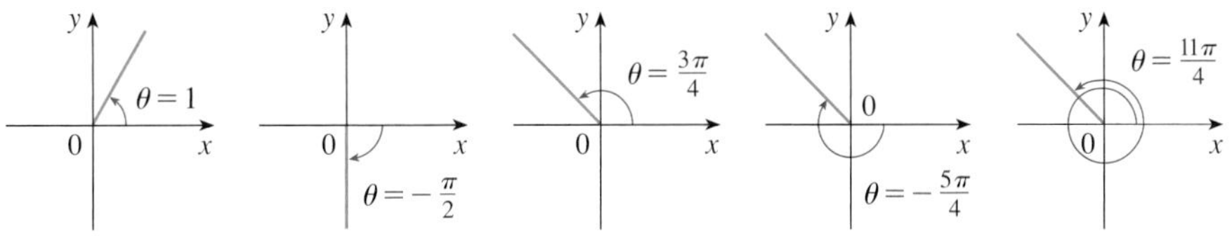

FIGURE 5
Angles in standard position

The Trigonometric Functions

For an acute angle θ the six trigonometric functions are defined as ratios of lengths of sides of a right triangle as follows (see Figure 6).

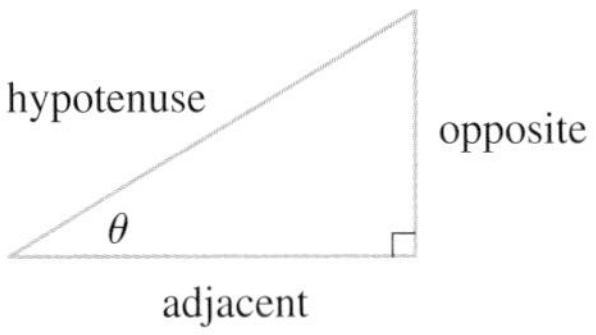

FIGURE 6

(4)

$$\sin\theta = \frac{\text{opp}}{\text{hyp}} \qquad \csc\theta = \frac{\text{hyp}}{\text{opp}}$$

$$\cos\theta = \frac{\text{adj}}{\text{hyp}} \qquad \sec\theta = \frac{\text{hyp}}{\text{adj}}$$

$$\tan\theta = \frac{\text{opp}}{\text{adj}} \qquad \cot\theta = \frac{\text{adj}}{\text{opp}}$$

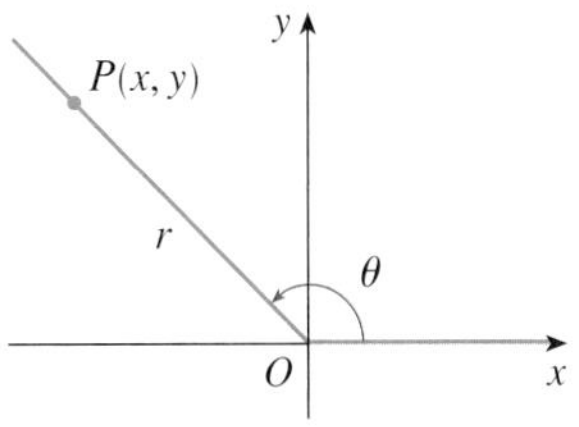

FIGURE 7

This definition doesn't apply to obtuse or negative angles, so for a general angle θ in standard position we let $P(x, y)$ be any point on the terminal side of θ and we let r be the distance $|OP|$ as in Figure 7. Then we define

(5)

$$\sin\theta = \frac{y}{r} \qquad \csc\theta = \frac{r}{y}$$

$$\cos\theta = \frac{x}{r} \qquad \sec\theta = \frac{r}{x}$$

$$\tan\theta = \frac{y}{x} \qquad \cot\theta = \frac{x}{y}$$

Since division by 0 is not defined, $\tan\theta$ and $\sec\theta$ are undefined when $x = 0$ and $\csc\theta$ and $\cot\theta$ are undefined when $y = 0$. Notice that the definitions in (4) and (5) are consistent when θ is an acute angle.

If θ is a number, the convention is that $\sin\theta$ means the sine of the angle whose *radian* measure is θ. For example, the expression $\sin 3$ implies that we are dealing with an angle of 3 rad. When finding a calculator approximation to this number, we must remember to set our calculator in radian mode, and then we obtain

$$\sin 3 \approx 0.14112$$

If we want to know the sine of the angle 3° we would write sin 3° and, with our calculator in degree mode, we find that

$$\sin 3° \approx 0.05234$$

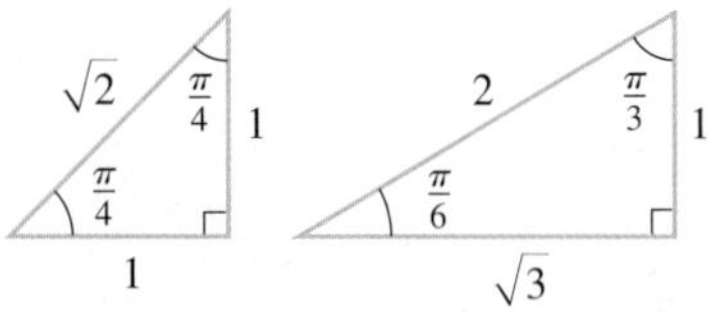

FIGURE 8

The exact trigonometric ratios for certain angles can be read from the triangles in Figure 8. For instance,

$$\sin\frac{\pi}{4} = \frac{1}{\sqrt{2}} \qquad \sin\frac{\pi}{6} = \frac{1}{2} \qquad \sin\frac{\pi}{3} = \frac{\sqrt{3}}{2}$$

$$\cos\frac{\pi}{4} = \frac{1}{\sqrt{2}} \qquad \cos\frac{\pi}{6} = \frac{\sqrt{3}}{2} \qquad \cos\frac{\pi}{3} = \frac{1}{2}$$

$$\tan\frac{\pi}{4} = 1 \qquad \tan\frac{\pi}{6} = \frac{1}{\sqrt{3}} \qquad \tan\frac{\pi}{3} = \sqrt{3}$$

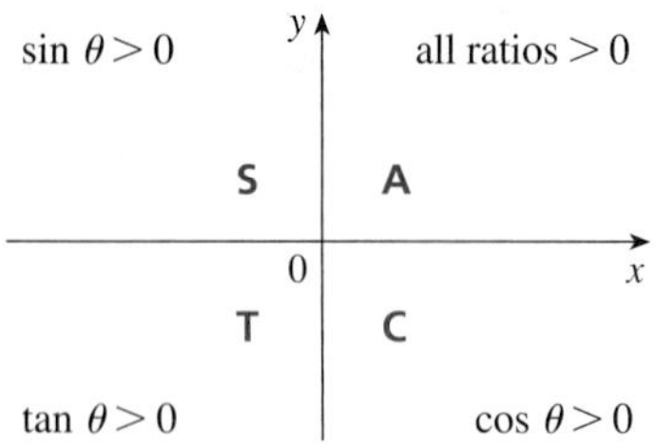

FIGURE 9

The signs of the trigonometric functions for angles in each of the four quadrants can be remembered by means of the rule "All Students Take Calculus" shown in Figure 9.

EXAMPLE 3 | Find the exact trigonometric ratios for $\theta = 2\pi/3$.

SOLUTION From Figure 10 we see that a point on the terminal line for $\theta = 2\pi/3$ is $P(-1, \sqrt{3})$. Therefore, taking

$$x = -1 \qquad y = \sqrt{3} \qquad r = 2$$

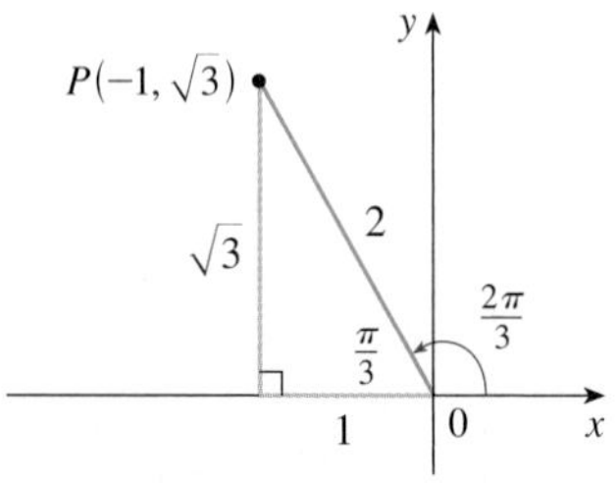

FIGURE 10

in the definitions of the trigonometric ratios, we have

$$\sin\frac{2\pi}{3} = \frac{\sqrt{3}}{2} \qquad \cos\frac{2\pi}{3} = -\frac{1}{2} \qquad \tan\frac{2\pi}{3} = -\sqrt{3}$$

$$\csc\frac{2\pi}{3} = \frac{2}{\sqrt{3}} \qquad \sec\frac{2\pi}{3} = -2 \qquad \cot\frac{2\pi}{3} = -\frac{1}{\sqrt{3}}$$ ■

The following table gives some values of $\sin\theta$ and $\cos\theta$ found by the method of Example 3.

θ	0	$\frac{\pi}{6}$	$\frac{\pi}{4}$	$\frac{\pi}{3}$	$\frac{\pi}{2}$	$\frac{2\pi}{3}$	$\frac{3\pi}{4}$	$\frac{5\pi}{6}$	π	$\frac{3\pi}{2}$	2π
$\sin\theta$	0	$\frac{1}{2}$	$\frac{1}{\sqrt{2}}$	$\frac{\sqrt{3}}{2}$	1	$\frac{\sqrt{3}}{2}$	$\frac{1}{\sqrt{2}}$	$\frac{1}{2}$	0	-1	0
$\cos\theta$	1	$\frac{\sqrt{3}}{2}$	$\frac{1}{\sqrt{2}}$	$\frac{1}{2}$	0	$-\frac{1}{2}$	$-\frac{1}{\sqrt{2}}$	$-\frac{\sqrt{3}}{2}$	-1	0	1

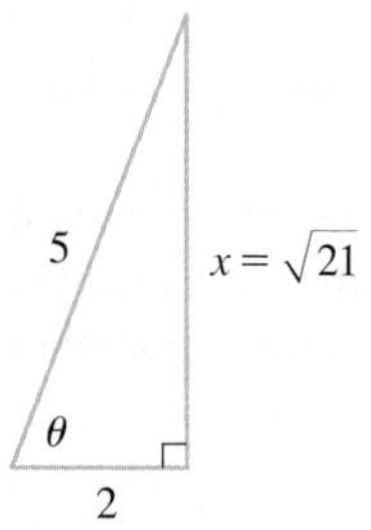

FIGURE 11

EXAMPLE 4 | If $\cos\theta = \frac{2}{5}$ and $0 < \theta < \pi/2$, find the other five trigonometric functions of θ.

SOLUTION Since $\cos\theta = \frac{2}{5}$, we can label the hypotenuse as having length 5 and the adjacent side as having length 2 in Figure 11. If the opposite side has length x, then the

Pythagorean Theorem gives $x^2 + 4 = 25$ and so $x^2 = 21$, $x = \sqrt{21}$. We can now use the diagram to write the other five trigonometric functions:

$$\sin\theta = \frac{\sqrt{21}}{5} \qquad \tan\theta = \frac{\sqrt{21}}{2}$$

$$\csc\theta = \frac{5}{\sqrt{21}} \qquad \sec\theta = \frac{5}{2} \qquad \cot\theta = \frac{2}{\sqrt{21}}$$

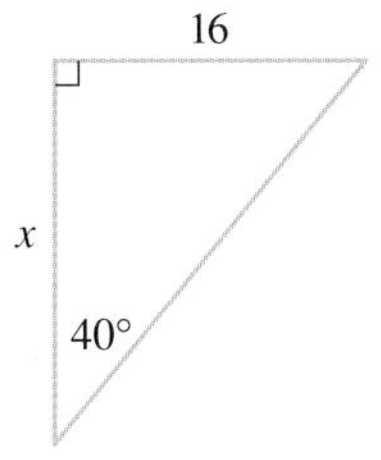

FIGURE 12

EXAMPLE 5 | Use a calculator to approximate the value of x in Figure 12.

SOLUTION From the diagram we see that

$$\tan 40° = \frac{16}{x}$$

Therefore
$$x = \frac{16}{\tan 40°} \approx 19.07$$

■ Trigonometric Identities

A trigonometric identity is a relationship among the trigonometric functions. The most elementary are the following, which are immediate consequences of the definitions of the trigonometric functions.

(6)
$$\csc\theta = \frac{1}{\sin\theta} \qquad \sec\theta = \frac{1}{\cos\theta} \qquad \cot\theta = \frac{1}{\tan\theta}$$
$$\tan\theta = \frac{\sin\theta}{\cos\theta} \qquad \cot\theta = \frac{\cos\theta}{\sin\theta}$$

For the next identity we refer back to Figure 7. The distance formula (or, equivalently, the Pythagorean Theorem) tells us that $x^2 + y^2 = r^2$. Therefore

$$\sin^2\theta + \cos^2\theta = \frac{y^2}{r^2} + \frac{x^2}{r^2} = \frac{x^2 + y^2}{r^2} = \frac{r^2}{r^2} = 1$$

We have therefore proved one of the most useful of all trigonometric identities:

(7)
$$\sin^2\theta + \cos^2\theta = 1$$

If we now divide both sides of Equation 7 by $\cos^2\theta$ and use Equations 6, we get

(8)
$$\tan^2\theta + 1 = \sec^2\theta$$

Similarly, if we divide both sides of Equation 7 by $\sin^2\theta$, we get

(9)
$$1 + \cot^2\theta = \csc^2\theta$$

The identities

(10a) $$\sin(-\theta) = -\sin\theta$$

(10b) $$\cos(-\theta) = \cos\theta$$

Odd functions and even functions are discussed in Section 1.1.

show that sine is an odd function and cosine is an even function. They are easily proved by drawing a diagram showing θ and $-\theta$ in standard position (see Exercise 19).

Since the angles θ and $\theta + 2\pi$ have the same terminal side, we have

(11) $$\sin(\theta + 2\pi) = \sin\theta \qquad \cos(\theta + 2\pi) = \cos\theta$$

These identities show that the sine and cosine functions are periodic with period 2π.

The remaining trigonometric identities are all consequences of two basic identities called the **addition formulas**:

(12a) $$\sin(x + y) = \sin x \cos y + \cos x \sin y$$

(12b) $$\cos(x + y) = \cos x \cos y - \sin x \sin y$$

The proofs of these addition formulas are outlined in Exercises 45, 46, and 47.

By substituting $-y$ for y in Equations 12a and 12b and using Equations 10a and 10b, we obtain the following **subtraction formulas**:

(13a) $$\sin(x - y) = \sin x \cos y - \cos x \sin y$$

(13b) $$\cos(x - y) = \cos x \cos y + \sin x \sin y$$

Then, by dividing the formulas in Equations 12 or Equations 13, we obtain the corresponding formulas for $\tan(x \pm y)$:

(14a) $$\tan(x + y) = \frac{\tan x + \tan y}{1 - \tan x \tan y}$$

(14b) $$\tan(x - y) = \frac{\tan x - \tan y}{1 + \tan x \tan y}$$

If we put $y = x$ in the addition formulas (12), we get the **double-angle formulas**:

(15a) $$\sin 2x = 2 \sin x \cos x$$

(15b) $$\cos 2x = \cos^2 x - \sin^2 x$$

Then, by using the identity $\sin^2 x + \cos^2 x = 1$, we obtain the following alternate forms of the double-angle formulas for $\cos 2x$:

(16a) $$\cos 2x = 2\cos^2 x - 1$$

(16b) $$\cos 2x = 1 - 2\sin^2 x$$

If we now solve these equations for $\cos^2 x$ and $\sin^2 x$, we get the following **half-angle formulas**, which are useful in integral calculus:

(17a) $$\cos^2 x = \frac{1 + \cos 2x}{2}$$

(17b) $$\sin^2 x = \frac{1 - \cos 2x}{2}$$

Finally, we state the **product formulas**, which can be deduced from Equations 12 and 13:

(18a) $$\sin x \cos y = \tfrac{1}{2}[\sin(x + y) + \sin(x - y)]$$

(18b) $$\cos x \cos y = \tfrac{1}{2}[\cos(x + y) + \cos(x - y)]$$

(18c) $$\sin x \sin y = \tfrac{1}{2}[\cos(x - y) - \cos(x + y)]$$

There are many other trigonometric identities, but those we have stated are the ones used most often in calculus. If you forget any of them, remember that they can all be deduced from Equations 12a and 12b.

EXAMPLE 6 | Find all values of x in the interval $[0, 2\pi]$ such that $\sin x = \sin 2x$.

SOLUTION Using the double-angle formula (15a), we rewrite the given equation as

$$\sin x = 2\sin x \cos x \qquad \text{or} \qquad \sin x(1 - 2\cos x) = 0$$

Therefore there are two possibilities:

$$\sin x = 0 \qquad \text{or} \qquad 1 - 2\cos x = 0$$

$$x = 0, \pi, 2\pi \qquad\qquad \cos x = \tfrac{1}{2}$$

$$x = \frac{\pi}{3}, \frac{5\pi}{3}$$

The given equation has five solutions: 0, $\pi/3$, π, $5\pi/3$, and 2π. ■

■ Graphs of the Trigonometric Functions

The graph of the function $f(x) = \sin x$, shown in Figure 13(a) on page 706, is obtained by plotting points for $0 \leqslant x \leqslant 2\pi$ and then using the periodic nature of the function (from Equation 11) to complete the graph. Notice that the zeros of the sine function occur at the integer multiples of π, that is,

$$\sin x = 0 \qquad \text{whenever } x = n\pi, \quad n \text{ an integer}$$

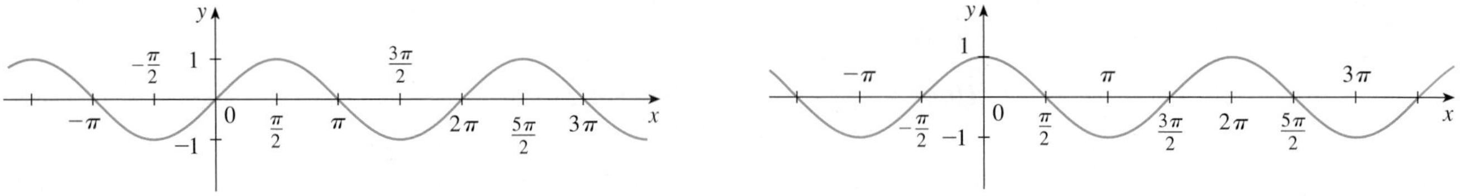

(a) $f(x) = \sin x$

(b) $g(x) = \cos x$

FIGURE 13

Because of the identity

$$\cos x = \sin\left(x + \frac{\pi}{2}\right)$$

(which can be verified using Equation 12a), the graph of cosine is obtained by shifting the graph of sine by an amount $\pi/2$ to the left [see Figure 13(b)]. Note that for both the sine and cosine functions the domain is $(-\infty, \infty)$ and the range is the closed interval $[-1, 1]$. Thus, for all values of x, we have

$$-1 \leqslant \sin x \leqslant 1 \qquad\qquad -1 \leqslant \cos x \leqslant 1$$

The graphs of the remaining four trigonometric functions are shown in Figure 14 and their domains are indicated there. Notice that tangent and cotangent have range $(-\infty, \infty)$, whereas cosecant and secant have range $(-\infty, -1] \cup [1, \infty)$. All four functions are periodic: tangent and cotangent have period π, whereas cosecant and secant have period 2π.

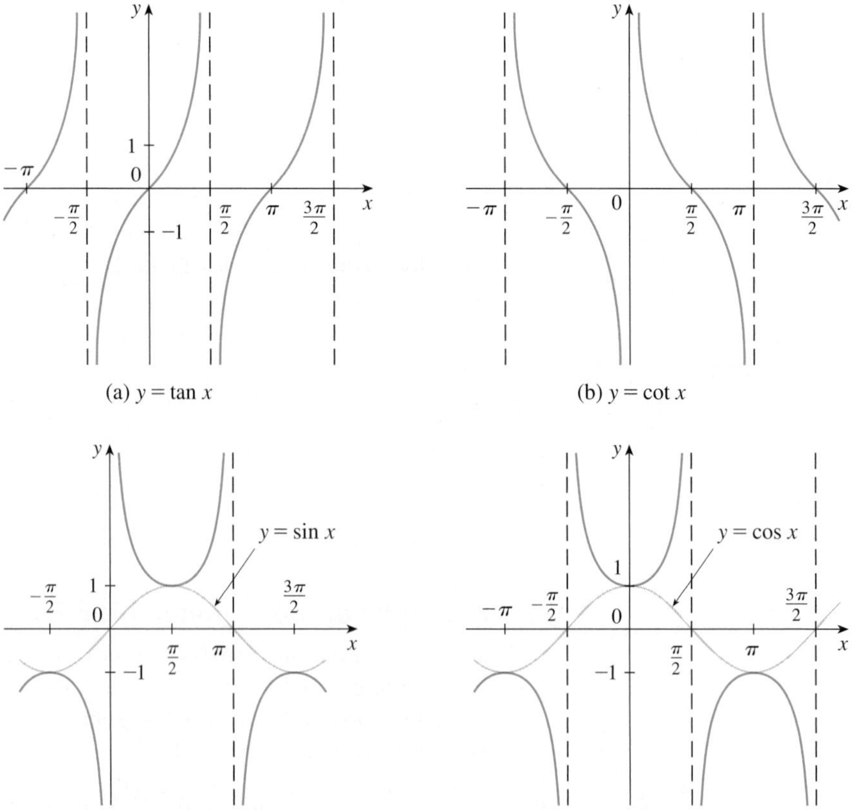

(a) $y = \tan x$

(b) $y = \cot x$

(c) $y = \csc x$

(d) $y = \sec x$

FIGURE 14

EXERCISES C

1–2 Convert from degrees to radians.

1. (a) $210°$ (b) $9°$

2. (a) $-315°$ (b) $36°$

3–4 Convert from radians to degrees.

3. (a) 4π (b) $-\dfrac{3\pi}{8}$

4. (a) $-\dfrac{7\pi}{2}$ (b) $\dfrac{8\pi}{3}$

5. Find the length of a circular arc subtended by an angle of $\pi/12$ rad if the radius of the circle is 36 cm.

6. If a circle has radius 10 cm, find the length of the arc subtended by a central angle of $72°$.

7. A circle has radius 1.5 m. What angle is subtended at the center of the circle by an arc 1 m long?

8. Find the radius of a circular sector with angle $3\pi/4$ and arc length 6 cm.

9–10 Draw, in standard position, the angle whose measure is given.

9. (a) $315°$ (b) $-\dfrac{3\pi}{4}$ rad

10. (a) $\dfrac{7\pi}{3}$ rad (b) -3 rad

11–12 Find the exact trigonometric ratios for the angle whose radian measure is given.

11. $\dfrac{3\pi}{4}$ **12.** $\dfrac{4\pi}{3}$

13–14 Find the remaining trigonometric ratios.

13. $\sin\theta = \dfrac{3}{5}, \quad 0 < \theta < \dfrac{\pi}{2}$

14. $\tan\alpha = 2, \quad 0 < \alpha < \dfrac{\pi}{2}$

15–18 Find, correct to five decimal places, the length of the side labeled x.

15.

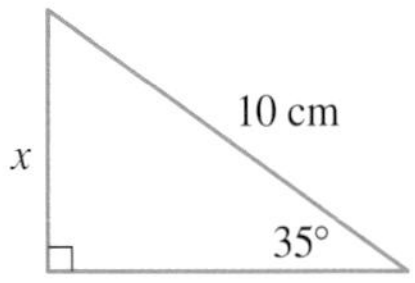

16.

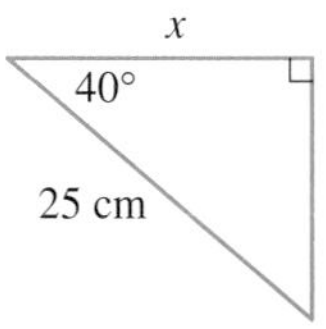

17.

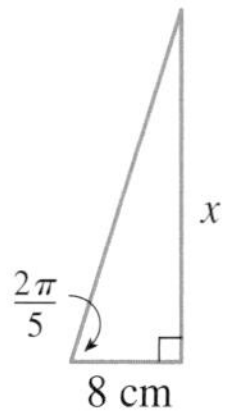

18.

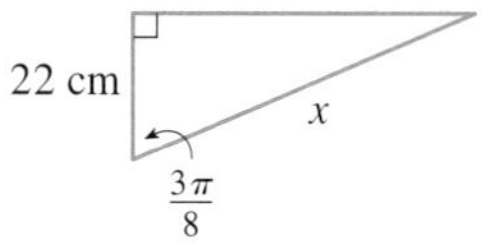

19–21 Prove each equation.

19. (a) Equation 10a (b) Equation 10b

20. (a) Equation 14a (b) Equation 14b

21. (a) Equation 18a (b) Equation 18b (c) Equation 18c

22–28 Prove the identity.

22. $\cos\left(\dfrac{\pi}{2} - x\right) = \sin x$

23. $\sin\left(\dfrac{\pi}{2} + x\right) = \cos x$ **24.** $\sin(\pi - x) = \sin x$

25. $\sin\theta\ \cot\theta = \cos\theta$ **26.** $(\sin x + \cos x)^2 = 1 + \sin 2x$

27. $\tan 2\theta = \dfrac{2\tan\theta}{1 - \tan^2\theta}$ **28.** $\cos 3\theta = 4\cos^3\theta - 3\cos\theta$

29–30 If $\sin x = \frac{1}{3}$ and $\sec y = \frac{5}{4}$, where x and y lie between 0 and $\pi/2$, evaluate the expression.

29. $\sin(x + y)$ **30.** $\cos 2y$

31–34 Find all values of x in the interval $[0, 2\pi]$ that satisfy the equation.

31. $2\cos x - 1 = 0$ **32.** $2\sin^2 x = 1$

33. $\sin 2x = \cos x$ **34.** $|\tan x| = 1$

35–38 Find all values of x in the interval $[0, 2\pi]$ that satisfy the inequality.

35. $\sin x \leqslant \frac{1}{2}$ **36.** $2\cos x + 1 > 0$

37. $-1 < \tan x < 1$ **38.** $\sin x > \cos x$

39–42 Graph the function by starting with the graphs in Figures 13 and 14 and applying the transformations of Section 1.3 where appropriate.

39. $y = \cos\left(x - \dfrac{\pi}{3}\right)$ **40.** $y = \tan 2x$

41. $y = \dfrac{1}{3}\tan\left(x - \dfrac{\pi}{2}\right)$ **42.** $y = |\sin x|$

43. Prove the **Law of Cosines**: If a triangle has sides with lengths a, b, and c, and θ is the angle between the sides with lengths a and b, then

$$c^2 = a^2 + b^2 - 2ab\cos\theta$$

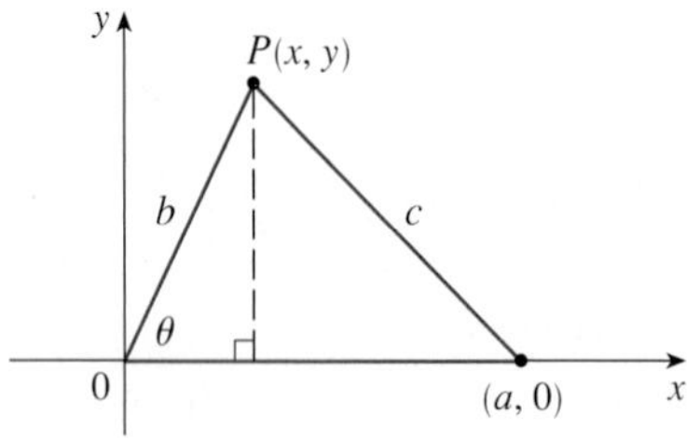

[*Hint:* Introduce a coordinate system so that θ is in standard position, as in the figure. Express x and y in terms of θ and then use the distance formula to compute c.]

44. In order to find the distance $|AB|$ across a small inlet, a point C is located as in the figure and the following measurements were recorded:

$\angle C = 103°$ $\quad |AC| = 820$ m $\quad |BC| = 910$ m

Use the Law of Cosines from Exercise 43 to find the required distance.

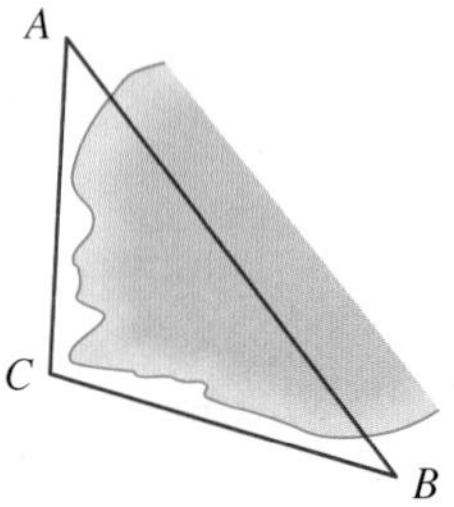

45. Use the figure to prove the subtraction formula

$$\cos(\alpha - \beta) = \cos\alpha\,\cos\beta + \sin\alpha\,\sin\beta$$

[*Hint:* Compute c^2 in two ways (using the Law of Cosines from Exercise 43 and also using the distance formula) and compare the two expressions.]

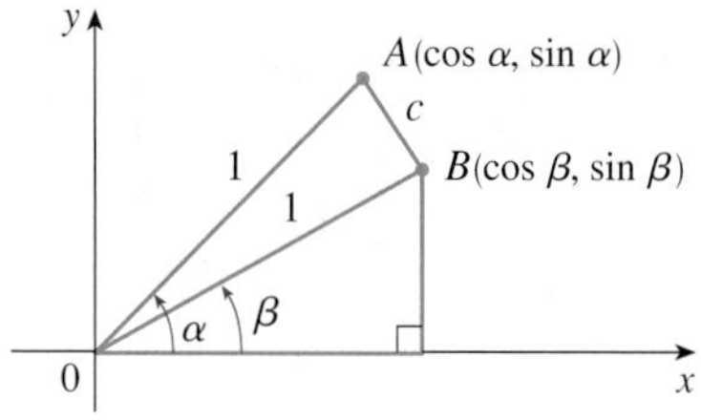

46. Use the formula in Exercise 45 to prove the addition formula for cosine (12b).

47. Use the addition formula for cosine and the identities

$$\cos\left(\frac{\pi}{2} - \theta\right) = \sin\theta \qquad \sin\left(\frac{\pi}{2} - \theta\right) = \cos\theta$$

to prove the subtraction formula for the sine function.

48. (a) Show that the area of a triangle with sides of lengths a and b and with included angle θ is

$$A = \tfrac{1}{2}ab\sin\theta$$

(b) Find the area of triangle ABC, correct to five decimal places, if

$|AB| = 10$ cm $\quad |BC| = 3$ cm $\quad \angle ABC = 107°$

D | Precise Definitions of Limits

The definitions of limits that have been given in this book are appropriate for intuitive understanding of the basic concepts of calculus. For the purposes of deeper understanding and rigorous proofs, however, the precise definitions of this appendix are necessary. In particular, the definition of a limit given here is used in Appendix E to prove some of the properties of limits.

■ Sequences

In Section 2.1 we used the notation $\lim_{n\to\infty} a_n = L$ to mean that the terms of the sequence $\{a_n\}$ approach the number L as n becomes large. Here we want to express, in quantitative form, that a_n can be made arbitrarily close to L by making n sufficiently large. This means that a_n can be made to lie within a prescribed distance from L (traditionally denoted by ε, the Greek letter epsilon) by requiring that n be made larger than a specified number N. The resulting precise definition of a limit is as follows.

(1) Definition A sequence $\{a_n\}$ has the **limit** L and we write

$$\lim_{n \to \infty} a_n = L \qquad \text{or} \qquad a_n \to L \quad \text{as} \quad n \to \infty$$

if for every $\varepsilon > 0$ there is a corresponding number N such that

$$\text{if} \quad n > N \qquad \text{then} \qquad |a_n - L| < \varepsilon$$

Definition 1 is illustrated by Figure 1, in which the terms $a_1, a_2, a_3, \ldots$ are plotted on a number line. No matter how small an interval $(L - \varepsilon, L + \varepsilon)$ is chosen, there exists an N such that all terms of the sequence from a_{N+1} onward must lie in that interval.

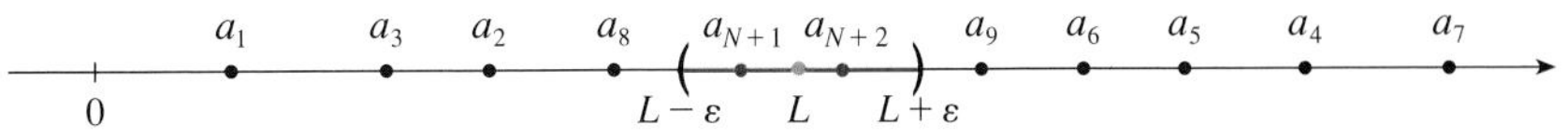

FIGURE 1

Another illustration of Definition 1 is given in Figure 2. The points on the graph of $\{a_n\}$ must lie between the horizontal lines $y = L + \varepsilon$ and $y = L - \varepsilon$ if $n > N$. This picture must be valid no matter how small ε is chosen, but usually a smaller ε requires a larger N.

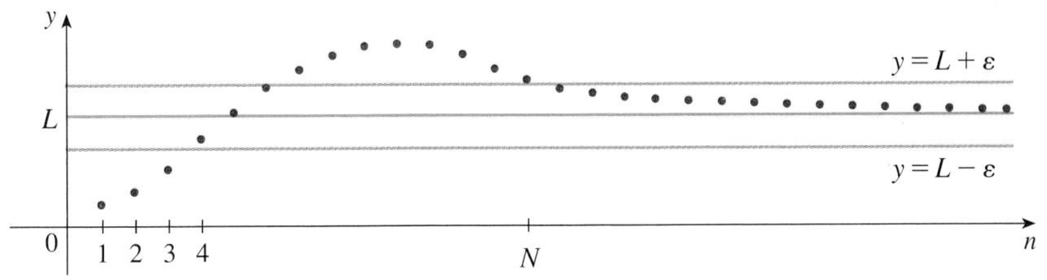

FIGURE 2

EXAMPLE 1 | Prove that $\displaystyle\lim_{n \to \infty} \frac{1}{n} = 0$.

SOLUTION Given $\varepsilon > 0$, we want to find a number N such that

$$\text{if} \quad n > N \qquad \text{then} \qquad \left| \frac{1}{n} - 0 \right| < \varepsilon$$

But $1/n < \varepsilon \iff n > 1/\varepsilon$. Let's choose $N = 1/\varepsilon$. So

$$\text{if} \quad n > N = \frac{1}{\varepsilon} \qquad \text{then} \qquad \left| \frac{1}{n} - 0 \right| = \frac{1}{n} < \varepsilon$$

Therefore, by Definition 1,

$$\lim_{n \to \infty} \frac{1}{n} = 0$$

■

■ Functions

The precise definition of the limit of a function at infinity, $\lim_{x \to \infty} f(x) = L$, is very similar to the definition of the limit of a sequence, $\lim_{n \to \infty} a_n = L$, given in Definition 1. So let's turn our attention to the definition of the limit of $f(x)$ as x approaches a finite number.

For a precise definition of $\lim_{x \to a} f(x) = L$ we want to express that $f(x)$ can be made arbitrarily close to L by making x sufficiently close to a (but $x \neq a$). This means that $f(x)$ can be made to lie within any preassigned distance ε from L by requiring that x be within a specified distance δ (the Greek letter delta) from a. That is, $|f(x) - L| < \varepsilon$ when

$|x - a| < \delta$ and $x \neq a$. Notice that we can stipulate that $x \neq a$ by writing $0 < |x - a|$. The resulting precise definition of a limit is as follows.

> **(2) Definition** Let f be a function defined on some open interval that contains the number a, except possibly at a itself. Then we say that the **limit of $f(x)$ as x approaches a is L**, and we write
>
> $$\lim_{x \to a} f(x) = L$$
>
> if for every number $\varepsilon > 0$ there is a number $\delta > 0$ such that
>
> $$\text{if} \quad 0 < |x - a| < \delta \qquad \text{then} \qquad |f(x) - L| < \varepsilon$$

Definition 2 is illustrated in Figures 3–5. If a number $\varepsilon > 0$ is given, then we draw the horizontal lines $y = L + \varepsilon$ and $y = L - \varepsilon$ and the graph of f. (See Figure 3.) If $\lim_{x \to a} f(x) = L$, then we can find a number $\delta > 0$ such that if we restrict x to lie in the interval $(a - \delta, a + \delta)$ and take $x \neq a$, then the curve $y = f(x)$ lies between the lines $y = L - \varepsilon$ and $y = L + \varepsilon$. (See Figure 4.) You can see that if such a δ has been found, then any smaller δ will also work.

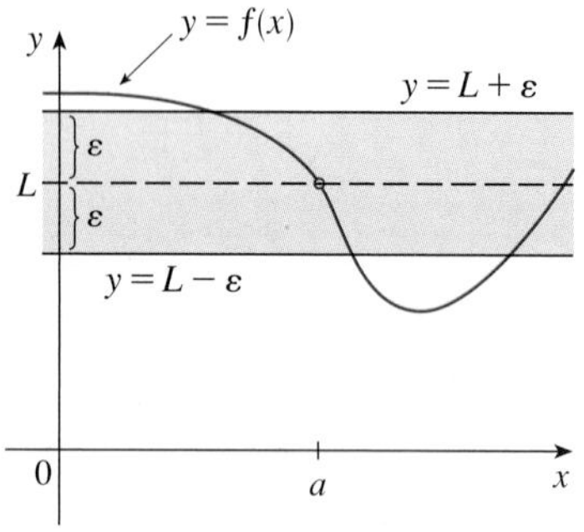

FIGURE 3

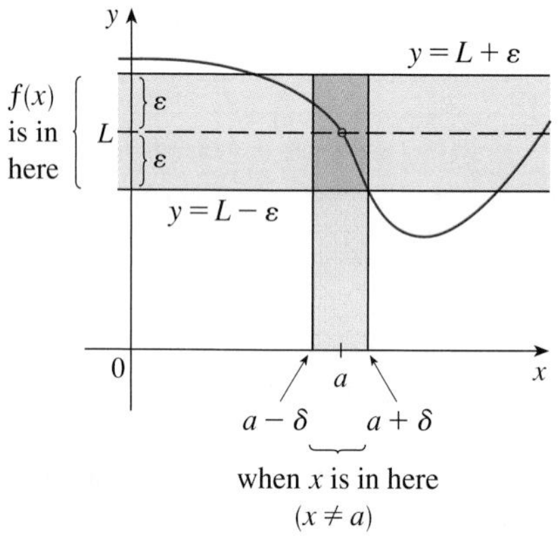

FIGURE 4

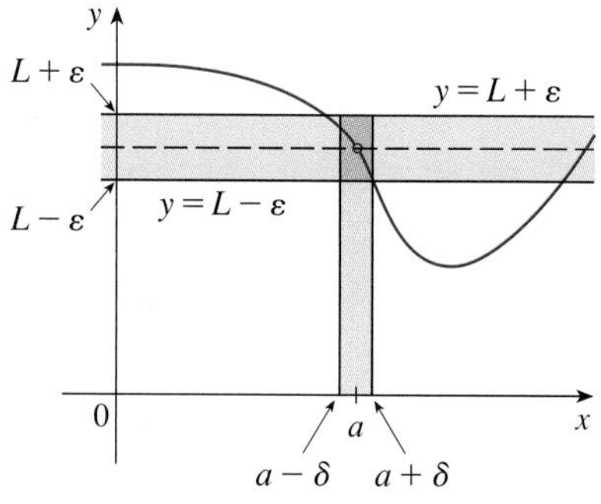

FIGURE 5

TEC In Module D you can explore the precise definition of a limit both graphically and numerically.

It's important to realize that the process illustrated in Figures 3 and 4 must work for *every* positive number ε no matter how small it is chosen. Figure 5 shows that if a smaller ε is chosen, then a smaller δ may be required.

EXAMPLE 2 | Use the ε, δ definition to prove that $\lim_{x \to 0} x^2 = 0$.

SOLUTION Let ε be a given positive number. According to Definition 2 with $a = 0$ and $L = 0$, we need to find a number δ such that

$$\text{if} \quad 0 < |x - 0| < \delta \qquad \text{then} \qquad |x^2 - 0| < \varepsilon$$

that is,

$$\text{if} \quad 0 < |x| < \delta \qquad \text{then} \qquad x^2 < \varepsilon$$

But, since the square root function is an increasing function, we know that

$$x^2 < \varepsilon \iff \sqrt{x^2} < \sqrt{\varepsilon} \iff |x| < \sqrt{\varepsilon}$$

So if we choose $\delta = \sqrt{\varepsilon}$, then $x^2 < \varepsilon \iff |x| < \delta$. (See Figure 6.) This shows that $\lim_{x \to 0} x^2 = 0$. ■

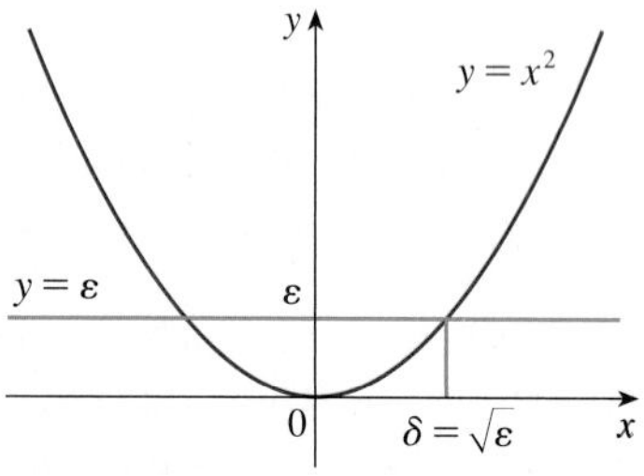

FIGURE 6

In proving limit statements it may be helpful to think of the definition of a limit as a challenge. First it challenges you with a number ε. Then you must be able to produce a suitable δ. You have to be able to do this for *every* $\varepsilon > 0$, not just a particular ε.

EXAMPLE 3 | Prove that $\lim_{x \to 3} (4x - 5) = 7$.

SOLUTION Let ε be a given positive number. According to Definition 2 with $a = 3$ and $L = 7$, we need to find a number δ such that

$$\text{if} \quad 0 < |x - 3| < \delta \quad \text{then} \quad |(4x - 5) - 7| < \varepsilon$$

Figure 7 shows the geometry behind Example 3.

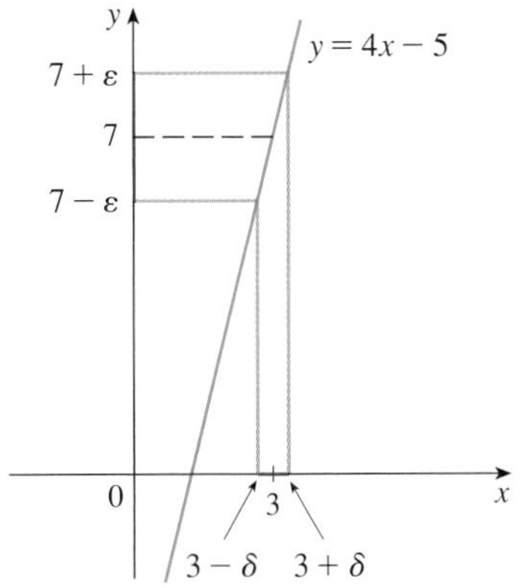

FIGURE 7

But $|(4x - 5) - 7| = |4x - 12| = |4(x - 3)| = 4|x - 3|$. Therefore we want δ such that

$$\text{if} \quad 0 < |x - 3| < \delta \quad \text{then} \quad 4|x - 3| < \varepsilon$$

Note that $4|x - 3| < \varepsilon \iff |x - 3| < \varepsilon/4$. So let's choose $\delta = \varepsilon/4$. We can then write the following:

$$\text{If} \quad 0 < |x - 3| < \delta \quad \text{then} \quad 4|x - 3| < \varepsilon \quad \text{so} \quad |(4x - 5) - 7| < \varepsilon$$

Therefore, by the definition of a limit,

$$\lim_{x \to 3} (4x - 5) = 7$$

This example is illustrated by Figure 7. ■

It's not always easy to prove that limit statements are true using the ε, δ definition. For a more complicated function such as $f(x) = (6x^2 - 8x + 9)/(2x^2 - 1)$, a proof would require a great deal of ingenuity. Fortunately, this is not necessary because the Limit Laws stated in Section 2.4 can be proved using Definition 2, and then the limits of complicated functions can be found rigorously from the Limit Laws without resorting to the definition directly.

■ Definite Integrals

In Section 5.2 we defined the definite integral of a function f on an interval $[a, b]$ as

$$\int_a^b f(x)\,dx = \lim_{n \to \infty} \sum_{i=1}^{n} f(x_i^*)\,\Delta x$$

where, at the nth stage, we have divided $[a, b]$ into n subintervals of equal width, $\Delta x = (b - a)/n$, and x_i^* is any sample point in the ith subinterval. The precise meaning of this limit that defines the integral is as follows:

For every number $\varepsilon > 0$ there is an integer N such that

$$\left| \int_a^b f(x)\,dx - \sum_{i=1}^{n} f(x_i^*)\,\Delta x \right| < \varepsilon$$

for every integer $n > N$ and for every choice of x_i^* in the ith subinterval.

This means that a definite integral can be approximated to within any desired degree of accuracy by a Riemann sum.

Functions of Two Variables

Here is a precise version of Definition 9.1.1:

> **(3) Definition** Let f be a function of two variables whose domain D includes points arbitrarily close to (a, b). Then we say that the **limit of $f(x, y)$ as (x, y) approaches (a, b) is** L and we write
>
> $$\lim_{(x, y) \to (a, b)} f(x, y) = L$$
>
> if for every number $\varepsilon > 0$ there is a corresponding number $\delta > 0$ such that
>
> if $\quad (x, y) \in D \quad$ and $\quad 0 < \sqrt{(x-a)^2 + (y-b)^2} < \delta$
>
> then $\quad |f(x, y) - L| < \varepsilon$

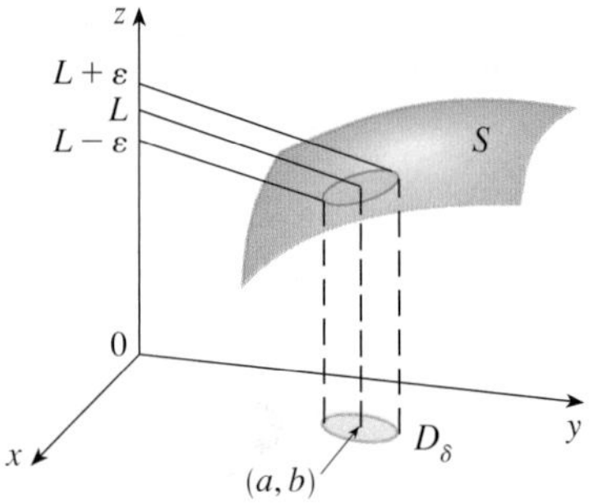

FIGURE 8

Notice that $|f(x, y) - L|$ is the distance between the numbers $f(x, y)$ and L, and $\sqrt{(x-a)^2 + (y-b)^2}$ is the distance between the point (x, y) and the point (a, b). Thus Definition 3 says that the distance between $f(x, y)$ and L can be made arbitrarily small by making the distance from (x, y) to (a, b) sufficiently small (but not 0). An illustration of Definition 3 is given in Figure 8 where the surface S is the graph of f. If $\varepsilon > 0$ is given, we can find $\delta > 0$ such that if (x, y) is restricted to lie in the disk D_δ with center (a, b) and radius δ, and if $(x, y) \neq (a, b)$, then the corresponding part of S lies between the horizontal planes $z = L - \varepsilon$ and $z = L + \varepsilon$.

EXAMPLE 4 | Prove that $\displaystyle\lim_{(x, y) \to (0, 0)} \frac{3x^2 y}{x^2 + y^2} = 0$.

SOLUTION Let $\varepsilon > 0$. We want to find $\delta > 0$ such that

$$\text{if} \quad 0 < \sqrt{x^2 + y^2} < \delta \quad \text{then} \quad \left| \frac{3x^2 y}{x^2 + y^2} - 0 \right| < \varepsilon$$

that is,

$$\text{if} \quad 0 < \sqrt{x^2 + y^2} < \delta \quad \text{then} \quad \frac{3x^2 |y|}{x^2 + y^2} < \varepsilon$$

But $x^2 \leqslant x^2 + y^2$ since $y^2 \geqslant 0$, so $x^2/(x^2 + y^2) \leqslant 1$ and therefore

$$\frac{3x^2 |y|}{x^2 + y^2} \leqslant 3|y| = 3\sqrt{y^2} \leqslant 3\sqrt{x^2 + y^2}$$

Thus if we choose $\delta = \varepsilon/3$ and let $0 < \sqrt{x^2 + y^2} < \delta$, then

$$\left| \frac{3x^2 y}{x^2 + y^2} - 0 \right| \leqslant 3\sqrt{x^2 + y^2} < 3\delta = 3\left(\frac{\varepsilon}{3}\right) = \varepsilon$$

Hence, by Definition 3,

$$\lim_{(x, y) \to (0, 0)} \frac{3x^2 y}{x^2 + y^2} = 0$$

■

EXERCISES D

1. (a) Determine how large we have to take n so that

$$\frac{1}{n^2} < 0.0001$$

(b) Use Definition 1 to prove that

$$\lim_{n \to \infty} \frac{1}{n^2} = 0$$

2. Use Definition 1 to prove that $\lim_{n \to \infty} \dfrac{1}{\sqrt{n}} = 0$.

3. Use the given graph of $f(x) = 1/x$ to find a number δ such that

$$\text{if} \quad |x - 2| < \delta \quad \text{then} \quad \left| \frac{1}{x} - 0.5 \right| < 0.2$$

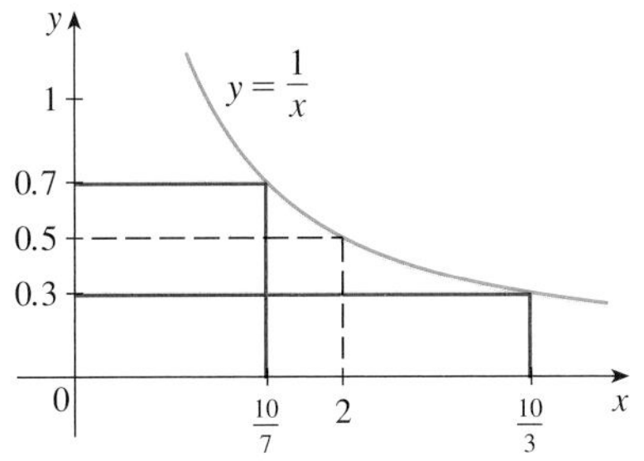

4. Use the given graph of f to find a number δ such that

$$\text{if} \quad 0 < |x - 5| < \delta \quad \text{then} \quad |f(x) - 3| < 0.6$$

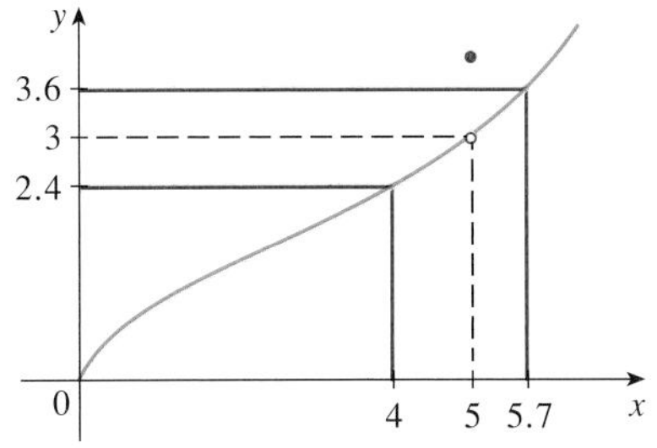

5. Use the given graph of $f(x) = \sqrt{x}$ to find a number δ such that

$$\text{if} \quad |x - 4| < \delta \quad \text{then} \quad |\sqrt{x} - 2| < 0.4$$

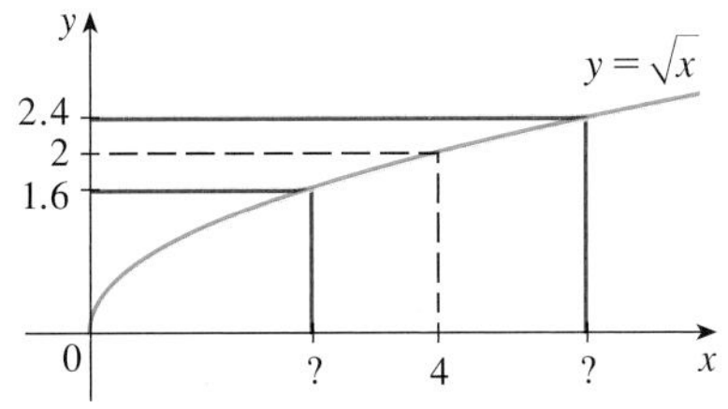

6. Use the given graph of $f(x) = x^2$ to find a number δ such that

$$\text{if} \quad |x - 1| < \delta \quad \text{then} \quad |x^2 - 1| < \tfrac{1}{2}$$

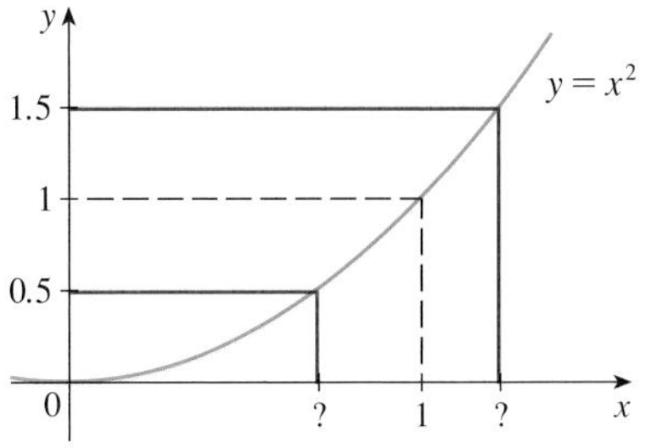

7. Use a graph to find a number δ such that

$$\text{if} \quad \left| x - \frac{\pi}{4} \right| < \delta \quad \text{then} \quad |\tan x - 1| < 0.2$$

8. Use a graph to find a number δ such that

$$\text{if} \quad |x - 1| < \delta \quad \text{then} \quad \left| \frac{2x}{x^2 + 4} - 0.4 \right| < 0.1$$

9. For the limit

$$\lim_{x \to 1} (4 + x - 3x^3) = 2$$

illustrate Definition 2 by finding values of δ that correspond to $\varepsilon = 1$ and $\varepsilon = 0.1$.

10. For the limit

$$\lim_{x \to 0} \frac{e^x - 1}{x} = 1$$

illustrate Definition 2 by finding values of δ that correspond to $\varepsilon = 0.5$ and $\varepsilon = 0.1$.

11. Use Definition 2 to prove that $\lim_{x \to 0} x^3 = 0$.

12. (a) How would you formulate an ε, δ definition of the one-sided limit $\lim_{x \to a^+} f(x) = L$?

(b) Use your definition in part (a) to prove that $\lim_{x \to 0^+} \sqrt{x} = 0$.

13. A machinist is required to manufacture a circular metal disk with area 1000 cm^2.

(a) What radius produces such a disk?

(b) If the machinist is allowed an error tolerance of ± 5 cm^2 in the area of the disk, how close to the ideal radius in part (a) must the machinist control the radius?

(c) In terms of the ε, δ definition of $\lim_{x \to a} f(x) = L$, what is x? What is $f(x)$? What is a? What is L? What value of ε is given? What is the corresponding value of δ?

14. A crystal growth furnace is used in research to determine how best to manufacture crystals used in electronic components for the space shuttle. For proper growth of the crystal, the temperature must be controlled accurately by adjusting the input power. Suppose the relationship is given by

$$T(w) = 0.1w^2 + 2.155w + 20$$

where T is the temperature in degrees Celsius and w is the power input in watts.

(a) How much power is needed to maintain the temperature at 200°C?

(b) If the temperature is allowed to vary from 200°C by up to ±1°C, what range of wattage is allowed for the input power?

(c) In terms of the ε, δ definition of $\lim_{x \to a} f(x) = L$, what is x? What is $f(x)$? What is a? What is L? What value of ε is given? What is the corresponding value of δ?

15. (a) Find a number δ such that if $|x - 2| < \delta$, then $|4x - 8| < \varepsilon$, where $\varepsilon = 0.1$.

(b) Repeat part (a) with $\varepsilon = 0.01$.

16. Given that $\lim_{x \to 2}(5x - 7) = 3$, illustrate Definition 2 by finding values of δ that correspond to $\varepsilon = 0.1$, $\varepsilon = 0.05$, and $\varepsilon = 0.01$.

17–18 Prove the statement using the ε, δ definition of a limit and illustrate with a diagram like Figure 7.

17. $\lim_{x \to -3} (1 - 4x) = 13$ **18.** $\lim_{x \to -2} \left(\frac{1}{2}x + 3\right) = 2$

19. Use Definition 1 to prove that

(a) $\lim_{n \to \infty} \dfrac{1}{2^n} = 0$

(b) $\lim_{n \to \infty} r^n = 0$ if $|r| < 1$

20. (a) For what values of x is it true that

$$\frac{1}{x^2} > 1{,}000{,}000$$

(b) The precise definition of $\lim_{x \to a} f(x) = \infty$ states that for every positive number M (no matter how large) there is a corresponding positive number δ such that if $0 < |x - a| < \delta$, then $f(x) > M$. Use this definition to prove that $\lim_{x \to 0}(1/x^2) = \infty$.

21. Use Definition 3 to prove that $\displaystyle\lim_{(x, y) \to (0, 0)} \frac{xy}{\sqrt{x^2 + y^2}} = 0.$

E | A Few Proofs

In this appendix we present proofs of some theorems that were stated in the main body of the text. We start by proving the Triangle Inequality, which is an important property of absolute value.

The Triangle Inequality If a and b are any real numbers, then

$$|a + b| \leqslant |a| + |b|$$

Observe that if the numbers a and b are both positive or both negative, then the two sides in the Triangle Inequality are actually equal. But if a and b have opposite signs, the left side involves a subtraction and the right side does not. This makes the Triangle Inequality seem reasonable, but we can prove it as follows.

Notice that

$$-|a| \leqslant a \leqslant |a|$$

is always true because a equals either $|a|$ or $-|a|$. The corresponding statement for b is

$$-|b| \leqslant b \leqslant |b|$$

Adding these inequalities, we get

$$-(|a| + |b|) \leqslant a + b \leqslant |a| + |b|$$

When combined, Properties 4 and 5 of absolute value (see Appendix A) say that

$$|x| \leqslant a \iff -a \leqslant x \leqslant a$$

If we now apply Properties 4 and 5 of absolute value from appendix A (with x replaced by $a + b$ and a by $|a| + |b|$), we obtain

$$|a + b| \leqslant |a| + |b|$$

which is what we wanted to show.

Next we use the Triangle Inequality to prove the Sum Law for limits.

The Sum Law was first stated in Section 2.4.

Sum Law If $\lim_{x\to a} f(x) = L$ and $\lim_{x\to a} g(x) = M$ both exist, then

$$\lim_{x\to a} [f(x) + g(x)] = L + M$$

PROOF Let $\varepsilon > 0$ be given. According to Definition 2 in Appendix D, we must find $\delta > 0$ such that

$$\text{if} \quad 0 < |x - a| < \delta \quad \text{then} \quad |f(x) + g(x) - (L + M)| < \varepsilon$$

Using the Triangle Inequality we can write

$$\textbf{(1)} \qquad \begin{aligned} |f(x) + g(x) - (L + M)| &= |(f(x) - L) + (g(x) - M)| \\ &\leqslant |f(x) - L| + |g(x) - M| \end{aligned}$$

We will make $|f(x) + g(x) - (L + M)|$ less than ε by making each of the terms $|f(x) - L|$ and $|g(x) - M|$ less than $\varepsilon/2$.

Since $\varepsilon/2 > 0$ and $\lim_{x\to a} f(x) = L$, there exists a number $\delta_1 > 0$ such that

$$\text{if} \quad 0 < |x - a| < \delta_1 \quad \text{then} \quad |f(x) - L| < \frac{\varepsilon}{2}$$

Similarly, since $\lim_{x\to a} g(x) = M$, there exists a number $\delta_2 > 0$ such that

$$\text{if} \quad 0 < |x - a| < \delta_2 \quad \text{then} \quad |g(x) - M| < \frac{\varepsilon}{2}$$

Let $\delta = \min\{\delta_1, \delta_2\}$, the smaller of the numbers δ_1 and δ_2. Notice that

$$\text{if} \quad 0 < |x - a| < \delta \quad \text{then} \quad 0 < |x - a| < \delta_1 \quad \text{and} \quad 0 < |x - a| < \delta_2$$

and so $$|f(x) - L| < \frac{\varepsilon}{2} \quad \text{and} \quad |g(x) - M| < \frac{\varepsilon}{2}$$

Therefore, by (1),

$$\begin{aligned} |f(x) + g(x) - (L + M)| &\leqslant |f(x) - L| + |g(x) - M| \\ &< \frac{\varepsilon}{2} + \frac{\varepsilon}{2} = \varepsilon \end{aligned}$$

To summarize,

$$\text{if} \quad 0 < |x - a| < \delta \quad \text{then} \quad |f(x) + g(x) - (L + M)| < \varepsilon$$

Thus, by the definition of a limit,

$$\lim_{x\to a} [f(x) + g(x)] = L + M$$ ■

Fermat's Theorem was discussed in Section 4.1.

Fermat's Theorem If f has a local maximum or minimum at c, and if $f'(c)$ exists, then $f'(c) = 0$.

PROOF Suppose, for the sake of definiteness, that f has a local maximum at c. Then, $f(c) \geqslant f(x)$ if x is sufficiently close to c. This implies that if h is sufficiently close to 0, with h being positive or negative, then

$$f(c) \geqslant f(c+h)$$

and therefore

$$f(c+h) - f(c) \leqslant 0 \tag{2}$$

We can divide both sides of an inequality by a positive number. Thus, if $h > 0$ and h is sufficiently small, we have

$$\frac{f(c+h) - f(c)}{h} \leqslant 0$$

Taking the right-hand limit of both sides of this inequality (using Theorem 2.4.2), we get

$$\lim_{h \to 0^+} \frac{f(c+h) - f(c)}{h} \leqslant \lim_{h \to 0^+} 0 = 0$$

But since $f'(c)$ exists, we have

$$f'(c) = \lim_{h \to 0} \frac{f(c+h) - f(c)}{h} = \lim_{h \to 0^+} \frac{f(c+h) - f(c)}{h}$$

and so we have shown that $f'(c) \leqslant 0$.

If $h < 0$, then the direction of the inequality (2) is reversed when we divide by h:

$$\frac{f(c+h) - f(c)}{h} \geqslant 0 \qquad h < 0$$

So, taking the left-hand limit, we have

$$f'(c) = \lim_{h \to 0} \frac{f(c+h) - f(c)}{h} = \lim_{h \to 0^-} \frac{f(c+h) - f(c)}{h} \geqslant 0$$

We have shown that $f'(c) \geqslant 0$ and also that $f'(c) \leqslant 0$. Since both of these inequalities must be true, the only possibility is that $f'(c) = 0$.

We have proved Fermat's Theorem for the case of a local maximum. The case of a local minimum can be proved in a similar manner. ■

This theorem was stated and used in Section 4.5.

(3) The Stability Criterion for Recursive Sequences Suppose that $\hat{x}$ is an equilibrium of the recursive sequence $x_{t+1} = f(x_t)$, where f' is continuous.

(a) If $|f'(\hat{x})| < 1$, the equilibrium is stable.

(b) If $|f'(\hat{x})| > 1$, the equilibrium is unstable.

PROOF OF PART (A) We want to show that if the sequence $\{x_t\}$ starts with x_0 sufficiently close to the equilibrium $\hat{x}$, then $x_t \to \hat{x}$ as $t \to \infty$.

By the Mean Value Theorem there is a number c between x_0 and $\hat{x}$ such that

$$f(x_0) - f(\hat{x}) = f'(c)(x_0 - \hat{x})$$

But $f(\hat{x}) = \hat{x}$, so we can write

$$x_1 - \hat{x} = f'(c)(x_0 - \hat{x})$$

Because f' is continuous and $|f'(\hat{x})| < 1$, we can guarantee that $|f'(x)| < 1$ when x is sufficiently close to $\hat{x}$. In fact there is a positive number α such that

$$|f'(x)| \leqslant \alpha < 1 \qquad \text{when } x \text{ is sufficiently close to } \hat{x}, \text{ say } |x - \hat{x}| < \delta$$

So if $|x_0 - \hat{x}| < \delta$, then

$$|x_1 - \hat{x}| = |f'(c)||x_0 - \hat{x}| \leqslant \alpha|x_0 - \hat{x}|$$

Using this argument t times, we get

$$|x_t - \hat{x}| \leqslant \alpha^t|x_0 - \hat{x}|$$

Because $0 < \alpha < 1$, we know that $\alpha^t \to 0$ as $t \to \infty$. Therefore $|x_t - \hat{x}| \to 0$ as $t \to \infty$. This means that $x_t \to \hat{x}$ as $t \to \infty$ and so $\hat{x}$ is a stable equilibrium. ■

Clairaut's Theorem was discussed in Section 9.2.

Clairaut's Theorem Suppose f is defined on a disk D that contains the point (a, b). If the functions f_{xy} and f_{yx} are both continuous on D, then

$$f_{xy}(a, b) = f_{yx}(a, b)$$

PROOF For small values of h, $h \neq 0$, consider the difference

$$\Delta(h) = [f(a + h, b + h) - f(a + h, b)] - [f(a, b + h) - f(a, b)]$$

Notice that if we let $g(x) = f(x, b + h) - f(x, b)$, then

$$\Delta(h) = g(a + h) - g(a)$$

By the Mean Value Theorem, there is a number c between a and $a + h$ such that

$$g(a + h) - g(a) = g'(c)h = h[f_x(c, b + h) - f_x(c, b)]$$

Applying the Mean Value Theorem again, this time to f_x, we get a number d between b and $b + h$ such that

$$f_x(c, b + h) - f_x(c, b) = f_{xy}(c, d)h$$

Combining these equations, we obtain

$$\Delta(h) = h^2 f_{xy}(c, d)$$

If $h \to 0$, then $(c, d) \to (a, b)$, so the continuity of f_{xy} at (a, b) gives

$$\lim_{h \to 0} \frac{\Delta(h)}{h^2} = \lim_{(c, d) \to (a, b)} f_{xy}(c, d) = f_{xy}(a, b)$$

Similarly, by writing

$$\Delta(h) = [f(a + h, b + h) - f(a, b + h)] - [f(a + h, b) - f(a, b)]$$

and using the Mean Value Theorem twice and the continuity of f_{yx} at (a, b), we obtain

$$\lim_{h \to 0} \frac{\Delta(h)}{h^2} = f_{yx}(a, b)$$

It follows that $f_{xy}(a, b) = f_{yx}(a, b)$. ■

The Second Derivatives Test was discussed in Section 9.6. Parts (b) and (c) have similar proofs.

> **Second Derivatives Test** Suppose the second partial derivatives of f are continuous on a disk with center (a, b), and suppose that $f_x(a, b) = 0$ and $f_y(a, b) = 0$ [that is, (a, b) is a critical point of f]. Let
>
> $$D = D(a, b) = f_{xx}(a, b)f_{yy}(a, b) - [f_{xy}(a, b)]^2$$
>
> (a) If $D > 0$ and $f_{xx}(a, b) > 0$, then $f(a, b)$ is a local minimum.
>
> (b) If $D > 0$ and $f_{xx}(a, b) < 0$, then $f(a, b)$ is a local maximum.
>
> (c) If $D < 0$, then $f(a, b)$ is not a local maximum or minimum.

PROOF OF PART (A) We compute the second-order directional derivative of f in the direction of $\mathbf{u} = [h, k]$. The first-order derivative is given by Theorem 9.5.3:

$$D_{\mathbf{u}} f = f_x h + f_y k$$

Applying this theorem a second time, we have

$$\begin{aligned} D_{\mathbf{u}}^2 f = D_{\mathbf{u}}(D_{\mathbf{u}} f) &= \frac{\partial}{\partial x}(D_{\mathbf{u}} f)h + \frac{\partial}{\partial y}(D_{\mathbf{u}} f)k \\ &= (f_{xx}h + f_{yx}k)h + (f_{xy}h + f_{yy}k)k \\ &= f_{xx}h^2 + 2f_{xy}hk + f_{yy}k^2 \qquad \text{(by Clairaut's Theorem)} \end{aligned}$$

If we complete the square in this expression, we obtain

(4) $$D_{\mathbf{u}}^2 f = f_{xx}\left(h + \frac{f_{xy}}{f_{xx}}k\right)^2 + \frac{k^2}{f_{xx}}(f_{xx}f_{yy} - f_{xy}^2)$$

We are given that $f_{xx}(a, b) > 0$ and $D(a, b) > 0$. But f_{xx} and $D = f_{xx}f_{yy} - f_{xy}^2$ are continuous functions, so there is a disk B with center (a, b) and radius $\delta > 0$ such that $f_{xx}(x, y) > 0$ and $D(x, y) > 0$ whenever (x, y) is in B. Therefore, by looking at Equation 4, we see that $D_{\mathbf{u}}^2 f(x, y) > 0$ whenever (x, y) is in B. This means that if C is the curve obtained by intersecting the graph of f with the vertical plane through $P(a, b, f(a, b))$ in the direction of $\mathbf{u}$, then C is concave upward on an interval of length 2δ. This is true in the direction of every vector $\mathbf{u}$, so if we restrict (x, y) to lie in B, the graph of f lies above its horizontal tangent plane at P. Thus $f(x, y) \geqslant f(a, b)$ whenever (x, y) is in B. This shows that $f(a, b)$ is a local minimum. ■

F | Sigma Notation

A convenient way of writing sums uses the Greek letter Σ (capital sigma, corresponding to our letter S) and is called **sigma notation**.

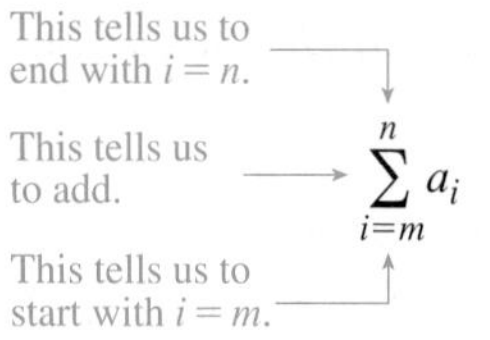
This tells us to end with $i = n$.
This tells us to add.
$\sum_{i=m}^{n} a_i$
This tells us to start with $i = m$.

> **(1) Definition** If $a_m, a_{m+1}, \ldots, a_n$ are real numbers and m and n are integers such that $m \leqslant n$, then
>
> $$\sum_{i=m}^{n} a_i = a_m + a_{m+1} + a_{m+2} + \cdots + a_{n-1} + a_n$$

With function notation, Definition 1 can be written as

$$\sum_{i=m}^{n} f(i) = f(m) + f(m+1) + f(m+2) + \cdots + f(n-1) + f(n)$$

Thus the symbol $\Sigma_{i=m}^{n}$ indicates a summation in which the letter i (called the **index of summation**) takes on consecutive integer values beginning with m and ending with n, that is, $m, m+1, \ldots, n$. Other letters can also be used as the index of summation.

EXAMPLE 1

(a) $\displaystyle\sum_{i=1}^{4} i^2 = 1^2 + 2^2 + 3^2 + 4^2 = 30$

(b) $\displaystyle\sum_{i=3}^{n} i = 3 + 4 + 5 + \cdots + (n-1) + n$

(c) $\displaystyle\sum_{j=0}^{5} 2^j = 2^0 + 2^1 + 2^2 + 2^3 + 2^4 + 2^5 = 63$

(d) $\displaystyle\sum_{k=1}^{n} \frac{1}{k} = 1 + \frac{1}{2} + \frac{1}{3} + \cdots + \frac{1}{n}$

(e) $\displaystyle\sum_{i=1}^{3} \frac{i-1}{i^2+3} = \frac{1-1}{1^2+3} + \frac{2-1}{2^2+3} + \frac{3-1}{3^2+3} = 0 + \frac{1}{7} + \frac{1}{6} = \frac{13}{42}$

(f) $\displaystyle\sum_{i=1}^{4} 2 = 2 + 2 + 2 + 2 = 8$ ■

EXAMPLE 2 | Write the sum $2^3 + 3^3 + \cdots + n^3$ in sigma notation.

SOLUTION There is no unique way of writing a sum in sigma notation. We could write

$$2^3 + 3^3 + \cdots + n^3 = \sum_{i=2}^{n} i^3$$

or

$$2^3 + 3^3 + \cdots + n^3 = \sum_{j=1}^{n-1} (j+1)^3$$

or

$$2^3 + 3^3 + \cdots + n^3 = \sum_{k=0}^{n-2} (k+2)^3$$ ■

The following theorem gives three simple rules for working with sigma notation.

(2) Theorem If c is any constant (that is, it does not depend on i), then

(a) $\displaystyle\sum_{i=m}^{n} ca_i = c\sum_{i=m}^{n} a_i$

(b) $\displaystyle\sum_{i=m}^{n} (a_i + b_i) = \sum_{i=m}^{n} a_i + \sum_{i=m}^{n} b_i$

(c) $\displaystyle\sum_{i=m}^{n} (a_i - b_i) = \sum_{i=m}^{n} a_i - \sum_{i=m}^{n} b_i$

PROOF To see why these rules are true, all we have to do is write both sides in expanded form. Rule (a) is just the distributive property of real numbers:

$$ca_m + ca_{m+1} + \cdots + ca_n = c(a_m + a_{m+1} + \cdots + a_n)$$

Rule (b) follows from the associative and commutative properties:

$$(a_m + b_m) + (a_{m+1} + b_{m+1}) + \cdots + (a_n + b_n)$$
$$= (a_m + a_{m+1} + \cdots + a_n) + (b_m + b_{m+1} + \cdots + b_n)$$

Rule (c) is proved similarly. ■

EXAMPLE 3 | Find $\sum_{i=1}^{n} 1$.

SOLUTION

$$\sum_{i=1}^{n} 1 = \underbrace{1 + 1 + \cdots + 1}_{n \text{ terms}} = n$$ ■

EXAMPLE 4 | Prove the formula for the sum of the first n positive integers:

$$\sum_{i=1}^{n} i = 1 + 2 + 3 + \cdots + n = \frac{n(n+1)}{2}$$

SOLUTION This formula can be proved by mathematical induction or by the following method used by the German mathematician Karl Friedrich Gauss (1777–1855) when he was ten years old.

Write the sum S twice, once in the usual order and once in reverse order:

$$S = 1 + 2 + 3 + \cdots + (n-1) + n$$
$$S = n + (n-1) + (n-2) + \cdots + 2 + 1$$

Adding all columns vertically, we get

$$2S = (n+1) + (n+1) + (n+1) + \cdots + (n+1) + (n+1)$$

On the right side there are n terms, each of which is $n + 1$, so

$$2S = n(n+1) \qquad \text{or} \qquad S = \frac{n(n+1)}{2}$$ ■

EXAMPLE 5 | Prove the formula for the sum of the squares of the first n positive integers:

$$\sum_{i=1}^{n} i^2 = 1^2 + 2^2 + 3^2 + \cdots + n^2 = \frac{n(n+1)(2n+1)}{6}$$

SOLUTION 1 Let S be the desired sum. We start with the *telescoping sum* (or collapsing sum):

Most terms cancel in pairs.

$$\sum_{i=1}^{n} [(1+i)^3 - i^3] = (2^3 - 1^3) + (3^3 - 2^3) + (4^3 - 3^3) + \cdots + [(n+1)^3 - n^3]$$
$$= (n+1)^3 - 1^3 = n^3 + 3n^2 + 3n$$

On the other hand, using Theorem 2 and Examples 3 and 4, we have

$$\sum_{i=1}^{n} [(1+i)^3 - i^3] = \sum_{i=1}^{n} [3i^2 + 3i + 1] = 3\sum_{i=1}^{n} i^2 + 3\sum_{i=1}^{n} i + \sum_{i=1}^{n} 1$$
$$= 3S + 3\,\frac{n(n+1)}{2} + n = 3S + \tfrac{3}{2}n^2 + \tfrac{5}{2}n$$

Thus we have

$$n^3 + 3n^2 + 3n = 3S + \tfrac{3}{2}n^2 + \tfrac{5}{2}n$$

Solving this equation for S, we obtain

$$3S = n^3 + \tfrac{3}{2}n^2 + \tfrac{1}{2}n$$

or

$$S = \frac{2n^3 + 3n^2 + n}{6} = \frac{n(n+1)(2n+1)}{6}$$

Principle of Mathematical Induction

Let S_n be a statement involving the positive integer n. Suppose that

1. S_1 is true.
2. If S_k is true, then S_{k+1} is true.

Then S_n is true for all positive integers n.

SOLUTION 2 Let S_n be the given formula.

1. S_1 is true because $\quad 1^2 = \dfrac{1(1+1)(2\cdot 1 + 1)}{6}$

2. Assume that S_k is true; that is,

$$1^2 + 2^2 + 3^2 + \cdots + k^2 = \frac{k(k+1)(2k+1)}{6}$$

Then

$$\begin{aligned}
1^2 + 2^2 + 3^2 + \cdots + (k+1)^2 &= (1^2 + 2^2 + 3^2 + \cdots + k^2) + (k+1)^2 \\
&= \frac{k(k+1)(2k+1)}{6} + (k+1)^2 \\
&= (k+1)\frac{k(2k+1) + 6(k+1)}{6} \\
&= (k+1)\frac{2k^2 + 7k + 6}{6} \\
&= \frac{(k+1)(k+2)(2k+3)}{6} \\
&= \frac{(k+1)[(k+1)+1][2(k+1)+1]}{6}
\end{aligned}$$

So S_{k+1} is true.

By the Principle of Mathematical Induction, S_n is true for all n. ■

We list the results of Examples 3, 4, and 5 together with a similar result for cubes (see Exercises 37–40) as Theorem 3. These formulas are needed for finding areas and evaluating integrals in Chapter 5.

(3) Theorem Let c be a constant and n a positive integer. Then

(a) $\displaystyle\sum_{i=1}^{n} 1 = n$

(b) $\displaystyle\sum_{i=1}^{n} c = nc$

(c) $\displaystyle\sum_{i=1}^{n} i = \frac{n(n+1)}{2}$

(d) $\displaystyle\sum_{i=1}^{n} i^2 = \frac{n(n+1)(2n+1)}{6}$

(e) $\displaystyle\sum_{i=1}^{n} i^3 = \left[\frac{n(n+1)}{2}\right]^2$

EXAMPLE 6 | Evaluate $\sum_{i=1}^{n} i(4i^2 - 3)$.

SOLUTION Using Theorems 2 and 3, we have

$$\sum_{i=1}^{n} i(4i^2 - 3) = \sum_{i=1}^{n} (4i^3 - 3i) = 4\sum_{i=1}^{n} i^3 - 3\sum_{i=1}^{n} i$$

$$= 4\left[\frac{n(n+1)}{2}\right]^2 - 3\,\frac{n(n+1)}{2}$$

$$= \frac{n(n+1)[2n(n+1) - 3]}{2}$$

$$= \frac{n(n+1)(2n^2 + 2n - 3)}{2}$$

■

The type of calculation in Example 7 arises in Chapter 5 when we compute areas.

EXAMPLE 7 | Find $\lim_{n\to\infty} \sum_{i=1}^{n} \frac{3}{n}\left[\left(\frac{i}{n}\right)^2 + 1\right]$.

SOLUTION

$$\lim_{n\to\infty} \sum_{i=1}^{n} \frac{3}{n}\left[\left(\frac{i}{n}\right)^2 + 1\right] = \lim_{n\to\infty} \sum_{i=1}^{n}\left[\frac{3}{n^3} i^2 + \frac{3}{n}\right]$$

$$= \lim_{n\to\infty}\left[\frac{3}{n^3}\sum_{i=1}^{n} i^2 + \frac{3}{n}\sum_{i=1}^{n} 1\right]$$

$$= \lim_{n\to\infty}\left[\frac{3}{n^3}\,\frac{n(n+1)(2n+1)}{6} + \frac{3}{n}\cdot n\right]$$

$$= \lim_{n\to\infty}\left[\frac{1}{2}\cdot\frac{n}{n}\cdot\left(\frac{n+1}{n}\right)\left(\frac{2n+1}{n}\right) + 3\right]$$

$$= \lim_{n\to\infty}\left[\frac{1}{2}\cdot 1\left(1 + \frac{1}{n}\right)\left(2 + \frac{1}{n}\right) + 3\right]$$

$$= \tfrac{1}{2}\cdot 1 \cdot 1 \cdot 2 + 3 = 4$$

■

EXERCISES F

1–10 Write the sum in expanded form.

1. $\sum_{i=1}^{5} \sqrt{i}$

2. $\sum_{i=1}^{6} \frac{1}{i+1}$

3. $\sum_{i=4}^{6} 3^i$

4. $\sum_{i=4}^{6} i^3$

5. $\sum_{k=0}^{4} \frac{2k-1}{2k+1}$

6. $\sum_{k=5}^{8} x^k$

7. $\sum_{i=1}^{n} i^{10}$

8. $\sum_{j=n}^{n+3} j^2$

9. $\sum_{j=0}^{n-1} (-1)^j$

10. $\sum_{i=1}^{n} f(x_i)\,\Delta x_i$

11–20 Write the sum in sigma notation.

11. $1 + 2 + 3 + 4 + \cdots + 10$

12. $\sqrt{3} + \sqrt{4} + \sqrt{5} + \sqrt{6} + \sqrt{7}$

13. $\frac{1}{2} + \frac{2}{3} + \frac{3}{4} + \frac{4}{5} + \cdots + \frac{19}{20}$

14. $\frac{3}{7} + \frac{4}{8} + \frac{5}{9} + \frac{6}{10} + \cdots + \frac{23}{27}$

15. $2 + 4 + 6 + 8 + \cdots + 2n$

16. $1 + 3 + 5 + 7 + \cdots + (2n - 1)$

17. $1 + 2 + 4 + 8 + 16 + 32$

18. $\frac{1}{1} + \frac{1}{4} + \frac{1}{9} + \frac{1}{16} + \frac{1}{25} + \frac{1}{36}$

19. $x + x^2 + x^3 + \cdots + x^n$

20. $1 - x + x^2 - x^3 + \cdots + (-1)^n x^n$

21–35 Find the value of the sum.

21. $\sum_{i=4}^{8} (3i - 2)$

22. $\sum_{i=3}^{6} i(i + 2)$

23. $\sum_{j=1}^{6} 3^{j+1}$

24. $\sum_{k=0}^{8} \cos k\pi$

25. $\sum_{n=1}^{20} (-1)^n$

26. $\sum_{i=1}^{100} 4$

27. $\sum_{i=0}^{4} (2^i + i^2)$

28. $\sum_{i=-2}^{4} 2^{3-i}$

29. $\sum_{i=1}^{n} 2i$

30. $\sum_{i=1}^{n} (2 - 5i)$

31. $\sum_{i=1}^{n} (i^2 + 3i + 4)$

32. $\sum_{i=1}^{n} (3 + 2i)^2$

33. $\sum_{i=1}^{n} (i + 1)(i + 2)$

34. $\sum_{i=1}^{n} i(i + 1)(i + 2)$

35. $\sum_{i=1}^{n} (i^3 - i - 2)$

---

36. Find the number n such that $\sum_{i=1}^{n} i = 78$.

37. Prove formula (b) of Theorem 3.

38. Prove formula (e) of Theorem 3 using mathematical induction.

39. Prove formula (e) of Theorem 3 using a method similar to that of Example 5, Solution 1 [start with $(1 + i)^4 - i^4$].

40. Prove formula (e) of Theorem 3 using the following method published by Abu Bekr Mohammed ibn Alhusain Alkarchi in about AD 1010. The figure shows a square $ABCD$ in which sides AB and AD have been divided into segments of lengths $1, 2, 3, \ldots, n$. Thus the side of the square has length $n(n + 1)/2$ so the area is $[n(n + 1)/2]^2$. But the area is also the sum of the areas of the n "gnomons" $G_1, G_2, \ldots, G_n$ shown in the figure. Show that the area of G_i is i^3 and conclude that formula (e) is true.

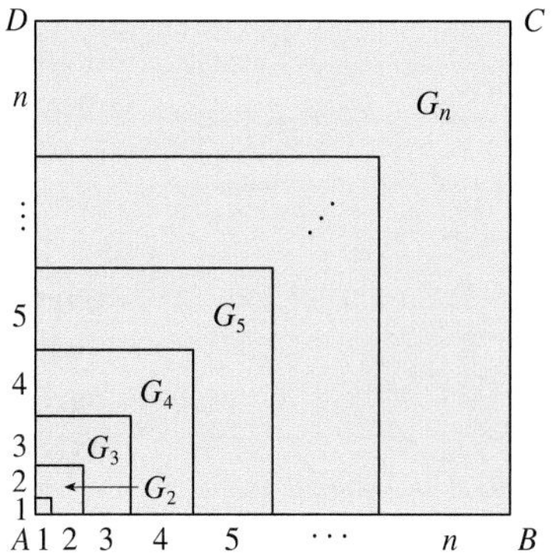

41. Evaluate each telescoping sum.

(a) $\sum_{i=1}^{n} [i^4 - (i - 1)^4]$

(b) $\sum_{i=1}^{100} (5^i - 5^{i-1})$

(c) $\sum_{i=3}^{99} \left(\frac{1}{i} - \frac{1}{i + 1}\right)$

(d) $\sum_{i=1}^{n} (a_i - a_{i-1})$

42. Prove the generalized triangle inequality:

$$\left|\sum_{i=1}^{n} a_i\right| \leqslant \sum_{i=1}^{n} |a_i|$$

43–46 Find the limit.

43. $\lim_{n\to\infty} \sum_{i=1}^{n} \frac{1}{n}\left(\frac{i}{n}\right)^2$

44. $\lim_{n\to\infty} \sum_{i=1}^{n} \frac{1}{n}\left[\left(\frac{i}{n}\right)^3 + 1\right]$

45. $\lim_{n\to\infty} \sum_{i=1}^{n} \frac{2}{n}\left[\left(\frac{2i}{n}\right)^3 + 5\left(\frac{2i}{n}\right)\right]$

46. $\lim_{n\to\infty} \sum_{i=1}^{n} \frac{3}{n}\left[\left(1 + \frac{3i}{n}\right)^3 - 2\left(1 + \frac{3i}{n}\right)\right]$

---

47. Prove the formula for the sum of a finite geometric series with first term a and common ratio $r \neq 1$:

$$\sum_{i=1}^{n} ar^{i-1} = a + ar + ar^2 + \cdots + ar^{n-1} = \frac{a(r^n - 1)}{r - 1}$$

48. Evaluate $\sum_{i=1}^{n} \frac{3}{2^{i-1}}$.

49. Evaluate $\sum_{i=1}^{n} (2i + 2^i)$.

50. Evaluate $\sum_{i=1}^{m} \left[\sum_{j=1}^{n} (i + j)\right]$.

G | Complex Numbers

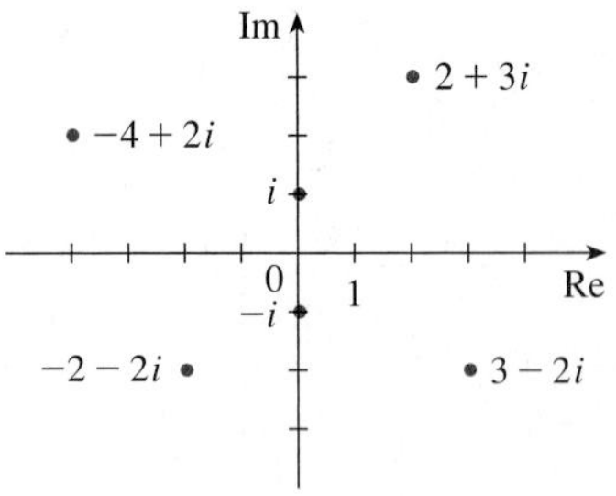

FIGURE 1
Complex numbers as points in the Argand plane

A **complex number** can be represented by an expression of the form $a + bi$, where a and b are real numbers and i is a symbol with the property that $i^2 = -1$. The complex number $a + bi$ can also be represented by the ordered pair (a, b) and plotted as a point in a plane (called the Argand plane) as in Figure 1. Thus the complex number $i = 0 + 1 \cdot i$ is identified with the point $(0, 1)$.

The **real part** of the complex number $a + bi$ is the real number a and the **imaginary part** is the real number b. Thus the real part of $4 - 3i$ is 4 and the imaginary part is -3. Two complex numbers $a + bi$ and $c + di$ are **equal** if $a = c$ and $b = d$, that is, their real parts are equal and their imaginary parts are equal. In the Argand plane the horizontal axis is called the real axis and the vertical axis is called the imaginary axis.

The sum and difference of two complex numbers are defined by adding or subtracting their real parts and their imaginary parts:

$$(a + bi) + (c + di) = (a + c) + (b + d)i$$

$$(a + bi) - (c + di) = (a - c) + (b - d)i$$

For instance,

$$(1 - i) + (4 + 7i) = (1 + 4) + (-1 + 7)i = 5 + 6i$$

The product of complex numbers is defined so that the usual commutative and distributive laws hold:

$$\begin{aligned}(a + bi)(c + di) &= a(c + di) + (bi)(c + di)\\ &= ac + adi + bci + bdi^2\end{aligned}$$

Since $i^2 = -1$, this becomes

$$(a + bi)(c + di) = (ac - bd) + (ad + bc)i$$

EXAMPLE 1

$$\begin{aligned}(-1 + 3i)(2 - 5i) &= (-1)(2 - 5i) + 3i(2 - 5i)\\ &= -2 + 5i + 6i - 15(-1) = 13 + 11i\end{aligned}$$ ■

Division of complex numbers is much like rationalizing the denominator of a rational expression. For the complex number $z = a + bi$, we define its **complex conjugate** to be $\bar{z} = a - bi$. To find the quotient of two complex numbers we multiply numerator and denominator by the complex conjugate of the denominator.

EXAMPLE 2 | Express the number $\dfrac{-1 + 3i}{2 + 5i}$ in the form $a + bi$.

SOLUTION We multiply numerator and denominator by the complex conjugate of $2 + 5i$, namely, $2 - 5i$, and we take advantage of the result of Example 1:

$$\frac{-1 + 3i}{2 + 5i} = \frac{-1 + 3i}{2 + 5i} \cdot \frac{2 - 5i}{2 - 5i} = \frac{13 + 11i}{2^2 + 5^2} = \frac{13}{29} + \frac{11}{29}i$$ ■

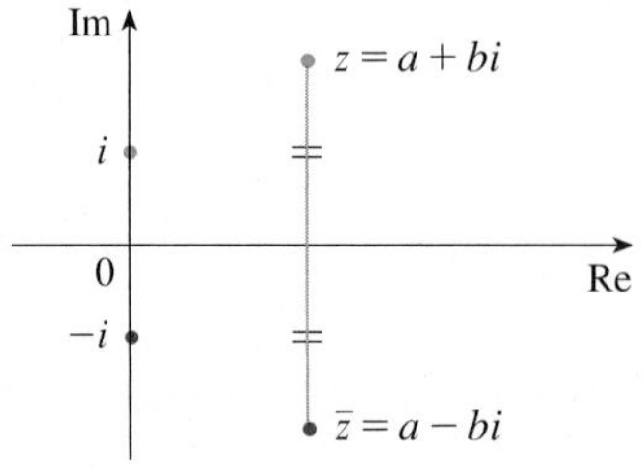

FIGURE 2

The geometric interpretation of the complex conjugate is shown in Figure 2: $\bar{z}$ is the reflection of z in the real axis. We list some of the properties of the complex conjugate in the following box. The proofs follow from the definition and are requested in Exercise 18.

Properties of Conjugates

$$\overline{z + w} = \bar{z} + \bar{w} \qquad \overline{zw} = \bar{z}\,\bar{w} \qquad \overline{z^n} = \bar{z}^n$$

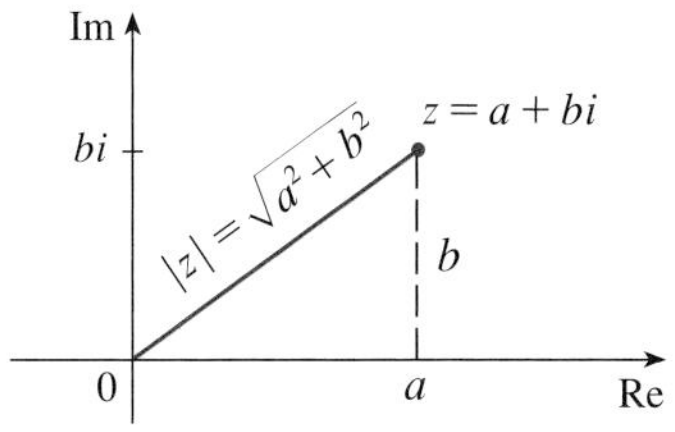

FIGURE 3

The **modulus**, or **absolute value**, $|z|$ of a complex number $z = a + bi$ is its distance from the origin. From Figure 3 we see that if $z = a + bi$, then

$$|z| = \sqrt{a^2 + b^2}$$

Notice that

$$z\bar{z} = (a + bi)(a - bi) = a^2 + abi - abi - b^2i^2 = a^2 + b^2$$

and so

$$z\bar{z} = |z|^2$$

This explains why the division procedure in Example 2 works in general:

$$\frac{z}{w} = \frac{z\bar{w}}{w\bar{w}} = \frac{z\bar{w}}{|w|^2}$$

Since $i^2 = -1$, we can think of i as a square root of -1. But notice that we also have $(-i)^2 = i^2 = -1$ and so $-i$ is also a square root of -1. We say that i is the **principal square root** of -1 and write $\sqrt{-1} = i$. In general, if c is any positive number, we write

$$\sqrt{-c} = \sqrt{c}\, i$$

With this convention, the usual derivation and formula for the roots of the quadratic equation $ax^2 + bx + c = 0$ are valid even when $b^2 - 4ac < 0$:

$$x = \frac{-b \pm \sqrt{b^2 - 4ac}}{2a}$$

EXAMPLE 3 | Find the roots of the equation $x^2 + x + 1 = 0$.

SOLUTION Using the quadratic formula, we have

$$x = \frac{-1 \pm \sqrt{1^2 - 4 \cdot 1}}{2} = \frac{-1 \pm \sqrt{-3}}{2} = \frac{-1 \pm \sqrt{3}\, i}{2}$$ ■

We observe that the solutions of the equation in Example 3 are complex conjugates of each other. In general, the solutions of any quadratic equation $ax^2 + bx + c = 0$ with real coefficients a, b, and c are always complex conjugates. (If z is real, $\bar{z} = z$, so z is its own conjugate.)

We have seen that if we allow complex numbers as solutions, then every quadratic equation has a solution. More generally, it is true that every polynomial equation

$$a_n x^n + a_{n-1} x^{n-1} + \cdots + a_1 x + a_0 = 0$$

of degree at least one has a solution among the complex numbers. This fact is known as the Fundamental Theorem of Algebra and was proved by Gauss.

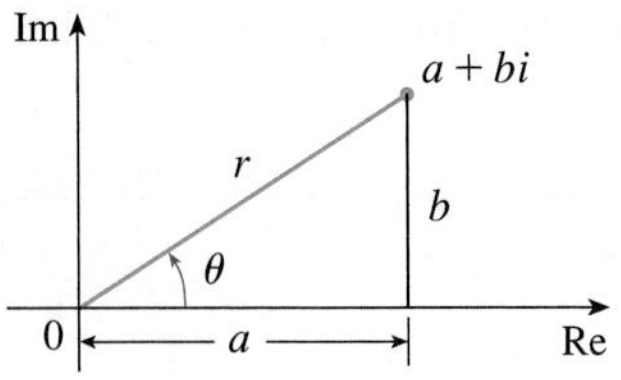

FIGURE 4

Polar Form

Any complex number $z = a + bi$ can be considered as a point (a, b) and any such point can be represented by **polar coordinates** (r, θ) with $r \geqslant 0$. In fact,

$$a = r\cos\theta \qquad b = r\sin\theta$$

as in Figure 4. Therefore we have

$$z = a + bi = (r\cos\theta) + (r\sin\theta)i$$

Thus we can write any complex number z in the form

$$\boxed{z = r(\cos\theta + i\sin\theta)}$$

where $\qquad r = |z| = \sqrt{a^2 + b^2} \qquad$ and $\qquad \tan\theta = \dfrac{b}{a}$

The angle θ is called the **argument** of z and we write $\theta = \arg(z)$. Note that $\arg(z)$ is not unique; any two arguments of z differ by an integer multiple of 2π.

EXAMPLE 4 | Write the following numbers in polar form.

(a) $z = 1 + i$ (b) $w = \sqrt{3} - i$

SOLUTION

(a) We have $r = |z| = \sqrt{1^2 + 1^2} = \sqrt{2}$ and $\tan\theta = 1$, so we can take $\theta = \pi/4$. Therefore the polar form is

$$z = \sqrt{2}\left(\cos\frac{\pi}{4} + i\sin\frac{\pi}{4}\right)$$

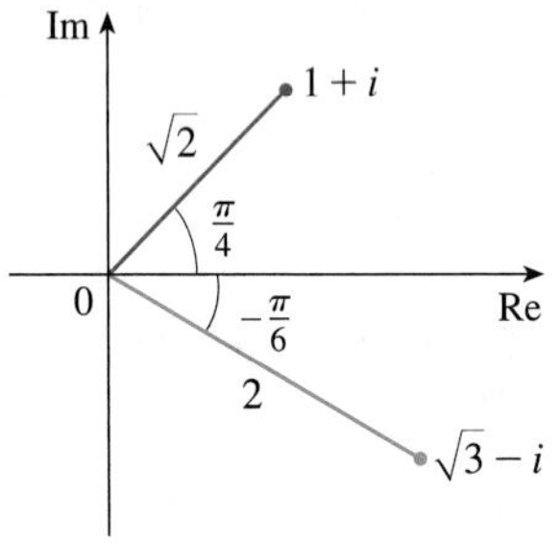

FIGURE 5

(b) Here we have $r = |w| = \sqrt{3 + 1} = 2$ and $\tan\theta = -1/\sqrt{3}$. Since w lies in the fourth quadrant, we take $\theta = -\pi/6$ and

$$w = 2\left[\cos\left(-\frac{\pi}{6}\right) + i\sin\left(-\frac{\pi}{6}\right)\right]$$

The numbers z and w are shown in Figure 5. ■

The polar form of complex numbers gives insight into multiplication and division. Let

$$z_1 = r_1(\cos\theta_1 + i\sin\theta_1) \qquad z_2 = r_2(\cos\theta_2 + i\sin\theta_2)$$

be two complex numbers written in polar form. Then

$$\begin{aligned} z_1 z_2 &= r_1 r_2(\cos\theta_1 + i\sin\theta_1)(\cos\theta_2 + i\sin\theta_2) \\ &= r_1 r_2[(\cos\theta_1\cos\theta_2 - \sin\theta_1\sin\theta_2) + i(\sin\theta_1\cos\theta_2 + \cos\theta_1\sin\theta_2)] \end{aligned}$$

Therefore, using the addition formulas for cosine and sine, we have

(1) $$\boxed{z_1 z_2 = r_1 r_2[\cos(\theta_1 + \theta_2) + i\sin(\theta_1 + \theta_2)]}$$

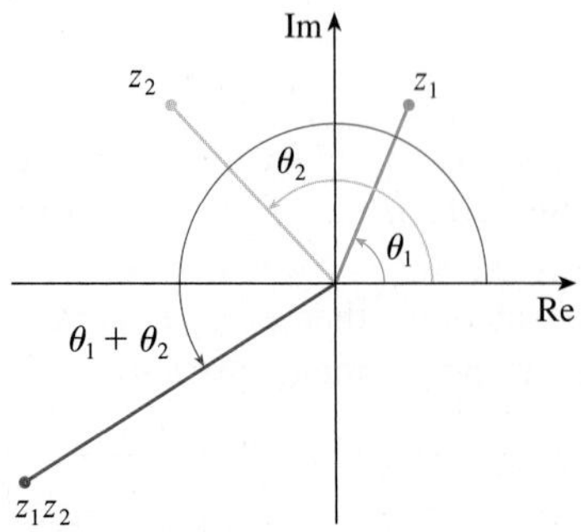

FIGURE 6

This formula says that to *multiply two complex numbers we multiply the moduli and add the arguments.* (See Figure 6.)

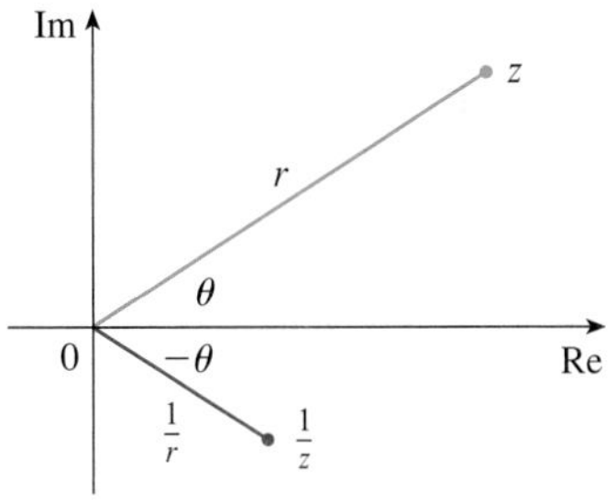

FIGURE 7

A similar argument using the subtraction formulas for sine and cosine shows that to *divide two complex numbers we divide the moduli and subtract the arguments.*

$$\frac{z_1}{z_2} = \frac{r_1}{r_2}[\cos(\theta_1 - \theta_2) + i\sin(\theta_1 - \theta_2)] \qquad z_2 \neq 0$$

In particular, taking $z_1 = 1$ and $z_2 = z$ (and therefore $\theta_1 = 0$ and $\theta_2 = \theta$), we have the following, which is illustrated in Figure 7.

$$\text{If} \quad z = r(\cos\theta + i\sin\theta), \quad \text{then} \quad \frac{1}{z} = \frac{1}{r}(\cos\theta - i\sin\theta).$$

EXAMPLE 5 | Find the product of the complex numbers $1 + i$ and $\sqrt{3} - i$ in polar form.

SOLUTION From Example 4 we have

$$1 + i = \sqrt{2}\left(\cos\frac{\pi}{4} + i\sin\frac{\pi}{4}\right)$$

and

$$\sqrt{3} - i = 2\left[\cos\left(-\frac{\pi}{6}\right) + i\sin\left(-\frac{\pi}{6}\right)\right]$$

So, by Equation 1,

$$(1 + i)(\sqrt{3} - i) = 2\sqrt{2}\left[\cos\left(\frac{\pi}{4} - \frac{\pi}{6}\right) + i\sin\left(\frac{\pi}{4} - \frac{\pi}{6}\right)\right]$$

$$= 2\sqrt{2}\left(\cos\frac{\pi}{12} + i\sin\frac{\pi}{12}\right)$$

This is illustrated in Figure 8. ■

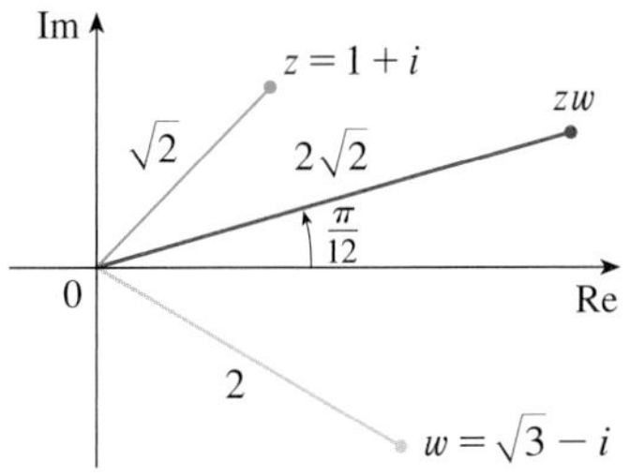

FIGURE 8

Repeated use of Formula 1 shows how to compute powers of a complex number. If

$$z = r(\cos\theta + i\sin\theta)$$

then

$$z^2 = r^2(\cos 2\theta + i\sin 2\theta)$$

and

$$z^3 = zz^2 = r^3(\cos 3\theta + i\sin 3\theta)$$

In general, we obtain the following result, which is named after the French mathematician Abraham De Moivre (1667–1754).

(2) De Moivre's Theorem If $z = r(\cos\theta + i\sin\theta)$ and n is a positive integer, then

$$z^n = [r(\cos\theta + i\sin\theta)]^n = r^n(\cos n\theta + i\sin n\theta)$$

This says that *to take the nth power of a complex number we take the nth power of the modulus and multiply the argument by n.*

EXAMPLE 6 | Find $\left(\frac{1}{2} + \frac{1}{2}i\right)^{10}$.

SOLUTION Since $\frac{1}{2} + \frac{1}{2}i = \frac{1}{2}(1 + i)$, it follows from Example 4(a) that $\frac{1}{2} + \frac{1}{2}i$ has the polar form

$$\frac{1}{2} + \frac{1}{2}i = \frac{\sqrt{2}}{2}\left(\cos\frac{\pi}{4} + i\sin\frac{\pi}{4}\right)$$

So by De Moivre's Theorem,

$$\left(\frac{1}{2} + \frac{1}{2}i\right)^{10} = \left(\frac{\sqrt{2}}{2}\right)^{10}\left(\cos\frac{10\pi}{4} + i\sin\frac{10\pi}{4}\right)$$

$$= \frac{2^5}{2^{10}}\left(\cos\frac{5\pi}{2} + i\sin\frac{5\pi}{2}\right) = \frac{1}{32}i$$

■

De Moivre's Theorem can also be used to find the nth roots of complex numbers. An nth root of the complex number z is a complex number w such that

$$w^n = z$$

Writing these two numbers in polar form as

$$w = s(\cos\phi + i\sin\phi) \qquad \text{and} \qquad z = r(\cos\theta + i\sin\theta)$$

and using De Moivre's Theorem, we get

$$s^n(\cos n\phi + i\sin n\phi) = r(\cos\theta + i\sin\theta)$$

The equality of these two complex numbers shows that

$$s^n = r \qquad \text{or} \qquad s = r^{1/n}$$

and $$\cos n\phi = \cos\theta \qquad \text{and} \qquad \sin n\phi = \sin\theta$$

From the fact that sine and cosine have period 2π, it follows that

$$n\phi = \theta + 2k\pi \qquad \text{or} \qquad \phi = \frac{\theta + 2k\pi}{n}$$

Thus $$w = r^{1/n}\left[\cos\left(\frac{\theta + 2k\pi}{n}\right) + i\sin\left(\frac{\theta + 2k\pi}{n}\right)\right]$$

Since this expression gives a different value of w for $k = 0, 1, 2, \ldots, n - 1$, we have the following.

(3) Roots of a Complex Number Let $z = r(\cos\theta + i\sin\theta)$ and let n be a positive integer. Then z has the n distinct nth roots

$$w_k = r^{1/n}\left[\cos\left(\frac{\theta + 2k\pi}{n}\right) + i\sin\left(\frac{\theta + 2k\pi}{n}\right)\right]$$

where $k = 0, 1, 2, \ldots, n - 1$.

Notice that each of the nth roots of z has modulus $|w_k| = r^{1/n}$. Thus all the nth roots of z lie on the circle of radius $r^{1/n}$ in the complex plane. Also, since the argument of each successive nth root exceeds the argument of the previous root by $2\pi/n$, we see that the nth roots of z are equally spaced on this circle.

EXAMPLE 7 | Find the six sixth roots of $z = -8$ and graph these roots in the complex plane.

SOLUTION In polar form, $z = 8(\cos \pi + i \sin \pi)$. Applying Equation 3 with $n = 6$, we get

$$w_k = 8^{1/6}\left(\cos \frac{\pi + 2k\pi}{6} + i \sin \frac{\pi + 2k\pi}{6}\right)$$

We get the six sixth roots of -8 by taking $k = 0, 1, 2, 3, 4, 5$ in this formula:

$$w_0 = 8^{1/6}\left(\cos \frac{\pi}{6} + i \sin \frac{\pi}{6}\right) = \sqrt{2}\left(\frac{\sqrt{3}}{2} + \frac{1}{2}i\right)$$

$$w_1 = 8^{1/6}\left(\cos \frac{\pi}{2} + i \sin \frac{\pi}{2}\right) = \sqrt{2}\, i$$

$$w_2 = 8^{1/6}\left(\cos \frac{5\pi}{6} + i \sin \frac{5\pi}{6}\right) = \sqrt{2}\left(-\frac{\sqrt{3}}{2} + \frac{1}{2}i\right)$$

$$w_3 = 8^{1/6}\left(\cos \frac{7\pi}{6} + i \sin \frac{7\pi}{6}\right) = \sqrt{2}\left(-\frac{\sqrt{3}}{2} - \frac{1}{2}i\right)$$

$$w_4 = 8^{1/6}\left(\cos \frac{3\pi}{2} + i \sin \frac{3\pi}{2}\right) = -\sqrt{2}\, i$$

$$w_5 = 8^{1/6}\left(\cos \frac{11\pi}{6} + i \sin \frac{11\pi}{6}\right) = \sqrt{2}\left(\frac{\sqrt{3}}{2} - \frac{1}{2}i\right)$$

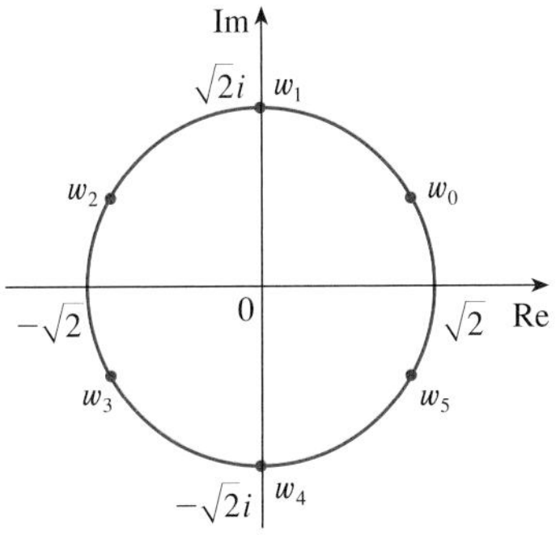

FIGURE 9
The six sixth roots of $z = -8$

All these points lie on the circle of radius $\sqrt{2}$ as shown in Figure 9. ■

■ Complex-Valued Functions and Their Derivatives

For a complex-valued function of one real variable t we can write

$$z(t) = x(t) + iy(t)$$

where $x(t)$ and $y(t)$ are the real and imaginary parts of $z(t)$. We can define its derivative $z'(t)$ by differentiating its real and imaginary parts:

$$z'(t) = x'(t) + iy'(t)$$

In particular, if we look at a complex number z on the unit circle in Figure 10 we can write it in polar form

$$z = \cos \theta + i \sin \theta$$

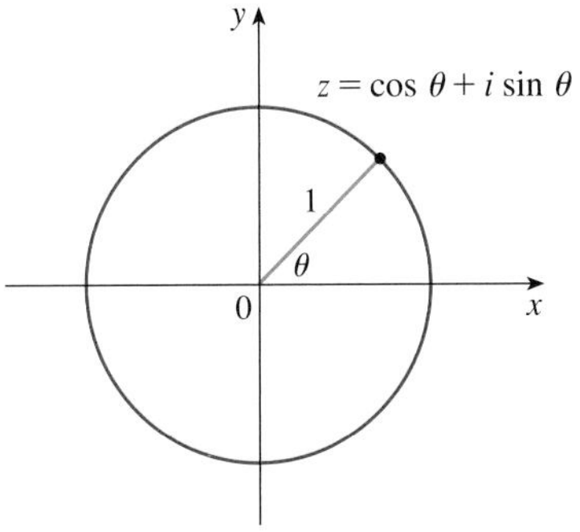

FIGURE 10

and consider the corresponding complex-valued function of the real variable θ

$$f(\theta) = \cos \theta + i \sin \theta$$

Then

$$f'(\theta) = \frac{d}{d\theta}(\cos \theta) + i \frac{d}{d\theta}(\sin \theta) = -\sin \theta + i \cos \theta$$

Because $i^2 = -1$, we could rewrite this as

$$\begin{aligned} f'(\theta) &= i\cos\theta + i^2\sin\theta \\ &= i(\cos\theta + i\sin\theta) = if(\theta) \end{aligned}$$

The equation $f'(\theta) = if(\theta)$ shows that the derivative of f is proportional to the function itself, the proportionality constant being the complex number i. In Section 7.4 we showed that the only solution of the differential equation $dy/dt = ky$ is the exponential function $y(t) = Ae^{kt}$. Assuming that this is also true for complex functions, we get

$$f'(\theta) = if(\theta) \qquad \Rightarrow \qquad f(\theta) = Ae^{i\theta}$$

where A is a constant. Putting $\theta = 0$, we get $f(0) = Ae^0 = A$. But

$$f(0) = \cos 0 + i\sin 0 = 1$$

and so $A = 1$ and therefore $f(\theta) = e^{i\theta}$. We have thus arrived at a famous formula:

(4) Euler's Formula

$$e^{i\theta} = \cos\theta + i\sin\theta$$

We could write the result of Example 8 as

$$e^{i\pi} + 1 = 0$$

This equation relates the five most famous numbers in all of mathematics: 0, 1, e, i, and π.

EXAMPLE 8 | Evaluate $e^{i\pi}$.

SOLUTION From Euler's Formula (4) with $\theta = \pi$, we get

$$e^{i\pi} = \cos\pi + i\sin\pi = -1 + i(0) = -1$$ ■

We could define the general complex exponential function e^z for any complex number $z = x + iy$ by writing

(5) $$e^z = e^{x+iy} = e^x e^{iy} = e^x(\cos y + i\sin y)$$

EXAMPLE 9 | Evaluate $e^{-1+i\pi/2}$.

SOLUTION Using Equation 5 we get

$$e^{-1+i\pi/2} = e^{-1}\left(\cos\frac{\pi}{2} + i\sin\frac{\pi}{2}\right) = \frac{1}{e}[0 + i(1)] = \frac{i}{e}$$ ■

Finally, we note that Euler's equation provides us with an easier method of proving De Moivre's Theorem:

$$[r(\cos\theta + i\sin\theta)]^n = (re^{i\theta})^n = r^n e^{in\theta} = r^n(\cos n\theta + i\sin n\theta)$$

EXERCISES G

1–14 Evaluate the expression and write your answer in the form $a + bi$.

1. $(5 - 6i) + (3 + 2i)$

2. $\left(4 - \frac{1}{2}i\right) - \left(9 + \frac{5}{2}i\right)$

3. $(2 + 5i)(4 - i)$

4. $(1 - 2i)(8 - 3i)$

5. $\overline{12 + 7i}$

6. $\overline{2i\left(\frac{1}{2} - i\right)}$

7. $\dfrac{1 + 4i}{3 + 2i}$

8. $\dfrac{3 + 2i}{1 - 4i}$

9. $\dfrac{1}{1 + i}$

10. $\dfrac{3}{4 - 3i}$

11. i^3

12. i^{100}

13. $\sqrt{-25}$

14. $\sqrt{-3}\sqrt{-12}$

15–17 Find the complex conjugate and the modulus of the number.

15. $12 - 5i$

16. $-1 + 2\sqrt{2}\,i$

17. $-4i$

18. Prove the following properties of complex numbers.

(a) $\overline{z + w} = \overline{z} + \overline{w}$ (b) $\overline{zw} = \overline{z}\,\overline{w}$

(c) $\overline{z^n} = \overline{z}^n$, where n is a positive integer

[*Hint:* Write $z = a + bi$, $w = c + di$.]

19–24 Find all solutions of the equation.

19. $4x^2 + 9 = 0$

20. $x^4 = 1$

21. $x^2 + 2x + 5 = 0$

22. $2x^2 - 2x + 1 = 0$

23. $z^2 + z + 2 = 0$

24. $z^2 + \frac{1}{2}z + \frac{1}{4} = 0$

25–28 Write the number in polar form with argument between 0 and 2π.

25. $-3 + 3i$

26. $1 - \sqrt{3}\,i$

27. $3 + 4i$

28. $8i$

29–32 Find polar forms for zw, z/w, and $1/z$ by first putting z and w into polar form.

29. $z = \sqrt{3} + i, \quad w = 1 + \sqrt{3}\,i$

30. $z = 4\sqrt{3} - 4i, \quad w = 8i$

31. $z = 2\sqrt{3} - 2i, \quad w = -1 + i$

32. $z = 4\left(\sqrt{3} + i\right), \quad w = -3 - 3i$

33–36 Find the indicated power using De Moivre's Theorem.

33. $(1 + i)^{20}$

34. $\left(1 - \sqrt{3}\,i\right)^5$

35. $\left(2\sqrt{3} + 2i\right)^5$

36. $(1 - i)^8$

37–40 Find the indicated roots. Sketch the roots in the complex plane.

37. The eighth roots of 1

38. The fifth roots of 32

39. The cube roots of i

40. The cube roots of $1 + i$

41–46 Write the number in the form $a + bi$.

41. $e^{i\pi/2}$

42. $e^{2\pi i}$

43. $e^{i\pi/3}$

44. $e^{-i\pi}$

45. $e^{2+i\pi}$

46. $e^{\pi+i}$

47. Use De Moivre's Theorem with $n = 3$ to express $\cos 3\theta$ and $\sin 3\theta$ in terms of $\cos\theta$ and $\sin\theta$.

48. Use Euler's formula to prove the following formulas for $\cos x$ and $\sin x$:

$$\cos x = \frac{e^{ix} + e^{-ix}}{2} \qquad \sin x = \frac{e^{ix} - e^{-ix}}{2i}$$

49. If $u(x) = f(x) + ig(x)$ is a complex-valued function of a real variable x and the real and imaginary parts $f(x)$ and $g(x)$ are differentiable functions of x, then the derivative of u is defined to be $u'(x) = f'(x) + ig'(x)$. Use this together with Equation 5 to prove that if $F(x) = e^{rx}$, then $F'(x) = re^{rx}$ when $r = a + bi$ is a complex number.

50. (a) If u is a complex-valued function of a real variable, its indefinite integral $\int u(x)\,dx$ is an antiderivative of u. Evaluate

$$\int e^{(1+i)x}\,dx$$

(b) By considering the real and imaginary parts of the integral in part (a), evaluate the real integrals

$$\int e^x \cos x\,dx \qquad \text{and} \qquad \int e^x \sin x\,dx$$

(c) Compare with the method used in Example 5.5.4.

H Statistical Tables

Table 1 $t_{C,n-1}$ values

d.f. $(n - 1)$	90%	95%	99%
1	6.31	12.71	63.66
2	2.92	4.30	9.92
3	2.35	3.18	5.84
4	2.13	2.78	4.60
5	2.02	2.57	4.03
6	1.94	2.45	3.71
7	1.89	2.36	3.50
8	1.86	2.31	3.36
9	1.83	2.26	3.25
10	1.81	2.23	3.17
15	1.75	2.13	2.95
20	1.72	2.09	2.85
25	1.71	2.06	2.79
30	1.70	2.04	2.75
40	1.68	2.02	2.70
50	1.68	2.01	2.68
∞	1.64	1.96	2.58

Table 2 *P*-Values for the *t*-Distribution

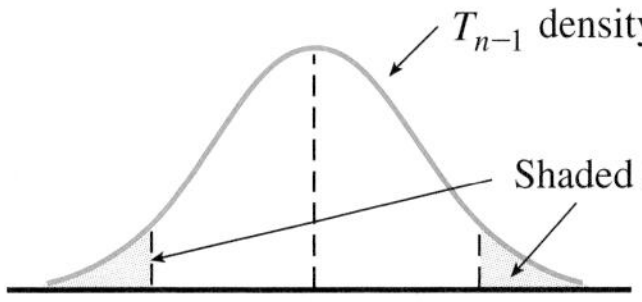

d.f. $(n-1)$	*P*-value .20	.10	.05	.01	.001
1	3.078	6.314	12.706	63.657	636.62
2	1.886	2.920	4.303	9.925	31.598
3	1.638	2.353	3.182	5.841	12.924
4	1.533	2.132	2.776	4.604	8.610
5	1.476	2.015	2.571	4.032	6.869
6	1.440	1.943	2.447	3.707	5.959
7	1.415	1.895	2.365	3.499	5.408
8	1.397	1.860	2.306	3.355	5.041
9	1.383	1.833	2.262	3.250	4.781
10	1.372	1.812	2.228	3.169	4.587
11	1.363	1.796	2.201	3.106	4.437
12	1.356	1.782	2.179	3.055	4.318
13	1.350	1.771	2.160	3.012	4.221
14	1.345	1.761	2.145	2.977	4.140
15	1.341	1.753	2.131	2.947	4.073
16	1.337	1.746	2.120	2.921	4.015
17	1.333	1.740	2.110	2.898	3.965
18	1.330	1.734	2.101	2.878	3.922
19	1.328	1.729	2.093	2.861	3.883
20	1.325	1.725	2.086	2.845	3.850
21	1.323	1.721	2.080	2.831	3.819
22	1.321	1.717	2.074	2.819	3.792
23	1.319	1.714	2.069	2.807	3.767
24	1.318	1.711	2.064	2.797	3.745
25	1.316	1.708	2.060	2.787	3.725
26	1.315	1.706	2.056	2.779	3.707
27	1.314	1.703	2.052	2.771	3.690
28	1.313	1.701	2.048	2.763	3.674
29	1.311	1.699	2.045	2.756	3.659
30	1.310	1.697	2.042	2.750	3.646
40	1.303	1.684	2.021	2.704	3.551
50	1.299	1.676	2.009	2.678	3.496
60	1.296	1.671	2.000	2.660	3.460
∞	1.282	1.645	1.960	2.576	3.291

Glossary of Biological Terms

agar a gelatinous substance made of seaweed used in the culturing of microbes

Allee effect a phenomenon by which the per capita growth rate of a population is negative for small population sizes (for example, because of difficulty finding mates for reproduction)

allele a particular variant of the DNA that is found at a specific gene; the terms *gene* and *allele* are often used interchangeably but technically it is not correct to do so

allometry the relationship between the size of one part of an organism and the size of the organism itself; allometric growth refers to this relationship during the growth of an individual

amino acid the basic chemical subunit that makes up proteins; there are 20 different possible amino acids that are specified by codons during the process of protein synthesis

anesthesiology a branch of medicine dealing with the administration of drugs and anesthetics for controlling pain and sensation, as well as other life sustaining factors, particularly in the context of surgery

antibiotic concentration profile a graph depicting the concentration of an antibiotic in the body as a function of the time since it was administered

antigen any foreign substance or molecule that stimulates an immune response specific to it; often a surface molecule or cellular component of a pathogen

antigenic cartography the construction of a plot in which the binding strength of a virus (or other pathogen) to each of a set of n immunity molecules is depicted as a point in n-dimensional space

antigenic evolution evolution of the antigenic properties of a pathogen; it allows the pathogen to escape any previous immune response and thereby to reproduce until an immune response specific to the new form of its antigen develops

antigenic space an n-dimensional space whose coordinate axes represent the binding strength to each of n different immunity molecules

antigenicity the capacity of a substance or molecule to induce an immune response

antiserum serum containing immune system molecules against specific antigens

biomechanics the study of animal form in terms of mechanical structure and physical composition, especially as it relates to movement

CT/CAT scan computerized tomography or computerized axial tomography; an X-ray technique used for generating cross-sectional images of the internal parts of a body

carrying capacity the population size at which the per capita growth rate of the population is zero

chemotaxis a phenomenon by which an organism moves in the direction of a chemical gradient

codon a specific sequence of three consecutive nucleotides that codes for a particular amino acid during the process of protein synthesis

compensatory growth a phenomenon by which an individual exhibits an increase in growth rate after a period of food deprivation so as to "catch up" to the size that is typical for their age

cross-feeding a process by which one species or genotype feeds on the waste products of another

diploid a cell or organism that has two copies of each gene

DNA acronym for DeoxyriboNucleic Acid; a double-stranded nucleic acid molecule that contains the genetic information (that is, the genetic code) for many living organisms.

dose response relationship (curve) the relationship between the concentration or amount of a drug and its effect

dynamics a broad term referring to the change in a variable of interest over time; for example, "the dynamics of N" refers to the way that N changes as a function of time

Einthoven's triangle an imaginary equilateral triangle superimposed on a person's chest, and used in vectorcardiography

electrocardiogram (ECG) a recording of the electrical activity of the heart over time

embolism a clot or blockage in a blood vessel

enzyme a type of molecule that catalyzes chemical reactions

fitness a measure of the survivorship and/or reproductive output of an organism

gait a pattern of animal locomotion such as walking, running, or galloping

gametocyte a specific kind of cell that is capable of producing either the male (sperm) or female (egg) cells required for sexual reproduction

gene the basic unit of heredity, typically identified as a specific location on the DNA; the terms *gene* and *allele* are often used interchangeably but technically it is not correct to do so

gene regulation a feedback process by which the expression of a gene is regulated by the products it generates

genome the entire set of genes of an organism

genome expression profile the measured levels of expression of a set of genes

genotype the collection of specific alleles within an organism

haploid a cell or organism that has one copy of each gene

Hardy-Weinberg law the statement that allele frequencies in a population remain constant over time in the absence of evolutionary processes such as natural selection or mutation

homeostatis the tendency of a biological system to maintain itself in a relatively stable state

inbreeding the interbreeding of genetically similar individuals

incidence in epidemiology, the rate at which individuals are getting a disease

infectious the state in which an infected individual is able to transmit a disease

island biogeography the study of the geographic distribution of organisms, particularly the species richness of isolated areas such as caves, islands, or lakes

kill curve a plot of the size of a bacterial population as a function of time after a dose of antibiotic is given

merozoite in malaria, a specific kind of cell formed during the stage of the life cycle when asexual cell division occurs

metamorphosis a large-scale change in the form and structure of an organism's body (and often its way of life) during development

metapopulation a group of spatially separated but interacting subpopulations

methylation (DNA) the attachment of chemical methyl groups to DNA

microarray a grid of microscopic gene fragments attached to a solid surface, and used to quantify the expression level of different genes

morphology a term referring to the form or structure of an organism

mRNA acronym for messenger RiboNucleic Acid; a type of RNA that serves as an intermediate molecule in the process of protein synthesis

mutate to change from one allele to another

mutation a term used to describe a new type of allele that arises when the "normal" allele mutates

neuron an electrically excitable cell that conducts electrical impulses (called nerve impulses) from one location in an organism to another

nucleotide a single letter of the genetic code that makes up the chemical structure of DNA and RNA

parasite an organism that obtains resources by living in or on another host organism (typically to the detriment of the host); the terms parasite and pathogen are sometimes used interchangeably

pathogen an organism that is capable of causing disease or illness in its host organism; the terms pathogen and parasite are sometimes used interchangeably

pathogenesis the development of a disease within an individual over time

per capita growth rate (per unit rate) the rate at which a population is growing divided by the size of the population at that time

pharmacokinetics the change in drug concentration within a patient during treatment

phenotype the observable properties of an organism in contrast to genotype, which is the collection of specific alleles within an organism and cannot be directly observed. An organism's genotype, in part, determines the organism's phenotype

photosynthesis the synthesis of organic carbon compounds from carbon dioxide, water, and sunlight

population genetics the study of how allele frequencies in a population change

prevalence in epidemiology, the proportion (or sometimes number) of individuals having a disease

renewable resource a biological resource that is self-reproducing, such as a forest or a fish population (as compared with nonrenewable resources such as oil)

reproduction number a term used in the study of infectious diseases; it is the number of new infections generated by a single infected individual when introduced into a population entirely susceptible to infection

RNA acronym for RiboNucleic Acid; a single-stranded nucleic acid molecule that plays an important role in translating the genetic code of DNA into proteins and also serves as the molecule of inheritance for some viruses

serum the clear portion of blood

sporozoite in malaria, a specific kind of cell involved in the stage of the life cycle when sexual cell division occurs; it is the cell type that is introduced into the human bloodstream by mosquitoes

substrate a substance upon which a chemical reaction occurs (often as a result of binding with an enzyme)

transcription an initial step in protein synthesis whereby the DNA sequence of a gene is transcribed into mRNA

translation a secondary step in protein synthesis whereby mRNA is read as a series of codons, each of which specifies an amino acid that goes into the synthesis of the protein

urea a chemical waste product produced by many organisms

vaccination a form of preventative medicine in which an individual is given an inactive form of a pathogen, called a vaccine, to stimulate an immune response that protects against subsequent infection

vaccine escape the antigenic evolution of a virus to the point where a vaccine no longer provides protection against it

vectorcardiography a method that records the direction and magnitude of electrical impulses generated by the heart

virulence a term referring to the severity, in terms of an individual's health, of an infectious disease

zygote the cell formed by the fusion of a male (sperm) and female (egg) cell during sexual reproduction

CHAPTER 1

EXERCISES 1.1 | page 13

1. Yes
3. (a) 3 (b) -0.2 (c) 0, 3 (d) -0.8
(e) $[-2, 4], [-1, 3]$ (f) $[-2, 1]$
5. No **7.** Yes, $[-3, 2], [-3, -2) \cup [-1, 3]$
9. (a) $\approx 13.8°C$ (b) ≈ 1992 (c) 1910; 2006
(d) $[13.5, 14.5]$
11. [12:23 AM, 12:52 AM]
13. (a) $\approx 100, 134$ (b) 30°N or 30°S (c) Even function
15. Diet, exercise, or illness
17.
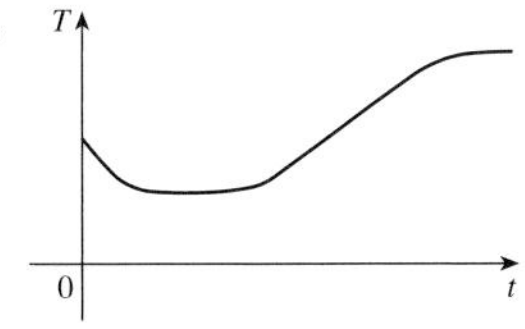

19. Nutrient consumption; growth; carrying capacity reached; death
21.
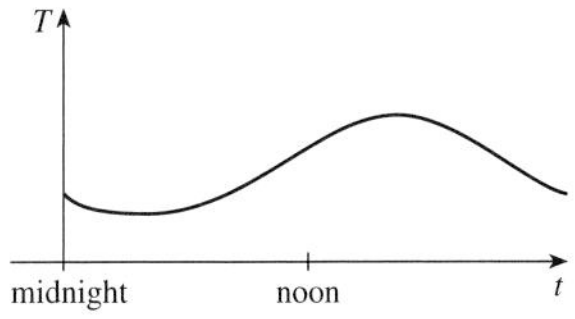

23.
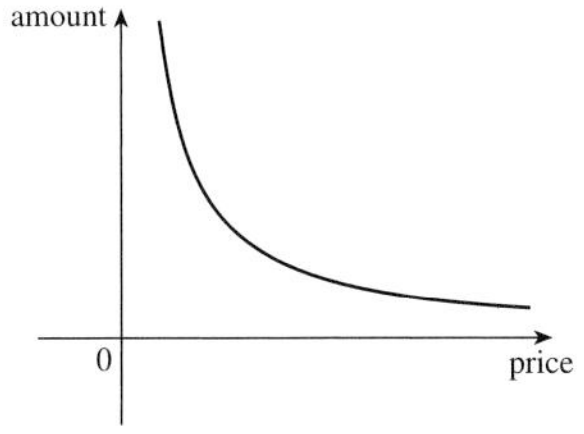

25. (a)
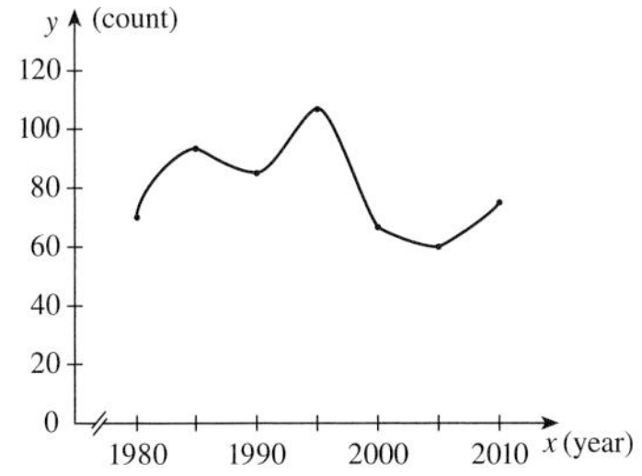

(b) $\approx 92{,}000$

27. $12, 16, 3a^2 - a + 2, 3a^2 + a + 2, 3a^2 + 5a + 4,$
$6a^2 - 2a + 4, 12a^2 - 2a + 2, 3a^4 - a^2 + 2,$
$9a^4 - 6a^3 + 13a^2 - 4a + 4, 3a^2 + 6ah + 3h^2 - a - h + 2$
29. $-3 - h$ **31.** $-1/(ax)$
33. $(-\infty, -3) \cup (-3, 3) \cup (3, \infty)$ **35.** $(-\infty, \infty)$
37. $(-\infty, 0) \cup (5, \infty)$ **39.** $[0, 4]$
41. $(-\infty, \infty)$
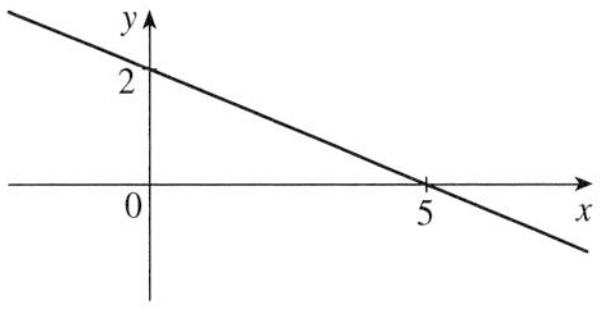

43. $(-\infty, \infty)$
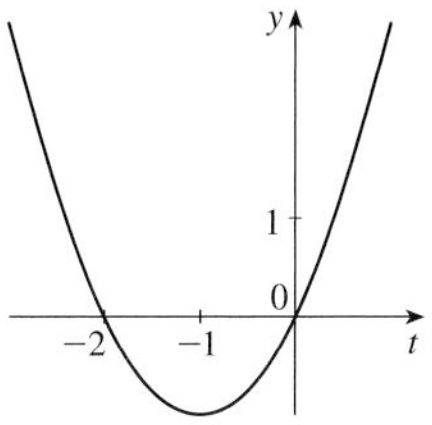

45. $[5, \infty)$
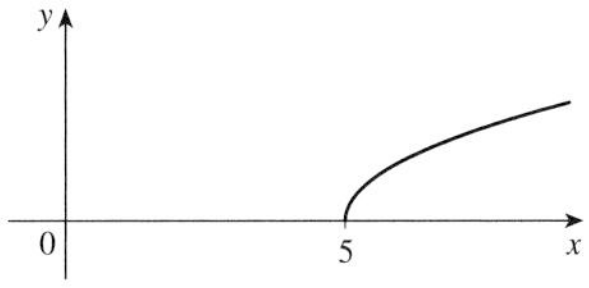

47. $(-\infty, 0) \cup (0, \infty)$
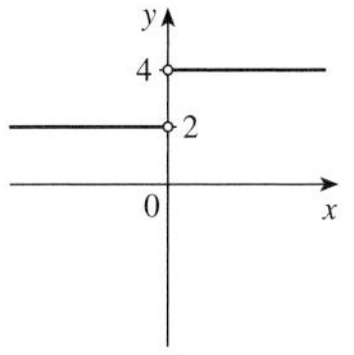

49. $(-\infty, \infty)$
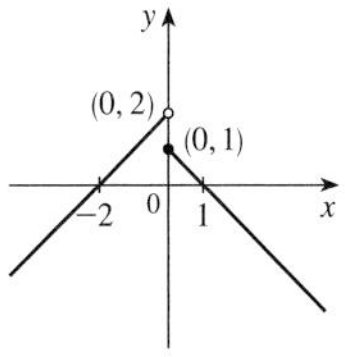

51. $(-\infty, \infty)$
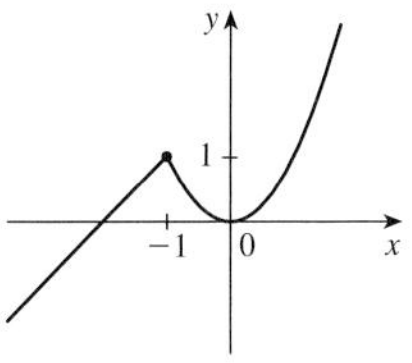

53. $A(L) = 10L - L^2, 0 < L < 10$
55. $A(x) = \sqrt{3}x^2/4, x > 0$ **57.** $S(x) = x^2 + (8/x), x > 0$
59. $T(x) = \begin{cases} 75x & \text{if } 0 < x \leq 2 \\ 150 + 50(x - 2) & \text{if } x > 2 \end{cases}$
61. ≈ 77 hours
63. f is odd, g is even **65.** (a) $(-5, 3)$ (b) $(-5, -3)$
67. Odd **69.** Neither **71.** Even
73. Even; odd; neither (unless $f = 0$ or $g = 0$)

EXERCISES 1.2 | page 28

1. (a) Logarithmic (b) Root (c) Rational
(d) Polynomial, degree 2 (e) Exponential (f) Trigonometric
3. (a) h (b) f (c) g

5. (a) $y = 2x + b$, where b is the y-intercept.

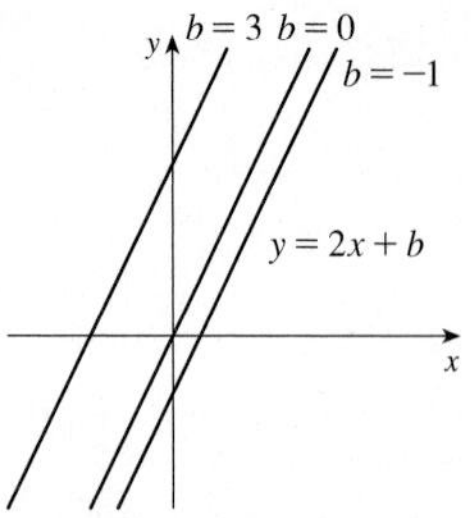

(b) $y = mx + 1 - 2m$, where m is the slope.
(c) $y = 2x - 3$

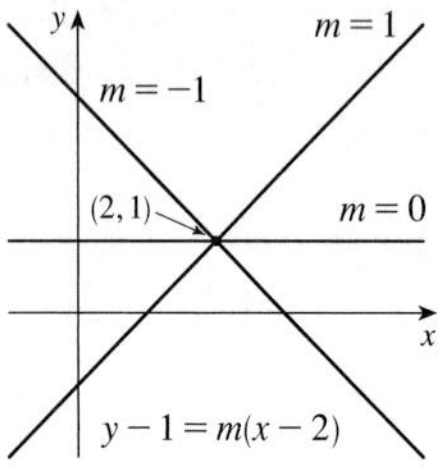

7. Their graphs have slope -1.

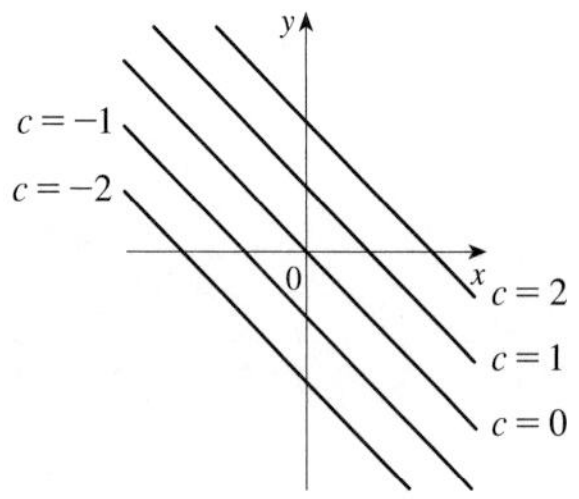

9. $f(x) = -3x(x + 1)(x - 2)$

11. (a) 8.34, change in mg for every 1 year change
(b) 8.34 mg

13. (a)

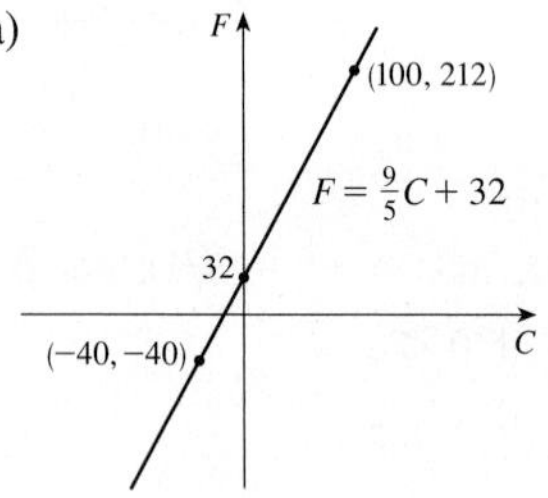

(b) $\frac{9}{5}$, change in °F for every 1°C change; 32, Fahrenheit temperature corresponding to 0°C

15. (a) $T = \frac{1}{6}N + \frac{307}{6}$ (b) $\frac{1}{6}$, change in °F for every chirp per minute change (c) 76°F

17. (a) Cosine (b) Linear

19. (a)

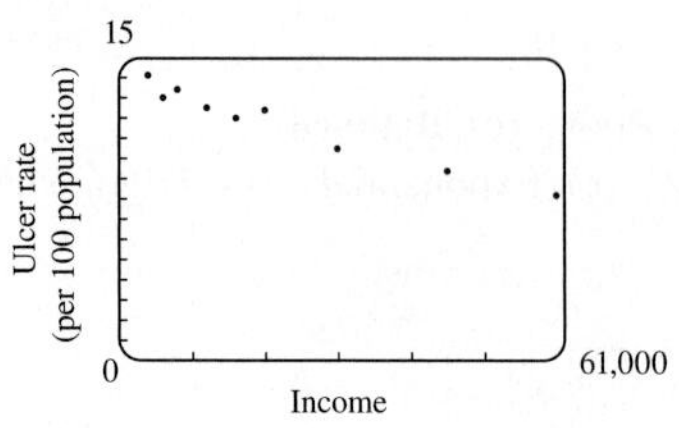

Linear model is appropriate.

(b) $y = -0.000105x + 14.521$

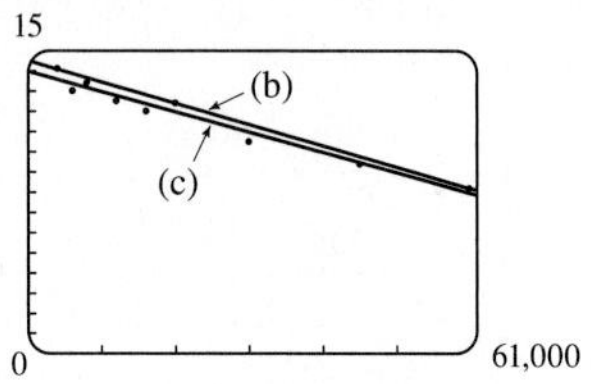

(c) $y = -0.00009979x + 13.951$ [See graph in part (b).]
(d) About 11.5 per 100 population (e) About 6% (f) No

21. (a)

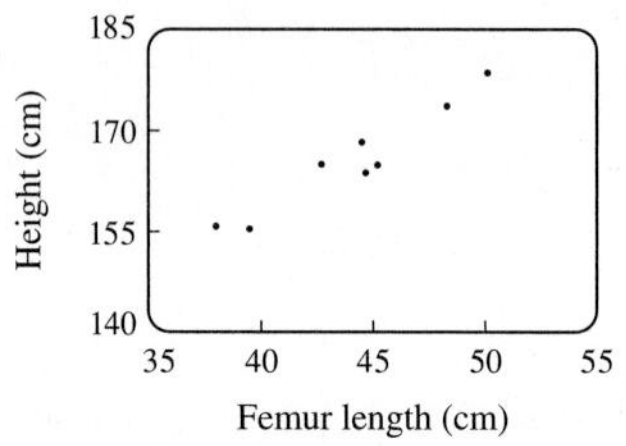

(b) $H = 1.8807L + 82.6497$, where H is the height in centimeters and L is the femur length in centimeters

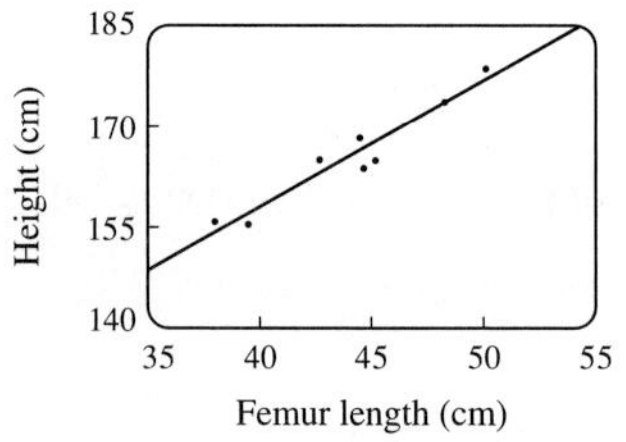

(c) $\approx$182.3 cm

23. Four times as bright

25. (a) $N = 3.1046A^{0.308}$ (b) 18

27. (a) $L = 0.0155A^3 - 0.3725A^2 + 3.9461A + 1.2108$

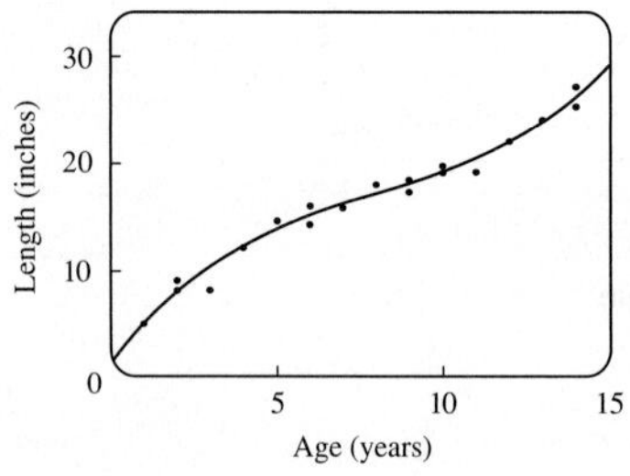

(b) $\approx$13.6 in (c) $\approx$10.88 years

EXERCISES 1.3 | page 38

1. (a) $y = f(x) + 3$ (b) $y = f(x) - 3$ (c) $y = f(x - 3)$
(d) $y = f(x + 3)$ (e) $y = -f(x)$ (f) $y = f(-x)$
(g) $y = 3f(x)$ (h) $y = \frac{1}{3}f(x)$

3. (a) 3 (b) 1 (c) 4 (d) 5 (e) 2

5. (a)

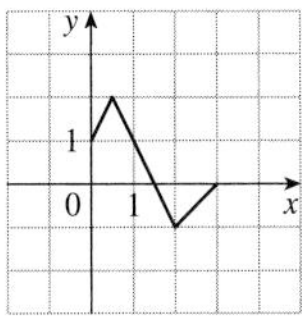

(b)

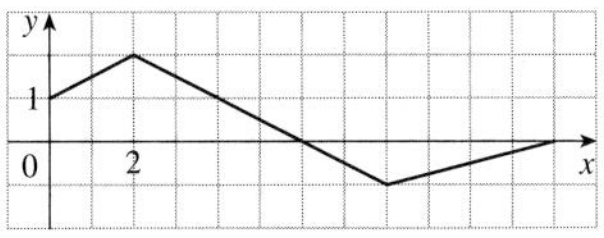

(c)

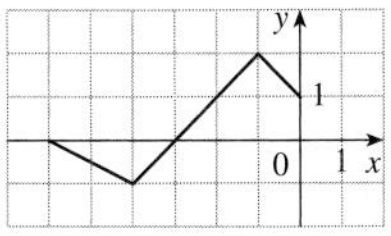

(d)

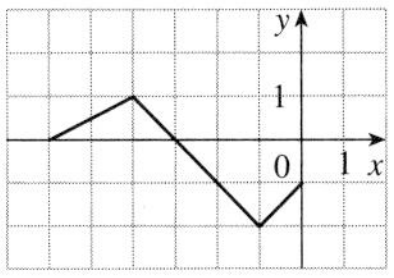

7.

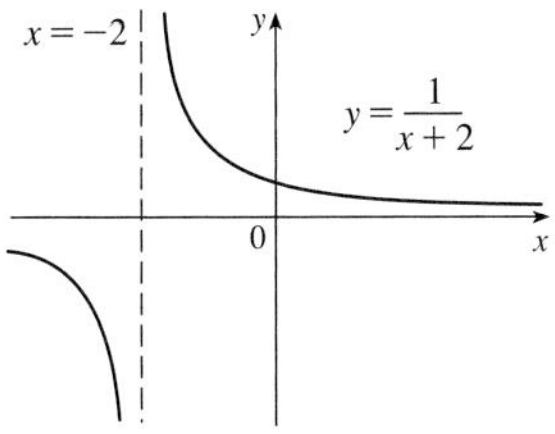

9.

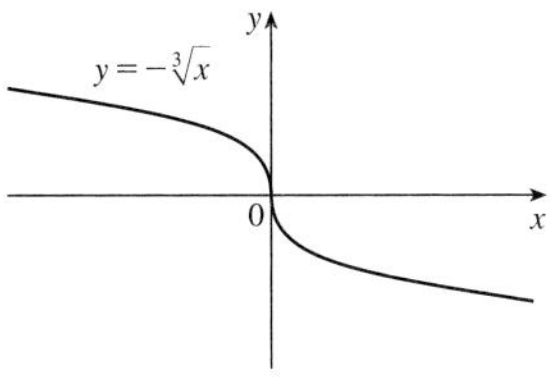

11.

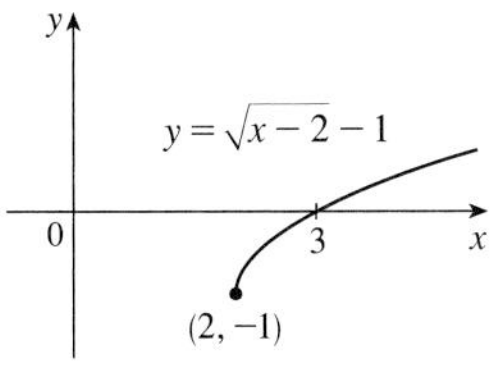

13.

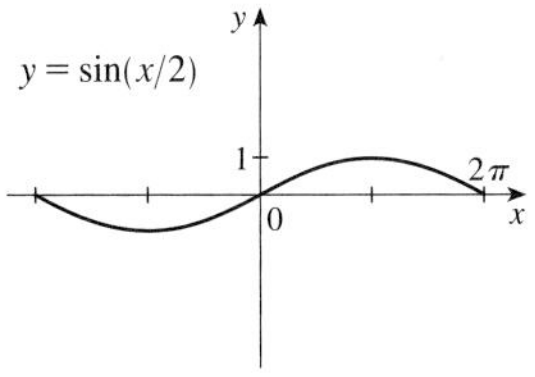

15.

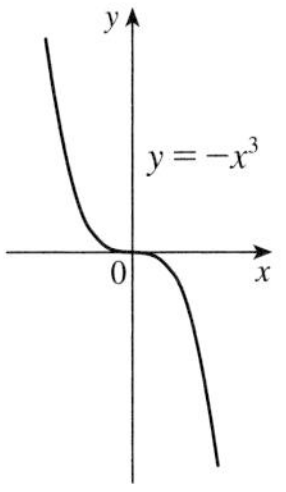

17.

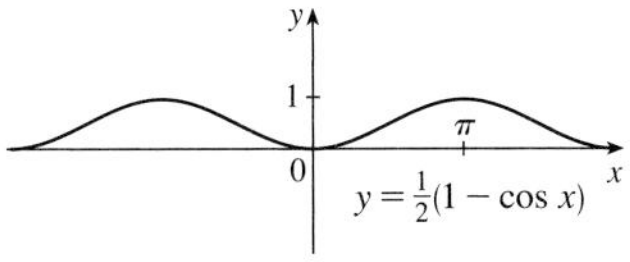

19.

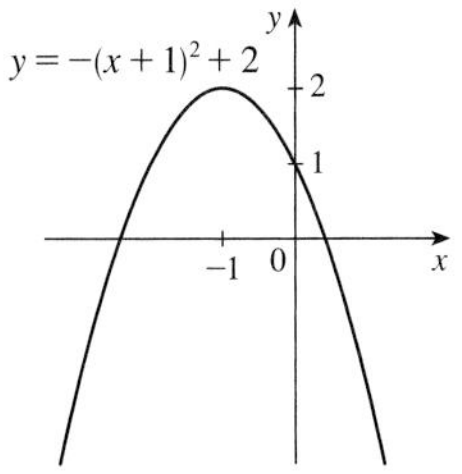

21. $L(t) = 12 + 2\sin\left[\dfrac{2\pi}{365}(t - 80)\right]$

23. $D(t) = 5\cos\left(\dfrac{\pi}{6}(t - 6.75)\right) + 7$

25. $f(t) = 30\sin\left(\dfrac{2\pi}{3}t\right) + 50$

27. (a) $(f + g)(x) = x^3 + 5x^2 - 1, (-\infty, \infty)$
(b) $(f - g)(x) = x^3 - x^2 + 1, (-\infty, \infty)$
(c) $(fg)(x) = 3x^5 + 6x^4 - x^3 - 2x^2, (-\infty, \infty)$
(d) $(f/g)(x) = \dfrac{x^3 + 2x^2}{3x^2 - 1}, \{x \mid x \neq \pm 1/\sqrt{3}\}$

29. (a) $(f \circ g)(x) = 4x^2 + 4x, (-\infty, \infty)$
(b) $(g \circ f)(x) = 2x^2 - 1, (-\infty, \infty)$
(c) $(f \circ f)(x) = x^4 - 2x^2, (-\infty, \infty)$
(d) $(g \circ g)(x) = 4x + 3, (-\infty, \infty)$

31. (a) $(f \circ g)(x) = 1 - 3\cos x, (-\infty, \infty)$
(b) $(g \circ f)(x) = \cos(1 - 3x), (-\infty, \infty)$
(c) $(f \circ f)(x) = 9x - 2, (-\infty, \infty)$
(d) $(g \circ g)(x) = \cos(\cos x), (-\infty, \infty)$

33. (a) $(f \circ g)(x) = \dfrac{2x^2 + 6x + 5}{(x + 2)(x + 1)}, \{x \mid x \neq -2, -1\}$
(b) $(g \circ f)(x) = \dfrac{x^2 + x + 1}{(x + 1)^2}, \{x \mid x \neq -1, 0\}$
(c) $(f \circ f)(x) = \dfrac{x^4 + 3x^2 + 1}{x(x^2 + 1)}, \{x \mid x \neq 0\}$
(d) $(g \circ g)(x) = \dfrac{2x + 3}{3x + 5}, \{x \mid x \neq -2, -\frac{5}{3}\}$

35. $(f \circ g \circ h)(x) = 3\sin(x^2) - 2$

37. $(f \circ g \circ h)(x) = \sqrt{x^6 + 4x^3 + 1}$

39. $g(x) = 2x + x^2, f(x) = x^4$

41. $g(x) = \sqrt[3]{x}, f(x) = x/(1 + x)$

43. $g(t) = t^2, f(t) = \sec t \tan t$

45. $h(x) = \sqrt{x}, g(x) = x - 1, f(x) = \sqrt{x}$

47. $h(x) = \sqrt{x}, g(x) = \sec x, f(x) = x^4$

49. (a) 4 (b) 3 (c) 0 (d) Does not exist; $f(6) = 6$ is not in the domain of g. (e) 4 (f) −2

51. (a) $r(t) = 60t$ (b) $(A \circ r)(t) = 3600\pi t^2$; the area of the circle as a function of time

53. (a) $s = \sqrt{d^2 + 36}$ (b) $d = 30t$
(c) $(f \circ g)(t) = \sqrt{900t^2 + 36}$; the distance between the lighthouse and the ship as a function of the time elapsed since noon

55. (a) $d(t) = gt$ (in mm)
(b) $(P \circ S \circ d)(t) = kg^2\pi t^2$; the rate of enzyme production as a function of time

57. Yes; $m_1 m_2$

59. No; h is odd; h is even

EXERCISES 1.4 | page 50

1. (a) 4 (b) $x^{-4/3}$ **3.** (a) $16b^{12}$ (b) $648y^7$

5. (a) $f(x) = b^x, b > 0$ (b) $\mathbb{R}$ (c) $(0, \infty)$
(d) See Figures 5(c), 5(b), and 5(a), respectively.

7.

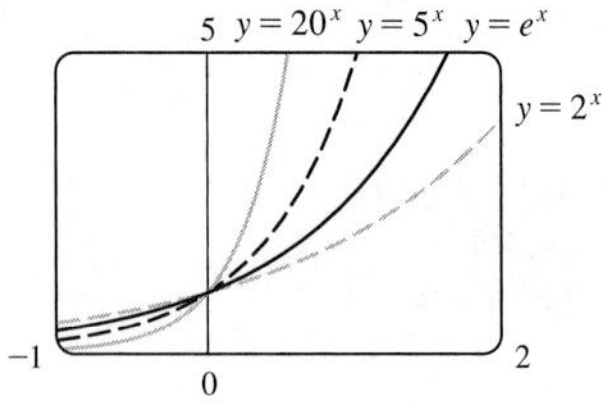

All approach 0 as $x \to -\infty$, all pass through (0, 1), and all are increasing. The larger the base, the faster the rate of increase.

9.

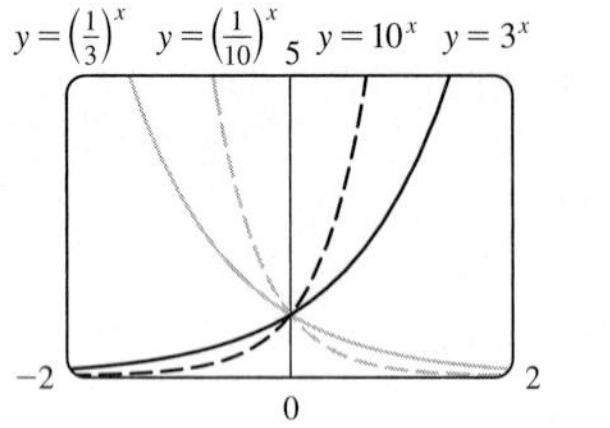

The functions with base greater than 1 are increasing and those with base less than 1 are decreasing. The latter are reflections of the former about the y-axis.

11.

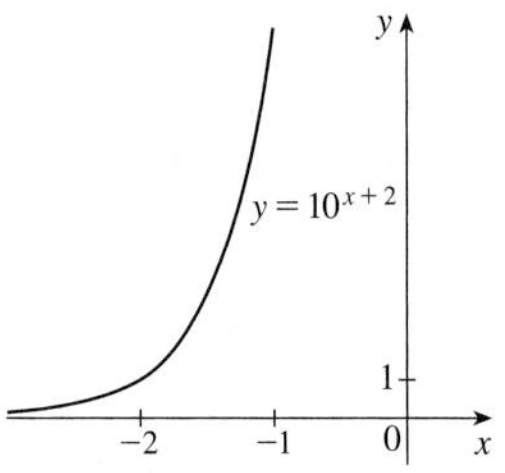

13.

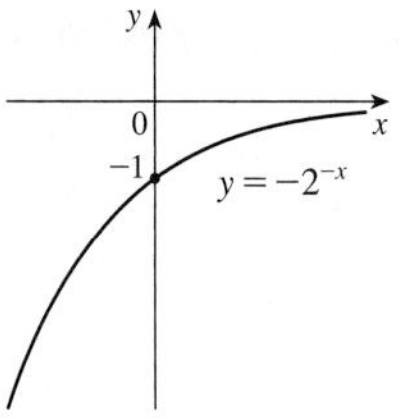

15.

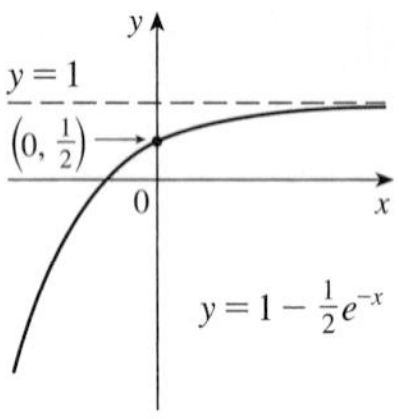

17. (a) $y = e^x - 2$ (b) $y = e^{x-2}$ (c) $y = -e^x$
(d) $y = e^{-x}$ (e) $y = -e^{-x}$

19. (a) $(-\infty, -1) \cup (-1, 1) \cup (1, \infty)$ (b) $(-\infty, \infty)$

21. $f(x) = 3 \cdot 2^x$ **27.** At $x \approx 35.8$

29. (a)

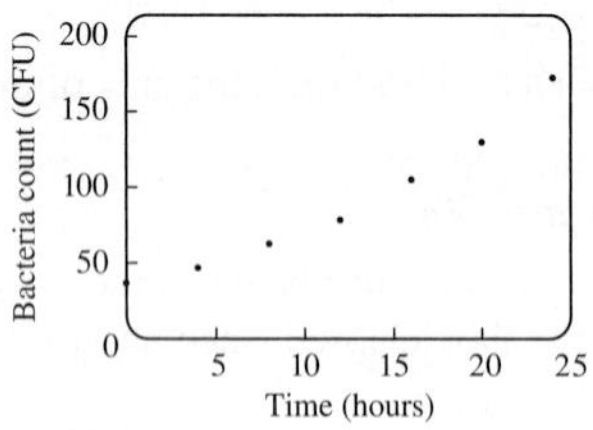

(b) $f(t) = (36.78) \cdot (1.07)^t$

(c)

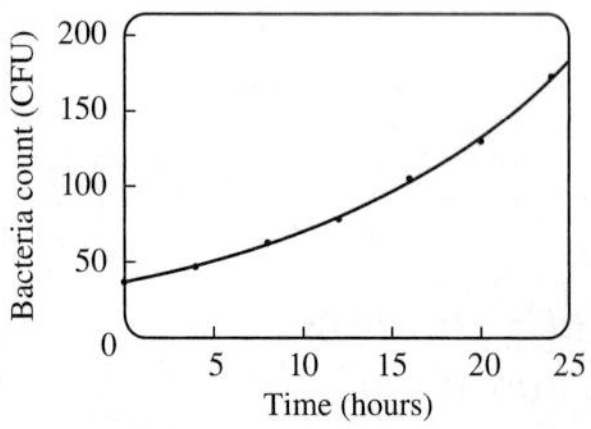

$\approx$10.8 hours for bacteria count to double

31. (a) 25 mg (b) $200 \cdot 2^{-t/5}$ (c) 10.9 mg (d) 38.2 days

33. $\approx$3.5 days

35. $P = 2614.086(1.01693)^t$, where $t = 0$ in 1950; 1993: $P \approx 5381$ million; 2020: $P \approx 8466$ million

EXERCISES 1.5 | page 65

1. (a) See Definition 1.
(b) It must pass the Horizontal Line Test.

3. No **5.** No **7.** Yes **9.** No

11. Yes **13.** No

15. (a) 6 (b) 3 **17.** 0

19. $F = \frac{9}{5}C + 32$; the Fahrenheit temperature as a function of the Celsius temperature; $[-273.15, \infty)$

21. $y = \frac{1}{3}(x - 1)^2 - \frac{2}{3}, x \geqslant 1$

23. $y = \frac{1}{2}(1 + \ln x)$

25. $y = e^x - 3$

27. $f^{-1}(x) = \sqrt[4]{x - 1}$

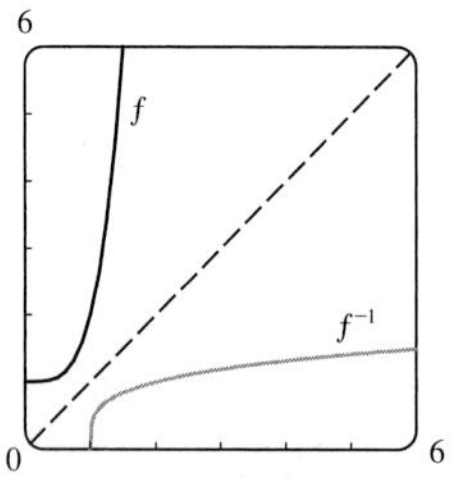

29.

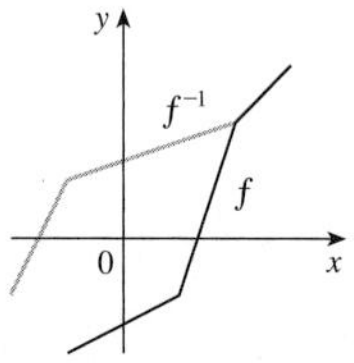

31. (a) $f^{-1}(x) = \sqrt{1 - x^2}, 0 \leqslant x \leqslant 1$; f^{-1} and f are the same function. (b) Quarter-circle in the first quadrant

33. (a) It's defined as the inverse of the exponential function with base b, that is, $\log_b x = y \iff b^y = x$.
(b) $(0, \infty)$ (c) $\mathbb{R}$ (d) See Figure 11.

35. (a) 3 (b) -3

37. (a) 3 (b) -2 **39.** ln 1215

41. $\ln \dfrac{\sqrt{x}}{x + 1}$ **43.** About 1,084,588 mi

45. (a)

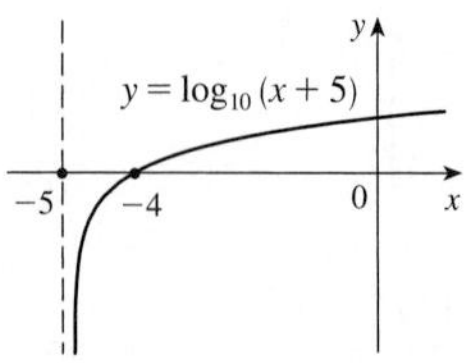

(b)

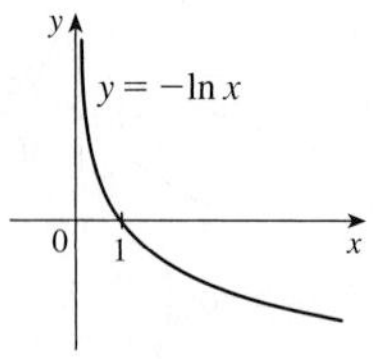

47. (a) $\frac{1}{4}(7 - \ln 6)$ (b) $\frac{1}{3}(e^2 + 10)$

49. (a) $5 + \log_2 3$ or $5 + (\ln 3)/\ln 2$ (b) $\frac{1}{2}\left(1 + \sqrt{1 + 4e}\right)$

51. (a) $0 < x < 1$ (b) $x > \ln 5$

53. (a) $-\dfrac{32941}{340}\ln\left(\dfrac{0.60}{1.65}\right) \approx 98.0$ minutes

(b) $T = -\dfrac{V}{K}\ln\left(\dfrac{c(T)}{c_0}\right)$

55. (a) $(\ln 3, \infty)$ (b) $f^{-1}(x) = \ln(e^x + 3)$; $\mathbb{R}$

57. (a) $t = \dfrac{\log(n/500)}{\log 4}$; the inverse function gives the number of hours that have passed when the population size reaches n

(b) $\dfrac{\log 20}{\log 4} \approx 2.16$ h

59. (a)

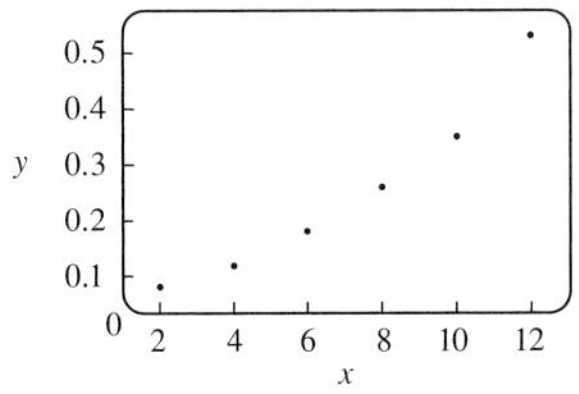

(b)

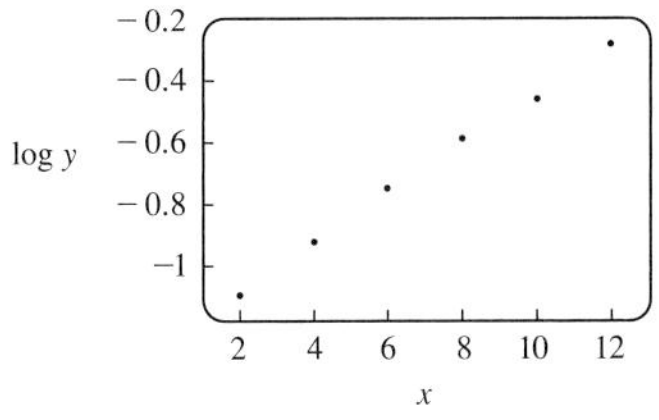

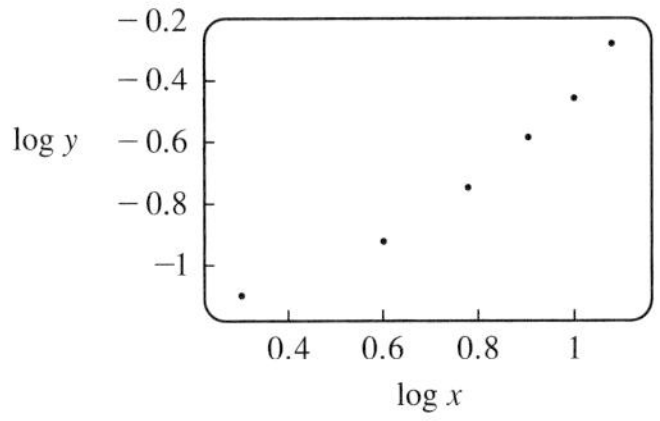

(c) Exponential

(d) $y = 0.056769(1.204651)^x$

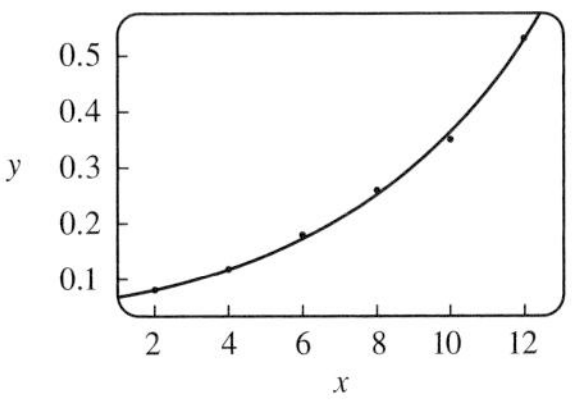

61. (a)

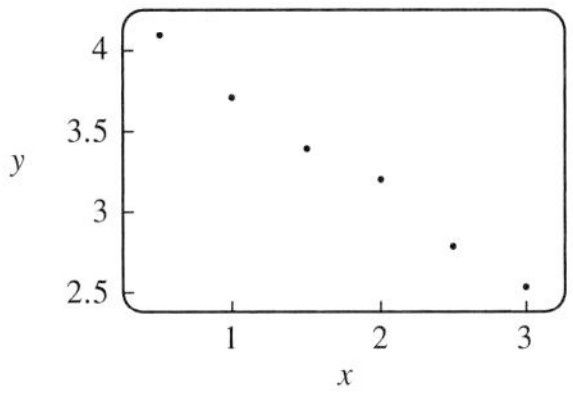

(b)

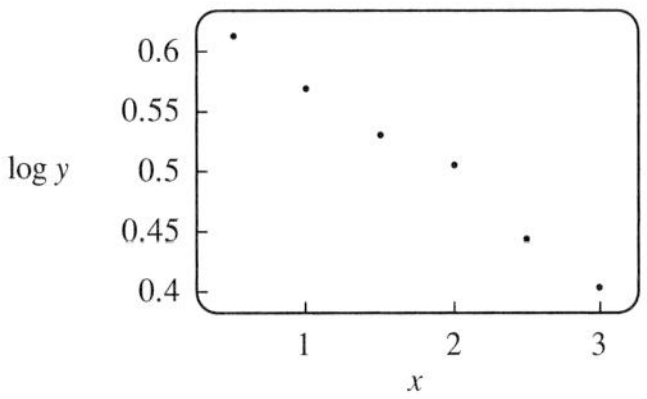

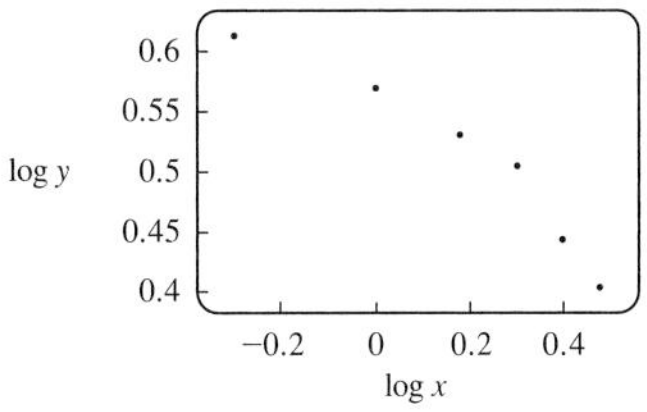

(c) Linear

(d) $y = -0.618857x + 4.368000$

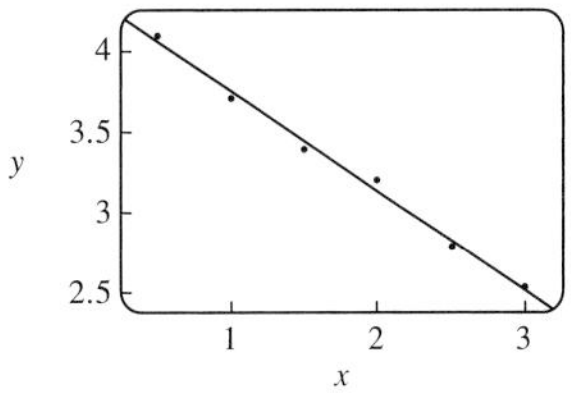

63. (a)

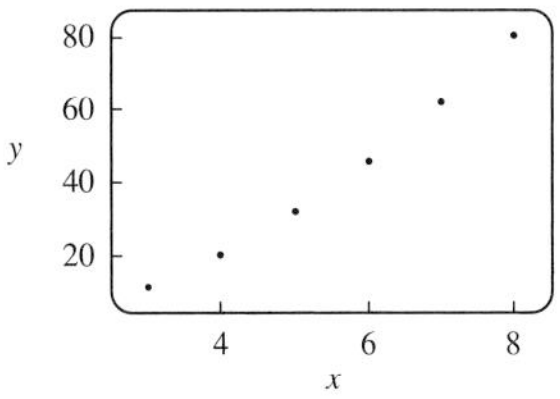

(b)

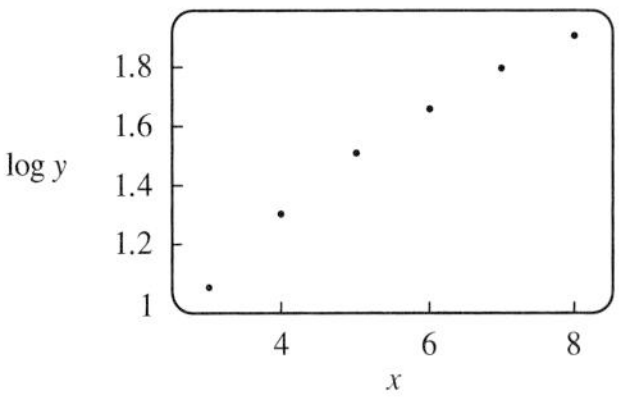

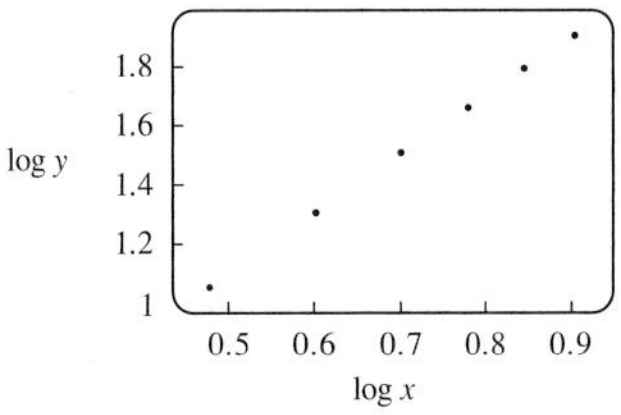

(c) Power model is appropriate.

(d) $y = 1.260294x^{2.002959}$

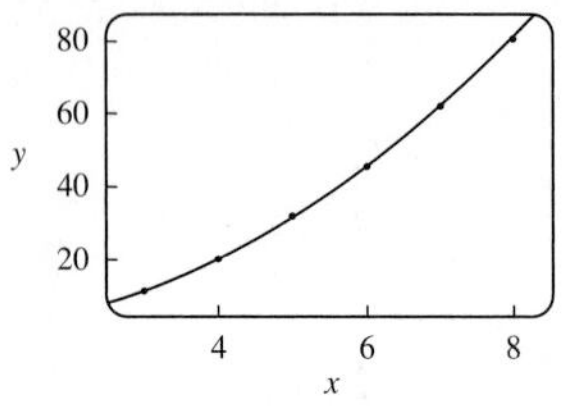

65. (a)

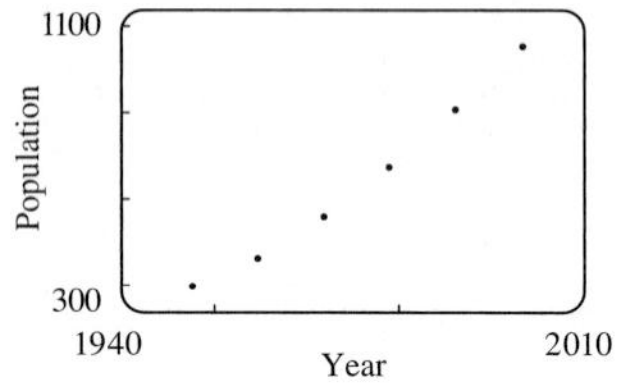

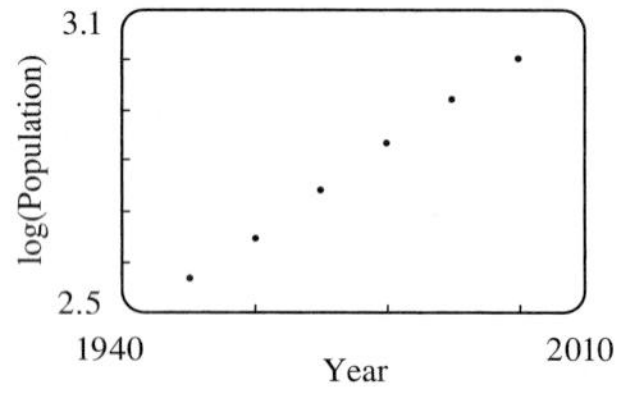

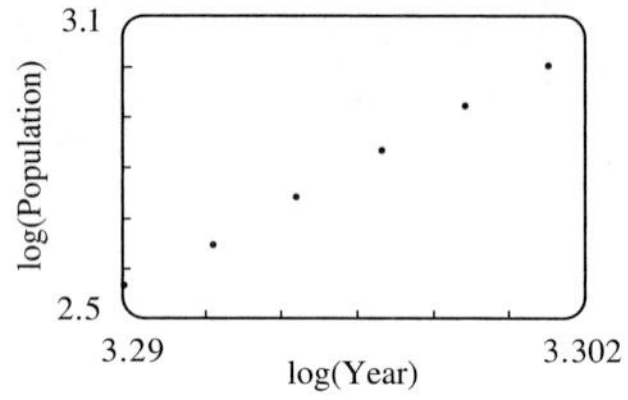

Exponential model is appropriate.

(b) $P = (2.276131 \cdot 10^{-15}) \cdot (1.020529)^Y$, where P is the population in millions and Y is the year.

(c) $P \approx 1247$ million. Model overestimates actual population by 74 million. It does not generalize well to future population growth.

67. (a)

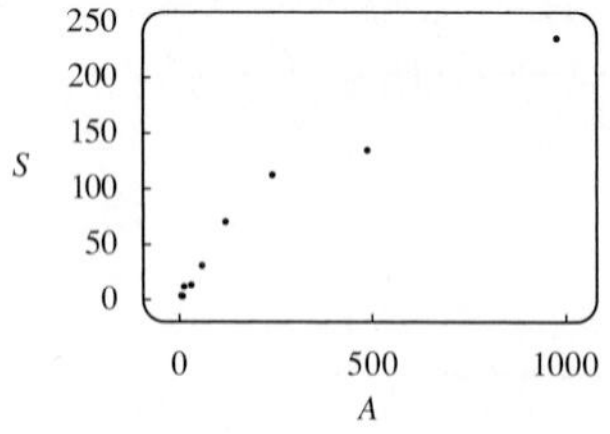

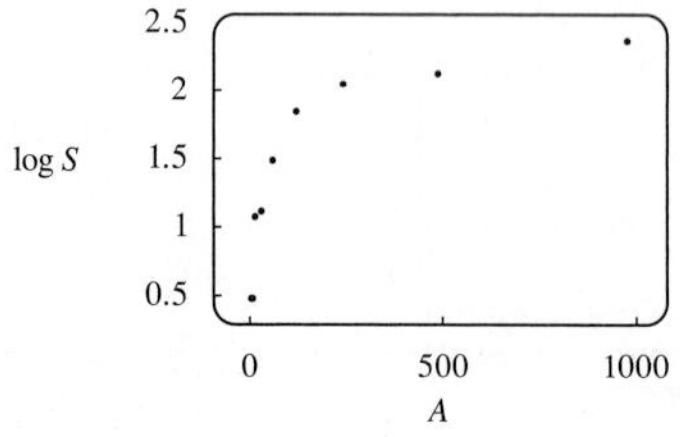

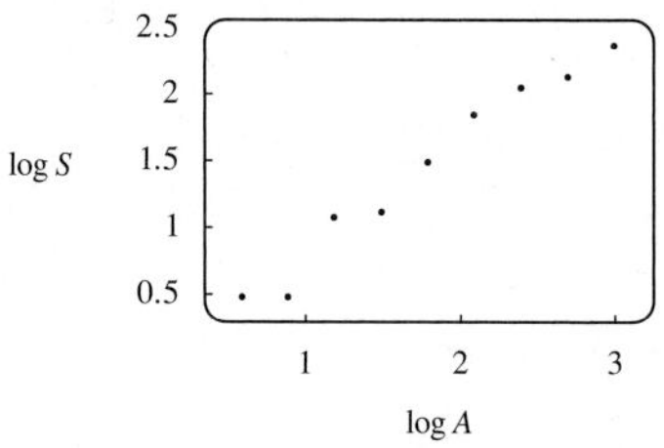

Power model is appropriate.

(b) $S = 0.881518A^{0.841701}$

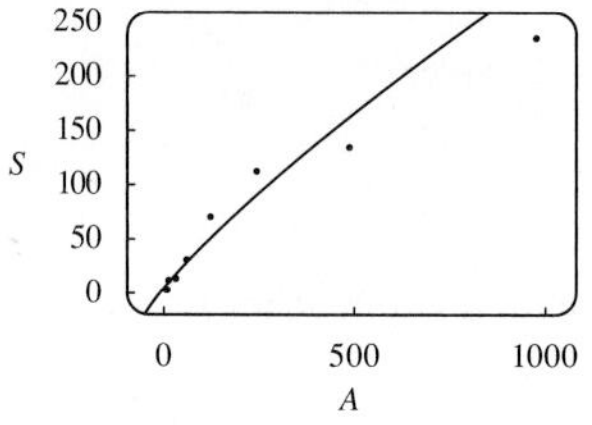

69. (a)

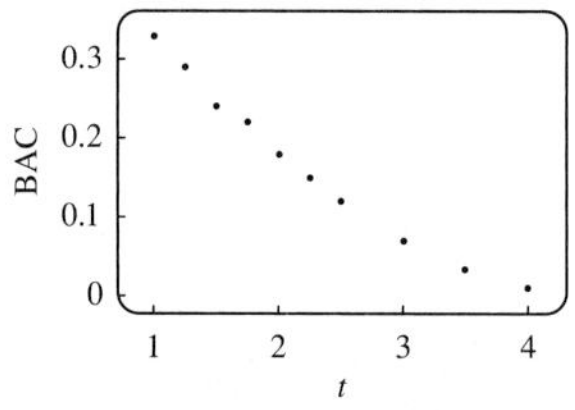

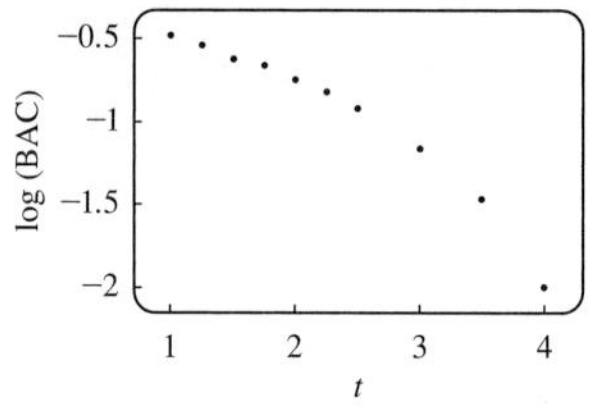

(b) $\text{BAC} = 1.343328(0.338676)^t$

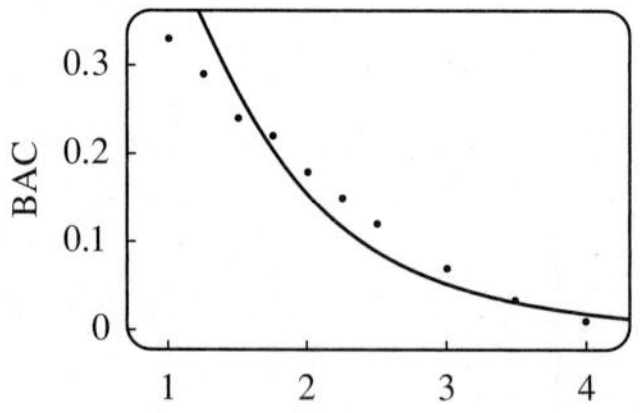

The exponential function overestimates BAC for small values of t.

(c) After 2.6 h

EXERCISES 1.6 | page 76

1. $\left\{1, \frac{4}{5}, \frac{3}{5}, \frac{8}{17}, \frac{5}{13}, \ldots\right\}$

3. $\left\{\frac{1}{5}, -\frac{1}{25}, \frac{1}{125}, -\frac{1}{625}, \frac{1}{3125}, \ldots\right\}$

5.

n	$a_n = \dfrac{3n}{1 + 6n}$
1	0.4286
2	0.4615
3	0.4737
4	0.4800
5	0.4839
6	0.4865
7	0.4884
8	0.4898
9	0.4909
10	0.4918

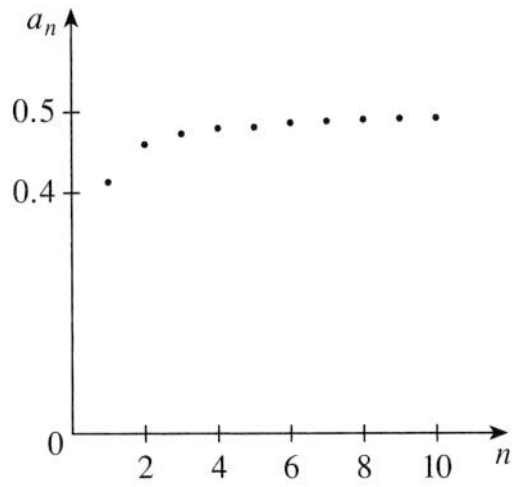

7.

n	$a_n = 1 + \left(-\frac{1}{2}\right)^n$
1	0.5000
2	1.2500
3	0.8750
4	1.0625
5	0.9688
6	1.0156
7	0.9922
8	1.0039
9	0.9980
10	1.0010

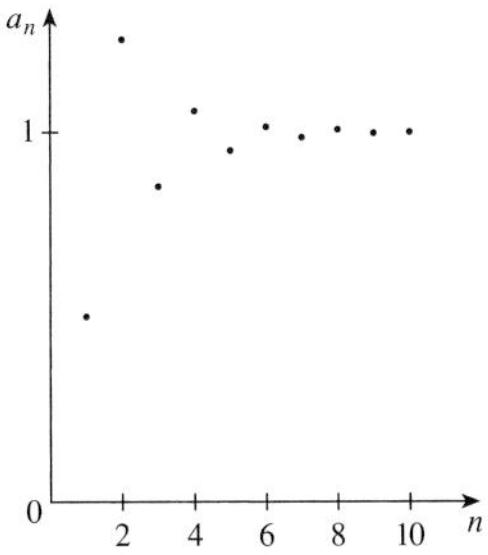

9. $a_n = 1/(2n - 1)$ **11.** $a_n = -3\left(-\frac{2}{3}\right)^{n-1}$

13. $a_n = (-1)^{n+1}\dfrac{n^2}{n + 1}$ **15.** $\{1, 2, 7, 32, 157, 782, \dots\}$

17. $\left\{2, \dfrac{2}{3}, \dfrac{2}{5}, \dfrac{2}{7}, \dfrac{2}{9}, \dfrac{2}{11}, \dots\right\}$

19. $\{1, 3^{1/2}, 3^{3/4}, 3^{7/8}, 3^{15/16}, 3^{31/32}, \dots\}$

21. $\{2, 1, -1, -2, -1, 1, \dots\}$

25. (a) As t increases, N_t approaches 0. (b) $N_t = 1$ for all t. (c) The sequence grows indefinitely as t increases.

27.

t	x_t
0	0.5000
1	0.3750
2	0.3516
3	0.3419
4	0.3375
5	0.3354
6	0.3344
7	0.3338
8	0.3336
9	0.3335
10	0.3334

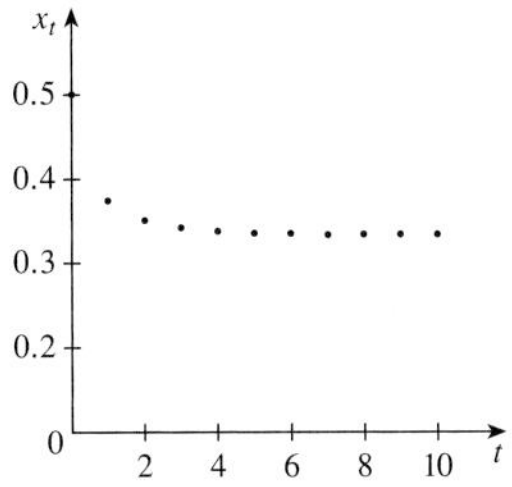

Approaches 1/3

29.

t	x_t
0	0.8750
1	0.3741
2	0.8008
3	0.5456
4	0.8479
5	0.4411
6	0.8431
7	0.4523
8	0.8472
9	0.4427
10	0.8438

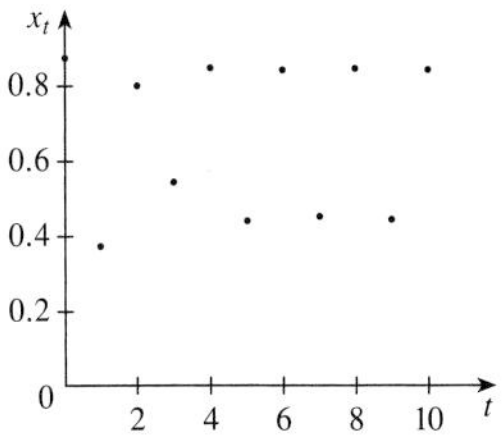

Cycles between values near 0.84 and 0.45

31.

t	x_t
0	0.5000
1	0.9250
2	0.2567
3	0.7060
4	0.7681
5	0.6591
6	0.8313
7	0.5189
8	0.9237
9	0.2608
10	0.7134

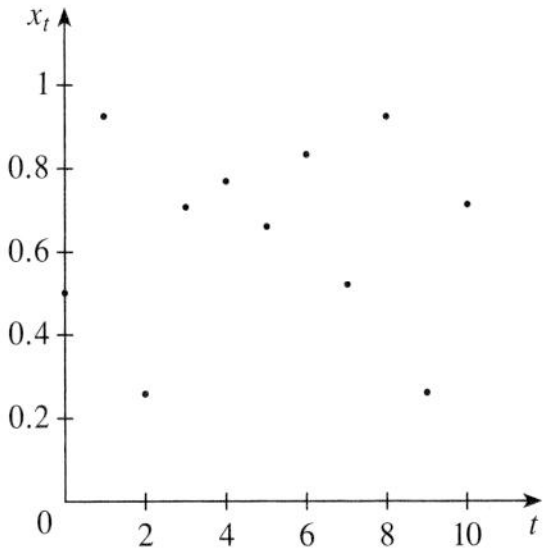

Cycles irregularly among a range of values between 0.26 and 0.92

33.

t	x_t
0	0.8750
1	1.2475
2	1.2254
3	1.2306
4	1.2294
5	1.2297
6	1.2296
7	1.2296
8	1.2296
9	1.2296
10	1.2296

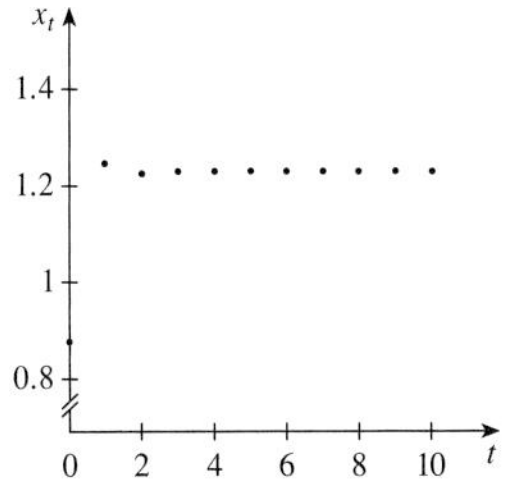

Ricker: approaches 1.2296;
Logistic: cycles between values

35. (a) $N_{t+1} = N_t + RK\sqrt{N_t}$; R and K constants

(b)

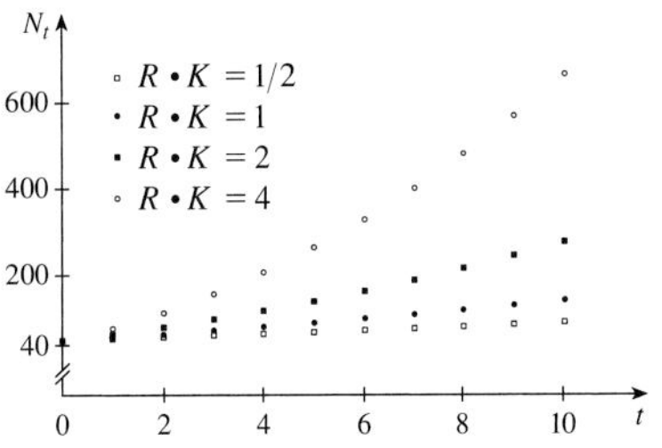

37. (a) $n_{t+1} = b(1 - d)n_t + m$
(b) $n_{t+1} = (1 - d)(bn_t + m)$
(c) Part (a) recursion formula

39. $p_{t+1} = \dfrac{\alpha p_t}{\alpha p_t + 1 - p_t}$

CHAPTER 1 REVIEW | page 81

True-False Quiz

1. False **3.** False **5.** True **7.** False **9.** True **11.** False

Exercises

1. (a) 2.7 (b) 2.3, 5.6 (c) $[-6, 6]$ (d) $[-4, 4]$
(e) $[-4, 4]$ (f) No; it fails the Horizontal Line Test.
(g) Odd; its graph is symmetric about the origin.
3. (a) ≈ -36 m
(b) 18,000 years ago; 121,000 years ago
(c) $[-114, 8]$
(d) Sea level drops correspond to periods of glaciation.
5. $2a + h - 2$
7. $\left(-\infty, \frac{1}{3}\right) \cup \left(\frac{1}{3}, \infty\right)$, $(-\infty, 0) \cup (0, \infty)$
9. $(-6, \infty)$, $\mathbb{R}$
11. (a) Shift the graph 8 units upward.
(b) Shift the graph 8 units to the left.
(c) Stretch the graph vertically by a factor of 2, then shift it 1 unit upward.
(d) Shift the graph 2 units to the right and 2 units downward.
(e) Reflect the graph about the x-axis.
(f) Reflect the graph about the line $y = x$ (assuming f is one-to-one).

13.

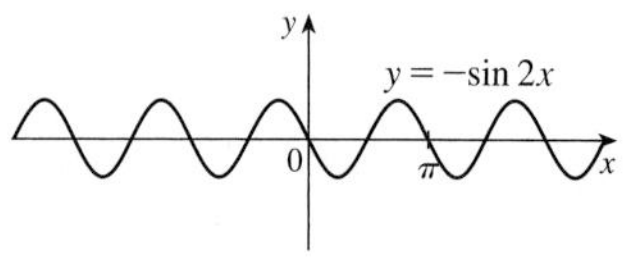

15.

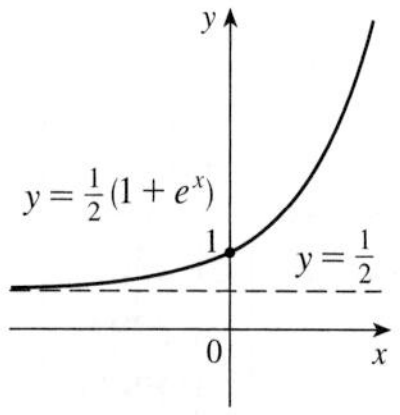

17.

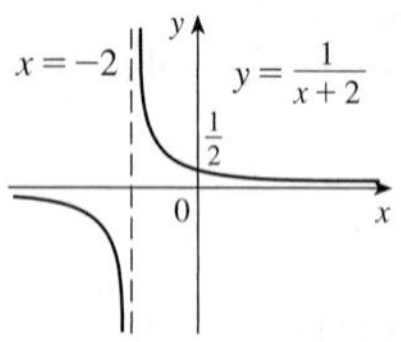

19.

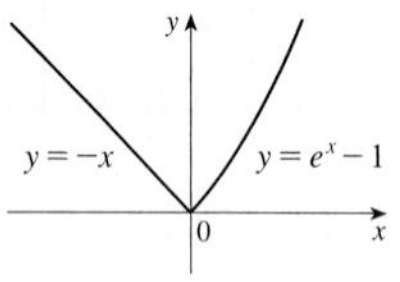

21. (a) $(f \circ g)(x) = \ln(x^2 - 9)$, $(-\infty, -3) \cup (3, \infty)$
(b) $(g \circ f)(x) = (\ln x)^2 - 9$, $(0, \infty)$
(c) $(f \circ f)(x) = \ln \ln x$, $(1, \infty)$
(d) $(g \circ g)(x) = (x^2 - 9)^2 - 9$, $(-\infty, \infty)$
23. $y = 0.2493x - 423.4818$; about 77.6 years
25. 1 **27.** (a) 9 (b) 2

29. (a) $\frac{1}{16}$ g (b) $m(t) = 2^{-t/4}$
(c) $t(m) = -4 \log_2 m$; the time elapsed when there are m grams of ^{100}Pd
(d) About 26.6 days

31.

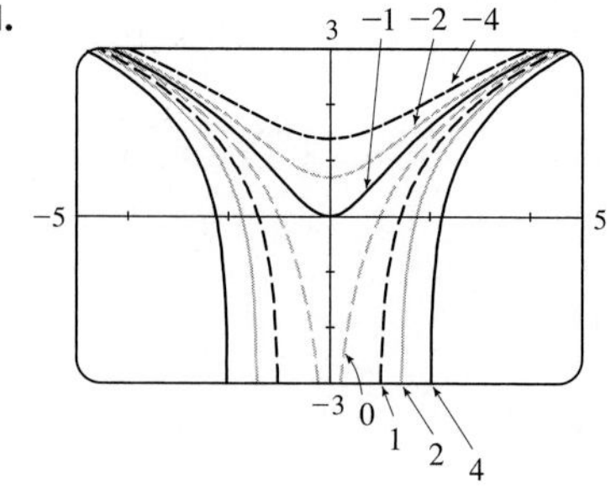

For $c < 0$, f is defined everywhere. As c increases, the dip at $x = 0$ becomes deeper. For $c \geqslant 0$, the graph has asymptotes at $x = \pm\sqrt{c}$.

33. (a)

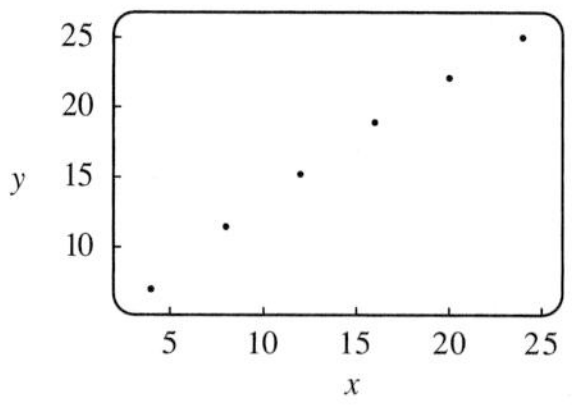

(b)

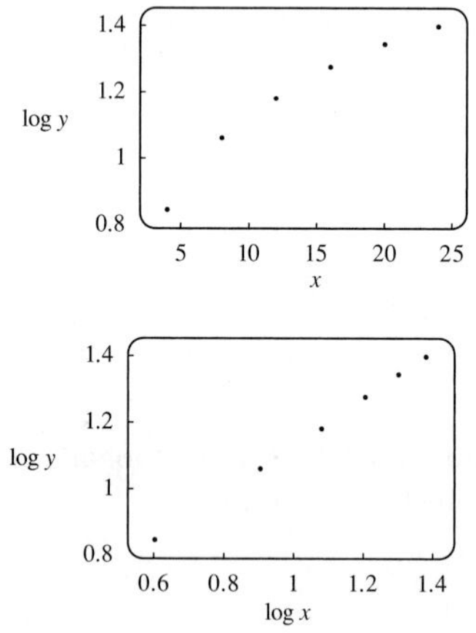

(c) Power
(d) $y = 2.608377x^{0.712277}$

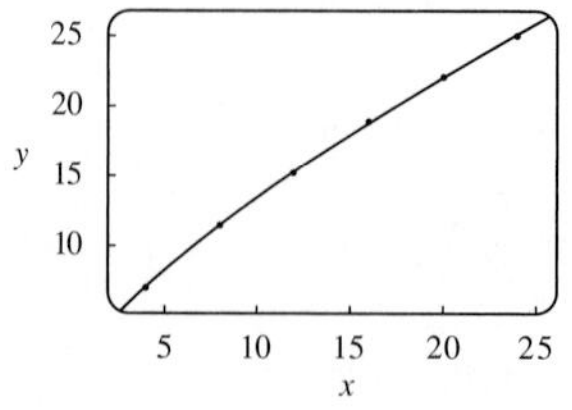

35. (a)

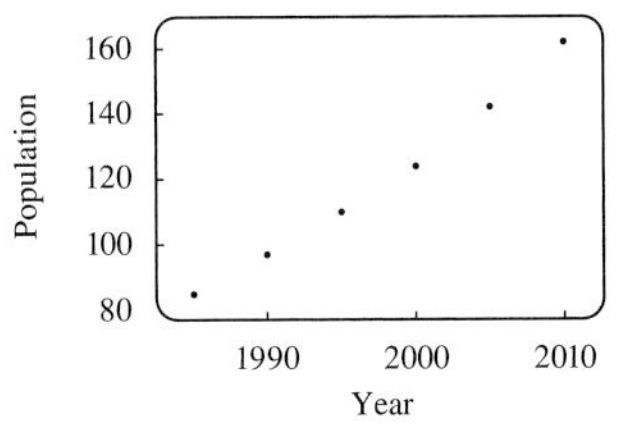

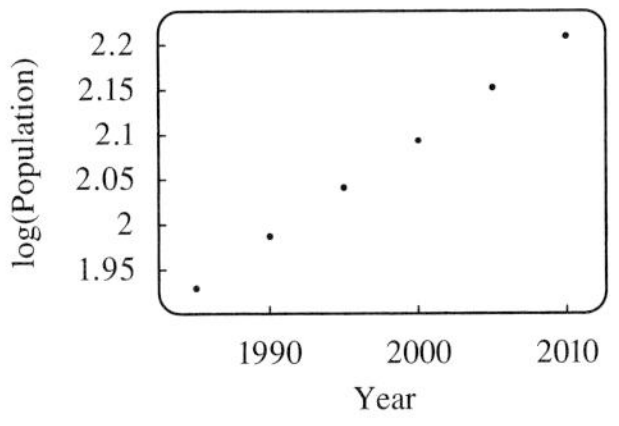

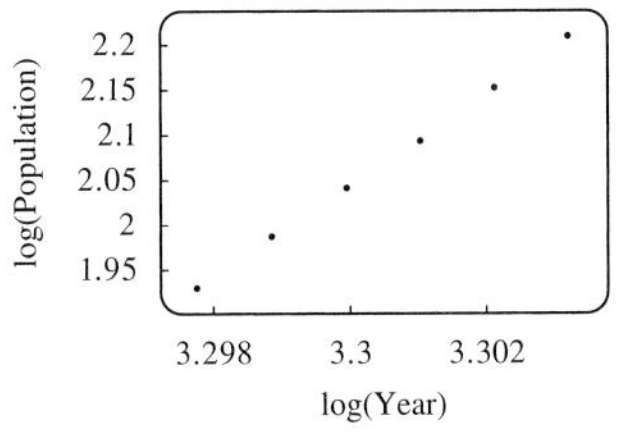

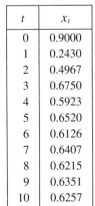

Exponential or power model
(b) $P = (6.6326 \cdot 10^{-21}) \cdot (1.025977)^Y$, where P is the population in millions and Y is the year.
(c) 153 million; 209 million

37. $\{3, 6, 13, 28, 59, 122, \ldots\}$

39.

t	x_t
0	0.9000
1	0.2430
2	0.4967
3	0.6750
4	0.5923
5	0.6520
6	0.6126
7	0.6407
8	0.6215
9	0.6351
10	0.6257

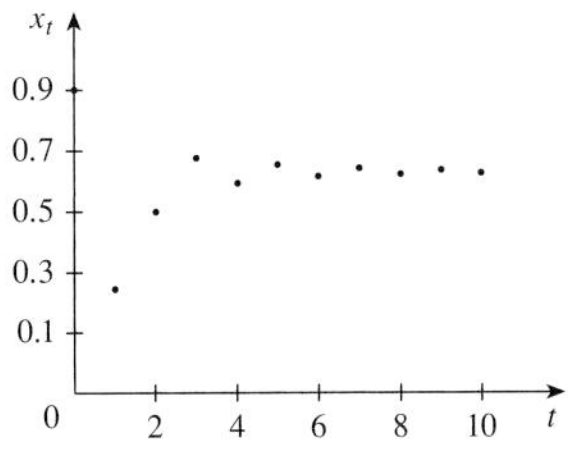

Approaches 0.63

CHAPTER 2

EXERCISES 2.1 | page 99

1. (a) A sequence is an ordered list of numbers. It can also be defined as a function whose domain is the set of positive integers.
(b) The terms a_n approach 8 as n becomes large.
(c) The terms a_n become large as n becomes large.

3. The sequence likely has a nonzero limit as $t \to \infty$ because human physiology will ultimately limit how fast a human can sprint 100 meters. This means that there is a certain world record time that athletes will never surpass.

5.

n	a_n
1	0.2000
2	0.2500
3	0.2727
4	0.2857
5	0.2941
6	0.3000
7	0.3043
8	0.3077
9	0.3103
10	0.3125

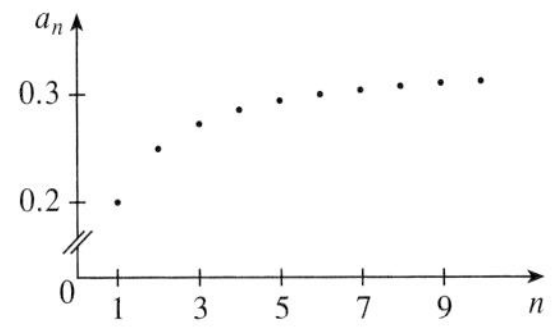

The sequence appears to converge to a number between 0.3 and 0.35.

$\lim_{n \to \infty} a_n = \frac{1}{3}$. This agrees with value predicted from the data.

7.

n	a_n
1	2.3333
2	3.4444
3	2.7037
4	3.1975
5	2.8683
6	3.0878
7	2.9415
8	3.0390
9	2.9740
10	3.0173

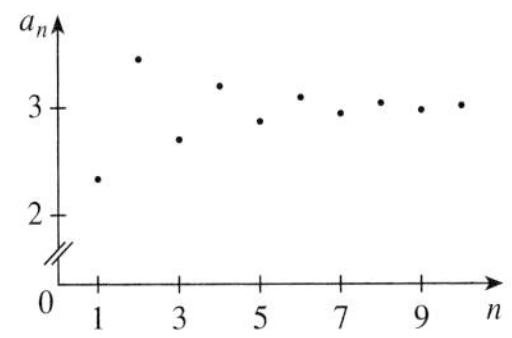

The sequence appears to converge to a number near 3.

$\lim_{n \to \infty} a_n = 3$. This agrees with value predicted from the data.

9. 0 **11.** 2 **13.** $\frac{5}{7}$
15. 1 **17.** Diverges
19. Diverges **21.** Diverges
23. ln 2 **25.** 0

27.

n	a_n
1	1.0000
2	1.5000
3	1.7500
4	1.8750
5	1.9375
6	1.9688
7	1.9844
8	1.9922

$\lim_{n \to \infty} a_n = 2$

29.

n	a_n
1	2.0000
2	3.0000
3	5.0000
4	9.0000
5	17.0000
6	33.0000
7	65.0000
8	129.0000

Divergent

31.

n	a_n
1	1.0000
2	3.0000
3	1.5000
4	2.4000
5	1.7647
6	2.1702
7	1.8926
8	2.0742

$\lim_{n\to\infty} a_n = 2$

33.

n	a_n
1	1.0000
2	1.7321
3	1.9319
4	1.9829
5	1.9957
6	1.9989
7	1.9997
8	1.9999

$\lim_{n\to\infty} a_n = 2$

35. (a) 120 mg, 124 mg (b) $Q_{n+1} = 100 + (0.20)Q_n$
(c) $Q_n = 125(1 - 0.20^n)$ (d) $\lim_{n\to\infty} Q_n = 125$ mg

37. (a) 157.875 mg; $\frac{3000}{19}(1 - 0.05^n)$ (b) 157.895 mg

39. (a) $x = 1$ (b) 1 (c) 2
(d) All rational numbers with a terminating decimal representation, except 0.

41. $\frac{8}{9}$ **43.** $\frac{838}{333}$ **45.** 5063/3300

47.

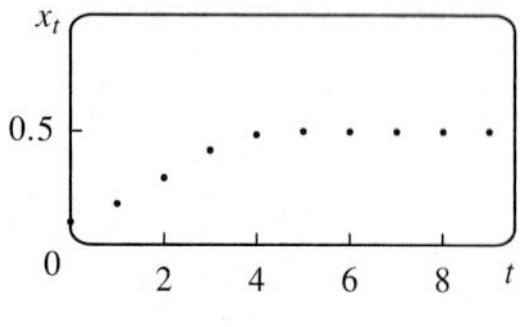

Converges to 1/2. $\lim_{t\to\infty} x_t = 1/2$.

49.

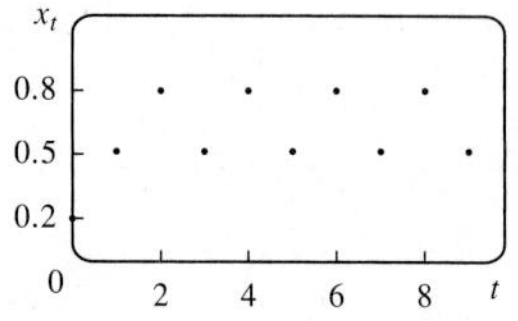

Diverges

51.

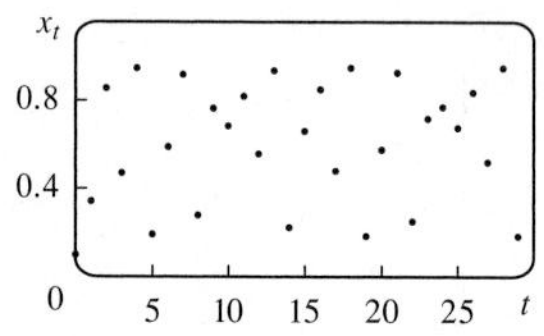

Diverges

53.

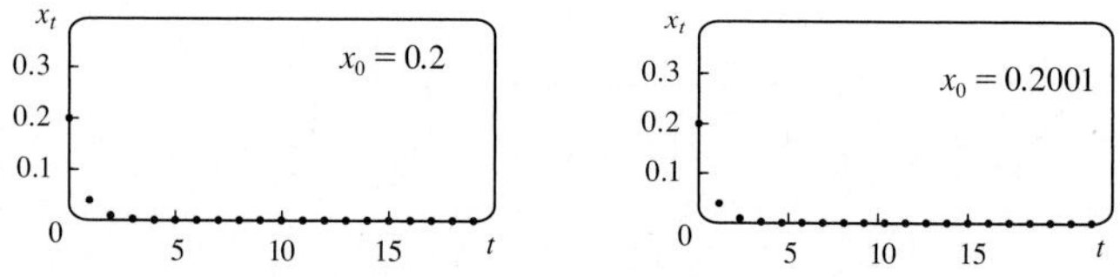

The plots show that the solutions are nearly identical, converging to zero as t increases.

55.

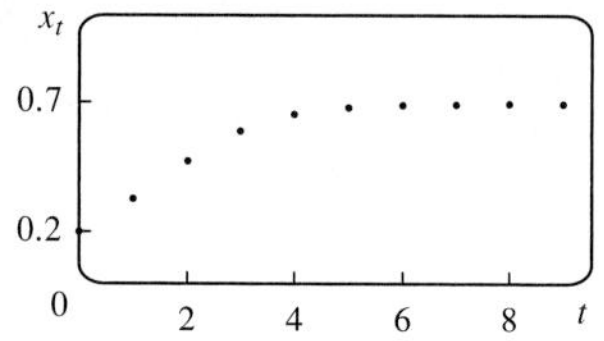

Converges to approximately 0.7; $\lim_{t\to\infty} x_t = \ln 2$

57.

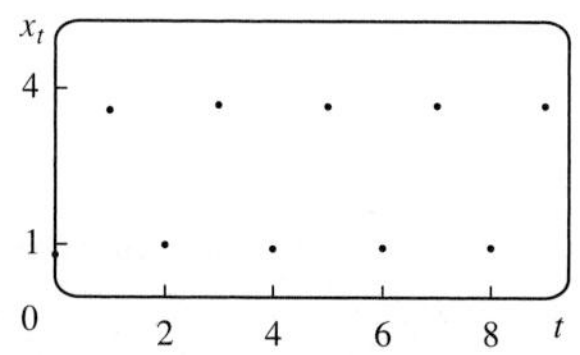

Diverges

59. The removed area of the Sierpinski carpet after the nth step of construction is $A_n = 1 - \left(\frac{8}{9}\right)^n$ so, $\lim_{n\to\infty} A_n = 1$ implying that the Sierpinski carpet has zero area.

■ EXERCISES 2.2 | page 109

1. (a) As x becomes large, $f(x)$ approaches 5.
(b) As x becomes large negative, $f(x)$ approaches 3.

3. 0 **5.** 0 **7.** $\frac{3}{2}$ **9.** $-\frac{1}{2}$ **11.** ∞ **13.** -1
15. 4 **17.** $\frac{1}{6}$ **19.** 2 **21.** ∞ **23.** $-\infty$ **25.** 1
27. $-\frac{1}{2}$

29. c is the concentration at which the growth rate is half that of the maximum possible value. This is often referred to as the half-saturation coefficient.

31. 0. As the mortality rate increases, the number of new infections approaches zero.

33. 8×10^7. In the long run the biomass of the Pacific halibut will tend to 8×10^7 kg.

35. $x > 9.21$

37. (a) v^* (b) 3.52 s

■ EXERCISES 2.3 | page 122

1. As x approaches 2, the value of $f(x)$ approaches 5; yes

3. (a) $\lim_{x\to -3} f(x) = \infty$ means that the values of $f(x)$ can be made arbitrarily large (as large as we please) by taking x sufficiently close to -3 (but not equal to -3).
(b) $\lim_{x\to 4^+} f(x) = -\infty$ means that the values of $f(x)$ can be made arbitrarily large negative by taking x sufficiently close to 4 through values larger than 4.

5. (a) 2 (b) 1 (c) 4 (d) Does not exist (e) 3
7. (a) 260 (b) 254 (c) Does not exist (d) 254
(e) 258 (f) Does not exist (g) 258
(h) On June 3 ($t = 2$), the population decreased by 6. This could have been a result of deaths, emigration, or a combination of the two. On June 5 ($t = 4$), the population increased by 4. This could have been a result of births, immigration, or a combination of the two.
9. (a) $-\infty$ (b) $-\infty$ (c) ∞ (d) 2 (e) -1
(f) Vertical: $x = 0$, $x = 2$; horizontal: $y = -1$, $y = 2$

11.

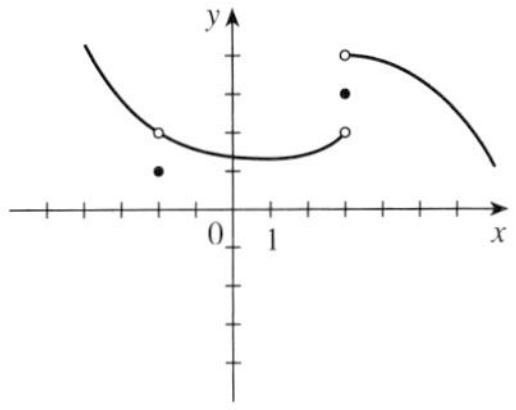

13.

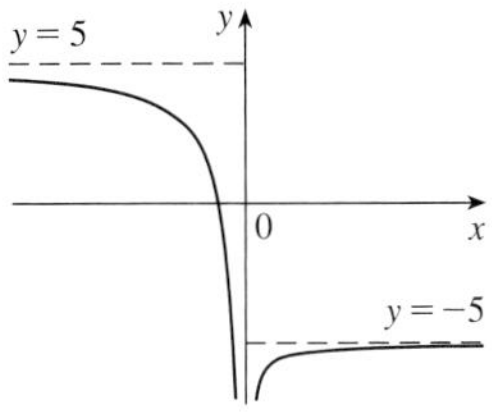

15.

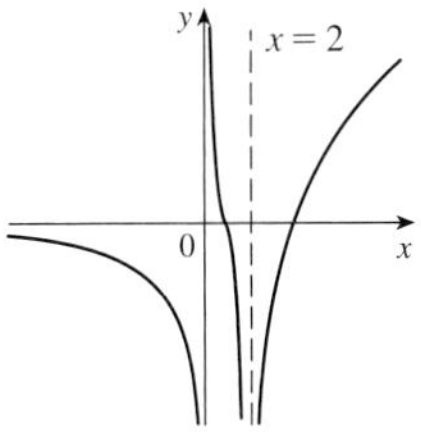

17.

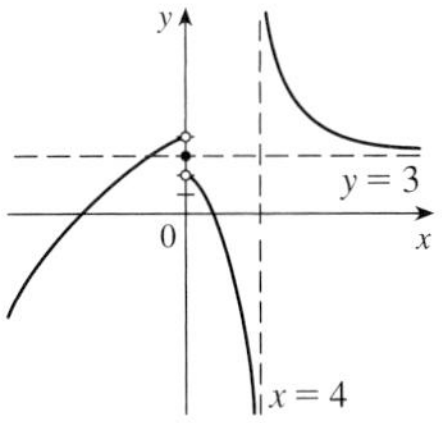

19. $\frac{2}{3}$ **21.** 5 **23.** $\frac{1}{4}$ **25.** $\frac{3}{5}$
27. (a) -1.5 (b)

x	$f(x)$
± 0.1	-1.493759
± 0.01	-1.499938
± 0.001	-1.499999
± 0.0001	-1.500000

29. $-\infty$ **31.** ∞ **33.** $-\infty$ **35.** $-\infty$ **37.** ∞
39. (a) $\lim_{x\to 1^-} f(x) = -\infty$ and $\lim_{x\to 1^+} f(x) = \infty$
(c)

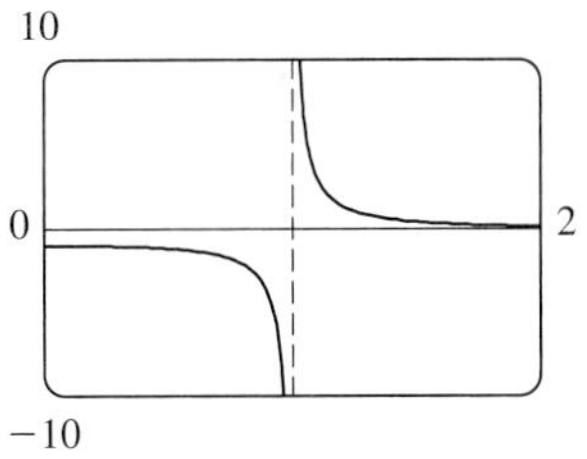

41. (a) 2.71828 (b)

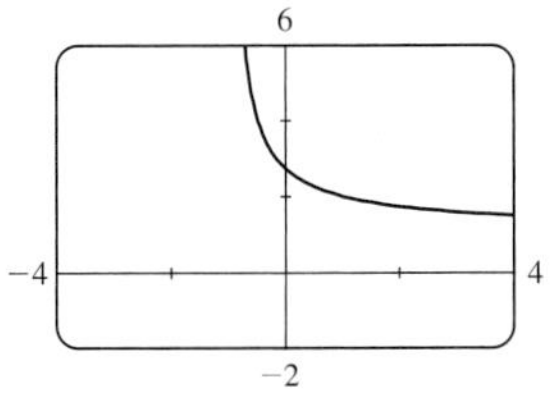

43. No matter how many times we zoom in toward the origin, the graph appears to consist of almost-vertical lines. This indicates more and more frequent oscillations as $x \to 0$.

EXERCISES 2.4 | page 135

1. (a) -6 (b) -8 (c) 2 (d) -6
(e) Does not exist (f) 0
3. 59 **5.** $\frac{3}{2}$ **7.** $\pi/2$
9. 4 **11.** Does not exist
13. $\frac{6}{5}$ **15.** 8 **17.** $\frac{1}{12}$ **19.** $-\frac{1}{16}$
21. $\frac{1}{128}$ **23.** $-\frac{1}{2}$ **25.** (a), (b) $\frac{2}{3}$
29. 7 **33.** 6 **35.** Does not exist
37. (a) (i) 5 (ii) -5 (b) Does not exist
(c)

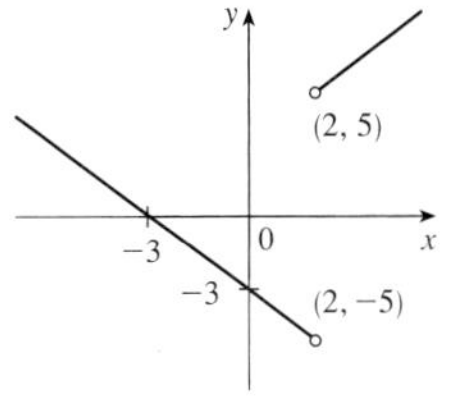

39. 3 **41.** 3 **43.** $\frac{1}{2}$ **49.** 8 **51.** 15; -1

EXERCISES 2.5 | page 147

1. $\lim_{x\to 4} f(x) = f(4)$
3. (a) $f(-4)$ is not defined and $\lim_{x\to a} f(x)$ does not exist [for $a = -2, 2,$ and 4]
(b) -4, neither; -2, left; 2, right; 4, right
5.

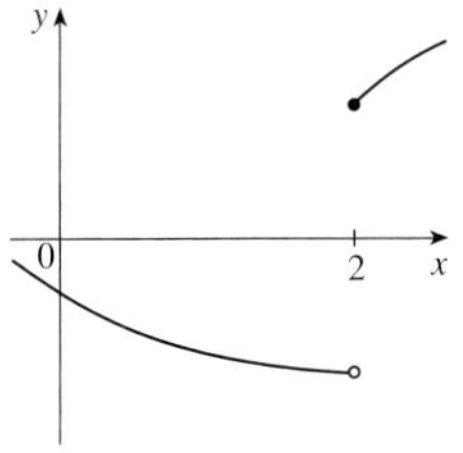

7.

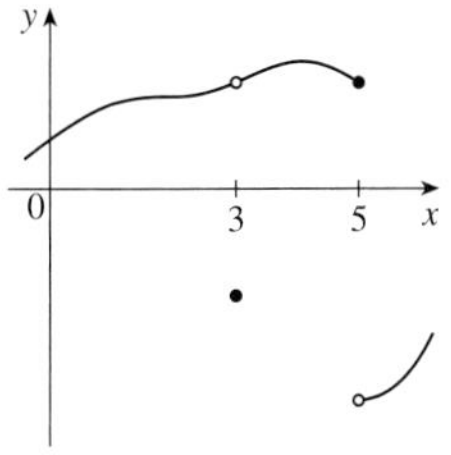

9. (a) Discontinuous at $t = 12, 24, 36$ (b) Jump discontinuities
11. (a)

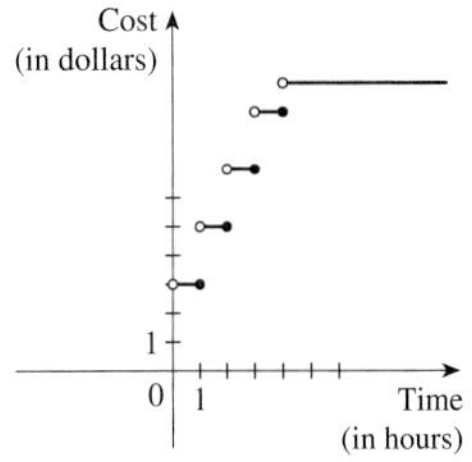

(b) Discontinuous at $t = 1, 2, 3, 4$
13. 6
17. $\lim_{x\to 0} f(x)$ does not exist.

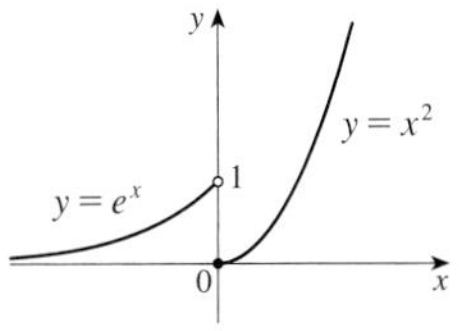

19. $\lim_{x\to 0} f(x) \neq f(0)$

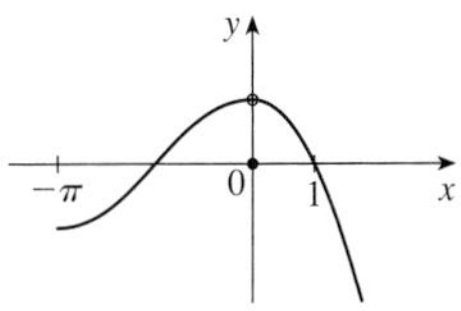

21. $\left[\frac{1}{2}, \infty\right)$ **23.** $(-\infty, \infty)$
25. $(-\infty, -1) \cup (1, \infty)$

27. $x = 0$

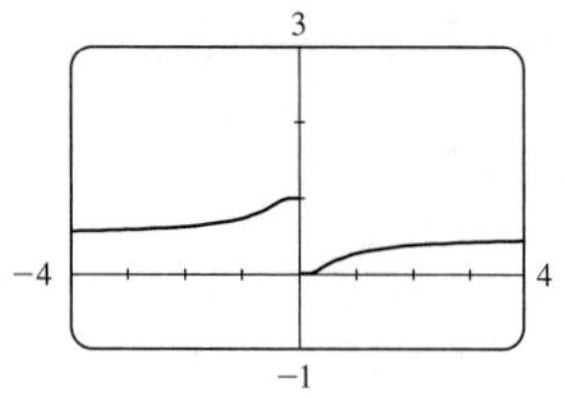

29. $\frac{7}{3}$ **31.** 1

35. 0, right; 1, left

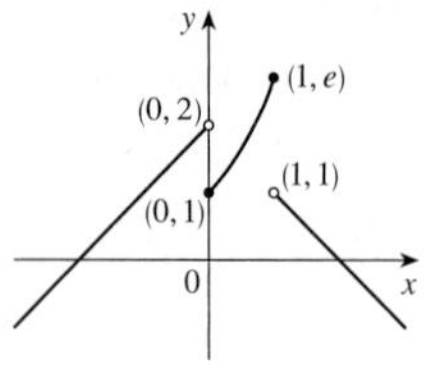

37. $\frac{2}{3}$ **45** (b) (0.86, 0.87) **47.** (b) 70.347 **49.** Yes

■ CHAPTER 2 REVIEW | page 150

True-False Quiz

1. True **3.** False **5.** True **7.** False **9.** True
11. True **13.** False **15.** True

Exercises

1. $\frac{1}{2}$ **3.** Diverges **5.** $\frac{9}{2}$ **7.** $\frac{4111}{3330}$

9. (a) (i) 3 (ii) 0 (iii) Does not exist (iv) 2
(v) ∞ (vi) $-\infty$ (vii) 4 (viii) -1
(b) $y = 4, y = -1$ (c) $x = 0, x = 2$ (d) $-3, 0, 2, 4$

11. $-\frac{1}{5}$ **13.** 1 **15.** $\frac{3}{2}$ **17.** 3 **19.** ∞ **21.** $\frac{4}{7}$

23. $-\infty$ **25.** $\frac{1}{2}$ **27.** 2

29. 0.50. As the concentration grows larger, the enzymatic reaction rate will approach 0.50.

31. (a) (i) 3 (ii) 0 (iii) Does not exist (iv) 0 (v) 0 (vi) 0
(b) At 0 and 3 (c)

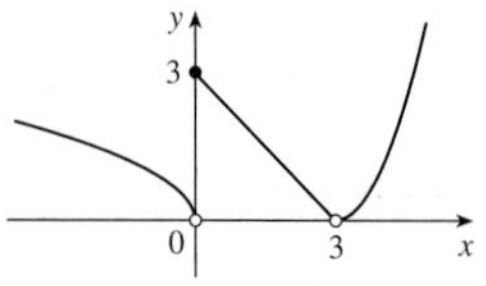

CHAPTER 3

■ EXERCISES 3.1 | page 165

1. (a) $\dfrac{f(x) - f(3)}{x - 3}$ (b) $\displaystyle\lim_{x \to 3} \frac{f(x) - f(3)}{x - 3}$

3. (a) 2 (b) $y = 2x + 1$ (c)

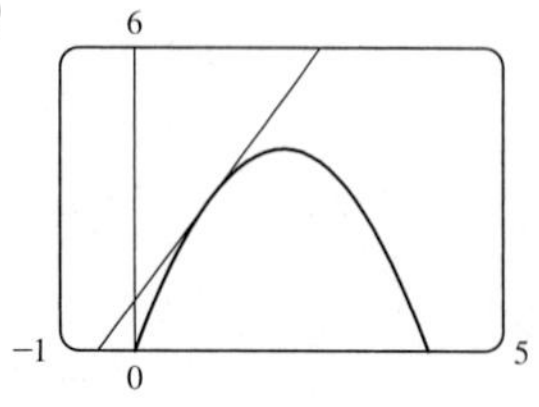

5. $y = -8x + 12$ **7.** $y = \frac{1}{2}x + \frac{1}{2}$

9. (a) $8a - 6a^2$ (b) $y = 2x + 3, y = -8x + 19$
(c)

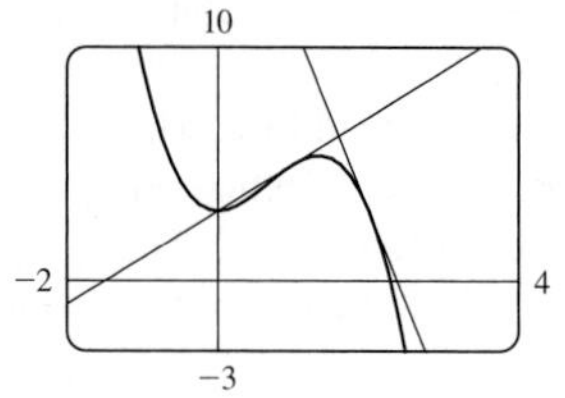

11. $g'(0), 0, g'(4), g'(2), g'(-2)$

13. $f(2) = 3$; $f'(2) = 4$

15.

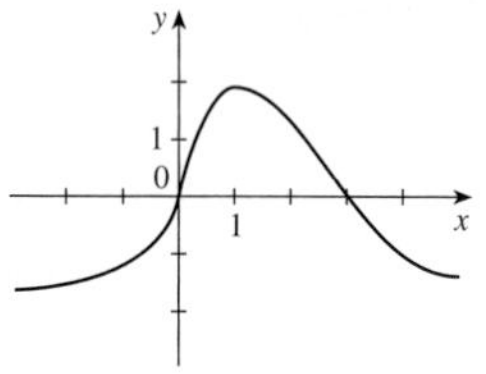

17. $y = 3x - 1$

19. (a) $-\frac{3}{5}$; $y = -\frac{3}{5}x + \frac{16}{5}$
(b)

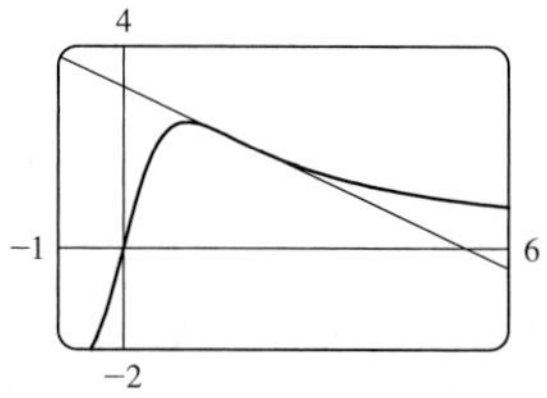

21. $6a - 4$ **23.** $\dfrac{5}{(a + 3)^2}$

25. $-\dfrac{1}{\sqrt{1 - 2a}}$ **27.** -24 ft/s

29. $-2/a^3$ m/s; -2 m/s; $-\frac{1}{4}$ m/s; $-\frac{2}{27}$ m/s

31. (a) (i) 3.26 million/year (ii) 3.18 million/year
(iii) 2.72 million/year (iv) 2.61 million/year
(b) $P'(2000) \approx 2.95$ million/year; the US population was growing at a rate of 2.95 million people per year in 2000.

33. (a) (i) -0.15 (mg/mL)/h (ii) -0.12 (mg/mL)/h
(iii) -0.12 (mg/mL)/h (iv) -0.11 (mg/mL)/h
(b) $C'(2.0) \approx -0.12$ (mg/mL)/h, meaning that the blood alcohol concentration was decreasing at a rate of 0.12 mg/mL per hour after 2.0 hours.

35. $T'(12) \approx 2.75$°F/h; $T'(12)$ is the rate of change of temperature 12 hours after midnight on May 7, 2012.

37. Greater (in magnitude)

39. (a) The rate at which the cost is changing per ounce of gold produced; dollars per ounce
(b) When the 800th ounce of gold is produced, the cost of production is \$17/oz.
(c) Decrease in the short term; increase in the long term

41. (a) The rate at which the oxygen solubility changes with respect to the water temperature; (mg/L)/°C
(b) $S'(16) \approx -0.25$; as the temperature increases past 16°C, the oxygen solubility is decreasing at a rate of 0.25 (mg/L)/°C.

EXERCISES 3.2 | page 178

1. (a) -0.2 (b) 0 (c) 1 (d) 2
(e) 1 (f) 0 (g) -0.2

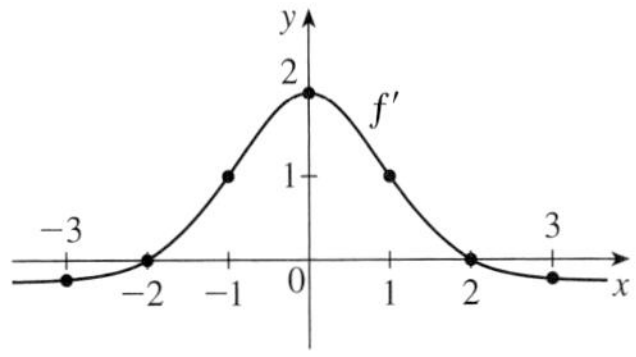

3. (a) II (b) IV (c) I (d) III

5.

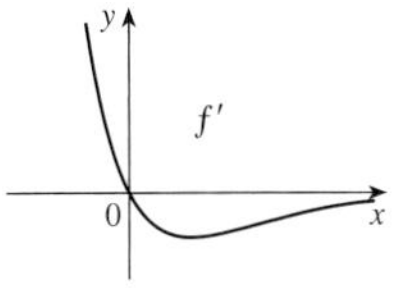

7.

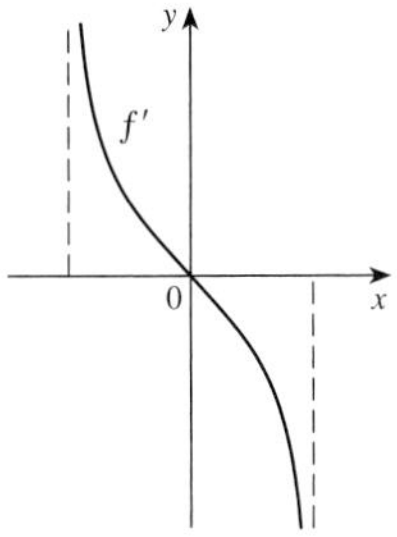

9.

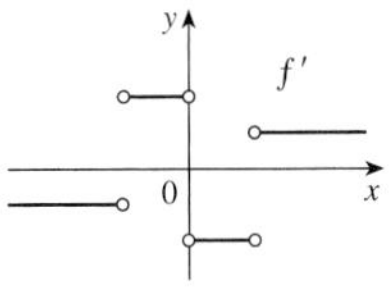

11.

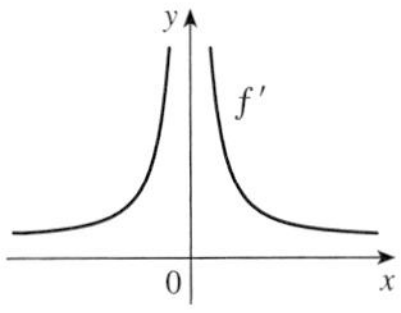

13. (a) $W'(t)$ is the rate of change of average body weight with respect to time for tadpoles raised in a density of 80 tadpoles/L.
(b)

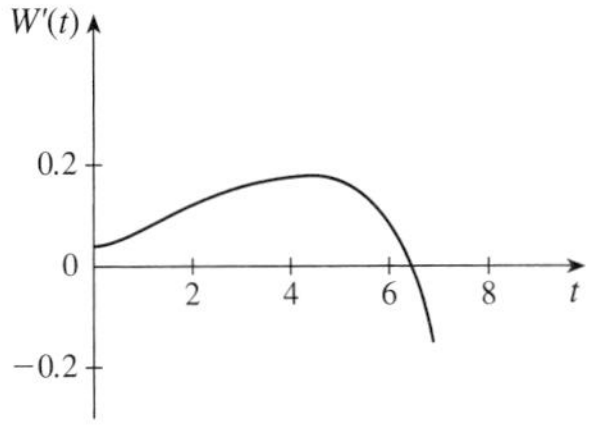

15.

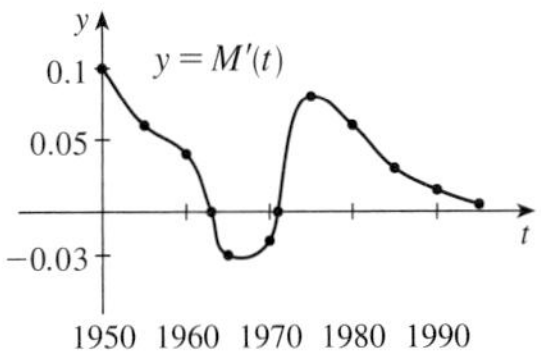

1963 to 1971

17.

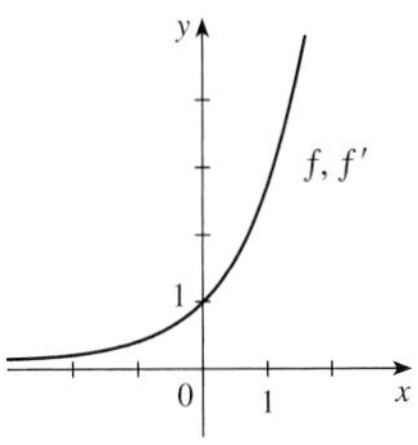

$f'(x) = e^x$

19. (a) 0, 1, 2, 4 (b) $-1, -2, -4$ (c) $f'(x) = 2x$

21. $f'(x) = \frac{1}{2}$, $\mathbb{R}$, $\mathbb{R}$ **23.** $f'(t) = 5 - 18t$, $\mathbb{R}$, $\mathbb{R}$

25. $f'(x) = 2x - 6x^2$, $\mathbb{R}$, $\mathbb{R}$

27. $g'(x) = 1/\sqrt{1 + 2x}$, $\left[-\frac{1}{2}, \infty\right)$, $\left(-\frac{1}{2}, \infty\right)$

29. $G'(t) = \dfrac{4}{(t + 1)^2}$, $(-\infty, -1) \cup (-1, \infty)$, $(-\infty, -1) \cup (-1, \infty)$

31. $f'(x) = 4x^3$, $\mathbb{R}$, $\mathbb{R}$

33. (a) $f'(x) = 4x^3 + 2$ (b)

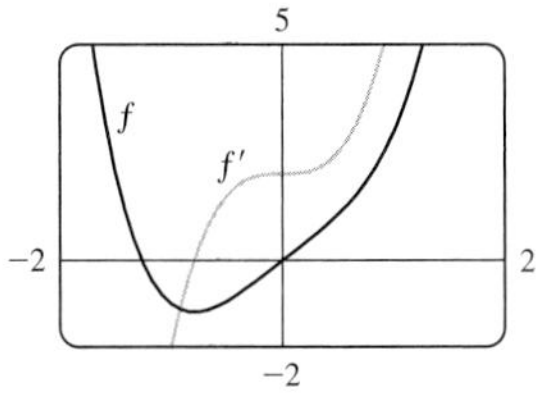

35. (a) $N'(t)$ is the rate of change of the number of malarial parasites with respect to time. Its units are (number of parasites/μL)/day.
(b)

t	$N'(t)$
2	6,261
3	12,152
4	179,791
5	1,095,390
6	3,188,035

37. -4 (corner); 0 (discontinuity)

39. -1 (vertical tangent); 4 (corner)

41.

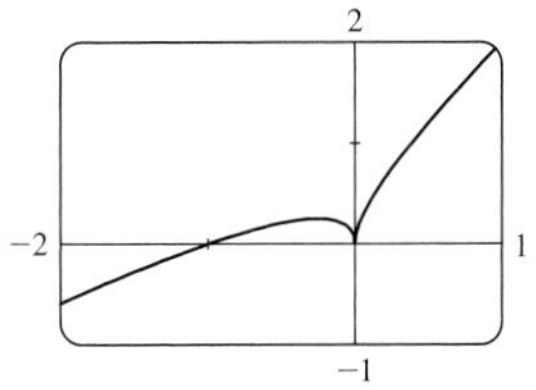

Differentiable at -1; not differentiable at 0

43. $a = f$, $b = f'$, $c = f''$

45. $6x + 2$; 6

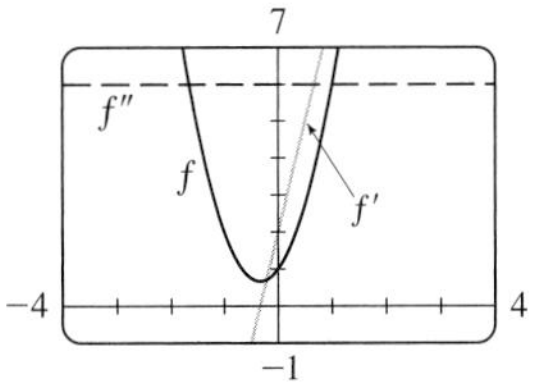

47. (a) Increasing on (0, 1), (4, 5); decreasing on (1, 4)
(b)

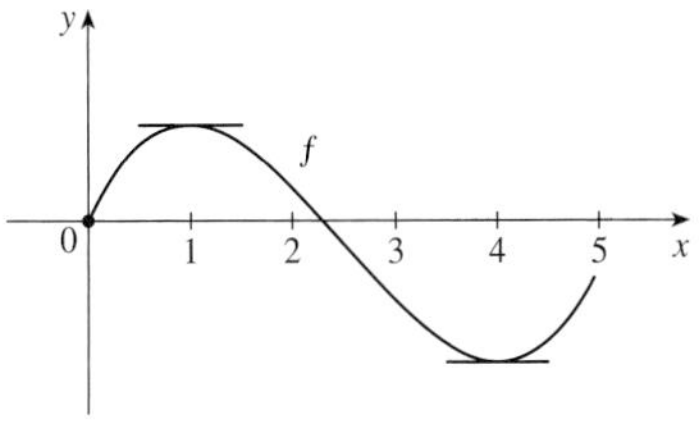

EXERCISES 3.3 | page 192

1. (a) See Definition of the number e (page 189).
(b) 0.99, 1.03; $2.7 < e < 2.8$

3. $f'(x) = 0$ **5.** $f'(x) = 5$ **7.** $f'(x) = 3x^2 - 4$

9. $1 - 3\cos x$ **11.** $f'(t) = t^3$ **13.** $A'(s) = 60/s^6$

15. $g'(t) = -\frac{3}{2}t^{-7/4}$ **17.** $y' = 3e^x - \frac{4}{3}x^{-4/3}$

19. $F'(x) = \frac{5}{32}x^4$ **21.** $y' = \frac{3}{2}\sqrt{x} + \dfrac{2}{\sqrt{x}} - \dfrac{3}{2x\sqrt{x}}$

23. $y' = 0$ **25.** $g'(y) = -10A/y^{11} - B \sin y$

27. $f'(x) = k(2x - a - b)$ **29.** $u' = \frac{1}{5}t^{-4/5} + 10t^{3/2}$

31. $G'(y) = -10A/y^{11} + Be^y$ **33.** $y = \frac{1}{4}x + \frac{3}{4}$

35. $y = -3\sqrt{3}x + 3 + \pi\sqrt{3},\ y = x/(3\sqrt{3}) + 3 - \pi/(9\sqrt{3})$

37. Tangent: $y = 2x + 2$; normal: $y = -\frac{1}{2}x + 2$

39. $y = \frac{3}{2}x + \frac{1}{2}$

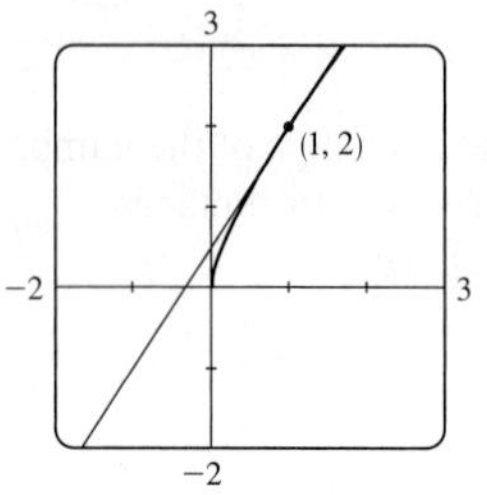

41. $45x^{14} - 15x^2$

43. $f'(x) = 4x^3 - 9x^2 + 16,\ f''(x) = 12x^2 - 18x$

45. $g'(t) = -2 \sin t - 3 \cos t,\ g''(t) = -2 \cos t + 3 \sin t$

47. $\left.\dfrac{dL}{dA}\right|_{A=12} = 1.718$ means that a 12-year old rock bass grows at a rate of 1.718 inches/year.

49. (a) 0.926 cm/s; 0.694 cm/s; 0
(b) 0; −92.6 (cm/s)/cm; −185.2 (cm/s)/cm
(c) At the center; at the edge

51. (a) $v(t) = 3t^2 - 9t - 7;\ a(t) = 6t - 9$
(b) $t = 4$ s (c) $t = 1.5$ s; the velocity has an absolute minimum

53. (a) 30 mm²/mm; the rate at which the area is increasing with respect to side length as x reaches 15 mm
(b) $\Delta A \approx 2x\,\Delta x$

55. (a) (i) 5π (ii) 4.5π (iii) 4.1π
(b) 4π (c) $\Delta A \approx 2\pi r\,\Delta r$

57. $-\cos x$ **59.** $(2n + 1)\pi \pm \frac{1}{3}\pi$, n an integer

63. $y = 12x - 15,\ y = 12x + 17$ **65.** $y = \frac{1}{3}x - \frac{1}{3}$

67. $(\pm 2, 4)$ **71.** $y = 2x^2 - x$

73. $a = -\frac{1}{2},\ b = 2$ **75.** 1000

EXERCISES 3.4 | page 200

1. $1 - 2x + 6x^2 - 8x^3$ **3.** $f'(x) = e^x(x^3 + 3x^2 + 2x + 2)$

5. $g'(t) = 3t^2 \cos t - t^3 \sin t$ **7.** $F'(y) = 5 + \dfrac{14}{y^2} + \dfrac{9}{y^4}$

9. $f'(x) = \cos x - \frac{1}{2}\csc^2 x$

11. $h'(\theta) = \csc\theta - \theta\csc\theta\cot\theta + \csc^2\theta$

13. $y' = (x - 2)e^x/x^3$ **15.** $g'(x) = 5/(2x + 1)^2$

17. $y' = \dfrac{x^2(3 - x^2)}{(1 - x^2)^2}$ **19.** $y' = \dfrac{2t(-t^4 - 4t^2 + 7)}{(t^4 - 3t^2 + 1)^2}$

21. $y' = (r^2 - 2)e^r$ **23.** $f'(\theta) = \dfrac{\sec\theta\tan\theta}{(1 + \sec\theta)^2}$

25. $y' = (x\cos x - 2\sin x)/x^3$ **27.** $y' = 2v - 1/\sqrt{v}$

29. $f'(t) = \dfrac{4 + t^{1/2}}{(2 + \sqrt{t})^2}$ **31.** $f'(x) = \dfrac{-ACe^x}{(B + Ce^x)^2}$

33. $f'(x) = \dfrac{2cx}{(x^2 + c)^2}$ **35.** $y = \frac{1}{2}x + \frac{1}{2}$

37. $y = 2x;\ y = -\frac{1}{2}x$ **39.** $(x^4 + 4x^3)e^x;\ (x^4 + 8x^3 + 12x^2)e^x$

41. $\theta\cos\theta + \sin\theta;\ 2\cos\theta - \theta\sin\theta$

43. $f^{(n)}(x) = e^x(n + x)$

49. (a) −16 (b) $-\frac{20}{9}$ (c) 20

51. (a) 0 (b) $-\frac{2}{3}$

53. (a) $y' = xg'(x) + g(x)$
(b) $y' = \dfrac{g(x) - xg'(x)}{[g(x)]^2}$ (c) $y' = \dfrac{xg'(x) - g(x)}{x^2}$

55. $\dfrac{dv}{d[S]} = \dfrac{0.0021}{(0.015 + [S])^2}$ is the rate of change of the enzymatic reaction rate with respect to the concentration of the substrate.

57. −0.2436 K/min **59.** Two, $\left(-2 \pm \sqrt{3}, \frac{1}{2}(1 \mp \sqrt{3})\right)$

61. (c) $3e^{3x}$ **63.** (b) $y' = -2x(2x^2 + 1)/(x^4 + x^2 + 1)^2$

EXERCISES 3.5 | page 212

1. $\dfrac{4}{3\sqrt[3]{(1 + 4x)^2}}$ **3.** $\pi\sec^2\pi x$ **5.** $\dfrac{e^{\sqrt{x}}}{2\sqrt{x}}$

7. $F'(x) = 10x(x^4 + 3x^2 - 2)^4(2x^2 + 3)$

9. $F'(x) = -\dfrac{1}{\sqrt{1 - 2x}}$ **11.** $f'(z) = -\dfrac{2z}{(z^2 + 1)^2}$

13. $y' = -3x^2\sin(a^3 + x^3)$ **15.** $h'(t) = 3t^2 - 3^t \ln 3$

17. $y' = e^{-kx}(-kx + 1)$

19. $y' = 8(2x - 5)^3(8x^2 - 5)^{-4}(-4x^2 + 30x - 5)$

21. $y' = (\cos x - x\sin x)e^{x\cos x}$ **23.** $y' = \dfrac{-12x(x^2 + 1)^2}{(x^2 - 1)^4}$

25. $y' = 4\sec^2 x\tan x$ **27.** $y' = (r^2 + 1)^{-3/2}$

29. $y' = 2\cos(\tan 2x)\sec^2(2x)$ **31.** $y' = 2^{\sin\pi x}(\pi\ln 2)\cos\pi x$

33. $y' = -2\cos\theta\cot(\sin\theta)\csc^2(\sin\theta)$

35. $y' = \dfrac{-\pi\cos(\tan\pi x)\sec^2(\pi x)\sin\sqrt{\sin(\tan\pi x)}}{2\sqrt{\sin(\tan\pi x)}}$

37. $y' = -2x\sin(x^2);\ y'' = -4x^2\cos(x^2) - 2\sin(x^2)$

39. $e^{\alpha x}(\beta\cos\beta x + \alpha\sin\beta x);\ e^{\alpha x}[(\alpha^2 - \beta^2)\sin\beta x + 2\alpha\beta\cos\beta x]$

41. $y = 20x + 1$ **43.** $y = -x + \pi$

45. 24 **47.** (a) 30 (b) 36

49. (a) $F'(x) = e^x f'(e^x)$ (b) $G'(x) = e^{f(x)}f'(x)$

51. 120 **53.** $-2^{50}\cos 2x$

55. $v(t) = \frac{5}{2}\pi\cos(10\pi t);\ a(t) = -25\pi^2\sin(10\pi t)$

57. $m'(t) = e^{-t}\cos t$

59. (a) $C'(10) \approx 0.00752$ (mg/mL)/min
(b) The BAC is decreasing at a rate of about 0.00316 (mg/mL)/min.

61. (a) 1 (b) $\dfrac{dp}{dt} = \dfrac{kae^{-kt}}{(1 + ae^{-kt})^2}$

(c)

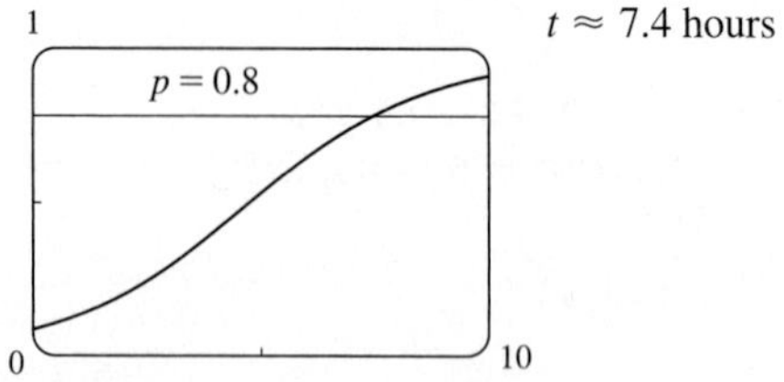

63. (a) $y' = -(y + 2 + 6x)/x$
(b) $y = (4/x) - 2 - 3x$, $y' = -(4/x^2) - 3$

65. $y' = -\dfrac{x^2}{y^2}$ **67.** $y' = \dfrac{2x + y}{2y - x}$

69. $y' = \dfrac{3y^2 - 5x^4 - 4x^3y}{x^4 + 3y^2 - 6xy}$ **71.** $y' = \tan x \tan y$

73. $y' = \dfrac{y(y - e^{x/y})}{y^2 - xe^{x/y}}$ **75.** $y' = \dfrac{e^y \sin x + y \cos(xy)}{e^y \cos x - x \cos(xy)}$

77. $y = -x + 2$ **79.** $y = x + \frac{1}{2}$

81. $\dfrac{dA}{dp} = \dfrac{1}{pq - e^{qA}}$ **83.** $dV/dt = 3x^2\, dx/dt$

85. 48 cm^2/s **87.** 80 cm^3/min

89. (a) $\dfrac{dm}{dt} = \pi r^2 L(1 - k^2)\dfrac{d\rho}{dt}$

(b) $\dfrac{dm}{dt} = -2\pi r^2 L(\rho - 1)k\dfrac{dk}{dt}$

91. $\dfrac{dr}{da} = \dfrac{2}{\sqrt{1 + 8s}}\dfrac{ds}{da}$ **93.** $\dfrac{dP}{dx} = -\dfrac{8\eta l v}{R^3}R'(x)$

EXERCISES 3.6 | page 220

1. About 235
3. (a) $100(4.2)^t$ (b) ≈ 7409 (c) $\approx 10{,}632$ bacteria/h
(d) $(\ln 100)/(\ln 4.2) \approx 3.2$ h
5. (a) 1508 million, 1871 million (b) 2161 million
(c) 3972 million; wars in the first half of century, increased life expectancy in second half
7. (a) $100 \times 2^{-t/30}$ mg (b) ≈ 9.92 mg
(c) ≈ -0.229 mg/year (d) ≈ 199.3 years

9. (a) $-\dfrac{\ln 2}{\ln 0.945} \approx 12.25$ years (b) $-\dfrac{\ln 5}{\ln 0.945} \approx 28.45$ years

11. $\approx 57{,}104$ years
13. (a) ≈ 137°F (b) ≈ 116 min
15. (a) $13.\overline{3}$°C (b) ≈ 67.74 min
17. (a) ≈ 64.5 kPa (b) ≈ 39.9 kPa

EXERCISES 3.7 | page 229

1. The differentiation formula is simplest.

3. $f'(x) = \dfrac{\cos(\ln x)}{x}$ **5.** $f'(x) = \dfrac{3}{(3x - 1)\ln 2}$

7. $f'(x) = \dfrac{1}{5x\sqrt[5]{(\ln x)^4}}$ **9.** $f'(x) = \dfrac{\sin x}{x} + \cos x \ln(5x)$

11. $F'(t) = \dfrac{6}{2t + 1} - \dfrac{12}{3t - 1}$ **13.** $g'(x) = \dfrac{2x^2 - 1}{x(x^2 - 1)}$

15. $y' = \dfrac{10x + 1}{5x^2 + x - 2}$ **17.** $y' = \dfrac{-x}{1 + x}$

19. $y' = \dfrac{1}{\ln 10} + \log_{10} x$

21. $y' = x + 2x \ln(2x)$; $y'' = 3 + 2 \ln(2x)$

23. $f'(x) = \dfrac{2x - 1 - (x - 1)\ln(x - 1)}{(x - 1)[1 - \ln(x - 1)]^2}$; $(1, 1 + e) \cup (1 + e, \infty)$

25. $y = 3x - 9$ **27.** 1

29. $\dfrac{dt}{dc} = \dfrac{3 + (9c - 4)/\sqrt{9c^2 - 8c}}{3c + \sqrt{9c^2 - 8c}}$ is the rate of change of dialysis duration as the initial urea concentration increases.

31. $\dfrac{da}{dN} = -\dfrac{5370}{(\ln 2)N}$ is the rate of change of the estimated age with respect to an increase in the measured amount of ^{14}C.

33. $y' = (2x + 1)^5(x^4 - 3)^6\left(\dfrac{10}{2x + 1} + \dfrac{24x^3}{x^4 - 3}\right)$

35. $y' = \dfrac{\sin^2 x \tan^4 x}{(x^2 + 1)^2}\left(2 \cot x + \dfrac{4 \sec^2 x}{\tan x} - \dfrac{4x}{x^2 + 1}\right)$

37. $y' = x^x(\ln x + 1)$ **39.** $y' = (\cos x)^x(-x \tan x + \ln \cos x)$

41. $y' = (\tan x)^{1/x}\left(\dfrac{\sec^2 x}{x \tan x} - \dfrac{\ln \tan x}{x^2}\right)$ **43.** $y' = \dfrac{2 \tan^{-1} x}{1 + x^2}$

45. $y' = -\dfrac{\sin\theta}{1 + \cos^2\theta}$ **47.** $y' = \dfrac{1}{2(1 + x^2)}$ **49.** $\pi/2$

51. $y' = \dfrac{2x}{x^2 + y^2 - 2y}$ **53.** $f^{(n)}(x) = \dfrac{(-1)^{n-1}(n - 1)!}{(x - 1)^n}$

EXERCISES 3.8 | page 237

1. $L(x) = -10x - 6$ **3.** $L(x) = -x + \pi/2$

5. $\sqrt{1 - x} \approx 1 - \frac{1}{2}x$;
$\sqrt{0.9} \approx 0.95$,
$\sqrt{0.99} \approx 0.995$

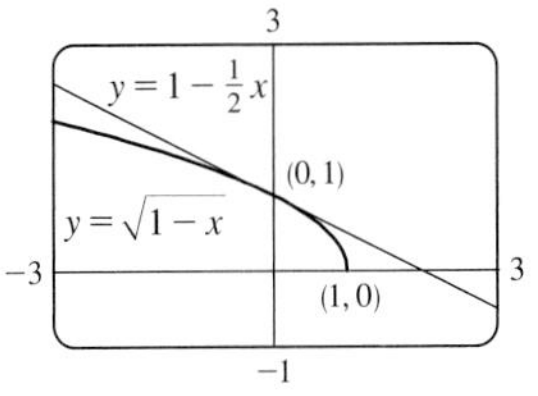

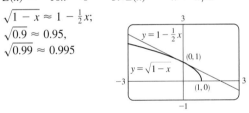

7. $-1.204 < x < 0.706$ **9.** $-0.045 < x < 0.055$
11. 32.08 **15.** $L(s) = p + p(1 - p)s$
17. A 5% increase in the radius corresponds to a 20% increase in blood flow.
19. (a) $x_2 \approx 2.3$, $x_3 \approx 3$ (b) No
21. 1.1797 **23.** -0.724492, 1.220744
25. 1.412391, 3.057104
27. -1.93822883, -1.21997997, 1.13929375, 2.98984102
29. -1.97806681, -0.82646233
31. 0.21916368, 1.08422462
35. $T_3(x) = 1 + x + \frac{1}{2}x^2 + \frac{1}{6}x^3$
37. $T_3(x) = 1 - (x - 1) + (x - 1)^2 - (x - 1)^3 + (x - 1)^4$
39. $L(x) = \frac{1}{4}x + \frac{7}{4}$
$P(x) \approx 2 + \frac{1}{4}(x - 1) - \frac{1}{64}(x - 1)^2$

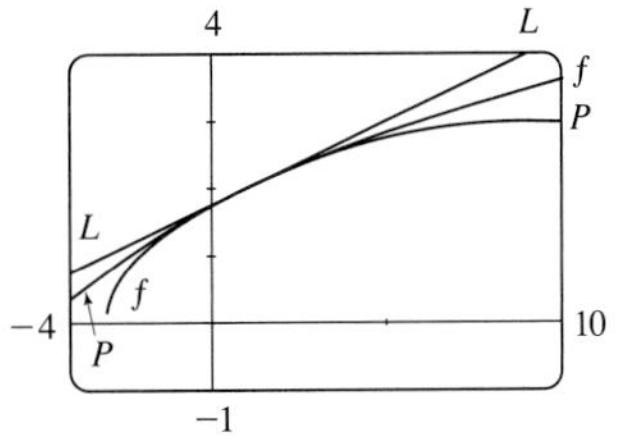

$P(x)$ is a better approximation than $L(x)$.

41. $T_1(x) = x$
$T_2(x) = x$
$T_3(x) = x - \frac{1}{6}x^3$
$T_4(x) = x - \frac{1}{6}x^3$
$T_5(x) = x - \frac{1}{6}x^3 + \frac{1}{120}x^5$

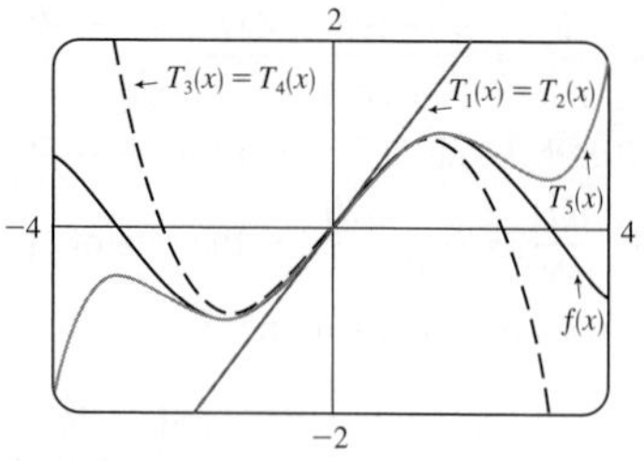

43. (a) $T_1(s) = 1 + 2s$ (b) $T_2(s) = 1 + 2s - 4s^2$

CHAPTER 3 REVIEW | page 240

True-False Quiz

1. False **3.** True **5.** True **7.** False **9.** False
11. True **13.** True

Exercises

1. $f''(5), 0, f'(5), f'(2), 1, f'(3)$

3. (a) The rate at which the cost changes with respect to the interest rate; dollars/(percent per year)
(b) As the interest rate increases past 10%, the cost is increasing at a rate of \$1200/(percent per year).
(c) Always positive

5.

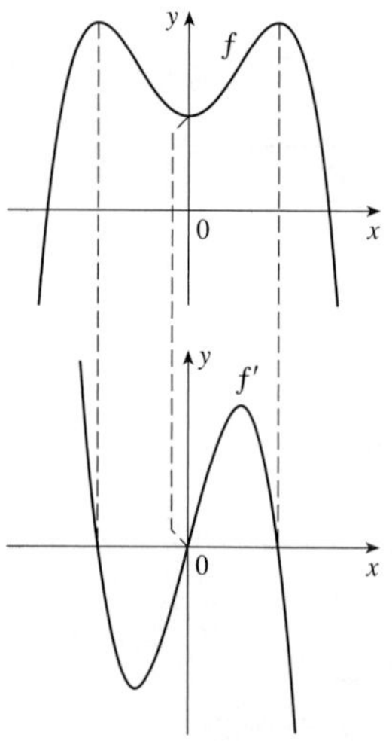

7.

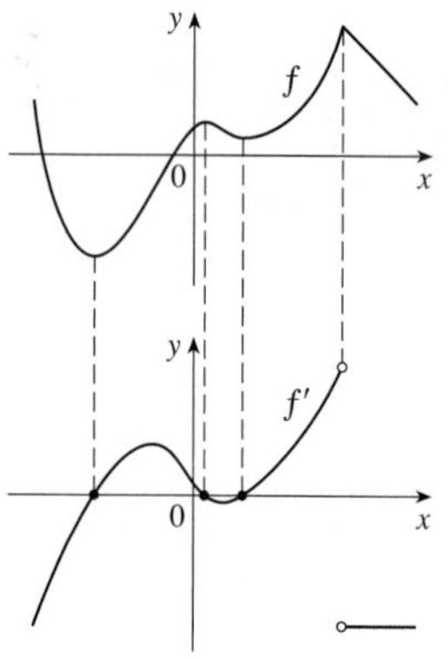

9. (a) Rate of change of heart rate with respect to time.
(b)

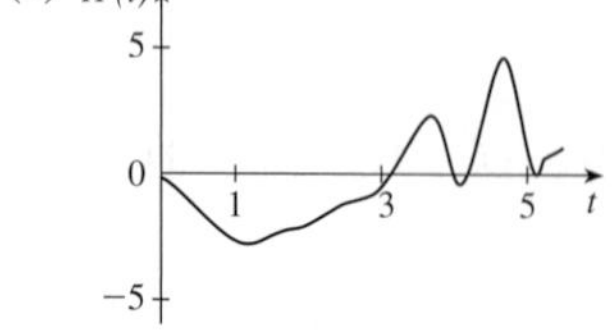

11. (a) $f'(x) = -\frac{5}{2}(3 - 5x)^{-1/2}$ (b) $\left(-\infty, \frac{3}{5}\right], \left(-\infty, \frac{3}{5}\right)$
(c)

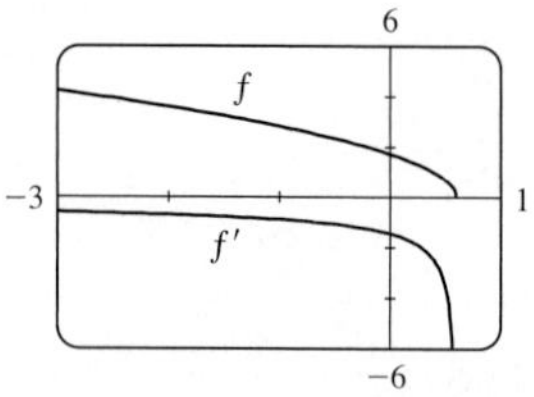

13. −4 (discontinuity), −1 (corner), 2 (discontinuity), 5 (vertical tangent)

15. $6x(x^4 - 3x^2 + 5)^2(2x^2 - 3)$

17. $\dfrac{1}{2\sqrt{x}} - \dfrac{4}{3\sqrt[3]{x^7}}$ **19.** $\dfrac{2(2x^2 + 1)}{\sqrt{x^2 + 1}}$ **21.** $2\cos 2\theta e^{\sin 2\theta}$

23. $\dfrac{t^2 + 1}{(1 - t^2)^2}$ **25.** $-\dfrac{e^{1/x}(1 + 2x)}{x^4}$ **27.** $\dfrac{1 - y^4 - 2xy}{4xy^3 + x^2 - 3}$

29. $\dfrac{2\sec 2\theta(\tan 2\theta - 1)}{(1 + \tan 2\theta)^2}$ **31.** $(1 + c^2)e^{cx}\sin x$

33. $\dfrac{2}{(1 + 2x)\ln 5}$ **35.** $\dfrac{2x - y\cos(xy)}{x\cos(xy) + 1}$

37. $3^{x\ln x}(\ln 3)(1 + \ln x)$ **39.** $\cot x - \sin x \cos x$

41. $\dfrac{4x}{1 + 16x^2} + \tan^{-1}(4x)$ **43.** $5\sec 5x$

45. $2\cos\theta\tan(\sin\theta)\sec^2(\sin\theta)$

47. $\cos\left(\tan\sqrt{1 + x^3}\right)\left(\sec^2\sqrt{1 + x^3}\right)\dfrac{3x^2}{2\sqrt{1 + x^3}}$

49. $\dfrac{-3\sin\left(e^{\sqrt{\tan 3x}}\right)e^{\sqrt{\tan 3x}}\sec^2(3x)}{2\sqrt{\tan 3x}}$

51. $-\frac{4}{27}$ **53.** $2^x(\ln 2)^n$

55. $y = 2\sqrt{3}x + 1 - \pi\sqrt{3}/3$

57. $y = -x + 2$; $y = x + 2$

59. (a) $\dfrac{10 - 3x}{2\sqrt{5 - x}}$ (b) $y = \frac{7}{4}x + \frac{1}{4}$, $y = -x + 8$
(c)

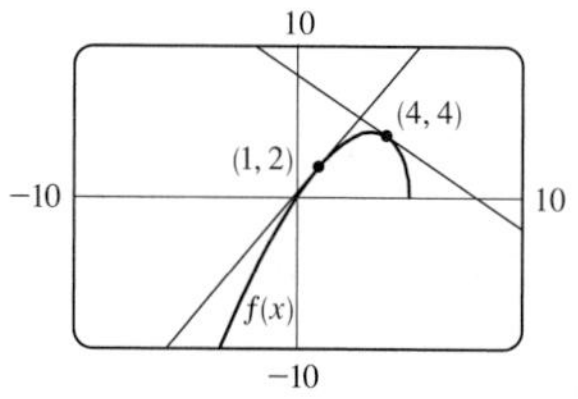

61. (a) 2 (b) 44

63. $2xg(x) + x^2g'(x)$ **65.** $2g(x)g'(x)$

67. $g'(e^x)e^x$ **69.** $g'(x)/g(x)$ **71.** $(-3, 0)$

73. $\left(\pm 2/\sqrt{6}, \mp 1/\sqrt{6}\right)$

75. $v(t) = -Ae^{-ct}[c\cos(\omega t + \delta) + \omega\sin(\omega t + \delta)]$,
$a(t) = Ae^{-ct}[(c^2 - \omega^2)\cos(\omega t + \delta) + 2c\omega\sin(\omega t + \delta)]$

77. (a) $dV/dh = \frac{1}{3}\pi r^2$ (b) $dV/dr = \frac{2}{3}\pi rh$

79. 1980: $E'(10) \approx$ \$21.56 billion/year;
2000: $E'(30) \approx$ \$125.64 billion/year

81. (a) $200(3.24)^t$ (b) $\approx$22,040
(c) $\approx$25,910 bacteria/h (d) $(\ln 50)/(\ln 3.24) \approx 3.33$ h

83. (a) C_0e^{-kt} (b) $\approx$100 h

85. $\frac{4}{3}$ cm²/min **87.** 1.297383

89. (a) $L(x) = 1 + x$; $\sqrt[3]{1 + 3x} \approx 1 + x$; $\sqrt[3]{1.03} \approx 1.01$
(b) $-0.235 < x < 0.401$

91. (a) $T_1(c) = \ln 2 + 2(c - 1)$
(b) $T_2(c) = \ln 2 + 2(c - 1) - 4(c - 1)^2$

93. $(\cos\theta)'\big|_{\theta=\pi/3} = -\sqrt{3}/2$

CHAPTER 4

EXERCISES 4.1 | page 256

Abbreviations: abs, absolute; loc, local; max, maximum; min, minimum

1. Abs min: smallest function value on the entire domain of the function; loc min at c: smallest function value when x is near c

3. Abs max at s, abs min at r, loc max at c, loc min at b and r

5. Local max: $f(0.18) \approx 0.36$ mV, $f(0.30) \approx 1.2$ mV, $f(0.57) \approx 0.31$ mV
Abs max: $f(0.30) \approx 1.2$ mV
Local min: $f(0.29) \approx -0.01$ mV, $f(0.32) \approx -0.76$ mV
Abs Min: $f(0.32) \approx -0.76$ mV

7. Local max: D(Oct 23, 1918) ≈ 91, D(Jan 22, 1919) ≈ 15
Abs max: D(Oct 23, 1918) ≈ 91
Local min: D(Nov 27, 1918) ≈ 4, D(Feb 13, 1919) ≈ 7
Abs min: D(Sept 11, 1918) ≈ 0

9.
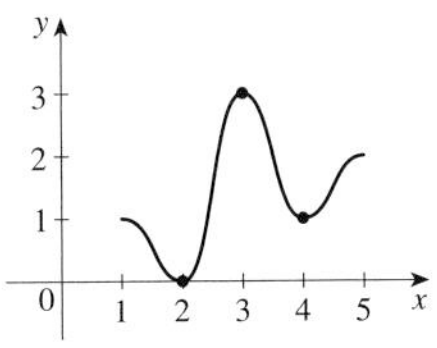

11.
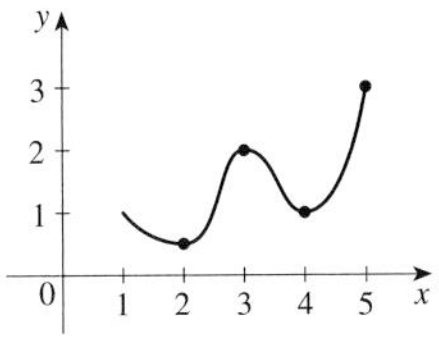

13. (a)
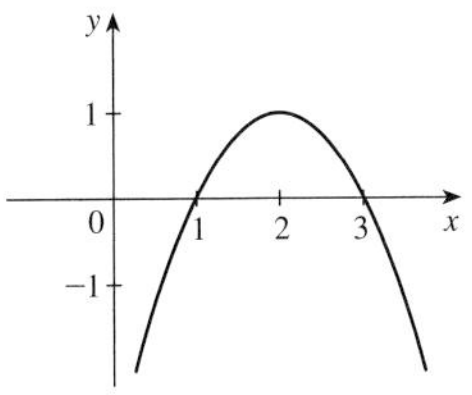

(b)
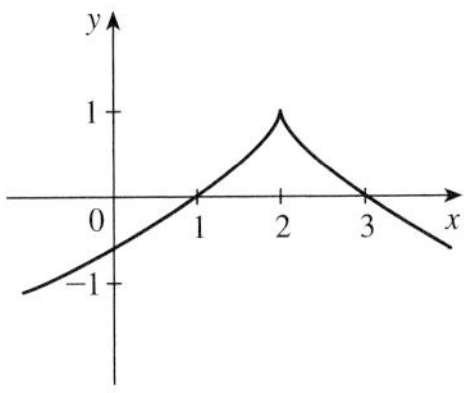

(c)
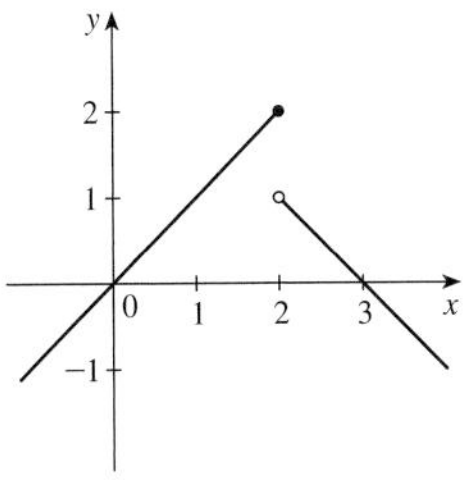

15. (a)
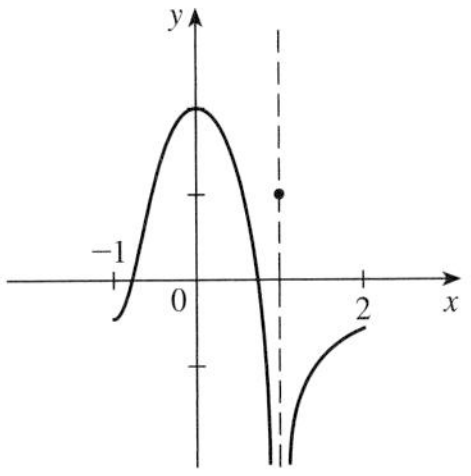

(b)
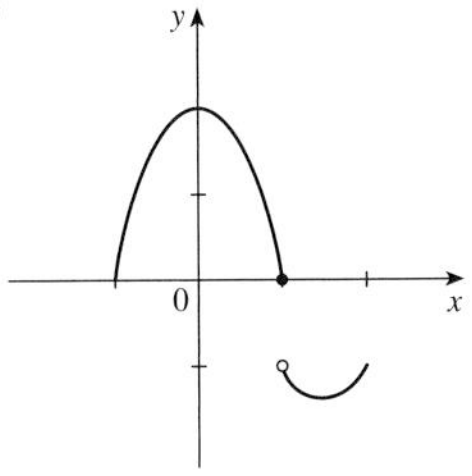

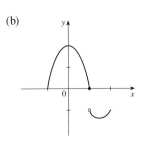

17. Abs max $f(3) = 4$ **19.** None

21. Abs max $f(2) = \ln 2$ **23.** Abs max $f(0) = 1$

25. $\frac{1}{3}$ **27.** $-4, 2$

29. $0, \frac{1}{2}(-1 \pm \sqrt{5})$ **31.** $0, 2$

33. $0, \frac{4}{9}$ **35.** $0, \frac{8}{7}, 4$ **37.** $n\pi$ (n an integer) **39.** $0, \frac{2}{3}$

41. $f(2) = 16, f(5) = 7$ **43.** $f(-1) = 8, f(2) = -19$

45. $f(3) = 66, f(\pm 1) = 2$ **47.** $f(\sqrt{2}) = 2, f(-1) = -\sqrt{3}$

49. $f(2) = 2/\sqrt{e}, f(-1) = -1/\sqrt[8]{e}$

51. $f(1) = \ln 3, f(-\frac{1}{2}) = \ln \frac{3}{4}$

53. $f(\pi/6) = \frac{3}{2}\sqrt{3}, f(\pi/2) = 0$

55. $f\left(\dfrac{a}{a+b}\right) = \dfrac{a^a b^b}{(a+b)^{a+b}}$

57. Smallest: $P(2.16) \approx 5.32$; largest: $P(0) = P(9) = 10$

59. (a) $r = \frac{2}{3}r_0$ (b) $v = \frac{4}{27}kr_0^3$
(c)
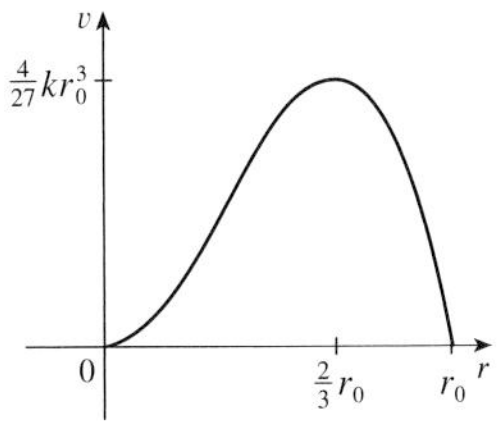

61. ≈ 3.9665°C

EXERCISES 4.2 | page 271

Abbreviations: inc, increasing; dec, decreasing; CD, concave downward; CU, concave upward; HA, horizontal asymptote; VA, vertical asymptote; IP, inflection point(s)

1. 0.8, 3.2, 4.4, 6.1

5. (a) (1, 3), (4, 6) (b) (0, 1), (3, 4) (c) (0, 2)
(d) (2, 4), (4, 6) (e) (2, 3)

7. (a) See the First Derivative Test
(b) See the Second Derivative Test and the note that precedes Example 9.

9. (a) 3, 5 (b) 2, 4, 6 (c) 1, 7

11. (a) Inc on $(-\infty, -3), (2, \infty)$; dec on $(-3, 2)$
(b) Loc max $f(-3) = 81$; loc min $f(2) = -44$
(c) CU on $(-\frac{1}{2}, \infty)$; CD on $(-\infty, -\frac{1}{2})$; IP $(-\frac{1}{2}, \frac{37}{2})$

13. (a) Inc on $(-1, 0), (1, \infty)$; dec on $(-\infty, -1), (0, 1)$
(b) Loc max $f(0) = 3$; loc min $f(\pm 1) = 2$
(c) CU on $(-\infty, -\sqrt{3}/3), (\sqrt{3}/3, \infty)$;
CD on $(-\sqrt{3}/3, \sqrt{3}/3)$; IP $(\pm\sqrt{3}/3, \frac{22}{9})$

15. (a) Inc on $(0, \pi/4), (5\pi/4, 2\pi)$; dec on $(\pi/4, 5\pi/4)$
(b) Loc max $f(\pi/4) = \sqrt{2}$; loc min $f(5\pi/4) = -\sqrt{2}$
(c) CU on $(3\pi/4, 7\pi/4)$; CD on $(0, 3\pi/4), (7\pi/4, 2\pi)$;
IP $(3\pi/4, 0), (7\pi/4, 0)$

17. (a) Inc on $(-\frac{1}{3}\ln 2, \infty)$; dec on $(-\infty, -\frac{1}{3}\ln 2)$
(b) Loc min $f(-\frac{1}{3}\ln 2) = 2^{-2/3} + 2^{1/3}$ (c) CU on $(-\infty, \infty)$

19. (a) Inc on $(0, e^2)$; dec on (e^2, ∞)
(b) Loc max $f(e^2) = 2/e$
(c) CU on $(e^{8/3}, \infty)$; CD on $(0, e^{8/3})$; IP $(e^{8/3}, \frac{8}{3}e^{-4/3})$

21. Loc max $f(\frac{3}{4}) = \frac{5}{4}$

23. (a) f has a local maximum at 2.
(b) f has a horizontal tangent at 6.

25. (a) Inc on $(-\infty, -1)$, $(2, \infty)$; dec on $(-1, 2)$
(b) Loc max $f(-1) = 7$; loc min $f(2) = -20$
(c) CU on $(\frac{1}{2}, \infty)$; CD on $(-\infty, \frac{1}{2})$; IP $(\frac{1}{2}, -\frac{13}{2})$
(d) See graph at right.

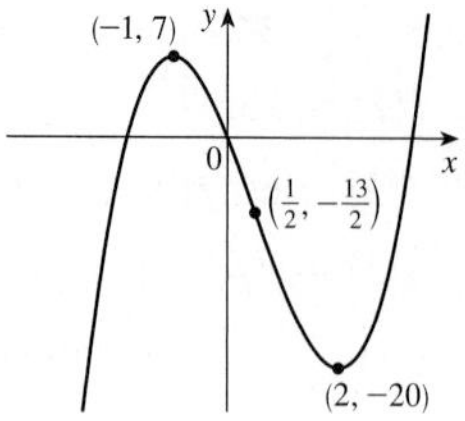

27. (a) Inc on $(-\infty, -1)$, $(0, 1)$; dec on $(-1, 0)$, $(1, \infty)$
(b) Loc max $f(-1) = 3, f(1) = 3$; loc min $f(0) = 2$
(c) CU on $(-1/\sqrt{3}, 1/\sqrt{3})$; CD on $(-\infty, -1/\sqrt{3})$, $(1/\sqrt{3}, \infty)$; IP $(\pm 1/\sqrt{3}, \frac{23}{9})$
(d) See graph at right.

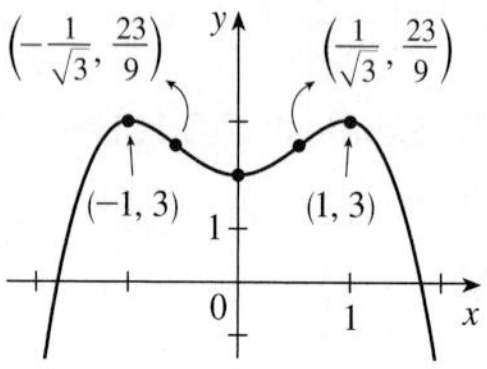

29. (a) Inc on $(-\infty, -2)$, $(0, \infty)$; dec on $(-2, 0)$
(b) Loc max $h(-2) = 7$; loc min $h(0) = -1$
(c) CU on $(-1, \infty)$; CD on $(-\infty, -1)$; IP $(-1, 3)$
(d) See graph at right.

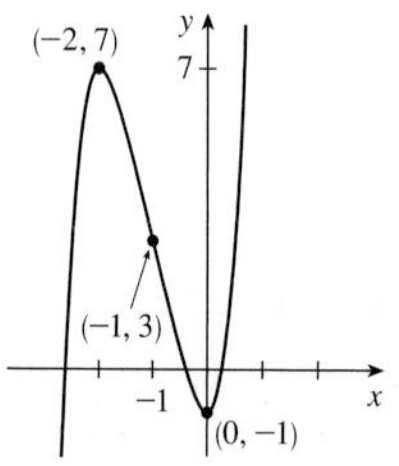

31. (a) Inc on $(-2, \infty)$; dec on $(-3, -2)$
(b) Loc min $A(-2) = -2$
(c) CU on $(-3, \infty)$
(d) See graph at right.

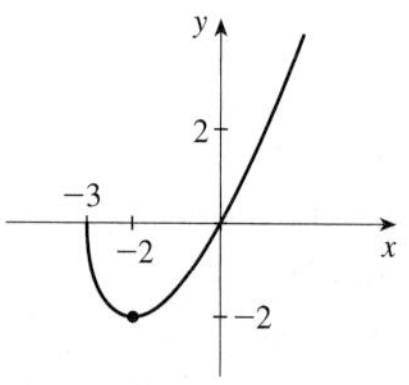

33. (a) Inc on $(-1, \infty)$; dec on $(-\infty, -1)$
(b) Loc min $C(-1) = -3$
(c) CU on $(-\infty, 0)$, $(2, \infty)$; CD on $(0, 2)$; IP $(0, 0)$, $(2, 6\sqrt[3]{2})$
(d) See graph at right.

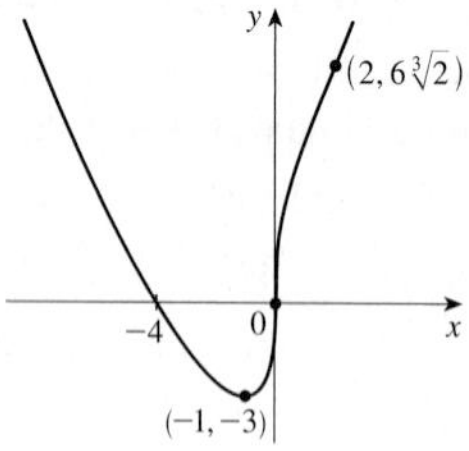

35. (a) Inc on $(\pi, 2\pi)$; dec on $(0, \pi)$
(b) Loc min $f(\pi) = -1$
(c) CU on $(\pi/3, 5\pi/3)$; CD on $(0, \pi/3)$, $(5\pi/3, 2\pi)$; IP $(\pi/3, \frac{5}{4})$, $(5\pi/3, \frac{5}{4})$
(d) See graph at right.

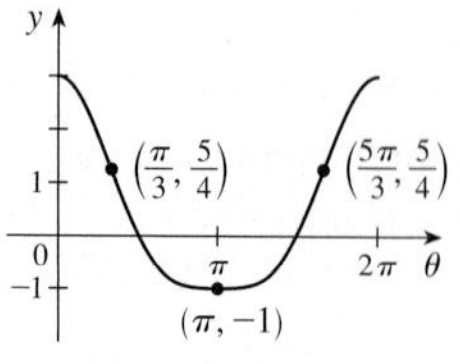

37. (a) HA $y = 1$, VA $x = -1, x = 1$
(b) Inc on $(-\infty, -1)$, $(-1, 0)$; dec on $(0, 1)$, $(1, \infty)$
(c) Loc max $f(0) = 0$
(d) CU on $(-\infty, -1)$, $(1, \infty)$; CD on $(-1, 1)$
(e) See graph at right.

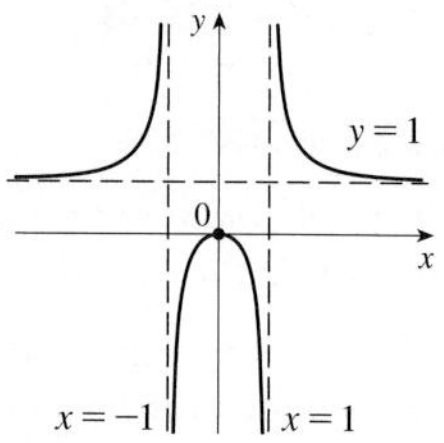

39. (a) HA $y = 0$
(b) Dec on $(-\infty, \infty)$
(c) None
(d) CU on $(-\infty, \infty)$
(e) See graph at right.

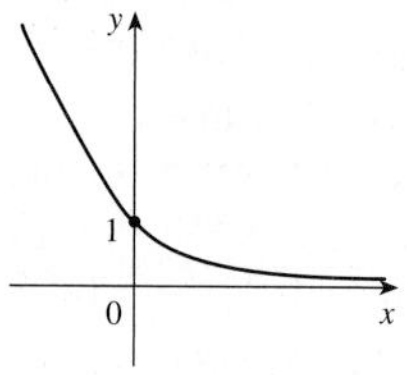

41. (a) VA $x = 0, x = e$
(b) Dec on $(0, e)$
(c) None
(d) CU on $(0, 1)$; CD on $(1, e)$; IP $(1, 0)$
(e) See graph at right.

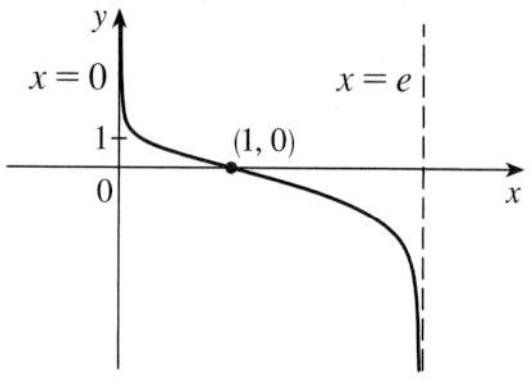

43. (a) HA $y = 1$, VA $x = -1$
(b) Inc on $(-\infty, -1)$, $(-1, \infty)$
(c) None
(d) CU on $(-\infty, -1)$, $(-1, -\frac{1}{2})$; CD on $(-\frac{1}{2}, \infty)$; IP $(-\frac{1}{2}, 1/e^2)$
(e) See graph at right.

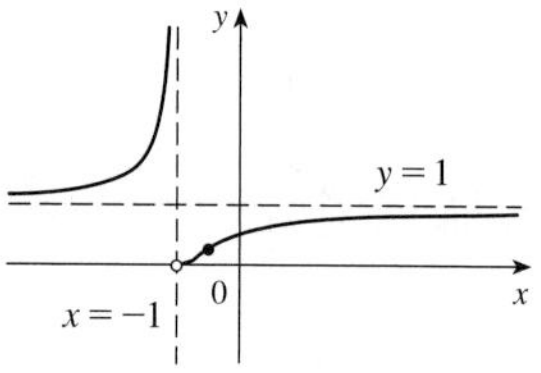

45. $(3, \infty)$

47. (a) Very unhappy (b) Unhappy (c) Happy (d) Very happy

49.

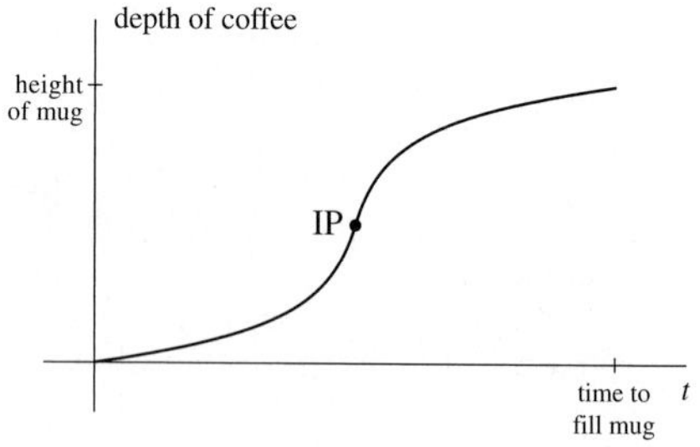

51. 28.57 min, when the rate of increase of drug level in the bloodstream is greatest; 85.71 min, when rate of decrease is greatest

53. (a) $c = 1$
(b) Constant concentration treatment yields a better response.

55.

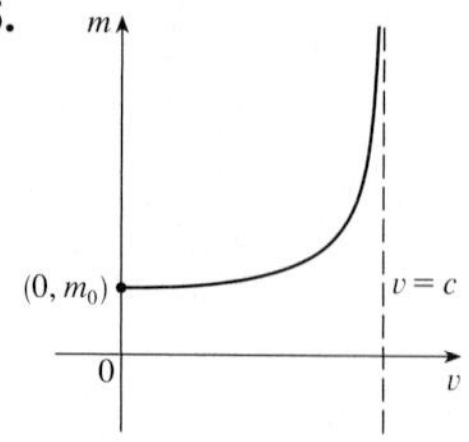

57. $f(x) = \frac{1}{9}(2x^3 + 3x^2 - 12x + 7)$

59. Inc on (0.92, 2.5), (2.58, ∞); dec on ($-\infty$, 0.92), (2.5, 2.58); loc max $f(2.5) = 4$; loc min $f(0.92) \approx -5.12$, $f(2.58) \approx 3.998$; CU on ($-\infty$, 1.46), (2.54, ∞); CD on (1.46, 2.54); IP (1.46, -1.40), (2.54, 3.999)

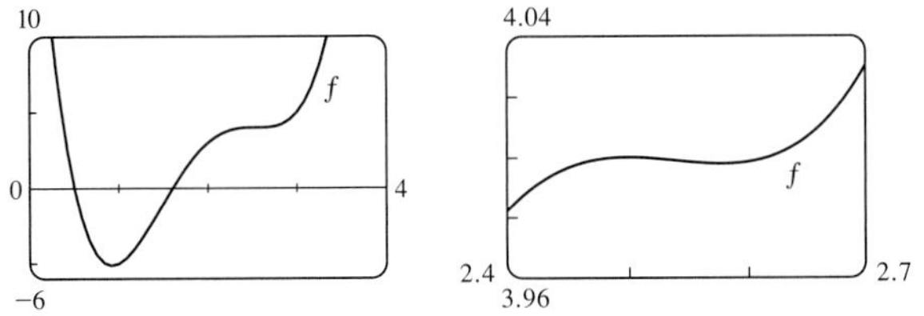

61. Inc on (-1.49, -1.07), (2.89, 4); dec on (-4, -1.49), (-1.07, 2.89); loc max $f(-1.07) \approx 8.79$; loc min $f(-1.49) \approx 8.75$, $f(2.89) \approx -9.99$, CU on (-4, -1.28), (1.28, 4); CD on (-1.28, 1.28); IP (-1.28, 8.77), (1.28, -1.48)

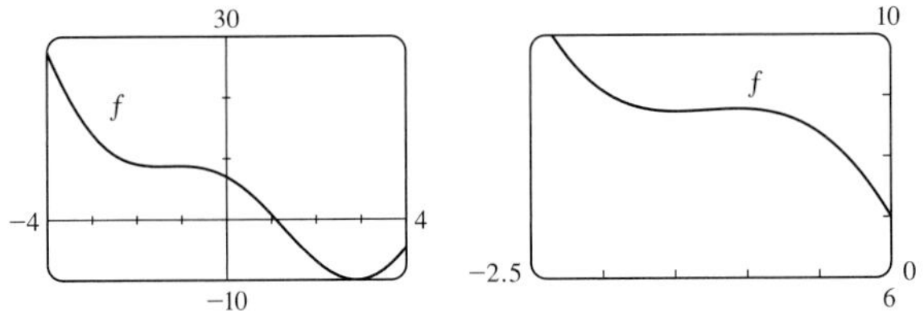

■ EXERCISES 4.3 | page 282

1. 2 **3.** $-\infty$ **5.** ∞ **7.** 0 **9.** $-\infty$ **11.** 3 **13.** $\ln\frac{5}{3}$
15. $\frac{1}{2}$ **17.** $-1/\pi^2$ **19.** $\frac{1}{2}a(a-1)$ **21.** $\frac{1}{24}$ **23.** π
25. 3 **27.** 0 **29.** $\frac{1}{2}$ **31.** $\frac{1}{2}$ **33.** ∞

35. HA $y = 0$

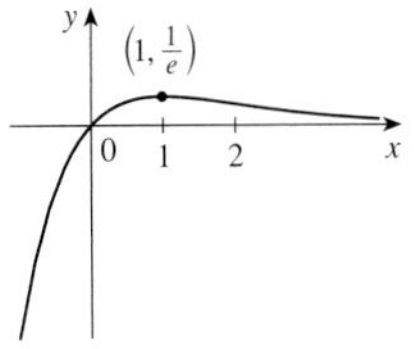

37. HA $y = 0$, VA $x = 0$

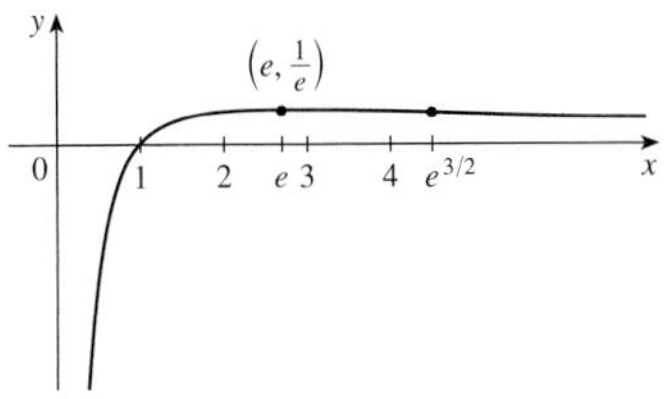

39. Fastest to slowest: $y = e^{3x}$, $y = e^{2x}$, $y = x^5$, $y = \ln(x^{10})$

41. Fastest to slowest: $y = \sqrt{x}$, $y = \sqrt[3]{x}$, $y = (\ln x)^3$, $y = (\ln x)^2$

43. ∞

45. Fastest to slowest: $y = e^{-x}$, $y = \dfrac{1}{x^2}$, $y = \dfrac{1}{x}$, $y = x^{-1/2}$

47. 1

49. (a) r (b) r (c) $f'(0)$

51. (a) 62.3%
(b) 80.1%; Answer is larger due to Stiles-Crawford effect. Light entering closer to center of pupil measures brighter.
(c) $\lim_{r\to 0} P = 100\%$;
Result expected: all light entering at center of pupil is sensed at retina.

53. $te^{-\beta t}$ **57.** $\frac{16}{9}a$

■ EXERCISES 4.4 | page 293

1. (a) 11, 12 (b) 11.5, 11.5
3. 10, 10 **5.** 25 m by 25 m
7. $N = 1$
9. (a)

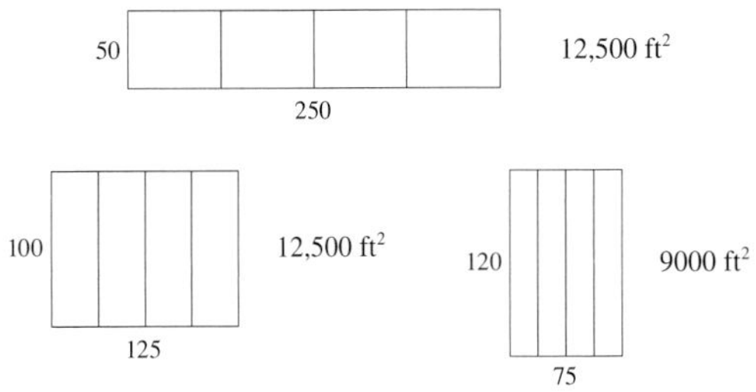

(b)

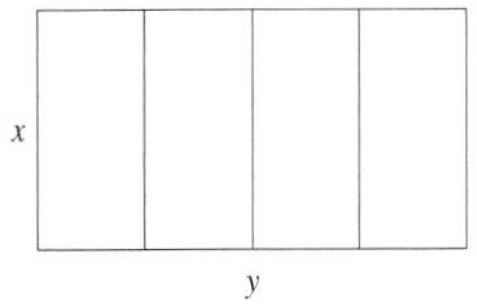

(c) $A = xy$ (d) $5x + 2y = 750$ (e) $A(x) = 375x - \frac{5}{2}x^2$
(f) 14,062.5 ft^2

11. 4000 cm^3 **15.** $\left(-\frac{6}{5}, \frac{3}{5}\right)$

17. (a) $a = e$ (b) $a = 1/\mu$

19. (b) (i) \$342,491; \$342/unit; \$390/unit (ii) 400
(iii) \$320/unit

21. (a) $h = r/2$
(b) $h = \dfrac{r}{2} + \dfrac{\alpha}{2p}$; $N = \dfrac{K}{2}\left(1 - \dfrac{\alpha}{rp}\right)$
(c) Part (b) gives a larger harvest and smaller stabilized fish population. Extra harvesting results in reduced unit costs making it profitable to harvest more.

23. (a) Function starts at zero and increases with time; concave down implies diminishing returns.

27. (a) $\frac{3}{2}s^2 \csc\theta\,(\csc\theta - \sqrt{3}\cot\theta)$
(b) $\cos^{-1}(1/\sqrt{3}) \approx 55°$ (c) $6s\left[h + s/(2\sqrt{2})\right]$

29. (a) About 5.1 km from B
(b) C is close to B; C is close to D;
$W/L = \sqrt{25 + x^2}/x$, where $x = |BC|$
(c) ≈ 1.07; no such value (d) $\sqrt{41}/4 \approx 1.6$

EXERCISES 4.5 | page 305

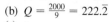

1. Stable: $x = 1$ **3.** Stable: $x = 1$; Unstable $x = 0, 2$
5. Stable: $x = 0$; Unstable: $x = 2$
7. Stable: $x = 0.8$; Unstable: $x = 0$
9. Unstable: $x = 0, \frac{1}{2}\ln(10)$
11. Stable: $x = 0, 3$; Unstable: $x = 1$
$\lim_{t\to\infty} x_t = 0$ when $x_0 = 0.5$
$\lim_{t\to\infty} x_t = 3$ when $x_0 = 2$
13. $x = 0$ is stable equilibrium when $|c| < 1$.
$x = c - 1$ is stable equilibrium when $c < -1$ or $c > 1$.
15. (a) $Q_{n+1} = 0.1Q_n + 200$
(b) $Q = \frac{2000}{9} = 222.\overline{2}$
(c)

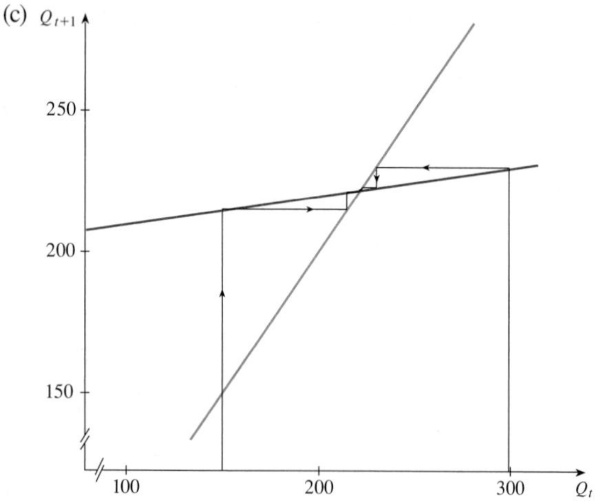

17.

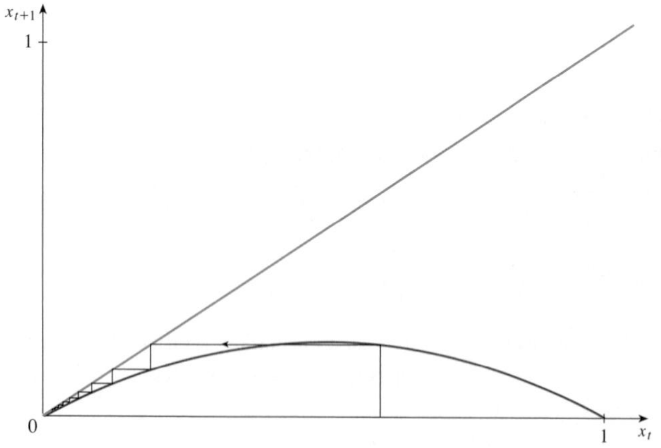

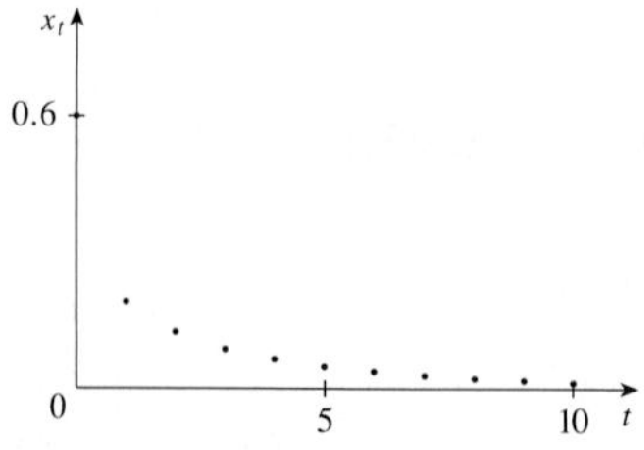

$\hat{x} = 0$ is a stable equilibrium.

19.

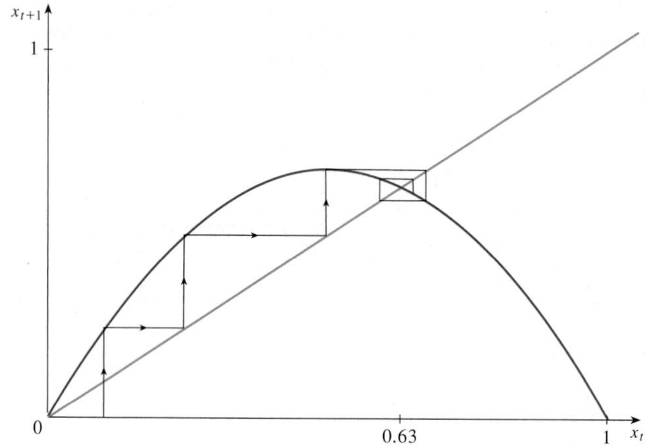

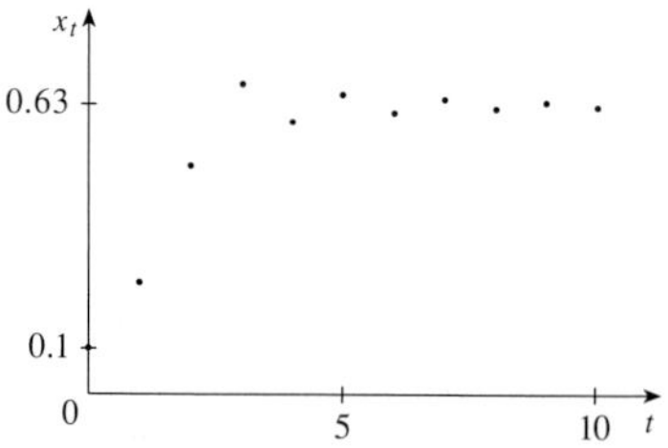

$\hat{x} = \frac{17}{27} = 0.\overline{629}$ is a stable equilibrium; $\hat{x} = 0$ is unstable.
21. $N = 0$ is a stable equilibrium when $r < h$.
$N = K\left(1 - \dfrac{h}{r}\right)$ is a stable equilibrium when
$h < r < h + 2$.
23. $d = 1$ is a stable equilibrium.
25. (a) $R = \dfrac{\theta K}{\theta + d}$ is a stable equilibrium when $0 < d + \theta < 2$.
(b) $R = 0$ is an unstable equilibrium;
$R = \sqrt{\dfrac{a - db}{d}}$ is a stable equilibrium.

EXERCISES 4.6 | page 311

1. $F(x) = \frac{1}{2}x + \frac{1}{4}x^3 - \frac{1}{5}x^4 + C$
3. $F(x) = \frac{2}{3}x^3 + \frac{1}{2}x^2 - x + C$
5. $F(x) = 4x^{5/4} - 4x^{7/4} + C$ **7.** $F(x) = 4x^{3/2} - \frac{6}{7}x^{7/6} + C$
9. $F(x) = \sqrt{2}x + C$
11. $C(t) = -\dfrac{3}{t} + K$, where K is a constant.
13. $G(\theta) = \sin\theta + 5\cos\theta + C$ **15.** $V(s) = 2s^2 + 3e^s + C$
17. $F(u) = \frac{1}{3}u^3 - 6u^{-1/2} + C$ **19.** $F(t) = \frac{1}{3}t^3 - t - \dfrac{1}{t} + C$
21. $y(t) = \frac{1}{3}t^3 + t + 6, t \geqslant 0$ **23.** $P(t) = \frac{2}{3}e^{3t} + \frac{1}{3}, t \geqslant 0$
25. $r(\theta) = \sin\theta + \sec\theta + 2 - \dfrac{\sqrt{3}}{2}, 0 < \theta < \pi/2$
27. $u(t) = \frac{2}{3}t^{3/2} + 4t^{1/2} + \frac{1}{3}, t > 0$ **29.** $x^3 + x^4 + Cx + D$
31. $\frac{3}{20}x^{8/3} + Cx + D$ **33.** $x - 3x^2 + 8$

35. $4x^{3/2} + 2x^{5/2} + 4$

37. $-\sin\theta - \cos\theta + 5\theta + 4$

39. $x^2 - 2x^3 + 9x + 9$ **41.** ≈ 44 bacteria

43. $s(t) = 1 - \cos t - \sin t$

45. (a) $s(t) = 450 - 4.9t^2$ (b) $\sqrt{450/4.9} \approx 9.58$ s
(c) $-9.8\sqrt{450/4.9} \approx -93.9$ m/s

47.

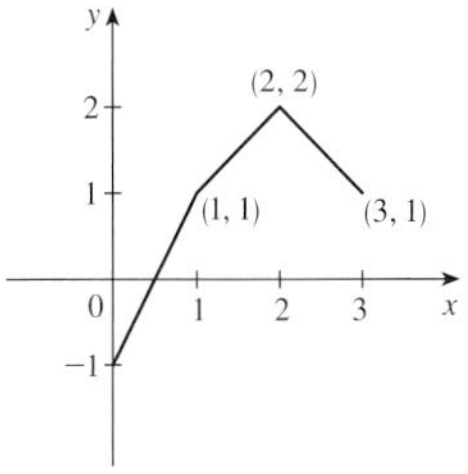

■ CHAPTER 4 REVIEW | page 312

True-False Quiz

1. False **3.** False **5.** True **7.** False **9.** True
11. True **13.** False **15.** True **17.** True

Exercises

1. Abs max $f(4) = 5$, abs and loc min $f(3) = 1$

3. Abs max $f(2) = \frac{2}{5}$, abs and loc min $f\left(-\frac{1}{3}\right) = -\frac{9}{2}$

5. Abs max $f(\pi) = \pi$; abs min $f(0) = 0$;
loc max $f(\pi/3) = (\pi/3) + \frac{1}{2}\sqrt{3}$;
loc min $f(2\pi/3) = (2\pi/3) - \frac{1}{2}\sqrt{3}$

7. (a) None
(b) Dec on $(-\infty, \infty)$
(c) None
(d) CU on $(-\infty, 0)$; CD on $(0, \infty)$;
IP $(0, 2)$
(e) See graph at right.

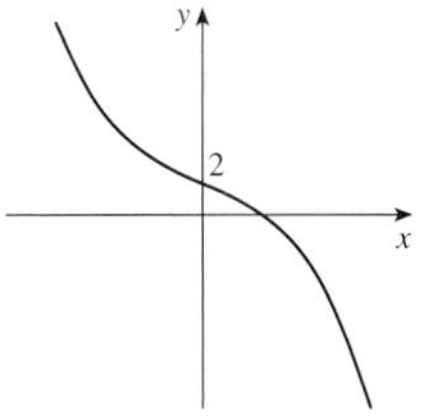

9. (a) None
(b) Inc on $\left(-\infty, \frac{3}{4}\right)$; dec on $\left(\frac{3}{4}, 1\right)$
(c) Loc max $f\left(\frac{3}{4}\right) = \frac{5}{4}$
(d) CD on $(-\infty, 1)$;
(e) See graph at right.

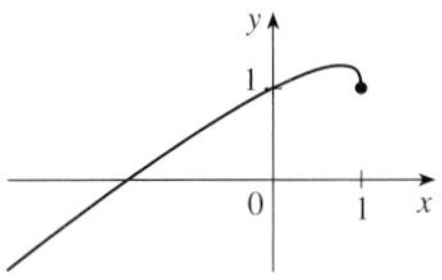

11. (a) None
(b) Inc on $(2n\pi, (2n+1)\pi)$,
n an integer;
dec on $((2n+1)\pi, (2n+2)\pi)$
(c) Loc max $f((2n+1)\pi) = 2$
loc min $f(2n\pi) = -2$
(d) CU on $(2n\pi - (\pi/3), 2n\pi + (\pi/3))$;
CD on $(2n\pi + (\pi/3), 2n\pi + (5\pi/3))$; IPs $\left(2n\pi \pm (\pi/3), -\frac{1}{4}\right)$
(e) See graph at right.

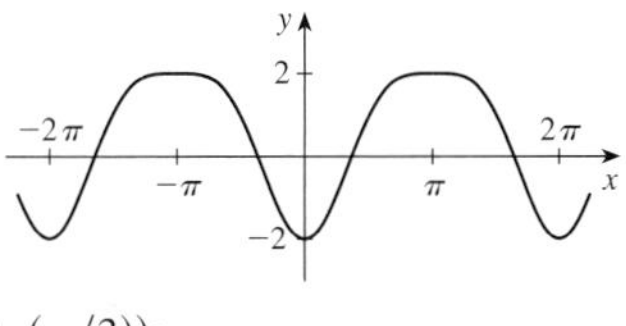

13. (a) None
(b) Inc on $\left(\frac{1}{4}\ln 3, \infty\right)$;
dec on $\left(-\infty, \frac{1}{4}\ln 3\right)$
(c) Loc min $f\left(\frac{1}{4}\ln 3\right) = 3^{1/4} + 3^{-3/4}$
(d) CU on $(-\infty, \infty)$
(e) See graph at right.

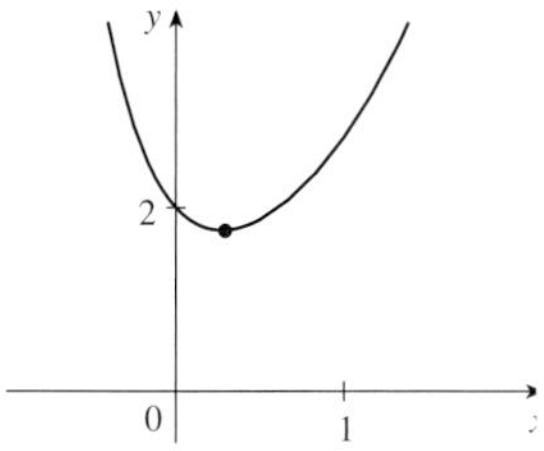

15. (a) Max: $C(2.118) \approx 0.7567$ μg/mL
(b) IP: $t = \frac{5}{2}\ln(49/9) \approx 4.236$ h
Rate of change of concentration begins to increase after 4.236 h.

17. Population $\leqslant 900$ after five weeks

19. π **21.** 8 **23.** 0

25. $\frac{1}{2}$ **27.** 500 and 125

29. $L = C$ **31.** Δp is largest when $p = 1/2$

33. Stable: $x = 3/5$; unstable: $x = 0$

35. Stable: $x = 0, 4$; unstable: $x = 2$
$\lim_{t\to\infty} x_t = 0$ when $x_0 = 1$; $\lim_{t\to\infty} x_t = 4$ when $x_0 = 3$

37. $F(x) = -\cos x + \sec x + C, 0 \leqslant x \leqslant \pi/2$

39. $Q(t) = \frac{1}{4}t^4 + \frac{1}{3}t^3 - \frac{1}{2}t^2 + t + C$

41. $y(t) = t - \dfrac{1}{\pi}e^{\pi t} + \dfrac{1}{\pi}$

43. $f(x) = \frac{1}{2}x^2 - x^3 + 4x^4 + 2x + 1$

45. $s(t) = -\sin t - 3\cos t + 3t + 3$

CHAPTER 5

■ EXERCISES 5.1 | page 327

1. (a) $L_4 = 33$, $R_4 = 41$

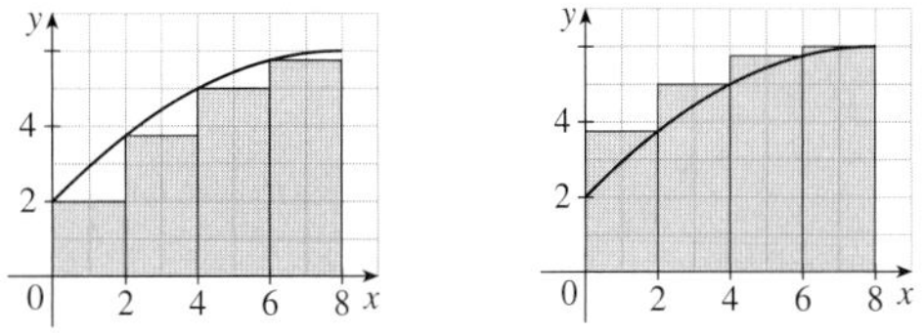

(b) $L_8 \approx 35.2$, $R_8 \approx 39.2$

3. (a) 0.7908, underestimate

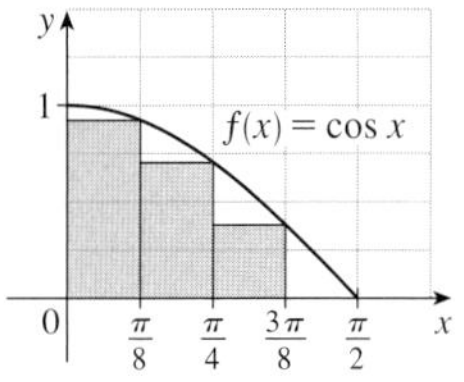

(b) 1.1835, overestimate

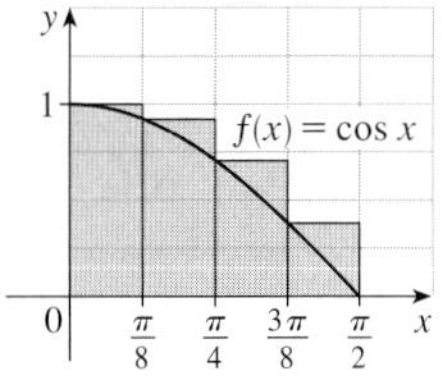

5. (a) 8, 6.875

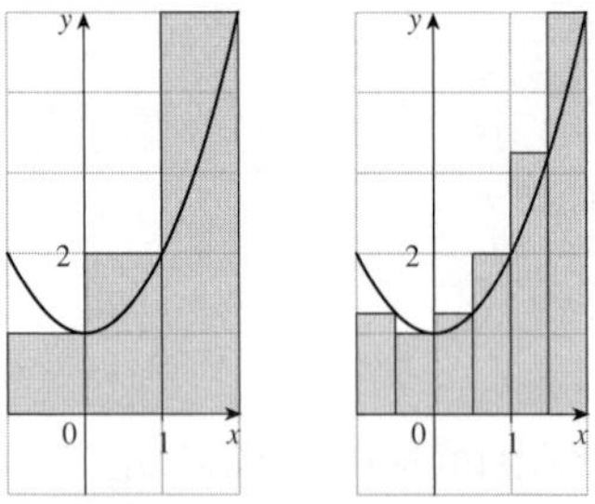

(b) 5, 5.375

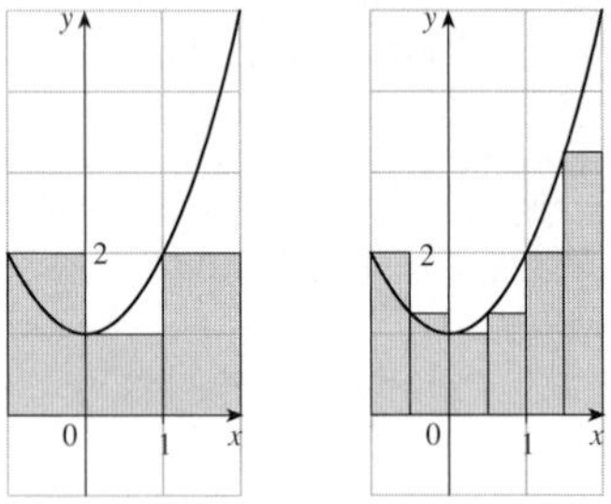

(c) 5.75, 5.9375

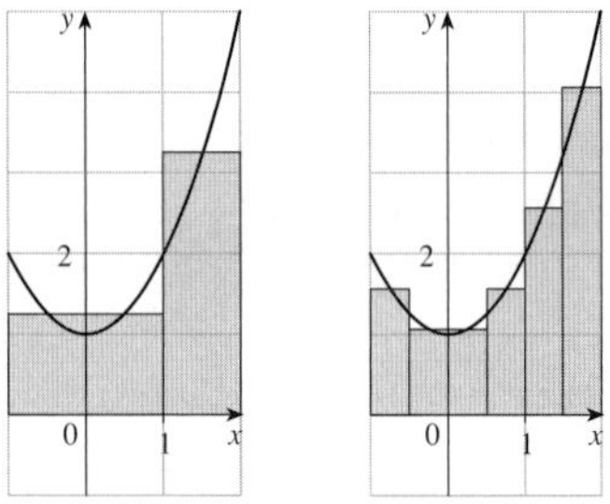

(d) M_6

7. 34.7 ft, 44.8 ft

9. $M_6 = 7840$ (infected cells/mL) · days

11. $L_6 \approx 24$ people, $R_6 \approx 28$ people **13.** 155 ft

15. $\lim_{n\to\infty} \sum_{i=1}^{n} \dfrac{2(1 + 2i/n)}{(1 + 2i/n)^2 + 1} \cdot \dfrac{2}{n}$ **17.** $\lim_{n\to\infty} \sum_{i=1}^{n} \left(\dfrac{i\pi}{2n} \cos \dfrac{i\pi}{2n}\right) \dfrac{\pi}{2n}$

19. (a) $L_n < A < R_n$

21. (a) $\lim_{n\to\infty} \dfrac{64}{n^6} \sum_{i=1}^{n} i^5$ (b) $\dfrac{n^2(n + 1)^2(2n^2 + 2n - 1)}{12}$ (c) $\frac{32}{3}$

23. $\sin b$, 1

EXERCISES 5.2 | page 339

1. −6

The Riemann sum represents the sum of the areas of the two rectangles above the x-axis minus the sum of the areas of the three rectangles below the x-axis; that is, the net area of the rectangles with respect to the x-axis.

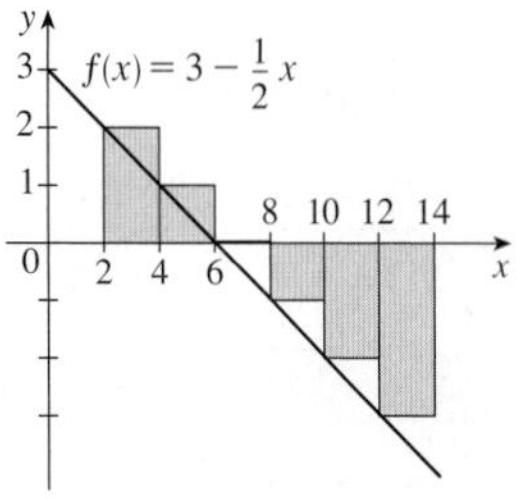

3. 2.322986

The Riemann sum represents the sum of the areas of the three rectangles above the x-axis minus the area of the rectangle below the x-axis.

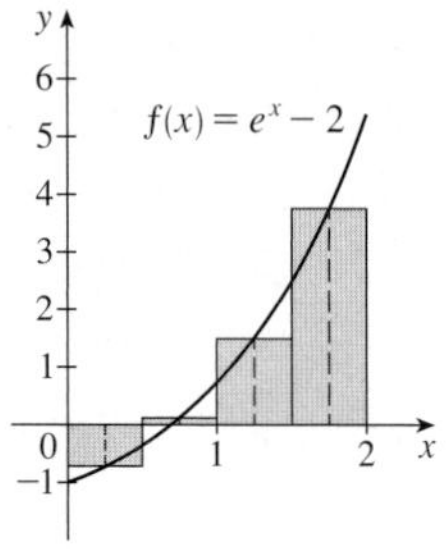

5. (a) 4 (b) 6 (c) 10

7. Lower, $L_5 = -64$; upper, $R_5 = 16$

9. 124.1644 **11.** 0.3084 **13.** 198 (μg/mL) · min

15. $\int_2^6 x \ln(1 + x^2)\, dx$ **17.** $\int_1^8 \sqrt{2x + x^2}\, dx$ **19.** 42

21. $\frac{4}{3}$ **23.** 3.75 **25.** $\lim_{n\to\infty} \sum_{i=1}^{n} \dfrac{2 + 4i/n}{1 + (2 + 4i/n)^5} \cdot \dfrac{4}{n}$

27. (a) 4 (b) 10 (c) −3 (d) 2

29. $-\frac{3}{4}$ **31.** $3 + 9\pi/4$ **33.** 2.5 **35.** 0

37. $\int_{-1}^{5} f(x)\, dx$ **39.** 122

41. $B < E < A < D < C$ **43.** 15 **47.** $\int_0^1 x^4\, dx$

EXERCISES 5.3 | page 350

1. $-\frac{10}{3}$ **3.** $\frac{56}{15}$ **5.** $\frac{5}{9}$ **7.** $-2 + 1/e$ **9.** $\frac{49}{3}$ **11.** $\frac{40}{3}$

13. $\frac{55}{63}$ **15.** 1 **17.** ln 3 **19.** $\dfrac{1}{e + 1} + e - 1$

21. $e^2 - 1$ **23.** $\ln 2 + 7$ **25.** $1 + \pi/4$ **27.** $\pi/6$

29. The function $f(x) = 1/x^2$ is not continuous on the interval $[-1, 3]$, so the Evaluation Theorem cannot be applied.

31. 2

33. 3.75

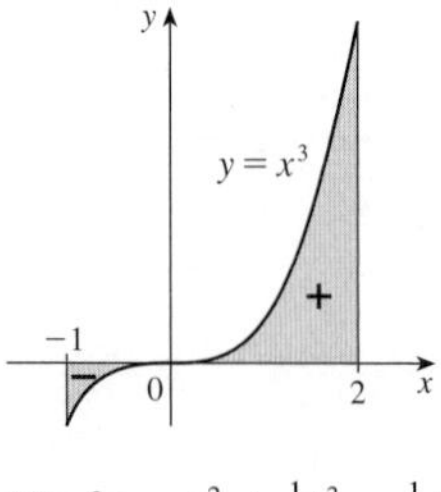

37. $\sin x + \frac{1}{4}x^2 + C$

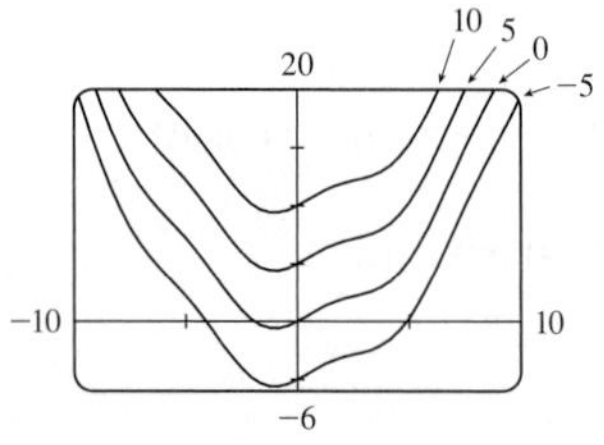

39. $2t - t^2 + \frac{1}{3}t^3 - \frac{1}{4}t^4 + C$ **41.** $\tan \alpha + C$

43. $\sec x + C$ **45.** 7848 (number of cells per mL) · days

47. The increase in the child's weight (in pounds) between the ages of 5 and 10

49. The total number of sea urchins between points a and b

51. The change in concentration between time t_1 and t_2

53. The number of gallons of oil leaked in the first two hours

55. 1800 liters **57.** 29.8 cm

59. (a) $\int_0^L A(x)\, dx$ (b) Upper: βL; lower: αL

61. (a) $P(t) = P_{\max}(1 - e^{-akt})$ (b) $P_{\max}\left[5 + \dfrac{1}{ak}e^{-5ka} - \dfrac{1}{ak}\right]$

(c) $P_{\max}\left[t + \dfrac{1}{ak}e^{-akt} - \dfrac{1}{ak}\right]$ (d) $P_{\max}(1 - e^{-akt})$

65. (a) 0, 2, 5, 7, 3
(b) (0, 3)
(c) $x = 3$
(d)

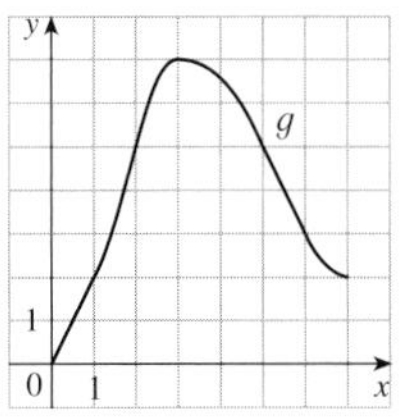

67.

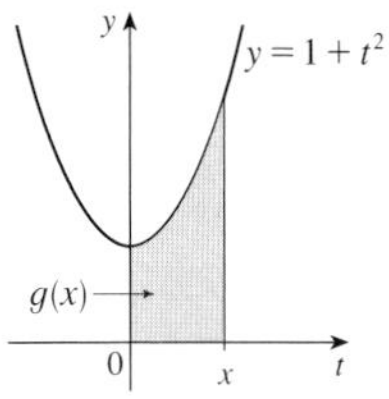

$g'(x) = 1 + x^2$

69. $g'(x) = 1/(x^3 + 1)$ **71.** $g'(y) = y^2 \sin y$

73. $F'(x) = -\sqrt{1 + \sec x}$ **75.** $h'(x) = -\dfrac{\arctan(1/x)}{x^2}$

77. $y' = \sqrt{\tan x + \sqrt{\tan x}}\,\sec^2 x$ **79.** 29 **81.** 3 **83.** $\frac{1}{4}$

85. $f(x) = x^{3/2}$, $a = 9$

EXERCISES 5.4 | page 360

1. $-e^{-x} + C$ **3.** $\frac{2}{9}(x^3 + 1)^{3/2} + C$ **5.** $-\frac{1}{4}\cos^4\theta + C$

7. $-\frac{1}{2}\cos(x^2) + C$ **9.** $\frac{1}{63}(3x - 2)^{21} + C$

11. $-(1/\pi)\cos \pi t + C$ **13.** $\frac{1}{3}(\ln x)^3 + C$

15. $-\frac{1}{3}\ln|5 - 3x| + C$ **17.** $\frac{2}{3}\sqrt{3ax + bx^3} + C$

19. $\frac{2}{3}(1 + e^x)^{3/2} + C$ **21.** $-1/(\sin x) + C$

23. $\frac{1}{15}(x^3 + 3x)^5 + C$ **25.** $-\frac{2}{3}(\cot x)^{3/2} + C$

27. $-\frac{1}{2}\cos(e^{2r}) + C$ **29.** $\frac{1}{3}\sec^3 x + C$

31. $\frac{1}{40}(2x + 5)^{10} - \frac{5}{36}(2x + 5)^9 + C$ **33.** $-\ln(1 + \cos^2 x) + C$

35. $\tan^{-1}x + \frac{1}{2}\ln(1 + x^2) + C$ **37.** $2/\pi$ **39.** $\frac{45}{28}$ **41.** $\frac{182}{9}$

43. $2(e^2 - e)$ **45.** 0 **47.** $\frac{16}{15}$ **49.** $\ln(e + 1)$ **51.** $\frac{1}{6}$

53. 6π **55.** All three areas are equal. **57.** ≈ 4512 L

59. $\dfrac{5}{4\pi}\left(1 - \cos\dfrac{2\pi t}{5}\right)$ L

61. $\approx 1.0\ \text{mm}^3$ **63.** ≈ 515 degree-days **65.** 5

EXERCISES 5.5 | page 367

1. $\frac{1}{3}x^3 \ln x - \frac{1}{9}x^3 + C$ **3.** $\frac{1}{5}x \sin 5x + \frac{1}{25}\cos 5x + C$

5. $2(r - 2)e^{r/2} + C$

7. $-\dfrac{1}{\pi}x^2 \cos \pi x + \dfrac{2}{\pi^2}x \sin \pi x + \dfrac{2}{\pi^3}\cos \pi x + C$

9. $x \ln \sqrt[3]{x} - \frac{1}{3}x + C$ **11.** $\frac{1}{13}e^{2\theta}(2 \sin 3\theta - 3 \cos 3\theta) + C$

13. $\pi/3$ **15.** $\frac{1}{2} - \frac{1}{2}\ln 2$ **17.** $\frac{1}{4} - \frac{3}{4}e^{-2}$

19. $2(\ln 2)^2 - 4 \ln 2 + 2$ **21.** $2\sqrt{x}\sin\sqrt{x} + 2\cos\sqrt{x} + C$

23. $-\frac{1}{2} - \pi/4$ **25.** $\frac{1}{2}(x^2 - 1)\ln(1 + x) - \frac{1}{4}x^2 + \frac{1}{2}x + \frac{3}{4} + C$

27. (c) $-\frac{1}{4}\cos x \sin^3 x + \frac{3}{8}x - \frac{3}{16}\sin 2x + C$

31. $x[(\ln x)^3 - 3(\ln x)^2 + 6 \ln x - 6] + C$

33. $\approx 10.3560\ (\mu\text{g/mL})\cdot\text{h}$

35. $\frac{1}{2}e^{-1}\cos 1 \approx 0.09938$ **37.** 2

EXERCISES 5.6 | page 370

1. (a) $\dfrac{1}{x^2 - 1} = \dfrac{A}{x + 1} + \dfrac{B}{x - 1}$ (b) $\dfrac{2}{x^2 + x} = \dfrac{A}{x} + \dfrac{B}{x + 1}$

3. $x + 6 \ln|x - 6| + C$ **5.** $2 \ln|x + 5| - \ln|x - 2| + C$

7. $\frac{1}{2}\ln\frac{3}{2}$ **9.** $a \ln|x - b| + C$ **11.** $2 \ln \frac{3}{2}$

13. $\frac{27}{5}\ln 2 - \frac{9}{5}\ln 3$ $\left(\text{or } \frac{9}{5}\ln\frac{8}{3}\right)$ **15.** $2 + \ln\frac{25}{9}$

17. $\ln[(e^x + 2)^2/(e^x + 1)] + C$

19. $2 \ln|x| + 3 \ln|x + 2| + (1/x) + C$

21. $A = 1, B = 1, C = 1$; $\ln|x| + \tan^{-1}x + \frac{1}{2}\ln|x^2 + 1| + C$

EXERCISES 5.7 | page 375

1. $\dfrac{1}{2\pi}\tan^2(\pi x) + \dfrac{1}{\pi}\ln|\cos(\pi x)| + C$

3. $-\sqrt{4x^2 + 9}/(9x) + C$ **5.** $\frac{1}{2}(e^{2x} + 1)\arctan(e^x) - \frac{1}{2}e^x + C$

7. $\pi^3 - 6\pi$ **9.** $-\frac{1}{2}\tan^2(1/z) - \ln|\cos(1/z)| + C$

11. $\frac{1}{9}\sin^3 x\,[3 \ln(\sin x) - 1] + C$

13. $\dfrac{1}{2\sqrt{3}}\ln\left|\dfrac{e^x + \sqrt{3}}{e^x - \sqrt{3}}\right| + C$ **15.** $\frac{1}{5}\ln|x^5 + \sqrt{x^{10} - 2}| + C$

17. $\frac{1}{2}(\ln x)\sqrt{4 + (\ln x)^2} + 2 \ln\left[\ln x + \sqrt{4 + (\ln x)^2}\right] + C$

21. $\frac{1}{3}\tan x \sec^2 x + \frac{2}{3}\tan x + C$

23. $\frac{1}{10}(1 + 2x)^{5/2} - \frac{1}{6}(1 + 2x)^{3/2} + C$

25. $-\ln|\cos x| - \frac{1}{2}\tan^2 x + \frac{1}{4}\tan^4 x + C$

27. $\frac{2}{5}(8 + 3\sqrt[3]{x^2} - 4\sqrt[3]{x})\sqrt{1 + \sqrt[3]{x}} + C$

EXERCISES 5.8 | page 380

1. $\frac{1}{2} - 1/(2t^2)$; 0.495, 0.49995, 0.4999995; 0.5

3. 2 **5.** Diverges **7.** $2e^{-2}$ **9.** Diverges **11.** 0

13. Diverges **15.** $\frac{1}{25}$ **17.** Diverges **19.** $\pi/9$ **21.** $\frac{1}{2}$

23. e

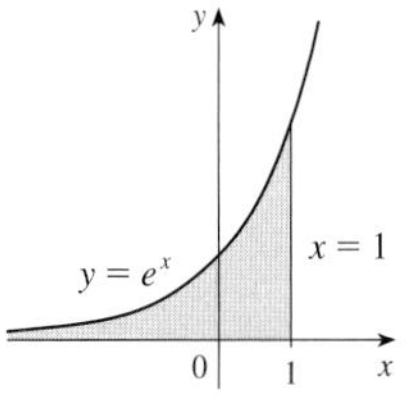

25. (a) $\approx 4.23\ \mu\text{g/mL}$; 0.5 h
(b) 5.75 (μg/mL) × hours; this is the long-term “availability” of a single drug dose.

27. $\int_0^\infty P(I(x))\,dx = \dfrac{a}{k}$ is the rate of photosynthesis per unit area in a water column of infinite depth. In practice, this is a good approximation for an entire water column of unit area in the deep ocean.

29. (a)

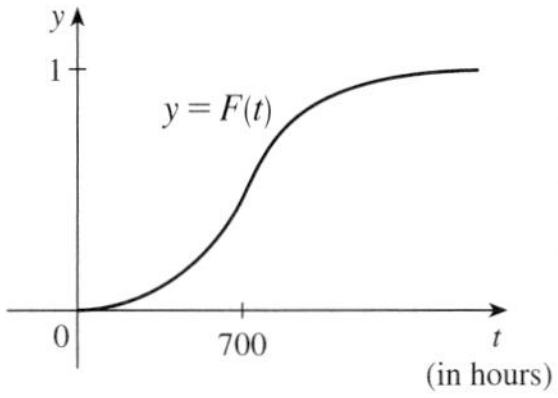

(b) The rate at which the fraction $F(t)$ increases as t increases
(c) 1; all bulbs burn out eventually

33. $\sqrt{\dfrac{\pi}{2}}$ **35.** $\frac{1}{2}\sqrt{\pi}$

CHAPTER 5 REVIEW | page 381

True-False Quiz

1. True **3.** True **5.** False **7.** True **9.** True
11. False **13.** False **15.** False **17.** False

Exercises

1. (a) 8 (b) 5.7

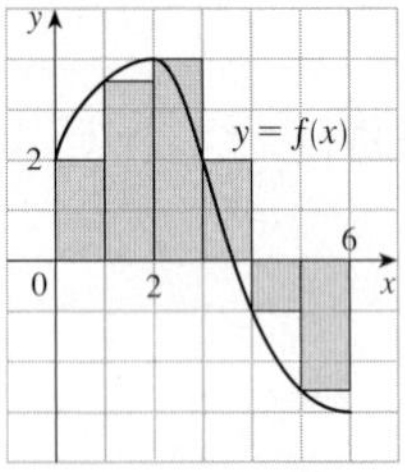

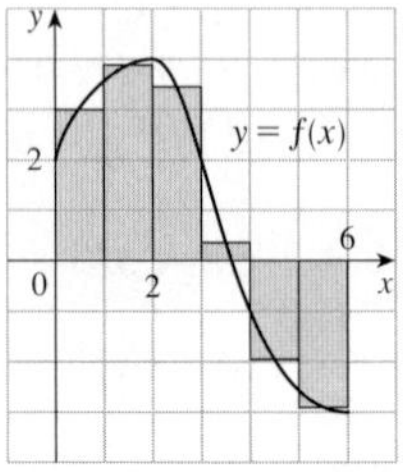

3. $\frac{1}{2} + \pi/4$ **5.** 3 **7.** f is c, f' is b, $\int_0^x f(t)\,dt$ is a
9. 37 **11.** $\frac{9}{10}$ **13.** $-(1/x) - 2\ln|x| + x + C$ **15.** $\frac{1}{2}\ln 2$
17. $\frac{1}{3}\sin 1$ **19.** $(1/\pi)(e^{\pi} - 1)$ **21.** $\sqrt{x^2 + 4x} + C$
23. $5 + 10\ln\frac{2}{3}$ **25.** 0 **27.** $\frac{64}{5}\ln 4 - \frac{124}{25}$
29. $3e^{\sqrt[3]{x}}(x^{2/3} - 2x^{1/3} + 2) + C$ **31.** $\ln|1 + \sec\theta| + C$
33. $\frac{64}{5}$ **35.** $\dfrac{\cos^3 x}{1 + \sin^4 x}$ **37.** $\frac{1}{2}\left[e^x\sqrt{1 - e^{2x}} + \sin^{-1}(e^x)\right] + C$
39. $4 \leqslant \int_1^3 \sqrt{x^2 + 3}\,dx \leqslant 4\sqrt{3}$ **41.** $\frac{1}{36}$ **43.** Diverges
45. The number of barrels of oil consumed from Jan. 1, 2000, through Jan. 1, 2015
47. 750 gallons
49. (a) $\int_a^b r(t)\,dt$
(b) $\int_a^b r(t)\,dt$ = (number of births) − (number of deaths)
51. $k - k\left(\dfrac{be^{-a}\sin b - ae^{-a}\cos b + a}{a^2 + b^2}\right)$
53. $e^{2x}(1 + 2x)/(1 - e^{-x})$

CHAPTER 6

EXERCISES 6.1 | page 392

1. $\frac{32}{3}$ **3.** $e - (1/e) + \frac{4}{3}$ **5.** $\frac{1}{3}$ **7.** 72 **9.** $e - 2$
11. 0, 0.90; 0.04
13. $\frac{1}{2}$

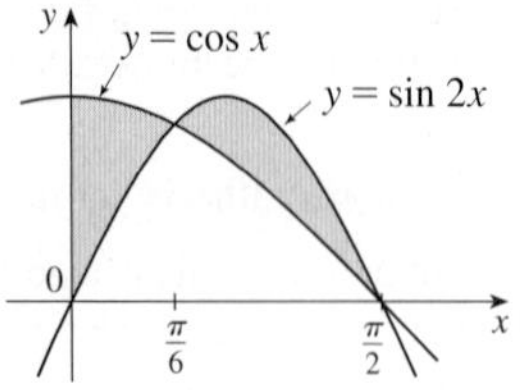

15. (a)

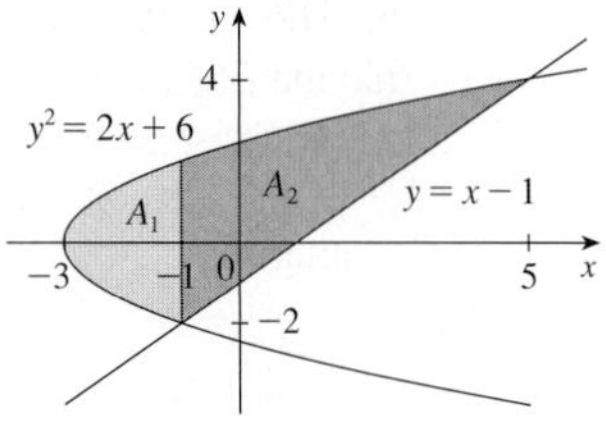

(b)

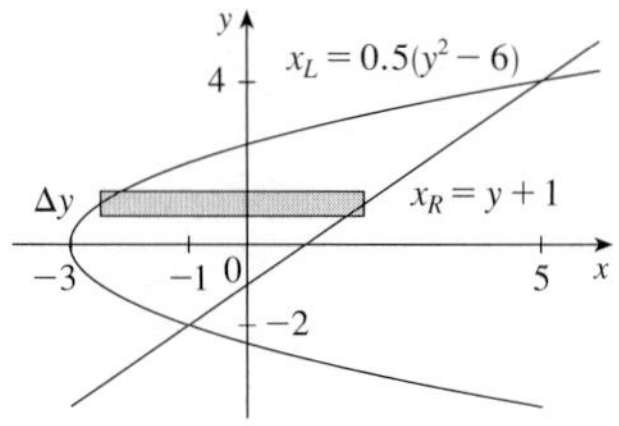

(c) 18

17. 13 cm^2 **19.** 4232 cm^2
21. (a) $M_5 = 0.09$ (mL/mL) · min (b) 711 mL/min
23. $117\frac{1}{3}$ ft
25. 8868; increase in population over a 10-year period
27. ± 6

EXERCISES 6.2 | page 399

1. $\frac{8}{3}$ **3.** $\frac{45}{28}$ **5.** $2/(5\pi)$
7. (a) 1 (b) 2, 4
(c)

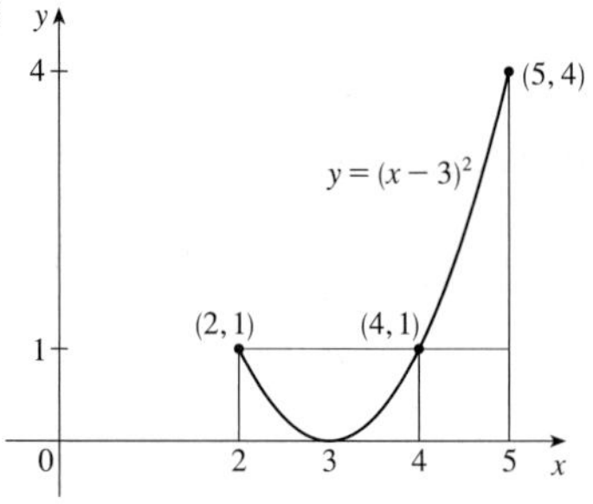

9. (a) $4/\pi$ (b) $\approx 1.24, 2.81$
(c)

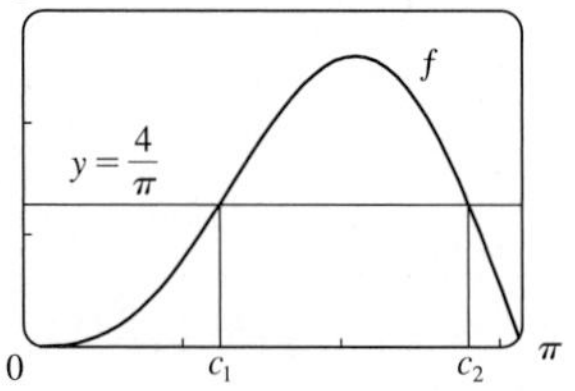

11. $\frac{9}{8}$ **13.** $(50 + 28/\pi)$°F $\approx$ 59°F
15. 74,719 million people
17. 845.25 cells/mL
19. $1/\pi \approx 0.32$ L/s

EXERCISES 6.3 | page 404

1. (a) 1480 members (b) 9440 members
(c) Not all of the 9440 new members survive.
3. 21,046 insects **5.** 48.3 mg **7.** 12,417 gallons
9. 1.19×10^{-4} cm^3/s **11.** 6.59 L/min **13.** 5.69 L/min

EXERCISES 6.4 | page 411

1. $19\pi/12$

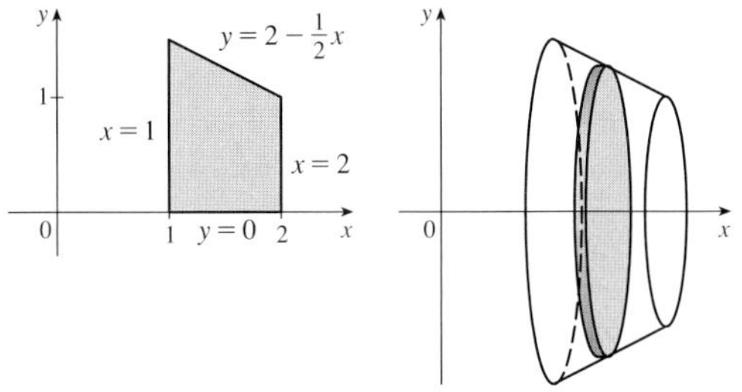

3. 8π

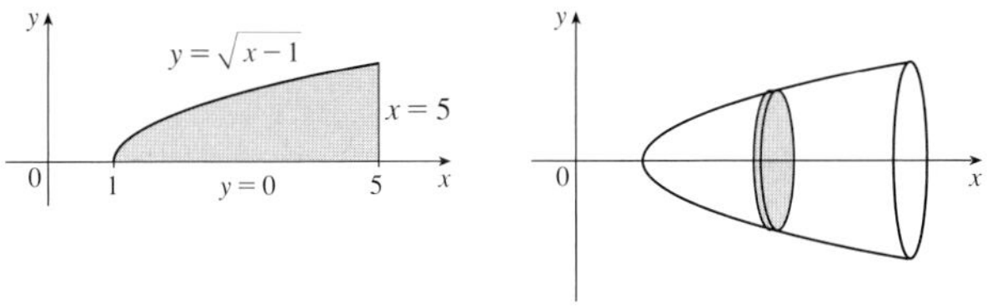

5. $4\pi/21$

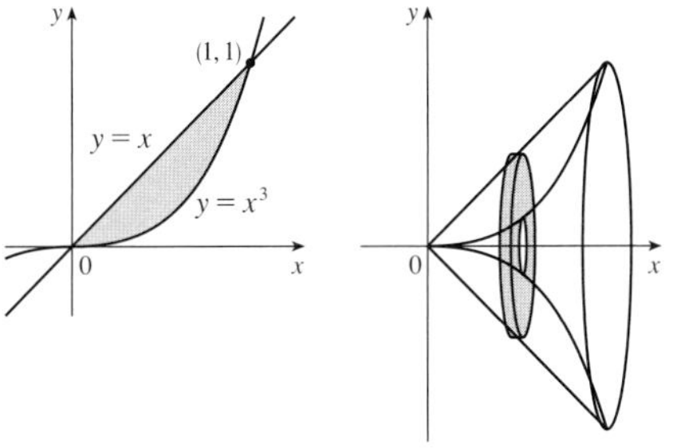

7. 162π

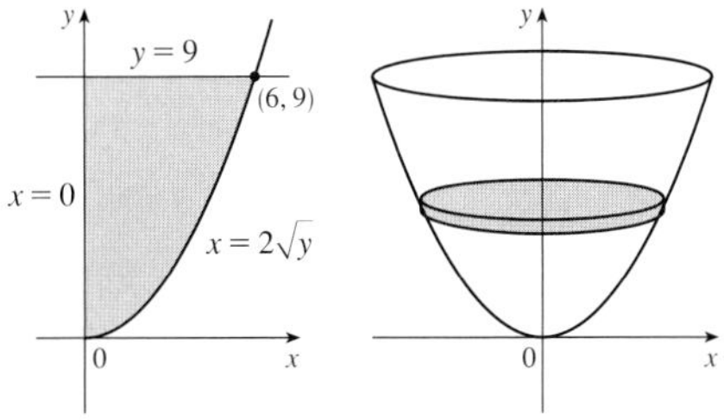

9. $V \approx M_6 = 106.8 \text{ cm}^3$ **11.** (a) 196 (b) 838
13. $\frac{1}{3}\pi r^2 h$ **15.** 24 **17.** $\frac{5}{12}\pi r^3$

CHAPTER 6 REVIEW | page 412

Exercises

1. $\frac{8}{3}$ **3.** $\frac{7}{12}$ **5.** $M_5 = 180.6 \text{ cm}^2$ **7.** ≈ 0.007
9. ≈ 0.848 µg/mL **11.** 54,916
13. 0.070 L/s
15. (a) 0.38 (b) 0.87
17. (a) $2\pi/15$ (b) $\pi/6$
19. 36

CHAPTER 7

EXERCISES 7.1 | page 428

1. Nonautonomous **3.** Pure-time
5. (a) It must be either 0 or decreasing.
(c) $y = 0$ (d) $y = 1/(x + 2)$
7. (a) $0 < P < 4200$ (b) $P > 4200$ (c) $P = 0, P = 4200$
11. (a) III (b) I (c) IV (d) II
13. Pure-time; rate of change of drug concentration is a positive constant; $c(t) = kt$
15. Nonautonomous; rate of change of concentration is proportional to the difference between the concentration and c_s, with a constant of proportionality k/t^b, which decreases over time

EXERCISES 7.2 | page 436

1. (a) (i) is locally stable.
(b) (i) and (ii) are locally stable.
3. (a)

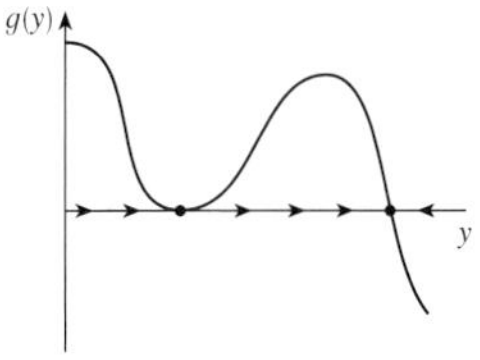

(b)

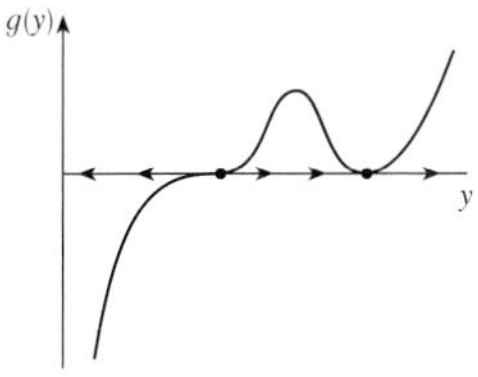

(c)

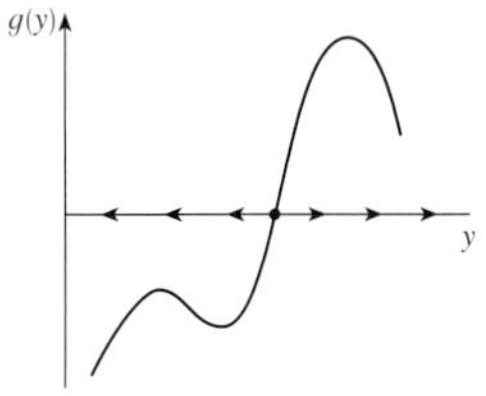

(d)

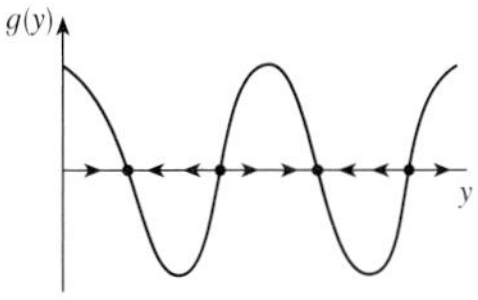

5. (a) $\hat{y} \approx 0.7035$

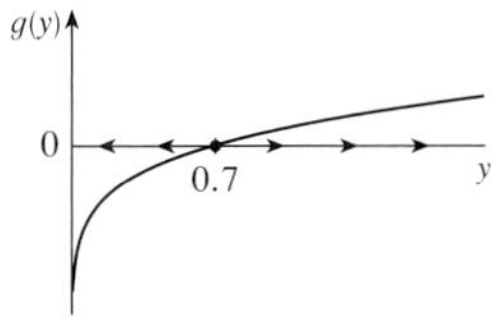

(b) $\hat{y} = \sqrt[3]{a}$

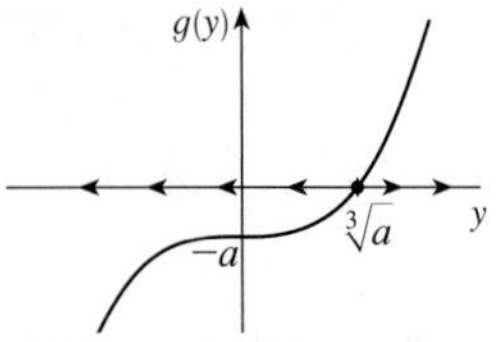

(c) No equilibria

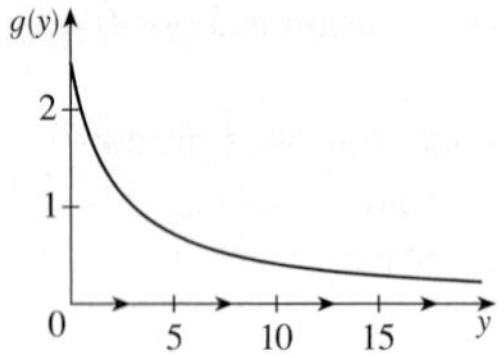

7. (a) $\hat{y} = \frac{5}{3}$ is locally stable

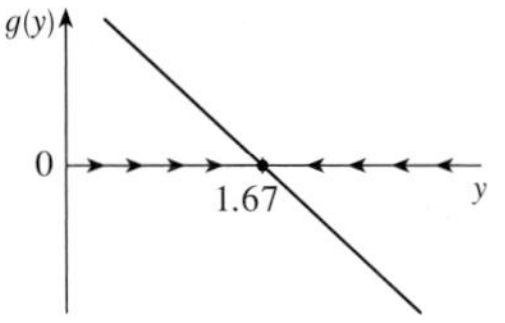

(b) $\hat{y} = 0$ is unstable and $\hat{y} = \frac{2}{3}$ is locally stable

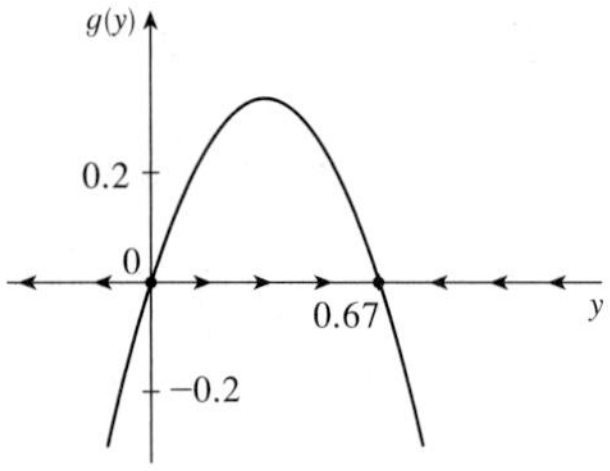

9. (a) $\hat{y} = -1/a$; locally stable when $a < 0$

(b) $\hat{y} = 0$; locally stable when $a < 0$

(c) $\hat{y} = \pi/2$; locally stable when $a > 0$

(d) $\hat{y} = 0$ is locally stable when $a < 0$;
$\hat{y} = a$ is locally stable when $a > 0$

11. (b) $r > 0$

(c) Slope of the phase plot at $N = K$ is zero (horizontal tangent line). Equilibrium may be locally stable or unstable.

(d) $\hat{N} = K$ is unstable in both cases.

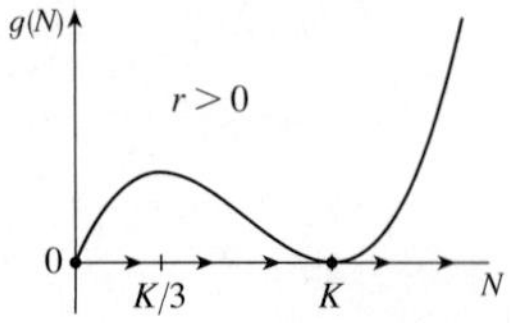

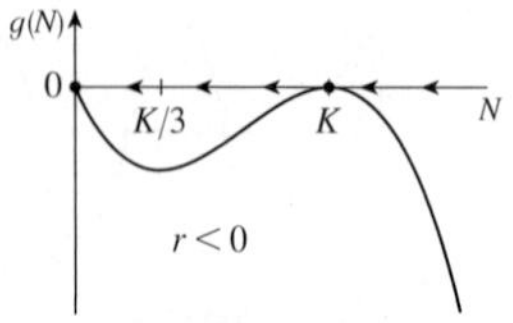

13. (a) $\hat{N} = 0, \hat{N} = 500(2 - h)$ (b) $h \geqslant 2$ (c) $h < 2$

15. (a) $\hat{p} = 0, \hat{p} = 1 - (m/c)$

(b) $m < c$; colonization rate greater than extinction rate

(c) $m < c$

19. (a) $dp/dt = sp(1 - p) - \mu p$, where $s = r_1 - r_2$

(b) $\hat{p} = 0, \hat{p} = 1 - \dfrac{\mu}{s}$

(c) $\hat{p} = 0$ is locally stable when $s < \mu$;
$\hat{p} = 1 - \dfrac{\mu}{s}$ is locally stable when $s > \mu$.

■ EXERCISES 7.3 | page 447

1. (a)

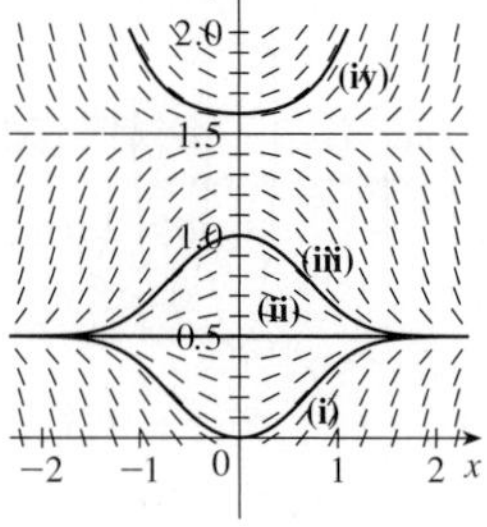

(b) $y = 0.5, y = 1.5$

3. III **5.** IV

7.

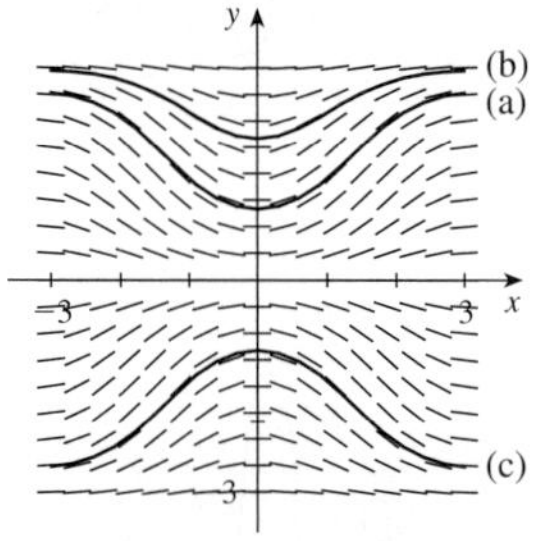

9.

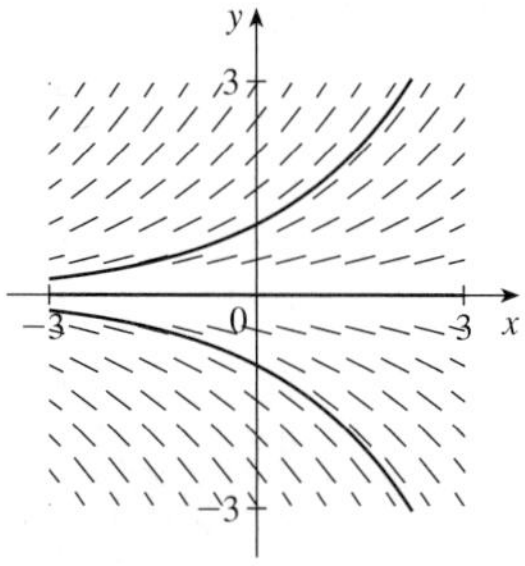

11.

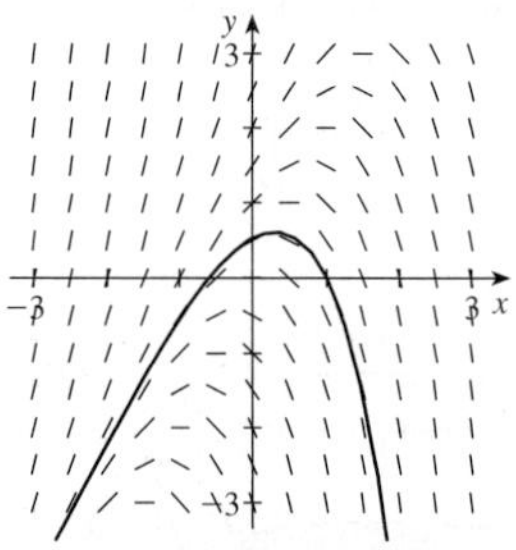

13.

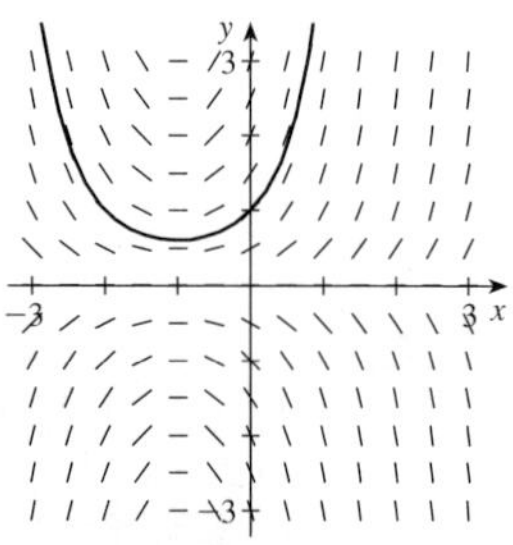

15.

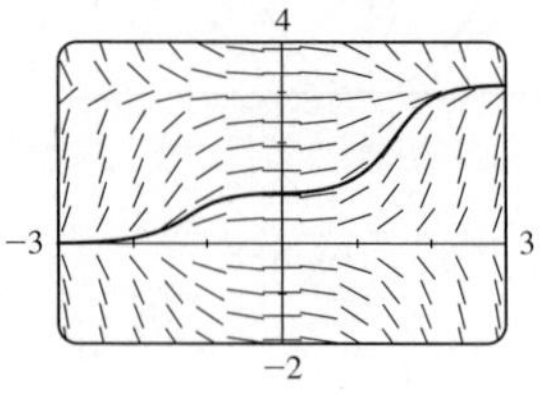

17.

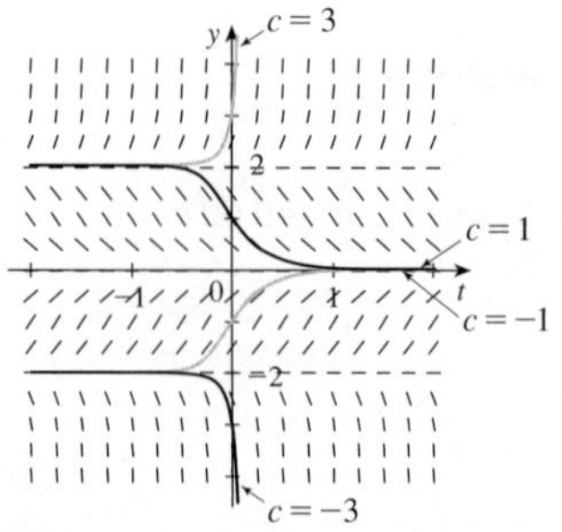

$-2 \leqslant c \leqslant 2$; $-2, 0, 2$

19. (a) (i) 1.4 (ii) 1.44 (iii) 1.4641

(b) Underestimates

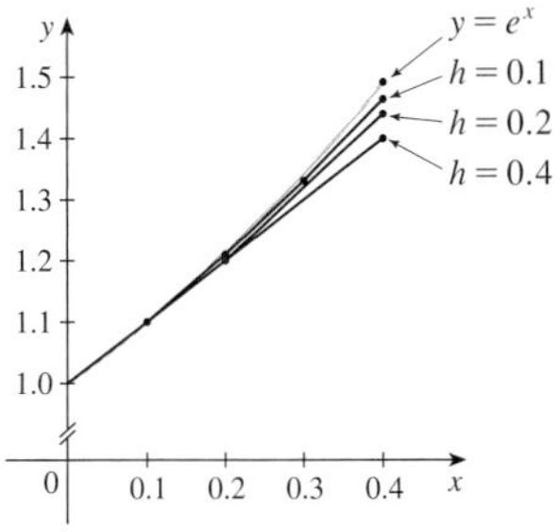

(c) (i) 0.0918 (ii) 0.0518 (iii) 0.0277

It appears that the error is also halved (approximately).

21. $N(4) \approx 1.3906$; close agreement with Table 7.1.1

23. $-1, -3, -6.5, -12.25$ **25.** 1.7616

27. (a) (i) 3 (ii) 2.3928 (iii) 2.3701 (iv) 2.3681

(c) (i) -0.6321 (ii) -0.0249 (iii) -0.0022 (iv) -0.0002

It appears that the error is also divided by 10 (approximately).

■ EXERCISES 7.4 | page 455

1. $y = \dfrac{2}{K - x^2}$, $y = 0$ **3.** $y = K\sqrt{x^2 + 1}$

5. $\frac{1}{2}y^2 - \cos y = \frac{1}{2}x^2 + \frac{1}{4}x^4 + C$

7. $y = \pm\sqrt{[3(te^t - e^t + C)]^{2/3} - 1}$

9. $u = Ae^{2t+t^2/2} - 1$

11. $y = -\sqrt{x^2 + 9}$ **13.** $u = -\sqrt{t^2 + \tan t + 25}$

15. $\frac{1}{2}y^2 + \frac{1}{3}(3 + y^2)^{3/2} = \frac{1}{2}x^2 \ln x - \frac{1}{4}x^2 + \frac{41}{12}$

17. $y = \dfrac{4a}{\sqrt{3}} \sin x - a$

19. $y = e^{x^2/2}$ **21.** $y = Ke^x - x - 1$

23. (a) $\sin^{-1}y = x^2 + C$ for $-\pi/2 \leq x^2 + C \leq \pi/2$; $y = 1$ is also a solution

(b) $y = \sin(x^2)$, $-\sqrt{\pi/2} \leq x \leq \sqrt{\pi/2}$ (c) No

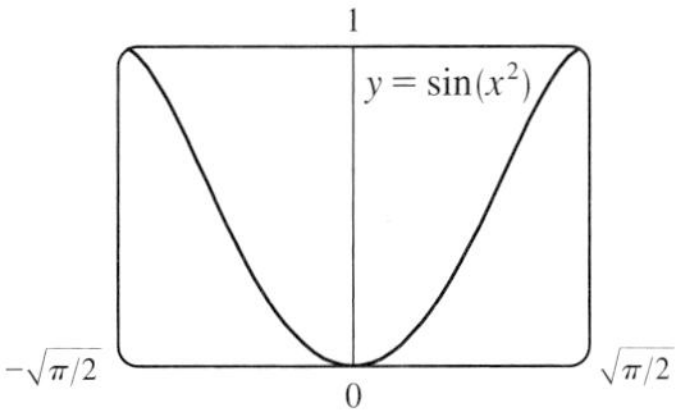

25. $\cos y = \cos x - 1$

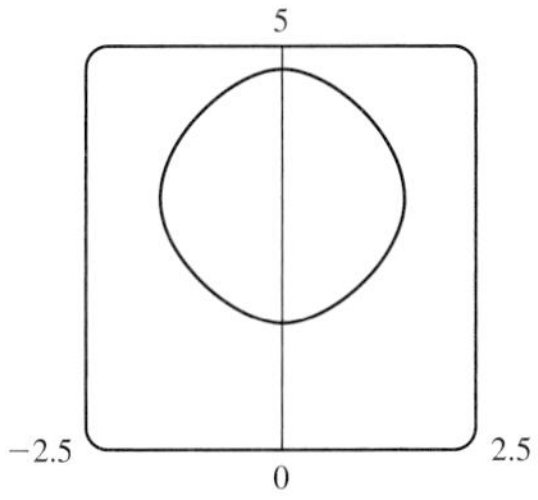

27. (a), (c)

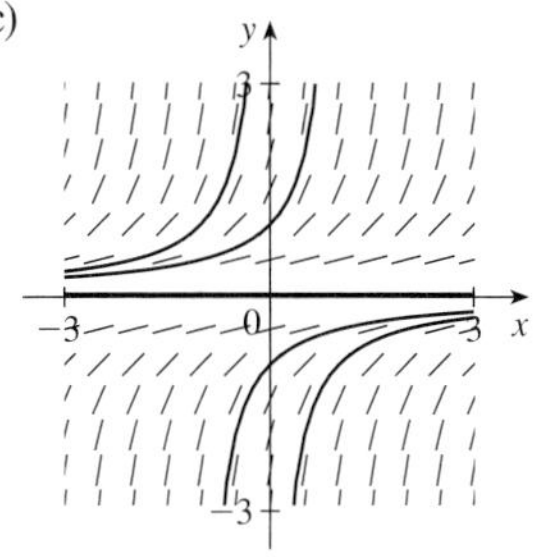

(b) $y = \dfrac{1}{K - x}$, $y = 0$

29. $y = 1 + e^{2-x^2/2}$ **31.** $y = \left(\frac{1}{2}x^2 + 2\right)^2$

33. $n(t) = n_0 e^{1-(e^{-t}+t)}$; population goes extinct

35. $c(t) = c_s - c_s e^{-kt}$ **37.** $n(t) = \frac{1}{4}(kt + 2)^2$

39. (a) $M(t) = \mu^{-3}\left[1 - (1 - \mu)e^{-\mu t/3}\right]^3$

(b) Approaches μ^{-3} grams

(c) $\dfrac{dD}{dt} = \dfrac{\mu}{3}\left(\dfrac{ka}{\mu} - D\right)$

41. $p(t) = \dfrac{p_0 e^{st}}{1 - p_0 + p_0 e^{st}}$

43. (a) $C(t) = (C_0 - r/k)e^{-kt} + r/k$

(b) r/k; the concentration approaches r/k regardless of the value of C_0

45. (a) $15e^{-t/100}$ kg (b) $15e^{-0.2} \approx 12.3$ kg

47. About 4.9% **49.** g/k

51. (a) $dA/dt = k\sqrt{A}\,(M - A)$

(b) $A(t) = M\left(\dfrac{Ce^{\sqrt{M}\,kt} - 1}{Ce^{\sqrt{M}\,kt} + 1}\right)^2$, where $C = \dfrac{\sqrt{M} + \sqrt{A_0}}{\sqrt{M} - \sqrt{A_0}}$ and $A_0 = A(0)$; $A(t) = 0$

53. $S(A) = CA^k$, where S is number of species, A is the area, and C and k are constants

■ EXERCISES 7.5 | page 465

1.

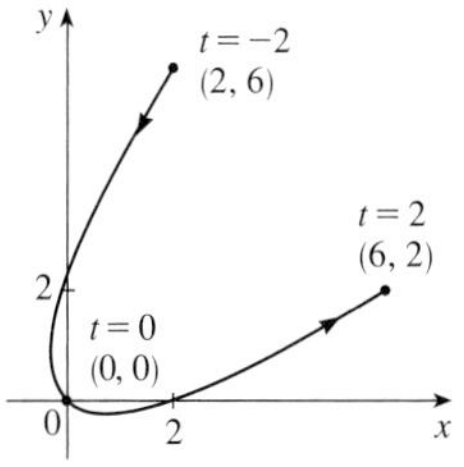

3.

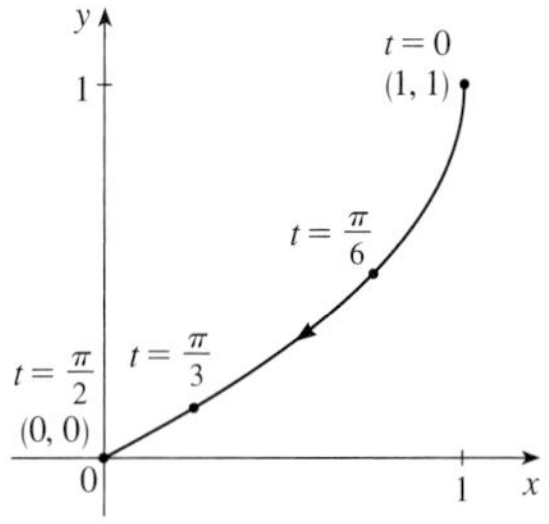

5. (a)

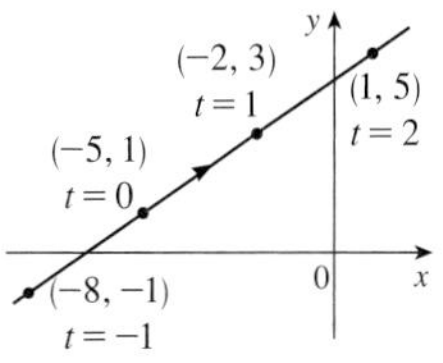

(b) $y = \frac{2}{3}x + \frac{13}{3}$

7. (a)

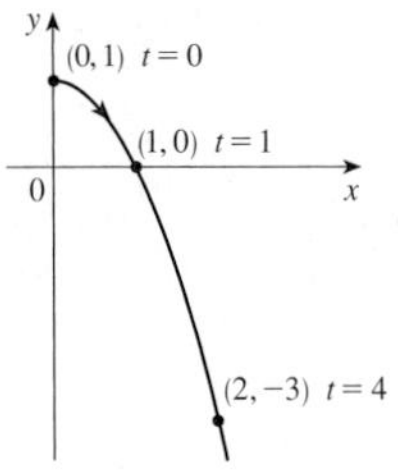

(b) $y = 1 - x^2, x \geqslant 0$

9. Moves counterclockwise along the circle $(x - 3)^2 + (y - 1)^2 = 4$ from (3, 3) to (3, −1)

11. Moves 3 times clockwise around the ellipse $(x^2/25) + (y^2/4) = 1$, starting and ending at (0, −2)

13. (a) x = predators, y = prey; growth is restricted only by predators, which feed only on prey.
(b) x = prey, y = predators; growth is restricted by carrying capacity and by predators, which feed only on prey.

15. Competition

17. (a) The rabbit population starts at about 300, increases to 2400, then decreases back to 300. The fox population starts at 100, decreases to about 20, increases to about 315, decreases to 100, and the cycle starts again.
(b)

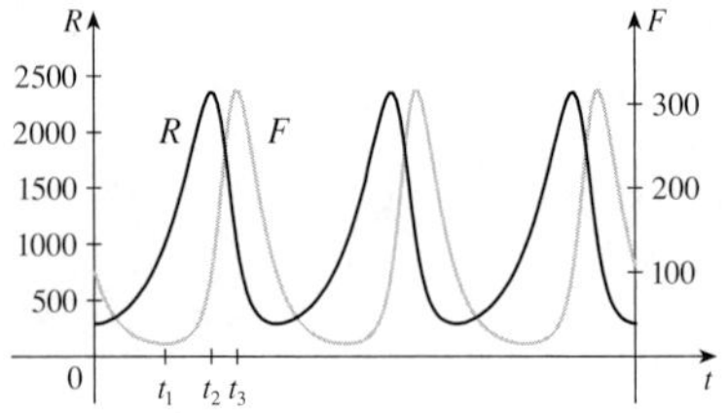

19.

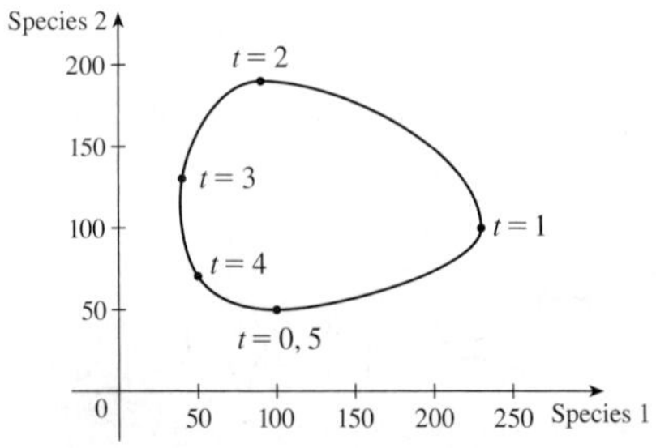

23. (a) Population stabilizes at 5000.
(b) The populations stabilize at 1000 rabbits and 64 wolves.
(c)

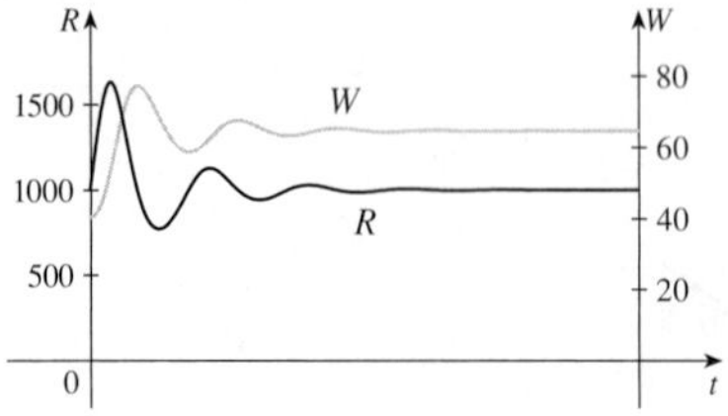

■ EXERCISES 7.6 | page 476

1.

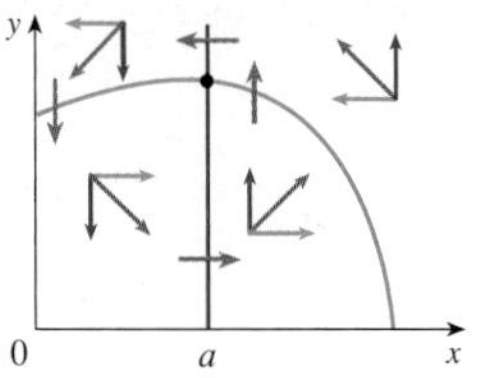

Cannot tell

3.

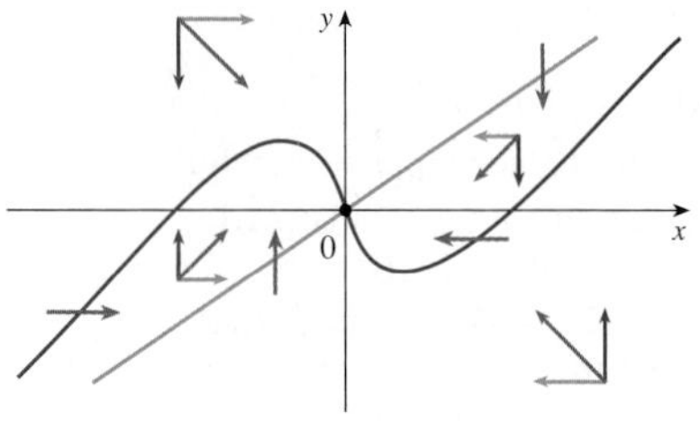

Cannot tell

5.

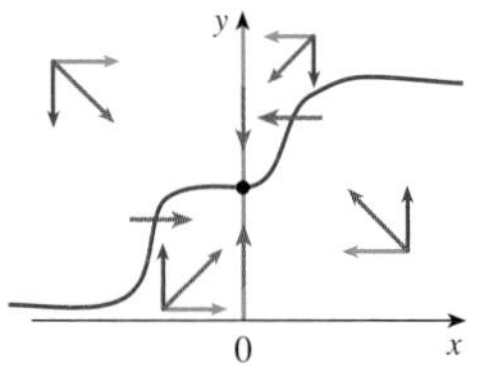

Locally stable

7. (a)

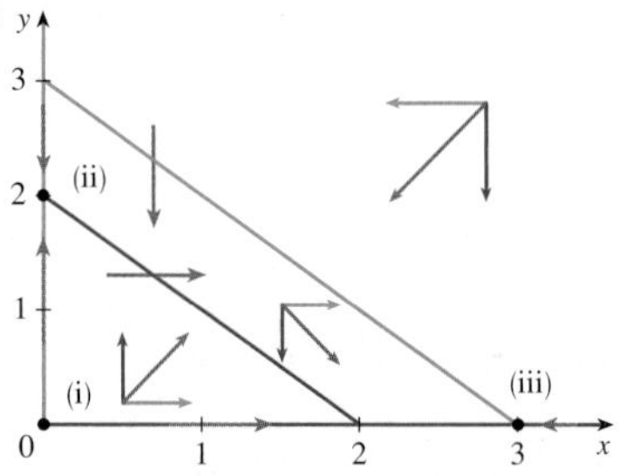

(b) (i) $\hat{x} = 0, \hat{y} = 0$ (ii) $\hat{x} = 0, \hat{y} = 2$ (iii) $\hat{x} = 3, \hat{y} = 0$

9. (a)

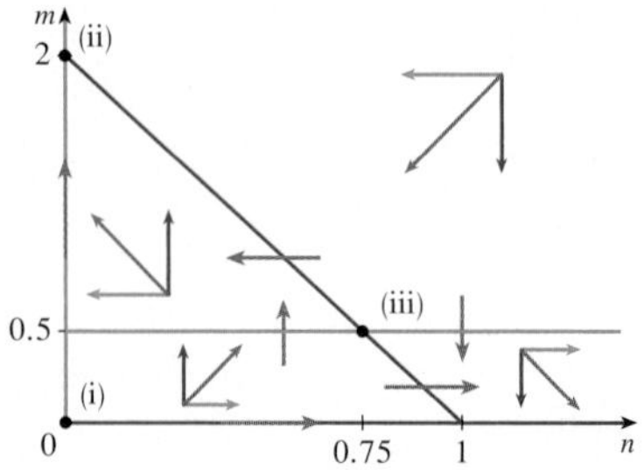

(b) (i) $\hat{n} = 0, \hat{m} = 0$ (ii) $\hat{n} = 0, \hat{m} = 2$ (iii) $\hat{n} = \frac{3}{4}, \hat{m} = \frac{1}{2}$

11. (a)

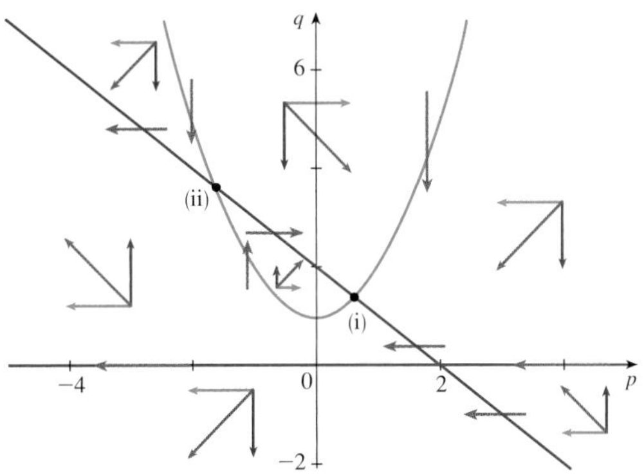

(b) (i) $\hat{p} \approx 0.618, \hat{q} \approx 1.382$ (ii) $\hat{p} \approx -1.618, \hat{q} \approx 3.618$

13. (a)

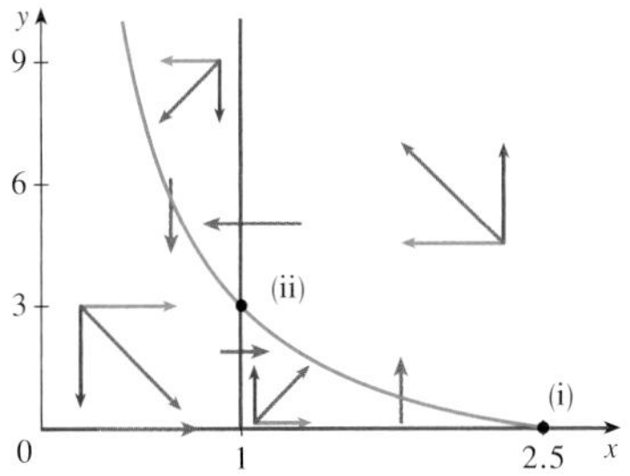

(b) (i) $\hat{x} = 2.5, \hat{y} = 0$ (ii) $\hat{x} = 1, \hat{y} = 3$

15. (a)

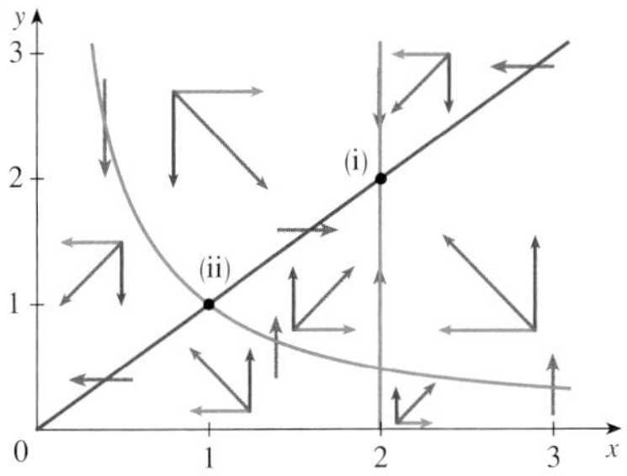

(b) (i) $\hat{x} = 2, \hat{y} = 2$ (ii) $\hat{x} = 1, \hat{y} = 1$

17. (a) Locally stable when $a > 1$ (b) $\hat{x} = 0, \hat{y} = 0$

19. (a) Locally stable for all values of a (b) $\hat{x} = a + 1, \hat{y} = 1$

21. (b)

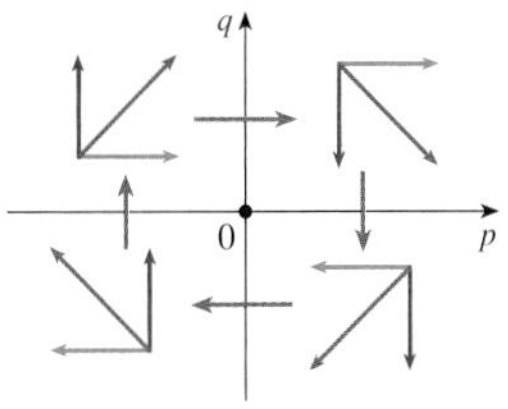

(c) The mass will cycle in front of and behind its rest position, $p = 0$, as time passes.

(d) The velocity of the mass will cycle from positive to negative values and back as time passes.

23. (a) $-\beta SI$: rate of decrease of the number of susceptible people due to transmission of the disease

βSI: rate of increase of the number of infected people due to transmission of the disease

$-\mu I$: rate of decrease of the infected population due to recovery

(b)

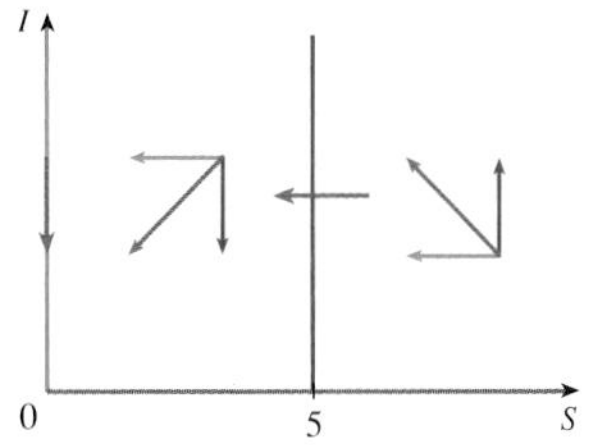

(c)

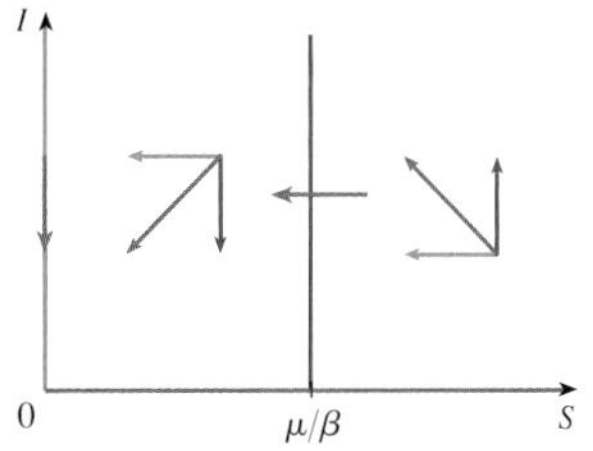

25. (a) $-k_f xyM$: rate of decrease of substrate molecules due to binding with enzymes, where k_f is the fractional rate of enzyme binding

$k_r(1 - y)M$: rate of increase of substrate molecules due to dissociation of the enzyme-substrate complex, where k_r is the fractional rate of dissociation

$k_{cat}(1 - y)M$: rate of increase of free enzymes due to the catalyzed reaction and rate at which product molecules are generated, where k_{cat} is the fractional rate of the forward reaction

(b) dx/dt and dy/dt are independent of z.

(c)

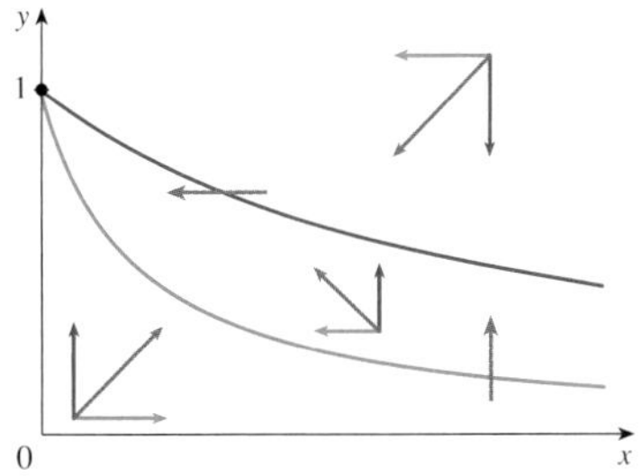

27. (a)

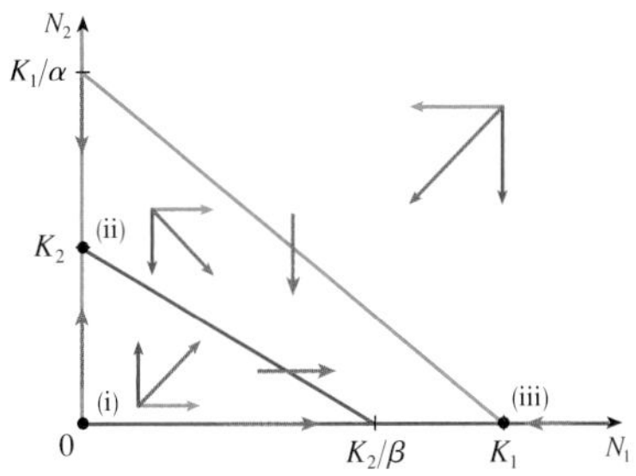

(b)

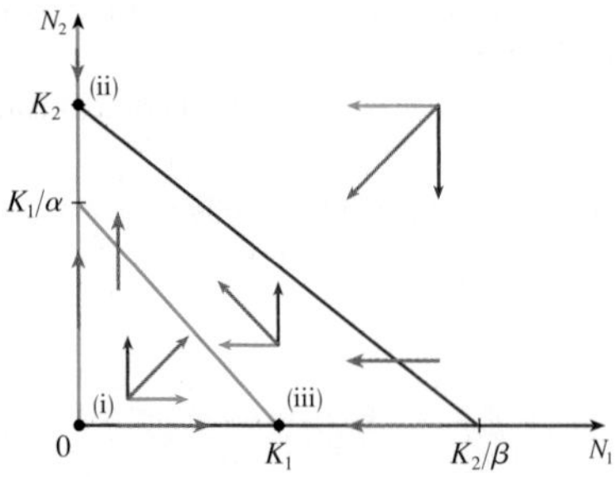

(c)

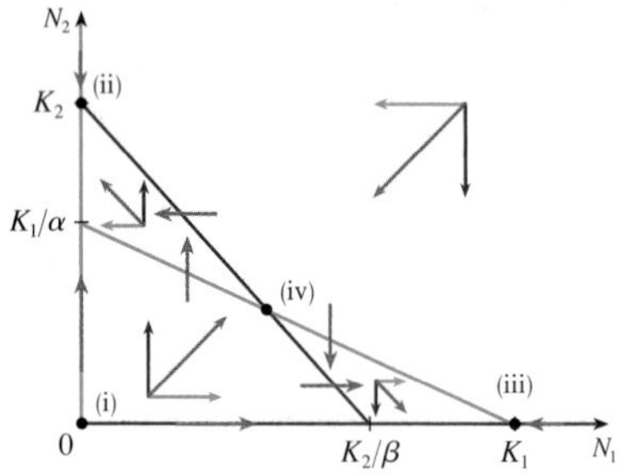

(d)

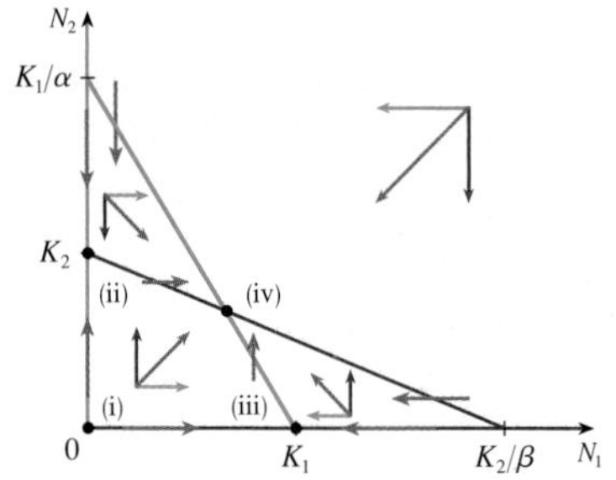

29.

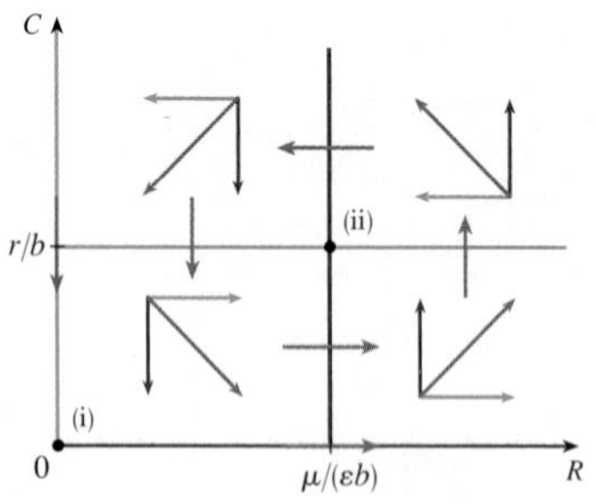

Oscillatory dynamics are predicted.

31. (a) $-\dfrac{K}{V}c$: rate of decrease of urea concentration in the blood due to dialysis, where K is the rate of flow through the dialyzer
ap: rate of increase of urea concentration in the blood due to urea outflow from the inaccessible pool
bc: rate of change of urea concentration due to urea flow from the blood back to the inaccessible pool

(b)

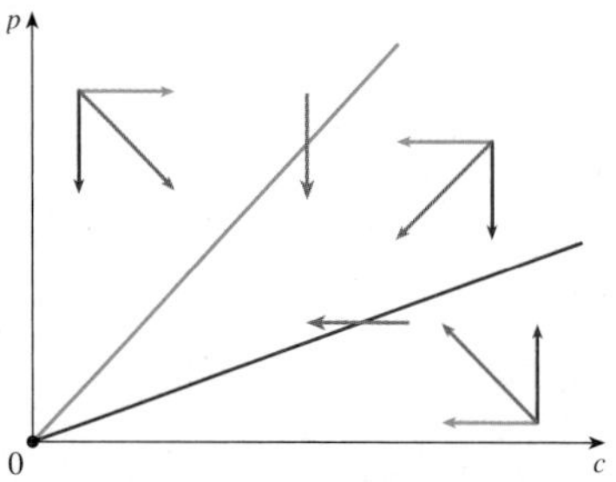

Urea concentration approaches zero in blood and pool.

33.

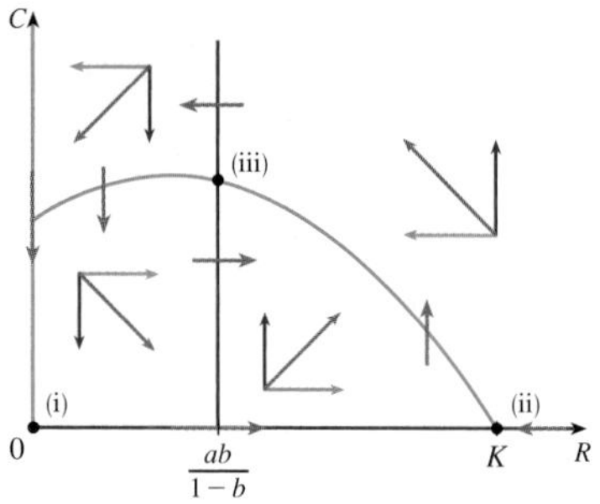

CHAPTER 7 REVIEW | page 481

True-False Quiz

1. True **3.** True **5.** True

Exercises

1. (a) $\hat{x} = 0, a$

(b)

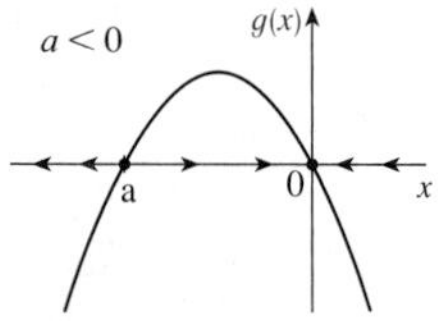

(i) $\hat{x} = 0$ is stable, $\hat{x} = a$ is unstable

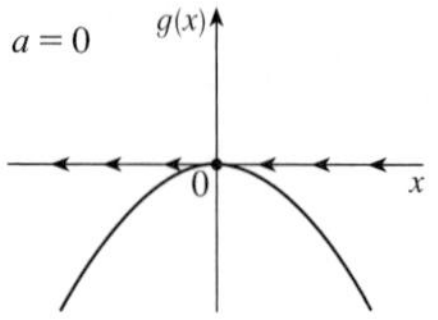

(ii) $\hat{x} = a = 0$ is unstable

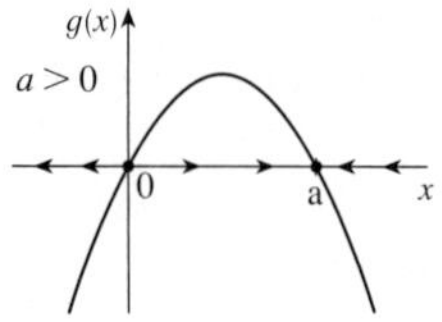

(iii) $\hat{x} = 0$ is unstable, $\hat{x} = a$ is stable

3. (a) $\hat{x} = 0, \pm\sqrt{a}$ (provided $a \geqslant 0$)

(b)

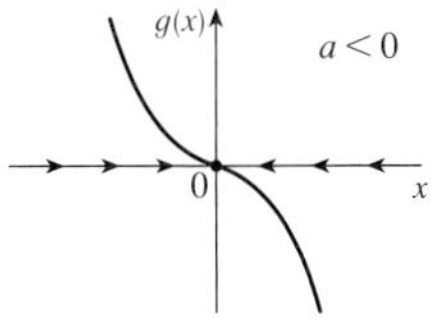

(i) $\hat{x} = 0$ is stable

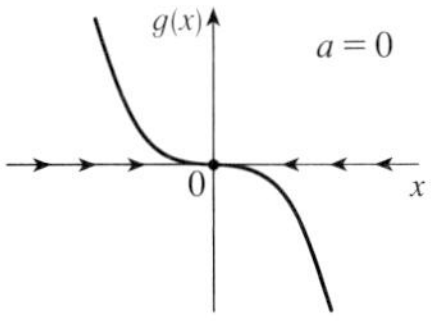

(ii) $\hat{x} = 0$ is stable

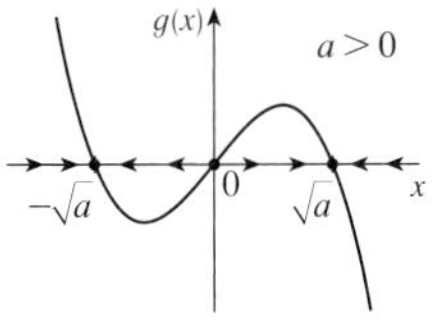

(iii) $\hat{x} = \pm\sqrt{a}$ are stable, $\hat{x} = 0$ is unstable

5. (a)

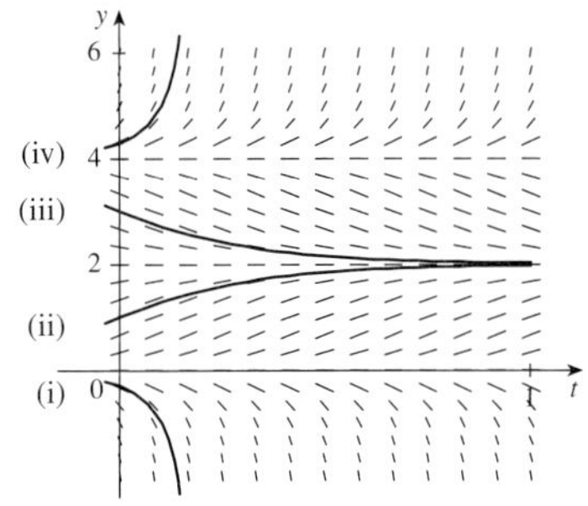

(b) $0 \leqslant c \leqslant 4$; $y = 0$, $y = 2$, $y = 4$ (c) Inconclusive

7. (a) $y(0.3) \approx 0.8$

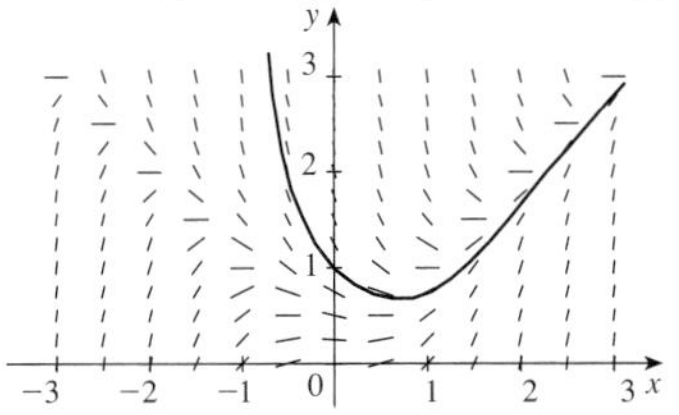

(b) 0.75676

(c) $y = x$ and $y = -x$; there is a loc max or min

9. $y = \pm\sqrt{\ln(x^2 + 2x^{3/2} + C)}$

11. $r(t) = 5e^{t-t^2}$

13. $n(t) = n_0 e^{(365/2\pi)\sin(2\pi t/365)}$

15. (a) $p(t) = \dfrac{p_0(c-m)e^{(c-m)t}}{c - m - cp_0 + p_0ce^{(c-m)t}}$ when $c \neq m$;

$p(t) = \dfrac{p_0}{cp_0t + 1}$ when $c = m$

(b) $c \leqslant m$

17. (a) $V(t) = 3000 - 2400e^{-(1/300)t}$

(b) Approximately 416 s or 6.9 h

19. $k \ln h + h = (-R/V)t + C$

21. (a) Stabilizes at 200,000

(b) (i) $x = 0$, $y = 0$: zero populations

(ii) $x = 200{,}000$, $y = 0$: In the absence of birds, the insect population is always 200,000.

(iii) $x = 25{,}000$, $y = 175$: Both populations are stable.

(c) The populations stabilize at 25,000 insects and 175 birds.

(d)

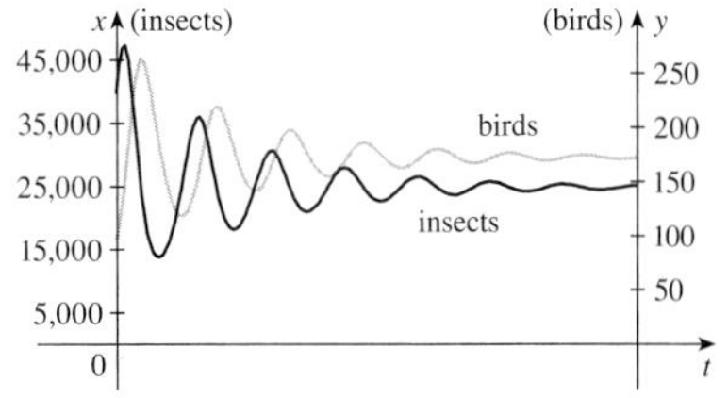

23. dp_1/dt has no term with p_2 so species 1 growth is unaffected by species 2; $-c_1p_1p_2$ term in dp_2/dt means species 2 population decreases as species 1 population grows.

(b)

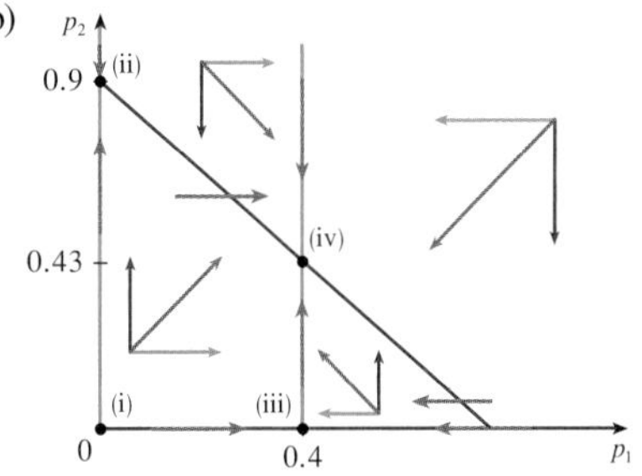

25. (a)

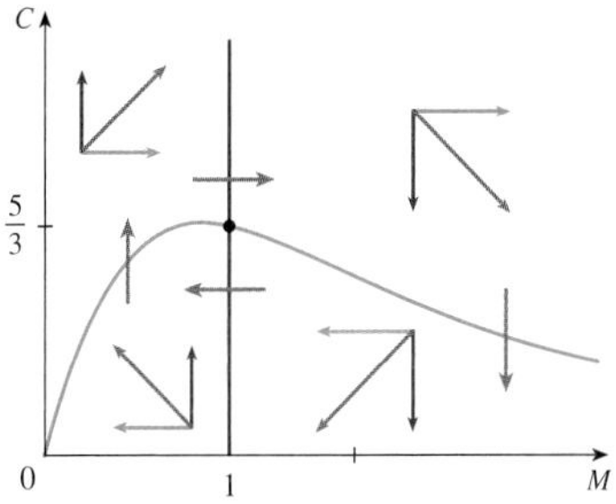

(b) M cycles between values above and below 1 mg/mL; cell division is a periodic process.

CHAPTER 8

EXERCISES 8.1 | page 493

1. (4, 0, −3) **3.** Q; R

5. A vertical plane that intersects the xy-plane in the line $y = 2 - x$, $z = 0$ (see graph at right)

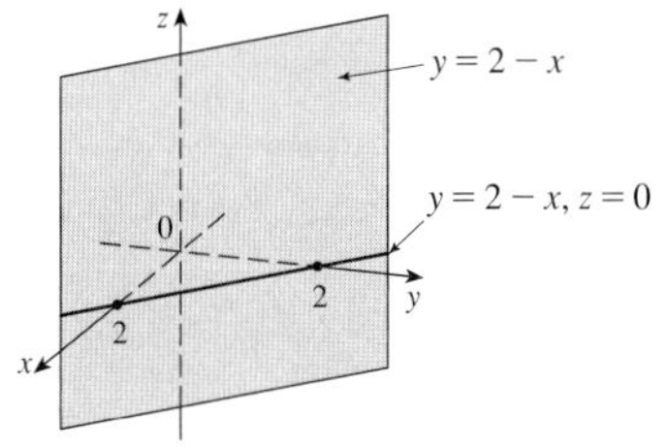

7. (a) $|PQ| = 6$, $|QR| = 2\sqrt{10}$, $|RP| = 6$; isosceles triangle
(b) $|PQ| = 3$, $|QR| = 3\sqrt{5}$, $|RP| = 6$; right triangle
9. (a) No (b) Yes
11. $(x - 3)^2 + (y - 8)^2 + (z - 1)^2 = 30$
13. $(3, -2, 1)$, 5 **15.** $(2, 0, -6)$, $9/\sqrt{2}$
17. (b) $\frac{5}{2}, \frac{1}{2}\sqrt{94}, \frac{1}{2}\sqrt{85}$
19. (a) $(x - 2)^2 + (y + 3)^2 + (z - 6)^2 = 36$
(b) $(x - 2)^2 + (y + 3)^2 + (z - 6)^2 = 4$
(c) $(x - 2)^2 + (y + 3)^2 + (z - 6)^2 = 9$
21. A plane parallel to the yz-plane and 5 units in front of it
23. A half-space consisting of all points to the left of the plane $y = 8$
25. All points on or between the horizontal planes $z = 0$ and $z = 6$
27. All points on a circle with radius 2 and center on the z-axis that is contained in the plane $z = -1$
29. All points on or inside a sphere with radius $\sqrt{3}$ and center O
31. $0 < x < 5$ **33.** $r^2 < x^2 + y^2 + z^2 < R^2$
35. (a) $(2, 1, 4)$ (b)

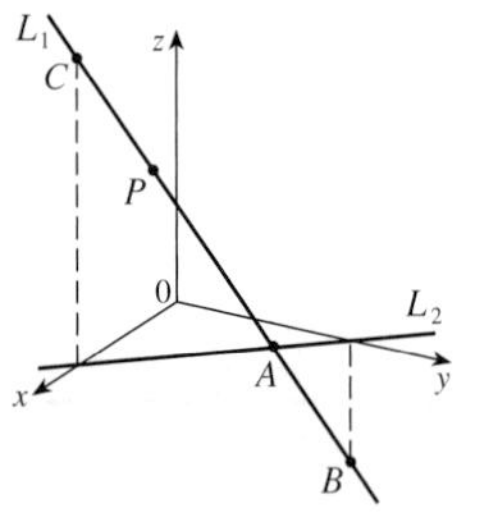

37. (a) 0.2 cm (b) 1.0 cm (c) ≈2.24 cm
39. (a) Yes (b) No (c) The circles in part (a) are the projections of the spheres in part (b) onto the xy-plane.

EXERCISES 8.2 | page 502

1. (a) Scalar (b) Vector (c) Vector (d) Scalar
3. $\overrightarrow{AB} = \overrightarrow{DC}$, $\overrightarrow{DA} = \overrightarrow{CB}$, $\overrightarrow{DE} = \overrightarrow{EB}$, $\overrightarrow{EA} = \overrightarrow{CE}$
5. (a)

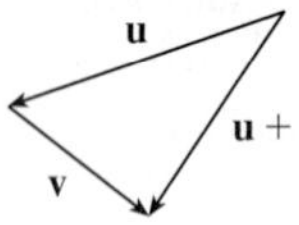

(b)

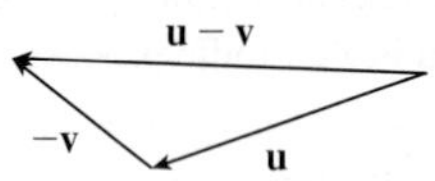

(c)

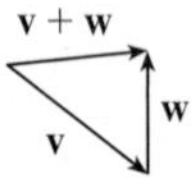

(d)

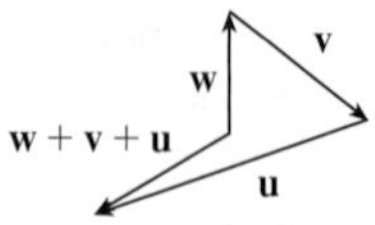

7. $\mathbf{a} = [3, -1]$

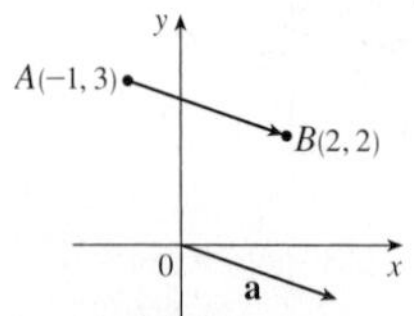

9. $\mathbf{a} = [2, 0, -2]$

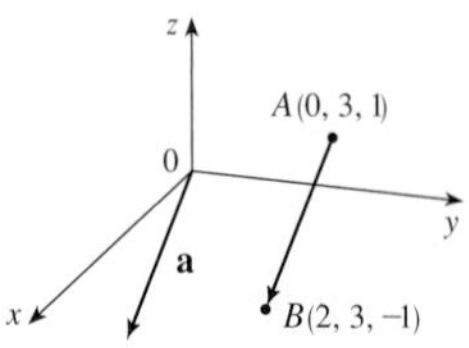

11. $[5, 2]$

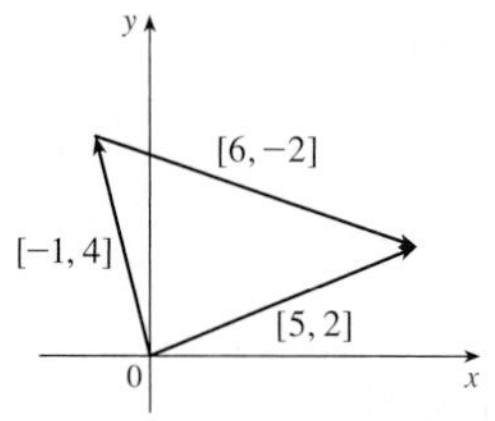

13. $[0, 1, -1]$

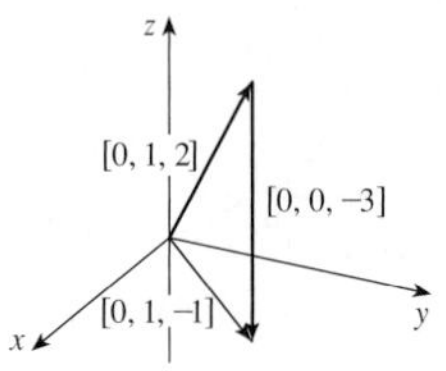

15. $[2, -18]$, $[1, -42]$, 13, 10
17. $[-1, 1, 2]$, $[-4, 1, 9]$, $\sqrt{14}$, $\sqrt{82}$
19. $\left[-\dfrac{3}{\sqrt{58}}, \dfrac{7}{\sqrt{58}}\right]$ **21.** $\left[\frac{8}{9}, -\frac{1}{9}, \frac{4}{9}\right]$
23. $\left[2, 2\sqrt{3}\right]$ **25.** ≈45.96 ft/s, ≈38.57 ft/s
27. $100\sqrt{7} \approx 264.6$ N, ≈139.1°
29. $\sqrt{493} \approx 22.2$ mi/h, N8°W
31. (a) $5\mathbf{i} + 2\mathbf{j} + 4\mathbf{k}$ (b) $-3\mathbf{i} + 2\mathbf{j} - 10\mathbf{k}$
(c) $14\mathbf{i} + 4\mathbf{j} + 15\mathbf{k}$ (d) $-23\mathbf{i} + 10\mathbf{j} - 64\mathbf{k}$
33. $\pm[1, 4]/\sqrt{17}$
35. (a), (b)

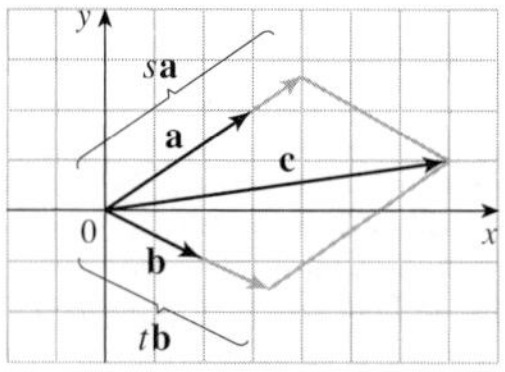

(c) $s \approx 1.3$, $t \approx 1.6$
(d) $s = \frac{9}{7}$, $t = \frac{11}{7}$

37. $\mathbf{a} \approx [0.50, 0.31, 0.81]$
39. (a) Observing a resultant voltage vector that points upwards and has a smaller (horizontal) magnitude compared to a healthy individual
(b) Observing a resultant voltage vector that points to the right of the patient
41. A sphere with radius 1, centered at (x_0, y_0, z_0)

EXERCISES 8.3 | page 511

1. (b), (c), (d) are meaningful **3.** -15 **5.** 14 **7.** 19
9. 1 **11.** $\mathbf{u} \cdot \mathbf{v} = \frac{1}{2}$, $\mathbf{u} \cdot \mathbf{w} = -\frac{1}{2}$
15. $\cos^{-1}\left(\dfrac{9 - 4\sqrt{7}}{20}\right) \approx 95°$ **17.** $\cos^{-1}\left(\dfrac{-1}{2\sqrt{7}}\right) \approx 101°$
19. 45°, 45°, 90°
21. (a) Neither (b) Orthogonal (c) Orthogonal
(d) Parallel
23. Yes **25.** $\left[\dfrac{1}{\sqrt{3}}, -\dfrac{1}{\sqrt{3}}, -\dfrac{1}{\sqrt{3}}\right]$ or $\left[-\dfrac{1}{\sqrt{3}}, \dfrac{1}{\sqrt{3}}, \dfrac{1}{\sqrt{3}}\right]$
27. $x - 2y + 5z = 0$
29. 3, $\left[\frac{9}{5}, -\frac{12}{5}\right]$ **31.** $1/\sqrt{21}$, $\left[\frac{2}{21}, -\frac{1}{21}, \frac{4}{21}\right]$
35. $\left[0, 0, -2\sqrt{10}\right]$ or any vector of the form $\left[s, t, 3s - 2\sqrt{10}\right]$, $s, t \in \mathbb{R}$
37. (a) Magnitude is about twice as large in North American viruses (b) Directions are perpendicular

39. (a) $\sqrt{115} \approx 10.7$
(b) Drug A: $37/115 \approx 0.32$ Drug B: $46/115 = 0.4$
(c) Drug B

41. $\cos^{-1}(1/\sqrt{3}) \approx 55°$

43. Center: $\left[\frac{1}{2}(a_1 + b_1), \frac{1}{2}(a_2 + b_2), \frac{1}{2}(a_3 + b_3)\right]$
Radius: $\frac{1}{2}\sqrt{(a_1 - b_1)^2 + (a_2 - b_2)^2 + (a_3 - b_3)^2}$

EXERCISES 8.4 | page 518

1. (a) 2×2 (b) Not defined (c) 2×3 (d) 3×3 (e) Not defined (f) 3×2 (g) Not defined (h) 3×3

3. (a) $\begin{bmatrix} -25 & -1 \\ -20 & -23 \end{bmatrix}$

(b) $\begin{bmatrix} 3x + 13 & 5 & 26 \\ 16 & 3y + 3 & 51 \\ -2 & -10 & -4 \end{bmatrix}$

(c) $\begin{bmatrix} 32 & 9 & 68 \\ 135 & 5y & 23 \end{bmatrix}$ (d) Not defined

(e) $\begin{bmatrix} 33 & 5x - 5 \\ 5a - 1 & 18 \end{bmatrix}$ (f) Not defined

(g) $\begin{bmatrix} 2x & 4 & 18 \\ 12 & 2y & 26 \\ 0 & 2 & 0 \end{bmatrix}$ (h) $\begin{bmatrix} 8 & 3 & 20 \\ 45 & -y & 5 \end{bmatrix}$

5. (a) $[3X\ 1\ 2]$ (b) $\begin{bmatrix} 3 \\ 3 \\ 9 \end{bmatrix}$ (c) $\begin{bmatrix} 2 & 8 \\ 1 & 3 \\ 7 & 6 \end{bmatrix}$ (d) $\begin{bmatrix} 2 & 1 \\ 1 & 3 \end{bmatrix}$

7. (b) d_{ii}^k

9. (a) $C^2 = \begin{bmatrix} 1 & 0 \\ 0 & 1 \end{bmatrix}, C^3 = \begin{bmatrix} 0 & 1 \\ 1 & 0 \end{bmatrix}, C^4 = \begin{bmatrix} 1 & 0 \\ 0 & 1 \end{bmatrix}, C^5 = \begin{bmatrix} 0 & 1 \\ 1 & 0 \end{bmatrix}$

(b)
$$C^k = \begin{cases} \begin{bmatrix} 0 & 1 \\ 1 & 0 \end{bmatrix} & \text{when } k \text{ is odd} \\ \begin{bmatrix} 1 & 0 \\ 0 & 1 \end{bmatrix} & \text{when } k \text{ is even} \end{cases}$$

17. $a_{ij} = \begin{cases} 0 & \text{when } i = j \text{ (diagonal terms)} \\ -a_{ji} & \text{when } i \neq j \text{ (off-diagonal terms)} \end{cases}$

EXERCISES 8.5 | page 525

1. (a) $\begin{bmatrix} X & 0 \\ X & 0 \end{bmatrix}$ (b) $\begin{bmatrix} 0 & X & 0 \\ X & 0 & X \\ 0 & X & 0 \end{bmatrix}$ (c) $\begin{bmatrix} X & X & 0 & 0 \\ X & 0 & X & 0 \\ 0 & X & X & X \\ 0 & 0 & X & 0 \end{bmatrix}$

(d) $\begin{bmatrix} 0 & X & 0 \\ X & 0 & 0 \\ 0 & 0 & X \end{bmatrix}$ (e) $\begin{bmatrix} X & X \\ X & X \end{bmatrix}$

3. (a) $\begin{bmatrix} 1 & 0 \\ 2 & 3 \end{bmatrix}$ (b) $\begin{bmatrix} 0 & 0 & 4 \\ 1 & 2 & 0 \\ 0 & 3 & 0 \end{bmatrix}$

(c) $\begin{bmatrix} 0 & 0 & 0 & 1 \\ 0 & 0 & 2 & 0 \\ 0 & 3 & 0 & 0 \\ 4 & 0 & 0 & 0 \end{bmatrix}$ (d) $\begin{bmatrix} 0 & 1 \\ 0 & 2 \end{bmatrix}$

5. (a)
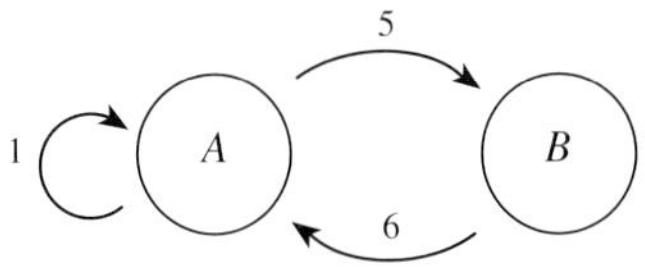

(b)
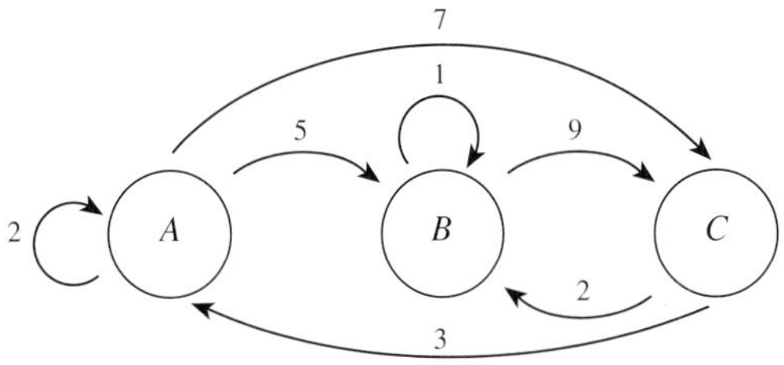

(c)
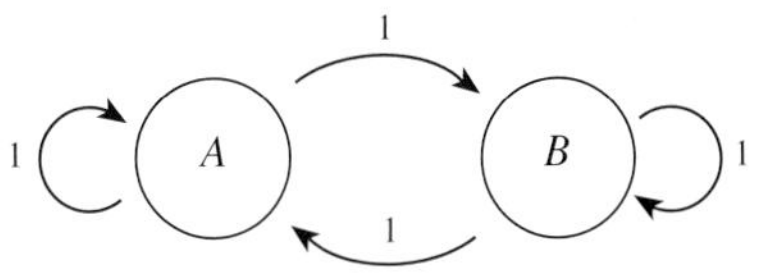

(d)
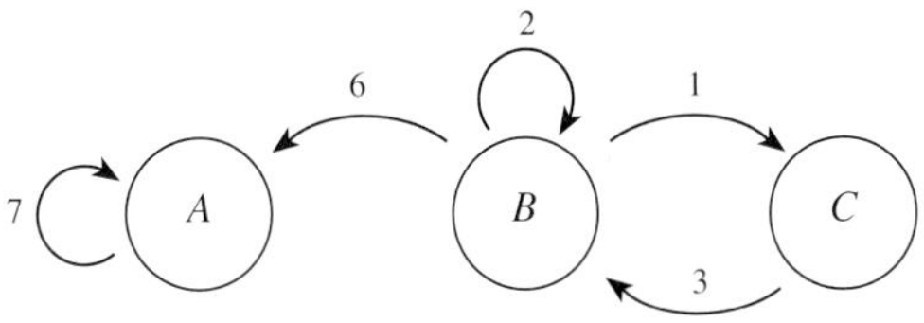

7. No contributions to the variable having a row of zeroes; a matrix diagram would have no incoming arrows to the circle representing this variable.

9.
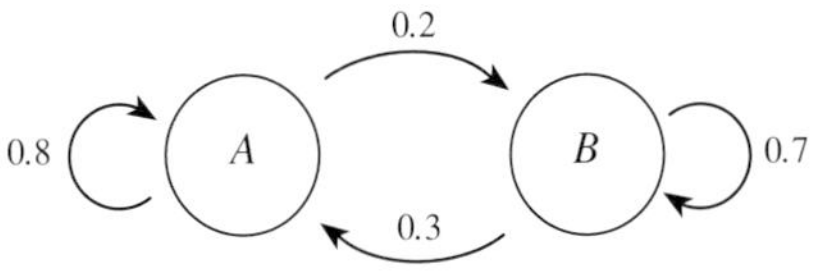

11.
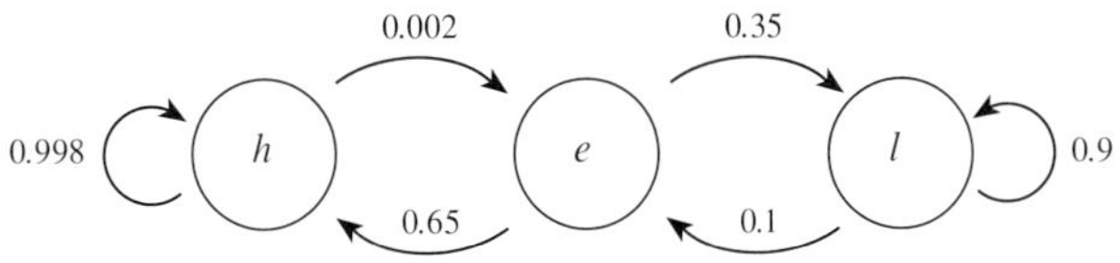

13.
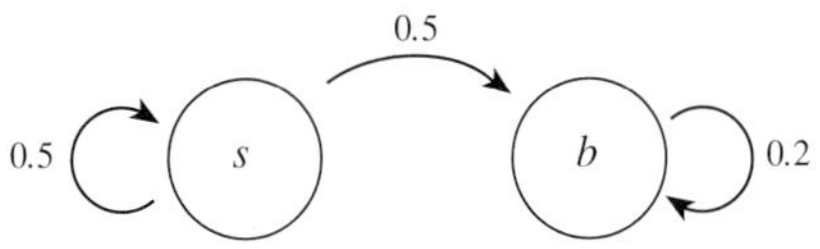

15.

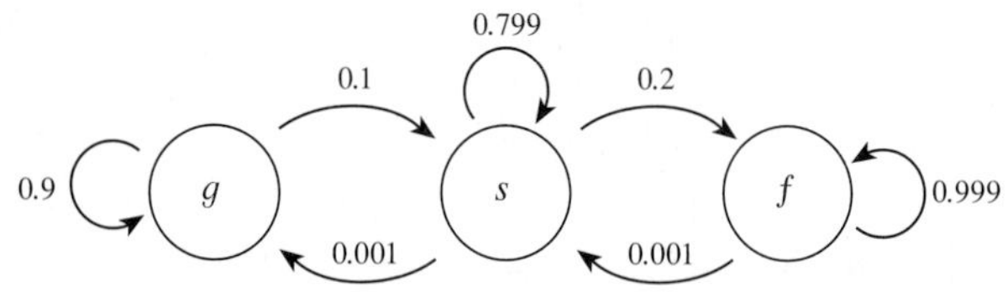

17. $\begin{Bmatrix} x_{t+1} = 2x_t + 3y_t \\ y_{t+1} = 0.9y_t \end{Bmatrix}$, $\begin{bmatrix} x_{t+1} \\ y_{t+1} \end{bmatrix} = \begin{bmatrix} 2 & 3 \\ 0 & 0.9 \end{bmatrix} \begin{bmatrix} x_t \\ y_t \end{bmatrix}$

19. $\begin{Bmatrix} d_{t+1} = 1d_t + \frac{1}{4}h_t \\ h_{t+1} = \frac{1}{2}h_t \\ r_{t+1} = \frac{1}{4}h_t + 1r_t \end{Bmatrix}$, $\begin{bmatrix} d_{t+1} \\ h_{t+1} \\ r_{t+1} \end{bmatrix} = \begin{bmatrix} 1 & \frac{1}{4} & 0 \\ 0 & \frac{1}{2} & 0 \\ 0 & \frac{1}{4} & 1 \end{bmatrix} \begin{bmatrix} d_t \\ h_t \\ r_t \end{bmatrix}$

21. $\begin{Bmatrix} X_{t+1} = 0.92X_t + 0.9Z_t \\ Y_{t+1} = 0.05X_t + 0.999Y_t \\ Z_{t+1} = 0.03X_t + 0.001Y_t + 0.1Z_t \end{Bmatrix}$,

$$\begin{bmatrix} X_{t+1} \\ Y_{t+1} \\ Z_{t+1} \end{bmatrix} = \begin{bmatrix} 0.92 & 0 & 0.9 \\ 0.05 & 0.999 & 0 \\ 0.03 & 0.001 & 0.1 \end{bmatrix} \begin{bmatrix} X_t \\ Y_t \\ Z_t \end{bmatrix}$$

EXERCISES 8.6 | page 534

1. (a) Yes (b) Yes (c) No (d) Yes (e) No (f) Yes

5. $\begin{bmatrix} \pm 1 & 0 \\ 0 & \pm 1 \end{bmatrix}$

11. 1 **13.** 1, 2 **15.** $\det A^k \neq 0$, so A^k is nonsingular.

19. One solution when $k \neq 6$; infinite number of solutions when $k = 6$

21. $x_1 = \dfrac{a_{22}b_1 - a_{12}b_2}{a_{11}a_{22} - a_{12}a_{21}}$, $x_2 = \dfrac{-a_{21}b_1 + a_{11}b_2}{a_{11}a_{22} - a_{12}a_{21}}$

25. $x_1 = \frac{11}{4}, x_2 = -\frac{3}{4}$ **27.** $x_1 = -1, x_2 = \frac{5}{6}$

29. Infinite number of solutions satisfying $x_2 = -2x_1 + 1$

31. $\mathbf{x}_t = \begin{bmatrix} 1 \\ 2 \end{bmatrix}$

33. $\hat{\mathbf{n}} = \begin{bmatrix} b & 2 \\ \frac{1}{2} & 0 \end{bmatrix} \hat{\mathbf{n}} \;\Rightarrow\; \hat{\mathbf{n}} - \begin{bmatrix} b & 2 \\ \frac{1}{2} & 0 \end{bmatrix} \hat{\mathbf{n}} = \mathbf{0}$

$\Rightarrow \begin{bmatrix} 1 & 0 \\ 0 & 1 \end{bmatrix} \hat{\mathbf{n}} - \begin{bmatrix} b & 2 \\ \frac{1}{2} & 0 \end{bmatrix} \hat{\mathbf{n}} = \mathbf{0}$

$\Rightarrow \left(\begin{bmatrix} 1 & 0 \\ 0 & 1 \end{bmatrix} - \begin{bmatrix} b & 2 \\ \frac{1}{2} & 0 \end{bmatrix} \right) \hat{\mathbf{n}} = \mathbf{0}$

$\Rightarrow \underbrace{\begin{bmatrix} 1-b & -2 \\ -\frac{1}{2} & 1 \end{bmatrix}}_{A} \hat{\mathbf{n}} = \begin{bmatrix} 0 \\ 0 \end{bmatrix}$

So finding $\hat{\mathbf{n}}$ requires solving the homogeneous system above. The matrix A will be invertible when $\det A \neq 0 \;\Rightarrow$ $(1 - b)(1) - (-\frac{1}{2})(-2) \neq 0 \;\Rightarrow\; b \neq 0$.

(a) $\hat{\mathbf{n}} = \begin{bmatrix} 0 \\ 0 \end{bmatrix}$

(b) $\hat{\mathbf{n}} = \begin{bmatrix} 2k \\ k \end{bmatrix}$, where k is any real number (infinite number of equilibria)

35. $\hat{\mathbf{v}} = \begin{bmatrix} k \\ 0 \end{bmatrix}$, where k is any real number (infinite number of equilibria)

EXERCISES 8.7 | page 544

1. (a) Reflects about the x-axis (b) Reflects about the y-axis
(c) Reflects about the x-axis and the y-axis
(d) Reflects about the x- and y-axes and the line $y = x$ (swaps x- and y-values).

3. Vertical segment of L rotates clockwise and increases in length
(b) Horizontal segment of L rotates clockwise and increases in length
(c) Vertical segment of L rotates counterclockwise and increases in length
(d) Horizontal segment of L rotates counterclockwise and increases in length

5. (a) $\begin{bmatrix} -1 & 0 \\ 0 & 1 \end{bmatrix}$ (b) $\begin{bmatrix} 2 & 0 \\ 0 & 1 \end{bmatrix}$ (c) $\begin{bmatrix} 1 & 1.5 \\ 0 & 1 \end{bmatrix}$

(d) $\begin{bmatrix} 1/\sqrt{2} & -1/\sqrt{2} \\ 1/\sqrt{2} & 1/\sqrt{2} \end{bmatrix}$

7. Denote the voltage vector of a healthy heart by $\mathbf{v} = [0.3, -0.2]$.

(a) $\begin{bmatrix} 1 & 0 \\ 0 & -1 \end{bmatrix}$ (b) $\begin{bmatrix} -1 & 0 \\ 0 & 1 \end{bmatrix}$ (c) $\begin{bmatrix} -1 & 0 \\ 0 & -1 \end{bmatrix}$

(d) $\begin{bmatrix} 1 & 1 \\ 0 & 0.3335 \end{bmatrix}$

9. (a) Yes (b) Yes (c) No (d) Yes (e) No (f) Yes

11. (a) 0, 2 (b) 1, −1 (c) 2, 3 (d) $i, -i$ (e) 0, 2
(f) 1, 2, 3

13. (a) $\lambda = 1, \mathbf{v} = \begin{bmatrix} 1 \\ 0 \end{bmatrix}$; $\lambda = -1, \mathbf{v} = \begin{bmatrix} 0 \\ 1 \end{bmatrix}$

(b) $\lambda = 3, \mathbf{v} = \begin{bmatrix} 1 \\ 1 \end{bmatrix}$; $\lambda = -1, \mathbf{v} = \begin{bmatrix} 1 \\ -1 \end{bmatrix}$

(c) $\lambda = 1 + 2i, \mathbf{v} = \begin{bmatrix} i \\ 1 \end{bmatrix}$; $\lambda = 1 - 2i, \mathbf{v} = \begin{bmatrix} 1 \\ i \end{bmatrix}$

(d) $\lambda = 2, \mathbf{v} = \begin{bmatrix} 1 \\ 0 \end{bmatrix}$; $\lambda = 5, \mathbf{v} = \begin{bmatrix} 7 \\ 3 \end{bmatrix}$

(e) $\lambda = -1 + \sqrt{10}, \mathbf{v} = \begin{bmatrix} -2 \\ 2 - \sqrt{10} \end{bmatrix}$;

$\lambda = -1 - \sqrt{10}, \mathbf{v} = \begin{bmatrix} -2 \\ 2 + \sqrt{10} \end{bmatrix}$

(f) $\lambda = 1 + \sqrt{31}, \mathbf{v} = \begin{bmatrix} -6 \\ 1 - \sqrt{31} \end{bmatrix}$;

$\lambda = 1 - \sqrt{31}, \mathbf{v} = \begin{bmatrix} -6 \\ 1 + \sqrt{31} \end{bmatrix}$

(g) $\lambda = 1, \mathbf{v} = \begin{bmatrix} 0 \\ 1 \\ 0 \end{bmatrix}; \lambda = 1 + \sqrt{3}, \mathbf{v} = \begin{bmatrix} \sqrt{3} \\ 2 \\ 3 \end{bmatrix};$

$\lambda = 1 - \sqrt{3}, \mathbf{v} = \begin{bmatrix} -\sqrt{3} \\ 2 \\ 3 \end{bmatrix}$

(h) $\lambda = 1, \mathbf{v} = \begin{bmatrix} 1 \\ 0 \\ 0 \end{bmatrix}; \lambda = 1 + \sqrt{14}, \mathbf{v} = \begin{bmatrix} 3 + \sqrt{14} \\ 7 \\ \sqrt{14} \end{bmatrix};$

$\lambda = 1 - \sqrt{14}, \mathbf{v} = \begin{bmatrix} -3 + \sqrt{14} \\ -7 \\ \sqrt{14} \end{bmatrix}$

15. (a) Free to choose any values for the components of the eigenvectors associated with the repeated eigenvalue $\lambda = 1$.

Two distinct eigenvectors $\begin{bmatrix} 1 \\ 0 \end{bmatrix}$ and $\begin{bmatrix} 0 \\ 1 \end{bmatrix}$ are associated with $\lambda = 1$.

(b) Only one eigenvector $\begin{bmatrix} 1 \\ 0 \end{bmatrix}$ is associated with the repeated eigenvalue $\lambda = 1$.

(c) Two distinct eigenvectors $\begin{bmatrix} 1 \\ 0 \\ 0 \end{bmatrix}$ and $\begin{bmatrix} 0 \\ 1 \\ 0 \end{bmatrix}$ are associated with the repeated eigenvalue $\lambda = 1$. One eigenvector $\begin{bmatrix} 0 \\ 0 \\ 1 \end{bmatrix}$ is associated with $\lambda = 2$.

25. (a) $\lambda_A + \lambda_B$ (b) $\lambda_A \lambda_B$

27. $\lambda = 2, \mathbf{v} = \begin{bmatrix} 1 \\ 0 \end{bmatrix}; \lambda = \frac{9}{10}, , \mathbf{v} = \begin{bmatrix} 30 \\ -11 \end{bmatrix}$

29. $\lambda = \frac{1}{2}b + \frac{1}{2}\sqrt{b^2 + 4}, \mathbf{v} = \begin{bmatrix} 4 \\ -b + \sqrt{b^2 + 4} \end{bmatrix};$

$\lambda = \frac{1}{2}b - \frac{1}{2}\sqrt{b^2 + 4}, \mathbf{v} = \begin{bmatrix} -4 \\ b + \sqrt{b^2 + 4} \end{bmatrix}$

31. $-\lambda^3 + \lambda^2 + \lambda + \frac{2}{3}$

EXERCISES 8.8 | page 556

1. $D = \begin{bmatrix} 1 & 0 \\ 0 & 2 \end{bmatrix}, P = \begin{bmatrix} 0 & 1 \\ 1 & 0 \end{bmatrix}, P^{-1} = \begin{bmatrix} 0 & 1 \\ 1 & 0 \end{bmatrix}$

3. $D = \begin{bmatrix} 3 & 0 \\ 0 & -1 \end{bmatrix}, P = \begin{bmatrix} 1 & 1 \\ 1 & -1 \end{bmatrix}, P^{-1} = \begin{bmatrix} \frac{1}{2} & \frac{1}{2} \\ \frac{1}{2} & -\frac{1}{2} \end{bmatrix}$

5. $D = \begin{bmatrix} 2 + \sqrt{5}\,i & 0 \\ 0 & 2 - \sqrt{5}\,i \end{bmatrix}, P = \begin{bmatrix} 2 & 2 \\ 1 + \sqrt{5}\,i & 1 - \sqrt{5}\,i \end{bmatrix},$

$P^{-1} = \begin{bmatrix} \frac{1}{4\sqrt{5}}i + \frac{1}{4} & -\frac{1}{2\sqrt{5}}i \\ -\frac{1}{4\sqrt{5}}i + \frac{1}{4} & \frac{1}{2\sqrt{5}}i \end{bmatrix}$

7. $D = \begin{bmatrix} 0 & 0 & 0 \\ 0 & 1 & 0 \\ 0 & 0 & 2 \end{bmatrix}, P = \begin{bmatrix} 1 & 0 & 1 \\ 0 & 1 & 0 \\ 0 & 0 & 2 \end{bmatrix}, P^{-1} = \begin{bmatrix} 1 & 0 & -\frac{1}{2} \\ 0 & 1 & 0 \\ 0 & 0 & \frac{1}{2} \end{bmatrix}$

9. $D^2 = \begin{bmatrix} 1 & 0 \\ 0 & 16 \end{bmatrix}, P = \begin{bmatrix} 1 & 0 \\ 0 & 1 \end{bmatrix}, P^{-1} = \begin{bmatrix} 1 & 0 \\ 0 & 1 \end{bmatrix}$

11. $D^2 = \begin{bmatrix} 1 & 0 \\ 0 & 4 \end{bmatrix}, P = \begin{bmatrix} 2 & 1 \\ -1 & 0 \end{bmatrix}, P^{-1} = \begin{bmatrix} 0 & -1 \\ 1 & 2 \end{bmatrix}$

13. $D^2 = \begin{bmatrix} a^2 & 0 \\ 0 & b^2 \end{bmatrix}, P = \begin{bmatrix} 1 & 0 \\ 0 & 1 \end{bmatrix}, P^{-1} = \begin{bmatrix} 1 & 0 \\ 0 & 1 \end{bmatrix}$

15. $\mathbf{n}_t = \begin{bmatrix} 0 \\ 1 \end{bmatrix} + \begin{bmatrix} 1 \\ 0 \end{bmatrix} 2^t$ **17.** $\mathbf{n}_t = \begin{bmatrix} 1 \\ 1 \end{bmatrix} 2^t$

19. $\mathbf{n}_t = \begin{bmatrix} 1 - a \\ 0 \end{bmatrix} + \begin{bmatrix} a \\ 1 \end{bmatrix} 2^t$

21–25 Denote the arbitrary initial condition by $\mathbf{n}_0 = \begin{bmatrix} x_0 \\ y_0 \end{bmatrix}$.

21. $\mathbf{n}_t = x_0 \begin{bmatrix} 1 \\ 0 \end{bmatrix} a^t + y_0 \begin{bmatrix} 0 \\ 1 \end{bmatrix} b^t$

23. $\mathbf{n}_t = -y_0 \begin{bmatrix} 1 \\ -1 \end{bmatrix} + (x_0 + y_0) \begin{bmatrix} 1 \\ 0 \end{bmatrix} 2^t$

25. $\mathbf{n}_t = \frac{1}{2}(x_0 + y_0) \begin{bmatrix} 1 \\ 1 \end{bmatrix} (a + b)^t + \frac{1}{2}(x_0 - y_0) \begin{bmatrix} 1 \\ -1 \end{bmatrix} (a - b)^t$

31. $\mathbf{n}_t = (\sqrt{2})^t \begin{bmatrix} \cos\left(\frac{\pi}{4}t\right) & -\sin\left(\frac{\pi}{4}t\right) \\ \sin\left(\frac{\pi}{4}t\right) & \cos\left(\frac{\pi}{4}t\right) \end{bmatrix} \begin{bmatrix} 1 \\ 1 \end{bmatrix}$

33. $\mathbf{n}_t = (\sqrt{6})^t \begin{bmatrix} \sqrt{5}\sin(\theta t) \\ \cos(\theta t) \end{bmatrix}$, where $\theta = \tan^{-1}\sqrt{5}$

35. (a) No (b) $\mathbf{v}_t = \begin{bmatrix} 0.3 \\ 0 \end{bmatrix} + \begin{bmatrix} 0 \\ -0.2 \end{bmatrix} (-1)^t$

(c) $\mathbf{v}_t$ cycles between $\begin{bmatrix} 0.3 \\ -0.2 \end{bmatrix}$ and $\begin{bmatrix} 0.3 \\ 0.2 \end{bmatrix}$.

37. (a) $\lambda_1 = 1, \lambda_2 = \frac{7}{10}$

(b) $\mathbf{y}_t$ approaches steady state in long term.

(c) $\mathbf{y}_t = \frac{1}{3}\left(\begin{bmatrix} 2 \\ 1 \end{bmatrix} + \begin{bmatrix} 1 \\ -1 \end{bmatrix}\left(\frac{7}{10}\right)^t\right)$

(d) $\lim_{t\to\infty} \mathbf{y}_t = \frac{1}{3}\begin{bmatrix} 2 \\ 1 \end{bmatrix}$

39. (a) $\lambda_1 = 1, \lambda_2 = \frac{3}{10}$

(b) $\mathbf{x}_t$ approaches steady state in long term.

(c) $\mathbf{x}_t = \frac{1}{7}\begin{bmatrix} 5 \\ 2 \end{bmatrix} + \left(a - \frac{5}{7}\right)\begin{bmatrix} 1 \\ -1 \end{bmatrix}\left(\frac{3}{10}\right)^t$

(d) $\lim_{t\to\infty} \mathbf{x}_t = \frac{1}{7}\begin{bmatrix} 5 \\ 2 \end{bmatrix}$

41. (b) $\lambda_1 = \dfrac{1 + \sqrt{5}}{2}$, $\lambda_2 = \dfrac{1 - \sqrt{5}}{2}$

(c) Yes; components of $\mathbf{z}_n$ grow by a factor λ_1 each time step and asymptotically approach the line defined by the associated eigenvector $\mathbf{v}_1$.

(d) $F_n = \dfrac{1}{\sqrt{5}}\left[\left(\dfrac{1+\sqrt{5}}{2}\right)^n - \left(\dfrac{1-\sqrt{5}}{2}\right)^n\right]$

CHAPTER 8 REVIEW | page 561

True-False Quiz

1. True **3.** False **5.** False **7.** False **9.** True
11. True **13.** False **15.** True **17.** True

Exercises

1. (a) $(x + 1)^2 + (y - 2)^2 + (z - 1)^2 = 69$
(b) $(y - 2)^2 + (z - 1)^2 = 68$, $x = 0$
(c) Center $(4, -1, -3)$, radius 5

3. $(x_1 - c_1)^2 + (x_2 - c_2)^2 + \cdots + (x_n - c_n)^2 = r^2$ for a hypersphere centered at $P(c_1, c_2, \ldots, c_n)$ with radius r

5. $-2, -4$ **7.** (a) 45° (b) 74.7° **9.** $\frac{13}{5}$

11. (a) $\sqrt{8} \approx 2.83$ (b) $\cos^{-1}\left(\dfrac{1}{\sqrt{5}}\right) \approx 1.11$ (or 63.4°)

13. $\begin{bmatrix} 0 & 0 & 0 \\ 0.5 & 1 & 2 \\ 0 & 0.5 & 0 \end{bmatrix}$

15. (a) Not defined (b) $\begin{bmatrix} 9 & 4 \\ 19 & 21 \end{bmatrix}$ (c) Not defined
(d) Not defined (e) $\begin{bmatrix} 16 & 1 & 7 \\ 4 & 5 & 11 \end{bmatrix}$ (f) $\begin{bmatrix} \frac{13}{7} & 4 \\ \frac{1135}{63} & \frac{107}{9} \end{bmatrix}$

21. $x = \frac{9}{11}$, $y = \frac{5}{11}$

23. Infinite number of solutions given by $y = -7x$

25. $\hat{\mathbf{n}} = \begin{bmatrix} 0 \\ 0 \\ 0 \end{bmatrix}$; population extinct

27. $\lambda = 2$, $\mathbf{v} = \begin{bmatrix} 1 \\ -1 \end{bmatrix}$; $\lambda = 4$, $\mathbf{v} = \begin{bmatrix} 1 \\ 1 \end{bmatrix}$

29. $\lambda = 2 + i$, $\mathbf{v} = \begin{bmatrix} 1 \\ i \end{bmatrix}$; $\lambda = 2 - i$, $\mathbf{v} = \begin{bmatrix} i \\ 1 \end{bmatrix}$

33. $\mathbf{n}_t = -\frac{1}{2}\begin{bmatrix} 1 \\ -2 \end{bmatrix}(-4)^t$ for $t \geq 1$

35. $\mathbf{n}_t = r^t\left(\begin{bmatrix} \cos(\theta t) - \sin(\theta t) \\ \sin(\theta t) + \cos(\theta t) \end{bmatrix}\right)$

where $r = \sqrt{1 + a^2}$ and $\theta = \tan^{-1}\left(\dfrac{1}{a}\right)$

37. Yes

39. (a) $A = \begin{bmatrix} \frac{1}{2} & \frac{1}{2} \\ 1 & 0 \end{bmatrix}$ (c) $\lambda_1 = 1$ and $\lambda_2 = -\frac{1}{2}$
(d) $\mathbf{z}_n$ approaches steady state in long term.
(e) $\lim_{n \to \infty} a_n = c$ (component 1 of $\mathbf{v}_1$), where c is constant

CHAPTER 9

EXERCISES 9.1 | page 580

1. (a) −27°C; a temperature of −15°C with wind blowing at 40 km/h feels equivalent to about −27°C without wind.
(b) When the temperature is −20°C, what wind speed gives a wind chill of −30°C? 20 km/h
(c) With a wind speed of 20 km/h, what temperature gives a wind chill of −49°C? −35°C
(d) A function of wind speed that gives wind-chill values when the temperature is −5°C
(e) A function of temperature that gives wind-chill values when the wind speed is 50 km/h

3. (a) ≈20.5; the surface area of a person 70 in tall who weighs 160 lb is approximately 20.5 square feet.
(b) Answers vary depending on height and weight.

5. (a) ≈94.2; When the manufacturer invests \$20 million and the number of labor hours is 120,000, its yearly production is about \$94.2 million.

7. Snake with $R = 3$ and $S = 1$ is likely to survive longer.

9. (a) 1 (b) $\mathbb{R}^2$ (c) $[-1, 1]$

11. (a) 3 (b) $\{(x, y, z) \mid x^2 + y^2 + z^2 < 4, x \geq 0, y \geq 0, z \geq 0\}$, interior of a sphere of radius 2, center the origin, in the first octant

13. $\{(x, y) \mid y \leq 2x\}$

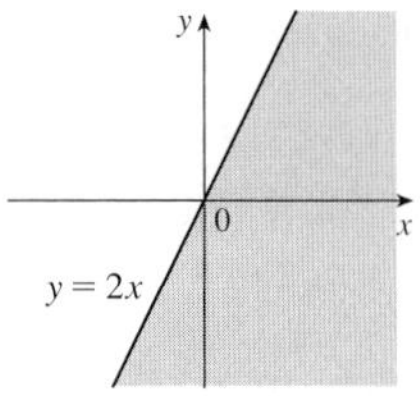

15. $\{(x, y) \mid -1 \leq x \leq 1, -1 \leq y \leq 1\}$

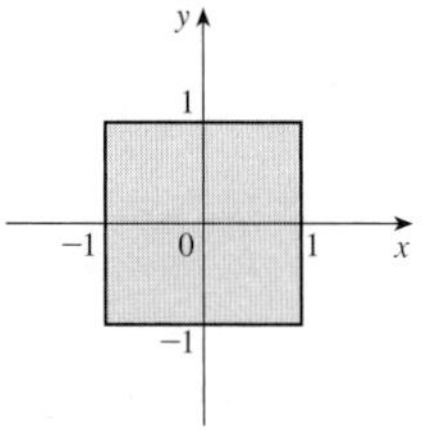

17. $\{(x, y) \mid y \geq x^2, x \neq \pm 1\}$

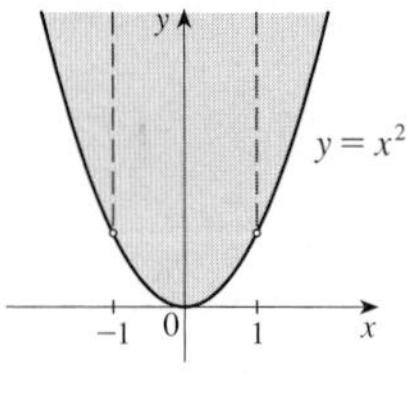

19. $\{(x, y, z) \mid x^2 + y^2 + z^2 \leq 1\}$

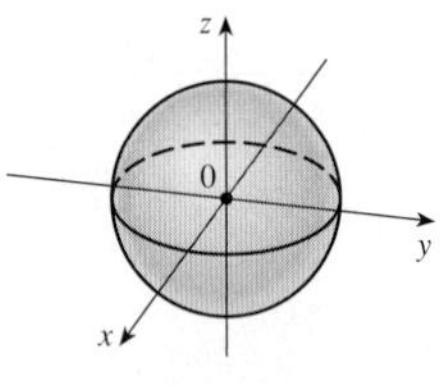

21. $z = 3$, horizontal plane

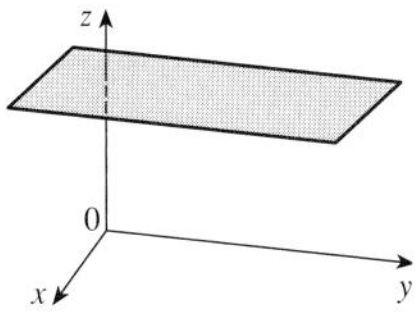

23. $4x + 5y + z = 10$, plane

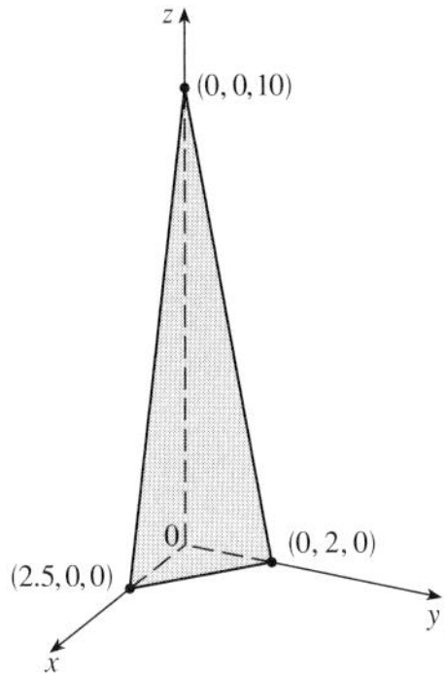

25. $z = y^2 + 1$, parabolic cylinder

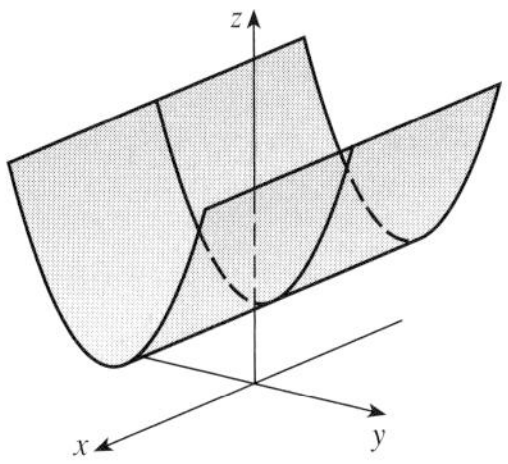

27. $\approx 56, \approx 35$ **29.** 11°C, 19.5°C

31. Steep; nearly flat

33. $y = 2x - k$

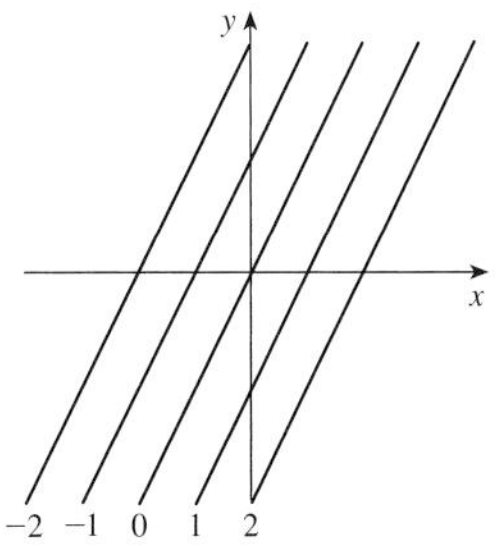

35. $y = k/x$

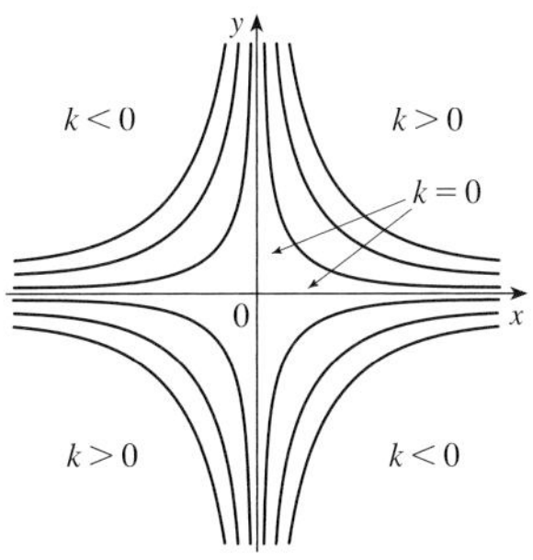

37. $y = ke^{-x}$

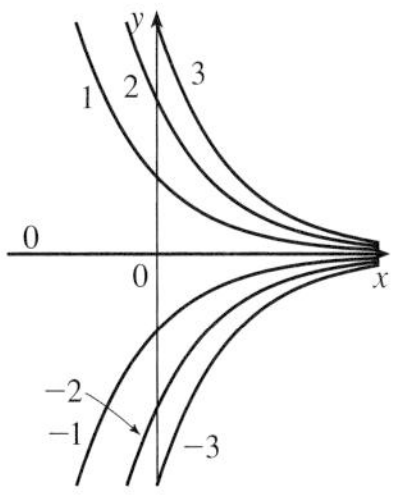

39. $m = kh^2$

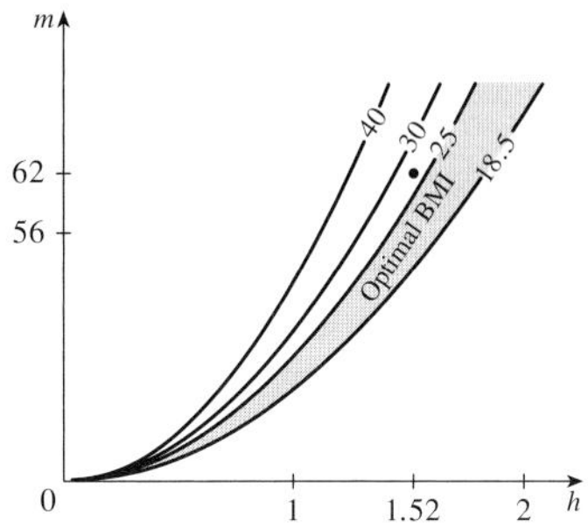

No

41. (a) C (b) II **43.** (a) F (b) I

45. (a) B (b) VI

47.

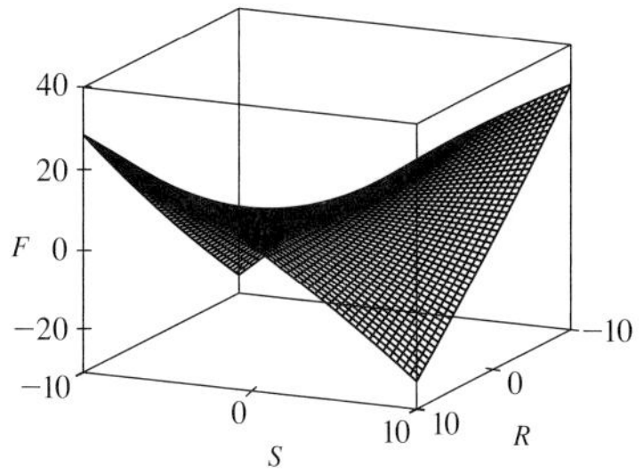

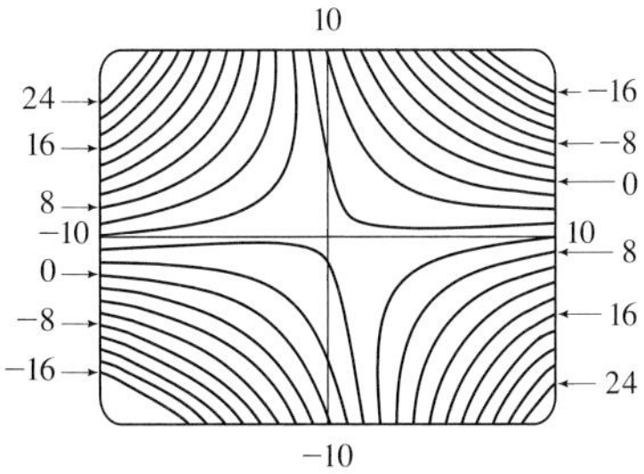

$F(R, S)$ is shaped like a saddle.

49. 1 **51.** $\frac{2}{7}$ **53.** Does not exist **55.** Does not exist

57. 0 **59.** $\{(x, y) \mid y \geqslant 0\}$ **61.** $\{(x, y) \mid x^2 + y^2 > 4\}$

63. $\{(x, y, z) \mid y \geqslant 0, y \neq \sqrt{x^2 + z^2}\}$

65. $\{(x, y) \mid (x, y) \neq (0, 0)\}$

EXERCISES 9.2 | page 593

1. (a) The rate of change of temperature as longitude varies, with latitude and time fixed; the rate of change as only latitude varies; the rate of change as only time varies

(b) Positive, negative, positive

3. (a) $f_T(-15, 30) \approx 1.3$; for a temperature of $-15°$C and wind speed of 30 km/h, the wind-chill index rises by 1.3°C for each degree the temperature increases. $f_v(-15, 30) \approx -0.15$; for a temperature of $-15°$C and wind speed of 30 km/h, the wind-chill index decreases by 0.15°C for each km/h the wind speed increases.
(b) Positive, negative (c) 0
5. (a) Positive (b) Negative
7. $f_x(1, 2) = -8 =$ slope of C_1, $f_y(1, 2) = -4 =$ slope of C_2

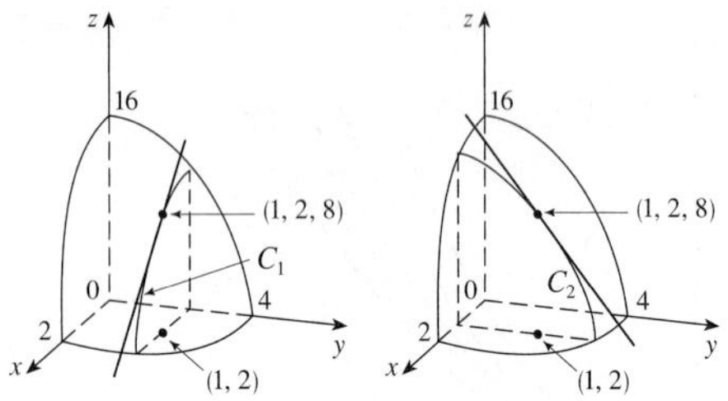

9. $f_x(x, y) = -3y$, $f_y(x, y) = 5y^4 - 3x$
11. $f_x(x, t) = -\pi e^{-t}\sin \pi x$, $f_t(x, t) = -e^{-t}\cos \pi x$
13. $\partial z/\partial x = 20(2x + 3y)^9$, $\partial z/\partial y = 30(2x + 3y)^9$
15. $f_x(x, y) = 2y/(x + y)^2$, $f_y(x, y) = -2x/(x + y)^2$
17. $\partial w/\partial \alpha = \cos \alpha \cos \beta$, $\partial w/\partial \beta = -\sin \alpha \sin \beta$
19. $f_r(r, s) = \dfrac{2r^2}{r^2 + s^2} + \ln(r^2 + s^2)$, $f_s(r, s) = \dfrac{2rs}{r^2 + s^2}$
21. $\partial u/\partial t = e^{w/t}(1 - w/t)$, $\partial u/\partial w = e^{w/t}$
23. $f_x = z - 10xy^3z^4$, $f_y = -15x^2y^2z^4$, $f_z = x - 20x^2y^3z^3$
25. $\partial w/\partial x = 1/(x + 2y + 3z)$, $\partial w/\partial y = 2/(x + 2y + 3z)$, $\partial w/\partial z = 3/(x + 2y + 3z)$
27. $\partial u/\partial x = e^{-t}\sin\theta$, $\partial u/\partial t = -xe^{-t}\sin\theta$, $\partial u/\partial\theta = xe^{-t}\cos\theta$
29. $f_x = yz^2\tan(yt)$, $f_y = xyz^2t\sec^2(yt) + xz^2\tan(yt)$, $f_z = 2xyz\tan(yt)$, $f_t = xy^2z^2\sec^2(yt)$
31. $\partial u/\partial x_i = x_i/\sqrt{x_1^2 + x_2^2 + \cdots + x_n^2}$ **33.** $\frac{1}{5}$ **35.** $\frac{1}{4}$
37. $\dfrac{\partial z}{\partial x} = \dfrac{3yz - 2x}{2z - 3xy}$, $\dfrac{\partial z}{\partial y} = \dfrac{3xz - 2y}{2z - 3xy}$
39. $\dfrac{\partial z}{\partial x} = \dfrac{1 + y^2z^2}{1 + y + y^2z^2}$, $\dfrac{\partial z}{\partial y} = \dfrac{-z}{1 + y + y^2z^2}$
41. (a) ≈ 0.0545 ft²/lb; rate at which body surface area increases with respect to weight when an individual weighs 160 lb and has a height 70 in.
(b) ≈ 0.213 ft²/in; rate at which body surface area increases with respect to height when an individual weighs 160 lb and has a height 70 in.
43. $\dfrac{\partial R}{\partial L} = \dfrac{C}{r^4}$, rate of increase of resistance of blood when artery length increases (with constant artery radius);
$\dfrac{\partial R}{\partial r} = -4C\dfrac{L}{r^5}$, rate of change of resistance of blood when artery radius increases (with constant artery length)
45. $\dfrac{\partial P}{\partial v} = 3Av^2 - \dfrac{B(mg/x)^2}{v^2}$, rate of change in required power that occurs from an increase in the bird's velocity;
$\dfrac{\partial P}{\partial x} = -2\dfrac{Bm^2g^2}{vx^3}$, rate of change in required power that occurs from an increase in the fraction of time the bird spends flapping;
$\dfrac{\partial P}{\partial m} = 2\dfrac{Bmg^2}{vx^2}$, rate of change in required power that occurs from an increase in the bird's mass
47. $E_m(400, 8) \approx 0.301$ kcal/g, an increase in mass of 1 g requires approximately 0.3 kcal of additional energy when $m = 400$ and $v = 8$.
$E_v(400, 8) \approx -4.89$ kcal/(km/h), an increase in speed of 1 km/h results in approximately 4.9 kcal of energy saved when $m = 400$ and $v = 8$.
49. $\dfrac{\partial T}{\partial P} = \dfrac{V - nb}{nR}$; $\dfrac{\partial P}{\partial V} = \dfrac{2n^2a}{V^3} - \dfrac{nRT}{(V - nb)^2}$
51. $f_{xx} = 6xy^5 + 24x^2y$, $f_{xy} = 15x^2y^4 + 8x^3 = f_{yx}$, $f_{yy} = 20x^3y^3$
53. $w_{uu} = v^2/(u^2 + v^2)^{3/2}$, $w_{uv} = -uv/(u^2 + v^2)^{3/2} = w_{vu}$, $w_{vv} = u^2/(u^2 + v^2)^{3/2}$
55. $z_{xx} = -2x/(1 + x^2)^2$, $z_{xy} = 0 = z_{yx}$, $z_{yy} = -2y/(1 + y^2)^2$
59. $12xy$, $72xy$
61. $24\sin(4x + 3y + 2z)$, $12\sin(4x + 3y + 2z)$
63. $\theta e^{r\theta}(2\sin\theta + \theta\cos\theta + r\theta\sin\theta)$
65. $6yz^2$ **73.** No

EXERCISES 9.3 | page 601

1. $z = -7x - 6y + 5$ **3.** $x + y - 2z = 0$ **5.** $z = y$
7. $2x + \frac{1}{4}y - 1$ **9.** $\frac{1}{9}x - \frac{2}{9}y + \frac{2}{3}$ **11.** $4x + 13y + 4z - 28$
15. 6.3 **17.** $\frac{3}{7}x + \frac{2}{7}y + \frac{6}{7}z$; 6.9914
19. $4T + H - 329$; 129°F **21.** $\approx 0.301m - 4.89v + 96$

EXERCISES 9.4 | page 608

1. $(2x + y)\cos t + (2y + x)e^t$
3. $[(x/t) - y\sin t]/\sqrt{1 + x^2 + y^2}$
5. $e^{y/z}[2t - (x/z) - (2xy/z^2)]$
9. $\partial z/\partial s = 2xy^3\cos t + 3x^2y^2\sin t$, $\partial z/\partial t = -2sxy^3\sin t + 3sx^2y^2\cos t$
11. 62 **13.** $\dfrac{2x + y\sin x}{\cos x - 2y}$ **15.** $\dfrac{\sin(x - y) + e^y}{\sin(x - y) - xe^y}$
17. $\dfrac{3yz - 2x}{2z - 3xy}$, $\dfrac{3xz - 2y}{2z - 3xy}$
19. $\dfrac{1 + y^2z^2}{1 + y + y^2z^2}$, $-\dfrac{z}{1 + y + y^2z^2}$
21. (a) $\dfrac{dW}{dt} = \dfrac{\partial W}{\partial F}\dfrac{dF}{dt} + \dfrac{\partial W}{\partial C}\dfrac{dC}{dt}$
(b) $\partial W/\partial F$ is positive, $\partial W/\partial C$ is negative.
(c) Wolf population increases.
(d) Nothing
23. $\dfrac{dB}{da} = \dfrac{1}{h^2}\dfrac{dm}{da} - 2\dfrac{m}{h^3}\dfrac{dh}{da}$
25. $\dfrac{\partial A}{\partial \rho} = \dfrac{1}{\rho q - e^{qA}}$ **27.** ≈ -0.27 L/s

EXERCISES 9.5 | page 617

1. ≈ -0.08 mb/km **3.** ≈ 0.778 **5.** $2 + \sqrt{3}/2$
7. (a) $\nabla f(x, y) = [5y^2 - 12x^2y, 10xy - 4x^3]$
(b) $[-4, 16]$ (c) $172/13$
9. (a) $\nabla f(x, y) = [2\cos(2x + 3y), 3\cos(2x + 3y)]$
(b) $[2, 3]$ (c) $\sqrt{3} - \frac{3}{2}$

11. $23/10$ **13.** $-8/\sqrt{10}$ **15.** $-6/\sqrt{5}$ **17.** $2/5$
19. $4\sqrt{2}, [-1, 1]$ **21.** $1, [0, 1]$
23. (b) $[-12, 92]$ **25.** All points on the line $y = x + 1$
27. Deeper **29.** $[-1, -2]$
31. (a) It is beneficial
(b)

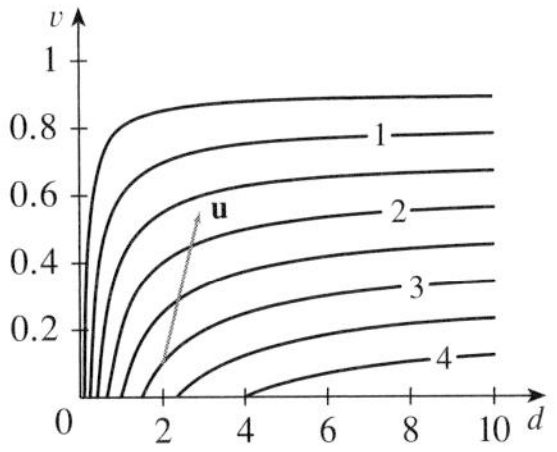

33. $[2, 3]$, $2x + 3y = 12$

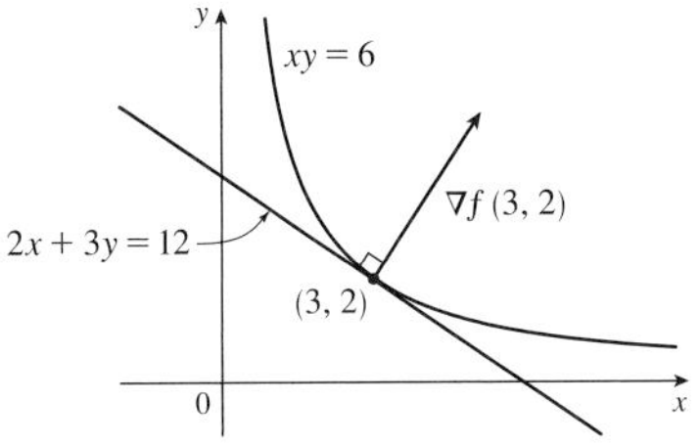

35. $\frac{774}{25}$

■ EXERCISES 9.6 | page 626

1. (a) f has a local minimum at (1, 1).
(b) f has a saddle point at (1, 1).
3. Local minimum at (1, 1), saddle point at (0, 0)
5. Minimum $f\left(\frac{1}{3}, -\frac{2}{3}\right) = -\frac{1}{3}$
7. Minima $f(1, 1) = 0, f(-1, -1) = 0$, saddle point at (0, 0)
9. Minimum $f(2, 1) = -8$, saddle point at (0, 0)
11. None **13.** Minimum $f(0, 0) = 0$, saddle points at $(\pm 1, 0)$
15. Minima $f(0, 1) = f(\pi, -1) = f(2\pi, 1) = -1$, saddle points at $(\pi/2, 0)$, $(3\pi/2, 0)$
17. Maximum $f(2, 0) = 9$, minimum $f(0, 3) = -14$
19. Maximum $f(\pm 1, 1) = 7$, minimum $f(0, 0) = 4$
21.

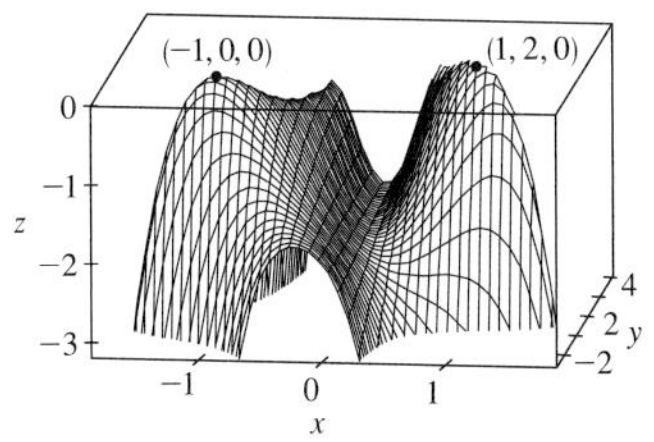

23. $\sqrt{3}$ **25.** $(2, 1, \sqrt{5}), (2, 1, -\sqrt{5})$ **27.** $\frac{100}{3}, \frac{100}{3}, \frac{100}{3}$
29. $8r^3/(3\sqrt{3})$ **31.** $\frac{4}{3}$ **33.** Cube, edge length $c/12$
35. Square base of side 40 cm, height 20 cm
37. (a) $H(p_1, p_2) = \ln\left[\dfrac{(1 - p_1 - p_2)^{p_1+p_2-1}}{p_1^{p_1} p_2^{p_2}}\right]$
(b) $\{(p_1, p_2) \mid p_1 \geqslant 0, p_2 \geqslant 0, p_1 + p_2 \leqslant 1\}$
(c) $\ln 3, p_1 = p_2 = p_3 = \frac{1}{3}$
39. $\frac{100}{21}$

■ CHAPTER 9 REVIEW | page 628

True-False Quiz

1. True **3.** False **5.** False **7.** True **9.** False
11. True

Exercises

1. $\{(x, y) \mid y > -x - 1\}$

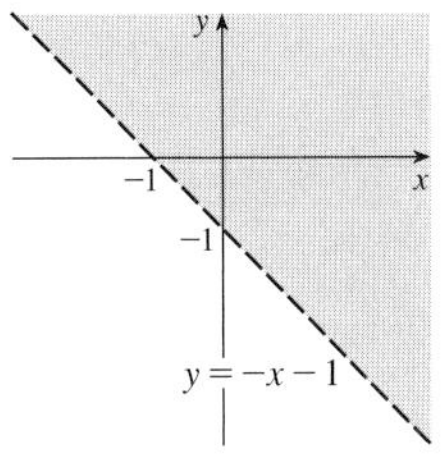

3.

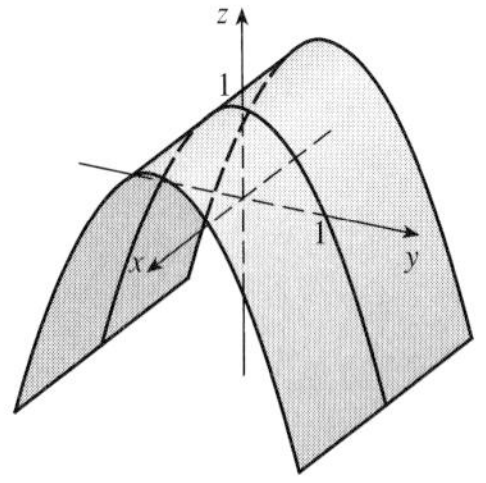

5.

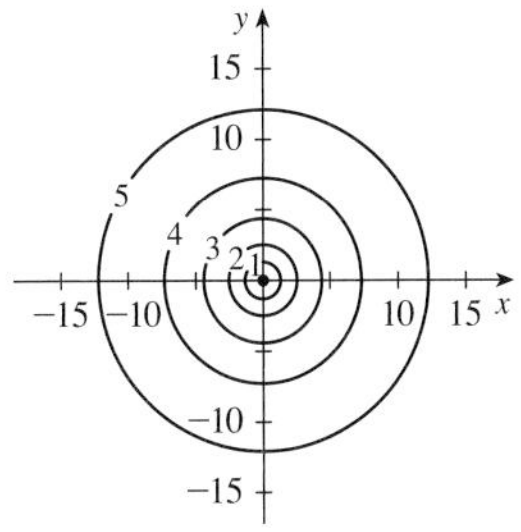

7. $\frac{2}{3}$
9. (a) $\approx 3.5°C/m$, $-3.0°C/m$ (b) $\approx 0.35°C/m$ by Equation 9.5.9 (Definition 9.5.2 gives $\approx 1.1°C/m$.)
(c) ≈ -0.25
11. $f_x = 1/\sqrt{2x + y^2}$, $f_y = y/\sqrt{2x + y^2}$
13. $g_u = \tan^{-1}v$, $g_v = u/(1 + v^2)$
15. $T_p = \ln(q + e^r)$, $T_q = p/(q + e^r)$, $T_r = pe^r/(q + e^r)$
17. (a) $\partial P/\partial m \approx 4.87$, $\partial P/\partial T \approx -0.013$
(b) Worsen (c) Yes
19. $f_{xx} = 24x$, $f_{xy} = -2y = f_{yx}$, $f_{yy} = -2x$
21. $f_{xx} = k(k - 1)x^{k-2}y^l z^m$, $f_{xy} = klx^{k-1}y^{l-1}z^m = f_{yx}$, $f_{xz} = kmx^{k-1}y^l z^{m-1} = f_{zx}$, $f_{yy} = l(l - 1)x^k y^{l-2}z^m$, $f_{yz} = lmx^k y^{l-1}z^{m-1} = f_{zy}$, $f_{zz} = m(m - 1)x^k y^l z^{m-2}$
25. $z = -8x - 2y$ **27.** $x - 2y + z = 4$ **29.** $x + 5y - 4$
31. $2xy^3(1 + 6p) + 3x^2y^2(pe^p + e^p) + 4z^3(p\cos p + \sin p)$
35. $[q^2e^{-pq} - pq^3e^{-pq}, 2pqe^{-pq} - p^2q^2e^{-pq}]$ **37.** $-\frac{4}{5}$
39. $\approx \frac{5}{8}$ knot/mi
41. Maximum $f(1, 1) = 1$; saddle points (0, 0), (0, 3), (3, 0)
43. Maximum $f(1, 2) = 4$, minimum $f(2, 4) = -64$

CHAPTER 10

■ EXERCISES 10.1 | page 638

1. Linear, homogeneous, nonautonomous
3. Nonlinear, autonomous
5. Linear, nonhomogeneous, nonautomonous
7. $\begin{bmatrix} dx/dt \\ dy/dt \end{bmatrix} = \begin{bmatrix} 5 & -3 \\ -1 & 2 \end{bmatrix}\begin{bmatrix} x \\ y \end{bmatrix}$

9. $\begin{bmatrix} dx/dt \\ dy/dt \end{bmatrix} = \begin{bmatrix} 0 & 3t \\ 2 & -3 \end{bmatrix}\begin{bmatrix} x \\ y \end{bmatrix} + \begin{bmatrix} -7 \\ 0 \end{bmatrix}$

11. $\begin{bmatrix} dx/dt \\ dy/dt \end{bmatrix} = \begin{bmatrix} 2 & 0 \\ 3 & 7 \end{bmatrix}\begin{bmatrix} x \\ y \end{bmatrix} + \begin{bmatrix} -5 \\ 0 \end{bmatrix}$

13. $\begin{bmatrix} dx/dt \\ dy/dt \end{bmatrix} = \begin{bmatrix} 1 & 4 \\ -1 & 1 \end{bmatrix}\begin{bmatrix} x \\ y \end{bmatrix} + \begin{bmatrix} -3t \\ 0 \end{bmatrix}$

15. Node

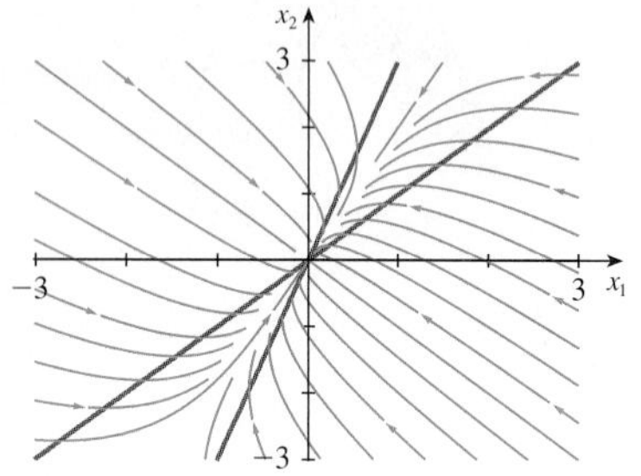

17. Spiral

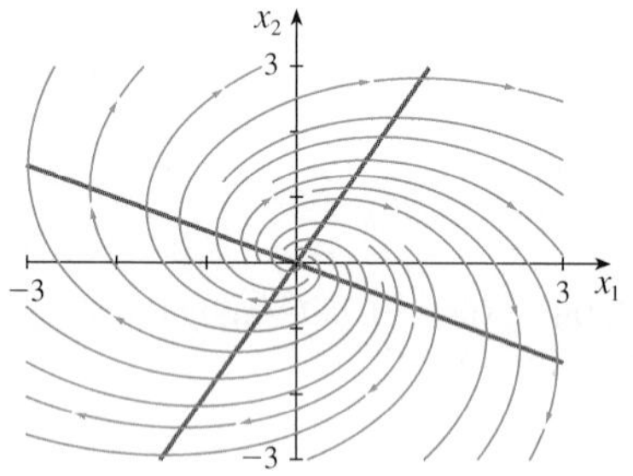

19. Spiral

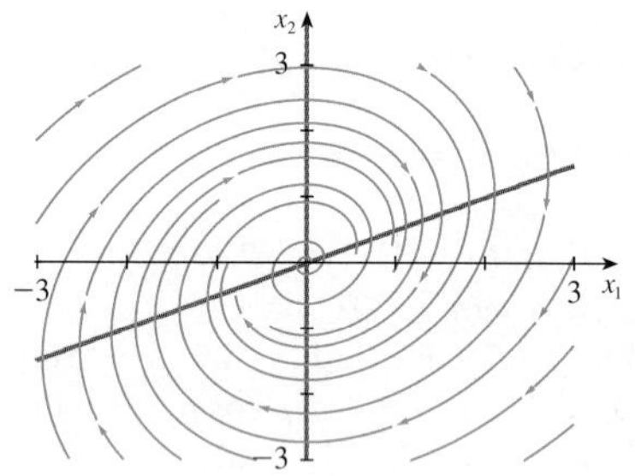

21. Node

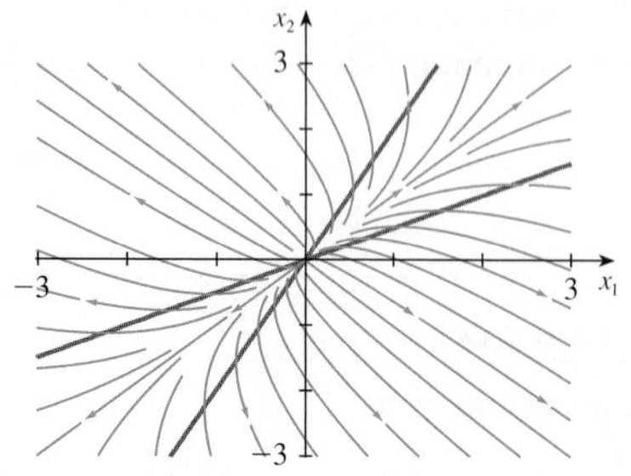

23. (a) $\hat{\mathbf{x}} = -A^{-1}\mathbf{g}$

(b) The distance between the vector $\mathbf{x}$ and the equilibrium vector $\hat{\mathbf{x}}$ along each coordinate dimension

31.

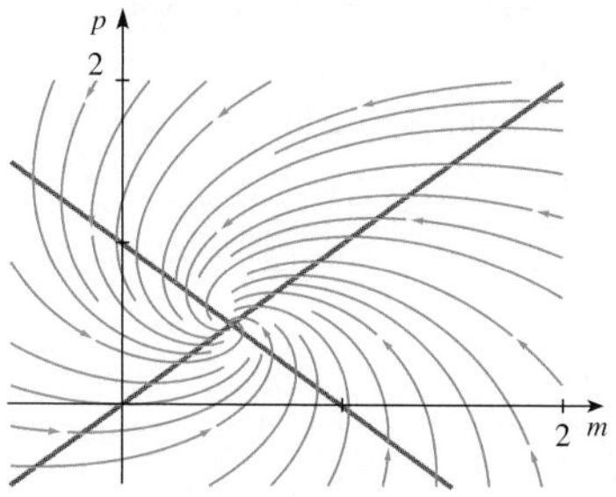

33. (a)

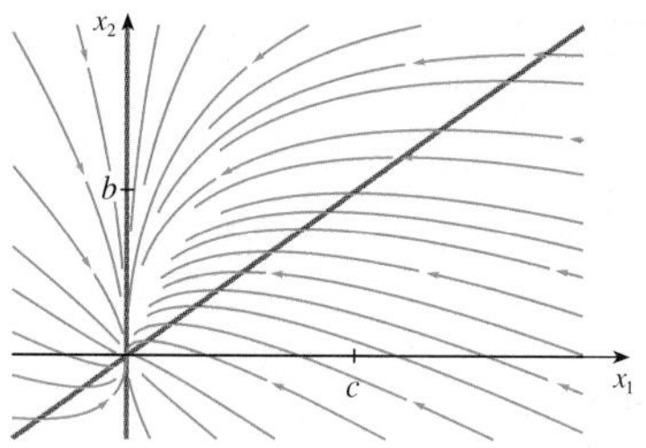

(b) Amount of antibody in blood and tumor approaches zero.

■ EXERCISES 10.2 | page 650

11.

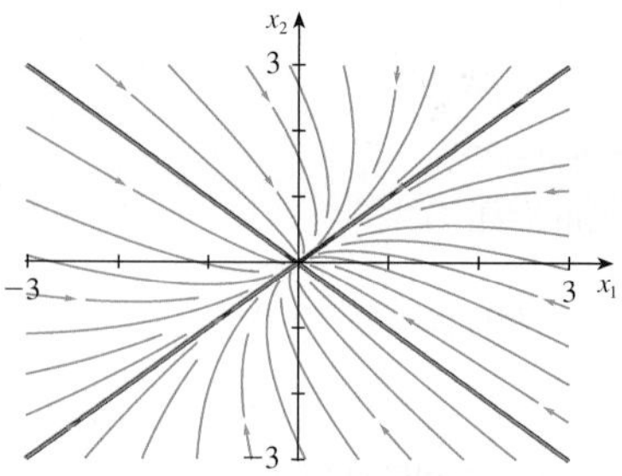

13.

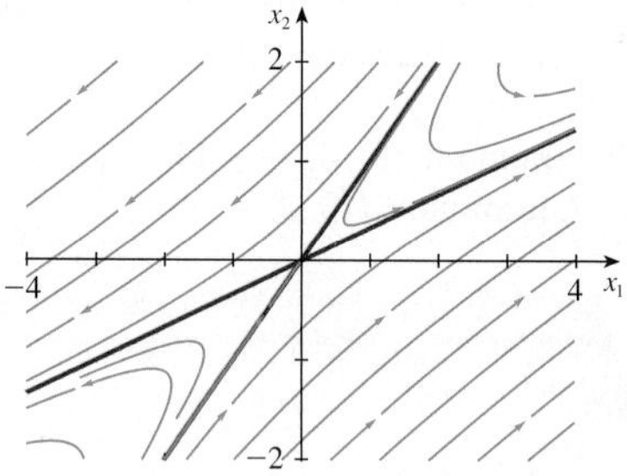

15.

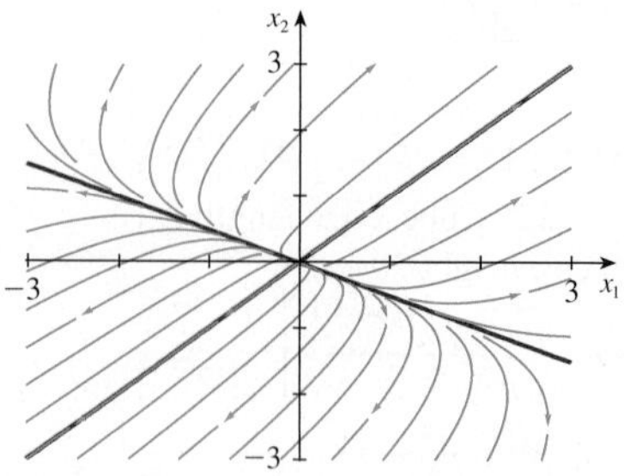

17. $\mathbf{x}(t) = \frac{3}{2}e^{-t}\begin{bmatrix} 1 \\ 1 \end{bmatrix} - \frac{1}{2}e^{-2t}\begin{bmatrix} 1 \\ -1 \end{bmatrix}$

19. $\mathbf{x}(t) = 3e^{t}\begin{bmatrix}1\\2\end{bmatrix} - 4e^{-t}\begin{bmatrix}0\\1\end{bmatrix}$

21. $\mathbf{x}(t) = -9e^{t}\begin{bmatrix}1\\1\end{bmatrix} + 4e^{3t}\begin{bmatrix}2\\3\end{bmatrix}$

23. $\mathbf{x}(t) = e^{-t}\begin{bmatrix}2\cos\sqrt{6}\,t\\-\sqrt{6}\sin\sqrt{6}\,t\end{bmatrix}$

25. $\mathbf{x}(t) = \frac{1}{2}e^{t}\begin{bmatrix}1\\-1\end{bmatrix} + \frac{3}{2}e^{-t}\begin{bmatrix}1\\1\end{bmatrix}$

27. $\mathbf{x}(t) = e^{3t/2}\begin{bmatrix}\cos\frac{\sqrt{39}}{2}t - \frac{3\sqrt{39}}{13}\sin\frac{\sqrt{39}}{2}t\\ \cos\frac{\sqrt{39}}{2}t + \frac{\sqrt{39}}{13}\sin\frac{\sqrt{39}}{2}t\end{bmatrix}$

29. $\mathbf{x}(t) = c_1\begin{bmatrix}1\\-2\end{bmatrix} + c_2e^{-t}\begin{bmatrix}1\\-1\end{bmatrix}$

31. (b) No restrictions on the components of the eigenvectors

33. $a < 0$: stable spiral; $0 \leqslant a < 1$: stable node;
$a = 1$: unstable (infinite number of equilibria);
$a > 1$: saddle (unstable)

35. $a < -1$: stable node;
$a = -1$: unstable (infinite number of equilibria);
$-1 < a < 1$: saddle (unstable);
$a = 1$: unstable (infinite number of equilibria);
$a > 1$: unstable node

■ EXERCISES 10.3 | page 661

1. $\mathbf{z}(t) = e^{-5\times10^{-4}t}\cdot$
$\begin{bmatrix}k_1\cos 0.1t + k_2\sin 0.1t\\(-5\times10^{-4}k_1 + 0.1k_2)\cos 0.1t - (0.1k_1 + 5\times10^{-4}k_2)\sin 0.1t\end{bmatrix}$

3. (a) $\dfrac{d\mathbf{x}}{dt} = A\mathbf{x}$, where $\mathbf{x}(t) = \mathbf{y}(t) - \begin{bmatrix}0\\d/2\end{bmatrix}$

(b) $\mathbf{x}(t) = c_1e^{-2t}\begin{bmatrix}0\\1\end{bmatrix} + c_2e^{-4t}\begin{bmatrix}-2\\1\end{bmatrix}$

(c) $\mathbf{x}(t) = \left(\frac{1}{2} - d/2\right)e^{-2t}\begin{bmatrix}0\\1\end{bmatrix} - \frac{1}{2}e^{-4t}\begin{bmatrix}-2\\1\end{bmatrix}$

(d) $y_2(t) = \left(\frac{1}{2} - d/2\right)e^{-2t} - \frac{1}{2}e^{-4t} + d/2$

(e) $-\frac{1}{2}\ln\left[\frac{1}{2}\left(1 - d + \sqrt{(d-1)(d+3)}\right)\right]$

5. (a) $\mathbf{x}(t) = c_1e^{(\rho-\delta)t}\begin{bmatrix}0\\1\end{bmatrix} + c_2e^{-(\alpha+\beta)t}\begin{bmatrix}\rho-\delta+\alpha+\beta\\-\alpha\end{bmatrix}$

(b) $\rho > \delta$: saddle (unstable); $\rho < \delta$: stable node

(c) $\mathbf{x}(t) = \dfrac{\alpha C_0}{\rho-\delta+\alpha+\beta}e^{(\rho-\delta)t}\begin{bmatrix}0\\1\end{bmatrix}$
$\qquad + \dfrac{C_0}{\rho-\delta+\alpha+\beta}e^{-(\alpha+\beta)t}\begin{bmatrix}\rho-\delta+\alpha+\beta\\-\alpha\end{bmatrix}$

7. (a) $x_0' = -u_0x_0,\ x_1' = u_0x_0 - u_1x_1,\ x_2' = u_1x_1$

(b) $\mathbf{x}(t) = \dfrac{k}{u_0-u_1}e^{-u_0t}\begin{bmatrix}u_0-u_1\\-u_0\end{bmatrix} + \dfrac{u_0k}{u_0-u_1}e^{-u_1t}\begin{bmatrix}0\\1\end{bmatrix}$

(c) $\dfrac{dx_2}{dt} = -\dfrac{u_0u_1k}{u_0-u_1}e^{-u_0t} + \dfrac{u_0u_1k}{u_0-u_1}e^{-u_1t}$

(d) $x_2(t) = \dfrac{u_1k}{u_0-u_1}(e^{-u_0t}-1) - \dfrac{u_0k}{u_0-u_1}(e^{-u_1t}-1)$

9. (a) Water temperature affected only by room temperature; coin temperature affected only by water temperature

(b) $\dfrac{d\mathbf{x}}{dt} = A\mathbf{x}$, where $\mathbf{x} = \begin{bmatrix}w(t)-R\\p(t)-R\end{bmatrix}$

(c) $\mathbf{x}(t) = c_1e^{-k_wt}\begin{bmatrix}k_w-k_p\\-k_p\end{bmatrix} + c_2e^{-k_pt}\begin{bmatrix}0\\1\end{bmatrix}$

(d) $\begin{bmatrix}w(t)\\p(t)\end{bmatrix} = \dfrac{w_0-R}{k_w-k_p}e^{-k_wt}\begin{bmatrix}k_w-k_p\\-k_p\end{bmatrix}$
$\qquad + \left(k_p\dfrac{w_0-R}{k_w-k_p} + p_0 - R\right)e^{-k_pt}\begin{bmatrix}0\\1\end{bmatrix} + \begin{bmatrix}R\\R\end{bmatrix}$

11. (a) $\hat{x}_1 = \hat{x}_2 = cV$

(b) $\dfrac{d\mathbf{y}}{dt} = A\mathbf{y}$, where $\mathbf{y} = \begin{bmatrix}x_1(t)\\x_2(t)\end{bmatrix} - \begin{bmatrix}cV/2\\0\end{bmatrix}$

(c) $\mathbf{y}(t) = cVe^{-at/V}\begin{bmatrix}1\\1\end{bmatrix} - \dfrac{cV}{2}e^{-2at/V}\begin{bmatrix}1\\0\end{bmatrix}$

(d) $\mathbf{x}(t) = cVe^{-at/V}\begin{bmatrix}1\\1\end{bmatrix} - \dfrac{cV}{2}e^{-2at/V}\begin{bmatrix}1\\0\end{bmatrix} + \begin{bmatrix}cV/2\\0\end{bmatrix}$

■ EXERCISES 10.4 | page 673

1. $\dfrac{d\boldsymbol{\varepsilon}}{dt} = \begin{bmatrix}4&0\\0&-2\end{bmatrix}\boldsymbol{\varepsilon}$, where $\boldsymbol{\varepsilon} = \begin{bmatrix}x_1\\x_2\end{bmatrix}$

3. $\dfrac{d\boldsymbol{\varepsilon}}{dt} = \begin{bmatrix}1&0\\-1&7\end{bmatrix}\boldsymbol{\varepsilon}$, where $\boldsymbol{\varepsilon} = \begin{bmatrix}x_1\\x_2\end{bmatrix}$

5. $\dfrac{d\boldsymbol{\varepsilon}}{dt} = \begin{bmatrix}-1&1\\0&2\end{bmatrix}\boldsymbol{\varepsilon}$, where $\boldsymbol{\varepsilon} = \begin{bmatrix}x_1\\x_2\end{bmatrix}$

7. (i) $\hat{x}_1 = 0,\ \hat{x}_2 = 0$; $\dfrac{d\boldsymbol{\varepsilon}}{dt} = \begin{bmatrix}-5&0\\0&1\end{bmatrix}\boldsymbol{\varepsilon}$, where $\boldsymbol{\varepsilon} = \begin{bmatrix}x_1\\x_2\end{bmatrix}$

(ii) $\hat{x}_1 = \frac{1}{5},\ \hat{x}_2 = 5$; $\dfrac{d\boldsymbol{\varepsilon}}{dt} = \begin{bmatrix}0&\frac{1}{5}\\-25&0\end{bmatrix}\boldsymbol{\varepsilon}$, where $\boldsymbol{\varepsilon} = \begin{bmatrix}x_1\\x_2\end{bmatrix} - \begin{bmatrix}\frac{1}{5}\\5\end{bmatrix}$

9. (i) $\hat{x}_1 = 0,\ \hat{x}_2 = 0$; $\dfrac{d\boldsymbol{\varepsilon}}{dt} = \begin{bmatrix}1&0\\8&0\end{bmatrix}\boldsymbol{\varepsilon}$, where $\boldsymbol{\varepsilon} = \begin{bmatrix}x_1\\x_2\end{bmatrix}$

(ii) $\hat{x}_1 = -24,\ \hat{x}_2 = -2$;
$\dfrac{d\boldsymbol{\varepsilon}}{dt} = \begin{bmatrix}-1&0\\0&-96\end{bmatrix}\boldsymbol{\varepsilon}$, where $\boldsymbol{\varepsilon} = \begin{bmatrix}x_1\\x_2\end{bmatrix} - \begin{bmatrix}-24\\-2\end{bmatrix}$

11. (i) $\hat{x}_1 = 0, \hat{x}_2 = 0$; $\dfrac{d\boldsymbol{\varepsilon}}{dt} = \begin{bmatrix} 1 & -1 \\ 1 & 0 \end{bmatrix}\boldsymbol{\varepsilon}$, where $\boldsymbol{\varepsilon} = \begin{bmatrix} x_1 \\ x_2 \end{bmatrix}$

(ii) $\hat{x}_1 = -1, \hat{x}_2 = -1$;

$\dfrac{d\boldsymbol{\varepsilon}}{dt} = \begin{bmatrix} e & -e \\ -1 & 0 \end{bmatrix}\boldsymbol{\varepsilon}$, where $\boldsymbol{\varepsilon} = \begin{bmatrix} x_1 \\ x_2 \end{bmatrix} - \begin{bmatrix} -1 \\ -1 \end{bmatrix}$

13. (i) unstable, (ii) locally stable

15. (i) inconclusive, (ii) unstable

17. (i) locally stable, (ii) unstable

19. (i) $\hat{x} = 0, \hat{y} = 0$; unstable (ii) $\hat{x} = 0, \hat{y} = 2$; unstable (iii) $\hat{x} = 3, \hat{y} = 0$; locally stable

21. (i) $\hat{n} = 0, \hat{m} = 0$; unstable (ii) $\hat{n} = 0, \hat{m} = 2$; locally stable (iii) $\hat{n} = \frac{3}{4}, \hat{m} = \frac{1}{2}$; unstable

23. (i) $\hat{p} = -\frac{1}{2} + \sqrt{5}/2, \hat{q} = \frac{5}{2} - \sqrt{5}/2$; locally stable (ii) $\hat{p} = -\frac{1}{2} - \sqrt{5}/2, \hat{q} = \frac{5}{2} + \sqrt{5}/2$; unstable

25. (i) $\hat{x} = 0, \hat{y} = 0$; locally stable when $a < 1$ (ii) $\hat{x} = \dfrac{1-a}{a}, \hat{y} = \dfrac{1-a}{a}$; locally stable when $a > 1$

27. (a) (i) $\hat{p}_1 = 0, \hat{p}_2 = 0$, (ii) $\hat{p}_1 = 0, \hat{p}_2 = \frac{9}{10}$, (iii) $\hat{p}_1 = \frac{2}{5}, \hat{p}_2 = 0$, and (iv) $\hat{p}_1 = \frac{2}{5}, \hat{p}_2 = \frac{13}{30}$

(b) $J(p_1, p_2) = \begin{bmatrix} 2 - 10p_1 & 0 \\ -35p_2 & 27 - 35p_1 - 60p_2 \end{bmatrix}$

(c) (i), (ii), and (iii) are unstable; (iv) is locally stable.

(d) Yes

29. $\hat{R} = 1, \hat{C} = 2$; locally stable

31. (i) $\hat{R} = 0, \hat{C} = 0$; unstable (ii) $\hat{R} = 5, \hat{C} = 0$; unstable (iii) $\hat{R} = 1, \hat{C} = \frac{8}{5}$; locally stable

33. (a) dx/dt and dy/dt do not contain any terms with z (b) $\hat{x} = 0, \hat{y} = 1$

(c) $J(x, y) = \begin{bmatrix} -k_f yM & -M(k_f x + k_r) \\ -k_f yM & -M(k_f x + k_r + k_{cat}) \end{bmatrix}$

(d) Locally stable

35. (b) $J(v, w) = \begin{bmatrix} -3v^2 + 2(1 + a)v - a & -1 \\ \varepsilon & -\varepsilon \end{bmatrix}$

(c) Locally stable for all positive constants a and ε

CHAPTER 10 REVIEW | page 676

True-False Quiz

1. False **3.** True **5.** True **7.** True **9.** False

Exercises

1. Nonlinear **3.** Linear

9. $\mathbf{x}(t) = \begin{bmatrix} \cos t + \sin t \\ \cos t - \sin t \end{bmatrix}$

11. $\mathbf{x}(t) = \frac{1}{3}e^{t}\begin{bmatrix} 1 \\ 0 \end{bmatrix} + \frac{2}{3}e^{-2t}\begin{bmatrix} 1 \\ -3 \end{bmatrix}$

13. (a) $\dfrac{d}{dt}\begin{bmatrix} x_1 \\ x_2 \end{bmatrix} = \begin{bmatrix} -\frac{4}{25} & 0 \\ \frac{4}{25} & -\frac{8}{25} \end{bmatrix}\begin{bmatrix} x_1 \\ x_2 \end{bmatrix} + \begin{bmatrix} 16 \\ 0 \end{bmatrix}$ and $\mathbf{x}(0) = \begin{bmatrix} 8 \\ 0 \end{bmatrix}$

(b) $\begin{bmatrix} x_1(t) \\ x_2(t) \end{bmatrix} = -92e^{-4t/25}\begin{bmatrix} 1 \\ 1 \end{bmatrix} + 42e^{-8t/25}\begin{bmatrix} 0 \\ 1 \end{bmatrix} + \begin{bmatrix} 100 \\ 50 \end{bmatrix}$

(c) 50 g

19. Cannot tell **21.** Locally stable

23. (i) $\hat{x} = 0, \hat{y} = 0$, unstable (ii) $\hat{x} = 0, \hat{y} = 4$, unstable (iii) $\hat{x} = -\frac{1}{2}, \hat{y} = 5$, locally stable

25. (i) $\hat{R} = 0, \hat{W} = 0$, unstable (ii) $\hat{R} = \frac{1}{2}, \hat{W} = 0$, locally stable (iii) $\hat{R} = 1, \hat{W} = -8$, unstable

27. (a) (i) $\hat{N}_1 = 0, \hat{N}_2 = 0$, (ii) $\hat{N}_1 = 0, \hat{N}_2 = K_2$, (iii) $\hat{N}_1 = K_1, \hat{N}_2 = 0$, (iv) $\hat{N}_1 = \dfrac{\alpha K_2 - K_1}{\alpha\beta - 1}, \hat{N}_2 = \dfrac{K_1\beta - K_2}{\alpha\beta - 1}$

(b) $J(N_1, N_2) = \begin{bmatrix} 1 - \dfrac{2N_1 + \alpha N_2}{K_1} & -\dfrac{\alpha N_1}{K_1} \\ -\dfrac{\beta N_2}{K_2} & 1 - \dfrac{2N_2 + \beta N_1}{K_2} \end{bmatrix}$

(c) (i) and (ii) are unstable; (iii) is locally stable; (iv) is not biologically feasible

(d) (i) and (iii) are unstable; (ii) is locally stable; (iv) is not biologically feasible

(e) (i) and (iv) are unstable; (ii) and (iii) are locally stable

(f) (i), (ii), and (iii) are unstable; (iv) is locally stable

29. $\dfrac{dp_1}{dt} = f_1(p_1, p_2) = c_1p_1(h - p_1) - m_1p_1$;

$\dfrac{dp_2}{dt} = f_2(p_1, p_2) = c_2p_2(h - p_1 - p_2) - m_2p_2 - c_1p_1p_2$

(a) (i) $\hat{p}_1 = 0, \hat{p}_2 = 0$, (ii) $\hat{p}_1 = 0, \hat{p}_2 = h - \dfrac{m_2}{c_2}$,

(iii) $\hat{p}_1 = h - \dfrac{m_1}{c_1}, \hat{p}_2 = 0$, and

(iv) $\hat{p}_1 = h - \dfrac{m_1}{c_1}, \hat{p}_2 = \dfrac{m_1}{c_1} - \dfrac{m_2}{c_2} - \dfrac{c_1h}{c_2} + \dfrac{m_1}{c_2}$

(b)

$J(p_1, p_2) = \begin{bmatrix} c_1(h - 2p_1) - m_1 & 0 \\ -c_2p_2 - c_1p_2 & c_2(h - p_1 - 2p_2) - m_2 - c_1p_1 \end{bmatrix}$

(c) $hc_1 < m_1$ and $hc_2 < m_2$

CHAPTER 11

EXERCISES 11.1 | page 691

1. Numerical and continuous; positive values

3. Numerical and discrete; nonnegative integer values

5. Categorical and nominal

7. $\bar{x} = 953.1$, Med $= 954$, s.d. ≈ 22.6405, IQR $= 31.5$

9. $\bar{x} = 27.45$, Med $= 25.5$, s.d. ≈ 19.966, IQR $= 34$

11. Min = 0.05 $Q_1 = 0.125$ Med = 0.21 $Q_3 = 0.33$ Max = 0.48

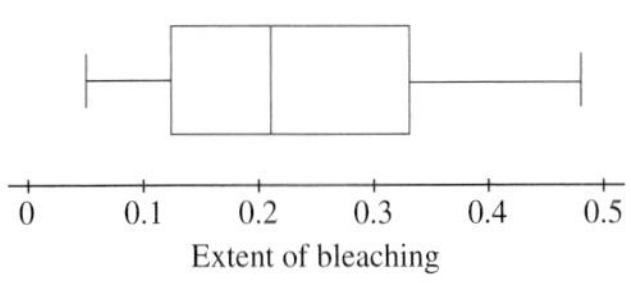

13. Min = 0.2 $Q_1 = 0.4$ Med = 0.95 $Q_3 = 1.1$ Max = 1.3

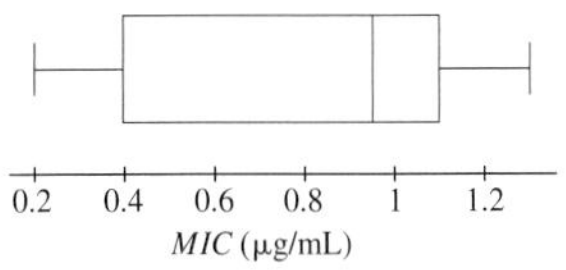

15.

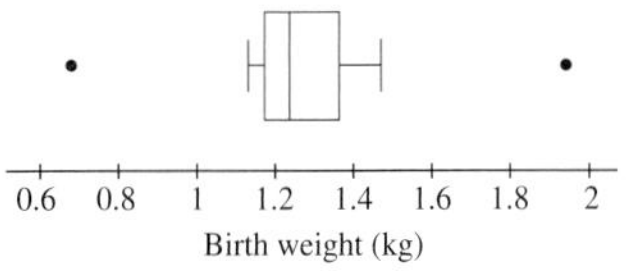

17.

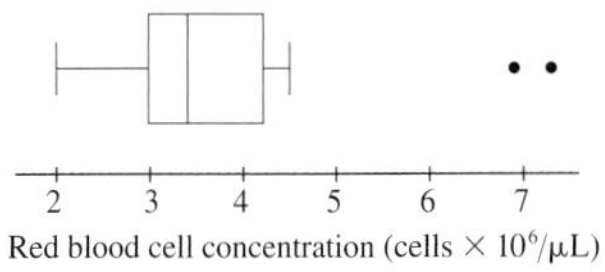

19. $\bar{x} \approx 0.3429$ L/min, s.d. ≈ 0.1116 L/min

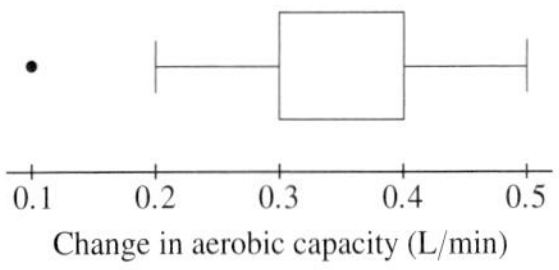

On average, an individual's aerobic capacity increases by about 0.34 L/min after EPO use.

EXERCISES 11.2 | page 700

1.

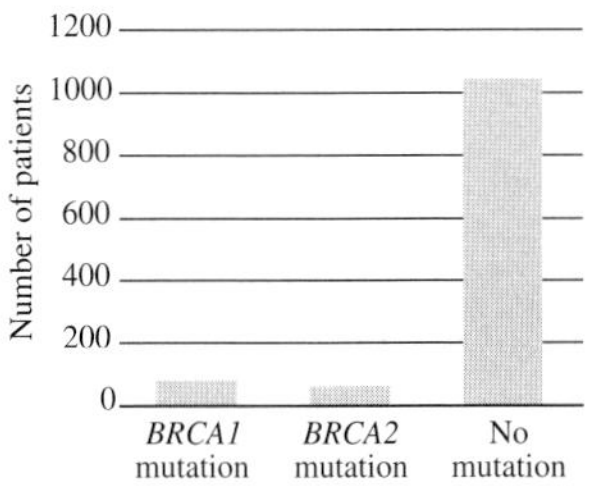

3.

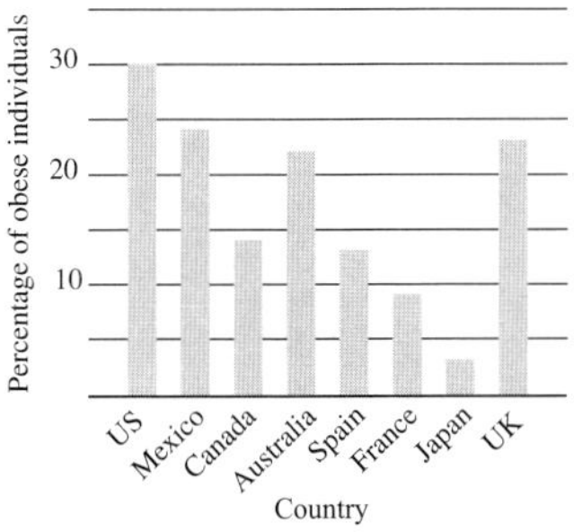

5.

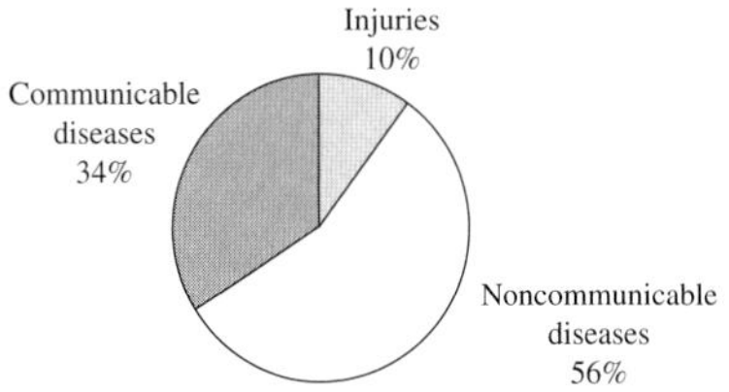

7.

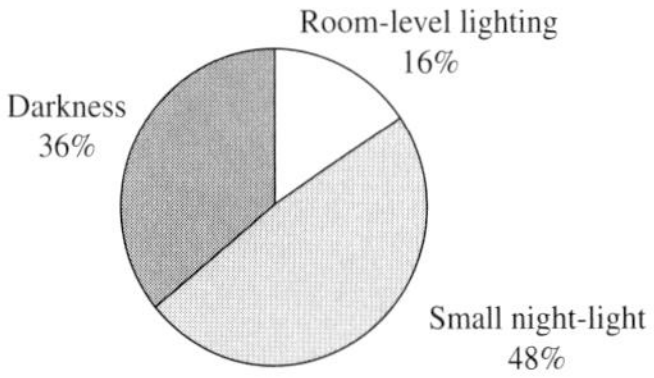

9–17. *Note:* Your histograms may differ slightly from those presented depending on the placement of the first bin and, in some cases, the choice of bin width.

9.

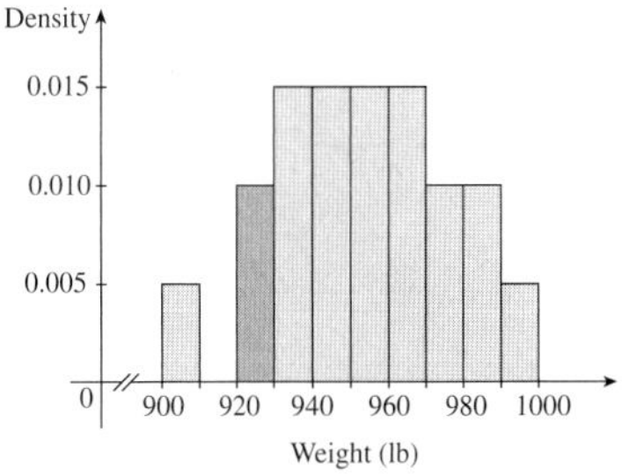

11.

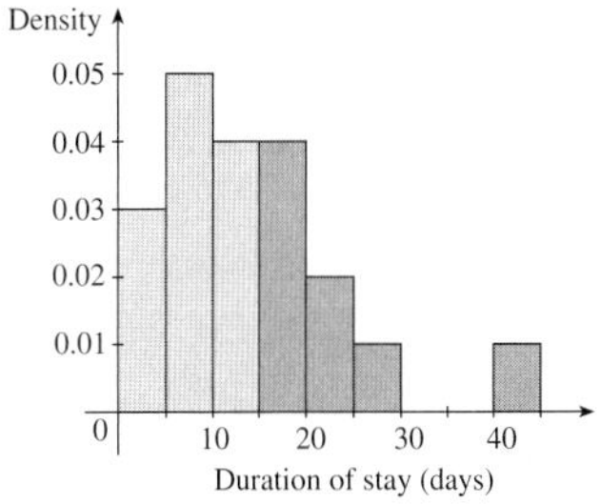

13.

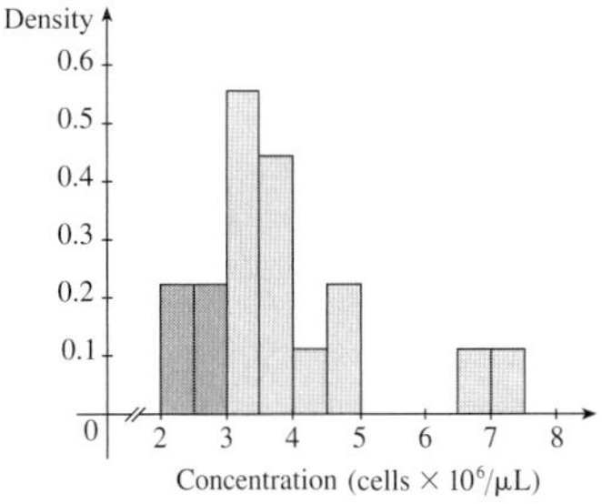

15.

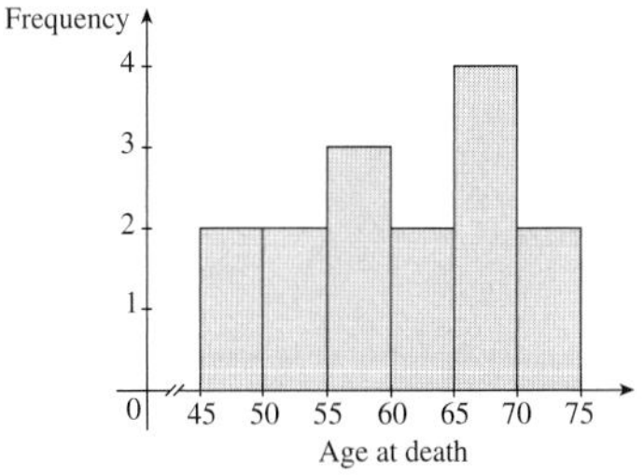

17. 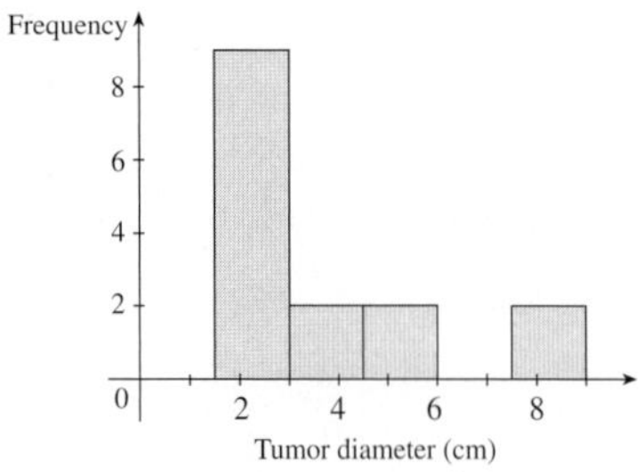

21. 0.682 (68.2%) **23.** 0.0225 (2.25%)
25. 0.8185 (81.85%) **27.** 0.221 (22.1%)

29. $\begin{cases} \frac{1}{30}(a - 45) & \text{if } 45 \leqslant a \leqslant 75 \\ 1 & \text{if } a > 75 \end{cases}$

■ EXERCISES 11.3 | page 709

1.

	Survived	Died	Row total
Vaccinated	25	0	25
Unvaccinated	0	25	25
Column total	25	25	50

Vaccinated individuals always survive and unvaccinated individuals always die.

3.

	Developed myopia	No myopia	Row total
Darkness	9	163	172
Night-light	31	201	232
Room lighting	48	27	75
Column total	88	391	479

Yes

5.

	Species X present	Species X not present	Row total
Herbivore-dominated	9	3	12
Carnivore-dominated	4	11	15
Column total	13	14	27

Yes

7. 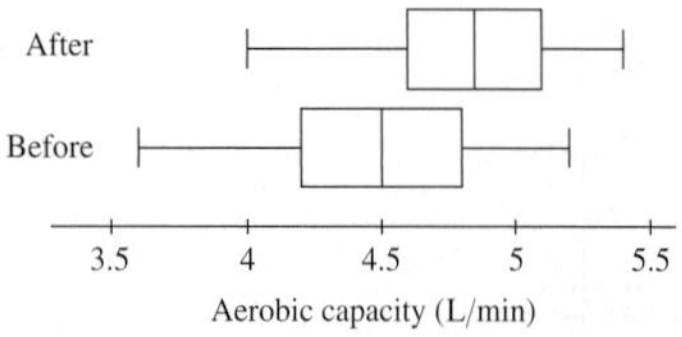

EPO use appears to increase aerobic capacity.

9. 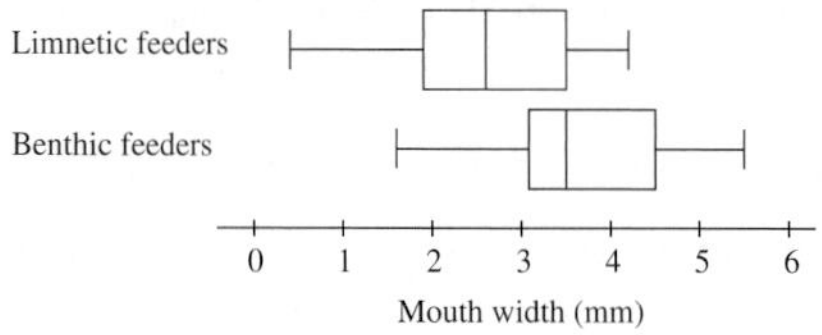

Limnetic feeders tend to have smaller mouth widths than benthic feeders.

11. 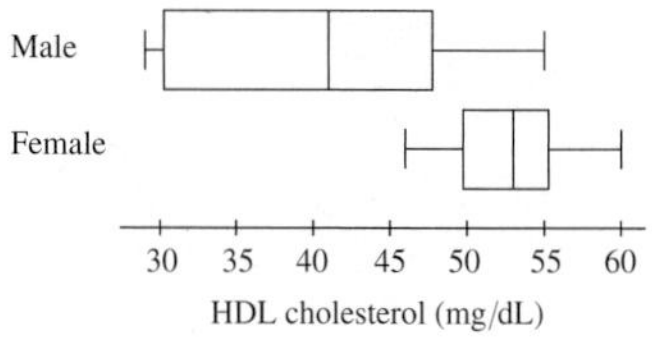

Females tend to have higher levels of HDL cholesterol than males.

13. 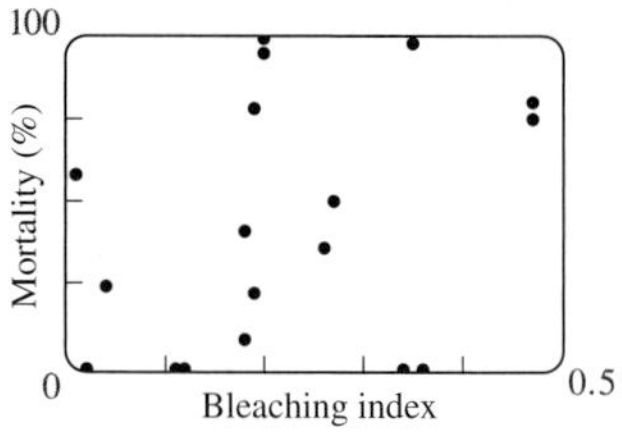

No apparent relationship between bleaching and mortality.

15. 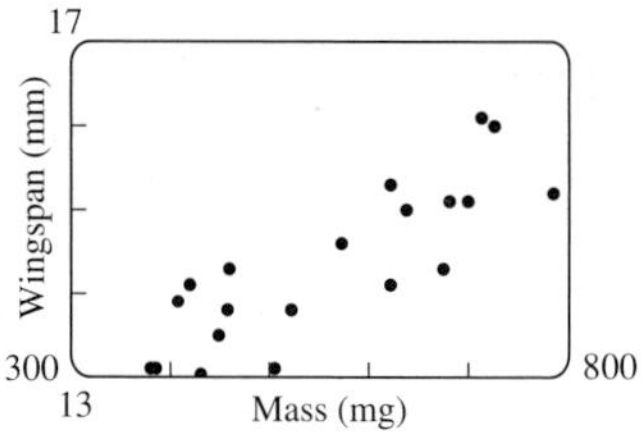

Moderate positive linear relationship: as mass increases, wingspan tends to increase.

19. $y = 0.49589x + 0.01529$

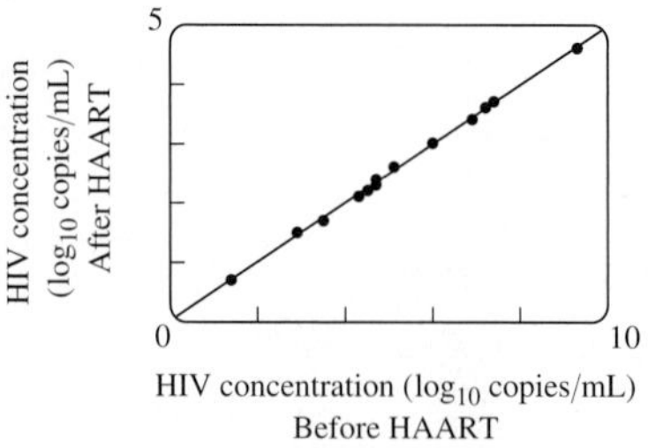

Positive linear relationship: individuals with high values of HIV concentration before HAART also tend to have high values after HAART, and vice versa.

21. $y = 92.6942x + 22.10727$

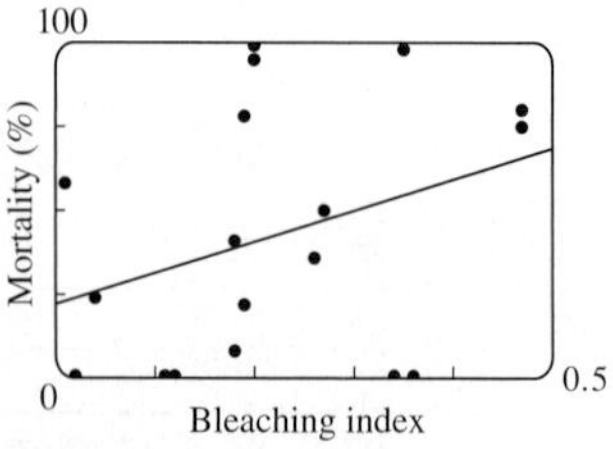

Data does not appear to exhibit a linear pattern: least-squares line is a poor approximation to the data.

23. $y = 0.366x - 617.33$

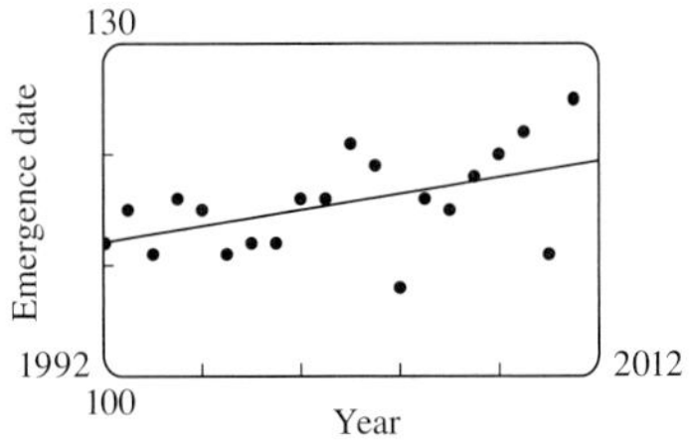

Weak positive linear relationship: after one year the emergence date will increase by about 0.366 day.

25. (a) The graph of $y(t)$ is a line having slope $-r$ and y-intercept $\ln A$.

(b)

Time x	$\ln(210N^{-1} - 1)$ y	Time x	$\ln(210N^{-1} - 1)$ y
0	6.9556	11	1.0361
1	6.4542	12	0.5671
2	6.0379	13	−0.1911
3	5.2465	14	−1.1632
4	5.0039	15	−1.2164
5	4.2008	16	−2.2513
6	4.0775	17	−2.4295
7	3.1061	18	−2.2513
8	2.9957	19	−5.3423
9	1.9834	20	−2.2513
10	1.9136		

(c) $y = -0.5527x + 7.0735$, $A \approx 1180.30$, and $r \approx 0.55267$

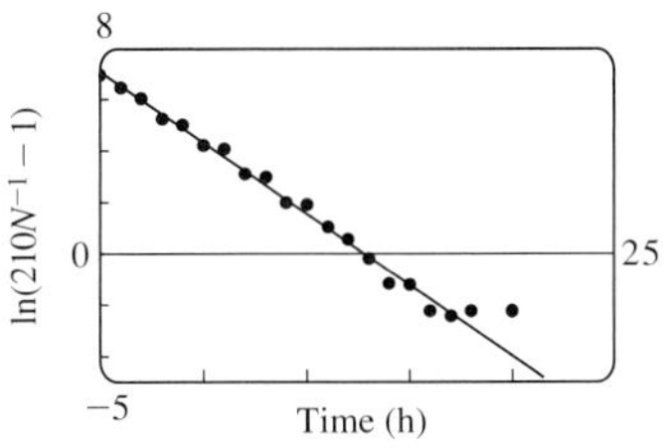

(d)

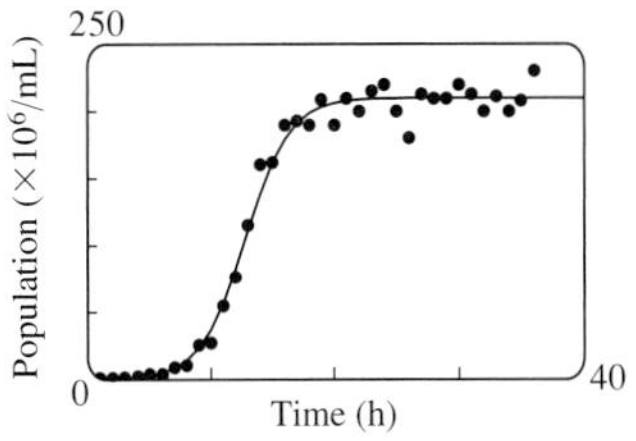

EXERCISES 11.4 | page 718

1. (a) Reproductive success of all yellow perch fish
(b) Experimental (c) Yes, provided the yellow perch in the lake are representative of the global perch population.

3. (a) Fraction of Nigerians with polio (b) Observational
(c) No, biased toward higher prevalence of polio because individuals who regularly visit health clinics are likely ill or prone to illness.

5. (a) Fraction of treated individuals with TB
(b) Experimental (c) Yes

7. (a) Reproductive success of all plants (b) Experimental
(c) Yes, provided the species *Arabidopsis thaliana* is representative of all plant species.

9. (a) Fraction of salmon longer than 30 cm
(b) Sample (2) (c) Sample (1)

11. (a) Fraction of individuals who support socialized health care
(b) Sample (1) (c) Sample (1)

13. (a)

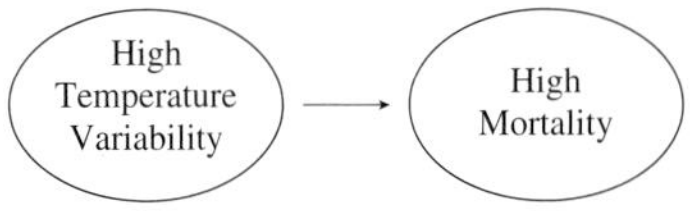

(b) Canopy cover: limited canopy cover (trees and leaves) means less food for insects, which increases mortality rates. This can also result in more sunlight, causing high temperature variability.

(c)

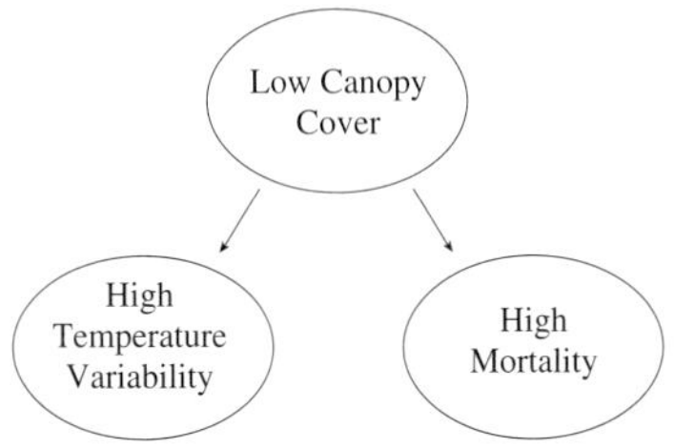

15. (a)

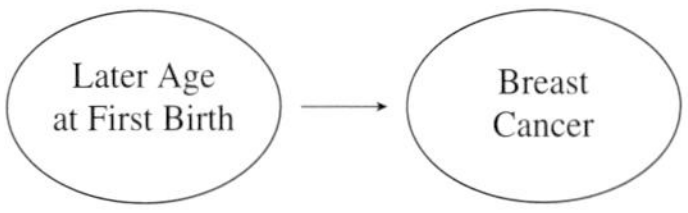

(b) Underlying health or fertility problem, which may result in later age at first birth due to longer duration attempting to conceive; this situation may also be a contributing factor to the development of breast cancer.

(c)

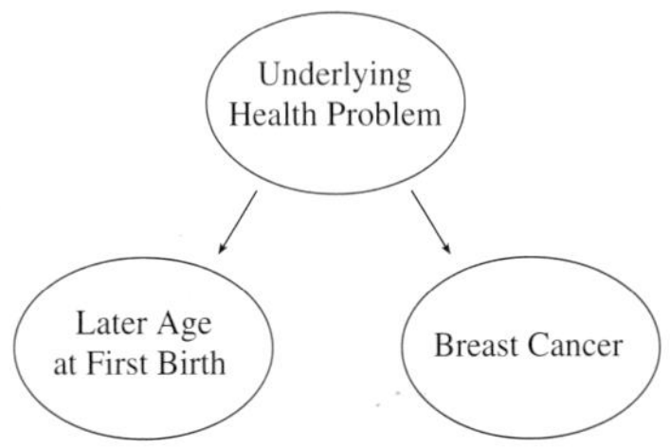

17. (a)

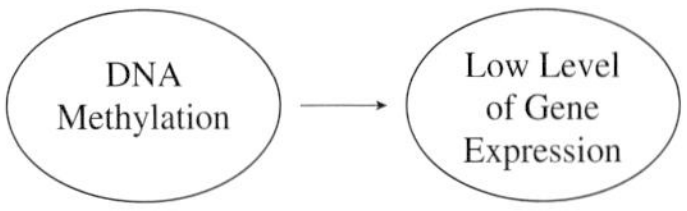

(b) Specialized inhibitor molecule, which may assist in the attachment of methyl groups to DNA and also limit gene expression.

(c)

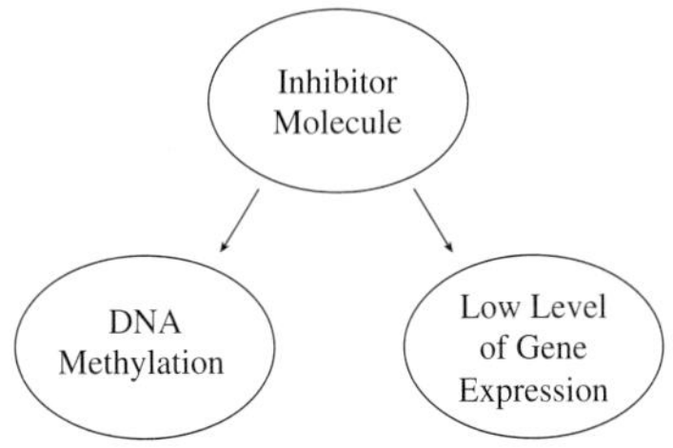

CHAPTER 11 REVIEW | page 722

True-False Quiz

1. False **3.** True **5.** True **7.** False **9.** False

Exercises

1. (a) 16.5 mg
(b) Min = 8 $Q_1 = 15$ Med = 18 $Q_3 = 20$ Max = 22
(c)

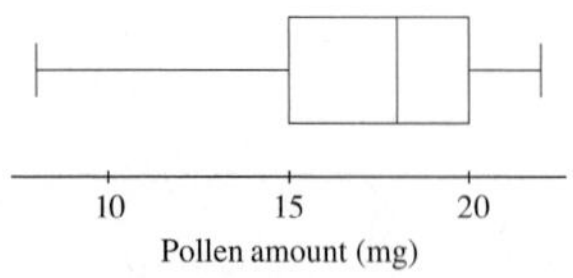

3. (a) 614 mg
(b) Min = 471 $Q_1 = 536.5$ Med = 587 $Q_3 = 698.5$ Max = 807
(c)

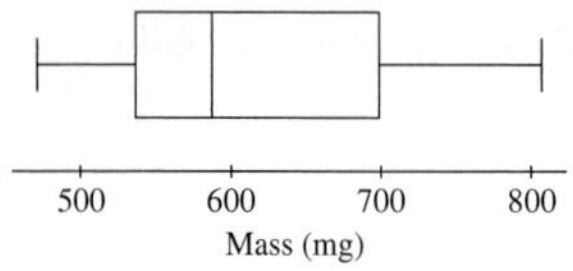

5. Categorical and nominal

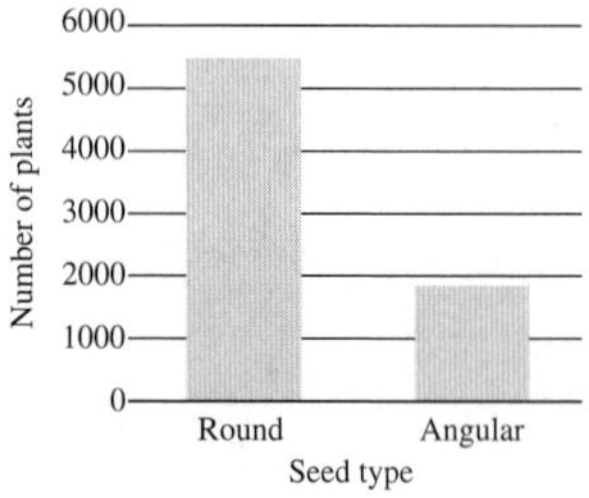

7.

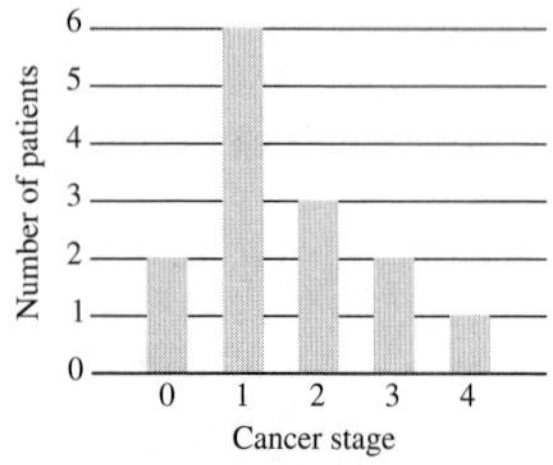

9.

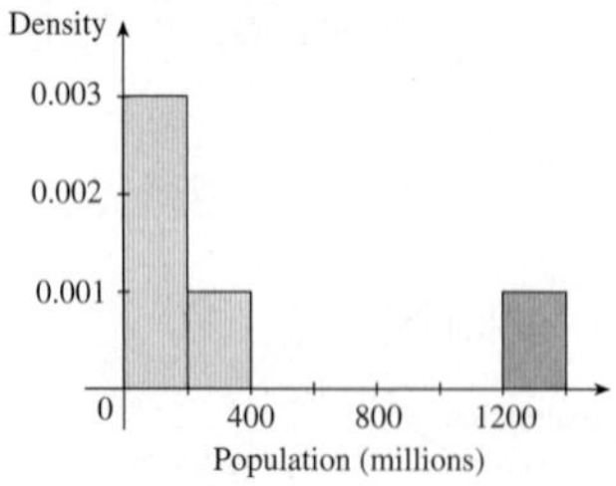

11. ≈0.8185 (81.85%) **13.** ≈0.20 (20%)

15.

	Positive test	Negative test	Row total
Cancer	0.75(490) ≈ 368	122	490
No cancer	0.55(346) ≈ 190	156	346
Column total	558	278	836

The PSA test may not be a reliable indicator of prostate cancer.

17.

	Positive test	Negative test	Row total
Male	13	8	21
Female	10	5	15
Column total	23	13	36

No apparent relationship between sex and test result.

19.

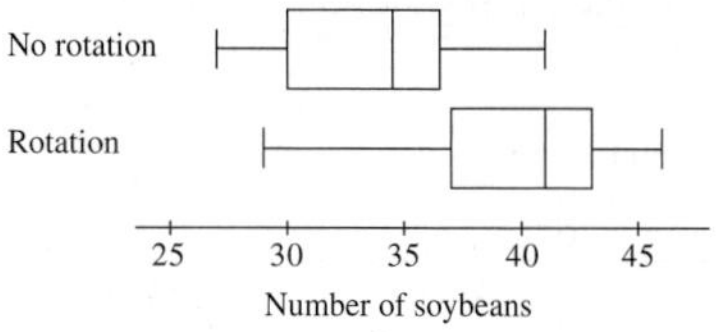

Crop rotation appears to have an effect on the soybean harvest in the sample.

21. $y = 0.08151x - 21.77342$

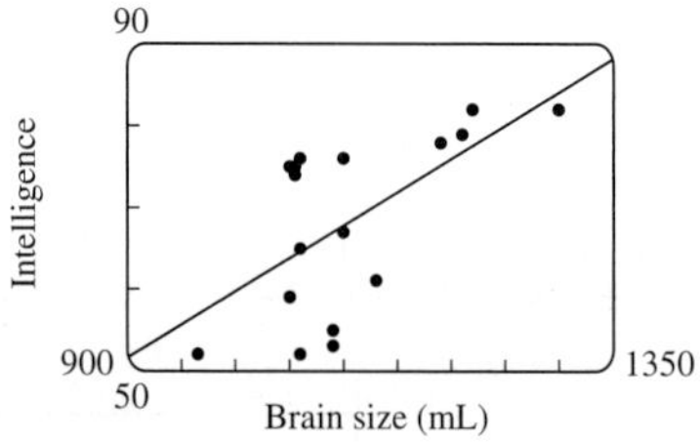

Weak positive linear relationship: moderately sized brains, around 1100 mL, have a wide range of intelligence.

23. $y = 1.3774x + 18.6662$

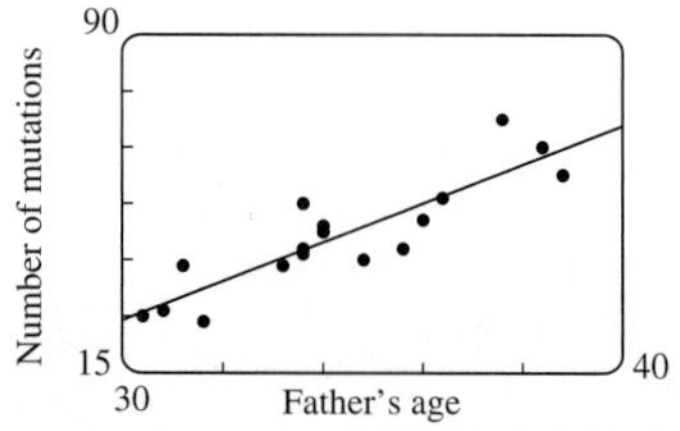

Positive linear relationship: as the father's age increases, the number of mutations increases.

25. (a)

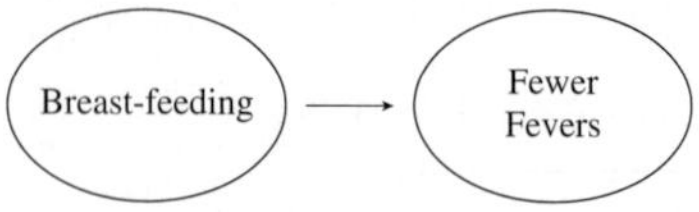

(b) Genetic condition: may make breast-feeding unsafe for the infant and may also cause fevers.

(c)

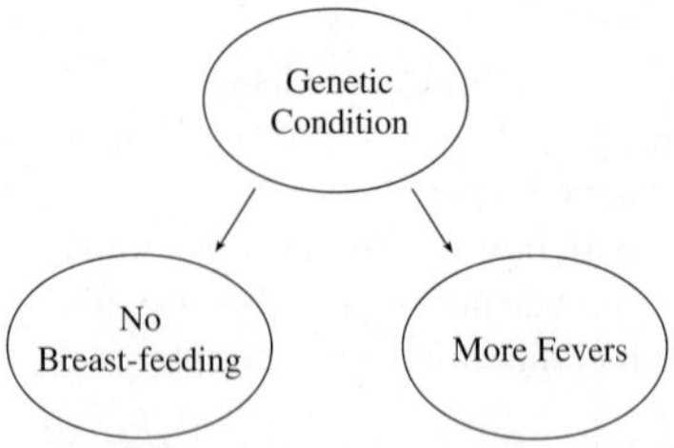

27. (a)

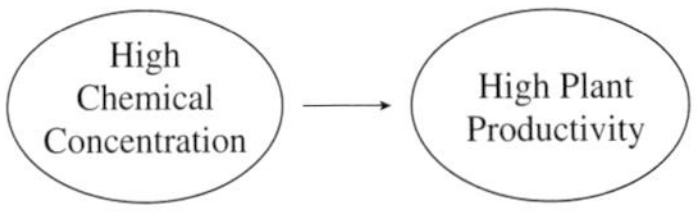

(b) Soil type: may have a higher concentration of the chemical of interest and also have a density or particle size that improves plant productivity.

(c)

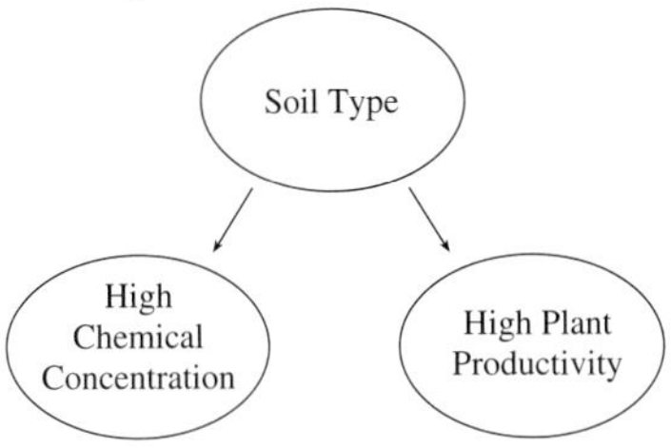

CHAPTER 12

EXERCISES 12.1 | page 734

1. 240 **3.** 8,000,000,000 **5.** (a) 6859 (b) 190
7. 17,576 **9.** 4^{800} **11.** 8 **13.** (a) 256 (b) 32
15. 3 **17.** 2520; 21 **19.** 24 **21.** (a) 120 (b) 60
23. $\dfrac{(n-1)!}{(n-1-k)!}$ **25.** 720 **27.** 24 **29.** 3,110,400
31. 3600 **33.** (a) 16 (b) 16 **35.** 120 **37.** 210
39. 125,970 **43.** 2520 **45.** (a) 495 (b) 70
47. (a) 3060 (b) 1155 **49.** 18,000

EXERCISES 12.2 | page 747

1. {0, 1, 2, 3, . . . , 20}

3.

Seat \ Row	1	2	3	⋯	29	30
A	1A	2A	3A	⋯	29A	30A
B	1B	2B	3B	⋯	29B	30B
C	1C	2C	3C	⋯	29C	30C
D	1D	2D	3D	⋯	29D	30D

5. (a) {1, 2, 3, 4, 5, 6} (b) {2, 4, 6} (c) {5, 6}

7. (a) 8
(b) {(P, P, P), (P, P, F), (P, F, P), (F, P, P), (P, F, F), (F, P, F), (F, F, P), (F, F, F)}
(c) {(P, P, F), (P, F, P), (F, P, P)}
(d) {(P, P, P), (P, P, F), (P, F, P), (F, P, P), (P, F, F), (F, P, F), (F, F, P)}
(e) {(P, P, P), (F, F, F)}

9. (a) 16
(b) F = infected with flu; N = not infected
{(F, F, F, F), (F, F, F, N), (F, F, N, F), (F, N, F, F), (N, F, F, F), (F, F, N, N), (F, N, F, N), (N, F, F, N), (N, F, N, F), (N, N, F, F), (F, N, N, F), (F, N, N, N), (N, F, N, N), (N, N, F, N), (N, N, N, F), (N, N, N, N)}
(c) {(F, F, F, F), (F, F, F, N), (F, F, N, F), (F, N, F, F), (N, F, F, F)}
(d) {(F, N, N, N), (N, F, N, N), (N, N, F, N), (N, N, N, F), (N, N, N, N)}
(e) {(F, F, F, F)}

11. {1, 2, 3, 4, 5, 6, 7, 8, 9} **13.** {1, 3, 4, 6, 7, 8, 9}
15. {2, 3, 4, 5, 6, 7, 8, 9}

17. (a)

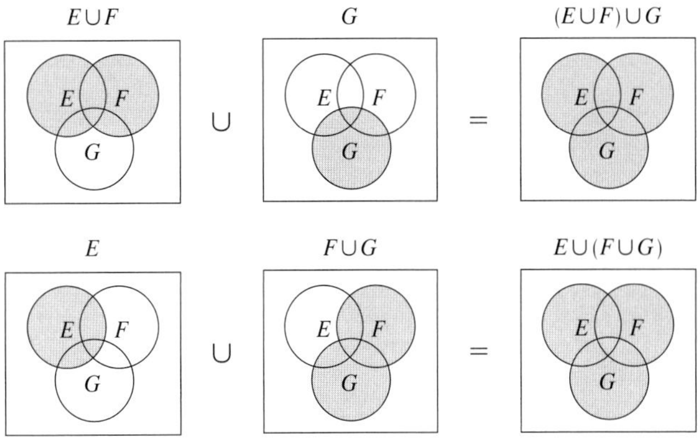

(b)

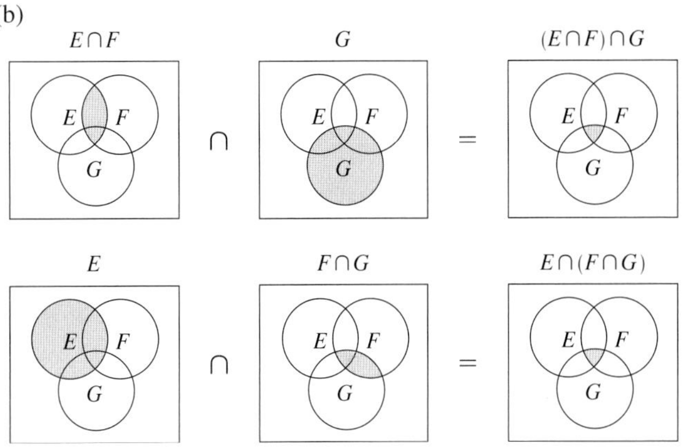

19. {1, 2, 3, 5, 7, 8, 9} **21.** ∅ **23.** {1, 3, 5, 7, 8, 9}

25.

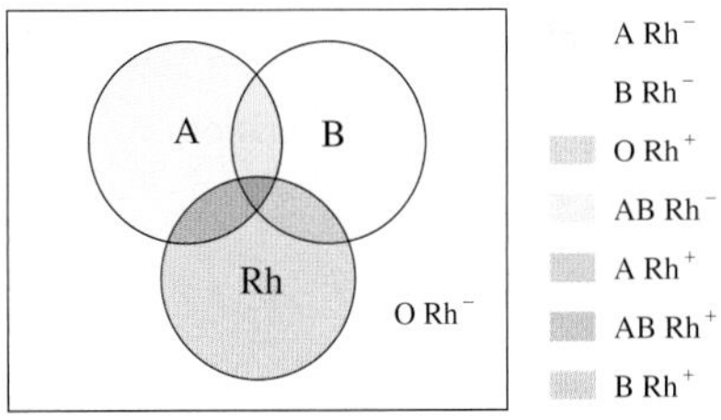

27. (a) Both foxes and wolves are present.
(b) Either foxes are absent or wolves are present (or both).
(c) Either wolves or chickadees are present and blue jays are present.
(d) No chickadees are present; that is, chickadees are absent.

29. (a) $\frac{1}{3}$ (b) $\frac{1}{2}$ (c) $\frac{1}{3}$

31. (a) $\frac{1}{4}$ (b) $\frac{1}{2}$ (c) $\frac{3}{4}$

33. Ω = {(H, H), (H, T), (T, H), (T, T)}
(a) $\frac{3}{4}$ (b) $\frac{1}{2}$

35. (a) $\frac{5}{8}$ (b) 1 (c) $\frac{3}{4}$

37. (a) {(1,1), (1, 2), (1, 3), (1, 4), (1, 5), (1, 6), (2, 1), (2, 2), (2, 3), (2, 4), (2, 5), (2, 6), (3, 1), (3, 2), (3, 3), (3, 4), (3, 5), (3, 6), (4, 1), (4, 2), (4, 3), (4, 4), (4, 5), (4, 6), (5, 1), (5, 2), (5, 3), (5, 4), (5, 5), (5, 6), (6, 1), (6, 2), (6, 3), (6, 4), (6, 5), (6, 6)}
(b) $\frac{1}{6}$ (c) $\frac{1}{9}$ (d) $\frac{1}{6}$ (e) $\frac{5}{6}$ (f) $\frac{5}{18}$

39. (a) $\frac{33}{66,640}$ (b) $\frac{33}{16,660}$ (c) $\frac{33}{108,290}$ (d) $\frac{1}{649,740}$

41. (a) $\frac{1}{2}$ (b) $\frac{1}{2}$

43. $\frac{3}{8}$ **45.** $\frac{182}{969}$ **47.** $\frac{109}{1292}$ **49.** $\frac{2011}{2300}$

55. (a) $\frac{70}{495} \approx 0.14$
(b) Yes, because the probability of the event is larger than 0.05.

57. (a) $\frac{1155}{3060} \approx 0.38$
(b) No, because the probability of the event is larger than 0.05.

59. (a) Not mutually exclusive; 1
(b) Mutually exclusive; $\frac{2}{3}$

61. (a) Not mutually exclusive; $\frac{4}{13}$
(b) Not mutually exclusive; $\frac{4}{13}$

63. $\frac{9}{20}$ (or 0.45) **65.** 0.12

EXERCISES 12.3 | page 762

1. (a) $\frac{1}{3}$ (b) $\frac{1}{3}$ **3.** (a) $\frac{1}{2}$ (b) $\frac{1}{2}$ (c) $\frac{1}{3}$

5. (a) $\frac{5}{12}$ (b) $\frac{7}{17}$

7. X^m denotes an X chromosome carrying a mutation.
(a) {XX, XY, X^mX, XX^m, X^mX^m, X^mY}
(b)

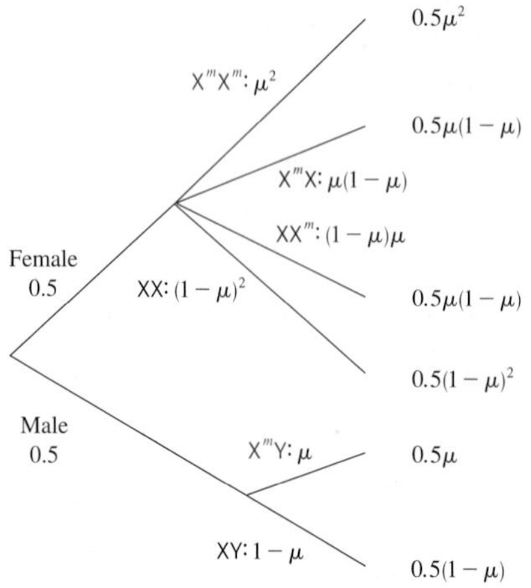

Red genotypes indicate individuals with disease.

(c) $0.5\mu(1 + \mu)$ (d) μ (e) μ^2

9. (a) $\frac{4}{11}$ (b) $\frac{5}{11}$ (c) $\frac{5}{11}$ (d) $\frac{4}{11}$

11. $\begin{cases} 0 & \text{if } k < 2 \text{ or } k > 7 \\ \dfrac{1}{k-1} & \text{if } 2 \leq k \leq 7 \end{cases}$

13. $\frac{1}{2}$ **15.** $\frac{0.32}{0.78} \approx 0.41$

17. $\frac{0.09}{0.78} \approx 0.12$ **19.** $\frac{0.01}{0.1} \approx 0.1$

21. (a) $\frac{1}{3}$ (b) $\frac{1}{2}$

23.

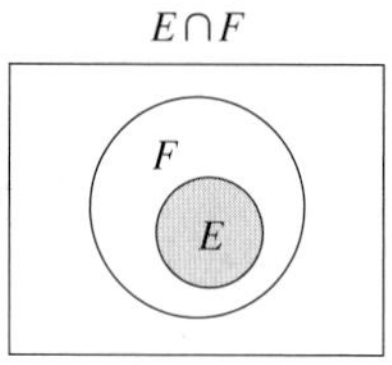

25. (a) $\frac{7}{30}$ (b) $\frac{7}{15}$

27. (a) $\frac{1}{36}$ (b) Yes

29.

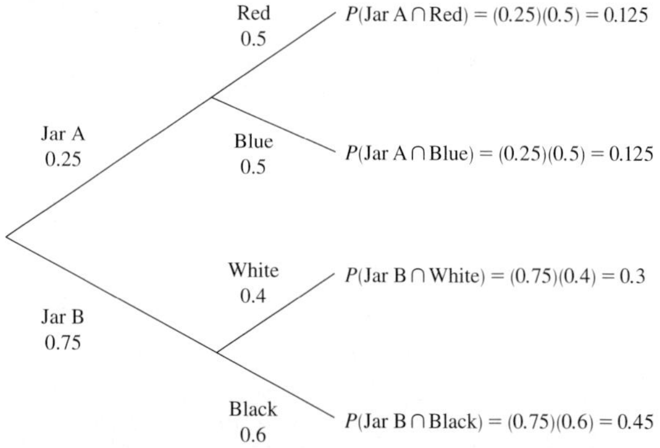

31. Not independent **33.** Independent

35. (a) $\frac{1}{4}$ (b) Yes

37.

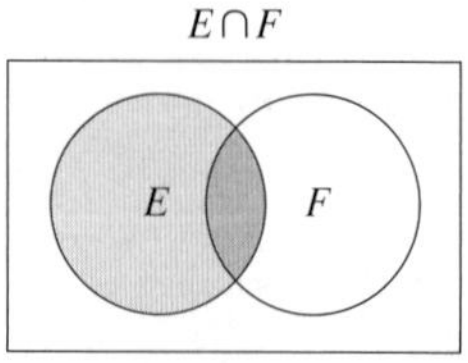

The fraction of the area of E that overlaps with F is the same as the area of F in the Venn diagram.

39. $\frac{1}{2}$ **41.** $\frac{1}{4}$ **43.** $\frac{3}{13}$ **45.** $\frac{1}{3}$ **47.** 0.00249

49. $\frac{3}{4}$ **51.** 0.075 **53.** 0.25 **55.** $\frac{2}{35}$

57. 0.06 **59.** $\approx$0.0094

EXERCISES 12.4 | page 778

1. There are five possible values for X with each value being equally likely. Therefore $p_i = \frac{1}{5}$ for $i = 1, 2, \ldots, 5$.

(a)

i	PDF (p_i)
1	0.2
2	0.2
3	0.2
4	0.2
5	0.2

(b)

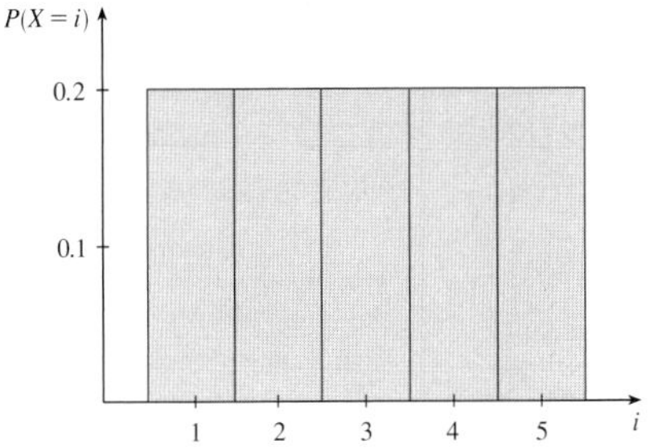

3. $P(X = 0) = \frac{1}{8}$
$P(X = 1) = \frac{3}{8}$
$P(X = 2) = \frac{3}{8}$
$P(X = 3) = \frac{1}{8}$

5. $P(X = 2) = \frac{1}{36}$ $\quad P(X = 8) = \frac{5}{36}$
$P(X = 3) = \frac{1}{18}$ $\quad P(X = 9) = \frac{1}{9}$
$P(X = 4) = \frac{1}{12}$ $\quad P(X = 10) = \frac{1}{12}$
$P(X = 5) = \frac{1}{9}$ $\quad P(X = 11) = \frac{1}{18}$
$P(X = 6) = \frac{5}{36}$ $\quad P(X = 12) = \frac{1}{36}$
$P(X = 7) = \frac{1}{6}$

7. $$P(X = i) = \begin{cases} \frac{1}{6} & \text{if } i = 0 \\ \dfrac{-2i + 12}{36} & \text{if } 1 \leqslant i \leqslant 5 \end{cases}$$

9. $P(X = 0) = \dfrac{38}{85}$

$P(X = 1) = \dfrac{36}{85}$

$P(X = 2) = \dfrac{132}{1105}$

$P(X = 3) = \dfrac{11}{1105}$

11. $P(X = 0) = \frac{7}{9}$, $\quad P(X = 1) = \frac{2}{9}$

13. $P(X = 0) = \frac{5}{12}$
$P(X = 1) = \frac{1}{2}$
$P(X = 2) = \frac{1}{12}$

15. (a) $\dfrac{1{,}048{,}576}{349{,}525}$ $\quad$ (b) $\dfrac{3}{1 - \left(\frac{1}{4}\right)^n}$

17. (a)

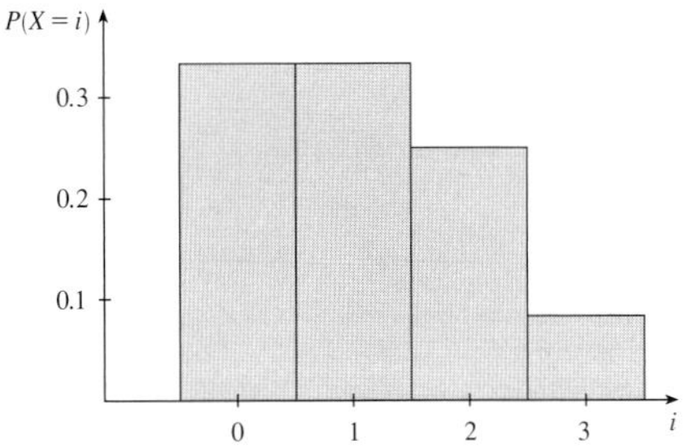

(b)

i	CDF (F_i)
0	$\frac{1}{3}$
1	$\frac{2}{3}$
2	$\frac{11}{12}$
3	1

(c)

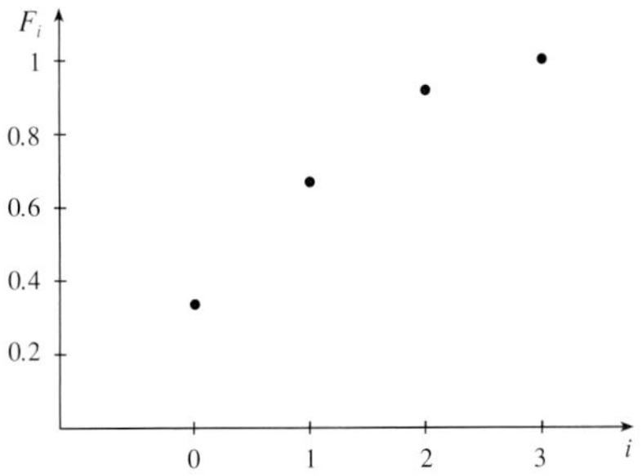

19. (a)

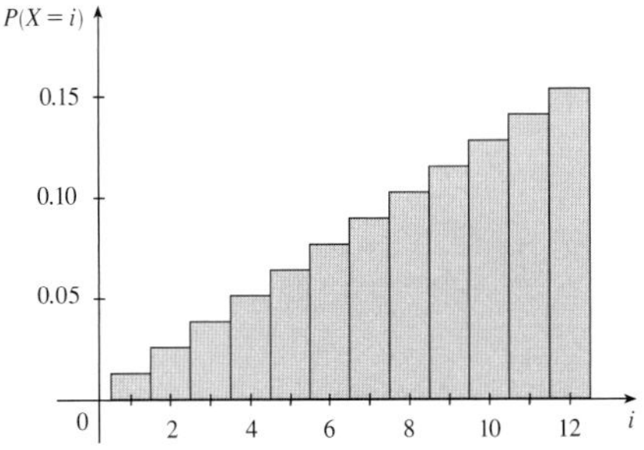

(b) $F_i = \dfrac{i(i + 1)}{156}$

(c)

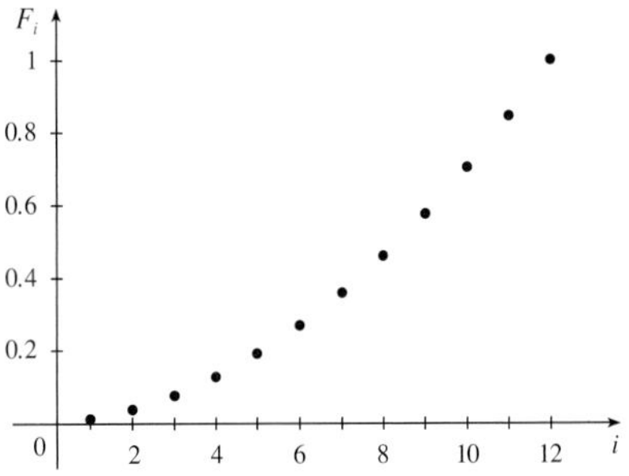

21.

i	PDF (p_i)
0	0.5
1	0
2	0
3	0.5

23. $\left(\frac{1}{3}\right)^9 - \left(\frac{1}{3}\right)^{20} \approx 0.000051$ **25.** $\dfrac{8}{55}$

27. (a) The probability of observing three or fewer different mammal species along a transect through a forest.

(b)

i	PDF (p_i)
0	0.01
1	0.15
2	0.20
3	0.35
4	0.25
5	0.04

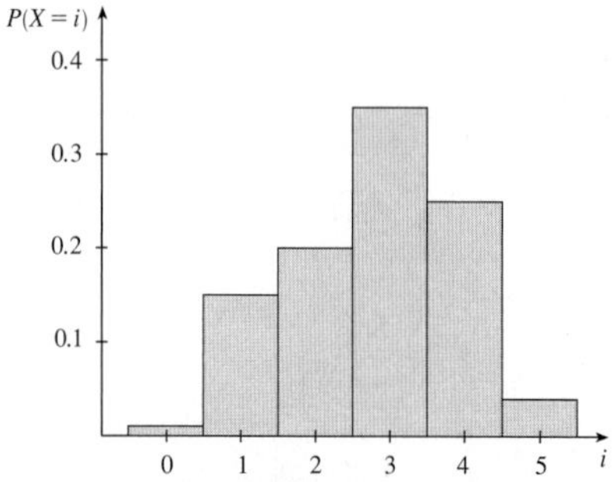

29. (a) $\frac{5}{6} \approx \$0.83$ (b) $\dfrac{605}{36}$

31. (a) \$2.50 (b) $\frac{1}{4}$

33. (a) \$0 (b) 4

35. (a) $-\frac{3}{10} = -\$0.30$ (b) 0.16

37. ≈ 5 **39.** ≈ 3

41. 1.85, 0.4275

43. 1.96, 1.1984

45. (a)

i	PDF (p_i)
0	0.0625
1	0.25
2	0.375
3	0.25
4	0.0625

(b)

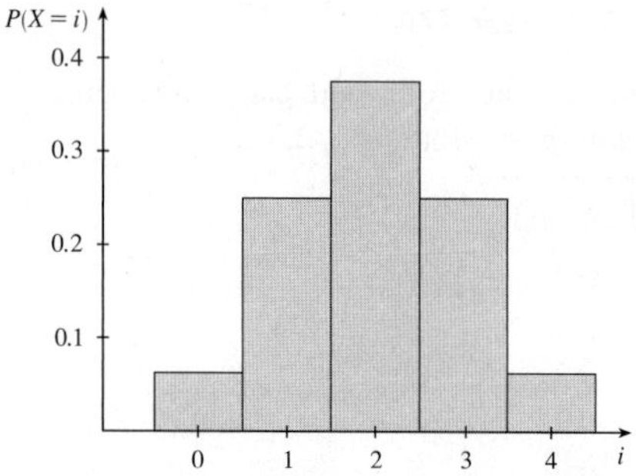

47. (a)

i	PDF (p_i)
0	0.2097152
1	0.3670016
2	0.2752512
3	0.1146880
4	0.0286720
5	0.0043008
6	0.0003584
7	0.0000128

(b)

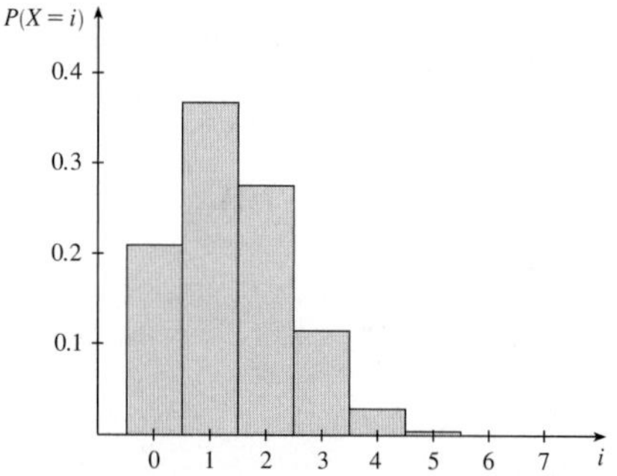

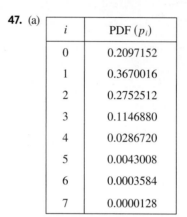

49. $\dfrac{1}{216} \approx 0.0046$

51. (a) $\dfrac{1}{78{,}125} = 1.28 \times 10^{-5}$ (b) $\dfrac{16{,}384}{78{,}125} \approx 0.210$

(c) $\dfrac{78{,}096}{78{,}125} \approx 0.9996$ (d) $\dfrac{13{,}312}{15{,}625} \approx 0.8520$

53. (a) ≈ 0.684 (b) 0.05078

55. ≈ 0.3439

57. (a) $\approx 3.32 \times 10^{-10}$ (b) 3.403×10^{-10}

59. (a) ≈ 0.00145 (b) ≈ 0.000649 (c) ≈ 0.244

61. (a) ≈ 0.484 (b) ≈ 0.123 (c) ≈ 0.996

63. (a) ≈ 0.0537 (b) ≈ 0.2397

(c) $220p^3(1-p)^9$ (d) $\frac{1}{4}$

69. ≈ 0.001726 **71.** (b) $F_i = 1 - (1-p)^i$

73. ≈ 0.0096

EXERCISES 12.5 | page 795

1. (a) The probability that a randomly chosen tire will have a lifetime between 30,000 and 40,000 miles
(b) The probability that a randomly chosen tire will have a lifetime of at least 25,000 miles
3. (a) The probability that a randomly selected fruit fly will die between 25 and 40 days old.
(b) The probability that a randomly selected fruit fly will die before it is 14 days old.
5. (b) $2/e - 3/e^2 \approx 0.33$
7. (a) $1/\pi$ (b) $\frac{1}{2}$
9. $F(x) = x; \frac{1}{3}$ **11.** $F(x) = 2 - x^{-1}; \frac{1}{6}$
13. $F(t) = t^{3/2}; \frac{3}{8}\sqrt{3} - \frac{1}{8} \approx 0.525$
15. $F(x) = \begin{cases} \frac{1}{2}e^x & \text{if } x \leq 0 \\ 1 - \frac{1}{2}e^{-x} & \text{if } x > 0 \end{cases}$;
$1 - \frac{1}{2}e^{-10} - \frac{1}{2}e^{-5} \approx 0.9966$
17. $F(x) = \dfrac{2}{\pi}\tan^{-1}x - \dfrac{1}{2}; \dfrac{2}{\pi}\tan^{-1}20 - \dfrac{2}{\pi}\tan^{-1}10 \approx 0.0316$
19. (a) $\dfrac{640}{15{,}676}(25 - 157e^{-2.4}) \approx 0.439$ (b) ≈ 0.725
21. (a) $\frac{9}{32} = 0.28125$ (b) $\frac{19}{32} = 0.59375$
23. $f(x) = 1$ **25.** $f(t) = \dfrac{2t}{(1 + t^2)^2}$
27. $f(t) = \dfrac{Ce^t}{(1 + Ce^t)^2}$ **29.** (a) $\frac{1}{2}$ (b) $\frac{1}{12}$
31. (a) $\ln 2$ (b) $\frac{1}{2} - (\ln 2)^2$
33. (a) 1 (b) 1
35. (a) 0 (b) 2
37. (a) $f(x) = \dfrac{1}{b - a}$ where $x \in [a, b]$ (b) $\dfrac{b + a}{2}$
(c) $\dfrac{(b - a)^2}{12}$
39. Yes, the mean and the variance are consistent with that of a uniform random variable on the interval $[0, 2]$.
41. (a) 40 min (b) $\frac{800}{3}$ min^2
43. $\frac{1}{2}$ **45.** $\frac{1}{3}$ **47.** ≈ 2.1555 **49.** $\frac{1}{2}$ **51.** 0
53. $\frac{1}{3}\sqrt{2}$ **55.** $\sqrt{3 - 4\ln 2 - (2\ln 2 - 1)^2}$
57. $e^{-2} \approx 0.135$
59. (a) $e^{-4/2.5} \approx 0.20$ (b) $1 - e^{-2/2.5} \approx 0.55$
(c) If you aren't served within 10 minutes, you get a free hamburger.
61. (a) $e^{-10} \approx 4.54 \times 10^{-5}$ (b) $1 - 1/e \approx 0.632$ (c) 1 year^2
63. (a) $1/\binom{n}{2}$ (b) $\left[\binom{n}{2}\right]^{-2}$ (c) $1 - e^{-20\binom{n}{2}}$
65. 0
67. (a) $e^{-2} \approx 0.135$ (b) $1 - 1/e \approx 0.632$
69. (a) 0.0668 (b) $\approx 5.21\%$
71. (a) ≈ 0.847 (b) $\approx 14.2\%$
73. (a) ≈ 0.157 (b) ≈ 0.520
75. μ

CHAPTER 12 REVIEW | page 800

True-False Quiz

1. True **3.** True **5.** True **7.** False **9.** False

Exercises

1. 4096 **3.** (a) 17,576,000 (b) 11,232,000
5. (a) 3,628,800 (b) 151,200
7. 35 **11.** $\{1, 2, 3, 4, 5, 7, 8, 9\}$ **13.** $\{1, 2, 3, 5, 6, 8, 9\}$
15. (a)

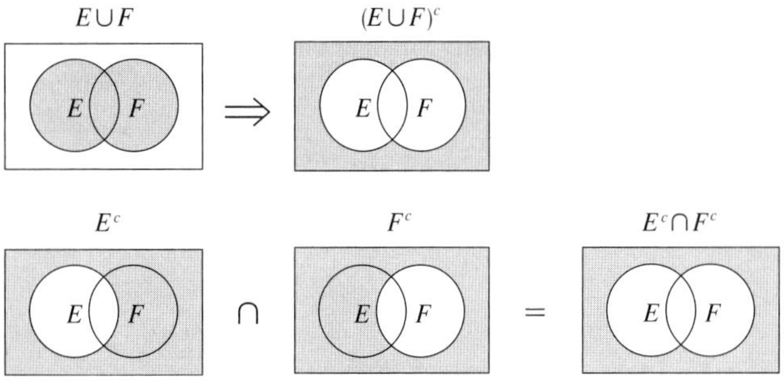

(b)

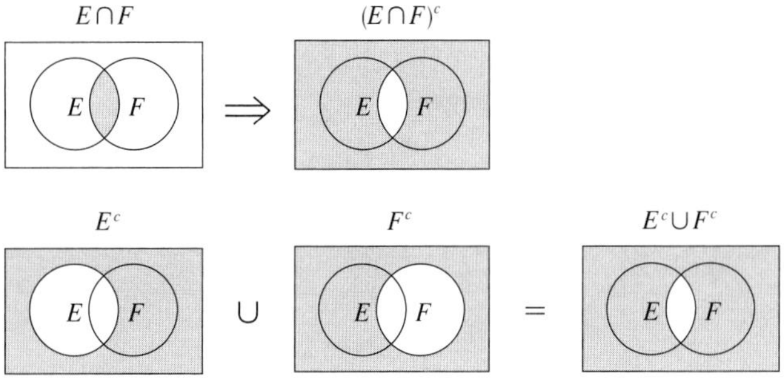

17. (a) $\dfrac{3}{13}$ (b) $\dfrac{9}{13}$ (c) $\dfrac{1}{4}$
19. (a) $\dfrac{4}{13}$ (b) 1 (c) $\dfrac{10}{13}$
21. (a) $1/5$ (b) $1/n$ **23.** $\dfrac{15{,}174{,}540}{146{,}746{,}831} \approx 0.1034$
25. (a)

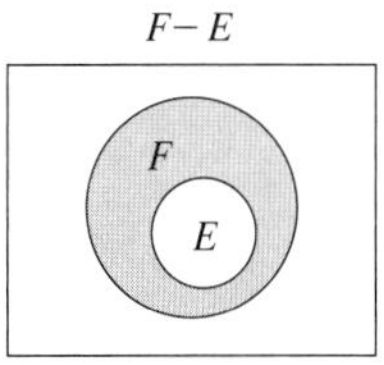

27. (a) $\dfrac{1}{4}$ (b) $\dfrac{1}{2}$
29. An HIV$^+$ individual may have a greater (or lesser) chance of displaying symptoms of tuberculosis than a randomly selected person.
33. (a) $\frac{255}{256} \approx 0.996$ (b) $\frac{243}{256} \approx 0.949$ (c) $\frac{81}{256} \approx 0.316$
35. (b) $1 - (\frac{1}{2})^{3/2} \approx 0.646$ (c) $(\frac{3}{4})^{3/2} - (\frac{1}{4})^{3/2} \approx 0.525$
(d) $\frac{3}{5}$ (e) $\frac{12}{175}$
37. (a) $1 - e^{-3/8} \approx 0.313$ (b) $e^{-5/4} \approx 0.287$
(c) $8\ln 2 \approx 5.55$ min
39. (b) The definite integral used to calculate $E[x]$ is divergent.

CHAPTER 13

EXERCISES 13.1 | page 810

1. (a) $F_1(10)\,F_2(8)$ (b) $F_1(5)\,[1 - F_2(15)]$
(c) $[F_1(10) - F_1(5)][F_2(8) - F_2(3)]$
3. $e^{-9/2}$
5. People living in a common household have similar influenza exposure; thus antibody measurements within a household will not be mutually independent.
7. Trees in a common region share nutrients, moisture, and so on; thus tree heights within a region will not be mutually independent.
9. ≈ 0.863 **11.** ≈ 0.0736 **13.** 7 **15.** ≈ 0.310
17. ≈ 0.844 **19.** ≈ 23.23 lb **21.** ≈ 9.72 days
23. ≈ 8.41 years
25. (a) $F(x_1)\,F(x_2)$
(b) $g(x_1, x_2) = \dfrac{1}{2\pi\sigma^2}\, e^{-(x_1-\mu)^2/(2\sigma^2)} e^{-(x_2-\mu)^2/(2\sigma^2)}$
(c) $g(160, 170) = \dfrac{1}{2\pi\sigma^2}\, e^{-(27{,}250-330\mu+\mu^2)/(\sigma^2)}$
(d) $\mu = 165$ cm

EXERCISES 13.2 | page 819

1. (943.2, 963.0) **3.** (9.1, 17.4)
5. (a) Not normal

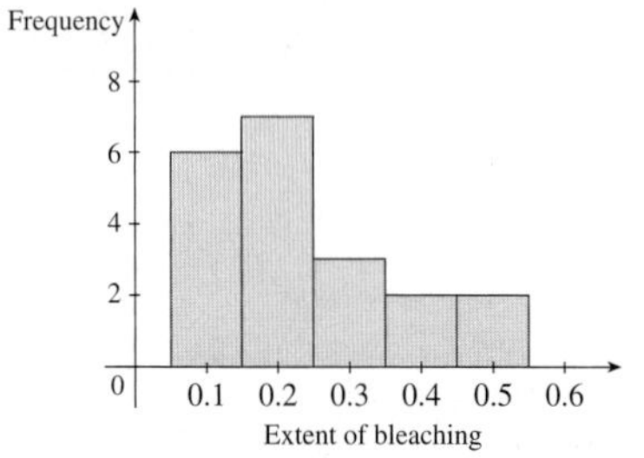

(b) (0.17, 0.27); Use of t-value assumes normally distributed data
7. (a) Not normal

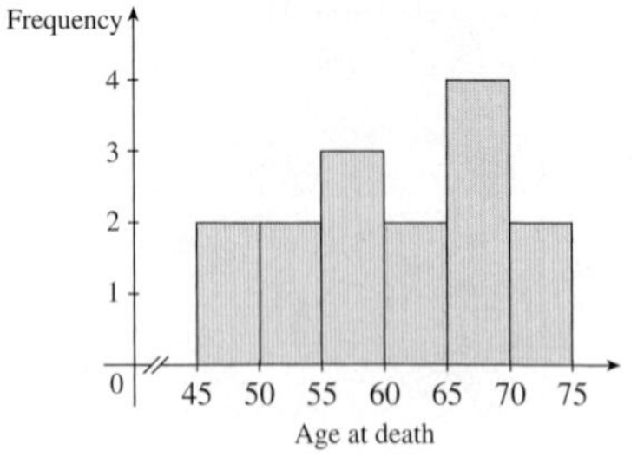

(b) (56.9, 63.7); Use of z_{90}-value assumes normally distributed data
9. (a) Not normal

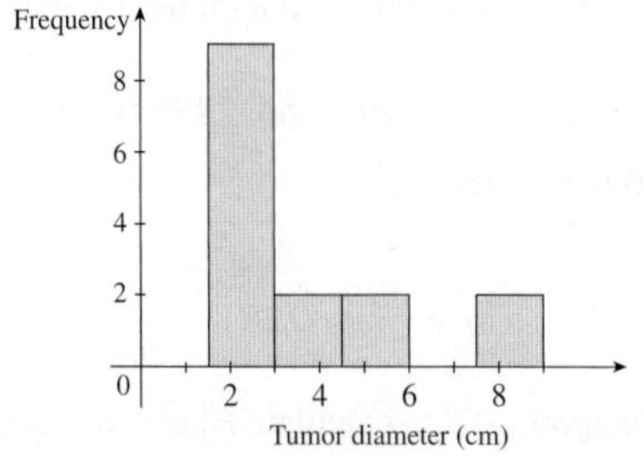

(b) Known σ: (2.41, 4.19); unknown σ: (2.33, 4.27);
Use of z-values and t-values assumes normally distributed data
11. (a)

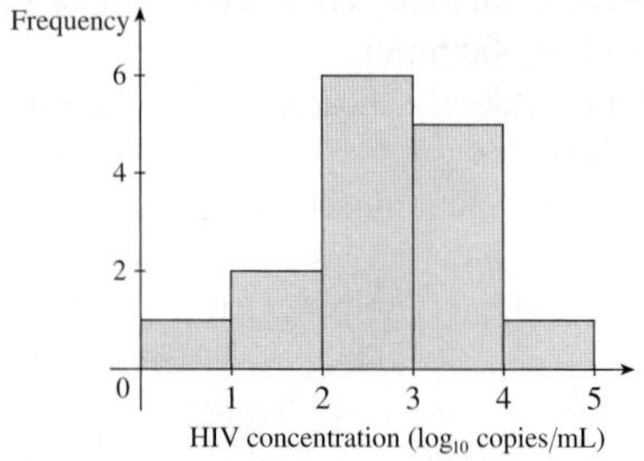

Approximately normally distributed
(b) Known σ: (2.27, 3.09); unknown σ: (2.22, 3.14);
CI for unknown σ is larger
13. Female: (50.4, 55.3); male (35.1, 44.5)

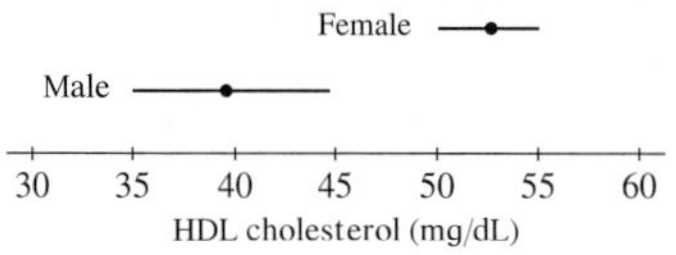

EXERCISES 13.3 | page 827

1. Yes (at both the 0.05 and 0.01 significance levels)
3. Mean hospital stay differs from 20 days (at both 0.05 and 0.01 significance levels); determination of P-value from t-distribution assumes normality of population data
5. No
7. $P < 0.001$ (or $P \approx 6.0 \times 10^{-12}$ using a calculator);
if the mean age of the population is 15, there is less than a 0.1% chance of having observed the given data; determination of P-value from t-distribution assumes normality of population data
9. $0.001 < P < 0.01$ (or $P \approx 0.0018$ using a calculator);
if the population mean wingspan is 10 cm, there is approximately a 0.18% chance of having observed the given data.
11. Yes, because the null hypothesis $\mu = 0$ is rejected at the 0.05, 0.01, and 0.001 significance levels.
13. $T_{12} \approx 2.567$; $\approx 6.1 \times 10^{-4}$

EXERCISES 13.4 | page 834

1.

	Survived	Died	Row total
Vaccinated	25 (12.5)	0 (12.5)	25
Unvaccinated	0 (12.5)	25 (12.5)	25
Column total	25	25	50

Vaccination status and survival are associated (at the 0.05, 0.01, and 0.001 significance levels).
3.

	Developed myopia	No myopia	Row total
Darkness	9 (17.03)	163 (154.97)	172
Night-light	31 (22.97)	201 (209.03)	232
Column total	40	364	404

Nighttime ambient light and myopia are associated (at the 0.05 and 0.01 significance levels).

5. Species X is a good indicator species (determined using a test with a 0.05 significance level).

7. A positive test is indicative of cancer (using a 0.05, 0.01, or 0.001 significance level).

CHAPTER 13 REVIEW | page 836

True-False Quiz

1. True **3.** True **5.** True **7.** True **9.** True

Exercises

1. ≈ 0.9791 **3.** ≈ 0.0614 **5.** (13.1, 19.9)

7. (548.1, 679.9)

9. Rotation (36.2, 42.8); no rotation: (31.0, 36.7)

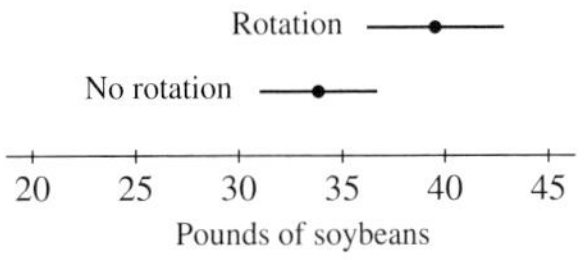

11. Data are consistent with a population mean pollen mass of 15 mg.

13. Population mean differs from 750 mg (using a 0.05, 0.01, or 0.001 significance level).

15.

	Positive test	Negative test	Row total
Male	13 (13.42)	8 (7.58)	21
Female	10 (9.58)	5 (5.42)	15
Column total	23	13	36

No association between sex and test outcome

APPENDIXES

APPENDIX A | page 844

1. 18 **3.** $5 - \sqrt{5}$ **5.** $2 - x$

7. $|x + 1| = \begin{cases} x + 1 & \text{for } x \geqslant -1 \\ -x - 1 & \text{for } x < -1 \end{cases}$ **9.** $x^2 + 1$

11. $(-2, \infty)$

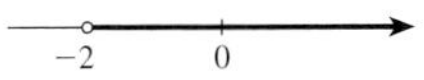

13. $[-1, \infty)$

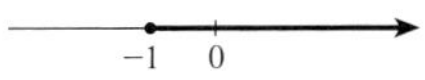

15. $(0, 1]$

17. $(-\infty, 1) \cup (2, \infty)$

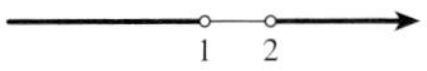

19. $(-\sqrt{3}, \sqrt{3})$

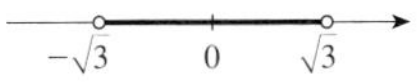

21. $(-\infty, 1]$

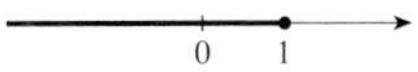

23. $(-1, 0) \cup (1, \infty)$

25. $(-\infty, 0) \cup (\frac{1}{4}, \infty)$

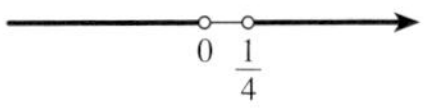

27. $10 \leqslant C \leqslant 35$

29. (a) $T = 20 - 10h,\ 0 \leqslant h \leqslant 12$
(b) $-30°\text{C} \leqslant T \leqslant 20°\text{C}$

31. $2, -\frac{4}{3}$ **33.** $(-3, 3)$

35. $(3, 5)$ **37.** $(-\infty, -7] \cup [-3, \infty)$ **39.** $[1.3, 1.7]$

41. $x \geqslant (a + b)c/(ab)$

APPENDIX B | page 854

1. 5 **3.** $-\frac{9}{2}$

7.

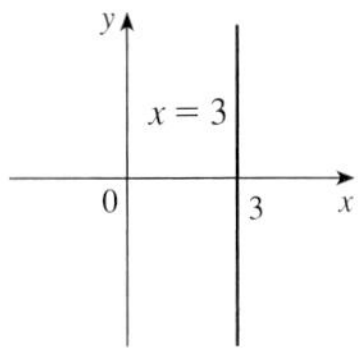

9.

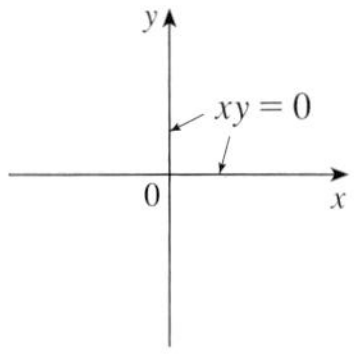

11. $y = 6x - 15$ **13.** $5x + y = 11$ **15.** $y = 3x - 2$

17. $y = 3x - 3$ **19.** $y = 5$ **21.** $x + 2y + 11 = 0$

23. $5x - 2y + 1 = 0$

25. $m = -\frac{1}{3},\ b = 0$

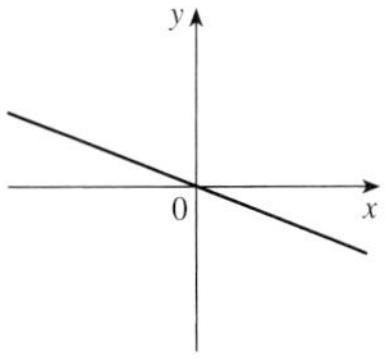

27. $m = \frac{3}{4},\ b = -3$

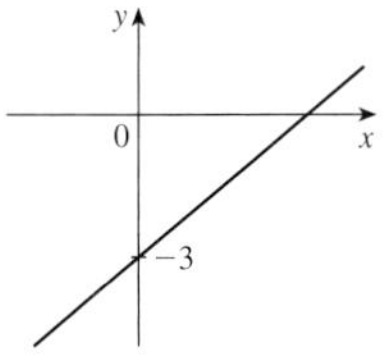

29.

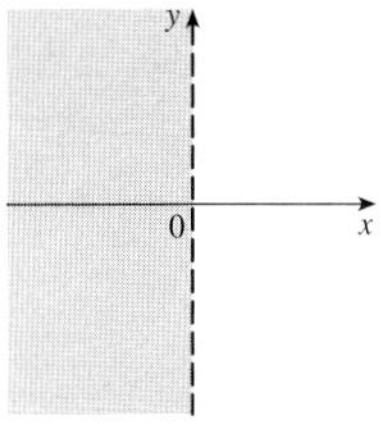

31.

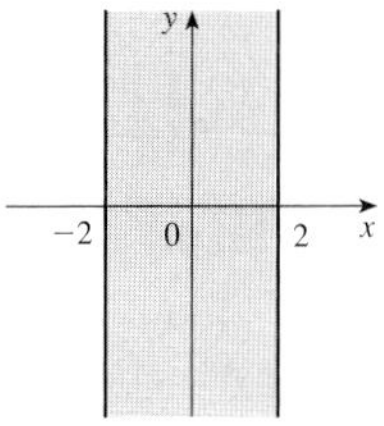

33.

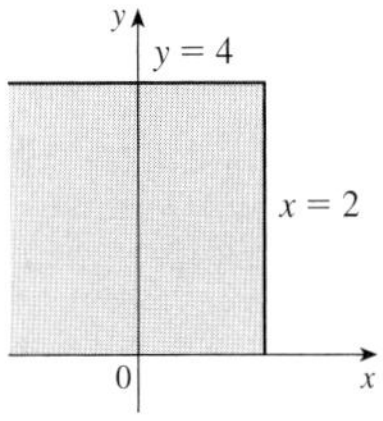

35.

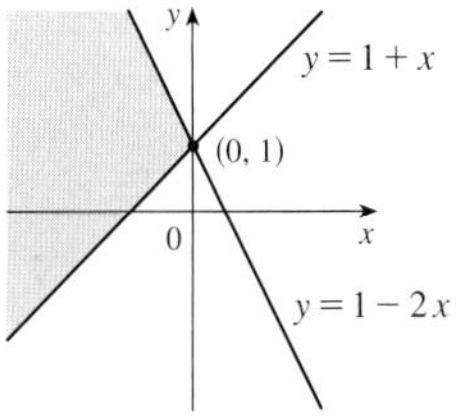

37. $(x - 3)^2 + (y + 1)^2 = 25$ **39.** $(2, -5), 4$ **41.** $(1, -2)$

45. $y = x - 3$

53.

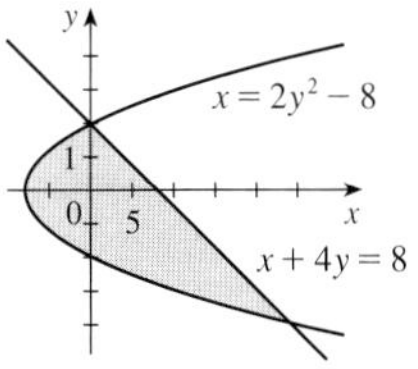

APPENDIX C | page 863

1. (a) $7\pi/6$ (b) $\pi/20$ **3.** (a) 720° (b) −67.5°

5. 3π cm **7.** $\frac{2}{3}$ rad $= (120/\pi)°$

9. (a)

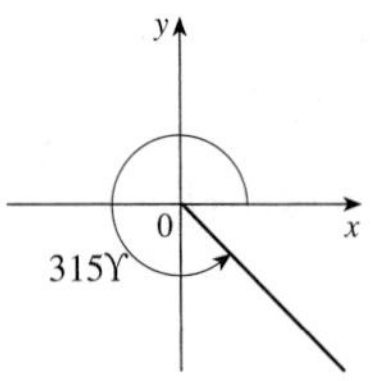

(b)

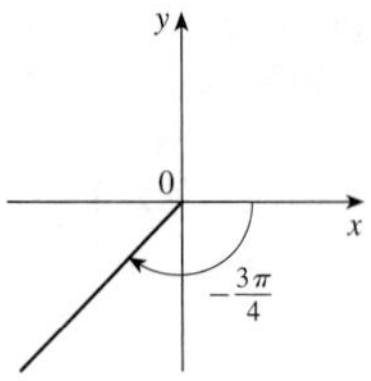

11. $\sin(3\pi/4) = 1/\sqrt{2}$, $\cos(3\pi/4) = -1/\sqrt{2}$, $\tan(3\pi/4) = -1$, $\csc(3\pi/4) = \sqrt{2}$, $\sec(3\pi/4) = -\sqrt{2}$, $\cot(3\pi/4) = -1$
13. $\cos\theta = \frac{4}{5}$, $\tan\theta = \frac{3}{4}$, $\csc\theta = \frac{5}{3}$, $\sec\theta = \frac{5}{4}$, $\cot\theta = \frac{4}{3}$
15. 5.73576 cm **17.** 24.62147 cm **29.** $\frac{1}{15}(4 + 6\sqrt{2})$
31. $\pi/3, 5\pi/3$ **33.** $\pi/6, \pi/2, 5\pi/6, 3\pi/2$
35. $0 \leqslant x \leqslant \pi/6$ and $5\pi/6 \leqslant x \leqslant 2\pi$
37. $0 \leqslant x < \pi/4$, $3\pi/4 < x < 5\pi/4$, $7\pi/4 < x \leqslant 2\pi$
39.

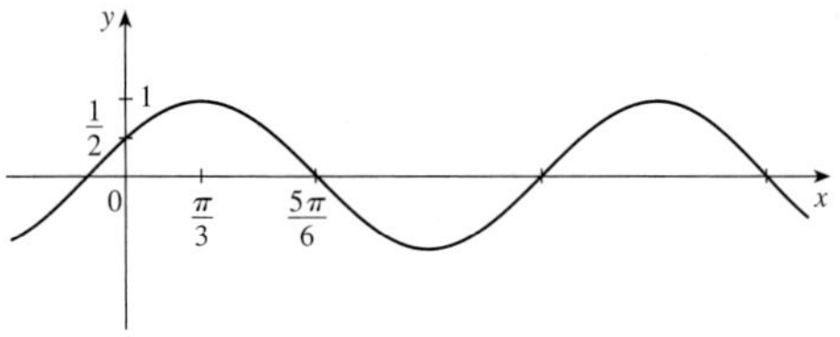

41.

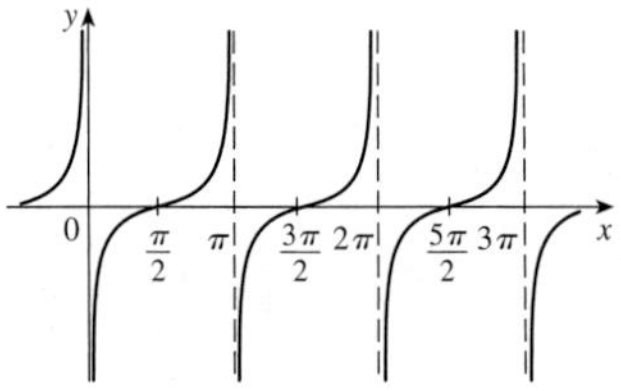

APPENDIX D | page 869

1. (a) $n > 100$
3. $\frac{4}{7}$ (or any smaller positive number)
5. 1.44 (or any smaller positive number)
7. 0.0906 (or any smaller positive number)
9. 0.11 or 0.012 (or smaller positive numbers)
13. (a) $\sqrt{1000/\pi}$ cm (b) Within approximately 0.0445 cm
(c) Radius; area; $\sqrt{1000/\pi}$; 1000; 5; ≈ 0.0445
15. (a) 0.025 (b) 0.0025

APPENDIX F | page 878

1. $\sqrt{1} + \sqrt{2} + \sqrt{3} + \sqrt{4} + \sqrt{5}$ **3.** $3^4 + 3^5 + 3^6$
5. $-1 + \frac{1}{3} + \frac{3}{5} + \frac{5}{7} + \frac{7}{9}$ **7.** $1^{10} + 2^{10} + 3^{10} + \cdots + n^{10}$
9. $1 - 1 + 1 - 1 + \cdots + (-1)^{n-1}$ **11.** $\sum_{i=1}^{10} i$
13. $\sum_{i=1}^{19} \frac{i}{i+1}$ **15.** $\sum_{i=1}^{n} 2i$ **17.** $\sum_{i=0}^{5} 2^i$ **19.** $\sum_{i=1}^{n} x^i$
21. 80 **23.** 3276 **25.** 0 **27.** 61 **29.** $n(n+1)$
31. $n(n^2 + 6n + 17)/3$ **33.** $n(n^2 + 6n + 11)/3$
35. $n(n^3 + 2n^2 - n - 10)/4$
41. (a) n^4 (b) $5^{100} - 1$ (c) $\frac{97}{300}$ (d) $a_n - a_0$
43. $\frac{1}{3}$ **45.** 14 **49.** $2^{n+1} + n^2 + n - 2$

APPENDIX G | page 887

1. $8 - 4i$ **3.** $13 + 18i$ **5.** $12 - 7i$ **7.** $\frac{11}{13} + \frac{10}{13}i$
9. $\frac{1}{2} - \frac{1}{2}i$ **11.** $-i$ **13.** $5i$ **15.** $12 + 5i$, 13
17. $4i$, 4 **19.** $\pm\frac{3}{2}i$ **21.** $-1 \pm 2i$
23. $-\frac{1}{2} \pm (\sqrt{7}/2)i$ **25.** $3\sqrt{2}\left[\cos(3\pi/4) + i\sin(3\pi/4)\right]$
27. $5\{\cos[\tan^{-1}(\frac{4}{3})] + i\sin[\tan^{-1}(\frac{4}{3})]\}$
29. $4[\cos(\pi/2) + i\sin(\pi/2)]$, $\cos(-\pi/6) + i\sin(-\pi/6)$, $\frac{1}{2}[\cos(-\pi/6) + i\sin(-\pi/6)]$
31. $4\sqrt{2}[\cos(7\pi/12) + i\sin(7\pi/12)]$, $(2\sqrt{2})[\cos(13\pi/12) + i\sin(13\pi/12)]$, $\frac{1}{4}[\cos(\pi/6) + i\sin(\pi/6)]$
33. -1024 **35.** $-512\sqrt{3} + 512i$
37. $\pm1, \pm i, (1/\sqrt{2})(\pm1 \pm i)$

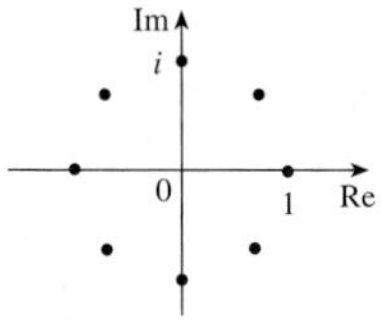

39. $\pm(\sqrt{3}/2) + \frac{1}{2}i, -i$

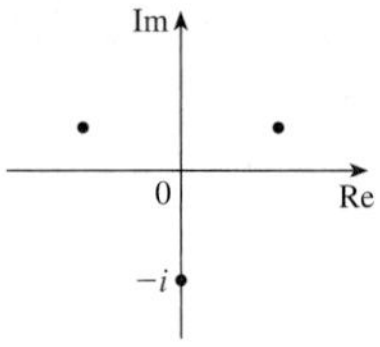

41. i **43.** $\frac{1}{2} + (\sqrt{3}/2)i$ **45.** $-e^2$
47. $\cos 3\theta = \cos^3\theta - 3\cos\theta\sin^2\theta$, $\sin 3\theta = 3\cos^2\theta\sin\theta - \sin^3\theta$

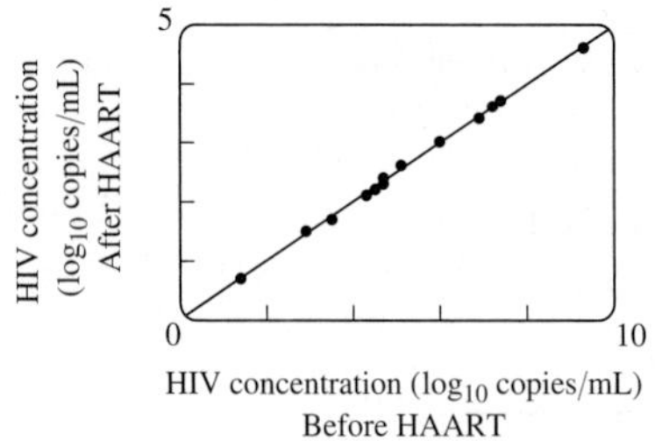

Biological Index

Index

RP denotes Reference Page numbers.

Cut here and keep for reference

1. (a) What is a function? What are its domain and range?

A function f is a rule that assigns to each element x in a set D exactly one element, called $f(x)$, in a set E. The set D is called the domain of the function. The range of f is the set of all possible values of $f(x)$ as x varies throughout the domain.

(b) What is the graph of a function?

If f is a function with domain D, then its graph is the set of ordered pairs $\{(x, f(x)) \mid x \in D\}$.

(c) How can you tell whether a given curve is the graph of a function?

Use the Vertical Line Test: A curve in the xy-plane is the graph of a function of x if and only if no vertical line intersects the curve more than once.

2. Discuss four ways of representing a function. Illustrate your discussion with examples.

The four ways to represent a function are: verbally, numerically, visually, and algebraically. An example of each is given below.

Verbally: An assignment of students to chairs in a classroom (a description in words)

Numerically: A tax table that assigns an amount of tax to an income (a table of values)

Visually: A graphical history of the Dow Jones average (a graph)

Algebraically: A relationship between distance, speed, and time: $d = rt$ (an explicit formula)

3. (a) What is an even function? How can you tell if a function is even by looking at its graph?

A function f is even if it satisfies $f(-x) = f(x)$ for every number x in its domain. If the graph of a function is symmetric with respect to the y-axis, then f is even. Examples of even functions: $f(x) = x^2$, $f(x) = x^4 + x^2$, $f(x) = |x|$.

(b) What is an odd function? How can you tell if a function is odd by looking at its graph?

A function f is odd if it satisfies $f(-x) = -f(x)$ for every number x in its domain. If the graph of a function is symmetric with respect to the origin, then f is odd. Examples of odd functions: $f(x) = x^3$, $f(x) = x^3 + x^5$, $f(x) = \sqrt[3]{x}$.

4. What is an increasing function?

A function f is called increasing on an interval I if $f(x_1) < f(x_2)$ whenever $x_1 < x_2$ in I.

5. What is a mathematical model?

A mathematical model is a mathematical description (often by means of a function or an equation) of a real-world phenomenon.

6. Give an example of each type of function.

(a) Linear function: $f(x) = 2x + 1, f(x) = ax + b$
(b) Power function: $f(x) = x^2, f(x) = x^a$
(c) Exponential function: $f(x) = 2^x, f(x) = b^x$
(d) Quadratic function: $f(x) = x^2 + x + 1$, $f(x) = ax^2 + bx + c \;\; (a \neq 0)$
(e) Polynomial of degree 5: $f(x) = x^5 + 2x^4 - 3x^2 + 7$
(f) Rational function: $f(x) = \dfrac{x}{x + 2}$, $f(x) = \dfrac{P(x)}{Q(x)}$ where $P(x)$ and $Q(x)$ are polynomials

7. Sketch by hand, on the same axes, the graphs of the following functions.

(a) $f(x) = x$ (b) $g(x) = x^2$
(c) $h(x) = x^3$ (d) $j(x) = x^4$

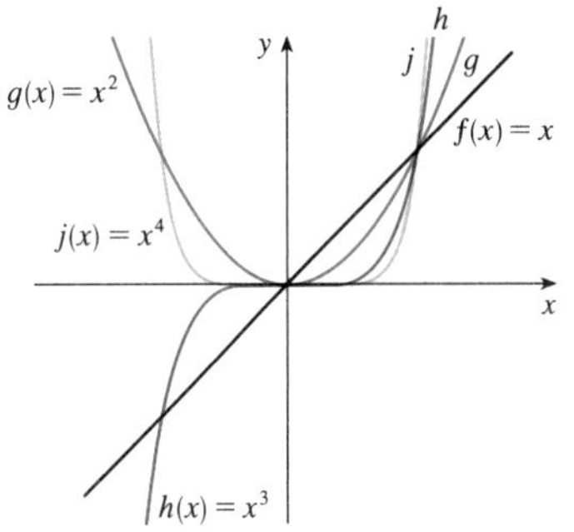

8. Draw, by hand, a rough sketch of the graph of each function.

(a) $y = \sin x$

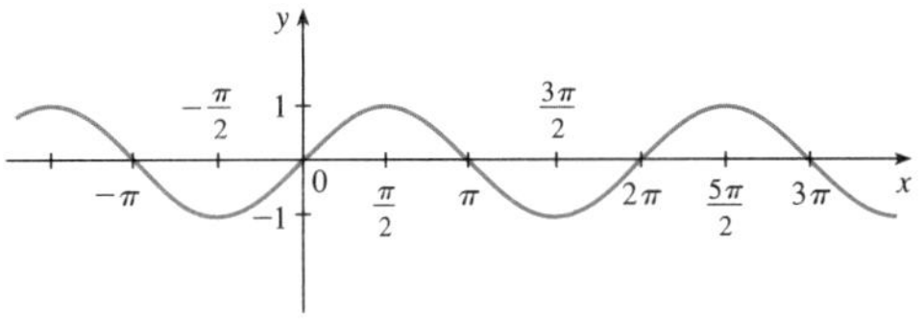

(b) $y = \cos x$

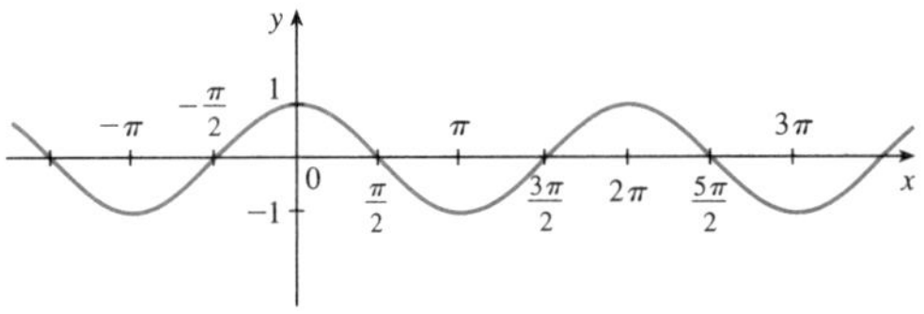

(c) $y = \tan x$

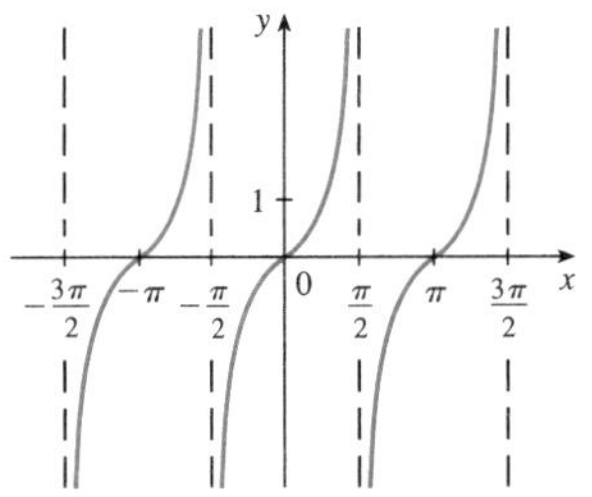

(d) $y = e^x$

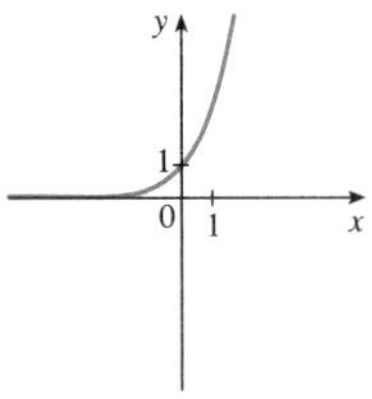

(e) $y = \ln x$

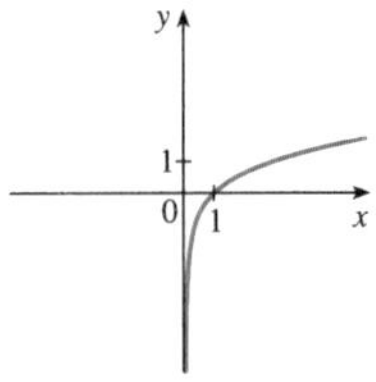

(f) $y = 1/x$

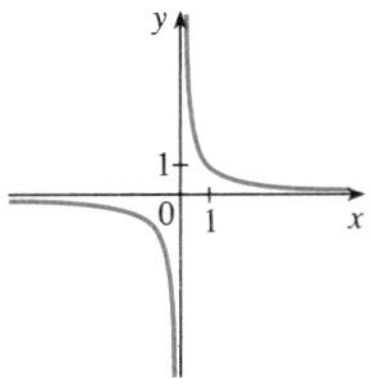

(g) $y = |x|$

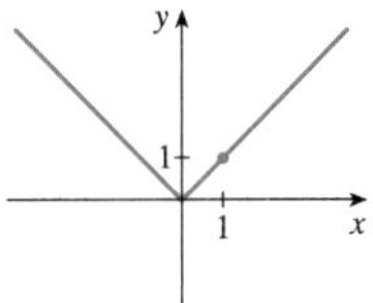

(continued)

(h) $y = \sqrt{x}$

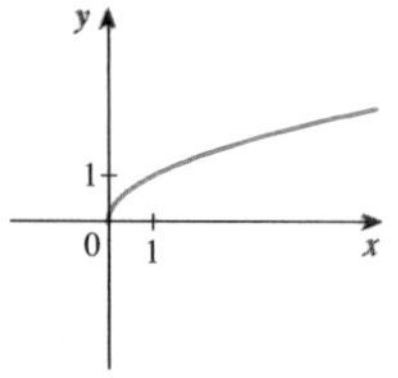

9. Suppose that f has domain A and g has domain B.

(a) What is the domain of $f + g$?

The domain of $f + g$ is the intersection of the domain of f and the domain of g; that is, $A \cap B$, the numbers that are in both A and B.

(b) What is the domain of fg?

The domain of fg is also $A \cap B$.

(c) What is the domain of f/g?

The domain of f/g must exclude values of x that make g equal to 0; that is, $\{x \in A \cap B \mid g(x) \neq 0\}$.

10. How is the composite function $f \circ g$ defined? What is its domain?

Given two functions f and g, the composite function $f \circ g$ is defined by $(f \circ g)(x) = f(g(x))$. The domain of $f \circ g$ is the set of all x in the domain of g such that $g(x)$ is in the domain of f.

11. Suppose the graph of f is given. Write an equation for each of the graphs that are obtained from the graph of f as follows.

(a) Shift 2 units upward: $y = f(x) + 2$

(b) Shift 2 units downward: $y = f(x) - 2$

(c) Shift 2 units to the right: $y = f(x - 2)$

(d) Shift 2 units to the left: $y = f(x + 2)$

(e) Reflect about the x-axis: $y = -f(x)$

(f) Reflect about the y-axis: $y = f(-x)$

(g) Stretch vertically by a factor of 2: $y = 2f(x)$

(h) Shrink vertically by a factor of 2: $y = \frac{1}{2}f(x)$

(i) Stretch horizontally by a factor of 2: $y = f\left(\frac{1}{2}x\right)$

(j) Shrink horizontally by a factor of 2 $y = f(2x)$

12. (a) What is a one-to-one function? How can you tell if a function is one-to-one by looking at its graph?

A function f is called a one-to-one function if it never takes on the same value twice; that is, if $f(x_1) \neq f(x_2)$ whenever $x_1 \neq x_2$. (Or, f is 1-1 if each output corresponds to only one input.)

Use the Horizontal Line Test: A function is one-to-one if and only if no horizontal line intersects its graph more than once.

(b) If f is a one-to-one function, how is its inverse function f^{-1} defined? How do you obtain the graph of f^{-1} from the graph of f?

If f is a one-to-one function with domain A and range B, then its inverse function f^{-1} has domain B and range A and is defined by

$$f^{-1}(y) = x \iff f(x) = y$$

for any y in B. The graph of f^{-1} is obtained by reflecting the graph of f about the line $y = x$.

13. (a) What is a semilog plot?

A semilog plot is a graph of the points $(x, \log y)$ given (x, y) data points.

(b) If a semilog plot of your data lies approximately on a line, what type of model is appropriate?

An exponential model is appropriate when a semilog plot of the data lies on a line.

14. (a) What is a log-log plot?

A log-log plot is a graph of the points $(\log x, \log y)$ given (x, y) data points.

(b) If a log-log plot of your data lies approximately on a line, what type of model is appropriate?

A power model is appropriate when a log-log plot of the data lies on a line.

15. (a) What is a sequence?

A sequence is an ordered list of numbers.

(b) What is a recursive sequence?

A recursive sequence is a sequence in which each term is defined using one or more preceding terms.

16. Discrete-time models

(a) If there are N_t cells at time t and they divide according to the difference equation $N_{t+1} = RN_t$, write an expression for N_t.

$N_t = N_0 R^t$

(b) If a population has carrying capacity K, write the logistic difference equation for N_t.

$$N_{t+1} = \left[1 + r\left(1 - \frac{N_t}{K}\right)\right] N_t$$

(c) Write the logistic difference equation for

$$x_t = \frac{r}{(1 + r)K} N_t$$

$x_{t+1} = R_{max} x_t (1 - x_t)$

1. (a) What is a convergent sequence?

A convergent sequence is an ordered list of numbers $\{a_n\}$ for which $\lim_{n\to\infty} a_n$ exists.

(b) What does $\lim_{n\to\infty} a_n = 3$ mean?

The terms of the sequence $\{a_n\}$ approach 3 as n becomes large.

2. What is $\lim_{n\to\infty} r^n$ in the following three cases?

(a) $0 < r < 1$

$\lim_{n\to\infty} r^n = 0$ when $0 < r < 1$

(b) $r = 1$

$\lim_{n\to\infty} r^n = 1$ when $r = 1$

(c) $r > 1$

$\lim_{n\to\infty} r^n = \infty$ when $r > 1$

3. (a) What is the sum of the finite geometric series $a + ar + ar^2 + \cdots + ar^n$?

$$a + ar + ar^2 + \cdots + ar^n = \frac{a(1 - r^{n+1})}{1 - r}$$

(b) If $-1 < r < 1$, what is the sum of the infinite geometric series $a + ar + ar^2 + \cdots + ar^n + \cdots$?

$$a + ar + ar^2 + \cdots + ar^n + \cdots = \frac{a}{1 - r} \text{ if } -1 < r < 1$$

4. Explain what each of the following means and illustrate with a sketch.

(a) $\lim_{x\to a} f(x) = L$

$\lim_{x\to a} f(x) = L$ means that the values of $f(x)$ approach L as the values of x approach a (from either side).

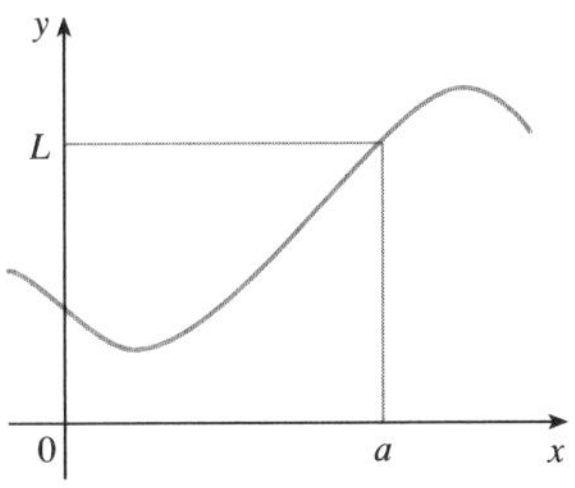

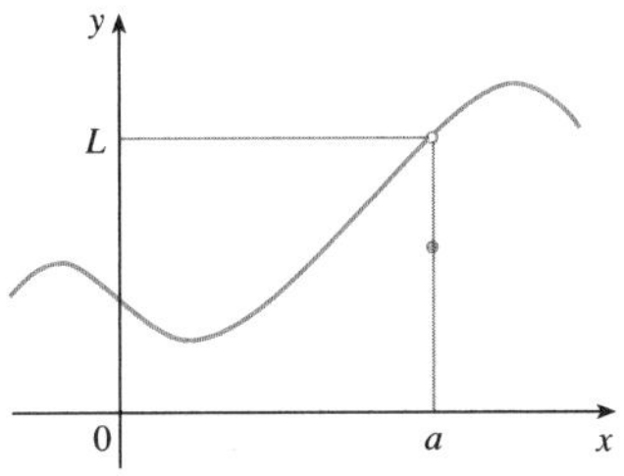

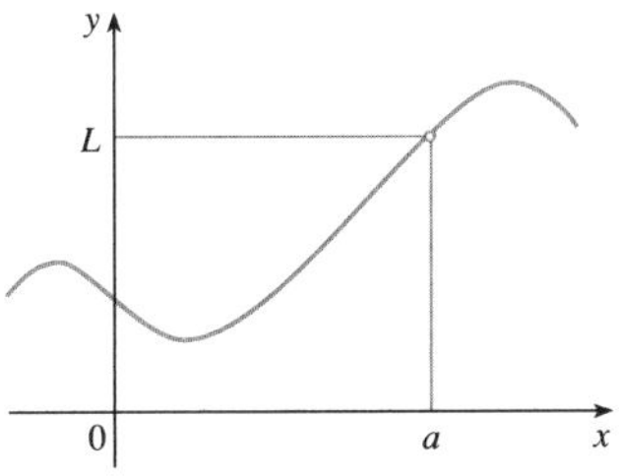

(b) $\lim_{x\to a^+} f(x) = L$

$\lim_{x\to a^+} f(x) = L$ means that the values of $f(x)$ approach L as the values of x approach a from the right side.

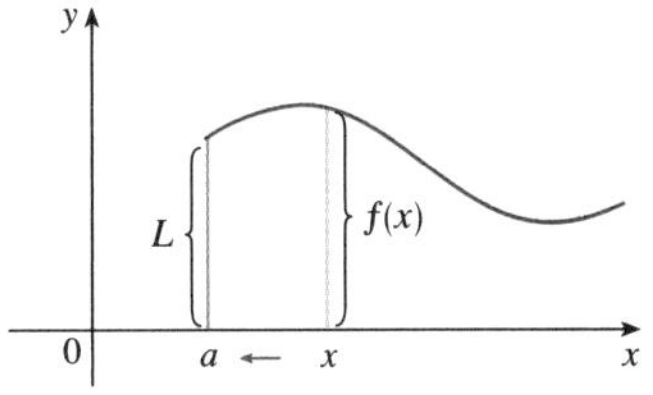

(c) $\lim_{x\to a^-} f(x) = L$

$\lim_{x\to a^-} f(x) = L$ means that the values of $f(x)$ approach L as the values of x approach a from the left side.

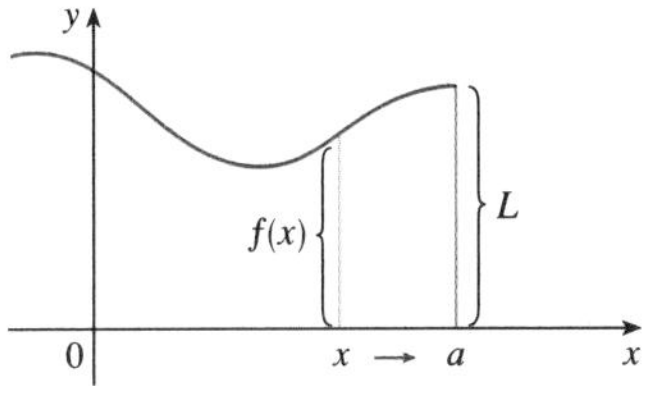

(d) $\lim_{x\to a} f(x) = \infty$

$\lim_{x\to a} f(x) = \infty$ means that the values of $f(x)$ become arbitrarily large as x approaches a (from either side)

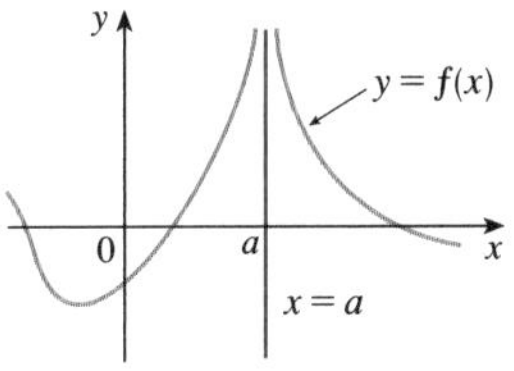

(e) $\lim_{x\to\infty} f(x) = L$

This means that the values of $f(x)$ approach L as x becomes large.

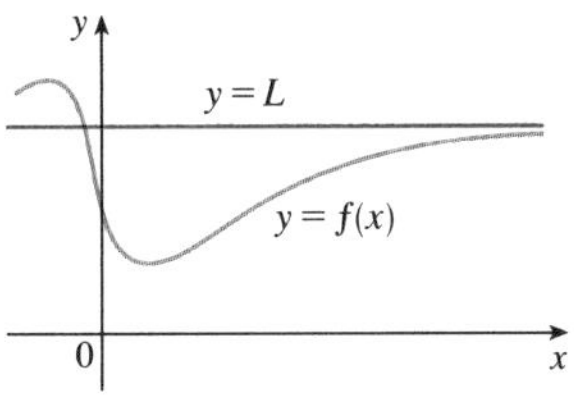

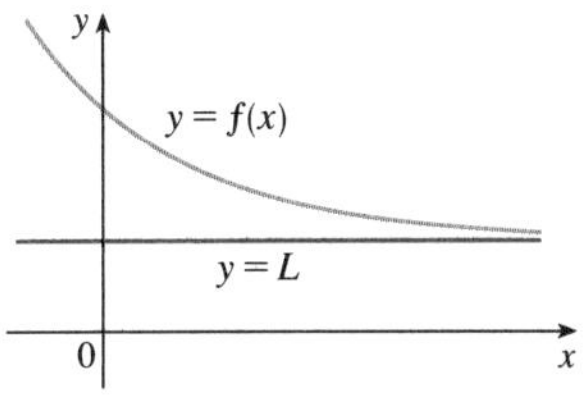

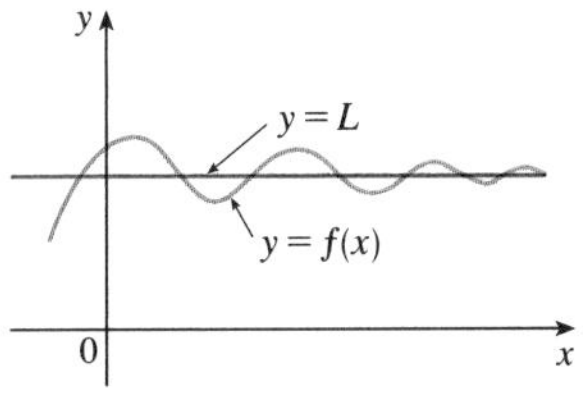

5. State the following Limit Laws for functions.

(a) Sum Law

The limit of a sum is the sum of the limits:
$\lim_{x\to a} [f(x) + g(x)] = \lim_{x\to a} f(x) + \lim_{x\to a} g(x)$

(continued)

(b) Difference Law

The limit of a difference is the difference of the limits:
$\lim_{x\to a}[f(x) - g(x)] = \lim_{x\to a} f(x) - \lim_{x\to a} g(x)$

(c) Constant Multiple Law

The limit of a constant times a function is the constant times the limit of the function: $\lim_{x\to a}[cf(x)] = c \lim_{x\to a} f(x)$

(d) Product Law

The limit of a product is the product of the limits:
$\lim_{x\to a}[f(x)g(x)] = \lim_{x\to a} f(x) \cdot \lim_{x\to a} g(x)$

(e) Quotient Law

The limit of a quotient is the quotient of the limits, provided that the limit of the denominator is not 0:

$$\lim_{x\to a}\frac{f(x)}{g(x)} = \frac{\lim_{x\to a} f(x)}{\lim_{x\to a} g(x)} \quad \text{if } \lim_{x\to a} g(x) \neq 0$$

(f) Power Law

The limit of a power is the power of the limit:
$\lim_{x\to a}[f(x)]^n = \left[\lim_{x\to a} f(x)\right]^n$ (for n a positive integer)

(g) Root Law

The limit of a root is the root of the limit:
$\lim_{x\to a}\sqrt[n]{f(x)} = \sqrt[n]{\lim_{x\to a} f(x)}$ where n is a positive integer
[If n is even, assume that $\lim_{x\to a} f(x) > 0$.]

6. What does the Squeeze Theorem say?

The Squeeze Theorem states that if $f(x) \leqslant g(x) \leqslant h(x)$ when x is near a (except possibly at a) and $\lim_{x\to a} f(x) = \lim_{x\to a} h(x) = L$, then $\lim_{x\to a} g(x) = L$.

7. (a) What does it mean to say that the line $x = a$ is a vertical asymptote of the curve $y = f(x)$? Draw curves to illustrate the various possibilities.

The line $x = a$ is a vertical asymptote of the curve $y = f(x)$ if at least one of the following is true:

$\lim_{x\to a} f(x) = \infty$

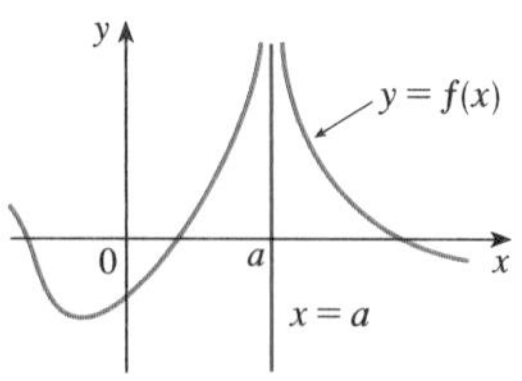

$\lim_{x\to a} f(x) = -\infty$

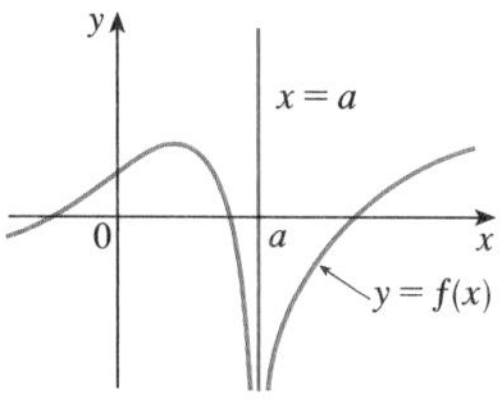

$\lim_{x\to a^-} f(x) = \infty$

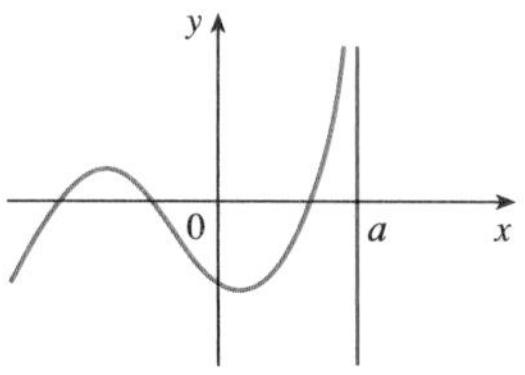

$\lim_{x\to a^+} f(x) = \infty$

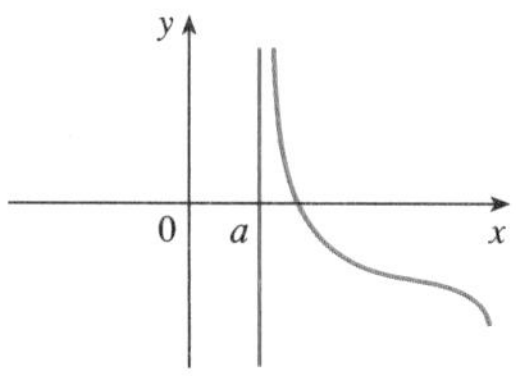

$\lim_{x\to a^-} f(x) = -\infty$

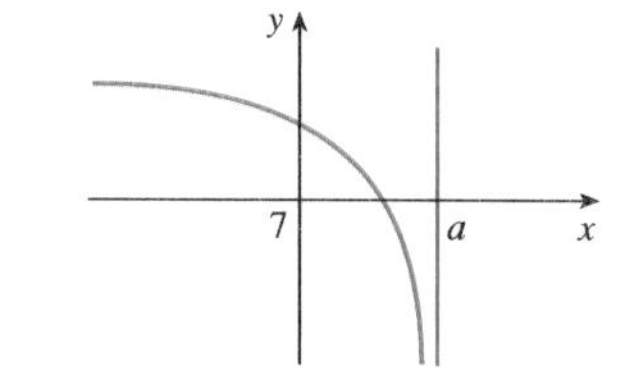

$\lim_{x\to a^+} f(x) = -\infty$

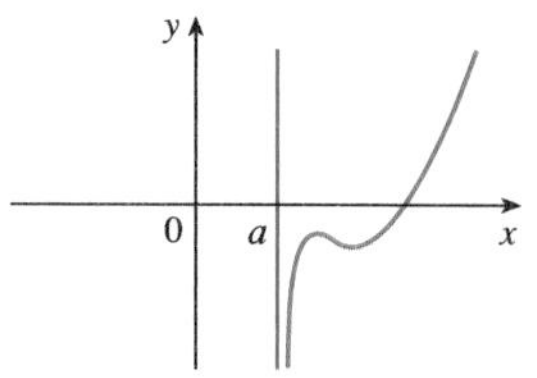

(b) What does it mean to say that the line $y = L$ is a horizontal asymptote of the curve $y = f(x)$? Draw curves to illustrate the various possibilities.

The line $y = L$ is a horizontal asymptote of the curve $y = f(x)$ if either $\lim_{x\to\infty} f(x) = L$ or $\lim_{x\to-\infty} f(x) = L$.

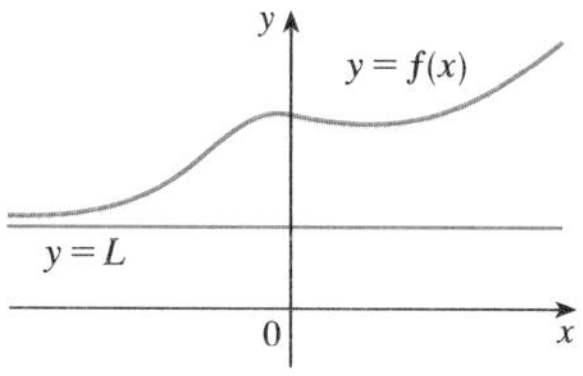

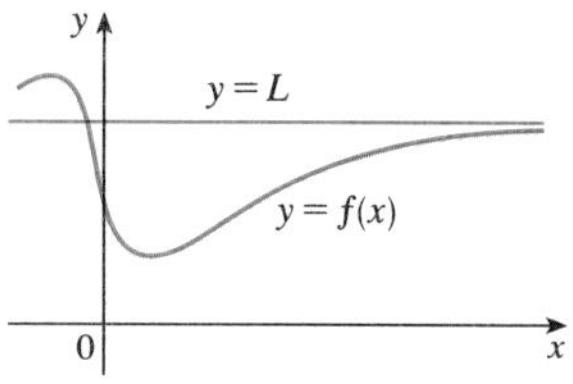

8. Which of the following curves have vertical asymptotes? Which have horizontal asymptotes?

(a) $y = x^4$: No asymptote

(b) $y = \sin x$: No asymptote

(c) $y = \tan x$: Vertical asymptotes $x = \frac{\pi}{2} + \pi n$, n an integer

(d) $y = e^x$: Horizontal asymptote $y = 0$ $\left(\lim_{x\to-\infty} e^x = 0\right)$

(e) $y = \ln x$: Vertical asymptote $x = 0$ $\left(\lim_{x\to 0^+} \ln x = -\infty\right)$

(f) $y = 1/x$:

Vertical asymptote $x = 0$, horizontal asymptote $y = 0$

(g) $y = \sqrt{x}$: No asymptote

9. (a) What does it mean for f to be continuous at a?

A function f is continuous at a number a if $f(x)$ approaches $f(a)$ as x approaches a; that is, $\lim_{x\to a} f(x) = f(a)$.

(b) What does it mean for f to be continuous on the interval $(-\infty, \infty)$? What can you say about the graph of such a function?

A function f is continuous on the interval $(-\infty, \infty)$ if f is continuous at every real number a. The graph of such a function has no break and every vertical line crosses it.

10. What does the Intermediate Value Theorem say?

The Intermediate Value Theorem states that if f is continuous on the closed interval $[a, b]$ and N is any number between $f(a)$ and $f(b)$, where $f(a) \neq f(b)$, then there exists a number c in (a, b) such that $f(c) = N$. In other words, a continuous function takes on every intermediate value between the function values $f(a)$ and $f(b)$.

1. Write an expression for the slope of the tangent line to the curve $y = f(x)$ at the point $(a, f(a))$.

The slope of the tangent line to the graph of $y = f(x)$ at the point $(a, f(a))$ is given by

$$\lim_{x \to a} \frac{f(x) - f(a)}{x - a} \quad \text{or} \quad \lim_{h \to 0} \frac{f(a + h) - f(a)}{h}$$

2. Define the derivative $f'(a)$. Discuss two ways of interpreting this number.

$f'(a) = \lim_{h \to 0} \dfrac{f(a + h) - f(a)}{h}$ or, equivalently,

$f'(a) = \lim_{x \to a} \dfrac{f(x) - f(a)}{x - a}$. The derivative $f'(a)$ is the instantaneous rate of change of f (with respect to x) when $x = a$ and also represents the slope of the tangent line to the graph of f at $x = a$.

3. If $y = f(x)$ and x changes from x_1 to x_2, write expressions for the following.

(a) The average rate of change of y with respect to x over the interval $[x_1, x_2]$.

The average rate of change of y with respect to x over the interval $[x_1, x_2]$ is $\dfrac{f(x_2) - f(x_1)}{x_2 - x_1}$.

(b) The instantaneous rate of change of y with respect to x at $x = x_1$.

The instantaneous rate of change of y with respect to x at $x = x_1$ is $\lim_{x_2 \to x_1} \dfrac{f(x_2) - f(x_1)}{x_2 - x_1}$.

4. Define the second derivative of f. If $f(t)$ is the position function of a particle, how can you interpret the second derivative?

The second derivative of f, f'', is the derivative of f'. If $f(t)$ is the position function of a particle, then the first derivative is velocity and the second derivative is the derivative of velocity, namely, acceleration.

5. (a) What does it mean for f to be differentiable at a?

A function f is differentiable at a number a if its derivative f' exists at $x = a$; that is, if $f'(a)$ exists.

(b) What is the relation between the differentiability and continuity of a function?

If f is differentiable at a, then f is continuous at a. So if f is *not* continuous at a, then f is *not* differentiable at a.

(c) Sketch the graph of a function that is continuous but not differentiable at $a = 2$.

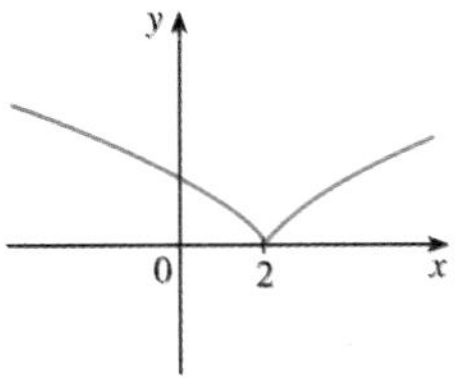

6. Describe several ways in which a function can fail to be differentiable. Illustrate with sketches.

A function is not differentiable at any value where the graph has a corner, where the graph has a discontinuity, or where it has a vertical tangent line.

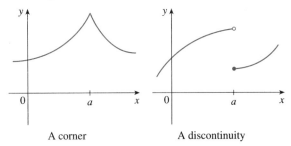

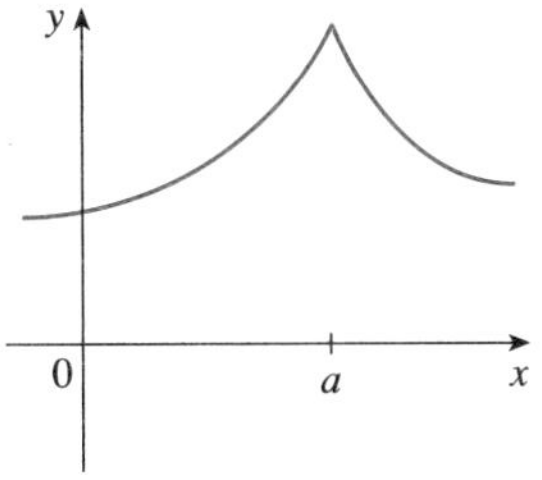

A corner

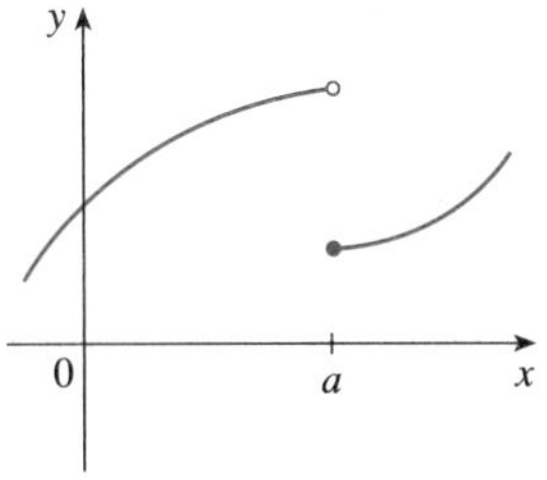

A discontinuity

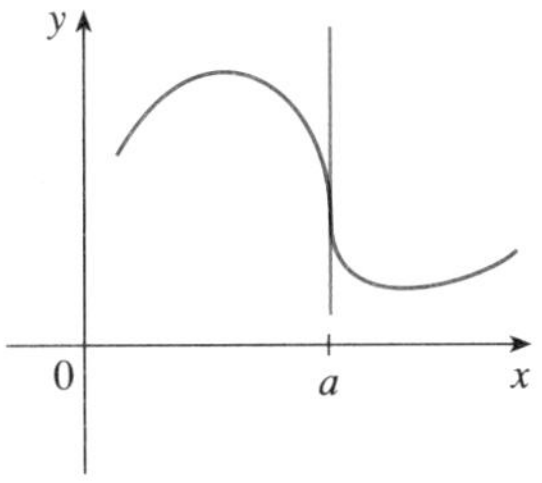

A vertical tangent

7. State each differentiation rule both in symbols and in words.

(a) The Power Rule

If n is any real number, then $\dfrac{d}{dx}(x^n) = nx^{n-1}$. The derivative of a variable base raised to a constant power is the power times the base raised to the power minus one.

(b) The Constant Multiple Rule

If c is a constant and f is a differentiable function, then $\dfrac{d}{dx}[cf(x)] = c\dfrac{d}{dx}f(x)$. The derivative of a constant times a function is the constant times the derivative of the function.

(c) The Sum Rule

If f and g are both differentiable, then $\dfrac{d}{dx}[f(x) + g(x)] = \dfrac{d}{dx}f(x) + \dfrac{d}{dx}g(x)$. The derivative of a sum of functions is the sum of the derivatives.

(d) The Difference Rule

If f and g are both differentiable, then $\dfrac{d}{dx}[f(x) - g(x)] = \dfrac{d}{dx}f(x) - \dfrac{d}{dx}g(x)$. The derivative of a difference of functions is the difference of the derivatives.

(e) The Product Rule

If f and g are both differentiable, then

$$\frac{d}{dx}[f(x)g(x)] = f(x)\frac{d}{dx}g(x) + g(x)\frac{d}{dx}f(x)$$

The derivative of a product of two functions is the first function times the derivative of the second function plus the second function times the derivative of the first function.

(continued)

(f) The Quotient Rule

If f and g are both differentiable, then

$$\frac{d}{dx}\left[\frac{f(x)}{g(x)}\right] = \frac{g(x)\dfrac{d}{dx}f(x) - f(x)\dfrac{d}{dx}g(x)}{[g(x)]^2}$$

The derivative of a quotient of functions is the denominator times the derivative of the numerator minus the numerator times the derivative of the denominator, all divided by the square of the denominator.

(g) The Chain Rule

If g is differentiable at x and f is differentiable at $g(x)$, then the composite function defined by $F(x) = f(g(x))$ is differentiable at x and F' is given by the product $F'(x) = f'(g(x))g'(x)$. The derivative of a composite function is the derivative of the outer function evaluated at the inner function times the derivative of the inner function.

8. State the derivative of each function.

(a) $y = x^n$: $y' = nx^{n-1}$

(b) $y = e^x$: $y' = e^x$

(c) $y = b^x$: $y' = b^x \ln b$

(d) $y = \ln x$: $y' = 1/x$

(e) $y = \log_b x$: $y' = 1/(x \ln b)$

(f) $y = \sin x$: $y' = \cos x$

(g) $y = \cos x$: $y' = -\sin x$

(h) $y = \tan x$: $y' = \sec^2 x$

(i) $y = \csc x$: $y' = -\csc x \cot x$

(j) $y = \sec x$: $y' = \sec x \tan x$

(k) $y = \cot x$: $y' = -\csc^2 x$

(l) $y = \tan^{-1} x$: $y' = 1/(1 + x^2)$

9. (a) How is the number e defined?

e is the number such that $\displaystyle\lim_{h\to 0} \frac{e^h - 1}{h} = 1$.

(b) Express e as a limit.

$e = \displaystyle\lim_{x\to 0} (1 + x)^{1/x}$

(c) Why is the natural exponential function $y = e^x$ used more often in calculus than the other exponential functions $y = b^x$?

The differentiation formula for $y = b^x$ $[y' = b^x \ln b]$ is simplest when $b = e$ because $\ln e = 1$.

(d) Why is the natural logarithmic function $y = \ln x$ used more often in calculus than the other logarithmic functions $y = \log_b x$?

The differentiation formula for $y = \log_b x$ $[y' = 1/(x \ln b)]$ is simplest when $b = e$ because $\ln e = 1$.

10. (a) Explain how implicit differentiation works. When should you use it?

Implicit differentiation consists of differentiating both sides of an equation involving x and y with respect to x, and then solving the resulting equation for y'. Use implicit differentiation when it is difficult to solve an equation for y in terms of x.

(b) Explain how logarithmic differentiation works. When should you use it?

Logarithmic differentiation consists of taking natural logarithms of both sides of an equation $y = f(x)$, simplifying, differentiating implicitly with respect to x, and then solving the resulting equation for y'. Use logarithmic differentiation when the calculation of derivatives of complicated functions involving products, quotients, or powers can be simplified by taking logarithms.

11. Write an expression for the linearization of f at a.

$L(x) = f(a) + f'(a)(x - a)$.

12. Write an expression for the nth-degree Taylor polynomial of f centered at a.

$$T_n(x) = f(a) + f'(a)(x - a) + \frac{f''(a)}{2!}(x - a)^2 + \cdots + \frac{f^{(n)}(a)}{n!}(x - a)^n$$

Cut here and keep for reference

1. Explain the difference between an absolute maximum and a local maximum. Illustrate with a sketch.

The function value $f(c)$ is the absolute maximum value of f if $f(c)$ is the largest function value on the entire domain of f, whereas f is a local maximum value if it is the largest function value when x is near c.

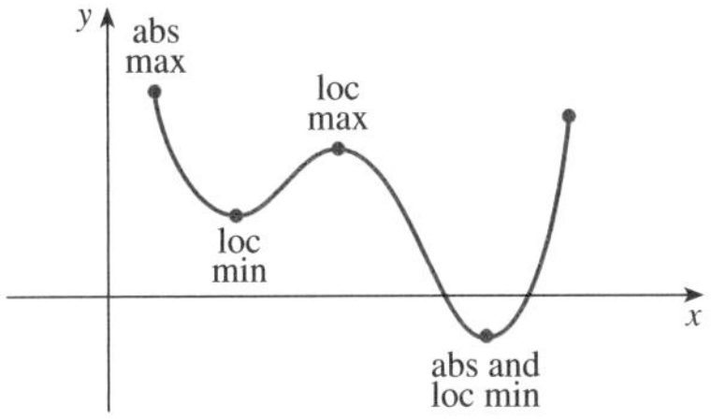

2. (a) What does the Extreme Value Theorem say?

The Extreme Value Theorem states that if f is a continuous function on a closed interval $[a, b]$, then it always attains an absolute maximum and an absolute minimum value on that interval.

(b) Explain how the Closed Interval Method works.

To find the absolute maximum and minimum values of a continuous function f on a closed interval $[a, b]$, we follow these three steps:

- Find the critical numbers of f in the interval (a, b) and compute the values of f at these numbers.
- Find the values of f at the endpoints of the interval.
- The largest of the output values from the previous two steps is the absolute maximum value; the smallest of these values is the absolute minimum value.

3. (a) State Fermat's Theorem.

Fermat's Theorem states that if f has a local maximum or minimum at c, and if $f'(c)$ exists, then $f'(c) = 0$.

(b) Define a critical number of f.

A critical number of a function f is a number c in the domain of f such that either $f'(c) = 0$ or $f'(c)$ does not exist.

4. State the Mean Value Theorem and give a geometric interpretation.

The Mean Value Theorem states that if f is a differentiable function on the interval $[a, b]$, then there exists a number c between a and b such that $f'(c) = \dfrac{f(b) - f(a)}{b - a}$.

Geometric interpretation: There is some point P on the graph of a function f [on the interval (a, b)] where the tangent line is parallel to the secant line that connects $(a, f(a))$ and $(b, f(b))$.

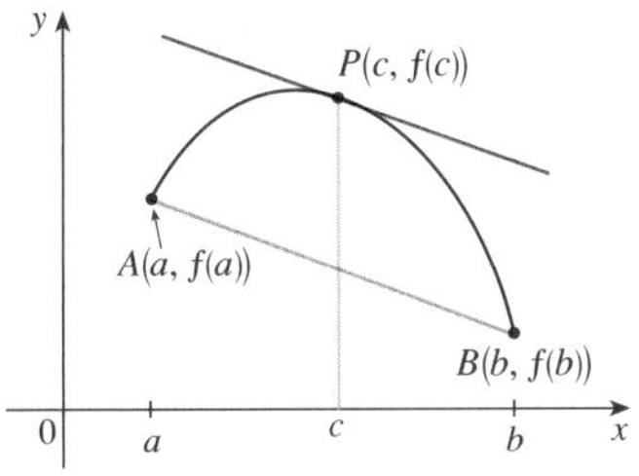

5. (a) State the Increasing/Decreasing Test.

If $f'(x) > 0$ on an interval, then f is increasing on that interval.

If $f'(x) < 0$ on an interval, then f is decreasing on that interval.

(b) What does it mean to say that f is concave upward on an interval I?

A function f is concave upward on an interval I if f' is an increasing function on I (or, equivalently, the graph of f lies above all of its tangent lines on I).

(c) State the Concavity Test.

If $f''(x) > 0$ on an interval, then the graph of f is concave upward on that interval.

If $f''(x) < 0$ on an interval, then the graph of f is concave downward on that interval.

(d) What are inflection points? How do you find them?

Inflection points on the graph of a continuous function f are points where the curve changes from concave upward to concave downward or from concave downward to concave upward. They can be found by determining the values at which the second derivative changes sign.

6. (a) State the First Derivative Test.

Suppose that c is a critical number of a continuous function f.

- If f' changes from positive to negative at c, then f has a local maximum at c.
- If f' changes from negative to positive at c, then f has a local mimimum at c.
- If f' does not change sign at c, then f has no local maximum or minimum at c.

(b) State the Second Derivative Test.

Suppose f'' is continuous near c.

- If $f'(c) = 0$ and $f''(c) > 0$, then f has a local minimum at c.
- If $f'(c) = 0$ and $f''(c) < 0$, then f has a local maximum at c.

(c) What are the relative advantages and disadvantages of these tests?

The Second Derivative Test is sometimes easier to use, but it is inconclusive when $f''(c) = 0$ and fails if $f''(c)$ does not exist. In either case the First Derivative Test must be used.

7. (a) What does l'Hospital's Rule say?

L'Hospital's Rule states that if f and g are differentiable functions, $g'(x) \neq 0$ near a (except possibly at a), and $\lim_{x \to a} f(x)/g(x)$ is an indeterminate form of type $\frac{0}{0}$ or ∞/∞, then

$$\lim_{x \to a} \frac{f(x)}{g(x)} = \lim_{x \to a} \frac{f'(x)}{g'(x)}$$ provided the right side limit exists.

(continued)

(b) How can you use l'Hospital's Rule if you have a product $f(x)\,g(x)$ where $f(x) \to 0$ and $g(x) \to \infty$ as $x \to a$?

Write fg as $\dfrac{f}{1/g}$ or $\dfrac{g}{1/f}$.

(c) How can you use l'Hospital's Rule if you have a difference $f(x) - g(x)$ where $f(x) \to \infty$ and $g(x) \to \infty$ as $x \to a$?

Convert the difference into a quotient using a common denominator, rationalizing, factoring, or some other method.

8. (a) What is an equilibrium of the recursive sequence $x_{t+1} = f(x_t)$?

An equilibrium of the recursive sequence $x_{t+1} = f(x_t)$ is a number $\hat{x}$ that is left unchanged by the function f, that is, $f(\hat{x}) = \hat{x}$.

(b) What is a stable equilibrium? An unstable equilibrium?

An equilibrium is stable if solutions that begin close to the equilibrium approach that equilibrium. An equilibrium is unstable if solutions that start close to the equilibrium move away from it.

(c) State the Stability Criterion.

If $\hat{x}$ is an equilibrium of the recursive sequence $x_{t+1} = f(x_t)$, where f' is continuous, and $|f'(\hat{x})| < 1$, then the equilibrium is stable. If $|f'(\hat{x})| > 1$, then the equilibrium is unstable.

9. (a) What is an antiderivative of a function f?

A function F is an antiderivative of f on an interval I if $F'(x) = f(x)$ for all x in I.

(b) Suppose F_1 and F_2 are both antiderivatives of f on an interval I. How are F_1 and F_2 related?

If F_1 and F_2 are both antiderivatives of f on an interval I, then they differ by a constant.

1. (a) Write an expression for a Riemann sum of a function f. Explain the meaning of the notation that you use.

$\Sigma_{i=1}^{n} f(x_i^*)\Delta x$ is an expression for a Riemann sum of a function f, where x_i^* is a point in the ith subinterval $[x_{i-1}, x_i]$ and Δx is the length of the subintervals.

(b) If $f(x) \geqslant 0$, what is the geometric interpretation of a Riemann sum? Illustrate with a diagram.

If f is positive, then a Riemann sum can be interpreted as the sum of areas of approximating rectangles, as shown in the figure.

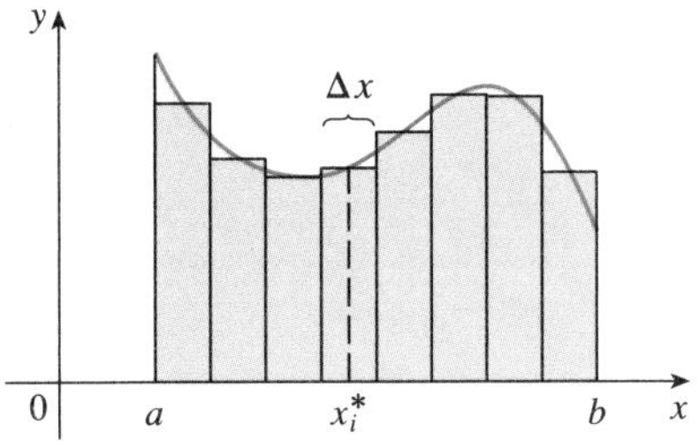

(c) If $f(x)$ takes on both positive and negative values, what is the geometric interpretation of a Riemann sum? Illustrate with a diagram.

If f takes on both positive and negative values, as in the figure below, then the Riemann sum $\Sigma f(x_i^*)\Delta x$ is the sum of the areas of the rectangles that lie above the x-axis and the *negatives* of the areas of the rectangles that lie below the x-axis (the areas of the blue rectangles *minus* the areas of the gold rectangles).

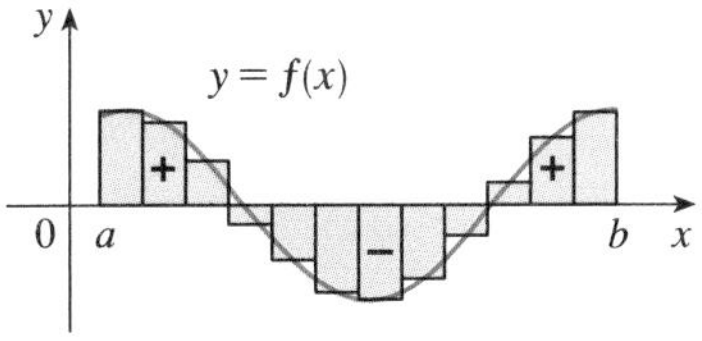

2. (a) Write the definition of the definite integral of a continuous function from a to b.

If f is a function defined for $a \leqslant x \leqslant b$, we divide the interval $[a, b]$ into n subintervals of equal width $\Delta x = (b - a)/n$. We let $x_0\ (= a), x_1, x_2, \ldots, x_n\ (= b)$ be the endpoints of these subintervals and we let $x_1^*, x_2^*, \ldots, x_n^*$ be any sample points in these subintervals, so x_i^* lies in the ith subinterval $[x_{i-1}, x_i]$. Then the definite integral of f from a to b is

$$\int_a^b f(x)\,dx = \lim_{n\to\infty} \sum_{i=1}^{n} f(x_i^*)\,\Delta x$$

provided that this limit exists.

(b) What is the geometric interpretation of $\int_a^b f(x)\,dx$ if $f(x) \geqslant 0$?

If f is positive, then $\int_a^b f(x)\,dx$ can be interpreted as the area under the graph of $y = f(x)$ and above the x-axis for $a \leqslant x \leqslant b$.

(c) What is the geometric interpretation of $\int_a^b f(x)\,dx$ if $f(x)$ takes on both positive and negative values? Illustrate with a diagram.

In this case $\int_a^b f(x)\,dx$ can be interpreted as a "net area," that is, the area of the region above the x-axis and below the graph of f (labeled "+" in the figure) minus the area of the region below the x-axis and above the graph of f (labeled "−").

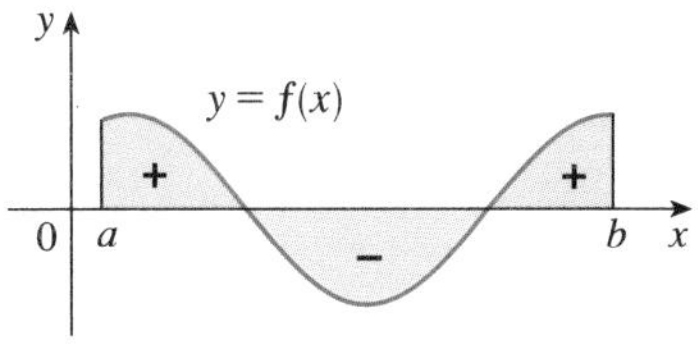

3. State the Midpoint Rule.

If f is a continuous function on the interval $[a, b]$ and we divide $[a, b]$ into n subintervals of equal width $\Delta x = (b - a)/n$, then

$$\int_a^b f(x)\,dx \approx \sum_{i=1}^{n} f(\bar{x}_i)\,\Delta x = \Delta x\,[f(\bar{x}_1) + \cdots + f(\bar{x}_n)]$$

where $\bar{x}_i = \frac{1}{2}(x_{i-1} + x_i) =$ midpoint of $[x_{i-1}, x_i]$.

4. (a) State the Evaluation Theorem.

The Evaluation Theorem says that if f is continuous on the interval $[a, b]$, then $\int_a^b f(x)\,dx = F(b) - F(a)$, where F is any antiderivative of f, that is, $F' = f$.

(b) State the Net Change Theorem.

The Net Change Theorem says that the integral of a rate of change is the net change: $\int_a^b F'(x)\,dx = F(b) - F(a)$.

5. If $r(t)$ is the rate of growth of a population at time t, where t is measured in months, what does $\int_6^{10} r(t)\,dt$ represent?

$\int_6^{10} r(t)\,dt$ is the change in population size from month 6 to month 10.

6. (a) Explain the meaning of the indefinite integral $\int f(x)\,dx$.

The indefinite integral $\int f(x)\,dx$ is another name for an antiderivative of f, so $\int f(x)\,dx = F(x)$ means that $F'(x) = f(x)$.

(b) What is the connection between the definite integral $\int_a^b f(x)\,dx$ and the indefinite integral $\int f(x)\,dx$?

The connection is given by the Fundamental Theorem:

$$\int_a^b f(x)\,dx = \int f(x)\,dx\Big]_a^b$$

if f is continuous.

7. State both parts of the Fundamental Theorem of Calculus.

Suppose f is continuous on $[a, b]$. The Fundamental Theorem of Calculus says

1. If $g(x) = \int_a^x f(t)\,dt$, then $g'(x) = f(x)$.

2. $\int_a^b f(x)\,dx = F(b) - F(a)$, where F is any antiderivative of f, that is, $F' = f$.

8. (a) State the Substitution Rule. In practice, how do you use it?

The substitution rule says that if $u = g(x)$ is a differentiable function whose range is an interval I and f is continuous on I, then $\int f(g(x))\,g'(x)\,dx = \int f(u)\,du$. In practice, we make the substitutions $u = g(x)$ and $du = g'(x)\,dx$ in the integrand to make the integral simpler to evaluate.

(continued)

(b) State the rule for integration by parts. In practice, how do you use it?

The rule for integration by parts states that $\int f(x)g'(x)\,dx = f(x)g(x) - \int g(x)f'(x)\,dx$. In practice, we try to choose $u = f(x)$ to be a function that becomes simpler when differentiated (or at least not more complicated) as long as $dv = g'(x)\,dx$ can be readily integrated to give v. Then the original integral $\int u\,dv$ becomes $uv - \int v\,du$.

9. Define the following improper integrals.

(a) $\int_a^\infty f(x)\,dx$

If $\int_a^b f(x)\,dx$ exists for every number $b \geqslant a$, then $\int_a^\infty f(x)\,dx = \lim_{b\to\infty} \int_a^b f(x)\,dx$ provided this limit exists (as a finite number).

(b) $\int_{-\infty}^b f(x)\,dx$

If $\int_a^b f(x)\,dx$ exists for every number $a \leqslant b$, then $\int_{-\infty}^b f(x)\,dx = \lim_{a\to-\infty} \int_a^b f(x)\,dx$ provided this limit exists (as a finite number).

(c) $\int_{-\infty}^\infty f(x)\,dx$

If both $\int_c^\infty f(x)\,dx$ and $\int_{-\infty}^c f(x)\,dx$ are convergent, then we define $\int_{-\infty}^\infty f(x)\,dx = \int_{-\infty}^c f(x)\,dx + \int_c^\infty f(x)\,dx$, where c is any real number.

10. Explain exactly what is meant by the statement that "differentiation and integration are inverse processes."

The Fundamental Theorem of Calculus (or, equivalently, the Net Change Theorem) states that $\int_a^b F'(x)\,dx = F(b) - F(a)$. This says that if we take a function F, first differentiate it, and then integrate the result, we arrive back at the original function, but in the form $F(b) - F(a)$. Also, the indefinite integral $\int f(x)\,dx$ represents an antiderivative of f, so $\dfrac{d}{dx}\int f(x)\,dx = f(x)$. Here we first integrate a function, then differentiate the result, and arrive back at the original function.

1. Draw two typical curves $y = f(x)$ and $y = g(x)$, where $f(x) \geqslant g(x)$ for $a \leqslant x \leqslant b$. Show how to approximate the area between these curves by a Riemann sum and sketch the corresponding approximating rectangles. Then write an expression for the exact area.

If we divide the interval $[a, b]$ into n subintervals of equal width Δx and label the endpoints of the subintervals $x_0, x_1, x_2, \ldots, x_n$, then the area between $f(x)$ and $g(x)$ can be approximated by the Riemann sum $\Sigma_{i=1}^{n} [f(x_i^*) - g(x_i^*)]\,\Delta x$, where x_i^* lies in the interval $[x_{i-1}, x_i]$. In the limit as $n \to \infty$ the Riemann sum approximation gives the exact area

$$A = \lim_{n\to\infty} \sum_{i=1}^{n} [f(x_i^*) - g(x_i^*)]\,\Delta x = \int_a^b [f(x) - g(x)]\,dx$$

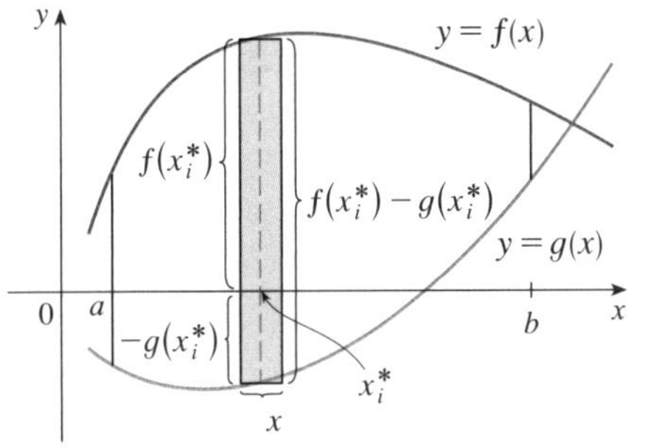

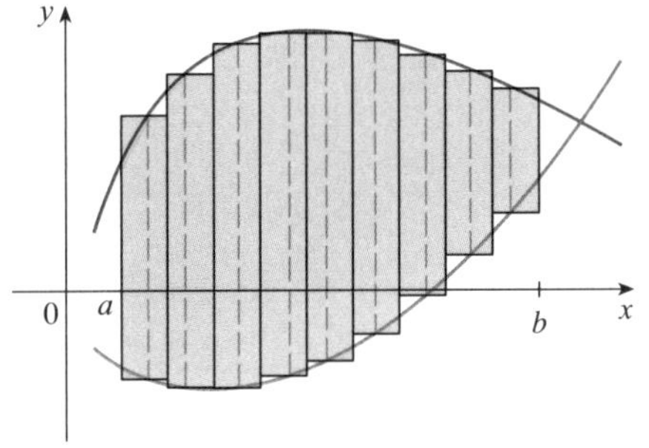

2. Suppose that Sue runs faster than Kathy throughout a 1500-meter race. What is the physical meaning of the area between their velocity curves for the first minute of the race?

It represents the number of meters by which Sue is ahead of Kathy after 1 minute.

3. (a) What is the average value of a function f on an interval $[a, b]$?

The average value of a function f on an interval $[a, b]$ is $f_{\text{ave}} = \dfrac{1}{b - a}\displaystyle\int_a^b f(x)\,dx$.

(b) What does the Mean Value Theorem for Integrals say? What is its geometric interpretation?

The Mean Value Theorem for Integrals says that if f is continuous on $[a, b]$, then there exists a number c in $[a, b]$ such that $\int_a^b f(x)\,dx = f(c)(b - a)$. The geometric interpretation is that, for *positive* functions f, there is a number c such that the rectangle with base $[a, b]$ and height $f(c)$ has the same area as the region under the graph of f from a to b.

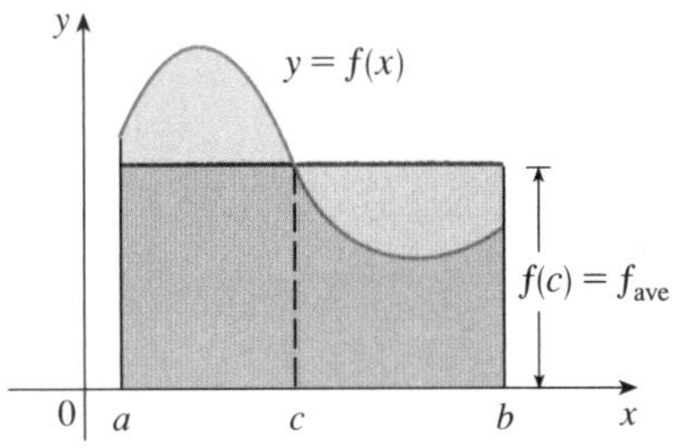

4. If we have survival and renewal functions for a population, how do we predict the size of the population T years from now?

If a population begins with P_0 members, new members are added at a rate given by the renewal function $R(t)$, where t is measured in years, and the proportion of the population that remains after t years is given by the survival function $S(t)$, then the population T years from now is given by

$$P(T) = S(T) \cdot P_0 + \int_0^T S(T - t)R(t)\,dt$$

5. (a) What is the cardiac output of the heart?

The cardiac output of the heart is the volume of blood pumped by the heart per unit time.

(b) Explain how the cardiac output can be measured by the dye dilution method.

Dye is injected into part of the heart and a probe measures the concentration of the dye leaving the heart over a time interval $[0, T]$ until the dye has cleared. If $c(t)$ is the concentration of the dye at time t, then the cardiac output is given by

$$F = \frac{A}{\int_0^T c(t)\,dt}$$

where A is the amount of dye used.

6. (a) Suppose S is a solid with known cross-sectional areas. Explain how to approximate the volume of S by a Riemann sum. Then write an expression for the exact volume.

If we divide S into n "slabs" of equal width Δx, label the endpoints of the slabs $x_0, x_1, x_2, \ldots, x_n$, and the cross-sectional area $A(x)$ of the solid is known, then the volume of S can be approximated by the Riemann sum $\sum_{i=1}^{n} A(x_i^*)\,\Delta x$, where x_i^* lies in the interval $[x_{i-1}, x_i]$. In the limit as $n \to \infty$ the Riemann sum approximation gives the exact volume

$$V = \lim_{n\to\infty} \sum_{i=1}^{n} A(x_i^*)\,\Delta x = \int_a^b A(x)\,dx$$

(b) If S is a solid of revolution, how do you find the cross-sectional areas?

If the cross-section is a disk, we find the radius in terms of x or y and use $A = \pi(\text{radius})^2$. If the cross-section is a washer, we find the inner radius r_{in} and outer radius r_{out} and use $A = \pi(r_{\text{out}}^2) - \pi(r_{\text{in}}^2)$.

1. (a) What is a differential equation?

A differential equation is an equation that contains an unknown function and one or more of its derivatives.

(b) What is the order of a differential equation?

The order of a differential equation is the order of the highest derivative that occurs in the equation.

(c) What is an initial condition?

An initial condition is a condition of the form $y(t_0) = y_0$.

(d) What are the differences between pure-time, autonomous, and nonautonomous differential equations?

Pure-time differential equations involve the derivative of the function but not the function itself and can be expressed in the form

$$\frac{dy}{dt} = f(t)$$

Autonomous differential equations involve both the derivative of the function and the function itself, but have no explicit dependence on the independent variable. They can be written as

$$\frac{dy}{dt} = g(y)$$

Nonautonomous differential equations are a combination of pure-time and autonomous differential equations. They involve the function, its derivative, and the independent variable explicitly.

2. What can you say about the solutions of the equation $y' = x^2 + y^2$ just by looking at the differential equation?

$y' = x^2 + y^2 \geqslant 0$ for all x and y. Also, $y' = 0$ only at the origin, so there is a horizontal tangent at $(0, 0)$, but nowhere else. The graph of the solution is increasing on every interval.

3. What is a phase plot for the differential equation $y' = g(y)$?

A phase plot helps to visualize the equilibria of $y' = g(y)$ and to determine the stability properties of the equilibria. A phase plot is constructed by graphing $g(y)$ as a function of y and drawing arrows on the horizontal axis to indicate the direction of change of y. Arrows point to the right when the curve lies above the horizontal axis and to the left when the curve lies below the horizontal axis. The points where the curve intersects the horizontal axis are the equilibrium points.

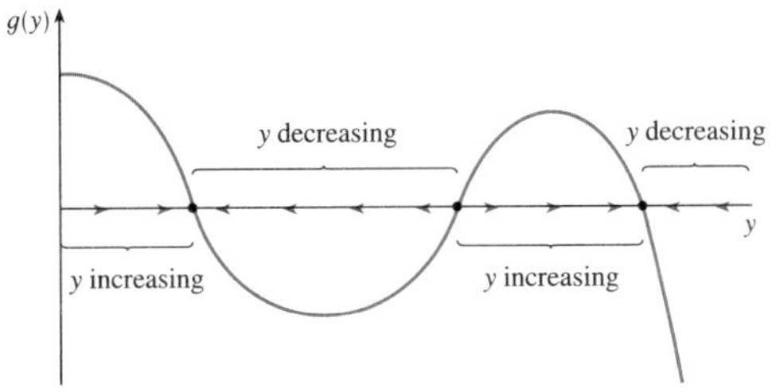

4. What is a direction field for the differential equation $y' = F(x, y)$?

A direction field is a plot that helps visualize the general shape of the solution curves for the differential equation $y' = F(x, y)$. The direction field can be constructed by drawing short line segments with slope $F(x, y)$ at several points (x, y). These line segments indicate the direction in which a solution curve is moving.

5. Explain how Euler's method works.

Euler's method is used to approximate values for the solution of the initial-value problem $y' = F(t, y)$, $y(t_0) = t_0$. The method involves starting at the initial point (t_0, y_0), moving a short distance in the direction indicated by the direction field, stopping and proceeding in the new direction given by the direction field, then repeat. More formally, if the step size is h, the approximate values at $t_{n+1} = t_n + h$ are $y_{n+1} = y_n + hF(t_n, y_n)$.

6. What is a separable differential equation? How do you solve it?

It is a first-order differential equation in which the expression for dy/dx can be factored as a function of x times a function of y, that is, $dy/dx = g(x)f(y)$. We can solve the equation by integrating both sides of the equation $dy/f(y) = g(x)\,dx$ and solving for y.

7. (a) Write the logistic equation.

$dN/dt = rN\left(1 - \dfrac{N}{K}\right)$, where K is the carrying capacity.

(b) Under what circumstances is this an appropriate model for population growth?

It is an appropriate model for population growth if the population grows at a rate proportional to the size of the population in the beginning, but eventually levels off and approaches its carrying capacity because of limited resources.

8. (a) Write Lotka-Volterra equations to model populations of sharks S and their food F.

$$\frac{dF}{dt} = rF - aFS \qquad \frac{dS}{dt} = -kS + bFS$$

(b) What do these equations say about each population in the absence of the other?

In the absence of sharks, the food grows exponentially, that is, $dF/dt = rF$, where r is a positive constant. In the absence of food, the shark population declines at a rate proportional to itself, that is, $dS/dt = -kS$, where k is a positive constant.

9. What is a nullcline?

A nullcline of a specific variable is a curve along which that variable does not change. For example, if $dx/dt = f(x, y)$ then the x-nullclines are the curves in the xy-plane that satisfy $f(x, y) = 0$. Along these curves, $dx/dt = 0$.

(continued)

10. (a) Write Lotka-Volterra competition equations for two competing fish species, x and y.

$$\frac{dx}{dt} = xr\left(1 - \frac{x + \alpha y}{K_x}\right) \qquad \frac{dy}{dt} = yr\left(1 - \frac{y + \beta x}{K_y}\right)$$

(b) What would the nullclines have to look like for species x to always outcompete species y?

The nullclines must result in movement away from equilibria (i) and (ii), but toward equilibrium (iii), as seen in the phase plane at the right.

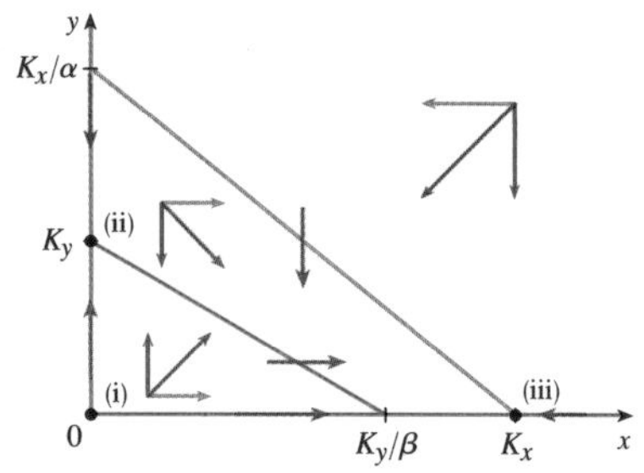

CHAPTER 8 CONCEPT CHECK ANSWERS

1. What is the difference between a vector and a scalar?

A scalar is a real number, whereas a vector is a quantity that has both a magnitude and a direction.

2. How do you add two vectors geometrically? How do you add them algebraically?

To add two vectors geometrically, we can use either the Triangle Law or the Parallelogram Law, as illustrated in the figures below.

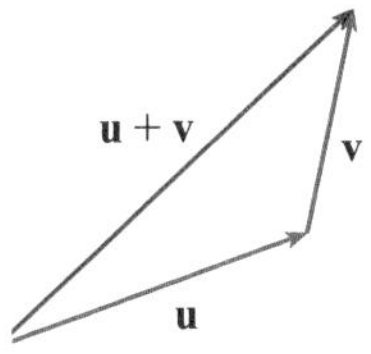

The Triangle Law

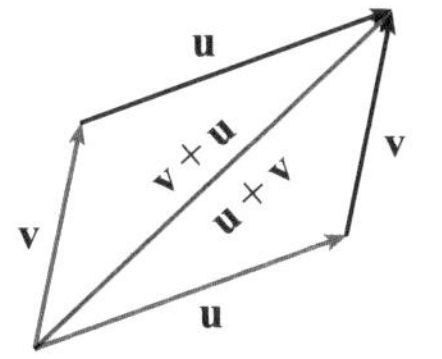

The Parallelogram Law

Algebraically, we add the corresponding components of the vectors.

3. If $\mathbf{a}$ is a vector and c is a scalar, how is $c\mathbf{a}$ related to $\mathbf{a}$ geometrically? How do you find $c\mathbf{a}$ algebraically?

$c\mathbf{a}$ is the vector whose length is $|c|$ times the length of $\mathbf{a}$ and whose direction is the same as $\mathbf{a}$ if $c > 0$ and is opposite to $\mathbf{a}$ if $c < 0$. If $c = 0$ or $\mathbf{a} = \mathbf{0}$, then $c\mathbf{a} = 0$. Algebraically, to find $c\mathbf{a}$ we multiply each component of $\mathbf{a}$ by c.

4. How do you find the vector from one point to another algebraically?

The vector from point $A(x_1, y_1, z_1)$ to point $B(x_2, y_2, z_2)$ is given by $\overrightarrow{AB} = [x_2 - x_1, y_2 - y_1, z_2 - z_1]$.

5. How do you find the dot product $\mathbf{a} \cdot \mathbf{b}$ of two vectors if you know their lengths and the angle between them? What if you know their components?

When the lengths of $\mathbf{a}$ and $\mathbf{b}$ and the angle θ between them are known, the dot product can be computed using

$$\mathbf{a} \cdot \mathbf{b} = |\mathbf{a}||\mathbf{b}| \cos \theta$$

If the components of $\mathbf{a} = [a_1, a_2, a_3]$ and $\mathbf{b} = [b_1, b_2, b_3]$ are known, then the dot product is given by

$$\mathbf{a} \cdot \mathbf{b} = a_1b_1 + a_2b_2 + a_3b_3$$

That is, we multiply corresponding components and add.

6. How are dot products useful?

The dot product can be used to find the angle between two vectors, to calculate the scalar or vector projection of one vector onto another, and to write equations of planes. In particular, the dot product can determine if two vectors are orthogonal. Also, the dot product can be used to analyze genome expression profiles for biological discovery and to diagnose cardiac problems using vectorcardiography.

7. Write expressions for the scalar and vector projections of $\mathbf{b}$ onto $\mathbf{a}$. Illustrate with diagrams.

Scalar projection of $\mathbf{b}$ onto $\mathbf{a}$: $\text{comp}_{\mathbf{a}}\,\mathbf{b} = \dfrac{\mathbf{a} \cdot \mathbf{b}}{|\mathbf{a}|}$

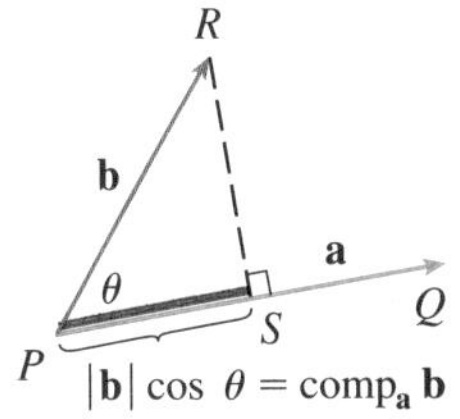

Vector projection of $\mathbf{b}$ onto $\mathbf{a}$: $\text{proj}_{\mathbf{a}}\,\mathbf{b} = \left(\dfrac{\mathbf{a} \cdot \mathbf{b}}{|\mathbf{a}|}\right)\dfrac{\mathbf{a}}{|\mathbf{a}|} = \dfrac{\mathbf{a} \cdot \mathbf{b}}{|\mathbf{a}|^2}\,\mathbf{a}$

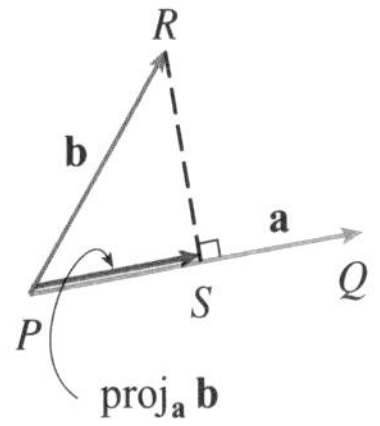

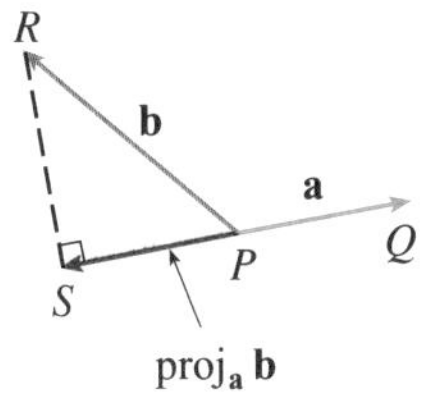

8. What is the equation of a sphere?

An equation of a sphere with center $C(h, k, l)$ and radius r is $(x - h)^2 + (y - k)^2 + (z - l)^2 = r^2$.

9. (a) How do you tell if two vectors are parallel?

Two (nonzero) vectors are parallel if and only if one is a scalar multiple of the other. In addition, two nonzero vectors are parallel if and only if their dot product is the product of their lengths.

(b) How do you tell if two vectors are perpendicular?

Two vectors are perpendicular if and only if their dot product is 0.

10. What is a symmetric matrix?

A matrix A is called symmetric if $A^T = A$.

11. If a matrix A rotates vectors counterclockwise by θ degrees, what does the matrix A^{-1} do?

The inverse matrix A^{-1} rotates vectors by θ degrees clockwise. In other words, it reverses the rotation of the matrix A.

12. If a 2×2 matrix has complex eigenvalues, what does this matrix do to vectors upon multiplication?

If a 2×2 matrix has complex eigenvalues $\lambda = a \pm bi$, it rotates vectors by an angle $\theta = \tan^{-1}(b/a)$ and stretches them by a factor $|\lambda| = \sqrt{a^2 + b^2}$.

13. Suppose A is a matrix and k is a positive integer. What does the notation A^k mean?

The notation A^k indicates that the matrix A is multiplied by itself k times, that is, $A^k = \underbrace{AA \cdots A}_{k \text{ times}}$.

(continued)

14. What is the relationship between the inverse of a matrix and the determinant of a matrix?

An $n \times n$ matrix A is invertible if and only if $\det A \neq 0$. That is, there exists a matrix inverse A^{-1} such that $AA^{-1} = I$, provided that $\det A \neq 0$.

15. Explain what eigenvalues and eigenvectors are.

An eigenvalue is a scalar quantity λ that, together with a nonzero vector $\mathbf{v}$ called an eigenvector, satisfies the equation $A\mathbf{v} = \lambda\mathbf{v}$ where A is an $n \times n$ matrix. Consequently, an eigenvalue must satisfy the equation $\det(A - \lambda I) = 0$.

16. Why does a 2×2 matrix have two eigenvalues?

The eigenvalues of a 2×2 matrix are determined by solving a quadratic characteristic polynomial. In general, there are two solutions to a quadratic equation, so there are two eigenvalues of a 2×2 matrix, though one of them may be repeated twice.

17. Suppose a 2×2 matrix with real entries has complex eigenvalues. Why must the eigenvalues be complex conjugates?

In general, the eigenvalues of a 2×2 matrix satisfy a second-degree characteristic polynomial of the form $\lambda^2 + b\lambda + c = 0$ with solutions given by the quadratic formula

$$x = \frac{-b \pm \sqrt{b^2 - 4c}}{2}$$

When the discriminant, $b^2 - 4c$, is negative, the solutions will be complex and have the form $x = -\frac{b}{2} \pm \frac{\sqrt{4c - b^2}}{2}i$. Hence, these solutions are complex-conjugate pairs.

18. Why is it sometimes useful to write a matrix A in the form $A = PDP^{-1}$, where D is a diagonal matrix?

The matrix decomposition $A = PDP^{-1}$ is useful when computing powers of the matrix A. For example, $A^k = PD^kP^{-1}$ and, since D is a diagonal matrix, D^k can be calculated by raising each entry of D to the power k. This provides a straightfoward method for raising a matrix to a power.

19. What does the Perron-Frobenius Theorem say, and why is it useful?

The Perron-Frobenius Theorem states that if A is a primitive $n \times n$ matrix whose entries are all nonnegative then

- There exists an eigenvalue of A, call it λ_1, that is real and positive.
- λ_1 is greater in magnitude than all other eigenvalues.
- The components of the eigenvector associated with λ_1 are all positive.

The Perron-Frobenius Theorem can be used to determine the long-term behavior of the recursion $\mathbf{n}_{t+1} = A\mathbf{n}_t$ without explicitly finding a solution.

Cut here and keep for reference

1. (a) What is a function of two variables?

It is a rule that assigns to each ordered pair of real numbers (x, y) in its domain a unique real number denoted by $f(x, y)$.

(b) Describe two methods for visualizing a function of two variables.

One way to visualize a function of two variables is by graphing it, resulting in the surface $z = f(x, y)$. Another method for visualizing a function of two variables is a contour map. The contour map consists of level curves of the function, with equations $f(x, y) = k$, which are horizontal traces of the graph of the function projected onto the xy-plane.

2. What does

$$\lim_{(x, y)\to(a, b)} f(x, y) = L$$

mean? How can you show that such a limit does not exist?

$\lim_{(x, y)\to(a, b)} f(x, y) = L$ means the values of $f(x, y)$ approach the number L as the point (x, y) approaches the point (a, b) along any path that is within the domain of f. We can show that a limit at a point does not exist by finding two different paths approaching the point along which $f(x, y)$ has different limits.

3. (a) What does it mean to say that f is continuous at (a, b)?

f is continuous at (a, b) if $\lim_{(x, y)\to(a, b)} f(x, y) = f(a, b)$.

(b) If f is continuous on $\mathbb{R}^2$, what can you say about its graph?

If f is continuous on $\mathbb{R}^2$, its graph will appear as a surface without holes or breaks.

4. (a) Write expressions for the partial derivatives $f_x(a, b)$ and $f_y(a, b)$ as limits.

$$f_x(a, b) = \lim_{h\to0} \frac{f(a + h, b) - f(a, b)}{h}$$

$$f_y(a, b) = \lim_{h\to0} \frac{f(a, b + h) - f(a, b)}{h}$$

(b) How do you interpret $f_x(a, b)$ and $f_y(a, b)$ geometrically? How do you interpret them as rates of change?

The partial derivative $f_x(a, b)$ is the slope of the tangent line at the point $(a, b, f(a, b))$ to the vertical trace in the plane $y = b$, and $f_y(a, b)$ is the slope of the tangent line to the trace in the plane $x = a$. Also $f_x(a, b)$ represents the rate of change of f with respect to x (at $x = a$) when y is fixed ($y = b$), and $f_y(a, b)$ is the rate of change of f with respect to y (at $y = b$) when x is fixed ($x = a$).

(c) If $f(x, y)$ is given by a formula, how do you calculate f_x and f_y?

To find f_x, regard y as a constant and differentiate $f(x, y)$ with respect to x. To find f_y, regard x as a constant and differentiate $f(x, y)$ with respect to y.

5. What does Clairaut's Theorem say?

Suppose f is defined on a disk D that contains the point (a, b). If the functions f_{xy} and f_{yx} are both continuous on D, then $f_{xy}(a, b) = f_{yx}(a, b)$.

6. How do you find an equation for the tangent plane to a surface $z = f(x, y)$?

If f has continuous partial derivatives, then an equation of the tangent plane to the surface $z = f(x, y)$ at the point $P(x_0, y_0, z_0)$ is $z - z_0 = f_x(x_0, y_0)(x - x_0) + f_y(x_0, y_0)(y - y_0)$.

7. Define the linearization of f at (a, b). What is the corresponding linear approximation? What is the geometric interpretation of the linear approximation?

The linearization of f at (a, b) is

$$L(x, y) = f(a, b) + f_x(a, b)(x - a) + f_y(a, b)(y - b)$$

and the corresponding linear approximation is

$$f(x, y) \approx f(a, b) + f_x(a, b)(x - a) + f_y(a, b)(y - b)$$

We can interpret the linearization of f at (a, b) geometrically as the linear function whose graph is the tangent plane to the graph of f at (a, b). Thus it is the linear function that best approximates f near (a, b).

8. (a) What does it mean to say that f is differentiable at (a, b)?

The function $f(x, y)$ is differentiable at (a, b) if

$$\lim_{(x, y)\to(a, b)} \frac{|f(x, y) - L(x, y)|}{\sqrt{(x - a)^2 + (y - b)^2}} = 0$$

(b) How do you usually verify that f is differentiable?

We check that the partial derivatives f_x and f_y exist and are continuous in some region D. If this is the case, then f is differentiable in D.

9. State the Chain Rule for the case where $z = f(x, y)$ and x and y are functions of a variable t.

$$\frac{dz}{dt} = \frac{\partial z}{\partial x}\frac{dx}{dt} + \frac{\partial z}{\partial y}\frac{dy}{dt}$$

10. If z is defined implicitly as a function of x and y by an equation of the form $F(x, y, z) = 0$, how do you find $\partial z/\partial x$ and $\partial z/\partial y$?

Provided $\partial F/\partial z \neq 0$, the partial derivatives are given by

$$\frac{\partial z}{\partial x} = -\frac{\dfrac{\partial F}{\partial x}}{\dfrac{\partial F}{\partial z}} \qquad \frac{\partial z}{\partial y} = -\frac{\dfrac{\partial F}{\partial y}}{\dfrac{\partial F}{\partial z}}$$

11. (a) Write an expression as a limit for the directional derivative of f at (x_0, y_0) in the direction of a unit vector $\mathbf{u} = [a, b]$. How do you interpret it as a rate? How do you interpret it geometrically?

$$D_{\mathbf{u}} f(x_0, y_0) = \lim_{h\to0} \frac{f(x_0 + ha, y_0 + hb) - f(x_0, y_0)}{h}$$

We can interpret the directional derivative as the rate of change of f at (x_0, y_0) in the direction of $\mathbf{u}$. Geometrically, if P is the point $(x_0, y_0, f(x_0, y_0))$ on the graph of f and C is the curve of intersection of the graph of f with the vertical plane that passes through P in the direction $\mathbf{u}$, the directional derivative of f at (x_0, y_0) in the direction of $\mathbf{u}$ is the slope of the tangent line to C at P. (See Figure 9.5.3.)

(continued)

(b) If f is differentiable, write an expression for $D_{\mathbf{u}}f(x_0, y_0)$ in terms of f_x and f_y.

$$D_{\mathbf{u}}f(x, y) = f_x(x, y)\,a + f_y(x, y)\,b \quad \text{where } \mathbf{u} = [a, b]$$

12. (a) Define the gradient vector ∇f for a function f of two variables.

$$\nabla f(x, y) = [f_x(x, y), f_y(x, y)]$$

(b) Express $D_{\mathbf{u}}f$ in terms of ∇f.

$$D_{\mathbf{u}}f(x, y) = \nabla f(x, y) \cdot \mathbf{u}$$

(c) Explain the geometric significance of the gradient.

The gradient vector of a function points in the direction of maximum rate of increase of the function. On a graph of the function, the gradient points in the direction of steepest ascent

13. What do the following statements mean?

(a) f has a local maximum at (a, b).

f has a local maximum at (a, b) if $f(x, y) \leqslant f(a, b)$ when (x, y) is near (a, b).

(b) f has an absolute maximum at (a, b).

f has an absolute maximum at (a, b) if $f(x, y) \leqslant f(a, b)$ for all points (x, y) in the domain of f.

(c) f has a local minimum at (a, b).

f has a local minimum at (a, b) if $f(x, y) \geqslant f(a, b)$ when (x, y) is near (a, b).

(d) f has an absolute minimum at (a, b).

f has an absolute minimum at (a, b) if $f(x, y) \geqslant f(a, b)$ for all points (x, y) in the domain of f.

(e) f has a saddle point at (a, b).

f has a saddle point at (a, b) if $f(a, b)$ is a local maximum in one direction but a local minimum in another.

14. (a) If f has a local maximum at (a, b), what can you say about its partial derivatives at (a, b)?

If f has a local maximum at (a, b) and the first-order partial derivatives of f exist there, then $f_x(a, b) = 0$ and $f_y(a, b) = 0$.

(b) What is a critical point of f?

A point (a, b) is a critical point of f if $f_x(a, b) = 0$ and $f_y(a, b) = 0$, or if one of these partial derivatives does not exist.

15. State the Second Derivatives Test.

Suppose that f_x and f_y are continuous on a disk with center (a, b), and that $f_x(a, b) = 0$ and $f_y(a, b) = 0$. Let

$$D = f_{xx}(a, b)\,f_{yy}(a, b) - [f_{xy}(a, b)]^2$$

- If $D > 0$ and $f_{xx}(a, b) > 0$, then $f(a, b)$ is a local minimum.
- If $D > 0$ and $f_{xx}(a, b) < 0$, then $f(a, b)$ is a local maximum.
- If $D < 0$, then f has a saddle point at (a, b).
- If $D = 0$, then the test gives no information.

16. (a) What is a closed set in $\mathbb{R}^2$? What is a bounded set?

A closed set in $\mathbb{R}^2$ is a set that contains all its boundary points. A bounded set in $\mathbb{R}^2$ is one that is contained within some disk. In other words, it is finite in extent.

(b) State the Extreme Value Theorem for functions of two variables.

If f is continuous on a closed, bounded set D in $\mathbb{R}^2$, then f attains an absolute maximum value $f(x_1, y_1)$ and an absolute minimum value $f(x_2, y_2)$ at some points (x_1, y_1) and (x_2, y_2) in D.

(c) How do you find the values that the Extreme Value Theorem guarantees?

To find the absolute maximum and minimum values of a continuous function f on a closed, bounded set D:

- Find the values of f at the critical points of f in D.
- Find the extreme values of f on the boundary of D.
- The largest of the values from steps 1 and 2 is the absolute maximum value; the smallest of these values is the absolute minimum value.

CHAPTER 10 CONCEPT CHECK ANSWERS

1. What is the difference between an autonomous and a non-autonomous system of differential equations?

A system of differential equations is autonomous if the equations do not explicitly contain the independent variable. If the differential equations depend on the independent variable, it is a non-autonomous system.

2. What is an equilibrium of a system of differential equations?

An equilibrium of the system of differential equations $d\mathbf{x}/dt = f(\mathbf{x})$ is a vector of values $\hat{\mathbf{x}}$ that satisfies the equation $d\mathbf{x}/dt = \mathbf{0}$.

3. Explain the difference between local and global stability in systems of differential equations.

An equilibrium $\hat{\mathbf{x}}$ of a system of differential equations is locally stable if $\mathbf{x}$ approaches the value $\hat{\mathbf{x}}$ as $t \to \infty$ for all initial values of $\mathbf{x}$ *sufficiently close* to $\hat{\mathbf{x}}$. The equilibrium is globally stable if the system approaches $\hat{\mathbf{x}}$ for *all* initial values of $\mathbf{x}$.

4. What is the difference between the solution of an initial-value problem and the general solution of a system of differential equations?

The general solution represents a *family* of solutions all of which satisfy the system of differential equations, whereas the solution of an initial-value problem is a *single* solution that satisfies the system of differential equations along with an initial condition $\mathbf{x}(t_0) = \mathbf{x}_0$.

5. What does the Existence and Uniqueness Theorem tell us about homogeneous systems of linear, autonomous differential equations?

The Existence and Uniqueness Theorem tell us that an initial-value problem involving a homogeneous systems of linear, autonomous differential equations will have one and only one solution $\mathbf{x}(t)$ and this solution is defined for all $t \in \mathbb{R}$.

6. Explain the Superposition Principle.

The Superposition Principle says that if $\mathbf{x}_1(t)$ and $\mathbf{x}_2(t)$ are solutions to the linear system of differential equations $d\mathbf{x}/dt = A\mathbf{x}$, then $\mathbf{x}(t) = c_1\mathbf{x}_1(t) + c_2\mathbf{x}_2(t)$, where c_1 and c_2 are scalar quantities, is also a solution.

7. Explain the difference between nullclines and eigenvectors in systems of linear autonomous differential equations.

The nullclines of a system of linear differential equations are straight lines in the phase plane along which the rate of change of one of the variables is zero. This means the solution curves move in either a vertical or horizontal direction along nullclines. Eigenvectors lie on straight lines that are themselves solution curves in the phase plane.

8. What is the linearization of a system of nonlinear differential equations?

The linearization of the system of nonlinear differential equations

$$\frac{dx_1}{dt} = f_1(x_1, x_2) \quad \text{and} \quad \frac{dx_2}{dt} = f_2(x_1, x_2)$$

is given by

$$\frac{d\boldsymbol{\varepsilon}}{dt} = \begin{bmatrix} \dfrac{\partial f_1(\hat{x}_1, \hat{x}_2)}{\partial x_1} & \dfrac{\partial f_1(\hat{x}_1, \hat{x}_2)}{\partial x_2} \\ \dfrac{\partial f_2(\hat{x}_1, \hat{x}_2)}{\partial x_1} & \dfrac{\partial f_2(\hat{x}_1, \hat{x}_2)}{\partial x_2} \end{bmatrix} \boldsymbol{\varepsilon}$$

where $\boldsymbol{\varepsilon}$ is the vector whose components are the deviations of x_1 and x_2 from their equilibrium values.

9. Explain what a Jacobian matrix is.

The Jacobian matrix of the system of differential equations

$$\frac{dx_1}{dt} = f_1(x_1, x_2) \quad \text{and} \quad \frac{dx_2}{dt} = f_2(x_1, x_2)$$

is defined as

$$J(x_1, x_2) = \begin{bmatrix} \dfrac{\partial f_1(x_1, x_2)}{\partial x_1} & \dfrac{\partial f_1(x_1, x_2)}{\partial x_2} \\ \dfrac{\partial f_2(x_1, x_2)}{\partial x_1} & \dfrac{\partial f_2(x_1, x_2)}{\partial x_2} \end{bmatrix}$$

The Jacobian matrix can be used to determine the local stability properties of equilibria.

10. What do the eigenvalues of a Jacobian matrix from a system of nonlinear differential equations tell us?

The eigenvalues of a Jacobian matrix can be used to determine the local stability properties of an equilibrium $(\hat{x}_1, \hat{x}_2)$. If r is the largest eigenvalue of $J(\hat{x}_1, \hat{x}_2)$ when they are real, or the largest real part of the eigenvalues when they are complex, then the equilibrium is stable when $r < 0$ and unstable when $r > 0$.

1. How do the goals of descriptive statistics differ from those of inferential statistics?

Descriptive statistics aims to summarize various properties of a data set using summary statistics such as means and standard deviations. Inferential statistics aims to quantify how well a summary statistic estimates a population parameter.

2. What is the difference between categorical and numerical variables?

Categorical variables describe membership in categories or groups, whereas numerical variables describe numeric measurements.

3. (a) What is the mode of a data set?

The mode of a data set is the element that appears most often in the data.

(b) What does it mean for the distribution of a variable to be bimodal?

If the distribution of a variable is bimodal, its histogram contains two peaks.

4. What is the relationship between the five-number summary and a box plot?

A box plot is a graphical display of the five-number summary. The box is drawn from Q_1 to Q_3 and divided by a line segment at the location of the median. Whiskers extend from both ends of the box to the location of the minimum and maximum values.

5. Describe the shape of a normal curve.

A normal curve is shaped like a bell and is symmetric about its center μ. As the spread σ increases, normal curves become wider.

6. What is a contingency table and what is it used for?

A contingency table is a way of summarizing the relationship between two categorical variables. The table rows represent one variable and the columns represent another variable. The number of data points falling into each combination of variables, along with the row and column totals, are recorded. A contingency table can be used to illustrate how membership in one group might be contingent on membership in another.

7. What is a scatter plot and what is it used for?

A scatter plot is a plot of two numerical variables. Each individual is represented by a single point in the plane, where the x-coordinate is the value of the x-variable and the y-coordinate is the value of the y-variable. A scatter plot is used to visualize the relationship between two numerical variables in a data set.

8. What is a confounding variable?

A confounding, or hidden, variable is one that complicates or confounds our ability to infer causation.

9. Explain the way in which a least-squares line best fits the data in a scatter plot.

The least-squares line $y = ax + b$ best fits the data $\{(x_1, y_1), \dots, (x_n, y_n)\}$ in a scatter plot when the constants a and b are chosen so as to minimize the total sum of squared deviations from the line; that is, a and b minimize the function $S(a, b) = (y_1 - ax_1 - b)^2 + \cdots + (y_n - ax_n - b)^2$.

10. What is a causal diagram?

A causal diagram consists of a set of nodes that represent the variables and arrows between the nodes that depict causal relationships. An arrow from variable A to variable B indicates that changes in the value of A can cause changes in the value of B, all else equal.

Cut here and keep for reference

CHAPTER 12 CONCEPT CHECK ANSWERS

1. What is the Fundamental Principle of Counting?

The Fundamental Principle of Counting states that if k selections can be made independently, the first having n_1 possible choices, the second having n_2 possible choices, . . . , and the kth having n_k possible choices, then, taken together, the total number of choices is $n_1 \cdot n_2 \cdot \cdots \cdot n_k$.

2. What is the difference between a permutation and a combination?

A permutation of a set of objects is an *ordered* arrangement of the objects without repetition, whereas a combination is an *unordered* subset of the objects without repetition. Order matters for permutations but does not matter for combinations.

3. Explain the relationship between an experiment, a trial, an outcome, and an event.

An experiment is any procedure, such as flipping a coin or rolling a die, that can be repeated an indefinite number of times under identical conditions (in principle, at least) and that has a set of distinct possible outcomes. Each repeated instance of such an experiment is called a trial, and an event is a particular set of outcomes from the experiment.

4. What is the difference between a simple and a compound event?

A simple event is an event consisting of a single outcome, whereas a compound event consists of more than one outcome.

5. In your own words explain what it means for two events to be independent.

Two events are independent if the probability of the second event is unaffected by the occurrence of the first event, and vice versa. Mathematically, events E and F are independent if $P(E \cap F) = P(E)\,P(F)$.

6. What is Bayes' Rule?

For any two events E and F with $P(E) > 0$,

$$P(F \mid E) = \frac{P(E \mid F)\,P(F)}{P(E)}$$

7. What is a random variable?

A random variable is a function whose domain is the sample space of some experiment and whose range is a subset of the real numbers.

8. (a) Name two specific examples of a discrete random variable.

The number of heads when two coins are tossed and the sum of two dice rolls

(b) Name two specific examples of a continuous random variable.

The lifetime of a randomly chosen bird and the cholesterol level of a randomly chosen male

9. (a) What is a probability density function (PDF)?

For a discrete random variable, the PDF $p_i = P(x = i)$ gives the probability that X takes on the value i.

For a continuous random variable, the PDF $f(x)$ is a measure of the probability that X takes on a given value.

Specifically, the probability of X falling within a subinterval $[a, b]$ is

$$P(a \leq X \leq b) = \int_a^b f(x)\,dx$$

(b) What is a cumulative distribution function (CDF)?

A cumulative distribution function gives the probability that a random variable X takes on a value no larger than x. Mathematically, $F(x) = P(X \leq x)$.

10. Explain the relationship between the Fundamental Theorem of Calculus, PDFs, and CDFs for continuous random variables.

The Fundamental Theorem of Calculus identifies the PDF of a continuous random variable as the derivative of its CDF, that is, $f(x) = F'(x)$. Additionally, the probability of X falling between any two values a and b can be obtained from the CDF as $P(a \leq X \leq b) = F(b) - F(a)$.

1. What are i.i.d. random variables?

Independent and identically distributed (i.d.d.) random variables are a set of random variables, all of which are mutually independent and have the same probability density function.

2. What is the sampling distribution of a statistic?

The sampling distribution of a statistic is the probability distribution of a random variable that represents the value of this statistic calculated from a random sample.

3. Explain what is meant by the precision of an estimate with reference to its sampling distribution.

The precision of an estimate refers to the variability of this estimate from sample to sample. An estimate is precise if it does not vary substantially from one sample to the next; that is, its sampling distribution is relatively narrow.

4. Explain what is meant by the bias of an estimate with reference to its sampling distribution.

Bias refers to any systematic over- or underestimation of a population parameter. An estimate is unbiased if the mean value of a statistic's sampling distribution coincides with the true population mean value. If this is not the case, then the estimate is biased.

5. What is the difference between the standard deviation of the data and the sample standard deviation?

The standard deviation of the data is calculated using the formula

$$\sigma = \sqrt{\frac{1}{n}\sum_{k=1}^{n}(x_k - \bar{x})^2}$$

and the sample standard deviation is given by

$$s = \sqrt{\frac{1}{n-1}\sum_{k=1}^{n}(x_k - \bar{x})^2}$$

The formulas differ by the scaling factor outside the summation. By dividing by $n - 1$ instead of n, the sample standard deviation provides an unbiased estimate of the population standard deviation.

6. Provide an interpretation of a 95% confidence interval.

A 95% confidence interval for a statistic is an interval that, if constructed repeatedly using samples of the same size, will contain the population parameter 95% of the time.

7. What is the difference between confidence intervals calculated when the population standard deviation is known versus when it is unknown?

When the population standard deviation is unknown, a confidence interval must be constructed using the sample standard deviation. This gives a confidence interval that is larger than one produced using a known population standard deviation because there is less certainty when estimating a parameter from a sample.

8. Explain what a null hypothesis and an alternative hypothesis are.

In hypothesis testing, the null hypothesis, denoted by H_0, is the hypothesis that there is no effect, and the alternative hypothesis, denoted by H_A, is the hypothesis that some effect is apparent. In the case of a two-tailed hypothesis test of a population mean, the null and alternative hypotheses are H_0: $\mu = \mu_0$ and H_A: $\mu \neq \mu_0$, where μ_0 is the value of the population mean being tested.

9. What is the t-statistic, and what is it used for?

For a sample of size n, the t-statistic is a measure of how far the sample mean Y_n lies from the hypothesized population mean μ_0, measured in units of the standard error in the mean $s/\sqrt{n}$. The t-statistic is given by

$$T_{n-1} = \frac{Y_n - \mu_0}{s/\sqrt{n}}$$

It is used during hypothesis testing to determine whether the sample data are compatible with the null hypothesis.

10. What is the chi-squared statistic, and what is it used for?

The chi-squared statistic is a test statistic that measures the departure of all observations in a study from what is expected under a null hypothesis. It is calculated using the formula

$$X = \sum_{i=1}^{4}\frac{(O_i - E_i)^2}{E_i}$$

where O_i are the observed values and E_i are the expected values. The chi-squared statistic is used to test for a relationship between categorical variables.